Aircraft
Powerplants

Aircraft
Powerplants

Eighth Edition

Thomas W. Wild

Michael J. Kroes

New York   Chicago   San Francisco   Athens   London   Madrid
Mexico City   Milan   New Delhi   Singapore   Sydney   Toronto

1 2 3 4 5 6 7 8 9 0   QVS/QVS   1 9 8 7 6 5 4 3

ISBN 978-0-07-179913-3
MHID 0-07-179913-3

Sponsoring Editor: Larry S. Hager
Editing Supervisor: Stephen M. Smith
Production Supervisor: Richard C. Ruzycka
Acquisitions Coordinator: Bridget L. Thoreson
Project Manager: Yashmita Hota, Cenveo® Publisher Services
Copy Editor: Megha Saini, Cenveo Publisher Services
Proofreader: Linda Manis Leggio
Indexer: Arc Films, Inc.
Art Director, Cover: Jeff Weeks
Composition: Cenveo Publisher Services

Printed and bound by Quad/Graphics.

McGraw-Hill Education books are available at special quantity discounts to use as premiums and sales promotions, or for use in corporate training programs. To contact a representative, please visit the Contact Us page at www.mhprofessional.com.

This book is printed on acid-free paper.

About the Authors

Thomas W. Wild is a professor in the Aviation Technology Department at Purdue University. He holds or has held several FAA certifications, including Aviation Maintenance Technician, Designated Mechanic Examiner, Flight Engineer, Inspection Authorization, and Sport Pilot. Professor Wild has earned numerous awards for his contributions to education during his over 32 years at Purdue. He has taught many courses dealing with reciprocating and gas-turbine engines, propellers, propeller control systems, and large aircraft systems. Professor Wild also serves as the managing editor of the *Aviation Technician Education Council Journal*. He has written and published many books and articles on several aviation-related subjects and served on boards of directors of aviation professional organizations.

Michael J. Kroes is an aviation practitioner and educator with more than 35 years of experience in the field. He holds or has held several FAA certifications, including Airframe and Powerplant Mechanic, Inspection Authorization, Designated Mechanic Examiner, Designated Engineering Representative, and Commercial Pilot. Mr. Kroes has worked for some of the top aviation companies, including Raytheon and Allied Signal, and spent 25 years as a professor and department head at Purdue University. Recognized as a leading expert on FAA technician certification, he authored a comprehensive study funded by the FAA. This study was used to develop new FAA technician certification content and guidelines.

Contents

Preface *xiii*
Acknowledgments *xv*

1. Aircraft Powerplant Classification and Progress   *1*

Engine Design and Classification   *5*
Review Questions   *27*

2. Reciprocating-Engine Construction and Nomenclature   *29*

The Crankcase   *29*
Bearings   *32*
The Crankshaft   *33*
Connecting-Rod Assemblies   *37*
Pistons   *40*
Cylinders   *45*
Valves and Associated Parts   *47*
The Accessory Section   *55*
Propeller Reduction Gears   *56*
Review Questions   *57*

3. Internal-Combustion Engine Theory and Performance   *59*

Science Fundamentals   *59*
Engine Operating Fundamentals   *60*
Valve Timing and Engine Firing Order   *62*
The Two-Stroke Cycle   *65*
Rotary-Cycle Engine   *66*
The Diesel Engine   *66*
Power Calculations   *68*
Engine Efficiency   *72*
Factors Affecting Performance   *73*
Review Questions   *78*

4. Lubricants and Lubricating Systems   *79*

Classification of Lubricants   *79*
Lubricating Oil Properties   *80*
The Need for Lubrication   *84*
Lubricant Requirements and Functions   *85*
Characteristics and Components of Lubrication Systems   *87*

Engine Design Features Related to Lubrication *95*
Typical Lubrication Systems *96*
Review Questions *100*

5. Induction Systems, Superchargers, Turbochargers, and Cooling and Exhaust Systems *101*

General Description *101*
Basic Induction System Components *101*
Principles of Supercharging and Turbocharging *104*
Internal Single-Speed Supercharger *110*
The Turbocharger *111*
Reciprocating-Engine Cooling Systems *121*
Reciprocating-Engine Exhaust Systems *124*
Review Questions *127*

6. Basic Fuel Systems and Carburetors *129*

Characteristics of Gasoline *129*
Fuel Systems *133*
Principles of Carburetion *136*
Float-Type Carburetors *144*
Carburetor Icing *154*
Inspection and Overhaul of Float-Type Carburetors *158*
Principles of Pressure Injection *161*
Pressure Carburetor for Small Engines *164*
Pressure Carburetors for Large Engines *164*
Water Injection *165*
Review Questions *167*

7. Fuel Injection Systems *169*

Definition *169*
Continental Continuous-Flow Injection System *169*
Bendix RSA Fuel Injection System *175*
Review Questions *189*

8. Reciprocating-Engine Ignition and Starting Systems *191*

Introduction *191*
Principles of Ignition *191*
Types of Magnetos *192*
Magneto Operational Theory *193*
Ignition Shielding *203*
Ignition Boosters and Auxiliary Ignition Units *205*
Continental Ignition High-Tension Magneto System for Light-Aircraft Engine *207*
Continental Dual-Magneto Ignition Systems *214*
Slick Series 4300 and 6300 Magnetos *215*
Other High-Tension Magnetos *218*
Low-Tension Ignition *218*
Low-Tension Ignition System for Light-Aircraft Engines *219*
FADEC System Description *219*
Compensated Cam *221*

Magneto Maintenance and Inspection *221*
Overhaul of Magnetos *223*
Spark Plugs *225*
Starters for Reciprocating Aircraft Engines *232*
Starters for Medium and Large Engines *236*
Troubleshooting and Maintenance *236*
Review Questions *237*

9. Operation, Inspection, Maintenance, and Troubleshooting of Reciprocating Engines *239*

Reciprocating-Engine Operation *239*
Engine Operation *241*
Cruise Control *243*
Engine Operating Conditions *245*
Reciprocating-Engine Operations in Winter *246*
Inspection and Maintenance *247*
Troubleshooting *258*
Review Questions *265*

10. Reciprocating-Engine Overhaul Practices *267*

Need for Overhaul *267*
Overhaul Shop *268*
Receiving the Engine *270*
Disassembly *270*
Visual Inspection *271*
Cleaning *273*
Structural Inspection *275*
Dimensional Inspection *281*
Repair and Replacement *286*
Reassembly *295*
Installation *299*
Engine Testing and Run-In *300*
Engine Preservation and Storage *302*
Review Questions *304*

11. Gas-Turbine Engine: Theory, Jet Propulsion Principles, Engine Performance, and Efficiencies *307*

Background of Jet Propulsion *307*
Basic Jet Propulsion Principles *309*
Types of Jet Propulsion Engines *310*
Gas-Turbine Engines *310*
Principles of Gas-Turbine Engines *311*
Types of Gas-Turbine Engines *311*
Gas-Turbine Engine Theory and Reaction Principles *314*
Airflow *315*
Gas-Turbine Engine Performance *317*
Efficiencies *321*
Turbine Engine *322*
Review Questions *323*

12. Principal Parts of a Gas-Turbine Engine, Construction, and Nomenclature *325*

The Inlet *325*
Types of Compressors *326*
Compressor Pressure Ratio *326*
Centrifugal-Flow Compressor *326*
Axial-Flow Compressor *327*
Multiple-Compressor Axial-Flow Engines *329*
Fan Bypass Ratio *330*
Compressor Stall *330*
Compressor Airflow and Stall Control *331*
Air-Bleed and Internal Air Supply Systems *331*
The Diffuser *333*
Combustion Chambers *333*
Turbine Nozzle Diaphragm *334*
Turbines *335*
Exhaust Systems *338*
Exhaust Nozzles *339*
Variable-Area Exhaust Nozzle *339*
Thrust Reversers *341*
Accessory Drive *342*
Reduction-Gear Systems *342*
Engine Noise *344*
Advanced Manufacturing Processes *345*
Review Questions *348*

13. Gas-Turbine Engine: Fuels and Fuel Systems *349*

Fuel Requirements *349*
Jet Fuel Properties and Characteristics *349*
Principles of Fuel Control *359*
Fuel Control Units for a Turboprop Engine *361*
Fuel Control Unit for a Large Turbofan Engine *364*
Hamilton Standard JFC68 Fuel Control Unit *364*
Fuel Control System for a Turboshaft Engine *370*
Electronic Engine Controls *373*
Review Questions *379*

14. Turbine-Engine Lubricants and Lubricating Systems *381*

Gas-Turbine Engine Lubrication *381*
Lubricating System Components *382*
Lubricating Systems *385*
Oil Analysis *391*
Review Questions *393*

15. Ignition and Starting Systems of Gas-Turbine Engines *395*

Ignition Systems for Gas-Turbine Engines *395*
Turbine-Engine Igniters *399*
Starting Systems for Gas Turbines *401*
Starting System for a Large Turbofan Engine *407*
Review Questions *411*

16. Turbofan Engines   413

Large Turbofan Engines   413
Small Turbofan Engines   463
Review Questions   478

17. Turboprop Engines   479

Large Turboprop Engines   480
Small Turboprop Engines   499
Review Questions   518

18. Turboshaft Engines   521

Auxiliary Power Unit   521
The Lycoming T53 Turboshaft Engine   524
The Allison Series 250 Gas-Turbine Engine   532
Helicopter Power Trains   536
Allison Series 250 Turboshaft Engine Operation in a Helicopter   537
Review Questions   539

19. Gas-Turbine Operation, Inspection, Troubleshooting, Maintenance, and Overhaul   541

Starting and Operation   541
Gas-Turbine Engine Inspections   546
Gas-Turbine Engine Maintenance   557
Gas-Turbine Engine Overhaul   568
Troubleshooting EGT System   575
Troubleshooting Aircraft Tachometer System   577
Gas-Turbine Engine Troubleshooting   577
Review Questions   578

20. Propeller Theory, Nomenclature, and Operation   581

Basic Propeller Principles   581
Propeller Nomenclature   581
Propeller Theory   582
Propeller Controls and Instruments   589
Propeller Clearances   590
General Classification of Propellers   590
Fixed-Pitch Propellers   592
Ground-Adjustable Propellers   593
Controllable-Pitch Propellers   594
Two-Position Propellers   594
Constant-Speed Propellers   594
McCauley Constant-Speed Propellers   598
Hartzell Constant-Speed Propellers   599
Hamilton Standard Counterweight Propellers   604
The Hamilton Standard Hydromatic Propeller   606
Anti-Icing and Deicing Systems   607
Propeller Synchrophaser System   609
Review Questions   612

21. Turbopropellers and Control Systems *615*

Turbopropeller Horsepower Calculations *615*
Hartzell Turbopropellers *615*
The Dowty Turbopropeller *619*
Hamilton Standard Turbopropellers *621*
McCauley Turbopropeller *624*
Composite Propeller Blades *624*
PT6A Propeller Control Systems *626*
Garrett TPE331 Engine Turbopropeller Control System *631*
Allison 250-B17 Reversing Turbopropeller System *635*
PW124 and R352 Turbopropeller Engine Interface *639*
General Electric CT7 Propeller Control System *641*
The Allison Turbopropeller *643*
Review Questions *648*

22. Propeller Installation, Inspection, and Maintenance *649*

Propeller Installation and Removal *649*
Aircraft Vibrations *656*
Maintenance and Repair of Propellers *664*
Checking Blade Angles *675*
Inspections and Adjustments of Propellers *679*
Review Questions *681*

23. Engine Indicating, Warning, and Control Systems *683*

Engine Instruments *683*
Fire Warning Systems *700*
Fire Suppression Systems *706*
Engine Control Systems *709*
Mechanical Engine Control Functions for Small Aircraft *713*
Mechanical Engine Control Systems for Large Aircraft *714*
Inspection and Maintenance of Control Systems *716*
Review Questions *718*

Appendix *721*

Glossary *727*

Index *735*

Preface

Aircraft Powerplants, Eighth Edition, is designed to provide future aviation professionals with the academic, theoretical, and practical knowledge they will need for a career in aviation. This text will prepare students for certification as an FAA powerplant technician in accordance with the Federal Aviation Regulations (FAR). This edition is a revision designed to reflect not only the latest changes in FAR Part 147 but also the current and changing needs of the aircraft industry. Throughout the text, FAR Part 147 has been used for reference to ensure that FAA requirements have been met. The FAA Written Test Guide Advisory has been reviewed carefully to ensure that all technical data that students will need in order to prepare for FAA written and oral examinations are included. This text also expands the knowledge available to students beyond the requirements of Part 147 and allows them to learn the material at a higher level than most A&P school curriculums.

In this edition of *Aircraft Powerplants,* Chap. 11 from the previous edition has been split into two chapters, one covering expanded turbine-engine theory and the other expanded nomenclature. Additional current models of turbofan, turboprop, and turboshaft engines have been included. Information on turbine-engine fuel, oil, and ignition systems has been expanded and updated. Pictures of actual components have been used as much as possible. A color insert has been provided to clarify diagrams and systems. Review questions at the end of each chapter enable students to check their knowledge of the information presented.

Throughout the text, a special emphasis has been placed on the integration of information on how individual components and systems operate together. This text will provide maximum benefit when used in conjunction with the books *Aircraft Basic Science, Aircraft Maintenance and Repair,* and *Aircraft Electricity and Electronics,* which as a group encompass information on all phases of airframe and aircraft powerplant technology.

This book is designed to be used as both a classroom text and an on-the-job reference for the technician. Technical information contained in this book should not be substituted for that provided by manufacturers.

Thomas W. Wild
Michael J. Kroes

Acknowledgments

The authors wish to express appreciation to the following organizations and individuals for their generous assistance in providing illustrations and technical information for this text:

AiResearch Manufacturing Company, Division of Honeywell Aerospace Co., Torrance, California; Alcor Inc., San Antonio, Texas; American Hall of Aviation History, Northrop University, Inglewood, California; American Society of Mechanical Engineers, New York, New York; Beech Aircraft Corporation, Wichita, Kansas; Bell Helicopter Textron, Fort Worth, Texas; Bendix Corporation, Energy Controls Division, South Bend, Indiana; B.F. Goodrich, Akron, Ohio; Boeing AIRLINER Magazine, Seattle, Washington; Boeing Commercial Airplane Company, Seattle, Washington; Bray Oil Company, Los Angeles, California; Cessna Aircraft Company, Wichita, Kansas; Champion Spark Plug Company, Toledo, Ohio; Continental Motors, Mobile, Alabama; Dee Howard Company, San Antonio, Texas; Dowty Rotol Ltd., Gloucester, England; Dyna-Cam Engines, Redondo Beach, California; Dynamic Solutions Systems, Inc., San Marcos, California; Elcon Division, Icore International, Inc., Sunnyvale, California; Facet Aerospace Products Company, Jackson, Tennessee; C. A. Faulkner; Federal Aviation Administration, Washington, D.C.; Fenwall, Inc., Ashland, Massachusetts; General Electric Company, Aircraft Engine Group, Cincinnati, Ohio; General Electric Company, Commercial Engine Division, Lynn, Massachusetts; Hamilton Standard Division, United Technologies, Windsor Locks, Connecticut; Hartzell Propeller Division, TRW, Piqua, Ohio; Honeywell Engines, Phoenix, Arizona; Howell Instruments, Inc., Fort Worth, Texas; Howmet Turbine Components Corporation, Greenwich, Connecticut; Hughes Helicopters, Inc., Culver City, California; D. C. Latia; Lamar Technologies LLC, Marysville, Washington; Magnaflux Corp., Chicago, Illinois; McCauley Accessory Division, Cessna Aircraft Company, Dayton, Ohio; Northrop University, Inglewood, California; Olympus Corporation, IFD, Lake Success, New York; Piper Aircraft Company, Vero Beach, Florida; Prestolite Division, Eltra Corporation, Toledo, Ohio; Pratt & Whitney Aircraft Group, United Technologies, East Hartford, Connecticut; Pratt & Whitney Aircraft of Canada, Ltd., Longueuil, Quebec, Canada; Precision Airmotive LLC, Marysville, Washington; Purdue University, West Lafayette, Indiana; Rolls-Royce, Ltd., Derby, England; Rolls-Royce Allison, Indianapolis, Indiana; Santa Monica Propeller Services, Santa Monica, California; SGL Auburn Spark Plug Company, Auburn, New York; Slick Electro, Inc., Rockford, Illinois; Spectro Metries, Atlanta, Georgia; B. M. Stair; Systron Donner Corporation, Berkeley, California; Tempest Spark Plugs, Taylors, South Carolina; Textron Lycoming, Stratford, Connecticut; Textron Lycoming, Williamsport, Pennsylvania; Walter Kidde and Company, Belleville, New Jersey; Welch Allyn, Scaneateles Falls, New York; Woodward Governor Company, Rockford, Illinois.

Special thanks are given to Douglas C. Latia, Carol A. Kroes, and Louise K. Wild for their assistance in compiling this text.

In addition to the above, the authors wish to thank the many aviation technical schools and instructors for providing valuable suggestions, recommendations, and technical information for this revision.

Aircraft
Powerplants

Aircraft Powerplant Classification and Progress 1

INTRODUCTION

People have dreamed of flying ever since they first gazed into the sky and saw birds soaring overhead. Early attempts at flight often resulted in failure. This failure was not primarily due to airfoil design but instead was attributable to the lack of technology needed to produce a source of power sufficient to sustain flight.

The development of aviation powerplants has resulted from utilization of principles that were employed in the design of earlier internal-combustion engines. During the latter part of the nineteenth century, a number of successful engines were designed and built and used to operate machinery and to supply power for "horseless carriages."

Since the first internal-combustion engine was successfully operated, many different types of engines have been designed. Many have been suitable for the operation of automobiles and/or aircraft, and others have been failures. The failures have been the result of poor efficiency, lack of dependability (owing to poor design and to materials which could not withstand the operating conditions), high cost of operation, excessive weight for the power produced, and other deficiencies.

The challenge to aviation has been to design engines that have high power-to-weight ratios. This was accomplished first with lightweight piston engines and then, more effectively, with gas-turbine engines.

In this chapter we examine the evolution, design, and classification of various types of engines.

Early Engines

Development of the internal-combustion engine took place largely during the nineteenth century. One of the first such engines was described in 1820 by the Reverend W. Cecil in a discourse before the Cambridge Philosophical Society in England. This engine operated on a mixture of hydrogen and air. In 1838 the English inventor William Barnett built a single-cylinder gas engine which had combustion chambers at both the top and the bottom of the piston. This engine burned gaseous fuel rather than the liquid fuel used in the modern gasoline engine.

The first practical gas engine was built in 1860 by a French inventor named Jean Joseph Étienne Lenoir. This engine utilized illuminating gas as a fuel, and ignition of the fuel was provided by a battery system. Within a few years, approximately 400 of these engines had been built to operate a variety of machinery, such as lathes and printing presses.

The first four-stroke-cycle engine was built by August Otto and Eugen Langen of Germany in 1876. As a result, four-stroke-cycle engines are often called *Otto-cycle engines*. Otto and Langen also built a two-stroke-cycle engine.

In the United States, George B. Brayton, an engineer, built an engine using gasoline as fuel and exhibited it at the 1876 Centennial Exposition in Philadelphia. The first truly successful gasoline engine operating according to the four-stroke-cycle principle was built in Germany in 1885 by Gottlieb Daimler, who had previously been associated with Otto and Langen. A similar gasoline engine was built by Karl Benz of Germany in the same year. The Daimler and Benz engines were used in early automobiles, and the engines used today are similar in many respects to the Daimler and Benz engines.

The First Successful Airplane Engine

Inasmuch as the first powered flight in an airplane was made by the Wright brothers on December 17, 1903, it is safe to say that the first successful gasoline engine for an airplane was the engine used in the Wright airplane. This engine was designed and built by the Wright brothers and their mechanic, Charles Taylor. The engine had the following characteristics: (1) water cooling; (2) four cylinders; (3) bore, $4\frac{3}{8}$ inches (in) [11.11 centimeters (cm)], and stroke, 4 in [10.16 cm], displacement, 240 cubic inches (in^3) [3 932.9 cm^3]; (4) 12 horsepower (hp) [8.94 kilowatts (kW)]; (5) weight, 180 pounds (lb) [82 kilograms (kg)]; (6) cast-iron cylinders with sheet-aluminum water jackets; (7) valve-in-head with the exhaust valve mechanically operated and the intake valve automatically operated; (8) aluminum-alloy crankcase; (9) carburetion by means of fuel flow into a heated manifold; and (10) ignition by means of a high-tension magneto. A picture of an early Wright engine is shown in Fig. 1-1.

World War I Aircraft Engines

The extensive development and use of airplanes during World War I contributed greatly to the improvement of engines.

1

FIGURE 1-1   Early Wright engine.

Rotary-Type Radial Engines

One type of engine that found very extensive use was the air-cooled **rotary-type radial engine**. In this engine the crankshaft is held stationary, and the cylinders rotate about the crankshaft. Among the best-known rotary engines were the LeRhone, shown in Fig. 1-2, the Gnome-Monosoupape, shown in Fig. 1-3, and the Bentley, which has a similar appearance. In these engines, the crankshaft is secured to the aircraft engine mount, and the propeller is attached to the engine case.

Even though the rotary engines powered many World War I airplanes, they had two serious disadvantages: (1) the torque and gyro effects of the large rotating mass of the engines made the airplanes difficult to control; and (2) the engines used castor oil as a lubricant, and since the castor oil was mixed with the fuel of the engine in the crankcase, the exhaust of the engines contained castor-oil fumes which were often nauseating to the pilots.

FIGURE 1-2   LeRhone rotary engine.

FIGURE 1-3   Gnome-Monosoupape rotary engine.

FIGURE 1-4   Early Hispano-Suiza engine.

In-Line Engines

A number of **in-line engines** were also developed during World War I. Among these was the Hispano-Suiza engine, shown in Fig. 1-4.

The cylinders of an in-line engine are arranged in a single row parallel to the crankshaft. The cylinders are either upright above the crankshaft or inverted, that is, below the crankshaft. The inverted configuration is generally employed. A typical inverted in-line engine is shown in Fig. 1-5. The engine shown is a Menasco Pirate, model C-4. The number of cylinders in an in-line engine is usually limited to six, to facilitate cooling and to avoid excessive weight per horsepower. There are generally an even number of cylinders in order to provide a proper balance of firing impulses. The in-line engine utilizes one crankshaft. The crankshaft is located above the cylinders in an inverted engine. The engine may be either air-cooled or liquid-cooled; however, liquid-cooled types are seldom utilized at present.

FIGURE 1-5 Inverted in-line engine.

Use of the in-line-type engine is largely confined to low- and medium-horsepower applications for small aircraft. The engine presents a small frontal area and is therefore adapted to streamlining and a resultant low-drag nacelle configuration. When the cylinders are mounted in the inverted position, greater pilot visibility and a shorter landing gear are possible. However, the in-line engine has a greater weight-to-horsepower ratio than those of most other types. When the size of an aircraft engine is increased, it becomes increasingly difficult to cool it if it is the air-cooled in-line type; therefore, this engine is not suitable for a high-horsepower output.

V-Type Engines

World War I saw the development of several **V-type engines**, including the Rolls-Royce V-12 engine, the U.S.-made Liberty V-12 engine, shown in Fig. 1-6, and several German engines. The V-type engine has the cylinders arranged on the crankcase in two rows (or banks), forming the letter V, with an angle between the banks of 90, 60, or 45°. There are always an even number of cylinders in each row.

FIGURE 1-6 Liberty engine.

Since the two banks of cylinders are opposite each other, two sets of connecting rods can operate on the same crankpin, thus reducing the weight per horsepower as compared with the in-line engine. The frontal area is only slightly greater than that of the in-line type; therefore, the engine cowling can be streamlined to reduce drag. If the cylinders are above the crankshaft, the engine is known as the **upright-V-type engine**, but if the cylinders are below the crankshaft, it is known as an **inverted-V-type engine**. Better pilot visibility and a short landing gear are possible if the engine is inverted.

Post-World War I Engines

After World War I, many different engine designs were developed. Some of those with rather unusual configurations are shown in Fig. 1-7.

A popular U.S. engine was the Curtiss OX-5 engine manufactured during and after World War I. This engine powered the Curtiss Jennie (JN-4) trainer plane used for training U.S. military aviators. After the war, many were sold to the public, and the majority were used in the early barnstorming days for air shows and passenger flights. An OX-5 engine is shown in Fig. 1-8.

Other engines developed in the United States between World War I and World War II were the Wright Hisso (a U.S.-built Hispano-Suiza), the Packard V-12, the Curtiss D-12 (a V-12 engine), the Wright Whirlwind and radial engines, and the Pratt & Whitney Wasp and Hornet engines, which are air-cooled radial types. Numerous smaller engines were also designed and built, including radial, opposed-cylinder, and in-line types.

Radial Engines

The **radial engine** has been the workhorse of military and commercial aircraft ever since the 1920s, and during World War I radial engines were used in all U.S. bombers and transport aircraft and in most of the other categories of aircraft. They were developed to a peak of efficiency and dependability; and even today, in the jet age, many are still in operation throughout the world in all types of duty.

A **single-row radial engine** has an odd number of cylinders extending radially from the centerline of the crankshaft. The number of cylinders usually ranges from five to nine. The cylinders are arranged evenly in the same circular plane, and all the pistons are connected to a single-throw 360° crankshaft, thus reducing both the number of working parts and the weight.

A **double-row radial engine** resembles two single-row radial engines combined on a single crankshaft, as shown in Fig. 1-9. The cylinders are arranged radially in two rows, and each row has an odd number of cylinders. The usual number of cylinders used is either 14 or 18, which means that the same effect is produced as having either two seven-cylinder engines or two nine-cylinder engines joined on one crankshaft. A two-throw 180° crankshaft is used to permit the cylinders in each row to be alternately staggered on the common crankcase. That is, the cylinders of the rear

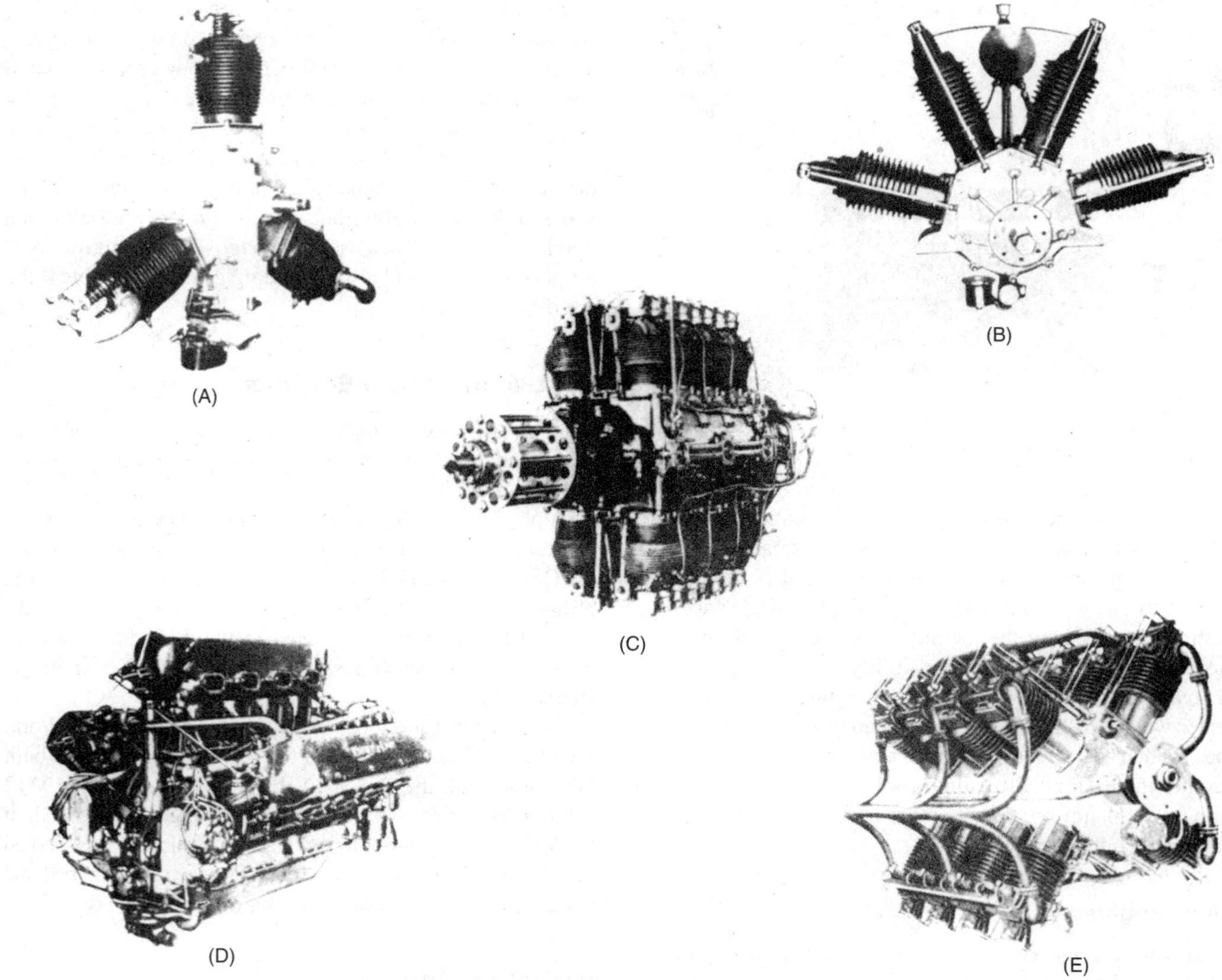

FIGURE 1-7 Different engine configurations developed after World War I. (A) Szekeley, 3-cylinder radial; (B) Italian MAB, 4-cylinder fan-type engine; (C) British Napier "Rapier," 16-cylinder H-type engine; (D) British Napier "Lion," 12-cylinder W-type engine; (E) U.S. Viking, 16-cylinder X-type engine.

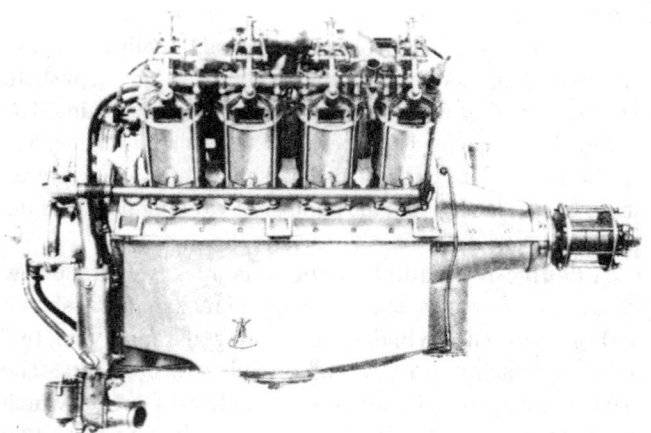

FIGURE 1-8 Curtiss OX-5 engine.

row are located directly behind the spaces between the cylinders in the front row. This allows the cylinders in both rows to receive ram air for the necessary cooling.

The radial engine has the lowest weight-to-horsepower ratio of all the different types of piston engines. It has the disadvantage of greater drag because of the area presented to the air, and it also has some problems in cooling. Nevertheless, the dependability and efficiency of the engine have made it the most widely used type for large aircraft equipped with reciprocating engines.

Multiple-Row Radial Engine

The 28-cylinder Pratt & Whitney R-4360 engine was used extensively at the end of World War II and afterward for both bombers and transport aircraft. This was the largest and most powerful piston-type engine built and used successfully in the United States. A photograph of this engine is shown in

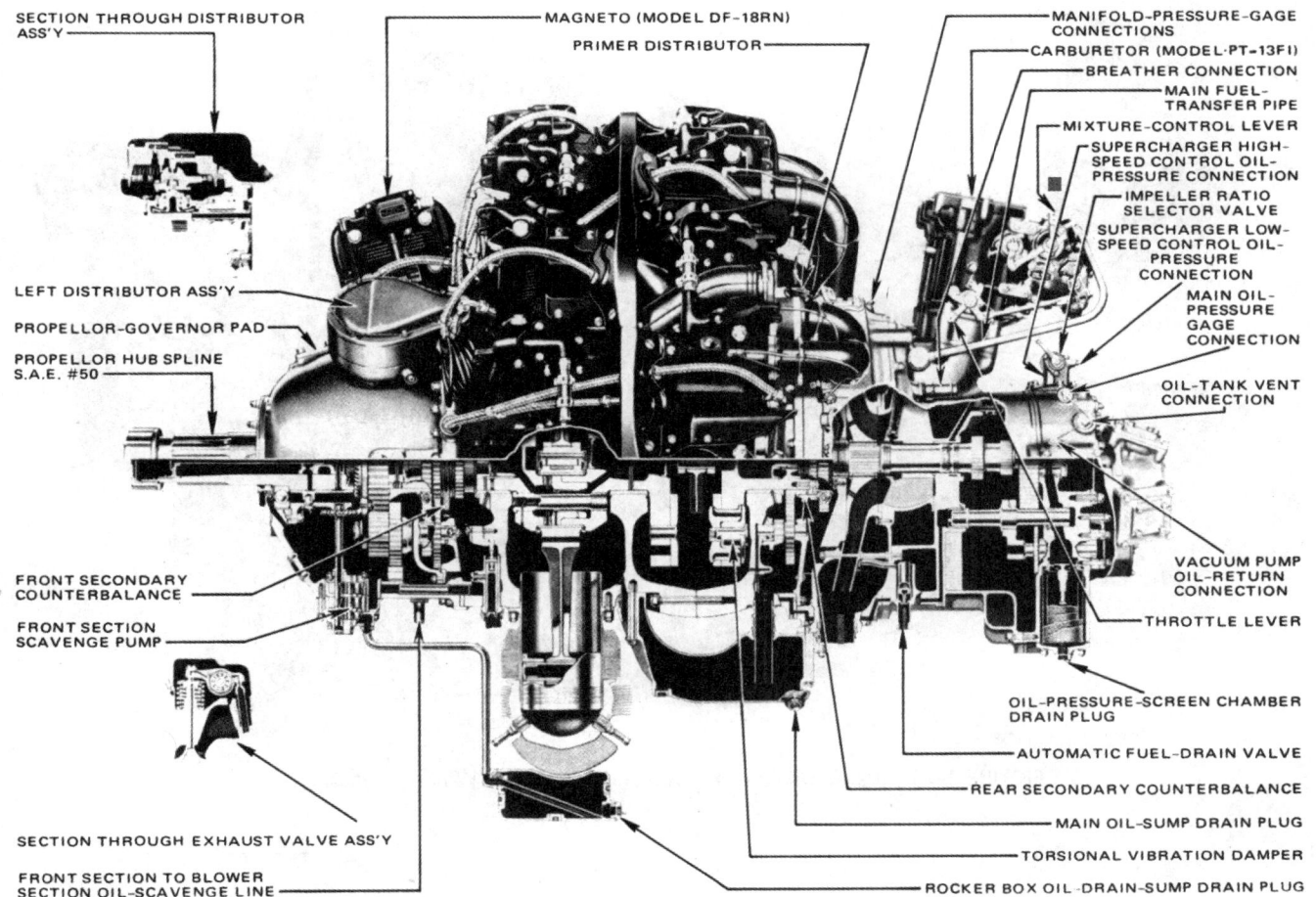

SECTION THROUGH DISTRIBUTOR ASS'Y

MAGNETO (MODEL DF-18RN)

PRIMER DISTRIBUTOR

MANIFOLD-PRESSURE-GAGE CONNECTIONS

CARBURETOR (MODEL PT-13FI)

BREATHER CONNECTION

MAIN FUEL-TRANSFER PIPE

MIXTURE-CONTROL LEVER

SUPERCHARGER HIGH-SPEED CONTROL OIL-PRESSURE CONNECTION

IMPELLER RATIO SELECTOR VALVE

SUPERCHARGER LOW-SPEED CONTROL OIL-PRESSURE CONNECTION

MAIN OIL-PRESSURE GAGE CONNECTION

OIL-TANK VENT CONNECTION

LEFT DISTRIBUTOR ASS'Y

PROPELLOR-GOVERNOR PAD

PROPELLOR HUB SPLINE S.A.E. #50

VACUUM PUMP OIL-RETURN CONNECTION

THROTTLE LEVER

FRONT SECONDARY COUNTERBALANCE

FRONT SECTION SCAVENGE PUMP

OIL-PRESSURE-SCREEN CHAMBER DRAIN PLUG

AUTOMATIC FUEL-DRAIN VALVE

REAR SECONDARY COUNTERBALANCE

MAIN OIL-SUMP DRAIN PLUG

TORSIONAL VIBRATION DAMPER

ROCKER BOX OIL-DRAIN-SUMP DRAIN PLUG

SECTION THROUGH EXHAUST VALVE ASS'Y

FRONT SECTION TO BLOWER SECTION OIL-SCAVENGE LINE

FIGURE 1-9 Double-row radial engine.

Fig. 1-10. Because of the development of the gas-turbine engine, the very large piston engine has been replaced by the more powerful and lightweight turboprop and turbojet engines. Since it has few moving parts compared with the piston engine, the gas-turbine engine is more trouble-free and its maintenance cost is reduced. Furthermore, the time between overhauls (TBO) is greatly increased.

Opposed, Flat, or O-Type Engine

The opposed-type engine is most popular for light conventional aircraft and helicopters and is manufactured in sizes delivering from less than 100 hp [74.57 kW] to more than 400 hp [298.28 kW]. These engines are the most efficient, dependable, and economical types available for light aircraft. Gas-turbine engines are being installed in some light aircraft, but their cost is still prohibitive for the average, private airplane owner.

The **opposed-type engine** is usually mounted with the cylinders horizontal and the crankshaft horizontal; however, in some helicopter installations the crankshaft is vertical. The engine has a low weight-to-horsepower ratio, and because of its flat shape it is very well adapted to streamlining and to horizontal installation in the nacelle. Another advantage is that it is reasonably free from vibration. Figure 1-11 illustrates a modern opposed engine for general aircraft use.

ENGINE DESIGN AND CLASSIFICATION

Conventional piston engines are classified according to a variety of characteristics, including cylinder arrangement, cooling method, and number of strokes per cycle. The most satisfactory classification, however, is by cylinder arrangement. This is the method usually employed because it is more completely descriptive than the other classifications. Gas-turbine engines are classified according to construction and function; these classifications are discussed in Chap. 11.

Cylinder Arrangement

Although some engine designs have become obsolete, we mention the types most commonly constructed throughout the history of powerplants. Aircraft engines may be classified according to cylinder arrangement with respect to the crankshaft as follows: (1) in-line, upright; (2) in-line, inverted; (3) V type, upright; (4) V type, inverted; (5) double-V or fan type; (6) X type; (7) opposed or flat type; (8) radial type, single-row; (9) radial type, double-row; (10) radial type, multiple-row or "corncob." The simple drawings in Fig. 1-12 illustrate some of these arrangements.

FIGURE 1-10   Pratt & Whitney R-4360 engine. (*Pratt & Whitney Canada.*)

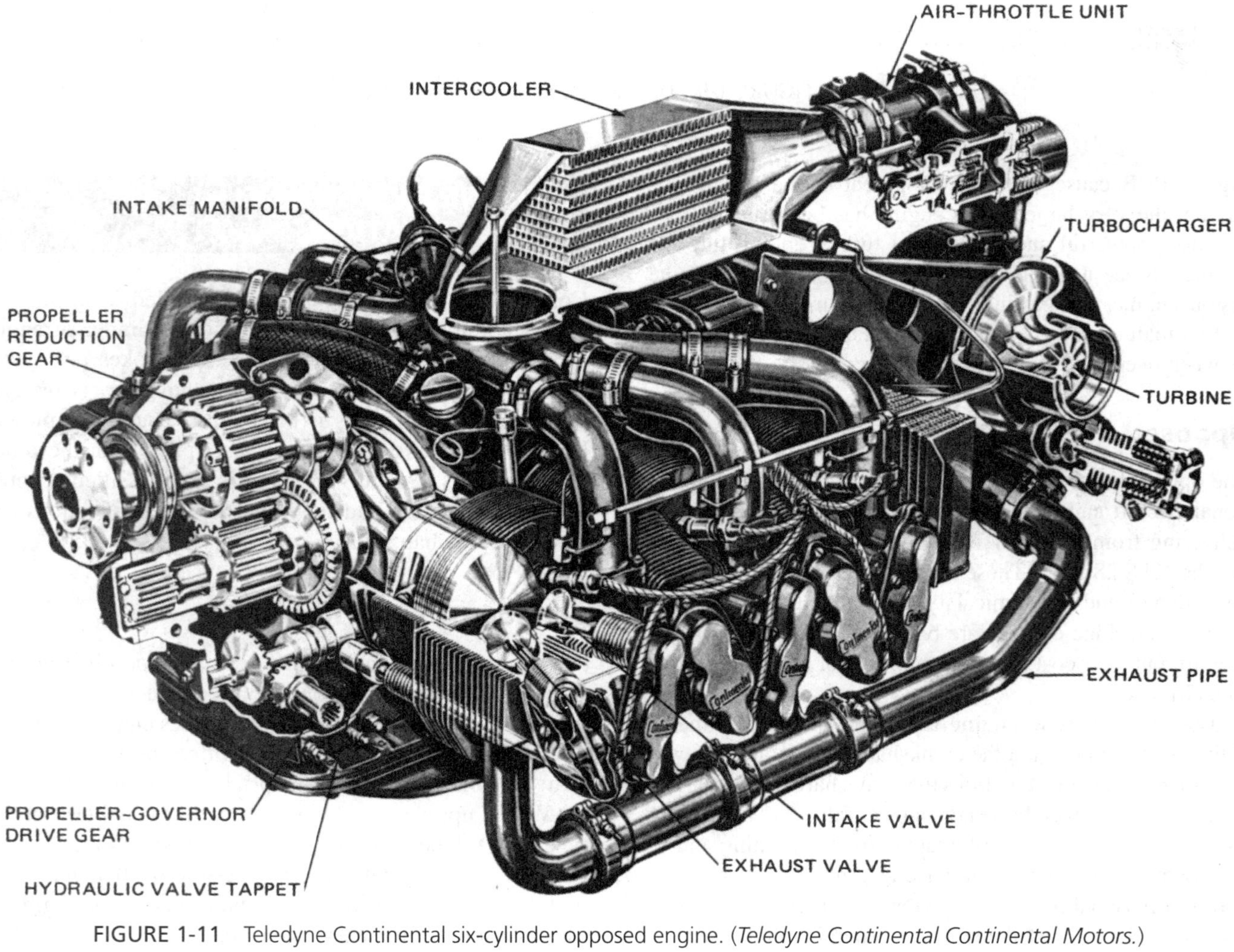

AIR-THROTTLE UNIT

INTERCOOLER

INTAKE MANIFOLD

TURBOCHARGER

PROPELLER
REDUCTION
GEAR

TURBINE

EXHAUST PIPE

PROPELLER-GOVERNOR
DRIVE GEAR

INTAKE VALVE

EXHAUST VALVE

HYDRAULIC VALVE TAPPET

FIGURE 1-11   Teledyne Continental six-cylinder opposed engine. (*Teledyne Continental Continental Motors.*)

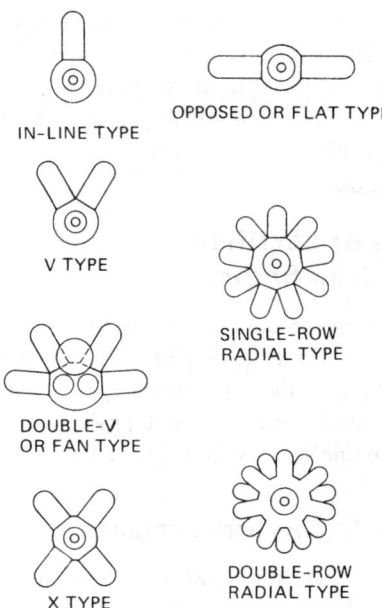

IN-LINE TYPE

OPPOSED OR FLAT TYPE

V TYPE

SINGLE-ROW RADIAL TYPE

DOUBLE-V OR FAN TYPE

X TYPE

DOUBLE-ROW RADIAL TYPE

FIGURE 1-12 Engines classified according to cylinder arrangement.

The double-V- or fan-type engine has not been in use for many years, and the only piston engines in extensive use for aircraft in the United States at present are the opposed and radial types. A few V-type and in-line engines may still be in operation, but these engines are no longer manufactured in the United States for general aircraft use.

Early Designations

Most of the early aircraft engines, with the exception of the rotary types, were water-cooled and were of either in-line or V-type design. These engines were often classified as liquid-cooled in-line engines, water-cooled in-line engines, liquid-cooled V-type engines, or water-cooled V-type engines. As air-cooled engines were developed, they were classified in a similar manner (air-cooled in-line, air-cooled V-type, etc.).

Classification or Designation by Cylinder Arrangement and Displacement

Current designations for reciprocating engines generally employ letters to indicate the type and characteristics of the engine, followed by a numerical indication of displacement. The following letters usually indicate the type or characteristic shown:

L   **Left-hand rotation** for counterrotating propeller

T   **Turbocharged** with turbine-operated device

V   **Vertical**, for helicopter installation with the crankshaft in a vertical position

H   **Horizontal**, for helicopter installation with the crankshaft horizontal

A   **Aerobatic**; fuel and oil systems designed for sustained inverted flight

I   **Fuel injected**; continuous fuel injection system installed

G   **Geared** nose section for reduction of propeller revolutions per minute (rpm)

S   **Supercharged**; engine structurally capable of operating with high manifold pressure and equipped with either a turbine-driven supercharger or an engine-driven supercharger

O   **Opposed cylinders**

R   **Radial engine**; cylinders arranged radially around the crankshaft

However, note that many engines are not designated by the foregoing standardized system. For example, the Continental W-670 engine is a radial type, whereas the A-65, C-90, and E-225 are all opposed-type engines. V-type engines and inverted in-line engines have such designations as V and I. In every case, the technician working on an engine must interpret the designation correctly and utilize the proper information for service and maintenance.

The two- or three-digit numbers in the second part of the engine designation indicate displacement to the nearest 5 in³. An engine with a displacement of 471 in³ [7.72 liters (L)] is shown as 470, as is the case with the Teledyne Continental O-470 opposed engine.

In some cases, the displacement number will end with a figure other than zero. In such a case, this is a special indication to reveal a characteristic such as an integral accessory drive.

Radial engines generally employ only the letter R followed by the displacement. For example, the R-985 is a single-row radial engine having a displacement of approximately 985 in³ [16.14 L].

An example of the standard designation for an engine is as follows.

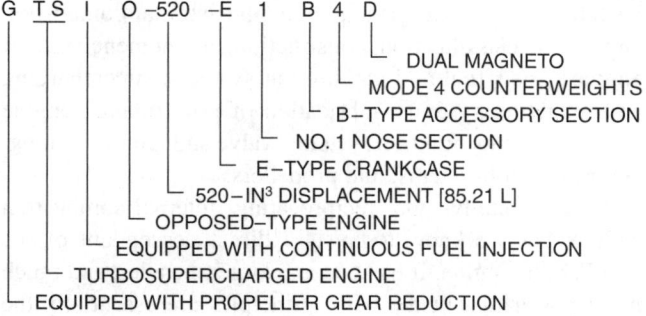

G T S I O –520 –E 1 B 4 D

DUAL MAGNETO
MODE 4 COUNTERWEIGHTS
B–TYPE ACCESSORY SECTION
NO. 1 NOSE SECTION
E–TYPE CRANKCASE
520–IN³ DISPLACEMENT [85.21 L]
OPPOSED–TYPE ENGINE
EQUIPPED WITH CONTINUOUS FUEL INJECTION
TURBOSUPERCHARGED ENGINE
EQUIPPED WITH PROPELLER GEAR REDUCTION

A system of suffix designations has also been established to provide additional information about engines. The first suffix letter indicates the type of power section and the rating of the engine. This letter is followed by a number from 1 to 9, which gives the design type of the nose section. Following the nose-section number is a letter indicating the type of accessory section, and after this letter is a number which tells what type of counterweight application is used with the crankshaft. This number indicates the mode of vibration, such as 4, 5, or 6. The mode number is found on the counterweights or dynamic balances on the crankshaft.

The final character in the designation suffix may be a letter indicating the type of magneto utilized with the engine. The letter D indicates a dual magneto.

Engine Classification by Cooling Method

Aircraft engines may be classified as being cooled either by air or by liquid; however, few liquid-cooled engines are in operation. Most aircraft engines are cooled by passing air over the engine's cylinders; through the convection process, excessive heat generated by the engine's combustion process is removed from the engine. In a liquid-cooled engine, the liquid is circulated through the engine areas that require heat removal. After the heat has been transferred to the liquid, the liquid passes through a heat exchanger which cools the liquid, and the cycle repeats. A complete discussion of engine cooling systems is presented in Chap. 5.

Progress in Design and Types of Current Reciprocating Engines

Engineers who specialize in the design of aircraft powerplants have used light alloy metals for construction of the engines and have adopted weight-saving cylinder arrangements, with the result that today the weight per horsepower on several engines is below 1.2 lb [0.54 kg] and on some less than 1 lb [0.45 kg].

Airplanes have increased in size, carrying capacity, and speed. With each increase has come a demand for more power, and this has been met by improvements in engine and propeller design and by the use of gas-turbine and turboprop engines. As piston engines increased in power, they became more complicated. The early powerplant engineers and mechanics had only a few comparatively simple problems to solve, but the modern powerplant specialist must be familiar with the principles of the internal-combustion engine; the classification, construction, and nomenclature of engines; their fuel and carburetion systems; supercharging and induction systems; lubrication of powerplants; engine starting systems; ignition systems; valve and ignition timing; engine control systems; and propellers.

Fundamentally, the reciprocating internal-combustion engine that we know today is a direct descendant of the first Wright engine. It has become larger, heavier, and much more powerful, but the basic principles are essentially the same. However, the modern reciprocating aircraft engine has reached a stage in its development where it is faced with what is commonly called the **theory of diminishing returns**. More cylinders are added to obtain more power, but the resulting increase in size and weight complicate matters in many directions. For example, the modern reciprocating engine may lose more than 30 percent of its power in dragging itself and its nacelle through the air and in providing necessary cooling.

The improvement in reciprocating engines has become quite noticeable in the smaller engines used for light aircraft. This has been accomplished chiefly with the opposed-type four- and six-cylinder engines. Among the improvements developed for light engines are geared propellers, superchargers, and fuel-injection systems. Whereas light airplanes were once limited to flight at comparatively low altitudes, today many are capable of cruising at altitudes of well over 20,000 feet (ft) [6 096 meters (m)].

Examples of Certified Reciprocating Engines

Many modern reciprocating engines for light certified aircraft (certificated under FAR part 33) are manufactured by Continental Motors, Inc. and Lycoming, a division of AVCO Corp. Although over time engine types can vary somewhat, some basic engine series will be presented.

Continental Motors Series Engines

200 Series. The first series for the continental engines is the 200 series shown in Fig. 1-13. This series of engines has been providing aircraft power for decades. The Continental's 200 series tuned induction system provides improved cylinder to cylinder intake air distribution for smoother operation and increased fuel efficiency. The lightweight O-200 D engine weighs 199 pounds and develops 100 continuous horsepower at 2750 rpm. Another version of the 200 series is the O-200-AF (alternative fuel) developed for use with lower octane/unleaded fuels for international markets.

300 Series. Some of Continental's 300 series engines top the horsepower charts at an impressive 225 hp in turbocharged (used to boost horsepower), and intercooled form. All 360s have six smooth-running cylinders. And every 360 engine is fuel injected for outstanding efficiency and range. At 283 pounds the TSIO-360-A, shown in Fig. 1-14, rates among the lightest of all six-cylinder aircraft powerplants. The Continental's 300 series designs are used in many different aircraft.

400 Series. The main engine in the 400 series is the O-470 powering many Cessna aircraft. Ranging in output from 225 to 260 hp, the 470 engines come equipped with either a carburetor or Continental's continuous-flow fuel-injection system. The 400 series uses a six-cylinder design and has many hours of successful operation and is used in several configurations. The O-470 series engines can be seen in Fig. 1-15.

500 Series. A family of engines that ranges in power from 285 to 375 hp is the Continental Motors 500 Series. Continental Motors introduced the first 500 series engine in the Beech Bonanza and the Cessna Centurion in 1964. The 500 series, shown in Fig. 1-16, includes both 520 and 550 in^3 models in either naturally aspirated or turbocharged configurations. There is even a geared variant that exceeds horsepower-to-displacement standards—a stunning 375 hp from 520 in^3. With the right combination of thrust and efficiency, the 500-series engines have powered many aircraft in general aviation.

FIGURE 1-13 Continental Motors 200 series engine.

Lycoming Series Engines

Lycoming four cylinder series. The Lycoming four-cylinder series engines, shown in Fig. 1-17, are four-cylinder, direct-drive, horizontally opposed, air-cooled models. The cylinders are of conventional air-cooled construction with heads made from an aluminum-alloy casting and a fully machined combustion chamber. Rocker-shaft bearing supports are cast integral with the head, along with housings to form the rocker boxes. The cylinder barrels have deep integral cooling fins, and the inside of the barrels are ground and honed to a specified finish. The IO-360 and TIO-360 series engines are equipped with a fuel-injection system, which schedules fuel flow in proportion to airflow. Fuel vaporization takes place at the intake ports. A turbo-charger is mounted as an integral part of the TIO-360 series.

Automatic waste-gate control of the turbocharger provides constant air density to the fuel-injector inlet from sea level to critical altitude. The following chapters will discuss all these engine components and details in great depth.

O-540 series. The Lycoming O-540 series engines are six-cylinder, direct-drive, horizontally opposed, air-cooled models. The cylinders are of conventional air-cooled construction with heads made from an aluminum-alloy casting and a fully machined combustion chamber. Rocker-shaft bearing supports are cast integral with the head, along with housings to form the rocker boxes as in the 0-360 series. The cylinder barrels are ground and honed to a specified finish and are equipped with cooling fins. The IO-540 (Fig. 1-18) and TIO-540 (turbocharged) series engines are

FIGURE 1-14 Continental Motors 300 series engine.

equipped with a fuel-injection system, which schedules fuel flow and delivers vaporized fuel at the intake ports in proportion to airflow. A turbocharger(s) is mounted as an integral part of the TIO-540 series. The turbocharger provides constant air density to the fuel-injector inlet. Some of the Lycoming series engines can be equipped with high compression heads, which increases the horsepower output.

IO-390 Series. The Lycoming IO-390 series engines are four-cylinder, direct-drive, horizontally opposed, air-cooled models. The engines are equipped with a fuel-injection system that schedules fuel flow in proportion to airflow. Fuel vaporization takes place at the intake ports. Implementing new technology in cylinder design proven by the performance of the 580 engine series to increase the displacement to 390 in^3, this model produces 210 hp at 2700 rpm and consumes 11.1 gallons per hour at 65 percent power. Designed to meet the growing demand for kit aircraft, the engine provides the required speed, payload, and low-fuel consumption.

IO-580 Series. The Lycoming IO-580 series engines are six-cylinder, direct-drive horizontally opposed, air-cooled models, see Fig. 1-19. The cylinders are of conventional

air-cooled construction with heads made from an aluminum-alloy casting and a fully machined combustion chamber. The engines are equipped with a fuel-injection system. The fuel injector meters fuel in proportion to induction airflow to air-bled nozzles at individual cylinder intake ports. Manual mixture control and idle cutoff are provided. This engine has a bore of 5.319 in, a stroke of 4.375 in and a piston displacement of 583 in^3. The IO-580 engine can be seen in Fig. 1-19.

IO-720 Series. The Lycoming IO-720 series engines, see Fig. 1-20, are eight-cylinder, direct-drive, horizontally opposed, air-cooled models. The cylinders are of conventional air-cooled construction with heads made from an aluminum-alloy casting and a fully machined combustion chamber. The engines are equipped with a fuel-injection system that schedules fuel flow in proportion to airflow. Fuel vaporization takes place at the intake ports.

Lycoming Integrated Electronic Engine

The Integrated Electronic Engine (IEE or iE^2) is shown in Fig. 1-21. The iE^2 electronics have been "integrated" throughout

FIGURE 1-15   Continental Motors 400 series engine.

FIGURE 1-16   Continental Motors 500 series engine.

Engine Design and Classification   **11**

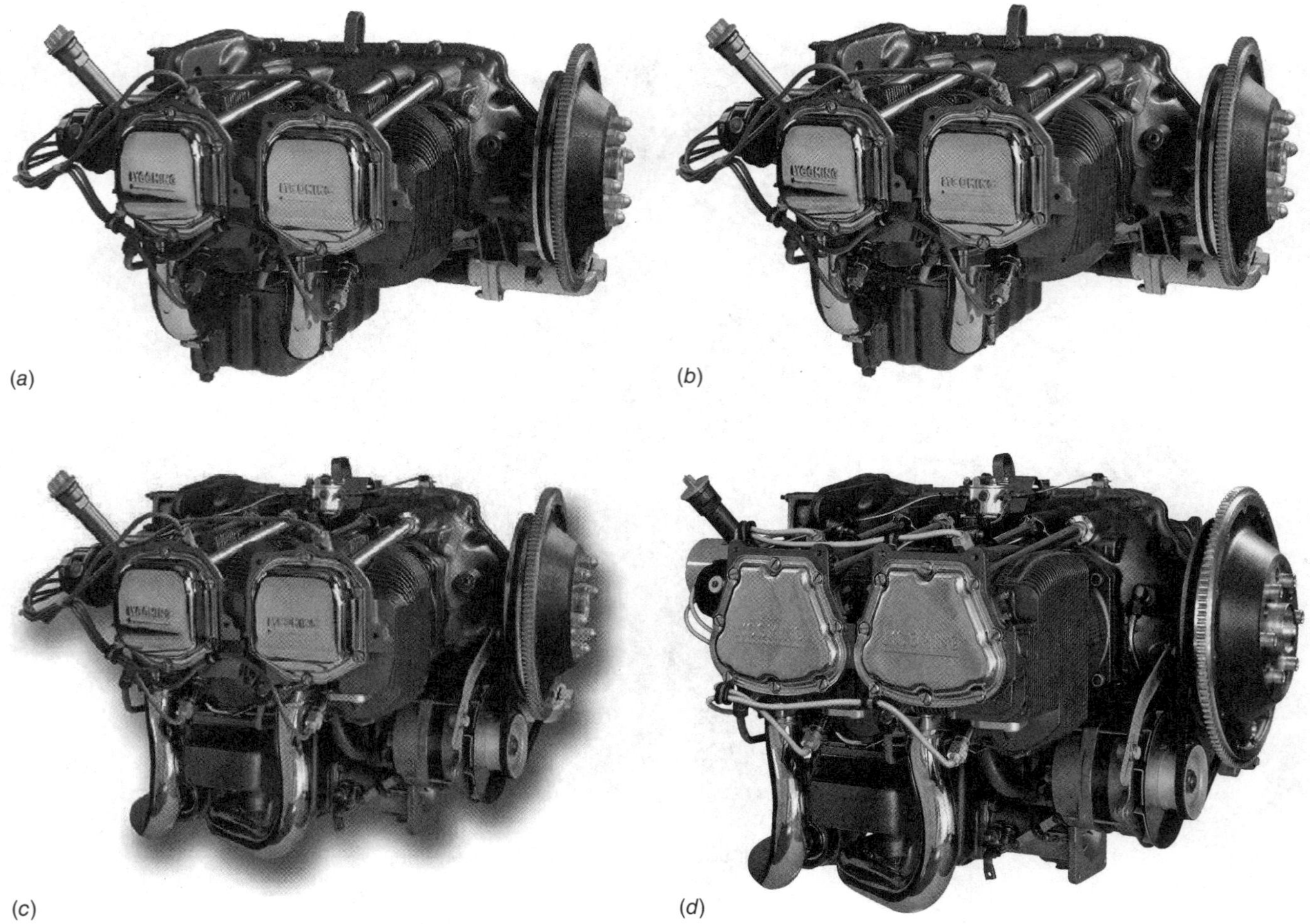

FIGURE 1-17 Four-cylinder Lycoming engines: (*a*) 0-235, (*b*) 0-320, (*c*) 0-360, (*d*) 0-390.

FIGURE 1-18 Six-cylinder IO-540 Lycoming engine.

FIGURE 1-19 Six-cylinder IO-580 Lycoming engine fuel injected.

FIGURE 1-20 IO-720 Lycoming eight-cylinder opposed engine.

FIGURE 1-21 Lycoming six-cylinder iE² engine.

the engine to optimize the systems operation, weight, and packaging. As a result, you have the benefit of an engine system that offers improved simplicity and reliability without sacrificing payload or performance.

In flight, the pilot can focus on flying the plane rather than managing the engine. Whether in takeoff, climb, or cruise, fuel leaning is automatic and optimized for each scenario. Monitoring CHTs, EGTs, and TITs is now a thing of the past as the engine condition is now managed electronically. Pilots need to set the desired power level to effectively control the engine.

Engine data recording capability gives the mechanic insights into engine operation that were never available before. Approved technicians with the proper tools and training can quickly review engine data to make sure your engine is operating at peak performance.

The most demanding aviation environments were the primary considerations throughout the process of developing the Lycoming iE² series engines. Advanced computer logic allows key engine parameters to have double and even triple redundancy without the weight and cost of extra sensors. The automated preflight check evaluates system components and signals the pilot if any anomalies exist prior to takeoff. During flight, the reduced pilot workload allows the pilot to focus on flying the plane and maintaining situational awareness rather than managing the engine. Electronic knock detection allows the engine to automatically adjust to prevent damage caused by engine detonation. Integrated electrical power generation allows the engine to maintain its own power supply to ensure that the engine will be able to run regardless of the airframe's power condition. The system even includes improved fuel consumption calculations to allow better estimates of fuel usage. In the event a problem is detected, the engine system alerts the pilot and keeps a record of the problem, allowing the technicians to quickly identify and remedy the issue.

System Operation

The first advantage of operation is the simplicity of the iE² single-lever engine controls. Mixture and propeller controls are now managed electronically, eliminating the need for additional cockpit levers. The engine starts easily and reliably, hot or cold, with a single button, bringing to mind today's modern automobiles. Engine preflight is automatic and starts with the push of a button. Within seconds, engine operations are checked then rechecked in the dual redundant system with the pilot receiving the "green light" when the sequence is successfully completed.

Conventional magneto and propeller control checks are now a "push of a button" process—set the engine to the appropriate rpm and push the "Preflight Test" button. A "Preflight Test" lamp will illuminate to let you know the process is underway. As the test progresses, you will notice variations in engine rpm and prop pitch as the system automatically checks the ignition, fuel, and turbo systems, as well as the propeller control system. The iE² electronics are designed to constantly monitor every sensor and actuator within the system. Each sensor and actuator is monitored for proper operation and even "cross-checked" with other sensors to verify accuracy. The data provided by each sensor is also monitored to identify potential issues that may have developed in other parts of the engine or aircraft. In the event a problem is found, the pilot is notified through one of the in-dash warning lamps. An electronic fault code is then stored in the system memory where it can be retrieved by a Lycoming Authorized Service Center to help in the repair process. The iE² engine uses a separate data logger to record a 30-minute detailed record of as many as 45 different engine parameters as well as a historic data trend of each parameter for the life of the engine. This data is available to mechanics with the appropriate training and equipment, and it can be used to greatly simplify the service and maintenance of the engine.

Types of Light-Sport and Experimental Engines

Note: All information in this text is for educational illustrational purposes and is not to be used for actual aircraft maintenance. This information is not revised at the same rate as the maintenance manual; always refer to the current maintenance information when performing maintenance on any engine.

Light-Sport Aircraft Engines

Light-sport/ultralight aircraft engines can be classified by several methods, such as by operating cycles, cylinder arrangement, and air- or water-cooled. An in-line engine generally has two cylinders, is two-cycle, and is available in several horsepower ranges. These engines may be either liquid-cooled, air-cooled, or a combination of both. They have only one crankshaft that drives the reduction gearbox or propeller directly. Most of the other cylinder configurations used are horizontally opposed, ranging from two to six cylinders from several manufacturers. These engines are either gear reduction or direct drive.

Two-Cycle, Two-Cylinder Rotax Engine Single Capacitor Discharge Ignition (SCDI) Dual Capacitor Discharge Ignition (DCDI)

Rotax 447 UL (SCDI) and Rotax 503 UL (DCDI). The Rotax in-line cylinder arrangement has a small frontal area and provides improved streamlining (Fig. 1-22). The two-cylinder, in-line two-stroke engine, which is piston ported with air-cooled cylinder heads and cylinders, is available in a fan or free air-cooled version. Being a two-stroke cycle

engine, the oil and fuel must be mixed in the fuel tank on some models. Other models use a lubrication system, such as the 503 oil injection lubrication system. This system does not mix the fuel and oil as the oil is stored in a separate tank.

As the engine needs lubrication, the oil is injected directly from this tank. The typical ignition system is a breakerless ignition system with a dual ignition system used on the 503, and a single ignition system used on the 447 engine series. Both systems are of a magneto capacitor discharge design.

The engine is equipped with a carburetion system with one or two piston-type carburetors. One pneumatic driven fuel pump delivers the fuel to the carburetors. The propeller is driven via a flange connected gearbox with an incorporated shock absorber. The exhaust system collects the exhaust gases and directs them overboard. These engines come with an integrated alternating current (AC) generator (12 V 170 W) with external rectifier-regulator as an optional extra.

Rotax 582 UL DCDI. The Rotax 582 is a two-stroke engine, two cylinders in-line with a rotary valve inlet, has liquid-cooled cylinder heads and cylinders that use an integrated water pump (Fig. 1-23). The lubrication system can be a fuel/oil mixture or oil injection lubrication. The ignition system is a dual ignition using a breakerless magneto capacitor discharge design. Dual piston type carburetors and a pneumatic fuel pump deliver the fuel to the cylinders. The propeller is driven via the prop flange connected gearbox with an incorporated torsional vibration shock absorber. This engine also uses a standard version exhaust system with an electric starter or manual rewind starter.

Description of Systems for Two-Stroke Engines

Cooling System of Rotax 447 UL SCDI and Rotax 503 UL DCDI. Two versions of air cooling are available for these engines. The first method is free air cooling, which is a process of engine cooling by an airstream generated

FIGURE 1-22   Rotax in-line cylinder arrangement.

FIGURE 1-23   Rotax 582 engine.

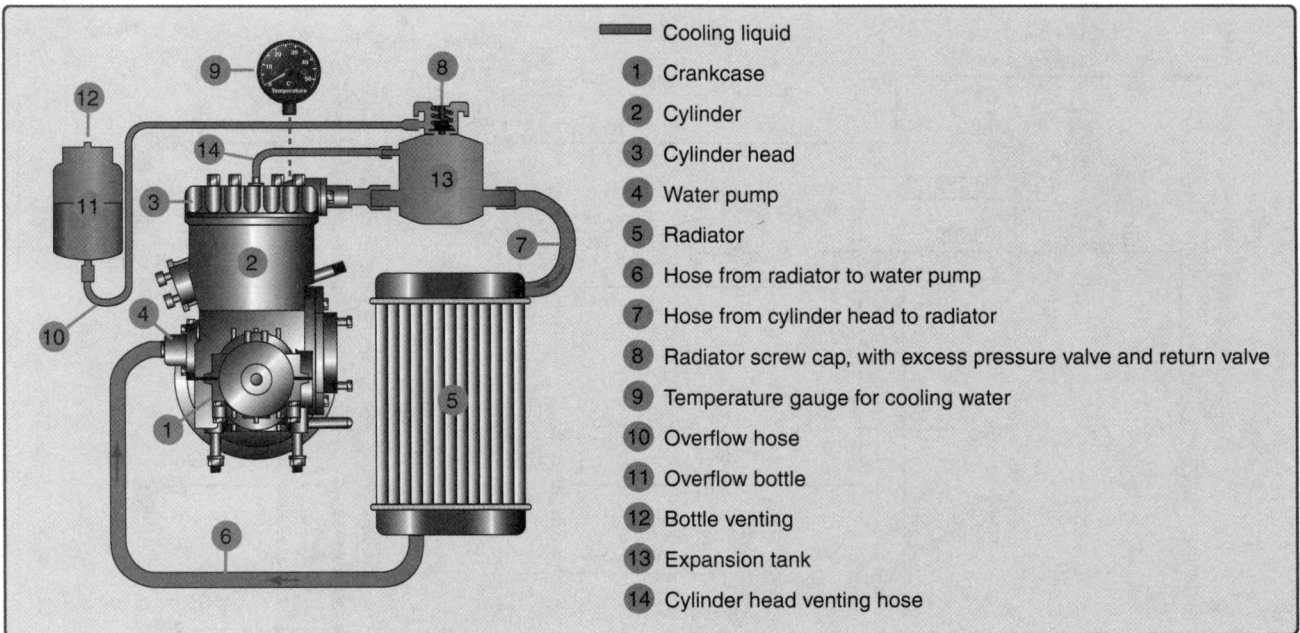

	Cooling liquid
1	Crankcase
2	Cylinder
3	Cylinder head
4	Water pump
5	Radiator
6	Hose from radiator to water pump
7	Hose from cylinder head to radiator
8	Radiator screw cap, with excess pressure valve and return valve
9	Temperature gauge for cooling water
10	Overflow hose
11	Overflow bottle
12	Bottle venting
13	Expansion tank
14	Cylinder head venting hose

FIGURE 1-24   Rotax 582 cooling system.

by aircraft speed and propeller. The second is fan cooling, which is cooling by an airstream generated by a fan permanently driven from the crankshaft via a V-belt.

Cooling System of the Rotax 582 UL DCDI.   Engine cooling for the Rotax 582 is accomplished by liquid-cooled cylinders and cylinder heads (Fig. 1-24). The cooling system is in a two-circuit arrangement. The cooling liquid is supplied by an integrated pump in the engine through the cylinders and the cylinder head to the radiator. The cooling system has to be installed, so that vapor coming from the cylinders and the cylinder head can escape to the top via a hose, either into the water tank of the radiator or to an expansion chamber. The expansion tank is closed by a pressure cap (with excess pressure valve and return valve). As the temperature of the coolant rises, the excess pressure valve opens, and the coolant flows via a hose at atmospheric pressure to the transparent overflow bottle. When cooling down, the coolant is sucked back into the cooling circuit.

Lubrication Systems

Oil Injection Lubrication of Rotax 503 UL DCDI and 582 UL DCDI.   Generally, the smaller two-cycle engines are designed to run on a mixture of gasoline and 2 percent oil that is premixed in the fuel tank. The engines are planned to run on an oil-gasoline mixture of 1:50. Other engines use oil injection systems that use an oil pump driven by the crankshaft via the pump gear that feeds the engine with the correct amount of fresh oil. The oil pump is a piston-type pump with a metering system. Diffuser jets in the intake inject pump supplied two-stroke oil with the exact proportioned quantity needed. The oil quantity is defined by the engine rotations per minute and the oil pump lever position. This lever is actuated via a cable connected to the throttle cable. The oil comes to the pump from an oil tank by gravity.

NOTE: In engines that use oil injection, the carburetors are fed with pure gasoline (no oil/gasoline mixture). The oil quantity in the oil tank must be checked before putting the engine into service as the oil is consumed during operation and needs to be replenished.

Electric System

The 503 UL DCDI and 582 UL DCDI engine types are equipped with a breakerless, SCDI unit with an integrated generator (Fig. 1-25). The 447 UL SCDI engine is equipped with a breakerless, SCDI unit with an integrated generator. The ignition unit is completely free of maintenance and needs no external power supply. Two charging coils are fitted on the generator stator, independent from each other, each feeding one ignition circuit. The energy supplied is stored in the ignition capacitor. At the moment of ignition, the external triggers supply an impulse to the control circuits and the ignition capacitors are discharged via the primary winding of the ignition coil. The secondary winding supplies the high voltage for the ignition spark.

Fuel System

Due to higher lead content in aviation gas (AVGAS), operation can cause wear and deposits in the combustion chamber to increase. Therefore, AVGAS should only be used if problems are encountered with vapor lock or if the other fuel types are not available. Caution must be exercised to use only fuel suitable for the relevant climatic conditions, such as using winter fuel for summer operation.

Fuel/Oil Mixing Procedure.   The following describes the process for fuel/oil mixing. Use a clean approved container of known volume. To help predilute the oil, pour a small amount of fuel into the container. Fill known amount of oil

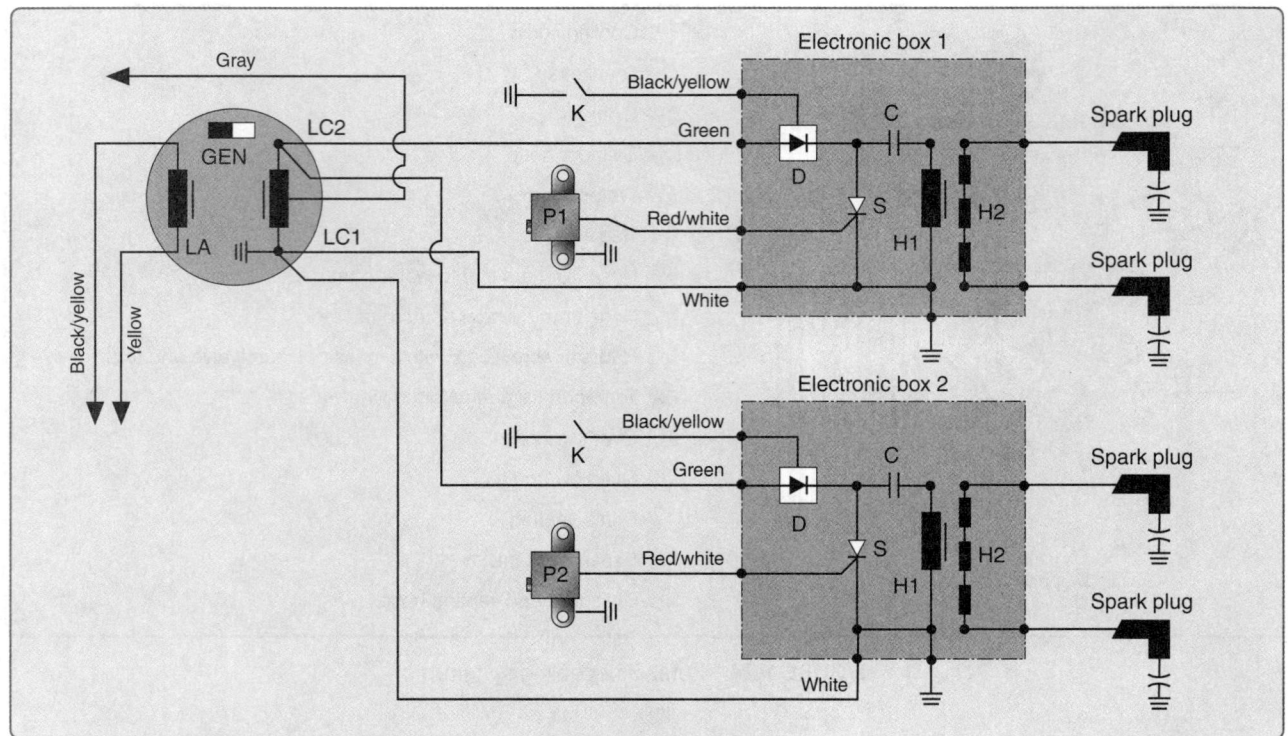

FIGURE 1-25   Rotax 503 and 582 electrical system.

[two-stroke oil ASTM/Coordinating European Council (CEC) standards], API-TC classification (e.g., Castrol TTS) mixing ratio 1:50 (2%), into the container. Oil must be approved for air-cooled engines at 50:1 mixing ratio. Agitate slightly to dilute oil with gasoline. Add gasoline to obtain desired mixture ratio; use a fine mesh screen. Replace the container cap and shake the container thoroughly. Then, using a funnel with a fine mesh screen to prevent the entry of water and foreign particles, transfer the mixture from the container into the fuel tank.

WARNING: To avoid electrostatic charging at refueling, use only metal containers and ground the aircraft in accordance with the grounding specifications.

Opposed Light-Sport, Experimental, and Certified Engines

Many certified engines are used with light-sport and experimental aircraft. Generally, cost is a big factor when considering this type of powerplant. The certified engines tend to be much more costly than the noncertified engines, and are not ASTM approved.

Rotax 912/914

Figure 1-26 shows a typical four-cylinder, four-stroke Rotax horizontally opposed engine. The opposed-type engine has two banks of cylinders directly opposite each other with a crankshaft in the center. The pistons of both cylinder banks are connected to the single crankshaft. The engine cylinder heads are both liquid-cooled and air-cooled; the air cooling is mostly used on the cylinder. It is generally mounted with the cylinders in a horizontal position. The opposed-type engine has a low weight to horsepower ratio, and its

FIGURE 1-26   Typical four-cylinder, four-stroke horizontally opposed engine.

narrow silhouette makes it ideal for horizontal installation on the aircraft wings (twin-engine applications). Another advantage is its low vibration characteristics. It is an ideal replacement for the Rotax 582 two-cylinder, two-stroke engine, which powers many of the existing light aircraft, as it is the same weight as the Rotax 582. These engines are ASTM approved for installation into light-sport category aircraft, with some models being FAA certified engines.

Description of Systems

Cooling System.   The cooling system of the Rotax 914 shown in Fig. 1-27 is designed for liquid-cooling of the

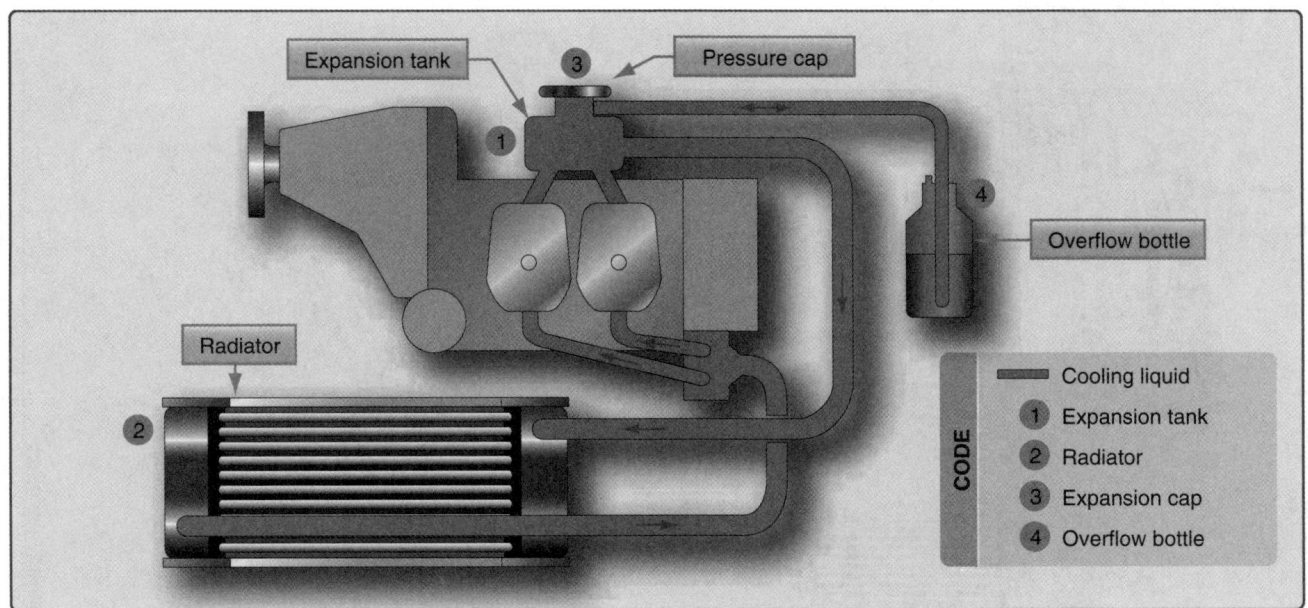

FIGURE 1-27 Rotax 914 cooling system.

FIGURE 1-28 Water-cooled heads.

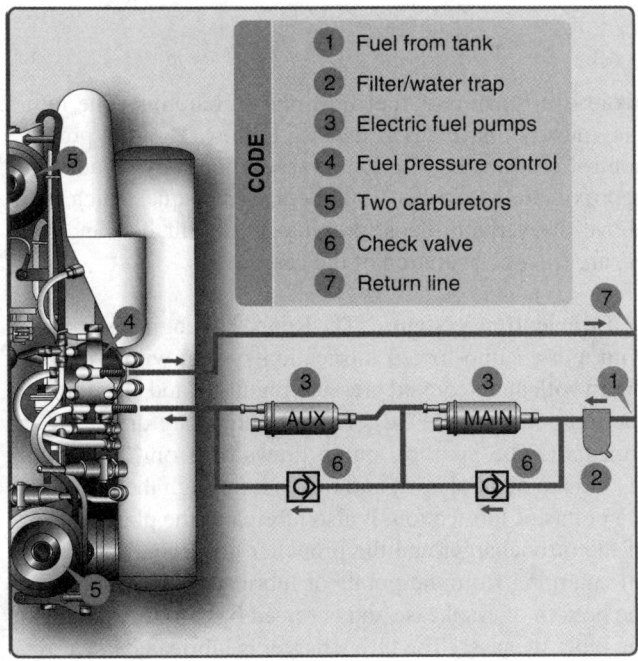

FIGURE 1-29 Fuel system components.

cylinder heads and ram-air cooling of the cylinders. The cooling system of the cylinder heads is a closed circuit with an expansion tank (Fig. 1-28). The coolant flow is forced by a water pump driven from the camshaft, from the radiator, to the cylinder heads. From the top of the cylinder heads, the coolant passes on to the expansion tank (1). Since the standard location of the radiator (2) is below engine level, the expansion tank located on top of the engine allows for coolant expansion.

The expansion tank is closed by a pressure cap (3) (with an excess pressure valve and return valve). As the temperature of the coolant rises, the excess pressure valve opens and the coolant flows via a hose at atmospheric pressure to the transparent overflow bottle (4). When cooling down, the coolant is sucked back into the cooling circuit. Coolant temperatures are measured by means of temperature probes installed in the cylinder heads 2 and 3. The readings

are taken on measuring the hottest point of cylinder head depending on engine installation (Fig. 1-28).

Fuel System. The fuel flows from the tank (1) via a coarse filter/water trap (2) to the two electric fuel pumps (3) connected in series (Fig. 1-29). From the pumps, fuel passes on via the fuel pressure control (4) to the two carburetors (5). Parallel to each fuel pump is a separate check valve (6) installed via the return line (7) that allows surplus fuel to flow back to the fuel tank. Inspection for possible constriction of diameter or obstruction must be accomplished to

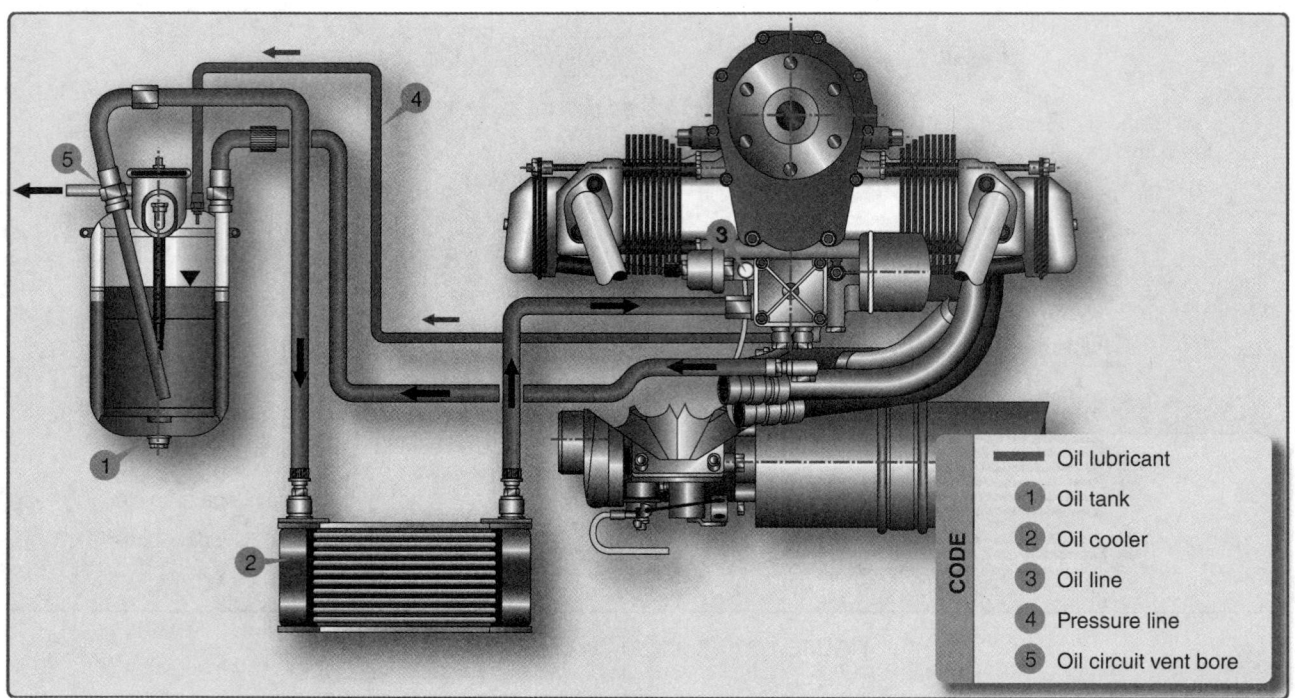

FIGURE 1-30 Lubrication system.

—		Oil lubricant
1		Oil tank
2		Oil cooler
3		Oil line
4		Pressure line
5		Oil circuit vent bore

CODE

avoid overflowing of fuel from the carburetors. The return line must not have any resistance to flow. The fuel pressure control ensures that the fuel pressure is always maintained approximately 0.25 bar [3.63 pounds per square inch (psi)] above the variable boost pressure in the air box and thus, ensures proper operation of the carburetors.

Lubrication System. The Rotax 914 engine is provided with a dry, sump-forced lubrication system with a main oil pump with an integrated pressure regulator and an additional suction pump (Fig. 1-30). The oil pumps are driven by the camshaft. The main oil pump draws oil from the oil tank (1) via the oil cooler (2) and forces it through the oil filter to the points of lubrication. It also lubricates the plain bearings of the turbocharger and the propeller governor. The surplus oil emerging from the points of lubrication accumulates on the bottom of crankcase and is forced back to the oil tank by the blow-by gases. The turbocharger is lubricated via a separate oil line (from the main oil pump). The oil emerging from the lower placed turbocharger collects in the oil sump by a separate pump and is pumped back to the oil tank via the oil line (3). The oil circuit is vented via the bore (5) in the oil tank. There is an oil temperature sensor in the oil pump flange for reading of the oil inlet temperature.

Electric System. The Rotax 914 engine is equipped with a dual ignition unit that uses a breakerless, capacitor discharge design with an integrated generator (Fig. 1-31). The ignition unit is completely free of maintenance and needs no external power supply. Two independent charging coils (1) located on the generator stator supply one ignition circuit each. The energy is stored in capacitors of the electronic modules (2). At the moment of ignition, two each of the four external trigger coils (3) actuate the discharge of the capacitors via the primary circuit of the dual ignition coils (4). The firing order is as follows: 1–4–2–3. The fifth trigger coil (5) is used to provide the revolution counter signal.

Turbocharger and Control System. The Rotax 914 engine is equipped with an exhaust gas turbocharger making use of the energy in the exhaust gas for compression of the intake air or for providing boost pressure to the induction system. The boost pressure in the induction system (air box) is controlled by means of an electronically controlled valve (waste gate) in the exhaust gas turbine. The waste gate regulates the speed of the turbocharger and consequently the boost pressure in the induction system. The required nominal boost pressure in the induction system is determined by the throttle position sensor mounted on the carburetor 2/4. The sensor's transmitted position is linear 0 to 115 percent, corresponding to a throttle position from idle to full power (Fig. 1-32). For correlation between throttle position and nominal boost pressure in the induction, refer to Fig. 1-33. As shown in the diagram, with the throttle position at 108–110 percent results in a rapid rise of nominal boost pressure.

To avoid unstable boost, the throttle should be moved smoothly through this area either to full power (115%) or at a reduced power setting to maximum continuous power. In this range (108–110% throttle position), small changes in throttle position have a big effect on engine performance and speed. These changes are not apparent to the pilot from the throttle lever position. The exact setting for a specific performance is virtually impossible in this range and has to be prevented, as it might cause control fluctuations or surging. Besides the throttle position, overspeeding of the

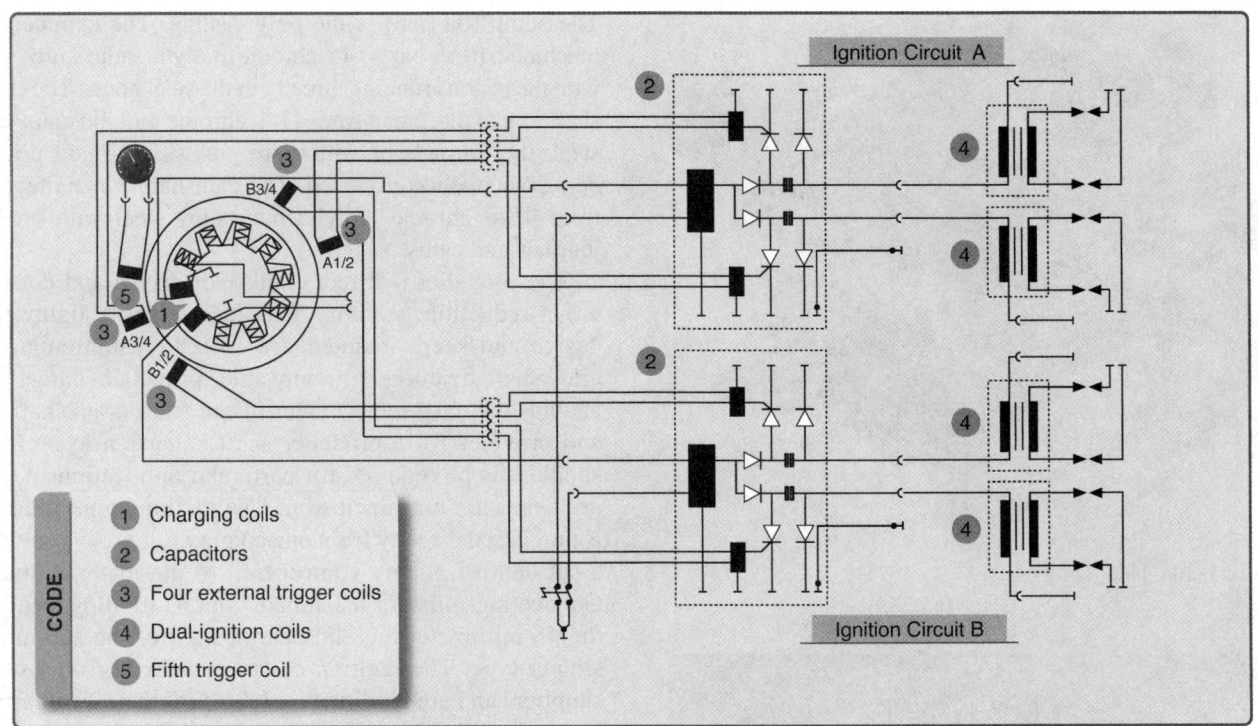

FIGURE 1-31 Electric system.

CODE
1 Charging coils
2 Capacitors
3 Four external trigger coils
4 Dual-ignition coils
5 Fifth trigger coil

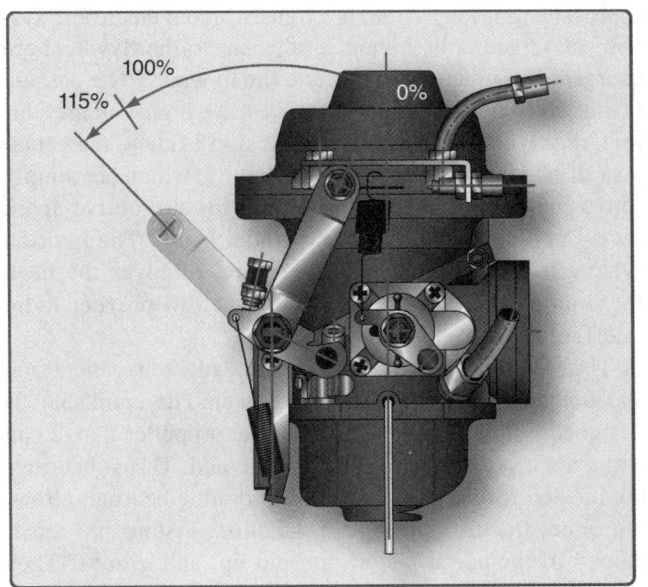

FIGURE 1-32 Turbocharger control system throttle range and position.

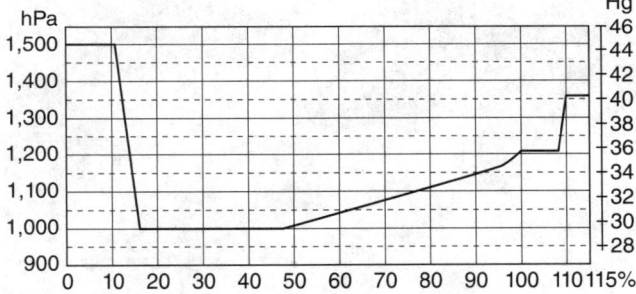

FIGURE 1-33 Correlation between throttle position and nominal boost pressure.

engine and too high intake air temperature have an effect on the nominal boost pressure. If one of the stated factors exceeds the specified limits, the boost pressure is automatically reduced, thus protecting the engine against overboost and detonation.

The turbo control unit (TCU) is furnished with output connections for an external red boost lamp and an orange caution lamp for indications of the functioning of the TCU. When switching on the voltage supply, the two lamps are automatically subject to a function test. Both lamps illuminate for 1 to 2 seconds, then they extinguish. If they do not, a check per the engine maintenance manual is necessary. If the orange caution lamp is not illuminated, then this signals that the TCU is ready for operation. If the lamp is blinking, this indicates a malfunction of the TCU or its periphery systems. Exceeding of the admissible boost pressure activates and illuminates the red boost lamp continuously. The TCU registers the time of full throttle operation (boost pressure). Full throttle operation for longer than 5 minutes, with the red boost light illuminated, makes the red boost lamp start blinking. The red boost lamp helps the pilot to avoid full power operation for longer than 5 minutes or the engine could be subject to thermal and mechanical overstress.

HKS 700T Engine

The HKS 700T engine is a four-stroke, two-cylinder turbocharged engine equipped with an intercooler (Fig. 1-34).

FIGURE 1-34   HKS 700T engine.

FIGURE 1-35   Jabiru engine.

The horizontally opposed cylinders house four valves per cylinder, with a piston displacement of 709 cc. It uses an electronic control fuel injection system. A reduction gearbox is used to drive the propeller flange at a speed reduction ratio of 2.13 to 1. The engine is rated at 77 hp continuous and 80 hp takeoff (3 minutes) at 4900 and 5300 rpm, respectively. A total, engine weight of 126 pounds provides a good power-to-weight ratio. The 700T has a TBO of 500 hours.

Jabiru Light-Sport Engines

Jabiru engines are designed to be manufactured using the latest manufacturing techniques (Fig. 1-35). All Jabiru engines are manufactured, assembled, and ran on a Dynometer, then calibrated before delivery. The crankcase halves, cylinder heads, crankshaft, starter motor housings, gearbox cover (the gearbox powers the distributor rotors), together with many smaller components are machined from solid material.

The sump (oil pan) is the only casting. The cylinders are machined from bar 4140 chrome molybdenum-alloy steel, with the pistons running directly in the steel bores. The crankshaft is also machined from 4140 chrome molybdenum-alloy steel, the journals of which are precision ground prior to being Magnaflux inspected. The camshaft is manufactured from 4140 chrome molybdenum-alloy steel with nitrided journals and cams.

The propeller is direct crankshaft driven and does not use a reduction gearbox. This facilitates its lightweight design and keeps maintenance costs to a minimum. The crankshaft features a removable propeller flange that enables the easy replacement of the front crankshaft seal and provides for a propeller shaft extension to be fitted, should this be required for particular applications. Cylinder heads are machined from a solid aluminum billet that is purchased directly from one company, thereby providing a substantive quality control trail to the material source. Connecting rods are machined from 4140 alloy steel and the 45-millimeter big end bearings are of the automotive slipper type. The ignition coils are sourced from outside suppliers and are modified by Jabiru for their own particular application.

An integral alternator provides AC rectification for battery charging and electrical accessories. The alternator is attached to the flywheel and is driven directly by the crankshaft. The ignition system is a transistorized electronic system; two fixed coils mounted adjacent to the flywheel are energized by magnets attached to the flywheel. The passing of the coils by the magnets creates the high-voltage current, that is transmitted by high-tension leads to the center post of two automotive type distributors, which are simply rotors and caps, before distribution to automotive spark plugs (two in the top of each cylinder head). The ignition system is fixed timing and, therefore, removes the need for timing adjustment. It is suppressed to prevent radio interference.

The ignition system is fully redundant, self-generating, and does not depend on battery power. The crankshaft is designed with a double bearing at the propeller flange end and a main bearing between each big end. Thrust bearings are located fore and aft of the front double bearing, allowing either tractor or pusher installation. Pistons are remachined to include a piston pin, circlip, and groove. They are all fitted with three rings, the top rings being cast iron to complement the chrome molybdenum cylinder bores. Valves are 7 mm (stem diameter) and are manufactured specifically for the Jabiru engine. The valve drive train includes pushrods from the camshaft followers to valve rockers. The valves are Computer Numerical Control (CNC) machined from steel billet, induction hardened, polished on contact surfaces, and mounted on a shaft through Teflon coated bronze-steel bush. Valve guides are manufactured from aluminum/bronze. Replaceable valve seats are of nickel steel and are shrunk into the aluminum cylinder heads. The valve train is lubricated from the oil gallery. Engines use hydraulic lifters that automatically adjust valve clearance. An internal gear pump is driven directly by the camshaft and provides engine lubrication via an oil circuit that includes an automotive spin-on filter, oil cooler, and built-in relief valve.

The standard engines are supplied with two ram-air cooling ducts, that have been developed by Jabiru to facilitate the cooling of the engine by directing air from the propeller to the critical areas of the engine, particularly the cylinder heads and barrels. The use of these ducts removes the need to design and manufacture baffles and the establishment of a plenum chamber, which is the traditional method of cooling air-cooled, aircraft engines. The fact that these baffles and plenum chamber are not required also ensures a cleaner engine installation, which in turn facilitates maintenance and inspection of the engine and engine components.

The engine is fitted with a 1.5 kW starter motor that is also manufactured by Jabiru and provides very effective starting. The engine has a very low vibration level; however, it is also supported by four large rubber shock mounts attached to the engine mounts at the rear of the engine. The fuel induction system uses a pressure compensating carburetor. Following the carburetor, the fuel/air mixture is drawn through a swept plenum chamber bolted to the sump casting, in which the mixture is warmed prior to entering short induction tubes attached to the cylinder heads.

An effective stainless-steel exhaust and muffler system is fitted as standard equipment ensuring very quiet operations. For owners wanting to fit vacuum instruments to their aircraft, the Jabiru engines are designed with a vacuum pump drive direct mounted through a coupling on the rear of the crankshaft.

Jabiru 2200 Aircraft Engine. The Jabiru 2200 cc aircraft engine is a four-cylinder, four-stroke horizontally opposed air-cooled engine. At 132 pounds [60 kgs] installed weight, it is one of the lightest four-cylinder, four-stroke aircraft engines. Small overall dimensions give it a small frontal area width (23.46 in, 596 mm) that makes it a good engine for tractor applications. The Jabiru engine is designed for either tractor or pusher installation. The Jabiru engine specifications are listed in Fig. 1-36.

The Jabiru 3300 (120 hp) engine features (Fig. 1-37):

- 4-stroke
- 3300 cc engine (200 in³)
- 6-cylinder horizontally opposed
- 1 central camshaft
- Fully machined aluminum-alloy crankcase
- Overhead valves (OHV)—pushrod operated
- Ram-air cooled
- Wet sump lubrication—4 liter capacity
- Direct propeller drive
- Dual transistorized magneto ignition
- Integrated AC generator
- Electric starter
- Mechanical fuel pump
- Naturally aspirated—1 pressure compensation carburetor

Aeromax Aviation 100 (IFB) Aircraft Engine. Aeromax Aviation produces a version of a 100-hp engine. The engine features a special made integral front bearing (Fig. 1-38). The engine uses an integral permanent magnet 35-amp alternator, lightweight starter, and dual ignition. The compact alternator and starter allow for a streamlined and aerodynamic cowl which improves the fuel efficiency of an experimental aircraft. The Aeromax aircraft engine is an opposed six-cylinder, air-cooled, and direct drive. Being a six-cylinder engine, it has smooth operation. The Aeromax engines are known for their heat dissipation qualities, provided the proper amount of cooling air is provided.

It features a crank extension supported by a massive integral front bearing (IFB) and bearing housing. These engines start out as a GM Corvair automobile core engine. These basic core engines are disassembled and each component that is reused is refurbished and remanufactured. The crankshaft in the Areomax 100 IFB aircraft engine is thoroughly inspected, including a magnaflux inspection. After ensuring the crank is free of any defects, it is extended by mounting the crank extension hub on its front. Then, the crank is ground true, with all five bearing surfaces (four original and the new-extended crank's front bearing), being true to each other and perpendicular to the crank's prop flange (Fig. 1-39).

All radiuses are smooth with no sharp corners where stress could concentrate. Every crankshaft is nitrated, which is a heat/chemical process that hardens the crank surfaces. The crank reinforcement coupled with the IFB is required to counter the additional dynamic and bending loads introduced on the crank in an aircraft application. The engine case is totally refurbished and checked for wear. Any studs or bolts that show wear are replaced. The engine heads are machined to proper specifications and all new valves, guides, and valve train components are installed. A three-angle valve grind and lapping ensure a good valve seal.

Once the engine is assembled, it is installed on a test stand, prelubricated, and inspected. The engine is, then, run several times for a total of 2 hours. The engine is carefully inspected after each run to ensure that it is in excellent operating condition. At the end of test running the engine, the oil filter is removed and cut for inspection. Its internal condition is recorded. This process is documented and kept on file for each individual engine. Once the engine's proper performance is assured, it is removed and packaged in a custom built crate for shipping. Each engine is shipped with its engine service and operations manual. This manual contains information pertaining to installation, break-in, testing, tune-up, troubleshooting, repair, and inspection procedures. The specifications for the Aeromax 100 engine are outlined in Fig. 1-40.

Direct Drive VW Engines

Revmaster R-2300 Engine

The Revmaster R-2300 engine maintains Revmaster's systems and parts, including its RM-049 heads that feature large

Specifications: Jabiru 2200 cc 85 HP Aircraft Engine	
Engine Features	Four stroke
	Four-cylinder horizontally opposed
Opposed	One central camshaft
	Pushrods
	Overhead valves (OHVs)
(OHV)	Ram-air cooled
	Wet sump lubrication
	Direct propeller drive
	Dual transistorized magneto ignition
Magneto Ignition	Integrated AC generator 20 amp
Generator 20 Amp	Electric starter
	Mechanical fuel pump
	Naturally aspirated - 1 pressure compensating carburetor
Pressure Compensating Carburetor	Six-bearing crankshaft
Displacement	2200 cc (134 cu. in.)
Bore	97.5 mm
Stroke	74 mm
Compression Ratio	8:1
Directional Rotation of Prop Shaft	Clockwise - pilot's view tractor applications
Ramp Weight	132 lb complete including exhaust, carburetor, starter motor, alternator, and ignition system
Ignition Timing	25° BTDC
Firing Order	1–3–2–4
Power Rating	85 hp @ 3300 rpm
Fuel Consumption at 75% Power	4 US gal/h
Fuel	AVGAS 100 LL or autogas 91 octane minimum
Oil	Aeroshell W100 or equivalent
Oil Capacity	2.3 quarts
Spark Plugs	NGK D9EA - automotive

FIGURE 1-36 Jabiru 2200 cc specifications.

fins and a hemispherical combustion chamber (Fig. 1-41). It maintains the earlier R-2200 engine's top horsepower (82) at 2950 rpm continuous (Fig. 1-42). Takeoff power is rated at 85 to 3350 rpm. The additional power comes from a bore of 94 mm plus lengthening of the R-2200's connecting rods, plus increasing the stroke from 78 to 84 mm. The longer stroke results in more displacement, and longer connecting rods yield better vibration and power characteristics. The lower cruise rpm allows the use of longer propellers, and the higher peak horsepower can be felt in shorter takeoffs and steeper climbs.

The Revmaster's four main bearing crankshaft runs on a 60-mm center main bearing, is forged from 4340 steel, and uses nitrided journals. Thrust is handled by the 55-mm #3 bearing at the propeller end of the crank. Fully utilizing its robust #4 main bearing, the Revmaster crank has built-in oil-controlled propeller capability, a feature unique in this horsepower range; nonwood props are usable with these engines.

Moving from the crankcase and main bearings, the cylinders are made by using centrifugally cast chilled iron. The pistons are forged out of high-quality aluminum-alloy,

Jabiru 3300 cc Aircraft Engine	
Displacement	3300 cc (202 cu. in.)
Bore	97.5 mm (3.838")
Stroke	74 mm (2.913")
Aircraft Engine	Jabiru 3300 cc 120 hp
Compression Ratio	8:1
Directional Rotation of Prop Shaft	Clockwise - Pilot's view tractor applications
Ramp Weight	178 lb (81 kg) complete including exhaust, carburetor, starter motor, alternator, and ignition system
Ignition Timing	25° BTDC fixed timing
Firing Order	1–4–5–2–3–6
Power Rating	120 hp @ 3300 rpm
Fuel Consumption at 75% Power	26 l/hr (6.87 US gal/h)
Fuel	AVGAS 100 LL or autogas 91 octane minimum
Oil	Aeroshell W100 or equivalent
Oil Capacity	3.5l (3.69 quarts)
Spark Plugs	NGK D9EA - automotive

FIGURE 1-37   Jabiru 3300 cc aircraft engine.

FIGURE 1-38   Aeromax direct-drive, air-cooled, six-cylinder engine.

FIGURE 1-39   Front-end bearing on the 1000 IFB engine.

machined and balanced in a set of four. There are two sizes of pistons, 92 mm and 94 mm, designed to be compatible with a 78- to 82-mm stroke crankshafts. The cylinder set also contains piston rings, wrist pins, and locks. The direct-drive R-2300 uses a dual CDI ignition with eight coil spark to eight spark plugs, dual 20-amp alternators, oil cooler, and its proprietary Rev-Flo carburetor, while introducing the longer cylinders that do not require spacers. The automotive-based bearings, valves, valve springs, and piston rings (among others) make rebuilds easy and inexpensive.

Great Plains Aircraft Volkswagen (VW) Conversions

Great Plains Aircraft is one company that offers several configurations of the Volkswagen (VW) aircraft engine conversion. One very popular model is the front-drive long block kits that offer a four-cycle, four-cylinder opposed engine with horsepower ranges from approximately 60–100 (Fig. 1-43). The long block engine kits, which are the complete engine kits that are assembled, in the field or can be shipped completely assembled, are available from 1600 cc up through 2276 cc. All the engine kits are built from proven time-tested

Engine Design and Classification **23**

Aeromax 100 Engine Specifications	
Power Output: 100 hp continuous at 3200 rpm	Air-cooled
Displacement: 2.7 L	Six cylinders
Compression: 9:1	Dual ignition–single plug
Weight: 210 lb	Normally aspirated
Direct Drive	CHT max: 475 F
Rear Lightweight. Starter and 45 amp alternator	New forged pistons
Counterclockwise rotation	Balanced and nitrated crankshaft
Harmonic balancer	New hydraulic lifters
Remanufactured case	New main/rod bearings
Remanufactured heads with new guides, valves, valve train, intake	New all replaceable parts
Remanufactured cylinders	New spark plug wiring harness
New lightweight aluminum cylinder - optional	Remanufactured dual-ignition distributor with new points set and electronic module
New high-torque cam	New oil pump
New CNC prop hub and safety shaft	New oil pan
New Aeromax top cover and data plate	Engine service manual

FIGURE 1-40   Aeromax 100 engine specifications.

FIGURE 1-41   Revmaster R-2300 engine.

FIGURE 1-43   Great Plain's Volkswagen conversion.

FIGURE 1-42   Hemispherical combustion chamber within the Revmaster R-2300 heads.

components and are shipped with a Type 1 VW Engine Assembly Manual. This manual was written by the manufacturer, specifically for the assembly of their engine kits. Also included are how to determine service and maintenance procedures and many tips on how to set up and operate the engine correctly. The crankshaft used in the 2180 to 2276 cc engines is an 82-mm crankshaft made from a forged billet of E4340 steel, machined and magnafluxed twice. The end of the crankshaft features a $\frac{1}{2}$-inch fine thread versus a 20-mm thread found on the standard automotive crank.

Teledyne Continental 0-200 Engine

The 0-200 Series engine has become a popular engine for use in light-sport aircraft. The 0-200-A/B is a four-cylinder, carbureted engine producing 100 bhp and has a

FIGURE 1-44   0-200 Continental engine.

four integral rear engine mounts. A crankcase breather port is located on the 1-3 side of the crankcase forward of the #3 cylinder. The engine lubrication system is a wet sump, high-pressure oil system. The engine lubrication system includes the internal engine-driven pressure oil pump, oil pressure relief valve, pressure oil screen mounted on the rear of the accessory case and pressure instrumentation. A fitting is provided at the 1-3 side of the crankcase for oil pressure measurement. The oil sump capacity is 6 quarts maximum. The 0-200-A/B induction system consists of an updraft intake manifold with the air intake and throttle mounted below the engine. Engine manifold pressure is measured at a port located on the 2-4 side of the intake air manifold. The 0-200-A/B is equipped with a carburetor that meters fuel flow as the flight-deck throttle and mixture controls are changed.

crankshaft speed of 2750 rpm (Fig. 1-44). The engine has horizontally opposed air-cooled cylinders. The engine cylinders have an overhead valve design with updraft intake inlets and downdraft exhaust outlets mounted on the bottom of the cylinder. The 0-200-A/B engines have a 201 cubic inch displacement achieved by using a cylinder design with a 4.06-inch diameter bore and a 3.88-inch stroke. The dry weight of the engine is 170.18 pounds without accessories. The weight of the engine with installed accessories is approximately 215 pounds. The engine is provided with

Lycoming 0-233 Series Light-Sport Aircraft Engine

Lycoming Engines, a Textron Inc. company, produces an experimental noncertified version of its 233 Series light-sport aircraft engine (Fig. 1-45). The engine is light and capable of running on unleaded automotive fuels, as well as AVGAS. The engine features dual CDI spark ignition, an optimized oil sump, a streamlined accessory housing, hydraulically adjusted tappets, a lightweight starter, and a lightweight alternator with integral voltage regulator. It has a dry weight of 213 pounds (including the fuel pump) and offers continuous power ratings up to 115 hp at 2800 rpm. In addition to

FIGURE 1-45   Lycoming 0-233 engine.

its multi-gasoline fuel capability, it has proven to be very reliable with a TBO of 2400 hours. The initial standard version of the engine is carbureted, but fuel-injected configurations of the engine are also available.

The Porsche PFM 3200 Aircraft Engine

Porsche has received Federal Aviation Administration (FAA) certification for its PFM 3200 engine, which is being supplemental type certificated for use in general aviation fixed- and rotary-wing aircraft in the 200- to 300-hp [149.14- to 223.71-kW] range. The engine is currently being produced in two versions, a 209-hp [155.85-kW] unit designed to burn autogas and avgas interchangeably and a 217-hp [161.82-kW] model for avgas only. Higher-horsepower versions are under development. The Porsche PFM 3200 engine is illustrated in Fig. 1-46.

The PFM 3200 engine is a derivative of the Porsche 911 sports-car engine. The 911 engine is a lightweight, six-cylinder horizontally opposed air-cooled engine. The 3200 engine incorporates many technologically advanced concepts such as a **single-power lever** that replaces the throttle; mixture, and propeller controls. The pilot operates the power lever much as the driver of a car operates the accelerator pedal. When the power lever is set, the fuel mixture and propeller speed adjust automatically to ensure maximum efficiency at all altitudes and power settings. The engine has a dual-ignition system with **variable ignition timing**. The correct timing is determined by the running speed, intake air temperature, and absolute intake manifold pressure. An automatic fuel injection system provides the correct air-fuel mixture at all altitudes. The engine is cooled by an engine-driven fan that provides cooling air in direct proportion to the power output.

FIGURE 1-46   Porsche PFM 3200 engine.

Jet Propulsion History

During World War II, the demand for increased speed and power expedited the progress which was already taking place in the development of jet propulsion powerplants. As a result of the impetus given by the requirements of the War and Navy departments in the United States and by similar demands on the part of Great Britain, engineers in England and the United States designed, manufactured, and tested in flight an amazing variety of jet propulsion powerplants. Note also that the German government was not trailing behind in the jet propulsion race, because the first flight by an airplane powered with a true jet engine was made in Germany on August 27, 1939. The airplane was a Heinkel He 178 and was powered by a Heinkel HeS 3B turbojet engine.

The first practical turbojet engines in England and the United States evolved from the work of Sir Frank Whittle in England. The success of experiments with this engine led to the development and manufacture of the Whittle W1 engine. Under agreement with the British government, the United States was authorized to manufacture an engine of a design similar to the W1. The General Electric Company was given a contract to build the engine because of its extensive experience with turbine manufacture and with the development of turbosuperchargers used for military aircraft in World War II. Accordingly, the General Electric GE I-A engine was built and successfully flown in a Bell XP-59A airplane.

The successful development of jet propulsion is beyond question the greatest single advance in the history of aviation since the Wright brothers made their first flight. The speed of aircraft has increased from below Mach 1, the speed of sound, to speeds of more than Mach 4. Commercial airliners now operate at more than 600 miles per hour (mph) [965.6 kilometers per hour (km/h)] rather than at speeds of around 350 mph [563.26 km/h], which is usually about the maximum for conventional propeller-driven airliners.

Gas-Turbine Engine Progress

The gas-turbine engine can be used as a **turbojet engine** (thrust developed by exhaust gases alone), a **turbofan engine** (thrust developed by a combination of exhaust gases and fan air), a **turboprop engine** (in which a gas-turbine engine turns a gearbox for driving propellers), and a **turboshaft engine** (which drives helicopter rotors). Small gas-turbine engines called **APUs** (auxiliary power units) have been developed to supply transport-category aircraft with electrical and pneumatic power. Although these units can be used both on the ground and in the air, they are mainly used on the ground.

Gas-turbine engines are used for propulsion in many different types of aircraft. These aircraft include airliners, business aircraft, training aircraft, helicopters, and agricultural aircraft. During the development of the gas-turbine engine, many challenges have faced designers and engineers. Some of the concerns which designers are constantly striving to improve on gas-turbine engines are performance, sound levels, fuel efficiency, ease of maintenance, dependability, and reliability. Many different types of gas-turbine engines have been developed to meet the overall propulsion needs

FIGURE 1-47 GE90 high bypass propulsion system. (*General Electric.*)

of the aviation community. An example of an engine which incorporates many of the learned technologies needed to provide excellent fuel consumption, low emissions, and many of the characteristics mentioned earlier, is the GE90, shown in Fig. 1-47. This engine is a **high bypass turbofan engine** which incorporates a fan with a diameter in excess of 10 ft [3 m] and has a thrust range of 72 000 to over 100 000 lb [32 660 to 43 090 kg] of takeoff thrust.

The largest increase in turbine technology has been the use of electronic engine control systems for full authority engine control. This also allowed for full integration with aircraft monitoring systems. The huge amount of thrust developed by each engine (well over 100 000 lbs of thrust) has allowed for heavier aircraft with fewer engines. The engine electronic control system automatically monitors and controls several separate systems during engine operation. Some examples of these systems are engine starting, relighting following flame-out detection, total control of fuel control as per pilot input, variable inlet guide vane/variable stator vanes position, compressor bleed valves, engine oil/fuel heat management, gearbox cooling airflow, turbine case cooling airflow, high-pressure valve for aircraft cabin air, probe heater control, thrust reverser, and turbine overheat data.

These systems and more will be explained in detail in the chapters on turbine engines later in this text.

REVIEW QUESTIONS

1. List the advantages of the in-line engine.
2. Describe the difference between upright-V-type and inverted-V-type engines.
3. What type of reciprocating engine design provides the best power-to-weight ratio?
4. List the advantages of the opposed-type engine design.
5. Conventional piston engines are classified according to what characteristics?
6. Name four common engine classifications by cylinder arrangement.
7. The letter "O" in an engine designation is used to denote what?
8. The greatest single advance in aircraft propulsion was the development of what type of engine?
9. What does the power lever on the Porsche engine operate?

Reciprocating-Engine Construction and Nomenclature 2

INTRODUCTION

Familiarity with an engine's components and construction is basic to understanding its operating principles and maintenance practices. In the construction of an aircraft reciprocating engine, reliability of the working parts is of primary importance. This need generally requires the use of strong, and at times heavy, materials which can result in a bulky and heavy engine. A major problem in the design of aircraft engine components lies in constructing parts strong enough to be reliable but light enough for use in an aircraft. Moving parts are carefully machined and balanced in an effort to minimize vibration and fatigue. In the construction of an aircraft engine, individual components must be designed and constructed in order to obtain a powerplant which has good reliability, weighs little, and is economical to operate.

This chapter describes the major components and construction principles of an aircraft reciprocating engine.

THE CRANKCASE

The **crankcase** of an engine is the housing that encloses the various mechanisms surrounding the crankshaft; therefore, it is the foundation of the engine. The functions of the crankcase are as follows: (1) it must support itself, (2) it contains the bearings in which the crankshaft revolves, (3) it provides a tight enclosure for the lubricating oil, (4) it supports various internal and external mechanisms of the powerplant, (5) it provides mountings for attachment to the airplane, (6) it provides support for the attachment of the cylinders, and (7) by reason of its strength and rigidity, it prevents the misalignment of the crankshaft and its bearings.

Crankcases come in many sizes and shapes and may be of one-piece or multipiece construction. Most aircraft engine crankcases are made of aluminum alloys because they are both light and strong. Although the variety of crankcase designs makes any attempt at classification difficult, they may be divided into three broad groups for discussion: (1) opposed-engine crankcases, (2) radial-engine crankcases, and (3) in-line and V-type engine crankcases.

Opposed-Engine Crankcase

The crankcase for a four-cylinder opposed engine is shown in Fig. 2-1. This assembly consists of two matching, reinforced aluminum-alloy castings divided vertically at the centerline of the engine and fastened together by means of a series of studs and nuts. The mating surfaces of the crankcase are joined without the use of a gasket, and the main-bearing bores are machined for the use of precision-type main-bearing inserts. Machined mounting pads are incorporated into the crankcase for attaching the accessory housing, cylinders, and oil sump, as shown in Fig. 2-2. Opposed engines with propeller reduction gearing usually incorporate a separate nose section to house the gears.

The crankcase of the opposed engine contains bosses and machined bores to serve as bearings for the camshaft. On the camshaft side of each crankcase half are the tappet bores which carry the hydraulic valve tappet bodies.

Essential portions of the lubricating system are contained in the crankcase. Oil passages and galleries are drilled in the sections of the case to supply the crankshaft bearings, camshaft bearings, and various other moving parts which require lubrication. During overhaul, the technician must make sure that all oil passages are free of foreign matter and that passages are not blocked by gaskets during assembly.

Radial-Engine Crankcase

Radial-engine crankcases (see Fig. 2-3) may have as few as three or as many as seven principal sections, the number depending on the size and type of the engine, although large engines usually have more sections than small ones. For the purpose of describing radial-engine crankcases, it is customary to assume that the typical radial-engine crankcase has four major sections, although this is not necessarily true.

The **front section**, or **nose section**, is usually made of an aluminum alloy, its housing is approximately bell-shaped, and it is fastened to the power section by studs and nuts or cap screws. In most cases, this section supports a propeller thrust bearing, a propeller-governor drive shaft, and a propeller reduction-gear assembly if the engine provides for propeller speed reduction. It may also include an oil scavenge pump and a cam-plate or cam-ring mechanism.

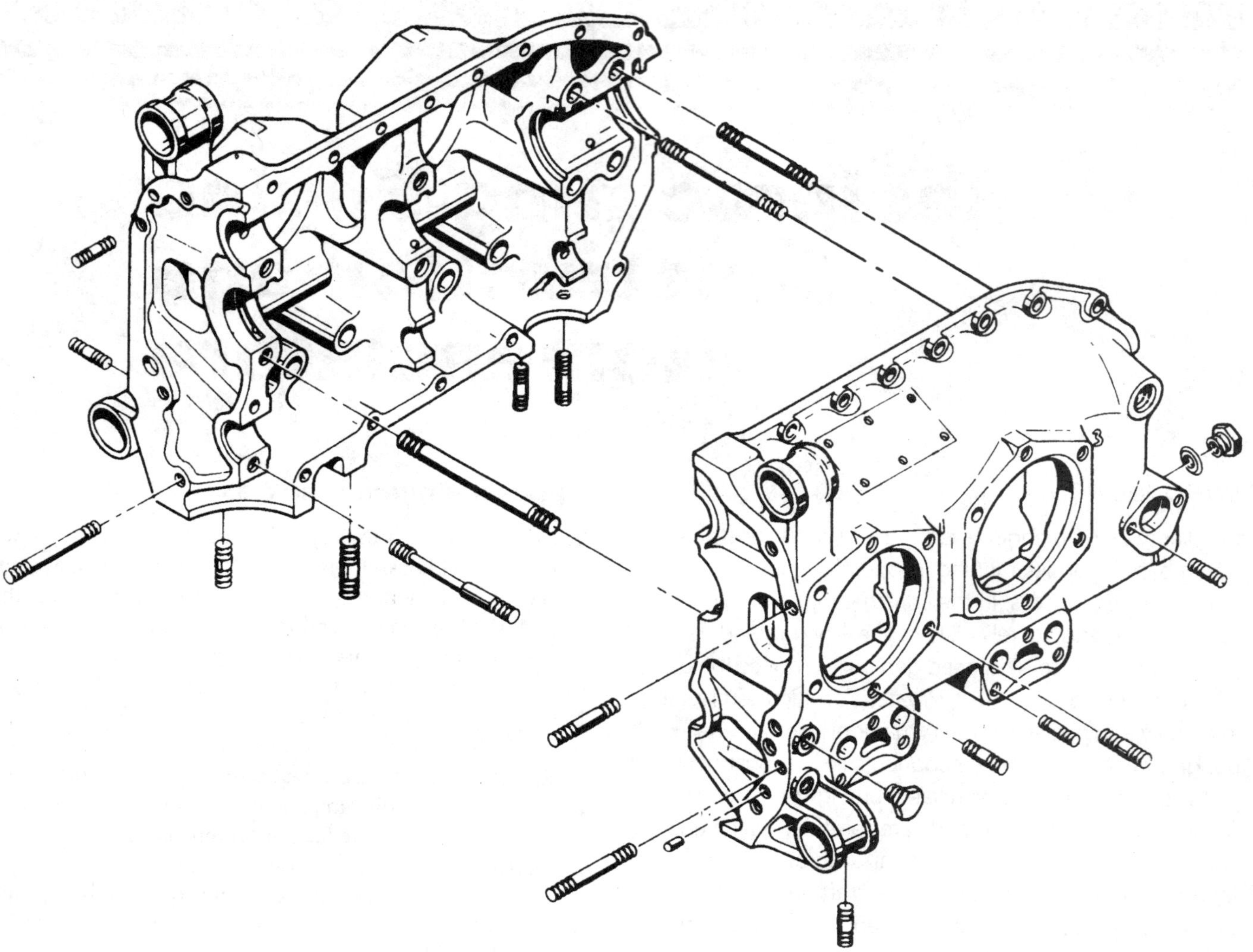

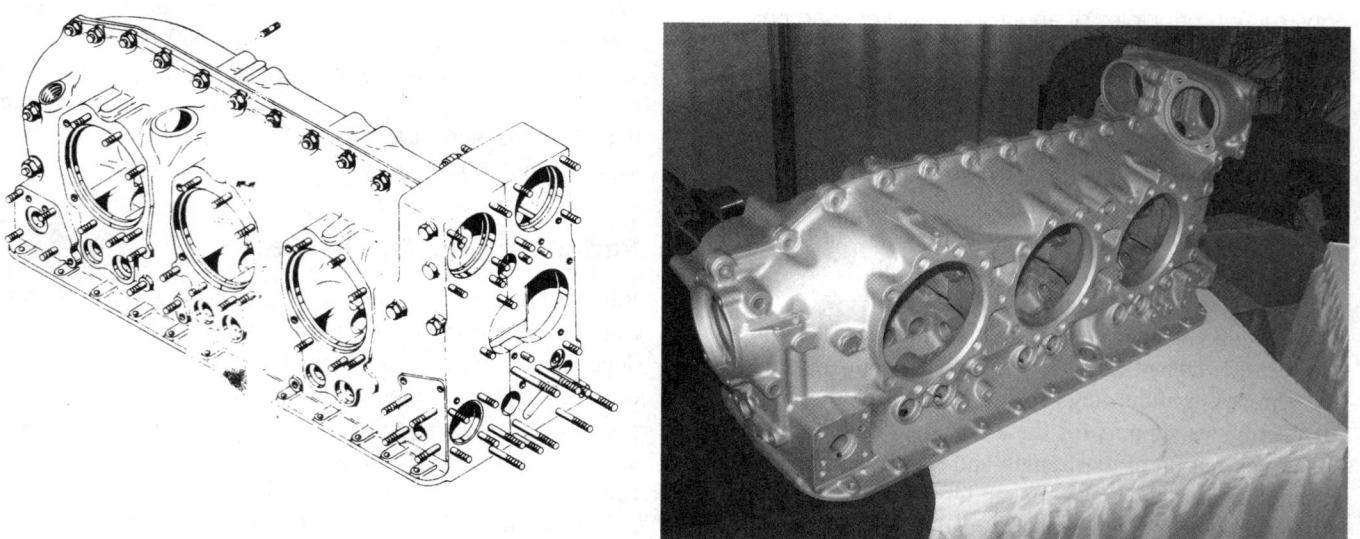

FIGURE 2-1 Crankcase for a four-cylinder opposed engine.

FIGURE 2-2 Crankcase machined mounting pads for a six-cylinder opposed engine.

This section may also provide the mountings for a propeller governor control valve, a crankcase breather, an oil sump, magnetos, and magneto distributors. The engines which have magnetos mounted on the nose case are usually of the higher power ranges. The advantage of mounting the magneto on the nose section is in cooling. When the magnetos are on the nose section of the engine, they are exposed to a large volume of ram air; thus, they are kept

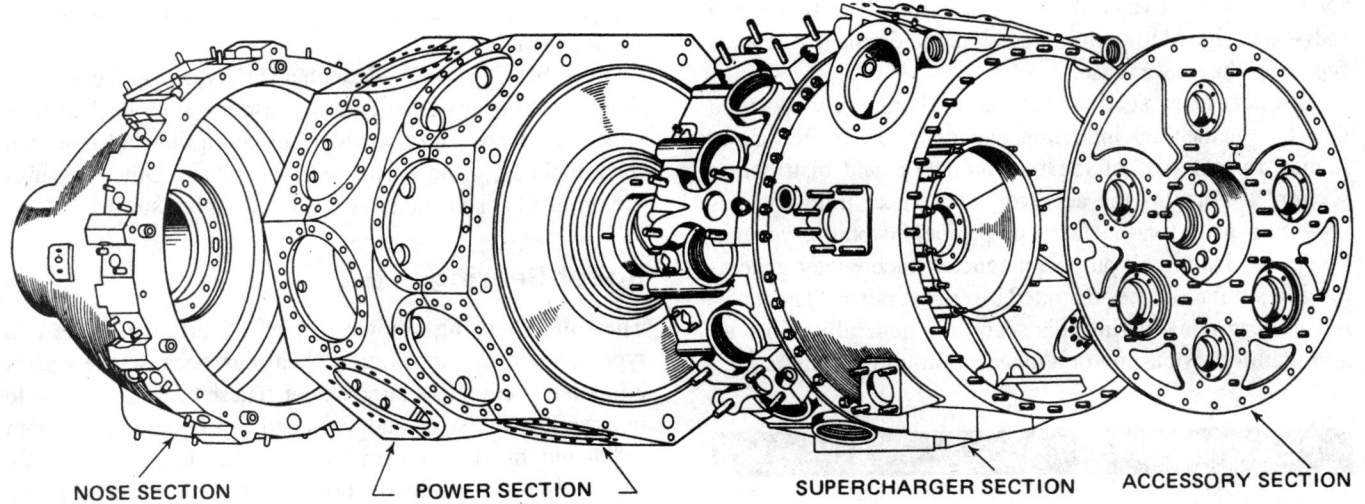

FIGURE 2-3 Crankcase for a twin-row radial engine.

NOSE SECTION — POWER SECTION — SUPERCHARGER SECTION — ACCESSORY SECTION

much cooler than when they are mounted on the accessory section.

The **main power section** usually consists of one, two, or possibly three pieces of high-strength heat-treated aluminum-alloy or steel forgings, bolted together if there is more than one piece. The use of a two-part main power section for a radial engine makes it possible to add strength to this highly stressed section of the engine. The cam-operating mechanism is usually housed and supported by the main crankcase section. At the center of each main crankcase web section are crankshaft bearing supports. Cylinder mounting pads are located radially around the outside circumference of the power section. The cylinders are fastened to the pads by means of studs and nuts or cap screws. Oil seals are located between the front section and the main section. Similar seals are installed between the power section and the fuel induction and distribution section.

The **fuel induction and distribution section** is normally located immediately behind the main (power) section and may be of either one- or two-piece construction. It is sometimes called the **blower section** or the **supercharger section** because its principal function is to house the blower or supercharger impeller and diffuser vanes. There are openings on the outside circumference of the housing for attaching the individual induction pipes, a small opening for the attachment of the manifold pressure line, and internal passages which lead to the supercharger drain valve.

The **accessory section** provides mounting pads for the accessory units, such as the fuel pumps, vacuum pumps, lubricating oil pumps, tachometer generators, generators, magnetos, starters, two-speed supercharger control valves, oil filtering screens, Cuno filters, and other items of accessory equipment. In some aircraft powerplants, the cover for the supercharger rear housing is made of an aluminum-alloy or a magnesium-alloy casting in the form of a heavily ribbed plate that provides the mounting pads for the accessory units; but in other powerplants, the housing for the accessory units may be mounted directly on the rear of the crankcase. Regardless of the construction and location of the accessory

housing, it contains the gears for driving the accessories which are operated by engine power.

In-Line and V-Type Engine Crankcases

Large in-line and V-type engine crankcases usually have four major sections: (1) the front, or nose, section; (2) the main, or power, section; (3) the fuel induction and distribution section; and (4) the accessory section.

The front, or nose, section is directly behind the propeller in most tractor-type airplanes. A **tractor-type airplane** is one in which the propeller "pulls" the airplane forward. The nose section may be cast as part of the main, or power, section, or it may be a separate construction with a dome or conical shape to reduce drag. Its function is to house the propeller shaft, the propeller thrust bearing, the propeller reduction-gear train, and sometimes a mounting pad for the propeller governor. In a very few arrangements where the nose section is not located close to the engine, the propeller is connected to the engine through an extension shaft and the reduction-gear drive has its own lubricating system. This same arrangement is found in some turboprop engines.

The main, or power, section varies greatly in design for different engines. When it is made up of two parts, one part supports one half of each crankshaft bearing and the other supports the opposite half of each bearing. The cylinders are normally mounted on and bolted to the heavier of the two parts of this section on an in-line engine, and the crankshaft bearings are usually supported by reinforcing weblike partitions. External mounting lugs and bosses are provided for attaching the engine to the engine mount.

The fuel induction and distribution section is normally located next to the main, or power, section. This section houses the diffuser vanes and supports the internal blower impeller when the engine is equipped with an internal blower system. The induction manifold is located between the fuel induction and distribution section and the cylinders. The housing of this section has an opening for the attachment of a manifold pressure gauge line, and it also has internal passages for the fuel drain valve of the blower case. The **fuel drain valve** is designed to permit the automatic drainage of excess fuel from the blower case.

The **accessory section** may be a separate unit mounted directly on the fuel induction and distribution section, or it may form a part of the fuel induction and distribution section. It contains the accessory drive-gear train and has mounting pads for the fuel pump, coolant pump, vacuum pump, lubricating oil pumps, magnetos, tachometer generator, and similar devices operated by engine power. The material used in constructing this section is generally either an aluminum-alloy casting or a magnesium-alloy casting.

BEARINGS

A **bearing** is any surface that supports or is supported by another surface. It is a part in which a journal, pivot, pin, shaft, or similar device turns or revolves. The bearings used in aircraft engines are designed to produce minimum friction and maximum wear resistance.

A good bearing has two broad characteristics: (1) it must be made of a material that is strong enough to withstand the pressure imposed on it and yet permit the other surface to move with a minimum of wear and friction, and (2) the parts must be held in position within very close tolerances to provide quiet and efficient operation and at the same time permit freedom of motion.

Bearings must reduce the friction of moving parts and also take thrust loads, radial loads, or combinations of thrust and radial loads. Those which are designed primarily to take thrust loads are called **thrust bearings**.

Plain Bearings

Plain bearings are illustrated in Fig. 2-4. These bearings are usually designed to take radial loads; however, plain bearings with flanges are often used as thrust bearings in opposed aircraft engines. Plain bearings are used for connecting rods, crankshafts, and camshafts of low-power aircraft engines. The metal used for plain bearings may be silver, lead, an alloy (such as bronze or babbitt), or a combination of metals. Bronze withstands high compressive pressures but offers more friction than babbitt. Conversely, **babbitt**, a soft bearing alloy, silver in color and composed of tin, copper, and antimony, offers less friction but cannot withstand high compressive pressures as well as bronze. Silver withstands compressive pressures and

is an excellent conductor of heat, but its frictional qualities are not dependable.

Plain bearings are made with a variety of metal combinations. Some bearings in common use are steel-backed with silver or silver-bronze on the steel and a thin layer of lead then applied for the actual bearing surface. Other bearings are bronze-backed and have a lead or babbitt surface.

Roller Bearings

The **roller bearings** shown in Fig. 2-5 are one of the two types known as "antifriction" bearings because the rollers eliminate friction to a large extent. These bearings are made in a variety of shapes and sizes and can be adapted to both radial and thrust loads. Straight roller bearings are generally used only for radial loads; however, tapered roller bearings will support both radial and thrust loads.

The bearing **race** is the guide or channel along which the rollers travel. In a roller bearing, the roller is situated between an inner and an outer race, both of which are made of case-hardened steel. When a roller is tapered, it rolls on a cone-shaped race inside an outer race.

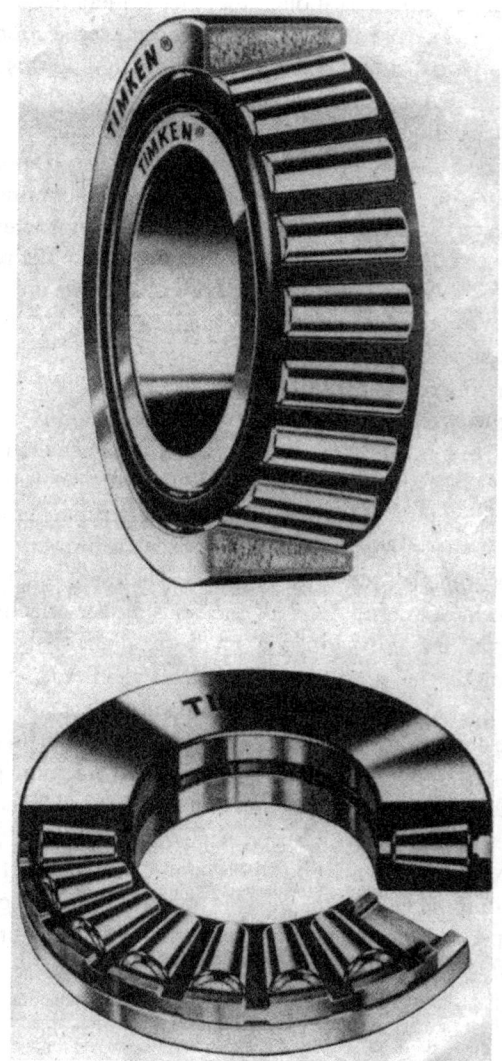

FIGURE 2-5   Roller bearings. (*Timken Roller Bearing Co.*)

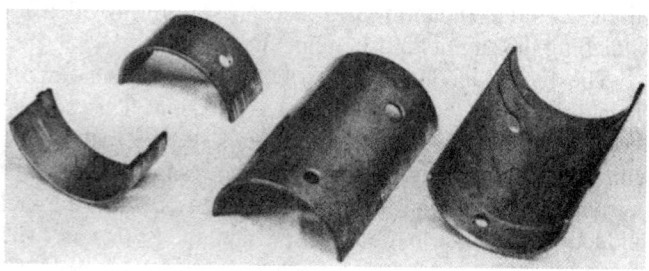

FIGURE 2-4   Plain bearings.

Roller bearings are used in high-power aircraft engines as main bearings to support the crankshaft. They are also used in other applications where radial loads are high.

Ball Bearings

Ball bearings provide less rolling friction than any other type. A ball bearing consists of an inner race and an outer race, a set of polished steel balls, and a ball retainer. Some ball bearings are made with two rows of balls and two sets of races. The races are designed with grooves to fit the curvature of the balls in order to provide a large contact surface for carrying high radial loads.

A typical ball-bearing assembly used in an aircraft engine is shown in Fig. 2-6. In this assembly, the balls are controlled and held in place by the ball retainer. This retainer is necessary to keep the balls properly spaced, thus preventing them from contacting one another.

Ball bearings are commonly used for thrust bearings in large radial engines and gas-turbine engines. Because of their construction, they can withstand heavy thrust loads as well as radial or centrifugal loads. They are also subject to gyroscopic loads, but these are not critical. A ball bearing designed especially for thrust loads is made with exceptionally deep grooves for the ball races. Bearings designed to resist thrust in a particular direction will have a heavier race design on the side which takes the thrust. It is important to see that this type of bearing is installed with the correct side toward the thrust load.

In addition to the large ball bearings used as main bearings and thrust bearings, many smaller ball bearings are found in generators, magnetos, starters, and other accessories used on aircraft engines. For this reason, the engine technician should be thoroughly familiar with the inspection and servicing of such bearings.

Many bearings, particularly for accessories, are prelubricated and sealed. These bearings are designed to function satisfactorily without lubrication service between overhauls. To avoid damaging the seals of the bearings, it is essential that the correct bearing pullers and installing tools be employed when the bearings are removed or installed.

THE CRANKSHAFT

The **crankshaft** transforms the reciprocating motion of the piston and connecting rod to rotary motion for turning the propeller. It is a shaft composed of one or more cranks located at definite places between the ends. These **cranks**, sometimes called **throws**, are formed by forging offsets into a shaft before it is machined. Since the crankshaft is the backbone of an internal-combustion engine, it is subjected to all the forces developed within the engine and must be of very strong construction. For this reason, it is usually forged from some extremely strong steel alloy, such as chromium-nickel-molybdenum steel (SAE 4340).

A crankshaft may be constructed of one or more pieces. Regardless of whether it is of one-piece or multipiece construction, the corresponding parts of all crankshafts have the same names and functions. The parts are (1) the main journal, (2) the crankpin, (3) the crank cheek or crank arm, and (4) the counterweights and dampers. Figure 2-7 shows the nomenclature of a typical crankshaft.

Main Journal

The **main journal** is the part of the crankshaft that is supported by and rotates in a **main bearing**. Because of this it may also properly be called a **main-bearing journal**. This journal is the center of rotation of the crankshaft and serves to keep the crankshaft in alignment under all normal conditions of operation. The main journal is surface-hardened by nitriding for a depth of 0.015 to 0.025 in [0.381 to 0.635 millimeter (mm)] to reduce wear. Every aircraft engine crankshaft has two or more main journals to support the weight and operational loads of the entire rotating and reciprocating assembly in the power section of the engine.

Crankpin

The **crankpin** can also be called a **connecting-rod bearing journal** simply because it is the journal for a connecting-rod bearing. Since the crankpin is off center from the main journals, it is sometimes called a **throw**. The crankshaft will rotate when a force is applied to the crankpin in any direction other than parallel to a line directly through the centerline of the crankshaft.

FIGURE 2-6 Ball-bearing assembly.

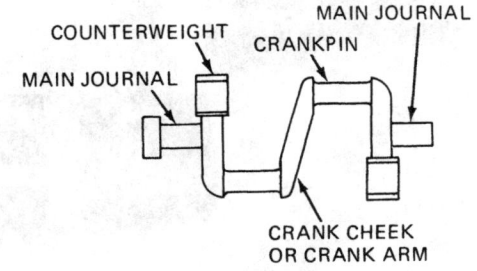

FIGURE 2-7 Nomenclature for a twin-row radial crankshaft.

The crankpin is usually hollow for three reasons: (1) it reduces the total weight of the crankshaft, (2) it provides a passage for the lubricating oil, and (3) it serves as a chamber for collecting carbon deposits, sludge, and other foreign substances which are thrown by centrifugal force to the outside of the chamber where they will not reach the connecting-rod bearing surface. For this reason the chamber is often called the **sludge chamber**. On some engines a drilled passage from the sludge chamber to an opening on the exterior surface of the connecting rod makes it possible to spray clean oil on the cylinder walls.

Lubrication of the crankpin bearings is accomplished by oil taken through drilled passages from the main journals. The oil reaches the main journals through drilled passages in the crankcase and in the crankcase webs which support the main bearings. During overhaul the technician must see that all oil passages and sludge chambers are cleared in accordance with the manufacturer's instructions.

Crank Cheek

The **crank cheek**, sometimes called the **crank arm**, is the part of the crankshaft which connects the crankpin to the main journal. It must be constructed to maintain rigidity between the journal and the crankpin. On many engines, the crank cheek extends beyond the main journal and supports a counterweight used to balance the crankshaft. Crank cheeks are usually provided with drilled oil passages through which lubricating oil passes from the main journals to the crankpins.

Counterweights

The purpose of the **counterweight** is to provide static balance for a crankshaft. If a crankshaft has more than two throws, it does not always require counterweights because the throws, being arranged symmetrically opposite each other, balance each other. A single-throw crankshaft, such as that used in a single-row radial engine, must have counterbalances to offset the weight of the single throw and the connecting-rod and piston assembly attached to it. This type of crankshaft is illustrated in Fig. 2-8.

Dynamic Dampers

The purpose of **dynamic dampers** is to relieve the whip and vibration caused by the rotation of the crankshaft. They are suspended from or installed in specified crank cheeks at locations determined by the design engineers. Crankshaft vibrations caused by power impulses may be reduced by placing floating dampers in a counterweight assembly. The need for counterweights and dampers is not confined to aircraft engines. Any machine with rotating parts may reach a speed at which so much vibration occurs in the revolving mass of metal that the vibration must be reduced or else the machine will eventually destroy itself.

Dampers or **dynamic balances** are required to overcome the forces which tend to cause deflection of the crankshaft and torsional vibration. These forces are generated principally by the power impulses of the pistons. If we compute the force exerted by the piston of an engine near the beginning of the power stroke, we find that 8000 to 10 000 lb [35 585 to 44 480 newtons (N)] is applied to the throw of a crankshaft. As the engine runs, this force is applied at regular intervals to the different throws of the crankshaft on an in-line or opposed engine and to the one throw of a single-row radial engine. If the frequency of the power impulses is such that it matches the natural vibration frequency of the crankshaft and propeller as a unit, or of any moving part of the engine, severe vibration will take place. The dynamic balances may be pendulum-type weights mounted in the counterweight (Fig. 2-9), or they may be straddle-mounted on extensions of the crank cheeks. In either case, the weight is free to move in a direction and at a frequency which will damp the natural vibration of the crankshaft. Dynamic balances are shown in Fig. 2-10.

The dynamic damper used in an engine consists of a movable slotted-steel counterweight attached to the crank cheek. Two spool-shaped steel pins extend into the slot and pass through oversized holes in the counterweight and crank cheek. The difference in the diameters between the pins and the holes provides a pendulum.

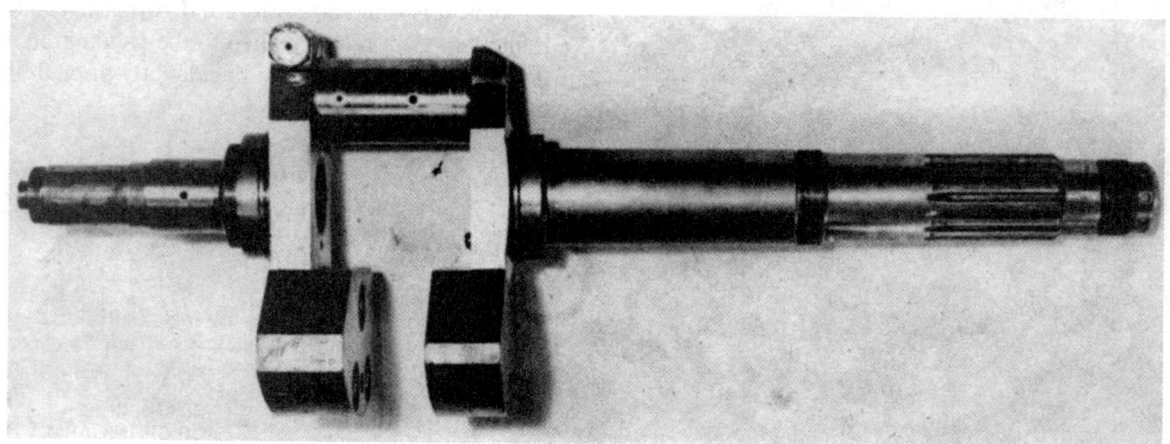

FIGURE 2-8   Single-throw crankshaft with counterweights.

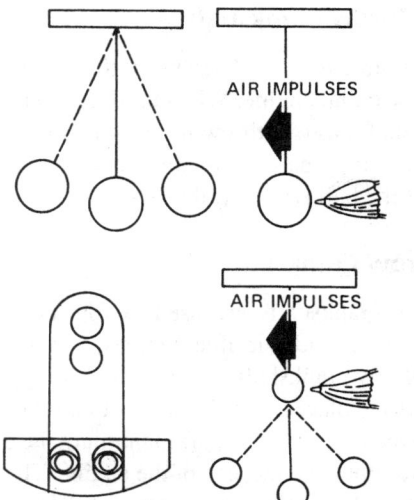

FIGURE 2-9 Dynamic balances and principles of operation.

FIGURE 2-10 Dynamic-balance weights.

The effectiveness of a dynamic damper can be understood by observing the operation of a pendulum. If a simple pendulum is given a series of regular impulses at a speed corresponding to its natural frequency, using a bellows to simulate a modified power impulse in an engine, it will begin swinging or vibrating back and forth from the impulses, as shown in the upper half of Fig. 2-9.

Another pendulum, suspended from the first, will absorb the impulse and swing itself, leaving the first pendulum stationary, as shown in the lower portion of Fig. 2-9. The dynamic damper, then, is a short pendulum hung on the crankshaft and tuned to the frequency of the power impulses to absorb vibration in the same manner as the pendulum illustrated in the lower part of the illustration. A **mode number** is used to indicate the correct type of damper for a specific engine.

Types of Crankshafts

The four principal types of crankshafts are (1) single-throw, (2) double-throw, (3) four-throw, and (4) six-throw.

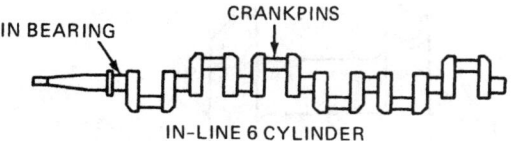
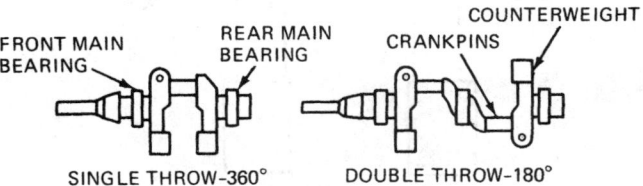

FIGURE 2-11 Three types of crankshafts.

Figure 2-11 shows the crankshafts for an in-line engine, a single-row radial engine, and a double-row radial engine. Each individual type of crankshaft may have several configurations, depending on the requirements of the particular engine for which it is designed. An engine which operates at a high speed and power output requires a crankshaft more carefully balanced and with greater resistance to wear and distortion than an engine which operates at lower speeds.

Single-Throw Crankshaft

The type of crankshaft and the number of crankpins it contains correspond in every case to the engine cylinder arrangement. The position of a crank on any crankshaft in relation to other cranks on the same shaft is given in degrees.

The single-throw, or 360°, crankshaft is used in single-row radial engines. It may be of single- or two-piece construction with two main bearings, one on each end. A single-piece crankshaft is shown in Fig. 2-12. This crankshaft must be used with a master rod which has the large end split.

Two-piece single-throw crankshafts are shown in Fig. 2-13. The first of these (Fig. 2-13A) is a clamp-type shaft, sometimes referred to as a **split-clamp crankshaft**. The front section of this shaft includes the main-bearing journal, the front crank-cheek and counterweight assembly, and the crankpin. The rear section contains the clamp by which the two sections are joined, the rear crank-cheek and counterweight assembly, and the rear main-bearing journal. The spline-type crankshaft shown in Fig. 2-13B has the same parts as the clamp type, with the exception of the device by which the two sections are joined.

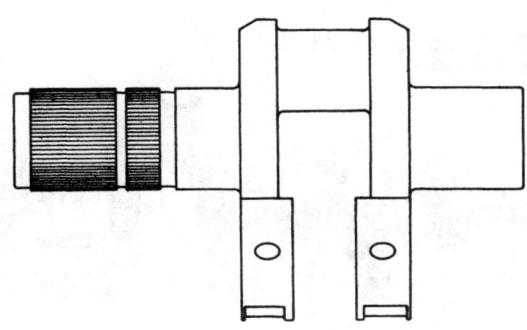

FIGURE 2-12 Single-piece crankshaft.

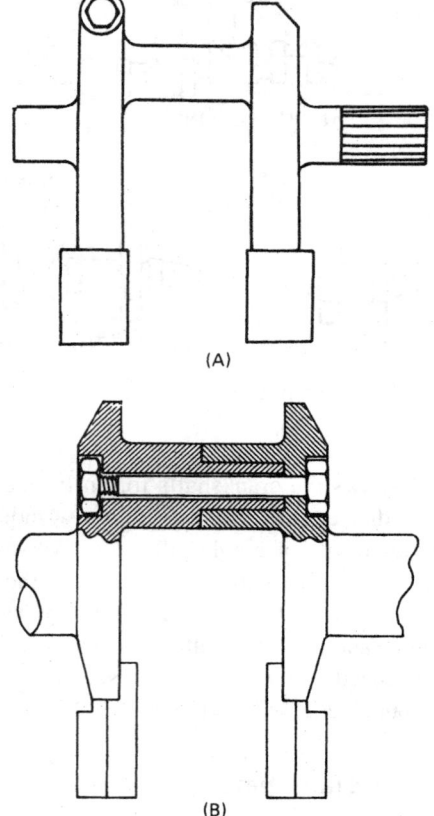

FIGURE 2-13 Two-piece single-throw crankshafts.

In this shaft the crankpin is divided, one part having a female spline and the other having a matching male spline. When the two parts are joined, they are held securely in place by means of a steel-alloy bolt.

Double-Throw Crankshaft

The double-throw, or 180°, crankshaft is generally used in a double-row radial engine. When used in this type of engine, the crankshaft has one throw for each row of cylinders. The construction may be one- or three-piece, and the bearings may be of the ball type or roller type.

Four-Throw Crankshaft

Four-throw crankshafts are used in four-cylinder opposed engines, four-cylinder in-line engines, and V-8 engines. In the four-throw crankshaft for an in-line or opposed engine, two throws are placed 180° from the other two throws. There may be three or five crankshaft main journals, depending on the power output and the size of the engine. The bearings for the four-cylinder opposed engine are of the plain, split-shell type. In the four-throw crankshaft, illustrated in Fig. 2-14, lubrication for the crankpin bearings is provided through passages drilled in the crank cheeks. During operation, oil is brought through passages in the crankcase webs to the main-bearing journals. From the main-bearing journals, the oil flows through the crank-cheek passages to the crankpin journals and the sludge chambers in the journals.

Six-Throw Crankshaft

Six-throw crankshafts are used in six-cylinder in-line engines, 12-cylinder V-type engines, and six-cylinder opposed engines. Since the in-line and V-type engines are not in general use in the United States, we limit our discussion to the type of shaft used in the six-cylinder opposed engine.

A crankshaft for a Continental six-cylinder opposed aircraft engine is shown in Fig. 2-15. This is a one-piece

FIGURE 2-14 Four-throw crankshaft with a tapered propeller shaft.

FIGURE 2-15 Crankshaft for a six-cylinder opposed engine with a splined propeller shaft.

six-throw 60° crankshaft machined from an alloy-steel (SAE 4340) forging. It has four main journals and one double-flanged main-thrust journal. The shaft is heat-treated for high strength and nitrided to a depth of 0.015 to 0.025 in [0.381 to 0.635 mm], except on the propeller splines, for maximum wear. The crankpins and main-bearing journals are ground to close limits of size and surface roughness. After grinding, nitriding, and polishing, the crankshaft is balanced statically and dynamically. Final balance is attained after assembly of the counterweights and other parts.

As shown in Fig. 2-15, the crankshaft is provided with dynamic counterweights. Since the selection of the counterweights of the correct mode is necessary to preserve the dynamic balance of the complete assembly, they cannot be interchanged on the shaft or between crankshafts. For this reason, neither counterweights nor bare crankshafts are supplied alone.

The crankshaft is line-bored the full length to reduce weight. Splined shafts have a threaded plug installed at the front end. The crankpins are recessed at each end to reduce weight. Steel tubes permanently installed in holes drilled through the crank cheeks provide oil passages across the lightening holes to all crankpin surfaces from the main journals. A U-shaped tube, permanently installed inside the front end of the shaft bore, conducts oil from the second main journal to the front main-thrust journal.

Propeller Shafts

Aircraft engines are equipped with one of three types of propeller mounting shafts: taper shafts, spline shafts, or flange shafts. In the past, low-power engines often were equipped with **tapered propeller shafts**. The propeller shaft is an integral part of the crankshaft (Fig. 2-14). The tapered end of the shaft forward of the main bearing is milled to receive a key which positions the propeller in the correct location on the shaft. The shaft is threaded at the forward end to receive the propeller retaining nut.

Crankshafts with **spline propeller shafts** are shown in Figs. 2-8 and 2-15. As can be seen in these illustrations, the splines are rectangular grooves machined in the shaft to mate with grooves inside the propeller hub. One spline groove may be blocked with a screw to ensure that the propeller will be installed in the correct position. A wide groove inside the propeller hub receives the blocked, or "blind," spline. The propeller is mounted on the spline shaft with front and rear cones to ensure correct positioning both longitudinally and radially. A retaining nut on the threaded front portion of the shaft holds the propeller firmly in place when the nuts are properly torqued. The installation of propellers is described in other chapters of this text.

Note that the propeller shaft in Fig. 2-8 is threaded about halfway between the forward end and the crank throw. This threaded portion of the shaft is provided to receive the thrust-bearing retaining nut, which holds the thrust bearing in the nose case of the engine. The shaft in Fig. 2-15 is not threaded aft of the splines because the design of the engine case eliminates the need for a thrust nut.

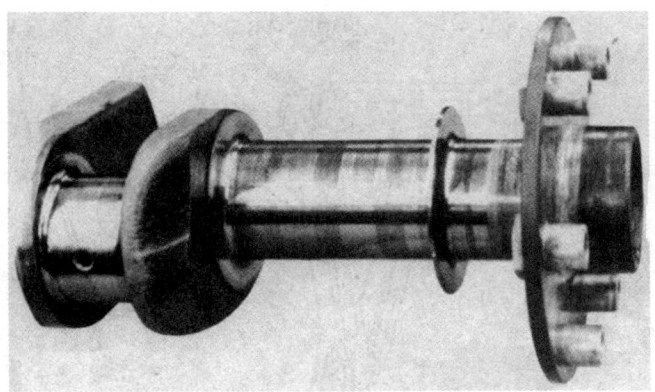

FIGURE 2-16 Flange-type propeller shaft.

Spline propeller shafts are made in several sizes, depending on engine horsepower. These sizes are identified as SAE 20, 30, 40, 50, 60, and 70. High-power engines have shafts from SAE 50 to SAE 70, and low-power engines are equipped with shaft sizes from SAE 20 to SAE 40.

Flange-type shafts are used with many modern opposed engines with power ratings up to 450 hp [335.57 kW]. Figure 2-16 shows a shaft of this type. A short stub shaft extends forward of the flange to support and center the propeller hub. Six high-strength bolts or studs are used to secure the propeller to the flange. In this type of installation, it is most important that the bolts or studs be tightened in a sequence which will provide a uniform stress. It is also necessary to use a torque wrench and to apply torque as specified in the manufacturer's service manual.

Some aircraft utilize **propeller-shaft extensions** to move the propeller forward, thus permitting a more streamlined nose design. Special instructions are provided by the aircraft manufacturer for service of such extensions.

Propeller-shaft loads are transmitted to the nose section of the engine by means of thrust bearings and forward main bearings. On some opposed-type engines, the forward main bearing is flanged to serve as a thrust bearing as well as a main bearing. The nose section of an engine either is a separate part or is integral with the crankcase. For either type of design, the nose section transmits the propeller-shaft loads from the thrust bearing to the crankcase, where they are applied to the aircraft structure through the engine mounts.

CONNECTING-ROD ASSEMBLIES

A variety of connecting-rod assemblies have been designed for the many different types of engines. Some of the arrangements for such assemblies are shown in Fig. 2-17. The **connecting rod** is the link which transmits forces between the piston and the crankshaft of an engine. It furnishes the means of converting the reciprocating motion of the piston to a rotating movement of the crankshaft in order to drive the propeller.

A tough steel alloy (SAE 4340) is the material used for manufacturing most connecting rods, but aluminum alloys

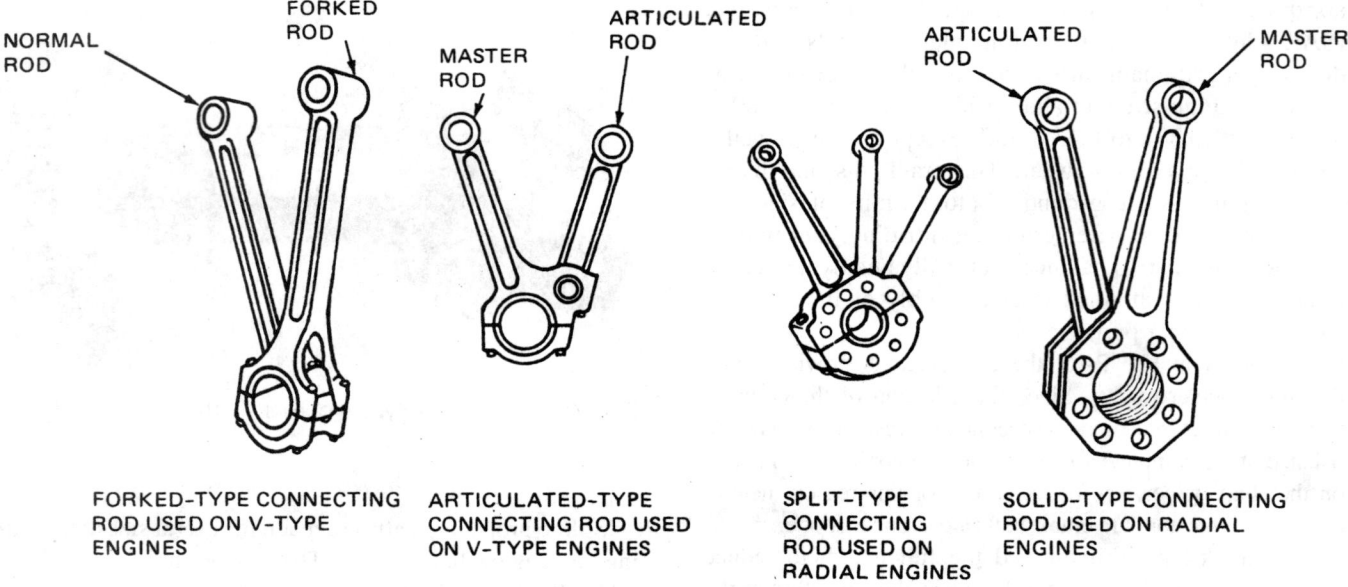

FORKED-TYPE CONNECTING ROD USED ON V-TYPE ENGINES

ARTICULATED-TYPE CONNECTING ROD USED ON V-TYPE ENGINES

SPLIT-TYPE CONNECTING ROD USED ON RADIAL ENGINES

SOLID-TYPE CONNECTING ROD USED ON RADIAL ENGINES

FIGURE 2-17 Connecting-rod assemblies.

have been used for some low-power engines. The cross-sectional shape of the connecting rod is usually like either the letter H or the letter I, although some have been made with a tubular cross section. The end of the rod which connects to the crankshaft is called the **large end**, or **crankpin end**, and the end which connects to the piston pin is called the **small end**, or **piston-pin end**. Connecting rods, other than tubular types, are manufactured by forging to provide maximum strength.

Connecting rods stop, change direction, and start at the end of each stroke; therefore, they must be lightweight to reduce the inertial forces produced by these changes of velocity and direction. At the same time, they must be strong enough to remain rigid under the severe loads imposed under operating conditions.

There are three principal types of connecting-rod assemblies: (1) the plain type, shown in Fig. 2-18, (2) the fork-and-blade type, shown in Fig. 2-19, and (3) the master and articulated type, shown in Fig. 2-20.

Plain Connecting Rod

The plain connecting rod is used in in-line engines and opposed engines. The small end of the rod usually has a bronze bushing that serves as a bearing for the piston pin. This bushing is pressed into place and then reamed to the proper dimension. The large end of the rod is made with a cap, and a two-piece shell bearing is installed. The bearing is held in place by the cap. The outside of the bearing flange bears against the sides of the crankpin journal when the rod assembly is installed on the crankshaft. The bearing inserts are often made of steel and lined with a nonferrous bearing material, such as lead bronze, copper lead, lead silver, or babbitt. Another type of bearing insert is made of bronze and has a lead plating for the bearing surface against the crankpin.

The two-piece bearing shell fits snugly in the large end of the connecting rod and is prevented from turning by dowel pins or by tangs which fit into slots cut into the cap and the connecting rod. The cap is usually secured on the end of the rod by bolts; however, some rods have been manufactured with studs for holding the cap in place.

During inspection, maintenance, repair, and overhaul, the proper fit and balance of connecting rods are obtained by always replacing the connecting rod in the same cylinder and in the same relative position as it was before removal.

The connecting rods and caps are usually stamped with numbers to indicate their positions in the engine. The rod assembly for the no. 1 cylinder is marked with a 1, the assembly for the no. 2 cylinder is marked with a 2, and so on. The caps are also marked and must be assembled with the numbers aligned, as shown in Fig. 2-18.

Fork-and-Blade Connecting-Rod Assembly

The fork-and-blade connecting rod, illustrated in Fig. 2-19, is generally used in V-type engines. The **forked rod** is split on the large end to provide space for the **blade rod** to fit between the prongs.

One two-piece bearing shell is fastened by lugs or dowel pins to the forked rod. Between the prongs of the forked rod,

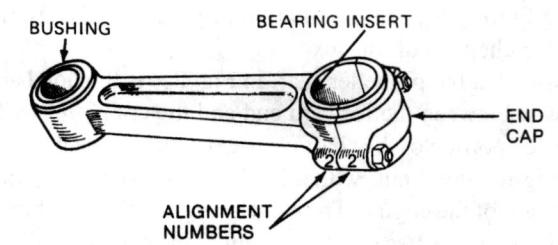

FIGURE 2-18 Plain-type connecting rod.

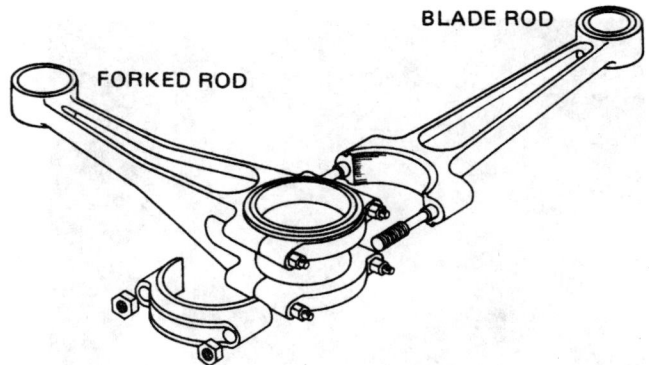

FIGURE 2-19   Fork-and-blade connecting rod.

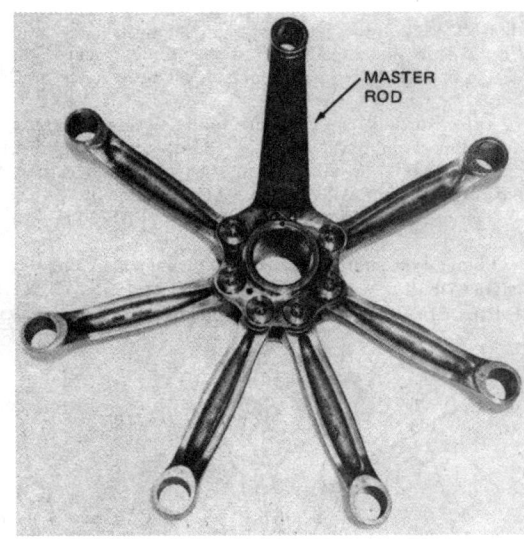

FIGURE 2-20   Master and articulated connecting-rod assembly.

the center area of the outer surface of this bearing shell is coated with a nonferrous bearing metal to act as a journal for the blade rod and cap.

During overhaul or maintenance, the fork-and-blade connecting rods are always replaced on the crankshaft in the same relative positions they occupied in their original installation. This ensures the proper fit and engine balance.

Master and Articulated Connecting-Rod Assembly

The **master and articulated connecting-rod assembly** is used primarily for radial engines, although some V-type engines have employed this type of rod assembly. The complete rod assembly for a seven-cylinder radial engine is shown in Fig. 2-20.

The **master rod** in a radial engine is subjected to some stresses not imposed on the plain connecting rod; therefore, its design and construction must be of the highest quality. It is made of an alloy-steel forging, machined and polished to final dimensions and heat-treated to provide maximum strength and resistance to vibration and other stresses. The surface must be free of nicks, scratches, or other surface

damage which may produce stress concentrations and ultimate failure.

The master rod is similar to other connecting rods except that it is constructed to provide for the attachment of the articulated rods (link rods) on the large end. The large end of the master rod may be a two-piece type or a one-piece type, as shown in Fig. 2-21.

If the large end of the master rod is made of two pieces, the crankshaft is one solid piece. If the rod is one piece, the crankshaft may be of either two- or three-piece construction. Regardless of the type of construction, the usual bearing surfaces must be supplied.

Master rod bearings are generally of the plain type and consist of a split shell or a sleeve, depending on whether the master rod is of the two- or one-piece type. The bearing usually has a steel or bronze backing with a softer nonferrous material bonded to the backing to serve as the actual

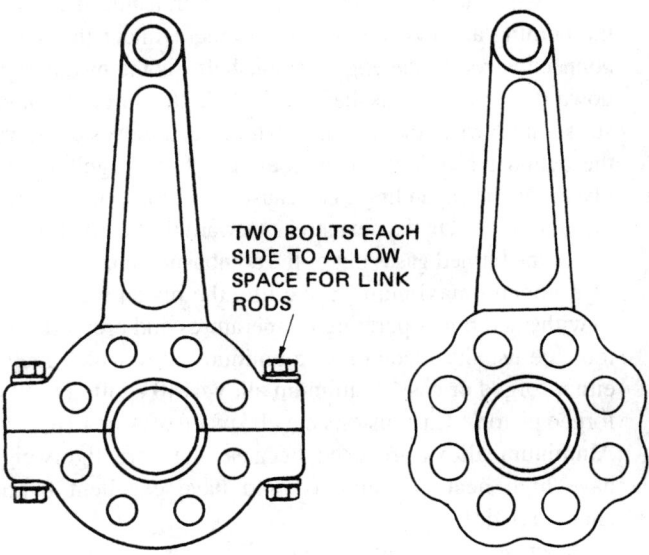

FIGURE 2-21   Types of master rods.

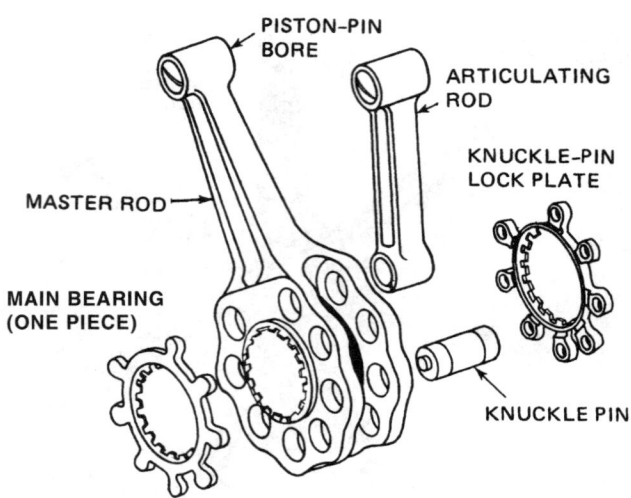

FIGURE 2-22 Master rod with full-floating knuckle-pin and lock-plate assembly.

bearing material. In low-power engines, babbitt material was suitable for the bearing surface, but it was found to lack the durability necessary for higher-power engines. For this reason, bronze, leaded bronze, and silver have been used in later engines. The actual bearing surface is usually plated with lead to reduce the friction as much as possible. During operation the bearing is cooled and lubricated by a constant flow of lubricating oil.

The **articulated rods** (link rods) are hinged to the master rod flanges by means of steel **knuckle pins**. Each articulated rod has a bushing of nonferrous metal, usually bronze, pressed or shrunk into place to serve as a knuckle-pin bearing. Aluminum-alloy link rods have been used successfully in some lower-power radial engines. With these rods, it is not necessary to provide bronze bushings at the ends of the rod because the aluminum alloy furnishes a good bearing surface for the piston pins and knuckle pins.

Articulated rods, when made of steel, are usually constructed in an I or H cross section. These configurations give the greatest strength and resistance to distortion with the lightest weight.

The knuckle pin resembles a piston pin. It is usually made of nickel steel, hollowed for lightness and for the passage of lubricating oil, and surface-hardened to reduce wear.

The articulated rod is bored and supplied with bushings at each end. One end receives the piston pin, and the other end receives the knuckle pin. The knuckle-pin bore in the articulated rod includes a bushing of nonferrous metal, which is usually bronze. It is pinned, pressed, shrunk, or spun into place. The bushing must be bored to precise dimension and alignment.

Knuckle pins installed with a loose fit so that they can turn in the master rod flange holes and also turn in the articulated rod bushings are called **full-floating knuckle pins**. Knuckle pins also may be installed so that they are prevented from turning in the master rod by means of a tight press fit. In either type of installation a lock plate on each side bears against the knuckle pin and prevents it from moving laterally (sideways).

Figure 2-22 shows a knuckle-pin and lock-plate assembly for a full-floating arrangement, and Fig. 2-23 shows a stationary knuckle-pin and lock-plate assembly.

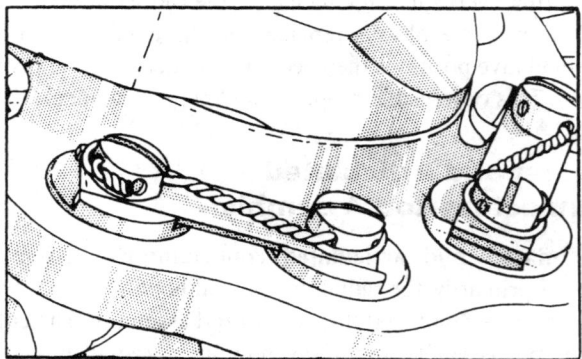

FIGURE 2-23 Stationary knuckle-pin and lock-plate assembly.

PISTONS

Construction

The **piston** is a plunger that moves back and forth or up and down within an engine cylinder barrel. It transmits the force of the burning and expanding gases in the cylinder through the connecting rod to the engine crankshaft. As the piston moves down (toward the crankshaft) in the cylinder during the intake stroke, it draws in the fuel-air mixture. As it moves up (toward the cylinder head), it compresses the charge. Ignition takes place, and the expanding gases cause the piston to move toward the crankshaft. On the next stroke (toward the head), the piston forces the burned gases out of the combustion chamber.

To obtain maximum engine life, the piston must be able to withstand high operating temperatures and pressures. Pistons are usually made of an aluminum alloy which may be either forged or cast. Aluminum alloy 4140 is often used for forged pistons. Cast pistons may be made of Alcoa 132 alloy. Aluminum alloys are used because they are lightweight, have high heat conductivity, and have excellent bearing characteristics.

A cross section of a typical piston is illustrated in Fig. 2-24. The top of the piston is the **head**. The sides form

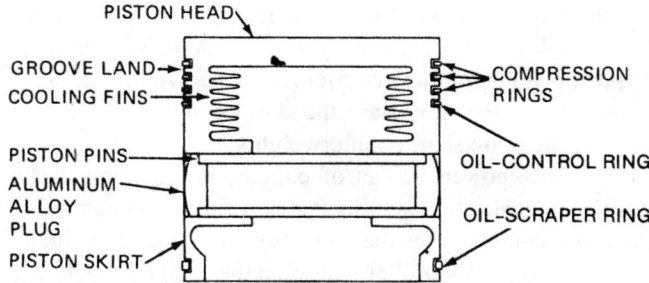

FIGURE 2-24   Cross section of an assembled piston.

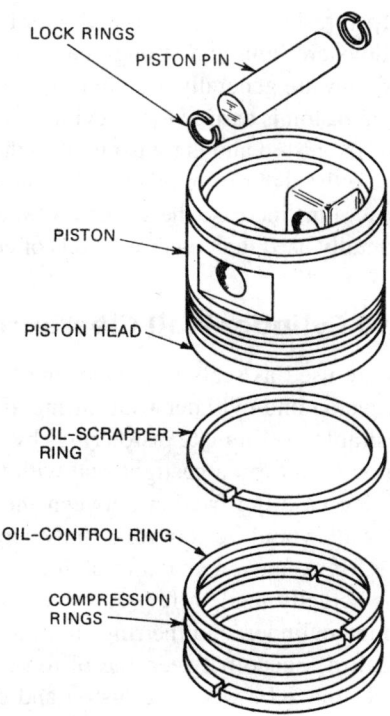

FIGURE 2-26   Components of a complete piston assembly.

the **skirt**. The underside of the piston head often contains ribs or other means of presenting maximum surface area for contact with the lubricating oil splashed on it. This oil carries away part of the heat conducted through the piston head.

Some pistons are constructed with a slightly oval cross section. The diameter perpendicular to the piston pin is greater, to allow for more wear of the piston due to additional side thrust against the cylinder walls and to provide a better fit at operating temperatures. Such a piston is called a cam ground piston and is shown in Fig. 2-25. The oval shape of the piston will hold the piston square in the cylinder when the engine is cold. This will reduce piston slap or the piston cocking in the cylinder during engine warm-up. As the piston reaches operating temperature, the oval shape will become round, providing the proper piston-cylinder-wall fit. This results from the greater mass of metal perpendicular to the piston pin.

Grooves are machined around the outer surface of the piston to provide support for the **piston rings**. The metal between the grooves is called a **groove land**, or simply a **land**. The grooves must be accurately dimensioned and concentric with the piston.

The piston and ring assembly must form as nearly as possible a perfect seal with the cylinder wall. It must slide along the cylinder wall with very little friction. The engine

lubricating oil aids in forming the piston seal and in reducing friction. All the piston assemblies in any one engine must be balanced. This means that each piston must weigh within $\frac{1}{4}$ ounce (oz) [7.09 grams (g)] of each of the others. This balance is most important to avoid vibration while the engine is operating. In any case, the weight limitations specified by the manufacturer must be observed.

The piston illustrated in the cross-sectional drawing of Fig. 2-24 has five piston rings; however, some pistons are equipped with four rings, and many operate with only three rings.

The parts of a complete piston assembly are shown in Fig. 2-26. This illustration shows the piston, piston pin, lock rings, oil scraper ring, oil control ring, and compression ring.

Piston Speed

To appreciate the loads imposed on a piston and connecting-rod assembly, it is helpful to consider the speed at which the piston must travel in the cylinder. To move at high speeds with a minimum of stress, the piston must be as light as possible. If an engine operates at 2000 rpm, the piston will start and stop 4000 times in 1 minute (min), and if the piston has a 6-in [15.24-cm] stroke, it may reach a velocity of more than 35 mph [56.32 km/h] at the end of the first quarter of crankshaft rotation and at the beginning of the fourth quarter of rotation.

Piston Temperature and Pressure

The temperature inside the cylinder of an airplane engine may exceed 4000°F [2204°C], and the pressure against the piston during operation may be as high as 500 pounds per

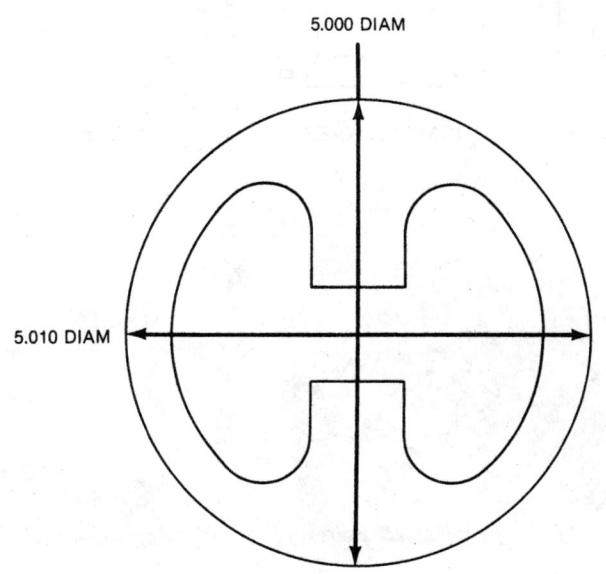

FIGURE 2-25   Cam ground piston.

square inch (psi) [3 447.5 kilopascals (kPa)] or higher. Since aluminum alloys are light and strong and conduct the heat away rapidly, they are generally used in piston construction. The heat in the piston is carried to the cylinder wall through the outside of the piston and is transmitted to the engine oil in the crankcase through ribs or other means on the inside of the piston head. Fins increase the strength of the piston and are more generally used than other methods of cooling.

Piston and Cylinder-Wall Clearance

Piston rings are used as seals to prevent the loss of gases between the piston and cylinder wall during all strokes. It would be desirable to eliminate piston rings by making pistons large enough to form a gastight seal with the cylinder wall, but in that case the friction between the piston and the cylinder wall would be too great and there would be no allowance for expansion and contraction of the metals. The piston is actually made a few thousandths of an inch smaller than the cylinder, and the rings are installed in the pistons to seal the space between the piston and cylinder wall. If the clearance between the piston and the cylinder wall became too great, the piston could wobble, causing piston slap.

Types of Pistons

Pistons may be classified according to the type of piston head used. The types of heads are flat, recessed, concave, convex, and truncated cone, all of which are illustrated in Fig. 2-27. Pistons in modern engines are usually of the flat-head type. The skirt of the piston may be of the trunk type, trunk type relieved at the piston pin, or slipper type. Typical pistons for modern engines are shown in Fig. 2-28. Note that some pistons have the skirt cut out at the bottom to clear the crankshaft counterweights.

The horsepower ratings of engines of the same basic design are changed merely by the use of different pistons. A domed piston increases the compression ratio and the brake mean effective pressure (bmep) when the engine is operating at a given rpm.

Piston-Ring Construction

Piston rings are usually made of high-grade gray cast iron that provides the spring action necessary to maintain a steady pressure against the cylinder wall, thus retaining the necessary seal. Cast-iron rings do not lose their elasticity even when they are exposed to rather high temperatures. The rings are split so that they can be slipped over the outside of the piston and into the ring grooves which are machined on the circumference of the piston. Some compression rings are given a chromium-plated surface on the face of the ring. *This type of ring must never be used in a chromium-plated cylinder.*

As shown in Fig. 2-29, the piston-ring gap may be a plain butt joint, a step joint, or an angle joint. The butt joint is commonly used in modern aircraft engines. When a piston ring is installed in a cylinder, there must be a specified gap clearance between the ends of the joint to allow for heat expansion during engine operation. This gap dimension is given in the Table of Limits for the engine. If a piston ring

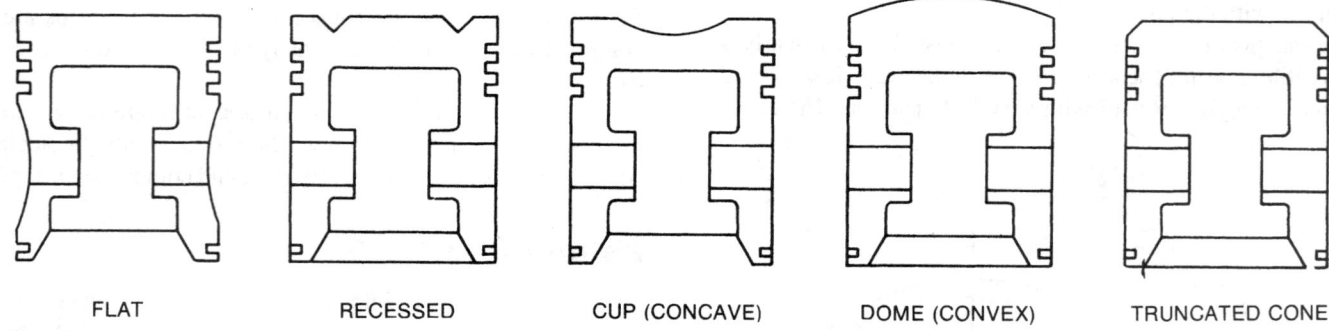

| FLAT | RECESSED | CUP (CONCAVE) | DOME (CONVEX) | TRUNCATED CONE |

FIGURE 2-27 Types of piston heads.

FIGURE 2-28 Several types of pistons.

FIGURE 2-29 Piston-ring joints.

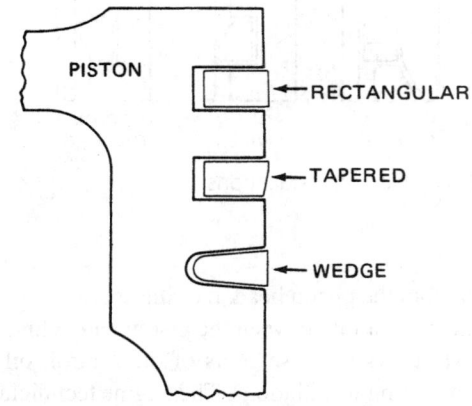

FIGURE 2-30 Cross sections of compression rings.

does not have sufficient gap clearance, the ring may seize against the wall of the cylinder during operation and cause scoring of the cylinder or failure of the engine. The joints of the piston rings must be staggered around the circumference of the piston in which they are installed at the time that the piston is installed in the cylinder. This staggering is done to reduce **blowby**—that is, to reduce the flow of gases from the combustion chamber past the pistons and into the crankcase. Blowby is evidenced by the emission of oil vapor and blue smoke from the engine breather. This same indication may occur as a result of worn piston rings; in this case, it is caused by oil entering and burning in the combustion chamber, and a part of the combustion gases blowing by the piston rings into the crankcase and out the breather. The greatest wear in a reciprocating engine usually occurs between the piston rings and the cylinder walls; excessive emission of blue smoke from the exhaust or from the engine breather indicates that repairs should be made.

The side clearances for the various rings are specified in the Table of Limits for the engine. Excessive side clearance will allow the rings to cock in their grooves, subjecting the rings to considerable wear and possible breakage.

Functions of Piston Rings

The importance of the piston rings in a reciprocating engine cannot be overemphasized. Their three principal functions are (1) to provide a seal to hold the pressures in the combustion chamber, (2) to prevent excessive oil from entering the combustion chamber, and (3) to conduct the heat from the piston to the cylinder walls. Worn or otherwise defective piston rings will cause loss of compression and excessive oil consumption. Defective piston rings will usually cause excessively high oil discharge from the crankcase breather and excessive emission of blue smoke from the engine exhaust during normal operation. The smoke is normal when an engine is first started, but it should not continue for more than a few moments.

Types of Piston Rings

Piston rings in general may be of the same thickness throughout the circumference or they may vary, but aircraft engine piston rings are almost always of the same thickness all the way around. Piston rings may be classified according to function as (1) compression rings and (2) oil rings.

The purpose of **compression rings** is to prevent gases from escaping past the piston during engine operation. The compression rings are placed in the ring grooves immediately below the piston head. The number of compression rings used on a piston is determined by the designer of the

engine, but most aircraft engines have two or three compression rings for each piston.

The cross section of the compression ring may be rectangular, tapered, or wedge-shaped. The rectangular cross section provides a straight bearing edge against the cylinder wall. The tapered and wedge-shaped cross sections present a bearing edge which hastens the seating of a new ring against the hardened surface of the cylinder wall. Figure 2-30 illustrates cross sections of rectangular, tapered, and wedge-shaped compression rings.

The principal purpose of **oil rings** is to control the quantity of lubricant supplied to the cylinder walls and to prevent this oil from passing into the combustion chamber. The two types of oil rings are **oil control rings** and **oil wiper rings** (sometimes called **oil scraper rings**).

Oil control rings are placed in the grooves immediately below the compression rings. There is generally only one oil control ring on a piston. However, there may be one or two oil scraper rings.

The purpose of the oil control ring is to control the thickness of the oil film on the cylinder wall. The groove of the oil control ring is often provided with holes drilled to the inside of the piston to permit excess oil to be drained away. The oil flowing through the drilled holes provides additional lubrication for the piston pins.

If too much oil enters the combustion chamber, the oil will burn and may leave a coating of carbon on the combustion-chamber walls, piston head, and valve heads. This carbon can cause the valves and piston rings to stick if it enters the valve guides and the ring grooves. In addition, the carbon may cause detonation and preignition. If the operator of an aircraft engine notices increased oil consumption and heavy blue smoke from the exhaust, probably the piston rings are worn and not providing the seal necessary for proper operation.

Oil wiper or oil scraper rings are placed on the skirts of the pistons to regulate the amount of oil passing between the piston skirts and cylinder walls during each of the piston strokes. The cross section is usually beveled, and the beveled edge is installed in either of two positions. If the beveled edge is installed nearest the piston heads, the ring scrapes oil toward the crankcase. If it is installed with the beveled

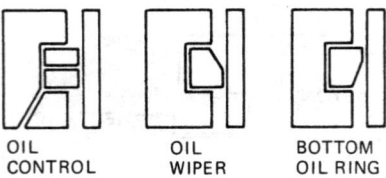

OIL CONTROL   OIL WIPER   BOTTOM OIL RING

FIGURE 2-31   Oil ring installations.

edge away from the piston head, the ring serves as a pump to keep up the flow of oil between the piston and cylinder wall. Figure 2-31 shows cross sections of oil control, oil wiper, and bottom oil ring installations. The engine technician must make sure the piston rings are installed according to the manufacturer's overhaul instructions.

Piston-Ring Cross Section

In addition to the cross sections of compression rings described previously, we must consider the shapes of oil rings. Oil control rings usually have one of three cross sections: (1) ventilated, (2) oil wiper (tapered with narrow edge up), or (3) uniflow-effect (tapered with wide edge up). The choice of the cross section to be used is normally determined by the manufacturer. In some cases the upper oil control ring is made up of two or more parts. **Ventilated-type oil control rings** are usually of two-piece construction and are made with a number of equally spaced slots around the entire circumference of the ring, to allow the oil to drain through to the holes in the ring groove and then into the crankcase. Another ring assembly consists of two thin steel rings, one on each side of a cast-iron ring. The cast-iron ring has cutouts along the sides to permit the flow of oil into the groove and through drilled holes to the inside of the piston. The choice of the particular ring to be used with an engine is normally made by the manufacturer. Where there is more than one position in which a ring may be installed, it should be marked to show the correct installation.

Narrow-surface rings are preferred to wide-surface ones because they adapt themselves better to the wall of the cylinder. Regardless of the cross section or the width, all modern piston rings are constructed to withstand wear and deterioration. Wear is generally caused by fine abrasive particles either in the lubricating oil or entering through the intake air and by the friction existing between the rings, ring grooves, and cylinder walls.

Wedge-shaped piston rings are fitted to beveled-edge grooves to obtain a sliding, self-cleaning action that will prevent sticking between the ring and groove. Also, the ring lands remain stronger when there are beveled-edge grooves between them. Since compression rings operate at the highest temperatures and are farthest from the oil source, they receive less lubrication than other rings and have the greatest tendency to stick. Therefore, wedge-shaped rings are installed as compression rings; that is, they are placed in the ring grooves immediately below the piston head.

On certain radial-type aircraft engines, the oil wiper rings on the upper cylinders face the piston head or dome to carry more oil to the top piston rings; therefore, the oil rings above the top ring serve as wiper rings. In that case, the oil control rings on the lower cylinders normally face the crankshaft to prevent overlubrication.

Piston Pins

A **piston pin**, sometimes called a **wrist pin**, is used to attach the piston to the connecting rod. It is made of steel (AMS 6274 or AMS 6322) hollowed for lightness and surface-hardened or through-hardened to resist wear. The pin passes through the piston at right angles to the skirt so that the piston can be anchored to the connecting-rod assembly. A means is provided to prevent the piston pin from moving sideways in the piston and damaging the cylinder wall. The piston pin is mounted in bosses and bears directly on the aluminum alloy of which the pistons are made. When the piston is made of an aluminum alloy, however, the bosses may or may not be lined with some nonferrous (no iron or steel) metal, such as bronze. Figure 2-32 shows a piston pin in a piston-pin boss. The piston pin passes through the piston bosses and through the small end of the connecting rod which rides on the central part of the pin.

Piston pins are usually classified as **stationary** (rigid), **semifloating**, or **full-floating piston pins**. The stationary type is not free to move in any direction and is securely fastened in the boss by means of a setscrew. The semifloating piston pin is securely held by means of a clamp screw in the end of the connecting rod and a half slot in the pin itself. The full-floating type is free to run or slide in both the connecting rod and the piston and is the most widely used in modern aircraft engines.

Piston-Pin Retainers

The three devices used to prevent contact between the piston-pin ends and the cylinder wall are circlets, spring rings, and nonferrous-metal plugs. Figure 2-33 shows these three retainers.

Circlets resemble piston rings and fit into grooves at the outside end of each piston boss.

Spring rings are circular spring-steel coils which fit into circular grooves cut into the outside end of each piston boss to prevent movement of the pin against the cylinder wall.

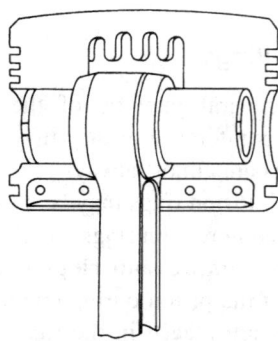

FIGURE 2-32   Piston pin in a piston-pin boss.

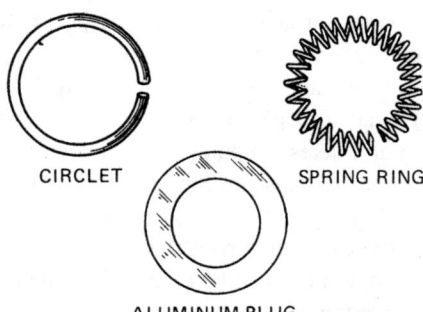

FIGURE 2-33 Piston-pin retainers.

Nonferrous-metal plugs, usually made of an aluminum alloy, are called **piston-pin plugs** and are used in most aircraft engines. They are inserted in the open ends of a hollow piston pin to prevent the steel pin end from bearing against the cylinder wall. The comparatively soft piston-pin plugs may bear against the cylinder walls without damage to either the plug or the wall.

Piston pins are fitted into the pistons and the connecting rod with clearances of less than 0.001 in [0.025 4 mm]. This is commonly called a "push fit" because the pin can be inserted in the piston boss by pushing with the palm of the hand. The proper clearances for piston pins, bosses, and connecting rods are listed in the Table of Limits for any particular engine.

CYLINDERS

The **cylinder** of an internal-combustion engine converts the chemical heat energy of the fuel to mechanical energy and transmits it through pistons and connecting rods to the rotating crankshaft. In addition to developing the power from the fuel, the cylinder dissipates a substantial portion of the heat produced by the combustion of the fuel, houses the piston and connecting-rod assembly, supports the valves and a portion of the valve-actuating mechanism, and supports the spark plugs.

The cylinder assembly used for present-day engines usually includes the following components: (1) cylinder barrel with an integral skirt, (2) cylinder head, (3) valve guides, (4) valve rocker-arm supports, (5) valve seats, (6) spark plug bushings, and (7) cooling fins. The cylinder assemblies, together with the pistons, connecting rods, and crankcase section to which they are attached, may be regarded as the main power section of the engine.

The two major units of the cylinder assembly are the cylinder barrel and the cylinder head. These are shown in Fig. 2-34. The principal requirements for this assembly are (1) sufficient strength to withstand the internal pressures developed during operation at the temperatures which are normally developed when the engine is run at maximum design loads, (2) lightweight, (3) the heat-conducting properties to obtain efficient cooling, and (4) a design which makes possible easy and inexpensive manufacture, inspection, and maintenance.

Cylinder Barrel

In general, the barrel in which the piston reciprocates must be made of a high-strength steel alloy, be constructed to save weight as much as possible, have the proper characteristics for operating under high temperatures, be made of a good bearing material, and have high tensile strength. The barrel is usually made of chromium-molybdenum steel or chromium nickel-molybdenum steel which is initially forged to provide maximum strength. The forging is machined to design dimensions with external fins and a smooth cylindrical surface inside. After machining, the inside surface is honed to a specific finish to provide the proper bearing surface for the piston rings. The roughness of this surface must be carefully controlled. If it is too smooth, it will not hold sufficient oil for the break-in period; if it is too rough, it will lead to excessive wear or other damage to both the piston rings and the cylinder wall. The inside of the cylinder barrel may be surface-hardened by means of nitriding, or it may be chromium plated to provide a long-wearing surface.

Chromium plating is usually done as a repair procedure and is accomplished by plating chromium on the cylinder wall. Chromium provides a very hard surface and possesses several advantages when compared to plain-steel or nitrided cylinder walls. Chromium is very resistant to corrosion and rusting. It has a uniform hardness and a very low coefficient of friction, and with the use of lubrication channels, it possesses excellent wear qualities.

Nitriding is a process whereby the nitrogen from anhydrous ammonia gas is forced to penetrate the surface of the steel by exposing the barrel to the ammonia gas for 40 h or more while the barrel is at a temperature of about 975°F [524°C].

In some cylinders, the cylinder is bored with a slight taper. The end of the bore nearest the head is made smaller than the skirt end to allow for the expansion caused by the greater operating temperatures near the head. Such a cylinder is said to be **chokebored**; it provides a nearly straight bore at operating temperatures.

The base of the cylinder barrel incorporates (as part of the cylinder) a machined mounting flange by which the cylinder is attached to the crankcase. The flange is drilled to provide holes for the mounting studs or bolts. The holes are reamed for accurate dimensioning. The cylinder **skirt** extends beyond the flange into the crankcase and makes it possible to use a shorter connecting rod and to reduce the external dimensions of the engine. The cylinders for inverted engines and the lower cylinders of radial engines are provided with extra-long skirts. These skirts keep most of the lubricating oil from draining into the cylinders. This reduces oil consumption and decreases the possibility of **hydraulic lock** (also termed **liquid lock**), which results from oil collected in the cylinder head.

The outer end of the cylinder barrel is usually provided with threads so that it can be screwed and shrunk into the cylinder head, which is also threaded. The cylinder head is heated in an oven to 575 to 600°F [302 to 316°C] and is then screwed onto the cool cylinder barrel.

As mentioned previously, cooling fins are generally machined directly on the outside of the barrel. This method provides the best conduction of heat from the inside of the barrel to the cooling air. On some cylinders, the cooling fins are on aluminum-alloy muffs or sleeves shrunk onto the outside of the barrel.

The barrel consists of a cylindrical, centrifugally cast sleeve of alloyed gray iron around which an aluminum-finned muff is die-cast. The aluminum muff (jacket) transmits heat from the liner to the cooling fins.

Cylinder Heads

The cylinder head encloses the combustion chamber for the fuel-air mixture and contains the intake and exhaust valves, valve guides, and valve seats. The cylinder head also provides the support for the rocker shafts upon which the valve rocker arms are mounted.

The openings into which the spark plugs are inserted are provided in the cylinder head at positions designed to provide the best burning pattern. The spark plug openings on older engines may contain bronze bushings shrunk and staked, or pinned, into the head. In most modern cylinders the threads are reinforced with steel inserts called **Heli-Coils**. The Heli-Coil inserts make it possible to restore the thread by replacement of the inserts.

Cylinder heads are usually made of a cast aluminum alloy (AMS 4220 or equivalent), to provide maximum strength with minimum weight. One disadvantage of using aluminum alloys for cylinder heads is that the coefficient of expansion of aluminum is considerably greater than that of steel. This disadvantage is largely overcome through the method by which the cylinder heads are attached to the cylinder barrels.

The cooling fins are cast or machined on the outside of the cylinder head in a pattern that provides the most efficient cooling and takes advantage of cylinder-head cooling baffles. The area surrounding the intake passage and valve does not usually have cooling fins because the fuel-air mixture entering the cylinder carries the head away. The intake side of the cylinder head can be quickly identified by noting which side is not finned.

The interior shape of the cylinder head may be flat, peaked, or hemispherical, but the latter shape is preferred because it is more satisfactory for scavenging the exhaust gases rapidly and thoroughly.

The three methods used for joining the cylinder barrel to the cylinder head are (1) the threaded-joint method, (2) the shrink-fit method, and (3) the stud-and-nut joint method. The method most commonly employed for modern engines is the threaded-joint method.

The threaded-joint method consists of chilling the cylinder barrel, which has threads at the head end, and heating the cast cylinder head to about 575°F [302°C], as previously explained. The cylinder head is threaded to receive the end of the barrel. A jointing compound is placed on the threads to prevent compression leakage, and then the barrel is screwed into the cylinder head. When the cylinder head cools, it contracts and grips the barrel tightly.

The cylinder head is provided with machined surfaces at the intake and exhaust openings for the attachment of the intake and exhaust manifolds. The manifolds are held in place by bolted rings which fit against the flanges of the

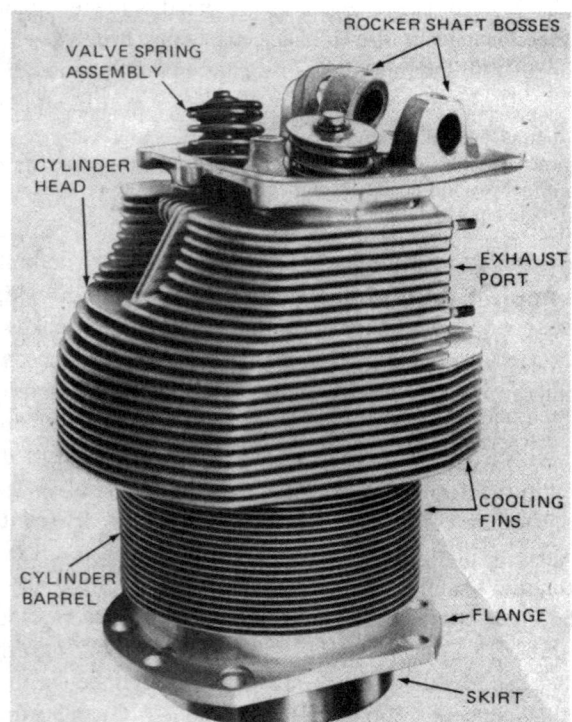

FIGURE 2-34 Cylinder assembly. (*Continental Motors.*)

manifold pipes. The intake pipes are usually provided with synthetic-rubber gaskets which seal the joint between the pipe and the cylinder. Exhaust pipes are usually sealed by metal or metal-and-asbestos gaskets. The mounting studs are threaded into the cast cylinder heads and are usually not removed except in case of damage.

The cylinder heads of radial engines are provided with fittings to accommodate rocker-box intercylinder drain lines (hoses) which allow for the evening of pressure and oil flow between cylinder heads. If oil flow is excessive in one or more rocker boxes. the excess will be relieved by flowing to other rocker boxes. The intercylinder drain lines also ensure adequate lubrication for all rocker boxes. If one or more of the rocker-box intercylinder drain lines becomes clogged, it is likely that excessive oil consumption will occur and that the spark plugs in the cylinders adjacent to the clogged lines will become fouled.

Cylinder Exterior Finish

Since air-cooled cylinder assemblies are exposed to conditions leading to corrosion, they must be protected against this form of deterioration. One method is to apply a coating of baked, heat-resistant enamel. In the past, this enamel was usually black; however, manufacturers have developed coatings which not only provide corrosion protection but also change color when over-temperature conditions occur. The Continental gold developed by Teledyne Continental Motors is normally gold but turns pink when subjected to excessive temperatures. The blue-gray enamel used on Lycoming engines also changes color when overheated, enabling the technician to detect possible heat damage during inspection of the engine.

In the past, manufacturers *metallized* engine cylinders, particularly for large radial engines, by spraying a thin layer of molten aluminum on the cylinders with a special metallizing gun. The aluminum coating, when properly applied, was effective in providing protection against the corrosive action

of saltwater spray, salt air, and the blasting effect of sand and other gritty particles carried by the cooling airstream.

VALVES AND ASSOCIATED PARTS

Definition and Purpose

In general, a **valve** is any device for regulating or determining the direction of flow of a liquid, gas, etc., by a movable part which opens or closes a passage. The word "valve" is also applied to the movable part itself.

The main purpose of valves in an internal-combustion engine is to open and close **ports**, which are openings into the combustion chamber of the engine. One is called the **intake port**, and its function is to allow the fuel-air charge to enter the cylinder. The other is called the **exhaust port** because it provides an opening through which burned gases are expelled from the cylinder.

Each cylinder must have at least one intake port and one exhaust port. On some liquid-cooled engines of high power output, two intake and two exhaust ports are provided for each cylinder. The shape and form of all valves are determined by the design and specifications of the particular engine in which they are installed.

Poppet-Type Valves

The word "poppet" comes from the popping action of the valve. This type of valve is made in four general configurations with respect to the shape of the valve head: (1) the flat-headed valve, (2) the semitulip valve, (3) the tulip valve, and (4) the mushroom valve. These are illustrated in Fig. 2-35.

Valves are subjected to high temperatures and a corrosive environment; therefore, they must be made of metals which resist these deteriorating influences. Since intake (or inlet) valves operate at lower temperatures than exhaust valves, they may be made of chromium-nickel steel. Exhaust valves, which

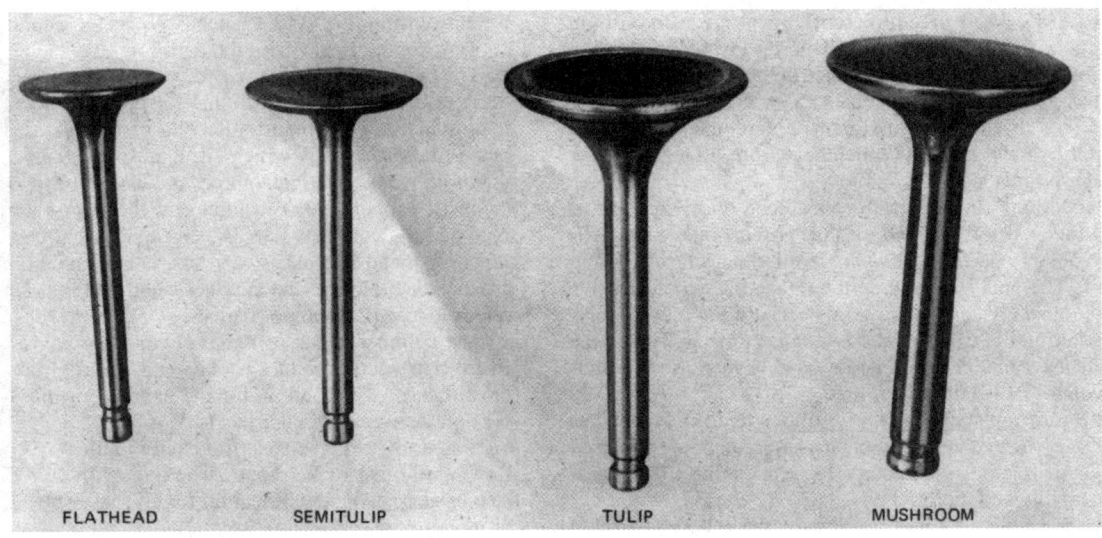

FLATHEAD   SEMITULIP   TULIP   MUSHROOM

FIGURE 2-35   Types of poppet valves.

operate at higher temperatures, are usually made of nichrome, silchrome, or cobalt-chromium steel. Poppet valves are made from these special steels and forged in one piece.

A **valve stem** is surface-hardened to resist wear. Since the **valve tip** must resist both wear and pounding, it is made of hardened steel and welded to the end of the stem. There is a machined **groove** on the stem near the tip to receive split-ring stem keys. These stem keys hold a split lock ring to keep the valve-spring retaining washer in place.

The stems of some radial-engine valves have a narrow groove below the lock-ring groove for the installation of **safety circlets** or **spring rings** which are designed to prevent the valves from falling into the combustion chambers if the tip should break during engine operation or during valve disassembly and assembly.

Exhaust Valves

Exhaust valves operate at high temperatures, and they do not receive the cooling effect of the fuel-air charge; therefore, they must be designed to dissipate heat rapidly. This is accomplished by making the exhaust valve with a hollow stem, and in some cases with a hollow mushroom head, and by partly filling the hollow portion with metallic sodium. The sodium melts at a little over 200°F [93.3°C], and during operation it flows back and forth in the stem, carrying heat from the head and dissipating it through the valve guides into the cylinder head. The cylinder head is cooled by fins, as explained previously. In some engines, exhaust valve stems contain a salt that serves as the cooling agent.

Not all lower-power engines are equipped with sodium-filled exhaust valves. It is important, however, for technicians to determine whether the exhaust valve stems upon which they may be working are of this type. *Under no circumstances should a sodium valve be cut open, hammered, or otherwise subjected to treatment which may cause it to rupture. Furthermore, sodium valves must always be disposed of in an appropriate manner.*

The faces of high-performance exhaust valves are often made more durable by the application of about 1/16 in [1.59 mm] of a material called **Stellite** (a very hard, heat-resisting alloy). This alloy is welded to the face of the valve and then ground to the correct angle. Stellite is resistant to high-temperature corrosion and withstands exceptionally well the shock and wear associated with valve operation.

The face of the valve is usually ground to an angle of either 30 or 45°. In some engines, the intake valve face is ground to an angle of 30°, and the exhaust valve face is ground to 45°. The 30° angle provides better airflow, and the 45° angle allows increased heat flow from the valve to the valve seat. This is of particular benefit to the exhaust valve.

The tip of the valve stem is often made of high-carbon steel or Stellite so it can be hardened to resist wear. Remember that the tip of the valve stem is continuously receiving the impact of the rocker arm as the rocker arm opens and closes the valve.

Intake Valves

Specially cooled valves are not generally required for the intake port of an engine because the intake valves are cooled by the fuel-air mixture. For this reason, the most commonly used intake valves have solid stems, and the head may be flat or of the tulip type. The valve is forged from one piece of alloy steel and then machined to produce a smooth finish. The stem is accurately dimensioned to provide the proper clearance in the valve guide. The intake valve stem has a hardened tip similar to that of the exhaust valve.

Intake valves for low-power engines usually have flat heads. Tulip-type heads are often used on the intake valves for high-power engines because the tulip shape places the metal of the head more nearly in tension, thus reducing the stresses where the head joins the stem.

Valve Guides

As shown in Fig. 2-36, the **valve guides** are positioned to support and guide the stems of the valves. The valve guides

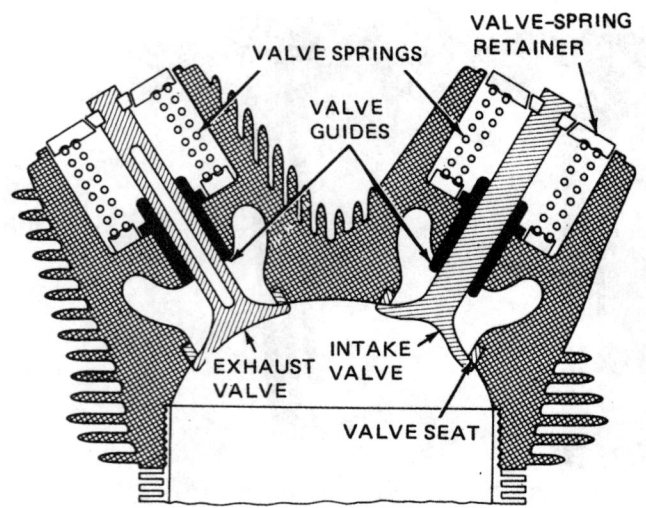

FIGURE 2-36 Installation arrangement for valve guides.

are shrunk into bored bosses with a 0.001- to 0.0025-in [0.025 4- to 0.063 5-mm] tight fit. Before the valve guides are installed, the cylinder head is heated to expand the holes in which the guides are to be installed. The guides are then pressed into place or driven in with a special drift. When the cylinder head cools, the guide is gripped so tightly that it will not become loose, even under severe heating conditions. It is common practice when valve guides are replaced to install new guides which are approximately 0.002 in [0.05 mm] larger than the holes in which they are to be installed. Valve guides are made of aluminum bronze, tin bronze, or steel, and in some cylinders the exhaust valve guide is made of steel while the intake valve guide is made of bronze.

Valve Seats

The aluminum alloy used for making engine cylinder heads is not hard enough to withstand the constant hammering produced by the opening and closing of the valves. For this reason, bronze or steel valve seats are shrunk or screwed into the circular edge of the valve opening in the cylinder head, as shown in Fig. 2-37. A typical six-cylinder opposed engine has forged aluminum-bronze intake valve seats and forged chromium-molybdenum steel seats for the exhaust valves.

In some cases the exhaust valve seat is made of steel with a layer of Stellite bonded on the seat surface to provide a more durable seat. Before the seat is installed, its outside diameter is from 0.007 to 0.015 in [0.178 to 0.381 mm] larger than the recess in which it is to be installed. To install the seats, the cylinder head must be heated to 575°F [301°C] or more.

The seat and mandrel can be chilled with dry ice, and the seat is pressed or drifted into place while the cylinder head is hot. Upon cooling, the head recess shrinks and grips the seat firmly. It is not always necessary to chill the seat before assembling it in the head; heating the cylinder head may provide sufficient clearance for installation.

When it is necessary to replace valve guides and valve seats, the valve guides should be replaced first, because the pilots for the seat tools are centered by the valve guides.

Valve seats are ground to the same angle as the face of the valve, or they may have a slightly different angle to provide an "interference fit." To obtain good seating at operating temperatures, valve faces are sometimes ground to an angle $\frac{1}{4}$ to 1° less than the angle of the valve seats. This provides a line

contact between the valve face and seat and permits a more positive seating of the valve, particularly when the engine is new or freshly overhauled. The importance of proper valve and seat grinding and lapping is discussed at length in the section on cylinder overhaul in Chap. 10.

Valve Springs

Valves are closed by helical-coil springs. Two or more springs, one inside the other, are installed over the stem of each valve. If only one spring were used on each valve, the valve would surge and bounce because of the natural vibration frequency of the spring. Each spring of a pair of springs is made of round spring-steel wire of a different diameter, and the two coils differ in pitch. Since the springs have different frequencies, the two springs together rapidly damp out all spring-surge vibrations during engine operation. A second reason for the use of two (or more) valve springs on each valve is that it reduces the possibility of failure by breakage from heat and metal fatigue.

The valve springs are held in place by steel **valve-spring retainers**, which are special washers shaped to fit the valve springs. The lower retainer seats against the cylinder head, and the upper retainer is provided with a conical recess into which the split stem keys (keepers) fit. The valve-spring retainers are sometimes called the upper and lower **valve-spring seats**.

Valve Operating Mechanism

The purpose of a valve operating mechanism in an aircraft engine is to control the timing of the engine valves so that each valve will open at the correct time, remain open for the required time, and close at the proper time. The mechanism should be simple in design, be ruggedly constructed, and give satisfactory service for a long time with a minimum of inspection and maintenance.

The two types of valve operating mechanisms most generally used today are the type found in the opposed engine and the type used in a typical radial engine. Since both these engines are equipped with overhead valves (valves in the cylinder head), the valve operating mechanisms for each are quite similar.

Valve Mechanism Components

A standard valve operating mechanism includes certain parts which are found in both opposed and radial engines. These parts may be described briefly as follows:

Cam. A device for actuating the valve lifting mechanism.

Valve lifter or **tappet.** A mechanism for transmitting the force of the cam to the valve pushrod.

Pushrod. A steel or aluminum-alloy rod or tube, situated between the valve lifter and the rocker arm of the valve operating mechanism for transmitting the motion of the valve lifter.

Rocker arm. A pivoted arm, mounted on bearings in the cylinder head, for opening and closing the valves. One end

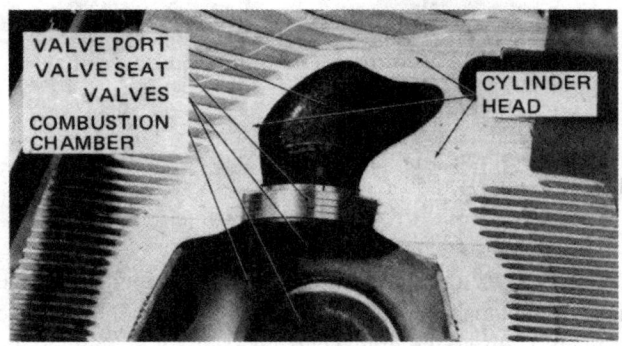

FIGURE 2-37 Valve seat in the cylinder head.

FIGURE 2-38 Camshaft for a six-cylinder opposed engine.

of the arm presses on the stem of the valve, and the other end receives motion from the pushrod.

The valve operating cam in an opposed or in-line engine consists of a shaft with a number of cam lobes sufficient to operate all the intake and exhaust valves of the engine. In a typical six-cylinder opposed engine, the camshaft has three groups of three cam lobes, as shown in Fig. 2-38. In each group, the center lobe actuates the valve lifters for the two opposite intake valves, whereas the outer lobes of each group actuate the lifters for the exhaust valves.

In a radial engine, the valve actuating device is a **cam plate** (or **cam ring**) with three or more lobes. In a five-cylinder radial engine, the cam ring usually has three lobes; in a seven-cylinder radial engine, the cam ring has three or four lobes; and in a nine-cylinder radial engine, the cam ring has four or five lobes.

Valve operating mechanisms for in-line and V-type engines utilize camshafts similar to those installed in opposed engines. The shaft incorporates single cam lobes placed at positions along the shaft which enable them to actuate the valve lifters or rocker arms at the correct times. Some in-line and V-type engines have overhead camshafts. The camshaft is mounted along the top of the cylinders and is driven by a system of bevel gears through a shaft leading from the crankshaft drive gear.

Valve Mechanism for Opposed Engines

A simplified drawing of a valve operating mechanism is shown in Fig. 2-39. The valve action starts with the crankshaft timing gear, which meshes with the camshaft gear. As the crankshaft turns, the camshaft also turns, but at one-half the rpm of the crankshaft. This is because a valve operates only once during each cycle and the crankshaft makes two revolutions (r) per cycle. A cam lobe on the camshaft raises the cam roller and therefore the pushrod to which the cam roller is attached. The ramp on each side of the cam lobe is designed to reduce opening and closing shock through the valve operating mechanism. In opposed engines, a cam roller is not employed, and in its place is a tappet or a hydraulic lifter. The pushrod raises one end of the rocker arm and lowers the other end, thus depressing the valve, working against the tension of the valve spring which normally holds the valve closed. When the cam lobe has passed by the valve lifter, the valve will close by the action of the valve spring or springs.

The valve-actuating mechanism starts with the drive gear on the crankshaft. This gear may be called the **crankshaft**

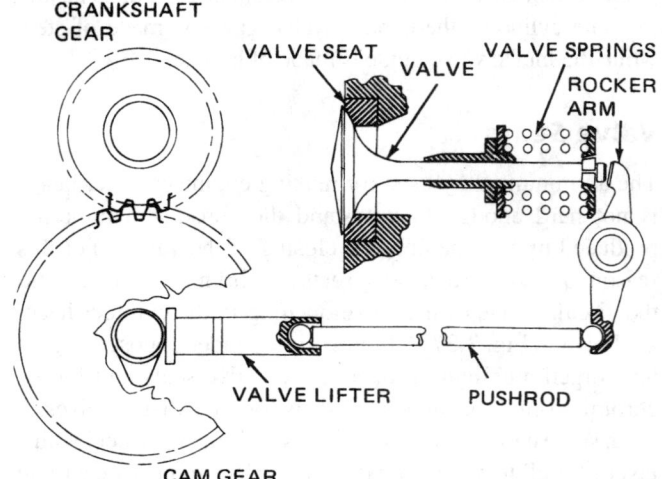

FIGURE 2-39 Valve operating mechanism for an opposed engine.

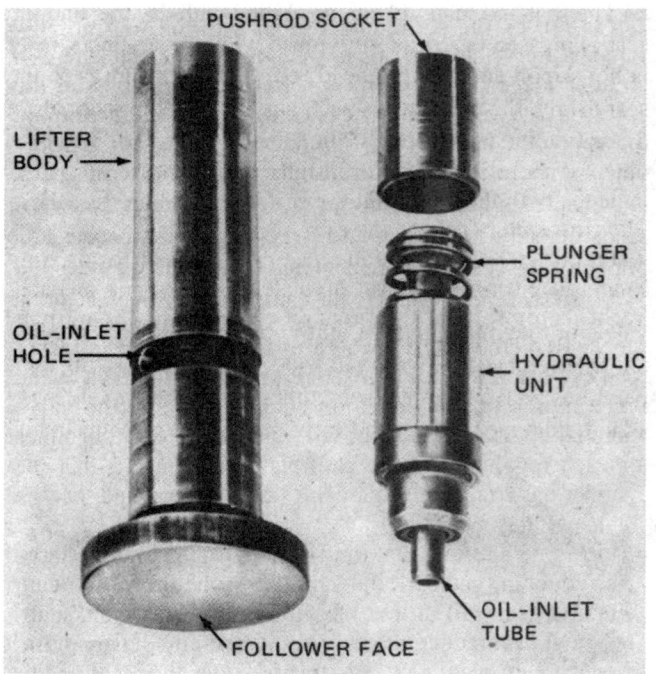

FIGURE 2-40 Hydraulic valve lifter assembly.

timing gear or the **accessory drive gear**. Mounted on the end of the camshaft is the **camshaft gear**, which has twice as many teeth as the crankshaft gear. In some engines the mounting holes in the camshaft gear are spaced so that they line up with the holes in the camshaft flange in only one position. In other engines, a dowel pin in the end of the crankshaft mates with a hole in the crankshaft gear to ensure correct position. Thus, when the timing marks on the camshaft and the crankshaft gear are aligned, the camshaft will be properly timed with the crankshaft.

Adjacent to each cam lobe is the **cam follower face**, which forms the base of the hydraulic valve lifter or lappet assembly. The outer cylinder of the assembly is called the **lifter body**. Inside the lifter body is the **hydraulic unit assembly**, consisting of the following parts: **cylinder, plunger, plunger springs, ball check valve**, and **oil inlet tube**. Figure 2-40 is an illustration of the complete lifter assembly. During operation, engine oil under pressure is supplied to the oil reservoir in the lifter body through an inlet hole in the side, as shown in Fig. 2-41. Since this oil is under pressure directly from the main oil gallery of the engine, it flows into the oil inlet tube, through the ball check valve, and into the cylinder. The pressure of the oil forces the plunger against the **pushrod socket** and takes up all the clearances in the valve operating mechanism during operation. For this reason, a lifter of this type has been called a "zero-lash lifter." When the cam is applying force to the cam follower face, the oil in the cylinder tends to flow back into the oil reservoir, but this is prevented by the ball check valve.

During overhaul of the engine, the hydraulic valve lifter assembly must be very carefully inspected. All the parts of a given assembly must be reassembled in order to ensure proper operation.

The ball end of the hollow valve pushrod fits into the pushrod socket, or cup, which bears against the plunger in the lifter. Both the socket and the ball end of the pushrod are drilled to provide a passage for oil to flow into the pushrod. This oil flows through the pushrod and out a hole at the end (the hole fits the pushrod socket in the **rocker arm**), thus providing lubrication for the rocker arm bearing (bushing) and valves. The rocker arm is drilled to permit oil flow to the bearing and valve mechanism.

Typical rocker arms are illustrated in Fig. 2-42. Rocker arms shown at (B) and (C) are designed for opposed-type engines. The rocker arm at (A) is used in a Pratt & Whitney R-985 radial engine.

The rocker arm is mounted on a steel shaft which, in turn, is mounted in rocker-shaft bosses in the cylinder head. The rocker-shaft bosses are cast integrally with the cylinder head and then are machined to the correct dimension and finish for installation of the rocker shafts. The rocker-shaft dimension provides a push fit in the boss. The shafts are held in place by the rocker-box covers or by covers over the holes through which they are installed. The steel rocker arms are fitted with bronze bushings to provide a good bearing surface. These bushings may be replaced at overhaul if they are worn beyond acceptable limits.

One end of each rocker arm bears directly against the hardened tip of the valve stem. When the rocker arm is rotated by the pushrod, the valve is depressed, acting against the valve spring pressure. The distance the valve opens and the time it remains open are determined by the height and contour of the cam lobe.

Valve Mechanism for Radial Engines

Depending on the number of rows of cylinders, the valve operating mechanism of a radial engine is operated by either one or two cam plates (or cam rings). Only one plate (or ring) is used with a single-row radial engine, but a double cam

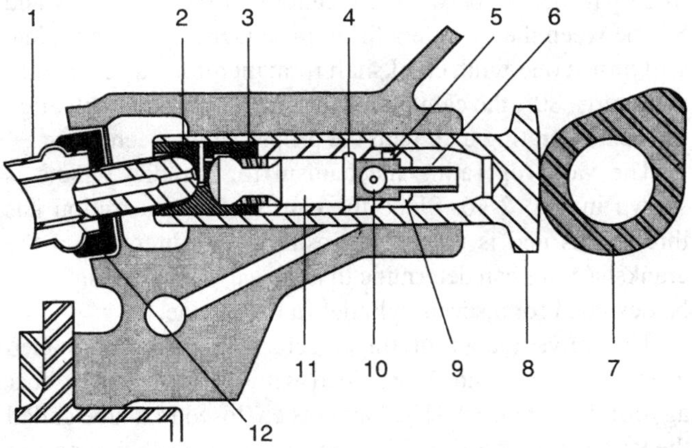

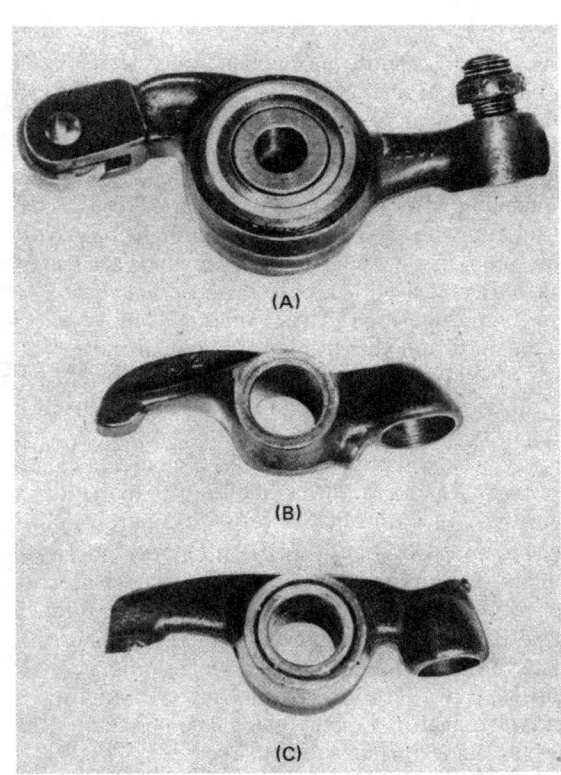

(A)

(B)

(C)

1. Shroud tube
2. Pushrod socket
3. Plunger spring
4. Oil pressure chamber
5. Oil hole
6. Oil supply chamber
7. Camshaft
8. Tappet body
9. Cylinder
10. Ball check valve
11. Plunger
12. Pushrod

FIGURE 2-41 Hydraulic valve lifter assembly cutaway view.

FIGURE 2-42 Typical rocker arms.

track is required. One cam track operates the intake valves, and the other track operates the exhaust valves. In addition, there are the necessary pushrods, rocker-arm assemblies, and tappet assemblies that make up the complete mechanism.

A cam ring (or cam plate), such as the one shown in Fig. 2-43, serves the same purpose in a radial engine as a camshaft serves in other types of engines. The cam ring is a circular piece of steel with a series of cam lobes on the outer surface. Each cam lobe is constructed with a **ramp** on the approach to the lobe, to reduce the shock that would occur if the lobe rise were too abrupt. The **cam track** includes both the lobes and the surfaces between the lobes. The **cam rollers** ride on the cam track.

Figure 2-44 illustrates the gear arrangement for driving a cam plate (or ring). This cam plate has four lobes on each track; therefore, it will be rotated at one-eighth crankshaft speed. Remember that a valve operates only once during each cycle and that the crankshaft makes 2 r for each cycle. Since there are four lobes on each cam track, the valve operated by one set of lobes will open and close four times for each revolution of the cam plate. This means that the cylinder

has completed four cycles of operation and that the crankshaft has made 8 r.

In Fig. 2-44, note that the crankshaft gear and the large cam reduction gear are the same size; therefore, the cam reduction gear will turn at the same rpm as the crankshaft. The small cam reduction gear is only one-eighth the diameter of the cam-plate gear, and this provides the reduction to make the cam plate turn at one-eighth crankshaft speed. The rule for cam-plate speed with respect to crankshaft speed may be given as a formula:

$$\text{cam-plate speed} = \frac{1}{\text{no. of lobes} \times 2}$$

In an arrangement of this type, the cam plate turns opposite to the direction of engine rotation. A study of the operation of the cam will lead us to the conclusion that a four-lobe cam turning in the opposite direction from the crankshaft will be used in a nine-cylinder radial engine. In the diagrams shown in Fig. 2-45, the numbers on the large outer ring represent the cylinders of a nine-cylinder radial engine. The firing order of such an engine is always 1-3-5-7-9-2-4-6-8. The small ring in the center represents the cam ring. In the first diagram we note that the no. 1 cam is opposite the no. 1 cylinder. We may assume that the cam is operating the no. 1 intake valve. In moving from the no. 1 cylinder to the no. 3 cylinder, the next cylinder in the firing order, we see that the crankshaft must turn 80° in the direction shown. Since the cam is turning at one-eighth crankshaft speed, a lobe on the cam will move 10° while the crankshaft is turning 80°. Thus, we see that the no. 2 cam lobe will be opposite the no. 3 cylinder, as shown by the second diagram. When the crankshaft has turned another 80° to the intake operation of the no. 5 cylinder, the no. 3 cam lobe is opposite the no. 5 cylinder.

If we draw a similar diagram for a nine-cylinder radial engine with a five-lobe cam, we will note that the cam must travel in the same direction as the crankshaft. This is because there will be 72° between the centers of the cam lobes and 80° between the cylinders firing in sequence. The cam plate will turn at one-tenth crankshaft rpm; therefore, as the crankshaft turns 80°, the cam plate will turn 8°, and this will align the next operating cam with the proper valve mechanism.

The valve operating mechanism for a radial engine is shown in Fig. 2-46. Since the cam in this illustration has three lobes and is turning opposite to the direction of the crankshaft, we can determine that the valve mechanism must be designed for a seven-cylinder radial engine.

The valve tappet in this mechanism is spring-loaded to reduce shock and is provided with a cam roller to bear against the cam track. The lappet is enclosed in a tube called the **valve-tappet guide**. The valve tappet is drilled to permit the passage of lubricating oil into the hollow pushrod and up to the rocker-arm assembly. The rocker arm is provided with a **clearance-adjusting screw** so that proper clearance can be obtained between the rocker arm and valve tip. This clearance is very important because it determines when the valve will start to open, how far it will open, and how long it will stay open.

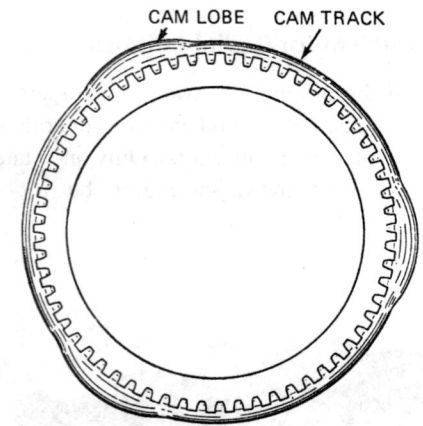

CAM LOBE   CAM TRACK

FIGURE 2-43   Cam ring.

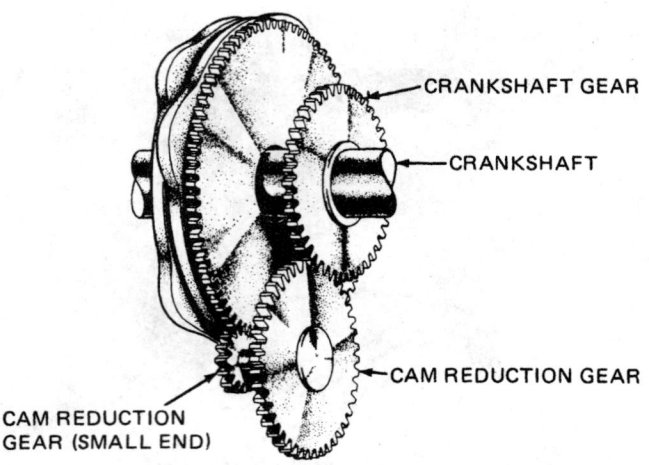

CRANKSHAFT GEAR

CRANKSHAFT

CAM REDUCTION GEAR

CAM REDUCTION GEAR (SMALL END)

FIGURE 2-44   Drive-gear arrangement for a radial-engine cam plate.

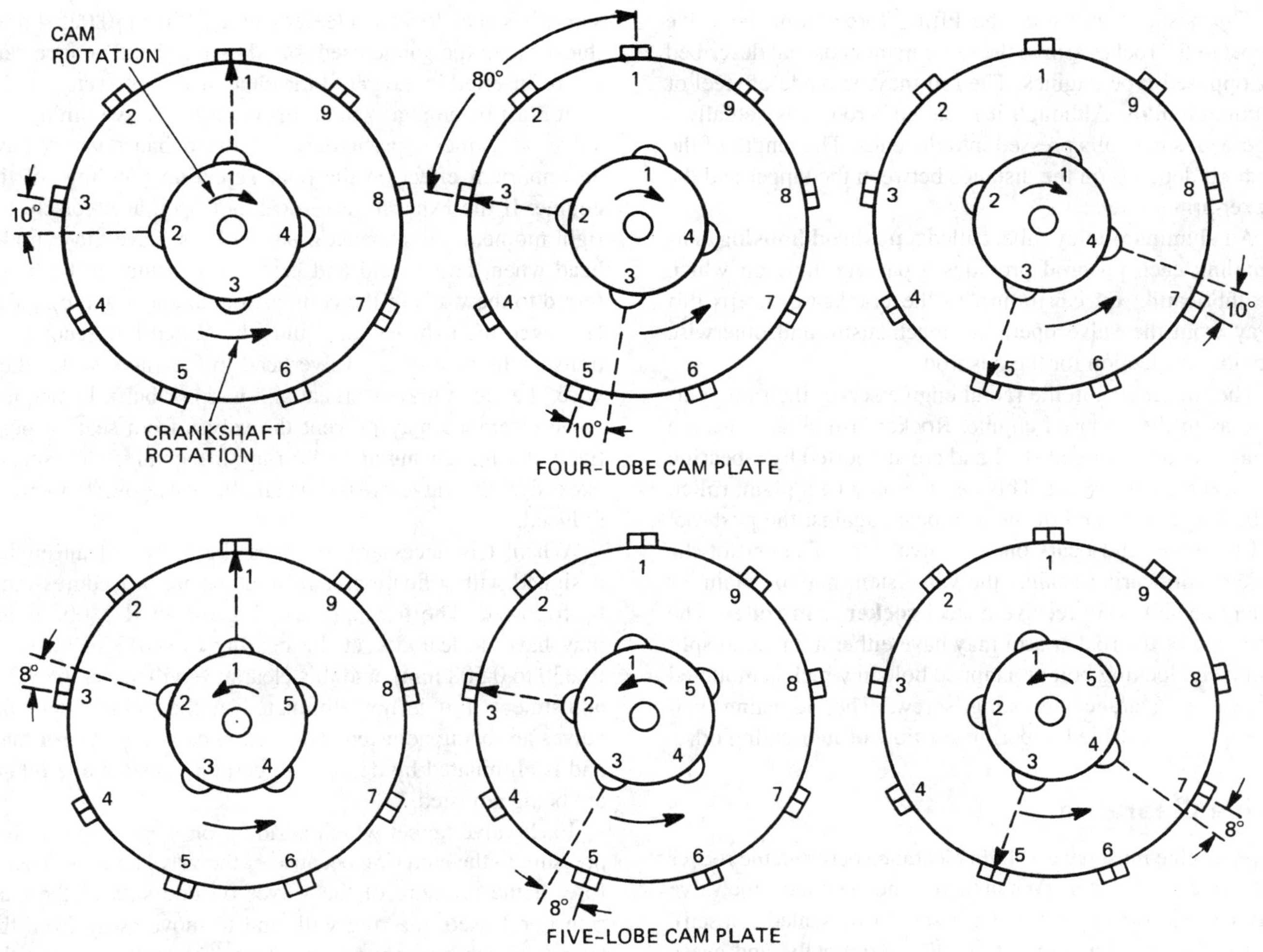

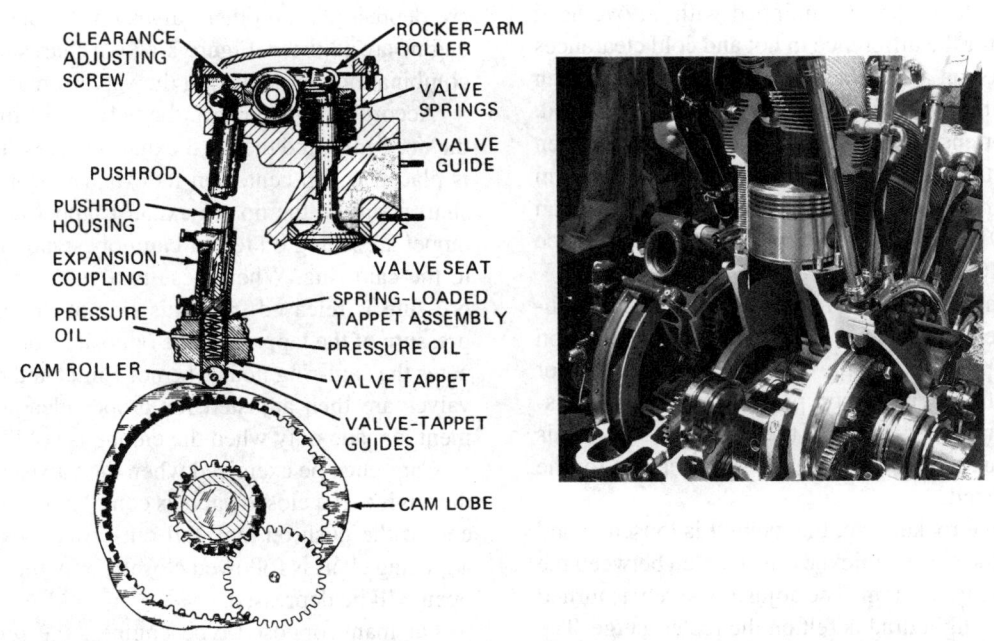

FIGURE 2-45 Diagrams showing cam-plate operation.

FOUR-LOBE CAM PLATE

FIVE-LOBE CAM PLATE

CAM ROTATION

CRANKSHAFT ROTATION

FIGURE 2-46 Valve operating mechanism for a radial engine.

CLEARANCE ADJUSTING SCREW

ROCKER-ARM ROLLER

VALVE SPRINGS

VALVE GUIDE

PUSHROD

PUSHROD HOUSING

EXPANSION COUPLING

PRESSURE OIL

CAM ROLLER

VALVE SEAT

SPRING-LOADED TAPPET ASSEMBLY

PRESSURE OIL

VALVE TAPPET

VALVE-TAPPET GUIDES

CAM LOBE

The pushrod transmits the lifting force from the valve tappet to the rocker arm in the same manner as that described for opposed-type engines. The rod may be made of steel or aluminum alloy. Although it is called a rod, it is actually a tube with steel balls pressed into the ends. The length of the pushrod depends on the distance between the tappet and the rocker-arm sockets.

An aluminum-alloy tube, called a **pushrod housing**, surrounding each pushrod provides a passage through which the lubricating oil can return to the crankcase, keeps dirt away from the valve operating mechanism, and otherwise provides protection for the pushrod.

The rocker arm in the radial engine serves the same purpose as in the opposed engine. Rocker-arm assemblies are usually made of forged steel and are supported by a bearing which serves as a pivot. This bearing may be a plain, roller, or ball type. One end of the arm bears against the pushrod, and the other end bears on the valve stem. The end of the rocker arm bearing against the valve stem may be plain, or it may be slotted to receive a steel **rocker-arm roller**. The other end of the rocker arm may have either a threaded split clamp and locking bolt or a tapped hole in which is mounted the valve clearance-adjusting screw. The adjusting ball socket is often drilled to permit the flow of lubricating oil.

Valve Clearance

Every engine must have a slight clearance between the rocker arm and the valve stem. When there is no clearance, the valve may be held off its seat when it should be seated (closed). This will cause the engine to operate erratically, and eventually the valve will be damaged. If, however, an engine is equipped with hydraulic valve lifters, there will be no apparent clearance at the valve stem during engine operation.

The **cold clearance** for the valves on an engine is usually much less than the "hot" (or operating) clearance. This is true except when the engine is equipped with an overhead cam. The reason for the difference in hot and cold clearances is that the cylinder on an engine becomes much hotter than the pushrod and therefore expands more than the pushrod. In effect, this shortens the pushrod and leaves a gap between the pushrod and the rocker arm or between the rocker arm and the valve stem. The hot valve clearance of engine can be as much as 0.070 in [1.778 mm], while the cold clearance may be 0.010 in [0.254 mm].

In adjusting the valve clearance of an engine, the technician must make sure that the cam is turned to a position where it is not applying any pressure on the pushrod. For any particular cylinder, it is good practice to place the piston in position for the beginning of the power stroke. At this point both cams are well away from the valve tappets for the valves being adjusted.

On an adjustable rocker arm, the locknut is loosened and a feeler gauge of the correct thickness is inserted between the rocker arm and the valve stem. The adjusting screw is turned to a point where a slight drag is felt on the feeler gauge. The lock screw or locknut is then tightened to the proper torque while the adjusting screw is held in place. After the adjusting screw has been locked, a feeler gauge 0.001 in [0.025 4 mm] thicker than the gauge used for adjusting the clearance cannot be inserted in the gap if the clearance is correct.

It must be emphasized at this point that valve timing and valve adjustment, particularly of the exhaust valve, have an important effect on the **heat rejection** (cooling) of the engine. If the exhaust valve does not open at precisely the right moment, the exhaust gases will not leave the cylinder head when they should and heat will continue to be transferred to the walls of the combustion chamber and cylinder. However, the exhaust valve must be seated long enough to transfer the heat of the valve head to the valve seat; otherwise, the valve may overheat and warp or burn. Inadequate valve clearance may prevent the valves from seating positively during starting and warm-up; if the valve clearance is excessive, the valve-open time and the valve overlap will be reduced.

When it is necessary to adjust the valves of an engine designed with a floating cam ring, special procedures must be followed. The floating cam ring for an R-2800 engine may have a clearance at the bearing of 0.013 to 0.020 in [0.330 to 0.508 mm], and this clearance will affect the valve adjustment if it is not eliminated at the point where the valves are being adjusted. The clearance is called **cam float** and is eliminated by depressing certain valves while others are being adjusted.

Each valve tappet which is riding on a cam lobe applies pressure to the cam ring because of the valve springs. Therefore, if the pressure of the valves on one side of the cam ring is released, the ring will tend to move away from the tappets which are applying pressure. This will eliminate the cam float on that side of the cam ring. The valves whose tappets are resting on the cam ring at or near the point where there is no clearance between the ring and the bearing surface, and which are between lobes, are adjusted, and then the crankshaft is turned to the next position. Certain valves are depressed, and other valves on the opposite side of the engine are adjusted. Figure 2-47 is a chart showing the proper combinations for adjusting the valves on an R-2800 engine.

According to the chart, the valve adjustment begins with the no. 1 inlet and the no. 3 exhaust valves. The no. 11 piston is placed at top center on its exhaust stroke. In this crankshaft position, the no. 15 exhaust tappet and the no. 7 inlet tappet are riding on top of cam lobes and applying pressure to the cam ring. When these two valves are depressed, the pressure is released from this side of the cam ring and the pressure of the tappets on the opposite side of the ring eliminates the cam-ring float. The no. 1 inlet and the no. 3 exhaust valves are then adjusted for proper clearance. The adjustment is made only when the engine is cold.

Care must be exercised when the valves are depressed on the engine. If a closed valve is completely depressed, the ball end of the pushrod may fall out of its socket. If the valve-adjusting chart is followed closely, only the valves which are open will be depressed.

On many opposed-type engines, the rocker arm is not adjustable and the valve clearance is adjusted by changing the pushrod. If the clearance is too great, a longer pushrod

Set Piston at Top Center of Its Exhaust Stroke	Depress Rockers		Adjust Valve Clearances	
	Inlet	Exhaust	Inlet	Exhaust
11	7	15	1	3
4	18	8	12	14
15	11	1	5	7
8	4	12	16	18
1	15	5	9	11
12	8	16	2	4
5	1	9	13	15
16	12	2	6	8
9	5	13	17	1
2	16	6	10	12
13	9	17	3	5
6	2	10	14	16
17	13	3	7	9
10	6	14	18	2
3	17	7	11	13
14	10	18	4	6
7	3	11	15	17
18	14	4	8	10

FIGURE 2-47 Valve-adjusting chart.

is used. When the clearance is too small, a shorter pushrod is installed. A wide range of clearances is allowable because the hydraulic valve lifters take up the clearance when the engine is operating. Valve clearance in these engines is normally checked only at overhaul.

THE ACCESSORY SECTION

The **accessory section** of an engine provides mounting pads for the accessory units, such as the fuel pressure pumps, fuel injector pumps, vacuum pumps, oil pumps, tachometer generators, electric generators, magnetos, starters, two-speed supercharger control valves, oil screens, hydraulic pumps, and other units. Regardless of the construction and location of the accessory housing, it contains and supports the gears for driving those accessories which are operated by engine power.

Accessory sections for aircraft engines vary widely in shape and design because of the engine and aircraft requirements.

Accessory Section for an Opposed Engine

The **accessory case** for a Lycoming opposed engine is shown in Fig. 2-48. This case is constructed from aluminum or magnesium and is secured to the rear of the crankcase. The accessory case conforms to the shape of the crankcase and forms part of the seal for the oil sump. The accessory case generally is for the purpose of housing and driving the engine accessories. To perform this function, it has mounting pads for the various accessories that the engine or the aircraft systems require. Some of the engine accessories that the case houses are the magnetos, oil filters, fuel pumps,

tachometer drives, and vacuum pumps. The accessory case shown in Fig. 2-48 is equipped with a dual magneto; therefore, this example contains only one magneto.

On many engines the accessory case has an internal pad to which the oil pump and its housing are bolted. There is generally a gasket between the accessory case and the engine crankcase. There is also a gasket between all engine-driven accessories and the accessory case. The accessory drive gears are usually mounted on the end of the crankcase. The accessory drive gears are housed in a cavity between the crankcase and the accessory case and are lubricated by engine oil.

In some engines the fuel pump is activated through the use of a plunger which is driven by an elliptical lobe on one of the accessory drive gears. Generally the accessory drive gears consist of a crankshaft gear, an idler gear, a camshaft gear, and various other gears which drive all the engine accessories.

The accessory case also serves as part of the lubrication system. As previously mentioned, the oil pump is housed with its drive gear and idler gear on the internal side of the accessory case. In some cases the oil pump and its housing are mounted externally on the accessory case.

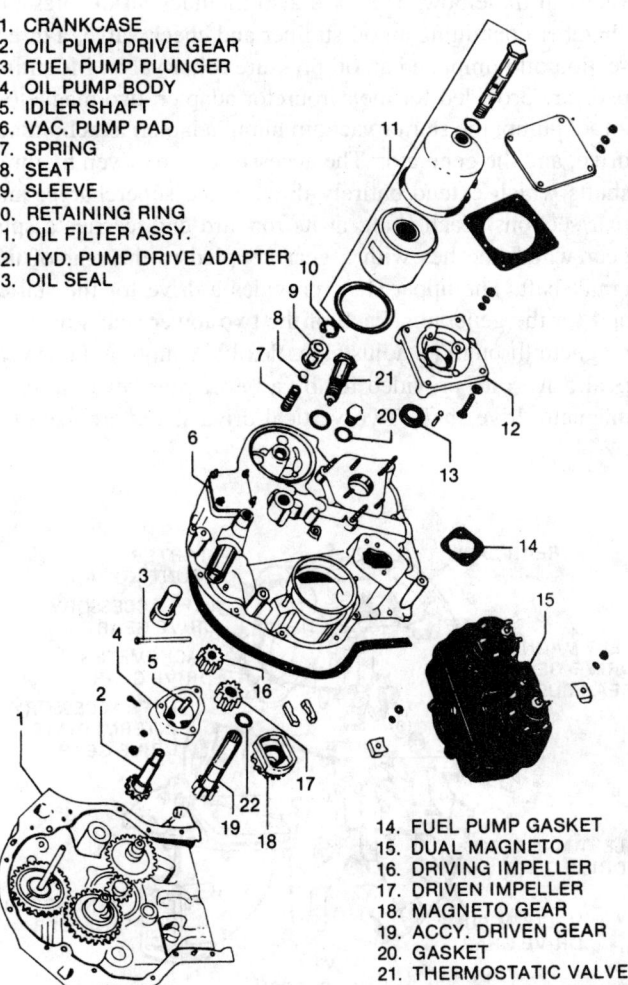

1. CRANKCASE
2. OIL PUMP DRIVE GEAR
3. FUEL PUMP PLUNGER
4. OIL PUMP BODY
5. IDLER SHAFT
6. VAC. PUMP PAD
7. SPRING
8. SEAT
9. SLEEVE
10. RETAINING RING
11. OIL FILTER ASSY.
12. HYD. PUMP DRIVE ADAPTER
13. OIL SEAL

14. FUEL PUMP GASKET
15. DUAL MAGNETO
16. DRIVING IMPELLER
17. DRIVEN IMPELLER
18. MAGNETO GEAR
19. ACCY. DRIVEN GEAR
20. GASKET
21. THERMOSTATIC VALVE
22. WASHER

FIGURE 2-48 Accessory case for a six-cylinder opposed engine. (*Textron Lycoming.*)

The accessory case has many oil passages drilled or cast into it during manufacture. Often, it serves as the mounting pad for the oil filter, oil screen, and oil cooler bypass valve. The accessory case's role in providing for engine ignition, fuel supply, oil filtering, and oil pressure makes this region of the engine very critical to engine operation. Although there are many differences in the types of accessories, the basic function of the accessory case and its drive gears is to supply the needs of the engine and other aircraft systems. Other accessories that can be located on the accessory case are the propeller governor and hydraulic pumps. Many of the accessory components have housings that contain an oil seal and a drive-gear mechanism between the accessory and the accessory case which aid in adapting the accessory drive to the accessory case.

Accessory Section for a Radial Engine

The accessory section which is shown in Fig. 2-49 is designed for the Pratt & Whitney R-985 Junior Wasp radial engine and is called the **rear case**. This case section attaches to the rear of the supercharger case and supports the accessories and accessory drives. The front face incorporates a vaned diffuser, and the rear face contains an intake duct with three vanes in the elbow. The case also includes an oil pressure chamber containing an oil strainer and check valve, a three-section oil pump, and an oil pressure relief valve. Mounting pads are provided for the carburetor adapter, two magnetos, a fuel pump, the starter vacuum pump adapter, a tachometer drive, and the generator. The accessories are driven by three shafts which extend entirely through the supercharger and rear sections. Each shaft, at its forward end, carries a spur gear which meshes with a gear coupled to the rear of the crankshaft. The upper shaft provides a drive for the starter and for the generator. Each of the two lower shafts drives a magneto through an adjustable, flexible coupling. Four vertical drives are provided for by a bevel gear keyed to each magneto drive shaft. Two vertical drive shafts are used for

operating accessories, and two tachometers are driven from the upper side of the bevel gears. The undersides of the bevel gears drive an oil pump on the right side and a fuel pump on the left. An additional drive, for a vacuum pump, is located at the lower left of the left magneto drive.

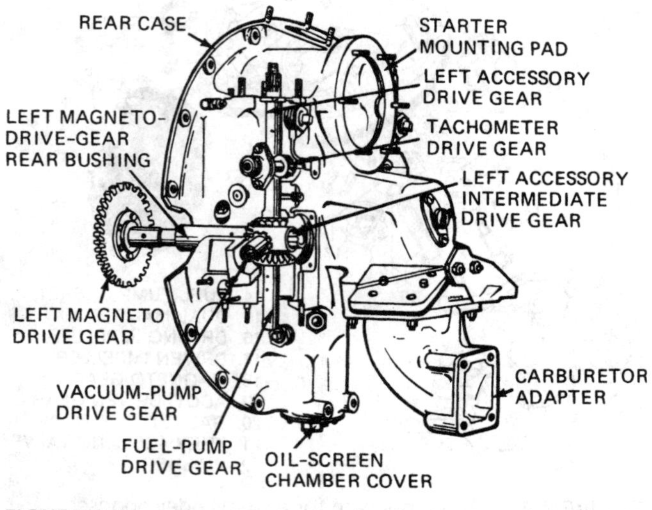

PROPELLER REDUCTION GEARS

Reduction gearing between the crankshaft of an engine and the propeller shaft has been in use for many years. The purpose of this gearing is to allow the propeller to rotate at the most efficient speed to absorb the power of the engine while the engine turns at a much higher rpm in order to develop full power. As noted in the previous chapter, the power output of an engine is directly proportional to its rpm. It follows, therefore, that an engine will develop twice as much power at 3000 rpm as it will at 1500 rpm. Thus, it is advantageous from a power-weight point of view to operate an engine at as high an rpm as possible so long as such factors as vibration, temperature, and engine wear do not become excessive.

A propeller cannot operate efficiently when the tip speed approaches or exceeds the speed of sound (1116 ft/s [340.16 m/s] at standard sea-level conditions). An 8-ft [2.45-m] propeller tip travels approximately 25 ft [7.62 m] in 1 r; therefore, if the propeller is turning at 2400 rpm (40 r/s), the tip speed is 1000 ft/s [304.8 m/s]. A 10-ft [3.05-m] propeller turning at 2400 rpm would have a tip speed of 1256 ft/s [382.83 m/s], which is well above the speed of sound.

Small engines that drive propellers no more than 6 ft [1.83 m] in length can operate at speeds of over 3000 rpm without creating serious propeller problems. Larger engines, such as the Avco Lycoming IGSO-480 and the Teledyne Continental Tiara T8-450, are equipped with reduction gears. The IGSO-480 operates at 3400 rpm maximum, and this is reduced to 2176 rpm for the propeller by means of the 0.64:1 planetary-reduction-gear system. The T8-450 engine operates at a maximum of 4400 rpm, and this is reduced to about 2200 rpm for the propeller by means of a 0.5:1 offset spur reduction gear. This ratio could also be expressed as 2:1.

FIGURE 2-49 Accessory section for the Pratt & Whitney R-985 radial engine.

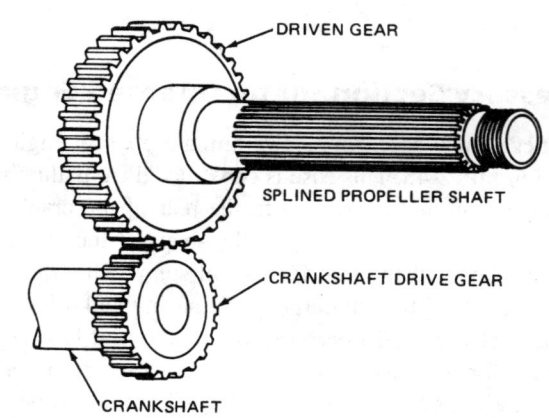

FIGURE 2-50 Spur-gear arrangement.

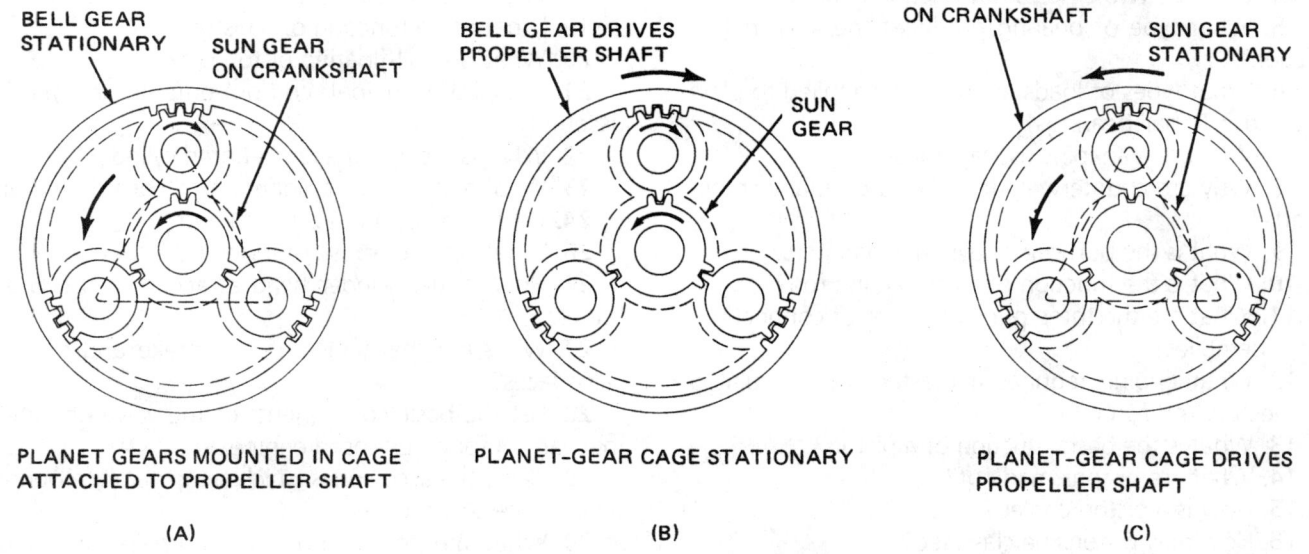

BELL GEAR STATIONARY   SUN GEAR ON CRANKSHAFT

BELL GEAR DRIVES PROPELLER SHAFT   SUN GEAR

BELL GEAR MOUNTED ON CRANKSHAFT   SUN GEAR STATIONARY

PLANET GEARS MOUNTED IN CAGE ATTACHED TO PROPELLER SHAFT

PLANET-GEAR CAGE STATIONARY

PLANET-GEAR CAGE DRIVES PROPELLER SHAFT

(A)   (B)   (C)

FIGURE 2-51   Different arrangements for planetary gears.

Remember that when reduction gears are employed, the propeller always rotates slower than the engine.

Reduction gears are designed as simple spur gears, planetary gears, bevel planetary gears, and combinations of spur and planetary gears. A **spur-gear** arrangement is shown in Fig. 2-50. The driven gear turns in a direction opposite that of the drive gear; therefore, the propeller direction will be opposite that of the engine crankshaft. The ratio of engine speed to propeller speed is inversely proportional to the number of teeth on the crankshaft drive gear and the number of teeth on the driven gear.

Arrangements for **planetary gears** are shown in Fig. 2-51. In Fig. 2-51A the outer gear, called the **bell gear**, is stationary and is bolted or otherwise secured to the inside of the engine nose case. The planet gears are mounted on a **carrier ring**, or **cage**, which is attached to the propeller shaft. The **sun gear** is mounted on the forward end of the crankshaft. When the crankshaft turns, the pinion (planet) gears rotate in a direction opposite that of the crankshaft. These gears are meshed with the stationary bell gear—therefore, they "walk" around the inside of the gear, carrying their cage with them. Since this assembly is attached to the propeller shaft, the propeller will turn in the same direction as the crankshaft and at a speed determined by the number of teeth on the reduction gears.

In Fig. 2-51B, the planet gears are mounted on stationary shafts so that they do not rotate as a group around the sun gear. When the crankshaft rotates, the sun gear drives the planet pinions which, in turn, drive the bell gear in a direction opposite the rotation of the crankshaft.

The arrangement where the sun gear is stationary is shown in Fig. 2-51C. Here the bell gear is mounted on the crankshaft, and the planet gear cage is mounted on the propeller shaft. The planet gears walk around the sun gear as they are rotated by the bell gear in the same direction as the crankshaft.

FIGURE 2-52   Bevel-planetary-gear arrangement.

In a **bevel-planetary-gear** arrangement (Fig. 2-52), the planet gears are mounted in a forged-steel cage attached to the propeller shaft. The bevel drive gear (sun gear) is attached to the forward end of the crankshaft, and the stationary bell gear is attached to the engine case. As the crankshaft rotates, the drive gear turns the pinions and causes them to walk around the stationary gear, thus rotating the cage and the propeller shaft. The bevel-gear arrangement makes it possible to use a smaller-diameter reduction-gear assembly, particularly where the reduction-gear ratio is not great.

REVIEW QUESTIONS

1. Describe the functions of a crankcase.
2. Of what material is a crankcase usually made?
3. Name the principal sections of the crankcase for a radial engine.

4. Describe two types of antifriction bearings.

5. What type of bearing produces the least rolling friction?

6. What types of loads are normally applied to plain bearings? To ball bearings?

7. Why are crankpins usually hollow?

8. Why are counterweights needed on many crankshafts?

9. What is the purpose of dynamic dampers?

10. What is the function of a connecting rod?

11. What are the three principal types of connecting-rod assemblies?

12. What engine requires a master and articulated connecting-rod assembly?

13. What is the basic function of a piston?

14. What are pistons made of?

15. How is a piston cooled?

16. How may pistons be classified?

17. Why is the piston-ring gap important?

18. What are the principal functions of piston rings?

19. What is the function of a piston pin?

20. What is a full-floating piston pin?

21. List the principal components in a cylinder assembly.

22. What is meant by a chokebored cylinder?

23. What material is a cylinder barrel constructed of?

24. What type of process is nitriding?

25. What is the purpose of a Heli-Coil insert?

26. How is the cylinder head attached to a cylinder barrel?

27. What are the angles of the intake and exhaust valve faces?

28. List the basic components of the valve operating mechanism for an opposed engine.

29. Name the accessories which are generally mounted on the accessory section.

30. What are the purposes of propeller reduction gears?

Internal-Combustion Engine Theory and Performance **3**

INTRODUCTION

The two most common types of aircraft engines used for the propulsion of almost all powered flights are the reciprocating engine and the gas-turbine engine. Both these engines are termed **heat engines** because they utilize heat energy to produce the power for propulsion.

Basically, an engine is a device for convening a source of energy to useful work. In heat engines, the source of energy is the fuel that is burned to develop heat. The heat, in turn, is converted to power (the rate of doing work) by means of the engine. The **reciprocating engine** uses the heat to expand a combination of gases (air and the products of fuel combustion) and thus to create a pressure against a piston in a cylinder. The piston, being connected to a crankshaft, causes the crankshaft to rotate, thus producing power and doing work. In the **gas-turbine engine**, the heat is used to expand the gas (air) as it moves through the engine, with the result that the velocity of the gases is greatly increased. The high-velocity flow of gases is directed through a turbine which rotates to produce shaft power. With a **turbojet engine**, the jet of gases from the engine exhaust results in thrust that is used to propel the aircraft.

The principles and operation of gas-turbine engines are discussed in Chap. 11.

SCIENCE FUNDAMENTALS

Conversion of Heat Energy to Mechanical Energy

Energy is the capacity for doing work. There are two kinds of energy: kinetic and potential. **Kinetic energy** is the energy of motion, such as that possessed by a moving cannon ball, falling water, or a strong wind. **Potential energy**, or **stored energy**, is the energy of position. A coiled spring has potential energy. Likewise, the water behind the dam of a reservoir has potential energy, and gasoline has potential energy.

Energy cannot be created or destroyed. A perpetual-motion machine cannot exist because even if friction and the weight of the parts were eliminated, a machine can never have more energy than that which has been put into it.

Energy cannot be created, but it can be transformed from one kind to another. When a coiled spring is wound, work is performed. When the spring unwinds, its stored (potential) energy becomes kinetic energy. When a mixture of gasoline and air is ignited, the combustion process increases the kinetic energy of the molecules in the gases. When the gas is confined, as in a reciprocating-engine cylinder, this results in increased pressure (potential energy), which produces work when the piston is forced downward.

Heat energy can be transformed to mechanical energy. mechanical energy can be transformed to electric energy, and electric energy can be transformed to heat, light, chemical, or mechanical energy.

The conversion of the potential energy in fuel to the kinetic energy of the engine's motion is controlled by certain laws of physics. These laws deal with pressure, volume, and temperature and are described in detail below.

Boyle's Law and Charles' Law

Boyle's law *states that the volume of any dry gas varies inversely with the absolute pressure sustained by it, the temperature remaining constant.* In other words, increasing the pressure on a volume of confined gas reduces its volume correspondingly. Thus, doubling the pressure reduces the volume of the gas to one-half, trebling the pressure reduces the volume to one-third, etc. The formula for Boyle's law is

$$\frac{V_1}{V_2} = \frac{P_2}{P_1}$$

Charles' law *states that the pressure of a confined gas is directly proportional to its absolute temperature.* Therefore, as the temperature of the gas is increased, the pressure is also increased as long as the volume remains constant. The formula for Charles' law is

$$\frac{V_1}{V_2} = \frac{T_1}{T_2}$$

These laws may be used to explain the operation of an engine. The mixture of fuel and air burns when it is ignited and produces heat. The heat is absorbed by the gases in the cylinder, and they tend to expand. The increase in pressure, acting on the head of the piston, forces it to move, and the motion is transmitted to the crankshaft through the connecting rod.

A further understanding of engine operation may be gained by examining the theory of the Carnot cycle. The Carnot cycle explains the operation of an "ideal" heat engine. The engine employs a gas as a working medium, and the changes in pressure, volume, and temperature are in accordance with Boyle's and Charles' laws. A detailed study of the Carnot cycle is not essential to the present discussion; however, if students desire to pursue the matter further, they can find a complete explanation in any good college text on physics.

ENGINE OPERATING FUNDAMENTALS

A **cycle** is a complete sequence of events returning to the original state. That is, a cycle is an interval of time occupied by one round, or course, of events repeated in the same order in a series—such as the cycle of the seasons, with spring, summer, autumn, and winter following each other and then recurring.

An **engine cycle** is the series of events that an internal-combustion engine goes through while it is operating and delivering power. In a four-stroke five-event cycle these events are intake, compression, ignition, combustion, and exhaust. An **internal-combustion engine**, whether it be a piston-type or gas-turbine engine, is so called because the fuel is burned inside the engine rather than externally, as with a steam engine. Since the events in a piston engine occur in a certain sequence and at precise intervals of time, they are said to be **timed**.

Most piston-type engines operate on the four-stroke five-event-cycle principle originally developed by August Otto in Germany. There are four strokes of the piston in each cylinder, two in each direction, for each engine operating cycle. The five events of the cycle consist of these strokes plus the ignition event. The four-stroke five-event cycle is called the **Otto cycle**. Other cycles for heat engines are the **Carnot cycle**, named after Nicolas-Leonard-Sadi Carnot, a young French engineer; the **Diesel cycle**, named after Dr. Rudolf Diesel, a German scientist; and the **Brayton cycle**, named for George B. Brayton, a U.S. engineer mentioned in Chap. 1. All the cycles mentioned pertain to the particular engine theories developed by the men whose names are given to the various cycles. All the cycles include the compression of air, the burning of fuel in the compressed air, and the conversion of the pressure and heat to power.

Stroke

The basic power-developing parts of a typical gasoline engine are the **cylinder**, **piston**, **connecting rod**, and **crankshaft**. These are shown in Fig. 3-1. The cylinder has a smooth surface such that the piston can, with the aid of piston rings and a lubricant, create a seal so that no gases can escape between the piston and the cylinder walls. The piston is connected to the crankshaft by means of the connecting rod so that the rotation of the crankshaft causes the piston to move

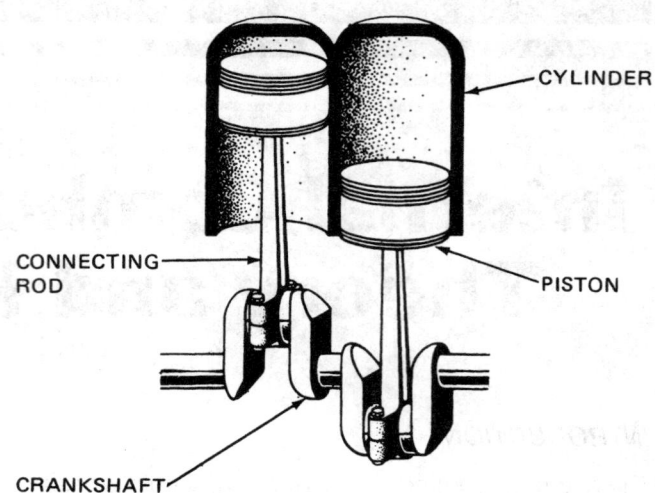

FIGURE 3-1   Basic parts of a gasoline engine.

with a reciprocating motion up and down in the cylinder. The distance through which the piston travels is called the **stroke**. During each stroke, the crankshaft rotates 180°. The limit of travel to which the piston moves into the cylinder is called **top dead center**, and the limit to which it moves in the opposite direction is called **bottom dead center**. For each revolution of the crankshaft there are two strokes of the piston, one up and one down, assuming that the cylinder is in a vertical position. Figure 3-2 shows that the stroke of the cylinder illustrated is 5.5 in [13.97 cm] and that its bore (internal diameter) is also 5.5 in [13.97 cm]. An engine having the bore equal to the stroke is often called a **square engine**.

It is important to understand top dead center and bottom dead center because these positions of the piston are used in setting the timing and determining the valve overlap. Top dead center (TDC) may be defined as the point which a piston has reached when it is at its maximum distance from the centerline of the crankshaft. In like manner bottom dead center (BDC)

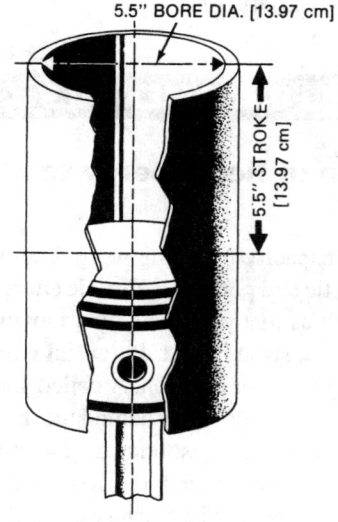

FIGURE 3-2   Stroke and bore.

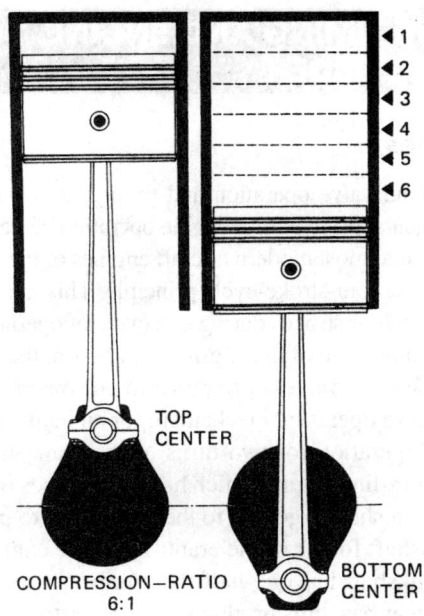

FIGURE 3-3 Top dead center and bottom dead center.

may be defined as the position which the piston has reached when it is at a minimum distance from the centerline of the crankshaft. Figure 3-3 illustrates the piston positions at TDC and at BDC.

Compression Ratio

The **compression ratio** of a cylinder is the ratio of the volume of space in the cylinder when the piston is at the bottom of its stroke to the volume when the piston is at the top of its stroke. For example, if the volume of the space in a cylinder is 120 in^3 [1.97 L] when the piston is at the bottom of its stroke and the volume is 20 in^3 [0.33 L] when the piston is at the top of its stroke, the compression ratio is 120:20. Stated in the form of a fraction, it is 120/20, and when the larger number is divided by the smaller number, the compression ratio is shown as 6:1. This is the usual manner for expressing a compression ratio. In Fig. 3-3, the piston and cylinder provide a compression ratio of 6:1.

The Four-Stroke Five-Event Cycle

The four strokes of a four-stroke-cycle engine are the intake stroke, the compression stroke, the power stroke, and the exhaust stroke. In a four-stroke-cycle engine, the crankshaft makes 2 r for each complete cycle. The names of the strokes are descriptive of the nature of each stroke.

During the **intake stroke**, the piston starts at TDC with the intake valve open and the exhaust valve closed. As the piston moves downward, a mixture of fuel and air, sometimes called the **working fluid**, from the carburetor is drawn into the cylinder. The intake stroke is illustrated in Fig. 3-4A.

When the piston has reached BDC at the end of the intake stroke, the piston moves back toward the cylinder head. The intake valve closes as much as 60° of crankshaft rotation after BDC in order to take advantage of the inertia of the incoming fuel-air mixture, thus increasing **volumetric efficiency**. (Volumetric efficiency is discussed later in this chapter.) Since both valves are now closed, the fuel-air mixture is compressed in the cylinder. For this reason the event illustrated in Fig. 3-4B is called the **compression stroke**. A few degrees of crankshaft travel *before* the piston reaches TDC on the compression stroke, **ignition** takes place. Ignition is caused by a spark plug which produces an electric spark in the fuel-air mixture. This spark ignites the fuel-air mixture, thus creating heat and pressure to force the piston downward toward BDC. The ignition is timed to occur a few degrees before TDC to allow time for complete combustion of the fuel. When the fuel-air mixture and the ignition timing are correct, the combustion process will be complete just after TDC at the beginning of the power stroke, producing maximum pressure. It the ignition should occur at TDC, the piston would be moving downward as the fuel burned and a maximum pressure would not be developed. Also, the burning gases moving down the walls of the cylinder would heat the cylinder walls, and the engine would develop excessive temperature.

The stroke during which the piston is forced down, as the result of combustion pressure, is called the **power stroke** because this is the time when power is developed in the engine. The movement of the piston downward causes the crankshaft to rotate, thus turning the propeller. The power

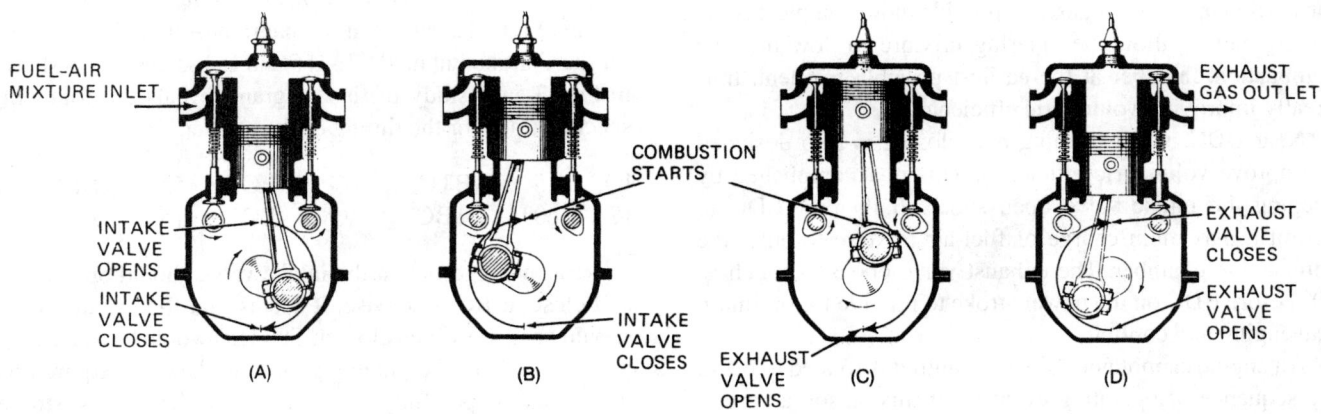

FIGURE 3-4 Operation of a four-stroke engine. (A) Intake stroke. (B) Compression stroke. (C) Power stroke. (D) Exhaust stroke.

stroke illustrated in Fig. 3-4C is also called the **expansion stroke** because of the gas expansion which takes place at this time.

Well before the piston reaches BDC on the power stroke, the exhaust valve opens, and the hot gases begin to escape from the cylinder. The pressure differential across the piston drops to zero, and the gases that remain in the cylinder are forced out the open exhaust valve as the piston moves back toward TDC. This is the **exhaust stroke** and is also called the **scavenging stroke** because the burned gases are scavenged (removed from the cylinder) during the stroke. The exhaust stroke is illustrated in Fig. 3-4D.

We may summarize the complete cycle of the four-stroke-cycle engine as follows: **intake stroke**—the intake valve is open and the exhaust valve closed, the piston moves downward, drawing the fuel-air mixture into the cylinder, and the intake valve closes; **compression stroke**—both valves are closed, the piston moves toward TDC, compressing the fuel-air mixture, and ignition takes place near the top of the stroke; **power stroke**—both valves are closed, the pressure of the expanding gases forces the piston toward BDC, and the exhaust valve opens well before the bottom of the stroke; **exhaust stroke**—the exhaust valve is open and the intake valve closed, the piston moves toward TDC, forcing the burned gases out through the open exhaust valve, and the intake valve opens near the top of the stroke.

The five-event sequence of intake, compression, ignition, power, and exhaust is a cycle which must take place in the order given if the engine is to operate at all, and it must be repeated over and over for the engine to continue operation. None of the events can be omitted, and each event must take place in the proper sequence. For example, if the gasoline supply is shut off, there can be no power event. The mixture of gasoline and air must be admitted to the cylinder during the intake stroke. Likewise, if the ignition switch is turned off, there can be no power event because the ignition must occur before the power event can take place.

Note at this point that each event of crankshaft rotation does not occupy exactly 180° of crankshaft travel. The intake valve begins to open substantially before TDC, and the exhaust valve closes after TDC. This is called **valve overlap** and is designed to take advantage of the inertia of the outflowing exhaust gases to provide more complete scavenging and to allow the entering mixture to flow into the combustion chamber at the earliest possible moment, thus greatly improving volumetric efficiency.

Near BDC, valve opening and closing is also designed to improve volumetric efficiency. This is accomplished by keeping the intake valve open substantially past BDC to permit a maximum charge of fuel-air mixture to enter the combustion chamber. The exhaust valve opens as much as 60° before BDC on the power stroke to provide for optimum scavenging and cooling.

An engine cannot normally start until it is rotated to begin the sequence of operating events. For this reason a variety of starting systems have been employed and are discussed Chap. 8 of this text.

VALVE TIMING AND ENGINE FIRING ORDER

Principles

To understand valve operation and timing, it is essential that the fundamental principles of engine operation be kept in mind. Remember that most modern aircraft engines of the piston type operate on the four-stroke-cycle principle. This means that the piston makes four strokes during one cycle of operation.

During one cycle of the engine's operation, the crankshaft makes 2 r and the valves each perform one operation. Therefore, the valve operating mechanism for an intake valve must make one operation for two turns of the crankshaft. On an opposed or in-line engine which has single lobes on the camshaft, the camshaft is geared to the crankshaft to produce 1 r of the camshaft for 2 r of the crankshaft. The cam drive gear on the crankshaft has one-half the number of teeth that the camshaft gear has, thus producing the 1:2 ratio.

On radial engines which utilize cam rings or cam plates to operate the valves, there may be three, four, or five cam lobes on the cam ring. The ratio of crankshaft to cam-ring rotation is then 1:6, 1:8, or 1:10, respectively.

Abbreviations for Valve Timing Positions

In a discussion of the timing points for an aircraft engine, it is convenient to use abbreviations. The abbreviations commonly used in describing crankshaft and piston positions for the timing of valve opening and closing are as follows:

After bottom center	ABC	Before top center	BTC
After top center	ATC	Exhaust closes	EC
Before bottom center	BBC	Exhaust opens	EO
		Intake closes	IC
Bottom center	BC	Intake opens	IO
Bottom dead center	BDC	Top center	TC
		Top dead center	TDC

Engine Timing Diagram

To provide a visual concept of the timing of valves for an aircraft engine, a valve timing diagram is used. The diagram for the Continental model E-165 and E-185 engines is shown in Fig. 3-5. A study of this diagram reveals the following specifications for the timing of the engine:

IO	15°	BTC	EO	55°	BBC
IC	60°	ABC	EC	15°	ATC

Reason suggests that the intake valve should open at TC and close at BC. Likewise, it seems that the exhaust valve should open at BC and close at TC. This would be true except for the inertia of the moving gases and the time required for the valves to open fully. Near the end of the exhaust stroke, the gases are still rushing out the exhaust valve. The inertia of the gases causes a low-pressure condition in the cylinder

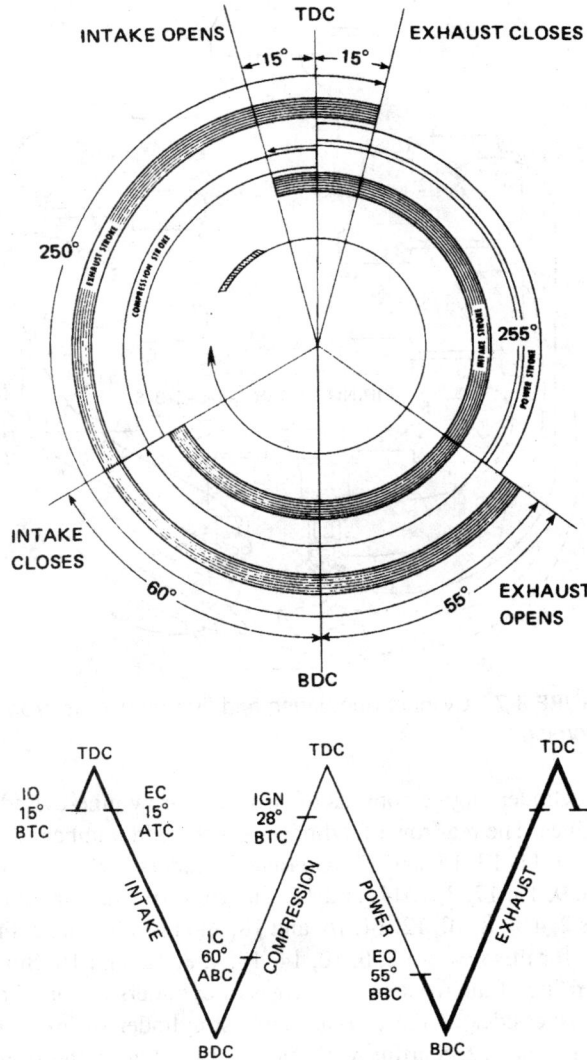

FIGURE 3-5 Diagram for valve timing.

intake valve is open is designed to permit the greatest possible charge of fuel-air mixture into the cylinder.

The exhaust valve opens before BC for two principal reasons: (1) more thorough scavenging of the cylinder and (2) better cooling of the engine. Most of the energy of the burning fuel is expended by the time the crankshaft has moved 120° past TC on the power stroke and the piston has moved almost to its lowest position. Opening the exhaust valve at this time allows the hot gases to escape early, and less heat is transmitted to the cylinder walls than would be the case if the exhaust valve remained closed until the piston reached BC. The exhaust valve is not closed until ATC because the inertia of the gases aids in removing additional exhaust gas after the piston has passed TC.

The opening or closing of the intake or exhaust valves after TC or BC is called **valve lag**. The opening or closing of the intake or exhaust valves before BC or TC is called **valve lead**. Both valve lag and valve lead are expressed in degrees of crankshaft travel. For example, if the intake valve opens 15° BTC, the valve lead is 15°.

Note from the diagrams of Fig. 3-5 that the valve lead and valve lag are greater in relation to the BC position than they are to the TC position. One reason for this is that the piston travel per degree of crankshaft travel is less near BC than it is near TC. This is illustrated in Fig. 3-6. In this diagram, the circle represents the path of the crank throw, point C represents the center of the crankshaft, TC is the position of the piston pin at top center, and BC is the position of the piston pin at bottom center. The numbers show the positions of the piston pin and the crank throw at different points through 180° of crankshaft travel. Note that the piston travels much farther

at this time. Opening the intake valve a little before TC takes advantage of the low-pressure condition to start the flow of fuel-air mixture into the cylinder, thus bringing a greater charge into the engine and improving volumetric efficiency. If the intake valve should open too early, exhaust gases would flow out into the intake manifold and ignite the incoming fuel-air mixture. The result would be **backfiring**. Backfiring also occurs when an intake valve sticks in the open position. The exhaust valve closes shortly after the piston reaches TC and prevents reversal of the exhaust flow back into the cylinder. The angular distance through which both valves are open is called **valve overlap**, or **valve lap**. When the intake valve opens 15° BTC and the exhaust valve closes 15° ATC, the valve overlap is 30°.

Figure 3-5 shows two diagrams that may be used as guides for valve timing. Either one may be employed to indicate the points in the cycle where each valve opens and closes. In the diagrams of Fig. 3-5, the intake valve remains open 60° ABC. This is designed to take advantage of the inertia of the fuel-air mixture rushing into the cylinder, because the mixture will continue to flow into the cylinder for a time after the piston has passed BC. The total period during which the

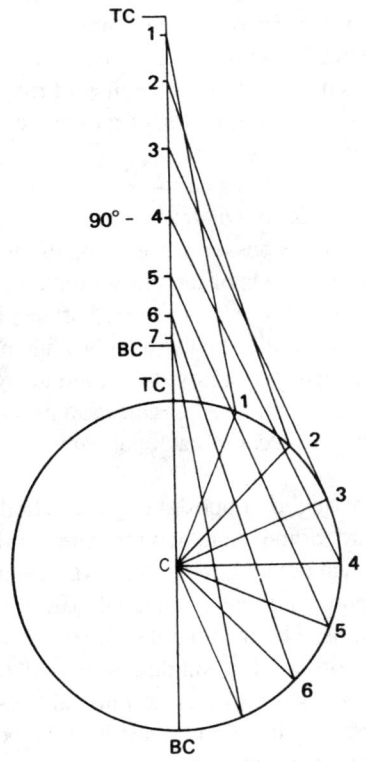

FIGURE 3-6 Relation between piston travel and crankshaft travel.

during the first 90° of crankshaft travel than it does during the second 90° and that the piston will be traveling at maximum velocity when the crank throw has turned 80 to 90° past TC.

By using the valve timing specifications for the diagrams of Fig. 3-5, it is possible to determine (1) the rotational distance through which the crankshaft travels while each valve is open and (2) the rotational distance of crankshaft travel while both valves are closed. Since the intake valve opens at 15° BTC and closes at 60° ABC, the crankshaft rotates 15° from the point where the intake valve opens to reach TC, then 180° to reach BC, and another 60° to the point where the intake valve closes. The total rotational distance of crankshaft travel with the intake valve open is therefore 15° + 180° + 60°, or a total of 255°. By the same reasoning, crankshaft travel while the exhaust valve is open is 55° + 180° + 15°, or a total of 250°. Valve overlap at TC is 15° + 15°, or 30°. The total rotational distance of crankshaft travel while both valves are closed is determined by noting when the intake valve closes on the compression stroke and when the exhaust valve opens on the power stroke. It can be seen from the diagram that the intake valve closes 60° ABC and that the crankshaft must therefore rotate 120° (180° – 60°) from intake-valve closing to TC. Since the exhaust valve opens 55° BBC, the crankshaft rotates 125° (180° – 55°) from TC to the point where the exhaust valve opens. The total rotational distance that the crankshaft must travel from the point where the intake valve closes to the point where the exhaust valve opens is then 120° + 125°, or 245°. The time the valves are off their seat is their **duration**. For example, the duration of the exhaust valve above is 250° of crankshaft travel.

Firing Order

In any discussion of valve or ignition timing, we must consider the firing orders of various engines because all parts associated with the timing of any engine must be designed and timed to comply with the engine's firing order. As the name implies, the **firing order** of an engine is the order in which the cylinders fire.

The firing order of in-line V-type and opposed engines is designed to provide for balance and to eliminate vibration to the extent that this is possible. The firing order is determined by the relative positions of the throws on the crankshaft and the positions of the lobes on the camshaft.

Figure 3-7 illustrates the cylinder arrangement and firing order for a six-cylinder opposed Lycoming engine. The cylinder firing order in opposed engines can usually be listed in pairs of cylinders, because each pair fires across the center main bearing.

The numbering of opposed-engine cylinders is by no means standard. Some manufacturers number their cylinders from the rear and others from the front of the engine. Always refer to the appropriate engine manual to determine the numbering system used by the manufacturer.

The firing order of a single-row radial engine which operates on the four-stroke cycle must always be by alternate cylinders, and the engine must have an odd number of cylinders. Twin-row radial engines are essentially two single-row engines joined together. This means that alternate cylinders in each row must fire in sequence. For example, an

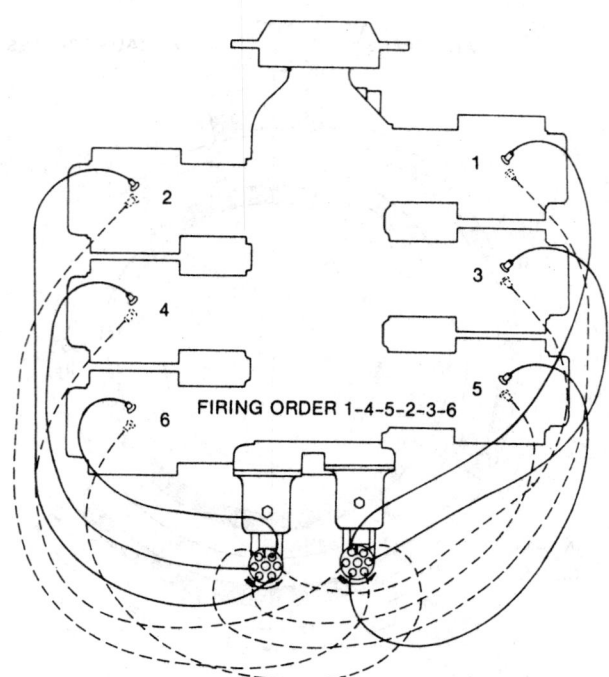

FIGURE 3-7 Cylinder numbering and firing order. (*Textron Lycoming.*)

18-cylinder engine consists of two single-row nine-cylinder engines. The rear row of cylinders has the odd numbers 1, 3, 5, 7, 9, 11, 13, 15, and 17. Alternate cylinders in this row are 1, 5, 9, 13, 17, 3, 7, 11, and 15. The front row has the numbers 2, 4, 6, 8, 10, 12, 14, 16, and 18, and the alternate cylinders for this row are 2, 6, 10, 14, 18, 4, 8, 12, and 16. Since the firing of the front and rear rows of cylinders is started on opposite sides of the engine, the first cylinder to fire after no. 1 is no. 12. Starting with the no. 12 cylinder, the front-row firing sequence is then 12, 16, 2, 6, 10, 14, 18, 4, and 8. By combining the rear-row firing with the front-row firing, we obtain the firing order for the complete engine: 1, 12, 5, 16, 9, 2, 13, 6, 17, 10, 3, 14, 7, 18, 11, 4, 15, and 8.

Figure 3-8 gives the firing orders for the majority of engine types. As an aid in remembering the firing order of large radial engines, technicians often use "magic" numbers. For a

Type	Firing Order
4-cylinder in-line	1-3-4-2 or 1-2-4-3
6-cylinder in-line	1-5-3-6-2-4
8-cylinder V-type (CW)	1R-4L-2R-3L-4R-1L-3R-2L
12-cylinder V-type (CW)	1L-2R-5L-4R-3L-1R-6L-5R-2L-3R-4L-6R
4-cylinder opposed	1-3-2-4 or 1-4-2-3
6-cylinder opposed	1-4-5-2-3-6
8-cylinder opposed	1-5-8-3-2-6-7-4
9-cylinder radial	1-3-5-7-9-2-4-6-8
14-cylinder radial	1-10-5-14-9-4-13-8-3-12-7-2-11-6
18-radial	1-12-5-16-9-2-13-6-17-10-3-14-7-18-11-4-15-8

FIGURE 3-8 Engine firing order.

14-cylinder radial engine, the numbers are +9 and −5, and for an 18-cylinder engine, the numbers are +11 and −7. To determine the firing order of a 14-cylinder engine, the technician starts with the number 1, the first cylinder to fire. Adding 9 gives the number 10, the second cylinder to fire. Subtracting 5 from 10 gives 5, the third cylinder to fire. Adding 9 to 5 gives 14, the fourth cylinder to fire. Continuing the same process will give the complete firing order. The same technique is used with an 18-cylinder engine by applying the magic numbers +11 and −7.

THE TWO-STROKE CYCLE

Although present-day aircraft engines of the reciprocating type usually operate on the four-stroke-cycle principle, a few small engines (such as those used on ultralight aircraft) operate on the two-stroke-cycle principle. The differences are in the number of strokes per operating cycle and the method of admitting the fuel-air mixture into the cylinder. The two-stroke-cycle engine is mechanically simpler than the four-stroke-cycle engine but is less efficient and is more difficult to lubricate; therefore, its use is restricted. The operating principle of the two-stroke-cycle engine is illustrated in Fig. 3-9.

Like the four-stroke-cycle engine, the two-stroke-cycle engine is constructed with a cylinder, piston, crankshaft, connecting rod, and crankcase; however, the valve arrangement and fuel intake system are considerably different. The upward movement of the piston in the cylinder of the engine creates low pressure in the crankcase. This reduced pressure causes a suction which draws die fuel-air mixture from the carburetor into the crankcase through a check valve. When the piston has reached TDC, the crankcase is filled with the fuel-air mixture and the inlet check valve is closed. The piston then moves downward in the cylinder and compresses the mixture in the crankcase. As the piston reaches the lowest point in its stroke, the intake port is opened to permit the fuel-air mixture which is compressed in the crankcase to flow into the cylinder. This is the **intake event**.

The piston then moves up in the cylinder, the intake port is closed, and the fuel-air mixture in the cylinder is compressed. While this is happening, a new charge of fuel and air is drawn into the crankcase through the check valve. This is the **compression event** and is shown in Fig. 3-9A.

The piston continues to move up in the cylinder, and when it is almost at the top of the stroke, a spark is produced at the gap of the spark plug, thus igniting the fuel-air mixture. This is the **ignition event**. As the fuel-air mixture burns, the gases of combustion expand and drive the piston down. This is the **power event** and is shown in Fig. 3-9B. During the power event, the fuel-air mixture in the crankcase is pressurized.

When the piston approaches the bottom point of its travel, the exhaust port is opened to allow the hot gases to escape. This occurs a fraction of a second before the intake port opens to allow the pressurized fuel-air mixture in the crankcase to flow through the intake port into the cylinder. As the exhaust gases rush out the exhaust port on one side of the cylinder, the fuel-air mixture flows into the other side. A baffle on the top of the piston reflects the incoming mixture toward the top of the cylinder, thus helping to scavenge the exhaust gases and reduce the mixing of the fuel-air mixture with the exhaust gases. Clearly the exhaust and intake events take place almost simultaneously, with the exhaust event leading by a small fraction of the piston stroke. This is illustrated in Fig. 3-9C.

Note that there are five events in the two-stroke engine cycle, but at one point, two of the events happen at approximately the same time. During the time that the combined exhaust and intake events are occurring, some of the fuel-air mixture is diluted with burned gases retained from the previous cycle, and some of the fresh mixture is discharged with the exhaust gases. The baffle on the top of the piston is designed to reduce the loss of the fresh mixture as much as possible.

It is important to understand that two strokes of the piston (one complete crankshaft revolution) are required to complete the cycle of operation. For this reason, all cylinders

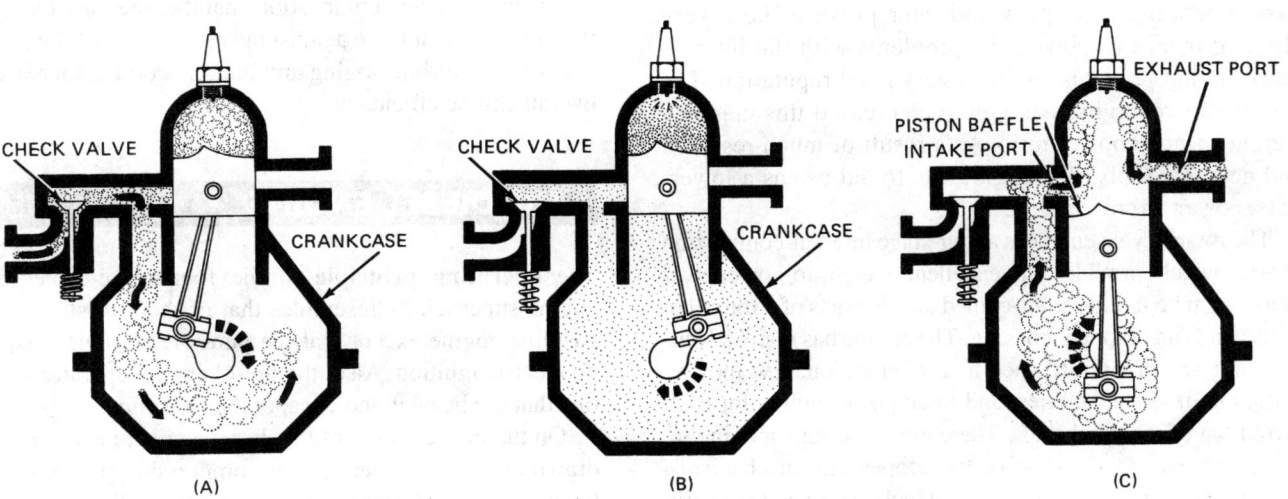

FIGURE 3-9 Operation of a two-stroke-cycle engine. (A) Compression event. (B) Ignition and power events. (C) Exhaust and intake events.

of a multicylinder two-stroke-cycle engine will fire at each revolution of the crankshaft. Remember that the four-stroke-cycle engine fires only once in two complete revolutions of the crankshaft.

The operation of the two-stroke-cycle engine may be summarized as follows: the piston moves upward and draws a fuel-air mixture into the crankcase through a check valve, the crankcase being airtight except for this valve; the piston moves downward and compresses the mixture in the crankcase; the intake port is opened, and the compressed fuel-air mixture enters the cylinder; the piston moves upward and compresses the mixture in the combustion chamber; near the top of the piston stroke, the spark plug ignites the mixture, thus causing the piston to move down; near the bottom of the stroke, the exhaust port is opened to allow the burned gases to escape, and the intake port opens to allow a new charge to enter the cylinder. Note that as the piston moves down during the power event, the fuel-air mixture is being compressed in the crankcase. As the piston moves upward during the compression event, the fuel-air mixture is being drawn into the crankcase.

The two-stroke cycle has three principal disadvantages: (1) there is a loss of efficiency as a result of the fuel-air charge mixing with the exhaust gases and the loss of some of the charge through the exhaust port; (2) the engine is more difficult to cool than the four-stroke-cycle engine, chiefly because the cylinder fires at every revolution of the crankshaft; and (3) the engine is somewhat difficult to lubricate properly because the lubricant must be introduced with the fuel-air mixture through the carburetor. This is usually accomplished by mixing the lubricant with the fuel in the fuel tank.

ROTARY-CYCLE ENGINE

A type of engine finding its way into general aviation is the **rotary cycle** (Wankel). The basic **Wankel cycle** was invented by Felix Wankel in 1957. The early versions of this engine had many problems with internal seals and high fuel consumption. Although the engine's basic operating concept would later prove to be a very efficient means of power, the problems with the internal seals did not give this engine a very good reputation. The use of supercharging has greatly decreased this engine's weight-to-horsepower ratio. As a result of much research and new materials, this engine has found use as a lower-horsepower aircraft engine.

The rotary-cycle engine is a four-stage internal-combustion engine which provides an excellent weight-to-horsepower ratio. It can be liquid- or air-cooled and consists of a rotor that turns inside an elliptical housing. The engine has many advantages for use in aircraft, such as low vibration, few moving parts, multifuel capabilities, and three power pulses for each revolution of the crankshaft. There are no pistons moving up and down and no camshaft or valve operating mechanisms. The advances made in seal design, which was one of the early problems with this type of engine, have greatly increased its reliability.

The basic theory of operation is that of a four-stage cycle similar to the reciprocating engines mentioned earlier. Intake, compression, power, and exhaust are the basic stages which are completed three times for each revolution of the triangular rotor. Since the rotor has three sides which contain three combustion chambers, each chamber is completing a different cycle simultaneously. This is illustrated in Fig. 3-10. The engine uses intake and exhaust ports, so valves are not needed. As the rotor turns past the intake port, it is uncovered and the fuel-air mixture is drawn into the combustion chamber on one side of the rotor. The rotor will turn until the next rotor tip passes over the original intake port, completing the intake stroke. Due to the eccentric shaft on which the rotor turns, the rotor tips are always in contact with the elliptical rotor housing. The rotor continues to turn, compressing the fuel-air charge due to the geometry of the engine housing and rotor. When the charge is compressed to its maximum, the spark plugs fire and the combustion of the fuel-air mixture drives the rotor in the direction of rotation. Due to the pressure of combustion acting off center of the eccentric, this pressure drives the rotor which is also attached to the output shaft. As the rotor continues to turn, it uncovers the exhaust port, allowing the exhaust gases to exit from the engine chamber. Because each side of the rotor is a separate and independent chamber, one rotor does the same work as a three-cylinder reciprocating engine.

Many times, two or more rotors are used together, as in a multicylinder reciprocating engine. With this configuration, a multirotor rotary engine can greatly increase the horsepower output. Many engine innovations have enhanced the fuel consumption qualities of rotary engines such as supercharging, stratified charge (a scheme for having two levels of fuel richness in a firing chamber), and multifuel capabilities.

The rotor has an internal gear that rotates about a stationary gear attached to the engine housing. The rotor then transmits its rotary motion to an output shaft. The ignition system incorporates two spark plugs which are fired by two separate ignition systems. One spark plug is designed to fire sooner than the other, making one the leading spark plug and the other the trailing spark plug. This ignition system design assists in producing combustion chamber pressure that gives the most efficient force against the rotor. Many of the aircraft versions use turbocharging and fuel injection to increase the overall engine efficiency.

THE DIESEL ENGINE

The **operating principle** of the four-stroke-cycle diesel engine superficially resembles that of the four-stroke-cycle gasoline engine except that the pure diesel engine requires no electric ignition. Also, the diesel engine operates on fuel oils that are heavier and cheaper than gasoline.

On the intake stroke of the diesel engine, only pure air is drawn into the cylinder. On the compression stroke, the piston compresses the air to such an extent that the air temperature is high enough to ignite the fuel without the use of an electric spark. As the piston approaches the top of its stroke,

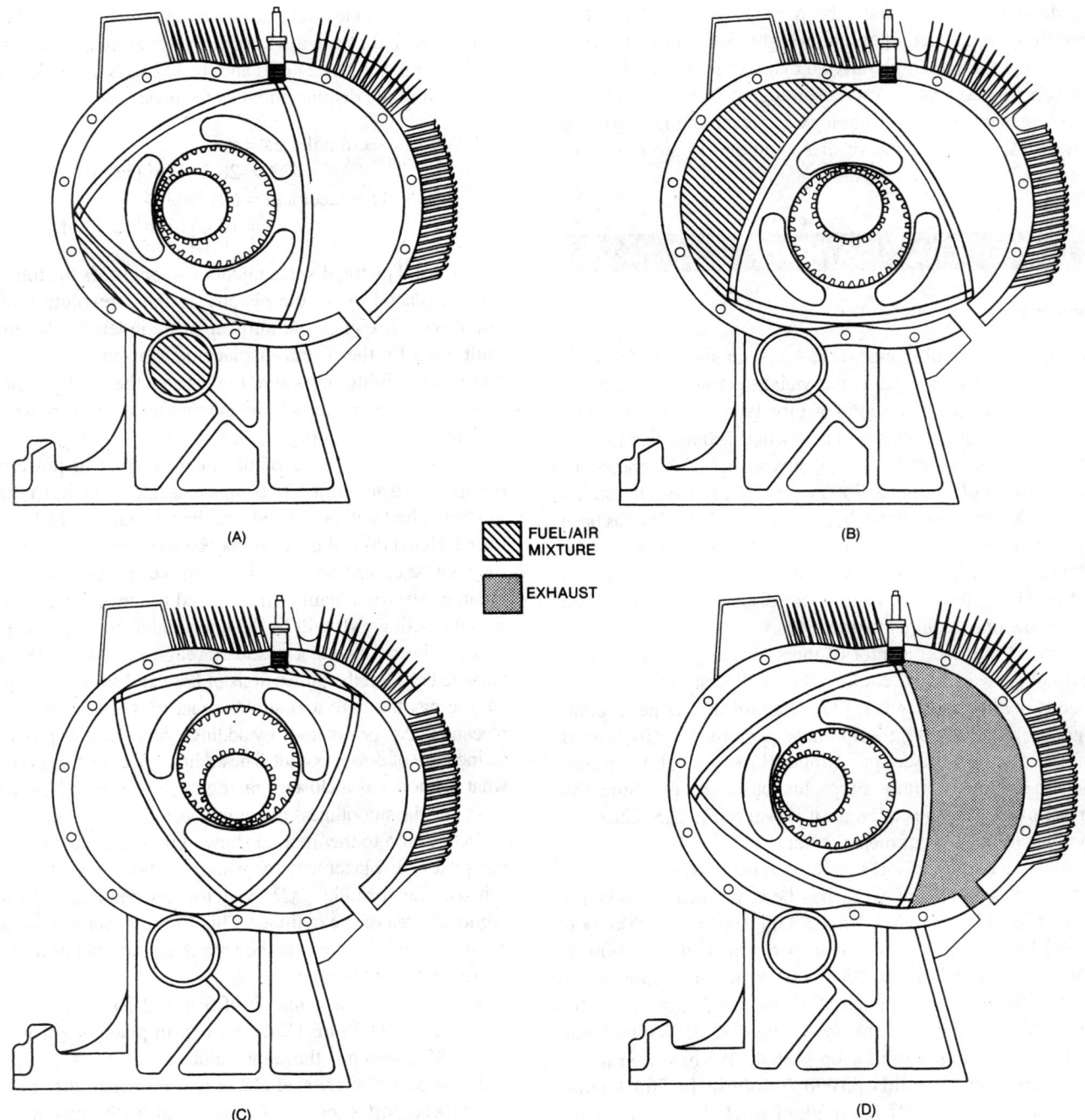

FUEL/AIR MIXTURE

EXHAUST

FIGURE 3-10 Rotary-cycle engine. (A) Intake stroke begins when rotor tip uncovers intake port. (B) Compression starts as intake port is closed and rotor reaches highest point in front of spark plug. (C) Combustion takes place when charge is most compressed. (D) Exhaust begins as rotor tip passes exhaust port.

the fuel is injected into the cylinder under a high pressure in a finely atomized state. The highly compressed hot air already in the cylinder ignites the fuel. The fuel burns during the **power stroke**, and the waste gases escape during the **exhaust stroke** just as they do in a gasoline engine. On many diesel engines, particularly those in automobiles, glow-plug igniters are installed to aid in starting the combustion of the fuel. These igniters are not in operation after the engine is running.

The **compression ratio**, discussed more fully later, is the ratio of the volume of space in a cylinder when the piston is at the bottom of its stroke to the volume when the piston is

at the top of its stroke. The compression ratio of a diesel engine may be as high as 14:1 as compared with a maximum of 10:1 or 11:1 for conventional gasoline engines. It is common for a gasoline engine to have a compression ratio of about 7:1; however, certain high-performance engines have higher ratios. The compression ratio of a conventional gasoline engine must be limited because the temperature of the compressed gases in the cylinder must not be high enough to ignite the fuel.

Like the gasoline internal-combustion engine, the diesel engine may be either a two-stroke-cycle or a four-stroke-cycle engine.

Many innovations have been made in diesel engines, especially in the area of engine weight. As described earlier, diesel engines have high internal cylinder pressures, because the compression ratio of a diesel engine may be as high as 14:1. Because of new technology in diesel-engine operating principles, the future use of diesel engines in aircraft is not only feasible but also probable.

POWER CALCULATIONS

Power

Power is the rate of doing work. A certain amount of work is accomplished when a particular weight is raised a given distance. For example, if a weight of 1 ton [907.2 kg] is raised vertically 100 ft [30.48 m], we may say that 100 ton-feet (ton•ft) [27 651 kilogram-meters (kg•m)] of work has been done. Since 1 ton [907.2 kg] is equal to 2000 lb [907.2 kg], we can also say that 200 000 foot-pounds (ft lb) [27 651 kg°m] of work has been done. When we speak of power, we must also consider the *time* required to do a given amount of work. Power depends on three factors: (1) the force extended, (2) the distance the force moves, and (3) the time required to do the work.

James Watt, the inventor of the steam engine, found that an English workhorse could work at the rate of 550 foot-pounds per second (ft•lb/s) [77 kilogram-meters per second (kg•m/s)], or 33 000 foot-pounds per minute (ft•lb/min) [4 563 kilogram-meters per minute (kg•m/min)], for a reasonable length of time. From his observations came the **horsepower**, which is the unit of power in the U.S. Customary System (USCS) of measurements.

When a 1-lb [0.45-kg] weight is raised 1 ft [0.304 8 m], 1 ft•lb [0.14 kg•m] of work has been performed. When a 1000-lb [450-kg] weight is lifted 33 ft [10.06 m], 33 000 ft•lb [4 563 kg•m] of work has been performed. If the 1000-lb [450-kg] weight is lifted 33 ft [10.06 m] in 1 min, 1 hp [0.745 kW] has been expended. If it takes 2 min to lift the weight through the same distance, $\frac{1}{2}$ hp [372.85 W] has been used. If it requires 4 min, $\frac{1}{4}$ hp [186.43 W] has been used. **One horsepower equals 33 000 ft•lb/min [4 563 kg•m/ min], or 550 ft•lb/s [77 kg•m/s], of work.** The capacity of automobile, aircraft, and other engines to do work is measured in horsepower. In the metric system, the unit of power is the **watt** (W). One kilowatt (kW) is equal to 1.34 hp.

Piston Displacement

To compute the power of an engine, it is necessary to determine how many foot-pounds of work can be done by the engine in a given time. To do this, we must know various measurements, such as cylinder bore, piston stroke, and piston displacement.

The **piston displacement** of one piston is obtained by multiplying the area of a cross section of the cylinder bore by the total distance that the piston moves during one stroke in the cylinder. Since the volume of any true cylinder is its cross-sectional area multiplied by its height, the piston displacement can be stated in terms of cubic inches of volume.

The piston displacement of one cylinder can be determined if the bore and stroke are known. For example, if the bore of a cylinder is 6 in [15.24 cm] and the stroke is 6 in [15.24 cm], we can find the displacement as follows:

$$\text{Cross-sectional area} = \pi r^2$$
$$= 28.274 \text{ in}^2 \text{ [182.41 cm}^2\text{]}$$
$$\text{Displacement} = 6 \times 28.274$$
$$= 169.644 \text{ in}^3 \text{ [2.779 L]}$$

The total piston displacement of an engine is the total volume displaced by all the pistons during 1 revolution of the crankshaft. It equals the number of cylinders in the engine multiplied by the piston displacement of one piston. Other factors remaining the same, the greater the total piston displacement, the greater will be the maximum horsepower that an engine can develop.

Displacement is one of the many factors in powerplant design which are subject to compromise. If the cylinder bore is too large, fuel will be wasted and the intensity of the heat and the restricted flow of the heat may be so great that the cylinder may not be cooled properly. If the stroke (piston travel) is too great, excessive dynamic stresses and too much angularity of the connecting rods will be the undesirable consequences.

It has been found that a "square" engine provides the proper balance between the dimensions of bore and stroke. (A **square engine** has the bore and stroke equal.) Increased engine displacement can be obtained by adding cylinders, thus producing an increase of power output. The addition of cylinders produces what is known as a **closer spacing of power impulses**, which increases the smoothness of engine operation.

In addition to the method shown previously for determining piston displacement by using the bore and stroke, we can use the formula $\frac{1}{4}\pi D^2 = A$ for determining the cross-sectional area of the cylinder. This formula can also be written $A = \pi D^2/4$, where A is the area in square inches and D is the diameter of the bore.

If a piston has a diameter of 5 in [12.70 cm], its area is $\frac{1}{4}\pi \times 25$, or 19.635 in^2 [126.68 cm^2]. In place of $\frac{1}{4}\pi$ we can use 0.7854, which is the same value.

If the piston mentioned above is used where the stroke is 4 in [10.16 cm], then the displacement of the piston is 4×19.635, or 78.54 in^3 [1.29 L]. If the engine has six cylinders, the total displacement of the engine is $78.54 \times 6 = 471.24$ in^3 [7.723 L]. This engine would be called an O-470 engine, where the O stands for "opposed."

One typical opposed engine has a bore of 5.125 in [13.01 cm] and a stroke of 4.375 in [11.11 cm]. The cross-sectional area of the cylinder is then $5.125^2 \times 0.7854 = 20.629$ in^2 [133.09 cm^2]. The displacement of one piston is $4.375 \times 20.629 = 90.25$ in^3 [1.479 L]. The engine has six cylinders; so the total displacement is $6 \times 90.25 = 541.51$ in^3 [8.875 L]. This engine is called an O-540 engine.

Indicated Horsepower

Indicated horsepower (ihp) is the horsepower developed by the engine, that is, the total horsepower converted from heat energy to mechanical energy.

If the characteristics of an engine are known, the ihp rating can be calculated. The total force acting on the piston in one cylinder is the product of the **indicated mean effective pressure** (imep) P and the area A of the piston head in square inches (found by the formula which states that the area of a circle is πr^2, or $\frac{1}{4}\pi D^2$).

The distance through which this total force acts in 1 min multiplied by the total force gives the number of foot-pounds of work done in 1 min. The work done in 1 min by one piston multiplied by the number of cylinders in operation gives the amount of work done in 1 min by the entire engine. This product is divided by 33 000 (the number of foot-pounds per minute in 1 hp) to obtain the indicated horsepower rating of the engine.

The length of the stroke in feet is represented by L, the area of the piston in square inches by A, the imep in pounds per square inch (psi) by P, the number of working strokes per minute per cylinder by N, and the number of cylinders by K. The ihp can then be computed by the formula

$$\text{ihp} = \frac{PLANK}{33\,000}$$

This formula can be made clear by remembering that **work** is equal to *force times distance* and that **power** is equal to *force times distance divided by time*. So *PLA* is the *product of pressure, distance, and area,* but *pressure times area equals force*; therefore, *PLA = FD*. In the formula, *PLANK* is the number of foot-pounds per minute produced by an engine because N represents the number of working strokes per minute for each cylinder, and K is the number of cylinders. To find horsepower, it is merely necessary to divide the number of foot-pounds per minute by 33 000 since 1 hp = 33 000 ft•lb/min [1 W = 6.12 kg•m/min].

Brake Horsepower

Brake horsepower (bhp) is the actual horsepower delivered by an engine to a propeller or other driven device. It is the ihp minus the friction horsepower. **Friction horsepower** (fhp) is that part of the total horsepower necessary to overcome the friction of the moving parts in the engine and its accessories. The relationship may be expressed thus: bhp = ihp − fhp.

Also, the bhp is that part of the total horsepower developed by the engine which can be used to perform work. On many aircraft engines it is between 85 and 90 percent of the ihp.

The bhp of an engine can be determined by coupling the engine to any power-absorbing device, such as an electric generator, in such a manner that the power output can be accurately measured. If an electric generator is connected to a known electric load and the efficiency of the generator is known, the bhp of the engine driving the generator can be determined. For example, assume that an engine is driving a generator producing 110 volts (V) and that the load on the generator is 50 amperes (A). Electric power is measured in watts and is equal to the voltage multiplied by the amperage. Therefore, the electric power developed by the generator is

50 × 110, or 5500 W. Since 1 hp = 746 W, 5500 W = 7.36 hp. If the generator is 60 percent efficient, the power required to drive it is equal to 7.36/0.60, or 12.27 hp [9.17 kW]. Therefore, we have determined that the engine is developing 12.27 bhp to drive the generator.

The Prony Brake

The **prony brake**, or dynamometer, illustrated in Fig. 3-11, is a device used to measure the **torque**, or turning moment, produced by an engine. The value indicated by the scale is read before the force is applied, and the reading is recorded as the **tare**. The force F produced by the lever arm equals the weight recorded on the scale minus the tare. The known values are then F, the distance L, and the rpm of the engine driving the prony brake. To obtain the bhp, these values are used in the following formula:

$$\text{bhp} = \frac{F \times L \times 2\pi \times \text{rpm}}{33\,000}$$

In this formula, the distance through which the force acts in 1 r is the circumference of the circle of which the distance L is the radius. This circumference is determined by multiplying the radius L by 2π. In the formula the force acts through a given distance a certain number of times per minute, and this gives us the foot-pounds per minute. When this value is divided by 33 000, the result is bhp.

If a given engine turning at 1800 rpm produces a force of 200 lb [889.6 N] on the scales at the end of a 4-ft [1.22-m] lever, we can compute the bhp as follows:

$$\text{bhp} = \frac{200 \times 4 \times 2\pi \times 1800}{33\,000}$$

$$= 274 \, [204.3 \, \text{kw}]$$

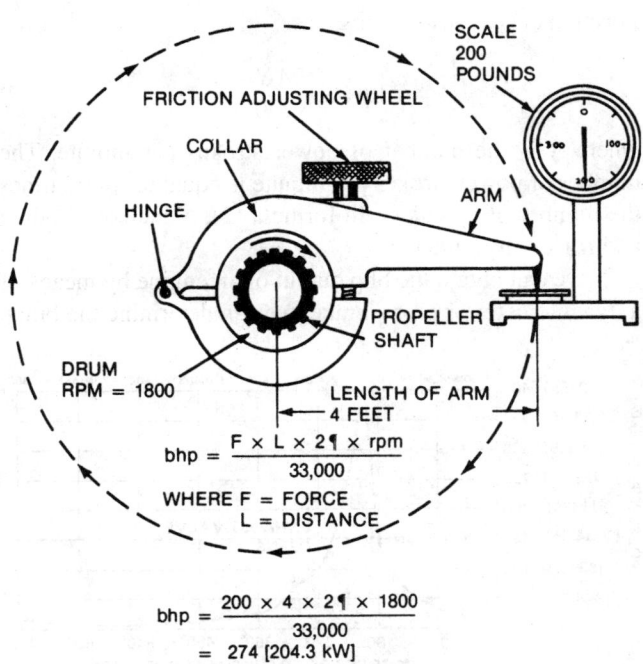

FIGURE 3-11 Prony brake.

Mean Effective Pressure

The **mean effective pressure** (mep) is a computed pressure derived from power formulas in order to provide a measuring device for determining engine performance. For any particular engine operating at a given rpm and power output, there will be a specific *indicated mean effective pressure* (imep) and a corresponding *brake mean effective pressure* (bmep).

Mean effective pressure may be defined as an average pressure inside the cylinders of an internal-combustion engine based on some calculated or measured horsepower. It increases as the manifold pressure increases. The imep is the mep derived from ihp and the bmep is the mep derived from bhp output.

The pressure in the cylinder of an engine throughout one complete cycle is indicated by the curve in Fig. 3-12. This curve is not derived from any particular engine; it is given to show the approximate pressures during the various events of the cycle. Note that ignition takes place shortly before TDC, and then there is a rapid pressure rise which reaches maximum shortly after TDC. Thus, the greatest pressure on the cylinder occurs during the first 5 to 12° after TDC. By the end of the power stroke, very little pressure is left, and this is being rapidly dissipated through the exhaust port.

The ihp of an engine is the result of the imep, the rpm, the distance through which the piston travels, and the number of cylinders in the engine. The formula for this computation was previously given as

$$\text{ihp} = \frac{PLANK}{33000} \qquad (1)$$

where P = imep
L = length of stroke, ft
A = area of piston
N = rpm divided by 2
K = no. of cylinders

Formula (1) can also be given as

$$\text{ihp} = \frac{PLAN}{33000} \qquad (2)$$

where N is the number of power strokes per minute. The number of power strokes per minute is equal to rpm/2 times the number of cylinders. In formula (2), N includes both N and K from formula (1).

If we can obtain the bhp output of an engine by means of a dynamometer or prony brake, we can determine the bmep by means of a formula derived from the power formula given above. By simple transposition, the formula setup for bhp becomes

$$P(\text{bmep}) = \frac{33000 \times \text{bhp}}{LAN} \qquad (3)$$

To simplify the use of the formula, we can convert the length of the stroke and the area of the piston to the displacement of one cylinder, and then multiply by the number of cylinders, to find the total displacement of the engine. In the formula, L is the length of the stroke in feet, A is the area of the piston (the area of the piston is calculated with the formula $A = \pi r^2$ or $\pi d^2/4$), and N is the number of cylinders times the rpm divided by 2. Since the area of the piston must be multiplied by the length of the stroke in inches to obtain piston displacement in cubic inches, S may be used for length of stroke in place of L and expressed in inches. Then $S/12$ is equal to L because L is expressed in feet. For example, if the stroke S is 6 in, then it is equal to $\frac{6}{12}$ or $\frac{1}{2}$ ft. With these adjustments in mind, we find that LAN becomes

$$\frac{SA}{12} \times \text{no. of cylinders} \times \frac{\text{rpm}}{2}$$

or

$$\frac{SA \times \text{no. of cylinders} \times \text{rpm}}{12 \times 2}$$

Since $SA \times$ no. of cylinders is the total displacement (disp) of the engine, we can express the above value as

$$\frac{\text{Displacement} \times \text{rpm}}{12 \times 2}$$

Substituting this value in formula (3) gives

$$\text{bmep} = \frac{33000 \times \text{bhp}}{(\text{disp} \times \text{rpm})/(12 \times 2)}$$
$$= \frac{24 \times 33000 \times \text{bhp}}{\text{disp} \times \text{rpm}}$$
$$= \frac{792000}{\text{disp}} \times \frac{\text{bhp}}{\text{rpm}} \qquad (4)$$

If an R-1830 engine is turning at 2750 rpm and developing 1100 hp [820.27 kW], we can find the bmep as follows:

$$\text{bmep} = \frac{792000}{1830} \times \frac{1100}{2750}$$
$$= 173 \text{ psi} [1192 \text{ kPa}]$$

For any particular engine computation, the value of 792 000/disp may be considered as a constant for that engine and given the designation K. Formula (4) then becomes

$$\text{bmep} = K \times \frac{\text{bhp}}{\text{rpm}}$$

To find the bhp of an engine, the foregoing formula is rearranged as follows:

$$\text{bhp} = \frac{\text{bmep} \times \text{rpm}}{K}$$

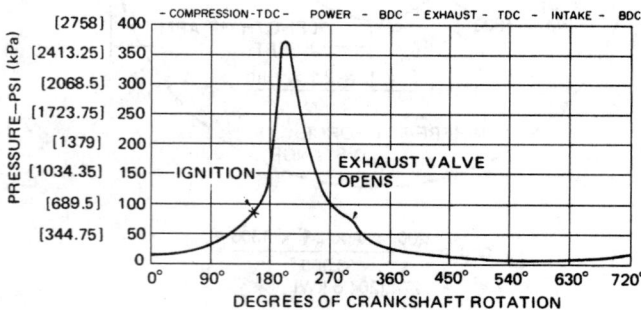

FIGURE 3-12 Curve showing cylinder pressure.

The constant K is often called the K factor of the engine.

Note that the foregoing formula may be used for imep as well as bmep if ihp is used in the formula instead of bhp. It is easy to determine the bhp of an engine by means of a dynamometer or prony brake; therefore, the bmep computation is more commonly employed to determine engine performance.

Brake mean effective pressure can be derived mathematically from the bhp, and vice versa. When the bhp has been determined by means of a dynamometer, the bmep can be computed by the formula

$$\text{bmep} = \text{bhp} \times \frac{33000}{LAN}$$

where L = stroke, ft
A = area of bore, in^2
N = number of working strokes per minute

In a four-stroke-cycle engine, $N = \frac{1}{2}$ the rpm of the engine multiplied by the number of cylinders.

The Indicator Diagram

Indicated horsepower (ihp) is based on the theoretical amount of work done according to calculations made from the actual pressure recorded in the form of a diagram on an indicator card, as illustrated in Fig. 3-13. This particular indicator diagram shows the pressure rise during the compression stroke and after ignition. It also shows the pressure drop as the gases expand during the power stroke. It clearly emphasizes the fact that the force acting on the piston during the combustion (power) stroke of the engine is not constant, because the fuel-air mixture burns almost instantaneously, with a resulting high pressure at the top of the stroke and a decreasing pressure as the piston descends.

Power computations for ihp are somewhat simplified by using the average pressure acting on the piston throughout the working stroke. This average pressure, often called the mean effective pressure, is obtained from the indicator diagram. The indicator diagram is drawn on the indicator card by a mechanical device attached to the engine cylinder. Modern engine manufacturers utilize much more sophisticated instrumentation with computers to determine the various engine measurements, or parameters. Figure 3-13 does, however, provide a graphic illustration of the process.

It has often been the practice with turboprop engines to equip one or more of the engines with a **torque-indicating system**. This system consists of a mechanism in the nose section of the engine which applies pressure to oil in a closed chamber in proportion to the engine torque. Since the planetary gears of a propeller reduction-gear system must work against a large stationary ring gear, the ring gear can be used to develop an indication of engine torque. The outside of the ring gear is constructed with helical teeth which fit into similar helical teeth in the nose case. When torque is developed, the ring gear tends to move forward. This movement is transmitted to hydraulic pistons which are connected to a torque gauge in the cockpit.

Power Ratings

The **takeoff power** rating of an engine is determined by the maximum rpm and manifold pressure at which the airplane engine may be operated during the process of taking off. The takeoff power may be given a time limitation, such as a period of 1 to 5 min. **Manifold pressure** is the pressure of the fuel-air mixture in the intake manifold between the carburetor or internal supercharger and the intake valve. The pressure is given in inches of mercury (inHg) [kilopascals (kPa)] above absolute zero pressure. Since sea-level pressure is 29.92 inHg [101.34 kilopascals (kPa)], the reading on the manifold pressure gauge may be either above or below this figure. As the manifold pressure increases, the power output of the engine increases, provided that the rpm remains constant. Likewise, the power increases as rpm increases, provided that the manifold pressure remains constant.

The takeoff power of an engine may be about 10 percent above the maximum continuous power-output allowance. This is the usual increase of power output permitted in the United States, but in British aviation the increase above maximum cruising power may be as much as 15 percent. It is sometimes referred to as the **overspeed** condition. The maximum continuous power is also called the **maximum except takeoff** (METO) power.

During takeoff conditions with the engine operating at maximum takeoff power, the volume of air flowing around the cylinders is restricted because of the low speed of the airplane during takeoff, and the initial carburetor air temperature may be very high in hot weather. For these reasons the pilot must exercise great care, especially in hot weather, to avoid overheating the engine and damaging the valves, pistons, and piston rings. The overheating may cause detonation or preignition, with a resultant loss of power in addition to engine damage.

The **rated power**, also called the **standard engine rating**, is the maximum horsepower output which can be obtained from an engine when it is operated at specified rpm and manifold pressure conditions established as safe for continuous operation. This is the power guaranteed by the manufacturer of the engine under the specified conditions and is the same as the METO power.

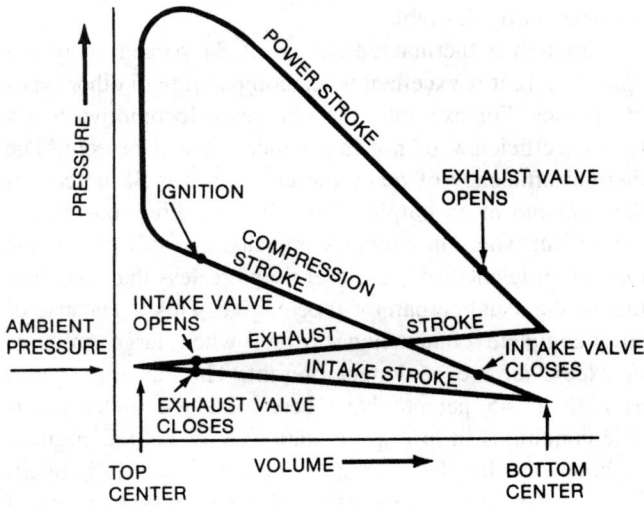

FIGURE 3-13 Cylinder pressure indicator diagram.

Maximum power is the greatest power output that the engine can develop at any time under any condition.

Critical Altitude

The **critical altitude** is the highest level at which an engine will maintain a given horsepower output. For example, an aircraft engine may be rated at a certain altitude which is the highest level at which rated power output can be obtained from the engine at a given rpm. Turbochargers and superchargers are employed to increase the critical altitude of engines. These applications are discussed in later chapters.

ENGINE EFFICIENCY

Mechanical Efficiency

The **mechanical efficiency** of an engine is measured by the ratio of the brake horsepower, or shaft output, to the indicated horsepower, or power developed in the cylinders. For example, if the ratio of the bhp to the ihp is 9:10. then the mechanical efficiency of the engine is 90 percent. In determination of mechanical efficiency, only the losses suffered by the energy that has been delivered to the pistons are considered. The word "efficiency" may be defined as the ratio of output to input.

Thermal Efficiency

Thermal efficiency is a measure of the heat losses suffered in converting the heat energy of the fuel to mechanical work. In Fig. 3-14, the heat dissipated by the cooling system represents 25 percent, the heat carried away by the exhaust gases represents 40 percent, the mechanical work on the piston to overcome friction and pumping losses represents 5 percent, and the useful work at the propeller shaft represents 30 percent of the heat energy of the fuel.

The thermal efficiency of an engine is the ratio of the heat developed into useful work to the heat energy of the fuel. It may be based on either bhp or ihp and is represented by

$$= \frac{\text{Indicated thermal efficiency}}{\text{wt/min of fuel burned} \times \text{heat value (Btu)} \times 778}$$
$$\frac{\text{ihp} \times 33000}{}$$

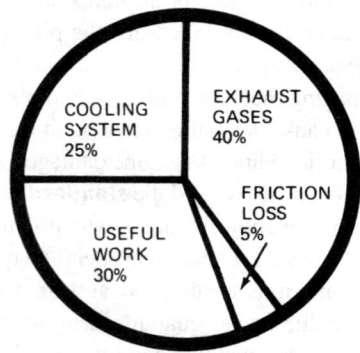

FIGURE 3-14 Thermal efficiency chart.

The formula for **brake thermal efficiency** (bte) is the same as that given above, with the word "brake" inserted in place of "indicated" on both sides of the equation.

If we wish to find the bte of a particular engine, we must first know the following quantities: the bhp, the fuel consumption in pounds per minute, and the heat value of the fuel in British thermal units (Btu). In this case, suppose that the engine develops 104 bhp at 2600 rpm and burns 6.5 gallons per hour (gal/h) [24.61 liters per hour (L/h)] of gasoline. The heat value of the fuel is 19 000 to 20 000 Btu [20 045 000 to 21 110 000 joules (J)].

First we must convert gallons [liters] per hour to pounds [kilograms] per minute. Since there are 60 min in 1 h, we divide 6.5 by 60 to obtain 0.108 gal/min [0.41 L/min]. Since each gallon of fuel weighs approximately 6 lb [2.72 kg], we multiply 0.108 by 6 to obtain 0.648 lb/min [0.29 kg/min]. The formula then becomes

$$\text{bte} = \frac{104 \times 33000}{0.648 \times 20000 \times 778}$$
$$= \frac{3432000}{10082880} = 0.34$$

Therefore, the bte is 34 percent.

To explain the formula, we must know only that the energy of 1 Btu is 778 ft•lb [107.6 kg•m]. The product of 104×33000 provides us with the total foot-pound output. The figures in the denominator give us the total input energy of the fuel. The fraction then represents the ratio of input to output.

In the foregoing problem, if the engine burns 100 gal [1378.54 L] of gasoline, only 34 gal [128.7 L] is converted to useful work. The remaining 66 percent of the heat produced by the burning fuel in the engine cylinders is lost by being exhausted through the exhaust manifold or through the cooling of the engine. This is an excellent value for many modern aircraft engines running at full power. At slightly reduced power, the thermal efficiency may be a little greater, and by the use of high compression with high-octane fuels, an engine may be made to produce as high as 40 percent bte. This is not normal, however, and for mechanical reasons is not necessarily desirable.

Although a thermal efficiency of 34 percent may not appear high, it is excellent when compared with other types of engines. For example, the old steam locomotive had a thermal efficiency of not much more than 5 percent. The thermal efficiency of many diesel engines is 35 percent if they are run at an output of one-half to three-fourths full power, but when the output is increased to full power, the thermal efficiency of the diesel drops to less than one-half that of the usual carburetor-type engine. This is because of an incomplete combustion of fuel when large amounts of excess air are no longer present. Thermal efficiencies as high as 45 percent have been obtained under favorable conditions in low-speed stationary or marine engines. Although the diesel engine has been used successfully in airplanes, in its present state of development it lacks many of the advantages of the carburetor-type aircraft engine.

Power and Efficiency

The efficiency of an engine is the ratio of output to input. For example, if the amount of fuel consumed should produce 300 hp [223.71 kW] according to its British thermal unit (Btu) rating and the output is 100 hp [174.57 kW], then the thermal efficiency is 100/300, or 33 $\frac{1}{3}$ percent.

An engine producing 70 hp [52.20 kW] burns about 30 lb/h [13.61 kg/h], or $\frac{1}{2}$ lb/min [0.23 kg/min], of gasoline. Since $\frac{1}{2}$ lb [0.23 kg] of gasoline has a heat value of about 10 000 Btu and since 1 Btu can do 778 ft•lb [107.60 kg•m] of work, the fuel being consumed should produce 778 × 10 000 ft•lb [1383 kg•m] of work per minute. Then

$$\text{Power} = \frac{778 \times 10\,000}{33\,000} = 235\,\text{hp}\,[175.25\,\text{kW}]$$

The fuel being consumed has a total power value of 235 hp but the engine is producing only 70 bhp. The thermal efficiency is then 70/235, or approximately 30 percent.

The question may be asked: What happens to the other 70 percent of the fuel energy? The answer is that the largest portion of the fuel energy is dissipated as heat and friction. The distribution of the fuel energy is approximately

Brake horsepower	30 percent
Friction and heat loss from engine	20 percent
Heat and chemical energy in exhaust	50 percent

Volumetric Efficiency

Volumetric efficiency is the ratio of the volume of the fuel-air charge burned by the engine at atmospheric pressure and temperature to the piston displacement. If the cylinder of an engine draws in a charge of fuel and air having a volume at standard atmospheric pressure and temperature which is exactly equal to the piston displacement of the cylinder, then the cylinder has a volumetric efficiency of 100 percent.

In a similar manner, if a volume of 95 in³ [1.56 L] of the fuel-air mixture is admitted into a cylinder of 100-in³ [1.64-L] displacement, the volumetric efficiency is 95 percent. Volumetric efficiency may be expressed as a formula thus:

$$\text{Vol eff} = \frac{\text{vol of charge at atmospheric pressure}}{\text{piston displacement}}$$

Factors that tend to decrease the mass of air entering an engine have an adverse effect on volumetric efficiency. Typical factors that have this effect are (1) improper valve timing, (2) high engine rpm, (3) high carburetor air temperature, (4) improper design of the induction system, and (5) high combustion chamber temperature. A combination of these factors can exist at any one time.

Improper timing of the valves affects volumetric efficiency because the degree of opening of the intake valve influences the amount of airflow into the cylinder and because the timing of the opening and closure of the exhaust valve affects the outflow of exhaust gases. The intake valve must be open as wide as possible during the intake stroke, and the exhaust valve must close precisely at the instant that exhaust gases stop flowing from the combustion chamber.

High engine rpm can limit volumetric efficiency because of the air friction developed in the intake manifold, valve ports, and carburetor. As the intake air velocity increases, friction increases and reduces the volume of airflow. At very high engine rpm, the valves may "float" (not close completely), thereby affecting airflow.

Carburetor air temperature affects volumetric efficiency because as the air temperature increases, the density decreases. This results in a decreased mass (weight) of air entering the combustion chambers. High combustion chamber temperature is a factor because it affects the density of the air.

Maximum volumetric efficiency is obtained when the throttle is wide open and the engine is operating under a full load.

A **naturally aspirated** (unsupercharged) **engine** always has a volumetric efficiency of less than 100 percent. However, the supercharged engine often is operated at a volumetric efficiency of more than 100 percent because the supercharger compresses the air before the air enters the cylinder. The volumetric efficiency of a naturally aspirated engine is less then 100 percent for two principal reasons: (1) the bends, obstructions, and surface roughness inside the intake system cause substantial resistance to the airflow, thus reducing air pressure below atmospheric in the intake manifold; and (2) the throttle and the carburetor venture provide restrictions across which a pressure drop occurs.

FACTORS AFFECTING PERFORMANCE

Earlier in this chapter, engine power was discussed, as well as mean effective pressure, rpm, displacement, and other factors involved in the measurement of engine performance. These areas will be explored in greater depth and applied to actual engine operation.

Manifold Pressure

As was explained previously, **manifold pressure**, or **manifold absolute pressure** (MAP), is the absolute pressure of the fuel-air mixture immediately before it enters the intake port of the cylinder. **Absolute pressure** is the pressure above a complete vacuum and is often indicated in *pounds per square inch absolute* (psia) or in *inches of mercury* (inHg). In the metric system, MAP may be indicated in kilopascals (kPa). The pressure we read on an ordinary pressure gauge is the pressure above ambient atmospheric pressure and is often called **gauge pressure**, or **pounds per square inch gauge** (psig). MAP is normally indicated on a pressure gauge in inches of mercury instead of a pounds per square inch gauge; therefore, the reading on a manifold gauge at sea level when an engine is not running will be about 29.92 inHg [101.34 kPa] when conditions are standard. When the engine is idling, the gauge may read from 10 to 15 inHg [33.87 to 50.81 kPa] because MAP will be considerably below atmospheric pressure owing to the restriction of the throttle valve.

MAP is of primary concern to the operator of a high-performance engine because such an engine will often be operating at a point near the maximum allowable pressure. It is essential, therefore, that any engine which can be operated at an excessive MAP be equipped with a MAP gauge so that the operator can keep the engine operation within safe limits.

The operator of an aircraft engine must take every precaution to avoid operating at excessive MAP or incorrect MAP-rpm ratios because such operation will result in excessive cylinder pressures and temperatures. Excessive cylinder pressures are likely to overstress the cylinders, pistons, piston pins, valves, connecting rods, bearings, and crankshaft journals. Excessive pressure usually is accompanied by excessive temperature, and this leads to detonation, preignition, and loss of power. Detonation usually results in engine damage if continued for more than a few moments. Damage may include piston failure by cracking or burning, failure of cylinder base studs, cracking of the cylinder head, and burning of valves.

Naturally aspirated engines using variable-pitch propellers *must be equipped with MAP gauges to ensure safe operation.*

Such engines, when equipped with fixed-pitch propellers, do not require the use of a MAP gauge because on these engines the MAP is a function of the throttle opening.

Detonation and Preignition

Detonation and preignition have already been mentioned; however, it is good to review these conditions in connection with a discussion of engine performance and engine parameter interaction.

Detonation is caused when the temperature and pressure of the compressed mixture in the combustion chamber reach levels sufficient to cause instantaneous burning (explosion) of the fuel-air mixture. Excessive temperatures and/or pressures can be caused by several different engine parameters, such as high inlet-air temperature, insufficient fuel octane rating, excessive engine load, overadvanced ignition timing, excessively lean fuel-air mixture, and excessive compression ratio. Factors that affect detonation are described in Fig. 3-15.

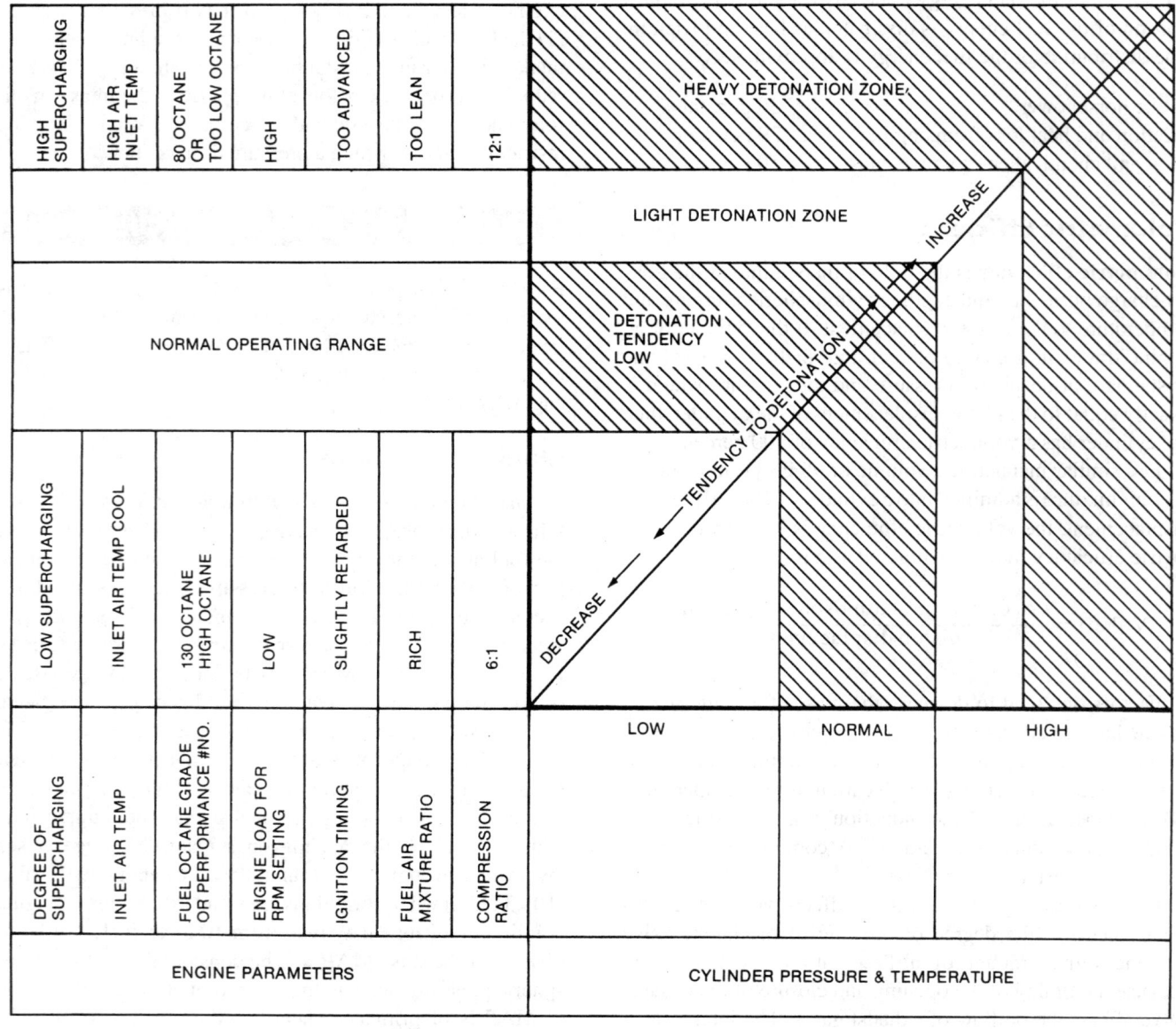

FIGURE 3-15 Factors that affect detonation.

Their relationship to detonation is shown with regard to cylinder pressure and temperature. A principal cause of detonation is operation of an engine with either a fuel whose octane rating is not sufficiently high for the engine or a high-combustion-rate fuel. A high-octane fuel can withstand greater temperature and pressure before igniting than can a low-octane fuel. When detonation occurs, the fuel-air mixture may burn properly for a portion of its combustion and then explode as the pressure and temperature in the cylinder increase beyond their normal limits, as illustrated in Fig. 3-16.

Detonation will further increase the temperature of the cylinders and pistons and may cause the head of a piston to melt. Detonation will generally cause a serious power loss. Instead of the piston receiving a smooth push, it gets a very short high-pressure push, much like the head of the piston being hit with a hammer. This high-pressure push occurs too quickly to be absorbed by the piston, with the result being a loss of power. This is shown in Figs. 3-17 and 3-18. Detonation will result whenever the temperature and pressure in the cylinder become excessive. A very lean mixture will lend to burn at a slower rate than will a rich mixture, allowing the cylinder to be subjected to high temperatures for a longer time than usual. If this condition is not corrected, the cylinder temperature will continue to climb until detonation occurs. Detonation can also be caused by excessive intake air temperature. This condition can be caused by the carburetor heat during high-power settings of the engine or excessive supercharging. Detonation cannot generally be detected in an aircraft engine as easily as preignition.

Preignition is caused when there is a hot spot in the engine that ignites the fuel-air mixture before the spark plug fires,

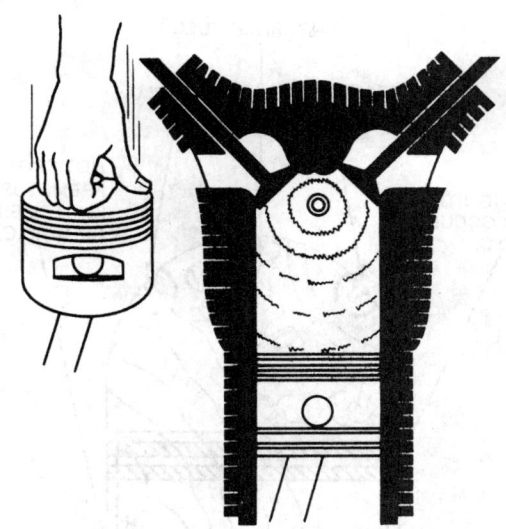

FIGURE 3-17 Normal combustion within a cylinder.

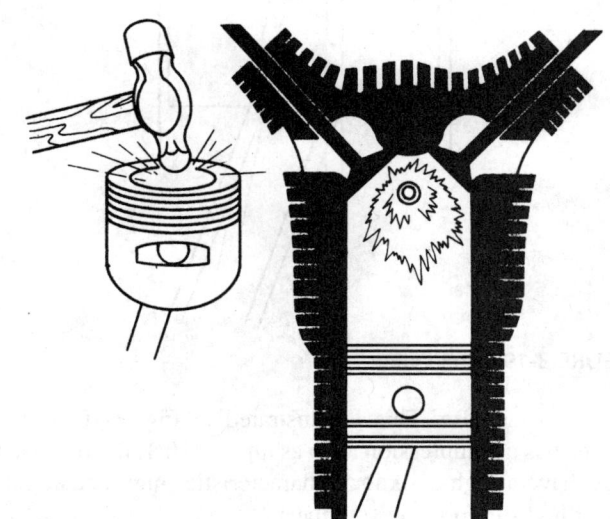

FIGURE 3-18 Detonation within a cylinder.

as illustrated in Fig. 3-19. The hot spot may be red-hot spark plug electrodes or carbon particles which have reached the burning temperature and are glowing. Preignition is indicated by roughness of engine operation, power loss, and high cylinder-head temperature. To prevent preignition and detonation, operate the engine with the proper fuel and within the correct limits of MAP and cylinder-head temperature.

Compression Ratio

As stated previously, the **compression ratio** of any internal-combustion engine is a factor that controls the maximum horsepower which can be developed by the engine, other factors remaining the same. Within reasonable limits, the maximum horsepower increases as the compression ratio increases. However, experience shows that if the compression ratio of the internal-combustion engine is much greater than 10:1, preignition or detonation may occur and cause overheating, loss of power, and damage to the engine. As the compression ratio increases, the pressure in the cylinder

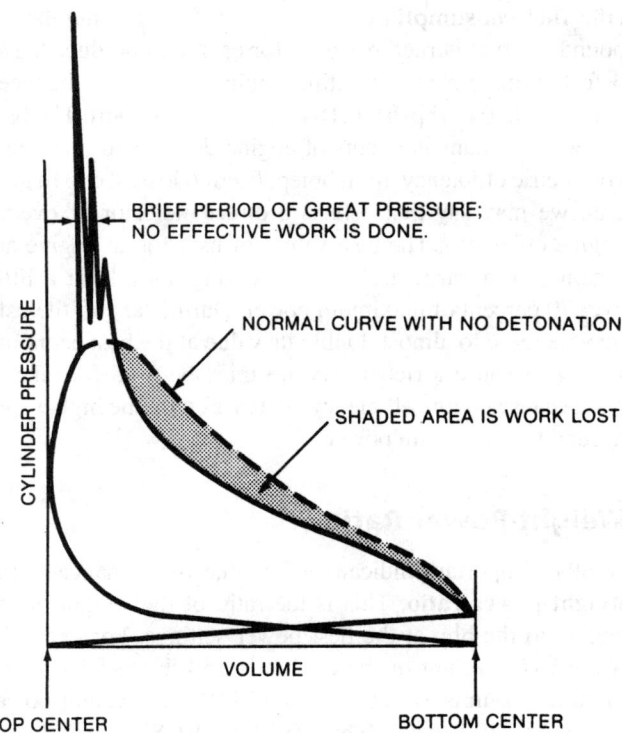

FIGURE 3-16 Detonation as illustrated by a pressure-volume diagram.

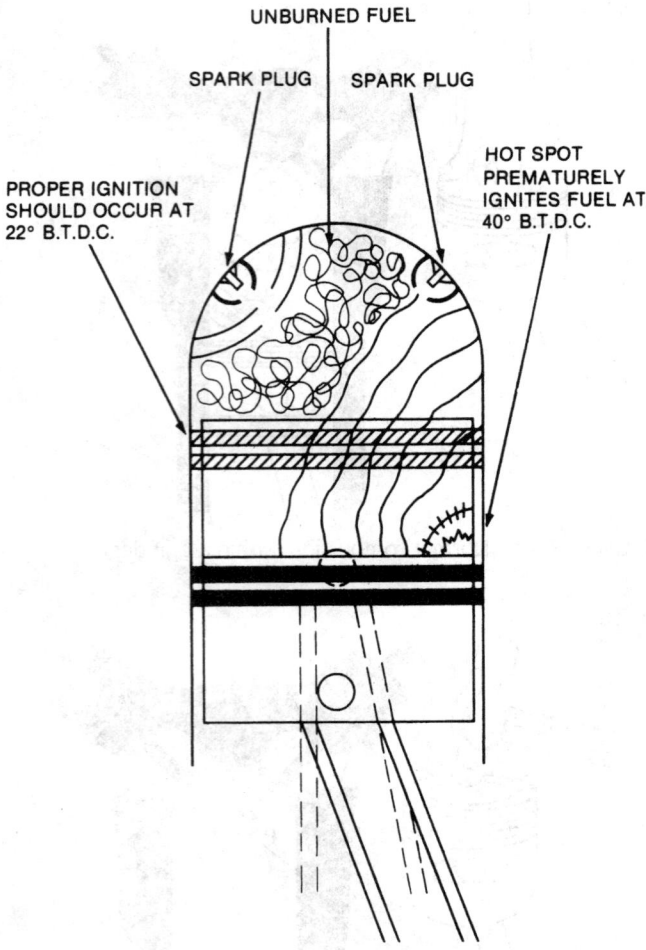

FIGURE 3-19 Preignition.

will also increase. This is illustrated in Fig. 3-20. If the engine has a compression ratio as high as 10:1, the fuel used must have a high-antiknock characteristic (high-octane rating or high-performance number).

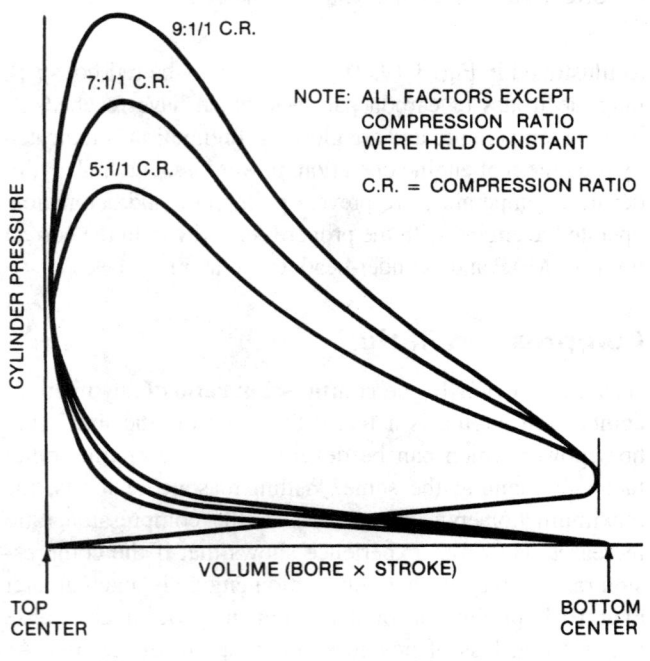

FIGURE 3-20 Compression ratio vs. pressure.

It is common for a gasoline engine to have a compression ratio of about 7:1; however, certain high-performance engines have higher ratios. The compression ratio of a conventional gasoline engine must be limited because the temperature or pressure of the compressed gases in the cylinder must not be high enough to ignite the fuel. The high-octane and high-performance-number fuels have made it possible to utilize higher compression ratios for conventional engines. If the compression ratio is too high for the fuel being used, detonation (explosion) of the fuel will occur, thus causing overheating, loss of power, and probable damage to the pistons and cylinders.

The **maximum compression ratio** of an engine, as indicated above, is limited by the detonation characteristics of the fuel used.

In addition to the detonation characteristics of the fuel used, the maximum compression ratio of an aircraft engine is affected by the design limitations of the engine, the availability of high-octane fuel, and the degree of supercharging.

The compression ratio of an engine may be increased by installing "higher" pistons, by using longer connecting rods, or by installing a crankshaft with a greater throw. In an engine which has a removable cylinder head, the combustion space in the head may be reduced by "shaving" the surface of the cylinder head where it mates with the top of the cylinder. This, of course, increases the compression ratio.

Increasing the compression ratio of an engine causes a lower specific fuel consumption (pounds of fuel burned per hour per horsepower) and a greater thermal efficiency.

Brake Specific Fuel Consumption

One of the measures of engine performance is **brake specific fuel consumption** (bsfc). The bsfc is the number of pounds of fuel burned per hour for each bhp produced. The bsfc for modern reciprocating engines is *usually* between 0.40 and 0.50 lb/(hp•h) [0.18 and 0.226 kg/kW•h]. The bsfc depends on many elements of engine design and operation, volumetric efficiency, rpm, bmep, friction losses, etc. In general, we may say that bsfc is a direct indicator of overall engine efficiency. The best values of bsfc for an engine are obtained at a particular cruising setting, usually at a little over 70 percent of maximum power. During takeoff the bsfc may increase to almost double its value at the best economy setting, because a richer mixture must be used for takeoff and because engine efficiency decreases with the higher rpm needed for maximum power.

Weight-Power Ratio

Another important indicator of engine performance is the **weight-power ratio**. This is the ratio of the weight of the engine to the bhp at the best power settings. For example, if the basic weight of the engine is 150 lb [68.04 kg] and the power output is 100 hp [74.57 kW], the weight-power ratio is 150:100, or 1.5 lb/hp [0.91 kg/kW]. Since weight is a prime consideration in the design of any aircraft, the weight-power ratio of the engine is always an important factor in the selection of the airplane powerplant. Weight-power ratios for reciprocating engines vary between 1.0 and 2.0 lb/hp [0.61 and 1.22 kg/kW], with the majority of high-performance engines having ratios between 1.0 and 1.5 lb/hp [0.61 and 0.91 kg/kW].

Effects of Altitude on Performance

The air density will affect the power output of an engine at a particular rpm and MAP. Since the air density depends on pressure, temperature, and humidity, these factors must be taken into consideration in determining the exact performance of an engine. To obtain the power of an engine from a power chart, it is necessary to find what corrections must be made. First, the **density altitude** must be determined by applying approximate corrections to the pressure altitude as shown on a standard barometer. A chart for converting pressure altitude to density altitude is shown in Fig. 3-21. If the temperature at a particular altitude is the same as the standard temperature at that altitude, no correction for density will be required unless humidity is a factor.

The charts in Fig. 3-22 are used to determine the power output of a Continental O-470-M engine at altitude. The chart at the left shows engine output at sea-level standard conditions with no ram air pressure applied to the carburetor intake. The chart at the right shows the effect of altitude and is used in conjunction with the first chart. The points corresponding to engine rpm and MAP are located on both charts. The horsepower indicated on the sea-level chart is transferred to an equivalent point C on the altitude chart. Then a straight line is drawn from point A to point C to establish

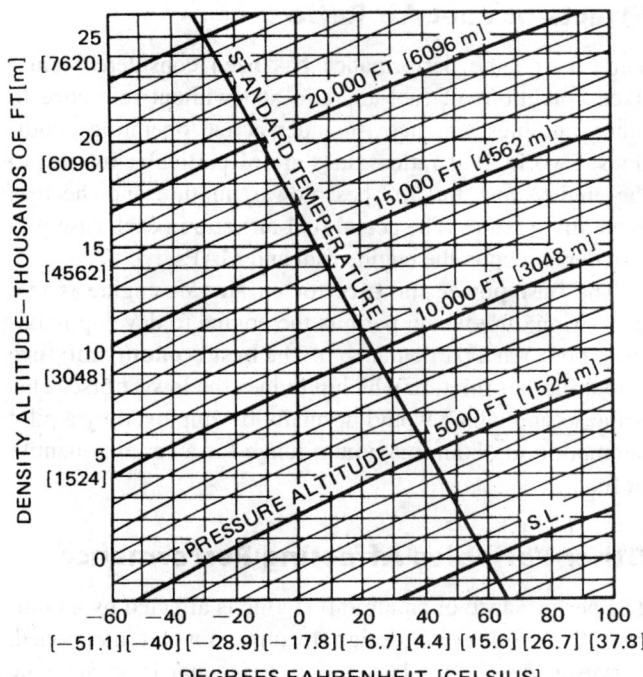

FIGURE 3-21 Chart for converting pressure altitude to density altitude.

the altitude correction. The intersection of this line with the density altitude line (D in the example) establishes the power output of the engine. The horsepower should be corrected by adding 1 percent for each 6°C [10.8°F] temperature decrease below T_S (standard temperature) and subtracting 1 percent for each 6°C [42.8°F] temperature increase above T_S.

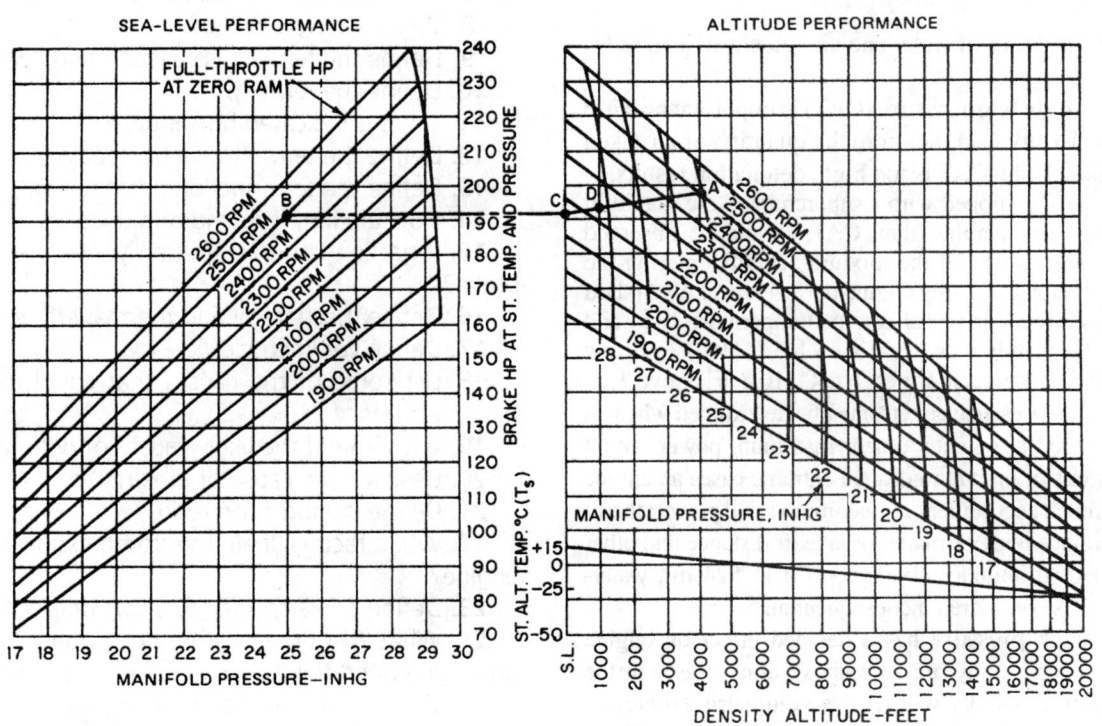

FIGURE 3-22 Finding actual horsepower from sea-level and altitude charts.

Effects of Fuel-Air Ratio

Thus far engine performance has been considered under fixed conditions of fuel-air ratio and without reference to other variables which exist under actual operating conditions. Two fuel-air ratio values are of particular interest to the engine operator: the best power mixture and the best economy mixture. The actual fuel-air ratio in each case will also depend upon the engine rpm and MAP.

The **best power mixture** for an aircraft engine is that fuel-air mixture which permits the engine to develop maximum power at a particular rpm. The **best economy mixture** is that fuel-air mixture which provides the lowest bfsc. This is the setting which would normally be employed by a pilot attempting to obtain maximum range for a certain quantity of fuel.

Other Variables Affecting Performance

The performance of an aircraft engine is affected by a number of conditions or design features not yet mentioned. However, these must be taken into account if an accurate evaluation of engine operation is to be made. Among these conditions are carburetor air intake ram pressure, carburetor air temperature, water-vapor pressure, and exhaust back pressure.

Ram air pressure at the carburetor air scoop is determined by the design of the scoop and the velocity of the air. Ram air pressure has the effect of supercharging the air entering the engine; therefore, the actual power output will be greater than it would be under standard conditions of rpm, pressure, and temperature. An empirical formula is

$$\text{Ram} = \frac{V^2}{2045} - 2$$

where ram is in inches of water and air velocity V is in miles per hour.

Carburetor air temperature (CAT) is important because it affects the density, and therefore the quantity, of air taken into the engine. If the CAT is too high, detonation results.

If an engine is equipped with a supercharger, the manifold mixture temperature, rather than CAT, should be observed because the temperature of the mixture actually entering the engine is the factor governing engine performance. A standard rule used to correct for the effects of temperature is to add 1 percent to the chart horsepower for each 6°C [10.8°F] below T_S and to subtract 1 percent for each 6°C [10.8°F] above T_S.

Water-vapor pressure effects must be determined when an engine is required to operate at near maximum power output under conditions of high humidity. In extreme cases an engine may lose as much as 5 percent of maximum rated power; therefore, an allowance must be made for takeoff distance and other critical factors. At altitudes above 5000 ft [1 524 m], water-vapor pressure is considered inconsequential.

Exhaust back pressure has a decided effect on engine performance because any pressure above atmospheric at the exhaust port of a cylinder will reduce volumetric efficiency. The design of the exhaust system is therefore one of the principal items to be considered by both the engine manufacturer and the manufacturer of the exhaust system. The exhaust-back-pressure effect begins at the cylinder with the exhaust port. Both the size and the shape of the opening and passages will affect the pressure. From the exhaust port onward, the exhaust stacks and sound reduction devices will produce varying amounts of back pressure, depending on their design.

Engineers have developed exhaust-augmenting systems to assist in reducing exhaust back pressure and to utilize the ejected exhaust gases for the production of additional thrust. These devices have proved effective for increasing engine performance on the airplane. Such systems usually consist of one or more tubes into which the exhaust stacks from the engine are directed. The engine's exhaust passing through the tubes through which ram air is also flowing results in a reduced pressure against the exhaust and an increased thrust, because the jet of exhaust gases is directed toward the rear of the aircraft. **Exhaust augmentors** with inlets inside the engine nacelle also increase airflow through the nacelle, thus improving cooling.

REVIEW QUESTIONS

1. Why are reciprocating engines for aircraft called heat engines?
2. Define the terms "bore" and "stroke."
3. What are the TDC and BDC positions of the piston?
4. List the four-stroke cycle of a piston engine.
5. What are the positions of the intake and exhaust valves during the power stroke?
6. At what point in the operating cycle of an engine does ignition take place?
7. Why is ignition timed to take place at this point?
8. What is valve overlap?
9. Define the terms "valve lead" and "valve lag."
10. Define the term "power."
11. What is indicated horsepower?
12. Define the term "brake horsepower."
13. Define the term friction horsepower."
14. How are ihp, fhp, and bhp related?
15. What is the purpose of a dynamometer or prony brake?
16. Define mechanical and thermal efficiencies.
17. Define *volumetric efficiency*.
18. List some of the causes that could reduce volumetric efficiency.
19. List some of the likely causes of detonation.
20. Describe the cause of preignition.
21. Define *compression ratio*.
22. What factors limit the compression ratio of an engine?
23. Define *brake specific fuel consumption*.
24. What effect does carburetor air temperature have on engine operation?
25. What effect does exhaust back pressure have on engine performance?

Lubricants and Lubricating Systems 4

Source of Engine Lubricating Oils

Petroleum, which is the source of volatile fuel gasoline, is also the source of engine lubricating oil. Crude petroleum is refined by the processes of distillation, dewaxing, chemical refining, and filtration.

In the process of distillation, crude petroleum is separated into a series of products varying from gasoline to the heaviest lubricating oils according to the boiling point of each. The dewaxing process essentially consists in chilling the waxy oil to low temperatures and allowing the waxy constituents to crystallize, after which the solid wax can be separated from the oil by filtration. After the removal of the wax, resinous and asphaltic materials are removed from the lubricating oil by chemical refining. The oil is then treated with an absorbent which removes the last traces of the chemical refining agents previously used, improves the color, and generally prepares the oil for shipment and use.

CLASSIFICATION OF LUBRICANTS

A **lubricant** is any natural or artificial substance having greasy or oily properties which can be used to reduce friction between moving parts or to prevent rust and corrosion on metallic surfaces. Lubricants may be classified according to their origins as animal, vegetable, mineral, or synthetic.

Animal Lubricants

Examples of lubricants having an animal origin are tallow, tallow oil, lard oil, neat's-foot oil, sperm oil, and porpoise oil. These are highly stable at normal temperatures, so they can be used to lubricate firearms, sewing machines, clocks, and other light machinery and devices. Porpoise oil, for example, is used to lubricate expensive watches and very delicate instruments. However, animal lubricants cannot be used for internal-combustion engines because they produce fatty acids at high temperatures.

Vegetable Lubricants

Examples of vegetable lubricants are castor oil, olive oil, rape oil, and cottonseed oil. These oils tend to oxidize when exposed to air. Vegetable and animal oils have a lower coefficient of friction than most mineral oils, but they wear away steel rapidly because of their ability to loosen the bonds of iron on the surface.

Castor oil, like other vegetable oils, will not dissolve in gasoline. For this reason it was used in rotary engines where the crankcase was used as a part of the induction system. It oxidizes easily and causes gummy conditions in an engine.

Mineral Lubricants

Mineral lubricants are used to a large extent in the lubrication of aircraft internal-combustion engines. They may be classified as solids, semisolids, and fluids.

Solid Lubricants

Solid lubricants, such as mica, soapstone, and graphite, do not dissipate heat rapidly enough for high-speed machines, but they are fairly satisfactory in a finely powdered form on low-speed machines. Solid lubricants fill the low spots in the metal on a typical bearing surface to form an almost perfectly smooth surface, and at the same time they provide a slippery film that reduces friction. When a solid lubricant is finely powdered and is not too hard, it may be used as a mild abrasive to smooth the surface previously roughened by excessive wear or by machine operations in a factory. Some solid lubricants can carry heavy loads, and therefore they are mixed with certain fluid lubricants to reduce the wear between adjacent surfaces subjected to high unit pressures. Powdered graphite is used instead of oils and greases to lubricate firearms in extremely cold weather, because oils and greases become thick and gummy, rendering the firearms inoperative.

Semisolid Lubricants

Extremely heavy oils and greases are examples of semisolid lubricants. Grease is a mixture of oil and soap. It gives good service when applied periodically to certain units, but its consistency is such that it is not suitable for circulating or continuous-operation lubrication systems. In general, sodium soap is mixed with oil to make grease for gears and hot-running equipment, calcium soap is mixed with oil to make cup grease, and aluminum soap is mixed with oil to make grease for ball-bearing and high-pressure applications.

Fluid Lubricants (Oils)

Fluid lubricants (oils) are used as the principal lubricants in all types of internal-combustion engines because they can be pumped easily and sprayed readily and because they absorb and dissipate heat quickly and provide a good cushioning effect.

Summary of Advantages of Mineral-Base Lubricants

In general, lubricants of animal and vegetable origin are chemically unstable at high temperatures, often perform poorly at low temperatures, and are unsuited for aircraft engine lubrication. However, lubricants having a mineral base are chemically stable at moderately high temperatures, perform well at low temperatures, and are widely used for aircraft engine lubrication.

Synthetic Lubricants

Because of the high temperatures required in the operation of gas-turbine engines, it became necessary for the industry to develop lubricants which would retain their characteristics at temperatures that cause petroleum lubricants to evaporate and break down into heavy hydrocarbons. Synthetic lubricants do not evaporate or break down easily and do not produce coke or other deposits. These lubricants are called synthetics because they are not made from natural crude oils. Typical synthetic lubricants are **Type I**, **alkyl diester oils** (MIL-L-L7808), and **Type II**, **polyester oils** (MIL-L-23699).

LUBRICATING OIL PROPERTIES

The most important properties of an aircraft engine oil are its flash point, viscosity, pour point, and chemical stability. Various tests for these properties can be made at the refinery and in the field. In addition, there are tests which are of interest principally to the petroleum engineers at the refinery, although all personnel interested in aircraft engine lubrication should have some familiarity with such tests so that they can intelligently read reports and specifications pertaining to petroleum products.

Some of the properties tested at the refinery are the gravity, color, cloud point, carbon residue, ash residue, oxidation, precipitation, corrosion, neutralization, and oiliness of the oil.

Gravity

The **gravity** of a petroleum oil is a numerical value which serves as an index of the weight of a measured volume of the product. Two scales are generally used by petroleum engineers. One is the specific-gravity scale, and the other is the American Petroleum Institute (API) gravity scale. *Gravity is not an index to quality*. It is a *property* of importance only to those operating the refinery but it is a convenient term for use in figuring the weights and measures and in the distribution of lubricants.

Specific gravity is the weight of any substance compared with the weight of an equal volume of a standard substance.

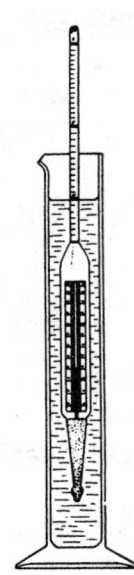

FIGURE 4-1   Hydrometer for determining API gravity.

When water is used as a standard, the specific gravity is the weight of a substance compared with the weight of an equal volume of distilled water.

A hydrometer is used to measure specific gravity, and it is also used in conducting the API gravity test.

Formerly, the petroleum industry used the Baumé scale, but it has been superseded by the API gravity scale which magnifies that portion of the specific-gravity scale which is of greatest interest for testing petroleum products. The test is usually performed with a hydrometer, a thermometer, and a conversion scale for temperature correction to the standard temperature of 60°F [15.56°C], as shown in Fig. 4-1.

Water has a specific gravity of 1.000, weighs 8.328 lb/gal [3.78 kg/gal], and has an API gravity reading of 10 under standard conditions for the test. An aircraft lubricating oil which has a specific gravity of 0.9340 and weighs 7.778 lb/gal [3.53 kg/gal] has an API gravity reading of 20 under standard conditions. If an aircraft lubricating oil has a specific gravity of 0.9042 and weighs 7.529 lb/gal [3.42 kg/gal], its API gravity reading is 24. These are merely examples of the relation between specific-gravity figures and API gravity readings. In actual practice, when the specific gravity and the weight in pounds per gallon at standard temperature are known, the API reading can be obtained from a table prepared for this purpose.

Flash Point

The **flash point** of an oil is the temperature to which the oil must be heated in order to give off enough vapor to form a combustible mixture above the surface that will momentarily flash or burn when the vapor is brought into contact with a very small flame. Because of the high temperatures at which aircraft engines operate, the oil used in such engines must have a high flash point. The rate at which oil vaporizes in an engine depends on the temperature of the engine and the grade of the oil. If the vaporized oil burns, the engine is not properly lubricated. The operating temperature of a

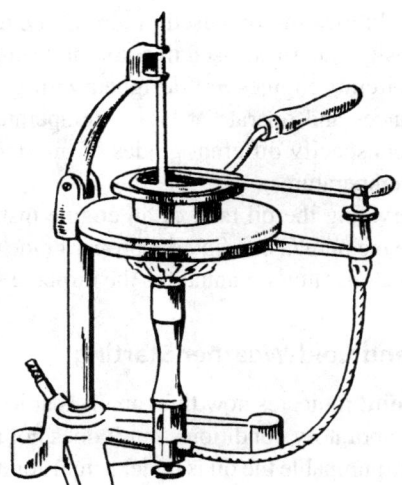

FIGURE 4-2   Cleveland open-cup tester.

particular engine determines the grade of oil which should be used.

The **fire point** is the temperature to which any substance must be heated in order to give off enough vapor to burn continuously when the flammable air–vapor mixture is ignited by a small flame. The fire-point test is mentioned occasionally in reports on lubricants, but it is not used as much as the flash-point test and should not be confused with it.

Lubricating oils can be tested by means of the **Cleveland open-cup tester** in accordance with the recommendations of the American Society for Testing and Materials (ASTM). This apparatus, shown in Fig. 4-2, is simple and adaptable to a wide range of products. It can be used for both flash-point and fire-point tests. When a test is made, the amount of oil, rate of heating, size of igniting flame, and time of exposure are all specified and must be carefully controlled to obtain accurate results.

In a test of stable lubricating oils, the fire point is usually about 50 to 60°F [28 to 33°C] higher than the flash point. Note that the determination of the fire point does not add much to a test, but the flash point of oil gives a rough indication of its tendency to vaporize or to contain light volatile material.

In a comparison of oils, if one has a higher or lower flash or fire point, this does not necessarily reflect on the quality of the oil—unless the fire point or flash point is exceptionally low in comparison with the fire or flash points of similar conventional oils.

If oil which has been used in an aircraft engine is tested and found to have a very low flash point, this indicates that the oil has been diluted by engine fuel. If the oil has been diluted only slightly with aviation-grade gasoline, the fire point is not lowered much because the gasoline in the oil ordinarily evaporates before the fire point is reached. If the oil has been greatly diluted by gasoline, the fire point will be very low.

In testing oil which has been used in an engine, it is possible to obtain more accurate results from the flash-point and fire-point tests if the sample of oil is obtained from the engine while both the engine and the oil are still hot.

Viscosity

Viscosity is technically defined as the fluid friction (or the body) of an oil. In simple terms, viscosity may be regarded as the resistance an oil offers to flowing. A heavy-bodied oil is high in viscosity and pours or flows slowly; it may be described as **viscous**. The lower the viscosity, the more freely an oil pours or flows at temperatures above the **pour point**, which indicates the fluidity of oil at lower temperatures. Oil that flows readily is described as having a low viscosity. The amount of fluid friction exhibited by the oil in motion is a measure of its viscosity.

The **Saybolt Universal viscosimeter**, illustrated in Fig. 4-3, is a standard instrument for testing petroleum products and lubricants. The tests are usually made at temperatures of 100, 130, and 210°F [38, 54, and 99°C].

This instrument has a tube in which a specific quantity of oil is brought to the desired temperature by a surrounding liquid bath. The time in seconds required for exactly 60 cm³ of the oil to flow through an accurately calibrated outlet orifice is recorded as **seconds Saybolt Universal viscosity**.

Commercial aviation oils are generally classified by symbols such as 80, 100, 120, and 140, which approximate the seconds Saybolt Universal viscosity at 210°F [99°C]. Their relation to Society of Automotive Engineers (SAE) numbers is given in Fig. 4-4.

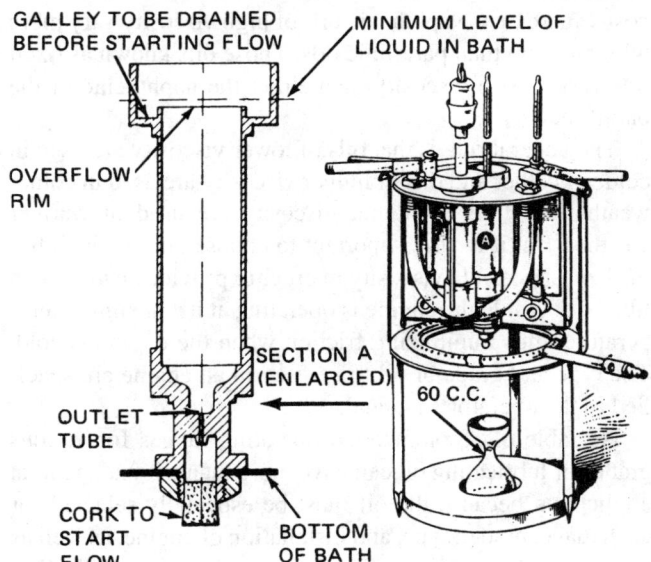

FIGURE 4-3   Saybolt Universal viscosimeter.

Commercial Aviation No.	Commercial SAE No.	AN Specification No.
65	30	1065
80	40	1080
100	50	1100
120	60	1120
140	70	

FIGURE 4-4   Grade designations for aviation oils.

Lubricating Oil Properties   **81**

Engineers use viscosity-temperature charts, published by the ASTM, to find quickly the variation of viscosity with the temperature of petroleum oils when the viscosities and any two temperatures are known. The two known temperatures are plotted on a chart and a straight line is drawn between these points. When the straight line is extended beyond the two known points, the viscosities at other temperatures can be read on the chart from that line.

The **viscosity index (VI)** is an arbitrary method of stating the rate of change in viscosity of an oil with changes of temperature. The VI of any specific oil is based on a comparative evaluation with two series of standardized oils: one has an assigned VI value of 100, which is somewhat typical of a conventionally refined Pennsylvania oil, and the other series has an assigned VI rating of 0, which is typical of certain conventionally refined naphthenic-base oils. The viscosity characteristics of these two series of standardized oils have been arbitrarily chosen and adopted by the ASTM.

Certain compounds can be added to the oil at the plant to raise the VI value above the value attained by any normal refining process. But this should not be interpreted by the reader to mean that it is safe to purchase compounds and dump them into the oil after it is received from the refinery or one of its agents.

The VI value is not fixed for all time when the oil is sold by the refinery or its distributors. If a lubricating oil is subjected to high pressure without any change in temperature, the viscosity increases. Naphthenic oils of high viscosity vary more with pressure than paraffinic oils. Those oils known as **fixed oils** vary less in viscosity than either the naphthenic or the paraffinic oils.

The general rule is that oils of lower viscosity are used in colder weather and oils of higher viscosity are used in colder weather and oils of higher viscosity are used in warmer weather. But it is also important to choose an oil which has the lowest possible viscosity in order to provide an unbroken film of oil while the engine is operating at its maximum temperature, thus minimizing friction when the engine is cold. The type and grade of oil to be used in an engine are specified in the operator's manual.

No table of recommended operating ranges for various grades of lubricating oil can have more than a broad, general application because the oil must be especially selected for each make, model, type, and installation of engine, as well as the operating conditions of both the engine and the airplane in which it is installed. However, a few recommendations may provide a starting point from which those selecting oils can proceed.

The grade of lubricating oil to be used in an aircraft engine is determined by the operating temperature of the engine and by the operating speeds of bearings and other moving parts. Commercial aviation grade no. 65 (SAE 30 or AN 1065) may be used at ground air temperatures of 4°C (40°F) and below. The **oil-in temperature** is the temperature of the oil before it enters the engine, as indicated by a thermometer bulb or other temperature-measuring device located in the oil system near the engine oil pump.

Note that lubricating oils used in aircraft engines have a higher viscosity than those used in automobile engines. This is because aircraft engines are designed with greater operating clearances and operate at higher temperatures. Some manufacturers specify different grades of oil, depending on outside air temperatures.

When servicing the oil tank of an engine installed in an airplane, the technician can find the proper grade of oil to be used from the operator's manual for the airplane.

Viscosity and Cold-Weather Starting

The **pour point** indicates how fluid an oil is at low temperatures under laboratory conditions, but it does not necessarily measure how pumpable the oil is under actual conditions. The viscosity is a far better indication of whether or not the oil will make it possible to start the engine at low temperatures and how well the oil can be pumped. At low temperatures it is desirable to have a combination of low pour point and low viscosity if the proper viscosity for operating temperatures is to be retained.

To thin the lubricating oil for starting engines in cold weather, engine gasoline may be added directly to the oil if provisions are made for this in the powerplant of the airplane. The cold oil diluted with gasoline circulates easily and provides the necessary lubrication. Then when the engine reaches its normal operating temperature, the gasoline evaporates and leaves the oil as it was before dilution.

When **oil dilution** is practiced, less power is needed for starting and the starting process is completed more quickly. The only important disadvantage is that the presence of ethyl gasoline in the oil may cause a slight corrosion of engine parts, but this disadvantage is outweighed by the advantages.

Color

The **color** of a lubricating oil is obtained by reference to transmitted light; that is, the oil is placed in a glass vessel and held in front of a source of light. The intensity of the transmitted light must be known in conducting a test because light intensity may affect the color.

The apparatus used for a color test is that approved by the ASTM and is called an **ASTM union colorimeter**. Colors are assigned numbers ranging from 1 (lily white) to 8 (darker than claret red). Oils darker than no. 8 color value are diluted with kerosene (85 percent kerosene by volume and 15 percent lubricating oil by volume) and then observed in the same manner as oils having color values from 1 to 8.

When reflected light (as distinguished from direct light) is used in a color test, the color is called the **bloom** and is used, among other things, to indicate the origin and refining method of the oil.

Cloud Point

The **cloud point** is the temperature at which the separation of wax becomes visible in certain oils under prescribed testing conditions. When such oils are tested, the cloud point is a temperature slightly above the **solidification point**. If the wax does not separate before solidification, or if the separation is not visible, the cloud point cannot be determined.

Pour Point

The pour point of an oil is the temperature at which the oil will just flow without disturbance when chilled. In practice, the pour point is the lowest temperature at which an oil will flow (without any disturbing force) to the pump intake. The fluidity of the oil is a factor of both pour test and viscosity. If the fluidity is good, the oil will immediately circulate when engines are started in cold weather. Petroleum oils, when cooled sufficiently, may become plastic solids as a result of either the partial separation of the wax or the congealing of the hydrocarbons comprising the oil. To lower the pour point, **pour-point depressants** are sometimes added to oils which contain substantial quantities of wax.

The general statement is sometimes made that the pour point should be within 5°F [3°C] of the average starting temperature of the engine. But this should be considered in connection with the viscosity of the oil, since the oil must be viscous enough to provide an adequate oil film at engine operating temperatures. Therefore, for cold-weather starting, the oil should be selected in accordance with the operating instructions for the particular engine, considering both the pour point and the viscosity.

Carbon-Residue Test

The purpose of the **carbon-residue** test is to study the carbon-forming properties of a lubricating oil. There are two methods: the Ramsbottom carbon-residue test and the Conradson test. The Ramsbottom test is widely used in Great Britain and is now preferred by many U.S. petroleum engineers because it seems to yield more practical results than the Conradson test, which was formerly more popular in the United States.

When doing the **Ramsbottom test**, a specific amount of oil is placed either in a heat-treated glass bulb well or in a stainless-steel bulb. The oil is then heated to a high temperature by a surrounding molten-metal bath for a prescribed time. The bulb is weighed before and after the test. The difference in weight is divided by the weight of the oil sample and multiplied by 100 to obtain the percentage of carbon residue in the sample.

The apparatus for the **Conradson test** allows oil to be evaporated under specified conditions. The carbon residue from the Conradson test should not be compared directly with the carbon residue from the Ramsbottom test, since the residues are obtained under different test conditions.

Tables have been prepared by engineers which give the average relation between the results of tests performed by the two methods.

Petroleum engineers advise those who are not experts in the field to be cautious in evaluating carbon-residue tests, since the carbon deposits from oil vary with type and mechanical condition of the engine, service conditions, cycle of operation, other characteristics of the oil, and method of carbureting the fuel. In the early days of internal-combustion engines, carbon-residue tests were more important as an indication of the carbon-forming properties of lubricating oil than they are today. The methods now used to refine petroleum products tend to make the carbon-residue tests less useful than before.

The Ash Test

The **ash test** is an extension of the carbon-residue test. If an unused (new) oil leaves almost no ash, it is regarded as pure. The **ash content** is a percentage (by weight) of the residue after all carbon and all carbonaceous matter have been evaporated and burned.

In a test of used lubricating oil, the ash is analyzed chemically to determine the content of iron, which shows the rate of wear; sand or grit, which come from the atmosphere; lead compounds, which come from leaded gasoline; and other metals and nonvolatile materials. The ash analysis tells something about the performance of the engine lubricating oil, but it is only one of many tests which are used to promote efficiency.

Oxidation Tests

Aircraft engine lubricating oils may be subjected to relatively high temperatures in the presence of both air and what the engineers call *catalytically active metals* or metallic compounds. This causes the oil to oxidize. It increases the viscosity, and it forms sludge, carbon residues, lacquers or varnishes (asphaltines), and sometimes inorganic acids.

There are several methods of testing for oxidation. The details do not interest most people outside the research laboratories, although the conclusions are important to aircraft engine personnel in general. The U.S. Air Force has its own oxidation test, the U.S. Navy has its work-factor test, and engine manufacturers often have their own tests.

When the carbon residue of engine oils is lowered below certain limits, the products of oxidation are soluble in hot oil. Deposits of lacquer form on the metallic surfaces, such as on the pistons, in the ring grooves, and on valve guides and stems and anywhere that the oil flows comparatively slowly in the engine. In addition, a sludge of carbon-like substance forms in various places. To overcome this situation, certain compounds known as **antioxidant** and **anticorrosion agents** have been used to treat lubricating oils before they are sold to the public.

Precipitation Number

The **precipitation number** recommended by the ASTM is the number of milliliters of precipitate formed when 10 milliliters (mL) of lubricating oil is mixed with 90 mL of petroleum naphtha under specified conditions and then centrifuged (subjected to centrifugal force) under prescribed conditions. The volume of sediment at the bottom of the centrifuge tube (container) is then taken as the ASTM precipitation number.

Corrosion and the Neutralization Number

Lubricating oils may contain acids. The **neutralization number** recommended by the ASTM is the weight in milligrams of potassium hydroxide required to neutralize 1 g of oil. A full explanation of this topic belongs in the field of elementary chemistry and is beyond the scope of this text.

The neutralization number does *not* indicate the corrosive action of the used oil that is in service. For example, in certain cases an oil having a neutralization number of 0.2 might have high corrosive tendencies in a short-operating period, whereas another oil having a neutralization number of 1.0 might have no corrosive action on bearing metals.

Oiliness

Oiliness is the property that makes a difference in reducing the friction when lubricants having the same viscosity but different oiliness characteristics are compared under the same conditions of temperature and film pressure. Oiliness, contrary to what might be expected, depends not only on the lubricant but also on the surface to which it is applied. Oiliness has been compared with metal wetting, but oiliness is a wetting effect that reduces friction, drag, and wear. It is especially important when the film of oil separating rubbing surfaces is very thin, when the lubricated parts are very hot, or when the texture (grain) and finish of the metal are exceedingly fine. When some oil films are formed, there may be almost no viscosity effects, and then the property of oiliness is the chief source of lubrication.

Extreme-Pressure (Hypoid) Lubricants

When certain types of gearing are used, such as spur-type gearing and hypoid-type gearing, the high tooth pressures and high rubbing velocities require the use of a class of lubricants called **extreme-pressure** (EP) **lubricants**, or **hypoid lubricants**. Most of these special lubricants are mineral oils containing loosely held sulfur or chlorine or some highly reactive material. If ordinary mineral oils were used by themselves, any metal-to-metal contact in the gearing would usually cause scoring, galling, and the local seizure of mating surfaces.

Chemical and Physical Stability

An aircraft engine oil must have **chemical stability** against oxidation, thermal cracking, and coking. It must have **physical stability** with regard to pressure and temperature.

Some of the properties discussed under other topic headings in this chapter are closely related to both chemical and physical stability. The oil must have resistance to emulsion; this characteristic is termed **demulsibility** and is a measure of the oil's ability to separate from water. Oil that is emulsified with water does not provide a high film strength or adequate protection against corrosion. Aircraft engine oil should also be nonvolatile, and there should be no objectionable compounds of decomposition with fuel by-products. The viscosity characteristics should be correct, as we have explained in detail.

If anything is added during the refining process, the resultant should be uniform in quality and purity.

When all the other factors are favorable, the oil should have a minimum coefficient of friction, maximum adhesion to the surfaces to be lubricated, good oiliness characteristics, and adequate film strength.

THE NEED FOR LUBRICATION

There are many moving parts in an aircraft engine. Some reciprocate and others rotate, but regardless of the motion, each moving part must be guided in its motion or held in a given position during motion. The contact between surfaces moving in relation to each other produces friction, which consumes energy. This energy is transformed to heat at comparatively low temperatures and therefore reduces the power output of the engine. Furthermore, the friction between moving metallic parts causes wear. If lubricants are used, a film of lubricant is applied between the moving surfaces to reduce wear and to lower the power loss.

Sliding Friction

When one surface slides over another, the interlocking particles of metal on each surface offer a resistance to motion known as **sliding friction**. If any supposedly smooth surface is examined under the microscope, hills and valleys can be seen. The smoothest possible surface is only relatively smooth. No matter how smooth the surfaces of two objects may appear to be, when they slide over each other, the hills in one catch in the valleys of the other.

Rolling Friction

When a cylinder or sphere rolls over the surface of a plane object, the resistance to this relative motion offered by the surfaces is known as **rolling friction**. In rolling contact, in addition to the interlocking of the surface particles which occurs when two plane objects slide over each other, there is a certain amount of deformation of both the cylinder or sphere and the plane surface over which it rolls. There is less rolling friction when ball bearings are used than when roller bearings are employed. Rolling friction is less than sliding friction and is always preferred by mechanical designers when the surface permits what they call **line** or **point contact**. A simple explanation of the reduction of friction obtained with rolling contact is that the interlocking of surface particles is considerably less than in the case of sliding friction. Therefore, even when the deformation is added, the total friction by rolling contact is less than it is by sliding contact.

Wiping Friction

Wiping friction occurs particularly between gear teeth. Gears of some designs, such as the hypoid gears and worm gears, have greater friction of the wiping type than do gears of other designs, such as the simple spur gear. Wiping friction involves a continually changing load on the contacting surfaces, both in intensity and in direction of movement, and it usually results in extreme pressure, for which special lubricants are required. Lubricants for this purpose are called EP lubricants.

Factors Determining the Amount of Friction

The amount of friction between two solid surfaces depends largely on the rubbing of one surface against the other, the condition and material of the surfaces, the nature of contact movement, and the load carried by the surfaces. The friction usually decreases at high speeds. When a soft bearing material is used in conjunction with hard metals, the softer metal can mold itself to the form of the harder metal, thus reducing friction. Increasing the load increases the friction.

The introduction of lubricant between two moving metallic surfaces produces a film which adheres to both surfaces. The movement of the surfaces causes a shearing action in the lubricant. In this manner the metallic friction between surfaces in contact is replaced by the smaller internal friction within the lubricant. Only fluid lubricants with a great tendency to adhere to metal are able to accomplish this purpose, since they enter where the contact between the surfaces is closest and where the friction would be the greatest if there were no lubrication. The adhesive quality of the lubricant tends to prevent actual metallic contact. The viscosity tends to keep the lubricant from being squeezed out by the pressure on the bearing surfaces.

Although the amount of friction between two solid surfaces depends on the load carried, the rubbing speed, and the condition and material of which the surfaces are made, the fluid friction of a lubricant is not affected in the same manner. The **internal friction of the lubricant** that replaces the metallic friction between moving parts is determined by the rubbing speed, the area of the surfaces in contact, and the viscosity of the lubricant. It is not determined by the load, by the condition of the surfaces, or by the materials of which they are made.

LUBRICANT REQUIREMENTS AND FUNCTIONS

Characteristics of Aircraft Lubricating Oil

The proper lubrication of aircraft engines requires the use of a lubricating oil which has the following characteristics:

1. It should have the proper **body (viscosity)** at the engine operating temperatures usually encountered by the airplane engine in which it is used; it should be distributed readily to the lubricated parts; and it must resist the pressures between the various lubricated surfaces.

2. It should have **high antifriction characteristics** to reduce the frictional resistance of the moving parts when separated only by boundary films. An ideal fluid lubricant provides a strong oil film to prevent metallic friction and to create a minimum amount of oil friction, or oil drag.

3. It should have **maximum fluidity at low temperatures** to ensure a ready flow and distribution when starting occurs at low temperatures. Some grades of oil become almost solid in cold weather, causing high oil drag and impaired circulation. Thus, the ideal oil should be as thin as possible and yet stay in place and maintain an adequate film strength at operating temperatures.

4. It should have **minimum changes in viscosity with changes in temperature** to provide uniform protection where atmospheric temperatures vary widely. The viscosities of oils are greatly affected by temperature changes. For example, at high operating temperatures, the oil may be so thin that the oil film is broken and the moving parts wear rapidly.

5. It should have **high antiwear properties** to resist the wiping action that occurs wherever microscopic boundary films are used to prevent metallic contact. The theory of fluid lubrication is based on the actual separation of the metallic surfaces by means of an oil film. As long as the oil film is not broken, the internal friction (fluid friction) of the lubricant takes the place of the metallic sliding friction which otherwise would exist.

6. It should have **maximum cooling ability** to absorb as much heat as possible from all lubricated surfaces—and especially from the piston head and skirt. One reason for using liquid lubricants is that they are effective in absorbing and dissipating heat. Another reason is that liquid lubricants can be readily pumped or sprayed. Many engine parts, especially those carrying heavy loads at high rubbing velocities, are lubricated by oil under direct pressure. Where direct-pressure lubrication is not practical, a spray mist of oil provides the required protection. Regardless of the method of application, the oil absorbs the heat and later dissipates it through coolers or heat exchangers.

7. It should offer the **maximum resistance to oxidation**, thus minimizing harmful deposits on the metal parts.

8. It should be **noncorrosive** to the metals in the lubricated parts.

Functions of Engine Oil

The aircraft engine must operate in a series of rapidly changing environments. At takeoff, the engine is running at maximum power output for several minutes. Then the power is gradually reduced until cruising power is established. The engine may operate for hours at cruising power which is usually about 70 percent of maximum power.

Aircraft piston engines do not use automotive engine oil because aircraft engines are air-cooled and run at much higher temperatures. Many of the additives which are used in automotive engine oils cannot be used in aircraft, because they would lead to preignition and possible engine failure.

Aviation oil may perform functions in addition to engine lubrication, such as serving as a hydraulic fluid to help the propeller function or as a lubricant for the propeller reduction gears located in the front of some engines.

Engine oil performs these functions:

1. It lubricates, thus reducing the friction between moving parts.

2. It cools various parts of the engine.

3. It tends to seal the combustion chamber by filling the spaces between the cylinder walls and piston rings, thus preventing the flow of combustion gases past the rings.

4. It cleans the engine by carrying sludge and other residues away from the moving engine parts and depositing them in the engine oil filter.

5. It aids in preventing corrosion by protecting the metal from oxygen, water, and other corrosive agents.

6. It serves as a cushion between parts where impact loads are involved.

Straight Mineral Oil

Several types of oils are used in reciprocating aircraft engines today. One such type is **straight mineral oil** blended from selected high-viscosity-index (high-VI) base stocks. These oils do not contain any additives, except for a small amount of pour-point depressant for improved fluidity at cold temperatures. This type of oil is used primarily during break-in for most four-cycle aviation piston engines. These oils are approved and generally used for piston engines when an ashless-dispersant oil is not required.

Ashless-Dispersant Oil

Most aircraft oils other than straight mineral oils contain a **dispersant** that suspends contaminants such as carbon, lead compounds, and dirt. The dispersant helps prevent these contaminants from gathering into clumps and forming sludge or plugging oil passages. Contaminants may then be filtered out or drained with the oil rather than deposited in the engine. This offers several advantages during engine operation, such as keeping the piston-ring grooves free of deposits, assisting the rings in maintaining their effectiveness, and providing a good compression seal. Oil consumption is also reduced because the oil-ring drain holes tend to stay clean.

Along with the dispersant, many oils contain an **additive** combination to provide for high VI, dispersancy, oxidation stability, and antiwear and antifoam properties. These additives are unusual because they leave no metallic ash, thus the term "ashless." A high ash content in the oil can cause preignition, spark plug fouling, and other engine problems. Most oils used in aircraft reciprocating engines are **ashless-dispersant** (AD) **oil**.

Multiviscosity Oils

In certain circumstances, all **single-grade** oils, which have been the industry's standard for several decades, have shortcomings. In cold-weather starts, single-grade oil generally flows slowly to the upper reaches and vital parts of the engine. This can lead to excessive wear and premature overhauls. Even when a single-grade oil does get to the engine's upper end, it may still provide poor lubrication for several minutes because it is too thick.

Multigrade oils, with viscosity characteristics that enable them to flow quickly in the cold, allow better lubrication during starting. Even when the ground temperature is warm, a multigrade oil will flow more quickly into the engine gallery at start-up than will a single-grade oil, and oil pressure will be achieved faster.

Another disadvantage of single-grade oils is that they react badly to temperature changes. When airplanes fly between hot and cold climates in just a few hours, a single-grade 65 (30 weight) or 80 (40 weight) oil gets too hot and is prone to thin out excessively. If a single-grade 100 (50 weight) oil gets too cold, it will not flow well. In either case, there is some risk that the oil will fail to lubricate properly.

The difference between a multiviscosity oil and a straight-grade oil, with regard to VI, is that multiviscosity oil can provide adequate lubrication at a wider range of temperatures than a straight-weight oil. This can be seen graphically in Fig. 4-5. At low temperatures, the thicker oil can take longer to reach critical bearing areas, especially during engine starting. This is important because most engine bearing wear occurs during starting. The time it takes for lubrication to arrive at the engine bearings is shown in Fig. 4-6. One caution generally agreed upon in the field is that multiviscosity oil should be used in high time engines with caution. Instances of high oil consumption have been reported when multiviscosity oils are used in high time engines due to some of the extra oil additives in the multiviscosity oils.

There are major differences in composition and performance among the available types of multigrade oil. The mineral-oil base is less expensive, but tests have shown that engines may end up with carbon deposits on the valves and considerable low-temperature sludge, which could lead to screen and filter plugging. Companies have also tested a full-synthetic oil. Synthetic oils tested did not properly disperse the lead by-products of combustion.

One manufacturer's multigrade 15W50 blend has a composition of 50 percent high-quality synthetic hydrocarbon oil and 50 percent mineral oil, thus the term "semisynthetic." The additives are exactly like those in the single-grade oil

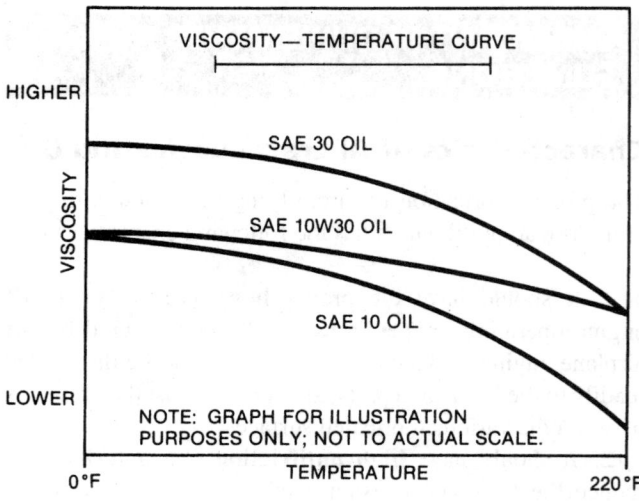

FIGURE 4-5 Multigrade vs. straight-grade oil.

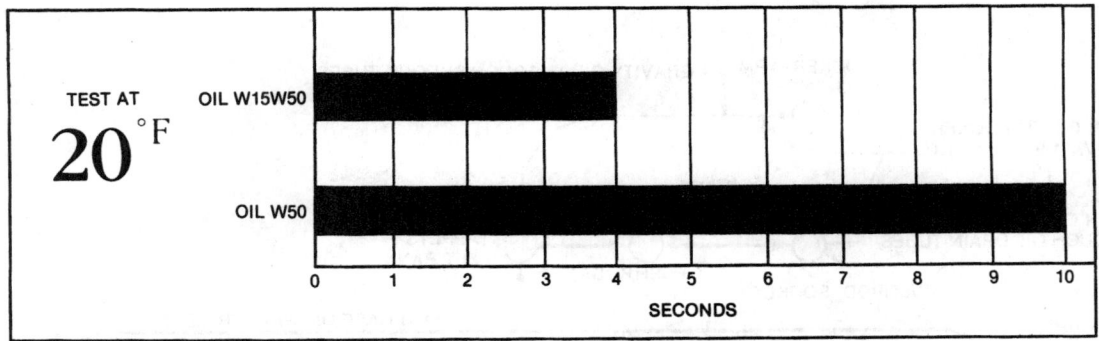

FIGURE 4-6 Time needed for oil to reach the gallery.

plus there is an antiwear additive to reduce engine wear and corrosion. The antiwear performance is provided by tricresyl phosphate (TCP), which has been used as an effective antiwear agent for over 20 years in a wide array of applications. In addition, the base-stock combination of mineral oil and synthetic hydrocarbons provides a very low pour point and gives excellent high-temperature protection.

Sport Aviation Oils

Aeroshell oil sport plus 2 is a two-stroke engine oil developed for ultrallight and light-sport engines (noncertified). Most of the two-stroke aviation engines require oil that can withstand intense operating conditions; full power takeoff, altitude cruise, and descent and idle conditions. Normal two-stroke oil tended to have a limited ability when dealing with these extreme operating parameters. A new type of two-stroke oil was developed with the engine manufacturers which was specifically suited for light-sport and ultralight aircraft. This oil also reduced engine wear, corrosion, and dealt with fuels such as 80 or 100 LL (leaded fuels) used in aviation better than other two-stroke oils.

Aeroshell oil sport plus 4 is a four-stroke engine oil developed for light-sport engines (noncertified). This is a multigrade oil with additive technology which meets the requirements of integrated gearbox and overload clutches. This oil can be used with both unleaded fuels and avgas 100 LL. It also contains advanced anticorrosion and antiwear additives, which are specific to reduce wear to internal parts, especially during starting helping the engine reach its time before overhaul. Different engines require different types of oils for various reasons. Before using any type of oil always consult the operator's handbook /manual to confirm the correct lubricant specification before use.

CHARACTERISTICS AND COMPONENTS OF LUBRICATION SYSTEMS

The purpose of a lubrication system is to supply oil to the engine at the correct pressure and volume to provide adequate lubrication and cooling for all parts of the engine which are subject to the effects of friction. The oil tank must have ample capacity, the oil pump volume and pressure must be adequate, and the cooling facilities for the oil must be such that the oil temperature is maintained at the proper level to keep the engine cool. Several typical systems are described in this section.

The lubricant is distributed to the various moving parts of a typical internal-combustion engine by pressure, splash, and spray.

Pressure Lubrication

In a **pressure lubrication system**, a mechanical pump supplies oil under pressure to the bearings. A typical lubrication system schematic is shown in Fig. 4-7. The oil flows into the inlet (or suction) side of the pump, which is usually located higher than the bottom of the oil sump so that sediment which falls into the sump will not be drawn into the pump. The pump may be of either the eccentric-vane type or the gear type, but the gear type is more commonly used. The pump forces oil into an oil manifold, which distributes the oil to the crankshaft bearings. A pressure relief valve is usually located near the outlet side of the pump.

Oil flows from the main bearings through holes drilled in the crankshaft to the lower connecting-rod bearings. Each of these holes through which the oil is fed is located so that the bearing pressure at that point will be as low as possible.

Oil reaches a hollow camshaft through a connection with the end bearing or the main oil manifold and then flows out of the hollow camshaft to the various camshaft bearings and cams.

Lubrication for overhead-valve mechanisms on reciprocating engines, both in conventional airplanes and in helicopters, is supplied by the pressure system through the valve pushrods. Oil is fed from the valve tappets (lifters) into the pushrods. From there, it flows under pressure to the rocker arms, rocker-arm bearings, and valve stems.

The engine cylinder surfaces and piston pins receive oil sprayed from the crankshaft and from the crankpin bearings. Since oil seeps slowly through the small crankpin clearances before it is sprayed on the cylinder walls, considerable time is required for enough oil to reach the cylinder walls, especially on a cold day when the oil flow is more sluggish. This situation is one of the chief reasons for diluting engine oil with engine fuel for starting in very cold weather.

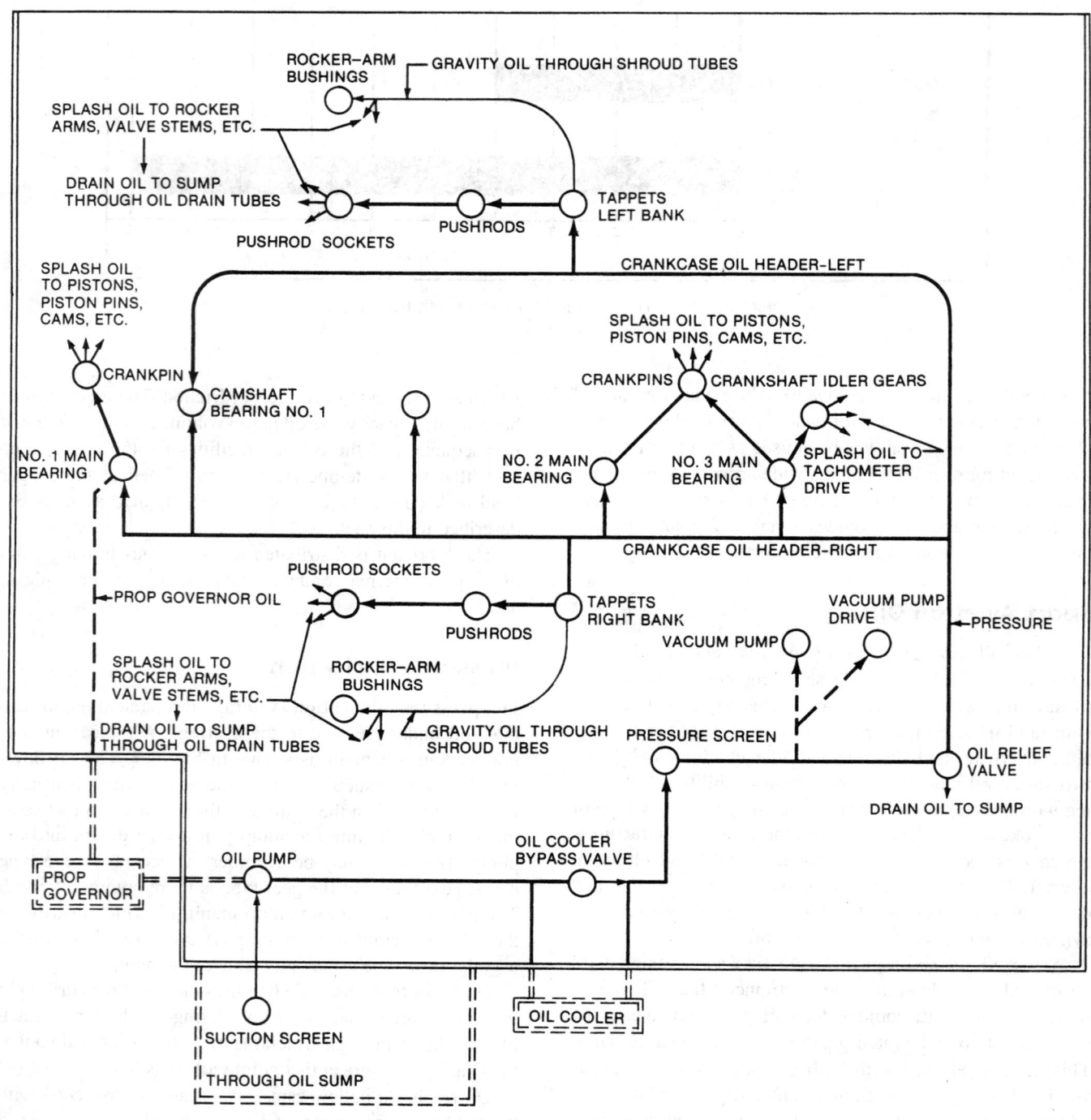

FIGURE 4-7 Lubrication for four-cylinder engines.

Splash Lubrication and Combination Systems

Pressure lubrication is the principal method of lubrication used on all aircraft engines. **Splash lubrication** may be used in addition to pressure lubrication on aircraft engines, but it is never used by itself. Therefore, aircraft engine lubrication systems are always of either the pressure type or the combination pressure, splash, and spray type—usually the latter. The lubrication system illustrated in Fig. 4-7 is an example of a combination pressure-splash system. This text discusses the pressure type of lubrication system but calls attention

to those units or parts which are splash-lubricated or spray-lubricated. The bearings of gas-turbine engines are usually lubricated by means of oil jets that spray the oil under pressure into the bearing cavities. Further information on the lubrication systems for gas-turbine engines is provided in a later chapter of this text.

Principal Components of a Lubrication System

An aircraft engine lubrication system includes a pressure oil pump, an oil pressure relief valve, an oil reservoir (either as a part of the engine or separate from the engine), an oil pressure gauge, an oil temperature gauge, an oil filter, and the necessary piping and connections. In addition, many lubrication systems include oil coolers and/or temperature-regulating devices. Oil dilution systems are included when they are deemed necessary for cold-weather starting.

Oil Capacity

The capacity of the lubrication system must be sufficient to supply the engine with an adequate amount of oil at a temperature not in excess of the maximum established as safe for the engine. On a multiengine airplane, the lubrication system for each engine must be independent of the systems for the other engines.

The usable tank capacity must not be less than the product of the endurance of the airplane under critical operating conditions and the maximum oil consumption of the engine under the same conditions, plus an adequate margin to ensure satisfactory circulation, lubrication, and cooling. In lieu of a rational analysis of airplane range, a fuel-oil ratio of 30:1 by volume is acceptable for airplanes not provided with a reserve or transfer system. If a transfer system is provided, a fuel-oil ratio of 40:1 is considered satisfactory.

Plumbing for the Lubrication System

The plumbing for the oil system is essentially the same as that required for fuel systems or hydraulic systems. Where lines are not subject to vibration, they are constructed of aluminum-alloy tubing and connections are made with approved tubing fittings, AN or MS type. In areas near the engine or between the engine and the fire wall where the lines are subject to vibration, synthetic hose of an approved type is used. The hose connections are made with approved hose fittings which are securely attached to the hose ends. Fittings of this type are described in the text *Aircraft Basic Science*.

Hose employed in the engine compartment of an airplane should be fire-resistant to minimize the possibility of hot oils being discharged into the engine area if a fire occurs. To aid in protecting oil hoses in high-temperature areas, a protective **fire sleeve**, such as that illustrated in Fig. 4-8 may be installed on the hose.

The size of oil lines must be such that they will permit flow of the lubricant in the volume required without restriction. The size for any particular installation is specified by the manufacturer of the engine.

Temperature Regulator (Oil Cooler)

As indicated by its name, the oil temperature regulator is designed to maintain the temperature of the oil for an

FIGURE 4-9   Oil cooler.

operating engine at the correct level. Such regulators are commonly called **oil coolers** because cooling of the engine oil is one of their principal functions. An oil cooler can be seen in Fig. 4-9. Oil temperature regulators are manufactured in a number of different designs, but their basic functions remain essentially the same.

One type of oil temperature regulator is illustrated in Fig. 4-10. The outer cylinder of this particular unit is about 1 in [12.54 cm] larger in diameter than the inner cylinder. This provides an oil passage between the cylinders and enables the oil to bypass the core when the oil is either at the correct operating temperature or too cold. When the oil from the engine is too hot for proper engine operation, the oil is routed through the cooling tubes by the viscosity valve. Note that the oil which passes through the core is guided by baffles which force it to flow around these tubes and thus to flow through the length of the core several times.

The oil flow through the cooling portion of the oil temperature regulator is controlled by some type of thermostatic valve. This valve may be called a **thermostatic control valve** or simply the **oil cooler bypass valve**. This valve is so designed that the temperature of the oil causes it to open or close, routing the oil lor little or no cooling when the oil is cold and for maximum cooling when the oil is hot. If the control valve should become inoperative or otherwise fail, the oil will still flow through or around the cooling portion of the unit.

Oil Viscosity Valve

The **oil viscosity valve**, illustrated in Fig. 4-11, is generally considered part of the oil temperature regulator unit and is employed in some oil systems. The viscosity valve consists essentially of an aluminum-alloy housing and a thermostatic control element. The valve is attached to the oil cooler valve. Together, the oil cooler valve and the oil viscosity valve, which form the **oil temperature regulator unit**, have the twofold duty of maintaining a desired temperature and keeping the viscosity within required limits by controlling the passage of oil through the unit.

Through its thermostatic control, the viscosity valve routes the oil through the cooling core of the cooler when

FIGURE 4-8   Silicone-coated asbestos fire sleeve.

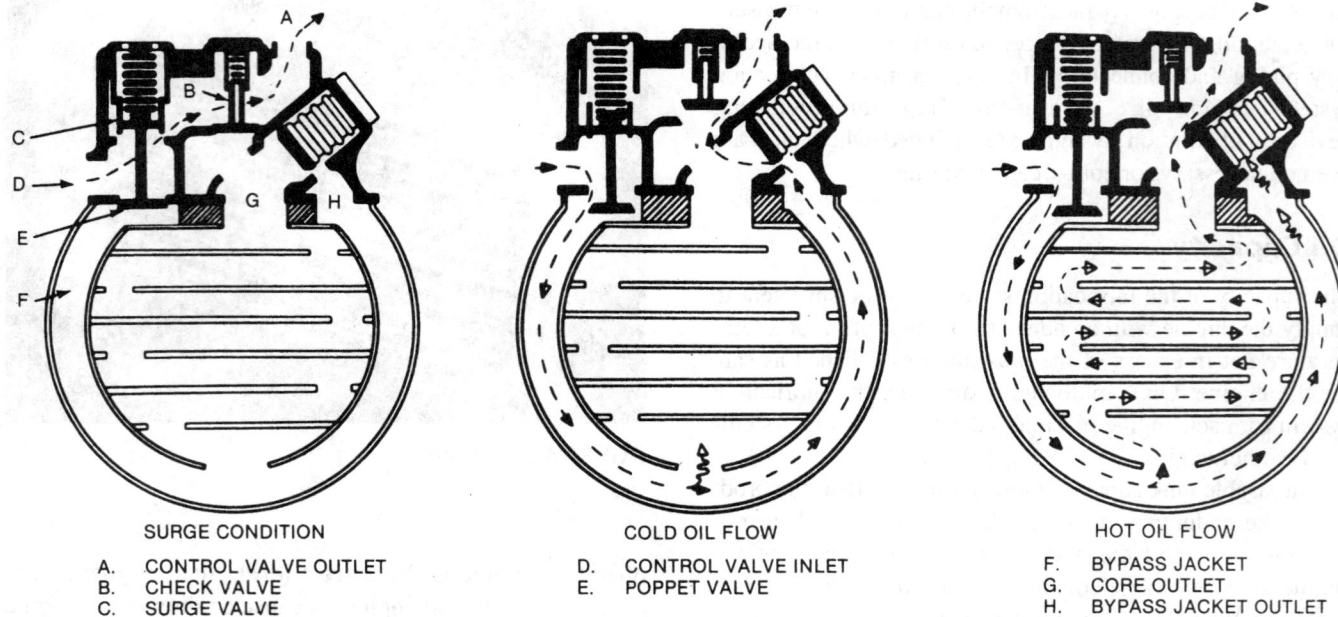

SURGE CONDITION

A. CONTROL VALVE OUTLET
B. CHECK VALVE
C. SURGE VALVE

COLD OIL FLOW

D. CONTROL VALVE INLET
E. POPPET VALVE

HOT OIL FLOW

F. BYPASS JACKET
G. CORE OUTLET
H. BYPASS JACKET OUTLET

FIGURE 4-10   Oil temperature regulator (oil cooler).

the oil is hot and causes the oil to bypass the core when the oil is not warm enough for correct engine lubrication.

When the oil is cold, the valve will be off its seat and oil can flow through the opening as shown on the left in Fig. 4-11. This passage permits the oil to flow from the area around the outside of the cooler; therefore, the oil does not become cooled. As the oil becomes heated, the valve closes, thus forcing the oil to flow through the opening on the right, which leads from the radiator section of the cooler. This, of course, exposes the oil to the cooling action of the radiator section.

Oil Pressure Relief Valves

The pressure of the oil must be great enough to lubricate the engine and its accessories adequately under all operating conditions. If the pressure becomes excessive, the oil system may be damaged and there may be leakage.

The purpose of an **oil pressure relief valve** is to control and limit the lubricating oil pressure, to prevent damage to the lubrication system itself, and to ensure that the engine parts are not deprived of adequate lubrication because of a

system failure. As noted previously and illustrated in Fig. 4-8, the oil pressure relief valve is located in the area between the pressure pump and the internal oil system of the engine; it is usually built into the engine. There are several types of oil pressure relief valves, a few of which are described below.

General Design of Relief Valves

Oil pressure relief valves utilized in modern light-airplane engines are comparatively simple in design and construction and usually operate according to the principle illustrated in Fig. 4-12. The relief valve assembly consists of a plunger and a spring mounted in a passage of the oil pump housing. When the oil pressure becomes too high, the pressure moves the plunger against the force of the spring to open a passage, allowing oil to return to the inlet side of the pump.

Oil pressure relief valves for large reciprocating engines are usually of the compensating type. This type of relief valve ensures adequate lubrication to the engine when the engine is first started by maintaining a high pressure until the oil has warmed sufficiently to flow freely at a lower pressure. The relief valve setting can usually be adjusted by means of a screw which changes the pressure on the spring or springs controlling the valve. Some of the simpler types of relief valves do not have an adjusting screw; in these cases, if the relief valve pressure setting is not correct, it is necessary to change the spring or to insert one or more washers behind the spring. The initial, or basic, oil pressure adjustment for an engine is made at the factory or engine overhaul shop.

Single Pressure Relief Valve

The typical single pressure relief valve, illustrated in Fig. 4-13, has a spring-loaded plunger, which has a tapered valve at one end, an adjusting screw for varying the spring

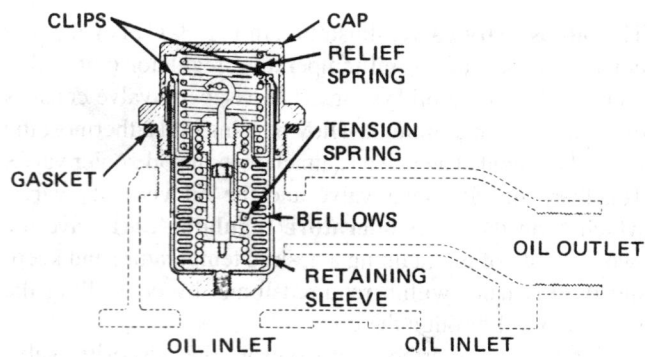

CLIPS   CAP
          RELIEF
          SPRING

          TENSION
          SPRING

GASKET

          BELLOWS

                    OIL OUTLET

          RETAINING
          SLEEVE
OIL INLET   OIL INLET

FIGURE 4-11   Oil viscosity valve.

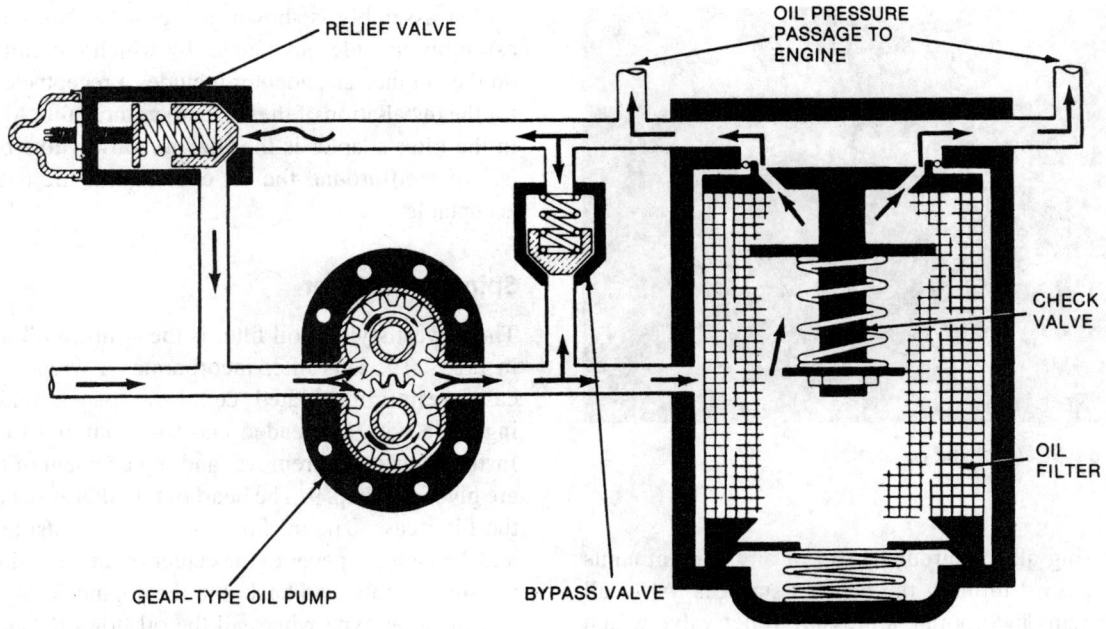

FIGURE 4-12   Engine oil pump and associated units.

tension, a locknut to keep the adjusting screw tight, a passage from the pump, a passage to the engine, and a passage to the inlet side of the pump. Normally, the valve is held against its seat by spring tension, but when the pressure from the pump to the engine becomes excessive, the increased

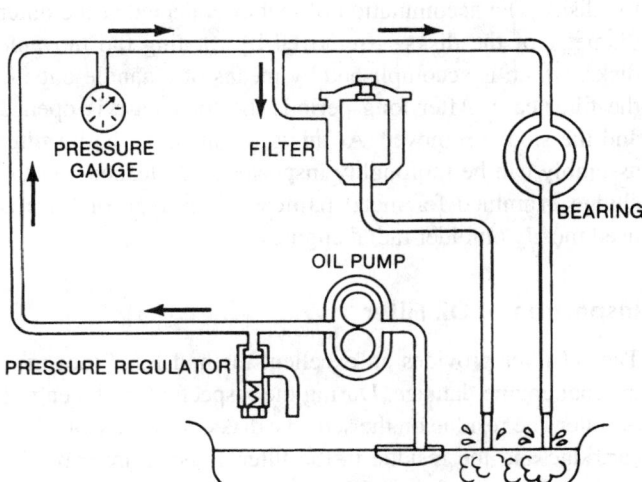

FIGURE 4-13   Single pressure relief valve.

pressure pushes the valve off its seat. The oil then flows past the valve and its spring mechanism and is thus bypassed to the inlet side of the oil pump.

Aircraft Engine Oil Filters

Most new aircraft engines are equipped with, or have provisions to accept, a **full-flow type** of oil filter system. However, some older-model engines do not have these provisions and have instead a **bypass system**, sometimes called a **partial-flow system**.

The bypass system filters only about 10 percent of the oil through the filtering element, returning the filtered oil directly to the sump. Note in Fig. 4-14 that the oil passing through the engine bearing is not filtered oil.

The newest type of oil filter is designed for a full-flow oil system. In this system, illustrated in Fig. 4-15, the filter is positioned between the oil pump and the engine bearings,

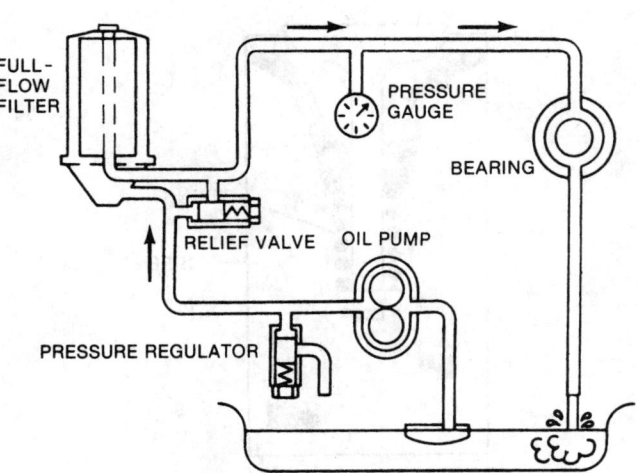

FIGURE 4-14   Bypass filter system.

FIGURE 4-15   Full-flow lubrication system.

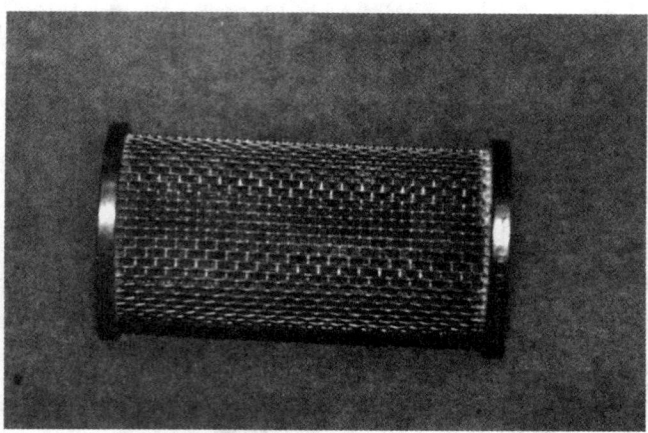

FIGURE 4-16 Strainer-type oil filter.

thereby filtering all the circulated oil of any contaminants before it is passed through the bearing surfaces. Also, all full-flow systems incorporate a pressure relief valve which opens by oil pressure at a predetermined differential pressure. If the filter becomes clogged, the relief valve will open, allowing the oil to bypass, preventing engine oil starvation.

Strainer-Type Filter

The purpose of any filter is to remove solid particles of foreign matter from the oil before it enters the engine. The **strainer-type oil filter** is simply a tubular screen, which is shown in Fig. 4-16. Some of these filters are designed so that they will collapse if they become clogged, thus permitting the continuation of normal oil flow. Other screens or filters are designed with relief valves which open if the screens become clogged (Fig. 4-17).

Disposable Filter Cartridge

Many modern engines for light aircraft incorporate oil systems which utilize external oil filters containing disposable filter elements in filter canisters. An exploded view of such

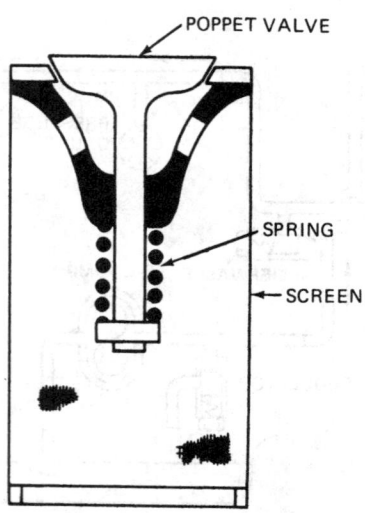

FIGURE 4-17 Oil screen with relief (bypass) valve.

a filter assembly is shown in Fig. 4-18. Note that the filter assembly includes an adapter by which the unit is mounted on the engine. The adapter includes a receptacle and fittings for the installation of the oil temperature bulb. Also, included in the filter adapter is the **thermostatic valve** by which oil is bypassed around the oil cooler until the temperature is acceptable.

Spin-On Oil Filter

The newest style of oil filter is the **spin-on oil filter**, shown in Fig. 4-19. This filter incorporates a wrench pad, a steel case, resin-impregnated cellulosic paper, and a mounting plate with a threaded end for mounting to the engine. Instructions on the removal and replacement of this oil filter are given in Chap. 9. The heart of the filter is the paper inside the filter case. The oil flows around the outside of the case and through the paper to the center of the filter down through the support tube and back into the engine. This type of filter is a full-flow type where all the oil flows through the filter. The filter medium (paper), shown in Fig. 4-20, provides both surface and depth filtration because the oil flows through many layers of locked-in fibers. There is no migration of filter material to clog engine oil passages or affect bearing surfaces. The filter also can contain an antidrain back valve and a pressure relief valve, all contained in the sealed disposable housing. A gasket on the mounting pad of the filter provides a seal between the filter and its adapter, which is bolted to the accessory case. This type of filter has proved very effective in providing engine protection for the entire service life of the engine.

Cuno Oil Filter

The Cuno oil filter has a series of laminated plates, or disks, with one set of disks rotating in the spaces between the other disks. The oil is forced through the spaces between the disks, flowing from the outside of the disks, between the disks and spacers, to the inside passage and then to the engine. Foreign-matter particles are stopped at the outer diameter of the disks. The minimum size of the particles filtered from the oil is determined by the thickness of the spacers between the disks. The accumulation of matter collected at the outer diameter of the disks is removed by rotating the movable disks, which is accomplished by means of a handle outside the filter case. After long periods the filter case is opened and the sludge removed. At this time, also, the entire filter assembly can be thoroughly inspected and cleaned and the sludge examined for metal particles. This type of filter is used mostly on older radial engines.

Inspection of Oil Filter

The oil filter provides an excellent method for discovering internal engine damage. During the inspection of the engine oil filter, the residue on the screens, disks, or disposable filter cartridge and the residue in the filter housing are carefully examined for metal particles. A new engine or a newly overhauled engine will often have a small amount of fine metal

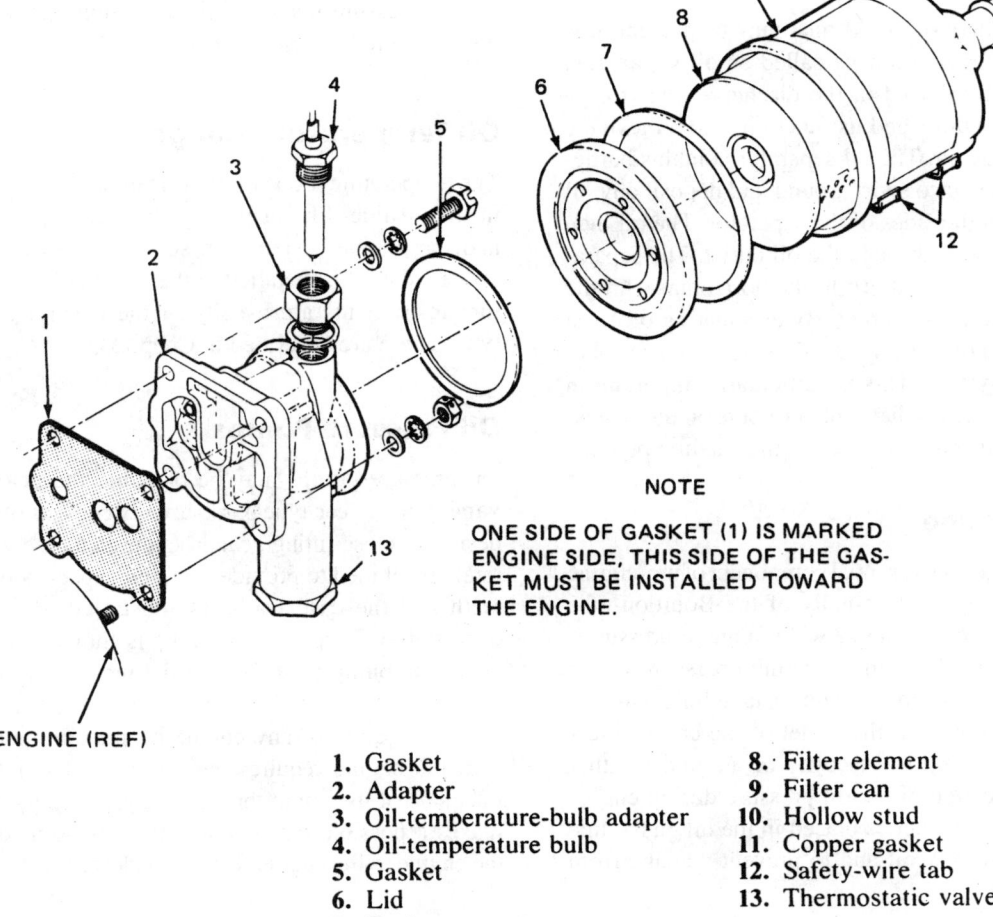

NOTE

ONE SIDE OF GASKET (1) IS MARKED ENGINE SIDE, THIS SIDE OF THE GASKET MUST BE INSTALLED TOWARD THE ENGINE.

ENGINE (REF)

1. Gasket
2. Adapter
3. Oil-temperature-bulb adapter
4. Oil-temperature bulb
5. Gasket
6. Lid
7. Gasket
8. Filter element
9. Filter can
10. Hollow stud
11. Copper gasket
12. Safety-wire tab
13. Thermostatic valve

FIGURE 4-18   External oil filter with disposable cartridge.

particles in the screen, or filter, but this is not considered abnormal. After the engine has been operated for a time and the oil has been changed one or more times, there should not be an appreciable amount of metal particles in the oil screen. If an unusual residue of metal particles is found in the oil screen, the engine should be taken out of service and disassembled to determine the source of the particles. This precaution will often prevent a disastrous engine failure in flight.

At oil changes, oil samples are often taken and sent to laboratories to be analyzed for wear metals. A complete discussion of oil analysis is given in Chap 14.

OIL FILTER WRENCH PAD

1 SAFETY WIRE TABS

3 RESIN IMPREGNATED CELLULOSIC MEDIA

2 CORRUGATED CENTER SUPPORT TUBE

4 FULL PLEAT MEDIA

FIGURE 4-19   Spin-on oil filter.

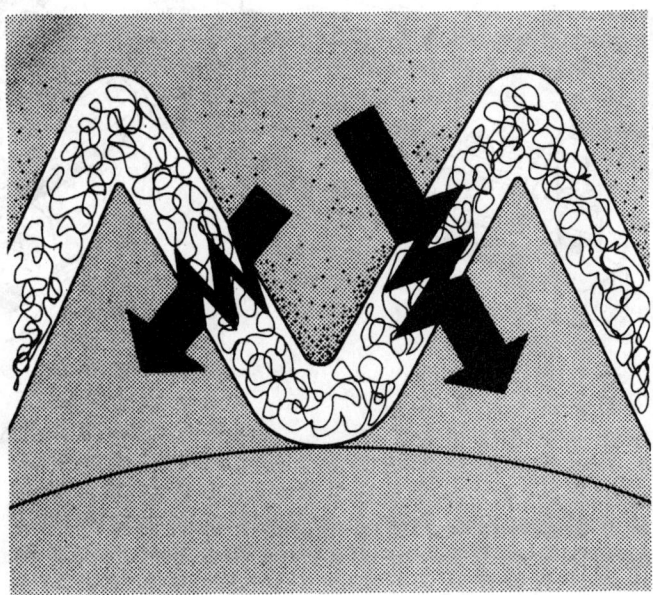

FIGURE 4-20   Oil filter medium (paper).

Oil Separator

In any air system where oil or oil mist may be present, it is often necessary to utilize a device called an **oil separator**. This device is usually placed in the discharge line from a vacuum pump or air pump, and its function is to remove the oil from the discharge air. The oil separator contains baffle-lates which cause the air to swirl around and deposit any oil on the baffles and on the sides of the separator. The oil then drains back to the engine through the oil outlet. The separator must be mounted at about 20° to the horizontal with the oil drain outlet at the lowest point. By eliminating oil from the air, the oil separator prevents the deterioration of rubber components in the system. This is particularly important in the case of deicer systems where rubber boots on the wings' leading edges are inflated with air from the vacuum pump.

Oil Pressure Gauge

An **oil pressure gauge** is an essential component of any engine oil system. These gauges are usually of the Bourdon-tube type and are designed to measure a wide range of pressures, from no pressure up to above the maximum pressure which may be produced in the system. The oil gauge line, which is connected to the system near the outlet of the engine pressure pump, is filled with low-viscosity oil in cold weather to obtain a true indication of the oil pressure during engine warm-up. A restricting orifice is placed in the oil gauge line to retain the low-viscosity oil and to prevent damage from

pressure surges. If high-viscosity oil is used in cold weather, the oil pressure reading will lag behind the actual pressure developed in the system.

Oil Temperature Gauge

The temperature probe for the oil temperature gauge is located in the oil inlet line or passage between the pressure pump and the engine system. On some engines the temperature probe (sensor) is installed in the oil filter housing. Temperature instruments are usually of the electrical or electronic type. These are described in Chap. 23.

Oil Pressure Pumps

Oil pressure pumps may be of either the **gear type** or the **vane type**. A gear-type pump usually consists of two specially designed, close-fitting gears rotating in a case which is accurately machined to provide minimum space between the gear teeth and the case walls, as illustrated in Fig. 4-21. The operation of a typical gear pump is shown in Fig. 4-22. The gear-type pump is utilized in the majority of reciprocating engines.

The capacity of any engine oil pressure pump is greater than the engine requires, and excess oil is returned to the inlet side of the pump through the pressure relief valve. This makes it possible for the pump to increase its oil delivery to the engine as the engine wears and clearances become greater.

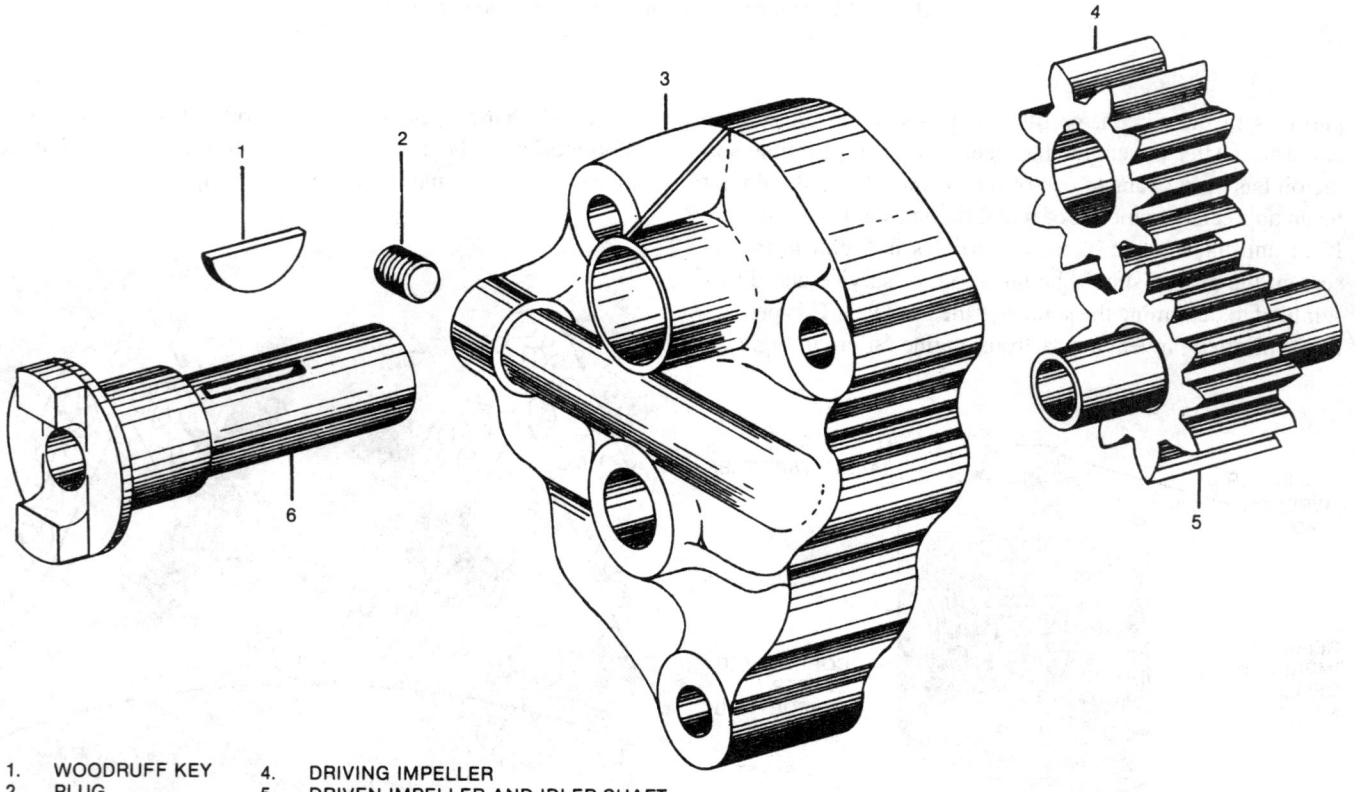

1.	WOODRUFF KEY	4.	DRIVING IMPELLER
2.	PLUG	5.	DRIVEN IMPELLER AND IDLER SHAFT
3.	OIL PUMP BODY	6.	OIL PUMP DRIVE SHAFT

FIGURE 4-21 Oil pump drive assembly.

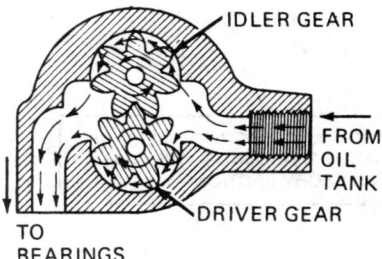

FIGURE 4-22   Gear-type oil pump.

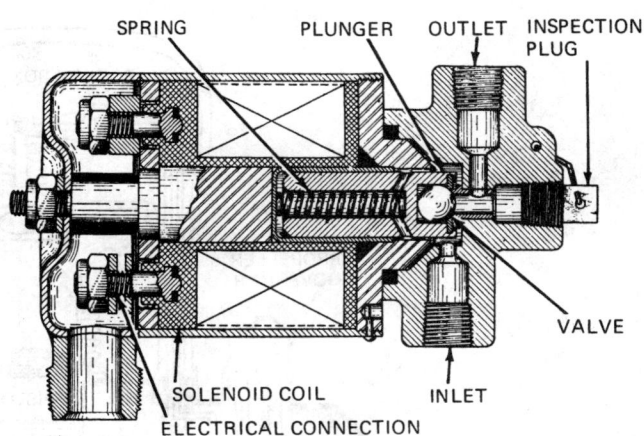

FIGURE 4-24   Solenoid valve.

If an engine oil pressure pump does not produce oil pressure within 30 s after the engine is started, this is an indication that the pump has lost its prime, probably due to wear. When the side clearance of the gears in the pump becomes too great, oil bypasses the gears and pressure cannot be developed. In this case the pump must be replaced.

Scavenge Pump

The scavenge pump or pumps for a dry-sump lubrication system or turbocharger are designed with a greater capacity than that of the pressure pump. In a typical engine, the gear-type scavenge pump is driven by the same shaft as the pressure pump, but the depth of the scavenge pump gears is twice that of the pressure pump gears. This gives the scavenge pump twice the capacity of the pressure pump. The reason for the higher capacity of the scavenge pump is that the oil which flows to the sump in the engine is somewhat foamy and therefore has a much greater volume than the air-free oil which enters the engine via the pressure pump. To keep the oil sump drained, the scavenge pump must handle a much greater volume of oil than the pressure pump.

Oil Dilution System

Figure 4-23 is a schematic diagram showing how the oil dilution system is connected between the fuel system and the oil system. In this diagram a line is connected to the fuel system on the pressure side of the fuel pump. This line leads to the oil dilution solenoid valve; and from the solenoid valve the line leads to the Y drain, which is in the engine inlet line of the oil system. If the system does not

include a Y drain, the oil dilution line may be connected at some other point in the engine inlet line before the pressure pump inlet. The oil dilution solenoid valve is connected to a switch in the cockpit so that the pilot can dilute the oil after flight and before shutting down the engine. A cutaway drawing of the solenoid valve is shown in Fig. 4-24.

If an oil dilution system's control valve becomes defective and leaks or remains open, gasoline will continue to be introduced to the engine oil during operation. This will result in low oil pressure, high oil temperature, foaming of the oil, high fuel consumption, and emission of excessive oil fumes from the engine breather.

ENGINE DESIGN FEATURES RELATED TO LUBRICATION

We have discussed the design of reciprocating engines and engine parts in general; however, at this point it is important to emphasize certain features directly related to lubrication.

Sludge Chambers

In some engines, the crankshaft is designed with chambers in the hollow connecting-rod journals by which carbon sludge and dirt particles are collected and stored. The chambers may be made by means of metal spools inserted in the hollow crankpins (journals) or by plugs at each end of the hollow journals. At overhaul it is necessary to disassemble or remove the chambers and to remove the sludge. Great care must be taken to ensure that the chambers are properly reassembled so that oil passages are not covered or plugged in any way. During the overhaul of crankshafts, all oil passages must be cleaned.

Intercylinder Drains

To provide lubrication for the valve operating mechanisms in many radial engines, the valve rocker-box cavities are interconnected with oil tubes called **intercylinder drains**. These drains ensure adequate lubrication for the valve mechanisms and provide a means whereby the oil can circulate and return

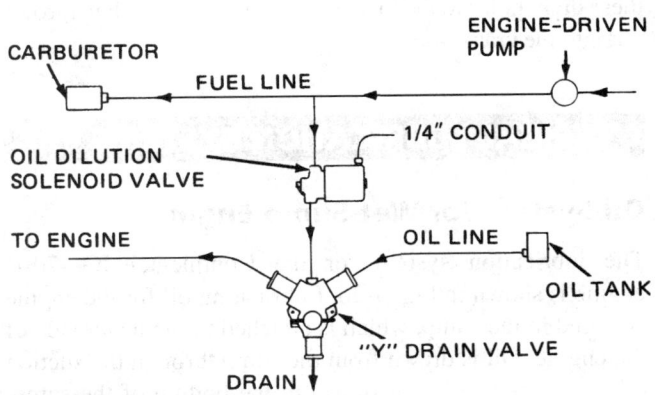

FIGURE 4-23   An oil dilution system.

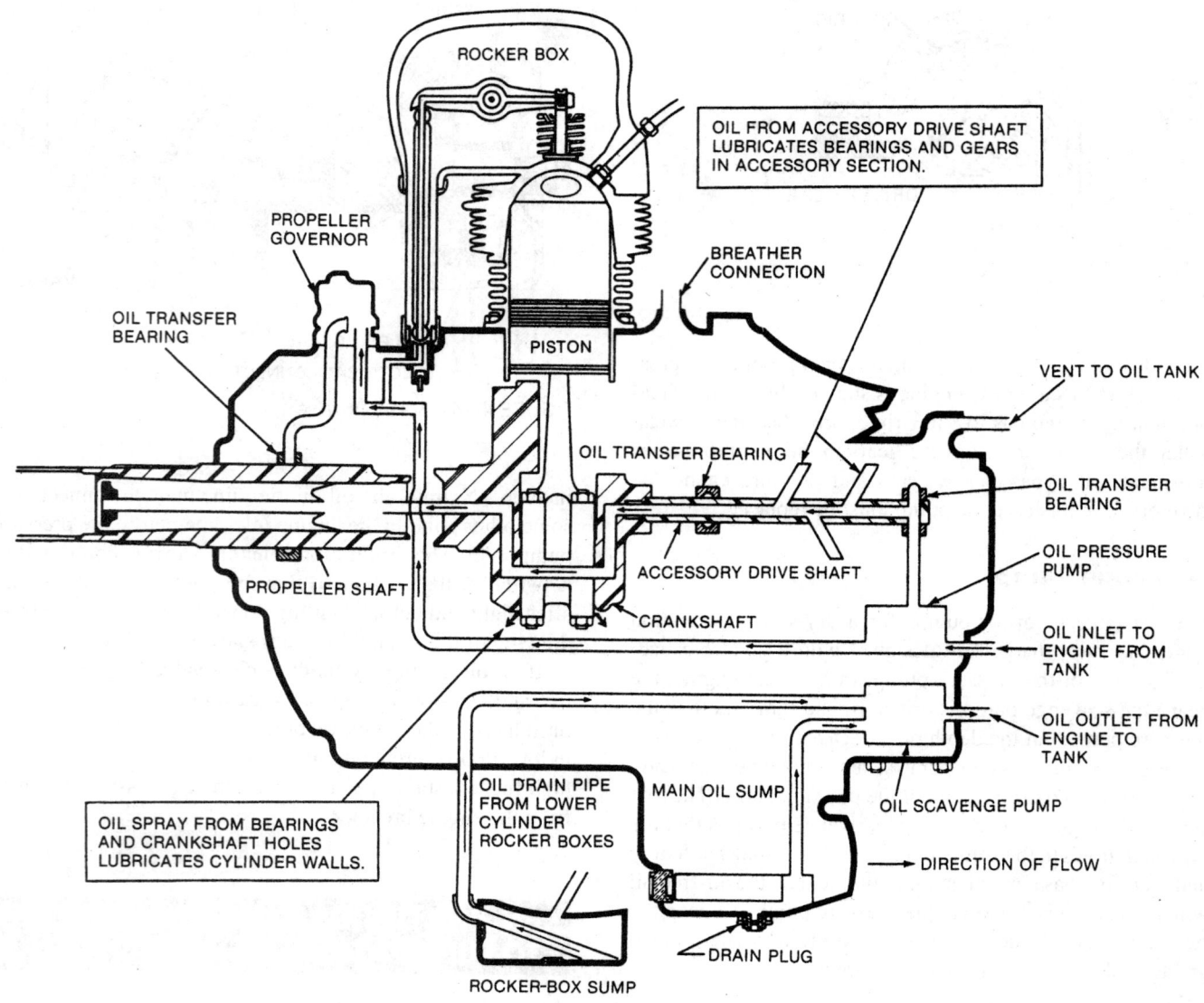

FIGURE 4-25 Schematic of radial-engine lubrication system.

to the sump. The drain tubes must be kept clear and free of sludge. If the drain tubes become partially or completely plugged, excess oil will build up in the rocker-box area and some of this oil will be drawn into the cylinder through the valve guide during the intake stroke. This will, of course, cause fouling of the spark plugs, particularly in the lower cylinders, and result in improper lubrication and cooling of the valve mechanism.

Oil Control in Inverted and Radial Engines

Some of the cylinders in a radial engine and all the cylinders in an inverted engine are located at the bottom of the engine. It is necessary, therefore, to incorporate features to prevent these cylinders from being flooded with oil. This is accomplished by means of long skirts on the cylinders and an effective scavenging system. During the operation of these engines, oil which falls into the lower cylinders is immediately thrown back out into the crankcase. The oil then drains downward and collects in the crankcase outside the cylinder skirts. From this point the oil drains into the sump, as illustrated in Fig. 4-25, so that there

is no buildup in the crankcase. The oil in the crankcase during operation is primarily in the form of a mist or spray. This oil lubricates the cylinder walls, pistons, and piston pins.

Excessive oil is prevented from entering the cylinder heads by means of oil control rings on the pistons, as explained previously. Some pistons incorporate drain holes under the oil control rings. Oil from the cylinder walls passes through these drain holes to the inside of the piston and is then thrown out into the crankcase.

TYPICAL LUBRICATION SYSTEMS

Oil System for Wet-Sump Engine

The lubrication system for the Continental IO-470-D engine is shown in Fig. 4-26. Lubricating oil for the engine is stored in the sump, which is attached to the lower side of the engine. Oil is drawn from the sump through the suction oil screen, which is positioned in the bottom of the sump. After passing through the gear-type oil pump, the oil is

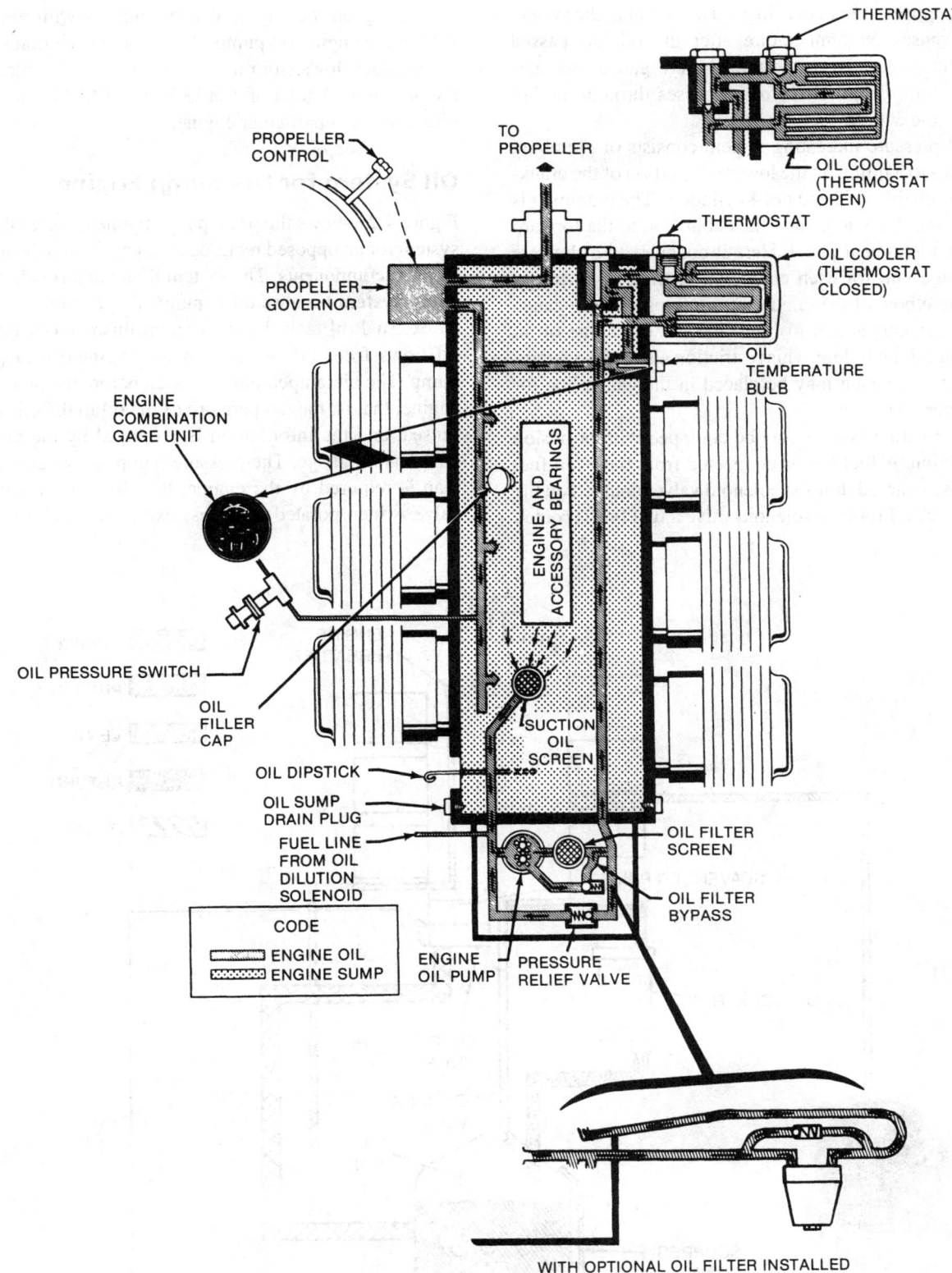

THERMOSTAT

OIL COOLER
(THERMOSTAT
OPEN)

THERMOSTAT

OIL COOLER
(THERMOSTAT
CLOSED)

TO
PROPELLER

PROPELLER
CONTROL

PROPELLER
GOVERNOR

OIL
TEMPERATURE
BULB

ENGINE
COMBINATION
GAGE UNIT

ENGINE AND
ACCESSORY BEARINGS

OIL PRESSURE SWITCH

OIL
FILLER
CAP

SUCTION
OIL
SCREEN

OIL DIPSTICK

OIL SUMP
DRAIN PLUG

FUEL LINE
FROM OIL
DILUTION
SOLENOID

OIL FILTER
SCREEN

OIL FILTER
BYPASS

CODE

▨ ENGINE OIL
▨ ENGINE SUMP

ENGINE
OIL PUMP

PRESSURE
RELIEF VALVE

WITH OPTIONAL OIL FILTER INSTALLED

FIGURE 4-26   Lubrication system for an opposed engine.

directed through the oil filter screen and along an internal gallery to the forward part of the engine where the oil cooler is located. A bypass check valve is placed in the bypass line around the filler screen to provide for oil flow in case the screen becomes clogged. A nonadjustable pressure relief valve permits excess pressure to return to the inlet side of the pump.

Oil temperature is controlled by a thermally operated valve which either causes the oil to bypass the externally mounted cooler or routes it through the cooler passages. Drilled and cored passages carry oil from the oil cooler to all parts of the engine requiring lubrication. Oil from the system is also routed through the propeller governor to the crankshaft and to the propeller for control of pitch and engine rpm.

The oil temperature bulb is located at a point in the system where it senses oil temperature after the oil has passed through the cooler. Thus, the temperature gauge indicates the temperature of the oil *before* it passes through the hot sections of the engine.

The oil pressure indicating system consists of plumbing that attaches to a fitting on the lower left portion of the crankcases between the no. 2 and no. 4 cylinders. The plumbing is routed through the wings, into the cabin, and to the forward side of the instrument panel. Here it connects to a separate engine gauge unit for each engine. A restrictor is incorporated in the elbow of the engine fitting to protect the gauge from pressure surges and to limit the loss of engine oil in case of a plumbing failure. This restriction also aids in retaining the light oil which may be placed in the gauge line for cold-weather operation.

This lubrication system may be equipped with provision for oil dilution. A fuel line is connected from the main fuel strainer case to an oil dilution solenoid valve mounted on the engine fire wall. From the solenoid valve a fuel line is routed to a fitting on the engine which connects with the suction side of the engine oil pump. When the oil dilution switch is closed, fuel flows from the fuel strainer to the inlet side of the oil pump. A total of 4 qt [3.79 L] of fuel is required for dilution in this particular engine.

Oil System for Dry-Sump Engine

Figure 4-27 shows the principal components of a lubrication system for an opposed reciprocating engine and the locations of these components. The system illustrated is called a **dry-sump system** because oil is pumped out of the engine into an external oil tank. In the system illustrated in Fig. 4-27, oil flows from the oil tank to the engine-driven pressure pump. The oil temperature is sensed before the oil enters the engine; that is, the temperature of the oil in the oil-in line is sensed, and the information is displayed by the engine oil temperature gauge. The pressure pump has greater capacity than is required by the engine; therefore, a pressure relief valve is incorporated to bypass excess oil back to the inlet

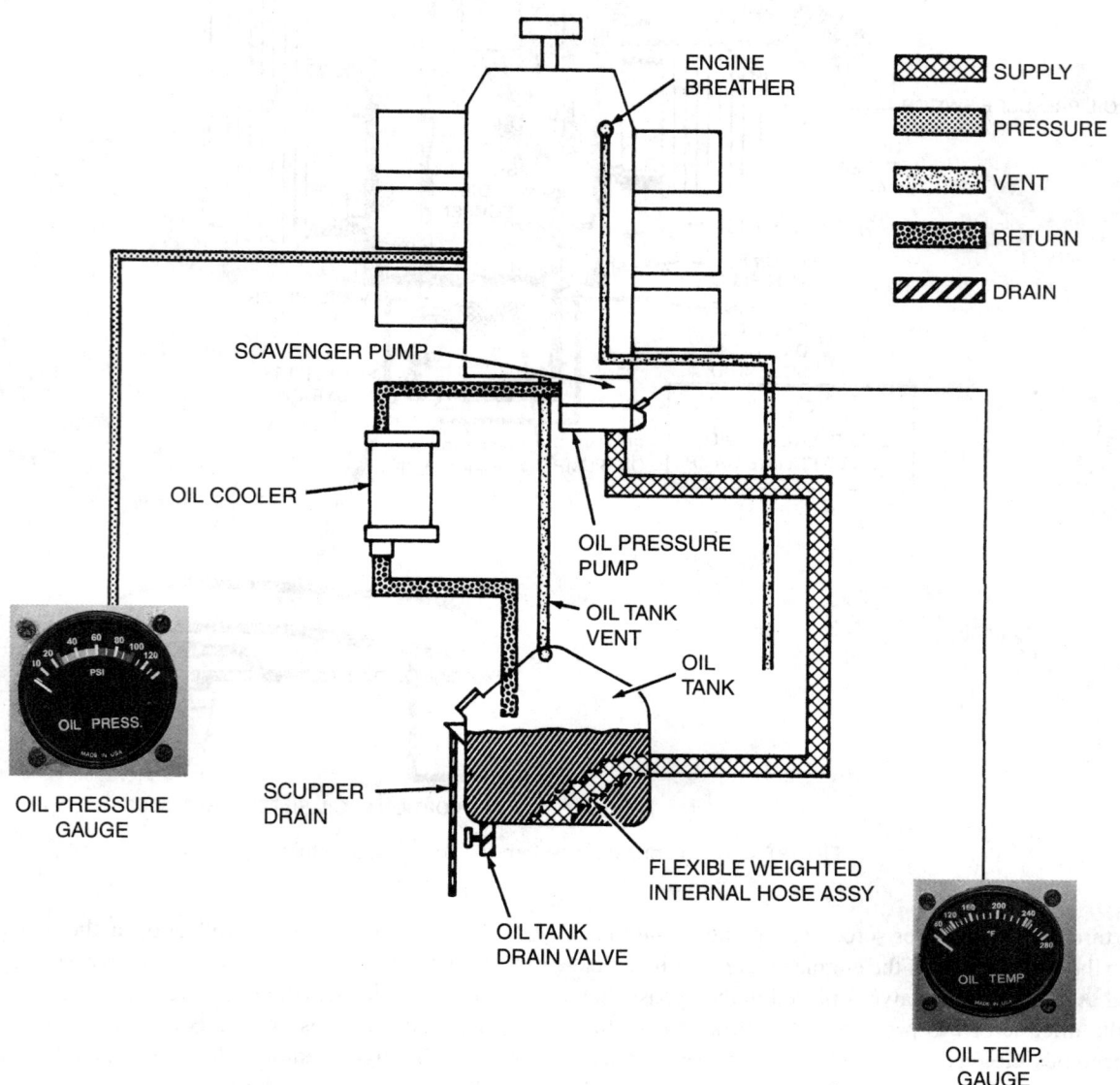

FIGURE 4-27 Schematic of dry-sump oil system.

side of the pump. A pressure gauge connection, or sensor, is located on the pressure side of the pressure pump to actuate the oil pressure gauge. The oil screen (oil filter) is usually located between the pressure pump and the engine system; oil screens are provided with bypass features to permit unfiltered oil to flow to the engine in case the screen becomes clogged, since unfiltered oil is better than no oil. After the oil has flowed through the engine system, it is picked up by the scavenge pump and returned through the oil cooler to the oil tank. The scavenge pump has a capacity much greater than that of the pressure pump, because the oil volume it must handle is increased as a result of the air bubbles and foam entrained during engine operation. The oil cooler usually incorporates a thermostatic control valve that bypasses the oil around the cooler until the oil temperature reaches a proper value. To prevent pressure buildup in the oil tank, a vent line is connected from the tank to the engine crankcase. This permits the oil tank to vent through the engine venting system. Check valves are employed in some systems to prevent oil from flowing by gravity to the engine when the engine is inoperative.

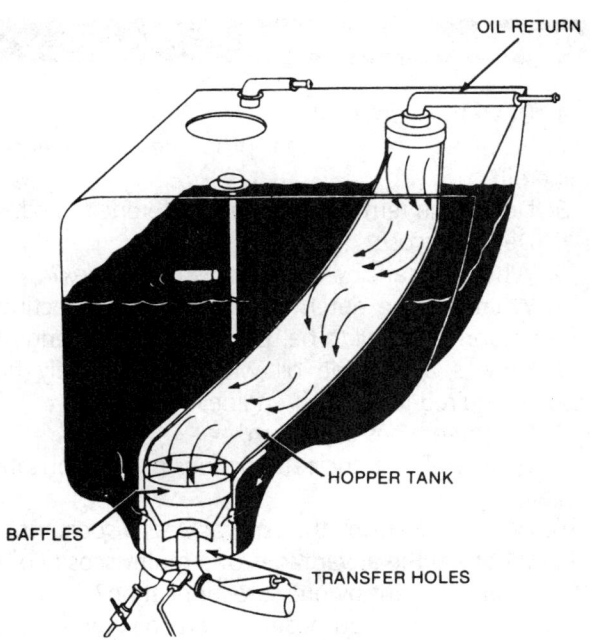

FIGURE 4-28   Oil tank with hopper.

Oil Tanks

Dry-sump engine lubrication systems require a separate oil tank for each engine system. These tanks are constructed of welded sheet aluminum, riveted aluminum, or stainless steel. Some aircraft are equipped with synthetic rubber tanks similar to fuel cells.

An outlet for the tank is normally located at the lowest section of the tank to permit complete drainage, either in the ground position or in a normal flight attitude. If the airplane is equipped with a propeller-feathering system, a reserve of oil must be provided for feathering, either in the main tank or in a separate reservoir. If the reserve oil supply is in the main tank, the normal outlet must be arranged so the reserve oil supply cannot be drawn out of the tank except when it is necessary to feather the propeller.

Some oil tanks, particularly those that carry a large supply of oil, are designed with a **hopper** that partially isolates a portion of the oil in the tank from the main body of the oil. Oil to the engine is drawn from the bottom of the hopper, and return oil feeds into the top of the hopper. This permits a small portion of the oil in the tank to be circulated through the engine, thus permitting rapid warm-up of the engine. The hopper is open to the main body of the oil through holes near the bottom, and in some cases the hopper is equipped with a flapper valve that allows oil to flow into the hopper as the oil in the hopper is consumed. The hopper is also termed a **temperature-accelerating well** and is illustrated in Fig. 4-28.

Provision must be made to prevent the entrance into the tank itself, or into the tank outlet, of any foreign object or material which might obstruct the flow of oil through the system. The oil tank outlet must not be enclosed by any screen or guard which would reduce the flow of oil below a safe value at any operating temperature condition. The diameter of the oil outlet must not be less than the diameter of the

inlet to the oil pump. That is, the pump must not have greater capacity than the outlet of the tank.

Oil tanks must be constructed with an expansion space of 10 percent of the tank capacity or $\frac{1}{2}$ gal [1.89 L], whichever is greater. It must not be possible to fill the expansion space when the tank is refilled. This provision is met by locating the filler neck or opening so that it is below the expansion space in the tank. Reserve oil tanks which have no direct connection to any engine must have an expansion space which is not less than 2 percent of the tank capacity.

Oil tanks designed for use with reciprocating engines must be vented from the top of the expansion space to the crankcase of the engine. The vent opening in the tank must be located so that it cannot be covered by oil under any normal flight condition. The vent must be designed so that condensed water vapor which might freeze and obstruct the line cannot accumulate at any point. Oil tanks for acrobatic aircraft must be designed to prevent hazardous loss of oil during acrobatic maneuvers, including short periods of inverted flight.

If the filler opening for an oil tank is recessed in such a manner that oil may be retained in the recessed area, a drain line must be provided to drain the retained oil to a point clear of the airplane. The filler cap must provide a tight seal and must be marked with the word "oil" and the capacity of the tank.

The oil tank must be strong enough to withstand a test pressure of 5 psi [34.48 kPa] and support without failure all vibration, inertia, and fluid loads imposed during operation.

The quantity of oil in the oil tank or in the wet sump of an engine can usually be determined by means of a dipstick. In some cases the dipstick is attached to the filler neck cap. Before any flight, it is standard practice to check visually the quantity of oil.

1. Describe a lubricant.
2. What are the principal properties of an aircraft engine oil?
3. Define the term *flash point* in reference to oil.
4. Define *viscosity*.
5. What is meant by the term *viscosity index*?
6. What are the determining factors in selecting a grade of lubricating oil to be used in an aircraft engine?
7. Why is an engine oil with comparatively high viscosity required for aircraft engines?
8. List three types of friction.
9. What are the principal functions of the lubricating oil in an engine?
10. What is meant by the term *ashless dispersant*?
11. What are the advantages of a multiviscosity oil?
12. What is meant by pressure lubrication?
13. Describe the purpose of an oil temperature regulator.
14. What is the function of an oil system relief valve?
15. What inspection should be made with respect to the engine when an oil filter is serviced?
16. What design feature is usually incorporated to prevent a clogged filter from restricting oil flow?
17. What is the purpose of an oil separator?
18. Why is low-viscosity oil used in the oil pressure gauge line?
19. Where is the oil temperature sensor placed in an engine lubrication system?
20. What will happen when the side clearance of the gears in a gear-type oil pressure pump becomes too great?
21. Why is the capacity of the scavenge pump in a dry-sump oil system greater than that of the pressure pump?
22. What happens to the excess oil supplied by the pressure pump?
23. Aircraft lubrication systems are normally classified into which two categories?
24. What is the purpose of a hopper in an oil tank?
25. Oil filler caps should contain what markings?

Induction Systems, Superchargers, Turbochargers, and Cooling and Exhaust Systems 5

GENERAL DESCRIPTION

The complete induction system for an aircraft engine includes three principal sections: (1) the air scoop and ducting leading to the carburetor; (2) the carburetor, or air control section, of an injection system; and (3) the intake manifold and pipes. These sections constitute the passages and controlling elements for all the air which must be supplied to the engine. A very basic induction system is shown in Fig. 5-1.

BASIC INDUCTION SYSTEM COMPONENTS

Air Scoop and Ducting

The ducting system for a nonsupercharged (naturally aspirated) engine comprises four principal parts; (1) air scoop, (2) air filter, (3) alternate air valve, and (4) carburetor air heater, or heater muff.

A typical **air scoop** is simply an opening facing into the airstream. This scoop receives ram air, usually augmented by the propeller slipstream. The effect of the air velocity is to "supercharge" (compress) the air a small amount, thus adding to the total weight of air received by the engine. The power increase may be as much as 5 percent. The design of the air scoop has a substantial effect on the amount of increased power provided by ram air pressure. Ducting is made from either solid molded parts or flexible hose used to seal and direct the intake air through the various components of the induction system.

Air Filters

The induction air filter is installed at or near the air scoop, as shown in Fig. 5-1, for the purpose of removing dirt, abrasive particles, sand, and even larger foreign materials before they are carried into the engine. Although the air filter reduces air pressure to the carburetor to some extent, thus reducing

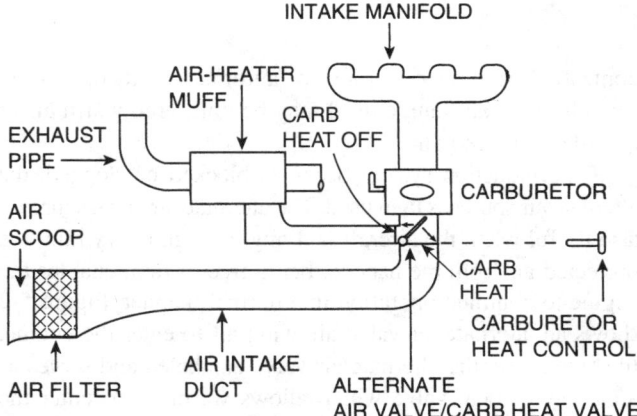

FIGURE 5-1 A simple induction system.

the power output, it prevents harmful dirt from entering the engine. There are basically three types of induction air filters: wetted-type mesh filters, dry paper filters, and polyurethane foam filters. The wetted mesh filter usually consists of a mat of metal filaments encased in a frame and dipped into oil. The oil film on the metal mesh filaments catches and holds dust and sand particles. The dry paper-type filter is similar to an automotive air filter. It is made of a pleated layer of paper filter elements through which the air must pass. The edges are sealed to prevent foreign material from entering the engine. The paper filter is usually replaced on a time-in-service basis, but if it becomes damaged or clogged, it should be replaced immediately. A new foam-type filter has been developed for aircraft use and utilizes polyurethane and a wetting agent. This type of filter is replacing many of the other filter types. Instructions on the cleaning, servicing, and replacement of air filters are given in Chap. 9.

Alternate Air Valve

The alternate air valve is designed to allow air to flow to the engine if the air filter or other parts of the induction system should become clogged. This valve can be manually

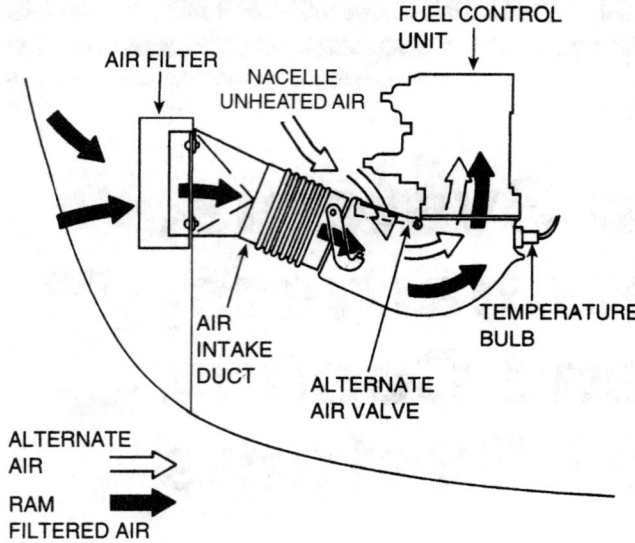

FIGURE 5-2 An alternate air valve system.

Intake Manifolds

The typical **opposed-type** (or **flat-type**) **aircraft engine** has an induction system with an individual pipe leading to each cylinder. On some models of this type, one end of each pipe is bolted to the cylinder by a flange, and the other end fits into a slip joint in the manifold. On other models of this type, the pipes are connected to the manifold by short sections of rubber (or synthetic rubber) hose held by clamps. In still other models of this type, the carburetor is mounted on the oil sump and the fuel-air mixture flows from the carburetor through passages in the oil sump and then out through each of the individual pipes leading to the engine cylinders. As the mixture of fuel and air flows through the passages in the oil sump, heat is transferred from the oil to the fuel-air mixture. This arrangement accomplishes two things: (1) it cools the oil slightly, and (2) it increases the temperature of the fuel-air mixture slightly for better vaporization of the fuel. An arrangement whereby heat is applied to the fuel-air mixture by means of heated oil or through proximity to the exhaust manifold is usually termed a **hot spot**.

The type of induction system used on a radial-type engine principally depends on the horsepower output desired from the engine. On a small radial engine of low output, the air is drawn through the carburetor, mixed with fuel in the carburetor, and then carried to the cylinders through individual intake pipes. In some engines, an intake manifold section is made a part of the main engine structure. The fuel-air mixture is carried from the outer edge of the manifold section to the separate engine cylinders by individual pipes, which are connected to the engine by a slip joint. The purpose of the slip joint is to prevent damage which would otherwise result from the expansion and contraction caused by changes in temperature.

controlled from the cockpit or be automatic in its operation. The alternate air source used may be unfiltered warm air or outside unfiltered air.

If the induction system becomes blocked or clogged, the alternate air source is then used. The alternate air valve source is also useful when the aircraft is flying through heavy rain; the protected air from the nacelle, being free of rain, enables the engine to continue operation in a normal manner. Figure 5-2 shows an alternate air valve allowing air to enter the engine. In some cases, the alternate air valve is labeled and serves as a carburetor heat valve which allows warm air to enter the engine's induction system to prevent carburetor ice build-up.

Carburetor Heat Valve and Heater Muff

The **carburetor heat air valve**, shown in Fig. 5-1, is operated by means of the carburetor heat control in the cockpit. The valve is simply a gate which closes the main air duct and opens the duct to the heater muff when the control is turned on. During normal operation, the gate closes the passage to the heater muff and opens the main air duct. The gate is often provided with a spring which tends to keep it in the normal position.

The **heater muff** is a shroud placed around a section of the exhaust pipe. The shroud is open at the ends to permit air to flow into the space between the exhaust pipe and the wall of the shroud. A duct is connected from the muff to the main air duct. During operation of the carburetor air heater system, **protected air** within the engine compartment flows into the space around the exhaust pipe where it is heated before being carried to the main air duct. Note that carburetor air heat should be applied only if necessary to prevent ice formation and to keep rain out of the carburetor. Since it is less dense than cool air, heated air results in a loss of power. For maximum power, therefore, it is desirable that a free flow of unheated air be provided for the engine. The use of heated air during periods of high-power operation is likely to cause detonation and will definitely cause a reduction in engine power output.

Importance of Gastight Seal

The portion of the intake system of an engine between the carburetor and the cylinders must be installed gastight for proper engine operation. When the manifold absolute pressure (MAP) is below atmospheric pressure, which is always the case with unsupercharged (naturally aspirated) engines, an air leak in the manifold system will allow air to enter and thin the fuel-air mixture. This can cause overheating of the engine, detonation, backfiring, or complete stoppage. Small induction system leaks will have the most noticeable effect at low rpm because the pressure differential between the atmosphere and the inside of the intake manifold increases as rpm decreases.

In a supercharged engine, a portion of the fuel-air mixture will be lost if leakage occurs in the intake manifold or pipes. This will result in a reduction in power and a waste of fuel.

One method of forming a gastight connection for intake pipes is to provide a synthetic-rubber packing ring and a packing retaining nut to form a slip-joint seal at the distribution chamber, thus allowing the intake pipes to slide in and out of the distribution chamber opening while the metal of the engine cylinder is expanding and contracting in response to changes in temperature. It is necessary to place a gasket at the cylinder intake port between the pipe flange and the cylinder port and to secure the flange rigidly with bolts and nuts.

Another method of forming a gastight connection for intake pipes is to use a packing ring and a packing retaining nut which screws into or over the intake port opening. Still another method is to have short stacks protruding from the intake ports, and to rubber couplings to connect the pipes to these protruding stacks.

Induction System for Six-Cylinder Opposed Engine

The principal assemblies of an induction system for a Continental engine are shown in Fig. 5-3. With this system, ram air enters the air box at the right rear engine baffle and is

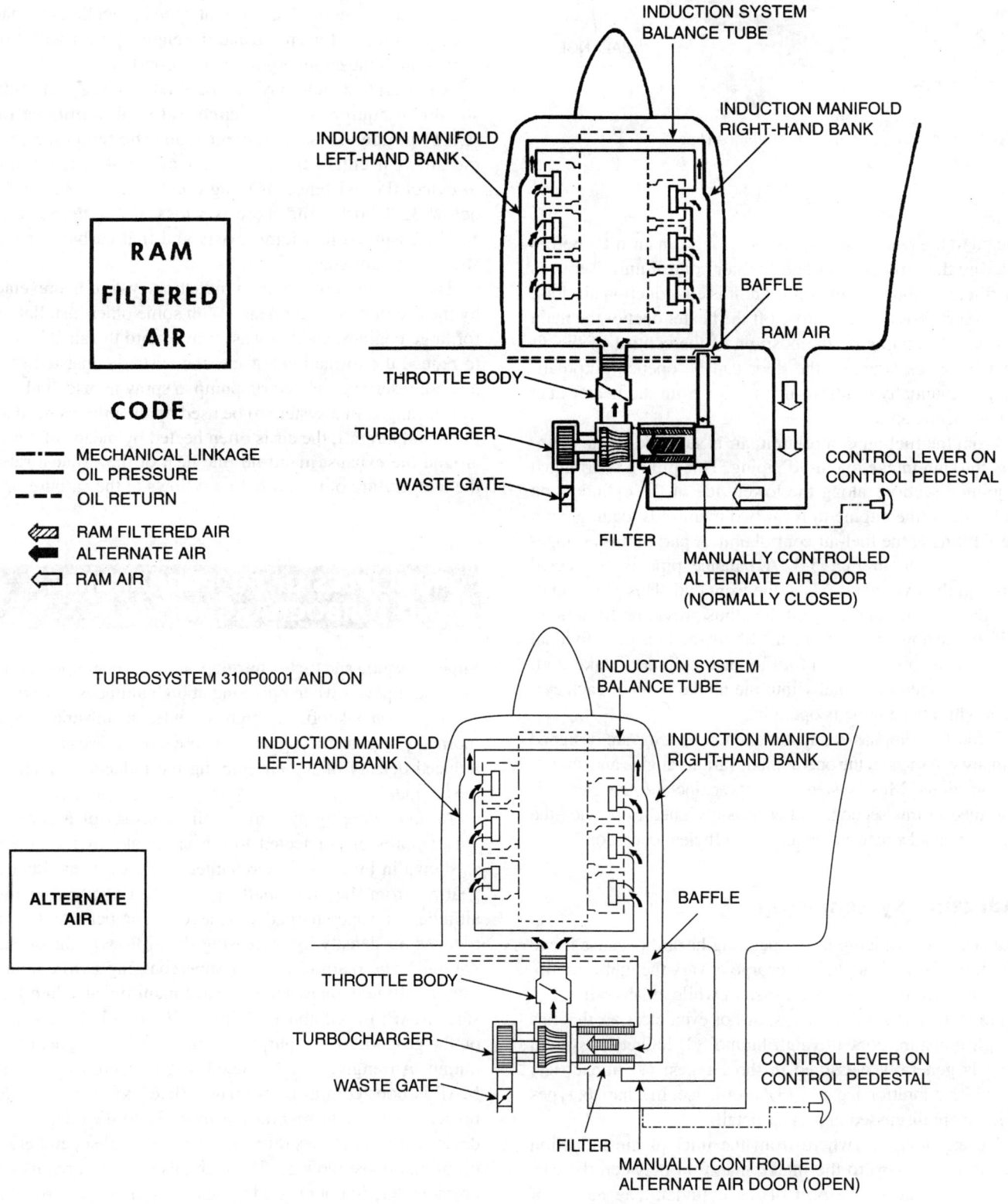

FIGURE 5-3 Induction system for a six-cylinder opposed engine.

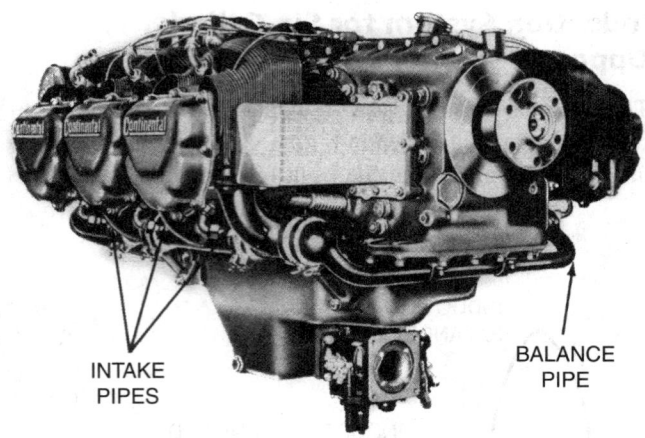

FIGURE 5-4 Intake pipes and balance pipe. (*Continental Motors.*)

ducted to the rear, where it passes through an air filter before entering the turbocharger and fuel-air control unit. Between the filter and the fuel-air control unit is an induction air door, which serves as a gate to close off the heater duct or the main air duct, depending on the position of the control. If the air filter becomes clogged, the door can be opened automatically or manually to allow air to enter from the heater duct or the nacelle area.

From the fuel-air control unit, air is supplied to the cylinders through intake manifold piping. This piping is arranged in jointed sections along the lower side of the cylinders on each side of the engine to form two manifolds leading from the Y fitting at the fuel-air control unit. A part of this arrangement is shown in Fig. 5-4. A balance pipe is connected between the two manifolds at the front end. This pipe equalizes the pressure in the manifolds, thus providing for a more uniform airflow to the cylinders. Short sections of individual piping lead from the manifold to each cylinder intake port. Fuel is injected continually into the intake ports of each cylinder while the engine is operating.

It must be emphasized that induction systems are designed in many ways to suit the operation of various engine and aircraft combinations. Most systems, however, include the elements described in this section, and each is designed to provide the engine with adequate air for the most efficient operation.

Induction System Icing

Induction system icing is an operating hazard because it can cut off the flow of the fuel-air charge or vary the fuel-air ratio. Ice can form in the induction system while an aircraft is flying in clouds, fog, rain, sleet, snow, or even clear air that has a high moisture content (high humidity). Induction system icing is generally classified in three types: (1) impact ice, (2) fuel evaporation ice, and (3) throttle ice. In Chap. 6, types of icing are discussed in greater detail.

Ice can form anywhere from the inlet of the induction system (air scoop) to the intake manifold between the carburetor and the intake port of the cylinder. The nature of the ice formation depends on atmospheric temperature and

humidity and the operating conditions of the engine. If the air scoop and ducting leading to the carburetor are at a temperature below the freezing point of water, impact ice will form when particles of water in the air strike the cold surfaces, particularly at the air screen and at turns in the intake duct. Any small protrusions in the duct are also likely to start the formation of ice. Icing can be detected by a reduction in engine power when the throttle position remains fixed. If the aircraft is equipped with a fixed-pitch propeller, the engine rpm will decrease. With a constant-speed propeller, the manifold pressure will decrease and the engine power will drop, even though the engine rpm remains constant.

An aircraft which may be operated in icing conditions should be equipped with a **carburetor air temperature (CAT) gauge**. This instrument reads the temperature of the air as it enters the carburetor and makes it possible to detect the existence of icing conditions. If the CAT is below 32°F [0°C] and there is a loss of engine power, it can be assumed that icing exists and that carburetor heat should be applied.

The formation of ice in an induction system is prevented by the use of carburetor heat. (With some older installations for large engines, alcohol was sprayed into the air inlet duct to reduce the formations of ice; the system consisted of an alcohol reservoir, an electric pump, a spray nozzle, and controls arranged in a system to be used by the pilot as needed.) For small aircraft, the air is often heated by means of a muff around the exhaust manifold; the heat of the exhaust raises the temperature of the air before it flows to the carburetor.

PRINCIPLES OF SUPERCHARGING AND TURBOCHARGING

Supercharging and **turbocharging** allows an engine to develop maximum power when operating at high altitudes or to boost its power on takeoff. At high altitude, an unsupercharged (normally aspirated) engine will lose power because of the reduced density of the air entering the induction system of the engine.

A supercharging system usually consists of a centrifugal compressor connected to the air intake of the engine, as shown in Fig. 5-5. Superchargers can be driven either by gearing from the crankshaft or by exhaust gases. At high altitudes, a supercharged engine can compensate for the reduced air density by increasing the airflow to the engine. Although the main reason for supercharging is to compensate for altitude, many engines have manifold absolute pressure (MAP) raised above 30 inHg [101.61 kPa] even when on the ground, for the purpose of boosting the engine power output. An engine is considered to be supercharged when MAP is boosted above 30 inHg [101.61 kPa]. Some high-power engines can be boosted from 34 to 48 inHg. These devices are sometimes referred to as sea-level superchargers or ground boost blowers. There are also engine-driven compressors that do not raise MAP above 30 inHg [101.61 kPa]. These compressors are called **normalizers**.

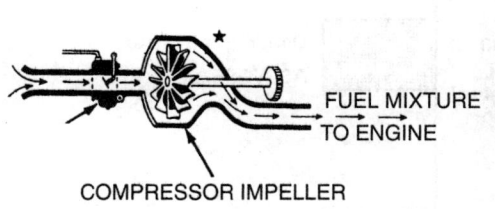

FUEL MIXTURE
TO ENGINE

COMPRESSOR IMPELLER

FIGURE 5-5 Compressor supercharger.

The turbocharger is so named because it is a supercharger driven by a turbine. The power to drive the turbine is furnished by the high-velocity gases from the engine exhaust. A turbocharger that boosts the intake air pressure above 30 inHg [101.61 kPa] is called a **turbosupercharger**.

As mentioned previously, the air density at high altitudes decreases. For the engine to develop maximum available power, the turbocharger will "normalize" or compensate for the decrease in air density. This normalizing turbocharging system is not used to boost MAP during takeoff, and it should never be used to boost MAP above 30 inHg [101.61 kPa]. The capacity of a supercharger, turbocharger, or turbosupercharger depends on the size of the impeller and the speed at which it is rotated. In the case of the turbocharger or turbosupercharger, its output also depends on the amount of exhaust gases exiting the engine.

Properties of Gases as Related to Supercharging

Understanding the principles of supercharging requires a knowledge of mass, volume, and density as applied to the properties of gases. All matter can be classified as solids and fluids. In turn, fluids include both liquids and gases. Solids, liquids, and gases all have weight, but the weight of gases is by no means a constant value under all conditions. For example, at sea-level pressure, about 13 ft³ [368.16 L] of air weighs 1 lb [0.45 kg], but at greater pressures the same volume would weigh more, and at lower pressures it would weigh less.

Mass is not the same as weight, but for an ordinary layperson's discussion it is often used as if they were the same. Mass should not be confused with volume, either, because volume designates merely the space occupied by an object and does not take density or pressure into consideration. The relation among these various factors is explained by the various laws pertaining to the behavior of gases.

Any gas can be compressed to some extent, and this compression is accomplished by exerting some force on the gas—that is, by increasing its pressure. **Boyle's law** expresses the relationship between pressure and volume as follows: *In any sample of gas, the volume is inversely proportional to the absolute pressure if the temperature is kept constant.*

With reference to the quantity of gas (air) in a closed cylinder fitted with a movable piston (as shown in Fig. 5-6), if it is assumed that the temperature is constant and that there is no leakage past the piston, then the volume and pressure are inversely related in accordance with Boyle's law.

At the left in Fig. 5-6, 15 ft³ [424.8 L] of gas weighs 1 lb [0.45 kg] at 10 psia [68.95 kPa] pressure. In the picture at the right, 15 ft³ [424.8 L] of gas weighs 2 lb [0.91 kg] at a pressure of 20 psia [137.9 kPa] because more gas is packed into the cylinders. The volume is the same in both pictures, but the pressure is changed, and therefore the mass (quantity) of air below the piston is changed.

In Fig. 5-7, cylinders are shown with their pistons. The one at the left has a volume of 10 in³ [0.17 L] under the piston, the one in the middle has a volume of 5 in³ [0.08 L] under the piston, and the one at the right has a volume of 20 in³ [0.33 L] under the piston.

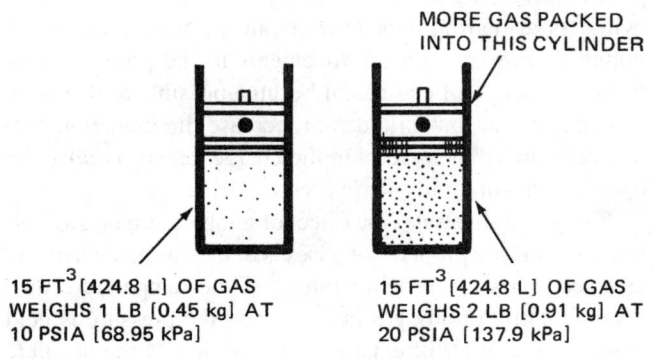

MORE GAS PACKED
INTO THIS CYLINDER

15 FT³ [424.8 L] OF GAS
WEIGHS 1 LB [0.45 kg] AT
10 PSIA [68.95 kPa]

15 FT³ [424.8 L] OF GAS
WEIGHS 2 LB [0.91 kg] AT
20 PSIA [137.9 kPa]

FIGURE 5-6 Quantity of gas (air) charge.

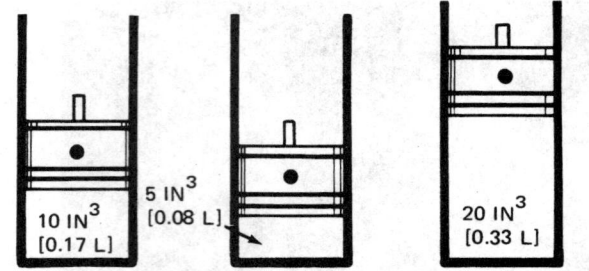

10 IN³ [0.17 L]   5 IN³ [0.08 L]   20 IN³ [0.33 L]

15 PSIA [103.43 kPa]   30 PSIA [208.85 kPa]   7.5 PSIA [52.71 kPa]

FIGURE 5-7   Relative volumes and pressures.

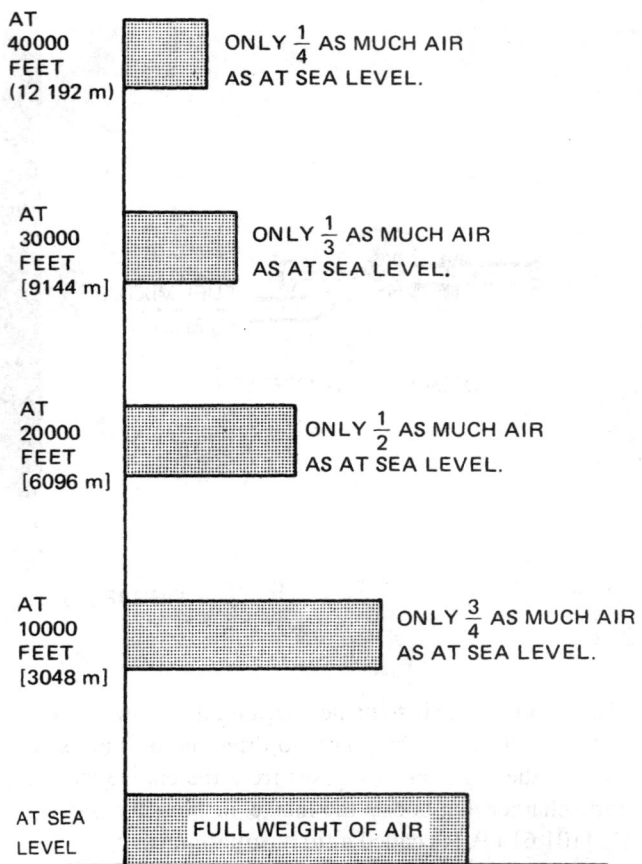

AT 40000 FEET (12 192 m)   ONLY $\frac{1}{4}$ AS MUCH AIR AS AT SEA LEVEL.

AT 30000 FEET [9144 m]   ONLY $\frac{1}{3}$ AS MUCH AIR AS AT SEA LEVEL.

AT 20000 FEET [6096 m]   ONLY $\frac{1}{2}$ AS MUCH AIR AS AT SEA LEVEL.

AT 10000 FEET [3048 m]   ONLY $\frac{3}{4}$ AS MUCH AIR AS AT SEA LEVEL.

AT SEA LEVEL   FULL WEIGHT OF AIR

FIGURES ARE APPROXIMATE AND MEAN WEIGHT OF AIR.

FIGURE 5-8   Effect of altitude on density of air.

If the density of the air in the cylinder at the left is accepted as standard, the air in the middle cylinder has a density of 2 and the air in the cylinder at the right has a density of only $\frac{1}{2}$.

The weight of the air surrounding the earth is sufficient to exert considerable pressure on objects at sea level. At altitudes above sea level, the pressures, density, and temperatures of the air are lower. At an altitude of 20 000 ft [6096 m], the pressure and density of the atmosphere are only about one-half their values at sea level.

Unless something were done to offset the decreasing density, at sea level the engine would receive the full weight of air, at 10 000 ft [3048 m] it would receive only three-fourths as much air by weight as at sea level, at 20 000 ft [6096 m] it would get only one-half as much air by weight as at sea level, at 30 000 ft [9144 m] it would obtain only one-third as much air by weight as at sea level, and at 40 000 ft [12 192 m] it would receive only one-fourth as much air by weight as at sea level. These relations are illustrated in Fig. 5-8.

These relations are important in any discussion of an internal-combustion engine because the power developed by the engine depends principally on the mass of the induced charge. An engine that is not supercharged can induce only a charge of a given mass, according to its volumetric efficiency and piston displacement. To increase the mass of the charge, it is necessary to increase the pressure and the density of the incoming charge by means of a supercharger or turbocharger. Therefore, it is possible to say that the function of a supercharger or turbocharger is to increase the quantity of air (or fuel-air mixture) entering the engine cylinders.

Superchargers were developed originally for the sole purpose of increasing the density of the air taken into the engine cylinders at high altitudes, to obtain the maximum power output. However, with improvements in the production of fuels and design of engines, it became possible to operate a supercharger at low altitudes to increase the induction system pressure (thus increasing the charge density) *above* the normal value of atmospheric pressure.

Figure 5-9 illustrates the effect of temperature on gas volume. The elastic property of gases is demonstrated when any change occurs in the temperature. If the temperature of a given quantity of any gas is raised and the pressure is held constant, the gas will expand in proportion to the absolute temperature. This is expressed by

$$\frac{V_1}{V_2} = \frac{T_1}{T_2}$$

This equation is known as **Charles' law** and is attributed to the French mathematician and physicist Jacques A. C. Charles (1746–1823).

In Fig. 5-9, the temperature of the gas in the first cylinder is 0°C [273 K (kelvins)]. When the gas temperature is raised to 273°C [546 K], the absolute temperature is

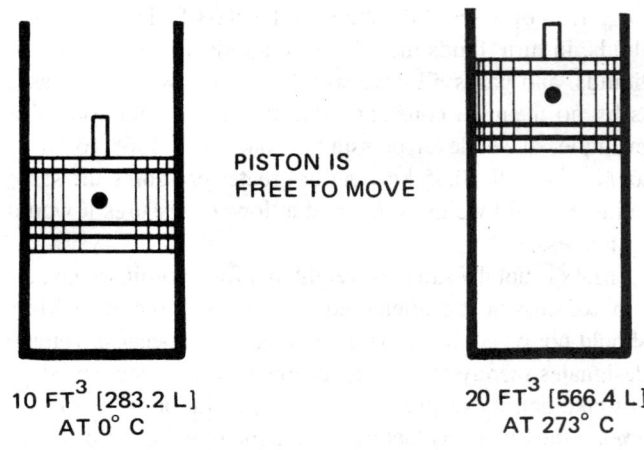

PISTON IS FREE TO MOVE

10 FT³ [283.2 L] AT 0° C   20 FT³ [566.4 L] AT 273° C

FIGURE 5-9   Effect of temperature on gas volume.

doubled; therefore, the volume is doubled. To illustrate further, if a quantity of gas has a volume of 10 ft³ [283.2 L] at 10°C [283 K] and we wish to find the volume at 100°C [373 K], we must convert the temperature indications to absolute (kelvin) values. To do this, we merely add 273 to the Celsius value. Then 10°C becomes 283 K, and 100°C becomes 373 K. Applying the formula gives

$$\frac{10}{V_2} = \frac{283}{373} \quad \text{pressure constant}$$

$$283\,V_2 = 3730$$

$$V_2 = \frac{3730}{283} = 13.18 \text{ ft}^3 [373.26\,L]$$

Thus, the increase in temperature has increased the volume of the gas from 10 to 13.18 ft³ [283.2 to 373.26 L], the pressure remaining constant.

The pressure of a gas varies in proportion to the absolute temperature if the volume is held constant. This is expressed by the equation

$$\frac{P_1}{P_2} = \frac{T_1}{T_2} \quad \text{volume constant}$$

This principle is illustrated in Fig. 5-10, where the gas temperature has been doubled; that is, it has been raised from 0°C [273 K] to 273°C [546 K]. Since the absolute temperature has been doubled with the volume held constant, the pressure has doubled.

Manifold Pressure

Manifold absolute pressure (MAP) is the pressure in the intake manifold of the engine. The weight of the fuel-air mixture entering the engine cylinders is measured by MAP and the temperature of the mixture. In a normally aspirated engine, MAP is less than outside atmospheric pressure because of the air friction losses in the air induction system. In a supercharged engine, however, MAP may be higher than the pressure of the atmosphere. When the supercharger is operating, MAP may be greater or less than atmospheric

pressure, depending on the settings of the supercharger control and throttle.

It must be pointed out that MAP is of prime importance in high-performance engines equipped with constant-speed propellers. If MAP is too high, detonation and overheating will occur. These conditions will damage the engine and cause engine failure if permitted to exist for an appreciable amount of time. During operation of a supercharged engine, strict attention must be paid to the power settings (rpm and MAP) of the engines.

If an engine has a particularly high compression ratio, it is quite likely that the supercharger cannot be used at all until an altitude of 5000 ft [11 524 m] or more is reached. If the supercharger on such an engine is operated at low altitudes, the pressures and temperatures in the combustion chambers will cause detonation and preignition.

Another factor governing the MAP to be used with an engine is the octane rating or performance number of the fuel. If a fuel has very high antidetonation characteristics, the maximum MAP may be higher than it would be with a fuel having a lower antiknock rating.

Purposes of Supercharging

The main purpose of supercharging an aircraft engine is to increase MAP above the pressure of the atmosphere to provide high power output for takeoff and to sustain maximum power at high altitudes.

Increased MAP increases the power output in two ways:

1. *It increases the weight of the fuel-air mixture (charge) delivered to the cylinders of the engine.* At a constant temperature, the weight of the fuel-air mixture that can be contained in a given volume is dependent on the pressure of the mixture. If the pressure on any given volume of gas is increased, the weight of that gas is increased because the density is increased.

2. *It increases the compression pressure.* The compression ratio for any given engine is constant; therefore, the greater the pressure of the fuel-air mixture at the beginning of the compression stroke, the greater will be the **compression pressure**, the latter being the pressure of the mixture at the end of the compression stroke. Higher compression pressure causes a higher mean effective pressure (mep) and consequently higher engine output.

Figure 5-11 illustrates the increase of the compression pressure. In the cylinder at the extreme left the pressure is only 36 inHg [121.93 kPa] at the beginning of the compression stroke. At the end of the compression stroke, the compression pressure in the second cylinder from the left is 270 inHg [914.49 kPa]. In this case, the intake pressure is comparatively low.

The pressure is 45 inHg [154.42 kPa] in the third cylinder from the left in Fig. 5-11, and the compression pressure in the cylinder at the extreme right is 405 inHg [1371.74 kPa]. Since the pressure is relatively high at the beginning of the compression stroke, it is still higher at the end. In this case

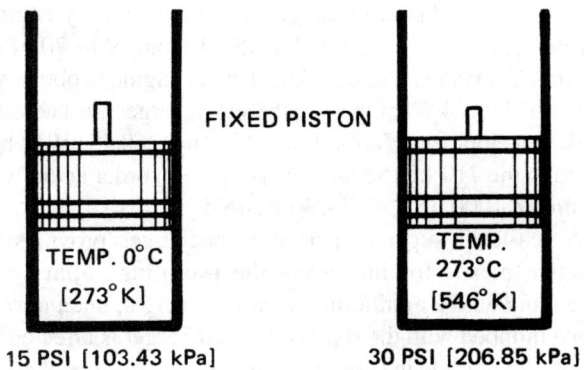

FIXED PISTON

TEMP. 0°C [273°K]

TEMP. 273°C [546°K]

15 PSI [103.43 kPa]   30 PSI [206.85 kPa]

FIGURE 5-10   Effect of temperature on gas pressure.

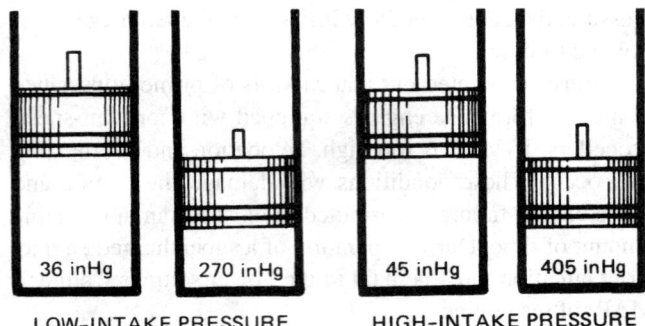

| 36 inHg | 270 inHg | 45 inHg | 405 inHg |

LOW-INTAKE PRESSURE   HIGH-INTAKE PRESSURE

FIGURE 5-11 Effects of air pressure entering the cylinder.

the intake pressure is high in comparison with that represented by the two drawings to the left. Increased temperature as a result of the compression also adds to the pressure.

Relation Between Horsepower and MAP

The relation between MAP and the engine power output for a certain engine at maximum rpm is shown in the chart in Fig. 5-12, where MAP in inches of mercury is plotted horizontally and horsepower is plotted vertically. By referring to a MAP of 30 inHg [101.61 kPa] and then following the vertical line upward until it intersects the curve, we see that the curve intersects the 30-inHg [101.61-kPa] line at about 550 hp [410.14 kW]. Thus, when the engine is not supercharged, the theoretical maximum pressure in the intake manifold is assumed to be almost 30 inHg [101.61 kPa], which is atmospheric pressure at sea level, and the power developed by the engine is about 550 hp [410.14 kW]. However, in actual practice it is impossible in an unsupercharged engine to obtain an intake MAP of 30 inHg [101.61 kPa] because of friction losses in the manifold.

Refer to a MAP reading of 45 inHg [154.42 kPa], and follow the vertical line upward until it intersects the curve. This shows that if MAP is increased to 45 inHg [154.42 kPa] by supercharging, the engine output is then about 1050 hp [782.99 kW]. MAP cannot be increased indefinitely to

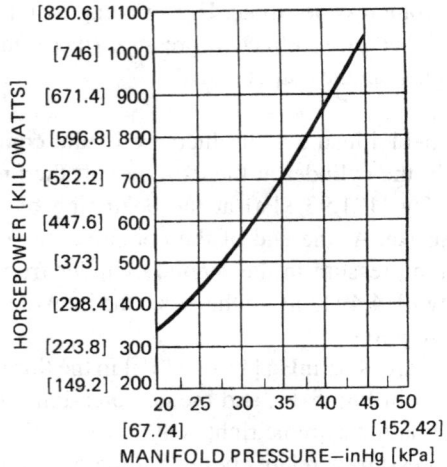

FIGURE 5-12 Relation between MAP and horsepower.

obtain more power. Excessive MAP has an adverse affect on engine operation and may ultimately damage the engine permanently as a result of high stresses, detonation, and high temperatures.

Supercharging Limitations

In supercharging an engine, if other factors remain the same, the gain in power output from the engine is proportional to the increase of pressure. However, the other factors do not remain constant, and there are limits beyond which safe operation is not possible. One of these factors is temperature. When air is compressed, its temperature is raised. This reduces the efficiency of the supercharger because heated air expands and increases the amount of power required to compress it and push it into the cylinders. The increased air temperature also reduces engine efficiency because any gas engine operates better if the intake mixture of fuel and air is kept cool. When the fuel-air mixture reaches an excessively high temperature, preignition and detonation may take place, resulting in a loss of power and often a complete mechanical failure of the powerplant.

The temperature increase resulting from supercharging is in addition to the heat generated by the compression in the engine cylinders. For this reason, the combined compression of the supercharger and of the cylinders must be kept within the correct limits determined by the antiknock qualities or octane ratings of the engine fuel.

If a supercharger were designed with enough capacity to raise the pressure of air at sea level from 14.7 to 20 psi [101.36 to 137.9 kPa], it would be possible to obtain about 40 percent more power than would be generated if there were no supercharging to increase the air pressure. If this supercharger were installed with a 1000-hp [745.7-kW] engine, the piston displacement of the 1000-hp [745.7-kW] engine would not need to be any greater than the piston displacement of a 710-hp [529.45-kW] engine that was not supercharged.

However, note that an engine which is to be provided with a supercharger must be designed to withstand the higher stresses developed by the increased power. It is not simply a matter of adding a supercharger to a 710-hp [529.45-kW] engine to obtain a 1000-hp [745.7-kW] output.

The difference between 710 and 1000 hp [529.45 and 745.7 kW] is 290 hp [216.25 kW], but it requires about 70 hp [52.22 kW] to operate the supercharger for this imaginary engine; therefore, not merely 290 hp [216.25 kW], but 290 + 70 = 360 hp [268.45 kW] must be developed in the engine to obtain the 1000-hp [745.7-kW] output. So the supercharger must account for the development of an additional 360 hp [268.45 kW] when added to the 710-hp [529.45-kW] engine in order actually to obtain the 1000 hp [745.7 kW] desired.

A sea-level supercharger or turbocharger provides an effective means for increasing the **pumping capacity** of the engine with a minimum increase in weight, but a powerplant equipped with the sea-level supercharger is affected by changes in altitude in the same manner as an unsupercharged engine, as shown in Fig. 5-13.

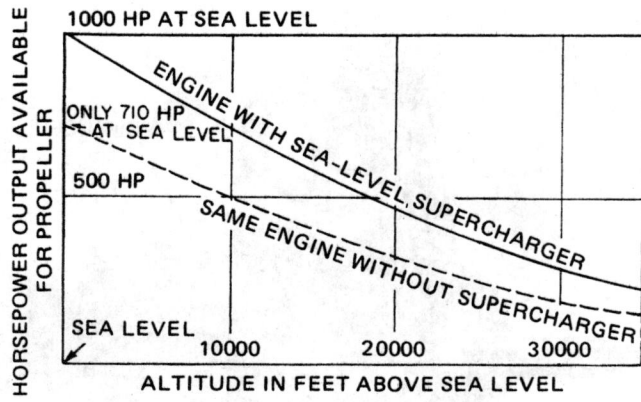

FIGURE 5-13   Effect of altitude with a sea-level supercharger.

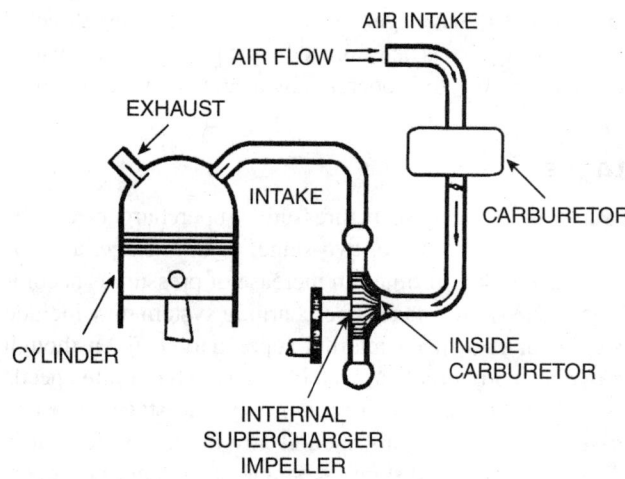

FIGURE 5-14   Location of an internal-type supercharger.

Some engines currently being manufactured employ turbochargers to **turbonormalize** the engine. The turbocharger provides the intake air compression that makes it possible to maintain sea-level pressure at high altitudes.

Factors Considered in Designing Altitude Superchargers

If the power to run the supercharger is taken from the engine crankshaft, the net gain in horsepower obtained by supercharging is reduced.

The net gain in horsepower obtained from supercharging is not fully reflected in the overall airplane performance because the supercharging equipment requires additional space and increases the weight of the airplane.

The degree of supercharging must be restricted within definite limits to avoid preignition and detonation which result from excessive temperatures and pressures.

A special cooling apparatus must be used to reduce the temperature of the fuel-air mixture because of the excessive heat resulting from the extra compression required at extreme altitudes. This apparatus requires space, adds to the weight of the airplane, and complicates operation, inspection, and maintenance. The special radiators used for such cooling are called **intercoolers** or **aftercoolers**, depending on their location with reference to the carburetor or fuel control.

Internal and External Superchargers

Most superchargers used on conventional airplanes are alike in that an impeller (blower) rotating at high speed is used to compress either the air before it is mixed with the fuel in the carburetor or the fuel-air mixture which leaves the carburetor. It is therefore possible to classify any supercharger, according to its *location* in the induction system of the airplane, as either an internal or external type.

When the supercharger is located between the carburetor and the cylinder intake ports, it is an **internal-type super-charger**, as shown in Fig. 5-14. Air enters the carburetor at atmospheric pressure and is mixed with the fuel. The fuel-air mixture leaves the carburetor at near-atmospheric pressure, is compressed in the supercharger to a pressure greater

than atmospheric, and then enters the engine cylinders. The power required to drive the supercharger impeller is transmitted from the engine crankshaft by a gear train. Because of the high gear ratio, the impeller rotates much faster than the crankshaft. If the gear ratio is adjustable for two different speeds, the supercharger is described as a **two-speed super-charger**. In general, the internal-type, supercharger may be used with an engine which is not expected to operate at very high altitudes or, in any event, where it is not necessary for air to be delivered under pressure to the carburetor intake.

An **external-type supercharger** delivers compressed air to the carburetor intake, as shown in Fig. 5-15. The air is compressed in the supercharger and then delivered through an air cooler to the carburetor, where it is mixed with the fuel. Since the power required to drive the ordinary type of external supercharger is obtained from the action of the engine exhaust gases against a bucket wheel, or turbine, the external type is also called a turbosupercharger or turbocharger, depending on whether it supercharges the air or merely maintains sea-level pressure. The speed of the impeller depends only on the quantity and pressure of the exhaust gases directed against the bucket wheel; therefore, the turbosupercharger is also a

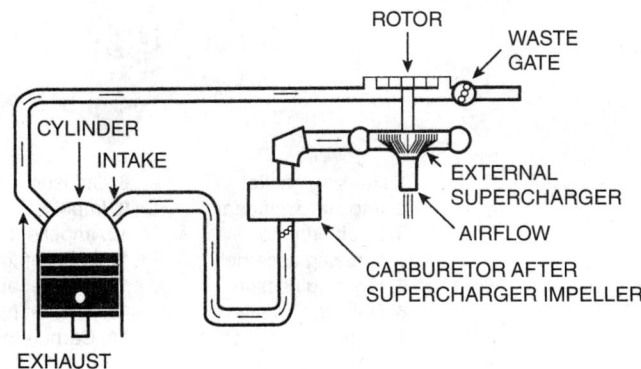

FIGURE 5-15   Location of an external-type supercharger.

multispeed supercharger. The volume of exhaust directed through the turbine is determined by the position of the **waste gate**. The waste gate is operated by a control in the cockpit.

Stages

A **stage** is an increase in pressure. Superchargers can be classified as single-stage, two-stage, or multistage, according to the number of times an increase of pressure is accomplished. The single-stage supercharging system may include a single- or two-speed internal supercharger. Even though the two-speed internal supercharger has two definite speeds at which the impeller can rotate, only one stage (boost in pressure) can be accomplished at any time; therefore, it is still a single-stage system. A single-stage internal supercharger system is shown in Fig. 5-14.

INTERNAL SINGLE-SPEED SUPERCHARGER

System for Six-Cylinder Opposed Engine

A type of supercharger still used on older opposed engines is an internal single-speed supercharger. It consists of a gear-driven impeller placed between the carburetor and the intake ports of the cylinders.

The exploded view in Fig. 5-16 shows the supercharger components. From the arrangement and design of the parts, we see that the fuel-air mixture from the carburetor is drawn into the center of the impeller; from there it is thrown outward (by centrifugal forces) through the diffuser and into the supercharger housing. From the housing it flows into the portion of the induction system surrounded by the oil sump and then into the individual intake pipes to each cylinder.

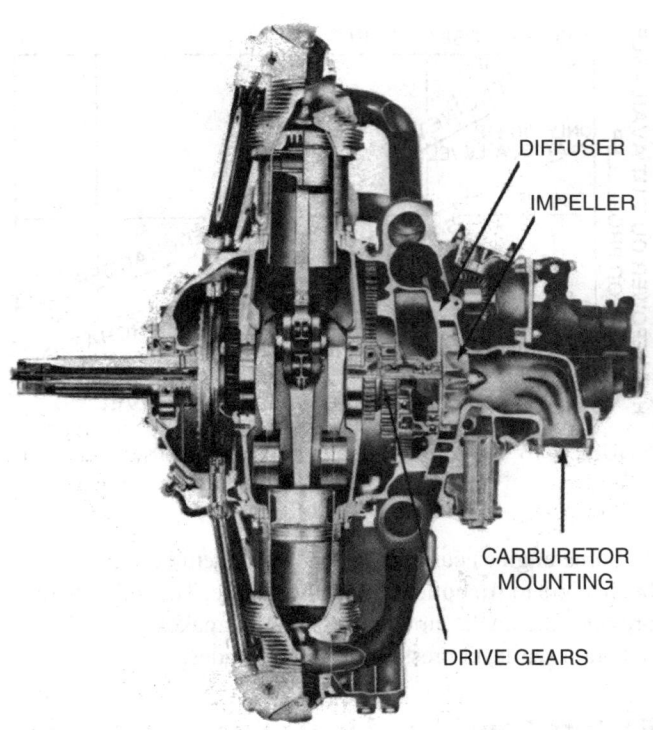

FIGURE 5-17 Single-speed internal supercharger for a radial engine.

Single-Speed Supercharger for a Radial Engine

In a typical **high-output radial engine**, an internal blower or a supercharger is located in the rear section of the engine. The fuel-air mixture passes from the carburetor through the supercharger or blower and then flows out through the diffuser section and individual intake pipes to the engine cylinders. In addition to providing some supercharging, the internal impeller (blower) helps to atomize and vaporize the fuel and to ensure an equal distribution of the fuel-air mixture to all cylinders.

The principal components of the supercharger section for the Pratt & Whitney R-985 engine are shown in Fig. 5-17.

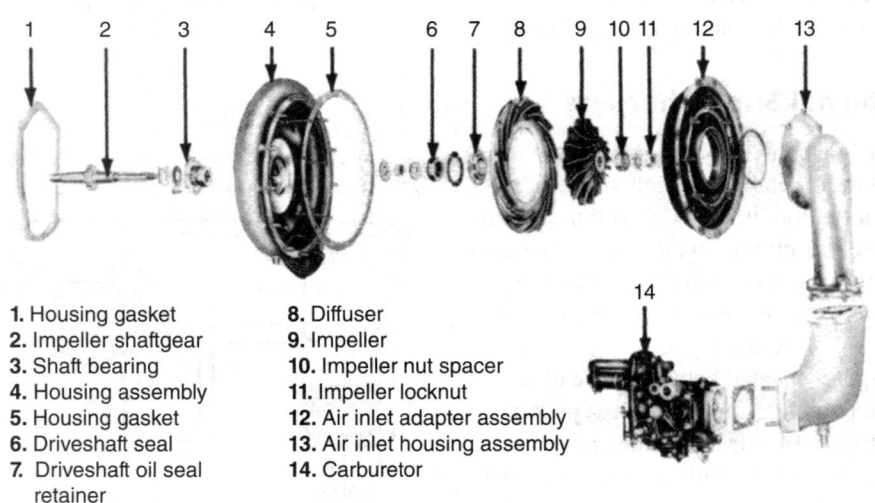

1. Housing gasket
2. Impeller shaftgear
3. Shaft bearing
4. Housing assembly
5. Housing gasket
6. Driveshaft seal
7. Driveshaft oil seal retainer
8. Diffuser
9. Impeller
10. Impeller nut spacer
11. Impeller locknut
12. Air inlet adapter assembly
13. Air inlet housing assembly
14. Carburetor

FIGURE 5-16 Exploded view of a supercharger.

The supercharger consists of an impeller mounted in the supercharger (blower) case of the engine, immediately forward of the rear case. The rear case contains the **diffuser vanes** which distribute the fuel-air mixture evenly to the nine intake pipes attached by seals and packing nuts to the openings around the outside of the case.

The impeller is driven by the engine at a speed of 10 times the crankshaft speed. This provides for a maximum MAP of 37.5 inHg [127.01 kPa] and a power output of 450 hp [335.57 kW].

Two-Speed Internal Supercharger

Some older radial-engine aircraft, were equipped with a two-speed internal supercharger which was designed to permit a certain degree of supercharging for altitude operation. A typical two-speed system is that installed on the Pratt & Whitney R-2800 Double Wasp series CB engines. The impeller for this supercharger is driven at a ratio of 7.29:1 by the low-ratio clutch and at 8.5:1 by the high-ratio clutch. In some models of the engine, the high gear ratio may be as much as 9.45:1.

THE TURBOCHARGER

A turbocharger is an externally driven device designed to be driven by a turbine wheel which receives its power from the engine exhaust. Ram air pressure or alternate air pressure is applied to the inlet side of the turbocompressor (blower), and the output is supplied to the inlet side of the carburetor or fuel injector. If a high degree of air compression is achieved by the compressor, it may be necessary to pass the compressed air through an intercooler to reduce the air temperature, as illustrated in Fig. 5-18. As mentioned previously, if carburetor air temperature (CAT) is too high, detonation will occur.

The exhaust gases are usually diverted from the main exhaust stack by means of a **waste gate**. The waste gate is

FIGURE 5-18   Arrangement of a turbosupercharger system.

closed to direct the exhaust gases through the turbine. The degree of closing determines the amount of air pressure boost obtained from the supercharger.

The general arrangement of a turbosupercharger system is shown in Fig. 5-18. A turbosupercharger system can be used to maintain a given MAP to the design altitude of the system. Above this altitude, called the *critical altitude,* MAP will begin to fall off as the altitude is increased. We may therefore define **critical altitude** as the altitude above which a particular engine-supercharger combination will no longer deliver full power.

Turbochargers and Normalizers for Light-Aircraft Engines

A turbocharger installation for a light-airplane engine is shown in Fig. 5-19. We see from the drawing that the exhaust pipes from each cylinder are all connected to one main exhaust stack. A waste gate is placed near the outlet of the stack to block the exit of the exhaust gases and to direct them through a duct into the turbine. The turbine drives the compressor, which increases the air pressure to the carburetor inlet.

This particular turbocharger is designed for use at altitudes above 5000 ft [1 524 m] because maximum engine power is available without supercharging below that altitude.

Description

The unit consists of a precision-balanced rotating shaft with a radial inflow turbine wheel on one end and a centrifugal compressor impeller on the other end, each with its own housing.

The principal components of the turbocharger for this system are illustrated in Fig. 5-20. The housings are machined castings, whereas the turbine and compressor impellers are formed by investment casting, machining, and grinding. The turbine, driven by engine exhaust gases, powers the impeller, which supplies air under pressure to the carburetor inlet. This higher pressure supplies more air by weight to the engine with the advantage of a proportionally higher power output and a minimum increase in weight.

A photograph of the turbocharger and its ducting to the carburetor air box is shown in Fig. 5-21. This photograph may be compared with the diagram in Fig. 5-22 to gain a clear understanding of the operating principles.

In the carburetor air box is a **swing check valve** which is open during **naturally aspirated operation**—that is, when the turbocharger is not in operation. The cheek-valve operation is automatic, and the valve will close when turbocharger boost pressure is greater than ram air pressure. When the waste gate in the exhaust stack outlet is closed or partially closed, exhaust gases are directed to the turbine. The rotation of the turbine causes the compressor to draw air from the air box and deliver it under pressure through the duct forward of the carburetor.

We see in Fig. 5-22 that carburetor heat may be obtained through the alternate air duct, regardless of whether the

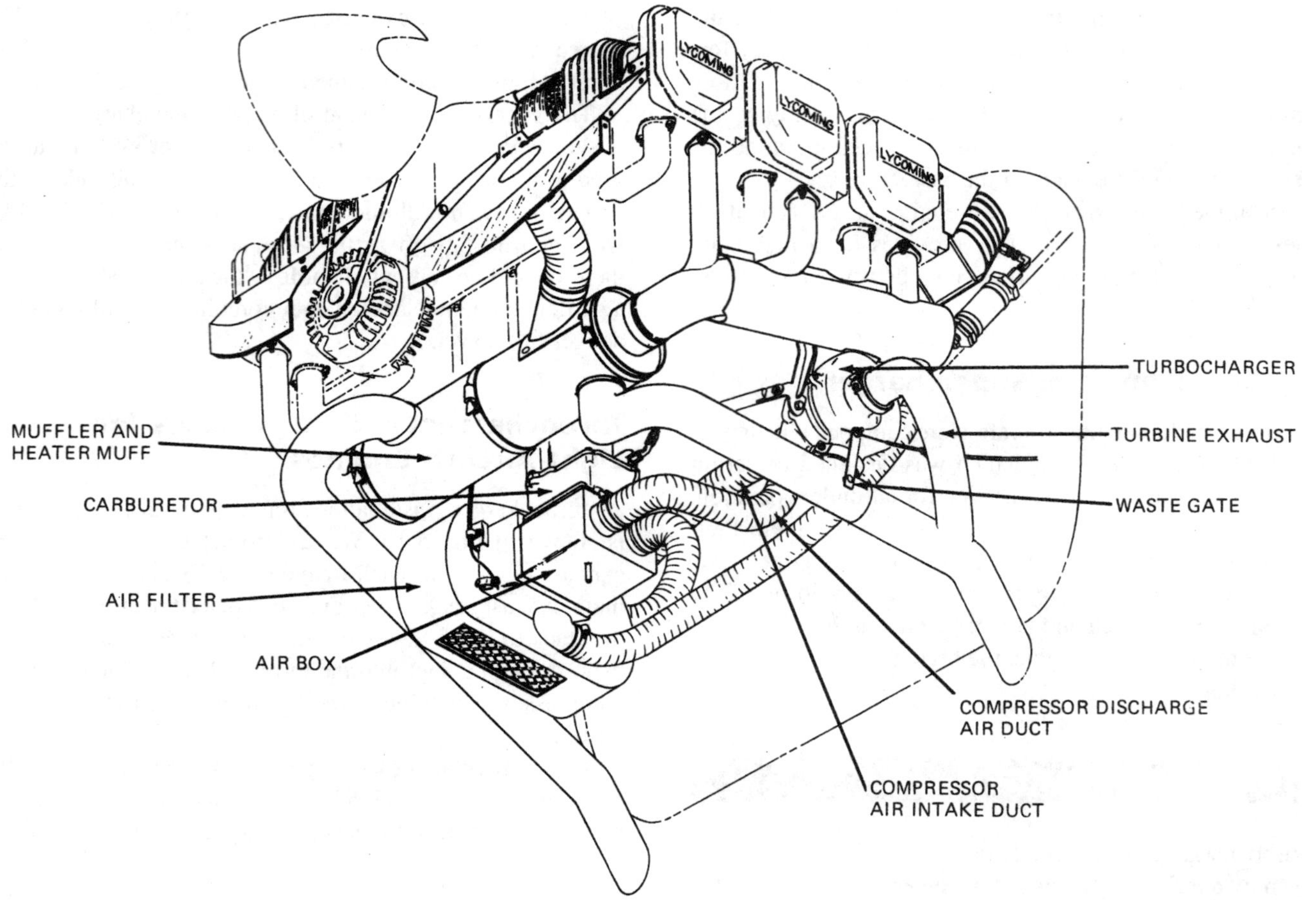

FIGURE 5-19   Installation of a light-aircraft turbonormalizing system.

MUFFLER AND
HEATER MUFF

CARBURETOR

AIR FILTER

AIR BOX

TURBOCHARGER

TURBINE EXHAUST

WASTE GATE

COMPRESSOR DISCHARGE
AIR DUCT

COMPRESSOR
AIR INTAKE DUCT

COMPRESSOR HOUSING

BEARING HOUSING

TURBINE HOUSING

BEARING

COMPRESSOR IMPELLER

TURBINE

FIGURE 5-20   Components of a turbocharger.

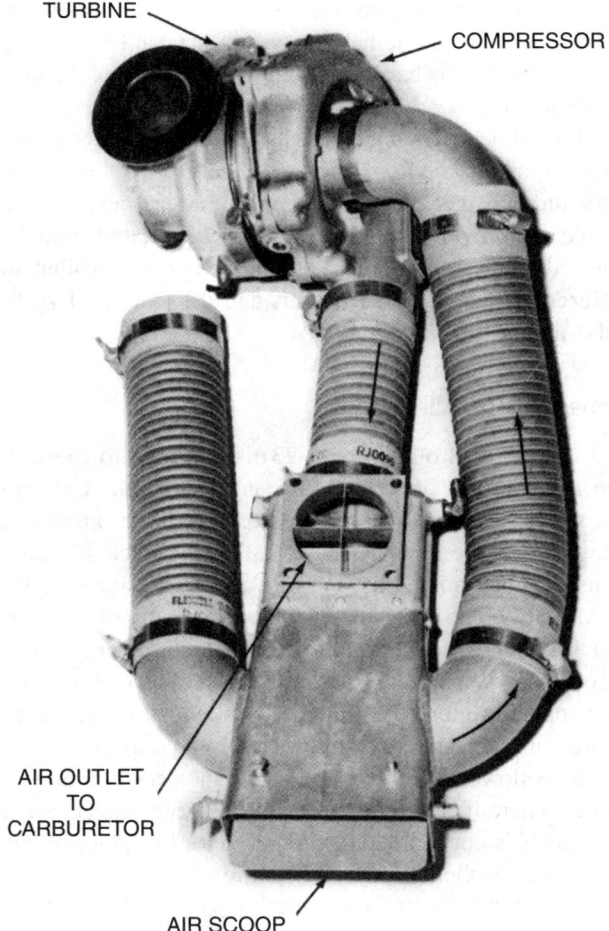

TURBINE

COMPRESSOR

AIR OUTLET
TO
CARBURETOR

AIR SCOOP

FIGURE 5-21 Turbocharger and ducting.

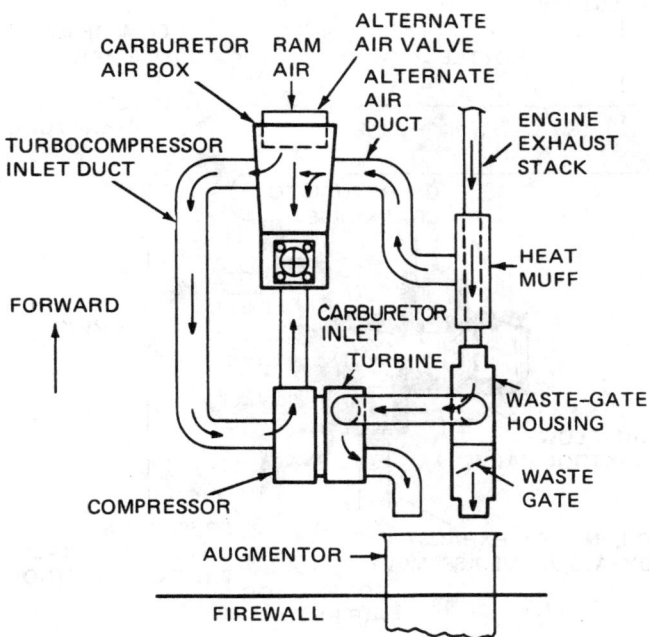

CARBURETOR
AIR BOX

RAM
AIR

ALTERNATE
AIR VALVE

ALTERNATE
AIR
DUCT

ENGINE
EXHAUST
STACK

TURBOCOMPRESSOR
INLET DUCT

HEAT
MUFF

FORWARD

CARBURETOR
INLET

TURBINE

WASTE–GATE
HOUSING

COMPRESSOR

WASTE
GATE

AUGMENTOR

FIREWALL

FIGURE 5-22 Turbocharger system.

supercharger is operating. If carburetor heat is desired, the alternate air valve is closed, shutting off ram air and allowing air to be drawn through the heater muff. With this installation, carburetor heat should not be used when the turbocharger boost is more than 5 inHg [16.94 kPa].

The bearings of the turbocharger are of the sleeve-journal type and utilize pressure lubrication from the engine. This type of bearing costs little and has a high degree of reliability.

The turbine and turbine housing are cast of high-temperature alloys, with the central main housing, compressor housing, and impeller cast of aluminum alloy for light weight and good thermal characteristics. The design and construction result in a unit which is completely air-cooled.

Lubrication System

Lubricant for the turbocharger is supplied by a line connected to a fitting on the engine-governor, fuel-pump, dual-drive pad. A fitting included in this lubricant supply line incorporates a pressure regulator poppet valve to reduce engine gallery oil pressure from the normal 60- to 80-psi [414- to 552-kPa] range to a pressure of 30 to 50 psi [207 to 345 kPa], which is required for the turbocharger. At this pressure range, between 1 and 2 qt/min [0.95 and 1.89 L/min] of lubricating oil will be supplied to the unit. This quantity of oil is but a small percentage of the total capacity of the engine oil pump. The oil supplied to the turbocharger is normally returned to the engine sump by way of the bypass pressure relief valve.

Incorporated in the turbocharger lubricant supply line is a pressure switch which activates a red warning light if turbocharger oil pressure goes below 27 to 30 psi [186 to 207 kPa]. If the oil pressure is lost, the pilot simply removes the turbocharger from service by opening the waste gate and returns the engine to naturally aspirated operation to save the turbocharger bearings. The turbocharger lubricating-oil sump is scavenged by means of the fuel-pump drive gears contained in the dual-drive unit for the pump.

Controls

As previously explained, the principal factor in turbocharger operation is the degree of waste-gate closure. This determines the amount of the total engine exhaust-gas flow through the turbine and the resulting level of boost. A separate push-pull control with a precise vernier adjustment is installed for actuation of the waste gate. This permits convenient, exact matching of MAPs for both engines on the airplane.

Operation

Since this particular turbocharger is designed for operation at altitudes above 4000 to 6000 ft [1219 to 1829 m], the ground operation of the engine is the same as that required for an unsupercharged engine. During climb, when the airplane reaches an altitude where power begins to fall off, the pilot places the throttle in the wide-open position and begins to close the waste gate with the separate control located on the fuel-valve selector console. The operation of the turbocharger makes it possible to maintain a sea-level MAP of 28 inHg [94.83 kPa] absolute during the climb of the aircraft to an altitude of 20 000 ft [6096 m].

Typical engine operating conditions for climb are as follows:

1. 2400 rpm maximum and 2200 rpm minimum with turbocharger operative.
2. 25- to 28-inHg [84.67- to 94.83-kPa] maximum MAP.
3. 400 to 475°F [204.44 to 246.11°C] maximum cylinder-head temperature with turbocharger operating. A cylinder-head temperature gauge is required.
4. Carburetor inlet air temperature 100 to 160°F [37.78 to 71.12°C], depending on the power setting and the temperature of outside air. A performance number for the fuel will have been specified to preclude any possibility of detonation.

After the airplane has attained the desired cruising altitude, power should be reduced to 23- to 25-inHg [77.9- to 84.67-kPa] MAP and the rpm to the 2200 to 2300 cruising range. The aircraft is then trimmed for cruising speed, and the fuel-air mixture is adjusted for best economy. When the engine is operating below 75 percent power, leaning can be accomplished by pulling the mixture control back slowly until there is a slight drop in MAP. The mixture control is then moved forward until smooth, steady engine operation is attained. Note that manual leaning must not be done when an engine is operating at more than 75 percent of power or when MAP exceeds a certain value specified in the operator's handbook.

When it is desired to let down and prepare for landing, the turbocharger should be shut off by opening the waste gate and the carburetor heat should be turned on. The throttle should not be closed suddenly, because rapid cooling of the cylinder heads may cause cracking or other damage. During letdown, the throttle should not be entirely closed and should occasionally be opened sufficiently to clear the engine.

Textron Lycoming Automatic Turbosupercharger

In this section we describe the operation of a typical automatic turbocharger manufactured by the AiResearch Manufacturing Co., a division of the Allied-Signal Aerospace Co., and used on the Textron Lycoming TIO-540-A2A engine which powers the Piper Navajo airplane. This is a **sea-level boosted engine**; that is, the turbocharger can operate at sea level and above to produce the selected power from the engine until an altitude of approximately 19 000 ft [5791.20 m] is attained. Thus, the engine can produce sea-level power up to the critical altitude of approximately 19 000 ft [5791.20 m].

At sea level, without turbocharging, the engine named above can produce 290 hp [216.25 kW]. With turbocharging, the same engine will have MAP boosted by 10 inHg [33.87 kPa], and the horsepower will be 310 hp [231.17 kW]. In addition to the increased horsepower at sea level, the engine can develop rated power up to approximately 19 000-ft [5791.20-m] altitude as stated above. At this altitude, a normally or naturally aspirated engine (one without supercharging) can develop only about one-half the power that can be developed at sea level. This is because of the lower density of the air at higher altitudes.

The turbocharger under discussion utilizes a centrifugal compressor attached to a common shaft with the turbine. This turbocharger has three control components, which enable it to provide automatically the power that the pilot has selected for operation. These are the **density controller**, the **differential-pressure controller**, and the **exhaust bypass-valve assembly** (waste gate).

Density Controller

The density controller (Fig. 5-23) is designed to sense the *density* of the air after it has passed through the compressor, that is, between the compressor and the throttle valve. The air pressure in this area is called the **deck pressure**. Density is determined by pressure and temperature, so that the density controller is equipped with a bellows assembly which contains dry nitrogen for temperature sensitivity, a valve, springs, and a housing with oil inlet and outlet ports. As temperature increases, the bellows tends to expand; and as pressure increases, the bellows tends to contract.

The bellows assembly extends into the deck pressure air-stream where it is exposed to the air entering the engine. If the density is not equal to that required for full-power operation, the controller will call for additional supercharging.

The bellows controls a metering valve which permits oil to escape back into the engine or restrains the oil and causes a pressure buildup to the exhaust bypass valve. When oil pressure builds up, the bypass valve closes and directs exhaust gases to the turbine, thus increasing compressor rpm,

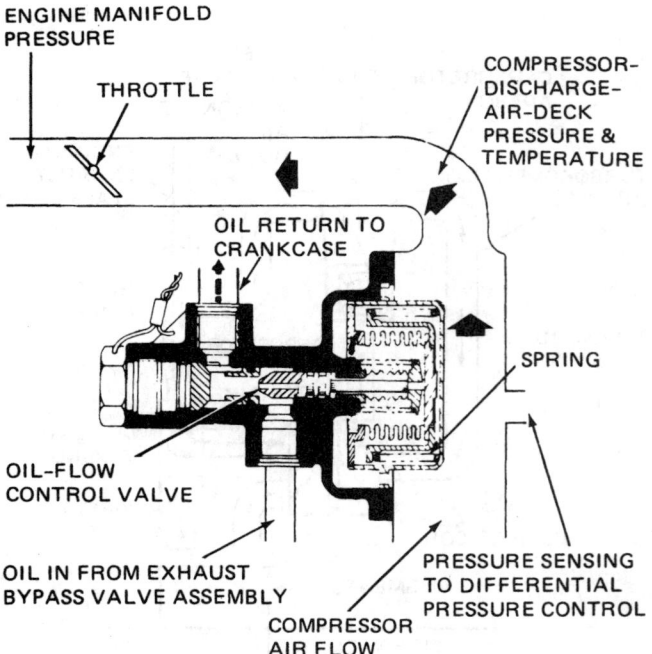

FIGURE 5-23 Density controller. (*Textron Lycoming.*)

which in turn increases compression of the incoming air. Pressurized oil in the exhaust bypass-valve actuator causes the butterfly valve to close an amount proportional to the oil pressure.

The density controller operates to restrain oil only during full-power operation. At partial-throttle settings, the control of the turbocharger is assumed by the differential-pressure controller.

Remember that the density controller is designed to maintain deck pressure and density at levels required for full-power operation of the engine. If the power selected is less than full power, the deck pressure will be less than that required for full power and the controller will not permit any oil to bypass back to the engine. The control of the oil pressure to the exhaust bypass valve will therefore be accomplished by the differential-pressure controller.

Differential-Pressure Controller

The differential-pressure controller is illustrated in Fig. 5-24. Note that the unit incorporates a diaphragm which is exposed on one side to deck pressure and on the other side to inlet MAP. The diaphragm is connected to a valve and is subjected to spring pressure, which holds the valve closed until the differential pressure across the diaphragm is sufficient to open the valve. The unit is set to provide approximately 2- to 4-inHg [6.77- to 13.55-kPa] pressure drop across the throttle at a specified MAP; however, the usual differential is approximately 2 inHg [6.77 kPa]. If the required pressure differential is exceeded, the oil valve will open sufficiently to reduce supercharging until the desired pressure differential is restored.

Exhaust Bypass-Valve Assembly

The exhaust bypass-valve assembly, often referred to as a waste gate, is illustrated in Fig. 5-25. When this valve is open to its maximum design limit, almost all the exhaust

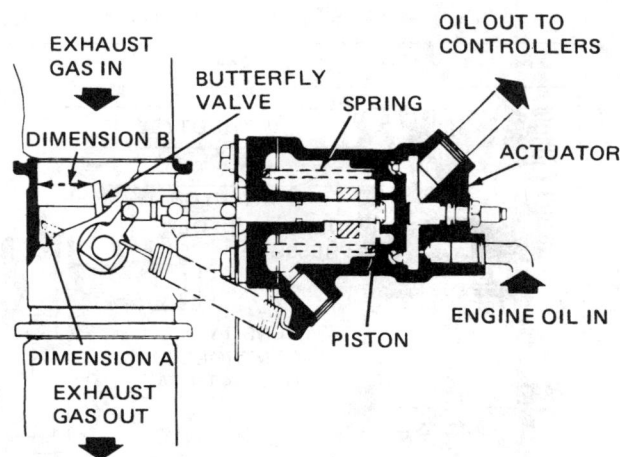

FIGURE 5-25   Exhaust bypass-valve assembly. (*Textron Lycoming.*)

gas is dumped overboard and is not directed to the turbine. Therefore, minimum supercharging is applied. When the valve is in the fully closed position, there is still a clearance of 0.023 to 0.037 in [0.58 to 0.94 mm], shown as dimension A in the drawing. By keeping the butterfly from completely closing and touching the walls of the valve housing, the danger of the valve's sticking is eliminated. Furthermore, the possibility of turbocharger overspeeding is reduced when all the exhaust gas is supplied to the turbine at high altitude. When the butterfly is fully open, there is a clearance of 0.73 to 0.75 in [18.54 to 19.05 mm] between the edge of the butterfly and the wall of the valve housing. This is shown as dimension B in the drawing. Remember that the dimensions given here are for one model of valve only; others will have different dimensions. The manufacturer's manual should be consulted to establish the correct dimension when the valve is overhauled or adjusted.

The butterfly moves in proportion to the amount of oil pressure built up by the controller. Therefore, the butterfly can be at any position between fully open and fully closed. Thus, the engine can deliver constant power for any power setting selected by the pilot as long as the aircraft is below the critical altitude for the selected power setting. The critical altitude is the point at which the butterfly is fully closed and the engine is delivering full-rated power. At any altitude above the critical altitude, the engine cannot deliver full rated power because the air density decreases and the amount of air entering the engine is not sufficient to produce full power. During operations below the critical altitude, such as cruise conditions and power-on letdown, the butterfly will be positioned somewhere between fully open and fully closed.

Operation of the Turbocharger Control System

A diagram of the automatic turbocharger control system under discussion is given in Fig. 5-26. Note that air enters the turbocharger compressor through the air filter and inlet duct. From the compressor the air passes the density controller (where

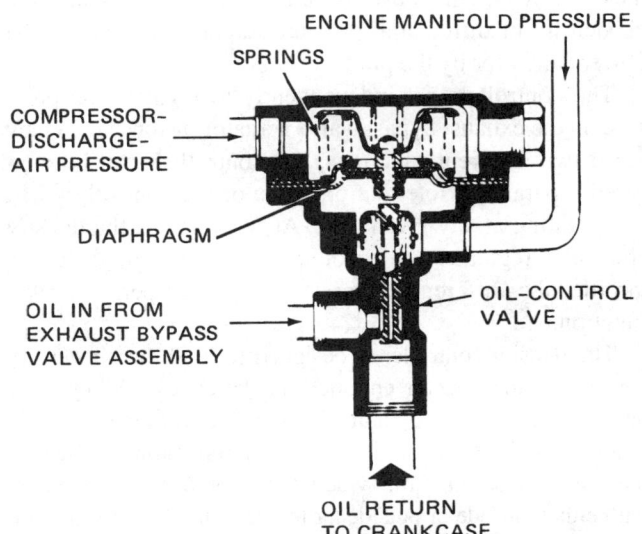

FIGURE 5-24   Differential-pressure controller. (*Textron Lycoming.*)

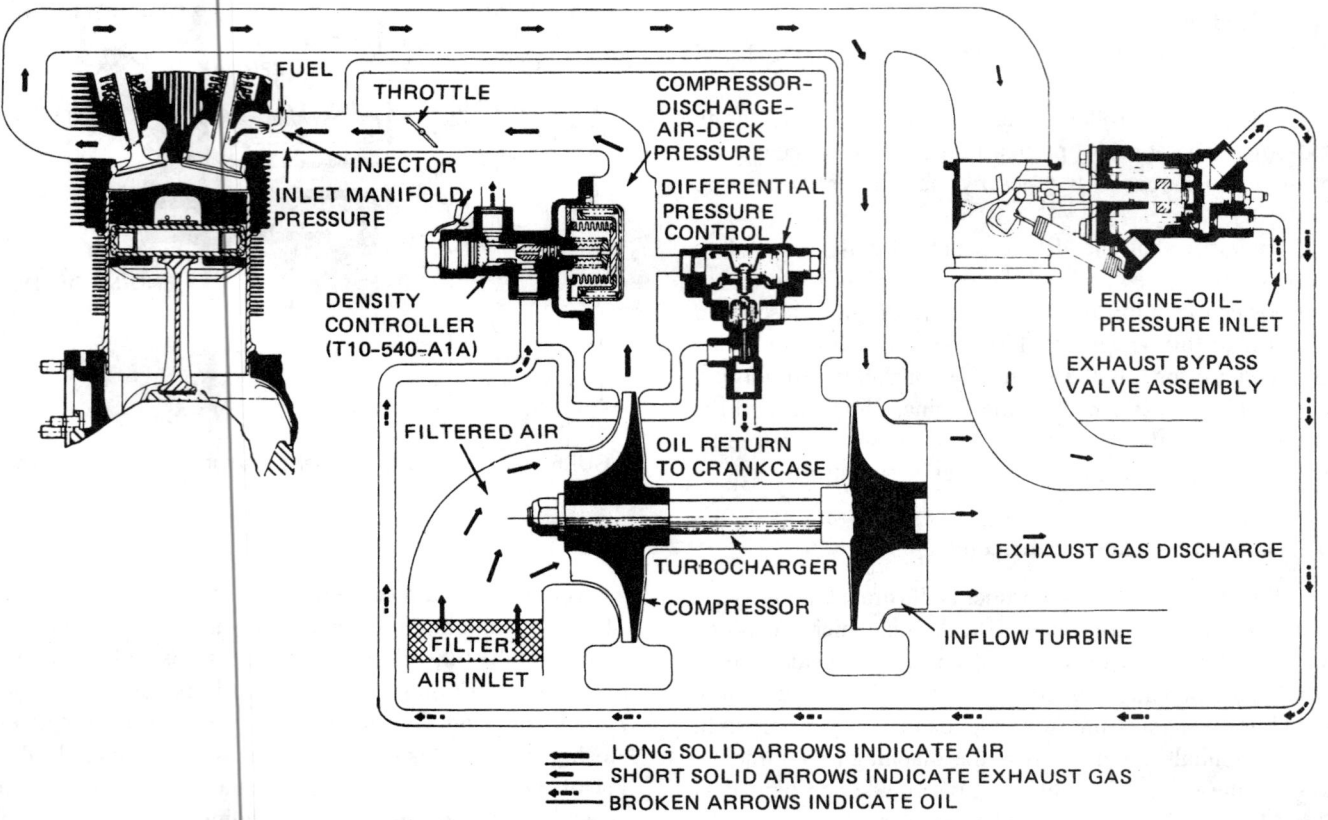

EXHAUST MANIFOLD

LONG SOLID ARROWS INDICATE AIR
SHORT SOLID ARROWS INDICATE EXHAUST GAS
BROKEN ARROWS INDICATE OIL

FIGURE 5-26 Schematic diagram of automatic turbocharger control system. (*Textron Lycoming.*)

the density is sensed) and continues through the throttle valve to the intake valve of the engine. From the exhaust valve, the exhaust gases enter the exhaust manifold and flow to either the exhaust bypass valve or the turbine, depending on the position of the butterfly valve. If the system is calling for maximum supercharging, the major portion of the exhaust gases must pass through the turbine because the butterfly valve is closed. If no supercharging is required, the butterfly valve will be open and the gases will pass out through the exhaust stacks.

Remember, the position of the butterfly valve of the exhaust bypass depends on the oil pressure in the actuator. Engine oil pressure is fed to the area above the piston in the actuator, and an oil outlet is connected to the same area. If oil can flow freely from the cylinder through the return line, the piston will not be moved and the butterfly valve will remain open. The oil flow out of the oil pressure line is controlled by the density controller and the differential-pressure control. If one or the other of these units permits the oil to return to the crankcase freely, pressure will not build up in the exhaust bypass actuating cylinder and the butterfly valve will remain open. If one of the controlling units is restraining the oil return, the bypass valve in the other unit will be completely closed and pressure will be built up in the actuating cylinder of the exhaust bypass valve. The piston will then move to close the butterfly valve, and exhaust gases will be directed to the turbine. This will increase the rpm of the turbine and compressor, and the incoming air will be supercharged or pressurized.

Since the exhaust bypass-valve assembly contains O-rings, seals, springs, and a piston, there is a certain amount of friction, which may result in slightly different valve positions for the same oil pressure, depending on whether the pressure is increasing or decreasing at the time the setting is made. Engine oil under pressure is fed through an orifice to the top of the piston and is bled back to the engine through the controllers, as previously explained. In all cases, the controllers will adjust the butterfly valve to establish the air density or differential pressure required to produce the power called for by the pilot.

The controllers act independently to regulate the pressure on the exhaust bypass-valve actuator piston. Since they serve two different functions, the controllers can be analyzed separately. Note that only one of the controllers will be operating at any given time. At full throttle, the density controller regulates the amount of supercharging; at part throttle, the differential-pressure controller performs this function.

The density controller is designed to hold the air density constant at the injector entrance (intake area in the cylinder head). As the air temperature is increased, a higher MAP is required, and this results in a greater temperature rise across the compressor. In turn, wide-open-throttle MAP increases with either altitude or outside air temperature. In a full-throttle climb, the gain is 3- to 4-inHg [10.16- to 13.55-kPa] MAP between sea level and critical altitude.

As deck pressure and temperature change, the bellows in the density controller is either expanded or contracted, which in turn repositions the oil-bleed valve, changes the amount of bleed oil, establishes a new oil pressure on the exhaust bypass-valve actuator piston, and repositions the butterfly valve. Controllers, engines, and turbochargers have individual differences; and with the length of time required to establish temperature equilibriums, two powerplant packages will show somewhat different MAPs.

The density controller can be adjusted, but it must be done under controlled conditions with ample stabilization time. Since this unit regulates wide-open-throttle power and MAP only, adjustments should not be made to correct any part-throttle discrepancies. The controller should be adjusted by authorized personnel and must be adjusted to the curve found in the engine operation manual.

If the differential-pressure controller were not used, the density controller would attempt to position the exhaust bypass valve so that the air density at the injector entrance would always be that which is required for maximum power. Since this high air density is not required during part-throttle operation, the differential-pressure controller reduces the air pressure to the correct level and causes the exhaust bypass valve to modulate over as high an operating range as possible.

As explained previously, the differential-pressure controller contains a diaphragm connected to a bleed valve in the oil line between the exhaust bypass valve and the engine crankcase. One side of the diaphragm senses air pressure before the throttle valve, and the other side senses air pressure after the throttle valve. During operation at wide-open throttle, the air pressure drop across the throttle (and across the diaphragm) is at a minimum, and the spring in the differential-pressure controller holds the oil-bleed valve in the closed position. Therefore, at wide-open throttle, the differential-pressure controller is not working, and the density controller determines the position of the exhaust bypass valve.

As the throttle is partially closed, the air pressure drop across the throttle valve and across the diaphragm is increased. This pressure differential on the diaphragm opens the bleed valve and allows oil to bleed back to the crankcase. This establishes a different oil pressure on the exhaust bypass-valve actuator piston and changes the position of the exhaust bypass valve. This cycle repeats itself until equilibrium is established.

The differential-pressure controller regulates the amount of supercharging to a point just slightly higher than that required for part-throttle operation and therefore reduces the time required for the system to seek and find equilibrium. The unit performs a necessary function, but it has a side effect that must be understood. When the pilot changes the position of the throttle, a long chain of events is triggered. Since heat transfers and turbocharger inertia are involved, many cycles may be required to reach a new equilibrium.

The sequence of events which occur as the result of a change in throttle position are as follows:

1. The pilot moves the throttle and establishes a different pressure drop across the throttle valve.

2. The differential-pressure controller diaphragm senses the change and repositions the bleed valve.

3. The new bleed-valve setting changes the oil flow which establishes a new pressure on the exhaust bypass-valve actuating cylinder.

4. The changed pressure on the actuating cylinder piston causes the piston to reposition the butterfly valve.

5. The new butterfly-valve position changes the amount of exhaust gas flow to the turbine.

6. This changes the amount of supercharging, which, in turn, changes the air pressure at the injector entrance.

7. The new pressure then changes the pressure drop across the throttle valve, and the sequence returns to step 2 and repeats until equilibrium is established.

The net result of these events is an effect called **throttle sensitivity**. When this operation is compared with that of naturally aspirated engines, the supercharged engine's MAP setting will require frequent resetting if the pilot does not move the throttle controls slowly and wait for the system to seek its stabilization point before making corrective throttle settings. Variations of temperature can also cause the turbocharger system to fluctuate. For instance, the bellows in the density controller responds relatively slowly to temperature changes. Changes in oil temperature have some effect on the system, too, because of corresponding changes in oil viscosity. A "thin" oil requires greater restriction than a "thick" oil to maintain the same back pressure on the exhaust bypass-valve actuating piston.

During acceleration, the supercharging system discussed in this section is noticeably subject to variations because all the components of the control system are continually resetting themselves and little time is available for stabilization. An **overboost** condition occurs when MAP exceeds the limits at which the engine was tested and certified by the FAA. This can be detrimental to the life and performance of the engine. Overboost can be caused by malfunctioning controllers or by an improperly operating exhaust bypass valve. The control system on the engine is designed to prevent overboost, but owing to the many components and the response time involved, another related condition known as **overshoot** must be considered. Overshoot occurs when automatic controls not having the ability to respond quickly enough to check the inertia of the turbocharger speed increase with rapid engine throttle advance.

Overshoot differs from overboost in that the high MAPs last for only a few seconds. This condition can usually be overcome by smooth throttle operation. With unstabilized temperatures and during transient conditions, a certain amount of overshoot must be expected. But overshoot can be held to an acceptable level by avoiding a too violent opening of the throttles and by changing the oil and filter at regular intervals specified by the manufacturer. It may also be necessary to disassemble and clean certain assemblies in accordance with manufacturers' instructions.

Variable-Pressure Controllers

Another turbocharging system utilized on some Lycoming engines such as the TSIO-541 utilizes a variable-pressure controller in place of the density and differential-pressure controller. A variable-pressure controller is used to react to the pressure of the air between the compressor and the throttle valve (deck pressure). The controller illustrated in Fig. 5-27 is equipped with a bellows assembly, springs, a poppet (servo) valve, a cam follower, a housing with oil inlet and outlet ports, and a cam which is connected to the throttle plate with a cable.

The bellows assembly extends into the deck pressure airstream. Variation in air pressure (and therefore, variation in the configuration of the bellows assembly) causes the poppet valve to move. The poppet valve, which serves as a metering jet, permits oil to escape into the engine oil sump; or it restrains the oil which causes a pressure buildup in the oil line to the exhaust bypass-valve actuator. Pressurized oil in the actuator causes the butterfly valve to close in proportion to the pressure. The controller cam is linked to the throttle valve. As the throttle valve is moved, the cam moves, causing a resultant movement of the cam follower. A spring is compressed between the cam follower and the poppet. As the cam follower is moved, the spring modifies the motion of the poppet, thereby affecting the amount of oil pressure in the controller and actuator.

Some controllers are equipped with two adjusting screws. The low-setting adjustment is a screw which prohibits cam follower travel away from the poppet. The high-setting adjustment is a screw on the movable cam arm.

Turbosupercharging Systems for Continental Engines

The turbocharging system for some Continental engines utilizes engine oil pressure which flows through the waste-gate actuator and then through the controllers. The system's controllers are the **absolute-pressure controller**, the **rate-of-change controller**, and the **pressure-ratio controller**. All the controllers either restrict or bypass oil from the waste-gate actuator. Each controller is vented to upper deck pressure which is used to control oil flow through each controller. This is illustrated in Fig. 5-28.

The absolute-pressure controller will unseat and bypass oil at a predetermined upper deck pressure. The rate-of-change controller will unseat when the upper deck pressure increases at too rapid a rate (approximately 6.5 inHg/s). At high altitudes, the turbocharger must work much harder to provide the amount of intake air needed to maintain a set MAP. As the turbocharger compresses the air, it will gain heat which could become excessive enough to cause detonation. At about 16 000 ft [4876.8 m], the ratio controller will unseat, decreasing the amount of turbo boost available, which will reduce the tendency toward detonation. The pressure-ratio controller will limit the upper deck pressure to 2.2 times the ambient air pressure in the nacelle. An intercooler (an air-to-air heat exchanger) can help reduce this problem,

but the extra weight of such a device sometimes reduces the benefits received.

The operation of the system begins with the waste-gate actuator, which is spring-loaded to the open position. When the engine is started, oil pressure is fed into the waste-gate actuator through the capillary tube and on through the lines to all the controllers. As the controllers are seated, blocking the flow of oil, the oil pressure builds up a force sufficient to close the waste gate. When the waste gate closes, exhaust gases are routed through the turbocharger turbine, which causes the compressor to speed up and increase upper deck pressure. Upper deck pressure will rise until the absolute-pressure controller aneroid bellows causes the controller to unseat, decreasing oil pressure.

When the oil pressure in the waste-gate actuator is lowered sufficiently, the waste gate will begin to open, bypassing exhaust gases and allowing the turbine to slow down. This will slow the compressor and reduce upper deck pressure to a preset value. If the compressor discharge pressure (upper deck pressure) increases more rapidly than 6.5 inHg/s, the rate-of-change controller will unseat, decreasing the oil pressure to the waste-gate actuator. This prevents the turbocharger from increasing the upper deck pressure too quickly, which prevents engine overboost. This is most important during rapid opening of the throttle.

The pressure-ratio controller, which senses both nacelle ambient pressure and compressor discharge pressure, begins the function of controlling turbocharger discharge pressure at a ratio of 2.2 times the ambient pressure. This generally occurs at an altitude of about 16 000 ft [4876.8 m].

Some of the turbocharging systems use a variable-orifice fuel pump. Its purpose is to reduce fuel flow during rapid opening of the throttle. With the engine at idle, MAP is about 14 inHg [47.39 kPa] and upper deck pressure is ambient. If the throttle is opened rapidly, the turbocharger and compressor will lag behind, allowing the upper deck pressure to drop. This condition can cause an overly rich mixture; therefore, during this time the variable-orifice fuel pump will bypass fuel until the turbocharger raises the upper deck pressure sufficiently to provide the correct fuel-air ratio. This is shown in Fig. 5-29.

A variation of this Continental turbocharging system uses a variable absolute-pressure controller which combines the functions of the controllers previously discussed into one unit. A variable absolute-pressure controller is illustrated in Fig. 5-30. This controller's operation is very similar to that of an absolute-pressure controller except for the movable seat. The movable seat can vary the amount of oil pressure at the waste gate, which can vary the upper deck pressure. As the cam releases the pressure against the seat, it will decrease the upper deck pressure needed for the controller to unseat and reduce the pressure on the waste-gate actuator. This action will decrease the turbocharger compressor output. The position of the cam can be adjusted by a linkage connected to the throttle. As the throttle is opened, more boost will be needed. The cam presses on the seat, increasing the amount of boost needed to unseat the controller, thereby increasing the turbocharger output.

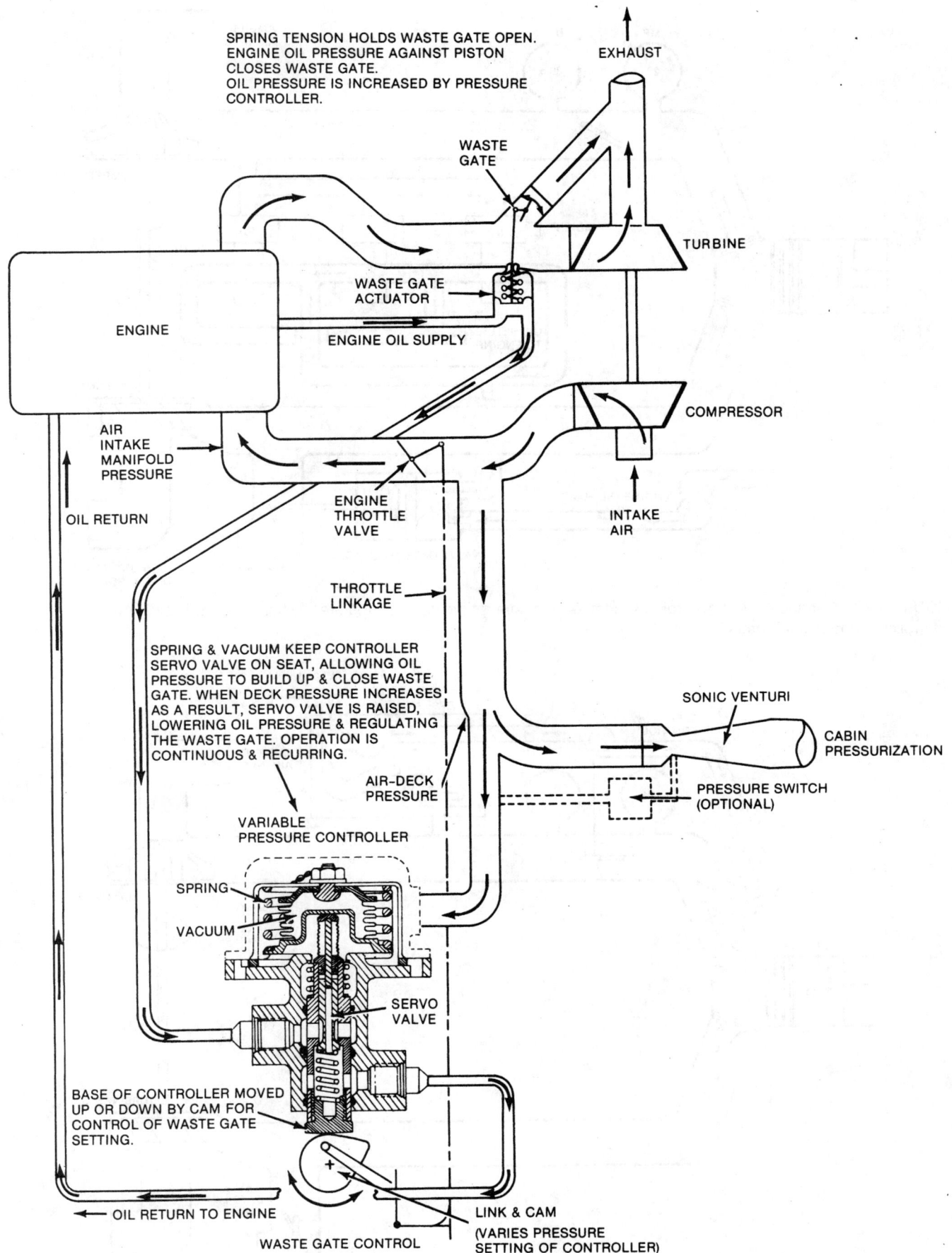

SPRING TENSION HOLDS WASTE GATE OPEN.
ENGINE OIL PRESSURE AGAINST PISTON
CLOSES WASTE GATE.
OIL PRESSURE IS INCREASED BY PRESSURE
CONTROLLER.

EXHAUST

WASTE
GATE

TURBINE

WASTE GATE
ACTUATOR

ENGINE

ENGINE OIL SUPPLY

COMPRESSOR

AIR
INTAKE
MANIFOLD
PRESSURE

OIL RETURN

ENGINE
THROTTLE
VALVE

INTAKE
AIR

THROTTLE
LINKAGE

SPRING & VACUUM KEEP CONTROLLER
SERVO VALVE ON SEAT, ALLOWING OIL
PRESSURE TO BUILD UP & CLOSE WASTE
GATE. WHEN DECK PRESSURE INCREASES
AS A RESULT, SERVO VALVE IS RAISED,
LOWERING OIL PRESSURE & REGULATING
THE WASTE GATE. OPERATION IS
CONTINUOUS & RECURRING.

SONIC VENTURI

CABIN
PRESSURIZATION

AIR-DECK
PRESSURE

PRESSURE SWITCH
(OPTIONAL)

VARIABLE
PRESSURE CONTROLLER

SPRING

VACUUM

SERVO
VALVE

BASE OF CONTROLLER MOVED
UP OR DOWN BY CAM FOR
CONTROL OF WASTE GATE
SETTING.

OIL RETURN TO ENGINE

LINK & CAM
(VARIES PRESSURE
SETTING OF CONTROLLER)

WASTE GATE CONTROL

FIGURE 5-27   Variable-pressure controller. (*Textron Lycoming.*)

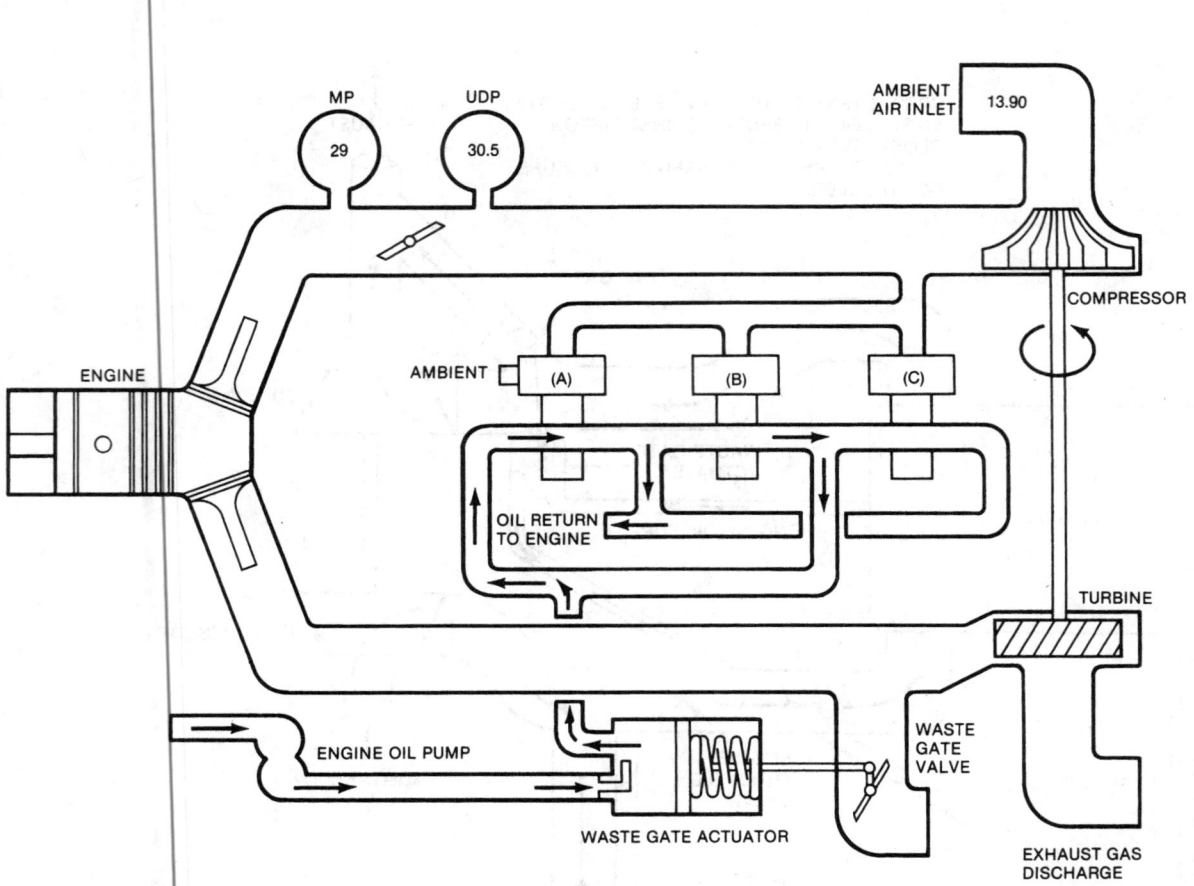

FIGURE 5-28 Continental turbocharging system. (A) Pressure-ratio controller. (B) Rate-of-change controller. (C) Absolute-pressure controller. (*Continental Motors.*)

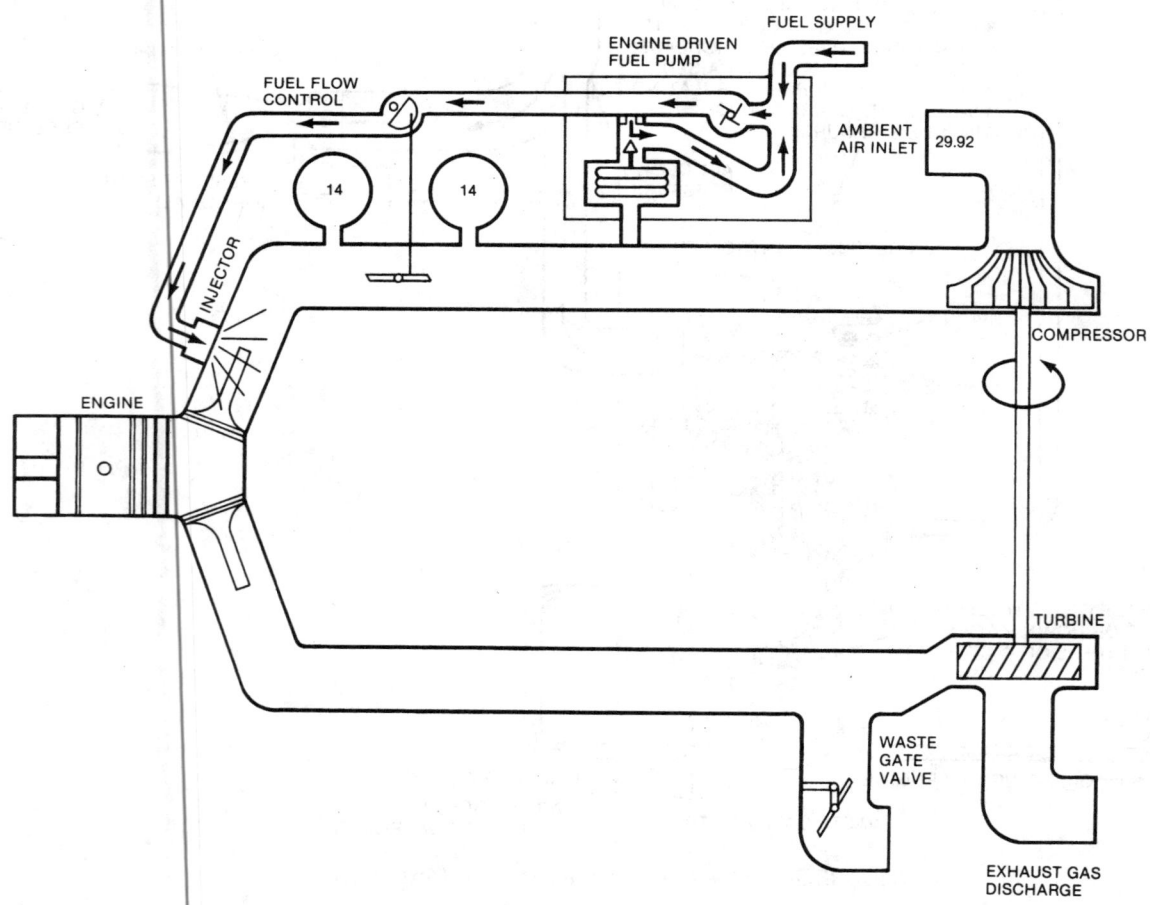

FIGURE 5-29 Variable orifice open to bypass excess fuel. (*Continental Motors.*)

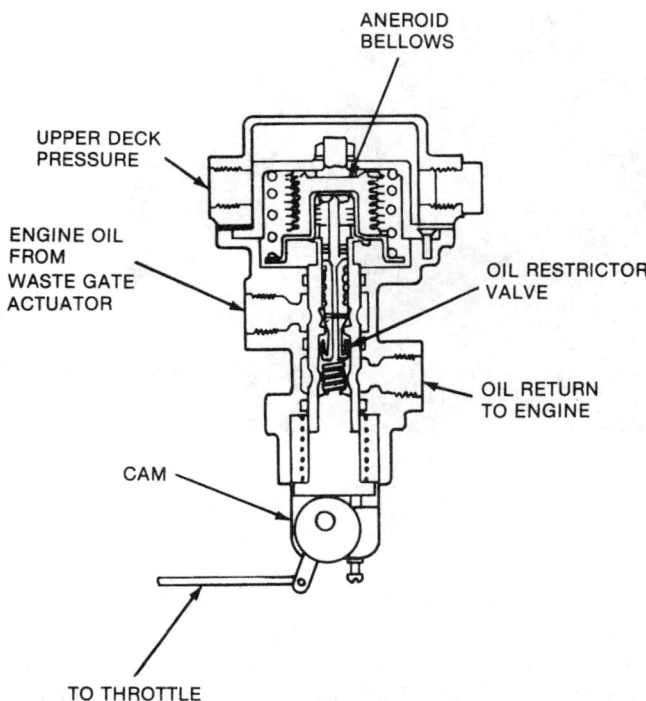

FIGURE 5-30   Variable absolute-pressure controller.
(*Continental Motors.*)

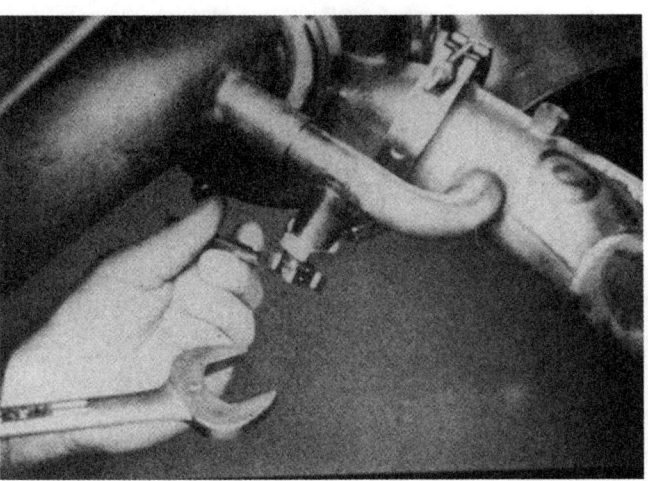

FIGURE 5-32   Waste gate with adjustable screw. (*Continental Motors.*)

Usually used in conjunction with the variable absolute-pressure controller is an overboost relief valve or MAP relief valve which is a spring-loaded pop-off valve that operates as a safety valve. In Fig. 5-31 the cutaway view of the overboost relief valve shows a bellows assembly that is evacuated to absolute pressure and compensates for any altitude changes. A spring and the bellows hold the plate and seat against the

mounting flange to provide a seal. If an overboost condition occurs, the valve will open, relieving the excess upper deck pressure until the deck pressure is within normal limits. The valve is located in the upper deck pressure manifold close to the compressor outlet.

Another Continental engine turbocharger system is installed on the TSIO-360 engine. This engine incorporates a waste-gate system that consists of an adjustable screw rather than a gate-type valve. As shown in Fig. 5-32, adjustment can be made by rotating the adjustment screw. Turning the screw into the pipe causes more exhaust gases to be sent to the turbine section, increasing the amount of engine boost. This engine system also uses an overboost relief valve, and its operation is very similar to that of the overboost relief valve mentioned earlier.

RECIPROCATING-ENGINE COOLING SYSTEMS

Excessive heat is undesirable in any internal-combustion engine for three principal reasons: (1) it adversely affects the behavior of the combustion of the fuel-air charge, (2) it weakens and shortens the life of the engine parts, and (3) it impairs lubrication.

If the temperature inside the engine cylinder is too great, the fuel mixture becomes preheated and combustion occurs before the proper time. Premature combustion causes detonation, "knocking," and other undesirable conditions. It also aggravates the overheated condition and is likely to result in failure of pistons and valves.

The strengths of many engine parts depend on their heat treatment. Excessive heat weakens such parts and shortens their lives. Also, the parts may become elongated, warped, or expanded to the extent that they freeze or lock together and stop the operation of the engine.

Excessive heat "cracks" the lubricating oil, lowers its viscosity, and destroys its lubricating properties.

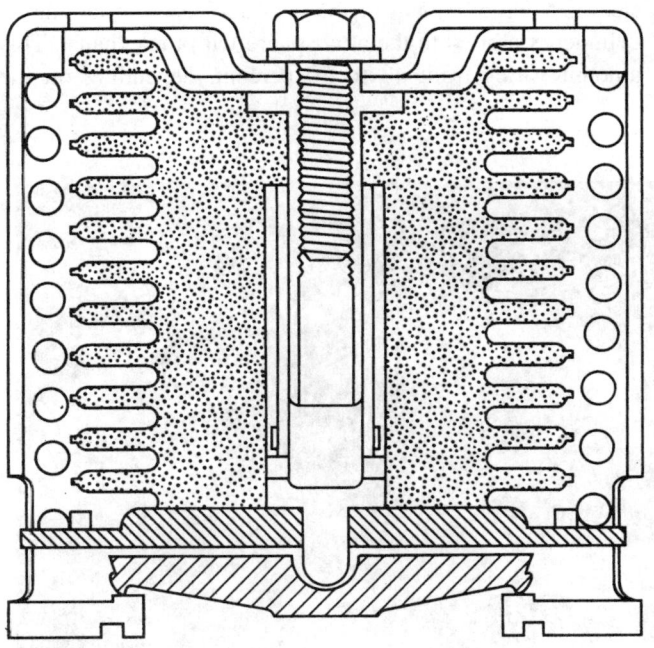

FIGURE 5-31   MAP relief valve (cutaway view).
(*Continental Motors.*)

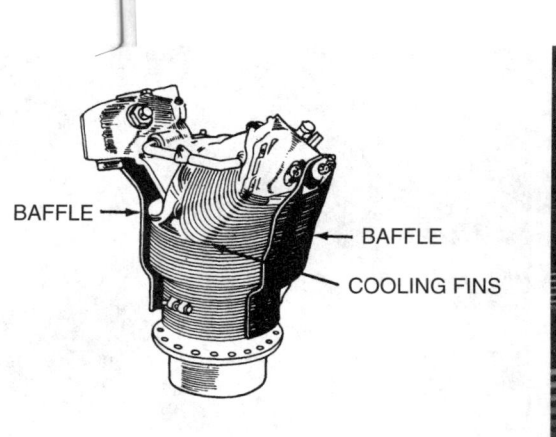

FIGURE 5-33 Cylinder with pressure baffles for cooling.

Air Cooling

In an air-cooled engine, thin metal fins project from the outer surfaces of the walls and heads of the engine cylinders, as illustrated in Fig. 5-33. When air flows over the fins, it absorbs the excess heat from the cylinders and carries it into the atmosphere. Deflector **baffles** fastened around the cylinders direct the flow of air to obtain the maximum cooling effect. The baffles are usually made of aluminum sheet. They are called **pressure baffles** because they direct airflow caused by ram air pressure. A cylinder with baffles is shown in Fig. 5-33. The operating temperature of the engine can be controlled by movable **cowl flaps** located on the engine cowling. On some airplanes, these cowl flaps are manually operated by means of a switch that controls an electric actuating motor. On other airplanes they can be operated either manually or by means of a thermostatically controlled actuator. Cowl flaps are illustrated in Fig. 5-34.

In the assembly of the engine baffling system, great care must be taken to ensure that the pressure baffles around the cylinders are properly located and secured. An improperly installed or loose baffle can cause a hot spot to develop, with

the result that the engine may fail. The proper installation of baffles around the cylinders of a twin-row radial engine is illustrated in Fig. 5-35. Baffling for an opposed-type engine is shown in Fig. 5-36. The baffling maintains a high-velocity airstream close to the cylinder and through the cooling fins. The baffles are attached by means of screws, bolts, spring hooks, or special fasteners.

Cylinder cooling is accomplished by carrying the heat from the insides of all the cylinders to the air outside the cylinders. Heat passes by conduction through the metal walls and fins of the cylinder assembly to the cooling airstream which is forced into contact with the fins by the baffles and cowling. The fins on the cylinder head are made of the same material as the head and are forged or cast as part of the head. Fins on the steel cylinder barrel air of the same metal as the barrel in most instances. In some cases, the inner part of the cylinder is a steel sleeve, and the cooling fins are made as parts of a muff or sleeve shrunk on the outside of the inner sleeve. A large amount of the heat developed in an engine cylinder is carried to the atmosphere with the exhaust. This amount varies from 40 to 50 percent, depending on the

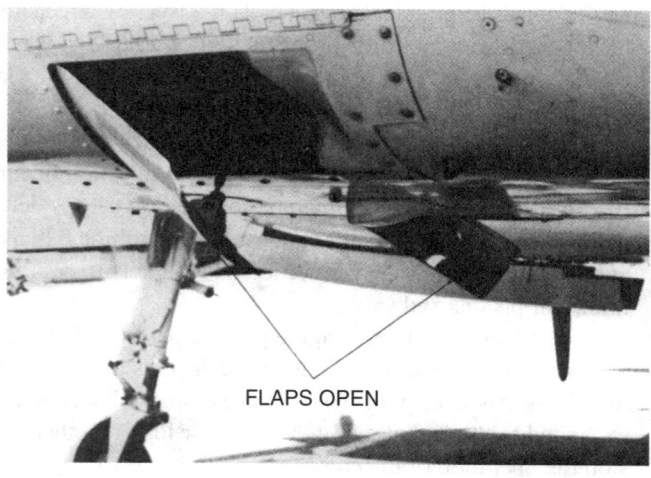

FIGURE 5-34 Cowl flaps. Reciprocating; radial cowl flaps.

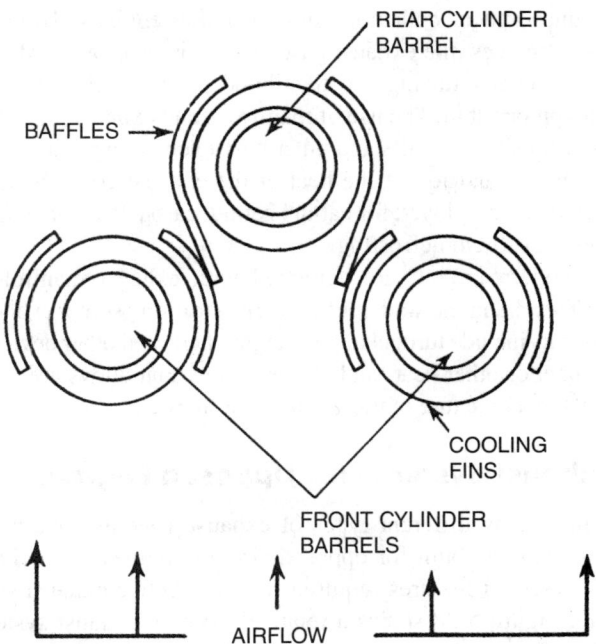

FIGURE 5-35 Baffles around the cylinders of a twin-row radial engine.

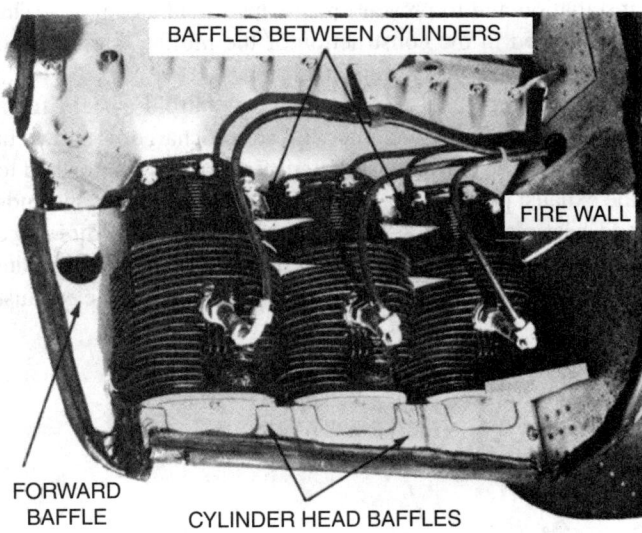

FIGURE 5-36 Baffling for an opposed-type engine.

design of the engine. The proper adjustment of valve timing is the most critical factor in heat rejection through the exhaust.

In the operation of a helicopter, the ram air pressure is usually not sufficient to cool the engine, particularly when the craft is hovering. For this reason, a large engine-driven fan is installed in a position to maintain a strong flow of air across and around the cylinders and other parts of the engine. Helicopters powered by turbine engines do not require the external cooling fan.

The principal advantages of air cooling are that (1) the weight of the air-cooled engine is usually less than that of a liquid-cooled engine of the same horsepower because the air-cooled engine does not need a radiator, connecting hoses and lines, and the coolant liquid; (2) the air-cooled engine is less affected by cold-weather operations; and (3) the air-cooled engine in military airplanes is less vulnerable to gunfire.

Liquid Cooling

Liquid-cooled engines are rarely found in U.S. aircraft today; however, powerplant technicians should have some understanding of the principal elements of such systems because new liquid-cooled engines are currently being developed. A liquid-cooling system consists of the liquid passages around the cylinders and other hot spots in the engine (see Fig. 5-37), a radiator by which the liquid is cooled, a thermostatic element to govern the amount of cooling applied to the liquid, a coolant pump for circulating the liquid, and the necessary connecting pipes and hoses. If the system is sealed, a relief valve is required to prevent excessive pressure and a sniffler valve is necessary to allow the entrance of air to prevent negative pressure When the engine is stopped and cooled off.

Water was the original coolant for liquid-cooled engines. Its comparatively high freezing point of 32°F [0°C] and its relatively low boiling point of 212°F [100°C] made it unsatisfactory for reciprocating engines used in military applications. The liquid most commonly used for liquid-cooled engines is either **ethylene glycol** or a mixture of ethylene glycol and water. Pure ethylene glycol has a boiling point of about 350°F [177°C] and a slush-forming freezing point of about 0°F [−17.78°C] at sea level. This combination of high boiling point and low freezing point makes it a satisfactory coolant for aircraft engines.

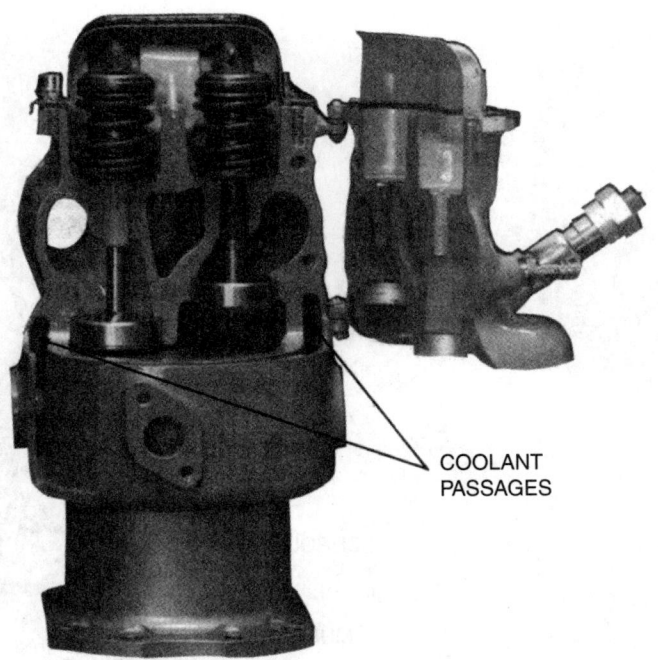

FIGURE 5-37 Liquid-cooled cylinder and jacket. (*Continental Motors.*)

RECIPROCATING-ENGINE EXHAUST SYSTEMS

One of the most critical systems employed in the engine's operation is the **exhaust system**, which removes the products of combustion from the engine safely and effectively. Since exhaust gases are both toxic and very hot, considerable care must be exercised in the design, construction, and maintenance of the exhaust system.

The maintenance and inspections of the engine exhaust system must be done on a regular basis and in accordance with the manufacturer's instructions. Poor maintenance and lack of inspections can lead to a nacelle fire, toxic gases entering the cockpit and cabin, damage to parts and structure in the nacelle, and poor engine performance.

Development of Exhaust Systems

Exhaust systems for the early aircraft engines were very simple. The engine exhaust was expelled from the exhaust port of each cylinder separately through short steel stacks attached to the exhaust ports. These systems were noisy and often permitted the exhaust gases to flow into the open cockpits of the aircraft. Pilots flying at night were often able to troubleshoot their engines by observing the color of the exhaust flame. A short, light-blue flame usually indicated that the mixture was correct and that the engine was operating satisfactorily. If the flame was shorter than normal for the engine, a lean mixture was indicated. A white or reddish flame indicated that the mixture was excessively rich. If only one cylinder produced a white or red flame, valve trouble or worn piston rings were likely causes.

The next step in the development of exhaust systems was the installation of **exhaust manifolds** for in-line and opposed engines and of collector rings for radial engines. Through these devices, the exhaust gases were directed outward and down, thus reducing the likelihood of gases entering the cockpit or cabin. The use of manifolds and exhaust pipes led to the design of muffs and other heat-exchanging equipment whereby a portion of the heat of the exhaust could be collected and employed for cabin heating, carburetor anti-icing, and windshield defrosting.

Modern aircraft are equipped with exhaust manifolds, heat exchangers, and mufflers. In addition, some exhaust systems include turbochargers, augmenters, and other devices. Inconel or other heat- and corrosion-resistant alloys are used in the manufacture of most exhaust systems.

Exhaust Systems for Opposed Engines

While many different types of exhaust systems have been designed and built for opposed aircraft engines, all include the essential features required for an effective exhaust system. Figure 5-38 shows a relatively simple exhaust system for a four-cylinder opposed aircraft engine. The exhaust manifold consists of risers leading from the muffler to the exhaust ports. The risers are attached by means of flanges, studs, and heat-resistant brass nuts. A copper-asbestos gasket is fitted between the flange and the cylinder to seal the installation and to prevent the escape of exhaust gases. The tubing used in the construction of the muffler, shroud, and risers is corrosion-resistant steel.

The arrangement of the exhaust manifold system for a six-cylinder opposed engine with a turbocharger is shown in Fig. 5-39. The corrosion-resistant steel risers are attached to the exhaust ports on the lower sides of the cylinders by studs and heat-resistant nuts. The risers are curved to direct the exhaust flow toward the rear in the manifold. The opposite side of the engine has a similar arrangement, and the exhaust

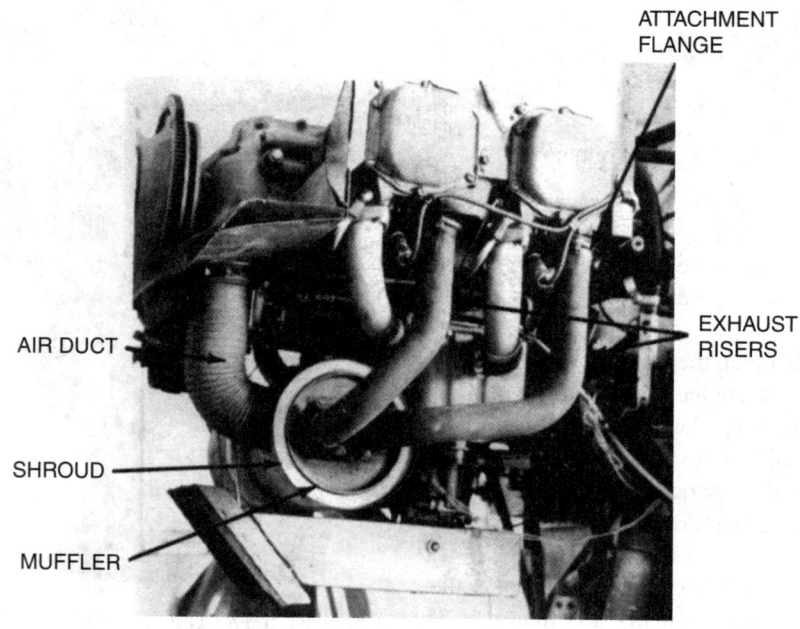

FIGURE 5-38   Exhaust system for an opposed engine.

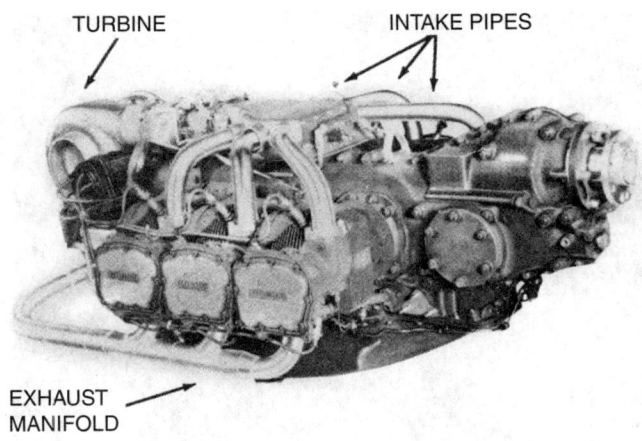

FIGURE 5-39 Exhaust manifold system for a six-cylinder opposed engine with a turbocharger. (*Textron Lycoming.*)

flow from both sides of the engine passes through exhaust pipes to the waste gate at the rear of the engine. When the engine is operated with turbocharging, the waste gate directs a portion of all the exhaust through the turbocharger turbine. Expansion joints are provided in the system to allow for uneven expansion and contraction due to changes in temperature.

An exhaust system for a light airplane, including the cabin heating system, is shown in Fig. 5-40. In this system, the exhaust is directed through the mufflers and out below the engine through a common tailpipe. A crossover pipe carries the exhaust from the left-hand cylinders to the tailpipe

of the right-hand muffler. The exhaust risers are attached to the mufflers by means of clamps which allow for expansion and contraction. Stainless-steel shells, or **shrouds**, are placed around the mufflers to capture the heat from the mufflers and direct it to the heater hoses. Shrouds of both mufflers are connected to a flexible duct system which routes outside air into the space between the mufflers and their shrouds. Heated air from the shrouds is carried through flexible ducts to plenum chambers in which the heat control valves are located. The shrouds are clamped around the mufflers by means of flanges joined by screws. This construction permits regular removal of the shrouds to inspect the mufflers for cracks and other signs of deterioration.

The exhaust system employed on a light twin-engine airplane is shown in Fig. 5-41. A combustion-type heating system is installed in this aircraft, so there is no need for a heat-exchanging system associated with the exhaust. The system on each side of the engine consists of three risers, a flexible joint between the aft riser and the muffler, and the tailpipe attached to the structure by means of a hanger with a spring-type **isolator** to allow for expansion and flexibility. Figure 5-42 illustrates the construction of the flexible joint between the aft riser and the muffler. Between the risers are slip joints which aid in alignment and allow for expansion.

Exhaust Systems for Radial Engines

The original radial engines usually disposed of gases through short stacks. The stacks prevented extreme, rapid temperature changes in the exhaust valves and the exhaust area of the

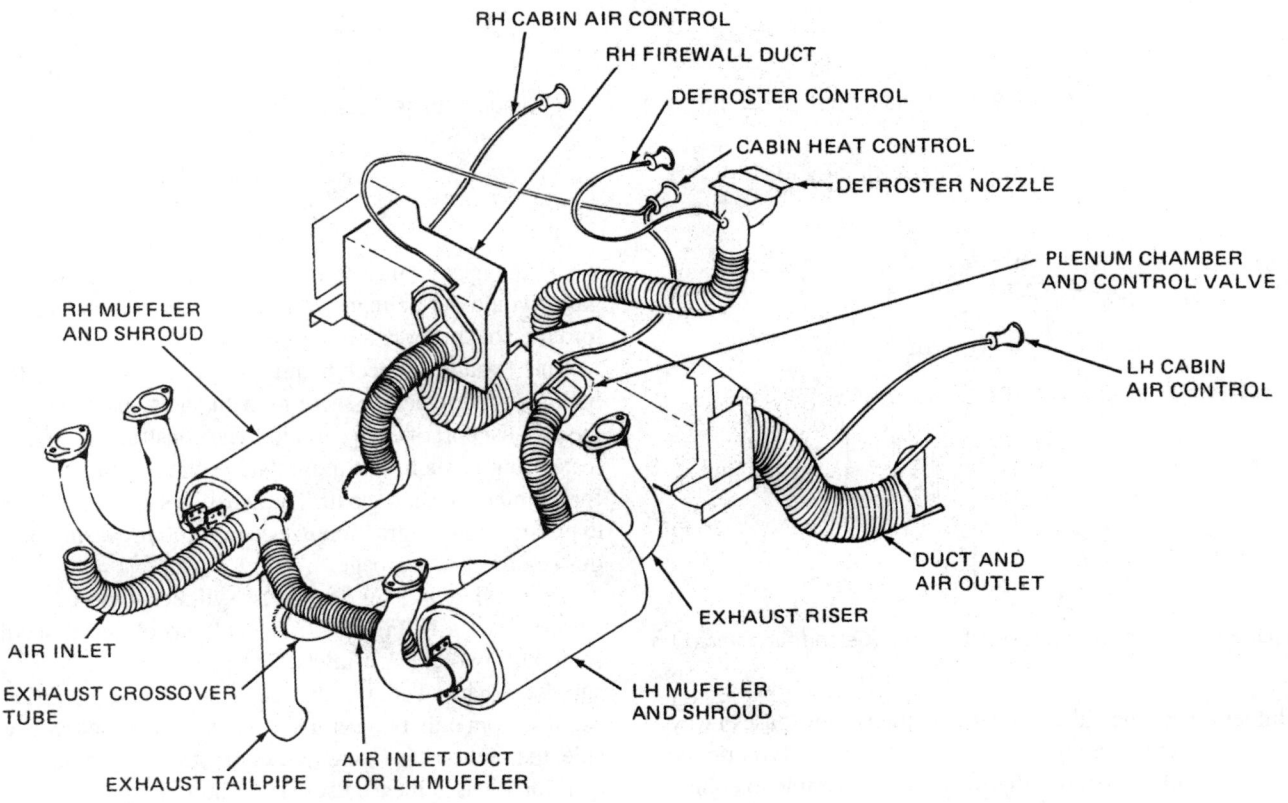

FIGURE 5-40 Exhaust system for a light airplane. (*Cessna Aircraft Co.*)

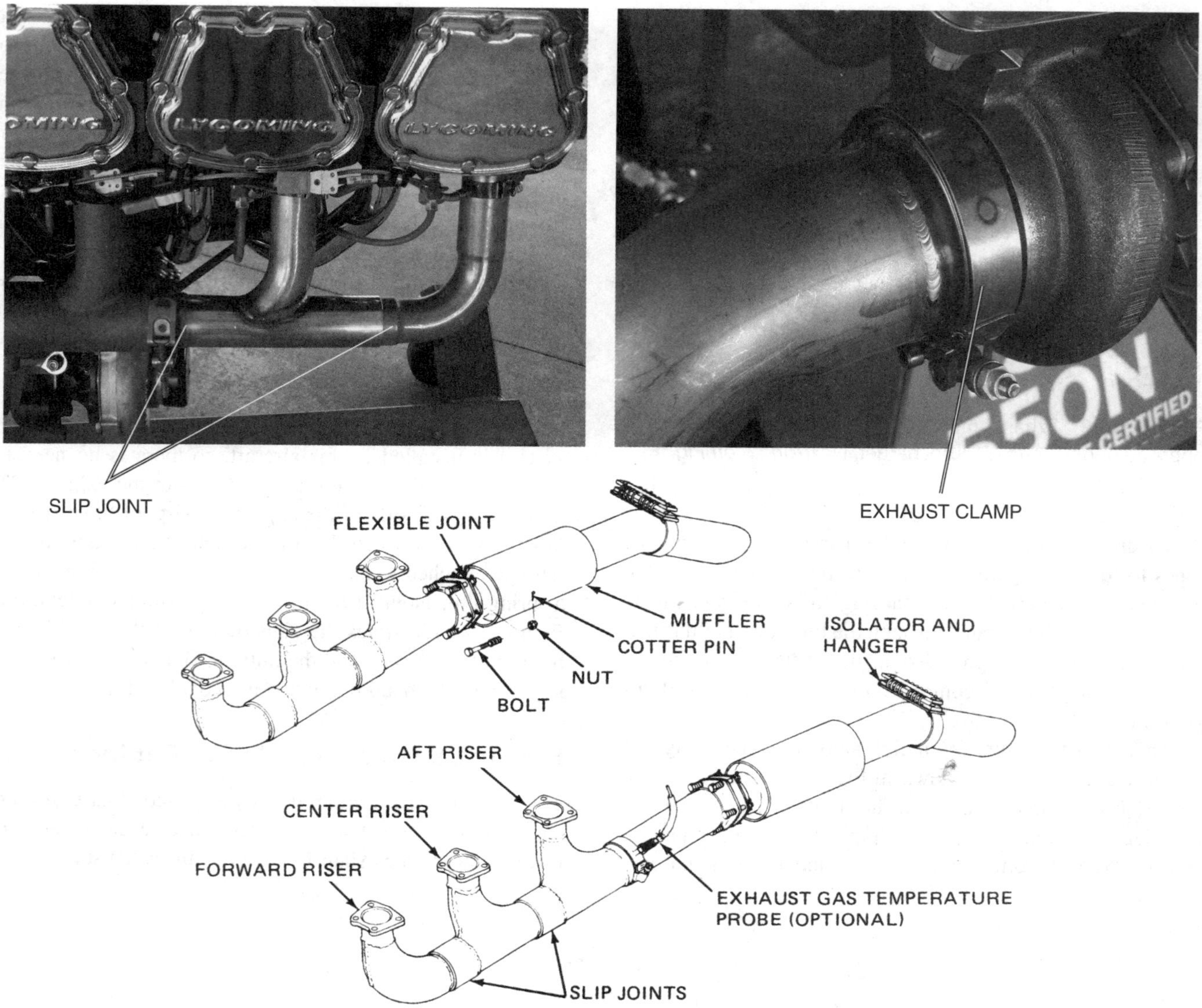

SLIP JOINT

FLEXIBLE JOINT

EXHAUST CLAMP

MUFFLER

COTTER PIN

NUT

BOLT

ISOLATOR AND
HANGER

AFT RISER

CENTER RISER

FORWARD RISER

EXHAUST GAS TEMPERATURE
PROBE (OPTIONAL)

SLIP JOINTS

FIGURE 5-41 Exhaust system components for a six-cylinder engine. (*Cessna Aircraft Co.*)

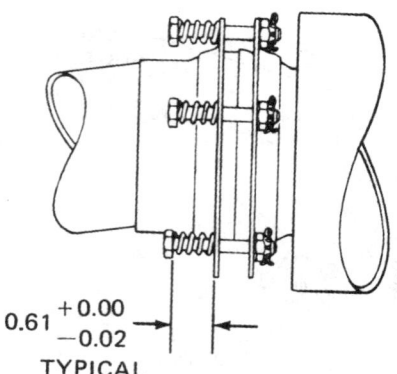

$0.61 \begin{array}{c} +0.00 \\ -0.02 \end{array}$

TYPICAL

FIGURE 5-42 Flexible-joint construction. (*Cessna Aircraft Co.*)

cylinder head. They also carried the hot exhaust gases away from the immediate area of the cylinder head. Experience revealed that for most installations it was desirable to collect the exhaust in a single manifold and discharge it at a point where the heat could not affect the aircraft structure and the gases would not be ingested into the engine intake or drawn into the cockpit or cabin.

The exhaust collector ring for a nine-cylinder R-985 engine is fabricated in sections with short stacks leading to the exhaust port of each cylinder. The construction of the collector ring is such that individual sections can be removed for maintenance and repair. The separate sections also permit expansion and contraction to occur without causing dangerous stresses and warpage.

The collector ring for a 14-cylinder twin-row radial engine, shown in Fig. 5-43, is made up of seven sections, each with two exhaust inlets. The section of the ring opposite the exhaust outlet is the smallest because it carries the exhaust from only two cylinders. From that point to the outlet side, the sections are increased in diameter to provide for the additional gases they must carry. The longer exhaust stacks connected to the ports of the collector ring reach forward

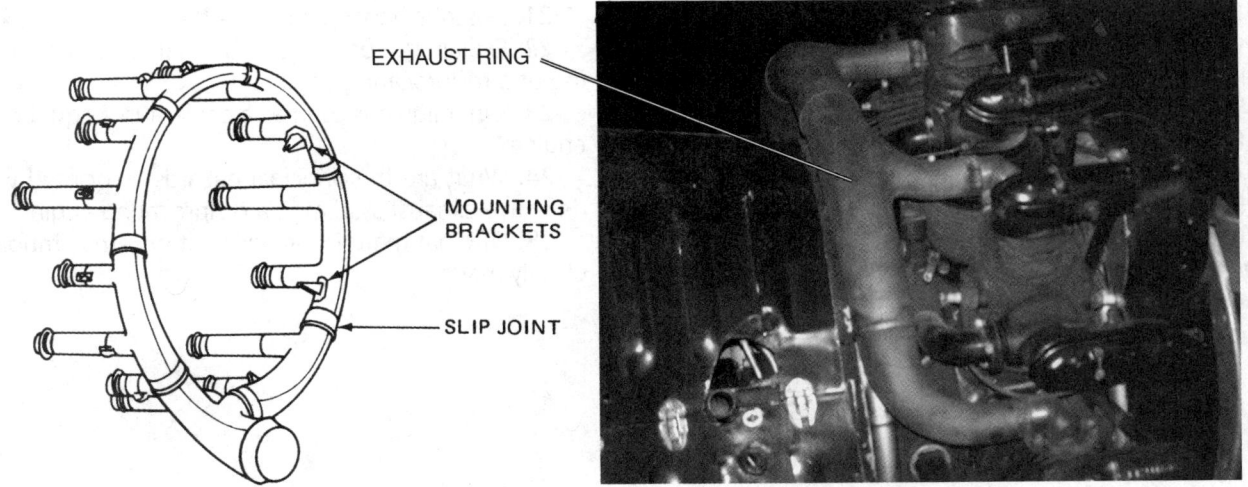

FIGURE 5-43   Exhaust collector ring and stacks for a 14-cylinder twin-row engine.

to connect with the front cylinders. The exhaust stacks from the cylinders are joined to the collector-ring ports by means of sleeve connections that allow for expansion and flexibility. Each section of the collector ring is attached to the blower section of the engine by a bolted bracket.

Exhaust Augmentors

Some aircraft exhaust systems include **exhaust augmentors**, also referred to as **ejectors**. In an augmentor-equipped 18-cylinder engine such as the R-2800, the exhaust is collected from the right side of the engine and discharged into the bellmouth of the right-hand augmentor and from the left side of the engine and discharged into the left augmentor. Four of the exhaust pipes on each side of the engine handle the exhaust from two cylinders each. The firing of the two cylinders feeding into each of the exhaust stacks is separated as much as possible to provide for maximum exhaust flow without excessive back pressure.

A simplified drawing of an augmentor is presented in Fig. 5-44. The augmentor produces a venturi effect which increases the flow of air from the engine nacelle. This increased airflow through the nacelle aids in cooling the engine and provides a small amount of "jet" thrust. Augmentor

tubes must be in perfect alignment with the exhaust flow to produce maximum effect.

The augmentor tubes are constructed of corrosion-resistant steel and sometimes contain an adjustable vane which can be controlled from the cockpit. In case the engine is operating at a temperature below that desired, the pilot can close the vane to reduce the cross-sectional area of the augmentor as much as 45 percent, to increase the operating temperature of the engine.

REVIEW QUESTIONS

1. Name the principal sections of an induction system.
2. What is the function of an alternate air valve?
3. How does an air filter affect engine power?
4. Why should carburetor heat not be used continuously?
5. What would happen if carburetor heat were turned on during high-power operation?
6. What is the effect of an air leak in the manifold between the carburetor and the cylinder for an unsupercharged (naturally aspirated) engine?
7. What is the function of the balance pipe connected between the forward ends of the intake manifolds on some engines?
8. What is the first indication of carburetor icing in an engine equipped with a fixed-pitch propeller?
9. What instrument indication will be observed when carburetor ice forms in an engine equipped with a constant-speed propeller?
10. What instrument is useful in determining whether icing may occur in the carburetor?
11. What are the advantages of supercharging?
12. What is the energy source for a turbocharger?
13. Compare the density of the atmosphere at sea level with the density at an altitude of 20 000 ft [6096 m].
14. Explain the significance of MAP in the operation of an aircraft engine.

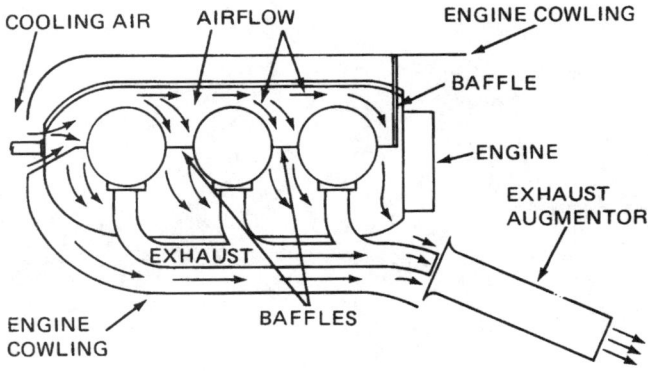

FIGURE 5-44   Operation of an exhaust augmentor.

15. What is the effect of excessive MAP?

16. What is the effect of supercharging on the density of air? How does it affect air temperature?

17. What is an internal supercharger?

18. Explain the function of the waste gate in a turbo-charger system.

19. What is meant by the term *critical altitude*?

20. Describe some of the effects of excessive heat in an aircraft engine.

21. Describe how cylinders are designed for air cooling.

22. Explain the use of pressure baffles in the air cooling of a reciprocating engine.

23. What liquid is used as a coolant in a liquid-cooled engine?

24. What are the principal hazards associated with a defective exhaust system in a reciprocating engine?

25. Of what material are exhaust system components usually made?

Basic Fuel Systems and Carburetors 6

INTRODUCTION

This chapter primarily explains basic aircraft fuels and fuel systems and their relation to both the engine and the fuel supply to the engine through the carburetor. The theory, operation, construction, and maintenance of float-type carburetors used on small and medium-size reciprocating engines are covered in detail. Also, pressure carburetors which were used on many older engines are included for your familiarization. Other types of fuel metering and fuel control devices are described in later chapters. Additional information on aircraft and engine fuel systems is provided in the text *Aircraft Maintenance and Repair*.

CHARACTERISTICS OF GASOLINE

Gasoline possesses desirable characteristics that make it suitable for use as an aviation fuel. These characteristics, compared with those of other fuels, are high heat value and the ability to evaporate when exposed to air at ordinary temperatures. A fuel such as gasoline evaporates readily at low temperatures and is said to have high volatility. For engine-starting purposes, high volatility is a desirable characteristic. However, a gasoline that evaporates too readily is apt to "boil" and form air bubbles in the fuel lines, resulting in vapor locks. A good aircraft fuel must have **volatility** that is high enough to start an engine easily but not so high as to readily form excessive vapors in the fuel system.

Testing Fuels

To determine the volatility of an aircraft fuel, a fractional-distillation test is made. Figure 6-1 illustrates this test and shows the type of apparatus used. A measured quantity of gasoline to be tested is placed in a glass flask, and then glass tubing is connected from the flask through a condenser unit to a calibrated receiver. Heat is applied under the flask and the amount of fuel condensed in the receiver at various temperatures is recorded. The data can be plotted on a graph as shown in Fig. 6-2, and a temperature range found where 10, 50, and 90 percent of the fuel condenses.

The temperature at which 10 percent of a test fuel is boiled off is indicative of the lowest atmospheric temperature at

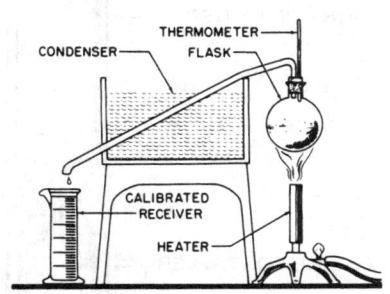

FIGURE 6-1 Fractional-distillation test.

which an engine will start when primed with this fuel. The temperature at which 50 percent of the fuel condenses determines the engine's acceleration ability; 90 percent condensation determines overall engine performance.

The volatility of a gasoline is also important because of its effect on carburetor icing. Fuel engineers speak of the **latent heat of vaporization**, which is simply the amount of heat necessary to vaporize a given amount of fuel. Vaporization cannot take place without heat. In a carburetor, this heat for vaporizing the fuel is taken from the air and from the metal. If too much heat is used in vaporization, ice may form. A highly volatile fuel extracts heat from its surroundings more rapidly than a less volatile fuel does. Carburetor icing has been practically eliminated from all aircraft except the small types equipped

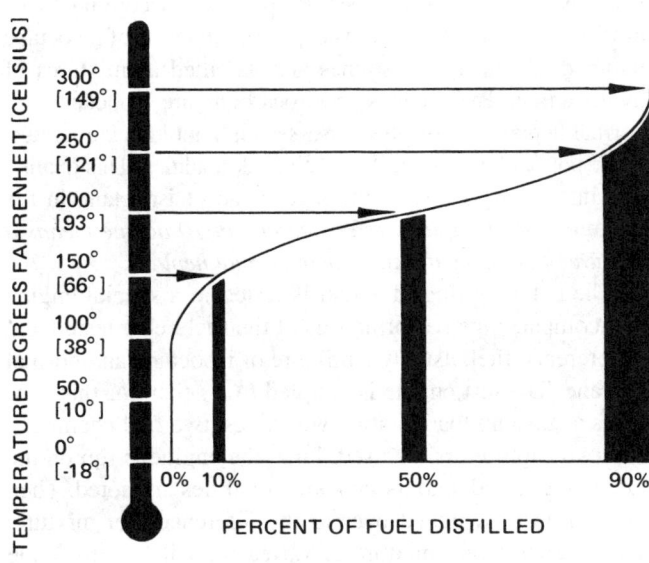

FIGURE 6-2 Fuel distillation at different temperatures.

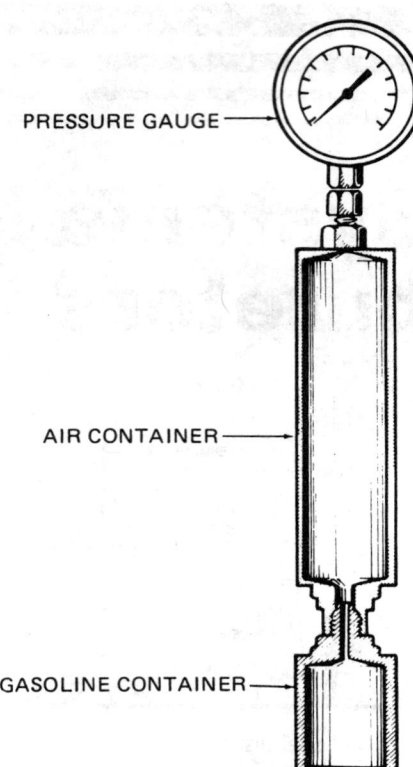

PRESSURE GAUGE

AIR CONTAINER

GASOLINE CONTAINER

FIGURE 6-3   Reid vapor pressure bomb.

with float-type carburetors. This has been done through the use of pressure injection and fuel injection systems, with the fuel injection occurring in locations not conducive to ice formation.

For this reason, aviation fuels that, in general, are blends of a number of different gasolines are checked carefully for vaporizing properties. An instrument known as the **Reid vapor pressure bomb** is used. In this apparatus, shown in Fig. 6-3, a pressure gauge attached to one end of a sealed container registers the amount of vapor pressure that a given fuel creates at various temperatures.

Octane Number

Gasoline is rated for engine fuel purposes according to its **antiknock value**; this value is expressed in terms of an octane number. Chemically, gasolines are classified as mixtures of hydrocarbon. Two of these hydrocarbons are isooctane and normal heptane. Isooctane possesses high antiknock qualities, while normal heptane has low antiknock qualities. This quality, or value, is expressed as the percentage of isooctane in the mixture. *For example, a reference fuel of 70 octane means a mixture of 70 percent isooctane in normal heptane.*

The octane rating of a fuel is tested in a special engine that compares the performances of the fuel being tested and of reference fuel, usually a mixture of isooctane and normal heptane. The test engine is coupled to a generator that provides a constant load factor. Two valves, two fuel chambers, and two carburetors are used. First, the engine is run on the fuel being tested, and its knocking qualities are noted. Then the engine is switched over to the reference-fuel mixture. The reference-fuel mixture is varied until it has the same knock qualities as the fuel being tested. The tested fuel is

given a number that is determined by the percentage of isooctane in the reference-fuel mixture.

Performance Number

Aviation fuels have been developed that possess greater antiknock qualities than 100 octane, or pure isooctane. To classify these fuels, a special **performance number** is used. This scale, or rating, is based on isooctane to which measured quantities of **tetraethyl-lead** (TEL) are added. For example, isooctane has an octane number of 100; likewise, it has a performance number of 100. If TEL is added to it, a performance number above 100 is obtained. The performance numbers obtained by mixing various amounts of TEL with octane are shown in the chart in Fig. 6-4.

When ordinary gasoline is rated, the octane number is generally used, but since, with the addition of lead, the rating may run over the fixed 100 rating of isooctane, it is preferable to rate aviation gasoline by giving it a performance number. Since the antiknock qualities of a fuel will vary according to the fuel-air (F/A) ratio, the performance numbers are expressed with two numbers—one rating for a lean mixture and the other for a rich mixture. The performance numbers are expressed as follows: 100/130, 115/145. The first number is the lean-performance number; the second number is the rich-performance number.

Use of Lead in Aviation Fuels

Lead, in the form of TEL, is used in relatively small quantities in aviation gasoline to improve antiknock qualities. The standard method in this country for expressing the quantity of lead is in terms of milliliters per gallon. The maximum lead concentration used is 630 volumes of gasoline to 1 volume of TEL. This corresponds to 6 milliliters (mL) of TEL to 1 gal [3.79 L] of gasoline.

Lead, if added alone to gasoline, will burn to form lead oxide, which is a solid with a very high boiling point. For this reason the lead remains as a residue in the cylinders to a large extent. To prevent this, a gasoline-soluble bromine compound is added to the lead. The mixture forms lead bromide,

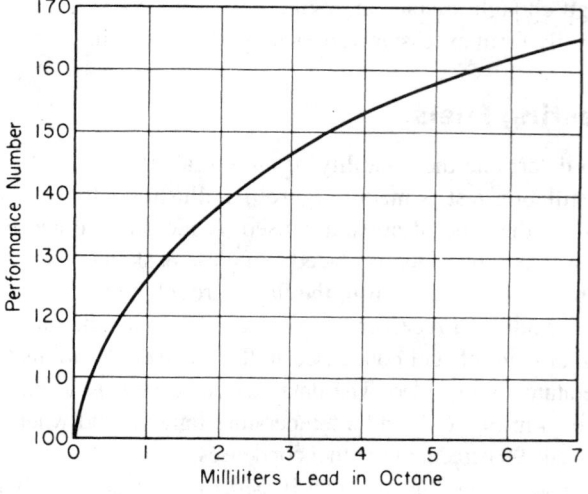

FIGURE 6-4   Chart of performance numbers.

which has a much lower boiling point than lead oxide, and therefore a large portion is expelled from the cylinders with the exhaust gases. Another chemical compound, **tricresyl phosphate** (TCP), is now being added to leaded fuel to further reduce the effects of the lead bromide deposits in the cylinder and on spark plug electrodes. The TCP converts the lead residue to lead phosphate rather than lead bromide. Lead phosphate is nonconductive and is easier to dispose of than lead bromide.

Aviation Gasoline: Grades and Color Codes

Technicians and refueling personnel should be familiar with grades of aviation gasoline (avgas) and their color codes, to ensure proper servicing of engines. Three grades of avgas are now produced for civil use: 80, 100LL (low lead), and 100. These grades replaced 80/87, 91/96, 100/130, and 115/145 avgas.

The lead quantity, or concentration of lead, in avgas is expressed in milliliters $\frac{1}{1000}$ per 1 gal [3.79 L] of avgas. The Standard Specification for Aviation Gasolines, Specification D910-75, developed by ASTM (American Society for Testing and Materials), established that grade 80 should be red and contain a maximum of 0.5 mL of TEL per 1 gal [3.79 L] of avgas. Grade 100LL is blue and contains a maximum of 2.0 mL/gal. Grade 100 is green and contains a maximum of 3.0 mL/gal.

Grades 100LL and 100 represent two avgases which are identical in antiknock quality but differ in maximum lead content and color. The color identifies the difference for engines with a low tolerance to lead.

Limited availability of grade 80 in some U.S. geographic areas has forced owners and operators to use the next higher grade of avgas. Specific use of higher grades is dependent on the applicable manufacturer's recommendations. Continuous use of higher-lead fuels in low-compression engines designed for low-lead fuels can cause erosion or necking of the exhaust valve stems and spark plug lead fouling.

Engine Design and Fuel Performance

Aircraft engines are specifically designed to operate with fuels having certain octane ratings, or performance numbers. The minimum octane rating, or performance number, is usually more critical than the maximum, although the use of a fuel with either too high or too low a rating may cause engine failure. FAA Type Certificate Data Sheets for Aircraft specify the minimum octane rating of fuel for each engine installation. Using a fuel with too low a rating usually leads to detonation, with accompanying damage to the pistons and cylinders, and eventual engine failure.

The principal factors governing the grade of fuel required for an engine are the compression ratio and the manifold air pressure (MAP). Supercharged engines require a higher grade of fuel than unsupercharged engines having the same compression ratio. As the compression ratio and MAP increase, the pressure of the F/A mixture in the cylinder also increases. Higher pressures lead to higher temperatures, which in turn increase the possibility of detonation.

Note that a fuel with a high performance number contains the same energy as a fuel with a low performance number. However, since the higher-rated fuel makes it possible for the engine to be operated at a higher compression ratio, a higher MAP, and higher temperatures, an engine designed to operate on higher-rated fuels can develop more power for the amount of fuel consumed.

High compression ratios enable economical long-range operation and aid in cylinder cooling. The cooling effect is a result of the fact that more of the heat in the fuel is converted to useful work at the crankshaft and that less goes into the cylinder walls. High compression is also advantageous in that a reduced weight of air is required for a given horsepower; this is a particular advantage at high altitudes.

An increase in the compression ratio improves fuel economy at the expense of reducing the power for takeoff and emergencies. Attempts to increase the range by employing a higher compression ratio are limited because this may so reduce takeoff power that the airplane cannot develop the power required for takeoff, since high compression ratios increase the possibility of detonation. In the ideal situation, a maximum compression ratio compatible with the fuel used is designed into the engine. Any increase in compression ratio above that at which the fuel will burn satisfactorily under full-power conditions will cause detonation and loss of power.

Corrosive Effects of Leaded Fuels

As previously stated, TEL is used with aviation fuels to improve the antiknock quality. Also, the use of TEL requires the addition of a bromide compound to help reduce the accumulation of lead deposits inside the combustion chamber. The bromide compound used commonly is ethylene dibromide.

The burning of TEL and ethylene dibromide produces the compound of lead bromide, most of which is carried out of the cylinder with the exhaust gases. However, a certain amount will remain, even under the best conditions. Lead bromide in the presence of water and metals, particularly aluminum, produces corrosive liquids, particularly hydrobromic acid, which cause rusting of steel and cast iron.

Great damage has been caused to engines by the corrosive action of hydrobromic acid and other chemical residues. The damage is especially great when an engine is allowed to stand unattended for several weeks. If an engine will not be operated for an extended period, it should be given a preservation treatment. This is done by operating the engine on unleaded fuel for at least 15 min and then spraying the interior of the cylinders with a rust-preventing oil while the engine is being rotated.

Aromatic and Alcohol Aviation Fuels

Gasoline, the aromatics, and alcohol are the ideal fuels for internal-combustion engines. In the following paragraphs we point out the relative merits of aromatics and alcohol as aviation fuels.

The **aromatics** are hydrocarbon compounds acquired either from coal (as a by-product during the manufacture of coke) or from oil. They are also present in straight-run gasoline as a natural product of the fractional-distillation process to an extent of about 7 percent or less. Aromatic fuel is so named because the atomic arrangement of its molecular structure is identical with that in perfumes.

When two fuels have equal performance numbers, both lean and rich, the fuel not containing aromatics is somewhat preferable to the fuel containing 15 percent aromatics. Also, aromatic fuel blends cause trouble with rubber parts and require the use of synthetic rubber for fuel hoses, pump packings, carburetor diaphragms, etc.

Benzol is the best known of the aromatic group. It has a high compression point at rich mixtures, and tests show that it will withstand a compression pressure of 175 psi [1206.63 kPa] within the combustion chamber of an engine before knocking occurs. But benzol has certain undesirable characteristics, among them a slow burning rate. It is also a powerful solvent of rubber. This objection to benzol was overcome to some extent by the development of synthetic-rubber fuel lines. However, for various reasons, the amount of benzol in aviation fuel at present is limited to 5 percent by volume.

Three other important aromatics are **toluene, xylene**, and **cumene**. Some of the characteristics of toluene, such as low freezing point, good volatility, and rubber-solvent properties less powerful than those of benzol, make it suitable for blending in aviation fuel up to 15 percent by volume. Xylene also has desirable qualities for blending in aviation fuels. However, it can be used only in limited quantities owing to its relatively high boiling point. Cumene is made from benzol. This limits the amount that can be used for blending purposes. Cumene has an extremely high boiling point, which tends to cause uneven distribution of the fuel-air charge to the various cylinders of the aircraft engine.

Since all aromatics are rubber solvents to some extent, the use of aviation fuels containing large amounts of aromatics requires aromatic-resistant materials in the fuel system. Parts such as flexible hoses, pump packings, and carburetor diaphragms must be made of special synthetic rubber.

Purity of Aviation Fuel

Aviation fuel must be free of any such impurities as water, dirt, microorganisms, acid, or alkali. It must also have low sulfur and gum contents. Sulfur has a corrosive effect on the various metal parts of the fuel system and engine. Gum may cause valves to stick, and it will clog fuel metering jets and restrictions.

Excessive water in an engine fuel system constitutes a serious problem. A small amount of water will pass through the system and carburetor jets without radically affecting engine performance, but any significant amount of water in the system will stop the flow of fuel through the metering jets and result in engine failure.

Water in the fuel poses a serious threat to high-altitude aircraft such as jet transport types. Fuel becomes cooled well below the freezing point of water, and since fuel cannot contain as much dissolved water at low temperatures as it can when it is comparatively warm, the water is precipitated out and forms minute ice crystals. These crystals adhere to fuel system screens and filters and effectively shut off the flow of fuel.

Effective water removers are used in airplane servicing equipment to eliminate the danger of water being pumped into the tanks during refueling. However, condensation will form within fuel tanks during periods of variable atmospheric temperatures, the amount of this condensation being proportional to the extent of airspace in the tank above the fuel line. To eliminate any possibility of this trouble, the tanks should be kept filled, especially at night or when extreme changes in temperature are likely.

When it is suspected that gasoline contains more water than is allowable [approximately 30 parts per million (ppm)], various tests can be made. A water-soluble coloring agent can be added to a sample of the fuel; if the fuel becomes tinted, water is present. Another testing method involves passing the fuel through a special filter paper; if the amount of water in the fuel is excessive, the yellow filter paper turns blue. Fuel cannot be tested for water content if the fuel is below the freezing temperature of water, for the water would be in the form of ice crystals and would not react with the testing agents. Aircraft that have been flying at high altitudes, especially in winter, are likely to contain fuel that has been "cold-soaked" to the extent that water tests cannot be made. When the fuel temperature is below the freezing temperature of water, the fuel must be allowed to warm up before tests for water are made.

As explained later in this section, fuel tanks are provided with sumps at the lowest points to collect water. Regular drainage of these sumps is an important step in preflight preparations.

Avgas vs. Automotive Gasoline

Because several of the major oil companies reduced or stopped the production of 80/87-octane avgas, many aircraft engines designed to operate on 80/87-octane avgas have been forced to use 100LL-octane (low-lead) avgas. This caused many problems since 100LL contains four times the lead of 80/87 avgas. These problems became apparent in the valves and spark plugs where excess lead deposits accumulated and other lead-related problems arose. Another problem was the increasing cost of 100LL avgas. To remedy some of these problems, several supplemental-type certificates (STCs) have been granted by the FAA to use automotive gasoline (autogas) in specific aircraft. Although the use of unleaded autogas has decreased or eliminated the problem with valves and spark plug deterioration, the use of autogas in aircraft still remains very controversial.

The aircraft manufacturers, engine manufacturers, major oil companies, and some aviation professional organizations remain staunchly opposed to the use of autogas. The four main areas of concern involve its volatility, combustion characteristics (antiknock), additives and blending components, and quality control. In the area of autogas volatility, a Reid vapor pressure of 9.0 to 15 is very common, compared to the vapor pressure of avgas, which is 5.5 to 7. At high temperatures and/or altitudes, because of the difference in volatility, autogas may vaporize in the fuel lines and pumps, preventing flow and possibly resulting in vapor lock. Due to the lack of stringent ASTM specifications for the combustion characteristics of autogases, there is much concern that not all unleaded autogas may meet the avgas rich rating of 87. Thus, autogas may not possess sufficient antiknock capability to prevent preignition and detonation in all applicable aircraft engines. Some blending agents used in autogas such as detergents and corrosion inhibitors could adversely affect the aircraft fuel system's components.

Quality control remains a concern in both the manufacture of autogas and its distribution to the aircraft. Although fuel contamination can occur in either avgas or autogas, the method of distribution of autogas remains somewhat questionable. Many times aircraft are fueled with autogas from 5-gal [18.9-L] cans that were filled at and transported from an automotive service station. Because the handling of autogas is not governed by quality control anywhere near that of avgas, there is a greater chance of fuel contamination. Another concern is the use of an autogas containing any type of alcohol. Alcohol is not compatible with the seals in most aircraft fuel systems. If the seals were to deteriorate and break loose, fuel starvation could occur, resulting in engine stoppage. Many people have used autogas very successfully in their aircraft by taking special handling precautions and closely adhering to the autogas STCs.

The avgas-vs.-autogas controversy in engines designed to use 80/87-octane avgas is far from resolved. Before using autogas in an aircraft, one should become very familiar with the special characteristics of autogas which could affect engine performance. It is also a good idea to check the engine manufacturer's warranty, since it may be void if autogas is used.

The future development of special engines and fuel systems designed to operate on autogas could increase its acceptance. Many view the widespread use of autogas as necessary to revive the general-aviation small-aircraft industry. Autogas is currently used widely in Europe, where valuable experience and data on its use are being gathered.

FUEL SYSTEMS

The complete fuel system of an airplane can be divided into two principal sections: the aircraft fuel system and the engine fuel system. The **aircraft fuel system** consists of the fuel tank or tanks, fuel boost pump, tank strainer (also called a *finger strainer*), fuel tank vents, fuel lines (tubing and hoses), fuel control or selector valves, main (or master) strainer, fuel flow and pressure gauges, and fuel drain valves. Fuel systems for different aircraft vary in complexity and may or may not include all these components. The **engine fuel system** begins where the fuel is delivered to the engine-driven pump and includes all the fuel controlling units from this pump through the carburetor or other fuel metering devices.

Requirements

The complete fuel system of an aircraft must be capable of delivering a continuous flow of clean fuel under positive pressure from the fuel tank or tanks to the engine under all conditions of engine power, altitude, and aircraft attitude and throughout all types of flight maneuvers for which the aircraft is certificated or approved. To do this and provide for maximum operational safety, certain conditions must be met:

1. Gravity systems must be designed with the fuel tank placed far enough above the carburetor to provide such fuel pressure that the fuel flow can be 150 percent of the fuel flow required for takeoff.

2. A pressure, or pump, system must be designed to provide 0.9 lb/h [0.41 kg/h] of fuel flow for each takeoff horsepower delivered by the engine, or 125 percent of the actual takeoff fuel flow of the engine, at the maximum power approved for takeoff.

3. In a pressure system, a **boost pump**, usually located at the lowest point in the fuel tank, must be available for engine starting, for takeoff, for landing, and for use at high altitudes. It must have sufficient capacity to substitute for the engine-driven fuel pump in case the engine-driven pump fails.

4. Fuel systems must be provided with valves so that fuel can be shut off and prevented from flowing to any engine. Such valves must be accessible to the pilot.

5. In systems in which outlets are interconnected, it should not be possible for fuel to flow between tanks in quantities sufficient to cause an overflow from the tank vent when the airplane is operated in the condition most apt to cause such overflow on full tanks.

6. Multiengine airplane fuel systems should be designed so that each engine is supplied from its own tank, lines, and fuel pumps. However, means may be provided to transfer fuel from one tank to another or to run two engines from one tank in an emergency. This is accomplished by a **cross-flow system** and valves.

7. A gravity-feed system should not supply fuel to any one engine from more than one tank unless the tank airspaces are interconnected to ensure equal fuel feed.

8. Fuel lines should be of a size sufficient to carry the maximum required fuel flow under all conditions of operation and should have no sharp bends or rapid rises, which would tend to cause vapor accumulation and subsequent vapor lock. Fuel lines must be kept away from hot parts of the engine insofar as possible.

9. Fuel tanks should be provided with drains and sumps to permit the removal of the water and dirt which usually accumulate in the bottom of the tank. Tanks must also be vented with a positive-pressure venting system to prevent the development of low pressure, which will restrict the flow of fuel and cause the engine to stop. Fuel tanks must be able to withstand, without failure, all loads to which they may be subjected during operation.

10. Fuel tanks must be provided with baffles if the tank design is such that a shift in fuel position will cause an appreciable change in balance of the aircraft. This applies chiefly to wing tanks, where a sudden shift of fuel weight

can cause loss of aircraft control. Baffles also aid in preventing fuel sloshing, which can contribute to vapor lock.

Gravity-Feed Fuel Systems

A gravity-feed fuel system is one in which the fuel is delivered to the engine solely by gravity. (See items 1 and 7 in the previous list.) The gravity system does not require a boost pump because the fuel is always under positive pressure to the carburetor. A fuel quantity gauge must be provided to show the pilot the quantity of fuel in the tanks at all times. The system includes fuel tanks, fuel lines, a strainer and sump, a fuel cock or shutoff valve, a priming system (optional), and a fuel quantity gauge. The carburetor may also be considered part of the system. A gravity-feed fuel system is shown in Fig. 6-5.

Pressure Systems

For aircraft in which the fuel tanks cannot be placed the required distance above the carburetors or other fuel metering devices and in which a greater pressure is required than can be provided by gravity, fuel boost pumps and engine-driven fuel pumps are needed.

For a fuel system which relies entirely on pump pressure, the **fuel boost pump** is located at the bottom of the fuel tank and may be either inside or outside the tank. In many systems the boost pump is submerged in the fuel at the bottom of the tank. In some systems (Fig. 6-6), gravity feeds the fuel to reservoir tanks and then through the fuel selector valve to the auxiliary (boost) fuel pump. This system utilizes a fuel injection metering unit and requires more pressure than can be supplied by gravity alone.

In a pressure system, the **engine-driven fuel pump** is in series with the boost pump, and the fuel must flow through the engine-driven pump to reach the fuel metering unit. The pump must be designed so that fuel can bypass it when the engine is not running. This is normally accomplished by a bypass valve.

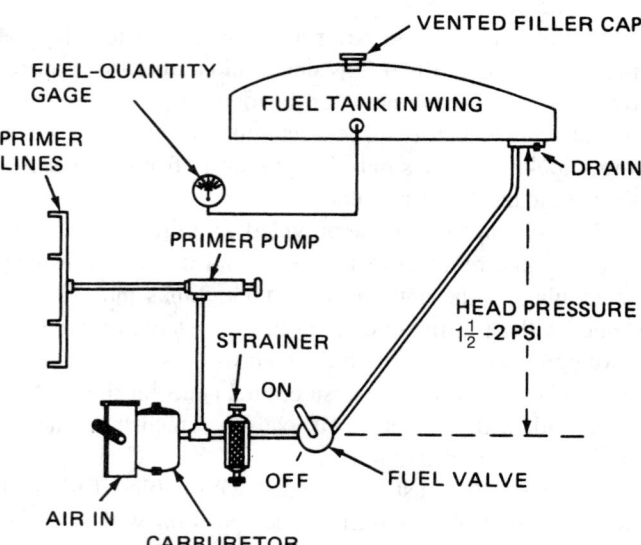

FIGURE 6-5   Gravity-feed fuel system.

The pump must also include a **relief valve** or similar unit to permit excess fuel to return to the inlet side of the pump. The engine-driven pump must be capable of delivering more fuel to the engine than is required for any mode of operation.

The fuel boost pump supplies fuel for starting the engine, and the engine pump supplies the fuel pressure necessary for normal operation. During high-altitude operation, takeoff, and landing, the boost pump is operated to ensure adequate fuel pressure. This is particularly important during landing and takeoff in case of engine pump failure.

Fuel Strainers and Filters

All aircraft fuel systems must be equipped with strainers and/or filters to remove dirt particles from the fuel. Strainers are often installed in fuel tank (cell) outlets, or they may be integral with the fuel boost pump assembly. Fuel tank strainers have a comparatively coarse mesh, some being as coarse as eight mesh to 1 in [2.54 cm]. **Fuel sump strainers**, also called **main strainers** or **master strainers**, are located at the lowest point in the fuel system between the fuel tank and engine and are much finer in mesh size, usually being 40 or more mesh per inch. The fuel filters installed in carburetors and other fuel metering units may be screens or sintered (heat-bonded) metal filters. Many of these are designed to remove all particles larger than 40 micrometers. A **micrometer** is one one-thousandth of a millimeter (1 μm = 0.001 m).

Fuel strainers and filters should be checked and cleaned as specified in the aircraft service manual.

Fuel System Precautions

In servicing fuel systems, remember that fuel is flammable and that the danger of fire or explosion always exists. The following precautions should be taken:

1. Aircraft being serviced or having the fuel system repaired should be properly grounded.
2. Spilled fuel should be neutralized or removed as quickly as possible.
3. Open fuel lines should be capped.
4. Fire-extinguishing equipment should always be available.
5. Metal fuel tanks must not be welded or soldered unless they have been adequately purged of fuel fumes. Keeping a tank or cell filled with carbon dioxide will prevent explosion of fuel fumes. Additional information on aircraft fueling procedures is provided in the text *Aircraft Basic Science*.

Vapor Lock

The condition known as **vapor lock** is caused by fuel vapor and air collecting in various sections of the fuel system. The fuel system is designed to handle liquid fuel rather than a gaseous mixture. When a substantial amount of vapor collects, it interferes with the operation of pumps, valves, and the fuel metering section of the carburetor. Vapor lock is caused to form by the low atmospheric pressure of high altitude, by excessive fuel temperature, and by turbulence (or sloshing) of the fuel.

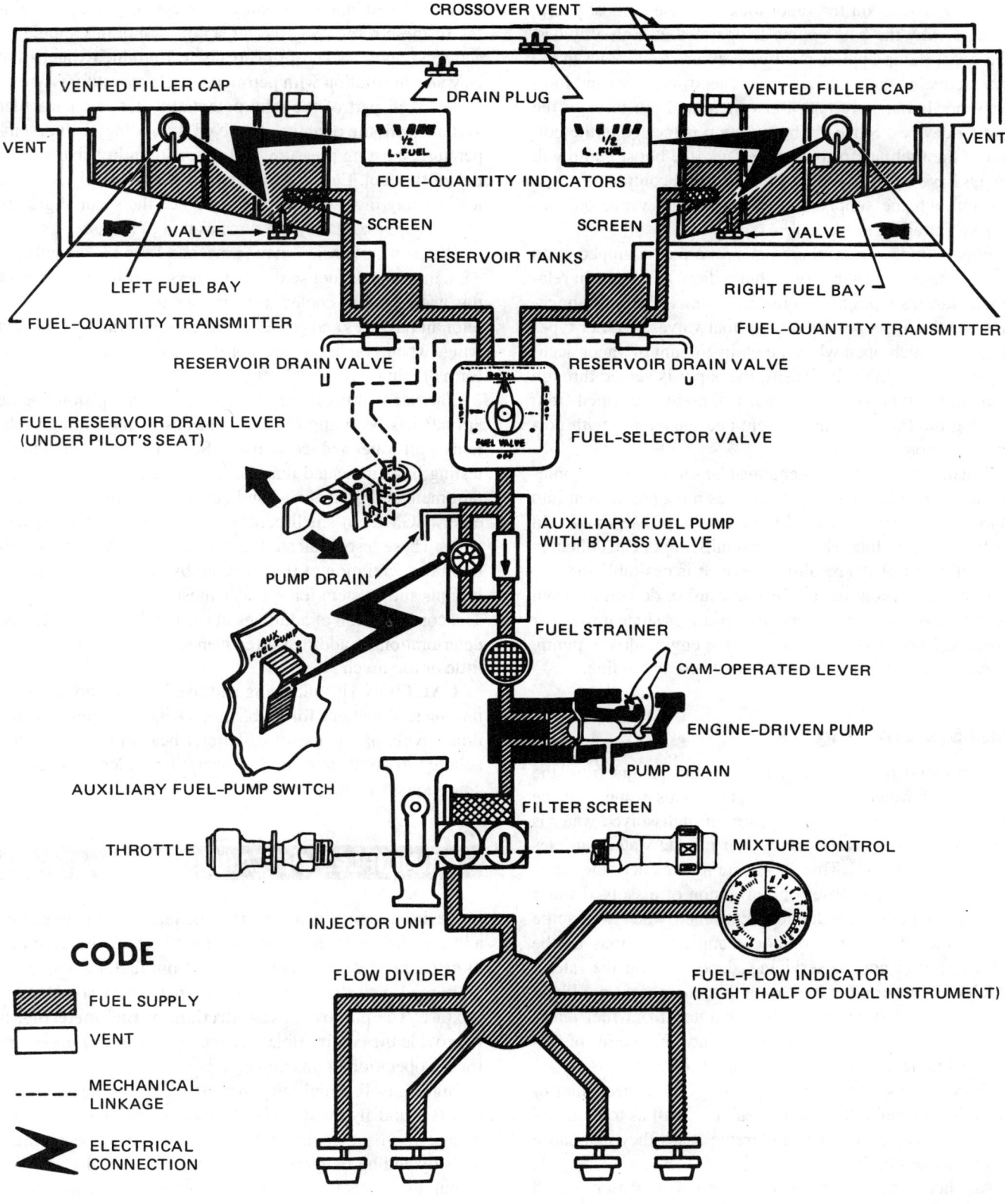

CROSSOVER VENT

VENTED FILLER CAP

DRAIN PLUG

VENTED FILLER CAP

VENT

½ L. FUEL

½ L. FUEL

VENT

FUEL-QUANTITY INDICATORS

VALVE → SCREEN

SCREEN

VALVE

RESERVOIR TANKS

LEFT FUEL BAY

RIGHT FUEL BAY

FUEL-QUANTITY TRANSMITTER

FUEL-QUANTITY TRANSMITTER

RESERVOIR DRAIN VALVE

RESERVOIR DRAIN VALVE

FUEL RESERVOIR DRAIN LEVER
(UNDER PILOT'S SEAT)

FUEL-SELECTOR VALVE

AUXILIARY FUEL PUMP
WITH BYPASS VALVE

PUMP DRAIN

FUEL STRAINER

AUX FUEL PUMP

CAM-OPERATED LEVER

ENGINE-DRIVEN PUMP

PUMP DRAIN

AUXILIARY FUEL-PUMP SWITCH

FILTER SCREEN

THROTTLE

MIXTURE CONTROL

INJECTOR UNIT

CODE

FLOW DIVIDER

FUEL-FLOW INDICATOR
(RIGHT HALF OF DUAL INSTRUMENT)

FUEL SUPPLY

VENT

MECHANICAL
LINKAGE

ELECTRICAL
CONNECTION

FUEL-INJECTION NOZZLES

TO ENSURE DESIRED FUEL CAPACITY WHEN REFUELING, PLACE THE
FUEL-SELECTOR VALVE IN EITHER LEFT OR RIGHT POSITION TO PRE-
VENT CROSS-FEEDING.

FIGURE 6-6 Pressure fuel system. (*Cessna Aircraft Company.*)

The best solution for vapor lock is to use a boost pump, which is why the boost pump is operated at high altitudes. The boost pump applies positive pressure to the fuel in the lines, reducing the tendency of the fuel to vaporize and forcing vapor bubbles through the system and out through the venting devices. Since the boost pump is located at the bottom of the fuel tank or below the tank, the boost pump will always have a good supply of fuel and will continue to force fuel through the supply lines, even though vapor bubbles may be entrained in the fuel.

Fuel pumps and carburetors are often equipped with vapor-separating devices (which are discussed later, in relation to various types of fuel metering units). **Vapor separators** are chambers provided with float valves or other types of valves which open when a certain amount of vapor accumulates. When the valve opens, the vapor is vented through a line to the fuel tank. The vapor is thereby prevented from entering the fuel metering system and interfering with normal operation.

The design of a fuel system must be such that an accumulation of vapor is not likely. Fuel lines must not be bent into sharp curves; neither should there be sharp rises or falls in the line. If a fuel line rises and then falls, vapor can collect in the high point of the resulting curve. It is desirable that the fuel line have a continuous slope upward or downward from the tank to the boost pump and a continuous slope upward or downward from the boost pump to the engine-driven pump, to prevent vapor from collecting anywhere in the line.

Fuel System Icing

Ice formation in the aircraft fuel system results from the presence of water in the fuel system. This water may be undissolved or dissolved. One form of undissolved water is **entrained water**, which consists of minute water particles suspended in the fuel. This may occur as a result of mechanical agitation of free water or conversion of dissolved water through temperature reduction. Entrained water will settle out in time under static conditions and may or may not be drained during normal servicing, depending on the rate at which it is converted to free water. In general, it is not likely that all entrained water can be separated from fuel under field conditions. The settling rate depends on a series of factors, including temperature and droplet size.

The droplet size will vary depending on the mechanics of formation. Usually, the particles are so small as to be invisible to the naked eye, but in extreme cases they can cause slight haziness in the fuel.

Another form of undissolved water is **free water**, which may be introduced as a result of refueling or the settling of entrained water at the bottom of a fuel tank. Free water is usually present in easily detectable quantities at the bottom of the tank. It can be drained from a fuel tank through the sump drains provided for that purpose. Free water frozen on the bottoms of reservoirs, such as the fuel tanks and fuel filter, may render drains useless and can later melt, releasing water into the system and thereby causing engine malfunction or stoppage. If such a condition is detected, the aircraft may be placed in a warm hangar to reestablish proper draining of these reservoirs, and all sumps and drains should be activated and checked prior to flight. Entrained water (that is, water in solution with petroleum fuels) constitutes a relatively small part of the total potential water in a particular system, with the quantity dissolved depending on fuel temperature, existing pressure, and water solubility characteristics of the fuel. Entrained water will freeze in cold fuel and tend to stay in suspension longer, since the specific gravity of ice is approximately the same as that of avgas.

Water in suspension may freeze and form ice crystals big enough to block fuel screens, strainers, and filters. Some of this water may be cooled further when the fuel enters carburetor air passages and causes carburetor metering-component icing, when conditions are not otherwise conducive to this form of icing.

The use of **anti-icing additives** for some piston-engine aircraft has been approved as a means of preventing problems with water and ice in avgas. Some laboratory and flight testing have indicated that the use of hexylene glycol, certain methanol derivatives, and ethylene glycol monomethyl ether (EGME) in small concentrations inhibits fuel system icing. These tests indicate that the use of EGME at a maximum concentration of 0.15 percent by volume substantially inhibits fuel system icing under most operating conditions. The concentration of additives in the fuel is critical. Marked deterioration in additive effectiveness may result from too little or too much additive.

CAUTION The anti-icing additive is in no way a substitute or replacement for carburetor heat. Operating instructions involving the use of carburetor heat should be strictly adhered to at all times during operation under atmospheric conditions conducive to icing.

PRINCIPLES OF CARBURETION

In the discussion of heat engine principles, we explained that a heat engine converts a portion of the heat of a burning fuel to mechanical work. To obtain heat from fuel, the fuel must be burned, and the burning of fuel requires a combustible mixture. The purpose of **carburetion**, or **fuel metering**, is to provide the combustible mixture of fuel and air necessary for the operation of an engine.

Since gasoline and other petroleum fuels consist of carbon (C) and hydrogen (H) chemically combined to form hydrocarbon molecules (CH), it is possible to burn these fuels by adding oxygen (O) to form a gaseous mixture. The carburetor mixes the fuel with the oxygen of the air to provide a combustible mixture which is supplied to the engine through the induction system. The mixture is ignited in the cylinder; then the heat energy of the fuel is released, and the fuel-air mixture is converted to carbon dioxide (CO_2), water (H_2O), and possibly some carbon monoxide (CO).

The carburetors used in aircraft engines are comparatively complicated because they play an extremely important part in engine performance, mechanical life, and the general efficiency of the airplane, due to the widely diverse conditions under which airplane engines are operated. The carburetor must deliver an accurately metered fuel-air mixture for engine loads and speeds between wide limits and must provide for automatic or manual mixture correction under changing conditions of temperature and altitude, all the while being subjected to a continuous vibration that tends to upset the calibration and adjustment. Sturdy construction is essential for all parts of an aircraft carburetor, to provide durability and resistance to the effects of vibration. A knowledge of the functions of these parts is essential to an understanding of carburetor operation.

Fluid Pressure

In the carburetor system of an internal-combustion engine, liquids and gases, collectively called **fluids**, flow through various passages and orifices (holes). The volume and density of liquids remain fairly constant, but gases expand and contract as a result of surrounding conditions.

The atmosphere surrounding the earth is like a great pile of blankets pressing down on the earth's surface. **Pressure** may be defined as force acting on an area. It is commonly measured in pounds per square inch (psi), inches of mercury (inHg), centimeters of mercury (cmHg), or kilopascals (kPa). The **atmospheric pressure** at any place is equal to the weight of a column of water or mercury a certain number of inches, centimeters, or millimeters in height. For example, if the cube-shaped box shown in Fig. 6-7, each side of which is 1 in^2 [6.45 cm^2] in area, is filled with mercury, that quantity of mercury will weigh 0.491 lb [2.18 N]; therefore, a force of 0.491 lb [2.18 N] is acting on the bottom 1 in^2 [6.45 cm^2] of the box. If the same box were 4 in [10.16 cm] high, the weight of the mercury would be 4 × 0.491, or 1.964 lb [8.74 N]. Therefore, the downward force on the bottom of the box would be 1.964 lb [8.74 N]. Therefore, each 1 in [2.54 cm] of height of a column of mercury represents a 0.491-psi [3.385-kPa] pressure. To change inches of mercury to pounds

per square inch, simply multiply by 0.491. For example, if the height of a column of mercury is 29.92 inHg [101.34 kPa], multiply 29.92 by 0.491, and the product is 14.69 psi, which is standard atmospheric pressure at sea level.

Refer again to Fig. 6-7. A glass tube about 36 in [91.44 cm] long, with one end sealed and the other end open, is filled completely with mercury. The tube is then placed in a vertical position with the open end submerged in a small container partly filled with mercury. If this is done at sea level under standard conditions, the mercury will sink and come to rest 29.92 in [76 cm] above the mercury in the container. There is then a vacuum above the mercury in the tube; therefore, there is no atmospheric pressure above the mercury in the tube.

The atmospheric pressure acts on the surface of the mercury in the container in Fig. 6-7. The weight of the mercury column above the surface of the mercury in the container must therefore equal the weight of the air column above the same surface. The length of the mercury column in the tube indicates the atmospheric pressure and is measured by a scale placed beside the tube or marked on its surface. Atmospheric pressure is expressed in *pounds per square inch, inches of mercury, kilopascals,* or *millibars.*

Standard sea-level pressure is 14.7 psi, 29.92 inHg, 101.34 kPa, or 1013 millibars (mbar). NASA and the International Committee for Aeronautical Operations (ICAO) have established a standard atmosphere for comparison purposes. The standard atmosphere table can be found in the Appendix of this text. **Standard atmosphere** is defined as a pressure of 29.92 inHg [101.34 kPa] at sea level and at a temperature of 15°C [59°F] when the air is perfectly dry at latitude 40°N. This is a purely fictitious and arbitrary standard, but it has been accepted and should be known to pilots, technicians, and others engaged in aircraft work.

The pressure of the atmosphere varies with the altitude.

- At 5000 ft [1524 m] it is 24.89 inHg [84 kPa].
- At 10 000 ft [3048 m] it is 20.58 inHg [69.7 kPa].
- At 20 000 ft [6098 m] it is 13.75 inHg [46.57 kPa].
- At 30 000 ft [9144 m] it is 8.88 inHg [30 kPa].
- At 50 000 ft [15 240 m] it is only 3.346 inHg [11.64 kPa].

Expressing the same principle in different terms, we say that the pressure of the atmosphere at sea level is 14.7 psi [101.34 kPa], but at an altitude of 20 000 ft [6096 m] the pressure is only about 6.74 psi [46.47 kPa].

The pressure exerted on the surface of the earth by the weight of the atmosphere, **absolute pressure**, can be measured by a barometer. A **relative pressure** assumes that the atmospheric pressure is zero. Relative or differential pressures are usually measured by fuel pressure gauges, steam gauges, etc. This means that when a pressure is indicated by such a gauge, the pressure actually shown is so many pounds per square inch above atmospheric pressure. This pressure is often designated as **psig**, meaning psi gauge. When absolute pressure is indicated, it is designated as **psia**, meaning psi absolute.

The effect of atmospheric pressure is important to the understanding of aircraft fluids, including fuel, oil, water, hydraulic fluid, etc. The effect of atmospheric pressure on

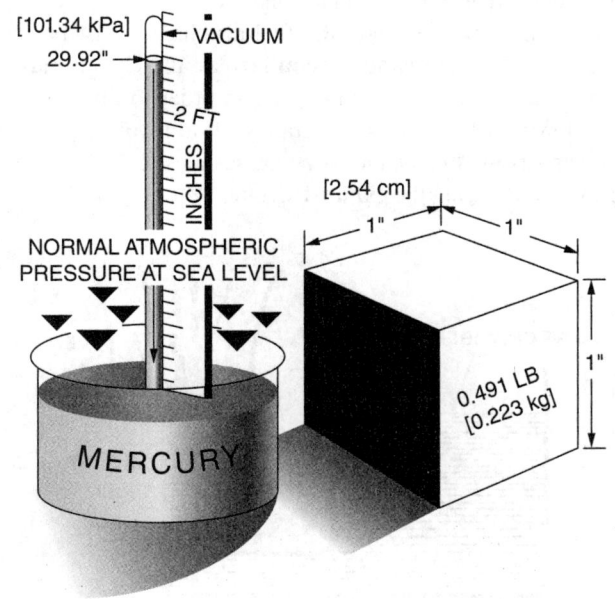

FIGURE 6-7   Measuring atmospheric pressure.

liquids can be shown by a simple experiment. Place a tube in a glass of water, place your finger over the open end of the tube, and take the tube out of the water, keeping your finger over the end. The water will not run out of the tube until you remove your finger. This shows the importance of providing and maintaining open vents to the outside atmosphere for tanks, carburetor chambers, and other parts or units which depend on atmospheric venting for their operation.

Venturi Tube

Figure 6-8 shows the operation of a **venturi tube**, which was originally used for the measurement of the flow of water in pipes. This device consists of a conical, nozzle-like reducer through which the air enters, a narrow section called the **throat**, and a conical enlargement for the outlet, which attains the same size as the inlet but more gradually.

The quantity of air drawn through the inlet is discharged through the same-size opening at the outlet. The velocity of the fluid must therefore increase as the fluid passes through the inlet cone, attain a maximum value in the throat, and thereafter gradually slow down to its initial value at the outlet. The pressure in the throat is consequently less than that at either the entrance or the exit.

Figure 6-8 shows manometers connected to the venturi. These are gauges for measuring pressure and are similar in principle to barometers.

The operation of the venturi is based on **Bernoulli's principle**, which states that the total energy of a particle in motion is constant at all points on its path in a steady flow; therefore, at a higher velocity the pressure must decrease. The pressure in the throat of the venturi tube is less than the pressure at either end of the tube because of the increased velocity in the constricted portion. This is explained by the fact that the same amount of air passes all points in the tube in a given time.

The venturi illustrates the relation between pressure (force per unit area) and velocity in a moving column of air. In equal periods, equal amounts of air flow through the inlet, which has a large area; through the throat, which has a small cross-sectional area; and then out through the outlet, which also has a large area.

If any body, fluid or solid, is at rest, force must be applied to set that body in motion. If the body is already in motion, force must be applied to increase its velocity. If a body in motion is to have its velocity decreased, or if the body is to be brought to a state of rest, an opposing force must be applied.

In Fig. 6-8, if the cross-sectional area of the inlet is twice that of the throat, the air will move twice as fast in the throat as it does in the inlet and outlet. Since the velocity of any moving object cannot be decreased without applying an opposing force, the pressure of the air in the outlet portion of the tube must be greater than it is in the throat. From this we see that the pressure in the throat must be less than it is at either end of the tube.

Venturi in a Carburetor

Figure 6-9 shows the venturi principle applied to a simplified carburetor. The amount of fluid which flows through a given passage in any unit of time is directly proportional to the velocity at which the fluid is moving. The velocity is directly proportional to the difference in applied forces. If a fuel discharge nozzle is placed in the venturi throat of a carburetor, the effective force applied to the fuel will depend on the velocity of air going through the venturi. The rate of flow of fuel through the discharge nozzle will be proportional to the amount of air passing through the venturi, and this will determine the supply of the required fuel-air mixture delivered to the engine. The ratio of fuel to air should be varied within certain limits; therefore, a mixture control system is provided for the venturi-type carburetor.

Review of the Engine Cycle

The conventional aircraft internal-combustion engine is a form of heat engine in which the burning of the fuel-air mixture occurs inside a closed cylinder and in which the heat energy of the fuel is converted to mechanical work to drive the propeller.

The **engine cycle** (Otto cycle) must be understood and remembered in order to understand the process of carburetion. The fuel and air must be mixed and inducted into the cylinder during the **intake stroke**; the fuel-air charge must be compressed during the **compression stroke**; the charge must be ignited and must burn and expand in order to drive the piston downward and cause the crankshaft to revolve during the **power stroke**; finally, the burned gases must be exhausted (or scavenged) during the **exhaust stroke**.

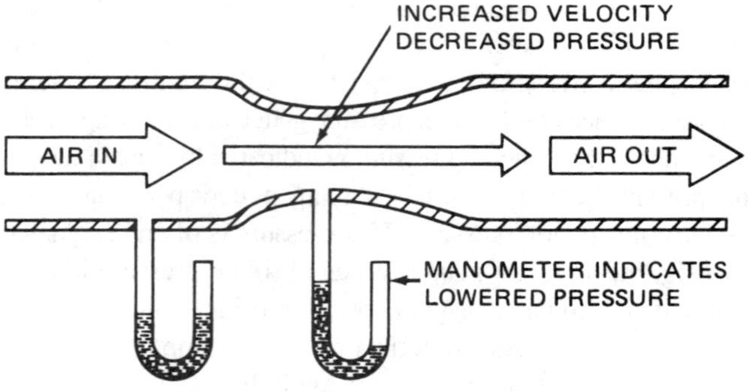

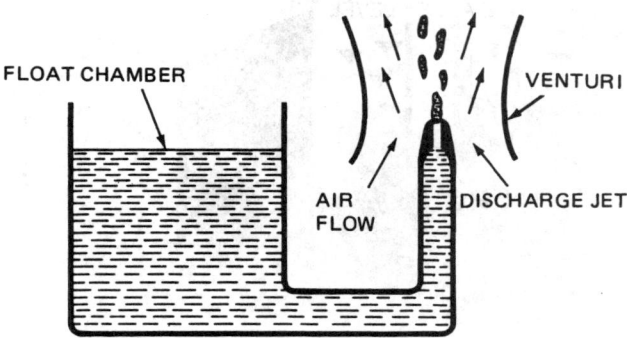

FIGURE 6-8 Operation of a venturi tube.

FIGURE 6-9 Venturi principle applied to a carburetor.

The quantity and the nature of the charge of fuel and air inducted into the engine cylinder must be given considerable attention because the power, speed, and operating efficiency of the engine are governed largely by this charge.

Fuel-Air Mixtures

Gasoline and other liquid fuels will not burn in the liquid state, but must be vaporized and combined with correct amounts of oxygen to form a combustible mixture. The mixture of fuel and air is described as **chemically correct** when there is just enough oxygen present in the mixture to burn the fuel completely. If there is not quite enough air, combustion may occur but will not be complete. If there is either too much or too little air, the mixture will not burn.

As mentioned previously, burning is a chemical process. Gasoline is composed of carbon and hydrogen, and a gasoline called isooctane has the formula C_8H_{18}. During the burning process, this molecule must be combined with oxygen to form carbon dioxide (CO_2) and water (H_2O). The equation for the process may be written

$$2C_8H_{18} \times 25O_2 = 16CO_2 + 18H_2O$$

Thus, we see that two molecules of this particular gasoline require 50 atoms of oxygen for complete combustion. The burning of fuel is seldom as complete as this, and the resulting gases would likely contain carbon monoxide (CO). In such a case the equation could be

$$C_8H_{18} + 12O_2 = 7CO_2 + 9H_2O + CO$$

Air is a mechanical mixture containing by weight about 75.3 percent nitrogen, 23.15 percent oxygen, and a small percentage of other gases. Nitrogen is a relatively inert gas which has no chemical effect on combustion. Oxygen is the only gas in the mixture which serves any useful purpose as far as the combustion of fuel is concerned.

Gasoline will burn in a cylinder if mixed with air in a ratio ranging between 8 parts of air to 1 part of fuel and 18 parts of air to 1 part of fuel (by weight). That is, the air-fuel (A/F) ratio would be from 8:1 to 18:1 for combustion. This means that the A/F mixture can be ignited in a cylinder when the ratio is anywhere from as *rich* as 8 parts of air by weight to 1 part of fuel by weight to as *lean* as 18 parts of air by weight to 1 part of fuel by weight. In fuel-air (F/A) mixtures, the proportions are expressed on the basis of weight, because a ratio based on volumes would be subject to inaccuracies resulting from variations of temperature and pressure.

The proportions of fuel and air in a mixture may be expressed as a ratio (such as 1:12) or as a decimal fraction. The ratio 1:12 becomes 0.083 (which is derived by dividing 1 by 12). The decimal proportion is generally employed in charts and graphs to indicate F/A mixtures.

Fuel-air mixtures employed in the operation of aircraft engines are described as **best-power mixture**, **lean best-power mixture**, **rich best-power mixture**, and **best-economy mixture**. The graph in Fig. 6-10 illustrates the effects of changes in the F/A ratio for a given engine operating at a given rpm. The mixture at A is the lean best-power mixture

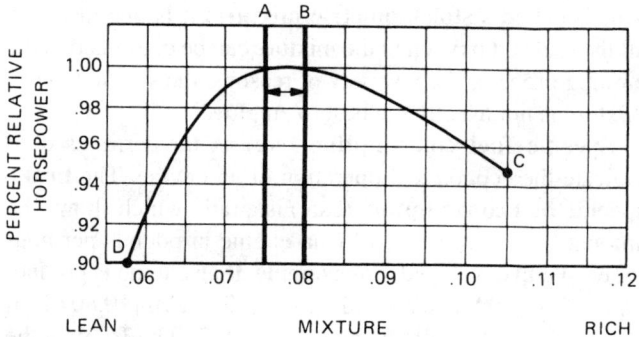

FIGURE 6-10 Effects of F/A ratios on power at constant rpm.

and is the point below which any further leaning of the mixture will rapidly reduce engine power; in other words, the lean best-power mixture is the leanest mixture that can be used and still obtain maximum power from the engine. The mixture indicated by B in Fig. 6-10 is the rich best-power mixture and is the richest mixture that can be used and still maintain maximum power from the engine. The mixtures from A to B in the graph therefore represent the best-power range for the engine. Points C and D in Fig. 6-10 are the limits of flammability for the F/A mixture; that is, the F/A mixture will not burn at any point richer than that represented by C or at any point leaner than that represented by D.

The chart in Fig. 6-11 illustrates how the best-power mixture will vary for different power settings. Note that there is a very narrow range of F/A ratios for the best-power mixture. For example, the setting for 2900 rpm is 0.077, for 3000 rpm is 0.082, and for 3150 rpm is 0.091. Any F/A ratios other than those given will result in a rapid falling off of power. Internal-combustion engines are so sensitive to the proportioning of fuel and air that the mixture ratios must be maintained within a definite range for any given engine. A perfectly balanced F/A mixture is approximately 15:1, or 0.067.

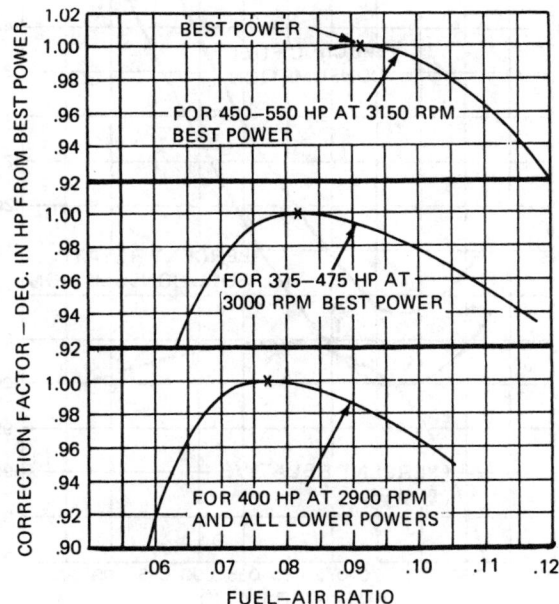

FIGURE 6-11 Chart of best-power mixtures for different power settings.

This is called a **stoichiometric mixture**; it is one in which all the fuel and oxygen in the mixture can be combined in the burning process. For a variety of reasons, the stoichiometric mixture is not usually the best to employ.

Specific fuel consumption (sfc) is the term used to indicate the economical operation of an engine. The **brake specific fuel consumption** (bsfc) is a ratio which shows the amount of fuel consumed by an engine in pounds per hour for each bhp developed. For example, if an engine is producing 147 hp [109.62 kW] and burns 10.78 gal/h [40.807 L/h] of fuel, the fuel weight being 6 lb/gal [0.719 kg/L], then the sfc is 0.44 lb/(hp•h) [0.15 kg/(kW•h)].

The chart in Fig. 6-12 was derived from a test run to determine the effects of F/A ratios. Note that the lowest sfc in this case occurred with an F/A ratio of approximately 0.067 and that maximum power was developed at F/A ratios between 0.074 and 0.087. For this particular engine, we may say that lean best power is at point A on the chart (F/A ratio of 0.074) and that rich best power is at point B (F/A ratio of 0.087).

Note further that the sfc increases substantially as the mixture is leaned or enriched from the point of lowest sfc. From this observation it is clear that excessive leaning of the mixture in flight will not produce maximum economy. Furthermore, excessive leaning of the mixture is likely to cause detonation, as mentioned previously.

If detonation is allowed to continue, the result will be mechanical damage or failure of the tops of the pistons and rings. In severe cases, cylinder heads may be fractured. It is therefore important to follow the engine operating instructions regarding mixture control settings, thereby avoiding detonation and its unfavorable consequences. A careful observance of cylinder-head temperature and/or exhaust gas temperature (EGT) in most cases will enable the pilot to take corrective action before damage occurs. A reduction of power and an enrichment of the mixture usually suffice to eliminate detonation.

We have already stated that if there is too much air or too much fuel, the mixture will not burn. In other words, when the mixture is excessively rich or excessively lean, it approaches the limit of flammability; as it approaches this limit, the rate of burning decreases until it finally reaches zero. This is much more pronounced on the lean side than it is on the rich side of the correct proportion of fuel and air.

We noted that the best-power mixtures of fuel and air for the operation of an engine are those which enable the engine to develop maximum power. There is, however, a mixture of fuel and air which will produce the greatest amount of power for a given consumption of fuel. This is called the **best-economy mixture**, attained by leaning the mixture below the lean best-power mixture. As the mixture is leaned, both power and fuel consumption drop, but fuel consumption decreases more rapidly than engine power until the best-economy mixture is reached. The point is reached with a mixture somewhere between 0.055 and 0.065, depending on the particular engine and operating conditions. Remember that leaning of the mixture below the best-power mixture is not practiced with the engine developing its maximum power. Usually the engine is operating at less than 75 percent power. The operator's manual should be consulted for specific information on the operation of a particular powerplant and aircraft combination.

The chart in Fig. 6-13 provides a graphic illustration of the difference between best-economy mixture and best-power mixture.

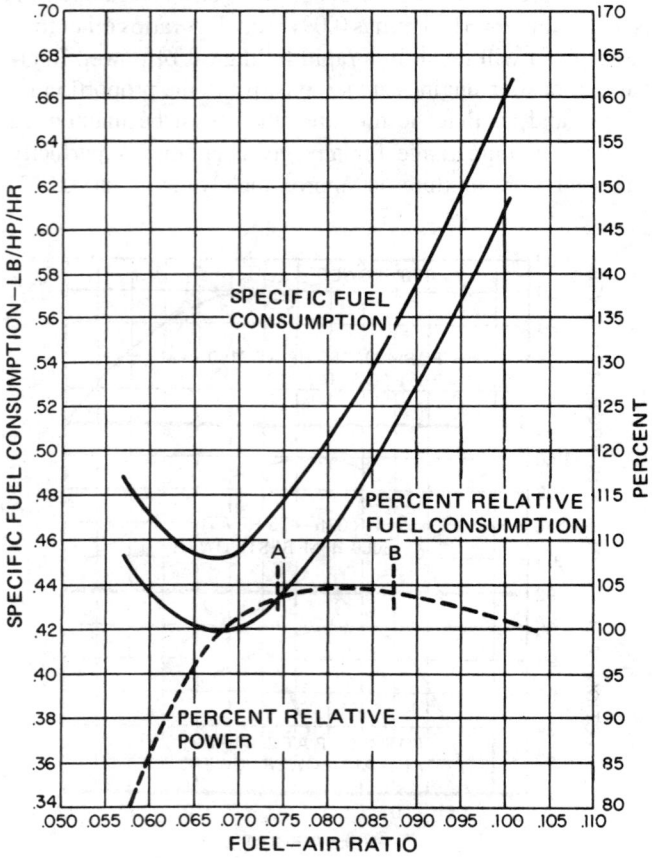

FIGURE 6-12 Effects of F/A ratios and power settings on fuel consumption.

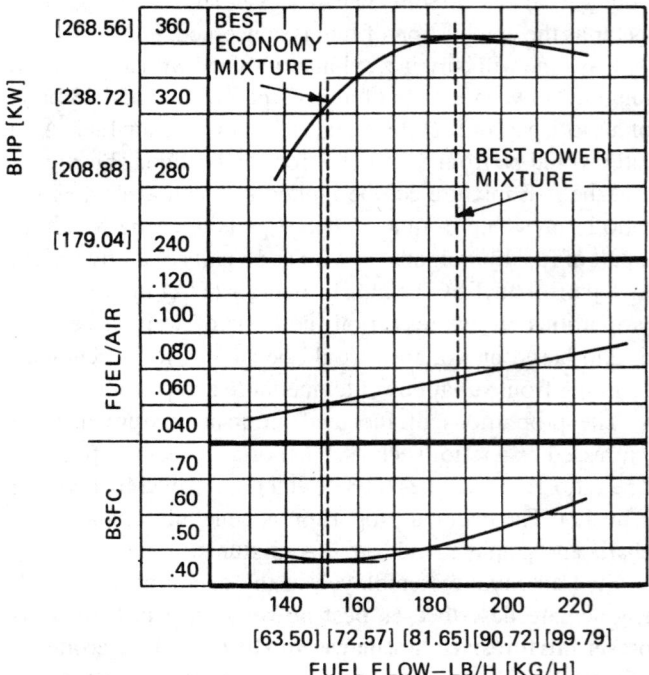

FIGURE 6-13 Best-economy mixture and best-power mixture at constant throttle and constant rpm.

This chart is based on a constant-throttle position with a constant rpm. The only variable is the F/A ratio. With a very lean F/A ratio of about 0.055, the engine delivers 292 bhp [217.7 kW] with a fuel flow of 140 lb/h [63.50 kg/h], and the bsfc is about 0.48 lb/(hp•h) [0.218 kg/(kW•h)]. The best-economy mixture occurs when the F/A ratio is approximately 0.062. At this point the bhp is 324, and the fuel flow is 152 lb/h [68.95 kg/h]. The bsfc is then 0.469 lb/(hp•h) [0.213 kg/(kW•h)]. As the strength of the F/A mixture is increased, a point is reached where the engine power has reached a peak and will begin to fall off. This is the best-power mixture, and it is approximately 0.075, or 1:13.3. At this point the bhp is 364, and the bsfc is 0.514 with a fuel flow of 187 lb/h [84.82 kg/h].

We have now established the effects of F/A ratio when other factors are constant, and we can see that the mixture burned during the operation of the engine will have a profound effect on the performance. We must, however, explore the matter further because the engine operating temperature must be considered. If an engine is operated at full power and at the best-power mixture, as shown in the upper curve in Fig. 6-11, it is likely that the cylinder-head temperature will become excessive and detonation will result. For this reason, at full-power settings the mixture will be enriched beyond the best-power mixture. This is the function of the **economizer**, or **enrichment valve**, in the carburetor or fuel control, as explained later. The extra fuel will not burn but will vaporize and absorb some of the heat developed in the combustion chamber. At this time the manual mixture control is placed in the FULL RICH or AUTO RICH position, and the F/A ratio will be at or above the rich best-power mixture.

When operating under cruising conditions of rpm and manifold pressure (MAP), it is possible to set the mixture at the lean best-power value to save fuel and still obtain a maximum value of cruising power from the engine. If it is desired to obtain maximum fuel economy at a particular cruise setting, the manual mixture control can be used to lean the mixture to the best-economy F/A ratio. This will save fuel but will result in a power reduction of as much as 15 percent.

The chart in Fig. 6-14 illustrates graphically the requirements of an aircraft engine with respect to F/A ratio and power output. As shown, a rich mixture is required for very low-power settings and for high-power settings. When the power is in the 60 to 75 percent range, the F/A ratio can be

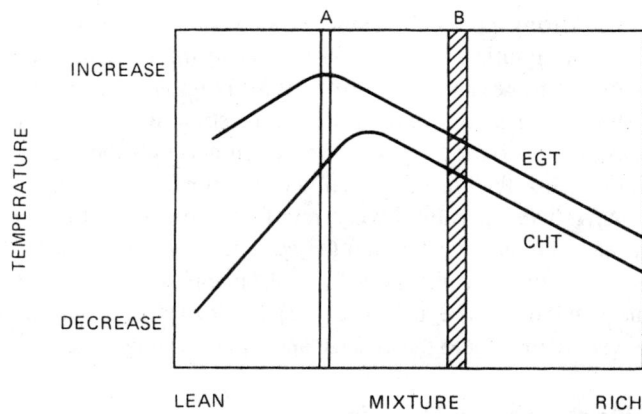

FIGURE 6-15 Effects of F/A mixture on cylinder-head and exhaust gas temperatures.

set for lean best power or for best economy. The curve shown in Fig. 6-14 will vary from engine to engine.

The effect of the F/A mixture on cylinder-head and exhaust gas temperatures is illustrated in Fig. 6-15. The temperatures rise as the mixture is leaned to a certain point; however, continued leaning leads to a drop in the temperatures. This is true if the engine is not operating at high power settings. An excessively lean mixture may cause an engine to **backfire** through the induction system or to stop completely. A backfire is caused by slow flame propagation. This happens because the F/A mixture is still burning when the engine cycle is completed. The burning mixture ignites the fresh charge when the intake valve opens, the flame travels back through the induction system, the combustible charge is burned, and often any gasoline that has accumulated near the carburetor is burned. This occurs because the flame propagation speed decreases as the mixture is leaned. Thus, a mixture which is lean enough will still be burning when the intake valve opens.

The **flame propagation** in an engine cylinder is the rate at which the flame front moves through the mixture of fuel and air. The flame propagation is most rapid at the best-power setting and falls off substantially on either side of this setting. If the mixture is too lean, the flame propagation will be so slow that the mixture will still be burning when the intake valve opens, thus igniting the mixture in the intake manifold and causing a backfire. The effect of the F/A ratio on flame propagation is illustrated in Fig. 6-16.

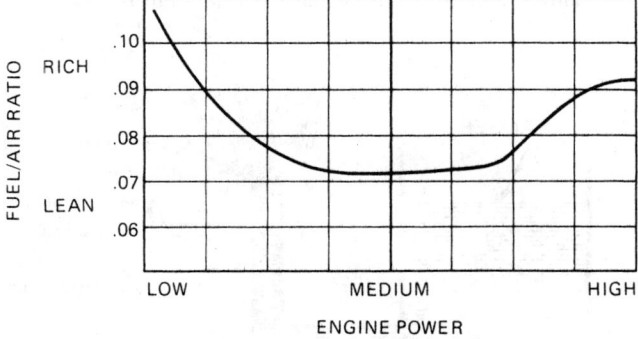

FIGURE 6-14 Fuel-air ratios required for different power settings.

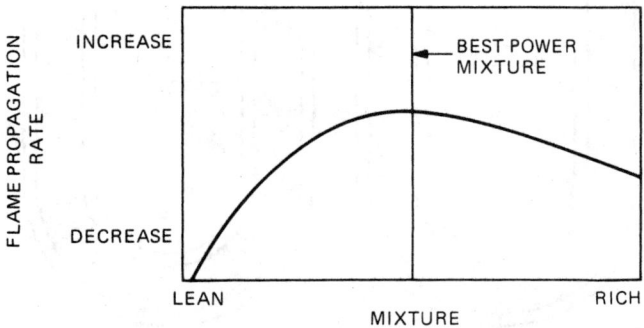

FIGURE 6-16 Effect of F/A ratio on flame propagation.

Principles of Carburetion **141**

Backfiring is *not* the same as **kickback**, which occurs when the ignition is advanced too far at the time that the engine is to be started. If the mixture is ignited before the piston reaches top center, the combustion pressure may cause the piston to reverse its direction and turn the crankshaft against the normal direction of rotation.

Afterfiring is caused when raw fuel is permitted to flow through the intake valve into the cylinder head, then out the exhaust valve into the exhaust stack, manifold and muffler, and heater muff. The fuel can cause a fire or explosion that can be very damaging to the exhaust and cabin-heating systems.

Effects of Air Density

Density may be defined simply as the *weight per unit volume of a substance.* The weight of l cubic foot (ft³) [28.32 L] of dry air at standard sea-level conditions is 0.076475 lb [0.034 7 kg]. A pound [0.453 6 kg] of air under standard conditions occupies approximately 13 ft³ [368.16 L].

The density of air is affected by pressure, temperature, and humidity. An increase in pressure will *increase* the density of air, and an increase in humidity will *decrease* the density. Therefore, the F/A ratio is affected by air density. For example, an aircraft engine will have less oxygen to burn with the fuel on a warm day than on a cold day at the same location; that is, the mixture will be richer when the temperature is high, and the engine cannot produce as much power as when the air is cool. The same is true at high altitudes, where air pressure decreases and density decreases. For this reason, pilots usually "lean out" the mixture at higher altitudes to avoid an overrich mixture and waste of fuel.

Water vapor in the air (humidity) decreases the density because a molecule of water weighs less than a molecule of oxygen or a molecule of nitrogen. Therefore, pilots must realize that an engine will not develop as much power on a warm, humid day as it will on a cold, dry day. This is because the less-dense air provides less oxygen for fuel combustion in the engine.

Effect of Pressure Differential in a U-Shaped Tube

Figure 6-17 shows two cross-sectional views of a U-shaped glass tube. In the left view, the liquid surfaces in the two arms of the tube are even because the pressures above them are equal. In the right view, the pressure in the right arm of the tube is reduced while the pressure in the left arm of the tube remains the same as before. This causes the liquid in the left arm to be pushed down while the liquid in the right arm is raised, until the difference in weights of liquid in the two arms are exactly proportional to the difference in forces applied on the two surfaces.

Pressure Differential in a Simple Carburetor

The principle explained in the preceding paragraph is applied in a simple carburetor, such as the one shown in Fig. 6-9. The rapid flow of air through the venturi reduces the pressure at the discharge nozzle so that the pressure in the fuel chamber can force the fuel out into the airstream. Since the airspeed in the tube is comparatively high and there is a relatively great reduction in pressure at the nozzle during medium and high engine speeds, there is a reasonably uniform fuel supply at such speeds.

When the engine speed is low and the pressure drop in the venturi tube is slight, the situation is different. This simple nozzle, otherwise known as a **fuel discharge nozzle**, in a carburetor of fixed size does not deliver a continuously richer mixture as the engine suction and airflow increase. Instead, a plain discharge nozzle will give a fairly uniform mixture at medium and high speeds; but at low speeds and low suction, the delivery falls off greatly in relation to the airflow.

This occurs partly because some of the suction force is consumed in raising the fuel from the float level to the nozzle outlet, which is slightly higher than the fuel level in the fuel chamber to prevent the fuel from overflowing when the engine is not operating. It is also caused by the tendency of the fuel to adhere to the metal of the discharge nozzle and to break off intermittently in large drops instead of forming a fine spray. The discharge from the plain fuel nozzle is therefore retarded by an almost constant force, which is not important at high speeds with high suction but which definitely reduces the flow when the suction is low because of reduced speed.

Figure 6-18 shows how the problem is overcome in the design and construction of the venturi-type carburetor. Air is bled from behind the venturi and passed into the **main discharge nozzle** at a point slightly below the level of the fluid, causing the formation of a finely divided F/A mixture which is fed into the airstream at the venturi. A metering jet between the fuel chamber and the main discharge nozzle controls the amount

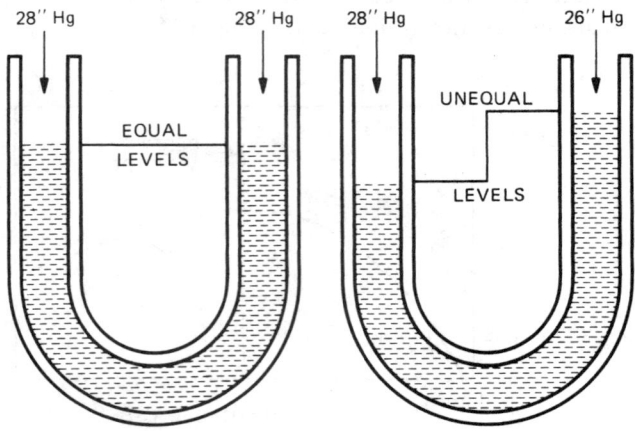

FIGURE 6-17 Pressure effects on fluid in a U-shaped tube.

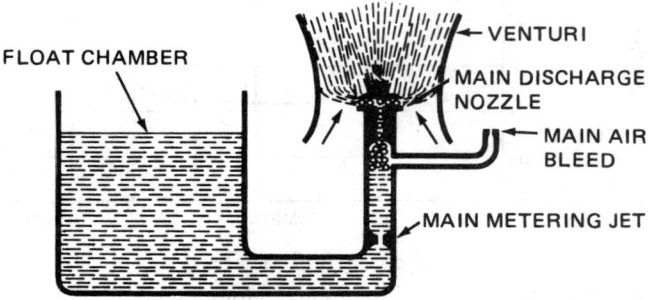

FIGURE 6-18 Basic venturi-type carburetor.

of fuel supplied to the nozzle. A **metering jet** is an orifice, or opening, which is carefully dimensioned to meter (measure) fuel flow accurately in accordance with the pressure differential between the float chamber and discharge nozzle. The metering jet is an essential part of the main metering system.

Air Bleed

The **air bleed** in a carburetor lifts an emulsion of air and liquid to a higher level above the liquid level in the float chamber than would be possible with unmixed fuel. Figure 6-19 shows a person sucking on a straw placed in a glass of water. The suction is great enough to lift the water above the level in the glass without that person drawing any into the mouth. In Fig. 6-20 a tiny hole has been pricked in the side of the straw above the surface of the water in the glass, and the same suction is applied as before. The hole causes bubbles of air to enter the straw, and liquid is drawn up in a series of small drops or slugs.

In Fig. 6-21 the air is taken into a tube through a smaller tube which enters the main tube below the level of the water. There is a restricting orifice at the bottom of the main tube; that is, the size of the main tube is reduced at the bottom. Instead of a continuous series of small drops or slugs being drawn up through the tube when the person sucks on it, a finely divided mixture of air and water is formed in the tube.

Since there is a distance through which the water must be lifted from its level in the glass before the air begins to pick it up, the free opening of the main tube at the bottom prevents

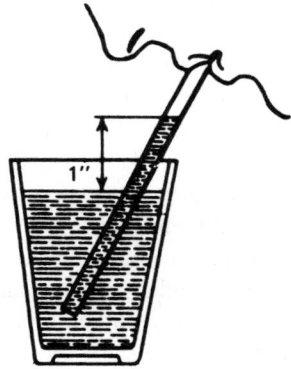

FIGURE 6-19 Suction lifts a liquid.

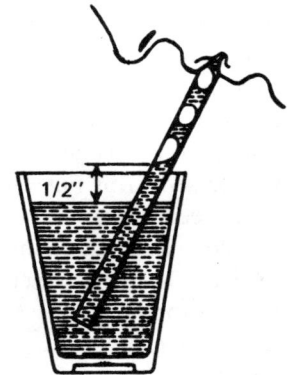

FIGURE 6-20 Effects of air bleed.

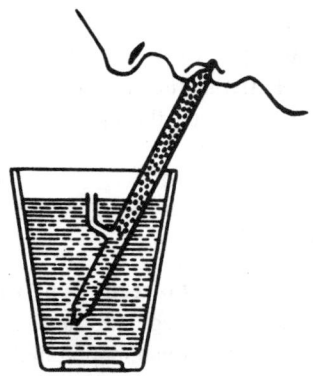

FIGURE 6-21 Air bleed breaking up a liquid.

a very great suction from being exerted on the air-bleed hole or vent. If the air openings were too large in proportion to the size of the main tube, the suction available to lift the water would be reduced.

In Fig. 6-21 the ratio of water to air could be modified for high and low airspeeds (produced by sucking on the main tube) by changing the dimensions of the air bleed, the main tube, and the opening at the bottom of the main tube.

The carburetor nozzle in Fig. 6-18 has an air bleed, as explained previously. We can summarize our discussion by stating that the purpose of this air bleed in the discharge nozzle is to assist in the production of a more uniform mixture of fuel and air throughout all operating speeds of the engine.

Vaporization of Fuel

The fuel leaves the discharge nozzle of the carburetor in a stream which breaks up into drops of various sizes suspended in the airstream, where they become even more finely divided. Vaporization occurs on the surface of each drop, causing the very fine particles to disappear and the large particles to decrease in size. The problem of properly distributing the particles would be simple if all the particles in each drop vaporized completely before the mixture left the intake pipe. But some particles of the fuel enter the engine cylinders while they are still in a liquid state and thus must be vaporized and mixed in the cylinder during the compression stroke.

The completeness of vaporization depends on the volatility of the fuel, the temperature of the air, and the degree of atomization. Volatile means readily vaporized; therefore, the more volatile fuels evaporate more readily. Higher temperatures increase the rate of vaporization; therefore, carburetor air intake heaters are sometimes provided. Some engines are equipped with "hot-spot" heaters which utilize the heat of exhaust gases to heat the intake manifold between the carburetor and the cylinders. This is usually accomplished by routing a portion of the engine exhaust through a jacket surrounding the intake manifold. In another type of hot-spot heater, the intake manifold is passed through the oil reservoir of the engine. The hot oil supplies heat to the intake manifold walls, and the heat is transferred to the F/A mixture.

The degree of atomization is the extent to which fine spray is produced; the more fully the mixture is reduced to fine spray and vaporized, the greater is the efficiency of the combustion process. The air bleed in the main discharge nozzle passage aids in the atomization and vaporization of the fuel. If the fuel is not fully vaporized, the mixture may run lean even though an abundance of fuel is present.

Throttle Valve

A **throttle valve**, usually a **butterfly-type valve**, is incorporated in the fuel-air duct to regulate the fuel-air output. The throttle valve is usually an oval-shaped metal disk mounted on the throttle shaft in such a manner that it can completely close the throttle bore. In the closed position, the plane of the disk makes an angle of about 70° with the axis of the throttle bore. The edges of the throttle disk are shaped to fit closely against the sides of the fuel-air passage. The arrangement of such a valve is shown in Fig. 6-22. The amount of air flowing through the venturi tube is reduced when the valve is turned toward its closed position. This reduces the suction in the venturi tube, so that less fuel is delivered to the engine. When the throttle valve is opened, the flow of the F/A mixture to the engine is increased. Opening or closing the throttle valve thus regulates the power output of the engine. In Fig. 6-23, the throttle valve is shown in the OPEN position.

FIGURE 6-22   Throttle valve.

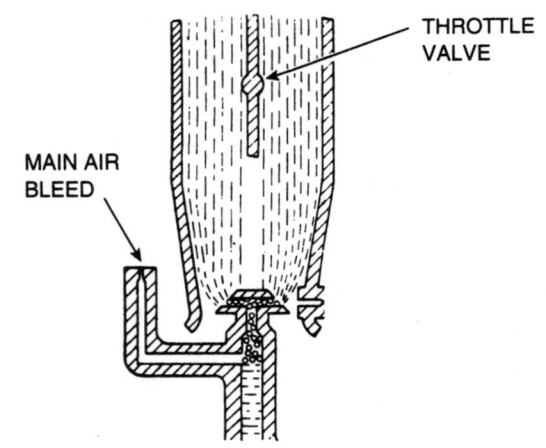

THROTTLE VALVE

MAIN AIR BLEED

FIGURE 6-23   Throttle valve in open position.

FLOAT-TYPE CARBURETORS

Essential Parts of a Carburetor

The carburetor consists essentially of a main air passage through which the engine draws its supply of air, mechanisms to control the quantity of fuel discharged in relation to the flow of air, and a means for regulating the quantity of F/A mixture delivered to the engine cylinders.

The essential parts of a float-type carburetor are (1) the float mechanism and its chamber, (2) the strainer, (3) the main metering system, (4) the idling system, (5) the economizer (or power enrichment) system, (6) the accelerating system, and (7) the mixture control system.

In the float-type carburetor, atmospheric pressure in the fuel chamber forces fuel from the discharge nozzle when the pressure is reduced at the venturi tube. The intake stroke of the piston reduces the pressure in the engine cylinder, thus causing air to flow through the intake manifold to the cylinder. Thus flow of air passes through the venturi of the carburetor and causes the reduction of presure in the venturi which, in turn, causes the fuel to be sprayed from the discharge nozzle.

Float Mechanism

As previously explained, the float in a carburetor is designed to control the level of fuel in the float chamber. This fuel level must be maintained slightly below the discharge-nozzle outlet holes to provide the correct amount of fuel flow and to prevent leakage of fuel from the nozzle when the engine is not running. The arrangement of a float mechanism in relation to the discharge nozzle is shown in Fig. 6-24. Note that the float is attached to a lever which is pivoted and that one end of the lever is engaged with the float needle valve. When the float rises, the needle valve closes and stops the flow of fuel into the chamber. At this point, the fuel level is correct for proper operation of the carburetor, provided that the needle valve seat is at the correct level.

As shown in Fig. 6-24, the float valve mechanism includes a needle and a seat. The needle valve is constructed of hardened steel, or it may have a synthetic-rubber section which fits the seat. The needle seat is usually made of bronze. There must be a good fit between the needle and seat to prevent fuel leakage and overflow from the discharge nozzle.

During operation of the carburetor, the float assumes a position slightly below its highest level, to allow a valve opening sufficient for replacement of the fuel as it is drawn out through the discharge nozzle. If the fuel level in the float chamber is too high, the mixture will be rich; if the fuel is too low, the mixture will be lean. To adjust the fuel level for the carburetor shown in Fig. 6-24, washers are placed under the float needle seat. If the fuel level (float level) needs to be raised, washers are removed from under the seat. If the level needs to be lowered, washers are added. The specifications for the float level are given in the manufacturer's overhaul manual.

For some carburetors, the float level is adjusted by bending the float arm.

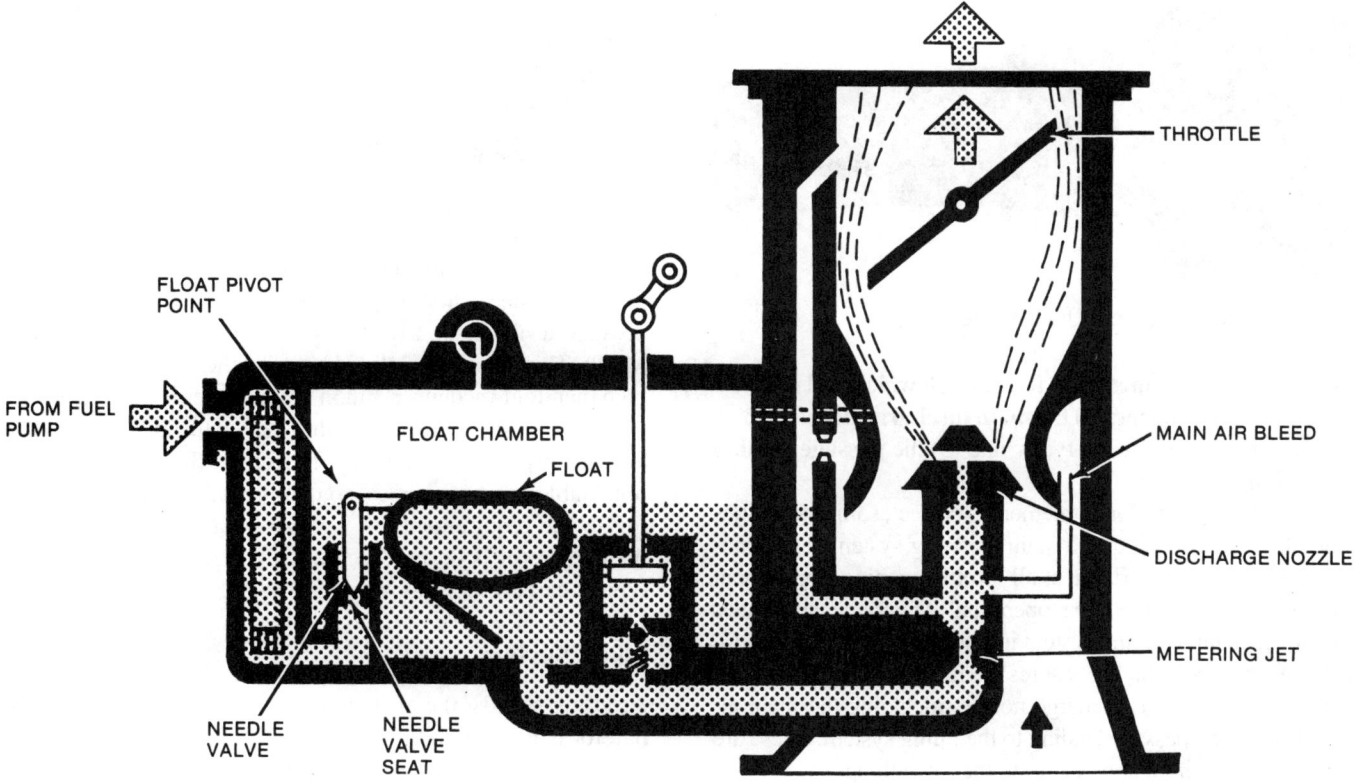

FIGURE 6-24 Float and needle valve mechanism in a carburetor.

Figure 6-25 shows a cutaway float carburetor. The inlet screen area and float chamber area are removed to show the float, inlet screen, mixture control, and the needle valve and seat.

Fuel Strainer

In most carburetors, the fuel supply must first enter a strainer chamber, where it passes through a strainer screen. The **strainer** consists of a fine wire mesh or other type of filtering device, cone-shaped or cylindrically shaped, located so that it will intercept any dirt particles which might clog the needle valve opening or, later, the metering jets. The strainer is usually removable so that it can be taken out and thoroughly drained and flushed. A typical strainer is shown in Fig. 6-26.

Main Metering System

The **main metering system** controls the fuel feed in the upper half of the engine speed range as used for cruising and full-throttle operations. It consists of three principal divisions, or

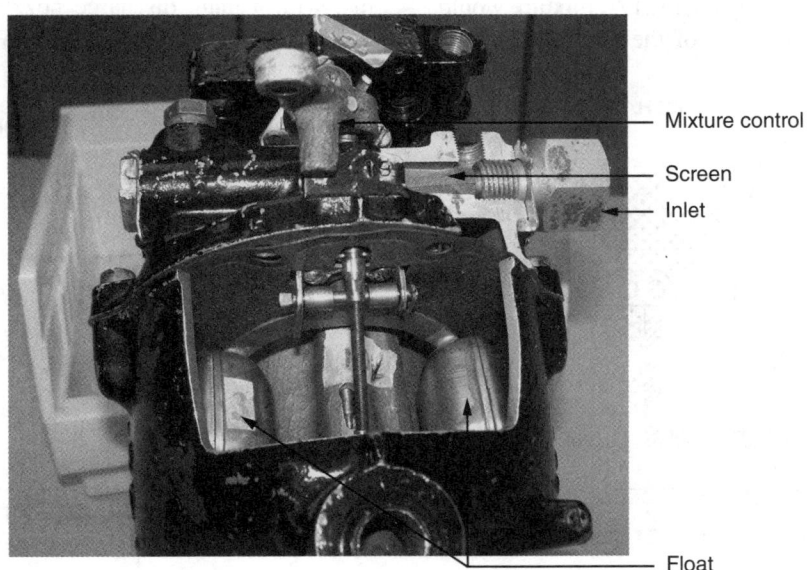

FIGURE 6-25 Concentric and eccentric float mechanisms.

FIGURE 6-26 Carburetor fuel strainer.

units: (1) the **main metering jet**, through which fuel is drawn from the float chamber; (2) the **main discharge nozzle**, which may be any one of several types; and (3) the **passage leading to the idling system**.

Although the previous statement is correct, some authorities state that the purpose of the main metering system is to maintain a constant F/A mixture at all throttle openings throughout the power range of engine operation. The same authorities divide the main metering system into four parts; (1) the venturi, (2) a metering jet which measures the fuel drawn from the float chamber, (3) a main discharge nozzle, including the main air bleed, and (4) a passage leading to the idling system. These are merely two different approaches to the same thing.

The three functions of the main metering system are (1) to proportion the F/A mixture, (2) to decrease the pressure at the discharge nozzle, and (3) to control the airflow at full throttle.

The airflow through an opening of fixed size and the fuel flow through an air-bleed jet system respond to variations of pressure in approximately equal proportions. If the discharge nozzle of the air-bleed system is located in the center of the venturi, so that both the air-bleed nozzle and the venturi are exposed to the suction of the engine in the same degree, it is possible to maintain an approximately uniform mixture of fuel and air throughout the power range of engine operations. This is illustrated in Fig. 6-27, which shows the air-bleed principle and the fuel level of the float chamber in a typical carburetor. If the main air bleed of a carburetor should become restricted or clogged, the F/A mixture would be excessively rich because more of the available suction

would be acting on the fuel in the discharge nozzle and less air would be introduced with the fuel.

The full-power output from the engine makes it necessary to have, above the throttle valve, a manifold suction (reduced pressure, or partial vacuum) which is between 0.4 and 0.8 psi [2.8 and 5.5 kPa] at full engine speed. However, more suction is desired for metering and spraying the fuel and is obtained from the venturi. When a discharge nozzle is located in the central portion of the venturi, the suction obtained is several times as great as the suction found in the intake manifold. Thus, it is possible to maintain a relatively low manifold vacuum (high MAP). This results in high volumetric efficiencies. In contrast, high manifold vacuums result in low volumetric efficiencies.

We have stated previously that the venturi tube affects the air capacity of the carburetor. Therefore, the tube should be obtainable in various sizes, so that it can be selected according to the requirements of the particular engine for which the carburetor is designed. The main metering system for a typical carburetor is shown in Fig. 6-28.

Idling System

At idling speeds, the airflow through the venturi of the carburetor is too low to draw sufficient fuel from the discharge nozzle, so the carburetor cannot deliver enough fuel to keep the engine running. At the same time, with the throttle nearly closed, the air velocity is high and the pressure is low between the edges of the throttle valve and the walls of the air passages. Furthermore, there is very high suction on the intake side of the throttle valve. Because of this situation, an idling system with an outlet at the throttle valve is added. This idling system delivers fuel only when the throttle valve is nearly closed and the engine is running slowly. An **idle cutoff** valve stops the flow of fuel through this idling system on some carburetors, and this is used for stopping the engine. An increased amount of fuel (richer mixture) is used in the idle range because at idling speeds there may not be enough air flowing around its cylinders to provide proper cooling.

Figure 6-29 shows a three-piece main discharge assembly, with a main discharge nozzle, main air bleed, main-discharge-nozzle stud, idle feed passage, main metering jet,

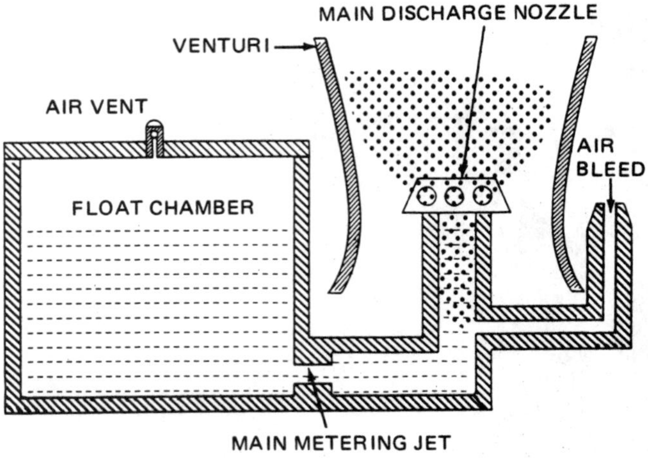

FIGURE 6-27 Location of air-bleed system and main discharge nozzle.

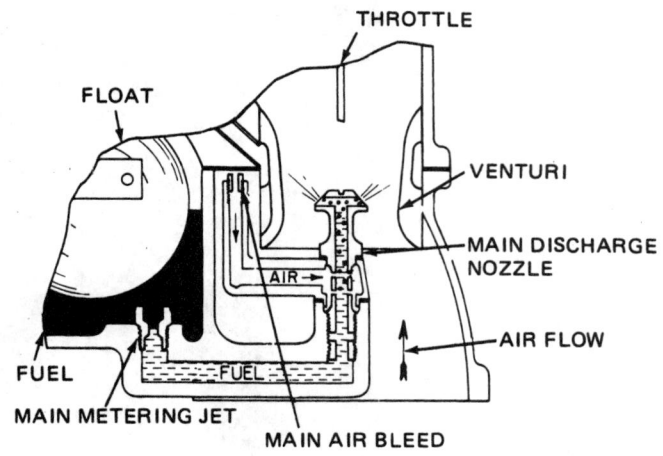

FIGURE 6-28 Main metering system in a carburetor. (*Continental Motors.*)

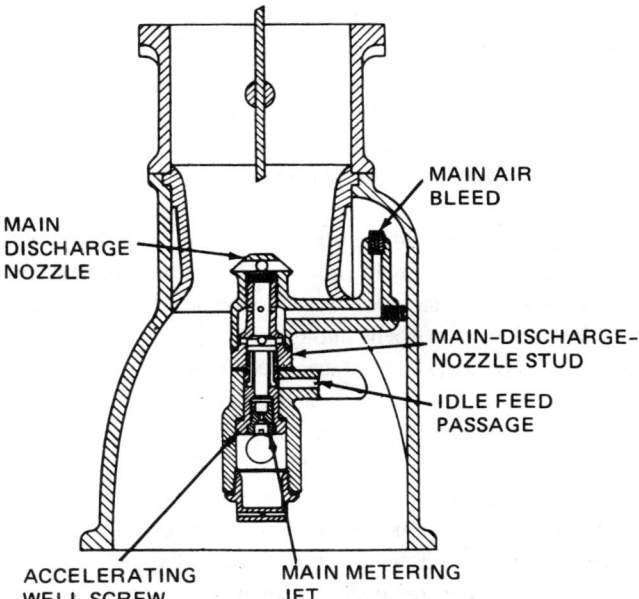

FIGURE 6-29 Three-piece main discharge assembly.

and accelerating well screw. This is one of the two types of main discharge nozzle assemblies used in updraft, float-type carburetors. An **updraft carburetor** is one in which the air flows upward through the carburetor to the engine. In the other type, the main discharge nozzle and the main-discharge-nozzle stud are combined in one piece that is screwed directly into the discharge-nozzle boss, which is part of the main body casting, thereby eliminating the necessity of having a discharge-nozzle screw.

Figure 6-30 is a drawing of a conventional idle system, showing the idling discharge nozzles, the mixture adjustment, the idle air bleed, the idle metering jet, and the idle

metering tube. Note that the fuel for the idling system is taken from the fuel passage for the main discharge nozzle and that the idle air-bleed air is taken from a chamber outside the venturi section. Thus the idle air is at air inlet pressure. The idle discharge is divided between two discharge nozzles, and the relative quantities of fuel flowing through these nozzles are dependent on the position of the throttle valve. At very low idle, all the fuel passes through the upper orifice, since the throttle valve covers the lower orifice. In this case, the lower orifice acts as an additional air bleed for the upper orifice. As the throttle is opened further, exposing the lower orifice, additional fuel passes through this opening.

Since the idle-mixture requirements vary with climatic conditions and altitude, a needle valve type of mixture adjustment is provided to vary the orifice in the upper idle discharge hole. Moving this needle in or out of the orifice varies the idle fuel flow accordingly, to supply the correct F/A ratio to the engine.

The idling system described above is used in the Bendix-Stromberg NA-S3A1 carburetor and is not necessarily employed in other carburetors. The principles involved are similar in all carburetors, however.

Figure 6-31 A, B, and C show a typical float-type carburetor at idling speed, medium speed, and full speed, respectively. The greatest suction (pressure reduction) in the intake

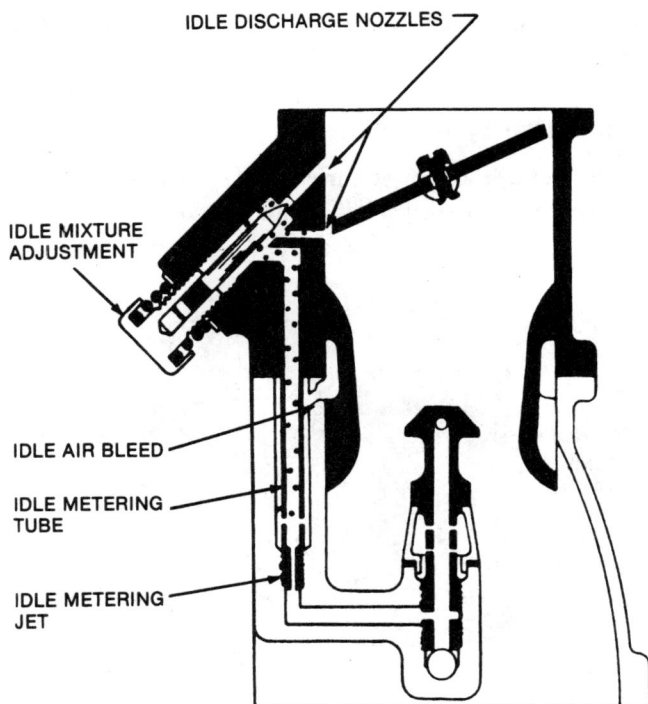

FIGURE 6-30 Conventional idle system.

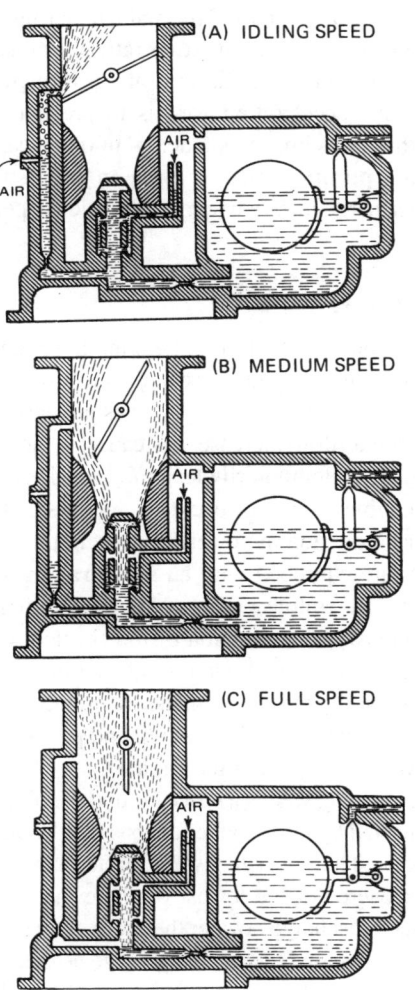

FIGURE 6-31 Float-type carburetor at different engine speeds.

manifold above the throttle is at the lowest speeds, when the smallest amount of air is received, which is also the condition requiring the smallest amount of fuel. When the engine speed increases, more fuel is needed, but the suction in the manifold decreases. For this reason, the metering of the idling system is not accomplished by the suction existing in the intake manifold. Instead, the metering is controlled by the suction existing in a tiny intermediate chamber, or slot, formed by the idling discharge nozzle and the wall of the carburetor at the edge of the throttle valve. This chamber has openings into the barrel of the carburetor, both above and below the throttle.

In Figs. 6-30 and 6-31, note that there is a small chamber surrounding the main discharge-nozzle passage just below the main air-bleed inlet. This chamber serves as an **accelerating well** to store extra fuel that is drawn out when the throttle is suddenly opened. If this extra supply of fuel were not immediately available, the fuel flow from the discharge nozzle would be momentarily decreased and the mixture entering the combustion chamber would be too lean, thereby causing the engine to hesitate or misfire.

In the carburetor shown in Fig. 6-31, when the engine is operating at intermediate speed, the accelerating well still holds some fuel. However, when the throttle is wide open, all the fuel from the well is drawn out. At full power, all fuel is supplied through the main discharge and economizer system, and the idling system then acts as an auxiliary air bleed to the main metering system. The main metering jet provides an approximately constant mixture ratio for all speeds above idling, but it has no effect during idling. Remember that the purpose of the accelerating well is to prevent a power lag when the throttle is opened suddenly. In many carburetors, an **accelerating pump** is used to force an extra supply of fuel from the discharge nozzle when the throttle is opened quickly.

Accelerating System

When the throttle controlling an engine is suddenly opened, there is a corresponding increase in the airflow; but because of the inertia of the fuel, the fuel flow does not accelerate in proportion to the airflow increase. Instead, the fuel lags behind, which results in a temporarily lean mixture. This, in turn, may cause the engine to miss or backfire, and it is certain to cause a temporary reduction in power. To prevent this condition, all carburetors are equipped with an **accelerating system**. This is either an accelerating pump or an accelerating well, mentioned previously. The function of the accelerating system is to discharge an additional quantity of fuel into the carburetor airstream when the throttle is opened suddenly, thus causing a temporary enrichment of the mixture and producing a smooth and positive acceleration of the engine.

The accelerating well is a space around the discharge nozzle and is connected by holes to the fuel passage leading to the discharge nozzle. The upper holes are located near the fuel level and are uncovered at the lowest pressure that will draw fuel from the main discharge nozzle; therefore, they receive air during the entire time that the main discharge nozzle operates.

Very little throttle opening is required at idling speeds. When the throttle is opened suddenly, air is drawn in to fill

the intake manifold and whichever cylinder is on the intake stroke. This sudden rush of air temporarily creates a high suction at the main discharge nozzle, brings into operation the main metering system, and draws additional fuel from the accelerating well. Because of the throttle opening, the engine speed increases and the main metering system continues to function.

The accelerating pump illustrated in Fig. 6-32 is a sleeve-type piston pump operated by the throttle. The piston is mounted on a stationary hollow stem screwed into the body of the carburetor. The hollow stem opens into the main fuel passage leading to the discharge nozzle. Mounted over the stem and piston is a movable cylinder, or sleeve, which is connected by the pump shaft to the throttle linkage. When the throttle is closed, the cylinder is raised and the space within the cylinder fills with fuel through the clearance between the piston and the cylinder. If the throttle is quickly moved to the open position, the cylinder is forced down, as shown in Fig. 6-33, and the increased fuel pressure also forces the piston partway down along the stem. As the piston moves down, it opens the pump valve and permits the fuel to flow through the hollow stem into the main fuel passage. With the throttle fully open and the accelerating pump cylinder all the way down, the spring pushes the piston up and forces most of the fuel out of the cylinder. When the piston reaches its highest position, it closes the valve and no more fuel flows toward the main passage.

There are several types of accelerating pumps, but each serves the purpose of providing extra fuel during rapid throttle opening and acceleration of the engine.

When a throttle is moved slowly toward the OPEN position, the accelerating pump does not force extra fuel into the discharge system, because the spring in the pump holds the valve closed unless the fuel pressure is great enough to overcome the spring pressure. When the throttle is moved slowly, the trapped fuel seeps out through the clearance between the piston and the cylinder, and the pressure does not build up enough to open the valve.

Economizer System

An **economizer**, or **power enrichment system**, is essentially a valve which is closed at low engine and cruising speeds but

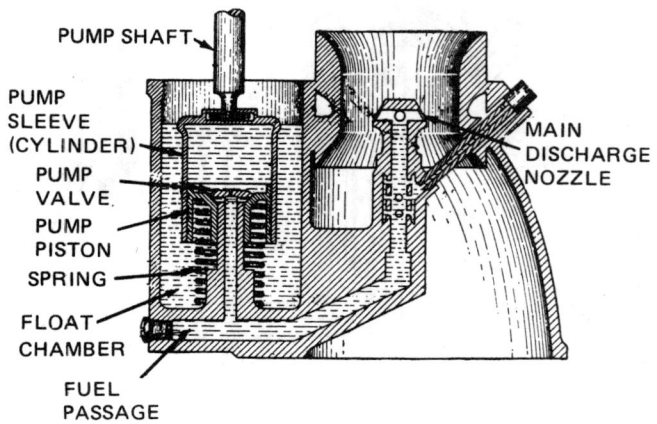

FIGURE 6-32 Movable-piston-type accelerating pump.

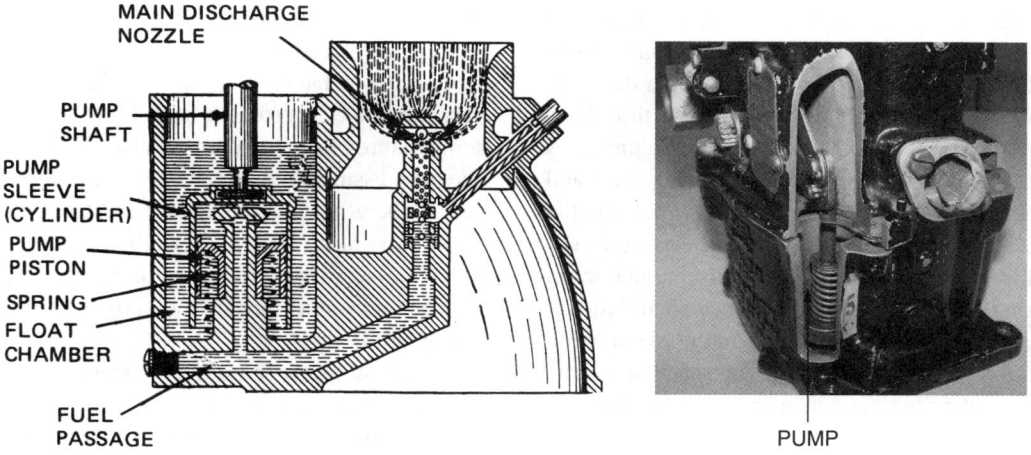

FIGURE 6-33 Accelerating pump in operation.

is opened at high speeds to provide an enriched mixture to reduce burning temperatures and prevent detonation. In other words, this system supplies and regulates the additional fuel required for all speeds above the cruising range. An economizer is also a device for enriching the mixture at increased throttle settings. It is important, however, that the economizer close properly at cruising speed; otherwise, the engine may operate satisfactorily at full throttle but will "load up" at and below cruising speed because of the extra fuel being fed into the system. The extra-rich condition is indicated by rough running and by black smoke emanating from the exhaust.

The economizer gets its name from the fact that it enables the pilot to obtain maximum economy in fuel consumption by providing for a lean mixture during cruising operation and a rich mixture for full-power settings. Most economizers in their modern form are merely enriching devices. The carburetors equipped with economizers are normally set for their leanest practical mixture delivery at cruising speeds, and enrichment takes place as required for higher power settings.

Three types of economizers for float-type carburetors are (1) the needle valve type, (2) the piston type, and (3) the MAP operated type. Figure 6-34 illustrates the **needle valve economizer**. This mechanism utilizes a needle valve which is opened by the throttle linkage at a predetermined throttle position.

This permits a quantity of fuel, in addition to the fuel from the main metering jet, to enter the discharge-nozzle passage. As shown in Fig. 6-34, the economizer needle valve permits fuel to bypass the cruise-valve metering jet.

The **piston-type economizer**, illustrated in Fig. 6-35, is also operated by the throttle. The lower piston serves as a

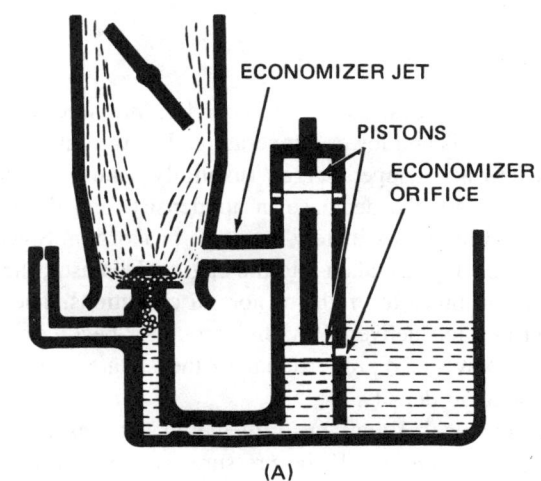

(A)

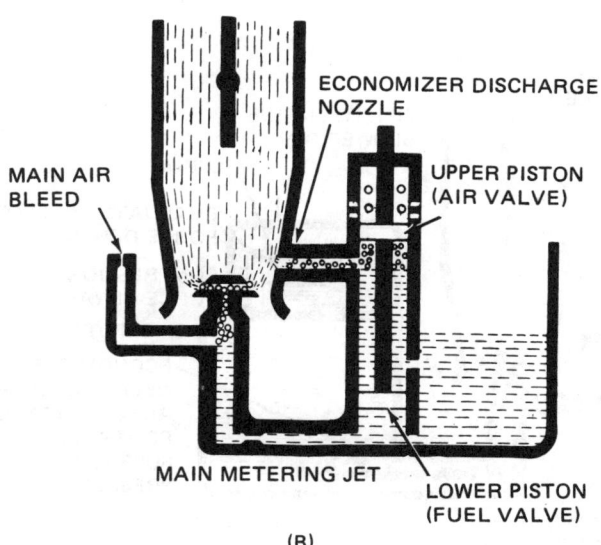

(B)

FIGURE 6-35 Piston-type economizer.

FIGURE 6-34 Needle valve economizer.

Float-Type Carburetors 149

fuel valve, preventing any flow of fuel through the system at cruising speeds. (See view A.) The upper piston functions as an air valve, allowing air to flow through the separate economizer discharge nozzle at part throttle. As the throttle is opened to higher power positions, the lower piston uncovers the fuel port leading from the economizer metering valve and the upper piston closes the air ports (view B). Fuel fills the economizer well and is discharged into the carburetor venturi where it adds to the fuel from the main discharge nozzle. The upper piston of the economizer permits a small amount of air to bleed into the fuel, thus assisting in the atomization of the fuel from the economizer system. The space below the lower piston of the economizer acts as an accelerating well when the throttle is opened.

The **MAP-operated economizer**, illustrated in Fig. 6-36, has a bellows which is compressed when the pressure from the engine blower rim produces a force greater than the resistance of the compression spring in the bellows chamber. As the engine speed increases, the blower pressure also increases. This pressure collapses the bellows and causes the economizer valve to open. Fuel then flows through the economizer metering jet to the main discharge system. The operation of the bellows and spring is stabilized by means of a dashpot, as shown in the drawing.

Mixture Control System

At higher altitudes, the air is under less pressure, has less density, and is at a lower temperature. The weight of the air taken into an unsupercharged (naturally aspirated) engine decreases with the decrease in air density, and the power is reduced in approximately the same proportion. Since the quantity of oxygen taken into the engine decreases, the F/A mixture becomes too rich for normal operations. The mixture proportion delivered by the carburetor becomes richer at a rate inversely proportional to the square root of the increase in air density.

Remember that the density of the air changes with temperature and pressure. If the pressure remains constant, the density of air will vary according to temperature, increasing

as the temperature drops. This will cause leaning of the F/A mixture in the carburetor because the denser air contains more oxygen. The change in air pressure due to altitude is considerably more of a problem than the change in density due to temperature changes. At an altitude of 18 000 ft [5 486.4 m], the air pressure is approximately one-half the pressure at sea level. Therefore, to provide a correct mixture, the fuel flow must be reduced to almost one-half what it would be at sea level. Adjustment of fuel flow to compensate for changes in air pressure and temperature is a principal function of the mixture control.

Briefly, a **mixture control system** can be described as a mechanism or device through which the richness of the mixture entering the engine during flight can be controlled to a reasonable extent. This control should be maintainable at all normal altitudes of operation.

The functions of the mixture control system are (1) to prevent the mixture from becoming too rich at high altitudes and (2) to economize on fuel during engine operation in the low-power range where cylinder temperature will not become excessive with the use of the leaner mixture.

Mixture control systems may be classified according to their principles of operation as (1) the **back-suction** type, which reduces the effective suction on the metering system; (2) the **needle** type, which restricts the flow of fuel through the metering system; and (3) the **air-port** type, which allows additional air to enter the carburetor between the main discharge nozzle and the throttle valve. Figure 6-37 shows two views of a back-suction-type mixture control system. The left view shows the mixture control valve in the closed position. This cuts off the atmospheric pressure from the space above the fuel in the fuel chamber. Since the float chamber is connected to the low-pressure area in the venturi of the carburetor, the pressure above the fuel in the float chamber will be reduced until fuel is no longer delivered from the discharge nozzle. This acts as an idle cutoff and stops the engine. In some carburetors, the end of the back-suction tube is located where the pressure is somewhat higher than that at the nozzle, thus making it possible for the mixture control valve to be completely closed without stopping the flow of fuel. The fuel flow is varied by adjusting the opening of the mixture control valve. To lean the mixture, the valve is moved toward the closed position; to enrich the mixture, the valve is moved toward the open position. The right-hand drawing in Fig. 6-37 shows the valve in the full rich position.

To reduce the sensitivity of the back-suction mixture control, a disk-type valve is sometimes used. This valve is constructed so that a portion of the valve opening can be closed rapidly at first and the remainder of the opening can be closed gradually. A disk-type mixture control valve is shown in Fig. 6-38. This assembly is called an altitude-control-valve **disk and plate**. The arrangement of the mixture control for an NA-S3A1 carburetor is shown in Fig. 6-39.

A needle-type mixture control is shown in Fig. 6-40. In this control, the needle is used to restrict the fuel passage through the main metering jet. When the mixture control is in the full rich position, the needle is in the fully raised position and the fuel is accurately measured by the main metering jet.

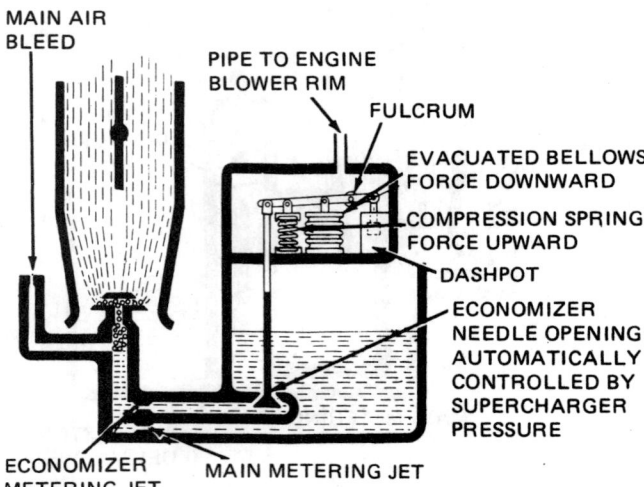

MAIN AIR
BLEED

PIPE TO ENGINE
BLOWER RIM

FULCRUM

EVACUATED BELLOWS
FORCE DOWNWARD

COMPRESSION SPRING
FORCE UPWARD

DASHPOT

ECONOMIZER
NEEDLE OPENING
AUTOMATICALLY
CONTROLLED BY
SUPERCHARGER
PRESSURE

ECONOMIZER
METERING JET

MAIN METERING JET

FIGURE 6-36   MAP-operated economizer.

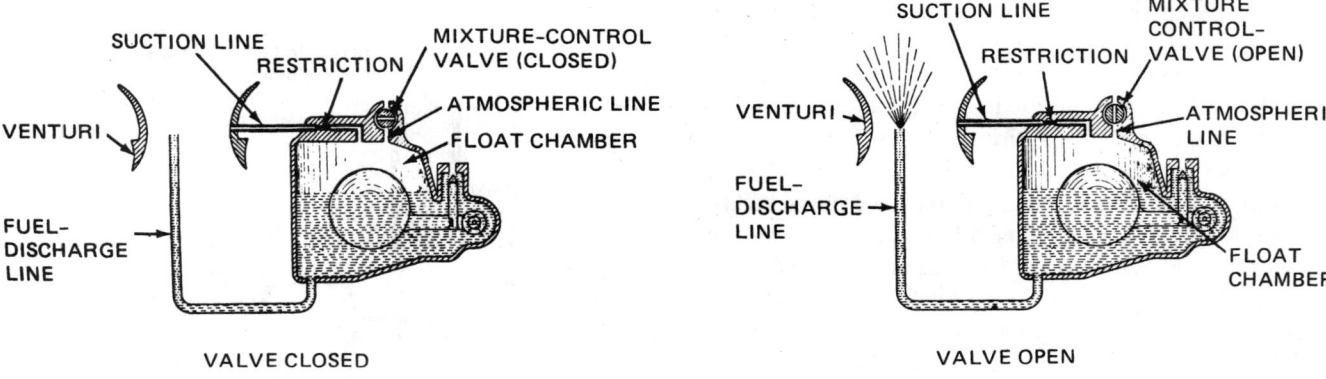

FIGURE 6-37   Back-suction-type mixture control.

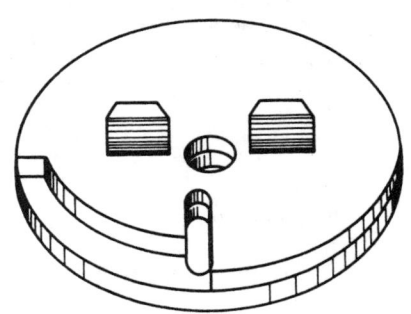

FIGURE 6-38   Disk-type mixture control valve.

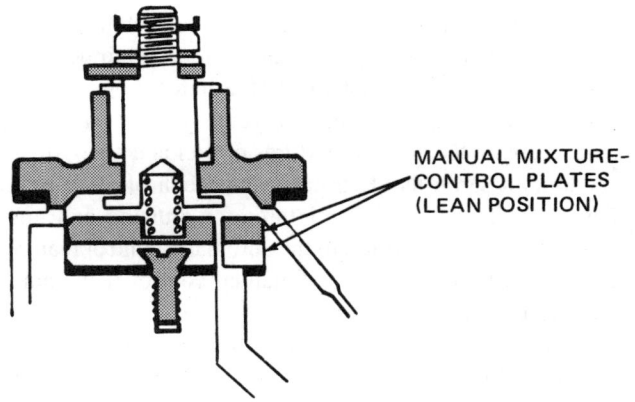

FIGURE 6-39   Mixture control for NA-S3A1 carburetor.

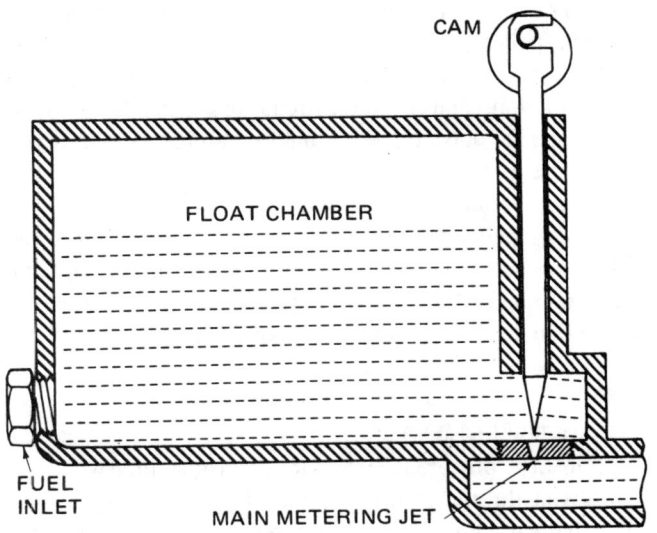

FIGURE 6-40   Needle-type mixture control.

The needle valve is lowered into the needle valve seat to lean the mixture, thus reducing the supply of fuel to the main discharge nozzle. Even though the needle valve is completely closed, a small bypass hole from the float chamber to the fuel passage allows some fuel to flow; therefore, the size of this bypass hole determines the control range.

The air-port type of mixture control, illustrated in Fig. 6-41, has an air passage leading from the region between the venturi tube and the throttle valve to atmospheric pressure. In the air passage is a butterfly valve which is manually controlled by the pilot in the cockpit. Obviously, when the pilot opens the butterfly valve in the air passage, air which has not been mixed with fuel will be injected into the F/A mixture. At the same time, the suction in the intake manifold will be reduced, thereby reducing the velocity of the air coming through the venturi tube. This will further reduce the amount of fuel being drawn into the intake manifold.

Idle Cutoff

The term "idle cutoff" has been mentioned previously; it describes the position of certain mixture controls in which the control is enabled to stop the flow of fuel into the intake airstream. Some float-type carburetors and the majority of pressure-type carburetors incorporate the IDLE CUTOFF position in the mixture control system.

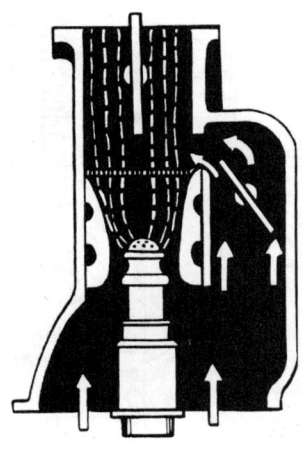

FIGURE 6-41   Air-port-type mixture control.

Float-Type Carburetors   **151**

Essentially, the idle cutoff system stops the flow of fuel from the discharge nozzle and is therefore used to stop the engine. This provides an important safety factor, because it eliminates the combustible mixture in the intake manifold and prevents the engine from firing as a result of a hot spot in one or more cylinders. In some cases, engines which have been stopped by turning off the ignition switch have kicked over after stopping, thus creating a hazard to someone who may move the propeller. In an engine equipped with the idle cutoff feature, the engine ignition switch is turned off *after* the engine is first stopped by means of moving the mixture control to the idle cutoff position. This procedure also eliminates the possibility of unburned fuel entering the cylinder and washing the oil film from the cylinder walls.

Automatic Mixture Control

Some of the more complex aircraft carburetors are often equipped with a device for automatically controlling the mixture as altitude changes. **Automatic mixture control (AMC) systems** may be operated on the back-suction principle and the needle valve principle or by throttling the air intake to the carburetor. In the latter type of AMC, the control regulates power output within certain limits in addition to exercising its function as a mixture control.

In AMC systems operating on the back-suction and needle valve principles, the control may be directly operated by the expansion and contraction of a pressure-sensitive evacuated bellows through a system of mechanical linkage. This is the simplest form of AMC and is generally found to be accurate, reliable, and easy to maintain. Some mixture control valves, such as the one illustrated in Fig. 6-42, are operated by a sealed bellows in a compartment vented to the atmosphere; therefore, the fuel flow is proportional to the atmospheric pressure. Figure 6-43 shows the bellows type of mixture control valve installed on a carburetor

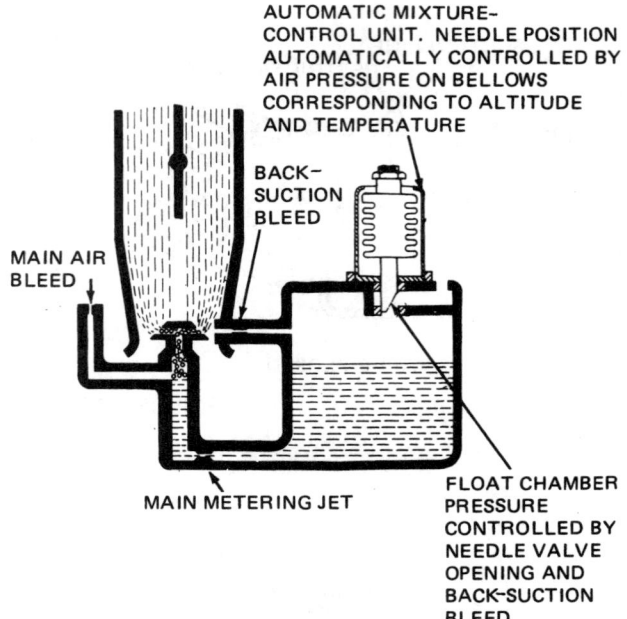

FIGURE 6-43   AMC in operation.

as a back-suction control device. As atmospheric pressure decreases, the bellows will expand and begin to close the opening into the fuel chamber. This will cause a reduction of pressure in the chamber, resulting in a decreased flow of fuel from the discharge nozzle. In some systems equipped with external superchargers (not illustrated here), both the fuel chamber and the bellows may be vented to the carburetor intake to obtain the correct mixtures of fuel and air.

Automatic controls often have more than one setting in order to obtain the correct mixtures for cruising and high speed operation. In addition to the automatic control feature, there is usually a provision for manual control if the automatic control fails.

When the engine is equipped with a fixed-pitch propeller which allows the engine speed to change as the mixture changes, a manually operated mixture control can be adjusted by observing the change in engine rpm as the control is moved. Obviously, this will not work with a constant-speed propeller.

If a constant-speed propeller cannot be locked into fixed-pitch position, and if the extreme-pitch positions cause engine speeds outside the normal operating range in flight, it is necessary to have an instrument of some type that indicates the F/A ratio or power output.

If the propeller can be locked in a fixed-pitch position, and if this does not lead to engine speeds outside the normal flight operating range, the following expressions may be employed to describe the manual adjustments of the mixture control.

Full rich. The mixture control setting in the position for maximum fuel flow.

Rich best power. The mixture control setting which, at a given throttle setting, permits maximum engine rpm with the mixture control as far toward full rich as possible without reducing the rpm.

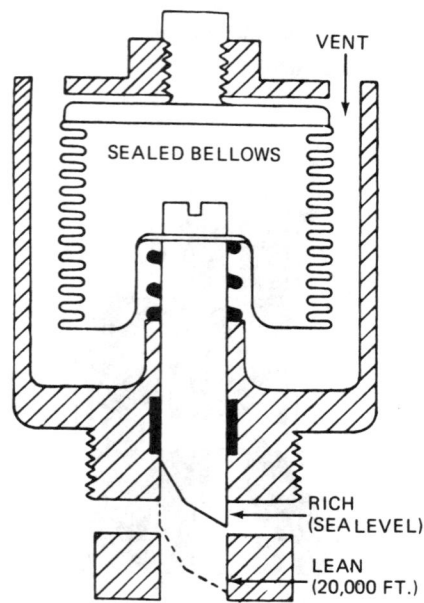

FIGURE 6-42   AMC mechanism.

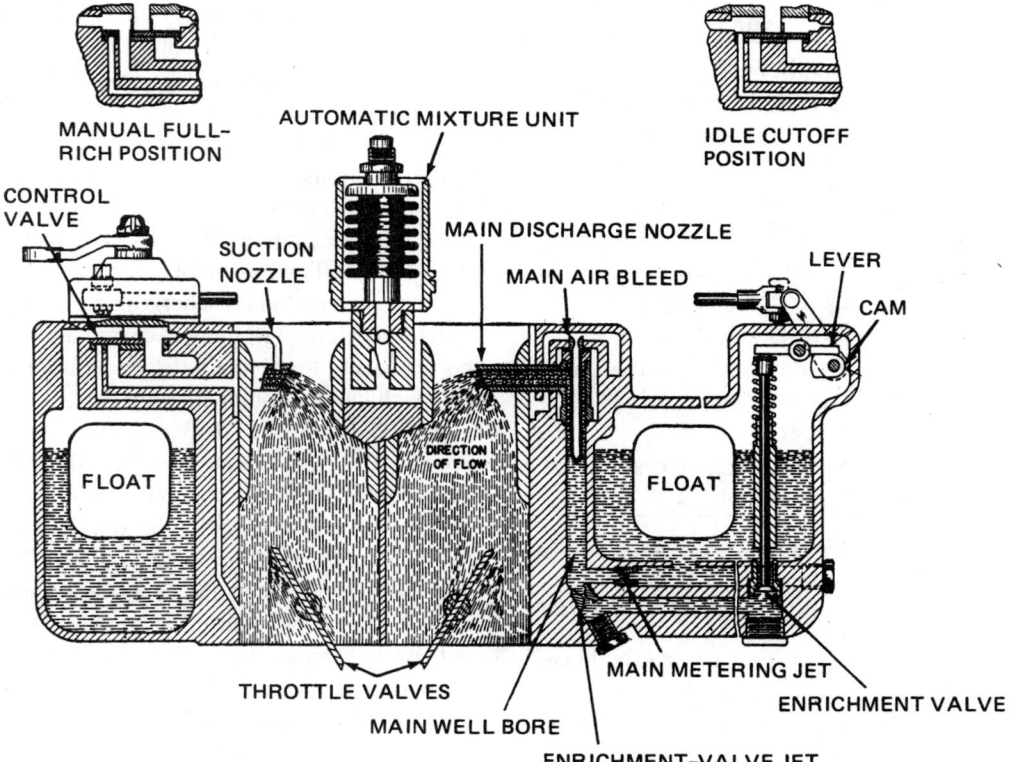

FIGURE 6-44   Downdraft carburetor.

Lean best power. The mixture control setting which, at a given throttle setting, permits maximum engine rpm with the mixture control as far toward lean as possible without reducing the rpm.

Downdraft Carburetors

So far we have principally dealt with updraft carburetors. Updraft means that the air through the carburetor is flowing upward. A **downdraft carburetor** (Fig. 6-44) takes air from above the engine and causes the air to flow down through the carburetor. Those who favor downdraft carburetors claim that they reduce fire hazard, provide better distribution of mixture to the cylinders of an upright engine, and have less tendency to pick up sand and dirt from the ground.

Downdraft carburetors are very similar in function and systems to the updraft types. Figure 6-44 illustrates one of several types of downdraft carburetors used for aircraft. This particular model has two float chambers, two throttle valves, and an AMC unit.

Figure 6-45 illustrates a portion of a downdraft carburetor and its idling system, emphasizing the path of the fuel leaving the float chamber and the position of the idle air bleed. When the engine is not operating, this air bleed, in addition to its other functions, prevents the siphoning of fuel. An average intake pressure to the mixture control system is supplied by the series of vents at the entrance to the venturi.

Model Designation

All Bendix-Stromberg float-type aircraft carburetors carry the general model designation NA followed by a hyphen; the next letter indicates the type, as shown in Fig. 6-46.

Following the type letter is a numeral that indicates the nominal rated size of the carburetor, starting at 1 in [2.54 cm], which is no. 1, and increasing in $\frac{1}{4}$-in [6.35-mm] steps. For example, a 2-in [5.08-cm] carburetor in rated no. 5. The actual diameter of the carburetor barrel opening is $\frac{3}{16}$ in [4.76 mm] greater than the nominal rated size, in accordance with the standards of the Society of Automotive Engineers (SAE). A final letter is used to designate various models of a given type.

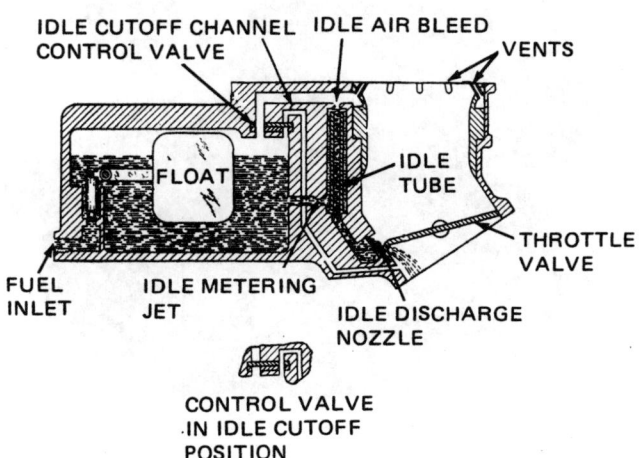

FIGURE 6-45   Downdraft carburetor and its idling system.

Type Letter	Type Description
S and R	Single barrel
D	Double barrel, float chamber to rear (obsolescent)
U	Double barrel, float chamber between barrels
Y	Double barrel, double chamber fore and aft of barrels
T	Triple barrel, double float chamber fore and aft of barrels
F	Four barrel, two separate float chambers

FIGURE 6-46 Basic Bendix-Stromberg model designation system for carburetors.

This system of model designation applies to inverted or downdraft as well as updraft carburetors. The model designation and serial number are found on an aluminum tag riveted to the carburetor. There are so many Bendix-Stromberg carburetors that it is necessary to consult the publications of the manufacturer to learn the details of the designation system, but the explanation above is ample for ordinary purposes.

Typical Float-Type Carburetors

A carburetor commonly used for light-aircraft engines is the Marvel-Schebler MA3, shown in Fig. 6-47. The MA3 carburetor has a double-float assembly hinged to the upper part of the

FIGURE. 6-47 Marvel-Schebler MA3 carburetor.

carburetor. This upper part is called the **throttle body** because it contains the throttle assembly. The fuel inlet, the float needle valve, and two venturis are also contained in the throttle body.

The carburetor body-and-bowl assembly contains a crescent-shaped fuel chamber surrounding the main air passage. The main fuel discharge nozzle is installed at an angle in the air passage with the lower end leading into the main fuel passage. An accelerating pump is incorporated in one side of the body-and-bowl assembly. The pump receives fuel from the fuel chamber and discharges accelerating fuel through a special accelerating-pump discharge tube into the carburetor bore adjacent to the main discharge nozzle. The carburetor also includes an altitude mixture control unit.

In addition to the MA3 carburetor, the Facet Aerospace Products Co. manufactures the MA3-SPA, MA4-SPA, MA4-5, MA4-5AA, MA5, MA6-AA, and HA6 carburetors. These models are essentially the same as the MA3 in design and operation, but they are designed for larger engines. The HA6 carburetor employs the same float and control mechanism; however, the airflow is horizontal rather than updraft. The essential features of these carburetors are the same, but size and minor variations account for the different model numbers.

The HA6 carburetor, having horizontal airflow, is mounted on the rear of the oil sump of the engine for which it is designed. The F/A mixture then passes through the oil sump where it picks up heat from the oil, which provides better vaporization of the fuel.

A simplified drawing of the arrangement and operation of an MA-type carburetor is given in Fig. 6-48.

The principal features of a more complex carburetor are illustrated in Fig. 6-49. This is a Bendix-Stromberg NAR-type carburetor, which includes all the systems explained previously. The operation of these systems can be easily understood through a careful study of the drawing. Observe particularly the float and needle valve, needle-type mixture control, economizer system, accelerating pump, idle system, main metering system, and air bleeds. The names of all the principal parts are included in the illustration.

Disadvantages of Float-Type Carburetors

Float-type carburetors have been improved steadily by their manufacturers, but they have two important disadvantages or limitations: (1) The fuel-flow disturbances in aircraft maneuvers may interfere with the functions of the float mechanism, resulting in erratic fuel delivery, sometimes causing engine failure. (2) When icing conditions are present, the discharge of fuel into the airstream ahead of the throttle causes a drop of temperature and a resulting formation of ice at the throttle valve.

CARBURETOR ICING

Water Vapor in Air

In addition to gases, the air always contains some water vapor, but there is an upper limit to the amount of water vapor (as an invisible gas) that can be contained in air at a given temperature. The capacity of air to hold water increases

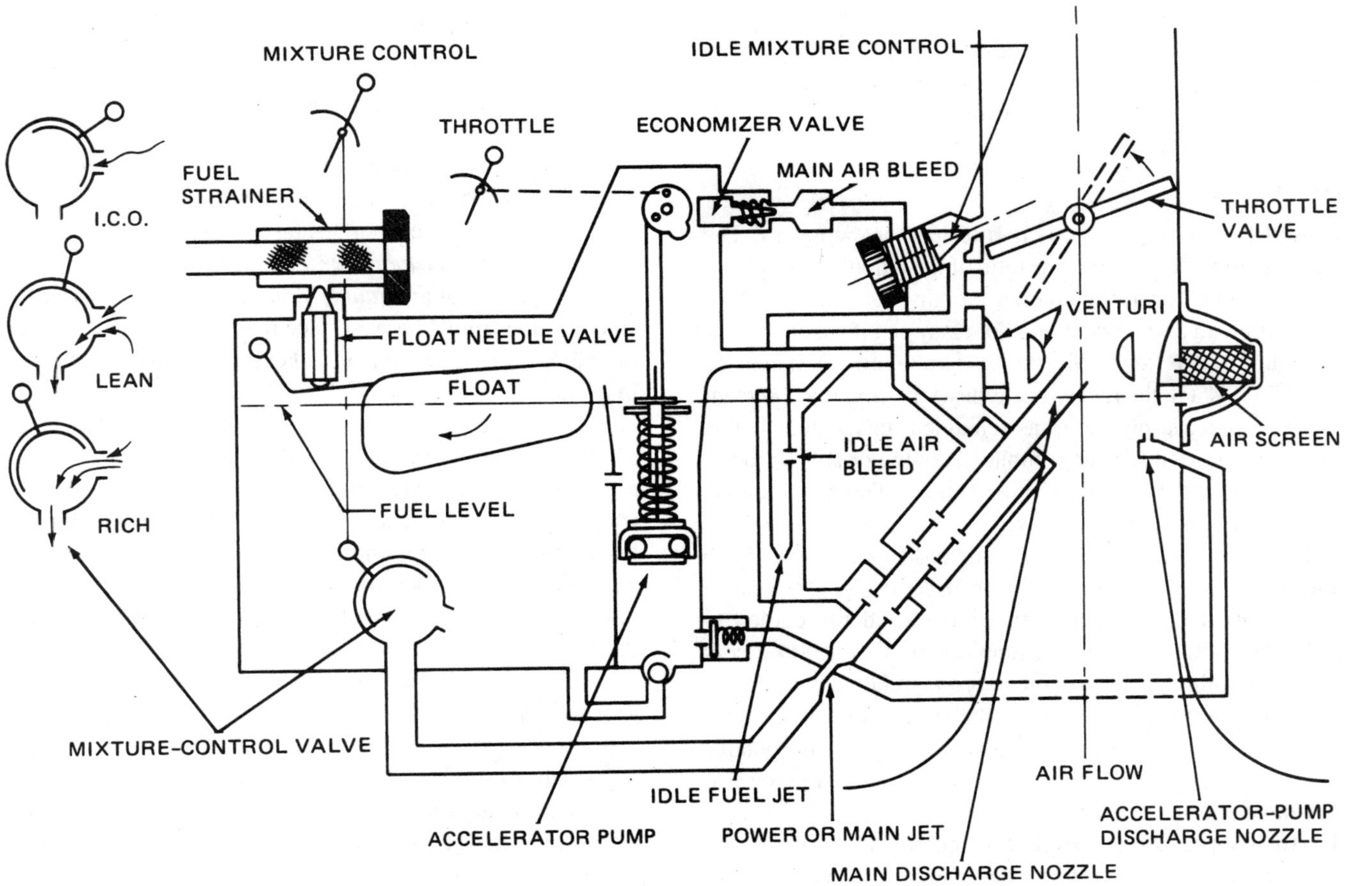

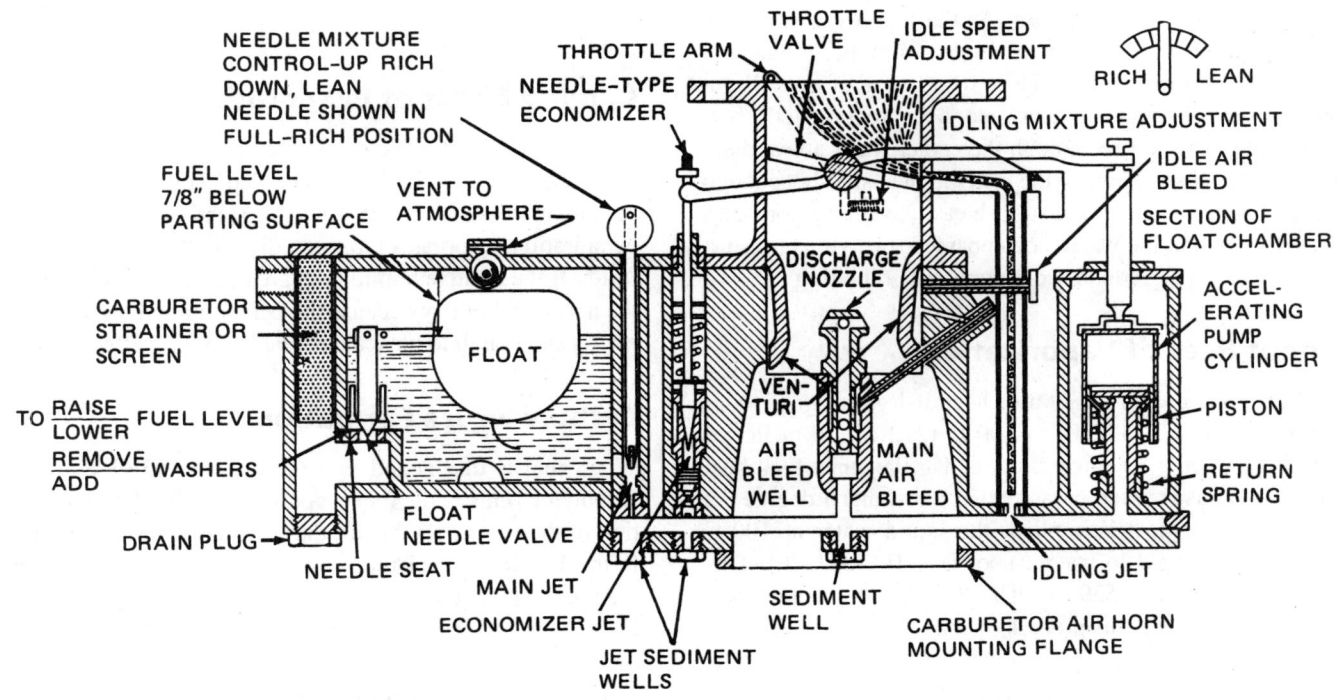

FIGURE 6-48 Simplified drawing of Marvel-Schebler carburetor.

MIXTURE CONTROL

THROTTLE

I.C.O.

LEAN

RICH

FUEL STRAINER

FLOAT NEEDLE VALVE

FLOAT

FUEL LEVEL

MIXTURE-CONTROL VALVE

ACCELERATOR PUMP

IDLE FUEL JET

POWER OR MAIN JET

MAIN DISCHARGE NOZZLE

IDLE MIXTURE CONTROL

ECONOMIZER VALVE

MAIN AIR BLEED

THROTTLE VALVE

VENTURI

IDLE AIR BLEED

AIR SCREEN

AIR FLOW

ACCELERATOR-PUMP DISCHARGE NOZZLE

FIGURE 6-49 Drawing of Bendix-Stromberg series NAR carburetor.

NEEDLE MIXTURE CONTROL-UP RICH DOWN, LEAN NEEDLE SHOWN IN FULL-RICH POSITION

THROTTLE ARM

THROTTLE VALVE

IDLE SPEED ADJUSTMENT

RICH   LEAN

NEEDLE-TYPE ECONOMIZER

IDLING MIXTURE ADJUSTMENT

FUEL LEVEL 7/8" BELOW PARTING SURFACE

VENT TO ATMOSPHERE

IDLE AIR BLEED

SECTION OF FLOAT CHAMBER

CARBURETOR STRAINER OR SCREEN

FLOAT

DISCHARGE NOZZLE

ACCEL-ERATING PUMP CYLINDER

TO RAISE/LOWER FUEL LEVEL

REMOVE/ADD WASHERS

VEN-TURI

AIR BLEED WELL

MAIN AIR BLEED

PISTON

RETURN SPRING

DRAIN PLUG

FLOAT NEEDLE VALVE

NEEDLE SEAT

MAIN JET

ECONOMIZER JET

JET SEDIMENT WELLS

SEDIMENT WELL

IDLING JET

CARBURETOR AIR HORN MOUNTING FLANGE

with temperature. When air contains the maximum possible amount of moisture at a given temperature in a given space, the pressure exerted by the water vapor is also at a maximum and the space is said to be **saturated**.

Humidity

Humidity, in simple terms, is moisture or dampness. The **relative humidity** of air is the ratio of the amount of moisture actually in the air to the amount the air could contain, usually expressed as a percentage. For example, if we have saturated air at 20°F [−6.67°C] and the temperature is increased to 40°F [4.44°C], the relative humidity will drop to 43 percent if the barometric pressure has remained unchanged. If this same air is heated further without removing or adding moisture, its capacity for holding water vapor will increase and its relative humidity will decrease.

Lowering the temperature of air reduces its capacity to hold moisture. For example, if air at 80°F [26.67°C] has a relative humidity of 49 percent and is suddenly cooled to 62°F [16.67°C], the relative humidity is then 100 percent because cooling the air has increased its relative humidity.

In the case just mentioned, a relative humidity of 100 percent means that the air is saturated; that is, the air contains all the moisture that it can hold. If the air should be cooled still more, further decreasing its moisture capacity, some of the water vapor will condense. The temperature at which the moisture in the air begins to condense is called the **dew point**.

Vaporization

The addition of heat can change a solid into a liquid or a liquid into a gas or vapor. The process of converting a liquid to a vapor is called **vaporization**. As the liquid is heated, the more rapidly moving molecules escape from the surface in a process called **evaporation**. Thus, when a pan of water is put on a hot stove, bubbles of water vapor begin to form at the bottom of the pan and rise through the cooler water above them and then collapse. When all the water in the pan is hot enough for the bubbles to reach the surface easily, vaporization takes place throughout the water, accompanied by a violent disturbance. This vaporization is commonly called **boiling**.

Latent Heat of Vaporization

When a liquid evaporates, it uses heat. If 1 gram (g) of water is raised from 0°C [32°F] to 100°C [212°F], 100 calories (cal) is required. (A **calorie** is defined as the quantity of heat required to raise the temperature of 1 g of water 1 degree C.)

When more heat is applied, the liquid water at 100°C [212°F] is changed to water vapor at 100°C [212°F]. This evaporation requires 539 cal. The temperature of the water does not rise during this process of changing from a liquid to a gas; therefore, 539 cal is required for the change of state. The reverse is also true; that is, when 1 g of water vapor is condensed to liquid water, 539 cal are given off. This energy (539 cal) required for the change of state from liquid to vapor or from vapor to liquid is called the **latent heat of vaporization** or **condensation**.

Laws of Evaporation

There are six **laws of evaporation** which have a bearing on the formation of ice in carburetors:

1. *The rate of evaporation increases as the temperature increases.* Heat increases the rate at which molecules move; therefore, hot water evaporates faster than cold water.

2. *The rate of evaporation increases with an increase of the surface area of the liquid.* More molecules can escape in a given time from a large surface than from a small surface; therefore, water in a big pan will evaporate faster than it will in a small vase.

3. *The rate of evaporation increases when the atmospheric pressure decreases.* As the weight of the air above a body of water decreases, the escaping water molecules encounter less resistance. In other words, evaporation is faster under low pressure than under high pressure. For example, it is possible to freeze water by the cooling effect of evaporation if a stream of dry air is passed over water in a partial vacuum.

4. *The rate of evaporation varies with the nature of the exposed liquid.* For example, alcohol evaporates faster than water.

5. *The rate of evaporation of water is decreased when the humidity of the air is increased.* The escaping molecules of water can get away more easily if there are only a few molecules of water already in the air above the water. Conversely, if the humidity is high, the escaping molecules encounter more resistance to their departure from the water.

6. *The rate of evaporation increases with the rate of change of the air in contact with the surface of the liquid.* Wet clothes dry faster on a windy day than on a day when the air is calm.

Cooling Effect of Evaporation

Every gram of water that evaporates from the skin takes heat from the skin, and the body is cooled. This is the function of human perspiration. If a person stands in a breeze, the perspiration evaporates more rapidly and the body is cooled faster. If the relative humidity is great, a person suffers from the heat on a hot day because the perspiration does not evaporate fast enough to have a noticeable cooling effect.

Carburetor Ice Formation

When fuel is discharged into the low-pressure area in the carburetor venturi, the fuel evaporates rapidly. This evaporation of the fuel cools the air, the walls, and the water vapor. If the humidity (moisture content) of the air is high and the metal of the carburetor is cooled below 32°F [0°C], ice forms and interferes with the operation of the engine. The fuel-air passages are clogged, the mixture flow is reduced, and the power output drops. Eventually, if the condition is not corrected, the drop in power output may cause engine failure. Ice formation in the carburetor may be indicated by a gradual loss of engine speed, a loss of MAP, or both, without change in the throttle position.

It is extremely important that a pilot recognize the symptoms of carburetor icing and the weather conditions which may be conducive to icing. The principal effects of icing are the loss of power (drop in MAP without a change in throttle position), engine roughness, and backfiring. The backfiring is caused when the discharge nozzle is partially blocked, which causes a leaning of the mixture.

Standard safety procedures with respect to icing involve (1) checking carburetor heater operation before takeoff, (2) turning on the carburetor heater when power is reduced for gliding or landing, and (3) using carburetor heat whenever icing conditions are believed to exist.

If the mixture temperature is slightly above freezing, there is little danger of carburetor icing. For this reason, a **mixture thermometer** [carburetor air temperature (CAT) gauge] is sometimes installed between the carburetor and the intake valve. This instrument not only indicates low temperatures in the carburetor venturi; which lead to icing conditions, but also is used to monitor temperature when a carburetor air intake heater is used to avoid excessively high temperatures, which causes detonation, preignition, and loss of power.

In general, ice may form in a carburetor system by any one of three processes: (1) The cooling effect of the evaporation of the fuel after being introduced into the airstream may produce fuel ice or fuel evaporation ice. (2) Water in suspension in the atmosphere coming in contact with engine parts at a temperature below 32°F [0°C] may produce impact ice or atmospheric ice. (3) Freezing of the condensed water vapor of the air at or near the throttle forms throttle ice or expansion ice.

Throttle Ice

Throttle ice is most likely to form when the throttle is partially closed, such as during letdown for a landing. The air pressure is decreased, and the velocity is increased as the air passes the throttle.

The rate of ice accumulation within and immediately downstream from the carburetor venturi and throttle butterfly valve is a function of the amount of entrained moisture in the air. If this icing condition is allowed to continue, the ice may build up until it effectively throttles the engine. Visible moisture in the air is not necessary for this type of icing, sometimes making it difficult for pilots to believe that it occurs unless they are fully aware of its effect. The result of throttle icing is a progressive decline in the power delivered by the engine. With a fixed-pitch propeller, this is evidenced by a decrease in engine rpm and a loss of altitude or airspeed unless the throttle is slowly advanced. With a constant-speed propeller, there will normally be no change in rpm, but a decrease in MAP or EGT will occur before any noticeable decrease in engine and airplane performance. If the pilot fails to note these signs and no corrective action is taken, the decline in engine power will continue and engine roughness will occur, probably followed by backfiring. Beyond this stage, insufficient power may be available to maintain flight, and complete stoppage may occur, especially if the throttle is moved abruptly.

Fuel Vaporization Ice

Fuel vaporization ice usually occurs in conjunction with throttle icing. It is most prevalent with conventional float-type carburetors. It occurs with pressure carburetors, to a lesser degree, when the F/A mixture reaches a freezing temperature as a result of the cooling of the mixture during the expansion process between the carburetor and engine manifold. This does not present a problem for systems which inject fuel at a location beyond which the passages are kept warm by engine heat. Thus the injection of fuel directly into each cylinder intake port, or air heated by a supercharger, generally precludes such icing. Vaporization icing may occur at temperatures from 32°F [0°C] to as high as 100°F [37.8°C] when relative humidity is 50 percent or above.

Impact Ice

Impact ice is formed by moisture-laden air at temperatures below freezing when the air strikes and freezes on elements of the induction system which are at temperatures of 32°F [0°C] or below. Under these conditions, ice may build up on components such as the air scoops, heat or alternate air valves, intake screens, and protrusions in the carburetor. The ambient temperature at which impact ice can be expected to build up most rapidly is about 25°F [−3.9°C], at which the supercooled moisture in the air is still in a semiliquid state. This type of icing affects an engine with fuel injection as well as carbureted engines. It is usually preferable to use carburetor heat or alternate air as an ice prevention means, rather than as a deicer, because fast-forming ice, which is not immediately recognized by the pilot, may significantly lower the heat available from the carburetor heating system. Impact icing is unlikely under extremely cold conditions, because the relative humidity is usually low in cold air and because the moisture present usually consists of ice crystals which pass through the air system harmlessly.

Icing Prevention Procedures

Remember that **induction system icing** is possible, particularly with float-type carburetors, at temperatures as high as 100°F [37.8°C] and relative humidity as low as 50 percent. It is more likely, however, at temperatures below 70°F [21.1°C] and the relative humidity above 80 percent. The likelihood of icing increases as the temperature decreases (down to 32°F [0°C]) and as the relative humidity increases.

When no carburetor air or mixture temperature instrumentation is available, the general practice for smaller engines should be to use full heat whenever carburetor heat is applied. With higher-output engines, however, especially those with superchargers, discrimination in the use of heat should be exercised because of the possible engine overheating and detonation hazard involved. In any airplane, the excessive use of heat during full-power operations, such as takeoffs or emergency go-arounds, may result in a serious reduction in the power developed as well as engine damage. Note that carburetor heat is rarely needed for brief high-power operations.

Carburetor Air Intake Heaters

The exhaust type of carburetor air intake heater is essentially a jacket or tube through which the hot exhaust gases of the engine are passed to warm the air flowing over the heated surface before the air enters the carburetor system. The principal value of carburetor air heat is to eliminate or prevent carburetor icing. The amount of warm air entering the system can be controlled by an adjustable valve.

The **alternate air inlet heating system** has a two-position valve and an air scoop. When the passage from the scoop is closed, warm air from the engine compartment is admitted to the carburetor system. When the passage from the scoop is open, cold air comes from the scoop.

In a third type of carburetor air intake heater, the air is heated by the compression which occurs in the external supercharger, but the air becomes so hot that it is passed through an **intercooler** to reduce its temperature before it enters the carburetor. Shutters at the rear of the intercooler can be opened or closed to regulate the degree of cooling to which the air warmed by the supercharger is subjected. Air entering the engine at too high a temperature can lead to detonation.

Excessively High CAT

At first thought, it seems foolish to first heat the air and then cool it when the purpose is to raise its temperature to avoid icing the carburetor. However, as stated earlier, excessively high CAT values are not wanted. Air expands when heated, and its density is reduced. Lowering the density of the air reduces the mass; that is, it cuts down the *weight* of the fuel-air charge in the engine cylinder, thus reducing volumetric efficiency. This results in a loss of power, because power depends on the weight of the F/A mixture burned in the engine. Note also that since the weight of the air is decreased while the fuel weight remains essentially unchanged, the mixture is enriched and power is decreased when carburetor heat is turned on.

Another danger inherent in a high fuel-air temperature is detonation. If the air temperature is such that further compression in the cylinder raises the temperature to the combustion level of the fuel, detonation will occur.

INSPECTION AND OVERHAUL OF FLOAT-TYPE CARBURETORS

Inspection in Airplane

Remove the carburetor strainer and clean it frequently. Flush the strainer chamber with gasoline to remove any foreign matter or water. Inspect the fuel lines to make certain that they are tight and in good condition. Inspect the carburetor to be sure that all safety wires, cotter pins, etc., are in place and that all parts are tight. On those models having economizers or accelerating pumps, clean the operating mechanism frequently, and put a small quantity of oil on the moving parts.

When you are inspecting the carburetor and associated parts, it is particularly important to examine the mounting flange closely for cracks or other damage. The mounting studs and safety devices should be checked carefully for security. If there is an air leak between the mounting surfaces or an air leak because of a crack, the F/A mixture may become so lean that the engine will fail. A very small leak can cause overheating of the cylinders and power loss.

Disassembly

Great care must be taken in disassembling a float-type carburetor to make sure that parts are not damaged. The sequence described in the manufacturer's overhaul manual should be followed, if a manual is available. This sequence is designed to ensure that parts still on the assembly will not interfere with parts to be removed. As parts are removed, they should be placed in a tray with compartments, to keep the components of each assembly together. When this is done, there is much less likelihood of installing parts in the wrong position when the carburetor is reassembled.

The tools used in the disassembly of a carburetor should be of the proper type. Screwdrivers should have the blades properly ground to avoid slipping in screw slots and damaging the screw heads or gouging the aluminum body of the carburetor. Metering jets and other specially shaped parts within the carburetor should be removed with the tools designed for the purpose. A screwdriver should not be used for prying, except where specific instructions are given to do so.

Cleaning

The cleaning of carburetor parts is usually described in the manufacturer's overhaul manual, but certain general principles may be followed. The first step is to remove oil and grease by using a standard petroleum solvent, such as Stoddard Solvent (Federal Specification P-S-661 or the equivalent). The parts to be cleaned should be immersed in the solvent for 10 to 15 min, rinsed in the solvent, and then dried.

To remove carbon and gum from the carburetor parts, a suitable carbon remover should be employed. Carbon remover MIL-C-5546A or the equivalent may be used. The remover should be heated to about 140°F [60°C] and the parts immersed in it for 30 min. The parts should then be rinsed thoroughly in hot water (about 176°F [80°C]) and dried with clean, dry compressed air, with particular attention paid to internal passages and recesses.

Wiping cloths or rags should never be used to dry carburetor parts because of the lint which will be deposited on the parts. Small particles of lint can obstruct jets, jam close-fitting parts, and cause valves to leak.

If aluminum parts are not corroded but still have some deposits of carbon, these deposits may be removed with No. 600 wet-or-dry paper used with water. After this, the parts should be rinsed with hot water and air dried.

Aluminum parts that are corroded can be cleaned by immersion in an alkaline cleaner such as Formula T or an equivalent agent inhibited against attack on aluminum.

Inspection of Parts

Before assembly of the carburetor, all parts should be inspected for damage and wear. Inspections for a typical carburetor are as follows:

1. Check all parts for bends, breaks, cracks, or crossed threads.
2. Inspect the fuel strainer assembly for foreign matter or a broken screen.
3. Inspect the float needle and seat for excessive wear, dents, scratches, or pits.
4. Inspect the mixture control plates for scoring or improper seating.
5. Inspect the float assembly for leaks by immersing it in hot water. Bubbles will issue from a point of leakage.
6. Inspect the throttle shaft's end clearance and the play in the shaft bushings.

In addition, certain assemblies must be checked for fits and clearances. Among these are the fulcrum bushing in the float, the fulcrum pin, the slot in the float needle, the pin in the float assembly, the bushing in the cover assembly, the mixture control stem, and the throttle shaft and bushings. The limits for these assemblies are given in the Table of Tolerance Values in the manufacturer's overhaul manual.

The foregoing inspections are specified for the Bendix-Stromberg NA-S3A1 carburetor. For other carburetors, such as the Marvel-Schebler MA3, MA4, and MA6 series, additional inspections are specified in the overhaul manual. In each case the manufacturer's instructions should be followed carefully.

Inspection of Metering Jets

The sizes of the metering jets in a carburetor are usually correct because these sizes are established by the manufacturer. Sometimes a jet may be changed or drilled to increase its size, so it is always a good idea to check the sizes when the carburetor is overhauled. The correct sizes are given in the specifications of the manufacturer in the overhaul manual, and the size numbers are usually stamped on the jet. The number on the jet corresponds to a numbered drill shank; therefore, it is possible to check the size of the jet by inserting the *shank* of a numbered drill into the jet, as shown in Fig. 6-50. If the drill shank fits the jet without excessive play, the jet size is correct. The number of the jet should also be checked against the specifications in the overhaul manual to see that the correct jet is installed. Metering jets should be examined closely to see that there are no scratches, burrs, or other obstructions in the jet passages, because these will cause local turbulence, which interferes with normal fuel flow. If a metering jet is defective in any way, it should be replaced by a new one of the correct size.

Repair and Replacement

The repair and replacement of parts for a carburetor depend on the make and model being overhauled and should be

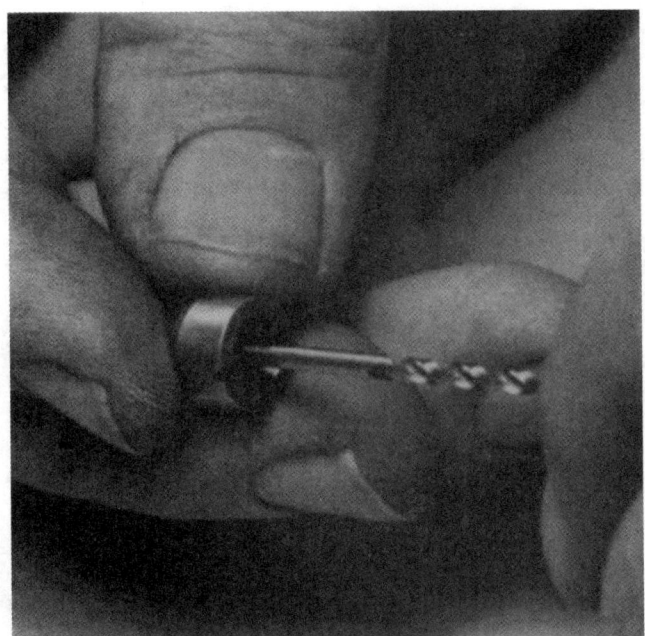

FIGURE 6-50 Checking metering jet for size.

performed in accordance with the manufacturer's instructions. It is always good practice to replace gaskets and fiber washers and any other part which shows substantial signs of wear or damage. When clearances and other dimensions are not within the specified limits, the parts involved must be repaired or replaced.

A carburetor float is usually made of formed brass sheet and can be checked for leaks by immersing it in hot water. The heat will cause the air and any fuel fumes in the float to expand, thus making a stream of bubbles emerge from the leak.

If the float has leaks, the leaks should be marked with a pencil or other means which will not cause damage. A small hole may then be drilled in the float to permit the removal of any fuel which may have been trapped inside. After the hole is drilled, the fuel should be drained and the float then immersed in boiling water until all fuel fumes have evaporated from the inside. This will permit soldering of the leaks without risking the danger of explosion. As a further precaution, the float should never be soldered with an open flame. The small leaks should be soldered before the drilled hole is sealed. Care must be taken to apply only a minimum of solder to the float, because its weight must not be increased more than necessary. An increase in the weight of the float will cause an increase in the fuel level which, in turn, will increase fuel consumption.

After the float is repaired, it should be immersed in hot water, to determine that all leaks have been sealed properly.

Checking the Float Level

As explained previously, the fuel in the float chamber of a carburetor must be maintained at a level which will establish the correct fuel flow from the main discharge nozzle while the carburetor is in operation. The fuel level in the discharge nozzle is usually from $\frac{3}{16}$ to $\frac{1}{8}$ in [4.76 to 3.18 mm] below the opening in the nozzle.

After the carburetor is partially assembled according to the manufacturer's instructions, the float level may be checked. In a Bendix-Stromberg carburetor where the float needle seat is in the lower part of the carburetor, the float level may be tested as follows:

1. Mount the assembled main body in a suitable fixture so that it is level when checked with a small spirit level.
2. Connect a fuel supply line to the fuel inlet in the main body, and regulate the fuel pressure to the value given on the applicable specification sheet. This pressure is $\frac{1}{2}$ psi [3.45 kPa] for an NA-S3A1 carburetor used in a gravity-feed fuel system. When the fuel supply is turned on, the float chamber will begin to fill with fuel and the flow will continue until it is stopped by the float needle on its seat.
3. Using a depth gauge, measure the distance from the parting surface of the main body to the level of the fuel in the float chamber approximately $\frac{1}{2}$ in [12.7 mm] from the side wall of the chamber, as shown in Fig. 6-51. If the measurement is taken adjacent to the side of the float chamber, a false reading will be obtained. The fuel level for the NA-S3A1 carburetor should be $\frac{13}{32} \pm \frac{1}{64}$ in [10.32 ± 0.397 mm] from the parting surface.
4. If the level of the float is not correct, remove the needle and seat and install a thicker washer under the seat to lower the level or a thinner washer to raise the level. A change in washer thickness of $\frac{1}{64}$ in [0.397 mm] will change the level approximately $\frac{5}{64}$ in [1.98 mm] for the NA-S3A1 carburetor.

Two different test procedures are used to establish the correct float level and float valve operation for the Marvel-Schebler MA series carburetors. The first is carried out during assembly after the float and lever assembly is installed. The throttle body is placed in an upside-down position,

FIGURE 6-51 Checking float level.

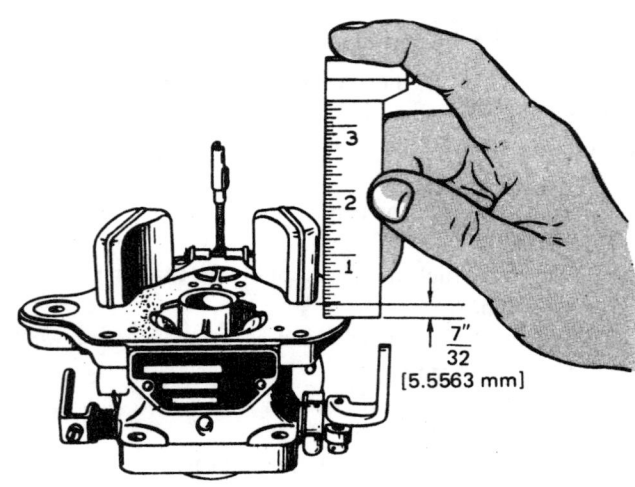

FIGURE 6-52 Measuring float distance on an MA3 carburetor.

as shown in Fig. 6-52. The height of the lower surface of each float above the gasket and screen assembly is then measured. For the MA3 and MA4 carburetors, this distance should be $\frac{7}{32}$ in [5.56 mm]. For the MA4-5 carburetor, the distance is $\frac{13}{64}$ in [5.16 mm]. When the throttle body is placed in the upside-down position, the float needle is bearing against the float valve and holding it in the closed position. This is the same position taken by the float when the carburetor is in the normal operating position and the float chamber is filled with fuel.

The method for testing float valve operation is illustrated in Fig. 6-53 and is performed after the carburetor is completely assembled. The procedure is as follows:

1. Connect the inlet fitting of the carburetor to a fuel pressure supply of 0.4 psi [2.76 kPa].
2. Remove the bowl drain plug, and connect a glass tube to the carburetor drain connection with a piece of rubber hose. The glass tubing should be positioned vertically beside the carburetor.
3. Allow the fuel pressure at 0.4 psi [2.76 kPa] to remain for at least 15 min, and then raise the fuel pressure to 6.0 psi [41.37 kPa]. (There will be a slight rise in the fuel level as the pressure is increased.) Allow the 6.0-psi [41.37-kPa] pressure to remain for at least 5 min after the fuel level has stabilized.
4. If the fuel does not rise to the level of the parting surface of the castings or run out of the nozzle, which can be observed through the throttle bore, the float valve and seat are satisfactory. If fuel is observed running out the nozzle, the bowl and throttle body must be separated and the float valve and seat cleaned or replaced.

In Fig. 6-53 the fuel level, shown as DISTANCE "A," will automatically be correct if the float height is correct and the float valve does not leak.

The foregoing procedures are given as typical operations in the inspection and overhaul of float-type carburetors; however, it is not possible in this text to give complete, detailed overhaul instructions for specific float-type carburetors. When faced with the necessity of overhauling a particular carburetor, the technician should obtain the correct manufacturer's manual and all special bulletins pertaining to the carburetor.

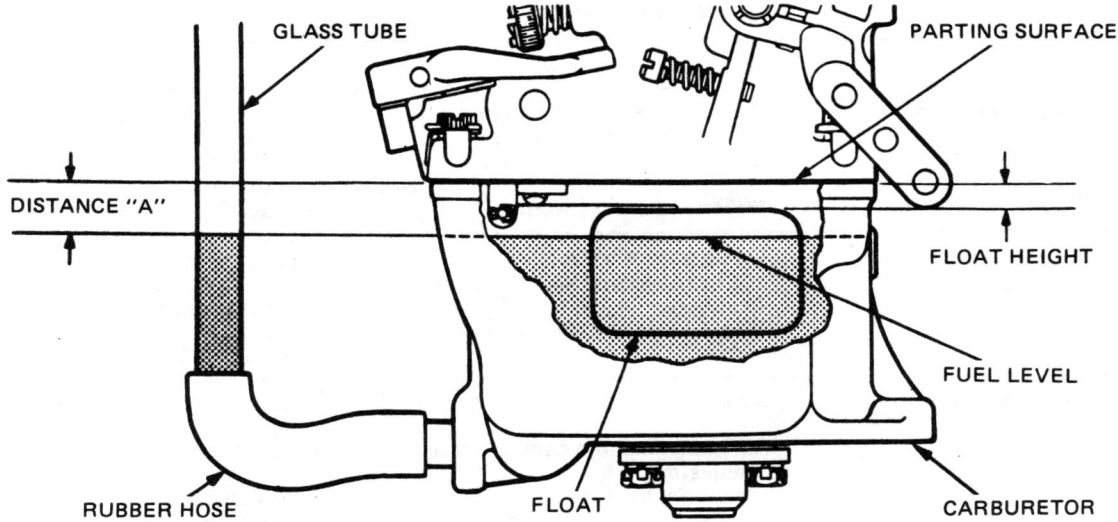

FIGURE 6-53 Testing float valve operation.

Also, the technician should check the FAA Engine Type Certificate Data Sheet for the particular engine on which the carburetor is to be used for any parts changes or modifications. The overhaul procedure should then be carried out according to the applicable instructions.

Troubleshooting

The troubleshooting chart presented as Fig. 6-54 provides some typical procedures for determining and correcting float-type carburetor malfunctions; however, it is not intended to cover all possible problems with the carburetor system. There are many types of carburetors, and among them are numerous variations in operational characteristics.

Installation of Carburetor

Before installing a carburetor on an engine, check it for proper lockwiring, and be sure that all shipping plugs have been removed from the carburetor openings. Put the mounting flange gasket in position. On some engines, bleed passages are found in the mounting pad. Install the gasket so that the bleed hole in the gasket aligns with the passage in the mounting flange.

Using the proper maintenance manual as a guide, tighten the carburetor mounting bolts to the value specified in the Table of Torque Limits found in the manual. Tighten securely the other nuts and bolts for the carburetor before connecting the throttle and mixture control levers.

After bolting the carburetor to the engine, check the throttle and mixture control lever on the unit for freedom of movement. Then connect the control cables or linkage.

Connect and adjust the carburetor or throttle controls of the fuel metering equipment so that full movement of the throttle corresponds to full movement of the control in the cockpit. Check and adjust the throttle control linkages so that springback on the throttle quadrant in the aircraft is equal for both FULL OPEN and FULL CLOSED positions. Finally, safety all controls properly to eliminate loosening during operation.

Adjusting Idle Speed and Idle Mixture

The correct idle speed and idle mixture are essential for the most efficient operation of an engine, particularly on the ground. The idle speed is established by the manufacturer at a level designed to keep the engine running smoothly, reduce overheating, and avoid spark plug fouling. A typical idle speed is 600 ± 25 rpm; however, this will vary somewhat among different aircraft.

The idle speed for an engine with a float-type carburetor is adjusted by turning the screw which bears against the throttle stop. Thus, the idle speed is established by varying the degree of throttle opening when the throttle lever is completely retarded. Usually, turning the screw to the right will increase the idle speed.

To adjust the idle mixture, do as follows:

1. Run the engine until it is operating at normal operating temperature.
2. Operate the engine at IDLE, and adjust for the correct idle speed.
3. Turn the idle mixture adjustment toward LEAN until the engine begins to run roughly.
4. Turn the mixture adjustment toward RICH until the engine is operating smoothly and the rpm has dropped slightly from its peak value.
5. Using the manual mixture control in the cockpit, move the control slightly toward LEAN. The rpm should increase slightly (about 25 rpm) before it begins to fall off and the engine starts to misfire. Returning the mixture control to FULL RICH should make engine operation smooth.

PRINCIPLES OF PRESSURE INJECTION

The **pressure injection carburetor** is a radical departure from float-type carburetor designs and takes an entirely different approach to aircraft engine fuel metering. It employs the simple method of metering the fuel through fixed orifices according to air venturi suction and air impact

Trouble	Cause	Correction
Carburetor leaks when engine is stopped.	Float needle valve not seated properly because of dirt on seat.	Tap carburetor body with soft mallet while engine is running. Remove and clean carburetor. Check float level.
	Float needle valve worn.	Replace float needle valve.
Mixture too lean at idle.	Fuel pressure too low.	Adjust fuel pressure to correct level.
	Idle mixture control out of adjustment.	Adjust idle mixture control.
	Obstruction in idle metering jet.	Disassemble and clean carburetor.
	Air leak in the intake manifold.	Check intake manifold for tightness at all joints. Tighten assembly bolts.
Mixture too lean at cruising speed.	Air leak in the intake manifold.	Check intake manifold for tightness at all joints. Tighten assembly bolts.
	Automatic mixture control out of adjustment.	Adjust automatic mixture control.
	Float level too low.	Check and correct float level.
	Manual mixture control not set correctly.	Check setting of manual mixture control. Adjust linkage if necessary.
	Fuel strainer clogged.	Clean fuel strainer.
	Fuel pressure too low.	Adjust fuel-pump relief valve.
	Obstruction in fuel line.	Check fuel flow and clear any obstructions.
Mixture too lean at full-power setting.	Same causes as those for lean cruise.	Make corrections the same as those for lean cruise.
	Economizer not operating correctly.	Check economizer system for operation. Adjust or repair as required.
Mixture too rich at idle.	Fuel pressure too high.	Adjust fuel pressure to correct level.
	Idle mixture control out of adjustment.	Adjust idle mixture.
	Primer line open.	See that primer system is not feeding fuel to engine.
Mixture too rich at cruising speed.	Automatic mixture control out of adjustment.	Adjust automatic mixture control.
	Float level too high.	Adjust float level.
	Manual mixture control not set correctly.	Check setting of manual mixture control. Adjust linkage if necessary.
	Fuel pressure too high.	Adjust fuel pump relief valve for correct pressure.
	Economizer valve open.	Check economizer for correct operation. Quick acceleration may clear.
	Accelerating pump stuck open.	Quick acceleration of engine may remove foreign material from seat.
	Main air bleed clogged.	Disassemble carburetor and clean air bleed.
Poor acceleration. Engine backfires or misses when throttle is advanced.	Accelerating pump not operating properly.	Check accelerating pump linkage. Remove carburetor, disassemble, and repair accelerating pump.

FIGURE 6-54   Troubleshooting chart for float carburetors.

pressure, combined with the additional function of atomizing the fuel spray under positive pump pressure. Although pressure carburetors are not used on modern aircraft, they are discussed so that readers have a basic understanding of their operating principles. Many older aircraft still incorporate various types of pressure carburetors. Pressure carburetors do have some advantages over float-type carburetors; for instance, they operate during all types of flight maneuvers (including acrobatics), and carburetor icing is less of a problem.

Principles of Operation

The basic principle of the pressure injection carburetor can be explained briefly by stating that mass airflow is utilized to regulate the pressure of fuel to a metering system which in turn governs the fuel flow. The carburetor therefore increases fuel flow in proportion to mass airflow and maintains a correct F/A ratio in accordance with the throttle and mixture settings of the carburetor.

The fundamental operation of a pressure injection carburetor may be illustrated by the simplified diagram in Fig. 6-55.

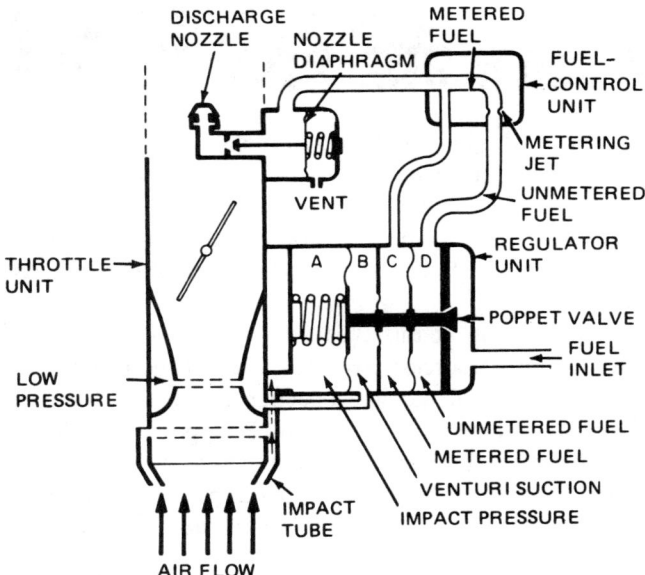

FIGURE 6-55 Simplified diagram of a pressure injection carburetor.

Shown in this diagram are four of the main parts of a pressure carburetor system: (1) the **throttle unit**, (2) the **regulator unit**, (3) the **fuel control unit**, and (4) the **discharge nozzle**.

When the carburetor is operating, the air flows through the throttle unit in an amount governed by the opening of the throttle. At the entrance to the air passage are impact tubes which develop a pressure proportional to the velocity of the incoming air. This pressure is applied to chamber A in the regulator unit. As the air flows through the venturi, a reduced pressure is developed in accordance with the velocity of the airflow. This reduced pressure is applied to chamber B in the regulator unit. The comparatively high pressure in chamber A and the low pressure in chamber B will create a differential of pressure across the diaphragm between the two chambers. The force of this pressure differential is called the **air metering force**, and as this force increases, it opens the poppet valve and allows fuel under pressure from the fuel pump to flow into chamber D. This unmetered fuel exerts force on the diaphragm between chamber D and chamber C and thus tends to close the poppet valve. The fuel flows through one or more metering jets in the fuel control unit and then to the discharge nozzle. Chamber C of the regulator unit is connected to the output of the fuel control unit to provide **metered fuel** pressure to act against the diaphragm between chambers C and D. Thus, unmetered fuel pressure acts against the D side of the diaphragm, and metered fuel pressure acts against the C side. The fuel pressure differential produces a force called the **fuel metering force**.

When the throttle opening is increased, the airflow through the carburetor is increased and the pressure in the venturi is decreased. Thus, the pressure in chamber B is lowered, the impact pressure to chamber A is increased, and the diaphragm between chambers A and B moves to the right because of the differential of pressure (air metering force). This movement opens the poppet valve and allows more fuel to flow into chamber D. This increases the pressure in chamber D and tends to move the diaphragm and the poppet valve to the left against the air metering force; however, this movement is modified by the pressure of metered fuel in chamber C. The pressure differential between chambers C and D (fuel metering force) is balanced against the air metering force at all times when the engine is operating at a given setting. The chamber C pressure is established at approximately 5 psi [34.48 kPa] by the spring-loaded, diaphragm-operated main discharge nozzle valve. This valve prevents leakage from the nozzle when the engine is not operating.

When the throttle opening is reduced, the air metering force decreases and the fuel metering force starts to close the poppet valve. This causes a decrease in the fuel metering force until it is again balanced by the air metering force.

Note particularly that an increase in airflow through the carburetor results in an increase in the fuel metering pressure across the metering jets in the fuel control section, and this increase causes a greater flow of fuel to the discharge nozzle. A decrease in airflow has the converse effect.

It must be understood that the regulator section of a pressure injection carburetor cannot regulate fuel pressure accurately at idling speeds because the venturi suction and the air impact pressure are not effective at low values. Therefore, it is necessary to provide **idling valves** which are operated by the throttle linkage to meter fuel in the idling range and which have springs in the pressure regulators to keep the poppet valve, from closing completely.

Types of Pressure Injection Carburetors

A number of different types of pressure injection carburetors have been manufactured, the majority having been developed by the Energy Controls Division of the Bendix Corporation. Models have been manufactured for almost all sizes of reciprocating engines.

The carburetor for small engines has a single venturi in a single barrel and is designated by the letters PS, meaning a pressure-type single-barrel carburetor.

A pressure carburetor for larger engines has a double barrel with boost venturis and is designated by the letters PD (for pressure-type, double-barrel). The triple-barrel carburetor is designated by the letters PT, and the rectangular-barrel carburetor is designated PR.

The numbers following the letter designation generally indicate the bore size of a carburetor or injection unit. Nominal bore sizes are designated in increments of $\frac{1}{4}$ in [6.35 mm], beginning with 1 in (2.54 cm) for the no. 1 bore size. The actual bore diameter is $\frac{3}{16}$ in [4.762 5 mm] larger than the nominal size. For example, the nominal diameter of the no. 10 size is 3.25 in [8.26 cm] $(9 \times \frac{1}{4} + 1 = 3.25$ in), but the actual bore diameter is $\frac{3}{16}$ in [4.762 5 mm] larger than the nominal diameter, or 3.4375 in [8.731 3 cm].

PRESSURE CARBURETOR FOR SMALL ENGINES

To provide the benefits of pressure injection carburetion for small aircraft engines, the Bendix Products Division (now Energy Controls Division) developed the PS series of carburetors. Figure 6-56 illustrates the PS-5C carburetor used on the Continental O-470 series engines.

The PS-5C carburetor utilizes the principles previously explained in this chapter. It includes a throttle unit, regulator section, fuel control unit, discharge nozzle, manual mixture control, accelerating pump, and idle system.

General Description

The PS-5C injection carburetor is a single-barrel updraft unit that provides a closed fuel system from the engine fuel pump to the carburetor discharge nozzle. Its function is to meter fuel through a fixed jet to the engine in proportion to mass airflow. The discharge nozzle is located downstream of the throttle valve to prevent ice from forming in the carburetor. This carburetor provides positive fuel delivery regardless of aircraft altitude or attitude and maintains proper F/A ratios regardless of engine speed, propeller load, or throttle lever position.

PRESSURE CARBURETORS FOR LARGE ENGINES

A number of different models of pressure injection carburetors have been designed for large engines; however, they all utilize the principles explained previously. These carburetors vary in size, type of mixture control, type of enrichment valves, shape of throttle body, and type of discharge nozzle.

One example of this type of carburetor is the Bendix PR-58. This carburetor possesses most of the characteristics of the typical pressure injection carburetors for large engines, including both the PD and PT types. A drawing of this carburetor system is shown in Fig. 6-57.

Principal Units. To understand the operation of the complete carburetor, one should note the construction and operation of each unit and its function in relation to the other units. There are four basic units of the carburetor: (1) the throttle unit, (2) the pressure regulator unit, (3) the fuel control unit, and (4) the AMC unit. In addition, this particular model of PR-58 carburetor is equipped to operate with a water-alcohol ADI (antidetonant injection) system which includes the derichment valve in the fuel control unit and a separate water-alcohol (W/A) regulator.

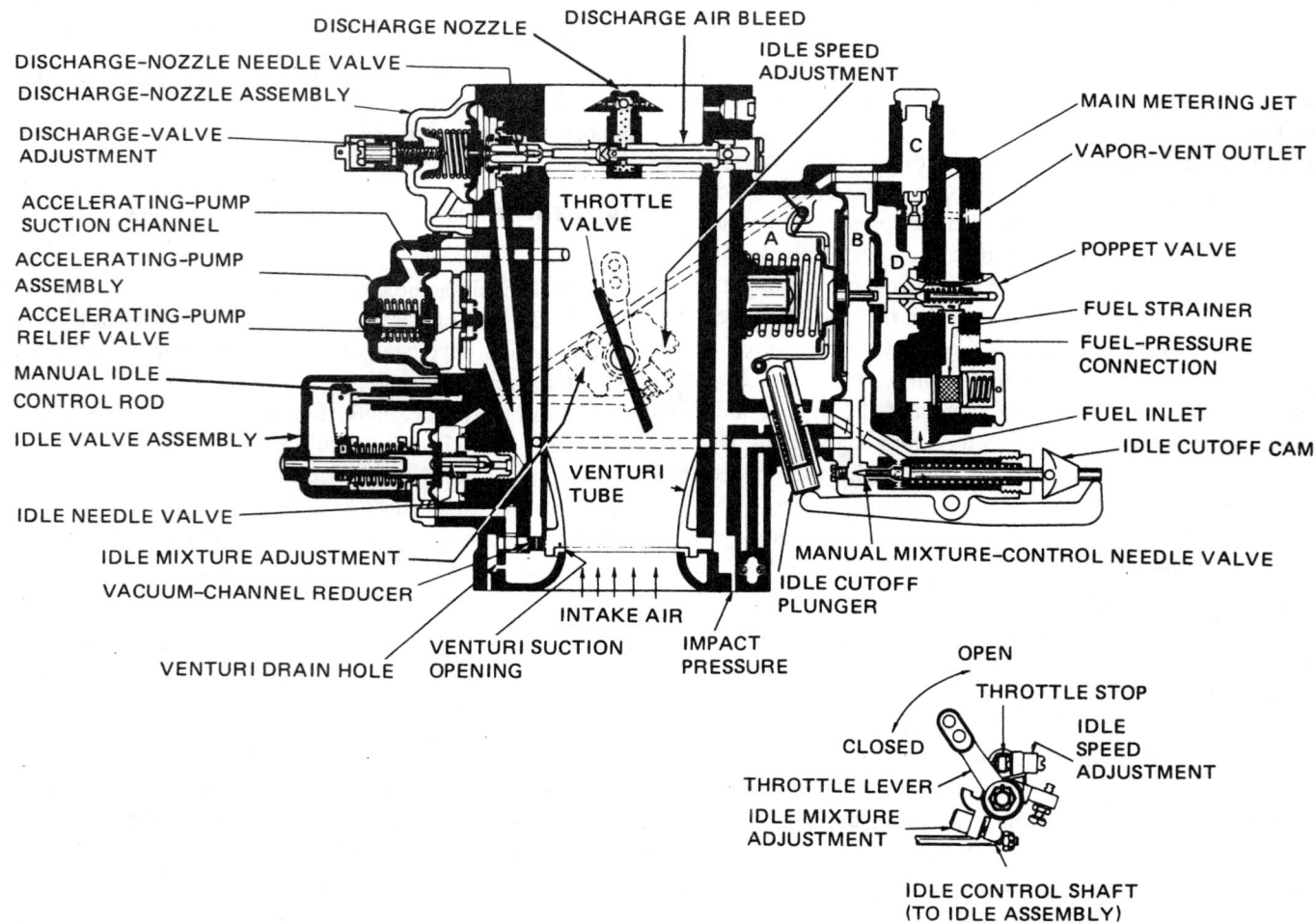

FIGURE 6-56   Drawing of the PS-5C carburetor. (*Continental Motors.*)

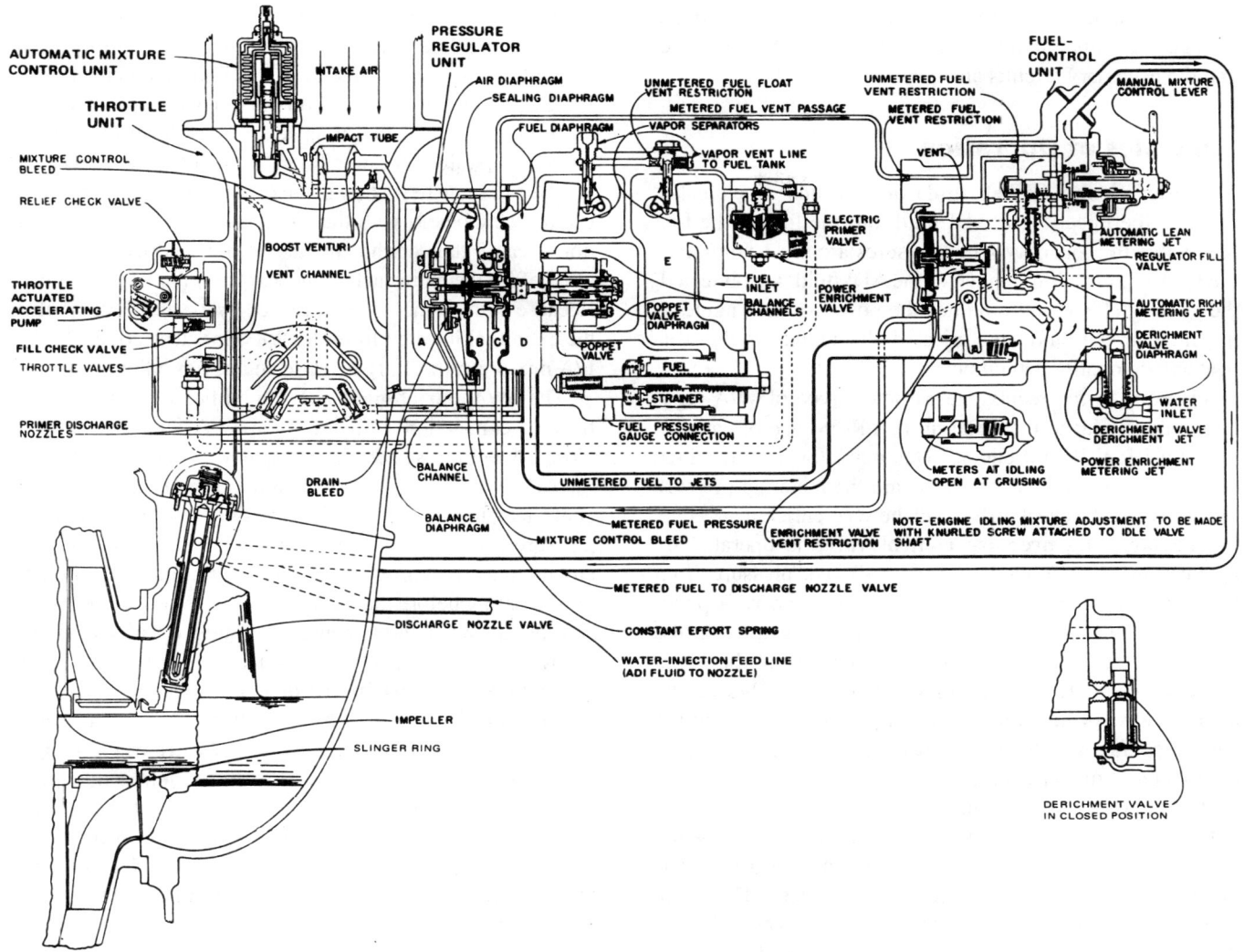

FIGURE 6-57 Drawing of a Bendix PR-58 type of carburetor system. (*Continental Motors.*)

WATER INJECTION

Water injection, also called **antidetonant injection** (ADI), is the use of water with the F/A mixture to provide cooling for the mixture and the cylinders so that additional power can be drawn from the engine without danger of detonation.

Instead of using pure water for the ADI system, it is necessary to use a water-alcohol mixture (methanol) with a small amount of water-soluble oil added. The alcohol prevents freezing of the water during cold weather and at high altitudes. The water-soluble oil is added to prevent the corrosion which would occur in the units of the system if they lacked oil. The water-alcohol-oil mixture is called **antidetonant injection fluid,** or simply **ADI fluid.** In servicing the ADI system, the technician must be sure that the correct mixture of fluid components is used. This information is contained in the manufacturer's service instructions.

Advantages

It is often necessary to use the maximum power which an engine can produce, such as for taking off from short fields and for go-arounds. A "dry" engine—that is, one without water injection—is limited in its power output by the detonation which results when operating limits are exceeded. The injection of water into the F/A mixture has the same effect as the addition of antiknock compounds in that it permits the engine to deliver greater power without danger of detonation.

The average engine operating without water injection requires a rich mixture of approximately 10 parts air to 1 part fuel by weight (F/A ratio of 0.10). With this mixture, a portion of the fuel is unburned and acts as a cooling agent. The additional unburned fuel subtracts from the power of the engine. But when water is added to the F/A mixture in proper quantities, the power of the engine can be increased. The water cools the F/A mixture, thus permitting a higher manifold pressure to be used. In addition, the F/A ratio can be reduced to the rich best-power mixture, thus deriving greater power from the fuel consumed. When water injection is employed, the F/A ratio can be reduced to approximately 0.08, which is a much more efficient mixture than the 0.10 ratio required otherwise.

The use of water injection permits an increase of 8 to 15 percent in takeoff horsepower.

The equipment required for water injection includes a storage tank, a pump, a water regulator, a derichment valve, and the necessary circuits and controls.

Principles of Operation

The **water-alcohol (W/A) regulator** is the unit which makes possible the injection of ADI fluid into the fuel at the fuel feed valve in a quantity which ensures a correct volume of the W/A mixture. If too much of the ADI fluid were injected, the cooling effect would reduce the power of the engine. If insufficient ADI fluid were injected, the engine would overheat and detonation would occur.

Figure 6-58 represents one particular type of W/A regulator, one similar to that used in the PR-58 type of carburetor on a P&W R2800 engine. This regulator includes three diaphragm-operated valves. These are the metering pressure control valve, the check valve, and the W/A enrichment valve.

The **metering pressure control valve** is operated by application of chamber D (unmetered) fuel pressure from the carburetor to one side of the diaphragm and W/A pump pressure to the opposite side. When the pilot turns on the ADI switch in the cockpit, ADI fluid flows from the pump to the regulator. The metering pressure control valve will be open because of the unmetered fuel pressure applied to the valve diaphragm. When pump pressure builds up to the level of unmetered fuel pressure, the valve will begin to close.

The **check valve** is normally closed when the system is not operating; however, it will begin to open slowly when pump pressure is applied to the diaphragm. The valve cannot open immediately because of the delay bleed. The **delay bleed** provides time for the derichment valve to close before the ADI fluid starts to discharge into the fuel feed valve.

When the system is *not* operating, the fuel backs up into the W/A feed line. Therefore, when the system is turned on, fuel will be the first substance injected into the fuel feed valve. If the carburetor is set for takeoff at FULL RICH or EMERGENCY RICH and additional fuel is injected from the W/A line, the overrich mixture will cause the engine to lose power and there will be a definite hesitation in the operation of the engine. The use of delay bleeds with the check valve prevents this situation because the derichment valve closes and leans the mixture before the extra fuel is injected from the W/A feed line. Because of the leaner mixture caused by the derichment valve, the extra fuel injected from the W/A line does not enrich the F/A mixture sufficiently to cause engine hesitation.

The **W/A enrichment valve** modifies the flow of ADI fluid in connection with the main W/A jet. This valve is closed when the system is not operating. The operation of the ADI system is initiated when the pilot turns on the ADI control switch in the cockpit. This switch is in series with a pressure switch operated by engine oil pressure or MAP; the engine must therefore be operating at a comparatively high power setting before the electric power can be directed to the ADI pump. When the system is operating, a **water pressure transmitter** connected to the regulator sends electric signals to the **water pressure indicator** in the cockpit. Water pressure from the regulator is also directed to a pressure warning switch which controls the **water pressure warning light** in the cockpit. The light is on while the system is operating. If the ADI fluid supply should become exhausted while the system is operating, the pressure switch will open and the warning light in the cockpit will turn off. At the same time, the derichment valve will open and permit enrichment fuel to flow through the fuel control unit, thus providing the necessary cooling to avoid detonation.

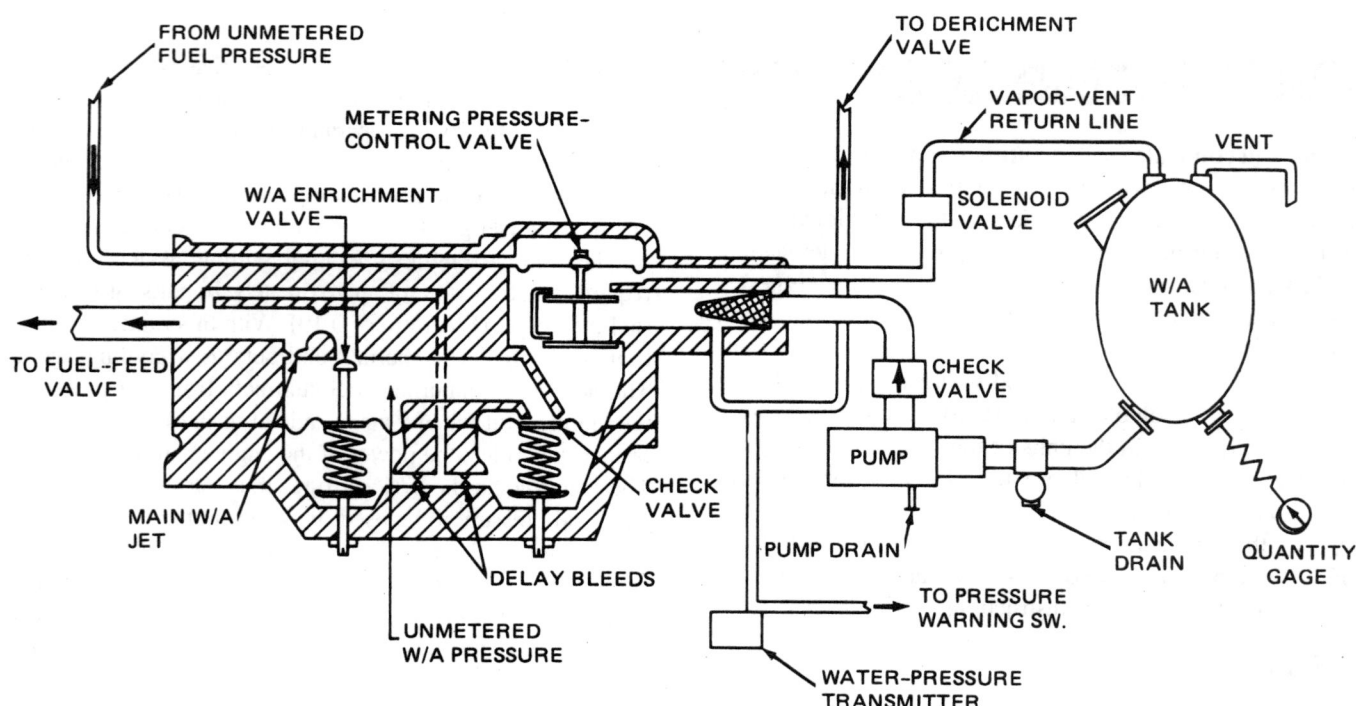

FIGURE 6-58 Schematic diagram of water-alcohol (ADI) regulator.

Some systems utilize a float-operated switch to turn off the fluid pump before the fluid supply is exhausted. This is particularly important if a vane-type pump is employed. Such a pump can maintain pressure on the derichment valve even though the fluid supply is exhausted. In such a case, the valve would not open to allow enrichment fuel to flow, and the engine would still be operating on the best-power mixture. This would allow detonation to occur.

The ADI system is particularly advantageous under conditions of high humidity. The water vapor in humid air displaces oxygen, so that a particular F/A ratio will increase in richness as humidity increases. Therefore, when an aircraft taking off is using the emergency rich mixture, high humidity will further enrich the mixture and cause a substantial loss of power. When an aircraft takes off wet (with ADI), the F/A ratio is set for best power and the enrichment caused by high humidity is not great enough to cause an appreciable loss of power. The water injected into the fuel does not have an appreciable effect on the F/A ratio because the water does not displace the oxygen in the air.

REVIEW QUESTIONS

1. Explain the meanings of *octane number* and *performance number*.
2. What is added to aviation fuel to improve its anti-knock qualities?
3. Into what two principal sections may the fuel system of an aircraft be divided?
4. Why does an engine-driven fuel pump require a bypass system?
5. Explain vapor lock and list three primary causes.
6. What is the purpose of carburetion or fuel metering?
7. What are standard sea-level pressure and temperature?
8. How is the venturi tube utilized in a carburetor?
9. In what terms are F/A mixtures generally expressed?
10. Explain the best-economy mixture of fuel and air.
11. What is meant by the brake specific fuel consumption of an engine?
12. Define *flame propagation*.
13. What atmospheric conditions affect air density?
14. How is the fuel discharge nozzle in a float-type carburetor located with respect to the fuel level in the float chamber?
15. Explain the use of an air bleed in a float-type carburetor.
16. Describe a throttle valve and its operation.
17. Explain the importance of float level in a carburetor.
18. Describe the main metering system in a carburetor.
19. What is idle cutoff?
20. Why is the F/A ratio enriched during the idle operation?
21. Why is an accelerating system needed?
22. What is the purpose of the economizer system in a carburetor?
23. What are the principal functions of the mixture control system?
24. Describe the purpose of an AMC.
25. Differentiate between updraft and downdraft carburetors.
26. What are the disadvantages of a float-type carburetor?
27. In what different ways may ice form in a carburetor system?
28. Describe methods for heating the carburetor intake air.
29. What are the effects of excessively high CAT?
30. How is the size of a metering jet checked?
31. Name the principal units of a pressure-type carburetor.
32. What is the purpose of water injection?

Fuel Injection Systems 7

DEFINITION

Fuel injection is the introduction of fuel or a fuel-air (F/A) mixture into the induction system of an engine or into the combustion chamber of each cylinder by means of a pressure source other than the pressure differential created by airflow through the venturi of a carburetor. The usual pressure source is an injection pump, which comes in several types. A **fuel injection carburetor** discharges the fuel into the airstream at or near the carburetor. A **fuel injection system** discharges the fuel into the intake port of each cylinder just ahead of the intake valve or directly into the combustion chamber of each cylinder.

Fuel injection systems have a number of advantages:

1. Freedom from vaporization icing, thus making it unnecessary to use carburetor heat except under the most severe atmospheric conditions
2. More uniform delivery of F/A mixture to each cylinder
3. Improved control of F/A ratio
4. Reduction of maintenance problems
5. Instant acceleration of engine after idling, with no tendency to stall
6. Increased engine efficiency

Fuel injection systems have been designed for all types of reciprocating aircraft engines. They are presently used on a wide variety of light-engine airplanes, large commercial aircraft, helicopters, and military aircraft.

CONTINENTAL CONTINUOUS-FLOW INJECTION SYSTEM

The Continental fuel injection system is a multinozzle, *continuous-flow* type which controls fuel flow to match engine airflow. The fuel is discharged into the intake port of each cylinder. Any change in air throttle position or engine speed, or a combination of both, causes changes in fuel flow in the correct relation to engine airflow. A manual mixture control and a pressure gauge, indicating metered fuel pressure, are provided for precise leaning at any combination of altitude and power. Since fuel flow is directly proportional to metered fuel pressure, the settings can be predetermined and the fuel consumption accurately predicted. The continuous-flow system permits the use of a typical rotary-vane fuel pump in place of a much more complex and expensive plunger-type pump. There is no need for a timing mechanism because each cylinder draws fuel from the discharge nozzle in the intake port as the intake valve opens.

The Continental fuel injection system consists of four basic units: the fuel injection pump, the fuel-air control unit, the fuel manifold valve, and the fuel discharge nozzle. These units are shown in Fig. 7-1.

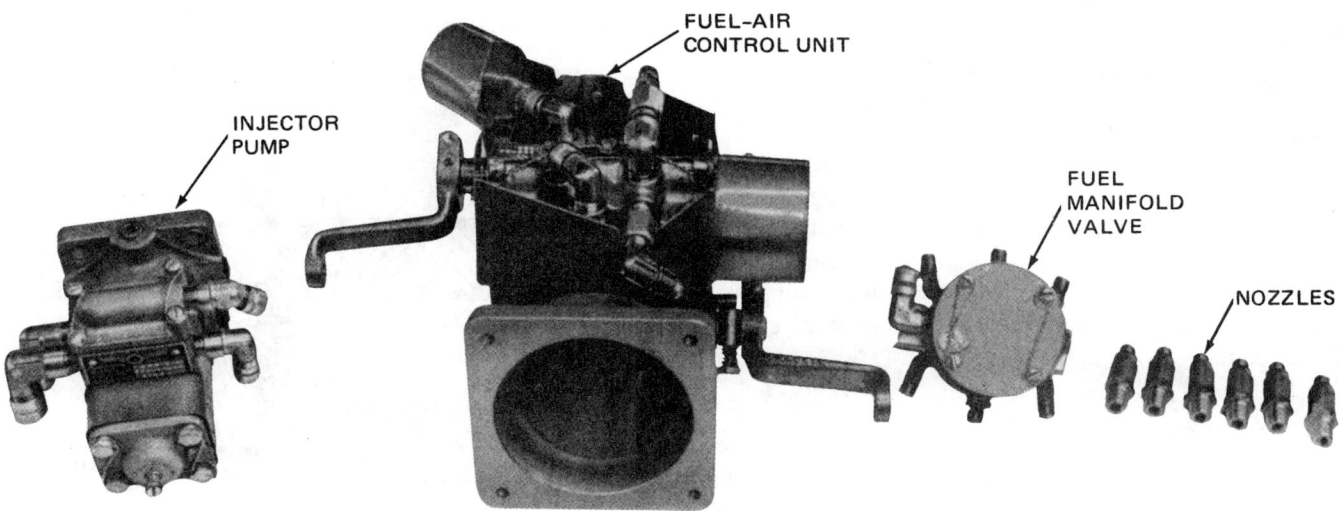

FIGURE 7-1 Units of the Continental fuel injection system. (*Continental Motors.*)

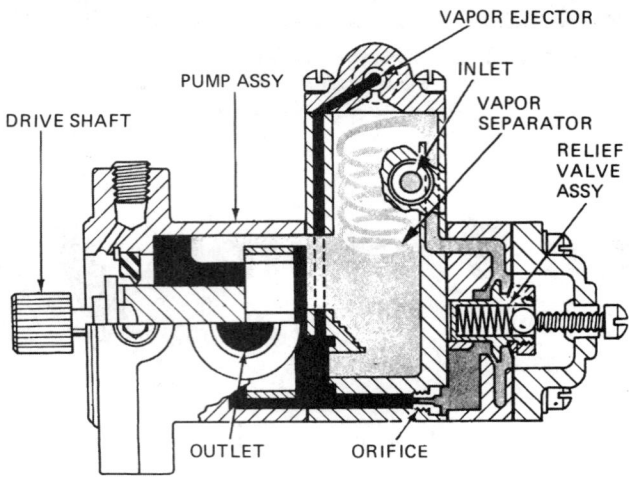

FIGURE 7-2   Fuel injection pump. (*Continental Motors.*)

Fuel Injection Pump

The **fuel injection pump** shown in Fig. 7-2 is a positive-displacement rotary-vane type with a splined shaft for connection to the accessory drive system of the engine. A spring-loaded diaphragm-type relief valve, which also acts as a pressure regulator, is provided in the body of the pump. Pump outlet fuel pressure passes through a calibrated orifice before entering the relief valve chamber, thus making the pump delivery pressure proportional to engine speed.

Fuel enters the pump assembly at the swirl well of the **vapor separator**, as shown in the drawing. At this point, any vapor in the fuel is forced up to the top of the chamber, where it is drawn off by the **vapor ejector**. The vapor ejector is a small pressure jet of fuel which feeds the vapor into the vapor return line, where it is carried back to the fuel tank. There are no moving parts in the vapor separator, and the only restrictive passage is used in connection with vapor removal; therefore, there is no restriction of main fuel flow.

Disregarding the effects of altitude or ambient air conditions, the use of a positive-displacement engine-driven pump means that changes in engine speed affect total pump flow proportionally. The pump provides greater capacity than is required by the engine; therefore, a recirculation path is required. When the relief valve and orifice are placed in this path, fuel pressure proportional to engine speed is provided. These provisions ensure proper pump pressure and delivery for all engine operating speeds.

A check valve is provided so that boost pressure to the system from an auxiliary pump can bypass the engine-driven pump during start-up. This feature is also available to suppress vapor formation under high ambient fuel temperatures. Furthermore, this permits use of the auxiliary pump as a source of fuel pressure in the event of failure of the engine-driven pump.

Fuel-Air Control Unit

The **fuel-air control unit**, shown in Fig. 7-3, occupies the position ordinarily used for the carburetor at the intake manifold inlet. The unit includes three control elements, one for

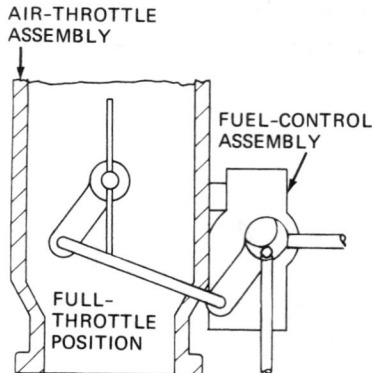

FIGURE 7-3   Fuel-air control unit. (*Continental Motors.*)

air in the **air throttle assembly** and two for fuel in the **fuel control assembly**, which is mounted on the side of the air throttle assembly.

The air throttle assembly is an aluminum casting which contains the shaft and butterfly valve. The casting bore size is tailored to the engine size, and no venturi or other restriction is employed. Large shaft bosses provide an adequate bearing area for the throttle shaft so that there is a minimum of wear at the shaft bearings. Wave washers provide protection against vibration. A conventional idle-speed adjusting screw is mounted in the air throttle shaft lever and bears against a stop pin in the casting.

The fuel control unit, shown in Figs. 7-3 and 7-4, is made of bronze for best bearing action with the stainless-steel valves. The central bore contains a metering valve at one end and a mixture control valve at the other end. These rotary valves are carried in oil-impregnated bushings and are sealed against leakage by O-rings. Loading springs are installed in the center bore of the unit between the end bushings and the large end of each control shaft, to force the valve ends against a fixed plug installed in the middle of the bore. This arrangement ensures close contact between the valve faces and the metering plug. The bronze metering plug has one passage that mates with the fuel return port and one through

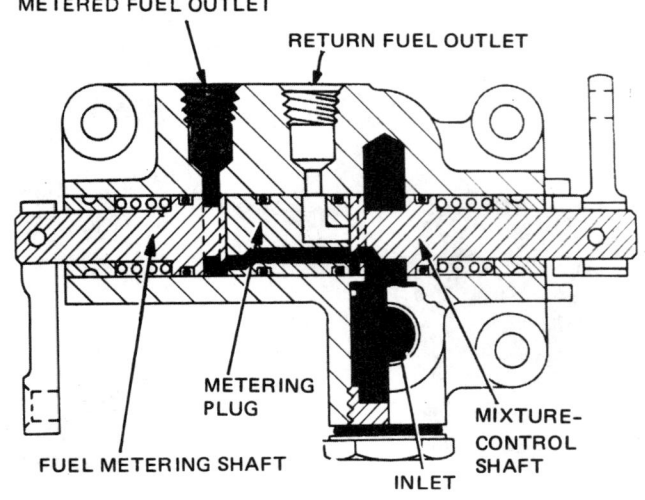

FIGURE 7-4   Fuel control unit. (*Continental Motors.*)

a passage that connects the mixture control valve chamber with the metering valve chamber. O-rings are used to seal the metering plug in the body.

Each stainless-steel rotary valve includes a groove which forms a fuel chamber. A contoured end face of the mixture control valve aligns with the passages in the metering plug, to regulate the fuel flow from the fuel chamber to the metering valve or to the return fuel outlet. A control lever is mounted on the mixture control valve shaft for connection to the cockpit mixture control. If the mixture control is moved toward the LEAN position, the mixture control valve in the fuel control unit causes additional fuel to flow through the return line to the fuel pump. This reduces the fuel flow through the metering plug to the metering valve. Rotation of the fuel control valve toward the RICH position causes more fuel to be delivered to the metering valve and less to the return fuel outlet.

On the metering valve, a cam-shaped cut is made on one outer part of the end face which bears against the metering plug. As the valve is rotated, the passage from the metering plug is opened or closed in accordance with the movement of the throttle lever. As the throttle is opened, the fuel flow to the metered fuel outlet is increased. Thus, fuel is measured to provide the correct amount for the proper F/A ratio. The linkage from the throttle to the fuel metering valve is shown in Fig. 7-5.

Fuel Manifold Valve

The **fuel manifold valve** is illustrated in Fig. 7-6. The fuel manifold valve body contains a fuel inlet, a diaphragm chamber, a valve assembly, and outlet ports for the lines to the individual fuel nozzles. The spring-loaded diaphragm carries the valve plunger in the central bore of the body. The diaphragm is enclosed by a cover which retains the diaphragm loading spring.

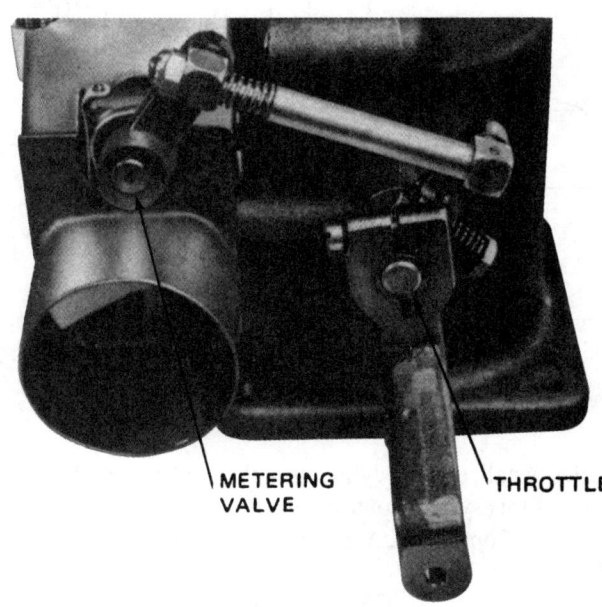

FIGURE 7-5   Linkage from throttle to fuel metering valve.

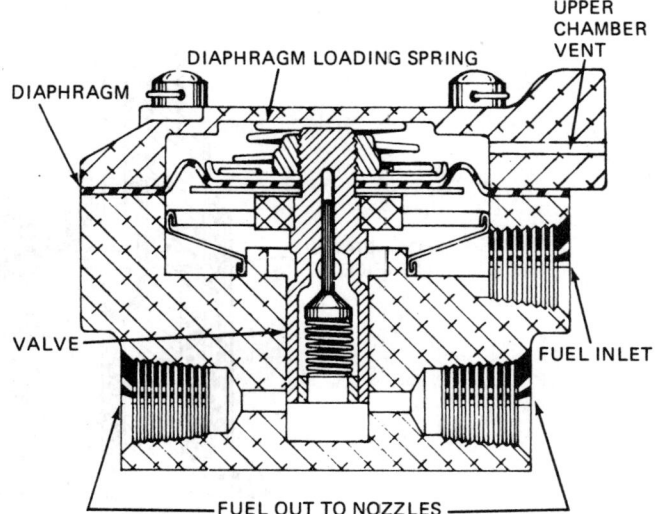

FIGURE 7-6   Fuel manifold valve.

When the engine is not running, there is no pressure on the diaphragm to oppose the spring pressure; therefore, the valve will be closed to seal off the outlet ports. Furthermore, the valve in the center bore of the valve plunger will be held on its seat to close the passage through the plunger. When fuel pressure is applied to the fuel inlet and into the chamber below the diaphragm, the diaphragm will be deflected and will raise the plunger from its seat. The pressure will also open the valve inside the plunger and allow fuel to pass through to the outlet ports.

A fine screen is installed in the diaphragm chamber so that all fuel entering the chamber must pass through the screen, where any foreign particles are filtered out. During inspection or troubleshooting of the system, this screen should be cleaned.

Fuel Discharge Nozzle

The **fuel discharge nozzle**, shown in Fig. 7-7, is mounted in the cylinder head of the engine with its outlet directed into the intake port. The nozzle body contains a drilled central passage with a counterbore at each end. The lower end of the nozzle is used for a fuel-air mixing chamber before the spray leaves the nozzle. The upper bore contains a removable orifice for calibrating the nozzles.

Near the top of the nozzle body, radial holes connect the upper counterbore with the outside of the nozzle body to provide for air bleed. These holes enter the counterbore above the orifice and draw outside air through a cylindrical screen fitted over the nozzle body. The screen keeps dirt and other foreign material out of the interior of the nozzle. A press-fitted shield is mounted on the nozzle body and extends over the greater part of the filter screen, leaving an opening near the bottom. This provides both mechanical protection and an air path with an abrupt change in direction as an aid to cleanness. Nozzles are calibrated in several ranges, and all nozzles furnished for one engine are of the same range. The range is identified by a letter stamped on the hexagon of the nozzle body.

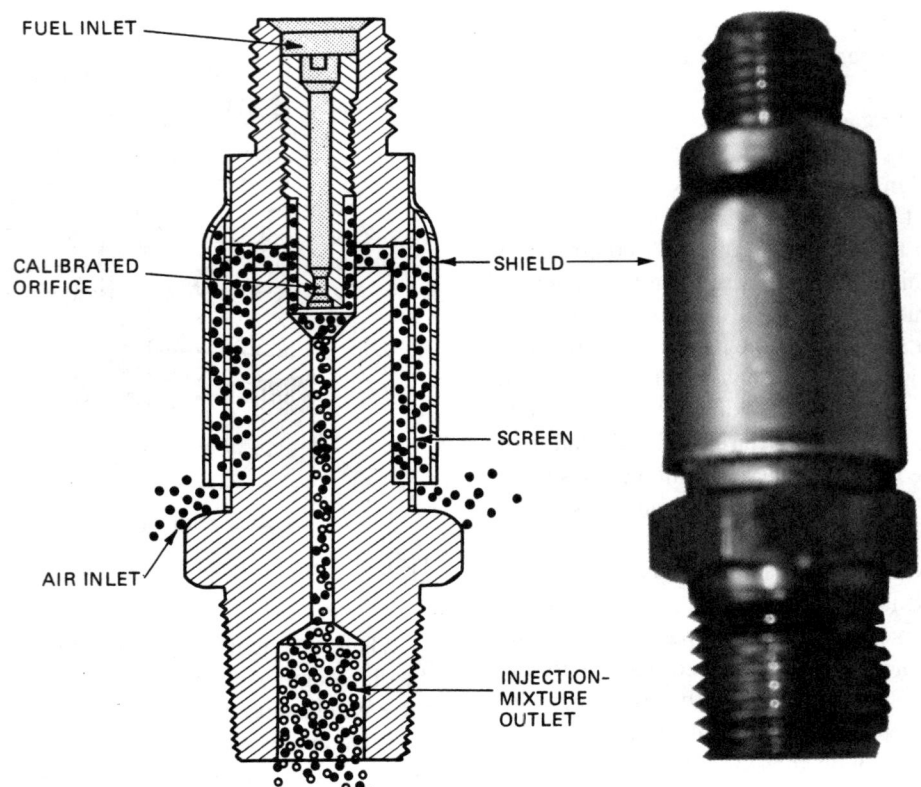

FIGURE 7-7 Fuel discharge nozzle.

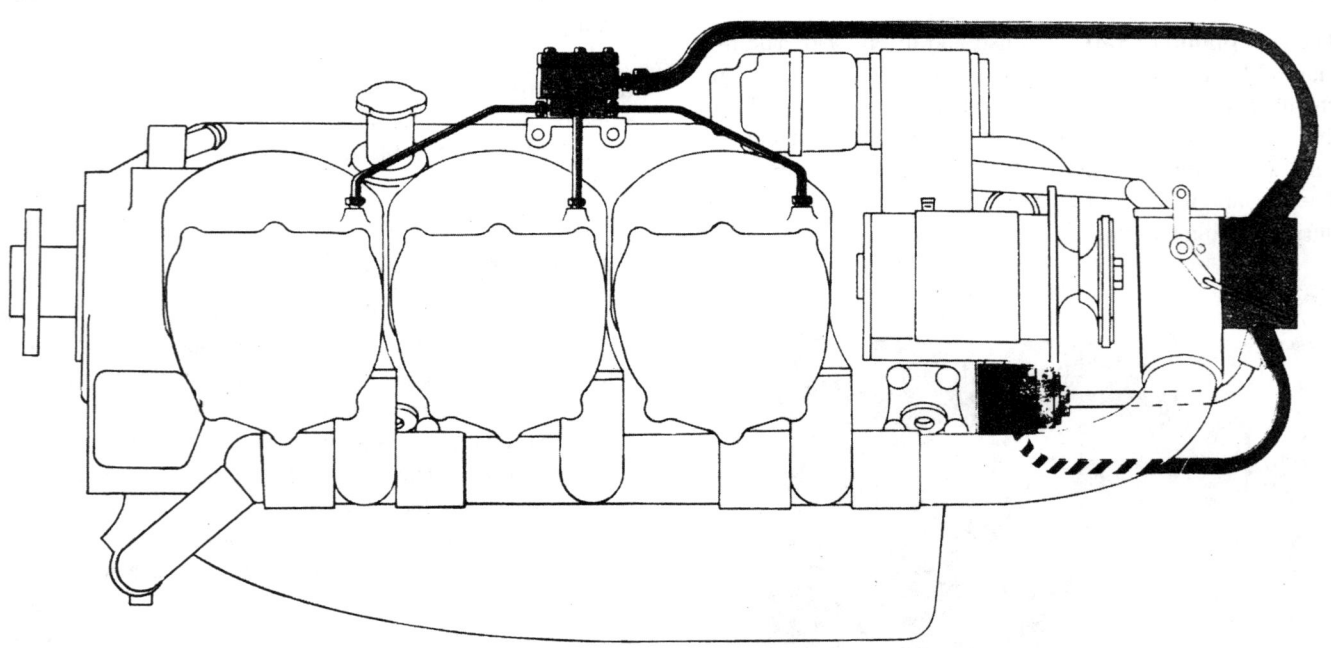

FIGURE 7-8 Continental fuel injection system installed on an engine. (*Continental Motors.*)

Complete System

The complete Continental fuel injection system installed on an engine is shown in Fig. 7-8. This diagram shows the fuel-air control unit installed at the usual location of the carburetor, the pump installed on the accessory section, the fuel manifold valve installed on the top of the engine, and the nozzles installed in the cylinders at the intake ports. The simplicity of this system contributes to ease of maintenance and economy of operation. A diagram of the complete system is shown in Fig. 7-9.

To summarize, the Continental fuel injection system utilizes fuel pressure (established by the engine rpm and the

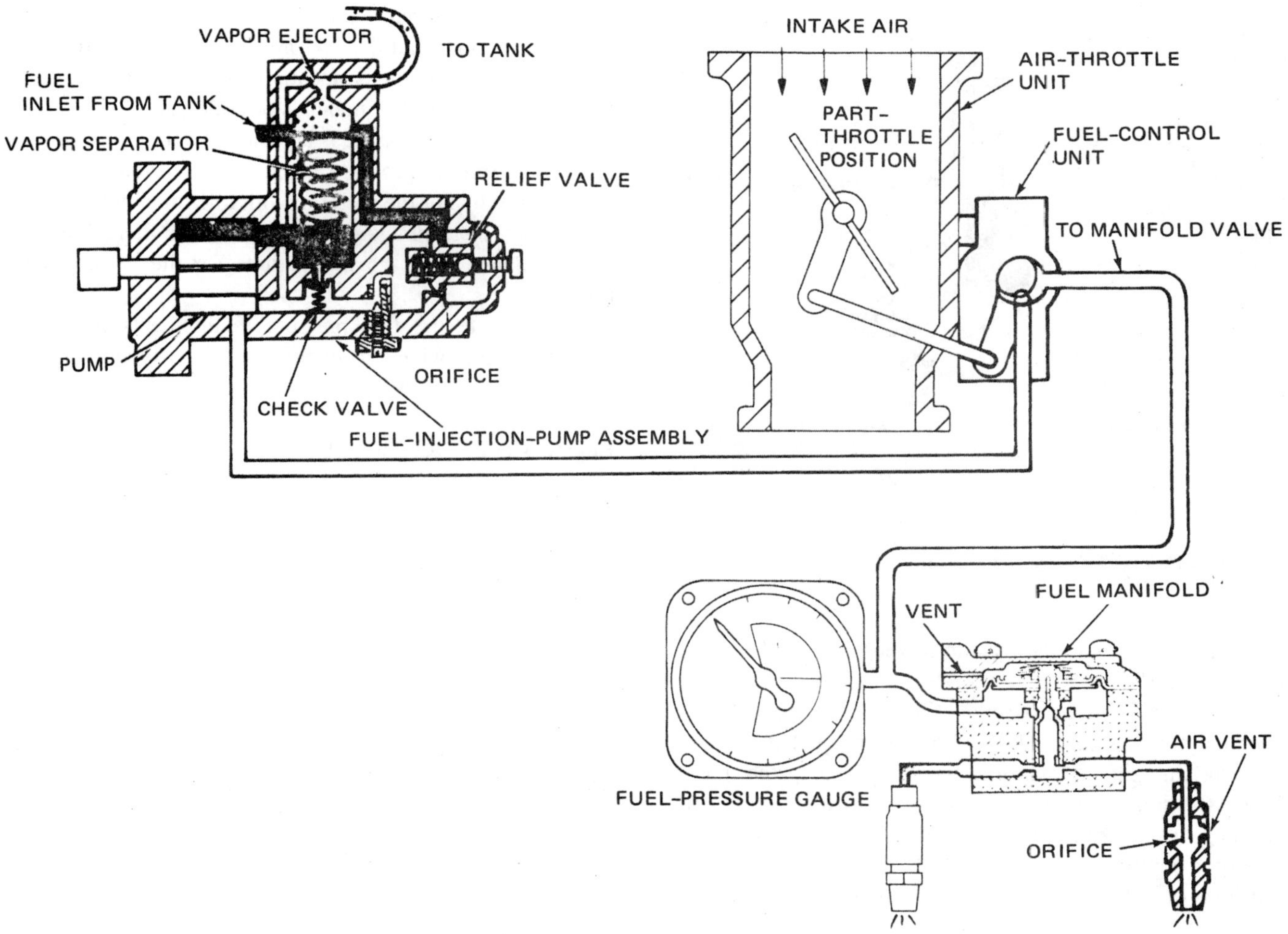

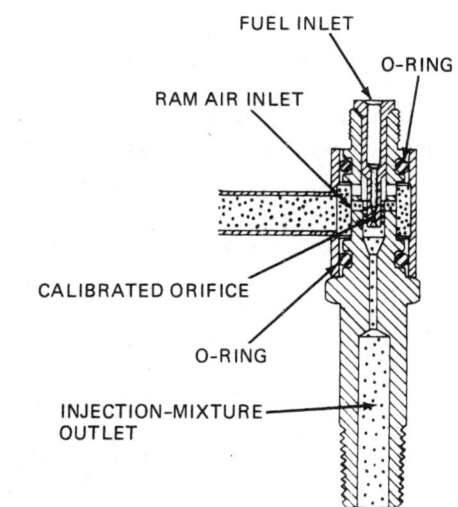

FIGURE 7-9 Schematic diagram of Continental fuel injection system. (*Continental Motors.*)

relief valve) and a variable orifice (controlled by throttle position) to meter the correct volume and pressure of fuel for all power settings. The mixture is controlled by adjusting the quantity of fuel returned to the pump inlet.

Installations for Turbocharged Engines

To permit the Continental fuel injection system to operate at the high altitudes encountered by turbocharged engines, the fuel nozzles are modified by incorporating a shroud that directs pressurized air to the air bleeds. The low air pressure at high altitudes is not sufficient for air to enter the bleeds and mix with the fuel prior to injection. Ram air is therefore applied to the nozzle shrouds, which direct it to the air bleeds. A shrouded fuel nozzle is illustrated in Fig. 7-10.

Another modification required for high-altitude operation is an **altitude compensating valve** in the fuel pump. This valve is operated by an aneroid bellows which moves a tapered plunger in an orifice. The plunger's movement varies the amount of fuel bypassed through the relief valve and thus compensates for the effect of lower pressures at high altitude. A pump which incorporates the aneroid valve is shown in Fig. 7-11.

FIGURE 7-10 Shrouded fuel nozzle.

Adjustments

The **idle speed adjustment** for the Continental fuel injection system is made with a conventional spring-loaded screw (Fig. 7-12) on the throttle lever of the fuel control unit.

Continental Continuous-Flow Injection System **173**

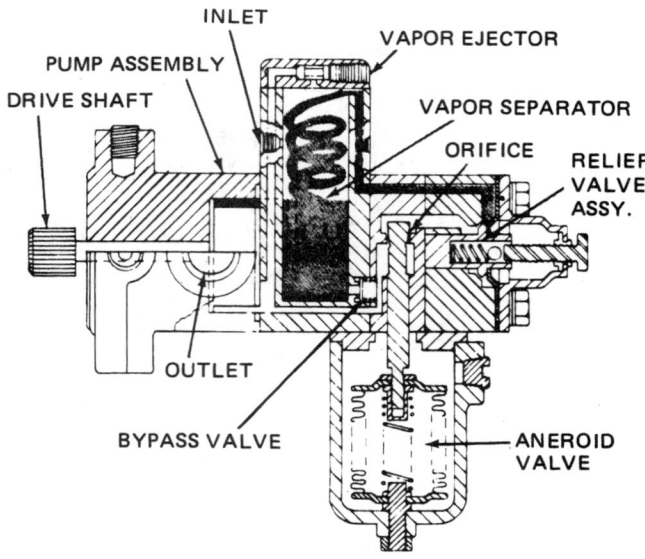

FIGURE 7-11 Fuel pump incorporating the aneroid valve.

The screw is turned to the right to increase and to the left to decrease the idle speed.

The **idle mixture adjustment** (Fig. 7-12) consists of the locknut at the metering-valve end of the linkage between the metering valve and the throttle lever. Tightening the nut to shorten the linkage provides a richer mixture, and backing off the nut leans the mixture. The mixture should be adjusted slightly richer than best power, as is the case with all systems. This is checked after adjustment by running the engine at idle and slowly moving the mixture control toward IDLE CUTOFF. Idle speed should increase slightly just before the engine begins to stop. When the manifold pressure gauge is used for the check, the MAP should be seen to decrease slightly as the mixture is leaned with the manual mixture control.

Pump Pressure for the Injection System

Since the flow of a liquid through a given orifice will increase as the pressure increases, the fuel pressure delivered by the

engine-driven fuel pump must be correct if the flow through the metering unit and the fuel nozzles is to be correct. The fuel pump adjustment is accomplished as directed by the manufacturer in service manuals or bulletins.

The fuel pump pressure is adjusted at both idle speed and full-power rpm. A pressure gauge is connected to the fuel pump outlet line or the metering unit inlet line by means of a tee fitting. The technician can then make the fuel adjustment in accordance with the instructions for the system being tested.

To illustrate the fuel pressure requirements for typical engines, the following information is quoted from the *Teledyne Continental Service Bulletin*:

Engine Model—IO-346-A,B

600 rpm		
Pump pressure	7–9 psi	48–62 kPa
Nozzle pressure	2.0–2.5 psi	13.8–17.24 kPa
2700 rpm		
Pump pressure	19–21 psi	131–145 kPa
Nozzle pressure	12.5–14.0 psi	86–96.5 kPa
Fuel flow	78–85 lb/h	13–14 gal/h
		35.4–38.5 kg/h

Engine Model—GTS1O-520-C

450 rpm		
Pump pressure	5.5–6.5 psi	38–45 kPa
Nozzle pressure	3.5–4.0 psi	24–28 kPa
2400 rpm		
Pump pressure	30–33 psi	207–228 kPa
Nozzle pressure	16.5–17.5 psi	114–121 kPa
Fuel flow	215–225 lb/h	36–38 gal/h
		98–102 kg/h

Inspections

To avoid any difficulty with the fuel injection system, it is important to perform certain inspections and checks, even though no operating discrepancies have been noted. The following inspections are recommended:

1. Check all attaching parts for tightness. Check all safetying devices.
2. Check all fuel lines for leaks and for evidence of damage, such as sharp bends, flattened tubes, or chaffing from metal-to-metal contact.
3. Check the control connections, levers, and linkages for tight-attaching parts, for safetying, and for lost motion owing to wear.
4. Inspect nozzles for cleanness, with particular attention to air screens and orifices. Use a standard $\frac{1}{2}$-in [12.70-mm] spark plug wrench (deep socket) to remove the nozzles. Do not remove the shields to clean the air screen in the nozzles. Do not use wire or other objects to clean the orifices. To clean the nozzles, remove them from the engine and immerse them in fresh cleaning solvent. Use compressed air to dry.

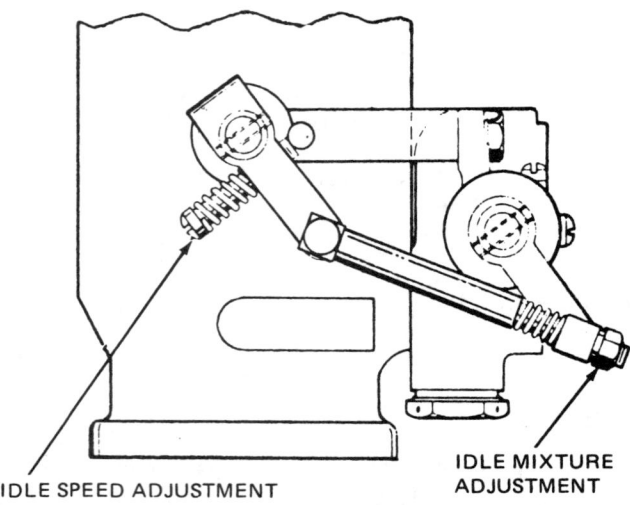

FIGURE 7-12 Idle speed and mixture adjustment.

5. Unscrew the strainer plug from the fuel injection control valve, and clean the screen in fresh cleaning solvent. Reinstall, safety, and check for leaks.

6. During periodic lubrication, apply a drop of engine oil on each end of the air throttle shaft and at each end of the linkage between the air throttle and the fuel metering valve. No other lubrication is required.

7. In the event that a line fitting in any part of the injection system must be replaced, only a fuel-soluble lubricant (such as engine oil) should be used on the fitting threads during installation. Do not use any other form of thread compound.

8. If a nozzle is damaged and requires replacement, it is not necessary to replace the entire set. Each replacement nozzle must match the one removed, as marked.

Operation

For the operation of an aircraft engine equipped with the Continental fuel injection system, certain facts must be remembered by the technician or pilot:

1. When the engine is being started, it is easy to flood the system if the timing of the starting events is not correct. When the mixture control is in any position other than IDLE CUTOFF, when the throttle is open (even slightly), and when the auxiliary fuel pump is operating, fuel will be flowing into the intake ports of the cylinders. Therefore, the engine should be started within a few seconds after the auxiliary fuel pump is turned on.

2. The engine cannot be started without the auxiliary fuel pump because the engine-driven pump will not supply adequate pressure until the engine is running.

3. The auxiliary fuel pump should be turned off during flight. It may be left on during takeoff as a safety measure.

4. For takeoff, the throttle should be fully advanced, and the mixture control should be set at FULL RICH.

5. For cruising, the engine rpm should be set according to instructions in the operator's handbook. The mixture control may be set for best power or an economy cruising condition, depending on the judgment of the operator. Care must be exercised not to lean the mixture too much.

6. Before reducing power for descent, set the mixture to best power. Once the traffic pattern is entered, the mixture control must be set to FULL RICH and kept in this position until after landing.

7. The engine is stopped by moving the mixture control to IDLE CUTOFF after the engine has been idled for a short time. All switches should be turned off immediately after the engine is stopped.

Troubleshooting

Before assuming that the fuel control system for an engine is at fault when the engine is not operating properly, the technician should make sure that other factors are not involved. A rough-running engine may have ignition problems which should be checked before dismantling the fuel control system.

The **troubleshooting chart** provided in Fig. 7-13 is a guide for determining what may be at fault. The items are arranged in order of the approximate ease of checking and not necessarily in the order of probability.

BENDIX RSA FUEL INJECTION SYSTEM

Principles of Operation

The **Bendix RSA** fuel injection system is a continuous-flow system and is based on the same principles as those of the pressure-type carburetors described in Chap. 6. A pictorial view of the RSA injector unit is shown in Fig. 7-14. Bendix fuel injection systems are designed to meter fuel in direct ratio to the volume of air being consumed by the engine at a given time. This is accomplished by sensing **venturi suction** and **impact air pressures** in the throttle body. Opening or closing the throttle valve results in a change in the velocity of air passing across the impact tubes and through the venturi, as illustrated in Fig. 7-15. When the air velocity increases, the pressure at the impact tubes remains relatively constant depending on the inlet duct configuration, air filter location, etc. The pressure at the venturi throat decreases. This decrease creates a differential (impact minus suction) which is used over the entire range of operation of the Bendix fuel injection system as a measurement of the volume of air consumed.

All reciprocating engines operate most efficiently in a very narrow range of air-fuel (A/F) [or fuel-air (F/A)] ratios. The Bendix injection system uses the measurement of air volume flow to generate a force which can be used to regulate the flow of fuel to the engine in proportion to the amount of air being consumed. This is accomplished by channeling the impact and venturi suction pressures to opposite sides of a diaphragm, as shown in Fig. 7-16. The difference between these two pressures then becomes a usable force which is equal to the area of the diaphragm times the pressure difference.

Fuel is supplied to the engine from the aircraft fuel system. This system usually includes a boost pump located in either the fuel tank or the fuel line between the tank and the engine. The engine-driven fuel pump receives fuel from the aircraft system (including the boost pump) and supplies that fuel at a relatively constant pressure to the fuel injector servo inlet. The engine manufacturer specifies the fuel pump pressure setting applicable to the specific fuel injector installation. The fuel injectors are calibrated at that inlet pressure. The settings are checked to ensure that metered fuel flow will not be affected by changes in inlet fuel pressure caused by normal ON or OFF operation of the boost pump.

The Bendix fuel injection system, if properly assembled and calibrated, will meet all performance requirements over an extremely wide range of inlet fuel pressures. Its heart is the **servo pressure regulator**.

Trouble	Probable Cause	Corrective Action
Engine will not start.	No fuel to engine.	Check mixture control for proper position, auxiliary pump ON and operating, feed valves open, fuel filters open, tank fuel level.
No gauge pressure.	Engine flooded.	Reset throttle, clear engine of excess fuel, and try another start.
With gauge pressure.	No fuel to engine.	Loosen one line at nozzle. If no fuel flow shows with metered fuel pressure on gauge, replace the fuel manifold valve.
Rough idle.	Nozzle air screens restricted.	Remove nozzles and clean.
	Improper idle mixture adjustment.	Read just as described under *Adjustments*, page 173.
Poor acceleration.	Idle mixture too lean.	Read just as described under *Adjustments*, page 173.
	Linkage worn.	Replace worn elements of linkage.
Engine runs rough.	Restricted nozzle.	Remove and clean all nozzles.
	Improper mixture.	Check mixture-control setting. Improper pump pressure; replace, or call an authorized representative to adjust.
Low gauge pressure.	Restricted flow to metering valve.	Check mixture control for full travel. Check for clogged fuel filters.
	Inadequate flow from pump.	May be worn fuel pump or sticking relief valve. Replace the fuel pump assembly.
	Mixture-control-level interference.	Check for possible contact with cooling shroud.
High gauge pressure.	Restricted flow beyond metering valve.	Check for restricted nozzles or fuel manifold valve. Clean or replace as required.
	Restricted recirculation passage in pump.	Replace pump assembly.
Fluctuating gauge pressure.	Vapor in system; excess fuel temperature.	If not cleared with auxiliary pump, check for clogged ejector jet in the vapor-separator cover. Clean only with solvent, no wires.
	Fuel in gauge line. Leak at gauge connection.	Drain the gauge line and repair the leak.
Poor idle cutoff.	Engine getting fuel.	Check mixture control to be in full IDLE CUTOFF. Check auxiliary pump to be OFF. Clean nozzle assemblies (screens) or replace. Replace manifold valve.

FIGURE 7-13 Continental fuel injection troubleshooting chart.

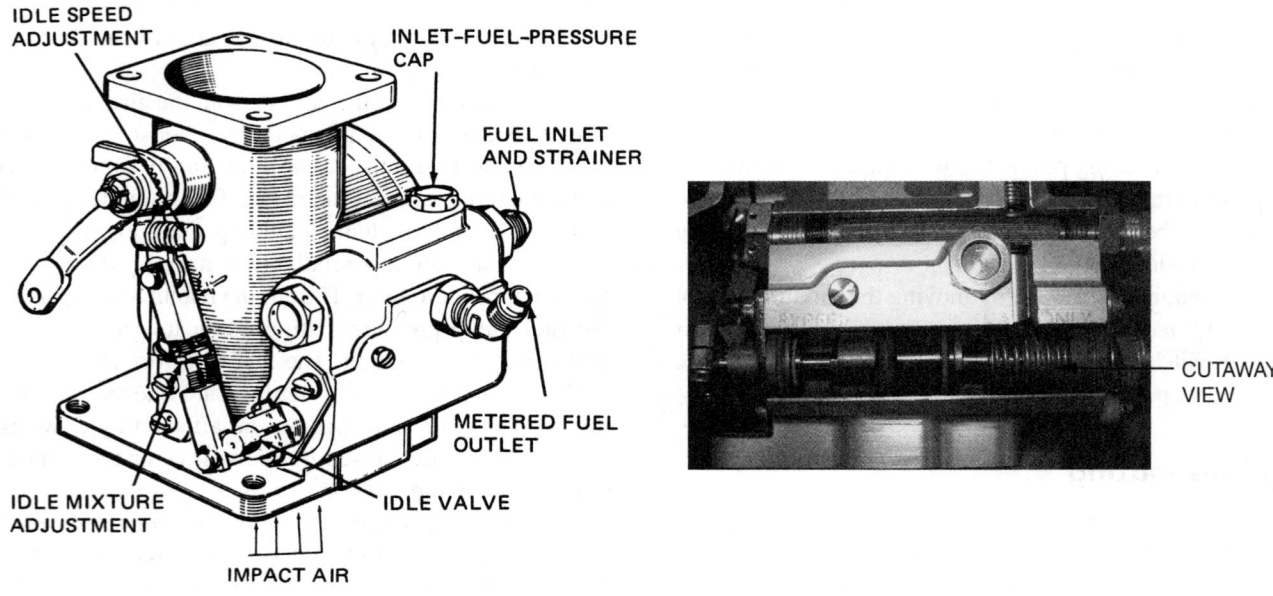

FIGURE 7-14 Bendix RSA injector unit. (*Continental Motors.*)

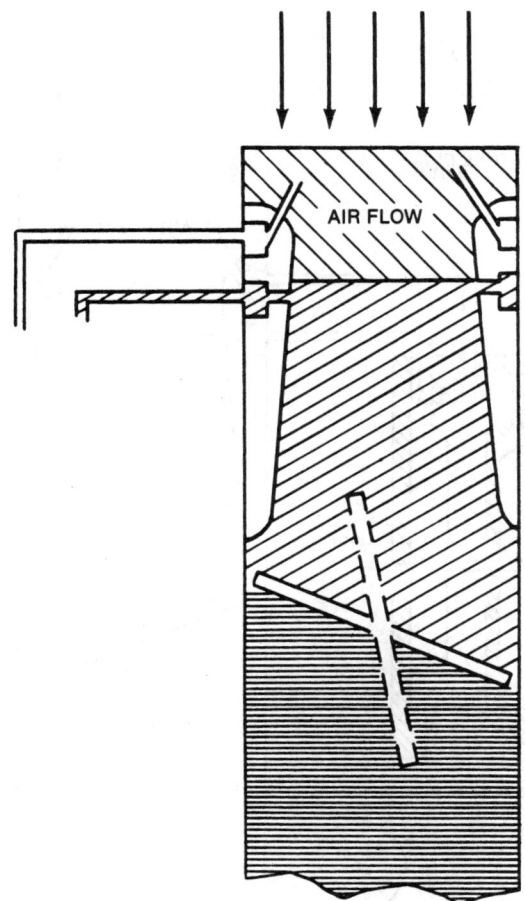

FIGURE 7-15 Airflow through the venturi. (*Continental Motors.*)

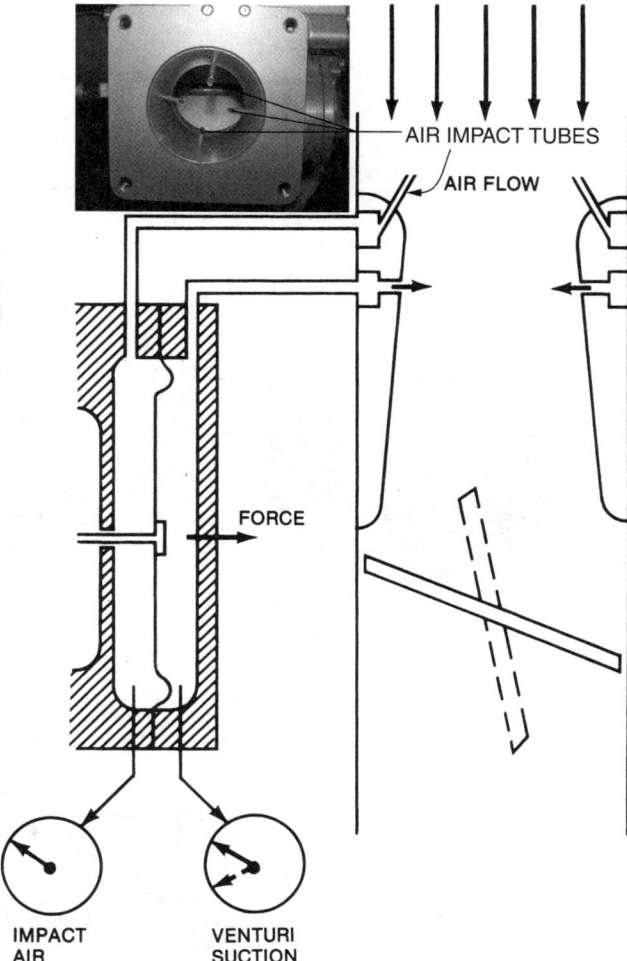

FIGURE 7-16 Impact and venturi suction pressures. (*Continental Motors.*)

The easiest way to explain the operation of this regulator and its relationship to the main metering jet is to describe a power change which requires a fuel flow change.

We start from a cruise condition where air velocity through the throttle body is generating an impact pressure minus venturi suction pressure differential at a theoretical value of 2. This air pressure differential of 2 is exerting a force to the right, as shown in Fig. 7-17.

Fuel flow to the engine, passing through the metering jet, generates a fuel pressure differential (unmetered minus metered fuel pressure). This pressure differential, applied across a second (fuel) diaphragm, is also creating a force with a value of 2. This value of 2 is exerting a force to the left, as shown in Fig. 7-17.

The two opposing forces (fuel and air differentials) are equal, and the **regulator servo valve** (which is connected to both diaphragms by a stem) is held at a fixed position that allows discharge of just enough metered fuel to maintain pressure balance. If the throttle is opened to increase power, airflow immediately increases. This results in an increase in the pressure differential across the air diaphragm to a theoretical value of 3. The immediate result is a movement of the regulator servo valve to the right. This increased servo valve opening causes both a decrease in pressure in the metered fuel chamber and an increase in fuel pressure differential across the main metering jet. When this increasing

fuel differential-pressure force reaches a value of 3 (equaling the air diaphragm force), the regulator stops moving and the servo valve stabilizes at a position which will maintain the balance of pressure differentials (that is, air and fuel, each equaling 3).

Fuel flow to the engine is increased to support the higher power level requested. The fuel diaphragm force being generated by the pressure drop across the main metering jet is equal to the air diaphragm force being generated by the venturi.

This sequence of operations is maintained over all regimes of power operation and all power changes. The regulator servo valve responds to changes in effective air diaphragm differential-pressure forces and adjusts the position of the servo valve to regulate unmetered to metered fuel pressure-differential forces accordingly. Fuel flow through the metering jet is a function of its size and the pressure differential across the jet. The servo valve does not meter fuel. It only controls the pressure differential across the metering jet.

The idle valve is connected to the throttle linkage. It effectively reduces the area of the main metering jet for accurate metering of fuel in the idle range. Figure 7-18 shows the function and operation of the idle valve. It is externally adjustable and allows the technician to properly tune the fuel injector to the engine installation for proper idle mixture.

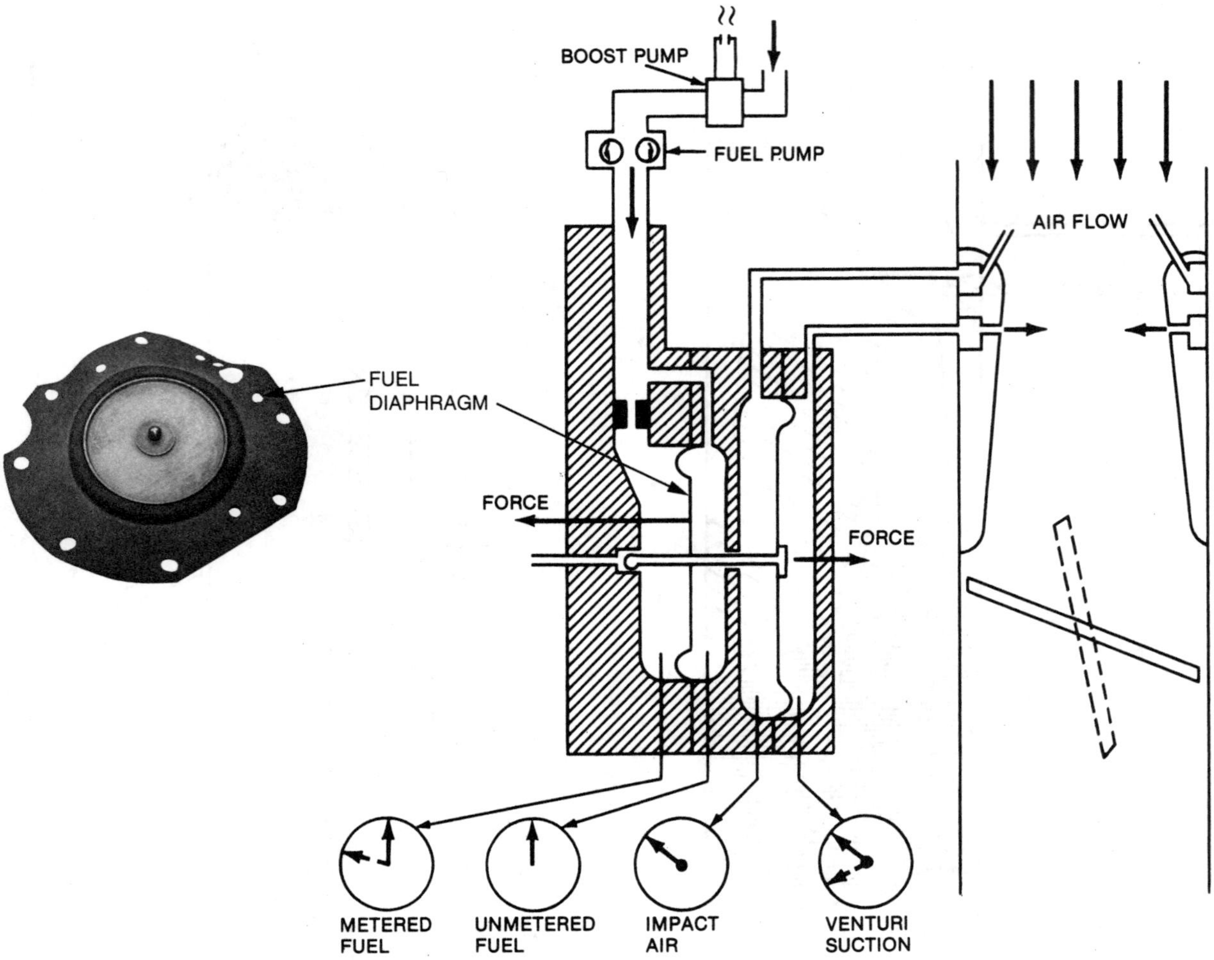

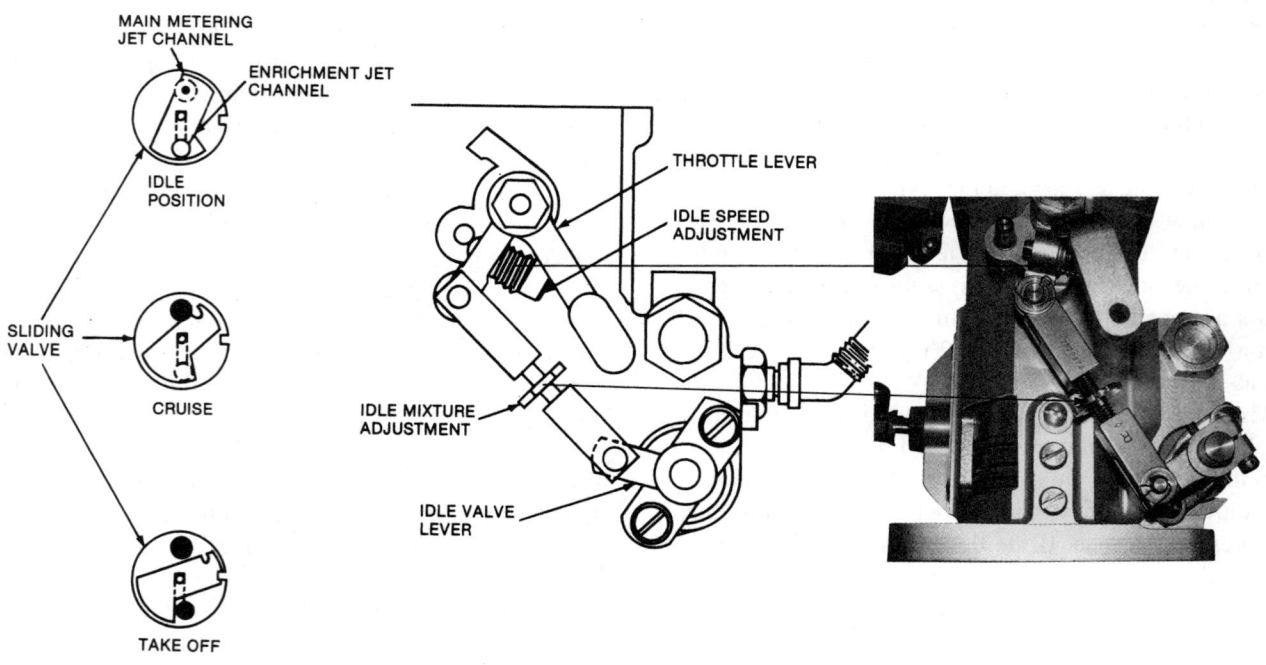

BOOST PUMP

FUEL PUMP

AIR FLOW

FUEL
DIAPHRAGM

FORCE

FORCE

METERED
FUEL

UNMETERED
FUEL

IMPACT
AIR

VENTURI
SUCTION

FIGURE 7-17 Servo pressure regulator. (*Continental Motors.*)

MAIN METERING
JET CHANNEL

ENRICHMENT JET
CHANNEL

IDLE
POSITION

THROTTLE LEVER

IDLE SPEED
ADJUSTMENT

SLIDING
VALVE

CRUISE

IDLE MIXTURE
ADJUSTMENT

IDLE VALVE
LEVER

TAKE OFF

IDLE VALVE POSITIONS

FIGURE 7-18 Idle valve and manual mixture control. (*Continental Motors.*)

The idle mixture is correct when the engine gains approximately 25 to 50 rpm from its idle speed setting as the mixture control is placed in IDLE CUTOFF. Manual control of the idle mixture is necessary because at the very low airflow through the venturi in the idle range, the air metering force is not sufficient to accurately control the fuel flow.

On some engines, according to specific installation requirements, an enrichment jet is added in parallel with the main metering jet. On these installations, the **sliding (rotation) idle valve** begins to uncover the enrichment jet at a preset throttle position, as shown in Fig. 7-18. This parallel-flow path increases the F/A mixture strength to provide for **fuel cooling** of the engine in the high-power range. In simple terms, increased fuel consumption is traded for added engine life.

The manual mixture control, shown in Fig. 7-18 as a sliding valve, can be used by the pilot to effectively reduce the size of the metering jet. With the servo pressure regulator functioning to maintain a differential pressure across the metering jet in proportion to the volume of airflow, the flow through the jet may be varied by changing its effective size. This gives the pilot the option of manually leaning the mixture for best cruise power or best SFC. It also provides the means to shut off fuel flow to the engine at engine shutdown.

The **constant head idle spring** augments the force of the air diaphragm in the idle range when the air pressure differential is not sufficient to open the servo valve. The idle spring shown in Fig. 7-19 ensures that the regulator servo valve is open sufficiently to allow fuel being metered by the idle valve to flow out to the flow divider. As airflow increases above idle, the air diaphragm will begin to move to the right in response to increasing air pressure differential. It will compress the constant head idle spring until its retainer and guide contact the diaphragm plate. From this point onward, in terms of airflow, fuel flow, or power, the constant head idle spring assembly is a solid member moving with the air diaphragm and exerts no force of its own. The constant head spring is furnished in various strengths so that the overhaul technician can properly calibrate the injector for idle fuel flow and for the transition to servo regulator controlled fuel flow.

In most installations, the transition from idle to servo regulator controlled fuel flow has to be supplemented with a **constant-effort spring**, pictured in Fig. 7-20. This spring also assists the air diaphragm in moving smoothly from the low-airflow idle range to the higher-power range of operation. It also comes in a selection of strengths to be utilized by the overhaul technician for proper calibration of the unit.

The fuel section of the servo pressure regulator is separated from the air section by a center body seal assembly, as shown in Fig. 7-20. Leakage through the center body seal causes extremely rich operation and poor cutoff. Leakage of raw fuel out of the impact tubes may indicate possible seal leakage. Failure of this seal requires repair in an overhaul shop. The seal cannot be replaced in the field.

Automatic Mixture Control

The AMC adjusts the F/A ratio to compensate for the decreased air density as the aircraft climbs to altitude. Figure 7-21 shows the function and operation of the AMC and expands on the description of the manual mixture control and idle valves. The mixture control is shown in the FULL RICH position and with the idle valve fully open, as it would be at cruise power or above. In the cutaway view, the two rotating valve assemblies are spring-loaded together, back to back, with an O-ring seal between them. Fuel flows through the mixture control valve, through the idle valve, and out to the regulator servo valve. The inlet strainer is located underneath the fuel inlet fitting and is installed spring end first, so the open end is mated to the inlet fitting. If the screen is blocked by contaminant material, inlet pressure will force the material away from the fitting, compressing the spring to allow fuel to bypass the screen if necessary. This screen filter is replaced at overhaul. There is no approved method of cleaning this screen for reuse.

When calculating the F/A ratio, both the fuel and the air are measured in pounds per hour. The fuel injector meters fuel on a pounds-per-hour basis, referenced to the volume of airflow, which, when converted to velocity passing through the venturi, produces the air metering signal previously discussed.

An engine pumps air on the basis of volume, not weight. This volume is determined by the engine displacement—that is, 540 in^3 per complete four-stroke cycle (intake, compression, power, and exhaust for all six cylinders). Therefore an IO-540 engine running at 2500 rpm would be consuming (pumping) air at a rate of

$$540 \times \frac{2500}{2} = 675\,000 \text{ in}^3/\text{min } [11 \text{ m}^3/\text{min}]$$

$$\frac{675\,000}{1728} = 390 \text{ ft}^3/\text{min } [11 \text{ m}^3/\text{min}]$$

$$390 \times 0.0765 = 30 \text{ lb/min } [13.6 \text{ kg/min}]$$

or

$$30 \text{ lb/min} \times 60 \text{ min} = 1800 \text{ lb/h } [816 \text{ kg/h}]$$

This would be equivalent to cruise power at sea level. An F/A ratio of 0.08 would result in a fuel flow rate of

$$1800 \times 0.08 = 144 \text{ lb/h } [65 \text{ kg/h}]$$

As the aircraft climbs to altitude, the specific weight of air decreases from 0.0765 lb/ft^3 [1.22 kg/m^3] until, at 15 000 ft, air weighs only 0.0432 lb/ft^3 [0.692 kg/m^3]. The engine operating at 2500 rpm would still be consuming 390 ft^3/min [11 m^3/min], resulting in an airflow rate of

$$390 \times 0.0432 \times 60 = 1011 \text{ lb/h } [459 \text{ kg/h}]$$

This 1011-lb/h [459-kg/h] airflow will produce the same air metering signal across the venturi that 1800 lb/h [816 kg/h]

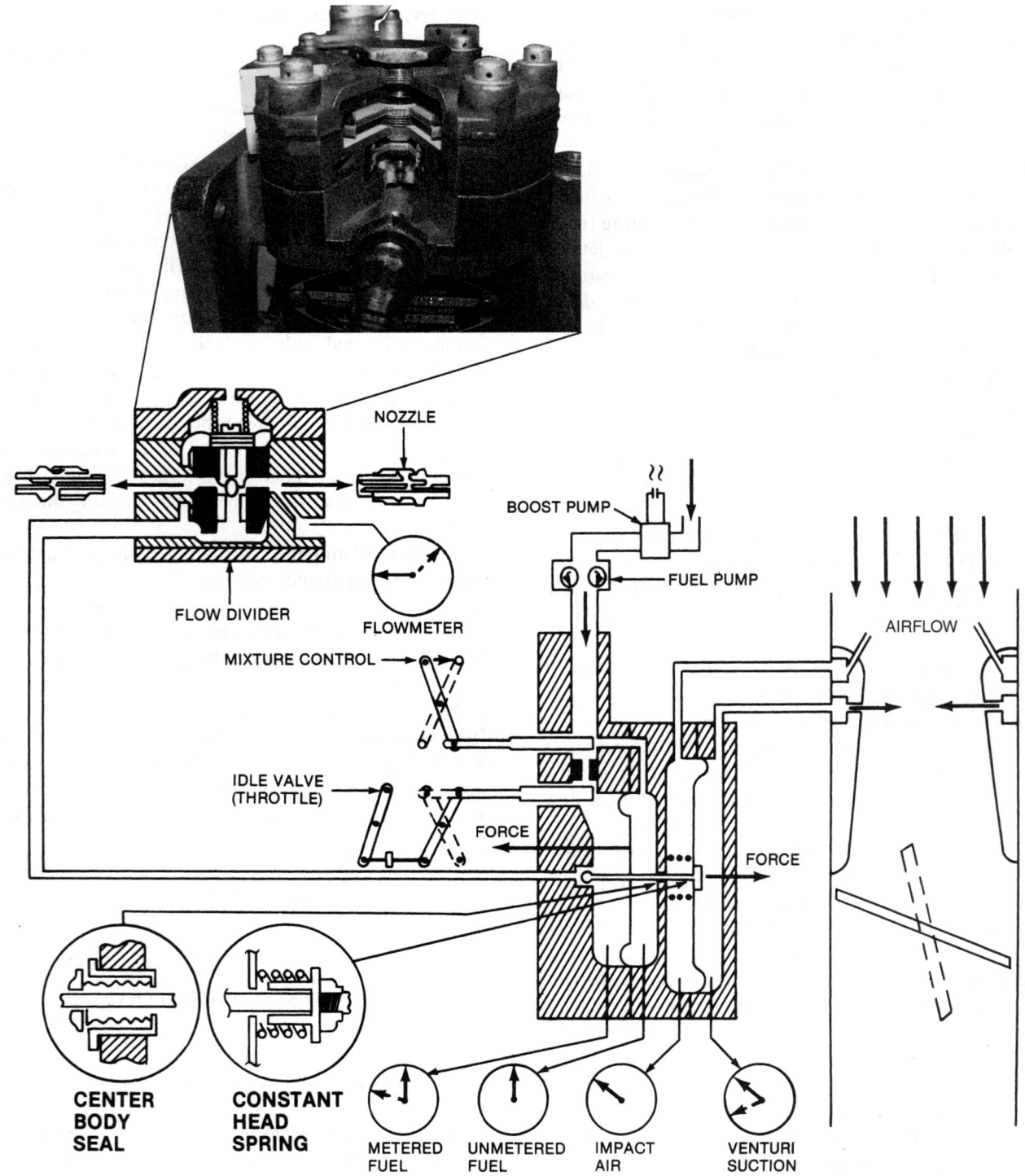

FIGURE 7-19 Constant head idle spring. (*Continental Motors.*)

did at sea level. This air metering signal will maintain the 144-lb/h [65-kg/h] fuel flow, which results in

$$\frac{144}{1011} = 0.142 \text{ F/A ratio}$$

Without an AMC, it would be necessary for the pilot to continually lean the mixture manually to maintain the

desired 0.08 F/A ratio. The AMC works independently of, and in parallel with, the manual mixture control by providing a variable orifice between the two air pressure signals (impact and suction) to modify the air metering signal force.

The AMC assembly consists of a contoured needle that is moved in and out of an orifice by a bellows assembly.

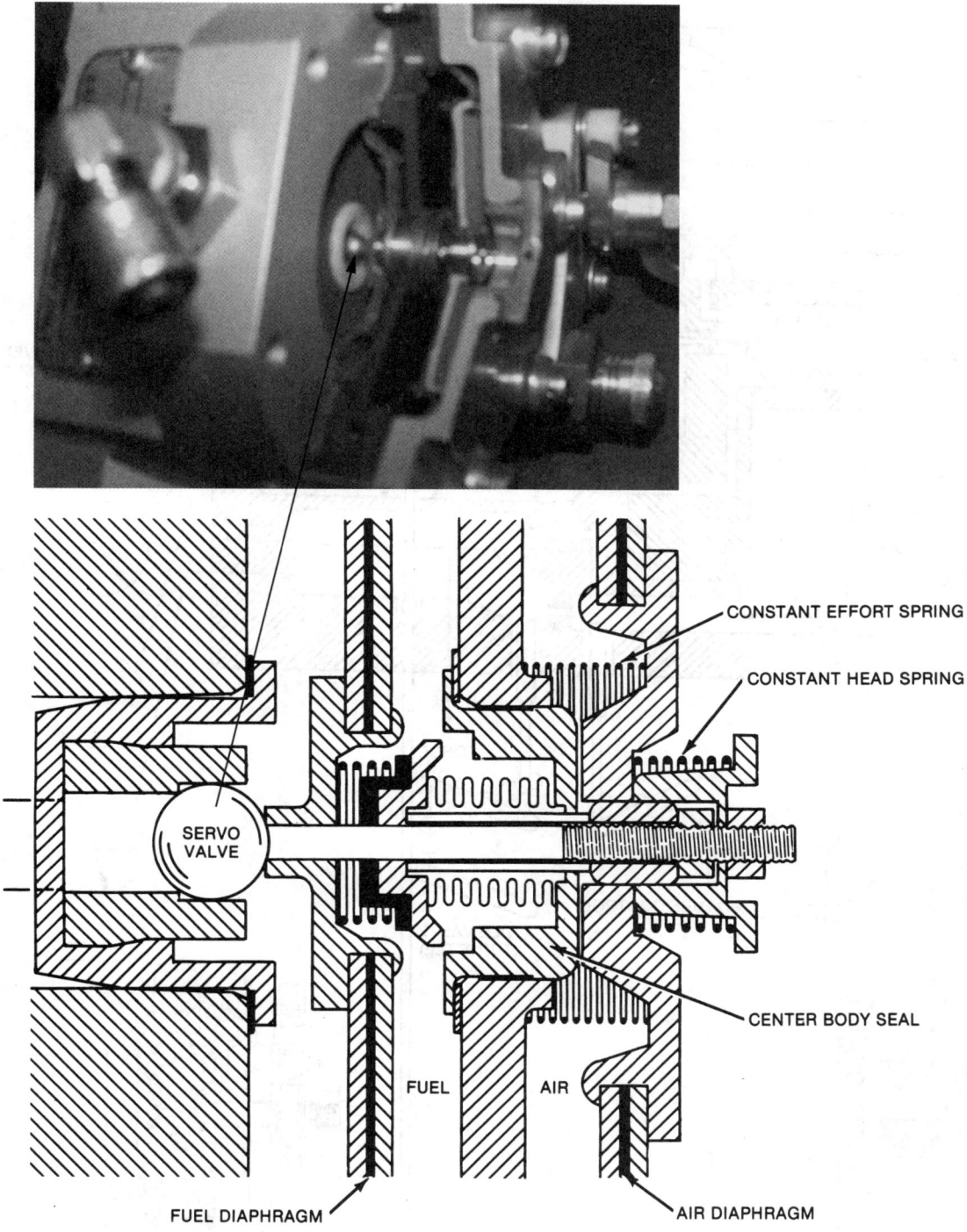

FIGURE 7-20 Constant-effort spring. (*Continental Motors.*)

This bellows reacts to changes in air pressure and temperature, increasing in length as pressure altitude increases. At ground level, the needle is positioned in the AMC orifice so that the orifice is closed, or nearly closed, to allow the maximum impact pressure on the impact pressure side of the air diaphragm.

When the aircraft flies at increased altitude, the AMC bellows elongates with the air pressure decrease and the needle is moved into its orifice. This increases the orifice opening between the impact air and venturi suction and allows impact air to bleed into the venturi suction channel. This reduces the air metering force across the air diaphragm.

The needle is contoured so that regardless of altitude (or air density) the correct air metering signal is established across the air diaphragm to maintain a relatively constant F/A ratio as the air density changes with altitude.

Current production-type fuel injectors use a **bullet-type AMC**, illustrated in Fig. 7-22, which is mounted in the bore of the throttle body. The exterior diameter of this unit is contoured to perform the function of the venturi. The function

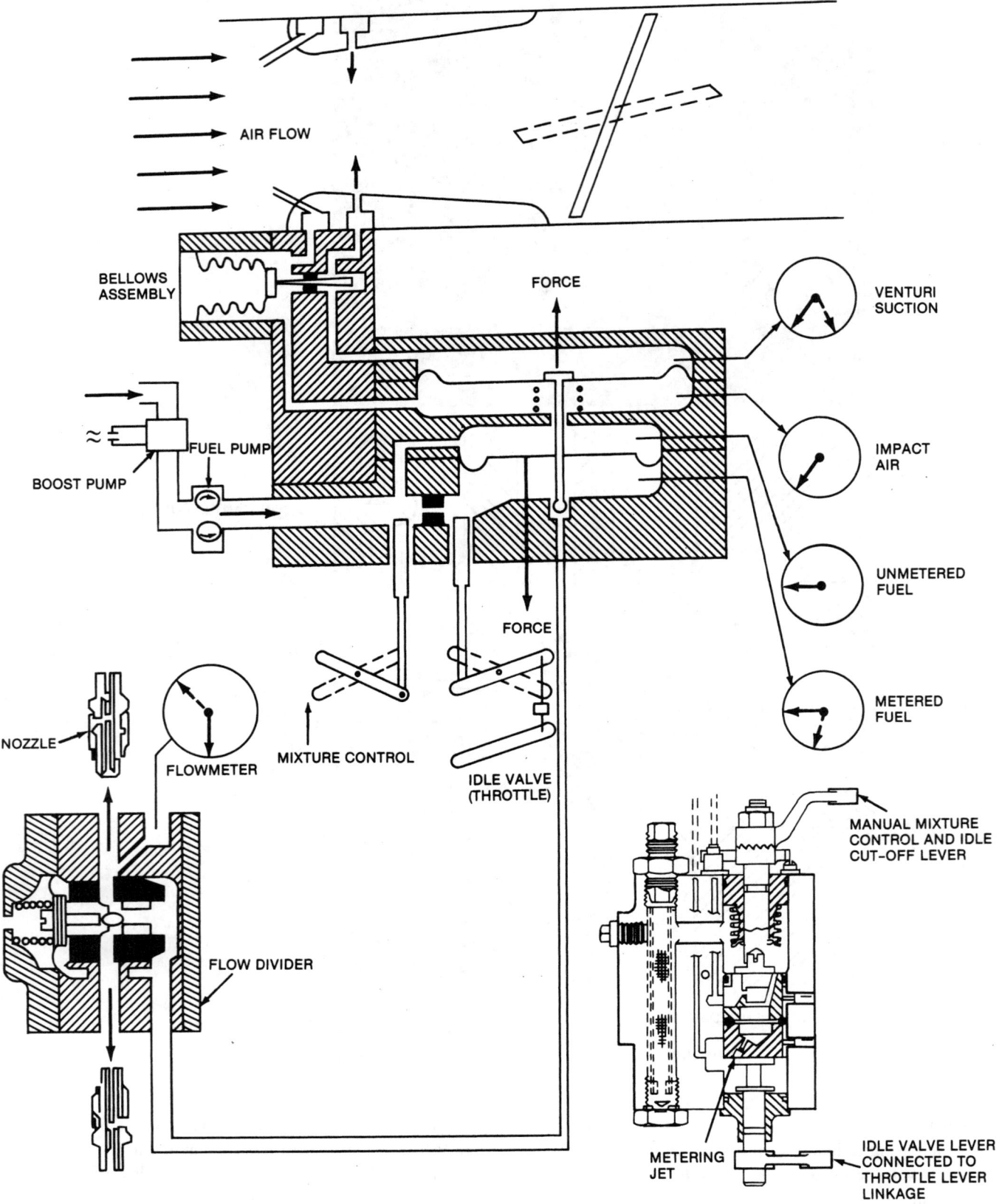

AIR FLOW

BELLOWS ASSEMBLY

FORCE

VENTURI SUCTION

IMPACT AIR

FUEL PUMP

BOOST PUMP

UNMETERED FUEL

METERED FUEL

FORCE

NOZZLE

FLOWMETER

MIXTURE CONTROL

IDLE VALVE (THROTTLE)

MANUAL MIXTURE CONTROL AND IDLE CUT-OFF LEVER

FLOW DIVIDER

METERING JET

IDLE VALVE LEVER CONNECTED TO THROTTLE LEVER LINKAGE

FIGURE 7-21   AMC. (*Continental Motors.*)

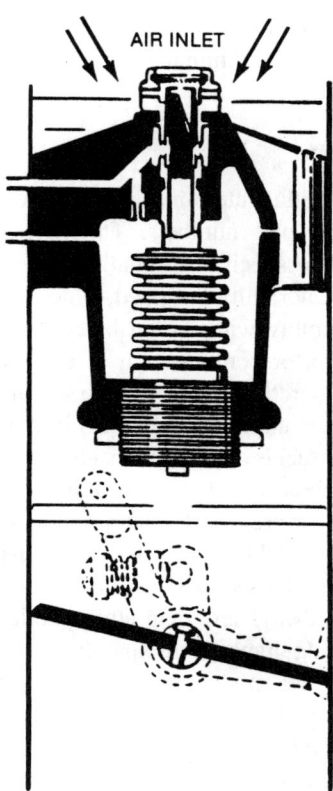

FIGURE 7-22 Bullet-type AMC. (*EContinental Motors.*)

performed and the principle of operation are exactly the same as for the externally mounted unit. There are two basic differences:

1. The bellows assembly is exposed to venturi suction rather than impact pressure.
2. As the needle is moved into its orifice, impact air pressure to the servo regulator is restricted, thus causing a reduction of the air metering force across the air diaphragm exactly as described above.

Flow Dividers

Metered fuel flow is delivered from the fuel injector servo unit to the engine through a system which usually includes a **flow divider** and a set of **discharge nozzles** (one nozzle per cylinder). A few engine installations do not use a flow divider. On these engines, the fuel flow is divided by either a single four-way fitting (four-cylinder engines) or a tee which divides the fuel flow into two separate paths.

The flow divider illustrated in Fig. 7-23 consists of a valve, sleeve, diaphragm, and spring. The valve is spring-loaded to the closed position in the sleeve. This effectively closes the path of fuel flow from the fuel injector servo to the nozzles and at the same time isolates each nozzle from all the others at engine shutdown. The two primary functions of the flow divider are

1. To ensure equal distribution of metered fuel to the nozzles at and just above idle
2. To provide isolation of each nozzle from all the others for clean engine shutdown

The area of the fuel discharge jet in the fuel nozzles is sized to accommodate the maximum fuel flow required at rated horsepower without exceeding the regulated fuel pressure range capability of the servo pressure regulator.

The area of the jet in the nozzle is such that metered fuel pressure at the nozzle is negligible at the low fuel flows required at and just above idle. Metered fuel from the injector servo enters the flow divider and is channeled to a chamber beneath the diaphragm. At idle, fuel pressure is only sufficient to move the flow divider valve slightly open, exposing the bottom of a V slot in the exit to each nozzle, as illustrated in Fig. 7-24. This position provides the accuracy of fuel distribution needed for smooth idle. As the engine is accelerated, metered fuel pressure increases at the flow divider inlet and in the nozzle lines. This pressure gradually moves the flow divider valve open against the spring pressure until the area of the V slot opening to each nozzle is greater than the area of the fuel restrictor in the nozzle. At this point, responsibility for equal distribution of metered fuel flow is assumed by the nozzles. Since metered fuel pressure (nozzle pressure) increases in direct proportion to metered fuel flow, a simple pressure gauge can be used as a flowmeter indicator. If the fuel restrictor in one or more nozzles becomes partially plugged, the total exit path for metered fuel flow is reduced. The fuel injector servo will continue to deliver the same amount of total flow. Therefore, the nozzle pressure will increase, giving an indication of fuel flow increase on

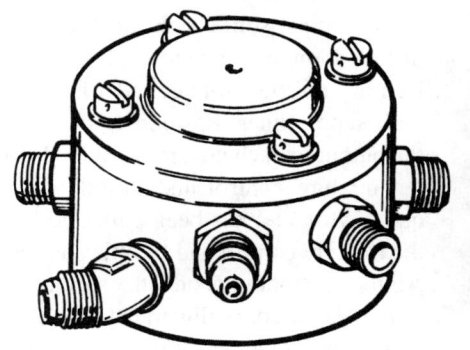

FIGURE 7-23 External view of flow divider and cutaway view.

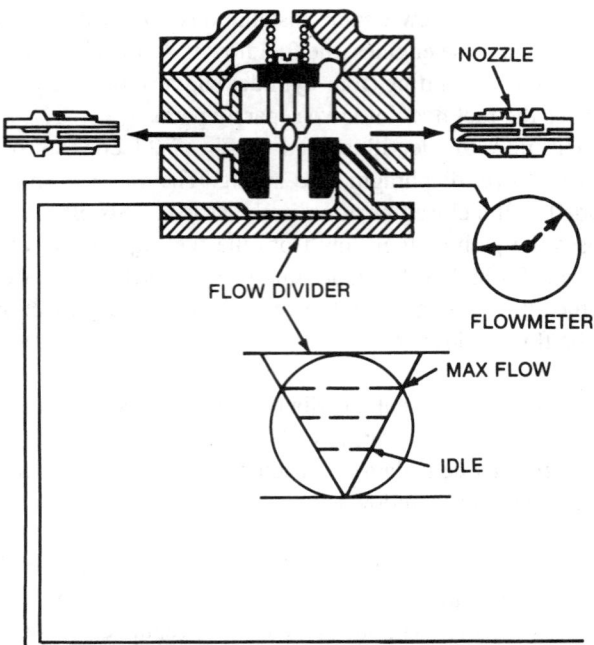

FIGURE 7-24 Fuel flow divider. (*Continental Motors.*)

the flowmeter gauge. An example of a fuel flow gauge is seen in Fig. 7-25.

The cylinder(s) having restricted nozzles will be running lean, and the remaining cylinders will be rich. The result is rough engine operation accompanied by a high-fuel-flow indication.

When the mixture control is placed in CUTOFF, fuel pressure to the flow divider drops to zero. The spring forces the flow divider valve to the CLOSED position and immediately interrupts the flow of fuel to each nozzle. This breaks the path of capillary flow, which would allow manifold suction to continue to draw fuel in dribbles from one or more nozzle lines as the engine coasts down. Without the flow divider,

FIGURE 7-25 Fuel pressure gauge (*flow indicator*).

this dribbling of fuel into one or more cylinders could keep the engine running for a minute or more.

Fuel Nozzles

Bendix nozzles, illustrated in Fig. 7-26, are furnished with several different part numbers. The part number generally identifies the specific installation requirement for the engine—that is, normally aspirated, which requires the simple nozzle assembly with the air-bleed screen and shroud pressed in place, or configuration of the shroud assembly to accept the supercharger air pressure signal to the nozzle.

All nozzles are of the air-bleed type illustrated in Fig. 7-27. This means that fuel is discharged inside the nozzle body into a chamber which is vented to either atmospheric air pressure or supercharger air pressure (injector top deck pressure). The nozzle is mounted in the intake valve port of the cylinder head. Its exit is always exposed to MAP, which on a normally aspirated engine is always less than atmospheric pressure. This results in air being drawn in through the air bleed and mixed with fuel in the fuel-air chamber to provide for fuel atomization. This is particularly important in the idle and low-power ranges where MAP is lowest and bleed-air intake is greatest. A plugged air bleed in this range allows the exit of the fuel restrictor to be exposed to manifold suction, which effectively increases the pressure differential across the restrictor and causes an increase in fuel flow through that nozzle. Since this nozzle is, in effect, stealing fuel from the other nozzles (injector servo output flow will remain the same), this cylinder will run rich and the other cylinders will be correspondingly lean. A net decrease in metered fuel pressure will result and show up on the flowmeter as a lower-fuel-flow indication. This would result in a rough idle with a low-fuel-flow indication and a higher than normal rpm rise going into cutoff. The engine will also have very poor cutoff, tending to continue chugging for several seconds following movement of the mixture control to CUTOFF.

CAUTION The following test is intended only as a troubleshooting aid and should not be construed as a calibration check of the nozzle assemblies. Should a question arise regarding serviceability of a given nozzle assembly, the unit must be sent to a certified overhaul and repair facility.

Most nozzles which are installed in cylinder heads are calibrated alike. For example, with exactly 12-psig [82.8-kPa] inlet pressure, applied flow should be 32 lb/h [14.53 kg/h] plus or minus 2 percent.

A comparison check of fuel flow from the nozzles installed on any given engine can be made as follows: Remove nozzles from cylinder heads and reconnect them to their supply lines. Position equal-size containers to capture the output of each nozzle. Turn on the boost pump, and open the mixture control and throttle. When a good reference quantity of fuel has been collected in each container, close the mixture control and throttle and turn off the boost pump. Align the containers on a flat surface, and compare the level of fuel captured, as illustrated in Fig. 7-28. A noticeably low volume of fuel in one or more containers indicates restriction in the nozzle fuel restrictor, flow divider, or lines.

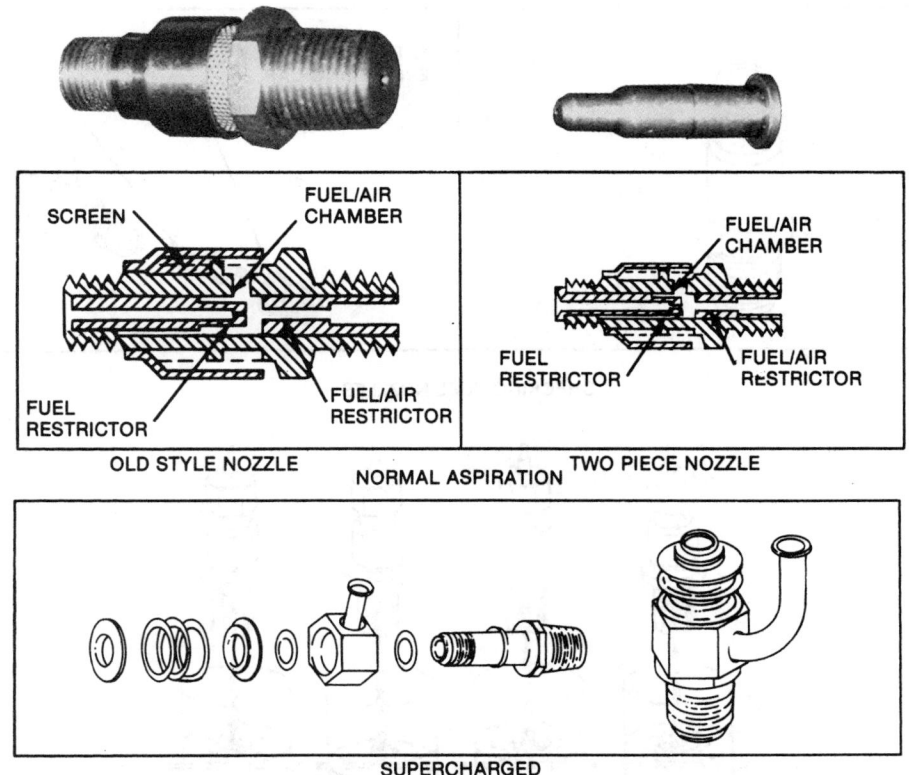

FIGURE 7-26   Bendix fuel nozzles. (*Continental Motors.*)

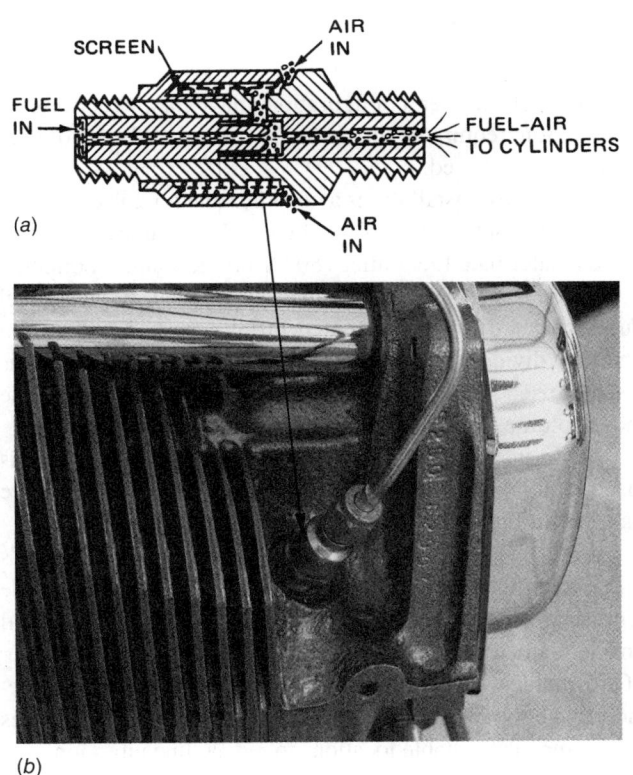

FIGURE 7-27   (*a*) Schematic diagram of air-bleed fuel nozzle. (*b*) nozzle installed in cylinder.

All current production nozzles have two-piece configurations. The fuel restrictor is a flanged insert which can easily be removed for cleaning. Also, it can be easily lost during handling. A lost or damaged insert will require a new nozzle assembly because they are flow-matched assemblies. In older production nozzles, the restrictor was pressed into the nozzle body and was not removable.

Prior to installing a nozzle assembly, always refer to the engine manufacturer's instruction manuals for the proper torque values of nozzles and lines.

Overtorque of the nut connecting the fuel line to the nozzle can press the insert deeper into the body and close off the air bleed on the older nozzles. Overtorque can crack the flange off the insert on the newer types. In either case, the nozzle is destroyed. Overtorque of the nozzle into the cylinder head distorts the base of the nozzle and upsets its calibration and spray pattern.

Although it is generally good practice to maintain a common configuration of a full set of one type of nozzle, installing one or more new-type nozzles on an engine equipped with the older type should create no problem. Also, where possible, individual two-piece nozzles should be kept as matched assemblies. Intermixing these parts may not cause a problem; however, if or when a problem occurs as a result of intermixing, a new nozzle will probably be the only recourse.

The **letter A** found stamped on one flat of the wrenching hexagon is located 180° from the air-bleed hole in the nozzle body. After final installation torquing, the air bleed should be positioned upward (A on the bottom) so that residual fuel in the line cannot drip out after engine shutdown. A thorough periodic cleaning of the nozzle is recommended to ensure continued satisfactory operation until overhaul is due.

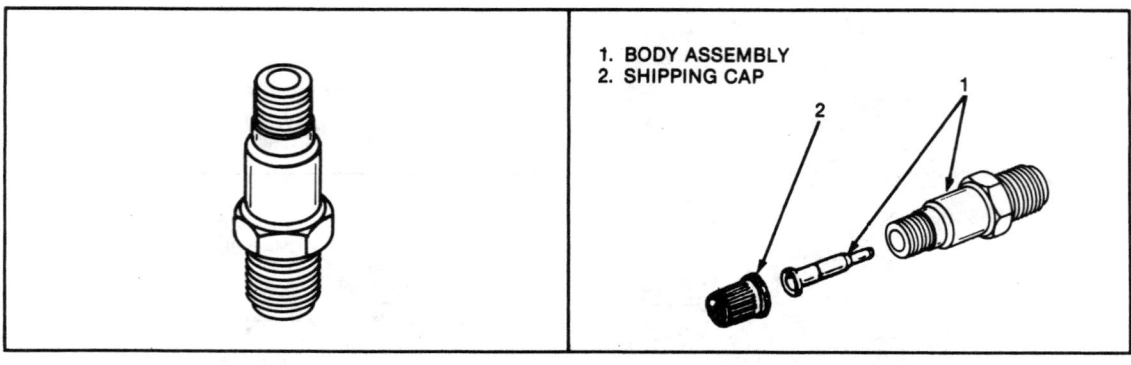

1. BODY ASSEMBLY
2. SHIPPING CAP

CHECKING FUEL NOZZLES

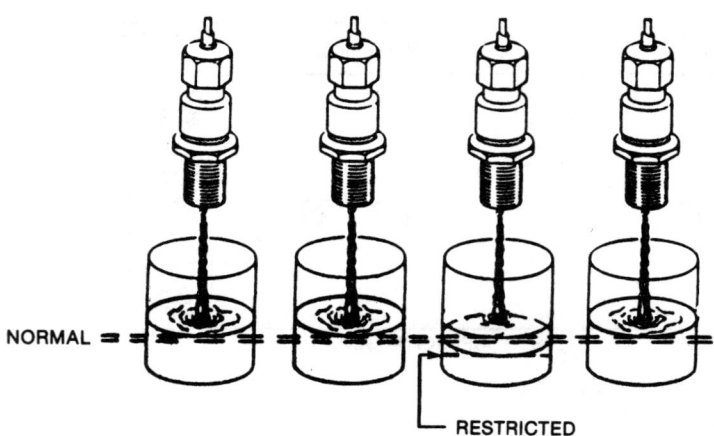

NORMAL

RESTRICTED

FIGURE 7-28   Testing fuel nozzle output. (*Continental Motors.*)

Installation of the Injection System

The following information about installation of the RSA fuel injection system was provided by the Energy Controls Division of the Bendix Corporation and is given here as an example of the data supplied to aircraft operators and maintenance personnel:

1. The injector can be mounted on the inlet flange of the engine's intake manifold at any attitude that facilitates engine-to-airframe combination installation, taking into consideration that the throttle linkage and the manual mixture control linkage must be attached to the unit.

2. An allowance should be made for adequate ventilation to the injector because of possibly high ambient temperatures within the engine nacelle.

3. The flow divider can be mounted at an optimum location with a predetermined bracket configuration; however, it must be mounted with the nozzle-line fittings in a horizontal plane.

4. On engines where the nozzle is installed horizontally, particular attention should be paid to the identification mark stamped on one of the hexagonal flats on the nozzle body. This mark is located 180° from the air-bleed hole and must appear on the lower side of the nozzle. This ensures that the air-bleed hole is on top—to reduce fuel bleeding from this opening just after shutdown.

5. A flexible hose is used from the engine-driven fuel pump to the injector fuel inlet. The size of this hose may vary according to the installation.

6. Fuel strainer configuration may vary according to installation requirements. In most cases a 74-micrometer (μm) screen is used.

7. In most installations a no. 4 flexible hose is used from the injector outlet to the flow divider. Later-model injectors have an alternate fuel outlet 180° from the standard outlet.

8. A $\frac{1}{8}$-in [3.175-mm] (outside-diameter) stainless-steel tube is routed from a restricted fitting (marked GAUGE) on the flow divider to the firewall. A no. 3 low-pressure hose is usually used from the firewall to the gauge. In all cases, the hose volume should be held to a minimum.

9. In installations where an inlet fuel pressure gauge is used, a no. 4 flexible hose is connected to the fuel pressure takeoff fitting on the fuel metering system.

10. The nozzle line length will depend on the engine's installation and the location of the flow divider. The nozzle lines are formed from 0.085- to 0.090-in [2.159- to 2.286-mm] inside-diameter (ID) × $\frac{1}{8}$-in [3.175-mm] outside-diameter (OD) stainless-steel tubing, with suitable fittings for connection to the top of the nozzle and to the flow divider. The lines are clamped at suitable locations to reduce line vibration.

Troubleshooting

The troubleshooting chart in Fig. 7-29 may be used to troubleshoot the Bendix RSA fuel injection system. It is important, however, to keep in mind any modifications of the units that may change the troubleshooting procedures.

Problem	Probable Cause	Remedy
Hard starting.	Technique.	Refer to aircraft manufacturer's recommended starting procedure.
	Flooded.	Clear engine by cranking with throttle open and mixture control in ICO.
	Throttle valve opened too far.	Open throttle to a position at approximately 800 rpm.
	Insufficient prime (usually accompanied by a backfire).	Increase amount of priming.
Rough idle.	Mixture too rich or too lean.	Confirm with mixture control. A too-rich mixture will be corrected and roughness decreased during lean-out, while a too-lean mixture will be aggravated and roughness increased. Adjust idle to give a 25 to 50 rpm rise at 700 rpm.
	Plugged nozzle(s) (usually accompanied by high takeoff fuel-flow readings).	Clean nozzles.
	Slight air leak into induction system through manifold drain check valve. (Usually able to adjust initial idle, but rough in 1000 to 1500 rpm range.)	Confirm by temporarily plugging drain line. Replace check valves as necessary.
	Slight air leak into induction system through loose intake pipes or damaged O rings. (Usually able to adjust initial idle, but rough in 1000 to 1500 rpm range.)	Repair as necessary.
	Large air leak into induction system. Several cases of $\frac{1}{8}$-in [3.175-mm] pipe plugs dropping out. (Usually unable to throttle engine down below 800 to 900 rpm.)	Repair as necessary.
	Internal leak in injector. (Usually unable to lean out idle range.)	Replace injector.
	Unable to set and maintain idle.	Replace injector.
	Fuel vaporizing in fuel lines or distributor. (Encountered only under high ambient temperature conditions or following prolonged operation at low idle rpms.)	Cool engine as much as possible before shutoff. Keep cowl flaps open when taxiing or when the engine is idling. Use fast idle to aid cooling.
Low takeoff fuel flow.	Strainer plugged.	Remove strainer and clean in a suitable solvent. Acetone or MEK is recommended.
	Injector out of adjustment.	Replace injector.
	Faulty gauge.	In a twin-engine installation, crisscross the gauges. Replace as necessary. In single-engine, change the gauge.
	Sticky flow divider valve.	Clean the flow divider valve.
High fuel flow indicator reading.	Plugged nozzle if high fuel flow is accompanied by loss of power and roughness.	Remove and clean.
	Faulty gauge.	Crisscross the gauges and replace if necessary.
	Injector out of adjustment.	Replace injector.
Staggered mixture-control levers.	If takeoff is satisfactory, do not be too concerned about staggered mixture-control levers because some misalignment is normal with twin-engine installation.	Check rigging.
Poor cutoff.	Improper rigging of aircraft linkage to mixture control.	Adjust.
	Mixture-control valve scored or not seating properly.	Eliminate cause of scoring (usually burr or dirt) and lap mixture-control valve and plug on surface plate.
	Vapor in lines.	Cool engine as much as possible before shutoff. Keep cowl flaps open when taxiing or when the engine is idling. Use fast idle to aid cooling.
Rough engine (turbo-charged) and poor cutoff.	Air-bleed hole(s) clogged.	Clean or replace nozzles.

FIGURE 7-29 Bendix RSA fuel injection troubleshooting chart.

Operation

A principal consideration in the ground operation of an engine equipped with an RSA fuel injection system is the temperature within the engine nacelle. In flight operation, the engine and nacelle are adequately cooled by the rush of air that results both from the ram effect and from the volume of air moved by the propeller. On the ground, however, there is little or no ram effect, and the volume of air from the propeller is not adequate for cooling at idling speeds during hot-weather conditions.

High temperatures in the engine nacelle cause fuel vaporization that, in turn, affects the way the engine responds in its starting, idling, and shutdown. The person operating the engine should be aware of the effects of high temperatures and should adjust the operating procedures accordingly.

In hot weather, after the engine is shut down, the high temperatures in the engine nacelle cause the fuel in the nozzle feed lines to vaporize and escape through the nozzles into the intake manifold. For this reason, it is not necessary to prime the engine if it is restarted within 20 to 30 min after shutdown. The engine is started with the mixture control in the IDLE CUTOFF position; as soon as the engine fires, the control is advanced to the FULL RICH position. The fuel feed lines quickly fill and supply fuel at the nozzles before the engine can stop. If an engine has been stopped for more than 30 min, it is likely that the fuel vapor in the manifold has dissipated and that some priming may be needed to restart the engine.

During ground operations, particularly in hot weather, every effort should be made to keep the engine and nacelle temperatures as low as possible. To do this, keep ground operations at a minimum, keep the engine rpm as high as practical, and keep the cowl flaps open. Upon restarting a hot engine, operate the engine at 1200 to 1500 rpm for a few minutes to dissipate residual heat. A higher rpm also aids in cooling the fuel lines by increasing fuel pressure and flow.

The idle speed and mixture should be adjusted to effect a compromise which will best serve the engine's operating requirements for both cool and hot weather. A comparatively high idling speed (700 to 750 rpm) is best for hot weather, but it should not be so high that it makes the aircraft difficult to operate on the ground.

The idle mixture for hot-weather operation should be set to provide a 50-rpm rise when the mixture control is moving to IDLE CUTOFF. The richer setting increases pressure and helps to dissipate any vapor which may form in the system.

Before a hot engine is stopped, it should be run at increased rpm for a few minutes to eliminate as much heat as possible. The mixture control is then moved slowly to IDLE CUTOFF. When the engine has stopped, the ignition switch is turned off. If the engine is quite warm, it will likely idle roughly for several seconds before stopping because of vaporized fuel feeding from the fuel nozzles. This is likely even though the idle cutoff will completely stop the flow of fuel to the flow divider.

The operator is cautioned that great care must be taken in moving the propeller on an engine that has just been shut down. Sometimes hot spots may exist in one or more combustion chambers, and fuel vapor can be ignited and cause the engine to kick over. It is best to avoid moving the propeller until the engine has cooled for several minutes.

Field Adjustments

As with the pressure discharge carburetor, field adjustments for the fuel injection system are generally limited to idle speed and idle mixture. The procedure for making these adjustments is outlined below; however, the procedure given in the airplane operator's manual should be followed for a particular type and model of aircraft:

1. Perform a magneto check in accordance with instructions. If the rpm drop for each magneto is satisfactory, proceed with the idle mixture adjustment.

2. Retard the throttle to the IDLE position. If the idle speed is not in the recommended range, adjust the rpm with the idle-speed adjusting screw adjacent to the throttle lever (see Fig. 7-14).

3. When the engine is idling satisfactorily, slowly move the mixture control lever in the cockpit toward the IDLE CUTOFF position. Observe the tachometer or MAP gauge. If the engine rpm increases slightly or MAP decreases, this indicates that the mixture is on the rich side of best power. An immediate decrease in rpm or increase in MAP indicates that the idle mixture is on the lean side of best power. An increase of 25 to 50 rpm or a decrease in MAP of $\frac{1}{4}$ inHg [0.85 kPa] should provide a mixture rich enough to provide satisfactory acceleration under all conditions and lean enough to prevent spark plug fouling or rough operation.

4. If the idle mixture is not correct, turn the idle adjustment (see Fig. 7-14) one or two notches in the direction required for correction and recheck, as explained previously. Make additional adjustments, if necessary, until the mixture is satisfactory.

5. Between adjustments, clear the engine by running it up to 2000 rpm before making the mixture check.

6. The mixture is adjusted by lengthening the linkage between the throttle lever and idle valve to enrich the mixture and by shortening the linkage to lean the mixture. The center screw has right-hand threads on both ends, but one end has a no. 10-24 thread and the other has a no. 10-32 thread. For easy reference, consider only the coarse-threaded end. When this end of the link is backed out of its block, the mixture is enriched. Leaning is accomplished by screwing the coarse-threaded end of the link into the block.

7. If the center screw bottoms out in one of the blocks before a satisfactory mixture is achieved, an additional adjustment must be made. First, measure the distance between the blocks. Then disconnect one of the blocks from its lever by removing the link pin. Turn the block and adjustment screw until the adjusting wheel is centered and the distance between the blocks is the same as previously measured. There is now additional adjustment range, and the reference point is retained.

8. Make the final idle-speed adjustment to obtain the desired idling rpm with the throttle closed.

9. If the setting does not remain stable, check the idle linkage for looseness. In all cases, allow for the effect of weather conditions. The prevailing wind can add to or reduce the propeller load and affect the engine rpm. During the idle mixture and rpm checks, the airplane should be placed crosswind.

Inspection and Maintenance

When a new injector unit is installed, the injector inlet strainer should be removed and cleaned after 25 h of operation. Thereafter, the strainer should be cleaned at each 50-h inspection.

If an aircraft engine is equipped with a fuel injector that includes an AMC, the operator should be alert for signs of problems with the unit. Dirt can build up on the needle and cause rich operation and possible sticking of the needle, with resultant loss of altitude compensation. The following instructions are given for cleaning the unit:

1. Carefully remove the AMC unit. If the gasket is damaged, replace it with a new gasket with the appropriate Bendix part number.
2. Remove the 9/16-24 plug and immerse the unit in clean naphtha or other approved petroleum solvent. Invert the unit so that it will fill completely with the solvent. Exercise the AMC needle with a hardwood or plastic rod to facilitate cleaning. Shake the unit vigorously while allowing the solvent to drain. Repeat several times to wash out all traces of contaminants.
3. Drain the unit, and allow the solvent to evaporate completely. Do not dry with compressed air.
4. Replace the plug, and install the unit on the injector. Torque to 50 to 60 pound inches (lb•in) [5.65 to 6.78 Newton-meters (N•m)].

The injector should be lubricated in accordance with the approved lubrication chart for the particular installation. The clevis pins used in connection with the throttle and the manual mixture control should be checked for freedom of movement and lubricated if necessary.

Lubricate the throttle shaft bushings by placing a drop of engine oil on each end of the throttle shaft so that the oil can work into the bushings.

Use care in cleaning and oiling the air filter element. If the element is replaced with excessive oil clinging to it, some of the oil will be drawn into the injector and will settle on the venturi. This can greatly affect the metering characteristics of the injector.

1. What is meant by *fuel injection*?
2. List the advantages of a fuel injection system.
3. What are the four basic units of the Continental continuous-flow fuel injection system?
4. How many fuel control units are included in the fuel control assembly? What are they?
5. At what locations on the engine are the fuel discharge nozzles installed?
6. Describe the construction of a fuel discharge nozzle for the Continental fuel injection system.
7. What special feature is required for the fuel discharge nozzles installed on a turbocharged engine?
8. Explain the operation of the altitude compensating valve.
9. How would you adjust the idle speed for the Continental fuel injection system?
10. How would you check the idle mixture for the Continental fuel injection system?
11. What unit is considered the heart of the Bendix RSA fuel injection system.
12. What means are provided in the RSA system to provide for adequate fuel flow at idling engine speeds?
13. What is the purpose of the AMC on the Bendix RSA fuel injection system?
14. Describe the operation of the AMC on the RSA fuel injector.
15. Describe the operation of the flow divider.
16. What factors determine the size of the calibrated jet in the fuel nozzle?
17. What is meant when a fuel nozzle is classified as an air-bleed type?
18. In what position must the flow divider valve be installed on the RSA system?
19. Why must horizontally installed fuel nozzles be installed with the identification mark on the lower side?
20. What methods are employed to minimize engine heat during ground operation?
21. What causes a warm engine to continue to idle for a few seconds after the mixture control is placed in IDLE CUTOFF?
22. List the items that are generally considered field adjustable on the RSA system.
23. Describe the procedure for checking the idle mixture for an RSA fuel injection system.

Reciprocating-Engine Ignition and Starting Systems 8

INTRODUCTION

The maintenance, service, and troubleshooting of aircraft engines require a thorough understanding of ignition systems on the part of the maintenance technician. The service and maintenance of ignition systems, starters, and other electric devices that are, or may be, associated with engines require knowledge of the principles of electricity and magnetism. It is not the purpose of this chapter to provide material in electrical theory and practice. Complete coverage of electrical and electronic theory is provided in the text *Aircraft Electricity and Electronics*.

PRINCIPLES OF IGNITION

Ignition Event in the Four-Stroke Cycle

During the first event of the four-stroke, five-event cycle, the piston moves downward as a charge of combustible fuel and air is admitted into the cylinder. This is the intake stroke. During the second event, which is the compression stroke, the crankshaft continues to rotate and the piston moves upward to compress the fuel-air (F/A) mixture.

As the piston approaches the top of its stroke within the cylinder, an electric spark jumps across the points of the spark plugs and ignites the compressed F/A mixture. This is the ignition event, or the third event. Having been ignited, the F/A mixture burns and expands, and the resulting gas pressure drives the piston downward. This causes the crankshaft to rotate. Since this is the only stroke and event that furnishes power to the crankshaft, it is usually called the power stroke, and it is the fourth event. The exhaust stroke is the fifth event. These facts have been presented before, but they must be reviewed briefly in order to understand aircraft engine ignition.

The electric spark jumps between the electrodes (points) of a spark plug installed in the cylinder head or combustion chamber of the engine cylinder. The ignition system furnishes sparks periodically to each cylinder at a certain position of piston and valve travel.

Battery Ignition System

A few aircraft and most automobiles use a **battery ignition** system which has a battery or generator rather than a magneto as its source of energy. In the battery ignition system, a cam, which is driven by the engine, opens a set of points to interrupt the flow of current in a primary circuit. The resulting collapsing magnetic field induces a high voltage in the secondary of the ignition coil, which is directed by a distributor to the proper cylinder. A simplified schematic of a battery ignition system is shown in Fig. 8-1.

Magneto Ignition

Magneto ignition is superior to battery ignition because it produces a hotter spark at high engine speeds and it is a self-contained unit, not dependent on any external source of electric energy.

The magneto is a special type of alternating-current (ac) generator that produces electric pulsations of high voltage for purposes of ignition.

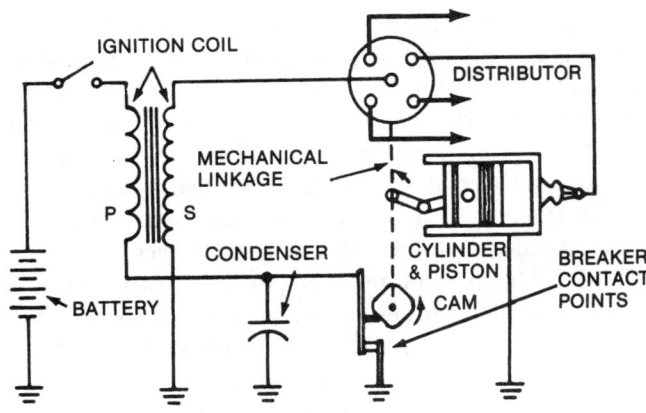

FIGURE 8-1 Battery ignition system.

There are many ways of classifying magnetos. They may be (1) low-tension or high-tension, (2) rotating-magnet or inductor-rotor, (3) single or double, and (4) flange-mounted or base-mounted.

Low- and High-Tension Magnetos

A **low-tension magneto** delivers current at a low voltage by means of the rotation of an armature, wound with only one coil, in the field of a permanent magnet. Its low-voltage current must be transformed to a high-tension (high-voltage) current by a transformer at or near the spark plug.

A **high-tension magneto** delivers a high voltage to the spark plug and has both a primary winding and a secondary winding. An outside induction coil is not needed because the double winding accomplishes the same purpose. The low voltage generated in the primary winding induces a high-voltage current in the secondary winding when the primary circuit is broken.

Rotating-Magnet and Inductor-Rotor Magnetos

In a **rotating-magnet magneto**, the primary and secondary windings are wound on the same iron core. This core is mounted between two poles, or inductors, which extend to shoes on each side of the rotating magnet. The rotating magnet is usually made with four poles, which are arranged alternately north and south in polarity.

The **inductor-rotor magneto** has a stationary coil (armature) just as the rotating-magnet type does. The difference lies in the method of inducing a magnetic flux in the core of the coil. The inductor-rotor magneto has a stationary magnet or magnets. As the rotor of the magneto turns, the flux from the magnets is carried through the segments of the rotor to the pole shoes and poles, first in one direction and then in the other.

Single- and Double-Type Magnetos

Two single-type magnetos are commonly used on piston-type engines. The **single-type magneto** is just what its name implies—one magneto.

The **double-type magneto** is generally used on different models of several types of engines. When made for radial engines, it is essentially the same as the magneto made for in-line engines except that two compensated cams are employed. The compensated cam is explained later in this chapter.

The double-type magneto is essentially two magnetos having one rotating magnet common to both. It contains two sets of breaker points, and the high voltage is distributed either by two distributors mounted elsewhere on the engine or by distributors forming part of the magneto proper. Since there are two sets of breaker points, an equal number of sparks will be produced by each coil assembly per revolution of the magneto drive shaft.

Flange- and Base-Mounted Magnetos

A **flange-mounted magneto** is attached to the engine by means of a flange on the end of the magneto. The mounting holes in the flange are not circular; instead, they are slots that permit a slight adjustment, by rotation, in timing the magneto with the engine.

The single-type magneto may be either base-mounted or flange-mounted. The double-type magneto is always flange-mounted.

A **base-mounted magneto** is attached to a mounting bracket on the engine by means of cap screws, which pass through holes in the bracket and enter tapped holes in the base of the magneto.

Symbols Used to Describe Magnetos

Magnetos are technically described by letters and figures which indicate make, model, type, etc., as shown in Fig. 8-2.

Example 1 The DF18RN is a double-type flange-mounted magneto for use on an 18-cylinder engine designed for clockwise rotation and made by Bendix.

Example 2 The SF14LU-7 is a single-type flange-mounted magneto for use on a 14-cylinder engine, designed for counterclockwise rotation and made by Bosch, seventh modification.

NOTE: Each manufacturer uses a dash followed by letters or numerals after the symbol for the product to indicate the series or particular model. The letters F and B are no longer used for civilian magnetos because all are flange-mounted.

Order of Designation	Symbol	Meaning
1	S	Single type
	D	Double type
2	B	Base-mounted
	F	Flange-mounted
4, 6, 7, 9, etc.		Number of distributor electrodes
4	R	Clockwise rotation as viewed from drive-shaft end
	L	Counterclockwise rotation as viewed from drive-shaft end
5	G	General Electric
	N	Bendix
	A	Delco Appliance
	U	Bosch
	C	Delco-Remy (Bosch design)
	D	Edison-Splitdorf

FIGURE 8-2 Magneto code designation.

MAGNETO OPERATIONAL THEORY

The magneto consists of three circuits: **magnetic**, **primary**, and **secondary circuits**. They work together to produce the high-tension spark at the spark plug. The magnetic circuit includes the permanent magnet, coil core, pole shoes, and pole shoe extensions. The primary circuit consists of the primary winding of the coil, the breaker points or contacts, and the condenser or capacitor. The secondary circuit contains the secondary windings, the distributor and rotor, the high-tension ignition lead, and the spark plug. These basic magneto circuits are illustrated in Fig. 8-3.

Magnetic Circuit

The magnetic circuit of the magneto may be designed in different ways. One type of design uses rotating permanent magnets having two, four, and even eight magnetic poles. These magnets are often made of **alnico**, an alloy of aluminum, iron, nickel, and cobalt that retains magnetism for an indefinite period. The magnets rotate under pole pieces, which complete a magnetic circuit through a coil core.

Another type of magneto design makes use of stationary permanent magnets and a system of rotating inductors which complete the magnetic circuit through the coil core by two different paths, thereby providing two directions through the coil core for the magnetic flux.

The first type of design is called the rotating-magnet type, and the second is called the inductor-rotor type. The rotating-magnet type is more widely used in aircraft ignition.

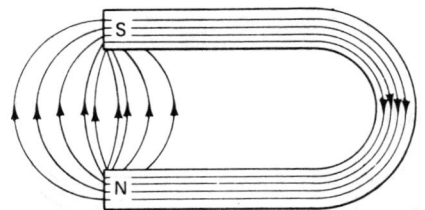

FIGURE 8-4   Permanent magnet and field.

Principles of the Rotating-Magnet Magneto

The horseshoe-shaped permanent magnet shown in Fig. 8-4 has a magnetic field that is represented by many individual paths of invisible magnetic flux, commonly known as "lines" of flux. Each of these lines of flux within the magnet itself extends from the north pole of the magnet (marked with the letter N) through the intervening airspace to the south pole (marked with the letter S), thereby forming the closed loop indicated in the drawing. These lines of magnetic flux are invisible, but their presence can be verified by placing a sheet of paper over the magnet and sprinkling iron filings on the paper. The iron filings will arrange themselves in definite positions along the lines of flux which compose the magnetic field represented by the solid lines containing arrows.

The lines of flux repel one another; therefore, they tend to spread out in the airspace between the poles of the magnet, as shown in Fig. 8-4. They also tend to seek the path of least resistance between the poles of the magnet. A laminated soft-iron bar provides an easier path for the lines of flux flowing between the poles than air does; consequently, the lines will

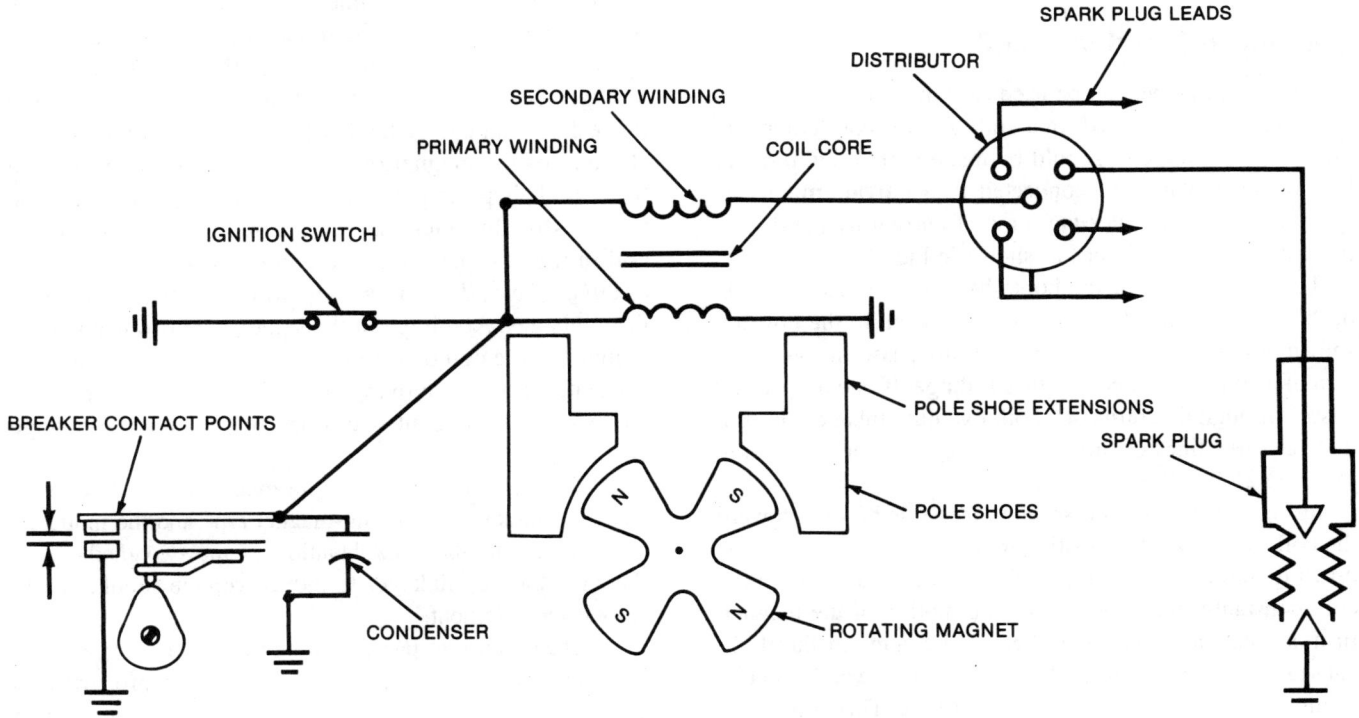

FIGURE 8-3   Basic magneto circuits.

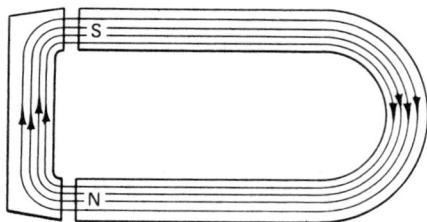

FIGURE 8-5 Permanent magnet with flux passing through an iron bar.

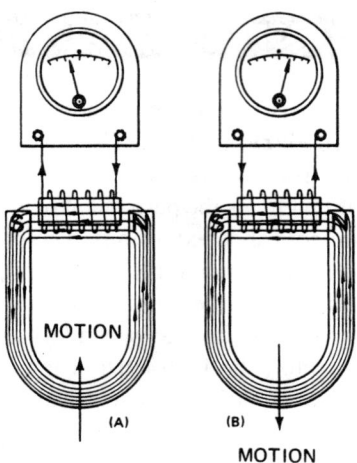

FIGURE 8-6 Inducing current with a magnetic field.

crowd together if such a bar (sometimes called a **keeper**) is put near the magnet. The lines of flux then assume the locations and directions shown in Fig. 8-5, where they are concentrated within the bar instead of being spread out as in the airspace of Fig. 8-4. This heavy concentration of lines of flux within the bar is described as a condition of high "density."

The direction taken by the lines of flux in the laminated soft-iron bar placed in a magnetic field is determined by the polarity of the permanently magnetized horseshoe magnet. For example, in Figs. 8-4 and 8-5, the direction of flow is from the north pole to the south pole. The direction would be reversed if the north pole were the upper pole in the illustrations and the south pole were the lower pole.

If the permanent magnet is made of alnico, Permalloy, or some hardened steel, it can retain a large portion of the magnetism induced in it when it was originally magnetized. Since the laminated iron bar is made of magnetically "soft" iron, it does not retain much of the magnetism when magnetic lines of flux pass through it. This makes it possible to change the direction of the lines of flux by turning the magnet over so that the north pole of the magnet in Figs. 8-4 and 8-5 is at the top, instead of the bottom, of the picture.

Current Induced in a Coil

A horseshoe magnet can be used to show that a current can be generated, or **induced**, in a coil of wire. For demonstration purposes, the coil should be made with a few turns of heavy copper wire and connected to a **galvanometer**, an instrument that indicates any flow of current by the deflection of its needle (pointer), as shown in Fig. 8-6.

The lines of flux of the horseshoe magnet pass through, or "link," the turns of wire in the coil when in the position shown in Fig. 8-6. When one line of flux passes through one turn of a coil, it is called one **flux linkage**. If one line of flux passes through five turns of a coil, five flux linkages are produced. If five lines of flux pass through five turns of a coil, there are 25 flux linkages, and so on.

In Fig. 8-6A, if the horseshoe magnet is brought toward the coil from a remote position to the position shown in the drawing, the number of lines of flux, which are linking the coil, constantly increases during the motion of the magnet. In more technical language, there is a change in flux linkages as the horseshoe magnet moves toward the coil, and this change induces a voltage in the coil of wire. This voltage, or electromotive force (emf), causes an electric current to flow

through the circuit, and this is indicated by the deflection of the galvanometer needle.

In Fig. 8-6B, if the horseshoe magnet is moved away from the coil, the flux linkages occur in the opposite direction during this movement, and this change in flux linkages induces a current in the coil of wire *in the opposite direction*, as indicated by the movement of the galvanometer needle.

The voltage induced in the coil of wire is proportional to the *rate of change of flux linkages*. The flux linkages can be increased by adding more turns in the coil of wire or by using a stronger magnet having more lines of flux. The **rate** involves an element of time and can be increased by moving the horseshoe magnet near the coil faster, thus increasing the speed of flux change. When any of these methods for increasing the rate of change of flux linkages is tried during an experiment, the galvanometer needle deflection indicates the magnitude of the induced current.

There must be a change of flux linkages to induce a voltage. Voltage is not induced in the coil of wire if the horseshoe magnet is held stationary, regardless of the strength of the magnet or the number of turns of wire in the coil. This principle is applied to the magneto because the lines of flux must have a magnetic path through the coil in the first place and then *there must be a movement of either the coil or the magnet* to produce the change in flux linkages. A voltage in the same proportions could be induced in the coil of wire by holding the horseshoe magnet stationary and moving the coil. This would provide the relative movement necessary to produce the change of flux linkages.

We have previously stated that magnetos can be divided into two classes: the rotating-magnet type and the inductor-rotor type. The above explanation of the two methods of changing the flux linkages to induce voltage should clarify this earlier statement.

Whenever current passes through a coil of wire, a magnetic field is established, which has the same properties as the magnetic field of the horseshoe or permanent magnet previously described.

Lenz' Law

Lenz' law can be stated in terms of induced voltage thus: *An induced voltage, whether caused by self-inductance or mutual inductance, always operates in such a direction as to oppose the source of its creation.*

The same thing can be stated in simpler terms: When a change in flux linkages produces a voltage which establishes a current in a coil or wire, the direction of the current is always such that its magnetic field opposes the motion or change in flux linkages which produced the current. Lenz' law is very important to operation of the magneto, as explained further in this text.

In Fig. 8-6A, when the magnet is moved toward the coil, the current flows up the left-hand wire, through the galvanometer, and down the right-hand wire, as shown by the arrows. When the magnet is moved away from the coil, the current flows up the right-hand wire, through the galvanometer, and down the left-hand wire, as shown in Fig. 8-6B.

Left-Hand Rule

The polarity of a magnetic field can be determined when the directions of the current and the winding of a coil are known. If the wire is grasped with the left hand, and if the fingers of the left hand extend around the coil in the direction of the current, the thumb will always point in the direction of the flux, or the north end of the field. This is called the **left-hand rule**.

If the left-hand rule is applied to the current in Fig. 8-6, the field which the current establishes opposes the *increase* or change of flux linkages. While the magnet is being moved up toward the coil in Fig. 8-6A, the usual tendency is to *increase* the flux through the coil core in the direction from right to left of the illustration, as shown by the arrows. However, as soon as the flux starts to increase, a current begins to flow in the coil. It establishes a field of direction from left to right, and this field opposes the increase of magnetic flux and actually exerts a small mechanical force that tends to push the magnet away from the coil.

In Fig. 8-6B, when the magnet is moving, away from the coil, the current flows up the right-hand wire, through the galvanometer, and down the left-hand wire. In accordance with the statement of the left-hand rule, the field of the coil is now helping the field of the magnet. As the magnet is moved away from the coil, the flux linkages occur in the opposite direction and induce a current in the coil which sets up a magnetic field that opposes the change, following the principle of Lenz' law. However, the change is now a decrease; therefore, the field of the coil now helps the magnetic field, aiding it in its effort to oppose the change. A small mechanical pull is actually exerted on the magnet by the coil's tending to resist the motion of the magnet away from the coil.

If the circuit of the coil is opened, no current can flow in the wire because there is not enough voltage to force the current across the gap where the wire in the circuit is broken or disconnected. To jump the gap requires high voltage, and this would make it necessary to increase greatly the rate of change of flux linkages. We have already discussed the methods of increasing the rate of change of flux linkages. In the simple horseshoe magnet and single coil shown in Fig. 8-6, if the size of the coil, the size of the magnet, and the rate of movement of the magnet were increased, the power required to move the magnet rapidly enough to produce the high rate of change of flux linkages would be so great that this device would not be practical. Therefore, the basic design must be changed to provide the compact, efficient source of high voltage required for igniting the fuel-air charge in the engine cylinder.

Rotating Magnet

Figure 8-7 shows a four-pole rotating magnet, similar to those used in some models of Bendix aircraft magnetos. The lines of flux of the rotating magnet, when it is not installed in the magneto, pass from its north pole through the airspace to its south pole, in a manner similar to the flow of the lines of flux in the simple horseshoe magnet in Fig. 8-4.

In Fig. 8-8, the pole shoes and their extensions are made of soft-iron laminations cast in the magneto housing. The coil core is also made of soft-iron laminations and is mounted on top of the pole-shoe extensions. The pole shoes D and their extensions E, together with the coil core C, form a magnetic path similar to that made by the laminated soft-iron bar (keeper) shown with the ordinary horseshoe magnet in Fig. 8-5. When the magnet is in the position shown in Fig. 8-8, the magnetic path produces a concentration of flux in the core of the coil.

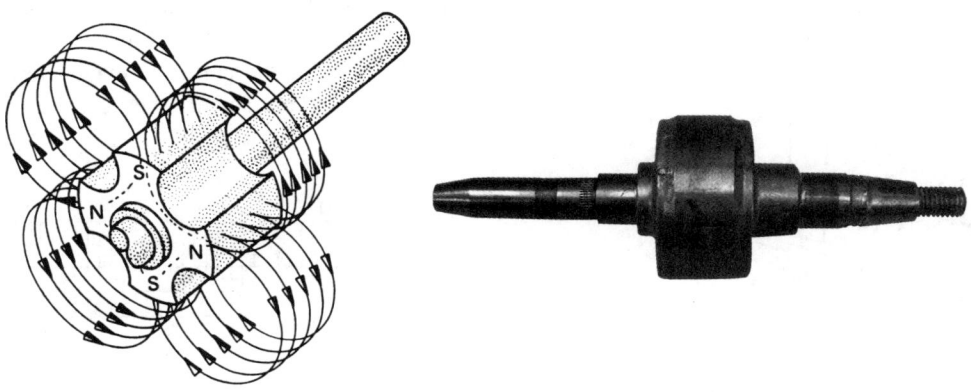

FIGURE 8-7   Four-pole rotating magnet shown with actual shaft and magnet.

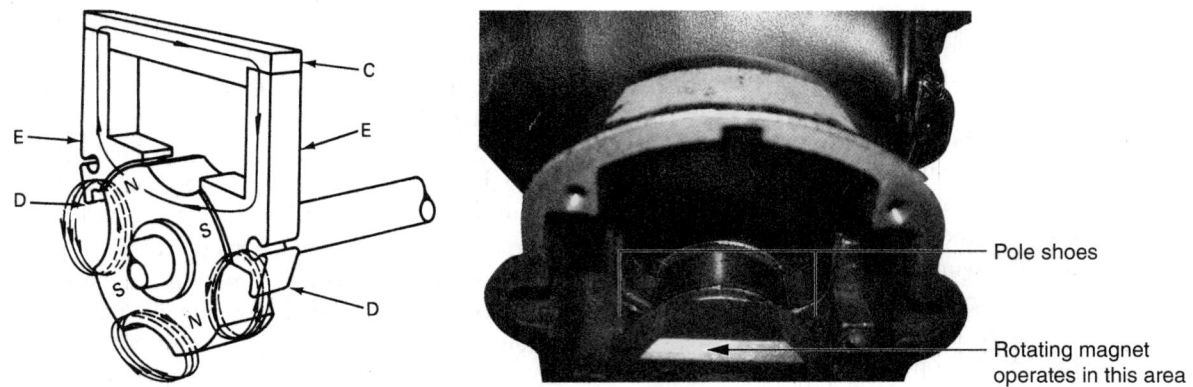

FIGURE 8-8 Arrangement of rotating magnet, pole shoes, and core of the coil in a magneto.

In Fig. 8-9, notice that the rotating magnet has rotated from the position it had in Fig. 8-8. The neutral position of any rotating magnet is that position where one of the pole pieces is centered between the pole shoes in the magneto housing, as shown in Fig. 8-9. When the rotating magnet is in its neutral position, the lines of flux do not pass through the coil core because they are short-circuited by the pole shoes. When a two-pole magnet moves into the neutral position, the poles of the magnet are positioned too far from the pole shoes to allow flux to flow through the coil core.

Notice especially that primary and secondary windings are *not* shown in the coil core in Figs. 8-8 and 8-9. They are omitted to make it easier to understand the magnetic action. Having learned the action without windings, the reader will find it easier to understand the functions of the windings explained later in this text.

The curve in Fig. 8-10 shows how the flux in the coil core changes when the magnet is turned with no windings present. This curve is called the **static-flux curve** because it represents the stationary or normal condition of the circuit. If the magnet is turned with no windings on the coil core, the flux will build up through the coil core in first one direction and then the other, as indicated by the curve.

This curve represents both the direction of the flux and its concentration. When the curve is above the horizontal line, the flux is passing through the coil in one direction, and the higher the curve above the line, the greater the number of lines of flux in the core.

When the curve is below the horizontal line, the flux is passing through the coil in the opposite direction, and the lower the curve below the line, the greater the number of lines of flux passing through the core in this other direction.

Whenever the magnet passes through a neutral position, the flux in the coil core falls to zero, and this is shown by the point where the curve touches the horizontal line. Having fallen to zero, the flux then builds up again in the opposite direction. Therefore, the greatest change in flux occurs when the magnet is passing through the neutral position. Note that the curve in Fig. 8-10 has a steep slope at the points corresponding to the neutral positions of the magnet—that is, wherever the curve crosses the horizontal line.

Primary-Circuit Components

Coil Assembly

The typical coil assembly consists of a laminated soft-iron core around which is placed a **primary winding** and a **secondary winding**. The primary winding consists of a comparatively few turns of insulated copper wire, and the secondary winding consists of several thousands of turns of very fine wire. The coil is covered with a case of hard rubber, Bakelite, varnished cambric, or plastic, according to the design requirements of the manufacturer. A primary condenser (capacitor) may be built into the coil between the primary winding and the secondary winding, or it may be connected in the external circuit.

The ends of the coil core extend beyond either end of the coil assembly so that they can be secured to the pole

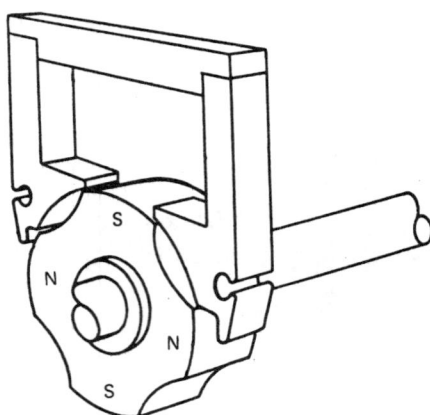

FIGURE 8-9 Rotating magnet in neutral position.

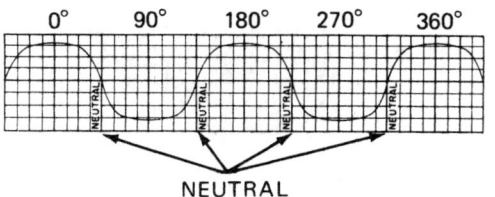

FIGURE 8-10 Static-flux curve.

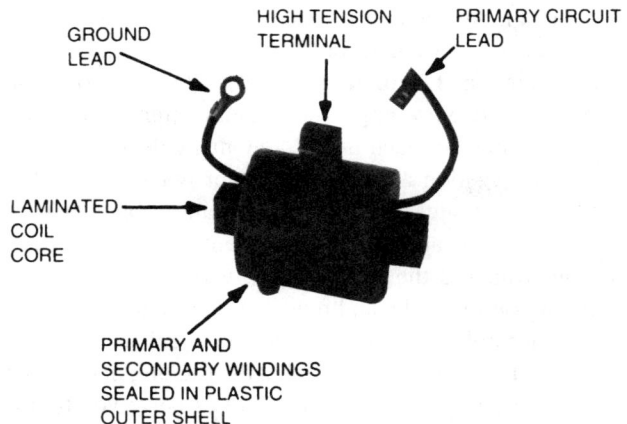

FIGURE 8-11   Magneto coil assembly.

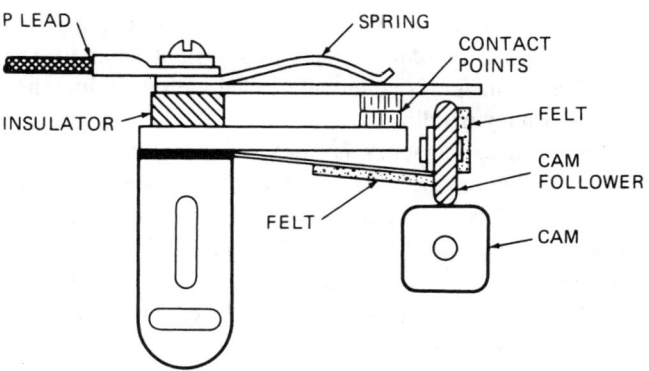

FIGURE 8-12   One type of pivotless breaker assembly.

shoe extensions. One end of the primary winding is usually grounded to the core, and the other end is brought out to a terminal connection or to a short length of connecting wire having a terminal on the end. This end of the primary coil is connected in the magnet to the **breaker points**, and a provision is also made for a connection to the ignition-switch lead. Note carefully that when the breaker points are closed, current flows from the coil to ground and back from ground to the coil in a complete circuit. This direction of flow alternates with the rotation of the magnet.

One end of the secondary winding is grounded inside the coil, and the other end is brought to the outside of the coil to provide a contact through which the high-tension (high-voltage) current can be carried to the distributor.

This description of a magneto coil does not necessarily apply to all magnetos, but it does give an overall explanation of coil-assembly design and construction. A magneto coil assembly is illustrated in Fig. 8-11.

Breaker Assembly

The **breaker assembly** of a magneto, also referred to as the **contact breaker**, consists of contact points actuated by a rotating cam. Its function is to open and close the circuit of the primary winding as timed to produce a buildup and collapse of the magnetic field.

Some early magnetos have **lever-type** or **pivot-type breaker points**. These are designed with a movable contact at one end of a lever or arm that is mounted on a pivot. A **cam follower** that rides on the surface of the breaker cam is attached to the breaker arm. Later models of magnetos have pivotless-type breaker assemblies with the movable contact point mounted on a spring-type lever. A leaf spring is often used for additional force. Pivotless breaker assemblies are not affected by the wear occurring at the pivot bearings; therefore, they stay in adjustment better than pivot types. One type of pivotless breaker assembly is illustrated in Fig. 8-12. The breaker contact points for a magneto are made of a platinum-iridium alloy or some other heat- and wear-resistant material. The cam and cam follower are lubricated by a felt pad on the cam follower. This pad is saturated with lubricating oil at regular service periods.

The breaker contact points are electrically connected across the primary coil so that there is a complete circuit through the coil when the points are closed and the circuit is broken when the points open. The magneto is timed so that the breaker points close at the position where magnetic flux through the coil core is at a maximum. At this time there is a minimum of **flux change**.

Primary Capacitor

During the operation of the magneto, voltage and current are induced in both the primary and the secondary windings of the coil. When the breaker points open, there is a tendency for the primary current to arc across the points. This results in burning of the contact points, and the rate of the collapse of the field is reduced, resulting in a weak spark from the secondary. To overcome this problem, a **primary capacitor** condenser is connected across the breaker points. The operation of the primary capacitor is illustrated in Fig. 8-13. The capacitor acts as a storage chamber to absorb the sudden rise of voltage in the primary coil when the breaker points begin to open. The primary capacitor prevents arcing between the contact points as they open by absorbing the "inertia" current induced in the primary coil by the collapse of the

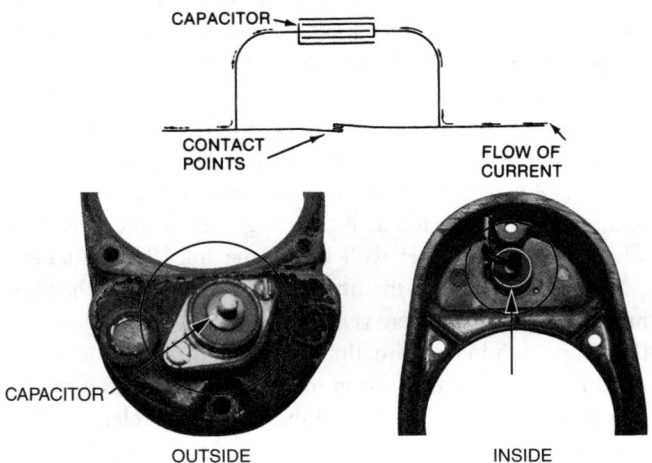

FIGURE 8-13   Operation of the primary capacitor, and picture of capacitor mounted in housing.

electromagnetic field established by the primary current. This field "cuts" the turns of the primary winding as it collapses. The underlying principle is commonly called the **self-inductance of the primary winding**.

The primary capacitor is always connected across the points, but its shape and location vary among different magneto models. It may be round, square, or of some other shape. It may be located in the coil housing, in the breaker housing with the breaker points, or on top of the coil. Briefly stated, its function is to absorb self-induced current flowing in the primary circuit.

The primary capacitor must have the correct *capacitance*. Too low a capacitance will permit arcing and burning of the breaker points plus a weakened output from the secondary. If the capacitance is too high, the mismatch between the coil and the capacitor will reduce the voltages developed and result in a weakened spark.

Elements of Magneto Operation

An understanding of the operation of a typical magneto may be enhanced by a study of the several electrical elements or factors involved in its operation. The graphic representation in Fig. 8-14 illustrates the strengths and direction of the electrical and magnetic factors in an operating magneto. The factors indicated in the graph are *static flux*, *breaker timing*, *primary current*, *resultant flux*, *primary voltage*, and *secondary voltage*.

The static flux curve is shown at the top of the graph in Fig. 8-14 and illustrates what the nature of the magnetic field (flux) through the core of the coil would be if there were no windings on the core. As the rotating magnet of the magneto is turned, the flux increases in one direction and then decreases to zero before increasing to a maximum in the opposite direction. The static-flux curve is shown as a dotted line with the curve for the resultant flux.

When primary and secondary windings are placed on the coil core and the magneto is rotated, a primary current is induced when the flux begins to reduce from maximum. This current flows during the time that the breaker points are closed, as shown in the graph. The direction of the current is such that it produces a magnetic field, as shown in Fig. 8-14, that *opposes* the change in magnetic flux produced by the rotating magnet. This is in accordance with Lenz' law, as explained previously. This effect of the primary current flow produces the resultant flux shown in the graph. Note that the resultant flux remains very strong in its original direction (above the zero line) until the breaker points open to stop the flow of primary current. Immediately before the breaker points open, the resultant flux is high above the zero line while the static flux is well below the line. This situation causes a high stress in the magnetic circuit so that when the breaker points open, there is an extremely rapid change in the flux from high in one direction to high in the opposite direction. The effect of this action may be compared to a spring that is stretched to its limit and then suddenly released. The rapid change in magnetic flux through the core of the magneto coil induces the high voltage in the secondary winding of the coil that produces the spark for ignition.

In a magneto, the number of degrees of rotation between the neutral position and the position where the contact points open is called the **E-gap angle**, usually shortened to simply **E gap**, or **efficiency gap**. The manufacturer of the magneto determines for each model how many degrees beyond the neutral position a pole of the rotor magnet should be to obtain the strongest spark at the instant of breaker-point separation. This angular displacement from the neutral position, which is the E-gap angle, varies from 5 to 17°, depending on the make and model; but for a representative type of four-pole magneto, such as the one discussed here, the correct E gap is 11°. Notice that the rotating magnet will be in the E-gap position as many times per revolution as there are poles.

When the magnet has reached the position where the contact points are about to open, a few degrees past the neutral position, the primary current is maintaining the original field in the coil core while the magnet has already turned past neutral. The primary current is now attempting to establish a field through the coil core in the opposite direction.

When the contact points are opened, the primary circuit is broken. This interrupts the flow of the primary current and causes an extremely rapid change in flux linkages. In an exceedingly short time, the field established by the primary current falls to zero and is replaced by the field of the opposite direction established by the magnet. This process is represented by the almost vertical portion of the resultant-flux curve in Fig. 8-14.

The secondary winding of a magneto coil consists of many turns of fine wire, often over 10 000, wound over the primary on the coil core. The large number of turns in the secondary winding and the very rapid change in flux together cause a high rate of change of flux linkages, which in turn produces the high voltage in the secondary winding.

The extremely rapid flux change represents the dissipation of the energy involved in the stress that existed in the magnetic circuit before the contact points opened. The flux in the coil core would normally change as represented by the static-flux curve in the illustration, but the primary current prevents this change and holds back the flux change while the magnet turns.

The factors illustrated by the curves in Fig. 8-14 are representative of the conditions that exist in a rotating magneto when the secondary circuit is open—that is, when there is not a complete circuit for the secondary winding. Under this condition, both secondary and primary voltages build to a maximum, as shown in the lower part of the graph. Note that secondary voltages are 100 times greater than the primary voltages. This is because the secondary coil has at least 100 times as many windings as the primary coil.

When a magneto is operated with a complete path for secondary current, such as a spark plug or a test gap, the secondary voltage rises only high enough to cause the current to jump the gap as a spark. The values and directions of the factors of operation in a magneto connected to spark plugs in a normally operating aircraft engine are shown in Fig. 8-15. When a magneto is operated on a test stand with open air spark gaps, the values are as shown in Fig. 8-15.

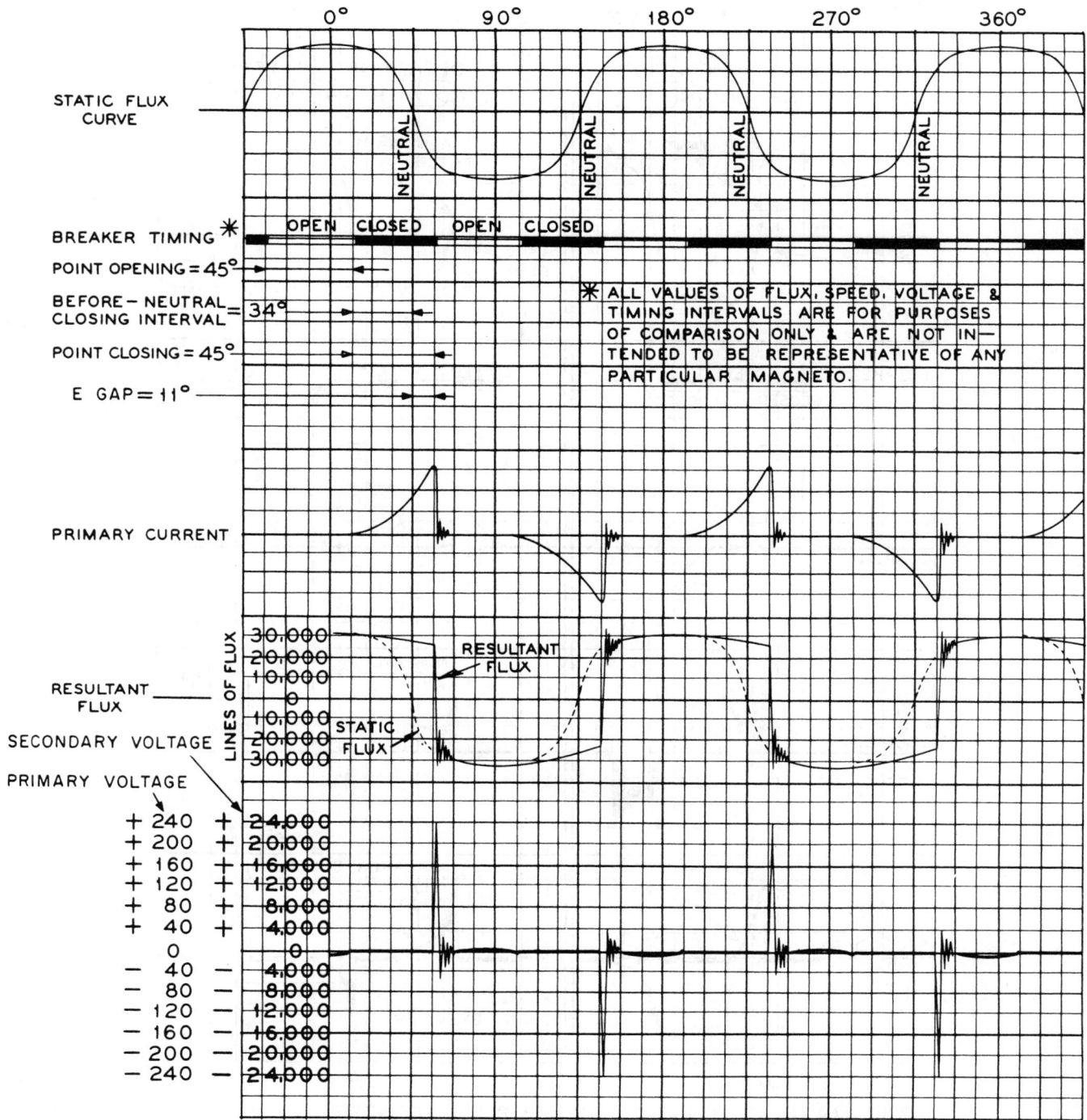

FIGURE 8-14 Graphic representation of the electrical and magnetic factors involved in the operation of a magneto, with the secondary circuit open. (*Continental Motors.*)

These values would exist for an eight-pole magneto in which the angular distance through which the breaker points are either open or closed is $22\frac{1}{2}°$.

In Fig. 8-15, where the magneto is operating normally on an aircraft engine, the pressure in the cylinder affects the level of secondary voltage necessary to cause the current to jump across the spark plug gap. In this case, secondary emf may reach only 5000 V before the spark plug fires, and then the voltage diminishes in an oscillating pattern, as shown in the graph. The initial oscillations are due to the sudden current load placed on the coil when the secondary current starts to flow.

The increasing "quench" oscillations are caused by the effect of turbulence and pressure on current flowing across the spark plug gap. The resulting flux change is decreasing at this time, and all energy is dissipated just before the breaker points close to begin the next cycle.

Distributor

The distributor and rotor gear are shown in Fig. 8-16. The large distributor gear which is driven by the smaller gear located on the drive shaft of the rotating magnet is used to

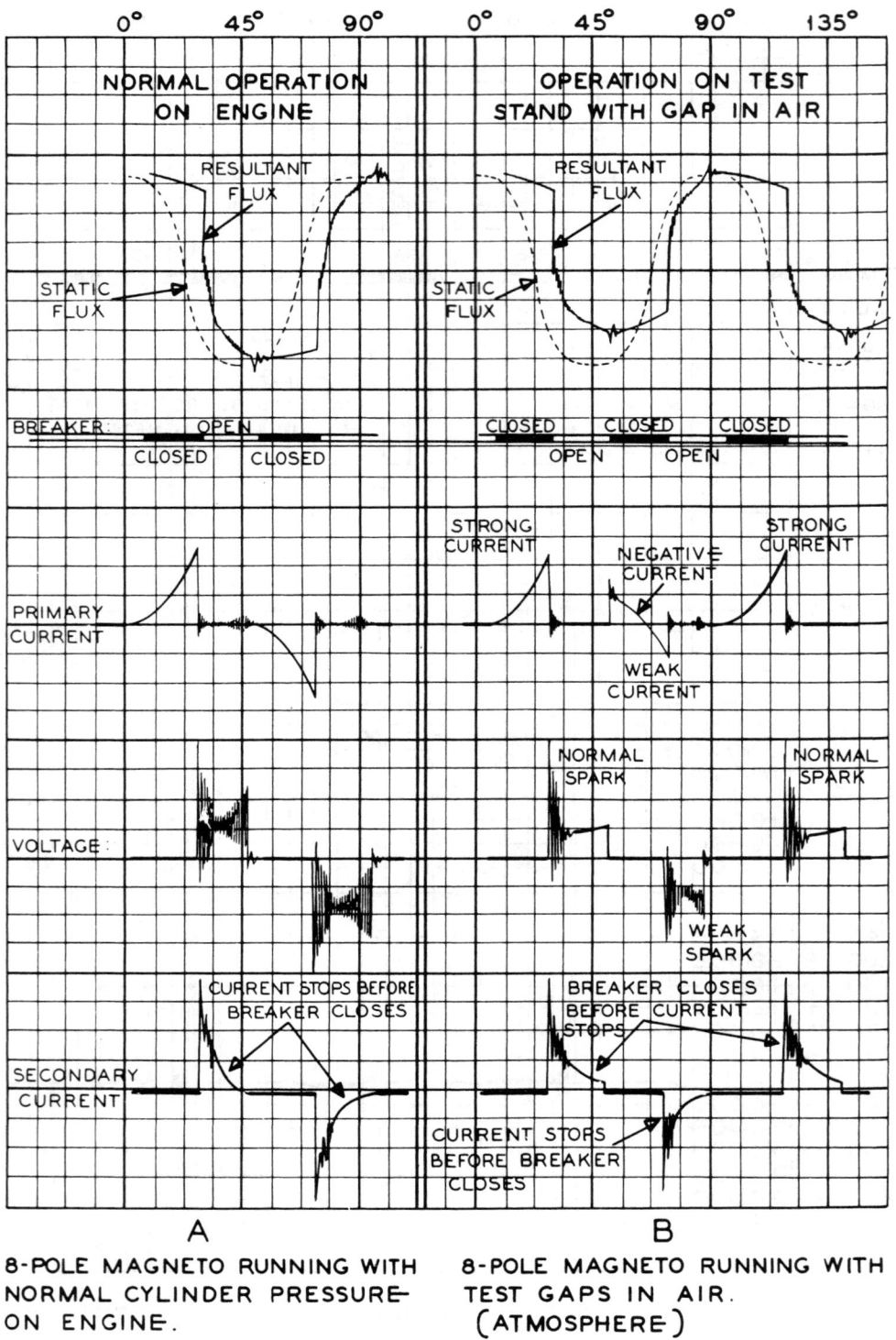

FIGURE 8-15 Graphic representation of the electrical and magnetic factors in a magneto operating with a spark gap in the secondary circuit. (*Continental Motors.*)

distribute the high-tension voltage to the different cylinder terminals on the distributor block. The ratio between these gears is always such that the distributor gear electrode (rotor) is driven at *one-half engine crankshaft speed*. This ratio of the gears ensures the proper distribution of the high-tension current to the spark plugs in accordance with the firing order of the particular engine.

In general, the **distributor rotor** of the typical aircraft magneto is a device that distributes the high-voltage current from the coil secondary terminal to the various connections of the distributor block. This rotor may be in the form of a finger, disk, drum, or other shape, depending on the design of the magneto manufacturer. In addition, the distributor rotor on very old magnetos may have been designed with

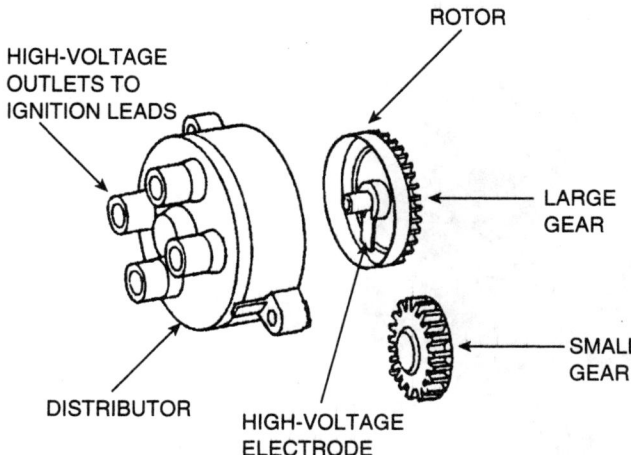

HIGH-VOLTAGE
OUTLETS TO
IGNITION LEADS

ROTOR

LARGE
GEAR

SMALL
GEAR

DISTRIBUTOR

HIGH-VOLTAGE
ELECTRODE

FIGURE 8-16 Magneto distributor, rotor gear, and magneto rotor shaft gear.

either one or two distributing electrodes. When there are two distributing electrodes, the leading electrode, which obtains high voltage from the magneto coil secondary, makes its connection with the coil secondary through the shaft of the rotor, whereas the trailing electrode obtains a high-tension voltage from the booster by means of a collector ring mounted either on the stationary distributor block or on the rotor itself.

It must be explained that the distributors with the trailing finger are not employed on late-model aircraft magnetos, although they may be encountered from time to time on magnetos used on older radial engines. The early systems utilized booster magnetos or high-tension booster coils to provide a strong spark during engine start-up, and the trailing finger of the distributor provided a retarded spark to prevent the engine from kicking back (trying to run backward).

Construction and Other Characteristics of Magnetos

Materials used in the construction of magneto components are selected primarily for their effects in the control of magnetic forces. Strength and durability with respect to mechanical stresses and heat are also considered. Dielectric materials (insulators) must be able to provide adequate insulation to withstand high voltages under all operating conditions.

The case of a magneto must be constructed of a nonmagnetic alloy, such as an aluminum alloy, so that it will not affect the magnetic circuit. The case supports and protects the operating mechanisms and provides mounting attachments. It completely covers the operating parts of the magneto to prevent the entrance of water, oil, or other contaminants. Screened vents permit ventilation and cooling. Some magnetos are provided with forced-air cooling to remove heat from inside the case.

As explained previously, the pole shoes and the coil core of the magneto are constructed of laminated soft iron. Soft iron has high permeability (ability to carry magnetic lines of force) but will not retain magnetism. For this reason, the magnetism in the pole shoes and coil core can change rapidly as required in the operation of the magneto.

The pole shoes and coil core are laminated with insulation between the laminations to reduce the effect of **eddy currents** which develop as a result of the rapidly changing magnetic forces in these parts. Eddy currents interfere with the proper changes in magnetic force and also generate heat.

The magnets employed in magnetos are constructed of very hard alloys, such as alnico or Permalloy. These have proved to be much stronger than hardened steel. The alloys are shaped to meet the requirements of the magneto and are magnetized by means of a strong electromagnetic field. If the design of the magneto requires a rotating magnet, the magnet is mounted on a steel shaft with suitable bearing surfaces for rotation.

The distributor of the magneto consists of a rotor made of a durable dielectric material, usually fabricated by molding. The material may be Formica, Bakelite, or some of the more recent thermosetting plastics. In any event, the rotor must be able to withstand high temperatures and high electric stresses. The high-tension output from the distributor is carried through a distributor cap or blocks in which the high-tension leads for the spark plugs are mounted. Inside the distributor cap or blocks are electrodes which pick up the high-tension current from the rotor electrode. The distributor rotor and cap or blocks are usually coated with a high-temperature wax to prevent moisture absorption and the possibility of high-voltage leakage.

Magneto Speed

Figure 8-17 is a schematic illustration of an aircraft ignition system using a rotating-magnet magneto. Notice the cam on the end of the magnet shaft. It is not a compensated cam; therefore, it has as many lobes as there are poles on the magnet, four in this case. The number of high-voltage impulses produced per revolution of the magnet is equal, therefore, to the number of poles. The number of cylinder firings per complete revolution of the engine is equal to one-half the number of engine cylinders. Therefore, the ratio of the magneto shaft speed to that of the engine crankshaft is equal to the number of cylinders divided by twice the number of poles on the rotating magnet. This can be stated as a formula in this manner:

$$\frac{\text{No. of cylinders}}{2 \times \text{no. of poles}} = \frac{\text{magneto shaft speed}}{\text{engine crankshaft speed}}$$

For example, if the uncompensated cam has four lobes (since there are four poles on the magnet) and if the engine has 12 cylinders, then

$$\frac{12 \text{ cylinders}}{2 \times 4 \text{ poles}} = \frac{12}{8} = 1\frac{1}{2}$$

Therefore, the magneto speed is $1\frac{1}{2}$ times the engine crankshaft speed.

Remember that in a four-stroke-cycle engine, each cylinder fires once for each two turns of the crankshaft. Therefore, we know that a 12-cylinder engine will fire six times for each

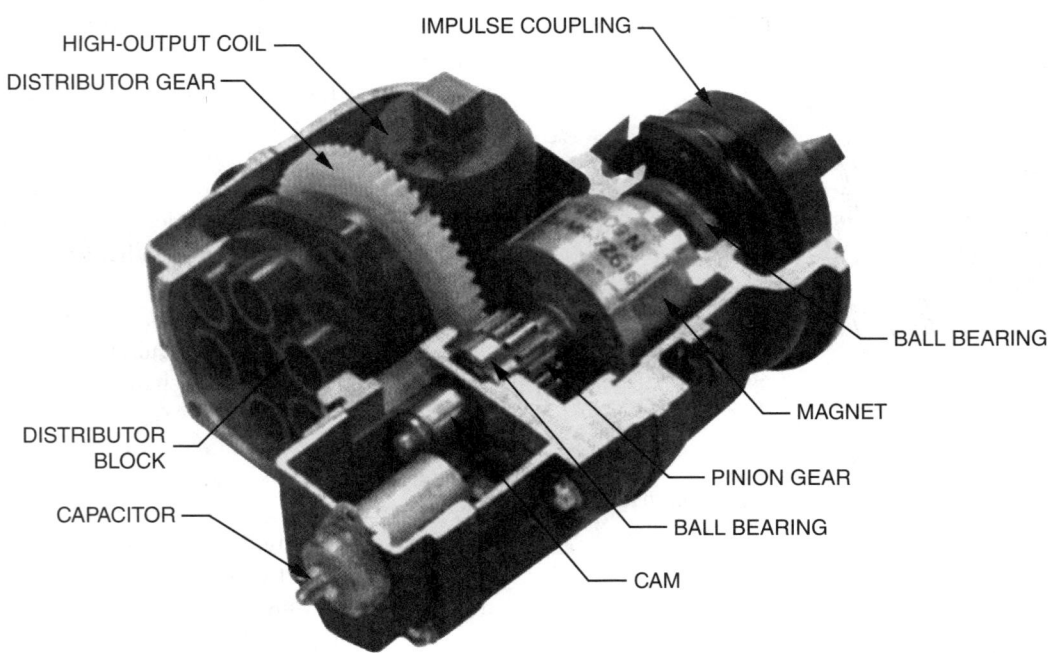

FIGURE 8-17   High-tension magneto primary ignition system.

revolution of the crankshaft. Also, a magneto having four lobes on the cam will produce four sparks for each turn of the cam. In the magneto under discussion, the cam is mounted on the end of the magneto shaft, so we know that the magneto produces four sparks for each revolution of the magneto shaft. Then, to produce the six sparks needed for each revolution of the crankshaft, the magneto must turn $1\frac{1}{2}$ times.

A nine-cylinder radial engine requires $4\frac{1}{2}$ sparks per revolution, and a four-pole magnet produces 4 sparks per revolution. The ratio of engine speed to magneto speed must therefore be $4:4\frac{1}{2}$, or 8:9. The engine requires 36 sparks for 8 r, and the magneto produces 36 sparks in 9 r.

Polarity or Direction of Sparks

Fundamentally the magneto is a special form of ac generator, modified to enable it to deliver the high voltage required for ignition purposes. In Fig. 8-14, the high rate of change of flux linkages represented by the almost vertical portion of the resultant-flux curve is responsible for the high voltage which produces the secondary spark. From the curves, clearly the rapid flux change is downward, then upward, alternating in direction at each opening of the contacts. Since the direction of an induced current depends on the direction of flux change which produced it, the sparks produced by the magneto are of alternating polarity; that is, they jump one way and then the other, as represented by the secondary-voltage current curve in Fig. 8-14, which is first above and then below the line, indicating alternating polarity.

Dual-Magneto Ignition

An arrangement in which two magnetos fire at the same or approximately the same time through two sets of spark plugs is known as a double-, or dual-, magneto ignition system. The principal advantages are the following: (1) If one magneto or any part of one magneto system fails to operate, the other magneto system will furnish ignition until the disabled system functions again. (2) Two sparks, igniting the F/A mixture in each cylinder simultaneously at two different places, provide quicker and more complete combustion than a single spark; therefore, the power of the engine is increased. All certificated reciprocating engines must be equipped with dual ignition.

Dual-ignition spark plugs may be set to fire at the same instant (synchronized) or at slightly different times (staggered). When staggered ignition is used, the two sparks occur at different times. The spark plug on the exhaust side of the cylinder always fires first because the lower rate of burning of the expanded and diluted F/A mixture at this point in the cylinder makes it desirable to have an advance in the ignition timing.

Magneto Sparking Order

Almost all piston-type aircraft engines operate on the principle of the four-stroke, five-event cycle. For this reason, the number of sparks required for each complete revolution of the engine is equal to one-half the number of cylinders in the engine. The number of sparks produced by each revolution of the rotating magnet is equal to the number of its poles. Therefore, the ratio of the speed at which the rotating magnet is driven to the speed of the engine crankshaft is always one-half the number of cylinders on the engine divided by the number of poles on the rotating magnet, as explained before.

The numbers on the distributor block show the **magneto sparking order, not the firing order of the engine**. The distributor block position marked 1 is connected to the no. 1

cylinder, the distributor block position marked 2 is connected to the second cylinder to be fired, the distributor block position marked 3 is connected to the third cylinder to be fired, and so on.

Some distributor blocks or housings are not numbered for all high-tension leads. In these cases, the lead socket for the no. 1 cylinder is marked and the others follow in order according to direction of rotation.

Coming-In Speed of Magneto

To produce sparks, the rotating magnet must be turned at or above a specified rpm, at which speed the rate of change in flux linkages is sufficiently high to induce the required primary current and the resultant high-tension output. This speed is known as the **coming-in speed** of the magneto; it varies for different types of magnetos but averages about 100 to 200 rpm.

Magneto Safety Gap

Magnetos are sometimes equipped with a **safety gap** to provide a return ground when the external secondary circuit is open. One electrode of the safety gap is screwed into the high-tension brush holder, while the grounded electrode is on the safety-gap ground plate. Thus, the safety gap protects against damage from excessively high voltage in case the secondary circuit is accidentally broken and the spark cannot jump between the electrodes of the spark plugs. In such a case, the high-tension spark jumps the safety gap to the ground connection, thereby relieving the voltage in the secondary winding of the magneto.

Blast Tubes

Blast tubes are used on some aircraft to cool the magnetos. The blast tubes are attached to the engine's baffling which collects ram air and directs it onto the magneto housing, thus providing a cooling effect.

IGNITION SHIELDING

Since the magneto is a special form of high-frequency generator, it acts as a radio transmitting station while it is in operation. Its oscillations are called **uncontrolled oscillations** because they cover a wide range of frequencies. The oscillations of a conventional radio transmitting station are waves of a **controlled frequency**. For this reason the ignition system must be shielded.

Shielding is difficult to define in general terms. Aircraft **radio shielding** is the metallic covering or sheath used for all electric wiring and ignition equipment, grounded at close intervals, and provided for the purpose of eliminating any interference with radio reception.

If the high-tension cables and switch wiring of the magneto are not shielded, they can serve as antennas from which the uncontrolled frequencies of the magneto oscillations are radiated. The receiving antenna on an airplane is comparatively close to the ignition wiring; therefore, the uncontrolled frequencies are picked up by the antenna along with the controlled frequencies from the aircraft radio station, thus causing interference (noise) to be heard in the radio receiver in the airplane.

Design of the Ignition Shielding

The magneto has a metallic cover made of a nonmagnetic material. The cover joints are fitted tightly to prevent dirt and moisture from entering. Since it is necessary to cover the cables completely, fittings are provided on the magneto for attachment of a shielded ignition harness. Provision is made for ventilation to remove condensation and the corrosive gases formed by the arcing of the magneto within the housing.

Shielding of high-tension leads for the ignition system installed on an opposed engine is accomplished by means of a woven wire sheath placed around each spark plug lead. This sheath is electrically connected to the magneto case and to the spark plug shells to provide a continuous grounded circuit.

Ignition Wiring System

The **low-tension wiring** on a high-tension magneto consists of a single shielded conductor from the primary coil to the engine ignition switch. Its circuit passes through the fire wall by means of a connector plug, frequently of a special design, which automatically grounds the magnetos when the plug is disconnected.

High-tension cable has a conductor of small cross section and insulation of comparatively large cross section, whereas low-tension cable has a conductor of large cross section and insulation of comparatively small cross section. The reason for this difference is that the capacity to carry current is the primary requisite of low-tension cable, whereas dielectric strength (insulating property) is the most important requirement of high-tension cable.

High-tension cable may consist of several strands of small wire; a layer of rubber, synthetic rubber, or plastic; a glass braid covering; and a neoprene, or plastic, sheath. It is available in several sizes, the most common being 5 and 7 mm.

High-tension wiring is placed in the special conduit arrangement known as the ignition harness or enclosed in a woven wire sheath to provide for radio shielding.

Ignition Switch and the Primary Circuit

All units in an aircraft ignition system are controlled by an **ignition switch** in the cockpit. The type of switch used varies with the number of engines on the aircraft and the type of magnetos used. All switches, however, turn the system off and on in a similar manner. The normal electric switch is closed when it is turned on. The magneto ignition switch is closed when it is turned off, because its purpose is to short-circuit the breaker points of the magneto and to prevent collapse of the primary circuit required for production of a spark.

In the ignition switch, one terminal is connected to the primary electric circuit between the coil and the breaker contact points. The ignition switch lead that connects the primary

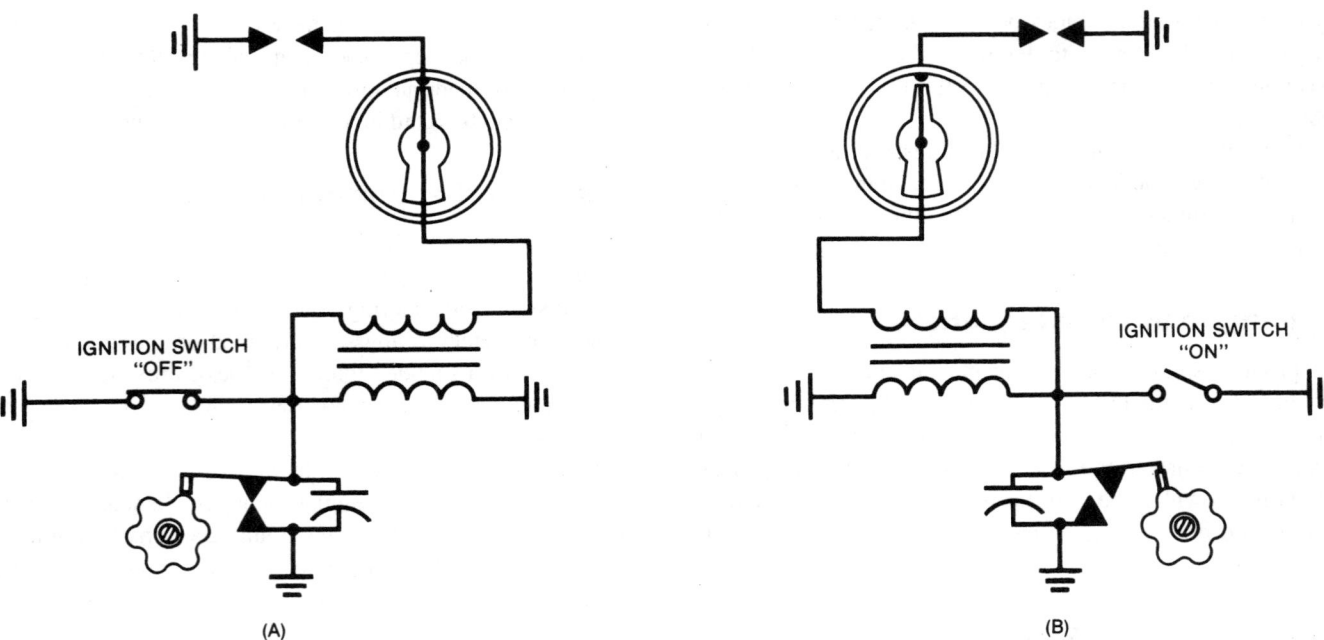

FIGURE 8-18 Typical ignition switch circuit. (A) Circuit in OFF position. (B) Circuit in ON position.

circuit and the switch is commonly referred to as the P-lead. The other terminal is connected to the aircraft ground (structure). As shown in Fig. 8-18A, there are two ways to complete the primary circuit: through the closed breaker points to ground or through the closed ignition switch to ground.

In Fig. 8-18A, the primary current will not be interrupted when the breaker contacts open, since there is still a path to ground through the closed (off) ignition switch. Since the primary current is not stopped when the contact points open, there can be no sudden collapse of the primary coil flux field and no high voltage induced in the secondary coil to fire the spark plug.

When the ignition switch is placed in the ON position (switch open), as shown in Fig. 8-18B, the interruption of primary current and the rapid collapse of the primary coil flux field are once again controlled or triggered by the opening of the breaker contact points. When the ignition switch is in the ON position, the switch has absolutely no effect on the primary circuit.

Many single-engine aircraft ignition systems employ a dual-magneto system in which the left magneto supplies the electric spark for the top plugs in each cylinder and the right magneto fires the bottom plugs. One ignition switch is normally used to control both magnetos.

The switch illustrated in Fig. 8-19 has four positions: **OFF**, **LEFT**, **RIGHT**, and **BOTH**. In the OFF position, both magnetos are grounded and thus are inoperative. When the switch is in the LEFT position, only the left magneto operates; in the RIGHT position, only the right magneto operates. In the BOTH position, both magnetos operate. The RIGHT and LEFT positions are used to check dual-ignition systems, allowing the magnetos to be turned off one at a time.

After the installation of an ignition switch or after the switch circuit has been rewired, the operation of the circuit

must be tested. This can be done with an ohmmeter or a continuity test light. Disconnect the P-lead from the magneto, and connect it to one terminal of the test unit. Connect the other terminal of the test unit to ground at or near the engine. When the ignition switch is turned to OFF, the test light should burn or the ohmmeter should indicate a complete circuit (little or no resistance). When the ignition switch is turned to ON, the test light should go out or the ohmmeter should show infinite resistance (an open circuit).

In working with the magneto system, the technician must always keep in mind that the magneto will be "hot" when the P-lead is disconnected or when there is a break in the circuit leading to the ignition switch. If the switch circuit is being repaired, it is important that the primary terminal of the magneto be connected to ground or that the spark plug leads be disconnected.

FIGURE 8-19 Ignition-starter switches. (*Continental Motors Ignition Systems.*)

IGNITION BOOSTERS AND AUXILIARY IGNITION UNITS

When attempting to start an engine, often the engine starter will not rotate the crankshaft fast enough to produce the required coming-in speed of the magneto. In these instances, a source of external high-tension current is required for ignition purposes. The various devices used for this purpose are called **ignition boosters** or **auxiliary ignition units**.

An ignition booster may be in the form of a booster magneto, a high-tension coil to which primary current is supplied from a battery, or a vibrator which supplies intermittent direct current from a battery directly to the primary of the magneto. Another device used for increasing the high-tension voltage of the magneto for start-up is called an impulse coupling.

Impulse Coupling

When an aircraft engine is started, the engine turns over too slowly to permit the magneto to operate. The impulse coupling installed on the drive shaft of a magneto is designed to give the magneto a momentary high rotational speed and to provide a retarded spark for starting the engine. This coupling is a spring-like mechanical linkage between the engine and magneto shaft which "winds up" and "lets go" at the proper moment for spinning the magneto shaft, thus supplying the high voltage necessary for ignition. The coupling consists of a shell, spring, and hub. The hub is provided with **flyweights** which enable the assembly to accomplish its purpose. These are illustrated in Fig. 8-20. In some manuals, the shell is referred to as the *body*, and the hub is called the *cam*.

When the impulse coupling is installed on the drive shaft of the magneto, the shell of the coupling may be rotated by the engine drive for a substantial portion of l r while the rotating magnet remains stationary, as shown in Fig. 8-21. While this is taking place, the spring in the coupling is being wound up. At the point where the magneto must fire, the flyweights are released by the action of the body contacting the trigger ramp. This action causes the flyweights to rotate on the pivot point and disengage from the stop pin, as shown in Fig. 8-22. This allows the spring to unwind giving the rotating magnet a

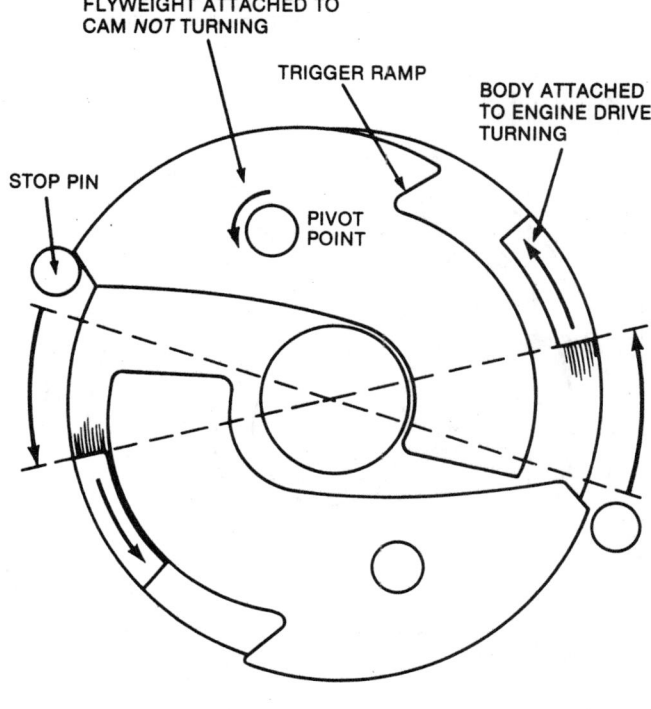

FIGURE 8-21 Impulse coupling in START position.

rapid rotation in the normal direction. This, of course, causes the magneto to produce a strong spark at the spark plug. As soon as the engine begins to run, the flyweights are held in the release position by centrifugal force (see Fig. 8-23), and the magneto fires in its normal advanced position. During engine

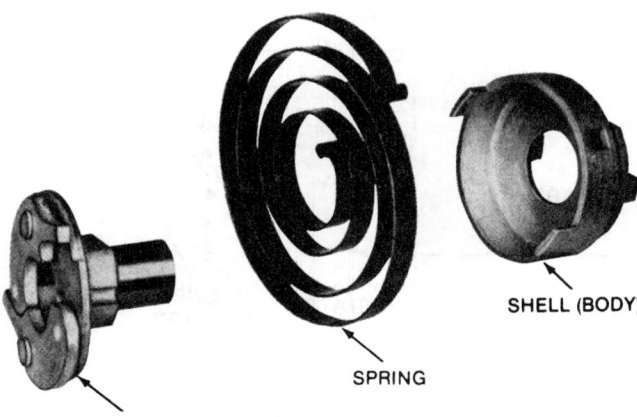

FIGURE 8-20 Components of an impulse coupling.

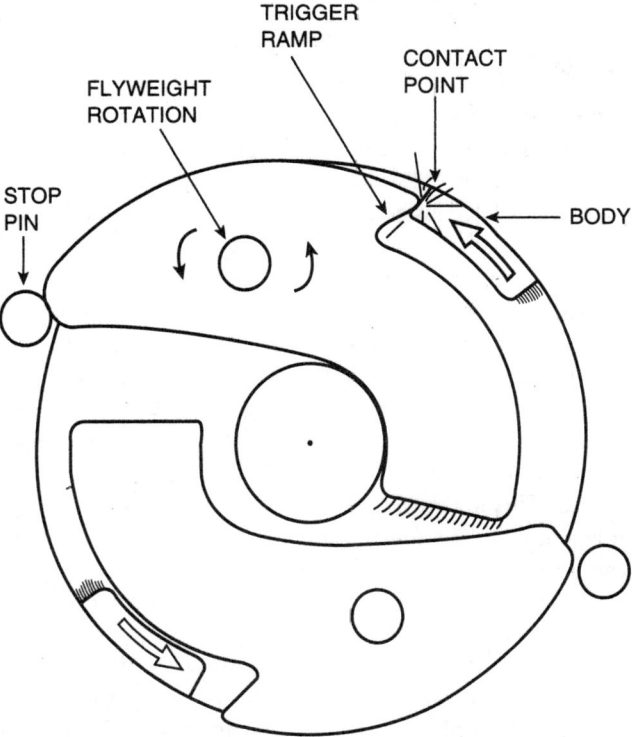

FIGURE 8-22 Impulse coupling in RELEASE position.

Ignition Boosters and Auxiliary Ignition Units **205**

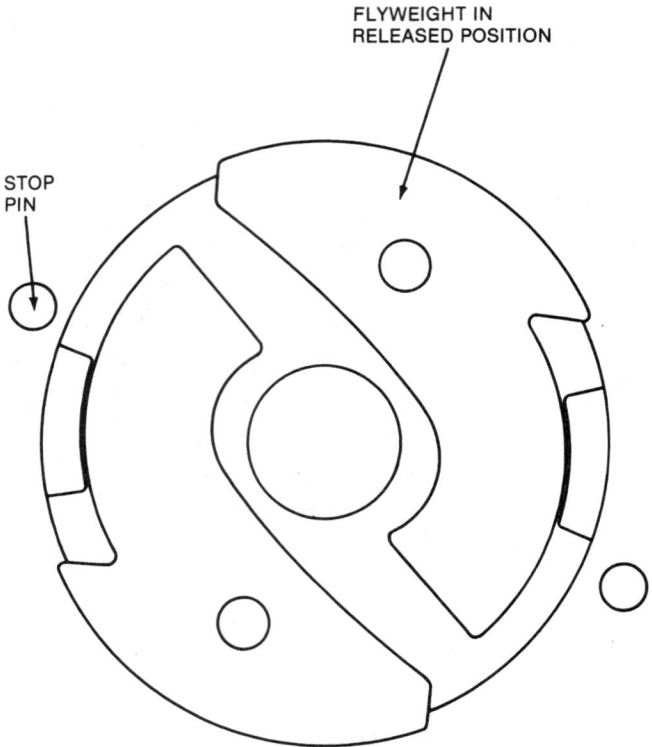

FIGURE 8-23   Impulse coupling in RUNNING position.

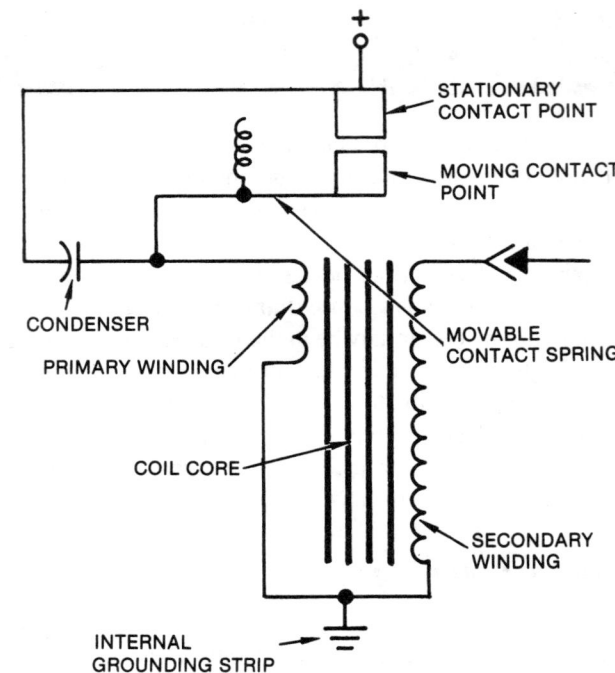

FIGURE 8-24   Booster coil schematic.

start-up, the retard spark is produced when the magneto rotation is held back by the impulse coupling. During normal operation, the impulse coupling spring holds the magneto in the advance spark position. If the spring were to break, the magneto would continue to rotate but would be in the retard spark position. The spark plugs fired by this particular magneto would be firing late.

Booster Coil

A **booster coil** is a small induction coil. Its function is to provide a **shower of sparks** to the spark plugs until the magneto fires properly. The booster coil is usually connected to the starter switch. When the engine has started, the booster coil and the starter are no longer required; therefore, they can be turned off together.

When voltage from a battery is applied to the booster coil, illustrated in Fig. 8-24, magnetism is developed in the core until the magnetic force on the soft-iron armature mounted on the vibrator overcomes the spring tension and attracts the armature toward the core. When the armature moves toward the core, the contact points and the primary circuit are opened. This demagnetizes the core and permits the spring to again close the contact points and complete the circuit. The armature vibrates back and forth rapidly, making and breaking the primary circuit as long as the voltage from the battery is applied to the booster coil.

The use of booster coils as described here is limited to a few older aircraft which are still operating. Most modern aircraft employ the induction vibrator or an impulse coupling.

Induction Vibrator

The function of the **induction vibrator** is to supply interrupted low voltage (pulsating direct current) for the magneto primary coil, which induces a sufficiently high voltage in the secondary for starting. A schematic diagram of the circuit for an induction vibrator designed for use with light-aircraft engine magnetos is shown in Fig. 8-25. Observe that when the starter switch is closed, battery voltage is applied to the vibrator coil through the vibrator contact points and through the retard contact points in the left magneto. As the coil is energized, the breaker points open and interrupt the current flow, thus de-energizing the coil, VC. Through spring action the contact points close and again energize the coil, causing the points to open. Thus the contact points of the vibrator continue to make and break contact many times per second,

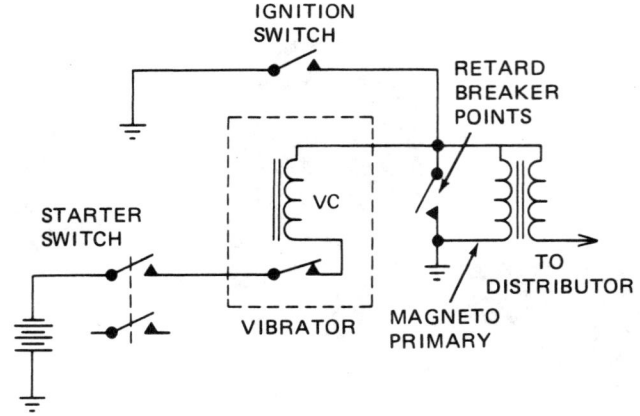

FIGURE 8-25   Induction vibrator circuit for light-aircraft engine.

sending an interrupted current through both the main and retard contact points of the magneto.

The vibrator sends an interrupted battery current through the primary winding of the regular magneto coil. The magneto coil then acts as a battery ignition coil and produces high-tension impulses, which are distributed through the distributor rotor, distributor block, and cables to the spark plugs. These high-tension impulses are produced during the entire time that both sets of magneto contact points are open. When the contact points are closed, sparks cannot be generated. Although the vibrator continues to send interrupted current impulses through the magneto contact points, the interrupted current will flow through the contact points to ground. This is the path of least resistance for the current.

A circuit for an induction vibrator as used with a Continental Shower of Sparks high-tension magneto ignition is shown in Fig. 8-26A. This circuit applies to one engine only, but a similar circuit would be used with each engine of a multiengine airplane. The induction vibrator is energized from the same circuit which energizes the starting solenoid. It is thus energized only during the time that the engines are being started.

When the ignition switch is in the ON position and the engine starter is engaged, the current from the battery is sent through the coil of a relay which is normally open. The battery current causes the relay points to close, thus completing the circuit to the vibrator coil and causing the vibrator to produce a rapidly interrupted current (pulsating direct current). This can be seen in Fig. 8-26B. In Fig. 8-26C, the engine has advanced far enough to open the advanced set of breaker points, but the retard contact points are still closed, preventing flow of the interrupted current through the primary magneto winding. As the engine continues to turn, at about top center (TC) engine position, the retard contact points open, as shown in Fig. 8-26D.

The rapidly interrupted current produced by the vibrator is sent through the primary winding of the magneto coil. This action can be seen in Fig. 8-26D. The path through the primary winding is now the easiest and only path to ground. By induction, high voltage is created in the secondary winding of the magneto coil, and this high voltage produces high-tension sparks which are delivered to the spark plugs through the magneto distributor block electrodes during the time that the magneto contact points are open.

This process is repeated each time the magneto contact points are separated, because the interrupted current once more flows through the primary of the magneto coil. The action continues until the engine is firing because of regular magneto sparks, and the engine starter is released. As can be seen in Fig. 8-26B, the ignition switch to the right (R) magneto is closed during starting to prevent firing of the right magneto. This is done to eliminate the possibility of the right magneto firing in the advanced position, causing kickback. Note that the vibrator starts to operate automatically when the engine ignition switch is turned to the ON position and the starter is engaged. The vibrator stops when the starter is disengaged, as seen in Fig. 8-26E, which is the normal engine running position.

CONTINENTAL IGNITION HIGH-TENSION MAGNETO SYSTEM FOR LIGHT-AIRCRAFT ENGINE

General Description

A typical ignition system for a light-aircraft engine consists of two magnetos, a starter vibrator, a combination ignition and starter switch, and a harness assembly. These parts are illustrated in Fig. 8-27. This illustration shows the components of the system associated with the Continental ignition series S-200 magneto. However, it is quite similar to other Continental ignition systems, and the principles are the same.

Magneto

The S-200 magneto is a completely self-contained unit incorporating a two-pole rotating magnet, a coil unit containing the primary and secondary windings, a distributor assembly, main breaker points, retard breaker points, a two-lobe cam, a feed-through type of capacitor, housing sections, and other components necessary for assembly.

The rotating magnet turns on two ball bearings, one located at the breaker end and the other at the drive end. A two-lobe cam is secured to the breaker end of the rotating magnet. In a six-cylinder magneto, the rotating magnet turns $1\frac{1}{2}$ times engine speed. Thus, six sparks are produced through 720° of engine rotation—that is, 2 r of the crankshaft. In a four-cylinder magneto, the rotating magnet turns at engine speed, thus producing four sparks through 2 r of the crankshaft.

As mentioned previously, the dual-breaker magneto incorporates a retard breaker. This breaker is actuated by the same cam as the main breaker and is positioned so that its contacts open a predetermined number of degrees after the main breaker contacts open. A battery-operated starting vibrator used with this magneto provides retarded ignition for starting, regardless of engine cranking speed. The retarded ignition is in the form of a shower of sparks instead of a single spark like that produced by an impulse coupling. Remember that the slow cranking speed of an engine during starting makes it necessary to retard ignition to prevent kickback. At starting speed, if advanced ignition is supplied, the full force of the combustion will be developed before the piston reaches TDC (top dead center) and the piston will be driven back down the cylinder, thus rotating the crankshaft in reverse of the normal direction.

Operation of the System

The operation of the S-200 magneto system can be understood by studying Fig. 8-28. In this circuit, the starting vibrator unit includes both a **vibrator** and a **control relay**. The interrupted battery current supplied by the vibrator is controlled by the retard breaker points in the left magneto. The control relay grounds the right magneto during the time that the starter switch is turned on, thus preventing an advanced spark from being applied to the spark plugs.

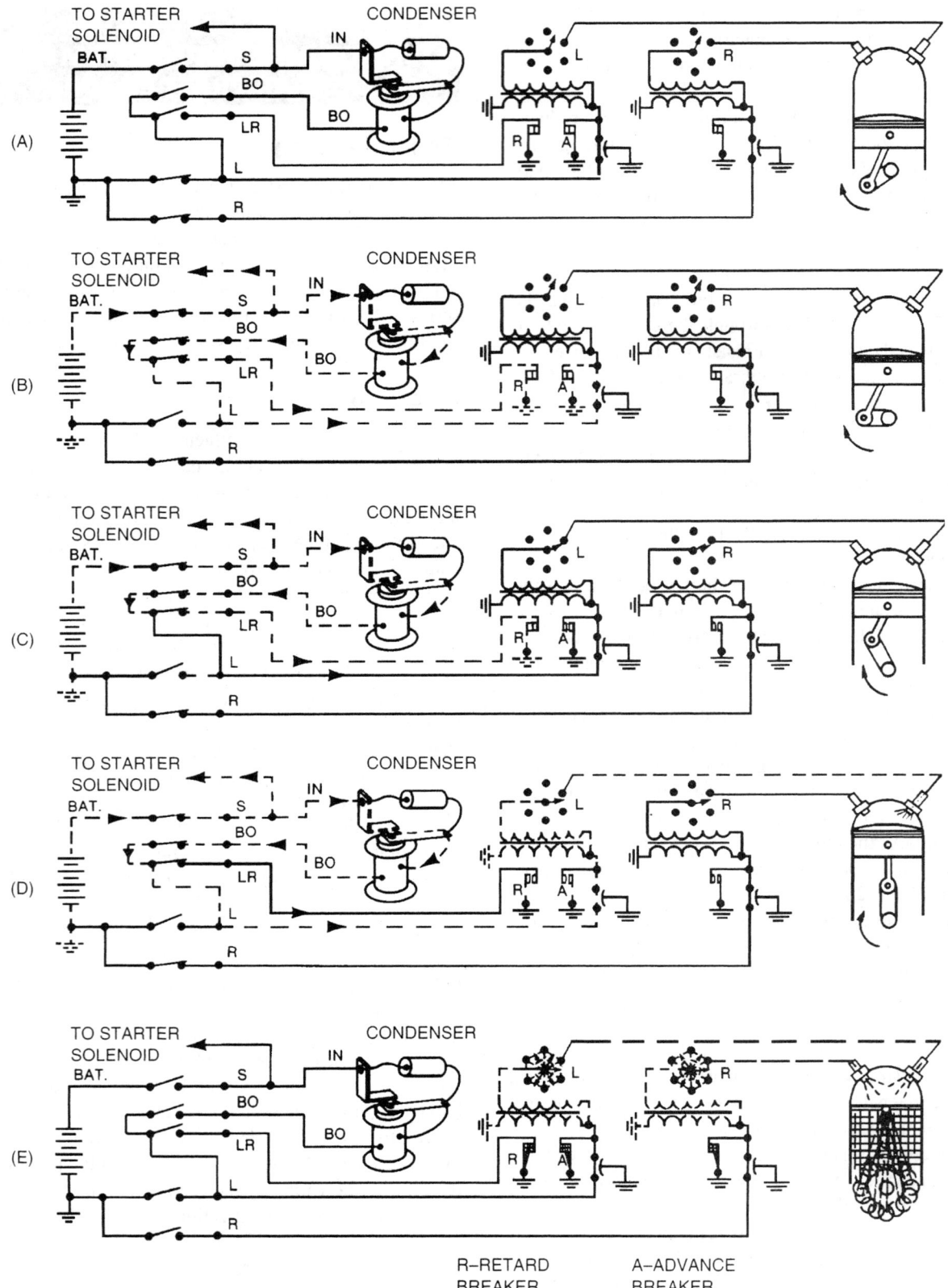

FIGURE 8-26 Continental ignition Shower of Sparks induction vibrator circuit. (A) All switches off; circuit not energized. (B) Switch in START position. (C) Advance breaker points open. (D) Retard breaker points open. (E) Switches released to BOTH position. (*Continental Motors.*)

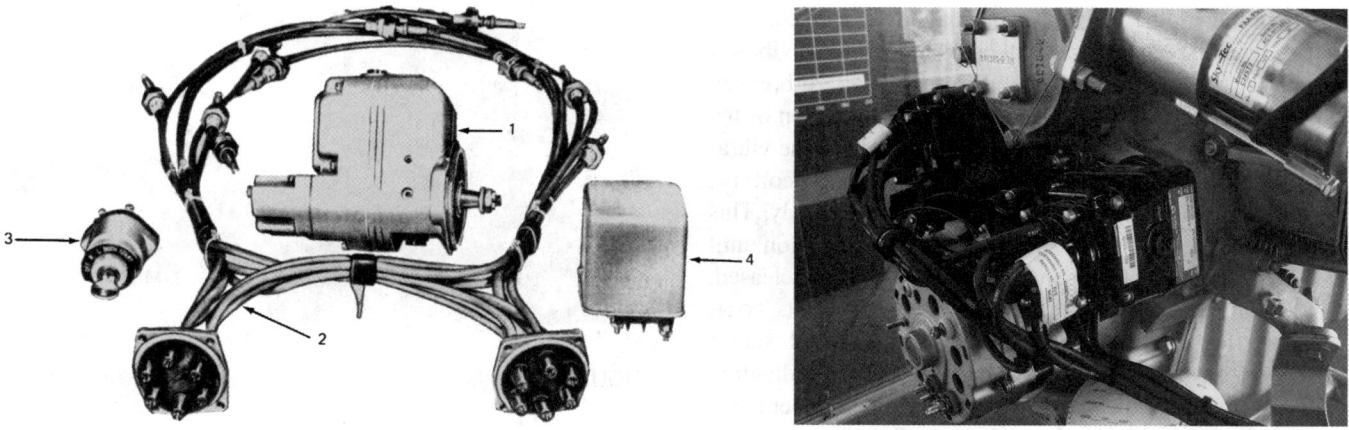

FIGURE 8-27 Components of a high-tension ignition system for a light-aircraft engine: (1) magneto, (2) harness assembly, (3) combination switch, and (4) vibrator. (*Continental Motors.*)

The vibrator assembly, which includes the control relay, is used with a *standard* ignition switch. The electrical operation of this switch can be seen in Fig. 8-28. In the diagram, all the switches and contact points are shown in their normal OFF positions.

With the standard ignition switch in its BOTH position and the starter switch turned on, the starter solenoid L3 and the relay coil L1 are energized, thus causing them to close their relay contacts R4, Rl, R2, and R3. Relay contact R3 connects the right magneto to ground, rendering it inoperative during starting procedures. Battery current flows through relay contact Rl, vibrator points V1, and coil L2, and then through the retard breaker of the left magneto to ground, through the relay contact R2, and through the main breaker to ground. The flow of current through coil L2 establishes a magnetic field which opens the vibrator points V1 and starts the vibrating cycle. The interrupted battery current thus produced is carried to ground through both sets of breaker points in the left magneto.

When the engine reaches its normal advance firing position, the main breaker opens. However, the current is still carried to ground through the retard breaker, which does not open until the starting *retard position* of the engine is reached. When the retard breaker opens (the main breaker is still open), the vibrator

current flows through the primary of transformer T1 (magneto coil), producing a rapidly fluctuating magnetic field around the coil. This causes a high voltage to be induced in the secondary, thus firing the spark plug to which the voltage is directed by the distributor. A shower of sparks is therefore produced at the spark plug owing to the opening and closing of the vibrator points while the main and retard breaker points are open.

When the engine fires and begins to pick up speed, the starter switch is released, thus deenergizing relay coil L2 and starter relay L3. This opens the vibrator circuit and the retard breaker circuit, thus rendering them inoperative. The single-breaker magneto (right magneto) is no longer grounded; therefore, both magnetos are firing in the full-advance position.

The schematic diagram presented as Fig. 8-29 illustrates the operation of a system utilizing a **combination starter-ignition switch** and a starting vibrator which does not include the control relay. When the combination switch is placed in the START position, the right magneto is grounded, starter solenoid L1 is energized, and current flows through vibrator L2 to both magneto breaker points and then to ground if the points are closed. Note that all contacts in the combination switch are moved to the START position; therefore, they will not be in the position shown in the diagram.

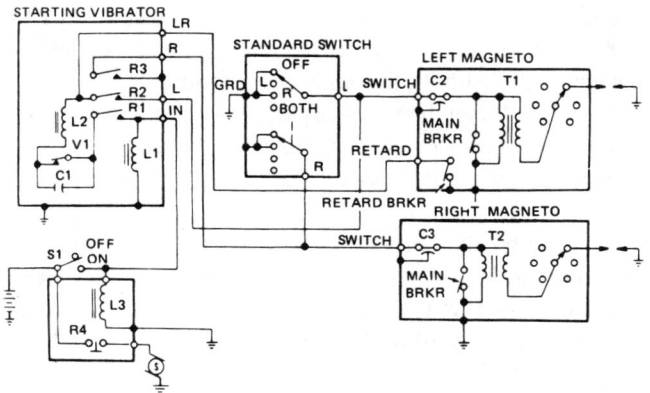

FIGURE 8-28 Circuit diagram for Continental ignition S-200 magneto system. (*Continental Motors.*)

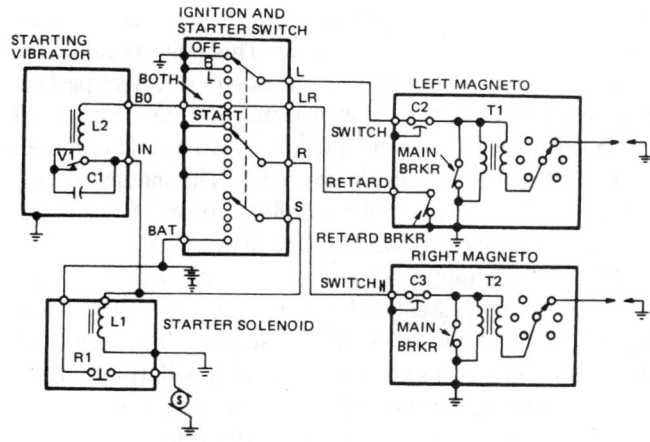

FIGURE 8-29 Ignition system with a combination starter-ignition switch.

When the engine reaches its normal advance firing position, the main breaker points will open; however, the vibrator current is still carried to ground through the retard breaker, which does not open until the starting retard position of the engine is reached. When the retard breaker opens, the vibrator current flows through the primary of magneto coil T1, thus inducing a high voltage, as explained previously. This voltage provides the retard spark necessary for ignition until the engine speed picks up and the starter switch is released. The combination switch automatically returns to its BOTH position, thus removing the starting vibrator and starter solenoid from the circuit. A study of the switch diagram will also show that the switch circuits of both magnetos are ungrounded; therefore, both magnetos will be firing.

The combination switch used with a magneto system has five positions and is actuated by either a switch or a key. The five positions are: (1) OFF—both magnetos grounded and not operating; (2) R—right magneto operating, left magneto off; (3) L—left magneto operating, right magneto off; (4) BOTH—both magnetos operating; and (5) START—starter solenoid operating and vibrator energized, causing an intermittent current to flow through the retard breaker on the left magneto while the right magneto is grounded to prevent advanced ignition. The START position on the switch is a *momentary* contact and is *on* only while being held in this position. When the switch is released, it automatically reverts to the BOTH position.

In Fig. 8-29, the magnetos are equipped with flow-through-type capacitors C2 and C3, which reduce arcing at the breaker contacts, help to eliminate radio interference, and cause a more rapid collapse of the magnetic field when the breaker points open during normal operation. A capacitor C1 is also necessary in the starter vibrator circuit to produce similar results during the production of the intermittent vibrator current.

Continental Ignition Internal Magneto Timing

As explained in the general discussion of magnetos, every magneto must be internally timed to produce a spark for ignition at a precise instant. Furthermore, the magneto breaker points must be timed to open when the greatest magnetic field stress exists in the magnetic circuit. This point is called the E gap, or efficiency gap, and it is measured in degrees past the neutral position of the magnet. The magneto distributor must be timed to deliver the high-tension current to the proper outlet terminal of the distributor block. The internal timing procedure varies somewhat for different types of magnetos; however, the principles are the same in every case.

Distributor timing is usually accomplished while the magneto is being assembled. Figure 8-30 shows the matching of the chamfered tooth on the distributor drive gear with the marked tooth on the driven gear. In this illustration, the magneto is being assembled for right-hand (clockwise) rotation. The direction of rotation refers to the direction in which the magnet shaft rotates, facing the drive end. When the teeth of the gears are matched as shown, the distributor will be in

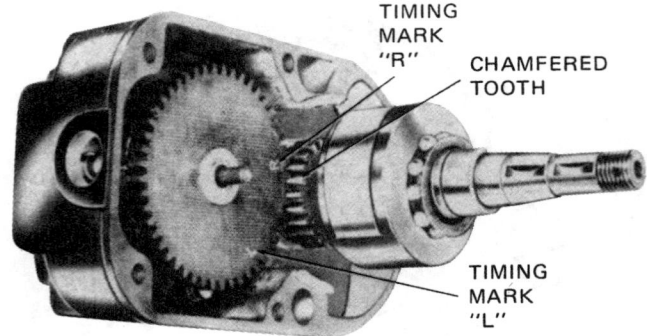

FIGURE 8-30 Matching marks on gears for distributor timing.

the correct position with respect to the rotating magnet and breaker points at all times. The large distributor gear also has a marked tooth, which can be observed through the timing window on top of the case to indicate when the distributor is in position for firing the no. 1 cylinder. This mark is not sufficiently accurate for timing the opening of the points, but it does show the correct position of the distributor and rotating magnet for timing to the no. 1 cylinder.

The following steps are taken to check and adjust the timing of the breaker points for the S-200 magneto, which does not have timing marks in the breaker compartment:

1. Remove the timing inspection plug from the top of the magneto. Turn the rotating magnet in its normal direction of rotation until the painted, chamfered tooth on the distributor gear is approximately in the center of the inspection window. Then turn the magnet back a few degrees until it is in its neutral position. Because of its magnetism, the rotating magnet will hold itself in the neutral position.

2. Install the timing kit as shown in Fig. 8-31, and place the pointer in the zero position. If the manufacturer's timing kit is not available, a substitute can be fabricated by using a protractor to provide accurate angular measurement.

3. Connect a suitable timing light across the main breaker points, and turn the magnet in its normal direction of rotation 10° as indicated by the pointer. This is the E-gap position. The main breaker points should be adjusted to open at this point.

4. Turn the rotating magnet until the cam follower is at the high point on the cam lobe, and measure the clearance between the breaker points. This clearance must be 0.018 ± 0.006 in [0.46 ± 0.15 mm]. If the breaker-point clearance is not within these limits, the points must be adjusted for correct setting. It will then be necessary to recheck and readjust the timing for breaker opening. If the breaker points cannot be adjusted to open at the correct time, they should be replaced.

On dual-breaker magnetos (those having retard breakers), the retard breaker is adjusted to open a predetermined number of degrees after the main breaker opens, within +2 to 0°. The amount of retard in degrees for any particular magneto is stamped on the bottom of the breaker compartment. To set the retard breaker points correctly, it is necessary to add the degrees of retard indicated in the breaker compartment to the reading of the timing pointer when the main breaker points

FIGURE 8-31   Installation and use of timing kit.

are opening. For example, if the main breaker points open when the timing pointer is at 10° and the required retard is 30°, then 30° should be added to 10°. The rotating magnet should thus be turned until the timing indicator reads 40°. The retard breaker points should be adjusted to open at this time.

If an engine is designed for ignition at 20° BTC (before top center) under normal operating conditions, the retard ignition should be set at least 20° later than the normal ignition. At this time the piston is close enough to TDC that it is not likely to kick back when ignition occurs.

Timing for Magneto with "Cast-In" Timing Marks

Some models of the S-200 magneto and other Continental ignition magnetos such as the S-1200 have timing marks cast in the breaker compartment. These marks are illustrated in Fig. 8-32. On each side of the breaker compartment, timing marks indicate E-gap position and various degrees of retard breaker timing. The marks on the left-hand side, viewed from the breaker compartment, are for clockwise-rotating magnetos, and the marks on the right-hand side are for counterclockwise-rotating magnetos. The rotation of the magneto is determined by viewing the magneto from the drive end.

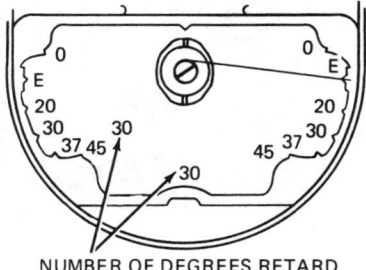

NUMBER OF DEGREES RETARD

FIGURE 8-32   Timing marks in breaker compartment.

The point in the center of the E-gap boss, shown at E in the drawing, indicates the exact E-gap position if the indicator is first set to zero with the magnet in the neutral position. The width of the boss on either side of the point is the allowable tolerance of ±4°. In addition to these marks, the cam has an indented line across its end for locating the E-gap position of the rotating magnet. This position is indicated when the mark on the cam is aligned with the mark at the top of the breaker housing.

Engine Timing Reference Marks

Most reciprocating engines have **timing reference marks** built into the engine. On an engine which has no propeller reduction gear, the timing mark will normally be on the propeller flange edge, as shown in Fig. 8-33. The TC mark stamped on the edge will align with the crankcase split line when the no. 1 piston is at TDC. Other flange marks indicate degrees BTC. Timing marks, also displayed on the starter ring gear, are aligned with a small hole located on the top face of the starter housing, as shown in Fig. 8-34.

On some engines, there are degree markings on the propeller reduction drive gear, as shown in Fig. 8-35. To time these engines, the plug provided on the exterior of the reduction-gear housing must be removed to view the timing marks. On other engines, the timing marks are on a crankshaft flange and can be viewed by removing a plug from the crankcase. In every case, the engine manufacturer's instructions give the location of built-in timing reference marks.

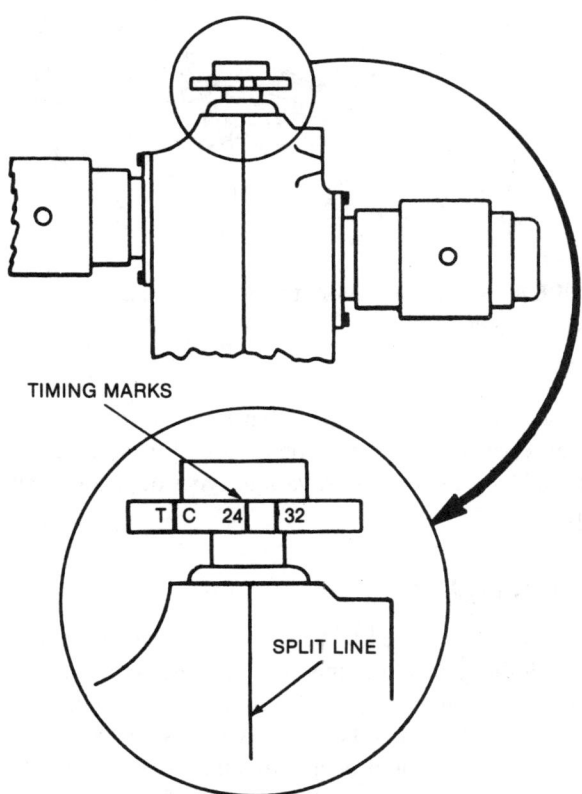

TIMING MARKS

SPLIT LINE

FIGURE 8-33   Propeller-flange timing marks.

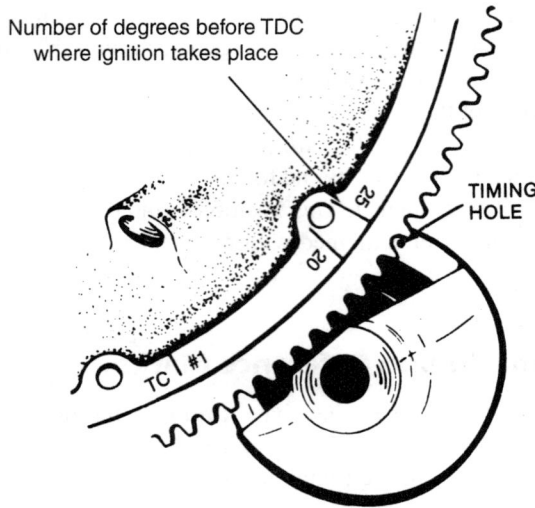

FIGURE 8-34 Engine timing marks on starter ring.

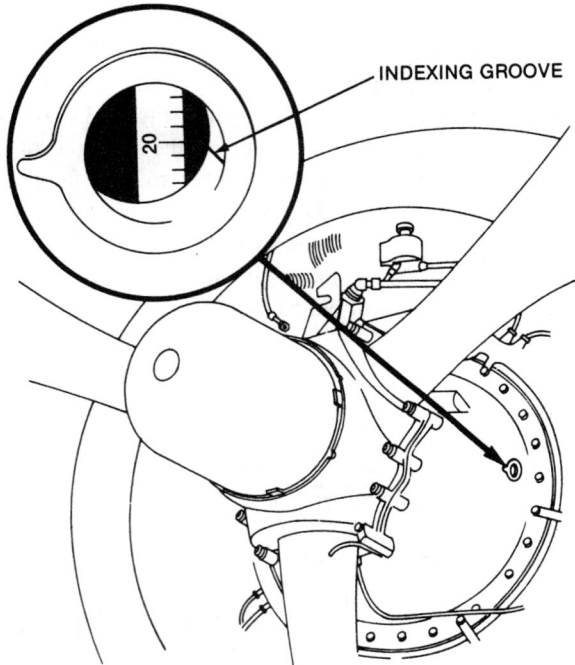

FIGURE 8-35 Typical built-in timing mark on propeller reduction gear.

Some older engines do not have built-in timing marks. The exact position of the piston on one of these engines can be found by using one of the various types of piston-position indicators.

Timing Lights

Timing lights are used to help determine the exact instant at which the magneto points open. Several types of timing lights are in common use today. In some, two lights go off when the points open; other lights work just the opposite and light up when the points open. Still other timing lights utilize an audible signal. With the wide variety of timing devices in use today, it is very important that the technician be familiar

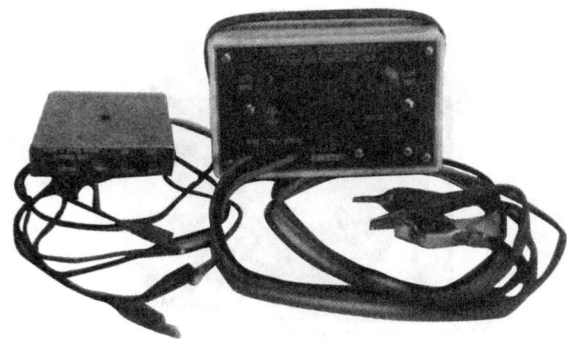

FIGURE 8-36 Timing lights.

with the instructions for the type being used. Two types of timing lights are illustrated in Fig. 8-36. Three wires come out of the top of the timing-light box.

There are also two lights on the front face of the unit and a switch to turn the unit on and off. To use the timing light, the center lead, marked "ground lead," is connected to the case of the magneto being tested. The other leads are connected to the primary leads of the breaker-point assembly of the magnetos being timed. With the leads connected in this manner, one can easily determine whether the points are open or closed by turning on the switch and observing the two lights.

Before the magneto is installed on the engine, check that it has the correct direction of rotation. Then proceed as follows.

Timing the Continental Ignition S-200 Magneto to the Engine

1. Remove the timing inspection plug from the top of the magneto, and turn the magneto in the normal direction of rotation until the painted chamfered tooth on the distributor gear is approximately in the center of the window, as in Fig. 8-37. The magneto is now in the correct E-gap position for firing the no. 1 cylinder.

2. Turn the engine to the no. 1 cylinder full-advance firing position (compression stroke) with the use of a suitable piston-position indicator or a timing disk and TC indicator.

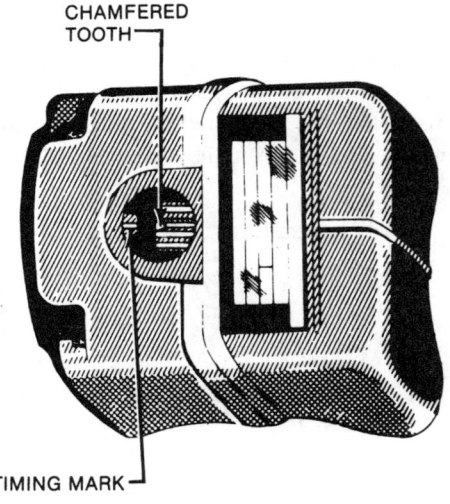

FIGURE 8-37 Magneto timing marks.

3. Install the magneto on the engine, and tighten the mounting bolts sufficiently to hold the magneto in position, but loosely enough that it can be rotated.

4. Connect the timing light to the magneto switch terminal, using the previously prepared terminal connection. When a direct-current (dc) continuity light is used for checking breaker-point opening time, the primary lead from the coil should be disconnected from the breaker points. This will prevent the flow of current from the battery in the tester through the primary winding.

5. If the timing light is out, rotate the magneto housing in the direction of its magnet rotation a few degrees beyond the point where the light comes on. Then slowly turn the magneto in the opposite direction until the light just goes out. Secure the magneto to the engine in this position by tightening the mounting bolts. Recheck the timing of the breaker points by turning the engine in reverse and then rotating it forward until the light goes out. The light should go out when the engine reaches the advance firing position as shown on the timing disk.

6. Remove the timing-light connection, and install the switch wire (P-lead) connection to the switch terminal of the magneto.

WARNING It is most important to note that the magneto is in the "switch on" condition whenever the P-lead wire is disconnected. It is therefore necessary to disconnect the spark plug wires when timing the magneto to the engine; otherwise, the engine could fire and cause injury to personnel.

Installation of High-Tension Harness

The high-tension spark plug leads are secured to the proper outlets in the magneto by means of the *high-tension outlet plate* and a rubber *grommet* or *terminal block*. The shielding of the cables is secured in the outlet plate by means of a ferrule, sleeve, and coupling nut. The ferrule and sleeve are crimped on the end of the shielding to form a permanent coupling fitting.

The high-tension cables are inserted through the outlet plate and into the grommet after the insulation has been stripped about $\frac{1}{2}$ in [12.7 mm] back from the end of the wire. The bare wires are extended through the grommet and secured by a small brass washer, as shown in Fig. 8-38. Another suitable method for securing copper high-tension cable is illustrated in Fig. 8-39. In this method, the wires are cut off even with the insulation, and then the cable is inserted into the grommet. A metal-piercing screw is used with a washer to hold the cables in place. The screws penetrate the ends of the stranded copper cable and form threads. The screws must not be turned too tight, or else they will strip.

Methods for securing ignition leads in distributor caps and for attaching spark plug terminals are described in maintenance manuals for specific magnetos. In all cases, follow the manufacturer's instructions.

During assembly of the high-tension harness, it is essential to note that the high-tension leads are installed in the outlet plate in the order of engine firing. The order of magneto firing for different magnetos is shown in Fig. 8-40;

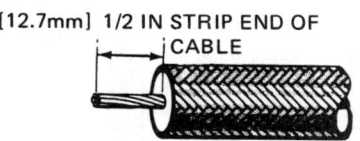

[12.7mm] 1/2 IN STRIP END OF CABLE

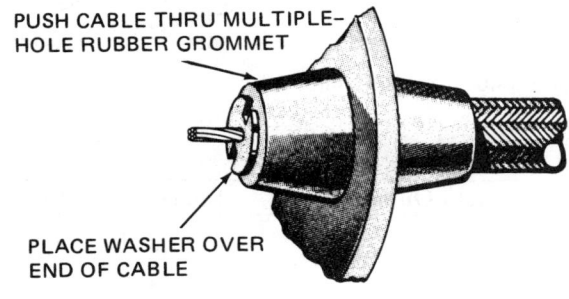

PUSH CABLE THRU MULTIPLE-HOLE RUBBER GROMMET

PLACE WASHER OVER END OF CABLE

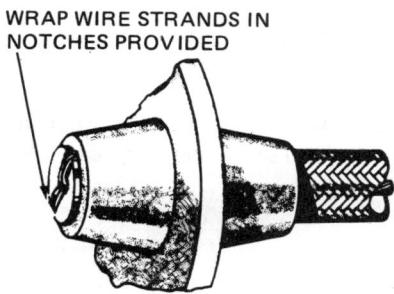

WRAP WIRE STRANDS IN NOTCHES PROVIDED

CROSS-SECTION VIEW OF WIRE STRANDS IN FINAL POSITION.

FIGURE 8-38 Connecting high-tension leads to the magneto.

however, the spark plug leads must not be connected in the same order. Since the firing order of a typical six-cylinder opposed engine is 1-4-5-2-3-6, the magneto outlets must be connected to spark plug leads as shown in Fig. 8-41.

In the practice of connecting the leads for dual-magneto systems, the right magneto fires the top spark plugs on the right-hand side of the engine and the bottom spark plugs on the left-hand side of the engine. The left magneto is connected

CUT OFF CABLE *SQUARELY* AND INSERT FIRMLY INTO SOCKET OF GROMMET.

INSERT SCREW THROUGH WASHER AND SECURE INTO CABLE. TIGHTENING FIRMLY BUT NOT EXCESSIVELY.

FIGURE 8-39 Use of screws for attaching copper high-tension cable.

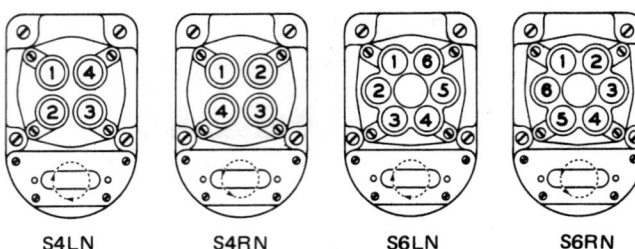

S4LN   S4RN   S6LN   S6RN

FIGURE 8-40   Magneto firing order.

Magneto Outlet	Spark Plug Lead
1	1
2	4
3	5
4	2
5	3
6	6

FIGURE 8-41   Magneto outlets with corresponding spark plug leads for a six-cylinder engine.

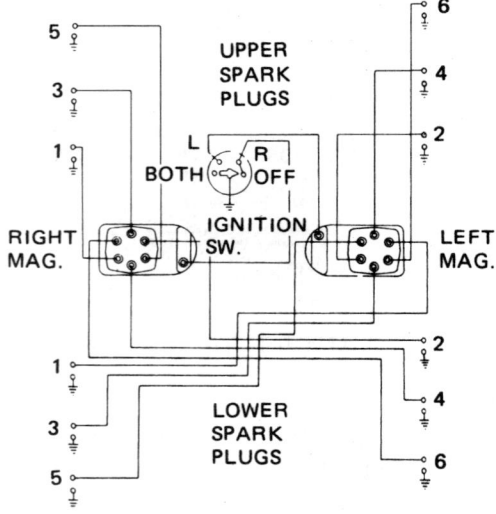

FIGURE 8-42   Wiring diagram for a high-tension magneto system on a six-cylinder opposed engine.

to fire the top spark plugs on the left side of the engine and the bottom spark plugs on the right side of the engine. A circuit diagram for this arrangement is shown in Fig. 8-42.

Maintenance of the Continental Ignition S-200 Magneto

It is recommended that S-200 magnetos be inspected after the first 25 h of operation and every 50 h thereafter. A typical inspection and check are performed as follows:

1. Remove the screws which hold the breaker cover and loosen the cover sufficiently to allow removal of the feed-through capacitor and retard lead terminals from the breakers.

The feed-through capacitor and retard leads will remain in the breaker cover when the cover is removed from the magneto.

2. Examine the breaker contact points for excessive wear or burning. Points which have deep pits or excessively burned areas should be discarded. Examine the cam follower felt for proper lubrication. If the felt is dry, apply two or three drops of an approved lubricant. Blot off any excess oil. Clean the breaker compartment with a clean, dry cloth.

3. Check the depth of the spring contact in the switch and retard terminals. The spring depth from the outlet face should not be more than $\frac{1}{2}$ in [12.7 mm].

4. Visually check the breakers to see that the cam follower is securely riveted to its spring. Check the screw that holds the assembled breaker parts together for tightness.

5. Check the capacitor mounting bracket for cracks or looseness. Test the capacitor for a minimum capacitance of 0.30 microfarad (μF) with a suitable capacitor tester.

6. Remove the harness outlet plate from the magneto, and inspect the rubber grommet and distributor block. If moisture is present, dry the block with a soft, dry, clean, lint-free cloth. Do not use gasoline or any solvent for cleaning the block. The solvent will remove the wax coating and could cause electrical leakage.

7. Reassemble all parts carefully.

The foregoing directions are indicative of the checks and inspections that should be made, especially if there is any sign of magneto trouble. If possible, use the manufacturer's manual to make sure that no important details are omitted.

CONTINENTAL DUAL-MAGNETO IGNITION SYSTEMS

The Continental D-2000 and D-3000 magneto ignition systems were designed to provide dual ignition for aircraft engines with only one magneto. These systems are available for four-, six-, and eight-cylinder engines. A complete system includes the dual magneto, the harness assembly, a starting vibrator (for the D-2200 and D-3200 systems), and an ignition switch. The D-3000 magneto is identical to the D-2000 series with the exception of a few structural changes.

Magneto

The dual magneto consists of a single driveshaft and rotating magnet that supplies the magnetic flux for two electrically independent ignition circuits. Each ignition circuit includes pole shoes, primary winding, secondary winding, primary capacitor, breaker points, and distributor. Figure 8-43 is a photograph of a D-3000 series magneto showing the block and bearing assembly. This system utilizes either a starting vibrator or impulse coupling to provide adequate starting ignition.

The D-3000 series magneto equipped with a starting vibrator has two separate breaker cams mounted on the same shaft. The lower cam operates the main breaker points for both magneto circuits, while the upper cam operates the

FIGURE 8-43 Continental ignition D-3000 magneto. (*Continental Motors Ignition Systems.*)

left magneto retard breaker and the tachometer breaker. The tachometer breaker provides electric impulses to operate the type of tachometer that utilizes such impulses for rpm indication. If the airplane is equipped with any other type of tachometer, the tachometer breaker is not used. The tachometer breaker is located above the right main breaker, and the retard breaker is above the left main breaker. On the D-3000 magneto that employs an impulse coupling, the upper cam is installed only if a tachometer breaker is required.

The primary capacitors are feed-through types and are mounted in the magneto cover, which is part of the harness assembly. When the harness is assembled to the magneto, the capacitor leads are attached to the breaker-point tabs before the cover is installed over the distributor block.

Starting Vibrator

The starting vibrator operates on the same principle as the induction vibrator described earlier. Two types of starting vibrators are used with D-2200 and D-3200 ignition systems, depending on the type of starter and ignition switches used in the system. One type of starting vibrator includes a relay to ground out the right magneto primary circuit during starting. If this were not done, the right magneto would produce an advanced spark and this would cause the engine to kick back. When the ignition switch incorporates the starter switch, the starter vibrator does not require a relay because the combination switch grounds out the right magneto when the switch is in the START position.

Harness

As mentioned previously, the harness assembly includes the magneto cover. Each harness is designed for a particular make and model of engine. The assembly is fully shielded to prevent electromagnetic emanations that would interfere with radio and other electronic equipment. The harness shielding consists of tinned copper braid that is impregnated with a silicone-base material. The harness is designed so that any part can be replaced in the field.

SLICK SERIES 4300 AND 6300 MAGNETOS

The Slick series magnetos manufactured for use with four and six-cylinder opposed engines are quite similar in operation to the Continental magneto systems previously discussed. The Slick 4300 and 6300 magnetos are improved designs over earlier manufactured models. Unlike some earlier models, they can be overhauled in the field according to the manufacturer's instructions.

The 4300 and 6300 magnetos have a common frame and rotor assembly, but the distributor housing, block, and gear differ between the two models simply because the 4300 model is designed for four-cylinder engines and the 6300 model for six-cylinder engines. The parts of a 4300 series magneto are illustrated in Fig. 8-44. The 4300 and 6300 magnetos are pictured in Fig. 8-45.

These magnetos utilize a two-pole magnetic rotor that revolves on two ball bearings located on opposite sides of the rotating magnet. The rotor and bearing assembly is contained within the drive and frame. Bearing preloading is provided by a loading spring, thus eliminating the need for selective shimming. The other components contained within the drive end frame are a high-tension coil that is retained by wedge-shaped keys in the contact breaker assembly which is secured to the inboard bearing plate with two screws. The contact breaker is actuated by a two-lobe cam at the end of the rotor shaft. The cam also serves to key the rotor gear to the shaft.

To provide a retarded spark for engine starting, the series 4300 and 6300 magnetos employ an impulse coupling or retard points.

Installation and Timing Procedure for the Slick 4300/6300 Series

The following is an example of the procedure for installing and timing Slick magnetos on a Textron-Lycoming engine. Always refer to the magneto or aircraft maintenance manual when performing any type of magneto repair or maintenance.

1. Remove the top spark plug from the no. 1 cylinder. Place a thumb over the spark plug hole, and turn the engine crankshaft in the normal direction of rotation until the compression stroke is reached. The compression stroke is indicated by positive pressure inside the cylinder which tends to lift the thumb off the spark plug hole. In this position, both valves of the no. 1 cylinder are closed. Turn the crankshaft opposite to its normal direction of rotation until it is approximately 35° BTC on the compression stroke of the no. 1 cylinder. Rotate the crankshaft in its normal direction

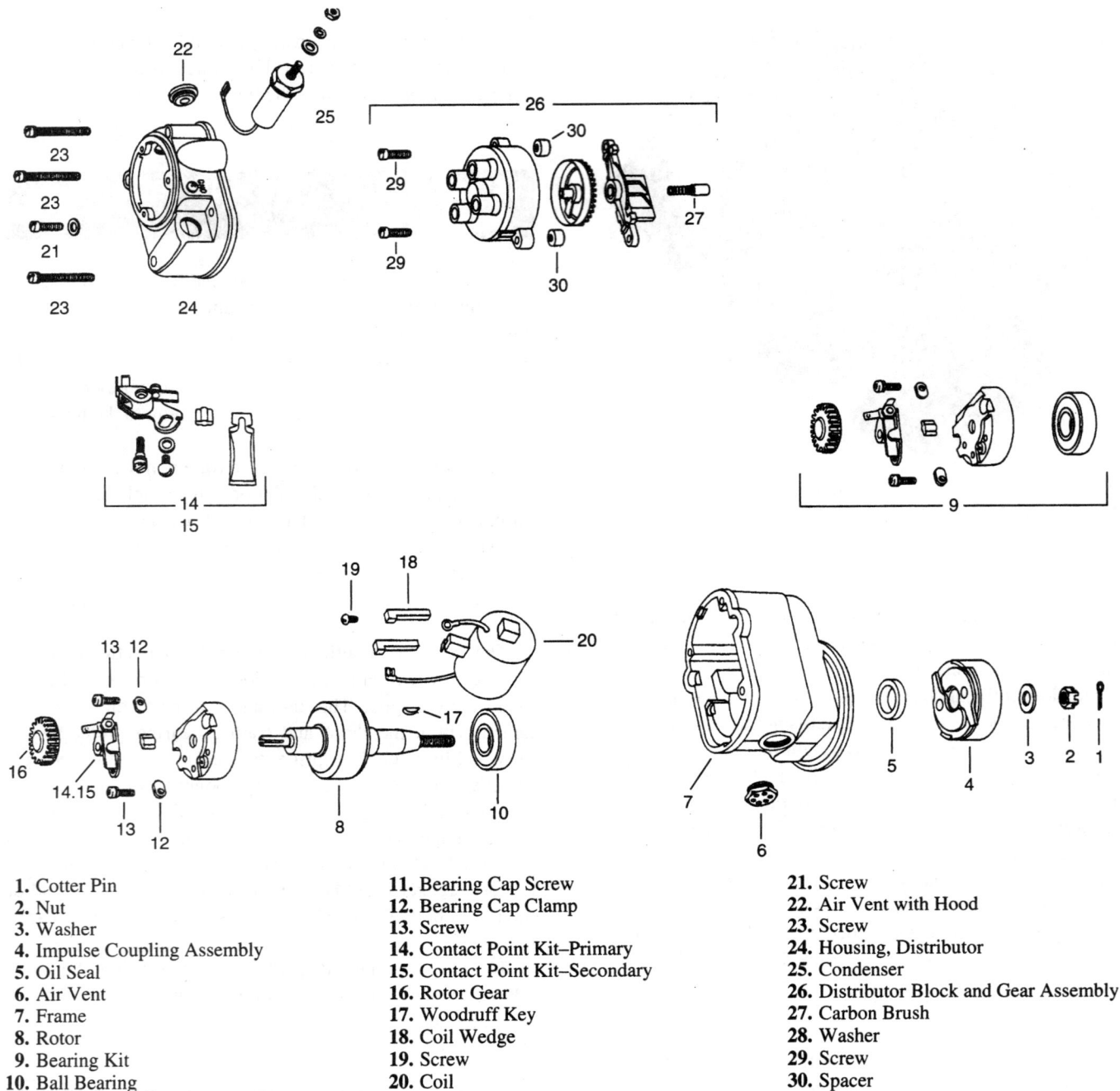

1. Cotter Pin
2. Nut
3. Washer
4. Impulse Coupling Assembly
5. Oil Seal
6. Air Vent
7. Frame
8. Rotor
9. Bearing Kit
10. Ball Bearing

11. Bearing Cap Screw
12. Bearing Cap Clamp
13. Screw
14. Contact Point Kit–Primary
15. Contact Point Kit–Secondary
16. Rotor Gear
17. Woodruff Key
18. Coil Wedge
19. Screw
20. Coil

21. Screw
22. Air Vent with Hood
23. Screw
24. Housing, Distributor
25. Condenser
26. Distributor Block and Gear Assembly
27. Carbon Brush
28. Washer
29. Screw
30. Spacer

FIGURE 8-44 Exploded view of Slick 4300 series magneto. (*Unison Industries.*)

of rotation until the 25° mark on the starter ring gear and the hole in the starter housing align.

2. Insert a **timing pin** in the L or R hole (depending on the rotation of the magneto) in the distributor block, as shown in Fig. 8-46. Turn the rotor opposite the rotation of the magneto until the pin engages the gear. Install the magneto and gasket on the mounting pad of the accessory housing, and *remove the timing pin*. Tighten the mounting nuts fingertight.

3. Connect a standard timing light between engine ground and the left magneto condenser terminal. *The ignition switch must be in the ON position.*

4. Rotate the complete magneto opposite to normal rotation of the magneto on the engine mounting until the timing light indicates that the contact breaker points are just opening. Secure the magneto in this position. Turn the ignition switch off.

5. Turn on the switch of the timing light. Turn the crankshaft very slowly in the direction of normal rotation until the timing mark on the front face of the starter ring gear aligns with the drill hole in the starter housing, at which point the light should come on. This type of timing light comes on when the points in the magneto open. Some timing lights go out when the points open. The type of timing light you

FIGURE 8-45   Slick 4300 and 6300 magnetos. (*Unison Industries.*)

are using should be determined before the timing procedure is begun. If the light does not come on, turn the magneto in its mounting flange slots and repeat the procedure until the light comes on at 25° before TDC. Tighten the two mounting bolts.

6. Connect the other positive wire of the timing light to the right magneto condenser terminal, and time the magneto in the same manner as the left magneto.

7. After both magnetos have been timed, leave the timing-light wires connected and recheck the magneto timing to make sure both magnetos are set to fire together. If the timing is correct, both timing lights will come on simultaneously when the 25° mark on the ring gear aligns with the drill hole in the starter housing. If the magnetos are not timed correctly, readjust the magneto timing until both magnetos fire at the same time and at 25° before TDC. Secure and torque

the bolts using the correct torque from the maintenance manual, remove the timing light, and make sure the ignition switch is off.

When checking the timing, scribe or paint a reference mark on the magneto mounting flange and engine accessory case before moving the magneto. After resetting the timing, check the mark made earlier and measure the distance from the original installed mark on the accessory case. If this dimension is greater than $\frac{1}{8}$ in [3.18 mm], the magneto must be removed and the contact breaker points must be inspected or adjusted; refer to the maintenance manual.

Maintenance Procedures for the Slick 4300 Series

The following information consists of examples of items that should be checked during an inspection of a magneto. Always refer to the maintenance manual when performing this type of inspection.

After 100 h of operation and every 100 h afterward, the magneto-to-engine timing should be checked. Other items should also be checked; refer to the maintenance manual.

At 500-h intervals, the contact-point assemblies should be checked for burning or wear. If the points are not discolored and have a white, frosty surface around the edges, the points are functioning properly and can be reused. Apply cam grease sparingly to each lobe of the cam if needed before reassembly.

If the points are blue (indicating excessive arcing) or pitted, they should be discarded.

At the 500-h inspection, it is necessary to check and replace the carbon brush in the distributor gear if it is worn, cracked, or chipped. The distributor block should also be

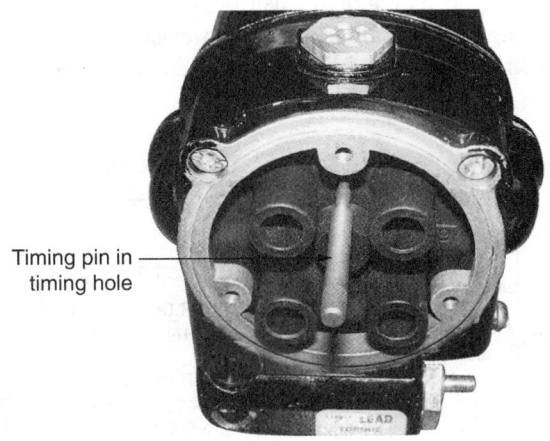

Timing pin in timing hole

FIGURE 8-46   Timing pin installed in Slick 4300 series magneto. (*Unison Industries.*)

checked for cracks and/or signs of carbon tracking and should be replaced if necessary.

Inspect and put a drop of SAE no. 20 nondetergent machine oil in each oilite bearing in the distributor block and bearing bar before reassembly.

Inspect the high-tension lead from the coil to make sure it makes contact with the carbon brush on the distributor gear shaft. Clean residue from the high-tension lead before reassembly, taking care not to scratch the surface of the lead.

At the 500-h inspection, visually inspect the impulse coupling shell and hub for cracks, loose rivets, or rounded flyweights that may slip during latching up on the stop pin. If any of these conditions is evident, the coupling should be replaced.

Pressurized Magnetos

Many magnetos that operate at high altitudes are pressurized by a regulated air source from the aircraft engine. The jumping of high voltage inside the distributor, called *flashover*, can occur, especially when the aircraft is operating at high altitudes. At high altitudes the air is less dense, allowing the high-tension spark to jump to ground more easily. To prevent this, air is pumped into the housing with a controlled bleed of air exiting the magneto at all times. The ventilation air passing through the magneto is necessary for proper venting of heat and other gases produced by the arcing between the distributor and rotor in the magneto. Most pressurized magnetos are gray or dark blue and are used on turbocharged engines.

OTHER HIGH-TENSION MAGNETOS

Numerous types of high-tension magnetos have been designed for use on aircraft engines; however, it is not essential to describe all types. The basic principles are the same for all such magnetos, and it is only necessary to determine how each is timed internally and to the engine. With a good understanding of the principles of operation and timing, the technician can usually adjust any magneto for satisfactory operation. If there is any question concerning a particular magneto and its installation, the manufacturer's manuals for the magneto and the engine should be consulted.

LOW-TENSION IGNITION

Reasons for Development

Several very serious problems are encountered in the production and distribution of the high-voltage electricity used to fire the spark plugs of an aircraft engine. High-voltage electricity causes corrosion of metals and deterioration of insulating materials. Electricity also has a marked tendency to escape from the routes provided for it by the designer of the engine.

There are four principal causes of the troubles encountered in the use of high-voltage ignition systems: (1) flashover, (2) capacitance, (3) moisture, and (4) high-voltage corona.

Flashover is the jumping of the high voltage inside a distributor when an airplane ascends to a high altitude. This occurs because the air is less dense at high altitudes and therefore has less dielectric, or insulating, strength.

Capacitance is the ability of a conductor to store electrons. In the high-tension ignition system, the capacitance of the high-tension leads from the magneto to the spark plugs causes the leads to store a portion of the electric charge until the voltage is built up sufficiently to cause the spark to jump the gap of a spark plug. When the spark has jumped and established a path across the gap, the energy stored in the leads during the rise of voltage is dissipated in heat at the spark plug electrodes. Since this discharge of energy is in the form of a relatively low voltage and high current, it burns the electrodes and shortens the life of the spark plug.

Moisture, wherever it exists, increases conductivity. Thus, it may provide new and unforeseen routes for the escape of high-voltage electricity.

High-voltage corona is a term often used to describe a condition of stress across any insulator (dielectric) exposed to high voltage. When the high voltage is impressed between the conductor of an insulated lead and any metallic mass near the lead, an electrical stress is set up in the insulation. Repeated application of this stress to the insulation will eventually result in insulation failure.

Low-tension ignition systems are designed so that the high voltage necessary to fire the spark plugs is confined to a very small portion of the entire circuit. The greater part of the circuit involves the use of low voltage; therefore, the term **low-tension ignition** is used to describe such a system.

Many of the problems associated with high-tension systems in the past have been overcome by the use of new insulating materials for high-tension leads. Most engine ignition systems today are of the high-tension type because of the high cost and added weight of low-tension systems.

Operation of Low-Tension Ignition System

The low-tension ignition system consists of (1) a low-tension magneto, (2) a carbon brush distributor, and (3) a transformer for each spark plug. Figure 8-47 shows the principal parts of a simple low-tension system. Because only one spark plug appears in this diagram, the distributor is not shown.

During the operation of the low-tension system, surges of electricity are generated in the magneto generator coil.

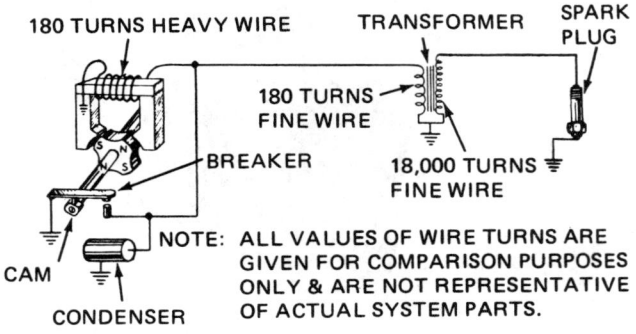

FIGURE 8-47 Low-tension ignition system.

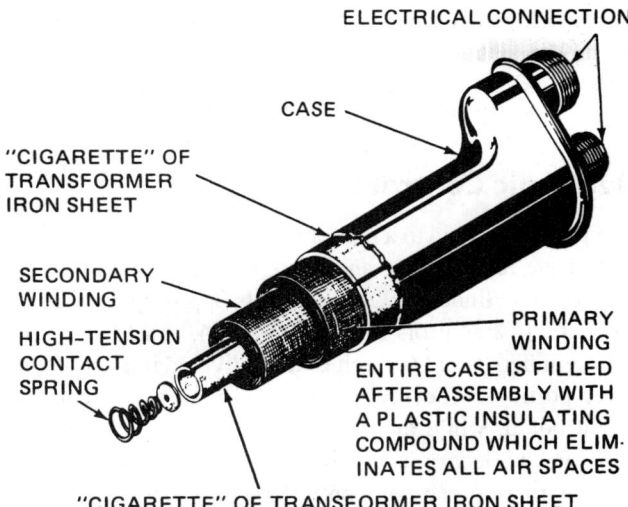

FIGURE 8-48 Transformer coil for a low-tension ignition system.

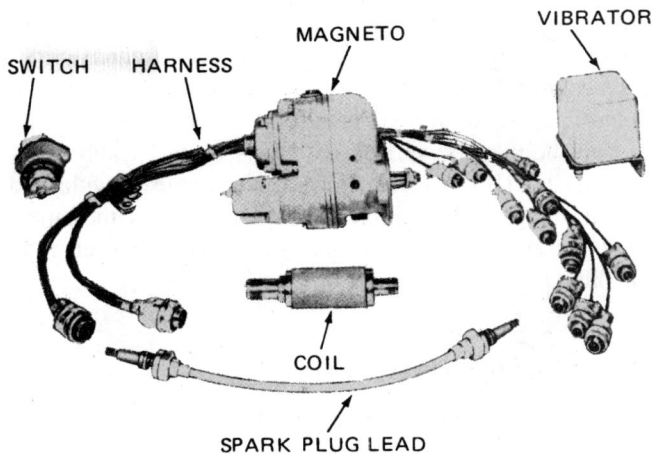

FIGURE 8-49 Components of Bendix S-600 low-tension ignition system. (*Continental Motors.*)

The peak-surge voltage is never in excess of 350 V and probably is nearer 200 V on most installations. This comparatively low voltage is fed through the distributor to the primary of the spark plug transformer.

Transformer Coil

Figure 8-48 is a drawing of a typical low-tension transformer coil "telescoped" (that is, pulled out of its case) to show its design. This coil consists of a primary and a secondary winding with a "cigarette" of transformer iron sheet in the center and another cigarette of transformer iron sheet surrounding the primary winding, which is on the outside of the secondary winding. Usually the transformer unit contains two transformers, one for each spark plug in the cylinder.

The complete transformer assembly provides a compact, lightweight unit convenient for installation on the cylinder head near the spark plugs. This permits the use of short high-tension leads from the transformer to the spark plugs, thus reducing to a large extent the opportunities for leakage of high-tension current.

An advantage of the low-tension system is that the failure of the primary or secondary of one transformer will affect only one spark plug. For example, if the primary winding is short-circuited, one spark plug will stop firing but the engine will continue to operate well. The "dead" coil and spark plug will be detected when the next magneto check is made on ground run-up.

LOW-TENSION IGNITION SYSTEM FOR LIGHT-AIRCRAFT ENGINES

General Description

A low-tension ignition system employed on some older types of engines for light aircraft is the series S-600 developed by the Continental Corporation. The model numbers are S6RN-600 for the dual-breaker magneto and S6RN-604 for the single-breaker magneto. These magnetos are also designed for left-hand rotation.

The components of the S-600 low-tension system are shown in Fig. 8-49. This system is designed for use on a six-cylinder opposed engine. Each installation consists of a retard-breaker magneto, single-breaker magneto, starting vibrator, harness assembly, transformer coils, high-tension leads, and either a combination ignition-starter switch or a standard ignition switch.

This system is designed to generate and distribute low-voltage current through low-tension cables to individual high-voltage transformer coils mounted on the engine crankcase. The low voltage is stepped up to a high voltage by the individual transformer coils and then conducted to the spark plugs by short lengths of high-tension cable. Both the low- and high-tension cables are shielded to prevent radio interference.

FADEC SYSTEM DESCRIPTION

A FADEC is a solid-state digital electronic ignition and electronic sequential port fuel injection system with only one moving part that consists of the opening and closing of the fuel injector. FADEC continuously monitors and controls ignition, timing, and fuel mixture/delivery/injection, and spark ignition as an integrated control system. FADEC monitors engine operating conditions (crankshaft speed, top dead center position, the induction manifold pressure, and the induction air temperature) and then automatically adjusts the fuel-to-air ratio mixture and ignition timing accordingly for any given power setting to attain optimum engine performance. As a result, engines equipped with FADEC require neither magnetos nor manual mixture control.

This microprocessor-based system controls ignition timing for engine starting and varies timing with respect to engine speed and manifold pressure.

PowerLink provides control in both specified operating conditions and fault conditions. The system is designed to prevent adverse changes in power or thrust. In the event of loss of primary aircraft-supplied power, the engine controls continue to operate using a secondary power source (SPS). As a control device, the system performs self-diagnostics to determine overall system status and conveys this information to the pilot by various indicators on the health status annunciator (HSA) panel. PowerLink is able to withstand storage temperature extremes and operate at the same capacity as a non-FADEC-equipped engine in extreme heat, cold, and high humidity environments.

Low-Voltage Harness

The low-voltage harness connects all essential components of the FADEC system.This harness acts as a signal transfer bus interconnecting the electronic control units (ECUs) with aircraft power sources, the ignition switch, speed sensor assembly (SSA), temperature and pressure sensors. The fuel injector coils and all sensors, except the SSA and fuel pressure and manifold pressure sensors, are hardwired to the low-voltage harness. This harness transmits sensor inputs to the ECUs through a 50-pin connector. The harness connects to the engine-mounted pressure sensors via cannon plug connectors. The 25-jpin connectors connect the harness to the speed sensor signal conditioning unit. The low-voltage harness attaches to the cabin harness by a firewall-mounted data port through the same cabin harness/bulkhead connector assembly. The bulkhead connectors also supply the aircraft electrical power required to run the system.

The ECU is at the heart of the system, providing both ignition and fuel injection control to operate the engine with the maximum efficiency realizable. Each ECU contains two microprocessors, referred to as a computer, that control two cylinders. Each computer controls its own assigned cylinder and is capable of providing redundant control for the other computer's cylinder.

The computer constantly monitors the engine speed and timing pulses developed from the camshaft gear as they are detected by the SSA. Knowing the exact engine speed and the timing sequence of the engine, the computers monitor the manifold air pressure and manifold air temperature to calculate air density and determine the mass airflow into the cylinder during the intake stroke. The computers calculate the percentage of engine power based on engine revolutions per minute (rpm) and manifold air pressure.

From this information, the computer can then determine the fuel required for the combustion cycle for either best power or best economy mode of operation. The computer precisely times the injection event, and the duration of the injector should be on time for the correct fuel-to-air ratio. Then, the computer sets the spark ignition event and ignition timing, again based on percentage of power calculation. Exhaust gas temperature is measured after the burn to verify that the fuel-to-air ratio calculations were correct for that combustion event. This process is repeated by each computer for its own assigned cylinder on every combustion/power cycle.

The computers can also vary the amount of fuel to control the fuel-to-air ratio for each individual cylinder to control both cylinder head temperature (CHT) and exhaust gas temperature (EGT).

Electronic Control Unit (ECU)

An ECU is assigned to a pair of engine cylinders. The ECUs control the fuel mixture and spark timing for their respective engine cylinders: ECU 1 controls opposing cylinders 1 and 2, ECU 2 controls cylinders 3 and 4, and ECU 3 controls cylinders 5 and 6. Each ECU is divided into upper and lower portions. The lower portion contains an electronic circuit board, while the upper portion houses the ignition coils. Each electronic control board contains two independent microprocessor controllers that serve as control channels. During engine operation, one control channel is assigned to operate a single engine cylinder. Therefore, one ECU can control two engine cylinders, one control channel per cylinder. The control channels are independent, and there are no shared electronic components within one ECU. They also operate on independent and separate power supplies. However, if one control channel fails, the other control channel in the pair within the same ECU is capable of operating both its assigned cylinder and the other opposing engine cylinder as backup control for fuel injection and ignition timing. Each control channel on the ECU monitors the current operating conditions and operates its cylinder to attain engine operation within specified parameters. The following transmit inputs to the control channels across the low-voltage harness:

1. Speed sensor that monitors engine speed and crank position
2. Fuel pressure sensors
3. Manifold pressure sensors
4. Manifold air temperature (MAT) sensors
5. CHT sensors
6. EGT sensors

All critical sensors are dually redundant with one sensor from each type of pair connected to control channels in different ECUs. Synthetic software default values are also used in the unlikely event that both sensors of a redundant pair fail. The control channel continuously monitors changes in engine speed, manifold pressure, manifold temperature, and fuel pressure based on sensor input relative to operating conditions to determine how much fuel to inject into the intake port of the cylinder.

PowerLink Ignition System

The ignition system consists of the high-voltage coils atop the ECU, the high-voltage harness, and spark plugs. Since there are two spark plugs per cylinder on all engines, a six-cylinder engine has 12 leads and 12 spark plugs. One end of each lead on the high-voltage harness attaches to a spark plug, and the other end of the lead wire attaches to the spark plug towers on each ECU. The spark tower pair is connected to opposite ends of one of the ECU's coil packs. Two coil packs are located in the upper portion of the ECU. Each coil pack generates a high-voltage pulse for two spark plug towers. One tower fires a positive polarity pulse and the other of the same coil fires a negative polarity pulse. Each ECU controls the ignition spark for two engine cylinders. The control channel within each ECU commands one of the two coil packs to control the ignition spark for the engine cylinders. The high-voltage harness carries energy from the ECU spark towers to the spark plugs on the engine.

For both spark plugs in a given cylinder to fire on the compression stroke, both control channels must fire their coil packs. Each coil pack has a spark plug from each of the two cylinders controlled by that ECU unit.

The ignition spark is timed to the engine's crankshaft position. The timing is variable throughout the engine's operating range and is dependent upon the engine load conditions. The spark energy is also varied with respect to the engine load.

NOTE: Engine ignition timing is established by the ECUs and cannot be manually adjusted.

During engine starting, the output of a magneto is low because the cranking speed of the engine is low. This is understandable when the factors that determine the amount of voltage induced in a circuit are considered.

To increase the value of an induced voltage, the strength of the magnetic field must be increased by using a stronger magnet, by increasing the number of turns in the coil, or by increasing the rate of relative motion between the magnet and the conductor.

Since the strength of the rotating magnet and the number of turns in the coil are constant factors in magneto ignition systems, the voltage produced depends upon the speed at which the rotating magnet is turned. When the engine is being cranked for starting, the magnet is rotated at about 80 rpm. Since the value of the induced voltage is so low, a spark may not jump the spark plug gap. To facilitate engine starting, an auxiliary device is connected to the magneto to provide a high ignition voltage.

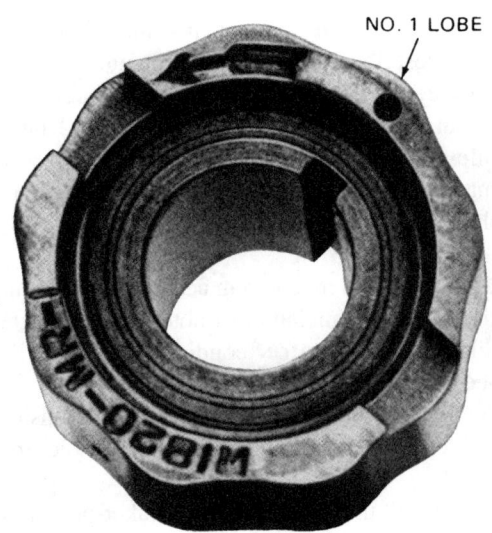

FIGURE 8-50 Compensated cam.

the TDC positions for some pistons and more than 40° for other pistons.

To obtain ignition at precisely 25° BTC, it is necessary to compensate the breaker cam in the magneto by providing a separate cam for each cylinder of the engine.

Design of Compensated Cam

A compensated cam for a nine-cylinder radial engine is shown in Fig. 8-50. The cam lobe for the no. 1 cylinder is marked with a dot, and the direction of rotation is shown by an arrow. A careful inspection of the cam shows slight differences among the distances between the various lobes. This variation is designed into the cam to compensate for the nonuniform movement of the pistons. The variation may be as much as 2.5° more or less than 40° for a nine-cylinder radial engine.

The compensated cam turns at one-half the crankshaft speed because it produces a spark for each cylinder during each complete revolution. Since the crankshaft must rotate through two turns to fire all the pistons, the cam can turn at only one-half crankshaft speed. The compensated cam is normally mounted on the same shaft that drives the distributor because the distributor also can turn at only one-half crankshaft speed.

COMPENSATED CAM

Reason for Compensated Cam

In a radial engine, because of the mounting of the link rods on the flanges of the master rod, the travel of the pistons connected to the link rods is not uniform. Normally we expect the pistons in a nine-cylinder radial engine to reach TC 40° apart. That is, for each 40° the crankshaft turns, another piston reaches TDC. Since the master rod tips from side to side while it is carried around by the crankshaft, the link rods follow an *elliptical path* instead of the circular path required for uniform movement. For this reason there is less than 40° of crankshaft travel between

MAGNETO MAINTENANCE AND INSPECTION

Some magnetos require inspections at regular intervals to ensure serviceability. Most magnetos should be inspected after every 100 h of operation or during the annual inspection. Some magnetos require a fairly detailed inspection after 500 h of service. If components are worn or damaged, they should be replaced. Generally no structural repairs are permissible on magneto components unless the repairs are specifically described in the overhaul manual.

The inspection and maintenance of a magneto are not difficult, but they require careful attention to detail. In any case, follow the instructions given in an approved maintenance manual. A complete inspection of a magneto generally includes the following:

1. Removal of the magneto in accordance with approved instructions. The technician must note that the magneto circuit is likely to be in an ON condition when the P-lead is disconnected.

2. Removal of the distributor blocks, high-tension plate, terminal block, or other part which may hold the high-tension leads.

3. Removal of the cover over the breaker-point assembly.

4. Examination and service of the breaker-point assembly and capacitor. If the breaker points have a smooth and frosty appearance, as shown in Fig. 8-51A, they are in good condition. This means that the points are worn in and mated to each other, thereby providing the best possible electrical contact and highest efficiency of performance. Minor irregularities, such as those illustrated in Fig. 8-51B, and roughness of point surfaces are not harmful. If the points have well-defined pits and mounds, such as in Fig. 8-51C, they should be rejected.

Tungsten oxide may occasionally form on the surfaces of contact breakers. This oxide acts as a **dielectric** and stops all current flow through the closed contacts. A fast and easy remedy is to take a piece of stiff, clean paper, such as a business card, and pull it through the closed contact surfaces. This should remove the oxide; if it does not, the breaker points should be replaced.

The breaker points should be checked for adequate lubrication. A drop or two of engine oil on the *oiler felt* attached to the point assembly will usually restore the lubrication. Care must be taken that no oil gets on the breaker points. Excess oil must be removed, or else it may cause the points to burn black.

Breaker-point spring tension can be checked by hooking a small spring scale to the movable point and applying sufficient force to open the points. The points should not be opened more than $\frac{1}{16}$ in [1.59 mm] because the spring may be weakened. A weak breaker-point spring will cause the points to "float," or fail to close in time to build up the primary field to full strength. This is most likely to occur at high speeds. The breaker-point area is checked for cleanness. With the exception of the oiler felt, all parts should be dry and clean. An approved solvent may be used to clean the metal parts, but the solvent must not get on the oiler felt.

The primary capacitor (condenser) should be tested with a suitable condenser tester. This instrument includes ranges for 0.1 to 0.4 μF, 0.4 to 1.6 μF, and 1.5 to 4.0 μF. The range is selected to accommodate the capacitor being tested by means of the selector and the MFD (microfarad) range switch. The MFD switch selects the correct capacitance range for the capacitor being tested. The unit of capacitance is the microfarad.

The manufacturer's instructions should be followed; however, these steps are generally taken:

a. See that the selector switch is in the OFF position.

b. Plug the instrument power cord into a power receptacle.

c. See that the test leads are not short-circuited or grounded.

d. Move the selector switch to SET, and adjust the instrument for proper setting with the SET knob.

e. Turn the selector switch off, and connect the test leads to the capacitor: one lead to the case and one lead to the insulated lead or terminal of the capacitor.

f. Set the instrument range for the capacitor being tested.

g. Check for capacitance, leakage, and series resistance by rotating the selector switch.

Do not handle the tester leads except when the selector switch is in the OFF position. The high voltage can cause a severe shock.

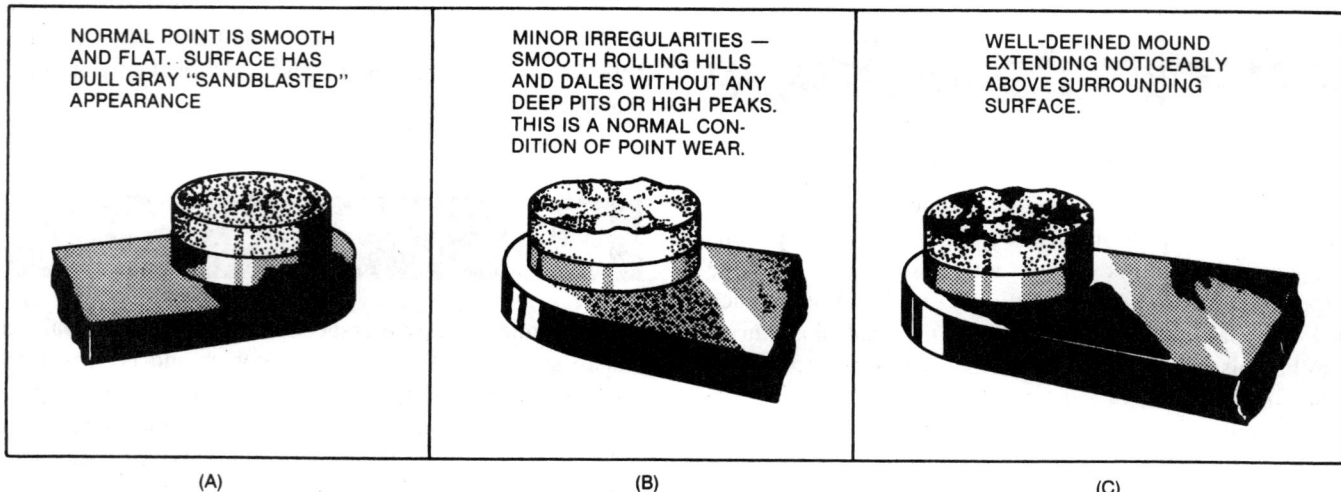

| NORMAL POINT IS SMOOTH AND FLAT. SURFACE HAS DULL GRAY "SANDBLASTED" APPEARANCE | MINOR IRREGULARITIES — SMOOTH ROLLING HILLS AND DALES WITHOUT ANY DEEP PITS OR HIGH PEAKS. THIS IS A NORMAL CONDITION OF POINT WEAR. | WELL-DEFINED MOUND EXTENDING NOTICEABLY ABOVE SURROUNDING SURFACE. |

(A)   (B)   (C)

FIGURE 8-51   Contact points. (A) Normal wear. (B) Serviceable condition. (C) Nonserviceable.

5. Checking of the magneto shaft and gears for excessive play and backlash. If these are beyond specified limits, the magneto must be overhauled.

6. Examination of high-tension parts, such as the distributor rotor, distributor block, and terminal block. These parts may be made of Formica, Bakelite, or another heat-resistant insulating material. These parts are sometimes called dielectric parts because they must have high dielectric strength to withstand the ignition voltages. Defects to look for are cracks, dark lines (carbon tracks) indicating flashover or leakage, and burning, also caused by leakage of high-tension current. Over time, a thin coating of dust may collect in the distributor area. This dust can absorb enough moisture from the air to make it conductive, so that the high-tension current can use the conductive dust as a bridge to ground. The current will often create a burned path referred to as a **carbon track**. This track becomes more conductive as the current flow continues and eventually acts as a short circuit. The hot spark burns the insulating material and dust and releases carbon, which is conductive. If carbon tracks are found, they should be removed if possible. If they cannot be removed, the part should be replaced. All the high-tension parts should be cleaned with recommended solvents and then dried and waxed with a high-temperature wax.

7. Thorough inspection of the magneto for corrosion. Corrosion problems can be attributed to a number of different factors—for instance, water ingestion, operation in salt-air environments (especially on magnesium housings), or unvented magnetos.

A magneto with a clogged vent plug or orifice plug (pressurized magnetos) will form a corrosive atmosphere as a result of electric arcing in the distributor and subsequent corona generation. This corrosive atmosphere, containing nitric acid, is very harmful to the life of internal components. Corrosion or contamination, if left unattended, can result in poor magneto performance with resultant rough engine operation.

8. Checking of internal timing. This is accomplished as explained in the discussion of magneto timing. By use of a timing light, it can be determined that the breaker points open at the E-gap position. If necessary, the points can be adjusted in position to establish the correct opening time. The timing of the distributor can usually be determined by noting the position of marks on the distributor gear with respect to matching marks on the distributor drive gear.

9. Reinstallation of the magneto. When this is done, the procedure for timing the magneto to the engine must be followed.

OVERHAUL OF MAGNETOS

It is not our intent to describe in detail the overhaul of any particular type of magneto, because such instructions can be found in the appropriate manufacturer's manual. We do, however, discuss the general requirements for overhaul of a typical magneto.

Overhaul of the magneto is generally recommended when the engine is overhauled, when operating conditions are unusual (for example, when there is an engine overspeed or sudden stoppage), and after 4 years regardless of how long the magneto has operated since the last overhaul or since its purchase. Magneto overhaul involves at least the following steps: disassembly, cleaning, inspection, repair, replacement, reassembly, and testing. Each step must be accomplished according to specifications in a current overhaul manual. During overhaul, many magneto parts are replaced. Some components are deemed 100 percent replacement parts, and they must be replaced during overhaul. Although which parts are to be replaced will vary somewhat with each magneto manufacturer, here is a general list of items to be replaced at overhaul:

Capacitor	Distributor block and
Ball bearings	gear assembly
Coil	Felt strips and washers
Impulse coupling	Lock washers
Oil seal	Gaskets
Contact-point assembly	Cotter pin
Rotor gear	Self-locking screws

Receiving and Cleaning

When a magneto is received for overhaul, all pertinent information such as make, type, and serial number should be recorded on the work order. In addition, the service record of the magneto should be noted, including the time of operation since its purchase or since the last overhaul.

The magneto should be cleaned thoroughly and disassembled according to instructions in the appropriate overhaul manual. The magnet should be handled carefully and should have a soft-iron keeper of the proper shape placed over the poles to prevent loss of magnetism. Care must be taken to ensure that the magnet is not dropped, jarred, or subjected to excessive heat, all of which can cause loss of magnetism.

It is good practice to place all parts of the magneto in a compartmented tray for protection and convenience of handling.

Inspection

The magnet and magnet shaft are inspected for physical damage and wear. The magnet should then be tested with a magnetometer (Gauss meter) to see that the magnetic strength is adequate for operation. Weak magnets can be returned to the manufacturer for remagnetization or can be remagnetized in the overhaul shop if the proper equipment is available.

The **magnetometer** is a device incorporating soft-iron shoes designed to fit the poles of the magnet. When the magnet is correctly positioned on the shoes, the indicator shows the level of magnetism.

The capacitor (condenser) for the primary circuit is often replaced at major overhaul to ensure maximum operational life. However, it is not usually necessary to replace mica capacitors if a capacitance test and leakage test reveal satisfactory condition.

The capacitance test is accomplished by a **capacity** (capacitance) **tester**, as explained previously. This device applies a carefully regulated alternating current to the capacitor, and the response of the capacitor is indicated in microfarads on an indicating dial. *Care must be observed in using the capacity tester, because the voltage is often at a level which can be injurious or even fatal.*

Leakage, indicating failure of the dielectric, should be tested in accordance with the manufacturer's recommendations. Usually this involves application of a direct current of specified voltage to the capacitor with a milliammeter hooked up in series. The amount of leakage is indicated by the milliammeter. Any appreciable current leakage is cause for rejection.

It is generally recommended that breaker-point assemblies be replaced at major overhaul. This will ensure best performance and maximum life. Worn points and worn cam followers, even though reconditioned, cannot provide the durability and performance of new assemblies. A cam follower worn beyond certain limits will make it impossible to adjust the breaker points for correct operation.

The breaker cam is inspected for wear and condition. If the wear is beyond specified limits, the cam must be replaced. The cam surface must be smooth and free of pits, corrosion, and other surface defects.

The distributor rotor is cleaned in an approved solvent and examined for cracks, carbon tracks, or other signs of failure. The solvent used for cleaning must be of a type which will not damage the finish of the rotor. Usually after inspection and any other processing specified by the manufacturer, the rotor is coated with a high-temperature wax to prevent high-voltage leakage and absorption of moisture.

Shaft bearings and distributor bearings are inspected and serviced just as are the bearings for other engine accessories. Since bearings are usually sealed, it is recommended that new bearings be installed at major overhaul. Overhaul manuals sometimes provide instructions for the reconditioning of sealed bearings.

The coil of a high-tension magneto includes both a primary and a secondary winding. Some coils include the primary capacitor in the coil. The coil should be tested for current leakage between the primary and secondary windings, for continuity of both windings, and for resistance of each winding. Resistance can be checked with an ohmmeter or multimeter—with primary lead to ground for the primary, and high-tension contact to ground for the secondary.

Assembly

After all parts are inspected, tested, and processed in accordance with instructions, the magneto is ready for assembly. The sequence of assembly procedures is determined by the make and type of magneto.

The principal factors in assembly are proper handling of parts to avoid damage; use of proper tools; correct torquing of screws, nuts, and bolts; and strict adherence to instructions relating to assembly and timing.

Testing Magnetos

Upon completion of a service inspection or an overhaul, a magneto should be tested on a **magneto test stand**. This stand includes a variable drive to permit operation of the magneto

FIGURE 8-52   A magneto test stand.

at any desired speed, a tachometer, a spark rack, and suitable controls. A typical magneto test stand is shown in Fig. 8-52.

The test stand is equipped for both base mounting and flange mounting. These mountings are adjustable to permit accurate alignment of the test stand drive to the magneto shaft. The test stand is reversible, so the magneto can be rotated to either the right or the left, depending on the requirement. The operator must take particular care that the magneto is rotated in the correct direction.

After the magneto is mounted on the stand, the high-tension leads are connected from the distributor to the **spark rack**. The spark rack is adjusted for the correct gap specified for the particular magneto being tested. Note that the gap used for the spark rack is much greater than the gap of a spark plug because the current at a particular voltage will jump a much greater distance in unpressurized air than in the compressed air in a cylinder.

The coming-in speed of a magneto is a good indication of the magneto's performance. With the spark gap set at the specified distance, the magneto rpm is slowly increased. When the coming-in speed is attained, there is a steady discharge of sparks at the spark gaps. The magneto speed is increased to the maximum specified rpm to test high-speed performance. Such testing is conducted for specified periods to ensure reliable performance under operating conditions.

If the coming-in speed of a magneto is too high, then the magnet is weak, the internal timing is not correct, the capacitor is defective, or there is some other defect in the magneto. The magnet can be tested with a magnetometer (Gauss meter); if weak, it can be recharged with a magnet charger of proper design. The capacitor can be tested with a capacitor tester, as explained previously.

During testing of a magneto on a test stand, the magneto must not be operated without the high-tension leads connected to the spark rack or without some other means whereby the high-tension current can flow to ground. If the current cannot discharge through normal paths, the voltage will build up to a level which may break down the insulation in the magneto coil and ruin the coil. The gap of the spark rack must not be increased to such a distance that the spark cannot jump, because the high-voltage current will seek another path to ground and this will damage the coil or distributor.

The same damage may occur during operation of the engine in flight if a spark plug lead should break or if high resistance should occur in an ignition lead for any other reason.

SPARK PLUGS

Function

The spark plug is the part of the ignition system in which the electric energy of the high-voltage current produced by the magneto, or other high-tension device, is converted to the heat energy required to ignite the F/A mixture in the engine cylinders. The spark plug provides an air gap across which the high voltage of the ignition system produces a spark to ignite the mixture.

Construction

An aircraft spark plug fundamentally consists of three major parts: (1) the electrodes, (2) the ceramic insulator, and (3) the metal shell.

Figure 8-53 shows the construction features of a typical aircraft spark plug. The assembly shown in the center of the spark plug is the inner electrode assembly consisting of the terminal contact, spring, resistor, brass cap and conductor (neither labeled in the illustration), and the nickel-clad copper electrode. The insulator, shown between the electrode assembly and the shell, is made in two sections. The main section extends from the terminal contact to a point near the electrode tip. The barrel-insulating section extends

from near the top of the shielding barrel far enough to overlap the main insulator.

The outer section of the spark plug illustrated in Fig. 8-53 is a machined-steel shell. The shell is often plated to eliminate corrosion and to reduce the possibility of thread seizure. To prevent the escape of high-pressure gases from the cylinder of the engine through the spark plug assembly, internal pressure seals, such as the cement seal and the glass seal, are used between the outer shell and the insulator and between the insulator and the center electrode assembly.

The shell of the spark plug includes the radio-shielding barrel. In some spark plugs, the shell and shielding barrel are made in two sections and are screwed together. The two parts should never be disassembled by the technician because during manufacture the correct pressure is applied to provide a gastight seal. Any disturbance of the seal may cause leakage.

The shell and the radio-shielding barrel complete the ground circuit for the radio shielding of the ignition harness. The shell is externally threaded on both ends so that it can be joined to the radio shielding of the ignition harness at the top and can be screwed into the cylinder head at the bottom.

Spark plugs are manufactured with many variations in construction to meet the demands of aircraft engines. **Resistor-type spark plugs** are designed to reduce the burning and erosion of electrodes in engines having shielded harnesses. The capacitance between the high-tension cable and the shielding is sufficient to store electric energy in quantities which produce a comparatively high-current discharge at the spark plug electrodes. The energy is considerably greater than is necessary to fire the F/A mixture; therefore, it can be reduced by means of a resistor in order to provide greater spark plug life.

Another improvement which leads to greater dependability and longer life is the use of iridium-alloy firing tips. A spark plug with this type of construction is illustrated in Fig. 8-54.

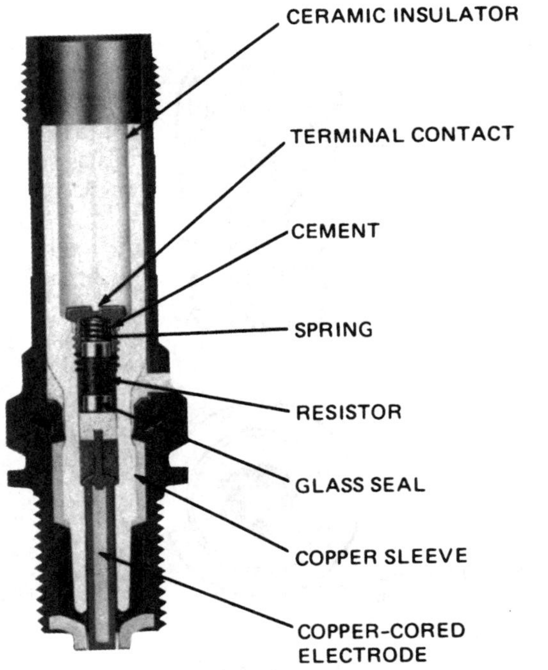

CERAMIC INSULATOR

TERMINAL CONTACT

CEMENT

SPRING

RESISTOR

GLASS SEAL

COPPER SLEEVE

COPPER-CORED ELECTRODE

FIGURE 8-53   Shielded spark plug. (*Champion Spark Plug Co.*)

SILVER-CORED CENTER ELECTRODE

IRIDIUM ELECTRODE

FIGURE 8-54   Spark plug with iridium electrodes. (*Champion Spark Plug Co.*)

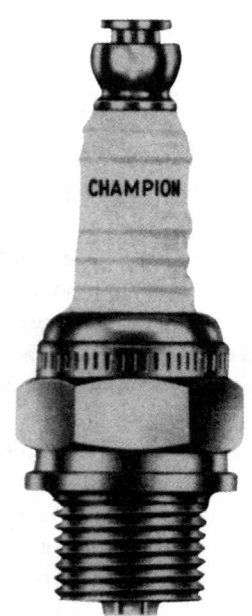

FIGURE 8-55 Unshielded spark plug. (*Champion Spark Plug Co.*)

Unshielded spark plugs are still used in a few light-aircraft engines. An unshielded spark plug is shown in Fig. 8-55.

The construction of spark plugs for aircraft engines is further illustrated in Fig. 8-56. The spark plug on the left is

the massive-electrode type, so named because of the size of the center and ground electrodes. This spark plug is a resistor type that reduces electrode erosion. Nickel seals are provided between the insulator and shell to effectively eliminate gas leakage. The center electrode consists of a copper core with a nickel-alloy sheath. The insulator tip is recessed to maintain the proper temperature to prevent fouling and lead buildup. The three ground electrodes are made of a nickel alloy and are designed to be cleaned with a three-blade vibrator tool. The center electrode is sealed against gas leakage by a metal-glass binder.

The spark plug on the right in Fig. 8-56 is a fine-wire type. It is similar in construction to the massive-electrode plug except for the electrodes. The center electrode is made of platinum, and the two ground electrodes are constructed of either platinum or iridium. The use of platinum and iridium ensures maximum conductivity and minimum wear.

In Fig. 8-57, four typical forms of electrode construction are illustrated: electrodes of the **projected core nose, two-prong fine-wire, two-prong ground**, and **push-wire** types are shown.

The projected core nose spark plugs have been developed for use in engines that have had problems with lead fouling of plugs. While the projected core nose does not necessarily prevent the accumulation of lead deposits, because of its design it is capable of firing despite a severe lead buildup.

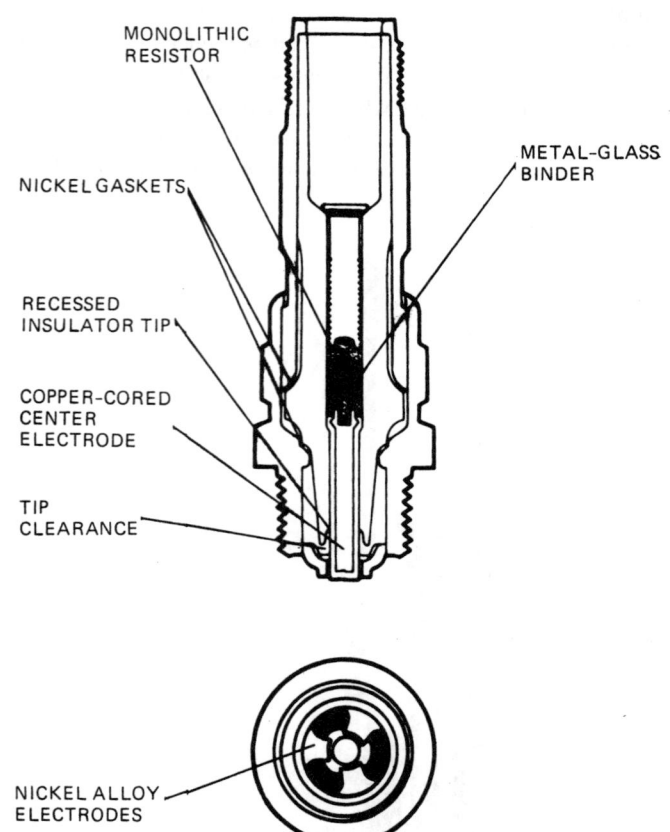

FIGURE 8-56 Massive-electrode and fine-wire spark plugs. (*SGL Auburn Spark Plug Co.*)

PROJECTED CORE NOSE   FINE WIRE   TWO-PRONG "E" GROUND ELECTRODES   PUSH-WIRE 90° TO CENTER

FIGURE 8-57   Typical types of electrode construction. (*Champion Spark Plug Co.*)

Spark Plug Reach

Reach is determined by the linear distance from the shell gasket seat to the end of the shell threads, commonly referred to as the shell skirt. An example of spark plug reach is shown in Fig. 8-58. The reach, or length, required for a given engine is determined by the cylinder-head design. The proper plug reach will ensure that the electrodes are positioned at the most satisfactory depth in the combustion chamber to ignite the fuel.

Classification of Shell Threads

Shell threads of spark plugs are classified as 14- or 18-mm diameter, **long reach** or **short reach**, thus:

Diameter	Long reach	Short reach
14 mm	$\frac{1}{2}$ in [12.7 mm]	$\frac{3}{8}$ in [9.53 mm]
18 mm	$\frac{13}{16}$ in [20.64 mm]	$\frac{1}{2}$ in [12.7 mm]

Terminal threads at the top of the radio-shielding spark plugs are either $\frac{5}{8}$ in [15.88 mm] 24 thread or $\frac{3}{4}$ in [19.05 mm] 20 thread. The latter type is particularly suitable for high-altitude flight and for other situations where flashover within the sleeve might be a problem. Examples of shielded terminal threads are shown in Fig. 8-59.

The designation numbers for spark plugs indicate the characteristics of the plug. The Champion Spark Plug Company utilizes letters and numbers to indicate whether the spark plug contains a resistor and to indicate the barrel style, mounting thread, reach, hexagon size, heat rating range, gap, and electrode style. The designations are as follows:

1. No letter or an R. The R indicates a resistor-type plug.
2. No letter, E, or H. No letter—unshielded; E—shielded $\frac{5}{8}$-in, 24 thread; H—shielded $\frac{3}{4}$-in, 20 thread.
3. Mounting thread, reach, and hexagon size.

a —18 mm, $\frac{13}{16}$-in reach, $\frac{7}{8}$-in stock hexagon
b —18 mm, $\frac{13}{16}$-in reach, $\frac{7}{8}$-in milled hexagon
d —18 mm, $\frac{1}{2}$-in reach, $\frac{7}{8}$-in stock hexagon
j —14 mm, $\frac{3}{8}$-in reach, $\frac{13}{16}$-in stock hexagon
l —14 mm, $\frac{1}{2}$-in reach, $\frac{13}{16}$-in stock hexagon
m —18 mm, $\frac{1}{2}$-in reach, $\frac{7}{8}$-in milled hexagon

4. Heat rating range. Numbers from 26 to 50 indicate coldest to hottest heat range. Numbers from 76 to 99 indicate special-application aviation plugs.
5. Gap and electrode style. E—two-prong aviation; N—four-prong aviation; P—platinum fine wire; B—two-prong massive, tangent to center; R—push wire, 90° to center.

Heat Range of the Spark Plug

The **heat range** of a spark plug is the principal factor governing aircraft performance under various service conditions. The term "heat range" refers to the classification of spark plugs according to their ability to transfer heat from the firing end of the spark plug to the cylinder head.

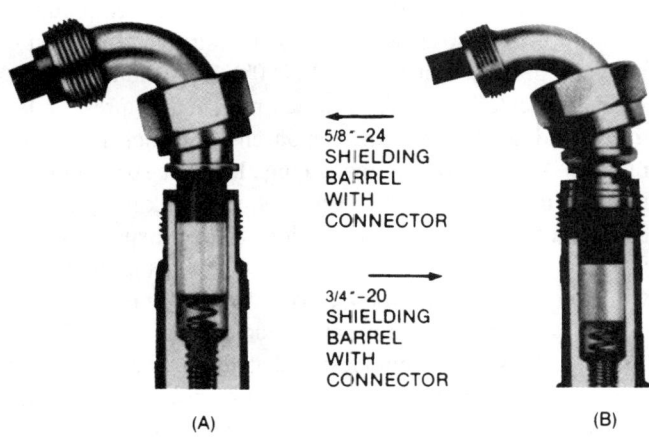

5/8"-24 SHIELDING BARREL WITH CONNECTOR

3/4"-20 SHIELDING BARREL WITH CONNECTOR

(A)   (B)

FIGURE 8-59   Shielded terminal thread designs. (A) $\frac{5}{8}$-in [15.88-mm], 24-thread standard design. (B) $\frac{3}{4}$-in [19.05-mm], 20-thread all-weather design which incorporates an improved seal to prevent entry of moisture.

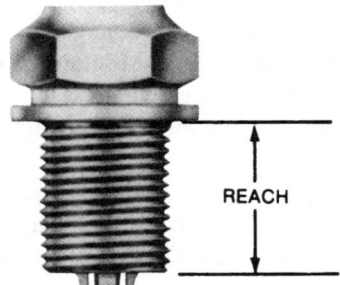

REACH

FIGURE 8-58   Spark plug reach.

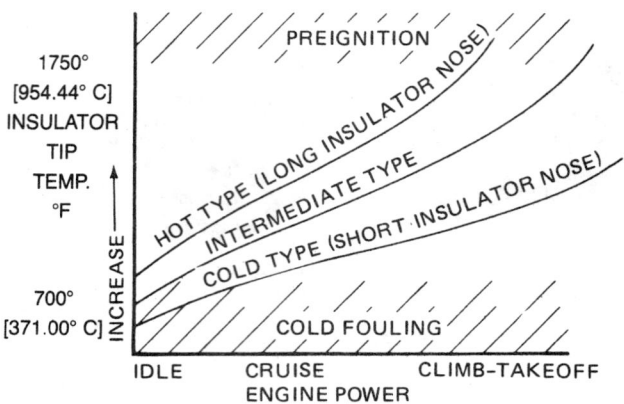

FIGURE 8-60 Chart of spark plug temperature ranges.

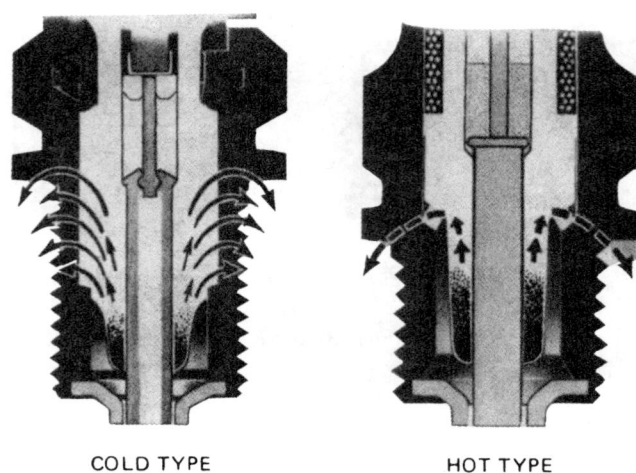

FIGURE 8-61 Construction of hot and cold spark plugs.

Spark plugs have been classified as "hot," "normal," and "cold." However, these terms may be misleading because the heat range varies through many degrees of temperature from extremely hot to extremely cold. Thus the word "hot" or "cold" or "normal" does not necessarily tell the whole story.

Since the insulator is designed to be the hottest part of the spark plug, its temperature can be related to the preignition and fouling regions, as shown in Fig. 8-60. Preignition is likely to occur if surface areas in the combustion chamber exceed critical limits or if the spark plug core nose temperature exceeds 1630°F [888°C]. However, fouling or short-circuiting of the plug due to carbon deposits is likely if the insulator tip temperature drops below approximately 800°F [427°C]. Thus, spark plugs must operate between fairly well-defined temperature limits, and so plugs must be supplied in various heat ranges to meet the requirements of different engines under a variety of operating conditions.

From the engineering standpoint, each individual plug must be designed to offer the widest possible operating range. This means that a given type of spark plug should operate as hot as possible at low speeds and light loads and as cool as possible under cruising and takeoff power. Plug performance therefore depends on the operating temperature of the insulator nose, with the most desirable temperature range falling between 1000 and 1250°F [538 and 677°C].

Fundamentally, an engine which runs hot requires a relatively cold spark plug, whereas an engine which runs cool requires a relatively hot spark plug. If a hot spark plug is installed in an engine which runs hot, the spark plug tip will be overheated and cause preignition. If a cold spark plug is installed in an engine which runs cool, the tip of the spark plug will collect unburned carbon, causing **fouling** of the plug.

A discussion of hot, normal, and cold plugs is technically correct, but we must emphasize that *different heat ranges of aircraft spark plugs cannot be substituted arbitrarily, as is common in automotive practice,* because the selection of aircraft spark plugs is governed by the aircraft engine manufacturers' and the FAA approvals governing the use of a particular spark plug in any aircraft engine.

The principal factors governing the heat range of aircraft spark plugs are (1) the distance between the copper sleeve around the insulator and the insulator tip, (2) the thermal conductivity of the insulating material, (3) the thermal conductivity of the electrode, (4) the rate of heat transfer between the electrode and the insulator, (5) the shape of the insulator tip, (6) the distance between the insulator tip and the shell, and (7) the type of outside gasket used. Hot- and cold-plug construction is illustrated in Fig. 8-61.

Other features of spark plug construction may affect the heat range to some extent. However, the factors just mentioned are those of primary consideration. In all cases, the technician should install the type of spark plug approved for the particular engine being serviced. Spark plugs approved for installation in an engine are listed in the engine Type Certificate Data Sheet. Some manufacturers use a color-code system on the cylinders to indicate the heat range of spark plugs to be installed.

Servicing Aircraft Spark Plugs

Scheduled servicing intervals are normally determined by the individual aircraft operator. These intervals will vary according to operating conditions, engine models, and spark plug types. The principal determining factor in the removal and servicing of spark plugs is the width of the spark gap—that is, the distance between the electrodes where the spark is produced. This spark plug gap widens with use until the distance becomes so great that the spark plug must be removed and either regapped or replaced. If the spark plug gap becomes too wide, a higher secondary voltage must be developed by the ignition system in order to create a spark at the gap. This higher voltage will tend to leak through the insulation of the ignition wiring, eventually causing failure of the high-tension leads.

The correct spark plug gap for a particular spark plug installation is established by the manufacturer. No spark plug should be operated with a gap greater than that specified in the manufacturer's instructions.

Servicing Procedure

In general, aircraft spark plugs are serviced in the following sequence:

1. **Removal.** Shielded terminal connectors are removed by loosening the elbow nut with the proper-size crowfoot or open-end wrench. Care must be taken to avoid damaging the elbow. Terminal sleeve assemblies must be pulled out in a straight line to avoid damaging either the sleeve or the barrel insulator.

The spark plug is loosened from the cylinder bushing by use of the proper-size deep-socket wrench. It is recommended that a six-point wrench be used because it provides a greater bearing surface than a 12-point wrench. The socket must be seated securely on the spark plug hexagon to avoid possible damage to the insulator or connector threads.

As each spark plug is removed, it should be placed in a tray with numbered holes so that the engine cylinder from which the spark plug has been removed can be identified. This is important, because the condition of the spark plug may indicate impending failure of some part of the piston or cylinder assemblies.

If a spark plug is dropped on a hard surface, cracks may occur in the ceramic insulation which are not apparent on visual examination. *Any spark plug which has been dropped should be rejected or returned to the manufacturer for reconditioning.*

When the threaded portion of a spark plug breaks off in the cylinder, great care must be exercised in the removal of the broken section. Normally it will be necessary to remove the cylinder from the engine. Before an attempt is made to remove the section, a liquid penetrant should be applied around the threads and allowed to stand for at least 30 min. If the broken part is tapped lightly with the end of a punch while the penetrant is working, the vibration will help the liquid to enter the space between the threads. The electrodes of the plug should be bent out of the way with a pin punch or similar tool to permit the insertion of a screw extractor (Easyout). A steadily increasing force on the screw extractor should be enough to remove the broken part.

If this process is not successful, then the part must be removed by cutting with a small metal saw. Such a saw can be made by cutting a hacksaw blade to a size which can be inserted inside the spark plug section. The saw must be carefully manipulated to cut three slots inside the section without cutting deeply enough to touch the threads of the cylinder. The three sections of the part can then be broken out by using a punch and a light hammer. The blows should be directed toward the center of the spark plug hole.

After the broken portion of the spark plug has been removed, the threads in the cylinder should be checked carefully for damage. Usually the threads can be cleaned by means of a thread chaser. If there is appreciable damage to the threads, a Heli-Coil insert should be installed. See instructions in Chap. 10 on overhaul practices.

2. **Preliminary inspection.** Immediately after removal, the spark plugs should be given a careful visual inspection, and all unserviceable plugs should be discarded. Spark plugs with cracked insulators, badly eroded electrodes, damaged shells, or damaged threads should be rejected.

3. **Degreasing.** All oil and grease should be removed from both the interior and exterior of the spark plugs according to the degreasing method approved for that particular type of spark plug. Either vapor degreasing or the use of solvents, such as Stoddard solvent, is usually recommended.

4. **Drying.** After they have been degreased, spark plugs should be dried, inside and out, to remove all traces of solvent. Dry compressed air can be used, or the spark plugs can be put in a drying oven.

5. **Cleaning.** During operation of an aircraft engine, lead and carbon deposits form on the ceramic core, the electrodes, and the inside of the spark plug shell. These deposits are most readily removed by an abrasive blasting machine especially designed for cleaning spark plugs.

The use of an abrasive blast spark plug cleaner is shown in Fig. 8-62. Instructions for the use of the cleaning unit follow:

a. Install the proper size rubber adapter in the cleaner, and press the firing end of the spark plug into the adapter hole.

b. Move the control lever to ABRASIVE BLAST, and slowly wobble the spark plug in a circular motion for 3 to 5 s. The wobbling-motion angle should be no greater than 20° from vertical to permit the abrasive materials to freely enter the firing end opening and to facilitate cleaning.

c. Continue the wobbling motion, and move the lever to AIR BLAST to remove the abrasive particles from the firing bore.

d. Remove the plug and examine its firing end. If cleaning is incomplete, repeat the cleaning cycle for 5 to 10 s. If cleaning is still incomplete, check the cleaner and replace the abrasive. For complete service information, refer to the manufacturer's service manual.

FIGURE 8-62 Use of abrasive blast spark plug cleaner.

Several companies offer spark plug cleaning machines with the abrasive supported in liquid. This is an excellent cleaning method, but the operator must make sure that the abrasive is of the **aluminum oxide** type. **Silica abrasive** can contribute to later plug fouling. Immediately after cleaning by the wet-blast method, the plugs should be oven-dried to prevent rusting and to ensure a satisfactory electrical test.

Excessive use of the abrasive blast should be avoided, to prevent too much wear of the electrodes and insulators. Spark plug threads are cleaned by using a wire wheel with soft bristles. Threads which are slightly nicked may be cleaned by using a chasing die.

The connector seat at the top of the shielding barrel must be cleaned to provide a proper seating surface for the gasket and nut at the end of the ignition harness. If necessary, fine-grained garnet paper or sandpaper may be used to smooth the seat. Emery paper should not be used because the emery compound conducts electricity and may establish a path for leakage of high-voltage current. After the seat is cleaned, the shielding barrel should be thoroughly blown out with an air blast.

6. **Regapping.** The tools and methods used to set spark plug gaps will vary with the shape, type, and arrangement of electrodes.

The gap in any spark plug is measured by **round wire gauges**. These gauges are supplied in two sizes for each gap, to measure both the minimum and maximum widths for the gap. For example, a spark plug gap gauge will have two wires, one 0.011 in [0.279 mm] in diameter and the other measuring 0.015 in [0.381 mm]. When the spark plug gap is being tested, the smaller-dimension gauge must pass through the gap and the larger-dimension gauge must be too large to pass through the gap. A gauge for checking the gap clearance of a spark plug and the correct method of insertion are shown in Fig. 8-63.

If a spark plug gap is too large, it is closed by means of a special gap-setting tool supplied by the manufacturer and used according to the manufacturer's instructions. A gap-setting tool is shown in Fig. 8-64.

If the gap of a four- or two-prong spark plug has been closed beyond limits, no effort should be made to open the gap and the plug should be discarded.

Single-electrode and two-prong wire electrode plugs are constructed so that the gap can be opened or closed without danger of cracking the ceramic insulator. A special tool for adjusting the gap of such a plug is shown in Fig. 8-65.

After spark plugs have been correctly gapped, check for erosion of the electrodes. The gauge shown in Fig. 8-66,

FIGURE 8-64   Gap-setting tool. (*Champion Spark Plug Co.*)

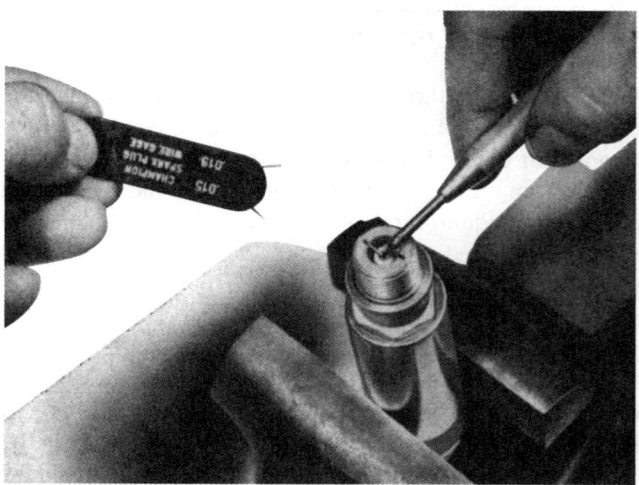

FIGURE 8-65   Setting the gap on a fine-wire electrode spark plug. (*Champion Spark Plug Co.*)

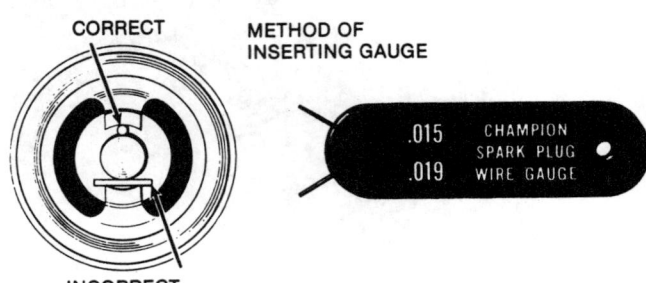

FIGURE 8-63   Spark plug gap gauge. (*Champion Spark Plug Co.*)

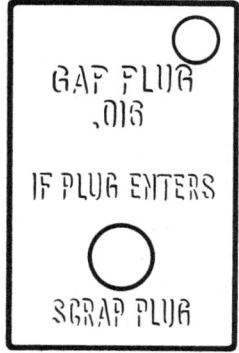
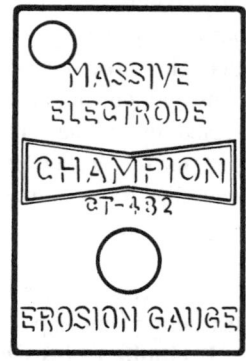

FIGURE 8-66   Spark plug erosion gauge. (*Champion Spark Plug Co.*)

Maintenance Guide
Aviation Spark Plug Resistance

OHM
METER

• **500 - 3000 Ω** = New - Good
• **500 - 5000 Ω** = Used - Good
• **5000 Ω & Up** = No Good - Hard Starting
  - Misfire - Poor Combustion - Pollution

FIGURE 8-67 Inspecting firing end and checking resistance of plug. (*Champion and Tempest Spark Plug Co.*)

which is used with massive-electrode plugs, eliminates the guesswork by identifying spark plugs that require replacement. If the electrode enters the center hole, too much erosion has taken place and the spark plug should be removed from service. Check using an OHM meter, as shown in Fig. 8-67.

7. **Inspection and testing.** The final steps in preparing a used spark plug for service are inspection and testing. Visual inspection is done with a magnifying glass, as shown in Fig. 8-67. Good lighting must be provided. The following items are examined: threads, electrodes, shell hexagons, ceramic insulation, and the connector seat.

Spark plugs are tested by applying high voltage, equivalent to normal ignition voltage, to the spark plug while the plug is under pressure. Testing devices have been designed by spark plug manufacturers to apply the correct pressure and voltage to the spark plug. Spark plugs are tested under pressure to simulate operating conditions to some extent. A spark plug that fires satisfactorily under normal atmospheric conditions may fail under pressure because of the increased resistance of the air gap under these conditions. The testing of a spark plug in a pressure tester is shown in Fig. 8-68. Instructions for operation are included with the test unit.

Spark plugs which fail to function properly during the pressure test should be baked in an oven for about 4 h

at 225°F [107°C]. This will dry out any moisture within the plug. After it has been baked, the plug should be tested again; if it fails under the second test, it should be rejected or returned to the manufacturer.

8. **Gasket servicing.** One of the most important essentials of spark plug installation involves the condition of the solid copper gasket used with the spark plug.

When spark plugs are installed, either new or reconditioned gaskets should be used.

Used spark plug gaskets should be annealed by being heated to a cherry red and immediately quenched in light motor oil. After the quenching, the oil should be removed with a solvent and the gaskets immersed in a solution of 50 percent water and 50 percent nitric acid to remove oxides. After the acid bath, the gaskets should be carefully rinsed in running water and air-dried.

Even though acceptable reconditioned spark plug gaskets may be available, it is recommended that new gaskets be installed with new or reconditioned spark plugs. The additional cost involved is so small that it cannot offset the advantage of new-gasket reliability.

9. **Plug rotation.** Excessive electrode erosion is caused by magneto constant-polarity firing and capacitance afterfiring. Constant polarity occurs with even-numbered-cylinder magnetos. One plug fires with positive polarity, causing excessive ground electrode wear, while the next plug fires negatively, causing excessive center electrode wear. Capacitance afterfiring wear is caused by the stored energy in the ignition shielded lead unloading after normal timed ignition.

To equalize this wear, keep the spark plugs in engine sets, placing them in trays identified by cylinder locations. After the plugs have been serviced, rotate them as indicated in Fig. 8-69. This will correct the polarity wear condition, and the capacitance wear will be corrected by swapping long-lead plugs with short-lead plugs.

10. **Installation.** Before installing spark plugs, the technician must have the proper type of plug for the engine. Two of the most critical factors are the *heat range* and the *reach*. The effect of improper heat range has been described previously. Briefly, if the plug is too hot, preignition and detonation may occur; if the plug is too cold, it will become fouled.

If a long-reach plug is installed in an engine designed for a short-reach plug, some of the threads of the plug will

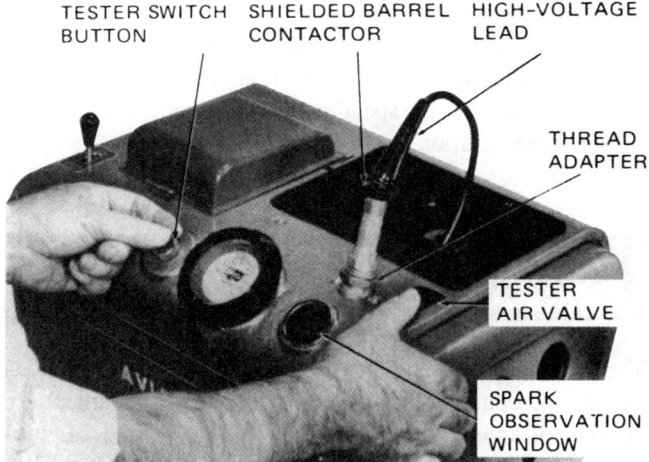

TESTER SWITCH   SHIELDED BARREL   HIGH-VOLTAGE
BUTTON   CONTACTOR   LEAD

THREAD
ADAPTER

TESTER
AIR VALVE

SPARK
OBSERVATION
WINDOW

FIGURE 8-68 Pressure-testing a spark plug. (*Champion Spark Plug Co.*)

Spark Plugs   **231**

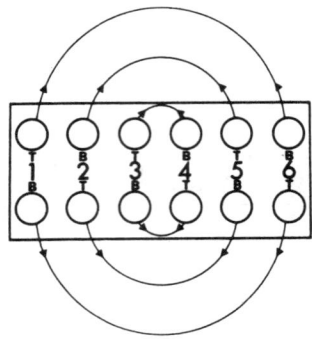 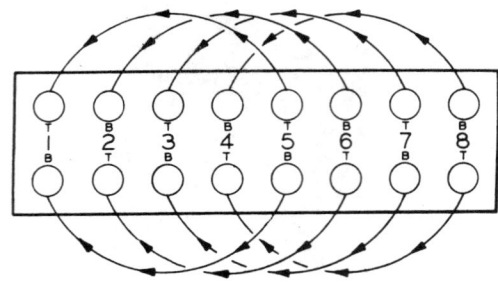

FIGURE 8-69 Spark plug rotation.

extend into the combustion chamber. In this position, the threads and the end of the plug may become overheated and thus cause preignition. This will result in loss of power, overheating, and possible detonation. The threads will be subject to high-temperature corrosion, and this will cause the plug to stick in the cylinder. The sticking tendency will be aggravated by the formation of carbon and lead residue at the end of the plug. When this condition exists, damage may be caused to the cylinder threads and to the spark plug at the time of removal.

Regardless of the care exerted by technicians in carrying out all the steps previously explained, their best efforts are in vain if they fail to install the spark plugs properly. The first step is to inspect the spark plug and the cylinder bushing. The threads of each should be clean and free of damage. A light coating of approved *antiseize compound* should be applied to the threads of the spark plug. It should not, however, be applied to the first two threads because the material may run down onto the electrodes when hot, thus short-circuiting the electrodes of the spark plug.

The spark plug is then installed together with the gasket in the cylinder bushing. It should be possible to screw the spark plug into the cylinder bushing by hand. The spark plug is tightened by means of a deep-socket wrench with a torque handle. It is very important that the spark plug be tightened to the torque specified by the manufacturer. The usual torque for 18-mm spark plugs is 360 to 420 in•lb [40.68 to 47.46 N•m], and the torque for 14-mm plugs is 240 to 300 in•lb [27.12 to 33.9 N•m]. Overtightening a spark plug may damage the threads and make the spark plug difficult to remove or, in extreme cases, change the gap setting. Overtightening will also cause the spark plug to stick in the cylinder. The ignition lead is connected to the spark plug by inserting the terminal sleeve in a straight line into the shielding barrel and tightening the terminal nut to the top of the shielding barrel. The terminal sleeve must be clean and dry and should not be touched by human hands. The entry of foreign matter or moisture into the terminal connector well can reduce the insulation value of the connector such that ignition system voltages at high power can cause flashover of the ignition current from the terminal or well surface to the ground, as shown in Fig. 8-70. This, in turn, can cause the plug to misfire.

The terminal nut at the top of the spark plug should be tightened as much as possible by hand and then turned about

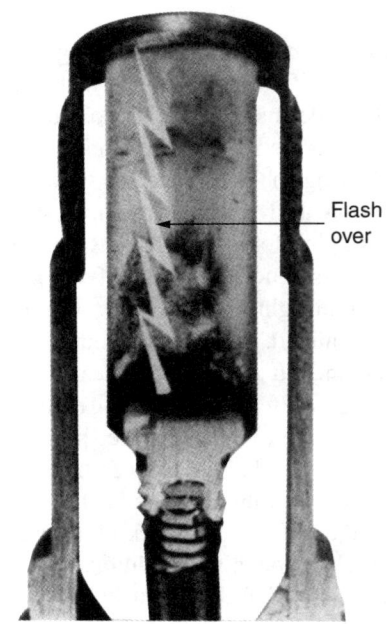

Flash over

FIGURE 8-70 Connector-well flashover. (*Champion Spark Plug Co.*)

one-quarter turn with the wrench. A good, snug fit is all that is required. Overtightening may cause damage to the elbow and the threads.

After a complete set of spark plugs has been installed, it is wise to run the engine and perform a magneto check to assess the general condition of the ignition system. A drop in engine rpm beyond the specified limit will require a check of the magneto, ignition cables, and spark plugs.

STARTERS FOR RECIPROCATING AIRCRAFT ENGINES

The direct-type starter, when electrically energized, provides instant and continuous cranking. This starter fundamentally consists of an electric motor, reduction gears, and an automatic engaging/disengaging mechanism, which is operated through an adjustable-torque overload release clutch. The engine is therefore cranked directly by the starter.

The motor torque is transmitted through the reduction gears to the adjustable-torque overload release clutch, which actuates a helically splined shaft. This, in turn, moves the starter jaw outward, along its axis, and engages the engine-cranking jaw. Then, when the starter jaw is engaged, cranking starts.

When the engine starts to fire, the starter automatically disengages. If the engine stops, the starter automatically engages again if current continues to energize the motor.

The automatic engaging/disengaging mechanism operates through the adjustable-torque overload release clutch, which is a multiple-disk clutch under adjustable spring pressure. When the unit is assembled, the clutch is set for a predetermined torque value. The disks in the clutch slip and absorb the shock caused by the engagement of the starter dogs; they also slip if the engine kicks backward. Since the engagement of the starter dogs is automatic, the starter disengages when the engine speed exceeds the starter speed.

The most prevalent type of starter used for light and medium engines is a series electric motor with an engaging mechanism, such as a Bendix drive or some other means of engaging and disengaging the starter gear from the engine gear. In all cases, the gear arrangement is such that there is a high gear ratio between the starter motor and the engine. That is, the starter motor turns at many times the rpm of the engine.

Starter Motor

A typical direct-cranking motor is illustrated in Fig. 8-71. The armature windings are made of heavy copper wire capable of withstanding very high amperage. The windings are insulated with special heat-resistant enamel, and after they are placed in the armature, the entire assembly is doubly impregnated with special insulated varnish. The leads from the armature coils are staked in place in the commutator bar and then soldered with high-melting-point solder. An armature constructed in this manner can withstand severe loads imposed for brief intervals during engine starting.

The field frame assembly is of cast-steel construction with the four field poles held in place by countersunk screws, which are threaded into the pole pieces. Since a motor of this type is series-wound, the field coils must be wound with heavy copper wire to carry the high starting current.

The end frames are of cast aluminum and are attached to the field frames by screws threaded into one end. The ball bearings are sealed and are pressed into recesses in the end frames. An oil seal is placed in the drive end frame to prevent engine oil from entering the motor.

The brush assemblies are attached to the commutator end frame. The brushes are not held in brush holders in the particular motor illustrated in Fig. 8-71 but are attached with screws to a pivoted brush arm. This arm is provided with a coil spring which holds the brush firmly against the commutator.

The field frame is slotted at the commutator end to provide access to the brushes for service and replacement. A cover band closes the slots to protect the motor from dirt and moisture.

The positive terminal extends through the field frame. It is insulated from the frame by a composition sleeve and washers of similar material. The negative side of the power supply comes through the field frame.

Overrunning Clutch

To engage and disengage a starter from an engine, it is necessary to employ some type of engaging mechanism. Various types of mechanisms have been designed for light-engine starters, one of which is called the **overrunning clutch**.

A typical overrunning-clutch arrangement is shown in Fig. 8-72. The clutch consists of an inner collar; a series of rollers, plungers, and springs enclosed in a shell assembly; and the shaft on which the assembly is mounted. The hardened-steel rollers are assembled into notches in the shell. The notches taper inward so that the rollers seize the collar when the shell

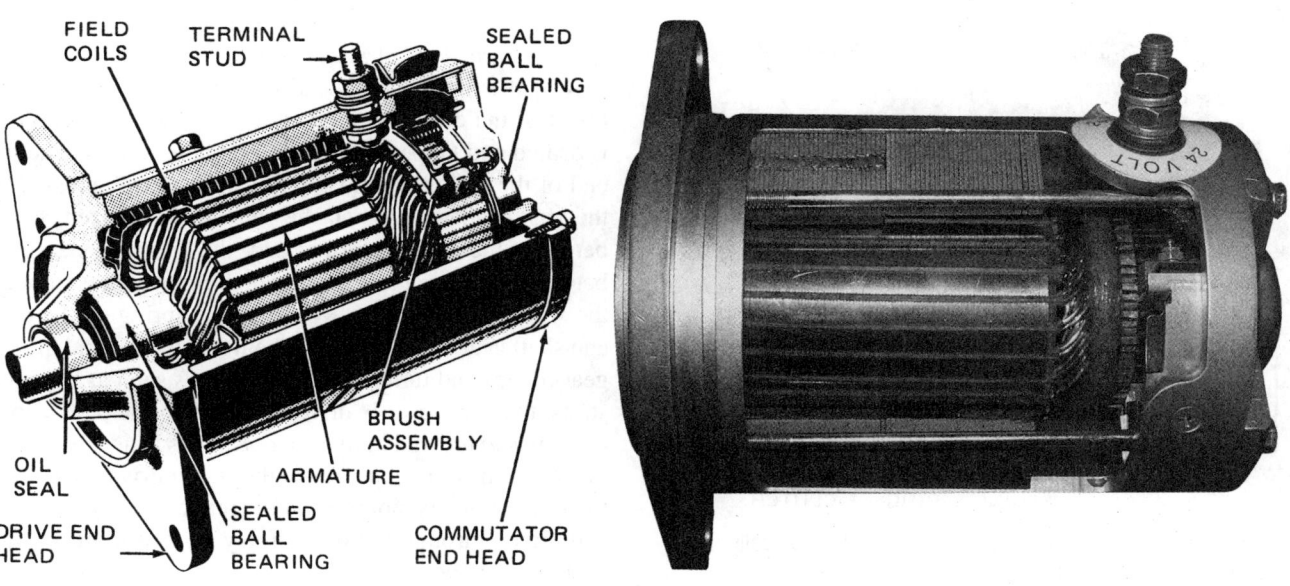

FIGURE 8-71 Direct-cranking starter motor cutaway.

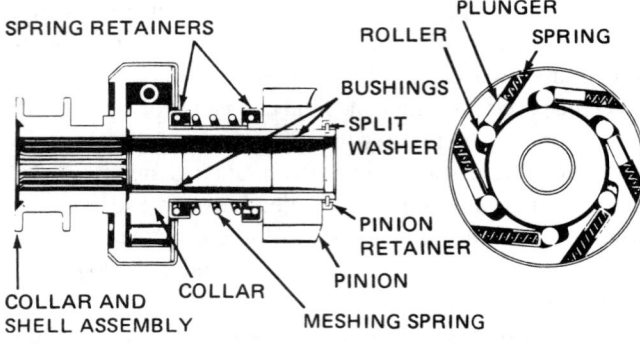

FIGURE 8-72 Overrunning clutch.

is turned in such a direction that the rollers are moved toward the small end of the notches. Thus, when the shell is turned in one direction, the collar must turn with it; however, when the shell is turned in the opposite direction, the collar can remain stationary.

Complete Starter Assembly

A starter assembly employing a manually operated switch and a shift lever is shown in Fig. 8-73. This starter assembly makes use of an overrunning clutch similar to that described above, and it incorporates a pair of gears so that there is a gear reduction between the motor armature and the drive pinion. This gear reduction provides an increase in cranking torque at the drive pinion.

The shift lever is operated by means of a cable or wire control from the airplane cockpit. The control has a return

spring with sufficient tension to bring the lever to the fully released position when the control is released.

When the control is operated to start the engine, the lower end of the shift lever thrusts against the clutch assembly and causes the overrunning-clutch drive pinion to move into mesh with the engine starter gear. If the drive pinion and starter gear teeth butt against each other instead of meshing, the **meshing spring** inside the clutch sleeve compresses to spring-load the drive pinion against the starter gear. Then, as soon as the armature begins to turn, engagement and cranking take place. In the unit illustrated, the overrunning-clutch drive pinion is supported on a stub shaft, which is part of the engine. As the drive pinion is moved into mesh, the stub shaft causes the **demeshing spring** inside the sleeve to be compressed. The demeshing spring produces demeshing whenever the shift lever is released. After the engine starts, overrunning-clutch action permits the pinion to overrun the clutch and gear during the brief period that the pinion remains in mesh. Thus, high speed is not transmitted to the cranking motor armature.

Starter Assembly with 90° Adapter Drive

The complete starter assembly, adapter, and clutch assembly for a six-cylinder opposed engine can be seen in the exploded view in Fig. 8-74. The starter is mounted on a right-angle drive adapter which is attached to the rear end of the crankcase. The tongue end of the starter shaft mates directly with the grooved end of the worm shaft. The worm shaft (43) is supported between a needle bearing (44) at its left end and a ball bearing (41) which is retained in the adapter by a snap ring. The worm (39) is driven by the shaft through a Woodruff key. The worm wheel (34) is attached by four bolts to a flange on the clutch drum (36) which bears on the shaft gear (35). Two dowels center the wheel on the drum and transmit the driving torque. A heavy helical spring (30) covers both the externally grooved drum and a similarly grooved drum machined on the shaft gear just ahead of the clutch drum. The spring is retained on the clutch drum by an in-turned offset at its rear end which rides in a groove around the drum, just ahead of the flange. The in-turned offset of the clutch spring is notched, and the clutch drum is drilled and tapped for a spring retaining screw. The front end of the spring fits closely in a steel sleeve, pressed into the starter adapter. When the starter is energized, friction between the clutch spring and the adapter sleeve and between the spring and the clutch drum, which is turned by the worm wheel, tends to wind up the spring on the clutch and shaft gear drums, locking them together so that the shaft gear rotates and turns the crankshaft. As soon as the engine starts, the shaft gear is driven faster than the clutch spring and tends to unwind it, thus increasing its inside diameter so that the shaft gear spins free of the starter drive. The generator drive pulley is mounted on the rear end of the shaft gear and driven through a Woodruff key, so that it always turns at shaft gear speed.

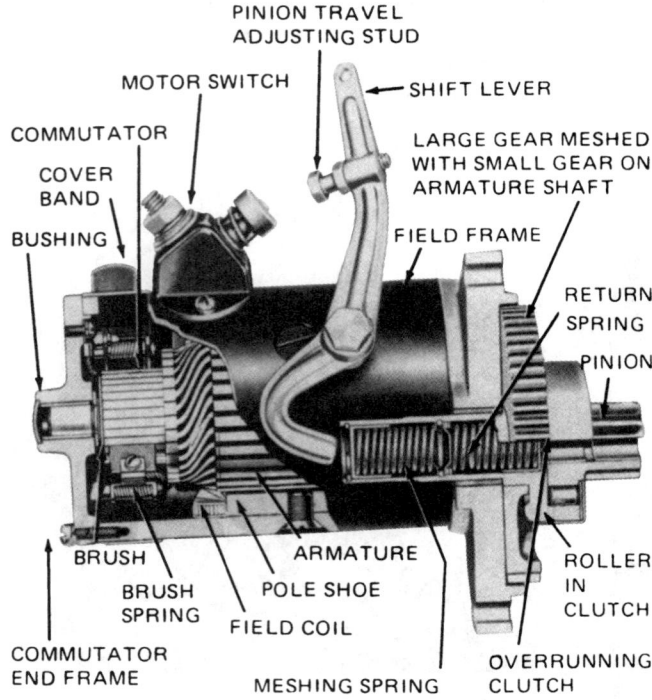

FIGURE 8-73 Starter with overrunning clutch, manually operated switch, and shift lever.

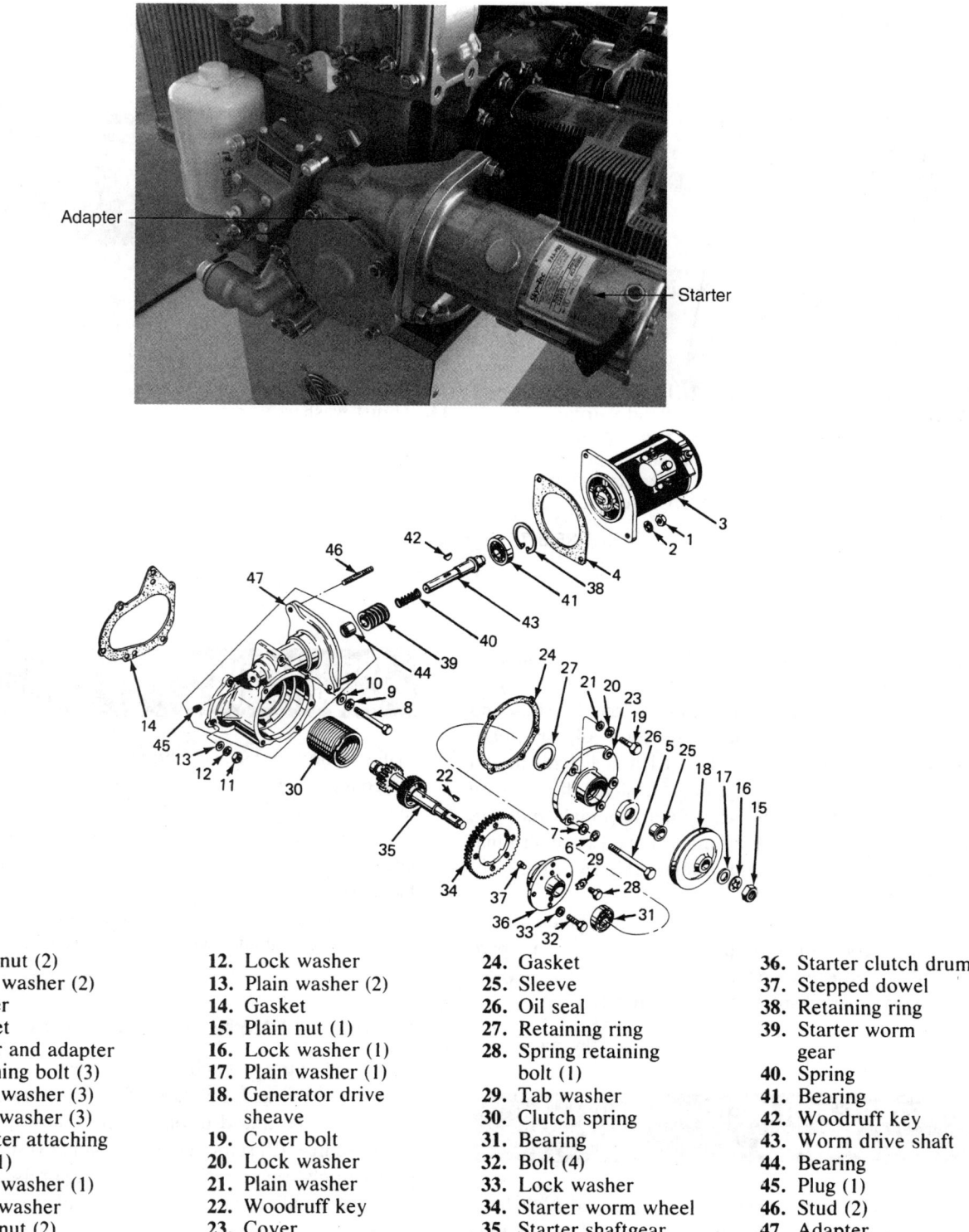

Adapter

Starter

1. Plain nut (2)	**12.** Lock washer	**24.** Gasket	**36.** Starter clutch drum
2. Lock washer (2)	**13.** Plain washer (2)	**25.** Sleeve	**37.** Stepped dowel
3. Starter	**14.** Gasket	**26.** Oil seal	**38.** Retaining ring
4. Gasket	**15.** Plain nut (1)	**27.** Retaining ring	**39.** Starter worm
5. Cover and adapter	**16.** Lock washer (1)	**28.** Spring retaining	gear
attaching bolt (3)	**17.** Plain washer (1)	bolt (1)	**40.** Spring
6. Lock washer (3)	**18.** Generator drive	**29.** Tab washer	**41.** Bearing
7. Plain washer (3)	sheave	**30.** Clutch spring	**42.** Woodruff key
8. Adapter attaching	**19.** Cover bolt	**31.** Bearing	**43.** Worm drive shaft
bolt (1)	**20.** Lock washer	**32.** Bolt (4)	**44.** Bearing
9. Lock washer (1)	**21.** Plain washer	**33.** Lock washer	**45.** Plug (1)
10. Plain washer	**22.** Woodruff key	**34.** Starter worm wheel	**46.** Stud (2)
11. Plain nut (2)	**23.** Cover	**35.** Starter shaftgear	**47.** Adapter

FIGURE 8-74   Direct-cranking starter with spring-type clutch and 90° drive. (*Continental Motors.*)

Starter with Bendix Drive

A typical Prestolite starter motor, used on many light-aircraft engines now in operation, is shown in Fig. 8-75. The Prestolite starter motor is quite similar to those described previously. An examination of Fig. 8-75 will show that the motor includes the conventional parts for a series motor—the end frames (end heads), field frame, field coils, brushes and brush plate assembly, armature, bearings, and assembly parts. This particular starter has the electric starter switch (starter relay) mounted on the field frame.

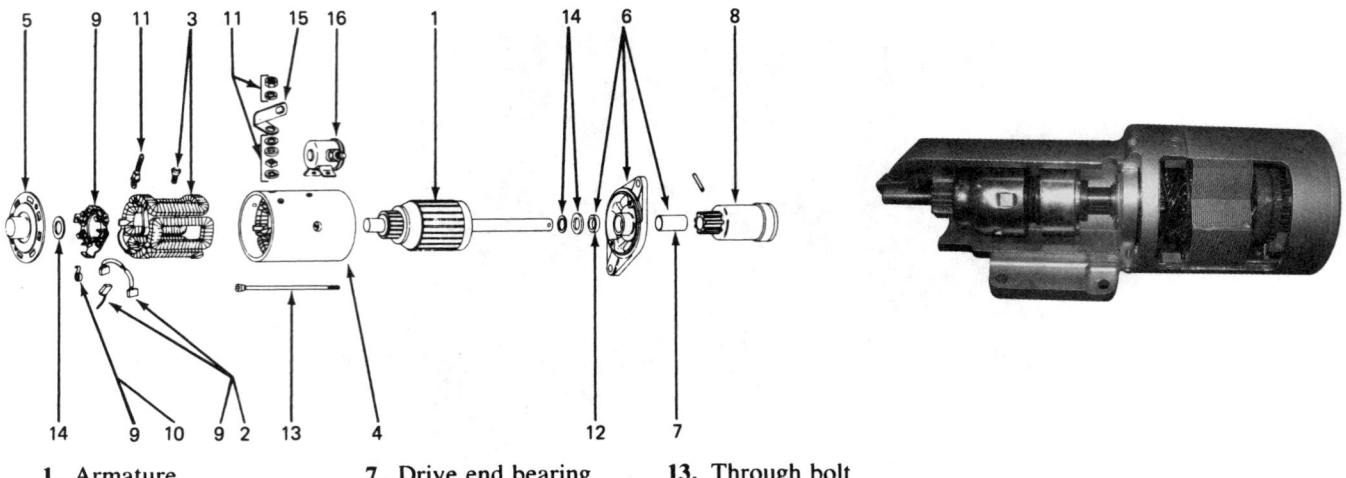

1. Armature
2. Brush set
3. Field coils
4. Field frame
5. Commutator end head
6. Drive end head
7. Drive end bearing
8. Bendix drive
9. Brush plate assembly
10. Brush spring
11. Terminal stud
12. Oil seal
13. Through bolt
14. Thrust washers
15. Connector
16. Starter relay

FIGURE 8-75 Exploded view of a starter.

The maintenance of the starter motor illustrated in Fig. 8-75 is the same as for any other motor of this type. Inspections include an examination of the brushes and commutator, a check for the presence of oil or grease inside the motor and the presence of lead particles in and near the brush plate assembly, and a test for the security of all electric connections. The main terminal studs are examined for tightness in their mountings. Loose terminal studs can result in arcing and ultimate failure of the electric connection.

During replacement of any starter, the technician must determine the correct part number for the replacement part. Part numbers are specified in the manufacturer's overhaul manual and the parts list for the aircraft involved.

STARTERS FOR MEDIUM AND LARGE ENGINES

Early-type radial engines were often equipped with hand inertia starters or electric hand inertia starters. These starters were effective. However, some inconvenience was associated with their use because of the necessity of accelerating the starter and then engaging it to the engine for the start. If the energy stored in the flywheel was dissipated before the engine started, then it was necessary to accelerate again. The inertia starter was therefore more complex in system, construction, and operation than a direct-cranking electric starter. However, designers and manufacturers eventually developed direct-cranking starters which were convenient and effective in operation.

TROUBLESHOOTING AND MAINTENANCE

In this section, the usual troubles found in the operation of typical, conventional starters installed on airplanes with reciprocating engines are discussed. Although the instructions are broad in their application, they cover the situations normally encountered. The study of manuals and handbooks issued by manufacturers and those responsible for inspection and maintenance is always recommended so that the equipment can be serviced in accordance with its own particular design characteristics.

Failure of Starter Motor to Operate

When the starter switch on an airplane is placed in the START position and the starter fails to operate, the trouble can usually be traced to one of the following: (1) electric power source, (2) starter control switch, (3) starter solenoid, (4) electric wiring, or (5) the starter motor itself. A check of these items will usually reveal the trouble.

The electric power source for a light airplane is the battery. If the starter fails to operate or if a click is heard, the battery charge may be low. This can be quickly checked by means of a hydrometer. A fully charged battery will give a reading of 1.275 to 1.300 on the hydrometer scale. When the battery is low, the starter solenoid may click or chatter and the starter motor will fail to turn or will turn very slowly. The solution to this problem is, of course, to provide a fully charged battery. In many cases the technician will merely connect a fully charged battery externally in parallel with the battery in the airplane. When this is done, the external battery must have the same voltage as the battery of the airplane and the terminals must be connected positive to positive and negative to negative.

If the battery of the airplane is found to be fully charged, the fault may lie in the control switch. This can be checked by connecting a jumper across the switch terminals. If the starter operates when this is done, the switch is defective and must be replaced.

If the starter control switch in the cockpit appears to be functional, the trouble may be in the starter solenoid. The solenoid can be checked in the manner described for the starter control switch. In this case, a heavy jumper must be used because the current flow across the main power circuit of the solenoid is much greater than that in the control circuit. If the starter operates when the jumper is connected across the main solenoid terminals, it is a sign that the solenoid is defective. Before the solenoid is replaced, however, the solenoid control circuit should be thoroughly checked. The circuit breaker or fuse in the control circuit may be failing, or the wiring may be defective.

If the power source, wiring, control switch, and starter solenoid are all functioning properly, the trouble lies in the starter motor. In this case, the band covering the brushes of the motor should be removed and the condition of the brushes and commutator checked. If the brushes are badly worn or the spring tension is too weak, the brushes will not make satisfactory contact with the commutator. This trouble can be corrected by replacing the brushes and/or the brush springs as required. If the commutator is black, dirty, or badly worn, it should be cleaned with no. 000 sandpaper or the starter should be removed for overhaul. Usually it is not necessary to overhaul a starter between engine overhauls.

Failure of Starter to Engage

If the starter motor turns but does not turn the engine when one attempts to start an engine, the overrunning clutch, disk clutch, or engaging mechanism has failed. Then it is necessary to remove the starter and correct the trouble. If the starter is engaged by a Bendix-type mechanism in which the engaging gear is moved into mesh by means of a heavy spiral thread on the screw shaft, then cold oil or grease on the thread will cause the gear to stick, thus preventing the gear from meshing. When this occurs, the problem can be solved by cleaning the spiral shaft and applying light oil for lubrication.

1. At what point in the four-stroke cycle does the ignition event take place?

2. Why is magneto ignition superior to battery ignition?

3. How may magnetos be classified?

4. Explain the difference between low-tension and high-tension magnetos.

5. Name the three circuits of a magneto.

6. Describe the difference between the primary and secondary winding in a coil.

7. What is the advantage of a pivotless breaker assembly?

8. Explain the function of the primary capacitor condenser in a magneto.

9. Define the term *E gap*.

10. Explain the events that occur at E gap.

11. Describe the function of the distributor rotor.

12. What are the advantages of a dual-magneto ignition?

13. Explain the term *coming-in speed*.

14. Explain the function of ignition shielding.

15. If the P-lead from the magneto to the ignition switch breaks or is disconnected, what effect will this have on operation of the system?

16. What is the function of an impulse coupling?

17. Describe the purpose of an induction vibrator.

18. During timing of a magneto to an engine, what device may be used to determine the exact location of point opening?

19. What problems associated with high-voltage ignition systems are reduced through the use of a low-tension ignition system?

20. Explain the reason for the use of a compensated cam.

21. Explain the purpose of spark plugs.

22. What are the major parts of a spark plug?

23. What is the purpose of the resistor in a spark plug?

24. Define *spark plug reach*.

25. What is meant by the heat range of a spark plug?

26. What is the purpose of spark plug rotation?

27. What is meant by the term *direct-type starter*?

28. What is the purpose of reduction gearing in a direct-cranking starter?

Operation, Inspection, Maintenance, and Troubleshooting of Reciprocating Engines 9

RECIPROCATING-ENGINE OPERATION

Starting Reciprocating Engines

The starting of an engine can be a relatively simple matter, or it can be a very complex and critical operation, depending on the size and type of engine. The following procedures are typical of those used to start reciprocating engines. There are, however, wide variations in procedures used for many reciprocating engines. **No attempt should be made to use the methods presented here for actually starting an engine.** Instead, always refer to the procedures contained in the applicable manufacturer's instructions.

Engine-Starting Precautions

Although the starting of an aircraft engine is a relatively simple procedure, certain precautions should be taken to obtain the best results and to avoid damage to the engine and injury to personnel.

Aircraft service personnel should acquire the following **safety habits**:

1. Treat all propellers as though the ignition switches were on.
2. Chock airplane or test stand wheels before working around the engine.
3. After an engine run and before the engine is shut down, perform an ignition switch test to detect a faulty ignition circuit.
4. Before moving a propeller or connecting an external power source to an aircraft, be sure that the aircraft is chocked, the ignition switches are in the OFF position, the throttle is closed, the mixture is in the IDLE CUTOFF position, and all equipment and personnel are clear of the propeller or rotor. Faulty diodes in aircraft electric systems have caused starters to engage when external power was applied, regardless of the switch position.

5. Remember, when you are removing an external power source from an aircraft, keep the equipment and yourself clear of the propeller or rotor.
6. Always stand clear of the rotor and propeller blade path, especially when you are moving the propeller. Be particularly cautious around warm engines.
7. The ground or pavement near the propeller should be checked for loose items which might be drawn into an operating propeller.

Ground support personnel who are in the vicinity of aircraft that are being run up need to wear proper eye and ear protection. Ground personnel must also exercise extreme caution in their movements about the ramp; a great number of very serious accidents have involved personnel in the area of an operating engine.

Ground Engine Fire

If an **engine fire** occurs while the engine is being started, move the fuel shutoff lever to the OFF position. Continue cranking or motoring the engine until the fire has been expelled from the engine. If the fire persists, **carbon dioxide** (CO_2) can be discharged into the inlet duct while the engine is being cranked. Do not discharge CO_2 directly into the engine exhaust, because it may damage the engine. If the fire cannot be extinguished, secure all switches and leave the aircraft.

If the fire is on the ground under the engine overboard drain, discharge the CO_2 on the ground rather than on the engine. This also is true if the fire is at the tailpipe and the fuel is dripping to the ground and burning.

Starting Procedures

Engine-starting procedures will vary for different fuel metering devices. Starting procedures for float carburetors, Bendix fuel injection, and Continental continuous-flow injection systems are described for familiarization purposes only.

Specific starting procedures are set forth in the operator's manual. It is extremely important that the operator be thoroughly familiar with the cockpit switches and the engine controls before attempting to start the engine. Figure 9-1 illustrates the standard knob shapes for the common engine controls, and a typical control console for a light twin-engine aircraft. It is important to remember, however, that these standard control knob shapes may not be found on many older aircraft.

Starting Procedure for Float Carburetors

1. Set the master switch to ON.
2. Turn on the boost pump if needed.
3. Open the throttle approximately $\frac{1}{2}$ in [1.27 cm].
4. If the engine is equipped with a constant-speed propeller, the propeller control should be set in the FULL INCREASE position.
5. Set the carburetor heat lever to the COLD (off) position.
6. Set the mixture lever in the FULL RICH position.
7. Clear the propeller.

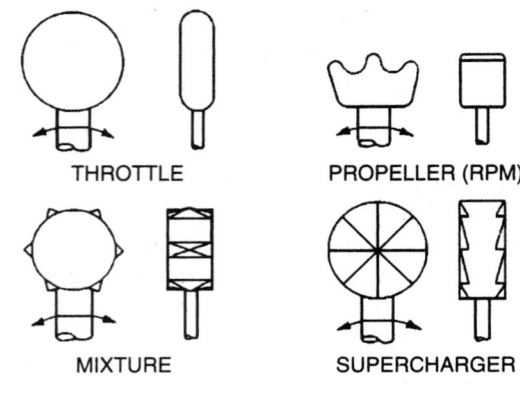

THROTTLE   PROPELLER (RPM)

MIXTURE   SUPERCHARGER

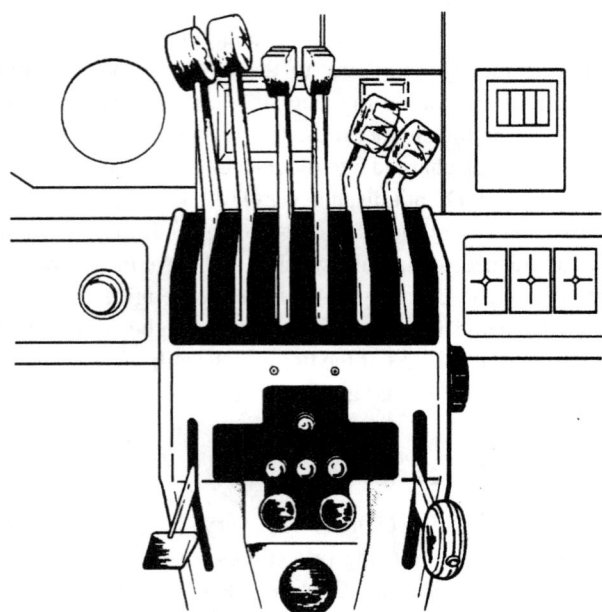

FIGURE 9-1 Engine control console and standard knob shapes.

8. Turn the ignition switch to the START position. (On most modern aircraft, this will also allow the magnetos to be energized.)
9. Release the ignition switch. (When the ignition switch is released from the START position, it is spring-loaded to return to the BOTH position. This action will de-energize the starter circuit and leave the magnetos in the ON position.)
10. Check for oil pressure.

Bendix Fuel Injection Starting Procedure

The starting procedure for an engine equipped with an RSA fuel injection system has been designed to avoid flooding the engine and to provide positive results. The normal steps for a cold start are as follows:

1. Put the mixture control in the IDLE CUTOFF position.
2. Adjust the throttle to $\frac{1}{8}$ in (3.18 mm) open.
3. Turn on the master switch.
4. Turn on the fuel boost pump switch.
5. Move the mixture control to FULL RICH until the fuel flow gauge reads 4 to 6 gal/h [15.14 to 22.7 L/h]; then immediately return the control to IDLE CUTOFF. If the aircraft does not have a fuel flow indicator, put the mixture control in the FULL RICH position for 4 to 5 s and return it to IDLE CUTOFF. Placing the mixture control in the FULL RICH position allows fuel to flow through the nozzles into the intake manifold to prime the engine.
6. Clear the propeller.
7. Turn the ignition switch to the START position.
8. As soon as the engine starts, move the mixture control to FULL RICH.
9. Release the ignition switch.
10. Check for oil pressure.

Continental Continuous-Flow Fuel Injection Starting Procedures

1. Turn the ignition switches to ON.
2. Open the throttle approximately $\frac{1}{2}$ in [1.27 cm].
3. Set the propeller pitch lever full forward to HIGH RPM.
4. Set the mixture lever full forward to FULL RICH.
5. Clear the propeller.
6. Turn the auxiliary fuel pump switch to the PRIME position. Avoid leaving the auxiliary fuel pump switch in either the PRIME or ON position for more than a few seconds unless the engine is running.
7. Turn the ignition switch to START when the fuel flow reaches 2 to 4 gal/h [7.57 to 15.14 L/h]. (Read the fuel pressure gauge.) If the engines are warm, first turn the ignition switch to START, then turn the auxiliary pump switch to PRIME.
8. Release the ignition switch as soon as the engine fires.
9. Turn off the auxiliary fuel pump switch when the engine runs smoothly. During very hot weather, if there is a sign of vapor in the fuel system (indicated by fluctuating fuel flow)

with the engine running, turn the auxiliary fuel pump switch to ON until the system is purged.

10. Check for an oil pressure indication within 30 s in normal weather and 60 s in cold weather. If no indication appears, shut off the engine and investigate.

11. Disconnect the external power source, if used.

12. Warm up the engine at 800 to 1000 rpm.

Starting Large Reciprocating Engines

Large radial engines installed on the DC-3, DC-6, Constellation, and other large aircraft should be started according to the manufacturer's instructions. The steps in starting are similar to those used for light-aircraft engines, but additional precautions are necessary. First, a fire guard should be placed to the rear and outboard of the engine being started, in case the engine backfires and fire burns in the engine induction system. The fire guard should have an adequate supply of carbon dioxide gas in suitable fire extinguisher bottles in order to immediately direct the gas into the engine induction system. The engine should be kept turning so that the fire will be drawn into the cylinders. Often it is not even necessary to use the extinguisher because the air rushing into the engine carries the fire with it and as the engine starts, the fire cannot continue to burn in the induction system.

Before any attempt to start a large reciprocating engine, the engine should be rotated several complete revolutions to eliminate the possibility of **liquid lock**, caused by oil in the lower cylinders. If the engine stops suddenly while being rotated by hand or with the starter, oil has collected in a lower cylinder, and the oil must be removed before the engine can be started. This is best accomplished by removing a spark plug from the cylinder. It is not recommended that the engine rotation be reversed to clear the oil. After the oil is drained from the cylinder, the spark plug can be replaced and the engine started.

For large reciprocating engines, priming is usually accomplished by means of a fuel pressure pump and an electrically operated priming valve. The fuel is carried from the primer to a spider (distributing fitting) and then to the top cylinders of the engine. This applies to a radial engine, either single- or twin-row. In a nine-cylinder radial engine, the top five cylinders of the engine receive priming. Priming is accomplished by pressing the priming switch while the fuel booster pump is turned on.

Large reciprocating engines may have direct-cranking starters similar to those used on light-aircraft engines but much more powerful, or they may have inertia starters in which the cranking energy is stored in a rapidly rotating flywheel. With the inertia starter, the flywheel must be energized by an electric motor or hand cranked until enough energy is stored to turn the engine for several revolutions. The engage switch is then turned on to connect the flywheel reduction gearing to the crankshaft through the starter jaws. A plate clutch, located between the flywheel and starter jaws, allows slippage to avoid damage due to inertial shock when the starter is first engaged.

With the engine properly primed, the throttle set, and the ignition switch on, the engine should start very soon after it is rotated by the starter. The throttle is then adjusted for proper warm-up speed.

Hand Cranking

Hand cranking of a starter-equipped engine with a low battery or defective starter, although convenient, can expose personnel to a potential safety hazard. For safety reasons, the replacement of the faulty starter and the use of a ground power source should be considered rather than hand cranking. *Only experienced persons should do the hand cranking, and a reliable person should be in the cockpit.* Hand cranking with the cockpit unoccupied has resulted in many serious accidents.

If the aircraft has no self-starter, the engine must be started by swinging the propeller. The person who is turning the propeller calls out, *"fuel on, switch off, throttle closed, brakes on."* The person operating the engine will check these items and repeat the phrase. The switch and throttle must not be touched again until the person swinging the prop calls *"contact."* The operator will repeat "contact" and then turn on the switch. Never turn on the switch and then call "contact."

When you are swinging the prop, a few simple precautions will help to avoid accidents. When you are touching a propeller, always assume that the ignition is on. The switches which control the magnetos operate on the principle of short-circuiting the current to turn off the ignition. If the switch is faulty, it can be in the OFF position and still permit current flow in the magneto primary circuit.

Be sure the ground is firm. Slippery grass, mud, grease, or loose gravel can lead to a fall into or under the propeller. Never allow any portion of your body to get in the way of the propeller. This applies even though the engine is not being cranked.

Stand close enough to the propeller to be able to step away as it is pulled down. Stepping away after cranking is a safeguard against brake failure. Do not stand in a position that requires leaning toward the propeller to reach it. This throws the body off balance and could cause you to fall into the blades when the engine starts. In swinging the prop, move the blade downward by pushing with the palms of the hand. Do not grip the blade with the fingers curled over the edge, since kickback may break them or draw your body into the blade path.

ENGINE OPERATION

Operating Requirements

The operation of any reciprocating engine requires that certain precautions be observed and that all operations be kept within the limitations established by the manufacturer. Among the conditions which must be checked during the operation of an engine are the following:

1. Engine oil pressure
2. Oil temperature
3. Cylinder-head temperature (CHT)

4. Engine rpm

5. Manifold pressure

6. Drop in rpm during switching to single-magneto operation

7. Engine response to propeller controls, if a constant-speed (controllable-pitch) propeller is used with the engine

8. Exhaust gas temperature (EGT)

Oil Pressure and Temperature Check

No engine should be operated at high-power settings unless its oil pressure and temperature are within satisfactory limits; otherwise, oil starvation of bearings and other critical parts will occur, thus potentially causing permanent engine damage. For this reason, a reciprocating engine must be properly warmed up before full-power operation is begun. When the engine is started, the oil pressure gauge should be observed to see that the oil pressure system is functioning satisfactorily. *If no oil pressure is indicated within 30 s after starting, the engine must be shut down and the malfunction located.* If the engine is operated without oil pressure for much more than 30 s, damage is likely to result.

Prior to takeoff, the reciprocating engine should be given an ignition check and a full-power test. This is usually done while the airplane is parked just off the end of the takeoff runway in a warm-up area. For the magneto check, the throttle is moved slowly forward until the engine rpm is at the point recommended by the manufacturer. This is usually from 1500 to 1800 rpm, although it may be outside this range. To make the check, the ignition switch is turned from the BOTH position to the LEFT magneto position and the tachometer is observed for rpm drop. The amount of drop is noted, and then the switch is turned back to the BOTH position for a few seconds until the engine is again running smoothly at the full-test rpm. The switch is then turned to the RIGHT magneto position for a few seconds so the rpm drop can be noted. The engine should not be allowed to operate for more than a few seconds on a single magneto because of possible plug fouling.

The permissible rpm drop during the magneto test varies, but it is usually between 50 and 125 rpm. In all cases the instructions in the operator's manual should be followed. Usually the rpm drop will be somewhat less than the maximum specified in the instructions.

When the magnetos are checked on an airplane having a constant-speed or controllable-pitch propeller, it is essential that the propeller be in the full HIGH-RPM (low-pitch) position; otherwise, a true indication of rpm drop may not be obtained.

Check of Constant-Speed Propeller Pitch

The propeller is checked to ensure proper operation of the pitch control and the pitch change mechanism. The operation of a **controllable-pitch propeller** is checked by the indications of the tachometer and manifold pressure (MAP) gauges when the propeller governor control is moved from one position to another. During this check, the propeller is cycled so as to circulate the cold oil from the propeller hub and to allow warmer oil to enter the hub. To cycle the propeller, the operator moves the propeller control in the cockpit rapidly to the full-decrease RPM position. As the engine rpm begins to slow down, the control is moved back to the full-increase RPM position. Cycling of the propeller is done during the run-up with the engine set at approximately 1600 to 2000 rpm. Usually the engine speed is not allowed to drop more than 500 rpm during this procedure. Because each type of propeller requires a different procedure, the applicable manufacturer's instructions should be followed.

The engine should be given a brief full-power check. This is done by slowly advancing the throttle to the full-forward position and observing the maximum rpm obtained. If the rpm level and MAP are satisfactory and the engine runs smoothly, the throttle is slowly retarded until the engine has returned to the desired idling speed.

When making the full-power check, the operator must make sure that the airplane is in a position which will not direct the propeller blast toward another airplane or into an area where damage to property, or injury or inconvenience to personnel, may be caused. The operator should also make sure that the brakes are on and that the elevator control is pulled back, if the airplane has conventional landing gear.

Power Settings and Adjustments

During operation of an airplane, the engine power settings must be changed from time to time for various types of operation. The principal power settings are for *takeoff, climb, cruises* (from maximum to minimum), *letdown,* and *landing.*

The methods for changing power settings differ according to the type of engine, type of propeller, whether the engine is equipped with a supercharger, type of carburetion, and other factors. The operator's manual will give the proper procedures for a particular airplane-engine combination.

The following rules generally apply to most airplanes and engines:

1. Always move the throttle slowly for a power increase or decrease. "Slowly" in this case means that the throttle movement from full open to closed, or the reverse, should require 2 to 3 s rather than the fraction of a second required to "jam" the throttle forward or "jerk" it closed.

2. Reduce the power setting to the climb value as soon as practical after takeoff if the specified climb power is less than maximum power. Continued climb at maximum power can produce excessive CHT and detonation. This is particularly true if the airplane is not equipped with a CHT gauge.

3. Do not reduce power suddenly when the CHT is high (at or near the red line on the gauge). The sudden cooling which occurs when power is reduced sharply will often cause the cylinder head to crack. When you are preparing to let down, reduce the power slowly by increments to allow for a gradual reduction of temperature.

4. When you are operating an airplane with a constant-speed propeller, *always reduce MAP with the throttle before*

reducing the rpm with the propeller control. Conversely, always increase the rpm with the propeller control before increasing MAP. If the engine rpm setting is too low and the throttle is advanced, it is possible to develop excessive cylinder pressures with the consequences explained previously. The operator of an engine should become familiar with the maximums allowable for the engine and then make sure that the engine is operated within these limits. Remember that a constant-speed propeller holds the engine rpm to a particular value in accordance with the position of the propeller control. When the throttle is moved forward, the propeller blade angle increases and MAP increases, but the rpm remains the same.

Follow these basic rules when you are changing power on an engine equipped with a constant-speed propeller:

a. To **increase power**, enrich the mixture, increase the rpm, then adjust the throttle.

b. To **decrease power**, reduce the throttle, reduce the rpm, and then adjust the mixture.

5. During a prolonged glide with power low (throttle near closed position) "clear the engine" occasionally to prevent spark plug fouling. This is done by advancing the throttle to a medium-power position for a few seconds. If the engine runs smoothly, the power may be reduced again.

6. Always place the manual mixture control in the FULL RICH position when the engine is to be operated at or near full power. This helps prevent overheating. The engine should be operated with the mixture control in a LEAN position only during cruise, in accordance with the instructions in the operator's manual. When power is reduced for letdown and in preparation for landing, the mixture control should be placed in the FULL RICH position. Some mixture controls and carburetors do not include a FULL RICH setting. In this case the mixture control is placed in the RICH position for high power and takeoff.

7. If there is any possibility of ice forming in the carburetor while power is reduced for a letdown preparatory to landing, it is necessary to place the carburetor heat control in the HEAT ON position. This is a precautionary measure and is common practice for all engines in which carburetor icing may occur.

8. At high altitude, adjust the mixture control to a position less rich than that used at low altitudes. The density of air at high altitudes is less than at lower altitudes; therefore, the same volume of air will contain less oxygen. If the engine is supercharged, the increase in altitude will not be of particular consequence until the capacity of the supercharger is exceeded. Usually the MAP gauge will provide information helpful for proper adjustment of the mixture control; however, an accurate EGT gauge is considered essential for leaning the mixture for cruise power at altitudes normally flown.

CRUISE CONTROL

Cruise control is the adjustment of engine controls to obtain the results desired in range, economy, or flight time. Since an engine consumes more fuel at high power settings than at lower settings, obviously maximum speed and maximum range or economy cannot be attained with the same power settings. If a maximum-distance flight is to be made, it is desirable to conserve fuel by operating at a low power setting. But if maximum speed is desired, it is necessary to use maximum power settings with a decrease in range capability.

Range and Speed Charts

The charts presented in Fig. 9-2 were developed for the operation of the Piper PA-23-160 aircraft. The chart on the left shows the effects of power settings on range, and the chart on the right shows how power settings and true airspeed (TAS) are related. From these charts we can easily determine the proper power settings for any flight within the range of the airplane, taking into consideration the flight altitude, flight distance, and desired flight time.

If we wish to make a flight of 900 mi [1 448.40 km] at an altitude of 6500 ft [1 981.20 m], we can determine the flight values for maximum speed or maximum range or choose a compromise setting. If we wish to make the flight in the shortest possible time, 75 percent of the engine power is used. With this setting (2400 rpm and full throttle), the TAS will be about 175 mph [281.63 km/h] and the flight will take 5.14 h, assuming no tail wind or head wind. At this setting the fuel consumption will be 18.8 gal/h [71.17 L/h]; therefore, the flight will require 96.7 gal (366.05 L) of fuel. If we wish to make this same flight with maximum economy, we may operate the engines at 45 percent of power with the mixture control leaned as far as good engine operation will permit. At this power setting, the TAS will be about 128 mph [205.99 km/h], and the fuel required for the trip will be about 77.7 gal [294.13 L]. The time required for the flight will be about 7 h.

We would seldom operate the engine at the extremes mentioned above, because the recommended power setting for cruise conditions is 65 percent of power. For flight at an altitude of 9500 ft [2895.6 m], this would provide a TAS of about 166 mph [267.14 km/h]. If we wished to operate more economically or with greater range, we would probably use a power setting about 55 percent of maximum.

Power Settings

To set the controls of an engine for a particular power output, the MAP and rpm are adjusted according to density altitude when the airplane is equipped with constant-speed propellers. Figure 9-3 shows the settings for the Lycoming O-320-B opposed engine. This table is adjusted for the use of pressure altitude at standard temperature T_s instead of density altitude. Observe the following facts regarding the settings for MAP and rpm:

1. At a given rpm and a given power setting, MAP must be decreased as altitude increases. This is because the T_s of the air decreases and the density therefore increases. Thus, a given volume of air at a certain pressure will have a greater weight as altitude increases, and MAP must be reduced to maintain constant power.

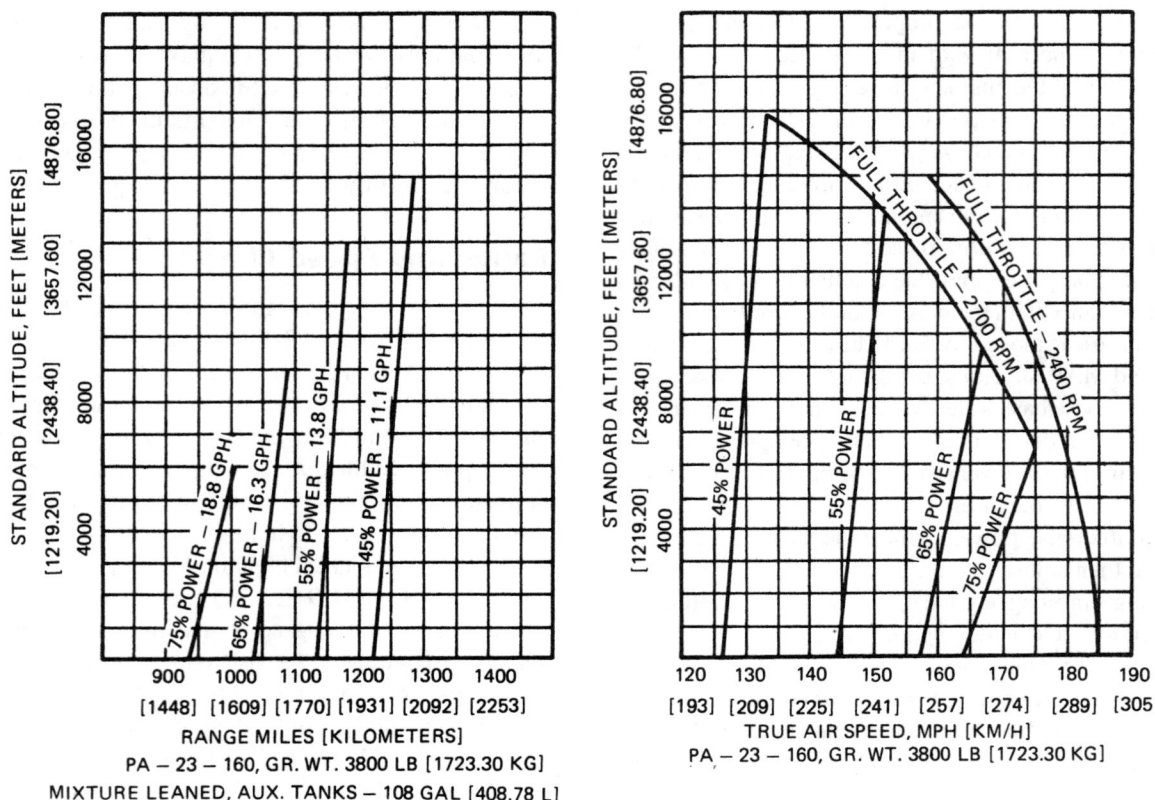

FIGURE 9-2 Charts showing range and airspeed in relation to power setting.

Press. alt. 1000 ft [304.80 m]	Std. alt. temp., °F [°C]	88 hp [65.62 kW]—55% rated Approx. fuel 7 gal/h [26.50 L/h] rpm & man. press.				104 hp [77.55 kW]—65% rated Approx. fuel 8 gal/h [30.28 L/h] rpm & man. press.				120 hp [89.48 kW]— 75% rated Approx. fuel 9 gal/h [34.07 L/h] rpm & man. press.		
		2100	2200	2300	2400	2100	2200	2300	2400	2200	2300	2400
SL	59 [15.0]	22.0	21.3	20.6	19.8	24.4	23.6	22.8	22.1	25.9	25.2	24.3
1	55 [12.8]	21.7	20.0	20.3	19.6	24.1	23.3	22.5	21.8	25.6	24.9	24.0
2	52 [11.1]	21.4	20.7	20.1	19.3	23.8	23.0	22.3	21.5	25.0	24.3	23.5
3	48 [8.9]	21.1	20.5	19.8	19.1	23.5	22.7	22.0	21.2	25.3	24.6	23.8
4	45 [7.2]	20.8	20.2	19.6	18.9	23.1	22.4	21.7	21.0	24.7	24.0	23.2
5	41 [5.0]	20.5	19.9	19.3	18.6	22.8	22.1	21.4	20.7	FT	23.7	23.0
6	38 [3.3]	20.2	19.6	19.0	18.4	22.5	21.8	21.2	20.5		FT	22.7
7	34 [1.1]	19.9	19.3	18.8	18.2	22.2	21.5	20.9	20.2			FT
8	31 [−0.56]	19.5	19.0	18.5	18.0	FT	21.2	20.6	19.9			
9	27 [−2.8]	19.2	18.8	18.3	17.7		FT	20.3	19.7			
10	23 [−5.0]	18.9	18.5	18.0	17.5			FT	19.4			
11	19 [−7.2]	18.6	18.2	17.8	17.3				FT			
12	16 [−8.9]	18.3	17.9	17.5	17.0							
13	12 [−11.1]	FT	17.6	17.3	16.8							
14	9 [−12.8]		FT	17.0	16.6							
15	5 [−15.0]			FT	16.3							

To maintain constant power, correct manifold pressure approximately 0.15 in Hg for each 10°F variation in carburetor air temperature from standard altitude temperature. Add manifold pressure for air temperatures above standard; subtract for temperatures below standard.

FIGURE 9-3 Power setting chart for Lycoming Model O-320-B, 160-hp [119.31-kW], engine.

2. When the engine is operated at higher rpm, a lower MAP is used to maintain the same power.

3. At a certain level of altitude, MAP can no longer be maintained because of the reduction in atmospheric pressure. This is the point in the chart shown as FT, meaning full throttle.

4. At 55 percent of rated power, the power can be maintained up to 15 000-ft [4 572-m] pressure attitude. An output of 75 percent power can be maintained only up to about 7000-ft [2 133.6-m] pressure altitude.

5. MAP settings must be adjusted to maintain a particular power output if the outside air temperature is above or below the standard given in the chart.

Stopping Procedure

Usually an aircraft engine has cooled sufficiently for an immediate stop because of the time required to move the airplane into the parking area. It is good practice, however, to observe the CHT gauge to see that the CHT is somewhat under 400°F [204.44°C] before stopping. If the engine is equipped with an idle cutoff on the mixture control, the engine should be stopped by placing the control in the IDLE CUTOFF position. Immediately after the engine stops, the ignition switch must be turned off. If the airplane is equipped with cowl flaps, the flaps should be left in the OPEN position until after the engine has cooled.

After stopping the engine, check that all switches in the cockpit are set to OFF. This is especially important for the ignition switches and the master battery switch. Check that all wheel chocks are installed, and release the parking brake to prevent undue stress on the brake system.

ENGINE OPERATING CONDITIONS

Leaning the Mixture

With the mixture in the FULL RICH position, a predetermined mixture of fuel and air is used. For takeoff, a mixture setting of FULL RICH is used. This setting ensures the best combination of power and cooling.

As an aircraft climbs, the air becomes less dense. On the FULL RICH setting, the carburetor is putting out about the same amount of fuel, but there is less air to mix with it, so the mixture gets richer. If the aircraft climbs high enough, the F/A ratio becomes too great for smooth operation. Not only will the engine run roughly, but also fuel will be wasted. The purpose of the fuel metering device is to establish the optimum F/A ratio for all operating conditions.

The two basic types of fuel metering devices discussed are the float carburetor and fuel injection. The general procedures for **leaning** at the manufacturer's recommended cruise power are as follows:

1. Float-type carburetor
   a. Fixed-pitch propeller. Lean to a **maximum increase in rpm** and airspeed or to the point just before engine

roughness occurs. Engine roughness is not detonation at cruise power, but is caused because the leanest cylinder does not fire due to a very lean F/A mixture which will not support combustion in that cylinder.

   b. Controllable-pitch propeller. Lean the mixture until engine roughness is encountered, and then enrich slightly until roughness is eliminated and engine runs smoothly. There may be a slight increase of airspeed noted in smooth air when the mixture is properly leaned at cruise compared to full rich.

2. Fuel injection. Because of the various models of fuel injectors used, the operator must consult the operating handbook for specific leaning instructions.

However, as a basic technique, at the *manufacturer's recommended cruise power limitation,* with a manual mixture control, lean initially by reference to the fuel flow (if available) for the percentage of cruise power, without exceeding the manufacturer's recommended limits. Then, for more precise leaning, if an EGT reading is available, find **peak EGT** without exceeding limits, and operate there.

If the EGT and fuel flow are not available, then lean to just before engine roughness, or to a slight airspeed loss.

The EGT method of mixture control relies on a thermocouple in the exhaust stack not far from the exhaust valve. To see the effect of mixture control, the operator may watch the EGT gauge as the mixture is leaned from the FULL RICH position. This is illustrated in Fig. 9-4. At FULL RICH, a large amount of excess fuel is unburned, which cools the exhaust gases and results in a lower EGT reading. As the mixture is leaned, the amount of excess fuel is reduced and the temperature climbs. At the point where there is complete burning of the F/A mixture, the peak EGT is realized. Leaning past this point results in a cooling effect caused by excess air, and the engine nears a condition of lean misfire. The mixture is then said to be on the lean side of peak EGT. Peak EGT is the key to the EGT method of mixture control.

Of course, the meter reading for peak EGT will vary with the power setting, altitude, outside air temperature, and whether or not the cylinder monitored is functioning normally.

As the mixture is leaned from the FULL RICH position, the airspeed will increase along with the EGT up to a point approximately 100°F [55.5°C] to the rich side of peak EGT. This is the mixture setting for maximum power. If leaning of the mixture is continued until the peak EGT is reached, the airspeed will decrease approximately 2 mph [3.2 km/h]. Fuel economy and the aircraft range, however, will be increased by about 15 percent.

Problems Caused by Spark Plug Lead Fouling

Many aircraft engines designed to operate on 80/87-octane avgas are forced to use 100LL (low-lead) avgas because of the reduction in the availability of 80/87 avgas. Although the engines are approved to operate on 100LL avgas, this presents the problem of **spark plug fouling** because 100LL

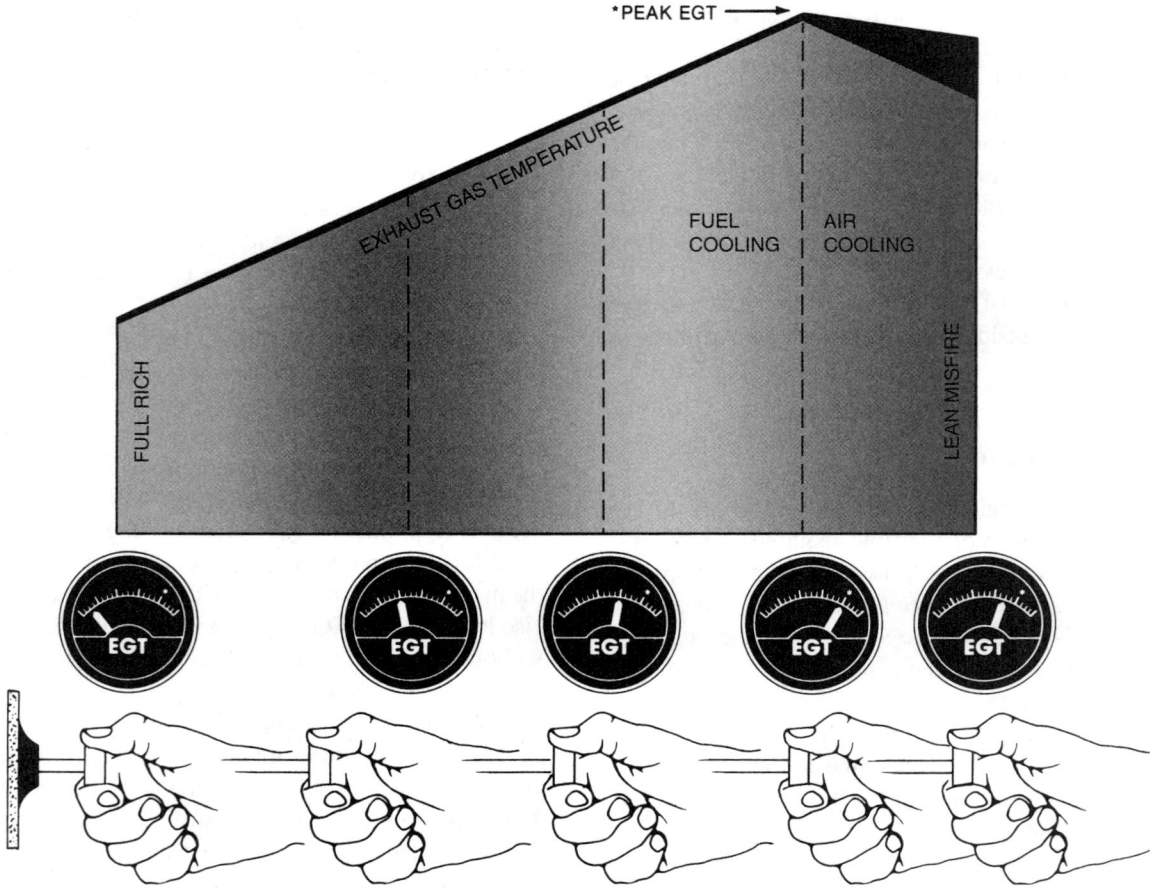

FIGURE 9-4 EGT changes with mixture leaning.

avgas contains four times the TEL (tetraethyl-lead) of 80/87-octane avgas. Normally 80/87 avgas has a lead content of 0.5 mL/gal whereas 100LL has 2.0 mL/gal. The engine's ability to scavenge the extra lead from the cylinder is greatly decreased while operating on 100LL avgas. As a result, lead deposits on the spark plugs, causing fouling of spark plug electrodes. This spark plug fouling can cause many operational problems and can increase the need for spark plug cleaning.

The solution to operating on 100LL avgas in an engine designed to operate on 80/87 is twofold. First, it is recommended that optimum mixture control be chosen to prevent excess lead buildup. Even with proper mixture control, it may not be possible to get the spark plug performance desired. To supplement mixture control procedures, a fuel treatment such as TCP (tricresyl phosphate) can be mixed with the fuel in the tank or mixed with fuel as it is dispensed into the tank. TCP reduces spark plug fouling by making the lead less conductive and less corrosive to the electrodes. It also softens the lead deposits formed, which can help in the scavenging of lead deposits from the combustion chamber. TCP is generally not used in turbocharged engines or engines that do not experience lead-related problems. Therefore, it is not used as a standard fuel additive.

Certain techniques of ground engine operation can aid in prevention of spark plug fouling when 100LL avgas is used in engines designed for 80/87 avgas: Let the engine idle

as little as possible because of the rich mixture used during idle; apply the power smoothly, and never open the throttle abruptly during normal takeoffs; run the engine to about 1000 rpm when shutting down the engine, and then move the mixture to IDLE CUTOFF. These are just a few examples of techniques for reducing spark plug fouling. The engine operator's handbook should always be consulted for specific instructions on engine operation.

RECIPROCATING-ENGINE OPERATIONS IN WINTER

Winterization Procedures

Cold-weather operation of an aircraft engine involves special preparations and precautions compared to normal-weather operation. Vaporization of the fuel becomes difficult, and the high viscosity of oil causes reduced cranking speed with accompanying high loads on the starter.

Often the engine accessories fail because of congealed oil. This is very clear from the increased number of oil cooler failures on reciprocating engines in cold weather. Excessive fuel priming washes the oil from the piston rings and cylinder walls, causing the piston to scuff and score the cylinders.

Some aircraft use winterizing kits to maintain desired engine operating temperatures and to prevent oil coolers and vapor vent lines from freezing. Attention to details, such as warming up the engine before takeoff and allowing the engine to cool down prior to shutting it off, pays dividends in reduced maintenance and extended engine life. Winter operation should include a check of the carburetor air heat system and the degree of heat rise available. At the same time, the engine idle rpm and mixture, with and without carburetor air heat, should be checked.

The crankcase breather should be checked in preparing for cold weather. Frozen breather lines have created numerous problems. Most of the water of combustion goes out of the exhaust; however, some water enters the crankcase and is vaporized. When the vapor cools, it condenses in the breather line, subsequently freezing it closed. Special preflight care is recommended to ensure that the breather system is free of ice.

Draining Sumps

Proper draining of the sump is very important during the preflight check. Sufficient fuel should be drawn off into a transparent container to see if the fuel is free of water and contaminants. This is especially important during changes in temperature, particularly near freezing. Ice, which may turn to water when the temperature rises, may be in the tanks and filter down into the carburetor or fuel controller, causing engine failure. Water can freeze in lines and filters, causing stoppage. A small amount of water, when frozen, can prevent proper operation of fuel pumps, selector valves, and carburetors.

Anti-Icing Additives

Although proper fuel sampling and proper draining of the sump are essential in preventing ice formation due to free water in the fuel, they will not eliminate the hazard of ice blockage of fuel flow. Under certain conditions, water in suspension or solution may form ice crystals. Since water in suspension or solution is not removed by the sump, the formation of ice crystals must be prevented by anti-icing additives, such as isopropyl alcohol or EGME (ethylene glycol monomethyl ether), in the fuel. Both additives absorb water and lower the freezing point of the mixture. When alcohol or EGME is used, instructions for proper use must be carefully followed.

Engine Preheating

Preheating an engine consists of forcing heated air into the engine area to heat the engine, lubricants, and accessories. Preheating is required for most aircraft reciprocating engines when outside temperatures are +10°F [−12.2°C] and below, as stipulated in the manufacturer's instructions. This does not mean that the engine will refuse to start after setting out in such an extremely cold environment, but starting without preheating has frequently caused engine damage.

Some types of damage are scored cylinders, scuffed piston skirts, and broken piston rings. The application of heat to only the cylinder area fails to ensure that the entire oil system is adequately heated. At temperatures of +10°F [−12.2°C] and below, preheating of the complete engine, oil supply tank, and oil system is required.

Engine Preheating Precautions

The following recommendations regard aircraft preheating:

1. Preheat the aircraft by storing in a heated hangar, if possible.
2. Use only heaters that are in good condition, and do not refuel the heater while it is operating.
3. During the heating process, do not leave the aircraft unattended, and keep a fire extinguisher handy.
4. Do not place heat ducting so that it will blow hot air directly on combustible parts of the aircraft, such as upholstery, canvas engine covers, or flexible fuel, oil, and hydraulic lines.

INSPECTION AND MAINTENANCE

Engines are designed and built to provide many years of service. For an engine to remain in airworthy condition, it should be operated in accordance with the recommendations of the manufacturer and cared for with sound inspection and maintenance practices. Many important points to be observed in the inspection and maintenance of an aircraft engine are explained in the following sections.

A visual inspection is needed to determine the current condition of the engine and its components. The repair of discrepancies is required to bring the engine back up to airworthy standards. To keep the aircraft in airworthy condition, the manufacturer may recommend that certain services be performed at various operating intervals. Also, before an engine is serviced, consult the handbooks and manuals issued by the manufacturers of the equipment for that particular make, model, and type of engine. The following general instructions apply broadly to all aircraft reciprocating engines.

During an inspection, a checklist must be used that meets the scope and detail of **FAR 43**, **Appendix D**. Most manufacturers have developed checklists that meet or exceed the scope and detail of Appendix D. According to FAR 43, Appendix D, each person performing an annual or 100-h inspection must inspect components of the engine and nacelle group as follows:

1. **Engine section**—for visual evidence of excessive oil, fuel, or hydraulic leaks as well as sources of such leaks.
2. **Studs and nuts**—for improper torquing and obvious defects.
3. **Internal engine**—for cylinder compression and for metal particles or foreign matter on screens and sump

drain plugs. If there is weak cylinder compression, check the internal condition and tolerances of the cylinder.

4. **Engine mount**—for cracks, looseness of mount, and looseness of engine to mount, and looseness of engine to mount.

5. **Flexible vibration dampers**—for poor condition and deterioration.

6. **Engine controls**—for defects, improper travel, and improper safetying.

7. **Lines, hoses, and clamps**—for leaks, improper condition, and looseness.

8. **Exhaust stacks**—for cracks, defects, and improper attachment.

9. **Accessories**—for apparent defects in security of mounting.

10. **All systems**—for improper installation, poor general condition, defects, and insecure attachment.

11. **Cowling**—for cracks and defects.

A manufacturer's inspection checklist is shown in Fig. 9-5. It shows the items to be inspected and the operational intervals for accomplishing each item. Notes at the bottom of the inspection list are for further information or contain references to other publications of the manufacturer. In the following text, we discuss the main items on the checklist in Fig. 9-5. Propeller inspection is covered in Chap. 22, and additional information on inspection programs may be found in the text *Aircraft Basic Science.*

Opening and Cleaning

FAR 43, Appendix D, begins by stating:

"Each person performing an annual or 100-h inspection shall, before that inspection, remove or open all necessary inspection plates, access doors, and cowling. He shall thoroughly clean the aircraft and aircraft engine."

When opening cowlings, the technician should note any accumulation of oil or other foreign material, which may be a sign of fluid leakage or other abnormal condition that should be corrected.

An engine and accessories wash-down should be done prior to each 100-h inspection to remove oil, grease, salt corrosion, or other residue that might conceal component defects during inspection. Precautions, such as wearing rubber gloves, an apron or coveralls, and a face shield or goggles, should be taken when working with cleaning agents. Use the least toxic of the available cleaning agents that will satisfactorily accomplish the work. These cleaning agents include: (1) Stoddard solvent; (2) a water-base alkaline detergent cleaner consisting of 1 part cleaner, 2 to 3 parts water, and 8 to 12 parts Stoddard solvent; or (3) a solvent-base emulsion cleaner comprising 1 part cleaner and 3 parts Stoddard solvent.

WARNING Do not use gasoline or other highly flammable substance for wash-down. Perform all cleaning operations in well-ventilated work areas, and ensure that adequate fire-fighting and safety equipment is available. Compressed air, used for cleaning agent application or drying, should be regulated to the lowest practical pressure. Use of a stiff-bristle fiber brush, rather than a steel brush, is recommended if cleaning agents do not remove excess grease and grime during spraying. Before cleaning the engine compartment, place a strip of tape on the magneto vents to prevent any solvent from entering these units. Place a large pan under the engine to catch waste. With the engine cowling removed, spray or brush the engine with solvent or a mixture of solvent and degreaser. To remove especially heavy dirt and grease deposits, it may be necessary to brush areas that have been sprayed.

WARNING Do not spray solvent into the alternator, vacuum pump, starter, or air intakes. Allow the solvent to remain on the engine for 5 to 10 min. Then rinse the engine clean with additional solvent and allow it to dry. Cleaning agents should never be left on engine components for extended periods. Failure to remove them may cause damage to components such as neoprene seals and silicone fire sleeves and could cause additional corrosion. Completely dry the engine and accessories, using clean, dry compressed air. If desired, the engine cowling may be washed with the same solvent. Remove the protective tape from the magnetos and lubricate the controls, bearing surfaces, etc., in accordance with the Lubrication Chart. Other parts of an airplane that often need cleaning prior to inspection are the landing gear and the underside of the aircraft. Most compounds used for removing oil, grease, and surface dirt from these areas are emulsifying agents. These compounds, when mixed with petroleum solvents, emulsify the oil, grease, and dirt. The emulsion is then removed by rinsing with water or by spraying with a petroleum solvent. Openings such as air scoops should be covered prior to cleaning.

Servicing Oil Screens and Filters

Most engines incorporate an oil suction screen which filters oil as it leaves the oil sump and before it enters the oil pressure pump. The oil suction screen is generally located on the bottom aft of the engine sump, installed horizontally. To remove the suction screen, cut the safety wire and remove the hexagonal-head plug. The screen should be checked and cleaned at each oil change or inspection to remove any accumulation of sludge and to examine for metal filings or chips. If metal particles are found in the screen, the engine should be examined for internal damage. After cleaning and inspection, replace the screen, tighten, and safety with safety wire.

Another type of oil screen is the oil pressure screen which is installed after the oil pump; because of the location of this screen, all oil passing through it is under pressure. It is located in a housing on the accessory case of the engine. The oil pressure screen should be cleaned and inspected at each oil change or aircraft inspection. After the pressure screen is removed, any accumulation of sludge should be removed and an inspection for metal particles should be made.

NOTE

Perform inspection or operation at each of the inspection intervals as indicated by a circle (O)

A. PROPELLER GROUP

Nature of Inspection	50	100	500	1000
1. Inspect spinner and back plate for cracks	O	O	O	O
2. Inspect blades for nicks and cracks	O	O	O	O
3. Check for grease and oil leaks		O	O	O
4. Lubricate propeller per Lubrication Chart		O	O	O
5. Check spinner mounting brackets for cracks		O	O	O
6. Check propeller mounting bolts and safety (Check torque if safety is broken)		O	O	O
7. Inspect hub parts for cracks and corrosion			O	O
8. Rotate blades of constant speed propeller and check for tightness in hub pilot tube			O	O
9. Remove constant speed propeller; remove sludge from propeller and crankshaft			O	O
10. Inspect complete propeller and spinner assembly for security, chafing, cracks, deterioration, wear and correct installation			O	O
11. Check propeller air pressure (at least once a month)	O	O	O	O
12. Overhaul propeller			O	O

B. ENGINE GROUP

CAUTION: Ground Magneto Primary Circuit before working on engine.

Nature of Inspection	50	100	500	1000
1. Remove engine cowl	O	O	O	O
2. Clean and check cowling for cracks, distortion and loose or missing fasteners		O	O	O
3. Drain oil sump (See Note 2)	O	O	O	O
4. Clean suction oil strainer at oil change (Check strainer for foreign particles)	O	O	O	O
5. Clean pressure oil strainer or change full flow (cartridge type) oil filter element (Check strainer or element for foreign particles)	O	O	O	O
6. Check oil temperature sender unit for leaks and security		O	O	O
7. Check oil lines and fitting for leaks, security, chafing, dents and cracks (See Note 4)		O	O	O
8. Clean and check oil radiator cooling fins		O	O	O
9. Remove and flush oil radiators			O	O
10. Fill engine with oil per information on cowl or Lubrication Chart		O	O	O
11. Clean engine	O	O	O	O

CAUTION: Use caution not to contaminate vacuum pump with cleaning fluid. Refer to Lycoming Service Letter 1221A.

Nature of Inspection	50	100	500	1000
12. Check condition of spark plugs (Clean and adjust gap as required; adjust per Lycoming Service Instruction No. 1042)		O	O	O
13. Check cylinder compression (Refer to AC 43.131A)		O	O	O
14. Check ignition harness and insulators (High tension leakage and continuity)		O	O	O
15. Check magneto points for proper clearance (Maintain clearance at .018 ± .006)			O	O
16. Check magneto for oil seal leakage			O	O
17. Check breaker felts for proper lubrication			O	O
18. Check distributor block for cracks, burned areas or corrosion and height of contact springs			O	O
19. Check magnetos to engine timing		O	O	O
20. Overhaul or replace magnetos (See Note 3)			O	O
21. Remove air filters and tap gently to remove dirt particles (Replace as required)		O	O	O
22. Clean fuel injector inlet line strainer (Clean injector nozzles as required) (Clean with acetone only)	O	O	O	O
23. Check condition of injector alternate air doors and boxes		O	O	O
24. Remove induction air box valve and inspect for evidence of excessive wear or cracks. Replace defective parts (See Note 7)	O		O	O
25. Inspect fuel injector attachments for loose hardware (See Note 8)		O	O	O
26. Check intake seals for leaks and clamps for tightness		O	O	O
27. Inspect all air inlet duct hoses (Replace as required)		O	O	O
28. Inspect condition of flexible fuel lines		O	O	O
29. Replace flexible fuel lines (See Note 3)			O	O
30. Check fuel system for leaks		O	O	O
31. Check fuel pumps for operation (Engine driven and electric)		O	O	O
32. Overhaul or replace fuel pumps (Engine driven and electric) (See Note 3)			O	O
33. Check vacuum pumps and lines		O	O	O
34. Overhaul or replace vacuum pumps (See Note 3)			O	O
35. Check throttle, alternate air, mixture and propeller governor controls for travel and operating condition		O	O	O

NOTE: Visually inspect the exhaust system per Piper Service Bulletin No. 373A at each 25 hours of operation. (See Note 10.)

Nature of Inspection	50	100	500	1000
36. Inspect exhaust stacks, connections and gaskets for cracks and loose mounting (Replace gaskets as required)	O	O	O	O
37. Inspect muffler, heat exchange, baffles and "augmentor" tube (See Note 6)	O	O	O	O
38. Check breather tubes for obstructions and security		O	O	O
39. Check crankcase for cracks, leaks and security of seam bolts		O	O	O
40. Check engine mounts for cracks and loose mountings		O	O	O
41. Check engine baffles for cracks and loose mounting		O	O	O
42. Check rubber engine mount bushings for deterioration (Replace as required)		O	O	O
43. Check fire wall seals		O	O	O
44. Check condition and tension of alternator drive belt		O	O	O
45. Check condition of alternator and starter		O	O	O
46. Check fluid in brake reservoir (Fill as required)	O	O	O	O
47. Inspect all lines, air ducts, electrical leads and engine attachments for security, proper routing, chafing, cracks, deterioration and correct installation		O	O	O
48. Lubricate all controls		O	O	O
49. Overhaul or replace propeller governor (See Note 3)			O	O
50. Complete overhaul of engine or replace with factory rebuilt (See Note 3)	O		O	O
51. Reinstall engine cowl	O	O	O	O

NOTES:

1. Both the annual and 100 hour inspections are complete inspections of the airplane, identical in scope, while both the **500 and 1000** hour inspections are extensions of the annual or 100 hour inspection, which require a more detailed examination of the airplane, and overhaul or replacement of some major components. Inspection must be accomplished by persons authorized by the FAA.

2. Intervals between oil changes can be increased as much as 100% on engines equipped with full flow (cartridge type) oil filters — provided the element is replaced each 50 hours of operation.

3. Replace or overhaul as required or at engine overhaul. (For engine overhaul, refer to Lycoming Service Instructions No. 1009.)

4. Replace flexible oil lines as required, but no later than 1000 hours of service.

5. Refer to Piper Service Letter No. 597 for flap control cable attachment bolt use.

6. Refer to Piper Service Bulletin No. 373A for exhaust system inspection.

7. Refer to Piper Service Bulletin No. 358.

8. Torque all attachment nuts to 135 to 150 inch-pounds. Seat "Pal" nuts finger tight against plain nuts and then tighten an additional 1/3 to 1/2 turn.

9. Inspect rudder trim tab for "free play," must not exceed .125 inches. Refer to Service Manual for procedure, Section V. Refer to Piper Service Bulletin No. 390A.

10. Compliance with Piper Service Letter No. 673 eliminates repetitive inspection requirements of Piper Service Bulletin No. 373A and FAA Airworthiness Directive No. 73-14-2.

11. Piper Service Letter No. 704 should be complied with.

FIGURE 9-5 Sample 100-h or annual inspection checklist.

FIGURE 9-6 Oil filter cutting tool and filter cutaway.

Most modern engines have a full-flow oil filter installed on the accessory case between the magnetos. This filter can be of the element type or the spin-on type. All the engine's oil flows through this filter under the pressure of the oil pump. The inspection procedure for each type of filter varies somewhat because of the different constructions. The element type is housed in a canister that is disassembled to allow inspection of the filter element.

The spin-on filter is generally cut with a special tool, as shown in Fig. 9-6, to allow inspection. Inspect the filter element by removing the outer perforated paper cover and using a sharp knife to cut through the folds of the element at both ends. Then carefully unfold the pleated element and examine for evidence of metal particles that could indicate internal engine damage.

In new or recently overhauled engines, some small particles of metallic shavings might be found. These are of little consequence and should not be confused with larger metal particles or chips. Evidence of internal engine damage found in the oil filter justifies further examination to determine the cause. The manufacturer's maintenance manual should always be consulted for proper determination of metal particles found in the engine oil filter. When these filters are reinstalled, they must be properly torqued and then safety-wired.

At oil changes, oil samples are often taken and sent away to laboratories to be analyzed for wear metals. A complete discussion of oil analysis can be found in Chap. 14.

Inspection of Oil System Lines

The inspection of the plumbing for an oil system is similar to the inspection of any other plumbing. The tubing, hose, tube fittings, hose fittings, hose clamps, and all other components of the system are inspected for cracks, holes, dents, bulges, and other signs of damage that might restrict the flow or cause a leak. All lines are inspected to ensure that they are properly supported and are not rubbing against a structure. Fittings should be checked for signs of improper installation, overtorquing, excessive tension, or other conditions which may lead to failure.

Compression Testing of Aircraft Engine Cylinders

The purpose of testing the cylinder compression is to determine the internal condition of the combustion chamber by ascertaining if any appreciable leakage is occurring.

Types of Compression Testers

The two basic types of compression testers currently in use are the **direct-compression tester** and the **differential-pressure tester**, shown in Fig. 9-7. Although it is common practice to use only the differential-type compression tester, ideally one would utilize both types in checking the compression of aircraft cylinders. In this respect, it is suggested that the direct-compression method be used first and the findings substantiated with the differential-pressure method. This yields a cross-reference to validate the readings obtained by each method and tends to ensure that the cylinder is defective before it is removed. Before a compression test is started, note the following points:

1. When the spark plugs are removed, identify the cylinders to which they belong. Close examination of the plugs will reveal the actual operating conditions and aid in diagnosing problems in the cylinders.

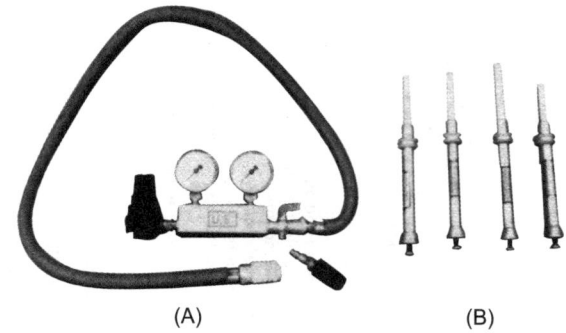

(A)   (B)

FIGURE 9-7 Compression testers. (A) Differential-pressure tester. (B) Direct-compression tester.

2. Review the operating and maintenance records of the engine. Records of previous compression tests reveal progressive wear conditions and help to establish the necessary maintenance approach.

3. Precautions should be taken to prevent the accidental starting of the engine. Remove all spark plug leads, and place them so that the spark plugs cannot fire.

4. The differential-pressure compression equipment must be kept clean and should be checked regularly for accuracy. Check equipment with the shutoff valve closed and regulated pressure at 80 psi [552 kPa]—the cylinder pressure gauge must indicate 80 ± 2 psi [552 ± 13.8 kPa]—and hold this reading for at least 5 s.

5. Combustion chambers with five piston rings tend to seal better than those with three or four, with the result that the differential-pressure tester does not consistently show excessive wear or breakage where five piston rings are involved.

6. If erratic readings are observed on the equipment, inspect the compression tester for water or dirt.

Direct-Compression Check

This type of compression test indicates the actual pressures within the cylinder. Although the particular defective component in the cylinder is difficult to determine by this method, the consistency of readings for all cylinders is an indication of overall engine condition. The following guidelines for performing a direct-compression test are suggested:

1. Thoroughly warm up the engine to operating temperature, and do the test as soon as possible after shutdown.

2. Remove the most accessible spark plug from each cylinder.

3. Rotate the engine with the starter to expel any excess oil or loose carbon in the cylinders.

4. If a complete set of compression testers is available, install one tester in each cylinder. If only one tester is being used, check each cylinder in turn.

5. Using the engine starter, rotate the engine at least three complete revolutions, and record the compression reading. *Note:* An external power source should be used, if possible, because a low battery will result in a low engine-turning rate and lower readings. This will noticeably affect the validity of the second engine test on a twin-engine aircraft.

6. Recheck any cylinder which shows an abnormal reading compared with the others. Any cylinder having a reading approximately 15 psi [103.4 kPa] lower than the others should be suspected of being defective.

7. If a compression tester is suspected of being defective, replace it with one known to be accurate and recheck the compression of the affected cylinders.

Differential-Pressure Compression Check

The differential-pressure tester is designed to check the compression of aircraft engines by measuring the leakage through the cylinders that is caused by worn or damaged components. The operation of the compression tester is based on this principle: For any given airflow through a fixed orifice, a constant pressure drop across that orifice will result. The dimensions of the restrictor orifice in the differential-pressure tester should be sized. A schematic diagram of the differential-pressure tester is shown in Fig. 9-8.

Since the regulated air pressure is applied to one side of the restrictor orifice with the air valve closed, there will be no leakage on the other side of the orifice and both pressure gauges will read the same. However, when the air valve is opened and leakage through the cylinder increases, the cylinder pressure gauge will record a proportionally lower reading.

The following procedures outline the principles involved in performing a differential-pressure compression test and are intended to supplement the manufacturer's instructions for the particular tester being utilized:

1. Perform the compression test as soon as possible after engine shutdown to ensure that the piston rings, cylinder walls, and other engine parts are well lubricated.

2. Remove the most accessible spark plug from each cylinder.

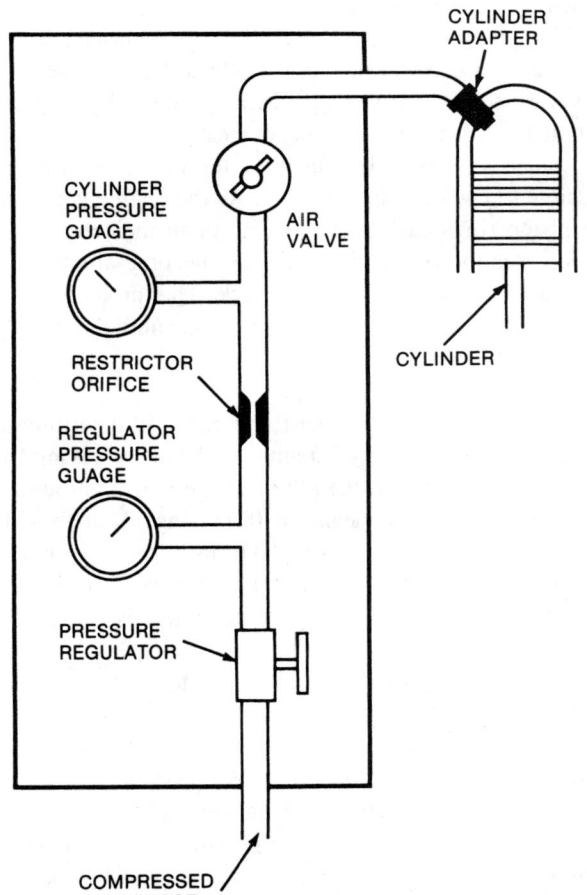

FIGURE 9-8 Schematic of typical differential-pressure compression tester.

3. With the air valve closed, apply an external source of clean air, approximately 100 to 120 psi [689 to 827 kPa], to the tester.

4. Install an adapter in the spark plug bushing, and connect the compression tester to the cylinder.

5. Adjust the pressure regulator to obtain a reading of 80 psi [552 kPa] on the pressure regulator gauge. At this time, the cylinder pressure gauge should also register 80 psi [552 kPa].

6. Turn the crankshaft by hand in the direction of rotation until the piston (in the cylinder being checked) is coming up on its compression stroke. Slowly open the air valve and pressurize the cylinder to approximately 20 psi [138 kPa]. *Caution:* Be careful in opening the air valve since sufficient air pressure will have built up in the cylinder to cause it to rotate the crankshaft if the piston is not at TDC.

Continue rotating the engine against this pressure until the piston reaches TDC. Reaching TDC is indicated by a flat spot or sudden decrease in force required to turn the crankshaft. If the crankshaft is rotated too far, back up at least 0.5 r and start over, to eliminate the effect of backlash in the valve operating mechanism and to keep the piston rings seated on the lower ring lands.

7. Open the air valve completely. Check the regulated pressure and adjust, if necessary, to 80 psi [552 kPa].

8. Observe the pressure indication on the cylinder pressure gauge. The difference between this pressure and the pressure shown by the regulator pressure gauge is the amount of leakage through the cylinder. A loss in excess of 25 percent of input air pressure is cause to suspect the cylinder of being defective; however, recheck the readings after operating the engine for at least 3 min to allow for sealing of the rings with oil.

9. If leakage is still occurring after a recheck, it may be possible to correct a low reading. As the engine is running, the piston rings can, over time, move in their grooves. In some cases the piston-ring gaps can become aligned. This will cause a low compression check. This problem can be corrected by simply running up the engine until the ring gaps become staggered again.

When leakage occurs, the technician can determine the source of the problem by listening. If the air is leaking from the crankcase breather, then the leakage is from around the piston rings or a hole in the piston. If spraying oil into the cylinder and rotating the engine several turns improve the reading on a compression test recheck, then the problem probably lies with the piston rings. This type of test is called a *wet check*.

If air is leaking from the valves, the technician will hear air exiting from the exhaust stacks or carburetor inlet. This leakage may be caused by a small piece of carbon stuck underneath the valve. Generally this problem can be corrected by placing a fiber drift on the rocker arm directly over the valve stem and tapping the drift several times with a hammer, to dislodge any foreign material or carbon between the valve face and seat. *Note:* When you are correcting a low reading in this manner, rotate the propeller so that the piston will not be at TDC. This will prevent the valve from striking the top of the piston in some engines. Rotate the engine

before rechecking the compression to reseat the valves in the normal manner.

Magneto Inspection

During a 100-h or annual inspection, magneto inspection is normally done with the magneto on the engine. Sometimes it is necessary to remove the magneto for inspection, such as with the Slick series magnetos for the 500-h check or if further inspection and disassembly are warranted by discrepancies. The following is an example list of magneto parts that should be inspected during a 100-h or annual inspection (see Fig. 9-9). (The numbers in the illustration refer to the following numbers in the text.) This inspection will require the removal of the ignition harness and breaker contact plate from the magneto.

1. Inspect the distributor block contact springs. If broken or corroded, they should be replaced.

2. Inspect the oil felt washer. It should be saturated with oil. If it is dry, check for worn bushing. Lubricate as needed with no. 30 oil.

3. Inspect the distributor block for cracks or burned areas. *Caution:* The wax coating on the block should not be removed.

4. Look for excess oil in the breaker compartment. It may mean a bad oil seal or oil seal bushing at the drive end.

5. Look for frayed insulation or broken wire strands in leads in back of the magneto. See that terminals are secure. Be sure the wires are properly positioned.

6. Inspect the capacitor visually. If possible, test for leakage, capacitance, and series resistance. Indiscriminate replacement of the capacitor each time the breaker points are replaced incurs an unnecessary expense.

7. Correct the adjustment of the breakers for proper internal timing of the magneto.

8. Check the breaker cam. It should be clean and smooth.

If the impulse coupling is accessible, it should be inspected in accordance with the proper service bulletin or maintenance manual.

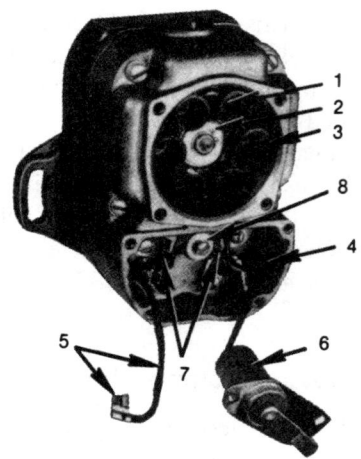

FIGURE 9-9 Magneto items to be inspected.

Inspection and Cleaning of Spark Plugs

As each spark plug is removed from the cylinder, the electrode end should be inspected for possible deposits. These deposits often reflect the internal condition of the cylinder, the operation of the fuel system, and the way in which the engine is being operated.

Visually inspect each spark plug for the following defects:

1. Severely damaged shell or shield threads nicked, stripped, or cross-threaded
2. Badly battered or rounded shell hexagons
3. Out-of-round or damaged shielding barrel
4. Chipped, cracked, or broken ceramic insulator parts
5. Badly eroded electrodes worn to approximately 50 percent of original size

After the spark plug has been inspected, it should be cleaned as required. After cleaning, it should be gapped and tested. Detailed information on spark plug servicing procedures can be found in Chap. 8.

Harness Testing and Inspection

As previously mentioned, high-tension cable for aircraft ignition systems consists of a few strands of stainless-steel wire covered with a thick layer of an insulating material such as silicone rubber. Over this is a layer of glass-fiber reinforcement, and over the reinforcement is another thick layer of insulating material.

The insulation of the ignition cable is designed to withstand very high voltage without breaking down. Over time, however, leakage of ignition current will occur. Even a new cable will leak somewhat, but this is not important until the leakage increases so much that the spark at the spark electrodes is weakened or stopped.

To ensure that the dielectric strength of ignition cable insulation is adequate and that excessive leakage is not occurring, a harness tester called a **megohmmeter**, or **megger**, is used. Typical testers are the Continental High-Tension Lead Tester and the Eastern Electronics Cable Tester.

A harness tester is an electric unit designed to produce dc voltages up to 15 000 V which can be applied to individual leads in an ignition harness. A typical unit may include gauges to measure the applied voltage and leakage current, a voltage control, input leads, output leads, and required control switches. These units include instructions for proper application.

To test ignition leads, all leads are disconnected from the spark plugs and all leads but the one being tested are grounded to the engine. With the leads grounded, the tester will show leakage between leads as well as from leads to ground. The high-voltage lead from the tester is connected to the spark plug terminal of the lead being tested. The ground lead is attached to the engine. The tester may also be grounded to earth through a water pipe or other means.

Manufacturer's instructions are provided for all harness testers and should be followed. Since such a unit produces very high voltage, it is essential that the operator be most careful when the unit is turned on.

The voltage of the tester is adjusted to the level given in the instructions, which is usually 10 000 V. When the control switch is turned on, this voltage is applied to the lead being tested. Leakage will show on the microammeter and should not exceed 50 microamperes (μA).

As testing of leads continues, one or more leads will likely show high leakage because of the position of the distributor rotor in the magneto. If the rotor finger is aligned with the electrode for the lead being tested, the current will jump the gap to the rotor and flow to ground through the magneto coil. When this occurs, the engine crankshaft should be rotated to change the alignment of the distributor rotor so that the lead can be retested. Using an ohmmeter, or **cable tester**, check each lead for **continuity**. If continuity does not exist, the lead is broken and must be replaced.

If the test shows excessive leakage in several cables, the distributor block or terminal block is probably defective. The block should therefore be thoroughly examined. If the test shows that the harness is faulty, all the cables should be replaced.

When a distributor block shows a modest amount of leakage, it can sometimes be restored to good condition by cleaning and waxing with an approved high-temperature wax. If leakage persists, the block should be replaced.

A comparatively simple cable tester is illustrated in Fig. 9-10. This unit operates from either 12 or 24 V dc. The instrument is set for the correct voltage by means of the selector switch. When the tester is properly connected to ignition cables, the indicator light will reveal excessive leakage.

During the inspection of an ignition harness, it is important to note the routing of individual spark plug cables with respect to engine parts and particularly to the exhaust manifold. Cables should be routed and supported so that they cannot

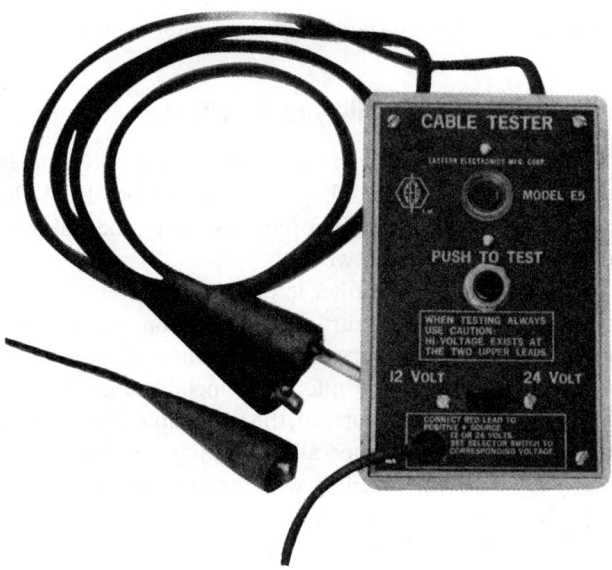

FIGURE 9-10 An ignition cable (harness) tester.

rub against engine parts or be located near hot parts which could burn the insulation. Sometimes it is necessary to adjust clamps and other supports to remove the cable from a position where it can become damaged by abrasion or heat.

Sharp bends should be avoided in ignition leads. If a cable is bent sharply or twisted, the insulation is under stress and can develop weak points through which high-tension current can leak.

Inspection and Maintenance of Induction System Air Filters and Ducting

The induction system air filter removes dirt and abrasive particles from the air before it enters the carburetor and/or the supercharger impeller. When the air filter has not been properly maintained, the result is the same as operating without a filter. The most common results of dirt or silicon entering the engine are worn piston rings and excessive ring groove wear. As ring groove wear progresses, the ring will eventually break.

It is imperative that the induction system air filter be installed properly. If it fits loosely so that the air can enter the induction system without being filtered, dirty air will enter the engine. This same problem will exist anywhere in the induction system where a leak is present.

There are several different types of air filters, and each has its own particular servicing procedures.

Dry Paper Filters

The **dry paper filter** must be cleaned daily when operation involves dusty conditions. If any holes or tears are noticed, the filter should be replaced immediately. To service the filter, remove the filter element and shake off the loose dirt by rapping on a hard, flat surface. Be especially careful not to crease or dent the sealing ends. When you are servicing a paper filter, never wash it in any liquid or soak it in oil, and never try to blow off dirt with compressed air. The filter housing can be cleaned by wiping with a cloth and a suitable solvent. When the housing is dry, reinstall and seal the filter element.

Wire Mesh Wetted Oil-Type Air Filter

The **wetted oil air filter** should be inspected daily for dirt accumulation and proper oiling. When dirt is found, the filter should be cleaned. If the filter requires oiling, the following procedure should be followed:

Thoroughly wash the filter in petroleum solvent.

Make certain that all dirt is removed from the filter and that the filter unit is in serviceable condition. If, after cleaning, the surfaces of the air filter show metallic wires through the remaining flocking material, the filter should be replaced. Dry the filter at room temperature, making certain it is thoroughly dry before proceeding with the next step. If the filter is not dry, the solvent will prevent the oil from adhering to the small surfaces of the filter and will thus decrease its efficiency. Next, immerse the filter in the recommended grade of oil for 5 min. After the filter is removed from the oil, allow

it to drain thoroughly before installing it in the aircraft. On many of these types of air filters, the cleaning instructions are printed on the filter housing.

Foam-Type Air Filters

Foam-type air filters are inspected daily, as are other types of air filters. However, unlike the other types of filters, there is no recommended cleaning procedure for foam-type filters. Instead, they are replaced at prescribed intervals, such as every 100 h.

Induction System Ducting

The inspection of the induction ducting is a visual inspection of the external surface and normally does not require duct removal. Inspect the external surface of the ducts for loose or broken strings, loose or displaced supporting wire, and signs of wear or perforation. Should any of these conditions exist, remove and replace the affected duct or ducts.

Inspection of Engine Fuel Systems and Carburetors

If possible, inspections should be carried out in accordance with manufacturer's or operator's instructions as set forth in appropriate manuals. If these are not available, the following general practices can be followed:

1. Remove cowling as necessary to gain access to the items to be inspected. Place cowling sections in suitable racks to avoid damage.

2. Examine all fuel line connections and fittings for signs of leakage. If fuel leakage is discovered, correct by tightening or replacing the fitting. If a leak cannot be stopped by applying the specified torque, the fitting or tube end should be replaced. Tubing fittings must not be overtorqued because of the danger of crushing the metal of the tubing and causing irreparable damage.

3. Observe the condition of the hoses. The outer surfaces should be smooth, firm, and free of blisters, bulges, collapsed bends, or deep cracks. Small blisters can be accepted, provided that there is no fuel leakage when the blister is punctured with a pin and when the hose is tested at $1\frac{1}{2}$ times the working pressure. Appreciable bulging at the hose fittings or clamps requires that the hose be replaced. Fine cracks which do not penetrate to the first fabric layer are acceptable.

4. Carefully examine the condition of the hoses and tubing at the clamps or brackets used for mounting. Both the mounting and the line should be checked for wear and looseness. A loose mounting will cause wear.

5. Check the metal fuel lines for wear, nicks, cuts, dents, and collapsed bends. Small nicks and cuts which do not extend deeper than 10 percent of the wall thickness and are not in the heel of a bend may be repaired by stoning and polishing with crocus cloth. Dents which are not deeper than 20 percent of the tubing diameter and are not in the heel of a bend are acceptable. Dents can be removed by drawing a

steel bullet through the tubing with an attached steel cable. Tubing which is not repairable must be replaced.

6. Remove the drain plugs in the carburetor and sumps to eliminate water and sediment. See that the plugs are reinstalled with proper torque and safetying. Install new washers with the plugs where required.

7. Remove all fuel screens and filters to clean them and to check their condition. Collapsed screens, and filters which do not provide free fuel flow, must be replaced. Main-line fuel screen sumps and tank drains should be opened briefly at preflight inspections to remove water and sediment.

8. If the fuel system includes an engine-driven fuel pump, check the pump for security, oil leakage from the mounting, and proper safetying of mounting nuts, bolts, or screws. Check the electric fuel boost pump for the operation and security of both fuel and electric connections. The brushes of the pump motor should be replaced in accordance with the schedule set forth in the service manual.

9. Check the carburetor for security of mounting, fuel leakage, and proper safetying. Check the gasket at the mounting flange or base to determine if there is a possibility of air leakage. Examine the throttle shaft bearings and the control arms for the throttle and mixture control for excessive play. Remember that excessive clearance at the throttle shaft can allow air to enter the carburetor and can lean the mixture. Apply lubricant to the bearings and moving joints in accordance with the service instructions or the approved lubrication chart.

Inspection and Maintenance of Fuel Injection System

The routine inspection and maintenance of an engine fuel system that includes a fuel injection system are similar to those of carburetors and fuel systems. The principal items to note are tightness and safetying of nuts and bolts, leakage from lines and fittings, and looseness in control linkages. Fuel strainers should be removed and cleaned as specified. Minor fuel stains at the fuel nozzles are normal and do not require repair.

When a new injector unit is installed, the injector inlet strainer should be removed and cleaned after 25 h of operation. Thereafter, the strainer should be cleaned at each 50-h inspection.

If an aircraft engine is equipped with a fuel injector that includes an AMC, the operator should be alert for signs of problems with the unit. Dirt can build up on the needle and cause rich operation and possible sticking of the needle, with resultant loss of altitude compensation.

Lubrication of the injector should be accomplished in accordance with the approved lubrication chart for the particular installation. The clevis pins used in connection with the throttle and the manual mixture control should be checked for freedom of movement and lubricated, if necessary.

Lubricate the throttle shaft bushings by placing a drop of engine oil on each end of the throttle shaft so that the oil can work into the bushings.

Use care in cleaning and oiling the air filter element. If the element is replaced while excessive oil is clinging to it, some of the oil will be drawn into the injector and will settle on the venturi. This can greatly affect the metering characteristics of the injector.

Inspection and Maintenance of Engine Controls

Engine controls, such as the throttle, mixture, propeller, and cowl flap controls, need to be checked during the course of a 100-h or annual inspection. The inspection of these controls should include the following steps:

1. Inspect push-pull controls for wear and smoothness of operation.

2. Operate the system slowly, and watch for signs of any strain on the rods and tubing that will cause bending or twisting.

3. Examine each rod end that is threaded, and observe whether the rod is screwed into the socket body far enough to be seen through the inspection hole.

4. Eliminate any play by making certain that all connections are tight.

5. Examine the guides to see if the rods bind too much on the guides, but do not mistake any binding for spring-back. Replace any guides that cause binding.

6. Adjust the lengths of screw-end rods by screwing them into or out of the control end. Retighten the locknuts.

7. If any rod is removed, label it to show its location on reassembly.

8. Replace any ball-bearing rod ends that cause lost motion.

Inspection and Maintenance of Exhaust Systems

The importance of proper inspection and maintenance of exhaust systems cannot be overemphasized. *Defective systems can lead to engine fire, engine failure, structural failure, or carbon monoxide poisoning.*

Approximately one-half of exhaust system failures occur in the exhaust gas-to-air heat exchanger, and as a result, carbon monoxide gas enters the cabin through the aircraft heater.

The exhaust system components are subjected to extreme temperatures, and the resulting expansion and contraction produce stresses which often lead to cracks and distortion resulting from warpage.

A primary reason for most exhaust system failures is inadequate and infrequent inspections and checks and the lack of routine and preventive maintenance between inspections. Exhaust systems deteriorate because of (1) engine operating temperatures, (2) vibration, which causes metal fatigue on areas of stress concentration and wear at joints or connections, and (3) engine backfiring and unburned fuel in the muffler. *Note:* These conditions begin to take effect the first hour of engine operation, and deterioration progresses through the life-span of the exhaust system components.

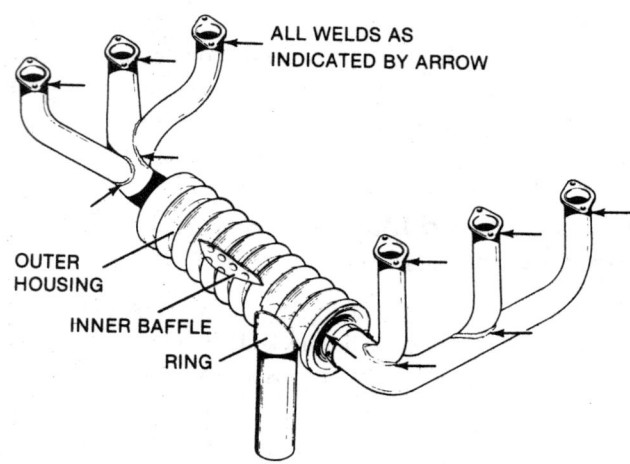

FIGURE 9-11 Exhaust system inspection points.

Indications of cracked or leaking exhaust systems can occur in any area of the system; however, the following are the most prominent problem areas:

1. Exhaust manifold and stack fatigue failures usually occur at welded or clamp joints (for example, exhaust stack flange, stack to manifold cross-pipes, or muffler connections). This is shown in Fig. 9-11.

2. Muffler failures and heat exchanger failures usually occur on the inner wall surface. Examples of fatigue areas are shown in Fig. 9-12. A proper inspection can be accomplished only when the outer heat shield is removed. This inspection should be done as recommended by the manufacturer or by a properly certificated technician or repair station.

Precautions

In the inspection and service of exhaust systems, certain precautions must be observed. Failure to employ adequate care in working with exhaust systems can result in their damage and deterioration.

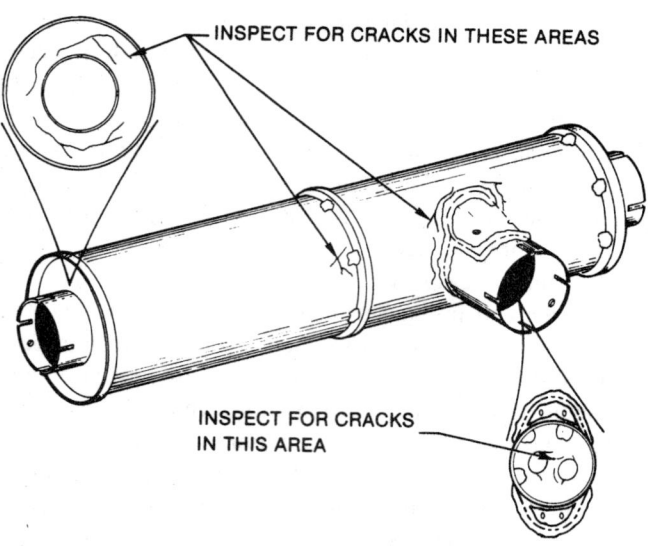

FIGURE 9-12 Typical muffler fatigue areas.

Corrosion-resistant exhaust system parts must be protected against contact with zinc-coated (galvanized) tools or any zinc-coated metal parts. Furthermore, lead pencils must not be used to mark exhaust system parts. At high temperatures, the metal of the exhaust system will absorb the zinc or lead of the lead pencil, and this will materially affect the molecular structure of the metal. Because of this, the softened metal will likely be subject to the development of cracks in the marked areas.

The exhaust system parts must be cleaned with care. This is particularly true of ceramic-coated parts. They must not be cleaned with harsh alkaline cleaners or by sandblasting. Degreasing with a suitable solvent will usually suffice. For a particular make and model of aircraft, the instructions of the manufacturer should be followed.

The reassembly of an exhaust system after inspection and repair is most critical. After the exhaust stacks or risers are secured to the cylinders, all other parts should be installed so that joints and other connections are in proper alignment to prevent exhaust leakage. Nuts, bolts, and clamp screws must be tightened to the correct torque. Overtorquing will probably result in failure.

Procedures

The procedures for performing exhaust system inspections are given in the manufacturer's maintenance manual for the aircraft. In general, the required inspections are similar in type and scope for most aircraft. The following steps are typical:

1. Remove the engine cowling sufficiently to see all parts of the exhaust system.

2. Examine all parts for cracks, wear, looseness, dents, corrosion, and any other apparent deterioration. Pay particular attention to attaching flanges, welded joints, slip joints, muffler shrouds, clamps, and attachment devices.

3. Check all joints for signs of exhaust leakage. Leakage can cause hot spots, in addition to being hazardous to passengers and crew. With supercharged engines operating at high altitudes, exhaust leaks assume the nature of blowtorches because of the sea-level pressures maintained in the system. These leaks are a fire hazard as well as being damaging to parts and structures. Evidence of leakage is a light gray or sooty spot at any slip joint or at any other point where pipes are joined. Leakage spots also reveal cracks in the system.

4. After a thorough visual inspection, the exhaust system should be pressure-checked. Attach the pressure side of an industrial vacuum cleaner to the tailpipe opening, using a suitable rubber plug to provide a seal. With the vacuum cleaner operating, check the entire system by feel or with a soap solution to reveal leaks. After the pressure test, remove the soap suds with water; dry the system components with compressed air.

5. For a complete inspection of the exhaust system, it may be necessary to disassemble the system and check individual components. Disassemble the system according

to the manufacturer's instructions, being careful to examine all attaching parts, such as clamps, brackets, bolts, nuts, and washers.

6. Remove the shrouds from the mufflers. Use rubber plugs to seal the openings, and apply $2\frac{1}{2}$-psi [17.24-kPa] air pressure while the muffler is submerged in water. Seal the exhaust stacks and pipes, and test in the same manner, using $5\frac{1}{2}$ psi [37.92 kPa]. Pressures used for testing may vary, but whatever the manufacturer recommends should be used.

7. After all components of the exhaust system have been examined and found satisfactory, reassemble the system on the engine loosely to allow for adjustment and proper alignment. Tighten the stack attachments to the cylinder exhaust ports first, using a torque wrench. Be sure that the proper type of heat-resistant nut is used and that new gaskets are installed. Next, tighten all other joints and attachments, making sure that all parts are in correct alignment.

8. After the exhaust system has been installed, run the engine long enough to bring it up to normal operating temperature. Shut down the engine and remove the cowling. Inspect each exhaust port and all joints where components are attached to one another. Look for signs of exhaust leaks, such as a light gray or sooty deposit. If a leak is found, loosen the connection and realign it.

9. If an exhaust system includes augmentors, inspect the augmentors in the same manner as the other components. Leaks in the augmentors can cause fires and escape of gases into the cockpit or cabin. The alignment of the augmentors is particularly critical. The manufacturer's specifications must be followed precisely.

10. On a system which includes a turbocharger, special inspections of the turbine and compressor assemblies must be made. Inspect the interior of all units for the buildup of coke deposits. These deposits can cause the waste gate to stick, causing excessive boost. In the turbine, carbon deposits will cause a gradual lessening of turbine efficiency, with a resulting decrease in engine power. Wherever coke buildup is found in any unit, remove it in accordance with the manufacturer's instructions.

Repairs

Exhaust system components which have become burned, cracked, warped, or so worn that leakage occurs should usually be replaced with new parts. In certain instances, cracks can be repaired by heliarc (inert-gas) welding with the proper type of welding rod. Care must be taken to avoid any repair which will cause a rough spot or protrusion inside an exhaust pipe or muffler. Any such area will create a hot spot and cause eventual burn-through.

Dents can sometimes be removed, provided that the dent has not caused a thin spot resulting from internal erosion and burning. Dents are removed by placing the exhaust pipe over a suitable mandrel and working out the dent with a soft hammer.

After any repair to a component of the exhaust system, a pressure test should be made.

Engine Mounts

Engine mounts are metal mounts of the type shown in Fig. 9-13, which connect the engine to the airframe. Rubber shock mounts isolate the engine from the metal engine mount. Most modern reciprocating-engine installations use **dyna-focal** engine mounts. The shock mounts on a dyna-focal engine mount point toward the center of gravity of the engine, as shown in Fig. 9-13. The bonded rubber shock mounts and metal mount are designed to reduce the transmission of engine vibrations to the airframe. This provides smoother aircraft operation and reduces the possibility of structural failure from vibration fatigue.

Cleaning and Inspection

All metal parts of the shock mount assemblies may be cleaned in a suitable cleaning solvent. The rubber pads (shock mounts) should be wiped clean with a dry, clean cloth. Do not clean the rubber parts with any type of solvent. Inspect the metal parts of the mounts for cracks, corrosion, dents, distortion, and excessive wear.

The rubber of the dynamic shock mount units can deteriorate with age and heat. Inspect the rubber pads for separation between pad and metal backing, swelling and cracking, or a pronounced set (distortion) of the pad. Replace all rubber pads that show evidence of damage.

The airframe manufacturer specifies the engine mount bolt torque for securing the engine to the mount and the mount to the airframe. The engine mount bolts should be checked for proper torque and any signs of rotation.

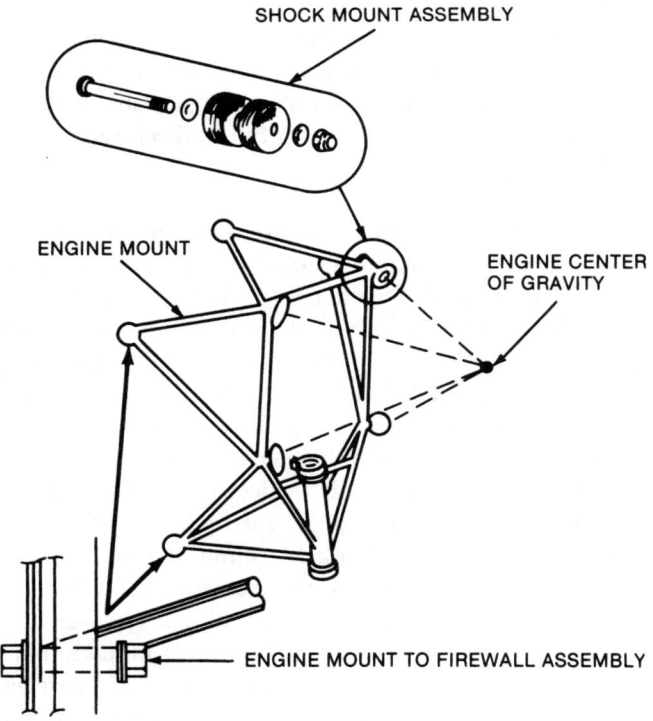

FIGURE 9-13 Dyna-focal engine mount.

Fire Wall Seals

During an inspection of the aircraft, the sealing compound and bushings that form seals around wiring and control cables that pass through the fire wall should always be inspected and repaired, if needed. Any type of openings in the fire wall area can allow exhaust gases and fumes to enter the cabin area.

Inspection, Maintenance, and Repair of Superchargers and Turbochargers

The same principles of inspection, maintenance, and repair that apply to other sections of the powerplant system apply to superchargers and turbochargers. Visual inspection of all visible parts should be performed daily to observe oil leaks, exhaust leaks, cracks in the metal of "hot sections," loose or insecure units, and other unacceptable conditions. Note that exhaust ducts, waste gates, nozzle boxes, and turbines are subjected to extremely high temperatures; thus, cracks develop because of the continual expansion and contraction of the metal as temperature changes occur.

The manufacturer's manual will specify the most important inspections to be accomplished and the service time established for periodic inspections. An inspection of a complete system should include the following, in addition to any other inspections specifically required by the company's operation manual or the manufacturer's maintenance manual:

1. Mounting of all units.
2. Oil leaks or dripping from any unit.
3. Security of oil lines.
4. Security and condition of electric wiring.
5. Cracks in ducting and other metal parts, including the turbine and housing.
6. Warping of metal ducts.
7. Operation of the complete system to determine performance, to discover undesirable sounds, and to note evidence of vibration. Unusual sounds and appreciable vibration require removal and replacement of the turbocharger to correct the faulty condition.

Improper lubrication or the use of an incorrect lubricant can cause serious malfunctions and the failure of units. Because of the high temperatures to which a turbine wheel is exposed, the turbine shaft is also subject to high temperatures. This can cause "coking" of the lubricant, with a subsequent buildup of carbon (coke) at turbine shaft seals and bearings. An appreciable amount of coking can cause failure of turbine shaft seals and bearings. Leaking shaft seals permit hot exhaust gases to reach the shaft bearings, where additional coking is likely to occur. Coking of the bearings is likely to limit the rpm that the turbine and compressor assembly can attain. In this case, the turbocharger will require removal and replacement. Because of the problems caused by coking in the turbine area, it is most essential that the proper type of lubricant be employed. The service manual for the aircraft will provide this information.

All overhauling and testing of superchargers and turbochargers should be accomplished at certified repair stations. This is particularly important because of the need to balance the turbine and compressor assembly accurately. These units rotate at speeds of up to 70 000 rpm; therefore, a slight unbalance can cause severe vibration and ultimate disintegration and failure.

TROUBLESHOOTING

With the increasing complexity of today's powerplants, maintenance technicians are more dependent on their ability to utilize published technical information in performing maintenance. Troubleshooting skills are increasingly needed by today's technicians. **Troubleshooting** is the step-by-step procedure used to determine the cause of a given fault and then select the best and quickest solution. When troubleshooting, the technician must evaluate the performance of the engine by comparing data on how the engine should operate with how it is currently performing. To troubleshoot, the technician needs thorough knowledge of the engine's theory of operation.

To pinpoint a fault, without wasting time and money, is not an easy job. Many times, faults can be intermittent, making the problem very difficult to isolate. Removing and replacing components on a trial-and-error basis can be viewed as "shotgun" troubleshooting in its worst form. To be effective, troubleshooting must be an analysis of the fault or faults. Probable causes, and the necessary actions to correct the problem, should be found through a logical and systematic approach.

A six-step troubleshooting procedure is illustrated in Fig. 9-14.

Symptom Recognition

The prerequisites for troubleshooting are to be familiar with the normal engine condition and to be able to recognize when an engine is not operating properly. Therefore, symptom recognition—the first step in troubleshooting—involves having knowledge of any engine condition that is not normal and knowing to what extent the fault is affecting the engine's performance.

Symptom Elaboration

Symptom elaboration is the next logical step, once a fault or malfunction has been detected. Test equipment, built-in or external, helps the technician to evaluate the performance of the engine and its components. The technician should use these aids to assess the effects of the symptoms and to provide additional information to further define the symptoms. Depending on the type of engine, the technician should ask some of the following questions to help elaborate the symptoms:

1. What components of the engine are not operating normally?
2. What operating rpm ranges are abnormal?

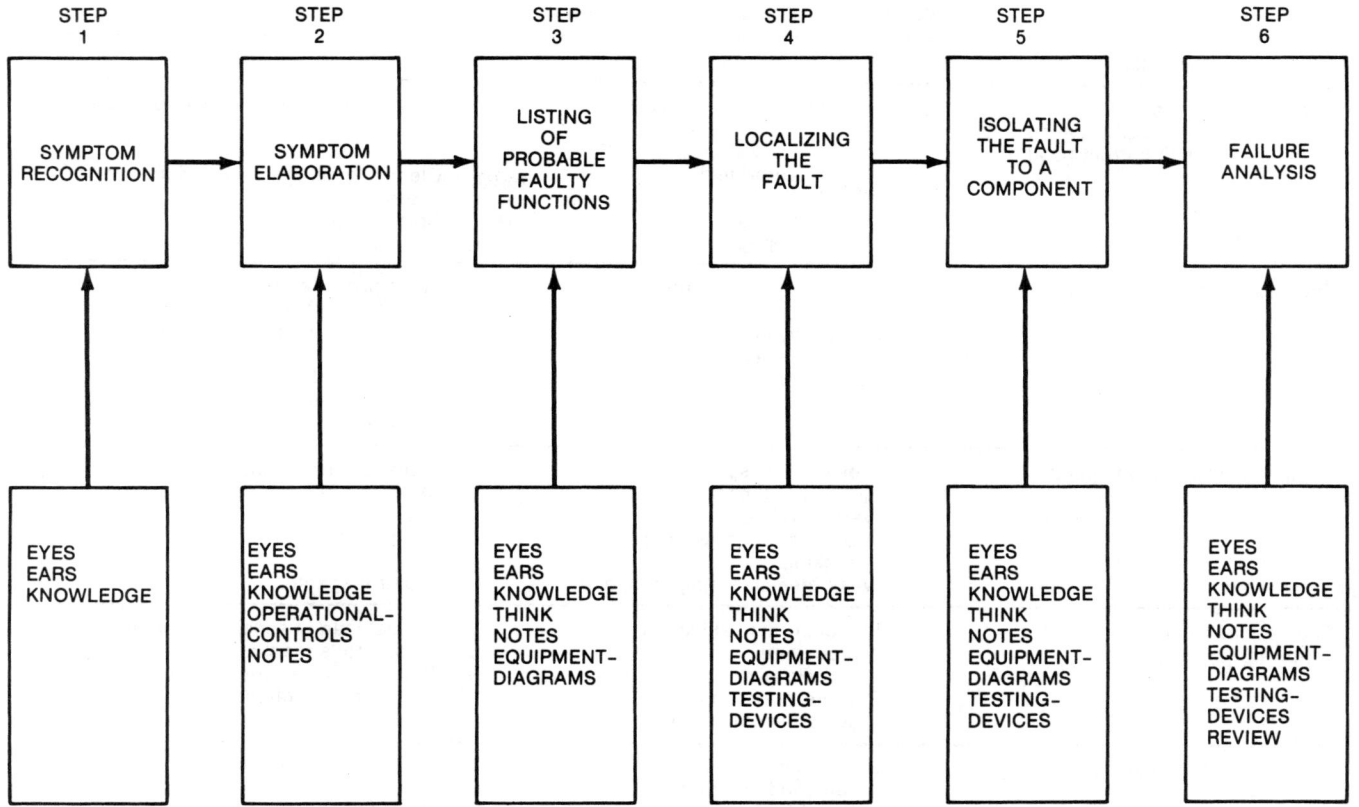

FIGURE 9-14   Six-step troubleshooting procedure.

3. Is this an intermittent fault or a continuous problem?

4. Is an engine parameter such as temperature or pressure out of limit?

5. Does the malfunction occur only under a specific set of circumstances?

These are a few examples of the types of questions that the technician may ask before continuing any maintenance action. The technician should try not to overlook any information. Sometimes the smallest detail can lead the way to the solution of the problem.

List of Probable Faulty Functions

When the technician has located all the symptoms of the malfunction or fault, the third step is to list, either mentally or on paper, the possible causes. To aid in this process, most manufacturer's technical service manuals list the "probable cause" for a certain engine symptom, along with the suggested corrective action. This information is often contained in a troubleshooting chart such as that in Fig. 9-15.

Localizing the Fault

Armed with a complete set of symptoms and their probable causes, the technician is ready for the fourth step, localizing the fault. Localizing the fault is an attempt to determine which functional system of the engine is actually creating the problem. The trouble may be traced by using the manufacturer's

troubleshooting charts and employing computers with special programs set up for troubleshooting engines.

Isolating the Fault to a Component

Once the malfunction is isolated to one system, additional testing is done to isolate the fault to a specific component. The technician often uses test equipment to measure or indicate the correct outputs for various system components.

Failure Analysis

Once the fault can be traced to a specific component or components, an attempt should be made to determine the cause of the failure. Substituting a new component into the system without analyzing the reason for the failure may just damage the new component. Sometimes many components have similar functions. In this case, all the components may need to be replaced or repaired.

To determine if there are multiple malfunctions, the technician should consider the effect of the component malfunction on engine operation. If the component is the probable cause of all the abnormal symptoms noted in earlier steps, then it can be assumed that the component is at fault.

Using the six-step procedure, we give an example of how to troubleshoot a fault on an engine that will not run consistently at idle speeds. The first step is symptom recognition.

Engine performance can be measured against several standards. The present performance of a particular engine

Indication	Cause	Remedy
Engine will not start. Engine cranking. All circuit breakers and switches in correct position.	Lack of fuel.	Check fuel valves. Service fuel tanks.
	Engine overprimed.	Clear engine. Follow correct starting procedure.
	Induction system leaks.	Correct leaks.
	Starter slippage.	Replace starter.
Engine will not run at idling.	Propeller lever set for DECREASE RPM.	Place propeller lever in HIGH RPM position for all ground operations.
	Improperly adjusted carburetor or fuel-injection system.	Readjust system as required.
	Fouled spark plugs.	Change spark plugs.
	Air leak in intake manifold.	Tighten loose connection or replace damaged part.
Engine misses at high speed.	Broken valve spring.	Replace valve spring.
	Plugged fuel nozzle.	Clean or replace.
	Warped valve.	Replace valve.
	Hydraulic tappet worn or sticking.	Replace tappet.
	Weak breaker spring in magneto.	Repair magneto.
Engine runs too lean at cruising power.	Improper manual leaning procedure.	Manual lean in accordance with operator's manual.
	Low fuel flow.	Check and clean fuel strainer.
	Carburetor or fuel-injection system malfunction.	Correct malfunction.
Engine runs rough at high speed.	Loose mounting bolts or damaged mount pads.	Tighten or replace mountings.
	Plugged fuel nozzle.	Clean or repair.
	Propeller out of balance.	Remove and repair propeller.
	Ignition system malfunction.	Troubleshoot ignition system and repair.
Engine idles rough.	Improperly adjusted carburetor or fuel-injection system.	Adjust system as required.
	Fouled spark plugs.	Clean or replace spark plugs.
	Improperly adjusted fuel controls.	Adjust fuel controls.
	Discharge-nozzle air vent manifold restricted or defective.	Clean or replace.
	Dirty or worn hydraulic lifters.	Clean or replace.
	Burned or warped exhaust valves, seats. Scored valve guides.	Repair or replace.
Engine runs rich at cruising power.	Restriction in air-intake passage.	Remove restriction.
Spark plugs continuously foul.	Piston rings worn or broken.	Overhaul engine.
	Spark plugs have wrong heat range.	Install proper range spark plugs.
Sluggish engine operation and low power.	Improper rigging of controls.	Rerig controls.
	Leaking exhaust system to turbo.	Correct exhaust system leaks.
	Restricted air intake.	Correct restriction.
	Turbo wheel rubbing.	Replace turbocharger.
	Ignition system malfunction.	Troubleshooting ignition system and correct malfunction.
	Carburetor or fuel-injection system malfunction.	Troubleshoot and correct malfunction.
	Engine valves leaking. Piston rings worn or sticking.	Overhaul engine.
High cylinder-head temperature.	Octane rating of fuel too low.	Drain fuel and fill with correct grade.
	Improper manual leaning procedure.	Use leaning procedure set forth in the operator's manual.
	Bent or loose cylinder baffles.	Inspect for condition and correct.
	Dirt between cooling fins.	Remove dirt.

FIGURE 9-15 General engine troubleshooting chart. (*Continued*)

Indication	Cause	Remedy
	Exhaust system leakage. Excessive carbon deposits in combustion chambers.	Correct leakage. Overhaul engine.
Oil pressure gauge fluctuates.	Low oil supply.	Determine cause of low oil supply and replenish.
Engine oil leaks.	Damaged seals, gaskets, O rings, and packings.	Repair or replace as necessary to correct leaks.
Low compression.	Excessively worn piston rings and valves.	Overhaul engine.
Engine will not accelerate properly.	Unmetered fuel pressure too high. Turbocharger waste gate not closing properly. Leak in turbocharger discharge pressure.	Adjust engine fuel pressure according to specifications. Refer to turbocharger and controls manual. Repair or replace as necessary.
Slow engine acceleration on a hot day.	Mixture too rich.	Lean mixture until acceleration picks up. Then return control to FULL RICH.
Engine will not stop at IDLE CUTOFF.	Manifold valve not seating tightly.	Repair or replace manifold valve.
Manifold pressure overshoot on engine acceleration.	Throttle moved forward too rapidly.	Open throttle about half way. Let manifold pressure peak, then advance throttle to full open.
Slow engine acceleration at airfields with ground elevation above 3500 ft [1066.80 m].	Mixture too rich.	Lean mixture with manual mixture contol until operation is satisfactory.
When climbing to 12,000 ft [3657.60 m], engine quits when power reduced.	Fuel vaporization.	Operate boost pump when climbing to high altitudes. Keep boost pump on until danger of vapor is eliminated.

FIGURE 9-15 General engine troubleshooting chart.

can be compared with its past performance, provided adequate records have been kept. Engine performance can also be compared with that of other engines installed on the same type of aircraft. Type Certificate Data Sheets and the engine operator's manual can be consulted for engine performance information.

Once the basic fault has been recognized (that the engine will not idle), the next step is to elaborate on the fault and ascertain the engine operating conditions at which it is exhibited. The following information may be of value in defining this problem:

1. Was any roughness noted? If so, under what conditions of operation?
2. How long have the engine and spark plugs been in use? How long has it been since the last inspection?
3. Were the ignition system (magneto) operational check and power check normal?
4. Did the problem change when the fuel boost pump was on?
5. When did the trouble first appear?
6. Was the full-throttle performance normal?

The next step is to list the probable faulty function. Refer to Fig. 9-15. Reasons for failure of the engine to idle could be

1. Propeller lever set to decrease rpm
2. Improperly adjusted carburetor or fuel injection system
3. Fouled spark plugs
4. Air leak in intake manifold

With a complete set of symptoms and probable causes, the technician is ready to do step 4 of troubleshooting, which includes testing various systems to localize the fault in one system. This can be done sometimes by eliminating systems that have been tested and found to be in good condition.

In most instances, assume that the trouble lies in one of the following systems:

1. Ignition system
2. Fuel metering system
3. Induction system
4. Power or mechanical system
5. Instrumentation system

Utilizing the manufacturer's troubleshooting manuals and charts to test and eliminate the various systems makes it possible to pinpoint the induction system as the problem here.

Let's go on to the next step—narrowing the problem to a specific component. Visual inspection of the induction system reveals that one cylinder intake pipe has fuel stains and a bad gasket. This is determined to be the cause of the fault, and with a new gasket, the engine can be repaired and its operation checked.

In troubleshooting step 6 (failure analysis), the technician should determine the cause of the gasket failure and whether other engine components have been affected by the malfunction. In this case, the cause of the gasket failure appears to be normal aging.

In assessing the possible damage, the technician notes that the cylinder with the leaking intake pipe has been operating with a very lean mixture, which could have caused the cylinder to run hot. Damage to other cylinders on the engine is highly unlikely, but damage could have occurred inside the leaking cylinder. The leaking cylinder should be thoroughly inspected for damage that could have resulted from the intake leak.

In all troubleshooting cases, the knowledge and experience of the technician will be needed along with a good logical approach to perform successful fault isolation.

Troubleshooting Examples

Engine operational malfunctions can usually be traced to one or more of three basic causes: (1) ignition malfunctions, (2) fuel system malfunctions, and (3) engine part malfunctions. Although in practice it is necessary to consult the specific manufacturer's manual, in the following text, each of these types of malfunctions is discussed separately.

Ignition Malfunctions

Ignition troubles may be traced to defective magnetos, defective transformers in a low-tension system, improper timing, spark plugs which are burned or otherwise damaged, poor insulation on the high-voltage leads, short-circuited or partially grounded primary or switch (P) leads, burned breaker points, or loose connections.

Missing at High Speeds. Misfiring of the engine at high speeds can be caused by almost all the foregoing defects in varying degrees. If the engine is operating at high speeds and high loads, the manifold pressure, and the cylinder pressure, will be high. As previously explained, more voltage is needed at the spark plug gap to fire the plug when the pressure at the gap is increased. This means that at high engine loads the ignition voltage will build up more than at low engine loads. This higher voltage will seek to reach ground through the easiest path, and if there is a path easier to follow than that through the spark plug gap, the spark plug will not fire and the spark will jump through a break in the insulation or follow a path where dampness has reduced the resistance.

If the airplane is operating at high altitudes, the high-voltage spark will be still more likely to leak off the high-tension leads instead of going through the spark plug. The lower air pressure at high altitudes permits the spark to jump a gap more readily than at the higher pressure near sea level.

A weak breaker-point spring will also cause misfiring at high speeds. This is because the breaker points do not close completely after they are opened by the cam. This condition is called **floating points** because the cam follower actually does not maintain contact with the cam but floats at some point between the cam lobes.

Engine Fails to Start. If the engine will not start, the trouble can be a defective ignition switch. Since aircraft engines have dual-ignition systems, it is rare that both systems fail at the same time. However, in some magneto switches, both magnetos could be grounded through a short circuit inside the switch.

If recommended practice is to start the engine on one magneto only and the engine will not start, try to start the engine on the other magneto. If the cause of the trouble is in the first magneto system, the engine will fire on the other magneto.

The checking of magnetos during the engine test will usually reveal malfunctions in one magneto or the other before a complete failure occurs. The defective magneto can then be removed and repaired before serious trouble occurs.

Defective Spark Plugs. Defective spark plugs are usually detected during the magneto check. If one spark plug fails, the engine rpm will show an excessive drop when it is checked on the magneto supplying the defective plug. The bad plug may be located by the **cold-cylinder check**. This is accomplished as follows: Start and run the engine for a few minutes on both magnetos. Perform a magneto check, and determine which magneto indicates a high rpm drop. Stop the engine, and let it cool until the cylinders can be touched without burning your hand. Start the engine again, and operate on the magneto for which the high rpm drop was indicated. Run the engine for about 1 min at 800 to 1000 rpm, and then shut it down. Immediately feel all cylinders with your hand, or use a cold-cylinder tester to determine which is the cold cylinder. This cylinder will have the defective spark plug. The time and expense involved in removing and replacing all the spark plugs to correct one defective plug can be avoided by pinpointing the defective plug through a cold-cylinder check.

Fuel System Troubles

Fuel systems, carburetors, fuel pumps, and fuel control units can cause a wide variety of engine malfunctions, some of which may be difficult to analyze. A thorough understanding of the system and its components is essential if the technician hopes to solve the problems of a particular system effectively. Figure 9-16 lists some of the most common problems encountered with fuel systems and suggests remedies.

Indication	Cause	Remedy
Engine will not start.	No fuel in tank.	Fill fuel tank.
	Fuel valves turned off.	Turn on fuel valves.
	Fuel line plugged.	Starting at carburetor, check fuel line back to tank. Clear obstruction.
	Defective or stuck mixture control.	Check carburetor for operation of mixture control.
	Pressure discharge-nozzle-valve diaphragm ruptured.	Replace discharge-nozzle valve.
	Primer system inoperative.	Repair primer system.
Engine starts, runs briefly, then stops.	Fuel tank vent clogged.	Clear the vent line.
	Fuel strainer clogged.	Clean fuel strainer.
	Water in the fuel system.	Drain sump and carburetor float chamber.
	Engine fuel pump inoperative or defective.	Replace engine-driven fuel pump.
Black smoke issues from exhaust. Red or orange flame at night.	Engine mixture setting too rich.	Correct the fuel–air mixture adjustment.
	Primer system leaking.	Replace or repair primer valve.
	At idling speed, idle mixture too rich.	Adjust idle mixture.
	Float level too high.	Reset carburetor float level.
	Defective diaphragm in pressure carburetor.	Replace pressure carburetor.

FIGURE 9-16 Fuel system troubleshooting chart.

Figure 9-16 does not cover all symptoms which may develop with fuel systems and carburetors because of the many different designs involved. The technician, in each case, should analyze the type of system upon which she or he is working and become familiar with the operation of the carburetor or fuel control unit used. Fuel injection systems involve some unique problems. Figure 9-17 lists problems that may be encountered with one particular fuel injection system.

Oil System Problems

Oil system troubles are usually revealed as leaks, absence of oil pressure, low oil pressure, fluctuating oil pressure, high oil pressure, and high oil consumption. The correction of oil leaks is comparatively simple in that it involves tracing the leak to its source and then making the indicated repair. If oil has spread over a large area of the engine, it is sometimes necessary to wash the engine with solvent and then operate it for a short period to find the leak.

A check for oil pressure when an engine is first started is always a standard part of the starting procedure. If the oil pressure does not show within about 30 s, the engine is shut down. Lack of oil pressure can be caused by any one of the following conditions: no oil in the tank; no oil in the engine oil pump (therefore, no prime); an air pocket in the oil pump; an oil plug left out of a main oil passage; an inoperative oil pump; an open pressure relief valve; a plugged oil supply line; or a broken oil line. If there is no oil pressure, the technician should start with the most likely cause and then check each possibility in turn until the trouble is located.

Low oil pressure can be caused by a variety of discrepancies including the following: oil pressure relief valve improperly adjusted, broken oil relief valve spring, sticking pressure relief valve, plug left out of an oil passage, defective gasket inside the engine, worn oil pressure pump, worn bearings and/or bushings, dirty oil strainer, excessive temperature, wrong grade of oil, and leaking oil dilution valve. The cause of low oil pressure is often more difficult to discover than the causes of some other oil problems; however, a systematic analysis of the problem by technicians will usually lead to a solution. One of the first questions technicians must ask is whether the condition developed gradually, or showed up suddenly. They should also check how many hours of operation the engine has had. Another most important consideration is the actual level of the oil pressure. If it is extremely low, technicians look for an "acute" condition, and if it is only slightly low, the cause will probably be different.

High oil pressure can result from only a few causes: an improper setting of the relief valve, a sticking relief valve, an improper grade of oil, low temperature of oil and engine, and a plugged oil passage. The cause can usually be located easily except in a newly overhauled engine where a relief valve passage may be blocked. Note that the oil pressure will be abnormally high when a cold engine is first started and is not yet warmed up.

High oil consumption is usually the result of wear or leaks. If blue oil smoke is emitted from the engine breather and exhaust, most likely the piston rings are worn, so that **blowby** occurs. In blowby, pressure built up in the crankcase causes the oil spray inside the crankcase to be blown out the breather. High breather pressures occur in some engines because of buildup of sludge in the breather tube; this may be detected by excessive leakage in the propeller shaft seal. Cleanness of the breather tube is usually checked at the

Indication	Cause	Remedy
Engine will not start. No fuel flow indication.	Fuel-selector-valve in wrong position.	Position fuel-selector-valve handle to main tank.
	Dirty metering unit screen.	Clean screen.
	Improperly rigged mixture control.	Correct rigging of mixture control.
Engine acceleration is poor.	Idle mixture incorrect.	Adjust fuel-air control unit.
Engine will not start. Fuel flow gauge shows fuel flow.	Engine flooded.	Clear engine of excessive fuel.
	No fuel to engine.	Loosen one line at fuel manifold nozzle; if no fuel shows, replace fuel manifold.
Engine idles rough.	Restricted fuel nozzle.	Clean nozzle.
	Improper idle mixture.	Adjust fuel-air control unit.
Very high idle and full-throttle fuel pressure present.	Relief valve stuck closed.	Repair or replace injector pump.
Engine runs rough.	Restricted fuel nozzle.	Clean nozzle.
	Improper pressure.	Replace pump.
Low fuel pressure at high power.	Leaking turbocharger discharge pressure.	Repair leaking lines and fittings.
	Check valve stuck open.	Repair or replace injector pump.
Low fuel flow gauge indication.	Restricted flow to metering valve.	Clean fuel filters and/or adjust mixture control for full travel.
	Inadequate flow from fuel pump.	Adjust fuel pump.
Fluctuating or erroneous fuel flow gauge indication.	Vapor in system.	Clear with auxiliary fuel pump.
	Clogged ejector jet in vapor-separator cover.	Clean jet.
	Air in fuel flow gauge line.	Repair leak and purge line.
High fuel flow gauge indication.	Altitude compensator stuck.	Replace fuel pump.
	Restricted nozzle or fuel manifold valve.	Clean or replace as required.
	Recirculation passage in pump restricted.	Replace fuel pump.
Fuel discharging into engine compartment. Relief valve probably not operating.	Leaking diaphragm.	Repair or replace injector pump.
No fuel pressure.	Check valve stuck open.	Repair or replace injector pump.
Unmetered fuel pressure.	If high, internal orifices are plugged.	Clean internal orifices in injector pump.
	If low, relief valve stuck open.	Repair or replace injector pump.

FIGURE 9-17   Fuel injection troubleshooting chart.

100-h inspection on engines where this may be a problem. The worn rings also allow oil to pass the piston and enter the combustion chamber, where it is burned. This, of course, produces blue smoke at the exhaust.

Another cause of high oil consumption is a worn master rod bearing in a radial engine; this permits too much oil to be sprayed from the bearing and into the cylinder bores. If the scavenger pump is defective, the oil will not be removed from the sump as rapidly as required, and this will also lead to excessive oil consumption.

Operating an engine at high power settings and high temperatures will increase the oil consumption. If an apparently normal engine is using more oil than it should, the pilot should be questioned regarding engine operation in flight. The pilot should observe the reading of the oil pressure gauge frequently during operation. If the gauge should begin to fluctuate, the flight should be terminated as soon as possible because there may be a low oil supply.

Induction System Problems

The designs of induction systems for reciprocating engines vary considerably for different aircraft-and-engine combinations. The simplest types of induction systems include an air filter in the forward-facing air scoop, a carburetor air heating system, ducting to the carburetor, the carburetor, and an intake manifold or intake pipes that carry the F/A mixture to the valves. Other systems include turbochargers, superchargers, alternate air systems, and carburetor deicing systems. For a particular aircraft-engine combination, the technician should consult the operator's and maintenance manuals for the aircraft. Induction systems are described in Chap. 5.

These problems may arise in a typical induction system:

1. Dirty and/or damaged air filter
2. Worn, loose, or damaged air ducting
3. Loose or defective air temperature bulb
4. Defective air heater valve
5. Defective alternate air valve
6. Loose carburetor mounting
7. Defective carburetor mounting gasket
8. Leaking packings or gaskets at intake pipes
9. Leaking intake manifold

Any crack or other opening that allows air to enter an intake manifold, in naturally aspirated engines, will cause the F/A mixture to be excessively lean and may cause engine damage and adversely affect engine performance. Leaks in the intake manifold of a supercharged or turbocharged engine will allow the F/A mixture to escape, thus reducing MAP and engine power output.

Inspection of a typical induction system includes the following:

1. Check the air filter for condition, cleanness, and security. Service the air filter according to instructions.
2. Check the air ducting to the carburetor for wear damage, cracks, and security of mounting.
3. Check the air heater valve and ducting for wear, cracks, and security of mounting. Check the valve door bearings for wear, and lubricate according to instructions.
4. Check the CAT bulb for security.
5. Check the carburetor mounting for security. Tighten any loose bolts or cap screws.
6. Check the carburetor mounting gasket for possible air leakage. Replace the gasket if it is damaged.
7. Check the intake pipes and/or manifold for condition and security.
8. Check the intake pipe packing nuts for tightness. Check the packings or gaskets for shrinkage or damage.
9. Check the alternate air system for condition.

Backfiring

Backfiring occurs when the flame from the combustion chamber burns back into the intake manifold and ignites the F/A mixture before the mixture enters the engine. It often occurs during starting of a cold engine because of poor (slow) combustion. The F/A mixture in the cylinder is still burning at the time the intake valve opens, and the flame burns back through the intake valve. This sometimes causes a fire in the induction system.

Any defect in the carburetor or fuel control system which causes an excessively lean mixture can lead to backfiring. If the condition persists after an engine is warmed up, follow a systematic procedure to locate the cause.

Another cause of backfiring is sticking intake valves. This does not usually occur with a new or recently overhauled engine, but it is likely to be encountered with an older engine operated at high temperatures. If a sticking intake valve remains open, it can cause the engine to stop and may cause considerable damage to the induction system.

Ignition troubles often cause backfiring. If high-tension current leaks at the distributor block, it can cause the plugs to fire out of time, so that the mixture may fire in a cylinder when the intake valve is open. If a newly overhauled engine is being started for the first time and backfiring persists, the technician should check for ignition timing and for proper connection of the spark plug leads. Ignition out of time can also cause afterfiring through the exhaust.

Afterfiring

Afterfiring is the burning of F/A mixture in the exhaust manifold after the mixture has passed through the exhaust valve. It is characterized by explosive sounds and large flames trailing outward from the exhaust stacks (torching). Usually excessive fuel (rich mixture) in the exhaust continues to burn after the mixture leaves the cylinder. The condition may be caused by overpriming, excessively rich mixture, poor ignition, and improper timing. Since there are comparatively few causes of afterfiring, it is usually easy to correct.

REVIEW QUESTIONS

1. If a fire occurs during starting of the engine, what should the operator do?
2. What fire extinguishing agent is commonly used on a reciprocating engine fire?
3. When starting an engine with a float-type carburetor, where should the mixture control be positioned?
4. What is meant by the term *liquid lock*?
5. What should be done if there is no indication of oil pressure shortly after starting the engine?
6. In the operation of an engine with a constant-speed propeller, what sequences should be followed in changing power settings?
7. What is meant by *cruise control*?
8. How is maximum range obtained in the operation of an aircraft engine?
9. When an engine is equipped with a carburetor which has an idle cutoff, what is the procedure for shutting down the engine?
10. Why is the mixture leaned out on a reciprocating engine?
11. What is the general procedure for leaning out the mixture on an engine equipped with a float-type carburetor?
12. What operating procedures can help reduce lead fouling of the spark plugs?

13. What is meant by preheating an engine?

14. What should be used to clean an aircraft engine?

15. What inspection should be made with respect to the engine when an oil screen is serviced?

16. How is an oil filter inspected when it is removed from an engine during inspection?

17. Name two types of compression testers utilized on reciprocating engines.

18. When using a differential-pressure compression tester, what percentage pressure loss is considered unacceptable?

19. What are the procedures for servicing a dry-paper induction air filter?

20. List the six steps that should be utilized in troubleshooting an engine.

Reciprocating-Engine Overhaul Practices 10

INTRODUCTION

Maintenance of airframes and powerplants is one of the most critical segments of the aviation industry. Individuals responsible for the maintenance or overhaul of a certificated aircraft engine or any of its parts bear a serious responsibility to the public, to themselves, and to their organizations. They must do their work in such a way that when they have completed a particular repair or overhaul job, they can say, "This part (or this engine) will not fail."

NEED FOR OVERHAUL

Someone entering the field of aviation maintenance may wonder why engine overhaul is necessary, what it consists of, and how often it must be done and may have a variety of other questions. These questions are not difficult, because experience has provided the answers.

After a certain number of hours of operation, an engine undergoes various changes which make an overhaul necessary. The most important of these changes are as follows:

1. Critical dimensions in the engine are changed as a result of wear and stresses, thus bringing about a decrease in performance, an increase in fuel and oil consumption, and an increase in engine vibration.
2. Foreign materials, including sludge, gums, corrosive substances, and abrasive substances, accumulate in the engine.
3. The metal in critical parts of the engine may be crystallized as a result of constant application of recurring stresses.
4. One or more parts may actually fail.

Experience with a particular make and model of engine establishes the average time this model may be operated without expectation of failure.

The **recommended overhaul time** is determined by the manufacturer for each model of engine based on experience and is made known to the aircraft owners and technicians through service publications. Operators of aircraft used in commercial air transportation must adhere to the overhaul time recommendations of the manufacturer or get approval from the FAA to operate an engine beyond this time.

However, the owner of a private aircraft has two other options when the aircraft engine has reached its recommended overhaul time, especially if the engine has been operated so that it has not been overheated or overstressed. The owner may disregard the recommended engine overhaul time until it is decided that a major overhaul is required, the decision being based on the performance of the engine. Or the owner may decide to have a **top overhaul** performed on the engine, which is the complete reconditioning of only the cylinders, pistons, and valve operating mechanism.

The owner must realize that whenever the overhaul period for an engine is extended, other factors may become important. The value of the aircraft may be depreciated because of the high engine time; or, when the engine is eventually overhauled, it may be worn beyond repairable limits, thus requiring extensive replacement of parts and greatly increasing the overhaul cost. The additional cost may be greater than the amount saved by extending the major overhaul period. The advice of a qualified engine specialist should be obtained before the decision is made to extend the overhaul time.

The actual time between overhauls (TBO) for an engine is determined largely by the manner of its operation. If the operator is careful and continually observes the rules for good engine operation, the life of the engine may be extended for several hundred hours. But careless operation may create a need for engine overhaul in much less than the normal period.

Overhauled vs. Rebuilt Engine

An **overhauled engine**, according to the FAA, is one that has been disassembled, cleaned, inspected, repaired as necessary, reassembled, and tested according to the manufacturer's instructions and specifications. A **rebuilt engine** must also undergo all the steps described above. In addition, the parts that are reused in the rebuilt engine must meet the same limits and tolerances specified for new parts by the engine manufacturer. The rebuilt engine must also be functionally tested and must meet the requirements of a new engine.

A rebuilt engine does not have to carry the previous operating history when it is returned to service. It can be said that the engine has been granted **zero time**. An aircraft engine may be labeled as rebuilt only by the manufacturer or an agency approved by the manufacturer. Many agencies

267

overhaul engines to the specifications of a rebuilt engine; however, since these agencies are not approved by the manufacturer for engine rebuilding, the engine must be labeled as overhauled and carry the previous operating time.

The term **remanufactured** is often used in regard to engine and accessory maintenance, but is not defined by the FAA. When an engine or accessory is labeled as remanufactured, it is important to obtain a detailed description of the maintenance performed.

Major Overhaul and Major Repairs

The process of taking apart, inspecting, repairing, reassembling, and testing an entire engine is referred to as a **major overhaul** and should not be confused with major repairs. A major overhaul can be performed or supervised by the powerplant mechanic and does not require FAA Form 337 unless the engine is equipped with an integral supercharger or has a propeller reduction system with other than spur-type gears. Repairs to structural engine parts such as welding, plating, and metallizing constitute major repairs and may be approved only by appropriately rated persons or agencies.

OVERHAUL SHOP

To provide for efficient overhaul operations so that engine overhaul can be accomplished at minimum cost to the aircraft owner, an overhaul shop must be well organized. The type of organization largely depends on the sizes of engines to be overhauled and the volume (number of engines to be overhauled in a given time). Overhaul shops may be established to handle comparatively few small engines, or they may be large companies which overhaul hundreds of small and large engines each year.

Regardless of the size of the operation, all engines go through the same steps, individually or several at a time on a production line. The basic steps of the overhaul process are as follows:

1. Receiving inspection
2. Disassembly
3. Visual inspection
4. Cleaning
5. Structural inspection
6. Dimensional inspection
7. Repair and replacement
8. Reassembly
9. Installation
10. Engine resting and run-in
11. Preservation and storage

Each of these steps will be discussed in detail. The processes involved in each step will also be described. Some of these processes require specialized equipment that may not be available to the individual powerplant technician. It is up to the powerplant technician to determine which functions

he/she can perform and which must be sent off to an appropriately rated repair station.

Tools and Equipment

Before attempting to overhaul an aircraft engine, the powerplant maintenance technician should make sure that all the required tools and equipment are available. Engine manufacturers design special tools and fixtures to be used for the overhaul of their engines; however, it is not necessary to have every tool listed in the overhaul manual. Many standard tools, such as wrenches, gear pullers, arbors, lifting slings, reamers, etc., can be used in place of similar tools which may be available from the manufacturer. However, if a particular overhaul shop is likely to receive many engines of the same type for overhaul, it is a good plan to get most (or all) of the tools designed for the overhaul of that type of engine. Among the tools and equipment needed for engine overhaul are the following:

Engine Shop Equipment

Arbors of several sizes	Drill press
Arbor press	Engine lathe
Chain hoist	Generator test stand
Cleaning equipment	Heating equipment
Crankshaft thread cap	Lifting sling
Crankshaft wrench	Penetrant inspection equipment
Magnetic inspection equipment	Test stand and propeller
Magneto test stand	Timing disk and TC indicator or Time-Rite equipment
Magneto timing tools	
Overhaul stand	Timing lights
Parts trays	Vises
Cylinder fixtures or holding blocks	

Engine Overhaul Tools

Connecting-rod alignment tools	Ignition harness tools
Cylinder base wrench	Piston ring compressor
Gear pullers	Reamers
Valve guide tools	Tap and the tools
Valve refacing equipment	Valve spring compressor
Valve seat grinding equipment	Valve spring tester

Hand Tools

Box wrenches	Safety wire tools
Bushing reamers	Screw extractors
Counterbores	Spanner wrenches
Diagonal cutters	Spot facers
Drifts of several sizes	Stud drivers
Drill motor	Stud pullers
End wrenches	Torque wrenches
Heli-Coil tools	Pliers

Precision Measuring Tools

Depth gauges	Small-hole gauges
Dial gauges	Surface plate
Height gauges	Telescope gauges
Micrometers	V blocks

Nondestructive Inspection and Testing Equipment

Magnaflux	Ultrasonic
Zyglo	Dye penetrant
Eddy current	

In addition to the tools and equipment listed above, it is likely that the engine will have certain features which require the manufacturer's special tools. It is therefore necessary to obtain these special tools for each different type of engine overhauled.

Manufacturer's Overhaul Manual

Every engine overhaul must be carried out in accordance with the instructions given in the manufacturer's overhaul manual. The overhaul manual provides information on all procedures of a special nature and on general procedures for the disassembly, cleaning, inspection, repair, modification, reassembly, and testing of engines. These procedures must be followed to ensure that the engine is overhauled in a manner which will provide the required reliability.

The manufacturer of an aircraft engine has all the information relating to the construction and design of engine parts and has made numerous tests of the assembled engine and its parts. The overhaul manual prepared by the manufacturer is therefore the most valid source of information for a particular make and model of engine.

The technician must be sure that the manual and other data referred to for overhaul information are the most current and apply to the specific model number, including the dash number, of the engine being overhauled.

Certification of Overhaul Station

To provide overhaul services for certificated aircraft engines, an overhaul agency should be an FAA certificated repair station with ratings covering all the types of overhaul work performed. A **certificated repair station** is defined by Federal Aviation Regulations as "*a facility for the maintenance, repair, and alteration of airframes, powerplants, propellers, or appliances, holding a valid repair station certificate with appropriate ratings issued by the Administrator.*" Certification by the FAA as a certificated repair station permits the repair station to perform major overhaul on all items for which appropriate ratings are held. This certification also permits the repair station to approve overhauled items for return to service. All agencies currently approved by the FAA are listed in the Consolidated Listings of FAA Certified Repair Stations. Further details regarding the responsibilities and privileges of a certificated repair station are given in Part 145 of the Federal Aviation Regulations.

Shop Safety and Preparation for Overhaul

In the preparation of an overhaul shop, all required equipment must be on hand. A listing of equipment is given later in this chapter. In the operation of the shop, it is important that adequate safety practices be exercised. Safety involves the design and arrangement of the shop, use of proper tools and equipment, and observance of safe practices by overhaul technicians and their helpers. To ensure that all reasonable safety precautions are taken and that all equipment meets safety requirements, it is advisable to hire a professional safety engineer during the establishment and organization of the shop. The safety engineer should be asked to make periodic inspections of the shop and shop operations to check for unsafe practices and conditions and to discover any equipment which has become unsafe.

The following are some of the requirements for safety and equipment in an overhaul shop:

1. The arrangement of the shop should be such that crowding does not exist.
2. Materials and parts must be stored in proper racks and containers.
3. Walkways should be painted on the floor and kept free of obstructions at all times except when necessary to move equipment.
4. Suitable engine overhaul stands must be provided for the types of engines to be overhauled. The design of engine stands varies with the type of engine.
5. Adequate work station and bench space should be provided.
6. All machinery must be in good repair and provided with approved safety guards.
7. Electric equipment and power outlets must be correctly identified for voltage and power handling capability.
8. Electric outlets must be designed so that it is not possible to plug in equipment other than that designed for the voltage and type of current supplied at the outlet.
9. Adequate lighting must be provided for all areas.
10. Cleaning areas must be adequately ventilated to remove all dangerous fumes from solvents, degreasing agents, and other volatile liquids.
11. Properly designed engine hoisting equipment must be available for lifting engines.
12. Operators should wear safety shoes to prevent injury to their feet in case heavy equipment or parts are dropped.
13. Operators of lathes, machine tools, drills, and grinders must wear safety goggles.
14. Operators of cleaning equipment must wear protective clothing, rubber gloves, and face shields.
15. The floor of the shop should have a surface which does not become slippery.
16. Grease, oil, and similar materials should be removed from the floor immediately after they are spilled.
17. Proper types of fire extinguishers should be available to shop personnel.
18. Personnel should take care to stay clear of an engine which is being hoisted except as necessary to prevent it from swaying or turning.
19. Cleanness is of prime importance in an engine overhaul shop. All areas must be kept free of sand, dust, grease, dirty rags, and other types of contaminants.
20. Greasy or oily rags must be stored in closely covered metal cans. The rag cans should be emptied frequently.

21. Flammable liquids must be stored in an approved fireproof area. The approval is normally granted by the local fire department.

22. The area where engines are run should be paved and kept free of all debris which could be picked up by air currents caused by a propeller or jet intake.

RECEIVING THE ENGINE

Receiving Inspection

The purpose of the **receiving inspection** is to determine the general condition of the engine when it is received, provide an inventory of the engine and all its accessories and associated parts, organize the engine records, and prepare the engine for the next step in the overhaul process.

Inspection

The general condition of the engine should be carefully noted on the receiving report. If the engine has been in a crash, fire, or other situation where damage may have been caused, such conditions have to be recorded. The cost of overhauling an engine which has suffered damage other than normal wear will be considerably greater than usual. In many cases it is not economically feasible to overhaul such an engine.

At this point the engine should be mounted on an overhaul stand and cleaned externally. In some cases the dirt-and-oil mixture is "baked" on to the extent that it is not entirely removable with an ordinary solvent. Usually, however, the oil and loose dirt can be removed by means of a petroleum solvent such as mineral spirits, kerosene, or other approved petroleum solvent. Some operators steam-clean the engine, thus removing oily deposits and caked dirt. When steam cleaning is done, the engine should be washed off with plain hot water afterward to remove all traces of the alkaline steam-cleaning solution. This is necessary to prevent the corrosion of aluminum and magnesium parts.

Inspection after cleaning may reveal cracks in the crankcase or cylinders or other damage which may have been hidden by dirt or oil. Damage found at this point in the overhaul process will increase the accuracy of the cost estimate and may save engine downtime if the parts in question have to be ordered or sent out for repair.

Inventory

It is essential that every part be accounted for, because an owner may not be aware that certain parts or accessories were removed before the engine was delivered to the overhaul facility. The receiving inspection report should be made on a standardized form, and a copy of this form should be given to the owner.

Included in the receiving inspection report should be the make, model, and serial number of the engine and the serial numbers of all its accessories. In addition, all parts, such as cylinder baffles, ducting, brackets, etc., should be listed so

that there will be no question regarding the responsibility for these parts after the overhaul has been completed.

Organization of Engine Records

Extensive records must be kept as the engine progresses through the overhaul process. This information will become part of the permanent records that must be kept by the overhauling agency. Even when a small number of engines are being processed, an accurate system for record keeping must be set up.

The FAA Type Certificate Data Sheet must be reviewed to ensure that the engine conforms to its **type certification**. For example, the magnetos, carburetor, fuel pump, and other components received with the engine must be the ones with which the engine was originally certificated. Also, notes and other information relating to the engine should be checked.

From time to time, engine manufacturers issue special **service bulletins** that require alterations or parts replacements designed to improve the performance and reliability of the engine. During the overhaul of an engine, bulletins must be complied with, and a statement of such compliance must be made in the overhaul record. The FAA issues **Airworthiness Directives** pertaining to aircraft and engines whenever it appears that certain changes should be made to correct discrepancies or to improve the reliability of the unit. During the engine overhaul, all Airworthiness Directives pertaining to the engine must be complied with, and a statement of compliance should be entered in the overhaul record.

The repair station should maintain a complete record of all repair operations and inspections performed on each engine or component overhauled. This record should contain an account of every repair operation, every inspection, and every part replacement. The inspection record should show the dimensions of each part measured and all fits and clearances. These measurements are compared with the manufacturer's **Table of Limits** (see Fig. 10-1) so that repairs and replacements can be made to restore worn parts to the required specifications.

The inspection record should show the types of inspection procedures employed, such as magnetic particle testing, liquid penetrant inspection, ultrasonic inspection, radiography, and any other approved process employed to determine the airworthiness of the engine and its parts. If any of these inspections are performed by another agency, the certificate of that agency must be included in the overhaul record.

DISASSEMBLY

Disassembly of a typical aircraft engine follows a sequence specified in the overhaul manual. During disassembly, the operator should identify and mark all parts. This is often done by attaching small metal or plastic tags to all parts with soft safety wire. The tags are stamped with the work order number. The overhaul manual specifies which parts must be replaced in their original locations. These parts may be

Ref. New	Ref. Old	Chart	Nomenclature	Dimensions		Clearances	
				Mfr. Min. & Max.	Serv. Max.	Mfr. Min. & Max.	Serv. Max.
500	501	A	All Main Bearings and Crankshaft			.0025L .0055L	.0060L
		B–D–G–J–S–T–Y–BD–BE–AF	Main Bearings and Crankshaft (Thin Wall Bearing - .09 Wall Approx.)			.0015L .0045L	.0060L
		B–G–J–S–T–Y–AF	Main Bearings and Crankshaft (Thick Wall Bearing - .16 Wall Approx.)			.0011L .0041L	.0050L
		A	Diameter of Main Bearing Journal on Crankshaft	2.3735 2.375	(E)		
		B–D–G–J–S–T–Y–BD–BE	Diameter of Main Bearing Journal on Crankshaft (2-3/8 In. Main)	2.3745 2.376	(E)		
		T1–T3–AF	Diameter of Main Bearing Journal on Crankshaft (2-5/8 In. Main)	2.6245 2.626	(E)		
		S8–S10	Diameter of Front Main Bearing Journal on Crankshaft (2-3/8 in. Main)	2.3750 2.3760	(E)		
		T1–T3–AF	Diameter of Front Main Bearing Journal on Crankshaft (2-5/8 in. Main)	2.6245 2.6255	(E)		

FIGURE 10-1 Sample Table of Limits. (*Lycoming Textron.*)

identified by attaching additional tags to show the locations. Nuts and bolts may be strung on safety wire in groups which are identified for location.

Cylinders, pistons, connecting rods, valve lifters, and other parts requiring exact location should be clearly marked to show their locations in the engine. For example, the piston from the no. 1 cylinder should be marked with the number 1. This number is stamped on top of the head or on the bottom side of the piston. Connecting rods have numbers stamped on the big end. Careful marking and identification of parts will ensure proper assembly and will provide for correct fits and clearances when measurements for wear are made. Organization of parts on a suitable parts rack will keep the overhaul process orderly and will help prevent part mixups and losses.

When the technician has carefully noted each step and carried out the operations as given, the engine is completely disassembled and ready for the first visual inspection.

VISUAL INSPECTION

All parts should be given a preliminary visual inspection either as they are removed from the engine or immediately at the completion of disassembly, before they are cleaned. Residue found in recesses and other areas of the engine may be indications of dangerous operating conditions.

Visual inspection is accomplished by direct examination and with the use of a magnifying glass. In each case a strong light should be used to aid in revealing all possible defects. During visual inspection, special attention is given to those areas of the engine and its parts where experience has shown damage to be most likely. Visual inspection will usually reveal cracks, corrosion, nicks, scratches, galling, scoring, and other disturbances of the metal surfaces. Parts that are obviously damaged beyond repair should be discarded and marked so that they will not be reused. A record of all parts thus discarded should be made for the overhaul file. The remaining parts should be grouped or labeled as apparently serviceable or repairable.

The various types of damage and defects in engine parts to be noted during visual inspection may be defined as follows:

Abrasion A roughened area where material has been eroded by foreign material being rubbed between moving surfaces.

Brinelling Indentations of a surface caused by high force pressing one material against another. A ball bearing pressed with sufficient force against a softer material will cause a Brinell mark, or indentation.

Burning Damage to a part due to excessive temperature. Figure 10-2 shows an example of burning in a cylinder head around the exhaust valve opening.

Burr A rough or sharp edge of metal, usually the result of machine working, drilling, or cutting.

Chafing The wear caused by two parts rubbing together under light pressure and without lubrication.

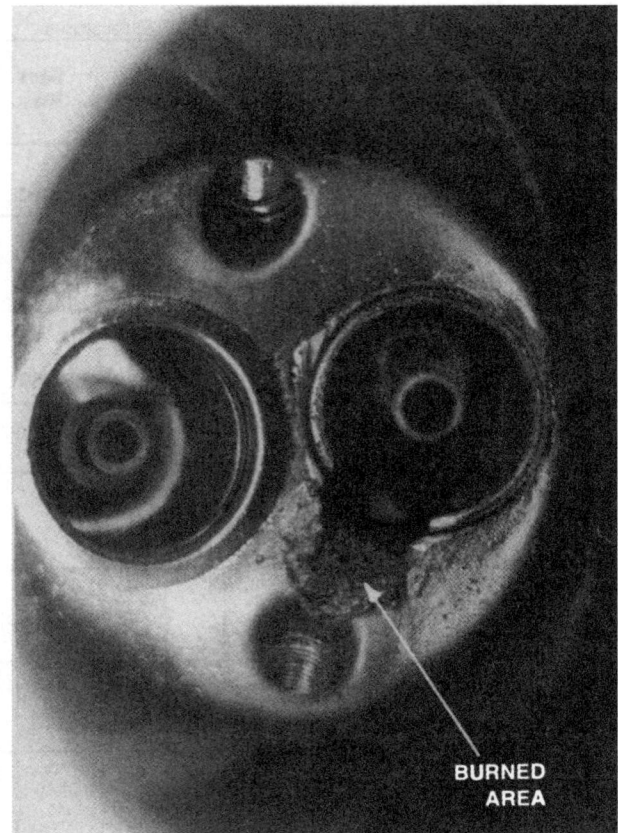

FIGURE 10-2   Burning around a valve seat in a cylinder head.

Chipping The breaking away of small pieces of metal from a part as a result of careless handling or excessive stress.

Corrosion Electrolytic and chemical decomposition of a metal, often caused by joining of dissimilar metals in a situation where moisture exists. Surface corrosion is caused by moisture in combination with chemical elements in the air.

Crack A separation of metal or other material, usually caused by various types of stress, including fatigue stresses resulting from repeated loads.

Dent Similar to a Brinell mark except that a dent is usually found in sheet metal where the metal is deformed on both the side where the force was applied and the opposite side, usually caused by the impact of a hard object against a surface.

Elongation Stretching or increasing in length.

Flaking The breaking away of surface coating or material in the form of flakes, caused by poor plating, poor coating, or severe loading conditions.

Fretting The surface erosion caused by very slight movement between two surfaces which are tightly pressed together.

Galling The severe erosion of metal surfaces pressed lightly together and moved one against the other. Galling can be considered a severe form of fretting.

Gouging The tearing away of metal by a hard object being moved across a softer surface under heavy force. A piece of hard material caught between two moving surfaces can cause gouging.

FIGURE 10-3   Spalling of a lifter body face.

Grooving The formation of a channel worn into metal as a result of poor alignment of parts.

Growth Elongation caused by excessive heat and centrifugal force on turbine-engine compressor and turbine blades.

Indentation Dent or depression in a surface caused by severe blows.

Nick A sharp-sided depression with a V-shaped bottom, caused by careless handling of tools or parts.

Oxidation Chemical combining of a metal with atmospheric oxygen. Aluminum oxide forms a tough, hard film and protects the surface from further decomposition. However, iron oxides do not form a continuous cover or protect underlying metal; thus, oxidation of steel parts is progressive and destructive.

Peening Depressions in the surface of metal caused by striking of the surface by blunt objects or materials.

Pitting, or spalling Small, deep cavities with sharp edges; may be caused in hardened-steel surfaces by high impacts or in any smooth part by oxidation. Figure 10-3 shows an example of spalling of a lifter body face.

Runout Eccentricity or wobble of a rotating part. Eccentricity of two bored holes or two shaft diameters. A hole or bushing out of square with a flat surface, usually measured with a dial indicator. Limits stated indicate the full deflection of the indicator needle in one complete revolution of the part or the indicator support.

Scoring Deep scratches or grooves caused by hard particles between moving surfaces. Similar to gouging. Figure 10-4 shows an example of scoring on a piston skirt.

Scratches Shallow, thin lines or marks, varying in degree of depth and width, caused by the presence of fine foreign particles during operation or by contact with other parts during handling.

Scuffing, or pickup The transfer of metal from one surface to a mating surface when the two are moved together under heavy force without adequate lubrication.

FIGURE 10-4   Scoring on a piston skirt.

All parts are to be inspected visually. The visual inspection of a few sample parts is as follows.

Crankcase

Crankcases are inspected for cracks, warping, damage to machine surfaces, worn bushings and bearing bores, loose or bent studs, corrosion damage, and other conditions which may lead to failure in service.

Studs

Studs are checked for bending, looseness, and erosion. All threaded parts are inspected for thread damage.

Gears

Accessory gears are examined for wear, overheating, scoring, pitting, and alteration of the tooth profile. Any appreciable damage is cause for rejection of the part.

Cam Surfaces

The surfaces of the cams must be examined with a magnifying glass to detect incipient failure. Any obvious wear, spalling, pitting, surface cracks, or other damage is cause for rejection of the camshaft. If examination with a magnifying glass reveals small pores in the surface, this also requires that the shaft be discarded.

Valves

Valves are inspected for burning, erosion, stretch, cracking, scoring, warping, and thickness of the head edge. Because of the high temperatures to which exhaust valves are exposed, these valves often need to be replaced even though the intake valves are still in serviceable condition. Valves are subjected to many thousands of recurring tension loads, because they are opened by cam action and closed by the valve springs. This, of course, tends to stretch the valves, and the effects of stretching are commonly noted in exhaust valves. Figure 10-5 shows how a valve is examined for stretch with a valve stretch gauge or contour gauge.

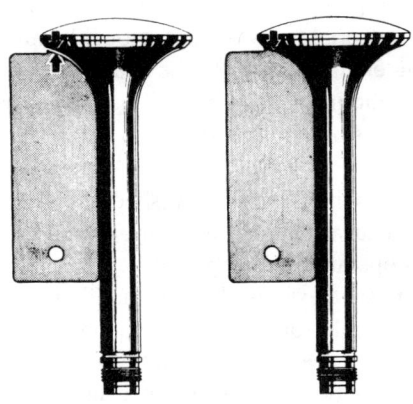

FIGURE 10-5 Inspection of valves for stretch with a contour gauge.

Valve Tappets

Valve tappets are inspected for wear, spalling, pitting, scoring, and other damage. Defective tappets must be replaced.

Hydraulic lifters or tappets are disassembled and inspected for wear. If any appreciable wear is noted, as evidenced by a feathered edge of worked metal at the shoulder in the tappet body, the entire tappet assembly must be discarded. The hydraulic tappet cylinder and plunger assembly must be checked to see that no burrs or binding exist and that the ball check valve is not leaking. A leaking check valve can be tested as follows: Make sure that the cylinder and plunger assembly are dry, and hold the lifter cylinder between the thumb and middle finger in a vertical position with one hand. Then place the plunger in position so that the plunger just enters the lifter cylinder. When the plunger is depressed quickly with the index finger, it should return to approximately its original position. If the plunger does not return but remains in a collapsed position, the ball check valve is not seating properly. The cylinder should be recleaned and then checked again. If the check valve still does not seat, the unit is defective and the entire cylinder and plunger assembly must be discarded.

Hydraulic cylinder and plunger parts cannot be interchanged from one assembly to another, and if a part of any assembly is defective, the entire assembly is discarded. The cylinder and plunger (hydraulic unit) assemblies can be interchanged in lifter bodies, provided that the clearances are within the approved limits.

CLEANING

It is necessary to clean the engine externally before disassembly and to clean the parts after disassembly. Great care must be taken during the cleaning processes because the engine parts can be seriously damaged by improper cleaning or application of the wrong types of cleaners to certain engine parts. In every case, the person in charge of the cleaning processes should study the engine manufacturer's recommendations regarding the cleaning procedures for various parts of the engine and comply with the directions provided by the manufacturer of the cleaning agent or process.

In general, two types of cleaning are required when an engine is overhauled: (1) degreasing, for removal of oil, soft types of dirt, and soft carbon (sludge); and (2) decarbonizing, or the removal of hard carbon deposits. The removal of grease and soft types of residues is relatively simple, but hard carbon deposits require the application of rather severe methods. It is during this type of cleaning that the greatest care must be exercised to avoid damaging the engine parts.

Degreasing after Disassembly

After the engine has been disassembled, all oil should be removed from the parts before further cleaning is attempted, because the additional cleaning processes are much more effective after the surface oil has been removed. Two of the

FIGURE 10-6 Solvent degreasing.

principal methods for removing the residual lubricating oil and loose sludge are washing in a petroleum solvent and employment of a vapor degreaser.

Cleaning with a petroleum solvent can be done in a special cleaning booth where the parts are supported on a grill and sprayed with a solvent gun using air pressure of 50 to 100 psi [345 to 690 kPa], as illustrated in Fig. 10-6. The spray booth should have a ventilating system to carry away the vapors left in the surrounding air, and the operator should wear adequate protective clothing. A drain should be located underneath the grill to collect the used cleaning solvent. During the cleaning process, particular care should be taken that all crevices, corners, and oil passages are cleaned.

Vapor degreasing is accomplished with equipment specially designed for this type of cleaning. A vapor degreaser consists of an enclosed booth in which a degreasing solution such as trichloroethylene is heated until it vaporizes. The engine parts are suspended above the hot solution, and the hot vapor dissolves the oil and soft residue to the extent that they flow off the parts and drop into the container below. The vapor degreaser should be operated only by a person thoroughly familiar with the equipment and the manufacturer's instructions regarding its use.

Decarbonizing

Parts contaminated with hard carbon, baked-on oil, resinous varnishes, and paint are ready for **decarbonizing** after the degreasing process has been completed. The most common methods of decarbonizing are stripping, grit blasting, and vapor blasting.

Stripping

The stripping process is used to remove paint and various resinous varnishes which have formed in the engine during its operation. Some stripping solutions are also effective in removing some of the harder carbon deposits. A number of solutions suitable for stripping have been developed by manufacturers

of chemical cleaning agents, but each solution must be used according to proper directions, or else the engine parts may be damaged. In all cases the person responsible for cleaning and stripping the parts must be thoroughly familiar with the solution used and must know what engine parts may be cleaned safely with it.

A typical cold-stripping process involves the immersion of parts, such as the engine crankcase, in a vat containing the solution for several hours or overnight. The parts are then removed and steam-cleaned to remove all traces of the stripping solution and all material loosened by the solution. After steam cleaning, the parts are washed with clear, hot water, and all openings, oil passages, and crevices are checked for cleanness. This is particularly important because the clogging of one oil passage can cause failure of the engine.

The stripping agents most commonly used are alkaline (caustic) solutions; they can be used cold, or they may be heated to expedite the cleaning process. Since stripping agents can be hazardous to the parts being cleaned as well as to the person performing the stripping operation, the following precautions should be observed.

1. The person stripping the engine parts must be careful to follow the instructions of the manufacturer of the stripping agent and the manufacturer of the engine.

2. **Dissimilar metals** should not be placed in the solution tank at the same time. Magnesium and aluminum parts can be completely destroyed by electrolytic action if they are placed in a cleaning vat with steel parts for more than a few minutes.

3. Monitor immersion times carefully. Alkaline (caustic) solutions will attack aluminum, magnesium, and some of the alloys employed in aircraft engines. Strong alkaline solutions will also remove the Alodine coating that is applied to many aluminum parts of engines. The Alodine coating is usually applied to the interior surfaces of crankcases, valve covers, accessory sections, and other surfaces that are not available for cleaning or treatment from the outside. Severe damage can result if parts are left in the stripping agent too long.

4. It is inadvisable to soak aluminum and magnesium parts in solutions containing soap, because some of the soap will become impregnated in the surface of the material even though the material is washed thoroughly after soaking. Then, during engine operation, the heat will cause the soap residue to contaminate the engine oil and cause severe foaming. This will result in loss of oil and possible damage to the engine.

5. Steel parts that have been washed with soap and water should be coated with a rust-inhibiting oil immediately after the cleaning operation is completed.

6. Care must be exercised in the handling of decarbonizing (stripping) solutions to prevent the material from contacting the skin, because these solutions will usually cause severe irritation or burning. The operator should therefore wear goggles, rubber gloves, and protective clothing while working with these solutions.

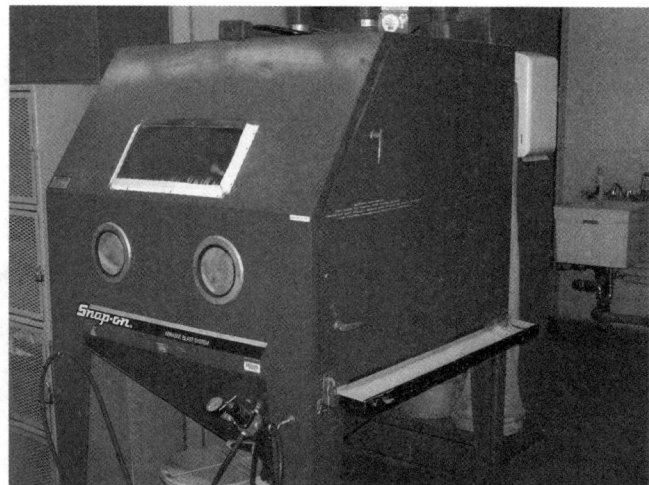

FIGURE 10-7 A grit-blasting machine. (*Pangborn.*)

Blasting with Grit and Other Materials

For the removal of hard carbon from the insides of cylinders and the tops of pistons, soft **grit blasting** offers one of the most satisfactory processes. The blasting material (grit) consists of ground walnut shells, ground fruit seeds, and other organic or plastic materials. The grit is applied by a high-pressure air gun in an enclosed booth, as shown in Fig. 10-7. The operator's hands and arms are extended into sleeves and gloves sealed into the machine for full protection while the gun and the parts being cleaned are manipulated. The work can be observed through a window in front of the booth.

In performing grit blasting for the removal of hard carbon, paint, and other residues, the operator must use care to avoid causing damage to highly polished surfaces and to avoid the plugging of oil passages. If crankcases are grit-blasted, all the small openings into which grit may enter must be plugged. Some operators do not use grit blasting on parts having oil passages.

In general, **sandblasting** is not employed in the cleaning of engine parts except the valve heads. This is because sandblasting will erode the metal so rapidly that serious damage may be done before the operator is aware of it. Furthermore, sand tends to embed itself in soft metals and will later become a source of abrasion in the operating engine. Under some circumstances, such as preparation for metallizing of cylinders, the outside of the cylinder is carefully sandblasted before application of the metal spray. Special uses of sandblasting may be approved for particular purposes; however, this may be done only by a thoroughly trained operator and then only within the limits specified for the process. After engine parts have been cleaned and dried, they should be coated with a preservative oil while awaiting inspection and further processing. This is done to prevent the rust and corrosion which are likely to occur.

In addition to the grit and other materials employed for cleaning by blasting, rice hulls, plastic pellets, baked wheat, and glass beads are used. Technicians must be sure to use an approved material for the blasting job they are performing.

When glass beads are used, the addition of aluminum oxide powder adds to the effectiveness of the process whether it is used as a cleaning procedure or as a mild form of "shot-peening." The size of the glass beads is important and should be checked in the instructions for the particular job.

Vapor Blasting

Vapor blasting is employed for special cleaning jobs and is accomplished by means of specially designed equipment and materials. The vapor solution is used with a fine abrasive and should be applied in an enclosed booth. The vapor is hot and is applied with a high-pressure air gun.

The use of vapor blasting is limited to parts and areas of parts which will not be damaged by a small amount of material erosion. It may be used on the tops of pistons, on the outsides of cylinders, and in almost any area which is not a bearing surface of some type. In all cases, the manufacturer's instructions must be followed.

Scraping

Except as permitted in the overhaul manual, wire wheels, steel scrapers, putty knives, or abrasives should not be used for cleaning parts or removing carbon. These items may leave scratches on the parts which will lead to stress concentrations and ultimate failure.

Hard carbon is removed from piston heads and from the insides of cylinder heads by soft grit blasting. The carbon which has collected inside the piston head and in the ring lands is the most difficult to remove. After the part has been soaked in the decarbonizing solution, the carbon can usually be removed by grit blasting, although it is sometimes necessary to use a soft scraper. The operator must be careful not to damage the piston-ring lands or to remove any metal from the small radii between the ring lands and the bottoms of the ring grooves. The glazed surfaces on the pistons and the piston pins should not be removed

STRUCTURAL INSPECTION

The processes and procedures for structural inspection of engine parts may vary somewhat from one overhaul shop to another. The function of the **structural inspection** is to determine the structural integrity of each part. Defective parts are removed from the overhaul process when they are found. This saves time and money and ensures that the parts going back into the engine are free from potentially hazardous defects.

Engine parts are structurally inspected by some or all of the following methods:

1. Magnetic particle testing
2. Liquid penetrant inspection
3. Eddy-current inspection
4. Ultrasonic inspection
5. Radiography

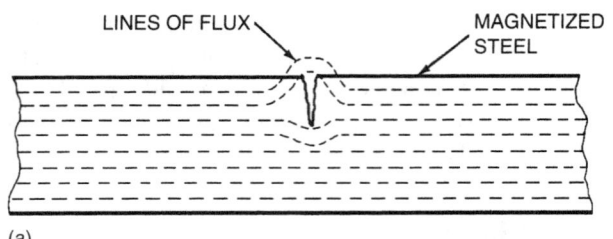

LINES OF FLUX — MAGNETIZED STEEL

(a)

(b)

FIGURE 10-8 (a) Magnetic particle machine. (b) Magnetic lines of flux leaving surface of metal.

Magnetic Particle Testing

Magnetic particle testing (see Fig. 10-8) is a **nondestructive** method for locating surface and subsurface discontinuities (cracks or defects) in ferromagnetic materials such as steel. The parts are magnetized by passing a strong electric current through or around them. An electric current is always accompanied by a magnetic field. The magnetic lines of flux travel through the part when it is magnetized. A crack or **discontinuity** will create a flux leakage. A **flux leakage** occurs where the lines of flux leave the surface of the material, resulting in a concentration of magnetic strength at the discontinuity, as shown in Fig. 10-8.

Magnetic particles are applied to the magnetized part and concentrate in the areas of flux leakage, giving a visible sign of a discontinuity.

Magnetic particle testing is done with special equipment, such as that manufactured by Magnaflux Corporation. Parts requiring magnetic inspection may be sent to an authorized inspecting agency, or the inspection may be performed by a person who is thoroughly trained in the use of magnetic inspection equipment.

Magnetic particle testing technicians must be trained and experienced to successfully find and interpret discontinuities. The major elements that must be understood and determined by the operator are

1. The direction in which the test part must be magnetized
2. How strong the magnetic field must be
3. What type of current to use
4. What type of particles to use
5. Interpretation of discontinuities

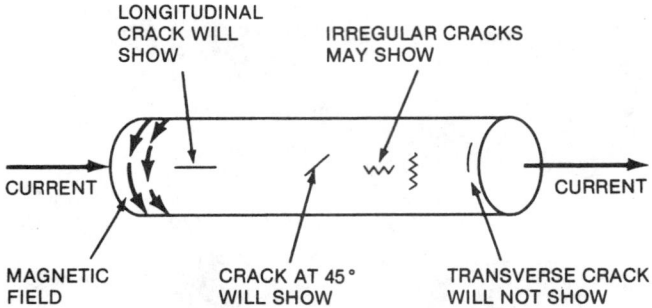

FIGURE 10-9 Effect of crack orientation in a circularly magnetized bar.

The magnetic field must intersect the discontinuities being sought at an appreciable angle, preferably 90°. The direction of magnetization is determined by the direction in which the current is passing through the part.

The direction of the magnetic field will always be perpendicular to current flow. If the discontinuity is expected to be approximately longitudinal, the part should be magnetized circularly, as illustrated in Fig. 10-9. **Circular magnetization** is achieved by passing the current directly through the part.

If the discontinuity is expected to run circularly or transversely through the part, the part must be magnetized longitudinally. **Longitudinal magnetization** is accomplished by wrapping heavy-gauge wire around the part, or by placing the part in the magnetizing coil, as illustrated in Fig. 10-10. In this case the current is passed through the coil, and not through the part itself. When a cable carrying current forms a loop, the flux lines surrounding the conductor all pass through the loop in one direction. This is shown in Fig. 10-11. Any ferromagnetic material placed in the center of the loop will take on the magnetic characteristics of the field within the loop.

It may be necessary to use a **central conductor**, as shown in Fig. 10-12, if the desired magnetization cannot be achieved by passing current through the part or by placing it in a coil.

FIGURE 10-10 Part in a magnetizing coil.

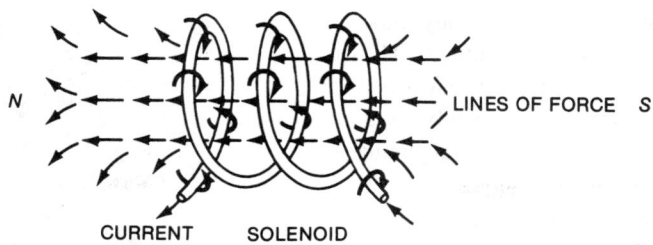

FIGURE 10-11 Field in and around a solenoid carrying direct current, and its direction.

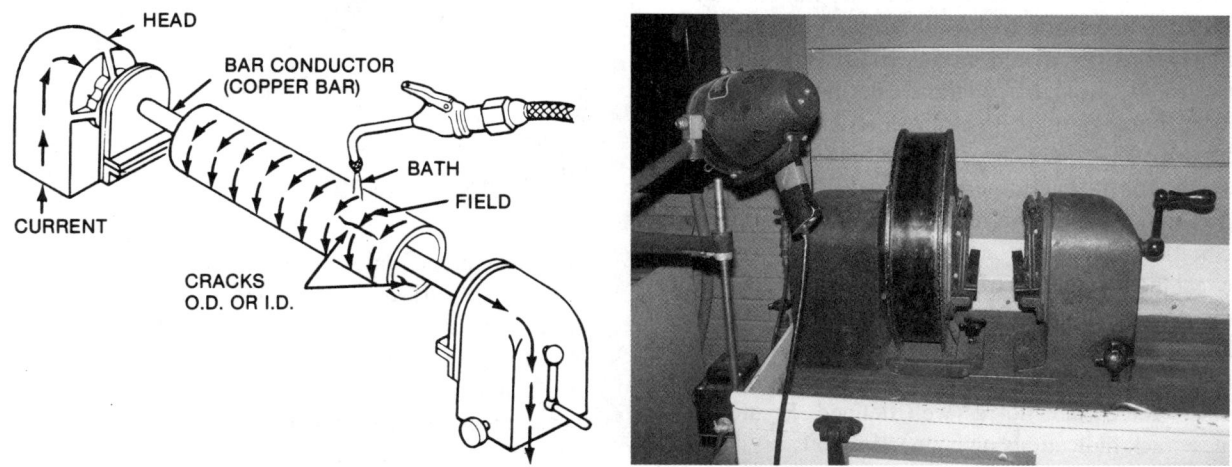

FIGURE 10-12 Use of a central conductor for magnetization. (*Magnaflux Corp.*)

The strength to which a part must be magnetized depends on several factors. As previously mentioned, the field strength must be great enough to cause a flux leakage at any discontinuity. However, the field strength must not be so great that a strong background is created which will hide indications.

The amount of magnetism is proportional to the amount of current passed through or around the part. The technician controls this by limiting the amperage applied. The general rule for determining how much current to use is 1000 A per 1 in [2.54 cm] of cross-sectional diameter. This works well for most small symmetric objects with diameters up to 3 or 4 in [7.6 or 10.2 cm]. Using too much current on a small part overheats the part and burns it at the contact ends when the part is circularly magnetized.

The **permeability** of a ferromagnetic material must be considered in determining the current strength. Permeability is the ease with which a magnetic field can be set up in a magnetic circuit. Hard steels have lower permeability and resist being magnetized. Soft steels have high permeability and can be magnetized easily.

The location of the defect will determine the type of current to be used. The magnetization process can be accomplished with either alternating or direct current. Direct current tends to penetrate the entire cross section of the material, making it more desirable for locating subsurface flaws. Alternating current has a tendency to flow along the surface of the material. This is referred to as the *skin effect*. Parts magnetized with alternating current must be limited to inspection for surface discontinuities.

Generally, magnetic particles are either visible or fluorescent. The visible magnetic particles are available in a variety of colors and can be applied dry or with a liquid vehicle. The dry powder is composed of gray-iron particles, and the particles used in a liquid vehicle are iron oxide. Both are treated with color pigments.

Dry powders are typically used for very local inspections and have proven superior to particles used in a liquid vehicle for locating subsurface discontinuities. However, the wet method offers ease of application, provides complete coverage, is more effective than the dry method in detecting surface cracks, and offers excellent visibility and contrast.

The fluorescent wet method along with dc magnetization is typically used for inspection of engine parts. However, the aircraft technician may use any of the other methods discussed that might be more effective for locating a particular type of defect.

When a fluid is used in which the magnetic particles are suspended, the fluid is applied to the part by means of a fluid nozzle. As the fluid flows over the part, the magnetic particles adhere to any area where the magnetic lines of flux leave the surface of the metal. There are two methods for applying the fluid bath to a part: the residual method and the continuous method. In the residual method, the part is magnetized first and then the bath is applied; in the continuous method, the part is flooded with the bath and then magnetized. The type of method used will depend largely on the operator and the permeability of the material.

The residual method is the less operator-dependent method; however, the materials being tested must be low in permeability and high in magnetic retentivity (they must hold their magnetism once the current is discontinued). The continuous method requires greater operator skill but is the more desirable method. The continuous method can be used on material with high or low permeability since magnetic particles will be present when the magnetizing current is passing through or around the part. Also, particle mobility is enhanced by the current flow, which makes it easier for particles to orient themselves at discontinuities.

The magnetic particle/liquid vehicle ratio, or particle concentration, is critical. The particles are mixed with the liquid vehicle according to a bath strength chart supplied by the manufacturer of the particles. The strength of the bath should be checked frequently. The most widely used method of checking the bath strength is gravity settling in a graduated ASTM pear-shaped centrifuge tube, shown in Fig. 10-13. The magnetic particles gradually settle to the bottom of the tube, where a reading may be taken by utilizing the marked gradations on the tube.

The technician must be trained and experienced to make accurate interpretations of the signs produced by magnetic particle inspection. Surface cracks, when inspected under a black light, will show up as sharp, bright, yellow-green lines. Subsurface defects are not as well defined. Figure 10-14 illustrates crack indications on connecting rods.

Many indications have no bearing on the structural strength or service usefulness of the part. These indications may be caused by any physical aspect of the part that interrupts the flux flow and are called *irrelevant indications*. Examples of irrelevant indications may occur at sharp edges of a part where there is a great concentration of flux. Tooling marks that may have no structural significance will produce

CRACK INDICATIONS

FIGURE 10-14 Crack indications revealed by magnetic particle inspection.

flux leakage. Internal keyways of splines will manifest themselves as subsurface flaws, as shown in Fig. 10-15. Irrelevant indications may also occur when the vehicle "runs away" from the particles on the horizontal tangents or bottom side of a cylinder, leaving "drainage lines." Indications such as these, which are not related to magnetic leakage fields, are called *false indications*. If false indications occur, the cylinder should be rotated and the bath reapplied to mobilize the particles.

After the parts have been inspected, they must be demagnetized. If this is not done, the parts will pick up and hold small steel particles, which can cause serious damage in the engine during operation. Demagnetization is accomplished by passing the parts slowly through a strong ac magnetic field and then moving the parts slowly out of the field until they are experiencing zero magnetic flux, as illustrated in Fig. 10-16.

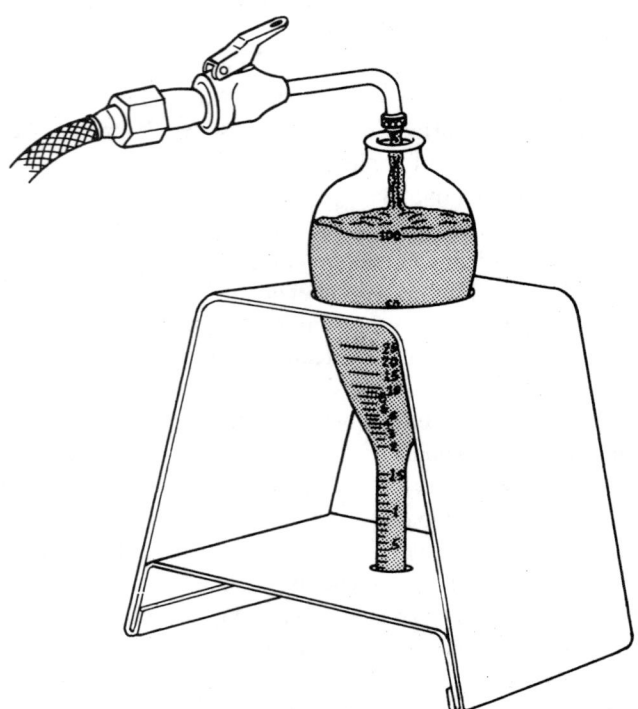

FIGURE 10-13 Checking of particle concentration with a centrifuge tube. (*Magnaflux Corp.*)

SUBSURFACE FLAWS

FIGURE 10-15 Gear and shaft showing irrelevant indications caused by internal splines.

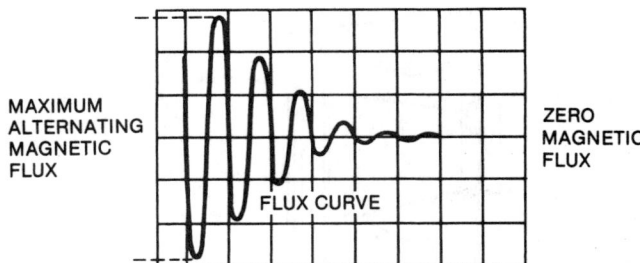

FIGURE 10-16 Slowly decaying ac magnetic field in a part during demagnetization.

The equipment for demagnetizing consists of a coil energized by alternating current. It may be the same coil in which the part was magnetized or a separate coil used solely for demagnetizing. Some machines feature an automatic demagnetizing cycle in which the part is held in the coil and the ac field slowly decays to zero. Regardless of the type of equipment used, the current must not be turned off until either the part is out of the coil or the cycle is complete.

Always consult manufacturers' maintenance manuals for specific instructions pertaining to magnetic particle testing of their products. Precautions might include plugging small oil passages or cavities that may collect and be obstructed by magnetic particles with wood plugs or hard grease.

Upon completion of the magnetic inspection, the engine parts should be washed in clean petroleum solvent and dried with compressed air. The parts should then be coated with a thin layer of corrosion-inhibiting oil.

Liquid Penetrant Inspection

Engine parts made of aluminum alloys, magnesium alloys, bronze, or any other metal which cannot be magnetized are usually inspected by means of a fluorescent penetrant, a dye penetrant, ultrasonic equipment, or eddy-current equipment. Parts on which **liquid penetrant inspection** is commonly used include crankcase halves, accessory cases, oil sumps, and cylinder heads.

Inspection with a penetrant should include the testing of the parts, the tabulation of the nature and extent of the discontinuities found, and the final decision regarding the suitability of parts for further service. The operator should be a specialist who has been thoroughly trained to evaluate correctly the various indications.

There are many manufacturers of liquid penetrant systems and several methods from which to choose. Determining which system and method to use depends on the materials being tested, work capacity of the shop, frequency of use, and type of discontinuities sought. The classification of liquid penetrant inspection is set forth by a military specification, MIL-I-25135. It is broken down by type, method, and level. Type classification indicates the type of penetrant, visible and/or fluorescent. The method specifies the process used for application of the materials. The sensitivity is indicated by the level number.

Regardless of the type, method, or level, the principles are the same. Parts are thoroughly cleaned and stripped of paint.

The penetrant is applied to the part and allowed to enter any surface discontinuities. The excess penetrant is cleaned off, and then the part is given a coat of developer. The developer draws out any penetrant that may have entered surface discontinuities, making them visible either under white light in the case of visible penetrant or black light in the case of fluorescent penetrant.

Liquid penetrant inspection can be performed simply with the use of aerosol cans. This method can be very effective, but it is limited to low-volume applications and is typically used for localized inspections. Liquid penetrant inspection materials are available in bulk quantities and can be applied with commercial spray equipment or by dipping the parts in vats. The system discussed here is an immersion system using fluorescent penetrant and an emulsifier. Because an **emulsifier** is used, this is called a postemulsification process. The equipment illustrated in Fig. 10-17 is manufactured by Magnaflux Corporation. This system would be used by an operator doing a high volume of liquid penetrant inspecting.

We must reemphasize that all parts must be completely clean. Avoid using any type of grit blasting to clean parts. The peening action of the blasting medium may close up surface discontinuities. Remember that liquid penetrant inspection will disclose only defects open to the surface. Should a part need to be grit-blasted, it must be etched to open up surface discontinuities before the inspection continues.

After being cleaned, the part is placed in the penetrant vat. It may stay in the penetrant from 10 to 45 min depending on the nature of the part, type and size of discontinuity sought, and temperature. The penetrant enters a discontinuity by capillary action, the same force that causes oil to rise in a lamp wick or sap to rise in trees.

The part is removed from the penetrant vat and placed on a drain rack. The part is given a prerinse with a water spray to remove the bulk of the penetrant and prevent excessive contamination of the emulsifier.

The penetrant used in this system is not water-soluble. Before the residual penetrant can be removed from the surface of the part, the penetrant must be emulsified, or made water-soluble. The part goes into the emulsifier vat. The emulsifier begins diffusing into the penetrant at the surface of the penetrant and works its way down to the surface of the part, as illustrated in Fig. 10-18. The dwell time in the emulsifier is critical. If the time is too short, penetrant will be left on the surface of the part and the bright background will disguise signs of discontinuities. If the dwell time is too long, the diffusion will continue to the bottom of the discontinuity and all penetrant will be washed away.

The part is given a postrinse after it is removed from the emulsifier. The emulsified penetrant is removed, leaving penetrant in the discontinuities. There is little danger of overrinsing or washing away indications with this system since the penetrant in the discontinuities is not water-soluble.

The part must be dry before it goes on to the developing stage. The part is placed in a small booth or tunnel where hot air is circulated around it. The part should be completely dry for best results.

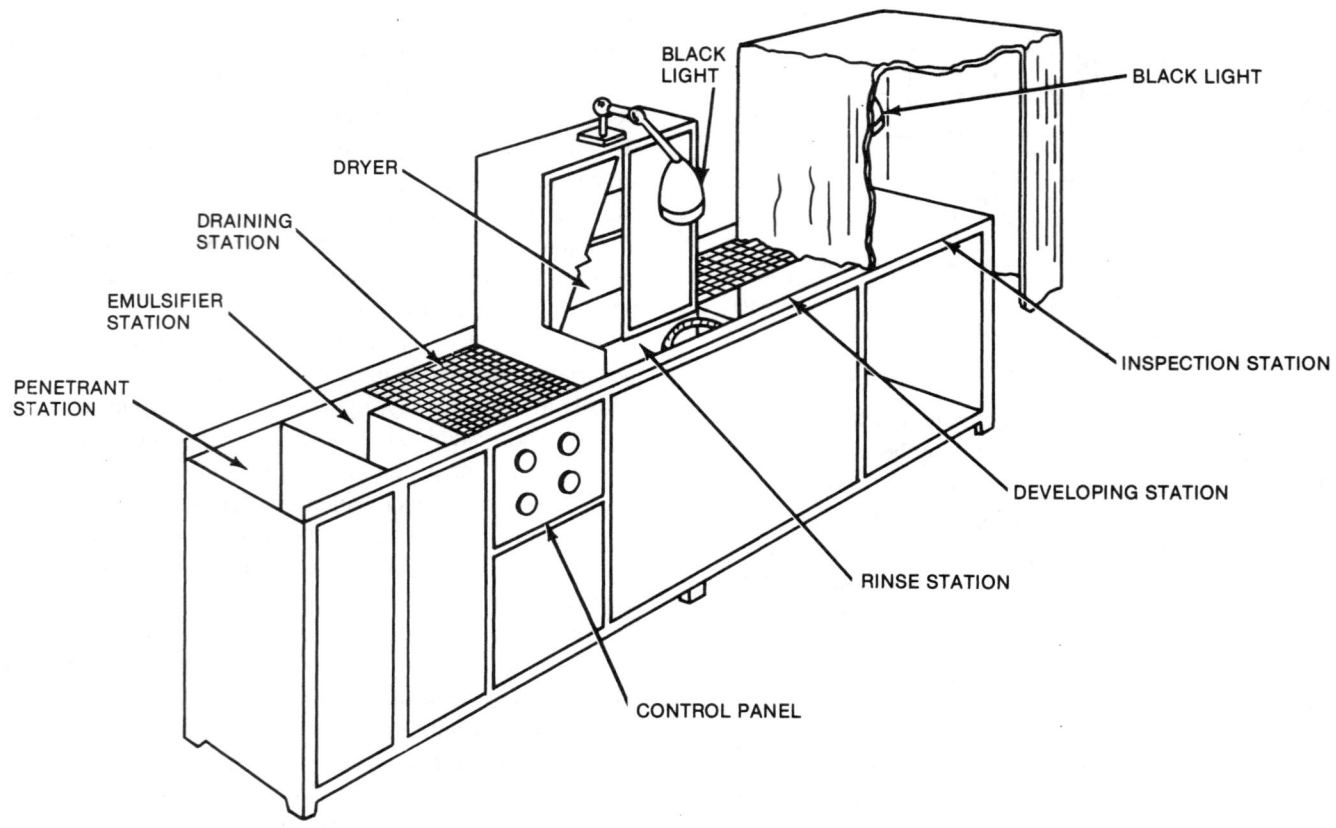

FIGURE 10-17 Liquid penetrant inspection unit. (*Magnaflux Corp.*)

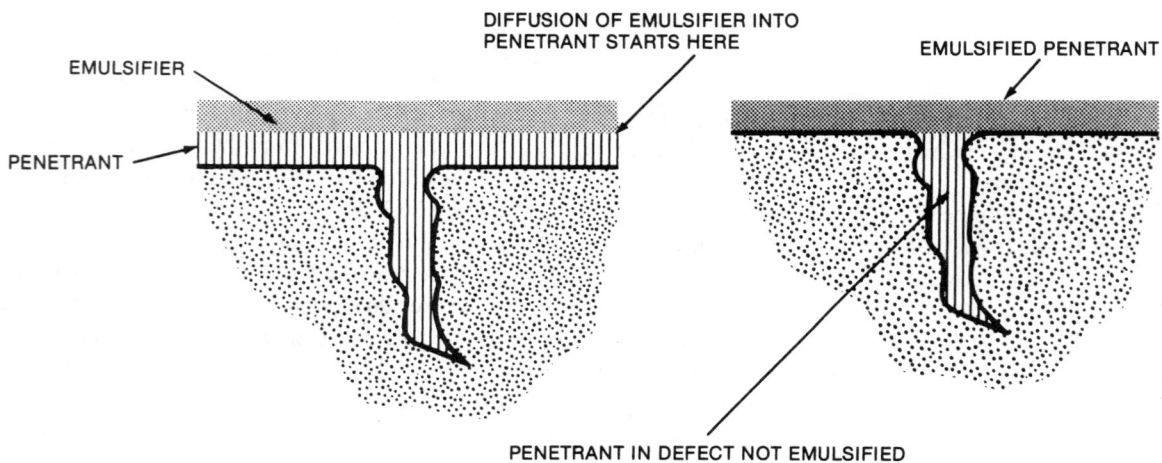

FIGURE 10-18 Action of emulsifier in liquid penetrant inspection.

The developers may be wet or dry. Dry developer is a powder which is dusted onto the part. Wet developers are usually applied by spraying or dipping. In this system, the part is dipped into the developer long enough to ensure that the entire part has been covered. Then the part goes back into the dryer until the developer has dried to a smooth, white film.

Inspection must be done in a darkened booth under a black light since fluorescent penetrant was used. Also, the proper development time must be observed. This may be from 5 to 30 min. Discontinuities will show up as bright, yellow-green indications. It is up to the inspector to determine if the indication represents a defect which requires that the part be rejected, if the part can be repaired, or if the indication is of no consequence. Acceptable parts are cleaned with a suitable cleaner and go on to the next phase in the inspection process.

Ultrasonic Inspection

Although it is not extensively used for overhaul of reciprocating engines, some manufacturers specify ultrasonic inspection for some engine parts. **Ultrasonic inspection** utilizes high-frequency sound waves to reveal flaws in metal parts.

The element transmitting the waves is placed on the part, and a reflected wave is received and registered on an oscilloscope. If there is a flaw in the part, the reflected wave will show a "blip" on the oscilloscope trace. The position of the blip indicates the depth of the flaw.

Ultrasonic inspection can be accomplished satisfactorily by a well-trained and experienced technician. The technician must be familiar with the responses of the equipment and be able to interpret all observed indications accurately.

X-ray Inspection

X-ray or **radiographic inspection**, though not employed on a routine basis for most engine overhauls, is specified from time to time to detect certain types of metal defects. The x-ray is particularly effective in detecting discontinuities inside castings, forgings, and welds. A powerful x-ray can penetrate metal for several inches and produce an image which will reveal defects within the metal. X-ray equipment is pictured in Fig. 10-19.

Only qualified personnel should attempt to perform x-ray examination of parts and materials. Particular attention must be paid to safety considerations to avoid injury from radiation. Qualified x-ray technicians are trained to observe all applicable safety precautions.

Eddy-Current Inspection

Eddy-current inspection is also effective in discovering defects inside metal parts. The eddy-current tester applies high-frequency electromagnetic waves to the metal, and these waves generate eddy currents inside the metal. If the metal is uniform in its structure, the eddy currents will flow in a uniform pattern and this will be shown by the indicator. If a discontinuity exists, the effect of the eddy currents will be changed and the indicator will produce a reading greater than normal for the particular test.

Like x-ray inspection, eddy-current inspection is not routinely used for engine overhaul. However, the eddy-current tester can be of great value to a technician who is experienced and knows what types of tests can be made effectively.

DIMENSIONAL INSPECTION

Dimensional inspection is employed to determine the degree of wear for parts of the engine where moving surfaces are in contact with other surfaces. If the wear between surfaces exceeds the amount set forth in the manufacturer's Table of Limits, the parts must be replaced or repaired in accordance with approved methods.

The Table of Limits, shown in Fig. 10-1, lists the name of the part, the serviceable or maximum limit, and the manufacturer's new part maximum and minimum limits. The serviceable limit sets forth the maximum amount of wear allowable and is the dimension that must be adhered to by the overhauler. The manufacturer's engineers have determined that if a part which is within the serviceable limits is installed in an engine, it should not wear to a point that jeopardizes safe operation of the engine prior to the next engine overhaul. The manufacturer's new part minimums and maximums are the tolerances to which the parts were manufactured. Engines that are "rebuilt" by the manufacturer must comply with these new-part dimensions. However, many overhauling agencies use the new-part tolerances to ensure a quality overhaul.

The inspector should know that parts are usually manufactured with dimensions in sixteenths of an inch in non-metric engines. This is in accordance with a practice established by SAE many years ago. A correctly dimensioned part will therefore measure $x/16$ in $+ 0.000$ minus an amount up to the tolerance. The decimals for sixteenths of an inch are as follows:

$\frac{1}{16} = 0.0625$	$\frac{7}{16} = 0.4375$	$\frac{13}{16} = 0.8125$
$\frac{2}{16} = 0.1250$	$\frac{8}{16} = 0.5000$	$\frac{14}{16} = 0.8750$
$\frac{3}{16} = 0.1875$	$\frac{9}{16} = 0.5625$	$\frac{15}{16} = 0.9375$
$\frac{4}{16} = 0.2500$	$\frac{10}{16} = 0.6250$	$\frac{16}{16} = 1.0000$
$\frac{5}{16} = 0.3125$	$\frac{11}{16} = 0.6875$	
$\frac{6}{16} = 0.3750$	$\frac{12}{16} = 0.7500$	

The dimensions of shafts, crankpins, main bearing journals, piston pins, and similar parts are measured with a micrometer caliper as shown in Fig. 10-20. The micrometer should be equipped with a vernier scale so that measurements can be taken to the nearest ten-thousandth. A micrometer having a ratchet sleeve or stem is recommended, to ensure that the measuring pressure between the anvil and stem is uniform for all measurements.

FIGURE 10-19   X-ray equipment.

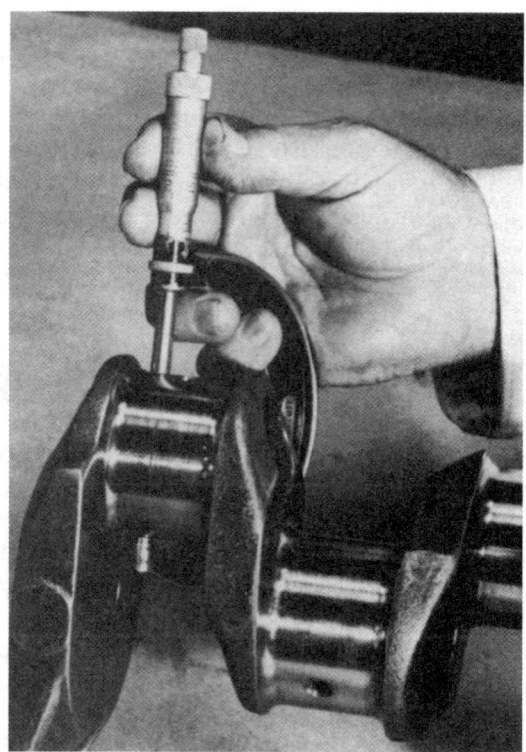

FIGURE 10-20 Measurement of a crankpin with a micrometer caliper.

The inside diameters (ID) of bushings, bearings, and similar openings are measured with telescoping gauges. While the gauge is in the opening, it is locked in place to preserve the dimension. The dimension of the telescoping gauge is then measured with a micrometer, as shown in Figs. 10-21 and 10-22.

The ID measurement of some bearings and bearing bores, such as crankshaft and camshaft bearing bores in the crankcase and large-end connecting-rod bores, requires partial assembly of the crankcase and connecting rods. Bolts and

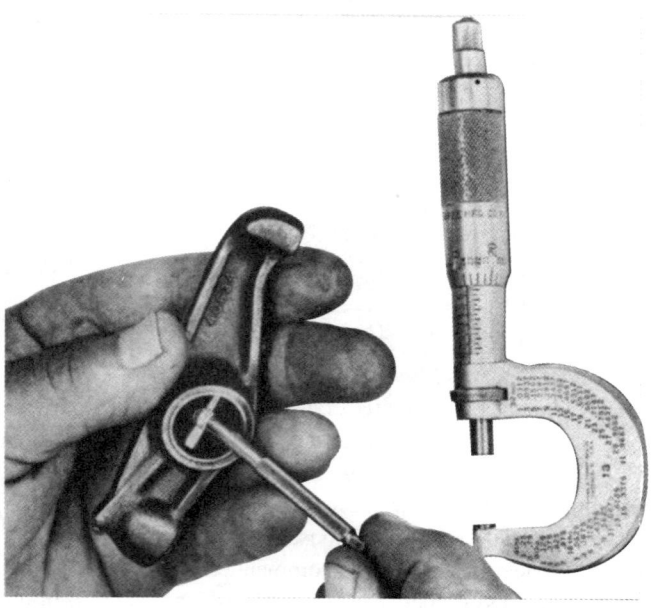

FIGURE 10-21 Measurement with a telescoping gauge.

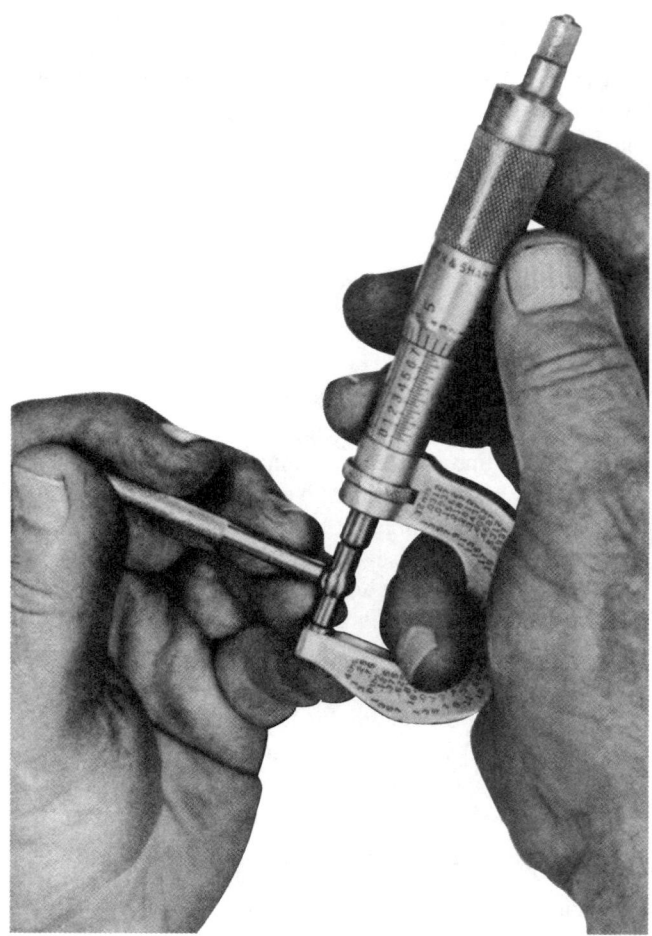

FIGURE 10-22 Measurement of the dimension of a telescoping gauge with a micrometer caliper.

studs must be installed and tightened down to remove all clearance between the parting surfaces before the measurement is taken with a telescoping gauge. The manufacturer may require that the measurement be taken at a specific angle from the parting surface or centerline of the part. These dimensions are compared with the outside diameters of the crankshaft main journals, crankpins, and camshaft journals. The difference between the two measurements is the **clearance**, and it must be within the values specified in the Table of Limits (Fig. 10-1).

If a hole is too small to receive a telescoping gauge, a **small-hole gauge** is used. Gauges of this type are also called **ball gauges**. The ball end of the gauge is inserted in the hole and expanded until it fits the hole snugly, as shown in Fig. 10-23. It is then removed and measured with a micrometer to obtain the dimension of the hole. Engine manufacturers often supply **plug gauges** or **go and no-go gauges** to measure the dimensions of certain holes or openings. These gauges are used according to the instructions given by the manufacturer.

Cylinder barrels are measured with a **cylinder bore gauge**, as shown in Fig. 10-24. This gauge will show the wear, out-of-roundness, and taper. The cylinder bore is measured by sliding the gauge from the top to the bottom of the cylinder in the direction of piston thrust and at 90° to this direction. In this way the out-of-roundness can be checked for the full length

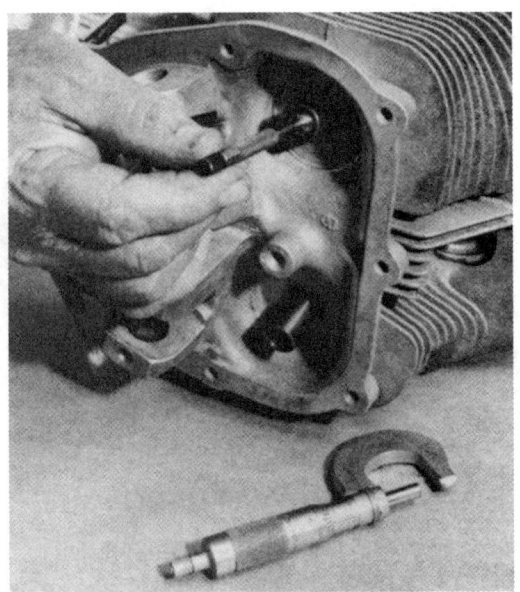

FIGURE 10-23 Measurement with a small-hole gauge.

FIGURE 10-24 Measurement with a cylinder bore gauge.

of the barrel. Before the gauge is placed in the cylinder barrel, the gauge needle is set at zero with the basic dimension of the barrel. Deviations from the basic dimension will be shown as the needle moves in a positive or negative direction.

Average cylinder measurements can be used to avoid extremes that can occur when single measurements are taken. Average measurements require that several measurements be taken and then averaged. Refer to the manufacturer's manual for measurement locations for each engine model.

Connecting rods must be measured for bearing and bushing dimensions and for alignment (twist and convergence of bearing with bushing). The requirements for a connecting rod to be used in a Continental O-470 engine are shown in Fig. 10-25. This drawing illustrates the dimensions as well as the tolerances for twist and convergence.

The twist of a connecting rod is checked by installing push-fit arbors in both ends and supporting the rod by means of the arbors on parallel steel bars resting on a surface plate. Measurements are then taken with a thickness gauge at each supporting point to determine the amount of twist. This check is shown in Fig. 10-26. The clearance between the support bar and the arbor in the small end of the rod is measured and divided by the number of inches between the center points of the supporting steel bars. The twist must not exceed 0.0005 in [0.012 7 mm] per inch of distance between the support point centerlines.

The arbor in the large end of the connecting rod may be supported by matched V blocks rather than on steel bars. Either method will give the desired results.

To check convergence, the difference in distance between the arbors is measured at a given distance on each side of the connecting-rod center. This is accomplished by installing a precision measuring arm with a ball end on one end of the arbor into the small end of the connecting rod. The distance of the measuring arm from the centerline of the rod is noted, and the measuring arm is adjusted so that the ball just touches the arbor in the big end of the rod. The measuring arm is then

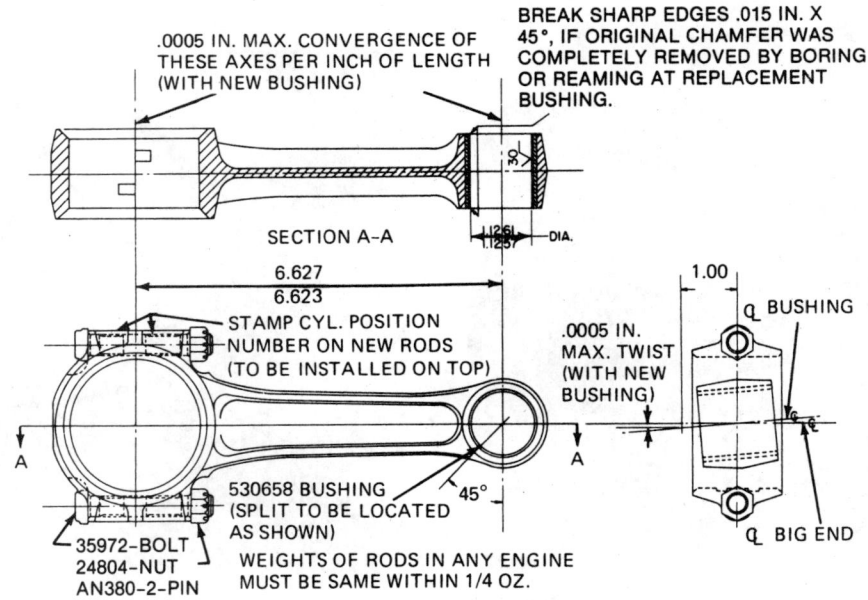

FIGURE 10-25 Dimensional requirements for a connecting rod.

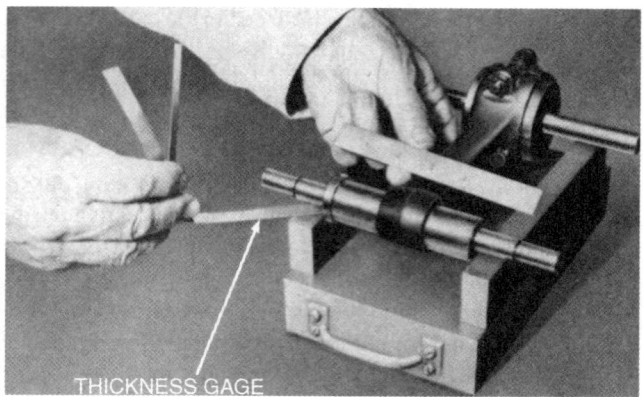

FIGURE 10-26 Checking the twist of a connecting rod.

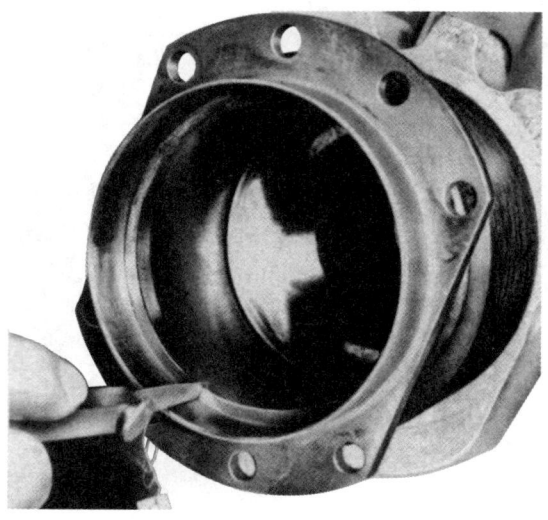

FIGURE 10-28 Measurement of a piston-ring gap.

moved to the opposite end of the arbor, and the difference in distance is checked with a thickness gauge, as shown in Fig. 10-27. If the distances are checked at points 3 in [7.62 cm] from the centerline of the connecting rod, the total distance between the measuring points is 6 in [15.24 cm]. Therefore, if the difference in distances between the arbors on each side of the connecting rod is less than 0.003 in [0.076 mm], the convergence is within limits.

The dimensions of small gaps, such as the clearances between piston rings and ring lands, is measured with a **thickness gauge**. If a gauge of the specified thickness will enter the gap without the use of undue force, it is evidence that the gap is at least as great as the gauge dimension. Thickness gauges are used to measure side clearances, end clearances, valve clearances, and other similar dimensions.

A thickness gauge is used to measure ring end gap and side clearances, when new rings are installed on the piston. The end clearance (gap) is checked by inserting the ring in the skirt of the cylinder and pushing it in with the head of a piston, to ensure that the ring is square with the bore. The ring must be positioned at a point inside the cylinder barrel specified by the manufacturer when the ring gap is measured.

See Fig. 10-28. This gap must be within the tolerance given in the Table of Limits. The piston-ring gap is necessary to prevent seizure of the rings in the cylinder as the pistons and rings expand with the high temperatures of operation.

After the piston-ring gap is measured, the rings are installed on the pistons in the proper grooves for inspection of side clearance. The side clearance of the piston rings is measured with a thickness gauge, as shown in Fig. 10-29. The specified ring clearance is necessary to ensure free movement of the rings in the ring grooves and a free flow of oil behind the rings. The correct fitting of piston rings to pistons and cylinders and installation of the approved type of piston rings in an engine are critical because the wear between piston rings and upper cylinder walls is usually greater than the wear in any other part of the engine.

Some manufacturers require measurement of piston-ring tension. Piston-ring tension is measured at a point 90° from the gap when the ring is compressed to the normal gap.

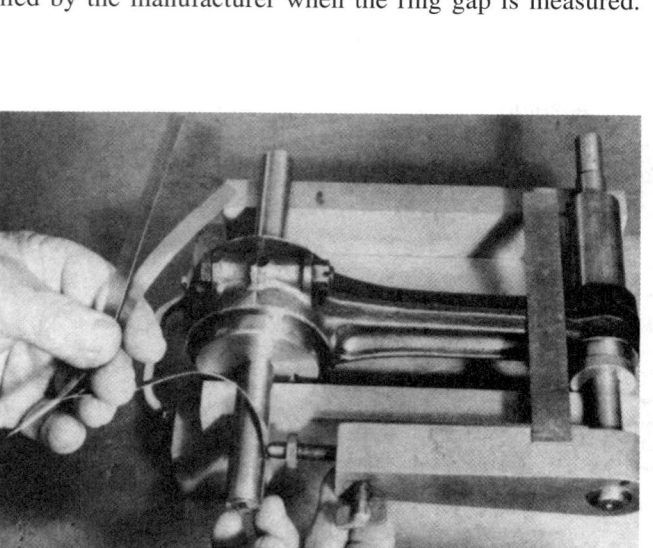

FIGURE 10-27 Checking the convergence of connecting-rod ends.

FIGURE 10-29 Measurement of piston-ring side clearance.

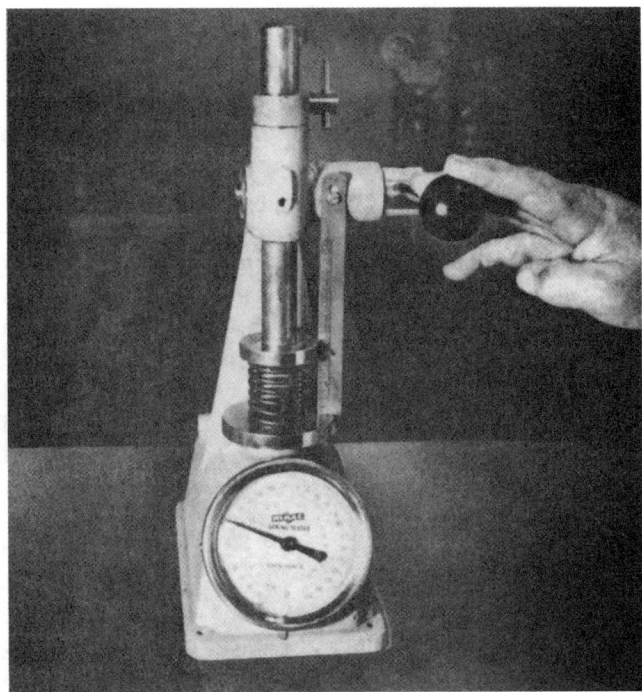

FIGURE 10-30   Testing the compression of a valve spring.

FIGURE 10-32   Checking crankshaft alignment (runout).

The tension is given in the Table of Limits in pounds. Smaller rings can be measured on a valve spring tester, as shown in Fig. 10-30. Larger rings must be measured on a suitable scale.

The **depth gauge** is used to provide an accurate indication of the distance between fixed surfaces, such as the distance from the parting surface of an oil pump housing to the end of the gear. The use of a depth gauge is shown in Fig. 10-31.

Dial gauges or indicators are particularly useful for checking the runout or out-of-roundness of rotating parts, such as crankshafts and camshafts. Crankshaft runout or bending is checked by mounting the shaft on V blocks placed on a level surface plate and rotating the shaft while reading the runout on the dial gauge. This operation is shown in Fig. 10-32. The crankshaft runout should be checked at the center main

journals while the shaft is supported at the thrust and rear journals. It should also be checked at the propeller flange or at the front propeller bearing seat. Permissible runout tolerances are given in the Table of Limits.

A crankshaft alignment and runout check should always be performed on an engine that has undergone a sudden stoppage, such as that caused when the propeller of the aircraft strikes the ground or a solid object. To perform the check with the engine still installed in the aircraft, the propeller is removed and a dial indicator is installed firmly on the front of the engine. The "finger" (actuating rod) of the dial indicator is placed so that it rests on the smooth part of the propeller shaft forward of the splines. The shaft is then rotated. Any eccentricity of the shaft will be indicated on the dial. If, during the runout check, a crankshaft is found to be bent, straightening is not recommended.

The camshaft of an opposed or in-line engine can be checked for alignment in the same manner as that described for a crankshaft. The shaft is mounted in the V blocks at the end bearings, and the runout is measured at the center bearings.

All the springs used in a reciprocating engine are checked for proper tension with the use of a valve spring compression tester, such as that shown in Fig. 10-30. The Table of Limits specifies a tension at a given spring height. Valve springs, oil pressure relief valve springs, oil filter bypass springs, and oil cooler bypass springs are all normally checked at overhaul.

Note here that the weight of some reciprocating-engine parts is critical. It is not uncommon to include the weighing of connecting rods and pistons in the dimensional inspection.

We have mentioned a variety of measuring instruments in this section; however, engine manufacturers often provide special gauges to be used with their particular engines. These gauges are described in the manufacturer's overhaul manual for the engine concerned and should be used according to the instructions given.

Many types of digital electronic measuring instruments are available. The dimensions are read directly from a liquid-crystal display. These instruments can be used alone, or they can be interfaced with digital processors or computers. Upper and lower dimensional limits can be programmed into

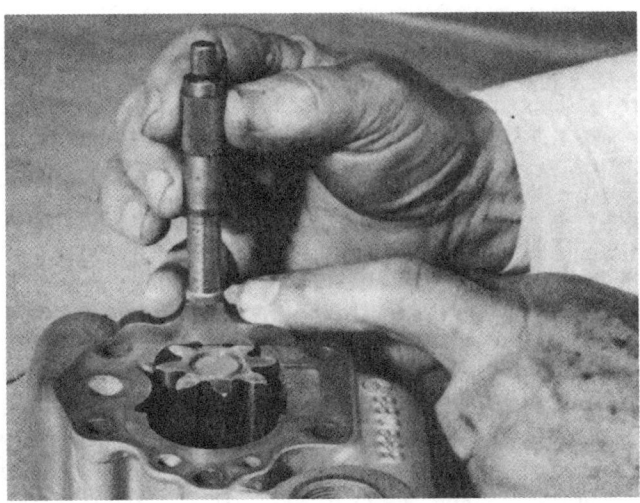

FIGURE 10-31   Measurement with a depth gauge.

the system. As the parts are measured by the operator, the measurements are automatically entered into the computer. The computer can provide the operator with a printout of all dimensions and indicate the parts that are out of limits.

REPAIR AND REPLACEMENT

At this point in the overhaul process, unrepairable parts and components should have been discarded, repairable parts labeled, and parts requiring no further action organized for reassembly on a parts rack. The discrepancy list and labels on repairable parts will serve as a guide for work to be done. In larger overhaul facilities, different parts and assemblies are routed to areas of the shop that do specific processes and repairs. For example, all cylinder work is done in one section of the shop, crankcase repair in another, and crankshaft and camshaft work in still another. Smaller shops may elect to send these same parts out to facilities that do specialized work, such as welding and grinding.

Repairs done on parts and assemblies by overhauling agencies must be done in accordance with the manufacturer's instructions or other industry practices that have been proved and accepted by the manufacturer and/or the FAA. In any case, the parts must be returned to airworthy condition before they go back in the engine.

Cylinders

The repair operations required for cylinders depend on the wear and damage found during inspection. Some operations are performed at each overhaul, and others are performed only if required to restore the cylinder to an airworthy condition.

If an engine is being overhauled for the first time, probably the only operations required on the cylinders will be routine inspections after cleaning, reseating the valves by grinding and lapping, and breaking the glaze on the cylinder walls by means of a cylinder hone.

When cylinders have been overhauled several times, the cylinder bore probably will have to be ground oversize or chromium-plated and ground to standard dimensions. Valve guides will need to be replaced, and if the valve seats have been ground a number of times, it may be necessary to replace the seats. Rocker shaft bosses which incorporate bushings must be rebushed from time to time to retain the dimension specified in the Table of Limits.

Cylinder Heads

Cracks often develop in cylinder heads after many hours of service. These cylinders should be replaced. However, some shops have approval to weld cracks in certain areas of the cylinder head. Cracks must be ground out completely and then filled in by welding. The weld material is ground or machined to match the contour of the surrounding area.

A small amount of pitting is not generally serious enough to render the cylinder unsatisfactory for use, but the rough edges of pits should be smoothed by a scraper, a burnishing tool, or a suitable file.

Fins

Inspection of the cylinder may reveal that some of the cooling fins have been damaged. The fins on the cylinder head are made as integral parts of the cast-aluminum-alloy head; therefore, they are brittle and easily cracked or broken. Broken head fins should be filed smooth at the broken edges to eliminate roughness and sharp edges. If it becomes necessary to cut out a V notch to stop a head fin crack, a slotted drill bushing that fits over the fin may be used with a $\frac{3}{16}$-in [4.76-mm] twist drill to cut the notch. The apex of the notch and the edges of the cut should be rounded to reduce the possibility of further cracking. The cylinder is considered beyond repair if an excessive amount of fin area is broken off or damaged. In any case, the manufacturer's overhaul manual should be consulted for limitations on fin repair.

Studs

The studs by which the exhaust pipes are attached to the cylinder head are usually subject to severe corrosion because of heat and the chemical effects of the exhaust gases. It is often necessary to replace these studs. During inspections, the threads of the studs should be carefully examined, and if appreciable erosion is noted, the studs should be replaced. Stud replacement is described in the section covering crankcase repair.

Valve Guides

If inspection reveals that valve guides are oversize or damaged in any way, they must be replaced. Valve guides should be removed according to the manufacturer's instructions, and the method usually requires driving with a special piloted drift or use of a valve guide puller. Since the valve guide has a flange on the outside end, it is necessary to drive it from the inside of the cylinder. Before it is driven or pulled, the inner end of the guide should be cleared of all hard carbon to prevent scoring of the guide hole. Some operators recommend heating the cylinder head to about 450°F [232°C] before removing the valve guides because this tends to expand the valve guide hole and permit easier removal. In all cases the cylinder should be properly supported on a cylinder-holding fixture while valve guides are being removed and replaced.

If equipment is not available to pull or drive the valve guide from inside the cylinder, it is possible to cut away the outer end and flange of the guide by means of a spot facer and then drive the guide to the inside of the cylinder. Care must be taken to avoid cutting into the guide flange seat.

After the valve guides are removed, the guide holes must be inspected for scoring, roughness, and diameter. If the hole size is not within specified limits, it will be necessary to ream or broach it to an oversize dimension. The actual size of the hole should be approximately 0.002 in [0.051 mm] smaller than the outside diameter of the guide to be installed. The Table of Limits for typical opposed engines establishes limits of 0.001- to 0.0025-in [0.025- to 0.064-mm] interference (T) fit. Valve guides are made in oversizes of 0.005, 0.010, 0.020, and 0.030 in [0.127, 0.254, 0.508, and 0.762 mm].

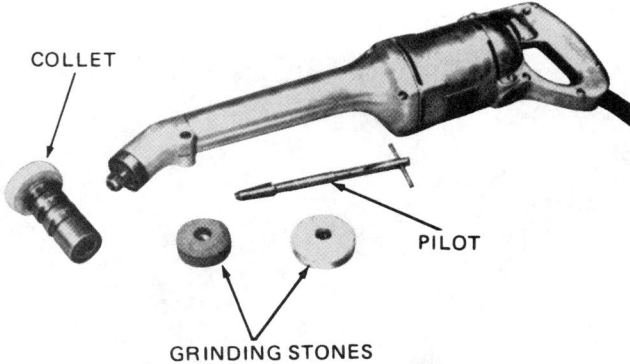

COLLET

PILOT

GRINDING STONES

FIGURE 10-33 Valve seat grinder.

Valve guides and valve guide holes are measured as shown in Fig. 10-23.

Guide holes which are within the required dimensional limits and are not scored or damaged in any way do not require repair. It is merely necessary to install new valve guides by driving with a suitable installing tool. The guide must be carefully aligned with the hole before driving. Valve seats must be reground *after* valve guides are replaced in every case.

Valve Seats

Valve seats can usually be repaired by grinding with specially designed seat grinding equipment. A typical valve seat grinder is shown in Fig. 10-33. As can be observed in the illustration, the valve seat grinder consists of an electric motor equipped with an angle drive to permit grinding of the seats at the angle at which they are installed in the cylinder. The grinding wheel is mounted on a collet equipped with a pilot which fits into the valve guide. The pilot holds the grinding wheel in exact alignment with the valve seat, as shown in Fig. 10-34.

The technician must choose the correct grinding wheel (angle, hardness, grit, and size). The stone angle must be correct for the particular seat being ground. Some engines have a 30° seat for the intake valve and a 45° seat for the exhaust valve. The 30° angle of the intake valve seat provides improved gas flow characteristics, whereas a 45° angle for the exhaust valve seat allows for better seat cooling. The hardness and size of the grinding wheel must be appropriate for the material and size of the seat being ground. A coarse-grit stone is used to make the initial cuts, and fine-grit stones, or finishing stones, are used to smooth out or polish the seat.

The grinding wheel must be dressed prior to the seat grinding operation. The grinding wheel is installed on the collet or stone holder, and the whole assembly is placed on a grinding wheel dresser. The grinding wheel is dressed by spinning the wheel at a high speed and moving a diamond-tipped dressing tool slowly across the cutting face at the correct angle, removing only enough material to true the wheel. The grinding stone must be dressed each time it is installed on the collet and whenever it begins to become grooved during grinding.

The motor must be centeured accurately on the wheel holder. If the motor is off center, chattering of the stone will result and a rough grind will be produced. It is very important that the grinding wheel be rotated at a speed that will permit grinding instead of rubbing. This speed is approximately 8000 to 10 000 rpm. Excessive pressure on the wheel can slow it down. It is not a good technique to let the wheel grind at low speed by putting pressure on the stone when starting or stopping the motor. The maximum pressure used on the stone at any time should be no more than that exerted by the weight of the motor.

Another practice often followed is to ease off on the stone every second or so to let the coolant wash away the chips on the seat; this rhythmic grinding action also helps keep the stone up to its correct speed. Since it is a critical job to replace a seat, remove as little material as possible during grinding. Inspect the job frequently to prevent unnecessary grinding.

It may be necessary to adjust the width and diameter of the seating surface after the 30° or 45° angle has been ground. The width of the seat is adjusted by using a 15° stone to reduce the outside diameter and a 75° stone to increase the inside diameter. The width and diameter of the seat after it is ground must be within the dimensions specified by the manufacturer.

If a valve seat has been reground so many times that the entire face of the 15° narrowing wheel must be brought into contact with the seat in order to grind to the required dimension, as illustrated in Fig. 10-35, then the seat is beyond limits and must be replaced.

A valve seat is typically removed in one of two ways. First, it can be cut out with a special counterbore suitable for the purpose, while the cylinder is mounted in a fixture which holds it at the proper angle under the spindle of a drill press. The drill press is used to rotate the counterbore, thus cutting the valve seat from its recess. This is a precision operation and must be performed by an experienced operator.

The second method does not require machining or a high level of skill. The cylinder is placed in an oven and heated

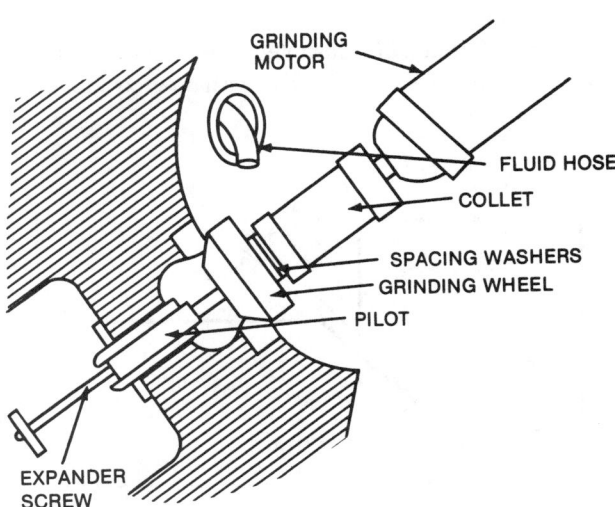

GRINDING MOTOR

FLUID HOSE

COLLET

SPACING WASHERS

GRINDING WHEEL

PILOT

EXPANDER SCREW

FIGURE 10-34 Alignment of pilot, grinding wheel, and grinding motor.

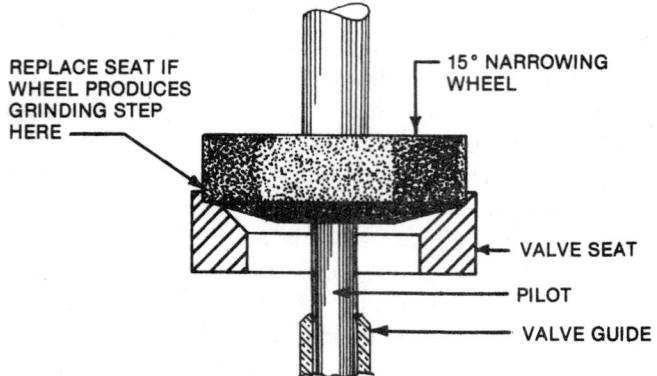

FIGURE 10-35   Cause for replacement of a valve seat.

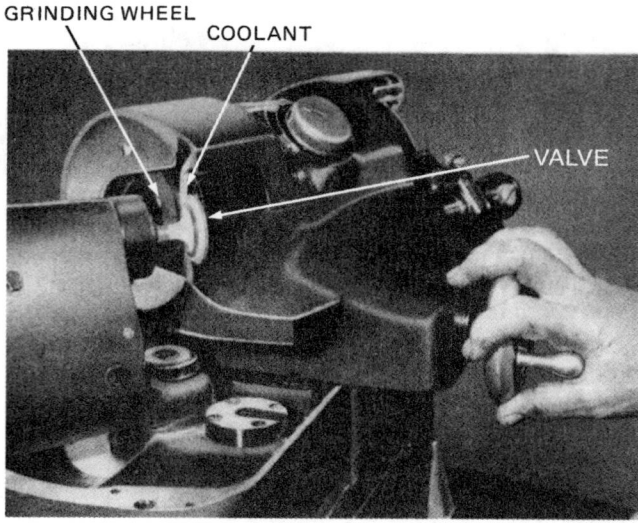

FIGURE 10-37   Refacing a valve.

to 600 to 650°F [316 to 343°C]. A special tool, shown in Fig. 10-36, is inserted into the valve seat and grasps the seat. The cold water used as a coolant shrinks the valve seat, relieving the interference fit and allowing the seat to be lifted out. Care must be taken in handling the hot cylinder, and protective clothing and equipment should be worn to prevent burns from hot water spattering off the cylinder.

After the seat is removed, the inside diameter of the valve seat recess must be measured to determine which oversize seat to install. The recess must then be cut to the correct oversize dimension as specified in the Table of Limits. This is accomplished by means of a valve recess cutter available from the manufacturer.

The cylinder is then heated to 600 to 650°F [316 to 343°C], and the new valve seat is driven into the recess with the replacement drift. The manufacturer's manual must be consulted to ascertain the correct temperature to be used.

After replacement, the new valve seats are ground to the proper face dimension as previously explained.

Valves

The valves are refaced by means of a standard valve refacing machine. This operation is illustrated in Fig. 10-37. The refacing machine rotates the valve while the grinding wheel is moved back and forth across the face.

Like many machine jobs, valve grinding is mostly a matter of setting up the machine. The following should be checked or performed before grinding is begun.

True the grinding wheel by means of a diamond-tipped dressing tool. Turn on the machine, and draw the diamond across the wheel, cutting just deep enough to true and clean the wheel.

Determine the face angle of the valve being ground, and set the movable head of the machine to correspond to this valve angle. Usually, valves are ground to the standard angles of 30 or 45°. However, in some instances, an interference fit of 0.5 or 1.5° less than the standard angle may be ground on the valve face.

The **interference fit**, illustrated in Fig. 10-38, is used to obtain a more positive seal by means of a narrow contact surface.

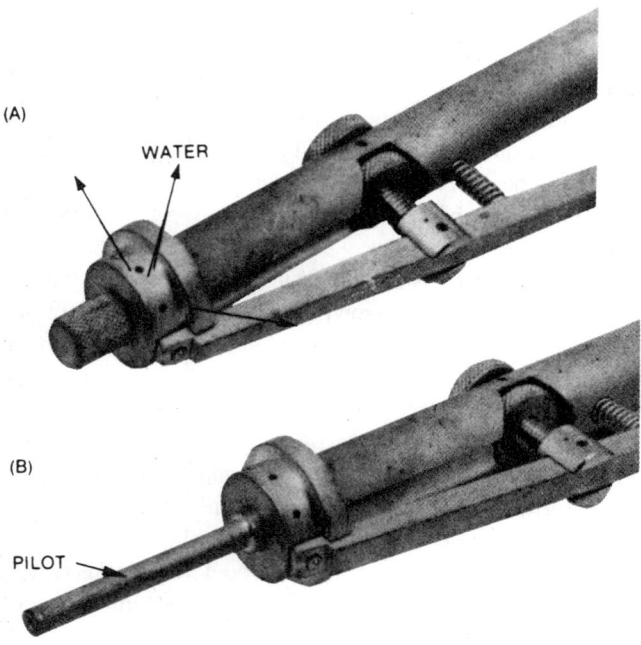

FIGURE 10-36   Tool for (A) removal and (B) replacement of valve seat.

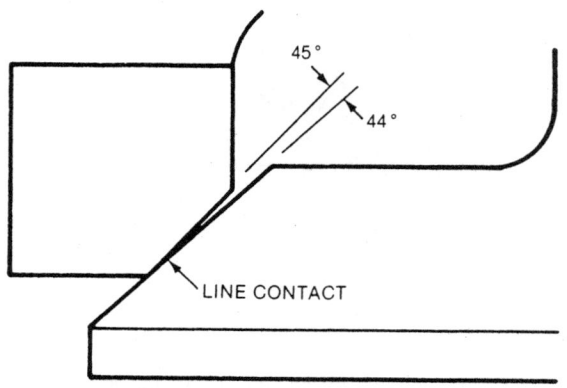

FIGURE 10-38   Interference fit of valve and valve seat.

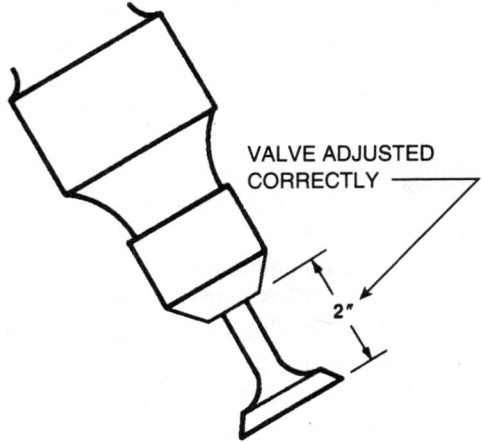

FIGURE 10-39   Proper installation of valve in chuck.

Theoretically, there is a line contact between the valve and seat. With this line contact, all the load that the valve exerts against the seat is concentrated in a very small area, thereby increasing the unit load at any one spot. The interference fit is especially beneficial during the first few hours of operation following an overhaul. The positive seal reduces the possibility of a burned valve or seat that a leaking valve might produce. After the first few hours of running, these angles tend to be pounded down and to become identical.

The interference angle is ground into the valve, not the seat. It is easier to change the angle of the valve grinder work head than to change the angle of a valve seat grinder stone. Do not use an interference fit unless the manufacturer approves it.

To grind a valve, first install the valve in the chuck, as shown in Fig. 10-39, and then adjust the chuck so that the valve face is approximately 2 in [5.08 cm] from the chuck. If the valve is chucked any farther out, there is danger of excessive wobble and a possibility of grinding into the stem.

Check the travel of the valve face across the stone. The valve should completely pass the stone on both sides and yet not travel far enough to grind the stem. There are stops, as illustrated in Fig. 10-40, on the machine which can be set to control this travel.

With the valve set correctly in place, turn on the machine and adjust the grinding fluid so that it splashes on the valve face. The grinding fluid is a water-soluble oil that is continuously run onto the valve face to provide cooling and to carry away grindings.

Back the valve away from the grinding wheel. Place the grinding wheel in front of the valve. Slowly bring the valve toward the grinding wheel until a light cut is made on the valve. The intensity of the grind is measured by sound more than anything else. Slowly move the wheel back and forth across the valve face without increasing the cut. Use the full face of the wheel, but always keep the wheel on the valve face. When the grinding sound diminishes, move the valve slightly closer to the grinding wheel, approximately 0.001 in [0.025 4 mm] each time, keeping the grinding pressure light. Heavy grinding will result in a rough finish unsuitable for proper valve sealing.

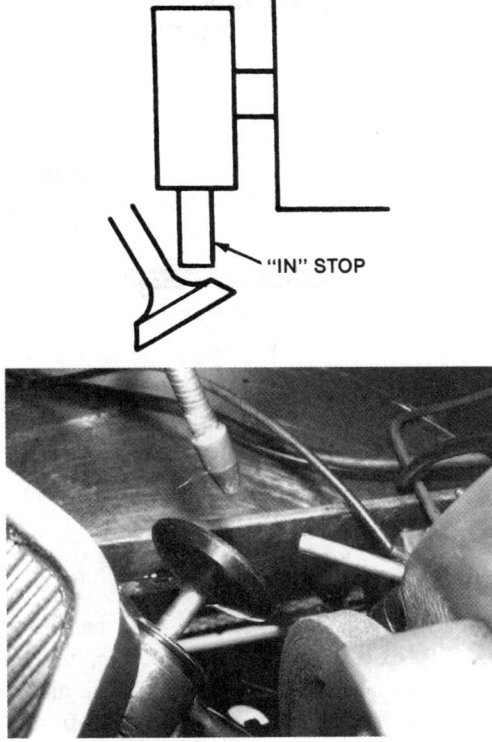

FIGURE 10-40   "In" stop prevents grinding the stem.

Back the valve away from the grinding wheel when grinding is complete. If inspection shows that more grinding is necessary, repeat the process described above.

After grinding, check the valve margin to be sure that the valve edge has not been ground too thin. A thin edge is called a **feather edge** and can lead to preignition. Such a valve edge would burn away in a short time, and the cylinder would have to be overhauled again. Figure 10-41 shows a valve with a normal margin and one with a feather edge.

After valves and seats are ground, they are **lapped** to provide a gastight and liquid-tight seal. Each valve is placed

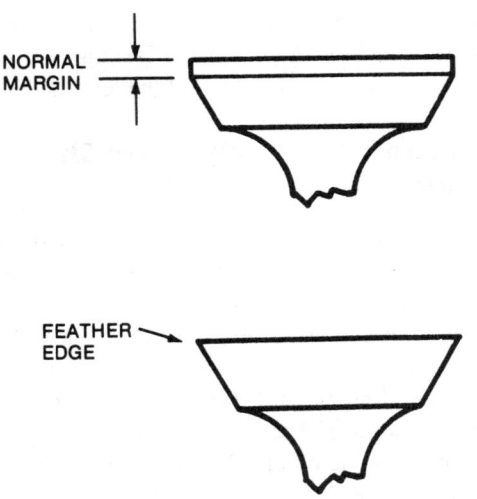

FIGURE 10-41   Engine valves showing normal margin and a feather edge.

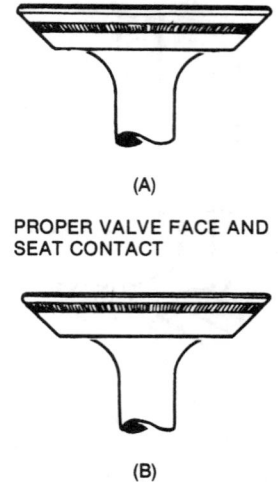

(A)

PROPER VALVE FACE AND
SEAT CONTACT

(B)

FIGURE 10-42 Correct (A) and incorrect (B) valve face conditions.

in its particular seat, one at a time, and the valve face is rotated against its seat with an approved lapping compound until there is a perfect fit between the valve and seat. All lapping compound is then carefully removed from both the seat and the valve. The valves are then placed in a numbered rack to make certain that they will be installed in the correct cylinder.

Lapped valves must be inspected to ensure that the face and seat of each valve make proper contact according to the limits specified in the overhaul manual. Figure 10-42A shows the lapped area of a properly ground valve. The lapped area is midway between the edge of the head and the bottom of the face. The area should have a frosty gray appearance. In Fig. 10-42B, the lapped area is too near the edge of the face at the top.

The final step is to check the mating surfaces for leaks, to see if they are sealing properly. This may be done by installing the valve in the cylinder, holding the valve by the stem with the fingers, and pouring kerosene or solvent into the valve port. While applying finger pressure on the valve stem, check whether the kerosene is leaking past the valve into the combustion chamber. If it is not, the valve reseating operation is finished. If kerosene is leaking past the valve, continue the lapping operation until the leakage is stopped.

Rocker Arms and Rocker-Arm Shaft Bushings

Frequently the rocker arms need no repair, and sometimes the only repairs required are replacement and reaming of the bushings. These operations are accomplished by means of a suitable arbor and arbor press by which the old bushings are pressed out and new bushings are pressed in. Each bushing hole should be examined for condition before the new bushing is pressed in. Special attention must be paid to the position of the oil hole in the bushing to make sure that it is aligned with the oil hole in the rocker arm.

Rocker shaft bushings in the cylinder head must be replaced if they are worn beyond limits. If the bushing is held

in place with a dowel pin, the pin must be drilled out before the bushing is removed. Removal of the bushing is accomplished with a drift or arbor. The cylinder must be properly supported while the rocker shaft bushings are removed to prevent damage.

After the rocker shaft bushings have been removed, each bushing hole must be checked for size with either a telescoping gauge or a special plug gauge. If the bushing hole dimension is above the maximum limit, it is necessary to install an oversize bushing. When it has been decided which oversize bushing is required, the hole is reamed to the correct size for the bushing. The new rocker shaft bushing is installed with an installation drift or similar tool in accordance with the manufacturer's instructions. If the bushing is held in place with a dowel pin, it will be necessary to drill a new hole in the bushing for the dowel pin and then to install a pin of the correct size. Special instructions relating to this operation are provided by the manufacturer for the engines in which dowel pins are employed.

After rocker shaft bushings are installed in the cylinder head, it is necessary to check them for dimension and ream them to size, if required. The final cut should be made with a finish reamer to produce a very smooth surface.

Cylinder Barrels

The condition of the cylinder barrels, pistons, and piston rings is a most vital factor in the performance of a reciprocating engine. If the dimensions and surface conditions of these parts are not satisfactory, combustion gases can escape past the piston rings into the crankcase, and oil from the crankcase can enter the combustion chambers. The engine will lose power, and the excess oil in the combustion chambers will foul the spark plugs and cause an accumulation of carbon. For these reasons, careful consideration must be given to the repair and servicing of cylinder barrels.

Cylinder barrels that have been determined to be structurally sound and within dimensional limits may only need to be deglazed before being returned to service. Cylinder barrels become very smooth after several hours of operation. Some intentional wearing of the new piston rings is necessary to allow the rings to conform to the cylinder walls. The smooth, or glazed, cylinder walls will not produce the friction needed for this wearing-in process.

Deglazing the cylinder walls is accomplished with the use of a deglazing hone such as the one illustrated in Fig. 10-43. The deglazing hone is turned by a suitable drill motor and moved in and out of the cylinder at a rate which will produce a crosshatched pattern, as shown in Fig. 10-44. This can also be done with no. 400 wet-or-dry sandpaper followed by the use of crocus cloth. This process will remove small amounts of corrosion or scoring as well. The dimensions of the barrel should be measured after deglazing; if they are not within limits, the barrel must be honed or ground to a standard oversize. Oversize dimensions are usually 0.010, 0.015, or 0.020 in [0.254, 0.381, or 0.508 mm], depending on how much metal must be removed to produce a uniform new surface on all parts of the cylinder wall. The manufacturer's overhaul

FIGURE 10-43 Cylinder deglazing hone.

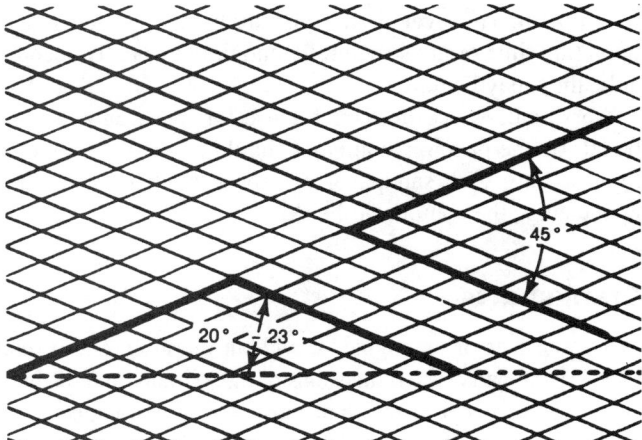

FIGURE 10-44 Crosshatched pattern.

manual should be consulted to determine the requirements for any particular make and model of engine. Note that the greatest wear of a cylinder barrel occurs near the top because of the high temperatures in this area during operation. For this reason, care must be taken to measure the cylinder barrel in this area during inspection.

When cylinder walls are reground, it is most important to consult the manufacturer's overhaul manual to determine whether the cylinder barrels are **nitrided** (surface-hardened) or chromium-plated. Nitrided barrels usually should not be ground to more than 0.010 in [0.254 mm] oversize because of the danger of grinding through the hardened surface. If worn beyond acceptable limits, chromium-plated barrels should be chemically stripped and replated to standard dimensions.

Grinding and chromium plating should be accomplished by an operator whose process has been approved by the FAA. The advantage of chromium plating is that the cylinder barrel will show very little wear between overhauls and will usually remain serviceable with standard dimensions for several thousand hours of operation.

When the cylinders of an engine have been chromium-plated, the piston rings used in these cylinders must not be chromium-plated. The overhaul technician must determine whether cylinders have been chromium-plated; if so, the piston rings must be unplated cast iron or steel. The plating facility employed to chromium-plate the cylinders can supply the proper piston rings to match its unique plating process.

If in a particular engine one or more cylinders must be ground to oversize, then all the cylinders in the engine should be given the same treatment. The crankshaft, piston rod, and piston assembly will be seriously out of balance if oversize pistons are installed in some cylinders while others are standard.

Cylinder grinding is accomplished by means of high-quality precision grinding equipment. The cylinder is firmly mounted on the grinding machine, and the grinding wheel is adjusted so that it will take the required cut from the cylinder wall to produce the correct dimension. The operator of the grinding machine must make sure to use the correct type of grinding wheel so that the finish will be as specified. Finish is specified in microinches, µin, usually from 10 to 30 [0.000 254 to 0.000 762 mm]. If a surface is ground to 10 µin, the depth of the grinding scratches will not exceed 10 millionths of an inch.

A **cylinder honing machine** is used to produce the final finish of the cylinder walls. The hone usually consists of four high-quality rectangular stones mounted on a fixture. When the cylinder is mounted on the machine and the hone is inserted in the cylinder, the stones make contact with the cylinder walls along their full length. The hone is rotated by means of an electric motor and is moved in and out of the cylinder at a uniform rate while being rotated. The stones remove surface roughness and irregularities, thus producing a smooth crisscross surface which the piston rings can wear in for a good working surface. In some cases it is recommended that piston rings be lapped to cylinder walls. A cylinder honing machine is shown in Fig. 10-45.

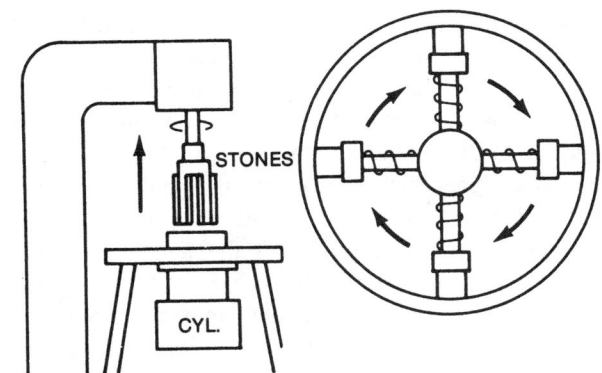

FIGURE 10-45 Cylinder honing machine.

Most maintenance technicians are not expected to grind cylinder barrels because this can be done most satisfactorily in a specially equipped shop set up for this type of work. Technicians must be able to inspect the cylinder barrel and determine what type of treatment is necessary to restore the barrel to satisfactory operating condition. They must check the top, middle, and bottom of the cylinder for out-of-round condition as set forth in the Table of Limits. While doing this they should also check for the **choke** of the barrel, provided that the barrel is designed with choke. A choked barrel is usually designed with a slightly smaller dimension at the top than at the skirt. Because of higher operating temperatures near the top of the cylinder, a choke bore provides a straight bore during engine operation.

Cylinders in operation for several hundred hours will usually have a step, or ridge, worn near the top of the barrel. This ridge is formed at the point where the top edge of the top piston ring stops when the piston is at TDC. When a cylinder is reground, the ridge is removed by the grinder; however, if the cylinder barrel is within limits and does not require grinding, the ridge should be removed or smoothed out by hand honing. If the ridge is not removed, the top piston ring will be damaged when the engine is assembled and operated.

When the cylinder barrels of an engine have been ground to a standard oversize or when the barrels have been chromium-plated to standard size, an oversize indicator must be provided on each cylinder. Oversizing of aircraft engine cylinders is limited because of the relatively thin cylinder walls. The color and location of the indicator may be specified by the manufacturer.

Skirt, Flange, and Fins

During the handling of cylinders, care must be exercised to avoid damaging the skirts. Approved practice calls for mounting the cylinder on a wooden cylinder block when it is not being worked on. If cylinder blocks are not available, the cylinder may be placed on its side on a wooden rack. If pushrod housings are attached, the cylinder must be placed on its side in a manner that does not put stress on the housings. The cylinder must not be lifted or carried by grasping the pushrod housings.

If the cylinder is handled properly, there is no reason why the skirt should become damaged. Usually such damage is caused by carelessly allowing the skirt to strike a hard metal object. If a small nick or scratch should be found on the skirt, it can be removed by careful stoning and polishing.

The cylinder mounting flange must be examined for cracks, warpage, damaged bolt holes, and bending. Warpage of the flange can be checked by mounting the cylinder on a specially designed surface plate and using a thickness gauge to determine the amount of warp. Another common method is to place a straightedge across the flange at locations about 45° apart and to check the gap under the straightedge with a thickness gauge. A very small amount of warp can be removed by lapping the bottom of the flange on the cylinder surface plate; otherwise, any appreciable defects require that the cylinder be discarded.

The cylinder flange should be given an especially careful examination if any of the cylinder hold-down bolts or nuts were found to be loose at the time of disassembly. A loose hold-down nut or bolt will cause exceptional stresses to be imposed on the flange and on the crankcase.

Steel barrel fins which have been bent can be straightened with a long-nose plier or a special slotted tool designed for the purpose.

It may be necessary, in some cases, to install a new cylinder barrel on the cylinder head. To do this, the cylinder is put in an oven and heated to 600 to 650°F [316 to 343°C]. The hot cylinder is then placed in a fixture which holds the cylinder base and sprays cold water on the inside of the cylinder at the threaded end. This shrinks the barrel and allows it to be unscrewed from the cylinder head. The new barrel is installed by heating the cylinder head and then threading the new barrel into the head. The cylinder head shrinks onto the threaded end of the barrel as it cools. Holes are drilled in the new cylinder flange, and final machining is done *after* the barrel is installed in the head.

Rebarreling is usually done in cases where cylinder assemblies are not available, and then only by well-equipped shops or by the manufacturer.

Pistons

Engine manufacturers recommend that pistons be replaced at overhaul. However, if the pistons have not seen much operating time or are structurally and dimensionally acceptable, they may be reused.

Very shallow scoring is not cause for rejection and may be left on the piston. No attempt should be made to remove light scoring with sandpaper or crocus cloth because this may change the contour of the skirt.

Crankshafts

The crankshaft of an engine is, without question, one of the most critical parts. The dimensions of the journals, and the balance and alignment of the shaft must be within tolerances, or else the engine will vibrate and may ultimately fail. It is easily understood that the crankshaft is subjected to extremely rigorous treatment during the operation of the engine because it must bear the constant hammering of the connecting rods as they transfer the force of the piston thrust to the connecting-rod journals. The repair of the crankshaft must therefore be accomplished with great care and precision if the crankshaft is to perform reliably for the hundreds of hours between overhauls.

Crankshaft main journals or crankpins found to be oval (out of round) may be ground undersize within the manufacturer's limits. Crankshafts that are ground undersize must be renitrided. If only a small amount of roughness is noted, the journal surfaces may be polished. It is best to do this while the shaft is rotated slowly in a lathe. A fine abrasive cloth is held against the journal or pin by means of a special block until all roughness is removed. The journals and pins must be remeasured and the dimensions recorded for comparison with bearing measurements at a later time to determine bearing clearances.

Some manufacturers allow limited straightening of the propeller flange. This may be done only on flanges that have not been nitrided. Consult the manufacturer's overhaul manual for instructions concerning propeller flange straightening.

No attempt should be made to straighten crankshafts that have an excessive amount of runout. Any bending of the shaft will fracture the nitrided surfaces and lead to complete failure of the crankshaft.

When a crankshaft is reground, the exact radii of the original journal ends must be preserved to avoid the possibility of failure during operation. If a small ledge or step is left in the metal at the radius location, the shaft is likely to develop cracks which may lead to failure of the journal.

Counterweights

The crankshafts of many engines are dynamically balanced by means of counterweights and dynamic balances mounted on extensions of the crank cheeks. The size and mounting of these counterweights and balances are such that they damp out (reduce) the torsional vibration which occurs as a result of the connecting-rod thrust. The proper method for removing and replacing the counterweights differs among various types and models of engines; therefore, the overhaul technician should make sure that the exact procedure outlined in the overhaul manual, for the model of engine on which work is being done, is followed.

Sludge Chambers and Oil Passages

Some crankshafts are manufactured with hollow crankpins which serve as sludge removers. Drilled oil passages through the crank cheeks carry the oil from inside the main journals to the chambers in the crankpins. The sludge chambers may be formed by means of spool-shaped tubes called **sludge tubes**, pressed into the hollow crankpins. The sludge tubes must be removed at overhaul and the soft-carbon sludge cleaned from the sludge chambers. New sludge tubes must be pressed back into the hollow crankpins. The overhaul technician must make certain that the tubes are reinstalled correctly to avoid covering the ends of the oil passages.

The front opening of the crankshaft may be sealed with an expansion plug if the engine is equipped with a fixed-pitch propeller. This plug must be removed at overhaul to clean the sludge from the flange end of the crankshaft.

The oil passages in the crankshaft should have been cleaned at the time that the shaft was originally cleaned; however, they should be checked again before the crankshaft is declared ready for reassembly in the engine. Soft-copper wire passed through the passages will verify that they are clear of dirt and other obstructions.

Connecting-Rod Bushing Replacement

If the piston-pin clearance in the connecting-rod bushing is excessive, it is necessary to replace the bushing. This is accomplished by pressing out the old bushing and pressing in a new bushing with an arbor press. The new bushing should be

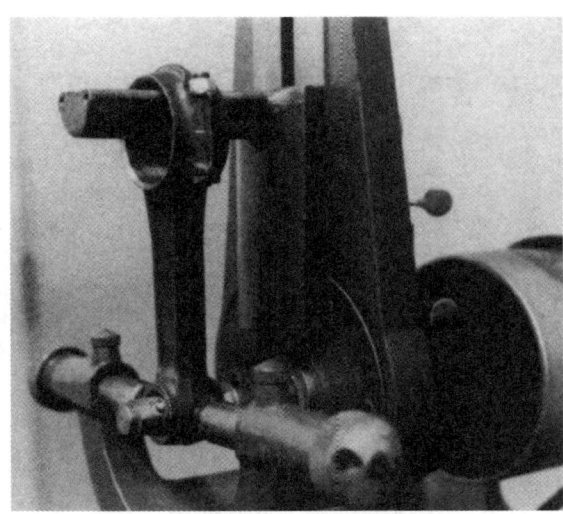

FIGURE 10-46 Hydrobore used for boring connecting-rod bushings.

lubricated before it is installed and must be perfectly parallel with the bore into which it is pressed. After a new piston-pin bushing is installed in the connecting rod, it is usually necessary to ream or bore the bushing to the correct size. This operation is particularly critical because the alignment of the bushing must be held within 0.0005 in [0.012 7 mm] per 1 in [25.4 mm], as previously explained. The bushing is usually bored with special equipment designed for this purpose. One such device is the **hydrobore** shown in Fig. 10-46. For this method the connecting rod is mounted on a faceplate, as shown, so that it is exactly perpendicular to the axis of the cutter bar rotation. The boring tool is then brought into contact with the inner surface of the bushing while turning at a slow rate. A small cut is taken with a smooth cutting tool for as many passes as necessary to provide the desired dimension.

Crankcases and Accessory Cases

Repairs of crankcases and accessory cases are generally limited to replacement of studs and dressings and fixing of small nicks, or scratches. Crankcase halves are manufactured as matched pairs. Both halves must be replaced if one half is rejected. Some facilities have FAA approval for making welding repairs of cracks and machining repairs to bring bearing bores back within limits.

Replacement of Studs

Studs or stud bolts are metal pins threaded on each end that are used for the attachment of parts to one another. One end is provided with coarse (NC) threads and the other with fine (NF) threads. The coarse-threaded end is designed to be permanently screwed into a casting such as the crankcase, and the fine-threaded end has a nut installed for holding an attached part.

Studs which are damaged, bent, or broken must be removed and replaced with new studs. Studs which are not broken can be removed with a **stud remover** or with a small pipe wrench. The stud should be turned slowly to avoid heating the casting. Broken studs which cannot be gripped by a stud remover or

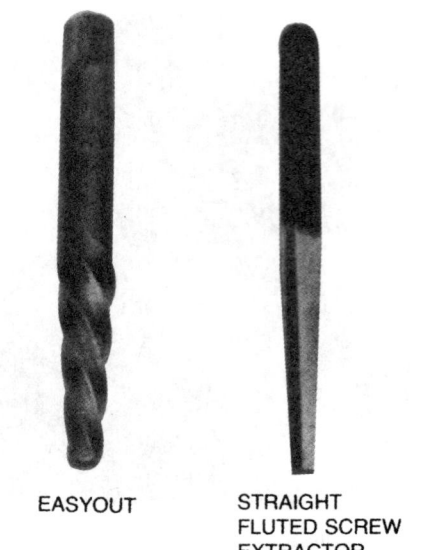

EASYOUT

STRAIGHT
FLUTED SCREW
EXTRACTOR

FIGURE 10-47   An Easyout and a straight-fluted screw extractor.

pipe wrench are removed by drilling a hole in the center of the stud and inserting a **screw extractor**. The screw extractor may have straight splines or helical flutes. The Easyout extractor has helical flutes, as shown in Fig. 10-47.

After a stud is removed, the coarse-threaded end should be examined to determine whether the stud is standard or oversize. This is indicated by machined or stamped markings on the coarse-threaded end. Identification markings are shown in Fig. 10-48. The replacement stud should be one size larger than the stud removed.

A **bottoming tap** of the same size as the internal threads should be used to clean the threaded hole. If the threads in the case are damaged, or if the old stud was maximum oversize, it will be necessary to retap the hole with a bottoming tap and install a Heli-Coil insert for a standard-size stud.

Before the new stud is installed, the coarse threads should be coated with a compound specified by the manufacturer. This compound lubricates and protects the threads and may also seal the threads to prevent the leakage of lubricating oil from inside the engine. The stud should be installed with a **stud driver** and screwed into the case by using the amount of torque specified by the manufacturer.

Heli-Coil Inserts

A **Heli-Coil insert** is a helical coil of wire having a diamond-shaped cross section. When this coil is properly installed in a threaded hole, it provides a durable thread to receive standard studs or screws.

When a threaded hole has been damaged or oversized beyond accepted limits, it can be repaired by retapping and installing a Heli-Coil insert. The tap and installing tools are provided by the manufacturer of the Heli-Coil inserts and must be used according to instructions. The inserts are used in cylinders for spark plug holes and stud holes and in other parts for stud holes and screw holes.

Heli-Coil inserts may be removed and replaced if they become worn or damaged. The special extracting tool provided by the manufacturer must be used for removal. Heli-Coil inserts and tools are illustrated in Fig. 10-49.

Treatment of Interior Engine Surfaces

The interior surfaces of some engine crankcases, accessory cases, and similar parts are provided with a protective coating to eliminate corrosion damage. During the overhaul of an engine, such coatings should be examined and restored if necessary. One commonly employed coating is Alodine, and it is easily restored by following the manufacturer's instructions. The restoration process involves cleaning the bare aluminum thoroughly with an approved, nongreasy solvent and then

Typical part no.	Oversize on pitch dia. of coarse thread, in.	Optional identification marks on coarse thread end		Identification color code
		Stamped	Machined	
XXXXXX	Standard	None		None
XXXXXXP003	0.003	 ⊙		Red
XXXXXXP006	0.006			Blue
XXXXXXP009	0.009			Green
XXXXXXP007	0.007			Blue
XXXXXXP012	0.012			Green

FIGURE 10-48   Identification data for oversize studs.

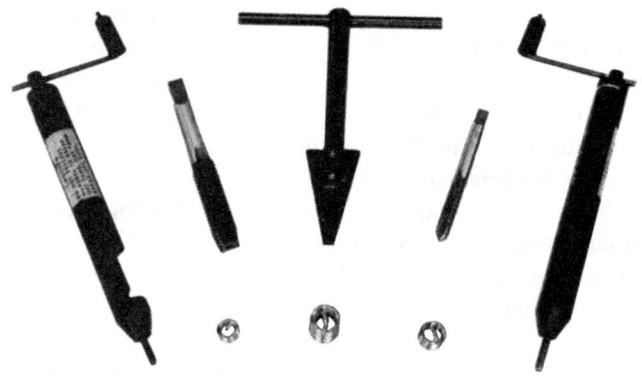

FIGURE 10-49 Heli-Coil inserts and tools.

applying the corrective solution as instructed. The solution is allowed to remain for a few minutes until the desired color is obtained. The area is then washed thoroughly and dried. In all cases, the manufacturer's overhaul instructions should be followed for each make and model of engine.

Recommended and Mandatory Replacement Parts

Many parts are rejected because of defects and wear found during the inspection phase of the overhaul, and others are tagged as serviceable only after repairs are made. Manufacturers have designated several parts as recommended or mandatory replacement items at overhaul regardless of their condition.

At Overhaul or Upon Removal

Any time the following parts are removed from any Lycoming reciprocating engine, it is mandatory that the following parts be replaced regardless of their apparent condition:

- All circlips, lockplates, retaining rings, and laminated shims
- All counterweight washers
- All lock washers and locknuts
- All main and connecting rod bearings (may also be referred to as "bearing inserts")
- All V-band coupling gaskets
- Stressed bolts and fasteners, such as
  - Stationary drive gear bolts (reduction gear)
  - Camshaft gear attaching bolts
  - Connecting-rod bolts and nuts
  - Crankshaft flange bolts
  - Crankshaft gear bolts

At Overhaul

During overhaul of any Lycoming reciprocating engine, it is mandatory that the following parts be replaced regardless of their apparent condition:

- All engine hoses
- All engine hose assemblies

- All oil seals
- All cylinder base seals
- All gaskets
- Piston rings
- Piston pins (thin wall)
- Piston pin plugs
- Propeller governor oil line elbow (aluminum)
- Propeller shaft sleeve rings
- Propeller shaft rollers (reduction gear pinion cage)
- Propeller shaft thrust bearings (all geared drive engines)
- Supercharger bearing oil seal (mechanically supercharged series)
- All exhaust valves (replace with current exhaust valves)
- All intake and exhaust valve guides
- All exhaust valve retaining rings
- Rocker arms and fulcrums
- Aluminum pushrod assemblies
- Hydraulic plunger assemblies
- Cylinder fin stabilizers
- Magneto drive cushions
- Magneto isolation drive bearings
- Thermostatic bypass valves
- Damaged ignition cables
- Crankshaft sludge tubes
- Counterweight bushings in crankshaft and in counterweights
- Accessory drive coupling springs
- AC diaphragm fuel pumps
- Fuel pump plunger for diaphragm fuel pumps
- Oil pump bodies
- Oil pump gears
- All V-band couplings and gaskets

REASSEMBLY

All serviceable and new engine parts should be organized on a parts rack prior to reassembly. These parts include those that replace parts found no longer serviceable, mandatory replacement items, and gaskets. All parts, including new ones, are cleaned and inspected for any damage that may have occurred during handling before the parts are installed on the engine.

In smaller overhaul shops, the total reassembly process may take place in one spot in the shop. In larger facilities and manufacturers' rebuild lines, the parts rack will follow the engine buildup stand through several stations. Each station is equipped with special tools for each phase of reassembly and stocked with miscellaneous hardware such as bolts, nuts, studs, washers, and lock washers. Most facilities have adopted the "one-person concept"—one person follows a particular engine along the assembly line and is solely responsible for assembling that engine.

The technician should be familiar with several accepted industry practices before beginning engine reassembly. A few of these practices will be reviewed before we describe the reassembly process.

Use of Safety Wire

During the final assembly of an engine and the installation of accessories, it is often necessary to **safety-wire** (lockwire) drilled head bolts, cap screws, fillister head screws, castle nuts, and other fasteners. The wire used for this purpose should be soft stainless steel or any other wire specified by the manufacturer.

The principal requirement for lockwire installation is to see that the tension of the wire tends to tighten the bolt, nut, or other fastener. The person installing safety wire must therefore see that the wire pull is on the correct side of the bolt head or nut to exert a tightening effect.

A length of safety wire is inserted through the hole in the fastener, and the two strands are twisted together by hand or with a special safety-wire tool. The length of the twisted portion is adjusted to fit the installation. One end of the wire is then inserted through the next hole, and the two ends are again twisted together. The wires are twisted tightly with pliers but not so tightly that the wire is weakened. After the job is completed, the excess wire is cut off to leave a stub end of about $\frac{1}{2}$ in [12.70 mm]. The stub should be bent back toward the nut. Typical examples of lockwiring are shown in Fig. 10-50.

Self-Locking Nuts

Self-locking nuts may be used on aircraft engines if all the following conditions are met:

1. Their use is specified by the manufacturer.
2. The nuts will not fall inside the engine should they loosen and come off.
3. There is at least one full thread protruding beyond the nut.
4. Cotter pin or lockwire holes in the bolt or stud have been rounded so they will not cut the fiber of the nut.
5. The effectiveness of the self-locking feature has been checked and found to be satisfactory prior to its use.

Engine accessories should be attached to the engine by the types of nuts furnished with the engine. On many engines, however, self-locking nuts are furnished for such use by the engine manufacturer for all accessories except the heaviest, such as starters and generators.

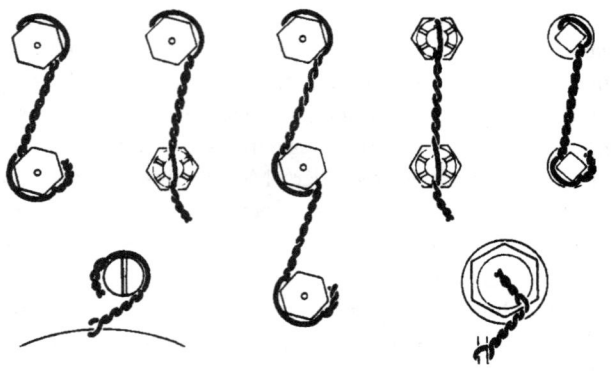

FIGURE 10-50   Examples of lockwiring.

On many engines, the cylinder baffles, rocker-box covers, drive covers and pads, and accessory and supercharger housings are fastened with fiber insert locknuts which are limited to a maximum temperature of 250°F [121°C] because above this temperature the fiber will char and consequently lose its locking characteristic.

Most engines require some specially designed nuts to provide heat resistance; to provide adequate clearance for installation and removal; to provide for the required degrees of tightening or locking ability which sometimes requires a stronger, specially heat-treated material, a heavier cross section, or a special locking means; to provide ample bearing area under the nut to reduce unit loading on softer metals; and to prevent loosening of studs when nuts are removed.

Washers

Flat washers (AN-960) are used under hexagonal nuts to protect the engine part and to provide a smooth bearing surface for the nut. Such washers may be reused, provided that they are inspected and found to be in good condition. Washers which are grooved, bent, scratched, or otherwise damaged should be discarded.

Lock washers may be used in some areas but only with the approval of the manufacturer. Lock washers are usually separated from aluminum or magnesium surfaces by flat washers to avoid damage to these soft metals. Where a part must be removed frequently, lock washers may not be used because of the damage which occurs each time the nut or bolt is loosened. Lock washers should be replaced each time they are removed.

Torque Values

One of the most important processes a technician must consider in the assembly of an engine or other parts of an aircraft is the **torque** applied to tighten nuts and bolts. Required torque values for various nuts and bolts in an engine are specified in the manufacturer's overhaul and maintenance manuals.

Torque wrenches are designed with a scale so that the technician can read the value of applied torque directly from the scale. The scale is marked in **inch-pounds** (in•lb) or **pound-inches** (lb•in) in the present system in the United States for small to medium bolts and nuts. For bolts and nuts of $\frac{3}{4}$-in [19.05-mm] diameter or larger, it is usually more convenient to employ **foot-pounds** (ft•lb) or **pound-feet** (lb•ft). In the metric system, the **newton-meter** (N•m) is used as a measure of torque. One foot-pound is equal to 1.356 N•m.

The value shown on the wrench in pound-inches is equal to the force applied in pounds at the handle multiplied by the number of inches (length) from the center of the handle to the center of the turning axis over the nut. This is illustrated in Fig. 10-51. If the torque wrench is used with an adapter, as shown in Fig. 10-52, the length of the adapter must be considered and the total torque computed.

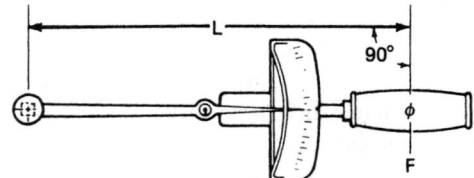

FIGURE 10-51   Measurement of torque.

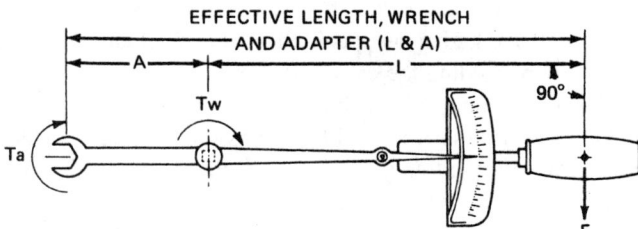

FIGURE 10-52   Torque wrench with adapter.

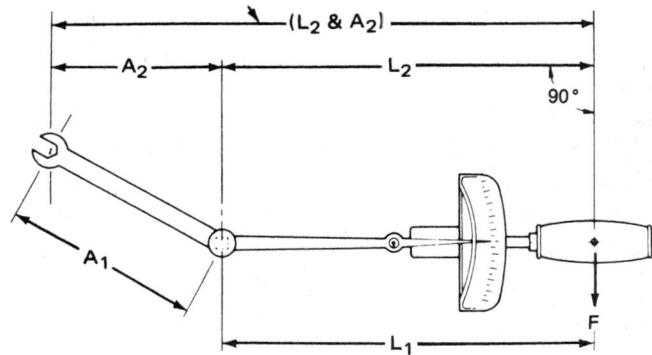

FIGURE 10-53   Torque wrench with offset adapter.

The following formula is used for finding the torque wrench setting T_w that will give a specified amount of applied torque T_a to the fastener:

$$T_w = \frac{T_a}{\dfrac{A}{L} + 1}$$

where T_w = torque read or set on torque wrench scale
T_a = torque to be applied to fastener
A = length of extension
L = length of torque wrench

The torque to be applied to the fastener T_a can be found in the manufacturer's table of torques. To arrive at the torque wrench setting or reading T_w, divide the length of the extension A by the length of the torque wrench L, add 1 to the quotient, and divide the result into the torque to be applied to the fastener T_a. When T_w is set or read on the torque wrench, the proper amount of torque will be applied to the fastener.

If the torque wrench is used with an offset adapter, then A is not the length of the adapter but the distance measured between two lines perpendicular to the axis of the wrench handle, one passing through the rotational axis of the wrench and the other passing through the center of the nut or bolt being turned. This is illustrated in Fig. 10-53. Excessive offsets should be avoided. A certain amount of error may be introduced as the angle increases even though the correct formula is used.

The torque range of a bolt or nut is critical, and an incorrect value of torque applied during assembly will often cause failure. In all assembly operations, the technician must consult the torque value charts supplied by the manufacturer.

Prelubrication of Parts

Manufacturer's recommendations for **prelubrication** of parts prior to assembly must be observed to avoid premature failure of engine parts. Proper lubrication of parts at assembly will protect the parts during the first few moments of operation, before engine oil can circulate to them. Manufacturers list the prelubricants to be used in either their overhaul manuals or their service publications.

Assembly

The assembly of an engine must follow a sequence recommended for the particular model of engine being assembled. Since the "core" of an engine is the crankshaft, assembly usually starts with the installation of connecting rods on the shaft.

Connecting Rods

New connecting-rod bearing inserts should be snapped into the rods and rod caps dry, and with the tangs of the inserts fitted into the cutouts provided. Each bearing is lubricated with approved oil. Then the rod and cap are installed on the crankpins according to the numbers stamped on the rods and caps. The manufacturer's overhaul manual specifies whether the rod and cap numbers face down or up.

Since different manufacturers of engines do not necessarily designate engine cylinder numbers from the same end of the engine, the overhaul technician must make sure to use the correct order when installing parts according to cylinder number. For example, Lycoming engines have the cylinders numbered with no. 1 being the right front cylinder, whereas the no. 1 cylinder of Continental engines is the right rear cylinder. In both cases the front of the engine is the propeller end.

The connecting rods are installed with new bolts, washers, nuts, and corrosion-resistant cotter pins (when used), and each nut is tightened with a suitable torque wrench to the torque specified. A nut must not be "backed up" in an effort to obtain a certain torque value but should always approach the correct value while being tightened. If the cotter-pin hole cannot be aligned with the nut when the nut has been torqued properly, it may be necessary to substitute nuts or bolts until the correct position can be attained. When the cotter pin is installed, one tang is bent back on the side of the nut and the other is bent out over the end of the bolt.

Some manufacturers specify the use of roll pins instead of cotter pins. When these are called for, the pin holes in the connecting-rod bolts must be of the proper diameter for the pins used.

The torque values for some connecting-rod bolts are given in bolt length rather than in foot- or inch-pounds. The bolts are tightened and then measured with a micrometer. Care must be taken not to overstretch the bolts. Bolts that are overstretched must be replaced.

After a connecting rod is installed on the crankpin, the side clearance should be checked with a thickness gauge to see that it is within approved limits. The rod should rotate freely on the crankpin, but there should be no noticeable play when the rod is tested manually.

Assembling the Crankcase

The crankcase of an opposed engine can be assembled to the crankshaft assembly either while the shaft is mounted in a vertical assembly stand or while the crankcase is supported on a workbench. The procedure depends on the manufacturer's recommendations and on the type of equipment available in the overhaul shop.

If the crankcase is assembled on a workbench, it must be supported so that the cylinder pads are about 6 in [15.24 cm] above the surface of the table. The right or left section of the case, as designated by the overhaul manual, should be placed on the supports, and all preliminary assembly operations specified should be completed.

The parting flange of the crankcase is coated with a thin layer of an approved sealing compound, care being taken not to apply so much that it will run inside the engine upon assembly. A single strand of no. 50 silk thread is then placed along the parting flange inside the bolt holes as specified by the manufacturer.

Prior to installation of the crankshaft and connecting-rod assembly in the crankcase, the front oil seal is installed on the crankshaft. The crankshaft gear can be installed either before or after the connecting rods.

The crankshaft is lifted carefully by two persons so that the correctly numbered connecting rods will be down and the others up. It is placed into the crankcase, care being used to ensure that the crankshaft seal fits into the seal recess without damage. The upper connecting rods are then laid gently to the side so that they rest on the crankcase flange.

If the engine construction is such that valve tappet bodies cannot be installed after the crankcase is assembled, these must be installed before the camshaft is placed in the assembly position.

The valve tappet bodies are lubricated on the outside and installed in the tappet bores of the opposite half of the crankcase. The camshaft with the cam gear installed is then placed in position and wired in place with brass or soft-steel wire. The camshaft holds the litter bodies in their bores while the opposite half of the crankcase is being lowered onto the crankcase half on the workbench.

Before the opposite half of the crankcase is placed in the assembly position, the end clearances of the crankshaft and camshaft are checked with a feeler gauge. This is done to ensure that the end play is within specified limits.

If required, the valve tappet bodies are installed in the crankcase section in which the crankshaft assembly has been placed. The other section of the crankcase in which the camshaft has been placed is now mated with the first section, care being taken to ensure that all parts fit together properly and that the cam-gear timing marks are aligned with the timing marks on the crankshaft gear. This automatically times the camshaft to the crankshaft. The two halves are then partially bolted together as specified in the overhaul manual, and the wire holding the camshaft is removed.

We must emphasize at this time that the foregoing procedures are usually followed for the assembly of the crankcase of an opposed engine. However, there is considerable variation in procedure for different makes and models, and the most desirable method is usually that described by the manufacturer in the overhaul manual.

When an assembly stand is employed which holds the crankshaft in a vertical position, the two halves of the crankcase are assembled simultaneously on the crankshaft and connecting-rod assembly. Regardless of which method is employed, care must be exercised to prevent damage to any part.

After the crankcase halves are bolted together, the connecting rods should be supported by means of rubber bands or special holding fixtures which fit over the cylinder hold-down studs, to prevent them from striking the edges of the cylinder pads.

Pistons and Cylinders

Valves are installed by inserting the stems through the valve guides from inside the cylinder and then holding them in place while the cylinder is placed over a cylinder post. The post bears against the valve heads and holds the valves on the valve seats. The lower spring seat is installed over the valve stem, and the valve springs are placed on the seat. Assembly instructions should be checked to see if there is a difference between the ends of the springs. The springs are compressed by means of a valve spring compressor. The valve retaining keys are installed in the groove around the valve stem, and the spring compressor is then released.

Piston rings are installed on the piston with a ring expander, or by hand. With respect to the installation of piston rings, it is especially important that the technician observe the instructions of the manufacturer. A particular make and model of engine requires a certain combination of piston rings as set forth in the parts manual or list for the engine. It is vital that some piston rings be installed top side up. The word "top" is usually etched on the side of the ring, indicating that this side of the ring must be nearest the top of the piston. Some piston rings which have symmetric cross sections may be installed with either side up.

Before installation, the piston and piston pin are generously coated with a preservative oil. This oil should be worked into the ring grooves so that all piston rings are thoroughly lubricated. Each piston is numbered and must be installed on the correspondingly numbered connecting rod. The piston is positioned so that the number on the head is in the location specified by the manufacturer, assuming that the engine is in its normal horizontal position. For radial engines, remember that the cylinder with the master rod is removed last during disassembly and installed *first* during assembly. This is done to provide adequate support for the master-rod assembly and to hold the link rods and pistons in such a position that the lower piston ring will not be below the skirt of the cylinder.

Before a piston is installed, the crankshaft should be turned so that the connecting rod for the cylinder being installed is in TDC position. The piston is installed by placing it in the proper position over the end of the connecting rod and pushing the piston pin into place through the piston and connecting rod. A piston-ring compressor is then hung over the connecting rod, ready for installation of the cylinder. Prior to installation, the cylinders should be checked for cleanness and the inside of the barrel coated with preservative oil. A base flange packing ring is installed around the skirt at the intersection of the skirt and flange. Check the rings for freedom of movement in the ring grooves, and stagger the ring gaps according to the manufacturer's instructions.

The correctly numbered cylinder is lifted into position, and the cylinder skirt is placed over the piston head. The ring compressor is then placed around the piston and upper piston rings, and the rings are compressed into the piston grooves. The cylinder can then be moved inward so that the skirt slides over the piston rings as the compressor is pushed back. When all the piston rings are inside the cylinder skirt, the compressor is removed and the cylinder flange stud holes are carefully moved into place over the studs. The base flange packing ring is then checked for position, and the cylinder is pushed into place. Cylinder hold-down nuts are screwed onto the studs and tightened lightly. The upper nuts are installed first to provide good support for the cylinder. After all the nuts are in place, they are tightened moderately but not torqued to full value. A torque handle is installed on the cylinder base wrench, and the nuts are torqued in the sequence specified in the overhaul manual. *It is very important that the cylinder hold-down nuts be tightened evenly and to the correct torque value, to prevent warping and undue strain on one side of the flange.* Some manufacturers require that all cylinders be installed before the final torquing of cylinder hold-down nuts.

Valve Mechanism

Since the valve mechanisms for engines of different makes and models vary considerably, no particular method of installation, assembly, or valve timing is discussed in this chapter. However, the assembly must be done in the sequence described by the manufacturer to avoid omitting any required operation. If this is done, the timing of valves will be correct. All parts should be perfectly clean before installation and should be coated with clean lubricant. This is particularly true of the valve lifter body and plunger assembly. It is especially important that the pushrod socket be in place in the lappet body before the pushrod is installed.

After the complete valve operating mechanism has been assembled, the clearance between the rocker arm and valve stem should be checked with the cylinder at TDC on the compression stroke. If the valve clearance is not within the limits specified, it must be adjusted by installing a pushrod of slightly different length or by adjusting the screw in the end of the rocker arm.

Upon completion of the valve mechanism installation, the rocker cover is installed with a gasket and torqued.

The **cam-gear backlash** with the crankshaft gear or idler gear (where used) should be checked before installation of the accessory case. If the backlash exceeds the value given in the Table of Limits, the gears must be replaced.

INSTALLATION

The assembled engine is ready for installation in a test cell where it will be tested, run in, and finally preserved for storage. This would be the order of events if the overhauling agency kept freshly overhauled engines in stock and the customers simply exchanged their engines for overhauled engines of the same model.

In many cases, the customer may receive an engine as soon as testing is complete and install it in the aircraft. In this instance, there is no need for preservation. It is also possible that the engine can be installed and tested in the aircraft provided that certain requirements are met. These requirements are discussed later in this chapter.

Installation in Test Stand

If an engine has just been overhauled but not tested and run in as required, it should be installed in a suitable test stand, such as that shown in Fig. 10-54, and run in according to

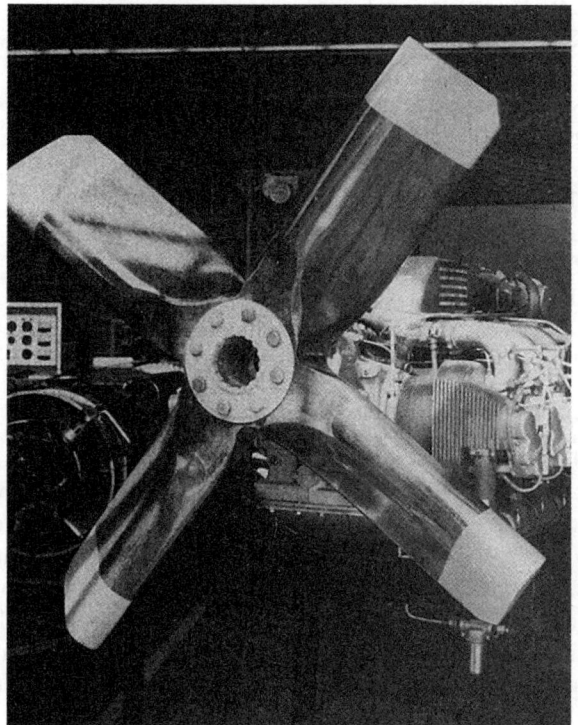

FIGURE 10-54 Engine with "test club" mounted on test stand.

the test schedule to make sure that it is operating in accordance with specifications. Small engines are sometimes run in on the airplane, but the standard run-in procedure must be modified to some extent if this is done.

The engine test stand should be mounted in a test cell, equipped with the necessary controls, instruments, and special measuring devices required for measurement of fuel consumption, power output, oil consumption, conduction of heat to the oil, and standard engine performance data. The following instruments and devices are usually required:

1. Fuel tank with adequate capacity at least 50 gal [189.27 L]
2. Oil tank with capacity of 10 gal [37.85 L] or more
3. Fuel flowmeter
4. Scales for weighing oil
5. Cylinder-head temperature gauges
6. Manifold pressure gauge
7. Tachometers (one which counts revolutions)
8. At least two oil temperature gauges (inlet and outlet)
9. Fuel pressure gauge
10. Oil pressure gauge
11. Manometer for testing crankcase pressure
12. A 12-V battery or other power source
13. Fuel pressure pump, either manual or electric
14. Engine test propeller (A test propeller, sometimes referred to as a *test club*, is utilized because it moves a large volume of air near the propeller hub for cooling purposes. A test propeller is illustrated in Fig. 10-54.)
15. Cooling shroud
16. Magneto switch
17. Suitable starter controls
18. Control panel for mounting instruments and controls
19. An accurate clock for checking run-in time
20. Throttle control
21. Mixture control

It is recommended that the following provisions be met if the engine is to be tested and run in on the airframe:

1. A test club should be used.
2. A cooling shroud or cowling should be installed.
3. There should be a means of monitoring the CHT of each cylinder.
4. The instruments used should be of known accuracy and independent of the aircraft instruments.

The control room of the test cell should be provided with a safety-glass window located so that the operator has a good view of the engine during the run-in procedure. The strength of the safety-glass window should be adequate to prevent any flying objects from entering the control room. The area in which the engine is installed should be protected by gates or doors to prevent personnel from entering the propeller area while the engine is running; however, provisions must be made for the operator to gain access to the rear part of the engine to make necessary adjustments.

The actual installation of the engine in the test stand depends partly on the design of the test stand. The following installation steps may be considered typical:

1. Make sure that the test stand is equipped with all items necessary for testing the make and model of engine to be installed.
2. Hoist the engine into place, and align mounting brackets.
3. Install mounting bolts, washers, lock washers, and nuts. Tighten the nuts to proper torque.
4. Install short exhaust stacks with gaskets to cylinder exhaust ports.
5. Connect the fuel supply line to the engine-driven pump inlet or to the carburetor as required. Make sure the supply pressure is correct for the carburetor or fuel unit. The pressure will vary among gravity-fed float-type carburetors, pump-fed float-type carburetors, injection carburetors, and direct fuel injection units.
6. Connect the oil supply and return lines if the engine is of the dry-sump type.
7. Attach CHT sensing units (thermocouples or temperature bulbs) as required.
8. Connect the magneto switch wires to the magnetos.
9. Connect the pressure line for the oil pressure gauge.
10. Connect the pressure line for the fuel pressure gauge.
11. Connect the oil temperature gauge line (electric or capillary).
12. Connect the manifold pressure line.
13. Connect the throttle control.
14. Connect the mixture control.
15. Connect the electric cable to the starter.
16. Connect the ground cable to the crankcase.
17. Connect the tachometer cable or electric lines as required.
18. Install a suitable cooling shroud for the engine.
19. Install the test propeller (test club). Make sure that the test propeller is of the correct rating for the engine being tested.
20. Service the engine with the proper grade of lubricating oil. If the engine is to be stored for a time after the test, use a preservative-type lubricating oil.
21. Perform a complete inspection of the installation to make sure that all required installation procedures have been completed.

ENGINE TESTING AND RUN-IN

Preoiling

Before the engine is actually started for the first time, it should be **preoiled** to remove air trapped in oil passages and lines and to ensure that all bearing surfaces are lubricated. Preoiling can be accomplished in several ways.

In one method of preoiling, one spark plug is removed from each cylinder. The crankcase or external oil tank is then filled with the oil to be used for run-in, and the engine is cranked with the starter until an oil pressure indication is read on the oil pressure gauge. Another method is to force oil, by means of a pressure oiler at a prescribed pressure, through the oil galleries until it comes out an oil outlet or the opposite end of an oil gallery. The engine manufacturer's overhaul manual should be consulted for the recommended procedure.

Run-In Test Schedule

The manufacturer's overhaul manual provides instructions and a run-in schedule for newly overhauled engines. The purpose of the run-in is to permit newly installed parts to burnish or "wear in," piston rings to seat against cylinder walls, and valves to become seated. The run-in also makes it possible to observe the engine's operation under controlled conditions and to ensure proper operation from idle to 100 percent power. The time during which an engine is operated at full power is referred to as a **power check**. The purpose of this check is to ensure satisfactory performance.

The engine run-in should be accomplished with the engine installed in a test cell equipped as specified in the manufacturer's overhaul manual. The engine should be equipped with a correctly designed and rated club propeller or a dynamometer which will apply the specified load to the engine. Calibrated instruments must be available in the test cell to measure such parameters as CHT, oil temperature, manifold pressure, intake air temperature, turbocharger intake air pressure, turbocharger air outlet pressure, turbocharger exhaust outlet pressure, and any other parameters specified by the manufacturer.

Slave (external) oil filters should be installed for both the engine and turbocharger oil systems. These filters are used to trap the metal particles often present in a newly overhauled engine. It is particularly important that metal particles be prevented from entering the turbocharger and turbocharger control units. For this purpose, the oil filter should have the capability of removing all particles having a dimension of 100 μm (0.1 mm) or greater and should have an area such that there is no restriction of oil flow.

A typical run-in schedule for a direct-drive (no propeller reduction gear) engine with the prescribed propeller load requires 10-min periods of operation at 1200, 1500, 1800, 2000, 2200, and 2400 rpm. Following this, the engine is operated at normal rated horsepower for 15 min. An oil consumption run is made after the standard run-in schedule has been completed.

In the run-in schedule shown in Fig. 10-55, the 5-min period during which the engine is operated at 3400 rpm is the power check. If the engine will not come up to normal operating speed when operated with a test club or fixed-pitch propeller, the engine is considered a "weak engine" and corrective action must be taken.

Oil Consumption Run

An **oil consumption run** is made at the end of the test in the following manner: Record the oil temperature. Stop the

Period	Time, min	rpm	Turbocharger outlet pressure, in Hg
1	5	1200	
2	10	1500	
3	10	2100	
4	10	2600	
5	10	2800	
6	10	3000	
7	10	3200	
8	5	3400 ± 25	42.0 – 43.0 (100% power)
9	5	3000	34.8 – 35.8 (68.5% power)
10	5	2600	33.5 – 34.5 (44.8% power)
11	10	600 ± 25	Cooling period (idle)

FIGURE 10-55 Sample run-in schedule.

engine in the usual manner. Place a previously weighed container under the external oil tank or engine sump, and remove the drain plug. Allow the oil to drain for 15 min. Replace the drain plug. Weigh the oil and the container. Record the weight of oil (that is, total weight less the weight of the container). Replace the oil in the tank or sump. Start the engine, warm up to the specified rpm ± 20 rpm and operate at this speed for 1 h. At the conclusion of 1 h of operation and with the oil temperature the same as that recorded at the time of previous draining (it is important to keep this oil temperature as constant as possible), again drain the oil as before. The difference in oil weights at the start and end of the run will give the amount of oil used during 1 h of operation.

The maximum amount of oil which can be used during the oil consumption run is determined by the manufacturer and is given in pounds of oil per hour. The result of the oil consumption run is an indication of how well the piston rings are sealing in the cylinders of a newly overhauled engine. If the amount of oil consumed is greater than that recommended by the manufacturer, an investigation should be made to determine the cause of the oil loss.

Starting Procedure

Before the engine is actually started, the engine area should be checked for loose objects which could be picked up by the propeller. The engine itself should be checked for tools, nuts, washers, and other small items which may be lying loose.

The following steps are typical of an engine starting procedure in the test cell:

1. With the magneto switch set to OFF and the mixture control in IDLE CUTOFF, turn the crankshaft 2 r with the propeller to check for liquid lock.
2. Turn the master power switch to ON.
3. Open the throttle about one-tenth of the total distance.
4. Turn on the fuel pump, and check the fuel pressure.
5. Turn the magneto switch to ON for the magneto having the impulse coupling or induction vibrator.
6. Place the mixture control in the FULL RICH or IDLE CUTOFF position, depending on the type of fuel control.

7. Check that the cell area is clear, and begin cranking the engine.

8. As soon as the engine starts running smoothly, adjust the throttle for the desired rpm, usually 1000 rpm or less for a newly overhauled engine. If the engine is equipped with Bendix or Continental fuel injection, the mixture control must be moved to FULL RICH as soon as the engine starts.

9. Check for oil pressure immediately. If oil pressure does not register within the prescribed time (10 to 30 s), shut down the engine and identify the problem.

10. If the engine operates properly, shut off the fuel boost pump.

11. As soon as the engine is operating smoothly, turn the magneto switch to OFF *momentarily* to determine whether the engine can be shut off with the switch in case of emergency.

12. If the engine is operating satisfactorily, continue with the test run specified by the manufacturer as described previously.

A log should be kept and the instrument readings recorded every 15 min. The log sheet should also include the date of the test, the engine number, and the type and nature of the test, along with the total number of hours of engine operation. All periods during the test run when the engine was not in operation should be recorded, along with the explanation. If for any reason it is necessary to replace any part, the complete reason for rejection of the part should also be recorded.

Preparation of Overhaul Records

Federal Aviation Regulations require that a permanent record of every maintenance (except preventive maintenance), repair, rebuilding, or alteration of any airframe, powerplant, propeller, or appliance be maintained by the owner or operator in a logbook or other permanent record satisfactory to the FAA administrator and contain at least the following information: (1) an adequate description of the work performed; (2) the date of completion of the work performed; (3) the name of the individual, repair station, manufacturer, or air carrier performing the work; and (4) the signature and the certificate number of the person, if a certificated mechanic or certificated repairer, approving as airworthy the work performed and authorizing the return of the aircraft or engine to service.

All major repairs and major alterations to an airframe, powerplant, propeller, or appliance must be entered on a form acceptable to the FAA administrator. This form must be executed in duplicate and must be disposed of in such manner as, from time to time, may be prescribed by the administrator.

All major alterations must be entered on FAA Form 337, the approved major repair and alteration form. This form must be executed in accordance with pertinent instructions, and the original copy given to the owner of the unit altered or repaired. The repair station should retain a copy for its permanent record, and one copy must be sent to the local FAA office within 48 h of the time that the powerplant or other unit is returned to service.

A certified repair station is allowed to subsitute the customer's work order upon which repairs are recorded in place of the Form 337. The original copy of the work order must be given to the owner or purchaser, and the duplicate copy must be kept for at least 2 years by the repair station.

The owner of an engine which has been overhauled by a certificated repair station should be supplied with a copy of a **maintenance release**. This release should accompany the engine until it is installed in the aircraft, and at that time the installing agency will make the release available to the owner for incorporation into the permanent record of the aircraft. The maintenance release may be included as a part of the work order, but it must contain the complete identification of the engine including the make, model, and serial number. The following statement must also be included:

The engine identified above was repaired and inspected in accordance with current Federal Aviation Regulations and was found airworthy for return to service.

Pertinent details of the repair are on file at this agency under Work Order No.———————————————

Date—————————————————————

Signed—————————————————————

(Signature of authorizing individual)

For—————————————————————

(Agency name)   (Certificate no.)

———————————————————————

(Address)

In addition to the formal records and statements required by FAA regulations, the repair station should maintain a complete record of all repairs and inspections performed. This record should contain an account of every repair operation and every replacement part. Details of all inspections made—dimensional inspection, structural inspection, service bulletin and airworthiness directive compliance, and engine specification or Type Certificate Data Sheet conformity—should be included in the record kept for each engine overhauled.

The preparation of adequate overhaul records cannot be overstressed. These records are particularly important for the protection of the overhaul agency in the event of an engine failure. If the record is complete and properly prepared, the overhaul agency can show that all overhaul work was accomplished in accordance with the manufacturer's overhaul manual and that all required operations were performed. This type of record will usually absolve the overhaul agency of responsibility in case of engine failure.

ENGINE PRESERVATION AND STORAGE

If an engine is to be stored for a time after having been run in, it should be preserved against corrosion. This is particularly important for the interiors of cylinders, where the products of combustion will initiate corrosion of the bare cylinder walls within a very short time.

Preservation Run-In

As previously mentioned, an engine which is to be stored should be run in with a preservation oil as the lubricant. In addition, if possible, the last 15 min of operation should be done with clear (unleaded) gasoline at about two-thirds of full rpm. This will tend to remove the accumulation of corrosive residues which are in the cylinders and combustion chambers.

Interior Treatment

When the engine is stopped, the preservative oil should be drained from the crankcase or sump. Spark plugs are then removed from the cylinders, and preservative oil is sprayed into the cylinders as the engine is rotated, several times for each cylinder. The rotation can be accomplished with the starter. Each cylinder is then sprayed one more time without further turning of the crankshaft.

After all cylinders have been sprayed and preservative oil has been sprayed in the crankcase through the oil filler neck or any other crankcase opening, dehydrator plugs containing silica gel are installed in the spark plug holes and in the sump drain. The dehydrator plugs absorb the moisture within the engine, thus reducing the tendency for the interior to corrode.

The short exhaust stacks should be removed and preservative oil sprayed into the exhaust ports. The ports should then be covered with airtight plugs.

Exterior Treatment

All openings into the engine should be sealed with airtight plugs or with waterproof tape. If the carburetor is removed for separate preservation, a **dehydrator bag** can be placed in the carburetor opening before it is sealed. The bag should be tied to an exterior fitting so that it can be easily removed when the engine is prepared for operation.

After the engine is completely sealed, it may be sprayed lightly with preservative oil or other approved coating. If the engine is to be stored for as long as 6 months, it should be sealed in a waterproof plastic bag. The bag is first placed over the mounting bolts in the engine case, and the engine is installed in the case with the mounting bolts sticking through the bag. The bag is sealed at the engine mounting bolts when the bolts are tightened.

After a number of dehydrator bags are placed in the waterproof bag with the engine, the waterproof bag should be sealed according to the directions furnished. An indicator is also placed in the bag with the desiccant exposed through a window to show when the humidity level in the bag has reached a point where it is necessary to represerve the engine. When the **desiccant** in the bags or dehydrator plugs loses its color and begins to turn pink, the preservation is no longer effective and must be redone.

The desiccant material in the bags and dehydrator plugs must be inspected frequently. When the desiccant material turns from blue to pink, it is no longer effective and must be replaced or dried in an oven. The interior of the engine must be inspected and resprayed periodically. The preservation process should be repeated if any signs of corrosion are found.

Inspection after Storage

When an engine is removed from storage after having been preserved, certain inspections should be made to ascertain that it has not been damaged by corrosion. The exterior inspection consists of a careful examination of all parts to see if corrosion has taken place on any bare metal part or under the enamel. Corrosion under enamel will cause the enamel to rise in small mounds or blisters above the smooth surface.

Interior inspections should be performed in all areas where it is possible to insert an inspection light. The most vulnerable area is inside the cylinders where the bare steel of the cylinders has been exposed to the combustion of fuel. Inspection of the cylinders is done by removing the spark plugs from the cylinders and inserting an inspection light in one of the spark plug holes. The inside of the cylinder can then be seen by looking through the other spark plug hole. If rust is observed on the cylinder walls, it is necessary to remove the cylinder and dispose of the rust. If the cylinder walls are badly pitted, it will be necessary to regrind the cylinders and install oversize pistons and rings.

The rocker-box covers should be removed to inspect for corrosion of the valve springs. Pitted springs will be likely to fail in operation owing to stress concentrations caused by the pitting.

When the exhaust port covers are removed to permit installation of the exhaust stacks, the ports and the valve stems can be examined for corrosion. A small amount of corrosion on the cast aluminum inside the exhaust port is not considered serious; however, if the valve stem is rusted, the rust must be removed or the valve replaced.

Installation in Aircraft

Overhauled or rebuilt engines may be crated and in "preserved" condition when they are received by the installer. Special steps must be taken to prepare a preserved engine for installation in the aircraft.

Preparation for Installation

If an engine has been stored in an engine case, special instructions for unpacking the engine will usually be included in the case. The case should be placed in the correct position (top side up) so that the engine case cover can be lifted off. After the attaching bolts are removed, the cover is carefully lifted to avoid damage to the engine.

If the engine has been properly preserved and packed to prevent corrosion damage while being moved from the factory overhaul shop to the purchaser, it will be sealed in a plastic envelope. The magnetos will probably be mounted on the engine together with the ignition harness. The carburetor may be in a separate package within the case.

The engine will be bolted to the supports built into the case, and it will be necessary to remove bolts and other attachments before the engine can be removed. The technician in charge of unpacking the engine must exercise great care to prevent damage and the loss of small parts. First, locate all paperwork, such as overhaul records and unpacking instructions, and then proceed according to instructions.

When the engine cover is removed, a hoist should be attached to the lifting eye, which is located along the top crankcase parting surface, and sufficient tension should be placed on the hoisting cable to remove most of the weight from the mounting brackets. The mounting bolts should then be removed, and the engine hoisted and placed on a suitable stand. This is necessary because several fittings and parts usually must be installed on the engine before the engine is ready to be installed in the airplane.

When the engine is firmly mounted on the stand, all shipping and preservative plugs are removed. These plugs, with the exception of drain plugs, should be replaced immediately with the proper fittings for engine operation. Fittings include the oil pressure fitting, manifold pressure fitting, crankcase vent fitting, oil temperature bulb fitting, and others.

When fittings having pipe threads are installed, the threads of the fittings should be lightly coated with an approved thread lubricant and the fittings should be installed with proper torque to prevent damage to the threads.

The desiccant plugs should be removed from the spark plug holes and from the oil sump. At this time the engine should be rotated a few times to permit drainage of the preservative oil.

New spark plugs and washers of the correct types should be installed, and the ignition harness elbows attached to the plugs.

If the carburetor or fuel injector has been preserved, it should be drained of the preservative and purged with clean fuel before installation. Care must be taken not to allow fuel to enter the air chambers of fuel injection units.

Before the engine is installed in the airplane, the engine should be inspected thoroughly. Both the engine manual and the airplane manual should be consulted to make sure that all fittings, baffles, and accessories are securely fastened and safetied as necessary. A careful check at this time may save much time and trouble later.

Installation in Airplane

The installation of the engine in the airplane should follow the directions given by the manufacturer. After an engine installation is completed, it is required that a complete installation inspection be made.

The following is an example of a post installation checklist.

1. Propeller mounting bolts safetied
2. Engine mounts secure
3. Oil-temperature-bulb electric connector secure and safetied; ground wire connection tight
4. Oil pressure relief valve plug safetied

5. Tachometer generator electric connector secure and safetied
6. Starter cable connection secure and insulating boot in place
7. CHT bulb installed and ground wire connection tight
8. Generator cable connections secure and cable shielding grounded
9. All wiring securely clamped in place
10. Fuel pump connections tight
11. Manifold pressure hose connections tight
12. Oil pressure connections clamped and tight
13. Fuel injection nozzles tight
14. Fuel injection lines clamped and tight
15. Fuel manifold secure
16. All flexible tubing in place and clamped
17. Crankcase breather-line connections secure
18. Air-oil separator exhaust line and return oil hose connections secure
19. Vacuum line and vacuum-pump outlet hose connections secure
20. Oil dilution hose connections tight
21. Propeller anti-ice hose connections tight
22. Engine controls properly rigged
23. Oil drain plugs tight and safetied
24. Oil quantity check, 12 qt [11.36 L] in each engine
25. Hoses and lines secure at fire wall
26. Fuel-air control unit and air intake box secure
27. Shrouds installed on engine-driven fuel pump, fuel filter, and fuel control unit; ram air tubes installed and clamped
28. Induction system clamps tight
29. Exhaust system secure
30. Spark plugs tight, ignition harness connections tight, and harness properly clamped
31. Magneto ground wires connected and safetied
32. Engine nacelle free of loose objects (tools, rags, etc.)
33. Cowling and access doors secure

The foregoing list of instructions for a specific engine and airplane is given to emphasize the many important details involved in engine installation. The installer and the inspector must follow this checklist to make sure that no operation has been left incomplete.

REVIEW QUESTIONS

1. Describe the changes which take place in an aircraft engine during operation and eventually make an overhaul necessary.
2. How is the term *overhauled engine* defined?
3. Describe the purpose of the receiving inspection.
4. Discuss the importance of reviewing manufacturers' bulletins and Airworthiness Directives during engine overhaul.
5. What are the method and purpose of the preliminary visual inspection?
6. Describe the difference between degreasing and decarbonizing.

7. Describe vapor degreasing.

8. List the most common methods of decarbonizing.

9. Describe vapor blasting.

10. For what purposes can sandblasting be employed in the overhaul of an aircraft engine?

11. Why should engine parts be coated with a preservative oil after they have been cleaned?

12. How are engine parts structurally inspected?

13. Briefly describe magnetic particle inspection.

14. Describe liquid penetrant inspection.

15. What is the function of a dimensional inspection?

16. What instrument is used to measure the cylinder bore?

17. Describe the use of a thickness gauge.

18. Discuss the repair of cooling fins and the limitations on the amount of damage which can be tolerated.

19. What is the advantage of a 30° valve face angle over a 45° angle?

20. How is the correct width of a valve seat obtained during the grinding process?

21. What is the purpose of an interference fit on valves?

22. What precautions should be observed with respect to grinding a cylinder which has been hardened by nitriding?

23. What is the advantage of a chromium-plated cylinder barrel?

24. What precautions must be observed with respect to the piston rings used in a chromium-plated cylinder barrel?

25. If one cylinder of an engine needs to be ground to an oversize dimension, what should be done to the other cylinders and pistons?

26. What is a choked cylinder barrel, and what is the purpose of choke in a cylinder barrel?

27. For what defects is a cylinder mounting flange inspected, and what procedures are followed?

28. What may be done if a crankshaft flange is found to be bent?

29. What is used to check that crankshaft oil passages are free from sludge?

30. Where may self-locking nuts be used on an engine?

31. Describe the preoiling procedure for a newly overhauled engine.

32. Describe a power check.

33. What material is used to absorb moisture inside an engine during storage?

34. What sections of the engine should be inspected for corrosion or rust after a long period of storage?

Gas-Turbine Engine: Theory, Jet Propulsion Principles, Engine Performance, and Efficiencies

11

BACKGROUND OF JET PROPULSION

Discovery of Jet Propulsion Principle

No one knows who first discovered the jet propulsion principle, but the honor is sometimes given to a man named Hero, who lived in Alexandria, Egypt in about 150 B.C. He invented a toy whirligig turned by steam, as illustrated in Fig. 11-1, and called his invention an **aeolipile**. But apparently he did not discover any very useful purpose for his discovery.

The historical records are not very definite in describing the aeolipile. If it resembled the picture in Fig. 11-1, it was a primitive form of a jet or reaction engine. But some authorities describe it as having been operated by hot air instead of steam. The heating of air in a vertical tube induced a flow of air in several tubes arranged radially around a horizontal wheel, and rotation resulted from the creation of an impulse effect. In that case, Hero's invention was a gas turbine.

About 1500, Leonardo da Vinci sketched a device that could be placed in a chimney where the upward movement

FIGURE 11-2   Newton's steam carriage.

of hot gases would turn a spit for roasting meat. In 1629, Giovanni Branca, another Italian, perfected a steam turbine that applied the jet principle and could be used to operate primitive machinery.

Figure 11-2 is a drawing of an invention called **Newton's carriage**, a jet-propelled steam carriage. Although Newton himself may have supplied the idea, there are authorities who attribute the design of the carriage to a Dutchman, Willem Jako Gravesande.

Turbine Development

The first patent covering a gas turbine was granted to John Barber of England in 1791. It included all the essential elements of the modern gas turbine except that it had a reciprocating-type compressor. In 1808, a patent was granted in England to John Bumbell for a gas turbine which had rotating blades but no stationary, guiding elements. Thus the advantages gained today by the multistage type of turbine were missed.

In 1837, a Frenchman named Bresson was granted a patent for a machine in which a fan delivered air under pressure to a combustion chamber, where the air was mixed with a gaseous fuel and burned. The hot products of combustion were then cooled by excess air and directed in the form of a jet against a turbine wheel. This was essentially a

FIGURE 11-1   Hero's aeolipile.

gas turbine, but there is apparently no record of its practical application.

In 1850, W. F. Fernihough was granted a patent in England for a turbine operated by both steam and gas, but as long as steam was used, the development of a true gas turbine was held back. However, a man named Stolze designed what was probably the first true gas turbine in 1872 and tested working models between 1900 and 1904. Stolze used both a multistage reaction gas turbine and a multistage axial compressor.

Sir Charles Parsons, the great English inventor, obtained a patent in 1884 for a steam turbine, in which he advanced the theory that a turbine could be converted to a compressor by driving it in an opposite direction with an external source of power. Parsons believed that compressed air could be discharged into a furnace or combustion chamber, fuel injected, and the products of combustion expanded through a turbine. This idea of a compressor was essentially the same as that which we have today except for the shape of the blades.

Charles G. Curtis is generally credited with the filing of the first patent application in the United States for a complete gas turbine. His application was filed in 1905, although previously, in 1902, he filed an application for a rotary compressor, blower, and pump combination and actually obtained the patent in 1914. There is some argument about how much Curtis did to develop the gas turbine, but he is credited, without dispute, with the invention of the Curtis steam engine, and he was one of the pioneers in the development of steam turbines.

Sanford A. Moss, who eventually became one of the leading engineers of the General Electric Company, completed his thesis on the gas turbine in 1900 and submitted it to the University of California in application for his master's degree. The contributions of Moss to the development of engines of all types are so extensive that to describe them completely would fill several volumes. However, a few of his outstanding contributions will be mentioned.

In 1902, experiments were conducted at Cornell University with what was probably the first gas turbine developed in the United States. A combustion chamber designed by Moss was used with a steam-turbine bucket wheel, which functioned as the gas-turbine rotor. A steam driven compressor supplied compressed air to the combustion chamber. The engine was not a success from the practical standpoint because the power required to drive the compressor was greater than the power delivered by the gas turbine. But from the experiments, Moss learned enough to enable him to start the General Electric Company's gas-turbine project the next year.

In the following years there were various turbine inventions and developments in the United States and in Europe, but the next outstanding one was the construction of the first General Electric turbosupercharger by Moss during World War I. The products of combustion of the engine exhaust drove a turbine wheel at constant pressure, and the turbine wheel, in turn, drove a centrifugal compressor that supplied the supercharging.

Strictly speaking, the first General Electric turbosupercharger was based on French patents by Rateau; therefore, Moss and the General Electric Company are entitled to credit for developing the running model, although the credit for the idea behind it belongs to Rateau of France.

It is interesting to consider that the turbosupercharger was developed as an offshoot of the gas turbine. The turbosupercharger then went through a long stage of development, and finally the engineers took the knowledge that they acquired from working with the turbosupercharger and applied it to jet propulsion.

Frank Whittle began work on gas turbines while he was still a Royal Air Force air cadet. He applied for a patent in England in 1930 for a machine having a blower compressor mounted at the forward end and a gas turbine at the rear end of the same shaft, supplied by energy from the combustion chamber. Discharge jets were located between the annular housings of the rotary elements and in line with several combustion chambers distributed around the circumference.

On May 14, 1941, flight trials began with a Gloster E28/39 experimental airplane equipped with Whittle's engine, shown in Fig. 11-3, which was known as the W1. The flight tests were successful, thus greatly increasing the interest of both the government and manufacturers and setting the stage for the tremendous progress to come.

While Whittle and his associates were working on the development of the W1 engine in England, the Heinkel Aircraft Company in Germany was also busy with a similar task. The German company was successful in making the first known jet-propelled flight on August 27, 1939, with a Heinkel He 178 airplane powered by a Heinkel HeS 3B turbojet engine having a thrust of 880 to 1100 lb [3914.24 to 4892.8 N].

The pioneer jet-propelled fighter planes built in England by the Gloucester (Gloster) Aircraft Company, Ltd., and in the United States by the Bell Aircraft Corporation, were powered by a combustion, gas-turbine, jet propulsion powerplant system developed from Frank Whittle's designs and built by the General Electric Company.

Only a few of the many important inventors and engineers who contributed to the modern jet engine program have been mentioned, but the work of every one of them has been based fundamentally on basic jet propulsion principles.

FIGURE 11-3   Whittle W1 engine.

BASIC JET PROPULSION PRINCIPLES

Jet Propulsion

To better understand the functioning of a turbojet engine, it is helpful to understand the basic principles of jet propulsion. Jet propulsion principles explain how jet aircraft and rockets are propelled. When one understands jet propulsion, it is easier to appreciate the fundamental simplicity of a turbojet engine.

A turbojet engine is a mechanical device which produces forward thrust by forcing the movement of a mass of gases rearward. This design is based on the principle that for every action there is an equal and opposite reaction. In the case of the jet engine, the action is the forcing of a large mass of exhaust gas out the rear of the engine. That is, the engine takes air in at the front or inlet at some velocity (depending on the aircraft speed) and forces it and combustion gases out the rear of the engine at a much higher speed. The reaction to the ejection of this mass of gas is a forward force on the engine and aircraft. The amount of force or thrust produced depends on the amount of mass of air moved through the engine and the extent to which this air can be accelerated and ejected.

A toy balloon may be used to demonstrate action and reaction as well as jet propulsion. When a balloon is inflated and then its mouth held closed, the balloon contains air, a gas, under pressure. The pressure within the balloon is exerted equally in all directions. The air presses with the same amount of force against the top, bottom, sides, front, and back of the balloon, as shown in Fig. 11-4. Since the pressure is equal in all directions, the total propulsive force is zero. When the balloon is released, it flies across the room. It loses its air and eventually falls to the floor. The short flight is due to jet propulsion. The pressures against the front and back were equal until the balloon was released.

When the balloon is released, there is nothing against which the air can exert a force at the open mouth or nozzle. All the other forces are still balanced, but there is an unbalanced force on the front of the balloon, and the balloon moves in that direction, as shown in Fig. 11-5. There is an

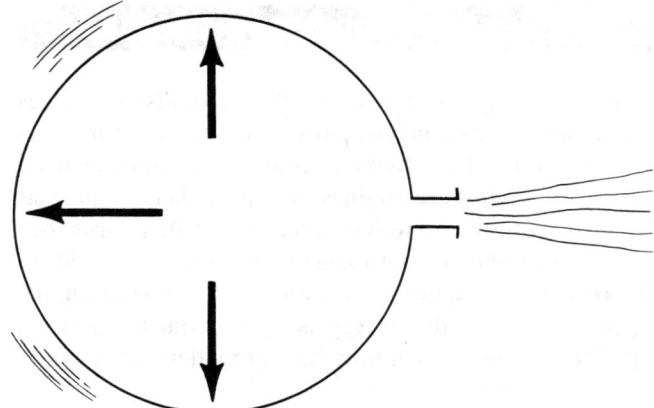

FIGURE 11-5 Unbalanced forces. (*General Electric Co., U.S.A.*)

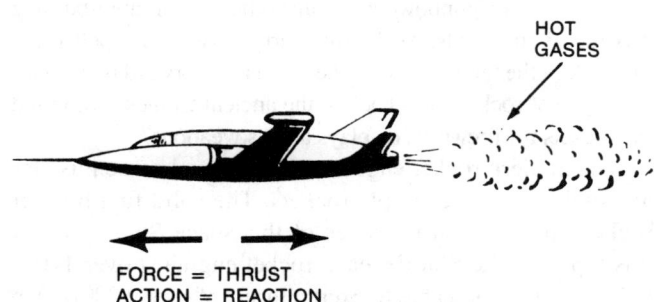

FIGURE 11-6 Newton's third law of motion.

equal and opposite force exerted by the balloon on the air it contains, and so the air is forced out the back of the balloon as exhaust. As the gases are accelerated out the rear of the balloon, the balloon is accelerated forward.

The acceleration and movement of a jet engine are caused by similar forces. In its simplest form, any jet engine draws in air, compresses the air, heats it, and forces it out through a jet nozzle at a very high velocity. As shown in Fig. 11-6, the exhaust gases are forced out the nozzle (action), and those gases exert a force or thrust (reaction) on the engine and aircraft in a forward direction. It is a common misunderstanding that the escaping gases push on the air behind the engine to move the engine forward. The thrust does not come from the jet engine pushing against the air behind it, but from the forces exerted by the hot expanding gases produced within the engine. The expanding gases push on the engine parts in sideways and forward directions, but since the back of the engine is just a large hole, no force can be exerted in a rearward direction. The engine moves in the direction of the unbalanced forces.

The amount of acceleration of an object is directly proportional to the force applied. This means that the greater the force, the faster the airplane will move. The amount of acceleration is inversely proportional to the mass of the object being moved. So if the mass or weight of an aircraft is increased, the acceleration of the aircraft will be decreased. These relationships are stated by the equation

$$\text{Acceleration} = \frac{\text{force}}{\text{mass}}$$

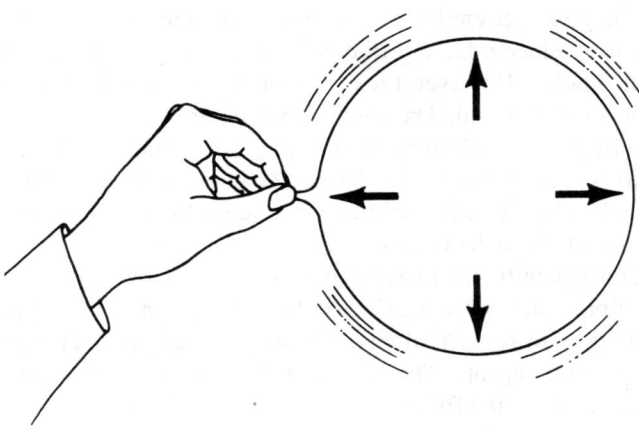

FIGURE 11-4 Balanced forces. (*General Electric Co., U.S.A.*)

There are many different types of jet propulsion engines, including rockets, fan jets, pulse jets, turbojets, turboprop jets, and turboshaft, bypass turbofan, and unducted fan jet engines. All these engines, except rockets, require air from the atmosphere to burn fuel. As the fuel burns, heat is produced and the hot expanding gases are forced out the rear of the engine. Depending on the type of engine, some or most of the energy is taken from the gases to operate turbines, which turn fans, propellers, or shafts.

Rockets

The simplest form of rocket is a tube made of metal, or even paper, filled with gunpowder or some other sort of rapid-burning mixture of chemicals. As the fuel burns, gases are expelled out the back of the tube and the rocket is pushed forward by the gas. This type of rocket dates back to the ancient Chinese who used the rockets for fireworks displays and as weapons.

The modern rocket engines which propel some missiles are similar to these simple rockets. The solid fuel booster rockets mounted on the sides of the Space Shuttle are of this type. On the Shuttle, each rocket engine is over 140 ft [42.67 m] long and able to exert a thrust of over 3 000 000 lb [13 350 kN].

Another type of rocket produces thrust by burning liquid fuel which has been mixed with a liquid oxidizer, usually liquid oxygen. The fuel or propellant is placed in one tank and the oxidizer in another tank. These are typically pumped to and mixed in the combustion chamber where the fuel is burned (see Fig. 11-7). As the gases rush out the nozzle at the back of the engine, thrust is produced. This nozzle has a definite shape and is known as a converging-diverging nozzle. This type of nozzle is required in rockets because of the desire for extremely high-velocity (highly accelerated) exhaust gases.

Ram Jet Engines

The **ram jet engine** is the simplest of the jet engines in that it contains no moving parts. In general, the ram jet construction resembles a pipe whose diameter enlarges in the combustion area and then gradually decreases toward the rear to reach its minimum size at the exit. Located in the combustion area are flame holders, a fuel nozzle, and an igniter. The ram jet must be assisted to attain a velocity in excess of about 250 mph [402.3 km/h] before it can be started. Once this speed is reached, there is sufficient combustion pressure to continue firing the engine. The flame holders located in the combustion chamber provide the necessary blockage in the passage to slow down the airflow so that the fuel and air can be mixed and ignited. A fuel metering system is also required to supply the proper amount of fuel to the engine.

Ram jets are most frequently used as augmenters or afterburners on turbojet engines. The augmenter is attached to the rear of the turbojet engine so that the jet exhaust passes through it.

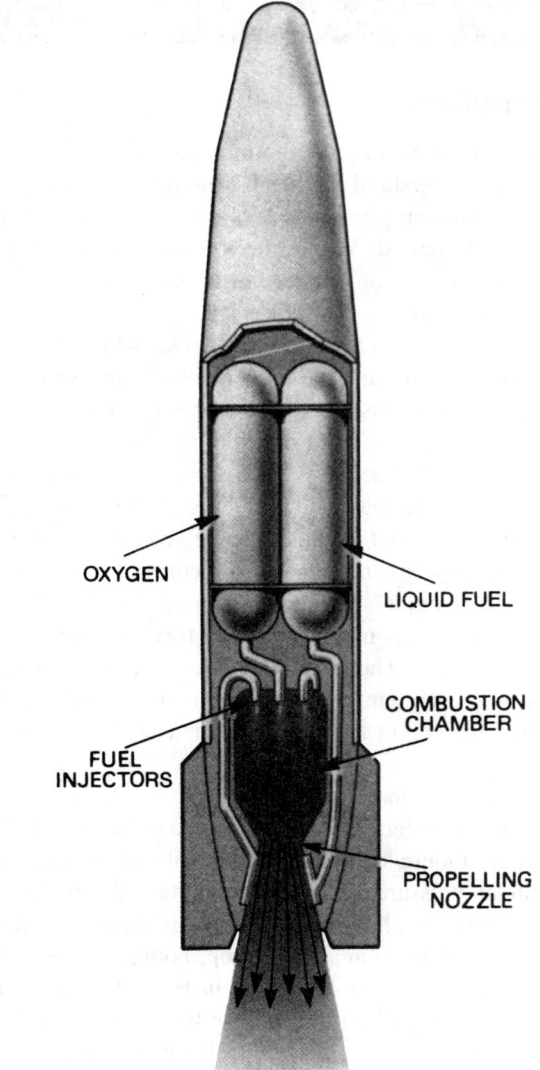

FIGURE 11-7   Liquid-fuel rocket engine. (*Rolls-Royce.*)

Additional fuel is sprayed into the exhaust gases which still contain some oxygen. The fuel burns, creating additional thrust.

GAS-TURBINE ENGINES

The **pulse jet engine** is somewhat more complex than the ram jet since it has a grill of shutters located at the inlet of the engine. However, the operation of the pulse jet is easier to understand. The shutters, which are kept open with springs, allow air to enter the combustion chamber. As the air is packed into the combustion chamber, it is mixed with fuel and ignited. The shutters are forced closed by the pressure of the exhaust gases when combustion takes place. Consequently, the exhaust gases can only move down the tailpipe and out the exhaust. Then the springs force the shutters to reopen, allowing more fresh air to enter, and the cycle repeats. The length of the tailpipe of the pulse jet regulates the frequency of the engine. Fuel flow is continuous, but flame propagation is intermittent, since the pulse jet operates in a step-by-step cycle. This is the only form of jet propulsion that operates by intermittent power surges, utilizing explosive rather than progressive or continuous combustion. However, in most pulse jet engines, the cycles per second are rather high and the net effect is practically continuous thrust. Pulse jet engines provide thrust for some guided missiles.

The basic operation of the gas-turbine or turbojet engine is relatively simple. Air is brought into the front of the turbine engine and compressed, fuel is mixed with this air and burned, and the heated exhaust gases rush out the back of the engine. The parts of a turbojet engine work together to change fuel energy to energy of motion, to cause the greatest thrust for the fuel used. A basic turbine engine is illustrated in Fig. 11-8.

A gas-turbine engine has three major sections: an air compressor, a combustion section, and a turbine section. The engine may also be divided into the cold section and the hot section. The forward or front part of the engine contains the air compressor, which is the cold section. The combustion and turbine sections make up the hot section of the engine. The compressor packs several tons of air into the combustion chamber every minute and works somewhat like a series of fans. Then fuel is forced into the combustion chamber through nozzles, a spark provides ignition, and the mixture burns in a process similar to a blowtorch, creating hot exhaust gases. These gases expand and are ejected from the rear of the engine. As the gases leave, they spin a turbine which is located just behind the combustion chamber. By means of an interconnecting shaft, the rotating turbine is connected to and turns the compressor, completing the cycle. After rushing by the turbine, the hot gases continue to expand and blast out through the exhaust nozzle at a high velocity, creating the force which propels a jet aircraft.

Gas turbines produce work in proportion to the amount of heat released internally. Therefore, it is necessary to study the production of heat in the engine, most of which is obtained by the burning of fuel (although some is obtained

by compressing the air in the compressor). An ordinary thermometer indicates the temperature of the gases, but does not tell the quantity of heat that is available. Heat cannot be measured directly, but must be calculated from three known quantities: temperature, mass (or weight), and specific heat. Although the exact nature of heat is not known, arbitrary standards have been internationally agreed upon by which changes in heat content can be calculated accurately. The standard in the British system, also accepted internationally, is the **British thermal unit** (Btu).

Imagine two turbojet engines, one using 10 000 lb/h [4540 kg/h] of fuel and the other using 1000 lb/h [454 kg/h]. Both engines are operating at the same turbine inlet temperature. However, the larger engine can do approximately ten times the work of the smaller because ten times more heat is released at the same temperature. This example illustrates that heat and temperature, although related, are not the same.

Both the reciprocating engine and the gas-turbine engine are considered **heat engines**. Both develop power or thrust by burning a combustible mixture of fuel and air. Both convert the energy of expanding gases to propulsive force. The reciprocating engine does this by changing the energy of combustion to mechanical energy which is used to turn a propeller. Aircraft propulsion is obtained as the propeller imparts a relatively small amount of acceleration to a large mass of air. The gas turbine, in its basic turbojet configuration, imparts a relatively large amount of acceleration to a smaller mass of air and thus produces thrust or propulsive force directly.

Because the turbojet engine is a heat engine, the higher the temperature of combustion, the greater the expansion of the gases. The combustion temperature, however, must not exceed a value that provides a turbine gas entry temperature suitable for the design and materials of the turbine assembly.

Gas-turbine engines come in various mechanical arrangements. Aircraft turbine engines can generally be classified into four types of engines: turbojet, turbofan, turboprop, and turboshaft. The basic components of all these engines are essentially the same: a compressor, a combustion chamber, a turbine to drive the compressor, and an exhaust nozzle. The difference lies in the type and arrangement of these components. The mechanical arrangements of various types of gas-turbine engines are shown in Fig. 11-9.

Turbojet Engine

A **turbojet engine** is a type of gas-turbine engine which produces thrust through the hot gases exiting the exhaust section of the engine. Although this engine was the first real successful gas-turbine engine, it has been replaced in most aircraft by the turbofan, because of its fuel efficiency. However, it is still used in some military applications.

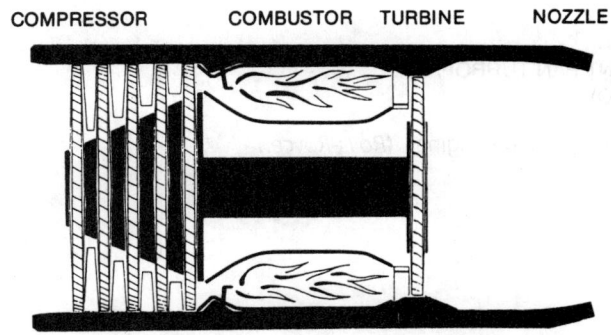

COMPRESSOR   COMBUSTOR   TURBINE   NOZZLE

FIGURE 11-8 Major parts needed to operate a gas-turbine engine. (*General Electric Co., U.S.A.*)

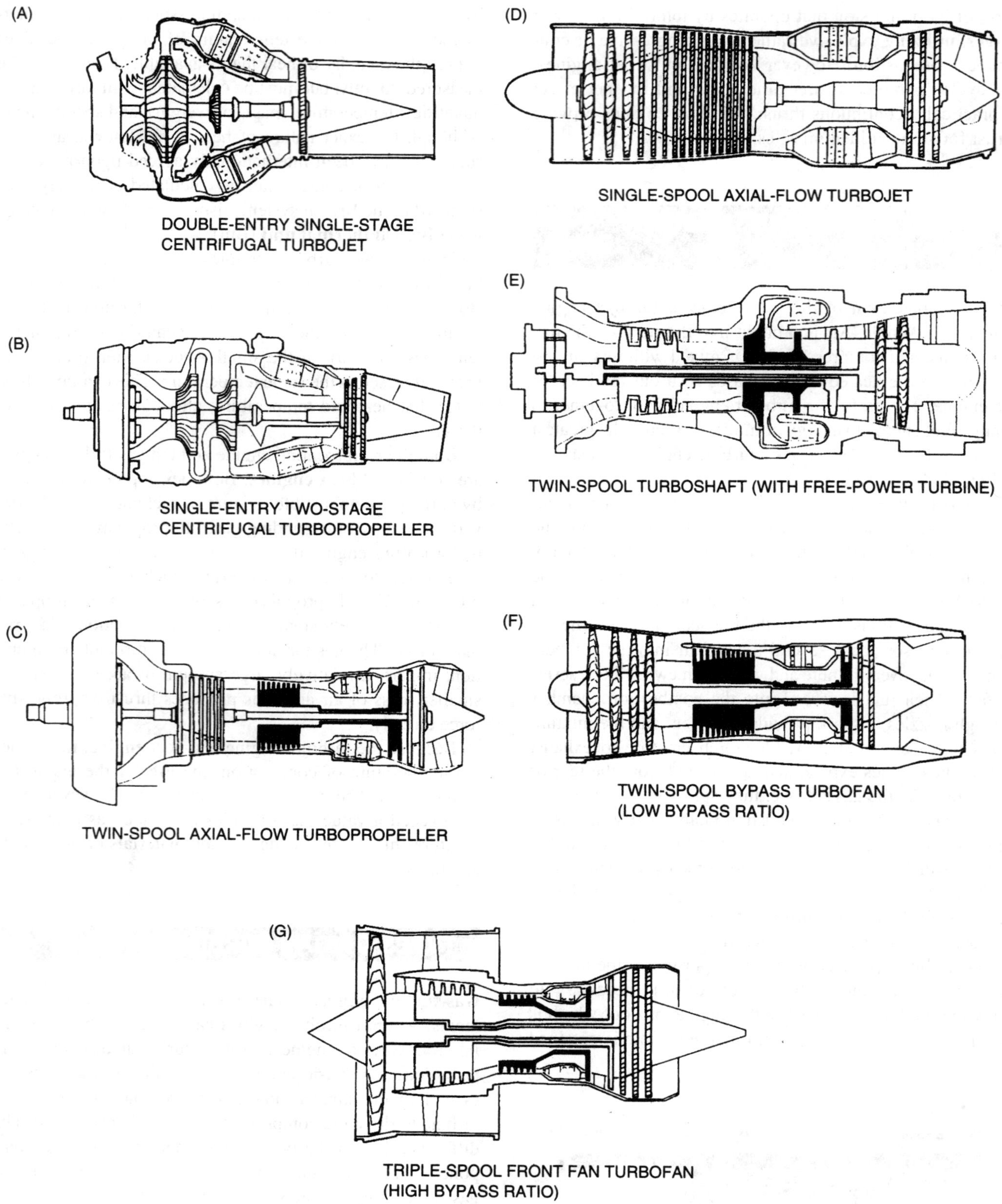

(A)

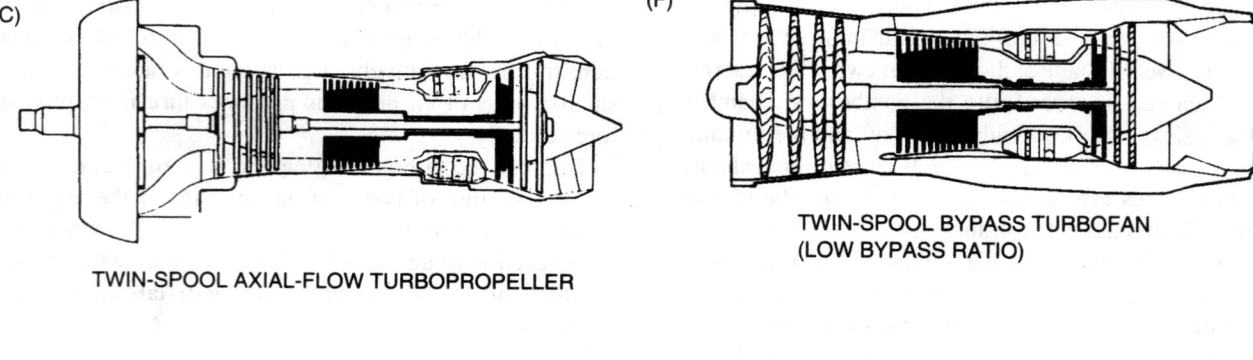

**DOUBLE-ENTRY SINGLE-STAGE
CENTRIFUGAL TURBOJET**

(D)

SINGLE-SPOOL AXIAL-FLOW TURBOJET

(B)

**SINGLE-ENTRY TWO-STAGE
CENTRIFUGAL TURBOPROPELLER**

(E)

TWIN-SPOOL TURBOSHAFT (WITH FREE-POWER TURBINE)

(C)

TWIN-SPOOL AXIAL-FLOW TURBOPROPELLER

(F)

**TWIN-SPOOL BYPASS TURBOFAN
(LOW BYPASS RATIO)**

(G)

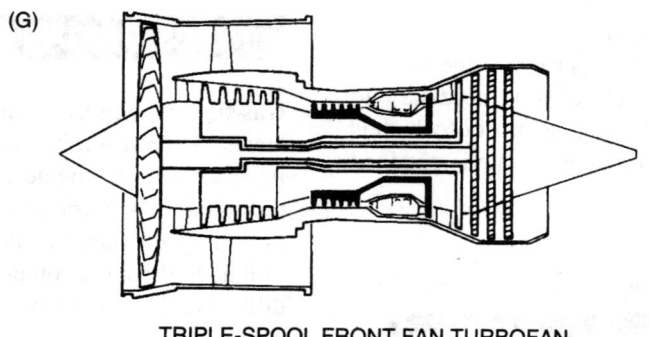

**TRIPLE-SPOOL FRONT FAN TURBOFAN
(HIGH BYPASS RATIO)**

FIGURE 11-9   Mechanical arrangements of gas-turbine engines. (*Rolls-Royce.*)

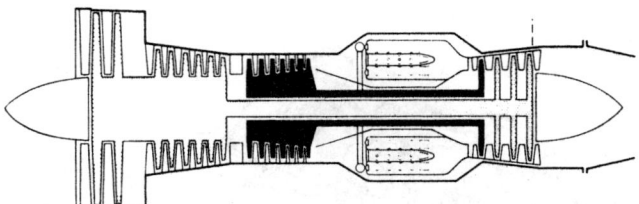

FIGURE 11-10 Arrangement of a forward turbofan engine. (*Pratt & Whitney Canada.*)

Turbofan Engine

A **turbofan engine** may be considered a cross between a turbojet engine and a turboprop engine. The turboprop engine drives a conventional propeller through reduction gears to provide a speed suitable for the propeller. The propeller accelerates a large volume of air in addition to that which is being accelerated by the engine itself. The turbofan engine accelerates a smaller volume of air than the turboprop engine but a larger volume than the turbojet engine.

The arrangement of a forward turbofan engine with a dual compressor is shown in Fig. 11-10. The fan's rotational speed on this engine is the same as the low-pressure (N_1) compressor speed. During operation, air from the fan section of the forward blades is carried outside to the rear of the engine, through ducting. The bypass engine has two gas streams: the cool bypass airflow and the hot turbine discharge gases which have passed through the core of the engine. The bypass air or fan air is cool because it has not passed through the actual gas-turbine engine. This fan air can account for around 80 percent of the engine's total thrust. The effect of the turbofan design is to greatly increase the power-weight ratio of the engine and to improve the thrust specific fuel consumption.

Turbofan engines may be high-bypass or low-bypass engines. The ratio of the amount of air that bypasses (passes around) the core of the engine to the amount of air that passes through the core is called the *bypass ratio*. A low-bypass engine does not bypass as much air around the core as a high-bypass engine.

Many different types of turbofan engines are in use and can be found on aircraft from small business jets to large transport-type aircraft. Most turbofan engines are constructed with a forward fan. The fan is driven by a set of core engine turbine stages designed to drive the fan only. These bypass fan blades extend into a coaxial duct which surrounds the main engine. Airflow enters the forward end of the duct and is expelled coaxially with the engine exhaust to produce additional thrust. Additional information on different types of turbofan engines is presented in Chap. 16.

High-Bypass Turbofan Engine

During recent years, the high-bypass turbofan engine has become one of the principal sources of power for large transport aircraft. Among such engines are the Pratt & Whitney JT9D,

the General Electric CF6, and the Rolls-Royce RB 211. These engines are used, respectively, in Boeing 747, Douglas DC-10, and Lockheed L-1011 aircraft.

A **high-bypass engine** utilizes the fan section of the compressor to bypass a large volume of air compared with the amount which passes through the engine. The bypass ratio for the Pratt & Whitney JT9D and the Rolls-Royce RB 211 is approximately 5:1. This means that the weight of the bypassed air is five times the weight of the air passed through the core of the engine. The bypass ratio for the General Electric CF6 engine is approximately 6.2:1; however, some models have a variable bypass ratio, and the amount of bypassed air may be more or less than stated above.

The principal advantages of the high-bypass engine are greater efficiency and reduced noise. The high-bypass engine has the advantages of the turboprop engine but does not have the problems of propeller control. The design is such that the fan can rotate at its most efficient speed, depending on the speed of the aircraft and the power demanded from the engine.

On some front fan engines, the bypass airstream is ducted overboard either directly behind the fan through short ducts or at the rear of the engine through longer ducts, thus the term **ducted fan**.

Very high bypass ratios, on the order of 15:1, are achieved by using Prop-Fans. These are variations on the turbopropeller theme but have advanced technology propellers capable of operating with high efficiency at high aircraft speeds.

Turboprop Engine

A **turboprop engine**, such as that illustrated in Fig. 11-11, is nothing more than a gas turbine or turbojet with a reduction gearbox mounted in the front or forward end to drive a standard airplane propeller. This engine uses almost all the exhaust-gas energy to drive the propeller and therefore provides very little thrust through the ejection of exhaust gases. The exhaust gases represent only about 10 percent of the total amount of energy available. The other 90 percent of the energy is extracted by the turbines that drive the compressor and a second turbine that drives the propeller. The basic components of the turboprop engine are identical to those of the turbojet—that is, the compressor, combustor, and turbine. The only difference is the addition of the gear-reduction box to reduce the rotational speed to a value suitable for propeller use. Further information on this type of engine is presented in Chap. 17.

Turboshaft Engine

A **turboshaft engine** is a gas-turbine engine which delivers shaft horsepower through an output shaft. This engine, like the turboprop, uses almost all the exhaust energy to drive the output shaft. This type of gas-turbine engine is used in aviation mainly on helicopters and for auxiliary power units on large transport aircraft. Additional information on this type of engine can be found in Chap. 18.

Types of Gas-Turbine Engines   **313**

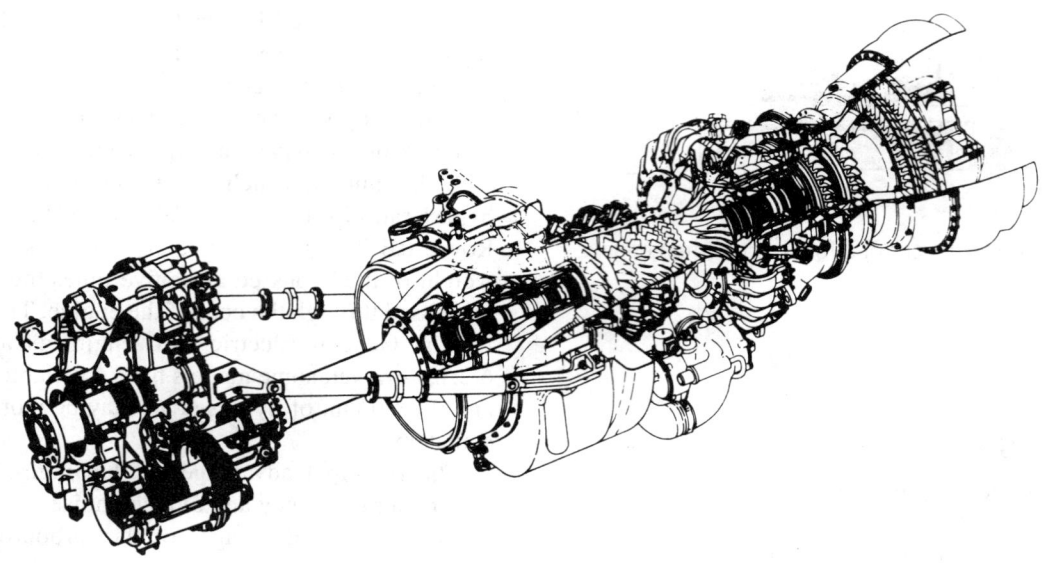

FIGURE 11-11   CT7 turboprop engine, cutaway drawing. (*General Electric Co., U.S.A.*)

GAS-TURBINE ENGINE THEORY AND REACTION PRINCIPLES

Newton's Laws of Motion

The thrust produced by a turbojet engine may be explained to a large extent by Newton's laws of motion. These may be stated as follows:

First Law. A body at rest tends to remain at rest, and a body in motion tends to continue in motion in a straight line unless caused to change its state by an external force.

Second Law. The acceleration of a body is directly proportional to the force causing it and inversely proportional to the mass of the body.

Third Law. For every action, there is an equal and opposite reaction.

The second and third laws of motion are the most applicable to the development of thrust. The equations for jet thrust may be derived directly from the second law. This law may be stated in words differing from those given above; however, the meaning is the same: **A force is created by a change of momentum, and this force is equal to the time rate of change of momentum.** The **momentum** of a body is defined as *the product of its mass and its velocity.*

From Newton's second law, clearly if we change the velocity of a body (or a mass), a force is required. In equation form, this principle is expressed as $F = dM/dt$, where F is the force, M is the momentum, and t is the time. This equation means that the force created by a change in velocity of a body is equal to the amount of change in velocity divided by the change in time, when appropriate units of measurement are used in the equation.

To understand the operation of equations dealing with acceleration, the nature of acceleration must be understood. **Acceleration** may be defined as a change in velocity. Generally, the word "acceleration" is used to indicate an increase in velocity. This is *positive acceleration.* A decrease in velocity is *negative acceleration,* or **deceleration**.

It is known through studies in the laws of physics that a free-falling body will accelerate at 32.2 feet per second per second, or feet per square second (ft/s²) [9.8 m/s²]. Thus, 32.2 ft/s² [9.8 m/s²] is called the **acceleration of gravity**, and the letter g is used to indicate this value.

If an automobile weighs 3000 lb [1360.8 kg] and we wish to know how much force is necessary to cause it to accelerate from 0 to 60 mph [96.56 km/h] in 5 s, we proceed as follows:

Since 60 mph [96.56 km/h] is equal to 88 ft/s [26.82 m/s], to reach this velocity, the automobile must accelerate at 17.6 ft/s [5.36 m/s] each second. Since the automobile would accelerate at 32.2 ft/s² [9.8 m/s²] if the force applied were equal to its weight, we know that the needed force is less than 3000 lb [1360.8 kg]. The force needed is then determined by multiplying 3000 by 17.6/32.2. The force needed for the required acceleration is then 1639.75 lb [7293.61 N].

The formula used for the foregoing problem may be written as

$$F = \frac{W \times A}{g}$$

where F = force, lb [kg]
    W = weight, lb [kg]
    A = acceleration, ft/s² [m/s²]
    g = 32.2

The problem can then be stated in equation form as

$$F = \frac{3000 \times 17.6}{32.2}$$

$$= 1639.75 \text{ lb } [7293.61 \text{ N}]$$

The basic equation which we may derive from Newton's second law is $F = MA$, where F is force, M is mass (weight), and A is acceleration. **Mass** is a basic property of matter, whereas **weight** is the effect of gravity on a mass. When we consider matter on the surface of the earth, mass and weight are approximately the same.

AIRFLOW

The path of the air through a gas-turbine engine varies according to the design of the engine. A straight-through flow system is the basic design, because it provides for an engine with a relatively small frontal area. In contrast, the reverse-flow system gives an engine with greater frontal area but a reduced overall length. The operation, however, of all engines is similar.

During the passage of air through the engine, aerodynamic and energy requirements demand changes in the velocity and pressure of the air. For instance, during compression, a rise in the pressure of the air, not an increase in its velocity, is required. After the air has been heated and its internal energy increased by combustion, an increase in the velocity of the gases is necessary to force the turbine to rotate. At the propelling nozzle, a high exit velocity is required, for it is the change in momentum of the air that provides the thrust on the aircraft. Local decelerations of airflow are also required, such as in the combustion chambers, to provide a low-velocity zone in which the flame can burn.

These various changes are effected by means of the size and shape of the ducts through which the air passes on its way through the engine. Where a conversion from velocity (kinetic) energy to pressure is required, the passages are divergent in shape, as shown in Fig. 11-12A. Conversely, where it is required to convert the energy stored in the combustion gases to velocity energy, a convergent passage or nozzle, as shown in Fig. 11-12B, is used. These shapes apply to the gas-turbine engine where the airflow velocity is subsonic or sonic (at the local speed of sound). Where supersonic speeds are encountered, such as in the propelling nozzle of the rocket or of some turbine engines, a **convergent-divergent nozzle** (see Fig. 11-13) is used to obtain the maximum conversion of energy in the combustion gases to kinetic energy.

Since the efficiency with which energy changes are effected depends on good design, the design of the passages and nozzles is very important. Any interference with the smooth airflow creates a loss in efficiency and could result in component failure due to vibration caused by eddies in or turbulence of the airflow.

When the air is compressed or expanded at 100 percent efficiency, the process is said to be **adiabatic**. Since such a change means there is no energy loss in the process

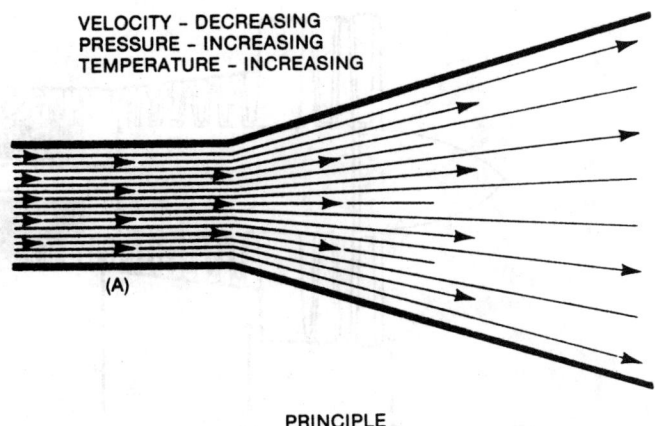

VELOCITY – DECREASING
PRESSURE – INCREASING
TEMPERATURE – INCREASING

(A)

PRINCIPLE

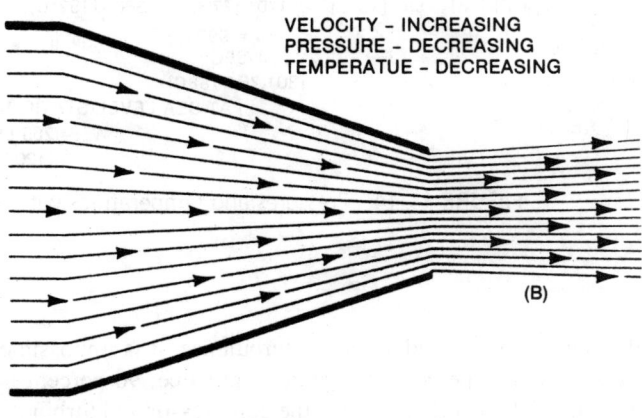

VELOCITY – INCREASING
PRESSURE – DECREASING
TEMPERATUE – DECREASING

(B)

PRINCIPLE

FIGURE 11-12 Airflow through (A) divergent and (B) convergent ducts. (*Rolls-Royce.*)

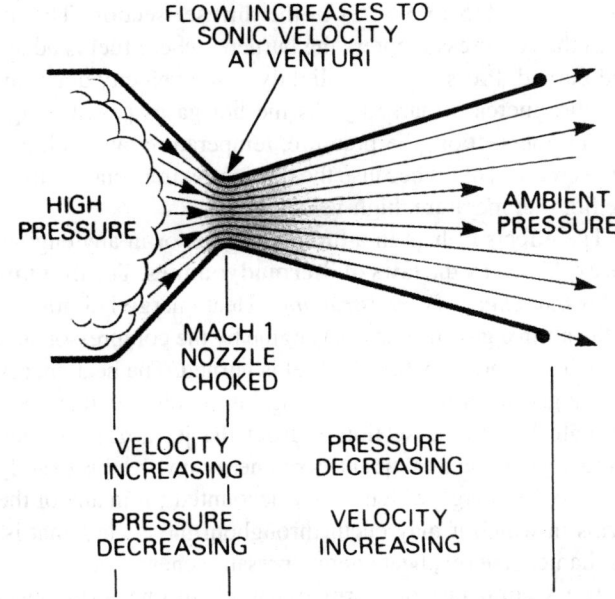

FLOW INCREASES TO
SONIC VELOCITY
AT VENTURI

HIGH
PRESSURE

AMBIENT
PRESSURE

MACH 1
NOZZLE
CHOKED

VELOCITY
INCREASING

PRESSURE
DECREASING

PRESSURE
DECREASING

VELOCITY
INCREASING

FIGURE 11-13 Supersonic airflow through a convergent-divergent nozzle or venturi. (*Rolls-Royce.*)

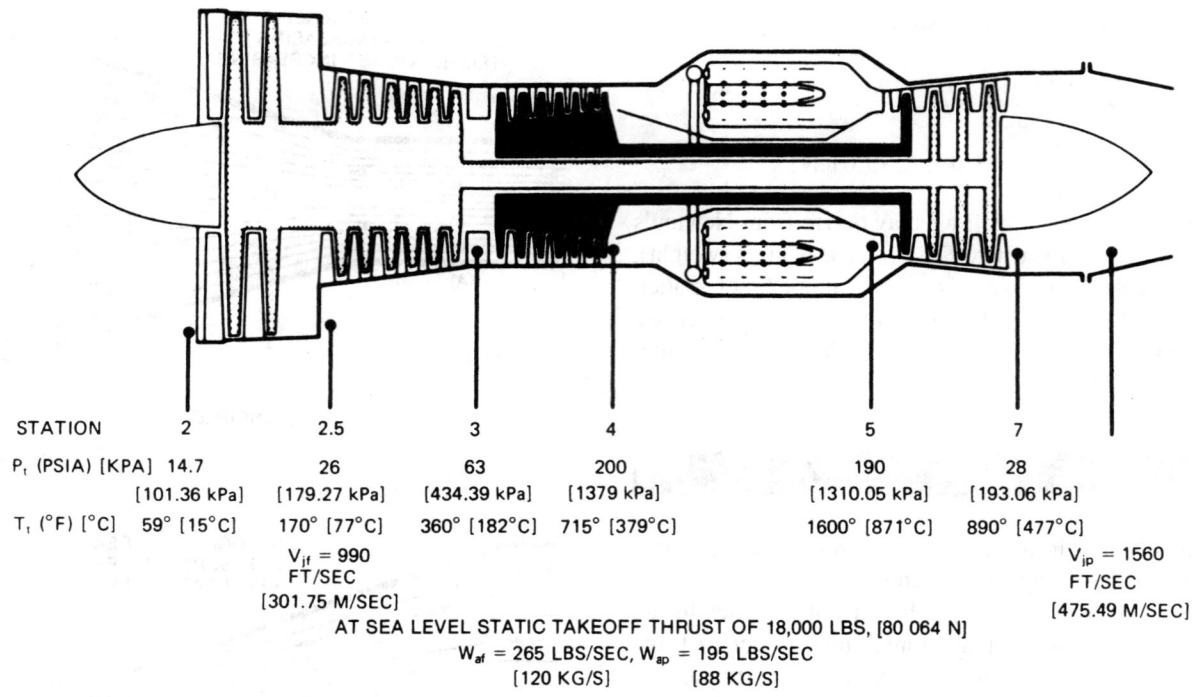

STATION	2	2.5	3	4	5	7	
P_t (PSIA) [KPA]	14.7	26	63	200	190	28	
	[101.36 kPa]	[179.27 kPa]	[434.39 kPa]	[1379 kPa]	[1310.05 kPa]	[193.06 kPa]	
T_t (°F) [°C]	59° [15°C]	170° [77°C]	360° [182°C]	715° [379°C]	1600° [871°C]	890° [477°C]	

V_{if} = 990 FT/SEC [301.75 M/SEC]

V_{ip} = 1560 FT/SEC [475.49 M/SEC]

AT SEA LEVEL STATIC TAKEOFF THRUST OF 18,000 LBS, [80 064 N]

W_{af} = 265 LBS/SEC, W_{ap} = 195 LBS/SEC
[120 KG/S] [88 KG/S]

FIGURE 11-14 Pressures and temperatures within a twin-spool turbofan engine. (*Pratt & Whitney Canada.*)

through friction, conduction, or turbulence, it is impossible to achieve 100 percent efficiency in practice; 90 percent is a good adiabatic efficiency for the compressor and turbine.

To better understand what takes place in a typical turbine engine, we examine the pressures and temperatures in the engine during operation. Figure 11-14 shows the values of pressure P_t (PSIA) and temperature, T_t (F°) at different stations, within a twin-spool turbofan engine. Note that there is a substantial rise in both pressure and temperature through the two compressor sections. Also, the pressure has increased more than 13:1, and the temperature has increased from 59 to 715°F [15 to 379°C] in the diffuser section. The air from the compressor enters the burners, where fuel is added and burned. Pressure drops slightly while velocity and temperature increase markedly. As the hot gases pass through the turbine section, the pressure, temperature, and velocity all decrease. This is because the turbines extract energy from the high-temperature, high-velocity gases.

The effects of heat in a turbine engine, or in any engine, are explained by the **laws of thermodynamics**. The **first law** states that *energy is indestructible*. Heat energy is imparted to the air in a gas-turbine (jet) engine by the compressor, and more heat is added when the fuel is burned. The heat energy is changed to thrust, and the gases are cooled as they pass through the turbine section and out the jet nozzle. In any energy cycle, the total quantity of energy involved is exactly equal to the energy which can be accounted for in any of the forms in which it may occur throughout the cycle—that is, mechanical energy, heat energy, pressure energy, etc.

The **second law** of thermodynamics in one form states that *heat cannot be transferred from a colder body to a hotter body*. The cooling of an engine involves this principle, in that heat is transferred from hotter bodies or substances to cooler bodies or substances.

The pressures and temperatures of gases follow the principles set forth in Boyle's law and Charles' law. **Boyle's law** states that *the volume of a confined body of gas varies inversely as its absolute pressure, the temperature remaining constant*.

Charles' law states that *the volume of a gas varies in direct proportion to the absolute temperature*. This law explains the expansion of gases that occurs when heat is added by the burning of fuel in the engine.

From the foregoing discussion, it is clear that many physical laws are involved in the operation of a gas-turbine engine. The operation of some of these laws is apparent when we examine the **Brayton cycle**, also known as the **constant-pressure cycle**, which defines the events that take place in the turbine engine. Figure 11-15A is a diagram that represents the events of the Brayton cycle. The actual values represented by the curves may vary considerably among different engines; however, the principle is the same for all. Point 1 in the drawing indicates the condition of the air in front of the engine before it is affected by the inlet duct of the engine. After the air enters the inlet duct, it is diffused and the static pressure increases. This is indicated by point 2, which represents the air condition at the entrance to the compressor. Through the compressor, the air volume is decreased and the pressure is increased substantially, as shown by the curve from point 2 to point 3. At point 3, fuel is injected and burned, causing a rapid increase in volume and temperature. Because of the design of the combustion chamber, the pressure drops slightly as the velocity of the hot gas mixture increases to the rear. At point 4, the heated

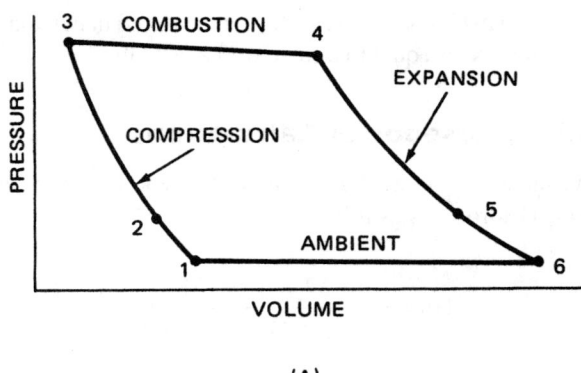

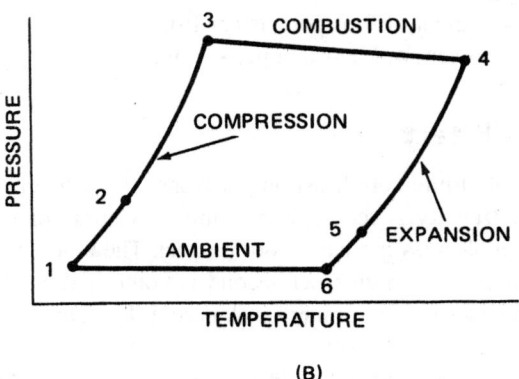

FIGURE 11-15   Brayton cycle.

gases enter the turbine where energy is extracted, causing decreases in both pressure and temperature. The curve from point 5 to point 6 represents the condition in the exhaust nozzle as the gases flow out to ambient pressure. The difference between the positions of point 1 and point 6 indicates the expansion of air caused by the addition of heat from the burning fuel.

The temperature diagram for the Brayton cycle is shown in Fig. 11-15B. Note that the temperature increases because of compression. During combustion, the pressure drops slightly and the temperature increases at a rapid rate as a result of the burning of fuel. During expansion of the gases through the turbine, the temperature and pressure of the gases are reduced to the point where the gases enter the atmosphere. At this time, the temperature is still considerably above the temperature of the ambient atmosphere, but it decreases rapidly as the gases leave the jet nozzle and go into the atmosphere behind the engine.

GAS-TURBINE ENGINE PERFORMANCE

The gas-turbine engine must operate efficiently from sea level to extreme altitudes, from a standstill to extreme speeds, and under all changes of atmospheric conditions, including temperatures from −80°F [−62°C] to 110°F [43.3°C] or more.

Jet Engine Symbols and Their Meanings

Maintenance, flight, and engineering personnel have found it very advantageous to use certain symbols to refer to particular areas or factors concerning gas-turbine engines. A complete listing of these symbols can be found in the Appendix of this text.

Thrust

Thrust is a reaction force which is measured in pounds. The **thrust** of a propeller, rocket, or gas-turbine engine depends on the acceleration of a mass (weight) in accordance with Newton's second law. A propeller accelerates a mass of air, a rocket accelerates the gases resulting from the burning of fuel, and a turbine engine accelerates both air and fuel gases. The quantity (mass) of air and gases accelerated and the amount of acceleration determine the thrust produced. Static thrust or gross thrust is the amount of thrust an engine is producing when the aircraft (engine) is not moving through the air. Net thrust is the amount of force or thrust being developed when the aircraft is in flight. To calculate the net thrust, the speed of the aircraft must be accounted for and used in the computations.

To determine the amount of thrust, the weight per unit time (normally in pounds per second) of the flow of air through the engine must be determined. This is calculated by multiplying the inlet volume (= $\pi r^2 \times$ flow velocity per second) by the weight of the air, measured in pounds per cubic foot, entering the engine. The weight of the air will vary with temperature. A conversion chart can be used to determine the weight of the air at a given temperature. For example, 1 ft³ [0.028 m³] of air at 59°F [15°C] will weigh 0.076 47 lb [0.034 69 kg]. To calculate mass airflow, the following formula can be used.

$$w_a = \pi r^2 \times \text{airflow velocity per second} \times 0.076\,47$$

The basic formula for the approximate thrust of a gas-turbine engine is:

$$F = \frac{w_a}{g}(V_2 - V_1)$$

where F = force, lb [kg]
  w_a = flow rate of air mass, lb/s [kg/s]
  g = acceleration of gravity, 32.2 ft/s² [9.8 m/s²]
  V_2 = final velocity of gases (velocity of exhaust gases at jet nozzle), ft/s [m/s]
  V_1 = initial velocity of gases (aircraft speed if calculating for net thrust; if calculating for gross or static thrust, this value is 0), ft/s [m/s]

The approximate thrust of an engine may be determined by considering the weight of the air only, because the fuel weight is a very small percentage of the air weight. However, to obtain an accurate indication of thrust produced by the acceleration of the fuel-air mixture, it is necessary to include both fuel and air in the equation. The equation then becomes

$$F = \frac{w_a}{g}(V_2 - V_1) + \frac{w_f}{g}(V_j)$$

where w_a = airflow through engine, lb/s [kg/s]
  w_f = fuel flow, lb/s [kg/s]
  V_j = velocity of gases at jet nozzle, ft/s [m/s]

In this equation, V_j represents the acceleration of the fuel because the initial velocity of the fuel is the same as that of the engine. In actual practice, not all the pressure of the gases flowing from the nozzle of a turbine engine can be converted to velocity. This is particularly true of a jet nozzle in which the velocity of the gases reaches the speed of sound (becomes choked). In these cases, the static pressure of the gases at the jet nozzle is above the ambient air pressure. This difference in pressure creates additional thrust proportional to the area of the jet nozzle. The thrust generated at the jet nozzle is given by

$$F_j = A_j(P_j - P_{amb})$$

where F_j = force (thrust, because of choked nozzle), lb [kg]
  A_j = area of jet nozzle, in² [m²]
  P_j = static pressure at jet nozzle, lb/in² [kg/m²]
  P_{amb} = static pressure of ambient air, lb/in² [kg/m²]

When the jet nozzle thrust is added to the reaction thrust created by acceleration of gases in the engine, the equation for the net thrust (F_n) becomes

$$F_n = \frac{w_a}{g}(V_2 - V_1) + \frac{w_f}{g}(V_j) + A_j(P_j - P_{amb})$$

By using calculations similar to those previously described, the thrust of a turbofan engine can also be determined. An example formula for an engine with an unchoked exhaust nozzle would be

$$F_n = (\text{Eng})\frac{w_a}{g}(V_2 - V_1) + \frac{w_f}{g}(V_j) + (\text{Fan})\frac{w_a}{g}(V_2 - V_1)$$

Converting Thrust to Horsepower

Because power is the product of a force and a distance, it is not possible to make a direct comparison of thrust and horsepower in a turbine engine. When the engine is driving an airplane through the air, however, we can compute the equivalent horsepower being developed. When we convert the foot-pounds per minute of 1 hp to mile-pounds per hour, we obtain the value 375. That is, 1 hp is equal to 33 000 ft • lb/min or 375 mi • lb/h. Thrust horsepower (thp) is then obtained by using the following formula:

$$thp = \frac{\text{thrust (lb)} \times \text{airspeed (mph)}}{375}$$

If a jet engine is developing a 10 000-lb [44 480-N] thrust and is driving an airplane at 600 mph [965.4 km/h], then

$$thp = \frac{10\,000 \times 600}{375} = 16\,000 \ [11\,931.2 \text{ kW}]$$

Note that **thrust** in the metric system is measured in newtons. One newton (N) is equal to a force of 0.224 82 lb.

Turbine Horsepower Calculations

The formula for finding the amount of horsepower a turbine wheel is absorbing is as follows:

$$\text{Hp} = \frac{Cp \times \Delta T \times w_a \times 778}{550}$$

where Cp = 0.24 (number of Btu's required to raise the temperature of 1 lb [0.45 kg] of air 1°F [0.55°C].
  ΔT = temperature drop from one side of the turbine wheel to the other or across the wheel
  w_a = mass airflow, lb/s
  778 = number of foot-pounds per Btu
  550 = horsepower constant per second

Altitude Effect

The effect of altitude on thrust output is actually a function of density. **Density** is the mass per unit of volume or the number of molecules per cubic foot (meter). The amount of air directed into the engine per second is controlled by the fixed engine inlet area, assuming constant rpm. Thus, the mass flow is determined by density.

In free air, a rise in temperature will cause the speed of the molecules to increase so that they run into each other more vigorously. When they are farther apart, a given number of molecules will occupy more space. When a given number of molecules occupy more space, fewer molecules pass through the engine inlet area. As the air temperature increases, thrust tends to decrease. When the pressure increases in tree air, it is because there are more molecules per cubic foot (meter). When there are more molecules per cubic foot (meter), more of them can be directed through the fixed inlet area. More molecules going through the inlet area per second means a greater weight, or mass, of air. As pressure goes up, density goes up. As density goes up, the weight of air goes up and consequently the thrust goes up. Density affects the weight of air w_a, and w_a affects thrust directly. When pressure goes up, density goes up; and when temperature goes up, density goes down. Density affects thrust proportionally. A 10 000-lb [44.5-kN] thrust engine on a hot day may put out no more than 8000 lb [35.6 kN] of thrust, whereas on a cold day the engine might produce as much as 12 000 lb [53.4 kN] of thrust.

As an aircraft climbs, air pressure and temperature decrease. As pressure decreases, thrust decreases; but as temperature decreases, thrust increases. However, the pressure drops off faster than the temperature so that there is actually a drop-off in thrust with an increase in altitude, as shown in Fig. 11-16. At approximately 36 000 ft [10 973 m], the temperature stops falling and remains constant while the pressure continues to fall. As a result, the thrust drops off more rapidly above 36 000 ft [10 973 m]. This makes 36 000 ft [10 973 m] the optimum altitude for long-range cruising, just before the rate of thrust fall-off increases.

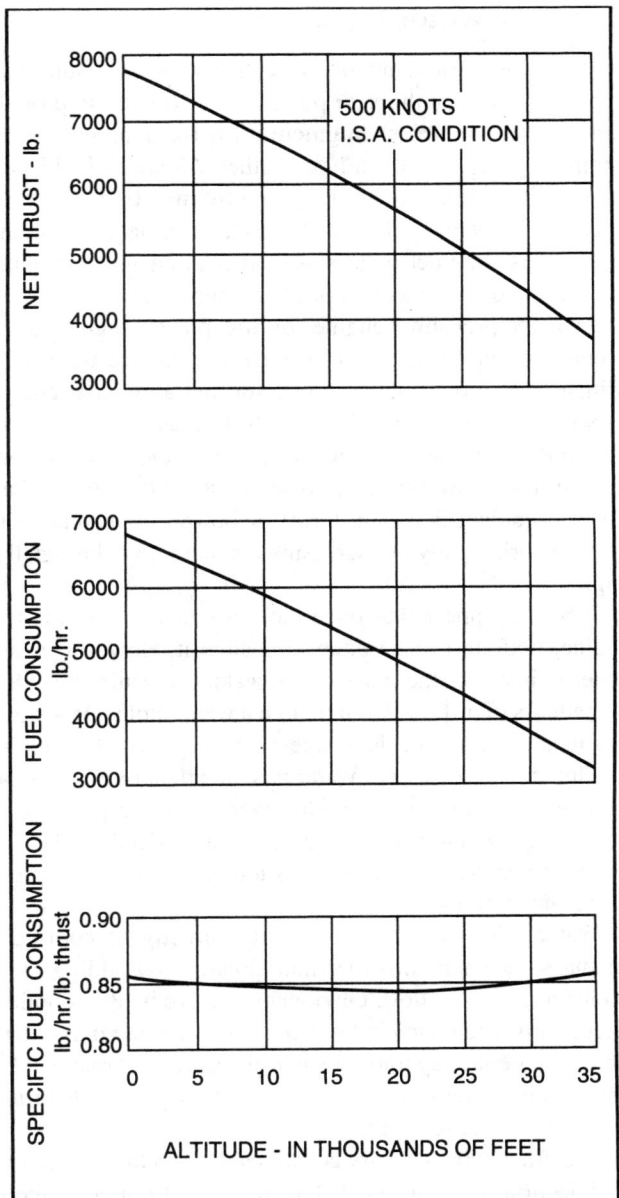

FIGURE 11-16 The effects of altitude on thrust and fuel consumption.

Ram Effect

As the aircraft picks up speed, the air moves by it at an increased speed, and as the air enters the divergent inlet duct, it fills up the added space. When this occurs, a drop in velocity and an increase in pressure occur. With the relative velocity of the air turned into pressure, the air molecules have developed their impact force. The restricted area of the inlet causes a pileup of molecules, thus increasing the density. When the density increases, there is an increase in thrust.

Any friction loss in the duct is a loss to the engine as far as ram effect is concerned. A quick summary of ram effect is presented in Fig. 11-17. Curve A represents the tendency for thrust to drop off with airspeed due to an increase in aircraft speed. Curve B represents the thrust generated by the ram effect or increased w_a. Curve C is the result of combining curves A and B.

Owing to the ram ratio effect from the aircraft forward speed, extra air is taken into the engine so that the mass airflow and also the engine velocity increase with aircraft speed. The effect of this tends to offset the extra intake momentum drag resulting from the forward speed so that the resultant loss of net thrust is partially recovered as the aircraft speed increases. The ram ratio effect, or the return obtained in terms of pressure rise at entry to the compressor in exchange for the unavoidable intake drag, is of considerable importance to the gas-turbine engine, especially at high speeds.

Effect of Engine Speed

At low engine speeds, the turbojet thrust increase is slight, even for a large increase in engine speed, and fuel consumption is high for the amount of thrust produced. For this reason, the cruise point is usually 85 to 90 percent of maximum rpm.

The input of heat energy needed to accomplish the required amount of work on a mass of air is controlled by the fuel control system. The variation in mass airflow on which the work is to be done is controlled by the engine's rpm. As a result, to increase thrust, the fuel control system must increase fuel flow, thus increasing rpm, and must do so in such proportions as to not overheat or overspeed the engine. Because the rpm controls only the mass airflow, the characteristics of the thrust line depend on the characteristics of the compressor as it pumps air at varying rpm values. At high engine speeds, even a small increase in rpm produces a large increase in thrust.

Piston engines have an almost opposite relationship between engine speed and propeller thrust. A turbojet engine's rpm should be kept high during an approach for landing. The acceleration time from a low idle speed to takeoff rpm is somewhat longer than that required for piston engines. However, this deficiency in turbine engines is gradually being overcome.

Effects of Humidity

It is interesting to compare the effects of humidity on output for turbine engines and piston-type engines. In a piston engine, an increase in the air humidity decreases the weight

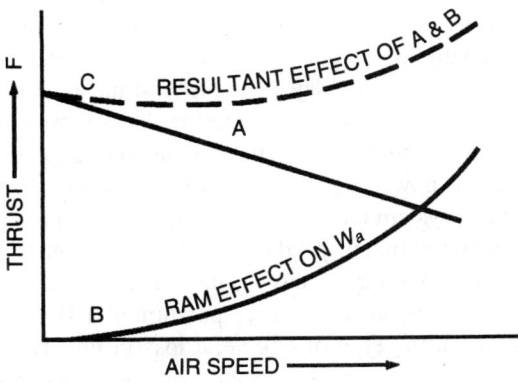

FIGURE 11-17 Ram effect.

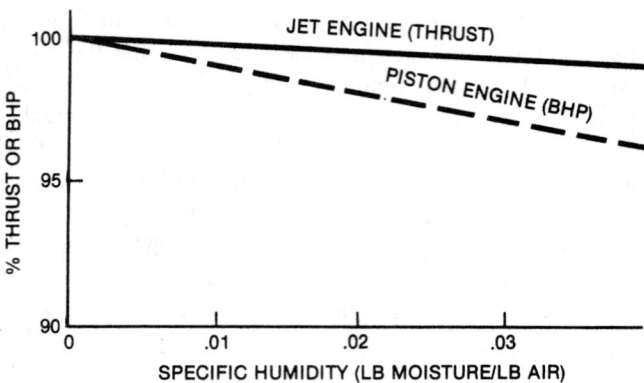

FIGURE 11-18   Effect of humidity on engine power output.

per unit volume of air. In high humidity, a piston engine will experience a fall-off in horsepower as a result of the fall-off in weight of airflow available for combustion under constant rpm, as illustrated in Fig. 11-18. Since the carburetor does not compensate fuel flow for humidity change, the air-fuel (A/F) ratio will drop, causing enrichment and loss of power.

The increased humidity also causes a decrease in weight per unit volume of air within the turbine engine. However, its reducing effect on output is almost negligible. Because the turbine engine operates with more air than is needed for complete combustion of all fuel, any lack of weight in the combustion air supply to give the proper A/F ratio will be made up from the cooling air supply. The engine will not be penalized by loss of heat energy from an improper A/F combustion ratio, as in the case of the piston engine, and therefore its output will not be reduced materially.

Temperature Effect

It was recognized early in the development and testing of gas-turbine engines that the atmospheric temperature at the time of takeoff affected a gas-turbine engine much more than it did a reciprocating engine. In some cases, the loss in power due to high atmospheric temperatures was twice the loss associated with piston engines.

On a cold day, the density of the air increases so that the mass of air entering the compressor for a given engine speed is greater, and therefore the thrust is higher, as shown in Fig. 11-19. The denser air does, however, increase the power required to drive the compressor or compressors; thus the engine will require more fuel to maintain the same engine speed. On a hot day, the density of the air decreases, thus reducing the mass of air entering the compressor and, consequently, the thrust of the engine for a given rpm. Because less power is required to drive the compressor, the fuel control system reduces the fuel flow to maintain a constant engine rotational speed or turbine entry temperature as appropriate; however, because of the decrease in air density, the thrust will be lower. At a temperature of 104°F (40°C), depending on the type of engine, a loss in thrust of up to 20 percent may occur. This means that some sort of thrust augmentation, such as water injection, may be required.

Effect of Water Injection

Since atmospheric temperatures above standard result in an appreciable loss of thrust or power, it is necessary to provide some means of **thrust augmentation** for nonafterburning engines during takeoff in hot weather. About 10 to 15 percent additional thrust can be gained by **injecting water**, or a mixture of water and alcohol, into the engine, either at the compressor air inlet or at some other point in the engine, such as the diffuser section or the burners.

In a reciprocating engine, during power augmentation brought about by means of water injection, the water acts primarily as a detonation suppressor and a cylinder-charge coolant. Higher takeoff power results because the engine can then operate at the best-power mixture without detonation. Gas turbines, however, have no detonation difficulties. When a liquid coolant is added, thrust or power augmentation is obtained principally by increasing the mass flow through the engine.

It would appear that once the mass of air is inside the engine, nothing can be done to change it. However, when water is injected, the mass of the water molecules is added to the mass of airflow. Even though a water molecule weighs less than an air molecule, whatever it weighs is added to the air already in the engine. Water has the effect of cooling the air mass inside the engine. However, the same pressure is maintained because water molecules are added to the air. Water molecules can be added as long as the constant pressure is maintained.

Water injection does two things directly. It cools the air mass and maintains the same pressure by adding molecules to the mass flow. Obviously, if there is little cooling effect, only a few molecules can be added to the mass flow. Thus, on a cold day, only a small increase in thrust can be obtained by water injection, but on a hot day a sizable thrust increase may be realized.

The effect will also vary considerably with the position of the injection. Water injected at the inlet will be affected more by outside temperatures than if it were injected somewhere

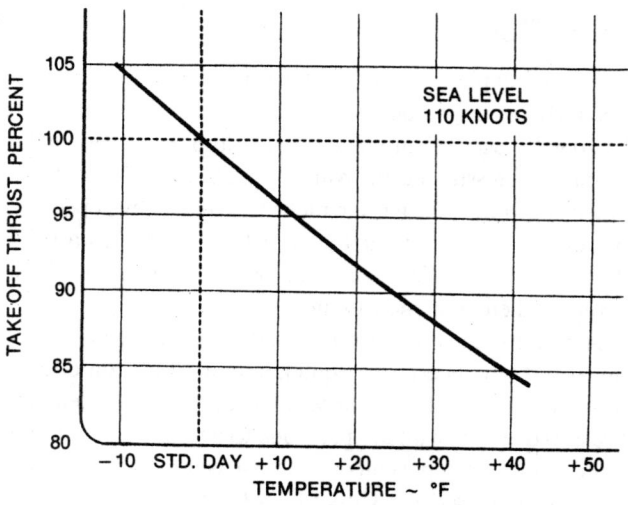

FIGURE 11-19   Relationship of temperature to thrust.

farther back in the engine where the heat generated by compression will counteract a cold outside temperature.

Not all the increase in thrust caused by water injection is due to the increase in mass flow. There is also a cycle effect. The increase in w_a causes a tendency to slow down the compressor. The fuel control system adds fuel to keep the compressor going at the same speed. The resulting increase in heat energy hits the turbines and the compressor is speeded up. When the compressor speeds up, it gives a greater weight of airflow. The end result is the realization of a greater thrust increase by increased w_a than is obtained by the addition of water molecules.

When pure water is used as the coolant and introduced into the compressor, a water-sensing line to the fuel control unit increases fuel supply to provide the added heat energy when water is injected. More heat energy introduced into the airflow will mean increased jet velocity (V_j), which means increased thrust.

Effect of Afterburning on Engine Thrust

Under takeoff conditions, the momentum drag of the airflow through the engine is negligible, so that the gross thrust can be considered to be equal to the net thrust. If afterburning is selected, an increase in takeoff thrust on the order of 30 percent is possible with the pure jet engine and considerably more with the bypass engine. This augmentation of basic thrust is of greater advantage for certain specific operating requirements. Under flight conditions, however, this advantage is even greater, since the momentum drag is the same with or without afterburning and, due to the ram effect, better utilization is made of every pound of air flowing through the engine.

EFFICIENCIES

The **efficiency** of any engine can be described as the output divided by the input. One of the main measures of turbine engine efficiency is the amount of thrust produced or generated, divided by the fuel consumption. This is called **thrust specific fuel consumption**, or tsfc. The tsfc is the amount of fuel required to produce 1 lb [0.004 45 kN] of thrust and can be calculated as follows:

$$\text{tsfc} = \frac{w_f}{F_n}$$

where w_f = fuel flow, lb/h [kg/h]
F_n = net thrust, lb [kg]

This leads to the conclusion that the more thrust obtained per pound of fuel, the more efficient the engine is.

Specific fuel consumption is made up of a number of other efficiencies. The two major factors affecting the tsfc are propulsive efficiency and cycle efficiency.

Propulsive Efficiency

Propulsive efficiency is the amount of thrust developed by the jet nozzle compared with the energy supplied to it in a usable form. In other words, the propulsive efficiency is the percentage of the total energy made available by the engine which is effective in propelling the engine. A comparison of the propulsive efficiencies of various jet engines is shown in Fig. 11-20.

Propulsive efficiency can also be expressed as:

$$\frac{\text{Work completed}}{\text{Work completed} + \text{work wasted in the exhaust}}$$

A simplified version of this formula for an unchoked engine is

$$\frac{2V}{V + V_j}$$

where V_j = jet velocity at propelling nozzle, ft/s [m/s]
V = aircraft speed, ft/s [m/s]

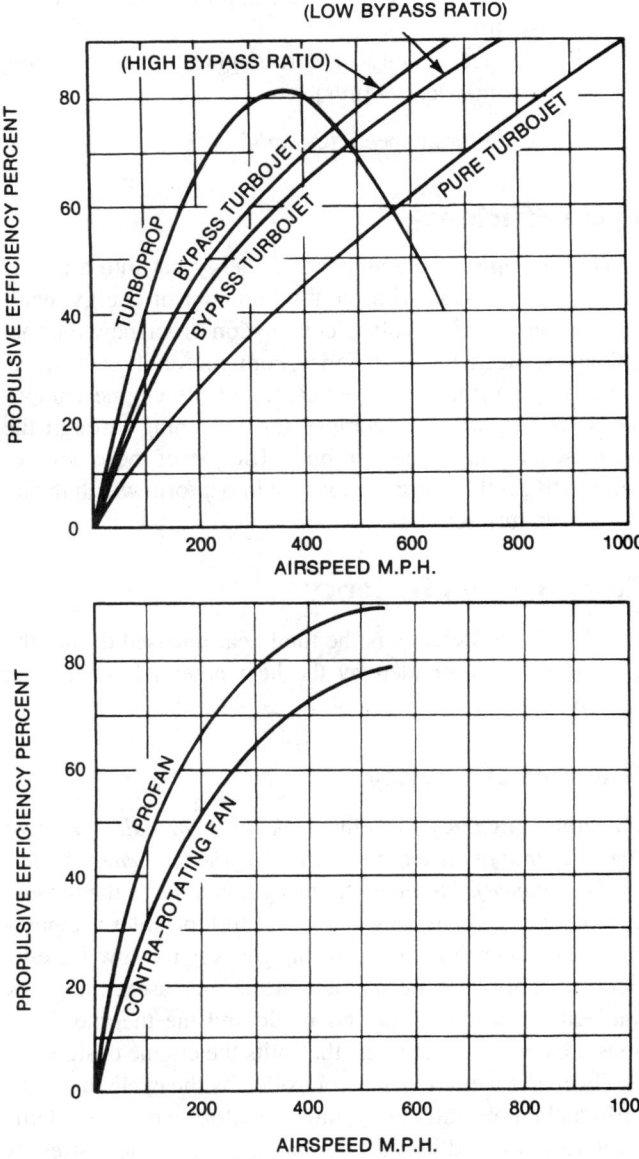

FIGURE 11-20 Comparative propulsive efficiencies. (*Rolls-Royce.*)

If an aircraft is traveling at a speed (V) of 400 mph [644 km/h] and its jet velocity (V_j) was 1150 mph [1851 km/h], the propulsive efficiency can be calculated as follows:

$$\frac{2 \times 400}{400 + 1150} = 52\%$$

The formula for calculating propulsive efficiency for a turbofan engine using separate exhaust nozzles is

$$\frac{w_{a_1} V(V_{j_1} - V) + w_{a_2} V(V_{j_2} - V)}{w_{a_1} V(V_{j_1} - V) + w_{a_2} V(V_{j_2} - V) + \frac{1}{2} w_{a_1}(V_{j_1} - V)^2 + \frac{1}{2} w_{a_2}(V_{j_2} - V)^2}$$

where w_{a_1} = mass of fan air passing through engine, lb/s [kg/s]

w_{a_2} = mass of core engine air passing through engine, lb/s [kg/s]

V_{j_1} = jet velocity of fan air at propelling nozzle, ft/s [m/s]

V_{j_2} = jet velocity of core engine air at propelling nozzle, ft/s [m/s]

V = aircraft speed, ft/s [m/s]

Cycle Efficiency

Cycle efficiency is the amount of energy put into a usable form in comparison with the total amount of energy available in the fuel. It involves combustion efficiency, thermal efficiency, mechanical efficiency, compressor efficiency, etc. It is, in effect, the overall efficiency of the engine components starting with the compressor and going through the combustion chamber and turbine. The job of these components is to get the energy in the fuel into a form which the jet nozzle can turn into thrust.

Combustion Efficiency

Combustion efficiency is the total heat released during the burning process, divided by the heat potential of the fuel burned.

Thermal Efficiency

Thermal efficiency is defined as *the heat value or heat energy output of the engine, divided by the heat energy input (fuel consumed)*. Thermal efficiency increases as the turbine inlet temperature increases. At low turbine inlet temperatures, the expansion energy of the gases is too low for efficient operation. As the temperature is increased, the gases (molecules) become more energetic and the thermal functions are performed at a rate that suits the engine design.

The thermal efficiency is controlled by the **cycle pressure ratio** and combustion temperature. Unfortunately, this temperature is limited by the thermal and mechanical stresses that can be tolerated by the turbine. The development of new

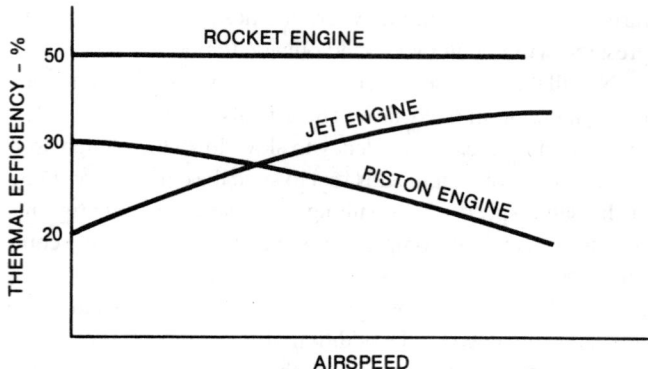

FIGURE 11-21   Ram effect on thermal efficiency.

materials and techniques to minimize these limitations is continually being pursued.

A turbine engine's thermal efficiency will tend to improve with airspeed due to the ram effect (see Fig. 11-21). Ram pressure, when multiplied across the compressor by the compressor ratio, can lead to improvements in w_a and the combustion chamber pressure, and thus increased thrust output, with little or no change in shaft energy input to the compressor. The average gas-turbine engine has a thermal efficiency under cruise conditions of 45 to 50 percent, whereas aircraft piston engines have 25 to 30 percent efficiency and rockets approximately 50 percent thermal efficiency.

Propeller Thrust Horsepower

Thrust horsepower can be considered the result of the engine and the propeller working together. If a propeller could be designed to be 100 percent efficient, the thrust and the bhp would be the same. However, the efficiency of the propeller varies with the engine speed, attitude, altitude, temperature, and airspeed. Thus, the ratio of the thrust horsepower and the bhp delivered to the propeller shaft will never be equal. For example, if an engine develops 1000 bhp, and it is used with a propeller having 85 percent efficiency, the thrust horsepower of that engine-propeller combination is 85 percent of 1000 or 850 thrust hp. Of the four types of horsepower discussed, it is the thrust horsepower that determines the performance of the engine-propeller combination.

TURBINE ENGINE

Thermal Efficiency Calculations

Any study of engines and power involves consideration of heat as the source of power. The heat produced by the burning of gasoline in the cylinders causes a rapid expansion of the gases in the cylinder, and this, in turn, moves the pistons and creates mechanical energy. It has long been known that mechanical work can be converted into heat and that a given amount of heat contains the energy equivalent of a certain amount of mechanical work. Heat and work are theoretically interchangeable and bear a fixed relation to each other.

Heat can therefore be measured in work units (for example, ft-lb) as well as in heat units. The British thermal unit (Btu) of heat is the quantity of heat required to raise the temperature of 1 pound of water by 1°F. It is equivalent to 778 ft-lb of mechanical work. A pound of petroleum fuel, when burned with enough air to consume it completely, gives up about 20 000 Btu, the equivalent of 15 560 000 ft-lb of mechanical work. These quantities express the heat energy of the fuel in heat and work units, respectively. A high thermal efficiency also means low specific fuel consumption and, therefore, less fuel for a flight of a given distance at a given power. Thermal efficiency is basically a comparison of how well the energy in the fuel is converted to direct power.

Thermal efficiency is expressed as a ratio of the net work produced by the engine to the fuel energy input.

$$TE = \frac{hp\ output\ of\ engine}{hp\ value\ of\ fuel\ consumed}$$

Example

A turboshaft engine is producing 750 shaft horsepower and is consuming 325 lb/h of fuel containing 18 730 Btu/lb. What is the engine's thermal efficiency?

Solution to finding thermal efficiency:

The fuel flow is 320 lb/h. This needs to be converted to lb/min by dividing by 60.

$$\frac{320\ lb/hr}{60} = 5.333\ lb/min$$

Using the heat value of the fuel in British thermal units (Btu's) and multiplying it times the lb/min of fuel flow, the Btu's/min can be obtained. Since 1 Btu is equal to 778 ft/lb, multiplying the Btu's/min by 778 converts to ft lb/min. By dividing this answer by 33 000 (amount of ft lb/min for 1 horsepower) the amount of fuel horsepower can be calculated. Fuel horsepower is the amount of horsepower the fuel would be capable of producing if all the fuel flowing to the engine was converted to horsepower (100% thermal efficiency).

Heat value = 18 730 Btu's/lb × 5.333 lb/min
Heat value = 99 877.09 Btu's/min (since 1 Btu = 778 ft lb)

$$Fuel\ horsepower = \frac{99\ 877.09 \times 778}{33\ 000\ ft\ lb/min/1\ horsepower}$$

Fuel horsepower = 2385.182 hp

The value of 800 shaft horsepower being developed by the engine represents the actual horsepower being developed by the engine. By dividing the fuel horsepower into the actual horsepower developed by the engine, the thermal efficiency can be calculated.

$$TE = \frac{800}{2385.182} = 0.33 \times 100 = 33\%$$

Example

A turbofan engine/produces 10 000 lb of net thrust in flight at 500 mph. The fuel being consumed is 5500 lb/h. What is the thermal efficiency if the fuel contains 18 730 Btu's?

Solution to thermal efficiency:

$$Fuel\ flow = \frac{5500\ lb/h}{60} = 91.66\ lb/min$$

Heat value = 18 730 × 91.66 = 1 716 791.8 Btu's/min

$$Fuel\ hp = \frac{1\ 716\ 791.8 \times 778}{33\ 000}$$

Fuel hp = 40 474.67

Since the engine is producing thrust instead of horsepower, the thrust must be converted to horsepower. By taking the thrust and multiplying it by the speed of the aircraft and dividing by 375, the thrust horsepower can be obtained. The value of 375 is obtained from 33 000 ft lb multiplied by 60 and divided by 5280 ft (feet in a statute mile).

$$thp = \frac{10\ 000 \times 500}{375}$$

thp = 13 333.33

$$TE = \frac{13\ 333.33}{46\ 474.67} = 0.32 \times 100 = 32\%$$

REVIEW QUESTIONS

1. What factors affect the amount of acceleration of an object?
2. Describe the basic operation of a gas-turbine engine.
3. What are the three major sections of a gas-turbine engine?
4. Aircraft gas-turbine engines are generally classified into what four types?
5. Describe the basic operation of a straight turbojet engine.
6. What are the two types of *turbofan engines*?
7. Describe the basic operation of a *turboprop engine*.
8. Describe the basic operation of a *turboshaft engine*.
9. When a conversion in the engine airflow from velocity (dynamic pressure) to static pressure is required, what type or shape of duct is used?
10. What is the relationship between turbine engine speed and thrust?

11. How does humidity affect turbine engine performance?

12. What is the effect on temperature of turbine engine performance?

13. What method may be used to increase engine thrust on high-temperature days?

14. Define the term *thrust specific fuel consumption*.

15. Define the term *propulsive efficiency*.

16. Define *thermal efficiency*.

17. List Newton's laws of motion.

18. What is the first law of thermodynamics?

19. What basic constant pressure cycle do turbine engines operate under?

20. At 600 mph airspeed, what type of engine has the highest propulsive efficiency?

Principal Parts of a Gas-Turbine Engine, Construction, and Nomenclature

12

THE INLET

The inlet of a turbine engine is typically located at the front of the compressor. It is not really a section of the engine defined by any one particular part. The inlet is formed by the structural support parts located forward of the compressor and has the purpose of admitting air to the forward end of the compressor. The opening of the inlet is usually of fixed size, but may be variable depending on the design of the compressor used in the particular engine.

The major engine part that may be located in this area of the engine is the compressor front frame. The front frame serves as the structural support member for the forward end of the engine and houses the bearing for the forward end of the compressor rotor. The inlet must have a clean aerodynamic design to ensure a smooth, evenly distributed airflow into the engine.

The air entrance is designed to conduct incoming air to the compressor with a minimum energy loss resulting from drag or ram pressure loss; that is, the flow of air into the compressor should be free of turbulence to achieve maximum operating efficiency. Proper inlet design contributes to aircraft performance by increasing the ratio of compressor discharge pressure to duct inlet pressure.

This is also referred to as the compressor pressure ratio. This ratio is the outlet pressure divided by the inlet pressure. The amount of air passing through the engine is dependent upon three factors:

1. The compressor speed (rpm)
2. The forward speed of the aircraft
3. The density of the ambient (surrounding) air

Basically turbine inlets are dictated by the type of gas-turbine engine. A high-bypass turbofan engine's inlet will be completely different than a turboprop or turboshaft. Large gas-turbine-powered aircraft are almost always a turbofan engine. The inlet on this type of engine is bolted to the front (A flange) of the engine. These engines are mounted on the

FIGURE 12-1 Turbofan engine air intake.

wings, or nacelles on the aft fuselage and a few are in the vertical fin. A typical turbofan inlet can be seen in Fig. 12-1. Since on most modern turbofan engines the huge fan is the first thing the incoming air comes into contact with icing protection must be provided. This prevents chunks of ice from forming on the leading edge of the inlet, breaking loose, and damaging the fan. Warm air is bled from the engine's compressor and is ducted through the inlet to prevent ice from forming. If inlet guide vanes are used to straighten the airflow they will also have anti-icing air flowing through them.

The inlet area can be controlled by a set of vanes known as the **inlet guide vanes**. The guide vanes in the axial-flow turbojet engine provide a change in direction of airflow so that air is directed on the first stage of the compressor at the proper angle. Controlling the amount of air flowing into the compressor in the axial-flow engine is necessary under some operating conditions, because at low engine speed the forward stages of the compressor could deliver more air than can be effectively handled by the rear stages of the compressor. When this condition exists, the engine may encounter **compressor stall**. To prevent this situation, the angles of the inlet guide vanes and some of the first stages of the stator vanes are varied to reduce the amount of air flowing through the engine. A less efficient way to reduce the amount of air reaching the rear stages is to bleed off some of the excess air partway through the compressor.

Air Inlet Icing

Axial compressor engines are seriously affected by the formation of ice on the compressor inlet guide vanes. All turbine engines equipped with nonretractable air inlet screens are very susceptible to icing. Ice forms on the guide vanes or inlet screen and restricts the flow of inlet air. This is indicated by a loss of thrust and a rapid rise in exhaust gas temperature (EGT). As the airflow decreases, the F/A ratio increases, which in turn raises the turbine inlet temperature. The fuel control attempts to correct any loss in engine rpm by adding more fuel, which aggravates the condition.

Centrifugal compressor engines, whether equipped with retractable screens or having no screens at all, are relatively free from the danger of ice collecting at the compressor inlet.

The inlet guide vanes can be heated to prevent the formation of ice. The inlet guide vanes and the inlet struts of axial compressor engines are usually hollow. Hot, high-pressure air is bled from the rear of the engine compressor and is ducted through an anti-icing system control valve to the hollow sections of the inlet struts and guide vanes. The heat provided prevents the adhesion of ice. Because such a system may not melt ice once it has formed, icing conditions should be anticipated in advance. Once ice has formed on the inlet struts or vanes, anti-icing air may cause large chunks of ice to enter the compressor, where they may damage the blades.

Anti-icing systems cause some reduction in thrust and are used only when needed. The engine specifications usually define the maximum allowable extraction of compressor bleed air at any one time. With all the anti-icing systems in use at one time on certain aircraft, the power loss can be as much as 30 percent.

TYPES OF COMPRESSORS

Since the more that air can be compressed, the more mass is available to do work when the air is heated and expanded, the gas-turbine engine must have some means of compressing air into the available space. There are three types of compressors with respect to airflow: the centrifugal type, combination of centrifugal and axial, and the axial type.

Even though each type of compressor has advantages and disadvantages, they all have a place in the type and size of engine that they are used with. The centrifugal-flow compressor's advantages are

1. High pressure rise per stage
2. Good efficiencies over wide rotational speed range

3. Simplicity of manufacture, thus low cost
4. Low weight
5. Low starting power requirements

The centrifugal-flow compressor's disadvantages are

1. Large frontal area for given airflow
2. More than two stages are not practical because of losses in turns between stages

The axial-flow compressor's advantages are

1. High peak efficiencies
2. Small frontal area for given airflow
3. Straight-through flow, allowing high ram efficiency
4. Increased pressure rise by increasing number of stages with negligible losses

The axial-flow compressor's disadvantages are

1. Good efficiencies over only narrow rotational speed range
2. Difficulty of manufacture and high cost
3. Relatively high weight
4. High starting power requirements (overcome somewhat by split compressors)

COMPRESSOR PRESSURE RATIO

An example of compressor pressure ratio is the ratio of the pressure of air at the compressor discharge to the compressor inlet air pressure. Normally the compressor inlet pressure will be approximately ambient pressure, or around 14.7 psia at sea level [101.04 kPa]. By knowing the compressor discharge pressure (280 psia), the pressure ratio can be calculated by dividing the discharge pressure by the inlet pressure (14.7 psia in this example).

$$CPR = \frac{280}{14.7} = 19{:}1$$

CENTRIFUGAL-FLOW COMPRESSOR

The centrifugal-flow compressor consists of three main parts: an impeller, a diffuser, and a compressor manifold (see Fig. 12-2). The term **centrifugal** means that the air is compressed by centrifugal force.

Centrifugal compressors operate by taking in outside air near the hub and rotating it by means of an impeller. The impeller, which is usually an aluminum-alloy forging, guides the air toward the outside of the compressor, building up the air velocity by means of high rotational speed of the impeller. The air then enters the **diffuser section**. The diffuser converts the kinetic energy of the air leaving the compressor to potential energy (pressure) by exchanging velocity for pressure. An advantage of the centrifugal-flow compressor is its high pressure rise per stage.

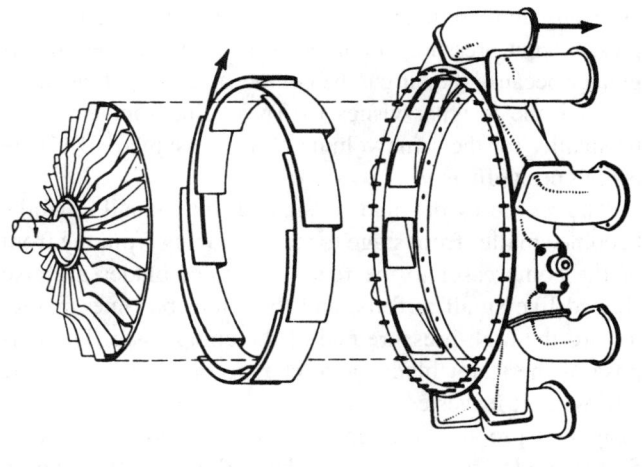

IMPELLER   DIFFUSER   COMPRESSOR MANIFOLD

FIGURE 12-2   Main parts of a compressor.

FIGURE 12-4   Cross-sectional drawing of a centrifugal turbo-prop engine.

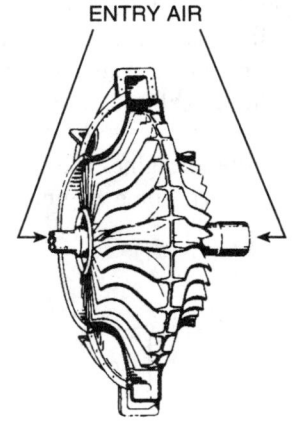

FIGURE 12-3   Drawing of a double-entry centrifugal-flow compressor.

A centrifugal compressor is either a double-entry type (Fig. 12-3) or a single-entry type (Fig. 12-2). In Fig. 12-3, a double-entry type compressor (double sided) is shown, with air inlets on both sides, front and rear. Air reaches the rear inlet of the compressor by flowing between the compressor outlet adapters.

Although the centrifugal compressor is not as expensive to manufacture as the axial-flow compressor, its lower efficiency eliminates the advantages of lower cost, except for some small turboprop engines. Among the successful

centrifugal engines manufactured in the United States was the TPE-331 engine. A cross-sectional drawing of this basic engine configuration is shown in Fig. 12-4.

AXIAL-FLOW COMPRESSOR

In an **axial-flow** jet engine, the air flows axially—that is, in a relatively straight path in line with the axis of the engine, as shown in Fig. 12-5. The axial-flow compressor consists of two elements: a rotating member called the rotor, and the stator, which consists of rows of stationary blades. The stator vanes are airfoil sections that are mounted in stationary casings. The compressor rotor and one-half of the stator case for an axial-flow turbojet engine are shown in Fig. 12-6. The rotor comprises the rotating components and castings that support the rotor blades which are attached to the rotor. The rotor is attached to a shaft which is driven by the turbine or turbine stages that drive this compressor. The rotor blades are attached to the rotor and are of an airfoil shape which maintains an axial air flow throughout the compressor. Methods of blade attachment are shown in Fig. 12-7.

The principle of operation of the axial-flow turbojet engine is the same as that of the centrifugal-flow engine; however, the axial-flow engine has a number of advantages: (1) The air flows in an almost straight path through the engine, and therefore less energy is lost as a result of the

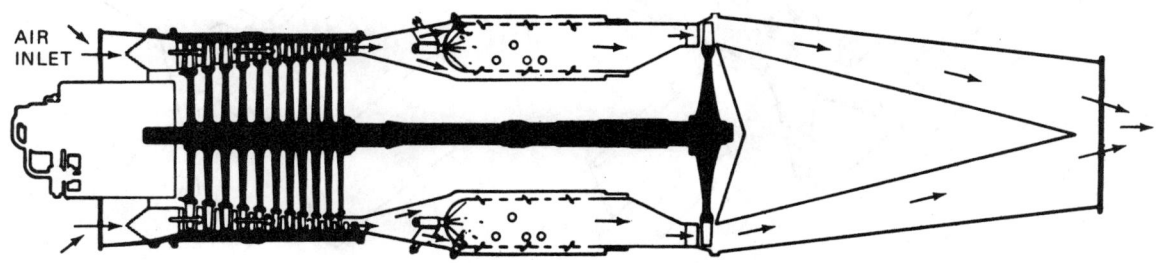

FIGURE 12-5   Drawing of an axial-flow turbojet engine.

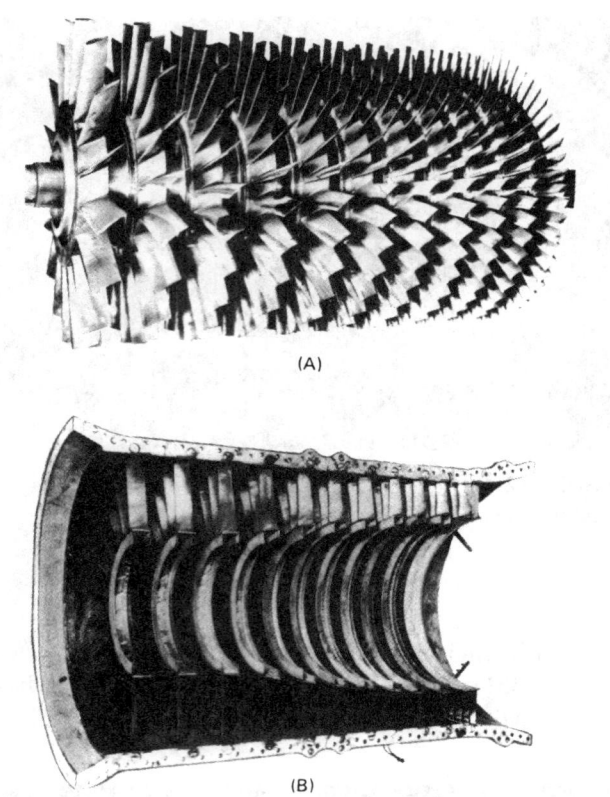

(A)

(B)

FIGURE 12-6 Rotor (A) and stator (B) of an axial-flow compressor.

air changing direction. (2) The pressure ratio (ratio of compressor inlet pressure to compressor discharge pressure) is greater because the air can be compressed through as many stages as the designer wishes. (3) The engine frontal area can be smaller for the same volume of air consumed. (4) There is high peak efficiency.

The compressor blades, shaped like small airfoils, become smaller from stage to stage, moving from the front of the compressor to the rear. The stator blades are also shaped like small airfoils, and they, too, become smaller toward the high-pressure end of the compressor. The purpose of the stator blades is to change the direction of the airflow as it leaves each stage of the compressor rotor and to give it proper direction for entry into the next stage. Stator blades also eliminate the turbulence that would otherwise occur between the compressor blades. The ends of the stator blades are fitted with shrouds to prevent the loss of air from stage to stage and to the interior of the compressor rotor.

During the operation of the compressor, the air pressure increases as it passes each stage, and at the outlet into the diffuser it reaches a value several times that of the atmosphere, the actual pressure being over 70 psi [482.65 kPa].

Sometimes gas-turbine engines use more than one axial-flow compressor; in fact, some engines use up to three

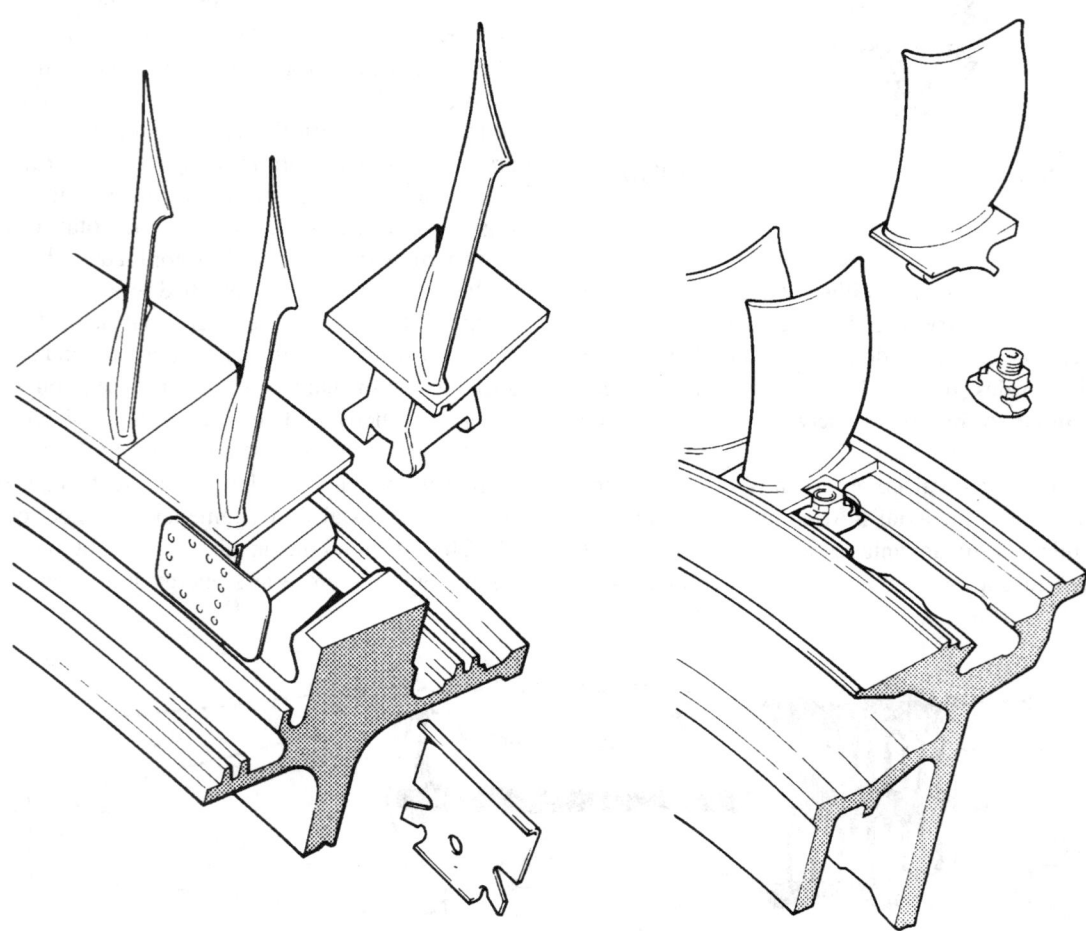

FIGURE 12-7 Methods of securing compressor blades to disk. (*Rolls-Royce.*)

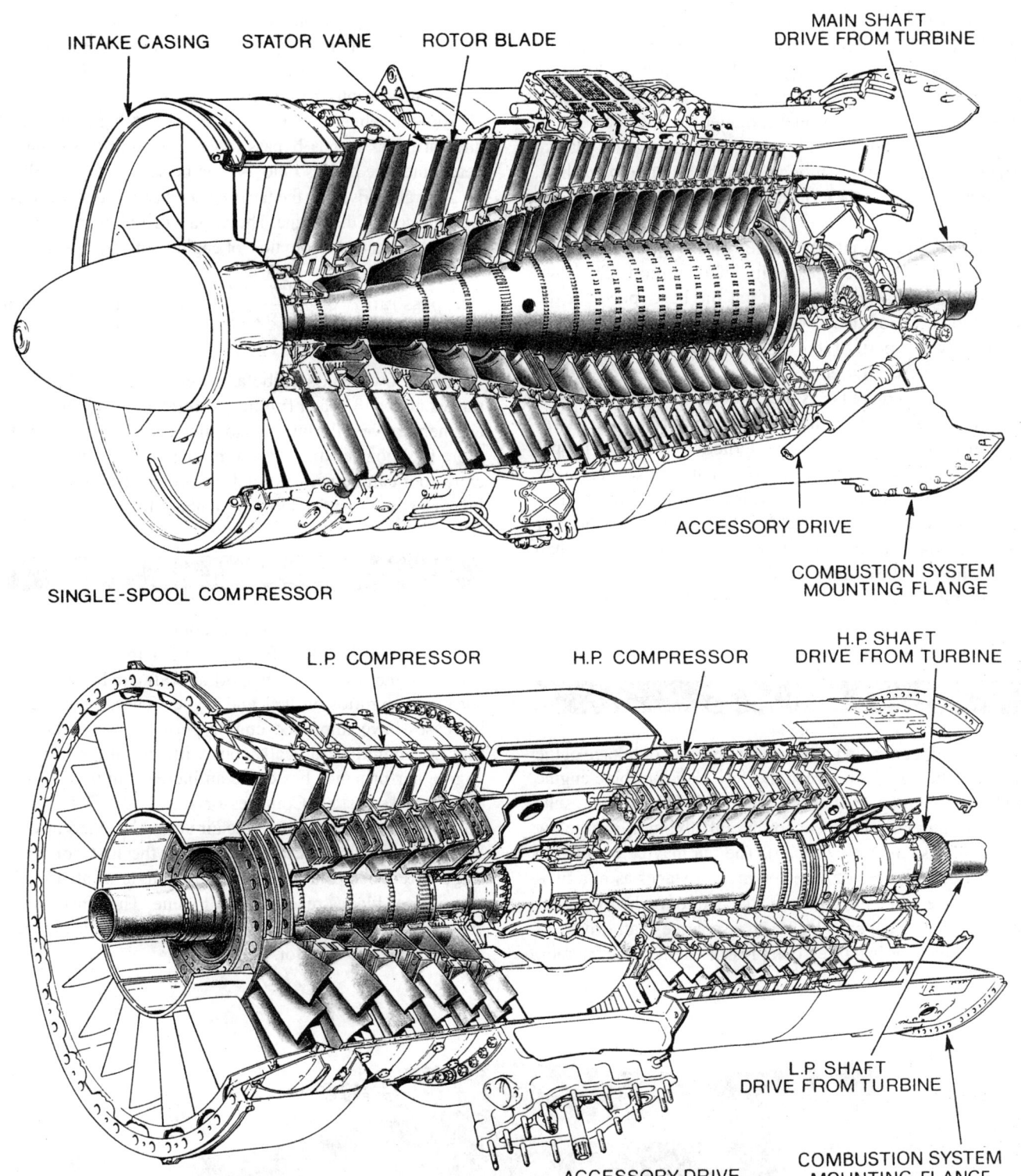

INTAKE CASING   STATOR VANE   ROTOR BLADE   MAIN SHAFT DRIVE FROM TURBINE

ACCESSORY DRIVE

COMBUSTION SYSTEM MOUNTING FLANGE

SINGLE-SPOOL COMPRESSOR

L.P. COMPRESSOR   H.P. COMPRESSOR   H.P. SHAFT DRIVE FROM TURBINE

L.P. SHAFT DRIVE FROM TURBINE

COMBUSTION SYSTEM MOUNTING FLANGE

ACCESSORY DRIVE

TWIN-SPOOL COMPRESSOR

FIGURE 12-8 Typical axial-flow compressors. (*Rolls-Royce.*)

separate compressors. The arrangement of a dual-axial (twin-spool) compressor is shown in Fig. 12-8. This compressor design makes it possible to obtain extremely high pressure ratios with reduced danger of compressor stall because the low-pressure compressor is free to operate at its best speed and the high-pressure compressor rotor is speed-regulated by the fuel control unit.

MULTIPLE-COMPRESSOR AXIAL-FLOW ENGINES

A dual-compressor turbine engine utilizes two separate compressors, each with its own driving turbine. This type of engine is also called a "twin-spool" or "split-compressor" engine.

The construction of the dual-compressor engine is shown in Fig. 12-8. The forward compressor section is called the *low-pressure compressor* (N_1) and the rear section the *high-pressure compressor* (N_2). The low-pressure compressor is driven by a two-stage turbine mounted on the rear end of the inner shaft, and the high-pressure compressor is driven by a single-stage turbine mounted on the outer coaxial shaft. The high-pressure rotor turns at a higher speed than the low-pressure rotor.

One of the principal advantages of the split-compressor arrangement is greater flexibility of operation. The low-pressure compressor can operate at the best speed for the accommodation of the low-pressure, low-temperature air at the forward part of the engine. During high-altitude operation where air density is low, the speed of the N_2 compressor will increase as the compressor load decreases. This makes N_1 in effect a supercharger for N_2. The high-pressure compressor is speed-governed to operate at proper speeds for the most efficient performance in compressing the high-temperature, high-pressure air toward the rear of the compressor section. The use of the dual compressor makes it possible to attain pressure ratios of more than 20:1, whereas the single axial-flow compressor produces pressure ratios of only 6:1 or 7:1 unless variable stator vanes are employed.

FAN BYPASS RATIO

The intake air generally is compressed by only one stage of the fan before being split between the core or gas-turbine engine and the bypass duct, as shown in Fig. 12-9. This design results in the optimum arrangement for aircraft flying at just below the speed of sound (.8–.9 Mach). The fan may be coupled to the front of a number of core compressor stages as in a two-compressor engine, or it may be attached to the low-pressure compressor. The fan can also be on a separate shaft driven by its own turbine, as in a three-compressor engine. The fan ratio is the amount of fan air (mass airflow) that flows through the fan duct compared to the airflow that flows through the core of the engine. The fan ratio can be calculated by dividing core-engine airflow into the fan-mass airflow.

Turbofan engines can be low bypass or high bypass. The amount of air that is bypass around the core of the engines determines the bypass ratio. As can be seen in Fig. 11-9G the air generally driven by the fan does not pass through the internal working core of the engine. The amount of airflow in lb/sec from the fan bypass to the core flow of engine is the bypass ratio.

$$\text{Bypass ratio} = \frac{100 \text{ lb/sec flow fan}}{20 \text{ lb sec flow core}} = 5\text{:}1 \text{ bypass ratio}$$

Some low-bypass turbofan engines are used in speed ranges above .8 Mach (military aircraft). These engines use augmenters or afterburners to increase thrust. By adding more fuel nozzles and a flame holder in the exhaust system extra fuel can be sprayed and burned which can give large increases in thrust for short amounts of time.

COMPRESSOR STALL

During the past several years, compressor ratios for gas turbines have increased from about 5:1 to more than 18:1. Some engines on large transport aircraft can have compressor pressure ratios of 30:1 and higher. These pressure ratio increases have improved engine performance and specific fuel consumption radically, but with this improvement in engine performance has come an increase in the likelihood of compressor stall. Compressor stall is the failure of the compressor blades to move the air at the designed flow rate. When this occurs, the air velocity in the first compressor stage is reduced to a level where the angle of attack of the compressor blades reaches a stall value. This unstable condition is often caused, in part, by piling up of air in the rear stages of the compressor.

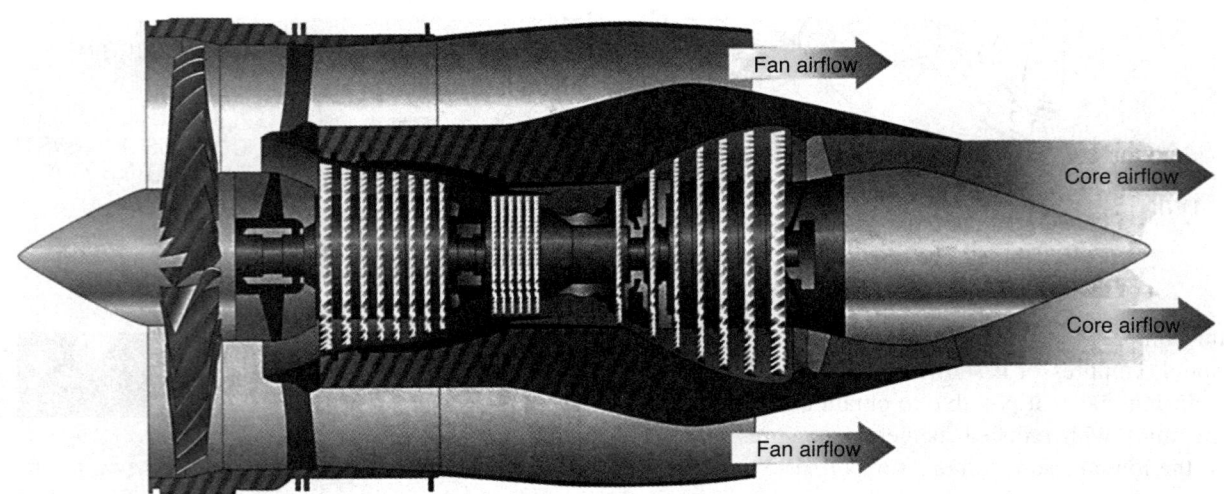

FIGURE 12-9 Core and fan flow.

Even though compressor rotor blades do not have variable angles, the effective angle of attack does not remain the same under all conditions. Compressor stall occurs most frequently whenever there is unusually high compressor speed and a low air-inlet velocity. Figure 12-10 shows how the effective angle of attack is changed by a combination of decreasing inlet air velocity and unchanged compressor speed. When the effective angle of attack increases because of the same high compressor speed together with a lower inlet velocity, the angle of attack reaches a stall condition.

Gas-turbine engine compressors are designed with margins adequate to prevent compressor stall from occurring under normal conditions, as shown in Fig. 12-11.

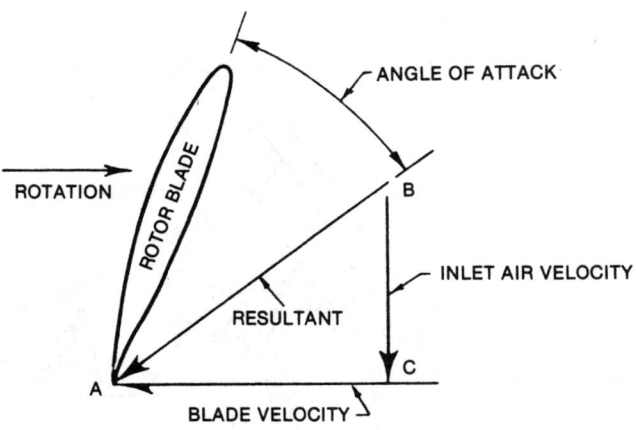

HIGH INLET AIR VELOCITY

COMPRESSOR AIRFLOW AND STALL CONTROL

Where high-pressure ratios on a single shaft are required, it becomes necessary to introduce airflow control into the compressor design. This may take the form of **variable inlet guide vanes** for the first stage, plus a number of stages incorporating **variable stator vanes**, as illustrated in Fig. 12-12. As the compressor speed is reduced from its design value, these stator vanes are progressively closed in order to maintain an acceptable air angle for the following rotor blades.

The variable vanes are automatically regulated in pitch angle by means of the fuel control unit. The regulating factors are compressor inlet temperature and engine speed. The effect of the variable vanes is to provide a means for controlling the direction of compressor interstage airflow, thus ensuring a correct angle of attack for the compressor blades and reducing the possibility of compressor stall.

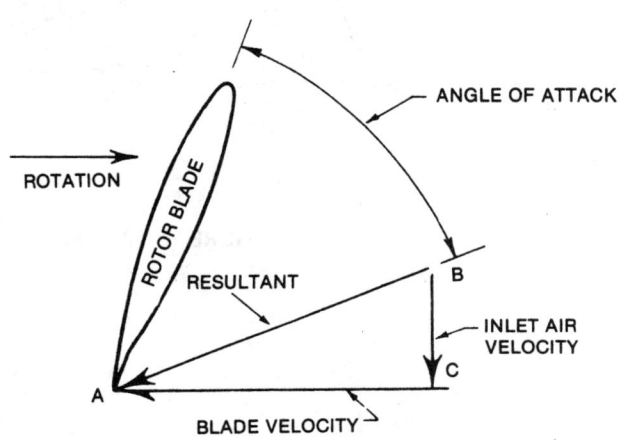

LOW INLET AIR VELOCITY

FIGURE 12-10 Compressor stall.

AIR-BLEED AND INTERNAL AIR SUPPLY SYSTEMS

Compressed air from the compressor section of the gas-turbine engine is used for a number of purposes. Compression of the air as it moves through the compressor causes a substantial rise in temperature. For example, air at the last stage of the compressor may reach a temperature of over 650°F [343.33°C] as a result of compression. This heated air is routed through the compressor inlet struts to prevent icing, and it is also used for various other heating tasks, such as operation of the fuel heater, aircraft heating, thermal anti-icing, etc.

Some engines are provided with automatic air-bleed valves which operate during engine starting or low-rpm conditions to prevent air from piling up at the high-pressure end of the compressor and "choking" (stalling) the engine. This permits easier starting and accelerating without the danger of compressor stall.

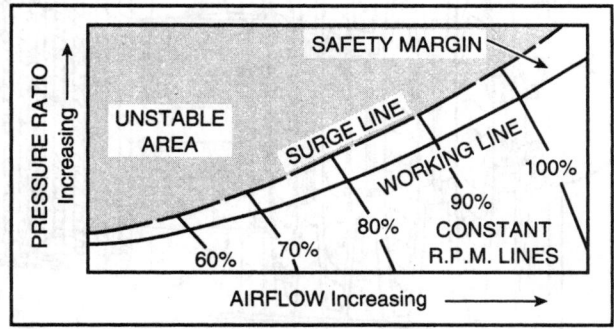

FIGURE 12-11 Limits of stable airflow.

Compressor air (see Fig. 12-13) is also utilized within the engine to provide cooling for the engine's internal hot section components, turbine wheel, and turbine inlet guide vanes. These vanes are hollow to provide passages for the cooling air, which is carried through the engine from the compressor to the area surrounding the nozzle diaphragm. Even though

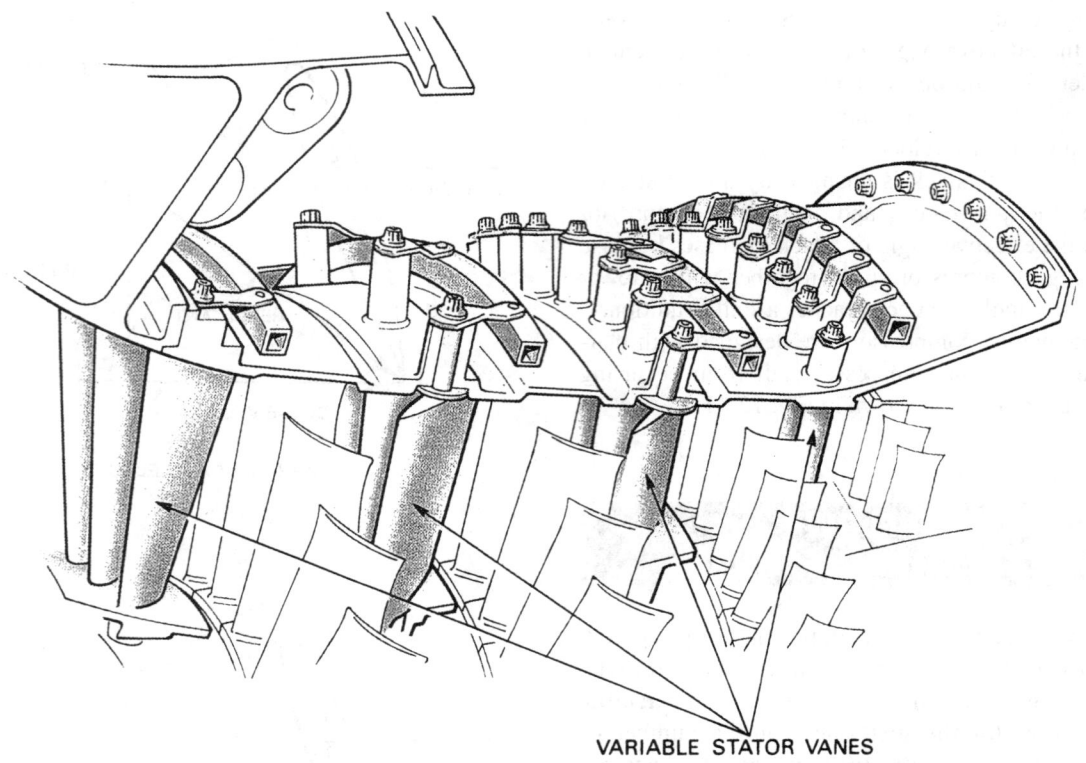

VARIABLE STATOR VANES

FIGURE 12-12 Typical variable stator vanes. (*Rolls-Royce.*)

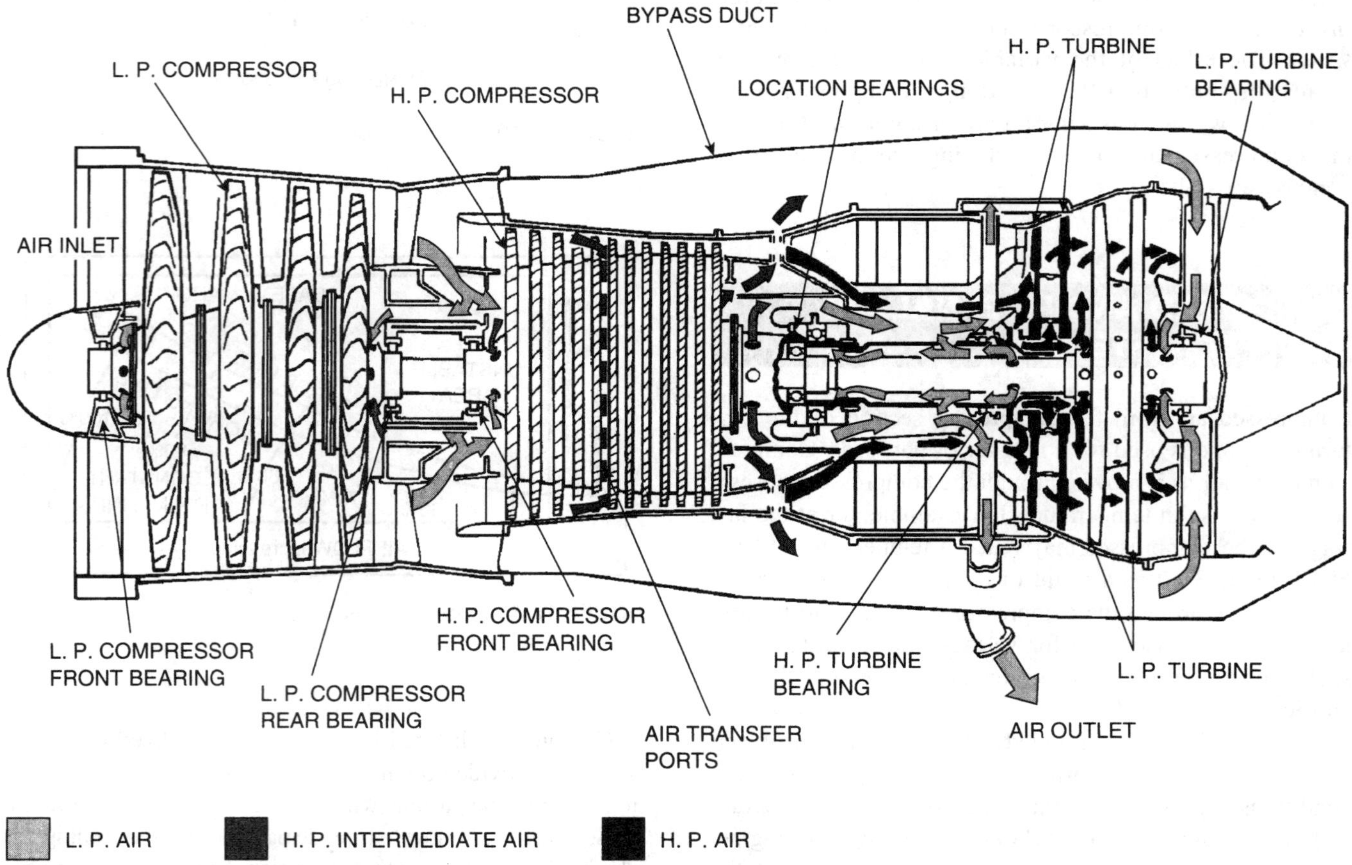

BYPASS DUCT

L. P. COMPRESSOR

H. P. COMPRESSOR

LOCATION BEARINGS

H. P. TURBINE

L. P. TURBINE BEARING

AIR INLET

L. P. COMPRESSOR FRONT BEARING

L. P. COMPRESSOR REAR BEARING

H. P. COMPRESSOR FRONT BEARING

AIR TRANSFER PORTS

H. P. TURBINE BEARING

AIR OUTLET

L. P. TURBINE

L. P. AIR   H. P. INTERMEDIATE AIR   H. P. AIR

FIGURE 12-13 General internal airflow pattern. (*Rolls-Royce.*)

the compressed air is heated by compression well above its initial temperature, it is still much cooler than the burning exhaust gases and can therefore provide cooling.

THE DIFFUSER

The diffuser for a typical gas-turbine engine is that portion of the air passage between the compressor and the combustion chamber or chambers. The purpose of the diffuser is to reduce the velocity of the air and prepare it for entry into the combustion area. As the velocity of the air decreases, its static pressure increases in accordance with Bernoulli's law. As the static pressure increases, the ram pressure decreases. The diffuser is the point of highest pressure within the engine.

COMBUSTION CHAMBERS

The combustion section of a turbojet engine may consist of individual combustion chambers ("cans"), an annular chamber which surrounds the turbine shaft, or a combination consisting of individual cans within an annular chamber. The latter type of combustor is called the **can-annular** type or simply the **cannular** type.

A typical can-type combustor, shown in Fig. 12-14, consists of an outer shell and a removable liner with openings to permit compressor discharge air to enter from the outer chamber. Approximately 25 percent of the air that passes through the combustion section is actually used for combustion, the remaining air being used for cooling. Located at the front end of the combustion chamber is a fuel nozzle through which fuel is sprayed into the inner liner. The flame burns in the center of the inner liner and is prevented from burning the liner by a blanket of excess air which enters through holes in the liner and surrounds the flame, as illustrated in

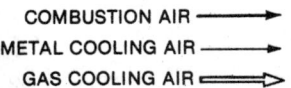
COMBUSTION AIR ⟶
METAL COOLING AIR ⟶
GAS COOLING AIR ⟹

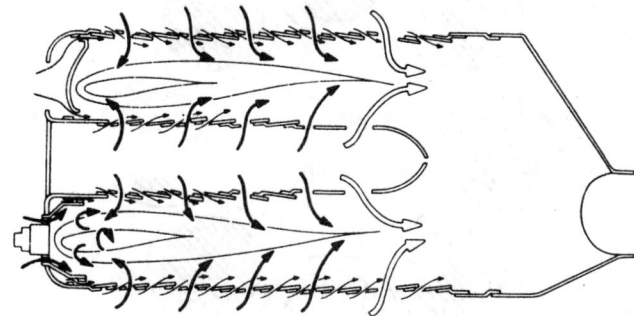

FIGURE 12-15   Airflow in a combustion liner.

Fig. 12-15. All burning is completed before the gases leave the chamber.

The high-bypass turbofan engines mentioned previously employ annular combustion chambers. These chambers have proven efficient and effective in producing smoke-free exhaust. The general configuration of the combustion chamber for the Pratt & Whitney JT9D engine is shown in Fig. 12-16. This is a two-piece assembly consisting of an inner and an outer liner. At the front are 20 fuel nozzle openings with **swirl vanes** to help vaporize the fuel. Two of

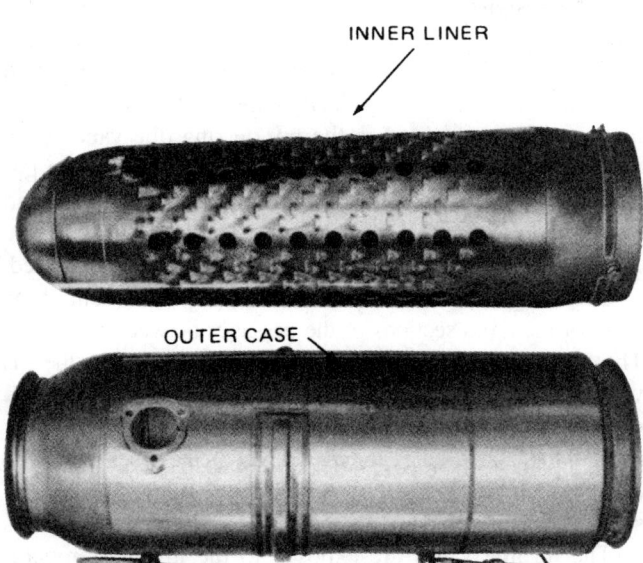

INNER LINER

OUTER CASE

FIGURE 12-14   Single can-type combustor.

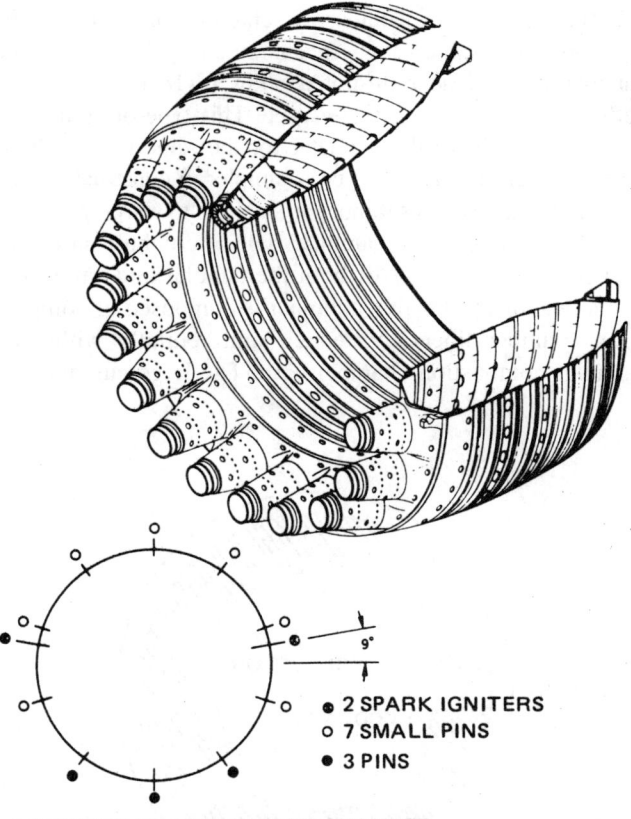

9°

● 2 SPARK IGNITERS
○ 7 SMALL PINS
● 3 PINS

LOCATION OF COMBUSTION–CHAMBER RETAINING PINS

FIGURE 12-16   Combustion chamber for the JT9D turbofan engine. (*Pratt & Whitney Canada.*)

Combustion Chambers   **333**

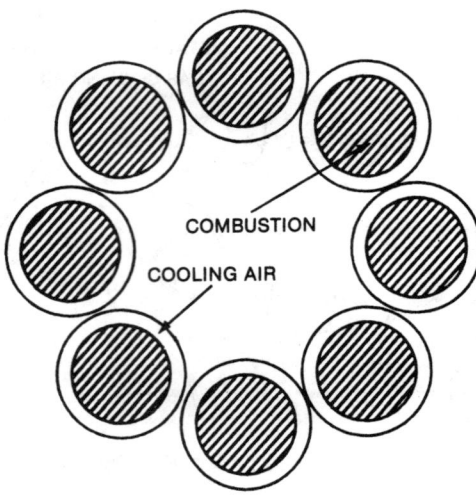

FIGURE 12-17   Can-type combustion chamber.

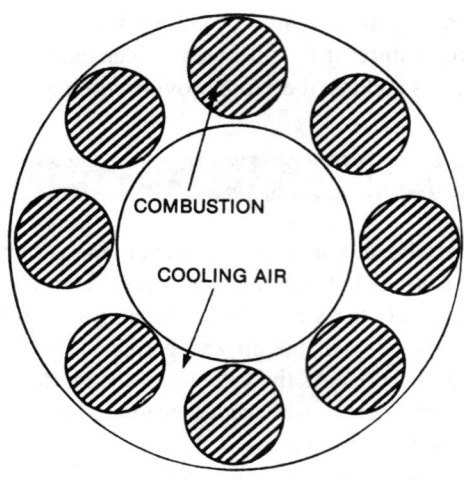

FIGURE 12-19   Can-annular-type combustion chamber.

the openings, on opposite sides of the combustion chamber, are designed to hold the igniter plugs.

One of the more important advantages of the can-type combustion chamber is that the liner surface has a large degree of curvature which results in high resistance to warpage. The main disadvantage of the can-type chamber illustrated in Fig. 12-17 is that it does not efficiently utilize the available space. Another disadvantage is found in the large area of metal required to enclose the required volume of gas flow.

The annular-type combustion chamber has some desirable advantages, including efficient air and gas handling. The annular-type chamber illustrated in Fig. 12-18 makes the most efficient use of the available space. This type of combustor requires about half the diameter for the same mass airflow. Even though the annular type is simpler in construction, the lower curvature makes it more susceptible to warping.

The can-annular or cannular type of combustion chamber has characteristics of both the annular and can types. This type of combustion chamber is composed of combustion chamber liners located circumferentially within an annular combustion chamber case. The large curvature of

liner surface is retained, thereby maintaining a high degree of resistance to warpage. Each liner has its own fuel nozzles. The space available is well utilized, although not to the same high degree as in the annular type. Individual liners tend to even out the air velocity distribution into the burner. The can-annular combustion chamber operates at a high pressure level, aiding efficient combustion at reduced power and high altitudes. Figure 12-19 illustrates the arrangement of the can-annular combustion chamber.

TURBINE NOZZLE DIAPHRAGM

The **turbine nozzle diaphragm** (turbine inlet guide vanes) is a series of airfoil-shaped vanes arranged in a ring at the rear of the combustion section of a gas-turbine engine. Its function is to control the speed, direction, and pressure of the hot gases as they enter the turbine. A nozzle diaphragm is shown in Fig. 12-20. The vanes of the nozzle diaphragm must be designed to provide the most effective gas flow for the particular turbine used in the engine.

The vanes in a turbine nozzle diaphragm are of airfoil shape to control the high-velocity gases in the most effective manner. When mounted in the nozzle ring, the vanes form convergent passages which change the direction of the gas flow, increase the gas velocity, reduce the gas pressure, and reduce the temperature of the gases. The heat and pressure energy of the gases is reduced as velocity energy is increased.

The total outlet area of the turbine nozzle is the sum of the areas of the cross sections of the passages between the vanes. The outlet area is less than the inlet area of the nozzle; the gas velocity is thus greater at the outlet than at the inlet. Note dimensions A, B, C, and D in Fig. 12-21, which shows the arrangement of the nozzle vanes and their effect on the gases. Note that the direction of the gas flow is changed to allow the gases to strike the turbine blades at the most effective angle.

The turbine vanes are exposed to the highest temperatures in the engine. Even though the gases are at a higher temperature during the fuel-burning process, the combustion

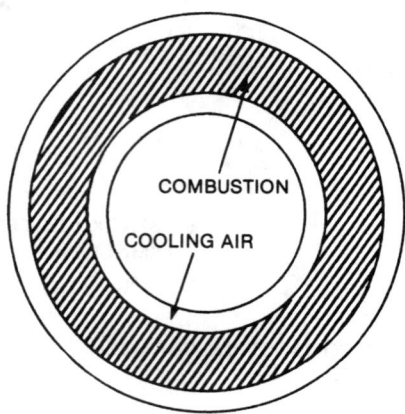

FIGURE 12-18   Annular-type combustion chamber.

FIGURE 12-20 Turbine nozzle diaphragm.

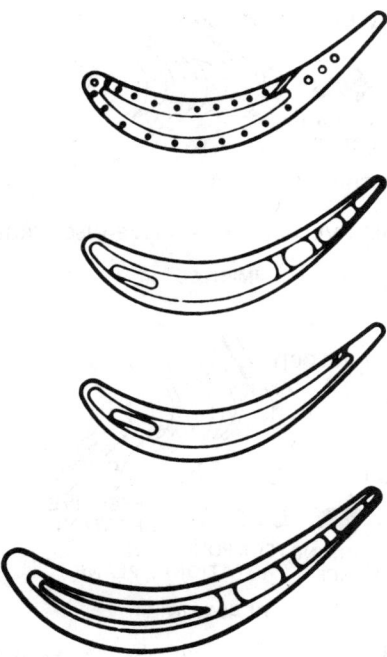

FIGURE 12-22 Cross sections of typical air-cooled vanes.

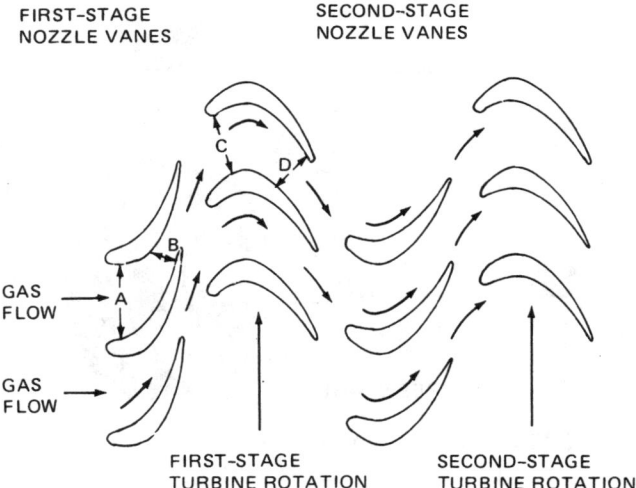

FIRST-STAGE
NOZZLE VANES

SECOND-STAGE
NOZZLE VANES

GAS FLOW

GAS FLOW

FIRST-STAGE
TURBINE ROTATION

SECOND-STAGE
TURBINE ROTATION

FIGURE 12-21 Arrangement of nozzle vanes and turbine blades.

chamber is protected from these high temperatures by a surrounding blanket of air. To withstand the extreme temperatures (1700 to 2000°F [927 to 1093°C]), the turbine nozzle vanes and support rings must be made of high-temperature alloys and must be provided with cooling. Furthermore, they must be mounted and assembled in a manner that permits expansion and contraction without causing warpage or cracking.

Cooling is accomplished by making the vanes hollow and flowing compressor bleed air through them. Cross sections of some typical air-cooled vanes are shown in Fig. 12-22. The air flows into the vanes and then out through holes in the leading and trailing edges, where it mixes with the exhaust gases. Air cooling of this type is called **convection** or **film cooling**.

In some engines, the nozzle vanes are constructed of sintered, high-temperature alloys to provide walls with a certain degree of porosity. Cooling air is directed to the insides of the vanes, after which it flows out through the porous walls. This is called **transpiration cooling**.

The high temperatures in the nozzle and turbine area increase the corrosion rates of even the most corrosion-resistant materials. For this reason, first-stage turbine vanes and turbine blades are often provided with a corrosion-resistant coating. One such treatment is called *Jo-Coating*.

Because of the expansion and contraction caused by the high temperature, nozzle vanes must be mounted in the inner and outer rings in a manner that prevents warping. This is accomplished by making the mounting holes in the support rings slightly larger than the ends of the vanes. In other cases, vanes are welded into the rings, but the rings are cut in sections to allow for expansion. In some modern engines, the turbine nozzle vanes are attached only to the outer ring; the vanes can thus expand and contract without warpage.

TURBINES

A turbojet engine may have a single-stage turbine or a multistage arrangement. The function of the turbine is to extract kinetic energy from the high-velocity gases leaving the combustion section of the engine. The energy is converted to shaft horsepower for the purpose of driving the compressor. Approximately three-fourths of the energy available from the burning fuel is required for the compressor. If the engine is used for driving a propeller or a power shaft, up to 90 percent of the energy of the gases will be extracted by the turbine section.

Turbines come in three types: the **impulse turbine**, the **reaction turbine**, and a combination of the two called a

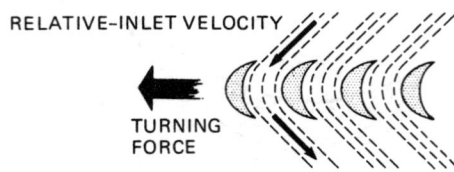

RELATIVE-INLET VELOCITY

TURNING FORCE

RELATIVE-DISCHARGE VELOCITY

RELATIVE-INLET VELOCITY · RELATIVE-DISCHARGE VELOCITY

IMPULSE

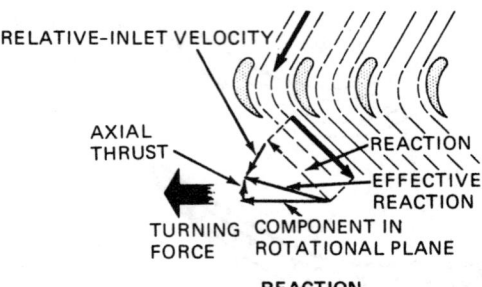

RELATIVE-INLET VELOCITY

AXIAL THRUST

REACTION

EFFECTIVE REACTION

TURNING FORCE   COMPONENT IN ROTATIONAL PLANE

REACTION

FIGURE 12-23 Comparison of impulse and reaction turbines.

reaction-impulse turbine. Turbojet engines normally employ the reaction-impulse type.

The difference between an impulse turbine and a reaction turbine is illustrated in Fig. 12-23. The pressure and speed of the gases passing through the impulse turbine remain essentially the same, the only change being in the direction of flow. The turbine absorbs the energy required to change the direction of the high-speed gases. A reaction turbine changes the speed and pressure of the gases. As the gases pass between the turbine blades, the cross-sectional area of the passage decreases and causes an increase in gas velocity. This increase in velocity is accompanied by a decrease in pressure according to Bernoulli's law. In this case the turbine absorbs the energy required to change the velocity of the gases. Typical turbines are illustrated in Fig. 12-24.

Since the nozzle vanes and turbine blades in gas-turbine engines are subjected to extremely high temperatures, they must be constructed of high-temperature alloys and some type of special cooling must be provided. If the vanes and blades in the turbine area cannot withstand the temperatures to which they are subjected, burning and stress-rupture cracks will develop. Remember that the efficiency of an engine becomes greater as the temperature of the gases at the burner outlet becomes greater. The development of high-temperature alloys containing cobalt, columbium (niobium), nickel, and other elements has made it possible to increase the operating temperature of, and thus the power available from, gas-turbine engines. A further development has been the application of coatings to the vanes and blades to enable them to withstand heat and to prevent high-temperature corrosion.

Since the temperature of the burning gases in a gas-turbine engine decreases substantially as the gases pass through each turbine stage, it is usually necessary to provide special cooling for the first stage only, although in some engines the second-stage nozzle vanes and turbine blades are also air-cooled.

TURBINE DISK AND BLADES

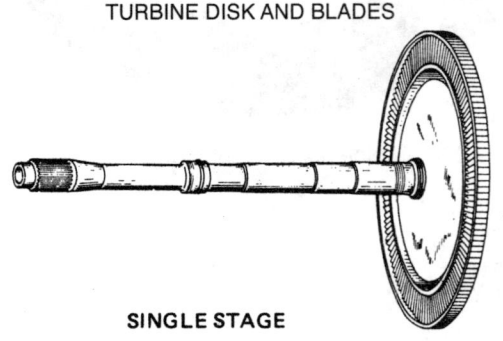

SINGLE STAGE

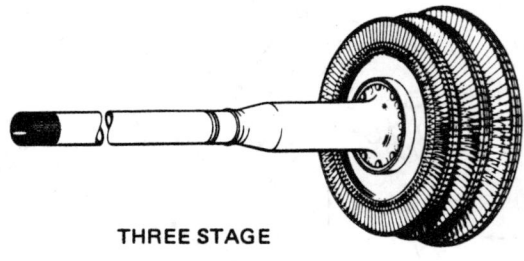

THREE STAGE

FIGURE 12-24 Typical turbines.

As mentioned previously, the first-stage nozzle vanes are made with interior passages through which air is passed for cooling. First-stage turbine blades are also made with air passages, as shown in Fig. 12-25. This illustration shows cross sections of typical turbine blades. The method by which cooling air is directed through the first-stage turbine blades of the Pratt & Whitney JT9D engine is shown in Fig. 12-26. The cooling, arrangement for the Rolls-Royce RB 211 turbine blades is shown in Fig. 12-27. Air for cooling is bled from the high-pressure compressor in each case.

Turbine blades are made in *shrouded* and *unshrouded* configurations. The shrouded blade has an extension cast on the tip to mate with the extension on the adjacent blade and form a continuous ring around the blade tips. This ring aids in preventing the escape of exhaust gases around the tips of the blades. The blades shown in Fig. 12-27 are of the shrouded type. Because of the added weight of the shrouds, shrouded

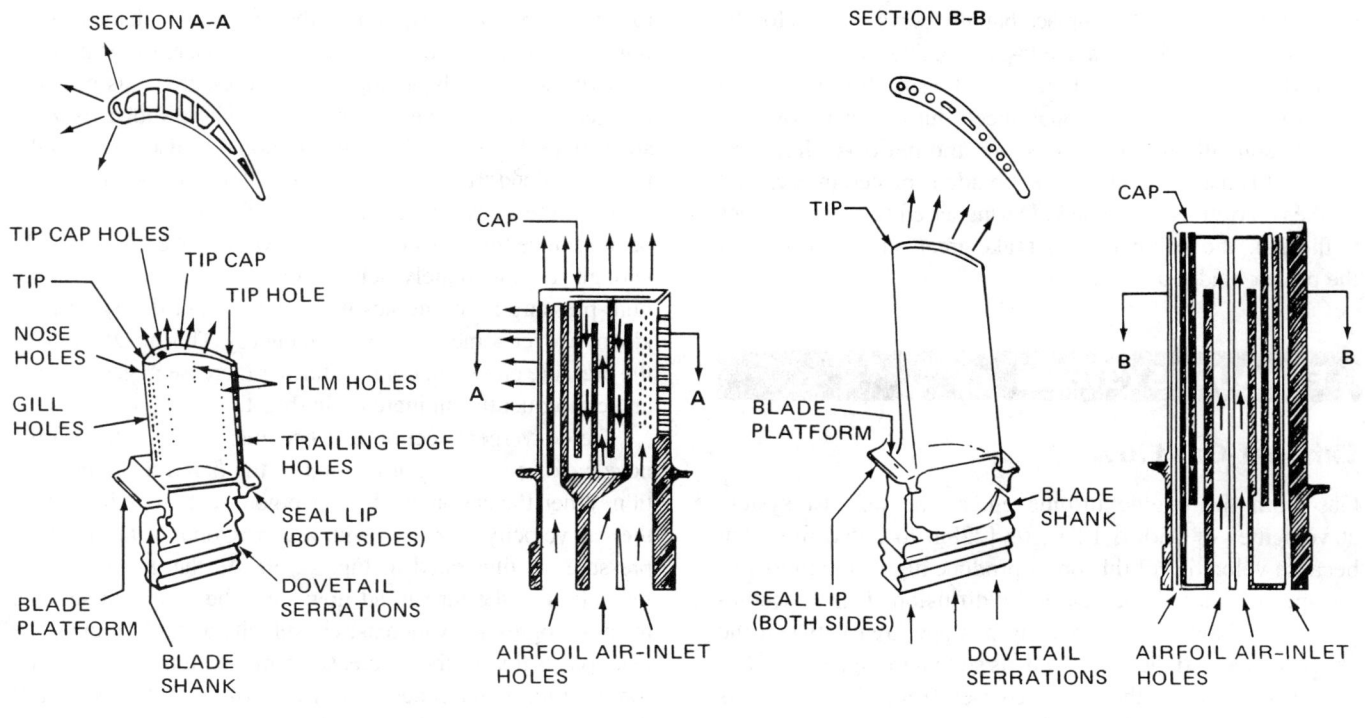

FIGURE 12-25 Air-cooled turbine blades.

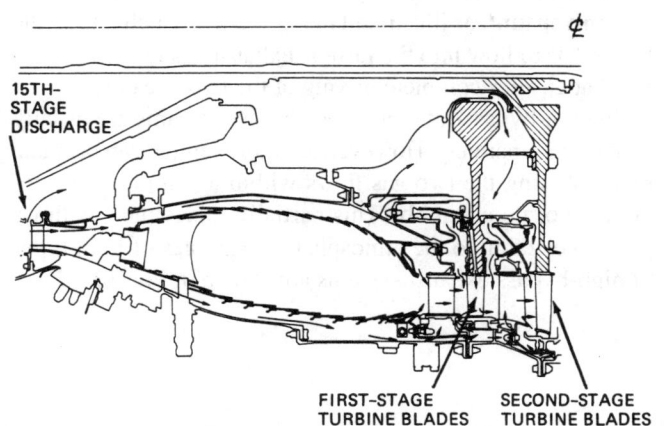

FIGURE 12-26 Cooling of first-stage turbine blades on a JT9D turbofan engine. (*Pratt & Whitney Canada.*)

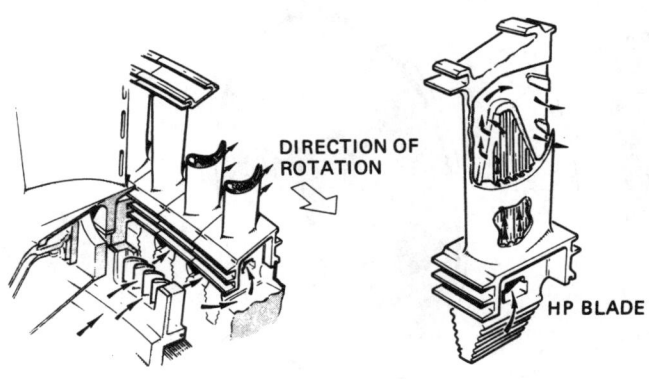

FIGURE 12-27 Cooling of turbine blades for an RB 211 turbofan engine. (*Rolls-Royce.*)

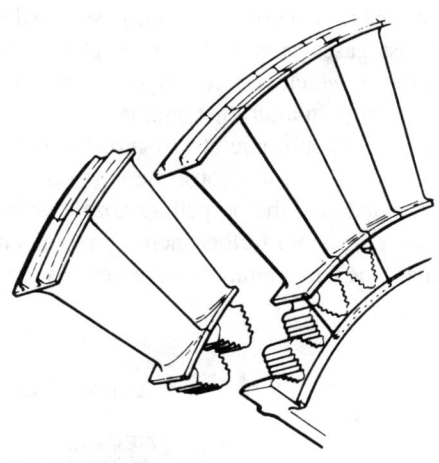

FIGURE 12-28 "Fir tree" method of attaching turbine blades to turbine disk.

turbine blades are particularly susceptible to **growth (creep)** caused by centrifugal force and overtemperature conditions.

The attachment of turbine blades to the turbine disk is usually accomplished by means of either "fir tree" slots (Fig. 12-28) or "dovetail" slots (Fig. 12-25) broached in the rim of the disk and matching bases cast or machined on the blades. After the blades are inserted in the slots on the rim of the turbine disk, they are held in place by means of rivets or metal tabs which are bent over the bases of the blades. In the General Electric CF6 engine, the turbine blades are held in the disk rim by means of a single-piece blade retainer bolted to the face of the disk. These retainers not only hold the blades in place but also serve as seals for the faces of the disks to prevent the loss of cooling air.

The balance of a turbine wheel assembly is critical because of the high rotational speed during operation. For this reason all turbine blades are **moment-weighed** and marked to ensure that the correct blade is placed in each slot on the rim of the turbine disk. During assembly of the blades to the disk, the technician must take great care to ensure that the proper blades are installed.

EXHAUST SYSTEMS

Exhaust-Gas Flow

Gas from the engine turbine enters the exhaust system at velocities of 750 to 1200 ft/s [228.6 to 365.8 m/s], but because velocities of this order produce high friction losses, the speed of flow is decreased by diffusion. This is accomplished by having an increasing passage area between the exhaust cone and the outer wall, as shown in Fig. 12-29. The cone also prevents the exhaust gases from flowing across the rear face of the turbine disk. It is usual to hold the velocity at the exhaust unit outlet to a Mach number of about 0.5, or approximately 950 ft/s [289.56 m/s]. Additional losses result from the residual whirl velocity in the gas stream from the turbine. To reduce these losses, the turbine rear struts (straightening vanes) in the exhaust unit are designed to straighten out the flow before the gases pass into the jet pipe.

The exhaust gases pass to the atmosphere through the propelling nozzle, which is a convergent duct, thus increasing the gas velocity. In a turbojet engine, the exit velocity of the exhaust gases is subsonic at low-thrust conditions only. During most operating conditions, the exit velocity reaches the speed of sound, and the propelling nozzle is then said to be "**choked**"—that is, no further increase in velocity can be obtained unless the temperature is increased. As the upstream total pressure is increased above the value at which the propelling nozzle becomes choked, the static pressure of the gases at the exit increases above atmospheric pressure. This pressure differential across the nozzle provides what is known as **pressure thrust**. Pressure thrust is additional thrust added to the thrust obtained from the momentum change of the gas stream.

With the convergent type of nozzle, a waste of energy occurs, since the gases leaving the exit do not expand rapidly enough to immediately achieve outside air pressure. Some high-pressure-ratio engines use a convergent-divergent nozzle to recover some of this wasted energy. This nozzle utilizes the pressure energy to obtain a further increase in gas velocity and consequently, an increase in thrust.

The convergent section exit becomes the throat, with the nozzle exit now being at the end of the flared divergent section. When the gas enters the convergent section of the nozzle, the gas velocity increases with a corresponding fall in static pressure, as illustrated in Fig. 12-30. As the gas leaves the restriction of the throat and flows into the divergent section, the gas progressively increases in velocity toward the exit. The reaction to this further increase in momentum is a pressure force acting on the inner wall of the nozzle. A component of this force acting parallel to the longitudinal axis of the nozzle produces the further increase in thrust.

The bypass engine has two gas streams to eject to the atmosphere: the cool bypass airflow and the hot turbine discharge gases.

In a low-bypass-ratio engine, the two flows are combined by a **mixer unit**, as illustrated in Fig. 12-31, which allows the bypass air to flow into the turbine exhaust gas flow in a manner that ensures complete mixing of the two streams.

In high-bypass-ratio engines, the two streams are usually exhausted separately. However, an improvement can be made by combining the two gas flows within a common, or integrated, nozzle assembly. This partially mixes the gas flows prior to ejection to the atmosphere. Examples of both types of high-bypass exhaust systems are shown in Fig. 12-32.

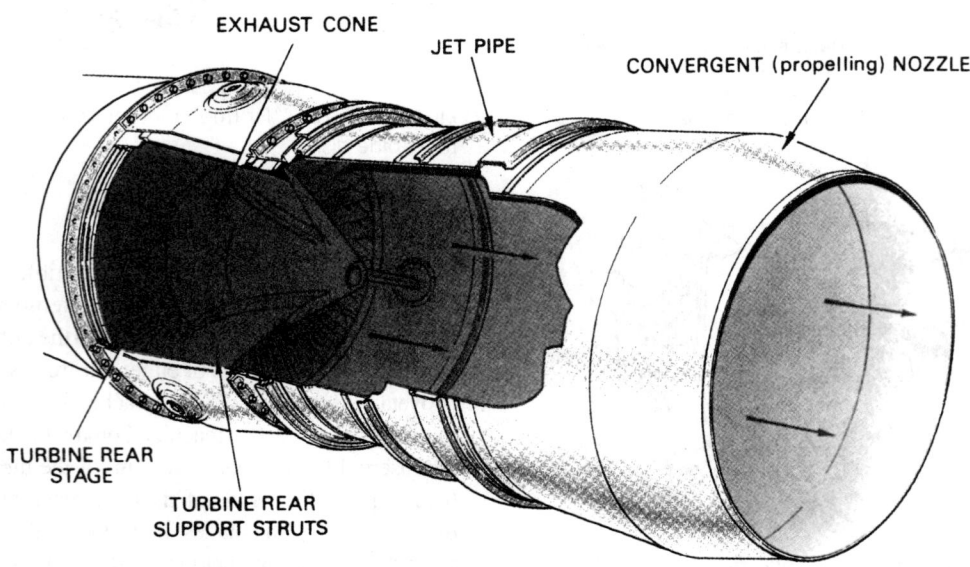

FIGURE 12-29 A basic exhaust system. (*Rolls-Royce.*)

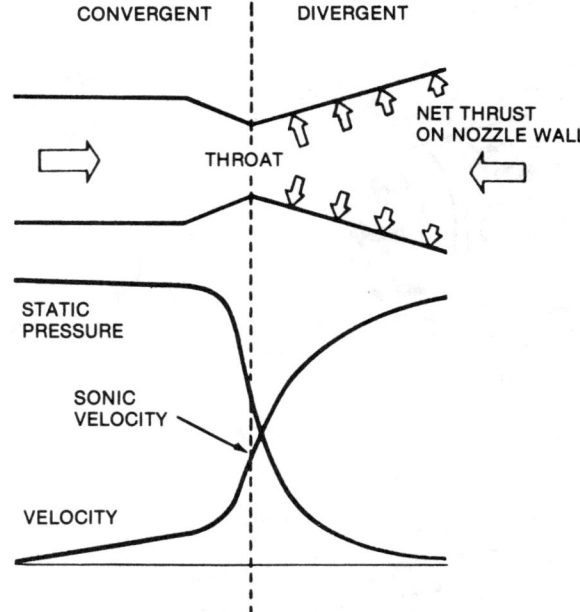

FIGURE 12-30 Gas flow through a convergent-divergent nozzle. (*Rolls-Royce.*)

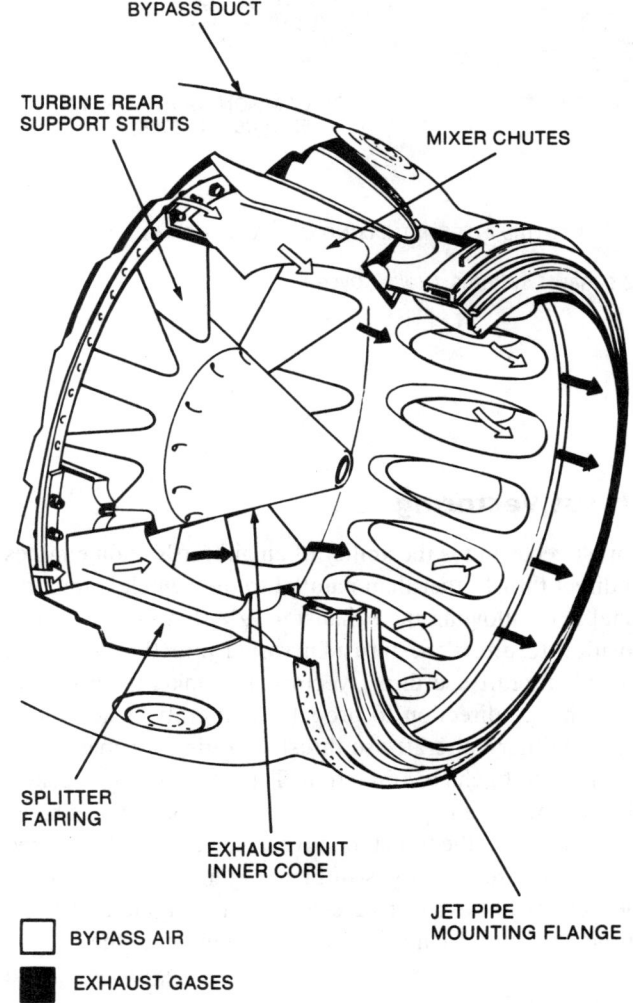

FIGURE 12-31 Low-bypass air mixer unit. (*Rolls-Royce.*)

EXHAUST NOZZLES

The normal function of the exhaust nozzle, or cone, is to control the velocity and temperature of the exhaust gases. Although a certain amount of thrust would be produced even if there were no exhaust nozzle, the thrust would be comparatively low and the direction of flow would not be properly controlled. When a convergent nozzle is used, the velocity of the gases is increased and the flow is directed so that the thrust is in line with the engine.

The cross-sectional area of the nozzle outlet is most critical. If the area is too large, the engine will not develop maximum thrust; and if the area is too small, the exhaust temperature will be excessive at full-power conditions and will damage the engine.

The exhaust system must be capable of withstanding high gas temperatures and is therefore manufactured from nickel or titanium. It is also necessary to prevent any heat from being transferred to the surrounding aircraft structure. This is achieved by passing ventilating air around the jet pipe, or by covering the exhaust system with an insulating blanket, as shown in Fig. 12-33. Each blanket has an inner layer of fibrous insulating material contained by an outer skin of thin stainless steel, which is dimpled to increase its strength. Acoustically absorbent materials are sometimes applied to the exhaust system to reduce engine noise.

Due to the wide variations of temperature to which the exhaust system is subjected, it must be mounted and have its sections joined together in such a manner as to allow for expansion and contraction without distortion or damage.

VARIABLE-AREA EXHAUST NOZZLE

The **variable-area exhaust nozzle** is generally used on engines which have some sort of thrust augmentation such as an afterburner. **Afterburners** are used to greatly increase the thrust and therefore the speed of an aircraft for relatively short periods by increasing the fuel flow. The afterburner portion of the engine is generally located immediately behind the turbine section and forward of the exhaust nozzle and operates similar to a ram jet. The afterburner system components, shown in Fig. 12-35, are a fuel manifold, an ignition system, and a flame holder. The manifold and flame holder provide the means for injecting fuel into the exhaust gases, and the ignition system ignites the F/A mixture.

A considerable amount of air passes through a core engine that is not needed for combustion (approximately 75 percent) but is used for cooling. Once the exhaust gases have passed through the turbine section, all this air is no longer needed for cooling and some of it may be mixed with fuel, ignited, and burned in the afterburner.

When combustion takes place in the afterburner, the gases are speeded up, increasing the temperature and thrust. With the afterburner in operation, the jet nozzle area must be increased. Otherwise, back pressure would increase at the discharge of the turbine and result in a temperature rise of the gas in the vicinity of the turbine beyond a permissible safe

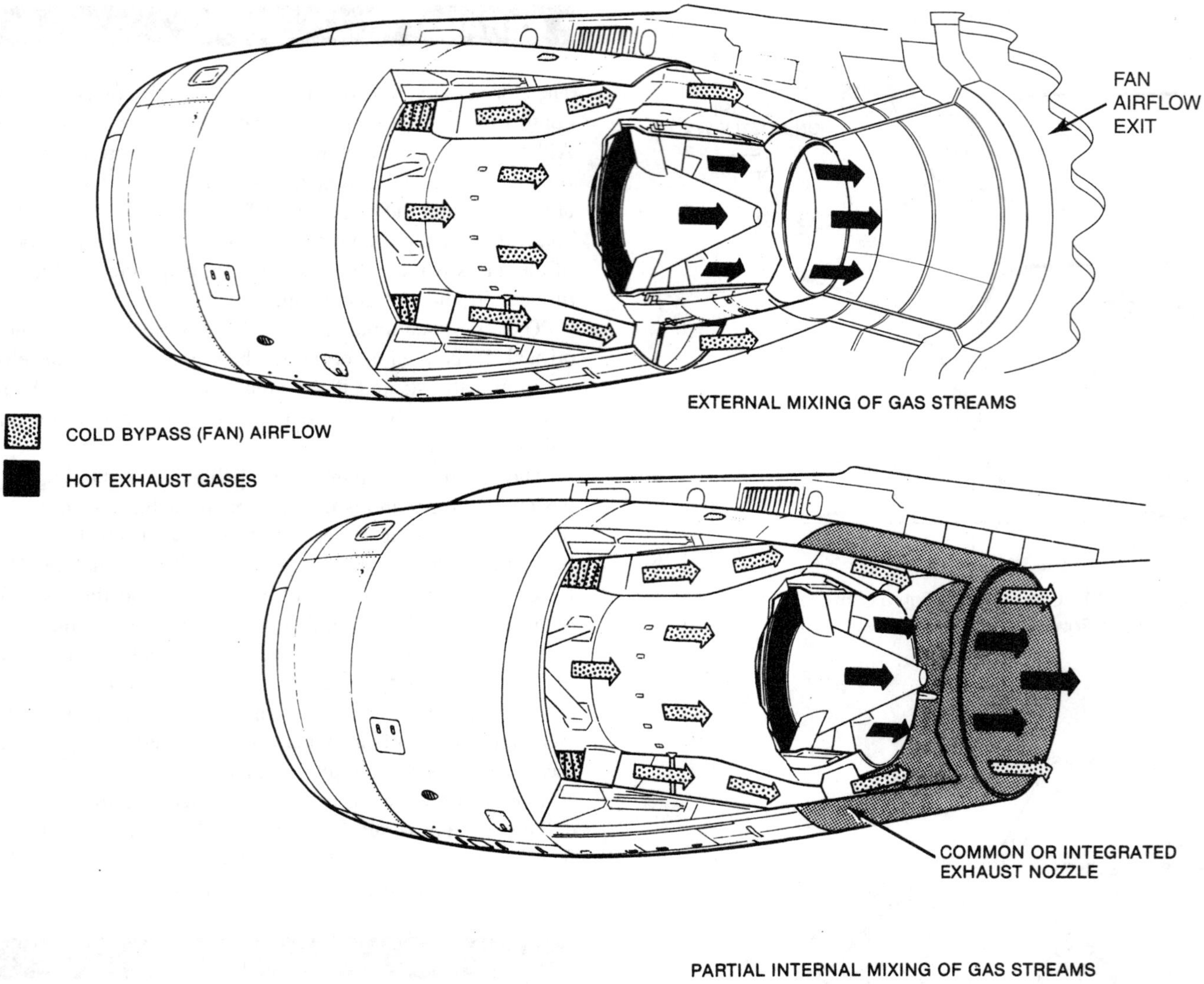

FAN
AIRFLOW
EXIT

EXTERNAL MIXING OF GAS STREAMS

☒ COLD BYPASS (FAN) AIRFLOW

■ HOT EXHAUST GASES

COMMON OR INTEGRATED
EXHAUST NOZZLE

PARTIAL INTERNAL MIXING OF GAS STREAMS

FIGURE 12-32   High-bypass-ratio engine exhaust systems. (*Rolls-Royce.*)

FIGURE 12-33   Insulating blanket. (*Rolls-Royce.*)

limit. With the variable-area nozzle, the size of the exhaust area can be changed. By opening the nozzle (increasing its size), the temperature of the exhaust gases can be reduced and kept within allowable limits.

The finger type of variable-area nozzle (see Fig. 12-35) has a number of flaps extending from the outer skin of the exhaust pipe. The angle of the flaps produces a larger or smaller exit area. The nozzle may be operated electrically, hydraulically, or pneumatically.

Thrust Vectoring

Thrust vectoring is the ability of an aircraft's main engines to direct thrust other than parallel to the vehicle's longitudinal axis, allowing the exhaust nozzle to move or change position to direct the thrust in varied directions. Vertical takeoff aircraft use thrust vectoring as takeoff thrust and then change direction to propel the aircraft in horizontal flight. Military aircraft use thrust vectoring for maneuvering in flight to change direction. Thrust vectoring is generally accomplished by relocating the direction of the exhaust nozzle to direct the thrust to move the aircraft in the desired path. At the rear of a gas-turbine engine, a nozzle directs the flow of hot exhaust gases out of the engine and afterburner. Usually, the nozzle points straight out of the engine. The pilot can move, or vector, the vectoring nozzle up and down by 20°. This makes the aircraft much more maneuverable in flight; see Fig. 12-34.

FIGURE 12-34 Vectoring nozzle providing directional thrust.

THRUST REVERSERS

Turbine aircraft include **thrust reversers** to assist the brake system in slowing the aircraft after landing. The two most commonly used types of thrust reversers are the aerodynamic

blockage system, illustrated in Fig. 12-36, and the mechanical blockage system, illustrated in Fig. 12-37.

Both the aerodynamic system and the mechanical system are subjected to high temperatures and to high gas loads. The components of both systems, especially the doors, are therefore constructed from heat-resisting materials and are of particularly heavy construction.

The thrust reverser shown in Fig. 12-36 is equipped with vanes and deflectors by which the exhaust gases and cold fan stream air are deflected outward and forward and are controlled through the thrust lever in the cockpit. This system consists of a number of cascade vanes and solenoids with a pneumatic motor operating through gears and shafts to move the cascade vanes to the deployed position.

After the aircraft has landed, the pilot moves the levers to the rear of the IDLE position. This causes the deflecting vanes to move into the mainstream of the gas flow through the engine and the cold airstream to reverse the flow direction. At the same time as the thrust levers are moved further rearward, the fuel flow to the engine is increased. When the thrust levers are in the FULL REVERSE position, the engine power output is approximately 75 percent of full-forward thrust capability.

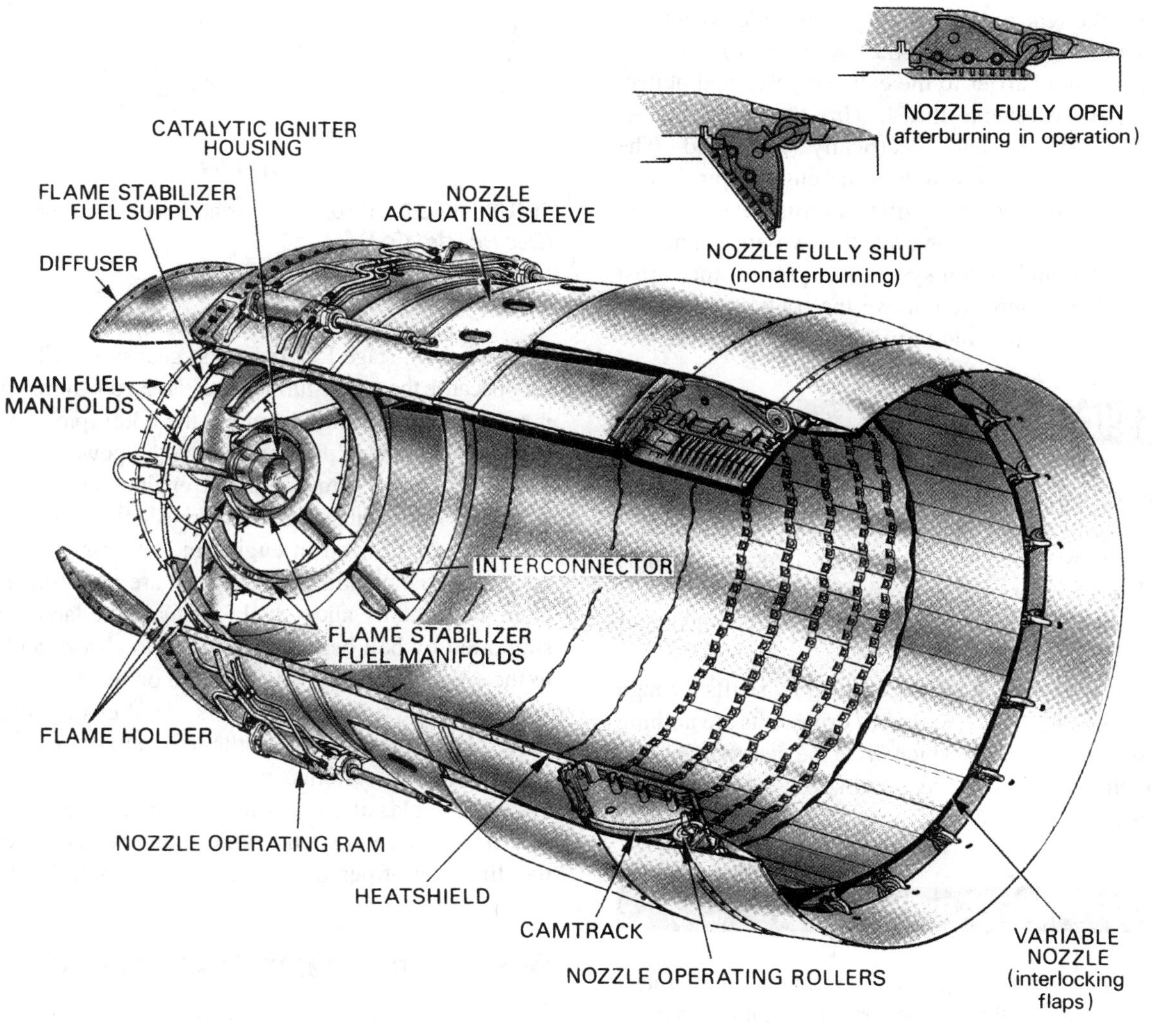

NOZZLE FULLY OPEN
(afterburning in operation)

NOZZLE FULLY SHUT
(nonafterburning)

CATALYTIC IGNITER
HOUSING

FLAME STABILIZER
FUEL SUPPLY

NOZZLE
ACTUATING SLEEVE

DIFFUSER

MAIN FUEL
MANIFOLDS

INTERCONNECTOR

FLAME STABILIZER
FUEL MANIFOLDS

FLAME HOLDER

NOZZLE OPERATING RAM

HEATSHIELD

CAMTRACK

NOZZLE OPERATING ROLLERS

VARIABLE
NOZZLE
(interlocking
flaps)

FIGURE 12-35 Typical afterburning jet pipe equipment. (*Rolls-Royce.*)

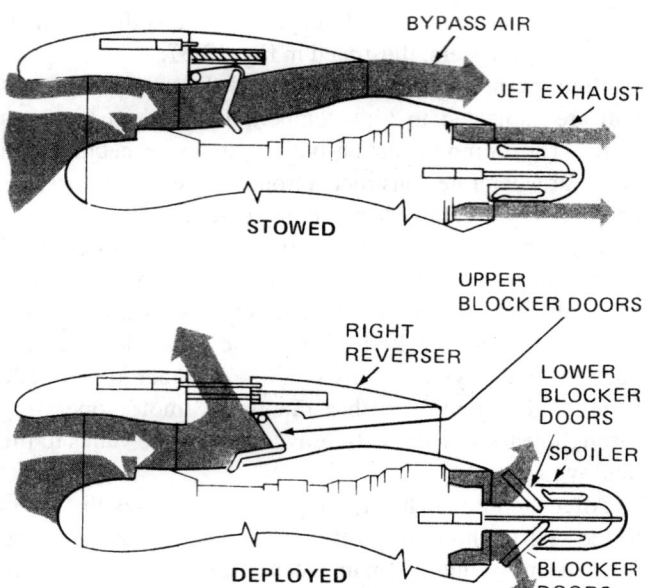

FIGURE 12-36   Thrust reverser action on a DC-10 airplane. (*McDonnell Douglas Corp.*)

The mechanical blockage reverser shown in Fig. 12-37 consists of two blocker doors, or "clamshells." which, when stowed, form the rear part of the engine nacelle. When the doors are deployed, as shown in the lower part of the illustration, they form a barrier to the exhaust gases and deflect them to produce a reverse thrust. This thrust reverser is hydraulically operated and electrically controlled. The reverser cannot be deployed unless the engine rpm is less than 65 percent. Some aircraft utilize a thrust reverser lever separate from the throttle; however, the two are connected to the same cable and clutch systems to prevent the thrust reverser from being deployed above the IDLE position. On other aircraft, the throttle and thrust lever are the same.

ACCESSORY DRIVE

The **accessory drive** section of a turbine engine is used to power engine and aircraft accessories such as fuel pumps, lubrication pumps, generators, alternators, hydraulic pumps, and any other necessary equipment. The accessory drive is usually located on the bottom forward part of the outside of the engine.

The accessory drive is a gearbox driven by shafts through the compressor front frame. Accessory drive units are nothing more than sets of gear trains which drive the necessary accessories through shafts at their required speeds. An accessory drive is illustrated in Fig. 12-38.

REDUCTION-GEAR SYSTEMS

Reduction-gear systems for gas-turbine engines have much higher reduction ratios than those for reciprocating engines. For example, the Pratt & Whitney of Canada PT6A-27 engine

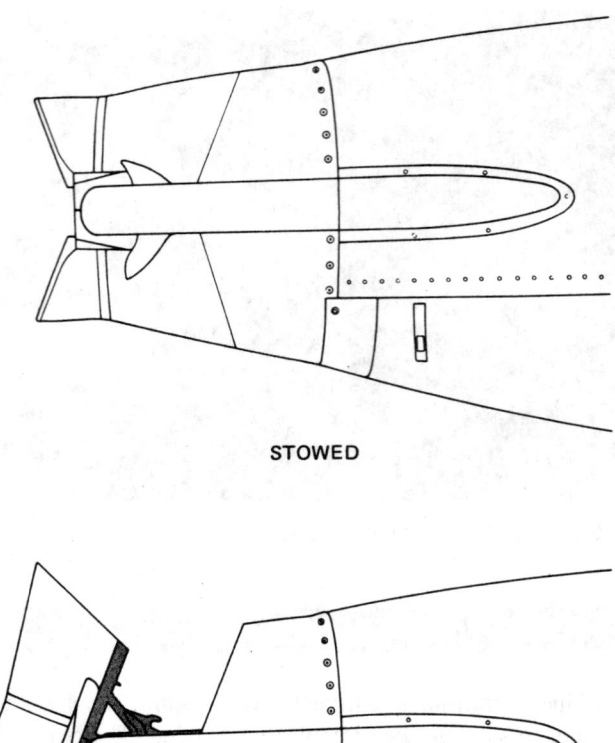

FIGURE 12-37   Thrust reverser action on a business aircraft. (*Dee Howard Co.*)

employs a two-stage 15:1 reduction-gear system. This is necessary because the power-turbine speed is 33 000 rpm. This engine is used for helicopters; therefore, the 6300 rpm of the output shaft can be reduced further through the power-train gearbox. Reduction-gear systems for turbine engines are described in the sections covering turboprop and turboshaft engines.

The gears utilized in engine reduction-gear systems are subjected to very high stresses and are therefore machined from high-quality alloy-steel forgings. The larger shafts are supported by ball bearings designed to absorb and transmit to the engine case all loads imposed on them.

Since reduction-gear systems are of critical importance in the reliability of engines, the overhaul of engines having reduction-gear systems, with the exception of spur-gear systems, is classified as a major repair. The overhaul and return to service of such engines must be carried out under the direction of persons suitably certificated by the FAA.

Gas-Turbine Engine Bearings and Seals

The main bearings have the critical function of supporting the main engine rotor. The number of bearings necessary for

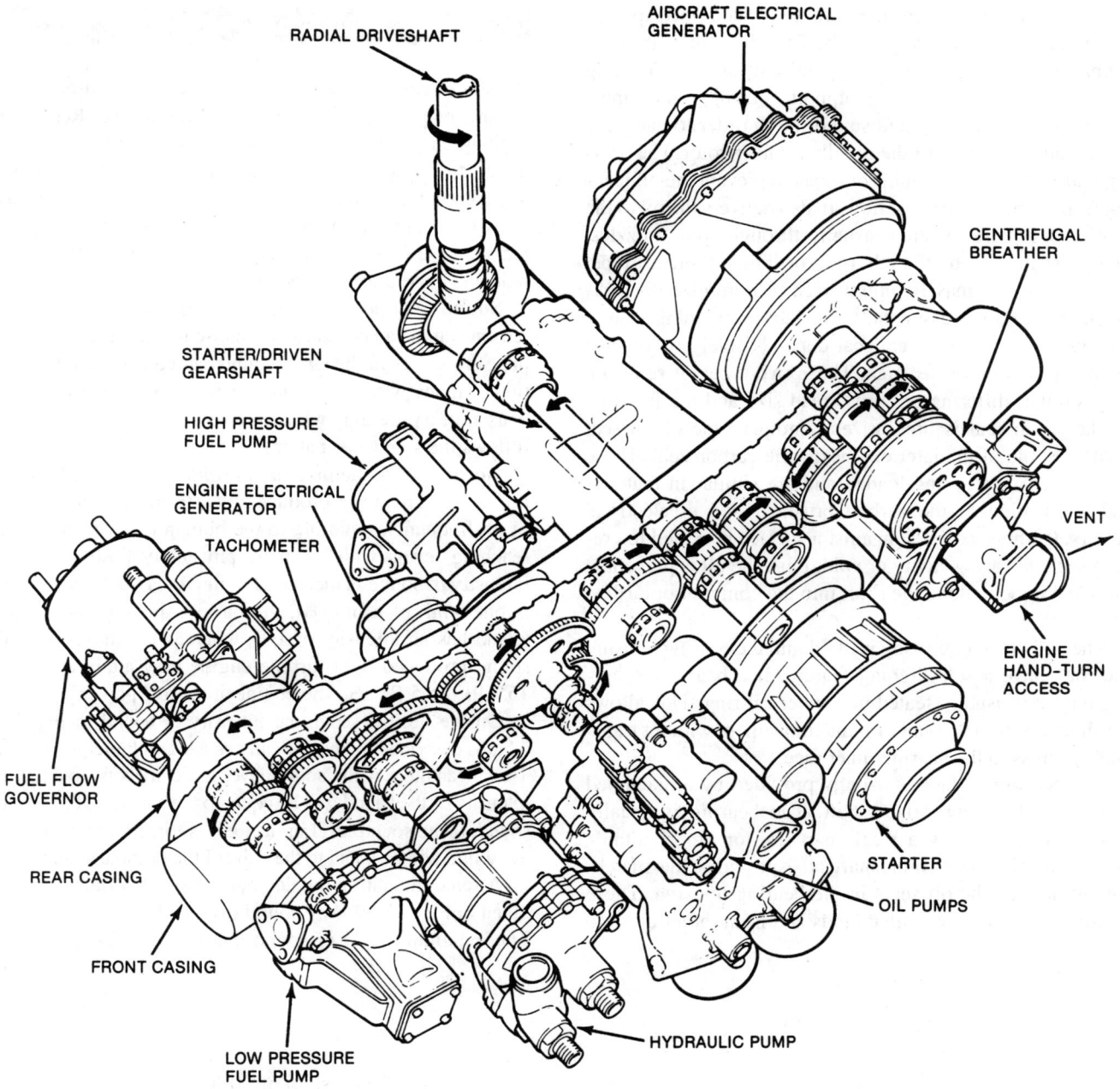

FIGURE 12-38 Accessory drive and accessory units. (*Rolls-Royce.*)

proper engine support will, for the most part, be decided by the length and weight of the engine rotor. The length and weight are directly affected by the type of compressor used in the engine. Naturally, a two-spool compressor will require more bearing support. The minimum number of bearings required to support one shaft is one deep groove ball bearing (thrust and radial loads) and one straight roller bearing (radial load only). Sometimes, it is necessary to use more than one roller bearing if the shaft is subject to vibration or its length is excessive. The gas-turbine rotors are supported by ball and roller bearings which are antifriction bearings. Many newer engines use hydraulic bearings where the outside race is surrounded by a thin film of oil. This reduces vibrations transmitted to the engine. These engines

1. Offer little rotational resistance
2. Facilitate precision alignment of rotating elements
3. Are relatively inexpensive
4. Are easily replaced
5. Withstand high momentary overloads
6. Are simple to cool, lubricate, and maintain
7. Accommodate both radial and axial loads
8. Are relatively resistant to elevated temperatures

The main disadvantages are their vulnerability to foreign matter and tendency to fail without appreciable warning. Usually the ball bearings are positioned on the compressor or turbine shaft so that they can absorb any axial (thrust) loads or radial loads. Because the roller bearings present a larger working surface, they are better equipped to support radial loads than thrust loads. Therefore, they are used primarily for this purpose. A typical ball or roller bearing assembly includes a bearing support housing, which must be strongly constructed and supported in order to carry the radial and axial loads of the rapidly rotating rotor. The bearing housing usually contains oil seals to prevent the oil leaking from its normal path of flow. It also delivers the oil to the bearing for its lubrication, usually through spray nozzles. The oil seals may be the labyrinth or thread (helical) type. These seals also may be pressurized to minimize oil leaking along the compressor shaft. The labyrinth seal is usually pressurized, but the helical seal depends solely on reverse threading to stop oil leakage. These two types of seals are very similar, differing only in thread size and the fact that the labyrinth seal is pressurized. Another type of oil seal used on some of the later engines is the carbon seal. These seals are usually spring-loaded and are similar in material and application to the carbon brushes used in electrical motors. Carbon seals rest against a surface provided to create a sealed bearing cavity or void; thus, the oil is prevented from leaking out along the shaft into the compressor airflow or the turbine section.

The ball or roller bearing is fitted into the bearing housing and may have a self-aligning feature. If a bearing is self-aligning, it is usually seated in a spherical ring. This allows the shaft a certain amount of radial movement without transmitting stress to the bearing inner race.

The bearing surface is usually provided by a machined journal on the appropriate shaft. The bearing is usually locked in position by a steel snap ring or other suitable locking device. The rotor shaft also provides the matching surface for the oil seals in the bearing housing. These machined surfaces are called lands and fit in rather close to the oil seal.

ENGINE NOISE

Because many complaints of excessive noise caused by gas-turbine engines have resulted in Federal Air Regulations limiting engine noise and because of the danger of physical injury from·such noise, engine and aircraft manufacturers together with government agencies have been actively engaged in experimenting with and modifying engines in an effort to reduce noise to an acceptable level. Some of the methods for reducing noise are described later in this section.

Sound may be defined as *that which can be heard*. The reason sound can be heard is because it consists of a series of pressure waves in the air. A sound can consist of a combination of many waves in a wide range of frequencies, or it can consist of a pure tone which is a single-frequency wave that follows the sine wave pattern.

Noise may be defined as *unwanted and usually irritating sound*. The noise produced by a turbine engine consists of all frequencies audible to the human ear with intensities reaching levels which can be physically destructive. The intensity of sound is measured in **decibels** (dB). One decibel is one-tenth of a **bel** (B), the basic unit. A barely audible sound has an intensity of 1 B, whereas the intensity of the sound produced by a turbine engine may attain a value of 155 dB (15.5 B) near the engine at takeoff power.

On the decibel scale, the intensity of sound increases on what is described mathematically as a logarithmic progression. This means that if the sound level in decibels is doubled, the intensity of the sound will be equal to the square of the original sound. If the sound level in decibels is tripled, the intensity of the sound will be equal to the cube of the original sound.

A scale indicating the decibel values of certain sounds is given in Fig. 12-39. Any sound over 100 dB is very intense. The maximum level of sound that can be evaluated by the

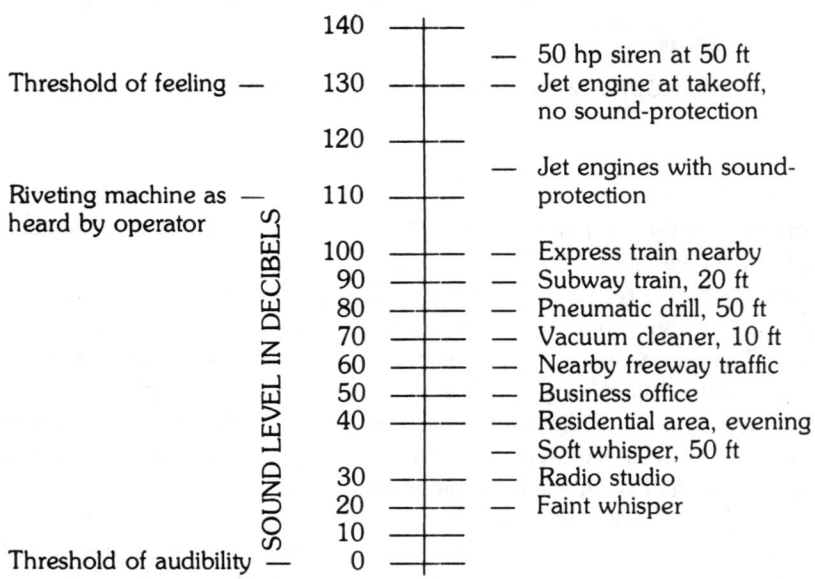

FIGURE 12-39 Scale of sound values.

human ear is approximately 120 dB. Above this level, the ear can feel increasing intensity but cannot hear the difference. Also, above this level, ear damage can occur.

Noise from a turbojet engine is caused by a number of forces, but it basically stems from the "torturing" of the air passing through the engine. Initially, the air is violently broken up and chopped into segments as it enters the inlet duct, passes through the inlet guide vanes, and encounters the compressor blades. Much of the sound thus created is in a wide range of frequencies, but a single frequency is also heard. This is the familiar "whine" caused by the compressor blades chopping the incoming air. A sound of this type is called "discrete" because it has an identifiable frequency and can be recognized in relation to other sounds. The most intense sound at high-power engine settings comes from the exhaust nozzle. This sound is caused by the shear turbulence between the relatively calm air outside the engine and the high-velocity jet of hot gases emanating from the nozzle. The noise caused by the jet exhaust is termed **broadband noise** because it includes many frequencies.

In a turbofan engine, considerable noise is caused by the secondary airflow from the fan section of the engine. This air has a lower velocity than the primary jet exhaust, and therefore the noise it produces is not as intense. An additional factor affecting the jet exhaust of the turbofan engine is that energy is extracted from the primary exhaust stream to drive the fan, and this decreases the velocity of the exhaust jet. Thus, the noise produced is lower in intensity.

Reduction of Noise

Experimentation has been going on for many years in an effort to reduce turbine engine noise. One of the first devices was a **multiple-tube jet nozzle** for commercial turbine aircraft instead of the single exhaust nozzles used by military aircraft and early commercial aircraft. The effect of this type of nozzle is to reduce the size of the individual jet streams and increase the frequency of the sound. The higher-frequency sound attenuates (reduces) more rapidly as distance from the source is increased; therefore, the sound at 500 or 1000 ft [152.4 or 304.8 m] from the aircraft is not as intense as it would be with the single-nozzle engine.

Another exhaust nozzle is the **corrugated-perimeter type**. This nozzle has an effect similar to that of the multiple-tube nozzle. In all modified nozzles, it is essential that the total outlet area be maintained to provide maximum jet efficiency. Multiple-tube and corrugated-perimeter nozzles are shown in Fig. 12-40.

The development of turbofan engines has made possible additional reduction of engine noise. In the turbofan engine, both the primary airflow and the secondary airflow are reduced in velocity compared with those in the turbojet. As explained previously, the reduced air velocity results in a decrease in noise intensity.

The high-bypass engines such as the General Electric CF6, the Pratt & Whitney JT9D, and the Rolls-Royce RB 211 produce noise of lower intensity than that of the turbojet or low-bypass

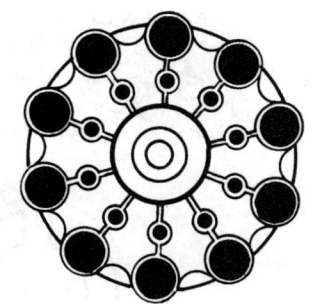

MULTIPLE-TUBE TYPE END VIEW

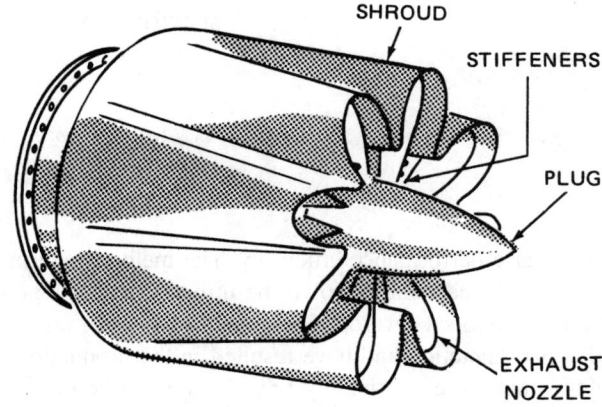

FIGURE 12-40   Noise suppressors.

turbofan engines for several reasons. The air discharge velocities are lower compared with those of other engines, there are no inlet guide vanes in the front of the fan section, and the engines are provided with sound-absorbing liners inside the fan ducts and exhaust nozzle. The arrangement of the noise-absorbent linings in the Rolls-Royce RB 211 engine is shown in Fig. 12-41. Advanced-design engines such as the Pratt & Whitney JT9D-7R4 and 2037 and the Garrett ATF3-6 have reduced noise further by the use of new fan blade design, improved exhaust systems, and more effective soundproofing materials.

Protection against Noise

It is essential that crew members working around turbine engines be provided with approved ear protectors. The most common types of protectors are over-the-ear devices in which earphones are installed for communication purposes. These protectors are muffs which completely enclose the ears, thus protecting them from noise while permitting the use of earphones for voice communication.

ADVANCED MANUFACTURING PROCESSES

Manufacturers of gas-turbine engines are continuously developing improved methods of manufacture to reduce costs, increase reliability, reduce weight, simplify maintenance,

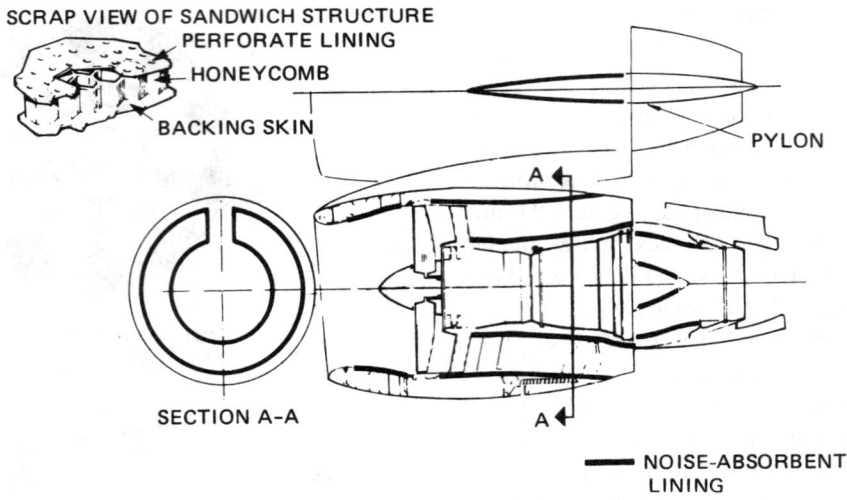

SCRAP VIEW OF SANDWICH STRUCTURE
PERFORATE LINING
HONEYCOMB
BACKING SKIN
PYLON
A
SECTION A-A
A
NOISE-ABSORBENT LINING

FIGURE 12-41 Noise-absorbent linings in a turbofan engine. (*Rolls-Royce.*)

and increase performance efficiency. The methods and processes employed today involve the utilization of technologies that have been developed as a result of years of research and experimentation and have resulted in the production of engines that are exceptionally reliable and can be operated for thousands of hours without the need for overhaul or repair.

Computer technology is very widely used in the manufacturing process to precisely control the operation of various machines. The specific requirements for each of the machine-generated operations are stored in a central computer. When initiated by the machine operators, the computer controls the machine so that it removes precise amounts of metal and generates the exact dimensions of the part.

Electrostream Drilling

Cooling holes are drilled in vanes and blades at precise locations and angles. **Electrostream drilling** is one of several processes employed to drill small, accurate holes. In this process, electrically charged acid is released through fine glass tubes onto the metal surface. Metal is dissolved at the point of contact, with the acid leaving a precise, small hole.

Electric-Discharge Machining

Another system for producing holes is **electric-discharge machining**. In this process, electric current is allowed to flow through an electrode which is held in close proximity to the part immersed in oil. As the current flows, a series of small sparks flash between the electrode and the part, creating a hole in the metal surface. The size and shape of the hole are dependent on the shape of the electrode. Operators are able to precisely adjust the feed rates and speed of the process.

Laser Drilling

The most cost-effective and state-of-the-art technique for drilling cooling holes is **laser drilling**. The part is loaded into

an appropriate fixture and placed in a machine that generates a pulsed beam of laser light, as shown in Fig. 12-42, which strikes the part at the location of a desired hole for 2 or 3 s. As a result of the heat generated, the metal exposed to the light melts and a hole is drilled. This process is very fast, and a large number of holes can be drilled efficiently and accurately in a very short time, keeping costs low.

Laser-Beam Welding

A **laser beam** is concentrated energy in the form of pure, coherent light. The diameter of the beam can be adjusted to the size necessary for a particular operation, even down to the size

FIGURE 12-42 Laser drilling. (*General Electric Co., U.S.A.*)

of a pinpoint. When used with automated controls, the laser beam can make welds much more rapidly and precisely than can be done with other methods.

Inertia Bonding

Inertia bonding is described as a solid-state welding technique which forms very strong weld joints with metallurgical properties equivalent to those of the base metal. The inertia bonds are more reliable than mechanical joints and have superior fatigue strength.

In inertia bonding, a rotating part is forced against its stationary mating part, and the resulting heat generated by friction and pressure causes the metal to bond without actually melting.

Powder Metallurgy

In powder metallurgy, powdered metal is formed into engine parts in appropriate dies under heat and pressure. The resulting parts are described as near-net-shape because they require very little machining and finishing. The process therefore greatly decreases the cost of production and the wastage of material. **Hot isostatic pressing** (HIP) is one of the most advanced processes for forming powdered alloys into finished parts by the application of extreme pressure and heat. Turbine disks of many modern engines are manufactured from powdered alloys by the HIP process.

Automated Investment Casting

Investment casting has been used for many years in a variety of applications; however, it is now being performed automatically, thus greatly reducing cost and ensuring the quality and uniformity of the cast products. This process is employed extensively for the production of turbine blades and vanes.

Investment casting is a process consisting of a number of distinct steps that result in a finished part that is an accurate copy of the original pattern in dimensions (Fig. 12-43). The pattern for a part to be produced by investment casting is constructed of a hard wax formed in the exact shape of the part to be cast. The pattern and mold are often designed so that many identical parts can be produced in one operation.

The wax pattern is dipped automatically by robot arms into a vat containing a suitable slurry. The pattern is then lifted out of the slurry and exposed to a spray of dry refractory material called "stucco" that adheres to the slurry. The dipping in the slurry and the application of the stucco are alternately performed by the robot devices until a mold of the desired thickness has been formed around the pattern. After the mold has solidified over the pattern, the mold and pattern are subjected to heat. The wax pattern melts and leaves the required cavity within the mold. The mold is "fired" by exposure to high temperature, to give it the strength and resistance to heat required for casting.

If the part to be cast requires internal passages, such as the cooling air passages for turbine blades and vanes, these are created by placing correctly shaped cores inside the mold in precisely located positions. The cores are made of a heat-resistant ceramic or similar material that can be removed from the part by leaching—that is, chemical dissolving.

Before casting, the finished mold is heated to a temperature at or near the temperature of the molten alloy to be cast. This is done to prevent rapid cooling and solidification of the alloy, which would result in unsatisfactory grain structure in the part being cast. The alloy is poured into the mold, after which the mold and alloy are cooled under

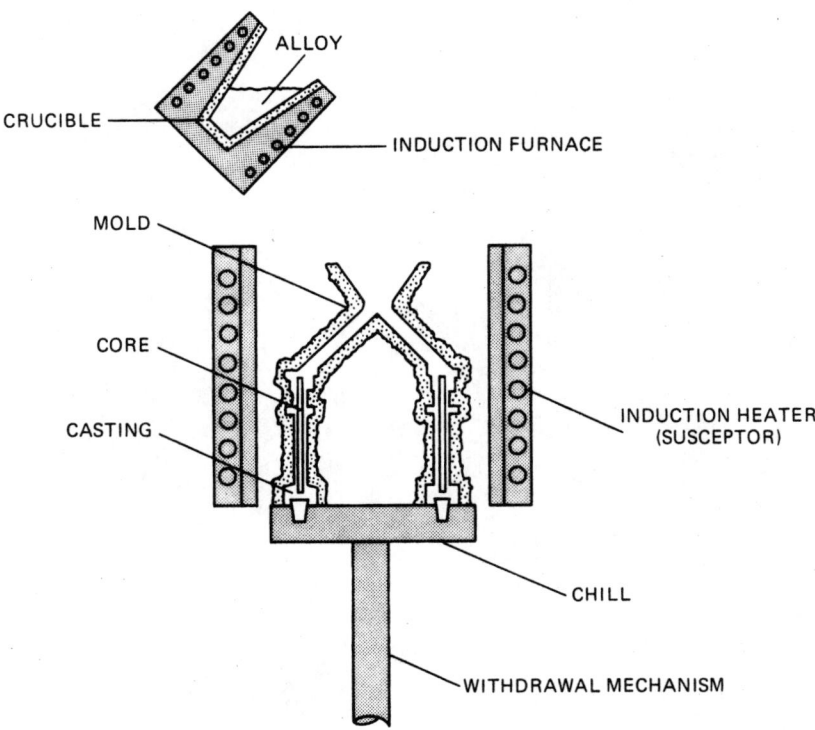

FIGURE 12-43   Drawing illustrating the investment casting process. (*Howmet Turbine Components Corp.*)

Advanced Manufacturing Processes   **347**

precisely controlled conditions. Upon cooling, the mold breaks away and leaves the cast part or parts ready for further finishing.

In recent years, directionally solidified and single-crystal castings have been developed by Pratt & Whitney and the Howmet Turbine Components Corporation. These castings are slowly cooled from the bottom of the mold. A **directionally solidified casting** develops crystals during cooling that extend from one end of the part to the other, thus eliminating the crosswise grain structure found in ordinary castings. Since the crystals in a turbine blade cast with the directionally solidified method are parallel to the centrifugal force loads applied during operation, the blade can withstand considerably more force than a blade with a granular structure.

Directionally solidified investment casting is carried out by means of a furnace, shown schematically in Fig. 12-43. The mold is placed on a heat sink (chill plate) within an induction furnace. The molten alloy is poured into the heated mold, and as heat is reduced, the molten alloy begins to solidify as the heat is drawn off by the water-cooled chill plate. The mold is then slowly withdrawn from the furnace, and solidification of the alloy continues upward. The effect of this procedure is to cause the metal crystallization to take place in a linear manner through the length of the casting.

Single-crystal castings are produced in a manner similar to that of the directionally stabilized casting. To produce a single-crystal casting, a "seed" crystal is placed at the bottom of each mold cavity. This causes a single crystal to be formed as the mold is withdrawn from the furnace. Single-crystal turbine blades provide maximum strength and are used in such engines as the Pratt & Whitney JT9D-7R4 and 2037.

The turbine blades are coated with alloys that resist heat, oxidation, sulfidation, and high-temperature corrosion.

REVIEW QUESTIONS

1. Describe the function of *inlet guide vanes*.
2. List the two types of compressors currently used in turbine engines.
3. Define the term *compressor pressure ratio*.
4. Describe the principle to operation of a *centrifugal-flow compressor*.
5. Describe the main components of an *axial-flow compressor*.
6. Discuss the advantages of axial-flow compressors.
7. Describe compressor stall.
8. What is the function of a *diffuser*?
9. List the three types of combustion chambers commonly used on turbine engines.
10. What is the function of the *turbine nozzle diaphragm*?
11. What is the purpose of the turbine?
12. List three types of turbine design.
13. Describe the function of *the exhaust nozzle*.
14. Describe the function and types of thrust reversers commonly being used on aircraft.
15. Why are anti-icing systems required on the inlet of some turbine engines?
16. What is normally used to heat the inlet of turbine engines for anti-icing purposes?
17. Where is the highest static pressure in the engine?
18. What prevents the metal of the combustion chamber from melting during engine operation?
19. Why do they make some turbine blades hollow?
20. What is *inertia bonding*?

Gas-Turbine Engine: Fuels and Fuel Systems 13

INTRODUCTION

Although aviation gasolines for piston engines have reached a high state of development and performance, these fuels are generally unsatisfactory for gas-turbine engines. Physical and thermal properties of aviation gasolines are such that extensive use of gasoline in turbines results in poor performance and possible engine damage. These factors generally limit aviation gasoline (avgas) to the role of an emergency fuel for turbine use.

Fuels suitable for gas-turbine engines have been in development for a number of years. The first jet fuels were prepared for military use. Later, the advent of gas-turbine engines in commercial aircraft brought new demands for jet fuels. These events generated extensive research to develop a fuel which embodied the properties and performance characteristics for optimum gas-turbine engine efficiency. In development of a production fuel, however, some compromise in desired features was necessary to achieve a proper balance among fuel availability, economy, handling, safety, and performance characteristics.

FUEL REQUIREMENTS

In general, a fuel for gas-turbine engines should have the following qualities:

1. Be pumpable and flow easily under all operating conditions
2. Permit engine starting under all ground conditions and give satisfactory flight relighting characteristics
3. Give efficient combustion under all conditions
4. Have as high a calorific (heat) value as possible
5. Produce minimal harmful effects on the combustion system or turbine blades
6. Produce minimal corrosive effects on fuel system components
7. Provide adequate lubrication for the moving parts of the fuel system
8. Reduce fire hazards to a minimum

The wide use of jet fuels has brought with it certain new problem areas which require diligent inspection and control to ensure safe and efficient aircraft performance. Among these are water and microbiological fuel contamination, static buildup and electric conductivity, and refueling hazards of intermixed fuels.

JET FUEL PROPERTIES AND CHARACTERISTICS

An essential property of any fuel is high calorific or heating value, since fuel is the basic source of energy. The heating value depends primarily on the type of hydrocarbons present in the fuel as determined in the refining process. Although the highest available heating value would be desirable, some compromise is necessary to maintain volatility, smoking, and other qualities within required ranges. Control of the heating value in jet fuels is based primarily on the limits set on specific gravity and volatility factors.

The sulfur content of jet fuels is held to low values since sulfur compounds cause corrosive deterioration of engine and fuel system components and produce objectionable odors.

The fuel freezing point is also an important factor in fuel specifications. When the freezing point of fuel is approached, waxy particles begin to form. Clogged filters and malfunctioning fuel control system components are the result. For turbine fuels, the freezing points range from −40 to −76°F [−40 to −60°C].

Combustion characteristics of fuel are carefully considered so that optimum benefit may be derived from the calorific value. Blowout limits, relighting, smoking, combustion efficiency, carbon formation, and life of high-temperature engine components are all directly or indirectly influenced by fuel hydrocarbon type and volatility. Clean and efficient combustion is more difficult with fuels having high molecular weights or high carbon-hydrogen ratios. Improvements in engine design have eased the earlier stringent requirements in this area.

Aside from its influence on engine performance, the **volatility** of a fuel is important in its relation to vapor-air mixtures which develop in the space above the fuel under various temperature and pressure conditions. Fuels with low volatility are desirable because of low boiling losses and the narrower range of temperatures in which ignitable F/A mixtures can be produced. Volatility is controlled by setting limits on specifications such as flash point, vapor pressure, distillation, and boiling point.

Types of Jet Fuels

Fuels generally available for use in turboprop and turbojet aircraft may be broadly categorized into two types: kerosene and wide-cut gasolines.

Kerosene-Type Jet Fuels (Jet A). Kerosene appears to be the more commonly available fuel for commercial aircraft. It is developed under a number of specifications, although differences among the various products are only slight.

Kerosene-type fuels consist essentially of the heavier hydrocarbon fractions and are more dense than the wide-cut gasolines. Because of its greater density, kerosene has a higher calorific or heating value per gallon (approximately 3 to 5 percent) than gasoline. However, the heating value of kerosene is slightly less than that of gasoline (approximately 0.3 to 0.5 percent) when measured in Btu per pound.

The freezing point of kerosene-type fuels varies somewhat with the type of hydrocarbon fraction present. Solid, waxy particles capable of clogging fuel filters begin to form at a temperature slightly above the pour point. For some kerosene, this occurs at approximately −40°F [−40°C] and for others as low as −58°F [−50°C].

The vapor pressure of kerosene-type fuels is extremely low, averaging about 0.125 psi [0.861 9 kPa]. As a result, fuel boiling and evaporation losses are negligible for flight conditions normally encountered. Also, the tendency to develop ignitable F/A mixtures in the space above the fuel is less than with the gasoline-type fuels.

Wide-Cut Gasolines (Jet B). The **wide-cut gasolines** are refined to contain various mixtures of gasoline and kerosene-type hydrocarbon fractions, therefore the term "wide-cut." Contributing to the ready availability of this fuel is the fact that crude oils may yield up to 50 percent of the wide-cut fractions.

Wide-cut gasolines are ordinarily less dense than kerosene-type fuels due to the presence of the lighter hydrocarbon fractions. The weight is 0.2 to 0.5 lb/gal [0.024 to 0.060 kg/L] less than the average kerosene (see Fig. 13-1), but the density range of wide-cut gasolines is quite wide, and a heavy fuel of this type could weigh more than a light kerosene. The heating value in Btu per pound for an average

wide-cut gasoline is slightly higher than that for an average kerosene.

An advantage of wide-cut gasoline is its relatively low freezing point, averaging less than −76°F [−60°C]. In addition, the viscosity of wide-cut gasoline stays low in colder temperatures. Jet B fuel is suited to operation at high altitudes where low temperatures prevail and in geographic areas subject to low-temperature ground conditions.

The wide-cut gasolines have a vapor pressure varying between 2 and 3 psi [13.8 and 20.7 kPa]. They are less volatile than aviation gasolines (6 to 7 psi [41.4 to 48.3 kPa] of vapor pressure) but somewhat more volatile than kerosene-type fuels. Compared with aviation gasolines and kerosene, the wide-cut gasolines generally produce flammable mixtures over a wider range of temperature and altitude conditions.

Water in Fuel

Although rigorous precautions are taken to ensure that fuel being pumped into an aircraft contains as little water as possible, aircraft fuel containing no water is an impossibility. The affinity that fuel has for water varies with its composition and temperature. The **saturation level** for a jet fuel in parts per million (ppm) by volume is approximately equivalent to the temperature in degrees Fahrenheit; that is, a jet fuel at 50°F [10°C] may contain approximately 50 ppm of dissolved water. When the fuel is cooled, that water which is above the saturation level is rejected as discrete water in minute particles. Until this water can collect and settle to the bottom of the tank, it will be carried in the fuel. At temperatures below the freezing point, these minute particles may be supercooled and will be deposited out only when they strike a solid obstruction and freeze. When low temperatures are encountered, water droplets combine with the fuel to form a frozen substance referred to as **gel**. The mass of gel, or "icing," that may be generated from moisture held in suspension in jet fuel can be much greater than that in gasoline.

An aircraft, after a long flight at high altitudes, will have fuel tank surfaces and fuel that are colder than the air being drawn into the tank during descent. When moisture-laden air enters the tank space, condensation may occur in the tank. Because of the higher viscosity of cold fuel, this water will not settle out readily and will be carried as dispersed water. Under these conditions, the dispersed water in the fuel may reach 100 ppm.

FUEL	AVERAGE SPECIFIC GRAVITY @ 15.6°C	AVERAGE WEIGHT IN LBS/GAL
AVIATION GASOLINE	0.696	5.7
KEROSENE	0.813	6.8
WIDE-CUT GASOLINE (JP-4)	0.773	6.5

FIGURE 13-1   Specific gravities and weights of typical aviation fuels.

Fuel Additives

Additives have been developed to combat problems with fuel icing. One such additive is **Prist**, manufactured and marketed by the Houston Chemical Co. Prist is designed to prevent ice and bacterial contamination in aviation fuel. It is sometimes referred to as PFA 55MB and is covered by specification MIL-I-27686D. Pure EGME (ethylene glycol monomethyl ether), the primary ingredient of Prist additive, is generally compatible with the components of turbine aircraft fuel systems. Prist is effective in preventing fuel ice because as the fuel temperature decreases, the additive combines with the water in greater concentrations. This combination will keep the freezing point of the water lower than the temperature of the fuel. Prist additive should be protected from contact with oxidizing agents, including long-term contact with air.

The recommended concentration of Prist in fuel is from 0.06 to 0.15 percent by volume. It must be incrementally blended into the fuel because its solubility in fuel is limited. It is completely soluble in water suspended in the fuel. Fuel-additive proportioners are engineered to blend the additive into fuel during refueling so that the turbulence of the flowing fuel stream will provide a mixing action. In all cases, additive must be used as directed by the manufacturer. No fuel additive should be used in a particular system unless the system has been approved for it.

Microorganisms in Fuel

Microorganisms of various types can grow in an aircraft fuel tank and in storage tanks if water is present. Microorganisms grow in fuel usually when the fuel is not disturbed frequently. If an airplane is stored with fuel and water in the fuel tank for an extended time, chances are that microorganisms will grow and appear as a slimy deposit in the water. The color will vary from brown to black, with all shades in between, including red.

When fuel has been stored for a lengthy period, either in an aircraft or in a fuel storage tank, the fuel and tank should be examined for signs of microbial contamination. If the fuel is discolored or has an abnormal smell, the tank should be drained and cleaned. Contaminated fuel should be discarded.

Turbine-engine fuels are more likely than gasoline to be contaminated with microorganisms. This is because turbine engine fuels can dissolve more water, and water is harder to remove because of the viscosity of turbine fuels. Technicians who have the responsibility of fueling and defueling turbine-powered aircraft must be alert to detect microorganism contamination and take the required measures to eliminate it. Fueling procedures established by airlines provide for protection against microorganism contamination and should be followed carefully.

The effects of microorganisms in fuel are serious regardless of whether the organisms are bacteria or fungi. Corrosive chemicals are formed which attack the metal walls and bottoms of fuel tanks. Because of this, many aircraft have suffered structural weakening of the wings since fuel tanks in large aircraft are usually integral with the wings. It has been the practice with integral wing tanks (wet wings) to coat the interior surfaces of the tank sections with a corrosion-resistant material. In some cases the microorganisms have penetrated the coating and caused severe corrosion. New coatings have been developed that resist not only corrosion but also attacks by the microorganisms. It is, therefore, important that the fuel tanks of aircraft which have been in service for a substantial time be thoroughly inspected inside. If evidence of microorganisms or corrosion exists, the interior of the tank should be cleaned, stripped, and recoated. If appreciable corrosion exists, replacing or repairing sections of the tank walls and bottoms may be necessary.

Refueling Precautions for Intermixed Jet Fuels

A major consideration in servicing an aircraft with a different type of jet fuel is the F/A mixture which develops in the space above the fuel. At normally encountered ground temperatures, the vapor pressure of kerosene is too low to develop an ignitable mixture. Under similar conditions, gasoline-type fuels produce an overrich mixture which will not ignite. However, when these two fuels are mixed, the fuel-vapor mixture is in the ignitable envelope throughout a much broader range of ground temperatures common at most operating locations. This condition exists when wide-cut gasoline (jet B) is added to a tank already containing kerosene (jet A) or vice versa.

The extra hazard of an ignitable mixture resulting from intermixing of fuels calls for special refueling procedures to minimize the buildup of electrostatic charges on the fuel. Precautionary procedures include underwing refueling only and reduction of flow rates by using a single hose attached to the fueling manifold.

Certain facts about electrostatic buildup in fuels should be explained. Although the exact nature of **electrostatic buildup** is not fully understood, recent tests and experience have yielded information to aid in its control. Studies show that a charge potential can develop and produce high-intensity sparks between the surface of the fuel and some point inside the tank structure, even though all structural parts of the tank are electrically bonded together. Turbulence within the fuel, high-velocity fuel flow, and fuel splashing tend to increase the static buildup.

A major factor in the development of charge potentials is the electric conductivity of the fuel. Clean fuel is a relatively poor conductor of electric charges. Therefore, a considerable charge potential may develop as a result of the slow dissipation or **bleed-off** rate of static charges in the fuel. The presence of water, dirt, and other contamination in the fuel improves fuel conductivity and aids in dissipating the static charge. However, use of contaminated fuel cannot be condoned even though it may be of some value in reducing static buildup.

The solution to the static buildup problem may lie in the development of fuel additives. Several patented additives are available which claim to reduce static buildup by improving fuel conductivity.

FIGURE 13-2 Submersible fuel boost pump.

Gas-Turbine Fuel System Components

The fuel supply is always carried in the airframe and may be contained in one or more fuel cells. Regardless of how many individual cells an aircraft system may contain, fuel is essentially fed to each engine from one main tank. Many transmitters may also be used to signal fuel quantity, pressure, flow, and temperature, which are read out on instruments in the cockpit.

Aircraft boost pumps, such as the one illustrated in Fig. 13-2, are located in the fuel tank. These electrically operated pumps are of the centrifugal or impeller type and maintain a constant fuel pressure in the line between the tank and the engine-driven fuel pump. An aircraft-mounted fuel shutoff valve, usually electrically operated, is installed in the line between the tank and the engine to allow fuel lines to be disconnected at the engine without danger of fuel spillage.

Located in the airframe structure downstream of the shutoff valve is a low-pressure fuel filter. This filter, illustrated in Fig. 13-3, is installed in the system to remove any large pieces of foreign material from the fuel which might cause damage to the engine-driven fuel pump.

In case the filter becomes clogged, a relief valve is provided to bypass fuel flow around the filter and continue to supply the engine with normal flow.

The most common type of fuel supply pump used on jet engines is the spur-gear type shown in Fig. 13-4. This is a positive-displacement type of pump which may contain one, two, or more individual pumping elements, depending on the fuel requirements of the engine. Since this is a positive-displacement type of pump, it will always supply a constant flow at a given rotational speed. A high-pressure relief valve is often incorporated in the housing to prevent damage to the pump in the event of a major line restriction

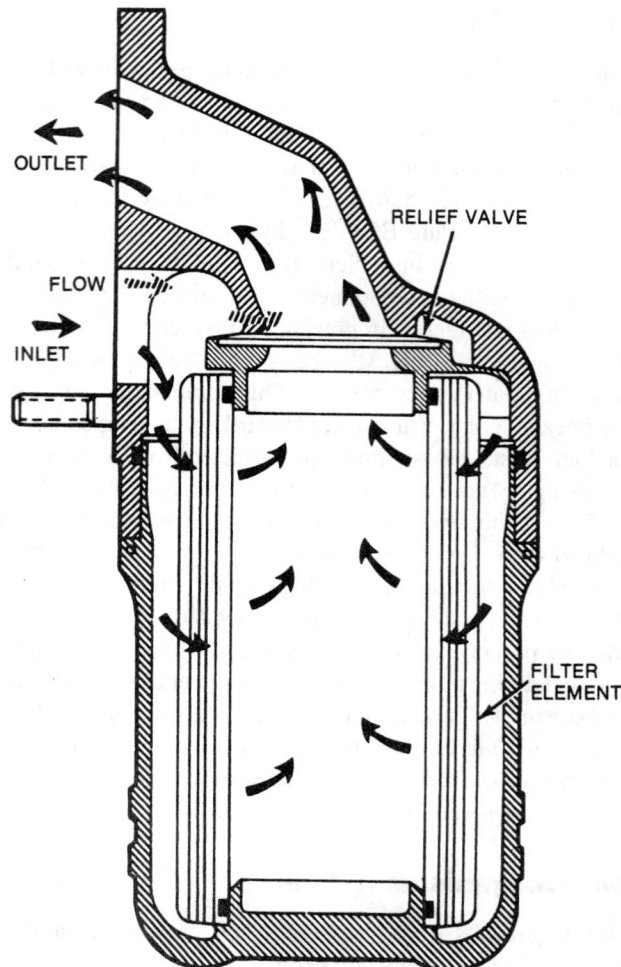

FIGURE 13-3 Fuel filter. (*General Electric Co., U.S.A.*)

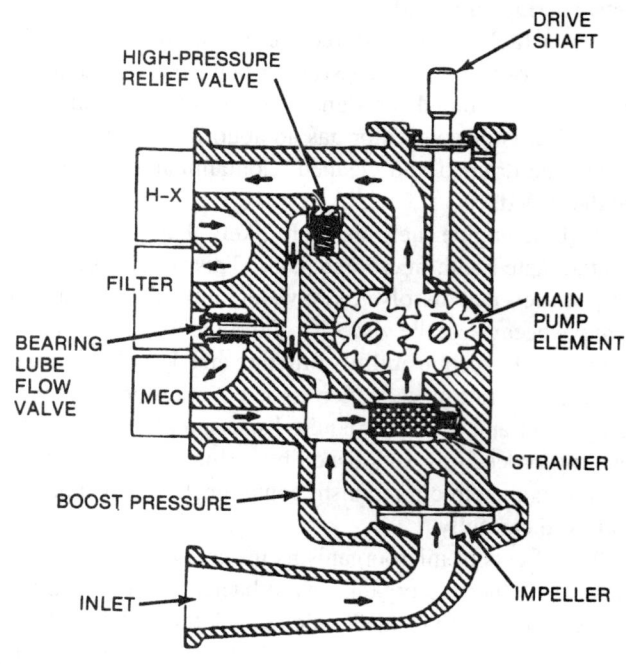

FIGURE 13-4 Spur-gear fuel pump. (*General Electric Co., U.S.A.*)

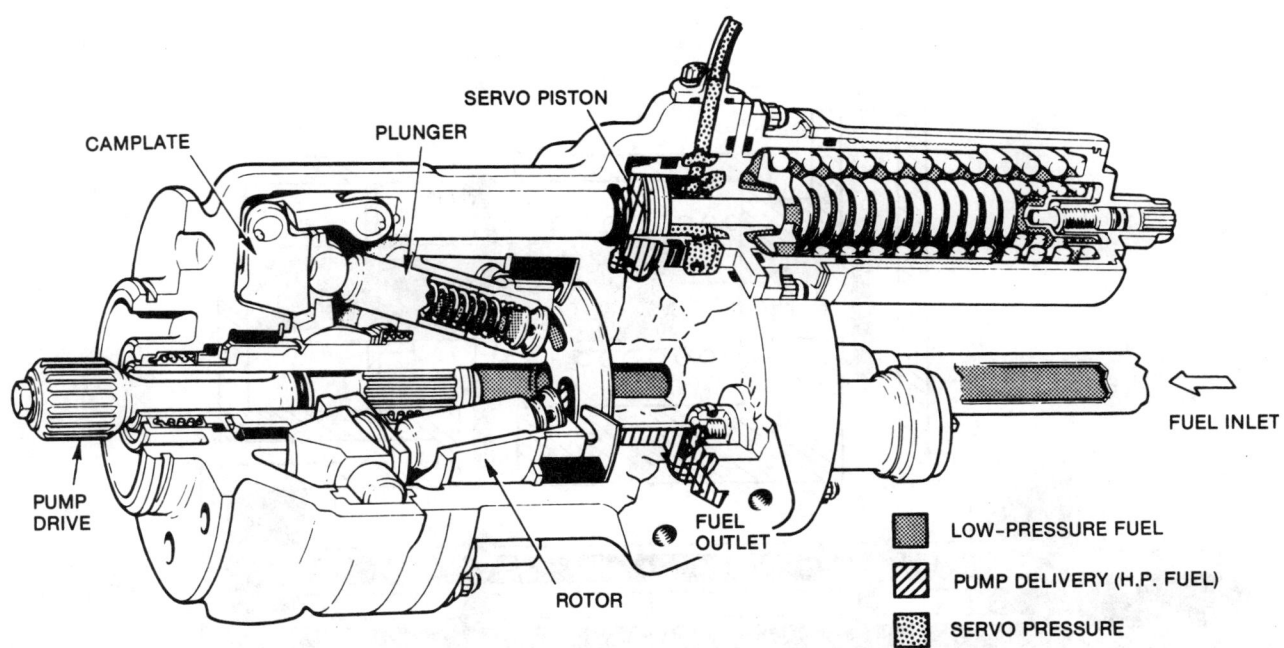

FIGURE 13-5   Plunger-type fuel pump. (*Rolls-Royce.*)

downstream of the pump. Another type of high-pressure pump is the plunger-type fuel pump, illustrated in Fig. 13-5. This fuel pump is driven by the engine gear train, and its output depends on its rotational speed and the stroke of the plungers. A single-unit fuel pump can deliver fuel at the rate of 100 to 2000 gal/h [378.5 to 7 570 L/h] at a maximum pressure of about 2000 psi [13 790 kPa]. To drive this pump, as much as 60 hp may be required.

From the engine-driven fuel pump, fuel is supplied to an engine-mounted high-pressure fuel filter such as that shown in Fig. 13-6. Any foreign material large enough to damage a fuel system component will be removed from the fuel by

this filter. At this point in the system, fuel flows through and is regulated by the main engine fuel control, which is discussed in detail later in this chapter.

After the fuel control, the fuel may pass through the fuel-cooled oil cooler if it has not done so earlier in the system. The purpose of a fuel-cooled oil cooler is to exchange the heat from the oil with that of the fuel. This accomplishes the purpose of warming the fuel and cooling the oil. A fuel-cooled oil cooler is illustrated in Fig. 13-7. Additional information on fuel system components and operation may be found in the text *Aircraft Maintenance and Repair*.

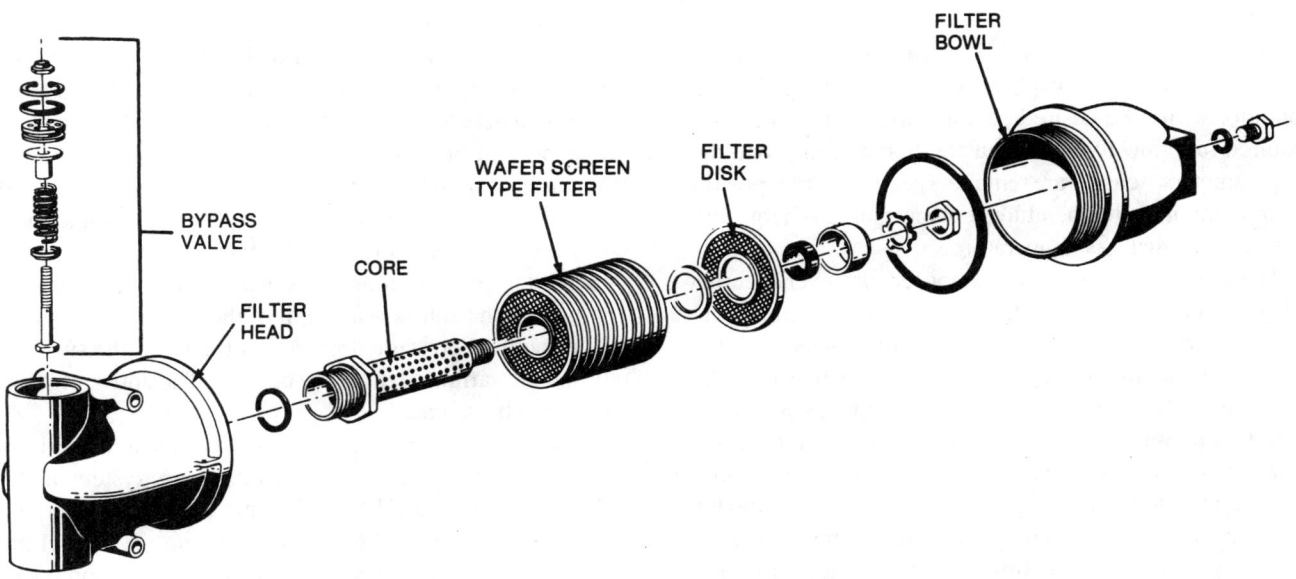

FIGURE 13-6   Typical mesh screen filter assembly.

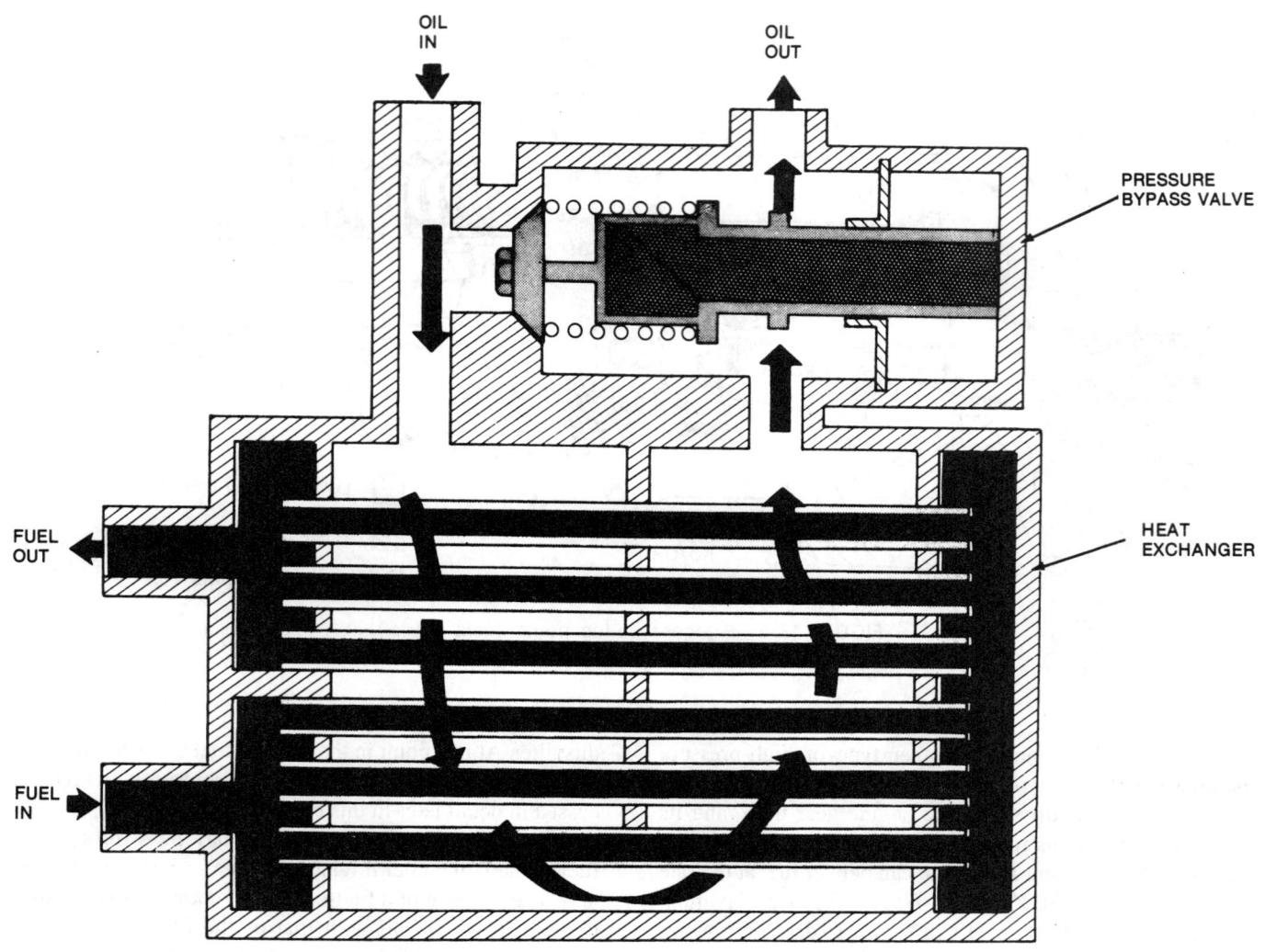

FIGURE 13-7 Fuel-cooled oil cooler.

Gas-Turbine Engine Fuel Systems

The purpose of a main fuel system is to provide a measured flow of fuel to the engine for a given set of engine operating conditions.

The fuel system of a turbojet engine senses a number of current engine and atmospheric conditions, compares these conditions with the desired conditions, and then meters the required fuel flow to the combustor. Some of the conditions or parameters sensed are engine speed, throttle position, compressor inlet temperature, compressor discharge pressure, and exhaust-gas temperature.

The main fuel system is composed of several components, mounted in the airframe and on the engine. The common airframe components are the fuel tanks, fuel boost pumps, fuel shutoff valve, and low-pressure fuel filter. The engine-mounted components in a typical system are the main fuel pump, fuel filter, main engine control, oil cooler, flow divider, and fuel nozzles. The fuel system can be divided into low- and high-pressure systems. The low-pressure system must supply the fuel to the engine at a suitable pressure, rate of flow, and temperature to ensure satisfactory engine operation. This system may include a

low-pressure pump to prevent vapor locking of the fuel and to provide a steady supply of fuel to the engine-driven pump. It also includes a fuel heater to prevent ice crystals from forming. A fuel filter is used in the system, and in some instances the flow passes through an oil cooler. On many engines, a fuel-cooled oil cooler is located between the fuel pump and the inlet to the fuel filter. The oil cooler is placed here to transfer the heat from the oil to the fuel and thus prevent blockage of the filter element by ice particles. When heat transfer by this means is insufficient, the fuel is passed through a fuel heater. There is usually some method to warn the pilot or flight engineer of this icing danger. One method uses a differential-pressure switch that obtains signals from the entrance and exit of the fuel filter. When there is a specified pressure drop across the filter, the switch will turn on a warning light in the cockpit. The anti-icing system can then be actuated.

The diagram in Fig. 13-8 represents a typical high-pressure fuel system. The high-pressure system includes the high-pressure (HP) fuel pump, fuel/oil cooler, filter, fuel control unit, fuel pressurizing/dump valve, fuel manifolds, fuel spray nozzles, and various other components. Fuel enters the system from the fuel supply and then passes

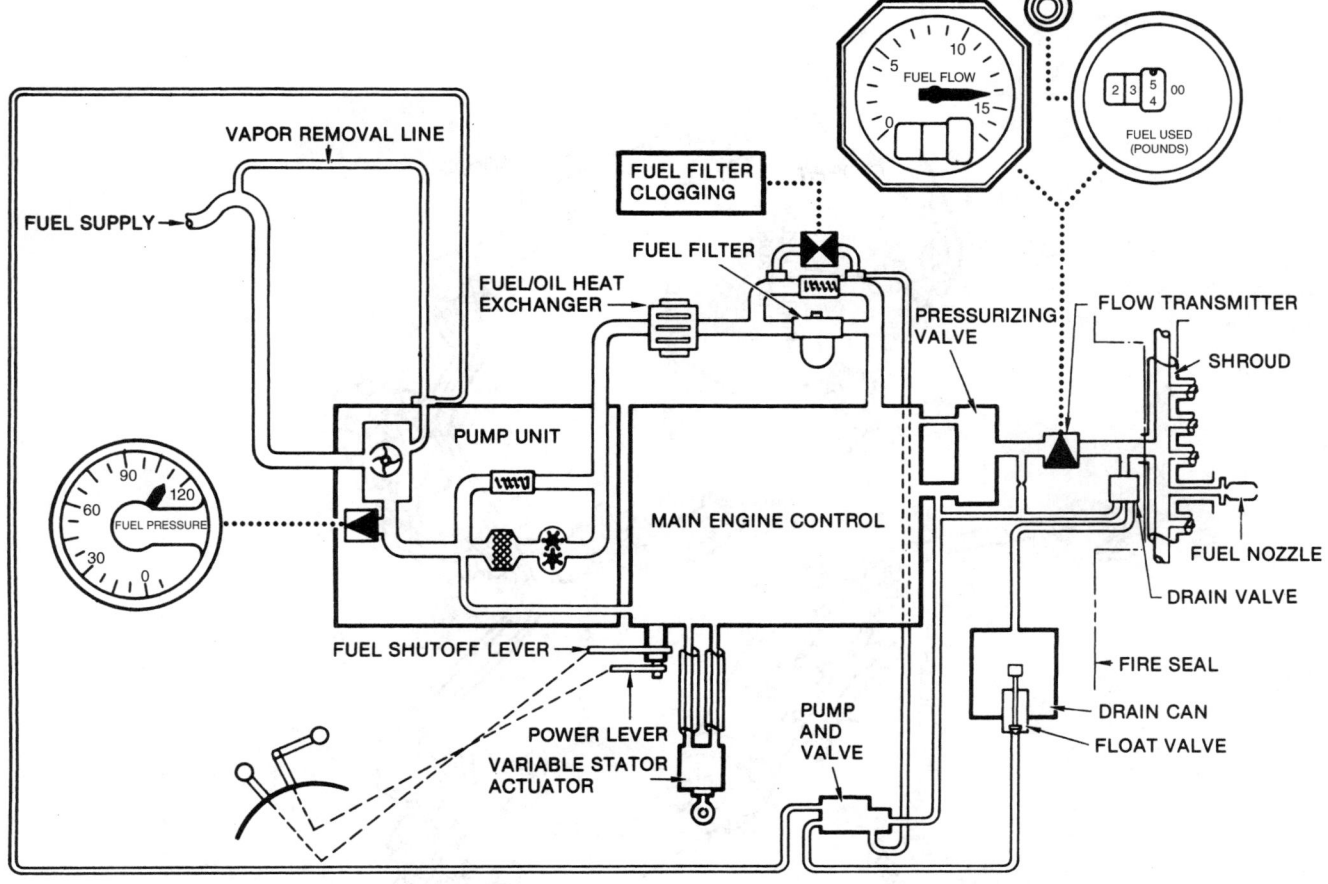

FIGURE 13-8 Turbine aircraft fuel system.

through a low-pressure pump and then through the high-pressure pump where it is pressurized up to several hundred psi. From the high-pressure pump the fuel enters the fuel/oil heat exchanger which is used to cool the oil. The fuel filter is next in the flow path to the engine. The filter has a bypass valve that will bypass fuel if the filter becomes clogged. As mentioned earlier, the filter inlet and outlet pressures are monitored to determine if the filter has become clogged. The fuel then passes into the main engine fuel control which meters the amount of fuel the engine requires for a given engine condition or rpm. Fuel on its way to the engine enters the pressurizing/dump valve which either directs fuel to the engine, such as in the engine running mode, or dumps fuel from the engine, as in the engine stop position. The fuel continues through the fuel flow transmitter and on to the fuel manifold, which distributes the fuel to each fuel nozzle.

Illustrated in Fig. 13-9 is another gas-turbine engine fuel system. This system has many of the same components as described earlier, but it uses a flow divider just before the fuel manifold which separates the fuel into primary and secondary fuel flow. The primary fuel flow is used during starting of the engine. As the engine requires more fuel at high engine speeds, the secondary flow begins.

Large Aircraft Turbofan Engine Fuel Systems

The fuel flow passes through several components which are all mounted around or on the engine, except for the fuel shutoff valves, which can be at the fuel tank or other places on the airframe. Fuel from the tank boost pumps is pressurized in the two stages of the engine-driven fuel pump, as shown in Fig. 13-10. The lower-pressure primary stage of the fuel pump supplies fuel through the fuel heater and fuel filter before the fuel passes into the secondary stage of the engine-driven pump. Fuel which passes through the fuel heater can be heated by allowing hot bleed air from the engine's compressors to pass through the fuel heater. The flow of bleed air is controlled by a switch on the flight engineer's panel. When the switch is turned on, the valve to the fuel heater is opened and the bleed air is allowed to flow through the fuel heater. After passing through the heater, the fuel flows next into the fuel filter, which is used to filter out impurities and ice crystals in the fuel. If enough ice crystals are present in the filter (caused by freezing of water droplets in the fuel tank), they can clog the filter. When the filter becomes clogged with ice crystals, the differential-pressure switch turns on the icing light, which informs the flight crew that icing is occurring in

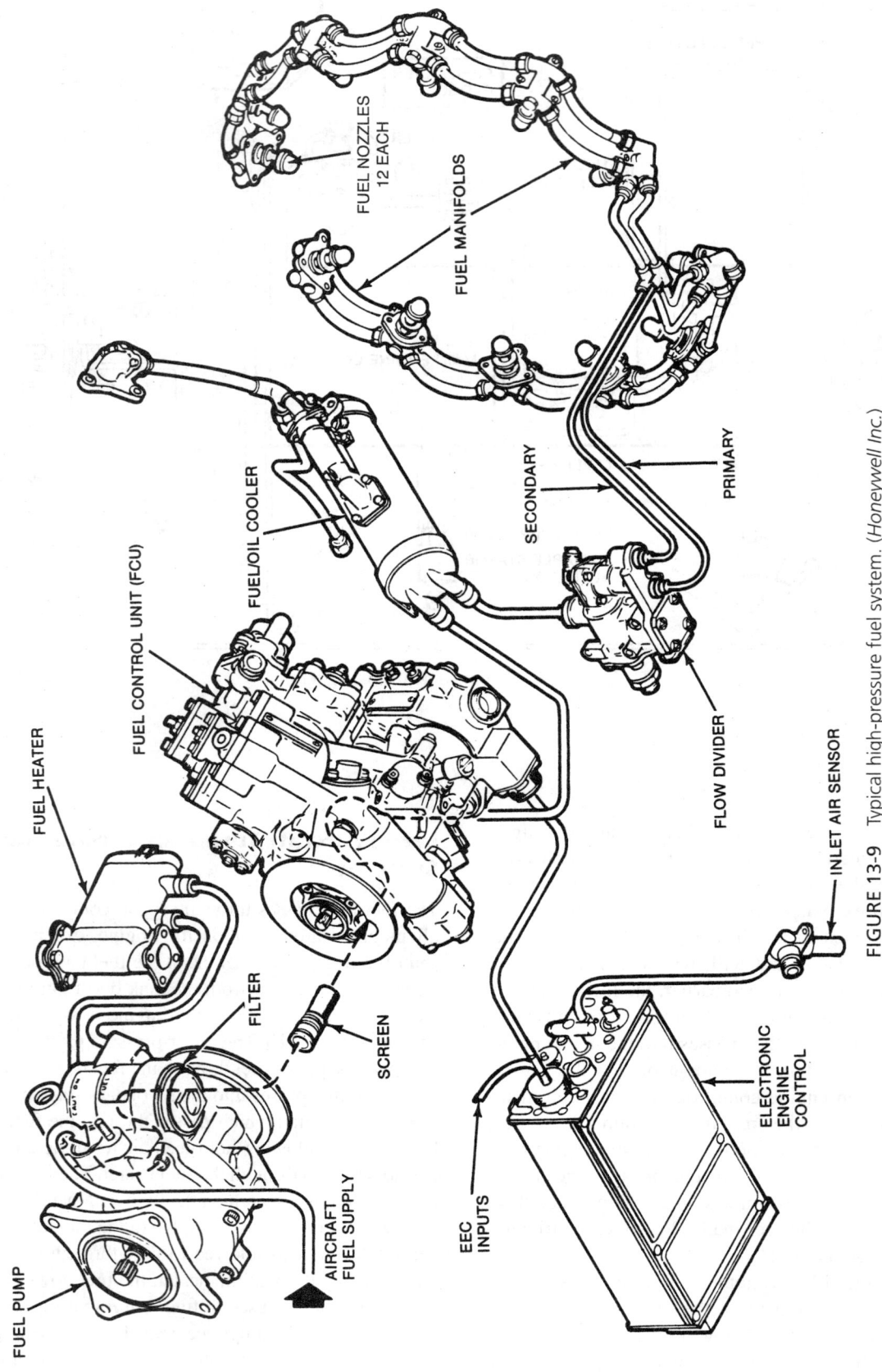

FUEL NOZZLES
12 EACH

FUEL MANIFOLDS

FUEL/OIL COOLER

FUEL CONTROL UNIT (FCU)

FUEL HEATER

SECONDARY

PRIMARY

FLOW DIVIDER

INLET AIR SENSOR

FILTER

SCREEN

ELECTRONIC
ENGINE
CONTROL

EEC
INPUTS

AIRCRAFT
FUEL SUPPLY

FUEL PUMP

FIGURE 13-9   Typical high-pressure fuel system. (*Honeywell Inc.*)

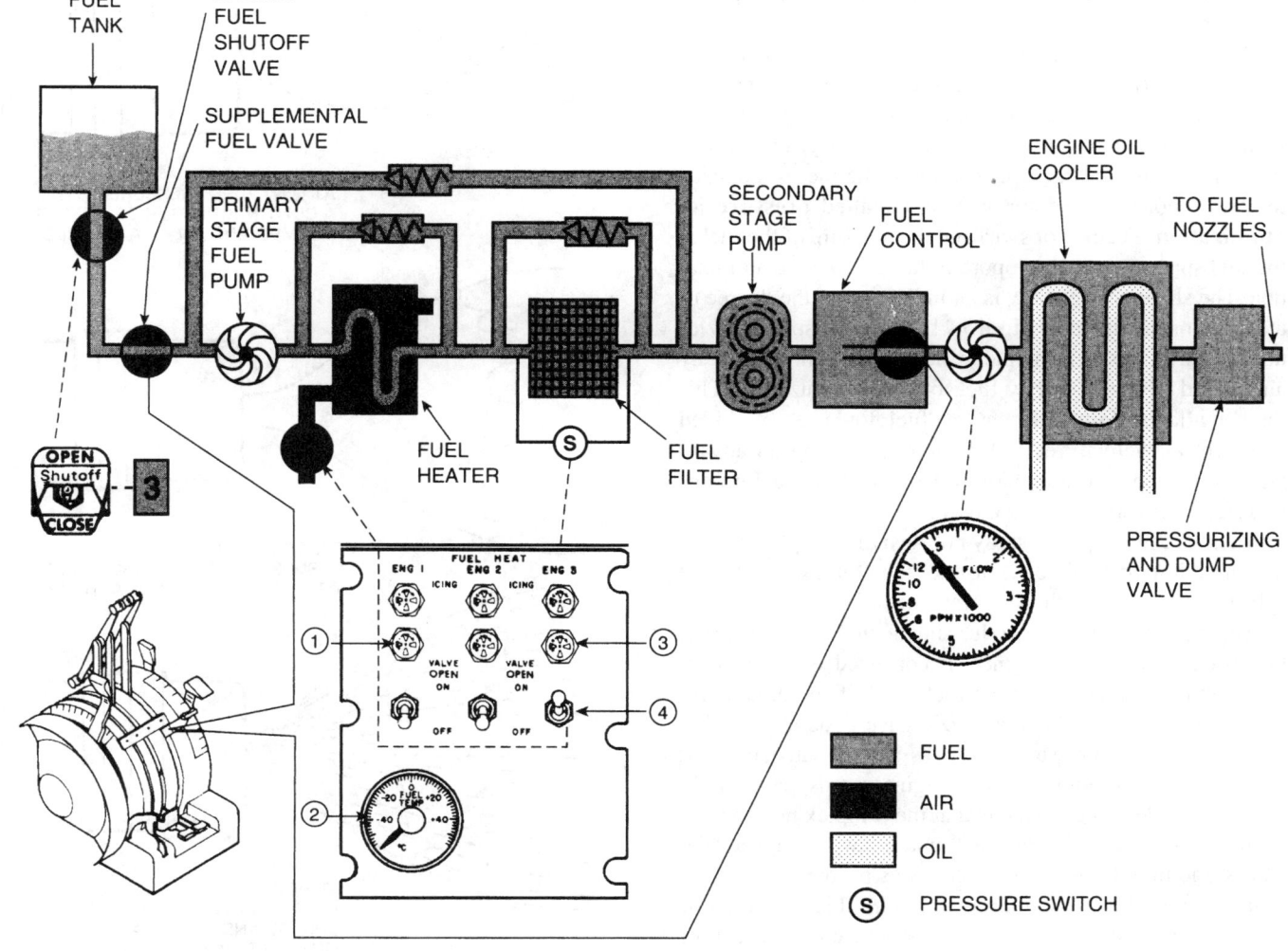

Fuel Heater

1. **Valve Open Light**
   Comes on when the fuel heat hot air valve is not closed.

2. **Fuel Temperature Indicator**
   Indicates the temperature (°C) in the No. 1 fuel tank.

3. **Fuel Heat Icing Light**
   Comes on when the fuel filter is blocked by ice crystals or other foreign matter.

4. **Fuel Heat Switch**
   When ON, opens fuel heat hot air valve to admit high temperature engine bleed air to the fuel heater.

FIGURE 13-10   Large turbofan engine fuel system.

the fuel filter. A fuel temperature indicator monitors the fuel temperature, and if the temperature of the fuel drops below freezing the fuel heating system can be activated. If for any reason the primary fuel pump, the fuel heater, or the fuel filter should become clogged, there are bypasses around each of the previously mentioned components. After passing through the fuel filter, the fuel flows next into the secondary-stage high-pressure pump, which increases the fuel pressure greatly. The hydromechanical fuel control meters the fuel needed and is operated, lubricated, and cooled by high-pressure fuel. Fuel is metered as a function of throttle position, N_2, compressor inlet temperature, and compressor discharge pressure.

After the fuel control meters the fuel, the fuel then flows through the fuel flow transmitter before entering the engine oil cooler. The engine oil cooler uses the fuel entering the engine to cool the oil and also warms the fuel. After exiting the oil cooler, the fuel passes through the pressurizing and dump valve and into the fuel manifold, where it is distributed to the fuel nozzles and sprayed into the combustion section of the engine.

Fuel Spray Nozzles

The final components of the fuel system are the **fuel spray nozzles**, which have as their essential function the task of atomizing or vaporizing the fuel to ensure its rapid burning. The difficulties involved in this process are the high velocity of the airstream from the compressor and the short length of the combustion system in which the burning must be completed.

An early method of atomizing the fuel was to pass it through a swirl chamber where holes or slots caused the fuel to swirl by converting its pressure energy to kinetic energy. In this state, the fuel is passed through the discharge orifice, which removes the swirl motion as the fuel is atomized to form a cone-shaped spray. This is called **pressure jet atomization**. The rate of swirl and the pressure of the fuel at the fuel spray nozzle are important factors in good atomization. The shape of the spray is an indication of the degree of atomization, as shown in Fig. 13-11. Later fuel spray nozzles utilized the air spray principle, which employs high-velocity air instead of high-velocity fuel to cause atomization. This method allows **atomization** at low fuel flow rates (provided sufficient air velocity exists), thereby providing an advantage over the pressure jet atomizer by allowing fuel pumps of a lighter construction to be used.

There are five types of spray nozzles: the Simplex nozzle, the variable-port (Lubbock) nozzle, the Duplex nozzle, the spill-type nozzle, and the air spray nozzle.

The **Simplex spray nozzle**, shown in Fig. 13-12, was first used on early jet engines. It consisted of a chamber. which induced a swirl into the fuel, and a fixed-area atomizing orifice. This fuel spray nozzle gave good atomization at the higher fuel pressures, but was very unsatisfactory at the low pressures required at low engine speeds, particularly at high altitudes. The reason is that the Simplex nozzle was, by the nature of its design, a "square-law" spray nozzle; that is, the flow through the nozzle was proportional to the square root of the pressure drop across it. This meant that if the minimum pressure for effective atomization was 30 psi [207 kPa], the pressure needed to give maximum flow would be about 3000 psi [20 700 kPa]. The fuel pumps available at that time were unable to cope with such high pressures, so the variable-port spray nozzle was developed in an effort to overcome the square-law effect.

The **variable-port** or **Lubbock nozzle**, illustrated in Fig. 13-13, made use of a spring-loaded piston to control the area of the inlet ports to the swirl chamber. At low fuel flows, the ports were partly uncovered by the movement of the piston; at high flows, they were fully open. By this method, the square-law pressure relationship was mainly overcome, and good atomization was maintained over a wide range of fuel flows. The matching of sets of spray nozzles and the sticking of the sliding piston due to dirt particles were, however, difficulties inherent in the design, and this type was eventually replaced by the Duplex fuel spray nozzles.

The **Duplex nozzle**, shown in Fig. 13-14, requires a primary and a main fuel manifold and has two independent orifices, one much smaller than the other. The smaller orifice handles the lower flows, and the larger orifice deals with the higher flows as the fuel pressure increases. A pressurizing valve may be employed with this type of spray nozzle to apportion the fuel to the manifolds. As the fuel flow and pressure increase, the pressurizing valve moves to progressively admit fuel to the main manifold and the main orifices. This gives a combined flow down both manifolds. In this way, the Duplex nozzle is

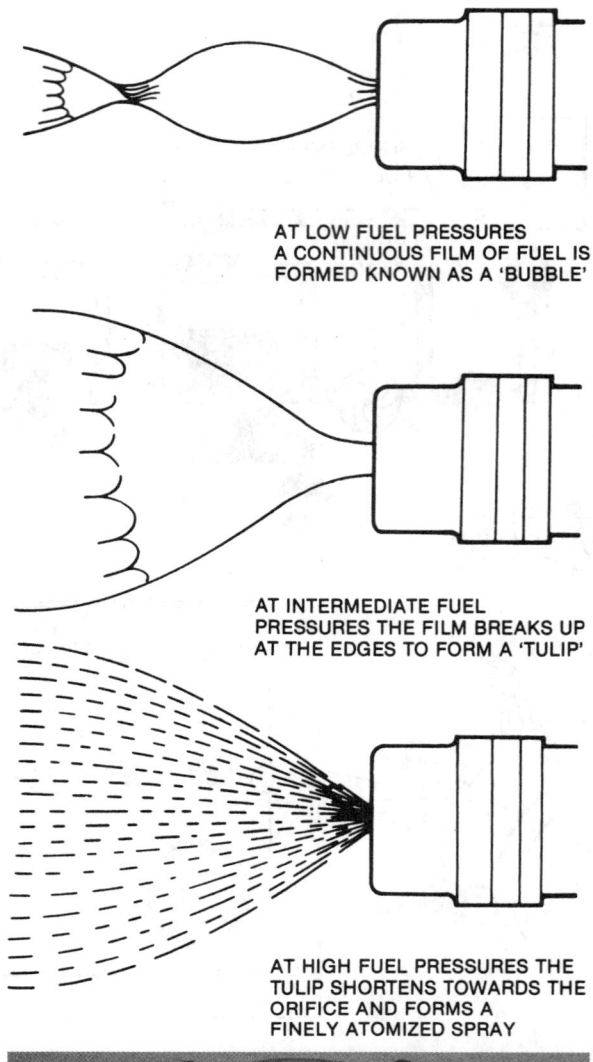

AT LOW FUEL PRESSURES A CONTINUOUS FILM OF FUEL IS FORMED KNOWN AS A 'BUBBLE'

AT INTERMEDIATE FUEL PRESSURES THE FILM BREAKS UP AT THE EDGES TO FORM A 'TULIP'

AT HIGH FUEL PRESSURES THE TULIP SHORTENS TOWARDS THE ORIFICE AND FORMS A FINELY ATOMIZED SPRAY

COMBUSTION AREA   FUEL NOZZLES

FIGURE 13-11   Full flow from all fuel nozzles into the combustion chamber.

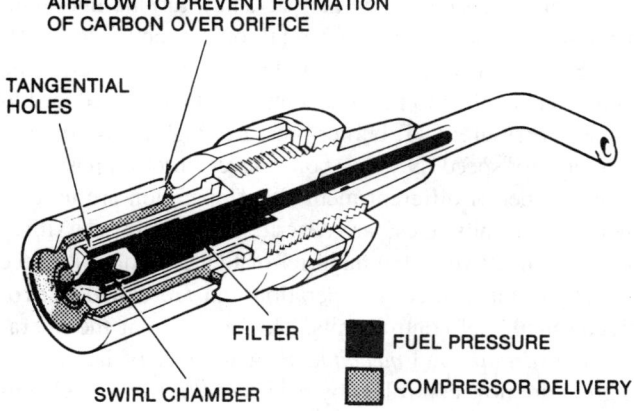

FIGURE 13-12   Simplex fuel spray nozzle. (*Rolls-Royce.*)

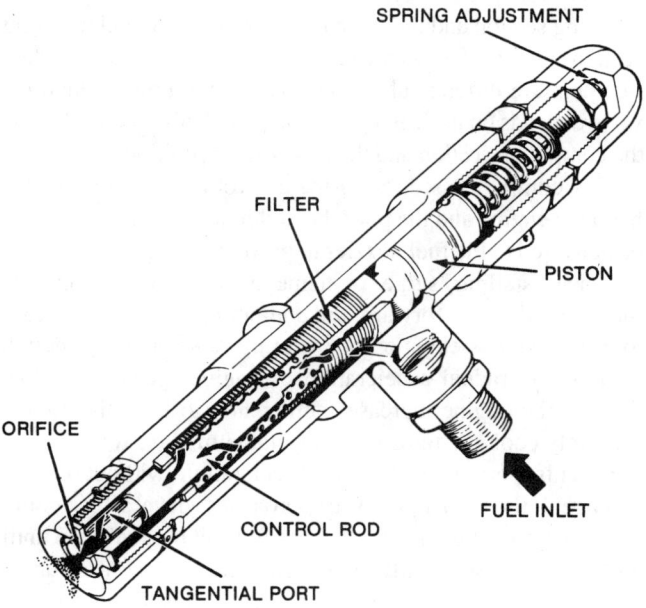

FIGURE 13-13   Variable-port or Lubbock fuel spray nozzle. (*Rolls-Royce.*)

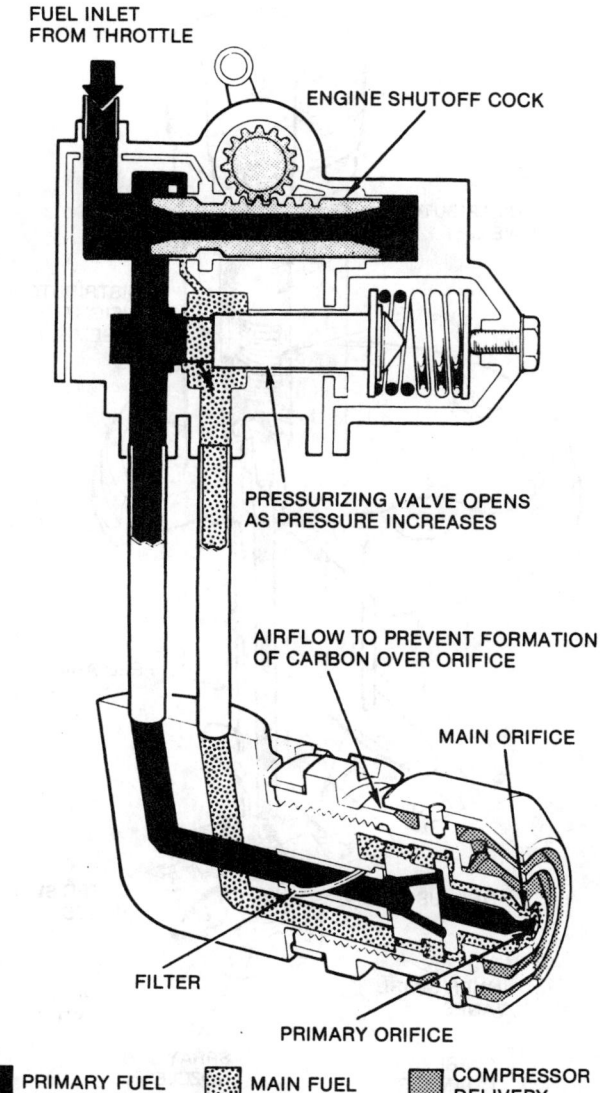

FIGURE 13-14   Duplex fuel spray nozzle and pressurizing valve. (*Rolls-Royce.*)

able to give effective atomization over a wider flow range than the Simplex spray nozzle for the same maximum fuel pressure. Also, efficient atomization is obtained at the low flows that may be required at high altitudes.

The **spill-type nozzle** can be described as being a Simplex spray nozzle with a passage from the swirl chamber for spilling fuel away. With this arrangement it is possible to supply fuel to the swirl chamber at a high pressure at all times. As the fuel demand decreases with altitude or reduction in engine speed, more fuel is spilled away from the swirl chamber, leaving less to pass through the atomizing orifice. Constant use of a relatively high pressure by the spill spray nozzle means that even at the extremely low fuel flows that occur at high altitudes there is adequate swirl to provide constant and efficient atomization of the fuel.

The spill spray nozzle system, however, involves a somewhat modified type of fuel supply and control system. A means has to be provided for removing the spill and

for controlling the amount of spill flow at various engine operating conditions. A disadvantage of this system is that excess heat may be generated when a large volume of fuel is being recirculated to the inlet. Such heat may eventually lead to a deterioration of the fuel.

The **air spray nozzle**, illustrated is Fig. 13-15, carries a portion of the primary combustion air with the injected fuel. By aerating the spray, the local fuel-rich concentrations produced by other types of spray nozzles are avoided, thus reducing both carbon formation and exhaust smoke. An additional advantage of the air spray nozzle is that the low pressures required for atomization of the fuel permit the use of the comparatively lighter gear-type pump.

PRINCIPLES OF FUEL CONTROL

Many different types and models of fuel control units have been designed and used for gas-turbine engines. The simplest

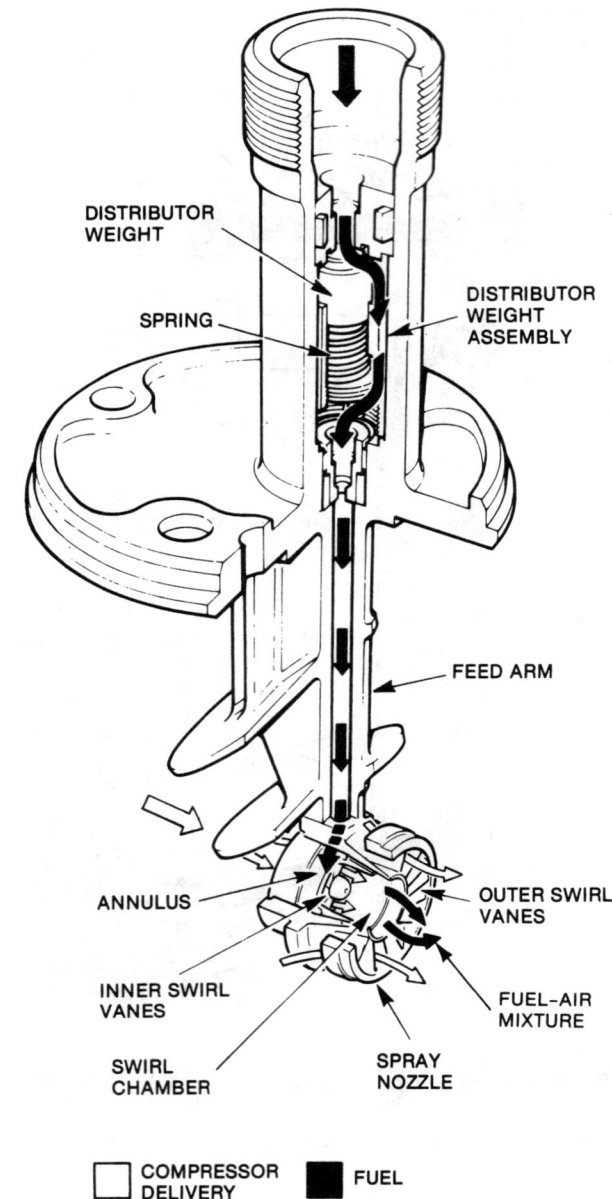

DISTRIBUTOR
WEIGHT

SPRING

DISTRIBUTOR
WEIGHT
ASSEMBLY

FEED ARM

ANNULUS

OUTER SWIRL
VANES

INNER SWIRL
VANES

FUEL-AIR
MIXTURE

SWIRL
CHAMBER

SPRAY
NOZZLE

☐ COMPRESSOR
DELIVERY   ■ FUEL

FIGURE 13-15   Air spray nozzle. (*Rolls-Royce.*)

type of control can be a manually controlled valve; however, this is not practical, because the pilot would have to watch several gauges and make frequent adjustments to keep the engine operating and to prevent damage. If the quantity of fuel is not correct for the velocity and pressure of the air, the engine cannot function. If too much fuel enters the combustion chamber or chambers, the turbine section may be damaged by excessive heat, the compressor may stall or surge because of back pressure from the combustion chambers, or a **rich blowout** may occur. A rich blowout takes place when the mixture is too rich to burn. If too little fuel enters the combustion chambers, a **lean die-out** occurs.

Among the various factors which must be considered in the regulation of fuel flow for a gas-turbine engine are ambient air pressure (P_{amb}), compressor inlet temperature (CIT), engine rpm, velocity of the air through the compressor, compressor inlet air pressure, compressor discharge pressure

(essentially the same as burner pressure), turbine inlet temperature, tailpipe temperature, and throttle or power lever setting. Since some of the foregoing factors are interrelated, the parameters applied to the fuel control unit can be reduced to ambient pressure, CIT, burner pressure, high-pressure compressor rotor speed (N_2), and power lever (throttle) position.

A number of different methods of operation are used in fuel control units; these are described in part by classification. For many years the majority of fuel control units have been hydromechanical in operation and are called **hydromechanical** fuel control units. This means that the operation is *hydraulic* and *mechanical*. A number of late-model engines are now controlled by means of electronic fuel control systems. These systems include computers that precisely measure all parameters and provide signals that result in maximum efficiency, reliability, and fuel economy.

The principal sections of a fuel control unit are the fuel metering section and the computing section. The fuel metering section consists of a fuel metering valve across which a constant fuel pressure differential is established. Fuel flow through the valve depends on its degree of opening, and this is controlled by the computing section and the power lever position.

The computing section of the control unit accepts signals from the engine and the pilot that tell it how much fuel should be delivered to the fuel nozzles to prevent damage from excessive heat, stalling, surge, or flameout. If the power lever is moved all the way forward, the computing section will begin to increase the opening of the metering valve but only enough to allow a gradual acceleration of engine speed as airflow through the engine increases sufficiently to keep the engine relatively cool and to provide a suitable mixture for combustion. This permits maximum acceleration without engine damage. If the burner pressure approaches a level which could cause surge or stall, the computer will limit the fuel flow until airflow is increased sufficiently. In some fuel control units, an air-bleed control system is included. This system causes bleed valves to open and allows excessive back pressure to be reduced in the compressor, thus avoiding stall or surge.

Figure 13-16 shows how one type of fuel control unit responds to engine conditions, air pressure and temperature,

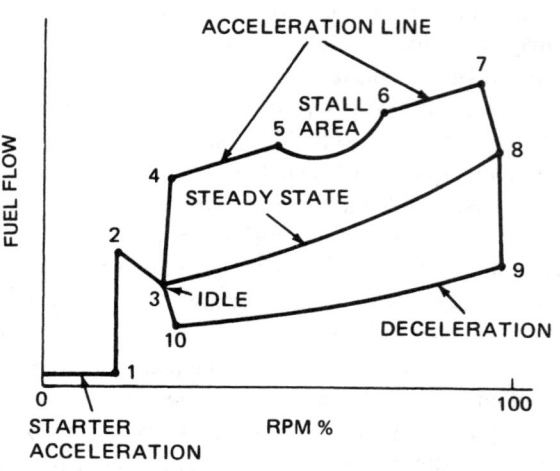

FIGURE 13-16   Curves showing operation of an FCU.

and power lever (throttle) position. In general, the following sequence takes place in a typical engine:

1. The engine is accelerated by the starter until the correct starting rpm is attained.

2. The power lever is advanced to the IDLE position to provide fuel (positions 1 to 2 on the curve).

3. As the engine rpm increases, fuel flow decreases from the acceleration level to the point where it is correct to sustain idle speed (position 3).

4. The power lever is advanced to the full-power position, and fuel flow is increased, as indicated by position 4.

5. As the engine rpm increases, fuel flow increases to position 5, and then it is decreased to avoid the stall and surge area. In the range of speeds between position 5 and position 6, the fuel must be reduced to prevent back pressure from the combustion chamber, which could cause compressor stall or surge.

6. At position 6, fuel flow is restored to the normal maximum acceleration level until the engine approaches maximum speed (position 7).

7. Shortly before maximum speed is reached, the governor reduces fuel flow to prevent overspeeding. The fuel flow is stabilized, as shown at position 8, at a quantity which will maintain 100 percent engine rpm.

8. When it is desired to reduce engine rpm to idle speed, the power lever is moved to the IDLE position. Fuel flow is immediately reduced to the value indicated by position 9 on the curve. This is a value which will permit maximum deceleration but will not let the F/A mixture drop to a ratio where a lean die-out may occur.

9. The engine rpm decreases, as indicated by the curve between positions 9 and 10, and the fuel flow decreases slowly.

10. When the engine rpm approaches idle speed, the fuel flow increases to the point where idle speed can be maintained. This is indicated by the curve between positions 10 and 3 in the diagram.

The curves in Fig. 13-16 will vary considerably as air pressure and temperature change. These values are applied to the computing section of the fuel control to modify fuel flow as conditions dictate.

FUEL CONTROL UNITS FOR A TURBOPROP ENGINE

An example of a **fuel control unit** (FCU) for a small turbine engine is shown in Fig. 13-17. This FCU is used on the Pratt & Whitney PT-6A-60/65 series engines. These engines are turboprop-type gas-turbine engines in the 1000-hp range. They are free turbine engines, which means that there is no direct mechanical link between the power turbine and the gas generator turbine. The gas generator turbine drives the compressor, and the power turbine turns a propeller shaft through a reduction gearbox.

The FCU is mounted on the rear flange of the fuel pump. A splined coupling between the pump and the FCU

transmits a speed signal, proportional to compressor rotor and turbine speed N_g, to the governing section in the FCU. The FCU determines the fuel schedule for the engine and provides the power required by controlling the speed of the compressor. Engine power output is directly dependent on N_g. Control of N_g is accomplished by regulating the amount of fuel to the combustion section of the engine. Compressor discharge pressure P_3 is sensed by the FCU and is used to establish acceleration fuel flow limits. This fuel-limiting function is used to prevent overtemperature conditions in the engine during starting and acceleration.

The FCU consists of the following major components: condition lever that selects the start, low-idle, and high-idle functions; power lever that selects gas generator speed between high idle and maximum through the three-dimensional cam, cam follower lever, and fuel valve; flyweight governor that controls fuel flow to maintain selected speed; and pneumatic bellows that controls the acceleration schedule and reduces gas generator speed in the event of propeller overspeed.

General Operation of FCU

As shown in Fig. 13-18, the FCU is supplied with fuel at pump pressure P_1. Fuel flow is established by the minimum flow orifice (4), fuel flow valve (7), and bypass valve system (1). The fuel pressure immediately after the minimum flow orifice and the fuel valve is called **metered fuel** (P_2). The bypass valve maintains an essentially constant fuel pressure differential ($P_1 - P_2$) across the fuel valve.

The orifice area of the fuel flow valve is changed by valve movement to meet specific engine fuel requirements, with fuel pump output P_1, in excess of these requirements returned via internal passages in the FCU and pump, to the pump inlet. Bypassed fuel is referred to as "P_0 fuel."

The bypass valve consists of a piston operating in a ported sleeve and is actuated by a spring. In operation, the spring force is balanced by the $P_1 - P_2$ differential operating on the piston. The valve is always in a position to maintain the $P_1 - P_2$ difference; excess P_1 pressure is bypassed to P_0. By turning the adjuster (3), the force on the spring (32) is varied. The adjuster is a manual compensation for fuels with different specific-gravity values.

An ultimate relief valve is incorporated in parallel with the bypass valve to prevent excessive P_1 pressure in the FCU. The valve is spring-loaded closed and remains closed until P_1 overcomes the spring force and opens the valve to P_0. As soon as the pressure is again reduced to an acceptable level, the valve closes.

A minimum pressurizing valve is incorporated in the output line to the flow divider. Its function is to ensure sufficient fuel pressure within the FCU to maintain correct fuel metering.

A fuel shutoff valve in the same line provides a positive means of stopping fuel flow to the engine. During normal operation, the valve is fully open and offers no restriction to flow of fuel to the flow divider. Engine shutdown is accomplished when the condition lever is moved to the shutdown position, depressing the plunger (34) on the

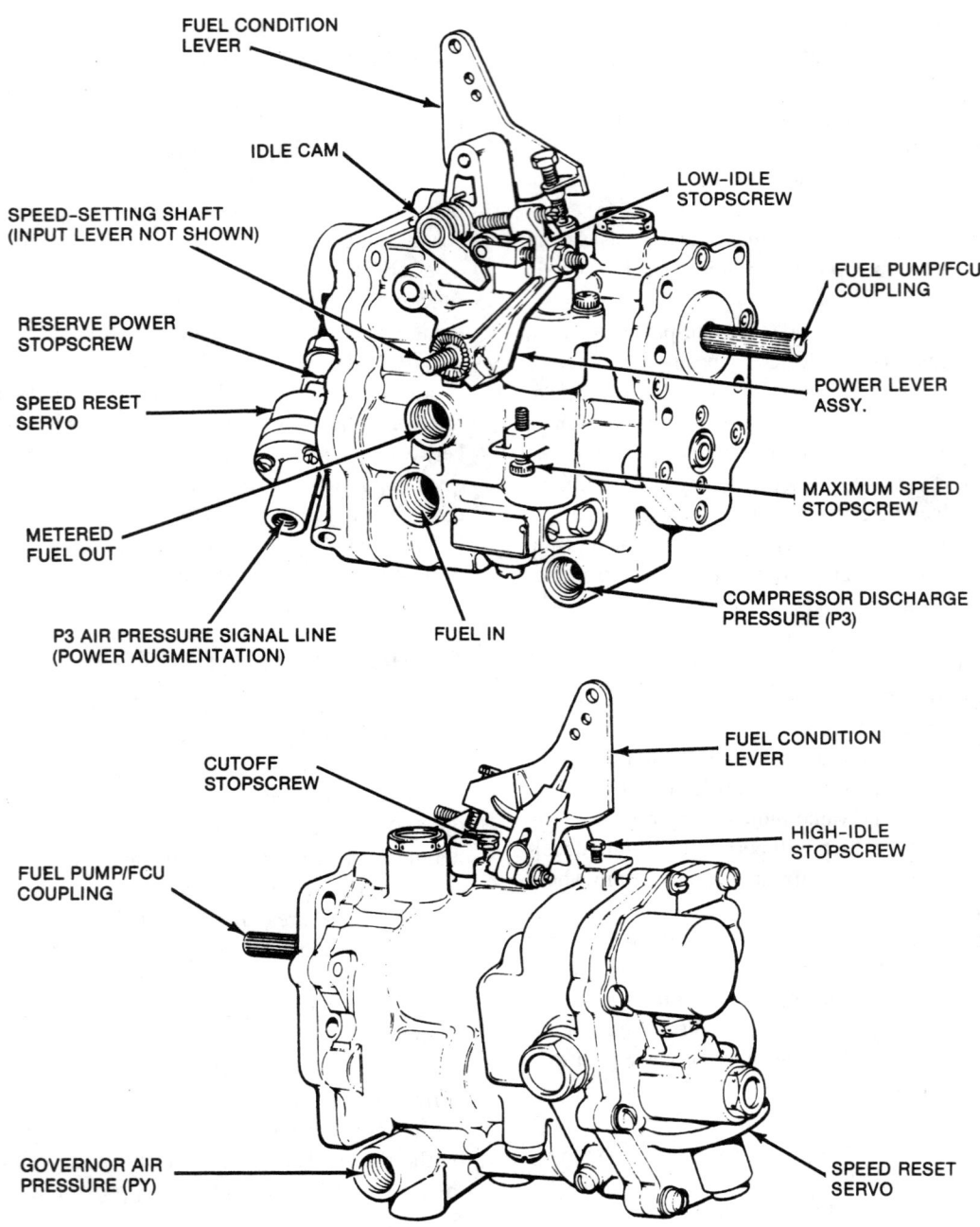

FIGURE 13-17 PT-6 fuel control unit. (*Pratt & Whitney Canada.*)

pump unloading valve, porting pressure P_2 to P_0. Decreasing pressure P_2 allows the pressurizing spring (33) to expand, moving the valve plunger (6) to the closed position and blocking the fuel flow.

Speed Control

Engine speed is controlled by the N_g speed sensor (tachometer). The tachometer contains two flyweights (10) mounted on a ball head driven by the engine. The centrifugal force generated by the flyweights is directly proportional to speed N_g and is transmitted to a valve (9) by the toes of the flyweights (10). The spring (23), controlled by the three-dimensional

cam (8), opposes this force. The valve (9) is positioned by the three-dimensional cam (8).

Fuel at pressure P_1 is supplied through an orifice (24) to the tachometer pilot-valve ports. Above the orifice (24), pressure P_1 is diverted to the area above the stationary piston (25) and exerts an upward force on the end of the three-dimensional cam (8). The valve (9) opening controls the pressure below the orifice. The three-dimensional cam is positioned at a point where the combination of the flyweight force and the spring sets the opening of the valve (9). By regulating the port area, an exact amount of pressure is bled off to maintain servo pressure (P_t) equal to $\frac{1}{2} P_1$. The three-dimensional cam is held stationary at this particular speed.

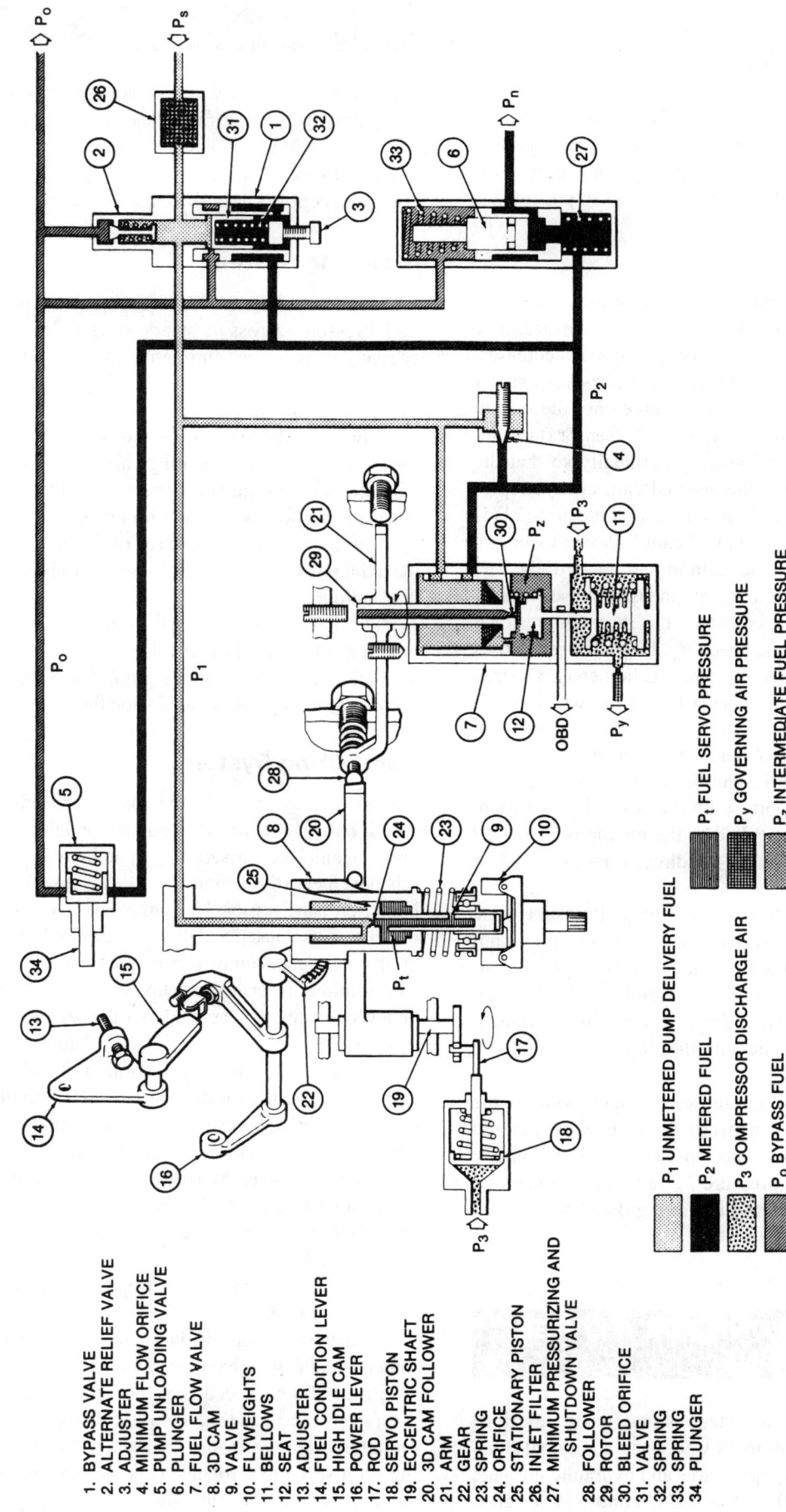

1. BYPASS VALVE
2. ALTERNATE RELIEF VALVE
3. ADJUSTER
4. MINIMUM FLOW ORIFICE
5. PUMP UNLOADING VALVE
6. PLUNGER
7. FUEL FLOW VALVE
8. 3D CAM
9. VALVE
10. FLYWEIGHTS
11. BELLOWS
12. SEAT
13. ADJUSTER
14. FUEL CONDITION LEVER
15. HIGH IDLE CAM
16. POWER LEVER
17. ROD
18. SERVO PISTON
19. ECCENTRIC SHAFT
20. 3D CAM FOLLOWER
21. ARM
22. GEAR
23. SPRING
24. ORIFICE
25. STATIONARY PISTON
26. INLET FILTER
27. MINIMUM PRESSURIZING AND
    SHUTDOWN VALVE
28. FOLLOWER
29. ROTOR
30. BLEED ORIFICE
31. VALVE
32. SPRING
33. SPRING
34. PLUNGER

P_t FUEL SERVO PRESSURE
P_y GOVERNING AIR PRESSURE
P_z INTERMEDIATE FUEL PRESSURE

P_1 UNMETERED PUMP DELIVERY FUEL
P_2 METERED FUEL
P_3 COMPRESSOR DISCHARGE AIR
P_o BYPASS FUEL

FIGURE 13-18   PT-6 fuel control unit schematic. (*Pratt & Whitney Canada.*)

Fuel Control Units for a Turboprop Engine   **363**

Overspeed

Increasing speed N_g increases the force of the flyweights, which forces the valve upward. The area of the metering port is decreased, and servo pressure (P_t) increases.

Increased servo pressure (P_t) forces the three-dimensional cam (8) downward. Force exerted by the spring (23) opposes the movement of the cam (8). A balance point is reached when the cam is held stationary at the new speed position.

Underspeed

Decreasing speed N_g decreases the force of the flyweights. The compressed spring moves the valve downward. The motoring port area increases, and servo pressure P_t decreases. Pressure P_1 above the piston moves the three-dimensional cam upward and decreases the spring force until the system is in equilibrium again. The action of the N_g sensor (tachometer) keeps these forces in balance continually so that the axial position of the three-dimensional cam always represents engine speed N_g. The follower (28) on the arm (21) is connected to the three-dimensional cam follower assembly (20). As the three-dimensional cam moves upward, the fuel valve port opens; fuel flow to the engine is increased and N_g increases. Downward movement of the three-dimensional cam decreases fuel flow and speed N_g. Engine speed is thus maintained by the N_g speed sensor (tachometer). Desired speed changes are set by the power lever (16), which determines the rotational position of the three-dimensional cam by means of a gear (22). When the power lever setting is changed, the positions of the cam followers (20 and 28) are moved. The area of the port in the fuel valve (7) is varied, and the amount of fuel supplied to the engine is increased or decreased. Engine speed N_g is directly proportional to fuel flow.

Limiting the Compressor Discharge Pressure. The compressor discharge pressure (CDP) (P_3) is a second input affecting the position of the fuel valve (7). The CDP P_3 sensor is a sealed, evacuated bellows assembly (11). Varying P_3 causes the bellows to expand or contract. This movement is transmitted by a hydraulic amplifier to move a rotor (29) axially.

Fuel at pressure P_1 is applied to the upper side of the rotor (29), imparting a downward force. Fuel is metered through an orifice in the rotor to the area immediately beneath it. Intermediate pressure P_2 in this area exerts an upward force on the rotor which is regulated by a bleed orifice (30).

FUEL CONTROL UNIT FOR A LARGE TURBOFAN ENGINE

The JFC68 FCU serves the functions previously described for other hydromatic controls as well as an additional function of supplying reference pressures and hydraulic pressure for the **engine vane control**. The JFC68 control is designed for use with the Pratt & Whitney JT9D turbofan engine.

HAMILTON STANDARD JFC68 FUEL CONTROL UNIT

The JFC68 fuel control unit is mounted on the main engine fuel pump and operates in connection with the engine vane control (EVC3) to regulate the thrust of the engine. The control utilizes the w_f/P_{s4} ratio as a control parameter, as do the other fuel controls described in this section.

Metering System

The JFC68 fuel control metering system applies regulated fuel pressure across a window-type throttle valve. The engine fuel is filtered through a coarse filter, and the servo fuel is filtered through a fine filter. The fuel metering system can be seen in the right center section of Fig. 13-19. The main units of the system are the pressure regulating valve sensor, the pressure regulating valve, and the throttle valve. Note that the throttle valve (near the center of the drawing) consists of a hollow sleeve with openings that allow fuel to flow out to the engine. The area of the openings depends on the axial position of the sleeve as determined by the computing section.

Fuel that is not necessary to maintain the pressure differential across the throttle valve is bypassed by the pressure regulating valve back to the pump interstage. Pump interstage pressure is maintained inside the case of the control.

Computing System

The computing system of the control unit utilizes engine N_2 speed, burner pressure P_{s4} compressor inlet temperature (T_{t2} or CIT), ambient pressure (P_{amb}), and power lever position to schedule fuel to the engine. Pressure P_{s4} is sensed by the engine burner pressure sensor. This unit consists of two bellows, one of which is evacuated and the other exposed to burner pressure on the outside and ambient pressure on the inside. This unit is shown in the lower right section of Fig. 13-19. Note that the bellows movement is applied to a flapper valve which directs servo pressure to and from each side of a servo. The servo acts on a lever which controls the position of rollers between the **ratio lever** and the **multiplying lever**. The multiplying lever moves the throttle-valve pilot valve to direct servo pressure, which moves the throttle valve. The force on the multiplying lever is balanced by the throttle-valve feedback spring as a function of actual fuel flow. The throttle-valve pilot valve and other spool-type valves in the fuel control are continuously rotated. This keeps them free from dirt particles which might become caught between operating surfaces and cause the valves to stick.

Acceleration control is provided through the three-dimensional cam, which is positioned axially by servo pressure from the N_2 speed governor through the operation of a servo piston inside the cam body. The cam is positioned radially by servo pressure from the T_{t2} (CIT) pilot valve acting on a servo that rotates a sector gear meshed with gear teeth on the cam. The three-dimensional cam is shown in the upper center portion of Fig. 13-19.

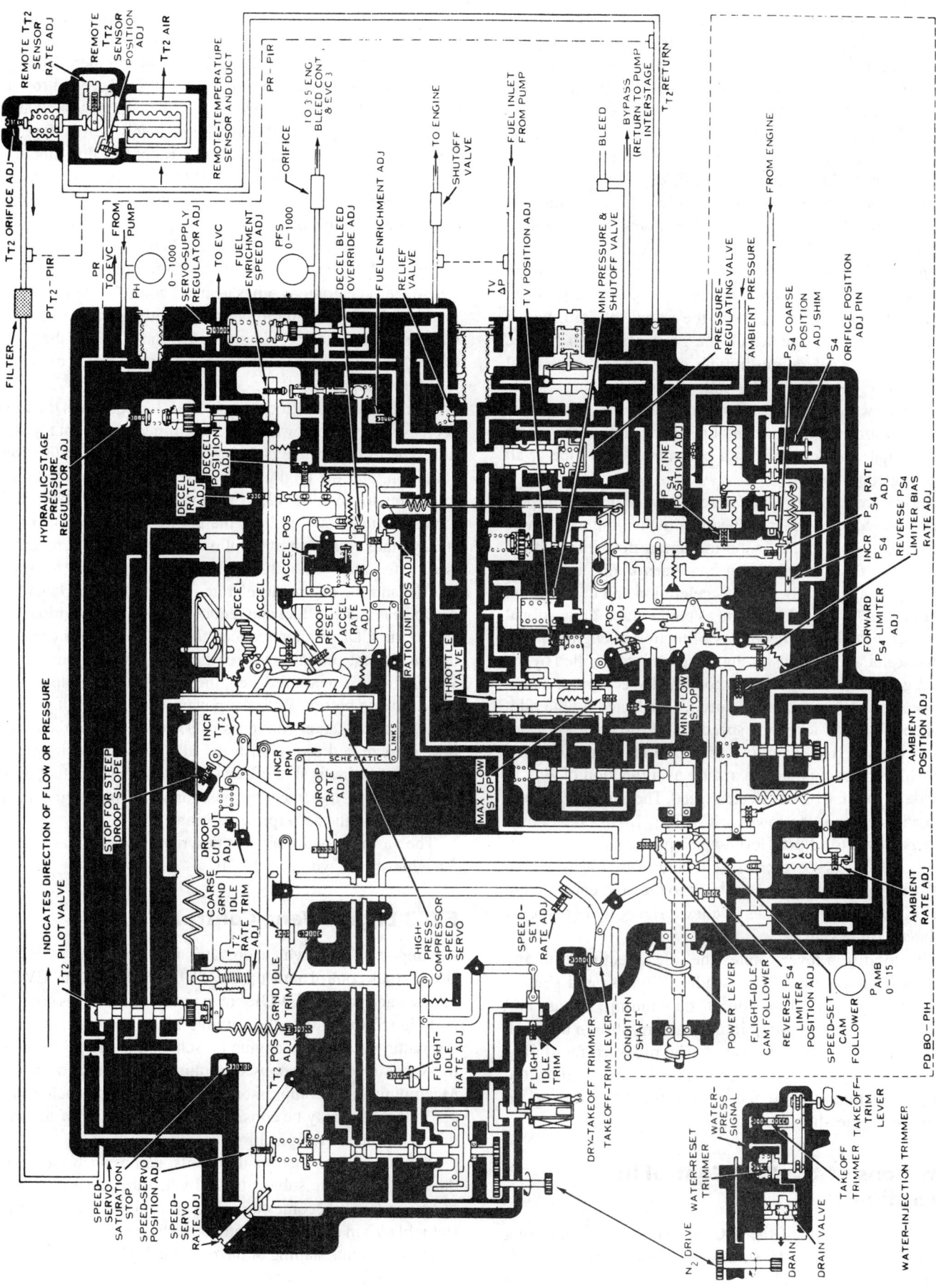

FIGURE 13-19 Schematic drawing of Hamilton Standard JFC63 FCU. (*Hamilton Standard.*)

The three-dimensional cam is contoured to define a schedule of w_f/P_{s4} vs. engine speed for each engine inlet temperature value. The operation of the cam is such that it permits engine accelerations which avoid the overtemperature and surge limits of the engine without adversely affecting engine accelerating time. The cam produces its effects through three cam followers: one for acceleration control, one for deceleration control, and one for droop reset. The outputs of these followers are fed through a series of linkages to the ratio-unit spring, which is connected to the throttle-valve actuating system. Also, the deceleration cam follower acts through a linkage and valve to control the compressor bleed actuator.

The power lever and the condition lever are shown in the lower left portion of the drawing above the ambient-pressure sensor. The power lever rotates the speed-set cam, which determines the point at which the governor droop linkage overrides the acceleration-limiting cam (three-dimensional cam) to decrease the w_f/P_{s4} ratio with increasing engine speed. This will continue until steady-state w_f/P_{s4} is attained. The governor droop-reset contour is provided on the three-dimensional cam to maintain a constant-droop slope at all operating conditions.

During acceleration, the three-dimensional cam provides a biased temperature schedule down to the new power lever setting droop line or the minimum flow line, whichever occurs first. Deceleration continues until steady-state operation is attained. Both the acceleration and deceleration schedules are provided by surfaces on the three-dimensional cam; therefore, both schedule as a function of engine speed.

As explained previously, engine speed is controlled by the speed governor. The governor continually compares actual engine speed N_2 with desired speed, as selected by the pilot through the power lever and speed-set cam. The power lever rotates the speed-set cam to set the desired governor droop line while the cam is moved axially as a function of P_{amb} to provide the biasing of the selected speed. The ambient-pressure sensor is shown at the lower left of Fig. 13-19. The sensor bellows moves to adjust a lever which positions a pilot valve. This valve directs servo pressure to and from a servo which moves the speed-set cam. Feedback to the ambient-pressure sensor is provided by means of a lever riding in a groove on the landing-idle cam.

The condition lever rotates the idle selection cam and the cam which actuates the windmill bypass and shutoff valve. The windmill function of the windmill bypass and shutoff valve directs pump interstage pressure to the back of the pressure-regulating valve. The pressure-regulating valve is thus permitted to open and bypass fuel to the pump interstage should the engine windmill during an in-flight engine shutdown.

Functions of the Fuel Control in Operation

The actual functions of the JFC68 fuel control unit during engine operation may be understood more clearly by following the diagram in Fig. 13-20. This diagram is similar in many respects to the fuel control curves shown previously;

however, this diagram of curves is based on the ratio w_f/P_{s4} and engine speed N_2, whereas the other diagram (Fig. 13-16) was based on fuel flow (w_f) and engine speed.

In Fig. 13-20, position 1 represents the engine condition at the time acceleration by the starter has proceeded to the point where fuel can be injected into the combustion chamber. At this time, the w_f/P_{s4} ratio immediately moves to position 2. The acceleration cam is in control of fuel flow, and the engine accelerates along line 2 to 3 to 4. If the power lever is in the IDLE position, the fuel ratio will decrease to points 5 and 6 as a result of governor droop, and the engine will now be in a steady-state condition. When the power lever is advanced, the fuel ratio increases from point 6 to point 7, where the acceleration cam is again in control. The F/A ratio increases to the maximum permitted and then remains constant as the engine accelerates to position 8, where the governor droop again takes effect until a steady-state condition is attained at position 9.

The line from position 9 to position 10 represents the fuel decrease when the power lever is returned to the IDLE position. The throttle valve closes to the minimum deceleration condition, and the engine decelerates to position 11. From this point to position 12, governor droop takes effect and increases fuel flow to allow the engine speed to reach the steady-state condition at position 12. If the thrust reverser is actuated at position 12, the fuel flow increases to position 13 and the engine accelerates to position 14. Governor droop again takes effect and reduces fuel flow to position 15, where the CDP limiter reduces fuel flow still further until the steady-state condition is reestablished at position 16 (9). When power is again reduced, the function line follows the same path as before. With the power lever in the IDLE position, the deceleration continues to position 18, and then fuel flow is increased to establish the idle speed at position 19. When the engine is shut off by the condition lever, fuel flow is stopped, as indicated by position 20. The engine then decelerates to the zero rpm condition.

The various conditions just described are summarized in the table shown with Fig. 13-20.

EVC3 Engine Vane Control

An important unit which operates in connection with the JFC68 fuel control unit on the JT9D engine is the EVC3 engine vane control (EVC) shown in Fig. 13-21. This control is designed to regulate the variable high-pressure compressor stator vanes of the engine by scheduling the position of the vanes in accordance with requirements dictated by the Mach number of compressor inlet airflow. The Mach number in this case may be considered as the velocity of the airflow adjusted for temperature.

The stator vanes are small airfoils, and as such they and the rotor blades are subject to stall when the angle of attack of the airstream becomes too great. Varying the angle of the stator blades in accordance with the airflow velocity and temperature prevents stalling of the rotor blades and stator vanes.

The EVC consists of a pressure ratio sensor, flapper valve, servo-operated three-dimensional cam, actuator pilot

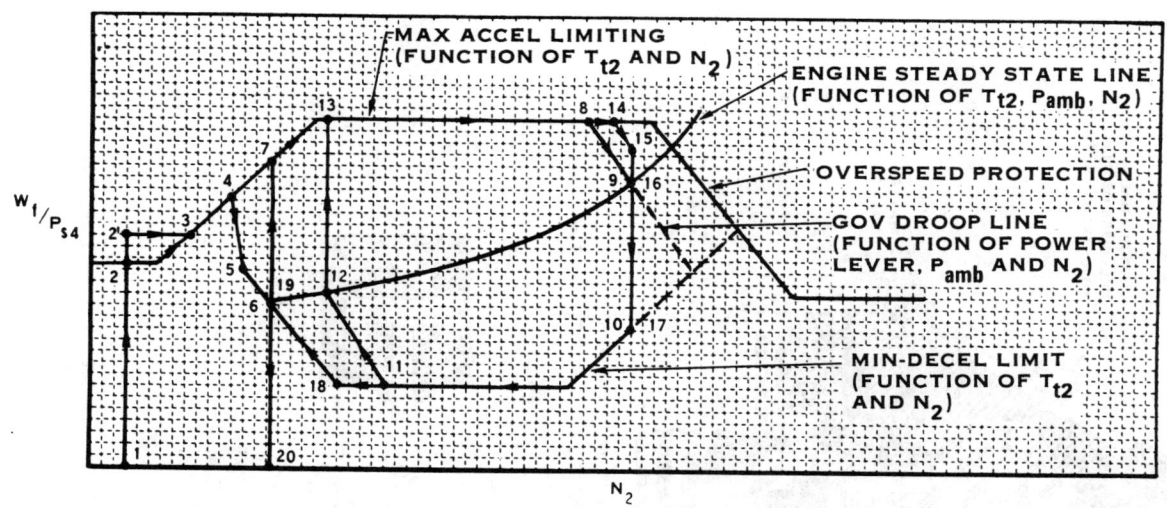

PATH	CONTROLLING FUNCTION	ENGINE OPERATING PARAMETER
1—2—3—4	ACCEL CAM	STARTING
1—2—3—4	ACCEL CAM AND SELECTOR SHAFT	COLD-START ENRICHMENT (MANUALLY SELECTED WHEN REQ'D.)
4—5—6	GOVERNOR DROOP	START TO MIN IDLE
6—7	POWER LEVER	STEP INPUT TO ACCEL
7—8	ACCEL CAM	ACCELERATION
8—9	GOVERNOR DROOP	ACCEL TO STEADY STATE
9—10	POWER LEVER	STEP INPUT TO DECEL
10—11	DECEL CAM	DECELERATION
11—12	GOVERNOR DROOP	DECEL TO STEADY STATE (LANDING IDLE)
12—13	POWER LEVER	STEP INPUT TO ACCEL (ACTUATE THRUST REVERSER)
13—14	ACCEL CAM	ACCEL (THRUST REVERSER ON)
14—15	GOVERNOR DROOP	ACCEL TO CDP LIMITING
15—16	CDP LIMITER	CDP LIMITING TO STEADY STATE
16—17	POWER LEVER	STEP INPUT TO DECEL
17—18	DECEL CAM	DECELERATION
18—19	GOVERNOR DROOP	DECEL TO MIN IDLE
19—20	SELECTOR LEVER	SHUT OFF

FIGURE 13-20 Curves illustrating operation of the Hamilton Standard JFC68 FCU. (*Hamilton Standard.*)

valve, and necessary linkage. In addition, the control contains two signal valves which supply hydraulic pressure, as a function of the pressure ratio, to two remotely mounted engine actuators. The control signal valves are also designed to supply a prescribed cooling flow through the supply lines to the engine bleed control valves.

The EVC schedules a hydraulic signal for operation of the engine stator vanes as a function of compressor inlet pressure ratio ($P_{t3} - P_{s3}/P_{t3}$). This is accomplished by a two-bellows null-type pressure ratio sensor of a force-vector design. This sensor is a torque-balance system arranged so that a ratio of two signal pressures may be determined, as shown in Fig. 13-22. In this application, the total pressure (P_{t3}) and the static pressure (P_{s3}) are measured by two bellows assemblies to provide force outputs that are proportional to the total pressure and the differential pressure ($P_{t3} - P_{s3}$) from which the pressure ratio ($P_{t3} - P_{s3}/P_{t3}$) is derived.

The bellows are arranged so that when they are in the null position, forces from the bellows act at right angles to one another through tension links that are attached to a common pivot. The resultant forces from the bellows are counteracted by a connecting link between the common pivot and the

feedback lever pivot. At the null position, the vector sum of the bellows' forces and the counteracting force in the connecting link are balanced.

At the null position of the system, the rotational axes of the common pivot and the feedback lever are on the same centerline. When the pressure ratio changes, the force balance is upset and the common pivot rotates from its null position. This rotation is constrained about the feedback lever pivot through the connecting link. The lateral part of the common pivot's rotation moves the flapper valve from the null position, thereby causing a high servo pressure on one side of the servo piston. The piston moves in proportion to flapper valve displacement. This movement of the servo piston, in turn, displaces the three-dimensional cam, which rotates the feedback lever, realigning the rotational axes of the common pivot and the feedback link. Realignment restores the null position of the flapper valve, restoring the force balance in the vector system.

Scheduling of the engine stator vanes is achieved by the three-dimensional cam that is contoured to a prescribed schedule. The cam displaces the actuator pilot valve, sending a high-pressure signal (P_{a1} or P_{a2}) to the remotely mounted

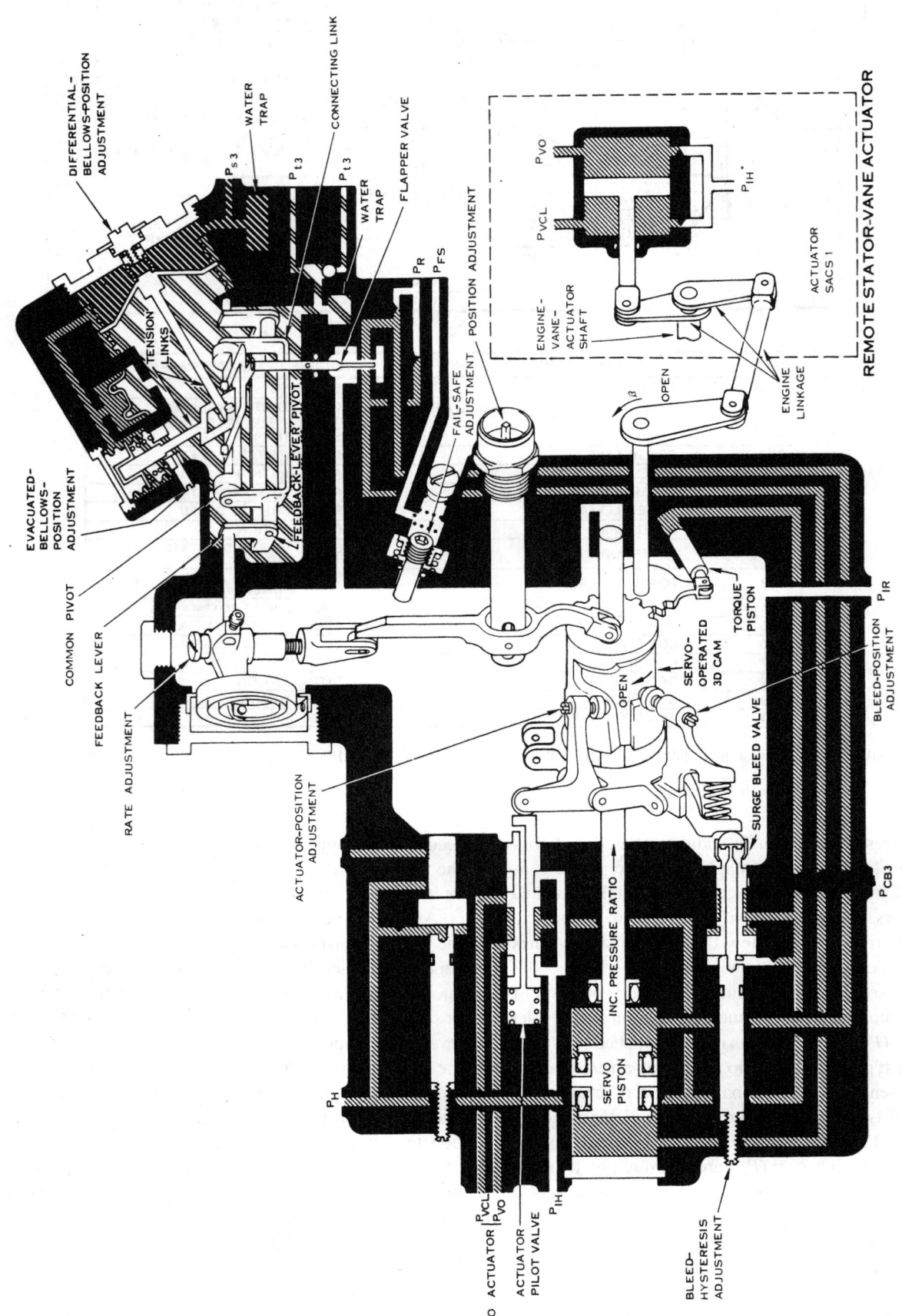

FIGURE 13-21 EVC3 engine vane control. (*Pratt & Whitney Canada*.)

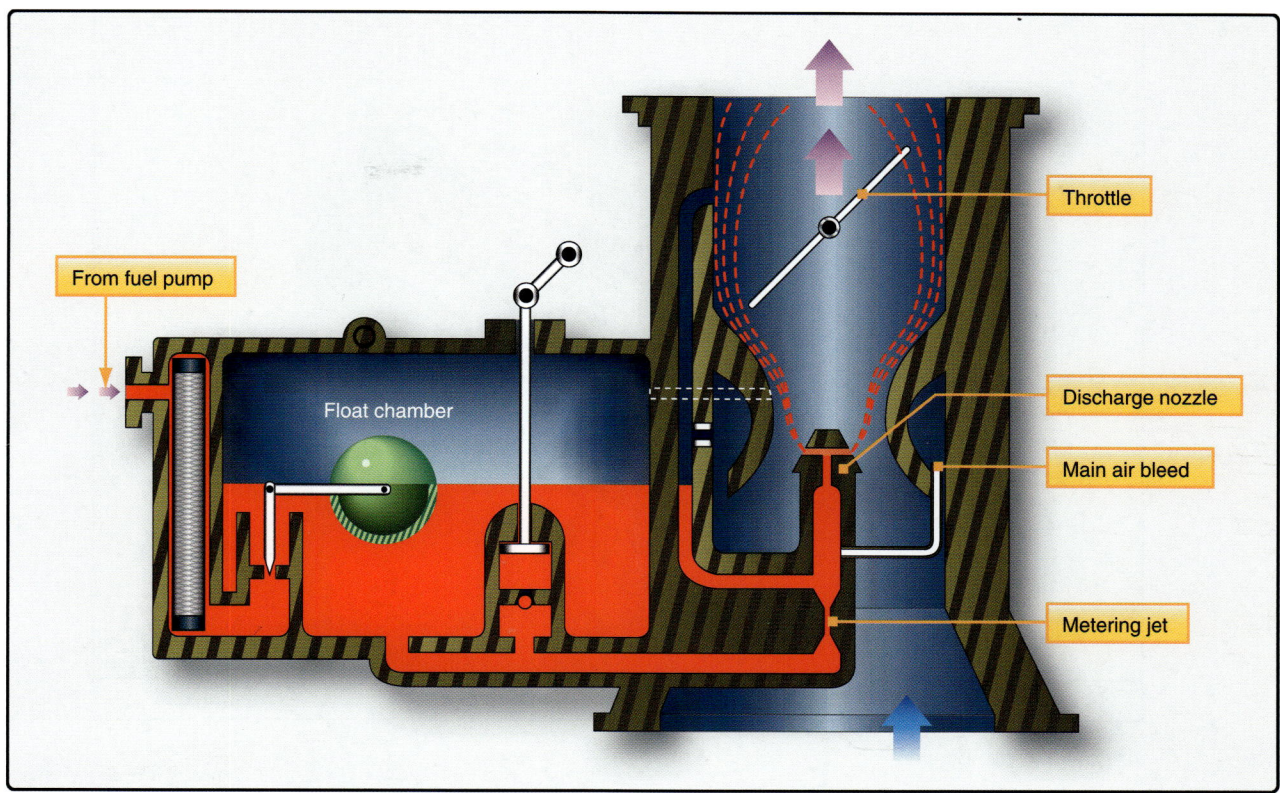

Fuel discharge.

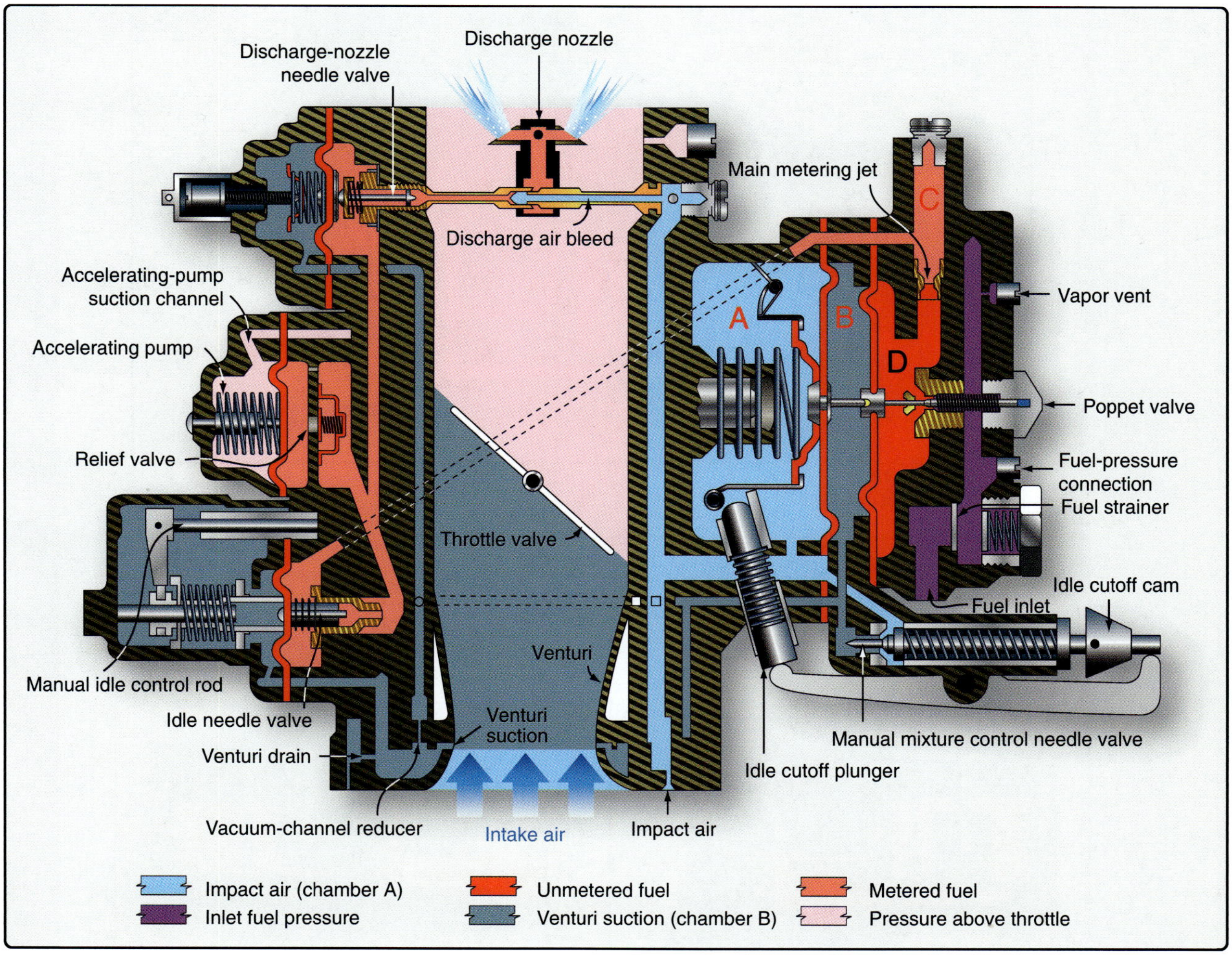

Discharge-nozzle
needle valve

Discharge nozzle

Main metering jet

C

Vapor vent

Accelerating-pump
suction channel

Discharge air bleed

A

B

D

Poppet valve

Accelerating pump

Fuel-pressure
connection

Relief valve

Fuel strainer

Idle cutoff cam

Fuel inlet

Throttle valve

Manual idle control rod

Venturi

Idle needle valve

Venturi
suction

Venturi drain

Manual mixture control needle valve

Vacuum-channel reducer

Intake air

Impact air

Idle cutoff plunger

| | Impact air (chamber A) | | Unmetered fuel | | Metered fuel |
| | Inlet fuel pressure | | Venturi suction (chamber B) | | Pressure above throttle |

Schematic of the PS series carburetor.

From fuel pump

Float chamber

Throttle

Discharge nozzle

Main air bleed

Metering jet

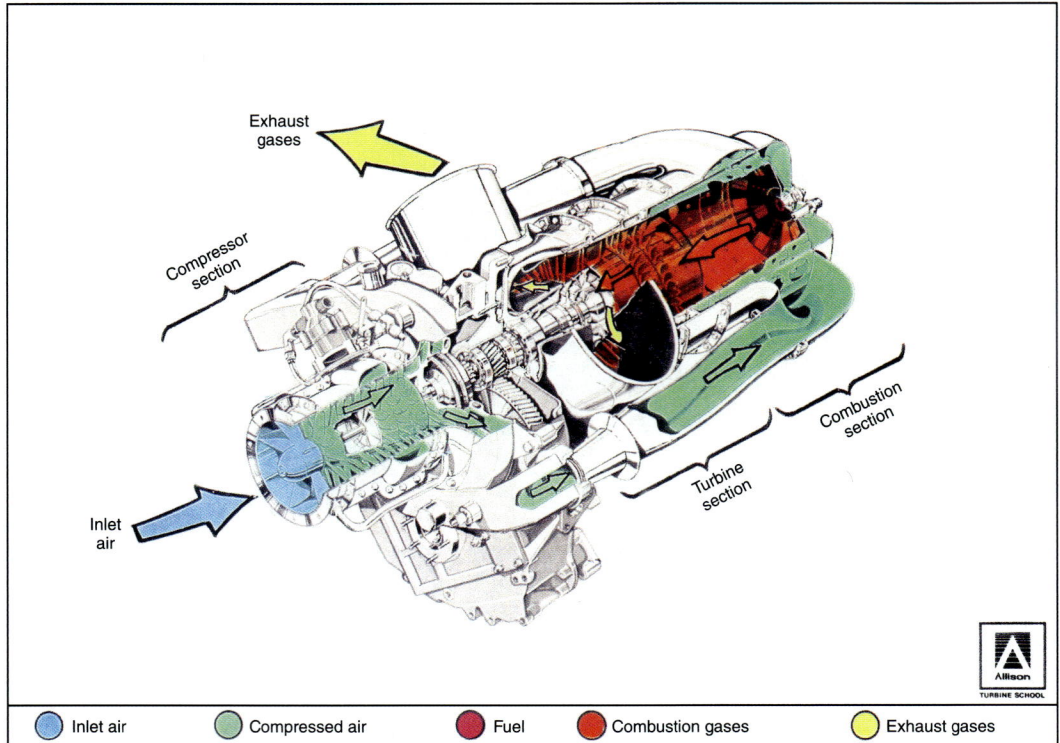

Inlet air · Compressed air · Fuel · Combustion gases · Exhaust gases

Airflow and combustion.

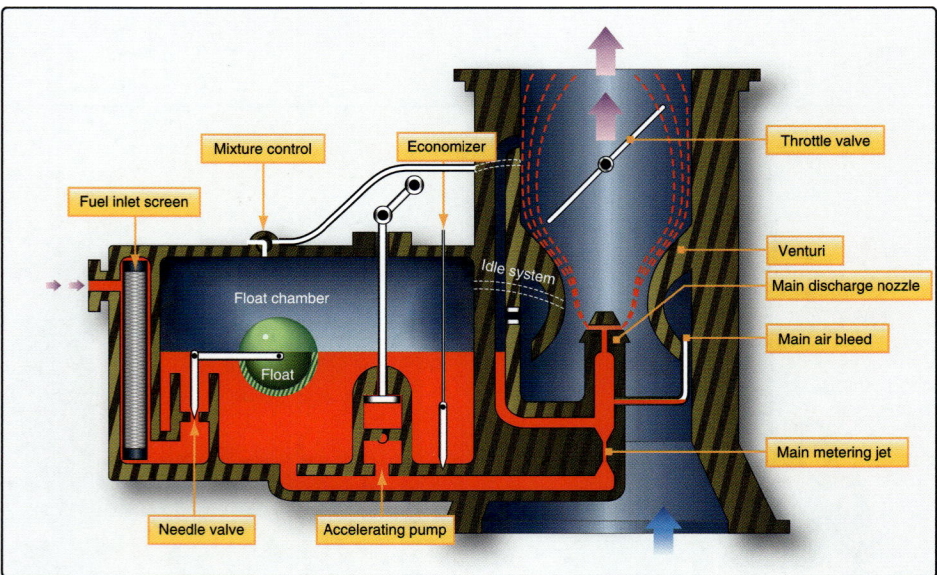

A float-type carburetor.

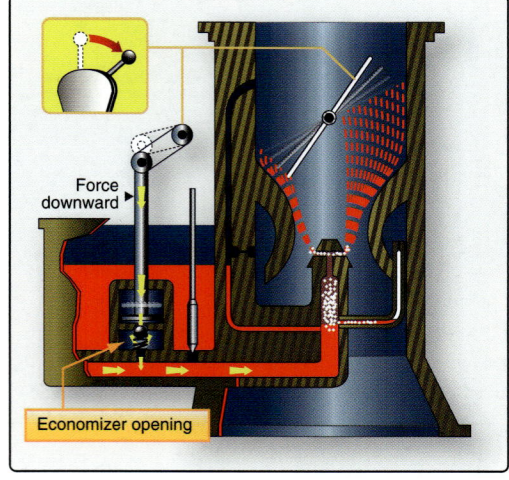

Accelerating system.

A needle-valve-type economizer system.

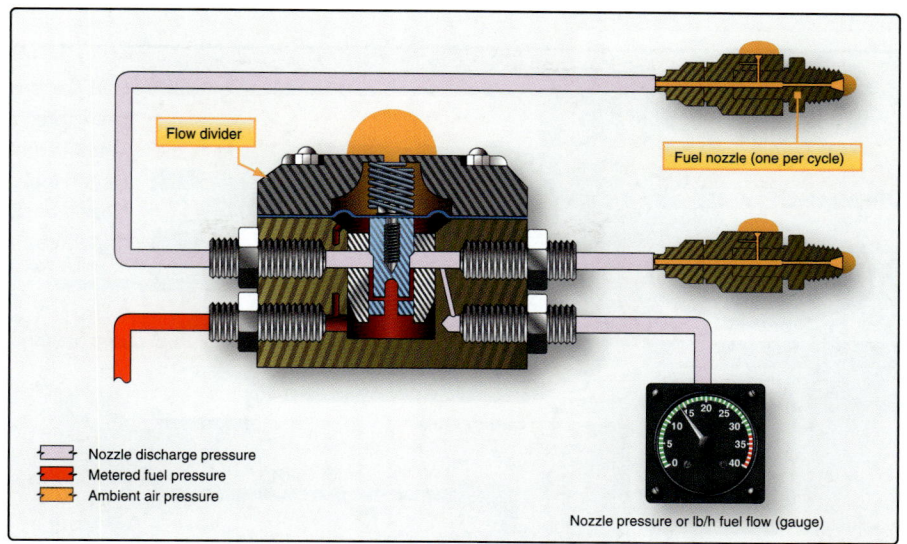

Nozzle discharge pressure
Metered fuel pressure
Ambient air pressure

Flow divider

Fuel nozzle (one per cycle)

Nozzle pressure or lb/h fuel flow (gauge)

Flow divider.

Flow divider cutaway.

Fuel nozzle assembly.

Fuel diaphragm with ball valve attached.

Impact tubes for inlet air pressure.

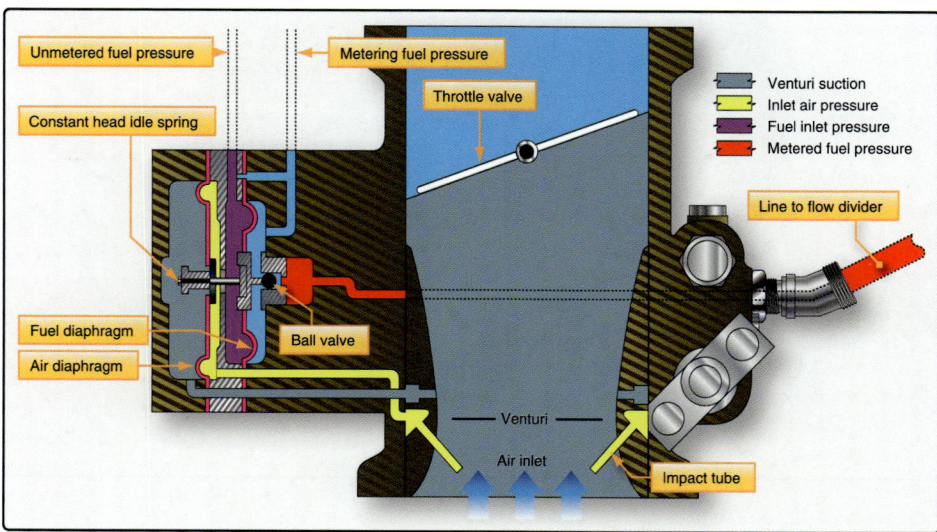

Unmetered fuel pressure

Metering fuel pressure

Constant head idle spring

Throttle valve

Venturi suction
Inlet air pressure
Fuel inlet pressure
Metered fuel pressure

Line to flow divider

Fuel diaphragm

Ball valve

Air diaphragm

Venturi

Air inlet

Impact tube

Airflow section of a fuel injector.

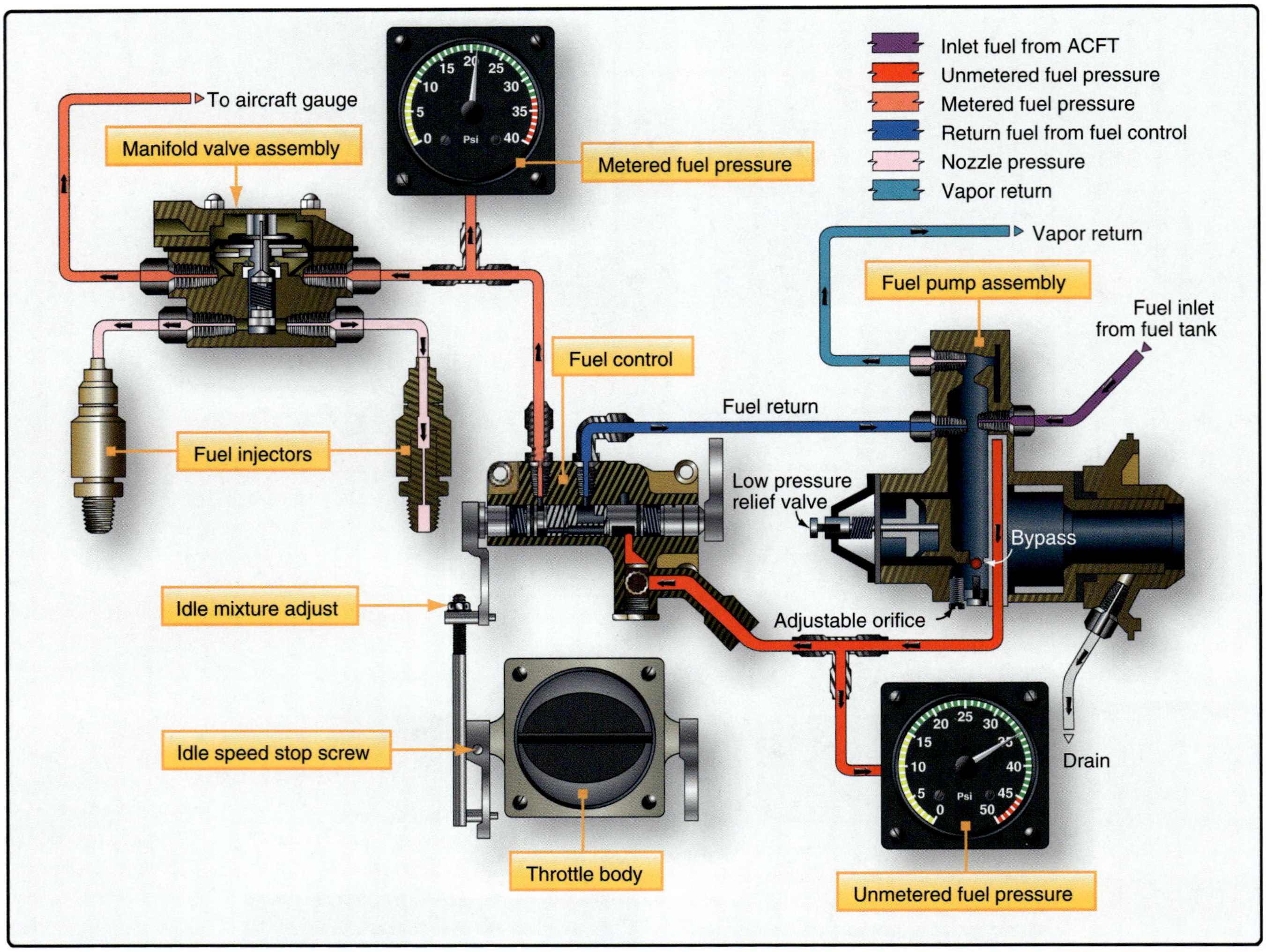

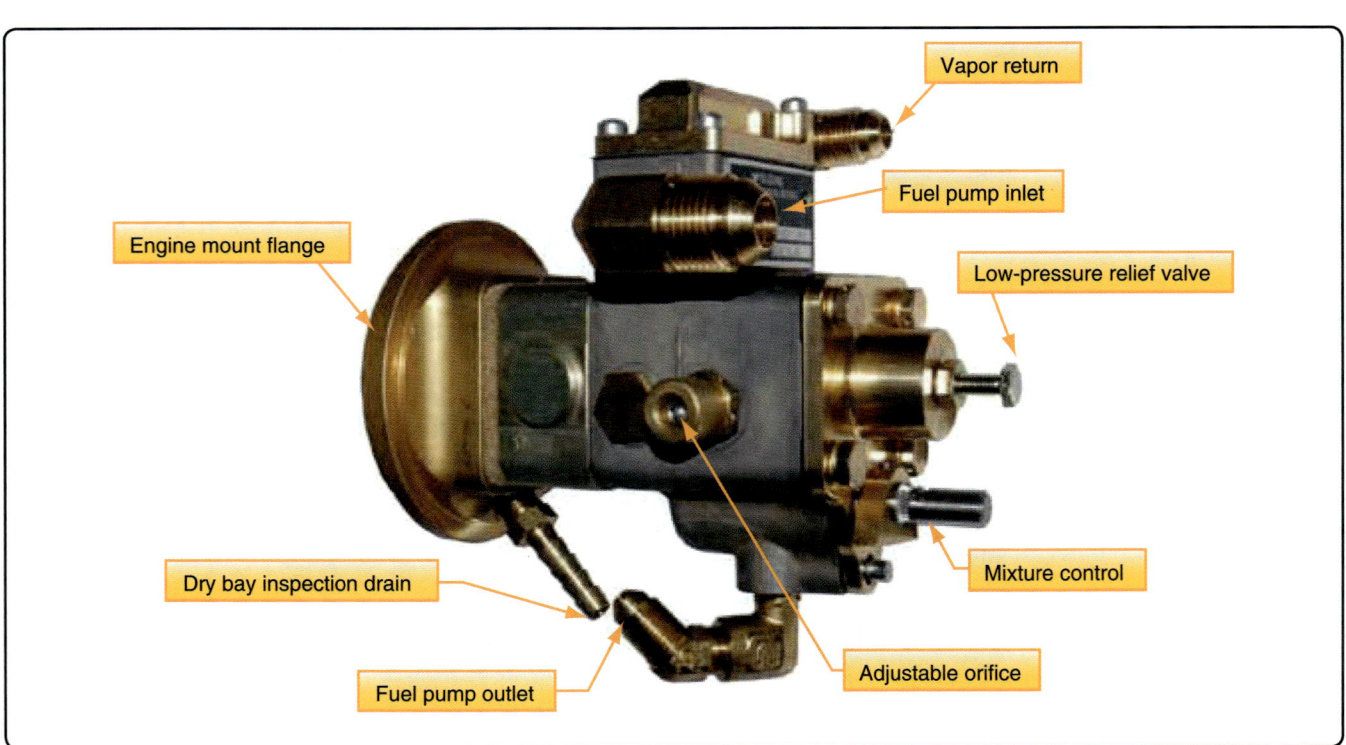

Legend:
- Inlet fuel from ACFT
- Unmetered fuel pressure
- Metered fuel pressure
- Return fuel from fuel control
- Nozzle pressure
- Vapor return

To aircraft gauge

Manifold valve assembly

Metered fuel pressure

Vapor return

Fuel pump assembly

Fuel inlet from fuel tank

Fuel control

Fuel return

Low pressure relief valve

Fuel injectors

Bypass

Idle mixture adjust

Adjustable orifice

Idle speed stop screw

Drain

Throttle body

Unmetered fuel pressure

Continental/TCM fuel injection system.

Vapor return

Fuel pump inlet

Engine mount flange

Low-pressure relief valve

Mixture control

Dry bay inspection drain

Adjustable orifice

Fuel pump outlet

Fuel pump.

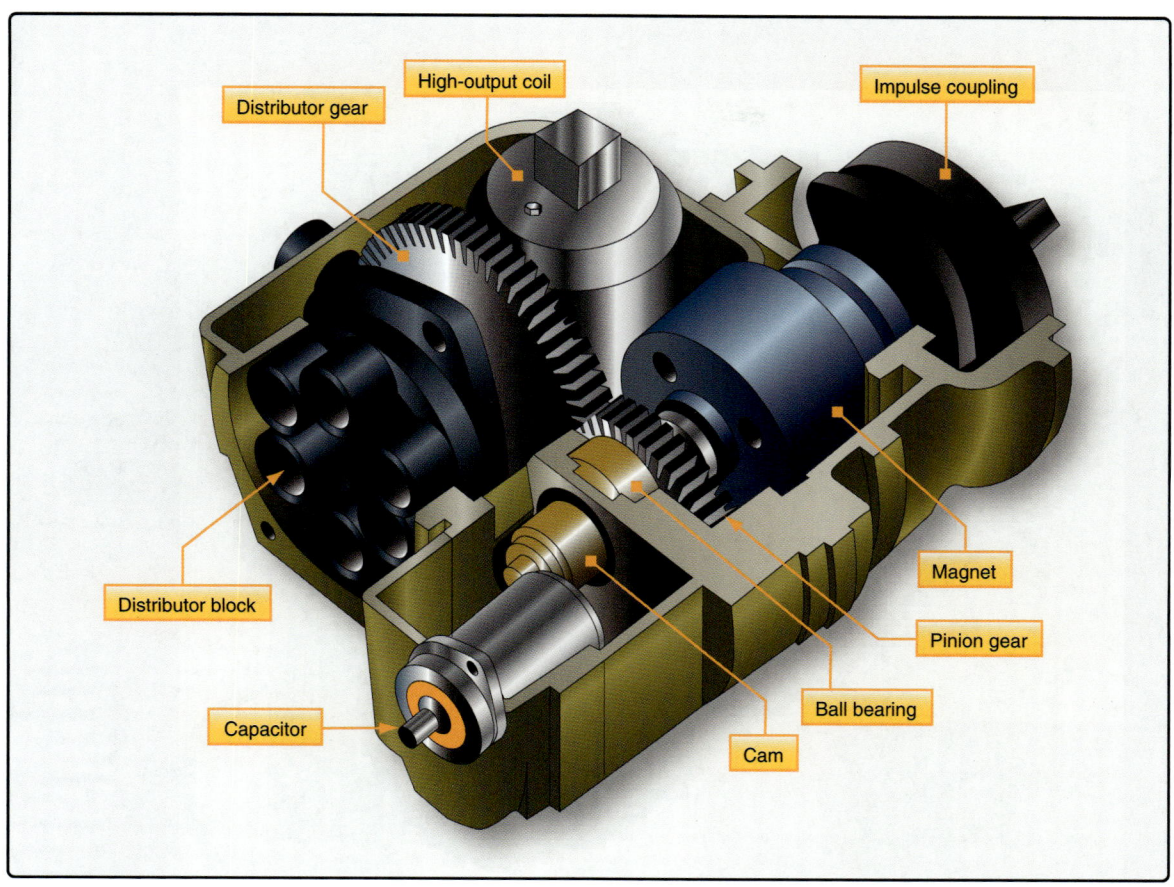

FIGURE 8-17   High-tension magneto primary ignition system.

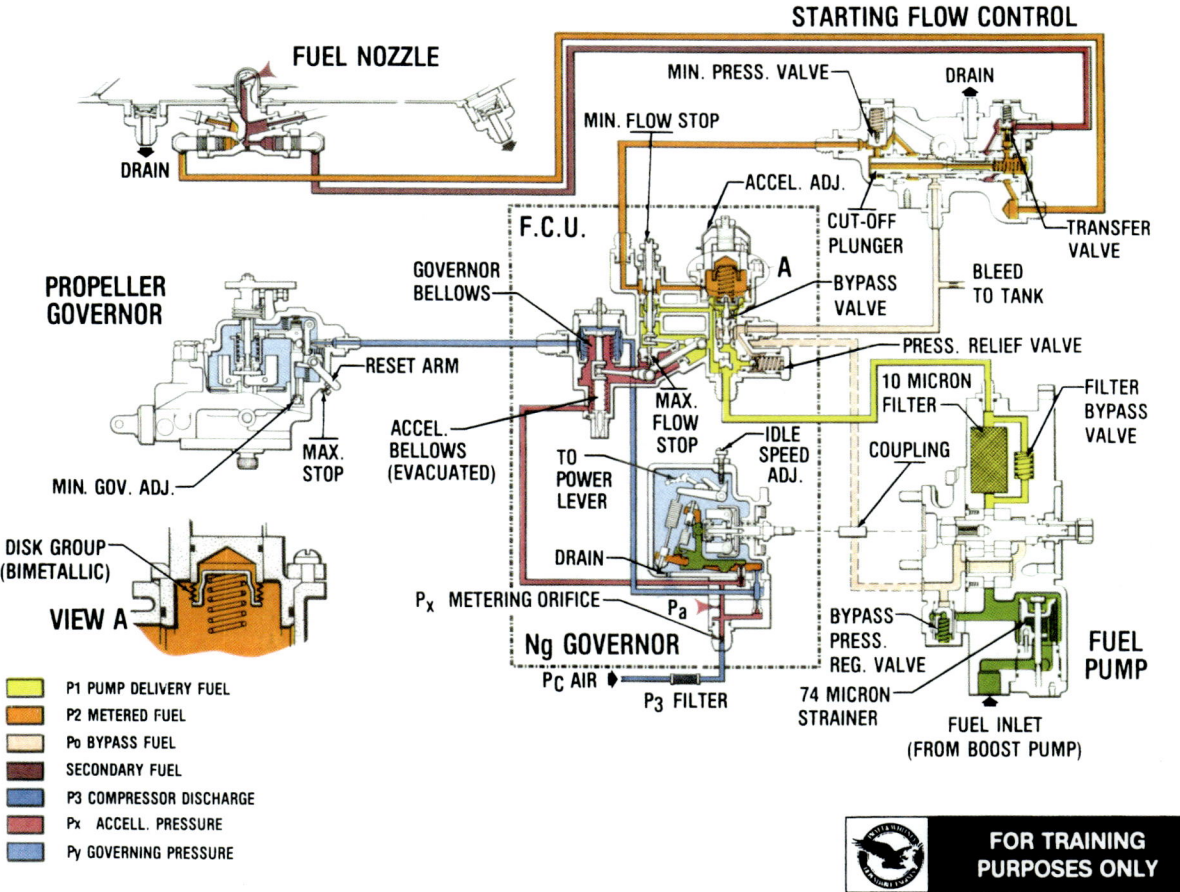

STARTING FLOW CONTROL

FUEL NOZZLE

MIN. PRESS. VALVE — DRAIN

MIN. FLOW STOP

DRAIN

ACCEL. ADJ.

F.C.U.

CUT-OFF PLUNGER

A

BLEED TO TANK

TRANSFER VALVE

PROPELLER GOVERNOR

GOVERNOR BELLOWS

BYPASS VALVE

PRESS. RELIEF VALVE

RESET ARM

10 MICRON FILTER

FILTER BYPASS VALVE

ACCEL. BELLOWS (EVACUATED)

MAX. FLOW STOP

IDLE SPEED ADJ.

COUPLING

MIN. GOV. ADJ.

MAX. STOP

TO POWER LEVER

DRAIN

DISK GROUP (BIMETALLIC)

VIEW A

P_x METERING ORIFICE

P_a

BYPASS PRESS. REG. VALVE

FUEL PUMP

Ng GOVERNOR

P_C AIR

P_3 FILTER

74 MICRON STRAINER

FUEL INLET (FROM BOOST PUMP)

▢	P1 PUMP DELIVERY FUEL
▢	P2 METERED FUEL
▢	Po BYPASS FUEL
▢	SECONDARY FUEL
▢	P3 COMPRESSOR DISCHARGE
▢	Px ACCELL. PRESSURE
▢	Py GOVERNING PRESSURE

FOR TRAINING PURPOSES ONLY

Engine fuel system schematic.

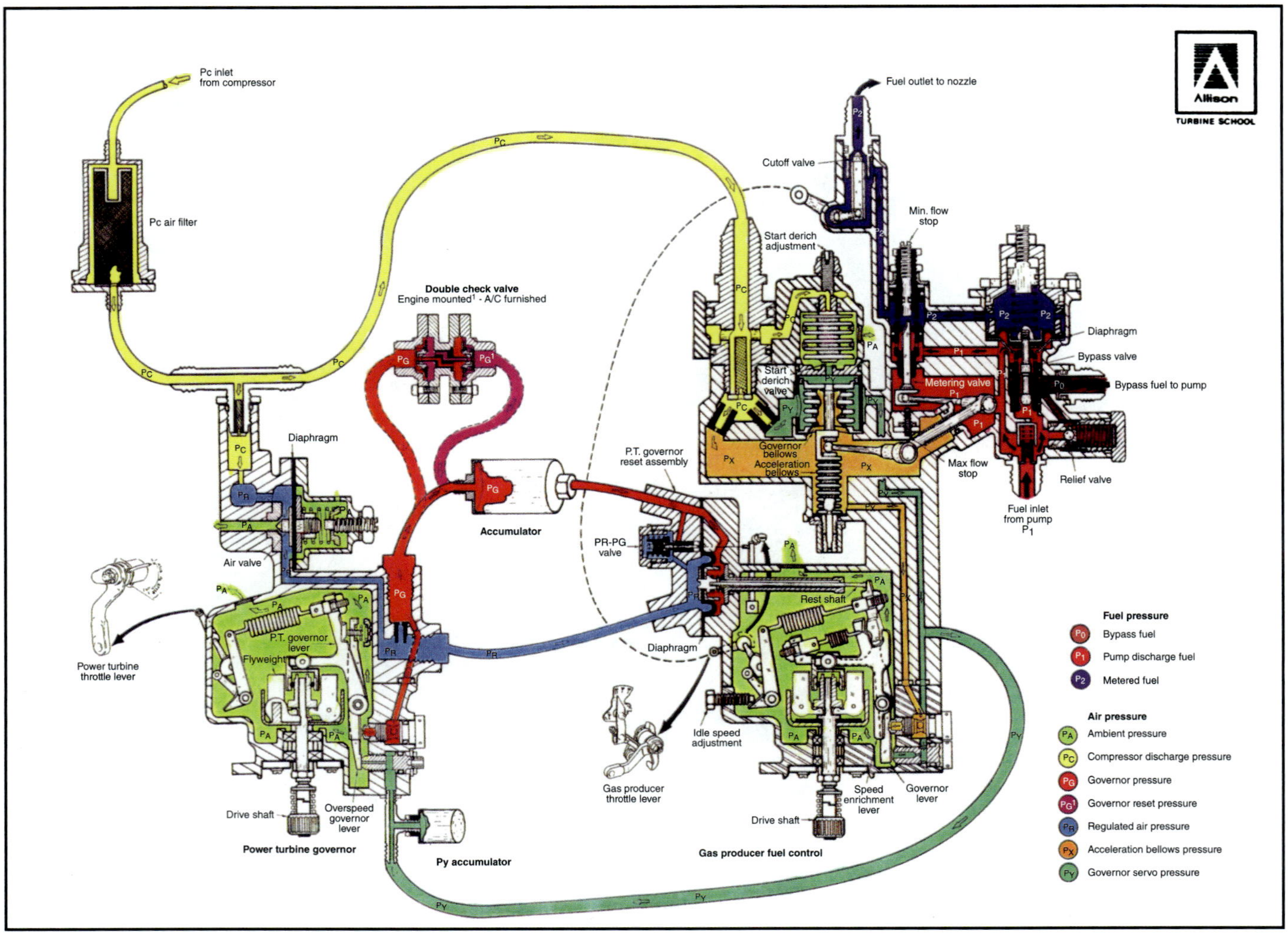

Fuel control system schematic.

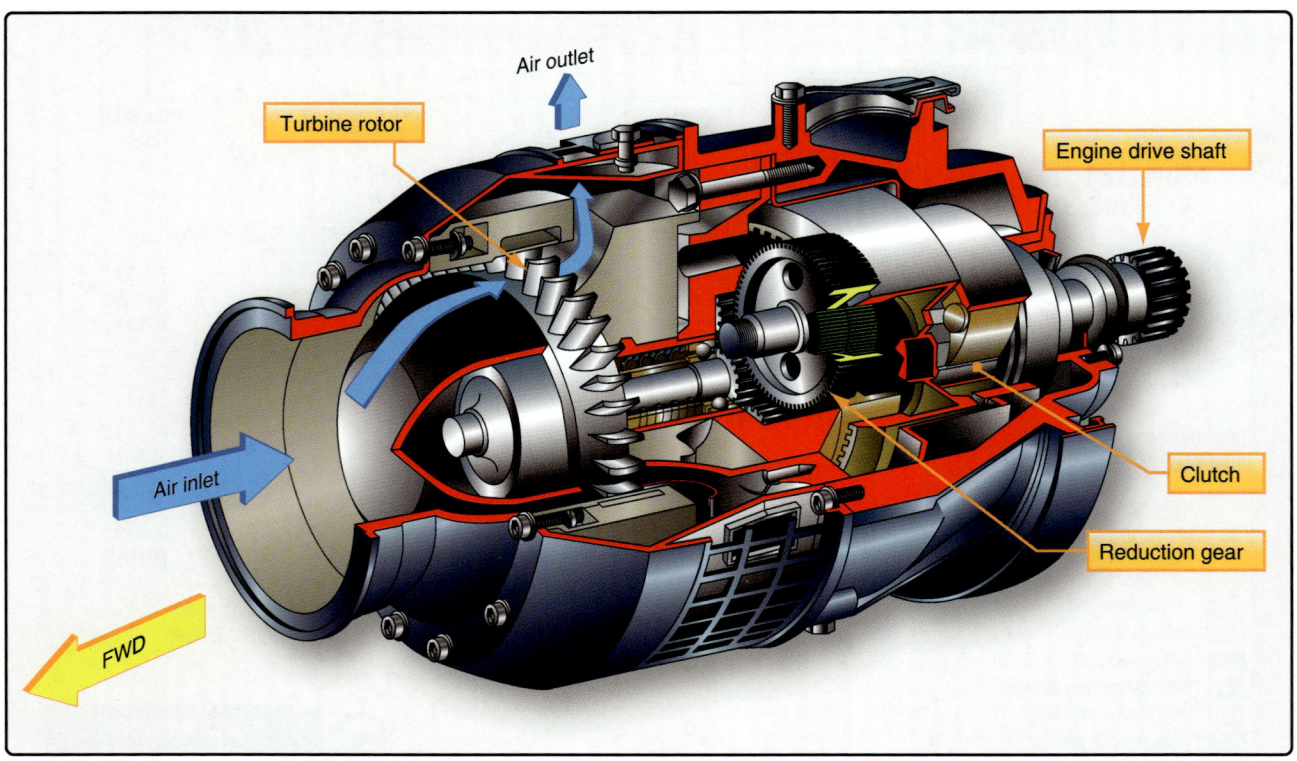

FIGURE 15-15 Air-turbine starter. (*Rolls-Royce.*)

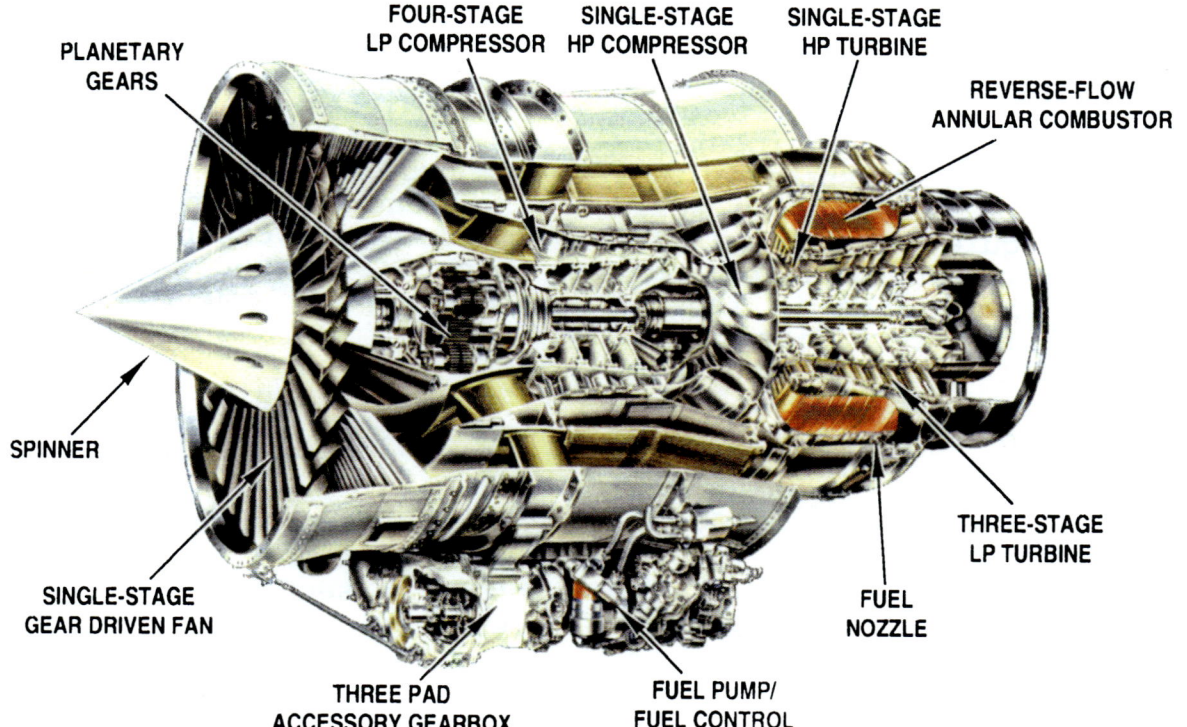

PLANETARY GEARS

FOUR-STAGE LP COMPRESSOR

SINGLE-STAGE HP COMPRESSOR

SINGLE-STAGE HP TURBINE

REVERSE-FLOW ANNULAR COMBUSTOR

SPINNER

SINGLE-STAGE GEAR DRIVEN FAN

THREE PAD ACCESSORY GEARBOX

FUEL PUMP/ FUEL CONTROL

FUEL NOZZLE

THREE-STAGE LP TURBINE

The TFE731 turbofan.

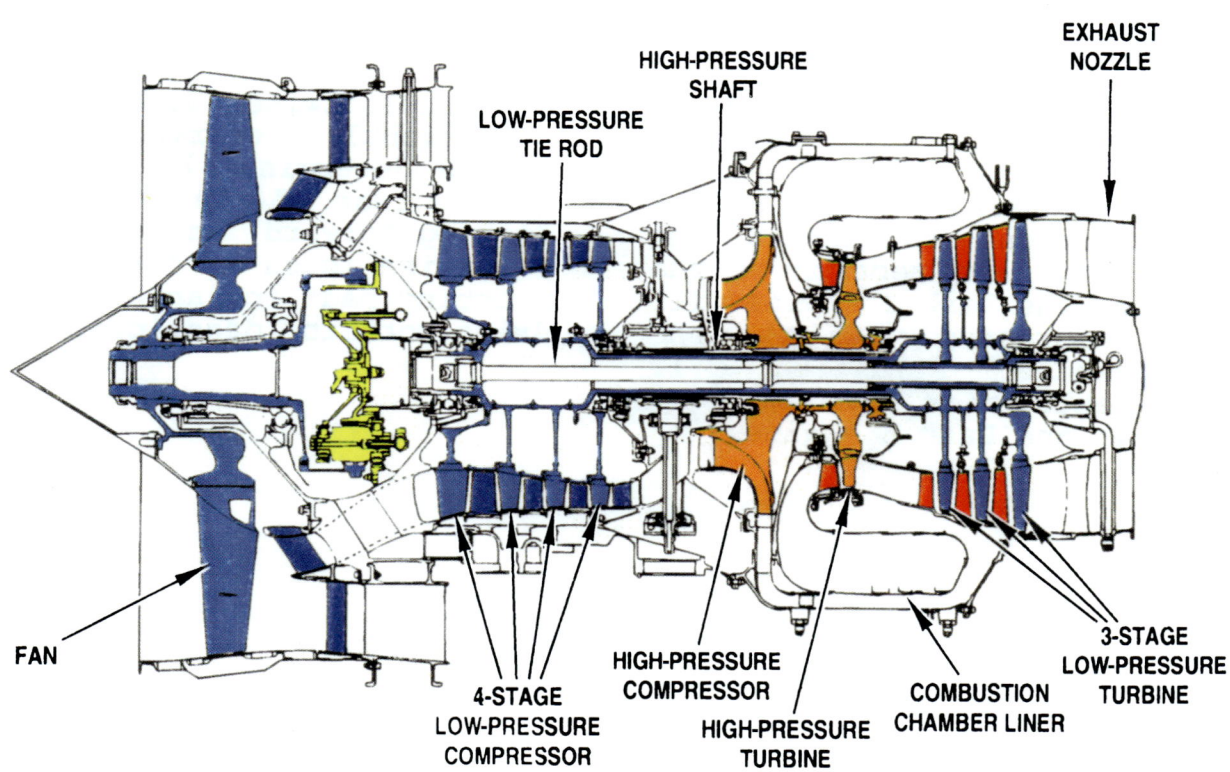

LOW-PRESSURE TIE ROD

HIGH-PRESSURE SHAFT

EXHAUST NOZZLE

FAN

4-STAGE LOW-PRESSURE COMPRESSOR

HIGH-PRESSURE COMPRESSOR

HIGH-PRESSURE TURBINE

COMBUSTION CHAMBER LINER

3-STAGE LOW-PRESSURE TURBINE

The TFE731 engine.

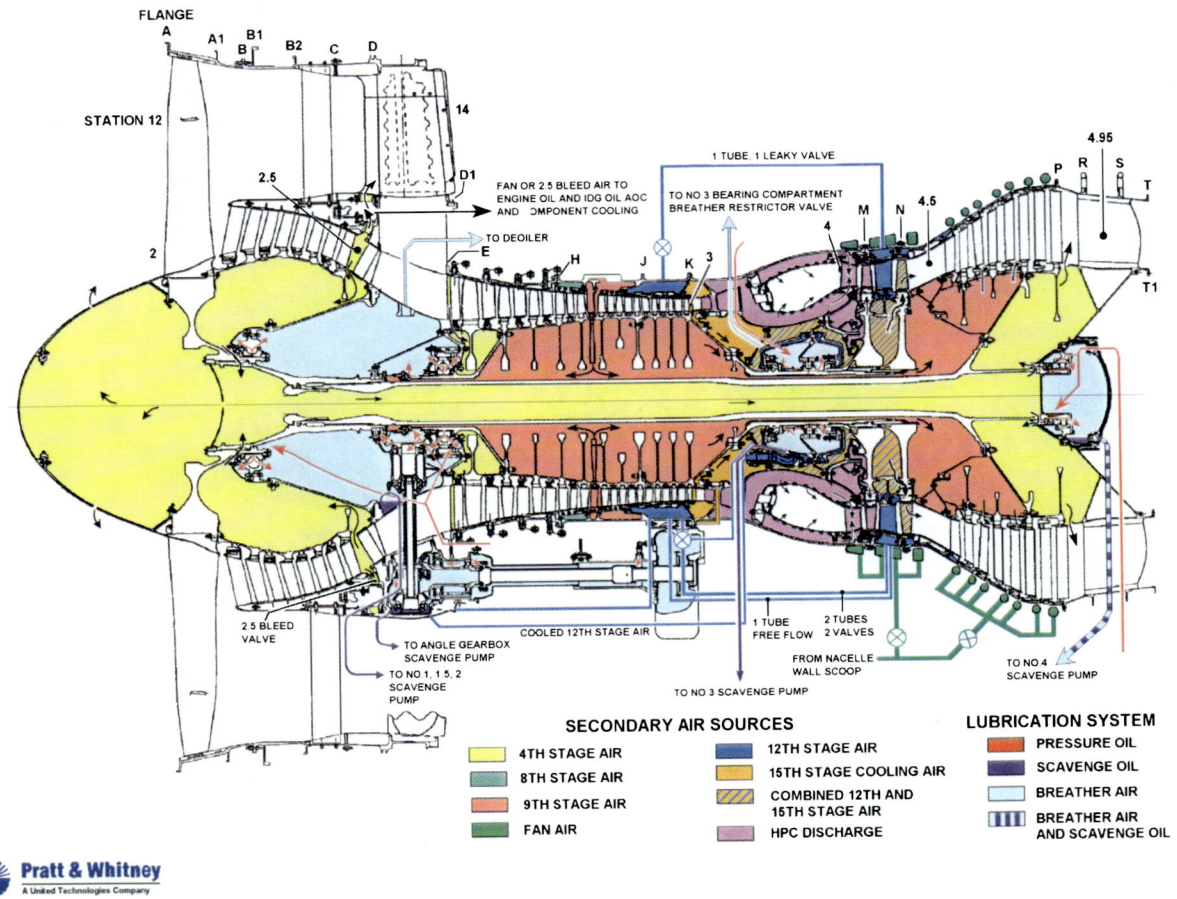

PW4000 (94-in) phase III.

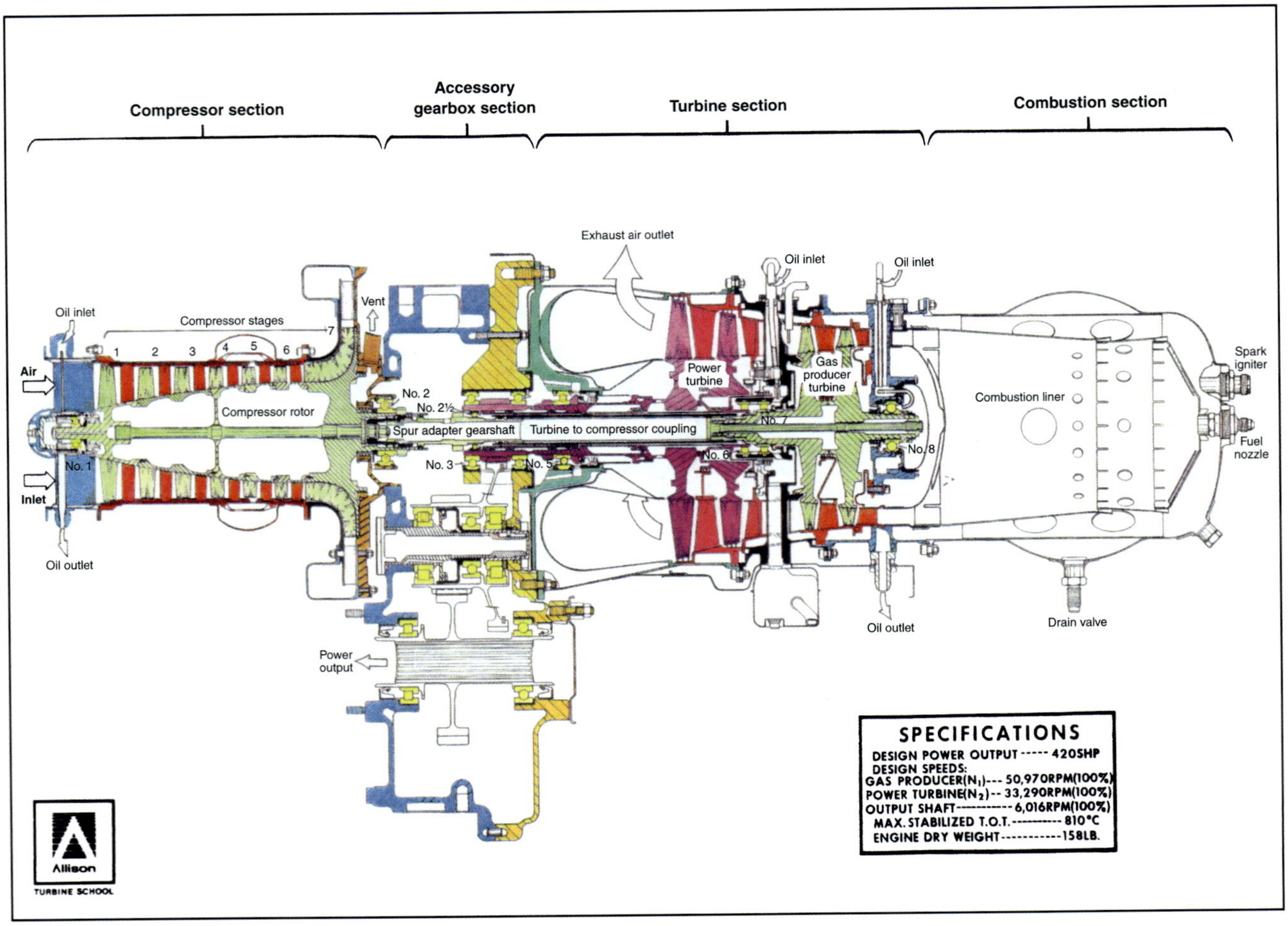

SECONDARY AIR SOURCES

(yellow)	4TH STAGE AIR
(teal)	8TH STAGE AIR
(orange)	9TH STAGE AIR
(green)	FAN AIR
(blue)	12TH STAGE AIR
(gold)	15TH STAGE COOLING AIR
(hatched gold)	COMBINED 12TH AND 15TH STAGE AIR
(pink)	HPC DISCHARGE

LUBRICATION SYSTEM

(red)	PRESSURE OIL
(purple)	SCAVENGE OIL
(light blue)	BREATHER AIR
(hatched blue)	BREATHER AIR AND SCAVENGE OIL

Labels on upper diagram:
FLANGE A, A1, B1, B, B2, C, D, STATION 12, 14, D1, 2.5, 2, TO DEOILER, FAN OR 2.5 BLEED AIR TO ENGINE OIL AND IDG OIL AOC AND COMPONENT COOLING, E, H, J, K, 3, M, N, 4, 4.5, 4.95, P, R, S, T, T1, 1 TUBE, 1 LEAKY VALVE, TO NO 3 BEARING COMPARTMENT BREATHER RESTRICTOR VALVE, 2.5 BLEED VALVE, TO ANGLE GEARBOX SCAVENGE PUMP, TO NO.1, 1.5, 2 SCAVENGE PUMP, COOLED 12TH STAGE AIR, 1 TUBE FREE FLOW, 2 TUBES 2 VALVES, FROM NACELLE WALL SCOOP, TO NO.4 SCAVENGE PUMP, TO NO 3 SCAVENGE PUMP

Labels on lower diagram:
Compressor section, Accessory gearbox section, Turbine section, Combustion section, Oil inlet, Compressor stages, Vent, Exhaust air outlet, Oil inlet, Oil inlet, Air, Inlet, No. 1, Compressor rotor, No. 2, No. 2½, Spur adapter gearshaft, Turbine to compressor coupling, No. 3, No. 5, No. 6, No. 7, No. 8, Power turbine, Gas producer turbine, Combustion liner, Spark igniter, Fuel nozzle, Oil outlet, Power output, Oil outlet, Drain valve

SPECIFICATIONS

DESIGN POWER OUTPUT ----- 420SHP
DESIGN SPEEDS:
GAS PRODUCER(N₁) --- 50,970RPM(100%)
POWER TURBINE(N₂) -- 33,290RPM(100%)
OUTPUT SHAFT --------- 6,016RPM(100%)
MAX. STABILIZED T.O.T. ---------- 810°C
ENGINE DRY WEIGHT ----------158LB.

Allison
TURBINE SCHOOL

Turboshaft engine cutaway schematic.

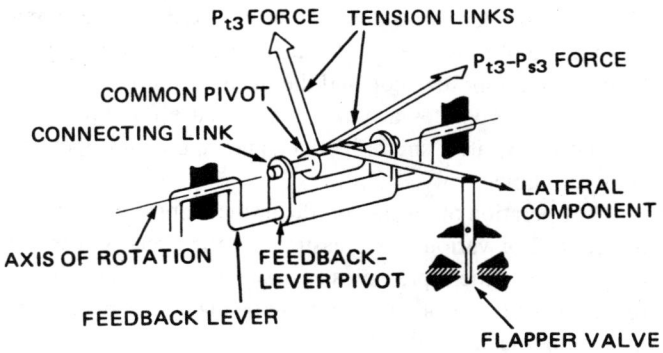

FIGURE 13-22 Drawing illustrating operation of pressure ratio sensor.

stator vane actuator (SACS1), which hydraulically positions the engine stator vanes. The SACS1 also provides a feedback signal, through a mechanical linkage, to rotate the

three-dimensional cam and reestablish the null position of the actuator pilot valve.

The three-dimensional cam also incorporates two additional contours that actuate the start and surge bleed control valves. These valves, in turn, provide high-pressure signals (P_{cb3} and $P_{cb3.5}$) to remotely mounted engine bleed valve actuators. The control valves are also designed to supply a prescribed cooling bleed flow through the supply lines to the engine bleed valve actuators.

Interrelation of Control Units

Figure 13-23 illustrates the interrelation between the main fuel control and the EVC and also the position of the stator vane actuator in the system. A careful study of this diagram and the various controlling and actuating forces will help the reader understand the many factors governing the operation of a large engine such as the JT9D.

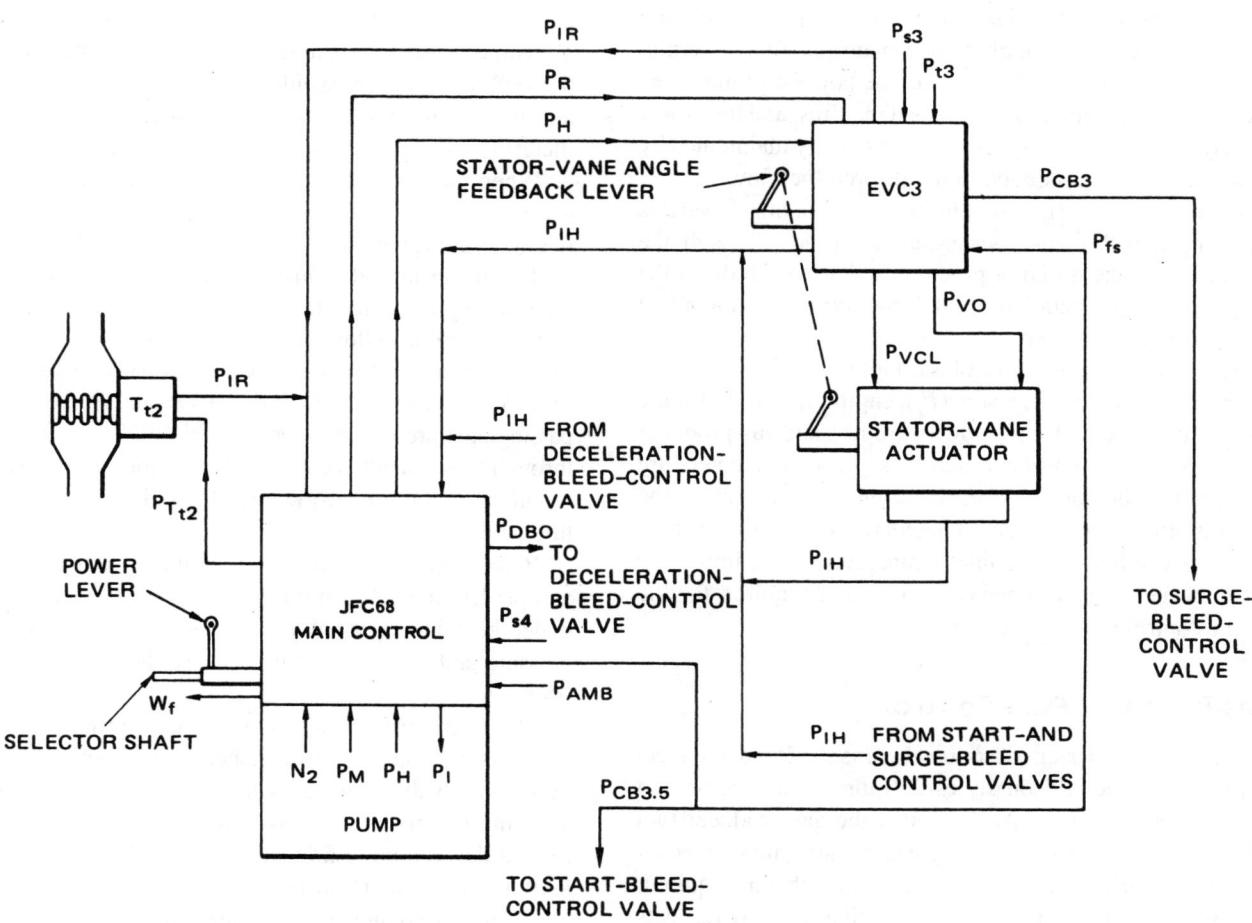

FIGURE 13-23 Drawing illustrating the interrelation between the FCU and the EVC. (*Hamilton Standard.*)

N_2—High-pressure compressor speed
P_{amb}—Ambient pressure
P_{DBO}—Deceleration-bleed override
P_F—Fine-filtered main-stage pressure
P_H—Controlled hydraulic-stage pressure

P_I—Controlled body pressure (pump-interstage pressure)
P_{IH}—Pump interstage from hydraulic supply
P_{IR}—Pump interstage from regulated supply
P_M—Main-stage inlet pressure

P_R—Regulated fine-filtered servo supply
P_S—Servo-control pressure
P_{s4}—Engine-burner pressure
T_{t2}—Engine-inlet total temperature
W_f—Metered fuel flow
P_{Tt2}—Signal pressure
T_{t2}— **Sensor**

P_{A1}—Metered vane-actuator pressure
P_{A2}—Metered vane-actuator pressure
P_{cb3}—Bleed-control pressure
$P_{cb3.5}$—Start-bleed pressure
P_{FS}—Speed-signal pressure

FUEL CONTROL SYSTEM FOR A TURBOSHAFT ENGINE

A typical turboshaft engine incorporates a gas producer (gas generator) and a power-turbine system within the engine. The gas producer consists of a compressor, the combustion chamber, fuel nozzle or nozzles, and the gas-producer turbine. This section produces the high-velocity, high-temperature gases which furnish the energy to drive the power turbine. The power turbine usually incorporates two or more stages (turbine wheels) which extract the energy from the gases and deliver power to the output shaft.

A fuel control system for a turboshaft engine is often comprised of two sections, one which senses and regulates the gas-producer part of the engine and the other which senses the operation and requirements of the power-turbine section. The complete system controls engine power output by controlling the gas-producer speed, which, in turn, governs the power output. The gas-producer speed levels are established by the action of the power-turbine fuel governor, which senses power-turbine speed. The power-turbine speed is selected by the operator as the load requires, and the power needed to maintain this speed is automatically maintained by power-turbine governor action on metered fuel flow.

The power-turbine governor incorporates rotating flyweights which continually sense power-turbine speed. Through the speed sensing, the governor produces actions which direct the gas-producer fuel control to schedule the correct amount of fuel for the required operation.

Fuel flow for engine control is established as a function of compressor discharge pressure (P_c), engine speed (N_1 for the gas producer and N_2 for the power turbine), and gas-producer throttle lever position. Note that these same parameters are employed in the control of other turbine engines, and some controls utilize additional parameters for fuel control. Turbojet engines utilize turbine inlet temperature as an important factor in fuel control; however, this is not required for the engine control under discussion.

Gas-Producer Fuel Control

The fuel control system described here is the Bendix system employed on the Allison series 250 turboshaft engine and is illustrated in Fig. 13-24. Note that the gas-producer fuel control and the power-turbine governor are interconnected so that each may affect the operation of the other as required.

The gas-producer fuel control is similar in many respects to other fuel controls described previously. Its primary function is the same. Fuel entering the control encounters a bypass valve which maintains a constant differential between fuel pump pressure P_1 and metered fuel pressure P_2. Excess fuel is bypassed back to the entering side of the pump. The constant pressure differential is applied across the metering valve; therefore, fuel flow will be in proportion to the opening of the metering valve. The degree to which the metering valve is open is controlled by the action of the governor bellows and the acceleration bellows and is modified by the action of the derichment valve during starting. The maximum range of movement of the metering valve is controlled by the minimum flow stop and the maximum flow stop. The unit also incorporates a maximum pressure relief valve, a manually operated shutoff valve, and a bellows-operated start-derichment valve.

The operation of the gas-producer fuel control is based on the control of various air pressures by the speed governors and the use of these pressures to move the metering valve as required. The control may therefore be classed as pneumatic or pneumomechanical.

The simplified drawing in Fig. 13-25 illustrates how air pressure may be controlled to operate a metering valve. In the drawing, air pressure P_c, which may be compared to CDP, is applied to the controller and flows through an air bleed. The rate of flow through the air bleed will determine the difference between P_c and modified pressure P_x. The rate of flow is determined by the position of the governor valve. If the governor valve is completely closed, there will be no flow through the air bleed and P_c will equal P_x. Since P_x would be at a comparatively high level, the pressure in the bellows chamber would cause the bellows to collapse and the metering valve to open through the linkage to the metering valve.

When the governor valve is closed, pressure P_x is much lower than P_c. This allows the bellows to expand and close the metering valve.

The metering valve in the gas-producer fuel control is operated by lever action in accordance with the movement of the governor bellows and the acceleration bellows. Note that the governor bellows and acceleration bellows are affected by variations in P_x and P_y. Pressures P_x and P_y are derived by passing pressure P_c through two air bleeds. The rate of airflow through these bleeds is controlled by action of the governor as modified by throttle position and the influence of the power-turbine governor.

Before lightoff and acceleration, the metering valve is set at a predetermined open position by the acceleration bellows under the influence of ambient pressure. At this point, ambient pressure and P_c are the same because the compressor is not operating.

The start-derichment valve is open during lightoff and acceleration until a preestablished P_c is reached. The open derichment valve vents pressure P_y to the atmosphere, thus allowing the governor bellows to move the metering valve toward the minimum flow stop. This keeps fuel flow at the lean fuel schedule required for starting and acceleration. As compressor rpm increases, the derichment valve is closed by P_c acting on the derichment bellows. When the derichment valve is closed, control of the metering valve is returned to the normal operating schedule in which the effects of P_x and P_y as regulated by the governor are operating through the governor bellows and acceleration bellows.

During acceleration, P_x and P_y are equal to the modified CDP P_c up to the point where the speed-enrichment orifice is opened by the governor flyweight action. This action bleeds pressure P_x while pressure P_y remains at a value equal to P_c. Under the influence of the $P_y - P_x$ pressure drop across the

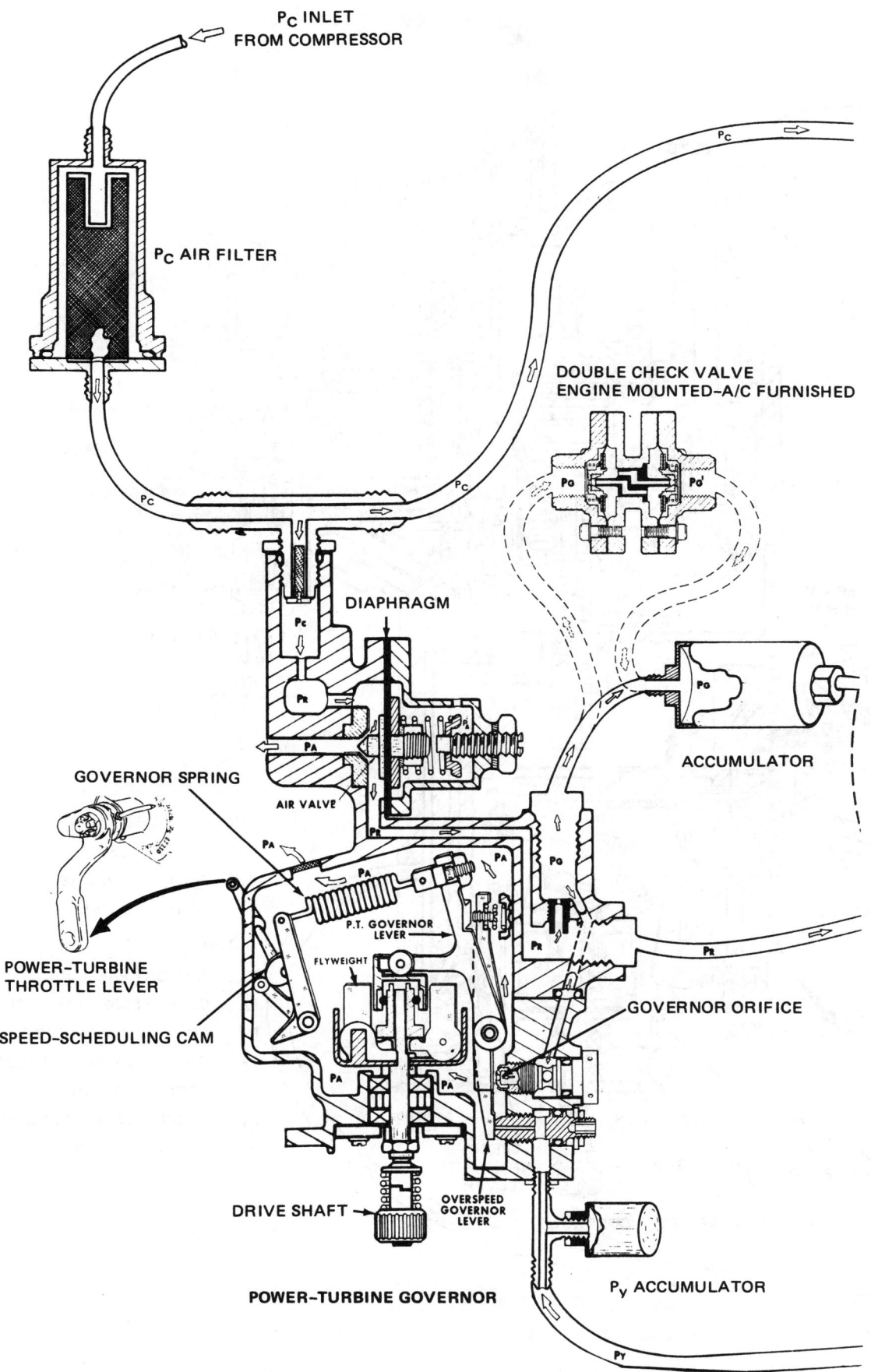

FIGURE 13-24 Schematic drawing illustrating a Bendix fuel control system for a turboshaft engine. (*Rolls-Royce.*)

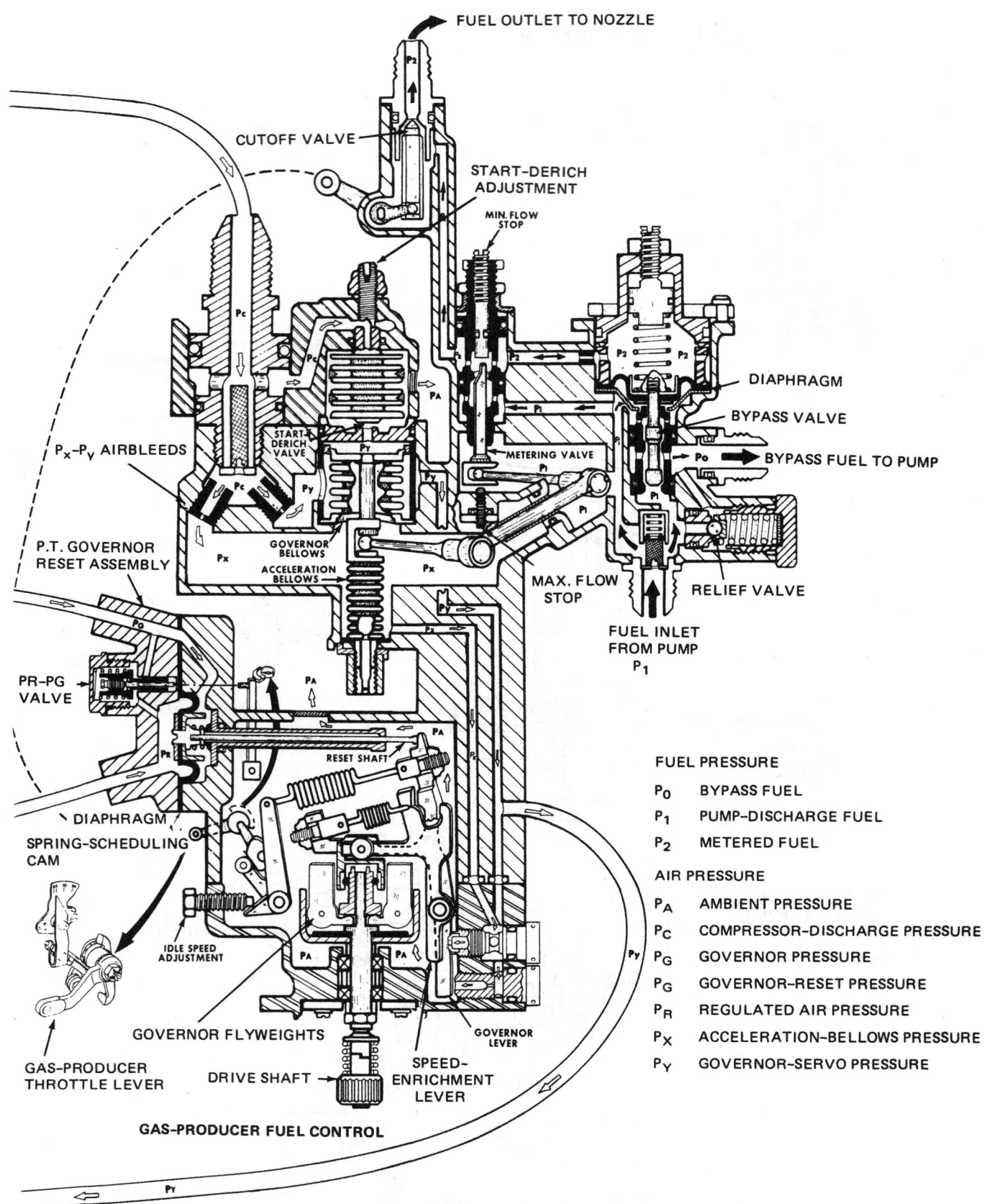

FIGURE 13-24 Schematic drawing illustrating a Bendix fuel control system for a turboshaft engine. (*Rolls-Royce.*) (*Continued*)

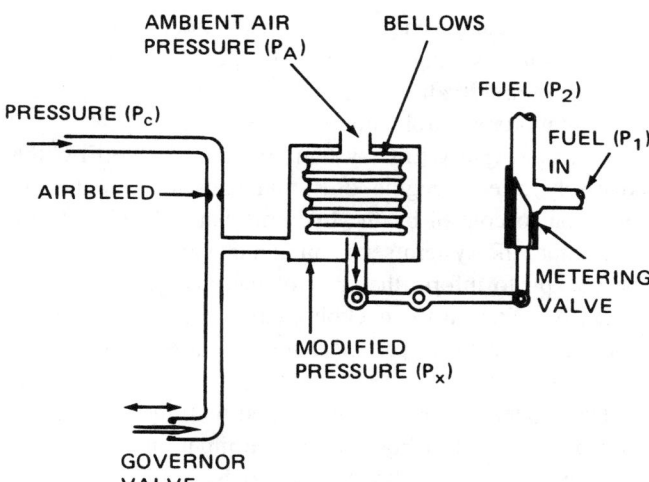

FIGURE 13-25 Simplified drawing showing the use of pneumatic control for a fuel metering valve.

governor bellows, the metering valve moves to a more open position, thus increasing fuel flow as required for acceleration.

Gas-producer rpm (N_1) is controlled by the gas-producer control governor. The governor flyweights operate the governor lever which controls the governor bellows (P_y) bleed at the governing orifice. The flyweight operation of the governor lever is opposed by a variable spring load which is changed in accordance with the position of the throttle acting through the spring-scheduling cam. Opening the governor orifice bleeds pressure P_y and allows pressure P_x to control the governor action on the bellows. The P_{xy} action on the bellows moves the metering valve to a more closed position until metered flow is at steady-state requirements.

The governor-reset section of the gas-producer fuel control is utilized by the power-turbine governor to override the speed-governing elements of the fuel control, to change the fuel schedule in response to load conditions applied to the power turbine. The diaphragm and spring in the governor-reset assembly apply force to the governor lever to modify the effect of the governor springs.

Power-Turbine Governor

The power-turbine section of the engine calls upon the gas generator section for more or less energy, depending on the load requirements. It is the function of the power-turbine fuel governor to provide the actuating force directed to the gas-producer fuel control, which responds by increasing or decreasing fuel as required to produce the needed gas energy.

As shown in Fig. 13-24, the power-turbine speed is scheduled by the power-turbine governor lever and the power-turbine speed-scheduling cam. The cam, operated by the throttle, sets a governor spring load which opposes the force of the speed flyweights. As the desired speed is approached, the speed weights, operating against the governor spring, move a link to open the power-turbine governor orifice. The speed flyweights also open the overspeed bleed (P_y) orifice but at a higher speed than that at which the regular governor orifice (P_g) is opened.

The power-turbine governor, like the gas-producer fuel control, utilizes controlled air pressure to accomplish its purposes. Compressor discharge pressure P_c enters the air valve, which is a pressure regulator. The output of the air valve is regulated pressure P_r, which is applied to one side of the diaphragm in the governor-reset section of the gas-producer fuel control. Governor pressure P_g, developed when pressure P_r passes through the P_g bleed, is applied to the opposite side of the diaphragm. When the governor orifice is closed, P_r and P_g are equal and produce no effect on the governor-reset diaphragm. When the governor orifice is opened by action of the flyweights, P_g is reduced. The effect of $P_r - P_g$ on the diaphragm is to produce force through the governor-reset rod to the gas-producer governor lever (power output link) to supplement the force of the flyweights in the gas-producer governor. This opens the P_y orifice and bleeds P_y, thus causing the gas-producer governor bellows to move the fuel metering valve to a more closed position. This, in turn, reduces gas-producer speed. Gas-producer speed cannot exceed the gas-producer fuel governor setting.

The governor-reset diaphragm is preloaded to establish the active $P_r - P_g$ range. This is accomplished by means of a spring, as shown in Fig. 13-24.

The overspeed orifice in the power-turbine governor bleeds P_y from the governing system of the gas-producer fuel control. This gives the system a rapid response to N_2 overspeed conditions.

ELECTRONIC ENGINE CONTROLS

Because of the need to control precisely the many factors involved in the operation of modern high-bypass turbofan engines, airlines and manufacturers have worked together to develop electronic engine control (EEC) systems that prolong engine life, save fuel, improve reliability, reduce flight crew workload, and reduce maintenance costs. The cooperative efforts have resulted in two types of EECs, one being the supervisory engine control system and the other the full-authority engine control system. The supervisory control system was developed and put into service first, and is used with the JT9D-7R4 engines installed in the Boeing 767.

Essentially, the **supervisory EEC** includes a computer that receives information regarding various engine operating parameters and adjusts a standard hydromechanical FCU to obtain the most effective engine operation. The hydromechanical unit responds to the EEC commands and actually performs the functions necessary for engine operation and protection.

The **full-authority EEC** is a system that receives all the necessary data for engine operation and develops the commands to various actuators to control the engine parameters within the limits required for the most efficient and safe engine operation. This type of system is employed on advanced-technology engines such as the Pratt & Whitney series 2000 and 4000.

Supervisory EEC System

The digital supervisory EEC system employed with the JT9D-7R4 turbofan engine includes a hydromechanical FCU such as the Hamilton Standard JFC68 described earlier, a Hamilton Standard EEC-103 unit, a hydromechanical air-bleed and vane control, a permanent-magnet alternator to provide electric power for the EEC separate from the aircraft electric system, and an engine inlet pressure and temperature probe to sense P_{t2} and T_{t2}. The hydromechanical units of the system control such basic engine functions as automatic starting, acceleration, deceleration, high-pressure rotor speed (N_2) governing, VSV compressor position, modulated and starting air-bleed control, and burner pressure (P_b) limiting. The EEC provides precision thrust management, N_2 and EGT limiting, and cockpit display information on engine pressure ratio (EPR) limit, EPR command, and actual EPR. It also provides control of modulated turbine-case cooling and turbine-cooling air valves and transmits information regarding parametric and control system condition for possible recording. Such recorded data are utilized by maintenance technicians in eliminating faults in the system.

The supervisory EEC, by measuring EPR and integrating thrust lever (throttle) angle, altitude data, Mach number, inlet air pressure P_{t2}, inlet air temperature T_{t2}, and total air temperature in the computation, is able to maintain constant thrust from the engine regardless of changes in air pressure, air temperature, and flight environment. Thrust changes occur only when the thrust lever angle is changed, and the thrust remains consistent for any particular position of the thrust lever. Takeoff thrust is produced in the full-forward position of the thrust lever. Thrust settings for climb and cruise are made by the pilot as the thrust lever is moved to a position that provides the correct EPR for the thrust desired. The EEC is designed so that the engine will quickly and precisely adjust to a new thrust setting without the danger of overshoot in N_2 or temperature. It adjusts the hydromechanical FCU through a torque motor electrohydraulic servo system.

In a supervisory EEC system, any fault in the EEC that adversely affects engine operation causes an immediate reversion to control by the hydromechanical FCU. At the same time, the system sends an annunciator light signal to the cockpit to inform the crew of the change in operating mode. A switch in the cockpit enables the crew to change from EEC control to hydromechanical control if it is deemed advisable.

The supervisory EEC is integrated with the aircraft systems as indicated in Fig. 13-26. The input and output signals are shown by the directional arrows. Although the EEC utilizes aircraft electric power for some of its functions, the electric power for the basic operation of the EEC is supplied by the separate engine-driven permanent-magnet alternator mentioned earlier.

The output signals of the supervisory EEC that affect engine operation are the adjustment of the hydromechanical FCU and commands to solenoid-actuated valves for control of modulated turbine-case cooling and turbine-cooling air.

Full-Authority EEC

A full-authority EEC performs all functions necessary to operate a turbofan engine efficiently and safely in all modes, such as starting, accelerating, decelerating, takeoff, climb, cruise, and idle. It receives data from the aircraft and engine systems, provides data for the aircraft systems, and issues commands to engine control actuators.

The information provided in this section is based on the Hamilton Standard EEC-104, an EEC designed for use with the

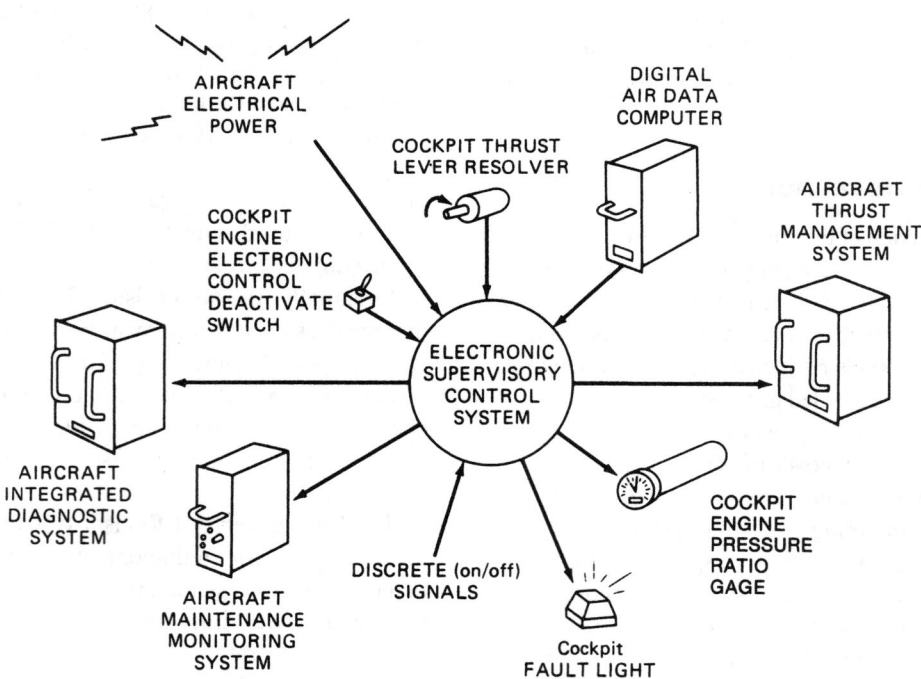

FIGURE 13-26 Integration of a supervisory EEC with aircraft system. (*ASME.*)

FIGURE 13-27 Hamilton Standard EEC-104 electronic engine control. (*Hamilton Standard.*)

Pratt & Whitney 2037 engine. The unit is shown in Fig. 13-27. This is a dual-channel unit having a "crosstalk" capability, so that either channel can utilize data from the other channel. This provision greatly increases reliability to the extent that the system will continue to operate effectively even though a number of faults may exist. Channel A is the primary channel, and channel B is the secondary, or backup, channel.

The following abbreviations and symbols are used in this section to identify functions, systems, and components:

ACC	Active clearance control
BCE	Breather compartment ejector
EEC	Electronic engine control
EGT	Exit (exhaust) gas temperature
EPR	Engine pressure ratio
FCU	Fuel control unit
LVDT	Linear variable differential transformer
N_1	Low-pressure spool rpm
N_2	High-pressure spool rpm
P_{amb}	Ambient air pressure
P_b	Burner pressure
PMA	Permanent-magnet alternator
P_{s3}	Static compressor air pressure
P_{t2}	Engine inlet total pressure
$P_{4.9}$	Exhaust-gas pressure
TCA	Turbine-cooling air
TLA	Throttle lever angle
TRA	Throttle resolver angle
T_{t2}	Engine inlet total air temperature
$T_{4.9}$	Exhaust-gas temperature
w_f	Fuel flow

Figure 13-28 is a block diagram showing the relationships among the various components of the EEC system. Input signals from the aircraft to the EEC-104 include throttle resolver angle (which tells the EEC the position of the throttle), service air-bleed status, aircraft altitude, total air pressure, and total air temperature. Information regarding altitude, pressure, and temperature is obtained from the air data computer as well as the P_{t2}/T_{t2} probe in the engine inlet.

Outputs from the engine to the EEC include overspeed warning, fuel flow rate, electric power for the EEC, high-pressure rotor speed N_2, stator vane angle feedback, position of the 2.5 air-bleed proximity switch, air/oil cooler feedback, fuel temperature, oil temperature, automatic clearance control (ACC) feedback, TCA position, engine tailpipe pressure $P_{4.9}$, burner pressure P_b, engine inlet total pressure P_{t2}, low-pressure rotor speed N_1, engine inlet total temperature T_{t2}, and exhaust-gas temperature (EGT or $T_{4.9}$). Sensors installed on the engine provide the EEC with measurements of temperatures, pressures, and speeds. These data are used to provide automatic thrust rating control, engine limit protection (overspeed, overheat, and overpressure), transient control, and engine starting.

Outputs from the EEC to the engine include fuel flow torque motor command, stator vane angle torque motor command, air/oil cooler valve command, 2.5 air-bleed torque motor command, ACC torque motor command, oil bypass solenoid command, breather compartment ejector solenoid command, and TCA solenoid command. The actuators that must provide feedback to the EEC are equipped with linear variable differential transformers (LVDTs) to produce the required signals.

During operation of the engine control system, fuel flows from the aircraft fuel tank to the centrifugal stage of the dual-stage fuel pump. The fuel is then directed from the pump through a dual-core oil/fuel heat exchanger which provides deicing for the fuel filter as the fuel is warmed and the oil is cooled. The filter protects the pump main-gear stage and the fuel system from fuel-borne contaminants. High-pressure fuel from the main-gear stage of the fuel pump is supplied to the FCU, which, through electrohydraulic servo valves, responds to commands from the EEC to position the fuel metering valve, stator vane actuator, and air/oil cooler actuator. Compressor air-bleed and ACC actuators are positioned by electrohydraulic servo valves that are controlled directly by the EEC, using redundant torque motor drivers and feedback elements. The word "redundant" means that units or mechanisms are designed with backup features so that a failure in one part will not disable the unit, and operation will continue normally. Actuator position feedback is provided to the EEC by redundant LVDTs for the actuators and redundant resolvers for the fuel metering valve. Fuel-pump discharge pressure is used to power the stator vane, 2.5 air-bleed, air/oil cooler, and ACC actuators. The EEC activates TCA, engine breather ejector, and the aircraft-provided thrust reverser through electrically controlled dual-solenoid valves.

The EEC and its interconnected components are shown in Fig. 13-29. Note that the EEC is mounted on the top left side of the engine fan case. The mounting is accomplished with vibration isolators (shock mountings) to protect the unit.

The benefits of employing a full-authority EEC result in substantial savings for the aircraft operator. Among these benefits are reduced crew workload, reduced fuel consumption, increased reliability, and improved maintainability.

Flight crew workload is decreased because the pilot utilizes the EPR gauge to set engine thrust correctly. An EPR

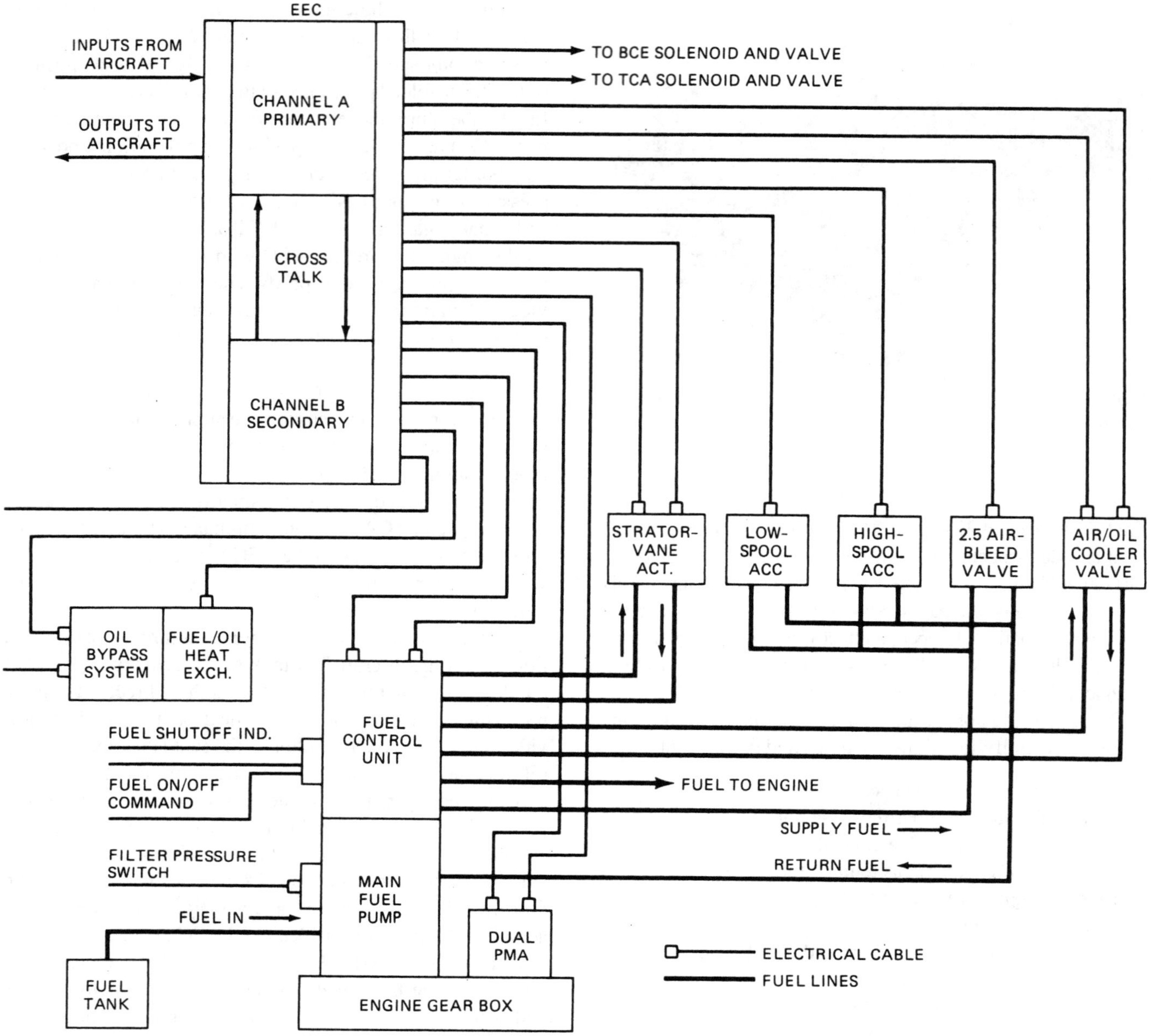

```
                         EEC
INPUTS FROM  ┌──────────────────┐
AIRCRAFT ──► │                  │ ─────────────► TO BCE SOLENOID AND VALVE
             │    CHANNEL A     │ ─────────────► TO TCA SOLENOID AND VALVE
             │    PRIMARY       │
OUTPUTS TO ◄─│                  │
AIRCRAFT     │                  │
             ├──────────────────┤
             │      CROSS       │
             │      TALK        │
             ├──────────────────┤
             │                  │
             │    CHANNEL B     │
             │    SECONDARY     │
             │                  │
             └──────────────────┘
```

FIGURE 13-28 Simplified block diagram of the EEC system with the Hamilton Standard EEC-104. (*Hamilton Standard.*)

Blocks and labels shown: OIL BYPASS SYSTEM, FUEL/OIL HEAT EXCH., FUEL SHUTOFF IND., FUEL ON/OFF COMMAND, FILTER PRESSURE SWITCH, FUEL IN, FUEL TANK, FUEL CONTROL UNIT, MAIN FUEL PUMP, DUAL PMA, ENGINE GEAR BOX, STRATOR-VANE ACT., LOW-SPOOL ACC, HIGH-SPOOL ACC, 2.5 AIR-BLEED VALVE, AIR/OIL COOLER VALVE, FUEL TO ENGINE, SUPPLY FUEL, RETURN FUEL.

Legend: ▭ ELECTRICAL CABLE, ━━ FUEL LINES

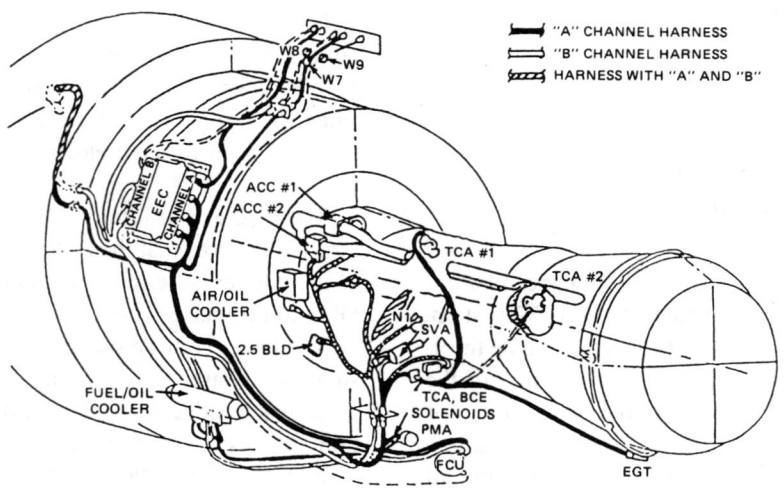

FIGURE 13-29 Drawing showing EEC units on an engine. (*Hamilton Standard.*)

Legend: "A" CHANNEL HARNESS, "B" CHANNEL HARNESS, HARNESS WITH "A" AND "B"

Labels: W8, W9, W7, CHANNEL B, CHANNEL A, EEC, ACC #1, ACC #2, TCA #1, TCA #2, AIR/OIL COOLER, N1, SVA, 2.5 BLD, FUEL/OIL COOLER, TCA, BCE SOLENOIDS PMA, FCU, EGT

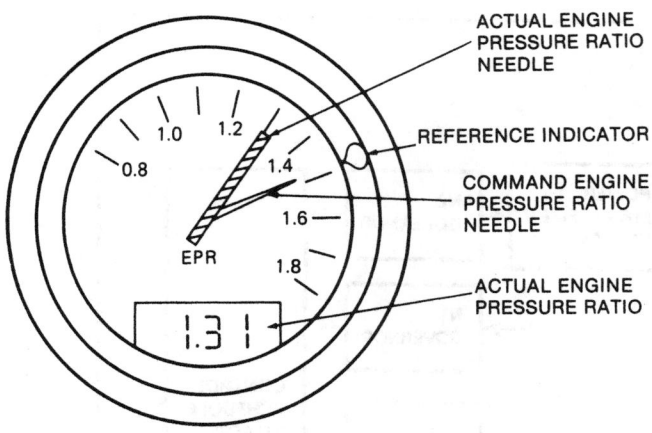

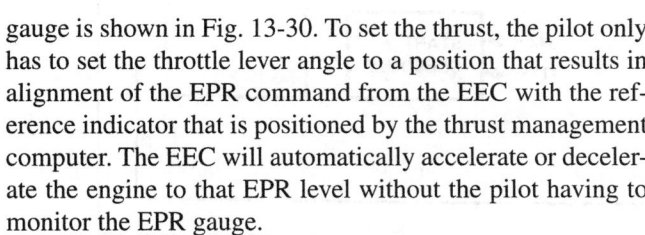

FIGURE 13-30 Drawing of an engine pressure ratio gauge.

FIGURE 13-31 Honeywell digital fuel controller. (*Honeywell, Inc.*)

gauge is shown in Fig. 13-30. To set the thrust, the pilot only has to set the throttle lever angle to a position that results in alignment of the EPR command from the EEC with the reference indicator that is positioned by the thrust management computer. The EEC will automatically accelerate or decelerate the engine to that EPR level without the pilot having to monitor the EPR gauge.

Reduced fuel consumption is attained because the EEC controls the engine operating parameters so that maximum thrust is obtained for the amount of fuel consumed. In addition, the ACC system ensures that compressor and turbine blade clearances are kept to a minimum, thus reducing pressure losses due to leakage at the blade tips. This is accomplished by the ACC system as it directs cooling air through passages in the engine case to control engine case temperature. The EEC controls the cooling airflow by sending commands to the ACC system actuator.

Engine trimming is eliminated by the use of the full-authority EEC. When an engine is operated with a hydromechanical FCU, it is necessary periodically to make adjustments on the FCU to maintain optimum engine performance. To trim the engine, it is necessary to operate the engine on the ground for extensive periods at controlled speeds and temperatures. This results in the consumption of substantial amounts of fuel plus work time for maintenance personnel and downtime for the aircraft. With the full-authority EEC, none of these costs is experienced.

The fault-sensing, self-testing, and correcting features designed into the EEC greatly increase the reliability and maintainability of the system. These features enable the system to continue functioning in flight and provide fault information that is used by maintenance technicians when the aircraft is on the ground. The modular design of the electronic circuitry saves maintenance time because circuit boards having defective components are quickly and easily removed and replaced.

Honeywell Digital Fuel Controller

The EEC designed for operation with the Honeywell TFE-731-5 turbofan engine is called a digital fuel controller

(DFC) and is a full-authority system. The DFC is shown in Fig. 13-31.

The DFC for the TFE-731-5 engine performs the following functions:

1. Maintains required thrust with varying altitude, airspeed, and inlet air temperature T_{t2}.
2. Maintains adequate surge margin throughout the operating range and during acceleration and deceleration of the engine.
3. Provides automatic fuel enrichment during starts.
4. Provides schedules for minimum and maximum fuel flow.
5. Provides temperature limiting at all times.
6. Automatically detects overspeed and actuates the fuel cutoff valve.
7. Provides for synchronizing engine speeds in multiengine applications.
8. Provides for automatic transfer to a backup mode if electric power is reduced below minimum or if critical failures are detected.
9. Provides for use of alternate values for noncritical faults.

The DFC utilizes a single power lever (throttle) position with dual (ground-flight) idle thrust. The power lever angle input to the controller establishes the engine thrust. The simplified block diagram in Fig. 13-32 shows the inputs to the DFC and the outputs to the engine. Inputs are power lever angle, engine inlet pressure, engine inlet temperature, engine spool speeds, and interturbine temperature. The discrete (on/off) inputs, such as the mode select switch, are derived from the cockpit and from the engine itself. The DFC outputs include a proportional drive for regulating fuel flow through a hydromechanical FCU, command to the air-bleed valve solenoids, and command to the overspeed solenoid.

The DFC includes an extensive, built-in test feature capable of isolating faults to a line replacement unit for interfacing components and for self-diagnosing the controller. It is capable of retaining fault history and annunciating faults on the cockpit panel display.

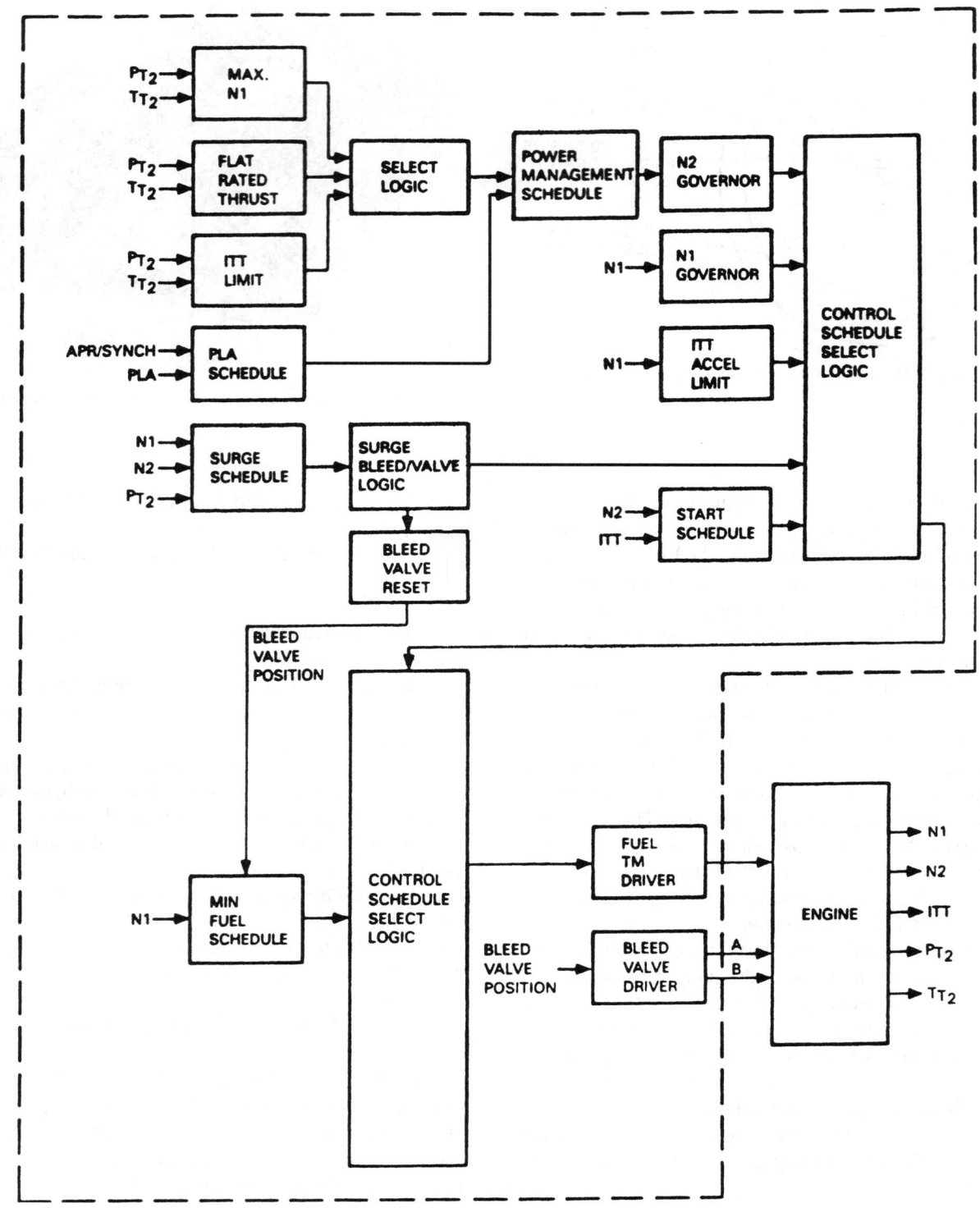

FIGURE 13-32 Simplified block diagram of a digital fuel control. (*Honeywell, Inc.*)

FADEC for an Auxiliary Power Unit

The first example system is an APU engine that uses the aircraft fuel system to supply fuel to the fuel control. An electric boost pump may be used to supply fuel under pressure to the control. The fuel usually passes through an aircraft shutoff valve that is tied to the fire detecting/extinguishing system. An aircraft furnished in-line fuel filter may also be used. Fuel entering the fuel control unit first passes through a 10-micron filter.

If the filter becomes contaminated, the resulting pressure drop opens the filter bypass valve and unfiltered fuel then is supplied to the APU. Shown in Fig. 13-33 is a pump with an inlet pressure access plug so that a fuel pressure gauge might be installed for troubleshooting purposes. Fuel then enters a positive displacement, gear-type pump. Upon discharge from the pump, the fuel passes through a 70-micron screen. The screen is installed at this point to filter any wear debris

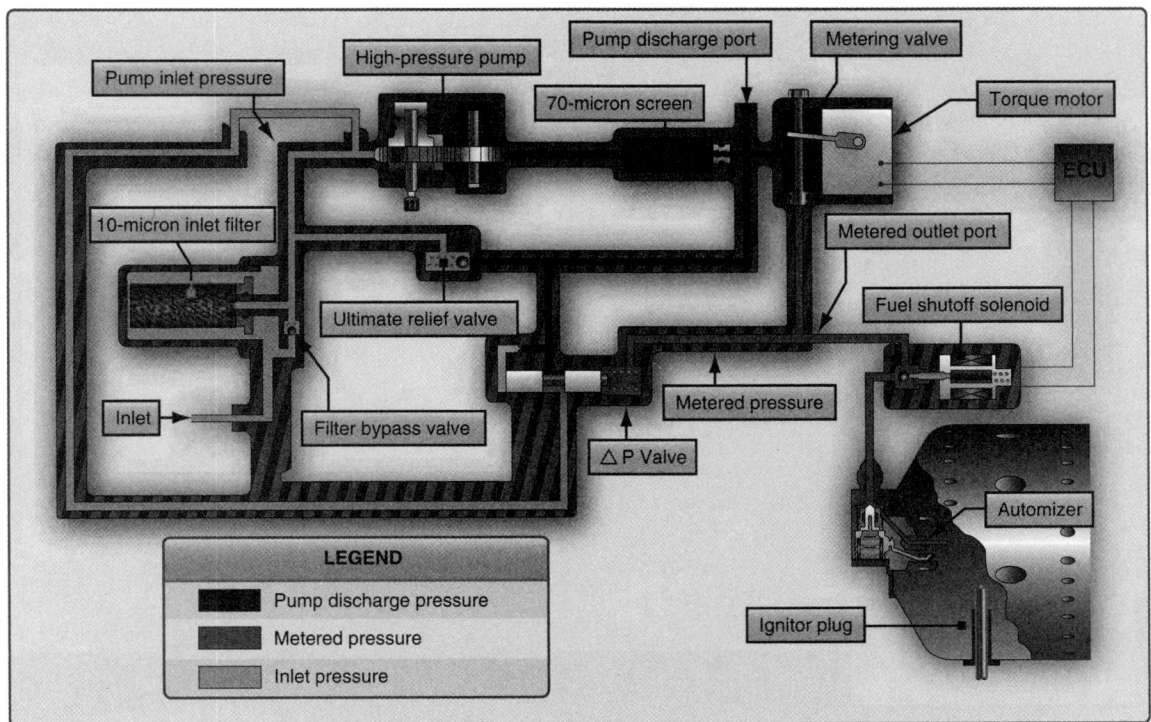

FIGURE 13-33 APU fuel system schematic.

that might be discharged from the pump element. From the screen, fuel branches to the metering valve, differential pressure valve, and the ultimate relief valve. Also shown at this point is the pump discharge pressure access plug, another point where a pressure gauge might be installed.

The differential pressure valve maintains a constant pressure drop across the metering valve by bypassing fuel to the pump inlet so that metered flow is proportional to the metering valve area. The metering valve area is modulated by the torque motor, which receives variable current from the ECU. The ultimate relief valve opens to bypass excess fuel back to the pump inlet whenever system pressure exceeds a predetermined pressure. This occurs during each shutdown since all flow is stopped by the shutoff valve and the differential pressure valve, is unable to bypass full pump capacity. Fuel flows from the metering valve out of the FCU, through the solenoid shutoff valve and on to the atomizer. Initial flow is through the primary nozzle tip only. The flow divider opens at higher pressure and adds flow through the secondary path.

REVIEW QUESTIONS

1. What qualities should turbine fuels possess?
2. What two types of turbine fuels are in common use today?
3. What are the differences between jet A and jet B fuels?
4. What is the danger in mixing jet A and jet B fuels?
5. List the principal components in a turbine-engine fuel system.

6. Where is the fuel-cooled oil cooler usually located in the system? Why is it located at this point?
7. List the different types of fuel spray nozzles.
8. Describe the principle of operation of the Duplex fuel nozzle.
9. Why is a manual fuel control not suitable for a gas-turbine engine?
10. What is a rich blowout? A lean die-out?
11. Name the engine operating conditions (parameters) which must be controlled to ensure efficient and safe performance of a gas-turbine engine.
12. Which of the parameters requested in the previous question are generally employed in the operation of an FCU?
13. What is a hydromechanical FCU?
14. Describe the function of a fuel metering section in a hydromechanical FCU.
15. What is the function of the computing section?
16. Describe the metering system for the JFC68 FCU.
17. What engine operating parameters are sensed by the computing section of the JFC68 FCU?
18. Describe the function of the EVC3 unit.
19. Name the principal parts of the EVC3 unit.
20. What engine operating parameters are utilized by the Honeywell DFC fuel control for the TFE-731-5 engine.
21. What feature allows channels A and B to communicate?
22. How is the VSV servo activated in the FCU?
23. Describe the basic design of a fuel control for a turboshaft engine.
24. What is the difference between a supervisory EEC and a full-authority EEC?
25. What benefits are derived from use of a full-authority EEC?

Turbine-Engine Lubricants and Lubricating Systems

14

GAS-TURBINE ENGINE LUBRICATION

Early gas turbines used oils that were thinner than those used in piston engines, but these oils were produced from the same mineral crude oil. When gas turbines that operated at higher speeds and temperatures were developed, these mineral oils oxidized and blocked the filters and oil passages. The development of low-viscosity **synthetic oils** overcame the major problems encountered with the early mineral oils.

Lubricating Oils

Lubricating oils for gas-turbine engines are usually of the synthetic type. This means that the oils are not manufactured in the conventional manner from petroleum crude oils. Petroleum lubricants are not suitable for modern gas-turbine engines because of the high temperatures encountered during operation. These temperatures often exceed 500°F [260°C], and at such temperatures, petroleum oils tend to break down. The lighter fractions of the oil evaporate, thus leaving carbon and gum deposits; the lubricating characteristics of the oil rapidly deteriorate, too.

Synthetic oils are designed to withstand high temperatures and still provide good lubrication. The first generally acceptable synthetic lubricating oil conformed to MIL-L-7808 and is known as type I, an aklyl diester oil. During recent years, type II oil, a polyester lubricant, has been found most satisfactory. This oil meets or exceeds the requirements of MIL-L-23699.

Lubricants for gas-turbine engines must pass a variety of exacting tests to ensure that they have the characteristics required for satisfactory performance. Among the characteristics tested are specific gravity, acid-forming tendencies, metal corrosion, oxidation stability, vapor-phase coking, gear scuffing, effect on elastomers, and bearing performance. These tests are designed to provide indications that the oil will supply the needed lubrication under all conditions of operation.

Viscosity of Synthetic Oils

The viscosity of the synthetic oils used in gas-turbine engines is generally expressed in units of the centimeter-gram-second (cgs) system. Under this system, the basic unit for the **coefficient of absolute viscosity** is the **poise** (P), named for the French physiologist Jean L. M. Poiseuille (1799–1869). If we imagine a flat plate being drawn across the surface of a layer of oil, the force necessary to move the plate at a given velocity is a measure of the viscosity of the oil. If the layer of oil is 1 cm thick and the plate is moved at the rate of 1 cm/s, the total number of **dynes** of force required to move the plate, divided by the area of the plate in square centimeters, will equal the coefficient of viscosity in poises. To express this in different terms, *when 1 dyne (dyn) will move a 1-cm² plate at a rate of 1 cm/s across the surface of a liquid with a thickness of 1 cm, the coefficient of absolute viscosity is 1 P.*

The viscosity of turbine-engine oil is considerably less than 1 P; therefore, the **centipoise** [1 cP = 0.01 P] is used to express the viscosity.

Because the density of oil is an important factor, it is common practice to employ the unit for **kinematic viscosity** in establishing the characteristic of gas-turbine lubricants. The unit for kinematic viscosity is the same as the poise when the density of a liquid is 1 gram per cubic centimeter [g/cm³]. Kinematic viscosity is expressed in **stokes** (St) [m²/s × 10⁻⁴] or **centistokes** (cSt), 1 cSt being equal to 0.01 St. Kinematic viscosity in stokes is equal to absolute viscosity in poises divided by the density of the liquid in grams per cubic centimeter. The Saybolt Universal viscosity of an oil having a kinematic viscosity of 5 cSt is approximately 42.6. This is roughly equivalent to what is known as 20-weight lubricating oil.

Type II synthetic lubricant is also described as a 5-centistoke (cSt) oil. This means that the oil must have a minimum kinematic viscosity of 5 cSt at a temperature of 210°F [99°C]. This specification is necessary because the oil must maintain sufficient body to carry all applied loads at operating temperatures.

Care in Handling Synthetic Lubricants

We must emphasize that the handling of synthetic lubricants requires precautions not needed for petroleum lubricants. Synthetic lubricants have a high solvent characteristic which causes them to penetrate and dissolve paints, enamels, and other materials. In addition, when synthetic oils are permitted to touch or remain on the skin, physical injury can result. It is therefore essential that the technician handling synthetic lubricants take every precaution to ensure that the lubricants are not spilled or allowed to be in contact with the skin. If a synthetic lubricant is spilled, it should be cleaned up immediately by wiping up, washing, or handling with a suitable cleaning agent. Safety precautions established by the aircraft operator should be observed carefully.

When changing or adding oil to a gas-turbine engine system, the technician must be certain that a lubricant of the correct type and grade is used.

Oil changes for turbine aircraft are governed by the approved service and maintenance procedures established by the airline operating the aircraft and the manufacturer.

LUBRICATING SYSTEM COMPONENTS

Oil Tank

Each engine is provided with an oil tank which is mounted on the engine and secured by a strap. The tank holds enough oil to lubricate the engine with some reserve for cooling and safety. A baffle serves to minimize sloshing of the oil in the tank and a deaerator in the tank separates most of the air from the returning oil, thus minimizing foaming. An example of a typical oil tank is shown in Fig. 14-1.

Pressure Oil Pump

The engine oil distribution system consists of a pressure system which supplies lubricant to the engine bearings, accessory drives, and other engine components. The pressure oil pump, which is a gear-type pump, develops oil pressure by trapping oil in the gear teeth as it is rotated by the engine. A pressure oil pump is illustrated in Fig. 14-2. Most turbine engines use this positive pressure (generally between 40 and 100 psi [275.8 and 689.5 kPa]) to spray oil on the engine's bearings.

Scavenge Oil System

Gas-turbine lubrication systems are usually of the dry-sump type, in that the oil is scavenged from the engine and stored in an oil tank. Scavenge pumps return oil from the engine's bearing cavities to a sump in an accessory drive gearbox or directly to the oil tank.

The scavenge system may consist of several stages—that is, individual pumps that draw oil from the different engine bearing cavities. Scavenge oil pumps are normally of higher

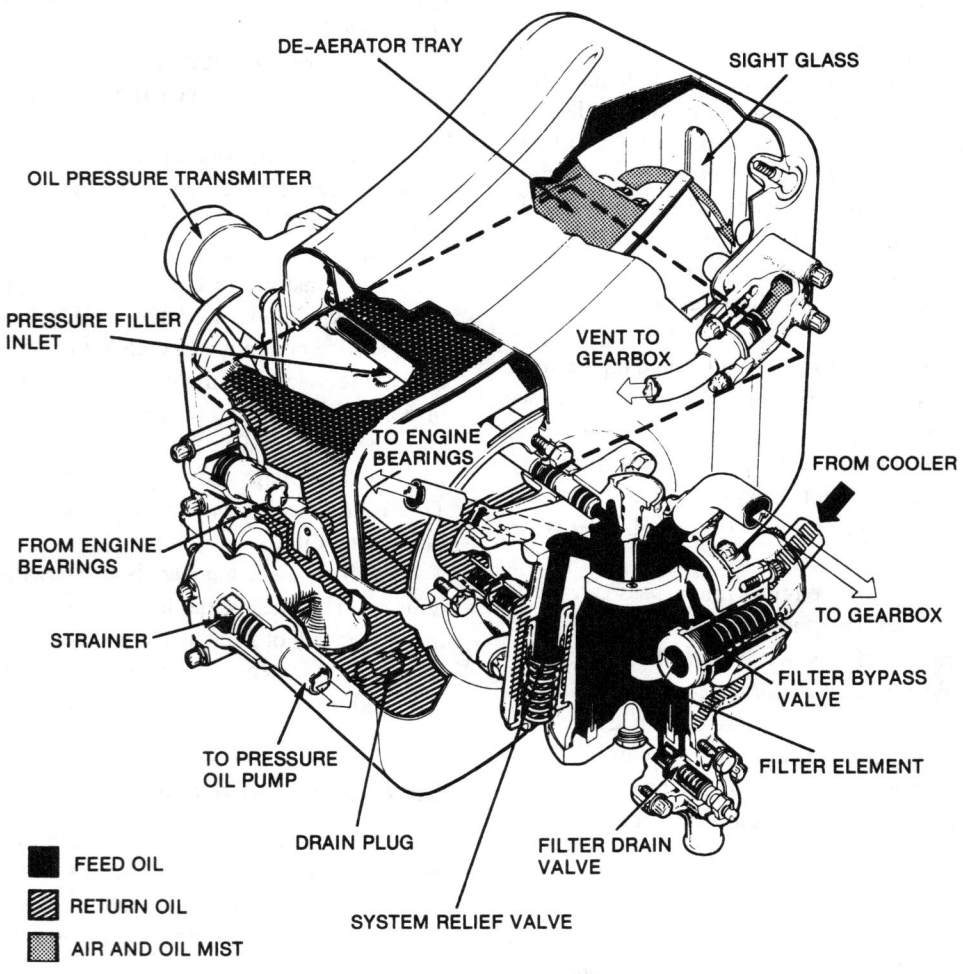

FIGURE 14-1 Oil tank. (*Rolls-Royce.*)

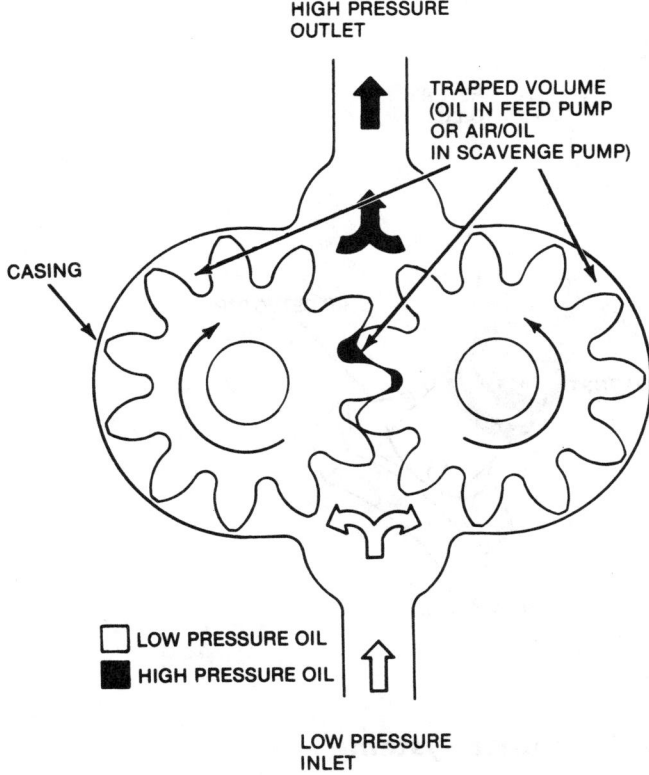

FIGURE 14-2　Principle of a pressure oil pump. (*Rolls-Royce.*)

capacity than engine-driven pumps because of the air that mixes with the oil (foaming) in the bearing cavities. A scavenge pump operates in much the same manner as a pressure pump. More information on oil pump design and operation can be found in Chap. 4.

Oil Filter

The pressure section of the main oil pump forces oil through the main oil filter located immediately downstream of the pump discharge. A typical pressure and scavenge oil filter for a gas-turbine engine is illustrated in Fig. 14-3. The oil enters the inlet port of the pressure filter, surrounds the filter cartridge, and flows through the cartridge to the inner oil chamber and out to the engine. If the filter becomes clogged, the oil is bypassed through the pressure relief valve to the discharge port. A differential pressure of 14 to 16 psi [96.53 to 110.32 kPa] is required to unseat the relief valve.

The size of the filter mesh is measured in **microns**, which is a very small mesh. A red blood cell is about 8 microns in size. Many of the contaminants in the oil that need to be removed are very small, requiring this type of filter. Additional information on filters is presented in Chap. 4.

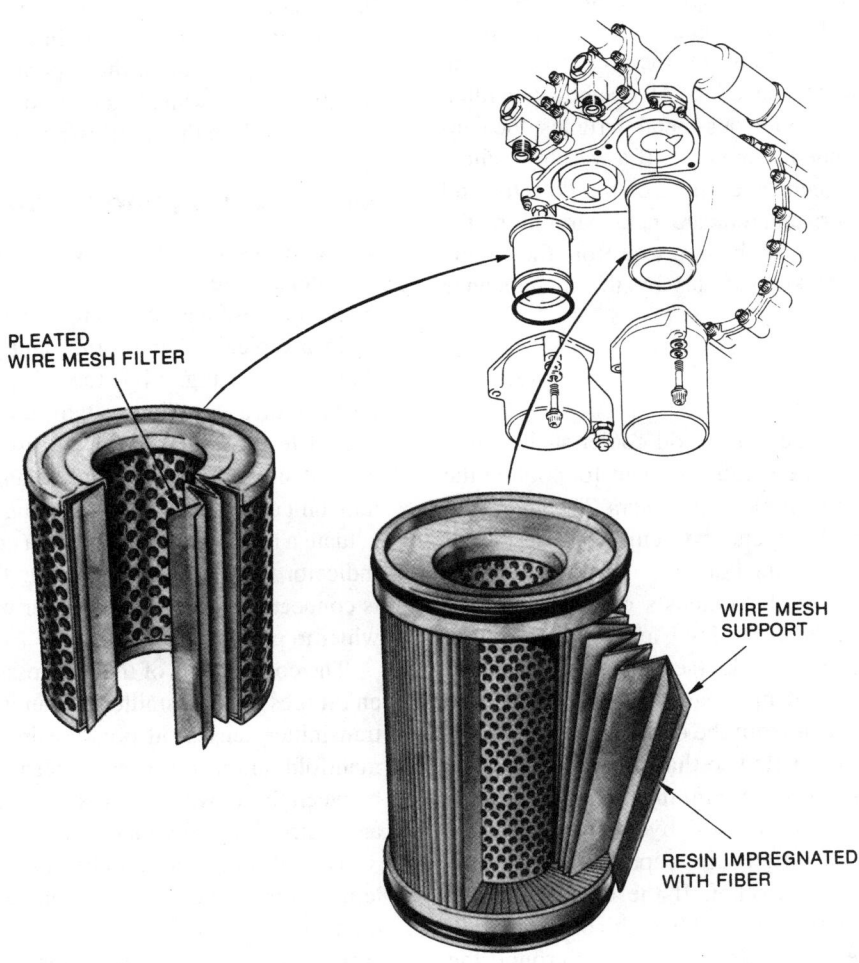

FIGURE 14-3　Typical pressure and scavenge oil filter. (*Rolls-Royce.*)

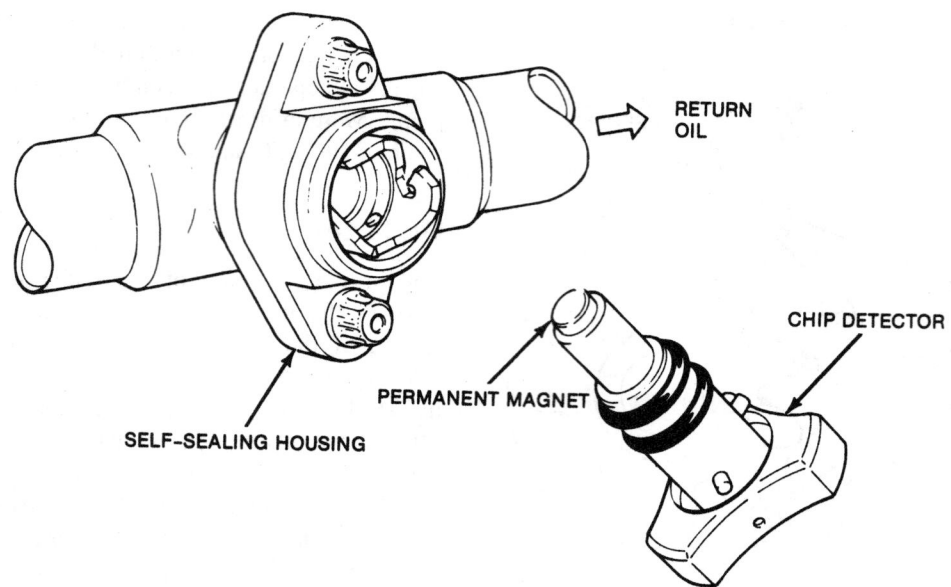

FIGURE 14-4 Magnetic chip detector. (*Rolls-Royce.*)

Magnetic Chip Detector

Magnetic chip detectors can be installed in the scavenge lines, oil tank, and accessory gearbox if the optional chip detector provisions are on the engine. A **magnetic chip detector**, illustrated in Fig. 14-4, is installed in the side of the filter case. This detector indicates the presence of metal contamination without the necessity of opening the filter. When the detector picks up ferrous-metal particles, the center plug becomes grounded to the case. If a warning light is connected between the center terminal of the detector and ground, the light will burn and indicate metal particles on the detector. The detector can also be removed from the engine and be inspected for metal particles by the maintenance technician.

Oil Coolers

Some systems employ a fuel-cooled oil cooler such as that illustrated in Fig. 14-5, others utilize ram air for cooling the oil, and still others do not employ oil coolers. The latter systems are referred to as "hot-tank" systems because the oil returning to the oil tank is quite hot.

The engine fuel oil cooler consists of an outer case which houses the cooler core. Fuel and oil enters and exits through passages in the cooler (see Fig. 14-5). Metered fuel from the fuel control unit passes through the core tubes and absorbs the heat from the oil. The hot oil passes around the tubes and is baffled so that it passes back and forth across the tubes to give maximum exchange of heat.

Although the fuel cools the oil by means of a heat exchanger, the oil and fuel are separate and never come in contact with each other. If the cooler were to become blocked, a bypass valve such as the one shown in Fig. 14-5 would unseat and allow oil to flow around the cooler.

Oil Breather System

An oil breather system connects the engine bearing cavities, the accessory drive gearbox, and the oil tank. Oil droplets and vapor are removed from the breather airstream by a centrifugal separator located in the accessory drive gearbox. After passing through the separator unit, the clean oil-free breather air is exhausted overboard through a vent pipe. A centrifugal breather is illustrated in Fig. 14-6.

Oil Indicating and Warning Systems

The temperature and pressure of the oil are critical to the correct and safe running of the engine. Provision is therefore made for these parameters to be indicated in the cockpit.

In a typical oil indicating and warning system such as that shown in Fig. 14-7, the oil quantity indicating system consists of a capacitance tank unit probe electrically connected to an indicator on the instrument panel to form a capacitance bridge circuit. A change in oil level alters the tank unit capacitance. The resulting flow of current is used to actuate a motor which positions a potentiometer wiper in the indicator to rebalance the circuit. The indicator dial pointer is connected to the potentiometer wiper and moves with the wiper to provide the oil quantity indication.

The components of the oil pressure indicating system are an oil pressure transmitter and an indicator. The oil pressure transmitter senses oil pressure in the external pressure oil manifold and also senses ambient pressure. The difference between these two pressures is measured and converted into an electrical signal which actuates the oil pressure indicator.

The oil temperature indicating system consists of an oil temperature indicator and a temperature-sensing bulb. The oil temperature bulb contains a resistance element which varies its resistance with temperature. This resistance of the bulb controls the current flowing through the indicator

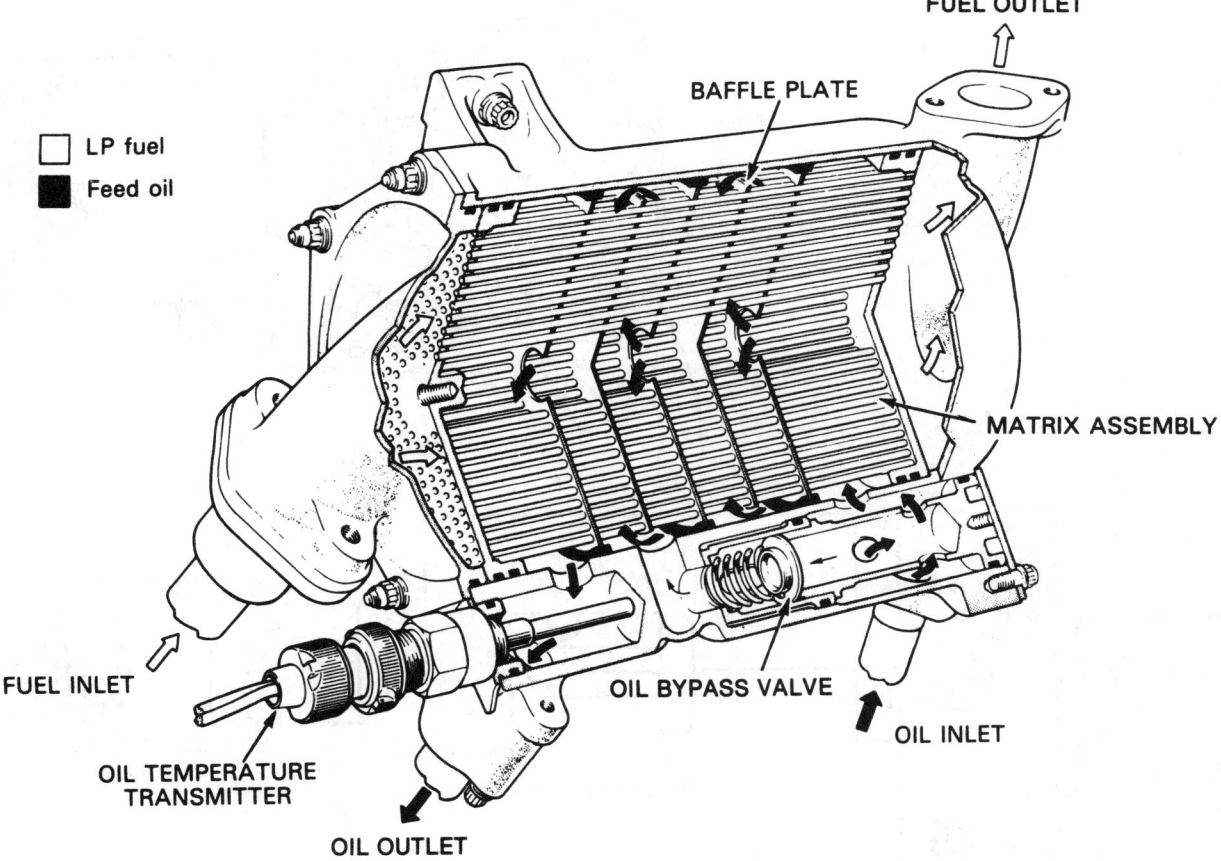

FIGURE 14-5 Low-pressure fuel-cooled oil cooler. (*Rolls-Royce.*)

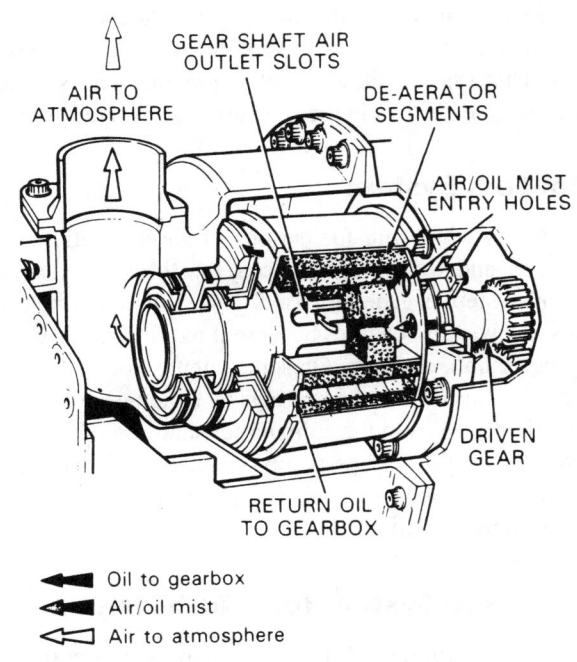

FIGURE 14-6 Centrifugal breather. (*Rolls-Royce.*)

deflection coil, and therefore controls the angular position of the pointer.

A warning light on the instrument panel is used to make the flight crew aware of a low oil pressure or oil filter bypass condition. The operation of this system can be seen in Fig. 14-7.

LUBRICATING SYSTEMS

Gas-turbine engines have been designed and manufactured in many different configurations; thus, there are correspondingly different designs for the lubrication systems of such engines.

There are three basic oil circulating systems, known as a pressure relief valve system, a full-flow system, and a total-loss system. The major difference lies in the control of oil flow to the bearings.

Pressure Relief Valve System

In the **pressure relief valve system**, the oil flow to the bearing chambers is controlled by limiting the pressure in the feed line to a given value. This is accomplished by the use of a spring-loaded valve which allows oil to be directly returned from the pressure pump outlet to the

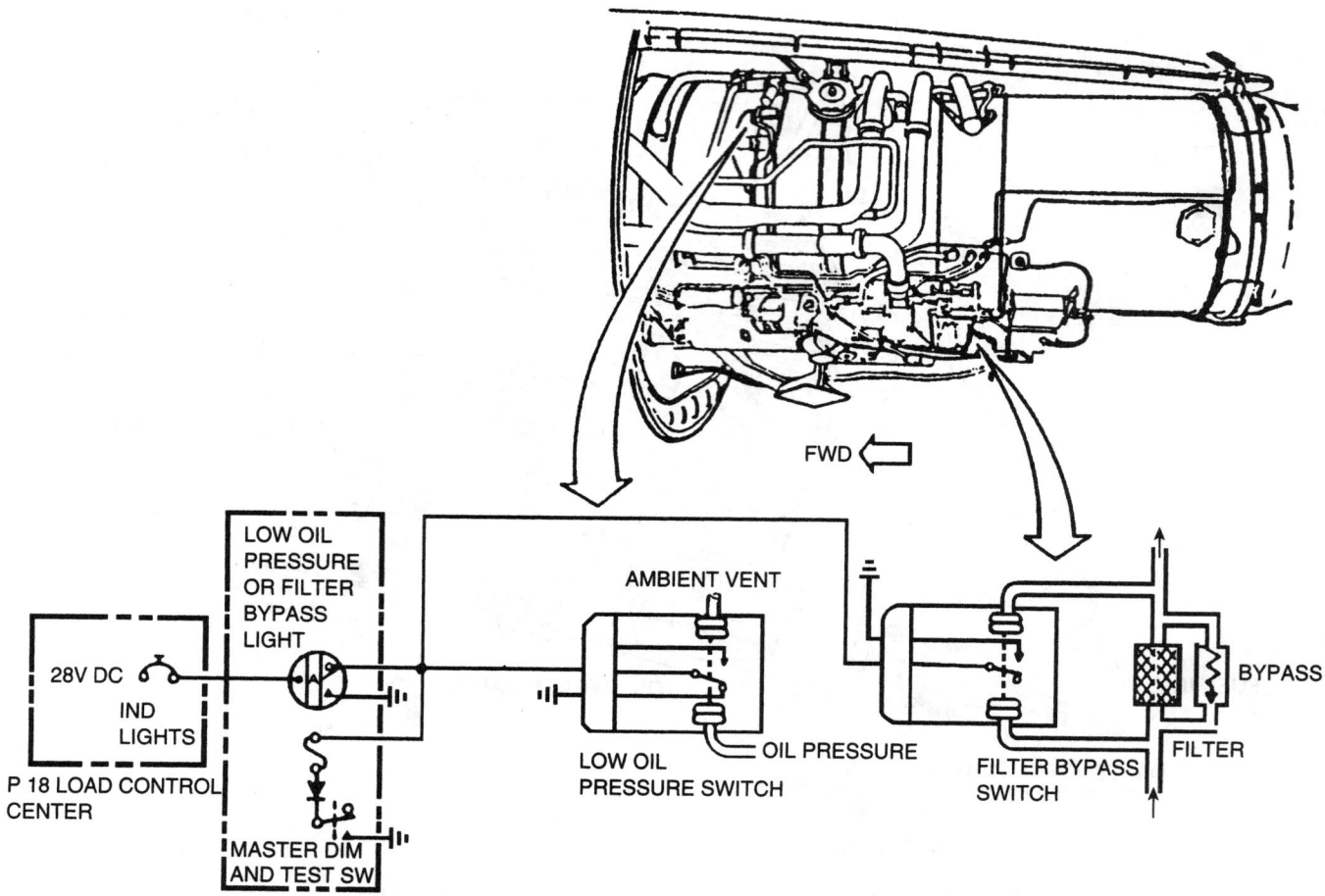

FIGURE 14-7 Low oil pressure and oil filter bypass warning system.

oil tank, or pressure pump inlet, when the design value is exceeded. The valve opens at a pressure which corresponds to the idling speed of the engine, thus giving a constant feed pressure over normal engine operating speeds. However, increasing engine speed causes the bearing chamber pressure to rise sharply. This reduces the pressure differential between the bearing chamber and feed jet, thus decreasing the oil flow rate to the bearings as the engine speed increases. To alleviate this problem, some pressure relief valve systems use the increasing bearing chamber pressure to augment the relief valve spring load. This maintains a constant flow rate at the higher engine speeds by increasing the pressure in the feed line as the bearing chamber pressure increases.

Figure 14-8 shows the pressure relief valve system and its location in a turbine-engine lubrication system.

Full-Flow System

Although the pressure relief valve system operates satisfactorily for engines which have low bearing chamber pressures that do not increase greatly with engine speed, it is an undesirable system for engines which have high bearing chamber pressures. For example, if a bearing chamber had a maximum pressure demand of 90 psi [620.55 kPa], it would require a pressure relief valve setting of 130 psi [896.35 kPa] to produce a pressure drop of 40 psi [275.8 kPa] at the oil feed jet. This would result in the

need for large pumps and would create difficulties in matching the required oil flow at lower speeds.

The **full-flow system** does not utilize a relief valve and achieves the desired oil flow rates throughout the complete engine speed range by allowing the pressure-pump delivery pressure to go directly to the oil feed jets. The pressure pump size is determined by the flow required at maximum engine speed. This system allows smaller pressure and scavenge pumps to be used than in the pressure relief valve system.

Total-Loss System

For engines which run for periods of short duration, such as booster and vertical-lift engines, the **total-loss** (expendable) **oil system** is generally used. This system is simple and incurs low weight penalties because it requires no oil cooler, scavenge pump, or filters. On some engines, oil is delivered in a continuous flow to the bearings by a plunger-type pump, indirectly driven from the compressor shaft; on others, oil is delivered by a piston-type pump operated by fuel pressure. Once the oil is used for bearing lubrication it is disposed of; there is no recirculation.

Lubrication System for a Turbofan Engine

A gas-turbine engine lubrication system that may be considered typical is the Pratt & Whitney JT8D, which powers

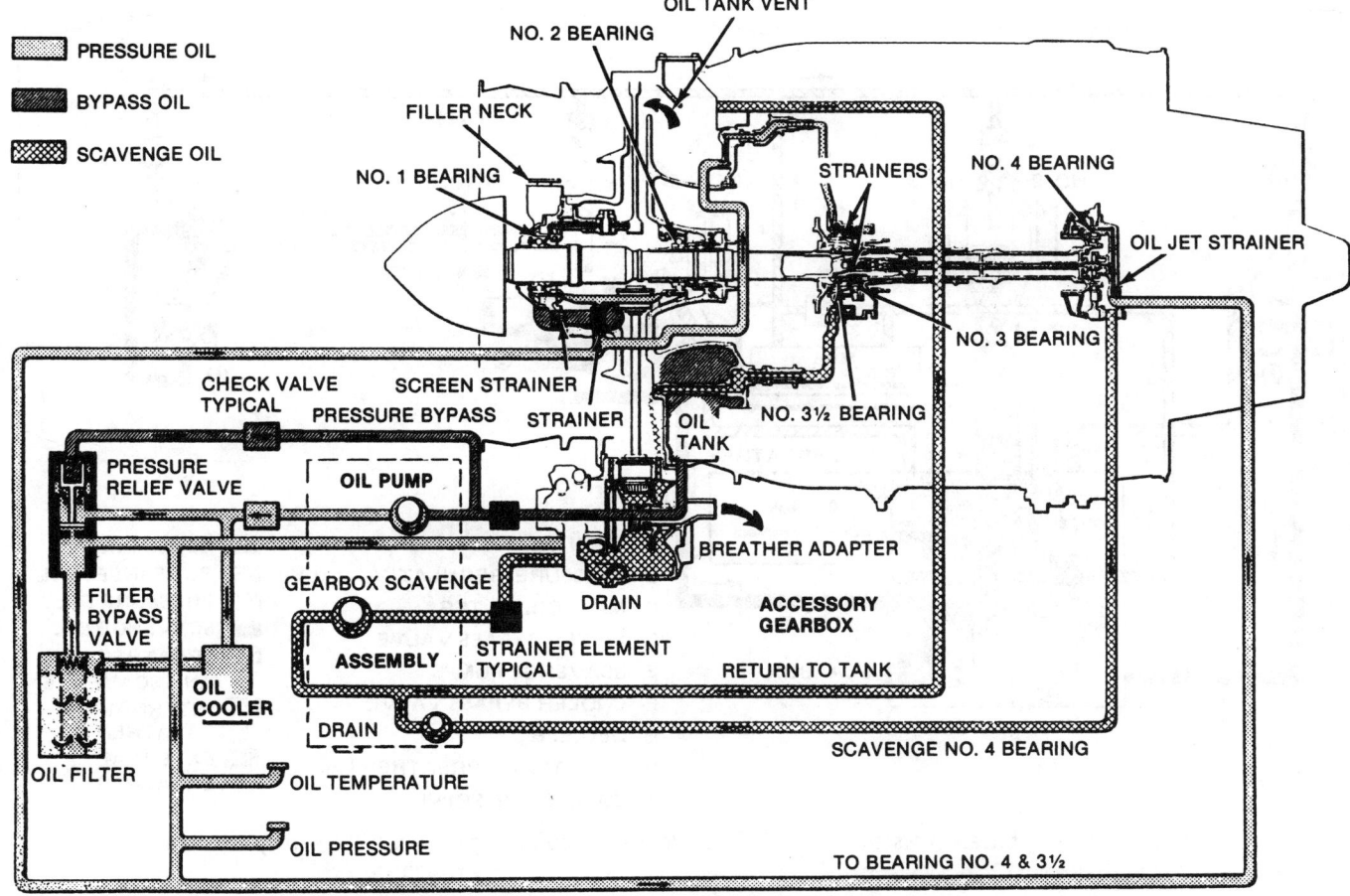

FIGURE 14-8 Engine lubrication system. (*Pratt & Whitney Canada.*)

the Boeing 727 and 737 aircraft, the Douglas DC-9, and many others. The system for the JT8D engine is shown in Fig. 14-9.

The JT8D lubrication system has a self-contained, high-pressure design consisting of a pressure system which supplies lubrication to the main engine bearings and to the accessory drives and a scavenge system by which oil is withdrawn from the bearing compartments and accessories and then returned to the oil tank. A breather system connecting the individual bearing compartments and the oil tank completes the lubrication system.

Oil is gravity-fed from the main oil tank into the main oil pump (A in Fig. 14-9) in the gearbox. The pressure section of the main oil pump forces oil through the main oil strainer located immediately downstream of the pump discharge.

The **filter cartridge** is composed of a stack of elements capable of filtering out particles larger than 46 microns. With an oil temperature at 150°F [65.56°C] and a flow of approximately 15 gal/min [56.8 L/min], the pressure drop across a clean filter is about 6 psi [41.4 kPa]. The estimated maximum pressure drop across a clogged filter is 23 psi [158.6 kPa]. A bypass valve is incorporated in the center of the filter element. If the filter element becomes clogged, the bypass valve will move off its seat and the oil will bypass around the filter.

Proper distribution of the total oil flow to various locations is maintained by metering orifices and clearances. The main oil pump is regulated by a pressure valve to maintain a specified pressure and flow. This valve is labeled B in Fig. 14-9. Note that the valve is located such that when oil pressure becomes too high, the valve will open and return a portion of the oil to the inlet side of the pump. Oil pressure, relative to internal engine breather pressure (tank pressure), and oil flow are essentially constant with changes in altitude and engine speed.

Oil leaves the gearbox and flows to the fuel-cooled oil cooler, where a portion of the oil's heat is transferred to the fuel flowing to the FCU. If the cooler is blocked, an oil cooler bypass valve (F in Fig. 14-9), opens to permit the continuous flow of oil. Oil leaves the cooler and flows into the oil pressure tubing to the main-bearing compartments. The pressure-sensor line maintains a constant oil pressure at the bearing jets, regardless of the pressure drop of the oil at the fuel-oil cooler. The pressure-sensor line can be seen in Fig. 14-9 leading from the outlet of the fuel-oil cooler back to the pressure regulating valve (B).

Oil for the no. 1 bearing enters the inlet case through a tube in the bottom inlet guide vane. For engines equipped with oil-damped no. 1 bearings, a transfer tube from the front accessory support leads back into the bearing support

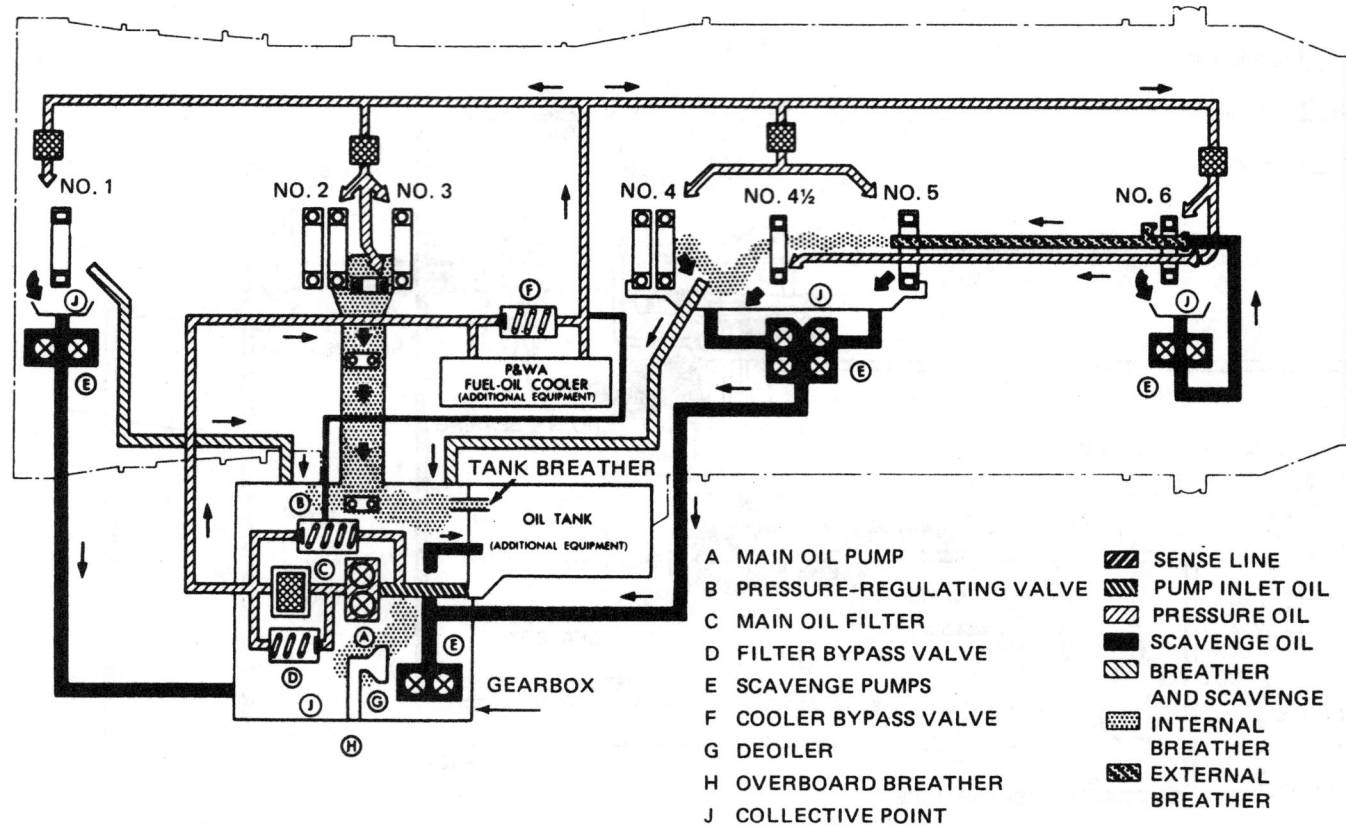

FIGURE 14-9 Lubrication system for the Pratt & Whitney turbofan engine. (*Pratt & Whitney Canada*)

A MAIN OIL PUMP
B PRESSURE-REGULATING VALVE
C MAIN OIL FILTER
D FILTER BYPASS VALVE
E SCAVENGE PUMPS
F COOLER BYPASS VALVE
G DEOILER
H OVERBOARD BREATHER
J COLLECTIVE POINT

SENSE LINE
PUMP INLET OIL
PRESSURE OIL
SCAVENGE OIL
BREATHER AND SCAVENGE
INTERNAL BREATHER
EXTERNAL BREATHER

to supply oil to a cavity around the bearing's outer race. The remainder of the oil moves up the tube and is then routed through a small strainer in the front accessory drive support into the accessory drive gear shaft. It moves to the outer wall holes in the front hub and in the inner race retaining nut to the front of the no. 1 bearing.

Oil enters the no. 2 and no. 3 bearing compartments through a small strainer and is sprayed onto the bearings through a three-legged oil nozzle assembly. A front leg, or nozzle, directs oil toward the no. 2 bearing, a second directs oil toward the no. 3 bearing, and a third directs oil toward the gearbox drive-shaft upper bearing. Oil flows through holes in the rear hub to the inside diameter (ID) of the no. 2 bearing. Flow through the gearbox drive bevel-gear holes carries oil to the ID of the no. 3 bearing.

Pressure oil for the no. 4 and no. 5 bearing locations flows into the engine through a tube at the eight o'clock location at the left side of the fan discharge diffuser's outer duct. Oil then flows upward around the diffuser case to the ten o'clock position and inward through the inner passage of the dual concentric pressure and breather tubing to the no. 4 bearing support. Here it is directed rearward through an elbow and flows into the multipassage no. 4 bearing oil nozzle assembly.

The no. 4 bearing oil nozzle assembly has an inlet passage, outlet holes at the bottom directing oil toward the no. 4 bearing, and an outlet passage toward the rear. An oil strainer

is positioned inside the inlet passage. The outlet passage toward the rear accommodates the long oil tube of the no. 5 bearing oil nozzle assembly. Oil passes rearward through this tube and is then directed through the no. 5 bearing oil nozzle assembly. From the oil nozzle it passes under the bearing race and through the seal plate to the no. 5 bearing compartment.

Oil flows to the no. 6 bearing area through a tube located in the upper turbine exhaust strut and down into the scavenge pump housing of the no. 6 bearing. In the scavenge pump housing it passes through a small strainer and then down into the outer passage of the no. 6 bearing oil nozzle assembly.

For engines having *oil-damped* no. 6 bearings, oil flows from the oil scavenge pump through a tube to the no. 6 bearing housing. The oil is then distributed to a cavity formed between the housing and the bearing's outer race. To minimize the effect of the dynamic loads transmitted from the rotating assemblies to the bearing housings, an **oil damped bearing (squeeze film bearing)** is used, as shown in Fig. 14-10. This oil film damps the radial motion of the rotating assembly and the dynamic loads transmitted to the bearing housing, thus reducing the vibration level of the engine and the possibility of damage by fatigue. Seal rings around the bearing's outer race help contain oil in the cavity.

The oil flows forward in the oil nozzle's outer passage and divides into two streams. One stream flows outward through

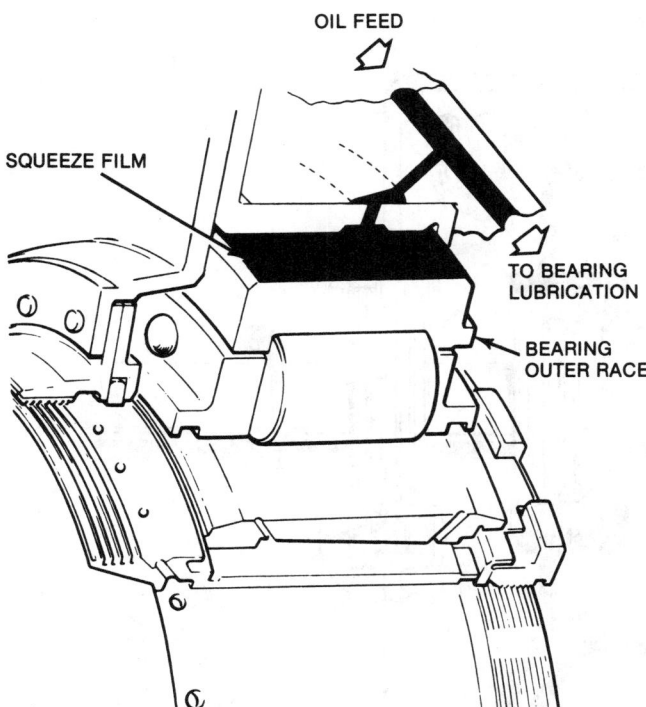

OIL FEED

SQUEEZE FILM

TO BEARING
LUBRICATION

BEARING
OUTER RACE

FIGURE 14-10 Squeeze film bearing. (*Rolls-Royce.*)

small holes on the outside diameter (OD) of the nozzle's outer passage tube to lubricate the no. 6 bearing area. From the same nozzle's outer passage tube, the other stream continues forward through holes on the nozzle's outer front face and into the outer passage of the turbine bearings' oil pressure and scavenge tubes assembly (trumpet) inside the front-compressor drive turbine rotor. The oil continues forward through the single (short) pressure tube in the oil trumpet to the no. $4\frac{1}{2}$ bearing area.

As previously mentioned, the outward stream for the no. 6 bearing area flows into the rear hub of the front-compressor drive turbine rotor. Through two sets of holes in the hub, it flows to the no. 6 bearing seals and to the no. 6 bearing's inner race.

The pressure oil in the oil trumpet flows forward and out through an oil baffle and then through holes in the long turbine shaft, to cool the no. $4\frac{1}{2}$ bearing seal spacers and to lubricate the no. $4\frac{1}{2}$ bearing.

The scavenge oil system of the engine includes four gear-type pumps (E in Fig. 14-9) with five pump stages. The pumps scavenge the main-bearing compartments and deliver the scavenged oil to the engine oil tank. Note that the scavenge pump for the no. 4 and no. 5 bearing areas has two pump stages.

The single-stage scavenge pump for the no. 1 bearing compartment is located in the cavity of the front accessory drive housing. The pump is driven by the front accessory drive gear shaft located in the front hub of the front-compressor rotor. The pump picks up the oil, sends it outward through a passage in the housing, and then sends it down a tube located in the bottom vane of the inlet case.

The second scavenge pump is located in the scavenge stage of the main oil pump assembly in the accessory drive gearbox. Scavenge oil from the gearbox drive-shaft bearings and from the no. 2 and no. 3 bearings is pumped from its collection point in the gearbox. Note that the scavenge oil from the no. 2 and no. 3 bearings drains down the outside of the accessory drive-shaft to the gearbox. One bevel gear drives both the pressure and scavenge stages of the pump.

The third pump, with two stages driven by the same gear, is the oil scavenge pump assembly for the no. 4 and no. 5 bearings, located inside the diffuser case. Together, the two stages of the pump scavenge oil from the no. 4 and no. 5 bearing areas. In addition, scavenge oil from the no. 6 bearing area, after flowing forward through the two long scavenge tubes in the oil trumpet, flows into this compartment. A tube in the combustion chamber heat shield allows passage of the oil forward from the no. 5 bearing cavity.

The discharge from the pump is carried forward into the scavenge adapter at just below the nine o'clock position in the no. 4 bearing support and then outboard. It flows through the inner tube of the dual concentric tubing to the outside of the diffuser case, then downward to the eight o'clock position where it is routed through the fairing to the outside of the diffuser's outer duct.

The fourth scavenge pump is located in the no. 6 bearing's scavenge pump housing, where it is driven by a gear shaft bolted to the rear of the turbine rotor's fourth-stage rear hub. It scavenges oil from the no. 6 bearing compartment and pumps it upward into the inner passage of the no. 6 bearing oil nozzle assembly.

The oil flows forward in the oil nozzle's inner passage and is discharged through the center hole in the front of the nozzle. Oil passes forward into the inner passage of the oil trumpet and continues forward through the two long scavenge tubes, as previously mentioned. At the front of the trumpet, the oil flows outward through holes in the front-compressor drive turbine shaft (inner shaft) and in the front of the no. $4\frac{1}{2}$ bearing's inner race retaining nut. The oil is then spun outward through holes in the rear-compressor drive turbine shaft (outer shaft) and into the no. 4 bearing cavity.

The return oil passed forward by the two rearmost pumps, as well as that from the front oil suction pump, is directed into the gearbox cavity. From here oil is pumped by the scavenge stage of the pump to the oil tank. Within the tank the oil passes through a deaerator where the major part of the entrapped air is removed.

To ensure proper oil flow and to maintain satisfactory scavenge pump performance during operation, the pressure in the bearing cavities is controlled by the **breather system**. The atmosphere of the no. 2 and no. 3 bearing cavities vents into the accessory gearbox. Breather tubes in the compressor inlet case and diffuser case discharge through external tubing into the accessory drive gearbox. Breather air from the no. 6 bearing compartment comes forward through the oil pressure and scavenge tube assembly (oil trumpet), along with the scavenge oil from that compartment, to the diffuser case cavity.

In a gearbox, vapor-laden atmosphere passes through centrifugal breather impellers (mounted on the starter drive

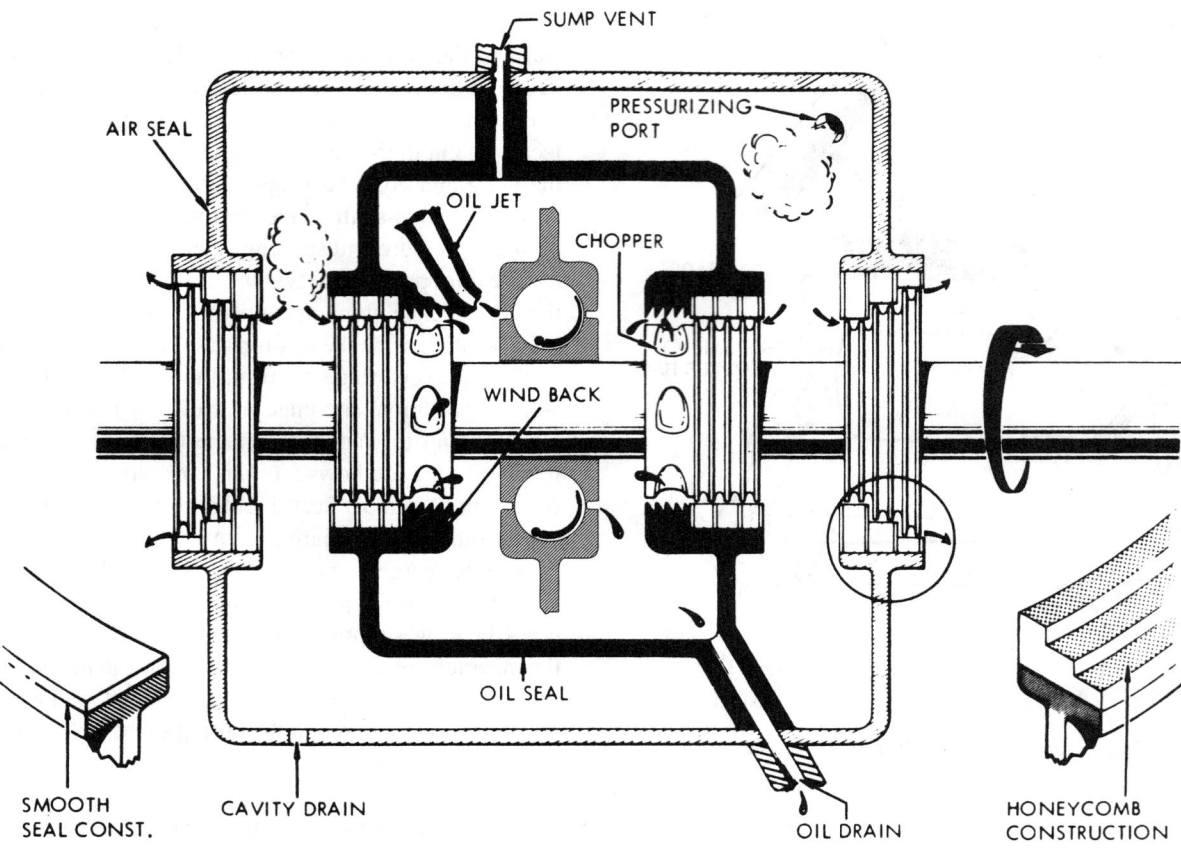

FIGURE 14-11 Typical bearing sump.

gear shaft), where the oil is removed. The relatively oil-free air reaching the center of the gear shaft is conducted overboard.

Oil Bearings and Seals

The airflow from the compressor is used within the engine to pressurize internal areas and to control oil flow through the **labyrinth seals** and around bearings. Figure 14-11 illustrates the type of bearing sump used on most modern turbine engines. Oil seals help retain the oil in the bearing cavity, and they may be a carbon type or the labyrinth type shown in Fig. 14-11. One great advantage of a labyrinth-type seal is that there is virtually no wear because the rotating and stationary members of the seal do not touch. As shown in the drawing, two seal chambers exist: the oil seal next to the bearing and the air seal. The close tolerances maintained in the air seal prevent the rapid loss of pressurized air and thereby creates a pressurized chamber between the oil and air seals. The oil seal must retain the oil in the sump with the aid of the air seal. Oil for lubrication of the bearing flows through the oil jet, over the bearing, and out through the oil drain. The pressurization chamber is pressurized by air supplied from the engine bleed-air system. The volume of air supplied must maintain an adequate flow of air inward across the oil seal to blow the oil inward to the sump while some of the air is leaking outward between the rotating and stationary members of the air seal. The theory is that if air is flowing inward, oil cannot flow outward. A vent is provided at the top of the sump to ensure inward airflow to the sump. An overboard drain is provided in the pressurized chamber to evacuate any oil that may leak past the oil seal. This prevents oil from getting past the air seal to the compressor, contaminating the bleed air used for cabin pressurization. Oil that flows from the oil drain is picked up by a scavenge pump and returned to the main tank. An external view of the JT8D engine's oil system and component location is shown in Fig. 14-12.

Lubricant Service for Gas-Turbine Engines

In the past, it has generally been the practice to drain and replace the lubricating oil in both reciprocating and gas-turbine engines at specified intervals. At these times, the oil filters or screens were examined for residues of metal to detect incipient failures in the engine. But experience and extensive investigations have shown that the periodic changes of oil are not necessary for gas-turbine engines using synthetic lubricants. The airlines have cooperated with oil companies and engine manufacturers in studies to determine the most effective and economical procedures with respect to the use of lubricants. The result is that engine oil is seldom, if ever, changed between engine overhaul periods.

To ensure continued effectiveness of the engine lubricant, two procedures are employed: **filtering** of the oil and **oil analysis**.

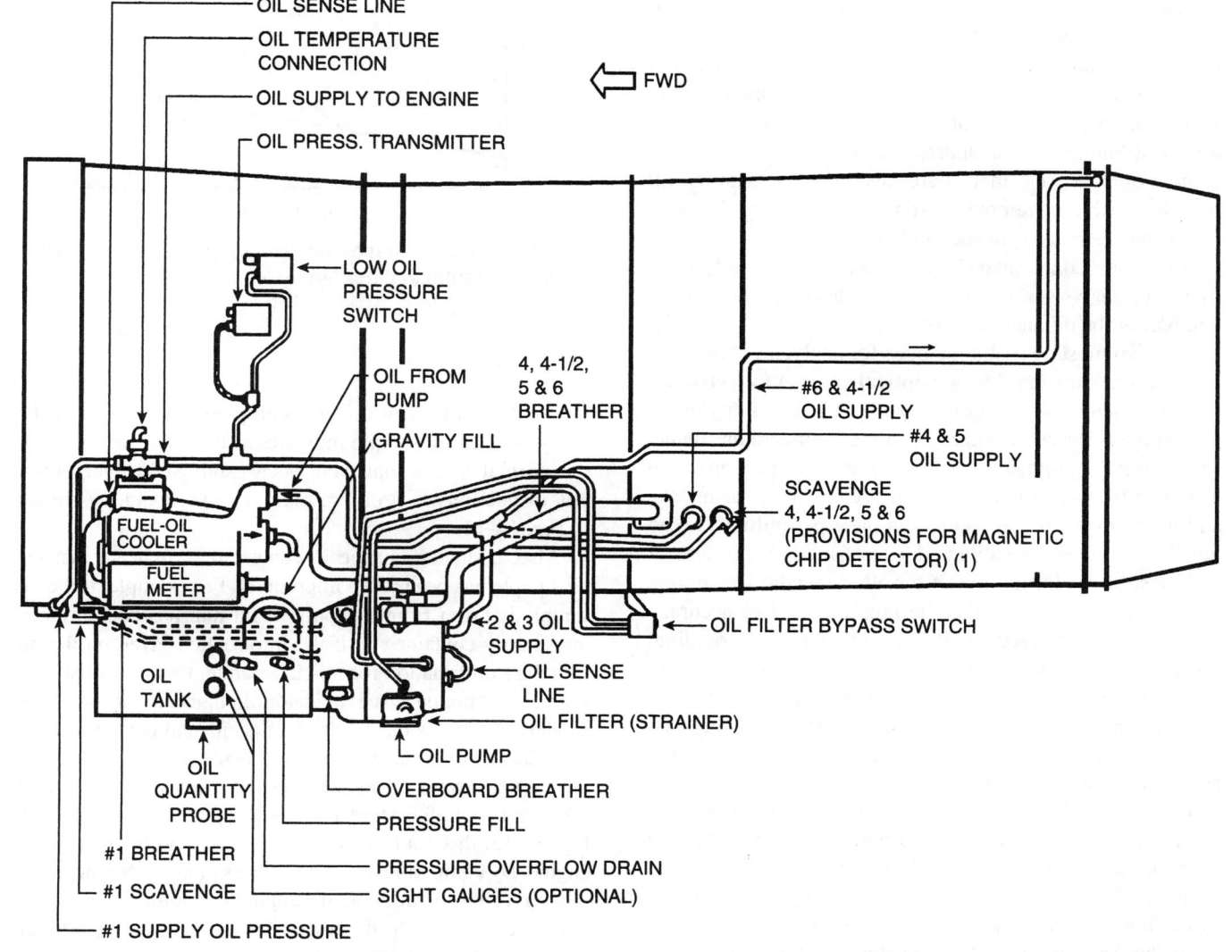

FIGURE 14-12 External oil distribution on a JT8D engine.

A typical procedure employed with an engine such as the Pratt & Whitney JT8D turbofan engine is the filtering of the oil every 250 h of operation. This is accomplished by removing the main oil screen (or filter) and installing a filter adapter. The filter adapter provides inlet and outlet connections by which a 15-micron filter is connected externally to the engine. When the filter has been connected, the engine is operated at medium speed for about 5 min. This causes all the engine oil to be passed through the filter, thereby removing even the smallest particles of suspended material.

This filtering method provides for thorough internal engine cleaning because the hot circulating oil flushes the inside of the engine while the engine is running. If the oil were drained and filtered, pockets or pools of unfiltered oil would remain in the engine along with residues which would normally settle in such pockets or pools.

After the filtering operation is completed, the 40-micron engine filter, having been examined for residues and cleaned, is reinstalled in the engine. Residues from the special 15-micron filter and from the engine oil filter are examined

for metal particles. The metals are identified by means of a magnet and by various chemicals to determine their type, in order to detect the possibility of incipient engine failure.

OIL ANALYSIS

Oil analysis is an analytical process which, when used within certain limits, can be used to identify some engine problems before major engine damage or failure occurs. Many oil analysis programs are available. The techniques and tests performed can vary widely from one laboratory to another.

The analysis of the oil condition in the engine can be divided into two basic areas. The first test determines the amount of **wear metals** in the oil sample, whereas the second test identifies the larger contaminants. Wear metals are very small particles of metal that have been worn away as the engine operates. By identifying each wear metal and noting the quantity present in the oil, the laboratory processing the oil sample can generally determine if this amount of wear metal is excessive and what part of the engine it has come from. The quantity of wear metals in the oil is measured in parts per million. Wear metals are generally smaller than 1 micron. To give an idea of the size of 1 micron, the diameter of a human hair is approximately 50 microns. These submicroscopic wear metals are much smaller than the oil filter is capable of removing from the system, so the wear metals will remain suspended in the oil.

The larger contaminants are trapped and retained by the filter. The analysis of these larger particles is another important part of the oil analysis process.

The two most common methods for analyzing wear metals are **atomic absorption** and **optical emission spectrometry**. In the atomic absorption method, a small quantity of the oil sample is burned (ionized) in a high-temperature flame. Then special equipment detects how much energy has been absorbed from a particular chemical element (wear metal), and the quantity of that element in parts per million is read out. The equipment is calibrated to test for a specific element and must be readjusted to test for other metals. The atomic absorption method provides the greatest level of accuracy for each metal analyzed, but the test is very time-consuming. For example, an analysis for 10 wear metals would require 10 equipment setups and 10 tests. Many times, several oil samples are analyzed for the same element by using one equipment setup and then resetting the equipment for the next element to be tested.

In the optical emission spectrometry procedure, a small quantity of the oil sample is also burned, but the detection device measures the different levels of light emitted. This equipment simultaneously measures the emitted light for as many as 18 different wear metals. The emission spectrometer offers somewhat lower accuracy than the atomic absorption method, but in little more than a minute it can complete its analysis of wear metals to within several parts per million.

The value of chemically analyzing the content of the oil supply in the engine relies heavily on information regarding the history of the engine prior to, and including, the time that the sample is taken. The curve in Fig. 14-13 illustrates the need for complete information regarding the oil sample. Figure 14-13 shows that as engine operating hours increase, the amounts of some materials (wear metals) in the oil tend to increase. A reading can be affected by either

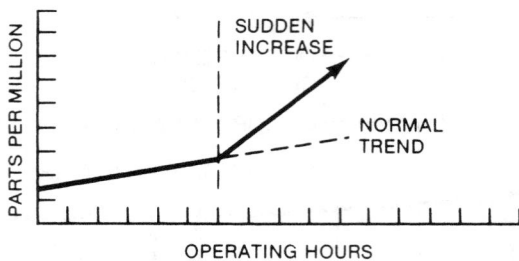

FIGURE 14-14 Illustration showing sudden increase in metal particle concentration. (*Honeywell Inc.*)

the addition of clean oil or a complete oil change. If the analysis of an oil sample indicates a sudden departure from the normal trends that have been established over previous inspections, a problem may be developing, as shown in Fig. 14-14.

Also, a sudden change in content may indicate that the oil sample has been taken improperly. For example, if the oil sample is taken from a supply of oil that has been drained into a dirty container, or if the information given on the oil sampling information form is inaccurate, then obviously the sample will not be a true indicator of engine condition. If the sudden increase is the result of an accidental contamination, the trend will normalize in the following oil analysis test. If the upward trend in contaminants continues in successive samples taken after short periods of operation, as shown in Fig. 14-15, this could verify a developing problem with the engine. By comparing the past analyses of many engines, predictions can be made about the engine's condition. This information can also help the technician choose further tests or inspections to more accurately determine engine condition.

Taking an Oil Analysis Sample

A sample kit will generally include a sample container and a form which contains sampling instructions as well as a place to record time on the oil and filter, engine model and serial number, time since the last oil sample was taken, amount of oil added since the last sample was taken, engine hours since the last oil change, and other information. The instruction sheet included in the oil analysis kit stresses the correct

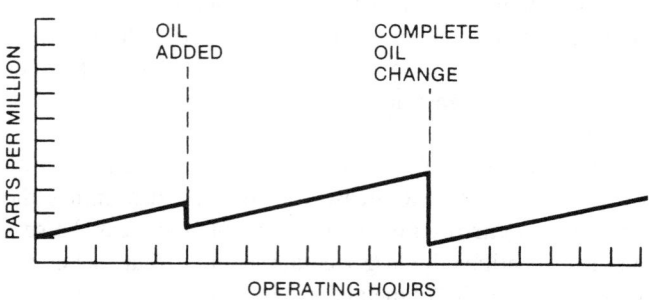

FIGURE 14-13 Effects of adding or changing oil on metal particle concentration. (*Honeywell Inc.*)

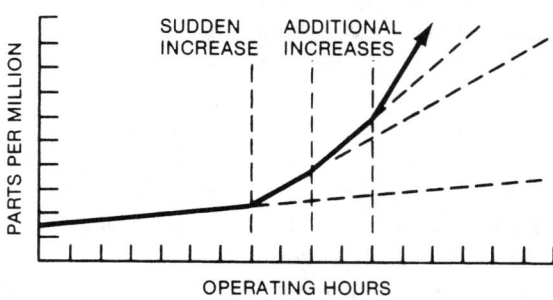

FIGURE 14-15 Oil analysis curve in which an engine problem is indicated by a continuing upward trend in metal particle concentration. (*Honeywell Inc.*)

method of taking an oil sample, how to complete the shipping form, and how to pack and ship the sample. These instructions are important, since on several occasions poor sampling technique or misinformation on the information form resulted in invalid testing. If the analysis reveals a condition that must be corrected quickly, the customer will generally be advised by telephone. If the situation is normal, the results will generally be reported by mail.

In addition to chemically analyzing the contents of the oil, many laboratories feel it is equally important to analyze the contents of the filter, because larger particles of wear material sometimes cannot be detected by chemically analyzing the oil. If a few pieces of a part have been trapped by the filter, there may not be enough wear metal in the oil to be detected. The filter is back-flushed so that the contents can be removed from the filter for examination. This includes weighing the samples to get a reference for the rate of contamination. The size and shape of the particles and the type of material will be analyzed so that the problem can be pinpointed to the specific area in the engine.

Oil analysis is a good example of preventive maintenance. The ability of an oil analysis program to locate impending failures is directly related to adherence to a complete program. It is important to take samples frequently enough that trends can be established. Although it is best to start a new engine on an oil analysis program, engines that have seen some use can also be started with a moderate degree of success. Results will not be accurate if an engine's analysis program is based on a single oil sample. A successful oil analysis program should be run over the engine's total operating life so that normal trends can be established. It is the engine's deviation from the normal trends which can reveal possible problems.

Oil Sample Analysis Report

An oil sample analysis report is shown in Fig. 14-16. The following example can help the technician understand the application of oil analysis and the interpretation of the results. This oil sample report provides feedback to the aircraft owner or operator and contains information about 18 different elements and/or wear metals.

The level of each element or wear metal is printed on the report in parts per million below the name of the element, in vertical columns (see Fig. 14-16). In the row immediately below the concentration levels, there are code letters. These code letters are part of a rating scale which is used to inform the aircraft's owner of the wear rates for each element and each wear metal. The letter N (normal) is used if the concentration of the wear metal or element is normal.

If the concentration is not normal, an A (abnormal) or an S (severe) is used to describe the condition.

The samples in Fig. 14-16 can be read by starting at the bottom of the report with the sample data dated 10-26-87. In this case, all wear metals and elements are in the normal category. In the report dated 12-22-87 the silica (dirt) has risen to the abnormal level, but this has not yet resulted in engine wear. The sample dated 03-02-88 shows that silica has risen to the severe category and two key engine wear metals are in the abnormal category. At this point the aircraft owner would be contacted by phone and informed of the potential engine problem. The most recent report, dated 03-24-88, shows that successful maintenance action was taken because all metals and elements are back to normal levels.

This hypothetical series of reports is a good example of the types of problems that can be detected and corrected before major engine damage occurs. In this example, if the problem of silicon entering the engine had not been corrected, engine wear would have continued to become excessive, resulting in premature engine overhaul or engine failure.

REVIEW QUESTIONS

1. Why was it necessary to develop synthetic lubricants for gas-turbine engines?
2. What types of synthetic lubricants are most commonly used for gas-turbine engines?
3. What should be done if synthetic lubricant is spilled during the servicing of an aircraft?
4. What is meant by the term "dry-sump oil system"?
5. What is the function of a scavenge pump?
6. What happens to system oil flow if the filter becomes clogged?
7. Describe the function of a magnetic chip detector.
8. In a turbine-engine oil cooler, what substances are used to cool the oil?
9. What is a "hot-tank" oil system on a gas-turbine engine?
10. What is the function of the oil breather system?
11. List the three general types of turbine-engine oil lubrication systems.
12. Describe the basic principle of operation of a pressure relief oil system.
13. Describe the basic principle of operation of a full-flow oil system.
14. Describe the basic principle of operation of a total-loss oil system.
15. Describe the two principle methods of oil analysis.

OIL SAMPLE ANALYSIS REPORT

Spectro/Metrics, Inc.
35 Executive Park Drive, N.E.
Atlanta, Georgia 30329
(404) 321-7909

AIRCRAFT MAKE: LEAR JET MODEL: 23 ENGINE MAKE: G.E. MODEL: CJ610-4 REGISTRATION: N68WM SERIAL: 241009RT

SAMPLE DATA	ALUMINUM	CHROMIUM	COPPER	IRON	LEAD	TIN	NICKEL	SILVER	SILICON (DIRT/ADDITIVE)	SODIUM (ADDITIVE/COOLANT)	POTASSIUM (COOLANT)	BORON (ADDITIVE/COOLANT)	MAGNESIUM (ADDITIVE/HOUSING)	MOLYBDENUM (ADDITIVE/RINGS)	BARIUM (ADDITIVE)	CALCIUM (ADDITIVE)	PHOSPHOROUS (ADDITIVE)	ZINC (ADDITIVE)	GLYCOL (ANTI-FREEZE)	WATER (CONDENSATION/COOLANT)	% SOOT (SOLIDS/CARBON)	% FUEL	VISCOSITY Cst 40°C	TAN
Sample Date: 03-24-88 Sample No.: 220477 Lab No.: 703801 Total Mls/Hrs: 4205 Mls/Hrs on Oil: 50 Mls/Hrs on Flt: 50	0 N	.8 N	.3 N	.2 N	0 N	0 N	0 N	.1 N	.4 N	.1 N	0 N	0 N	.3 N	0 N	0 N	0 N	2788 N	0 N	NEG N	-.05 N	.01 N	-1 N	26.3	.29 N
Sample Date: 03-02-88 Sample No.: 220472 Lab No.: 701266 Total Mls/Hrs: 4155 Mls/Hrs on Oil: 415 Mls/Hrs on Flt: 100	0 N	6.6 A	4.0 N	9.5 A	0 N	0 N	0 N	.8 N	38 S	.3 N	0 N	0 N	1.0 N	0 N	0 N	0 N	2419 N	0 N	NEG N	-.05 N	.06 N	-1 N	27.6	.36 N
Sample Date: 12-22-87 Sample No.: 220471 Lab No.: 299239 Total Mls/Hrs: 3950 Mls/Hrs on Oil: 210 Mls/Hrs on Flt: 100	0 N	2.0 N	1.9 N	3.8 N	0 N	0 N	0 N	.4 N	10 A	.3 N	0 N	0 N	.4 N	0 N	0 N	0 N	2693 N	0 N	NEG N	-.05 N	.02 N	-1 N	27.1	.30 N
Sample Date: 10-26-87 Sample No.: 220474 Lab No.: 289592 Total Mls/Hrs: 3850 Mls/Hrs on Oil: 110 Mls/Hrs on Flt: 110	0 N	1.2 N	.4 N	.9 N	0 N	0 N	0 N	.1 N	3.1 N	.2 N	0 N	0 N	.2 N	0 N	0 N	0 N	2700 N	0 N	NEG N	-.05 N	.01 N	-1 N	26.4	.30 N

Remarks:

CONSIDERING LOW HOURS ON OIL. ENGINE WEAR RATES AND CONTAMINANT LEVELS SATISFACTORY. CONDITION OF OIL SUITABLE FOR FURTHER SERVICE. RESAMPLE AS PER YOUR PROGRAM PLAN. — ANALYST JS (REC'D: 03-25-88)

CHROMIUM AND IRON LEVELS ABNORMAL. DIRT CONTENT HIGH. CHECK ALL DIRT ACCESS POINTS. CHANGE OIL AND PERFORM FILTER SERVICE. FLUSH THOROUGHLY. CHECK FOR VISIBLE METAL PARTICLES. RESAMPLE IN 45-50 HOURS TO MONITOR. — ANALYST CW (REC'D: 03-05-88) (REF. PHONE CALL, 03-05-88)

INCREASE IN ENGINE WEAR RATES NOTED. ALL ENGINE WEAR RATES NORMAL. SILICON LEVEL (DIRT/SEAL MATERIAL) ABNORMAL. CHECK FOR SOURCE OF CONTAMINANTS ENTRY. CHANGE OIL AND PERFORM FILTER SERVICE. RESAMPLE AS PER YOUR PROGRAM PLAN. — ANALYST JS (REC'D: 12-27-88)

ENGINE WEAR RATES AND CONTAMINANT LEVELS SATISFACTORY. CONDITION OF OIL SUITABLE FOR FURTHER SERVICE. RESAMPLE AT NEXT SERVICE INTERVAL TO MONITOR AND ESTABLISH WEAR TREND. — ANALYST CW (REC'D: 10-28-87)

AFS, INC.
ATTN: MITCH NOBLE
BOX 852, WILLOW RUN AIRPORT
YPSILANTI, MI 48198

LAST OVERHAUL:
SYSTEM CAPACITY: 5 QTS.
OIL MAKE & TYPE: EXXON 2380
HISTORY & REMARKS:

LEGEND:
N = Normal
A = Abnormal
S = Severe
+ = greater than
- = less than

FIGURE 14-16 Oil sample analysis report. (*Spectro/Metrics.*)

Ignition and Starting Systems of Gas-Turbine Engines 15

INTRODUCTION

Two separate systems are required to ensure that a gas-turbine engine will start satisfactorily. One system must be able to rotate the compressor and turbine at a speed at which adequate air passes into the combustion system to mix with fuel from the fuel spray nozzles. A second system must enable ignition of the fuel-air (F/A) mixture in the combustion system to occur. During engine start-up, these two systems must operate simultaneously.

The functioning of both systems is coordinated during the **starting cycle**, and their operation is automatically controlled after the initiation of the cycle by an electric circuit. A typical sequence of events during start-up of a turbojet engine is shown in Fig. 15-1.

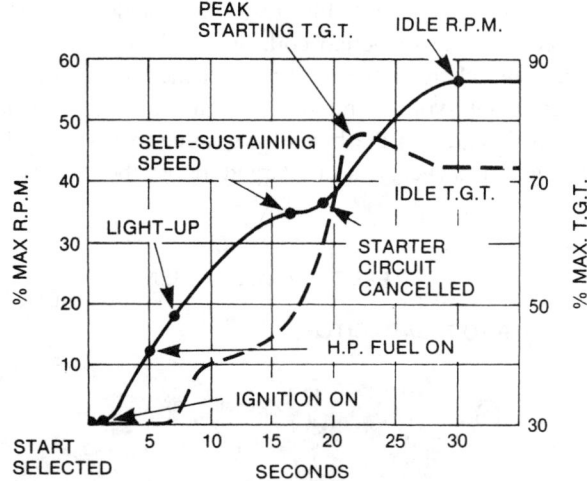

FIGURE 15-1 Typical starting sequence for a gas turbine engine. (*Rolls-Royce.*)

IGNITION SYSTEMS FOR GAS-TURBINE ENGINES

Ignition systems for gas-turbine engines consist of three main components: the exciter box, the ignition lead, and the igniter. The exciter box sends high-voltage current to the ignition lead, which transfers the high voltage to the igniter. The igniter is mounted in the engine in such a way that it protrudes into the combustion section of the engine. When the system is activated, the exciter creates a high voltage which is discharged across the igniter electrodes and ignites the fuel inside the engine's combustion section during starting.

Ignition systems for gas-turbine engines are required to operate for starting only; their total operating time is therefore almost insignificant in comparison with the operating time of an ignition system for a reciprocating engine. For this reason, the gas-turbine ignition system is almost trouble-free.

An important characteristic of a gas-turbine ignition system is the high-energy discharge at the igniter plug. This high-energy discharge is necessary because it is difficult to ignite the fuel-air (F/A) mixture under some conditions, particularly at the high altitudes that turbine aircraft operate when their engines have "flamed out." The high-energy discharge is accomplished by means of a storage capacitor in what is termed a **high-energy capacitor discharge system**. The effect of this system is to produce what appears to be a white-hot ball of fire at the electrodes of the igniter plug. In some designs, the igniter actually "shoots" the electric "flame" several inches. The technician must use great care when working on gas-turbine ignition systems because of the possibility of a lethal shock.

Ignition units are rated in **joules**. Ignition units are designed to provide variable output according to ignition requirements. A high-value output, such as 12 J, is necessary to ensure that the engine will obtain a satisfactory relight at high altitudes; also, a high-value output is sometimes necessary for starting the engine.

Turbine Ignition Systems and Components

The **spark ignition system** has been developed to provide turbine engines with an ignition system capable of quick light-ups over a wide temperature range. Most systems consist of one exciter unit, two individual high-tension cable assemblies, and two spark igniters. Systems can be energized from the aircraft's 28-V dc supply or from 115 V, 400 Hz ac, or both. The normal spark rate of a typical ignition system is between 60 and 100 sparks per minute.

Exciter Units

As shown in Fig. 15-2, the **ignition exciter unit** is a small box mounted on and secured to the engine. Flexible absorption mounts are often used in mounting the ignition exciter to isolate the exciter from the effects of engine-induced vibration.

Most ignition exciters are sealed units containing electronic components encased in an epoxy resin. The exciter transforms the input voltage to a pulsed high-voltage output through solid-state circuitry, transformers, and diodes.

DC Trembler Exciter Unit

The ignition unit shown in Fig. 15-3 is a typical dc trembler-operated unit. An induction coil, operated by the trembler mechanism, charges the reservoir capacitor (condenser) through a high-voltage rectifier. When the voltage in the capacitor is equal to the breakdown value of a sealed discharge gap, the energy is discharged across the face of the

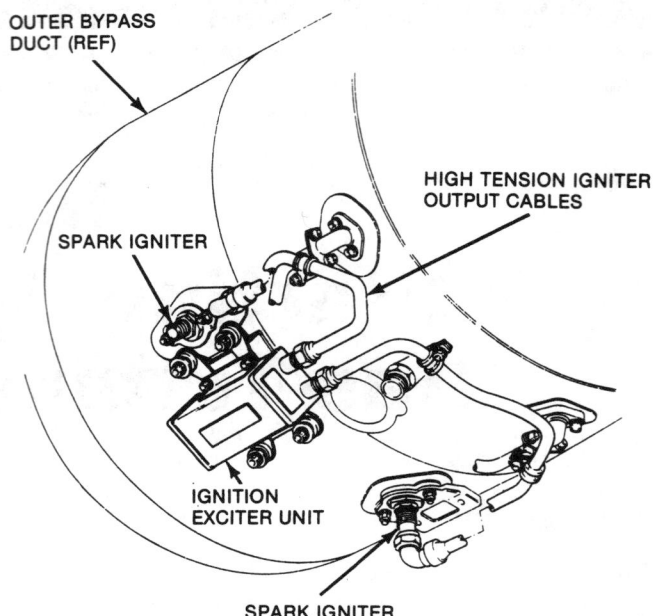

FIGURE 15-2 Ignition system components. (*Pratt & Whitney Canada.*)

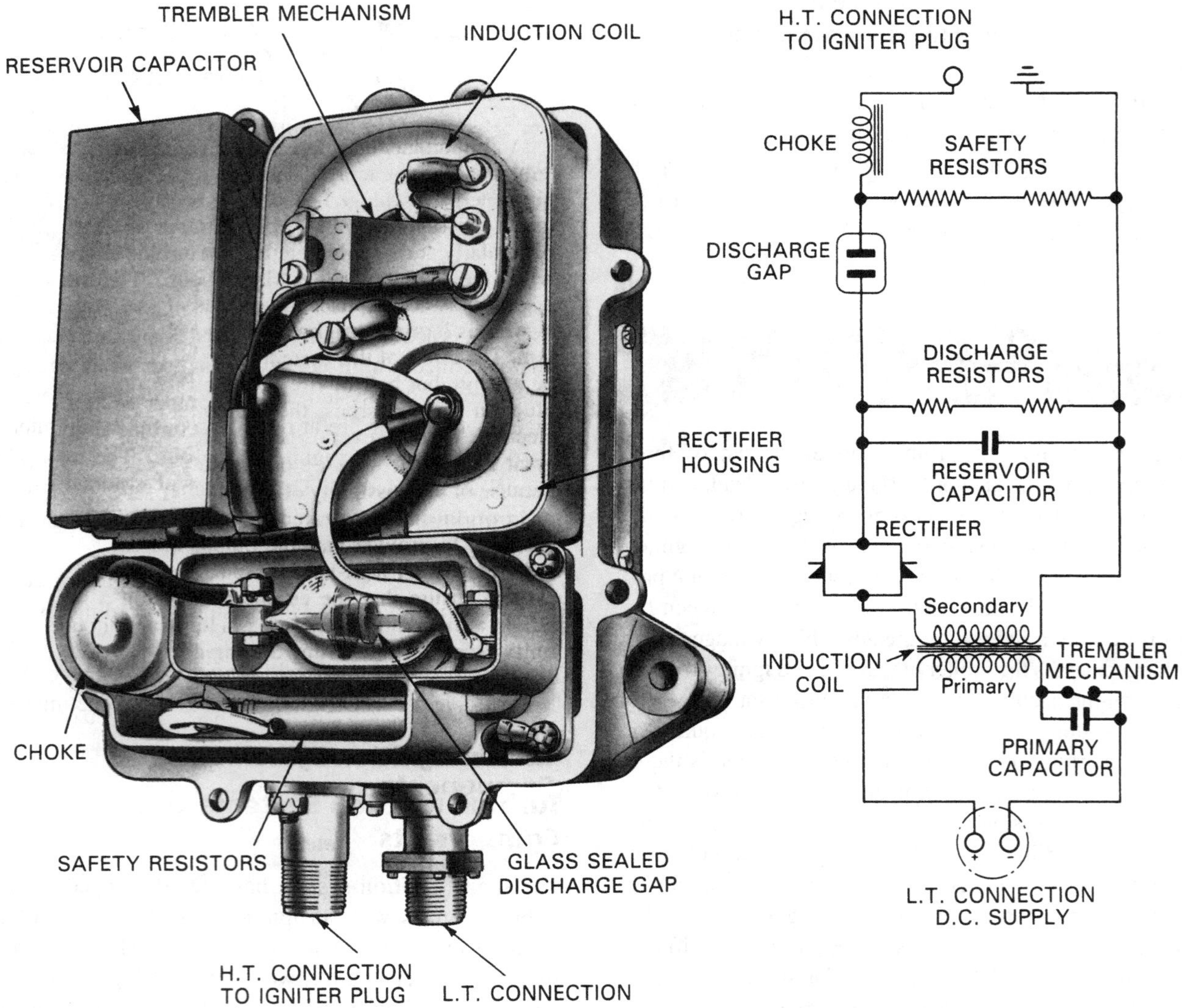

FIGURE 15-3 A dc trembler-operated ignition unit. (*Rolls-Royce.*)

igniter plug. A choke is fitted to extend the duration of the discharge and a discharge resistor is fitted to ensure that any residual stored energy in the capacitor is dissipated within 1 minute of the system being switched off. A safety resistor is fitted to enable the unit to operate safely, even when the high-tension lead is disconnected and isolated.

Transistor Ignition Exciter Unit

Operation of the transistorized ignition unit is similar to that of the dc trembler-operated unit, except that the trembler unit is replaced by a transistor chopper circuit. A typical transistorized unit is shown in Fig. 15-4; such a unit has many advantages over the trembler-operated unit because it has no moving parts and has a much longer operating life. The size and weight of the transistorized unit are less than those of the trembler-operated unit.

AC Ignition Exciter Unit

The ac ignition unit, shown in Fig. 15-5, receives an alternating current which is passed through a transformer and rectifier to charge a capacitor. When the voltage in the capacitor is equal to the breakdown value of a sealed discharge gap, the capacitor discharges the energy across the face of the igniter plug. Safety and discharge resistors are fitted as in the trembler-operated unit.

Continuous or Low-Intensity Ignition

Under certain flight conditions, such as takeoff, landing, heavy rain, snow, icing, compressor stall, or emergency descent, it may be necessary to have the ignition system operating continuously to provide automatic relight should a flameout occur. For this condition, a low-value output (from 3 to 6 J) is preferable, because it results in longer life for the igniter plug and the ignition unit. Consequently, to suit all engine operating conditions, a combined system with both high- and low-value outputs is needed. Such a system can consist of one ignition unit emitting a high output to one igniter plug and a second unit sending a low output to a second igniter plug, or it can consist of an ignition unit that is capable of supplying both high and low outputs, with the output value being preselected as required.

Glow Plug Ignition System

A glow plug system, used on some models of gas-turbine engines, provides current to a hot coil element in each glow plug which makes the plug glow red hot. This glow plug reaches an extremely high temperature very rapidly and ignites the fuel spray in the combustion section of the engine during starting.

Autoignition System

An autoignition system is generally used on turboprop engines. The ignition is activated when the system is armed and the engine torque drops below a certain value. If this condition occurs, the system will automatically activate the engine's ignition system, to help prevent the engine from flaming out or stopping. If the system is armed and the engine torque remains above the torque activation value, the autoignition system will remain inactive.

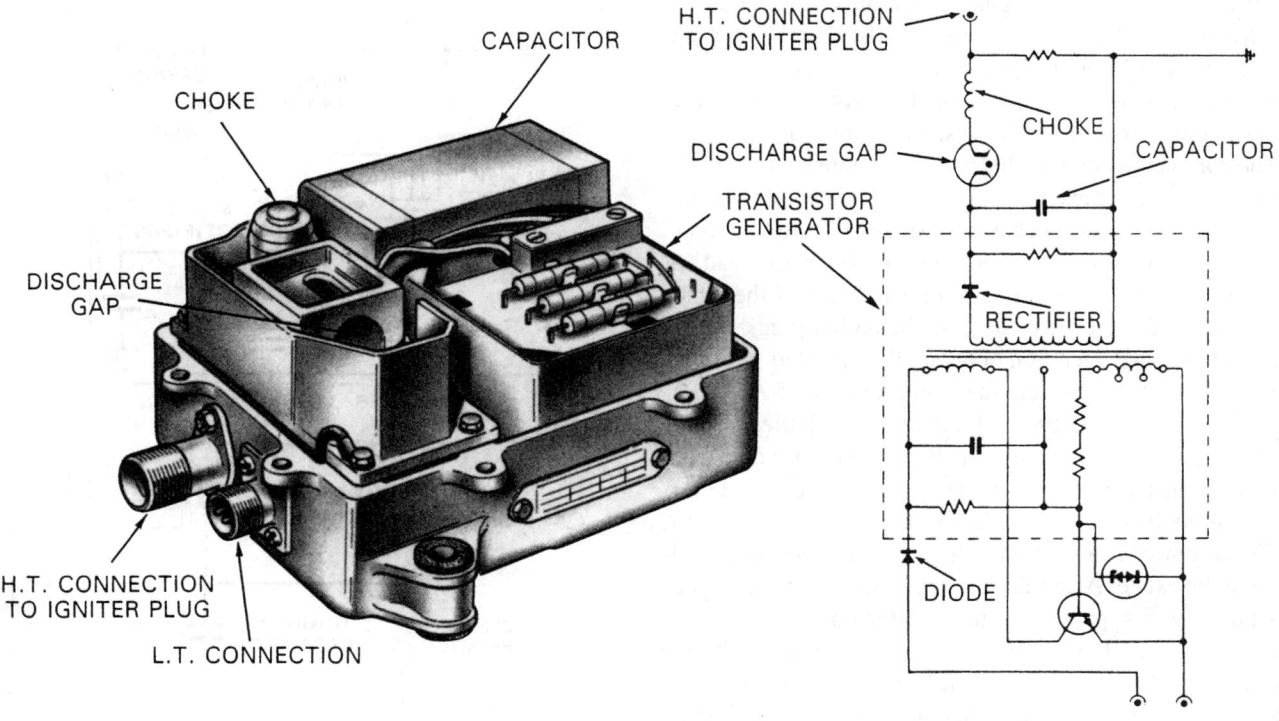

FIGURE 15-4   A transistorized ignition unit. (*Rolls-Royce.*)

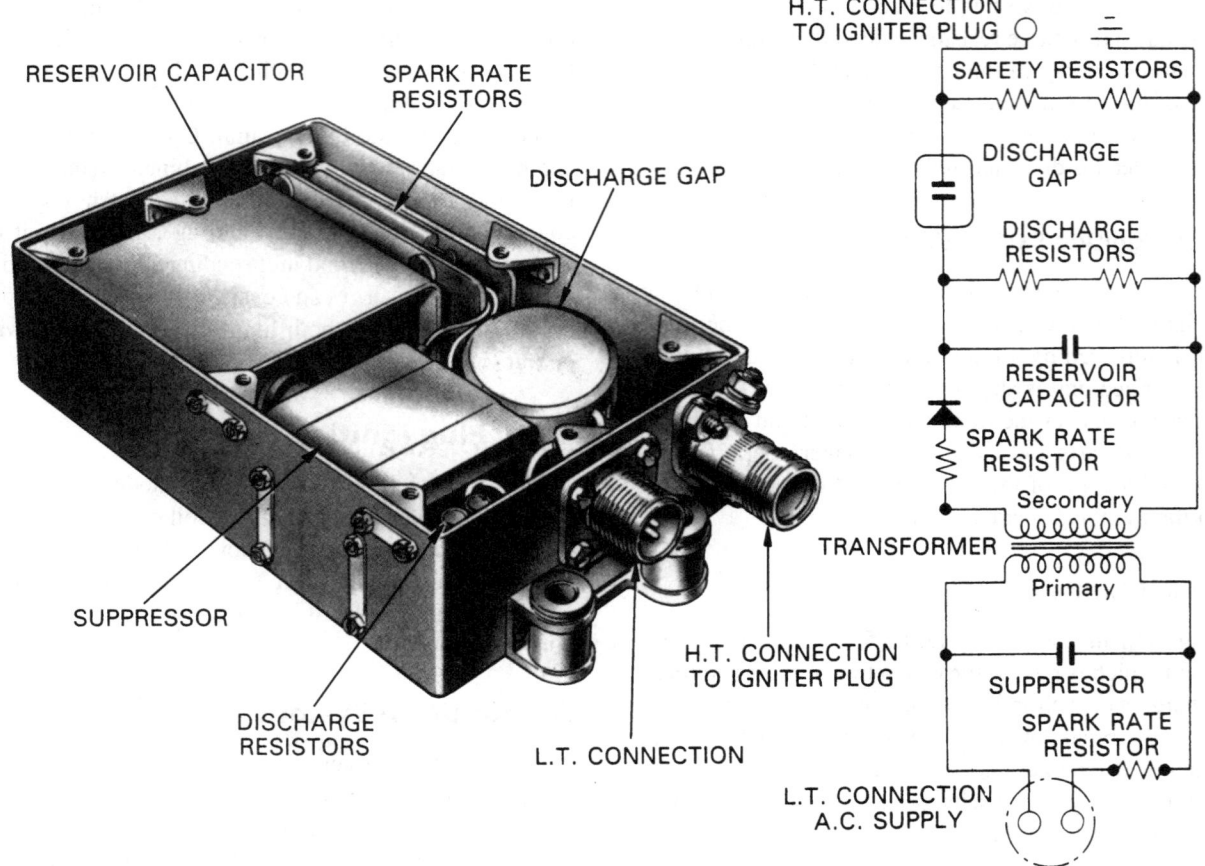

FIGURE 15-5 An ac ignition unit. (*Rolls-Royce.*)

JT9D Ignition System

A widely used gas-turbine engine is the Pratt & Whitney JT9D high-bypass turbofan, which powers the Boeing 747 aircraft and others. The ignition system for this engine is a good example of modern ignition systems for gas-turbine engines. A drawing of the circuit for this system is shown in Fig. 15-6.

The complete system includes two separate heat-shielded and shock-mounted exciters. Figure 15-6 shows the circuit for one of the exciters. Each exciter supplies ignition energy to a recess gap igniter plug through a high-tension lead. A small amount of engine-fan air is directed to cool the high-tension leads, the exciter boxes, and the igniter plugs.

Input power for operation of the ignition exciters is 115 V, 400 Hz ac, with an input current not in excess of 2.5 A to each exciter. The stored energy is 4 J nominal. One **joule**, defined previously, can also be defined as the amount of work done in maintaining a current of 1 A against a resistance of 1 Ω for 1 s; this is equivalent to 0.73746 ft·lb.

The ac power from the aircraft electric system is applied between the exciter A and B terminals at the input connector. This power is first passed through a **filter circuit** consisting of a reactor and a feed-through capacitor to prevent high-frequency feedback into the aircraft electrical system. The reactor also serves as a power choke to limit spark rate variations over the input voltage frequency range. From the filter, the voltage is applied across the primary of the power transformer.

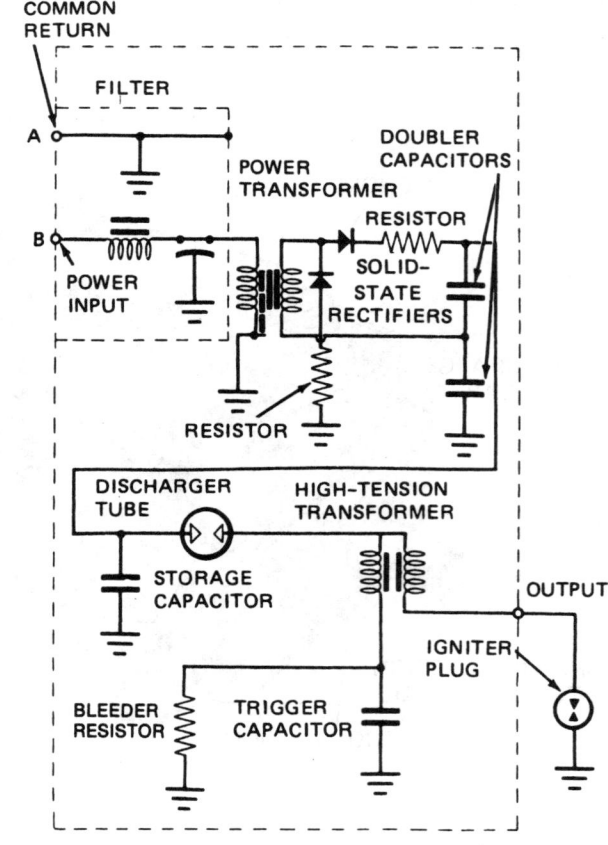

FIGURE 15-6 Ignition system for the JT9D turbofan engine.

The high voltage generated in the secondary of the power transformer is rectified in the **doubler circuit** by the two solid-state rectifiers and the doubler capacitors, so that with each change in polarity a pulse of dc voltage is sent to the storage capacitor. The resistors in the doubler circuit serve to limit the current passing through the rectifiers during those intervals of discharge of the storage capacitor when the voltage has reversed. With successive pulses, the storage capacitor assumes a greater and greater charge at increasing voltage.

When the voltage of the storage capacitor reaches the predetermined level for which the spark gap in the discharger tube has been calibrated, the gap breaks down and a portion of the charge accumulated on the storage capacitor flows through the primary of the high-tension transformer and to the trigger capacitor. This flow of current induces in the transformer secondary a voltage high enough to ionize the air gap in the igniter plug. With the gap thus made conductive, the remaining charge on the storage capacitor is delivered to the igniter plug as a high-current, low-voltage spark across the gap. This is the high-energy spark necessary to produce ignition under adverse conditions.

The **bleeder resistor** is provided to dissipate the energy in the circuit if the igniter plug is absent or fails to fire. It also serves to provide a path to ground for any residual charge on the trigger capacitor between cycles. When the storage capacitor has discharged all its accumulated energy, the cycle of operation recommences. Variations in input voltage or frequency will affect the spark repetition rate, but the stored energy will remain virtually constant.

Ignition Cable Assemblies

There are two individual **ignition cable assemblies** that carry the electrical output from the ignition exciter to the engine-mounted spark igniters. Each lead assembly consists of an electrical lead contained in a flexible metal braiding. Coupling nuts at each end of the assembly facilitate its connection to respective connectors on the ignition exciter and spark igniter. An ignition cable assembly is illustrated in Fig. 15-7.

TURBINE-ENGINE IGNITERS

Although either a gas-turbine engine igniter or a glow plug can serve the same purpose as a spark plug, the design and configuration of these units are considerably different from

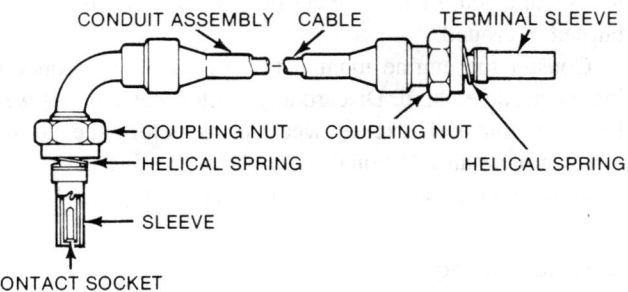

FIGURE 15-7   Ignition cable assembly. (*Pratt & Whitney Canada.*)

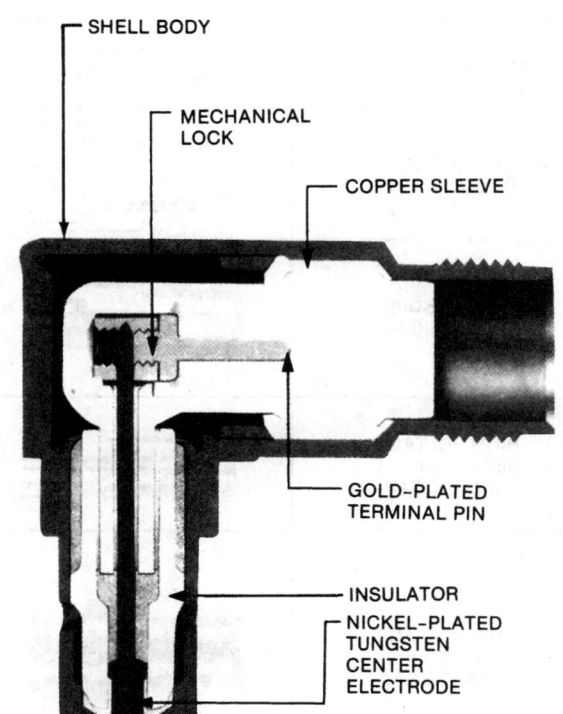

FIGURE 15-8   Turbine engine igniter. (*Champion Spark Plug Co.*)

the design and configuration of spark plugs used in reciprocating engines.

A cross-sectional view of a typical igniter can be seen in Fig. 15-8. The igniter is made up of four basic types of parts: insulators, electrodes, a shell body, and internal seals. Insulators are made of an aluminum-oxide ceramic with a diamond-like hardness that provides mechanical strength, high-voltage insulation, and rapid heat conductivity. Electrodes are made of Inconel Chromel D, tungsten, or any of various nickel alloys, depending on specific service requirements. The shell body is made of extremely high-quality stainless steel or Inconel to resist burner-can combustion temperatures. The internal seals are made of ceramic materials.

The sizes and shapes of igniters have not been standardized; therefore, there are no fixed rules for service and maintenance. In every case it is important to follow the manufacturer's recommendations for cleaning and reconditioning. As an example of servicing differences, it is desirable to clean the firing ends of some types of igniters, while cleaning the firing ends of certain other types of igniters will make the igniters completely unusable.

Figure 15-9 illustrates different types of igniter plugs and a glow plug. Since an igniter is designed to operate at a lower surrounding pressure than is a spark plug, the spark gaps in an igniter are greater. The power source for the igniter supplies a very high level of energy; therefore, the spark produced is of relatively high amperage and resembles a white-hot flame rather than a spark.

The igniter plug tip protrudes approximately 0.1 in [2.54 mm] into the flame tube. During operation, the spark penetrates a further 0.75 in [19.05 mm] into the flame tube. The fuel mixture is ignited in the relatively stable boundary layer and then spreads throughout the combustion system.

GAP DESCRIPTION	TYPICAL FIRING END CONFIGURATION
HIGH VOLTAGE AIR SURFACE GAP	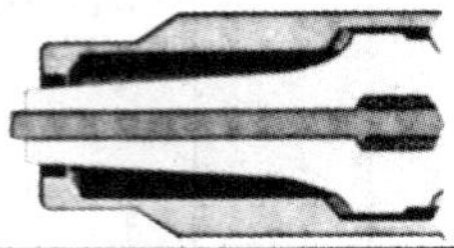
HIGH VOLTAGE SURFACE GAP	
HIGH VOLTAGE RECESSED SURFACE GAP	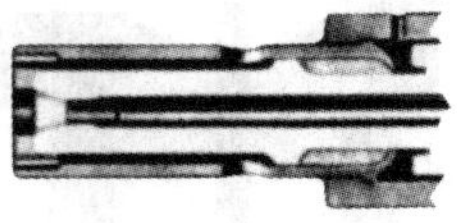
LOW VOLTAGE SHUNTED SURFACE GAP	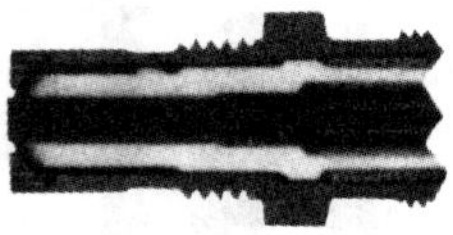
LOW VOLTAGE GLOW COIL ELEMENT	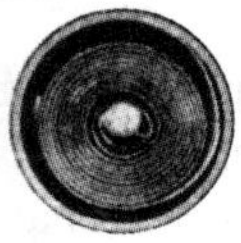

FIGURE 15-9 Various types of igniters and a glow plug. (*Champion Spark Plug Co.*)

The spark discharge of an igniter causes a much more rapid erosion of the electrodes than does the spark provided by a spark plug. However, since the igniter usually operates just long enough to start the engine, the total erosion over a long period of engine operation is not great.

In some cases the igniter must also operate during aircraft operational modes in order to ensure a restart if the engine should flame out. This ignition state, generally called **continuous ignition**, provides a low-intensity spark to one igniter plug, which continues to spark in one burner can as long as continuous ignition is selected by the pilot. Generally, low-intensity ignition (continuous ignition) should be used during aircraft operation under the following conditions: takeoff, icing, emergency descent, compressor stall in flight, approach and landing, and turbulent air.

Service and Inspection of Igniters and Glow Plugs

The procedures and techniques given in this section provide a general guide to ignition system maintenance, but they do not supersede instructions provided in approved maintenance manuals.

Igniters will need to be serviced, cleaned, and sometimes replaced because of erosion and carbon buildup. Specific operational, maintenance, and inspection procedures concerning

igniters are contained in the appropriate aircraft and service engine manuals. These manuals provide specific details that apply to the particular requirements of given models of engines or aircraft.

There are several steps involved in servicing igniters. With the exception of the operational check, the instructions that follow are generally confined to the servicing of an igniter once it has been removed from the engine.

Operational Check

All igniters, except the glow plug variety, emit a sharp snapping noise when firing. The higher the energy of the ignition system, the louder the noise of the spark. Engine cowling suppresses this noise considerably. In performing an operational check, one technician stands to one side of the tailpipe while another technician in the cockpit turns on the ignition. Igniters are operating satisfactorily if a steady pulsed firing is heard from both igniters.

Removal Precautions

When it becomes necessary to remove igniters for visual inspection, servicing, or replacement, extreme care should be exercised. The electric charge which may be stored in the condenser of a high-energy ignition unit is potentially lethal. It is essential that any safety precautions spelled out in the applicable engine manual be strictly followed. For example, some manuals call for disconnecting the low-voltage primary lead from the ignition exciter unit and waiting at least 1 minute before disconnecting the high-voltage cable from the igniter, to permit the stored energy to dissipate.

Inspection

Prior to igniter inspection, residue should be removed from the shell exterior using a dry cloth or fiber bristle brush. *Do not, under any circumstances, remove any deposits or residue from the firing ends of low-voltage igniters.* Only on glow plugs and high-voltage igniters can the firing end be cleaned to aid inspection.

Visually inspect the igniter for mechanical damage. An igniter should be rejected if it shows thread damage, cracked or loose ceramic in the terminal well, or chipped, cracked, or grooved ceramic in the firing-end insulator. Also, discard an igniter if the wrench hexagon or the mounting flange is physically damaged, or if the electrode or shell end is severely burned or eroded.

Consult the engine manual for specific limits concerning electrode erosion. Discard any igniter that exceeds wear limits or that will exceed wear limits before the aircraft can complete an additional scheduled operating period. An example of electrode erosion can be seen in Fig. 15-10.

Reconditioning

The terminal well insulator should be cleaned with a cotton or felt swab saturated with Stoddard solvent or

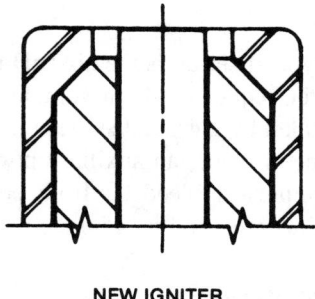
NEW IGNITER

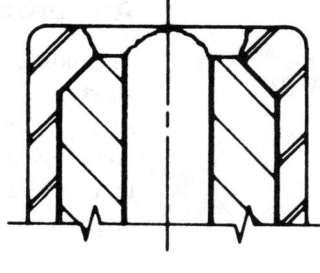
PARTIAL WEAR

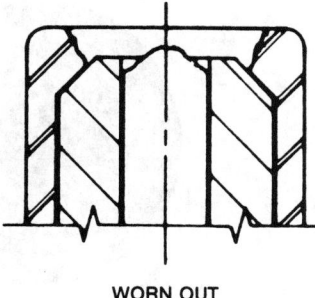
WORN OUT

FIGURE 15-10 Electrode erosion. (*Champion Spark Plug Co.*)

methyl-ethyl-ketone (MEK). Swabs should be approximately $\frac{5}{8}$ in. by $\frac{3}{16}$ in. in size and should project slightly beyond the end of a holder made of slotted-wood or plastic for safe cleaning of the terminal well and contact. Do not use a metal brush for cleaning.

If stains on the terminal well insulator cannot be removed with solvent alone, an abrasive such as the Carborundum Company's "Aloxite" (approximately 325 mesh sieve fineness), Bon Ami, or finely powdered flint may be used. Dip a swab in the solvent and then in the abrasive. Scrub the well insulator thoroughly with a twisting motion and a stroke long enough to remove the stain. Wet a second clean swab with solvent only and clean out all residue. Then blow the terminal well dry with an air blast. The connector seat at the top of the shielding barrel should be cleaned to ensure a satisfactory seal and shield bond when the ignition lead is installed. If solvent alone does not remove dirt and corrosion from the chamfered surface, use fine-grained garnet paper or sandpaper. Do not use emery cloth or paper. Hold the igniter in a partially inverted position to prevent abrasive particles from entering the terminal well. After cleaning thoroughly, blow out the terminal well with an air blast.

Examine the cleaned terminal well thoroughly for insulator cracks and terminal damage, and reject those igniters with physical faults.

The firing end of the igniter should be cleaned only if a cleaning procedure is approved by the manufacturer. Some igniters have semiconductor materials at the firing end and will be damaged if normal cleaning procedures are followed.

If inspection of a glow plug element reveals that it has carbon deposit buildup, it may be serviced as follows:

1. Immerse element end of glow plug in cold carbon remover to loosen carbon deposit (as required).
2. Brush off loosened carbon with a soft nylon or fiber brush. Never use a metal brush, because it will damage the oxide insulation coating on the element coils.
3. Rinse element thoroughly in hot running water. Blow element dry with air blast.
4. Inspect element for possible fused area, as illustrated in Fig. 15-11. The fusing of element coils is a random condition caused by a combustion deposit bridging the coils or an overwattage surge. Any plug that exceeds the manufacturer's acceptable fused limit should be replaced.

Electrical Testing

Some igniters are tested using a high-voltage source, with the firing end pressurized with compressed air, whereas others are tested in open air utilizing the engine ignition system. A steady firing across the gap indicates that the igniter is satisfactory. To test glow plugs, the technician should attach the plugs to the engine ignition system leads and turn the ignition switch on. The glow plug element should heat up to a bright yellow color within 30 s.

Installation Information

Igniter installation mountings vary extensively. Consult the appropriate instruction manuals for each particular engine model for specific igniter installation instructions.

Installation gaskets come in many sizes and shapes and are generally available only from the engine manufacturer. An installation gasket is considered to be an engine part. In a few instances, gaskets are supplied by the igniter manufacturer and are packaged with the igniter.

Some igniters require a specific mounting-gasket thickness to position the igniter firing end properly in the combustor can. The gaskets are usually supplied in a set containing gaskets of several thicknesses. If an igniter is mounted too deep, the firing-end temperature can increase considerably, resulting in both reduced life and reduced reliability of the igniter.

STARTING SYSTEMS FOR GAS TURBINES

Starters for gas-turbine engines may be classified as air-turbine (pneumatic) starters, electric starters, and fuel-air (F/A) combustion starters.

Electric Starters

The comparatively small gas-turbine engines (under 6000 lb [26 690 N] of thrust) are often equipped with heavy-duty **electric starters** or **starter generators**. These are simply electric motors or motor-generator units which produce very high starting torque because of the large amounts of electric power they consume.

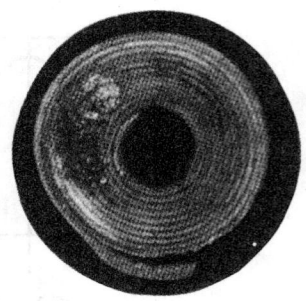

FIGURE 15-11 Glow plug with fused area on element coils. (*Champion Spark Plug Co.*)

Air-Turbine Starters

The most commonly used type of starter is the **air-turbine starter**. This type of starter requires a high-volume air supply, which may be provided by a ground starter unit, a compressed-air bottle on the airplane, an auxiliary power unit on the aircraft, or compressor bleed air from other engines on the aircraft, as illustrated in Fig. 15-14.

Low-Pressure Air-Turbine Starter

The **low-pressure air-turbine starter** is designed to operate with a high-volume, low-pressure air supply, usually obtained from an external turbocompressor unit mounted on a ground service cart or from the airplane's low-pressure air supply. The air supply must produce a pressure of about 35 psig [241.33 kPa] and a flow of more than 100 lb/min [45.36 kg/min].

A drawing of an air-turbine starter is shown in Fig. 15-15. This starter is a lightweight turbine air motor equipped with a rotating assembly, a reduction gear system, a splined output shaft, a cutout switch mechanism, an overspeed-switch scroll assembly, and a gear housing. A cross-sectional drawing of an AiResearch air-turbine starter is shown in Fig. 15-16. The low-pressure air is introduced into the scroll (5) through a 3-in [7.62-cm] duct, which is not shown in the illustration. From the scroll, the air passes through nozzle vanes to the outer rim of the turbine wheel (4). Since this is an inward-flow turbine design, the air expands radially inward toward the center of

The electric starter is coupled to the engine through a reduction gear and ratchet mechanism, or clutch, which automatically disengages after the engine has reached a self-sustaining speed. An example of an electric starter can be seen in Fig. 15-12.

The electric supply, which may be of a high or low voltage, is passed through a system of relays and resistances to allow the full voltage to be built up progressively as the starter gains speed. The electric supply also provides the power for the operation of the ignition system. The electric supply is automatically canceled when the starter load is reduced, either after the engine has satisfactorily started or when the time cycle is completed. A typical electric starting system is illustrated in Fig. 15-13.

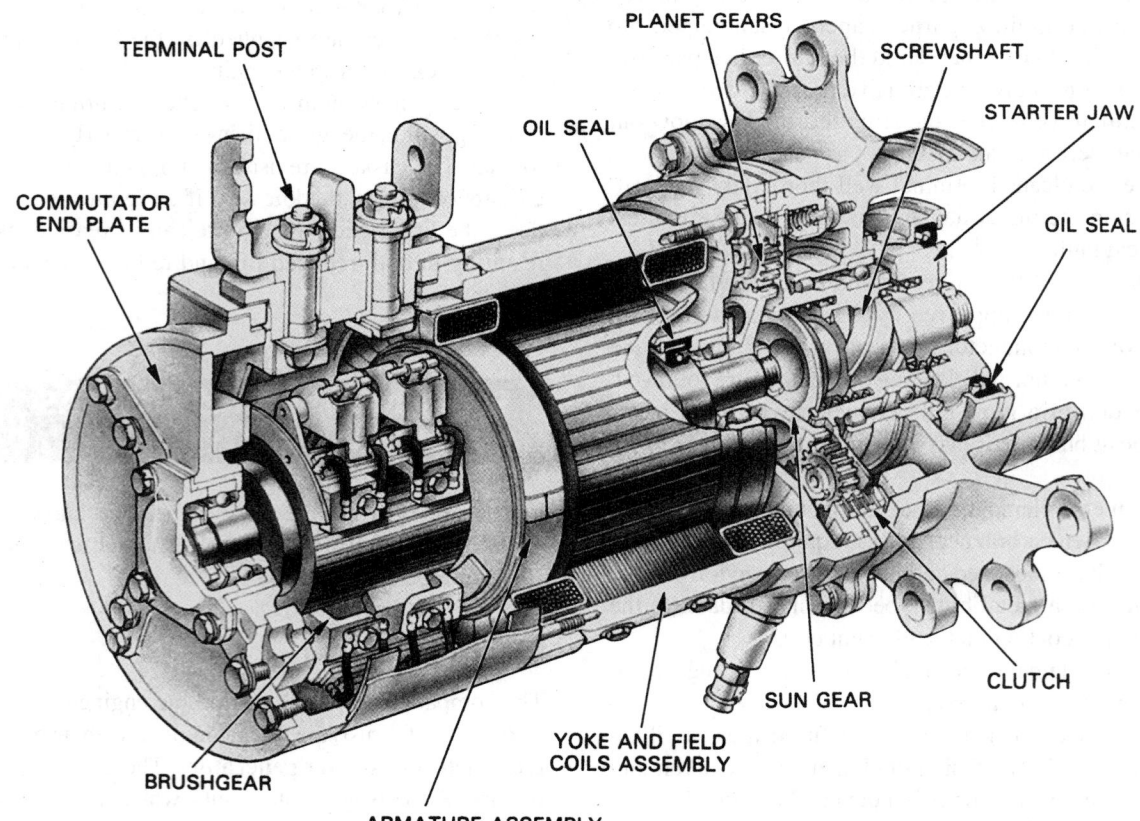

FIGURE 15-12 Electric starter. (*Rolls-Royce.*)

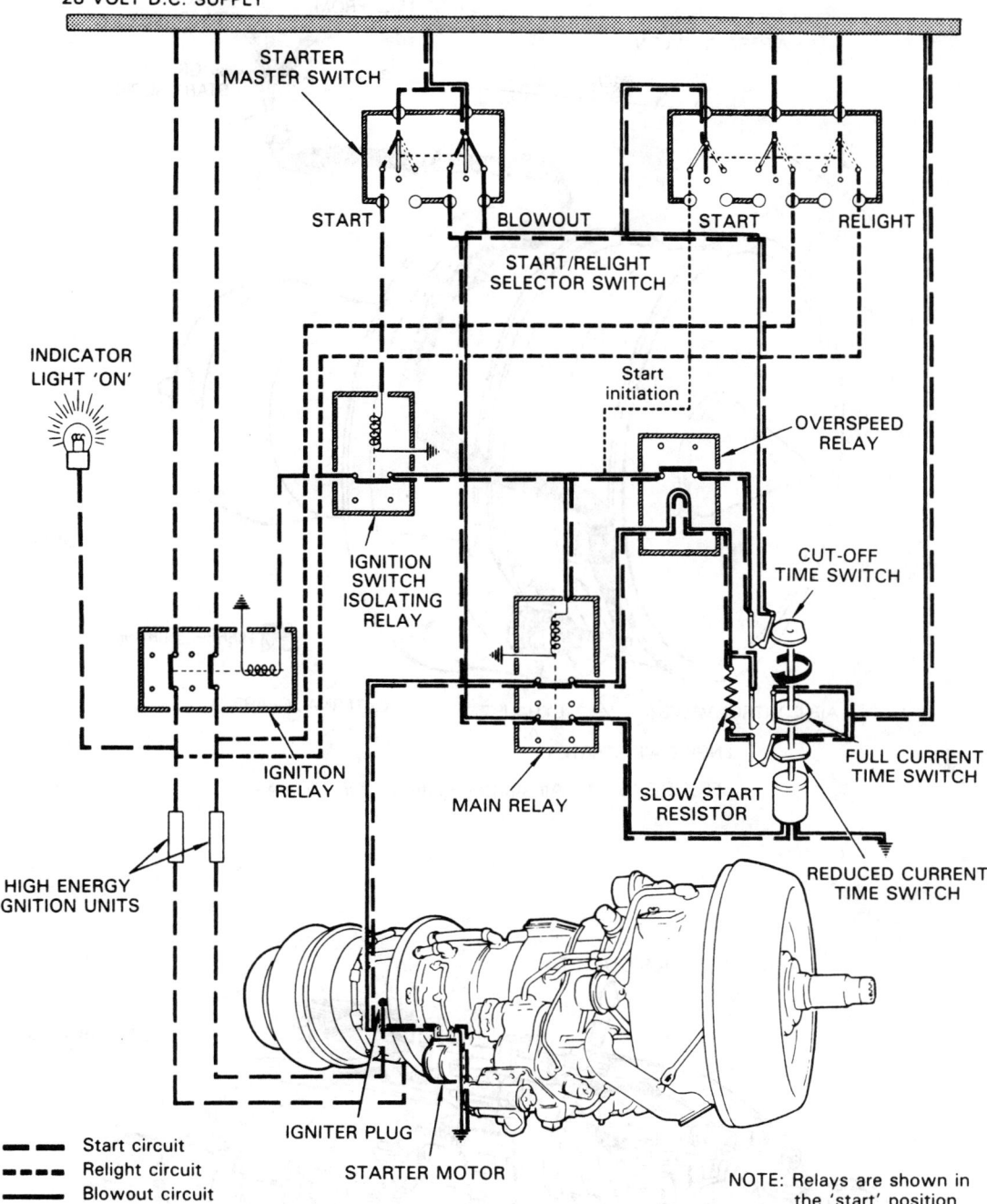

FIGURE 15-13 Low-voltage electric starting system. (*Rolls-Royce.*)

the wheel and is then expelled through the exducer (3). The exhausted air passes through the screen (1) and out to the atmosphere. The expansion of the air from a pressure of about 35 psig [241.33 kPa] to atmospheric pressure imparts energy to the turbine wheel, causing it to reach a speed of about 55 000 rpm. This low-torque high speed is converted to a high-torque low speed by means of the 23.2:1 reduction gearing.

The **rotating assembly** of the starter consists of the turbine wheel, a spacer, a spur gear (9), and a nut. The turbine wheel assembly is an integral wheel and shaft with an exducer pinned on the exhaust end of the shaft against the front face

of the wheel. The spacer, gear, and nut are installed on the opposite end of the shaft, which also provides for the installation of the two ball bearings in which the rotating assembly is mounted. The spacer, bearings, and gear are held on the shaft by the nut, which is secured by a roll pin through the shaft.

The **heat barrier and oil seal** (36) are positioned to provide the correct clearance between the turbine wheel assembly and the heat barrier. The heat barrier and the oil seal prevent the passage of compressed air from the scroll into the housing and the passage of lubricating oil from the housing into the scroll.

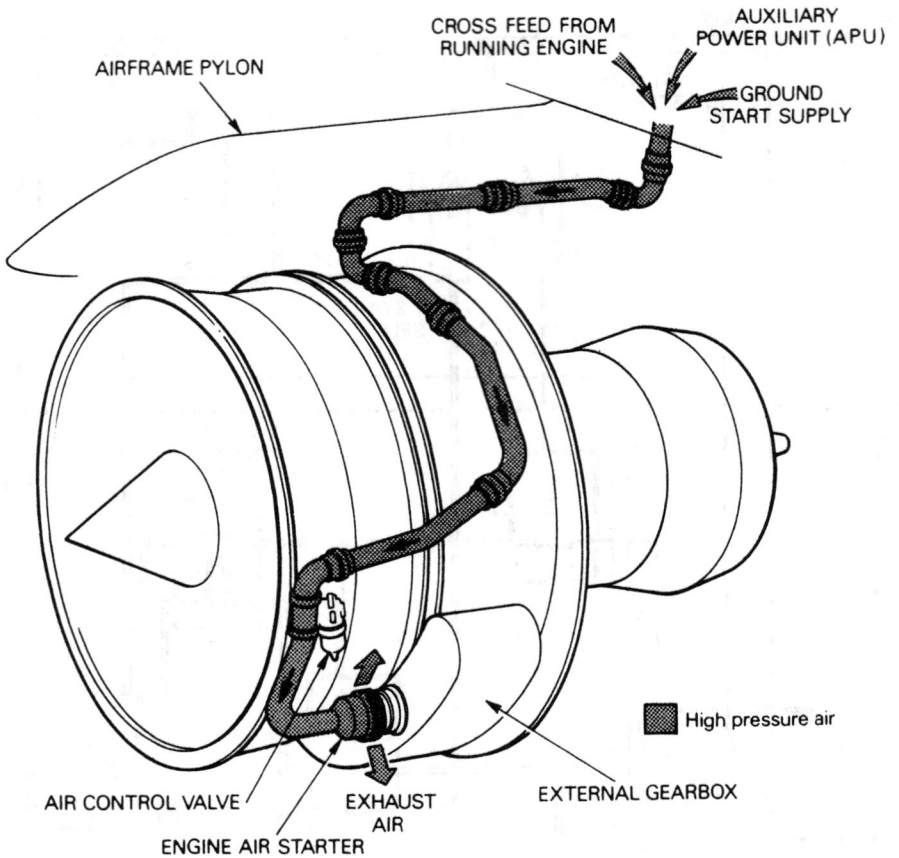

FIGURE 15-14 Air starting system. (*Rolls-Royce.*)

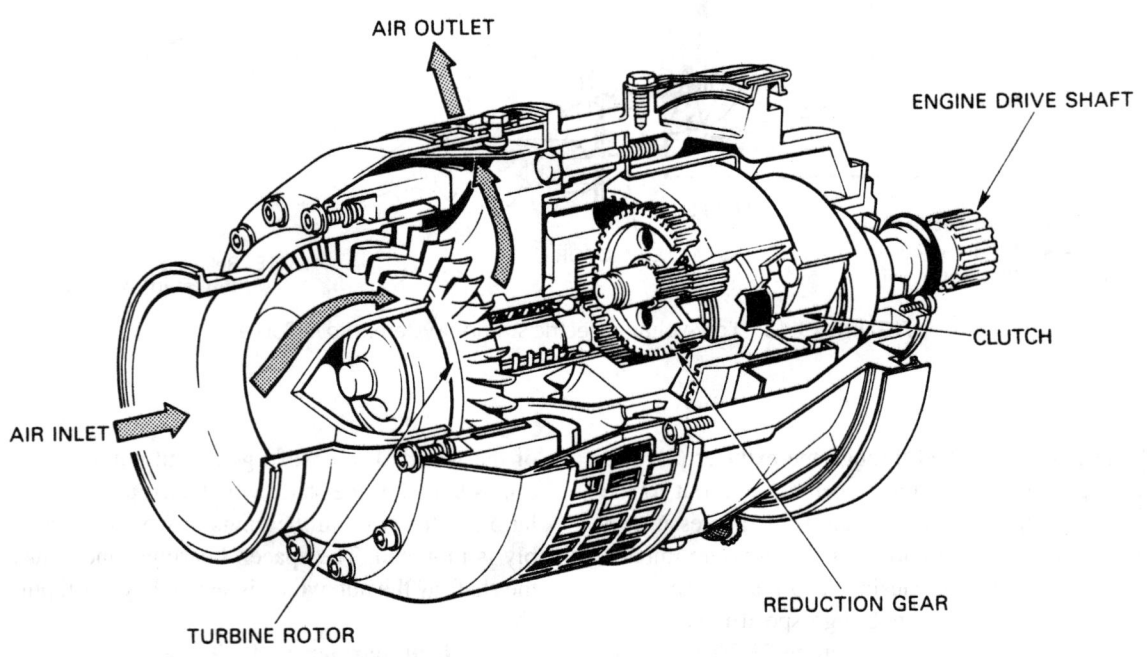

FIGURE 15-15 Air-turbine starter. (*Rolls-Royce.*)

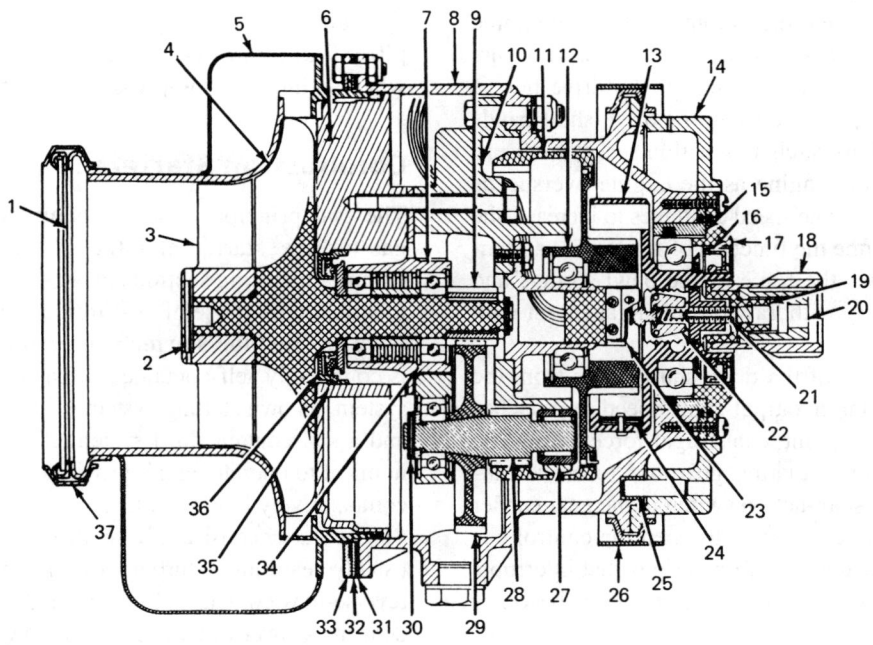

FIGURE 15-16 Cross-sectional drawing of an air-turbine starter. (*AiResearch.*)

The **oil-seal assembly** (36) consists of a rotor mounted on the turbine wheel shaft and a stator placed in the heat barrier. The rotor serves as an oil slinger. The carbon stator, containing an O-ring packing, is spring-loaded against the rotor, providing sealing against the passage of compressed air or lubricating oil.

The **bearing carrier** (7) supports the two ball bearings in which the rotating assembly is mounted. The bearing carrier and heat barrier with the oil seal are bolted to the gear carrier of the reduction gear system.

The **reduction-gear system** consists of a **gear carrier** (10), three spur-gear shaft assemblies (only one assembly shown at 28 and 29), and an internal gear (11). The **gear carrier assembly** is a matched pair of forgings brazed together and bolted inside the housing. The gear carrier provides for the installation of a ball bearing and a needle bearing in which each of the spur-gear shaft assemblies rotates and a ball bearing on which the internal gear hub rotates. Each of the three **spur-gear shaft assemblies** consists of an integral spur gear and tapered shaft (28), a planet spur gear (29), a bearing, and a nut. The planet spur gear, which has a tapered bore, is friction-mounted on the tapered shaft and held in position by the bearing and nut. The nut is secured on the shaft by a rollpin through the shaft. The planet spur gears mesh with and are driven by the spur gear of the rotating assembly. The internal gear, mounted over the spur-gear shaft assemblies, meshes with and is driven by the three spur gears of the gear-shaft assemblies. The internal gear hub is installed in and attached to the internal gear by means of a lock ring. The internal gear hub rotates on the ball bearing installed on the gear carrier assembly and is integral with the jaw of the engagement mechanism.

The **engagement mechanism** consists of a drive hub (12) and a drive-shaft assembly (13). The drive hub has a series of ratchet teeth equally spaced about the outside diameter, and these teeth engage the three pawls which are mounted inside the drum of the drive-shaft assembly. The arrangement of the pawl and spring drive assembly is shown in Fig. 15-17. The drive shaft is internally threaded for installation of the switch actuating governor (22 in Fig. 15-16) and internally splined for installation of the output shaft (18), and it provides a sealing surface for the drive-shaft seal (16) which is installed in the housing. Each pawl-spring assembly (24) is a series of leaf-type springs of varying length, as also shown in Fig. 15-17. Each spring assembly is riveted inside the drive-shaft drum.

The operation of the engagement mechanism is such that the drive-shaft pawls are disengaged from the teeth of the ratchet (pawl drive jaw) when the engine is running. Before and during starting, at low rotational speeds, the pawl springs in the drive shaft force the drive-shaft pawls into engagement with the ratchet teeth of the drive hub. This engagement transmits the rotation of the drive hub through

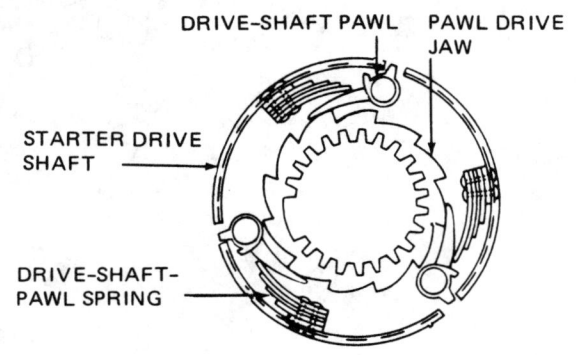

FIGURE 15-17 Engagement mechanism for the AiResearch air-turbine starter.

the drive-shaft assembly and the output shaft to the engine on which the starter is installed. When engine lightoff occurs and the drive-shaft speed exceeds that of the drive hub, a ratcheting action takes place between the drive-shaft pawls and the ratchet teeth. This ratcheting action serves to disengage the starter from the engine as the engine overspeeds the starter drive hub. As engine speed continues to increase, the starter drive pawls assume the function of flyweights as centrifugal force overcomes the force of the pawl springs and the pawls are completely withdrawn from engagement with the drive hub.

Before disengagement of the drive-shaft pawls from the drive hub and as the starter output-shaft speed approaches the predetermined cutoff point, centrifugal force causes the flyweights of the switch actuating governor to move outward. This actuates the snap-action switch in the gear carrier of the starter reduction-gear system to open the control circuit. This initiates the sequence of operations that interrupts the flow of compressed air, and the starter then becomes inoperative.

High-Pressure Air-Turbine Starters

A **high-pressure air-turbine starter** is essentially the same as a low-pressure starter except that it is equipped with an axial-flow turbine in place of the radial, inward-flow turbine previously described. The high-pressure starter is fitted with both low-pressure and high-pressure air connections to provide for operation from either type of air supply.

The usual air supply for the high-pressure starter operation is a high-pressure air bottle mounted in the airplane.

This air bottle is charged to a pressure of about 3000 psi [20 685 kPa] and is used for starting one of the engines when an internal low-pressure source is not available.

Combustion Starters

The two principal types of **combustion starters** are the gas-turbine starter and the cartridge-type starter. The turbine-operating sections of these starters are similar to or identical with those of air-turbine starters.

A **gas-turbine starter** is used for some jet engines and is completely self-contained. It has its own fuel and ignition system, its own starting system (usually electric or hydraulic), and a self-contained oil system. This type of starter is economical to operate, and it provides a high power output for a comparatively low weight.

The starter consists of a small, compact gas-turbine engine, usually featuring a turbine-driven centrifugal compressor, a reverse-flow combustion system, and a mechanically independent free-power turbine. The free-power turbine is connected to the main engine via a two-stage reduction gear, an automatic clutch, and an output shaft. A typical gas-turbine starter is shown in Fig. 15-18.

When the starting cycle is initiated, the gas-turbine starter is rotated by its own starter motor until it reaches a self-sustaining speed, at which point the starting and ignition systems automatically switch off. Acceleration then continues up to a controlled speed of approximately 60 000 rpm. While the gas-turbine starter engine is accelerating, the exhaust gas is being directed, via nozzle guide vanes, onto the free-power turbine to provide the drive to the main engine. Once the

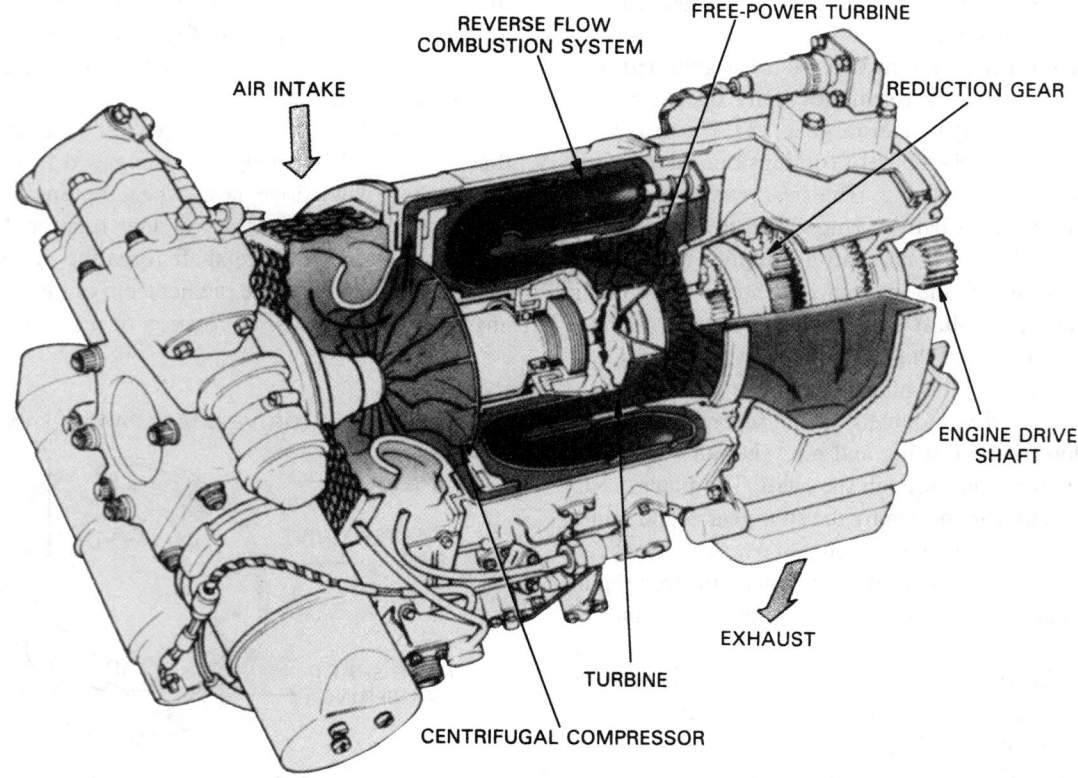

FIGURE 15-18 Gas-turbine starter. (*Rolls-Royce.*)

main engine reaches a self-sustaining speed, a cutout switch operates to shut down the gas-turbine starter. As the starter runs down, the clutch automatically disengages from the output shaft and the main engine accelerates up to idling rpm under its own power.

Other Types of Turbine Starters

The **cartridge-type starter** may be considered an air-turbine starter operated by means of hot gases from a solid fuel cartridge instead of compressed air. Some air-turbine starters can be adapted to cartridge operation merely by installing a cartridge combustion chamber and a gas duct.

The advantage of the cartridge-type starter is that it is a self-contained unit and does not require an external power source. The two principal disadvantages are the high cost of the fuel cartridges and the erosion of the turbine parts by solid particles in the hot gases.

Some turbojet engines are not fitted with starter motors but instead use air forced onto the turbine blades as a means of rotating the engine. The air is obtained from an external source or from a running engine, and it is directed through nonreturn valves and nozzles onto the turbine blades. This type of system is often referred to as **air impingement starting** and is illustrated in Fig. 15-19.

STARTING SYSTEM FOR A LARGE TURBOFAN ENGINE

The starting system for a large airliner, the McDonnell Douglas DC-10, is described in this section. The system is used with the General Electric CF6 engines which power the DC-10.

Auxiliary Power Unit

From the diagram shown in Fig. 15-20, it will be observed that the energy for the starting system may be supplied by

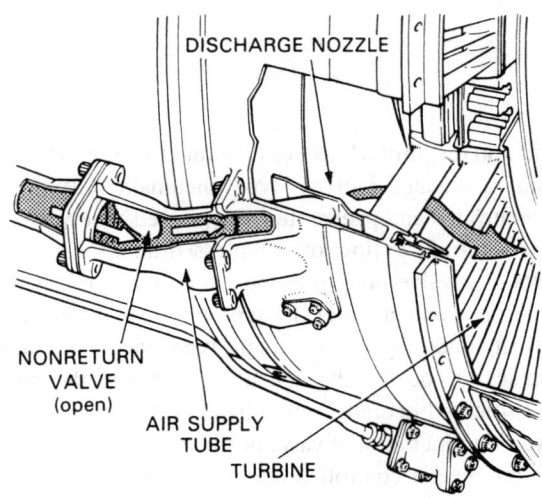

FIGURE 15-19   Air impingement starting. (*Rolls-Royce.*)

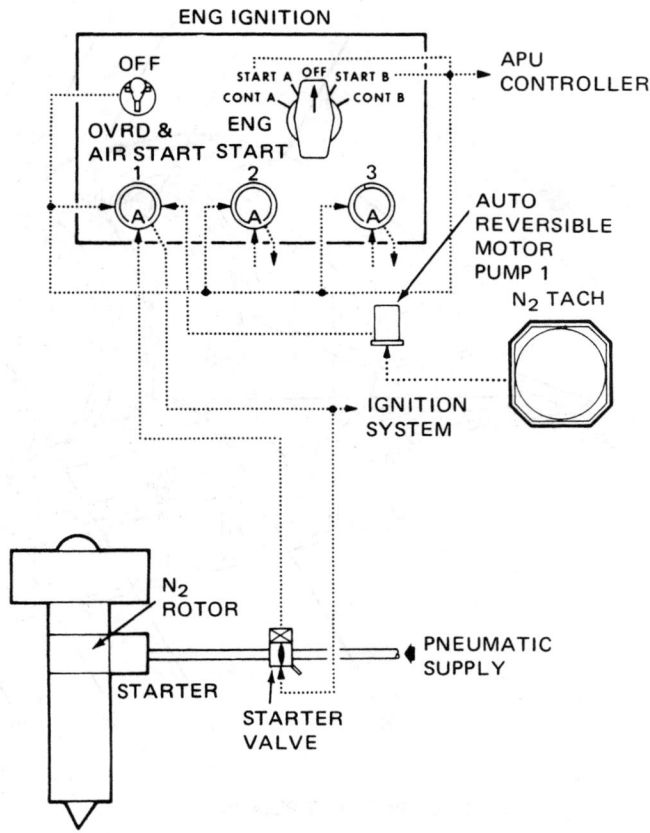

FIGURE 15-20   Schematic drawing of the starting system for the Douglas DC-10. (*McDonnell Douglas Corp.*)

an onboard **auxiliary power unit (APU)**, or by a ground air supply or an operating engine on the aircraft. The APU and the engines are connected through check valves and control valves to the common pneumatic system manifold. An APU, shown in Fig. 15-21, is a compact, self-contained unit which provides compressed air and aircraft electric power. Most APUs have automatic controls for starting, stopping, and maintaining operation within safe limits. These controls provide correct sequencing of the starting cycle, as well as protection against overspeed, loss of oil pressure, high oil temperature, or high turbine-gas temperature.

Fuel for the APU is taken from the aircraft fuel system. Electric power to start the APU comes from the aircraft electric system or battery. APUs are used to supply compressed air (compressor bleed air, or air from an external compressor), which is used for engine starting, air conditioning, and heating. The APU can also be used for generating aircraft electric power. Normally the APU is operated only on the ground, but some aircraft can utilize the APU in flight as a source of electric power, if needed.

Control and Indicating System

The control and indicating system provides means to actuate the starter shutoff valve, control, and pneumatic supply to the starter; to indicate the position of the starter shutoff valve; and to terminate the starting cycle. The system includes the engine start switch, the starter shutoff valve,

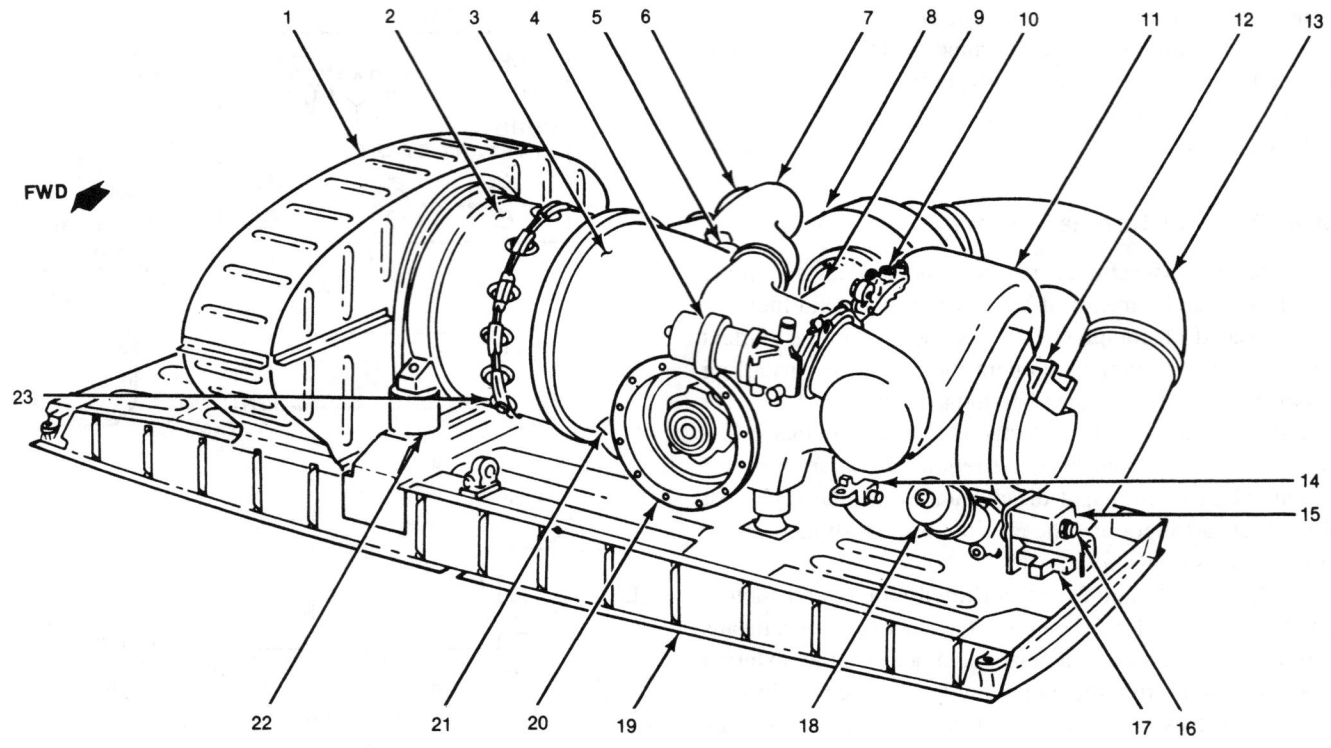

FIGURE 15-21 Auxiliary power unit.

1	ENGINE INLET PLENUM	13	FAN INLET AIR DUCT /2\
2	ENGINE	14	CURRENT TRANSFORMER
3	ENGINE INSULATION BLANKETS	15	INLET GUIDE VANE ACTUATOR ASSY /2\
4	SURGE VALVE ASSEMBLY /1\	16	FEEDBACK TRANSDUCER
5	FREE TURBINE SPEED SENSOR (TOP) /2\	17	LOAD COMPRESSOR CONTROLLER
6	EXHAUST MUFFLER DUCT /2\	18	INLET GUIDE VANE ACTUATOR
7	SURGE DUCT /2\	19	PLATFORM
8	COOLING FAN	20	GENERATOR MOUNTING PAD
9	LOAD COMPRESSOR GEARBOX	21	FREE TURBINE SPEED SENSOR (BOTTOM) /1\
10	SURGE VALVE CONTROLLER /2\	22	VIBRATION MOUNT (ONE OF THREE)
11	LOAD COMPRESSOR DUCT	23	IGNITER PLUG (ONE EACH SIDE) /1\
12	OIL TO AIR HEAT EXCHANGER ASSY		

and the pneumatic starter mounted on the engine accessory gearbox.

The **engine start switch** is located on the forward overhead panel in the flight compartment (cockpit) and controls the operation of the **starter shutoff valve**. The switch is a push-button type with an integral indicating light in the knob. The switch is actuated when depressed and is held in the ON position by a holding coil until it is released by a speed signal from the N_2 **rpm indicator speed switch**. The rotational speed of the high-pressure compressor rotor in the engine is N_2 rpm. The speed switch is operated in conjunction with the ignition system controls. Power to the switch is provided by the engine ignition switch.

The **starter shutoff valve**, illustrated in Fig. 15-22, is a diaphragm-actuated butterfly-type pneumatic valve, electrically controlled and pneumatically operated. The valve functions to control the flow of compressed air to the starter. It consists of a valve body housing with an integral, butterfly-type closure element (gate) and appropriate in-line end flanges for direct mounting; a diaphragm-type pneumatic actuator mechanically coupled through a lever arm to the butterfly shaft; a solenoid-operated single-ball selector valve with manual override for control of valve position; a rate-control orifice which provides a controlled opening time; a stainless-steel, sintered wire-mesh filter; and a mechanical pointer for visual indication of valve position. The lower end of the butterfly

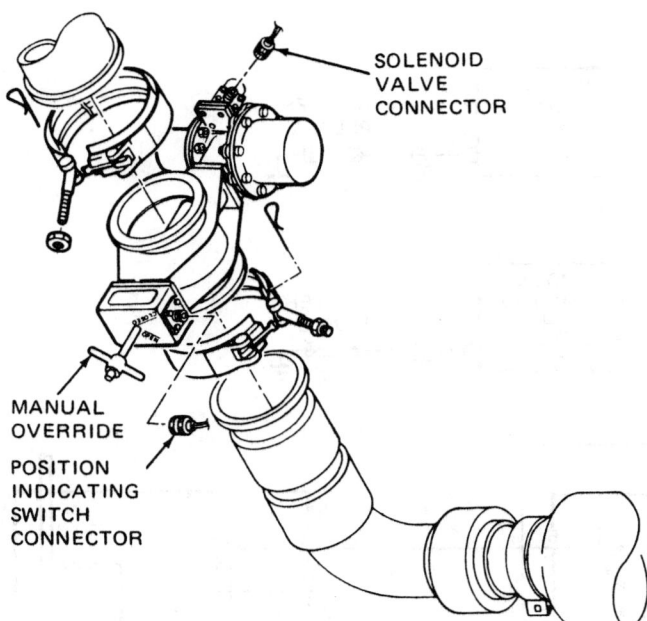

FIGURE 15-22 Starter shutoff valve. (*McDonnell Douglas Corp.*)

shaft is equipped with a handle to permit manual opening of the valve should the actuator supply pressure be lost.

As shown in Fig. 15-22, a **position indicating switch** is located on the lower end of the butterfly shaft and energizes the light in the engine start switch when the butterfly valve is not closed. As previously mentioned, the valve is equipped with a rate-control orifice so that the opening rate is controlled to limit the maximum starter impact torque developed during running starter engagements. **Running engagements** occur when the starter is engaged while the engine is rotating. These may occur during a restart in flight or a start on the ground when the engine is turning.

Because of the need for dry atmosphere in the solenoid- and servo-valve area of the starter shutoff valve, a pneumatic heater is made integral with the valve. Moisture supports corrosion and is detrimental to the operation of electric circuits.

A typical pneumatic starter is the AiResearch ATS 100-350 air turbine. A drawing of this starter and its mounting on the accessory gearbox is shown in Fig. 15-23. The starter is a single-stage turbine consisting of a stator, a turbine wheel, a reduction gear, a ring gear and hub jaw, a gear housing, an overrunning clutch, a splined output shaft, and an integral quick-attach-detach (QAD) mounting ring.

The starter gears and bearings are lubricated by a self-contained oil system using the same type of oil as that required by the engine (MIL-L-23699). Fill and drain ports are provided in the housing for servicing the oil system. A magnetic chip detector is incorporated in the oil drain plug.

The starter output shaft is splined to fit a matching internally splined shaft in the accessory gearbox. The shaft is lubricated by the engine oil system and is provided with a shear section to protect the engine in case of starter malfunction

or failure. The shear section is designed to shear at 1400 to 1600 lb•ft [1898.4 to 2169.6 N•m].

Operation

The operation of the DC-10 engine starting system involves energizing the starter with compressed air while the engine is provided with the proper amount of fuel and the ignition system is functioning. The procedure for starting can be understood by studying the diagram in Fig. 15-24.

The engine start and ignition control switch is turned on to provide electric power for the engine start switch and for ignition. The ignition comprises two independent systems for high-intensity start ignition or continuous ignition to prevent flameout. The systems are labeled A and B, with each containing its own ignition exciter unit lead and igniter. When the start switch is depressed, the holding-coil circuit is completed through the N_2 rpm indicator switch. The energized coil holds the start switch in the ON position until the N_2 compressor rotor speed is 45 percent of maximum, at which time the switch in the N_2 rpm indicator opens. The start switch can be opened manually at any time.

Actuation of the engine start switch energizes the starter shutoff-valve solenoid, as can be seen in Fig. 15-24. This allows inlet-air pressure to be ported to the OPEN chamber of the valve actuator. Since the effective area of the OPEN chamber is greater than the effective area of the CLOSE chamber, the actuator opens the butterfly valve and the compressed air can flow to the turbine starter. As the butterfly valve begins to open, the valve position indicating light switch closes and the indicating light in the starter switch comes on.

Air entering the starter inlet flows through the stator and is directed radially inward to propel the turbine wheel to high-speed rotation. Expended air is exhausted into the fan compartment of the engine and exited overboard through the compartment ventilating ports.

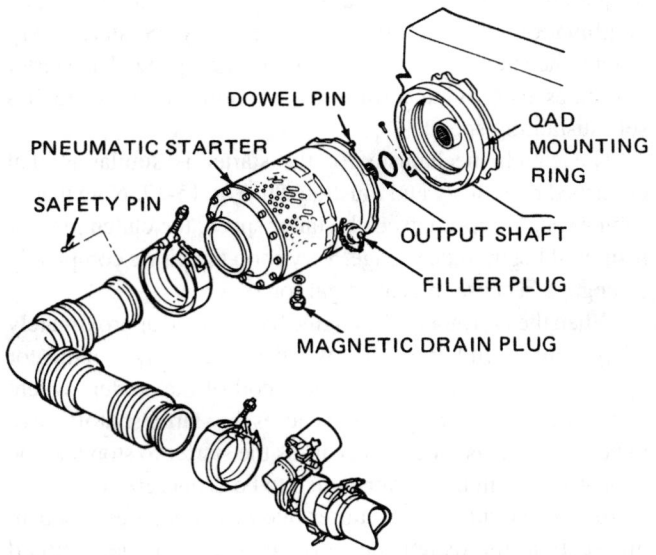

FIGURE 15-23 Pneumatic starter and its mounting. (*McDonnell Douglas Corp.*)

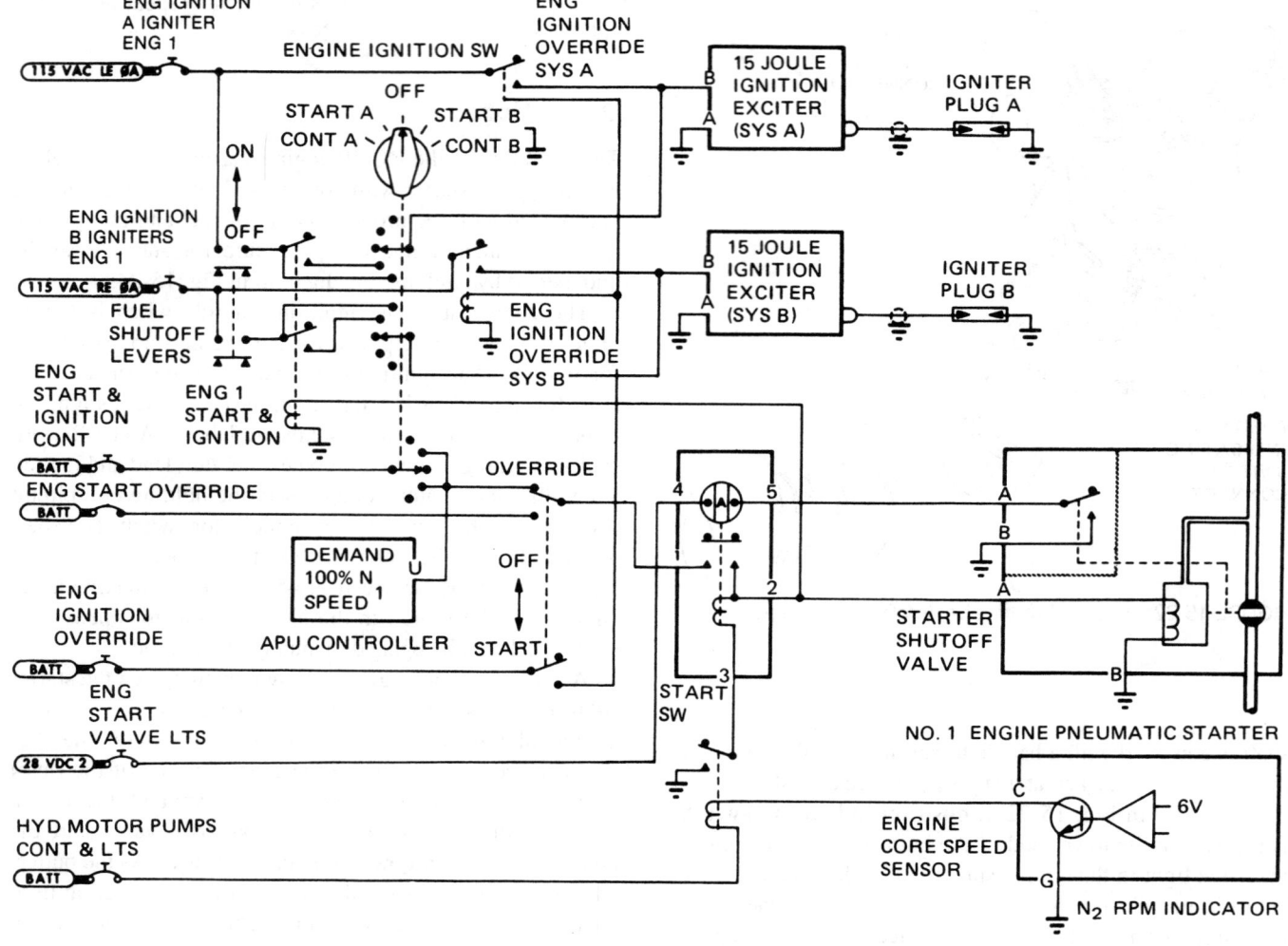

FIGURE 15-24 Electric circuit and controls for the air-turbine starter on the Douglas DC-10. *(McDonnell Douglas Corp.)*

The high rotational speed of the turbine is reduced from a high-speed, low-torque characteristic to a low-speed, high-torque characteristic by means of a planetary and spur-gear combination. The high torque delivered by the starter very quickly accelerates the engine to light off speed. The starter continues to provide torque to the engine until it reaches self-sustaining speed.

The clutch mechanism for the starter is similar to that described previously and illustrated in Fig. 15-17. Note that as soon as engine speed exceeds starter speed, the clutch mechanism will begin to disengage. Very soon the pawls completely disengage because of centrifugal force.

When the N_2 rotor of the engine has attained approximately 45 percent of full rpm, the switch in the N_2 rpm indicator opens and deenergizes the holding coil of the starter switch. The switch therefore opens and causes the starter shutoff valve to be closed. This, of course, causes the starter to stop and the indicator light in the starter switch to be deenergized.

In the operation of a starter such as those described in this section, the technician, pilot, or flight engineer should follow the procedure set forth by the manufacturer or the airline which operates the aircraft. Airlines have established

procedures which have been developed to produce the safest and most effective results.

Boeing 787 Starter Generator System

The variable frequency starter generator (VFSG) is shown in Fig. 15-25. The VFSG has a housing made of aluminum which contains a six-pole generator and is driven by the engine's gearbox. It is a variable frequency synchronous, three-phase brushless, and alternating current generator. Its rating is 235 volts alternating current (VAC), 250 KVA, and 360–800 Hz output. The VFSG delivers several advantages such as replacing the bleed air system used to feed the airplane's environmental control system with a weight savings by eliminating heavy bleed air system components. Components such as heavy regulation valves, ducting, and coolers are no longer needed. This also eliminates the energy loss of the bleed air from the engine's compression stages which on previous aircraft provided air for the aircraft's environmental system. The common motor start controllers (CMSCs) are used to control the VFSG start function and properly regulate torque during the start sequence. Once the engine

FIGURE 15-25   Variable frequency starter generator.

is started, the CMSC switches over to controlling the cabin air compressors, thereby performing a second function. The electric start system affords maximum flexibility from a variety of power sources: APU generators, external power cart, and cross engine (opposite engine VFSGs). The VFSG system provides full-maintenance diagnostics for the entire system.

Inspection and Maintenance of Turbine-Engine Starters

Routine inspections of turbine starters are similar to inspections of other accessories. These inspections include checks for security of mounting; freedom from oil leaks; and security of air and gas ducts, liquid lines, and electrical wiring. Special inspections required for particular units are described in the manufacturer's operation and service manuals.

Maintenance and service include the changing of the lubricant in the gear housing. Since the starter is attached to the engine, it is likely to be exposed to very high temperatures. For this reason, the lubricant used is of the high-temperature type, such as MIL-L-7808C or MIL-L-23699. Even this type of lubricant will lose its lubricating qualities after a time, and therefore it must be tested regularly. The lubricant is checked for the presence of metal particles whenever a change is made, to detect an incipient failure. Very small particles are normal, but particles which produce a sandy feel in the lubricant are a definite indication of internal damage in the starter.

Normally, overhaul of turbine starters should not be necessary at less than 2000 h of engine service; however, the life of a starter is largely dependent on the skill and knowledge of the flight engineer or pilot who operates it. Under no circumstances should anyone but a well-trained individual be permitted to operate the starting system.

REVIEW QUESTIONS

1. List the main components of a gas-turbine engine ignition system.
2. Why is it necessary to design gas-turbine ignition systems that have very high energy outputs, compared with the outputs of ignition systems for reciprocating engines?
3. Why must a technician exercise great care when working around a turbine-engine ignition system?
4. On what type of aircraft is autoignition used, and how is it activated?
5. Compare a gas-turbine igniter plug with a spark plug.
6. What is meant by the term "continuous ignition," and when is it generally used?
7. Where will the technician find specific information on service and maintenance of the igniters?
8. How are turbine-engine igniters operationally checked?
9. What should be done if inspection of a glow plug reveals that it has fused coils?
10. Name four types of starters for gas-turbine engines.
11. What pressure and flow of air are required for a low-pressure air-turbine starter?
12. Describe the engagement mechanism for an AiResearch air-turbine starter.
13. From what sources may pneumatic energy be obtained to operate the starter on a large turbofan engine?
14. Describe the starter shutoff valve on the starting system for a DC-10 aircraft.
15. What is the purpose of the position indicating switch?
16. What is the purpose of the shear section in the starter output shaft?
17. At what engine speed does the starter system disengage?

Turbofan Engines 16

INTRODUCTION

During the 70 years since gas-turbine engines first became a reality, many changes and improvements have been made in the design and performance of these engines. Among the first practical engines were the German Jumo, the British W1, and the U.S. GE I-A, which was an American version of the British engine. These were all pure jet engines in that they did not drive propellers; neither did they employ the bypass principle used almost universally with modern engines.

In this section, the construction and general configuration of some of the most commonly used engines for large and medium-size aircraft will be examined. Airliners in the United States are usually equipped with Pratt & Whitney JT8D, JT9D, 2000 series, or 4000 series engines; General Electric CF6 series or GE90 engines; or Rolls-Royce RB 211 series engines. Engines of the CFM56 and IAE 2500 series are also widely used on transport-category aircraft. Small and medium-size aircraft usually are equipped with turbofan engines in the 500- to 5000-lb [2224- to 22 240-N] thrust range.

The engines for large aircraft are all axial-flow, multiple-compressor turbofan engines with thrust ratings as high as 100 000 lb [444 800 N]. The turbofan engines of this type have proved to be the most efficient and economical for the operation of large airliners.

In recent years, makers of turbofan engines have improved their engines by changing the aerodynamic design of turbine and compressor blades and vanes, utilizing improved alloys for blades and vanes, modifying fan blade design, changing the internal engine-cooling systems, and making a number of other modifications. As a result, today's turbofan engines are more fuel-efficient, develop more thrust for their weight, and operate more quietly. Some of these improvements will be discussed later in this chapter.

LARGE TURBOFAN ENGINES

The Pratt & Whitney JT8D Gas-Turbine Engine

Introduction

The JT8D, illustrated in Fig. 16-1, powers about 4400 aircraft—more than half of the world's commercial transports—including the Boeing 727 and 737, the McDonnell Douglas DC-9, and two French aircraft—the Aerospatiale Super Caravelle and the Dassault Mercure. Since entering service in 1964, JT8D engines have flown more than 300 million h, and they continue to accumulate more than 17 million h each year.

The JT8D's initial rating of 14 000 lb [62 270 N] of takeoff thrust has been upgraded over the years to 17 400 lb [77 395 N]. This increase in thrust has permitted significant increases in aircraft payload and range, shorter takeoff distances, and higher rates of climb to reduce noise in communities surrounding airports.

To improve the engine's fuel efficiency, Pratt & Whitney has developed a series of engine performance improvement kits for the JT8D that provide significant reductions in engine fuel consumption.

FIGURE 16-1   The JT8D engine. (*United Technologies, Pratt & Whitney Aircraft.*)

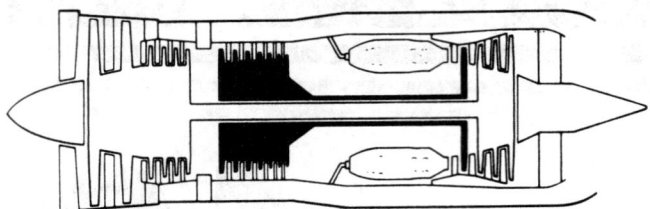

FIGURE 16-2 Arrangement of the Pratt & Whitney JT8D engine.

The Pratt & Whitney JT8D is a **twin-spool axial-flow gas-turbine engine** which is equipped with a secondary air duct that encases the full length of the engine. A diagram illustrating the general configuration of the JT8D engine is presented in Fig. 16-2.

Specifications

The specifications for a typical JT8D engine given below are for information only. As explained previously, the various models of a particular type of engine may have different specifications in some areas.

Length	120.0 in	304.80 cm
Diameter	43.0 in	109.22 cm
Frontal area	101 ft^2	9.38 m^2
Weight	3300 lb (approx.)	1497 kg
Takeoff thrust	14 500 lbt	64 496 N
Power-weight ratio	4.50 lbt/lb	43.08 N/kg
Fuel consumption	0.57 lb/lbt/h	58.1 g/N/h
Oil consumption	3.0 lb/h	1 360 g/h
Compressor ratio	16.9:1	
Bypass ratio	1.03:1	
Fan pressure ratio	1.91:1	
Air mass flow, fan	159 lb/s	72.1 kg/s
Air mass flow, core	163 lb/s	73.9 kg/s

General Configuration

The general configuration of the JT8D engine may be understood by examining Figs. 16-1 and 16-2. The bearing arrangement in the JT8D is similar to those of many turbofan engines. The no. 1 and no. 2 bearings support the low-pressure compressor (N_1) rotor, the no. 3 and no. 4 bearings support the high-pressure compressor (N_2) rotor, the no. $4\frac{1}{2}$ bearing is the intershaft bearing supporting the inner drive shaft within the hollow outer drive shaft, and the no. 5 and no. 6 bearings support the turbine.

A more complete understanding of the bearings and their functions can be obtained from Fig. 16-3.

The JT8D engine has six general sections. These are the **air inlet section**, the **compressor section**, the **combustion section**, the **turbine and exhaust section**, the **accessory drives**, and the **fan discharge section**. The different areas of the engine are shown in Fig. 16-4.

In modern gas-turbine engines it is common practice to identify the flanges by which the engine sections are bolted together. Figure 16-5 identifies the flanges for the JT8D engine.

Air Inlet Section

As shown in the drawing of Fig. 16-4, the **air inlet section** is the most forward section of the engine. This section forms the air inlet for the engine and houses the inlet guide vanes. The purpose of the vanes is to direct the incoming air at the proper angle to the first fan stage. The vane at the bottom of the air inlet is thicker than the others in order to accommodate engine tubing.

As indicated previously, the no. 1 bearing front support assembly is mounted in the center of the compressor inlet case. Behind the front support is the **no. 1 bearing rear support**. Mounted on the front of the inlet case, in the center, is the **front accessory drive support**. The air inlet, or fan inlet, case is shown in Fig. 16-6.

As can be seen in the drawing, the compressor inlet case contains 19 inlet guide vanes. Eighteen of these are identical, but the bottom vane is thicker to accommodate engine tubing, as mentioned previously. The vanes are brazed between the inner and outer shroud cases, which are made of titanium. The outer shroud case has a double wall; that is, a second ring is brazed outside the inner ring of the outer case to create an annular passage into which anti-icing air flows from the compressor. This air passes through the hollow vanes to the center of the assembly, where it discharges forward through the front of the inner shroud case.

The front accessory drive support is constructed of cast magnesium and includes a four-stud N_1 tachometer pad on the upper front face. A pressure oil passage in the support carries oil from the rear of the outer flange into the center, then rearward through the no. 1 bearing oil nozzle. An oil scavenge passage carries oil from a pump boss on the lower rear face of the support cavity back toward the outside of the support, then to another opening in the rear of the outer flange.

The N_1 tachometer-drive gear shaft and the scavenge pump gear shaft are driven by the **front accessory drive gear shaft** located in the front hub of the front-compressor rotor. The no. 1 bearing oil scavenge pump is mounted on the pump boss inside the front accessory drive support.

Fan and Compressor Sections

The **front-compressor section** and the **fan section** of the JT8D engine are both part of the same rotating assembly. The fan is actually the outer ends of the first two compressor blades. These two blades are enclosed by the front and rear fan cases. Figure 16-7 illustrates the arrangement of the JT8D engine with respect to position of units and airflow.

The N_1, or low-pressure, compressor partially compresses the air entering the engine before the air is delivered to the N_2, or high-pressure, compressor. While primary air flows through the core of the engine, secondary air from the fan is passing through the fan duct surrounding the engine. In this engine, the primary air and secondary air are almost equally divided by weight.

The front compressor is driven by the second, third, and fourth turbine stages through the inner or **front-compressor drive shaft**. Thus, the front compressor rotates independently of the rear compressor.

Location and Function of Bearings for JT8D Engine

Bearing Name	Location	Number	Type
Front compressor, front	Inlet case; front-compressor front hub	1	Roller
Front compressor, rear	Compressor case, front; front-compressor rotor rear hub	2	Duplex ball
Rear compressor, front	Compressor case, rear: main accessory drive gear	3	Ball
Rear compressor, rear	Diffuser case; rear-compressor rotor rear hub	4	Duplex ball
Turbine intershaft	In line with midpoint of combustion chamber case; outer race within rear-compressor drive-turbine shaft; inner race and rollers on front-compressor drive-turbine shaft	$4\frac{1}{2}$	Roller
Turbine front	In line with combustion-chamber-case rear flange; inner race on rear-compressor drive-turbine shaft	5	Roller
Turbine rear	Turbine exhaust case; front-compressor drive-turbine rear hub	6	Roller

FIGURE 16-3 Locations and functions of bearings for a JT8D engine.

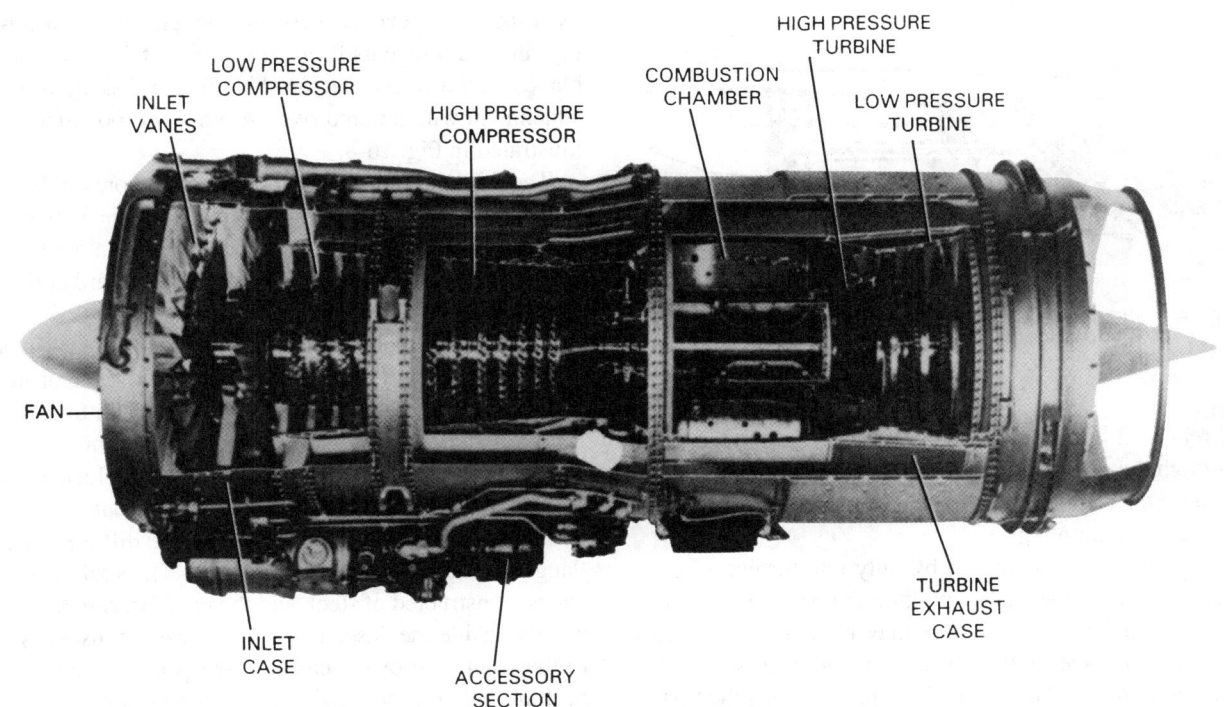

FIGURE 16-4 Sections of the JT8D engine. (*United Technologies, Pratt & Whitney Aircraft.*)

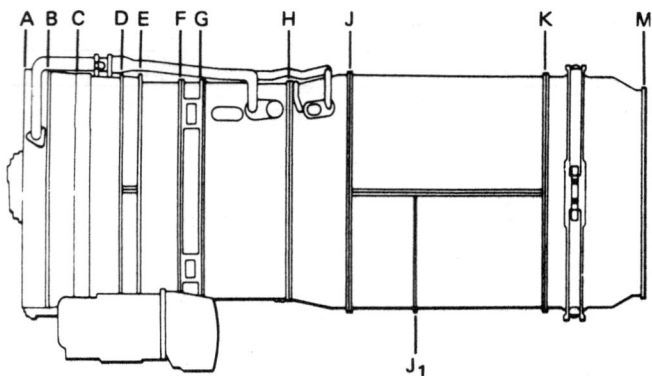

FIGURE 16-5 Flanges by which the JT8D is assembled.

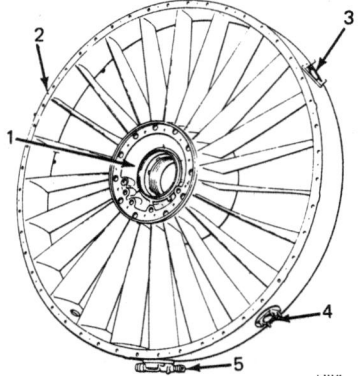

1. NO. 1 BEARING FRONT SUPPORT
2. FAN-INLET CASE
3. BOSS, INLET CASE
4. BOSS, INLET CASE
5. CONNECTOR ASSEMBLY

FIGURE 16-6 Air inlet case.

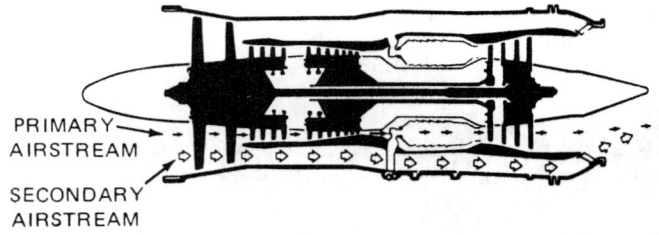

PRIMARY
AIRSTREAM

SECONDARY
AIRSTREAM

FIGURE 16-7 JT8D engine arrangement, showing airflow.

The rear (N_2) compressor is driven by the first-stage turbine through the **rear-compressor drive shaft**. This shaft is splined onto the rear-compressor rear hub and is retained by the turbine-shaft coupling.

The N_2 compressor is driven by only one turbine stage; however, it actually does more work on the primary air than does the N_1 compressor. This fact may raise a question as to why the first stage of the turbine can produce so much more power than the other stages. The answer is that the first-stage turbine is exposed to the hot gases from the engine at the time that the gases have their maximum energy in the forms of velocity, pressure, and heat. Moreover, the last three stages of the turbine not only must start the compression of the primary air but must also drive the fan section to accelerate and partially compress the secondary airflow to greatly increase engine thrust. In the JT8D turbofan engine, three turbine stages are used to drive the N_1 compressor and fan. Thus, it appears that the power developed by one extra turbine stage is necessary to drive the fan section.

The front-compressor rotor and stator assembly consists of the sixth-stage front-compressor rotor, the front-compressor front and rear cases, and the vanes and shrouds for stages 1 through 5. The sixth-stage vanes are in the compressor-intermediate case. The front-compressor rotor has a front hub which serves as the first-stage disk, a rear hub which serves as the fourth-stage disk, four rotor disks, six stages of blades secured in the hubs and disks, five rotor-disk spacers, and two sets of tie rods.

Each of the rotor-disk spacers has two knife-edge air seals on its outer diameter. These knife edges rotate just inside matching seal rings on the inside diameter of the vane and shroud assemblies. The knife-edge air seal of the second-to-third-stage spacer is incorporated in the rear flange of the spacer and matches the ring inside the second-stage vanes. As explained previously, the purpose of the knife-edge seals is to prevent loss of air between stages. Remember that there is a distinct increase in air pressure between successive stages and that the higher-pressure air will flow around the ends of the stator vanes and compressor blades through any available space to the preceding stage. The purpose of the air seals is to reduce the leak space to a minimum.

The first-stage compressor blades include the fan section and are dovetailed into matching grooves in the front-hub rim. They are retained at the leading edge of the blade root by a tab that prevents rearward movement and a positioning ring that prevents forward movement. The second-stage blades, and also the fan, are held in the disk by a pin-joint attachment with a flared rivet. A pin-joint root attachment is illustrated in Fig. 16-8.

The third through the sixth stages of compressor blades are fastened to the disks by dovetail-root sections which fit into broached slots in the disk rim. Tab locks fit in the bottom of the blade root and disk slots and are bent inward at the tab to effect blade retention.

The front hub and the second-stage disk are made of titanium. The third-stage disk is steel, and the rear hub and fifth- and sixth-stage disks are titanium. The first- through the sixth-stage compressor blades are titanium.

The **compressor-intermediate case** is located to the rear of the front compressor. It forms the outer wall of the inner engine from the fan discharge to the diffuser-case front flange, defined as flange H (Fig. 16-5). The sixth-stage stator vanes, constructed of steel, and the no. 3 bearing housing are welded inside the case. The no. 2 bearing housing is bolted to the front face of the case. A steel support and a support bushing below it are positioned at the bottom center of the case to accommodate the **main accessory drive gear shaft**

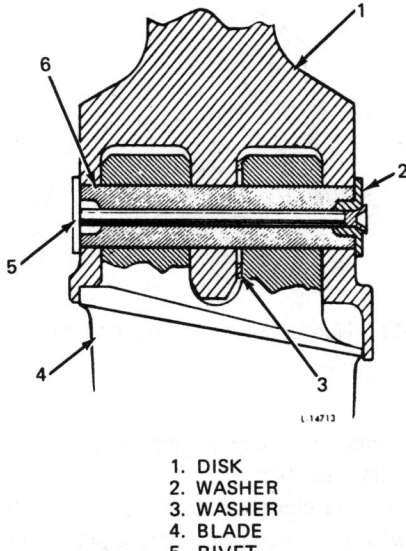

FIGURE 16-8 A pin-joint attachment. (*United Technologies, Pratt & Whitney Aircraft.*)

1. DISK
2. WASHER
3. WASHER
4. BLADE
5. RIVET
6. PIN

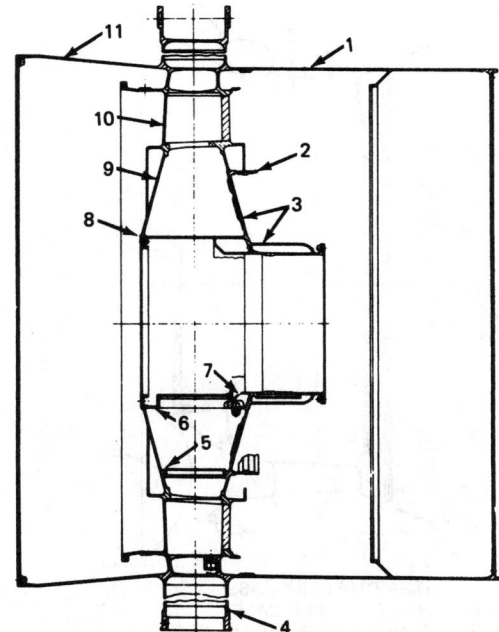

1. Fan-discharge-rear-compressor inner duct
2. Rear-compressor-rotor front-airseal ring
3. No. 3 bearing–housing support
4. Compressor-intermediate case
5. Gearbox-drive-bearing-housing lower support
6. Gearbox-drive-bearing-housing upper support
7. Seal-bleed manifold segment
8. Mounting flange for no. 2 bearing housing
9. No. 2 bearing-housing support
10. Sixth-stage vane
11. Fan-discharge-front-compressor inner duct

FIGURE 16-9 Cross section of the compressor-intermediate case and the compressor-intermediate fan case.

bearing housing. Inside the bearing housing is the **main accessory drive bevel-gear shaft** with a roller bearing at the top and a ball bearing at the bottom.

The **compressor-intermediate fan case** and the compressor-intermediate case are shown in Fig. 16-9. It will be noted that the sections labeled **fan-discharge rear-compressor inner duct** and **fan-discharge front-compressor inner duct** form the inner lining of the fan duct. The compressor-intermediate fan case incorporates streamlined struts between the intermediate case and the outer diameter of the engine. The larger six-o'clock-position strut accommodates the accessory gearbox main drive shaft.

The rear compressor utilizes a rotor having seven stages of disks and blades. These are separated by disk spacers and six steel stator-vane stages. The blade stages are numbered 7 through 13 from front to rear. The vane stages, behind their blade stages, are numbered correspondingly, 7 through 12, in the rotor and stator assembly. The thirteenth stator stage and the compressor exit vanes are located in the front end of the diffuser case.

The rear compressor (N_2) is held together by 12 tie rods. These hold the disks, spacers, and front and rear hubs together axially. A triple knife-edge air seal is secured to the front of the seventh-stage disk, and a four-edge air seal is integral with the rear of the thirteenth-stage disk. Blade attachment to the disks is accomplished by a dovetailed lock at the blade root, with the exception of the seventh stage, which uses a pin-joint attachment.

The seventh-, eighth-, and ninth-stage blades of the rear compressor rotor are titanium. The tenth- through thirteenth-stage blades are steel. Steel is used for these parts because of the high temperature of the compressed air at this point. The rear-compressor rotor disks are steel, except for the thirteenth-stage disk, which is made of nickel alloy. The thirteenth-stage disk incorporates an air-sealing configuration

on the rear with four knife edges. The knife edges match the steel thirteenth-stage air-sealing ring positioned inside the diffuser case.

Diffuser Section

The function of the diffuser section has been explained previously. In the JT8D engine, the air passes through the last row of rear-compressor blades at a high velocity. The motion is both rearward and circular in pattern around the engine. Two rows of radial, straightening exit-guide vanes at the entrance to the **diffuser case** slow the circular whirl pattern and convert the whirl-velocity energy to pressure energy. After passing through these straightening vanes, the air still has a strong rearward velocity. This velocity is so high that it would be nearly impossible to maintain a flame in its airstream. A gradually increasing cross section of the air passage decreases the velocity of the airflow and at the same time converts the velocity energy to pressure energy.

The forward part of the diffuser case houses the rearmost section of the rear compressor. The exit stator assembly of the compressor is bolted to flanges in the front openings of the case. This unit contains an inner shroud, outer shroud, and small vanes brazed in place.

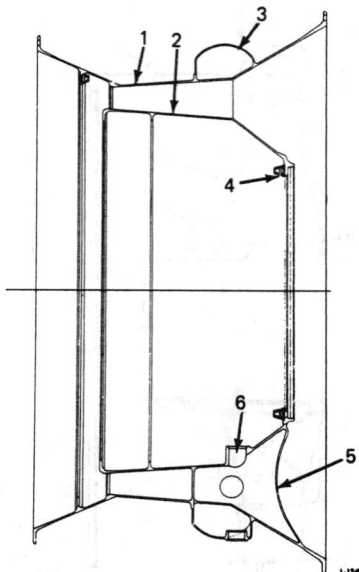

1. OUTER DIFFUSER CASE
2. INNER DIFFUSER CASE
3. DIFFUSER-CASE AIR MANIFOLD
4. NO. 4 BEARING-SUPPORT BOLT CIRCLE
5. DIFFUSER-CASE STRUT
6. NO. 4 BEARING AIR-BLEED-TUBE OPENING

FIGURE 16-10   Diffuser case.

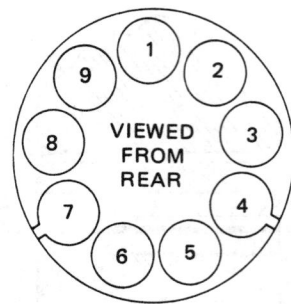

FIGURE 16-11   Numbering of combustion chambers.

In the divergent section of the diffuser case are nine hollow struts having small circular openings on either side which supply compressor discharge air to a manifold around the case. The manifold provides the discharge air for anti-icing and airframe use through two ports on its perimeter. The construction of the diffuser case is illustrated in Fig. 16-10.

Nine fuel nozzle support mounting pads are located radially near the rear of the case between the hollow struts. Behind the mounting pads are mounting lugs for the front of the individual combustion chambers. The pressure-sensing boss for P_{t4} pressure is located near the two o'clock position on the right-side outer surface of the diffuser case.

The no. 4 bearing compartment is located in the center of the diffuser case near the rear, as indicated in Fig. 16-10. Heat shields are bolted and lockwired in front of the bearing compartment to minimize the temperature inside the compartment. In addition, a tubing system brings eighth-stage discharge air to the annulus between the second and third labyrinth-seal units and bleeds air from the annulus between the first and second labyrinth units to the fan discharge path. The tubes, secured to the openings in the no. 4 bearing air seal-ring assembly, hold down the bearing compartment temperature by bleeding hot air before it can reach the compartment.

The scavenge pump for the no. 4 and no. 5 bearings is located inside the diffuser case in the rear portion of the no. 4 bearing housing. It has two stages and is driven by a gear mounted on the rear-compressor rear hub.

Combustion Section

The combustion section of the JT8D engine is of the cannular type, as explained previously. This means that individual combustion chambers are arranged around the engine inside a single annular chamber.

The combustion chambers in the JT8D engine are numbered clockwise as viewed from the rear of the engine. The no. 1 chamber is at the top of the engine, as shown in Fig. 16-11. In Fig. 16-12, the individual combustion chambers are located between the **fuel nozzles** and the **combustion-chamber rear support**. They are held in place by means of a mounting pin at the front. All the chambers have one male and one female flame tube, which interconnect to provide rapid flame propagation when the engine is first started. Chambers 4 and 7 each have a spark-igniter opening. Nine interconnector tubes connect the male and female flame tubes of the chambers.

The annular combustion chamber in which the individual chambers are located is formed by the concentric combustion-chamber outer case and combustion-chamber inner case. The **combustion-chamber inner case** is secured to the diffuser case inner rear flange and to the outer flange of the no. 5 bearing housing. It forms the inner wall of the combustion chamber and serves to position the no. 5 bearing through the bearing housing. The inner and outer cases are major structural members of the engine.

Inside the rear section of the combustion area is the welded **combustion-chamber rear support**. This is illustrated in Fig. 16-13. This unit is a large circular plate with nine openings around a simple central opening. The nine smaller openings support the rear ends of the individual combustion chambers. The rear outer flange is equipped with bolts attached to the support by stops and rivets.

Behind the combustion-chamber rear support are the **combustion-chamber inner and outer outlet ducts**, which form the passage through which the hot gases are carried to the first-stage turbine nozzle. Included are air deflector ducts which divide the cooling air in both the inner and outer ducts into dual streams. The hot gases pass through nine support openings as they are guided to the turbine nozzle.

Inside the combustion-chamber inner case are the **turbine-shaft heat shields**. These are bolted to the rear of the no. 4 bearing support. Also, inside the inner case are the oil scavenge pump heat-shield assembly, the no. $4\frac{1}{2}$ bearing heat-shield assembly, the no. 5 bearing oil-nozzle assembly, the oil scavenge pump shield, and the no. 5 bearing oil scavenge tube. These are illustrated in Fig. 16-14.

The **combustion-chamber outer case**, which is attached by bolts to the rear flange of the diffuser case and the front

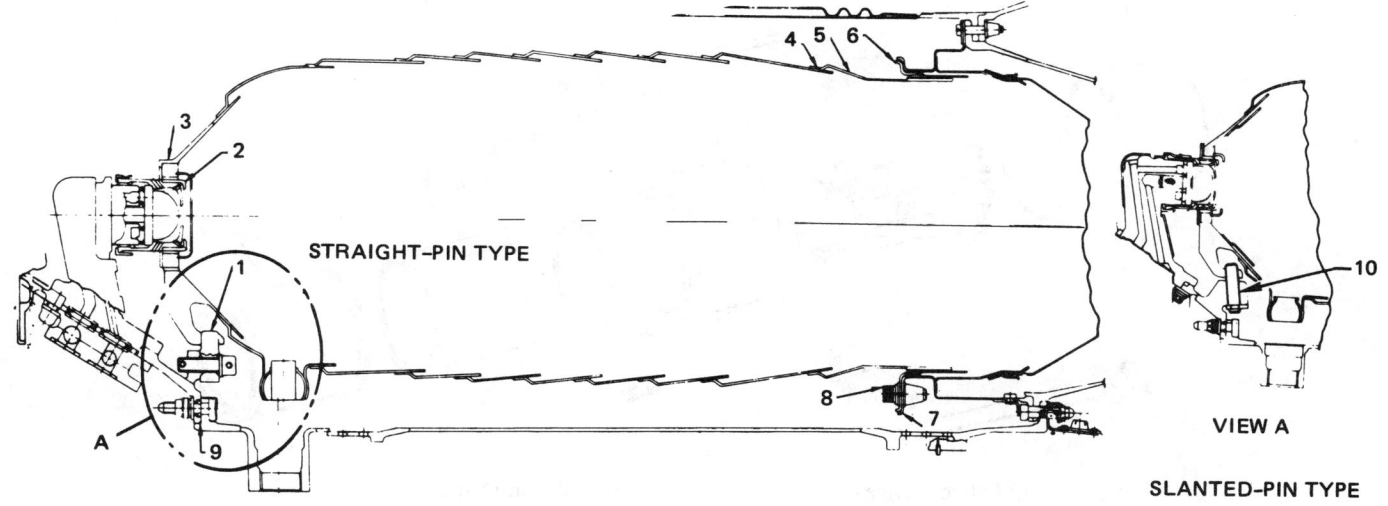

STRAIGHT-PIN TYPE

SLANTED-PIN TYPE

VIEW A

1. Mounting pin
2. Fuel-nozzle heatshield
3. Fuel-nozzle stator
4. Combustion-chamber detail rear liner
5. Diverging portion of rear liner
6. Combustion-chamber rear positioning guide
7. Combustion-chamber rear support
8. Combustion-chamber support bolt (two required at each location)
9. Diffuser-case rear flange
10. Mounting pin

FIGURE 16-12   Cross section of a combustion chamber. (*United Technologies, Pratt & Whitney Aircraft.*)

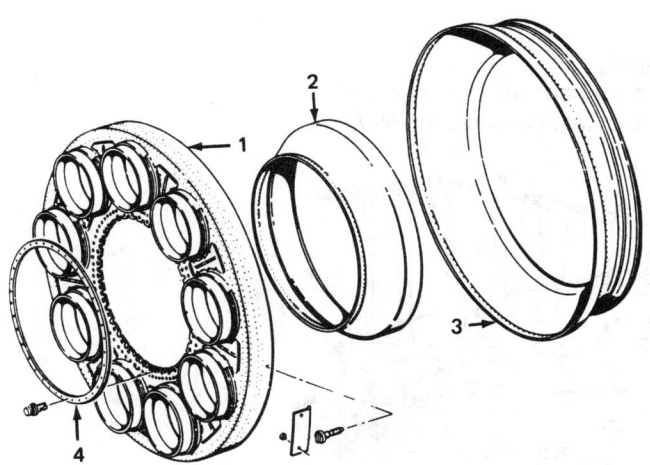

1. **COMBUSTION-CHAMBER REAR SUPPORT**
2. **COMBUSTION-CHAMBER INNER-OUTLET DUCT**
3. **COMBUSTION-CHAMBER OUTER-OUTLET DUCT**
4. **REAR-SUPPORT REINFORCING PLATE**

FIGURE 16-13   Combustion-chamber rear support. (*United Technologies, Pratt & Whitney Aircraft.*)

flange of the turbine nozzle case, encloses the combustion chamber. It forms the outer wall of the main engine and the inner wall of the fan annular duct.

Fuel drain valves are located on the bottom centerline of the outer combustion case, one at the front and one at the rear. A fuel drain manifold carries drain fuel outside the outer fan duct. Both flanges of the combustion-chamber inner duct turn inward, and the front flange is scalloped. The case is constructed of corrosion- and heat-resistant steel

with nickel-cadmium and baked-on aluminum enamel at the flanges to resist corrosion.

Turbine and Exhaust Section

The turbine section includes the turbine front case, the turbine rear case, an inner case and seal, four stages of turbine nozzle vanes, the two turbine rotors with four stages, and the coaxial drive shafts.

The **turbine front case** is a comparatively short section attached to the rear of the combustion-chamber outer case. It is of decreasing diameter and extends to the first-stage turbine. The first-stage turbine vanes are mounted in the rear of the turbine front case. Immediately to the rear of the nozzle vanes is the first-stage turbine. The case is constructed of corrosion- and heat-resistant steel.

Attached to the rear flange of the turbine front case is the **turbine rear case**. This section is also constructed of steel and increases in diameter from front to rear in order to accommodate the second, third, and fourth turbine vanes. The rear flange of this case is bolted with the **fourth-stage turbine outer seal ring** to the front flange of the **turbine exhaust case**.

The arrangement of the turbine and exhaust section can be made clear by a study of Fig. 16-15. The turbine nozzle vanes are mounted in the inside diameter of the turbine front case and turbine rear case. The blade shroud seal rings can be observed between the nozzle vanes. Remember that the turbine blade tips form a complete ring or shroud to reduce gas losses. The outer rim of the shroud forms a knife-edge seal which mates with the blade shroud seal rings.

Attached to the outer flange of the no. 5 bearing housing and extending rearward is the **turbine nozzle inner case**

Large Turbofan Engines   **419**

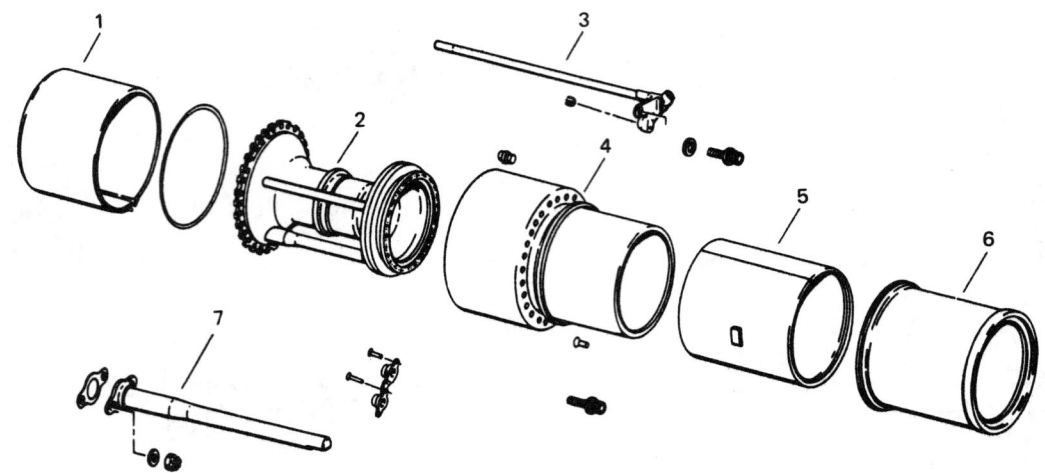

1. Oil-scavenge-
   pump heatshield assembly
2. No. $4\frac{1}{2}$ bearing
   heatshield assembly
3. No. 5 bearing oil-nozzle
   assembly
4. Oil-scavenge-pump
   shield
5. Turbine-shafts-
   bearing heatshield asssembly
6. Turbine-shafts-
   bearing heatshield-shield
   assembly
7. No. 5 bearing
   oil-scavenge tube

FIGURE 16-14 Heat shields for turbine shafts and bearings. (*United Technologies, Pratt & Whitney Aircraft.*)

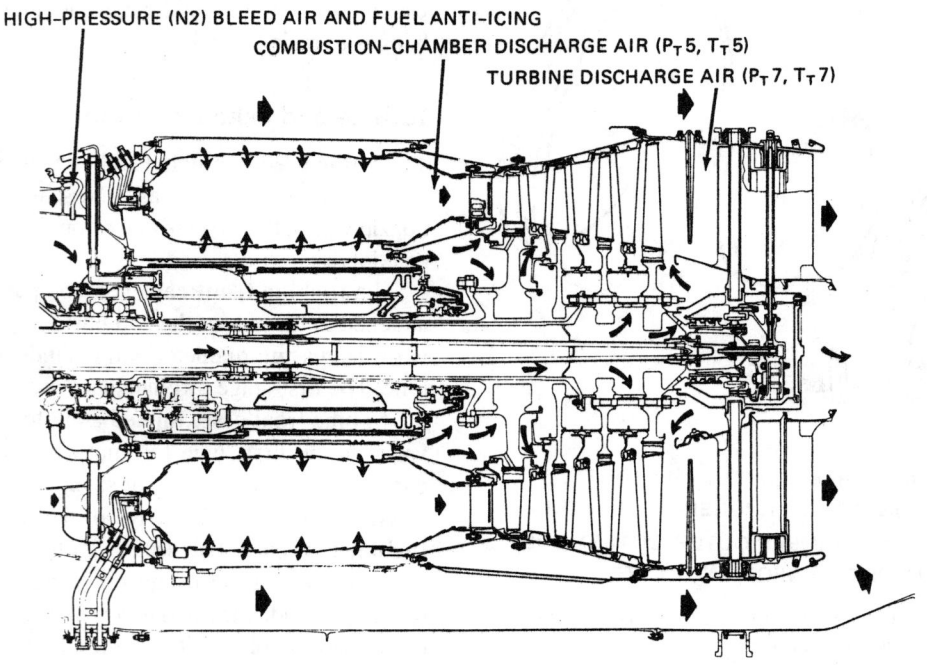

HIGH-PRESSURE (N2) BLEED AIR AND FUEL ANTI-ICING
COMBUSTION-CHAMBER DISCHARGE AIR (P_T5, T_T5)
TURBINE DISCHARGE AIR (P_T7, T_T7)

FIGURE 16-15 Combustion and turbine sections, showing airflow and gas flow. (*United Technologies, Pratt & Whitney Aircraft.*)

and seal assembly. Riveted to the rear inner flange of this assembly is the **turbine rotor inner first-stage air seal**, which matches the integral shoulders on the front of the first-stage turbine disk. A groove in the rear outer flange of the turbine nozzle inner case and seal assembly accommodates the inner rear shroud of the first-stage turbine nozzle vanes.

Segmented multiple turbine vane shroud nuts are riveted to the forward outer flange.

The first-stage **turbine nozzle vanes** are coated and air-cooled. In later models of the JT8D engine, the cooling air is passed inside the vanes and out through air-exit holes in the airfoil trailing edge on the concave side.

Figure 16-15 shows the installation of the second-, third-, and fourth-stage turbine nozzle vanes. All the nozzle vanes are attached to the inside of the turbine rear case.

As explained previously, the rear, or high-pressure, compressor (N_2) is driven by the first-stage turbine. In early engines, the turbine disk and the drive shaft were one unit, whereas in later engines the first-stage turbine is a separate unit bolted to the shaft flange, as shown in Fig. 16-16. As seen in the drawing, the turbine blades are attached to the turbine disk by means of fir-tree slots in the rim of the disk. They are held in the slots by means of rivets.

The front-compressor drive-turbine rotor includes the front-compressor drive-turbine shaft, three turbine stages, and the spacers and air seals between the disks. Twelve tie rods secure the disks and spacers to each other and to the rear flange of the rotor shaft. The turbine bearing pressure and scavenge oil tubes assembly is located inside the shaft.

The blades for the second, third, and fourth turbines are secured in the disks by fir-tree slots and rivets. The numbers of blades in the three disks are 88, 92, and 74, respectively, from the second stage to the fourth stage. Lugs on the second- and third-stage turbine rotor inner air seals mate with slots on the second- and third-stage disks to prevent rotation between the disks and seals.

The turbine disks for the front-compressor drive turbine can be assembled in only one position because one of the 12 tie-rod holes in each disk is offset. This ensures proper balance of the complete assembly. A drawing of the turbine assembly is shown in Fig. 16-17.

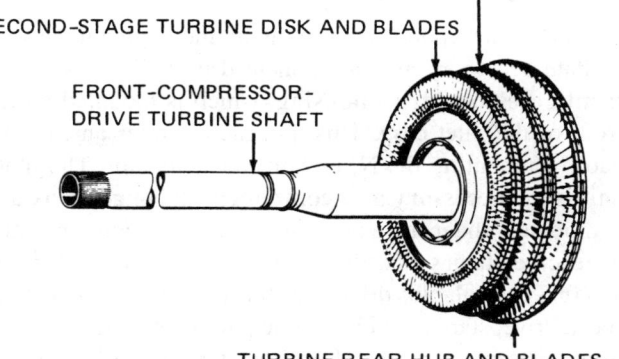

FIGURE 16-17 Front-compressor drive-turbine rotor assembly. (*United Technologies, Pratt & Whitney Aircraft.*)

In later engines of the JT8D type, the front-compressor drive-turbine rotor assembly is designed as a unit or module with the stator assembly. The complete module, which includes the turbine case, nozzle vanes, and turbines, is called the **front-compressor drive-turbine rotor and stator assembly**. This makes it possible to remove and replace the entire assembly without disturbing the other components of the engine. The change can be made with the engine installed in the aircraft.

The **turbine exhaust case**, constructed of welded steel, is the most rearward section of the basic inner engine. It is bolted to the rear flange of the turbine case and decreases in diameter from front to rear in order to increase the velocity of the exhaust gases. The turbine exhaust case is the principal support for the no. 6 bearing housing, the thermocouples for temperature sensing, and the pressure-sensing manifold. The **turbine exhaust case and fairing assembly** with the no. 6 bearing support rods acts as a structural member to support the no. 6 bearing and associated units, as shown in Fig. 16-18.

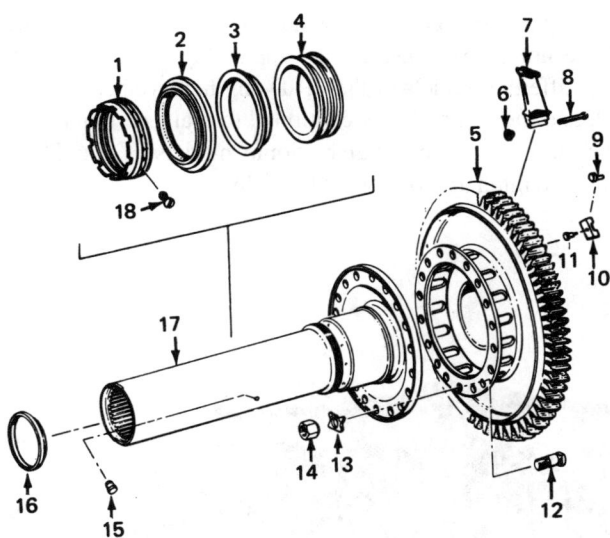

1. No. 5 bearing inner-race retaining nut
2. Seal seat
3. Bearing spacer
4. Labyrinth seal
5. First-stage turbine disk
6. Washer
7. First-stage turbine blade
8. Rivet
9. Counterweight
10. Counterweight
11. Rivet
12. Tiebolt
13. Keywasher
14. Tie-rod nut
15. Positioning plug
16. Turbine-shaft spacer
17. Rear-compressor-drive turbine shaft
18. Retaining screw

FIGURE 16-16 Rear-compressor drive-turbine rotor assembly. (*United Technologies, Pratt & Whitney Aircraft.*)

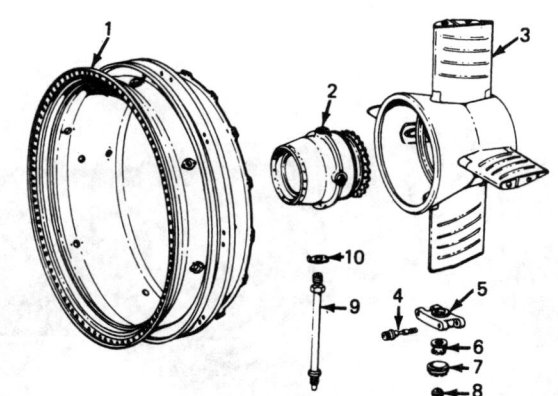

1. Turbine-exhaust case
2. No. 6 bearing housing
3. Turbine-exhaust duct and fairing assembly
4. No. 6 bearing-support rod boss
5. Bolt
6. Locking nut
7. Externally threaded ring
8. Plug
9. No. 6 bearing strut
10. Keywasher

FIGURE 16-18 Turbine exhaust case and fairing assembly. (*United Technologies, Pratt & Whitney Aircraft.*)

Accessory and Component Drives

The JT8D incorporates two units on the engine to accommodate accessory and component drives. The first is the **front accessory drive housing**, which is mounted on the front of the inlet case. This unit incorporates an external pad for mounting the N_1 tachometer generator. The other unit is the **accessory and component drive gearbox** and is mounted under the engine at the fan-discharge intermediate case. It has one drive pad for both the fuel pump and fuel control, plus drive pads for the alternator, constant-speed drive, starter, and hydraulic pump. Other units driven from this section are the N_2 tachometer, oil pressure pump, and oil scavenge pumps. An oil pressure relief valve, oil strainer, and bypass valve are included in this section.

Engine Systems

The engine systems for the JT8D engine are similar to those described previously in this text. The technician is advised to consult the overhaul and maintenance manuals for specific instructions regarding a particular system.

The Pratt & Whitney JT8D Series 200 Engine

The manufacturers and operators of all modern gas-turbine engines continuously seek to improve the performance, reliability, and cost-effectiveness of the engines. This is particularly true for the Pratt & Whitney JT8D turbofan, which is the most widely used engine in existence. The most recent version of this engine is the JT8D-200, a model of which is shown in Fig. 16-19. The original JT8D engine produced 14 000 lb [62 272 N] of thrust, and this performance was gradually increased to 17 400 lb [77 395 N]. The JT8D-200 engines produce from 19 250 lb [85 624 N] to 20 850 lb [92 741 N] of thrust.

The primary considerations in the improvement of gas-turbine engines have been reduced fuel consumption, increased reliability, and reduced noise. In the JT8D-200 engines, fuel consumption has been reduced through use of improved materials; more effective aerodynamic design of fan, compressor, and turbine blades and vanes; better cooling techniques; more effective gas seals; and other refinements. Noise reduction has been accomplished through improved fan design, better sound-absorbing materials, and the use of a multiple-tube exhaust discharge. The design of the exhaust discharge system can be seen in the cutaway view in Fig. 16-19.

The Pratt & Whitney JT9D Turbofan Engine

The Pratt & Whitney JT9D turbofan engine, shown in Fig. 16-20, is designed chiefly to power wide-body aircraft such as the Boeing 747 and the Douglas DC-10. It is designated a high-bypass engine because the fan section bypasses more than five times as much air as passes through the core, or main part, of the engine.

The JT9D is a twin-spool engine, driving the fan and low-pressure compressor by means of the four most rearward turbine stages through the center coaxial shaft. The high-pressure compressor is driven by the two forward-most turbine stages. The combustion chamber is of the full annular type, with 20 fuel nozzles at the front of the assembly.

Performance

The specifications given below are for the JT9D-3A engine and are not to be construed as applicable for all models. Many different models of the JT9D engine have been developed, and the technician is cautioned to make certain of the "dash-number" of the engine. Some models of the JT9D engine are rated at more than 50 000 lbt.

FIGURE 16-19 The JT8D series 200 engine. (*United Technologies, Pratt & Whitney Aircraft.*)

FIGURE 16-20 Pratt & Whitney JT9D turbofan engine. (*United Technologies, Pratt & Whitney Aircraft.*)

Length	154.2 in	391.67 cm
Diameter	95.6 in	242.82 cm
Frontal area	50 ft^2	4.6 m^2
Weight	8470 lb	3 842 kg
Takeoff thrust (wet)	45 000 lb	200160 N
Power-weight ratio	5.12 lbt/lb	52.09 N/kg
Fuel consumption	0.36 lb/lbt/h	36.7 g/N/h
Oil consumption	3.0 lb/h	1 360 g/h
Compressor ratio	22:1	
Bypass ratio	5.17:1	
Fan pressure ratio	1.51:1	
Air mass flow, fan	1271 lb/s	576 kg/s
Air mass flow, core	246 lb/s	111 kg/s

General Configuration

The basic design of the JT9D turbofan engine is depicted in Fig. 16-21, a simplified drawing that shows the principal sections of the engine. The engine comprises five rotating assemblies, or modules, that are independently balanced so that they may be used interchangeably with other engines of the same model. The five rotating modules are the fan assembly, low-pressure compressor (N_1) assembly, high-pressure compressor (N_2) assembly, high-compressor drive-turbine

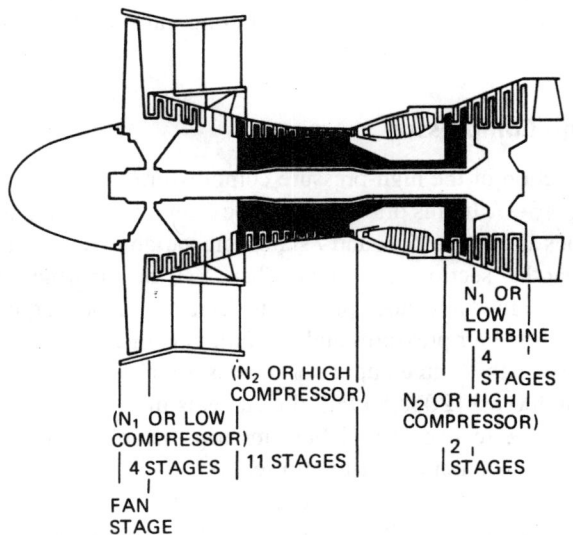

FIGURE 16-21 Basic arrangement of the JT9D engine.

assembly, and low-compressor drive-turbine assembly. When the engine is assembled, there are only two principal rotating assemblies. These may be termed the **fan and low-pressure compressor and drive turbine** and the **high-pressure compressor and drive turbine**.

The flanges by which the JT9D engine is assembled are shown in Fig. 16-22. This drawing provides views of mounting lugs, handling pads, accessory drives, oil tank, air bleeds, fuel lines, and other features.

A cross-sectional drawing of the JT9D engine is shown in Fig. 16-23. This drawing shows the internal construction of the engine. The low-pressure compressor consists of the fan rotor plus three additional stages mounted on a single disk which is attached to the drive shaft just forward of the no. 1 bearing.

The high-pressure inlet guide vanes and the no. 5, 6, and 7 stator vanes are variable to provide for adjustment of airflow direction in accordance with air velocity and pressure. The no. 2 bearing supports the forward end of the high-compressor shaft. As can be seen in the drawings, the section of the high-compressor shaft assembly is comprised of front and rear conical hubs between which are mounted the disks and spacers that support the compressor blades. The assembly thus has the appearance of a drum with conical ends.

As shown in the drawing, the high compressor is driven by the no. 1 and no. 2 turbine stages. The no. 3 bearing supports the assembly at the rear. The four rearmost stages of the turbine drive the low compressor and the fan. This assembly is supported at the rear by the no. 4 bearing.

Fan and Low-Compressor Assembly

As mentioned previously, the fan stage forms the first stage of the low compressor and also provides the high bypass of fan air. The fan has 46 blades balanced in pairs; they are thus replaceable in balanced sets. The blades have dovetail roots and are retained axially in the disk by means of shear locks.

The fan case and outlet duct are lined with sound-absorbing material to reduce noise. Additional noise-reduction factors are the relatively low tip speed of the fan blades and the fact that there are no inlet guide vanes ahead of the fan.

The fan-disk hub is splined to a coupling which is, in turn, splined to the low-turbine shaft. The fan disk is positioned axially by means of a spacer between the fan and the coupling. Axial positioning of the fan is an important factor in adjusting fan blade-tip clearance. The fan case is tapered so that the axial movement of the fan will vary the tip clearance.

The construction of the fan and fan hub (together with the low-pressure compressor hub and coupling) provides for easy removal of the fan, low turbine, and low compressor. The complete assembly is held in place by means of a large coupling nut at the forward end of the turbine shaft.

Each fan blade has a rubber air-sealing strip cemented to the butt end to reduce leakage of air from the airfoil section of the blade around the root end.

An enlarged drawing of a cross section of the low-pressure compressor is shown in Fig. 16-24, which shows the no. 2, 3, and 4 compressor blades between the stationary vanes. It also

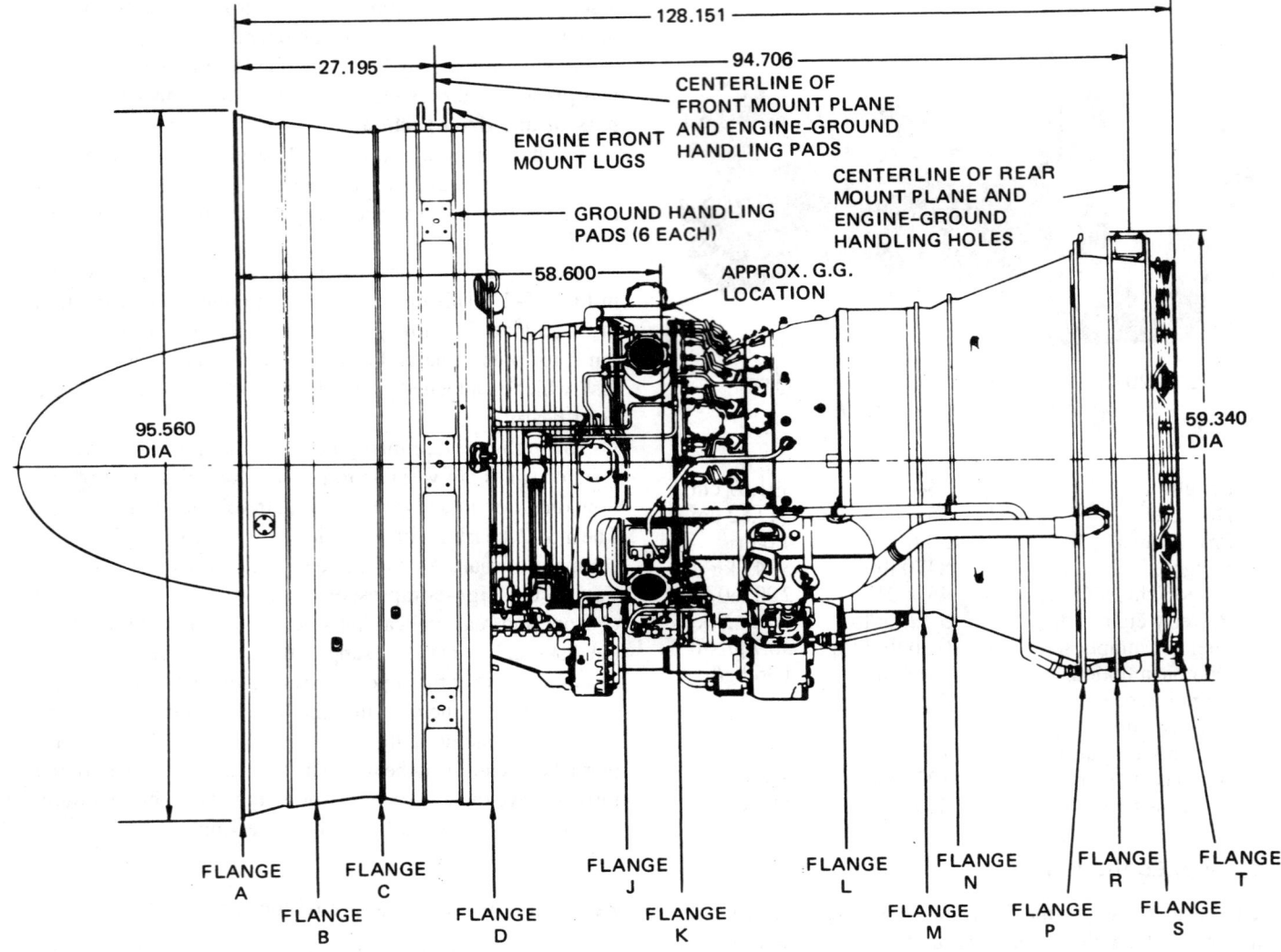

FIGURE 16-22 Assembly flanges for the JT9D engine. (*United Technologies, Pratt & Whitney Aircraft.*)

shows how the compressor blades are attached to the compressor hub. It must be remembered that the no. 1 compressor stage is the fan, which is not shown in the drawing.

A drawing of the anti-icing system for the first-stage stator vanes is provided in Fig. 16-25, which shows the hot-air duct that brings air from the ninth stage of the compressor forward to the first-stage vane. It can be seen that the air flows into the hollow vane and out through holes near the trailing edge of each vane.

Intermediate Case

The **intermediate case** is the section of the engine between the low compressor and the high compressor. It serves a major structural function and supports the no. 1 and no. 2 bearings. The **tower shaft**, which provides for accessory drives, engages the bevel ring gear, which is mounted on the forward end of the high-compressor front hub. The shaft passes through the lower strut of the case. This can be seen in Fig. 16-23.

A cross section of the bearing compartments in the intermediate case is shown in Fig. 16-26. This drawing reveals some of the details, including various types of oil and air seals.

High Compressor

A section of the high-pressure compressor (N_2) is shown in Fig. 16-27. In this drawing it can be seen that the inlet guide vanes and the no. 5, 6, and 7 stators are variable. As explained in another section of this text (Chap. 12), the **variable vanes** are designed to change pitch as necessary to avoid surge and stall as the air pressures and velocities change.

Since the high compressor increases air pressure to more than 300 psi [2 068.5 kPa], the air seals in the section must be very effective. Abradable rub strips are placed between rotors and stators to ensure tip seal.

Stages 8 through 11 are enclosed by two outer cases which serve as manifolds for bleed air from stages 7 through 9. Stages 12 through 15 are enclosed by the forward section

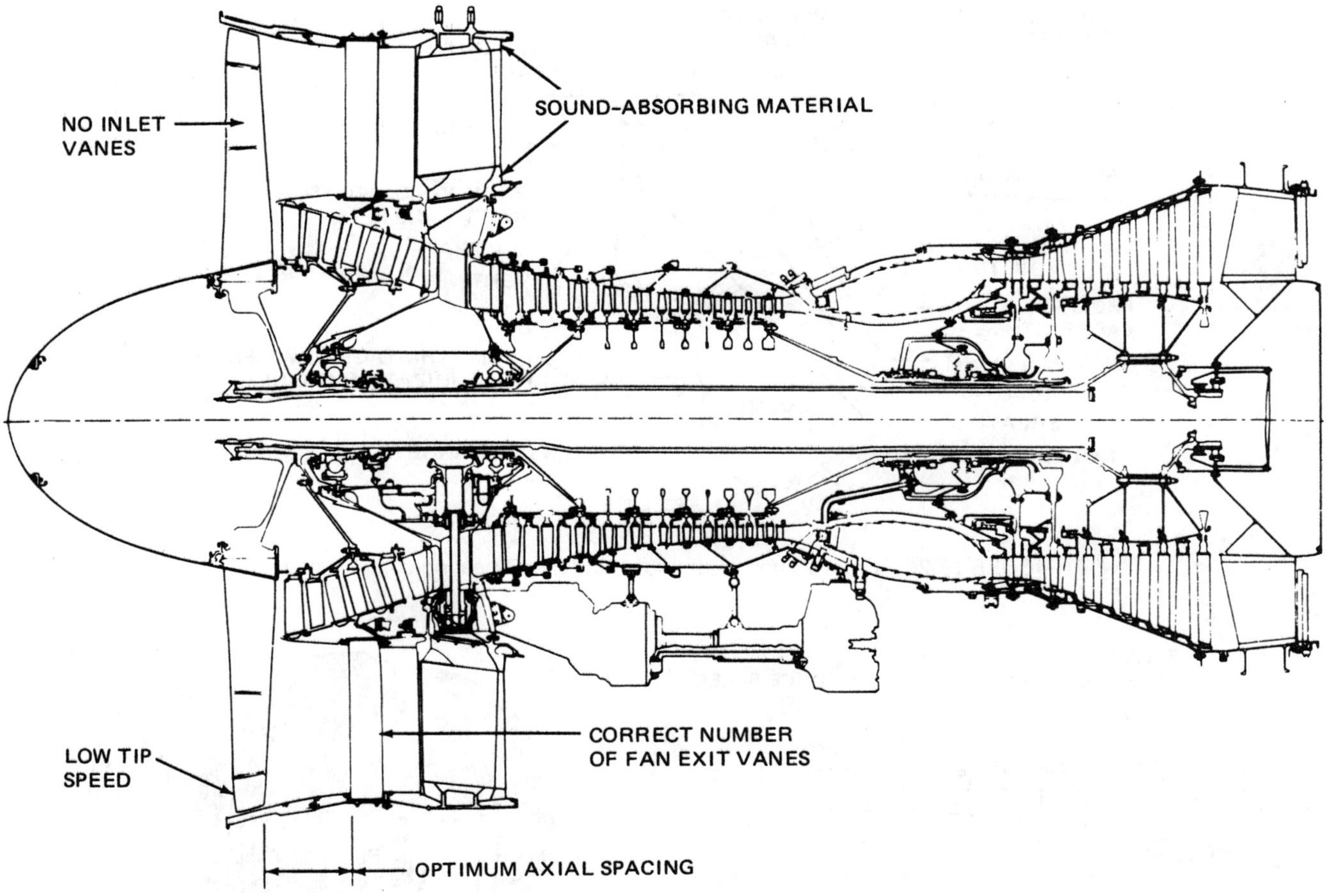

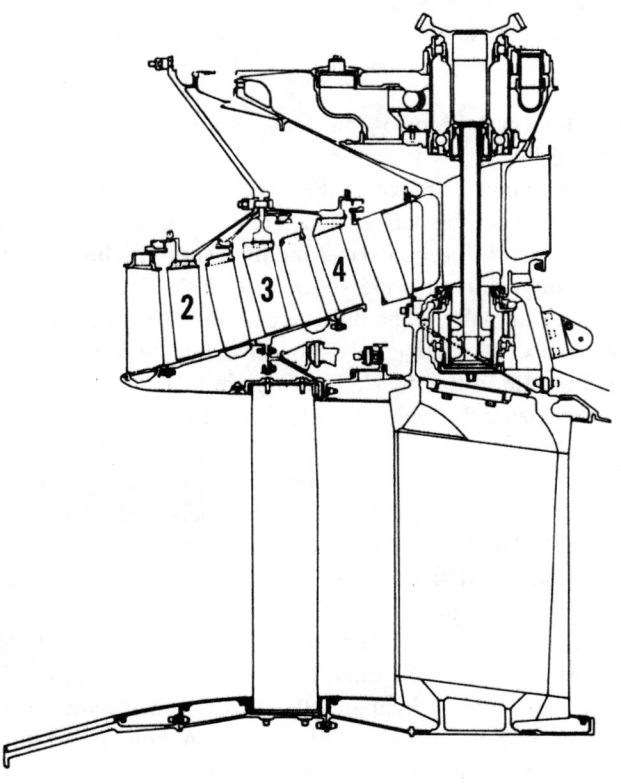

FIGURE 16-23 Cutaway drawing of the JT9D engine.

NO INLET VANES

SOUND-ABSORBING MATERIAL

LOW TIP SPEED

CORRECT NUMBER OF FAN EXIT VANES

OPTIMUM AXIAL SPACING

FIGURE 16-24 Cross section of the low-pressure compressor.

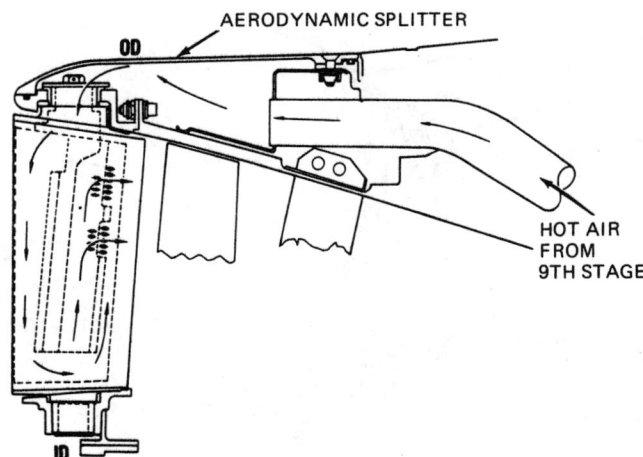

FIGURE 16-25 Anti-icing system for the first-stage stator.

AERODYNAMIC SPLITTER

OD

HOT AIR FROM 9TH STAGE

ID

of the **diffuser case**. This forms a manifold around the rear stages of the compressor for fifteenth-stage air bleed.

The arrangement for adjustment of the variable vanes is shown in Fig. 16-28. Each stage of vanes is operated through a unison ring which makes all vanes in the stage rotate the same amount. All four unison rings are connected to a bell crank rotated by means of a hydraulic cylinder. The hydraulic

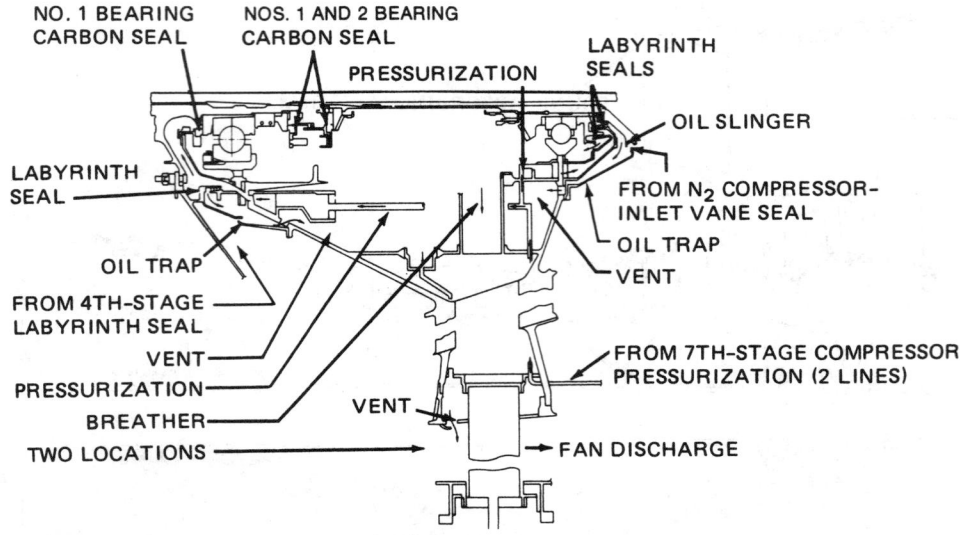

FIGURE 16-26 Cross section of the no. 1 and. no. 2 bearing compartment.

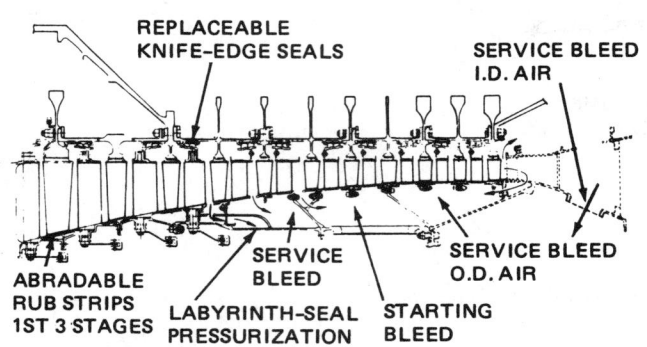

FIGURE 16-27 Section of the high-pressure compressor.

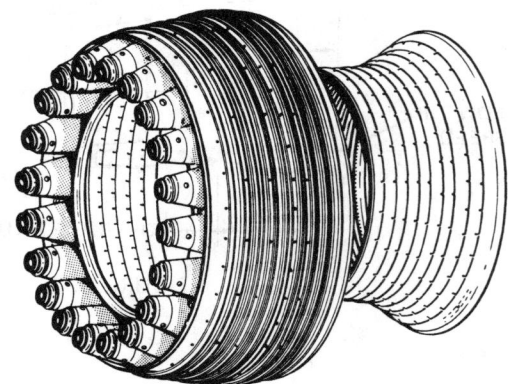

FIGURE 16-29 Combustion chamber with inner and outer liners separated. (*United Technologies, Pratt & Whitney Aircraft.*)

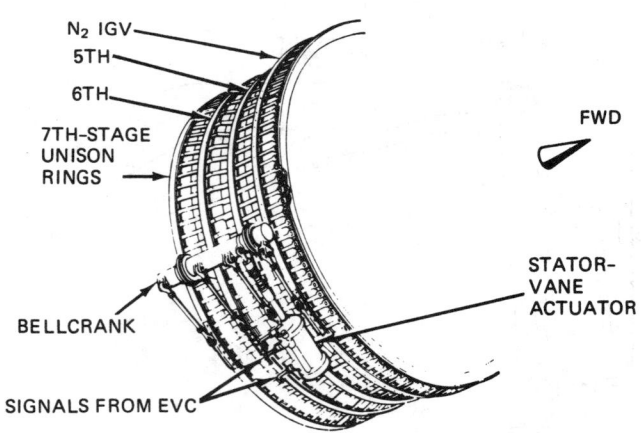

FIGURE 16-28 Variable vanes arrangement. (*United Technologies, Pratt & Whitney Aircraft.*)

cylinder receives operating energy in the form of hydraulic pressure from the engine vane control (EVC).

Combustion Section

The combustion chamber for the JT9D engine is a two-piece annular chamber having inner and outer liners. The two liners

which form the combustion chamber are shown separated in Fig. 16-29.

The cutaway drawing in Fig. 16-30 shows a cross section of one side of the combustion chamber together with a section of the diffuser case and the **first-stage turbine nozzle assembly**. This drawing shows one of the 20 fuel nozzles bolted externally to the diffuser case and extending into the forward end of the combustion chamber. Surrounding the fuel nozzle in the nozzle opening are swirl vanes to assist in vaporizing the fuel as it is discharged into the combustion chamber.

Each fuel nozzle is equipped with three fittings. One of these is for primary fuel, one for secondary fuel, and one for water. The fuel and water are carried in three steel manifolds, which surround the diffuser case. Individual lines and fittings carry the fuel and water to the nozzles.

At the rear of the combustion chamber in Fig. 16-30, the first-stage nozzle vane can be seen. There are 66 vanes in the nozzle assembly, which are classified for nozzle-area sizing and are individually replaceable. They are hollow and internally cooled with fifteenth-stage air. The vanes in their support can be removed from the engine as a complete assembly.

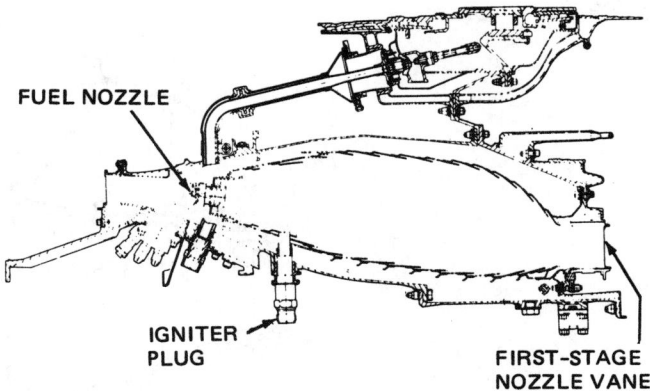

FIGURE 16-30   Cross section of the combustion chamber.

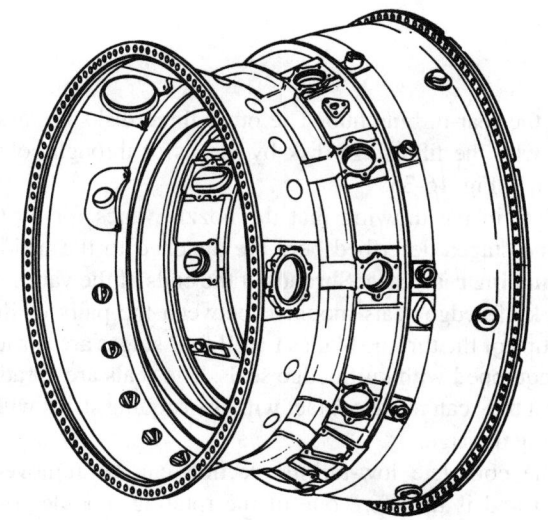

FIGURE 16-31   Diffuser case. (*United Technologies, Pratt & Whitney Aircraft.*)

The rear of the combustion section is designed with an outer combustion chamber which can be telescoped forward to permit individual nozzle vane replacement.

As shown in the drawings, the diffuser case comprises the forward end of the combustion section. Figure 16-31 shows the diffuser case and Fig. 16-32 shows some of the functions of the case.

The no. 3 bearing housing is supported inside the case by means of 10 struts. Four of the struts have openings inside the engine to bleed fifteenth-stage clean air to four openings on the case. Two of the struts carry seventh-stage cooling air into the cooling jacket around the no. 3 bearing and carry the bearing breather pressure out. Two other struts carry seventh-stage cooling air out of the bearing area. It will also be noted that oil pressure for the no. 3 bearing is earned through one of the struts and that scavenge oil is removed through the strut in the six o'clock position.

The diffuser case holds the mountings for the two spark igniters. These igniter plugs are mounted on each side of the case and extend into the inside of the combustion chamber on each side, 9° above the centerline. An igniter plug is pictured in Fig. 16-30.

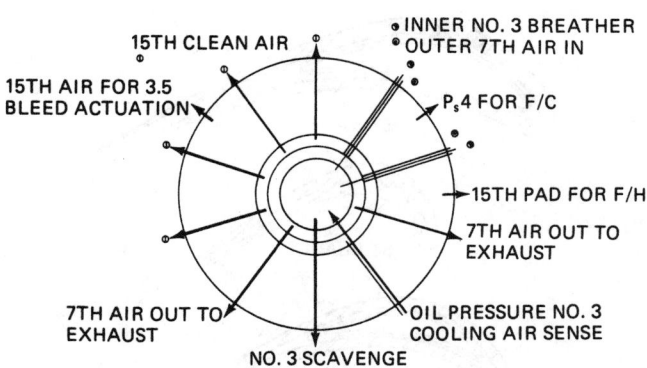

FIGURE 16-32   Some functions of the diffuser case.

Cooling of the first-stage nozzle and first-stage turbine is accomplished by fifteenth-stage air which is bypassed around the combustion chamber into a cooling duct. A series of jets at the aft end of the duct direct the air into 116 holes drilled tangentially through the first-stage disk at the broached end. The air then passes into the 116 hollow first-stage blades and exits from the blade tips and from small holes near the leading edges.

The entire turbine area is cooled by flowing air through numerous holes and passages, which eventually lead the air to openings where it can join the exhaust stream.

Turbine Section

As explained previously, the turbine section of the JT9D engine is comprised of two turbine assemblies. The first is the high-pressure compressor drive turbine and the second is the low-pressure compressor drive turbine. The **high-pressure compressor drive turbine** (also called the **high-compressor drive turbine**) consists of the first two stages of the turbine, which are located immediately to the rear of the combustion section. A cross section of the high-compressor drive-turbine assembly is shown in Fig. 16-33.

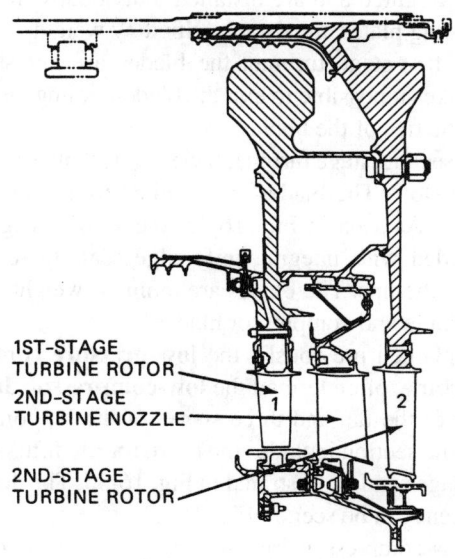

FIGURE 16-33   Cross section of a high-pressure compressor drive turbine.

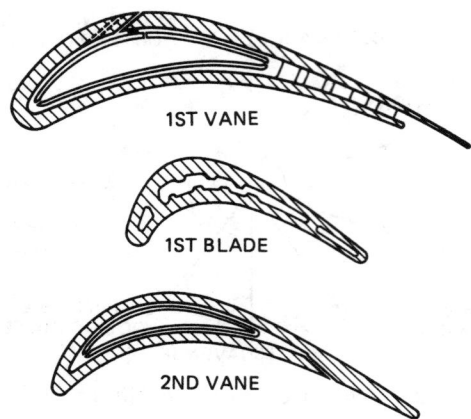

FIGURE 16-34   Cooling features of turbine nozzle vanes and blades.

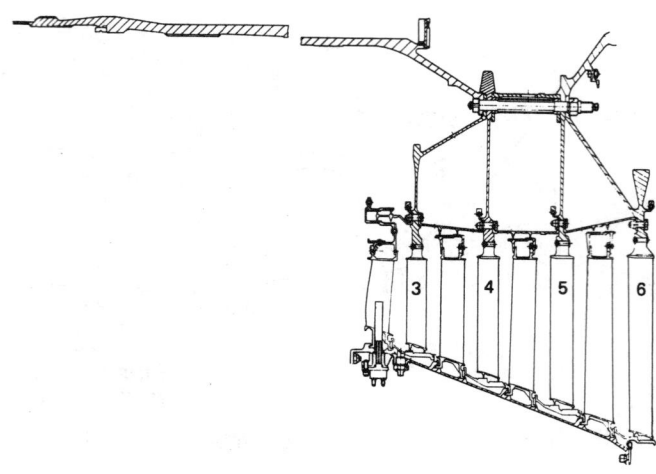

FIGURE 16-35   Section of the low-compressor drive turbine.

In the drawing it can be seen that the first-stage turbine disk is attached to the second-stage disk by means of a bolted flange. Knife-edge air seals are provided to prevent air leakage between the turbine blades and vanes. The air seals and blades operate against abradable shrouds at the outside diameter of the blades and the inside diameter of the vanes. Abradable shrouds permit rubbing without damage to the parts. The second-stage disk is splined to the high-compressor drive shaft (hub), as shown in the drawing.

The nozzle vanes and turbine blades are cooled by means of air which flows through interior passages, as shown in Fig. 16-34. The turbine blades are Jo-Coated and the vanes are chromallized to reduce gas erosion and high-temperature corrosion.

The first-stage turbine has 116 blades mounted in fir-tree slots broached in the rim of the disk. The blades are held in place by 29 blade-retaining plates riveted to the disk and to each other by 116 long rivet pins. The plates contain the cooling airflow from the disk so that it flows into the blades and not out the fir-tree slots. The turbine blades are moment-weighted and are installed individually in the disk. The retaining plates are also classified by weight.

In the first-stage turbine, the blades are not shrouded, which makes it possible to exit the blade-cooling air through holes in the tips of the blades.

In the second-stage turbine, there are 138 blades mounted in fir-tree slots. The blades are retained by means of flare-type rivets. As seen in Fig. 16-33, the second-stage blades are shrouded with integral knife-edge seals to reduce gas leakage at the tips. The blades are moment-weighted, as are all the turbine and compressor blades.

As explained previously, the **low-pressure compressor drive turbine**, often termed the **low-compressor drive turbine**, drives the fan and three stages of the N_1 compressor. This turbine section includes the third, fourth, fifth, and sixth turbine stages and is illustrated in Fig. 16-35. The drive-shaft arrangement can be seen in Fig. 16-23.

The low-compressor drive turbine incorporates four disks to which the turbine blades are attached in fir-tree slots. The blades are held in place by rivets. The fifth-stage disk is integral

with the rear-turbine hub. The other three disks are assembled with the fifth-stage disk by means of through-bolts, as shown in Fig. 16-35.

Note in the drawing that the nozzle vanes for the four turbine stages described here are attached to the inside of the **turbine rear case**. Shrouds at the ends of the vanes mate with knife-edge seals mounted between the pairs of disks. The tips of the turbine blades for all four stages are shrouded and equipped with knife-edge seals. The seals are abradable so that they can make contact with the rubbing strips without causing damage.

The complete low-turbine section can be removed as a unit and is therefore one of the rotating modules of the engine. The module is balanced and can be interchanged with similar modules on other engines of the same type.

The **turbine exhaust case** is illustrated in Fig. 16-36. This is a complete welded structure with 15 struts between the outer case and the no. 4 bearing housing. Within the bearing housing is the no. 4 bearing compartment surrounded by seventh-stage air for cooling and heat protection.

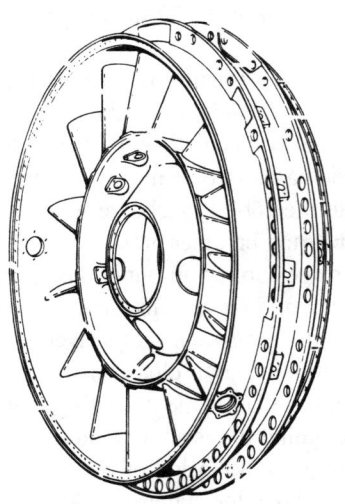

FIGURE 16-36   Turbine exhaust case.

Six pressure-sensing probes for measuring P_{t7} are manifolded together between the last two flanges of the case. Pressure oil for the no. 4 bearing flows through one of the struts, and scavenge oil flows out through another.

The rear mountings for the engine are located on the top of the turbine exhaust case. This requires that the case be designed and built for high structural integrity.

To reduce engine distortion resulting from heat and stress, a thrust frame is installed between the intermediate case and the turbine exhaust case. Through this frame, engine thrust is transmitted directly from the intermediate case to the aircraft structure. This is accomplished through a fitting called a "puck," which forms the attachment from the engine to the aircraft.

Accessory Section

The accessory section of the JT9D engine consists of two gearboxes mounted under the engine and driven through a tower shaft. The tower shaft incorporates a spur-gear pinion which meshes with the **accessory drive** mounted on the front hub of the high-compressor hub.

Power from the tower shaft is carried to the **angle gearbox** and from there through a horizontal shaft to the main gearbox located at the bottom of the engine near the forward end of the combustion section.

Conclusion

The foregoing description does not cover many details of the JT9D engine and its construction, but it does provide sufficient detail so that the technician can understand the general design and operation of the engine. To obtain complete information about any engine, the technician should consult the maintenance and overhaul manuals for the engine concerned.

Systems for the JT9D Turbofan Engine

Some of the systems for the JT9D engine have been partly described earlier. The fuel system includes the Hamilton Standard JFC68 fuel control, a TRW fuel pump with booster, an engine vane control, filters, a fuel-oil cooler, and other units common to the majority of gas-turbine fuel systems.

The lubrication system for the JT9D engine is similar to lubrication systems for other gas-turbine engines. A pressure pump delivers oil to the various bearings where the oil is sprayed on the bearings from jets. The oil and oil spray are contained in the bearing compartments by means of air pressure acting through the seals. The oil is returned to the oil tank by scavenge pumps in each area. Air is separated from the oil and is vented overboard. Relief valves provide protection against extreme pressure. The system is equipped with a main filter plus a number of "last-chance filters" to ensure that the oil delivered to the bearings is clean.

The ignition system has been described earlier. As mentioned previously, there are actually two systems of the high-energy type. The electrical energy is stored in capacitors and then released to provide an extremely high-energy spark for the ignition of fuel. The energy is required only for starting;

however, the ignition system may be turned on at times when there is a possibility of flameout.

The Pratt & Whitney JT9D-7R4 Turbofan Engine

The Pratt & Whitney JT9D-7R4 turbofan engine (shown in Fig. 16-37), which powers the Boeing 767, the Boeing 747-300, and the Airbus A310, is the product of years of research and development combined with experience gained in the operation of earlier engines for millions of flight hours. Advanced technology has been applied in many parts of the engine, resulting in substantial improvements in fuel consumption, reliability, engine life, maintainability, and noise emission. A cutaway view of the JT9D-7R4 engine is shown in Fig. 16-38. Some of the improved features are pointed out in the illustration.

The basic design of the-7R4 engine is essentially the same as that of the JT9D described in the foregoing section; however, the changes and additions made on the-7R4 have provided substantial gains for the operators of the engine. Fuel savings up to 8 percent and thrusts up to 56 000 lb [249 088 N] are possible with some models of this engine.

Fan Section

The fan of the -7R4 engine has been modified by the use of wide-chord single-shroud fan blades and the use of a reduced number of fan blades and exit guide vanes. These features, together with the convergent-divergent fan exit nozzle, contribute to significant fuel savings during cruise operation. The aft-positioned blade shroud reduces losses associated with operation of the fan at high rotational speeds. Figure 16-39 shows the difference between the fan blade used on earlier models of the JT9D and the blade used on the -7R4 model. The modification of the fan section has produced fuel savings, as well as reduced noise emission.

Low-Pressure Compressor

The JT9D-7R4 engine is equipped with the four-stage compressor utilized in the JT9D-59A, -70A and -7Q models of the engine. Previous models employed a three-stage low-pressure compressor, as illustrated in Fig. 16-40. The four-stage design increases the overall engine pressure ratio and airflow capacity, providing improved performance, lower turbine operating temperatures, growth capability, and improved surge margin.

High-Pressure Compressor Stators

The high-pressure compressor stators in the-7R4 engine are coated with an alloy with the trade name "SermeTel" that has proven very beneficial. This is a long-lasting, thin-film coating that provides improved protection against corrosion and erosion in the compressor section, thereby extending parts life and reducing maintenance and overhaul costs. The coating also contributes to lower fuel consumption.

FIGURE 16-37   Pratt & Whitney JT9D-7R4 turbofan engine. (*United Technologies, Pratt & Whitney Aircraft.*)

Burner Design

The bulkhead configuration burner design used in later series JT9D engines is also employed in the-7R4. This burner improves performance and durability. Because it provides lower temperatures at the outer diameter of the annulus, it reduces air seal expansion, thus retaining clearances and minimizing turbine rubbing. The fuel nozzles are of the aerating type; that is, air under pressure is mixed with the fuel before it is sprayed into the burner. This type of fuel nozzle reduces emission levels and eliminates coking.

Turbine Blades

The blades of the high-pressure turbine are single-crystal castings. The special Pratt & Whitney alloy used and the single-crystal feature substantially increase the mechanical properties of the blades—such as creep strength and resistance to low-cycle fatigue, oxidation, and high-temperature corrosion—and provide increased heat resistance. The use of these blades in the high-pressure turbine lowers maintenance costs and saves fuel by reducing cooling-air requirements.

Turbine Cooling

Cooling air is supplied to the turbine blades and vanes (airfoils) at high pressures through injection nozzles near the disk rim. As part of the-7R4 series turbine tailoring, the injection nozzles are sized to meet the cooling requirements of the particular engine model. By using only the amount of air from the engine airflow that is required for adequate cooling, maximum efficiency is attained and durability is maintained.

The-7R4 series engines employ a modulated turbine-case cooling system to improve performance and save fuel during cruise operation. The system starts operating after takeoff and climb because during these operations the turbine blade clearance must be greater than it is at cruise. During cruise operation, air flows through cooling ducts around the turbine case and the resultant cooling shrinks the case to minimize blade tip clearances, thus increasing the turbine's effectiveness and saving fuel. The case cooling system utilized to control blade tip clearance is termed **active clearance control** (ACC).

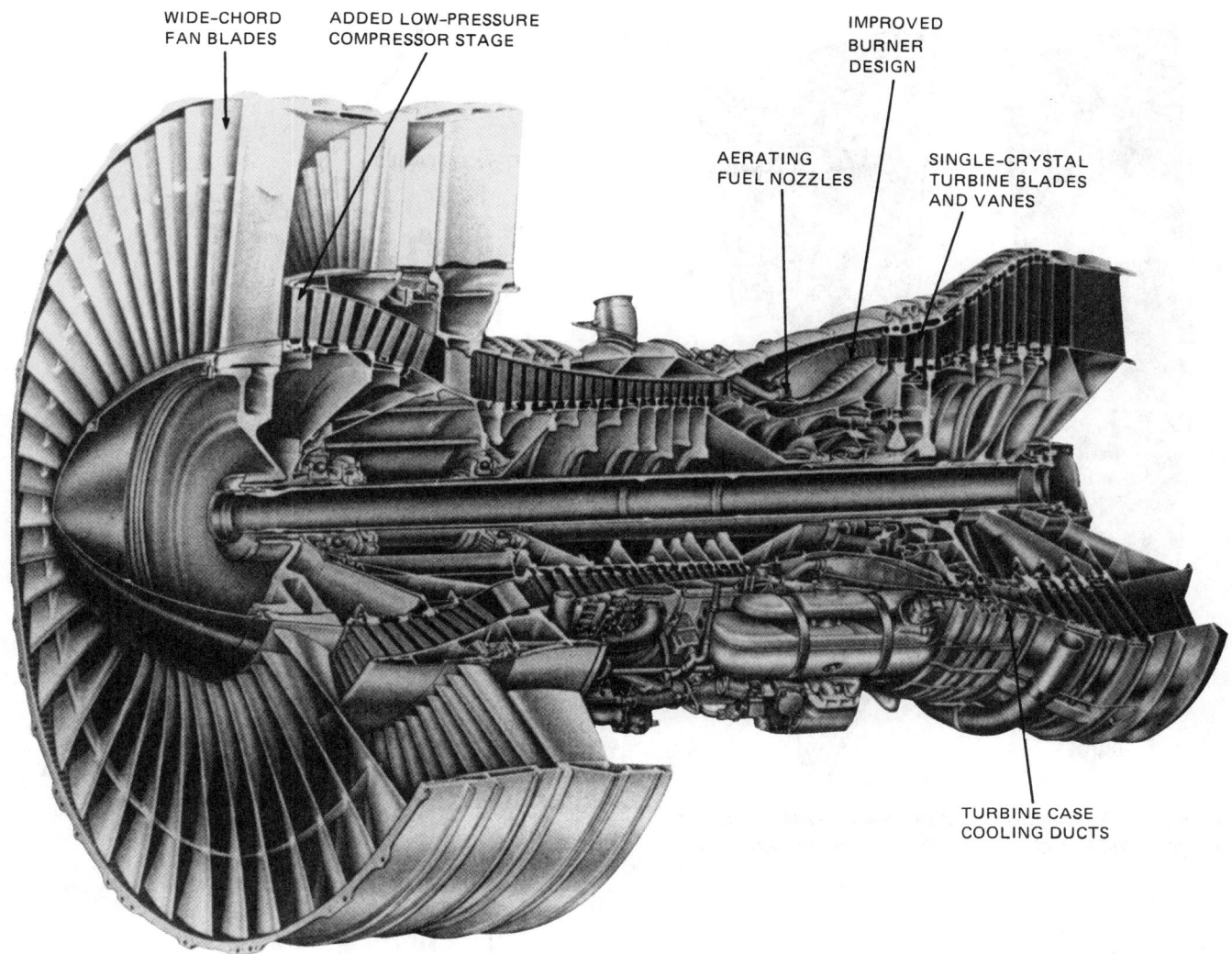

FIGURE 16-38 Cutaway view of the JT9D-7R4 turbofan engine. (*United Technologies, Pratt & Whitney Aircraft.*)

Labels on figure:
WIDE-CHORD FAN BLADES
ADDED LOW-PRESSURE COMPRESSOR STAGE
IMPROVED BURNER DESIGN
AERATING FUEL NOZZLES
SINGLE-CRYSTAL TURBINE BLADES AND VANES
TURBINE CASE COOLING DUCTS

Air Seals

Lightweight carbon air seals for the no. 2 and no. 3 bearing compartments in the-7R4 engine have replaced the labyrinth seals used in earlier engines. This change has resulted in a reduction of weight by eliminating the external plumbing required for seal-pressurization air ducting. The use of carbon seals improves engine performance and reduces oil consumption because there is less air leakage from the bearing compartments.

Electronic Engine Control

Although it is not a part of the engine, the **electronic engine control** (EEC) utilized with the-7R4 engine contributes substantially to the engine's performance. The engine control used with the-7R4 engine is the Hamilton Standard EEC-103. This is the **supervisory engine control** described in Chap. 12. The precise control of engine functions by the EEC-103 in conjunction with the hydromechanical fuel control unit increases engine efficiency significantly, thereby reducing fuel consumption.

The Pratt & Whitney 2037 Turbofan Engine

The Pratt & Whitney 2037 turbofan engine, shown in Fig. 16-41, is termed an "advanced-technology engine" because it incorporates all the advanced design features and materials that contribute to engine efficiency, reliability, low fuel consumption, and maintainability. The 2037 engine powers the Boeing 757 aircraft and is tailored to meet all the power requirements of that aircraft.

Specifications

The general specifications for the Pratt & Whitney 2037 turbofan engine are as follows:

Length	133.7 in	339.60 cm
Diameter, fan tip	78.5 in	199.39 cm
Diameter, fan case	85.0 in	215.90 cm
Diameter, turbine exhaust case	49.0 in	124.46 cm
Weight	6675 lb	3028 kg

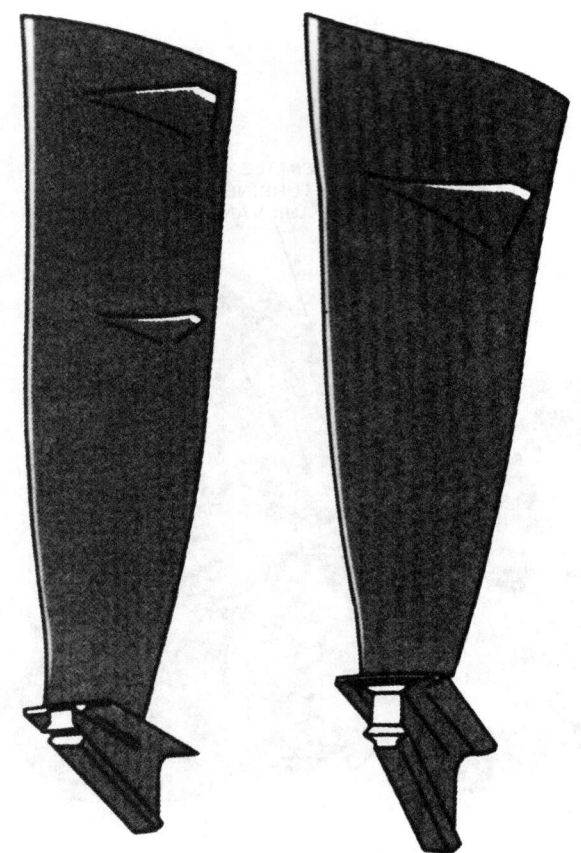

FIGURE 16-39 Comparison of fan blades. -7R4 is shown on the right. (*United Technologies, Pratt & Whitney Aircraft.*)

Takeoff thrust, sea-level static	37 000 lb	164 576 N
Fuel consumption, thrust specific, 35 000 ft [10 668 m], 0.8 Mach	0.536 lb/lbt/h	54.7 g/N/h
Overall pressure ratio	30:1	
Bypass ratio (cruise)	5.8:1	
Fan pressure ratio	1.70:1	
Total airflow	1193 lb/s	541 kg/s
Low-pressure compressor stages	Fan plus 4	
High-pressure compressor stages	12	
Low-pressure turbine stages	5	
High-pressure turbine stages	2	

Description

The Pratt & Whitney 2037 is a twin-rotor, high-bypass-ratio, axial-flow turbofan engine designed for use with commercial and military transport aircraft. Its configuration, shown in Fig. 16-41, is similar to that of the JT9D (Fig. 16-38), although the 2037 is somewhat smaller.

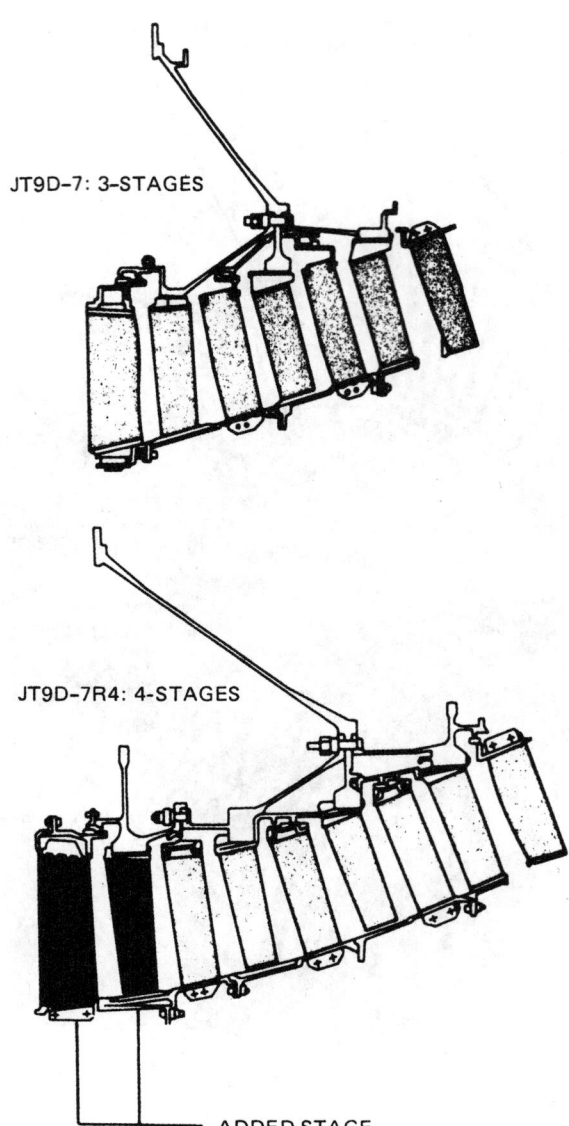

FIGURE 16-40 Comparison of low-pressure compressors.

The single-stage fan and four-stage low-pressure compressor are driven by a five-stage turbine at the rear of the engine. The 12-stage high-pressure compressor is driven by a two-stage turbine through a hollow shaft that is outside and coaxial with the low-pressure compressor drive shaft. The two rotors are supported by five main bearings.

The fan has 36 wide-chord, single-shroud blades similar to those on the JT9D-7R4 engine. The aft location of the shroud reduces airflow interference. The design of the fan section contributes to maximum efficiency and minimum noise.

The airfoils (blades and vanes) are described as "controlled-diffusion" airfoils. These airfoils permit higher Mach numbers without loss in efficiency and are used in both the compressors and turbines. The thicker leading and trailing edges of the airfoils, shown in Fig. 16-42, also enhance performance retention (continued serviceability) by increasing resistance to airborne particle erosion.

Variable stators are employed in the first five stages of the high-pressure compressor to increase compressor efficiency,

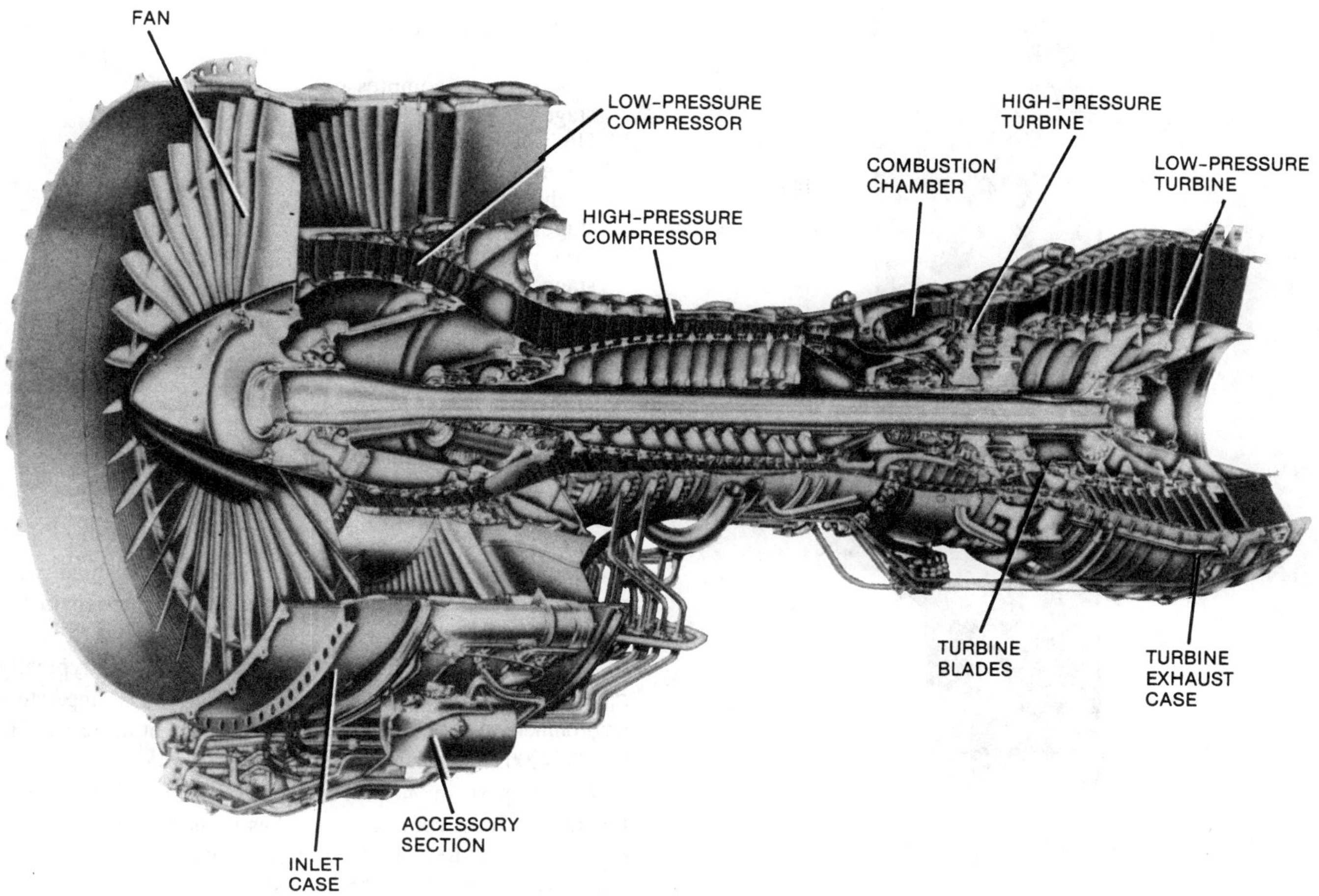

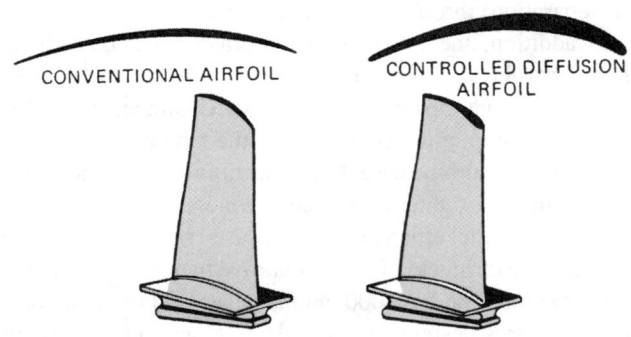

FAN

LOW-PRESSURE COMPRESSOR

HIGH-PRESSURE COMPRESSOR

COMBUSTION CHAMBER

HIGH-PRESSURE TURBINE

LOW-PRESSURE TURBINE

INLET CASE

ACCESSORY SECTION

TURBINE BLADES

TURBINE EXHAUST CASE

FIGURE 16-41   Cutaway view of Pratt & Whitney 2037 turbofan engine. (*United Technologies, Pratt & Whitney Aircraft.*)

CONVENTIONAL AIRFOIL

CONTROLLED DIFFUSION AIRFOIL

FIGURE 16-42   Comparison of airfoils.

control surge, and eliminate compressor stall. These stators provide stability over the entire operating envelope by controlling the aerodynamic loading on the compressor stages. The disks and spacers in the high-pressure rotor are joined by electron-beam welding. This technique, used instead of bolted joints, eliminates leakage and improves clearance control by stiffening the rotor.

The 2037 engine has an annular combustion chamber with 24 single-pipe air-blast fuel nozzles. The combustor design provides improved heat transfer from the lower lip, which reduces thermal stress. In addition, the design provides uniform discharge of the cooling air to the inner surface of the downstream burner wall.

The high-pressure turbine blades are single-crystal castings made from PW1480 alloy. A blade of this type is shown in Fig. 16-43. These blades provide high creep strength, high thermal limits and low-cycle fatigue life, and resistance to oxidation and high-temperature corrosion. The outer air seals for the high-pressure turbine are of an abradable, nonmetallic material. These seals improve performance by reducing blade tip clearances during operation. They also improve performance retention and reduce the risk of turbine blade damage.

The high-pressure turbine disks and the last high-pressure compressor disk are produced from fine-mesh PW1100 powder metal. Compared to conventional forged disks, the powder-metal disks offer higher tensile strength, greater stress-rupture resistance, and longer low-cycle fatigue life.

It will be noted in Fig. 16-41 that low-pressure turbines increase in diameter substantially from the first stage to the last stage. This flow-path configuration is designed to provide the best possible aerodynamic performance. The large mean diameter and smooth gas-flow transition from the high-pressure turbine is more efficient than the smaller mean diameter turbines used on earlier engines.

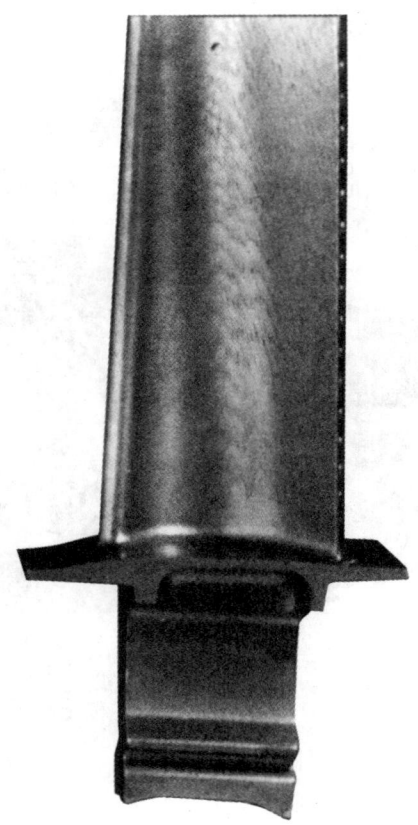

FIGURE 16-43 Single-crystal turbine blade.

Engine Control

The Pratt & Whitney 2037 engine is controlled by the Hamilton Standard EEC-104 **full-authority electronic engine control** described in Chap. 12. All of the engine variables—including starting, fuel flow, variable stator positioning, acceleration, and deceleration—are regulated by the EEC. The automated thrust thus provided eliminates the need to continually readjust throttle settings during takeoff and climb. The EEC reduces engine maintenance time by eliminating the requirement for engine trimming.

The Pratt & Whitney PW4000 Engine

Pratt & Whitney's PW4000 commercial jet engine, shown in Fig. 16-44, is a high-bypass-ratio turbofan engine. The PW4000 engine is a two-spool engine and has a nominal bypass ratio of 5:1.

Thrust of the engines in the PW4000 engine series ranges from 50 000 lb [222 400 N] to more than 60 000 lb [266 880 N], with growth capability to 70 000 lb [311 360 N], making the engine ideal for current and new versions of the Boeing 747 and 767; the Airbus A300, A310, A330; and McDonnell Douglas MD-11 wide-body aircraft. The PW4000 engine specifications are as follows:

Length	132.7 in	337 cm
Fan case diameter	97 in	246 cm
Fan tip diameter	93.4 in	237 cm
Weight	9200 1b	4173 kg

Turbine exhaust case diameter	53.4 in	136 cm
Low-pressure compressor stages	Fan plus 4	
High-pressure compressor stages	11	
High-pressure turbine stages	2	
Low-pressure turbine stages	4	
Combustor type	Annular	
Takeoff thrust, sea-level static	56 000 lb	249 088 N
Total airflow	1705 lb/s	773 kg/s
Fan pressure ratio	1.72:1	
Bypass ratio (at cruise)	5.0:1	
Overall pressure ratio (at cruise)	28.1	

The PW4000's higher rotational speed, coupled with a simplified design, permits a large-scale reduction in the number of engine parts.

The high-pressure turbine (HPT), an area which typically requires many replacement parts because of the high-temperature environment in which it operates, has 67 percent fewer parts in the PW4000 engine than it has in the JT9D.

Casting parts in one piece instead of welding components together to form complete parts has reduced the number of parts in the PW4000. Diffuser, turbine exhaust, and intermediate cases in the PW4000 are cast.

Three-dimensional airfoil designs are used in the PW4000 compressors and turbine to improve aerodynamic efficiency, as well as durability. Airfoils are designed with thicker leading and trailing edges to provide uniform aerodynamic flow (no separation) throughout the blade area.

In addition, the PW4000 incorporates an **active clearance control** (ACC), which controls the clearances between the turbine blade tips and turbine case. Clearance control is achieved by impinging cold air from the fan duct on the outside of the hot turbine case, thus controlling case dimensions and resulting in tighter clearances, reduced blade tip losses, and improved fuel efficiency. ACC is used on both HPTs and low-pressure turbines (LPTs) to improve fuel efficiency in the cruise regime. The PW4000 engine's patented thermal compressor rotor is designed to provide ACC without impinging cold air on the compressor case. The high-pressure compressor disk is uniformly expanded with hot engine air to reduce rotor-to-stator seal clearances and thus improve efficiency.

Ceramic abradable outer air seals are used in the high-pressure turbine to minimize operating clearances and further improve operating efficiency. **Single-crystal turbine blades** are designed to provide higher turbine inlet temperature capability and improved turbine life. **Radial gradient vanes** are designed to increase the low-pressure turbine aerodynamic efficiency by substantially reducing endwall losses. Controlled diffusion airfoils are designed to improve compression system efficiency and also reduce the number of airfoils. In addition, controlled diffusion airfoils are designed to have greater tolerance to erosion and therefore to have improved performance retention.

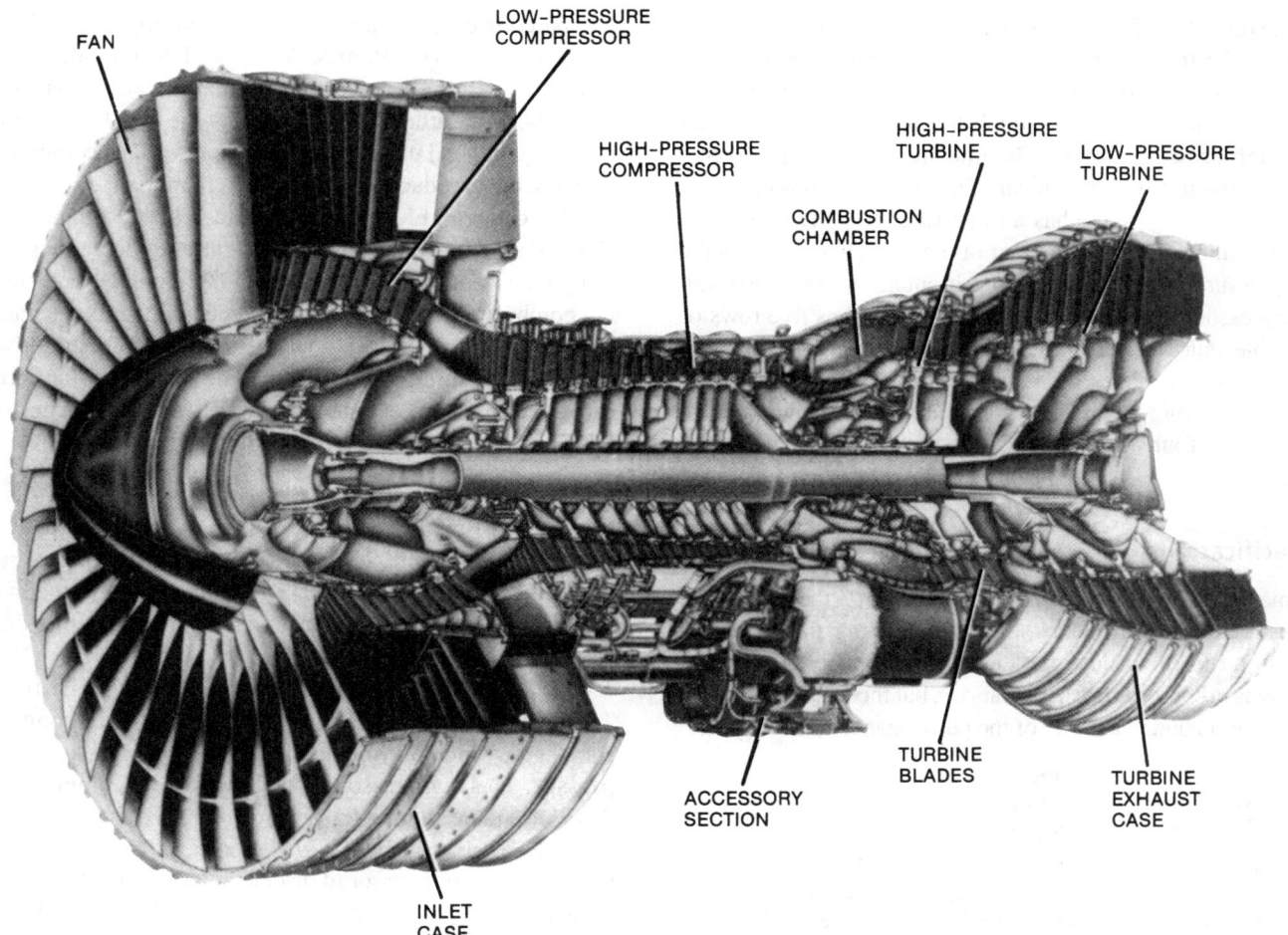

FIGURE 16-44 Cutaway view of Pratt & Whitney PW4000 turbofan engine. (*United Technologies, Pratt & Whitney Aircraft.*)

Cooling of the HPT blades is handled much more efficiently in the PW4000 than it is in earlier engines. These blades typically operate in temperatures of more than 2000°F [1093°C]. Improved internal cooling schemes minimize the amount of air diverted from the compressor to cool the blades. The less diversion of air from the compressor, the higher the engine efficiency.

Full-Authority Digital Electronic Control System

The PW4000 has a proven **full-authority digital electronic engine control (FADEC) system**. The FADEC control replaces previous hydromechanical units, producing significant savings by precise and uniform control of power settings and other critical engine operating functions. The control does not employ any hydromechanical computational elements.

The FADEC performs the following functions: basic engine operation (starting, acceleration, deceleration, speed governing, compressor vane and bleed scheduling), engine operating and rating data, fault detection, and indication of fault status for display and maintenance. The pilot sets the power by setting the thrust lever position. The FADEC uses the thrust lever position to determine the commanded thrust setting parameter and modulates fuel flow to make the actual and the commanded values equal. The FADEC system reduces crew workload, controls system costs, and extends engine life. In addition, the EEC system provides improved engine fuel efficiency and allows improved integration and coordination of engine control and aircraft systems.

The FADEC system assists in engine maintenance through its internal diagnostic capability, which is able to detect faults and generate maintenance messages that identify which control system components need to be repaired.

The FADEC system will allow the same basic engine configuration to produce from 50 000 lb [2 22 400 N] to over 70 000 lb [311 360 N] of thrust simply by changing the data entry plug for the control and the data plate for the engine.

The General Electric CF6 Turbofan Engine

The General Electric CF6 turbofan engine is also one of the new generation of high-bypass engines designed for the large, wide-body aircraft. The principal models at present are the CF6-6 and the CF6-50, which are designed for use in the Douglas DC-10 airliner. The CF6-50 is also installed in the Airbus Industrie A-300B airplane. A photograph of the CF6-50 is shown in Fig. 16-45.

The differences between the CF6-6 and the CF6-50 are in the numbers of compressor and turbine stages. The CF6-6 engine has a single-stage low-pressure booster immediately to the rear of the fan. This is called a *booster stage*. The compressor for the CF6-6 has six rows of variable stator blades and ten rows of fixed stator blades. The compressor rotor has 16 stages and is driven through the outer drive shaft from the high-pressure turbine. The low-pressure turbine has five stages, and it drives the fan and the one-stage booster.

The CF6-50 engine has a three-stage low-pressure compressor immediately to the rear of the fan. This increases the air pressure considerably before it enters the high-pressure compressor. The high-pressure compressor has five rows of variable stator vanes and ten rows of fixed vanes. The rotor consists of 15 stages of blades and is driven by the outer drive shaft from the high-pressure turbine. The low-pressure turbine has four stages instead of the five stages in the CF6-6 engine.

Specifications

Some of the differences in performance of the two types of engines can be understood by examining the specifications given below. Note that there have probably been recent changes in some of the specifications, but those given will provide a good understanding of the performance of each type.

	CF6-6	CF6-50
Length	193 in	190 in
	[490.22 cm]	[482.6 cm]
Diameter	94.0 in	94.0 in
	[238.76 cm]	[238.76 cm]
Frontal area	48.2 ft² [4.48 m²]	48.2 ft² [4.48 m²]
Weight	7450 lb [3379 kg]	8225 lb [3730 kg]
Takeoff thrust	40 000 lbt	50 000 lbt
	[177 920 N]	[220 400 N]
Power-weight		
ratio	5.45 lbt/lb	5.85 lbt/lb
	[52.65 N/kg]	[59.08 N/kg]
Fuel		
consumption	0.35 lb/lbt/h	0.39 lb/lbt/h
	[35.7 g/N/h]	[39.8 g/N/h]
Oil		
consumption	3.0 lb/h	3.0 lb/h
	[1360 g/h]	[1360 g/h]
Compressor ratio	26.6:1	29.9:1
Bypass ratio	6.2:1	4.4:1
Fan pressure ratio	1.64:1	1.69:1
Air mass flow, fan	1160 lb/s	1178 lb/s
	[562.18 kg]	[534.34 kg]
Air mass flow, core	183 lb/s	270 lb/s
	[83.01 kg/s]	[122.47 kg/s]

Note from the specifications above that both the CF6-6 and CF6-50 engines have very high pressure ratios. Note also that the CF6-50 utilizes a greater percentage of the air through the core of the engine and achieves a 25 percent increase in takeoff thrust. This is due in part to the greater pressure ratio. The greater total volume of air passing through the core and fan also increases the total thrust substantially.

General Configuration

As can be seen from the photograph in Fig. 16-45 and the drawings in Figs. 16-45 and 16-46, the CF6-50 engine

can be considered as a dual-compressor engine similar in many respects to the Pratt & Whitney JT9D previously discussed. It will be noted, however, that the performance of the CF6-50 is a little higher than that of the JT9D and is similar to that of the JT9D-7R4, largely because of the compressor design which produces the high pressure ratio of 29.9:1.

The General Electric CF6-6 and CF6-50 engines are classed as dual-rotor high-bypass-ratio turbofan engines. They incorporate variable stators in the compressors, annular combustors, an air-cooled core engine turbine, and a coaxial front fan with a low-pressure compressor driven by a low-pressure turbine. The engine includes a fan thrust reverser and a core engine thrust reverser or spoiler to produce reverse thrust during the landing roll.

Engine Description

The CF6 engine is basically divided into five principal sections, as shown in Fig. 16-46. These are similar to the main sections of other gas-turbine engines and are classified according to function.

The **fan section** of the CF6 engine is shown in Fig. 16-47. The CF6-50 engine fan section includes the three compressor stages, which may be defined as the **low-pressure compressor**. This compressor supercharges the incoming air so that the engine pumps 55 percent more air than the basic engine. Variable air-bypass valves are provided aft of the low-pressure compressor to discharge air into the fan stream to establish proper flow matching between the low- and high-pressure spools during transient operation.

The fan rotor assembly is made of forged titanium disks and blades with aluminum platforms and spacers. Details of the fan blade installation are shown in Fig. 16-48. Note that the blade roots are inserted into dovetail slots in the rims of the disks. Platforms are installed between the fan blades to provide a smooth airflow at the blade roots.

Construction of the fan rotor is shown in Fig. 16-49. The locations of the no. 1 and no. 2 bearings are shown at the front and rear of the **forward shaft**.

Compressor

The compressor rotor for the CF6-6 engine is shown in Fig. 16-50. The design and construction of this rotor are similar to those described previously.

The stator for the CF6 engine comprises two sections: the **compressor front-stator assembly** and the **compressor rear-stator assembly**. In the CF6-50 engine, the front section has stator blades in the first 11 stages of compression and the rear section mounts stator blades for the twelfth through fourteenth stages of compression. The inlet guide vane for the high-pressure compressor and the stator vanes for the first six stages are variable. As explained previously, the variable vanes are employed to provide for correct direction of airflow in the compressor under all operating conditions. This reduces the possibility of compressor surge and stall. Signals from the fuel control unit schedule high-pressure fuel to the **variable-stator-vane actuators** to adjust the vane angles as required.

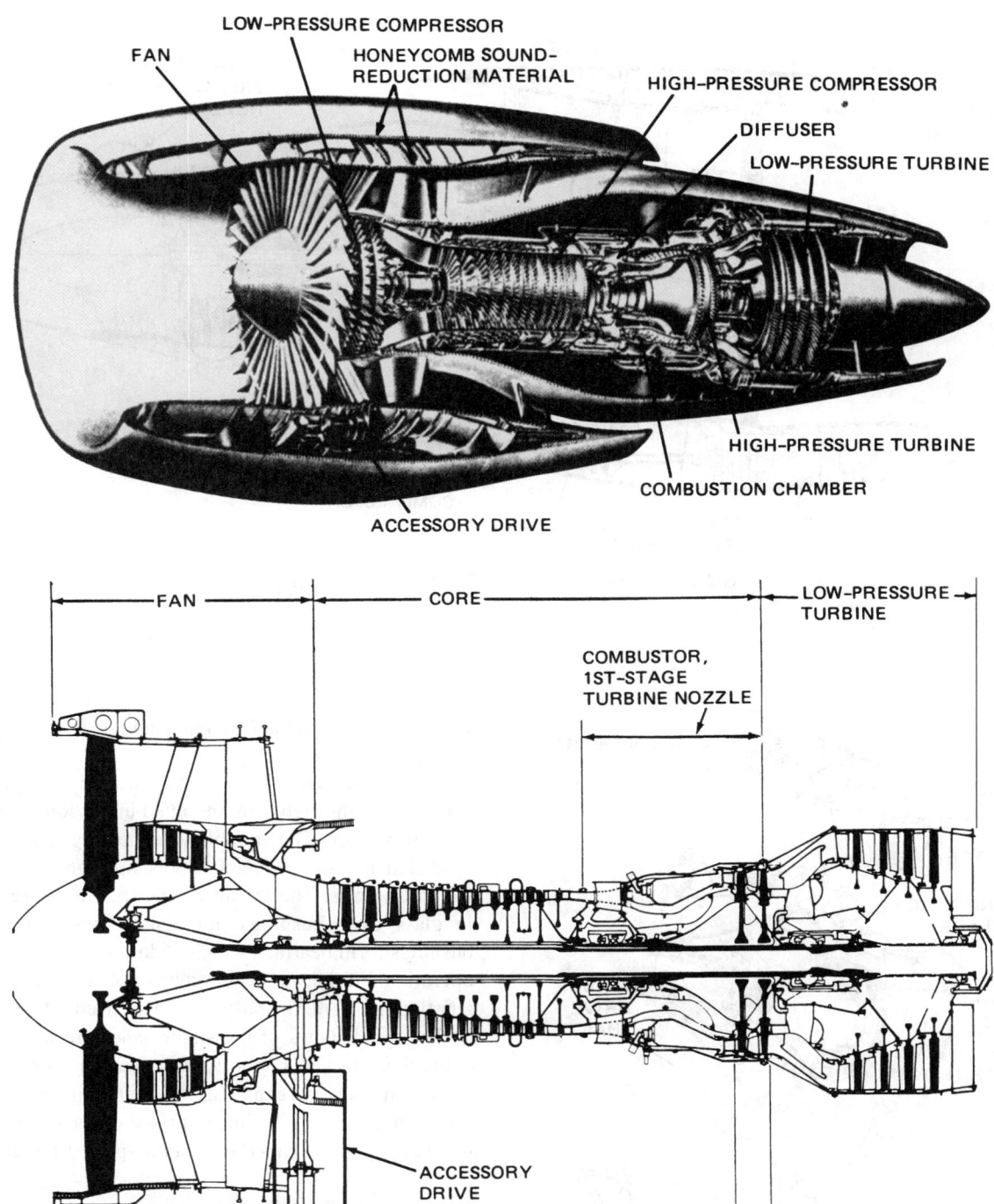

FIGURE 16-45   General Electric CF6-50 turbofan engine. (*General Electric.*)

The high-pressure compressor rotor is a combined spool-and-disk structure utilizing axial dovetails to retain the compressor blades in stages 1 and 2 and circumferential dovetails for the remaining stages. The blades for the first two stages are inserted axially in the slots cut into the disk rims and are held in place by tab-type **blade retainers**. The blades in the circumferential slots are inserted into the grooves through entry slots and then moved along the groove so that the dovetails' shoulders engage the locking lugs and hold the blades

in place. The blade dovetails are stronger than the blades; therefore, a failure will not cause as much damage if the blade and dovetail should come out of the slot as would be the case otherwise.

On the CF6-50 engine, the compressed air for aircraft use (customer air) is bled from the inside of the air passage through hollow stage-8 stator vanes. It passes through the vane bases and then through round holes in the casing skins into a pair of manifolds. Engine air is extracted at stage 7 for

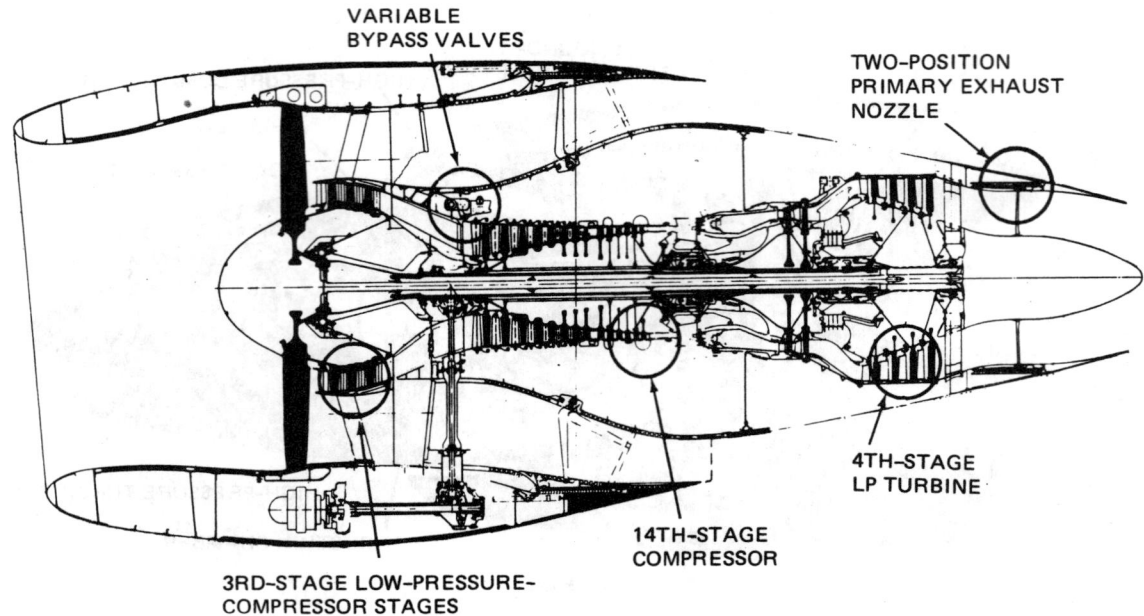

FIGURE 16-46 Cutaway view of the CF6-50 engine. (*General Electric.*)

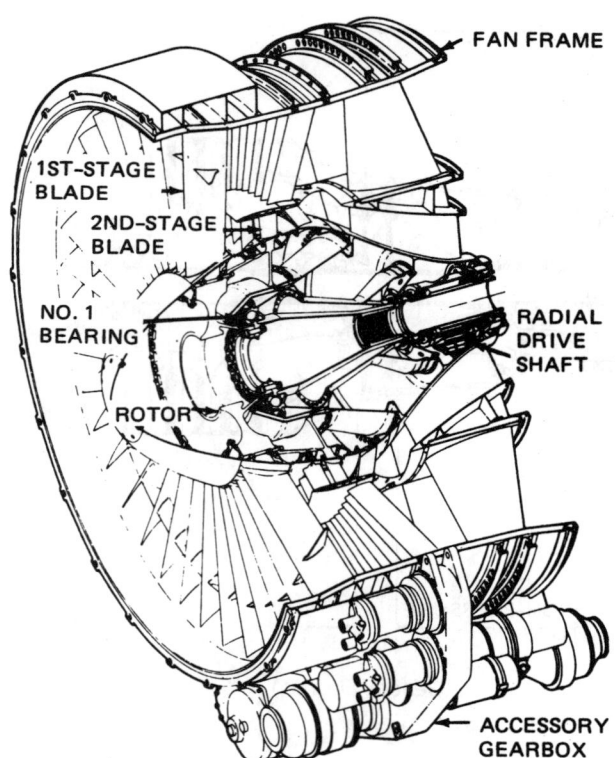

FIGURE 16-47 The CF6 fan section. (*General Electric.*)

turbine midframe cooling. This air passes through semicircular slots in adjacent stage-7 vane bases and then through round holes in the casing skins to the manifold.

Engine air is extracted at stage-10 stators for the second-stage high-pressure turbine cooling and nose cowl anti-icing. Similarly, this air passes through semicircular slots in the stage-10

vane bases and then through round holes in the casing skins to a pair of manifolds.

In the compressor casing, the variable inlet guide vanes and stages 1 through 6 are mounted in the conical part of the forward section. The variable vane bearing seats are formed by radial holes and counterbores through circumferential supporting ribs. The variable vane trunnions are supported in glass cloth bushings impregnated with Teflon. These bushings form bearings which endure thousands of hours of operation. Furthermore, Teflon has a self-lubricating characteristic; therefore, the bushings never need lubrication.

The fixed stages of the stator vanes are mounted in the cylindrical portion of the forward casing. These vanes are forged in one piece and are mounted in circumferential tracks machined in the inner surface of the casing. The forward casing is machined from a one-piece titanium forging.

Compressor Rear Frame

The compressor rear frame of the CF6 engine is a principal support structure for the engine. In the forward-center portion is a bearing housing containing two bearings for the rear support of the compressor rotor. The front flange of the compressor rear frame is bolted to the rear flange of the compressor stator. The bearing housing is supported by 10 streamlined struts extending outward through the outer shell of the frame. In addition to providing bearing support, the struts are utilized as passages for air and oil.

The compressor rear frame, which provides support for the rear of the compressor, houses the combustor, 30 fuel nozzles, and two igniters. The configuration of the compressor rear frame is shown in Fig. 16-51. Note that the unit is attached to the compressor by two forward flanges and to the high-pressure turbine section by one rear flange.

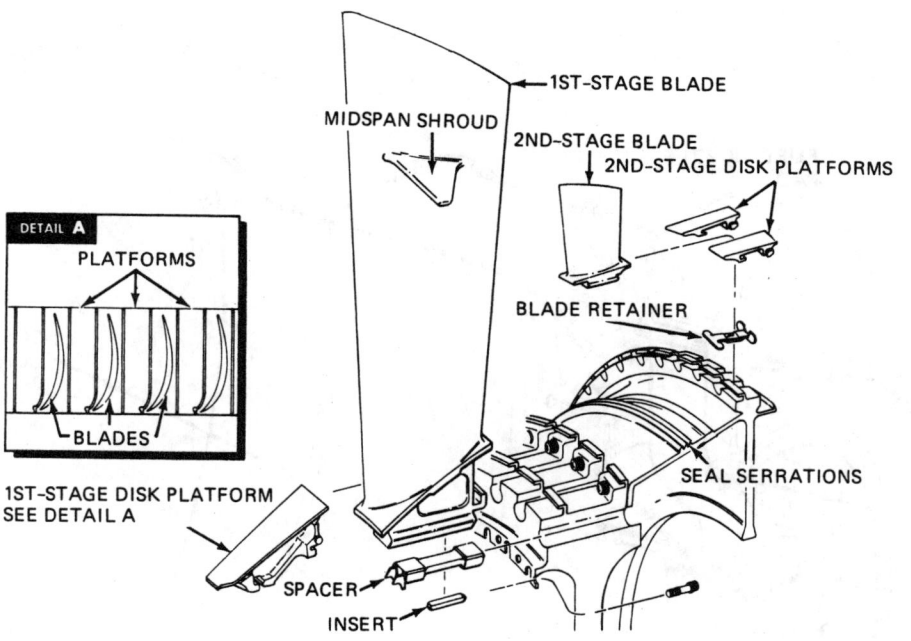

FIGURE 16-48 Fan blade installation. (*General Electric.*)

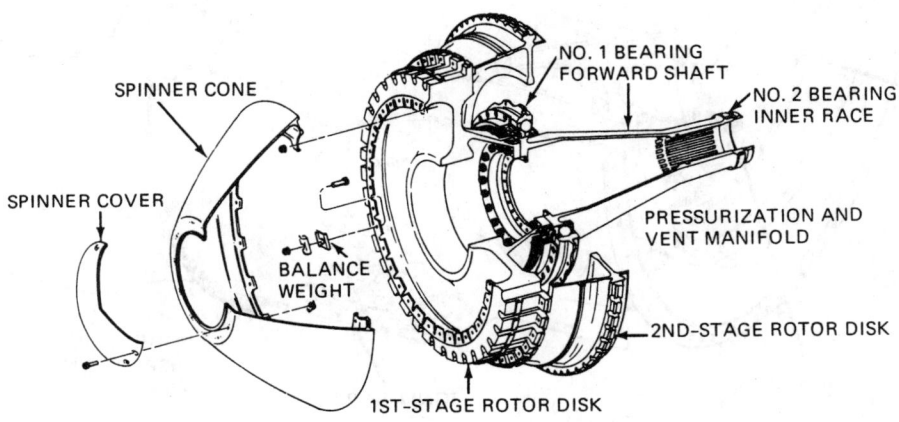

FIGURE 16-49 Construction of the fan rotor. (*General Electric.*)

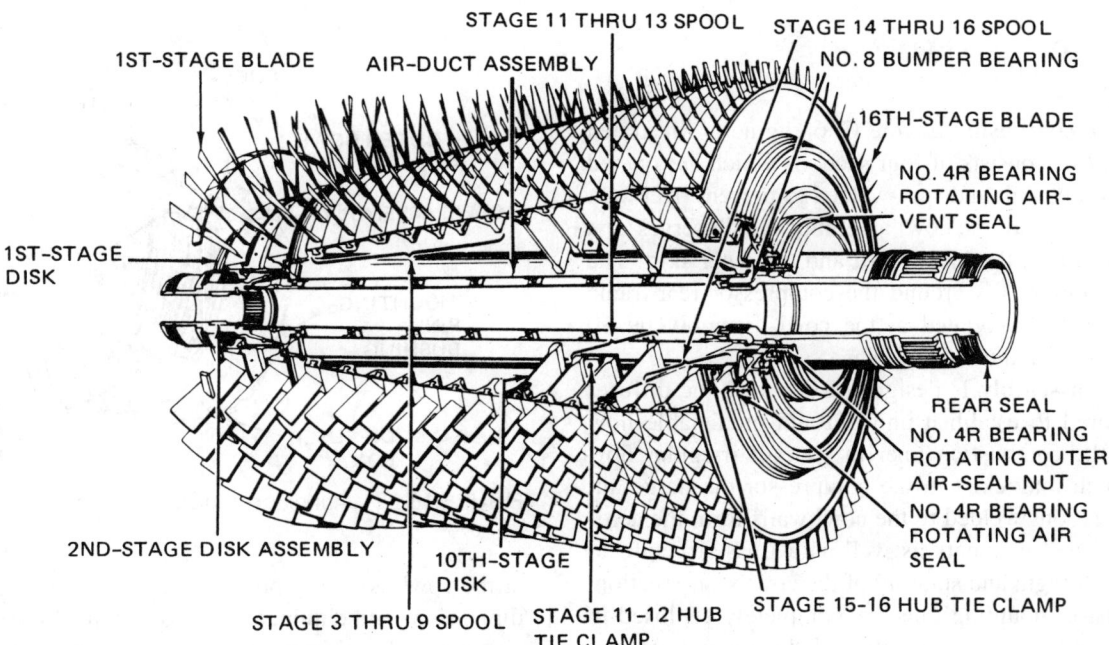

FIGURE 16-50 Compressor rotor, high pressure. (*General Electric.*)

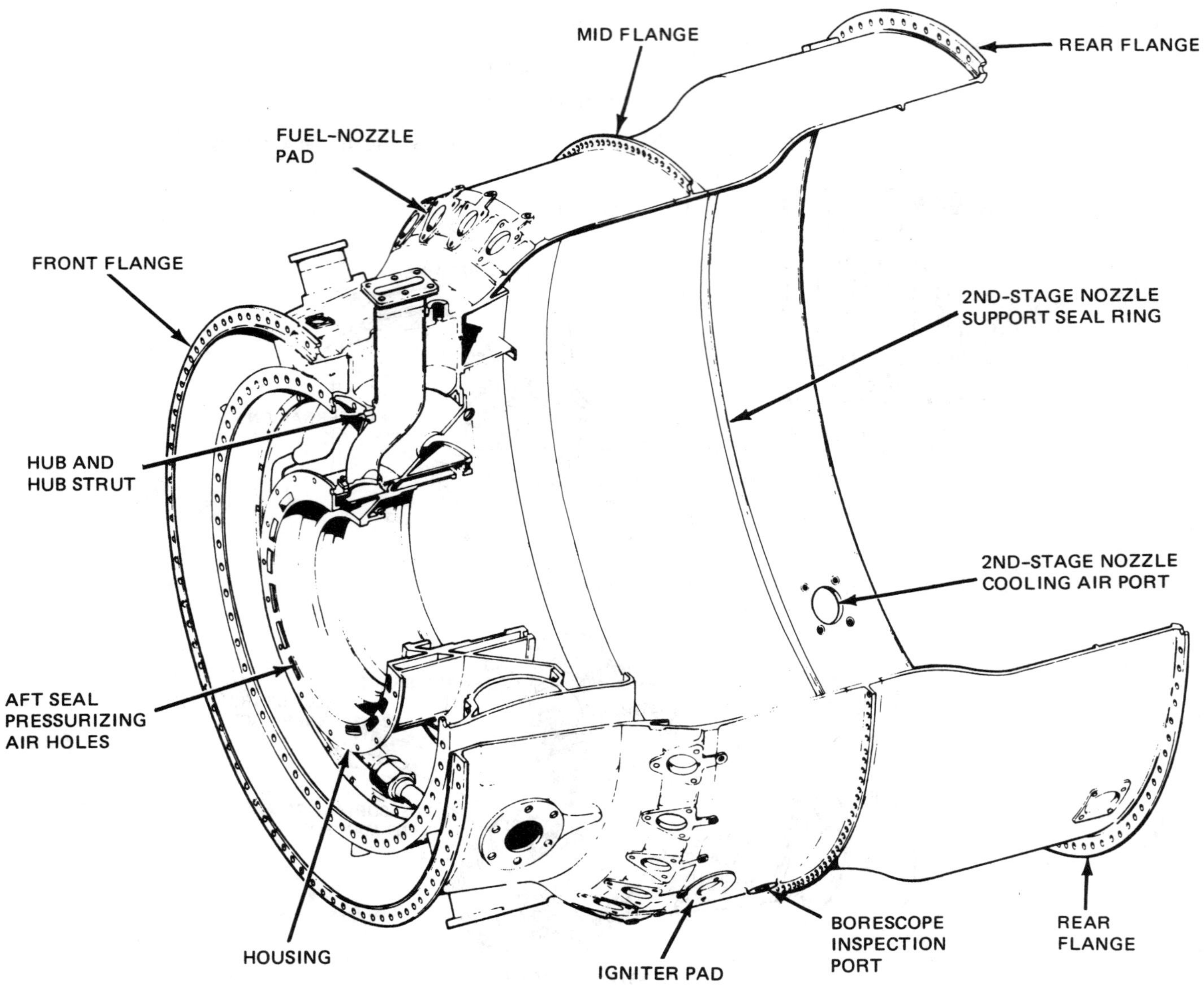

MID FLANGE

REAR FLANGE

FUEL-NOZZLE
PAD

FRONT FLANGE

2ND-STAGE NOZZLE
SUPPORT SEAL RING

HUB AND
HUB STRUT

2ND-STAGE NOZZLE
COOLING AIR PORT

AFT SEAL
PRESSURIZING
AIR HOLES

HOUSING

IGNITER PAD

BORESCOPE
INSPECTION
PORT

REAR
FLANGE

FIGURE 16-51   Compressor rear frame.

Combustor

The **annular combustor** for the CF6 engine is illustrated
in Fig. 16-52. It consists of four sections which are riveted
together into one unit and spot-welded to prevent rivet loss.
The four sections are the cowl assembly, which serves as a
diffuser; the dome; the inner skirt; and the outer skirt. The
complete assembly fits around the compressor rear-frame
struts, where it is mounted at the cowl assembly by 10
equally spaced radial mounting pins.

The **cowl assembly** is designed to provide the diffuser
action required to establish uniform and predictable flow
profiles to the combustion liner in spite of irregular flow
profiles which may exist in the compressor discharge air.
Forty box sections welded to the cowl walls form the aerodynamic diffuser elements as well as a truss structure to
provide the strength and stability of the cowl's ring section.
The combustor mounting pins are completely enclosed in
the compressor rear-frame struts and do not impose any
drag losses in the diffuser passage. Mounting the combustor

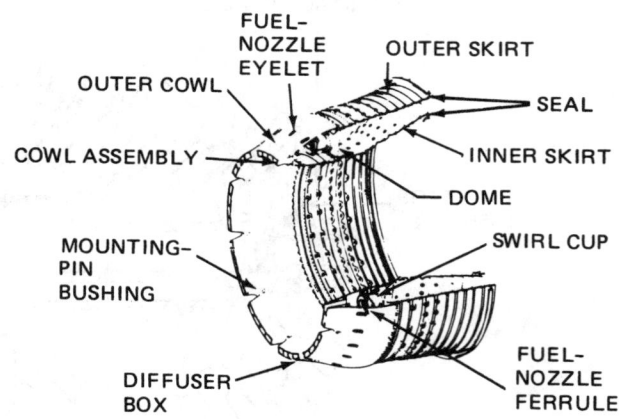

FUEL-
NOZZLE
EYELET

OUTER COWL

OUTER SKIRT

COWL ASSEMBLY

SEAL

INNER SKIRT

DOME

MOUNTING-
PIN
BUSHING

SWIRL CUP

DIFFUSER
BOX

FUEL-
NOZZLE
FERRULE

FIGURE 16-52   Annular combustor for the CF6 engine.

at the cowl assembly provides accurate control of diffuser
dimensions and eliminates changes in the diffuser flow pattern which could be caused by expansion of the assembly
because of heat.

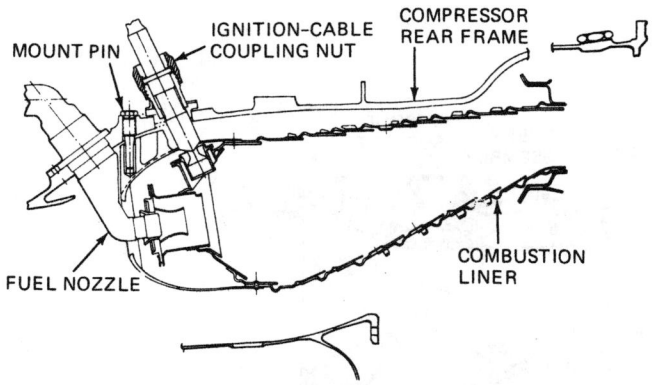

FIGURE 16-53 Cross section of the combustor.

A cross section of the combustor is shown in Fig. 16-53. The **dome** is the forward section inside the cowl assembly; it is the part which forms the actual forward unit of the combustion area. In the dome are 30 vortex-inducing axial swirl cups, one for each fuel nozzle. The swirl cups are designed to provide the airflow patterns required for flame stabilization and proper fuel-air mixing. The dome design and swirl cup geometry, coupled with the fuel-nozzle design, are the principal factors which contribute to smokeless combustion. The axial twirlers serve to lean out the fuel-air mixture in the primary zone of the combustor, and this helps to eliminate the formation of the high-carbon visible smoke which normally results from overrich burning in this zone. The dome is continuously film-cooled by the airflow.

The **combustor skirts** may be compared with combustion chamber liners, as described for some other engines.

Each skirt consists of circumferentially stacked rings which are joined by resistance-welded and brazed joints. The liners are continuously film-cooled by primary combustion air which enters each ring through closely spaced circumferential holes. The primary-zone hole pattern is designed to admit the balance of the primary combustion air and to augment the recirculation for flame stabilization. Three axial planes of dilution holes on the outer skirt and five planes on the inner skirt are employed to promote additional mixing and to lower the gas temperature at the turbine inlet. Combustion-liner and turbine-nozzle air seals provided on the trailing edge of the skirt allow for expansion due to heat (thermal growth) and accommodate manufacturing tolerances. The seals are coated with wear-resistant material.

High-Pressure Turbine

The high-pressure turbine (HPT) for the CF6 engine, shown in Fig. 16-54, consists of only two stages, located immediately to the rear of the combustor. This turbine drives the high-pressure compressor through the outer coaxial shaft.

The nozzle-vane arrangement for the first- and second-stage turbine stages is shown in Fig. 16-55. The first-stage vanes are cooled by convection, impingement, and film cooling. The details of vane construction and cooling features are provided in Fig. 16-56. The first-stage nozzle assembly is bolted at its inner diameter to the first-stage support and receives axial support at its outer diameter from the second-stage nozzle support.

The first-stage nozzle support is a sheet-metal and machined ring weldment. In addition to supporting the first-stage nozzle, it forms the inner flow path wall from the compressor

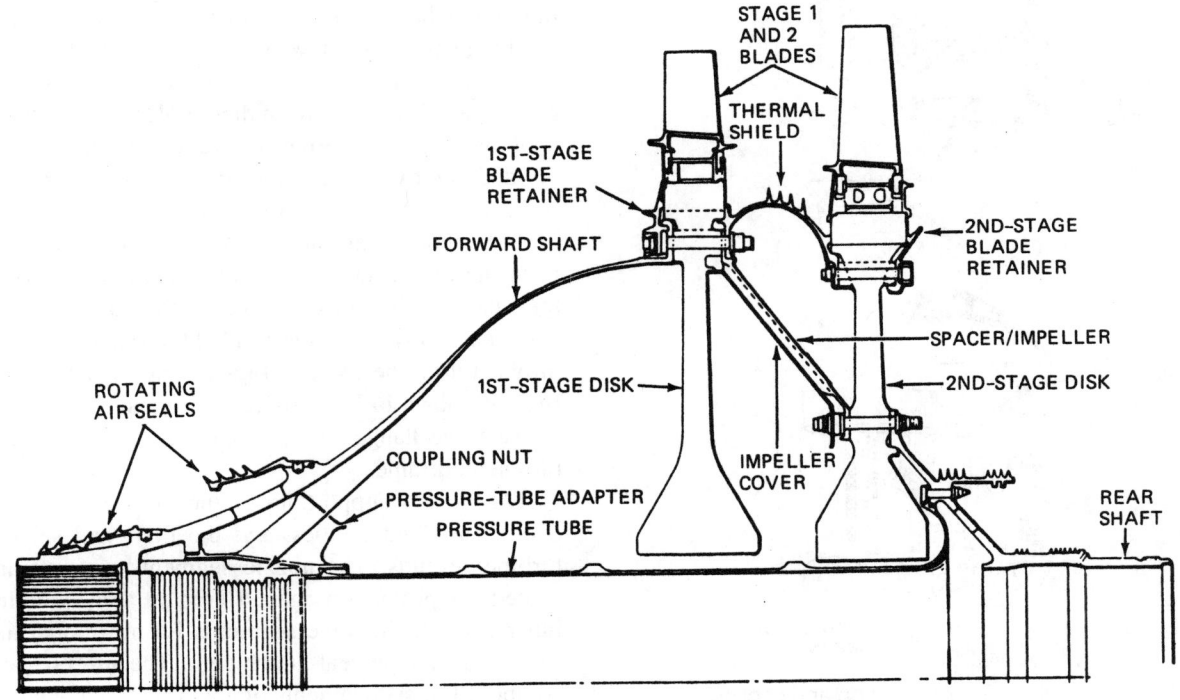

FIGURE 16-54 The CF6 high-pressure turbine. (*General Electric.*)

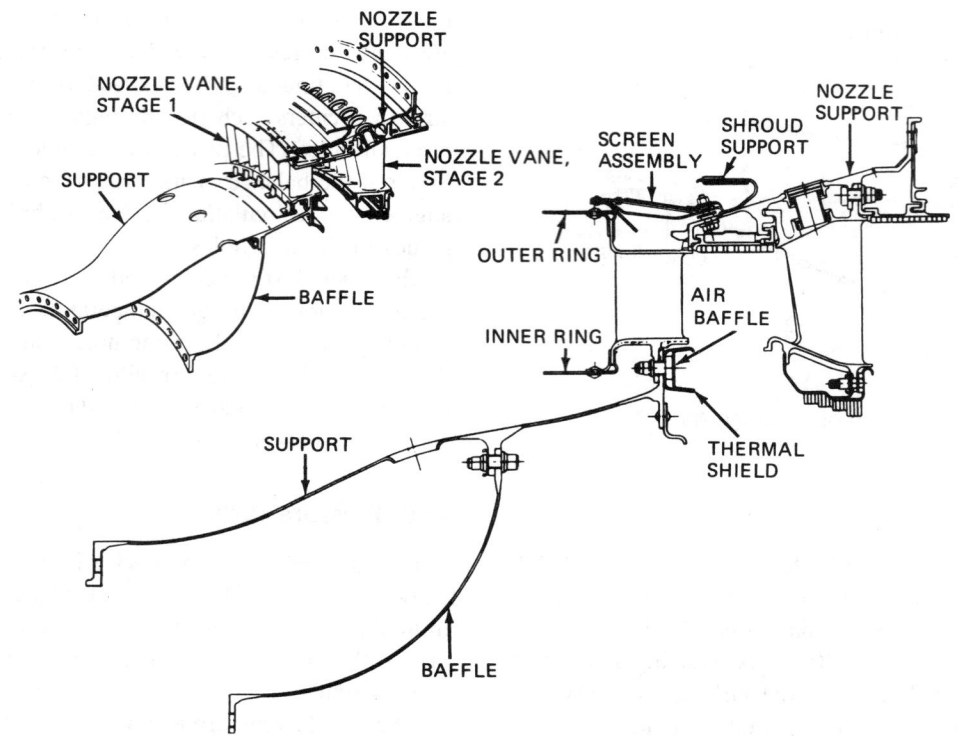

FIGURE 16-55 Nozzle-vane arrangement and supports for the first two stages of the turbine. (*General Electric.*)

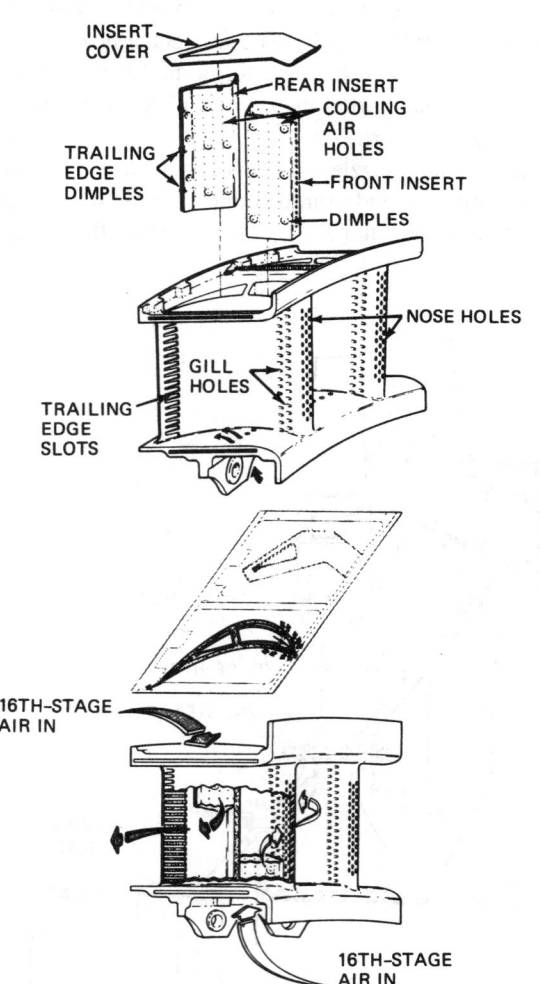

FIGURE 16-56 Nozzle-vane construction and cooling features. (*General Electric.*)

rear frame to the nozzle. The first-stage nozzle support also retains the baffle and the pressure balance seal support.

The first-stage vanes are coated to improve erosion and oxidation resistance. The vanes are cast individually and welded into pairs to decrease the number of gas leakage paths and to reduce the time required for field replacement. These welds are partial-penetration welds to allow easy separation of the two vanes for repair and replacement of individual halves. The vanes are cooled by compressor discharge air which flows through a series of leading-edge holes and fill holes located close to the leading edge on each side. Air flowing from these holes forms a thin film of cool air over the length of the vane. Internally, the vane is divided into two cavities and air flowing into the aft cavity is discharged through trailing-edge slots.

The second-stage nozzle is cooled by convection. Air from the thirteenth stage of the compressor flows through the vanes, exiting through holes in the trailing edges of the vanes and also out the inner end of the vanes to help cool the turbine rotor. The **second-stage nozzle support** is a conical ring, as shown in Fig. 16-55. The support is bolted rigidly between the flanges of the compressor rear frame and the turbine midframe.

The nozzle support holds the nozzle-vane segments, cooling-air feeder tubes, and the first- and second-stage turbine shrouds. The nozzle segments are cast and then coated for protection against erosion and oxidation. The inner ends of the vane segments form a mounting circle for the interstage-seal attachment. The **interstage seal** is composed of six segments of about 60° each which bolt to the vane segments. The function of the seal is to minimize

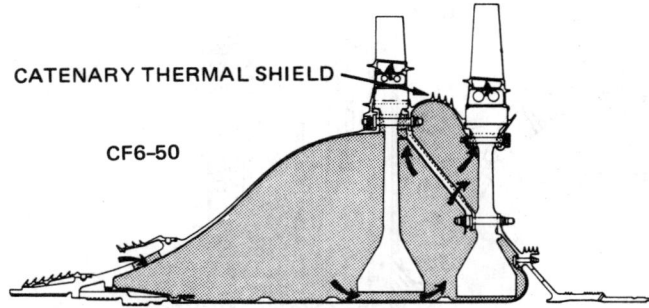

FIGURE 16-57 Cross section of the high-pressure turbine. (*General Electric.*)

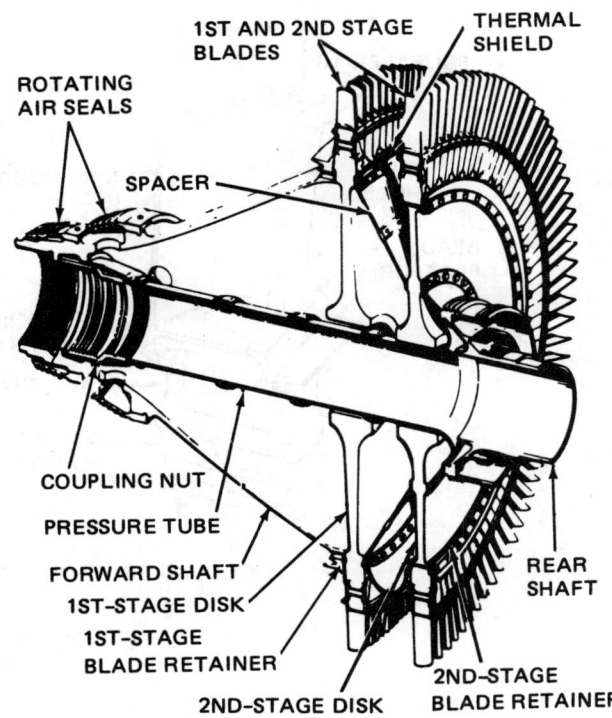

FIGURE 16-58 High-pressure turbine rotor.

the leakage of gases between the inside diameter of the second-stage nozzle and the turbine rotor. The interstage seal has four consecutive steps to provide for maximum effectiveness of each sealing tooth. These sealing teeth are shown in Fig. 16-57, which is a cross section of the HPT. The teeth are between the turbine disks on the **catenary thermal shield**.

The **turbine shrouds** form a portion of the outer aerodynamic flow path through the turbine. The shrouds are located axially in line with the turbine blades and form a pressure seal to prevent high-pressure gas leakage or bypass at the blade tip ends. The sealing surfaces are of conventional honeycomb filled with nickel aluminide compound. This construction of the sealing surfaces makes it possible for the blade tips to rub the surface without causing damage. The first- and second-stage turbine shrouds consist of 24 and 11 segments, respectively.

The HPT rotor consists of a conical forward turbine shaft, two turbine disks, two stages of turbine blades, a conical rotor spacer, a catenary-shaped thermal shield, the aft stub shaft, and precision-fit rotor bolts. Drawings of the HPT are shown in Figs. 16-57 and 16-58. The rotor of the turbine is cooled by a continuous flow of compressor discharge air drawn from holes in the nozzle support. This flow cools both the inside of the rotor and both disks before passing between the paired dovetails and out to the blades.

The conical **forward turbine shaft** transmits energy for rotation of the compressor. Torque is transmitted through the female spline at the forward end of the shaft. Two rotating air seals attached to the forward end of the shaft maintain compressor discharge pressure in the rotor combustion chamber plenum to furnish part of the corrective force necessary to minimize the unbalanced thrust load on the high-pressure rotor thrust bearing. The thrust bearing is a ball-type bearing designed to take the thrust loads of the high-pressure rotor system. This bearing does not take any of the axial bearing loads because these loads are handled by roller bearings.

The **turbine rotor spacer** can be seen between the turbine disks in Fig. 16-58. This spacer is a cone which serves as the structural support member between the turbine disks and also transmits the torque from the second-stage turbine to the forward shaft.

The turbine blades are brazed together in pairs with side-rail doublers added for structural integrity. Channel-shaped

squealer tip caps are inserted into the blade tips and are held in place by crimping the blade tip and brazing. Both stages of blades are cooled by compressor discharge air which flows through the dovetail and between the blade shanks into the blades. The construction of the first-stage blades to allow for cooling is shown in Fig. 16-59. First-stage blade cooling is a combination of internal convection and external film cooling. The convection cooling of the midchord region is accomplished through a labyrinth. The leading-edge circuit provides internal convection cooling by impingement of air against the inside surface and by flow through the leading-edge and gill holes. Convection cooling of the trailing edge is provided by air flowing through the trailing-edge exit holes. Second-stage blades are entirely cooled by convection. All the cooling air is discharged from the tips of these blades.

The rear shaft of the HPT bolts to the second-stage disk and supports the aft end of the turbine rotor. The shaft end is supported by a roller bearing.

The **blade retainers** serve two primary functions: they prevent the blades from moving axially under gas loads, and they seal the forward face of the first-stage rim dovetail and the aft face of the second-stage rim dovetail from the leakage of cooling air. An additional function is to cover the rotor bolt ends at the rotor rim and thereby prevent a substantial drag loss. These retainers are of single-piece construction and are held on by the same bolts that attach the forward shaft and thermal shield to the turbine disks.

The **pressure tube** serves to separate the high-pressure rotor internal-cooling air supply from the region of the fan midshaft which is concentric to the rotor. It is threaded into the front shaft and bolted to the rear shaft.

Large Turbofan Engines   **443**

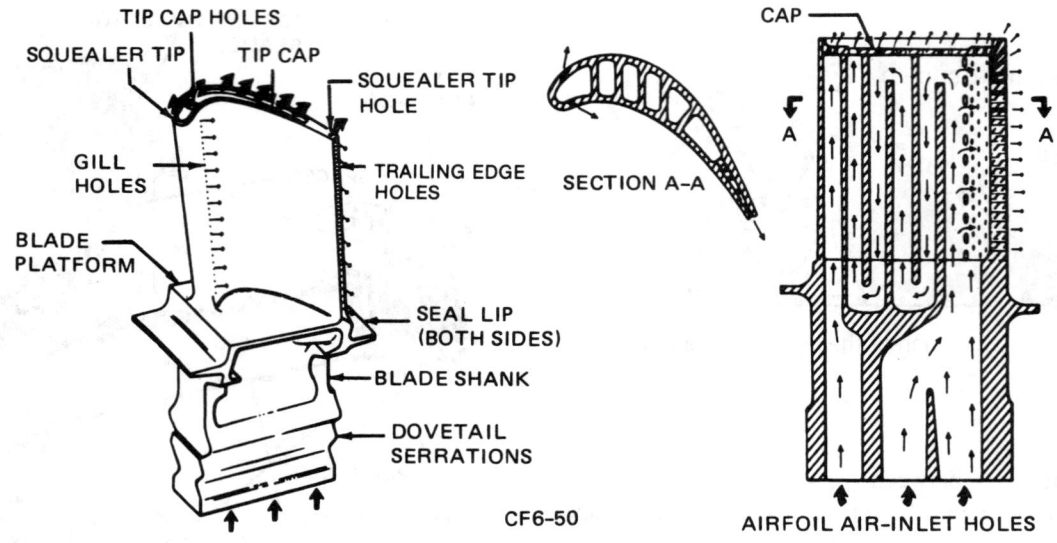

FIGURE 16-59 Construction of the first-stage turbine rotor.

Low-Pressure Turbine

The low-pressure turbine (LPT) for the CF6 engine has two different configurations. The CF6-6 model has five stages and the CF6-50 has four stages. In either case, the LPT utilizes a rotor supported between roller bearings mounted in the turbine midframe and the turbine rear frame. A horizontally split LPT casing containing stator vanes is bolted to these frames to complete the structural assembly. This provides a rigid, self-contained module which can be precisely and rapidly interchanged on the engine without requiring a subsequent engine test run. The **low-pressure turbine shaft** engages the long **fan drive shaft** through a spline drive and is secured by a lock bolt. The forward flange of the turbine midframe is bolted to the aft flange of the compressor rear frame, after installation of the HPT, to complete the engine assembly. A cross section of the CF6 low-pressure turbine is shown in Fig. 16-60.

Since the temperature of the exhaust gases is reduced considerably through the HPT, the LPT blades do not require cooling. Compressor-seal leakage air is used to cool the first two LPT disks to reduce thermal gradients.

The **first-stage low-pressure turbine nozzle** consists of 14 segments, each containing six vanes, as shown in Fig. 16-61. The segments are supported at their inner and outer ends by the turbine midframe and LPT casing, respectively. This provides low vane-bending stress and freedom to expand or contract thermally without thermal loads and stresses. The other stages of LPT vanes are cantilevered from the LPT stator casing. These can be seen in Fig. 16-62, which illustrates one-half of the LPT stator casing. The major parts of the **stator** are the nozzle stages, a split casing, shrouds, and interstage air seals. Each of the nozzle stages is composed of cast segments of six vanes each, any segment of which can be replaced with simple tools. The shrouds are in segments

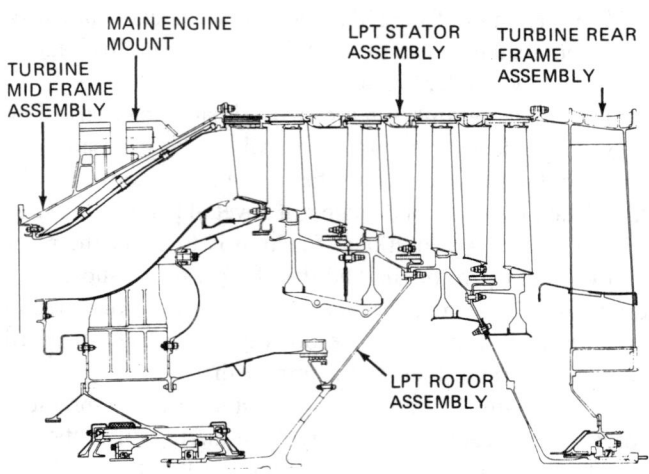

FIGURE 16-60 Cross section of the low-pressure turbine.

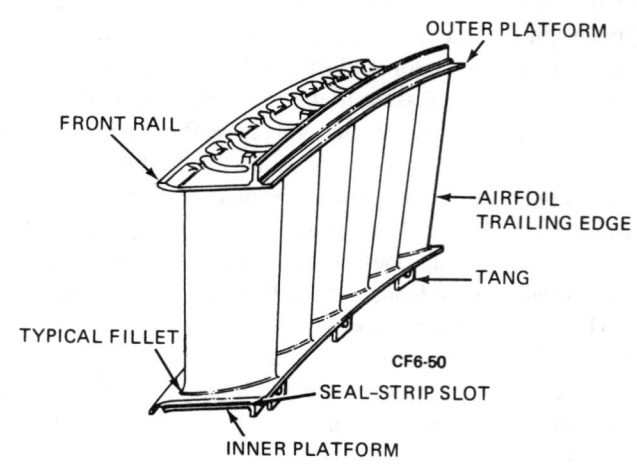

FIGURE 16-61 Segment of the first-stage low-pressure turbine nozzle.

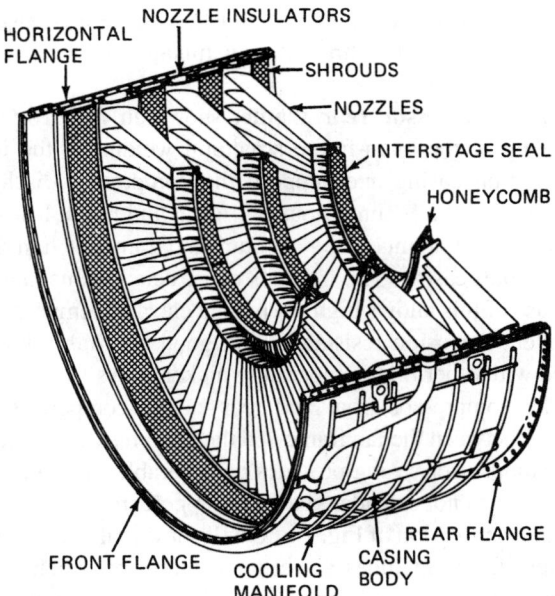

FIGURE 16-62 Low-pressure turbine stator casing and vanes. (*General Electric.*)

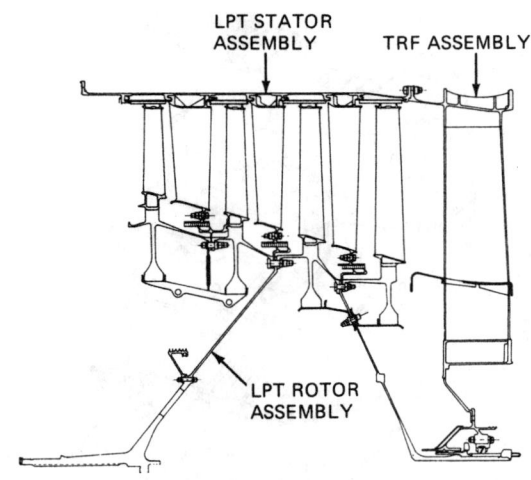

FIGURE 16-64 Attachment of low-pressure turbine blades.

held in place by projections mating with slots formed by the casing and nozzles. The interstage seals are bolted to the inside-diameter flange of the nozzles. Both shrouds and seals have abradable honeycomb sealing surfaces to allow close clearance without the risk of rotor damage caused by unusual rubs. The interstage seals partially restrain the inner vane ends to provide damping which results in low vane stresses.

The **low-pressure turbine rotor assembly** is shown in Fig. 16-63. Note that the turbine blades have interlocking blade shrouds to produce a continuous shrouded surface around each turbine stage. This reduces blade stress and helps to seal gases within the desired flow path. The shroud interlocks are in contact at all rotor speeds by virtue of an interference fit caused by pretwisting of the blades. This provides adequate damping at low speeds and damping proportional to rotor speed at higher speeds. Shroud interlocks are hard-coated for wear resistance.

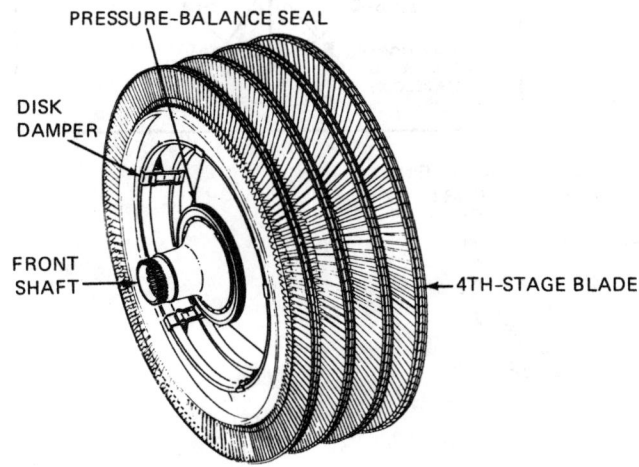

FIGURE 16-63 Low-pressure turbine rotor assembly. (*General Electric.*)

The disks for the LPT are made of Inconel 718 alloy, and each has integral torque-ring extensions. These are attached to the adjacent disks by close-fitting bolts. Bolt holes through the disk webs have been eliminated by locating them in the flanges where stresses are low.

Figure 16-64 shows how the front and rear shafts are attached to the disks. This arrangement forms a stiff rotor structure between bearings.

The turbine blades are attached to the disks by means of multitang dovetails. Replaceable rotating air seals mounted between the disk flanges mate with stationary seals to provide interstage air sealing.

Because of its low length-to-diameter ratio and the accurate fitting of bolts to fasten the disks together, the LPT rotor is highly stable. In the event of a midshaft failure, the LPT rotor would move aft until blade rows interfered with the stators. This would prevent any dangerous overspeed.

In the CF6-50 engine, an external impingement system is employed to cool the LPT rotor casing. This system employs fan-discharge air bled from the inner fan flow path. Utilizing fan air instead of the stage-9 air which is used in the CF6-6 engine results in a lower chargeable airflow and provides a further improvement of specific fuel control.

The LPT module, consisting of a turbine and case, is designed for easy removal and reinstallation, with the engine either in or out of the aircraft. The only quick-engine-change (QEC) items requiring removal are the thrust spoiler and drives, the EGT electrical harness, and condition-monitoring leads, if installed.

Support Structures

The supporting structures of the CF6 engine are the fan frame, compressor rear frame, turbine midframe, and turbine rear frame. These structures, bolted together, provide the rigid casing necessary to support the rotating and combustion elements of the engine.

The **fan frame** is illustrated in Fig. 16-65. It supports the forward end of the compressor, the fan rotor, fan stator, forward engine mount, radial drive shaft, transfer gearbox, horizontal

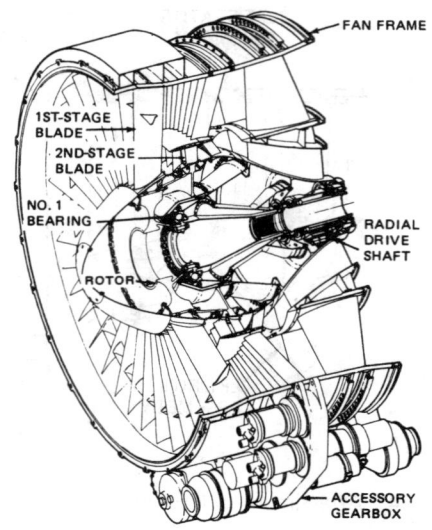

FIGURE 16-65 Drawing of the fan section. (*General Electric.*)

The forward engine mount attaches to the aft flange of the fan frame and a thrust mount linkage connects to the no. 1 strut casing.

The **compressor rear frame**, shown in Fig. 16-66, is made up of the main-frame structural weldment, the inner combustion casing and support, the compressor discharge air seal, and the B sump housing. Axial and radial loads are taken in the rigid inner-ring structure and transmitted in shear to the outer casing. The frame is constructed of Inconel 718 because of this alloy's high strength at elevated temperatures, excellent corrosion resistance, and good repairability compared with other high-temperature nickel alloys.

The compressor front and rear casing components are located between the fan frame and the compressor rear frame. The compressor casing provides considerable structural support, but it is not considered a main support structure.

An examination of Fig. 16-66 will show that the compressor rear frame includes a section at the front which serves partially as a diffuser. It also houses the combustion chamber described previously.

Bearing axial and radial loads and a portion of the high-pressure turbine first-stage nozzle loads are taken in the inner ring, or hub, and transmitted through the 10 radial struts to the outer shell. The inner ring of the frame is a casting which contains approximately half the radial strut length. The cross-sectional shape of the hub is a box, to provide structural rigidity.

The outer strut ends are castings which complete the formation of the struts when welded to the hub. Combining the hub and outer strut casting in this manner forms a

drive shaft, and accessory gearbox. The fan frame has 12 stainless-steel struts equally spaced at the leading edge. The struts bolt to the aft outer casing with eight bolts per strut. The six and twelve o'clock struts are the leading edges of the upper and lower pylons, shaped to flair into the pylon sidewalls. The radial shaft is enclosed within the lower pylon and is bolted to the strut in the six o'clock position. The fan frame houses the A sump, which includes the no. 1, 2, and 3 bearings. Sump pressuring air enters through the leading edges of struts 4, 5, 9, and 10 in the fan stream section.

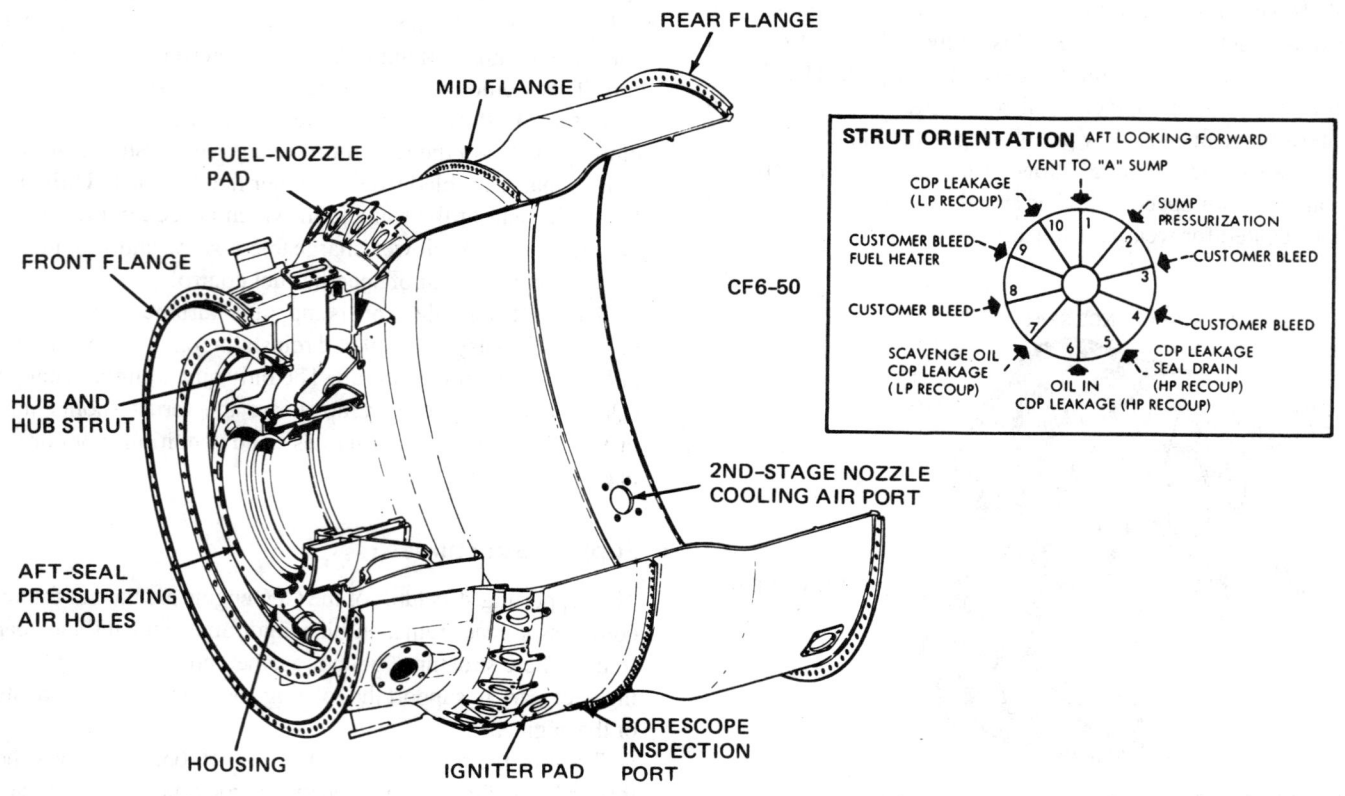

FIGURE 16-66 Compressor rear frame.

smooth strut-to-ring transition with the minimum concentration and no weld joints in the transition area. The 10 radial struts are airfoil-shaped to reduce aerodynamic losses and are sized to provide adequate internal area for sump service lines and bleed airflow. The hub and outer strut end assembly is then welded into the outer shell, which is a sheet-metal and machined ring weldment defining the outer-flow annulus boundary, as well as providing the structural load path between the high-pressure compressor casing and turbine midframe.

To provide for the differential thermal growth between sump service tubing and the surrounding structure, the tubes are attached only at the sump, and slip joints are used where tubes pass through the outer strut ends.

The **turbine midframe** is illustrated in Fig. 16-67. It consists of the outer casing reinforced with hat-section stiffeners, the link mount castings, the strut end castings, the cast hub, eight semitangential bolted struts, the C sump housing, and a one-piece flow path liner. The frame casing and cast hub operate coolly enough at all conditions to permit the use of Inconel 718. The operating temperature does not exceed 1100°F [593.33°C].

The struts are made of René 41, which is a high-temperature alloy developed by the General Electric Company. The selection of bolted structural joints at the inner and outer ends of the struts was made to satisfy the unique requirements arising from the location of the midframe between the high-pressure and low-pressure turbines. The liner and fairing assembly is a one-piece unit.

The turbine midframe supports the no. 5 and no. 6 bearings. The bearing support cone and sump housing is bolted to the forward flange of the frame's inner hub for ease of replacement, maintenance, and manufacture. Tubing through the frame's structural struts for sump service is secured to the sump by bolted flanges or B nuts to make the pump completely separable from the frame structure. The sump housing

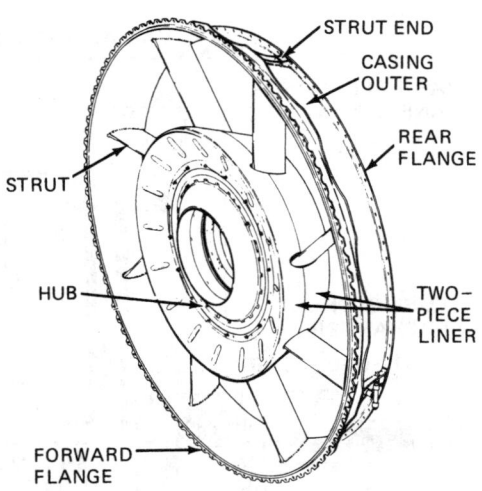

FIGURE 16-68   Turbine rear frame.

is of double-walled construction so that the wall of the sump bathed in oil is cooled by fan-discharge air, keeping its temperature below 350°F [176.67°C].

The **turbine rear frame**, illustrated in Fig. 16-68, is the rearmost structural member of the engine. This unit can be divided into the main-frame structure, the inner flow path liner, the D sump, and the sump service piping. It supports the no. 7 bearing in the D sump.

The assembly of the turbine rear frame is similar to that of the midframe but without the use of bolted struts. It is a welded Inconel 718 structure with eight equally spaced, partially tangential struts supported on two axially spaced rings at both the hub and outer ring. With the struts and with a gas temperature 500°F [278°C] lower than that of the turbine midframe, the rear frame can be designed without flow-path-liner or strut fairings. The main advantage of this type of construction is the accessibility of structural welds for visual inspection without disassembly.

Refer back to Fig. 16-46. The turbine rear frame can be seen at the extreme rear of the engine. It is separated from the turbine midframe by the low-pressure turbine stator assembly, which is illustrated in Fig. 16-62.

Engine Systems

The systems for the CF6 engine are similar to systems described previously. All systems utilize components of proven technology and reliable performance.

The fuel system is shown in Fig. 16-69. The principal units are the fuel pump, fuel-oil heat exchanger, main engine control, CIT sensor, variable-stator-vane actuator, pressurizing valve, and fuel nozzle.

The lubrication system is of the pressure type, with oil controlled in the various sumps by means of air seals and pressurization from the engine compressor. Scavenge oil is drained to the scavenge pump; from there it is returned to the oil tank.

The electric system is shown in Fig. 16-70. Note that the ignition system is energized by means of 400-Hz 115-V

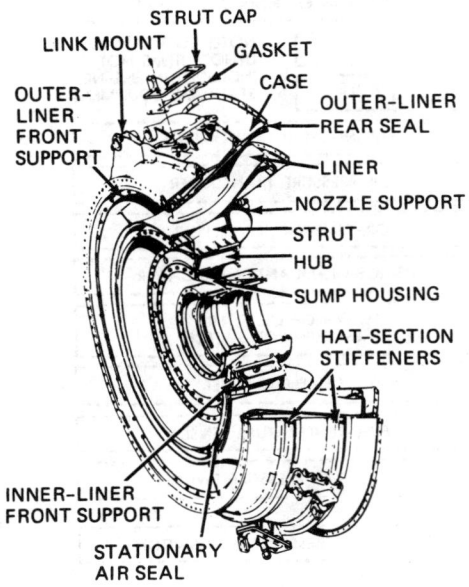

FIGURE 16-67   Turbine midframe.

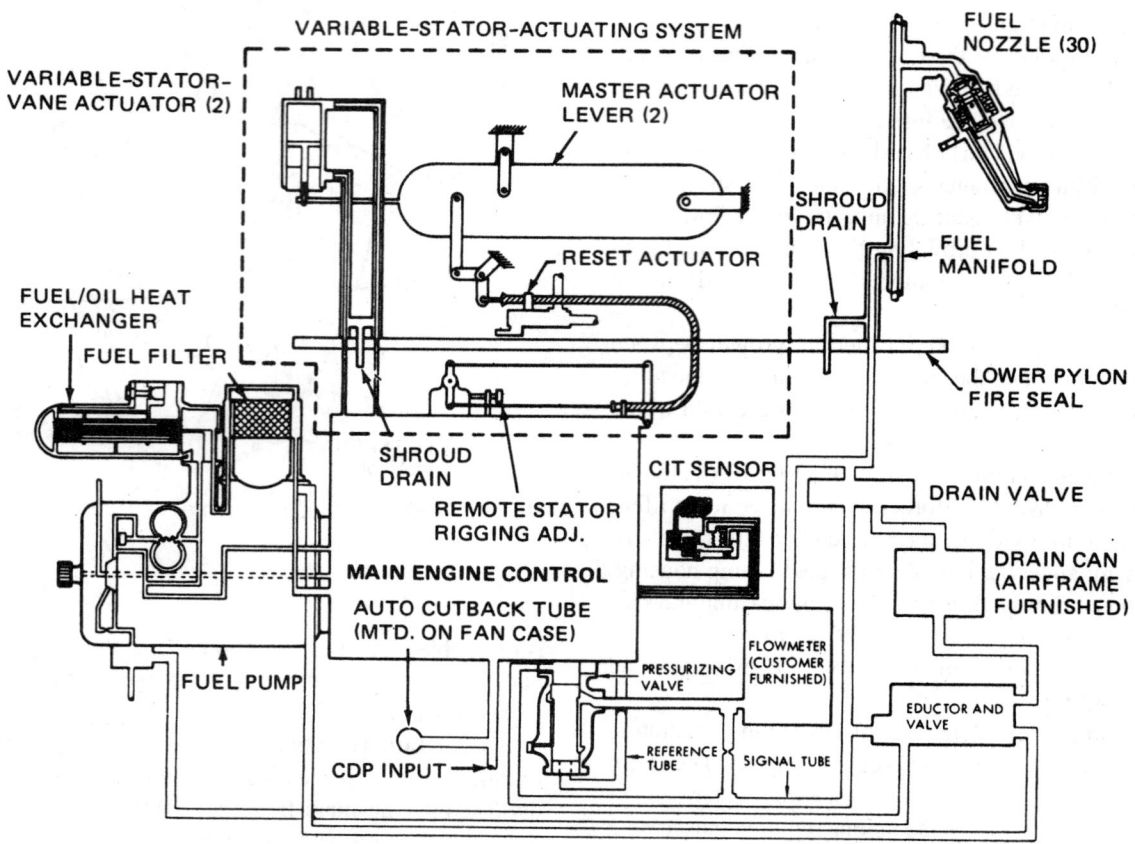

FIGURE 16-69 The CF6 fuel system. (*General Electric.*)

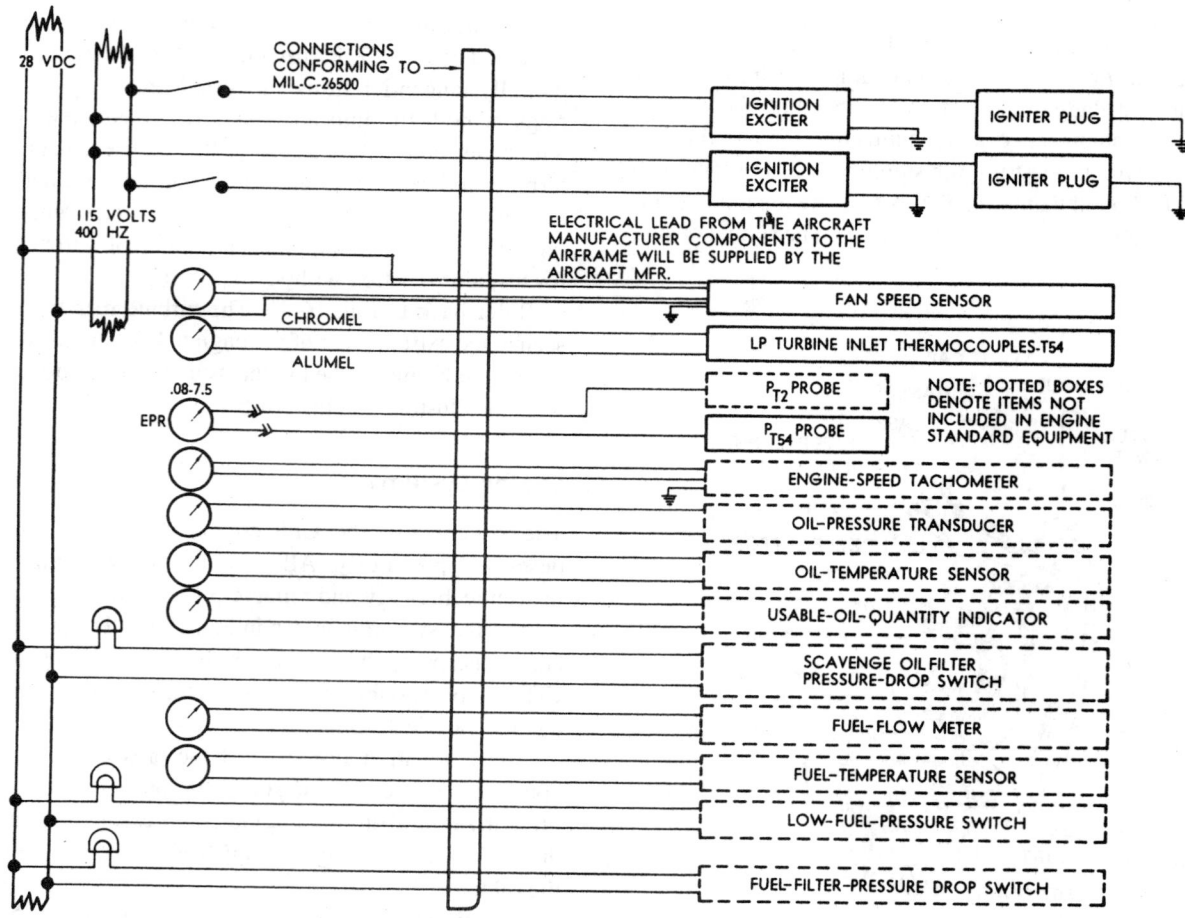

FIGURE 16-70 The CF6 electric system. (*General Electric.*)

alternating current. Other circuits are supplied from the 28-V dc source.

The General Electric CF6-80 Engine

The General Electric CF6-80 engine is essentially an upgraded version of the CF6-50 engine. Both engines are in approximately the same thrust class; however, design improvements have been made in length, weight, and maintainability to produce the CF6-80 model. The improvements incorporated in the CF6-80 design have resulted in greater dependability, durability, and efficiency.

The CF6-80A model engine is used in the Boeing 767 aircraft, and the CF6-80A1 engine is used to power the Airbus A310. The main difference between the CF6-80A and -80A1 models is in the location of the accessory gearbox. The CF6-80A has a core-mounted gearbox, which provides a smaller frontal area, thus improving the aerodynamic efficiency of the engine. The-80A1 model has the gearbox mounted on the fan stator case, which provides easier access for maintenance. Each model of engine was designed specifically for the aircraft in which it was to be used. General Electric manufactures both the engine and nacelle for the Airbus A310 but supplies the engine only for the Boeing 767.

The CF6-80 engine produces 48 000 lb [213 504 N] of thrust and has a bypass ratio of 4.66:1. The compressor pressure ratio is 28:1.

CFM56 Series Turbofan Engines

The CFM56 turbofan engine is produced by CFM International, a company owned jointly by General Electric of the United States and SNECMA (Société Nationale d'Étude et de Construction de Moteurs d'Aviation) of France. The CFM56 is a high-bypass, dual-rotor, axial-flow, advanced-technology turbofan engine, available in several thrust ratings for different aircraft applications. The CFM56-3, shown in Fig. 16-71, is rated at 18 500 to 23 500 lb [82 to 105 kN] thrust and is the exclusive powerplant for Boeing 737-300, -400, and -500 aircraft. The CFM56-2, rated at 22 000 lb [98 kN] thrust for commercial applications, powers the McDonnell Douglas DC-8 Super 71, 72, and 73 aircraft; the CFM56-2, rated at 22 000 to 24 000 lb [98 to 107 kN] thrust for military applications, powers the Boeing KC-135R and C-135FR tankers, E-3A airborne warning and control system (AWACS) aircraft, E-6A submarine communications aircraft, and KE-3A tanker. The CFM56-5A, rated at 25 000 to 26 500 lb [111 to 118 kN] thrust, powers the Airbus Industrie A320. The CFM56-5B series, rated at 22 000 to 31 000 lb [98 to 138 kN] thrust, powers the Airbus Industrie A319, A320, and A321. The CFM56-5C series, rated at 31 200 to 32 500 lb [139 to 144 kN] thrust, powers the Airbus Industrie A340.

Engine Structure

The external flanges of the CFM56 engine have been assigned letter designations, as shown in Fig. 16-72. The letter designations will be used for flange identification wherever it is necessary to be specific about flange location,

such as in discussions of the positioning of brackets, clamps, bolts, etc. The rotors are supported by five main bearings that are mounted in two engine sumps of the lubricating system. The forward sump in the fan frame contains bearing no. 1 (ball) and bearing no. 2 (roller), which supports the fan and booster assembly. The HP compressor front shaft is supported by bearing no. 3 (ball). The aft sump in the turbine frame contains the internal LPT shaft bearing no. 4 (roller), which supports the aft end of the HP turbine rear shaft. The turbine frame also contains bearing no. 5 (roller), which supports the aft end of the low-pressure turbine.

Modular Design

The CFM56-3 engine has three major modules that can be separated from the assembly to perform specific maintenance operations. The three major modules are the fan major module, the core engine major module, and the low-pressure turbine major module, as shown in Fig. 16-73.

The fan major module consists of the fan rotor, the fan outlet guide vanes (OGVs), the booster rotor and stator, the no. 1 and no. 2 bearing supports, the inlet gearbox (IGB) and no. 3 bearing support, the fan casing, and the fan frame. The principal function of the fan is to direct the major volume of accelerated incoming air to the engine secondary airflow nozzle. The result of the secondary airflow accounts for approximately 78 percent of the total engine thrust. The fan and low-pressure booster module consists of a single-stage fan rotor (38 blades) and a three-stage axial-flow compressor that is bolted to the rear outer flange of the fan disk. There are 38 titanium-alloy (Ti/TA6V) fan blades. The blades are approximately 14.5 in (368 mm) in length, and feature an integral midspan shroud. Each blade has a dovetail base that engages into the matching disk slot. The interlocking midspan shrouds provide increased blade assembly stiffness. Blades are individually retained in position by a blade retainer and a spacer, as shown in Fig. 16-74.

The core engine major module consists of the rotor and stator assemblies, the combustion casing and chamber, the HP turbine-nozzle and rotor assemblies, and the HP turbine shroud and stage 1 LP turbine-nozzle assembly. The compressor section of the core engine major module provides compressor air to the combustor and bleed air from stages 5 and 9 for engine and aircraft use. The HP compressor rotor, shown in Fig. 16-75, is an axial-flow, nine-stage, spool-and-disk structure. The stages 1 and 2 spool, front shaft, stage-3 disk, and stages 4 through 9 spool are assembled and bolted together at the stage-3 disk. The CDP seal is bolted between the stages 4 through 9 spool and HPT forward shaft. An air duct through the inside of the rotor forms a flow path for pressurizing and cooling air. The upper and lower stator case halves contain stator vanes, vane shrouds, liners, and variable vane and actuator assemblies. The inlet guide vanes (IGVs) and compressor stages 1 through 3 are variable stator vanes. The combustion case surrounds the combustion chamber and provides a path for compressor discharge airflow around the chamber. The compressor outlet guide vanes are part of the combustion case at the forward flange.

FIGURE 16-71 CFM56-3 high-bypass turbofan engine. (*CFM International.*)

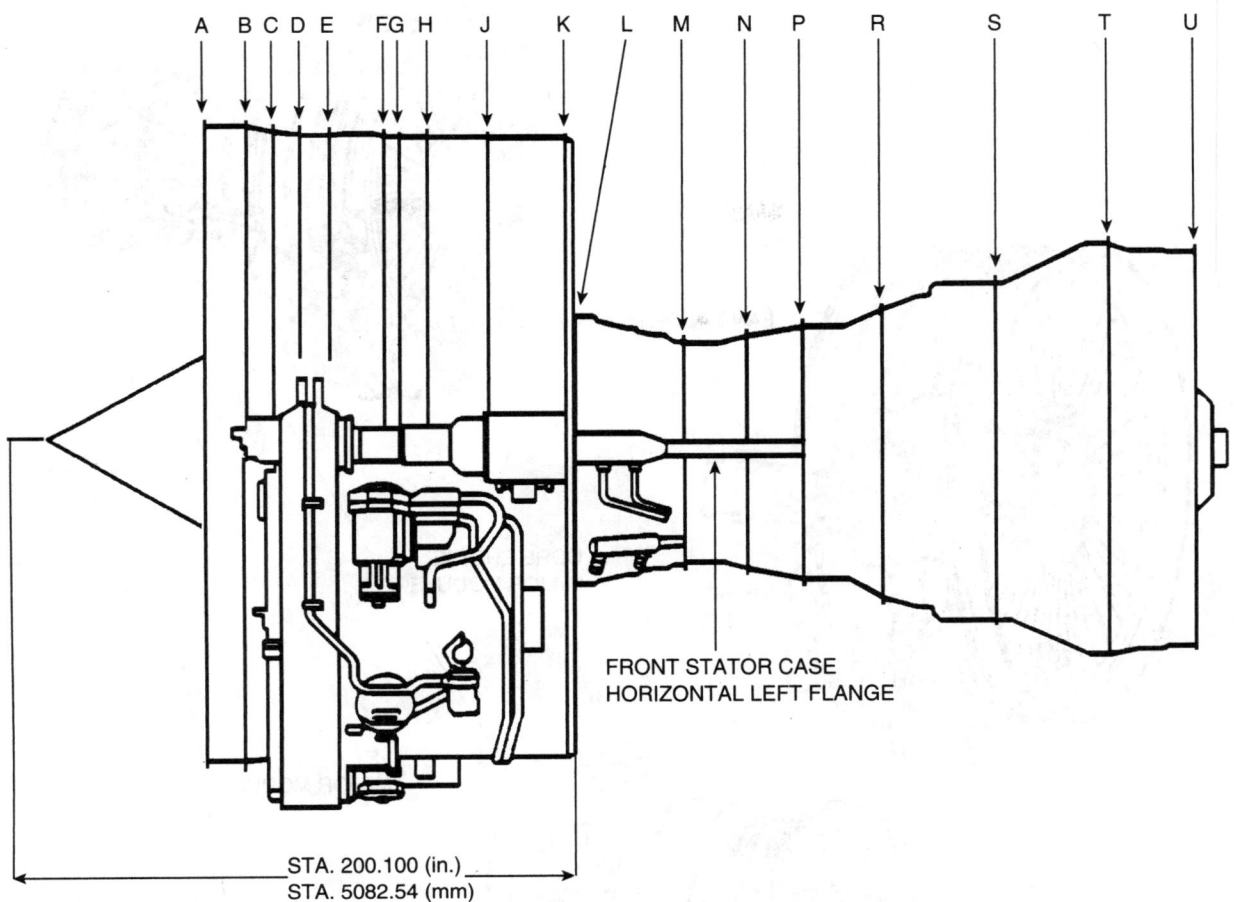

FIGURE 16-72 Flanges and station identification. (*CFM International.*)

The combustion chamber is of an annular design and contains an outer support ring, outer liner, outer cowl, inner support ring, inner liner, and inner cowl. Inside the outer and inner cowls, the primary swirl nozzle, venturi, secondary swirl nozzle, dome ring, sleeve, and deflector form a bolted assembly of 20 equally spaced positions for fuel nozzles, as shown in Fig. 16-76. The high-pressure turbine (HPT) section consists of the high-pressure turbine stators and the high-pressure turbine rotor. The HPT stator uses 23 nozzle segments comprised of two vanes each brazed into inner and outer platforms. The nozzle segments are internally cooled by CDP air that bypasses the combustor inner and outer liners. The high-pressure turbine rotor uses 72 blades which slide into individual disk dovetail slots and are internally cooled by CDP air by three different methods: convection, impingement, and film.

The low-pressure turbine rotor and stator module, shown in Fig. 16-77, consists of the following major assemblies: LPT stator, LPT rotor, and LPT case cooling manifold. The main function of these assemblies is to produce the necessary power to drive the fan and booster rotor. The main purpose of the LPT nozzle assembly (stator) is to direct the high-velocity gases from the HPT rotor onto the blades of LP turbine rotor stage l at the proper angle, as well as to expand and accelerate gases, resulting in decreases in pressure and temperature. Stage 1 of the LPT nozzle assembly forms the

interface between the core and the LP turbine. The LPT rotor disk has 174 dovetail slots and 174 tip-shrouded blades. The disk material is Inconel 718 and the blades are constructed of Inconel 100. The LPT case cooling manifold provides active clearance control for the LPT turbine and consists of two separate air cooling sources. Cooling is provided by a collection of secondary fan airflow from the engine, which also film-cools the outside of the LPT case. Each manifold is made up of six tubes with one end capped and the other end attached to the manifold for the distributor assembly. In each manifold tube there are small orifices which direct the air (supplied by the fan) toward the LPT case.

GE90 Turbofan Engine

The new GE90 high-bypass turbofan engine, shown in Fig. 16-78 at 87 400 lb [388 kN] thrust, but has the proven capability to "grow" in response to demand. In ground testing, the GE90 has already produced 105 400 lb [468 kN] of thrust, the highest thrust level ever attained by a gas turbine engine. Embodying unique technological innovations and the benefits of GE Aircraft Engines' extensive experience, particularly with the CF6 and CFM56 turbofan engines and the NASA/GE Energy Efficient Engine (E^3) program, the GE90 will provide airlines with a 9 percent improvement in specific fuel consumption while reducing

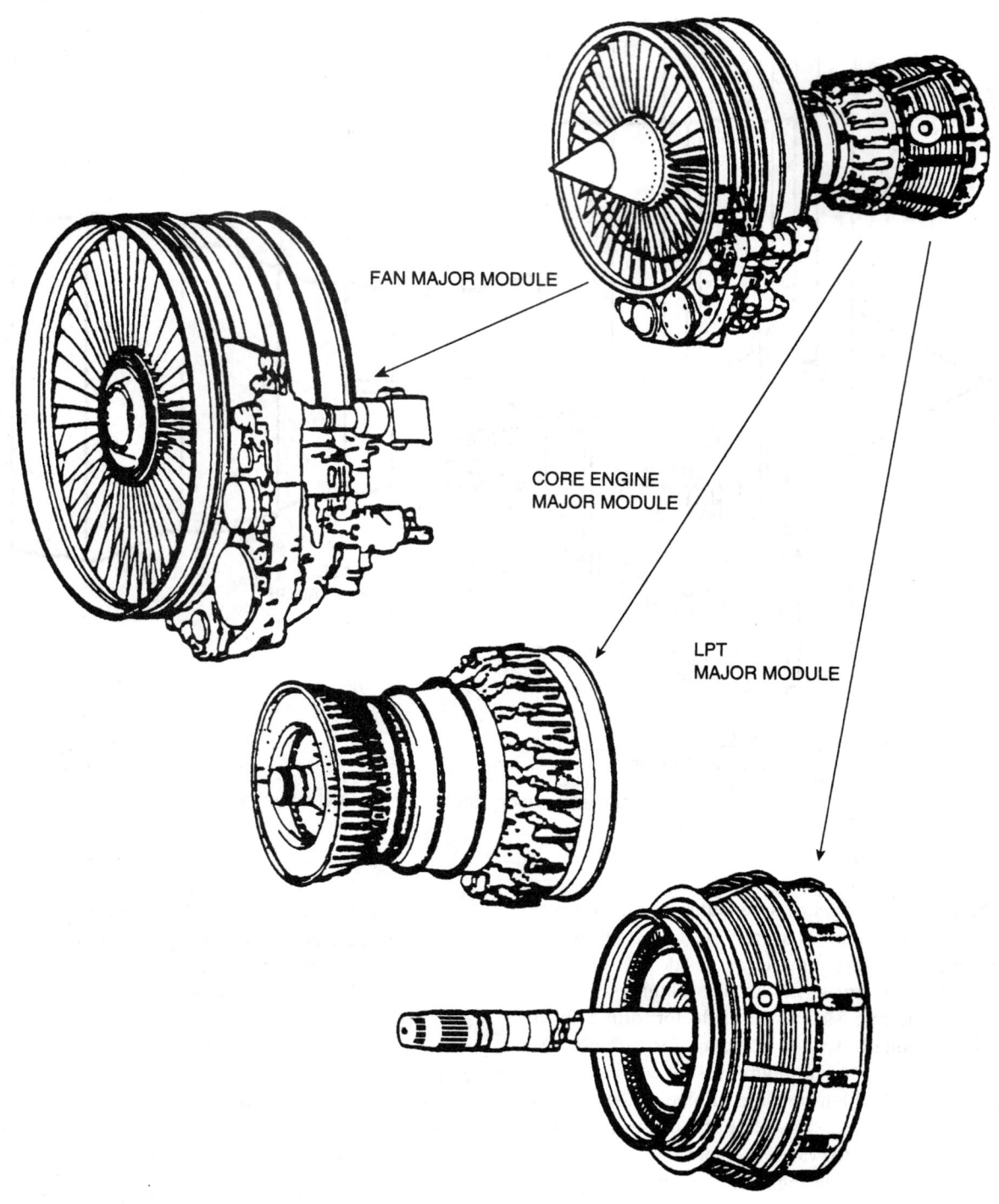

FAN MAJOR MODULE

CORE ENGINE
MAJOR MODULE

LPT
MAJOR MODULE

FIGURE 16-73 CFM56-3 modular design. (*CFM International.*)

noise and emissions levels to well within the limits imposed by new, more stringent environmental regulations.

The GE90 will feature a 123-in [312-cm] diameter fan with wide-chord composite blades and an energy-efficient compressor. The dual-dome combustor, which produces very low emissions, is also extremely durable. Likewise, the two-stage high-pressure turbine incorporates advanced materials and clearance control along with a six-stage low-pressure turbine. Airframes for which the GE90 may potentially be selected include the Boeing 777, the Airbus Industrie A330 derivatives, and McDonnell Douglas MD-12X.

Rolls-Royce RB 211 Turbofan Engine

The Rolls-Royce RB 211 turbofan series of engines is another example of the modern high-bypass high-performance engines designed for use on large, wide-body transport aircraft. The RB 211 engine, illustrated in Fig. 16-79, is a three-shaft engine, designed primarily to power the Lockheed L-1011 aircraft. Newer versions of the RB 211 family are used, or are options, on the Boeing 747 757, MD 11, and other current-production aircraft. One version of the new RB 211 family is the RB 211-524G/H, which is shown in Fig. 16-80.

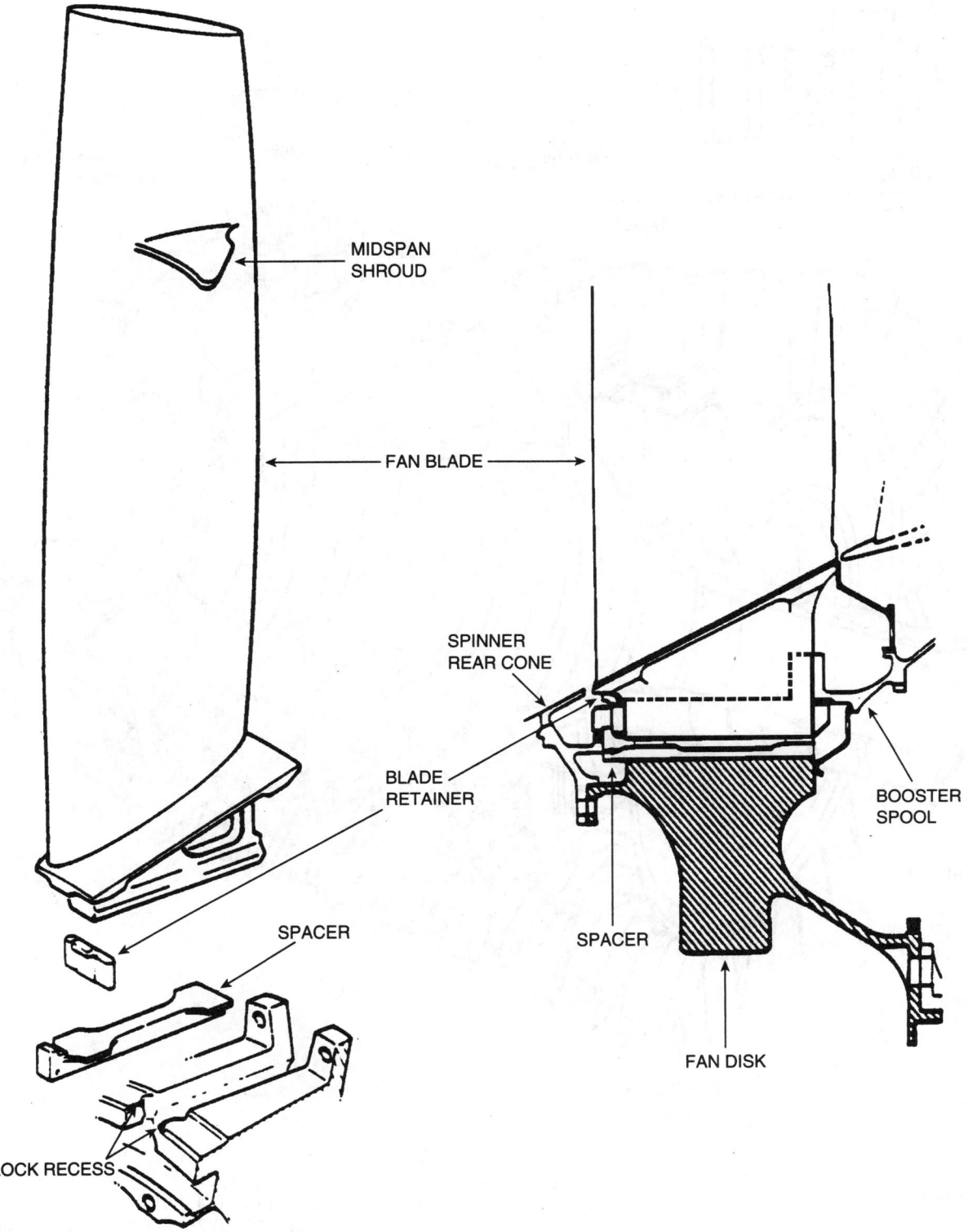

FIGURE 16-74 Fan blade retention. (*CFM International.*)

The three-shaft configuration is employed to permit the fan to rotate independently so that it will not limit the optimum rotational speed of the intermediate compressor. The fan is driven by the rearmost three stages of the turbine through the inside coaxial shaft, the intermediate compressor is driven by one stage of the turbine, and the high-pressure compressor is driven by the first stage of the turbine through the outermost of the three coaxial shafts. The drive arrangement can be seen in Fig. 16-81.

The combustion chamber is a short, annular-type chamber fitted with 18 fuel nozzles.

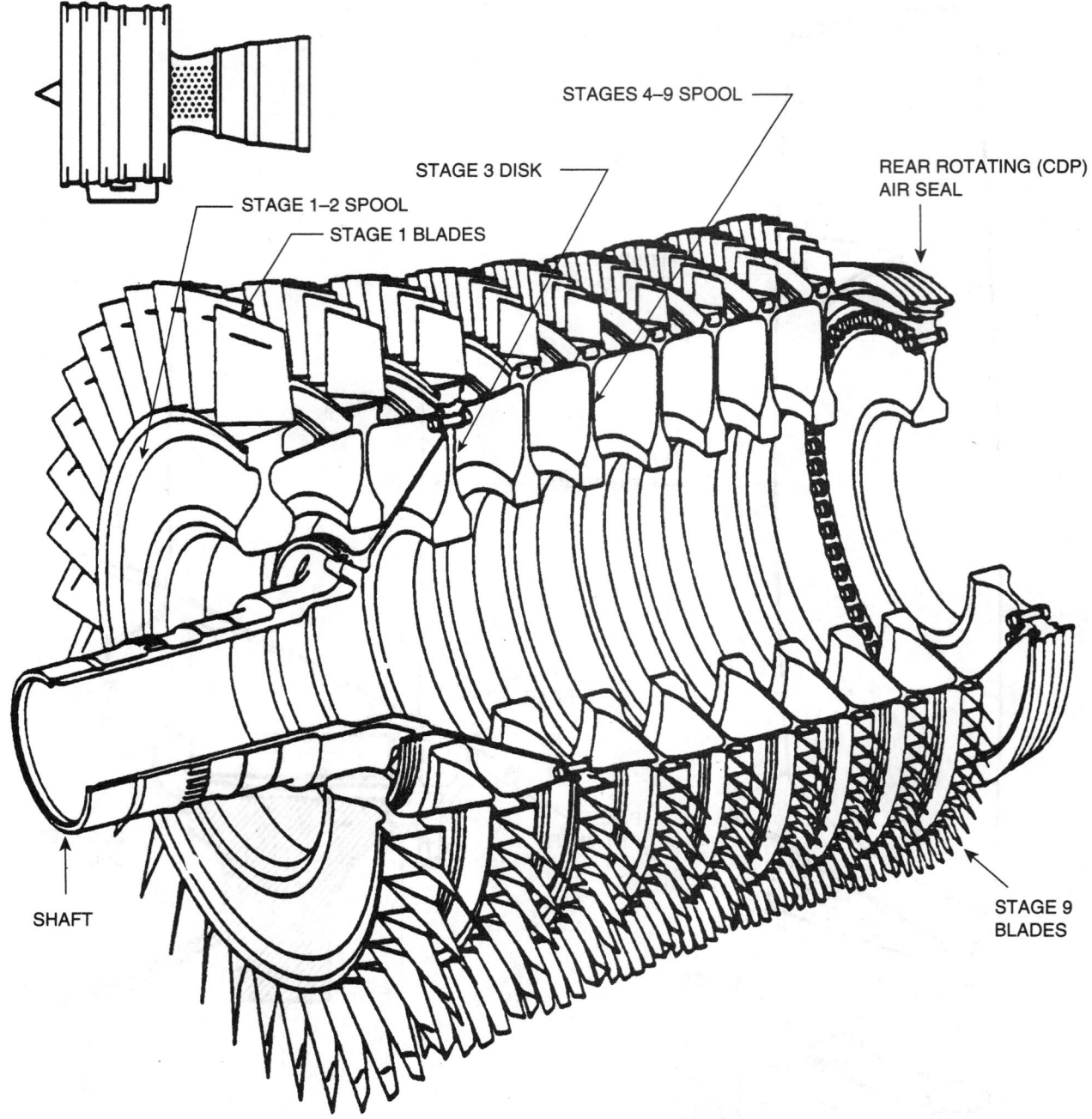

FIGURE 16-75 HP compressor rotor. (*CFM International.*)

Performance

Specifications for the RB 211 engine are given below. As is true of other engines, the specifications given for one model are not necessarily the same for others. The RB 211 engine was designed for growth; therefore, it is likely that later models will have higher thrust ratings as well as variations in fuel consumption and pressure ratios.

Length	119.4 in	303.28 cm
Diameter	85.5 in	217.17 cm
Frontal area	30.0 ft²	2.79 m²
Weight	6353 lb	2882 kg
Takeoff thrust	42 000 lb	186 816 N

Power-weight ratio	6.4 lbt/lb	64.82 N/kg
Fuel consumption	0.36 lb/lbt/h	36.7 g/N/h
Oil consumption	2.0 lb/h	907 g/h
Compressor ratio	25:1	
Bypass ratio	5:1	
Fan pressure ratio	1.42:1	
Air mass flow, fan	1095 lb/s	497 kg/s
Air mass flow, core	230 lb/s	104 kg/s

General Configuration

The general arrangement of the Rolls-Royce RB 211 engine can be understood by studying the illustrations in Figs. 16-79 to 16-82. The use of the three-shaft design makes it possible

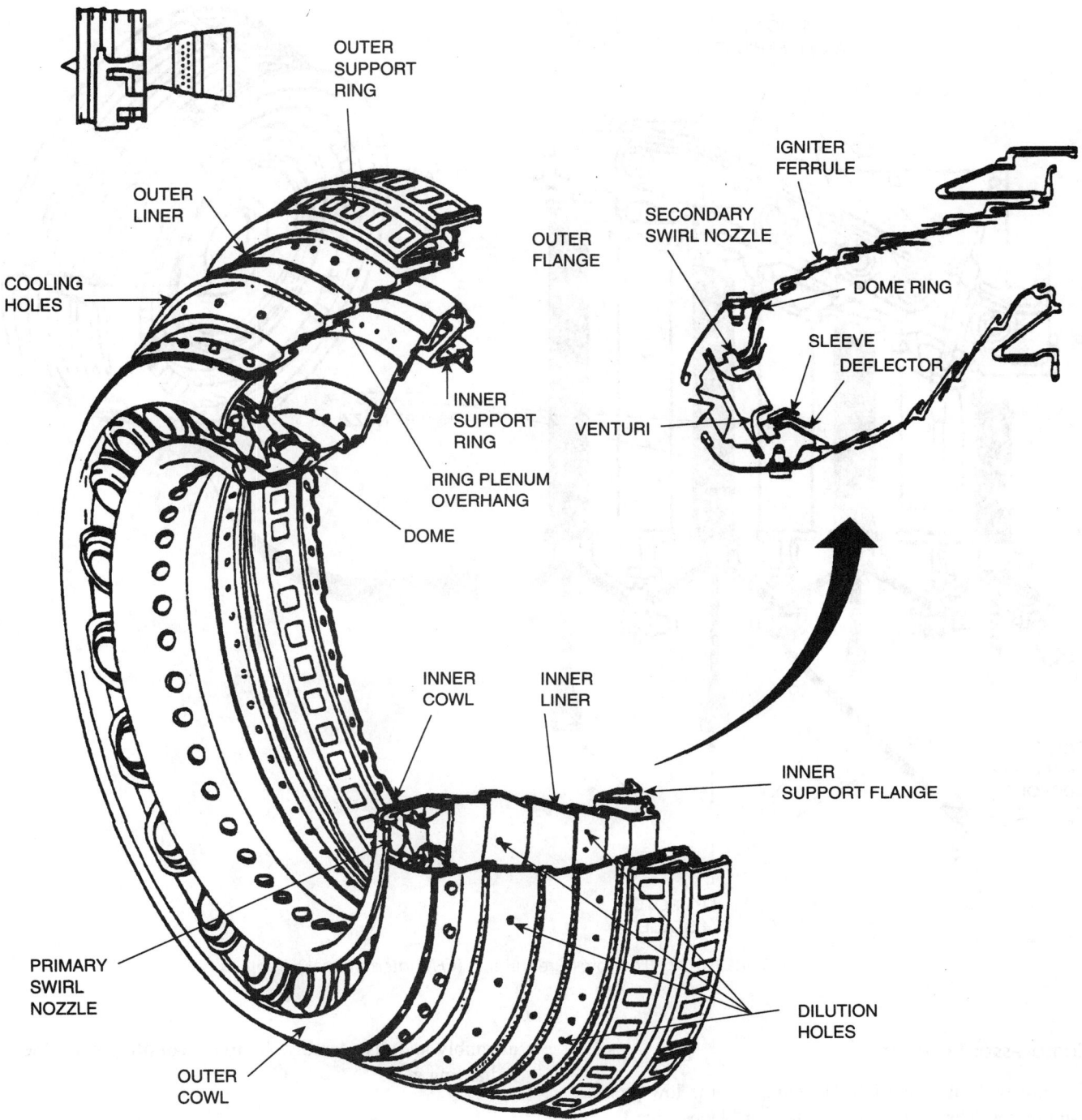

FIGURE 16-76 Combustion chamber. (*CFM International.*)

Labels on figure:
- OUTER SUPPORT RING
- OUTER LINER
- COOLING HOLES
- INNER SUPPORT RING
- RING PLENUM OVERHANG
- DOME
- OUTER FLANGE
- IGNITER FERRULE
- SECONDARY SWIRL NOZZLE
- DOME RING
- SLEEVE DEFLECTOR
- VENTURI
- INNER COWL
- INNER LINER
- INNER SUPPORT FLANGE
- PRIMARY SWIRL NOZZLE
- OUTER COWL
- DILUTION HOLES

to produce an engine which is shorter and more rigid and has fewer compressor and turbine stages.

A major consideration in the design of the RB 211 engine is stiffness or rigidity. Figure 16-82 shows the major load-carrying structures.

Bearing Arrangement

The RB 211 engine utilizes eight bearings to support the rotating assemblies (Fig. 16-83). The inner shaft through which the fan is driven is supported by a bearing at the front and a bearing at the rear of the engine. Note that there are two bearings near the front. The second of these bearings supports the forward end of the intermediate compressor shaft. The rear end of this same shaft is supported by the third bearing.

The three thrust ball bearings are grouped in the **intermediate casing**. The **low-pressure thrust bearing** is an intershaft bearing that forms a cluster with the **intermediate-pressure bearing**, as shown in the drawing. The fifth and sixth bearings support the high-pressure compressor system and are located at a distance from the hot combustion zone to keep the operating temperature at a satisfactory level.

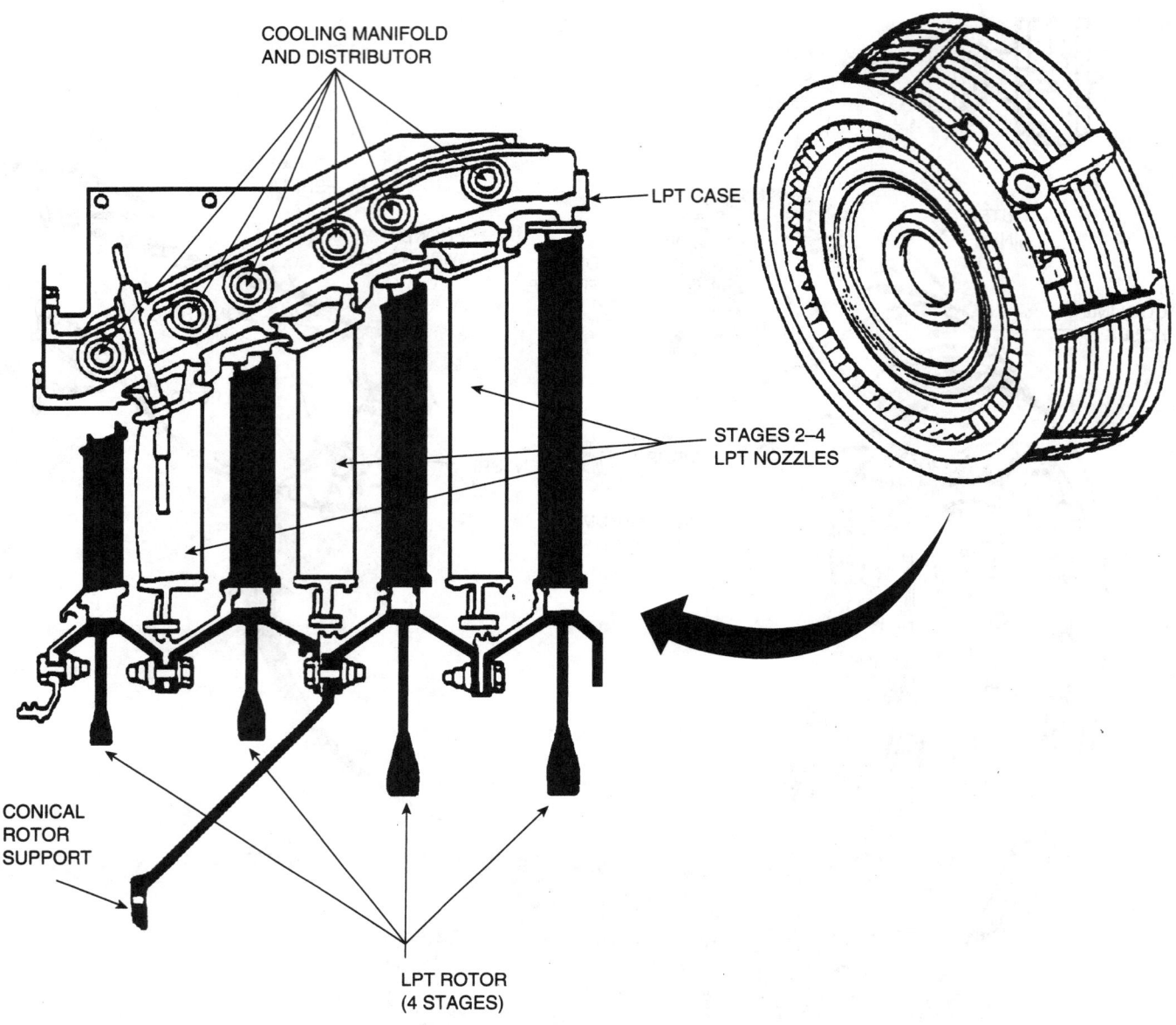

COOLING MANIFOLD
AND DISTRIBUTOR

LPT CASE

STAGES 2–4
LPT NOZZLES

CONICAL
ROTOR
SUPPORT

LPT ROTOR
(4 STAGES)

FIGURE 16-77   Low-pressure turbine. (*CFM International.*)

Compressor Sections

The **fan section** of the RB 211 engine is the low-pressure compressor. The inner ends of the fan blades give the air flowing into the core of the engine its first boost. The air then flows through a single stage of variable stator vanes into the intermediate-pressure section.

The fan section consists of 33 titanium blades held in the fan disk by means of fir-tree roots. The disk is faced with an anti-icing nose cone.

The **intermediate-pressure compressor** is a seven-stage unit built up of disks welded together by means of the electron-beam process. The blades are held in the disks by means of dovetail roots.

The **high-pressure compressor** has six stages. It consists of two electron-beam-welded drums bolted through the third-stage disk. The blades are dovetailed in the disks.

Compressor-to-shaft joints are made by means of curvic couplings, rather than spline couplings, to facilitate modular

assembly and disassembly. A **curvic coupling** resembles a radial ring gear.

Combustion Section

The **combustion chamber** for the RB 211 engine is of the annular design, similar to those described previously for other engines. A cross section of the combustion chamber is shown in Fig. 16-84. This particular chamber is shorter than most other designs, which makes a two-bearing support system possible, with no bearings located near the combustion area.

The annular combustion chamber provides a clean, aerodynamic extension of the compressor outlet casing and permits consistent cooling over a small surface area. The combustion cooling rings are of an advanced design which provides for high cooling efficiency.

The fuel is atomized in a high-velocity airstream by the fuel injection nozzles before it enters the combustion zone. This design gives good atomization of the fuel over the

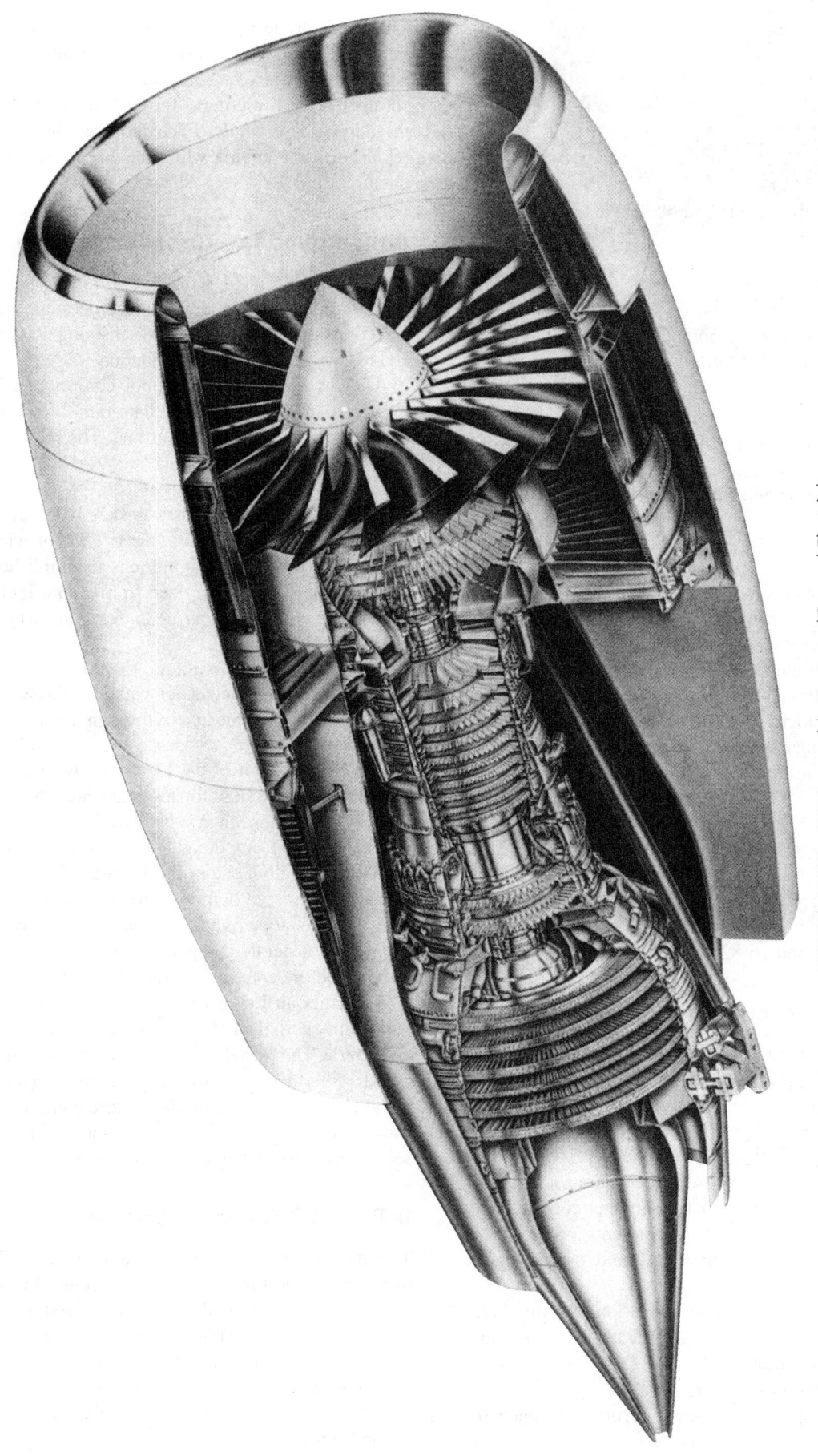

FIGURE 16-78   GE90 propulsion system. (*General Electric.*)

FIGURE 16-79 Cutaway photograph of the Rolls-Royce RB 211 turbofan engine. (*Rolls-Royce.*)

entire operating range and eliminates any tendency towards overrichness in the center of the combustion chamber, thus reducing heat radiation and smoke generation.

Turbine Section

The HPT blades of the RB 211 engine are impingement-cooled by air bled from the high-pressure compressor. Air is fed to the turbine blades through pre-swirl nozzles which accelerate the air in the direction of disk rotation. The cooling air enters the finned interior of the blades, after which it flows at high velocity through a slot to impinge against the leading-edge inner surface. This results in a high degree of heat transfer from the metal to the air. The air is then exhausted through the blade trailing edge. A discussion of this cooling system is given in Chap. 12.

As mentioned previously, the turbine section of the engine consists of three parts. These are (1) a one-stage HPT, (2) a one-stage intermediate-pressure (IP) turbine, and (3) a three-stage low-pressure and fan turbine. These can be seen in Figs. 16-81 and 16-82.

Engine Control System

The control system for the RB 211 engine is shown in Fig. 16-85. The system provides the following characteristics of performance:

1. Flat-rated takeoff thrust. Once the system is set for takeoff, no further power level adjustments are necessary.
2. An altitude schedule which maintains correct thrust during climb without power lever adjustments, provided that the required scheduled climb airspeed is held to a reasonable tolerance.
3. Accurate measurement and indication of thrust independent of air temperature. These indications reveal any engine damage or deterioration.
4. Surge-free engine handling.
5. Freedom to advance the power levers fully open for go-around or other emergency. Automatic limiters protect the engine from overboost, overtemperature, and overspeed.

Modular Construction

The modular design of the RB 211 engine is illustrated in Fig. 16-86. This construction permits sectional repair and maintenance of the engine, thus making it possible to set up service-life criteria for each module, rather than for the engine as a whole. Each module can be removed and replaced in the engine while the engine is still installed in the aircraft.

Rolls-Royce Tay Turbofan Engine

The Tay, shown in Fig. 16-87, is a high-bypass ratio, twin-spool turbofan, designed to achieve optimum fuel efficiency and component reliability, while satisfying noise requirements and reducing environmental pollution.

The low-pressure (LP) spool comprises a single-stage fan and a three-stage intermediate-pressure (IP) compressor driven by a three-stage LP turbine. The high-pressure (HP) spool consists of a 12-stage HP compressor driven by a two-stage HP turbine.

Combustion is effected by means of an annular system featuring 10 interconnected liners, each of which incorporates a fuel spray nozzle. Ignition is accomplished with high-energy igniter plugs energized from a dual-ignition system. The engine is started by an air starter motor which rotates the HP compressor shaft.

The fan bypass airstream and LP turbine exhaust are mixed in a 12-lobed forced mixer, shown in Fig. 16-88, before being discharged to the atmosphere through a common propelling nozzle.

The driving shaft of the LP assembly passes through the HP shaft, and each shaft rotates independently of the other in a clockwise direction when viewed from the front of the engine.

The output from the fan is divided into two separate airflows. One flow enters the IP and HP compressor to be further compressed before entering the combustion section, where fuel is added and the resultant mixture ignited. The hot gas, so generated, expands through both the HP and LP turbines before entering the forced mixer.

The other airflow is directed through the annular bypass duct to mix with the LP turbine exhaust in a common jet pipe and nozzle. The forced mixer produces a uniform low-velocity gas stream which expands through the propelling nozzle to the atmosphere. A portion of the core compressor airflow is used to cool certain parts of the engine, and to pressurize oil seals, before being vented to the atmosphere.

IAE V2500 Turbofan Engine

The internationally developed and manufactured V2500 turbofan engine for use in the McDonnell Douglas MD-90 aircraft represents another advanced design for fuel efficiency, low noise, and high reliability in turbine powerplants.

Blending the technology of five nations, the V2500, shown in Fig. 16-89, is the product of Pratt & Whitney division of United Technologies in the United States, Rolls-Royce in the United Kingdom, MTU of the Federal Republic of Germany, Fiat of Italy, and Japanese Aero Engines Corp.

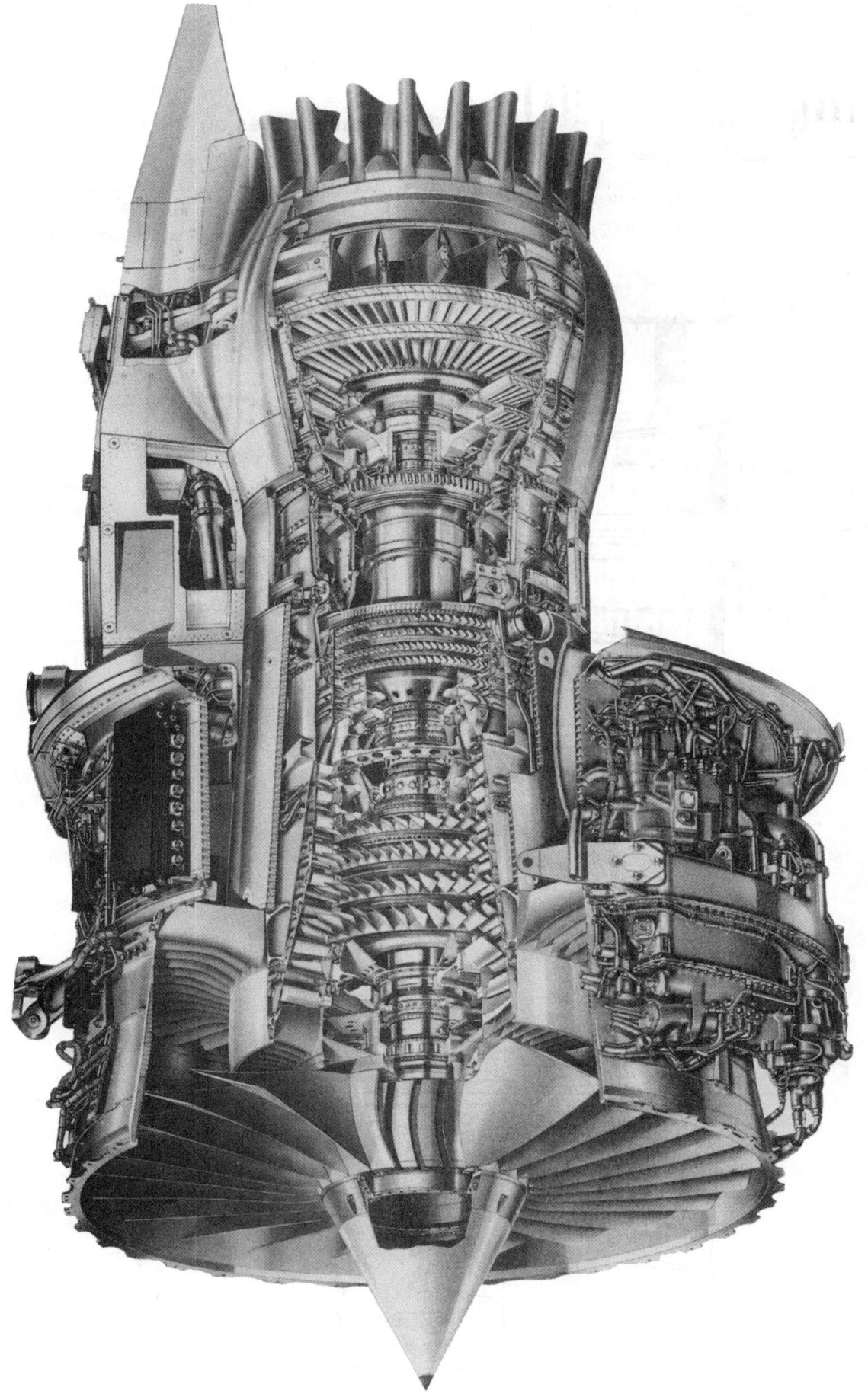

FIGURE 16-80   Rolls-Royce RB 211-524G/H. (*Rolls-Royce.*)

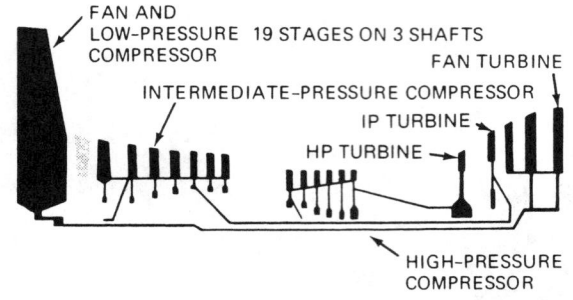

FIGURE 16-81 Drive arrangement for the RB 211 engine.

The engine is a high-bypass-ratio turbofan with a fan diameter of 5.25 ft [1.6 m], approximately 10.5 ft [3.2 m] in length. It is offered at thrust levels ranging from 22 000 lb [10 000 kg] to 28 000 lb [12 727 kg].

Turbofan Engines on the Boeing 787

The Boeing 787 can be equipped with either a Rolls-Royce Trent 1000 (Fig. 16-90), or a General Electric GEnx engine (Fig. 16-91). Both engines use the all electrical bleedless systems which eliminate the air transfer tubes and intercoolers

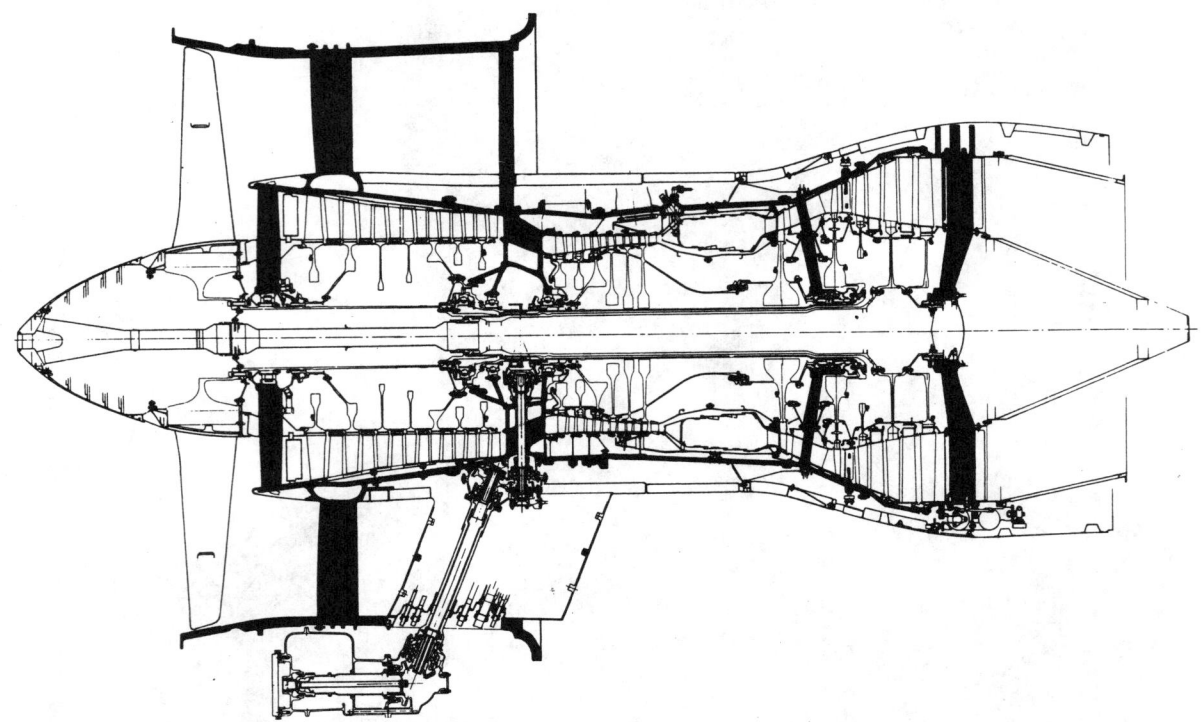

FIGURE 16-82 Arrangement of the RB 211 engine, emphasizing the principal load-carrying structures. (*Rolls-Royce.*)

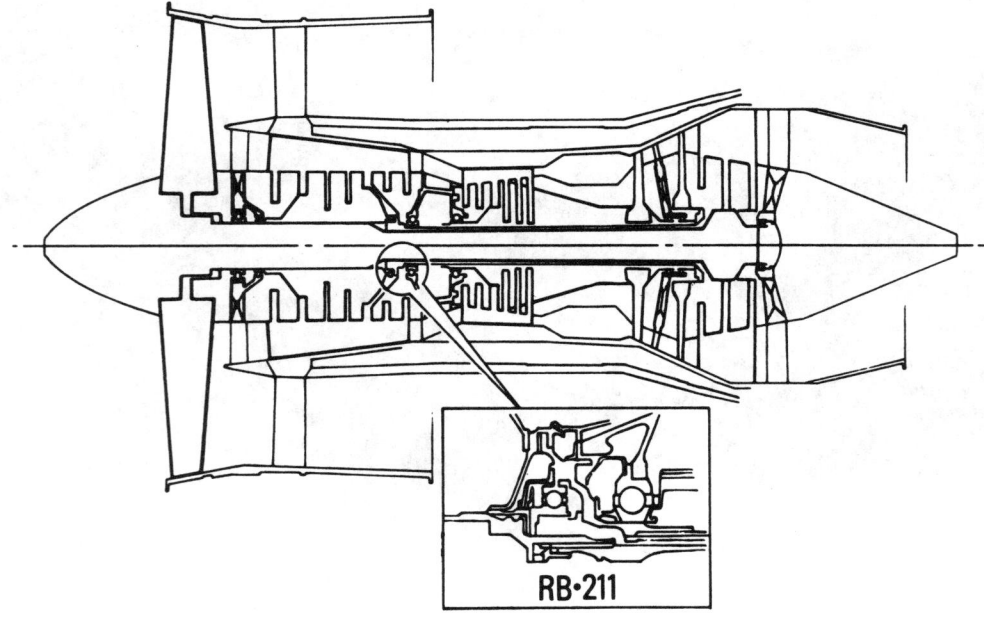

FIGURE 16-83 Bearing arrangement in the RB 211 engine. (*Rolls-Royce.*)

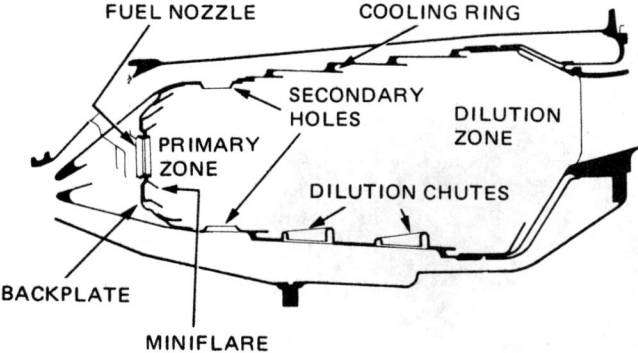

FIGURE 16-84 Cross section of the combustion chamber.

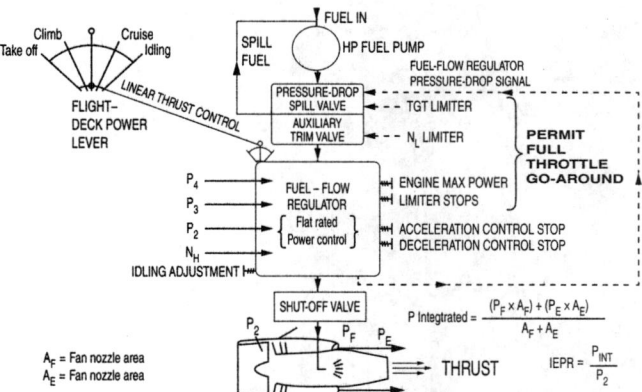

FIGURE 16-85 Control system for the RB 211 engine.

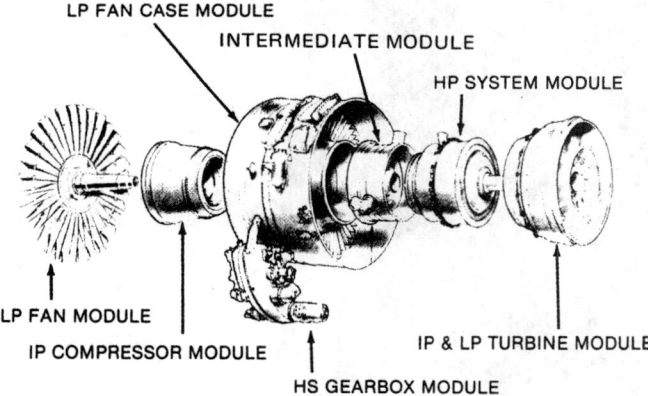

FIGURE 16-86 Modular design of the RB 211 engine.

throughout the aircraft. This accounts for major weight saving. Sound-absorbing quiet technology is used in the air inlet which uses new materials to reduce noise. The redesigned exhaust duct whose rims are typed in a toothed pattern called chevrons allow for mixing of the exhaust gases and the ambient air. The aircraft allows for flexibility by allowing either manufacturer's engine to be compatible with the aircraft.

The GEnx uses some technology from the GE90 turbofan, including composite fan blades, and the smaller core featured in earlier variants of the engine. The engine uses composite technology into the fan case, reducing weight. A first is the elimination of bleed air systems using high-temperature/

high-pressure air from the propulsion engines to power aircraft systems such as the starting, air-conditioning, and anti-ice systems. The GEnx and the Trent 1000 allow for a move toward the More Electric Airplane in the 787.

The GEnx produces thrust from 53 000 to 75 000 lbf (240 to 330 kN) with reduced fuel consumption of up to 20 percent and is significantly quieter than current turbofans. A 66 500 lbf (296 kN) thrust version (GEnx-2B67) will be used on the 747-8. Unlike the initial version, for the 787, this version has a traditional bleed air system to power internal pneumatic and ventilation systems. It will also have a smaller overall diameter than the initial model to accommodate installation on the 747. The GEnx as used on the 787, features a number of weight-saving innovations, the first of which is the fan diameter of 111 in [2.8 m] for the 787-9. The continued use of composite fan blades with titanium leading edges and a fan case of composite material reduces weight and thermal expansion. Several technological factors help in the fuel burn reduction such as a fan bypass ratio of 9.6:1, which also helps reduce noise. Also a high-pressure compressor, with a 23:1 pressure ratio and only 10 stages, adds to improved engine operation. Also, shrouded guide vanes reduce secondary flows counter-rotating spools for the reaction turbines used to reduce load on guide vanes. The engine uses a Lean TAPS (twin annular premixed swirler) combustor to reduce environmentally harmful emissions with improved airflow to prevent backflash in the combustor. The internal engine temperatures have been reduced by using more efficient cooling techniques with regard to the internal air cooling system. As with most large fan engines, the fan uses centrifugal force to extract debris ingested into the intake of the engine within the low-pressure compressor. This action guards the high-pressure compressor from damage. As items (birds or other items) flow into the engine the fan will cause the debris to flow out and through the fan air duct instead of flowing into the core of the engine or high-pressure compressor.

General Characteristics

Type: Turbofan

Length: 196 in (4.98 m)

Diameter: 144 in (3.66 m)

Dry weight: 12 822 lb (5816 kg)

Components

Compressor: Axial, 1-stage fan, 4-stage low-pressure compressor, 10-stage high-pressure compressor

Combustors: Annular

Turbine: Axial, 2-stage high-pressure turbine, 7-stage low-pressure turbine

Performance:

Maximum thrust: 63 800 lbf (284 kN)

Overall pressure ratio: 41:1

The Rolls-Royce Trent 1000 engine is a three-spool engine consisting of the fan, the intermediate spool, and the high-pressure spool. Thrust ranges from 53 000 lbf to 74 000 lbf. The 20 titanium aerodynamic fan blades measure 112 in. in diameter. The intermediate compressor has eight stages

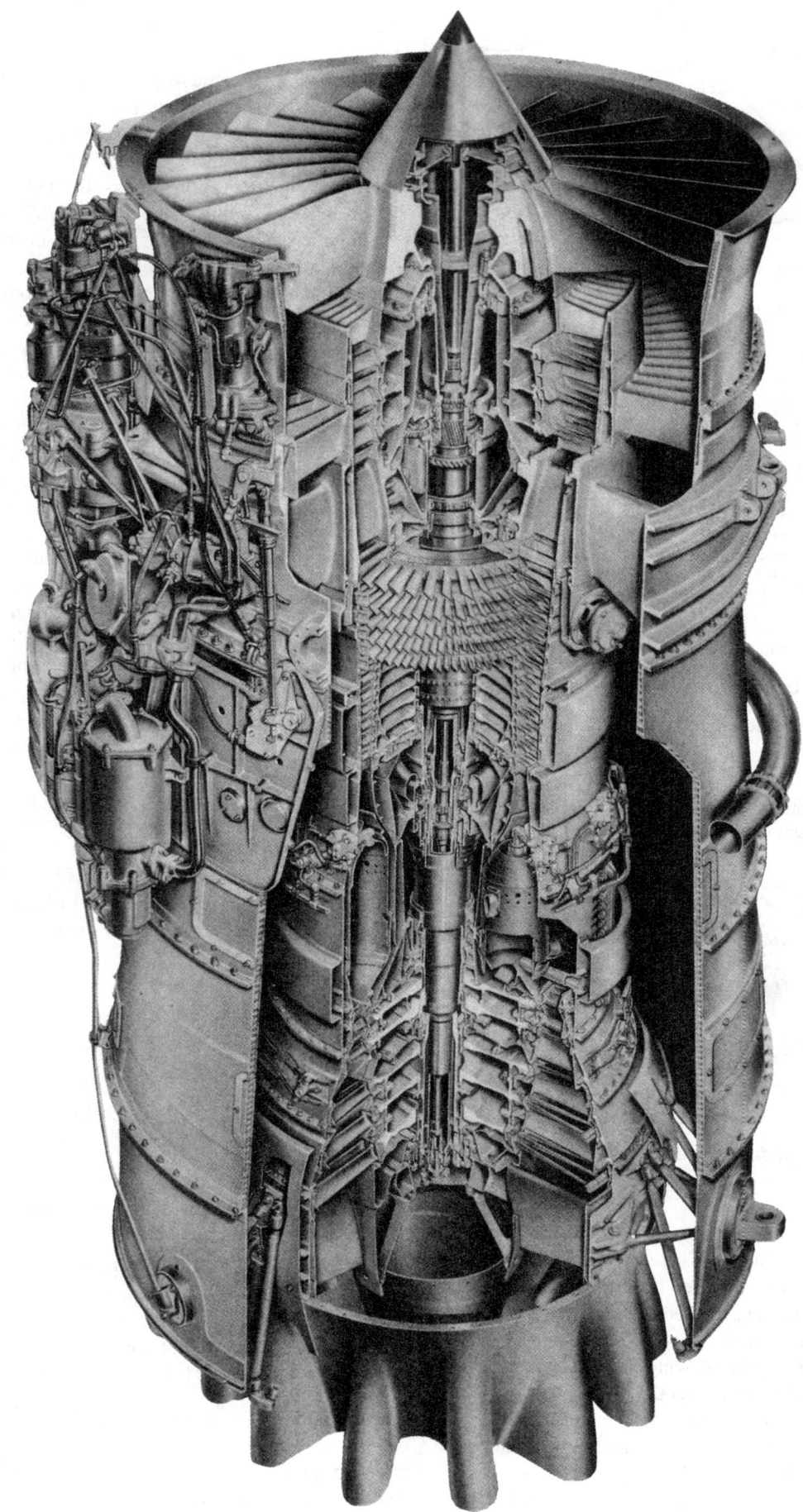

FIGURE 16-87 Rolls-Royce Tay engine. (*Rolls-Royce.*)

FIGURE 16-88   Hot and cold stream mixer. (*Rolls-Royce.*)

and is driven by a single-stage turbine. The high-pressure compressor has six stages which are driven by a one-stage turbine. The fan is driven by a low-pressure turbine with six stages. The engine has a bypass ratio of 10:1 and an overall pressure ratio of 50:1. Both the bypass and pressure ratio show a marked increase over earlier engines. This engine has increased fuel efficiency, monitoring capability, lower noise levels, lower emissions, and lower scheduled line maintenance.

General Motors Allison GMA 3007 Turbofan Engine

The GMA 3007, shown in Fig. 16-92, is a high-bypass-ratio turbofan engine in the 6000 - to 8000-lb [26688- to 35584-N] thrust class that was designed to provide the regional airline and business jet market with reliable high-performance propulsion.

It incorporates a single-stage, direct-drive fan which features wide-chord, clapperless blades. The fan provides excellent performance characteristics and resistance to foreign object damage (FOD). The fan is driven by a three-stage low-pressure turbine. Engine control and power management of the GMA 3007 are by full-authority digital electronic control (FADEC). The Embraer EMB-145 and the Cessna Citation X are aircraft that use the GMA 3007 turbofan engine.

The Garrett-AiResearch TFE731 Turbofan Engine

A medium-power turbofan engine for business and general aviation aircraft is the TFE731 engine manufactured by the Garrett Turbine Engine Company, a division of the Garrett Corporation. A photograph of this engine is shown in Fig. 16-93.

The TFE731 engine utilizes some rather innovative design features. The fan is driven through a reduction gear to avoid overspeeding and yet allow the low-pressure compressor to

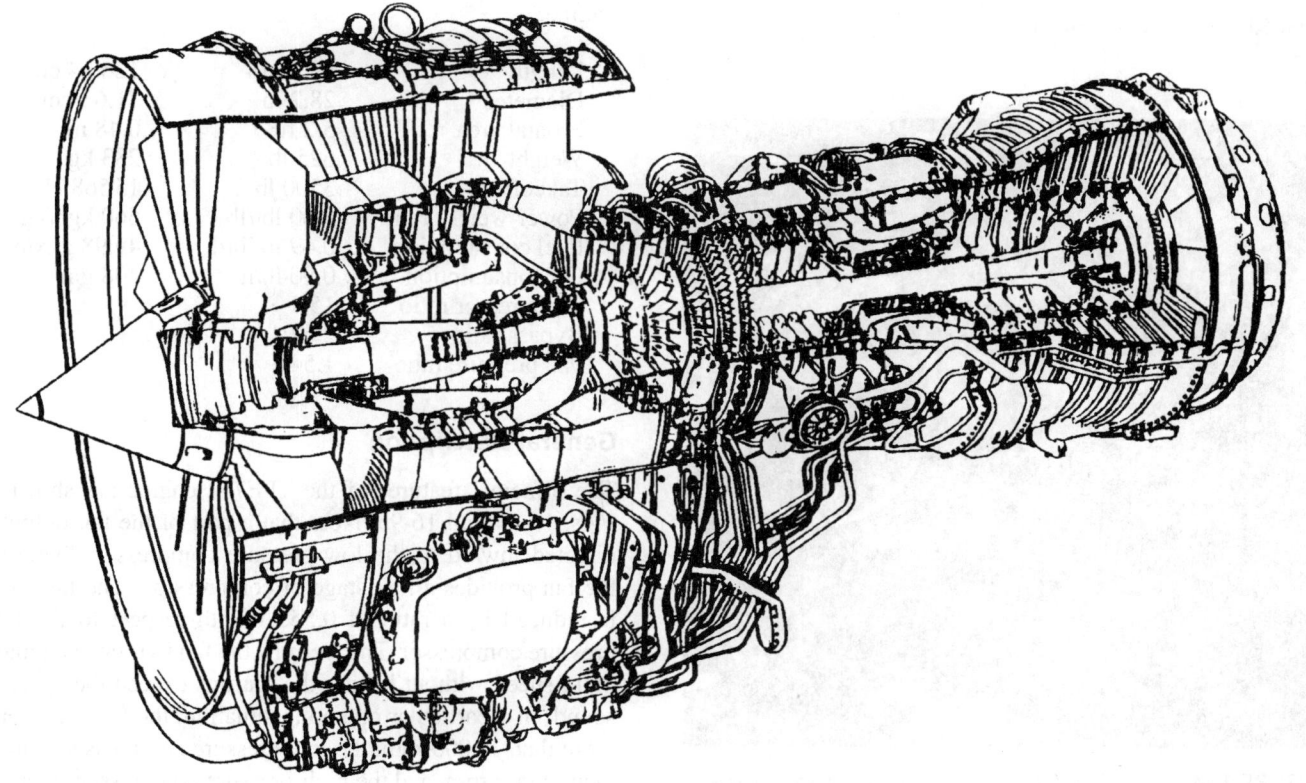

FIGURE 16-89   Cutaway drawing of the IAE V2500 turbofan engine.

FIGURE 16-90   Rolls-Royce Trent 1000.

FIGURE 16-93   The Garrett TFE731 engine. (*Honeywell, Inc.*)

FIGURE 16-91   GE GEnx-1B.

FIGURE 16-92   GMA 3007 turbofan engine. (*Allison Rolls-Royce.*)

operate at the most efficient speed. This feature is illustrated in Fig. 16-94. The low-pressure compressor is of the axial type, whereas the high-pressure compressor is a centrifugal unit. The total pressure ratio developed is 15:1. The combustion chamber is a reverse-flow annular type, which accounts for the reduced length of the engine.

Specifications

The general specifications of the TFE731 engine are listed below; remember, however, that these specifications may vary somewhat, depending on the particular model of the engine.

Length	49.7 in	126.24 cm
Diameter	28.2 in	71.63 cm
Frontal area	5.2 ft^2	0.48 m^2
Weight	625 lb	283 kg
Takeoff thrust	3500 lb	15 568 N
Power-weight ratio	5.60 lbt/lb	560 kgp/kg
Fuel consumption	0.49 lb/lbt/h	49.98 g/N/h
Oil consumption	0.90 lb/h	408 g/h
Compressor ratio	15:1	
Bypass ratio	2.67:1	
Fan pressure ratio	1.54:1	

General Description

The principal features of the TFE731 engine are shown in Figs. 16-94 and 16-95. Note that a part of the fan output is directed inward to the low-pressure compressor; therefore, the fan provides a first stage of compression. The fan speed is reduced by a ratio of 0.555:1 with respect to the low-pressure compressor. This permits the fan to rotate at a maximum speed without having the fan tips exceed the speed of sound. In a particular operating situation, the fan can turn at more than 10 000 rpm, the low-pressure compressor at more than 18 000 rpm, and the high-pressure compressor at almost 29 000 rpm.

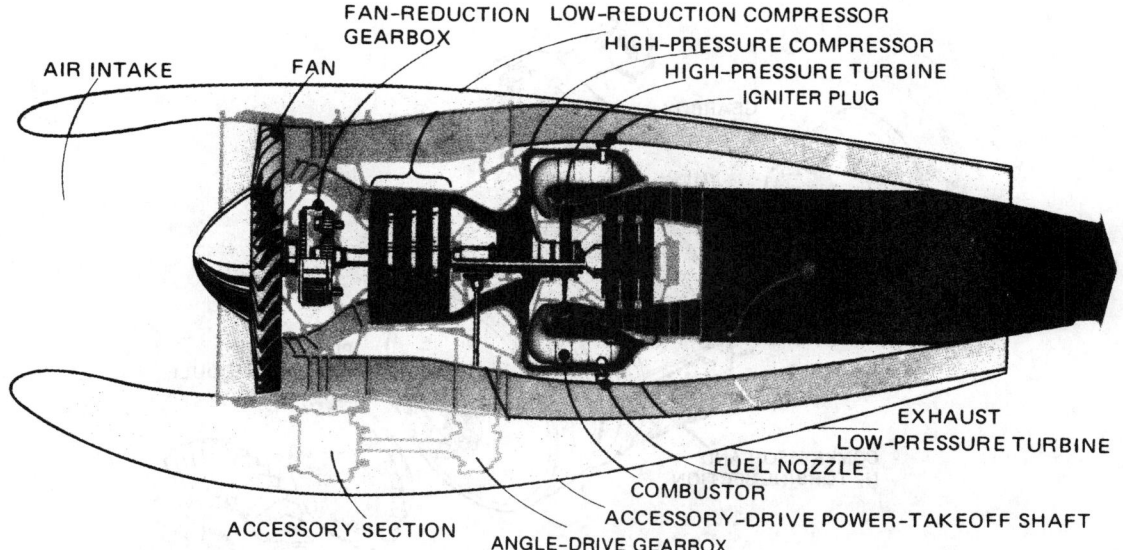

FIGURE 16-94 Operational drawing of the TFE731 engine. (*Honeywell, Inc.*)

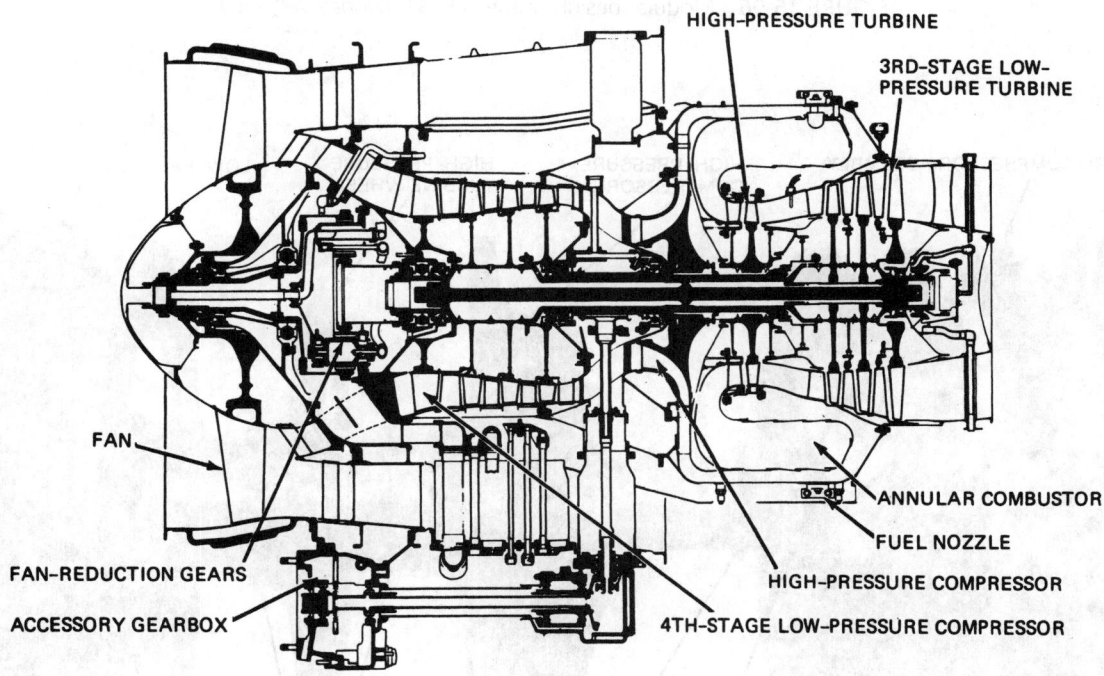

FIGURE 16-95 Cutaway view of the TFE731 engine. (*Honeywell, Inc.*)

Figure 16-95 shows all the principal components of the engine, including the fan, compressors, reduction gearing, bearings, combustion chamber, and turbine. The modular design of the engine, which allows for easy maintenance, is illustrated in Fig. 16-96.

The low-pressure compressor assembly, shown in Fig. 16-97, has four axial-flow compressor rotors shafted to, and driven by, a three-stage axial-flow LPT assembly, which also drives the bypass fan through the planetary gear assembly.

A single-stage centrifugal compressor impeller shafted to, and driven by, a single-stage axial-flow turbine wheel comprises the high-pressure compressor assembly.

The high-pressure compressor drives the accessory gear case through a transfer gear case. Accessory mounting is provided on the accessory gear case for such accessories as the fuel pump, fuel control, oil pumps, starter-generator, and other optional equipment.

Both the low-pressure and high-pressure compressors are mounted on coaxial shafts, but they are not mechanically connected together.

As illustrated in Fig. 16-96, the TFE731 is designed in modules. The **bypass fan assembly module** consists of the following: fan spinner, spinner support, fan rotor assembly, fan support assembly, planetary gear assembly, low-pressure

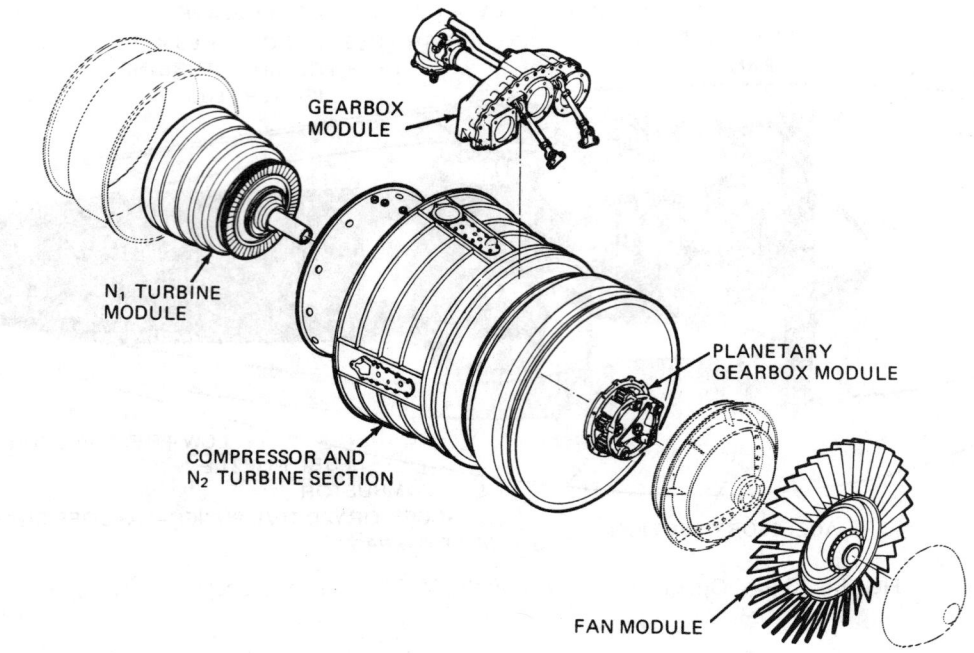

FIGURE 16-96 Modular design of the TFE731. (*Honeywell, Inc.*)

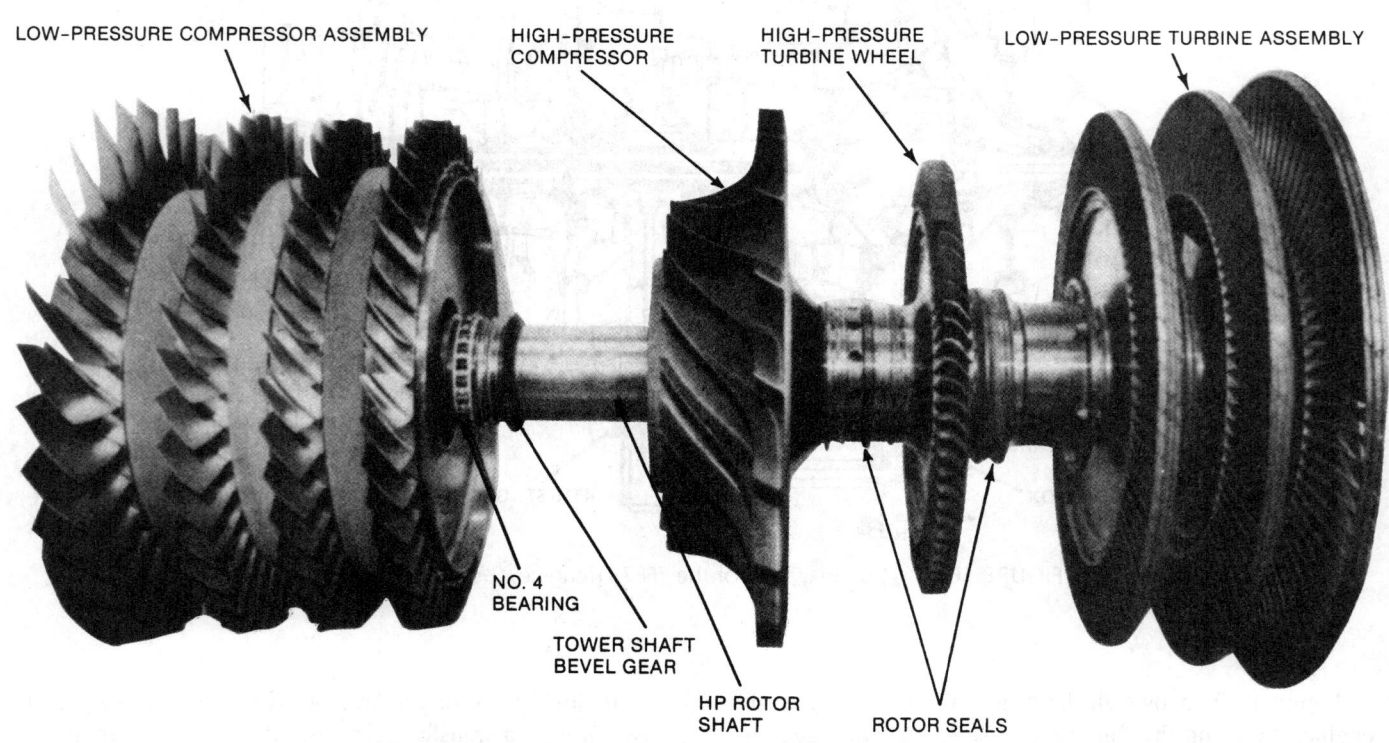

FIGURE 16-97 TFE731 rotating group. (*Honeywell, Inc.*)

compressor stator, bypass fan inlet housing, and fan bypass stator. The **turbine section modules** include the thrust nozzle and exhaust assembly, low-pressure turbine module, transition duct, ITT thermocouple harness, combustion liner, HPT wheel, and HPT nozzle assembly, as shown in Fig. 16-98.

The Garrett TFE731-5 Turbofan Engine

The Garrett TFE731-5 engine shown in Fig. 16-99 is an upgraded version of the TFE731. The engine has undergone changes in four major areas: fan section, fan gearbox, high-pressure turbine, and low-pressure turbine. The general

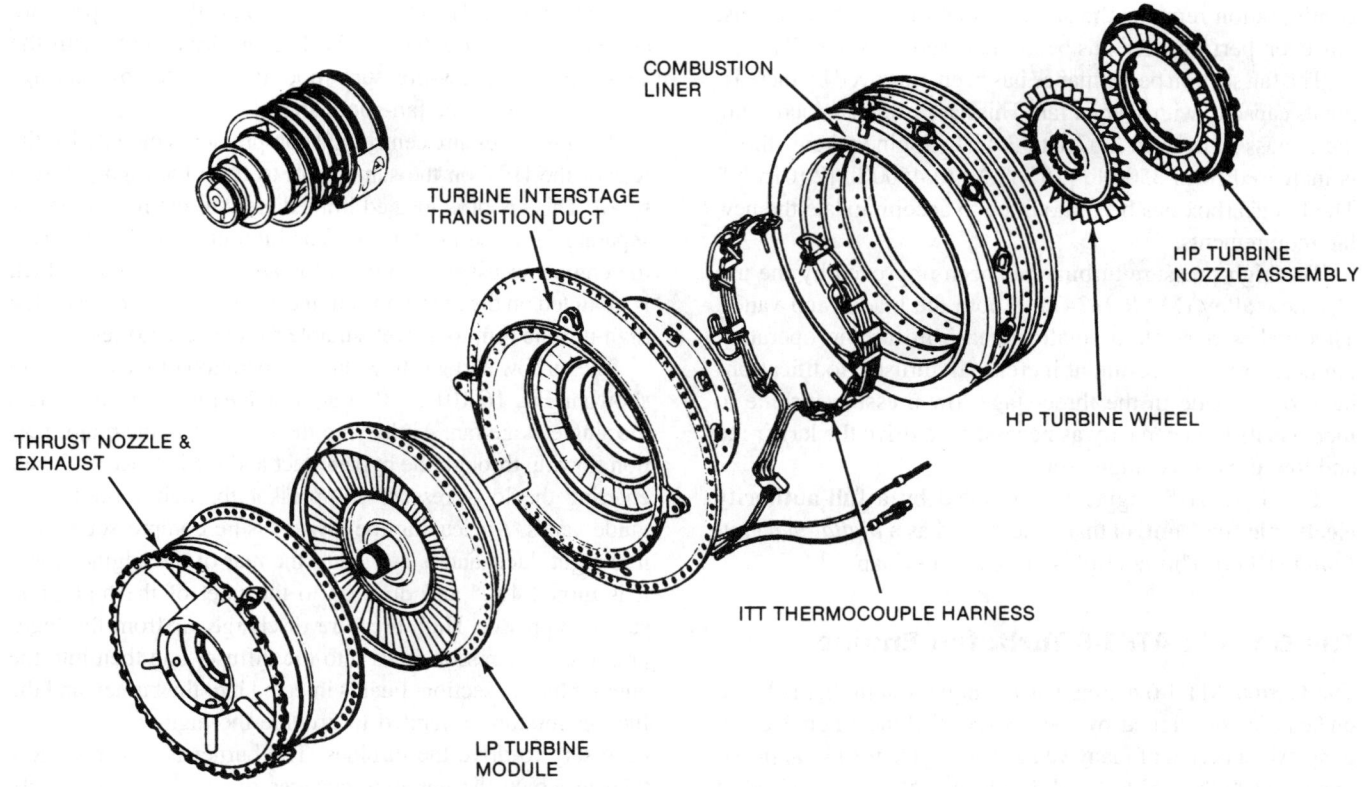

FIGURE 16-98 TFE731 turbine and combustion modules. (*Honeywell, Inc.*)

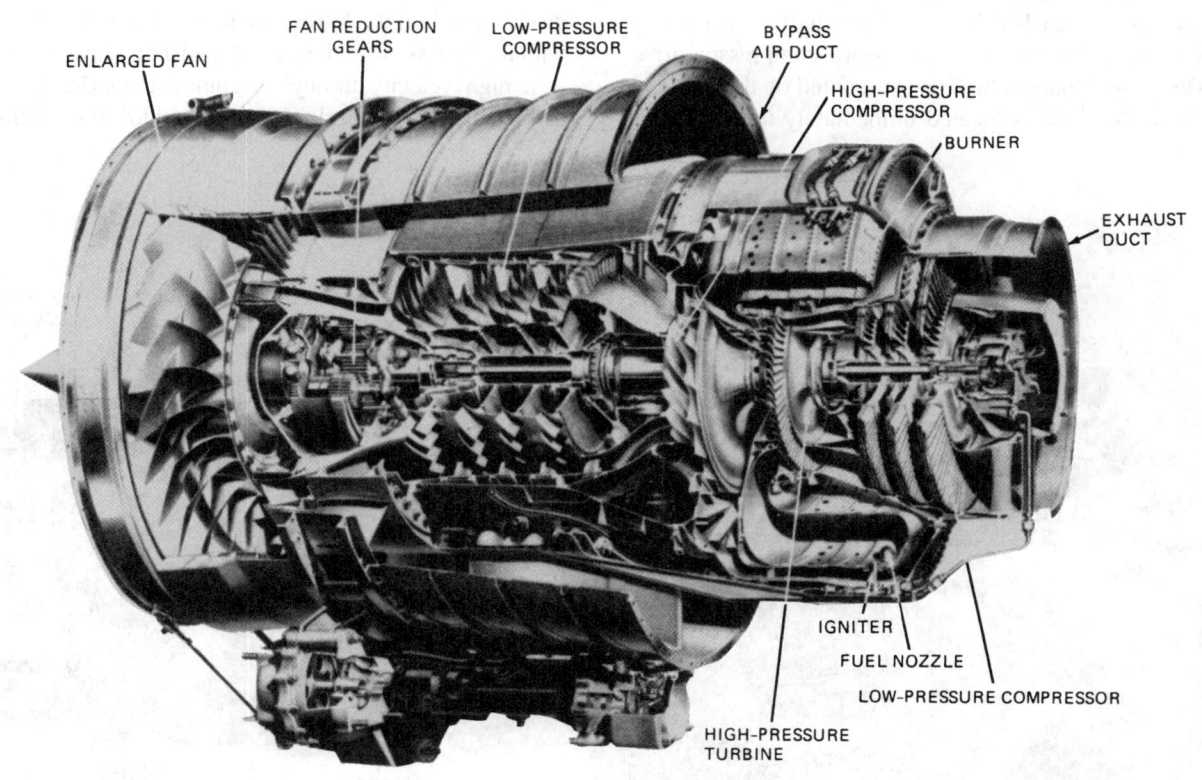

FIGURE 16-99 Cutaway view of the TFE731-5 engine. (*Honeywell, Inc.*)

configuration remains the same as that of previous models; however, performance has been increased substantially.

The fan section performance has been improved by increasing its capacity with a larger fan. This has resulted in increasing the bypass ratio from 2.67:1 to 3.48:1. Engine takeoff thrust is increased from 3500 lb [15 568 N] to 4500 lb [20 016 N]. The fan gearbox has been modified to accommodate the new fan requirements.

The high-pressure turbine has been improved by the use of a new alloy (MAR-M247) in both the blades and vanes. This makes possible a small increase in turbine operating temperature and a resultant increase in thrust. Modifications have been made in the three-stage low-pressure turbine to increase its load capacity as necessary to drive the larger fan and low-pressure compressor.

The TFE731-5 engine is controlled by a **full-authority electronic fuel control** that is described as a *digital fuel controller* (DFC). This control is discussed in Chap. 13.

The Garrett ATF3-6 Turbofan Engine

The Garrett ATF3-6 engine has a unique design that is based on features developed by the Garrett Turbine Engine Company over a period of many years. The engine weighs approximately 1100 lb [499 kg] and develops 5050 lb [22 462 N] of thrust at sea level. The bypass ratio is 2.67:1 and the thrust specific fuel consumption is 1020.9 g/N/h.

The ATF3-6 is a three-spool engine with separate turbines to drive the fan, low-pressure compressor, and high-pressure compressor. This configuration can be seen in the cutaway photograph in Fig. 16-100. The fan drive shaft is surrounded by and is coaxial with the low-pressure compressor drive shaft. The three-stage fan turbine is mounted on the rear end of the drive shaft and is located immediately to the rear of the two-stage LPT. The five-stage axial-flow low-pressure compressor is mounted on the hollow drive shaft with the LPT. The entire low-pressure spool is supported by bearings on the outside of the fan-spool shaft.

The high-pressure centrifugal compressor is mounted to the rear of the HPT on the same shaft to form the high-pressure spool. This spool is located immediately to the rear of and is separate from the fan turbine. The high-pressure spool drives the engine accessories through the accessory gearbox, which is mounted on the rear of the engine. Reduction gears lower the high-turbine rpm to a level suitable for the accessories.

The airflow and gas flow during operation of the engine are shown in Fig. 16-101. Airflow starts at the inlet cowl and enters the single-stage fan. Air leaving the fan is split, the major portion flowing through the bypass duct and the balance of the air entering the low-pressure compressor through variable-inlet guide vanes. Air leaving the low-pressure compressor is split into eight ducts and is carried to the rear of the engine where it is turned 180° and directed to the inlet of the high-pressure compressor. High-pressure discharge air from the high-pressure compressor flows into the diffuser and then into the annular burner section. Fuel is injected into the burner, and the fuel-air mixture is ignited to provide the high-velocity gases necessary to drive the turbines. The burner is a reverse-flow type in which the gas flow changes direction approximately 180°. The gases are directed inward from the burner to enter the HPT. Leaving this turbine, the gases flow forward and outward to deliver energy to the fan turbine and LPT.

Gas from the turbines flows forward to enter eight turning-vane modules. These modules turn the gas flow to the rear and discharge it into the bypass airstream. The exhaust gases mix with the bypass air, and the combined air and gas is discharged at a high velocity through a common nozzle. This design assures maximum nozzle efficiency and low noise emission.

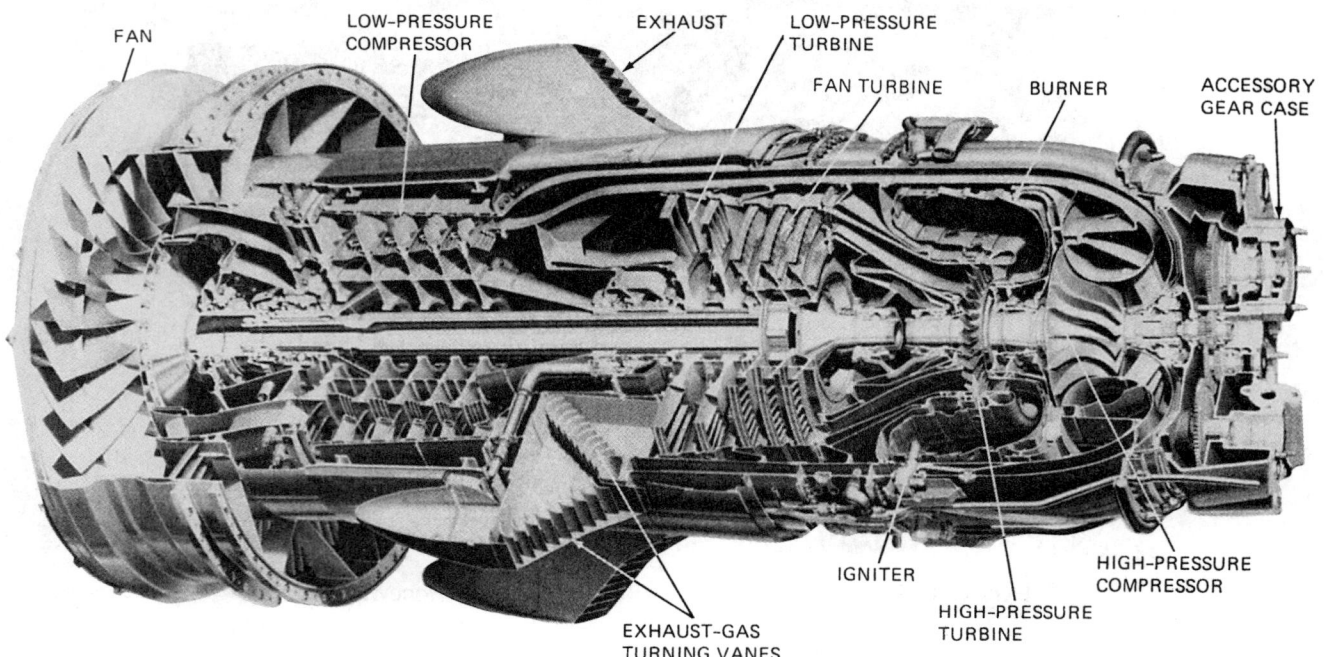

FIGURE 16-100 The Garrett ATF3-6 turbofan engine. (*Honeywell, Inc.*)

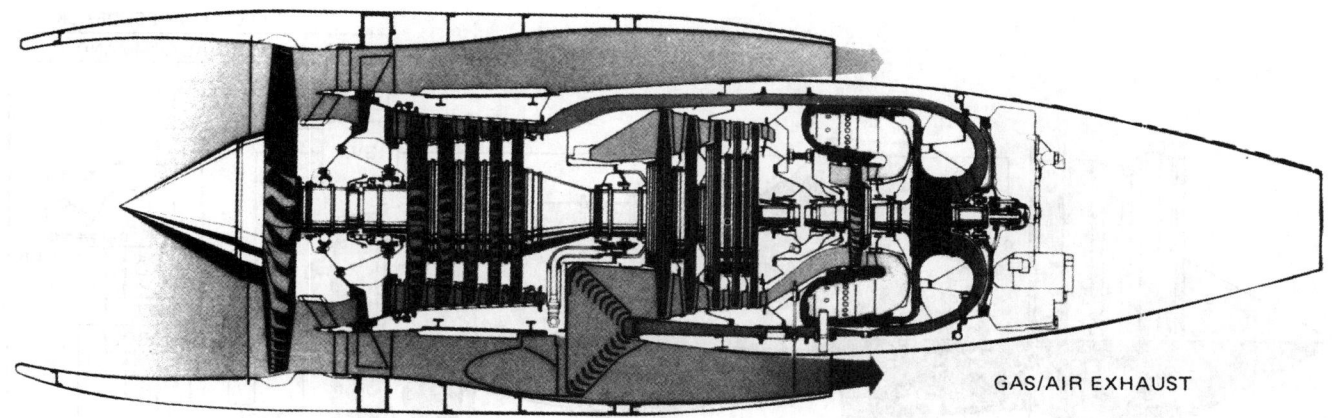

GAS/AIR EXHAUST

FIGURE 16-101 Airflow and gas flow in the ATF3-6 engine. (*Honeywell, Inc.*)

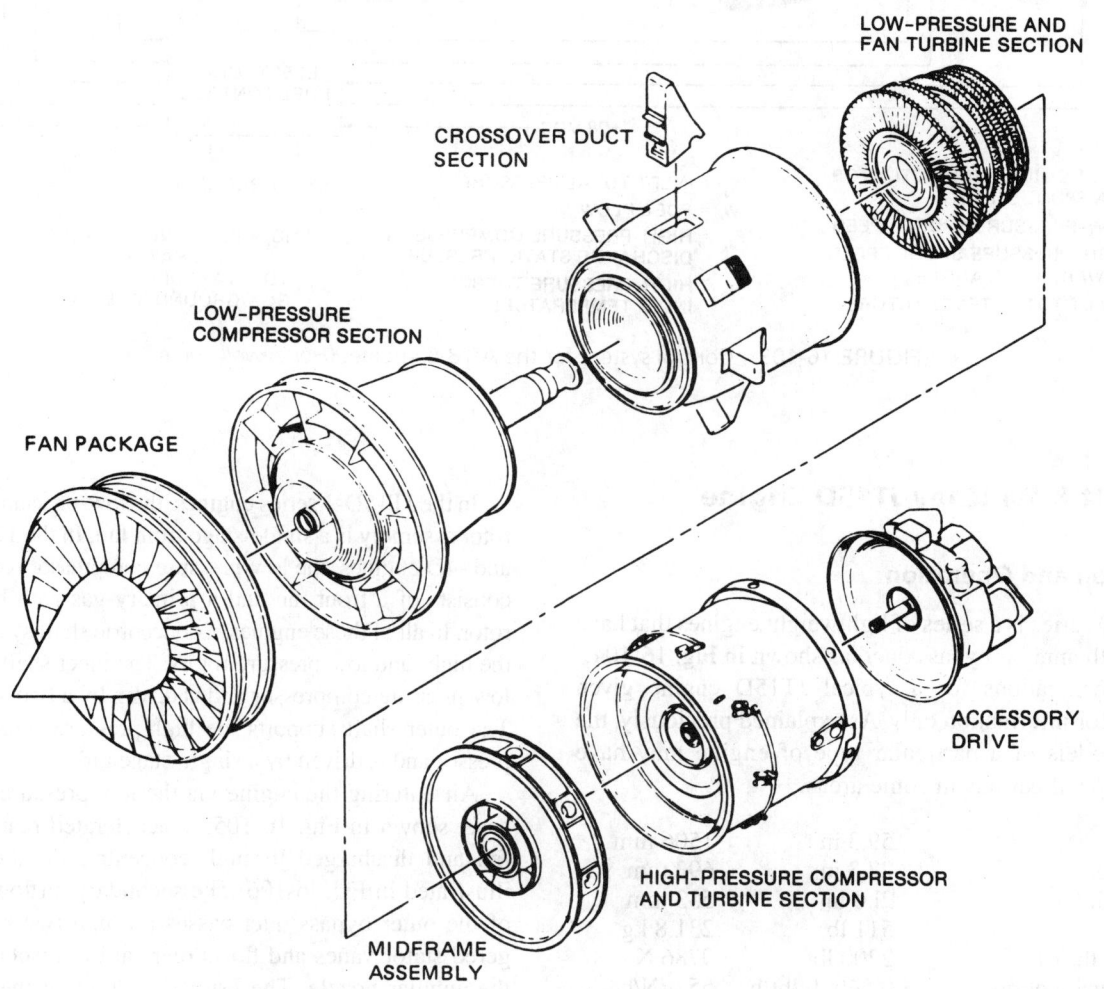

LOW-PRESSURE AND
FAN TURBINE SECTION

CROSSOVER DUCT
SECTION

LOW-PRESSURE
COMPRESSOR SECTION

FAN PACKAGE

ACCESSORY
DRIVE

HIGH-PRESSURE COMPRESSOR
AND TURBINE SECTION

MIDFRAME
ASSEMBLY

FIGURE 16-102 Modular construction of the ATF3-6 engine.

The ATF3-6 is of modular construction, as shown in Fig. 16-102. This type of design simplifies maintenance by making it possible to remove sections for repair or overhaul without having to disassemble the complete engine. While the engine is installed in the aircraft, those sections that require the most frequent service and repair, such as the high-pressure section and accessory drives, can be removed and overhauled without disturbing the other sections of the engine. With a modular engine design it is not necessary to perform complete engine overhaul on a periodic schedule. This permits module replacement or repair on an "as required" basis.

The ATF3-6 engine is controlled by a full-authority electronic fuel control similar to that employed with the TFE731-5 engine. The control system and interconnected units are shown in Fig. 16-103.

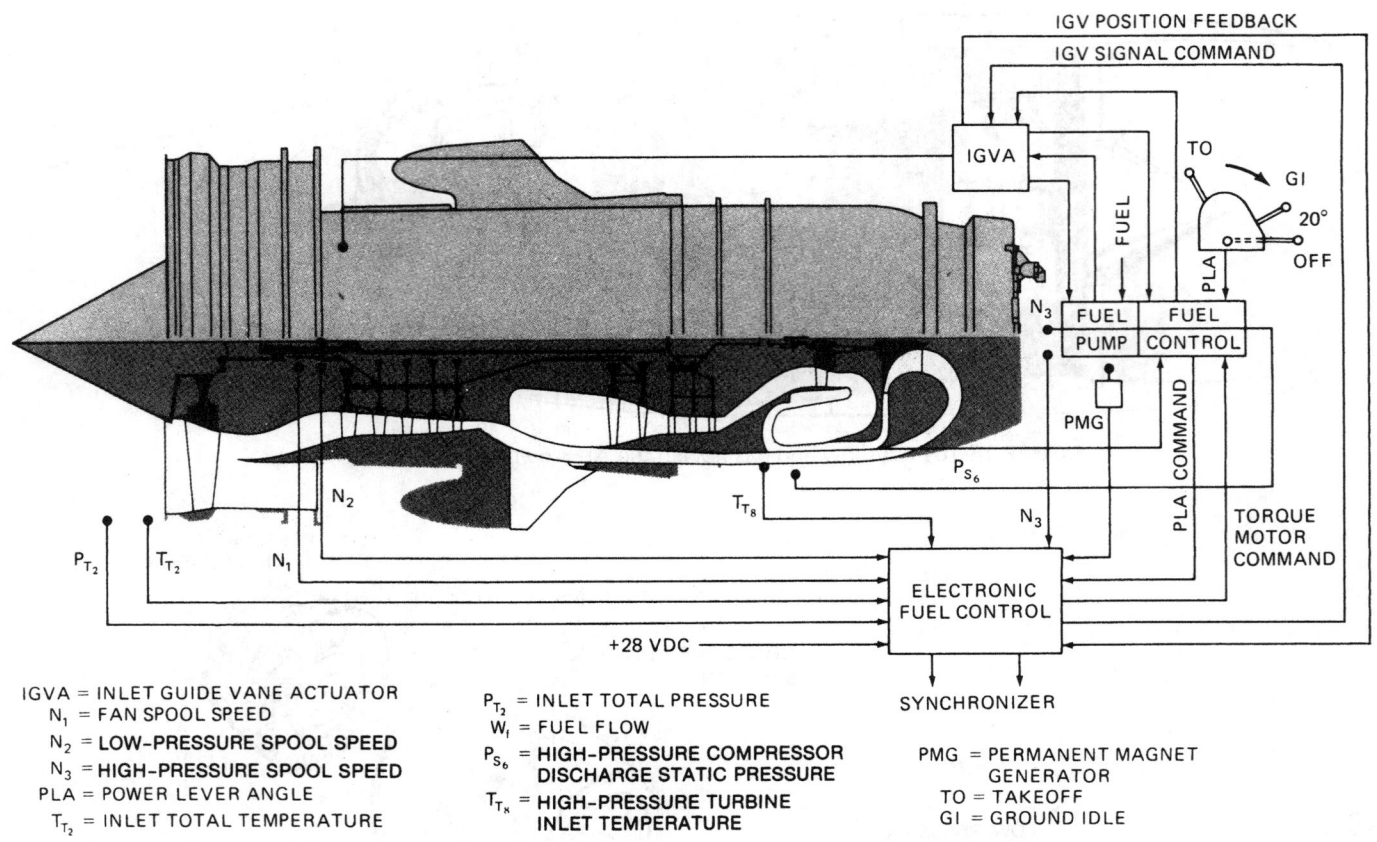

IGVA = INLET GUIDE VANE ACTUATOR
  N_1 = FAN SPOOL SPEED
  N_2 = LOW-PRESSURE SPOOL SPEED
  N_3 = HIGH-PRESSURE SPOOL SPEED
PLA = POWER LEVER ANGLE
  T_{T_2} = INLET TOTAL TEMPERATURE

P_{T_2} = INLET TOTAL PRESSURE
W_f = FUEL FLOW
P_{S_6} = HIGH-PRESSURE COMPRESSOR
     DISCHARGE STATIC PRESSURE
T_{T_x} = HIGH-PRESSURE TURBINE
     INLET TEMPERATURE

PMG = PERMANENT MAGNET
     GENERATOR
TO = TAKEOFF
GI = GROUND IDLE

FIGURE 16-103 Control system for the ATF3-6 engine. (*Honeywell, Inc.*)

The Pratt & Whitney JT15D Engine

Description and Operation

The JT15D series is a series of lightweight engines that have a full-length annular bypass duct, as shown in Fig. 16-104.

The specifications for a typical JT15D engine given below are for information only. As explained previously, the various models of a particular type of engine may have different specifications in some areas.

Length	59.3 in	1506 mm
Diameter	27.3 in	693 mm
Engine inlet	21.0 in	533 mm
Weight	511 lb	231.8 kg
Takeoff thrust	2200 lb	9786 N
Fuel consumption	0.540 lb/lbt/h	55 g/N/h
Oil consumption	0.5 lb/h	227 g/h
Compressor ratio	10:1	
Bypass ratio	3.3:1	
Fan pressure ratio	1.5:1	
Air mass flow, total	75 lb/s	34 kg/s
Type of combustion chamber	Annular reverse-flow type	
Fan rotation	Clockwise	

In the JT15D-1 series engines, the low-pressure compressor rotor assembly is a single-stage front fan. In the JT15D-4, -4B, and -4D engines, the low-pressure compressor rotor assembly consists of a front fan and a primary gas path booster stage rotor. In all of these engines, a concentric shaft system supports the high- and low-pressure rotors. The inner shaft supports the low-pressure compressor and is driven by a two-stage turbine. The outer shaft supports the high-pressure centrifugal compressor and is driven by a single-stage turbine.

Air entering the engine via the low-pressure compressor case, shown in Fig. 16-105, is accelerated rearward by the fan and discharged through concentric dividing ducts, as illustrated in Fig. 16-106. The secondary airflow at the inlet of the outer bypass duct passes through two rows of staggered stator vanes and flows rearward to discharge through the annular nozzle. The primary airflow at the inlet of the inner duct passes through a single row of stator vanes in the JT15D-1 series engines. In the JT15D-4, -4B, and -4D engines, the primary air is passed through a single row of stator vanes following the fan rotor and then through a second row of stator vanes following the booster rotor. In all of these engines, primary air is directed through an inlet guide stator vane assembly to the centrifugal impeller. The high-pressure air from the impeller passes through a diffuser assembly which returns the flow direction to the axial; the air then passes around the combustion chamber liner.

FIGURE 16-104   Pratt & Whitney JT15D, cutaway view. (*Pratt & Whitney Canada*)

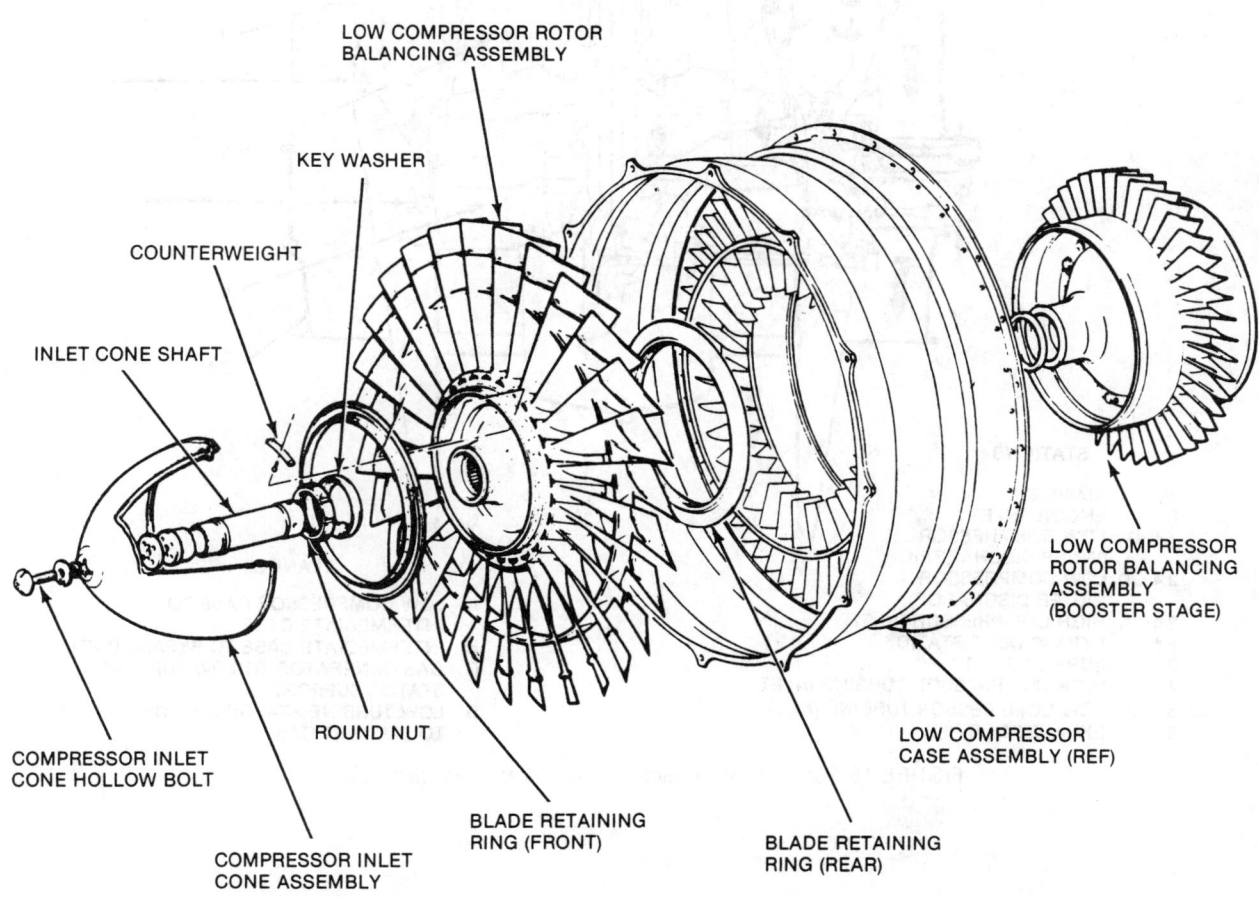

LOW COMPRESSOR ROTOR
BALANCING ASSEMBLY

KEY WASHER

COUNTERWEIGHT

INLET CONE SHAFT

LOW COMPRESSOR
ROTOR BALANCING
ASSEMBLY
(BOOSTER STAGE)

COMPRESSOR INLET
CONE HOLLOW BOLT

ROUND NUT

LOW COMPRESSOR
CASE ASSEMBLY (REF)

COMPRESSOR INLET
CONE ASSEMBLY

BLADE RETAINING
RING (FRONT)

BLADE RETAINING
RING (REAR)

FIGURE 16-105   JT15D low-pressure compressor assembly. (*Pratt & Whitney Canada*)

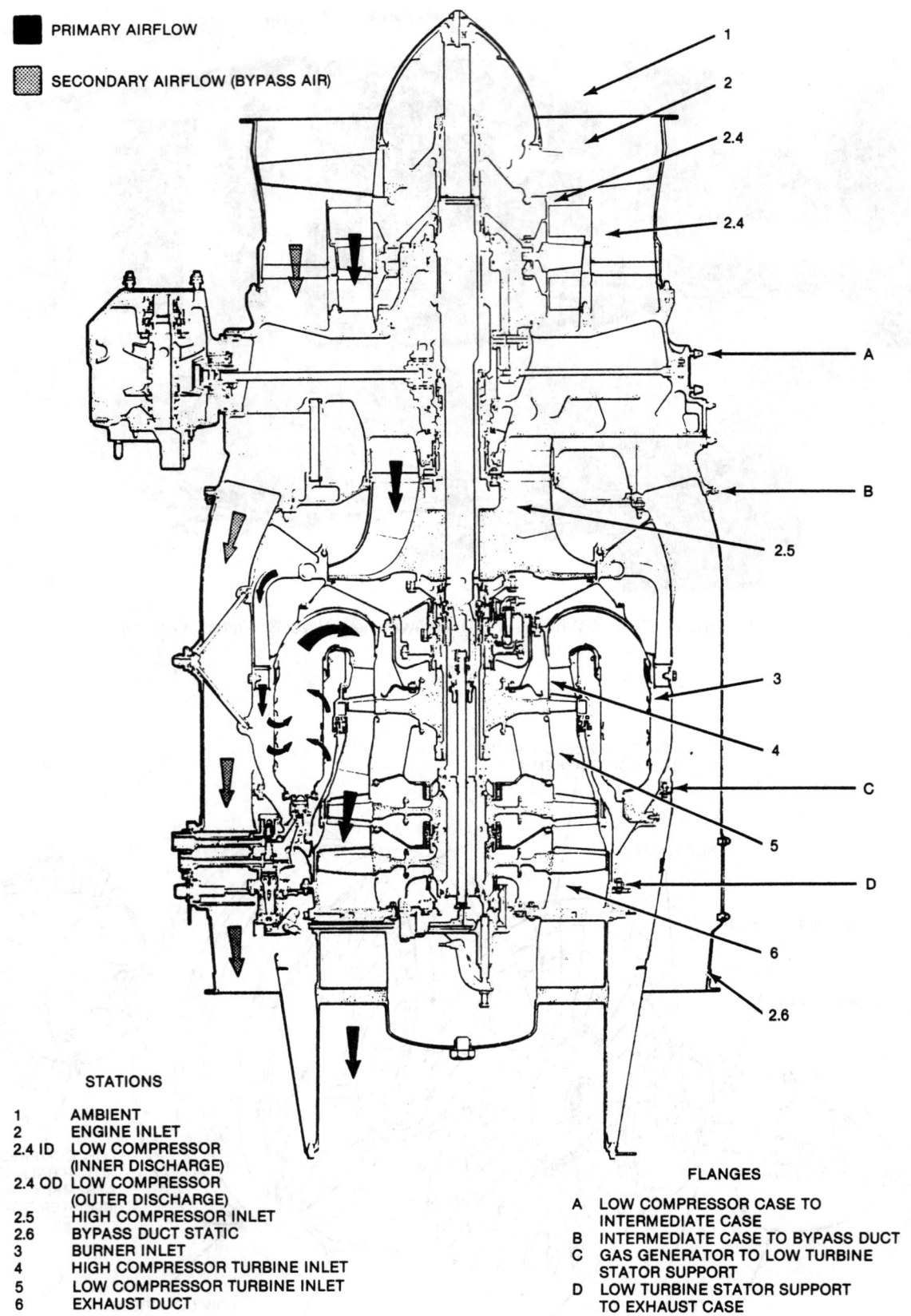

PRIMARY AIRFLOW

SECONDARY AIRFLOW (BYPASS AIR)

STATIONS

1   AMBIENT
2   ENGINE INLET
2.4 ID LOW COMPRESSOR
        (INNER DISCHARGE)
2.4 OD LOW COMPRESSOR
        (OUTER DISCHARGE)
2.5   HIGH COMPRESSOR INLET
2.6   BYPASS DUCT STATIC
3   BURNER INLET
4   HIGH COMPRESSOR TURBINE INLET
5   LOW COMPRESSOR TURBINE INLET
6   EXHAUST DUCT

FLANGES

A   LOW COMPRESSOR CASE TO
    INTERMEDIATE CASE
B   INTERMEDIATE CASE TO BYPASS DUCT
C   GAS GENERATOR TO LOW TURBINE
    STATOR SUPPORT
D   LOW TURBINE STATOR SUPPORT
    TO EXHAUST CASE

FIGURE 16-106   JT15D engine airflow. (*Pratt & Whitney Canada*)

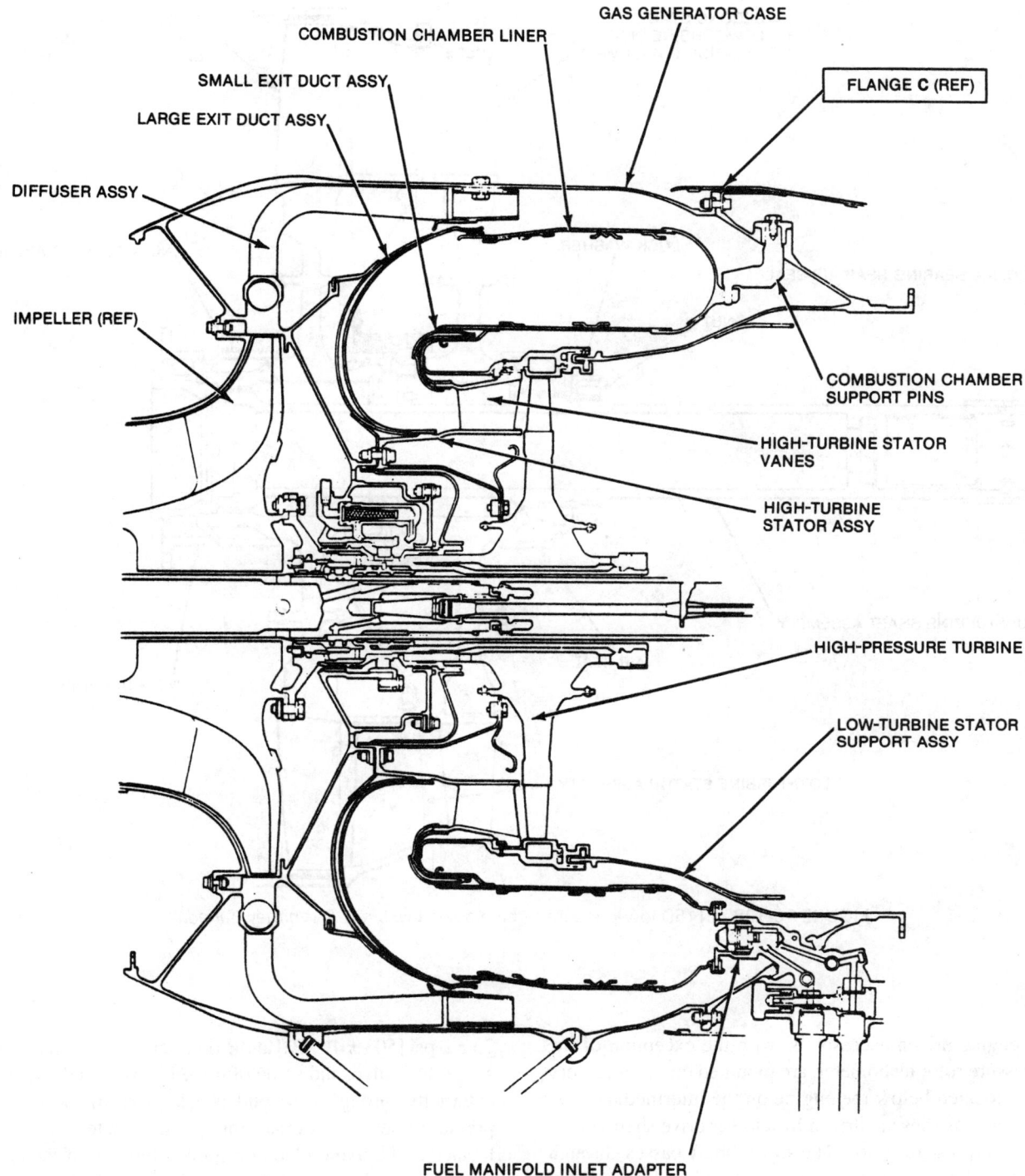

FIGURE 16-107 JT15D combustion chamber. (*Pratt & Whitney Canada*)

The combustion chamber liner, shown in Fig. 16-107, consists of an annular reverse-flow weldent made with varying sizes of perforations which allow entry of compressed air. The primary combustion air enters the combustion chamber liner and mixes with fuel. Secondary dilution air enters the liner downstream to drop the temperature peaks.

Fuel is injected into the combustion chamber by 12 dual-orifice-type nozzles supplied by a dual manifold. The mixture is ignited by two spark igniters which protrude into the combustion chamber liner. The resultant gases expand from the combustion chamber liner, reverse direction, and pass through the high-pressure compressor turbine guide vanes to the high-pressure compressor turbine. The turbine guide vanes ensure that the expanding gases impinge on the turbine blades at the proper angle, with maximum velocity and minimum loss of energy. The still-expanding gases pass rearward to the two-stage LPT, illustrated in Fig. 16-108, and to the associated guide vanes; the gases then pass to the atmosphere through the exhaust duct.

Small Turbofan Engines **473**

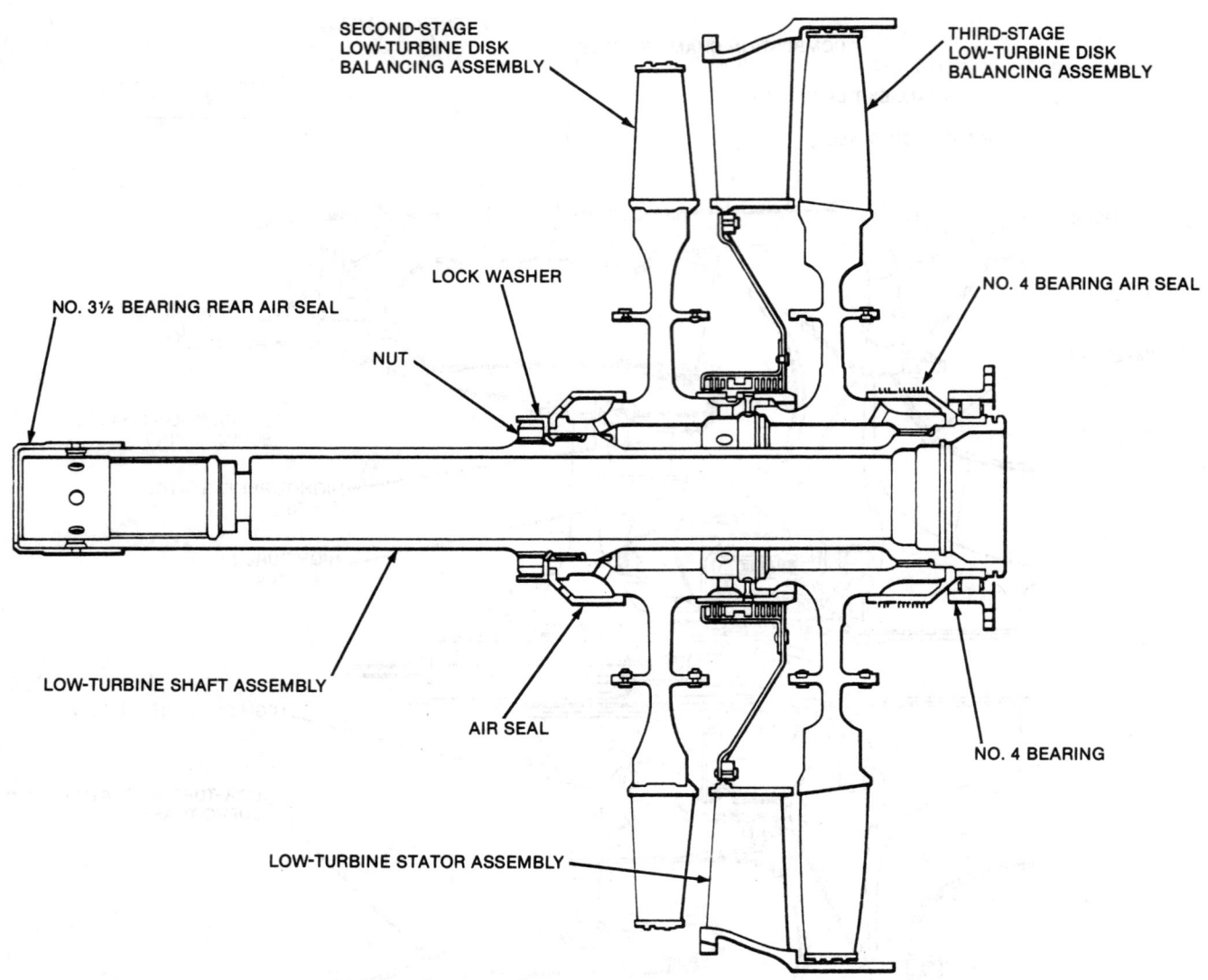

FIGURE 16-108 JT15D low-pressure turbine assembly. *(Pratt & Whitney Canada)*

All engine-driven accessories, with the exception of the low-pressure rotor tachometer, are mounted on the accessory gearbox located below the engine on the intermediate casing. The accessories are driven by a tower drive shaft geared to the high-pressure rotor. The tower shaft passes through the intermediate casing and meshes with a bevel gear on the starter-generator drive shaft to drive the accessories.

Pressure Oil System for JT15D

The **pressure oil system** for the JT15D consists basically of a pressure system, a scavenge system, and a breather system. The engine oil supply tank is contained within the intermediate casing and is located between the primary and bypass flow passages. Oil drawn from the tank by the pressure oil pump element is ducted through a check valve to the pressure relief valve inlet of the oil filter assembly. The oil is then passed through the oil cooler, which is mounted on the oil filter housing and the filter element, which, in the event of clogging, is bypassed by a valve. Oil pressure in excess of

73 ± 6 psi [503 ± 41 kPa] at the oil filter outlet opens the pressure relief valve, and some of the oil is bypassed and ducted externally through a second check valve to the oil pump pressure inlet. An external transfer tube routes oil to a boss located at the five o'clock position at the rear of the engine, and an internal transfer tube takes the oil to the bearings.

Scavenge Oil System

The function of the **scavenge oil system** is to return used oil to the tank. This return is achieved by allowing the oil from the no. 1, 2, 3, and $3\frac{1}{2}$ bearings to drain into the accessory gearbox, aided by the airflow from the bearing compartment labyrinth seals. The no. 4 bearing scavenge oil is pumped by a separate pump element in the oil pump assembly.

The scavenge oil returned to the accessory gearbox collects in a sump at the bottom of the housing. Sump oil is pumped out by a separate and larger scavenge pump element. This pump element returns both the bearing and gearbox scavenge oil to the oil tank.

JT15D Fuel System

The basic fuel system, shown in Fig. 16-109, consists of an engine-driven pump, a fuel control unit with a flexible hose leading to a temperature sensor unit, a flow divider valve, and a dual fuel manifold with 12 dual-orifice fuel nozzles. An airframe-supplied flowmeter may be installed in the high-pressure line (metered flow) between the fuel control unit and the flow divider valves, and a motive flow valve may also be installed, as an option. A drain hole in the low-turbine stator support, as well as two outlet bosses on the gas generator case connected to the single drain valve on the bypass duct, ensure drainage of residual fuel after engine shutdown or an aborted wet start.

Honeywell ALF 502 Engine

The Honeywell ALF 502 engine used on the British Aerospace 146 aircraft is shown in Fig. 16-110. This engine is a two-spool, high-bypass turbofan engine which makes use of a gear reduction unit to drive a single-stage fan off a two-stage LPT. The engine consists of four major modules: the fan module, the gas producer module, the combustor turbine module, and the accessory gearbox module. Each module has its own data plate, and is interchangeable with others of the same type. The core engine, which includes the gas producer module and combustor turbine module, are derivatives of the T55 series gas-turbine engine.

ALF 502 Specifications

Weight	1270 lb	576.1 kg
Diameter	41.7 in	106 cm
Length	56.8 in	144 cm
Thrust	6700 lb	29 800 N
Fan bypass ratio	5.7:1	
Gear reduction ratio	2.3019:1	
Fuels:		
Kerosene	Jet A, JP-5, Jet A-1, and JP-8	
Wide-cut	Jet B and JP-4	
Lubricating oils:		
Type I	MIL-L-7808	
Type II	MIL-L-23 699	
Oil tank capacity	3.6 gal (U.S.)	
Operating ambient temperature	−65°F to 122°F	

Williams/Rolls FJ44 Turbofan Engine

The FJ44 turbofan engine, shown in Fig. 16-111, is a two-spool, co-rotating, axial-flow turbofan engine with medium bypass ratio, mixed exhaust, and a high-cycle pressure ratio. It weighs 445 lb [202 kg] and produces 1900 lb [8451 N] of takeoff thrust at sea level and at ambient temperatures of up to 72°F [22°C]. Thrust is managed through power lever input to an engine-mounted hydromechanical fuel control. This fuel control is comprised of a hydromechanical metering unit (HMU) mounted on a fuel pump, and is driven by the engine gearbox. The HMU provides steady-state and transient scheduling of high-compressor rotor speed. HMU and fuel pump reliability and maintainability are enhanced by taking advantage of design elements and components from existing engine programs, where practical.

The low-pressure spool of the engine incorporates a single-stage, low-aspect-ratio, foreign-object-tolerant fan having integral blades, as can be seen in Fig. 16-112. The fan is followed by a single-stage axial intermediate-pressure (IP) compressor in the gas generator flow path. The fan and IP compressor are directly driven by two axial-flow inserted blade turbine rotors. The high-pressure spool consists of a single-stage, high-pressure-ratio centrifugal compressor driven by one uncooled axial turbine having replaceable blades. A folded annular combustor is provided which is fed by a rotating fuel slinger that atomizes and delivers fuel uniformly to the primary combustion zone. An accessory gearbox, which is driven through a tower shaft from the high-pressure spool, provides power for driving aircraft accessories, including a starter-generator and an additional accessory such as a hydraulic pump. Aircraft pressurization, cabin air-conditioning and/or heating, windshield defogging, and deice/anti-ice needs are supplied through ports that deliver high-pressure bleed air. A mechanical valve, actuated by the fuel control HMU, bleeds HP compressor inducer air into the bypass duct to improve transient response.

TURBOFAN ENGINE DEVELOPMENT

Turbofan engines are constantly being developed and improved with increased fuel economy and efficiency. Engines with conventional configurations have almost reached the maximum level of improved efficiency. Huge leaps in engine performance with current compression ratios and the current amount of internal bleed cooling air seem unlikely. The development of new engine configurations and strategies with much higher compression ratios, improved combustion techniques, higher internal temperatures (from new materials and design techniques), and reduction gearing for the fan will be the new architecture for turbofan engines in the near and foreseeable future. Engines of 5:1 compression ratio are considered high bypass engines, but the new generation of turbofan engines have compression ratios of 10:1 and even higher. These increased compression ratios will boost the output power and overall efficiency of future engines. Improved thermal efficiency from increased internal temperature and pressure, along with efficient combustors using a lean burn combustion approach, will reduce fuel burn and emissions from the engine. Engines will also have to meet new and rigid emission standards. The emissions from turbofan engines will always need to be lowered and will always continue to be somewhat of a challenge. Many of the emissions such as nitrogen oxide (NO_x) and carbon dioxide (CO_2) have been and will continue to be reduced by more and more efficient operating engines.

Much of the development and improvement in engine design comes from the use of techniques using three

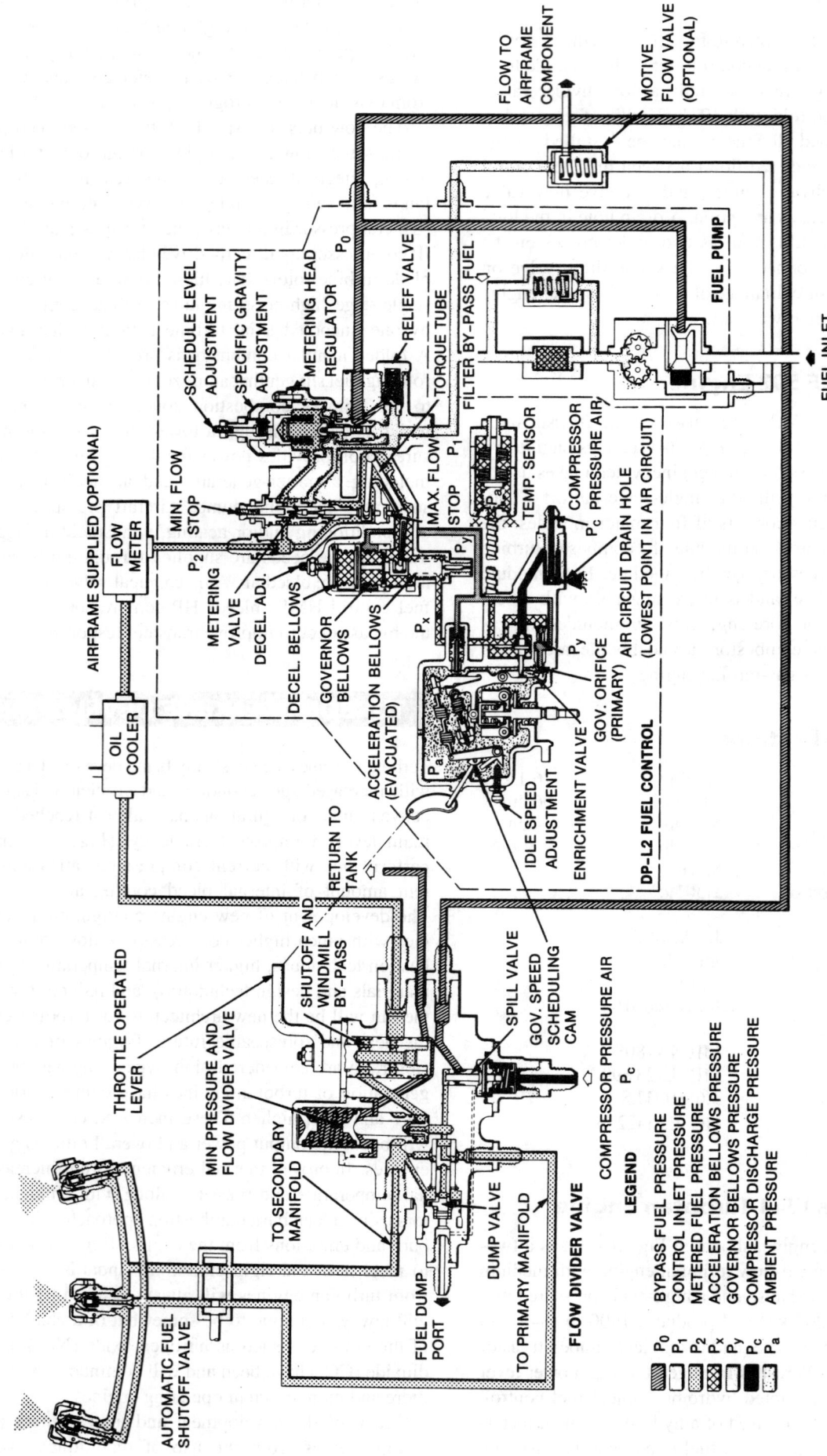

FIGURE 16-109 JT15D fuel control system schematc. (*Pratt & Whitney Canada*)

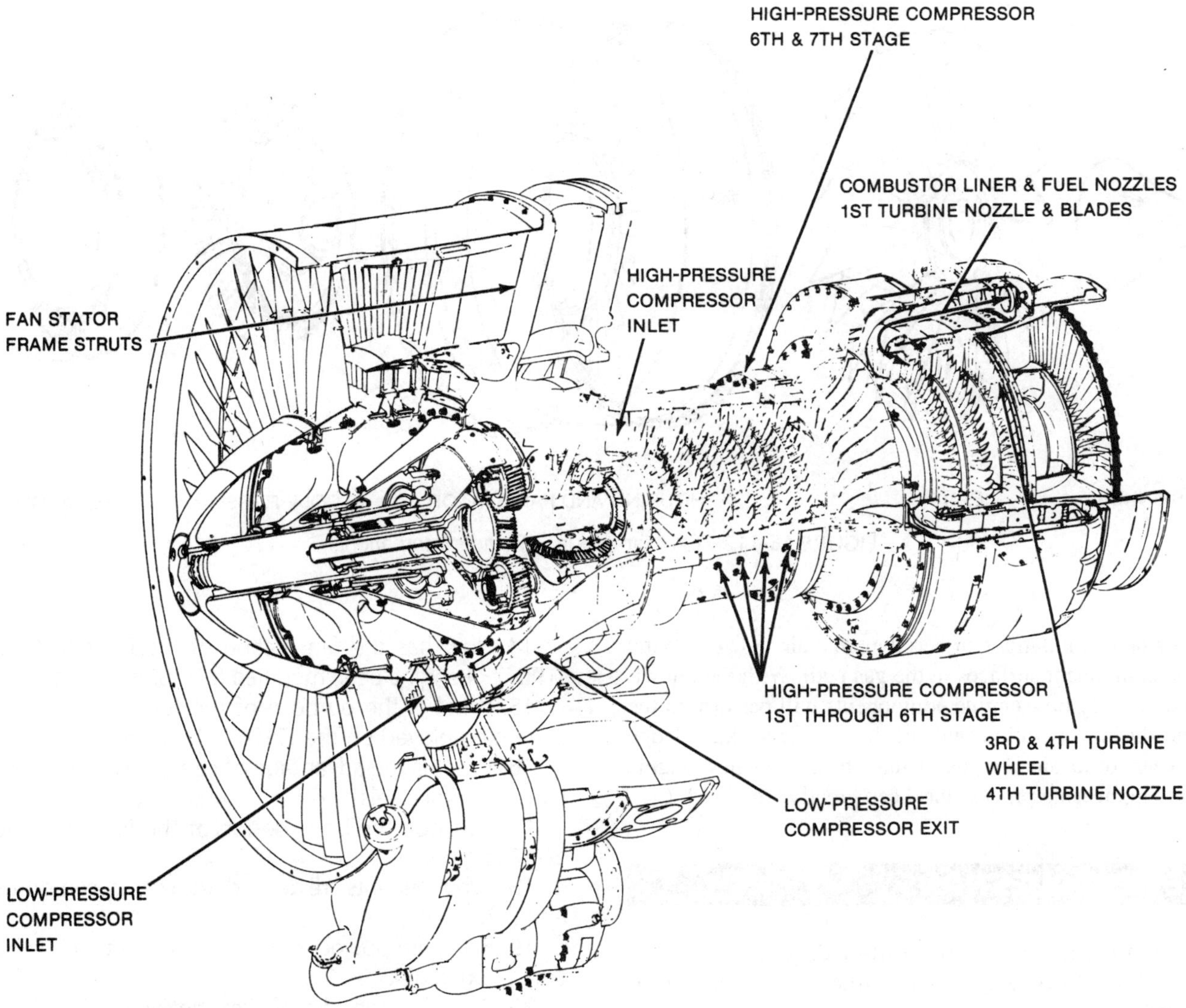

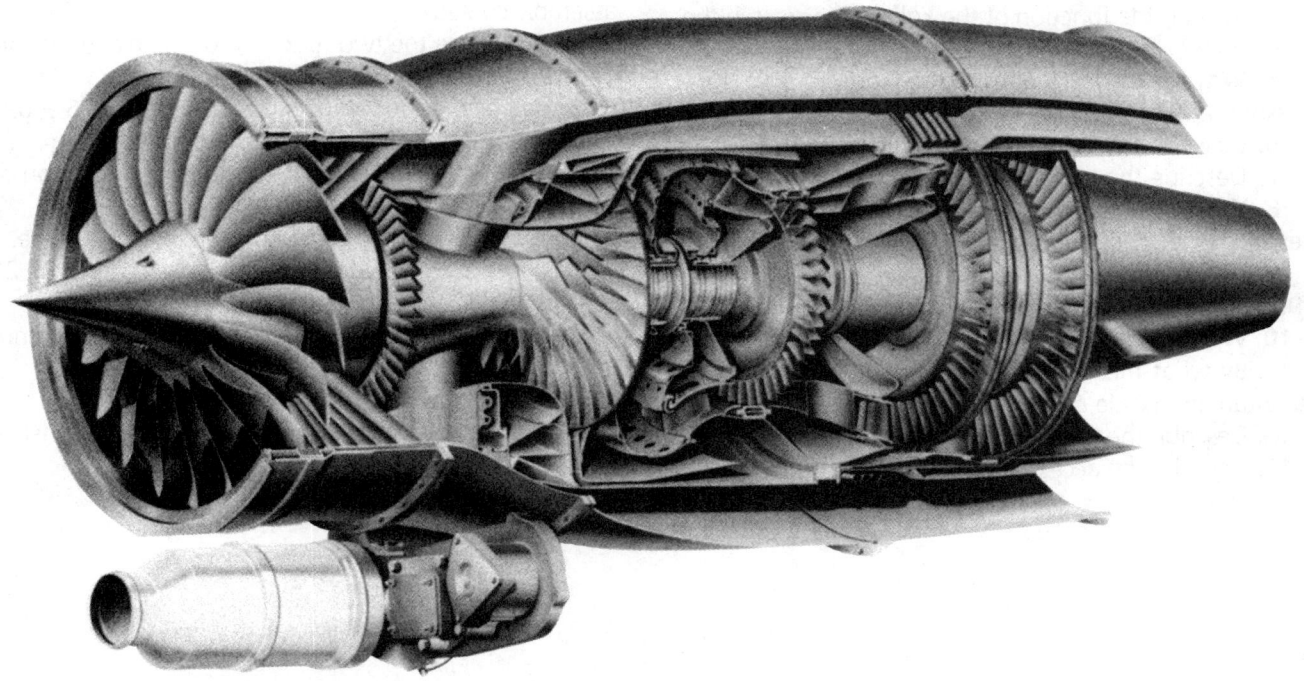

FAN STATOR
FRAME STRUTS

HIGH-PRESSURE
COMPRESSOR
INLET

HIGH-PRESSURE COMPRESSOR
6TH & 7TH STAGE

COMBUSTOR LINER & FUEL NOZZLES
1ST TURBINE NOZZLE & BLADES

HIGH-PRESSURE COMPRESSOR
1ST THROUGH 6TH STAGE

3RD & 4TH TURBINE
WHEEL
4TH TURBINE NOZZLE

LOW-PRESSURE
COMPRESSOR EXIT

LOW-PRESSURE
COMPRESSOR
INLET

FIGURE 16-110   Honeywell ALF 502 engine, cutaway view. (*Honeywell, Inc.*)

FIGURE 16-111   Williams/Rolls FJ44 engine. (*Williams International.*)

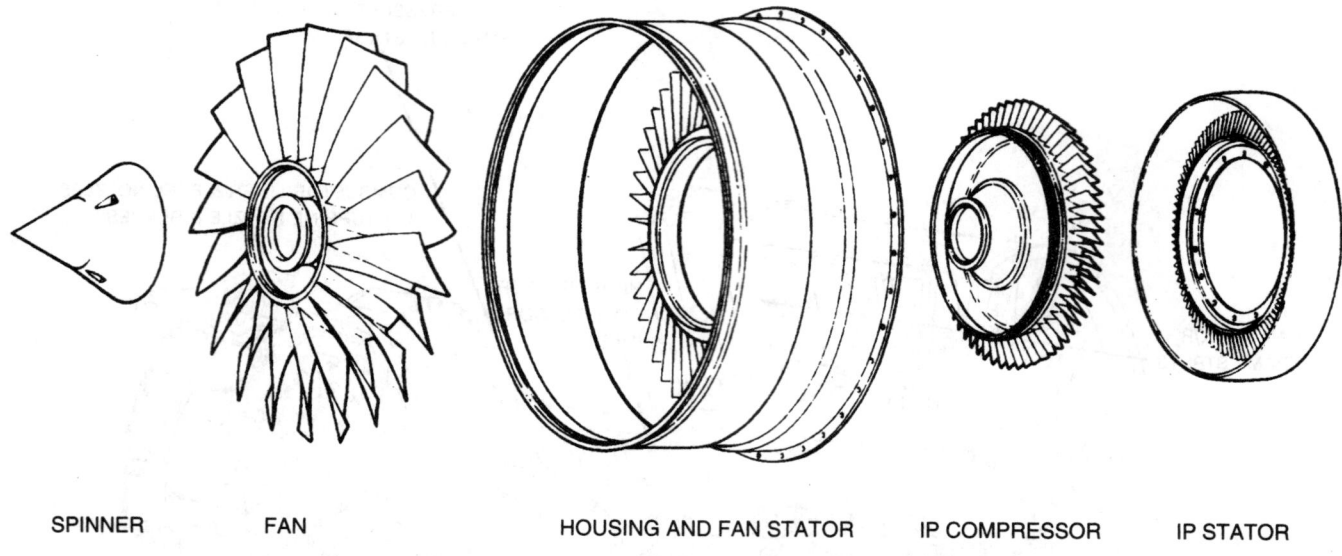

| SPINNER | FAN | HOUSING AND FAN STATOR | IP COMPRESSOR | IP STATOR |

FIGURE 16-112   FJ44 fan group. (*Williams International.*)

dimensional simulations of aerodynamic air flow over vital engine component surfaces in the gas path. With the knowledge of exactly how engine components will perform in the engine during operation and on the collective knowledge from years of turbine engine manufacturing, the new engine design/engineering process has been greatly enhanced.

REVIEW QUESTIONS

1. List the main sections of the JT8D engine.
2. Which turbine stages drive the front, or N_1, compressor?
3. Why can the first-stage turbine develop more power than other turbine stages?
4. What is the function of the knife-edge seals in the compressor?
5. Why are the last stages of the high-pressure compressor made of steel?
6. Describe the combustion section of the JT8D.
7. Describe the turbine section of the JT8D.
8. What turbine stages drive the high-pressure compressor of the JT9D?
9. List the rotating assemblies or modules of the JT9D engine.
10. Which compressor stator vanes are variable?
11. By what means is cooling provided for the first-stage turbine nozzle and turbine blades?
12. Describe the turbine section of the JT9D.
13. Describe the accessory drive section of the JT9D.

14. How has the fan section of the Pratt & Whitney JT9D-7R4 engine been modified to improve performance?
15. What is the purpose of the turbine case cooling system employed on the JT9D-7R4 engine?
16. What is the advantage of carbon air seals over the labyrinth-type seals?
17. Describe the basic design of the Pratt & Whitney 2037 engine.
18. What benefits are derived from controlled-diffusion airfoils?
19. Compare powder-metal turbine disks with conventional forged disks.
20. Describe the differences between the General Electric CF6-6 and CF6-50 engines.
21. How are the nozzle vanes for the first-stage nozzle assembly cooled?
22. What are the two functions of the blade retainers on the disks of the CF6 engine HPT?
23. Name the supporting structures of the CF6 engine.
24. List the main modules of the CFM56-3 engine.
25. Why is the three-shaft configuration used on the RB 211 engine?
26. Describe the fan section of the RB 211 engine.
27. How does the combustion chamber of the RB 211 engine compare with other combustion chambers?
28. What is unusual about the compressor arrangement in the Garrett-AiResearch TFE731 engine?
29 Describe the Garrett ATF3-6 engine.
30. Explain the advantages of the modular construction of the ATF3-6 engine.

Turboprop Engines 17

INTRODUCTION

The gas-turbine engine in combination with a reduction-gear assembly and a propeller has been in use for many years and has proved to be a most efficient power source for aircraft operating at speeds of 300 to 450 mph [482.70 to 724.05 km/h]. These engines provide the best specific fuel consumption of any gas-turbine engine, and they perform well from sea level to comparatively high altitude (over 20 000 ft [6096 m].

Although various names have been applied to gas-turbine engine/propeller combinations, the most widely used name is **turboprop**, which will generally be used in this section. Another popular name is "propjet."

The power section of a turboprop engine is similar to that of a turbojet engine; however, there are some important differences, the most important of which is found in the turbine section. In the turbojet engine, the turbine section is designed to extract only enough energy from the hot gases to drive the compressor and accessories. The turboprop engine, on the other hand, has a turbine section which extracts as much as 75 to 85 percent of the total power output to drive the propeller. For example, the Allison Model 501 engine extracts 3460 hp [2580.12 kW] for the propeller and produces 726 lb [3229.25 N] of thrust. The total **equivalent shp** (shaft horsepower plus thrust, or *eshp*) is given as 3750 eshp [2796.38 kW]. This means that the turbine section of the turboprop usually has more stages than that of the turbojet engine and that the turbine blade design of the turboprop is such that the turbines extract more energy from the hot gas stream of the exhaust. In a turboprop engine, the compressor, the combustion section, and the compressor turbine comprise what is often called the **gas generator** or **gas producer**. The gas generator produces the high-velocity gases which drive the **power turbine**. The gas generator section performs only one function: converting fuel energy into high-speed rotational energy.

In the turboprop engine, the primary effort is directed toward driving the propeller. One method of doing this is to use what is referred to as a **free turbine**. A free turbine is not mechanically connected to the gas generator; instead, an additional turbine wheel is placed in the exhaust stream from the gas generator. This extra turbine wheel, referred to as the **power turbine**, is shown in Fig. 17-1.

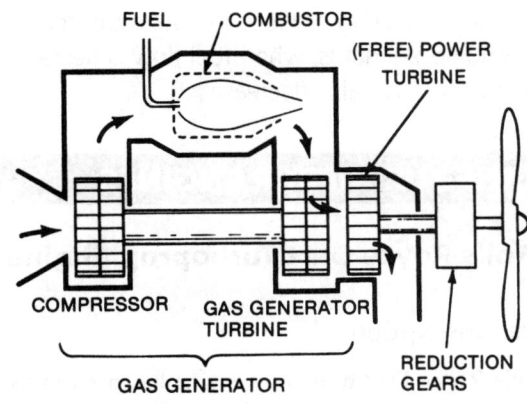

FIGURE 17-1   Free-turbine-type power conversion. (*Honeywell Inc.*)

A different method of converting the high-speed rotational energy from the gas generator into usable shaft horsepower is illustrated in Fig. 17-2. In this case, the **gas generator** (shown at right in the illustration) has an additional (third) turbine wheel. This additional turbine capability utilizes the excess hot gas energy (that is, energy in excess of that required to drive the engine's compressor section) to drive the propeller. In a **fixed shaft engine**, the shaft is mechanically connected to the gearbox so that the high-speed low-torque rotational energy transmitted into the gearbox from the turbine can then be converted to the low-speed high-torque power required to drive the propeller.

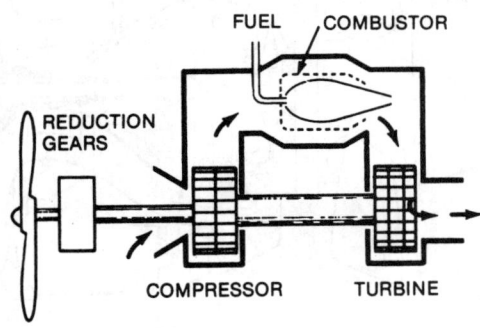

FIGURE 17-2   Fixed-shaft-type power conversion. (*Honeywell Inc.*)

The gear reduction from the engine to the propeller is of a much higher ratio than that used for reciprocating engines because of the high rpm of the gas-turbine engine. For example, the gear reduction for the Rolls-Royce Dart engine is 10.75:1, and the gear reduction for the Allison Model 501 engine is 13.54:1.

Because the propeller must be driven by the turboprop engine, a rather complex propeller control system is necessary to adjust the propeller pitch for the power requirements of the engine. At normal operating conditions, both the propeller speed and engine speed are constant. The propeller pitch and the fuel flow must then be coordinated in order to maintain the constant-speed condition—that is, when fuel flow is decreased, propeller pitch must also decrease.

LARGE TURBOPROP ENGINES

The Rolls-Royce Dart Turboprop Engine

General Description

The Rolls-Royce Dart turboprop engine has been in use for many years on a variety of aircraft, including the Vickers Viscount and the Fairchild F-27 Friendship. This engine has proven to be rugged, dependable, and economical, with overhaul periods extending to more than 2000 h.

The Dart engine utilizes a single-entry two-stage centrifugal compressor, a can-type through-flow combustion section, and a three-stage turbine. The general design of the engine is illustrated in Fig. 17-3. This drawing shows the arrangement of the propeller, reduction gear, air inlet, compressor impellers, combustion chambers, turbine, and exhaust. The engine is approximately 45 in [114.3 cm] in diameter and 98 in [248.92 cm] in length.

Engine Data

The general and performance data for the Dart Model 528 engine are as follows:

Power output	1825 shp plus 485 lbt	1368 kW plus 2157 N
Compression ratio	5.62:1	
Engine rpm	15 000	
Weight (without propeller)	1415 lb	642 kg
SFC (specific fuel consumption)	0.57 lb/eshp/h	346.72 g/kW/h
Power-weight ratio	1.51 eshp/lb	2.13 kW/kg

Internal Features

The cutaway photograph of the Dart engine shown in Fig. 17-4 reveals the internal construction of the engine. At the forward end is the **reduction-gear assembly**, which reduces the propeller-shaft speed to 0.093 of the speed of the engine. The reduction-gear housing is integral with the **air-intake casing**.

Immediately to the rear of the reduction-gear assembly is the compressor section, which includes two centrifugal impellers. Both impellers are clearly visible in the illustration. Accessory drives are taken from the reduction-gear assembly and through a train of gears aft of the second-stage compressor impeller.

Seven interconnected **combustion chambers** are located between the compressor section and the turbine. These combustion chambers are skewed, or arranged in a spiral configuration, to shorten the engine and take advantage of the direction of airflow as it leaves the compressor.

A **three-stage turbine** is located to the rear of the combustion chambers. As in other turboprop engines, this turbine is designed to extract as much energy as possible from the high-velocity exhaust gases.

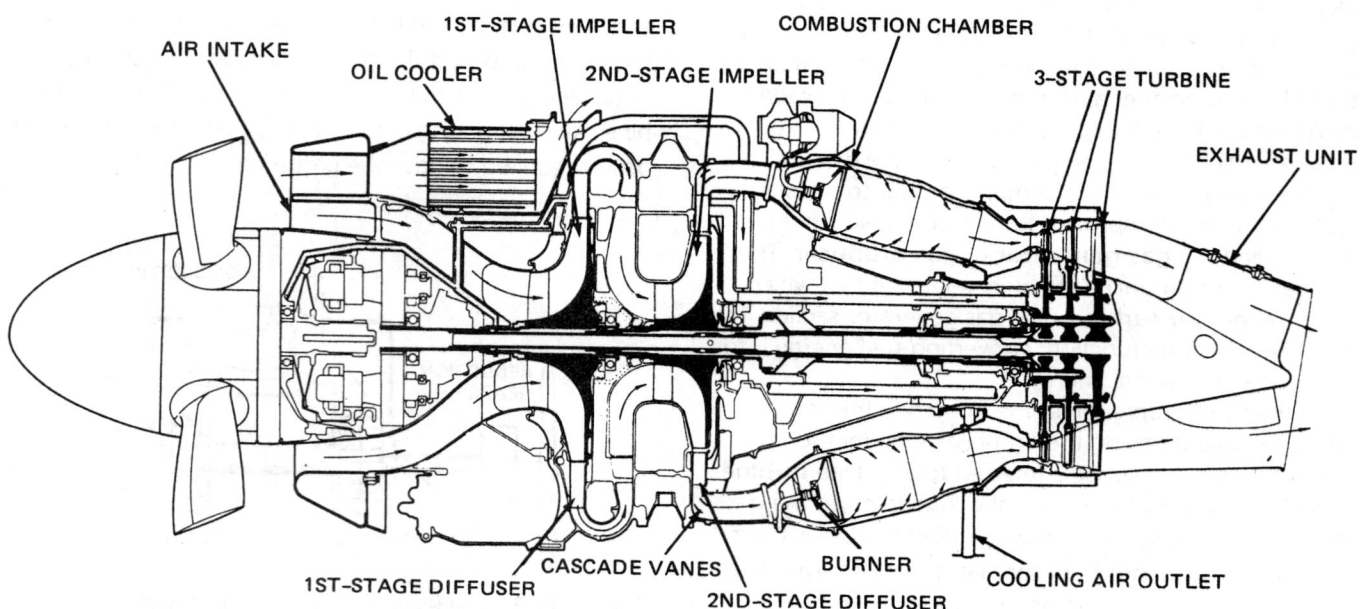

FIGURE 17-3 Arrangement of the Rolls-Royce Dart engine. (*Rolls-Royce.*)

FIGURE 17-4   Cutaway view of the Dart engine. (*Rolls-Royce.*)

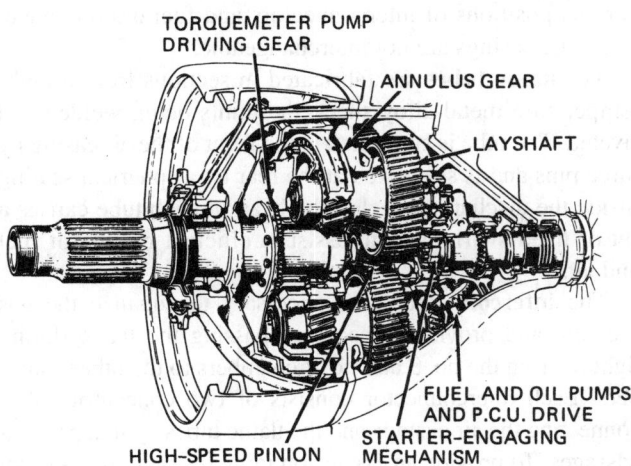

FIGURE 17-5   Reduction-gear assembly for the Dart engine.

Reduction-Gear Assembly

The reduction-gear assembly, shown in Fig. 17-5, is of the compound type having high-speed and low-speed gear trains. The **high-speed gear train** consists of a high-speed pinion connected to the main shaft that drives three layshafts through helical gear teeth. To isolate the main shaft couplings from propeller vibrations, a torsionally flexible shaft is used to couple the high-speed pinion to the main shaft. The three layshafts are mounted in roller bearings supported by panels in the gear casing.

The **low-speed gear train** consists of helical gears, formed on the front ends of the layshafts, which drive the internal, helically toothed **annulus gear**. This annulus gear is bolted to the propeller-shaft driving disk. As a result of driving through the helical gears, the layshafts tend to move axially. This movement is limited by limit shafts mounted coaxially within the layshafts. Each limit shaft is prevented from moving by a ball thrust race at the rear.

The **propeller shaft** is supported by roller bearings housed in the front panel and the domed front casing. Axial thrust is taken on a ball bearing mounted behind the front

roller bearing. A labyrinth-type seal assembly, pressurized by compressed air, surrounds the propeller shaft where it passes through the front cover and prevents loss of lubrication oil to the atmosphere.

To permit propeller oil to be transferred from the stationary casing to the rotating propeller shaft, a transfer seal assembly is used. It consists of babbitt lined with bronze bushings fitting closely around an adapter located inside the rear end of the propeller shaft. Tubes screwed into the adapter convey the oil to the pitch-control mechanism.

Torquemeter

Under normal operating conditions, the helical teeth of the gear train produce a forward thrust in each layshaft which is proportional to the propeller-shaft torque. This load is hydraulically balanced by oil pressure acting on a piston assembly incorporated in the forward end of each layshaft. The necessary oil pressure is obtained by boosting engine oil pressure with a gear pump mounted on the layshaft front-bearing housing and driven from a gear attached to the propeller shaft.

The forward thrust of the layshafts resulting from the greater torque of the low-speed gear train is partially balanced by the rearward thrust produced by the lesser torque of the high-speed gear train. The residual forward thrust is balanced by the torquemeter oil pressure. A gauge in the cockpit indicates this pressure, which is a measure of the torque transmitted by the gear. The engine power is calculated from the reading of the gauge.

Auxiliary Drives

The auxiliary drives receive power from a bevel gear, splined to the rear of the lower limit shaft, which meshes with another bevel gear supported in plain bearings in the rear panel of the reduction-gear case. Through the auxiliary drives, the oil pumps, fuel pumps, and propeller control unit are driven.

Large Turboprop Engines   **481**

Compressor

The compressor for the Dart engine comprises two stages, one immediately to the rear of the other, as shown in Fig. 17-4. The first-stage impeller is 20 in [50.8 cm] in diameter and has 19 blades, while the second-stage impeller is 17.6 in [44.7 cm] in diameter with 19 blades.

The compressor casings include the **front-compressor casing**, the **intermediate casing**, and the **second-stage outlet casing**. The front-compressor casing and the second-stage casing carry the diffuser-vane rings, and the intermediate casing carries the interstage guide vanes internally and the engine mounting points externally.

Each rotating assembly consists of an impeller and rotating guide vanes (RGVs). The assemblies are splined onto separate shafts and individually balanced. The split shaft facilitates bearing alignment and makes it unnecessary to disturb the balance during engine buildup. The guide vanes and impellers are locked to the shafts by nuts and cup washers.

Passages are machined through the first-stage rotating guide vanes and between the impeller vanes to permit water-methanol injection. The first-stage shaft is supported at the front by a roller bearing and at the rear by a ball bearing. The second-stage shaft is supported at the front by helical splines inside the rear of the first-stage shaft and at the rear by a ball bearing.

Surrounding each rotating assembly is a **diffuser-vane ring**. Each ring consists of a number of fixed vanes forming divergent channels.

Between the compressor stages is a set of guide vanes. Air leaving the first-stage compressor passes between these vanes before entering the second-stage RGVs. The vanes are so angled that they impart a whirling velocity to the airstream.

Combustion Section

The combustion section consists of seven individual combustion chambers such as that shown in Fig. 17-6, arranged in an inward spiral (skewed) with respect to the engine main shaft to shorten the engine and promote a smooth gas flow.

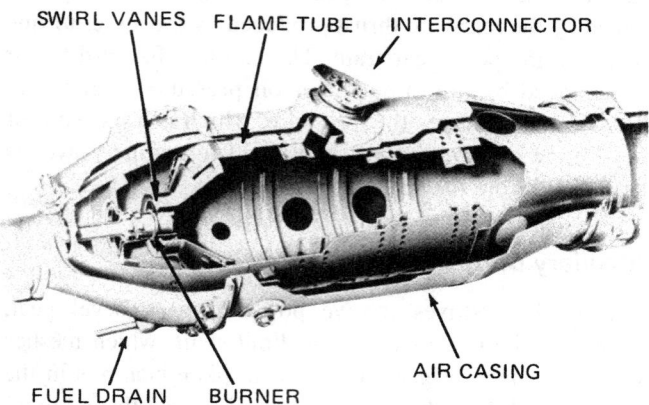

FIGURE 17-6 Combustion chamber for the Dart engine. (*Rolls-Royce.*)

The chambers are numbered counterclockwise, viewed from the rear, with no. 1 being at the top. Each combustion chamber consists of an expansion chamber, an air casing, the flame tube, and interconnectors.

The **expansion chambers**, forming the forward ends of the combustion chambers, are fitted to the compressor outlet elbows by two link bolts, the seating between the chamber and elbow being formed by a spherical joint ring. At the rear they are attached to the air casing by means of a bolted flange. Each expansion chamber provides the location for a fuel burner (nozzle), and provision is made for fuel drain connections where necessary. High-energy igniter plugs are carried in the no. 3 and no. 7 chambers.

The **air casings** are bolted to the expansion chambers at the front; however, they are inserted in the discharge nozzles at the rear with a slip fit sealed by piston rings. This permits expansion and contraction of the casings. Each air casing carries two interconnectors, three flame-tube locating pins, and fuel drain connections where necessary. Because of the various positions of interconnectors and fuel drain connections, the casings are not interchangeable.

The **flame tubes** are fabricated in sections from a high-temperature metal-alloy sheet, the joints being welded and riveted. The tube is supported at the front of the air casing by three pins and is supported at the rear by a spherical seating inside the discharge nozzle. The head of each tube carries a set of fixed swirl vanes to assist in efficient mixing of fuel and air.

The **interconnectors** are necessary to equalize the gas pressure and provide a means of passing the flame during lightoff from the no. 3 and no. 7 chambers to the other chambers. Each interconnector consists of two concentric tubes connecting the air casings and the flame tubes by independent passages. To provide an expansion joint, the outer tubes carry sealing rings seated in bores in the air casings. A three-bolt flange forms the joint between each interconnector connecting adjacent combustion chambers.

Turbine Section

The turbine section of the Dart engine consists of three turbine wheels fitted with blades and of the nozzle box assembly, which contains three sets of nozzle guide vanes (NGVs). The compressor drive shaft and the inner reduction-gear drive shaft are coaxial and are attached with bolted flanges to the three turbine wheels.

The **nozzle box** is a welded two-piece casing into which are fitted the seven combustion-chamber discharge nozzles. It is surrounded by a heat shield. On the front flange of the nozzle box is fitted the nozzle box mounting drum, which, together with the inner cone and turbine bearing housing, is bolted to the intermediate casing. Flanges on the inside of the nozzle box and inner cone, and interstage labyrinth-seal platforms, provide the location of the nozzle guide vanes.

The **nozzle guide vanes** form a series of nozzles in which the gases are accelerated. They are airfoil shaped and are cast hollow to maintain, as nearly as possible, a constant sectional thickness to reduce thermal stress.

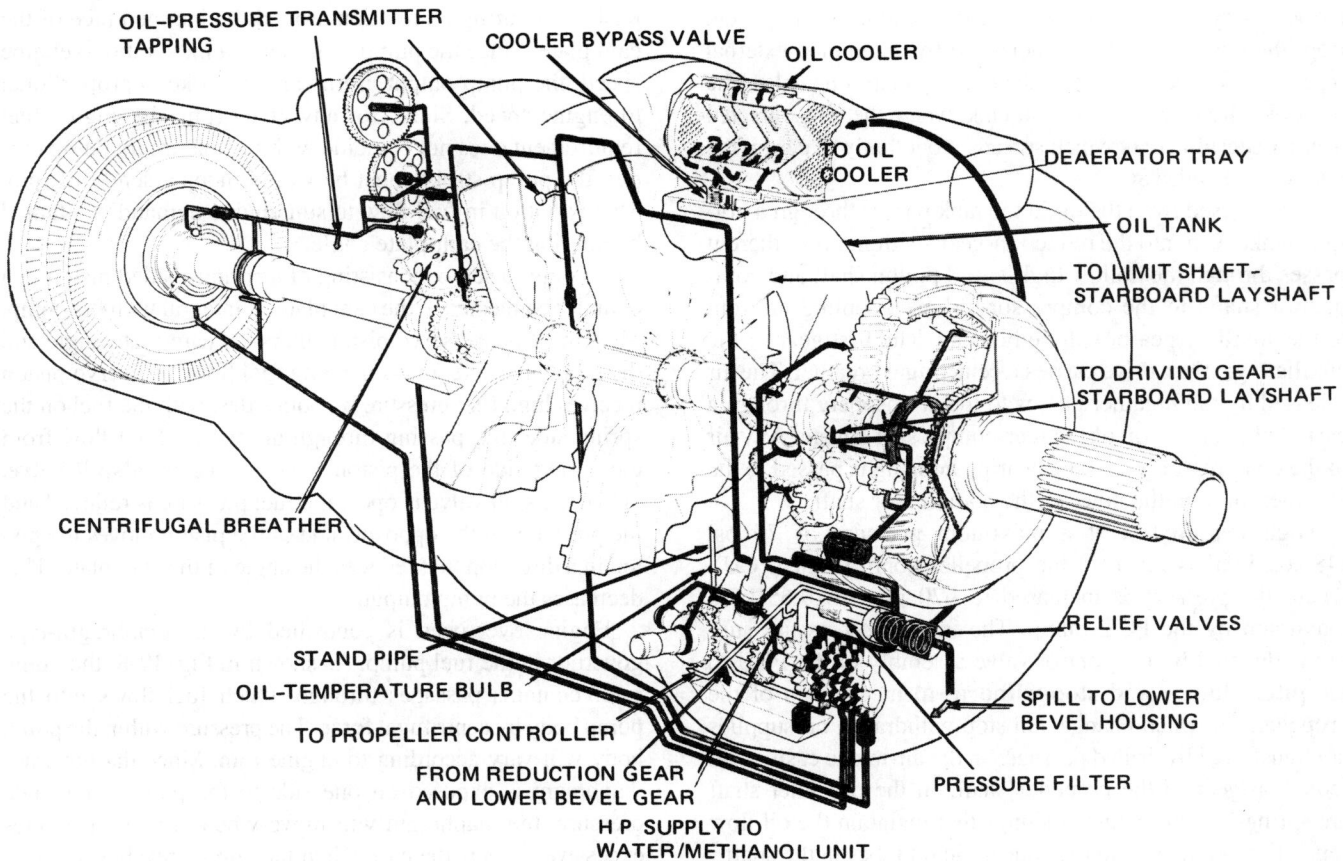

FIGURE 17-7 Oil system for the Dart engine. (*Rolls-Royce.*)

There are 70 high-pressure vanes hooked into flanges machined on the inner cone and nozzle box outer casing, and 14 of these are used as locators. The inner location is provided by slots in the flange of the inner cone, and the outer location is provided by locating pegs fitted through the nozzle box casing and engaging in the guide vane outer platforms.

Fifty-six intermediate-pressure vanes are supported in grooves in the nozzle box outer casing by the tongues on the outer platforms hooking into the grooves in the nozzle box. They are positioned axially by two rings, and the turbine interstage seal is carried on their inner platforms. At the leading edges of 12 of the vanes, provision is made for fitting the thermocouples. Twenty-eight of the vanes are used as locators.

The three **turbine wheels** are secured to the turbine and inner drive shaft by taper bolts. Each wheel consists of a steel disk to which is fitted Nimonic-alloy turbine blades, and each blade carries its own shroud. To reduce losses at the blade tips, seals are formed on the shrouds of the blades. The root of each blade is of fir-tree shape and fits into a corresponding slot broached in the rim of the disk. The blades are locked to the disk by locking tabs. Labyrinth-type seals are fitted between the stages of the turbine to control the disk-cooling airflows.

Exhaust Unit

The **exhaust unit**, which is bolted to the nozzle box, consists of two concentric cones joined by three support fairings. Each fairing is secured by setscrews to a sole plate

on the outer cone. The interior of the inner cone is vented to the exhaust-gas stream by three circumferentially positioned holes called **pressure balance holes**. Fuel drain holes are incorporated in the assembly to prevent the accumulation of fuel.

When the engine is installed, the exhaust unit is arranged within a conical shroud with its discharge end centrally located in the jet pipe inlet. An annular gap formed between the discharge end of the unit and the jet pipe inlet creates an ejector effect which draws air into the stream. This air is drawn from the combustion compartments between the exhaust-unit outer cone and its surrounding shroud. A flow of cooling air is thus provided over the whole combustion compartment.

Oil System

The oil system for the Dart engine is shown in Fig. 17-7. The oil tank is an integral part of the engine, consisting of the annular chamber surrounding the first-stage air inlet. The oil cooler is mounted at the top of the tank as shown. During operation, oil is drawn from the standpipe at the bottom of the tank and flows past an oil temperature bulb and then to the pressure pump. The pump applies pressure to the oil and forces it to all parts of the engine that require lubrication.

There are four scavenge pumps in the engine oil system. These pumps scavenge oil from the reduction-gear section, the interstage bearing, the second-stage compressor rear bearing, the accessory gearbox drive gears, and the turbine bearings. Oil from the scavenge pumps is delivered by a common external pipe on the left side of the air-intake casing to the oil cooler. The oil cooler discharges into the oil tank, where the oil is directed over a **deaerator tray** which spreads it out thinly to permit the release of included air.

Air released from the oil in the tank passes through a hollow intake web into the reduction-gear section. From there it passes through the hollow high-speed pinion shaft and compressor shafts to the compressor-turbine coupling and out to the auxiliary gearbox drive housing. The first gear of the auxiliary gearbox drive carries a centrifugal breather. The air released by the breather passes to the atmosphere through a cast pocket in the top of the rear-compressor casing. Any air in the compressor interstage bearing housing is passed to the breather through the holes in the compressor shaft.

High-pressure oil at a maximum pressure of 70 psi [482.65 kPa] is taken to the propeller control unit (PCU), where the pressure is increased to 670 psi [4619.65 kPa] maximum by the PCU pump. The increased-pressure supply is directed by the control valve assembly of the PCU to the **pitch-change and stop-withdrawal** mechanism of the propeller. The pitch-change and stop-withdrawal oil supplies are transferred by drilled passages in the air-intake casing and reduction gear to the propeller shaft. In the propeller shaft are spring-loaded sealing bushings that maintain the oil flow separation on transfer to the concentric oil tubes in the shaft.

The oil system includes features considered standard for engine oil systems, such as filters, pressure relief valves, oil quantity indicator (dipstick), scavenge oil filters, oil cooler, oil pressure transmitter, oil temperature bulb, and oil pressure warning light.

Fuel System

The fuel system for the Dart engine is designed to satisfy the basic requirements of the engine for all types of operation. The system must provide full atomization of the fuel over the complete range of fuel flow, control fuel flow according to engine demand, provide engine overspeeding control, ensure a specific flow for a given throttle position, compensate fuel flow for altitude conditions, limit flow to suit the engine power rating, provide a correct idling fuel flow, and provide for complete fuel shutoff when it is desired to stop the engine.

The operation of the fuel pump and fuel control unit can be understood by examining Fig. 17-8. The fuel pump consists of an engine-driven rotor carrying seven plungers spring-loaded against a circular cam plate. The output of the pump is varied by changing the angle of the cam plate relative to the rotor through the action of a servo piston. The piston assembly is carried in an alloy body which incorporates the inlet and outlet ports. These ports communicate with the revolving rotor through a fixed valve plate containing two kidney-shaped ports.

As the rotor of the pump revolves around the cam plate, each plunger in turn is extended and receives low-pressure fuel. It then delivers the fuel at high pressure as the plunger is pushed in during its rotation around the inclined face of the cam plate. Since the pump is driven at a fixed ratio to engine speed, the pump output at maximum stroke is proportional to engine speed. Since, for any given rpm, the engine fuel requirement does not coincide with the maximum pump output, the pump stroke must be varied independently of rpm. This variation in fuel flow to suit engine demand is attained by altering the cam-plate angle.

The pump servo, consisting of a spring-loaded piston in a cylinder connected to the cam plate, is integral with the pump. Movement of the servo piston alters the cam-plate angle and the plunger stroke, thus changing fuel flow. The servo piston receives high fuel pressure on both sides, with the fuel on the spring side first passing through an orifice. Fuel flow from the spring side of the piston is controlled by a **spill valve**. When the spill valve is open, the fuel pressure is relieved and the pressure on the opposite side of the piston moves the piston in a direction that reduces the angle of the cam plate. This decreases the pump output.

Engine overspeed is controlled by the diaphragm-type governor in the fuel pump. As shown in Fig. 17-8, the pump rotor contains passages through which fuel flows into the pump body by centrifugal force. The pressure within the pump body will vary according to engine rpm. Since the governor diaphragm is exposed on one side to the pump centrifugal pressure, the diaphragm will move when pressure becomes excessive. This is the case when the engine reaches an overspeed condition. As the diaphragm moves, it pushes a lever which releases a spill valve controlling fuel pressure on the spring side of the servo piston and thus reduces the pump cam-plate angle, which, in turn, reduces fuel flow. The reduction in fuel flow continues until the engine speed stabilizes at the predetermined overspeed rpm set by the tension spring.

A secondary function of the overspeed governor spill valve is to prevent excessive fuel pressures in the system. Thus, it acts as a relief valve. The overspeed governor spill-valve rocker arm is loaded by a spring which, through its leverage, maintains the spill valve in a closed position unless there is an excessive rise in pump delivery pressure. If this occurs, the spill valve opens and reduces pump delivery pressure.

In the fuel flow control unit, a spill valve controls the pump servo according to throttle position. This valve is kept informed of the throttle position by a spring-loaded control piston which senses the fuel flow via pressure signals from upstream and downstream of the throttle valve. Attenuators in the pressure-sensing lines damp out any pressure fluctuation from the fuel pump. The control piston movement is transmitted by a pushrod to the flexibly mounted lever housing the spill valve.

Under stabilized conditions the fuel-pressure differential across the control piston balances the control piston spring force. The spill-valve position is thus automatically adjusted so that the pump servo piston selects the correct pump stroke for fuel flow. When the throttle is opened, the pressure differential across the throttle valve decreases and the control piston senses this decrease. The piston moves to close the spill valve, thus causing the pump output to increase until

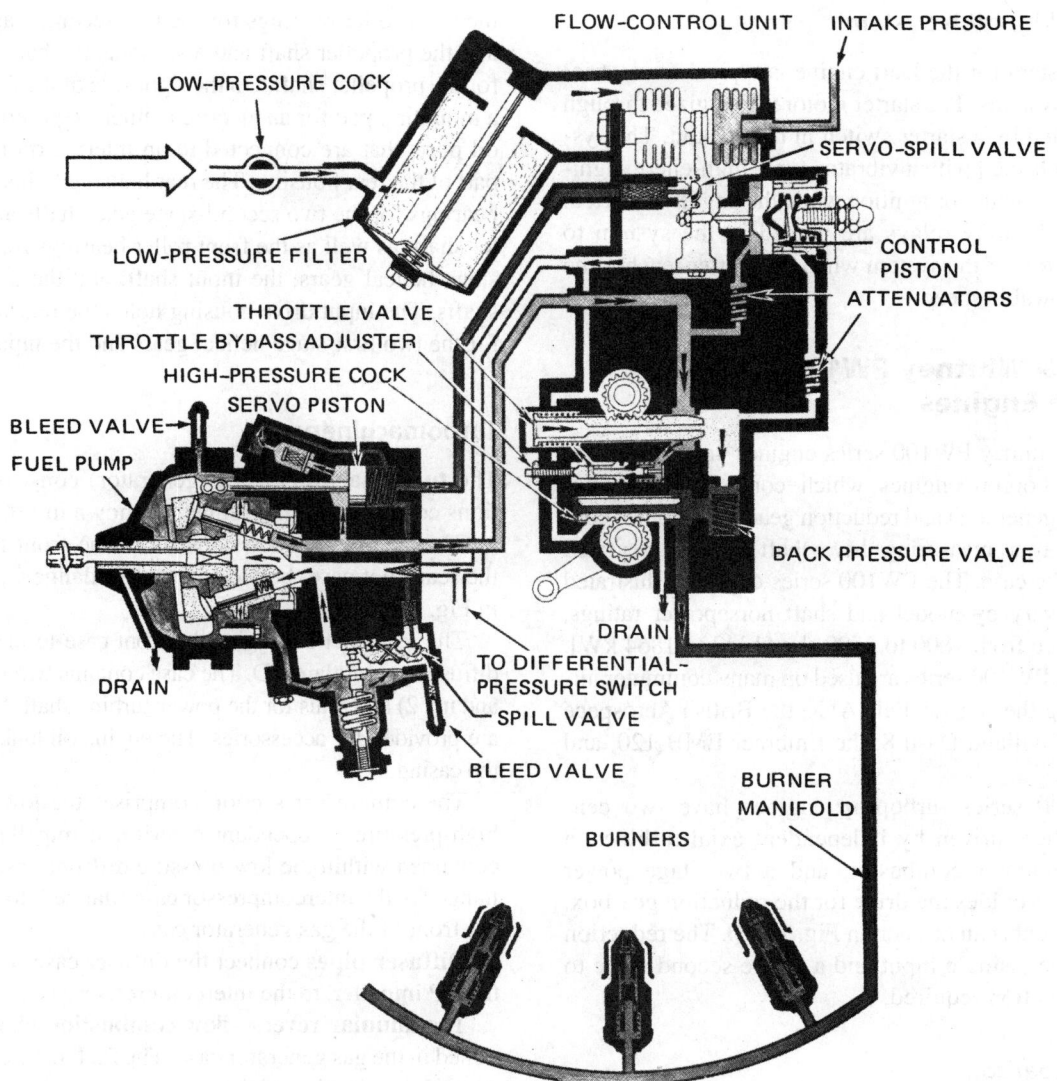

FIGURE 17-8 Fuel control unit and variable pump for the Dart engine. (*Rolls-Royce.*)

fuel flow is correct for the new throttle setting. The system then stabilizes at the new position.

Fuel flow adjustment for variations in altitude is accomplished through the action of the intake pressure aneroid bellows shown in Fig. 17-8. As altitude increases, the bellows exerts pressure on the spill valve, which reduces the pump output. The bellows is designed so that no further action of the bellows to increase fuel flow can take place when ambient pressure reaches 14.7 psi [101.36 kPa]. This is done to prevent the engine from being provided with excessive fuel.

Fuel flow from the control unit passes through the high-pressure cock and then passes to the burners in the combustion chambers. These burners, or nozzles, are designed to provide a hollow conical spray of fuel at the forward end of each combustion chamber. The burners include thread-type filters.

Water-Methanol System

An engine in operation under high-ambient-temperature conditions undergoes a reduction in engine mass airflow, and

the fuel flow is reduced by trimming in order to maintain the turbine working temperatures within acceptable limits. This results in a reduction of engine shaft horsepower (shp), which can be restored to takeoff level by injection of a water-methanol mixture into the first-stage compressor through drilled passages in the rotating guide vanes and impeller. Water and methanol from the aircraft tank are fed by a tank pump and electrically operated feed cock to the metering valve of the water-methanol unit. The cockpit selector switch operates both the feed cock and tank pump, and a cockpit light indicates when water and methanol are being supplied. The feed cock is interconnected with the propeller feathering system so that the water-methanol supply is automatically shut off when the propeller is feathered.

The water-methanol mixture used in the Dart engine consists of water containing between 36 and 38 percent of methanol (methyl alcohol) by weight. This is approximately equivalent to 43.8 volumes of methanol and 56.2 volumes of water. The water and methanol must meet rigid specifications of quality and purity.

Starting and Ignition System

The starter system for the Dart engine is typical of electric-motor starter systems. The starter motor is energized through relays controlled by a starter switch in the cockpit. The system is interconnected with a vibrator-type, high-energy ignition system to provide for ignition when the engine is started. Overspeed and safety relays are placed in the system to provide for cutoff of the system when the starter reaches the maximum allowable speed.

The Pratt & Whitney PW100 Series Turboprop Engines

The Pratt & Whitney PW100 series engines are free-turbine propulsion turboprop engines which consist of a **turbomachine** (gas generator) and reduction gearbox modules connected by a torque-measuring drive shaft and an integrated structural intake case. The PW100 series engines, illustrated in Fig. 17-9, vary by model and shaft horsepower ratings, which can range from 1800 to 2500 shp [1342 to 1864 kW]. Engines of the PW100 series are used on many commuter aircraft, including the Aerospatiale ATR, the British Aerospace ATP, the DeHavilland Dash 8, the Embraer EMB 120, and the Fokker 50.

The PW100 series turboprop engines have two centrifugal impellers driven by independent axial turbines, a reverse-flow annular combustor, and a two-stage power turbine which provides the drive for the reduction gearbox. These components can be seen in Fig. 17-10. The reduction gearboxes have a single input and a single second stage to obtain the reduction required.

Reduction Gearbox

The **reduction gearbox**, shown in Fig. 17-11, has an accessory drive cover and three housings; the front housing, the rear housing, and the input housing. The front housing holds the front roller bearings for the two second-stage gear shafts and the propeller shaft and also holds the ball thrust bearing for the propeller shaft. On the right side of the front housing is a mounting pad for an electric feathering pump. This pad has oil ports that are connected to an internal oil tank, which is part of the rear housing. The rear housing holds the rear roller bearings for the two second-stage gear shafts and the propeller shaft, as well as the front roller bearings for the two first-stage helical gears, the input shaft, and the accessory drive shafts. The input drive housing holds the rear roller bearings for the two first-stage helical gears and the input shaft.

Turbomachinery

The **turbomachinery** (gas generator) consists of four sections contained in six casings, as shown in Fig. 17-12.

The air inlet section consists of the front inlet case and the rear inlet case, bolted together at flange C, as illustrated in Fig. 17-13.

The rear inlet case joins the front case to the low-pressure diffuser case at flange D. The case contains two bearings (no. 1 and no. 2) and seals for the power-turbine shaft. Mounting pads are provided for accessories. The engine oil tank forms part of the casing.

The compressor section comprises the low-pressure and high-pressure independent centrifugal impellers. These are contained within the low-pressure diffuser case (flange D to flange E), the intercompressor case (flange E to flange F), and the front of the gas generator case.

Diffuser pipes connect the diffuser case, which contains the LP impeller, to the intercompressor case.

The **annular reverse-flow combustion chamber** is contained in the gas generator case. The fuel manifold is mounted around the exterior of the gas generator case and has spray nozzles which protrude into the combustion chamber liner. Two igniter plug bosses are provided on the gas generator case; there are corresponding bosses in the liner.

FIGURE 17-9   Pratt & Whitney Canada PW100 series engine. (*Pratt & Whitney Canada*)

FIGURE 17-10 PW100 series engine. Cutaway view. (*Pratt & Whitney Canada*)

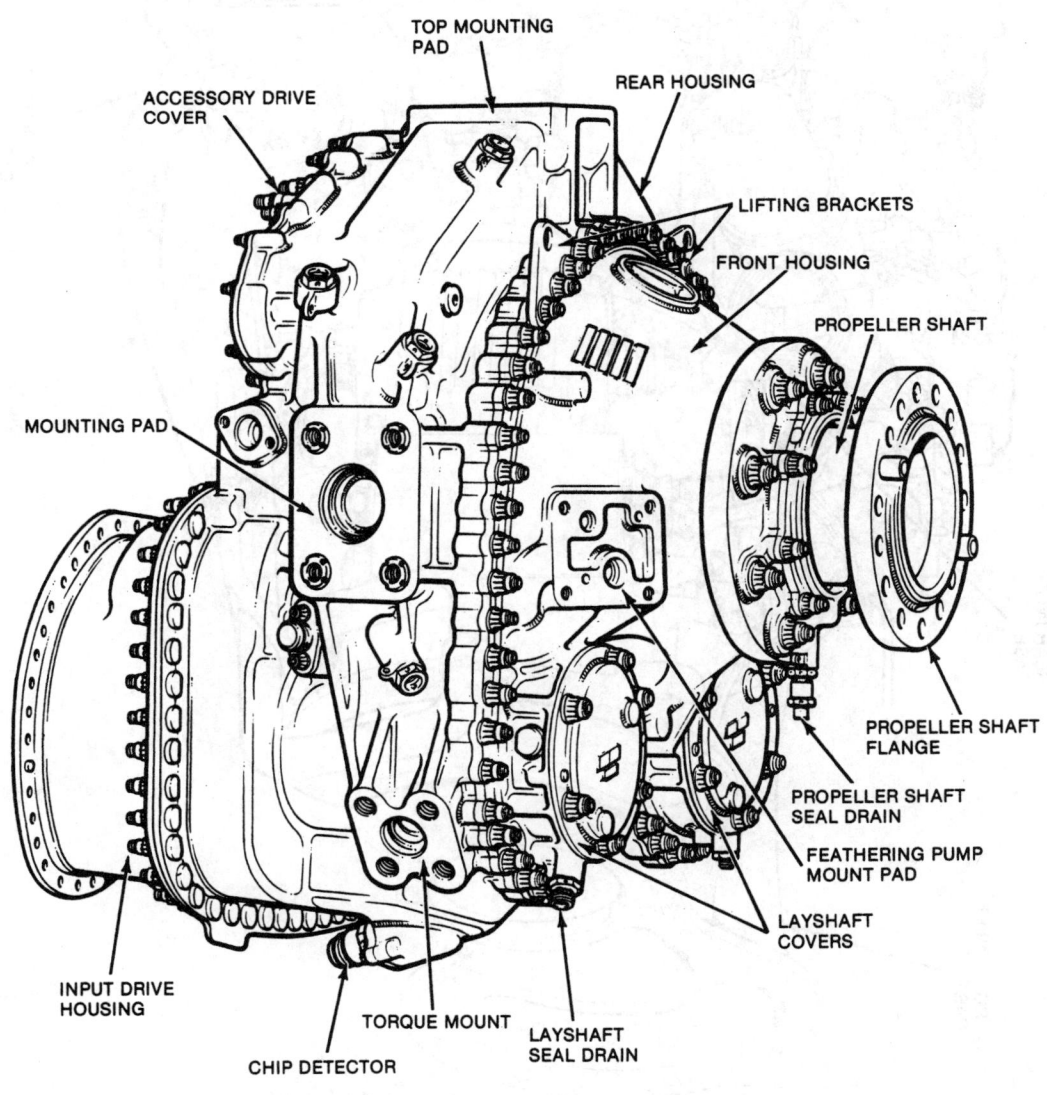

FIGURE 17-11 PW120 reduction gearbox. (*Pratt & Whitney Canada*)

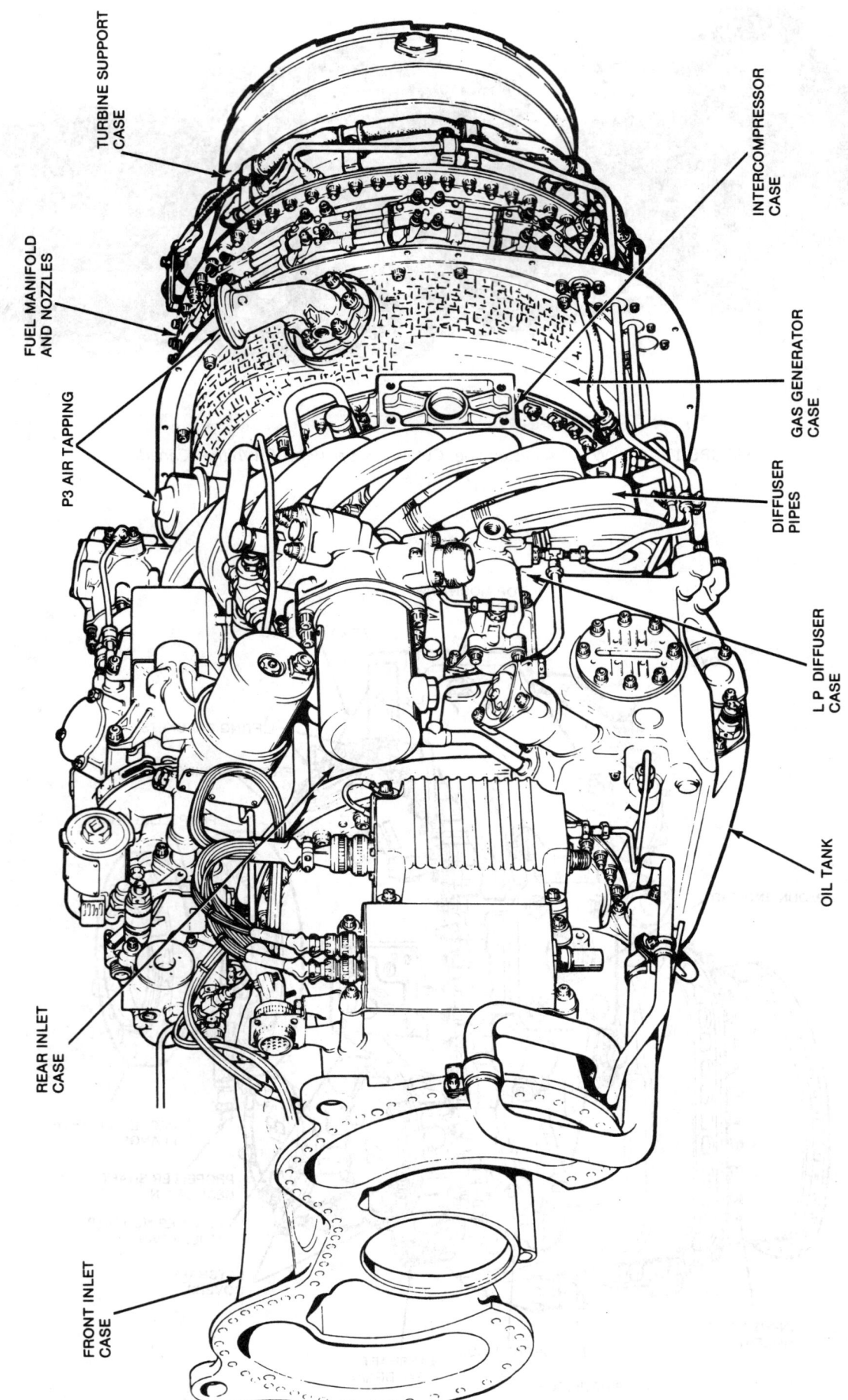

TURBINE SUPPORT CASE

FUEL MANIFOLD AND NOZZLES

P3 AIR TAPPING

INTERCOMPRESSOR CASE

GAS GENERATOR CASE

DIFFUSER PIPES

L P DIFFUSER CASE

OIL TANK

REAR INLET CASE

FRONT INLET CASE

FIGURE 17-12   PW120 turbomachinery. (Pratt & Whitney Canada)

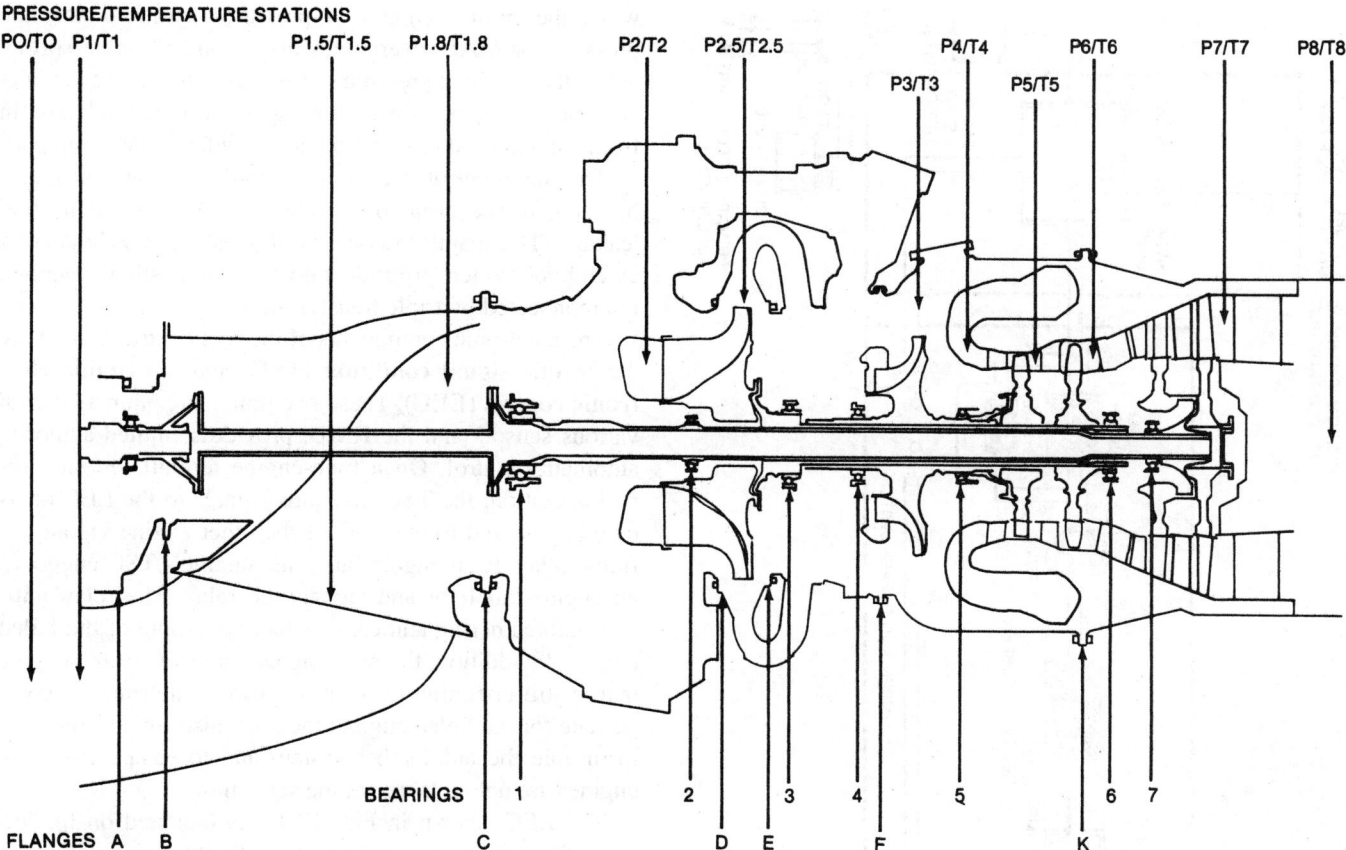

FIGURE 17-13 PW120 bearings, flanges, and stations. (*Pratt & Whitney Canada*)

The pressure turbines are housed in the rear of the gas generator case, and the power turbines are housed in the turbine support case. Concentric shafts connect the two-stage power turbine to the gearbox, and the single-stage LP and HP turbines to the impellers.

Oil System

The oil system is a wet-sump system, cooled by an externally mounted cooler. The oil is stored in a tank which is integral with the rear inlet case. The tank has a filler neck with a cap, a pressure oil strainer, an oil level indicator, and a scavenge oil chip detector. The single system supplies oil to the reduction gearbox and the turbomachinery. The oil system consists of two subsystems: the pressure system, which supplies oil to the engine; and the scavenge system, which returns the used oil to the tank.

Fuel and Control System

The engine fuel flow is controlled by the **power lever** and the **condition lever** through two integrated systems: the **hydromechanical control system** and the **electronic control system**. The hydromechanical control system, illustrated in Fig. 17-14, consists of a fuel pump and a **hydromechanical metering unit (HMU)** mounted on the accessory drive casing, a flow divider and dump valve, and a fuel nozzle manifold mounted on the gas generator case.

The fuel heater consists of a filter and a fin-type heater in two integral housings. The filter housing contains a bypass valve to ensure adequate fuel flow in the event of blockage, and an indicator to warn of impending blockage. The heater housing is divided into two circuits. Turbomachinery lubricating oil flows through one circuit, transferring heat to the fuel which flows through the other circuit.

Fuel Pump

The fuel pump, shown in Fig. 17-15, is a positive-displacement-type pump. The pump inlet contains a screen which, when blocked, lifts from its seat and allows fuel to pass by. The fuel passes through the inlet screen and is pumped through the outlet filter by a single-stage matched pair of spur gears. The outlet filter also has a valve that allows fuel to bypass the filter in the event of blockage and has an indicator to warn of impending blockage.

Flow Divider and Dump Valve

The flow divider and dump valve, illustrated in Fig. 17-16, is connected to the fuel manifold at the bottom of the gas generator case (flange F to flange K in Fig. 17-13). It comprises primary and secondary spool valves in a housing equipped with inlet and dump ports. The primary valve opens, giving access to the primary manifold, when the inlet fuel pressure overcomes the valve spring. The secondary valve opens

Large Turboprop Engines   **489**

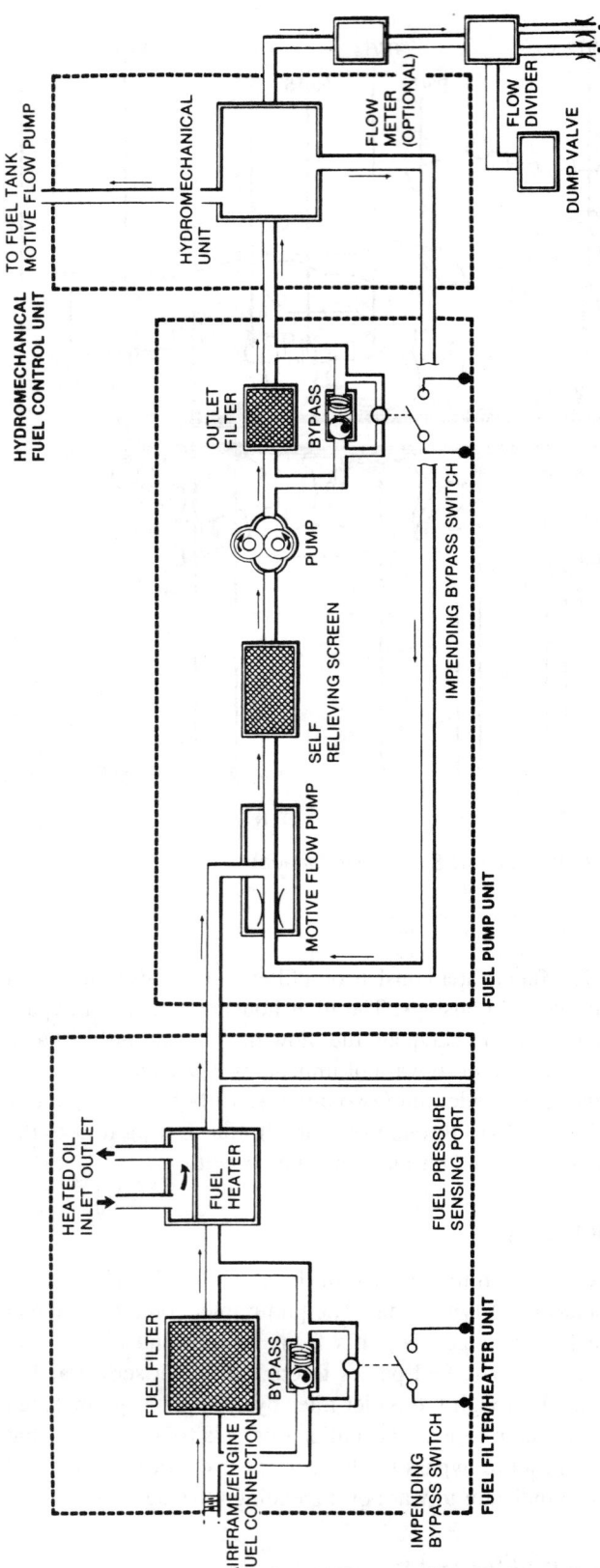

FIGURE 17-14 PW120 fuel system schematic.
(*Pratt & Whitney Canada*)

when the primary manifold pressure (manifold absolute pressure, or MAP) overcomes the secondary valve spring. When the fuel inlet pressure ceases, the valves close the inlet and open the dump ports, allowing any residual fuel to drain from the manifold through the flow divider to the dump port.

The fuel manifold delivers fuel to the combustion chamber and, in the event of a defective packing, drains fuel leakage. The manifold consists of sheathed nozzle adapter assemblies which protrude into the combustion chamber, interconnected by triple transfer tubes.

The electronic components of the fuel control system are the **torque signal condition (TSC)** and the **engine electronic control (EEC)**. These two units, in conjunction with various sensors and the HMU, provide a limited-authority automatic control. On a twin-engine aircraft, during normal operation, the TSC transmits signals to the EEC of its own engine and to the TSC of the other engine via an airframe relay. If an engine fails, that engine's TSC energizes an engine-fail light and signals the relay. The relay initiates autofeathering and cancels fuel governing of the failed engine. In addition, the relay signals the EEC of the engine that is still operating to increase power ("uptrim") to compensate for the failed engine; the relay also isolates the TSC to disable the autofeather system and to ensure that both engines are not feathered at the same time.

The EEC, shown in Fig. 17-17, is mounted on the left side of the front inlet case and works in conjunction with the HMU to control the fuel flow to the engine. The EEC, which requires 28 V of direct current, monitors the engine operating condition through analysis of various inputs from both the airframe and the engine. These inputs are processed by the circuits in the EEC and compared with reference data stored in the unit's memory. Based on these comparisons, commands are generated and transmitted to the torque motor in the HMU (telling it to adjust fuel flow) and also to a reference "bug" on the torque indicator. The torque needle is then matched to the bug by adjusting the power lever.

Engine Control System

In addition to the components of the fuel control system, the engine power output is also governed by the propeller control system. The propeller control system consists of the propeller control unit, the propeller overspeed governor, and, in the beta range (propeller operation on the ground), the EEC.

The **propeller control unit (PCU)** is mounted behind the propeller shaft on the rear face of the reduction gearbox. The unit has a power lever, which controls reverse pitch and beta scheduling, and a condition lever, which governs the propeller pitch range and thus the propeller speed. A switch is linked to the PCU condition lever to restrict the use of reverse pitch and fine pitch in the "quiet taxi" range. The PCU receives oil from a hydraulic pump mounted on the reduction gearbox. A transfer tube located in the propeller shaft conveys the oil from the PCU to the propeller pitch-change mechanism.

The **propeller overspeed governor** is a hydromechanical unit. In normal operation it routes pressure oil to operate a selector valve in the PCU. The governor monitors propeller

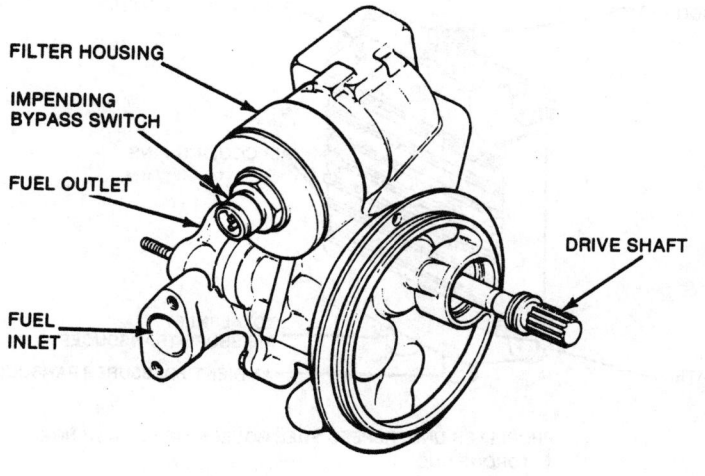

FILTER HOUSING

IMPENDING
BYPASS SWITCH

FUEL OUTLET

DRIVE SHAFT

FUEL
INLET

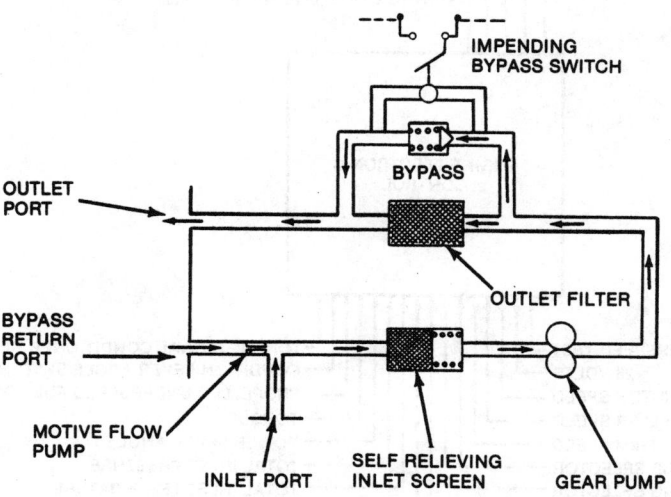

IMPENDING
BYPASS SWITCH

OUTLET
PORT

BYPASS

OUTLET FILTER

BYPASS
RETURN
PORT

MOTIVE FLOW
PUMP

INLET PORT

SELF RELIEVING
INLET SCREEN

GEAR PUMP

FIGURE 17-15 PW120 fuel pump, view and schematic. (*Pratt & Whitney Canada*)

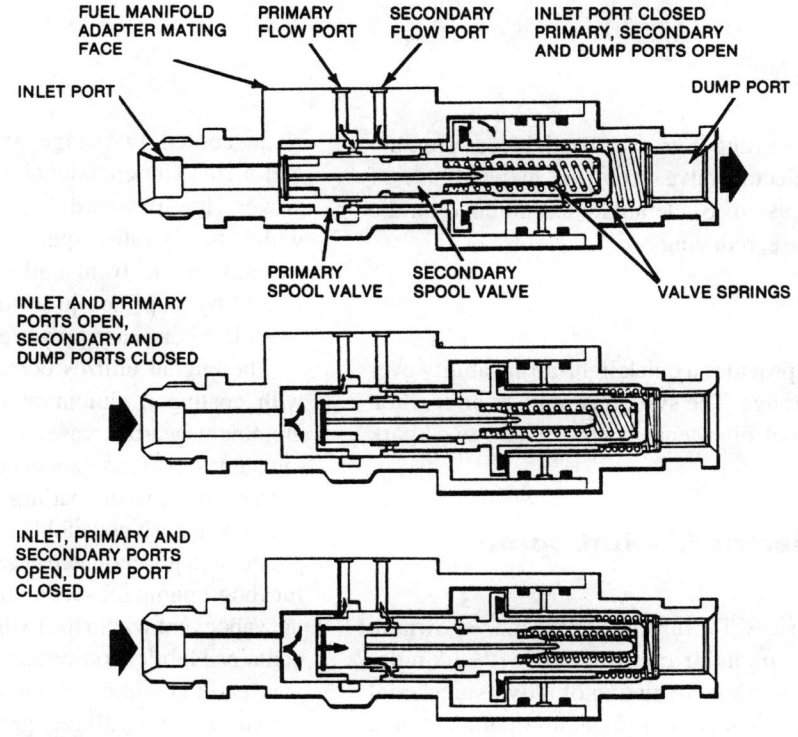

FUEL MANIFOLD
ADAPTER MATING
FACE

PRIMARY
FLOW PORT

SECONDARY
FLOW PORT

INLET PORT CLOSED
PRIMARY, SECONDARY
AND DUMP PORTS OPEN

INLET PORT

DUMP PORT

PRIMARY
SPOOL VALVE

SECONDARY
SPOOL VALVE

VALVE SPRINGS

INLET AND PRIMARY
PORTS OPEN,
SECONDARY AND
DUMP PORTS CLOSED

INLET, PRIMARY AND
SECONDARY PORTS
OPEN, DUMP PORT
CLOSED

FIGURE 17-16 PW120 flow divider and dump valve, schematic. (*Pratt & Whitney Canada*)

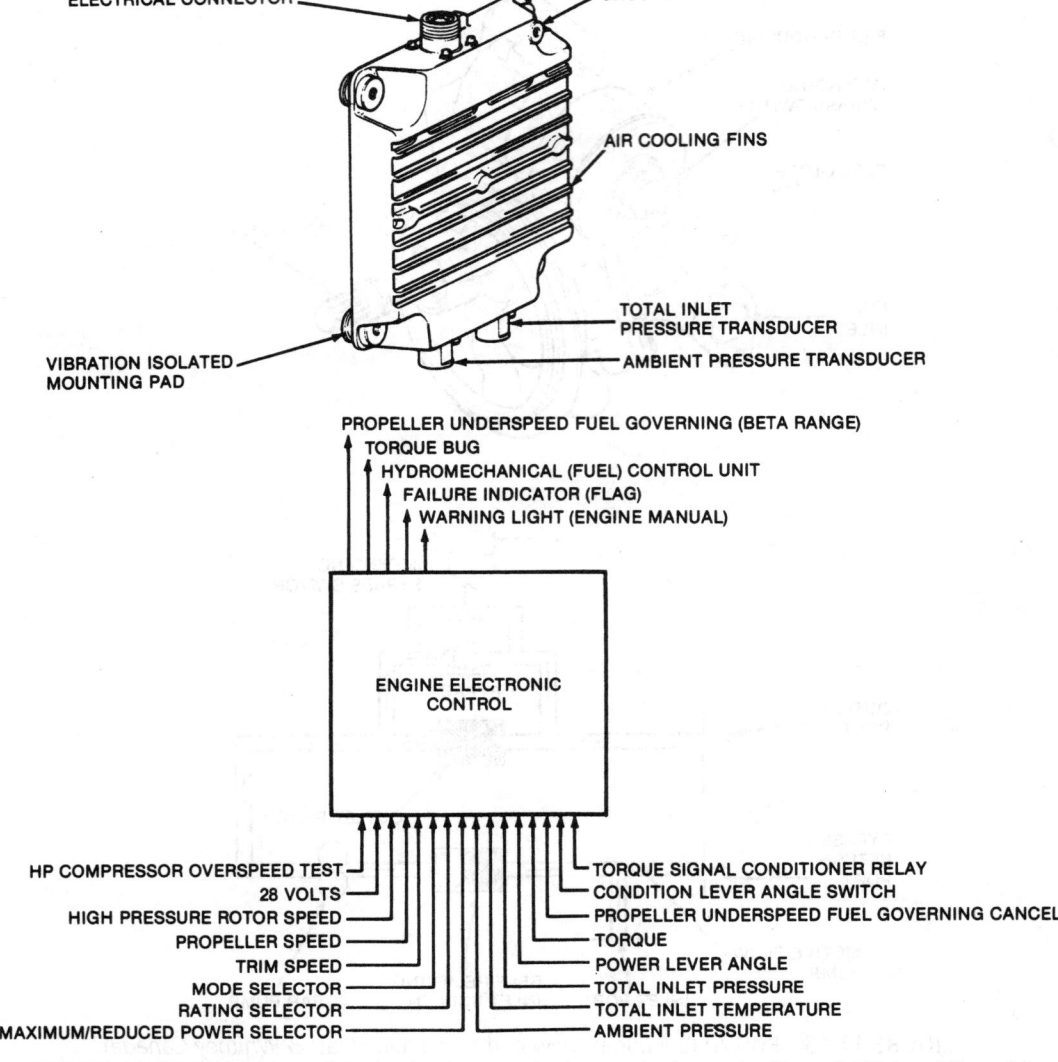

FIGURE 17-17 PW120 engine electronic control, view and schematic. (*Pratt & Whitney Canada*)

speed (N_p) and, in the event of an overspeed, bleeds pressure oil, via the PCU selector valve, from the metered side of the propeller servo piston. Blade angle and the load on the power turbine increase, reducing N_p.

Ignition System

The ignition system provides a quick light-up capability over a wide temperature range. The system comprises an ignition exciter, two individual high-tension cables, and two spark igniters.

The General Electric CT7 Turboprop Engine

The General Electric CT7 turboprop engine, shown in Fig. 17-18, features modular construction with a single-spool gas generator section consisting of a five-stage axial compressor and a single-stage centrifugal flow compressor; a low-fuel-pressure through-flow annular combustion chamber;

an air-cooled, two-stage, axial-flow high-pressure turbine; and a free (independent), two-stage, uncooled axial-flow power (low-pressure) turbine. The power-turbine shaft, which has a rated speed of 22 000 rpm, is coaxial and extends to the front end of the engine where it is connected by a splined joint to the output shaft assembly for propeller gear case power extraction.

The engine utilizes corrosion-resistant steel parts (some with coatings), aluminum inlet and main frames, and an aluminum gearbox case. There are four frames, three bearing sumps, two gas generator turbine rotor bearings, and four power-turbine rotor bearings. The engine incorporates an integral water-wash manifold; an integral foreign-object-damage protector; a top-mounted accessory package; an engine-driven fuel boost pump for suction fuel capability; a remote and manual vapor vent within the hydromechanical unit; separate, self-contained lubrication systems for the power unit and propeller gear case; condition monitoring and diagnostics provisions; a hydromechanical gas generator control system; and an electronic power control system that provides power-turbine

FIGURE 17-18   General Electric CT7 turboprop engine. (*General Electric.*)

speed bottoming governing, constant torque on takeoff, and overspeed protection.

The **module concept** (see Fig. 17-19) allows for the replacement of entire subsystems in a minimum amount of time. The CT7 turboprop engine power unit consists of four modules: the accessory module, the cold section module, the hot section module, and the power-turbine module. These modules, plus the propeller gear case, make up the CT7 turboprop engine. When modules are removed, there are no exposed sumps and no balance weights to remove and replace.

Major Power-Unit Components

The major components of the CT7 power unit, which can be seen in Fig. 17-20, consist of the following: the inlet frame, the main frame, the inlet guide vane casing, and the scroll case, which comprise the inlet section of the engine; a vertically split

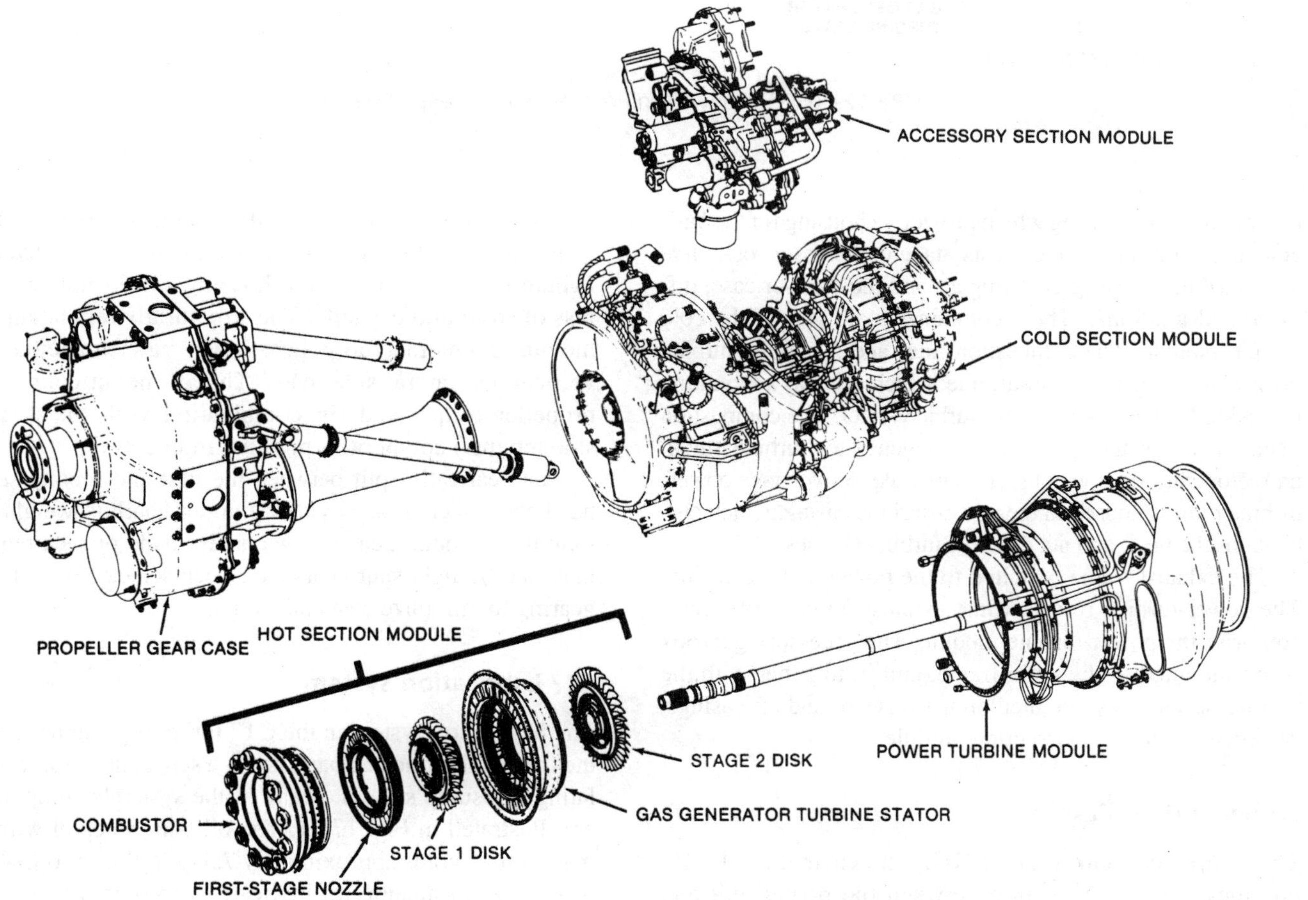

ACCESSORY SECTION MODULE

COLD SECTION MODULE

POWER TURBINE MODULE

PROPELLER GEAR CASE

HOT SECTION MODULE

STAGE 2 DISK

GAS GENERATOR TURBINE STATOR

COMBUSTOR

STAGE 1 DISK

FIRST-STAGE NOZZLE

FIGURE 17-19   CT7 modular breakdown. (*General Electric.*)

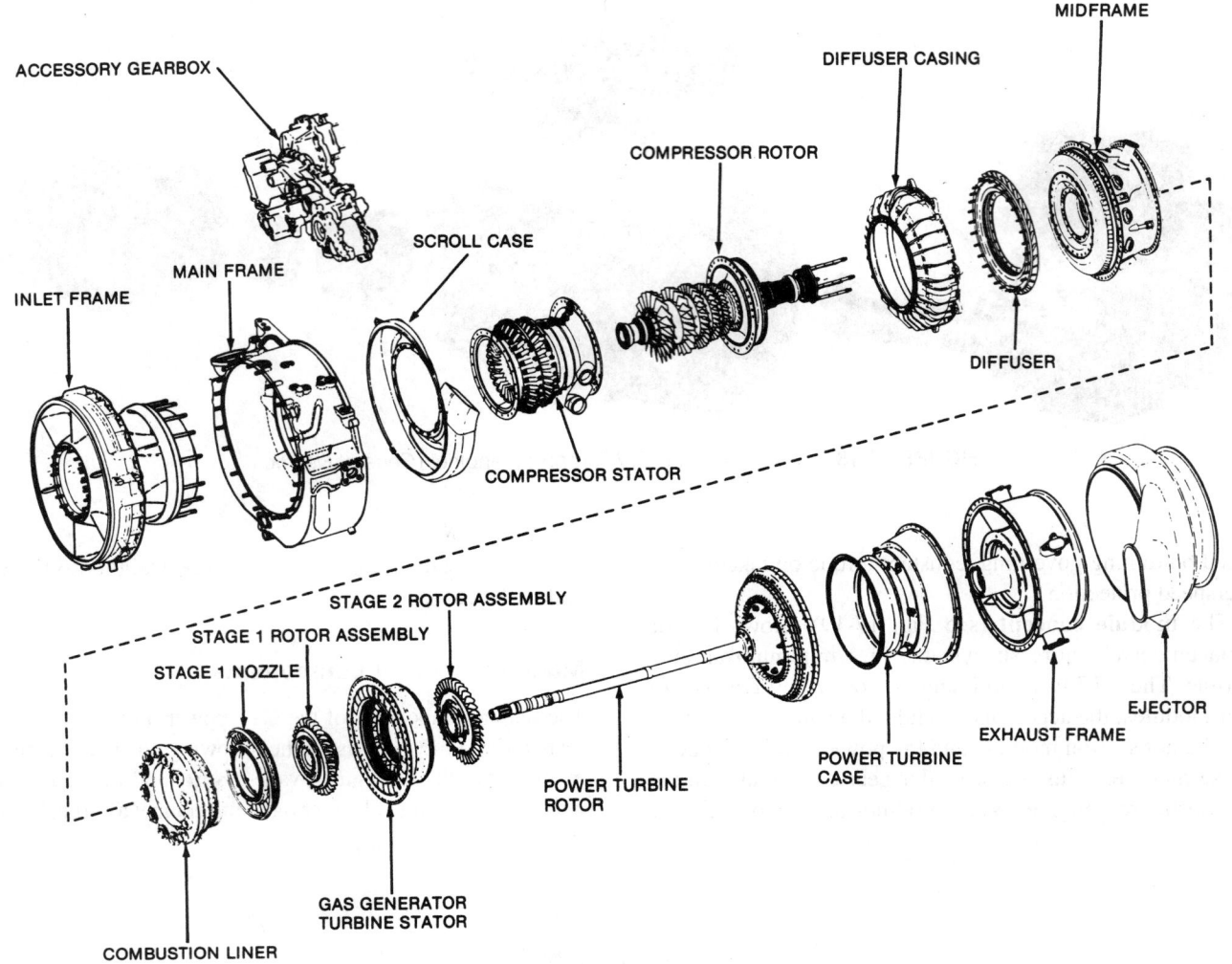

FIGURE 17-20 CT7 power-unit components. (*General Electric.*)

compressor stator casing which provides a housing for the variable and fixed stator vanes; a six-stage compressor rotor (five stages axial, one stage centrifugal); and the diffuser case, diffuser, and midframe. These components are part of the cold section module. The combustion liner and stage-one turbine nozzle are housed in the midframe, which also has a mounting provision for the gas generator turbine stator. The combustion liner, stage-one turbine nozzle, gas generator turbine stator, and rotor comprise the hot section module. A two-stage power-turbine rotor is housed in the power-turbine casing, which also contains the no. 3 and no. 4 power-turbine nozzles.

The exhaust frame is bolted to the power-turbine casing. The power-turbine rotor, casing, exhaust frame, and ejector comprise the power-turbine module. The accessory gearbox is top-mounted to the main frame, and it, together with the various accessories mounted on the forward and aft casings, comprise the accessory section module.

Propeller Gear Case

The **propeller gear case (PGC)**, shown in Fig. 17-19, provides the gear reduction between the power unit and the propeller. The PGC housing is an aluminum casting.

The main gear case forward side mounts are located at the gear case split line to provide one-mount-out redundancy. Failure of either half of the housing would not result in loss of structural capacity. The structurally redundant side mounts counteract propeller thrust, yaw moments, and vertical and lateral side loads. The aft mount counteracts propeller torque and (in combination with the forward side mounts) counteracts pitching moments.

The gear ratio split between the first and second stages has been selected to provide optimum gearbox weight and minimum frontal area. The gears are made of a carburized material. Straight spur gears have been selected over helical gearing to minimize cost and weight.

CT7 Lubrication System

The lubrication system in the CT7 turboprop engine distributes oil to all lubricated parts and is a self-contained, recirculating dry-sump system. Many of the system's components are illustrated in Fig. 17-21. The oil tank, integral with the main frame, holds approximately 7.3 qt [6.9 L] of oil, which is a sufficient quantity to lubricate the required power-unit parts without an external oil supply.

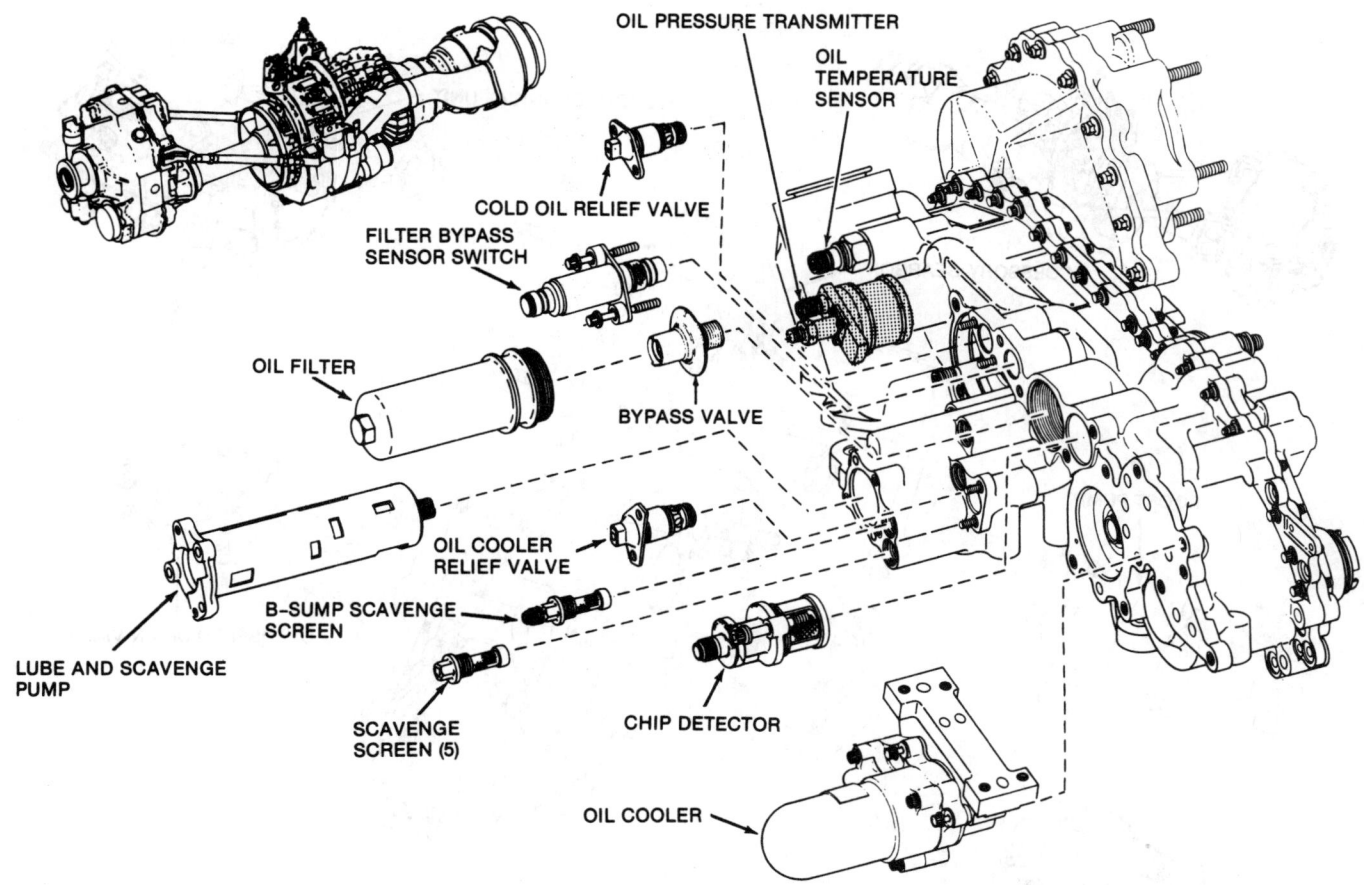

FIGURE 17-21 CT7 lubrication system components. (*General Electric.*)

Ignition System

The ignition system is an intermittent, ac-powered capacitor-discharge, low-voltage system. It includes a dual exciter unit mounted on the right-hand side and two igniter plugs. The spark rate of each ignition circuit is two sparks per s minimum, with a maximum output of 7000 V dc. The exciter is powered by one winding of the engine alternator. The ignition system is deactivated after engine start-up by shorting the alternator output through the ignition control circuit.

For normal starting, the ignition control circuit is tied in with the aircraft starting system to deenergize the ignition system at the starter-dropout speed. The ignition control circuit can be adjusted to the OFF position by the pilot when ignition is not desired. Additionally, the system can be armed to operate automatically in the event of an engine flameout. This feature monitors compressor discharge pressure (P_3) and energizes the ignition exciter if P_3 drops below 70 psi [483 kPa], indicating flameout.

Fuel System

The fuel system consists of the hydromechanical control unit, fuel boost pump, fuel filter, fuel flow transmitter, overspeed and drain valve, oil cooler, double-walled fuel manifold, and fuel injectors. These fuel system components, other than the fuel manifold and injectors, are mounted on the accessory gearbox and oriented as shown in Fig. 17-22.

Fuel enters the engine at the fuel boost pump. The boost pump pressurizes the fuel, which then flows through cored passages in the accessory drive gearbox to the fuel heater. From the heater, the fuel passes through the fuel filter, which has an impending bypass sensor that actuates a warning light in the cockpit when a differential pressure of 9 ± 1 psi [62 ± 6.9 kPa] occurs. The filtered, pressurized fuel then flows to the inlet of the high-pressure fuel pump within the hydro-mechanical metering unit (HMU). Fuel from the HMU passes through an external hose to the accessory gearbox. The flow of fuel continues through the fuel flow transmitter and then to the oil cooler. It enters the overspeed and drain valve (ODV) through other passageways in the accessory gearbox. The fuel is then routed through the main fuel manifold to the 12 fuel injectors for engine operation.

Control System

The **electrical control unit** (ECU) has four basic functions: power-turbine overspeed protection system, minimum power-turbine speed during ground operation, constant torque on takeoff system, and cockpit indication signals of propeller speed, torque, and temperature.

Power-turbine (N_p) overspeed protection system. The overspeed protection system receives a power-turbine rotor speed (N_p) signal from the torque and N_p speed sensor. If the N_p exceeds 25 000 ± 250 rpm (1560 propeller rpm),

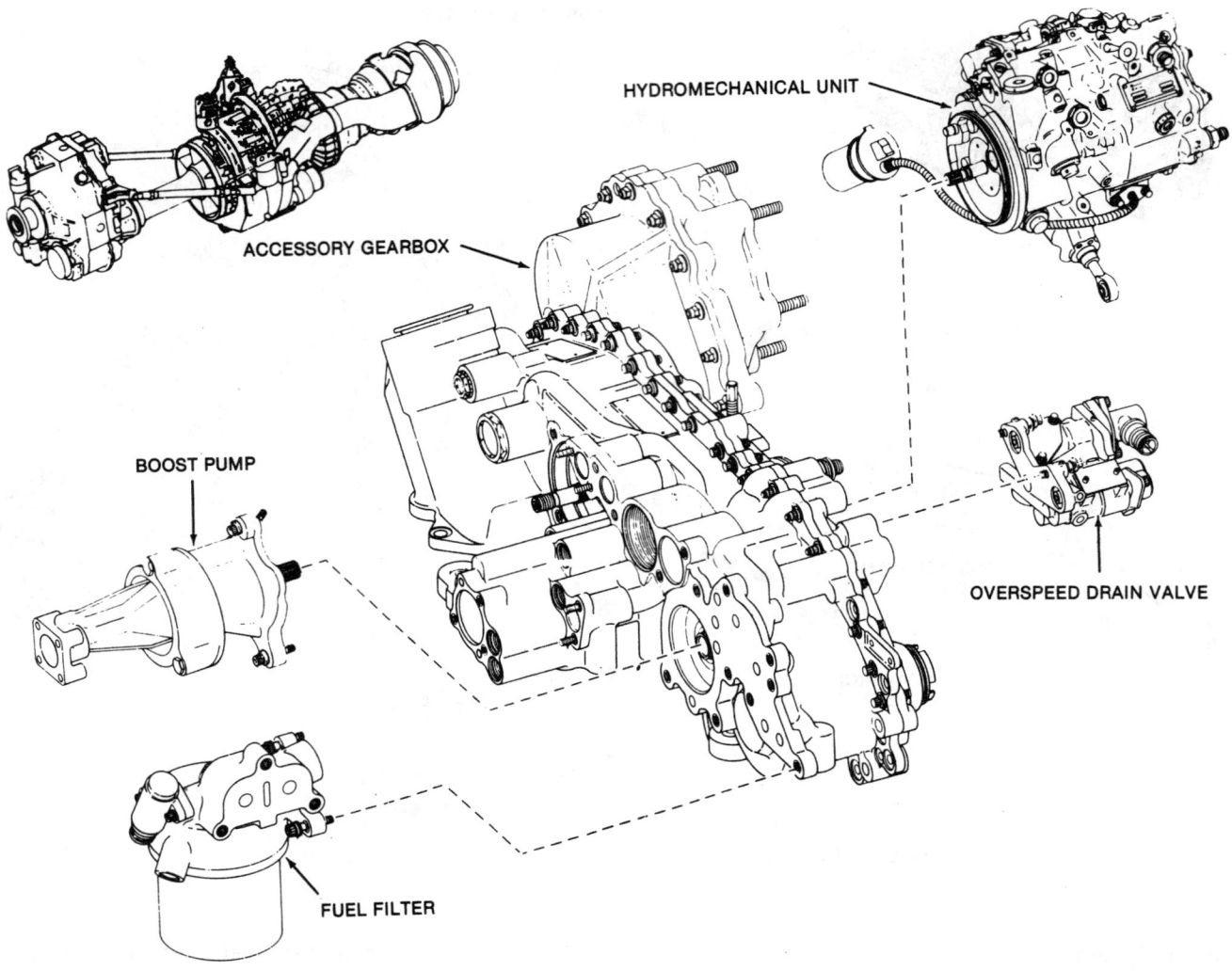

FIGURE 17-22 CT7 fuel system components. (*General Electric.*)

output from the protection system activates the overspeed solenoid in the overspeed and drain valve and energizes the auto relight circuit in the ECU. This shuts off fuel flow to the engine and starts the ignition circuit. The engine will flame out and start to decelerate. When the engine has decelerated to the reset speed, fuel flow is reestablished and the ignition system remains energized for 5 s to restart the engine.

Minimum power-turbine speed during ground operation (bottoming governor system). This system monitors the power-turbine rotor speed (N_p) signal sent from the power-turbine shaft speed sensor. If N_p drops below the N_p bottoming governor speed setting, the system increases fuel flow by actuating a torque motor in the HMU. The bottoming governor is enabled whenever propeller speed is greater than 35 percent N_p (485 rpm) and the condition lever is above the minimum propeller speed quadrant position which closes a switch located in the quadrant to activate the bottoming governing circuit on the ECU. When the condition lever is set to a point below the minimum propeller speed position, the switch is open and the bottoming governor function is disabled.

Constant torque on takeoff system (torque hold). To counteract ram effect, which creates a torque bloom as the aircraft picks up forward speed, the torque hold system adds torque motor trim to the power level (PL) control input to the HMU. By setting the power lever to establish a torque output from the engine that is lower than that required, the pilot can then set the torque hold system to compensate for the missing torque with a cockpit-mounted potentiometer.

Cockpit indication signals of propeller speed, torque, and temperature. The ECU supplies output signals to the torque and temperature indicators in the cockpit. The torque signal is processed by the ECU to provide a dc voltage to the cockpit instrument (1 V dc = 100 ft•lb [135.6 N•m] of torque). The temperature signal is processed in millivolts. The propeller speed signal is not processed by the ECU, but the signal is routed through the ECU to the cockpit indicator.

Propeller System

The airframe-furnished propeller system consists of a four-bladed, constant-speed, variable-pitch propeller.

The pitch-change mechanism is hydraulically operated by high-pressure oil from the PCU or by an auxiliary feathering pump, both of which are mounted on the aft propeller gearbox housing. Two concentric tubes are fitted in the bore of the gearbox-propeller shaft to direct operating oil from the PCU to the propeller pitch-change cylinder.

The propeller system operates in two modes: the **constant-speed mode**, for normal flight conditions, and the **beta control mode**, for taxiing and reverse thrust.

A high-pressure oil supply for propeller pitch changes is provided by a second stage in the main lube pump. An alternative supply for emergency feathering is available from an electric auxiliary feathering pump.

The design of the propeller control system is integrated with the engine's two-lever management system, which is illustrated in Fig. 17-23. The power lever controls engine power output in the constant-speed mode and propeller blade pitch in the beta control mode. Propeller rpm is controlled by the condition lever during the constant-speed mode, and the feathering valve is controlled by the condition lever during engine shutdown.

The power lever system provides fuel flow scheduling inputs to the HMU from the GROUND IDLE to MAX POWER settings. Additionally, the power lever functions to provide direct pitch control, through the PCU in the beta range. The beta range is from the MAX REVERSE position to the FLIGHT IDLE position.

The condition lever system provides the inputs to the HMU for mechanical stopcock of fuel, vapor venting of the engine fuel system, and mechanical lockout of the torque motor input from the ECU. A switch mounted in the condition lever quadrant "enables" the bottoming governor circuit in the ECU when the lever angle is above the MIN

PROP SPEED position. The lever, through the pitch control unit, feathers and unfeathers the propeller and sets propeller speed in the constant-speed mode through the constant-speed governor.

Allison T56/501D Series III

The T56/501D series III turboprop engine is used in over 60 countries around the globe on many different aircraft. Some of the aircraft that use the T56/501D engine are: the Lockheed L-100, which is a commercial version of the C-130 Hercules; the C-130 Hercules; the Grumman E-2; the Hawkeye; the Convair 580; the Lockheed P-3; and the Orion. A cutaway view of the T56/501D series III engine is shown in Fig. 17-24. As can be seen in the diagram, the engine consists of a gas producer turbine engine that drives a gearbox which in turn drives the propeller. The engine develops a shaft horsepower of up to 5000 shp [3729 kW].

Engine Data

General and performance data for the T56/501D series III engine are as follows:

Engine weight	1835 lb	832 kg
Pressure ratio	9.5:1	
Airflow	32.35 lb/sec	14.67 kg/sec
Compressor stages	14	
Turbine stages	4	
Allowable bleed air	8	
Engine speed	13 820 rpm	
Reduction gear ratio	13.54:1	
Max rated gas temperature	1970°F	1077°C

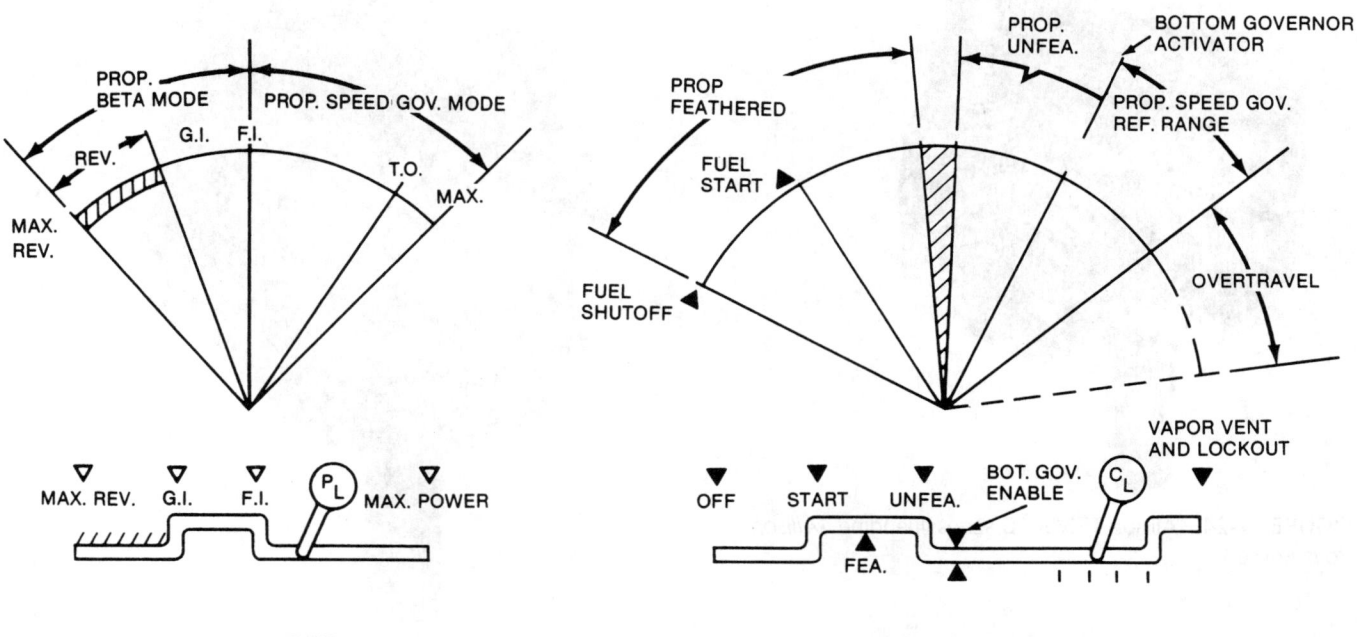

POWER LEVER   **CONDITION LEVER**

FIGURE 17-23   CT7 power and condition lever functions.

Allison T56/501D Series IV

The Allison T56/501D series IV engine, shown in Fig. 17-25, has many improvements over the series III engine—such as a 24 percent increase in horsepower, a 13 percent reduction in specific fuel consumption, and a 10 to 25 percent

FIGURE 17-24 Allison T56/501D series III engine. (*Allison Rolls-Royce.*)

FIGURE 17-25 Allison T56/501D series IV engine. (*Allison Rolls-Royce.*)

FIGURE 17-26 Air blast fuel nozzle. (*Allison Rolls-Royce.*)

improvement in average time between maintenance-related removals. Several modern features have been incorporated into the series IV engine, such as an engine monitoring system (EMS), air blast fuel nozzles, digital electronic control, and vane and blade coatings.

The engine monitoring system (EMS) is used to monitor the engine from before the aircraft leaves the ground until it is guided back into the hangar. Both the pilot and the ground crew know the exact status of the engine at all times and can use the information to keep ahead of trouble and to increase overall "up time." The system monitors the health of the engine by diagnostics and feeds this information into an onboard computer. The computer maintains a tape that can be downloaded to a ground station.

The single-entry-piloted air blast fuel nozzle, shown in Fig. 17-26, provides virtually smoke-free operation of the engine. Durability is also enhanced by replacing multiple parts with a single forging.

Perhaps the most significant control system change is the replacement of the temperature datum control with a modern technology digital electronic control. This new digital control monitors turbine temperature, calculates burner outlet temperature using other cycle measurements (rotor speed and compressor air inlet temperature), and controls engine fuel flow accordingly. The digital electronic control also limits maximum engine torque and performs the functions previously handled by the speed-sensitive control—i.e., energize and deenergize ignition, fuel manifold drain valve, starting fuel enrichment solenoid, fuel pump paralleling solenoid, and fuel cutoff actuator.

The coating on first- and second-stage turbine blades and vanes is CoCrAly overlay coating. This coating provides better bonding, better coverage, and greater hot corrosion resistance than the aluminum coatings previously used.

Allison T406-AD-400

The Allison T406 was chosen as the primary power source for the V22 Osprey Tiltrotor program. The V-22 Osprey aircraft is shown in Fig. 17-27.

The T406 compressor incorporates highly efficient aerodynamics and has variable geometry to permit free-turbine

operation and ease of starting. The annular combustor, lifted directly from the Allison T701-AD-700 Heavy-Lift Helicopter engine, features a new dump diffuser and has demonstrated that its flame pattern will ensure even temperatures into the turbine stages, which will improve their durability.

The high-pressure turbine design incorporates advanced cooling and material technology which further enhances the T406 low-risk design.

The T406 utilizes two independent full-authority digital electronic fuel controls to manage the engine fuel. This control "architecture" provides high redundancy and allows for multiple fail modes, thus maximizing flight safety. All controls are completely integrated with the V-22 aircraft control system.

Allison GMA 2100

The Allison GMA 2100 turboprop engine is shown in Fig. 17-28. The GMA 2100 utilizes the power section of the T406 engine developed for the V-22 Osprey Tiltrotor transport aircraft. It features a compressor which has demonstrated over 90 percent stage efficiency, an annular combustor, and a two-stage air-cooled turbine. The GMA 2100 prototype consists of the Allison T406 core engine, a newly designed reduction gearbox with flange mount, and the Dowty Aerospace six-bladed propeller. Among the aircraft slated for its use is the SAAB 2000 aircraft.

SMALL TURBOPROP ENGINES

The United Technologies Pratt & Whitney Canada PT6A Turboprop Engine

The PT6A turboprop engine, manufactured by United Technologies Pratt & Whitney Canada, is one of the most widely used engines in various configurations for a number of aircraft. The PT6A engine can provide 550 shp [410 kW] and higher, depending on the particular model and its application. A photograph of this engine is shown in Fig. 17-29. The PT6A engine is also produced in a turboshaft version.

An exploded view and a cut away view showing the principal components of the PT6A engine is given in Figs. 17-30 and 17-31. The air intake for this engine is at the rear, and the exhaust outlet is near the front. The engine includes the principal sections described for other gas-turbine engines: air inlet, compressor, diffuser, combustion chamber, turbine section, and exhaust. A cross section of the PT6A engine is shown in Fig. 17-32.

The PT6A engine is described as a lightweight, **free-turbine engine**, designed for use in fixed-wing or rotary-wing aircraft. The term "free turbine" means that the turbine which drives the output shaft is not mechanically connected to the turbine which drives the compressor.

Inlet air enters the engine through an annular plenum chamber formed by the compressor inlet case. From the inlet, the air is directed inward to the three-stage axial compressor and from there to the single-stage centrifugal compressor. The two compressor sections are constructed as one unit.

FIGURE 17-27 V-22 Osprey aircraft.

FIGURE 17-28 GMA 2100 series turboprop engine. (*Allison Rolls-Royce.*)

FIGURE 17-29 PT6A turboprop engine. (*Pratt & Whitney Canada*)

Air from the centrifugal compressor is thrown outward through diffuser pipes and turned 90° to a forward direction before being led through straightening vanes to the combustion chamber.

The **combustion chamber** is formed by the gas generator case and the rear end of the exhaust duct assembly. The **combustion chamber liner**, located in the combustion chamber, is an annular, reverse-flow unit with perforations of various sizes which provide for the entry of compressed air. The flow of air changes direction to enter the combustion chamber liner, where it reverses direction and mixes with the fuel.

Fuel is injected into the combustion chamber liner by way of 14 simplex nozzles supplied by a common manifold. The fuel-air mixture is initially ignited by two ignition glow plugs when the engine is being started. The glow plugs protrude into the combustion chamber liner, where they are exposed to the fuel-air mixture. The burned gases expand from the combustion chamber liner and reverse direction before passing through guide vanes to the compressor turbine. The vanes guide the expanding gases to ensure that they impinge on the turbine blades at the correct angle with minimum loss of energy. The gases continue to expand and pass forward through a second set of vanes to the power turbine.

The compressor and power turbines are located in the approximate center of the engine with their shafts extending in opposite directions. This provides for simplified installation and inspection procedures. Engine exhaust gases are discharged through an exhaust plenum to the atmosphere through exhaust ports.

The **accessory gear case** is located at the rear of the engine, immediately behind the air inlet. The gear case incorporates appropriate accessory drives and mounting pads. The accessories are driven from the compressor by means of a coupling shaft which extends the drive through a conical tube in the oil tank center section. The oil tank is integral with the compressor inlet case and has a capacity of 2.3 gal (U.S.) [8.71 L].

The propeller shaft is driven from the power turbine through a two-stage, planetary reduction gear having a ratio of 15:1. The power turbine turns at 33 000 rpm; therefore, the propeller turns at 2200 rpm.

Compressor Inlet Case

The **compressor inlet case**, shown in Fig. 17-33, is a circular aluminum-alloy casting, the front of which forms a plenum chamber for the passage of compressor inlet air. The rear portion consists of a hollow compartment which forms the integral oil tank. The intake is screened to preclude the ingestion of foreign objects.

FIGURE 17-30 PT6A turboprop cutaway.

The no. 1 bearing, bearing support, and air seal are contained within the compressor inlet-case centerbore. The bearing support is secured to the inlet-case centerbore flange by four bolts. A special nut and shroud washer retain the no. 1 bearing outer race in its support housing. A puller groove is provided on the rear face of the no. 1 bearing split inner race to facilitate its removal. The compressor assembly and the no. 1 bearing area are shown in Fig. 17-34. An oil nozzle, fitted at the end of a cored passage, supplies lubricant to the rear of the no. 1 bearing at approximately the one o'clock position. Other cored passages are provided for pressure and scavenge oil.

The oil pressure relief valve and the engine main oil filter, with check-valve and bypass-valve assemblies, are located on the right side of the inlet case at the one and three o'clock positions, respectively. A fabricated conical tube complete with preformed packings is fitted in the center of the oil tank compartment to provide a passage for the coupling shaft which extends the compressor drive to the rear accessories. The pressure oil pump, driven by an accessory drive, is located in the bottom portion of the integral oil tank and is secured by four bolts to the accessory diaphragm.

Compressor Rotor and Stator Assembly

The **compressor rotor and stator assembly**, shown in Fig. 17-34, consists of a three-stage axial rotor, three interstage spacers, three stators, and a single-stage centrifugal impeller and housing. The first-stage rotor blades are made of titanium to improve impact resistance, while the second and third stages are made of stainless steel. The rotor blades are dovetailed into their respective disks with a clearance between the blade and disk which causes a clicking sound during compressor rundown. The clearance is allowed to accommodate metal expansion due to the high temperatures.

Axial movement of the rotor disks is limited by the interstage spacers placed between the disks. The airfoil cross section of the first-stage blades differs from that of the second and third stages, the latter two being identical. The length of the blades differs in each stage, decreasing from the first to the third stage.

The first- and second-stage stator assemblies each contain 44 vanes, and the third stage contains 40 vanes. Each set of stator vanes is held in position by a circular ring with the vane outer ends protruding through and brazed to the ring. Part of each ring also provides the shrouds for the adjacent set of compressor blades.

The **compressor front stubshaft**, the **centrifugal impeller**, and the **impeller housing** are positioned in that order, followed alternately by an interstage spacer, a stator assembly, and a compressor rotor disk. These components are stacked and securely held together by six numbered tie rods. A series of slots and lugs on the impeller housing and compressor shroud assemblies interlock and prevent rotation of the assembly. The impeller housing is in turn secured in the gas generator case by eight eccentric bolts. The compressor front stubshaft consists of a hollow steel forging machined to

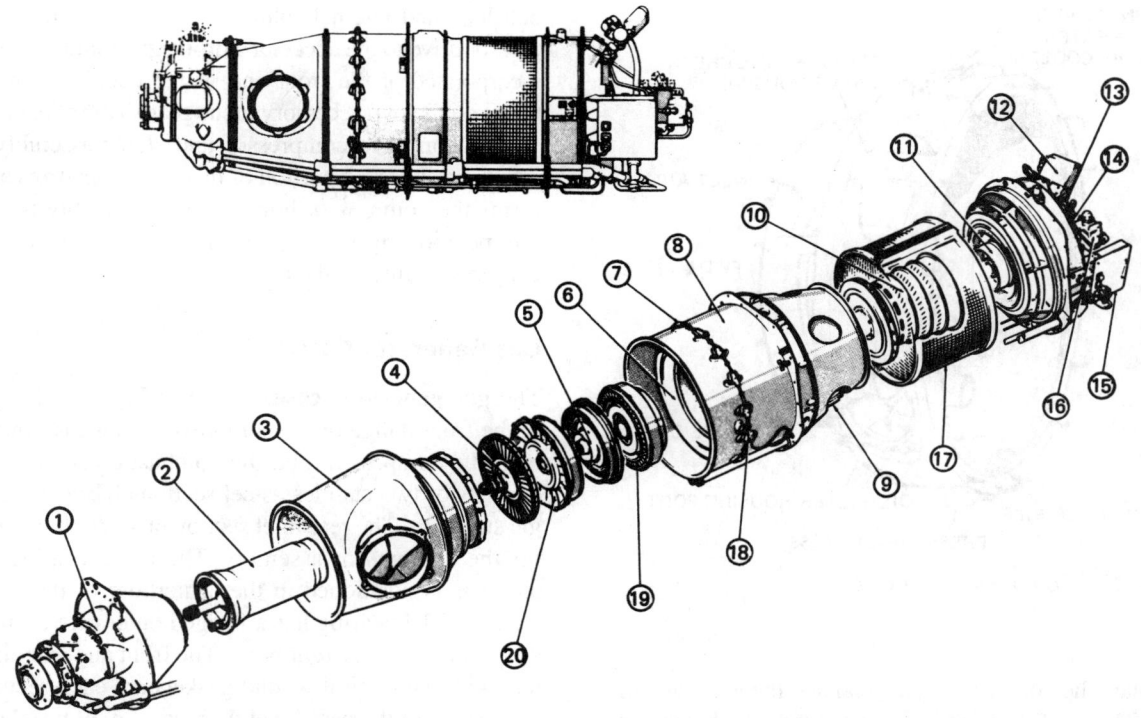

1. Propeller reduction gearbox
2. Power-turbine support housing
3. Exhaust duct
4. Power turbine
5. Compressor turbine
6. Combustion-chamber liner
7. Fuel manifold
8. Gas generator case
9. Compressor bleed valve
10. Compressor assembly
11. Compressor-inlet case
12. Oil to fuel heater
13. Dipstick and filler cap
14. Accessory gearbox
15. Ignition-current regulator
16. Fuel-control unit and pump
17. Air-inlet screen
18. Ignition glow plug
19. Compressor-turbine guide vanes
20. Power-turbine guide vanes

FIGURE 17-31   Exploded view of the PT6A engine.

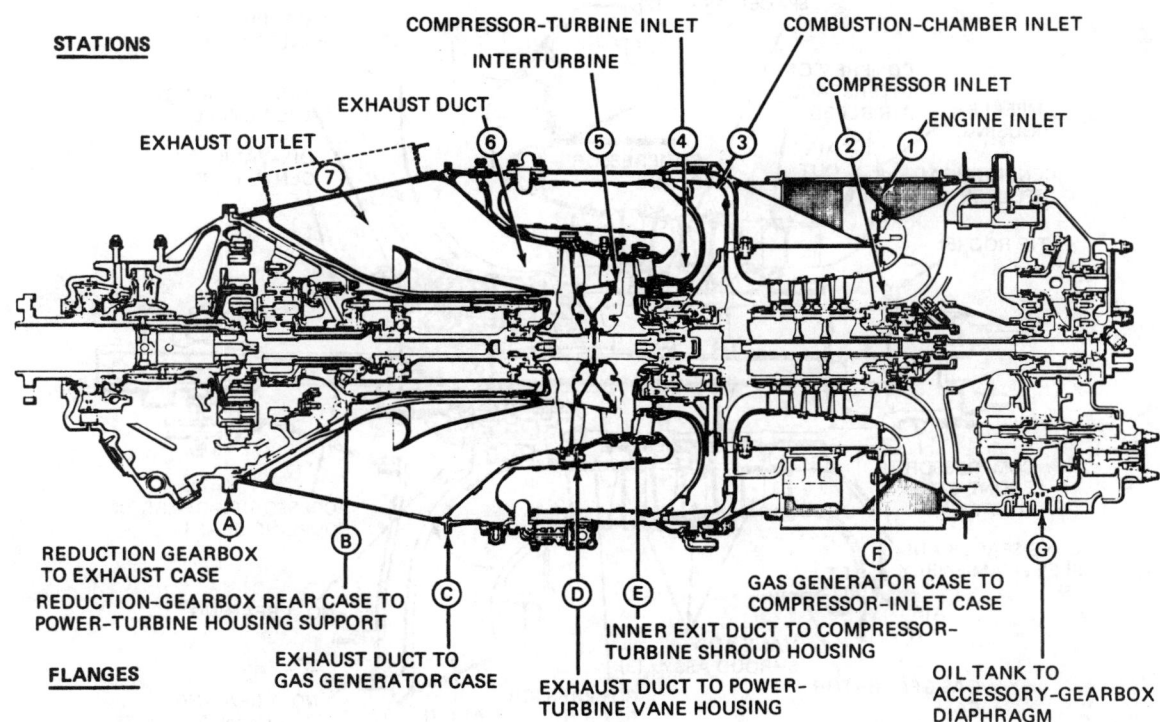

FIGURE 17-32   Cross section of the PT6A engine, showing flanges. (*Pratt & Whitney Canada*)

Small Turboprop Engines   **503**

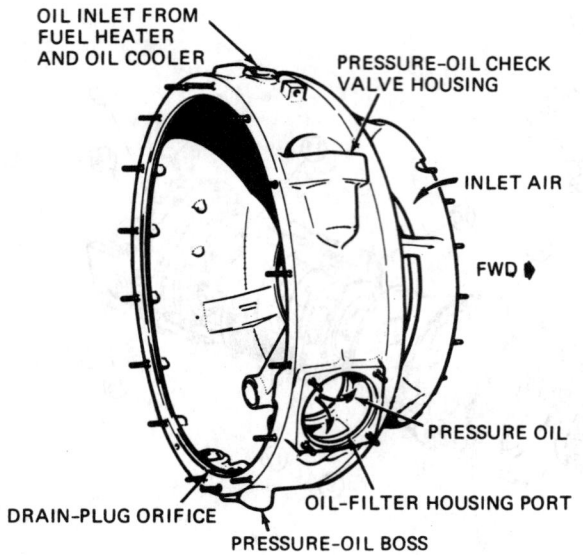

FIGURE 17-33 Compressor inlet case.

accommodate the no. 2 bearing air seal and the no. 2 bearing assembly. The no. 2 bearing is a roller type which supports the front of the compressor and the attached turbine in the gas generator case.

The **compressor rear hub** is an integral part of the first-stage compressor rotor disk. It consists of a steel forging machined with an extended hollow shaft to accommodate the no. 1 bearing air seal and the no. 1 bearing. The no. 1 bearing, which supports the rear of the compressor assembly in the inlet case, is a ball-type bearing. A short, hollow steel coupling, with

ball lock and internal splines at each end, extends the compressor drive to the **accessory input-gear shaft**. The ball lock, incorporated at the front end of the coupling, prevents end thrust on the two accessory input-gear shaft roller bearings.

The complete compressor and stator assembly is fitted into the center rear portion of the **gas generator case**, which forms the compressor housing. The assembly is secured in this position by the impeller housing at the front and the compressor inlet at the rear.

Gas Generator Case

The **gas generator case**, shown in Fig. 17-31, is attached to the front flange of the compressor inlet case and encloses both the compressor and the combustion section. The case consists of two stainless-steel sections fabricated into a single structure. The rear inlet section provides housing support for the compressor assembly. The no. 2 bearing with two air seals is positioned in the centerbore of the gas generator case. The bearing has a flanged outer race secured in the support housing by four bolts. The front and rear air-seal stators with their spiral-wound gaskets are each secured in the centerbore of the case by eight bolts. An oil nozzle with two jets, one in the front and the other in the rear of the bearing, provides lubrication. The 14 radial vanes brazed inside the double-skin center section of the gas generator case increase the pressure of the compressed air as it leaves the centrifugal impeller. The compressed air is then directed through 70 straightening vanes welded inside the gas generator case diffuser and out to the combustion chamber area through a slotted diffuser outlet baffle.

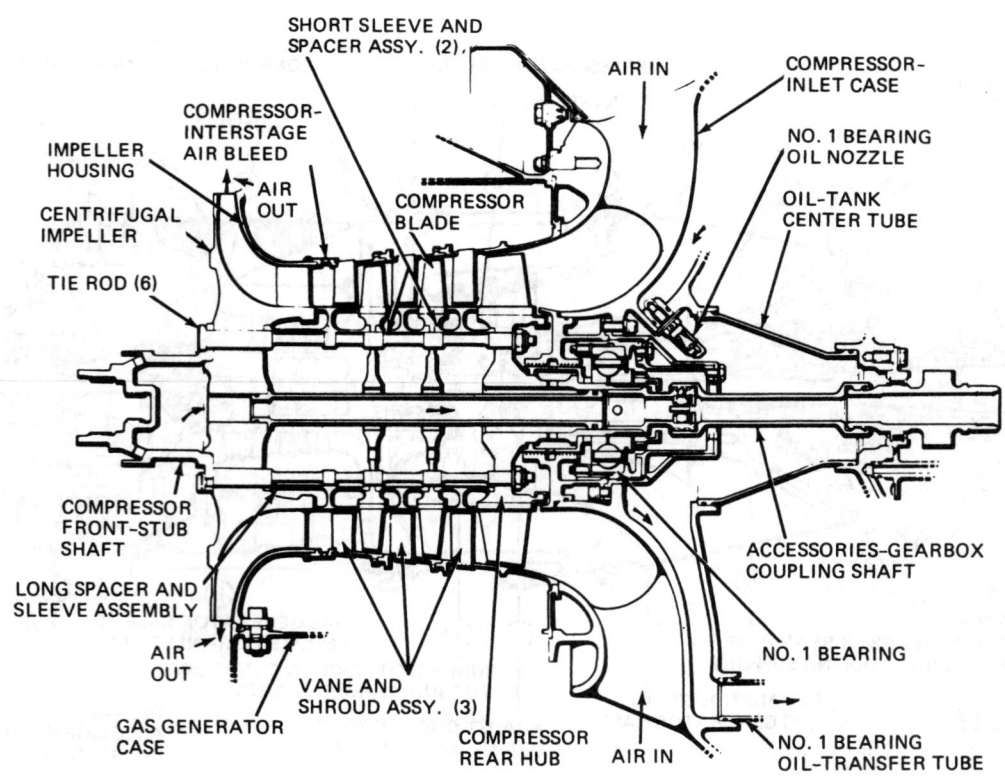

FIGURE 17-34 Compressor assembly and no. 1 bearing area.

The front section of the gas generator case forms an outer housing for the combustion chamber liner. It consists of a circular stainless-steel structure provided with mounting bosses for the 14 fuel nozzle assemblies and common manifold. Mounting bosses are also provided at the six o'clock position for the fuel dump valve and combustion chamber front and rear drain valves. Two ignition glow plugs are located at the four and eight o'clock positions. The plugs protrude into the combustion chamber liner to ignite the fuel-air mixture. Three equally spaced pads, located on the outer circumference of the gas generator case, provide accommodation for flexible-type engine mounts. The compressor bleed-valve outlet port is located at the eight o'clock position.

Combustion Chamber Liner

The **combustion chamber liner**, as shown in Fig. 17-35, is of the reverse-flow type and consists primarily of an annular, heat-resistant steel liner open at one end. A series of straight, plunged, and shielded perforations allow air to enter the liner in a manner designed to provide the best fuel ratios for engine starting and for sustained combustion. Direction of airflow is controlled by cooling rings especially located opposite the perforations. The perforations ensure an even temperature distribution at the compressor-turbine inlet. The domed front end of the combustion chamber liner is supported inside the gas generator case by seven of the 14 fuel-nozzle sheaths. The rear of the liner is supported by sliding joints which fit into the inner and outer exhaust duct assemblies. The duct assemblies form an envelope which effectively changes the direction of the gas flow by providing an outlet in close proximity to the compressor-turbine vanes. The outlet duct and heat-shield assembly is attached to the gas generator case by seven equally spaced support brackets located in the case.

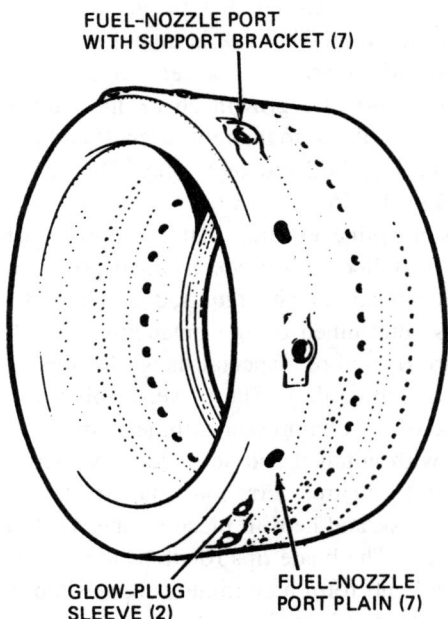

FUEL-NOZZLE PORT
WITH SUPPORT BRACKET (7)

GLOW-PLUG
SLEEVE (2)

FUEL-NOZZLE
PORT PLAIN (7)

FIGURE 17-35 Combustion chamber liner.

The heat shield forms a passage through which compressor discharge air is routed for cooling purposes. A scalloped flange on the rear of the outer duct locks in the support brackets and secures the assembly. The center section of the assembly is bolted with the compressor-turbine guide-vane support to the centerbore of the gas generator case.

Turbine Section

The turbine rotor section, shown in Fig. 17-36, consists of two separate single-stage turbines located in the center of the gas generator case and completely enveloped by the annular combustion chamber liner. The two turbines are mounted on shafts which extend in opposite directions. The rear shaft drives the compressor and the forward shaft drives the propeller through the reduction-gear assembly.

Compressor-Turbine Vanes

The compressor-turbine vane assembly (guide vanes) consists of 29 cast-steel vanes located between the combustion chamber exit ducts and the compressor turbine. The vanes are cast with individual dowel pins on the inner platforms which fit into the outer surface of the vane support. Sealing is ensured by a ceramic fiber cord packing located on the outer surface of the support, which is secured to the centerbore of the gas generator case by eight bolts. The outer platforms of the vanes are sealed with a chevron packing and fit into the shroud housing and exit duct. They are secured to the centerbore by 12 bolts. The shroud housing extends forward and forms a runner for two **interstage sealing rings**. The interstage sealing rings provide a power seal and an internal mechanical separation point for the engine. Fourteen compressor-turbine shroud segments located in the shroud housing act as a seal and provide a running clearance for the compressor turbine.

Compressor Turbine

The compressor turbine consists of a two-plane, balanced turbine disk with blades and classified weights. This turbine drives the compressor in a counterclockwise direction. The two-plane, balanced assembly is secured to the front stub-shaft by a simplified center lock bolt. A master spline is provided to ensure that the disk assembly is always installed in a predetermined position to retain proper balance. The disk has a reference circumferential groove for checking disk growth, when required. The 58 blades in the compressor-turbine disk are secured in fir-tree serrations machined in the outer surface of the disk and held in position by individual tubular rivets. The blades are made of cast-steel alloy and have **squealer tips**. A squealer tip is designed to cause a minimum amount of pickup if the blade should come into contact with the shroud segments during operation.

The required number of classified weights is determined during balancing procedures. These weights are riveted to the appropriate flanges machined on the turbine disk. A small rotor machined on the rear face of the turbine disk provides a sealing surface to control the flow of turbine disk cooling air.

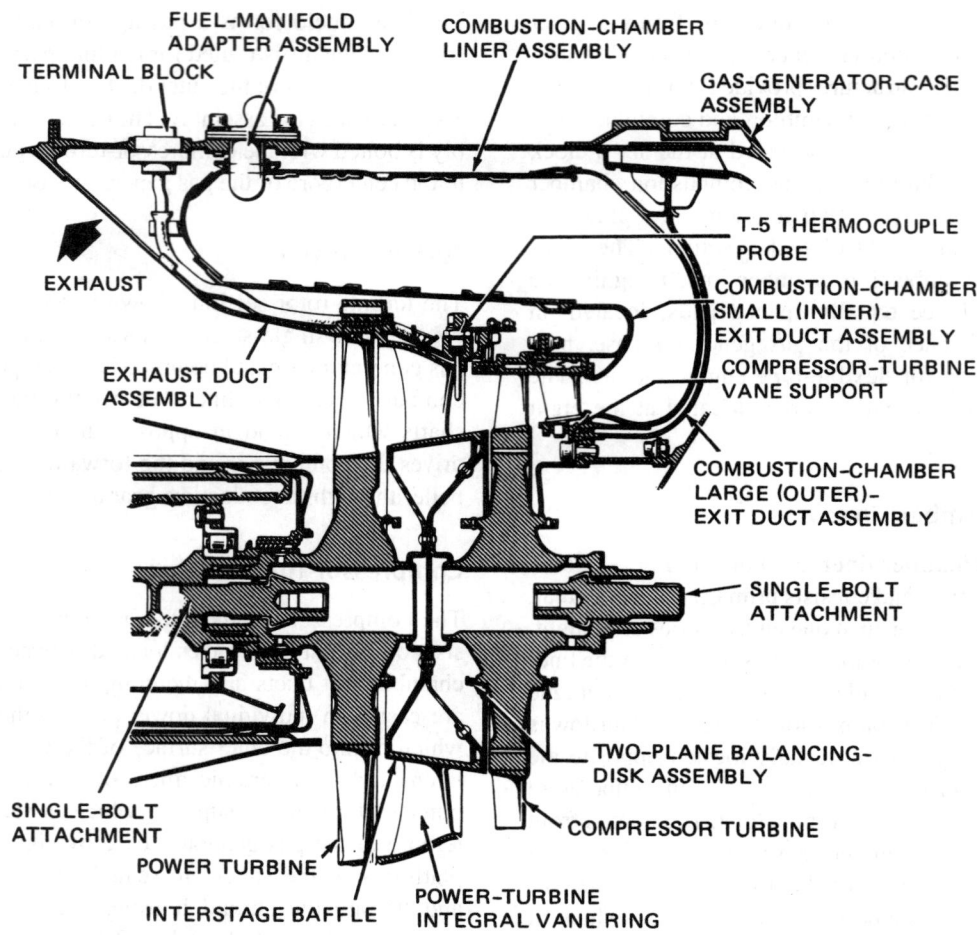

FuEL-MANIFOLD
ADAPTER ASSEMBLY

TERMINAL BLOCK

COMBUSTION-CHAMBER
LINER ASSEMBLY

GAS-GENERATOR-CASE
ASSEMBLY

EXHAUST

T-5 THERMOCOUPLE
PROBE

COMBUSTION-CHAMBER
SMALL (INNER)-
EXIT DUCT ASSEMBLY

EXHAUST DUCT
ASSEMBLY

COMPRESSOR-TURBINE
VANE SUPPORT

COMBUSTION-CHAMBER
LARGE (OUTER)-
EXIT DUCT ASSEMBLY

SINGLE-BOLT
ATTACHMENT

TWO-PLANE BALANCING-
DISK ASSEMBLY

SINGLE-BOLT
ATTACHMENT

COMPRESSOR TURBINE

POWER TURBINE

INTERSTAGE BAFFLE

POWER-TURBINE
INTEGRAL VANE RING

FIGURE 17-36 Turbine rotor section. (*Pratt & Whitney Canada*)

Interstage Baffle

The compressor turbine is separated from the power turbine by an **interstage baffle** which prevents dissipation of turbine gases and consequent transmission of heat to the turbine disk faces. The baffle, shown in Fig. 17-36, is secured to and supported by the power-turbine ring. The center section of the baffle includes small circular-lipped flanges on the front and rear faces. The flanges fit over mating rotor seals machined on the respective turbine disk faces to provide control of cooling airflow through the perforated center of the baffle.

Power-Turbine Guide Vanes

Between the compressor turbine and the power turbine is a ring which holds the **power-turbine guide vanes**. Depending on the model of the engine, either the vanes are separately cast and fitted into the interstage baffle rim by means of integral dowel-pin platforms or they are cast integrally with the turbine vane ring. The position of the vane ring can be seen in Fig. 17-36. The **stator housing** with the enclosed vane assembly is bolted to the exhaust duct and supports two sealing rings at flange D (see Fig. 17-32). The rings are self-centering and held in position by retaining plates bolted to the rear face of the stator housing.

Power Turbine

The **power-turbine disk assembly**, which consists of a turbine disk, blades, and classified weights, drives the reduction gearing through the power-turbine shaft in a clockwise direction. The disk is manufactured with close tolerances and incorporates a reference circumferential groove to permit disk-growth check measurements. The turbine disk is splined to the turbine shaft and secured by a single center lock bolt and keywasher. This arrangement is shown in Fig. 17-36.

A master spline ensures that the power-turbine disk is always installed in a predetermined position to retain the original balance. The required number of classified weights is determined during balancing procedures. The weights are riveted to a special flange located on the rear face of the turbine disk. The power-turbine blades differ from those of the compressor turbine in that they are cast complete with notched and shrouded tips. The 41 blades are secured by fir-tree serrations machined in the rim of the turbine disk and held in place by means of individual tubular rivets. The blade tips rotate inside a double knife-edge shroud and form a continuous seal when the engine is running. This reduces tip leakage and increases turbine efficiency.

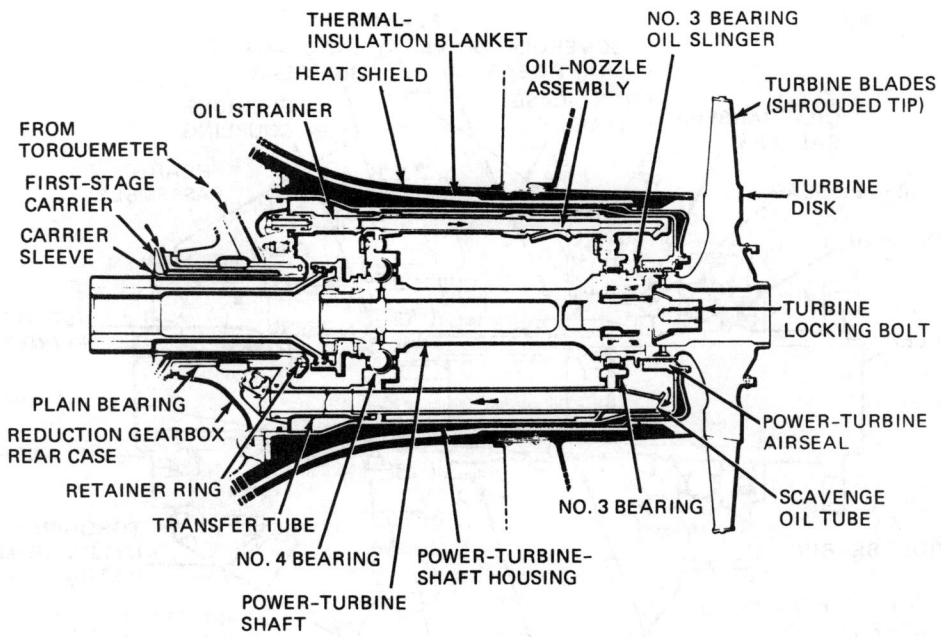

FIGURE 17-37 Power-turbine shaft and bearings.

Support Bearings

The power-turbine disk and shaft assembly is supported and secured in the **power-turbine shaft housing** by two bearings. The no. 3 bearing is a roller type and can be seen at the rear of the power-turbine shaft in Fig. 17-37. The no. 4 bearing is a ball type and is shown in the drawing at the forward end of the power-turbine shaft. The no. 3 bearing includes an inner race and a flanged outer race. The inner race is secured on the shaft together with the power-turbine air seal and power-turbine disk by the center lock bolt. The outer race is secured inside the power-turbine shaft housing by four bolts and tab-washers.

The no. 4 bearing includes a split inner race and a flanged outer race. The split inner race is secured on the front of the power-turbine shaft, together with the first-stage reduction-gear coupling and a coupling positioning ring, by a key-washer and spanner nut. A puller groove is incorporated on the front half of the split inner race to facilitate removal. The flanged outer race is secured inside the power-turbine shaft housing by four bolts and tab-washers.

Reduction Gearbox

The PT6A engine is equipped with a gearbox at the front for the reduction of engine speed to a level suitable for driving a propeller or a power shaft. In this section we are concerned only with the part of the engine which drives a propeller and in which the reduction gear ratio is 15:1. This means that the engine speed of 33 000 rpm is reduced to 2200 rpm at the propeller drive shaft.

The reduction gearbox, shown in Fig. 17-38, consists of two magnesium-alloy castings bolted to the front flange of the exhaust duct. The first stage of reduction is contained in the rear case.

The first-stage reduction sun gear consists of a short, hollow steel shaft which has an integral spur gear at the front end and is externally splined at the rear end. The external splines engage the retainer coupling by which the first-stage sun-gear shaft and the power-turbine shaft are joined.

The first-stage ring gear is located in helical splines provided in the first-stage **reduction-gearbox rear case**. The torque developed by the power turbine is transmitted through the sun gear and planet gears to the ring gear, which is opposed by the helical splines. Thus, the ring gear cannot rotate but the planet-gear carrier is caused to rotate. The ring gear moves axially a short distance because of the helical splines, and this movement is used to operate the torquemeter. The torquemeter will be described later.

The second stage of reduction gearing is contained in the **reduction-gearbox front case**. The first-stage planet carrier is attached to the second-stage sun gear by a flexible coupling, which also serves to damp any vibrations between the two rotating masses. The second-stage sun gear drives five planet gears in the second-stage carrier. A second-stage ring gear is fixed by splines to the reduction-gearbox front case and is secured by three bolted retaining plates. The second-stage carrier is in turn splined to the propeller shaft and secured by a retaining nut and shroud washer. A flanged bearing assembly, secured by bolts to the case, provides support for the second-stage carrier and the propeller shaft. An oil transfer tube and nozzle assembly fitted with preformed packings is secured within the propeller shaft to provide lubrication for the no. 4 bearing.

The accessories located on the reduction-gearbox front case are driven by a bevel gear mounted on the propeller shaft behind the thrust-bearing assembly. The propeller thrust loads are absorbed by a flanged ball bearing located in the front face of the reduction-gearbox centerbore. The bevel drive gear,

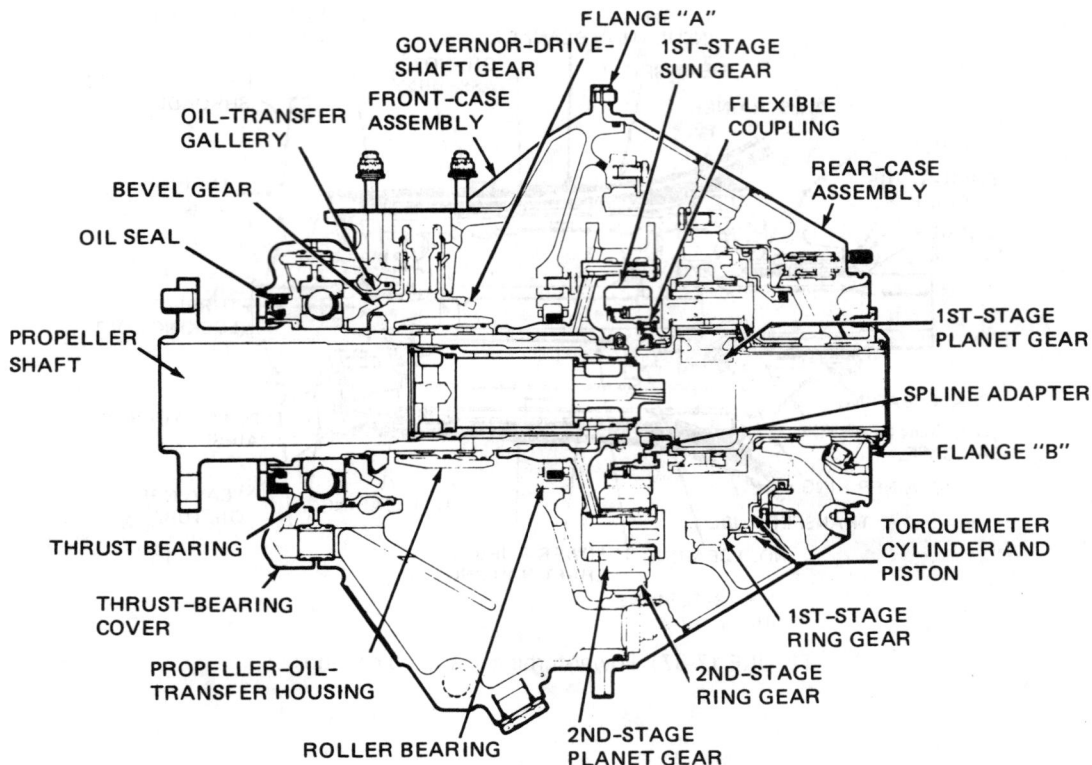

FIGURE 17-38 Two-stage reduction gearbox.

adjusting spacer, thrust bearing, and seal runner are stacked and secured to the propeller shaft by a single spanner nut and keywasher. The thrust-bearing cover is secured to the front of the reduction gearbox and incorporates a removable oil-seal retaining ring to facilitate removal of the oil seal.

Torquemeter

As mentioned previously, a **torquemeter** is used to determine the torque force being exerted by the engine. The value indicated by the torquemeter is used to determine power output.

The torquemeter for the PT6A engine operates in a manner similar to those previously described. The mechanism consists of a torquemeter cylinder, torquemeter piston, valve plunger, and spring (see Fig. 17-39).

Rotation of the reduction-gear first-stage ring gear is resisted by the helical splines, which impart an axial movement to the ring gear; this movement is transmitted to the torquemeter piston. This in turn moves a valve plunger against a spring, opening a metering orifice and allowing an increased flow of oil to enter the torquemeter chamber. This movement will continue until the oil pressure in the torque chamber balances the force on the ring gear caused by the torque being absorbed by the gear. Any change in the power control lever setting will recycle the sequence until a state of equilibrium is again achieved.

Hydraulic lock in the torquemeter is prevented by allowing the oil to bleed continuously from the pressure chamber into the reduction-gear casing through a small bleed hole provided in the top of the torquemeter cylinder.

Because the external oil pressure within the reduction gearbox may vary and affect the total oil pressure on the torquemeter piston, the internal pressure is measured. The difference between the torquemeter oil pressure and the reduction-gearbox internal oil pressure accurately indicates the torque being produced. The two pressures are internally routed to bosses located on the top

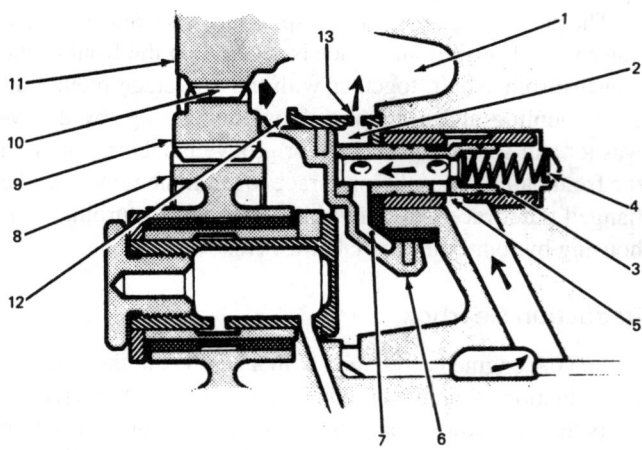

1. Gearbox pressure
2. Torquemeter pressure
3. Oil-control piston
4. Control spring
5. Metering orifice
6. Piston
7. Torquemeter chamber
8. First-stage planet gear
9. First-stage ring gear
10. Helical splines
11. Casting
12. Cylinder
13. Bleed hole

FIGURE 17-39 Torquemeter.

of the reduction-gearbox front case, where connections can be made to suit individual cockpit instrumentation requirements.

Power-Turbine Support Housing

The **power-turbine support housing** consists of a fabricated steel cylindrical unit attached to the reduction-gearbox rear case by 12 studs. The housing provides support for the power-turbine shaft assembly and two bearings, as shown in Fig. 17-37. A labyrinth-type seal, secured at the rear of the housing, prevents oil leakage into the power-turbine section. An internal oil transfer tube has four oil nozzles and provides front and rear lubrication to the no. 3 and no. 4 bearings. A scavenge tube, secured inside the housing at the six o'clock position, transfers bearing scavenge oil to the front of the engine.

Exhaust Duct

The **exhaust duct**, shown in Fig. 17-40, consists of a divergent, heat-resistant steel unit provided with two outlet ports, one on each side of the case. The duct is attached to the front flange of the gas generator case and consists of inner and outer sections. A reinforcing ring is provided at the rear of the exhaust duct at flange D. This consists of a scalloped stainless-steel ring machined in two halves and coupled to form a circle by two equally spaced clevis-pin joints. The power-turbine stator housing is secured to flange D by 12 bolts which screw into (and also secure) the reinforcing ring. The outer conical section, which has two flanged exhaust outlet ports, forms the outer gas path and also functions as a structural member to support the reduction gearbox. The inner section forms the inner gas path and provides a compartment for the reduction-gearbox rear case and the power-turbine support housing. A removable sandwich-type heat shield insulates the power-turbine support housing from the hot exhaust gases. A short no. 3 bearing cover and spacer are

secured at the rear of the power-turbine support housing by a retaining ring. A drain passage, located at the six o'clock position in the exhaust duct, enables residual fuel accumulation in the exhaust duct during engine shutdown to drain into the gas generator case, where it is discharged overboard through the front drain plug.

Accessory Gearbox

Accessories for the PT6A engine are driven from the **accessory gearbox** located at the rear of the engine. The gearbox, shown in Fig. 17-41, consists of two magnesium-alloy castings, both of which are attached to the rear flange of the compressor inlet case by 16 studs. The front casting, which incorporates front and rear preformed packings, forms an oil-tight diaphragm between the oil tank compartment of the inlet case and the accessory drives. The diaphragm also provides support for the accessory drive-gear bearings, seals, and the main pressure oil pump secured to the diaphragm by four bolts. The diaphragm is attached to the accessory-gearbox housing by four countersunk screws and nuts located at the fourth, eighth, fourteenth, and eighteenth positions in clockwise rotation, assuming that the first is at the twelve o'clock position.

The rear casting of the gearbox forms a cover and provides support bosses for the accessory drive bearings and seals. The internal oil scavenge pump is secured inside the housing, and a second scavenge pump is externally mounted. Mounting pads and studs are provided on the rear face for the combined starter-generator, the fuel control unit with the sandwich-mounted fuel pump, and the N_g tachometer. (N_g is the speed of the compressor turbine.) A large access plug located below the starter-generator mounting pad provides passage for the puller tool that is used to disengage and hold the ball-locked coupling shaft and input-gear shaft during disassembly. Three additional pads are available for

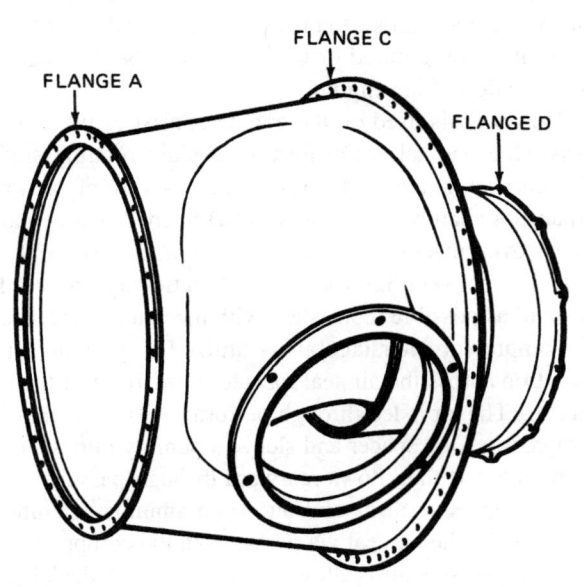

FIGURE 17-40 Exhaust duct assembly.

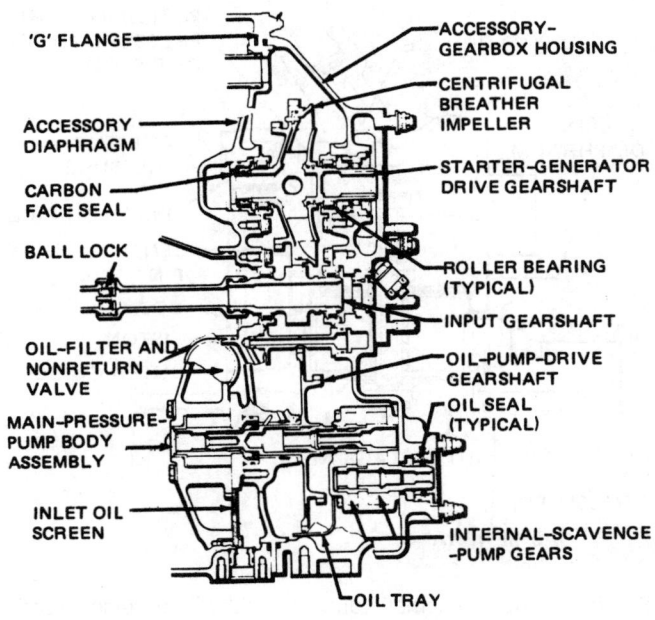

FIGURE 17-41 Cross section of the accessory gearbox.

optional requirements. Accessory drives are supported on similar roller bearings fitted with garter-type oil seals. An oil tank filler cap with an integral dipstick is located at the eleven o'clock position on the rear housing to facilitate servicing of the oil system. A centrifugal oil separator mounted on the starter-generator drive-gear shaft separates the oil from the engine breather air in the accessory-gearbox housing. A cored passage in the accessory diaphragm connects the oil separator to an external mounting pad located at the twelve o'clock position on the rear housing. A carbon face seal located on the front of the gear shaft in the accessory diaphragm prevents pressure leakage through the bearing assembly.

Air Systems

The PT6A engine has three separate air-bleed systems: a compressor bleed control, a bearing compartment air seal and bleed system, and a turbine-disk cooling system. A fourth system is available as an optional source of high-pressure air for use in operating auxiliary airframe equipment. A blanked mounting flange located on the gas generator case is provided for external connections.

Compressor bleed valve. The compressor bleed valve automatically opens a port in the gas generator case to spill interstage compressed air ($P_{2.5}$), thereby providing antistall characteristics at low engine speeds (less than 80 percent N_g). The port closes gradually as higher engine speeds are attained.

The compressor bleed valve, located on the gas generator case at the seven o'clock position and secured by two bolts, consists of a piston-type valve in a ported housing. This valve is illustrated in Fig. 17-42. The piston assembly

is supported in the bore of the housing and is guided by a seal support plate, guide pin, and guide-pin bolt, the latter holding the piston assembly together. A rolling diaphragm permits the piston full travel in either direction, to open or close the port, while at the same time effectively sealing the compartment at the top of the piston. A port in the gas generator case provides a direct passage for the flow of compressor interstage air ($P_{2.5}$) to the bottom of the bleed-valve piston.

Compressor discharge air (P_3) is tapped off and applied to the bleed valve through a nozzle (fixed orifice) in the bleed-valve cover, then passed through an intermediate passage and out to the atmosphere through a metering plug (convergent-divergent orifice). The control pressure (P_x) between the two orifices acts on the upper side of the bleed-valve piston, so that when P_x is greater than $P_{2.5}$, the bleed valve closes. In the closed position, the interstage air port is sealed by the seal support plate, which is forced against its seat by the effect of P_x. Conversely, when P_x is less than $P_{2.5}$, the bleed valve opens and allows interstage pressure ($P_{2.5}$) to be discharged to the atmosphere. The piston is prevented from closing off the P_x feed line to the upper section of the valve chamber and from damaging the piston assembly by the stop formed by the hexagon of the guide pin, which is screwed into the valve cover, and the hexagon of the piston guide-pin bolt.

For calibration purposes, P_x is measured by installing a suitable fitting in the valve cover, and adjustment is made by varying the diameter of the metering plug and/or nozzle orifice. The calibration should be such that P_x exceeds $P_{2.5}$ when the overall compressor ratio (P_3/P_2) is 3.70. P_2 is the compressor inlet pressure.

Bearing compartment seals, turbine cooling, and air-bleed systems. Pressure air is utilized to seal the first, second, and third bearing compartments and also to cool both the compressor and free turbines. The airflow for these purposes is shown in Fig. 17-43. The pressure air is used in conjunction with air seals which establish and control the required pressure gradients. Remember that air pressure is used in turbine engines to prevent oil from leaking into areas where it is not required or where it would be detrimental to the operation of the engine.

The air seals used on the engine consist of two separate parts. One part takes the form of a plain rotating surface. The corresponding part consists of a series of stationary expansion chambers (the **labyrinth**) formed by deep annular grooves machined in the bore or a circular seal. The clearance between the rotating and stationary parts is kept as small as possible (consistent with mechanical safety).

Compressor interstage air is utilized to provide a pressure drop across the air seal located in the front of the no. 1 bearing. The air is led through perforations in the rim of the compressor long spacer and sleeve assembly into the center of the rotor. It then flows rearward through passages in the three compressor disks and out to an annulus machined in the center of the air seal via passages in the compressor rear hub. The pressure air is allowed to leak through the labyrinth and thereby provides the required pressure seal.

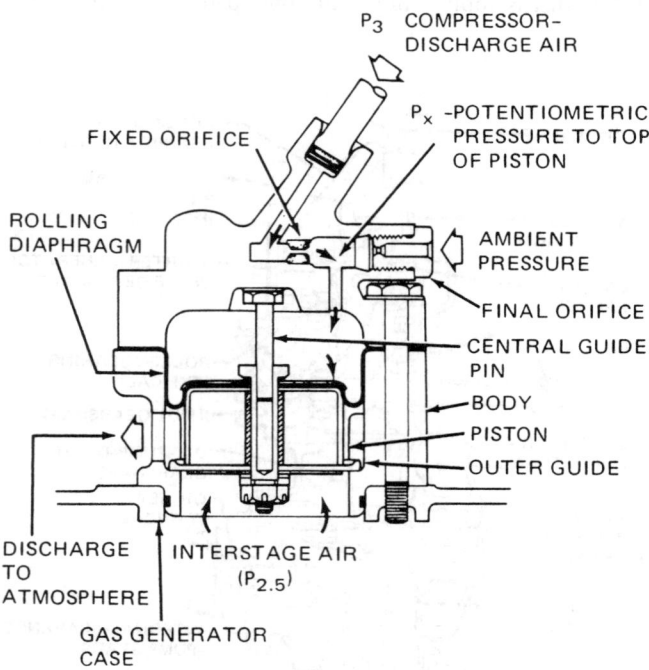

FIGURE 17-42 Schematic drawing of the compressor bleed valve.

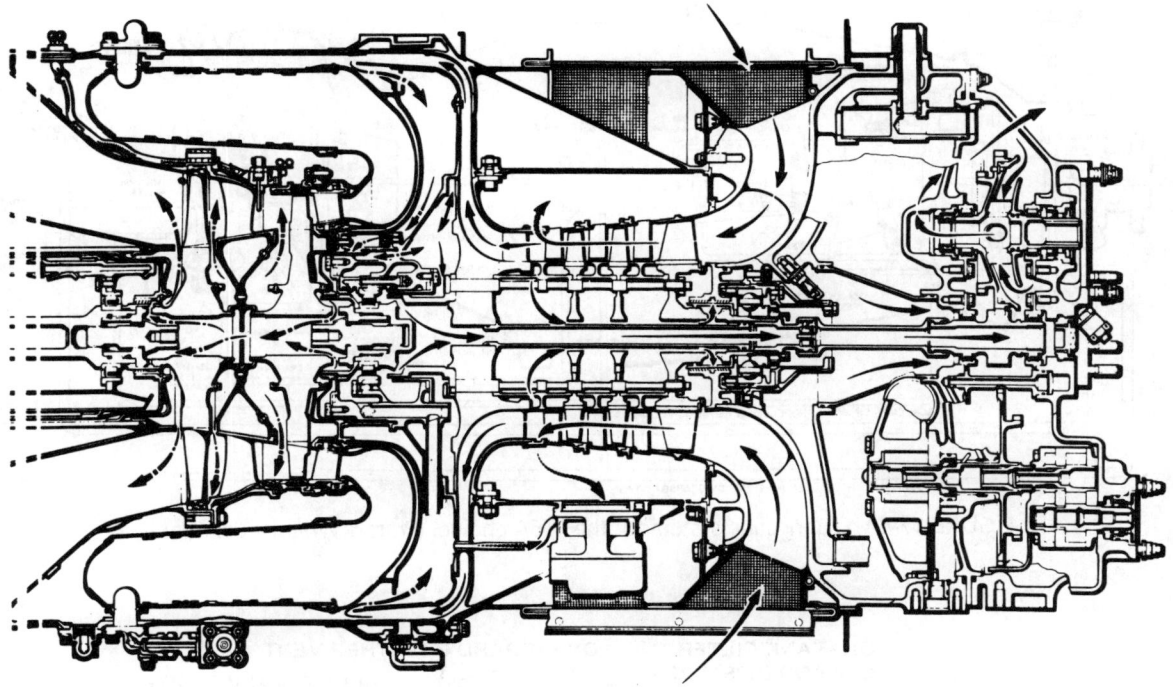

FIGURE 17-43 Bearing compartment seals, turbine cooling, and air-bleed systems. (*Pratt & Whitney Canada*)

The no. 2 bearing is protected by an air seal at the front and at the rear of the bearing. Pressure air for this system is bled either from the centrifugal impeller tip or, depending on engine speed, from the labyrinth seal connecting it to the turbine-cooling air system. The air flows through passages in the no. 2 bearing support, equalizing the air pressure at the front and rear of the bearing compartment and thereby ensuring a pressure seal in the front and rear labyrinths.

The compressor- and power-turbine disks are both cooled by compressor discharge air bled from the slotted diffuser baffle area down the rear face of the outer exit duct assembly. The air is then metered through holes in the compressor-turbine vane support into the turbine baffle hub, where it is divided into three paths. Some of the air is metered to cool the rear face of the compressor-turbine disk and some to pressurize the bearing seals. The balance is led forward through passages in the compressor-turbine hub to cool the front face of the compressor turbine. A portion of this cooling air is also led through a passage in the center of the interstage baffle, where the flow is divided. One path flows up the rear face of the power-turbine disk while the other is led through the center of the disk hub and out through drilled passages in the hub to the no. 3 bearing air seals and front face of the power-turbine disk.

The cooling air from both turbine disks is dissipated into the main gas stream flow to the atmosphere. The bearing cavity leakage air is scavenged with the scavenge oil into the accessory gearbox and vented to the atmosphere through the centrifugal breather.

Lubrication System

The lubrication system for the PT6A engine is shown in Fig. 17-44. As explained previously in this section, the oil tank is integral with the accessory gearbox. The lubricant used for the engine is type I, MIL-L-7808, or type II, MIL-L-23699. The Pratt & Whitney specification for the oil is PWA 521. The oil tank capacity is 2.3 gal (U.S.) [8.71 L] with an expansion space of 0.7 gal (U.S.) [2.65 L].

During operation, oil from the oil tank is picked up by the **main oil pressure pump** and forced through the **main oil filter**. The filter is a disposable cartridge type or a cleanable metal type. If the filter should be clogged, the oil flows through a bypass valve. A check valve is incorporated in the end of the filter housing, thus making it possible to change the filter without draining the oil tank.

The **oil pressure relief valve** is located on the gear case above the filter cover. The relief valve and filter are shown in Fig. 17-45. The pressure relief valve is adjusted at the factory, and no further adjustment is usually required. Oil released by the relief valve is returned to the tank. Figure 17-45 shows the location of the oil tank filler cap, oil tank drain plug, and other units on the accessory gear case.

As can be seen in Fig. 17-44, lubricating oil is carried to all parts of the engine through a system of tubes and passages. Thus, pressure oil reaches all bearings, gears, and other moving parts requiring lubrication. The oil reduces friction, cools the engine parts, and cleans the engine by removing foreign particles either deposited in the bottom of the oil tank or picked up by the oil filter. Scavenge oil is

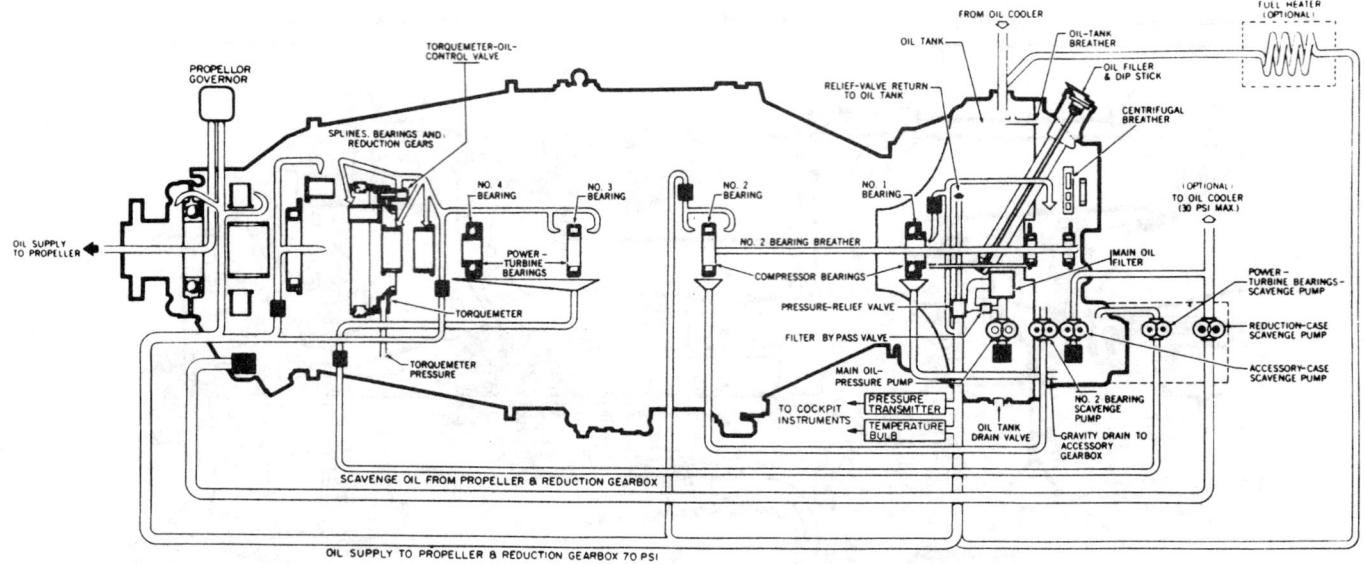

FIGURE 17-44 Lubrication system for the PT6A engine. (*Pratt & Whitney Canada*)

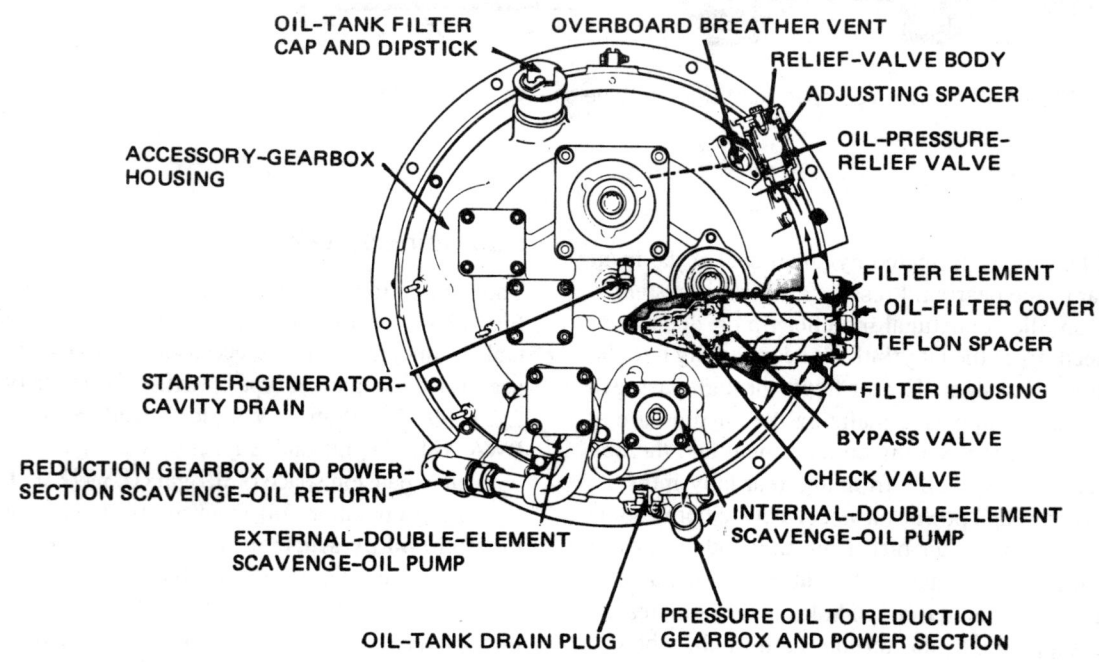

FIGURE 17-45 Accessory gearbox, showing oil pressure relief valve and filter.

drained into sumps and is pumped back to the oil tank. Oil from the reduction-case scavenge pump and the accessory-case scavenge pump is routed through an oil cooler if the engine installation in the aircraft requires such a cooler. The oil cooler is supplied by the airframe manufacturer.

Some installations include a fuel heater through which the fuel is heated by engine oil. The **fuel heater** is a simple heat exchanger with an automatic temperature control. A ball valve is closed by a vernatherm plunger when the fuel is below the desired temperature; this causes the hot oil from the engine to flow through the heater. Heat from the oil passes through the metal walls of the heater passages to the fuel.

Instruments for monitoring the oil system include a temperature gauge and a pressure gauge. The normal operating pressure is 65 to 85 psig [448.18 to 586.08 kPa] with a minimum of 40 psig [275.8 kPa]. Normal operating temperature of the oil is 166 to 176°F [74 to 80°C].

Ignition System

The PT6A engine employs **glow plugs** for ignition rather than the high-energy spark igniters employed by the majority of gas-turbine engines. The PT6A ignition system provides the engine with ignition capable of quick light-ups at extremely low ambient temperatures. The basic system consists of an ignition-current regulator with a selectable circuit to the two sets of ballast tubes, two shielded ignition cable and clamp assemblies, and two ignition glow plugs.

The **ignition-current regulator** is usually secured by three bolts to the accessory gearbox; however, it may be airframe-mounted if necessary. The regulator box has a removable cover and contains four ballast tubes. The circuit for the regulator system is shown in Fig. 17-46. Each ballast tube consists of a pure iron filament surrounded by helium and hydrogen gases enclosed in a glass envelope sealed to an octal base. The iron filament has a positive coefficient of resistance; that is, the resistance increases as temperature increases. Thus, when current flows through the filament, the temperature rises and the resistance increases to reduce the current flow. At low temperatures the ballast-tube resistance is low, which compensates for power losses due to low temperature. As the temperature rises, resistance increases, thus stabilizing the glow-plug current. Each glow plug is wired in series with two parallel-connected ballast tubes, and either one or both of the glow plugs can be selected for light-up. Ballast tubes provide an initial surge of current when the system is turned on and a lower stabilized current after approximately 30 s. This characteristic provides rapidly heated glow plugs for fast light-ups.

The ignition glow plug consists of a helical heating element fitted into a short plug body. The plugs are secured to the gas generator case in threaded bosses at the four and eight o'clock positions. The heating element lies slightly below the end of the plug body. Four holes, equally spaced on the periphery of the plug body, lead into an annulus provided below the coil. During starting, the fuel sprayed by the fuel nozzles runs down along the lower wall of the combustion chamber liner into the annulus. The fuel is vaporized and ignited by the hot coil element, which heats up to approximately 1316°C [2400°F]. This is the yellow-heat range, which is sufficient to ignite fuel vapors instantly. The four air holes bleed compressor discharge air from the gas generator case into the plug body and then past the hot coil into the combustion chamber to produce a "hot streak," or torching effect, which ignites the remainder of the fuel-air mixture in the chamber. The air also serves to cool the coils when the engine is running with the ignition system turned off.

Fuel System

The fuel system for the PT6A turboprop engine includes a gear-type or a vane-type fuel pump with a 74-μm inlet filter, a **fuel control unit** (FCU), a **temperature compensator**, a **power-turbine governor**, an **automatic fuel dump valve**, and a **fuel manifold adapter assembly**, which includes the fuel nozzles. The function of the system is to provide the correct amount of fuel for all operating conditions and to limit fuel as necessary to avoid damage to the engine, stall, and flameout.

The fuel pump is mounted on the accessory gearbox and is driven through a splined coupling. Another splined coupling shaft extends the drive to the FCU, which is bolted to the rear face of the fuel pump. Fuel from the aircraft boost pump enters the fuel pump through the 74-μm filter and flows into the pump chamber. Fuel from the pump is delivered at a high pressure to the FCU through a 10-μm filter. Both the inlet filter and the outlet filter have bypass valves to permit fuel flow if the filters should become clogged. The fuel control unit is classed as a pneumatic type since the primary actuating element is a pair of bellows which respond to varying air pressures. The controlling parameters for the fuel control unit are *power lever position, compressor discharge pressure* (P_3), *compressor inlet temperature* (T_2), *and compressor rpm* (N_g). An additional governor supplies an input signal to prevent overspeeding of the power turbine when the propeller governor is not in control of power-turbine speed (N_f).

The **temperature compensator** receives inputs from the inlet temperature and compressor discharge pressure. The effect of the temperature is to modify the P_3 pressure delivered to the fuel control **differential bellows** through the N_g governor. The N_g governor varies the pressures delivered to the differential bellows (pressures delivered depend on whether the N_g speed is correct, as required by the setting of the power-control lever).

The differential bellows moves the metering valve to deliver fuel to the fuel nozzles. The pressure differential across the metering valve is maintained at a constant value; therefore, fuel flow is proportional to the opening of the metering valve.

An **automatic fuel-dump valve** is provided to drain fuel from the fuel manifold in the combustion chamber when the engine is shut down. Drains are also provided to release fuel from the bottom of the combustion chamber. The drains eliminate the possibility of fuel collecting in the combustion chamber at shutdown.

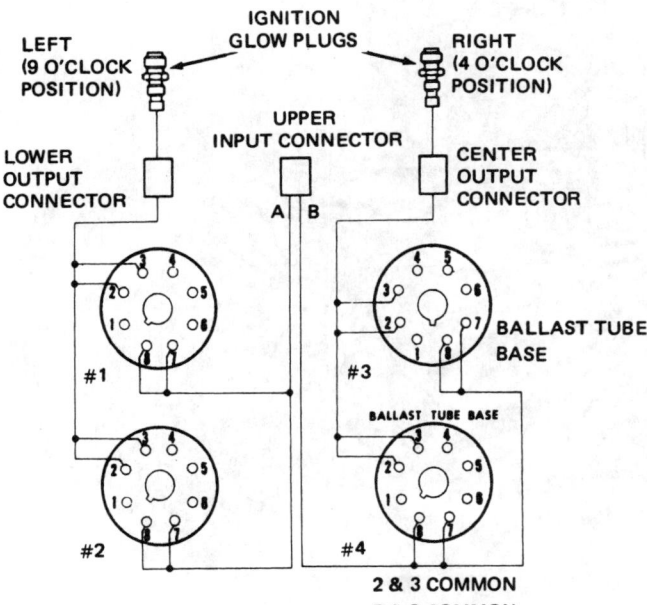

FIGURE 17-46 Ignition-current regulator circuit.

The Garrett TPE331 Turboprop Engine

The Garrett TPE331 turboprop engine, designated T76 for military purposes, is illustrated in Fig. 17-47. This engine has been in use for many years; however, improvements are constantly being made with resultant increases in power, efficiency, and reliability. These improvements include changes in compressor design, new airfoil configurations for turbine blades and vanes, and new superalloys for hot section parts. The most recent model can produce more than 1712 eshp [1277 ekW] for takeoff, whereas original models were rated at 655 shp [488.43 kW]. Specific fuel consumption (sfc) has improved from 0.556 to 0.49 lb/hp/h [338.2 to 298.12 g/kW/h].

Description

The TPE331 engine is a single-spool machine; that is, it has one main rotating assembly that includes both the compressor and the turbine. As can be seen in Fig. 17-48, at the forward end of the shaft is the gear that drives the propeller reduction and accessory gears. A two-stage centrifugal (radial) compressor is located on the main shaft in the forward section of the engine. To the rear of the compressor section is the three-stage turbine that extracts power from the hot, high-velocity gases and delivers the power through the main shaft to both the compressor and the propeller reduction gears.

The turbine is surrounded by the reverse-flow annular combustor. The engine is of modular construction to simplify repair and maintenance procedures.

Operation

During operation of the TPE331 engine, air flows from the inlet to the first stage of the compressor. From the first stage the air flows outward and then is routed through ducting back toward the center of the second-stage compressor. From the second stage the pressurized air flows outward and back around the outside of and into the annular combustor. Atomized fuel injected through nozzles in the rear of the combustor is ignited, and the resulting hot gases flow forward, then turn inward and flow to the rear through the three-stage turbine, where they are ejected out the rear of the engine.

Engine Systems

The propeller control system for the TPE331 engine is similar to those for other turboprop engines. During flight, the system automatically adjusts propeller pitch to maintain a constant propeller speed. As engine power is increased through movement of the power lever, propeller pitch increases as the propeller governor moves to release oil from the pitch-changing mechanism in the propeller hub. For ground operations, the propeller control system provides for reverse thrust

FIGURE 17-47   Cutaway view of the Garrett TPE331-14 turboprop engine. (*Honeywell Inc.*)

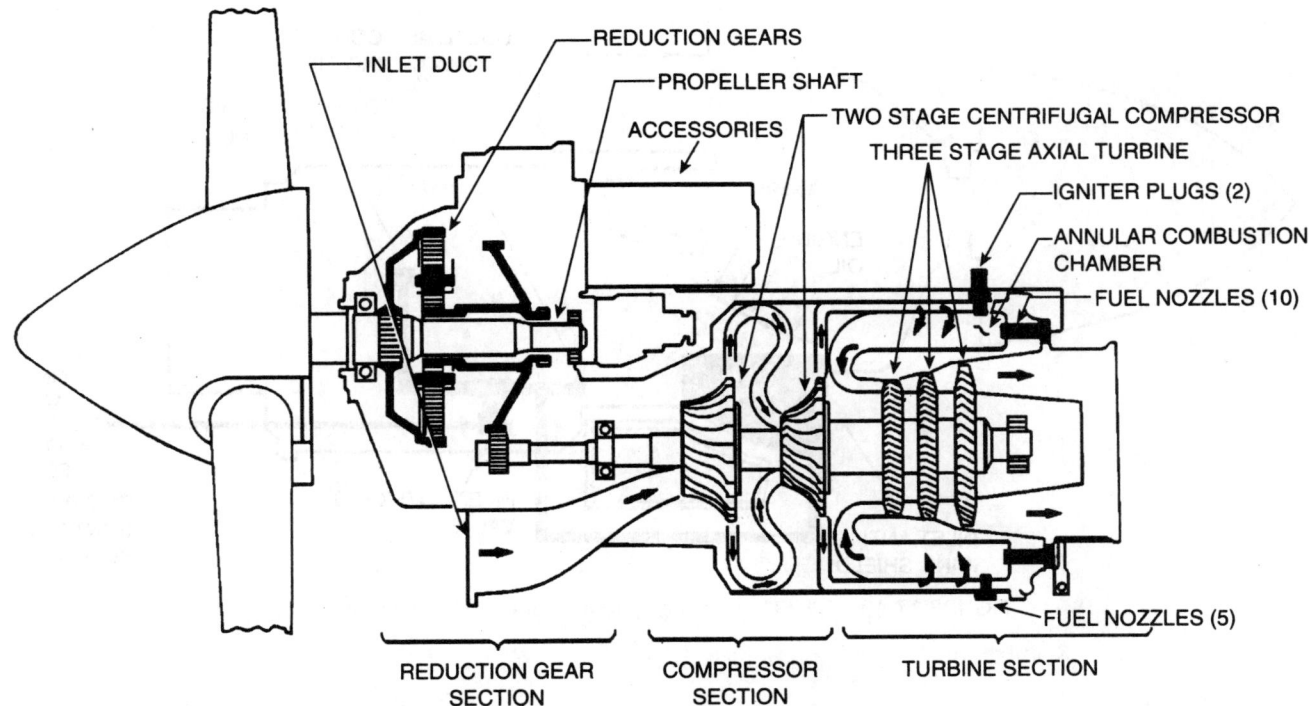

FIGURE 17-48 Cross-sectional view of TPE331 engine. (*Honeywell Inc.*)

and a beta mode of operation for taxiing. As explained earlier in this text, the **beta** mode is employed for low-power and reduced-rpm operations on the ground.

As mentioned before, the propeller pitch-changing mechanism is actuated hydraulically and is designed for both feathering and reverse thrust. Hydraulic pressure moves the propeller toward low pitch; however, high pitch and feathering are accomplished by means of coil springs and counterweight force.

Pilot-operated controls include the power lever, engine rpm lever, manual feather control, and unfeather switch. The system includes a torque sensor to detect negative torque—that is, a situation in which the engine is being rotated by the windmilling propeller. When a negative torque signal (NTS) is produced, the system automatically feathers the propeller to stop engine rotation and reduce drag.

Anti-icing for the TPE331 engine, illustrated in Fig. 17-49, involves an electrically actuated system controllable by the pilot. The inlet anti-icing system employs warm bleed air from the second-stage compressor. This air is directed through a forward manifold to an anti-icing shield surrounding the outer side of the air inlet. The air then flows rearward and is discharged into the engine nacelle. The inner wall of the air inlet is a part of the gearbox case and is warmed by engine oil.

The fuel system for the TPE331 engine, shown in Fig. 17-50, is designed to pressurize and regulate the fuel as it is directed to the nozzles that atomize it as it is injected into the rear of the combustor. The fuel control unit is coordinated with the propeller control to provide the correct amount of fuel to meet the speed and power requirements of the engine. The system includes a fuel control unit, solenoid valve, flow divider, fuel nozzle and manifold assembly, and oil/fuel heat exchanger.

The system automatically controls fuel flow for variations in power lever position, compressor discharge pressure, and inlet-air temperature and pressure. The solenoid valve is actuated automatically to open during starting and to close on engine shutdown. It is manually closed when the propeller is feathered.

Allison Model 250 Turboprop Engine

The Allison Model 250 turboprop engine is shown in Fig. 17-51. The Allison Model 250 is used more widely as a turboshaft engine on helicopters than as a turboprop engine. The gas generator (gas producer) section of the engine is explained in detail in Chap. 18.

Another arrangement in which a Model 250 engine drives a propeller is shown in Fig. 17-52. In this installation, a turboshaft engine drives a separate gearbox on which the propeller is installed. This arrangement is called a "**turbine pac.**"

Description and Operation

The Allison 250 series turboprop engine is an internal-combustion gas-turbine engine featuring a free-power turbine. The engine consists of a combination axial-centrifugal compressor; a single "can"-type combustor; a turbine assembly which incorporates a two-stage gas producer turbine; a two-stage power turbine; an exhaust collector; an accessory gearbox which incorporates a gas producer gear train and a power-turbine gear train; and a reduction gearbox that incorporates a planetary gear train.

Since the internal gas generator operation is much like that of the turboshaft version, the discussion here will be concerned with the unique characteristics of the turboprop version. The two sections of the engine that are the most

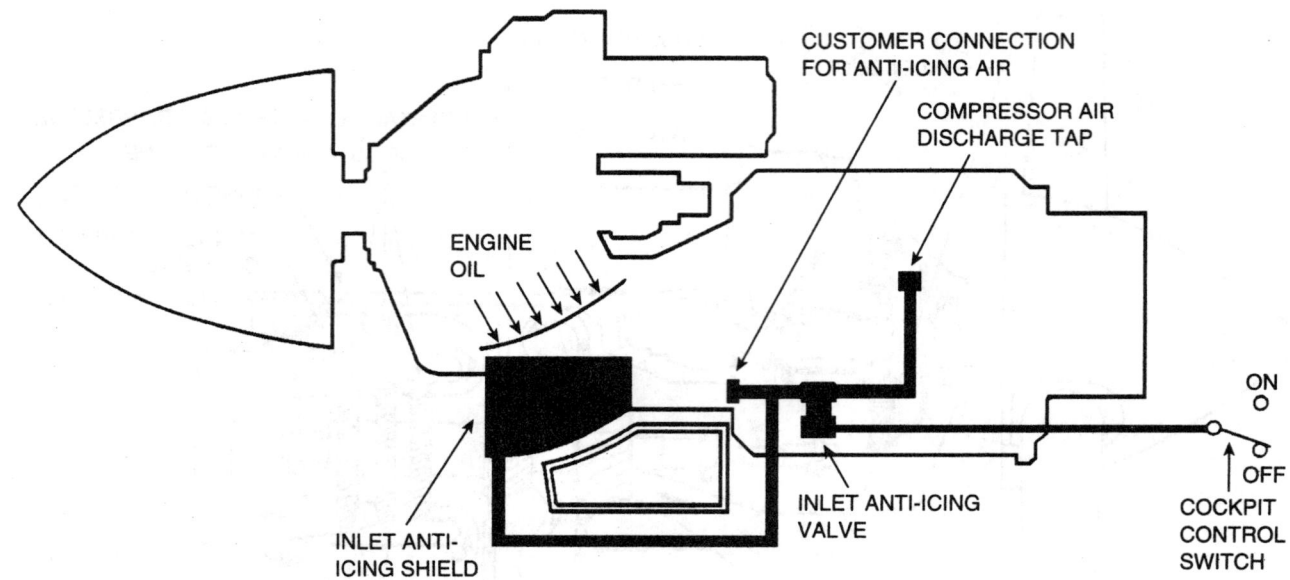

FIGURE 17-49 TPE331 anti-icing system diagram. (*Honeywell Inc.*)

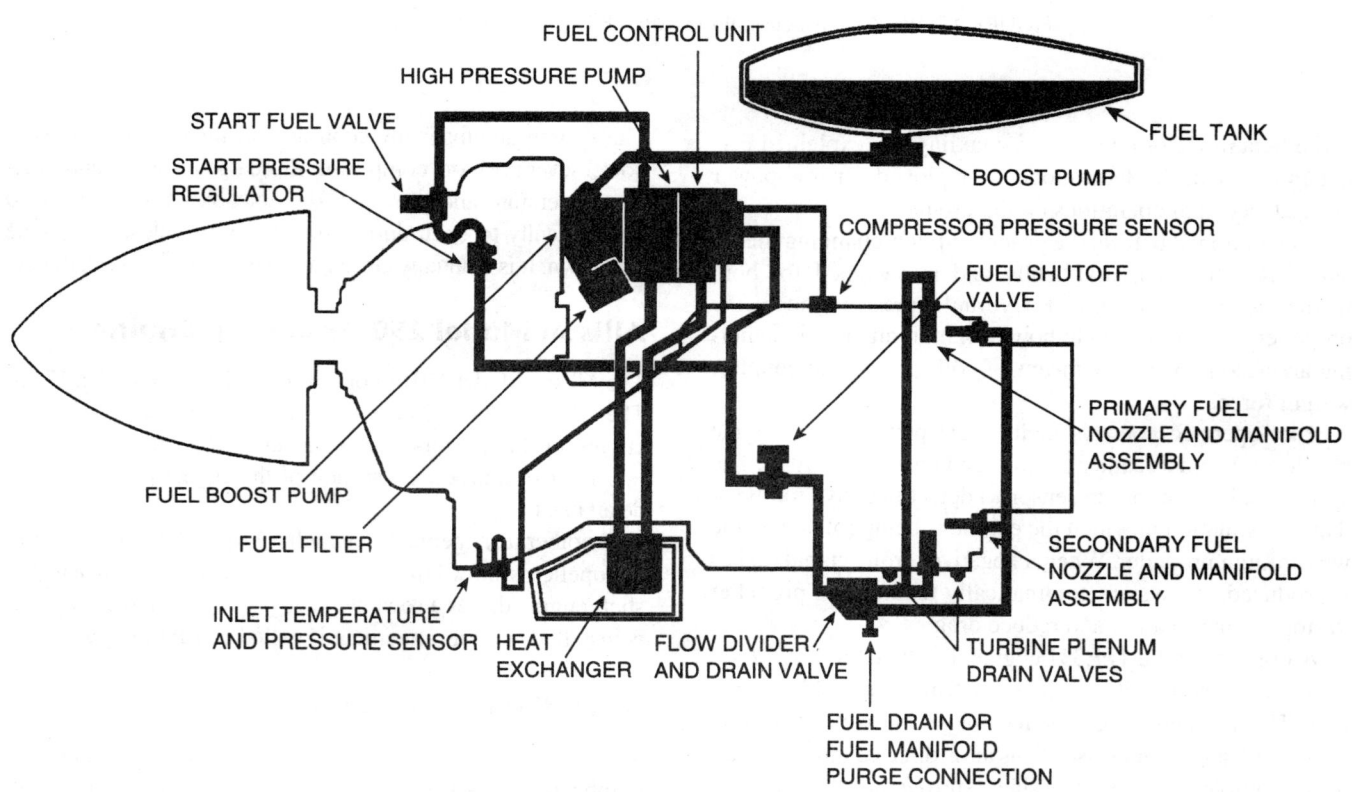

FIGURE 17-50 Simplified diagram of the TPE331 fuel system. (*Honeywell Inc.*)

different from their counterparts in the turboshaft version are the propeller reduction gearbox and the power control system.

The reduction-gearbox assembly mounts at the power output pad on the front of the accessory gearbox. The internal planetary-gear assembly reduces the input speed of 6016 rpm at the accessory gearbox to 2030 rpm at the propeller drive

flange. The propeller bolts to the propeller drive flange at the front end of the reduction gearbox.

The reduction-gearbox assembly, shown in Fig. 17-53, consists of a front housing, a rear housing, a sun gear, eight planet gears, a ring gear, a planet carrier, a propeller shaft, a propeller drive flange, and several attaching parts.

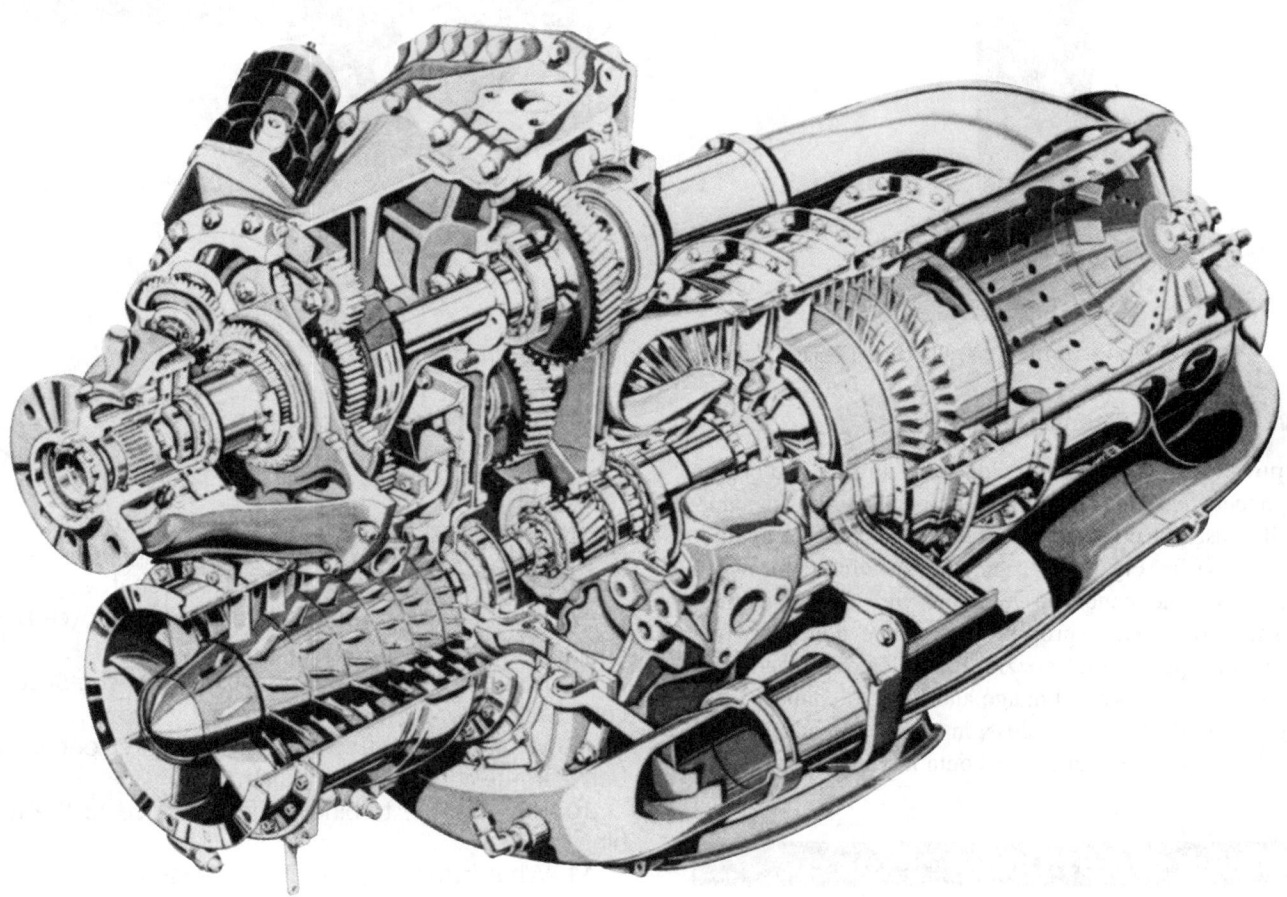

FIGURE 17-51 Two turboprop models of the Allison 250 engine. (*Rolls-Royce.*)

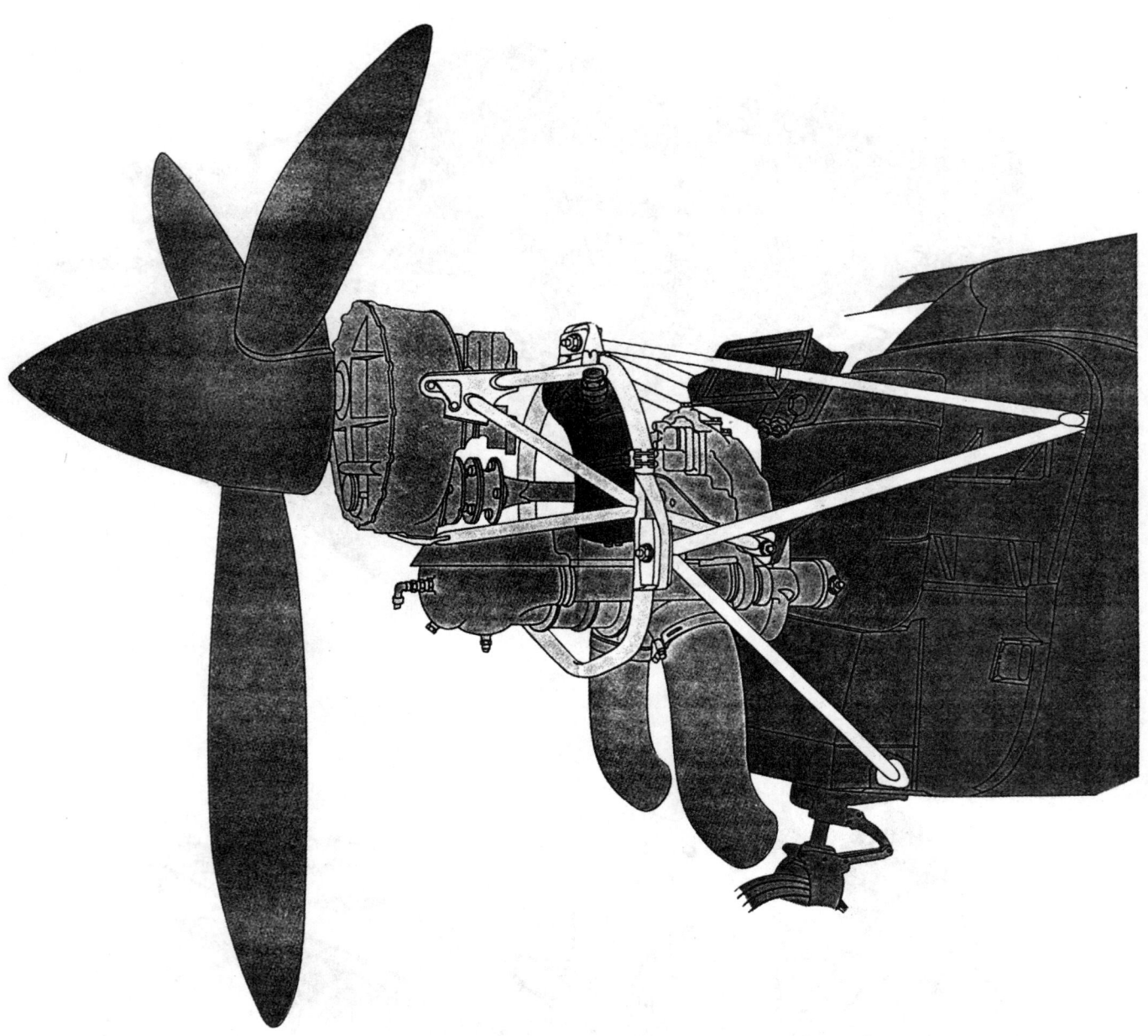

FIGURE 17-52   The Allison "turbine pac" arrangement.

Power Control System

The **power control system** includes the aircraft and engine components necessary to control engine power settings during all phases of operation. Some of this system's components, shown in Fig. 17-54, include the aircraft-furnished power lever and condition lever, as well as the engine-furnished coordinator, prop-power turbine governor, fuel control, fuel pump, fuel nozzle, propeller overspeed governor, and the necessary linkage and lines to connect these components together. A discussion of the operation of the propeller-related components is contained in Chap. 20.

REVIEW QUESTIONS

1. Compare the power section of a turboprop engine with that of a turbojet engine.

2. Define the term "eshp."

3. What components comprise the gas generator section of the engine?

4. What is the function of the gas generator?

5. What is meant by the term "free turbine"?

6. What is meant by the term "fixed shaft engine"?

7. Describe the compressor of the Rolls-Royce Dart engine.

8. Name the principal parts of the Dart combustion chamber.

9. What is the function of the interconnectors between the combustion chambers?

10. How are the turbine blades attached to the turbine disks?

11. What is the purpose of the deaerator tray in the oil system?

12. Through what unit is the water-methanol mixture injected into the engine?

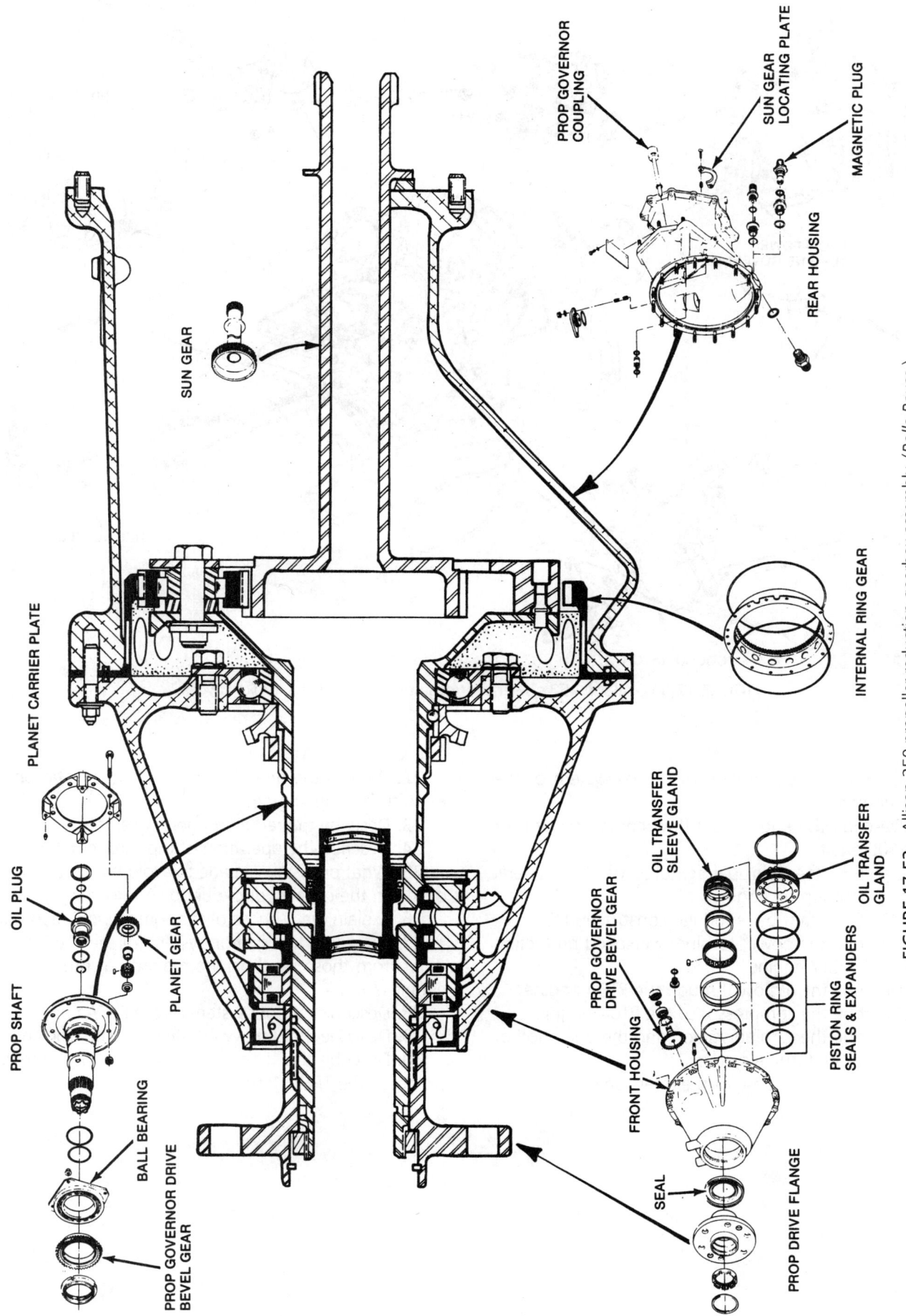

PROP GOVERNOR COUPLING

SUN GEAR LOCATING PLATE

MAGNETIC PLUG

REAR HOUSING

SUN GEAR

PLANET CARRIER PLATE

INTERNAL RING GEAR

OIL PLUG

OIL TRANSFER SLEEVE GLAND

PROP SHAFT

PROP GOVERNOR DRIVE BEVEL GEAR

OIL TRANSFER GLAND

PLANET GEAR

PROP GOVERNOR DRIVE BEVEL GEAR

BALL BEARING

PISTON RING SEALS & EXPANDERS

FRONT HOUSING

SEAL

PROP DRIVE FLANGE

FIGURE 17-53   Allison 250 propeller reduction-gearbox assembly. *(Rolls-Royce.)*

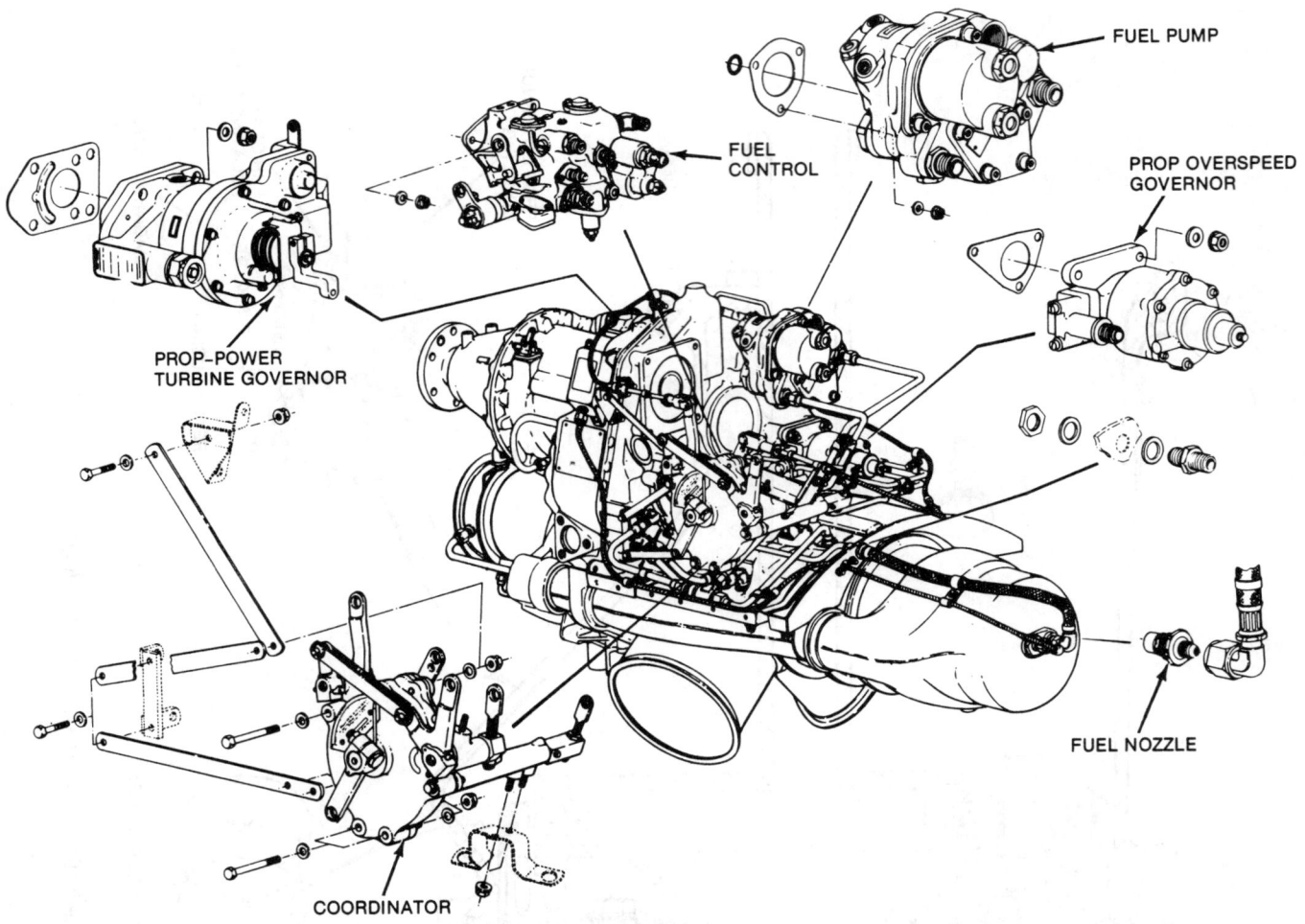

FIGURE 17-54 Allison 250 power control system components. (*Rolls-Royce.*)

13. Give a brief description of the oil system of the PW100 engine.

14. Describe the engine fuel flow control system for the PW100 engine.

15. Describe the principal features of the General Electric CT7 turboprop engine.

16. What is meant by "modular construction"?

17. Describe the power-turbine overspeed protection system of the CT7 engine.

18. Describe the airflow through the PT6A engine.

19. Describe the compressor of the PT6A engine.

20. Describe the turbine section and the operation of the two separate turbine systems.

21. What is the purpose of the interstage baffle?

22. What is the propeller reduction-gear ratio on the PT6A turboprop engine?

23. Describe the reduction-gear assembly.

24. Describe the operation of the torquemeter.

25. What provision is made to avoid the accumulation of fuel in the exhaust duct during shutdown?

26. Explain the purpose of the compressor bleed valve.

27. How does the ignition system for the PT6A engine differ from those used on the majority of other turbine engines?

28. Describe the fuel system of the PT6A engine.

29. Describe the Honeywell TPE331 turboprop engine.

30. Describe the function and operation of the NTS system in the TPE331 engine.

Turboshaft Engines 18

INTRODUCTION

A gas-turbine engine that delivers power through a shaft to operate something other than a propeller is referred to as a **turboshaft engine**. Turboshaft engines are similar to turboprop engines. Many of the turbofan and turboprop engines previously discussed in this text are also manufactured, with minor variations, in a turboshaft version. The **shaft turbine** may produce some thrust, but it is primarily designed to produce **shaft horsepower** (shp). The turboshaft engine has the same basic components found in a turbojet engine, with the addition of a turbine wheel or wheels to absorb the power of the escaping gases of combustion. The power takeoff may be coupled directly to the engine turbine, or the shaft may be driven by a turbine of its own (*free turbine*) located in the exhaust stream. Both types have been successfully used in helicopter applications; however, the free turbine is the most popular in use today. Another use of turboshaft engines is the **auxiliary power unit** or **APU**. These small gas-turbine engines are mostly used on large transport aircraft for providing auxiliary power either on the ground or in flight if needed. They are designed to provide the aircraft with electrical or pneumatic power for several on-board functions, making the aircraft more independent of ground support equipment.

AUXILIARY POWER UNIT

An airborne auxiliary power system, such as that shown in Fig. 18-1, supplies electrical and pneumatic power for most transport aircraft. On the ground, electrical and pneumatic power are available for such uses as lighting, ventilation, and engine starting. This power makes the airplane independent of ground support equipment. Pneumatic and electric power are available on most large aircraft in flight up to a specified altitude.

The Boeing 757 aircraft uses a Garrett GTCP (gas-turbine compressor powered) 331-200ER engine, which is electronically controlled. This auxiliary power unit (APU) is controlled by an electronic APU control unit (ECU) which supervises all operations of the APU. The ECU is located in the electrical and electronics compartment rack, which is located aft of the aft cargo door. This rack also contains the APU battery and charger, which are dedicated to starting the APU. The APU control unit coordinates the starting sequence, monitors the operation and pneumatic output of the APU, and ensures proper shutdown. The ECU features extensive built-in test equipment (BITE) that monitors many line-replaceable units and initiates protective shutdowns if damage to the APU is possible.

APU Operation

The airborne auxiliary power system is controlled from the APU control panel in the cockpit. This panel features a three-position rotary switch and fault and run annunciator lights. The APU can be started by turning this switch momentarily to the START position, after which the spring-loaded switch will return to the ON position. The white RUN light will illuminate when the APU is running at speeds greater than 95 percent. APUs normally operate at about 100 percent rpm. The amber FAULT light illuminates when the APU fuel shutoff valve is in disagreement or when there is a protective shutdown. To shut down the APU normally, the control switch is moved to the OFF position. The APU may be shut down in an emergency by pulling the APU fire handle or activating the APU shutdown switch on the APU remote control panel. The Engine Indicating and Crew Alert System (EICAS) displays APU exhaust gas temperature (EGT), rpm, and oil status. An APU hour meter is used to display APU hours of operation.

APU Description and Location

The APU is composed of three distinct modules: the power section, the load compressor, and the gearbox, as shown in Fig. 18-2. Air flowing into the APU can flow into the power section or the load compressor. The power section is a single-shafted gas-turbine engine which converts air and fuel into shaft horsepower. The shaft horsepower generated by the power section is used to drive the load compressor, gearbox, and accessories. The load compressor, which is driven by the power section, supplies compressed air for the airplane's pneumatic system. Inlet guide vanes regulate the amount of airflow through the compressor. The gearbox, which is also driven by the power section, contains gears and drive pads for the various APU accessories, including the APU generator.

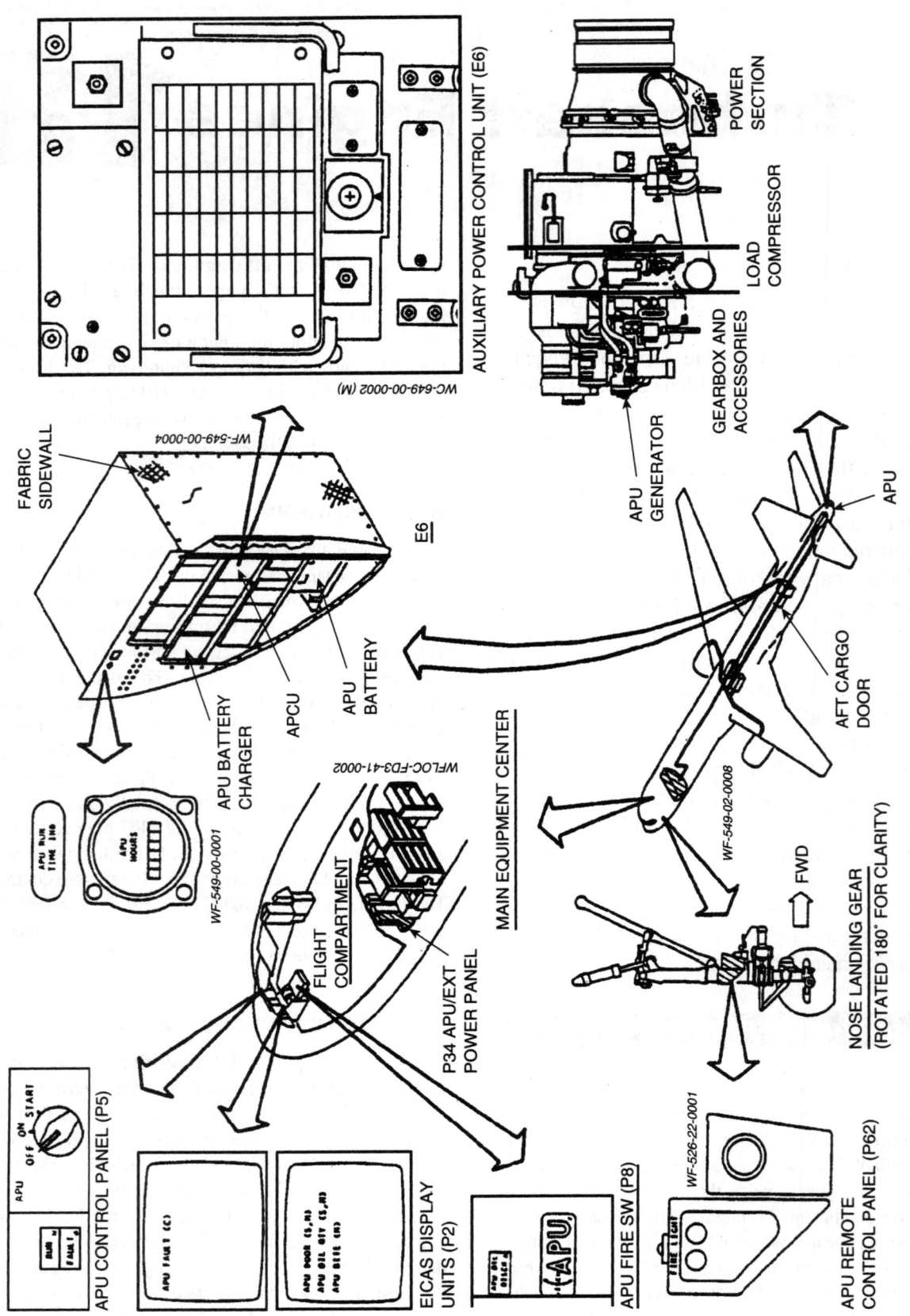

AUXILIARY POWER CONTROL UNIT (E6)

WC-649-00-0002 (M)

POWER SECTION

LOAD COMPRESSOR

GEARBOX AND ACCESSORIES

APU GENERATOR

FABRIC SIDEWALL

WF-549-00-0004

E6

APU BATTERY CHARGER

APCU

APU BATTERY

WF-549-00-0001

WFLOC-FD3-41-0002

FLIGHT COMPARTMENT

P34 APU/EXT POWER PANEL

MAIN EQUIPMENT CENTER

AFT CARGO DOOR

APU

WF-549-02-0008

FWD

NOSE LANDING GEAR
(ROTATED 180° FOR CLARITY)

APU CONTROL PANEL (P5)

APU
OFF   ON   START

RUN
FAULT

EICAS DISPLAY UNITS (P2)

APU FAULT (C)

APU DOOR (S,R)
APU OIL QTY (S,R)
APU EGT (R)

APU FIRE SW (P8)

APU

WF-526-22-0001

APU REMOTE CONTROL PANEL (P62)

FIGURE 18-1   Boeing 757 auxiliary power system. (Boeing.)

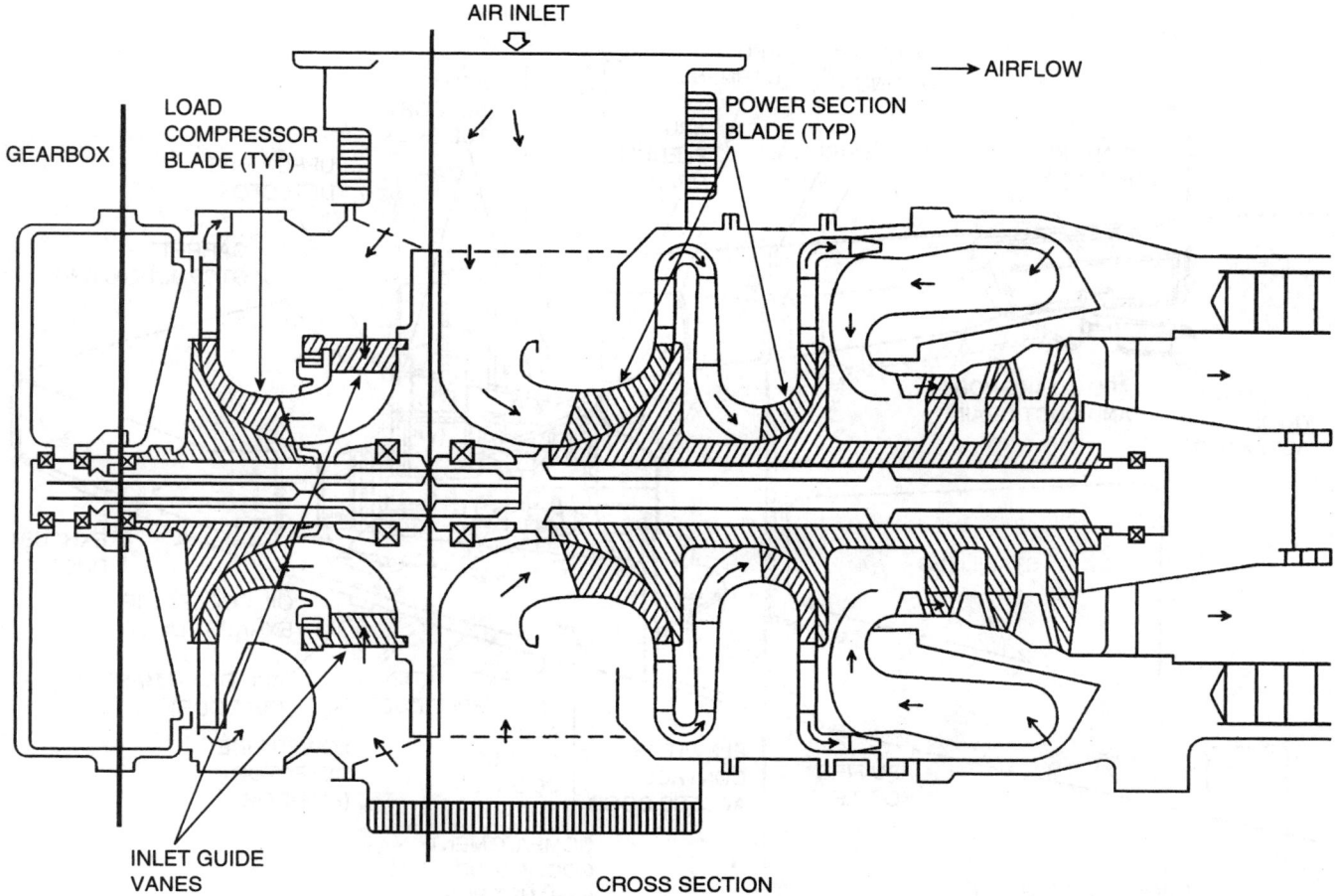

FIGURE 18-2 Diagram of auxiliary power unit engine airflow. (*Boeing.*)

The APU is located in the aft portion of the fuselage (tail cone area). It is suspended in its compartment from the APU air inlet plenum. Access into the APU compartment is through the APU access doors, which are shown in Fig. 18-3.

APU Systems

External ambient air enters the APU through the APU air intake door. This door is located on the upper right side of the fuselage next to the vertical stabilizer. An electrical actuator drives the door open to allow air into the APU air intake plenum, as shown in Fig. 18-3. Air that enters from the air intake system is used for cooling the APU compartment, for supporting combustion, and as a pneumatic power source. The APU uses a common oil system to cool and lubricate the bearings, gearbox, and generator of the APU.

The APU engine fuel system receives pressurized fuel from the airplane fuel tanks, increases this fuel pressure, and then regulates and distributes the fuel for engine combustion and pneumatic control. The fuel control unit pressurizes and regulates fuel for two functions: combustion which keeps the APU speed and EGT within limits, and power for the inlet guide vane actuator which controls pneumatic intake of the APU. The fuel control unit is electrically controlled by the ECU.

The APU is started by using a 28-volt dc electric motor. This motor is powered by the APU battery. Power is removed from the starter motor at 50 percent rpm of the APU. A single ignition unit sends a high voltage to the igniter plug that provides sparks to ignite the fuel-air mixture to start the combustion process. This ignition is controlled by the APU control unit. It begins ignition at 7 percent and automatically turns it off at 95 percent rpm. All operations of the APU are controlled and monitored by the APU control unit. The ECU is powered whenever the APU control panel switch is in the ON or START position or APU rpm is greater than 7 percent. Operating conditions of the auxiliary power unit (rpm, EGT, and oil status) are sent to the EICAS computers for display. APU EGT thermocouples measure exhaust gas temperature, which is provided to the APU control unit (ECU). The oil quantity transmitter signals oil level in the gearbox sump directly to the EICAS computers.

Honeywell GTCP85 Series APU

The GTCP85 series APU is used in many transport-type aircraft, such as the BAC 1-11, the DC-9 series, the Boeing 727 series, and the Boeing 737 series. Figure 18-4 is a cutaway view of the GTCP85 gas-turbine APU. This design incorporates a rotating group assembly of a single-piece first-stage impeller, next to a second-stage impeller connected to a turbine wheel which drives the assembly, as shown in Fig. 18-4. Although there are many different types and sizes of APUs, their basic function of providing electrical and pneumatic power remains the same.

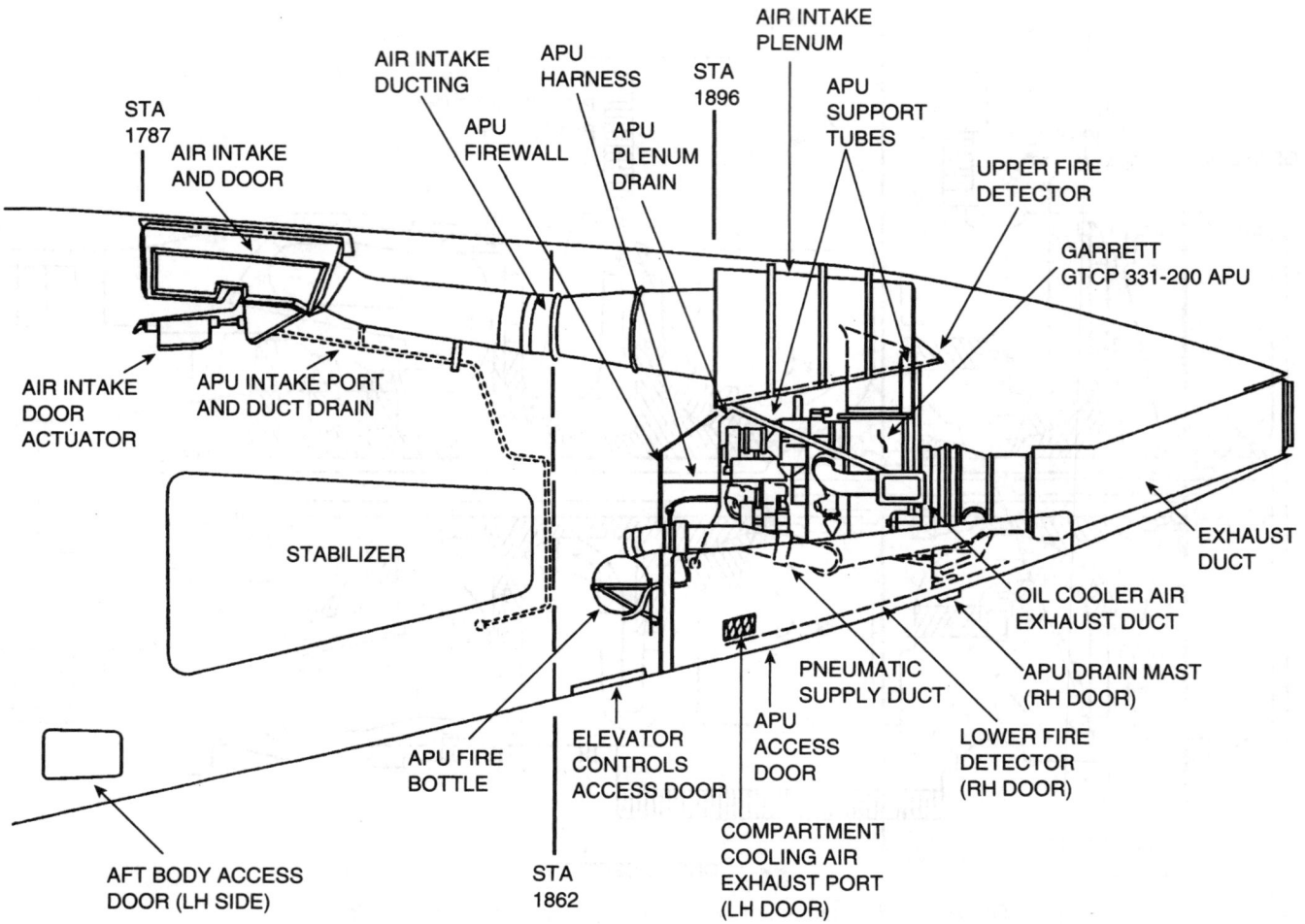

FIGURE 18-3 Boeing 757 auxiliary power unit installation. (*Boeing.*)

THE LYCOMING T53 TURBOSHAFT ENGINE

The Lycoming T53 turboshaft engine was initially manufactured in 1958 and, since that time, the family of T53 engines and others has grown in number as well as in performance capabilities without compromising reliability. T53 engines are manufactured for use as turboshaft engines for helicopters and as turboprop engines for fixed-wing aircraft. This engine is also used in marine and industrial fields.

In this section, we will discuss the T53-L-13B turboshaft engine illustrated in Fig. 18-5. Performance data and specifications for this engine are as follows:

Rated shaft power (30 min) at sea level, std. day (59°F) [15°C]	1400 hp	1044 kW
Maximum shaft power (continuous)	1250 hp	932 kW
Specific fuel consumption at rated power	0.58 lb/shp/h	

Maximum allowable oil consumption	0.14 gal/h	
Maximum gas producer rpm	25 600	
Maximum output rpm	6640	
Reduction-gear ratio (power-turbine output shaft)	3.2105:1	
Compression ratio	7.2:1	
Diameter	23.00 in	58.42 cm
Length	47.60 in	120.90 cm
Frontal area	2.88 ft^2	0.27 cm^2
Weight	540 lb	245 kg
Power-weight ratio	2.59 shp/lb	4 kW/kg
Oil specification	MIL-L-23699 and -7808, or equivalent	
Fuel specification	MIL-F-5264 and -46005A	

Engine Description

The T53-L-13B engine, shown in Fig. 18-6, is a free-turbine engine designed primarily for helicopter applications. However, because of its durability and reliability, other uses have developed. **Free turbine** means that the

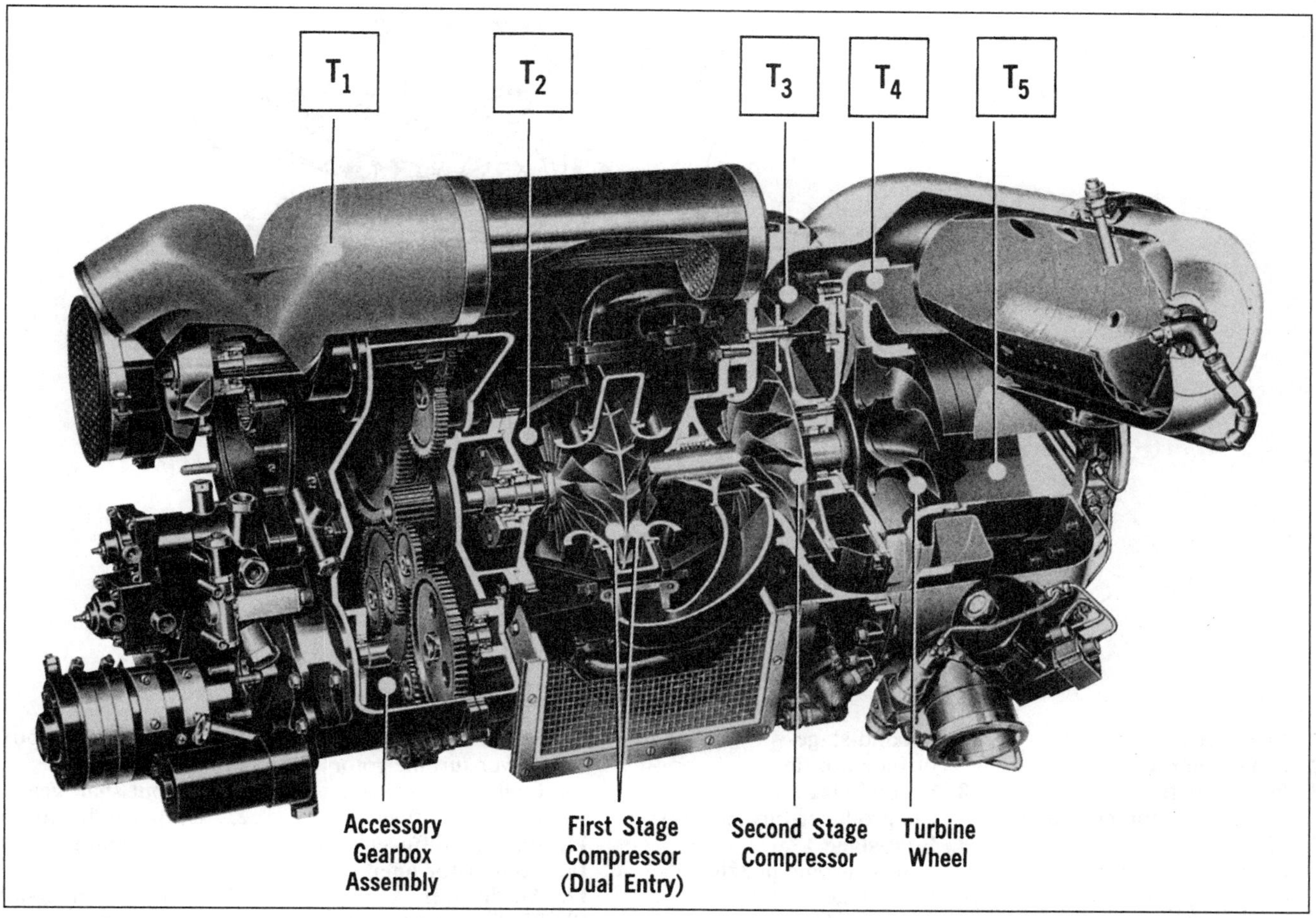

T_1 T_2 T_3 T_4 T_5

Accessory
Gearbox
Assembly

First Stage
Compressor
(Dual Entry)

Second Stage
Compressor

Turbine
Wheel

FIGURE 18-4   Cutaway diagram of the GTCP85 series auxiliary power unit. (*Boeing.*)

FIGURE 18-5   Lycoming T53-L-13B turboshaft engine. (*Textron Lycoming.*)

power turbines are not mechanically coupled to, or limited by the speed of, the compressor drive turbines. The engine consists of the air inlet housing, a carrier and gear assembly, a five-stage axial one-stage centrifugal compressor, an air diffuser, a reverse-airflow combustion chamber, a two-stage gas producer turbine, a two-stage power turbine, an exhaust diffuser, and a power shaft. The complete gas producer assembly, or N_1, system consists of five axial stages (first and second stages transonic), one centrifugal impeller, and the two gas producer turbine rotors. The gas producer turbine

rotors drive the compressor rotor, and, coaxial with it, the power-turbine rotors drive the power shaft, sun gear, output reduction carrier, and gear assembly which constitute the N_2 system.

Directional references for the engine are as follows:

Front	End of engine from which power is extracted.
Rear	End of engine from which exhaust gas is expelled.
Right and left	Determined by viewing the engine from the rear.
Bottom	Determined by the location of the accessory drive gearbox (six o'clock).
Top	Directly opposite, or 180° from, the accessory drive gearbox (twelve o'clock). (The engine lifting eyes are located at the top of the engine.)
Direction of rotation	Determined as viewed from the rear of the engine. The direction of rotation of the gas producer turbines is counterclockwise. The power turbines and the output gear shaft rotate in a clockwise direction.
O'clock	Position as viewed from the rear of the engine.

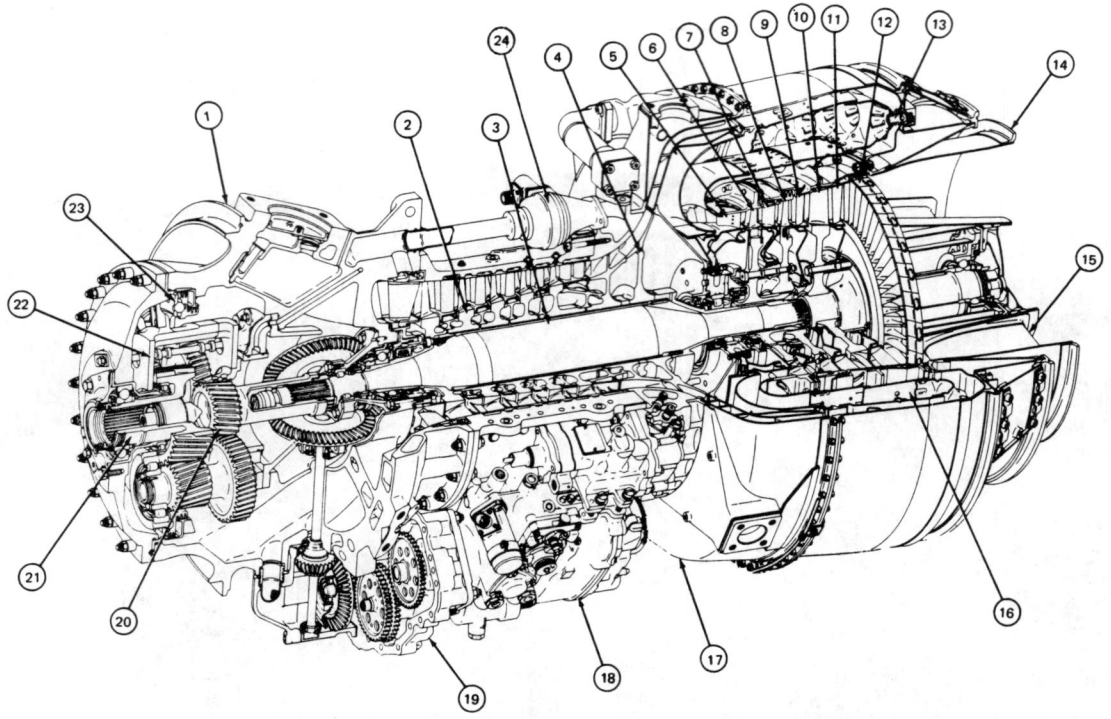

1. Inlet housing
2. Axial compressor
3. Power shaft
4. Centrifugal compressor
5. First-stage gas-producer-turbine nozzle
6. First-stage gas-producer-turbine rotor
7. Second-stage-turbine nozzle
8. Second-stage gas-producer-turbine rotor
9. First-stage power-turbine nozzle
10. First-stage power-turbine rotor
11. Second-stage power-turbine nozzle
12. Second-stage power-turbine rotor
13. Fuel-injector nozzle
14. Exhaust diffuser
15. Exhaust-diffuser strut
16. Combustor liner
17. Air diffuser
18. Fuel reg. and power-turbine governor
19. Accessory gearbox
20. Sun gear
21. Output shaft gear
22. Output-reduction carrier and gear assembly
23. Torquemeter system
24. Anti-icing valve

FIGURE 18-6 Cutaway drawing of the Lycoming T53 engine. (*Textron Lycoming.*)

Operational Description

The engine is started by an electric starter, geared to the compressor rotor shaft, which mechanically starts rotation of the compressor rotor. At the same time, the starting-fuel solenoid valve and the ignition system are also energized. This causes the fuel pump to drive fuel to the four starting nozzles at the two, four, eight, and ten o'clock positions in the combustor and the adjacent igniter plugs to ignite and sustain combustion of the fuel jets until the starting system is deenergized.

At 8 to 13 percent N_1 (compressor) speed, the distributor valve opens, allowing fuel to flow from the fuel regulator through 22 atomizers into the already inflamed combustor, where it accelerates both the compressor and power rotors. At approximately 40 percent N_1 speed, the flow of air and combustion are self-sustaining and the starter, starting-fuel valve, and igniters are deenergized. Control of power and speed is then taken over by the fuel regulator and power-turbine governor throughout the operating range.

Airflow and Gas Flow

Atmospheric air is drawn into the annular passageway of the inlet housing, passing rearward across the variable-inlet guide vanes, which direct the air to the engine compressor section.

The combination of one rotating member (compressor rotor) and one stationary member (stator) constitutes one stage of compression. The fifth-stage stator includes a second set of vanes which serve as exit guide vanes for the axial compressor and direct the air at the proper angle to the leading edge of the centrifugal compressor vanes.

The centrifugal compressor further accelerates the air as it passes radially to the diffuser housing passageway. Three sets of vanes in the air diffuser convert the air velocity into pressure and redirect the airflow rearward to the combustor.

At this point the air enters the combustor section, surrounds the combustor liner, and passes into the annular combustion area through slots, louvers, and holes in the liner. As the air enters the combustion area, its flow direction is reversed. At the same time, the air performs the dual function of cooling the combustor liner and supporting combustion. Combustion is made possible by the introduction of fuel into the combustion area through 22 atomizers (nozzles). The atomized fuel mixes with the air and burns.

The hot gas flows forward in the combustion area to the deflector, which reverses its flow. Flowing rearward again, the gas is directed across the two-stage gas producer turbine system. The first gas producer nozzle directs (impinges) the high-energy gas onto the first-stage turbine, through the second gas producer nozzle, and onto the second gas producer turbine. The power system also utilizes the two-stage turbine concept. On leaving the second gas producer turbine, the gas, still possessing a high work potential (approximately 40 percent), flows across the first power-turbine nozzle onto the first power turbine, through the second power-turbine nozzle, and onto the second power turbine. On passing from the second power turbine, the gas is exhausted into the atmosphere through the exhaust-diffuser passageway.

Cooling and Pressurization

As with other gas-turbine engines, the T53 engine requires controlled airflow for cooling internal components subject to high temperatures and for pressurizing the main-bearing seals and the intershaft oil seal at the forward end of the power shaft. The arrows in Fig. 18-7 indicate the flow of cooling air inside the engine. Compressor air is bled through ports at the periphery of the centrifugal compressor impeller, cooling the diffuser-housing front face and pressurizing the no. 2 main-bearing forward seal and aft oil seal, and is then ported through the rear-compressor shaft between the rotor assembly and the power shaft. This air flows both forward and aft. Some flows between the carbon elements of the no. 1 main-bearing seal, and then forward to the aft face of the intershaft seal. Some emerges at the aft end of the rear-compressor shaft, where it cools the rear face of the second gas producer rotor, the forward face of the first power-turbine rotor, and the first-stage power-turbine nozzle, and then flows into the airstream. The remainder flows through holes into the power shaft and aft into the interior of the second power-turbine rotor assembly. It is then ported through the turbine hub and spacer to cool the rear surface of the first power-turbine rotor, the forward surface of the second power-turbine rotor, and both faces of the second power-turbine nozzle. It is then discharged into the exhaust stream.

Compressed cooling air is directed through various ports and channels to all the moving components and the stationary nozzles subjected to the highest temperatures within the turbine section of the engine in a complex and fairly comprehensive system. The exhaust diffuser and the rear face of the second-stage power-turbine rotor are cooled only by circulation of ambient air through the hollow diffuser struts, impelled by the venturi action where it joins the exhaust.

Compressor and Impeller Housing Assemblies

The axial and centrifugal magnesium compressor housings each consist of two matched halves. Mounted within the axial housings are five stages of stator vanes and five rows of steel inserts that maintain radial blade clearances to resist erosion. Mounted externally on the axial housing are the starting-fuel solenoid valve and a two-piece steel band which is part of the interstage bleed system.

The centrifugal impeller housing provides mounting facilities for the anti-icing valve, ignition exciter, and interstage bleed-actuator assembly. In addition, the hollow housing is utilized as a manifold by which compressor-discharge-pressure bleed air may flow to the anti-icing valve and a customer air-bleed port located externally at the twelve o'clock position.

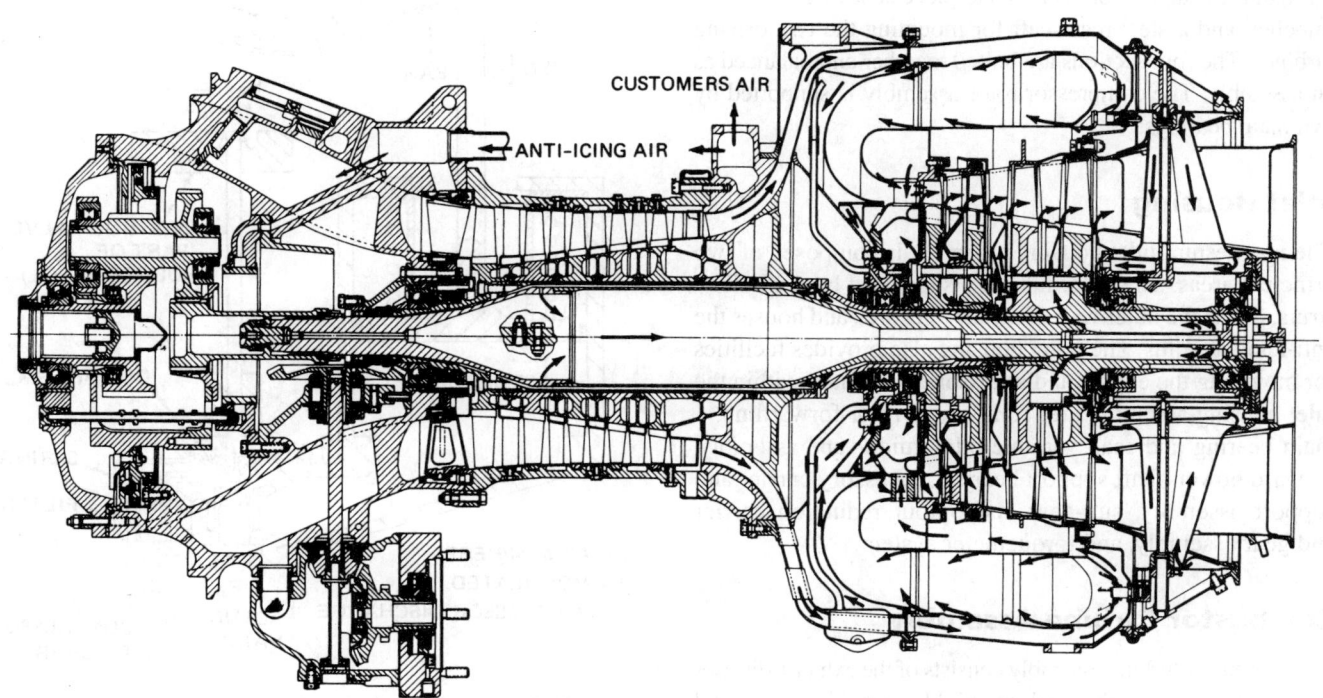

FIGURE 18-7 Cooling and pressurization airflow. (*Textron Lycoming.*)

Anti-Icing

Pressurized hot air from the annular manifold in the centrifugal compressor housing flows forward through the manually controlled airflow shutoff regulator into the hollow annulus on top of the inlet housing (see Fig. 18-7). This hot air is then directed through five of the six hollow inlet housing support struts to anti-ice the air inlet area. Hot scavenge oil, draining through the lower strut into the accessory drive gearbox, anti-ices the bottom of the air inlet area. Hot air also flows through the inlet guide vanes for anti-icing and then into the compressor area. In the event of electric power failure, anti-icing becomes continuous.

Air-Diffuser Housing

The air-diffuser housing is comprised of an inner and outer shell separated by three rows of vanes. The outer shell provides three additional engine mount pads and an aft lifting eye. The airflow, upon exiting the impeller, is turned, slowed, and smoothed out by the three vane stages prior to entering the combustor housing. A portion of the air exiting the third row of vanes is bled into an air manifold in the outer shell for anti-icing, customer bleed, and operation of the interstage air-bleed actuator. The inner shell supports the compressor rotor's rear no. 2 main bearing and seals, the deflector assembly, the first-stage gas producer nozzle, and the second-stage gas producer cylinder. Two oil tubes provide paths for pressure and scavenge oil into and out of the no. 2 bearing housing from pressure-scavenge ports on the outer shell.

Compressor Rotor Assembly

The compressor rotor assembly consists of a steel first-stage rotor disk, a welded titanium sleeve constituting the second-through fifth-stage rotor disks, a one-piece centrifugal titanium impeller, and a steel rear shaft for mounting the two driving turbines. The four sections are bolted together and balanced as an assembly. The compressor rotor assembly is supported by two main bearings.

Inlet Housing

The magnesium inlet housing assembly is composed of two principal areas. The outer housing, supported by six hollow struts, forms the outer wall of the air inlet area and houses the anti-icing annulus. The outer housing also provides facilities for mounting the engine and accessories. Located within the inlet housing assembly are: compressor rotor forward no. 1 main bearing and seal, variable-inlet guide-vane assembly, forward power-shaft support bearing, accessory carrier and support assembly, sun-gear and output reduction carrier and gear assembly, and torquemeter system.

Combustor Turbine Assembly

The combustor turbine assembly consists of the exhaust-diffuser support cone assembly, fuel manifold assembly, fire-shield assembly, exhaust-diffuser assembly, second power-turbine rotor and bearing housing assembly, V-band coupling, atomizing combustion chamber assembly, second-stage power-turbine nozzle, first-stage power-turbine rotor, and first-stage power-turbine nozzle. The second power-turbine rotor and bearing housing assembly consists of the turbine disks and blades, the no. 3 and no. 4 main bearings, the no. 3 main-bearing seal, and the bearing housing. The **exhaust diffuser** contains hollow struts through which cooling air is supplied to the no. 3 and no. 4 main-bearing housings and the rear face of the second-stage power-turbine disk. The combustion chamber assembly consists of the combustion chamber housing and the combustion chamber liner.

Description of Interstage Bleed System

The interstage bleed system is supplied with the engine to improve compressor acceleration characteristics. The system automatically relieves the compressor of a small amount of air to prevent compressor stall in the low-speed range and during compressor acceleration or deceleration. A schematic drawing of the system is shown in Fig. 18-8.

The air-bleed actuator operates by means of compressor discharge air extracted from a port on the right-hand side of the air diffuser. Compressor discharge air entering the

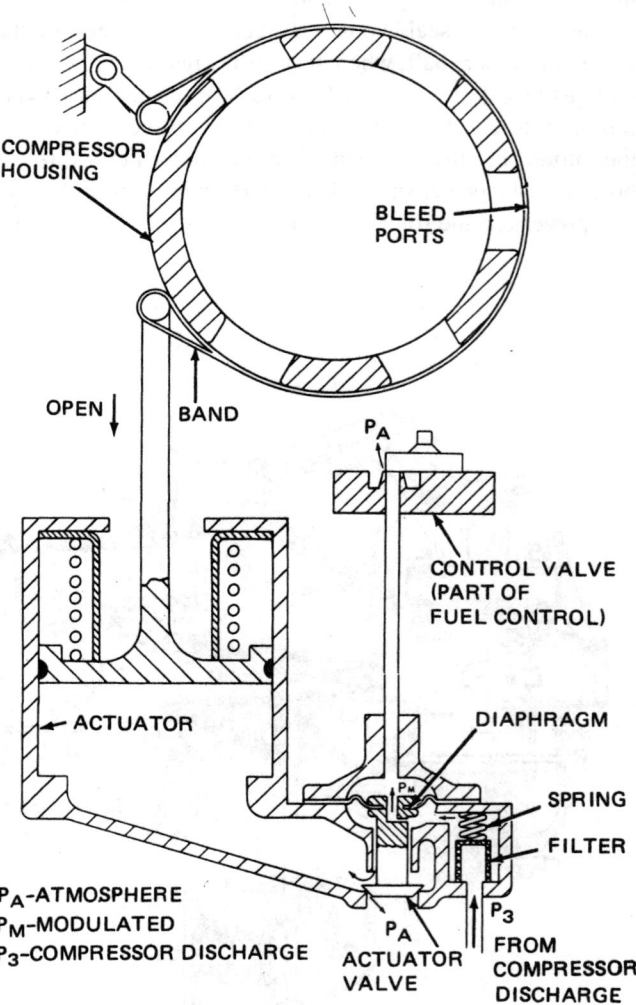

FIGURE 18-8 Schematic drawing of the interstage bleed system.

actuator assembly passes through a filter to the underside of the relay-valve diaphragm. A small portion of this air, which is under the diaphragm, is bled through an orifice in the base of the relay-valve assembly to an external line which directs it to a slide valve located on the fuel regulator housing. When the slide valve is in the open position, this air is vented overboard, reducing pressure at the top surface of the diaphragm. Simultaneously, air is being bled overboard through the open relay valve. This reduces pressure at the bottom surface of the diaphragm. This equalization of pressure on both surfaces of the diaphragm causes it to remain in a neutral position, holding the relay valve in its open position. With the relay valve held in the open position, the major portion of the compressor discharge air that enters the actuator assembly is vented to the atmosphere. When the compressor discharge actuating pressure is vented, the actuator spring, located on top of the actuator piston, expands and pushes the piston downward, causing the bleed band to open and remain open as long as the slide valve on the fuel regulator is in the open position.

It follows, therefore, that when the slide valve is closed, the bleed band will be closed. This is accomplished by a buildup of pressure on the top side of the relay-valve diaphragm, which forces the relay valve down, closing the overboard vent. With the overboard vent closed, the compressor discharge pressure is now routed into the actuator piston assembly, overcoming the spring load and forcing the piston assembly to move upward. This, in turn, causes the bleed band to close around the compressor bleed ports.

The entire sequence of operations is controlled by the fuel regulator, which senses gas producer speed, fuel flow, and pilot demand, thereby ensuring proper opening and closing of the interstage air bleed.

Power-Turbine Governor and Tachometer Drive Assembly

The power-turbine governor and tachometer drive assembly is mounted at the ten o'clock position on the exterior of the inlet housing and is driven through shafts and gearing from the power shaft. The drive assembly provides mounts and drives for the power-turbine tachometer generator and the power-driven rotary torquemeter boost scavenge pump. The drive assembly also drives the power-turbine governor and incorporates a strainer and metering cartridge for lubrication of the drive-gear train. A torquemeter relief valve, located on the upper portion of the housing, allows for adjustment of the torquemeter boost oil pressure.

Accessory Drive Gearbox Assembly

The accessory drive gearbox assembly is mounted at the six o'clock position on the exterior of the inlet housing. It is driven through a shaft and gearing from the compressor forward shaft. The power-driven rotary oil pump, oil filter, fuel regulator, compressor rotor tachometer generator, and starter-generator are mounted on the gearbox. A magnetic chip-detector drain plug is installed in the bottom of the gearbox.

Variable-Inlet Guide-Vane System

To provide the desired compressor surge margin, the angle of attack of the inlet air to the first compressor rotor must be within the stall-free operating range of the transonic airfoil (first two stages of the compressor). Since this stall-free operating range varies with compressor speed, it is necessary to vary the angle of attack as a function of compressor speed. This is accomplished by varying the angular position of the inlet guide vane. The variable-inlet guide-vane assembly is located in front of the first compressor rotor and consists of a series of hollow blades positioned by a synchronizing ring. Operation of the guide vanes is controlled by an actuator assembly mounted externally at the inlet to the axial compressor housing split line. The actuator is connected to the synchronizing ring by a control rod which attaches to a piston within the actuator. Depending on inlet air temperature and compressor speed, varying amounts of high-pressure fuel from the regulator assembly are ported through external hoses to one side of the piston which positions the control rod. The inlet guide-vane position is transmitted back to the regulator assembly by an external feedback rod to nullify the high-pressure-fuel signal.

Lubrication System

The lubrication system for the T53 engines is designed for operation up to 25 000 ft [7620 m] above sea level with oil conforming to MIL-L-23699 or -7808 specifications. It will provide satisfactory lubrication at oil inlet temperatures ranging from −65 to 200°F [−53.89 to 93.33°C].

The lubrication system consists of a vane-type pump incorporating a pressure and scavenge element, a main 40-μm oil filter with bypass capabilities, an oil manifold, and external oil hoses. A schematic diagram of the system is shown in Fig. 18-9.

Main Fuel Regulator Assembly

Control of fuel flow is a critical function for any gas-turbine engine and must be achieved with precision to get satisfactory and efficient turbine operation throughout the operating range. The critical objective is to maintain balance between combustion air and fuel uniting in the combustion chamber, but there are too many variables for even the most skilled pilot to achieve this with only a throttle lever. Too much fuel, relative to air, causes hot starts and surging. Too little fuel results in no start or, if the engine is running, a flameout. The fuel flow and fuel control system is shown in Fig. 18-10.

If the fuel regulator is to add the correct amount of fuel to the air in the combustor, it must continuously sense how much air there is in the combustor. This depends on compressor speed (N_1), air temperature (T_1), and air pressure (P_1), which vary with altitude.

The engine speed cannot change as rapidly as the throttle is moved, so the fuel regulator is governed by the gas producer governor (N_1) and the power-turbine governor (N_2) to damp the changes of fuel flow demanded by throttle movement. This delays the changes in fuel flow just long enough

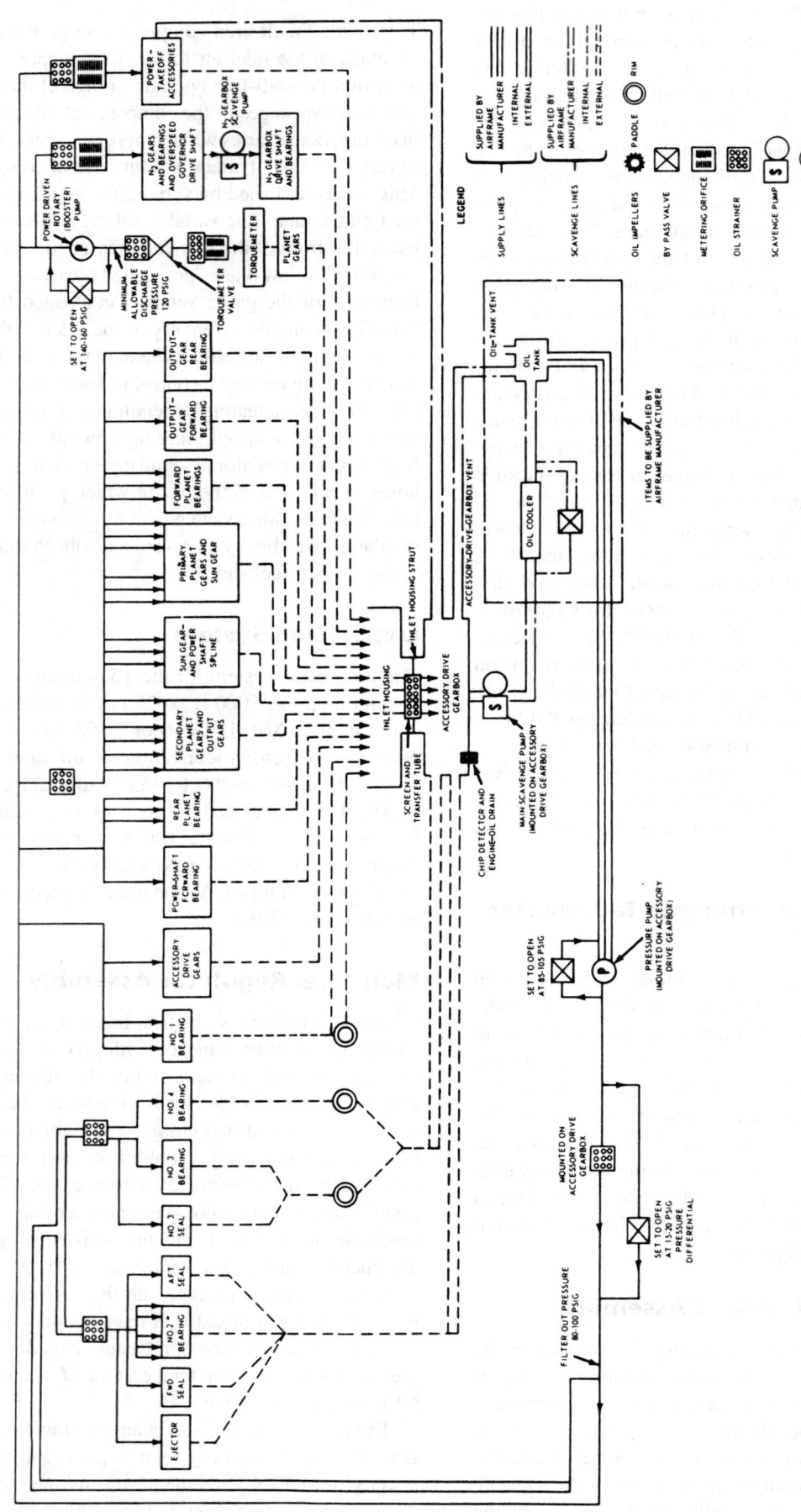

FIGURE 18-9 Schematic diagram of the lubrication system for the Lycoming T53 engine. (*Textron Lycoming.*)

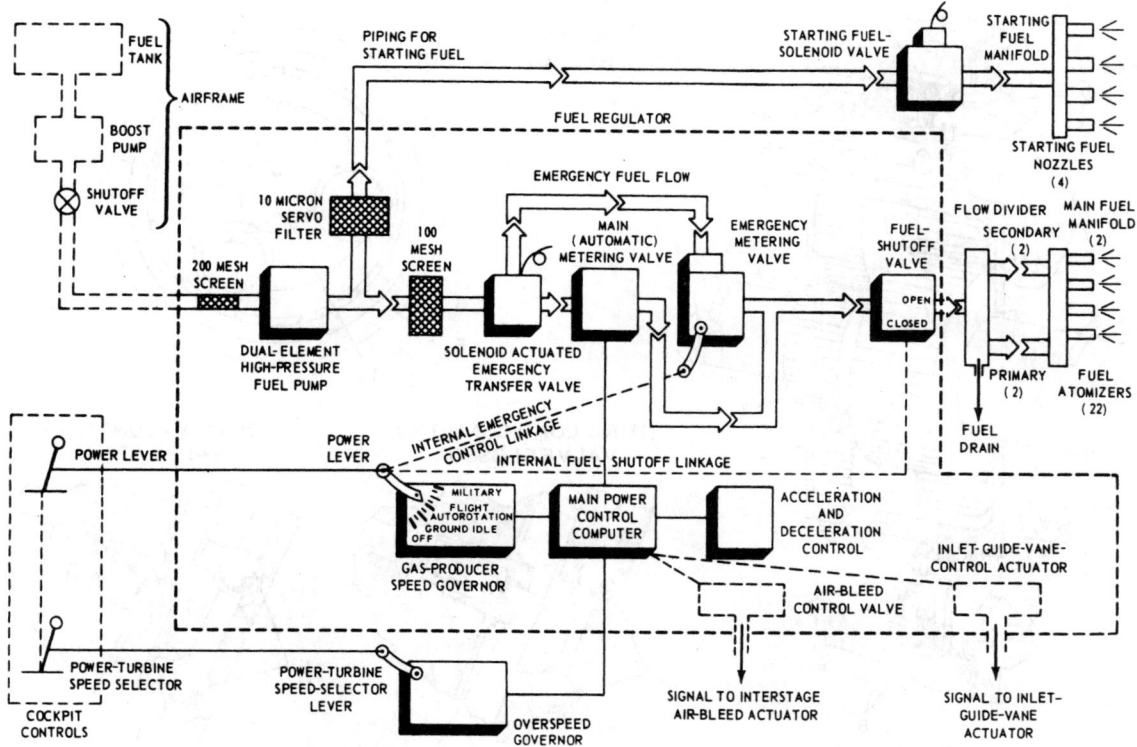

FIGURE 18-10 Schematic diagram of the fuel flow and fuel control system. (*Textron Lycoming.*)

for them to be acceptable at the changing engine speeds without causing the improper engine operation that results from an unbalanced fuel-air mixture.

The hydromechanical fuel regulator assembly is made up of a Model TA-2S fuel regulator and a Model PTG-3 power-turbine governor. It consists essentially of the following components:

1. Dual-element fuel pump
2. Speed input servo and gas producer governor (N_1)
3. Power-turbine speed governor (N_2)
4. Acceleration and deceleration fuel flow control
5. Manual (emergency) control system
6. P_1 multiplier assembly (altitude compensator)
7. T_1 motor assembly (temperature compensator)
8. Transient air-bleed control
9. Variable-inlet guide-vane actuating system

Because of the complexity of the fuel regulator, as indicated by the foregoing list, provision is made for last-resort manual control by the pilot in case of serious malfunction of the automatic control.

Torquemeter

The torquemeter (see Fig. 18-11) is a hydromechanical torque-measuring device located in the reduction-gear section of the engine inlet housing. It uses engine oil as the means for determining and measuring engine torque, which is read in the cockpit as oil pressure (psi). Although this system uses engine oil, it is not a part of the lubrication system.

The mechanical portion of the torquemeter consists of two circular plates. One is attached to the inlet housing and is identified as the stationary plate. The second, or movable, plate is attached to the reduction-gear assembly. The movable plate contains front and rear torquemeter sealing rings, which enable it to function as a piston in the rigidly mounted cylinder-chamber assembly of the torquemeter. The cylinder assembly houses the variable-opening torquemeter (poppet) valve, and the movable plate maintains the fixed-orifice metered bleed, which functions in relation with the poppet valve. The movable plate is separated from the stationary plate by steel balls positioned in matched conical sockets machined in the surfaces of both plates.

When the engine is not operating [section A-A (1) in Fig. 18-11], the torquemeter assembly's movable plate is in a position forward and clear of the torquemeter's valve plunger, allowing the spring-loaded valve to remain in the closed position. With the engine operating and a load applied to the output shaft, the torque developed in the engine to drive the shaft is transmitted from the sun gear through the reduction-gear assembly. The attached movable plate therefore tends to rotate with the assembly. However, this mechanically limited rotary movement positions the steel balls against the conical sockets of both plates, resulting in the movable plate being axially directed rearward in the assembly [section A-A (2), Fig. 18-11].

The plate, moving rearward, contacts the torquemeter-valve plunger, opening the valve and allowing oil to flow into the cylinder. This contact is maintained during all engine operation, and the size of the valve opening varies as the plate moves rearward or forward. As torque continues

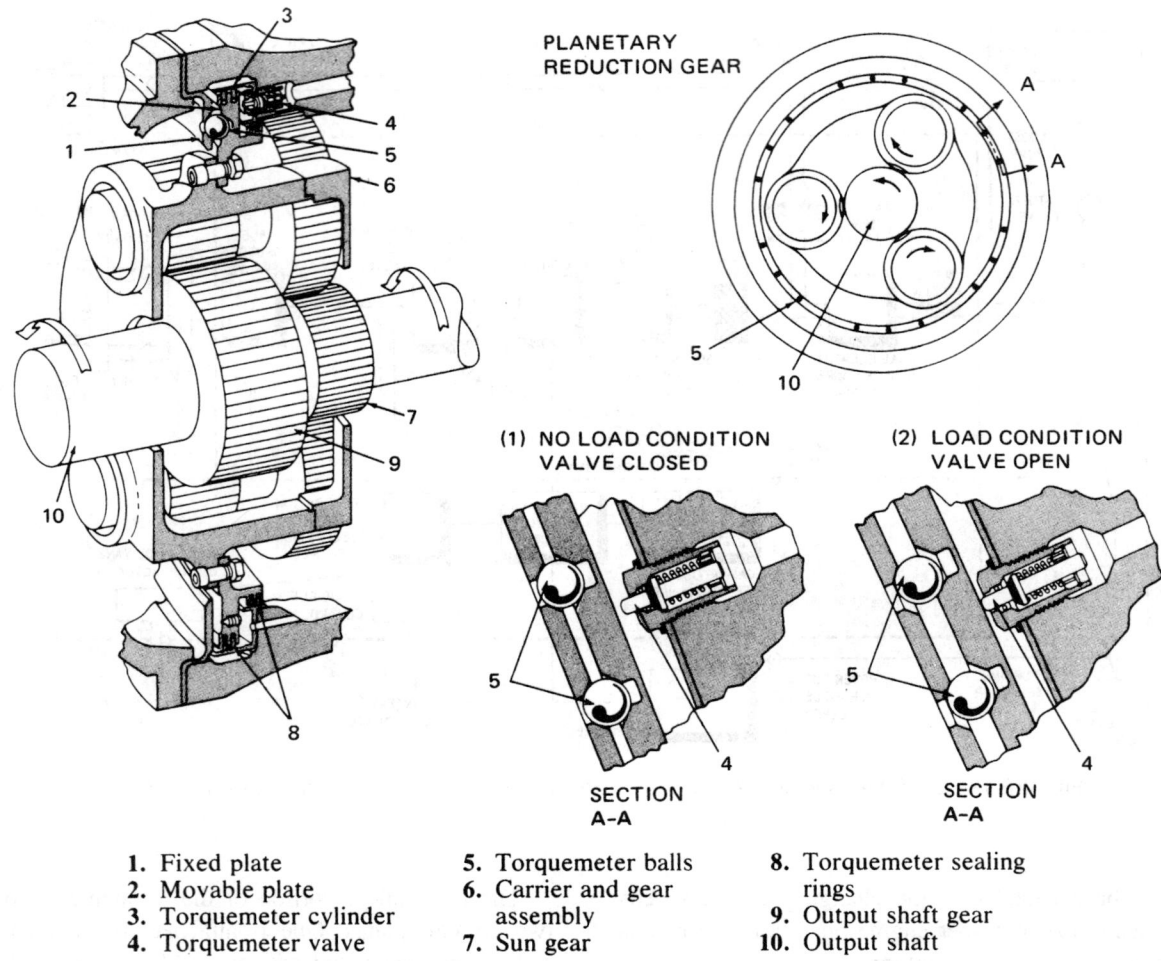

PLANETARY
REDUCTION GEAR

(1) NO LOAD CONDITION
VALVE CLOSED

(2) LOAD CONDITION
VALVE OPEN

SECTION
A–A

SECTION
A–A

1. Fixed plate
2. Movable plate
3. Torquemeter cylinder
4. Torquemeter valve
5. Torquemeter balls
6. Carrier and gear assembly
7. Sun gear
8. Torquemeter sealing rings
9. Output shaft gear
10. Output shaft

FIGURE 18-11 Hydromechanical torquemeter.

to increase and the torquemeter valve opens further, the oil pressure developed in the cylinder exerts pressure against the piston (movable plate), restraining its rearward movement. With the engine operating in a steady-state condition, the cylinder oil pressure and movement of the plate are held in an equalized position, maintaining a constant pressure in the cylinder. This pressure is proportional to engine torque, measured and read in the cockpit.

Main Electric Cable Assembly

The main electric cable assembly, shown in Fig. 18-12, furnishes all necessary interconnecting wiring between the main disconnect plug and the eight branched electric connectors. The eight electric accessories served by this cable are the gas producer tachometer generator, oil temperature bulb, manual fuel regulator solenoid valve, ignition exciter, anti-icing solenoid valve, starting-fuel solenoid valve, chip detector, and power-turbine tachometer generator. The main disconnect plug mates with an electric receptacle of the airframe wiring, establishing electric continuity of the various airframe components.

Ignition Exciter

Ignition of the four igniter plugs is powered by a coil which builds up the originally introduced 28-V primary voltage

to approximately 2500 V, on the same basis as automobile ignition.

Radio-frequency energy is generated within the exciter during normal operation. An inductive capacitive filter has been incorporated at the input to prevent this energy from being fed back onto the 28-V input line. Radio-frequency interference on this line could be detrimental to the operation of other electric accessories. The filter is tuned to radio frequencies and does not offer any appreciable opposition to the flow of 28-V direct current.

THE ALLISON SERIES 250 GAS-TURBINE ENGINE

The Allison Series 250 gas-turbine engine, developed and manufactured by Rolls-Royce, is manufactured in a number of different configurations, including turboshaft and turboprop models. Four of the turboshaft engines are shown in Fig. 18-13.

The Allison 250-C20 engine described in this section is composed of four major sections, or modules. These are the **compressor section**, the **turbine section**, the **combustion section**, and the **accessory gearbox section**. The compressor

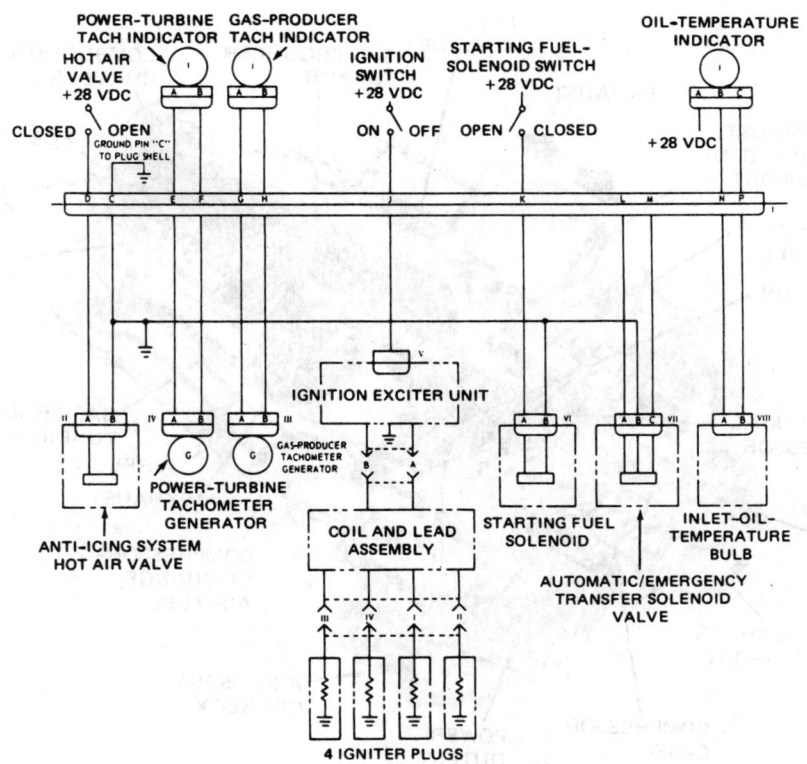

FIGURE 18-12 Schematic diagram of the main electric cable assembly. (*Textron Lycoming.*)

MODEL 250-C20B/F/J

MODEL 250-C28B

MODEL 250-C28C

MODEL 250-C30/P/R

FIGURE 18-13 Various models of the Allison 250 turboshaft engine. (*Rolls-Royce.*)

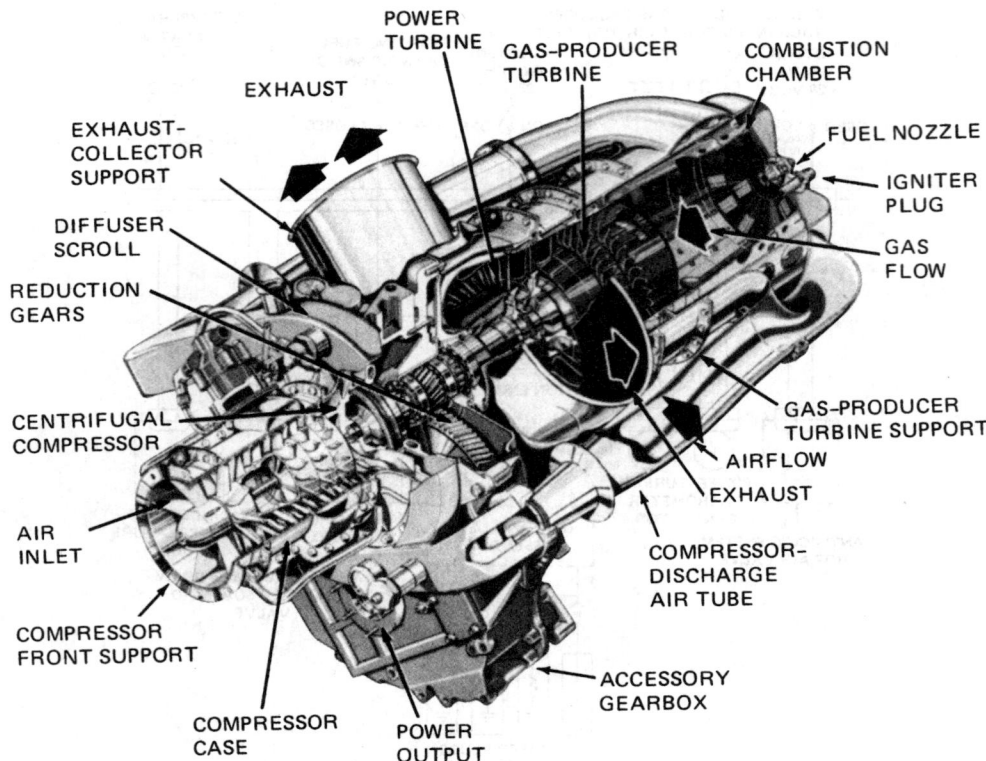

FIGURE 18-14 Cutaway view of the Allison 250-C20 turboshaft engine. (*Rolls-Royce.*)

section is at the front of the engine, the combustion section is at the rear, and the turbine section is near the center. Figure 18-14 is a cutaway drawing that illustrates the construction of the 250-C20.

Performance Data and Specifications

	250-C20	250-C20B
Takeoff power	400 hp [298 kW]	420 hp [313 kW]
Jet thrust	40 lb [178 N]	42 lb [187 N]
Gas producer rpm	52 000	53 000
Ouput shaft rpm	6016 rpm	Same
Weight	155 lb [70.3 kg]	Same
Specific fuel consumption	0.630 lb/shp/h	0.650 lb/shp/h
Length	40.7 in [103.4 cm]	Same
Height	23.2 in [58.9 cm]	Same
Width	19.0 in [48.3 cm]	Same

The foregoing specifications apply only to the engine models shown. Some models of the Series 250 engine can produce more than 700 hp [522 kW].

Compressor Section

The compressor section of the engine under discussion consists of a compressor front support, case assembly, rotor wheels with blades (for axial compressor section), centrifugal impeller, front diffuser assembly, rear diffuser assembly, diffuser-vane assembly, and diffuser scroll. Some models of

the Allison 250 engine do not include the six-stage axial compressor section; in these models all compression is accomplished by means of the centrifugal compressor.

Air enters the engine through the compressor inlet and is compressed by the axial and centrifugal compressor sections. It then passes through the scroll-type diffuser into two external ducts which convey it to the combustion section at the rear of the engine. The compressor is driven directly by the gas producer turbine at speeds of more than 50 000 rpm. The airflow and gas flow through the Allison 250-C20B engine are illustrated in Fig. 18-15.

Combustion Section

The combustion section, illustrated in Fig. 18-16, is located at the rear of the engine and consists of the **outer combustion case** and the **combustion liner**. A fuel nozzle and a spark igniter are mounted in the aft end of the outer combustion case. Compressed air from the two external ducts enters the combustion liner at the aft end through holes in the liner dome and skin. The air is mixed with the fuel sprayed from the fuel nozzle and combustion takes place. The hot gases move forward out of the combustion liner to the first-stage gas producer turbine nozzle.

Turbine Section

The **turbine** consists of a two-stage **gas producer rotor**, a two-stage **power-turbine rotor**, a **gas producer turbine support**, a **power-turbine support**, and a **turbine and exhaust**

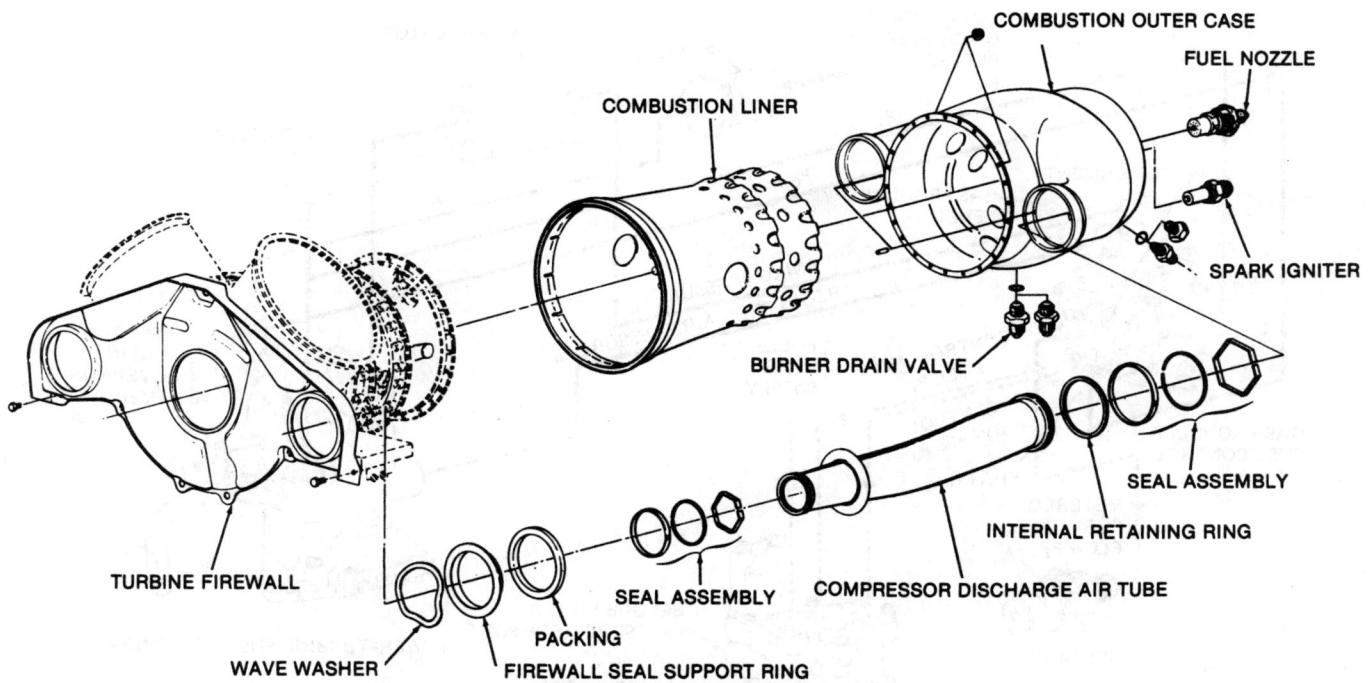

FIGURE 18-15 Airflow and gas flow through the Allison 250-C20B engine. (*Rolls-Royce.*)

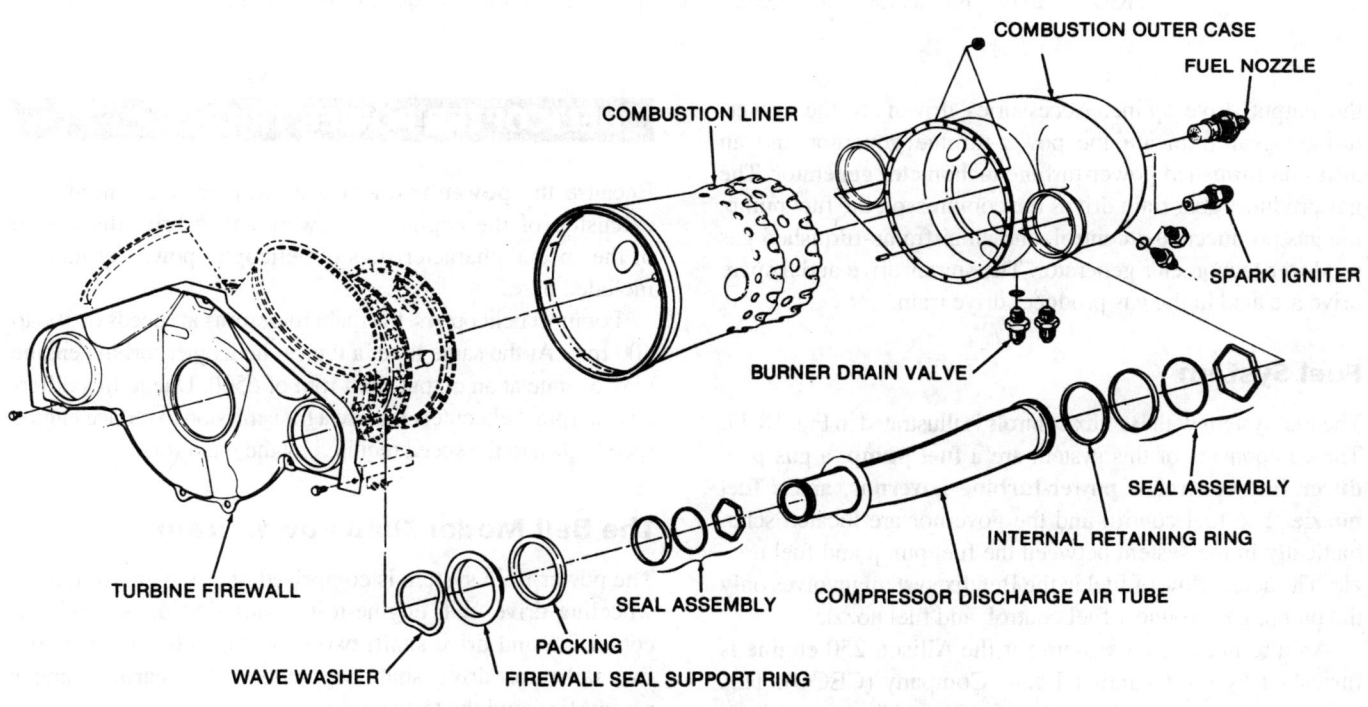

FIGURE 18-16 Allison 250 combustion assembly. (*Rolls-Royce.*)

collector support. The turbine is mounted between the combustion section and the power and accessory gearbox. The two-stage gas producer turbine drives the compressor and the accessory gear train. The power turbine drives the output shaft through the reduction-gear train. The expanded gas, having passed through the turbine stages, discharges in an upward direction through the twin ducts of the turbine and exhaust collector.

Power and Accessory Gearbox

The main power and accessory drive-gear trains are enclosed in a single gear case. This gear case serves as the structural support for the engine, and all engine components, including the engine-mounted accessories, are attached to it. A two-stage helical- and spur-gear set is used to reduce the rotational speed from 33 290 rpm at the power turbine to 6016 rpm at

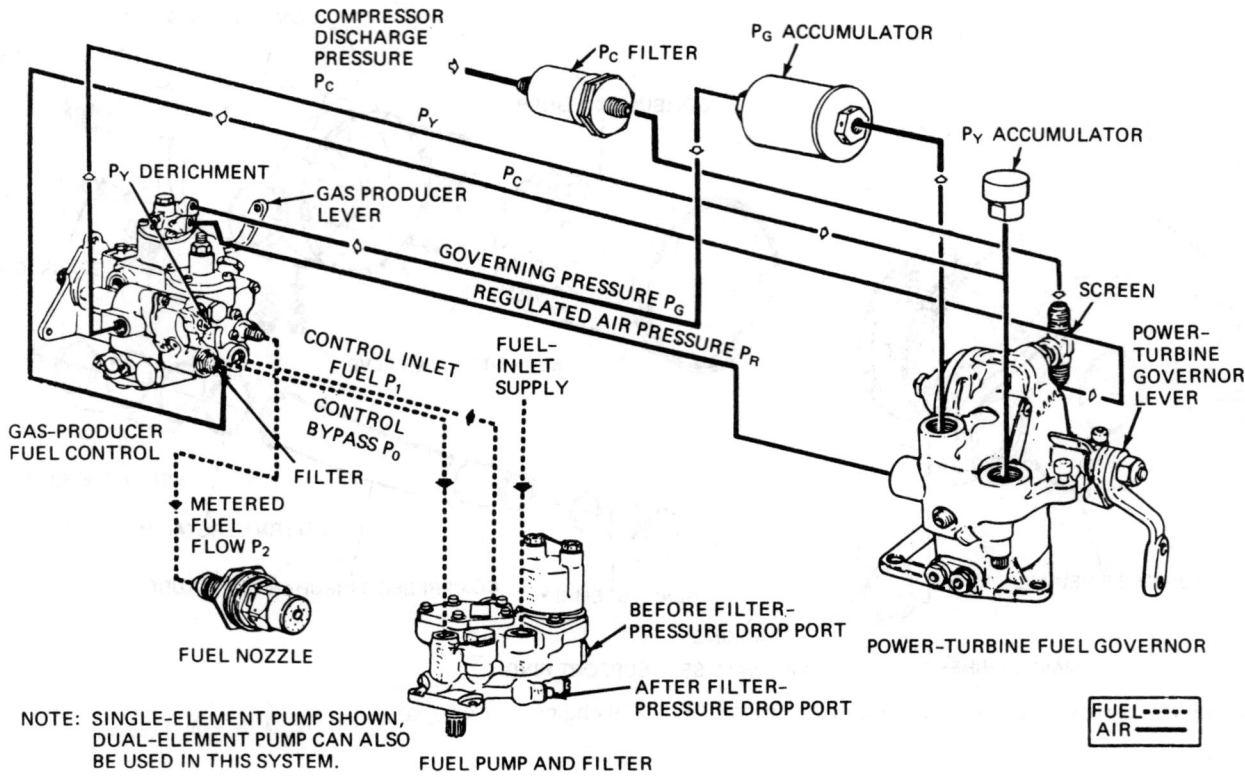

FIGURE 18-17 Bendix fuel control system for the Allison 250-C20 engine. (*Rolls-Royce.*)

the output drive spline. Accessories driven by the power-turbine gear train are the power-turbine governor and an airframe-furnished power-turbine tachometer generator. The gas producer gear train drives the compressor, the fuel pump, the gas producer fuel control, and an airframe-furnished gas producer tachometer generator. The starter drive and a spare drive are also in the gas producer drive train.

Fuel System

The fuel system with Bendix controls is illustrated in Fig. 18-17. The components of this system are a **fuel pump**, a **gas producer fuel control**, a **power-turbine governor**, and a **fuel nozzle**. The fuel control and the governor are located schematically in the system between the fuel pump and fuel nozzle. The actual flow of fuel in the Bendix system involves only the pump, gas producer fuel control, and fuel nozzle.

An alternative fuel system for the Allison 250 engine is furnished by the Chandler Evans Company (CECO). This system uses hydromechanical controls, and fuel enters both the gas producer fuel control and the power-turbine fuel governor.

Lubrication System

The lubrication system for the Allison 250-C20 turbine engine is shown in Fig. 18-18. This is a dry-sump system utilizing a pressure pump and scavenge pumps. Two metal chip detectors are included in the system to aid in discovering wear problems within the engine. The oil tank and the oil cooler are both airframe-furnished.

HELICOPTER POWER TRAINS

Because the **power train** of a helicopter is essentially an extension of the engine, or powerplant, a brief discussion of the special characteristics of helicopter power systems is included here.

For most helicopters, the main rotor turns at speeds of 300 to 400 rpm. At the same time, a typical helicopter turbine engine may operate at an output shaft rpm of 6500. Due to this difference in rpm, helicopters require a transmission to reduce engine speed down to the speed required by the main rotor.

The Bell Model 204B Power Train

The power-train system is comprised of a transmission freewheeling drive unit, engine-to-transmission drive shaft, oil cooler fan and drive shaft, two short-tail-rotor drive shafts, five tail-rotor drive shaft segments with bearing hanger assemblies, and the tail-rotor gearbox.

A portion of the power train for the Bell Model 204B helicopter is shown in Fig. 18-19. This figure shows how the main transmission is installed at the front of the turboshaft engine. In the drive coupling between the engine and the transmission is a **freewheeling** (sprag) **clutch**, which allows the rotor systems to continue to operate even if the engine stops, permitting **autorotational descent** in case of engine failure. The gearing in the transmission includes two planetary systems which have a total gear ratio of 20.37:1; that is, the engine drive shaft turns more than 20 times as fast as the main rotor.

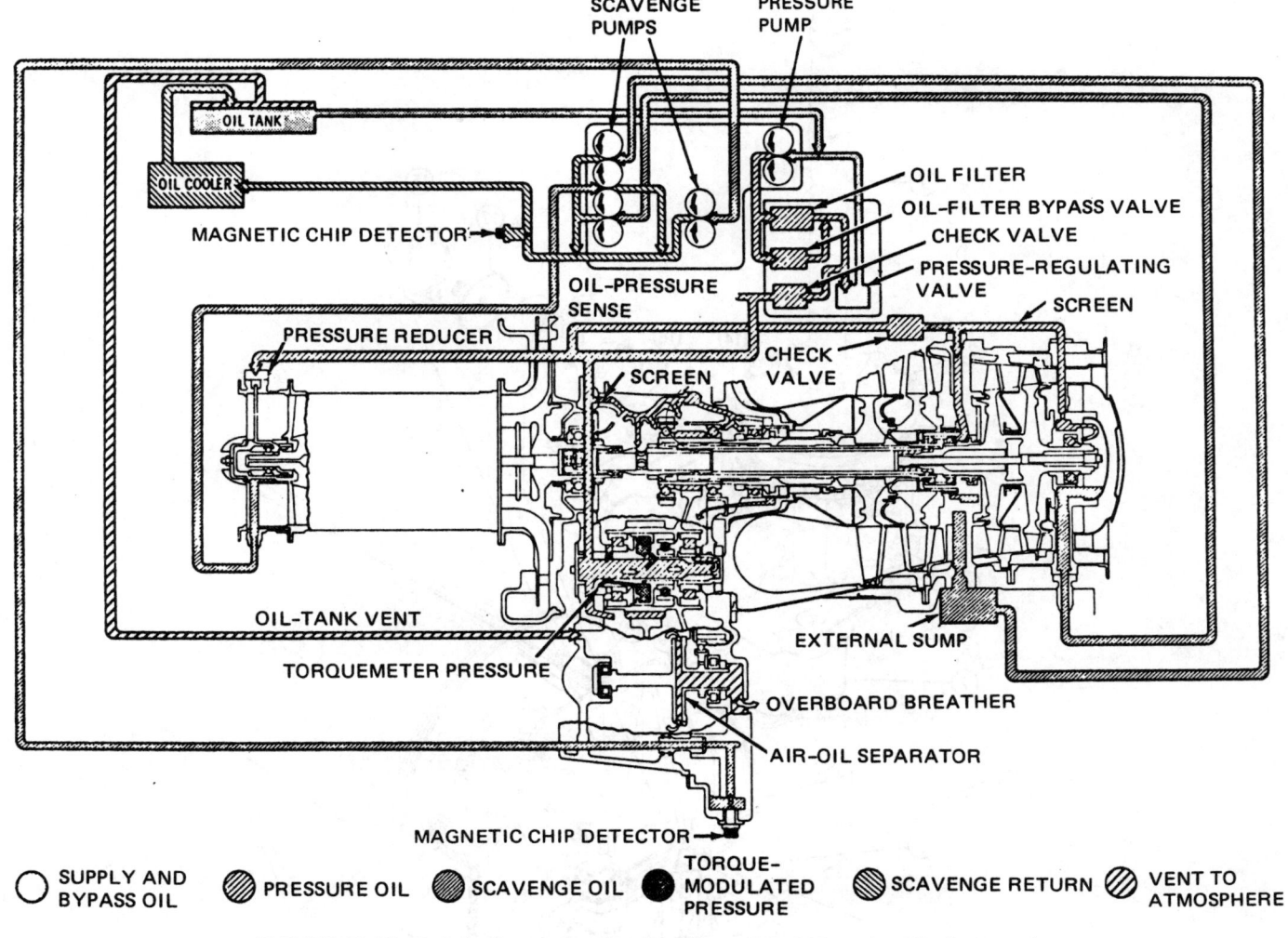

FIGURE 18-18 Lubrication system for the Allison 250-C20 engine. (*Rolls-Royce.*)

The gear reduction to the tail rotor, including the tail-rotor transmission gearing, provides a gear reduction of 4:1. The tail-rotor drive extends to the rear from the accessory drive and sump case, which is below the main transmission.

ALLISON SERIES 250 TURBOSHAFT ENGINE OPERATION IN A HELICOPTER

The rotor speed on a helicopter must be kept within certain limits. If rotor rpm is too high, the resulting centrifugal forces can overstress the rotating parts. If the rotor rpm is too low, excessive rotor blade coning (upward bending) will result. Therefore, the engine control system for helicopter installations must control the power output of the engine such that the rotor rpm (N_R) remains within established limits. The device which allows the engine to drive the rotor but prevents the rotor from driving the engine is generally called a **freewheeling unit**, or **overrunning clutch**.

When the engine is delivering power to the rotor system, the percentage of rotor rpm (N_R) and the percentage of power-turbine rpm (N_2) are the same. N_2 and N_R rpms

are indicated on the same instrument. When the N_2 and N_R percentages are the same, the tachometer indicator N_2 and N_R needles are said to be "locked." "Split needles" describes a condition in which the percentage of N_R is greater than the percentage of N_2; when the needles are split, the engine delivers no power to the helicopter rotor and the helicopter rotor delivers no power to the engine.

Starting

A minimum starter load is always desirable when an engine is being started. Thus, an engine must be able to be cranked without the helicopter rotor imposing any load on the starter. Helicopters powered by reciprocating engines incorporate a clutch system which enables the starter to crank the engine and not crank the rotor. This clutch system also allows for a gradual stress-free pickup of rotor momentum while the system is being engaged. Helicopters powered by the Series 250 engine do not incorporate a clutch system because the free-turbine design permits the starter to crank the gas producer system without any helicopter rotor load on the starter. When a Series 250 engine is started, N_2N_R speed will not begin to increase when the starter cranks the

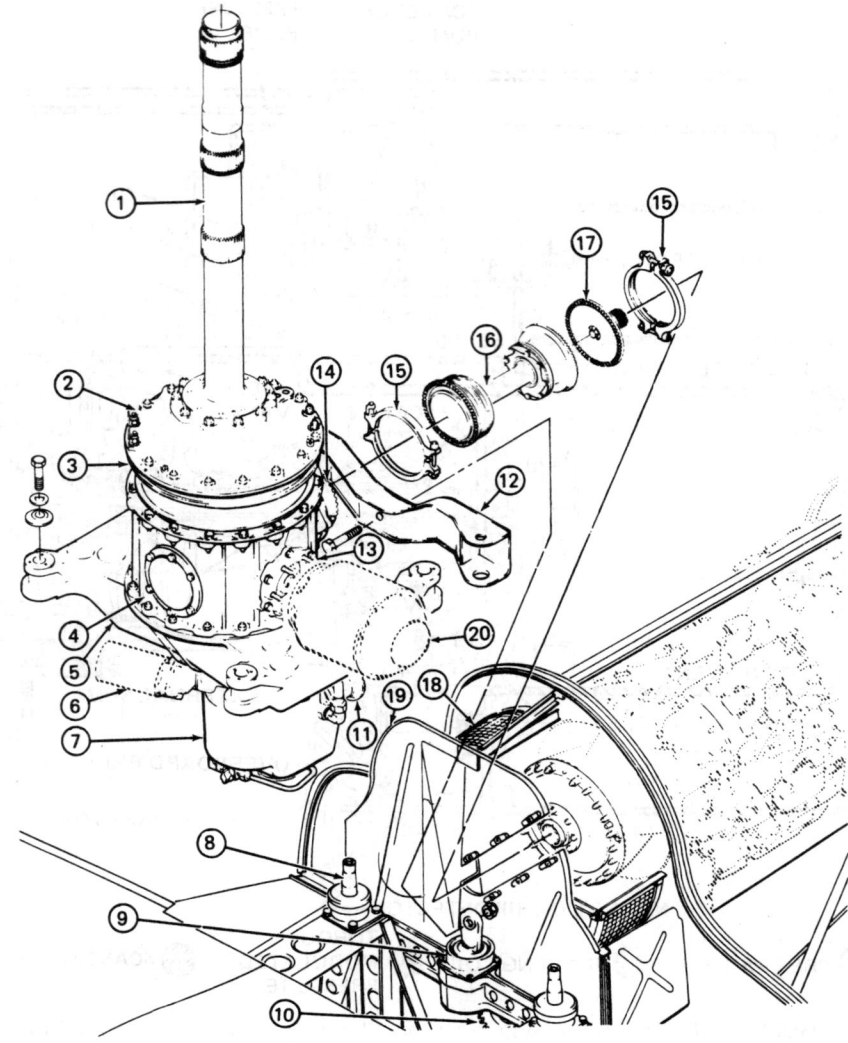

1. Mast
2. Top case
3. Planetary-ring-gear case
4. Main case
5. Support case
6. Hydraulic pump and tachometer drive quill
7. Accessory drive and sump case
8. Pylon main mounts
9. Fifth pylon mount
10. Tail-rotor drive shaft
11. Tail-rotor drive quill
12. Fifth mount support
13. Generator drive quill
14. Input drive quill
15. Coupling clamps
16. Input drive shaft
17. Engine coupling adapter
18. Screen
19. Baffle
20. DC generator

FIGURE 18-19   Main transmission and drive shafts for the Bell Model 204B helicopter. (*Bell Helicopter Co.*)

engine. The speed of N_2N_R gradually increases as N_1 speed increases to idle rpm. Thus, a free turbine allows for stress-free pickup of rotor momentum and permits cranking of an engine with the rotor imposing no load on the starter.

Collective-Pitch Increase

If the engine is running at stabilized ground idle, and takeoff power is required, the operator must move the twist grip from GROUND IDLE to FULL OPEN. When this is done, the gas producer fuel control governor spring is reset from the N_1 ground idle rpm setting to the N_1 overspeed governor setting. This setting results in an increase in N_1 rpm, an increase in

N_2 rpm to 100 percent, and an output of approximately 70 shp [52.2 kW] with the collective-pitch stick at its minimum setting. On free-turbine installations, it is not necessary for the operator to coordinate the twist grip with the collective-pitch stick. As the collective-pitch stick is pulled up, the rotor pitch changes such that the rotor power requirements increase. Thus, rotor rpm will tend to droop. As N_2 droops, the power-turbine governor senses the droop and initiates the necessary action to cause the gas producer fuel control to increase fuel flow. As the fuel flow increases, N_1 rpm increases and expansion through the power turbine increases. Thus, the power turbine develops more power, which is delivered to the rotor system to prevent excessive N_2 rpm droop.

The characteristics of the power-turbine governor are such that as the power requirements of the helicopter rotor system increase, N_2N_R speed tends to decrease, and if the rotor system power requirements decrease, N_2N_R tends to increase. On helicopters, it is highly desirable to vary rotor system power requirements without having a change in N_2N_R rpm, as described in the previous paragraph. Therefore, in order to prevent variation of N_2N_R rpm when a power change is made, the helicopter manufacturer provides a droop compensator which acts on the power-turbine governor in such a way that the N_2N_R rpm holds constant as power to the rotor system is varied. The droop compensator "resets" the power-turbine governor spring during power changes so that the resulting stabilized N_2N_R following a power change is the same as it was before the power change. Thus, when the operator increases collective, the power delivered to the rotor system increases and the stabilized N_2N_R rpm remains the same.

If the operator wishes to operate the engine at a different N_2N_R rpm, the power-turbine governor must be reset by some means other than the droop compensator. For this purpose, the helicopter manufacturer provides an electric beeper system. By means of a manually positioned beeper switch, the beeper system can reset the power-turbine governor so that N_2N_R rpm will be governed at a different speed.

Autorotation

In the event of an engine failure during flight, a helicopter can usually make a safe **autorotational** landing without damage to the helicopter or injury to its passengers. An **autorotation** is a condition of flight in which the helicopter rotor (N_R) speed and the resultant lift are derived entirely from the airflow up through the rotor system. If an engine fails or if a power loss occurs such that powered flight is no longer possible, the pilot must immediately initiate autorotation by moving the collective-pitch lever down to select minimum rotor pitch. As the helicopter descends, the airflow up through the rotor will maintain N_R speed, the overrunning clutch will prevent the rotor system from delivering power to the engine, and the N_2N_R tachometer indicator needles will (should) split. Additional information on helicopter flight controls may be found in the text *Aircraft Basic Science*.

Engine Failure Warning System

The natural pilot response to loss of altitude is to increase the collective pitch. If the loss of altitude is the result of an engine failure and the pilot increases collective pitch, N_2N_R rpm will rapidly decrease and a soft autorotational landing may be impossible.

When a reciprocating engine fails on a helicopter, there is a significant change in sound level, and the pilot is warned of the engine failure by this change. On gas-powered helicopters, however, an engine failure in flight is not easily detected, for there is very little sound-level variation at the time of power loss. For this reason, helicopters powered by turbine engines should be equipped with engine failure warning systems.

REVIEW QUESTIONS

1. What are the principal applications of the turboshaft engine?

2. Describe the function of an APU.

3. At what rotational speed do APUs normally operate?

4. How is power supplied to the Garrett GTCP 331-200ER for starting purposes?

5. Describe airflow through the compressor of the Lycoming T53 engine.

6. Describe the airflow through the combustion chamber of the T53 engine.

7. Explain how the turbine components of the T53 engine are cooled.

8. What materials are used in the construction of the compressor rotor assembly of the T53 engine?

9. What is the function of the interstage bleed system for the T53 engine?

10. What accessories for the T53 engine are driven from the accessory drive gearbox?

11. Describe the operation of the variable-inlet guide-vane system of the T53 engine.

12. Describe the components of the T53 engine lubrication system.

13. Describe the basic design and operation of the torquemeter for the T53 engine.

14. What are the principal modules of the Allison 250 engine?

15. Describe the principal units in the compressor section of the Allison 250 engine.

16. Describe the principal units in the turbine section of the Allison 250 engine.

17. What is the purpose of the metal chip detectors located in the lubrication system?

18. What are the principal units in the Bell 204B power train?

19. What are the principal functions of the main transmission?

20. What is the purpose of a freewheeling unit in a helicopter?

Gas-Turbine Operation, Inspection, Troubleshooting, Maintenance, and Overhaul

19

INTRODUCTION

Because of the great variety of gas-turbine engines, it is not possible to set forth standard procedures which will apply to all such engines. There are, however, certain common characteristics among gas-turbine engines and a number of features which lend themselves to accepted standards of procedure and workmanship. This chapter describes some general practices and provides specific examples of operation, inspection, and maintenance practices.

The most important considerations for the technician responsible for the operation and maintenance of gas-turbine engines are (1) an adherence to principles of good workmanship and attention to detail and (2) a consistent practice of following the instructions provided in the operator's and manufacturer's manuals.

STARTING AND OPERATION

Starting Gas-Turbine Engines

A gas-turbine engine should be started only when all condition required for the safety of the engine and nearby property and personnel are met. The engine pod or nacelle should be checked for loose material, tools, or other items which could be ingested by the engine and cause **foreign-object damage (FOD)** to vanes, blades, and other interior parts. The best type of surface on which to operate gas-turbine engines mounted in aircraft is smooth concrete that is free of all items or material which could be drawn into the engine intake. The aircraft must be positioned so that the exhaust heat and high-velocity gases will not cause damage to other aircraft, ground service equipment, or vehicles and will not cause injury to personnel.

Figure 19-1 shows the temperatures and velocities of the exhaust stream exiting a JT9D engine at both idle and takeoff speeds. Even at idle speed, the temperatures and gas velocities are hazardous. At takeoff power, the engine produces dangerous temperatures and high gas velocities more than 200 ft [60.96 m] to the rear of the jet nozzle. Observe in the drawings that a 25-ft [7.62-m] radius at the front of the engine is considered a danger zone because of the velocity of the air flowing to the engine inlet. This area should be carefully avoided by maintenance personnel and should be kept clean to prevent foreign objects from being drawn into the engine's intake. There have been many reported instances of inlet ingestion of personnel working around operating engines, demonstrating the serious consequences of violating the hazard zones delineated in most maintenance manuals. The inlet airflow velocity is low at the hazard boundary, but it rapidly increases to the point of no recovery as the inlet is approached. Personnel should not approach the engine's inlet by moving forward alongside the engine's cowling. The airflow that moves forward along the sides of the nacelle will increase greatly near the engine's inlet.

The principles involved in the starting of a gas-turbine engine are relatively simple. It is merely necessary to rotate the engine at a speed sufficient to provide adequate air volume and velocity for starting, provide high-intensity ignition in the combustion chamber, and introduce fuel through the fuel nozzles in an amount that will not produce excessive heat but will provide sustained combustion and further acceleration of the engine.

Small gas-turbine engines are often equipped with starter-generators. These units are electric motors which apply starting torque for the engine during starting; then, when the engine rpm attains a self-sustaining level, the starter motor becomes a generator to supply the aircraft electrical system. The starters draw a high level of current when starting and therefore heat up very rapidly; they are thus subject to damage by overheating if their operation is not limited to the time intervals specified.

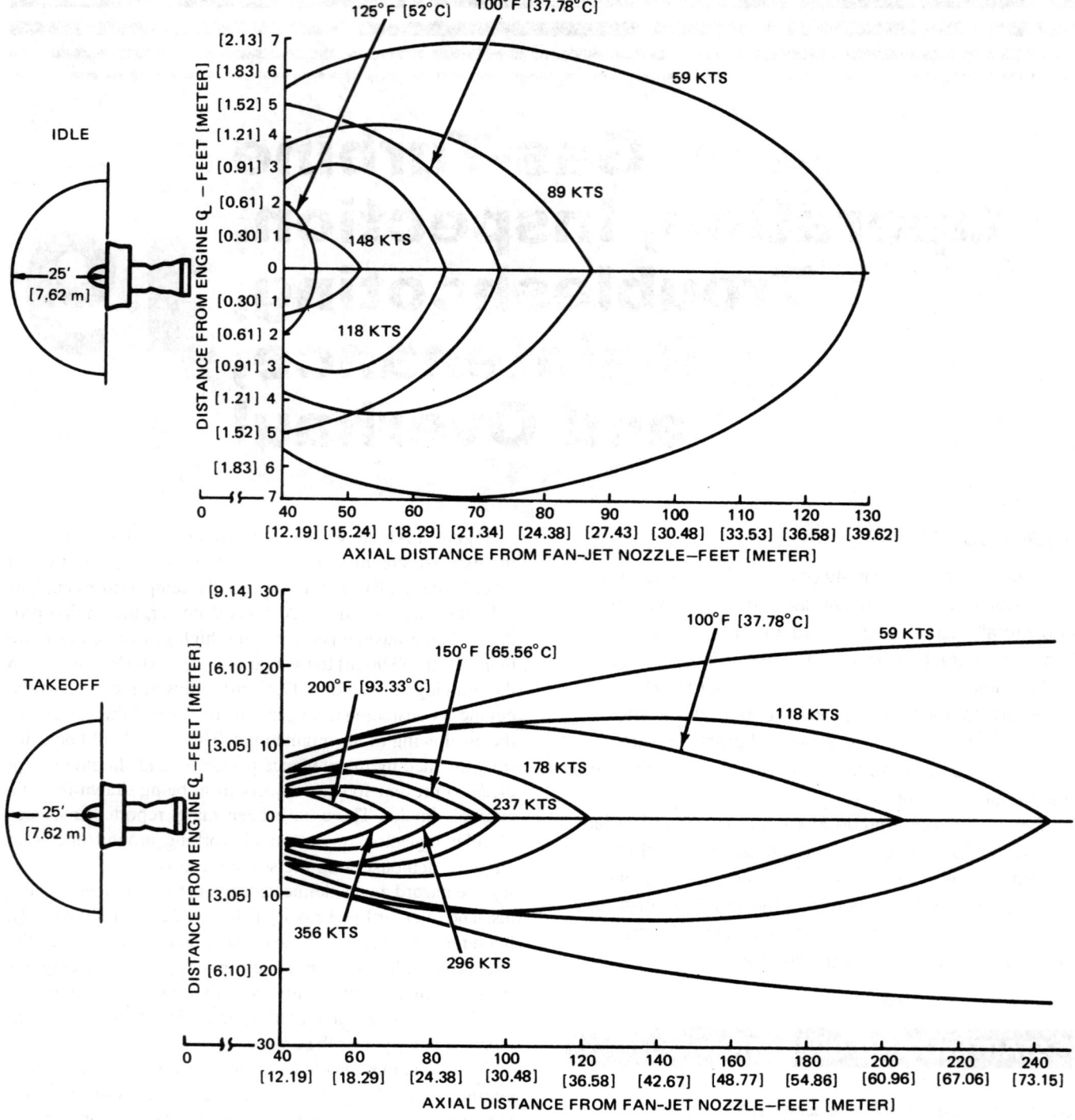

FIGURE 19-1 Temperatures and velocities of exhaust gases exiting an operating JT9D engine.

Large gas-turbines engines are generally equipped with air-turbine starters which receive their air supply from a ground service unit (air cart), another engine, or an auxiliary power unit (APU) installed in the aircraft.

The starting of gas-turbine engines can be manual or automatic, depending on the particular engine installation. The majority of starts for today's engines are accomplished automatically.

The manual start of a gas-turbine engine usually includes the following steps:

1. Connect the starting power unit to the aircraft, if used.
2. Provide for electrical power on the aircraft.
3. Turn on the fuel and fuel boost pumps.
4. Turn on or arm the ignition circuit. The ignition system is often energized by a switch connected to the start lever or power lever.
5. Check the starting system to ensure that there is enough starting energy for a complete start cycle. Engage the starter and accelerate the engine until the high-pressure compressor (N_2) or gas generator speed is between 10 and 25 percent rpm.

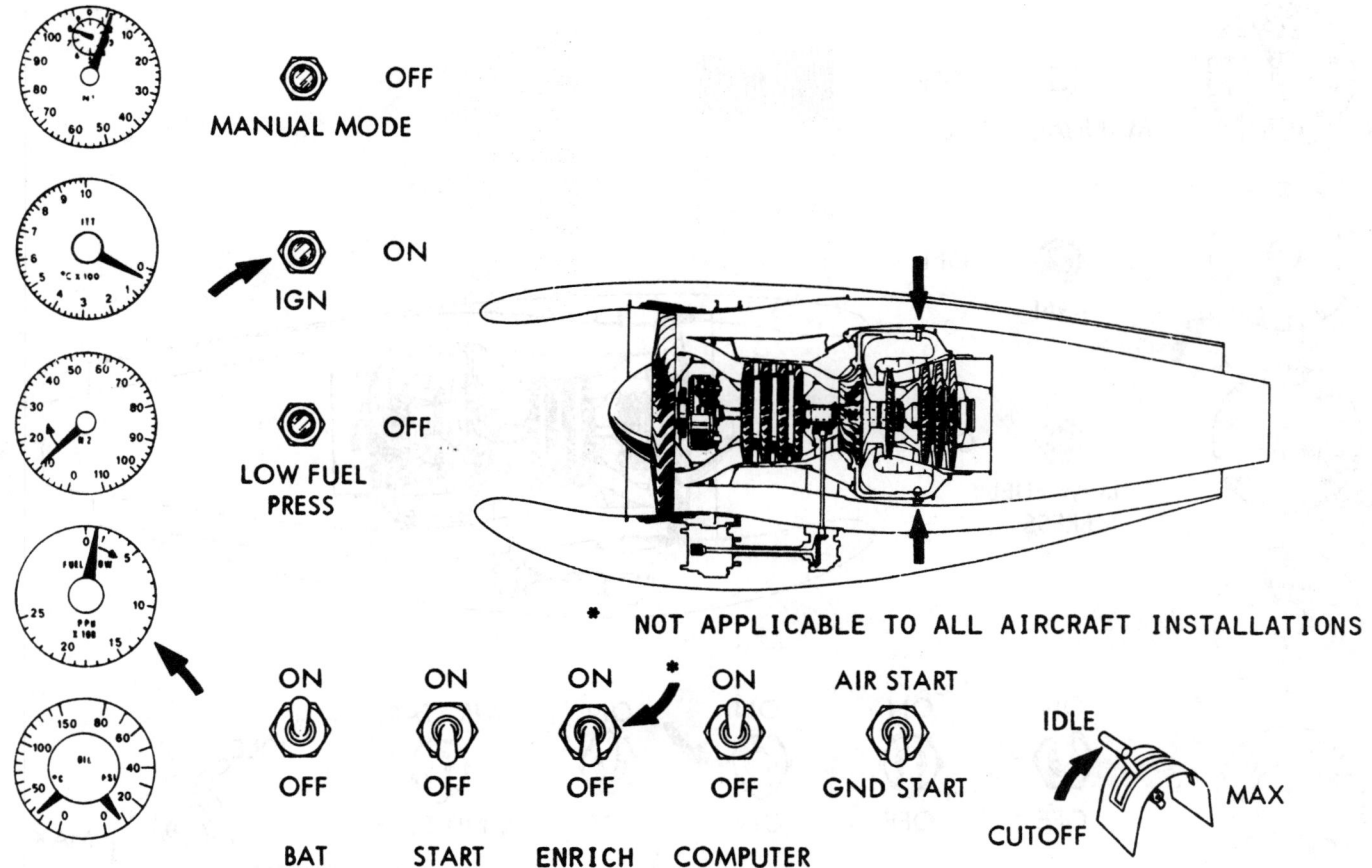

FIGURE 19-2 Engine accelerated to 10 percent N₂ rpm for starting. (*Honeywell Inc.*)

This speed will vary among different engines. In this example, the speed is 10 percent rpm, as shown in Fig. 19-2.

6. Advance the power lever or start lever depending on the type of engine being started. On some engines, the power lever is set in the IDLE position for starting and the start lever controls fuel flow to the engine. Note the fuel flow gauge and check for a stabilized fuel flow. Now watch the interstage turbine temperature (ITT) or exhaust-gas temperature (EGT) indicator until a rapid temperature rise is indicated, as shown in Fig. 19-3. This signifies a lightoff (fuel burning in the combustion section). When a start lever is used, it is generally moved to the START or IDLE position in one movement. If the power lever is used for starting, do not advance the power lever further until the temperature stabilizes. *The temperature must not be permitted to exceed the maximum allowable specified for the engine.* If the temperature climbs too high, the fuel flow must be shut off to prevent a **hot start condition.** A hot start can burn and destroy the engine's internal parts. If the EGT rises above the specified amount, the fuel flow should be stopped and the engine turned to draw cool air through it. If no rise in EGT is detected, then the condition is a **no-lightoff condition.** This is generally caused by a faulty ignition system or an incorrect fuel control start fuel-air mixture.

7. If lightoff has taken place and the engine's EGT has stabilized, the EGT will begin to decrease as engine rpm increases. Advance the power lever until the engine rpm reaches idle

speed, but do not permit the temperature to exceed the level specified.

8. The starter should be released after the operator is certain that the engine has reached a self-sustaining speed, as shown in Fig. 19-4. The starter cutout can be automatic or manual depending on the aircraft being started. It is important that the starter is cut out at the proper percent rpm to prevent overspeeding and damaging of the starter.

To start a gas-turbine engine with an automatic system, the operator must follow the procedure established for the particular aircraft and engine. An automatic start may be accomplished as follows, with some variations depending on the aircraft and engine combination:

1. See that an air supply is available from a ground service unit, an onboard APU, or an operating engine on the aircraft.

2. Turn on the aircraft electric power.

3. Turn on the fuel and fuel pump.

4. Place the power lever in the IDLE position.

5. Rotate the engine selector switch to the position for starting the engine selected.

6. Press the starter switch. It is normally equipped with a holding coil so that it will remain closed until the engine is rotating at a self-sustaining speed. At this point, current to the holding coil is cut off and the switch opens.

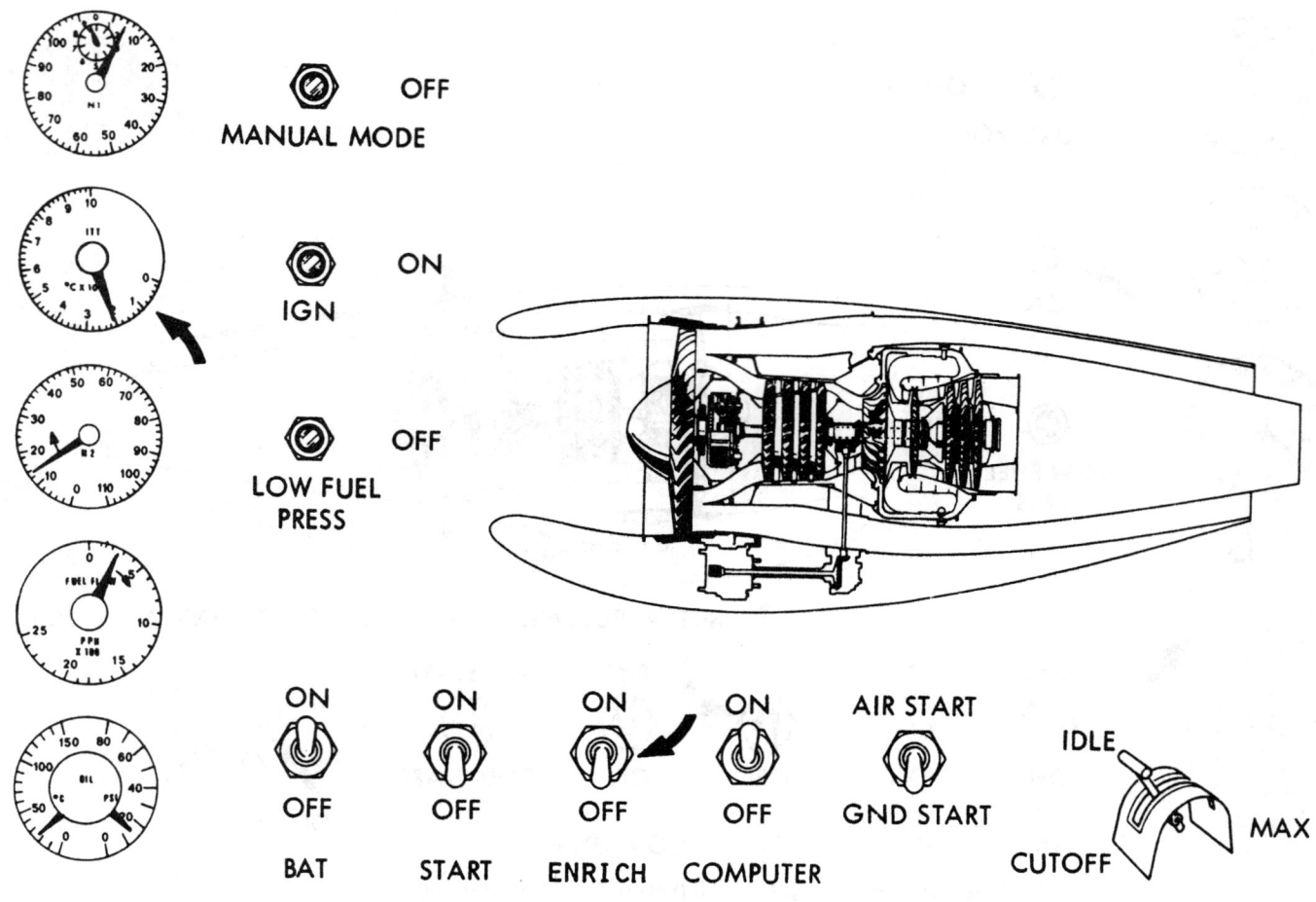

FIGURE 19-3 Engine lightoff indicated by a rise in interstage turbine temperature. (*Honeywell Inc.*)

The engine accelerates when it reaches a level of N_2 rpm. Somewhat above 10 percent of maximum, fuel will be supplied under pressure to the fuel nozzles. The ignition is turned on automatically at the time that the starter switch is depressed or shortly thereafter, but it is always on before fuel flow is supplied. Soon after fuel flow begins (within 20 s), fuel is ignited and the engine accelerates at an increased rate. Fuel flow is controlled automatically by the fuel control unit and is not permitted to exceed the amount required for the correct rate of acceleration. Normally an engine will not have a hot start or a hung start unless there is a defect in the starting system or fuel control.

When the engine has accelerated to 35 percent or more of maximum rated rpm, a centrifugal switch cuts off electric power to the holding coil of the start switch, and the switch opens. The engine accelerates to idle speed and remains in that condition until the power lever is moved by the operator.

It is important that the operator of a gas-turbine engine observe the instruments pertaining to engine operation while performing the starting procedure. A fuel pressure gauge should indicate correct fuel pressure before the engine is started. The EGT gauge should be watched closely at lightoff to ensure that the temperature has not exceeded the maximum value allowed. Oil pressure should register

on the oil pressure gauge shortly after the engine starts to rotate. If the engine does not accelerate properly or if the EGT exceeds the safe limit, the power-control lever should be retarded to cut off the fuel and stop the engine.

If, during the starting of a gas-turbine engine, the EGT exceeds the prescribed safe limit, the engine is said to have had a **hot start**. When this occurs, the engine fuel flow should be shut down immediately. Continue to rotate the engine with the starter to provide some engine cooling. Hot starts are usually caused by an excess of fuel entering the combustion chamber (malfunctioning fuel control) or insufficient starting power (low battery or air pressure). In an engine with an automatic fuel control unit, hot starts normally do not occur unless the unit is malfunctioning.

If the engine fails to accelerate properly or does not reach the idle rpm position, the starting attempt is called a **false start**, or a **hung start**. The automatic systems used on modern engines prevent this type of problem from occurring, as long as the fuel control unit is functioning properly and sufficient energy is supplied to the starter. Either a hung start or a hot start can be caused by an attempt to ignite the fuel before the engine has been accelerated sufficiently by the starter. Another type of problem is not having sufficient energy to turn the starter to complete the starting process.

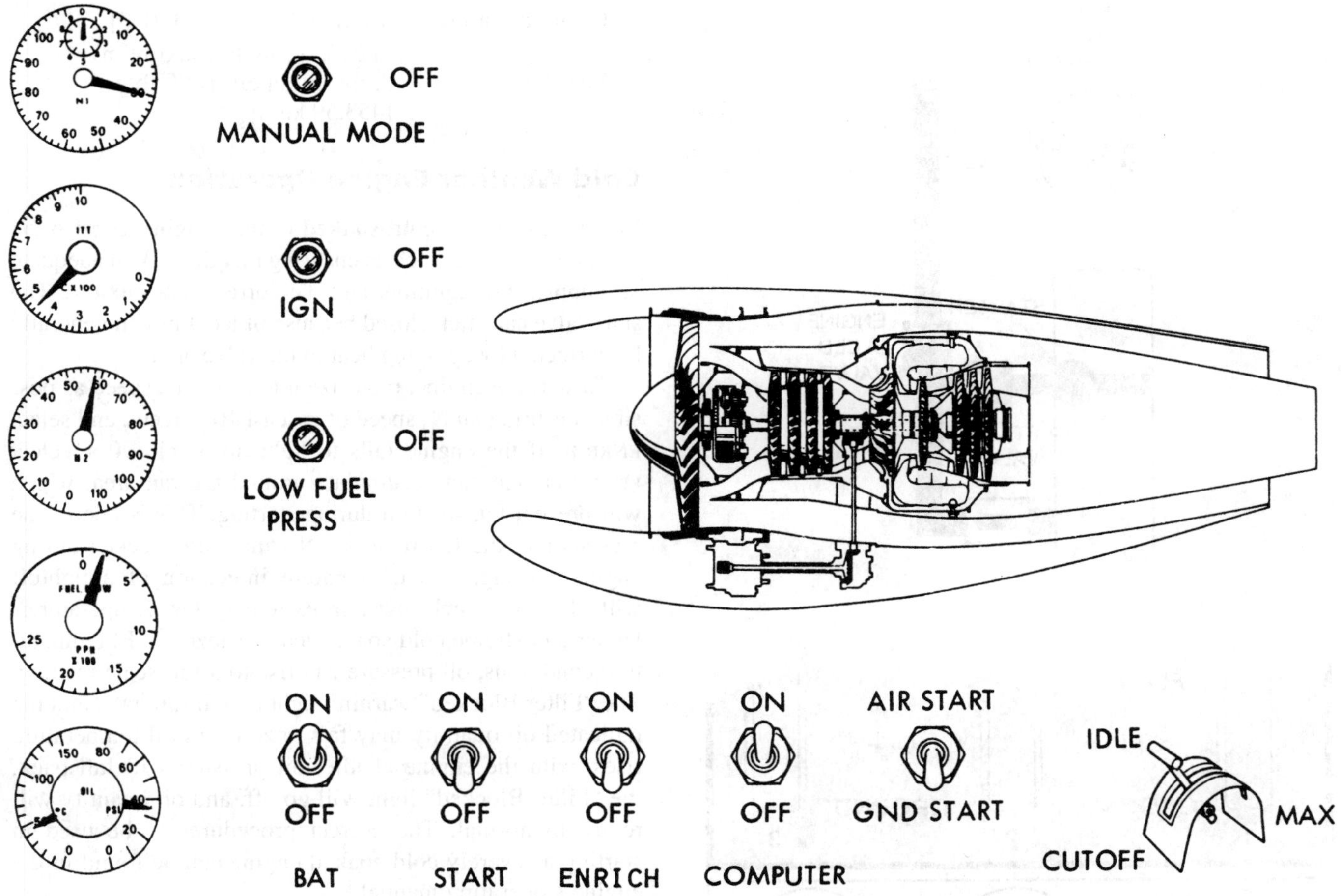

FIGURE 19-4 Engine at idle with N_1 and N_2 rpm stabilized. (*Honeywell Inc.*)

Typical Airline Start Procedure

Technicians and flight-crew members employed by an airline are given specific instructions and checklists which they must follow before and during an engine start. The procedures to be followed in the starting of a JT8D engine on a 727 aircraft are as follows:

1. Perform a *walk-around* inspection according to the appropriate checklist.

2. Perform a control cabin prestart check as set forth in the checklist.

3. Start the APU. (An APU control panel is shown in Fig. 19-5.)

    a. Turn essential bus-selector switch to APU.

    b. Switch ac meter to APU.

    c. Arm automatic fire shutdown switch.

    d. Reset APU fire detection system.

    e. Turn master switch to ON.

    f. After 10 s, turn master switch to START and release to ON position.

    g. Observe for lightoff and EGT max.

    h. Check APU frequency and voltage (400 Hz, 115 V/ac) and close APU generator breaker to power the aircraft.

4. Obtain "All Clear" signal.

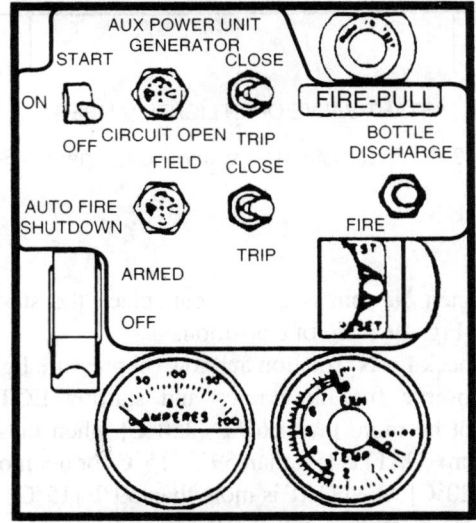

FIGURE 19-5 A typical APU control panel. (*Boeing.*)

5. Check pneumatic pressure (35 psi [241.33 kPa] minimum).

6. Set engine start switch to GROUND START, as shown in Fig. 19-6. This will allow the start valve to be opened and pneumatic pressure to turn the air-turbine starter which turns the N_2 compressor. The starter will stay engaged until it is released, later in the starting process.

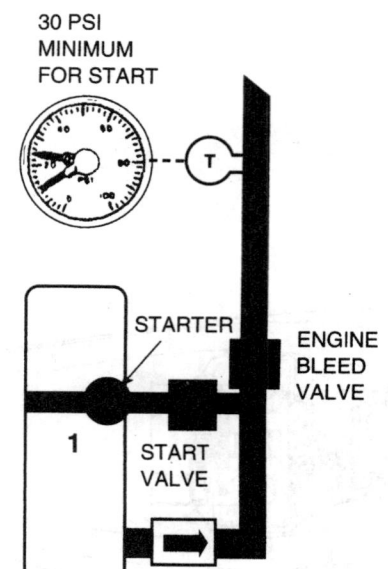

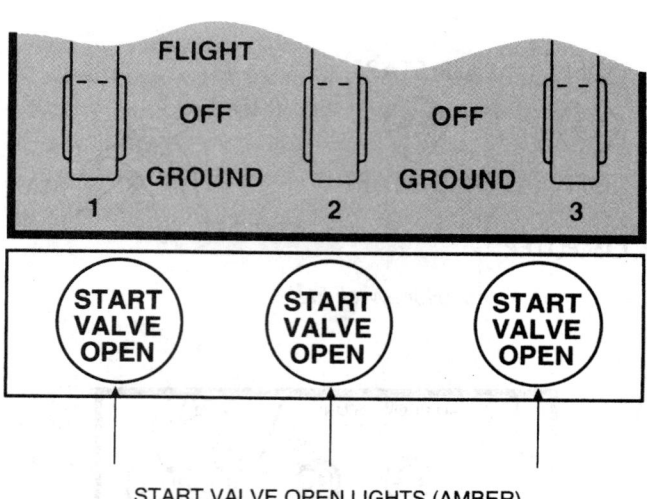

FIGURE 19-6 Pneumatic start system components. (*Boeing.*)

7. When N_2 rpm is 20 percent, place the start lever, shown in Fig. 19-7, in IDLE position.

8. Check for N_1 rotation and low oil pressure light out.

9. Observe fuel flow rates and starting EGT. EGT should not be more than 662°F [350°C] when outside air temperature (OAT) is less than 59°F [15°C] or not more than 788°F [420°C] when OAT is more than 59°F [15°C].

10. Release start switch at 40 percent N_2 speed.

11. Check the following engine parameters after idle speed stabilizes:

EGT	300 to 420°C [572 to 788°F] when air bleed or power extraction is used
N_2 rpm	54 to 59.4 percent
Oil pressure	40 to 55 psi [275.8 to 379.23 kPa] (44 to 46 psi [303.38 to 317.17 kPa] desired)

Oil temperature	40 to 60°C [104 to 140°F] (120°C [248°F] maximum)
Fuel flow	Approximately 1000 lb/h [453.59 kg/h]

Cold Weather Engine Operation

In the start-up of a **cold-soaked** turbine engine, some basic steps are essential, such as ensuring that there is an adequate air supply, dual ignition, and the correct fuel mixture. The start valve can stick closed because of ice; this can generally be corrected by applying heat to the valve area.

To start the engine, motorize it to as high a speed as possible, ensuring an N_2 speed of at least 10 percent, and select ENRICH. If the engine fails to light up within 90 s, select FUEL OFF. Ice can accumulate around the fan area, which will prevent fan rotation during starting. This is one of the reasons that the fan rotation (N_1) must be checked during engine starting. The temperature indication of a lightoff will likely be much lower in extremely low temperatures. Under an extreme cold soak, such as subzero (°F) temperature conditions, oil pressure can rise to a full-scale reading. The "Filter Blocked" warning light may illuminate, and the indicated oil quantity may fall to zero. As oil temperature rises, with the engine at idle, oil pressure will fall back, the "Filter Blocked" light will go off, and oil quantity will return to normal. The correct procedures to be used in starting a severely cold-soaked engine can be found in the engine's operating manual.

GAS-TURBINE ENGINE INSPECTIONS

Because of the great variety of gas-turbine engines in existence, no attempt will be made in this text to give instructions about inspection and maintenance procedures for specific engines. We shall, however, examine some of the conditions common to the majority of gas-turbine engines and provide information supplied by manufacturers and operators as examples of typical methods and processes. Remember that a particular method or process approved for one type of engine may not be satisfactory for another type of engine. *It is essential, therefore, that all inspection and maintenance practices be done in accordance with the manufacturer's and operator's maintenance manuals.*

The inspections established for gas-turbine engines fall into a number of classifications and are dependent on the types of operation to which the engines are subjected. For example, an airline whose routes are long will need an inspection schedule different from a local-service airline for which flights are short and takeoffs and landings are frequent. Inspections and maintenance procedures are scheduled, therefore, with consideration of the number of flight cycles an engine has experienced as well as the total hours of engine operation. A **flight cycle** is normally defined as one takeoff and landing.

To illustrate the difference in the number of flight cycles which may be imposed on an engine in different types of

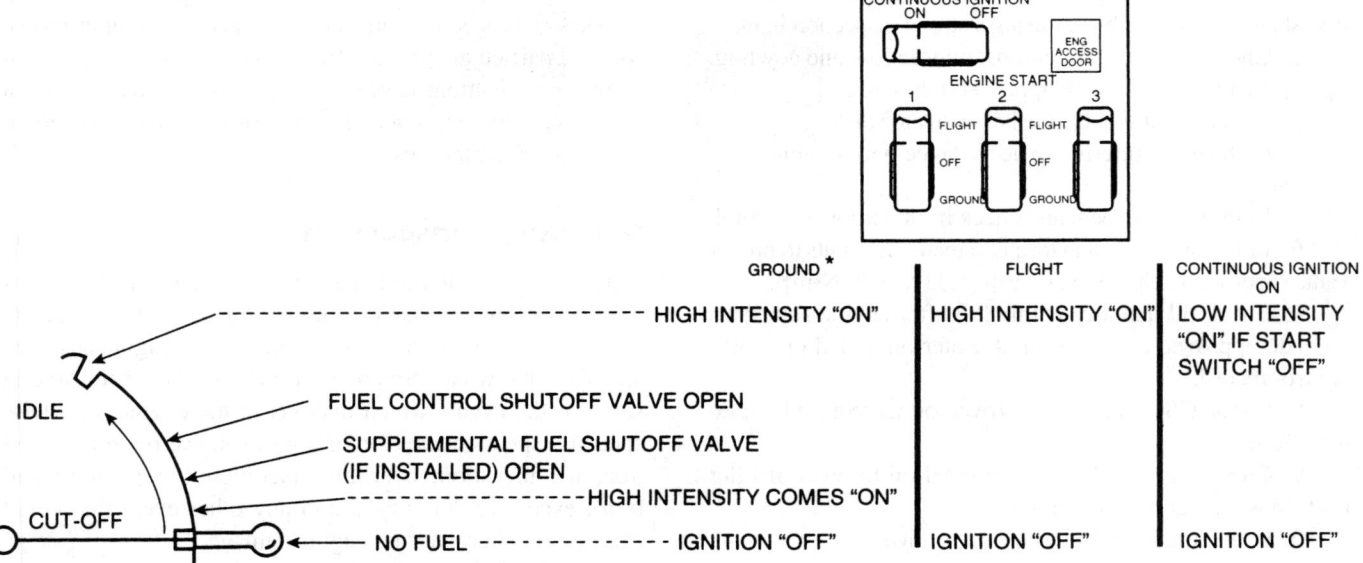

FIGURE 19-7 Engine start levers and switches. (*Boeing.*)

operation, a commuter operation from Los Angeles to San Francisco may be compared with an overseas flight to the Orient. A commuter airplane may accumulate 12 h of operation time while completing 15 flight cycles. On the other hand, a trans-Pacific flight to Hong Kong may put 18 h of operation time on the engines while only three flight cycles are completed. It is clear that the wear, erosion, and heat damage will be much greater on the commuter airplane engines than it will for the trans-Pacific engines. Accordingly, inspection and maintenance operations will have to be scheduled more often for the commuter airplane engines.

Periodic inspections are required after a given number of operation hours, flight cycles, or a combination of both. These inspections may be classified as routine, minor, or major. Scheduling of such inspections is established by the operator of the aircraft in accordance with the results of operational experience.

In addition to the periodic inspections performed on a regular basis, airlines often specify turnaround inspections, which may be called "line checks" or "A checks." Each airline has its own classifications for inspections; the only way for a person to know what is to be inspected and how it is to be inspected is to get the information from the check sheet furnished by the company.

Routine Operational Inspections

A typical airline may designate standard service operations and inspections by such names as "no. 1 service," "no. 2 service," "A" check, and "B" check. These various operations will include a number of standard operations plus special operations as needed.

A no. 1 service may be performed by station personnel each time the airplane lands or after several landings, depending on

the amount of time the aircraft is in flight. Usually the service will include correction of critical log items as well as regular service (fuel and resupply), plus a **walk-around** inspection. The walk-around inspection includes inspection of all items which can be observed from the ground. The engine inspection at this time includes a look at the engine inlet and fan, observation of any fuel or oil leakage from engine pods, and an examination of the tailpipe and turbine section with a flashlight.

The no. 2 service may include the following engine-related items:

1. Review of the flight log and cabin log
2. Check of engine oil quantity
3. Visual inspection of the engines with cowls open

The "A" check discussed here is performed after approximately 100 h of operation. Inspections and service relating to the engines are as follows:

1. Fill oil tanks. Enter in the inspection records the number of quarts added for each engine.
2. Service the constant-speed drive (CSD) as required.
3. Check engine inlet, cowling, and pylon for damage. Check for irregularities and exterior leakage.
4. Inspect the engine exhaust section for damage using a strong inspection light. Note condition of rear turbine.
5. Check the thrust-reverser ejectors and reverser buckets for security and damage.
6. Check the reverser system, with ejectors extended, for cracks, buckling, and damage.

The "B" check is more comprehensive than the "A" check; it includes the following:

Gas-Turbine Engine Inspections   **547**

1. Check engine nose cowl, inlet chamber, guide vanes, and first-stage compressor blades using a strong inspection light.

2. Check engine, installations, midsection, and cowling. Spray cowling latches with approved lubricant.

3. Check the fire extinguisher indicator disks.

4. Perform oil filtering in accordance with maintenance manual.

5. Remove oil screen, and check it for carbon and metal.

6. Install oil screen and torque screen-cover nuts to proper value (approximately 25 to 30 in•lb [2.83 to 3.39 N•m]).

7. Check oil quantity within 2 h after engine shutdown and add approved oil as required. Enter oil added on work-control record.

8. Check CSD oil. Add approved oil as required but do not overfill.

9. Check starter oil. Add approved oil to level of filler port. Make a record of oil added.

10. Check the ignition system as follows:

    a. Move four air-bleed switches on the air-conditioning panel to OFF.

    b. Move start lever to IDLE.

    c. Move start-control switch on overhead panel to FLIGHT position.

    d. See that the igniter at the no. 7 combustion chamber is firing.

    e. Return controls to the OFF position.

    f. Move start-control switch to GROUND position.

    g. Move start lever to the START position.

    h. Ensure that the igniters are firing by use of an approved tester.

    i. Return controls to the OFF position.

    j. Close engine cowling and check security of latches and inspection plates.

In checking the ignition systems of a modern gas-turbine engine, it is important that body contact not be made with the high-energy output. The voltage is such that a flow of current through the body could be fatal.

11. Check reversers and deflector doors as follows:

    a. Place reverser in reverse-thrust position. Install lock clamps and warning tags on reverse levers.

    b. Using a strong inspection light, check the tailpipe and fairing, the reverser clamshells, the turbine exit area, outlet guide vanes, and rear turbine blades.

    c. Check deflector doors and fittings for cracks.

    d. Check for delamination of the door inner and outer skin using an inspection light and testing by hand.

    e. Check the inner and outer skin for dents, cracks, and punctures.

    f. Check the deflector door forward link to the support pivot bolt for tightness. The bolt should not turn by hand.

    g. Check the bolts which secure the forward link support assembly to the reverser structure with a wrench.

    h. Check the deflector drive pivot bolt. It should not turn by hand.

    i. Check the deflector doorstops for excessive looseness and lubricate door-link and rod end bearings.

These notes are given only as examples of common inspection procedures and are not necessarily appropriate for any particular aircraft. Remember that all aircraft and engine combinations have specific procedures that have been established and approved. The approved procedures should be followed in all cases.

Nonroutine Inspections

During the operation of a gas-turbine engine, various events may occur which cause the engine to require an immediate special inspection to determine whether the engine has been damaged and what corrective actions must be taken. Among some of the events which may cause the engine to require special inspections are foreign-object ingestion, bird ingestion, ice ingestion, overlimit operation (temperature and rpm), excessive "G" loads, and any other event that could cause internal or external engine damage.

Nonroutine inspections require the same techniques as those used for daily and periodic inspections. These techniques include unaided visual inspection, inspection with lights, use of magnifiers, application of fluorescent or dye penetrants, use of a borescope or videoscope, and use of radiography techniques. Usually the maintenance manual for the engine will specify which technique is the most effective for a particular inspection.

Borescope

Fiberscope, and Electronic Imaging

The **borescope** was used for many years as a device for examining the insides of cylinder bores on reciprocating engines and is now extensively used on turbine engines.

The borescope is a rigid instrument that may be compared with a small periscope. At one end is an eyepiece with one or more lenses attached to the light-carrying tube. At the end of the tube are a mirror, a lens, and strong light. The tube is inserted through engine **borescope ports** located in the engine case at points necessary to allow for examination of all critical areas inside the engine. The ports are normally closed (with removable plugs). One type of borescope system is shown in Fig. 19-8.

When borescope inspections are to be performed, the technician should identify the plugs as they are removed to be sure that they are reinstalled in the same ports. Upon reinstallation, the threads and pressure faces of the plugs should be lightly coated with an antiseize compound such as MIL-T-5544 or, its equivalent.

A variation of the rigid borescope is the **fiberscope**, which is shown in Fig. 19-9. The flexible fiberscope usually has a controllable bending section near the tip so that the observer can direct the scope after it has been inserted into an engine inspection port. This bending action, illustrated in Fig. 19-10, allows the fiberscope to scan the area inside the engine once inside the port. Many times it is necessary to use the fiberscope to inspect around corners inside the engine when no inspection entry port is available to allow a direct line of sight.

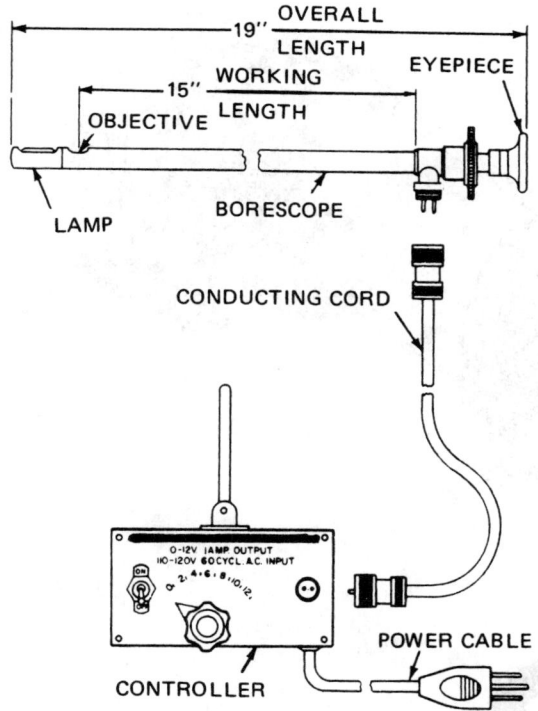

FIGURE 19-8 A borescope system.

The system uses a tiny charged-coupled device (CCD) sensor in the tip of the probe. The solid-state CCD sensor acts like a miniature TV camera to transmit the image electronically to a video monitor. First, light is transmitted to the inspection area, either by light-emitting diodes (LEDs) or by fiber-optic light guides, depending on the inspection probe selected. A fixed-focus lens in the tip of the probe gathers reflected light from the area and directs it to the surface of the CCD sensor. The signal then travels down the length of the probe through amplifiers. The video processor receives the signal, digitizes it, assembles it, and outputs it directly to a video monitor, videotape recorder, or computer enhancement equipment.

Video imaging lends itself to high-quality videotape and photographic documentation. The images can be viewed by several inspectors at different locations via multiple video monitors.

Foreign-Object Damage (FOD)

Foreign-object damage to a gas-turbine engine may consist of anything from small nicks and scratches to complete disablement or destruction of the engine. The flight crew of an aircraft may or may not be aware that FOD has occurred during a flight. If damage is substantial, however, it will be indicated by vibration and by changes in the engine's normal operating parameters. Damage to the compressors or turbines usually results in an increase in EGT, a decrease in engine pressure ratio (EPR), and a change in the rpm ratio between the core engine and the fan section (N_2/N_1 ratio).

When FOD has occurred, the inspections required depend on the nature of the foreign object or objects. If an external inspection indicates substantial damage to the fan section or to inlet guide vanes, the engine must be removed and overhauled. If the damage to the forward sections of the engine is slight, a borescope inspection of the interior of

Correct identification of cracks, stress, and corrosion is critical during maintenance inspections. Inspectors often find it difficult to differentiate between an actual defect and an unclear image. A new imaging technique, **electronic imaging**, is able to produce sharp, true-color, magnified images that can be seen on a video monitor. One such system, the Videoprobe 2000 (manufactured by the Welch Allyn Company), is shown in Fig. 19-11.

A video imaging system includes an inspection probe, a videoprocessor, and a video monitor for displaying the image.

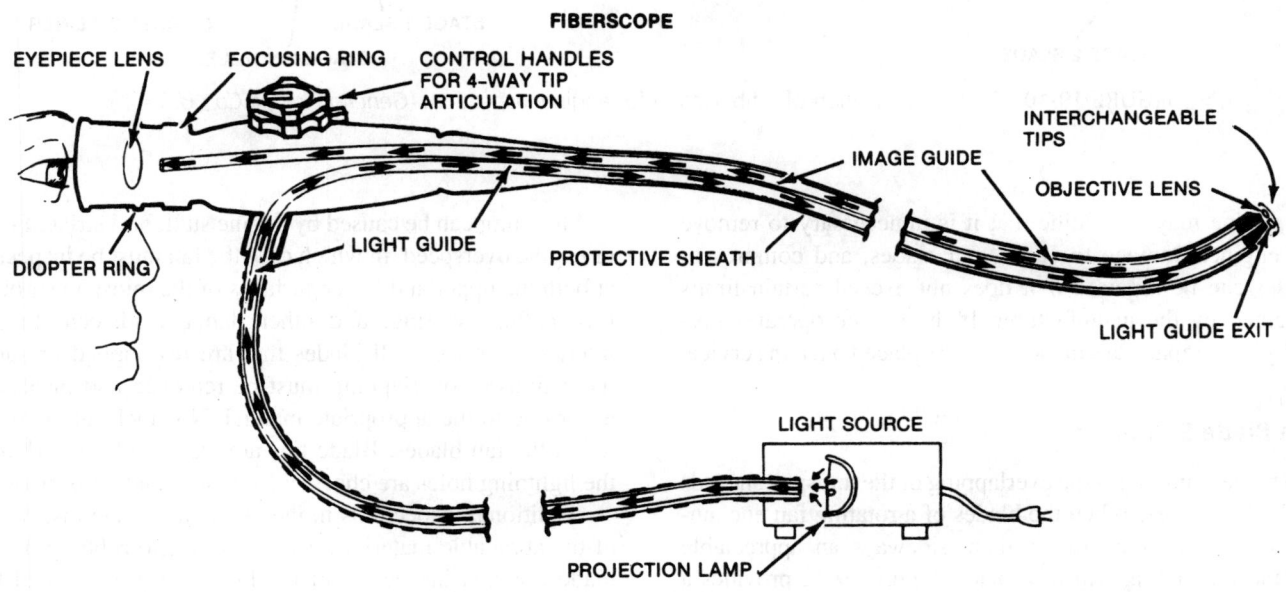

FIGURE 19-9 A fiberscope. (*Olympus Corp.*)

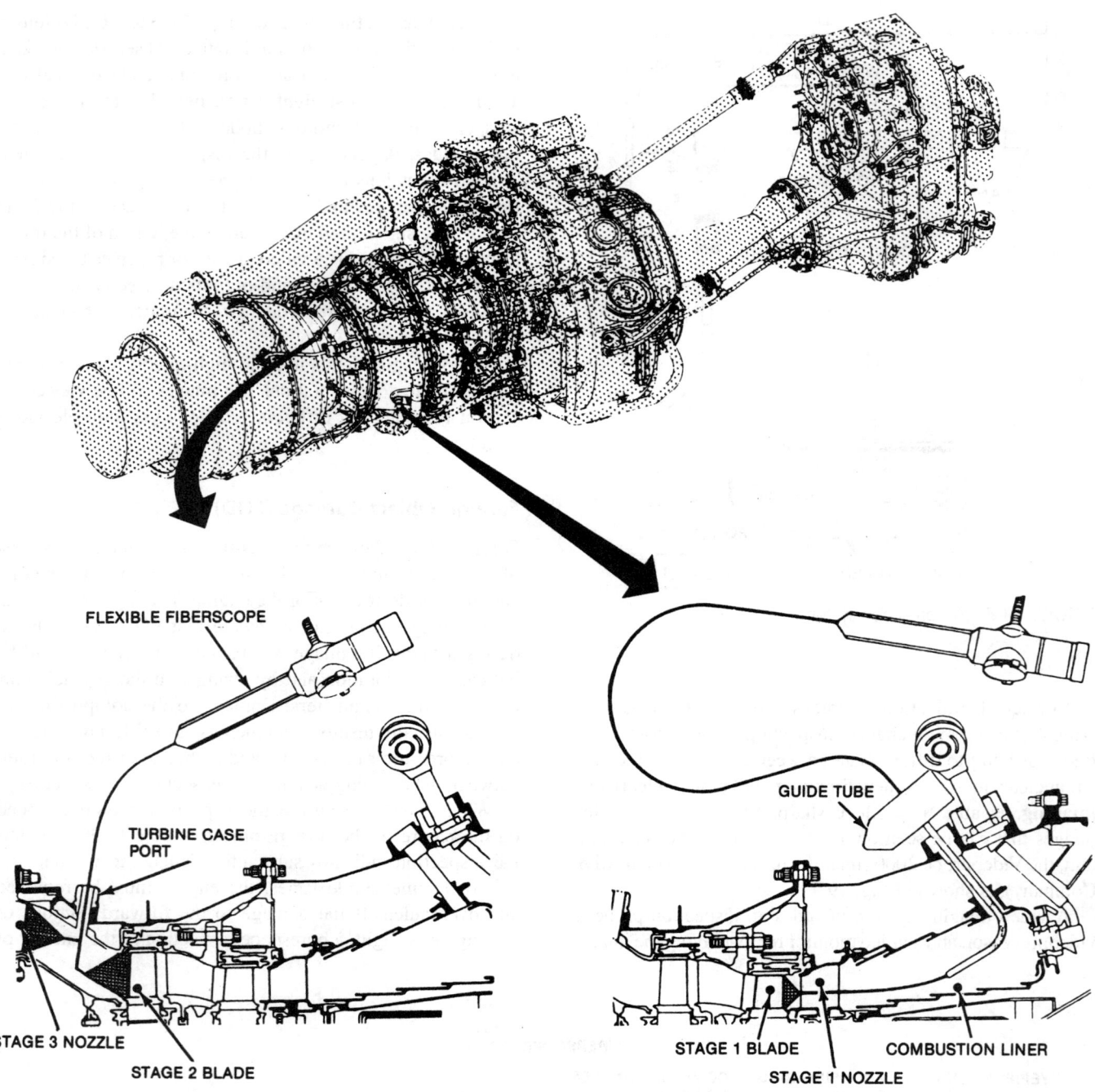

FIGURE 19-10 Typical application of a fiberscope for engine inspection. (*General Electric Co., U.S.A.*)

the engine may determine that it is unnecessary to remove the engine. Damage to vanes, fan blades, and compressor blades can be repaired if it does not exceed certain limits specified by the manufacturer. If the engine operates normally after repairs are made, it can be placed back in service.

Fan Blade Shingling

Fan blade shingling is the overlapping of the midspan shrouds of the fan blades. When the blades of a rotating fan encounter resistance which forces them sideways an appreciable distance, shingling will take place. Figure 19-12 provides a simplified illustration of the situation which causes shingling.

Shingling can be caused by engine stall, bird strike, FOD, or engine overspeed, in which case the fan must be inspected at both the upper and lower surfaces of the midspan shrouds for chafing, scoring, and other damage adjacent to the interlock surfaces. All blades that are overlapped or show indications of overlapping must be removed and inspected according to the appropriate manual. No cracks are permitted in the fan blades. Blade tips are examined for curl, and the lightning holes are checked for cracks and deformation.

Additional inspections in the fan area include inspection of the abradable material for damage due to rubbing of fan blade tips and inspection of the fan-speed sensor head for damage due to blade contact.

FIGURE 19-11 Videoscope used for engine inspection. (*Welch Allyn.*)

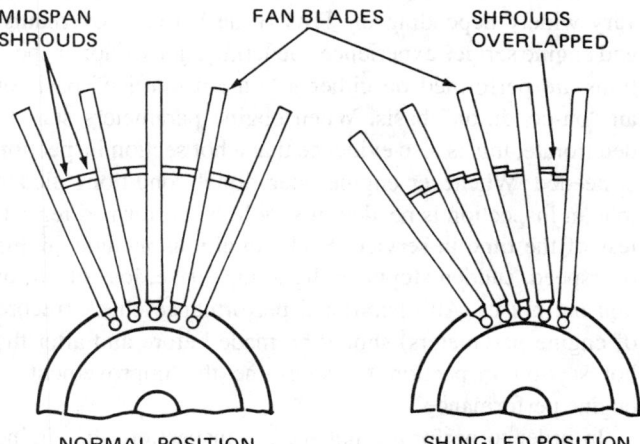

FIGURE 19-12 Shingling of fan blades.

Inspections for Overlimit Operation

Even though technicians and flight crews take every precaution possible to prevent overlimit operation of engines, such operation sometimes occurs. Often the cause is a malfunction of the engine fuel control or a malfunction in the engine. In any case, when overlimit operation does occur, it is necessary to perform certain inspections to determine what damage may have resulted.

At starting, the most critical parameter for the engine is EGT. The technician or crew member starting the engine must watch the EGT gauge carefully. As soon as lightoff occurs, there is a rapid rise in EGT; but if all systems are working properly, the EGT should not exceed limits. If it does, the person starting the engine should immediately retard the start lever to reduce fuel flow to the combustion chamber.

The technician who starts a gas-turbine engine should be familiar with the operating limitations. Figure 19-13 is a temperature-limit chart for starting a large, high-bypass engine.

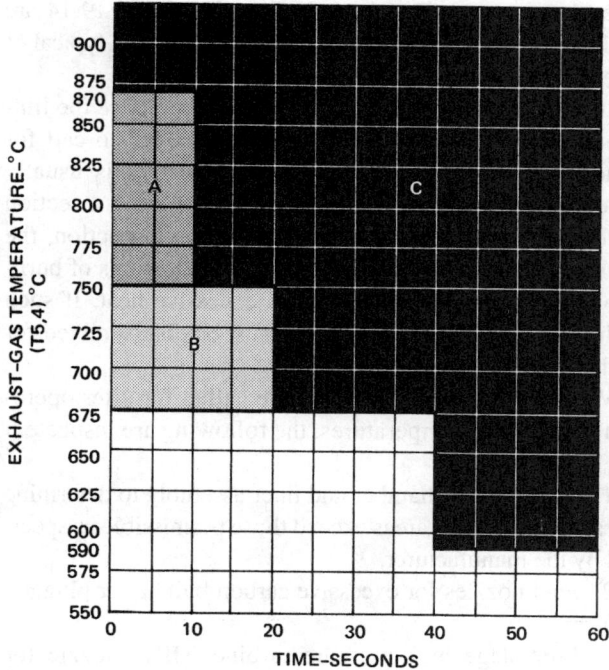

NOTE: ANY OPERATION IN AREAS "A," "B," OR "C" MUST BE VERIFIED AND INDICATION SYSTEM CALIBRATED'

Starts in area A must be recorded and immediate corrective action must be taken prior to further start attempts. A borescope inspection must be performed, prior to further start attempts.

Starts in area B must be recorded. Persistent starts in this area are cause for corrective action

Any operation in area C requires removal of engine to overhaul.

FIGURE 19-13 Temperature-limit chart for starting.

Note that any temperature above 675°C [1247°F] is cause for special attention. Temperatures that fall in area A require special inspections, and temperature-time values that fall in area C are cause for engine overhaul.

After an engine has been started and the operation is stabilized at ground idle, higher temperatures can be permitted during taxiing and preparation for takeoff. The chart in Fig. 19-14 shows temperature-time limitations for operations

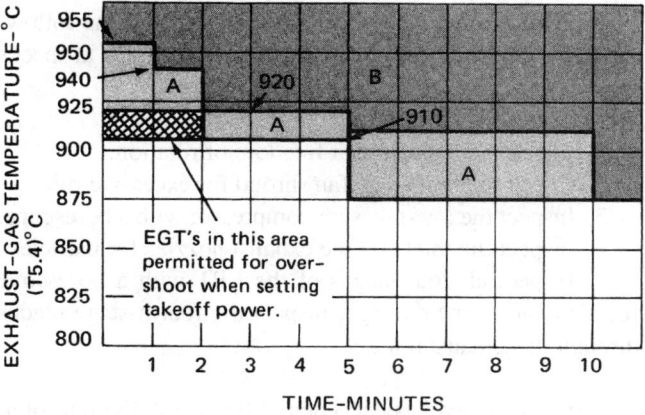

FIGURE 19-14 Temperature-time limits for operations other than starting.

other than starting. The charts in Figs. 19-13 and 19-14 are applicable to one particular engine only and are not typical of all engine limitations.

If a gas-turbine engine has been operated above the limits set for EGT but at a level not high enough to call for removal and overhaul, a borescope inspection is usually called for. An external visual inspection of the hot section of the engine should also be made. In this inspection, the hot section of the engine is checked for indications of burn-through or metal distortion due to excessive heat. If such indications are found, the section must be disassembled for further inspection and repair.

When borescope inspections are called for after operation at excessive temperatures, the following are inspected:

1. Combustion chamber and liner assembly to determine if cracks and burned areas exceed those permissible as specified by the manufacturer.
2. Fuel nozzles for excessive carbon buildup or plugged orifices.
3. First-stage high-pressure turbine (HPT) nozzle for cracks, burned areas, warping, and plugged cooling-air passages. Serviceable limits for defects specified in the maintenance manual must be met.
4. Second-stage HPT nozzles for defects as listed.
5. HPT rotor for cracks, tears, nicks, dents, and metal loss. Cracks in the turbine blades are cause for removal and replacement. Dents and nicks within certain limits may be permitted in the second-stage blades, as specified by the manufacturer.
6. Turbine midframe liner for cracks, nicks, dents, burns, bulges, and gouges. Limitations for these defects are specified by the manufacturer. Bulges associated with heat discoloration are cause for rejection.
7. First-stage low-pressure turbine (LPT) nozzle for cracks, nicks, dents, burns, etc., as for other turbine sections.
8. LPT stator assembly as above.
9. LPT rotor assembly as above. No cracks are permitted in any turbine blades. Limited dents and nicks are allowed.

Overspeed inspections. Overspeed inspection for a typical high-bypass fan engine is primarily concerned with rotating assemblies. One manufacturer specifies the following inspections if the fan section has been operated at speeds from 116 to 120 percent rpm:

1. Check the fan rotor for freedom of rotation.
2. Check the first-stage fan shroud for excessive rub.
3. Inspect the low-pressure compressor with a borescope.
4. Inspect the inlet and the exhaust nozzles for particles.
5. Inspect all four stages of the LPT with a borescope for blade and vane damage. Inspect the fourth-stage blades through the exhaust nozzle.

If the fan speed has exceeded 120 percent, the fan rotor, fan midshaft, and LPT rotor must be removed, disassembled, and inspected in accordance with instructions.

If the core-engine rotor (high-pressure compressor and high-pressure turbine) has been operated at speeds from 107 to 108.5 percent, the following inspections are specified:

1. Inspect the exhaust nozzle for particles.
2. Inspect the core compressor with a borescope for blade and vane damage.
3. With a borescope, inspect the HPT for blade damage

If the core engine rotor has been operated above 108.5 percent, the engine must be removed, disassembled, and inspected according to instructions.

Hot Section Inspections

A hot section inspection is needed to determine the integrity of the components in the hot section of the engine. The hot section of the engine consists of the combustion section (burner cans or liners), turbine inlet guide vanes, turbine wheels, and related parts. Hot section inspection intervals vary widely depending on TBO (time between overhauls) and engine service experience. Generally, hot section inspections are performed on either a "time-in-service" basis or an "on-condition" basis. When engine parameters start to deteriorate, this is also evidence that a hot section inspection is needed. Whenever engine operational conditions dictate that an inspection is needed, it should be performed regardless of the time in service. Such conditions include engine overspeed, sudden stoppage, lightning strike, loss of oil, or unusual noises. An operational performance check (record of engine parameters) should be made before and after the hot section inspection to determine the improvement in engine performance.

In order to perform a hot section inspection, the engine will have to be somewhat disassembled. However, before this can be done, parts of the ignition system and fuel system will need to be disassembled and removed from the engine. Proper references and special tools needed should always be obtained before the disassembly procedure begins. The condition of the engine's components as well as certain clearances and dimensions, which are outlined in the disassembly and inspection sections of the maintenance manual, may need to be recorded during the disassembly phase of the inspection. During the inspection phase, the condition of integral engine components such as the turbine and combustion chamber sections will be closely examined and their airworthiness evaluated. At this time, faulty parts are rejected and replaced or sent to a certified repair station for reworking. During engine reassembly, it is critical that the manufacturer's instructions be carefully followed. All clearances and torque values must be observed. After completion of the reassembly phase, all paperwork should be completed, and a complete inventory of all tools and materials should be made.

In this text it would be impossible to cover all the hot section inspections for the many different types of turbine engines used in aviation. However, an attempt will be made to cover one example of a typical hot section inspection.

PT6A Hot Section Inspection

The Pratt & Whitney PT6A turboprop engine, a common type of gas-turbine engine used in many turboprop aircraft, will be used to illustrate common hot section inspection procedures. Some of the objectives of a hot section inspection are to maintain the turbine blade tip clearance, ensure turbine vane and wheel disk blade integrity and combustion section integrity, and check various other engine components. The information contained here should not be applied to work on an actual engine because it is designed only to familiarize the student with the basic procedures of a hot section inspection. The current manufacturer's manual must always be used when performing such an important inspection. The PT6A hot section procedure is outlined in the following paragraphs, with excerpts from the manual.

Remove fuel manifold adapter assembly. Some procedures need to be accomplished before removal of the fuel nozzles and manifold (refer to Fig. 19-15). To facilitate access to the fuel transfer tubes (1) and the manifold adapters (2) adjacent to glow plug/spark igniters, disconnect the ignition leads. Also, release the ignition lead loop clamps from support brackets, at the center fireseal lower attachment brackets, and move the leads clear. Install blanking caps on the glow plugs or igniters and lead connectors. Using a suitable dye marker, number the position of each manifold adapter to identify the original location for reinstallation. The removal procedure consists of removing the nozzle bolts, transfer tube locking plate (6), and manifold inlet adapter (2) from the gas generator case. Remove bolts, locking plates (6), and adapters (5) on each side of the inlet adapter (2). Support all three adapters and slide interconnecting fuel transfer tubes (1) into bores of adapters (5), moving them away from inlet adapter bores. When these are clear, remove the inlet manifold adapter (2). Remove the gasket (4) from the fuel nozzle sheath (3) on the inlet manifold adapter (2). Then remove the remaining primary and secondary manifold adapters (5) progressively from the gas generator case by removing bolts and locking plates (6). As each adapter is removed, withdraw the interconnecting fuel transfer tubes (1) and remove preformed packings (13). The gaskets (4) from sheaths (3) on adapters can now also be removed. Remove sheaths (3) from the manifold inlet and all primary and secondary adapters, using a puller if the sheath is a tight fit on the adapter boss. To prevent exposure to dust and dirt, place the manifold components in a clean, covered container or polyethylene bag until required for use.

Remove ignition spark igniters/glow plugs. Disconnect the power input cable from the receptacle on the ignition current regulator (ignition exciter). Then disconnect the ignition cable from its respective glow plug (refer to Fig. 19-16). Remove the glow plugs from the gas generator ease and discard the gaskets.

Remove the power section. Place a suitable container under the reduction gearbox (2) (see Fig. 19-17), and remove oil drain plug or chip detector from boss at the 6 o'clock position on the front case of the reduction gearbox. Allow the oil to drain completely. Disconnect the coupling nut (1) from the governor pneumatic (P_y) line (3) at the center fireseal bulkhead coupling tube. Cap the line and coupling tube.

Disconnect electrical connections from the airframe and engine accessories mounted on the reduction gearbox, as necessary.

Remove the cotter pin, nut, bolt, and washer securing the clevis end (6) of the push-pull terminal to the propeller reversing lever (7). Remove the cotter pin, nut, washers, spacer, and bolt securing the propeller governor interconnecting rod end (9) to the reset arm (4) on the propeller governor. Release the locknut (17) that secures the front swivel joint of the propeller reversing linkage to the front lifting bracket (10). Continue by removing the inner nut and bolt (16) and slacken outer nut and bolt (19) that are securing the swivel joint retaining plate (18) to the front lifting bracket (10). Swing away the retaining plate and lift the forward end of the propeller reversing linkage clear of the front lifting bracket. Suitably support this linkage so as to prevent damage. Disconnect the T5 trim harness leads (14) and airframe leads from their terminal block (12) on the gas generator case. Release the terminal block from the gas generator case by removing two bolts (13). Attach the appropriate sling to the power section as follows: When a propeller is installed, position sling and secure the sling to the front lifting bracket using a pin. Position the hoist hook into sling eye and take up the slack on the hoist. Remove the nuts and bolts at flange C and separate power section from gas generator case. Using the sling, partially withdraw the power section. Cap all of the tubes and install plugs on the couplings. Withdraw the power section from the gas generator case and retain it in the sling or install it in a maintenance stand as applicable.

Remove combustion chamber liner. Withdraw the combustion chamber liner, as shown in Fig. 19-18, from the gas generator case. If difficulty is experienced in withdrawing the liner from the case, strike the inside wall of the liner with the palm of the hand to dislodge it from the small and large exit duct joints.

Remove compressor-turbine disk and blade assembly. Install the compressor-turbine disk wrench (2), as shown in Fig. 19-19, on flange C (1) of the gas generator case. Ensure that the slots in the wrench are correctly seated over the lugs (4) on the balancing rim of the turbine disk. Secure the wrench with two Tee-head bolts, washers, and knurled nuts. Insert a protector sleeve in the centerbore of the wrench. The wrench prevents disk rotation during removal of the stubshaft bolt (turbine wheel bolt) and keywasher. Insert the compressor-turbine disk keywasher spreader (6) into the centerbore and over the turbine disk stubshaft bolt. Unlock the keywasher that is locked to the stubshaft bolt and remove the spreader. Using a conventional socket wrench, break the torque on the stubshaft bolt and remove the bolt,

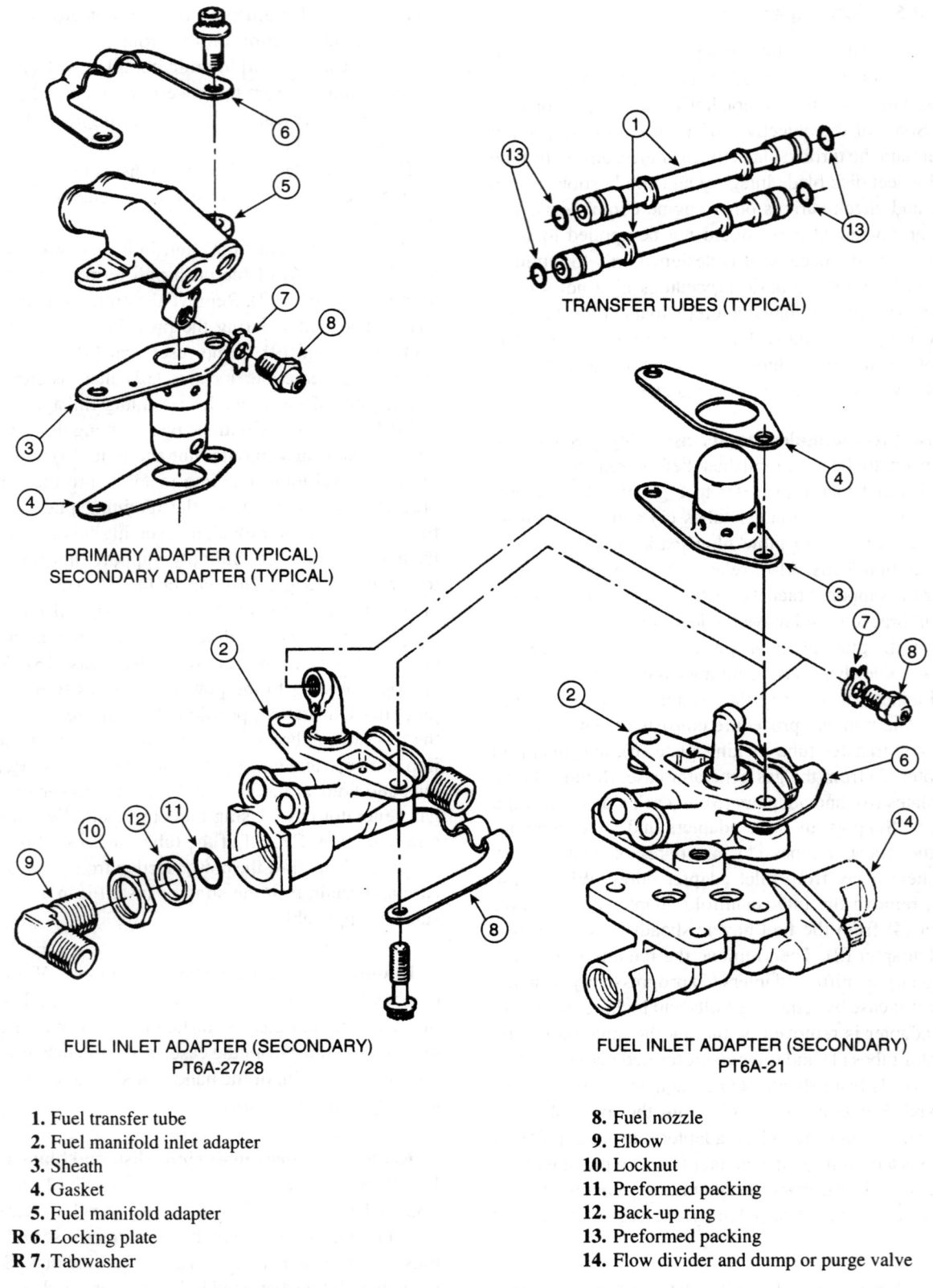

FIGURE 19-15 Fuel manifold installation. (*Pratt & Whitney Canada.*)

TRANSFER TUBES (TYPICAL)

PRIMARY ADAPTER (TYPICAL)
SECONDARY ADAPTER (TYPICAL)

FUEL INLET ADAPTER (SECONDARY)
PT6A-27/28

FUEL INLET ADAPTER (SECONDARY)
PT6A-21

1. Fuel transfer tube
2. Fuel manifold inlet adapter
3. Sheath
4. Gasket
5. Fuel manifold adapter
R 6. Locking plate
R 7. Tabwasher

8. Fuel nozzle
9. Elbow
10. Locknut
11. Preformed packing
12. Back-up ring
13. Preformed packing
14. Flow divider and dump or purge valve

keywasher, wrench, and socket wrench from the turbine disk. Insert the compressor-turbine disk puller into the centerbore of the turbine disk and release the turbine disk from the engine. Withdraw the turbine disk and blade assembly from the gas generator case. The puller will become free in its operation when the disk snap diameters (splines) are released.

Place the disk and blade assembly on a clean bench and remove the puller. Install the assembly in a suitable container to prevent any damage.

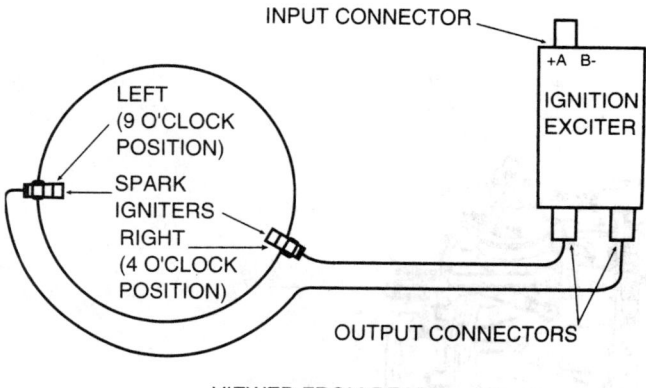

INPUT CONNECTOR

+A B-

IGNITION EXCITER

LEFT (9 O'CLOCK POSITION)

SPARK IGNITERS

RIGHT (4 O'CLOCK POSITION)

OUTPUT CONNECTORS

VIEWED FROM REAR

FIGURE 19-16   Spark ignition schematic. (*Pratt & Whitney Canada.*)

Inspection procedures. Following are procedures that should be performed during a PT6A hot section inspection.

A more detailed and expanded inspection procedure, in the manufacturer's maintenance manual, is needed to complete the hot section inspection completely.

Shroud segment grinding. To determine shroud segment concentricity and diameter, some special tools are needed. These tools are a radius gauge, gauge master, grinder, adapter, spacers, and grinding wheel. To maintain concentricity and a uniform turbine blade-tip-to-shroud-segment clearance, grinding should be carried out with the turbine vane ring, including the no. 2 bearing cover, fitted and installed in the engine.

Replacement of the compressor-turbine shroud segments may entail localized grinding to remove high spots and any slight eccentricity of the segments to obtain the correct turbine blade tip clearance. If tip clearance exceeds a specified limit after removal of the high spots and eccentricity, the next higher class of shroud segments must be selected, as shown in Fig. 19-20.

Preassemble the compressor-turbine vane ring with the shroud housing and bladed compressor-turbine disk. These assemblies are usually machine ground to provide minimum compressor-turbine blade-tip-to-shroud-segment clearance. Upon installation, a final grind may be necessary to ensure that prescribed clearances are maintained. To determine the shroud segment concentricity and diameter, it will be necessary to install an adapter (6) (see Fig. 19-21) on the compressor front stubshaft and secure it with a bolt (5). Next, install four spacers under the small exit duct, approximately 90° apart (see Fig. 19-22). Install the radius gauge (9 in Fig. 19-21) on the adapter (6) and secure it with the bolts (7).

Position the contact point (3) at the center of a segment (4) using the adjuster (1), as shown in Fig. 19-21. Take four readings 90° apart around the shroud segments. Adjust the spacers as necessary to obtain shroud concentricity, indicated by equal gauge readings at the four points.

Remove the radius gauge from the adapter (6) and install it in the gauge master (8). Set the dial indicator (10) to zero.

Reinstall the radius gauge on the adapter (6) mounted in the engine. Rotate the gauge and record the high and low readings. Adjust the gauge in and out to determine the segment's taper.

A low reading or excessive segment taper will require that the segment be replaced. To calculate the required grinding of the shroud segment, the following typical example can be used:

where A = true radius as stamped on gauge master
  B = radius of bladed disk, $\frac{1}{2}$ OD as measured with a micrometer
  C = dial indicator reading on shroud segments
  D = radius of shroud segment
  E = existing blade tip clearance, before grinding

Example:

where A = 4.282 in
  B = 4.266 in
  C = 0.013 in clockwise reading on minus side of gauge zero datum
  D = A − C = 4.282 in − 0.013 in = 4.269 in
  E = D − B = 4.269 in − 4.266 in = 0.003 in; then
  E = 0.003 in (turbine-blade-to-shroud-segment clearance)

In this example, 0.010 in must be ground from the shroud segments to provide the minimum blade tip clearance of 0.013 in (0.003 in + 0.010 in = the clearance needed, 0.013 in). Using a suitable material, block off all the cavities around the combustion chamber liner and compressor-turbine vane ring.

Install a grinding wheel and die grinder on the adapter.

Position the wheel (1) (Fig. 19-22) in line with and slightly overlapping segments, using the axial adjuster (6). Allow the wheel to touch the vane ring, then back off the adjuster (6) one turn and lock using the lockscrew (8). Rotate the grinder around the segments, adjusting the radial adjuster (9) until the wheel (1) contacts the high spots. Then back off the adjuster one turn. Connect the grinder air line (4) to shop air of 90 psig [620.6 kPa] with flow rate of 36 SCFM (std ft³/min) [1 m³/min] and operate the grinder. Rotate the grinder slowly and avoid removing an excessive amount of material in rotation. Adjust the wheel (1) with the radial adjuster (9) until it contacts the segments, and then grind the segment's ID to the required dimension. Install a radius gauge and recheck the shroud segment diameter. Check the segment taper by running the gauge contact point across the width of the segments. Using a suitable suction cleaner, remove all of the grinding residue from the gas generator case. Reassemble the engine following the maintenance manual and perform an operational performance check on the completed engine.

Other Special Inspections

In addition to the events requiring inspections discussed so far, there are certain events which happen occasionally that require special attention. Among these are tire damage, operation with no oil pressure, accident damage, and engine stall.

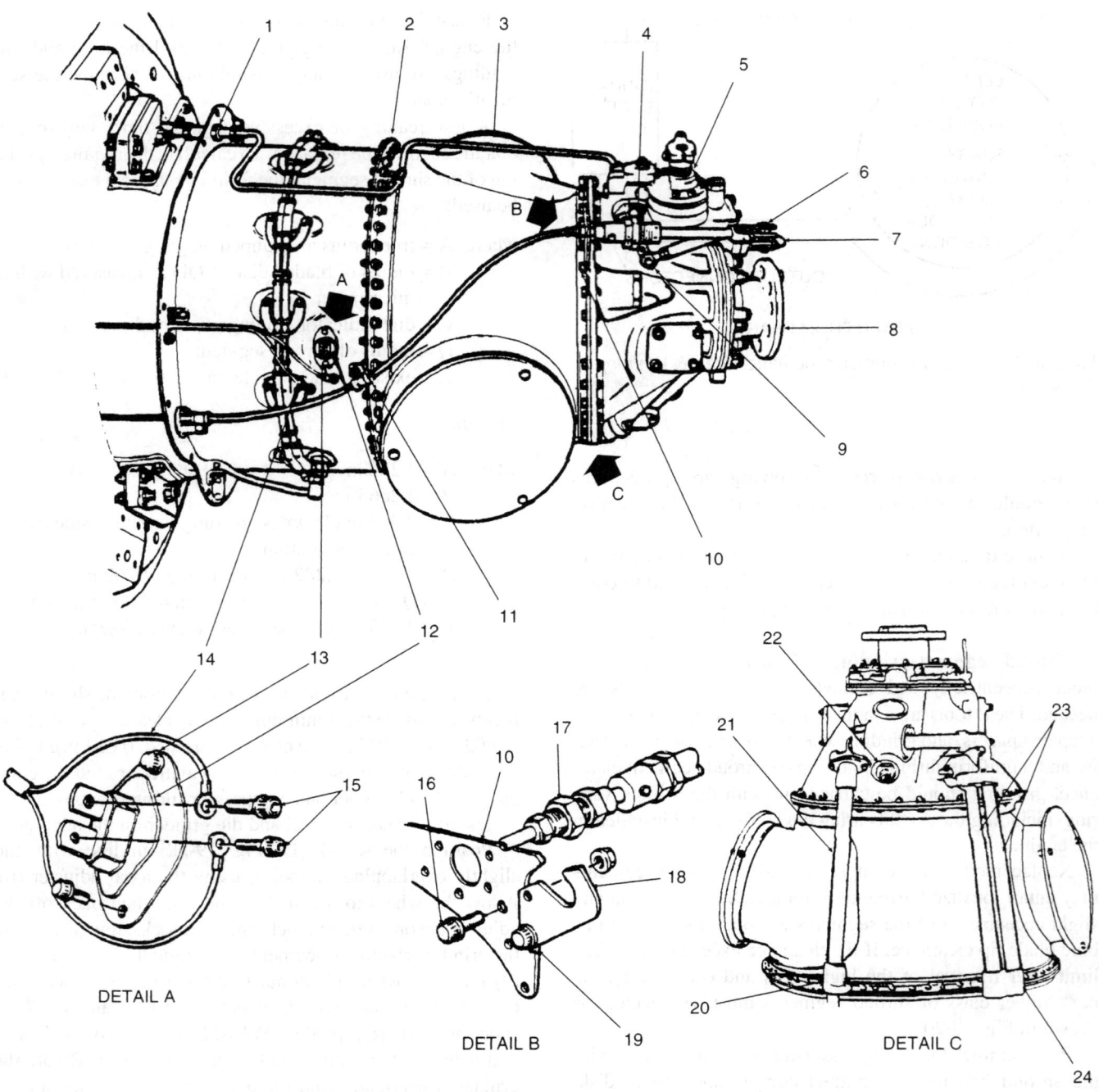

DETAIL A

DETAIL B

DETAIL C

1. Coupling nut at center fireseal bulkhead
2. Flange C
3. P$_y$ line to propeller governor
4. Reset arm (Propeller governor)
5. Propellor governor
6. Clevis end (Push-pull terminal)
7. Propeller reversing lever
8. Flange, propeller shaft
9. Rod end (interconnect rod)
10. Engine front lifting bracket
11. Bracket (Loop clamp)
12. Terminal block (T5)
13. Mounting bolt (Terminal block)
14. Wiring harness lead (T5)
15. Connecting bolts (Terminal leads)
16. Inner bolt (Retaining plate mounting)
17. Locknut (Front swivel joint)
18. Retaining plate (Front swivel joint)
19. Outer bolt (Retaining plate mounting)
20. Coupling (Pressure oil tube)
21. Pressure oil tube
22. Chip detector
23. Scavenge oil tubes
24. Coupling (Scavenge oil tubes)

FIGURE 19-17 Preparation for power section removal. (*Pratt & Whitney Canada.*)

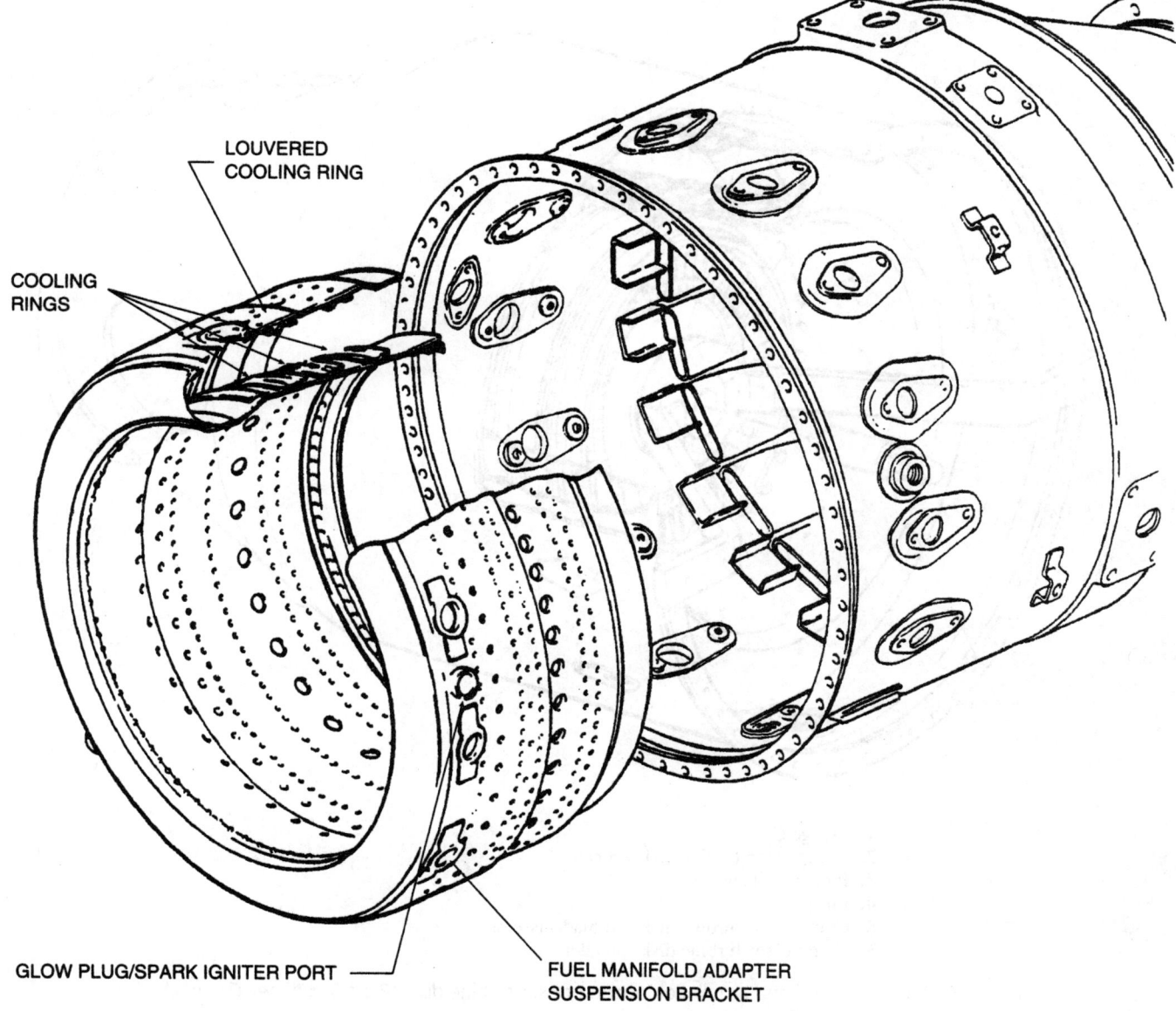

LOUVERED
COOLING RING

COOLING
RINGS

GLOW PLUG/SPARK IGNITER PORT

FUEL MANIFOLD ADAPTER
SUSPENSION BRACKET

FIGURE 19-18   Combustion chamber liner. (*Pratt & Whitney Canada.*)

Each inspection required depends on the nature and severity of the event, but these inspections often result in removal of the engine for disassembly and possible major overhaul.

When an engine has been operated with no oil pressure for more than 2 min, the engine must be removed for overhaul. If an engine is involved in an accident, the nature and severity of the accident are taken into account, and the matter is frequently referred to the engine manufacturer's technical representative.

GAS-TURBINE ENGINE MAINTENANCE

It is not practical (or perhaps even possible) to describe in a textbook all the practices and procedures for the repair and maintenance of all gas-turbine engines; these procedures vary from engine to engine and are covered in the various manuals available for each particular engine and its accessories. However, it is worthwhile to describe a few of the typical repair and maintenance practices which may be employed by the technician under certain conditions.

Maintenance covers both the work that is required to maintain an engine and its systems in an airworthy condition while it is installed in an aircraft (on-wing or line maintenance) and the work that is required to return an engine to airworthy condition once it has been removed from an aircraft (overhaul or shop maintenance). **On-wing maintenance** falls into two basic categories: **scheduled maintenance** and **unscheduled maintenance**.

Scheduled Maintenance

Scheduled maintenance includes the periodic and recurring inspections that must be made in accordance with the engine section of the aircraft maintenance schedule.

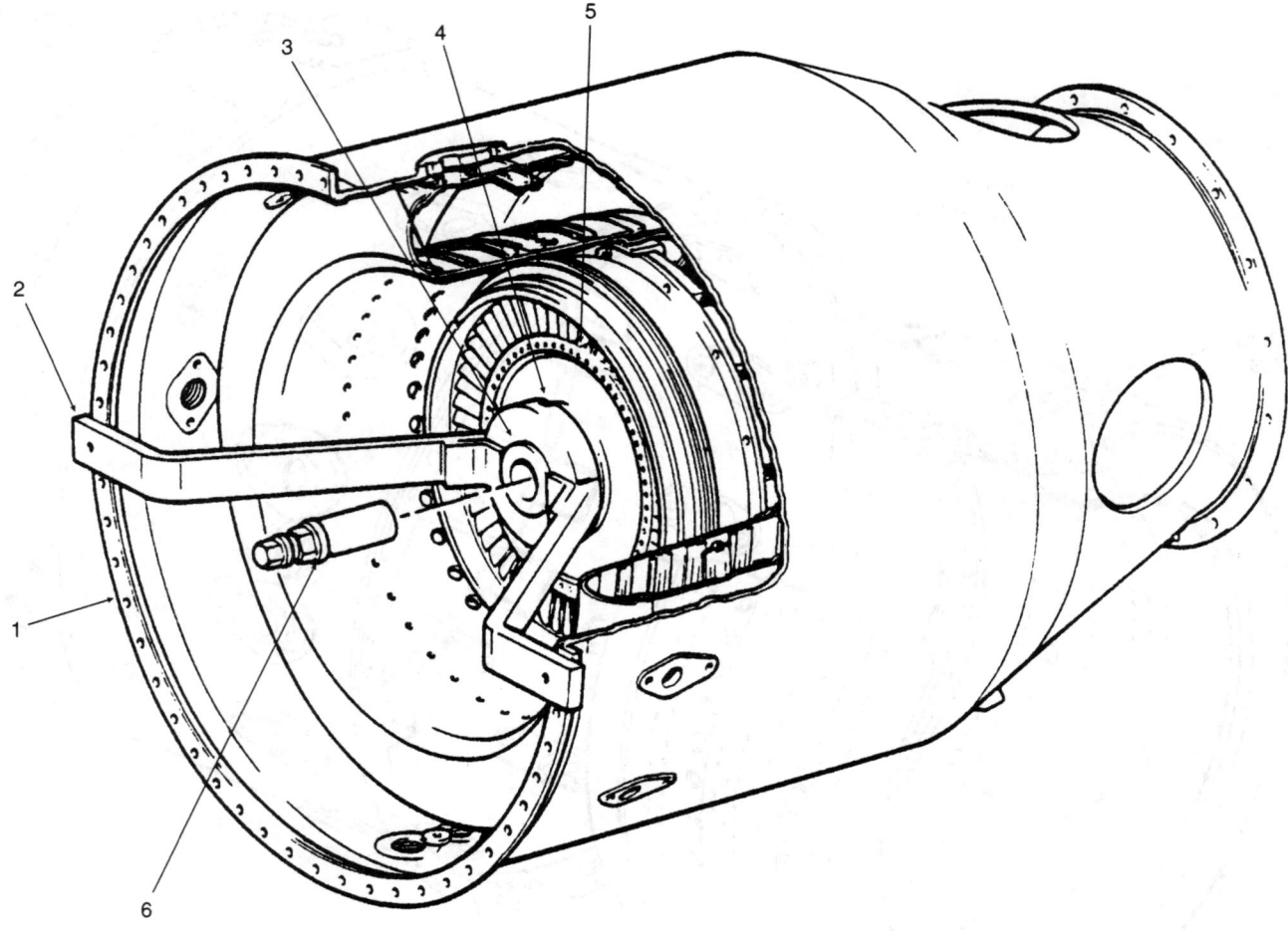

1. Flange C
2. Compressor-turbine disk wrench
3. Protector sleeve
4. Lugs
5. Compressor-turbine disk and blade assembly
6. Compressor-turbine disk spreader

FIGURE 19-19   Removal and installation of compressor turbine disk. (*Pratt & Whitney Canada.*)

These checks range from inspections which do not entail opening of cowls to more elaborate checks within specified time limits, usually calculated in aircraft flying hours or cycles.

A policy of **continuous maintenance**, whereby checks are carried out progressively and as convenient within given time limits, rather than at specific aircraft check periods, has been widely adopted. With the introduction of condition monitoring devices of increased efficiency and reliability, a number of traditionally accepted scheduled checks may become unnecessary.

Unscheduled Maintenance

Unscheduled maintenance covers work necessitated by occurrences that are not normally related to time limits—e.g., bird ingestion, a strike by lightning, or heavy landing. Unscheduled work may also result from malfunctions, troubleshooting, or scheduled maintenance inspections.

Line Maintenance

Maintenance functions also fall into the two categories of line maintenance and heavy maintenance. The scope of line maintenance consists of removal and installation of external components and engine accessories as well as hot section inspection. Much of the work considered to be line maintenance is removal and replacement of malfunctioning line replaceable units (LRUs). All procedures are considered line maintenance that do not fall under heavy maintenance.

Heavy Maintenance

Heavy maintenance entails removal, installation, and repair of components normally considered beyond the capabilities of the average line maintenance facility. Normally, procedures that are considered heavy maintenance are noted as such and require considerable equipment and engine knowledge. Heavy maintenance is normally performed at an overhaul facility.

Premachined Turbine Bladed Disk OD (Inches)	Recommended Class	Shroud Segment Dimension "A"	Grind Shroud Segment ID (Inches)	
			Minimum	Maximum
8.548 ± 0.001	Class 1	0.042–0.043	8.574	8.580
8.542	Class 2	0.045–0.046	8.568	8.574
8.536	Class 3	0.048–0.049	8.562	8.568
8.530	Class 4	0.051–0.052	8.556	8.562
8.524	Class 5	0.054–0.055	8.550	8.556
8.518	Class 6	0.057–0.058	8.544	8.550
8.512	Class 7	0.060–0.061	8.538	8.544
	Class 8	0.069–0.072	*	*
	Class 9	0.079–0.083	*	*

*Select class and grind to suit as necessary.

NOTE: If blade tip clearance exceeds specified limits after removing eccentricity on shroud segments, select next class (thicker) segments and grind as necessary. For bladed disk OD below 8.512 inches, use class 8 or 9.

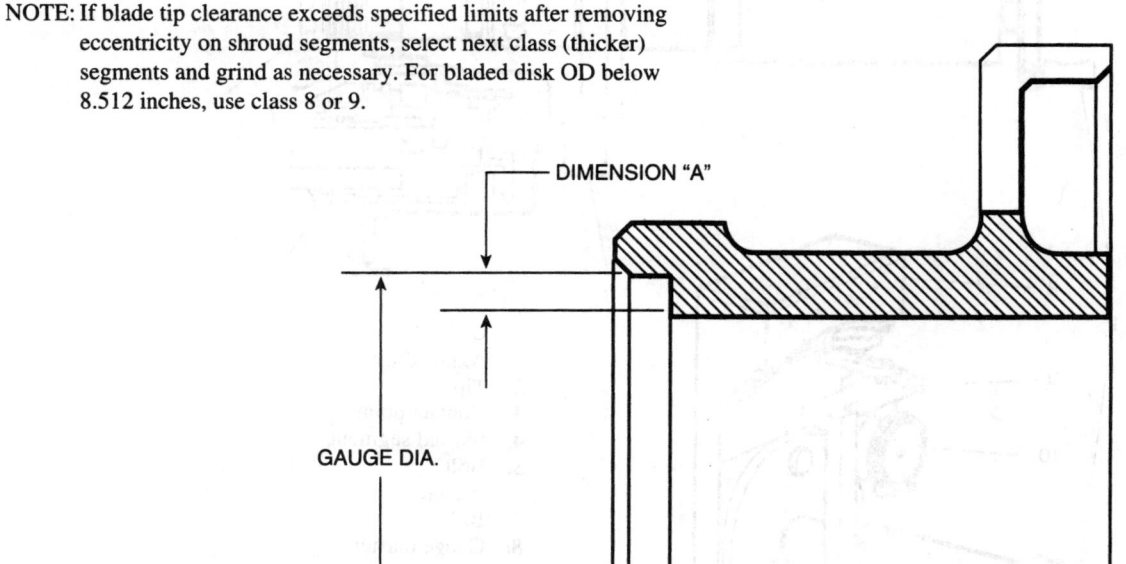

FIGURE 19-20 Compressor turbine shroud segment classification. (*Pratt & Whitney Canada.*)

Maintenance Precautions

Certain precautions must be observed during engine maintenance. The ignition system is potentially lethal; therefore, before any work is done on the high-energy ignition units, the igniter plugs or the harness must be disconnected and at least 1 min allowed to elapse before the high-tension lead is disconnected. Similarly, before carrying out work on units connected to the electrical system, make sure that the system is safe, either by switching off the power or by tripping and tagging the appropriate circuit breakers. With some installations, the isolation of certain associated systems may be required.

When the oil system is being replenished, care must be taken so that no oil is spilled. If any oil is accidentally spilled, clean it off immediately, because it is injurious to paint and to certain rubber compounds that may be found in the electrical harnesses or other components. Oil can also be toxic through absorption if it is allowed to come into contact with the human skin for prolonged periods. Care should be taken not to overfill the oil system. (Overfilling may easily

occur if the aircraft is not on level ground or if the engine has been stationary for a long period of time.)

Before making an inspection of the air intake or exhaust system, make sure that there is absolutely no possibility of the starter system being operated or the ignition system being energized.

Always make a final inspection of the engine, air intake, and exhaust system after any repair, adjustment, or component change to ensure that no loose items have been left inside.

Always observe fire safety precautions at all times when procedures involve the use of fuels or similar combustibles, especially during testing of fuel nozzles or fuel system components.

Fuel Nozzle Testing

Testing PT6A Fuel Nozzles

Because the information in this section is not totally complete, the manufacturer's maintenance manual as well as appropriate

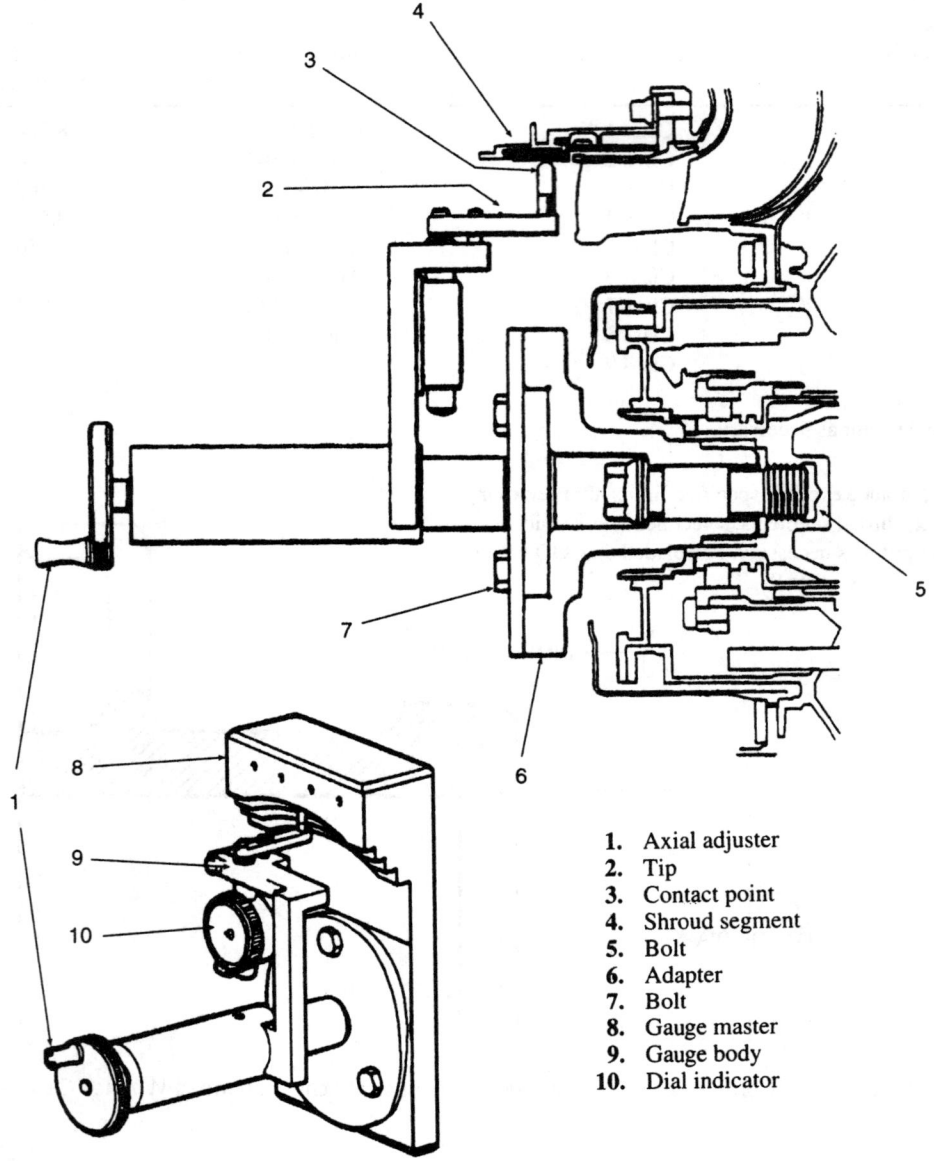

1. Axial adjuster
2. Tip
3. Contact point
4. Shroud segment
5. Bolt
6. Adapter
7. Bolt
8. Gauge master
9. Gauge body
10. Dial indicator

FIGURE 19-21 Shroud segment measurement of blade tip clearance. (*Pratt & Whitney Canada.*)

safety precautions should always be used when performing any type of maintenance on fuel nozzles. An example of a fuel nozzle for a PT6A engine is shown in Fig. 19-15. It consists of a fuel manifold adapter (5), a sheath (3), a locking plate (6), a tabwasher (7), and a fuel nozzle (8). Due to the important nature of the spray pattern of a nozzle, it is necessary to clean and test fuel nozzles at certain intervals.

The procedure for testing fuel nozzles begins with lubricating the nozzles with fuel, then installing the nozzles into the manifold adapters (5 or 2) using new tabwashers for each nozzle, as shown in Fig. 19-15. Tighten the nozzle assemblies and torque, but do not bend the lugs of the tabwashers until testing is completed.

Fuel nozzles are subjected to two separate tests, the **leakage test** and the **functional test**. The leakage test is used to make sure the nozzle is not leaking at the adapter connection. The functional test is used to observe the spray pattern.

To perform the leakage test, back off the torque screw (6) and the setscrew (4), loosen the pivot screw (8), and rotate the pivot block, as shown in Fig. 19-23. This will provide adequate space between the plugs in the pivot block and the plugs in the upright end of the fixture (2) to allow the installation of the adapter and nozzle assembly (3). Position the adapter and nozzle assembly (3) between the pivot block and the end of the fixture (2) with the nozzle facing toward the plastic pad (5).

Insert the plugs of the pivot block into the ports in the adapter and, by rotating the block, insert the plugs of the fixture (2) into the opposite ports in the adapter. Do not dislodge the preformed packings from the seal grooves in the plugs during this operation. With the plugs fully inserted into the adapter ports, hold the parts firmly and tighten the pivot screw (8).

Close off the nozzle orifice by turning the setscrew (4) until it just makes contact with the rear of the adapter behind

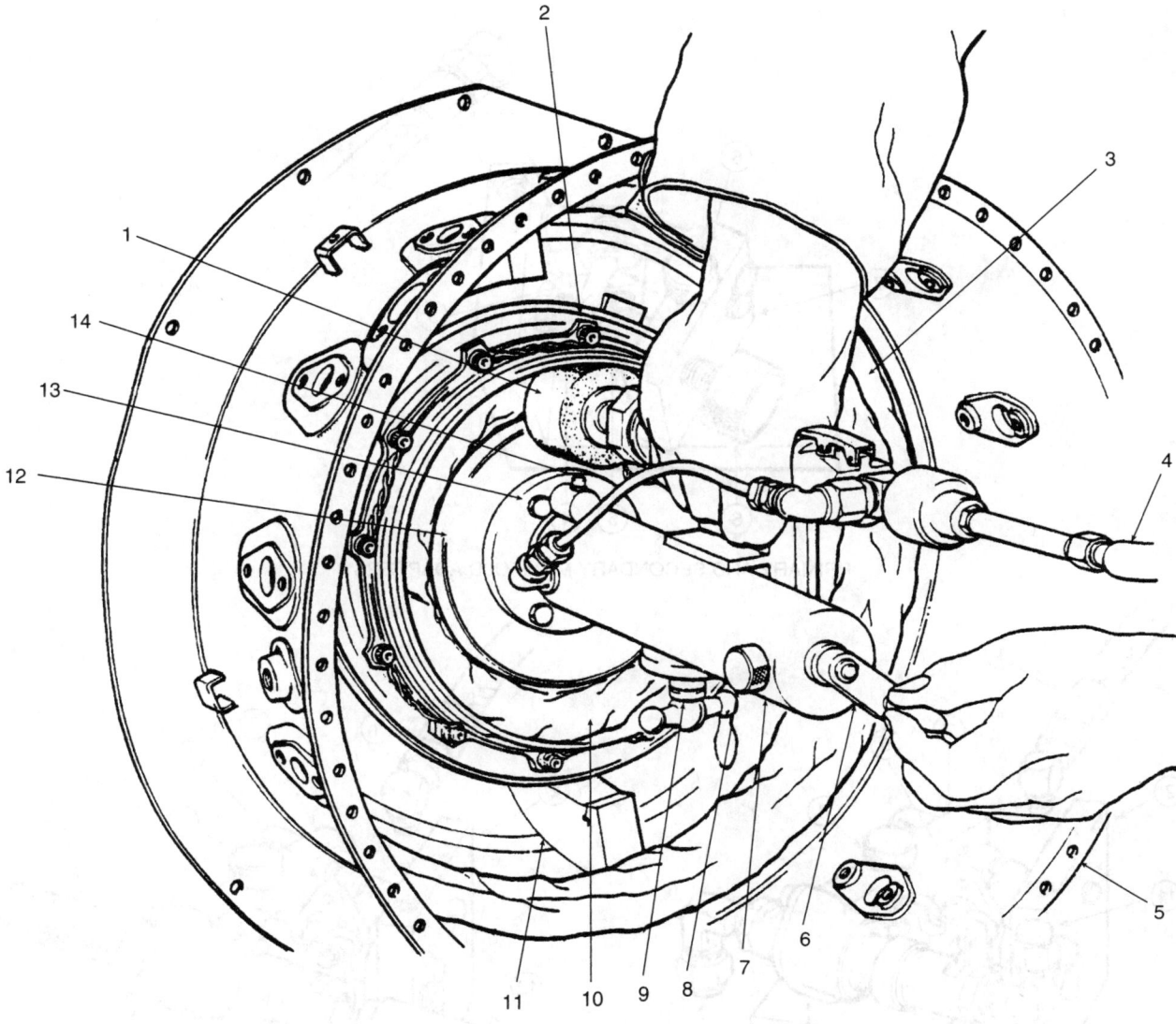

1. Grinding wheel	**8.** Axial adjustment lockscrew
2. Bolt (Shroud housing)	**9.** Radial adjuster
3. Blanking material (Exit duct)	**10.** Blanking material (Stator assembly)
4. Air hose	**11.** Spacer
5. Gas generator case	**12.** Adapter (Blanking area)
6. Axial adjuster	**13.** Bolt
7. Grinder	**14.** Diamond wheel dresser

FIGURE 19-22 Typical shroud segment grinding. (*Pratt & Whitney Canada.*)

the nozzle. Turn the torque screw (6) until the plastic pad (5) seats on the nozzle orifice. Tighten the setscrew (4) and torque the torque screw (6), simultaneously, to ensure that the plastic pad (5) closes the nozzle orifice without distortion of the adapter. Tighten the locknut on the setscrew (4).

When testing the inlet manifold adapter and nozzle assembly (9) with the attached flow divider and dump or purge valve (10), close off the elbows on the flow divider and dump or purge valve (10) with caps (11) (see Fig. 19-23A). If elbows are not installed, close off ports in the flow divider and dump or purge valve (10) with plugs (12), as shown in Fig. 19-23B.

Check for leakage between the nozzle and the adapter by connecting the hose assembly (1) to a supply of clean, dry compressed air or nitrogen and applying a pressure of 500 psig [3450 kPa] to the test fixture (2). Check for leakage using leak check fluid or by immersing in a petroleum solvent. No leakage is permitted.

The functional test consists of flowing fluid through the nozzle and observing the spray pattern. There are several terms used in this text that describe specified test conditions. For example, the term "onion" describes a spray condition which sometimes occurs at low flow rates when the spray exhibits a distinct "onion-bulb" shape. "Streakiness" is

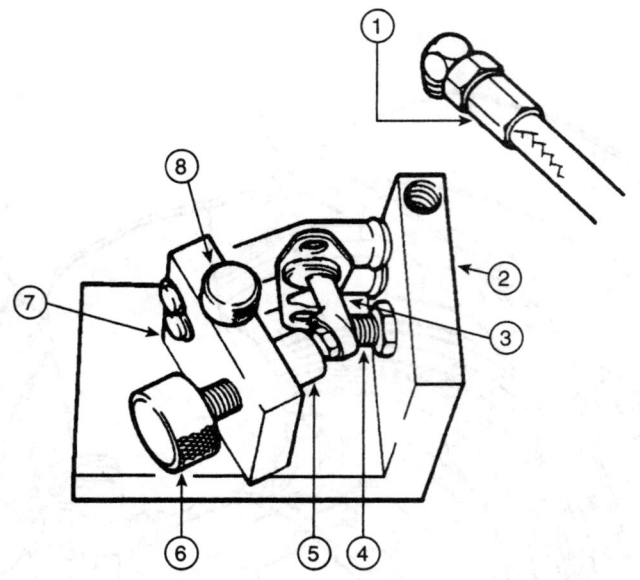

PRIMARY AND SECONDARY MANIFOLD ADAPTERS

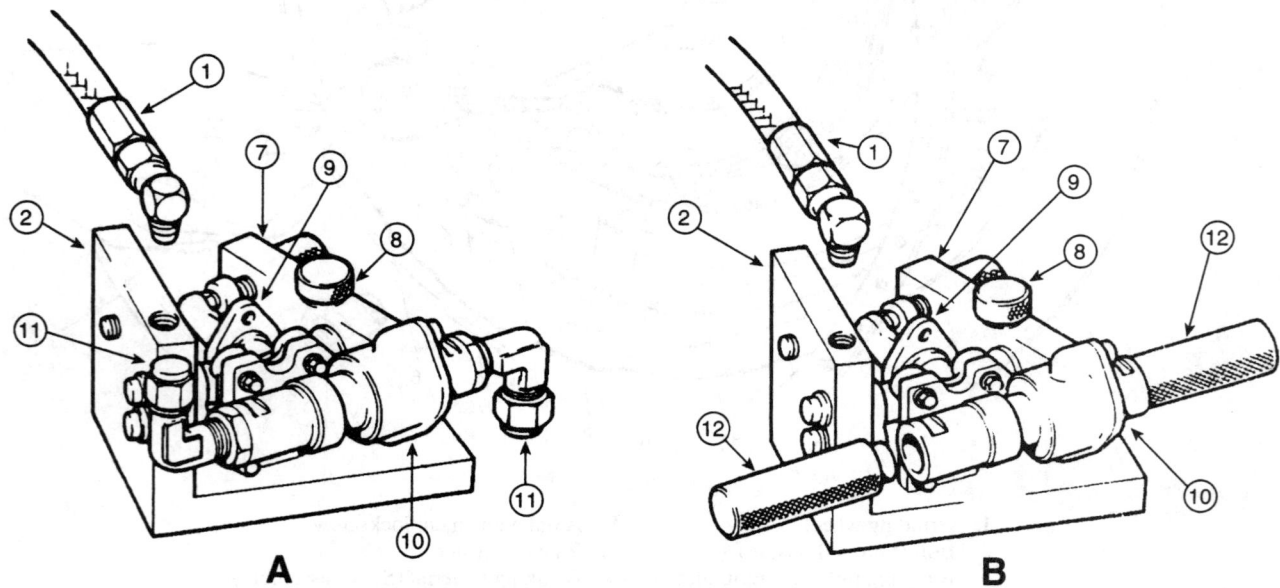

A

B

INLET MANIFOLD ADAPTER WITH FLOW DIVIDER AND DUMP VALVE

1. Hose assembly
2. Test fixture
3. Primary or secondary manifold
   Adapter and nozzle assembly
4. Setscrew (part of fixture)
5. Plastic pad (part of fixture)
6. Torque screw (part of fixture)

7. Pivot block (part of fixture)
8. Pivot screw (part of fixture)
9. Inlet manifold adapter and
   nozzle assembly
10. Flow divider and dump or purge valve
11. Blanking cap
12. Plug

FIGURE 19-23   Manifold adapter and nozzle assembly leakage test fixture. (*Pratt & Whitney Canada.*)

defined as variation in spray quantity among different parts of the spray cone and appears as a darker streak in the spray. It is specified as a percentage variation from nominal. "Spitting" is a condition which exists when large drops of unatomized fuel occur intermittently and usually on the outside of the spray cone. "Drooling" is a condition which occurs when large drops of unatomized fuel form on the nozzle face.

Figure 19-24 illustrates a functional test of the primary adapter and nozzle assemblies. By closing off the primary tube assembly using the blanking tubes (19) and connecting the tube assembly (10) of the flow fixture (11) into the ports as shown, the two remaining blanking tube assemblies (19) are not required for this test. Place the adapter (12) into the flow fixture (11) and ensure that the collars of the tube assemblies

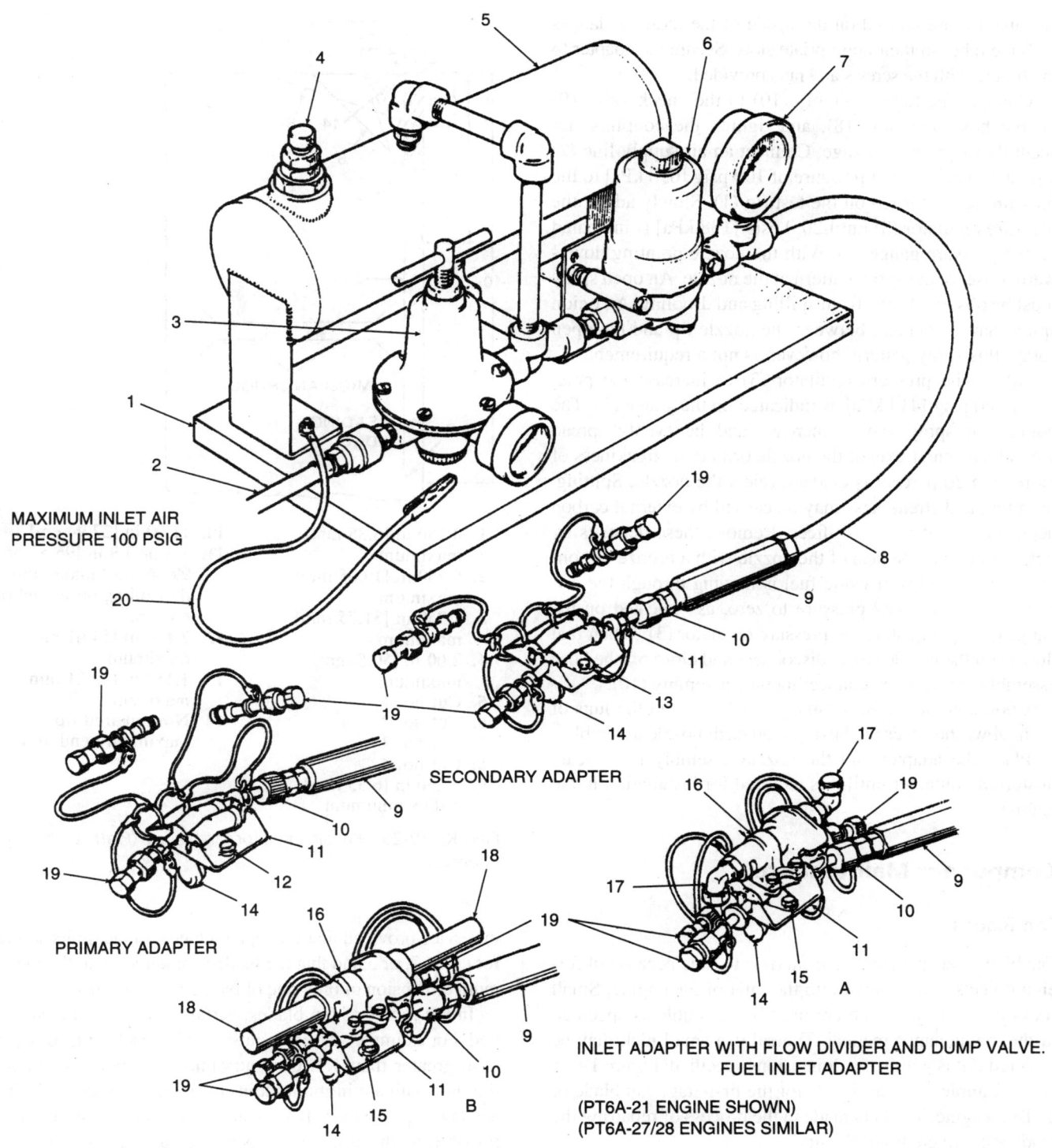

MAXIMUM INLET AIR
PRESSURE 100 PSIG

SECONDARY ADAPTER

PRIMARY ADAPTER

INLET ADAPTER WITH FLOW DIVIDER AND DUMP VALVE.
FUEL INLET ADAPTER

(PT6A-21 ENGINE SHOWN)
(PT6A-27/28 ENGINES SIMILAR)

1. Test rig
2. Air supply line
3. Pressure regulator (0 to 250 psig)
4. Relief valve (150 psig)
5. Reservoir
6. Filter (10 micron nom.)
7. Pressure gauge (0 to 100 psig)
8. Hose assembly
9. Check valve
10. Connecting tube assembly
11. Nozzle flow fixture

12. Primary adapter
13. Secondary adapter
14. Nozzle assembly
15. Inlet manifold adapter
16. Flow divider and dump or
    purge valve inlet adapter
17. Blanking cap
18. Plug
19. Blanking tube assembly
20. Ground cable

FIGURE 19-24 Manifold adapter and nozzle assembly functional test fixture. (*Pratt & Whitney Canada.*)

(19 and 10) are located on the inside of the fixture's flanges with the tubes in their appropriate slots. Secure the adapter to the fixture with the screws and nuts provided.

Connect the tube assembly (10) to the check valve (9) on the hose assembly (8), and tighten the coupling nut securely to prevent leakage. Connect an air supply line (2) with a maximum inlet pressure of 100 psig [690 kPa] to the pressure regulator (3) on the test rig (1). Slowly adjust the pressure regulator (3) until 20.0 psig [138 kPa] is indicated on the pressure gauge (7). With the nozzle pointing downward, observe the spray pattern at the nozzle. An open spray must be observed, free from spilling and drooling. An onion spray may be evident between the nozzle tip and the open spray; this spray pattern, however, is not a requirement.

Adjust the pressure regulator (3) to increase the pressure to 60 psig [414 kPa] as indicated on the gauge (7). The volume of spray should increase and be evenly spread around the center axis of the nozzle orifice. If streakiness of more than 20 percent is evident, reject the nozzle. Spitting, drooling, and streakiness may be caused by external carbon deposits around nozzle orifices. Remove these deposits by lightly brushing the lace of the nozzle with a bronze or nonmetallic bristle brush while fuel is flowing through the orifice. Reduce the flow pressure to zero, as indicated on the gauge (7), by adjusting the pressure regulator (3). When fuel flow from the nozzle stops, disconnect and close off the hose assembly (8) from the connecting tube assembly (10).

Upon completion of a satisfactory test, bend the lugs of each tabwasher over the hexagon on each nozzle assembly.

Place the adapter and the nozzle assembly in a clean, dustproof container until it is required for installation in the engine.

Component Maintenance

Fan Blades

Fan blades receive damage from time to time because of foreign objects being drawn into the inlet of the engine. Small rocks cause nicks which are usually repairable as specified in the maintenance manual. Typically, a small nick may be repaired if it is within the dimensions specified. Figure 19-25 is an example of repair limits for the first-stage fan blade of a JT8D engine. The cuts made in the process of repairing the blade are termed "flyback cuts."

If fan blade damage is such that all damaged sections can be removed within the limitations shown in Fig. 19-25, the blade can be continued in service for a maximum of 20 h. The repair must adhere to any combination of limits shown for cuts 1, 2, 3, and 4; and the blade may be repaired up to the maximum dimension defined by the envelope created by all four cuts.

In Fig. 19-25 it will be noted that the leading edge of the blade can be cut back a distance of 0.250 in [6.35 mm] for a distance of 11 in [27.94 cm] along the blade. Toward the tip, deeper cuts can be made as shown.

For blades with FOD (foreign-object damage) confined to the blade tip only, repair may be made and the blade continued

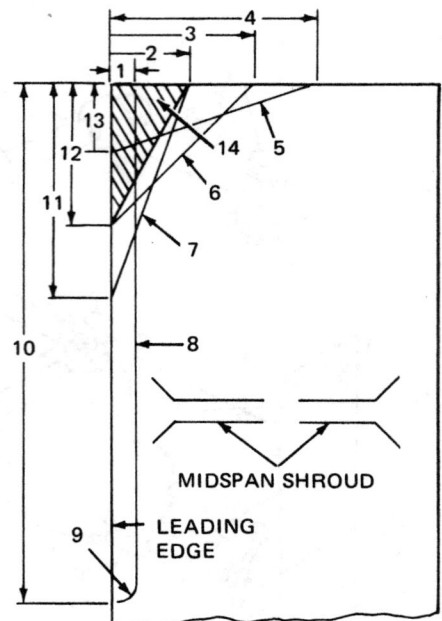

1. 0.250 in [6.35 mm] maximum
2. 0.750 in [19.05 mm] maximum
3. 1.250 in [31.75 mm] maximum
4. 2.00 in [50.8 mm] maximum
5. Cut no. 1
6. Cut no. 2
7. Cut no. 3
8. Cut no. 4
9. 0.250 in [6.35 mm] radius minimum
10. 11.00 in [27.94 cm]
11. 3.8 or 3.9 in [96.52 or 99.06 mm] maximum depending on model of engine.
12. 2.150 in [54.61 mm] maximum
13. 1.150 in [29.21 mm] maximum
14. Nonrepaired tip maximum bend area

FIGURE 19-25 Repair limits for fan blades. (*Pratt & Whitney Canada.*)

in service provided that the repair adheres to the limits shown for cut 1, 2, or 3 and that the blade is repaired up to the maximum dimension of only one of the permissible cuts.

In the repair of fan blades, certain conditions are specified. For example, all repair cuts must have a length-to-depth ratio greater than 4:1. Contours must be smooth and continuous, with a minimum radius of 0.250 in [6.35 mm]. The leading-edge contour after repair should conform as nearly as possible to the original. Repaired areas must be checked with a dye or fluorescent penetrant to ensure that there are no cracks.

The repair of fan blades while the fan rotor is installed in the engine requires that the area to be reworked be completely masked off to ensure that no metal splatter can strike any other blade or disk surface. Cutting is accomplished with a 2-in [5.08-cm] cutting wheel mounted in an air chuck operating at 18 000 rpm maximum. A minimum of 0.060 in [1.52 mm] of material must be left for hand filing and polishing to ensure removal of any heat-affected areas.

Shingled blades may be unshingled and continued in service for a maximum of 20 h provided that they can be unshingled without further damage and that inspection shows that the midspan shroud of a shingled blade has not hit the airfoil section of an adjacent blade or the radius between the airfoil and the midspan shroud of the adjacent blade. Blades showing evidence of having been hit in this manner must be removed from service before further flight. After 20 h of service, shingled blades should be removed and subjected to overhaul-type inspection.

Compressor Blades

Compressor blades are subject to the same type of damage encountered by fan blades, and the repair procedures are similar. Figure 19-26 is adapted from the maintenance manual for the Pratt & Whitney JT8D engine and shows some of the permissible repairs for compressor blades. Note that there are definite limits on the depth of a cut that is allowed in removing a nick, scratch, or other damage caused by the ingestion of a foreign object. The limits vary in accordance with the part of the blade where the damage is located. The portions of the blade which have higher stresses may not be cut as deeply as the portions subjected to lower stresses during operation. During blade repairs, care must be taken to maintain the original profile of the blade within reasonable limits.

The foregoing examples of blade repair are provided for information only, to illustrate typical practices. For a specific engine, the appropriate specifications given in the maintenance manual must be used.

Compressor and Turbine Wash

As an engine operates, deposits accumulate on the engine's internal gas path components such as the compressor and turbine blades. These deposits can accumulate to the point of deteriorating the engine's performance. To recover this performance loss, a type of compressor wash must be performed to remove the baked-on salt, dirt, or other types of contamination deposits. Cleaning of the engine can be divided into internal and external washing. Washing of the compressor section of the engine is accomplished by injecting the applicable fluid into the engine's intake using either an installed compressor wash ring or a hand-held wash wand. This provides the engine with the correct flow of fluid in the form of a spray. Turbine washes are done in much the same manner except that the wash tube is generally attached to the combustion section of the engine.

Internal engine washing can be done by two methods: the motoring wash and the running wash. The motoring wash is generally done by turning the engine using only the starter. This ensures that the wash fluid stays in a liquid form. The engine is run up by the starter to between 10 and 25 percent rpm. When the engine reaches about 5 percent rpm, the cleaning mixture can be sprayed into the engine's intake as the engine continues to accelerate. Spraying should be stopped as the engine slows back down to about 5 percent rpm. Sometimes it is necessary to do a cleaning wash and a rinse wash to remove the cleaning fluid. Before the internal wash is attempted, the compressor bleed air and any other components

that might become contaminated must be closed off or isolated to prevent contamination. Starter limits must also be adhered to so as to prevent overheating and damaging of the starter. As a general rule, the starter should not be operated for more than 30 s, and the correct cooldown time should be observed between runs. The engine's ignition must also be turned off during the motoring runs.

The running wash is performed with the engine running at idle speed and the cleaning fluid mixture and the rinse solution injected at the correct flow rate.

Some different types of internal engine washes are the compressor performance recovery wash, the compressor desalination wash, and the turbine desalination wash. The engine's maintenance manual will list the proper fluids and frequency of washes. In cold weather and in very contaminated environments, special frequency and fluids need to be used. A schematic of a compressor/turbine wash rig is shown in Fig. 19-27.

Turbine Nozzles and Vanes

First-stage high-pressure turbine nozzles and vanes are subjected to the highest temperatures during operation since they are exposed to the gases as they exit from the combustion chamber. The high temperatures lead to expansion cracks, stress-rupture cracks, some burning, and other damage. Stress-rupture cracks usually appear along the leading edge of the blade. Manufacturers and operators have determined what amounts of damage are acceptable and will not affect the safe operation of the engine. The following list gives examples of serviceable limits for a high-bypass engine in the first-stage nozzle vanes.

Inspection (Borescope)	Serviceability Limits
1. Axial cracks in the trailing edge (concave side only) or in slots adjacent to trailing edge	Any number 0.30 in [7.62 mm] in length allowed provided they are 0.06 in [1.52 mm] apart; or two per vane 0.80 in [20.32 mm] long, provided they are 0.30 in [7.62 mm] apart; or one 1.50 in [3.81 cm] long with two 0.50 in [12.70 mm] long provided they are 0.30 in apart and do not extend forward of the leading edge gill holes
2. Axial cracks in the leading edge	Any number 0.50 in [12.70 mm] long if separated by at least 0.25 in [6.35 mm] or any number interconnecting the cooling holes provided the total length of interconnecting cracks does not exceed 0.60 in [15.24 mm]
3. Radial cracks in the concave surface between the inner and outer platforms	Any number 0.50 in [12.7 mm] long; or two per vane 0.80 in [20.32 mm] long provided they are at least 0.30 in [7.62 mm] apart

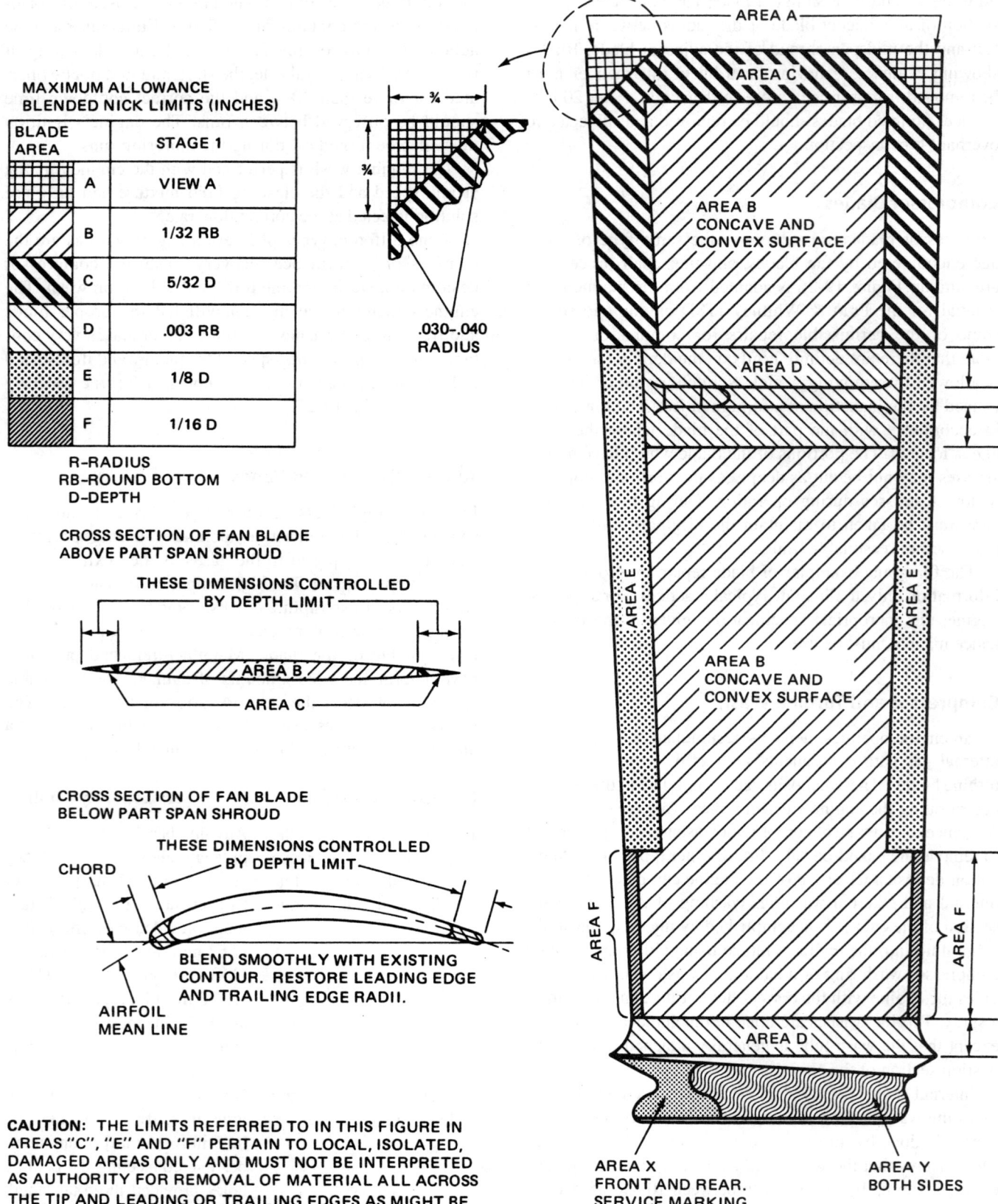

MAXIMUM ALLOWANCE
BLENDED NICK LIMITS (INCHES)

BLADE AREA		STAGE 1
	A	VIEW A
	B	1/32 RB
	C	5/32 D
	D	.003 RB
	E	1/8 D
	F	1/16 D

R–RADIUS
RB–ROUND BOTTOM
D–DEPTH

¾

¾

.030–.040
RADIUS

AREA A

AREA C

AREA B
CONCAVE AND
CONVEX SURFACE

AREA D

AREA E

AREA B
CONCAVE AND
CONVEX SURFACE

AREA E

AREA F

AREA F

AREA D

CROSS SECTION OF FAN BLADE
ABOVE PART SPAN SHROUD

THESE DIMENSIONS CONTROLLED
BY DEPTH LIMIT

AREA B

AREA C

CROSS SECTION OF FAN BLADE
BELOW PART SPAN SHROUD

THESE DIMENSIONS CONTROLLED
BY DEPTH LIMIT

CHORD

BLEND SMOOTHLY WITH EXISTING
CONTOUR. RESTORE LEADING EDGE
AND TRAILING EDGE RADII.

AIRFOIL
MEAN LINE

AREA X
FRONT AND REAR.
SERVICE MARKING
TIME ON REAR, ONLY.

AREA Y
BOTH SIDES

CAUTION: THE LIMITS REFERRED TO IN THIS FIGURE IN AREAS "C", "E" AND "F" PERTAIN TO LOCAL, ISOLATED, DAMAGED AREAS ONLY AND MUST NOT BE INTERPRETED AS AUTHORITY FOR REMOVAL OF MATERIAL ALL ACROSS THE TIP AND LEADING OR TRAILING EDGES AS MIGHT BE DONE IN A SINGLE MACHINING CUT.

FIGURE 19-26   Repair limits for compressor blades. (*Pratt & Whitney Canada.*)

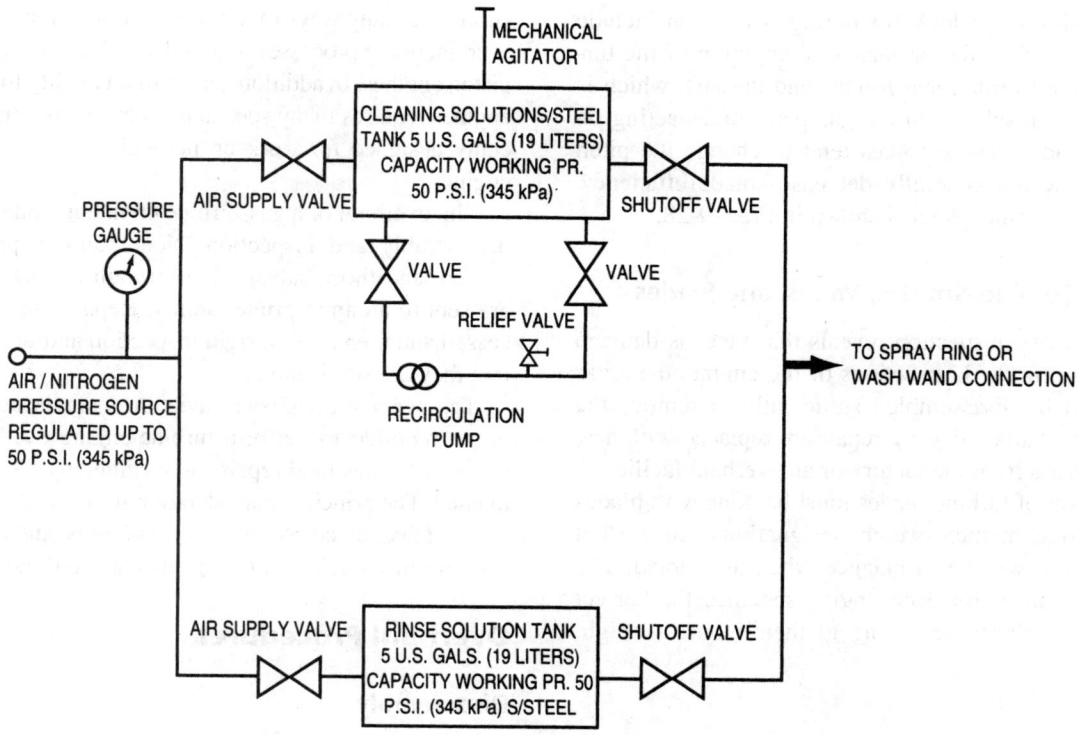

FIGURE 19-27   Compressor wash ring schematic. (*Pratt & Whitney Canada.*)

4. Radial cracks in the convex surfaces between the inner and outer platforms — One crack allowed 0.80 in [20.32 mm] long

5. Blocked cooling air passages — Five nose holes and four gill holes in each row. A minimum separation of one open hole shall exist between blocked holes. Three trailing-edge slots may be blocked provided that blocked slots are not adjacent

6. Nicks, scores, and scratches — Any number, any length allowed if not over 0.03 in [0.76 mm] deep. Nicks up to 0.10 in [2.54 mm] deep and 0.25 in [6.35 mm] long allowed on airfoil trailing edge

7. Buckling or bowing of the trailing edge — Any number up to 0.30 in [7.62 mm] from original contour

8. Axial cracks in concave surface — Two per vane extending aft from the first row of gill holes to (not through) the slot in the trailing edge

9. Axial cracks in convex surface — Two per vane between the midchord strut and the trailing edge, 0.25 in [6.35 mm] radially apart, total not to exceed 1.0 in [2.54 cm]. One per vane, length not to exceed 1.0 in; width not to exceed 0.50 in [12.70 mm] aft of the midchord strut. A maximum of three vanes per assembly, not adjacent

10. Burns in the trailing edge (loss of metal) — The total area removed from the trailing edge not to exceed 3.00 in² [19.35 cm²] per assembly. Cumulative area is determined by summing individual vane area, radial height by axial length

11. Burns and cracks on the convex and concave sides — Not to exceed an area of 1.50 in [3.81 cm] long and 1.0 in [2.54 cm] wide per vane. Maximum of four vanes per 90° arc

12. Burns or spalling on vane leading edge (charred only, no holes through airfoil) — 0.50 in [12.70 mm] diameter per vane, maximum of four vanes affected per 90° arc

13. Craze cracking (craze cracking is defined as uperficial surface cracks which have no visual width or depth) — Any amount

Turbine Blades

Serviceability limits for turbine blades are much more stringent than are those for nozzle vanes. This is particularly true for first-stage blades because of the high temperatures involved. The centrifugal stresses to which turbine blades are subjected require that the blades be free of cracks in any area and that no nicks or dents exist in the root area. A limited number of small nicks and dents can be permitted in the areas of the blade away from the root area. No burning or distortion is permitted.

Other conditions to look for during inspection include blade creep, which is the permanent elongation of the turbine blades due to rotational forces, and untwist, which is a condition that results from the gas path forces acting on the turbine blades. These forces tend to change the pitch of the blade, which generally decreases blade efficiency. A compressor-turbine wheel is shown in Fig. 19-28.

Repairs for Turbine Nozzles, Vanes, and Blades

When a borescope inspection reveals that there is damage or deterioration in the hot sections of the engine, the areas involved must be disassembled sufficiently to remove the defective parts. Parts requiring repair are replaced with new or reworked parts from the factory or an overhaul facility.

Replacement of turbine blades must be done with blades having the correct moment-weight designation to ensure that the turbine rotor will be in balance when assembled. The maintenance manual for each engine specifies the correct arrangement of blades according to their moment-weight markings.

GAS-TURBINE ENGINE OVERHAUL

In the past, most engines had specified numbers of hours they could operate before they needed to be overhauled. This period became known as the **time between overhauls (TBO)**.

The length of time between overhauls varies widely with different types of engines. When a new type of engine enters service, its TBO is fairly short, but as condition monitoring, the engine's service record, and inspections prove the engine to be reliable, the TBO is generally extended. Many engines have proven to be so reliable that they are overhauled only when they need major maintenance. This concept is a form of "**on condition**" maintenance or overhaul.

Because the TBO is actually determined by the life of one or two assemblies within the engine, during overhaul it is generally found that the other assemblies are mechanically sound and fit to continue in service for a much longer period. Therefore, with the introduction of modular engines and the improved inspection and monitoring techniques available, the TBO method of limiting the engine's life on-wing has been replaced by the "on condition" method. Basically this means that a life is not declared for the total engine but only for certain parts of the engine. On reaching their life limits, these parts are replaced and the engine continues in service, with the remainder of the engine being overhauled "on condition." Modularly constructed engines are particularly suited to this method, because the module containing a life-limited part can be replaced by a similar module and the engine returned to service with minimum delay. The module is then returned to the manufacturer or the overhaul shop and is disassembled for life-limited part replacement, repair, or complete overhaul, as required.

The overhaul of gas-turbine engines is accomplished by the manufacturer or at approved overhaul stations. The process is similar in many ways to the overhaul of piston engines; however, there are processes required which are not necessary for piston engines. In addition, an overhaul facility for gas-turbine engines requires many special tools and some equipment specially designed for work on particular types and models of engines.

The overhaul of a gas-turbine engine includes a complete disassembly and inspection. Nonrepairable parts are discarded, and those salvageable through rework or recycling are sent to an appropriate facility. Repairable parts are processed and then given a rigid inspection and/or test to ensure that they are serviceable.

The average certificated aviation maintenance technician is not required to perform turbine-engine overhaul but can perform various field repairs as specified by the maintenance manual. The principal consideration for the technician is to be sure to have the correct manuals, bulletins, and other instructions when servicing and repairing a gas-turbine engine.

Overhaul Procedures

Disassembly

The engine can be disassembled in the vertical or horizontal position. When it is disassembled in the vertical position, the engine is mounted, usually front end downward, on a floor fixture. To enable it to be disassembled horizontally, the engine is mounted in a special turnover stand.

The engine is disassembled into main **subassemblies** or **modules**, which are fitted in separate stands and sent to other areas where they are further disassembled into individual parts. The individual parts are taken to a cleaning area in preparation for inspection.

Cleaning

The cleaning agents used during overhaul range from organic solvents to acids, and other chemical cleaners, and extend to electrolytic cleaning solutions. Organic solvents include kerosene for washing, trichloroethane for degreasing, and paint-stripping solutions which can generally be used on the majority of components for carbon and paint removal. The more restricted and sometimes rigidly controlled acids and other chemical cleaners are used for removing corrosion, heat scale, and carbon from certain components. To achieve the highest degree of cleanness needed to perform the detailed inspection that is considered necessary for certain major rotating parts, such as turbine disks, electrolytic cleaning solutions are often used.

Inspections

After the components have been cleaned, they are visually and, when necessary, dimensionally inspected to establish their general condition and then further inspected for cracks. Inspection for cracks includes magnification, magnetic, or penetrant inspection techniques, used either alone or consecutively, depending on the components being inspected and the degree of inspection necessary.

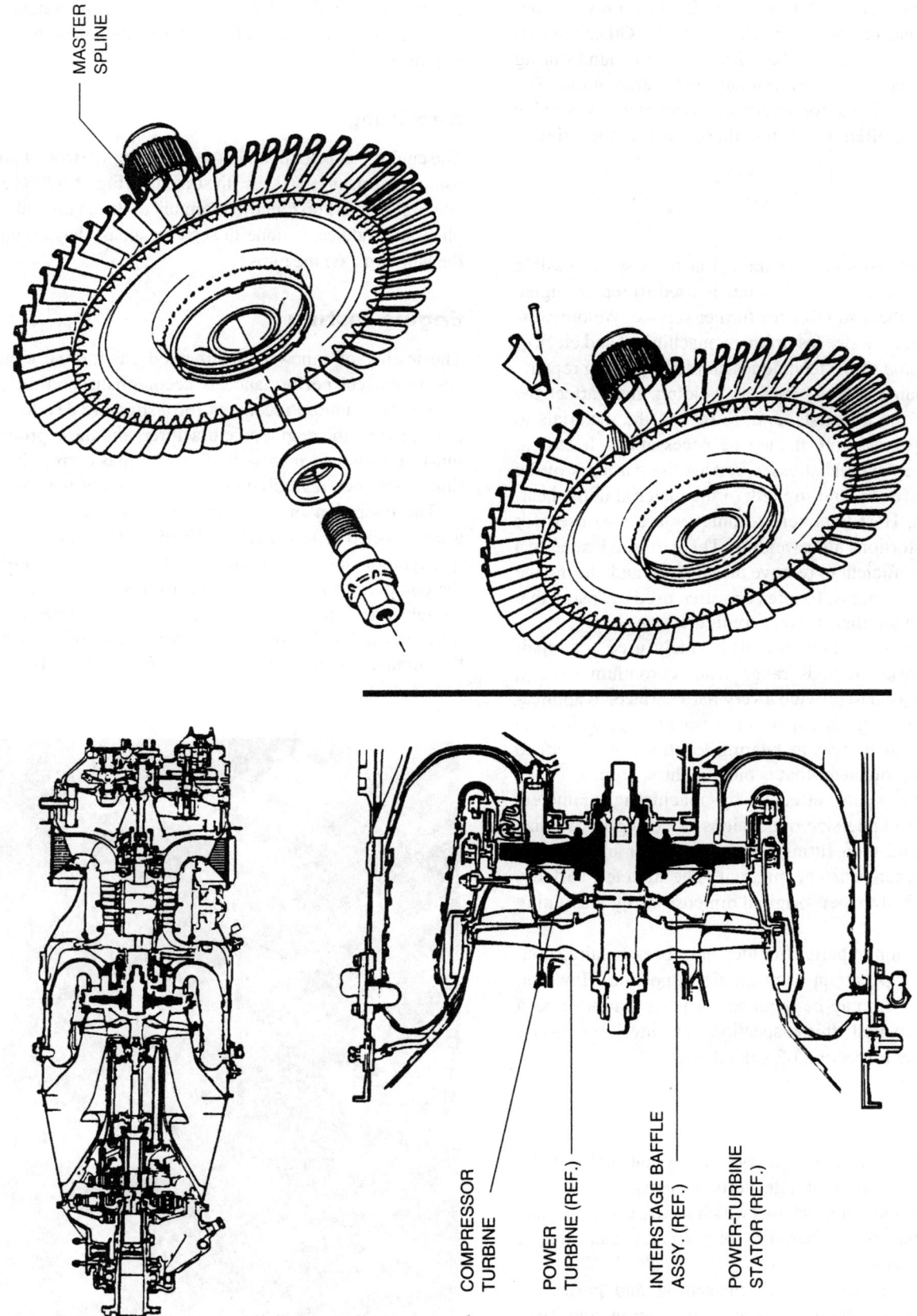

MASTER
SPLINE

COMPRESSOR
TURBINE

POWER
TURBINE (REF.)

INTERSTAGE BAFFLE
ASSY. (REF.)

POWER-TURBINE
STATOR (REF.)

FIGURE 19-28   Typical compressor turbine. (*Pratt & Whitney Canada.*)

Dimensional inspection consists of measuring specific components to ensure that they are within the limits and tolerances given in the Table of Limits. Some of the components are measured at each overhaul because only a small amount of wear or distortion is permissible. Other components are measured only when the condition found during visual inspection requires dimensional verification. The tolerances laid down for overhaul, supported by service experience, are often wider than those used during original manufacture.

Repair

To ensure that costs are maintained at the lowest possible level, a wide variety of techniques is used to repair engine parts to make them suitable for further service. Welding, fitting of interference sleeves or liners, machining, and electroplating are some of the techniques employed during repair.

Some repair methods, such as welding, may affect the properties of the materials, and, to restore the materials to a satisfactory condition, it may be necessary to heat-treat the parts to remove the stresses, reduce the hardness of the weld area, or restore the strength of the material in the heat-affected area. Heat-treatment techniques are also used for removing distortions after welding. The parts are heated to a temperature sufficient to remove the stresses, and, during the heat-treatment process, fixtures are often used to ensure that the parts maintain their correct configurations.

Electroplating methods are also widely used for repair purposes. These methods range from chromium plating, which can be used to provide a very hard surface, to application of thin coatings of copper or silver plating, which can be applied to such areas as bearing locations on a shaft to restore a fitting diameter that is only slightly worn.

Many repairs are effected by machining diameters and/or faces to undersize dimensions or boring to oversize dimensions and then fitting shims, liners, or metal spray coatings of wear-resistant material. The affected surfaces are then restored to their original dimensions by machining or grinding.

The inspection of parts after they have been repaired consists mainly of penetrant or magnetic inspection. However, further inspection may be required for parts that have been extensively repaired; this inspection may involve pressure testing or x-ray inspection of welded areas.

Balancing

Because of the high rotational speeds, any unbalance in the main rotating assembly of a gas-turbine engine is capable of producing vibrations and stresses which increase as the square of the rotational speed. Therefore, very accurate balancing of the rotating assembly is necessary.

The two main methods of measuring and correcting unbalance are **single-plane** (static) **balancing** and **two-plane** (dynamic) **balancing**. Single-plane balancing is used when the unbalance is in one plane only; that is, the unbalance goes centrally through the component at 90° to the axis.

The single-plane method is appropriate for components such as individual compressors and turbine disks. For compressor assemblies and turbine-rotor assemblies possessing appreciable axial length, unbalance may be present at many positions along the axis; therefore, two-plane balancing may be required.

Assembling

The engine can be built up in the vertical or horizontal position using a **ram**, or stand, as illustrated in Fig. 19-29 (vertical) and Fig. 19-30 (horizontal). Assembly of the engine subassemblies, or modules, is done in separate areas, thus minimizing the build time on the rams.

Engine Testing

The testing of a new or overhauled gas-turbine engine to ensure correct performance is accomplished on an instrumented test stand. Procedures for testing are developed and published by the engine manufacturer, and these procedures must be followed precisely to ensure that correct information is obtained regarding the performance of the engine.

The operation of an engine on a test stand is usually accomplished with a **bellmouth air inlet**. The purpose of this type of inlet is to eliminate any loss of air pressure at the compressor inlet. The reason for loss of pressure with a straight inlet and the effect of the bellmouth inlet are illustrated in Fig. 19-31. Since a large volume of air is drawn into the engine, a rapid increase in air velocity must take place as

FIGURE 19-29   Vertical engine assembly. (*Rolls-Royce.*)

FIGURE 19-30 Horizontal engine assembly. (*Rolls-Royce.*)

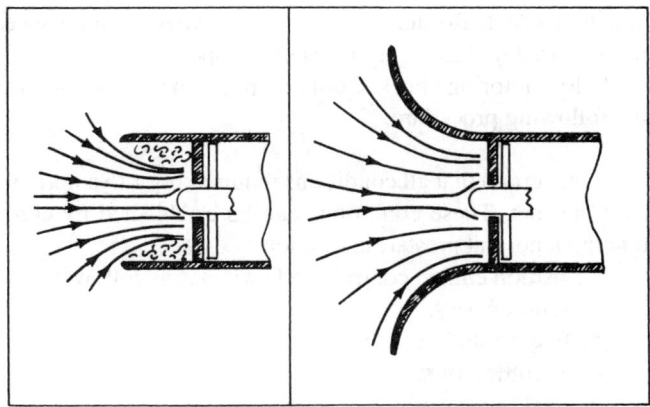

FIGURE 19-31 Effect of a bellmouth air inlet.

the air nears the inlet. Moreover, to supply the demand, air must flow from areas outside the area directly in front of the engine. Much of the airflow will have to change direction almost 90° as it comes from the sides of the inlet and enters the compressor. With a straight inlet duct, this directional change results in a pressure drop. However, the bellmouth duct guides the air in such a way that there is essentially no pressure drop at the compressor inlet. If the bellmouth duct is protected by a screen, a certain amount of pressure drop will occur and must be taken into consideration when the performance of the engine is measured.

Turbine Engine Calibration

Some of the most important factors affecting turbine engine life are EGT, engine cycles (a cycle is generally a takeoff and landing), and engine speed. Excess EGT of a few degrees reduces turbine component life. Low EGT materially reduces

turbine engine efficiency and thrust. So, to make the engine highly efficient, the exhaust temperatures need to be as high as possible while maintaining an EGT operating temperature that does not damage the turbine section of the engine. If the engine is operated at excess exhaust temperature, engine deterioration occurs. Since the EGT temperature is set by the EGT temperature gauge, it is imperative that it is accurate. Excessive engine speed can cause premature engine wear and, if extreme can cause engine failure.

In testing of a gas-turbine engine, it is common practice to measure certain essential parameters in order to evaluate the engine performance correctly. Among these parameters are the following:

1. Ambient air temperature (T_{amb})
2. Ambient air pressure (P_{amb})
3. Exhaust total pressure (P_{t7})
4. Low-pressure compressor rpm (N_1)
5. High-pressure compressor rpm (N_2)
6. Exhaust-gas temperature (EGT)
7. Fuel flow in pounds per hour (pph) (W_f)
8. Thrust (F_n)
9. Low-pressure compressor outlet pressure (P_{s3})
10. High-pressure compressor outlet pressure (P_{s4})

These parameters are usually adequate to determine engine performance, but others may be recorded if desired or necessary.

When an engine is assembled as a complete powerplant for a quick engine change (QEC), it is necessary to consider the equipment installed on the engine, because it may affect some of the performance measurements. Oil flow and temperature will be changed as a result of the engine oil cooler and the engine pump. Likewise, fuel flow and pressures will be affected by the engine-driven fuel pump.

Because standard performance of an engine occurs only under standard conditions, air pressure and temperature must be corrected to standard conditions. This is accomplished by means of correction factors designated by the Greek letters **delta** (δ) and **theta** (θ). Delta is used to correct for pressure and theta provides the correction for temperature.

The values for delta and theta may be found on an appropriate chart or they may be calculated as follows:

$$\delta = \frac{P}{P_0} = \frac{P}{29.92}$$

$$\theta = \frac{T}{T_0} = \frac{t(°F)+460}{519}$$

where P = observed barometric pressure (in HG abs)
P_0= standard-day barometric pressure
T = temperature, °R (°F + 460)
T_0= standard-day temperature, 519°R

If Kelvin degrees are used to indicate absolute temperature, then the Celsius or centigrade scale is used. Adding 273 converts centigrade to Kelvin. Standard-day temperature in degrees Kelvin is 288.

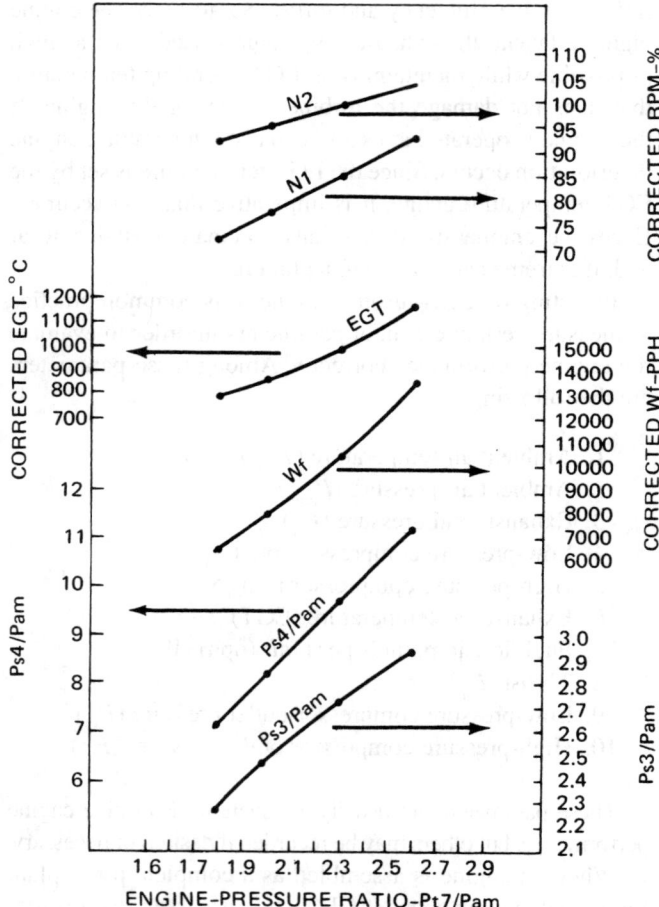

FIGURE 19-32 Typical operational curves for a gas generator.

To apply δ and θ to the correction or measurements, the following methods are employed:

$$N_2 \text{ (corrected)} = \frac{N_2 \text{(observed)}}{\sqrt{\theta_{t2}}}$$

$$EGT°R \text{ (corrected)} = \frac{EGT \text{ (observed)} + 460}{\sqrt{\theta_{t2}}}$$

$$W_f \text{(corrected)} = \frac{W_f \text{(observed)}}{\Delta_{r2}\sqrt{\theta_{t2}}}$$

$$F_n \text{(corrected)} = \frac{F_n \text{(observed)} + \text{Inst corrected} + \text{cell corrected}}{\delta_{t2}}$$

The values observed and calculated as shown for N_2 rpm, EGT, and fuel flow are recorded and plotted on a chart, as shown in Fig. 19-32. Note that the pressure ratios for the low-pressure compressor and high-pressure compressor are recorded. These pressure ratios are indicated as P_{s3}/P_{amb} for the low-pressure compressor and P_{s4}/P_{amb} for the high-pressure compressor. The net thrust (F_n) of an engine can be determined directly from the thrust meter in the cell and can also be found from the EPR (engine pressure ratio) and an EPR conversion table for the engine.

The foregoing discussion is presented as an example of how the performance of an engine can be determined. In actual practice, other tests may be performed and other parameters measured. For any particular type or model of engine, specific instructions are made available by the manufacturer for the testing of the engine in a test cell or on the aircraft.

Operational Checks

To ensure that a gas-turbine engine is in satisfactory operating condition, engine and aircraft manufacturers specify certain **operational checks** to be routinely performed by maintenance personnel. The particular types of checks and the procedures to be followed vary, depending on the type of engine and aircraft involved. In this section, to provide an example of typical checks, the checks recommended for the General Electric CF6-50 engine are described briefly.

Dry motoring check. The dry motoring check may be required during or after inspection or maintenance to ensure that the engine rotates freely, that instrumentation functions properly, and that starter operation meets speed requirements for successful starts. This check is also used to prime and leak-check the lubrication system when maintenance has required replacement of system components.

A dry motoring check should be performed according to the following procedure:

1. Ascertain that all conditions required prior to a normal start are met. These conditions can be established by conducting a normal prestart inspection.
2. Position engine controls and switches as follows:
   a. Ignition, OFF.
   b. Fuel shutoff lever, OFF.
   c. Throttle, IDLE.
   d. Fuel boost, ON.

3. Energize the starter and motor the engine as long as necessary to check instruments for positive indications of engine rotation and oil pressure.
4. Deenergize the starter and make the following checks during coastdown:

   a. Listen for unusual noises. Check for roughness. Normal noise consists of clicking of compressor and turbine blades, and gear noise.
   b. Inspect the lubricating system lines, fittings, and accessories for leakage.
   c. Check the oil level in the oil tank.

Wet motoring check. When it is necessary to check the operation of fuel-system components after removal and replacement or to perform a depreservation of the fuel system, the wet motoring check is employed. This is accomplished as follows:

1. Position the engine controls and switches as for a dry motoring check.
2. Energize the starter.

3. When core engine speed (N₂) reaches 10 percent, move the fuel shutoff lever to ON and check for oil pressure indication.

4. Continue motoring the engine until the fuel flow is 500 to 600 lb/h [226.80 to 272.16 kg/h] or for a maximum of 60 s. Observe the starter operating limits.

5. Move the fuel shutoff lever to OFF and continue motoring the engine for at least 30 s to clear the fuel from the combustion chamber. Check to see that fuel flow drops to zero.

6. Deenergize the starter and, during coastdown, check for unusual noises as mentioned for a dry motoring check.

7. Inspect the fuel system lines, fittings, and accessories for leakage.

8. Check the concentric fuel shroud for leakage. No leakage is permitted.

9. Inspect the lubrication system for leakage.

10. Check the oil level in the oil tank.

Idle check. The idle check consists of checking for proper engine operation as evidenced by leak-free connections, normal operating noise, and correct indications on engine-related instruments. Engine drain lines must be disconnected from drain cans to check for leakage.

After the engine is started according to approved procedure, the following steps should be taken:

1. Stabilize engine at ground idle.

2. Check fan speed (N₁), core engine speed (N₂), oil pressure, and exhaust-gas temperature (EGT) to see that they are within the proper ranges according to the ground idle speed chart and engine specifications. Engine speeds will vary according to compressor inlet temperature (T_{t2}).

3. Visually inspect fuel, lubrication, and pneumatic lines, fittings, and accessories for leakage.

4. Deenergize flight-idle solenoid. During operations above ground idle, do not exceed the open-cowling limitations imposed by the airframe manufacturer.

5. Stabilize at flight idle and check the same parameters checked for ground idle. See that they are within the limitations set forth on the flight idle speed charts.

Power assurance check. The power assurance check is performed to make sure that the engine will achieve takeoff power on a hot day without exceeding rpm and temperature limitations. During the tests, the engine being tested is not used to supply power for any aircraft systems—electric, hydraulic, or other. The engine is tested at 50 percent, 75 percent, and maximum power.

During engine operation for the power assurance check, EGT must be observed constantly to avoid the possibility of overtemperature. Should the temperature approach maximum allowable, the throttle must be retarded sufficiently to hold the EGT within limits. In the operation of the engine, the throttle should always be moved slowly.

To perform the power assurance test, follow these steps:

1. Set the engine power at nominal N₂ speed as indicated on the appropriate chart for the total air temperature (TAT).

For example, the nominal N₂ curve on the chart may coincide with the 91.8 percent line at 10°C for the 50 percent power setting. The throttle will therefore be adjusted to produce 91.8 percent rpm when the TAT is 10°C for a 50 percent power setting.

2. Four minutes after the throttle lever is set, record the average readings of TAT, N₁ speed, N₂ speed, EGT, EPR (engine pressure ratio), and fuel flow (W_f). Correct W_f for local barometric pressure in accordance with instructions.

$$\text{Corrected } W_f = \frac{\text{observed } W_f \times 29.92}{\text{actual barometric pressure}}$$

3. Using N₁ (where N₁ = target N₁ − observed N₁) as a correction factor, adjust readings according to the parameter adjustments set forth in the operations manual.

In the operation of gas-turbine engines of any type, it must be emphasized that temperatures and rpm for both N₁ and N₂ must be watched carefully. If it is expected that a beyond-limits condition is developing, the operator should take immediate action by retarding the throttle or shutting the engine down.

Before a hot engine is shut down, it should be operated at ground idle speed for about 3 min to permit temperature reduction and stabilization. As soon as the engine is shut down, the EGT gauge should be observed to see that EGT starts to decrease. If EGT does not decrease, an internal fire is indicated, and the engine should be dry-motored at once to blow out the fire.

During coastdown after the engine is shut down, a technician should listen for unusual noises in the engine such as scraping, grinding, bumping, and squealing.

Preparing Engine for Storage and Transportation

The preparation of the engine for storage and transportation is of major importance, because storage and transportation call for special treatment to preserve the engine. So that the fuel system will resist corrosion during storage, it is filled with a special oil and all openings are sealed off. The external and internal surfaces of the engine are also protected, either by special inhibiting powders or by paper impregnated with inhibiting powder. The engine is enclosed in a reusable bag or plastic sheeting into which a specific amount of desiccant is inserted. If the engine is to be transported, it is often packed in a crate or metal case.

Engine Trimming and Adjustment

Trimming a gas-turbine engine is the process of adjusting the fuel control unit so that the engine will produce its rated thrust at the designated rpm. The thrust is determined by measuring the **engine pressure ratio (EPR)**, which is the ratio of turbine discharge pressure to engine inlet pressure (P_{t6}/P_{t2}). On engines equipped with variable compressor vanes, it is necessary to check the vane angles and the operation of the **engine vane control** (EVC) during the trimming process. The trimming of a gas-turbine engine may be compared with the tuning of a piston engine for optimum performance. Gas-turbine engines with computer-controlled fuel systems do not require trimming because trimming adjustments are made automatically by the fuel control computer.

Gas-turbine engines manufactured by Pratt & Whitney are tested at the factory and adjusted to produce rated thrust. The engine speed (N_2) which is required for the engine to deliver rated thrust is stamped on the engine data plate or recorded on the engine data sheet of the engine log book. This information is supplied in both rpm and percent of maximum rpm. Because of manufacturing tolerances and slight variations which occur during the manufacture of engines, no two engines are exactly alike, and very rarely will two engines of the same model produce rated thrust at exactly the same rpm. The rpm for rated thrust stamped on the data plate will therefore vary from engine to engine.

Engine trimming is required from time to time because of changes that take place during the life of the engine. Dust and other particulate matter will adhere to the surfaces of the compressor rotor blades and stator vanes and lead to a slight resistance to airflow. Erosion of the leading edges of blades and vanes caused by dust, sand, and other material changes the characteristics and performance of the compressor. The turbine blades and vanes, which are exposed to very high temperatures, are subject to corrosion, erosion, and distortion. All the foregoing factors tend to cause the engine thrust to decrease over a period of time; therefore, trimming is necessary to restore the rated performance of the engine. Generally speaking, when an engine indicates high exhaust-gas temperature (EGT) for a particular EPR, it means that the engine is out of trim.

The following general principles for trimming are for information only:

1. Head the airplane as nearly as possible into the wind. Wind velocity should not be more than 20 mph [32.19 km/h] for best results. See that the area around the aircraft is clean and free from items which could enter the engine or cause other problems during the engine run.

2. Install the calibrated instruments required for trimming. One of these instruments is a pressure gauge for reading turbine discharge pressure (P_{t7}) or EPR. Another important instrument is the calibrated tachometer, which is used to read N_2 rpm.

3. Install a part-throttle stop or fuel control trim stop as specified in the trim instructions.

4. Record ambient temperature and barometric pressure. These values are necessary to correct performance readings to standard sea-level conditions. The pressure and temperature information is used to determine the desired turbine discharge pressure or EPR by means of the **trim curve** published for the engine.

5. Start the engine and operate it at idle speed for the time specified to ensure that all engine parameters have stabilized. Operate the engine at trim speed as established by the trim stop on the fuel control for about 5 min to stabilize all conditions. The overboard air-bleed valves should be fully closed and all accessory air bleed must be turned off.

6. Observe and record the P_{t7} or EPR to determine how much trimming (if any) is required. If trimming is required, adjust the fuel control unit to give the desired P_{t7} or EPR. When this is attained, record the engine rpm, the EGT, and the fuel flow.

7. The observed rpm is corrected for speed bias by means of a temperature-vs.-rpm curve to provide a new engine trim speed in percent corrected to standard conditions.

Note that these procedures will vary considerably, depending on the type and model of engine being trimmed. The purpose of trimming for all engines, however, remains the same: to provide optimum engine performance without exceeding the limits of rpm and temperature established for the engine. As explained previously, engines equipped with computer-controlled fuel controls do not require periodic trimming because the adjustments are made automatically by the computer.

Functional Check of Aircraft EGT Circuit

During the EGT system functional test and the thermocouple harness checks, the analyzer has a specific degree of accuracy at the test temperature, which is usually the maximum operating temperature of the turbine engine (Fig. 19-33). Each engine has its own maximum operating temperature, that can be found in applicable technical instructions.

The test is made by heating the engine thermocouples in the exhaust nozzle or turbine section to the engine test temperature. The heat is supplied by heater probes through the necessary cables. With the engine thermocouples hot, their temperature is registered on the aircraft EGT indicator. At the same time, the thermocouples embedded in the heater

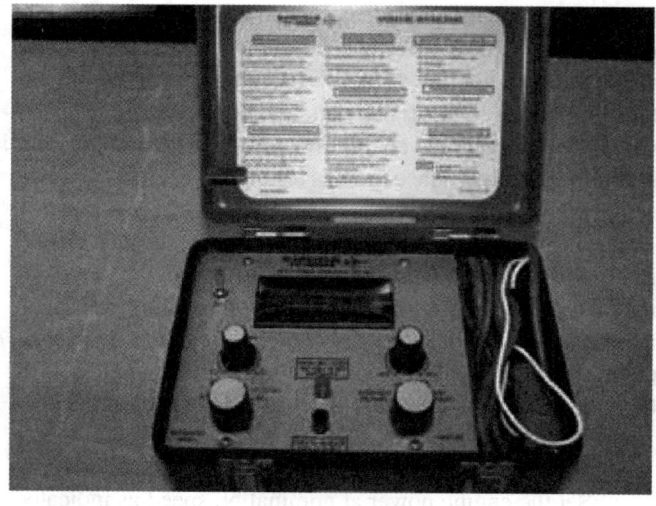

FIGURE 19-33 EGT analyzer.

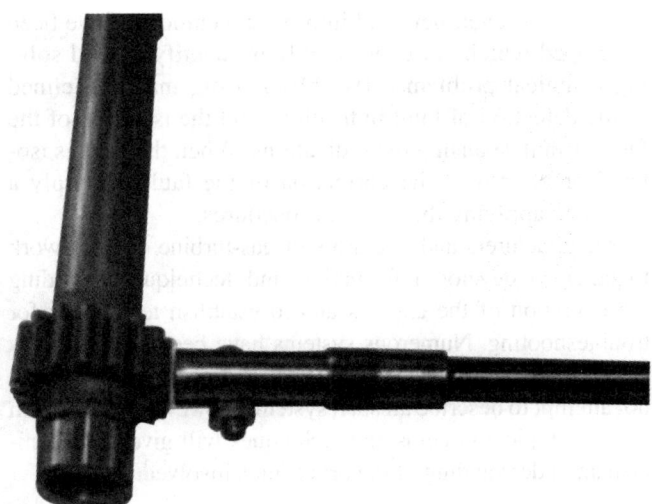

FIGURE 19-34 Magnetic pickup and gear.

probes, which are completely isolated from the aircraft system, are picking up and registering the same temperature on the test analyzer.

The temperature registered on the aircraft EGT indicator should be within the specified tolerance of the aircraft system and the temperature reading on the temperature analyzer. When the temperature difference exceeds the allowable tolerance, troubleshoot the aircraft system.

EGT Indicator Check

The EGT indicator is tested after being removed from the aircraft instrument panel and disconnected from the aircraft EGT circuit leads. Attach the instrument cable and EGT indicator adapter leads to the indicator terminals, and place the indicator in its normal operating position. Adjust the analyzer switches to the proper settings. The indicator reading should correspond to the readings of the analyzer within the allowable limits of the EGT indicator.

Correction for ambient temperature is not required for this test, as both the EGT indicator and analyzer are temperature compensated. The temperature registered on the aircraft EGT indicator should be within the specified tolerance of the aircraft system and the temperature reading on the analyzer readout. When the temperature difference exceeds the allowable tolerance, troubleshoot the aircraft system.

Resistance and Insulation Check

The thermocouple harness continuity is checked while the EGT system is being checked functionally. The resistance of the thermocouple harness is held to very close tolerances, since a change in resistance changes the amount of current flow in the circuit. A change of resistance gives erroneous temperature readings. The resistance and insulation check circuits make it possible to analyze and isolate any error in the aircraft system. How the resistance and

insulation circuits are used is discussed with troubleshooting procedures.

Tachometer Check

To read engine speed with an accuracy of ±0.1 percent during engine run, the frequency of the tachometer-generator (older style) is measured by the rpm check analyzer. The scale of the rpm check circuit is calibrated in percent rpm to correspond to the aircraft tachometer indicator, which also reads in percent rpm. The aircraft tachometer and the rpm check circuit are connected in parallel, and both are indicating during engine run-up. The rpm check circuit readings can be compared with the readings of the aircraft tachometer to determine the accuracy of the aircraft instrument.

Many newer engines use a magnetic pickup that counts passing gear teeth edges, which are seen electrically as pulses of electrical power as they pass by the pickup (Fig. 19-34). By counting the amount of pulses, the rpm of the shaft is obtained. This type of system requires little maintenance, other than setting the clearance between the gear teeth and the magnetic pickup.

TROUBLESHOOTING EGT SYSTEM

An appropriate analyzer is used to test and troubleshoot the aircraft thermocouple system at the first indication of trouble, or during periodic maintenance checks.

The test circuits of the analyzer make it possible to isolate the troubles listed below. Following the list is a discussion of each trouble mentioned.

1. One or more inoperative thermocouples in engine parallel harness
2. Engine thermocouples out of calibration
3. EGT indicator error
4. Resistance of circuit out of tolerance
5. Shorts to ground
6. Shorts between leads

One or More Inoperative Thermocouples in Engine Parallel Harness

This error is found in the regular testing of aircraft thermocouples with a hot heater probe and is a broken lead wire in the parallel harness, or a short to ground in the harness. In the latter case, the current from the grounded thermocouple can leak off and never be shown on the indicator. However, this grounded condition can be found by using the insulation resistance check.

Engine Thermocouples Out of Calibration

When thermocouples are subjected for a period of time to oxidizing atmospheres, such as encountered in turbine engines, they drift appreciably from their original calibration.

On engine parallel harnesses, when individual thermocouples can be removed, these thermocouples can be bench-checked, using one heater probe. The temperature reading obtained from the thermocouples should be within manufacturer's tolerances.

EGT Circuit Error

This error is found by using the EGT and comparing the reading of the aircraft EGT indicator with the analyzer temperature reading. The analyzer and aircraft temperature readings are then compared.

Resistance of Circuit Out of Tolerance

The engine thermocouple circuit resistance is a very important adjustment since a high-resistance condition gives a low indication on the aircraft EGT indicator. This condition is dangerous, because the engine is operating with excess temperature, but the high resistance makes the indicator read low. It is important to check and correct this condition.

Shorts to Ground/Shorts between Leads

These errors are found by doing the insulation check using an ohmmeter. Resistance values from zero to 550 000 ohms can be read on the insulation check ohmmeter by selecting the proper range.

TROUBLESHOOTING AIRCRAFT TACHOMETER SYSTEM

A function of the rpm check is troubleshooting the aircraft tachometer system. The rpm check circuit in the analyzer is used to read engine speed during engine run-up with an accuracy of ±0.1 percent. The connections for the rpm check are the instrument cable and aircraft tachometer system lead to the tachometer indicator. After the connections have been made between the analyzer rpm check circuit and the aircraft tachometer circuit, the two circuits, now classed as one, are a parallel circuit. The engine is then run-up as prescribed in applicable technical instructions. Both systems can be read simultaneously.

If the difference between the readings of the aircraft tachometer indicator and the analyzer rpm check circuit exceeds the tolerance prescribed in applicable technical instructions, the engine must be stopped, and the trouble must be located and corrected.

GAS-TURBINE ENGINE TROUBLESHOOTING

The troubleshooting of turbine engines follows, in general, the procedures traditionally employed for reciprocating

engines; however, new and improved techniques have been developed which aid considerably in identifying and solving technical problems. **Troubleshooting** may be defined as the detection of fault indications and the isolation of the fault or faults causing the indications. When the fault is isolated or identified, the correction of the fault is simply a matter of applying the correct procedures.

Manufacturers and operators of gas-turbine engines work together to develop information and techniques regarding the operation of the engines and to establish techniques for troubleshooting. Numerous systems have been developed by which faults are detected, analyzed, and corrected. We shall not attempt to describe all such systems; however, a discussion of some typical systems and techniques will give the technician an understanding of the procedures involved.

Fault Indicators

Fault indicators include any instruments or devices on an aircraft which can give a member of the crew information about a problem developing in the operation of the engine. These indicators may be divided into two groups: the standard engine instruments used to monitor the operation of the engines, and special devices designed to detect indications of trouble which may not be revealed by the engine instruments. Typical engine instruments for a gas-turbine engine are EGT gauges, percent rpm gauges (N_1 and N_2), EPR gauges, oil temperature gauges, oil pressure gauges, and fuel gauges. When turboshaft or turboprop engines are installed, torque-indicating gauges are often included. These instruments are all effective in detecting faults.

In addition to the standard instruments, built-in troubleshooting equipment (BITE) systems are often installed. These systems include special sensors and transducers which produce signals of vibration and other indications that are indicative of developing problems. By comparing the tendency of engine parameters to change up or down, the technician can troubleshoot engine problems. Engine speed (% rpm), exhaust-gas temperature (EGT), and fuel flow (W_f) are primary engine parameters for troubleshooting. Figure 19-35 gives some examples showing this concept. For long-term troubleshooting, these parameters are monitored over a period of time so that a trend can be noticed. This will allow corrective action to be taken as promptly as possible.

Condition Monitoring, Trend Monitoring, and Gas Path Health

Aviation maintenance and operations groups are continually striving to improve the service reliability of their gas-turbine engines and, at the same time, reduce operating costs. One tool which can aid both of these efforts is engine performance monitoring, through trend analysis. Trend analysis involves the recording and analysis of gas-turbine engine performance and certain mechanical parameters over a period of time.

The primary aim of trend analysis is to provide a means of detecting significant changes in the performance parameters

GRAPH	SYMPTOM	MOST PROBABLE SOLUTION

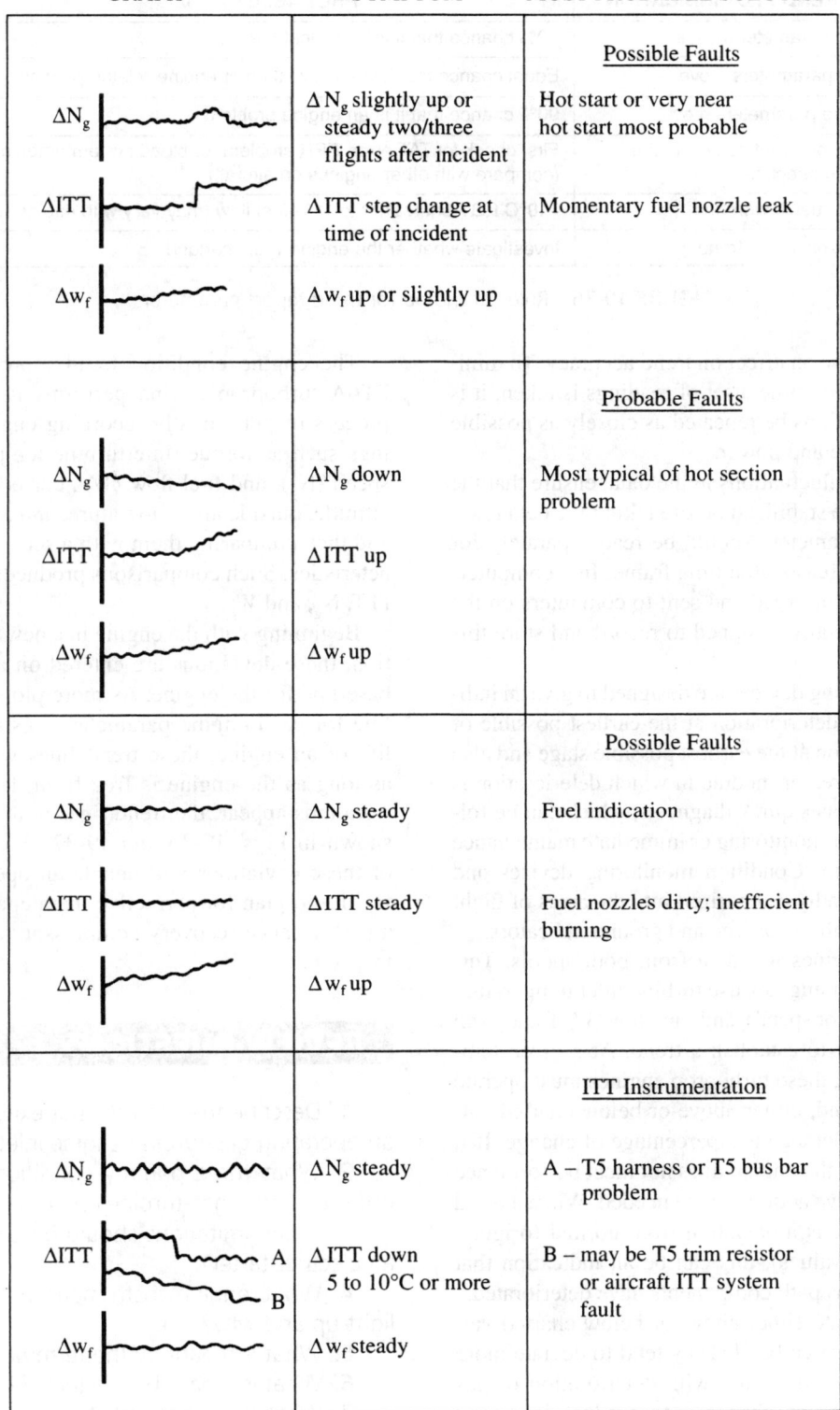

		Possible Faults
ΔN_g	ΔN_g slightly up or steady two/three flights after incident	Hot start or very near hot start most probable
ΔITT	ΔITT step change at time of incident	Momentary fuel nozzle leak
Δw_f	Δw_f up or slightly up	
		Probable Faults
ΔN_g	ΔN_g down	Most typical of hot section problem
ΔITT	ΔITT up	
Δw_f	Δw_f up	
		Possible Faults
ΔN_g	ΔN_g steady	Fuel indication
ΔITT	ΔITT steady	Fuel nozzles dirty; inefficient burning
Δw_f	Δw_f up	
		ITT Instrumentation
ΔN_g	ΔN_g steady	A – T5 harness or T5 bus bar problem
ΔITT	ΔITT down 5 to 10°C or more	B – may be T5 trim resistor or aircraft ITT system fault
Δw_f	Δw_f steady	

FIGURE 19-35 Turbine troubleshooting examples using trend analysis. (*Pratt & Whitney Canada.*)

resulting from changes in the mechanical condition of the engine. A gas-turbine engine operates in accordance with predetermined relationships among the various performance parameters at steady-state conditions. Once the initial relationships have been established for the various parameters, a specific engine will not vary significantly from this calibration unless some external forces effect it. Thus, abnormal performance of an engine will be indicated by parameter relationships deviating from the norm.

Data collection methods will vary depending on whether the data are collected manually or by an onboard computer, as with many airline-type aircraft. Data should normally be collected at regular intervals. Variable loads extracted from the engine, such as generator, hydraulic, air conditioning,

TREND PLOT INDICATION	PROBABLE CAUSE
One parameter moves	90% chance that it is an indicator error
Two parameters move	Equal chance that it is an indication or engine-related problem
Three parameters move	90% chance that it is an engine problem
Four parameters shift in the same direction	First check for TAT error, EPR problem, or bleed system problem (compare with other engines on aircraft)
EGT fuel flow trends	+10°C EGT equivalent to +1% fuel flow (may vary with engine type)
Unexplainable trend	Investigate whether the engine was changed

FIGURE 19-36   Rules of thumb for trend report parameters.

and bleed air, will have an effect on trend accuracy. To minimize these effects, each time a set of readings is taken, it is preferable that conditions be repeated as closely as possible with regard to altitude and power.

In order to reduce fluctuations in the data, ensure that the engine parameters are stabilized before taking the data readings. The engine parameters should be read separately for each engine and in a reasonable time frame. In a computerized system, the data are read and sent to computers on the ground that are especially designed to record and store this information.

Condition monitoring devices are designed to give an indication of any engine deterioration at the earliest possible of any engine deterioration at the earliest possible stage and also to help identify any area or module in which deterioration is occurring. This facilitates quick diagnosis, which can be followed by either further monitoring or immediate maintenance action on the problem. Condition monitoring devices and equipment can be broadly categorized into the areas of flight deck indicators, in-flight recorders, and ground indicators.

Most turbofan engines use data from both spools. Turboprop and turboshaft engines use turbine inlet temperature (TIT), N_g (gas generator speed), and fuel flow (W_f) for a given torque or horsepower to establish a trend. At a given temperature and pressure, these turboprop engines must operate within a tolerance band, either above or below charted values. Generally, the tolerance is a percentage of change. If it is too high or too low, the engine does not meet performance standards and corrective action will be needed. When a trend is established or there is a deviation from normal (original operating parameter values), this can be an indication that overoperating time gas path components have deteriorated.

New engines operate either above or below charted values and within a tolerance band. They tend to deviate more from these values over time and with deterioration of gas path components. Abrupt changes, or gradual increased rate of change of the normal deviations from charted values are critical indicators of gas path component conditions. As such, changes can be detected before any drastic failure occurs.

Since there are generally three to four parameters that are monitored, the number of parameters that show a shift or trend is very important in determining the cause of the engine malfunction. As shown in Fig. 19-36, the number of parameters that shift is of great importance to predicting the probable cause.

The engine condition trend monitoring system for a PT6A turboprop engine performs its function through a process of periodically recording engine instrument readings such as torque, interturbine temperature, compressor speed (N_g), and fuel flow (W_f), correcting the readings for altitude, outside air temperature, and airspeed if applicable, and then comparing them with a set of typical engine characteristics. Such comparisons produce a set of deviations in ITT, N_g, and W_f.

Beginning with the engine in a new or overhauled condition, these deviations are entered on a chart to establish a base line for the engine. As more plots are entered, a trend line for each engine parameter is established. During the life of an engine, these trend lines will remain stable for as long as the engine is free from deterioration. As deteriorations appear, the trend lines will gradually deviate, as shown in Figs. 19-36 and 19-37. A correct interpretation of these deviations will enable an operator's maintenance facility to plan for corrective maintenance actions, such as a performance recovery compressor wash or a hot section inspection.

REVIEW QUESTIONS

1. Describe the hazards that exist in the area around an operating gas-turbine engine inlet.
2. What three primary conditions are necessary in order to start a gas-turbine engine?
3. What limitations should be observed when using an electric starter?
4. What engine instrument must be observed at light-up and why?
5. What is meant by the term *hot start*?
6. What is meant by the term *hung start*?
7. What is a flight cycle?
8. What are periodic inspections?
9. List some events that could result in an immediate special inspection being performed on the engine.
10. Describe a borescope and explain its purpose.
11. During engine operation, what are the symptoms of foreign-object damage (FOD)?
12. Describe the condition known as fan blade shingling.
13. What type of inspection is likely to be required if a gas-turbine engine has been operated over temperature limits?

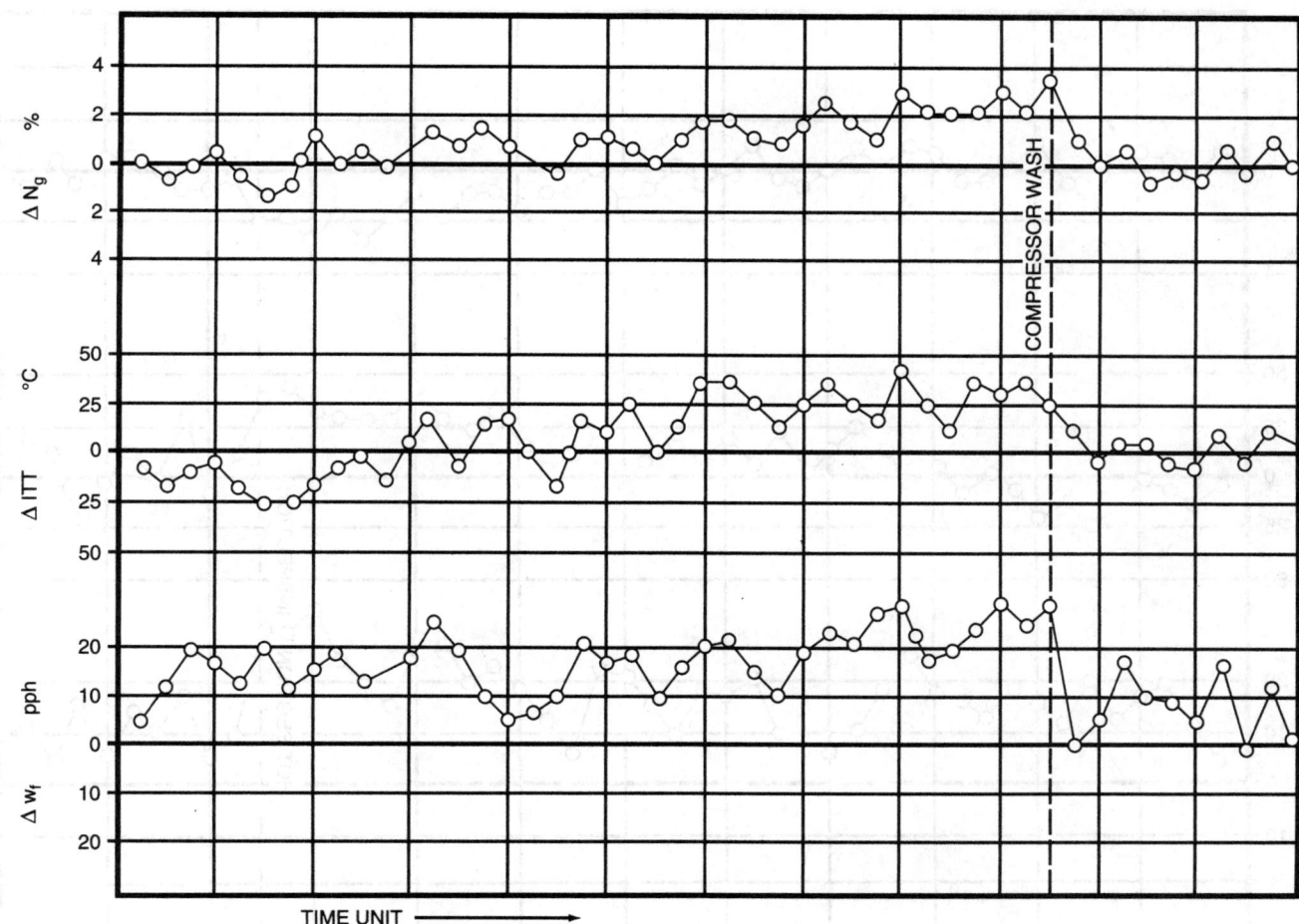

FIGURE 19-37 Trend analysis, compressor contamination. (*Pratt & Whitney Canada.*)

14. What engine assembly inspections may be called for after overspeed operation?

15. What engine parts comprise the hot section?

16. What is the purpose of performing a hot section inspection?

17. Discuss the difference between scheduled and unscheduled maintenance.

18. What two tests are performed in checking PT6A fuel nozzles?

19. What is the purpose of a compressor wash?

20. What are the two methods used to determine when turbine engines are overhauled?

21. Why are the rotating assemblies of turbine engines balanced?

22. What corrections must be applied to test results to ensure accurate evaluation of engine performance?

23. Describe a dry motoring check.

24. What is the purpose of a wet motoring check?

25. What is the purpose of a power assurance check?

26. What action should be taken by the operator if the EGT does not drop at shutdown?

27. What investigation should be made by the technician during coastdown after an engine is shut down?

28. What is meant by "trimming" an engine?

29. What engine instruments are useful as fault indicators?

30. Describe the purpose of trend analysis.

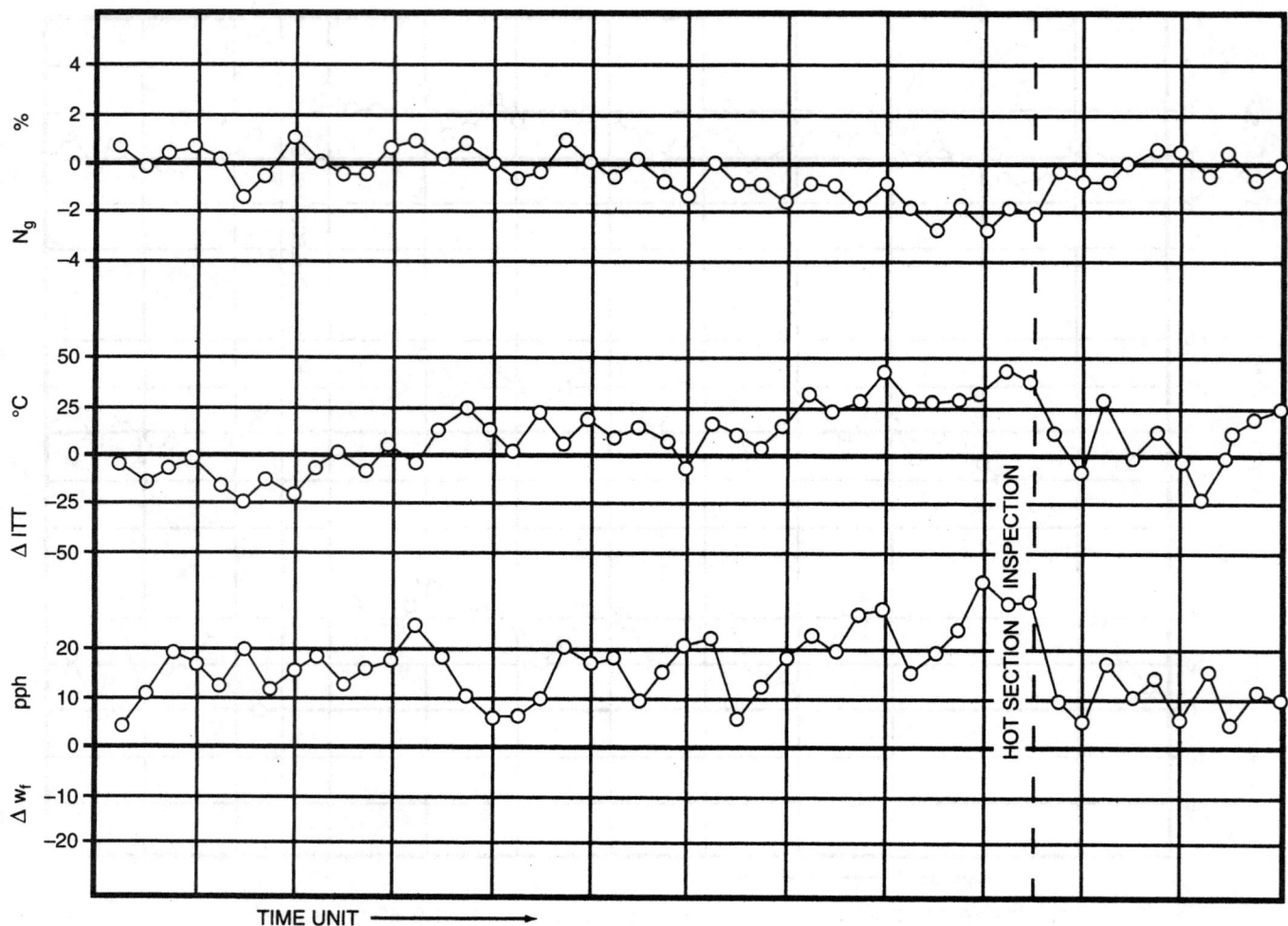

FIGURE 19-38 Trend analysis, hot section inspection. (*Pratt & Whitney Canada.*)

Propeller Theory, Nomenclature, and Operation 20

BASIC PROPELLER PRINCIPLES

The aircraft propeller consists of two or more blades and a central hub to which the blades are attached. Each blade of an aircraft propeller is essentially a rotating wing. As a result of their construction, propeller blades produce forces that create thrust to pull or push the airplane through the air.

Power to rotate an aircraft's propeller blades is furnished by the engine. On low-horsepower engines, the propeller is mounted on a shaft that is usually an extension of the crankshaft. On high-horsepower engines, the propeller is mounted on a propeller shaft that is geared to the engine crankshaft. In either case, the engine rotates the airfoils of the blades through the air at high speeds, and the propeller transforms the rotary power of the engine into thrust.

PROPELLER NOMENCLATURE

In order to explain the theory and construction of propellers, it is necessary first to define the parts of various types of propellers and give the nomenclature associated with propellers. Figure 20-1 shows a fixed-pitch one-piece wood propeller designed for light aircraft. Note the *hub, hub bore, bolt holes, neck, blade, tip,* and *metal tipping*. These are the common terms applied to a wood propeller.

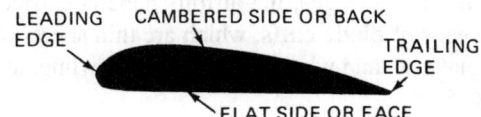

FIGURE 20-2   Cross section of a propeller blade.

The cross section of a propeller blade is shown in Fig. 20-2. This drawing is shown to illustrate the **leading edge** of the blade, the **trailing edge**, the cambered side, or **back**, and the flat side, or **face**. This illustration shows that the propeller blade has an airfoil shape similar to that of an airplane wing.

Since the propeller blade and the wing of an airplane are similar in shape, each blade of an aircraft propeller may be considered as a rotating wing. It is true that it is a small wing, which has been reduced in length, width, and thickness, but it is still a wing in shape. At one end this small wing is shaped into a shank, thus forming a propeller blade. When the blade starts rotating, air flows around the blade just as it flows around the wing of an airplane, except that the wing, which is approximately horizontal, is lifted upward, whereas the blade is "lifted" forward.

The nomenclature for an **adjustable propeller**, or **ground-adjustable propeller**, is illustrated in Fig. 20-3. This is a metal propeller with two blades clamped into a steel **hub assembly**. The hub assembly is the supporting unit for the blades, and it provides the mounting structure by which the propeller is attached to the engine propeller shaft.

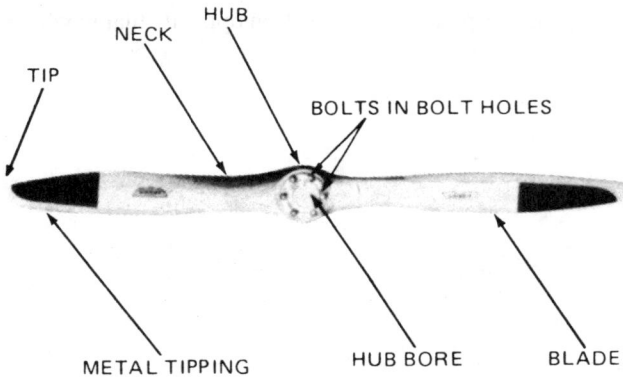

FIGURE 20-1   Fixed-pitch one-piece wood propeller.

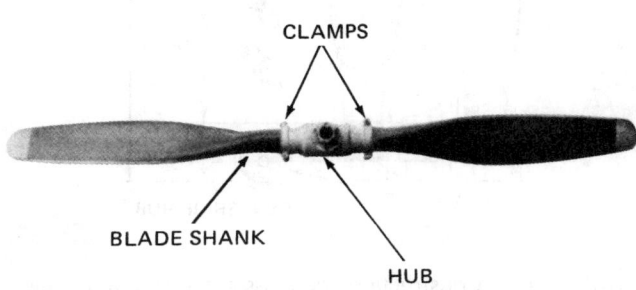

FIGURE 20-3   Nomenclature for a ground-adjustable propeller.

581

The propeller hub is split on a plane parallel to the plane of rotation of the propeller to allow for the installation of the blades. The **blade root** consists of machined ridges which fit into grooves inside the hub. When the propeller is assembled, the sections of the hub are held in place by means of **clamping rings** secured by means of bolts. When the clamping-ring bolts are properly tightened, the blade roots are held rigidly so that the blades cannot turn and change the blade angle.

Figure 20-4 shows two views and various cross sections of a propeller blade. The **blade shank** is that portion of the blade near the butt of the blade. It is usually made thick to give it strength, and it is cylindrical where it fits the hub barrel, but the cylindrical portion of the shank contributes little or nothing to thrust. In an attempt to obtain more thrust, some propeller blades are designed so that the airfoil section (shape) is carried all the way along the blade from the tip to the hub. In other designs, the airfoil shape is carried to the hub by means of **blade cuffs**, which are thin sheets of metal or other material, and which function like cowling, as shown in Fig. 20-5.

In Fig. 20-4 the tip section, the center of the hub, and the blade butt are shown. The **blade butt**, or **base**, is merely the end of the blade which fits into the hub.

PROPELLER THEORY

The Blade-Element Theory

The first satisfactory theory for the design of aircraft propellers was known as the **blade-element theory**. This theory was evolved in 1909 by a Polish scientist named Dryewiecki; therefore, it is sometimes referred to as the **Dryewiecki theory**.

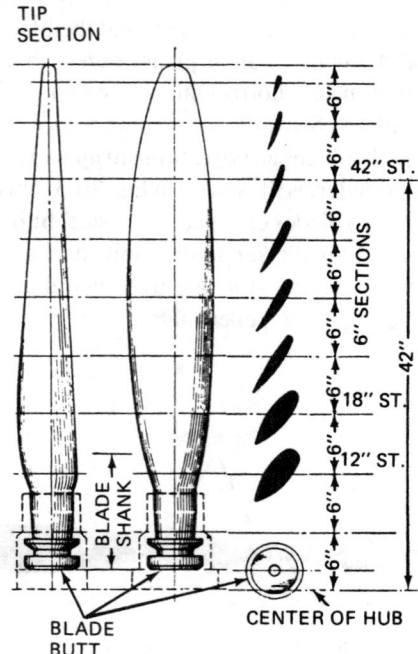

FIGURE 20-4 Construction and cross sections of a propeller blade.

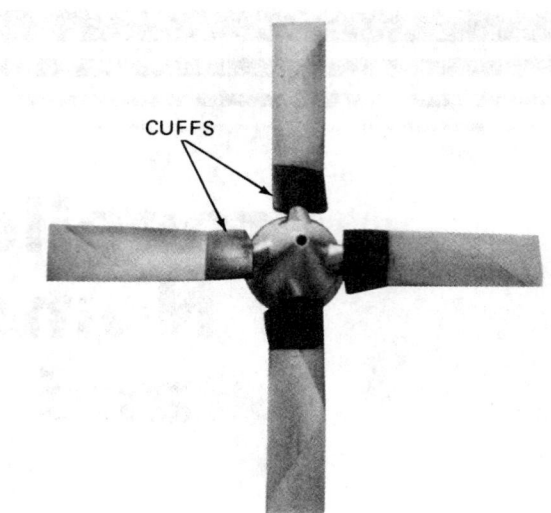

FIGURE 20-5 Propeller with blade cuffs.

This theory assumes that the propeller blade from the end of the hub barrel to the tip of the blade is divided into various small, rudimentary airfoil sections. For example, if a propeller 10 ft [3 m] in diameter has a hub 12 in [30.48 cm] in diameter, then each blade is 54 in [137.16 cm] long and can be divided into 54 1-in [2.54-cm] airfoil sections. Figure 20-6 shows one of these airfoil sections located at a radius r from the axis of rotation of the propeller. This airfoil section has a span of 1 in and a chord C. At any given radius r, the chord C will depend on the plan form or general shape of the blade.

According to the blade-element theory, the many airfoil sections, or **elements**, being joined together side by side, unite to form an airfoil (the blade) that can create thrust when revolving in a plane around a central axis. Each element must be designed as part of the blade to operate at its own best angle of attack to create thrust when revolving at its best design speed.

The thrust developed by a propeller is in accordance with Newton's third law of motion: *For every action, there is an equal and opposite reaction, and the two are directed along the same straight line.* In the case of a propeller, the first action is the acceleration of a mass of air to the rear of the airplane. This means that if a propeller is exerting a force of 200 lb [889.6 N] in accelerating a given mass of air, it is, at the same time, exerting a force of 200 lb in "pulling" the airplane in the direction opposite that in which

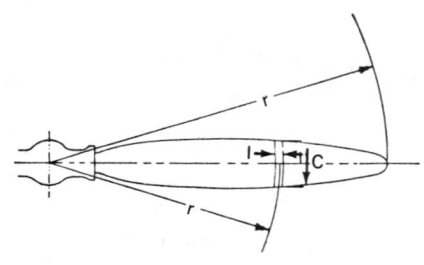

FIGURE 20-6 The blade element of a propeller.

the air is accelerated. That is, when the air is accelerated rearward, the airplane is pulled forward. The quantitative relationships among mass, acceleration, and force can be determined by the use of the formula for Newton's second law: $F = ma$. In words: *Force is equal to the product of mass and acceleration.* This principle is discussed further in the chapter on the theory of jet propulsion.

A **true-pitch propeller** is one that makes use of the blade-element theory. Each element (section) of the blade travels at a different rate of speed—that is, the tip sections travel faster than the sections close to the hub. When the elements (sections) are arranged so that each is set at the proper angle to the relative airstream, they all advance the same distance during any single revolution of the propeller.

Blade Stations

Blade stations are designated distances, in inches, measured along the blade from the center of the hub. Figure 20-4 shows the location of a point on the blade at the 42-in station.

This division of a blade into stations provides a convenient means of discussing the performance of the propeller blade, locating blade markings and damage, finding the proper point for measuring the blade angle, and locating antiglare areas.

Blade Angle

Technically, the **blade angle** is defined as the angle between the face or chord of a particular blade section and the plane in which the propeller blades rotate. Figure 20-7 is a drawing of a four-blade propeller, but only two blades are shown in order to simplify the presentation. The blade angle, the plane of rotation, the blade face, the longitudinal axis, and the nose of the airplane are all designated in this illustration. The plane of rotation is perpendicular to the crankshaft.

In order to obtain thrust, the propeller blade must be set at a certain angle to its plane of rotation, in the same manner that the wing of an airplane is set at an angle to its forward path. While the propeller is rotating in flight, each section of the blade has a motion that combines the forward movement

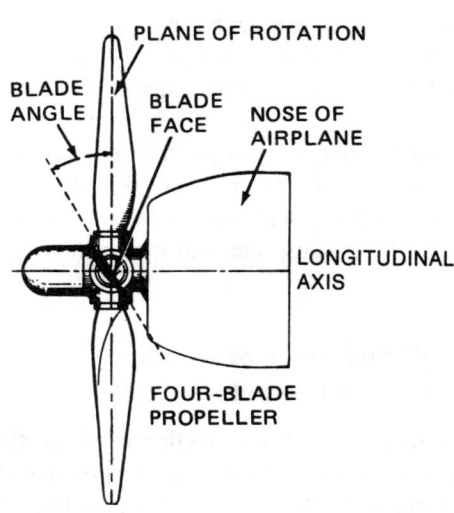

FIGURE 20-7   Four-blade propeller.

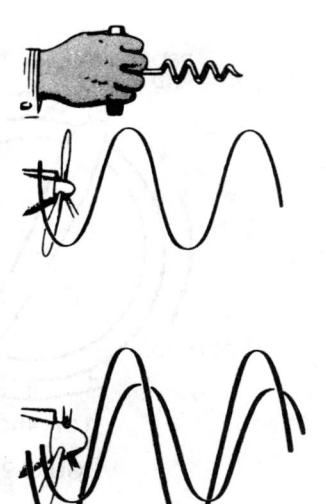

FIGURE 20-8   Path of a propeller through the air.

of the airplane with the circular or rotary movement of the propeller. Therefore, any section of a blade has a path through the air that is shaped like a spiral or a corkscrew, as illustrated in Fig. 20-8. The amount of bite (amount of air) taken by each blade is determined by its blade angle, as shown in Fig. 20-9.

An imaginary point on a section near the tip of the blade traces the largest spiral, a point on a section midway along the blade traces a smaller spiral, and a point on a section near the shank of the blade traces the smallest spiral of all. In one turn of the blade, all sections move *forward* the same distance, but the sections near the tip of the blade move a greater *circular* distance than the sections near the hub.

If the spiral paths made by various points on sections of the blade are traced, with the sections at their most effective angles, then each individual section must be designed and constructed so that the angles gradually decrease toward the tip of the blade and increase toward the shank. This gradual change of blade section angles is called **pitch distribution** and accounts for the pronounced *twist* of the propeller blade, as illustrated in Fig. 20-10. Since the blade is actually a twisted airfoil, the blade angle of any particular section of a particular blade is different from the blade angle of any other section of the same blade.

The blade angle is measured at one selected station when the blade is set in its hub, depending on the propeller diameter. If the blade angle at this station is correct, then all blade angles should be correct if the blade has been carefully designed and accurately manufactured.

The blade angle is so important that a change in blade angle of only 1° will decrease the rotational speed of a direct-drive engine by 60 to 90 rpm. The effect that it would have on an engine with propeller-gear reduction would depend on the gear ratio. In some installations, a deviation of 1° or less in blade-angle setting is occasionally permitted by specific instructions, but this is by no means a common practice.

The reason for the variation of the propeller blade angle from the blade hub to the blade tip is demonstrated in Fig. 20-11. The three triangles show the relative movement of the airplane and a

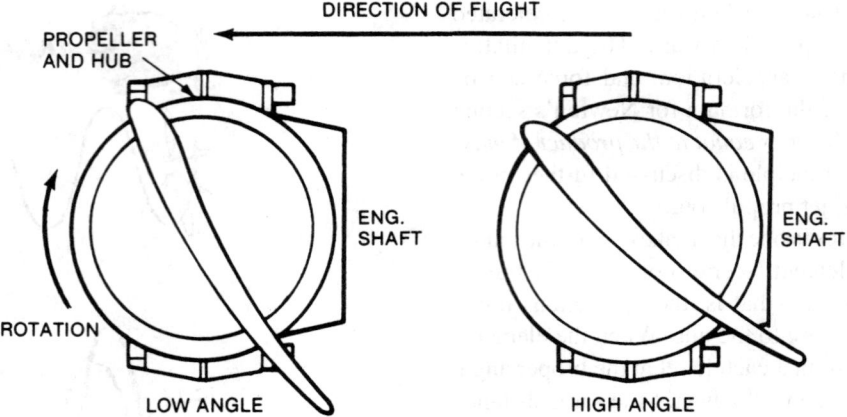

FIGURE 20-9   Propeller blade angle. (*Hartzell Propeller.*)

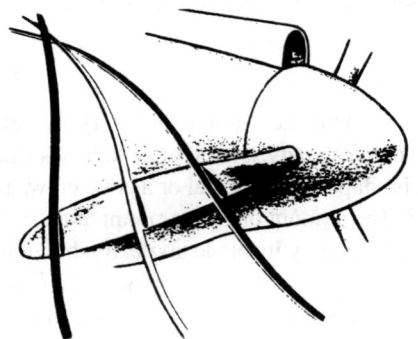

FIGURE 20-10   Pitch distribution.

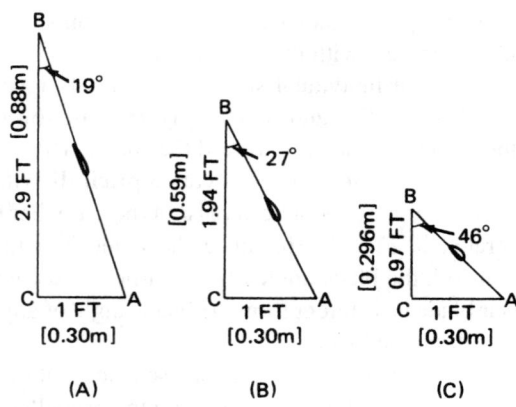

FIGURE 20-11   Demonstrating the reason for a variation of propeller blade angle from root to tip.

particular section of the propeller blade during flight at 150 mph [241.40 km/h] with the propeller turning at 2000 rpm. The triangle in Fig. 20-11A represents the movement of the blade section 36 in [91.44 cm] from the propeller hub.

The diameter of the circle traversed by the blade section at 36 in from the hub is $3 \times 2\pi$ ft, or 18.85 ft [5.75 m]. With the propeller turning at 2000 rpm, the section of the blade travels 37 700 ft/min [11 490.96 m/min] or about 628 ft/s [191.41 m/s].

If the airplane is traveling at a true airspeed of 150 mph [241.40 km/h], it is moving through the air at 220 ft/s [67.06 m/s]. This means, then, that the airplane travels 220 ft while the blade section is moving 628 ft. From these data we can determine that, while the airplane is moving 1 ft [0.30 m], the propeller blade section at 36 in from the hub is moving slightly less than 2.9 ft [0.88 m]. This is illustrated in Fig. 20-11A, which shows blade section distance in its plane of rotation as BC and shows the airplane distance as CA. The actual track of the propeller blade section is BA, and the relative wind direction is along the line AB. The angle of attack of the propeller blade is the difference between the angle of AB with respect to the plane of rotation and the propeller blade angle.

From a table of tangents or by measuring, we can find that the angle ABC is a little more than 19°. If the blade angle is set at 22°, the angle of attack of the blade will be somewhat more than 3°.

The triangle in Fig. 20-11B represents the travel of a blade section at 24 in [60.96 cm] from the hub when the airplane is moving at 150 mph TAS (true airspeed) and the propeller is turning at 2000 rpm. By using the same methods of computation, we find that the angle at B is about 27°, and that to provide an angle of attack of 3° the blade angle would have to be set at 30°. Under the same conditions, we will find that a blade section located 12 in [30.48 cm] from the hub will move at an angle of 46° from the plane of rotation, as shown in Fig. 20-11C.

It is apparent that a fixed-pitch propeller will be efficient only through a narrow range of operating conditions. A fixed-pitch propeller is therefore designed to operate most efficiently at the cruising speed of the airplane on which it is installed.

Blade Angle and Angle of Attack in Flight and on the Ground

The **angle of attack of a propeller blade section** is the angle between the face of the blade section and the direction of the relative airstream, as illustrated in Fig. 20-12. The direction of the relative airstream depends on the direction

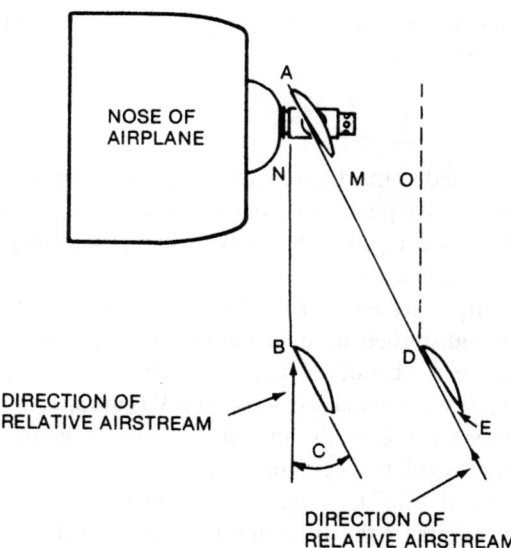

FIGURE 20-12 Blade angle with aircraft at rest and in flight.

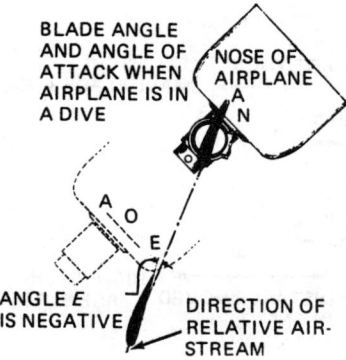

FIGURE 20-14 High speed results in a negative angle of attack.

that the airfoil moves through undisturbed air and the velocity of forward movement.

In Fig. 20-12, the blade airfoil section M of the rotating propeller travels from A to B when the airplane is parked on the ground. The trailing edge of the propeller determines the plane of rotation represented by the line AB.

The **relative wind** is the direction of the air with respect to the movement of the airfoil, as shown in Fig. 20-13. When the airplane is in the air, the relative wind results from the forward motion of the airplane and the circular motion of the propeller blade sections. When the engine is run on the ground, there is no forward movement of the airplane (on the assumption that the airplane is standing) but there is a relative wind caused by the air flowing through the propeller. There is also a certain amount of pitch angle of the air motion with regard to the blade sections. The angle of attack is therefore represented in Fig. 20-12 by the angle C. This is the angle at which the propeller section meets the relative airstream.

Also with respect to Fig. 20-12, in flight the airplane moves forward, and as it moves forward from N to O, the airfoil section M will travel from A to D. The trailing edge follows the path represented by the line AD, which represents the relative airstream. The angle of attack then becomes smaller and is represented by E in the illustration.

The normal in-flight angle of attack of the propeller blades for many airplanes varies from 0 to 15°. Referring to Fig. 20-14, we see that in a power dive the acceleration due to the force of gravity may give the airplane a speed greater than the speed which the propeller tends to reach. The angle of attack, represented by the letter E, is then negative and tends to hold back the airplane.

In a steep climb with the forward speed reduced, the angle of attack is increased, as shown in Fig. 20-15. Whether the airplane is in a power dive or in a steep climb, the aerodynamic efficiency of the propeller is low.

Note that the propeller blade angles in the foregoing illustrations appear to be high, because they have been exaggerated in these views for the purpose of illustration.

Pitch

Effective Pitch

The **effective pitch** is the actual distance the airplane moves forward during one revolution (360°) of the propeller in flight. "Pitch" is not a synonym for "blade angle," but the two terms are commonly used interchangeably because they are so closely related. Figure 20-16 shows two different

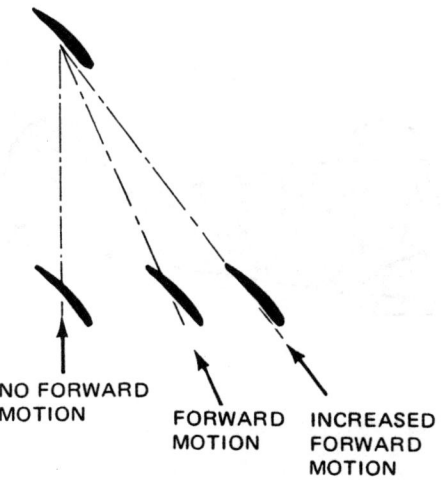

FIGURE 20-13 Relative wind with respect to propeller blade.

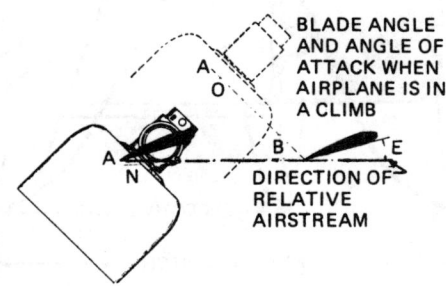

FIGURE 20-15 Increased angle of attack as airplane climbs.

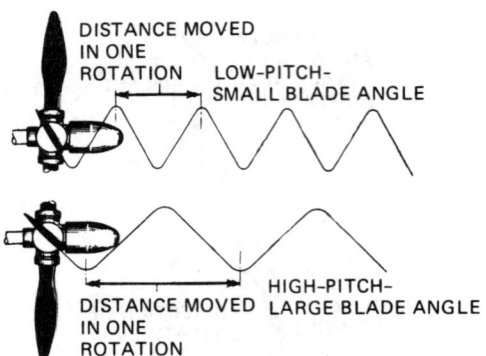

FIGURE 20-16 Low pitch and high pitch.

pitch positions. The heavy black airfoil drawn across the hub of each represents the cross section of the propeller to illustrate the blade setting. When there is a small blade angle, there is a low pitch and the airplane does not move very far forward in one revolution of the propeller. When there is a large blade angle, there is a high pitch and the airplane moves forward farther during a single revolution of the propeller.

Geometric Pitch

A distinction is made between effective pitch and other kinds of pitch. The **geometric pitch** is the distance an element of the propeller would advance in one revolution if it were moving along a helix (spiral) having an angle equal to its blade angle. Geometric pitch is a linear measurement, measured in units of inches. Geometric pitch can be calculated by multiplying the tangent of the blade angle by $2\pi r$, r being the radius of the blade station at which it is computed. For example, if the blade angle of a propeller is 20° at the 30-in [76.2-cm] station, we can apply the formula

$$GP = 2\pi \times 30 \times 0.364 = 68.61 \text{ pitch inches [174.27 cm]}$$

The tangent of 20° is 0.364. The geometric pitch of the propeller is therefore 68.61 in. This is the distance the propeller would move if it were going forward through a solid medium with no slippage.

Slip

Slip is defined as the difference between the geometric pitch and the effective pitch of a propeller (see Fig. 20-17). It may be expressed as a percentage of the mean geometric pitch or as a linear dimension.

The **slip function** is the ratio of the speed of advance through undisturbed air to the product of the propeller diameter and the number of revolutions per unit time. This may be expressed as a formula, $V(nD)$, where V is the speed through undisturbed air, D is the propeller diameter, and n is the number of revolutions per unit time.

The word "slip" is used rather loosely by many people in aviation to refer to the difference between the velocity of the air behind the propeller (caused by the propeller) and that of the aircraft with respect to the undisturbed air well ahead of the propeller. It is then expressed as a percentage of this difference in terms of aircraft velocity.

If there were no slippage of any type, and if the propeller were moving through an imaginary solid substance, then the geometric pitch would be the calculated distance that the blade element at two-thirds the blade radius would move forward in one complete revolution of the propeller (360°).

Zero-Thrust Pitch

The **zero-thrust pitch**, also called the **experimental mean pitch**, is the distance a propeller would have to advance in one revolution to produce no thrust.

The **pitch ratio** of a propeller is the ratio of the pitch to the diameter.

Terms Used in Describing Pitch Change

The principal terms used in describing propeller-pitch change are: (1) **two-position**, which makes available only two pitch settings; (2) **multiposition**, which makes any pitch

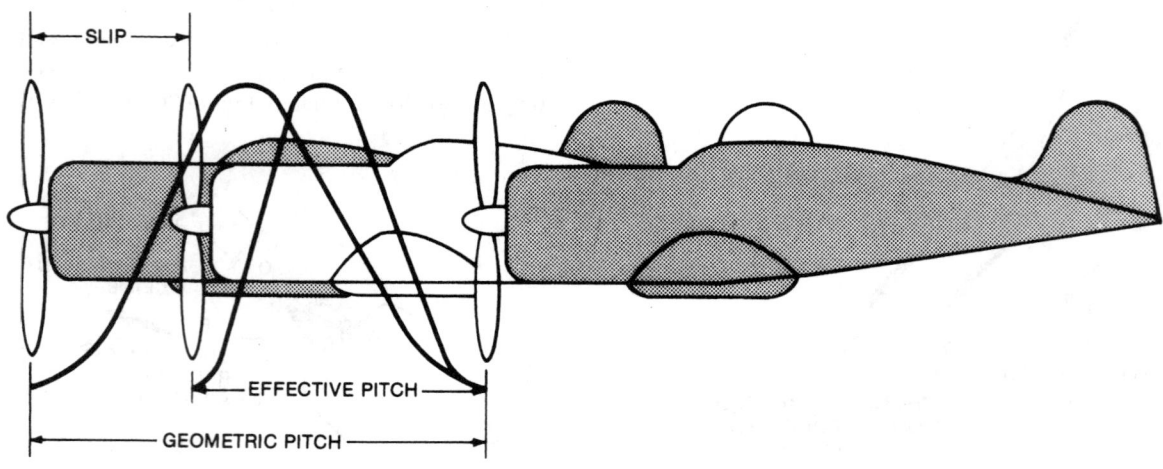

FIGURE 20-17 Effective and geometric pitch.

setting within limits possible; (3) **automatic**, which provides a pitch-setting control by some automatic device; and (4) **constant speed**, which enables pilots to select and control, during flight, the exact conditions at which they want the propeller to operate. A constant speed propeller uses a governor to maintain speed, regardless of aircraft altitude.

Forces Acting on a Propeller in Flight

The forces acting on a propeller in flight are: (1) **thrust**, which is the component of the total air force on the propeller and is parallel to the direction of advance and induces bending stresses in the propeller; (2) **centrifugal force**, which is caused by the rotation of the propeller and tends to throw the blade out from the central hub and produces tensile stresses; and (3) **torsion**, or **twisting forces**, in the blade itself, caused by the fact that the resultant air forces do not go through the neutral axis of the propeller, producing torsional stresses.

Forces to Which Propellers Are Subjected at High Speeds

Figure 20-18 illustrates the five general types of forces to which propellers rotating at high speeds are subjected. These stresses are: centrifugal force, torque bending force, thrust bending force, aerodynamic twisting force, and centrifugal twisting force.

Centrifugal force (see Fig. 20-18A) is a physical force that tends to throw the rotating propeller blades away from the hub. The hub resists this tendency, and therefore the blades "stretch" slightly.

Torque bending force (see Fig. 20-18B), in the form of air resistance, tends to bend the propeller blades in a direction that is opposite to the direction of rotation.

Thrust bending force (see Fig. 20-18C) is the thrust load that tends to bend propeller blades forward as the aircraft is pulled through the air. Bending forces are also caused by other factors, such as the drag caused by the resistance of the air, but these are of small importance in comparison with the bending stresses caused by the thrust forces.

Aerodynamic twisting force (see Fig. 20-18D) tends to turn the blades to a high blade angle. **Centrifugal twisting force** (see Fig. 20-18E), which is greater than the aerodynamic twisting force, tries to force the blades toward a low blade angle. In some propeller control mechanisms, this centrifugal twisting force is employed to aid in turning the blades to a lower angle when such an angle is necessary to obtain greater propeller efficiency in flight, thus putting a natural force to work. Torsional forces (aerodynamic and centrifugal twisting forces) increase with the square of the rpm. For example, if the propeller's rpm is doubled, the forces will be four times as great.

Tip Speed

Flutter or vibration may be caused by the tip of the propeller blade traveling at a rate of speed approaching the speed of sound, thus causing excessive stresses to develop. This condition can be overcome by operating at a lower speed

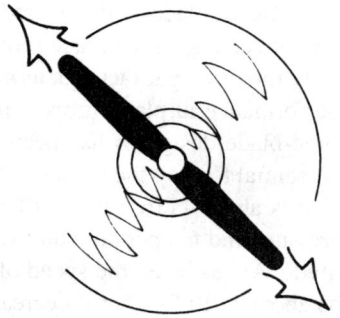

A. CENTRIFUGAL FORCE

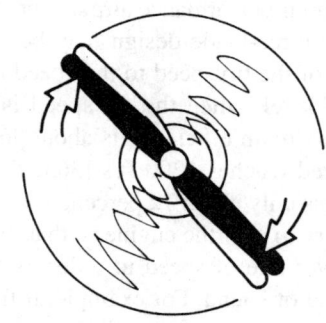

B. TORQUE BENDING FORCE

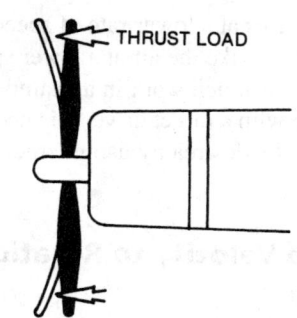

THRUST LOAD

C. THRUST BENDING FORCE

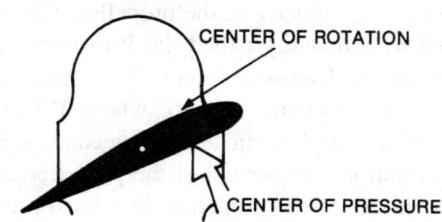

CENTER OF ROTATION

CENTER OF PRESSURE

D. AERODYNAMIC TWISTING FORCE

E. CENTRIFUGAL TWISTING FORCE

FIGURE 20-18 Forces acting on a rotating propeller.

or by *telescoping* the propeller blades—that is, reducing the propeller diameter without changing the blade profile.

Tip speed is actually the principal factor determining the efficiency of high-performance airplane propellers of conventional two- or three-blade design. It has been found by experience that it is essential to keep the tip speed below the speed of sound, which is about 1116.4 ft/s [340.28 m/s] at standard sea-level pressure and temperature and varies with temperature and altitude. At sea level, the speed of sound is generally taken to be about 1120 ft/s, but it decreases about 5 ft/s [152.4 cm/s] for each increase in altitude of 1000 ft [304.80 m].

The efficiency of high-performance airplane propellers of conventional two- or three-blade design may be expressed in terms of the ratio of the tip speed to the speed of sound. For example, at sea level, when the tip speed is 900 ft/s [274.32 m/s], the maximum efficiency is about 86 percent, but when the tip speed reaches 1200 ft/s [365.76 m/s], the maximum efficiency is only about 72 percent.

It is often necessary to gear the engine so that the propeller will turn at a lower rate of speed in order to obtain tip ratios below the speed of sound. For example, if the engine is geared in a 3:2 ratio, the propeller will turn at two-thirds the speed of the engine.

When the propeller turns at a lower rate of speed, the airfoil sections of the blades strike the air at a lower speed, and they therefore do not do as much work in a geared propeller as they would do in one with a direct drive. It is necessary in this case to increase the blade area by using larger-diameter or additional blades.

Ratio of Forward Velocity to Rotational Velocity

The efficiency of a propeller is also influenced by the ratio of the forward velocity of the airplane in feet per second to the rotational velocity of the propeller. This ratio can be expressed by a quantity called the *V*-over-*nD* ratio (or the slip function, as discussed previously), which is sometimes expressed as a formula, $V/(nD)$, where V is the forward velocity of the airplane in feet per second, n is the number of revolutions per second of the propeller, and D is the diameter in feet of the propeller. Any fixed-pitch propeller is designed to give its maximum efficiency at a particular aircraft speed, which is usually the cruising speed in level flight, and at a particular engine speed, which is usually the speed employed for cruising. At any other condition of flight where a different value of the $V/(nD)$ ratio exists, the propeller efficiency will be lower.

Propeller Load

A propeller being driven at a given speed will absorb a specific amount of power. It requires more power to drive a propeller at high speeds than at low speeds. Actually, the power required to drive a propeller varies as the cube of the rpm. This is expressed by the formula

$$hp = K \times rpm^3$$

where K is a constant whose value depends on the propeller type, size, pitch, and number of blades. Another formula that can be used to express the same principle is

$$hp_2[W_2] = hp_1[W_1] = \left(\frac{rpm_2}{rpm_1}\right)^3$$

It requires eight times as much power to drive a propeller at a given speed than to drive it at half that speed. If the speed of a propeller is tripled, it will require 27 times as much power to drive it as it did at the original speed.

Propeller load curves are shown in Fig. 20-19. This chart shows the manifold pressure, the power output, and the brake specific fuel consumption (bsfc) at different rpms when the engine is operated at full throttle with a particular fixed-pitch propeller.

At the top of the chart, it will be noted that MAP (manifold absolute pressure) decreases at full throttle as rpm increases. From the prop load curve at the top of the chart, we can see that the propeller can be turned at 1950 rpm with a manifold pressure of 20 inHg [67.72 kPa], at 2200 rpm at a manifold pressure of 22 inHg [74.49 kPa], and at 2600 rpm with a manifold pressure of 27.8 inHg [94.13 kPa]. This is the maximum output available with this propeller, because the load curve meets the manifold pressure curve at this rpm.

From the curves in the middle portion of the graph, we can see that the engine power output increases as rpm increases. The increase is not proportional because of the decrease in manifold pressure which takes place as rpm increases. We also note from the prop load curve that the propeller can be driven at 2100 rpm with 142 hp [105.89 kW], at 2400 rpm with

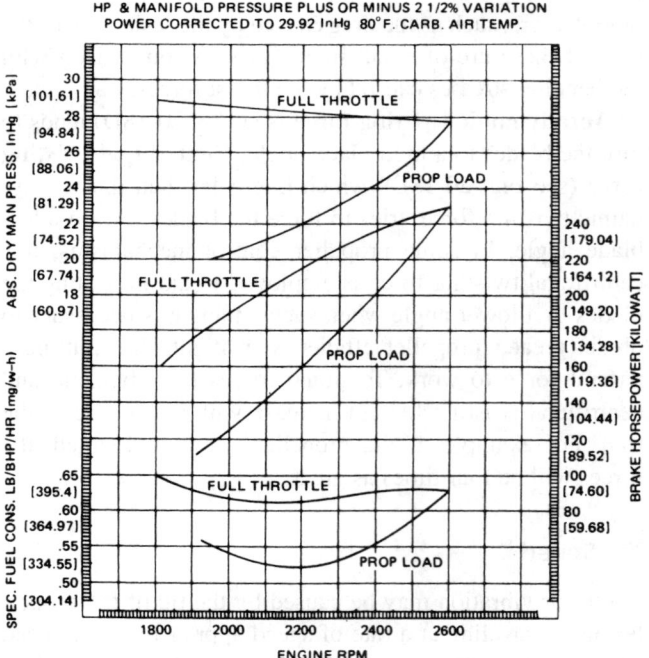

FIGURE 20-19 Propeller load curves.

202 hp [150.63 kW], and at 2600 rpm with 248 hp [184.93 kW]. Another way of saying the same thing is that the propeller absorbs 248 hp at 2600 rpm.

The curves at the bottom of the graph in Fig. 20-19 show the specific fuel consumption (sfc) under various conditions of rpm and prop load. It will be observed that the best fuel consumption takes place at approximately 2200 rpm when the propeller is absorbing 160 hp [119.31 kW]. The bsfc at this point is about 0.52 lb/hp/h [0.316 kg/kW/h]. If the engine were operated at full throttle with the rpm at 2200, the bsfc would be about 0.61 lb/hp/h [0.371 kg/kW/h].

Propeller Efficiency

Some of the work performed by the engine is lost in the slipstream of the propeller, and some is lost in the production of noise. The lost work cannot be converted to horsepower for turning the propeller. The effect of tip speed on propeller efficiency has already been examined. In addition, the maximum propeller efficiency that has been obtained in practice under the most ideal conditions, using conventional engines and propellers, has been only about 92 percent, and in order to obtain this efficiency it has been necessary to use thin airfoil sections near the tips of the propeller blades and very sharp leading and trailing edges. Such airfoil sections are not practical where there is the slightest danger of the propeller picking up rocks, gravel, water spray, or similar substances that might damage the blades.

The **thrust horsepower** is the actual amount of horsepower that an engine-propeller unit transforms into thrust. This is less than the **brake horsepower** developed by the engine, since propellers are never 100 percent efficient.

In the study of propellers, two forces must be considered: **thrust** and **torque**. The thrust force acts perpendicular to the plane of rotation of the propeller, and the torque force acts parallel to the plane of rotation of the propeller. The thrust horsepower is less than the torque horsepower. The **efficiency** of the propeller is the ratio of the thrust horsepower to the torque horsepower:

$$\text{Propeller efficiency} = \frac{\text{thrust horsepower}}{\text{torque horsepower}}$$

Propeller Torque Reaction

Torque reaction involves Newton's third law of physics: for every action there is an equal and opposite reaction. As applied to the airplane, this means that as the internal engine parts and propeller are revolving in one direction, an equal force is trying to rotate the airplane in the opposite direction.

Asymmetric Loading (*P* Factor)

When an airplane is flying at a high angle of attack (climbing), as illustrated in Fig. 20-20, the load of the downward-moving propeller blade is greater than the load on the upward-moving propeller blade. This moves the center of thrust to the right of the propeller center, thus causing a yawing moment toward the left around the vertical axis of the aircraft. With the airplane being flown at positive angles of attack, the right or down-swinging propeller blade is passing through an area of resultant velocity which is greater than that affecting the left or up-swinging blade. Since the propeller blade is an airfoil, increased velocity means increased lift. Therefore, the down-swinging blade, having more lift, tends to pull the aircraft's nose to the left (all directional references are from the pilot's seat looking forward).

PROPELLER CONTROLS AND INSTRUMENTS

The cockpit propeller controls are generally located on the center pedestal, which is between the pilot and copilot seats. Looking forward from the pilot's seat, the throttle is on the left side of the pedestal, and the propeller control is just to the right of the throttle, as shown in Fig. 20-21. The propeller controls are rigged so that the full INCREASE RPM

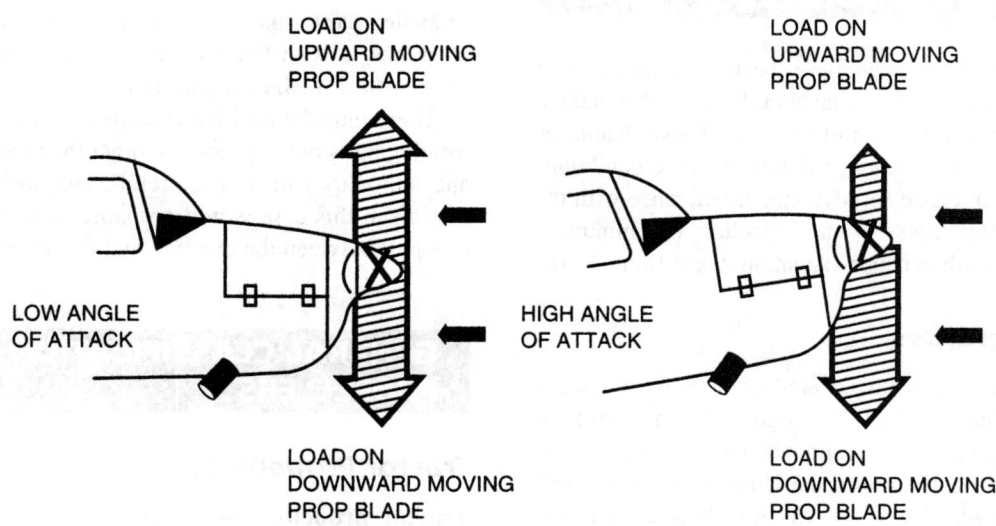

FIGURE 20-20   Asymmetrical loading of propeller (*P* factor).

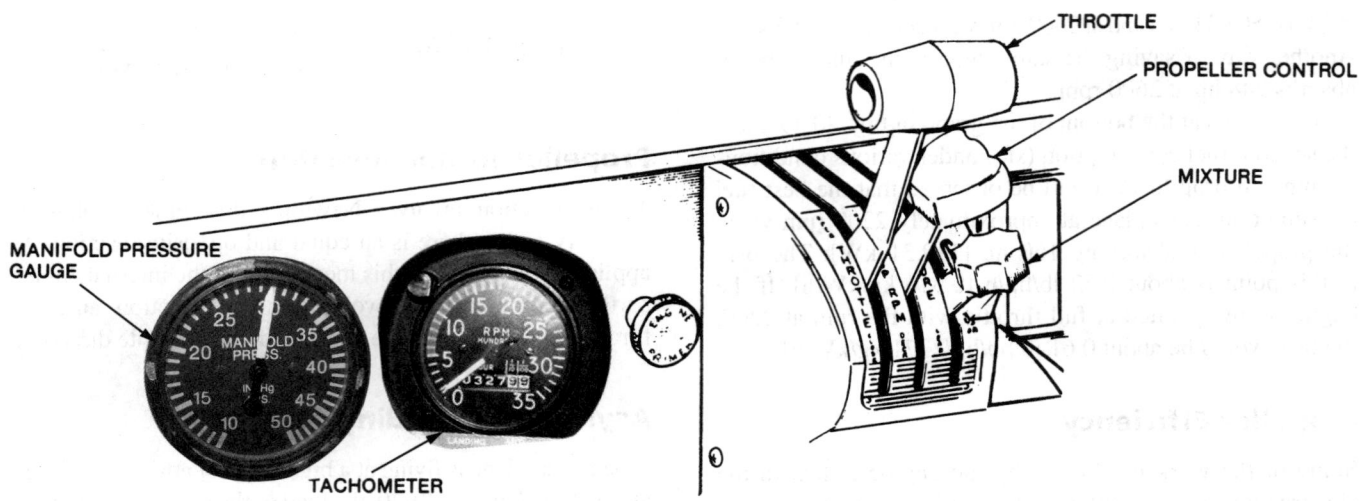

FIGURE 20-21 Cockpit instruments and propeller control.

position is forward (in) on the control and: the DECREASE RPM position is obtained by moving the control aft (out). Cockpit propeller controls are blue; their shapes can be seen in Fig. 20-21. Controls for turbopropeller engines will be discussed in Chap. 21.

The cockpit instruments that are mostly concerned with the control setting of the propeller are the tachometer and the MAP (manifold absolute pressure) gauge. The **tachometer** indicates the rpm of the engine's crankshaft while the **manifold pressure gauge** measures the absolute pressure in the intake manifold. If an aircraft is equipped with a constant-speed propeller, the aircraft will also have a MAP gauge to assist with setting the correct amount of engine power during climb and cruise. The MAP gauge and the tachometer are each marked with a green arc to indicate the normal operating range and a yellow arc for the takeoff and precautionary range. The tachometer is also sometimes marked with a red arc for critical vibration range and a red radial line for maximum operating limit. A typical tachometer and MAP gauge can be seen in Fig. 20-21.

PROPELLER CLEARANCES

Certain minimum clearances have been established with respect to the distances between an aircraft's propeller and the ground, the water, and the aircraft structure. These clearances are necessary to prevent damage during extreme conditions of operation and to reduce aerodynamic interference with the operation and effectiveness of the propeller. The minimum clearances are set forth in Federal Aviation Regulations (FAR).

Ground Clearances

Aircraft equipped with tricycle landing gear must have a minimum clearance of 7 in [17.78 cm] between the tips of the propeller blades and the ground when the aircraft is in the taxiing or takeoff position with the landing gear deflected. Aircraft with tail-wheel landing gear must have a minimum clearance of 9 in [22.86 cm] between the tips of the propeller

blades and the ground under the same conditions—that is, with the aircraft in the position where the propeller blade tips would come nearest the ground during operation. This would normally be during takeoff for an aircraft equipped with tail-wheel landing gear.

Water Clearance

Seaplanes or amphibious aircraft must have a clearance of at least 18 in [45.72 cm] between the tips of the propeller blades and the water unless it can be shown that the aircraft complies with the regulations regarding water-spray characteristics set forth in FAR 25.239.

Structural Clearances

The tips of the propeller blades must have at least 1 in [2.54 cm] of radial clearance from the fuselage or any other part of the aircraft structure. If this is not sufficient to avoid harmful vibrations, additional clearance must be provided.

Longitudinal clearance (fore and aft) of the propeller blades or cuffs must be at least $\frac{1}{2}$ in [12.70 mm] between propeller parts and stationary parts of the airplane. This clearance is with the propeller blades feathered or in the most critical pitch configuration.

There must be positive clearance between the spinner or rotating parts of the propeller, other than the blades or cuffs, and stationary parts of the aircraft. The stationary part of the aircraft in this case would probably be the engine cowling or a part between the cowling and the spinner.

GENERAL CLASSIFICATION OF PROPELLERS

Tractor Propellers

Tractor propellers are propellers mounted on the front end of the engine structure. Most aircraft are equipped with this type (location) of propeller. A major advantage of the tractor propeller is that relatively low stresses are induced in the propeller as it rotates in relatively undisturbed air.

Pusher Propellers

Pusher propellers are propellers mounted on the rear end of the engine behind the supporting structure. Seaplanes and amphibious aircraft use a greater percentage of pusher propellers than do other kinds of aircraft.

On land planes, where propeller-to-ground clearance is less than propeller-to-water clearance of watercraft, pusher propellers are subject to more damage than tractor propellers. Rocks, gravel, and small objects, dislodged by the wheels, quite often may be thrown or drawn into a pusher propeller. Similarly, seaplanes with pusher propellers are apt to encounter propeller damage from water spray thrown up by the hull during landing or takeoff. Consequently, the pusher propeller quite often is mounted above and behind the wings to prevent such damage.

Types of Propellers

In designing propellers, engineers try to obtain the maximum performance of an airplane from the horsepower delivered by the engine under all conditions of operation, such as takeoff, climb, cruise, and high speed.

Fixed Pitch

A **fixed-pitch propeller** is a rigidly constructed propeller on which the blade angles may not be altered without bending or reworking the blades. When only fixed blade-angle propellers were used on airplanes, the angle of the blade was chosen to fit the principal purpose for which the airplane was designed. The fixed-pitch propeller is made in one piece with two blades which are generally made of wood, aluminum alloy, or steel. Fixed-pitch propellers are in wide use on small aircraft.

With a fixed blade-angle propeller, an increase in engine power causes increased rotational speed, and this causes more thrust, but it also creates more drag from the airfoil and forces the propeller to absorb the additional engine power. In a similar manner, a decrease in engine power causes a decrease in rotational speed and consequently a decrease in both thrust and drag from the propeller.

When an airplane with a fixed blade-angle propeller dives, the forward speed of the airplane increases. Since there is a change in the direction of the relative wind, there is a lower angle of attack, thus reducing both lift and drag and increasing the rotational speed of the propeller. On the other hand, if the airplane goes into a climb, the rotational speed of the propeller decreases, the change in the direction of the relative wind increases the angle of attack, and there is more lift and drag and less forward speed for the airplane. The propeller can absorb only a limited amount of excess power by increasing or decreasing its rotational speed. Beyond this point, the engine will be damaged. For this reason, as aircraft engine

power and airplane speeds both increased, engineers found it necessary to design propellers with blades that could rotate in their sockets into different positions to permit changes in the blade-angle setting to compensate for changes in the relative wind brought on by the varying forward speed. This made it possible for the propeller to absorb more or less engine power without damaging the engine.

Ground-Adjustable

The pitch setting of a ground-adjustable propeller can be adjusted only with tools on the ground, when the engine is not operating. This older-type propeller usually has a split hub. On some airplanes it may be necessary to remove a ground-adjustable propeller from the engine when the pitch is being adjusted, but on other airplanes this is not necessary.

Two-Position Pitch

On this type of propeller, the blade angle may be adjusted during operation to either a preset low-angle setting or a high-angle setting. A low-angle setting is used for takeoff and climb, and then a shift is made to a high-angle setting for cruise. Only high-angle or low-angle settings may be selected on this older-type propeller.

Controllable Pitch

A **controllable-pitch** propeller is one provided with a means of control for adjusting the angle of the blades during flight. The pilot can change the pitch of a controllable-pitch propeller in flight or while operating the engine on the ground by means of a pitch-changing mechanism that may be mechanically, hydraulically, or electrically operated. Pitch may be set at any position between high and low pitch.

Automatic Pitch

With automatic-pitch propellers, the blade-angle change within a preset range occurs automatically as a result of aerodynamic forces acting on the blades. The pilot has no control over the angle changes. This type of propeller is not widely used.

Constant Speed

The constant-speed propeller utilizes a hydraulically or electrically operated pitch-changing mechanism controlled by a governor. The setting of the governor is adjusted by the pilot with the propeller rpm lever in the cockpit. During operation, the constant-speed propeller automatically changes its blade angle to maintain a constant engine speed. In straight and level flight, if engine power is increased, the blade angle is increased to make the propeller absorb the additional power while the rpm remains constant. The pilot may select the engine speed desired for any particular type of operation.

Feathering Constant Speed

A **feathering propeller** is a constant-speed propeller that has a mechanism for changing the pitch to an angle such that forward aircraft motion produces no windmilling. Feathering propellers are generally used on multiengine aircraft to reduce propeller drag under engine-failure conditions. The term **feathering** refers to the operation of rotating the blades of a propeller to an edge-to-the-wind position for the purpose of stopping the rotation of the propeller whose blades are thus "feathered" and thereby reducing drag. A **feathered blade** is in an approximate in-line-of-flight position, streamlined with the line of flight. Some, but not all, constant-speed propellers can be feathered.

Feathering is necessary when an engine fails or when it is desirable to shut off an engine in flight. The pressure of the air on the face and back of the feathered blade is equal, and the propeller will stop rotating. If it is not feathered when its engine stops driving it, the propeller will "windmill" and cause excessive drag, which may be detrimental to aircraft operation. This is the primary reason for feathering a propeller. Another advantage of being able to feather a propeller is that a feathered propeller creates less resistance (drag) and disturbance in the flow of air over the wings and tail of the airplane. Furthermore, a feathered propeller prevents additional damage to the engine if the failure has been caused by some internal breakage, and it also eliminates the vibration which might damage the structure of the airplane.

The importance of feathering the propeller of an engine which has failed on a multiengine airplane cannot be overemphasized. If the propeller cannot be feathered at low aircraft speeds, such as during takeoff, the aircraft could stall. Another problem that could occur during cruise flight if the propeller cannot be feathered is engine "runaway"—that is, overspeeding to the point where great damage may be caused. The lubrication system of the engine may fail because of the excessive speed, and this will cause the engine to "burn up." The heat generated may set the engine on fire, in which case the airplane itself may be destroyed. The excessive speed of the engine could result in the propeller losing a blade, thus bringing about an unbalanced condition which will cause the engine to be wrenched from its mounting. Numerous cases of runaway engines resulting in airplane crashes are on record.

Feathering a propeller when an engine failure occurs not only reduces drag but also allows for better performance on the part of the remaining engines and better aircraft control. Because of these advantages, an airplane suffering engine failure can usually be flown safely to a point where an emergency landing can be made.

Reverse Pitch

A **reverse-pitch propeller** is a constant-speed propeller for which the blade angles can be changed to a negative value during operation. The purpose of a reversible-pitch feature is to produce a high negative thrust at low speed by using engine power. A reverse-pitch propeller is used principally as an aerodynamic brake to reduce ground roll after landing.

Practically all feathering and reverse-pitch propellers are of the constant-speed type; however, some constant-speed propellers are not of the feathering and reversing type. When propellers are reversed, their blades are rotated below their positive angle (that is, rotated through "0 thrust" pitch) until a negative blade angle is obtained which will produce a thrust acting in the opposite direction to the forward thrust normally produced by the propeller.

This feature is helpful for landing multiengine turboprop airplanes because it reduces the length of the landing roll, which in turn reduces the amount of braking needed and substantially increases the life of the brakes and tires. Almost all turboprop-equipped aircraft use reversing propellers.

FIXED-PITCH PROPELLERS

Wood Propellers

In the early days of aviation, all propellers were made of wood, but the development of larger and higher-horsepower aircraft engines made it necessary to adopt a stronger and more durable material; therefore, metal is now extensively used in the construction of propellers for all types of aircraft. Some propeller blades have been made of plastic materials, specially treated wood laminations, and plastic-coated wood laminations. For most purposes, however, metal propellers have been most satisfactory where cost has not been a primary consideration.

The aviation technician today will seldom be required to repair wood propellers. For this reason, the details of wood propeller repair are not covered in this text, and we only give the technician basic information about wood propellers.

Construction

The first consideration in the construction of a wood propeller is the selection of the right quality and type of wood. It is especially important that all lumber from which the propeller laminae (layers) are to be cut be kiln-dried. A wood propeller is not cut from a solid block but is built up of a number of separate layers of carefully selected and well-seasoned hardwoods, as illustrated in Fig. 20-22.

Many types of wood have been used in making propellers, but the most satisfactory are sweet or yellow birch, sugar maple, black cherry, and black walnut. In some cases, alternate layers of two different woods have been used to reduce the tendency toward warpage. This is not considered necessary, however, because the use of laminations of the same type of wood will effectively reduce the tendency for a propeller to warp under ordinary conditions of use.

The spiral or diagonal grain of propeller wood should have a slope of less than 1 in 10 when measured from the longitudinal axis of the laminae.

Propeller lumber should be free from checks, shakes, excessive pinworm holes, unsound and loose knots, and decay. Sap stain is considered a defect. The importance of selecting a high grade of lumber to reduce the effect of the

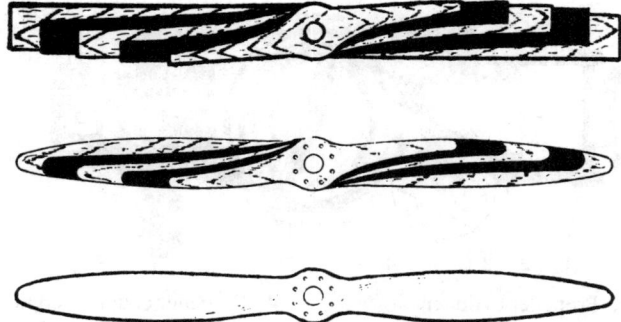

FIGURE 20-22   Construction of a typical wood propeller.

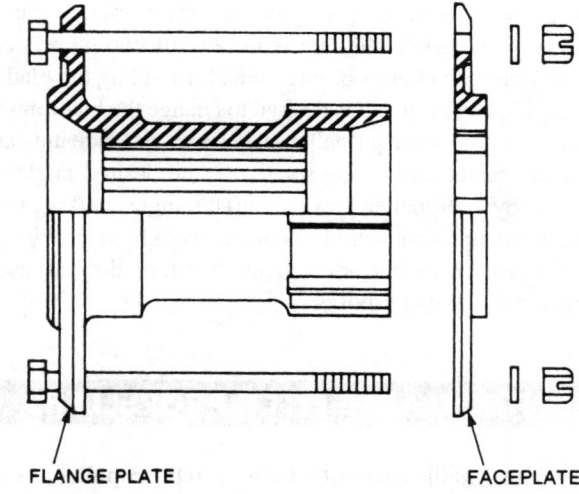

FLANGE PLATE   **FACEPLATE**

FIGURE 20-24   Hub assembly.

FIGURE 20-25   McCauley Met-L-Prop.

internal variations present in all wood cannot be too strongly emphasized.

As shown in Fig. 20-22, the laminations of wood are given a preliminary shaping and finishing and then are stacked together and glued with high-quality glue. Pressure and temperature are carefully controlled for the prescribed time. After the glue has set according to specifications, the propeller is shaped to its final form using templates and protractors to ensure that it meets design specifications.

After the propeller is shaped, the tip of each blade is covered with fabric to protect the tip from moisture and reduce the likelihood of cracking or splitting. The fabric is thoroughly waterproofed. Finally, the leading edge and tip of each blade are provided with a sheet-brass shield to reduce damage due to small rocks, sand, and other materials encountered during takeoff and taxiing. The metal tipping and leading-edge shield are shown in Fig. 20-23.

The centerbore of the hub and the mounting-bolt holes are very carefully bored to exact dimensions, which is essential to good balance upon installation. A hub assembly is inserted through the hub bore to facilitate the installation of the mounting bolts and faceplate. A hub assembly is illustrated in Fig. 20-24.

Metal Propellers

Description. A fixed-pitch metal propeller is usually manufactured by forging a single bar of aluminum alloy to the required shape. Typical of such propellers is the McCauley

Met-L-Prop shown in Fig. 20-25. The propeller shown in the illustration is provided with a centerbore for the installation of a steel hub or adapter to provide for different types of installation. The six hub bolt holes are dimensioned to fit a standard engine crankshaft flange. The following information should be printed on the propeller hub or the butts of the blades: builder's name, model designation, serial number, type certificate number, and production certificate number. The propeller is anodized to prevent corrosion.

Advantages

The advantages of a single-piece fixed-pitch metal propeller are (1) simplicity of maintenance, (2) durability, (3) resistance to weathering, (4) light weight, (5) low drag, and (6) minimum service requirements. Such a propeller is efficient for a particular set of operating conditions.

GROUND-ADJUSTABLE PROPELLERS

As previously mentioned, a ground-adjustable propeller is designed to permit a change of blade angle when the airplane is on the ground. This permits the adjustment of the propeller for the most effective operation under different conditions of flight. If it is desired that the airplane have a maximum rate of climb, the propeller blades are set at a comparatively low angle so that the engine can rotate at maximum speed to produce the greatest power. The propeller blade, in any case, must not be set at an angle which will permit the engine to overspeed. When it is desired that the engine operate efficiently at cruising speed and at high altitudes, the blade angle is increased.

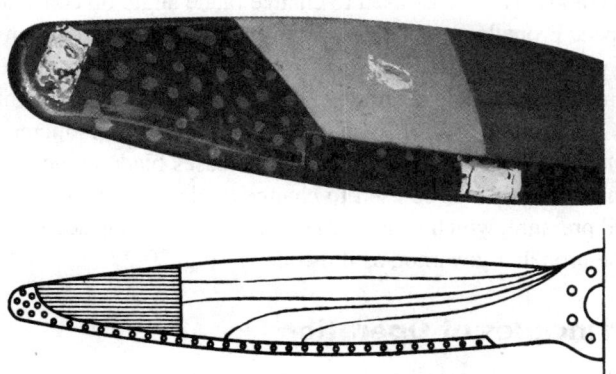

FIGURE 20-23   Metal-tipped propeller blade.

A ground-adjustable propeller may have blades made of wood or of metal. The hub is usually of two-piece steel construction with clamps or large nuts for holding the blades securely in place. When it is desired to change the blade angle of a ground-adjustable propeller, the clamps or blade nuts are loosened and the blades are rotated to the desired angle as indicated by a propeller protractor. The angle markings on the hub are not considered accurate enough to provide a good reference for blade adjustment; therefore, they are used chiefly for checking purposes.

CONTROLLABLE-PITCH PROPELLERS

As the name implies, a **controllable-pitch propeller** is one on which the blade angle can be changed while the aircraft is in flight. Propellers of this type have been used for many years on aircraft for which the extra cost of such propellers was justified by the improved performance obtained.

Advantages

The controllable-pitch feature makes it possible for the pilot to change the blade angle of the propeller at will in order to obtain the best performance from the aircraft engine. At takeoff, the propeller is set at a low blade angle so that the engine can attain the maximum allowable rpm and power. Shortly after takeoff, the angle is increased slightly to prevent overspeeding of the engine and to obtain the best climb conditions of engine rpm and airplane speed. When the airplane has reached the cruising altitude, the propeller can be adjusted to a comparatively high pitch for a low cruising rpm or to a lower pitch for a higher cruising rpm and greater speed.

TWO-POSITION PROPELLERS

A **two-position propeller** does not have all the advantages mentioned in the foregoing paragraph; however, it does permit a setting of blade angle for best takeoff and climb (low-pitch, high rpm) and for best cruise (high-pitch, low rpm). A schematic diagram of a two-position propeller pitch-changing mechanism is shown in Fig. 20-26. The principal parts of this assembly are the hub assembly, the counterweight and bracket assembly, and the cylinder and piston assembly. The blade angle is decreased by the action of the cylinder and piston assembly when engine oil enters the cylinder and forces it forward. The cylinder is linked to the blades by means of a bushing mounted on the cylinder base and riding in a slot in the counterweight bracket. As the cylinder moves outward, the bracket is rotated inward, and since the bracket is attached to the base of the blade, the blade is turned to a lower angle.

When the oil is released from the cylinder by means of a three-way valve, the centrifugal force acting on the counterweights moves the counterweights outward and rotates the blades to a higher angle. At the same time, the cylinder is pulled back toward the hub of the propeller.

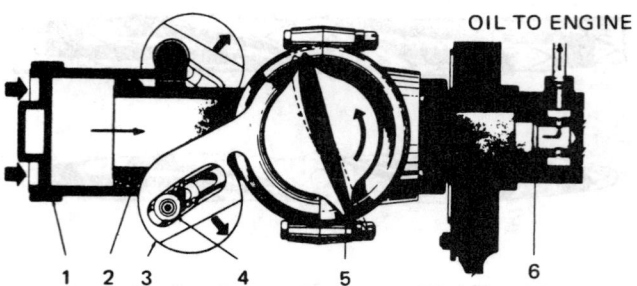

1. Propeller cylinder
2. Propeller piston
3. Propeller counterweight and bracket
4. Propeller counterweight shaft and bearing
5. Propeller blade
6. Engine propeller shaft

FIGURE 20-26 Drawing of a two-position propeller pitch-changing mechanism.

The basic high-pitch angle of the propeller is set by means of four blade-bushing index pins which are installed in aligned semicircular notches between the counterweight bracket and the blade bushing when the two are assembled. The pitch range is set by adjusting the counterweight adjusting screw nuts in the counterweight bracket.

A counterweight-type propeller may also be designed as a constant-speed propeller to be controlled by a propeller governor. In this case, the governor controls the flow of oil to and from the propeller cylinder in accordance with engine rpm. The governor is adjusted for the desired engine rpm by means of a control in the cockpit.

CONSTANT-SPEED PROPELLERS

As previously explained, a **constant-speed propeller** is controlled by a speed governor which automatically adjusts propeller pitch to maintain a selected engine speed. If the rpm of the propeller increases, the governor senses the increase and responds by causing the propeller blade angle to increase. If the propeller rpm decreases, the governor causes a decrease in propeller blade angle. An increase in blade angle will cause a decrease in engine rpm, and a decrease in blade angle will cause an increase in engine rpm.

The pitch-changing devices for constant-speed propellers include electric motors, hydraulic cylinders, devices in which centrifugal force acts on flyweights, and combinations of these methods. The forces used to change blade angle on constant-speed propellers can be divided into fixed and variable. Some types of **fixed forces** that are used to move propeller blades are counterweights, springs, centrifugal twisting moments (CTM), and air-nitrogen charges. All of these forces increase blade angle except CTM, which decreases blade angle. The main **variable force** used to change blade angle is governor oil pressure, which is metered by the speed-sensing section of the propeller governor, as illustrated in Fig. 20-27.

Principles of Operation

The blade-angle changes of the propeller are dependent on the balance between governor-boosted oil pressure and

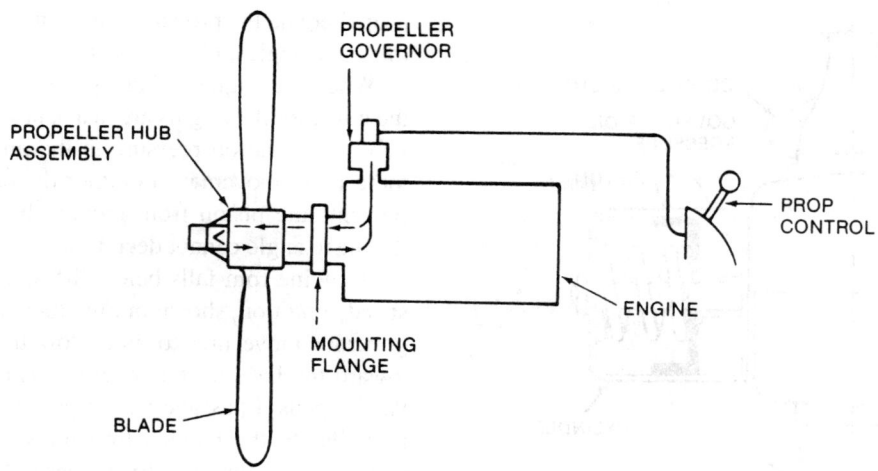

FIGURE 20-27 Propeller control mechanism (oil flow to and from engine). (*Hartzell Propeller.*)

the inherent centrifugal tendency of the propeller blades to maintain a low-pitch angle. The balance differential is maintained by the governor, which either meters oil pressure to, or allows oil to drain from, the propeller cylinder in the quantity necessary to maintain the proper blade angle for constant-speed operation. A drawing of the governor is shown in Fig. 20-28.

Within the governor, the L-shaped **flyweights** are pivoted on a disk-type **flyweight head** coupled to the engine gear train through a hollow drive-gear shaft. The **pilot-valve plunger** extends into the hollow shaft and is so mounted that the pivoting motion of the rotating flyweights will raise the plunger against the pressure of the **speeder spring** or allow the spring pressure to force the plunger down in the hollow shaft. The position assumed by the plunger determines the flow of oil from the governor to the propeller. Governor oil is directed to a transfer ring on the engine crankshaft and then into the crankshaft tube, which carries it into the rear side of the piston cylinder arrangement in the propeller hub. The linear motion of the piston is changed to rotary motion of the blades. Since the centrifugal twisting force of the propeller blades is transmitted to the propeller piston, the governor-boosted oil pressure must overcome this force to change the engine rpm. Forward motion of the piston increases pitch and decreases engine rpm, while rearward motion of the piston decreases pitch and increases engine rpm.

The action of the pitch-changing mechanism is shown in Fig. 20-29. As governor oil pressure enters the cylinder to the rear of the piston, the piston moves forward. This motion is transmitted through the piston shaft to each actuator bushing mounted on the butt of each blade, and when the bushings are moved forward, the blades are forced to rotate. During operation of the propeller in flight, the governor flyweights react to engine rpm. If the engine is turning faster than the selected rpm, the flyweights will move outward and cause the pilot valve in the governor to move upward, or toward the governor head. The resulting overspeed condition is illustrated in Fig. 20-30. With this valve position, the oil pressure from the governor pump is directed to the

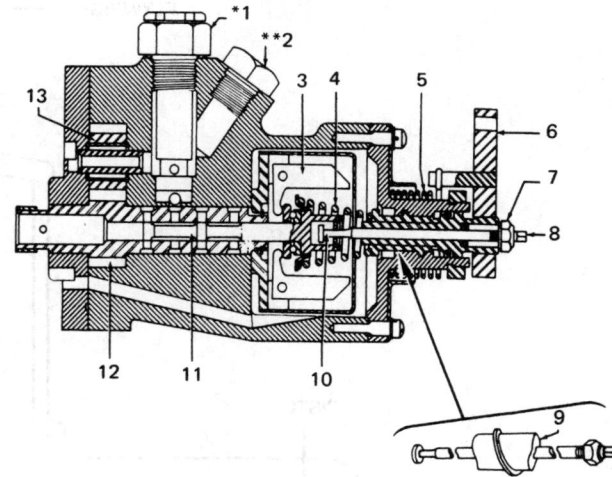

1. Differential-pressure-relief valve
2. High-pressure-relief valve
3. Flyweights
4. Speeder spring
5. Control-lever spring
6. Speed-adjusting control lever
7. Locknut
8. Lift-rod adjustment
9. Speed-adjusting worm
10. Pilot-valve lift rod
11. Pilot valve
12. Governor-pump drive gear
13. Governor-pump idler gear

FIGURE 20-28 Woodward propeller governor.

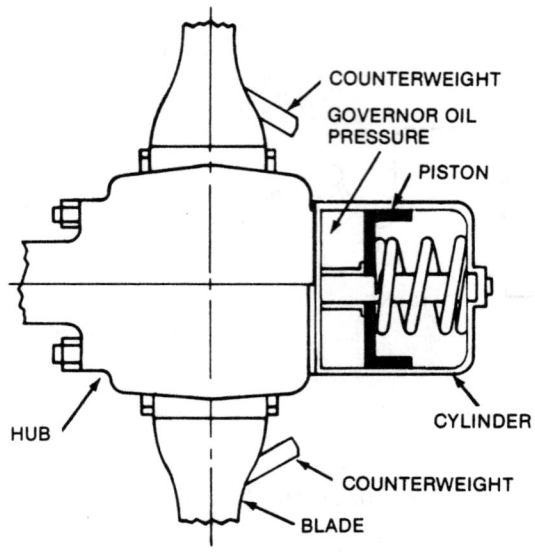

FIGURE 20-29 Typical constant-speed propeller operation. (*Hartzell Propeller.*)

propeller and the propeller piston moves forward to increase the blade angle and decrease the rpm.

When the engine is "on speed," as shown in Fig. 20-31, the governor flyweights are in a neutral position and the pilot valve seals the oil pressure in the propeller system so that there is no movement in either direction. The oil pressure prevents the piston from moving backward, and therefore the blade angle cannot decrease.

If engine rpm falls below the selected speed (an underspeed condition, shown in Fig. 20-32), the flyweights of the governor move inward and allow the pilot valve to move toward the base of the governor. This position of the pilot valve opens a passage which permits oil to flow from the propeller to the engine, thus allowing the blade angle to decrease and the rpm to increase. The blade angle tends to decrease because of the centrifugal twisting force, as explained previously.

Propeller Governor

We have discussed the propeller governor previously and described its operation to some degree in explaining the operation of constant-speed propellers; however, it will be

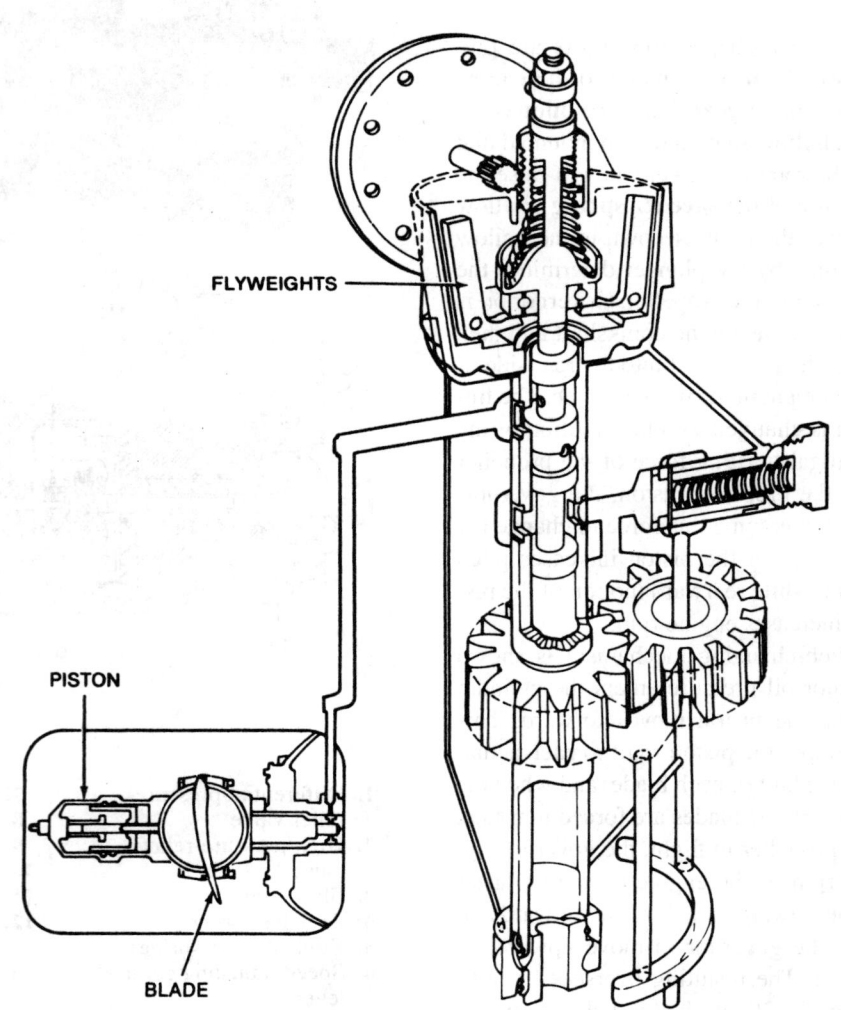

FIGURE 20-30 Position of governor flyweights in response to engine overspeeding. (*Hartzell Propeller.*)

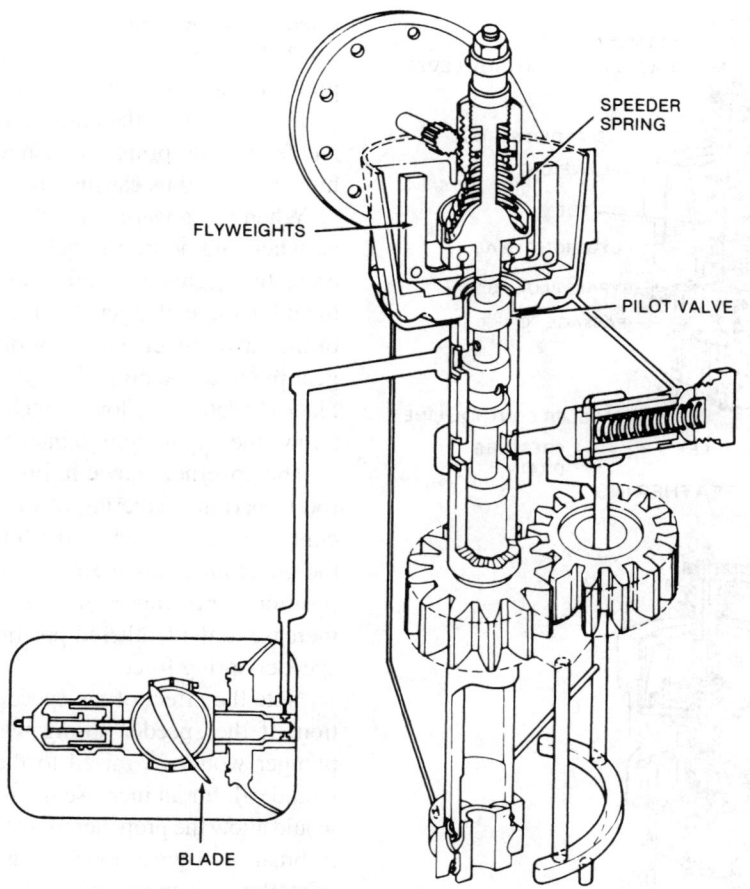

FIGURE 20-31   Position of governor flyweights during "on-speed" condition. (*Hartzell Propeller.*)

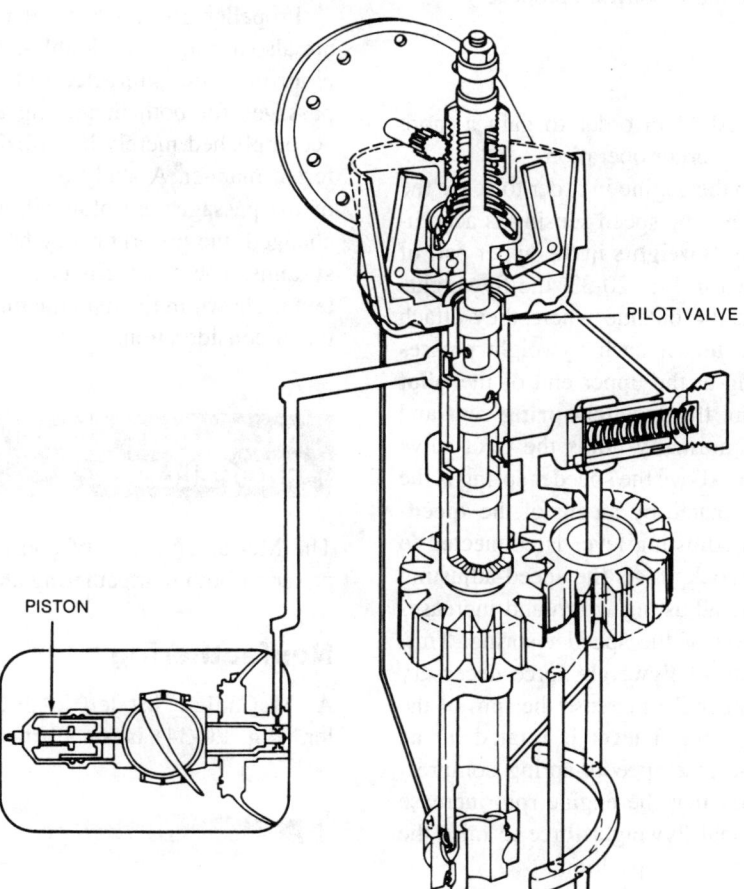

FIGURE 20-32   Position of governor flyweights in response to engine underspeeding. (*Hartzell Propeller.*)

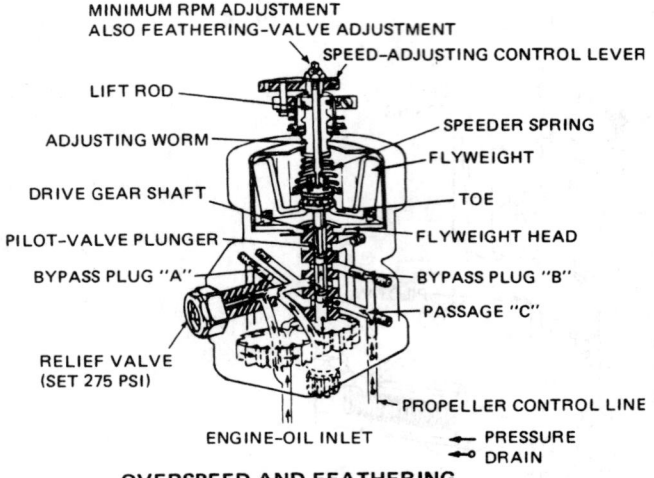

MINIMUM RPM ADJUSTMENT
ALSO FEATHERING-VALVE ADJUSTMENT
SPEED-ADJUSTING CONTROL LEVER
LIFT ROD
ADJUSTING WORM
DRIVE GEAR SHAFT
PILOT-VALVE PLUNGER
BYPASS PLUG "A"
RELIEF VALVE
(SET 275 PSI)
SPEEDER SPRING
FLYWEIGHT
TOE
FLYWEIGHT HEAD
BYPASS PLUG "B"
PASSAGE "C"
PROPELLER CONTROL LINE
ENGINE-OIL INLET
PRESSURE
DRAIN

OVERSPEED AND FEATHERING

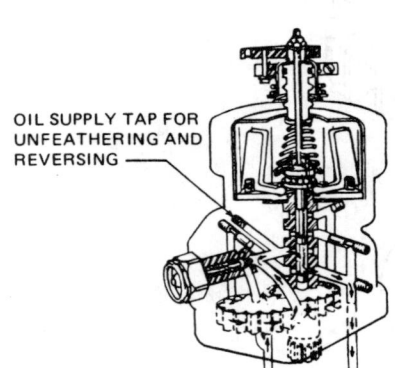

OIL SUPPLY TAP FOR
UNFEATHERING AND
REVERSING

FIGURE 20-33 Operation of the Woodward propeller governor.

beneficial to examine Fig. 20-33 in order to gain a more complete understanding of governor operation.

The governor is geared to the engine in order to sense the rpm of the engine at all times. The speed sensing is accomplished by means of rotating **flyweights** in the upper part of the governor body. As shown in Fig. 20-33, the flyweights are L-shaped and hinged at the outside where they attach to the **flyweight head**. The **toe** of each flyweight presses against the race of a bearing at the upper end of the **pilot valve**. Above the bearing are the **speeder-spring seat** and the **speeder spring**, which normally holds the pilot-valve plunger in the down position. Above the speeder spring is the **adjusting worm**, which is rotated by means of the **speed-adjusting lever**. The speed-adjusting lever is connected to the propeller control in the cockpit. As the speed-adjusting lever is moved, it rotates the adjusting worm and increases or decreases the compression of the speeder spring. This, of course, affects the amount of flyweight force necessary to move the pilot-valve plunger. To increase the rpm of the engine, the speed-adjusting control lever is rotated in the proper direction so as to increase speeder-spring compression. It is therefore necessary that the engine rpm increase in order to apply the additional flyweight force to raise the pilot-valve plunger to an "on-speed" position.

In the top drawing in Fig. 20-33, the governor is in the overspeed condition. The engine rpm is greater than that

selected by the control, and the flyweights are pressing outward. The toes of the flyweights have raised the pilot-valve plunger to a position which permits oil pressure from the propeller to return to the engine. The propeller counterweights and feathering spring can then rotate the propeller blades to a higher angle, thus causing the engine rpm to decrease.

When the governor is in an underspeed condition—that is, when engine rpm is below the selected value—the governor flyweights are held inward by the speeder spring and the pilot-valve plunger is in the down position. This position of the valve directs governor oil pressure from the governor gear pump to the propeller cylinder and causes the propeller blades to rotate to a lower-pitch angle. The lower-pitch angle allows the engine rpm to increase.

The governor shown in Fig. 20-33 is equipped with a lift rod to permit feathering of the propeller. When the cockpit control is pulled back to the limit of its travel, the lift rod in the governor holds the pilot-valve plunger in an overspeed position. This causes the blade angle of the propeller to increase to the feathered position regardless of flyweight or speeder-spring force.

Note the effect of the speeder spring on governor operation. If the speeder spring were to break, the pilot-valve plunger would be raised to the overspeed position, which would call for an increase in propeller pitch. This, of course, would allow the propeller to feather. If the speeder spring were to break in a governor for a nonfeathering, constant-speed propeller, the propeller blades would rotate to maximum high-pitch angle.

Propeller governors similar to the one described above are also arranged for double-acting operation in which governor pressure is directed to the propeller through different passages for both increasing and decreasing rpm. This is accomplished merely by utilizing the oil passages in a different manner. A study of Fig. 20-33 will show that some of the passages are plugged, and if the use of passages is changed, the governor may be adapted to different types of systems. The arrangement for any particular propeller system is shown in the manufacturer's manual for the propeller under consideration.

McCAULEY CONSTANT-SPEED PROPELLERS

The McCauley series of constant-speed propellers is comprised of both nonfeathering and feathering types.

Nonfeathering

A McCauley Model 2A36C18 nonfeathering propeller (Fig. 20-34) is an all-metal constant-speed propeller

FIGURE 20-34 McCauley Model 2A36C18 propeller. (*McCauley Propeller.*)

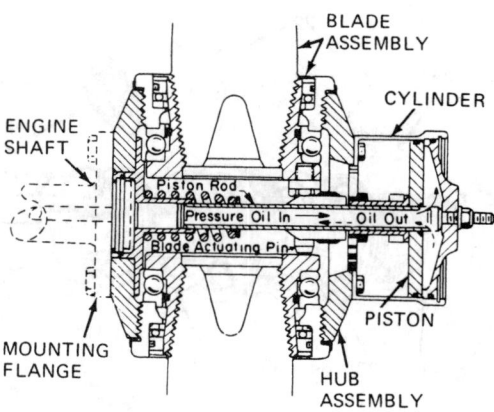

FIGURE 20-35 Drawing of the hub mechanism for the McCauley 2A36C18 propeller. (*McCauley Propeller.*)

controlled by a single-acting governor. The blades are made of forged aluminum alloy, and the hub parts are made of steel.

A schematic diagram of the propeller hub mechanism is shown in Fig. 20-35. A careful study of this drawing will reveal the *cylinder* at the front of the propeller hub, the *piston* inside the cylinder, the hollow *piston rod* through which oil flows to and from the cylinder, the *blade actuating pin,* the *low-pitch return boost spring,* the *hub assembly,* and the *blade assembly.* During operation, when the piston is fully forward, the blades are in the low-pitch position. If the engine overspeeds, the governor will direct governor oil pressure through the crankshaft into the hollow piston rod of the propeller. The oil flows through the piston rod and into the cylinder, forcing the piston to move back. The piston rod is linked to the blade butts through *link assemblies* and the blade actuating pins, and as the piston rod moves backward, the blades are forced to rotate in the hub. This increases the pitch and reduces the engine speed. If the engine rpm falls below the value selected by the governor control, the governor pilot valve will move downward and open the passages, which will allow the oil in the propeller piston to return through the piston rod to the engine. The piston is pushed forward by the low-pitch return boost spring and by the centrifugal twisting of the rotating blades. When the propeller is "on speed," the oil pressure in the cylinder is balanced against the two forces tending to turn the blades to low pitch.

The detailed construction of the hub assembly, the pitch-changing mechanism, and the blade assemblies are shown in the exploded view in Fig. 20-36.

Feathering

The McCauley feathering propeller, illustrated in Fig. 20-37, is of the constant-speed, full-feathering type. It is a single-acting unit in which hydraulic pressure opposes the forces of springs and counterweights to obtain the correct pitch. Hydraulic pressure moves the blades toward low pitch, while the springs and counterweights move the blades toward high pitch. An engine-mounted governor is required for operation of the propeller. No other external components are required,

although unfeathering accumulators may be installed on the aircraft.

The source of the hydraulic pressure for operation is oil from the engine lubricating system, boosted in pressure by the governor gear pump and supplied to the propeller hub through the engine shaft flange. Oil is metered to and from the propeller by the governor control valve as positioned by the flyweights. This either increases or decreases the blade angle (changes the pitch) as required when the propeller speed control setting is altered. Increases or decreases in blade angle can also occur with the propeller speed control remaining in a fixed setting in order to stabilize engine speed for varying flight attitudes. In flight, complete reduction of hydraulic pressure will cause the control springs and counterweights to automatically move the blades to the full-feathered position.

The complete pitch-changing mechanism is entirely enclosed in the hub structure. No operating or wearing surfaces are exposed to the elements. All functioning parts of the actuating mechanism are made from materials specifically selected to require no lubrication between overhauls.

The propeller has a ground stop mechanism installed within the hub structure which functions in response to centrifugal force acting on rotating latch weights. The mechanism includes latch weights which will engage a fixed stop, thus blocking movement of the piston in the direction of increased blade pitch beyond a predetermined setting and keeping the propeller from feathering. The latching movement is possible only when the engine is shut off, on the ground. If the engine were to be shut down in flight, propeller windmilling would provide sufficient centrifugal force to keep the latches disengaged, thus allowing the propeller to feather. Under all normal engine operating conditions, the weights are kept out of the latching position by centrifugal force and thus offer no resistance to feathering of the propeller.

HARTZELL CONSTANT-SPEED PROPELLERS

There are two major ways to categorize Hartzell constant-speed propellers. One is by the type of blade angle: nonfeathering, feathering, or reversing. The other is by the type of metal used in the propeller's construction: steel hub or compact hub (aluminum hub).

Steel Hub Propellers

The Hartzell steel hub propellers (nonfeathering, feathering, reversible) of the present series are similar in basic design and have many parts in common, with the exception that the feathering propeller has a greater blade angle range and a heavy spring for feathering the propeller.

Nonfeathering

An example of a Hartzell steel hub nonfeathering propeller is shown in Fig. 20-38. In order to control the pitch of

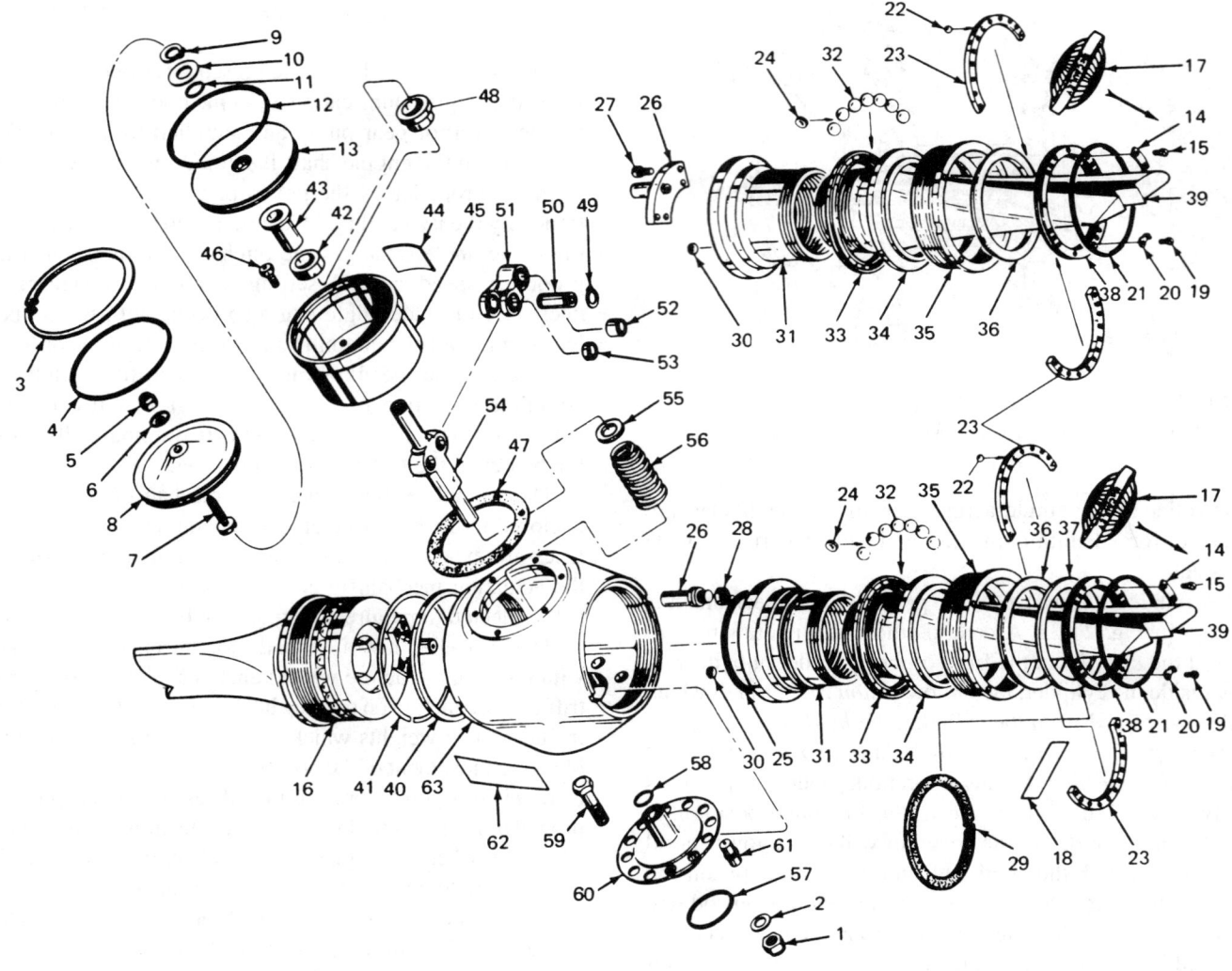

FIGURE 20-36 Exploded view of the McCauley constant-speed propeller. (*McCauley Propeller.*)

1. Nut	**20.** Preload nut lock	**37.** Outer preload bearing race	**52.** Blade actuating pin bearing
2. Plain washer	**21.** O-ring packing	**38.** Preload nut	**53.** Piston-rod-bearing
3. Internal retaining ring	**22.** Bearing ball	**39.** Blade	**54.** Piston rod
4. O-ring packing	**23.** Preload bearing retainer	**40.** Retention-nut lock ring	**55.** Plain washer
5. Self-locking nut	**24.** Ball separator	**41.** Balancing shim	**56.** Low-pitch return boost spring
6. Dyna seal	**25.** O-ring packing	**42.** High-pitch stop-spacer stock	**57.** O-ring packing
7. Low-pitch stop screw	**26.** Blade actuating pin	**43.** Piston-rod sleeve	**58.** O-ring packing
8. Cylinder head	**27.** Knurled-socket-head cap screw	**44.** Spring kit-installation decal	**59.** Hub mounting bolt
9. Bowed retaining ring	**28.** Actuating-pin washer	**45.** Cylinder assembly	**60.** Hub and piston guide flange
10. Piston washer	**29.** Gasket	**46.** Bolt	**61.** Hub-alignment dowel
11. O-ring packing	**30.** Ferrule-staking plug	**47.** Cylinder gasket	**62.** Propeller-installation-instructions decal
12. O-ring packing	**31.** Blade-retention ferrule	**48.** Cylinder bushing	
13. Piston	**32.** Bearing ball	**49.** External retaining ring	**63.** Propeller hub
14. Balance weight	**33.** Inner race	**50.** Piston-rod pin	
15. Screw	**34.** Outer race	**51.** Link assembly	
16. Blade assembly	**35.** Blade-retention nut		
17. Decal	**36.** Inner preload bearing race		
18. Decal			
19. Screw			

the blades, a hydraulic piston-cylinder element is mounted on the front of the hub spider. The piston is attached to the blade clamps by means of a sliding rod and fork system for the nonfeathering models. The piston is actuated in the forward direction by means of oil pressure supplied by the governor, which overcomes the opposing force exerted by the counterweights and decreases the pitch of the blades. The counterweights attached to the blade clamps utilize centrifugal force to increase the pitch of the blades. The centrifugal force, due to rotation of the propeller, tends to move the

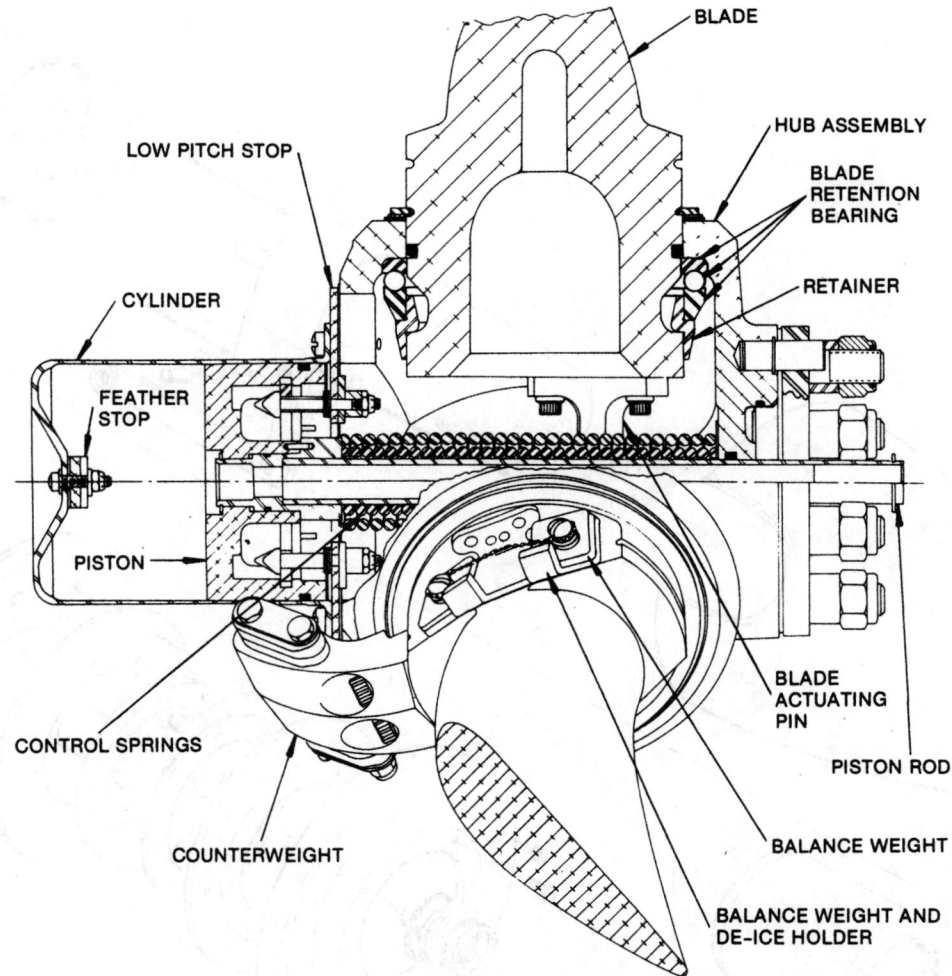

FIGURE 20-37 McCauley feathering propeller. (*McCauley Propeller.*)

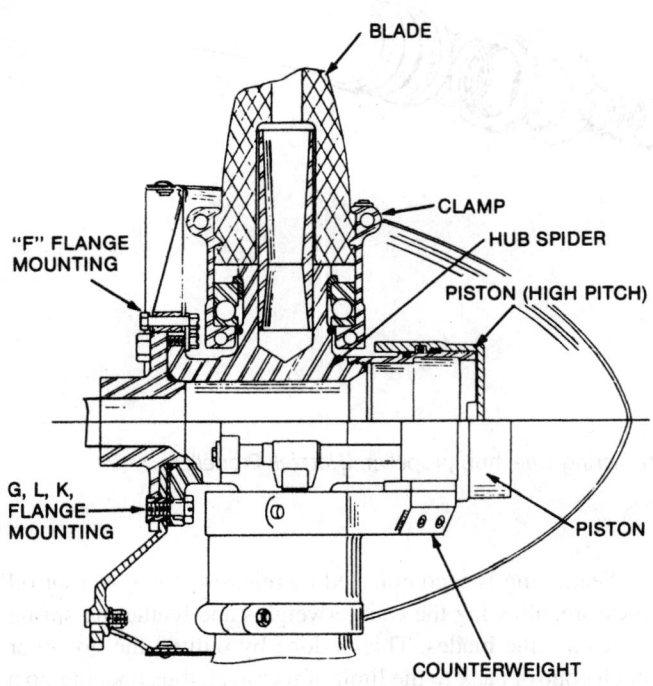

FIGURE 20-38 Hartzell steel hub nonfeathering propeller. (*Hartzell Propeller.*)

counterweights into the plane of rotation, thereby increasing the pitch of the blades.

Some of the steel hub propellers do not have counterweights; therefore, they have less total weight. In these propellers, the forces used to change pitch are reversed: blade centrifugal force reduces pitch and governor oil pressure increases pitch.

Full-Feathering

Hartzell constant-speed, full-feathering steel hub propellers utilize hydraulic pressure to reduce the pitch of the blades and a combination of spring and counterweight force to increase the pitch. If the pitch is increased to the limit, the blades are in the feathered position. A typical three-blade feathering steel hub propeller is shown in Fig. 20-39.

Operation. Figure 20-40 is a schematic drawing of the Hartzell Model HC-82XF-2 feathering propeller hub assembly that illustrates the pitch-changing mechanism. When the engine speed is below that selected by the pilot, the governor pilot valve directs governor oil pressure to the propeller. This pressure forces the cylinder forward and reduces the

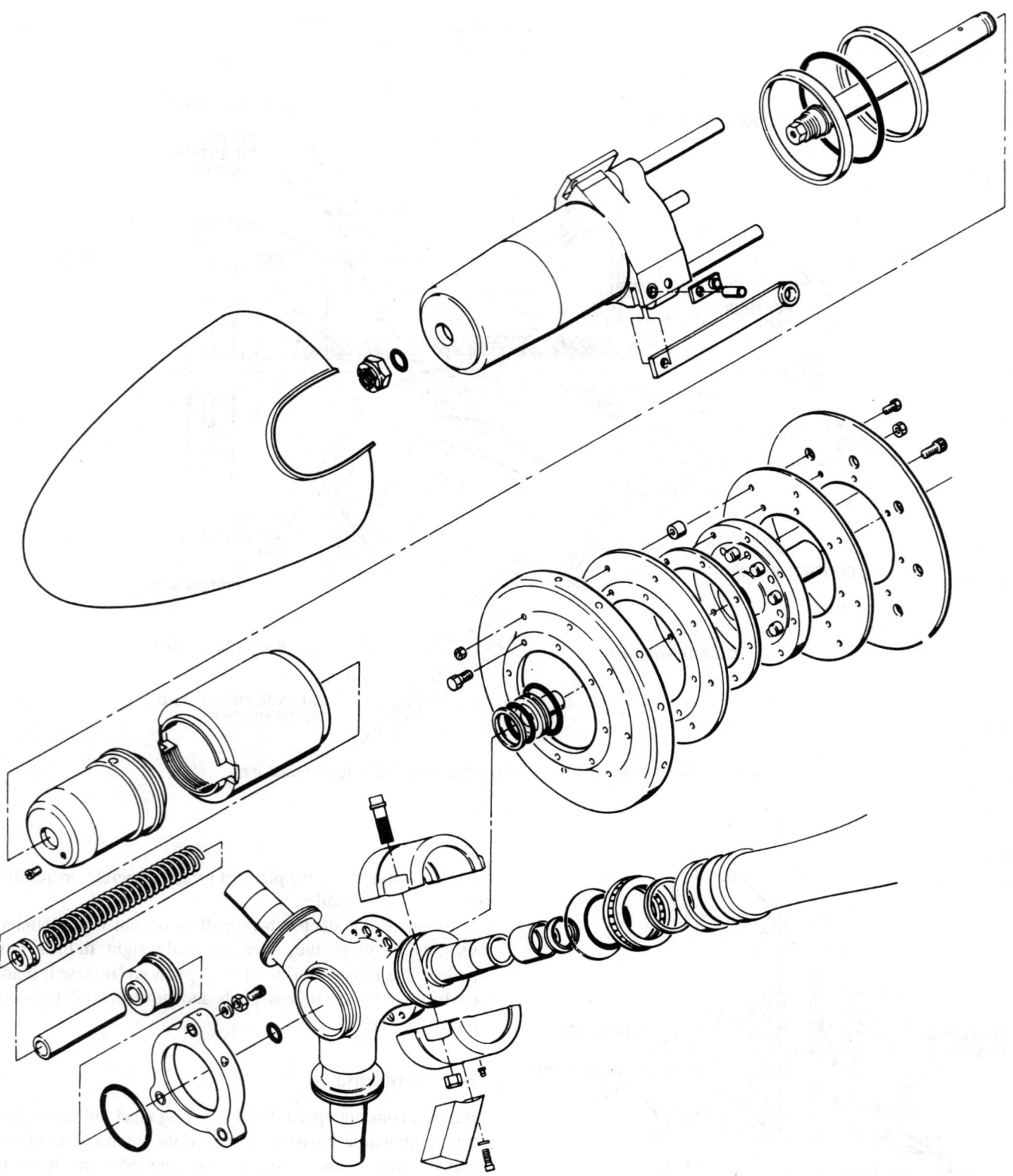

FIGURE 20-39 Typical three-blade constant-speed feathering steel hub propeller. (*Hartzell Propeller.*)

propeller pitch. When the cylinder moves forward, it also compresses the feathering spring. If engine speed increases above the rpm selected, the governor opens the oil passage to allow the oil in the propeller cylinder to return to the engine. The feathering spring and the counterweight force cause the blades to rotate to a higher-pitch position.

Feathering is accomplished by releasing the governor oil pressure, allowing the counterweights and feathering spring to feather the blades. This is done by pulling the governor pitch control back to the limit of its travel, thus opening up a port in the governor through which the oil from the propeller can drain back to the engine. The time required for feathering

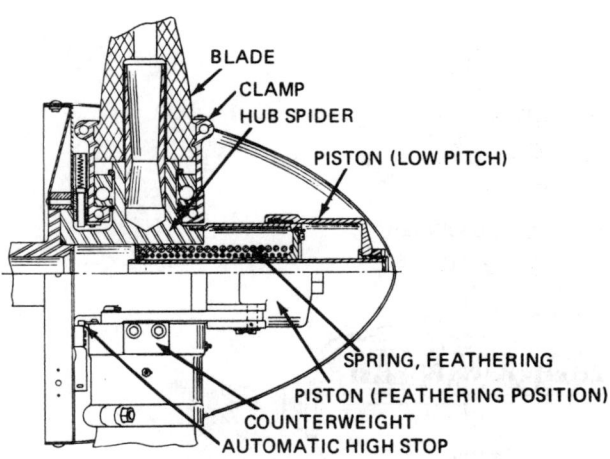

FIGURE 20-40 Drawing of the Hartzell HC-82XF-2 feathering propeller hub assembly. (*Hartzell Propeller.*)

depends on the size of the oil passage from the propeller to the engine and on the force exerted by the spring and counterweights. The larger the passages through the governor and the heavier the springs, the quicker the feathering action. The elapsed time for feathering is usually between 3 and 10 s.

Unfeathering the propeller is accomplished by repositioning the governor control within the normal flight range and restarting the engine. As soon as the engine cranks over a few turns, the governor starts to unfeather the blades and soon windmilling takes place, thus speeding up the process of unfeathering. In order to facilitate cranking of the engine, the feathering blade angle is set at 80 to 85° at the three-fourths station on the blades. In general, restarting and unfeathering can be accomplished within a few seconds.

Special unfeathering systems may be installed with the Hartzell propeller when it is desired to increase the speed of unfeathering. Such a system is shown in Fig. 20-41. During normal operation, the accumulator stores governor oil pressure; when the propeller is feathered, this pressure is trapped in the accumulator because the accumulator valve is closed at this time. When the propeller control is placed in the normal position, the pressure stored in the accumulator is applied to the propeller to rotate the blades to a low-pitch angle. Remember that when the propeller is feathered, there is no

pressure available from the governor because the engine is stopped. The pressure stored in the accumulator is used in place of the pressure which would normally be supplied by the governor.

Reversible

Steel hub reversible propellers, used primarily for turboprop installations, are similar to the feathering propellers, except that the pitch setting is extended into the reverse range and a hydraulic low-pitch stop is introduced. A typical Hartzell steel hub reversible propeller is illustrated in Fig. 20-42. In order to obtain greater pitch travel, the piston and cylinder are made longer. Detailed information on the Hartzell reversible propeller and its control systems is presented in Chap. 21.

Compact (Aluminum Hub) Propellers

Hartzell **compact propellers** incorporate many new concepts in basic design. They combine low weight with simplicity in design and rugged construction.

The hub is made as compact as possible, utilizing aluminum-alloy forgings for most of the parts. The hub shell is made in two halves that are bolted together along the plane of rotation. The hub shell carries the pitch-change mechanism and blade roots internally. The hydraulic cylinder which provides power for changing the pitch is mounted at the front of the hub. A compact propeller can be installed only on an engine with flanged crankshaft mounting provisions. Compact propellers are currently made in two- and three-blade configurations.

Nonfeathering

The constant-speed "dash-1" Hartzell propellers utilize oil pressure from a governor at pressures ranging from 0 to 300 psi [2068 kPa] to move the blades into high pitch (reduced rpm). The centrifugal twisting of the blades tends to move them into low pitch (high rpm) in the absence of governor oil pressure. The "dash-4" model routes the oil to the rear of the piston so that the oil pressure reduces the blade pitch. Counterweights are added to the blades to cause them to move to high pitch in the absence of oil pressure.

Feathering

Hartzell compact feathering propellers are currently manufactured in two configurations, air pressure–oil models and air pressure–counterweight–oil models. **Air pressure–oil propellers** utilize a combination of air pressure force and a mechanical spring to increase pitch and feather (see Fig. 20-43). These forces are opposed by governor-regulated oil pressure to reduce pitch. The increasing-pitch force is exerted by an air charge which is trapped in the cylinder head plus a coil spring located in the propeller shaft extension housing. Only two-blade propellers with extension shafts can be constructed in this manner. **Air pressure–counterweight–oil propellers** employ a combination

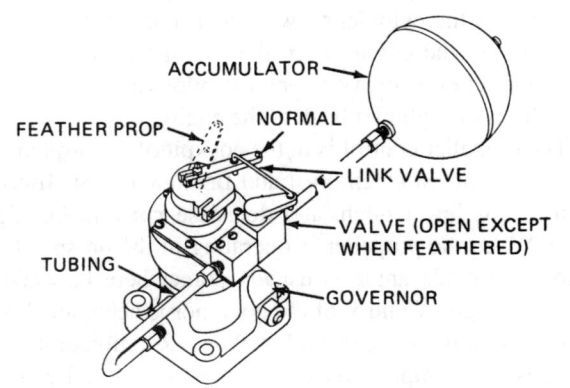

FIGURE 20-41 Unfeathering system for the Hartzell propeller.

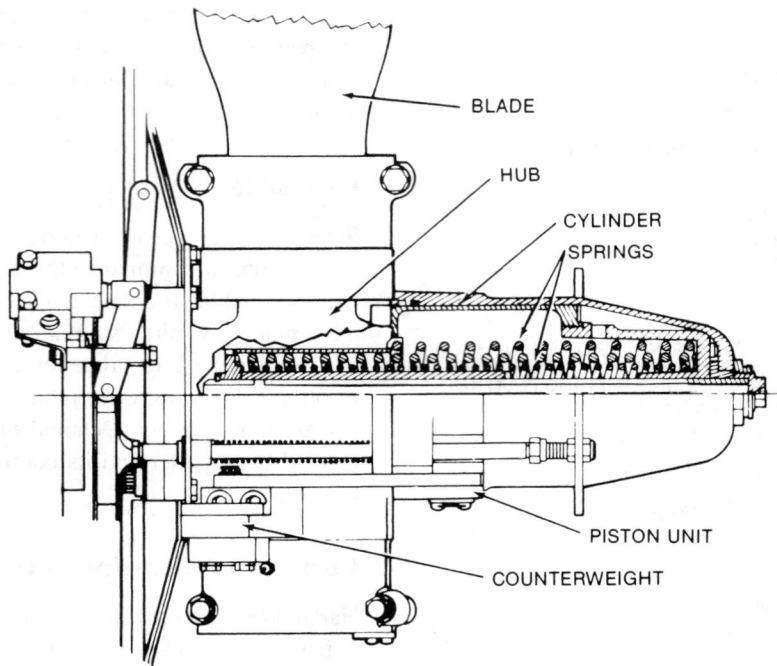

FIGURE 20-42 Steel hub reversible propeller. (*Hartzell Propeller.*)

of air pressure and blade counterweights to increase pitch and feather; these forces are opposed by governor-regulated oil pressure to reduce pitch. In both types of propellers, feathering is accomplished by the pilot by pulling the pitch-control lever back to the limit of its travel, which allows oil to drain out of the propeller and back to the engine sump. These propellers are unfeathered in a manner similar to the unfeathering of the steel hub models.

HAMILTON STANDARD COUNTERWEIGHT PROPELLERS

Two-Position Propeller

Although the Hamilton Standard counterweight propellers are no longer in production, there are still some aircraft which employ these propellers. For this reason, it is useful for the aviation technician to have some knowledge regarding their operation.

The two-position propeller is controlled from the cockpit by means of a three-way valve operated by the pilot. The valve directs engine oil pressure to the propeller cylinder to cause the pitch to decrease. When the valve position is changed to allow the oil to flow back to the engine, the counterweights rotate the blades to high pitch.

The range of pitch change for the two-position propeller is about 10°. The low-pitch position is used for takeoff so that the engine can develop its maximum rpm. After takeoff and climb, the propeller is placed in the high-pitch position for cruising.

The operation of the two-position propeller is illustrated in Fig. 20-44.

Constant-Speed Counterweight Propeller

The constant-speed counterweight propeller is essentially the same as the two-position propeller with the exception of the blade-angle range and the controlling mechanism. The constant-speed propeller may have a range of either 15 or 20° depending on the type of installation. This range is determined by the adjustment of the stops in the counterweight assembly. If the propeller is set for a 20° range, a return-spring assembly is installed in the piston to assist the counterweights in returning the cylinder to the rearward position when the governor calls for increased pitch.

Control of the constant-speed propeller is accomplished by means of a propeller governor similar to those described previously. The governor operates by means of flyweights which control the position of a pilot valve. When the propeller is in an underspeed condition, with rpm below that for which the governor is set, the governor flyweights move inward and the pilot valve then directs engine oil pressure to the propeller cylinder through the engine propeller shaft. This moves the cylinder forward and reduces the propeller pitch. When the engine is in the overspeed condition, the governor action is opposite and the pilot valve allows oil to drain from the cylinder back to the engine.

The propeller control is in the cockpit of the airplane and is marked for INCREASE RPM and DECREASE RPM. **Increase rpm** means lower pitch, and **decrease rpm** means higher pitch. When the propeller is operating in the on-speed condition, the blade angle is usually somewhere between the extreme ranges. Control of the governor is accomplished by rotating a shaft through a cable linkage. Rotation of the shaft changes the compression of the governor speeder spring, which controls the flyweight position.

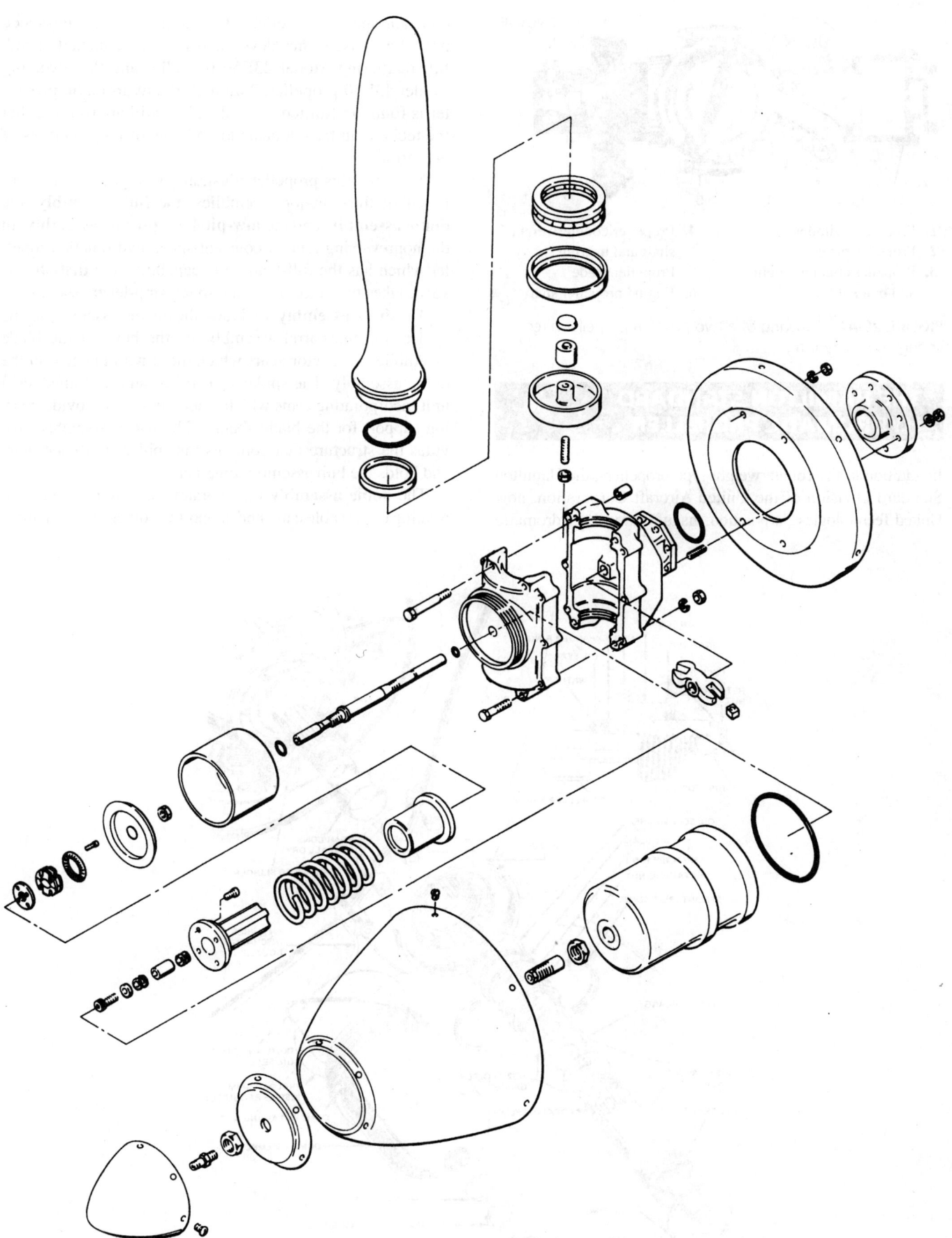

FIGURE 20-43 Typical three-blade constant-speed feathering aluminum hub propeller. (*Hartzell Propeller.*)

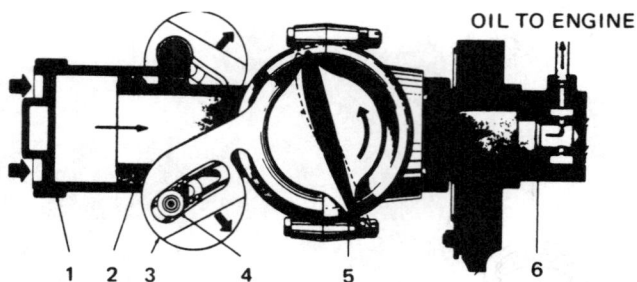

1. Propeller cylinder
2. Propeller piston
3. Propeller counterweight and bracket
4. Propeller counterweight shaft and bearing
5. Propeller blade
6. Engine propeller shaft

FIGURE 20-44 Drawing of a two-position propeller pitch-changing mechanism.

THE HAMILTON STANDARD HYDROMATIC PROPELLER

In addition to the counterweight-type propellers, the Hamilton Standard Division of the United Aircraft Corporation, now United Technologies Corporation, manufactured the hydromatic constant-speed propeller. The original constant-speed propeller was further developed into the constant-speed, full-feathering Model 23E50 propeller and the reversing Model 43E60 propeller. Although the hydromatic propeller is found in limited use today, it is still appropriate that the technician have a basic knowledge of its principles of operation.

Note that this propeller illustrated in Fig. 20-45 is composed of three major assemblies: the **hub assembly**, the **dome assembly**, and the **low-pitch stop-lever assembly**. In the nonreversing type of constant-speed hydromatic propeller which has the full-feathering capability, the **distributor valve** takes the place of the low-pitch stop-lever assembly.

The **hub assembly** includes the blade assemblies, the spider, and the barrel assembly. At the butts of the blade assemblies are sector gears which mesh with the gear in the dome assembly. The spider is a forged and machined steel unit incorporating arms which extend into and provide bearing support for the blade shanks. The barrel assembly provides the structure that contains the spider and blade butts and holds the hub assembly together.

The **dome assembly** incorporates the piston, fixed cam, rotating or movable cam, and support for other required units.

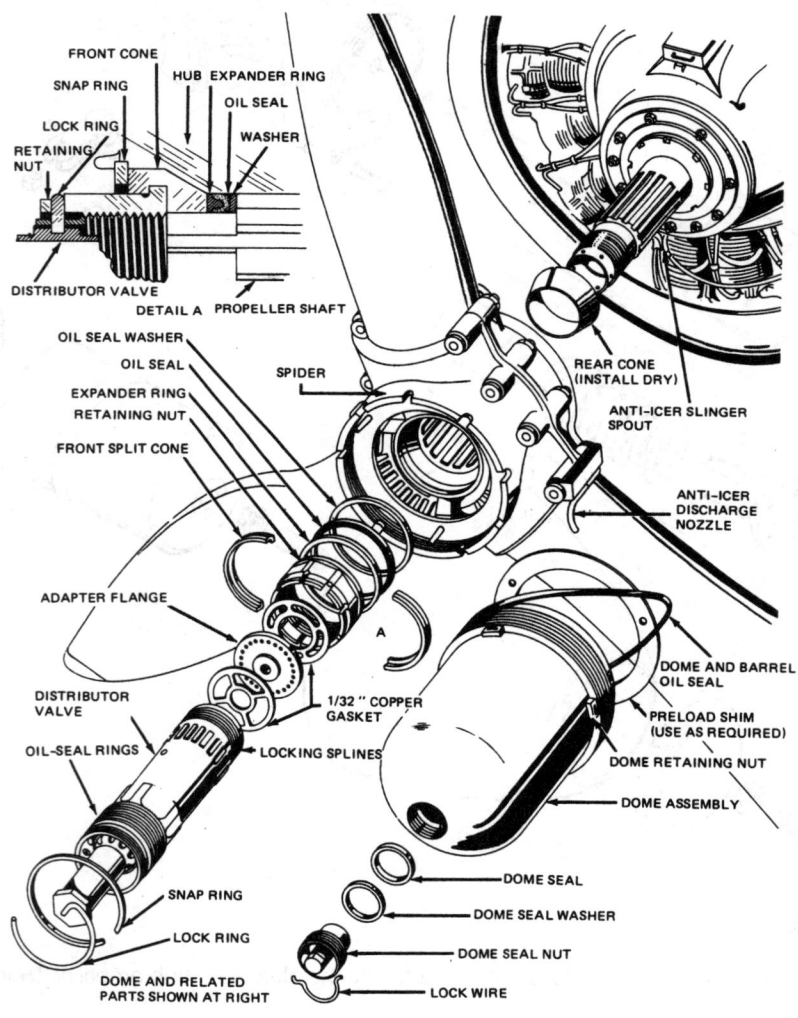

FIGURE 20-45 Typical hydromatic propeller assembly.

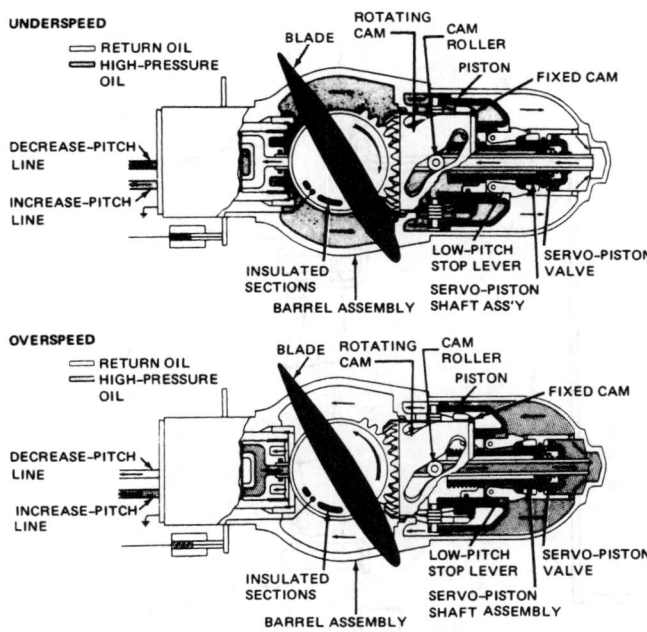

UNDERSPEED

OVERSPEED

FIGURE 20-46 Operation of the hydromatic pitch-changing mechanism.

The bevel pitch-changing gear which meshes with the sector gears on the blade butts is on the rotating cam.

Principles of Operation

The pitch-changing mechanism for both underspeed and overspeed conditions is shown in Fig. 20-46.

The forces acting to control the blade angle of the propeller are **centrifugal twisting moment** and **high-pressure oil**. The centrifugal twisting moment tends to turn the blades to a lower angle. The high-pressure oil is directed to the propeller to change the blade angle in either direction.

The pitch-changing mechanism consists of a piston which moves forward and backward in the dome cylinder. Rollers on the piston engage cam slots in the cams. The cams are so arranged, one within the other, that the rotation of one cam is added to the rotation of the other, thus doubling the movement that would be obtained from one cam alone. The outer cam is stationary, and the inner cam is rotated. This inner cam carries the bevel gear which meshes with the blade gears and in turn changes the blade angle.

The feathering process for the 23E50 propeller is nearly the same as that for the 43E60 propeller except that the piston moves *rearward* in the dome of the 23E50 propeller to decrease the pitch whereas it moves *forward* in the 43E60 propeller to decrease the pitch. When the feather button in the cockpit is depressed, a holding circuit for holding the button in is completed through the pressure cutout switch. Depressing the button also completes the circuit to the feather pump motor, which picks up oil from the main oil system and pumps it to the propeller governor through an external oil line. The feather pressure pump repositions the pressure transfer valve in the governor, allowing the high-pressure oil to bypass the governor and pass through

the propeller shaft to the distributor valve and into the rear side of the piston. Auxiliary pressure moves the piston forward until it reaches the feather stop. The feathering process is then completed when the system pressure rises above the pressure-cutoff-switch setting. When the cutout switch opens, the holding circuit to the switch is broken and the button automatically pops out, stopping the feather motor. Normally, when the system pressure increases to above 650 psi [4481.75 kPa], the switch will open.

To unfeather the propeller, the button is also pressed, but the button must be manually held in to overcome the pressure cutout switch. With the button held in, the feather pump pressure is approximately 650 psi [4481.75 kPa]. The high pressure that is exerted on the rear side of the distributor valve moves the distributor valve forward, rerouting the pressure to the front of the piston, moving the piston to the rear, and rotating the blades out of the feather position. As soon as the blades move from the full-feather position, the propeller begins to rotate (windmill) and the governor resumes its normal governing operation.

ANTI-ICING AND DEICING SYSTEMS

Propeller anti-icing may be accomplished by spraying isopropyl alcohol along the leading edges of the blades. A typical anti-icing system is illustrated in Fig. 20-47. The anti-icing fluid is carried in a reservoir in the airplane in sufficient quantities for normal demand. The fluid is pumped from the reservoir to the propeller by an electrically driven **anti-icing pump** which is controlled from the cockpit by a rheostat. The propeller is equipped with a **slinger ring** having nozzles aligned with the leading edge of each blade. When the pump is turned on, the fluid is forced out the nozzles of the slinger ring by centrifugal force and carried along the leading edges of the propeller blades. This prevents ice from accumulating on the blades. The length of time the system can provide icing protection depends on the amount of fluid carried and the rate at which it is used (controlled by the rheostat setting). This system is not used much on modern aircraft because the time limit on its use is controlled by the amount of fluid carried.

A deicing system which is used extensively employs electric heating elements to heat the blades. These elements are centered in boots attached to the leading edges of the blades or are mounted within the blades. Power for the heating elements is transferred through slip rings at the rear of the propeller hub.

The deicing system, shown in Fig. 20-48, applies heat to the surfaces of the propeller blades where ice normally would adhere. This heat, plus centrifugal force and the blast from the airstream, removes accumulated ice, as illustrated in Fig. 20-49. A minimum thickness or weight of ice must be formed before centrifugal force becomes a significant factor in propeller ice removal.

To conserve electric power while still providing effective ice removal, power is provided to the deicers in timed intervals

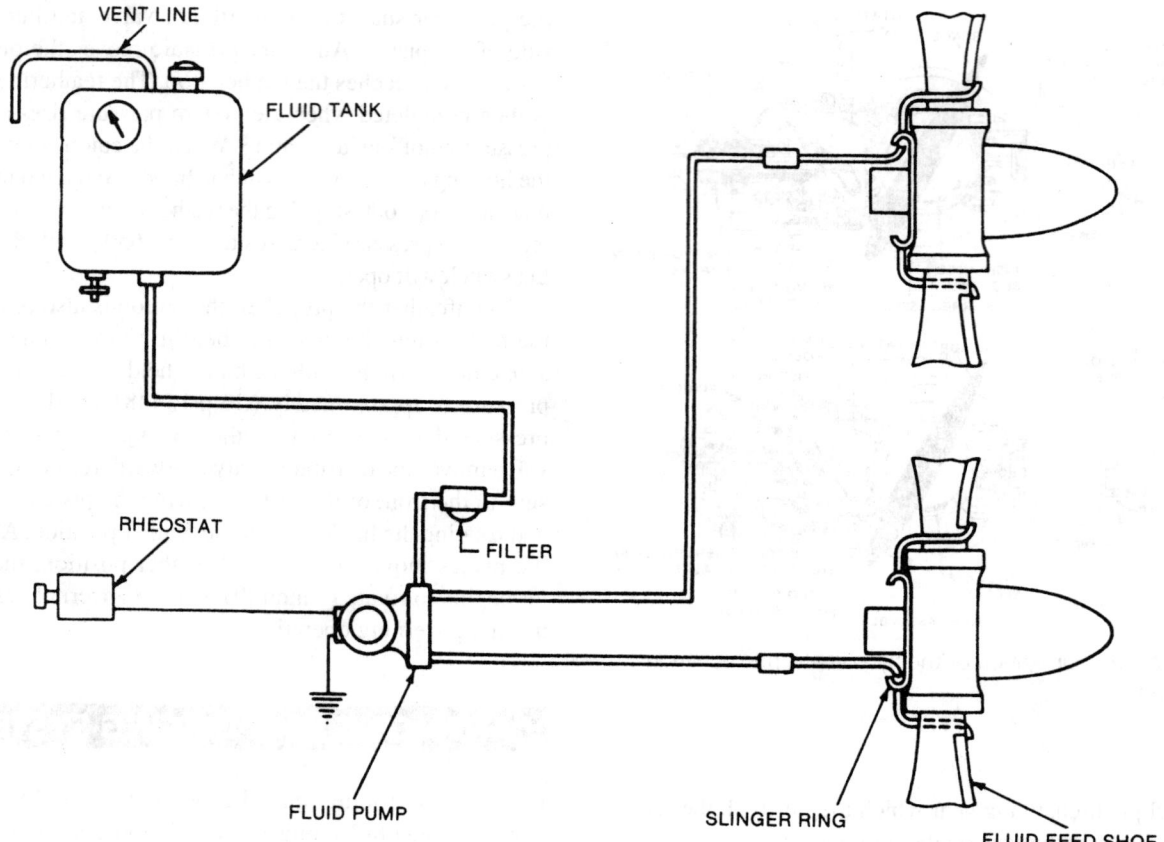

VENT LINE

FLUID TANK

RHEOSTAT

FILTER

FLUID PUMP

SLINGER RING

FLUID FEED SHOE

FIGURE 20-47 Typical propeller fluid anti-icing system.

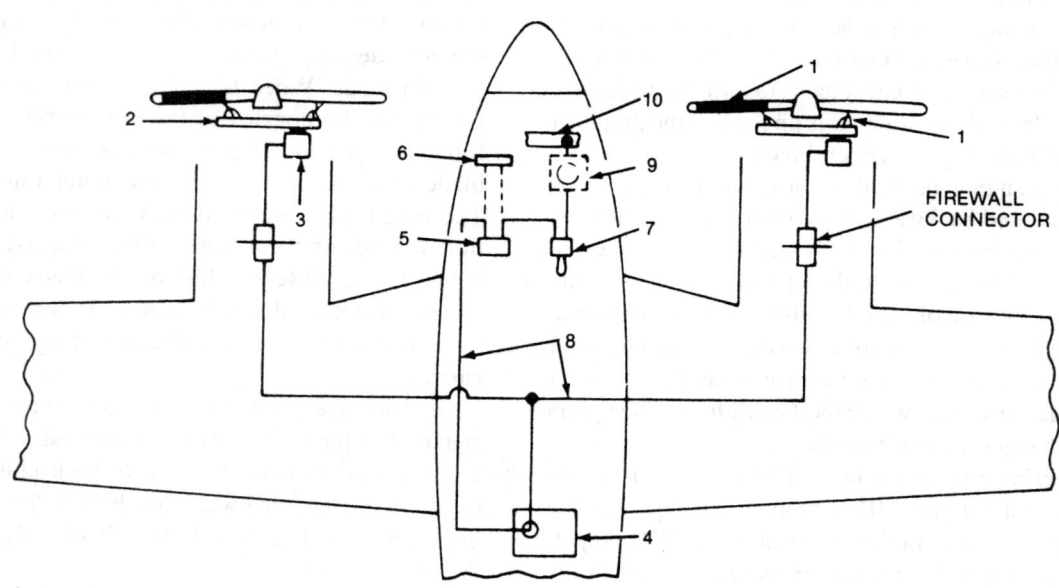

FIREWALL CONNECTOR

1. DE-ICER (AND WIRE HARNESS, WHEN USED)
2. SLIP RING
3. BRUSH BLOCK
4. TIMER
5. AMMETER (WHEN USED)

6. SHUNT (WHEN USED)
7. SWITCH (OR CIRCUIT BREAKER/SWITCH)
8. WIRING
9. CIRCUIT BREAKER (WHEN USED)
10. AIRCRAFT POWER SOURCE

FIGURE 20-48 Typical twin-engine deicing system. (*B.F. Goodrich.*)

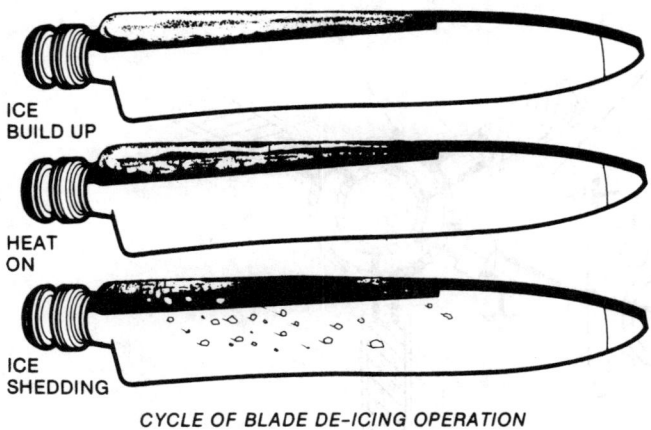

ICE BUILD UP

HEAT ON

ICE SHEDDING

CYCLE OF BLADE DE-ICING OPERATION

FIGURE 20-49 Cycle of blade deicing operation. (*Dowty.*)

rather than continuously. The time lapse between the heating cycles for any given deicer element will allow ice to accumulate slightly before removal. Rotational balance of the propeller or propellers is maintained by heating either the entire deicer on each blade of a propeller or corresponding sections of the deicers on each blade.

Deicer systems can be of the **single-** or **dual-element** type. In **dual-element deicer systems**, each deicer has two separate electrothermal heating elements: an inboard section and an outboard section. When the switch is turned on, the timer provides power through the brush block and slip ring to all outboard elements on one propeller for approximately 34 s. This will reduce ice adhesion in these areas, as illustrated in Fig. 20-50. Following this, the same procedure is applied to the inboard elements. The sequence in which the heating elements are activated on a dual-element, twin-engine, two-bladed propeller deicer system is as follows: outboards, inboards on one propeller; then outboards, inboards on the

other propeller. In the single-heating-element system, illustrated in Fig. 20-51, each deicer has one electrothermal element. During operation of this system, the timer switches power to all heating elements on one propeller for the same length of time. The timer then switches the power off for an equivalent length of time.

Servicing, inspection, maintenance, and repair of anti-icing and deicing systems for propellers will vary among different propellers and aircraft. It is therefore essential that the technician performing these operations follow the instructions furnished by the manufacturer of the aircraft and the manufacturer of the propeller.

PROPELLER SYNCHROPHASER SYSTEM

Some twin-engine aircraft with constant-speed propellers are equipped with an automatic synchronizer system that matches the rotational speeds of the two engines. Synchronization is merely the control of the propellers so that they operate at the same speed. Some earlier aircraft were equipped with a synchronization system only, but as technology advanced, control of the blade phase angle also was accomplished, hence the name synchrophasing. Synchrophasing is used to give optimum propeller performance and low noise and vibration levels.

Phasing has to do with the relative positions of the two propellers at any instant in time. Taking a vertical position of the front propeller as reference, if the rear propeller is exactly vertical, the relative phase angle between the two propellers at that time is 90°. If the rear propeller is at the angles shown in Fig. 20-52, the relative phase angles could be considered as (A) 0° or 180°, (B) 20° or 160°, and (C) 48°

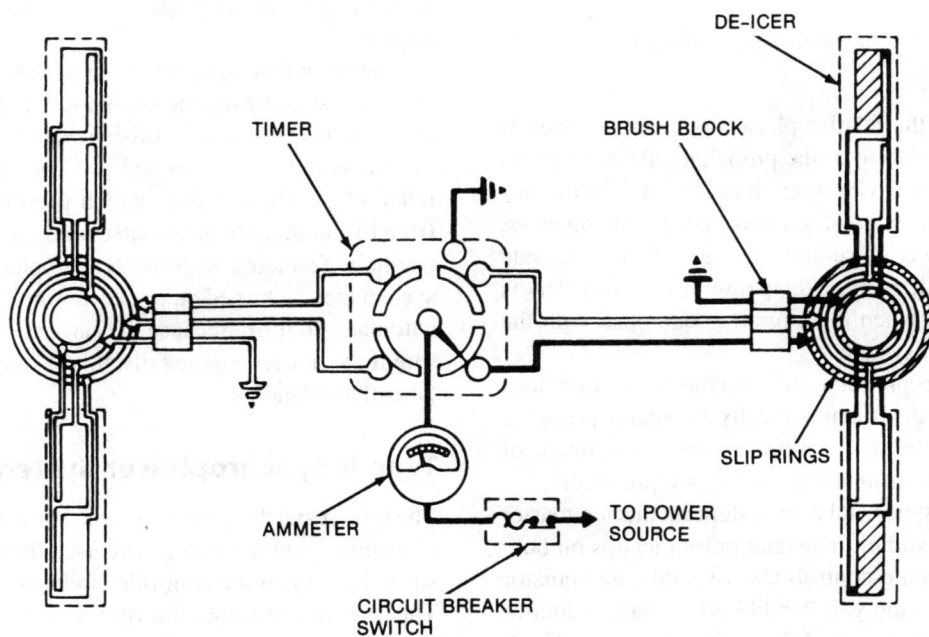

FIGURE 20-50 Dual-element, twin-engine deicer cycle sequence. (*B.F. Goodrich.*)

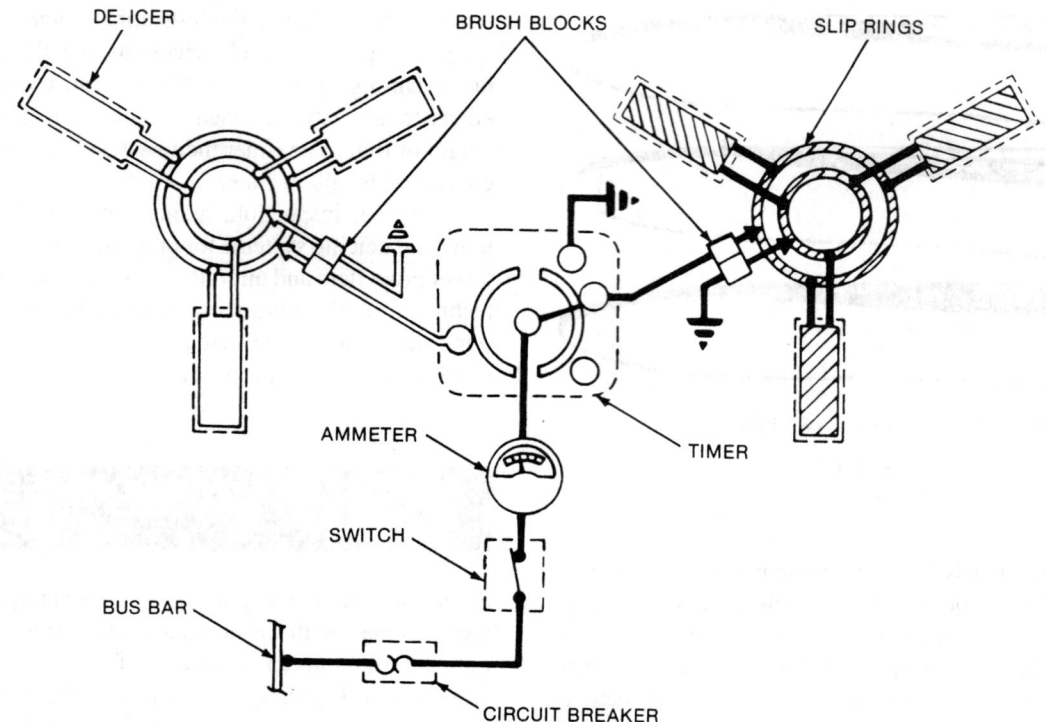

FIGURE 20-51   Single-element, twin-engine deicer cycle sequence. (*B.F. Goodrich.*)

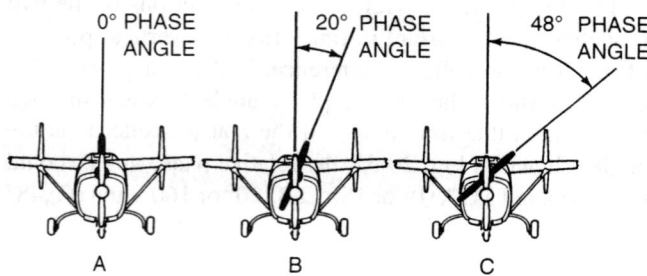

FIGURE 20-52   Phase angle relationships. (*Cessna.*)

or 132°. Of course, the relative phase angle can be taken at any point in the revolution of the propellers. Relative phase angle affects noise and vibration characteristics in the aircraft. Other dynamic factors, such as speed and rough vs. smooth air, also affect noise and vibration. To compensate for all of the dynamic factors, the phase of the propellers is manually adjustable when the aircraft is equipped with the type I synchrophaser system.

The type I synchrophaser system performs two functions in twin-engine aircraft. It automatically maintains propeller speed synchronization and it allows manual adjustment of the phase angle relationship between the two propellers.

A typical type I synchrophaser system includes a master governor, a slave governor, magnetic pulse pickups on both governors, an electronic control-box assembly, an actuator motor, a trimmer assembly, a flexible drive shaft, a control switch, and an indicator light. A schematic diagram of such a system is shown in Fig. 20-53.

During operation of the system, the two governors send pulse signals from the magnetic pickups to the control-box circuit. If the pulse signals are not exactly at the same frequency, the control circuit will rotate the actuator motor in a direction that equalizes the governor signals. As the actuator motor rotates, it turns the flexible shaft leading to the trimmer assembly which is attached to the control lever on the slave governor. The trimmer moves the governor control lever to make the appropriate rpm adjustment for the slave engine.

The automatic synchrophaser system is placed in operation by first adjusting the engine rpm manually for synchronization at the desired cruising value. The synchrophaser system switch is then turned on. The system has a limited range of synchronization, which prevents the slave engine from losing more than a limited amount of rpm if the master engine is feathered with the synchrophaser system on. This system may be turned on or off, thus providing for manual or automatic control. Because the slave engine will decrease its rpm if the master engine fails, this system is not used during takeoff and landing.

Type II Synchrophaser System

The type II synchrophaser system, shown in Fig. 20-54, is an electronic system which can be used for takeoff and landing, since the engines are controlled independently of each other. The system maintains the rpm and an established propeller blade phase relationship for each engine to reduce cabin noise and vibration.

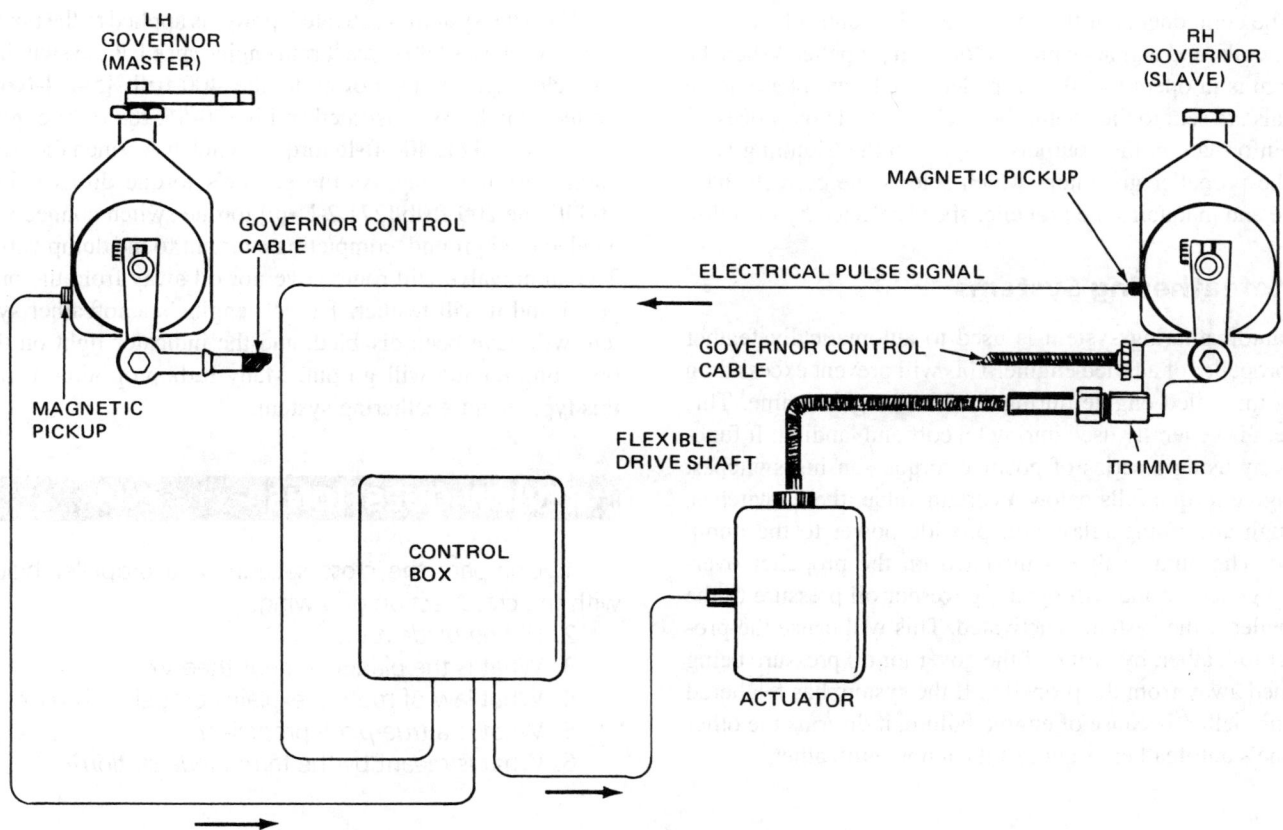

FIGURE 20-53 Schematic diagram of a synchrophaser system.

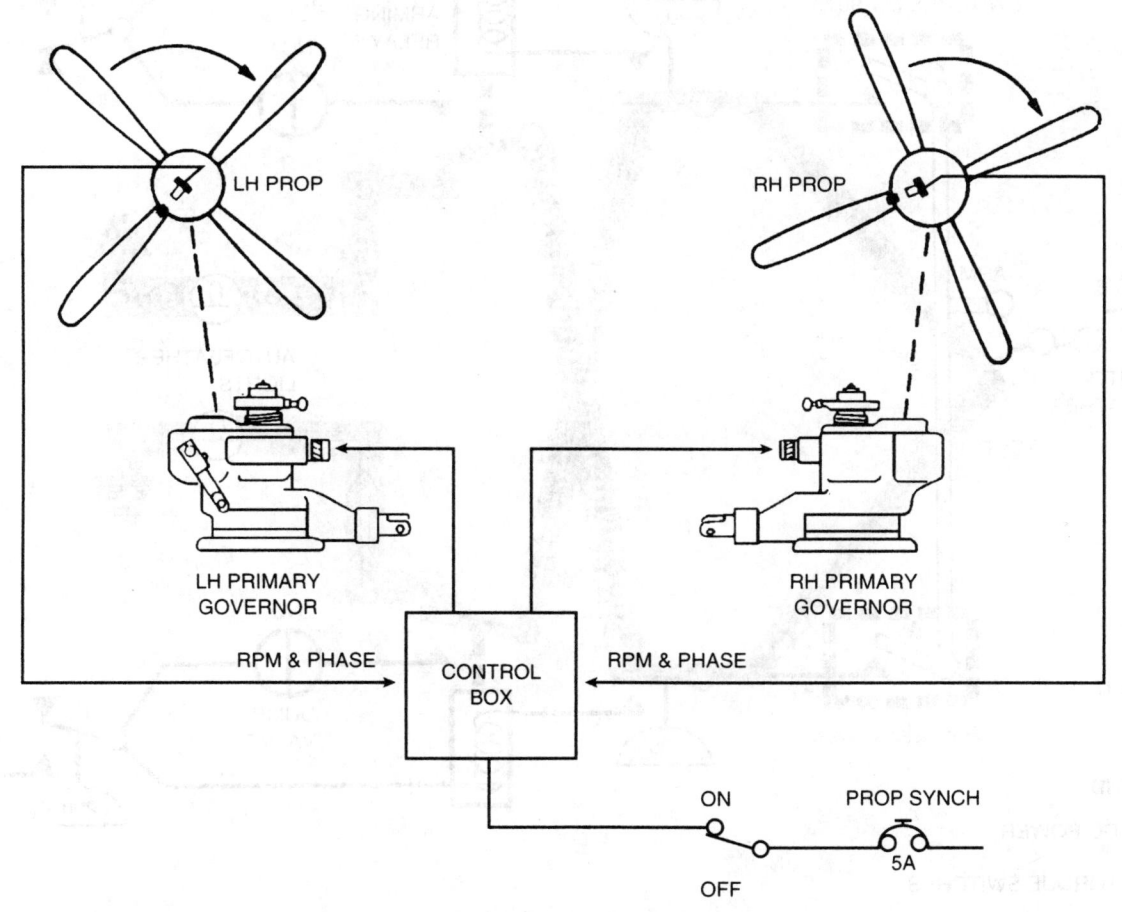

FIGURE 20-54 Type II propeller synchrophaser system. (*Beech Aircraft Corp.*)

The components of this system are the control box, trimming coils, and magnetic pickups on each propeller. When the system is in operation, the propeller speeds and phase angle signals are sent to the control box. The control box processes this information, then outputs a signal to the trimming coils on the propeller governors which selects the correct phase angle and maintains the propeller speed selected by the pilot.

Autofeathering Systems

An autofeathering system is used to automatically feather the propeller of a failed engine. This will prevent excess drag from the failed engine during a critical flight regime. This system is generally used during takeoff and landing. It functions by using a series of positive torque-sensing switches. If engine torque falls below a certain value, these switches, through an arming relay, will provide power to the dump valve. The dump valve is mounted on the propeller overspeed governor and will bypass governor oil pressure to the propeller if the system is activated. This will cause the propeller to feather, by virtue of the governor oil pressure being drained away from the propeller. If the system has feathered one propeller, because of engine failure, it disarms the other engine's autofeather circuit, so it cannot autofeather.

When the system is activated, power is applied to the power lever switches. At 90 percent rpm engine speed, these switches will close providing power to the 400-ft•lb [542.4-N•m] torque switch. As illustrated in Fig. 20-55, the right engine has failed and the 400-ft•lb torque switch has armed the right engine arming relay, As the engine's torque drops below 200 lb, the 200-ft•lb [271.2-N•m] torque switch connects to an electrical ground, completing a circuit to the dump valve. The dump valve will route governor oil away from the propeller and it will feather. The left engine's autofeather system will have been disabled, and the indicator light on the operating engine will go out. Many turboprop aircraft use this type of autofeathering system.

REVIEW QUESTIONS

1. Compare the cross section of a propeller blade with the cross section of a wing.
2. Define *blade root*.
3. What is the blade-element theory?
4. What law of motion explains propeller thrust?
5. What is a *true-pitch* propeller?
6. What is meant by the term *blade station*?

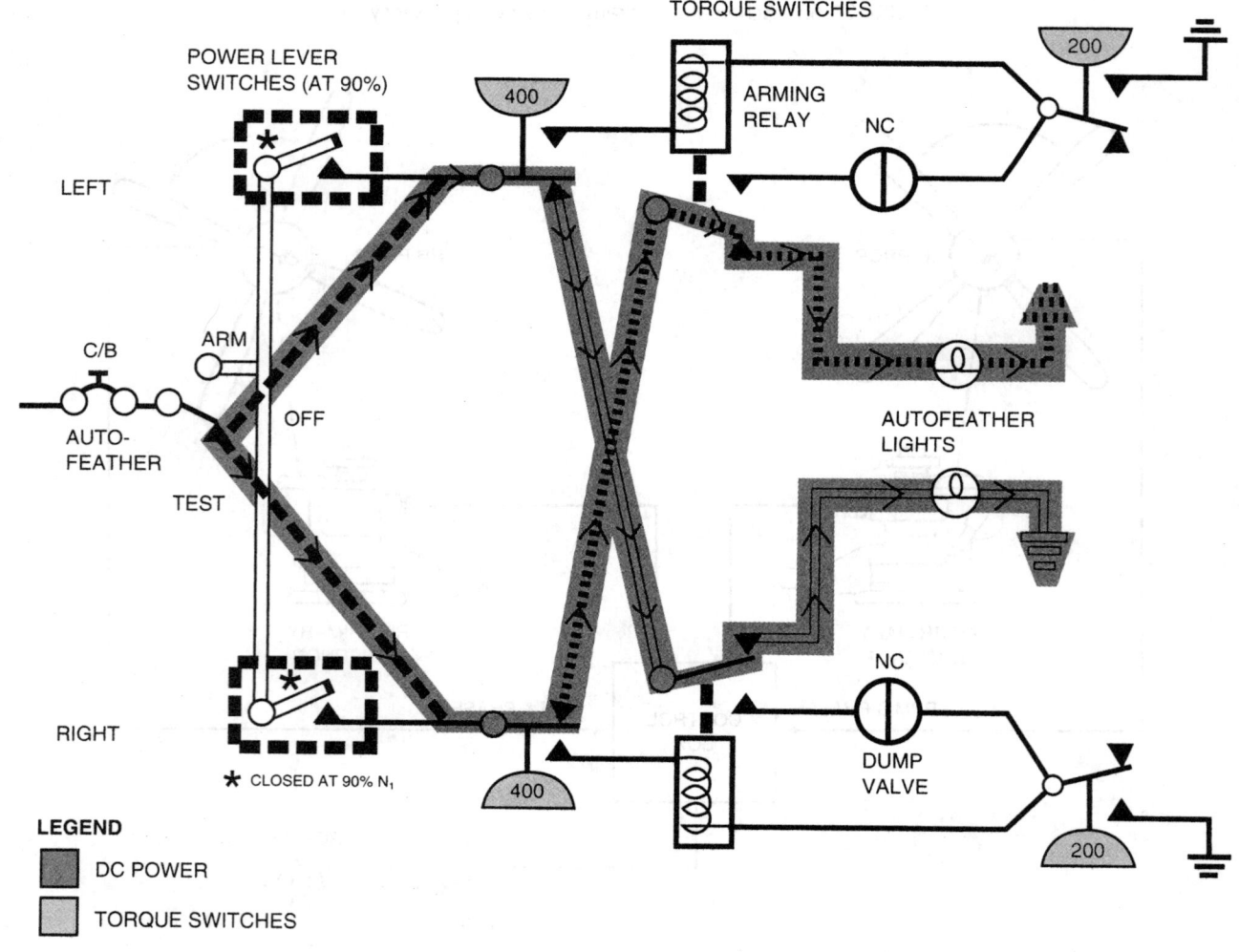

FIGURE 20-55 Autofeather operating schematic (right engine failure).

7. Define the term *blade angle*.

8. Discuss the term *angle of attack*, in regard to a propeller blade.

9. What is the effective pitch of a propeller?

10. What is the geometric pitch of a propeller?

11. Define propeller *slip*.

12. Explain the effect of each of the three principal forces acting on a propeller in flight.

13. What is the limiting factor with respect to the rpm at which a propeller may be operated?

14. What radial tip clearance is required between a propeller and the fuselage of an aircraft?

15. What are the two basic classifications of propellers?

16. Describe a fixed-pitch propeller.

17. Explain the difference between a ground-adjustable propeller and a controllable propeller.

18. Explain the operation of a constant-speed propeller.

19. What is the purpose of propeller feathering?

20. What is the purpose of reverse thrust?

21. What mechanism keeps a McCauley feathering propeller from feathering when it is shut off on the ground?

22. List the basic classifications of Hartzell constant-speed propellers.

23. Describe methods for propeller anti-icing and deicing.

24. What is meant by *propeller synchronizing*?

25. Define, and discuss the purpose of, *propeller synchrophasing*.

Turbopropellers and Control Systems 21

INTRODUCTION

Throughout the world, many midsize and commuter-size aircraft such as the Beechcraft King Air shown in Fig. 21-1 employ propellers powered by turbine engines for thrust. Technicians who perform maintenance on such aircraft should be familiar with turbopropeller operations and their control systems. In this chapter, some typical turbopropeller systems for gas-turbine engines will be examined. Among the turbopropeller engines discussed here are the Pratt & Whitney PT6A and PW100 series, the Garrett TPE331, the Allison 250, and the Allison turbopropeller used with the Allison Model 501-D13 engine and the General Electric CT7 engine.

In designing a propeller for a specific manufacturer's specific aircraft model, engineers give particular attention to that aircraft's design requirements, such as performance, noise, blade stresses, and vibration characteristics. Conversely, the aircraft manufacturers often incorporate propeller compatibility into their designs. Particular attention is given to such items as the propeller thrust line, the propeller vibration frequency, and a satisfactory engine mounting system.

Many different propeller manufacturers make turbopropellers for different engine installations. Each engine can generally use more than one manufacturer's propeller. Although different propellers may be used, the basic control system remains very similar. The principles explained in this chapter should provide the technician with a sufficient understanding of turbopropellers and control systems to be able to interpret and use operation and service instructions for any propeller.

TURBOPROPELLER HORSEPOWER CALCULATIONS

Two important aspects of turbopropeller operation are horsepower and torque. In the realm of turboprops, **brake horsepower** is usually referred to as **shaft horsepower** (shp), meaning the horsepower delivered to the propeller shaft. The following formula can be used to calculate shp:

$$\text{shp} = \text{actual propeller rpm} \times \text{torque [ft·lb]} \times K$$

where K is the torquemeter constant. The actual value of K is 0.0001904 ($K = 2\pi/33\,000$). This formula is used in test-stand calculations and for computing shp in a ground check of the engine in the aircraft.

With a turboprop engine, some jet velocity is left at the jet nozzle after the turbines have extracted the required energy for driving the compressor, reduction gear, accessories, etc. This velocity can be calculated as **net thrust** (F_n), which also aids in propelling the aircraft. If shaft horsepower and net thrust are added together, a new term, "equivalent shaft horsepower" (eshp), results. However, the net thrust must be converted to equivalent horsepower.

The formula for calculating eshp is

$$\text{eshp} = \text{shp} + \frac{F_n}{2.5}$$

The formula for calculating flight eshp is

$$\text{Flight eshp} = \text{shp} + \frac{F_n + V_p}{261}$$

where V_p = airplane velocity (knots).

Turboprop engine operation is quite similar to the operation of a turbojet engine except for the added feature of a propeller. A turboprop aircraft in flight chiefly requires attention to engine operating limits, such as rpm, turbine inlet temperature, exhaust-gas temperature, and torquemeter readings. Instruments for tracking engine performance, such as the torquemeter pressure gauge, are therefore very important. Although torquemeters indicate only the power being supplied to the propeller, rather than the eshp, torquemeter pressure is approximately proportional to the total power output and therefore is used as a measure of engine performance.

HARTZELL TURBOPROPELLERS

The HC-B3TN-2 propeller, illustrated in Fig. 21-2, is a constant-speed, feathering, three-blade model designed primarily for use

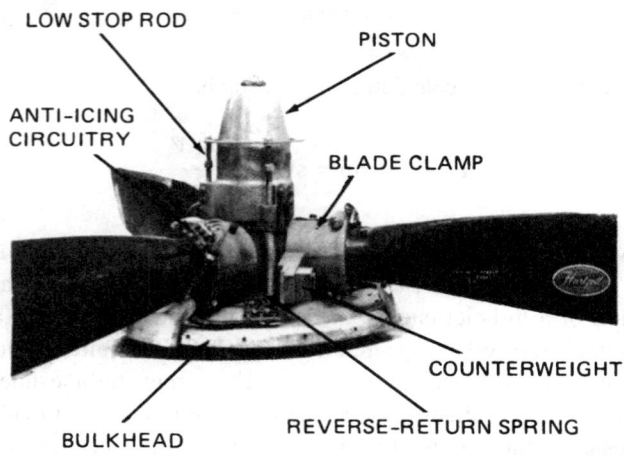

FIGURE 21-1   Beechcraft King Air. (*Beech Aircraft Corp.*)

FIGURE 21-2   Hartzell HC-B3TN-2 propeller. (*Santa Monica Propeller Service.*)

with the PT6A turboprop engine. The HC-B3TN-3 propeller is also designed for use with the PT6A engine; however, this model is reversible. The HC-B3TN-5 model is a constant-speed, feathering, reversible propeller designed for use with the Garrett-AiResearch TPE331 turboshaft (turboprop) engine.

The propellers mentioned above are all similar in design to the models described in Chap. 19. They operate on the same principles, and the same general maintenance practices

should be employed. The blade shanks of the HC-B3TN propellers each have two sets of needle bearings in order to reduce friction as shown in Fig. 21-3. The blade shanks, designated by the letter "T," can be used only with turboprop engines. The propeller flange, designated by "N," has eight $\frac{9}{16}$-in [14.29-mm] diameter bolt holes on a $4\frac{1}{2}$-in [11.43-cm] diameter bolt circle.

The reversible models of the propeller have lengthened piston cylinders in order to provide for the additional movement beyond the low-pitch position necessary to attain the required reverse-pitch angle. Low-pitch and reverse-pitch angles are brought about by means of governor oil pressure directed through a passage in the engine shaft to the propeller piston. This causes the piston to move outward (forward). The piston is connected to the blade shanks by means of **link arms**, thus producing rotation of the blades as the piston moves.

The nomenclature for the piston and cylinder in the propellers under discussion here is unusual in that the **piston** is outside the **cylinder**. This is the reverse of common terminology. Note in Fig. 21-3 that the piston forms the outer case of the operating assembly and the cylinder is inside.

In Fig. 21-3 it can be seen that there are three coaxial springs inside the piston-cylinder assembly. These constitute the **feathering-spring assembly**. When the piston moves

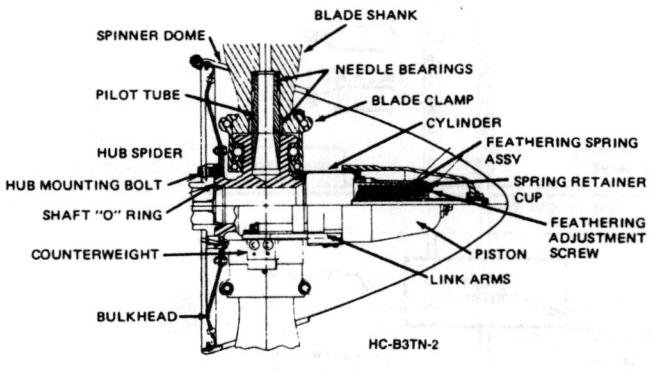

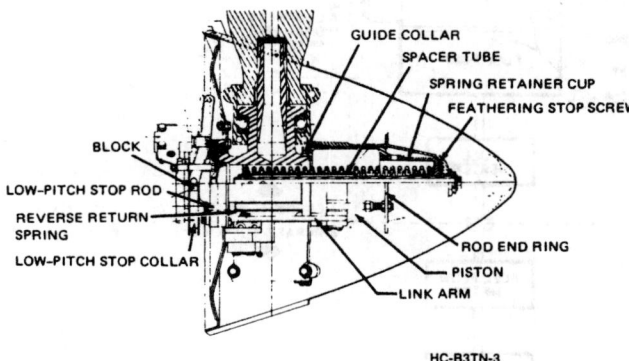

FIGURE 21-3 Cutaway drawings of the HC-B3TN-2 and HC-B3TN-3 propeller operating systems. (*Hartzell Propeller.*)

forward and the pitch is reduced, the springs are compressed. When the governor allows engine oil to flow from the propeller back to the engine, the springs and the force produced by the **blade counterweights** pull the piston back toward the hub and the blade pitch increases. If the propeller control is set for **feathering**, the feathering-spring assembly will pull the piston back to the position where the blades are feathered. The total pitch-change range is 110°.

As mentioned earlier, the propellers discussed in this section are controlled by a speed-sensing governor which maintains the engine rpm selected by the pilot. The governor is installed on the engine and provides oil pressure varying from 0 to 385 psi [2654.58 kPa]. Governor oil pressure is used to decrease propeller pitch and to place the blades in the reverse-pitch position.

Feathering of the propeller blades is accomplished by opening a valve in the governor to permit oil to return from the propeller piston to the engine. This allows the feathering spring to force the oil out of the propeller and move the piston rearward to the feathered position. The HC-B3TN-5 propeller has spring-loaded, centrifugal responsive latches which prevent feathering when the propeller is stationary but permit feathering when the propeller is rotating. The HC-B3TN-2 and -3 models have no latches, so they can be feathered when the engine is not running. Unfeathering, in the cases of the -2 and -3 propellers, is accomplished by starting the engine to allow the governor to pump oil into the propeller piston cylinder. In the case of the HC-B3TN-5 propeller, unfeathering is best accomplished by means of an electric pump.

The reversible propellers are provided with hydraulic low-pitch valves, commonly called "beta" (β) valves. The **beta valve** prevents the governor from moving the piston beyond a prescribed low-pitch position. This prevents the propeller from reversing unless the reverse position is selected by the pilot. Reversing is accomplished by manually moving the control which adjusts the low-pitch stop to the reverse-pitch position. During this operation, the governor is adjusted automatically to produce oil pressure for the reversing operation. Return from reverse is accomplished by manually moving the control to place the low-pitch stop in the normal low-pitch position. This allows the oil to drain from the piston cylinder as the forces of the springs and counterweights move the piston rearward.

The operation of the complete propeller control system is described later in this chapter. Figure 21-4 shows the control system as installed on the PT6A turboprop engine.

Hartzell Lightweight Turbine Propeller

The Hartzell **compact propeller**, named for its small hub size, has been in service for many years on reciprocating engines and has an outstanding record for performance and reliability. The split aluminum hub offers high strength, ease of maintenance, and low overhaul cost. A three-blade version and a four-blade version of this propeller have recently been developed for use on turbine engines. On the four-blade propeller, shown in Fig. 21-5, a significant weight savings was realized, compared with the Hartzell steel hub-type propeller. The light weight is achieved in part through the extensive use of aluminum and plastics, specially designed counterweights and feathering spring, and a reduced oil volume.

The compact propeller is compatible with external and internal beta systems in use on turbine engines today. Blade-angle adjustments for reverse, start locks, flight idle, and feather are external and conveniently accessed. Total blade pitch travel is sufficient to ensure full reverse and autofeathering capabilities. The low-friction pitch-change system makes for very smooth propeller operation.

Hartzell Six-Blade Propeller

A six-blade propeller, illustrated in Fig. 21-6, has been designed for use on turboprop commuter aircraft. This propeller, designed for a maximum diameter of 14 ft, utilizes a **dual-acting hydraulic pitch-control system** and is capable of handling 3000 hp [2237.1 kW]. High-pressure oil, supplied by the governor, is applied to one side of the piston while oil pressure on the other side is relieved in accordance with the pitch-change requirements. Either side of the piston can be subjected to the high-pressure oil, which is why the system is referred to as "dual-acting." This system offers a substantial weight savings over other propellers of its class. A dual-acting propeller system for 3000-hp [2237.1-kW] applications, complete with governor and oil, weighs less than 300 lb [136.1 kg]. The design includes an aluminum hub, which, in addition to being lightweight, also allows for easy accessibility for maintenance purposes. The propeller incorporates external adjustments for all major functions.

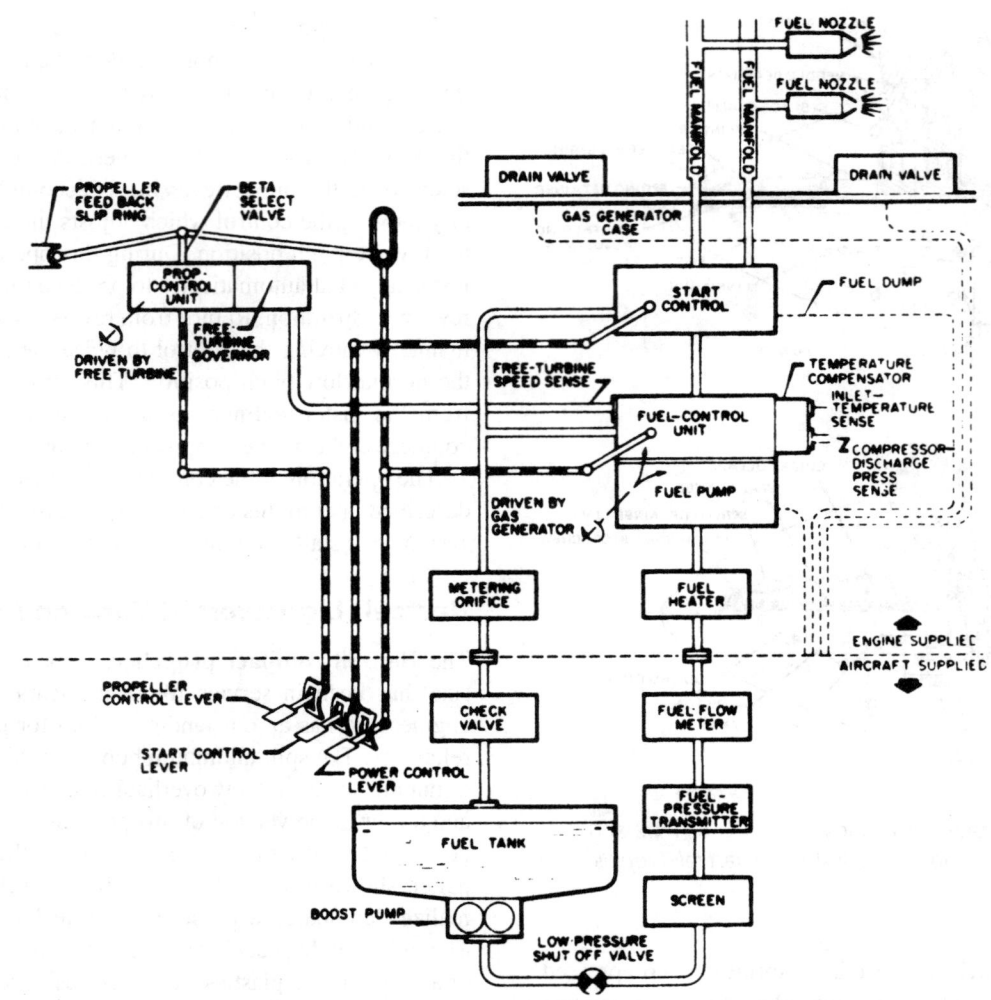

FIGURE 21-4 Propeller control system for the PT6A turboprop engine. (*Pratt & Whitney Canada.*)

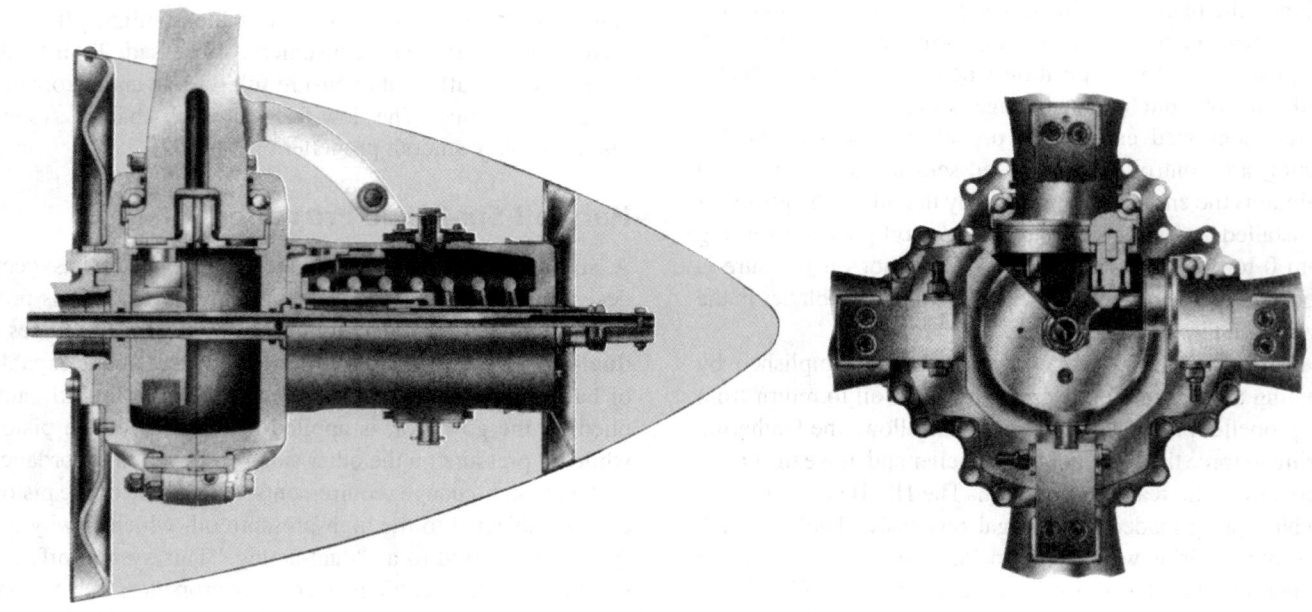

NUMBER OF BLADES: FOUR (ALUMINUM ALLOY)
HUB MATERIAL: ALUMINUM ALLOY
CURRENT INSTALLATION: GARRETT ENGINE POWERED
WEIGHT: 160 LBS. AIRCRAFT

FIGURE 21-5 Hartzell HV-D4N-5 four-blade propeller. (*Hartzell Propeller.*)

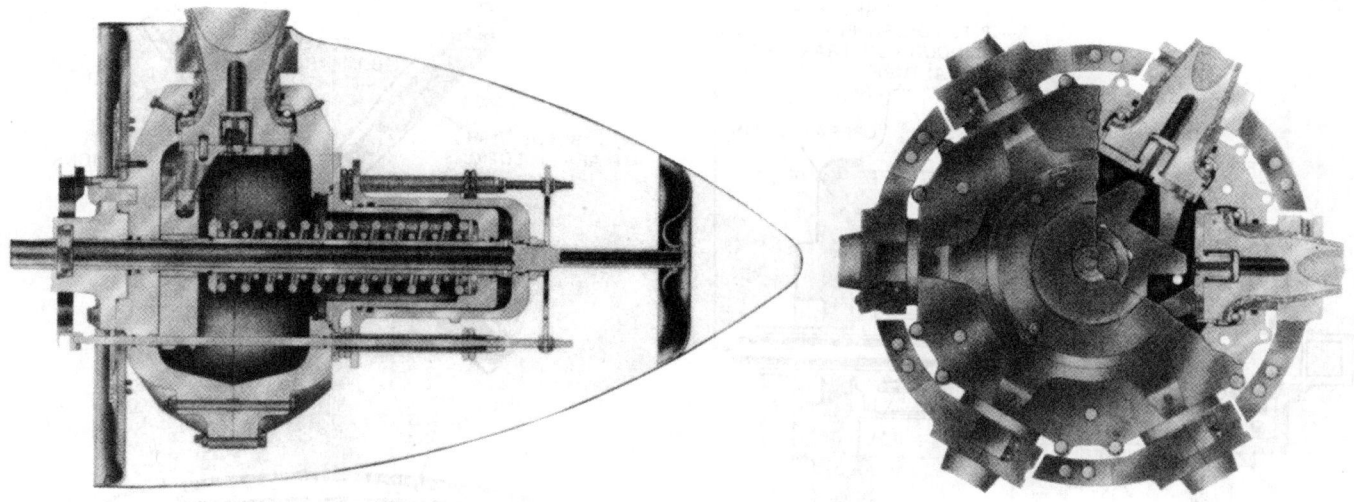

NUMBER OF BLADES: SIX (COMPOSITE)
HUB MATERIAL: ALUMINUM ALLOY
CURRENT INSTALLATION: SHORTS 360 W/PT-6 ENGINES
WEIGHT: 190 LBS.

FIGURE 21-6 Hartzell HC-A6A-3 six-blade propeller. (*Hartzell Propeller.*)

The propeller has a rapid control response rate and an available travel which is comparable to present engine applications. The system also includes a highly accurate propeller synchronizer system which aids in decreasing cabin noise levels.

The governor has energy-saving features which reduce maintenance and also save fuel. The entire propeller and control system is designed to have long maintenance intervals, in keeping with the requirements of the commuter industry. The deicing system has been completely redesigned for reduced maintenance and improved performance.

THE DOWTY TURBOPROPELLER

Dowty R321 Four-Bladed Turbopropeller

The Dowty R321 four-bladed turbopropeller is similar in operating principle to the Hartzell turbopropeller described previously. The propeller blades are rotated to low pitch and reverse pitch by means of oil pressure. When the oil pressure is released, the blades are rotated to the high-pitch and feathered positions by means of a combination of spring and counterweight forces. The propeller control system balances the opposing forces to provide the correct pitch for constant-speed operation in flight. The operating principle of the pitch-changing mechanism is an important safety factor because it ensures that the propeller will be feathered quickly in case of oil pressure loss during takeoff. This eliminates propeller drag and makes the aircraft easier to control.

A cutaway drawing that illustrates the operating mechanism and construction of the Dowty propeller is presented in

Fig. 21-7. The propeller hub consists of two identical steel forgings that bolt together to retain the propeller blades. A 21-ball thrust bearing between the hub shoulder and blade root shoulder provides for low-friction rotation of each blade and absorbs the centrifugal force.

An **operating pin** is mounted on the face of each blade root by means of a socket screw. The operating pin incorporates a needle bearing for reducing friction between the pin and the **crosshead**. The operating pins fit into the crosshead slot to provide blade rotation as the crosshead moves forward and backward.

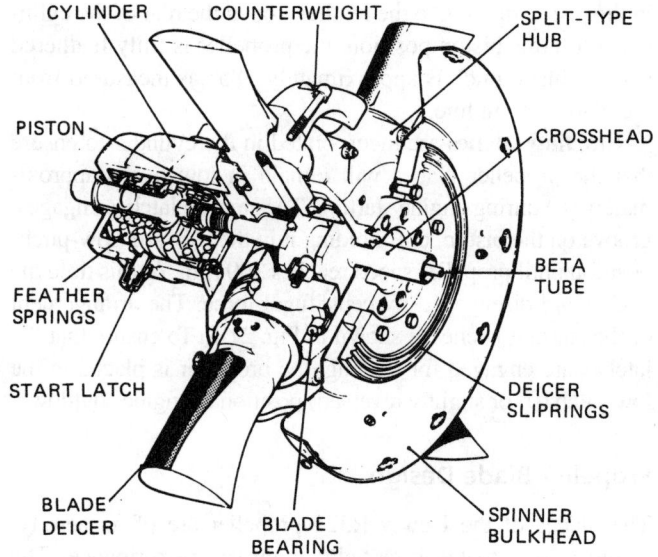

FIGURE 21-7 Cutaway illustration of the operating mechanism for a Dowty turbopropeller. (*Dowty.*)

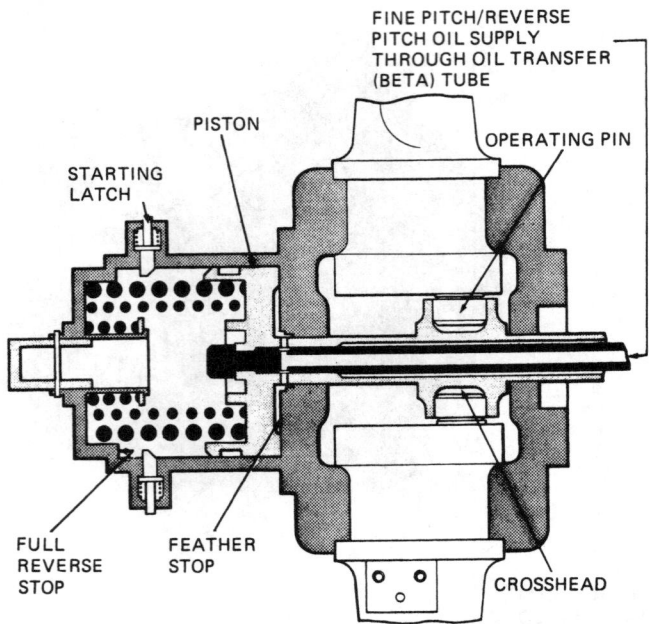

FIGURE 21-8 Simplified drawing of a pitch-changing mechanism. (*Dowty.*)

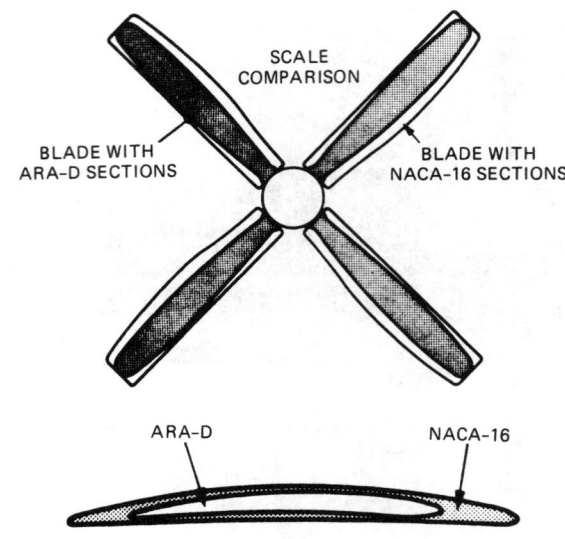

FIGURE 21-9 Comparison of propeller blade designs. (*Dowty.*)

An oil transfer tube, called the **beta tube**, screws into the center of the crosshead shaft, as shown in the simplified drawing in Fig. 21-8, and terminates in the **beta sleeve** in the engine. This tube carries oil pressure to the rear side of the piston that is attached to the forward end of the crosshead. The oil pressure forces the piston forward, causing the crosshead to rotate the propeller blades to low pitch and reverse pitch. When the piston is in the full-forward position, it bears against the full-reverse stop and the propeller blades are in the full-reverse position, which is –12.5° to –14.5°. When oil pressure is released, the oil flows out through the beta tube, the piston moves to the rear, and the propeller blades rotate toward high pitch because of feather spring pressure and counterweight force. Complete release of the pressure results in the piston moving to the feather stop at the rear of the cylinder. With this piston position, the propeller is fully feathered and the blade pitch is approximately +85° as measured from the blade setting line.

Starting latches are incorporated in the cylinder to ensure that the propeller blades will remain in low pitch (approximately 0°) during engine starting. The starting latches engage a groove on the piston, thus holding it in the forward (low-pitch) position until propeller speed reaches 300 rpm. At this time the latches are disengaged by centrifugal force. The arrangement of the starting latches is shown in Fig. 21-8. To ensure that the latches are engaged for starting, the propeller is placed in the low-pitch (0° or slightly reversed) position at engine shutdown.

Propeller Blade Design

The blades of the Dowty R321 propeller are of an aerodynamic design that substantially improves performance. The blades formerly used were of the NACA-16 design shown in Fig. 21-9. The new configuration is a Dowty development

designated ARA-D. The improved performance of the ARA-D design is shown in Fig. 21-10.

Propeller Control System

The control system for the Dowty R321 propeller, as installed on the Fairchild Swearingen Metroliner, is illustrated in Fig. 21-11. The **pitch control** mechanism is adjusted by means of two levers that select the operating mode for the propeller. The constant-speed governing mode is selected for in-flight operations, and the beta mode is selected for ground operations. A low-blade-angle stop for approach and landing is provided by the **beta valve**.

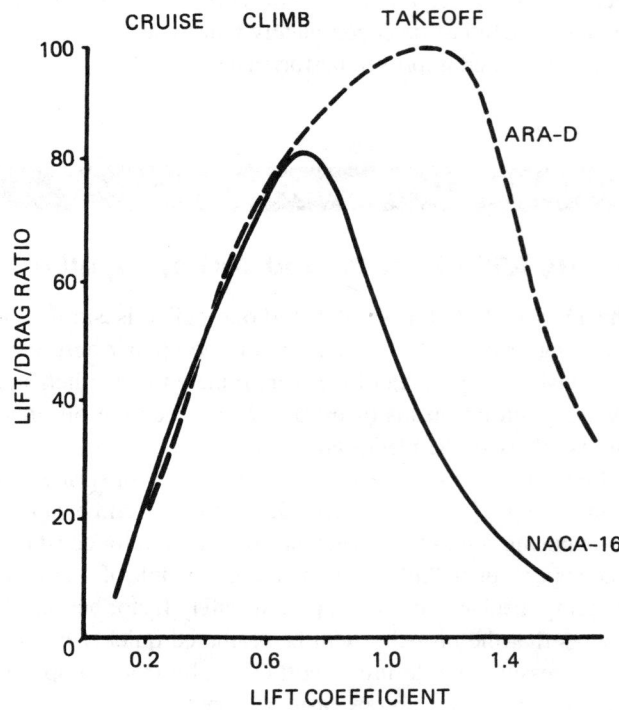

FIGURE 21-10 Curves showing propeller blade performance. (*Dowty.*)

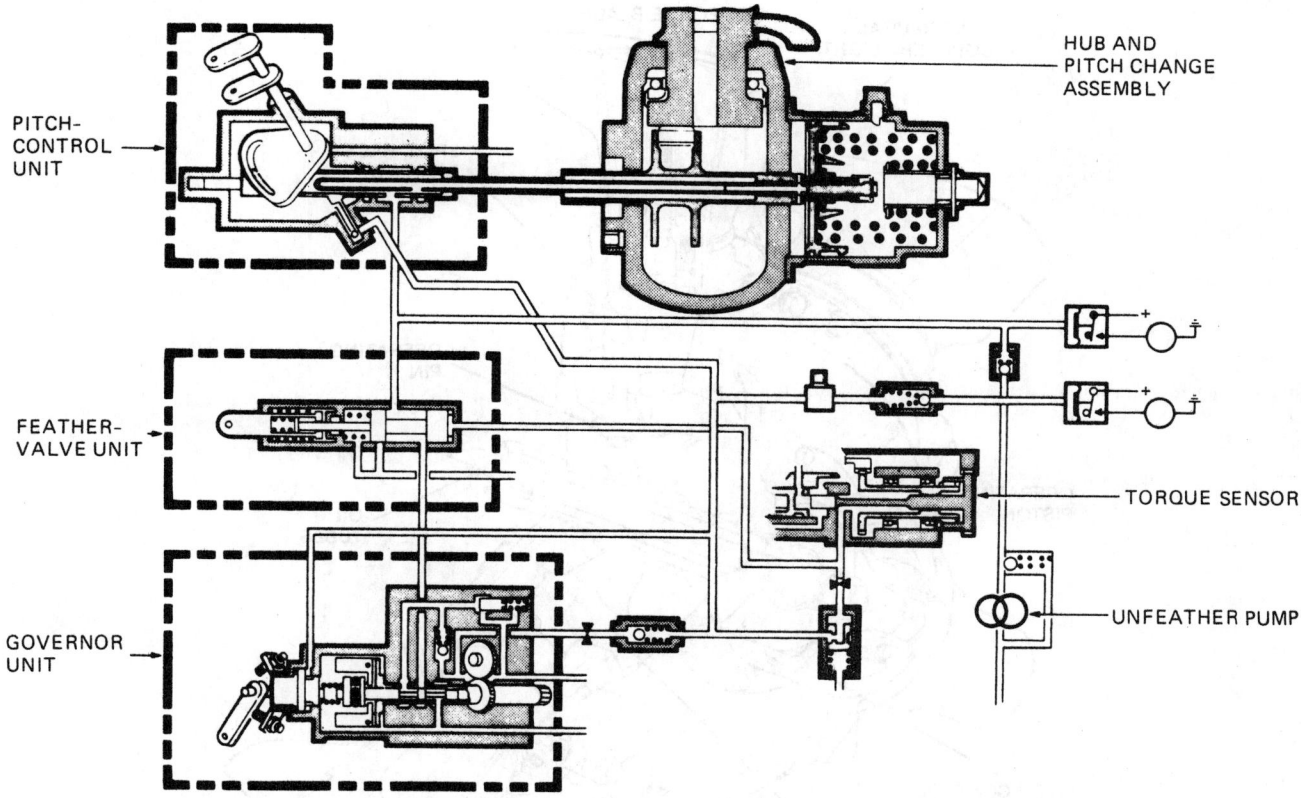

FIGURE 21-11 Propeller control system. (*Dowty.*)

The **feather valve** can be operated either manually or hydraulically. Thus, the pilot can feather the propeller when it is deemed necessary. In case of engine failure, a negative torque signal (NTS) is received by the **torque sensor**, which then moves to operate the feather valve hydraulically and feather the propeller. Negative torque is developed when the propeller is windmilling and driving the engine.

The **propeller governor** is a flyweight-controlled mechanism similar to the governors described previously. It is manually adjusted by the pilot to select the desired engine rpm. The selected rpm is maintained constantly during in-flight operations.

Dowty R352 Advanced Turbopropeller

In the Dowty R352 turbopropeller, illustrated in Fig. 21-12, the pitch-change mechanism is fitted within and in front of the hub. Engine oil, supplied by way of the propeller control system, is used to actuate the pitch-change mechanism. The oil is fed through two concentric tubes, as illustrated in Fig. 21-13, to either side of an operating piston. Low-pitch oil passes through the outer tube to the rear of the piston. High-pitch oil passes through the inner tube to the front of the piston.

The operating piston is connected to a crosshead unit (Fig. 21-12) which engages pins on the root ends of the six blades retained in the hub assembly. Thus, forward movement of the piston turns the blades toward low pitch, and rearward movement turns them toward high pitch. The oil tubes also constitute a mechanical feedback extending from

the operating piston to a beta control valve for approach and landing followed by on-ground reversing.

Turbopropeller blade pitch is controlled by the pitch control unit in flight, as illustrated in Fig. 21-14, giving constant speed; blade pitch is controlled by manual pitch selection via the beta valve for approach, landing, and ground maneuvering.

Blade counterweights, shown in Fig. 21-15, are offset masses bolted to the root of each blade. They consist of heavy metal slugs carried on light alloy arms that are secured to the blade roots.

The counterweights, which generate a force in a high-pitch direction, exceed normal blade centrifugal twisting moments toward low pitch. The counterweights therefore ensure blade movement toward operational high pitch in the event of pitch-change oil pressure failure. The use of counterweights as a simple fail-safe system to overcome the natural tendency of the blades to seek a lower pitch and to overspeed eliminates the need for complicated and heavy pitch locks within the turbopropeller mechanism.

HAMILTON STANDARD TURBOPROPELLERS

Model 24PF Propeller

The flange-mounted, engine-oil-actuated, single-acting Model 24PF propeller, shown in Fig. 21-16, was especially designed for the deHavilland DASH 7 and offers low-noise design

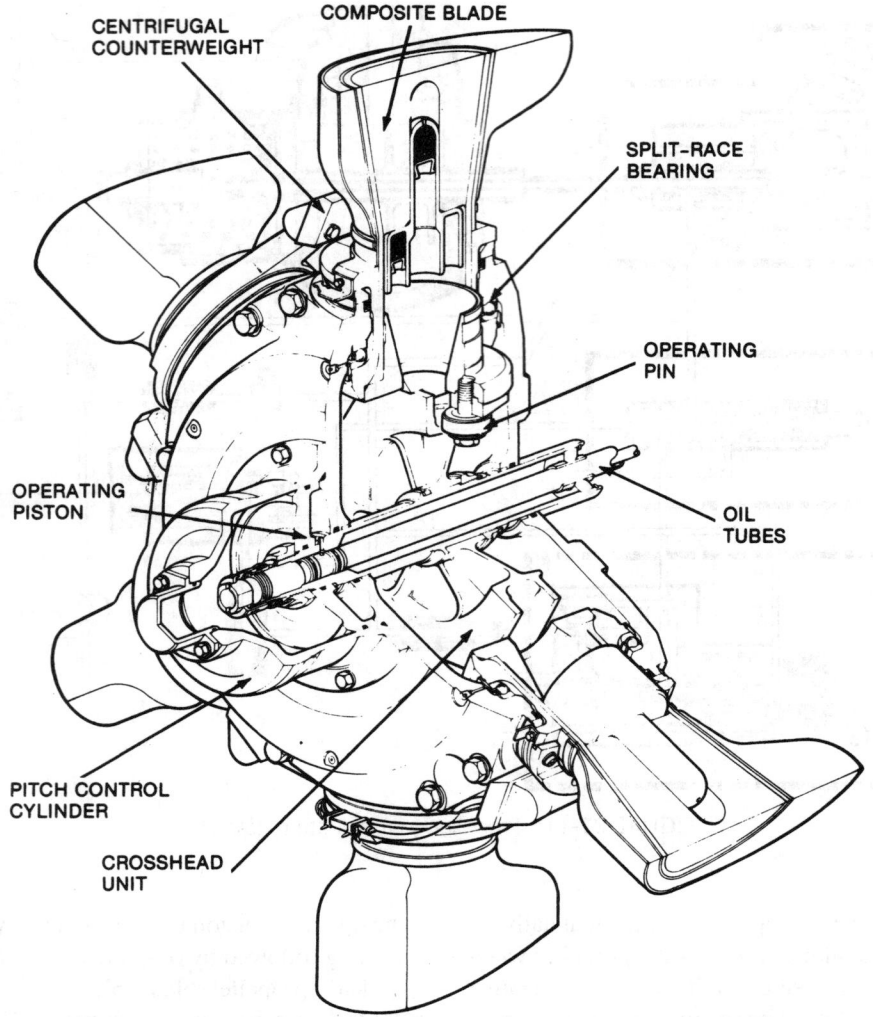

CENTRIFUGAL
COUNTERWEIGHT

COMPOSITE BLADE

SPLIT–RACE
BEARING

OPERATING
PIN

OPERATING
PISTON

OIL
TUBES

PITCH CONTROL
CYLINDER

CROSSHEAD
UNIT

FIGURE 21-12 Dowty R352 turbopropeller. (*Dowty.*)

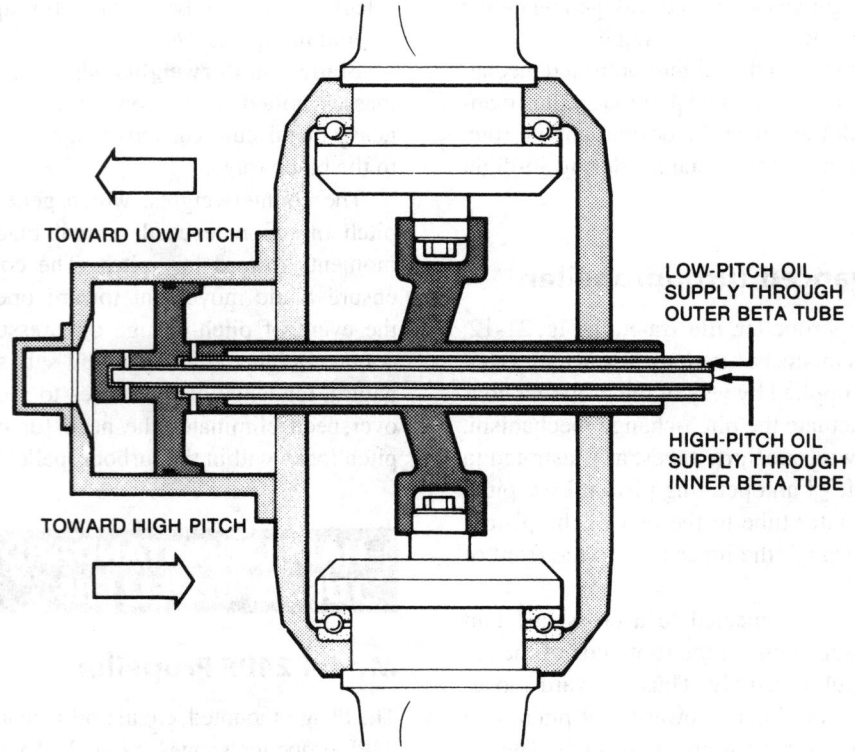

TOWARD LOW PITCH

LOW-PITCH OIL
SUPPLY THROUGH
OUTER BETA TUBE

HIGH-PITCH OIL
SUPPLY THROUGH
INNER BETA TUBE

TOWARD HIGH PITCH

FIGURE 21-13 Concentric oil tubes. (*Dowty.*)

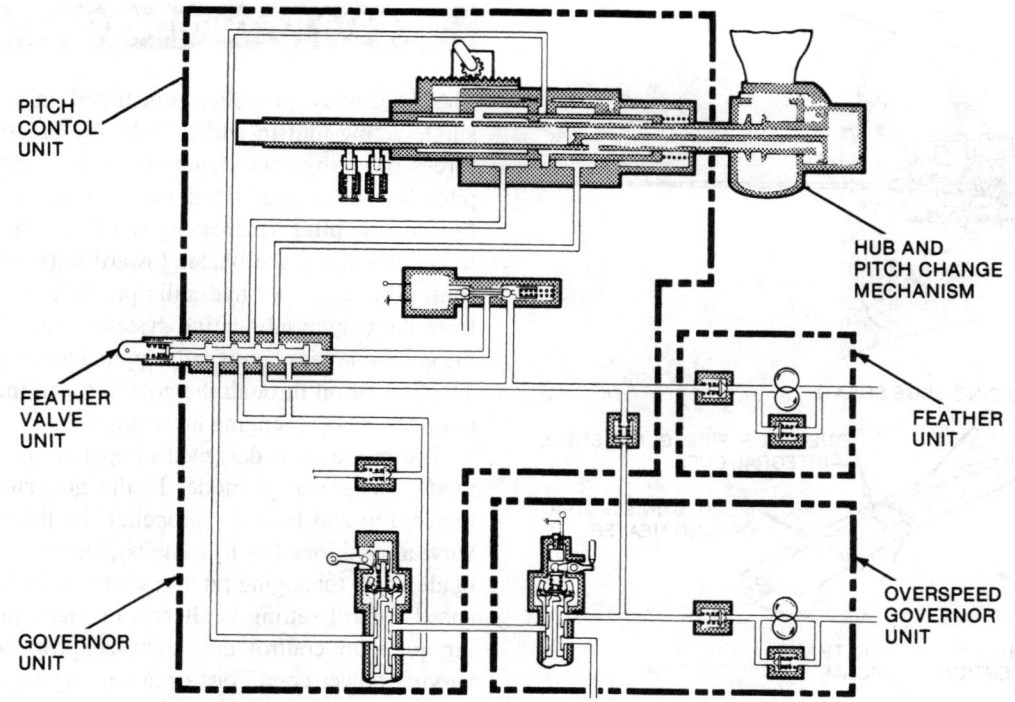

FIGURE 21-14 R352 pitch control unit. (*Dowty.*)

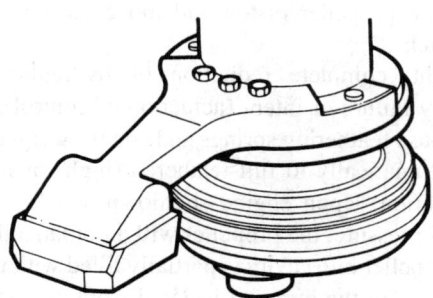

FIGURE 21-15 R352 blade counterweight. (*Dowty.*)

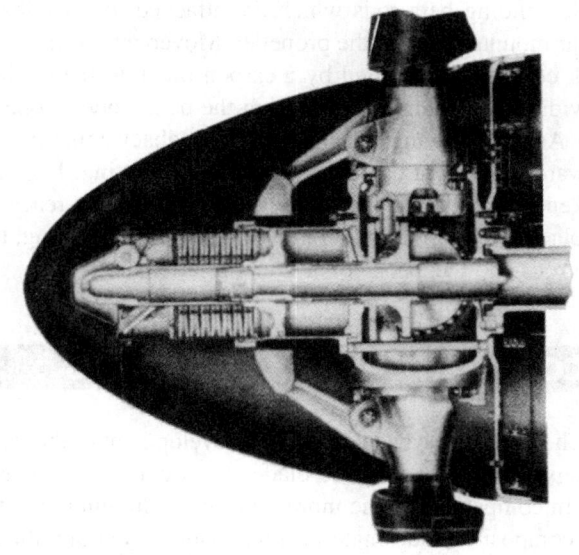

FIGURE 21-16 Hamilton Standard Model 24PF propeller. (*Hamilton Standard.*)

features such as low blade tip speed (715 fps) and a special blade tip shape for reduced aircraft noise. This propeller features the first Federal Aviation Administration- (FAA-) certified composite material blade for use in a fixed-wing aircraft. The spar/shell construction, illustrated in Fig. 21-17, permits weight savings for the entire aircraft (over an aircraft employing a comparable propeller with solid aluminum blades) equivalent to the weight of four passengers. The blades can be removed individually from the aircraft.

Model 14RF Propeller

The Model 14RF propeller is part of a new generation of flange-mounted, lightweight, high-performance, easily maintained propellers also commonly known as "commuter props." An illustration of the Model 14RF is presented in Fig. 21-18. These propellers were developed for regional aircraft that can carry 30 or more passengers. They incorporate many technological advances over earlier turbine-engine propellers beyond the normal feathering, reversing, synchrophasing, and electric blade-deicing features.

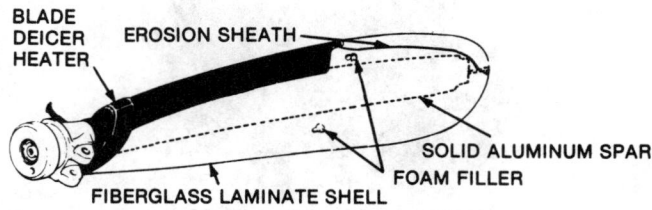

FIGURE 21-17 Model 24PF blade construction. (*Hamilton Standard.*)

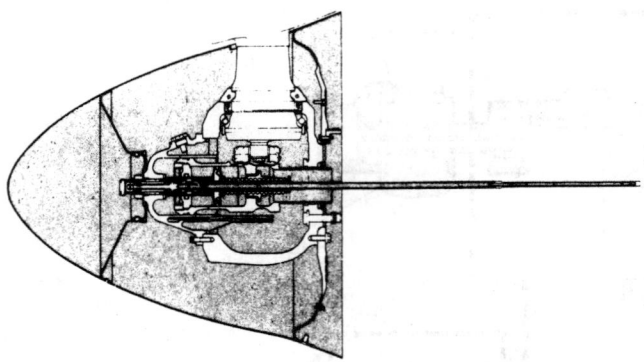

NICKEL LEADING EDGE SHEATH

LIGHTNING
STRIKE
TAB

ONE-PIECE, FIBERGLASS SHELL
—INTEGRAL CUFF

MOLDED-IN ELECTRIC
DEICING HEATER

ONE-PIECE,
SOLID ALUMINUM
SPAR-TIP TO RETENTION

URETHANE
FOAM

FIGURE 21-18 Hamilton Standard Model 14RF propeller.
(*Hamilton Standard.*)

A new airfoil (HS-I series) was especially developed by Hamilton Standard to cater to the unique mission profile of the regional operators. High-speed cruise efficiencies of earlier series airfoils have been maintained while low-speed efficiency has been improved by one to four percentage points. Shorter airfoil chord lengths permit narrower, low-weight blade plan forms.

Additional noise-reduction techniques are employed, such as selective use of sweep in the blade tip and improved power loading distributions. Pitch change is hydraulically actuated in both increase and decrease pitch directions. Pitch change is controlled by a compact, engine-gearbox-mounted propeller control unit (PCU), shown in Fig. 21-19. The PCU is modular and easily serviced or replaced. An active, in-place pitch lock screw provides instantaneous lockup of blade pitch-change motion in the event of a loss in supply oil. All major subsystems (blades, actuators, spinners, PCU) are modular, field replaceable, and interchangeable.

PROPELLER CONTROL UNIT (PCU)

FIGURE 21-19 Propeller control unit. (*Hamilton Standard.*)

McCAULEY TURBOPROPELLER

The McCauley propeller, illustrated in Fig. 21-20, is a single-acting unit in which hydraulic pressure opposes the forces of springs and counterweights to obtain the correct pitch for engine load. Hydraulic pressure moves the blades toward **low pitch** (increasing rpm), and springs and counterweights move the blades toward **high pitch** (decreasing rpm). The source of hydraulic pressure for operation is oil from the engine lubricating system, boosted in pressure by the governor gear pump and supplied to the propeller piston. The flow of oil through the governor and the propeller does not interfere with engine lubrication.

The propeller is designed to operate in two modes: beta mode and governor mode. In the **governor mode**, oil is metered to and from the propeller (by the governor control valve as positioned by flyweights), increasing and decreasing blade angle (changing pitch) as required when the propeller speed control setting is altered, or increasing and decreasing pitch to control and stabilize propeller speed under varying power conditions or at varying flight attitudes with a fixed speed setting. The pilot may select **beta mode** for ground reversing or taxi operation through aircraft-engine mechanical linkage which repositions the propeller-reversing lever and beta valve to provide access for high-pressure oil to reach the propeller piston and move the blades toward reverse pitch.

In flight, complete reduction of hydraulic pressure, whether by failure or intent (actuation of control), will permit and cause feathering springs and counterweights to move blades automatically to full-feathered (high- or maximum-pitch) position. Upon engine shutdown and loss of pitch-change oil pressure, the propeller will automatically feather.

The propeller hub cavity is partially filled with turbine oil that is sealed in the hub and isolated from engine oil. This turbine oil provides lubrication and corrosion protection for blade bearings and other internal parts.

The propeller piston linkage is connected to spring-loaded sliding beta rods which are attached to a feedback collar mounted behind the propeller. Movement of the feedback collar is transmitted by a carbon block to the engine, providing blade-angle input during the beta mode of operation. As the blade angle decreases, the feedback collar begins forward movement at the low-blade-angle setting. Forward movement continues until the reverse-pitch stop is reached. At blade angles higher than the low-blade-angle setting, the beta rods and feedback collar remain stationary.

COMPOSITE PROPELLER BLADES

Much research has been devoted to development of the technology required to produce blades from composite materials. In comparison to the more traditional aluminum alloys, the composite blade material offers not only a significant weight reduction, but also improved repairability, damage tolerance, vibration damping, and design flexibility.

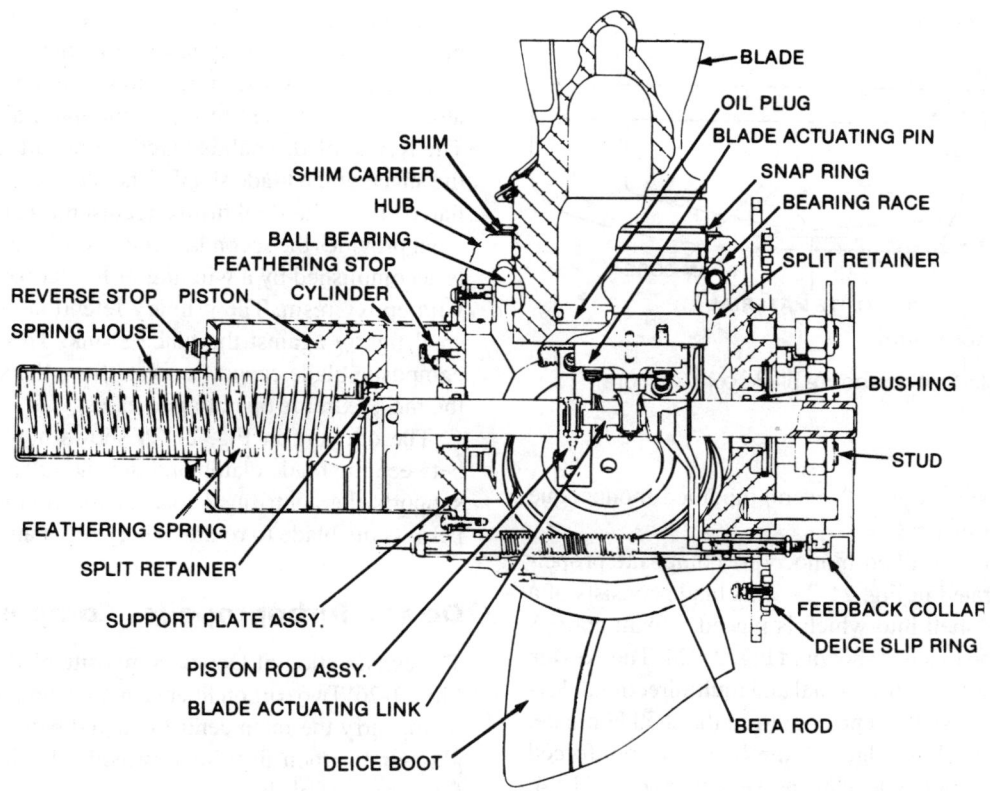

FIGURE 21-20   McCauley turbopropeller. (*McCauley Propeller.*)

The advantages this construction method offers in terms of reduced weight compared with light alloy propellers are illustrated in Fig. 21-21. Note that the weight advantage increases sharply with diameter. This sharp increase is due not only to the blade weight, but also to the impact the reduced blade weight has on hub and counterweight design. Counterweights become prohibitively heavy on large-diameter metal-blade propellers.

Composite blade construction involves the use of various plastic resins reinforced with fibers or filaments composed of glass, carbon, Kevlar, or boron. The resin matrix may be of epoxy, polyester, or polyamide. Glass fiber with epoxy was used extensively for many years to manufacture a wide variety of lightweight high-strength structures. Graphite or carbon filament with epoxy was developed more recently and has proven even stronger and more durable than glass fiber composites. Other combinations of fibers and resins have followed, with the result that there has been a continual improvement in propeller design, particularly with respect to weight.

Among propeller companies that manufacture propellers having composite blades are Hamilton Standard, Hartzell, and Dowty. Blade construction and design vary among the companies; however, similar results are achieved.

A composite propeller blade designed and manufactured by the Hamilton Standard Company consists of a solid aluminum-alloy spar around which a fiberglass shell with the correct airfoil shape is placed. The space between the spar and the shell is filled with a plastic foam that provides a firm support for the shell. The outer surface of the shell is given a coating of polyurethane. The term **spar/shell** is used to describe this type of construction. A cross section of the spar/shell propeller blade is shown in Fig. 21-22.

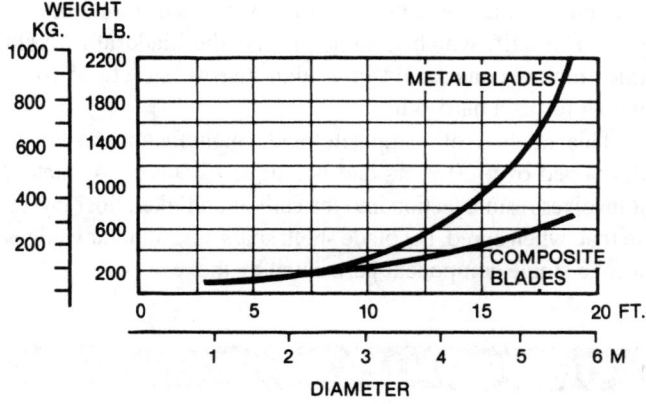

FIGURE 21-21   Propeller weight-diameter comparisons. (*Dowty.*)

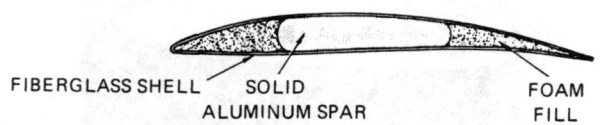

FIGURE 21-22   Cross section of a spar/shell propeller blade. (*Hamilton Standard.*)

Composite Propeller Blades   **625**

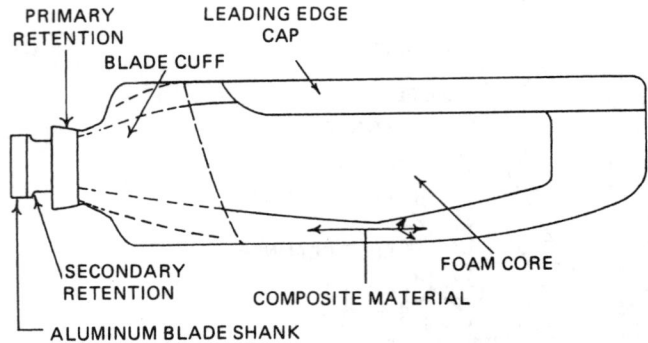

FIGURE 21-23   Hartzell composite blade construction.
(*Hartzell Propeller.*)

The spar/shell design is one of several modified-monocoque types of propeller blades.

Another type of modified-monocoque composite propeller blade is illustrated in Fig. 21-23. This blade consists of a laminated Kevlar shell into which is placed a foam core. A cross section of the blade is shown in Fig. 21-24. The **Kevlar shell** consists of both unidirectional and multidirectional layers of material bonded with epoxy to form the shell laminate. The leading and trailing edges of the blade are reinforced with solid unidirectional Kevlar, as shown in Fig. 21-24. Two unidirectional Kevlar shear webs are placed between the camber and face surfaces of the shell to provide resistance to flexing and buckling. The polyurethane foam that fills the spaces inside the shell supplies additional resistance to any distortion that could be caused by operating stresses.

The drawing in Fig. 21-23 illustrates the construction and configuration of the blade. This illustration shows the metal cap that is bonded to the leading edge of the blade. This cap serves to reduce the erosive and damaging effects of sand, gravel, rain, and other materials that may be encountered

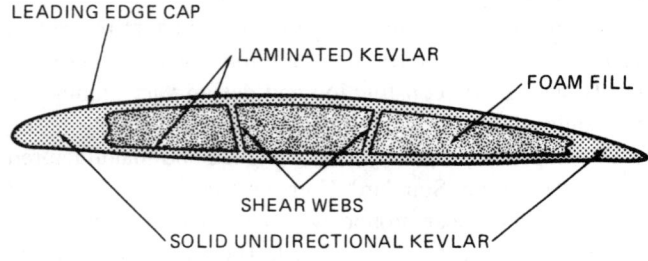

FIGURE 21-24   Cross section of composite blade. (*Hartzell Propeller.*)

during operation. A Kevlar cuff is attached at the base of the blade to improve aerodynamic efficiency.

Figure 21-25 illustrates how the composite blade is attached to (retained by) the aluminum-alloy blade shank. The Kevlar blade shell is flared at the butt end to conform to the shape of the blade shank. The blade clamp then holds the flared end of the shell firmly against the shank. This arrangement provides the secondary retention. The primary retention is accomplished by a winding of Kevlar roving impregnated with epoxy resin. The primary retention winding holds the shell tightly against the blade shank. This ensures that the composite blade cannot separate from the shank even under the most extreme operating stresses.

The drawing in Fig. 21-25 shows how the ball bearing between the blade clamp and the shoulder of the hub spider supports the centrifugal load of the rotating propeller and permits the blade to rotate axially for changes in pitch.

Dowty Turbopropeller Composite Blades

The construction of Dowty composite blades is illustrated in Fig. 21-26. Two carbon fiber spars, separated by polyurethane foam, carry the main centrifugal and bending loads, while a glass-and-carbon fiber shell provides torsional stiffness and forms the airfoil shape.

Additional layers of glass fiber around the leading edge strengthen this area so that it can withstand the impact of foreign objects. Lightning protection is provided by light alloy braiding laid into both faces of the blade, as shown.

As the carbon fiber spars approach the blade shank, their leading and trailing edges are gradually brought together until they join and enter the metal root components as a cylinder. At the inner end of this cylinder, the carbon fiber layers are opened out by the insertion of glass fiber wedges, as illustrated in Fig. 21-27.

These annular wedges, once formed, are trapped between the inner and outer metal sleeves. Structural integrity is ensured by this wedge formation, which is resin-injected in one operation. These wedges, which are pulled tight by centrifugal forces acting on the blade, can carry loads in any direction and do not depend on adhesion to the metal components for structural strength. The complete fiber content of the blade is laid up dry, and hot resin is injected under vacuum, so the complete structure is molded in one operation. The cuff, which is molded onto the blade after resin injection, is constructed of a high-density structural foam which forms a hard skin.

This method of composite-blade manufacture has been developed over 20 years and is unique to Dowty. As stated, it involves resin injection into the carbon-and-glass fiber lay-up so that, when cured, the blade shell, spars, and annular wedges form a single component permeated by the resin.

PT6A PROPELLER CONTROL SYSTEMS

The PT6A reversing-propeller installation utilizes a single-acting hydraulic reversing propeller controlled by a propeller governor.

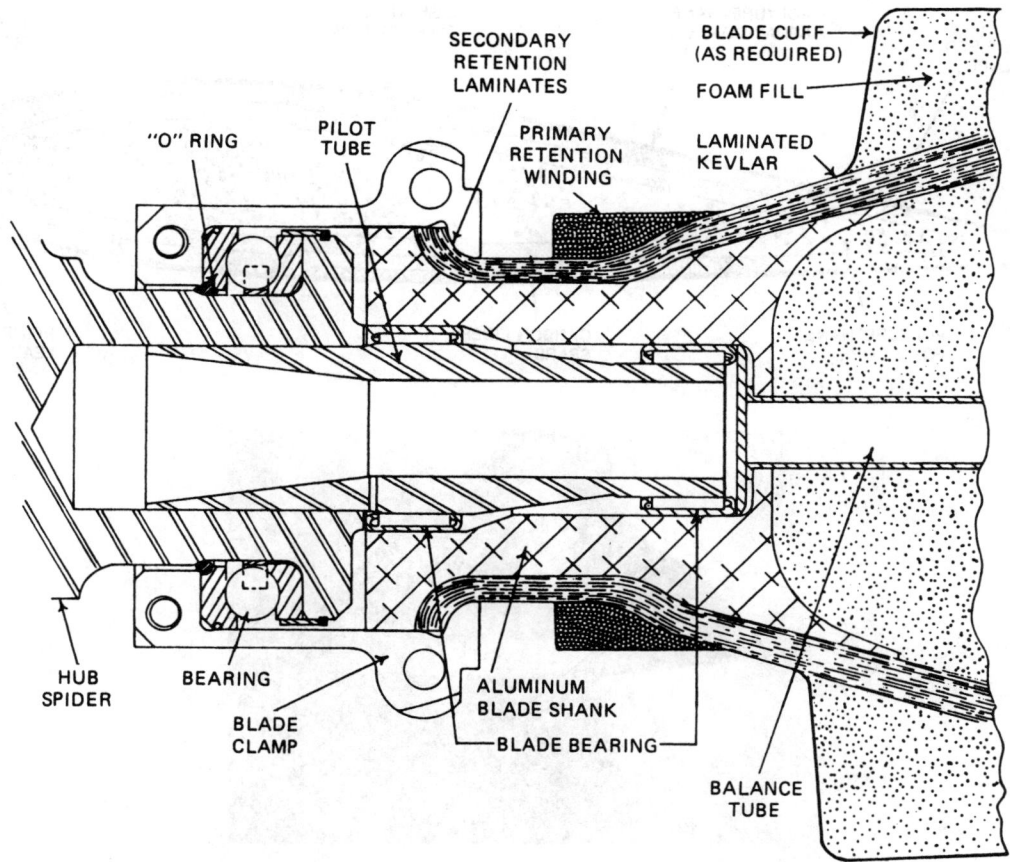

FIGURE 21-25 Blade retention system. (*Hartzell Propeller.*)

There are two propeller overspeed governors used to limit power-turbine speed in the event of failure of the propeller governor. The main overspeed governor bleeds governor oil pressure from the propeller if an overspeed should occur. The power-turbine governor has an air-bleed orifice that controls fuel flow to the gas generator. During an overspeed condition, it bleeds compressor discharge air from the fuel control, decreasing fuel flow.

Cockpit Controls

The integrated engine and propeller combination has three cockpit controls: a fuel cutoff lever (condition lever), a power control lever, and a propeller control lever. The fuel cutoff lever usually has two positions: CUTOFF and IDLE. An additional HI-IDLE position is provided on certain installations. This lever controls the cutoff function of the fuel control unit and resets the power control lever to IDLE STOP to provide 50 percent minimum compressor turbine speed (N_g) in the IDLE position and up to a minimum of 80 percent N_g in the HI-IDLE position, where applicable. The power control lever is used to select a maximum value of N_g, which is reached only if the power-turbine governor does not act to reduce N_g below the selected value. The power control lever is also used to select low blade angle (beta range). The propeller control lever is connected to the propeller governor through linkage. Movement of the linkage determines and limits the

rpm range of the propeller governor and so governs the propeller. When the control lever is moved to the minimum rpm position, the propeller will automatically feather.

Propeller Governor

The **propeller governor** performs two functions. Under normal flight conditions it acts as a constant-speed governing unit to regulate power-turbine speed (N_f) by varying the propeller blade pitch to match the load torque to the engine torque in response to changing flight conditions. During low-airspeed operations, the propeller governor can be used to select the required blade angle (beta control). While in the beta control range, engine power is adjusted by the fuel control unit and the power-turbine governor to maintain power-turbine speed at a speed slightly lower than the selected rpm.

The propeller governor, shown in Fig. 21-28, is installed at the 12 o'clock position on the front case of the reduction gearbox and is driven by the propeller shaft through an accessory drive shaft. The governor controls the propeller speed and pitch settings as dictated by the cockpit control settings and flight conditions.

Governor Description

The propeller governor is a base-mounted centrifugal type designed for use with hydraulic constant-speed propellers.

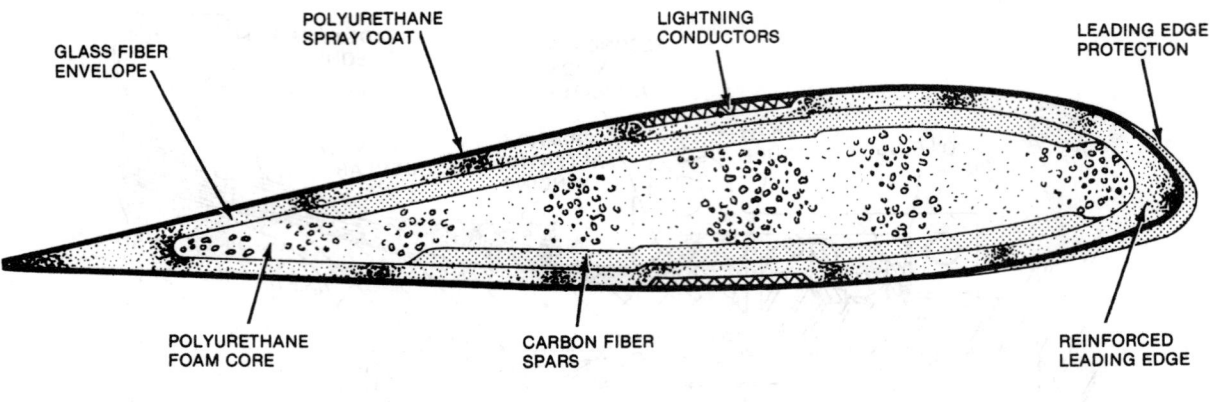

GLASS FIBER ENVELOPE

POLYURETHANE SPRAY COAT

LIGHTNING CONDUCTORS

LEADING EDGE PROTECTION

POLYURETHANE FOAM CORE

CARBON FIBER SPARS

REINFORCED LEADING EDGE

FIGURE 21-26 Dowty composite blade construction. (*Dowty.*)

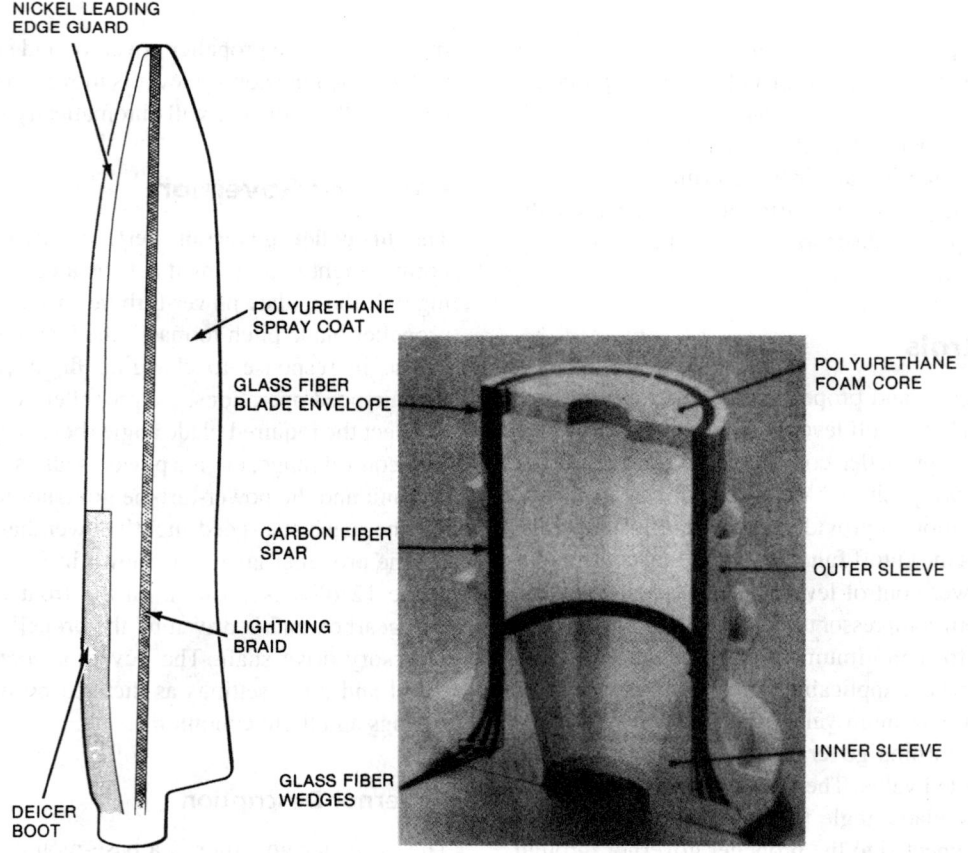

NICKEL LEADING EDGE GUARD

POLYURETHANE SPRAY COAT

GLASS FIBER BLADE ENVELOPE

CARBON FIBER SPAR

LIGHTNING BRAID

DEICER BOOT

GLASS FIBER WEDGES

POLYURETHANE FOAM CORE

OUTER SLEEVE

INNER SLEEVE

FIGURE 21-27 Dowty composite blade and root details. (*Dowty.*)

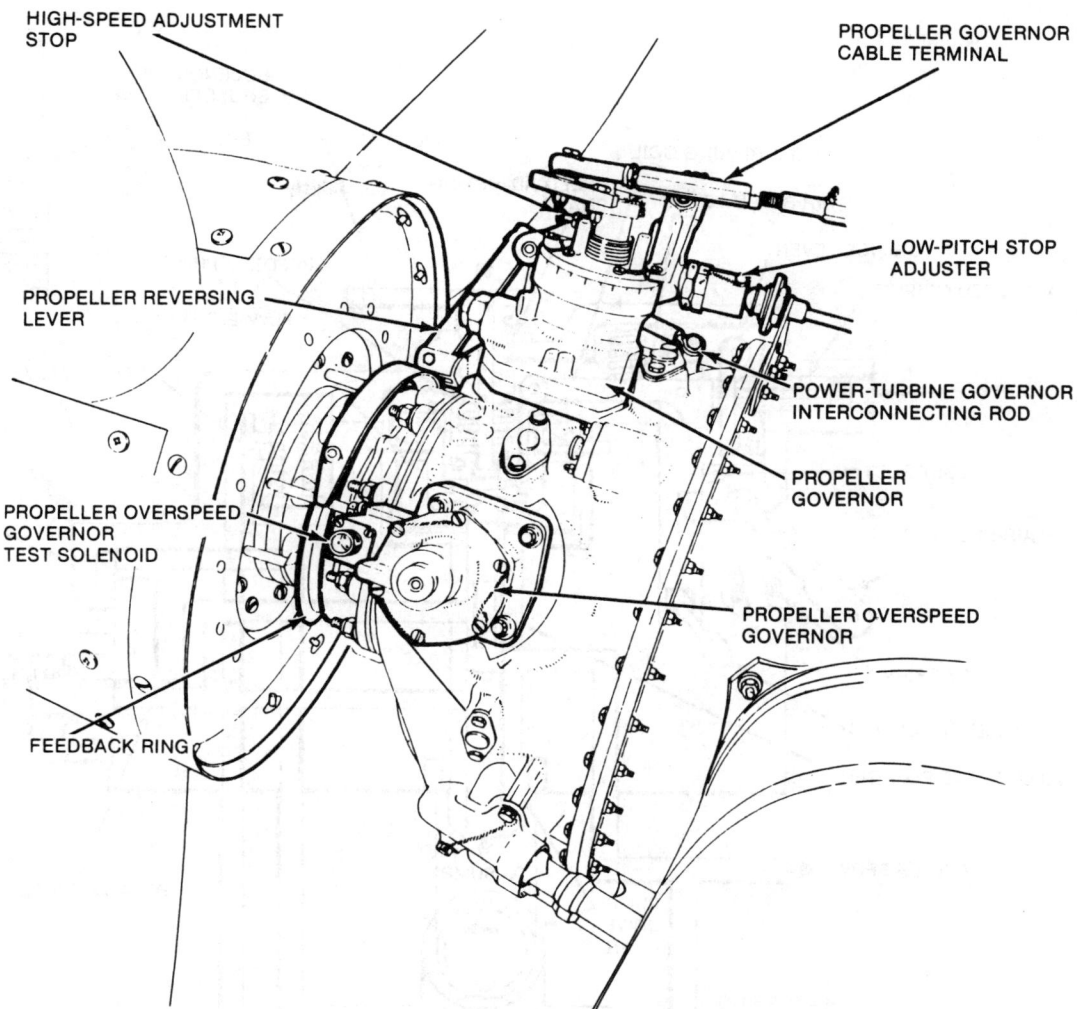

FIGURE 21-28 PT6A propeller governor installation.

Labels in figure:
- HIGH-SPEED ADJUSTMENT STOP
- PROPELLER GOVERNOR CABLE TERMINAL
- LOW-PITCH STOP ADJUSTER
- PROPELLER REVERSING LEVER
- POWER-TURBINE GOVERNOR INTERCONNECTING ROD
- PROPELLER GOVERNOR
- PROPELLER OVERSPEED GOVERNOR TEST SOLENOID
- PROPELLER OVERSPEED GOVERNOR
- FEEDBACK RING

Pitch of the propeller is varied by the governor to match load torque to engine torque in response to changing flight conditions.

Oil pressure is used to decrease propeller pitch. Pitch change in the increase-pitch direction is accomplished by the use of springs and propeller counterweights on single-acting propellers. The governor is equipped with an integral beta valve, as illustrated in Fig. 21-29, which selects propeller blade angle while the aircraft is operating in the beta control range. It also uses a pneumatic orifice that bleeds compressor discharge from the main fuel control unit, thereby limiting and controlling turbine speed while operating in the constant-speed range. This function is sometimes referred to as the N_f **governor**. On some governors, a speed-biasing coil raises the speed setting of the governor to synchronize the "slave" engine with the "master" engine. A feathering valve may be incorporated to allow rapid dumping of oil from the propeller servo when the propeller is feathered. Also, a lock-pitch solenoid (shutoff) valve may be fitted to prevent the propeller from moving into reverse pitch in flight. The valve is energized by switches on the landing gear.

The load imposed on an aircraft turboprop engine is the propeller. The amount of load varies with the angle or pitch of the propeller blades. A propeller in low pitch decreases the load on the engine and allows the engine to increase speed. Constant engine speed, therefore, is achieved by controlling the load on the engine by varying the pitch of the propeller blades.

Governor Operation

Engine oil is supplied to a gear-type engine-driven oil pump in the base of the governor, as shown in Fig. 21-29. Maximum oil pressure in the governor is limited by the governor relief valve. When oil pressure reaches the allowable maximum, the relief valve opens and the oil recirculates through the pump.

"On-speed" position. The pilot valve plunger moves up and down, as illustrated in Fig. 21-29, in the hollow drive-gear shaft to control the flow of oil to and from the propeller servo. When the pilot valve plunger is centered (i.e., when the plunger covers the oil discharge ports in the drive-gear shaft), no oil flows to or from the propeller servo, but oil is recirculated through the pump. The propeller blade angle and engine speed remain constant.

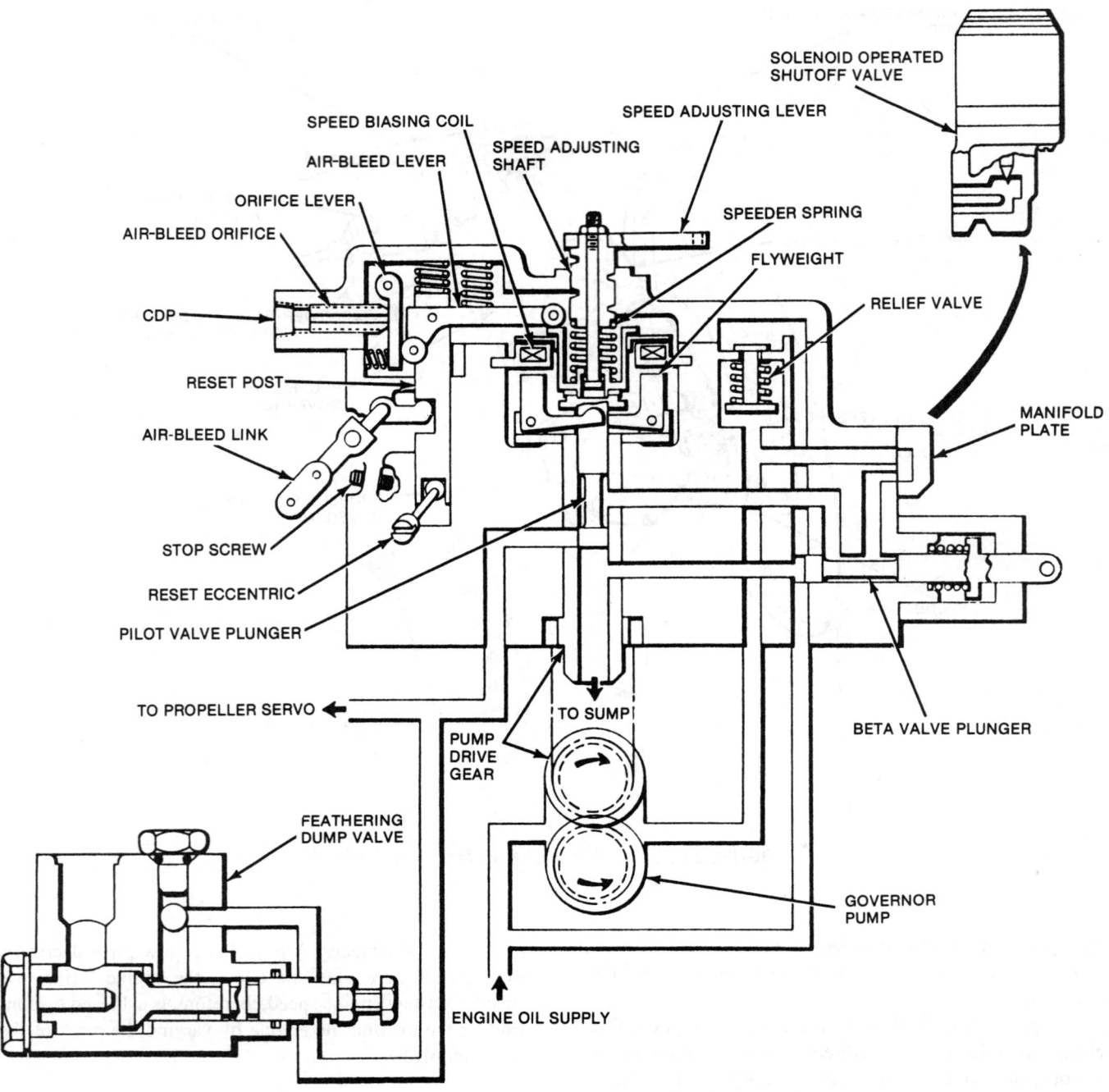

FIGURE 21-29 Schematic diagram of propeller governor operation. (*Pratt & Whitney Canada*)

This is the "on-speed" condition. The centrifugal force developed by the rotating flyweights is transformed into an upward force which tends to lift the plunger. This centrifugal force is opposed by the downward force of the speeder spring, which may be varied by means of the speed-adjusting lever. When the opposing forces are equal, the plunger is stationary.

Underspeed position. An underspeed condition occurs as the result of an increase in propeller load or movement of the control lever in the INCREASE RPM direction. The flyweight force thus becomes less than the speeder-spring

force. The flyweights move in, lowering the pilot valve plunger and uncovering the ports in the drive-gear shaft, thus allowing oil to flow to the propeller pitch-changing servo. The propeller blades move in the decrease-pitch direction against the force of the propeller counterweights. With load on the engine reduced, engine speed increases, and the centrifugal force developed by the rotating flyweights is increased. The flyweight toes lift the pilot valve plunger to cover the control ports and shut off the flow of oil. The forces acting on the engine-governor-propeller combination are again balanced, and the engine has returned to the speed called for by the governor control lever setting.

Overspeed position. An overspeed condition occurs as the result of a decrease in propeller load or movement of the control lever in the DECREASE RPM direction. The flyweight force thus becomes greater than the speeder-spring force. The flyweights move out, raising the pilot valve plunger and aligning the ports, which allows oil to drain from the propeller servo, through the governor, to the sump. As the propeller blades move in the increase-pitch direction, load on the engine is increased, and engine speed is reduced. The centrifugal force lowers the pilot valve plunger to its centered position. The pilot valve plunger once more covers the ports in the drive-gear shaft, blocking the flow of oil to or from the propeller servo, and the system is again in equilibrium.

Feathering. When the governor speed adjusting lever is pulled back against the "minimum rpm" stop, the pilot valve plunger is raised by the pilot valve lift rod. This allows the propeller counterweights and feathering spring (in the propeller hub) to force the oil out of the propeller servo into the engine sump. The blades then rotate to the feathered position.

Reversing Operation

The propeller governor beta valve plunger is operated by the propeller-reversing linkage (through the propeller control cam assembly) and the beta lever. Reverse pitch is obtained by moving the power lever aft of the FLIGHT IDLE position. The amount of reverse pitch applied is controlled by how far aft the pilot moves the power lever.

During reverse operation, the propeller governor supplies more oil, via the beta valve, to the propeller servo piston. When the amount of reverse pitch called for by the pilot is reached, the beta ring provides feedback through the carbon block to the beta lever. This movement of the beta lever pulls the beta valve plunger outward and blocks the propeller oil supply, thus preventing further pitch change. This action is illustrated in Fig. 21-30. As the blade angle reaches a reverse pitch position, engine speed is increased by the control cam assembly to provide enough power for the reverse blade angle selected.

When a higher blade angle is selected by the pilot, the propeller comes out of the beta range. The blade pitch will automatically increase to maintain the speed setting which has been established by the propeller governor speed-adjusting lever.

Overspeed Governor Air-Bleed Orifice

The air-bleed orifice serves to protect the engine against possible propeller overspeed. The air-bleed section of the propeller governor (N_f governor) is open to the compressor discharge pressure (CDP) sensor of the main fuel control. The air-bleed lever assembly is raised or lowered as flyweight action changes with speed variation, as shown in Fig. 21-29. The air-bleed lever assembly operates the orifice lever assembly which seals against the air-bleed orifice. Opening of the orifice bleeds off CDP from the fuel control, resulting in a lower CDP signal being received by the fuel control; fuel to the engine is accordingly reduced.

The speed at which the orifice opens is dependent on the setting of the propeller governor speed-adjusting lever and the setting of the air-bleed link. Normally the air-bleed orifice will be opened at approximately 106 percent above the propeller governor speed setting.

Propeller Overspeed Governor

The overspeed governor is a hydraulic-mechanical device which is driven by the aircraft's engine. It is supplied with oil from the propeller governor, as shown in Fig. 21-30. A pair of flyweights mounted on a flyweight head are rotated by the drive shaft. Centrifugal force of the flyweights is proportional to drive speed. This force is transmitted to a pilot valve and is opposed by the force of a spring (see Fig. 21-31). When the drive speed is below a specific value, the spring force holds the pilot valve down and the servo pressure port in the bottom of the governor is connected to the engine sump. If the drive speed exceeds a specific value, as in the case of an overspeed, the servo pressure port is supplied with oil at governor pump pressure, draining oil away from the propeller; this in turn causes an increase in blade angle, slowing down the prop.

Shutoff Valve

The solenoid-operated shutoff valve is a lock-pitch solenoid valve, shown in Fig. 21-29, and is attached to the side of the governor. It acts as an automatic safety device which prevents the propeller from going into reverse pitch during normal flight operation. The shutoff valve stops the flow of oil from the governor to the propeller servo if the propeller moves into the reverse-pitch range prior to the aircraft touching down on landing. The solenoid is actuated by a switch connected to the propeller slip-ring linkage. The blade angle becomes fixed, drifting slowly in the increase-pitch direction as oil pressure is lost to the propeller.

GARRETT TPE331 ENGINE TURBOPROPELLER CONTROL SYSTEM

The propeller most often used with the Garrett TPE331 is a hydraulic, single-acting, and reversing propeller. Governor oil pressure moves the propeller toward the low- and reverse-pitch positions, whereas a spring and counterweights move the propeller to the high-pitch and feather positions. Some of the control system components are the propeller pitch control, the feathering valve, the propeller governor, and the underspeed governor. The relationship among the cockpit controls, the control system components, and the propeller is shown in Fig. 21-32. In this engine system, the power turbines are connected directly to the reduction gearbox; therefore, the engine is a fixed-turbine arrangement.

Propeller Governor

The propeller governor for the TPE331 operates in very much the same manner as the propeller governors described previously.

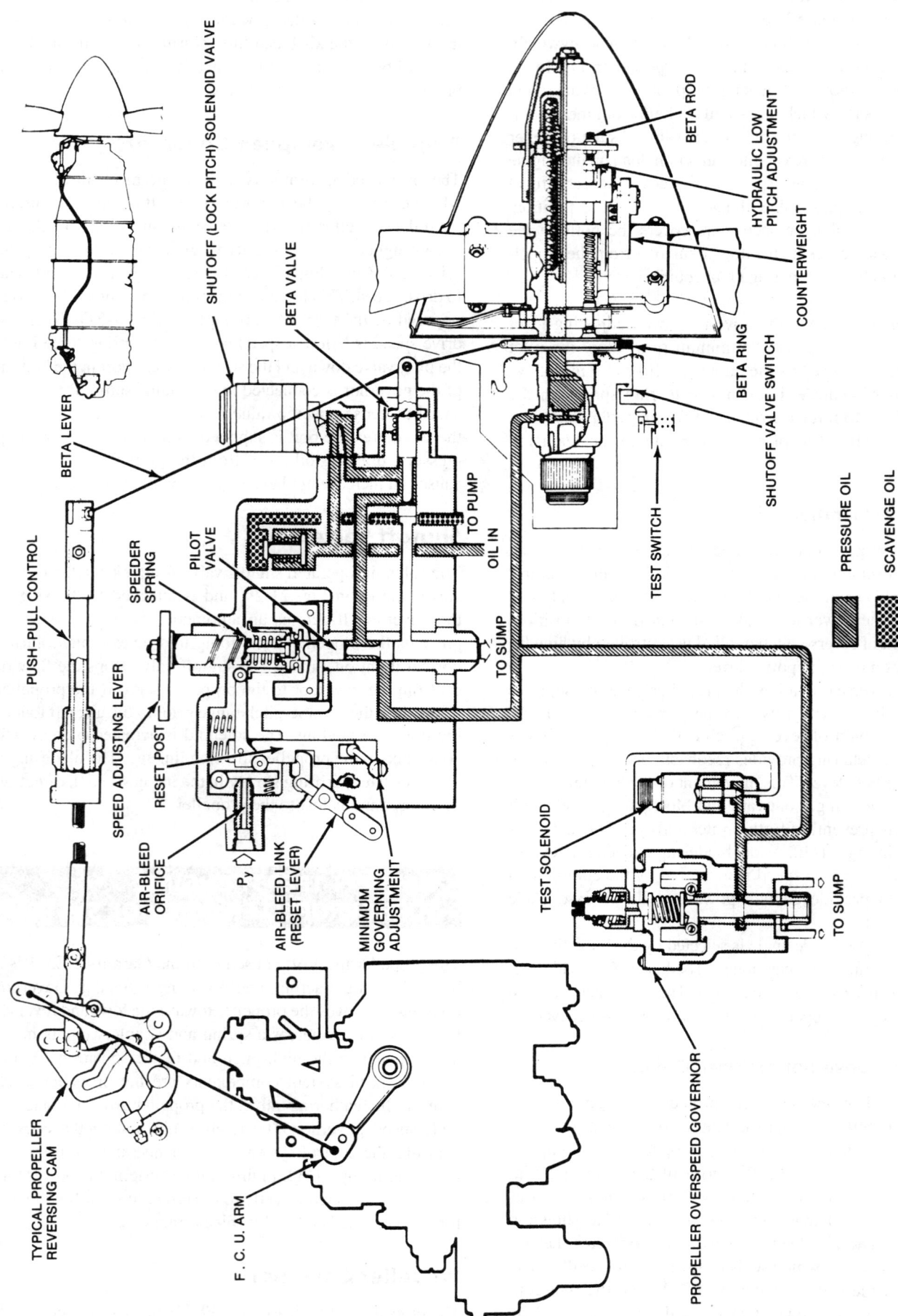

SHUTOFF (LOCK PITCH) SOLENOID VALVE

BETA VALVE

BETA ROD

HYDRAULIC LOW PITCH ADJUSTMENT

COUNTERWEIGHT

BETA RING

BETA VALVE SWITCH

SHUTOFF VALVE SWITCH

BETA LEVER

PUSH-PULL CONTROL

SPEEDER SPRING

PILOT VALVE

TO PUMP

OIL IN

TO SUMP

TEST SWITCH

PRESSURE OIL

SCAVENGE OIL

SPEED ADJUSTING LEVER

RESET POST

AIR-BLEED ORIFICE

P_y

AIR-BLEED LINK (RESET LEVER)

MINIMUM GOVERNING ADJUSTMENT

TEST SOLENOID

TO SUMP

TYPICAL PROPELLER REVERSING CAM

F. C. U. ARM

PROPELLER OVERSPEED GOVERNOR

FIGURE 21-30   Propeller governor reversing operation.

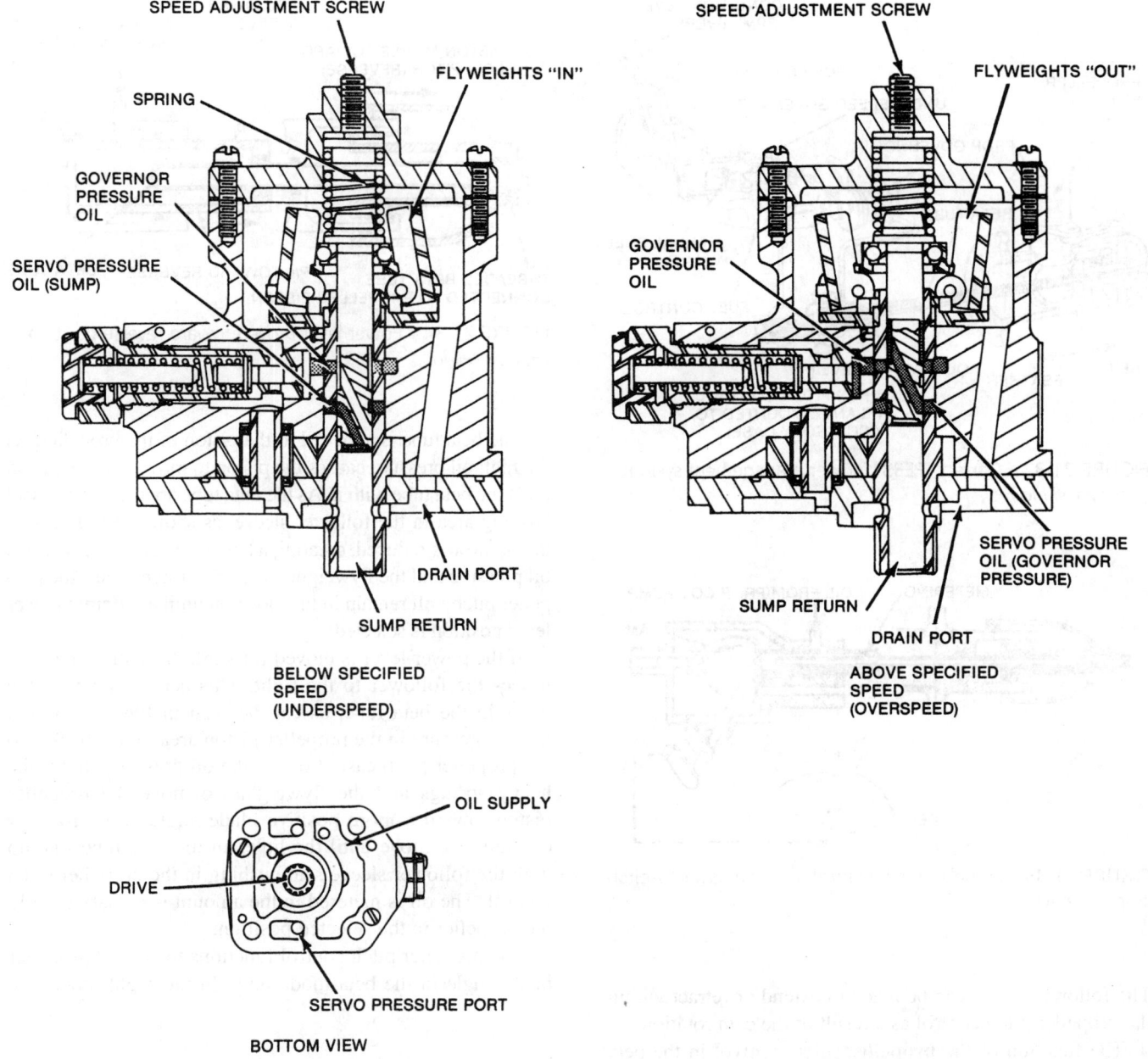

SPEED ADJUSTMENT SCREW

SPRING

FLYWEIGHTS "IN"

GOVERNOR PRESSURE OIL

SERVO PRESSURE OIL (SUMP)

DRAIN PORT

SUMP RETURN

BELOW SPECIFIED SPEED (UNDERSPEED)

SPEED ADJUSTMENT SCREW

FLYWEIGHTS "OUT"

GOVERNOR PRESSURE OIL

SERVO PRESSURE OIL (GOVERNOR PRESSURE)

SUMP RETURN

DRAIN PORT

ABOVE SPECIFIED SPEED (OVERSPEED)

OIL SUPPLY

DRIVE

SERVO PRESSURE PORT

BOTTOM VIEW

FIGURE 21-31 Propeller overspeed governor. (*Woodward Governor.*)

It is controlled by the **condition lever** (also known as the **rpm lever**) and is connected to the condition lever and underspeed governor through linkage, as shown in Fig. 21-32. The governor provides propeller control oil pressure and controls blade angle in the normal flight modes.

Underspeed Governor

In the beta mode (ground operation), the underspeed governor controls fuel flow to the engine, thereby controlling engine speed. As a speed-sensing device, the underspeed governor tries to maintain the speed setting selected by the condition lever in the cockpit. If, during the beta mode, something takes place that would tend to slow down the engine, such as an increase in propeller pitch,

the underspeed governor increases fuel flow to maintain the engine speed selected. Conversely, if propeller pitch decreases, the engine has a tendency to increase in speed, so the underspeed governor decreases fuel flow so as to maintain a set engine speed. The underspeed governor functions primarily during the beta mode with the condition lever set in the TAXI position.

Propeller Pitch Control

The **power lever** is mechanically connected to both the propeller pitch control and the main fuel control, as shown in Fig. 21-32. In the beta mode, movement of the power lever results in a rotation of the cam (see Fig. 21-33). The slot in the cam is attached (by virtue of a pin) to a follower sleeve.

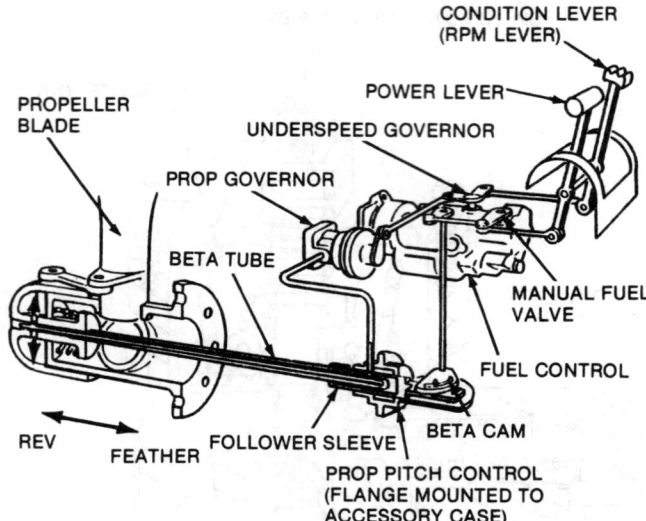

FIGURE 21-32 Garrett TPE331 power management system. (*Honeywell Inc.*)

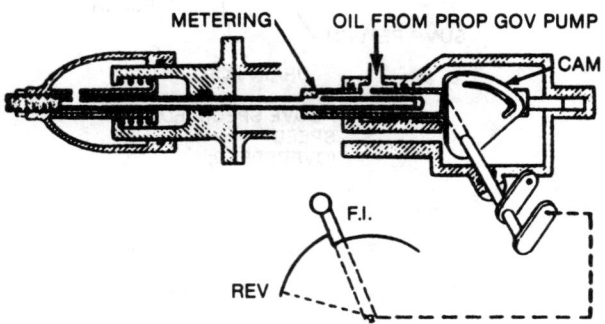

FIGURE 21-33 Propeller pitch control function as a low-pitch stop. (*Honeywell Inc.*)

The follower sleeve can be made to extend or retract within the propeller pitch control as a result of the cam rotation.

The function of the **propeller pitch control** in the beta mode is to meter the oil from the propeller governor pump into the propeller through the beta tube. This metering takes place at the point indicated in Fig. 21-33. Remember that a propeller governor cannot select reverse blade angle, because the governor senses rpm only. As the arrow in Fig. 21-34 notes, the power lever is in a reverse position and the cam has moved the follower sleeve in the propeller pitch control to the position shown. Propeller governor high-pressure oil is available to the propeller pitch control, as indicated by the arrow. The same pressure is extended into the inside of the follower sleeve and through holes within the beta tube.

The beta tube—acting as an oil transfer tube—carries the oil pressure to the piston area of the propeller. The increase in oil pressure causes the propeller piston to move to the left and rotate the blades toward a reverse pitch angle. This rotation would continue until the blades reached the full reverse position, as limited by the propeller internal reverse stop, if it were not for the feedback provided by the beta tube and propeller pitch control.

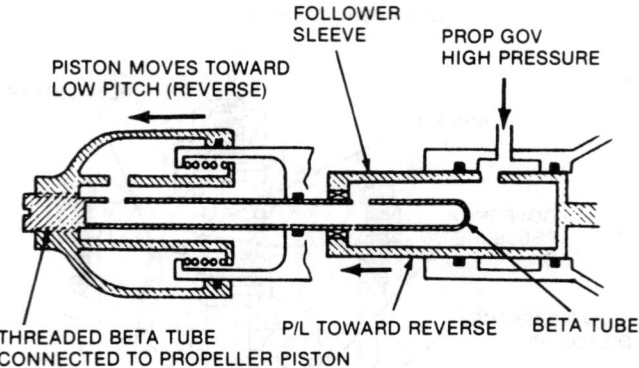

FIGURE 21-34 Power lever moved toward reverse position. (*Honeywell Inc.*)

The beta tube is attached to the piston of the propeller. As the high oil pressure causes the piston to move left, the piston pulls the beta tube with it. As the beta tube approaches the seal bushing area in the follower sleeve, as shown in Fig. 21-35, the oil flow is reduced, creating a balanced condition between oil pressure and the flyweight and spring force. Thus, the propeller pitch will remain in this position until a different power lever position is selected.

If the power lever is moved forward, the cam rotates and moves the follower to the right. This action uncovers the holes in the beta tube, as can be seen in Fig. 21-36, and the oil pressure in the propeller piston area drains back into the propeller pitch case. Loss of the oil pressure allows the heavy springs and the flyweights to move the propeller piston toward a more positive blade angle. The propeller continues to move until the holes in the beta tube line up with the follower sleeve seal bushing in the propeller pitch control. The oil is metered in the amount necessary to hold the propeller in the selected position.

The propeller pitch control functions to control propeller blade angle in the beta mode only. In the flight mode, the

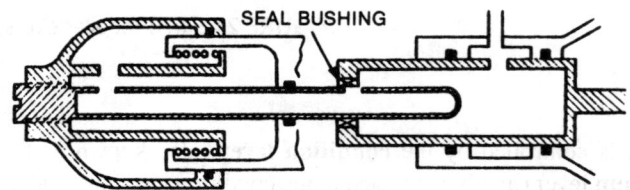

FIGURE 21-35 Beta tube in balanced position. (*Honeywell Inc.*)

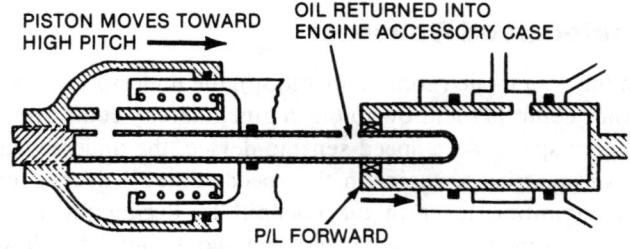

FIGURE 21-36 Beta tube, holes uncovered. (*Honeywell Inc.*)

propeller governor controls blade angle. During the flight mode, the propeller pitch control and beta tube serve only as an oil passage to the propeller piston area.

Feathering Valve

The basic function of the **feathering valve**, if actuated, is to open passages that allow oil pressure in the propeller to come back through the beta tube and the propeller pitch control, draining into the accessory case. If the oil pressure in the propeller piston is reduced, the propeller moves toward the feathered position by spring and flyweight force.

System Operation

System operation during the beta mode involves the propeller pitch control (controlled by the power lever), which controls blade angle, and the underspeed governor (set by the condition lever), which controls engine speed. In the flight mode, the propeller governor controls blade angle (set by the condition lever), and engine speed is controlled by the fuel control through the power lever.

Because this system is a fixed-turbine type of engine, the propeller cannot be allowed to feather upon engine shutdown. If the propeller were allowed to feather at engine shutdown, there would be an excessive load placed on the engine from the propeller during starting. Therefore, these propellers are equipped with start lockpins which lock the propeller blades at a 2° blade angle. This low blade angle reduces the load requirement on the electric starter motor and allows for easier engine acceleration during engine starting.

Figure 21-37 shows the major components of the start lock system. A plate is attached to the hub section of the propeller blade. The start lockpin is in a pin spring housing, and the pin is held against the plate by a spring within the housing. Each blade on the propeller has the same arrangement of the start lockpin in its housing and a plate attached to the hub of the blade. If the propeller is rotated to reverse, the plate rotates to the right until the pin can be extended in front of the edge of the plate. When the force that moves the propeller blade to reverse is removed, as with loss of oil pressure during shutdown, the heavy feather spring in the

piston causes the propeller to rotate toward the feathered position. The end of the plate contacts the pin, and the blade is held in this position.

In the flight mode during takeoff, the condition lever is in the TAKEOFF or FULL FORWARD position. The power lever is in the FLIGHT IDLE position. As the pilot moves the power lever forward for takeoff, the cam in the fuel control increases the fuel being sent to the engine. As the rpm increases from 97 percent, it approaches the propeller governor setting of 100 percent. When the propeller governor senses that the engine is at 100 percent rpm, the propeller governor meters oil pressure to the propeller. At this point, the change from the beta mode of operation to the propeller-governing mode of operation occurs. The underspeed governor in the fuel control now has no effect, since the fuel is now being controlled by the power lever positioning of the fuel control. The propeller governor positions its metering valve to reduce the pressure to the propeller, which results in a higher blade-angle position. Since the beta tube is attached to the piston, it will push the holes in the beta tube inside of the follower sleeve. From this point on, the propeller pitch control's only purpose is to move oil in and out of the propeller; it has no effect on the blade angle. Since the propeller governor metering is upstream of the beta switch, the beta light is turned off, signaling the transition from beta mode to propeller-governing mode (flight mode).

In the propeller-governing mode, the propeller governor is controlling, or metering, the oil pressure so as to maintain the right amount of oil in the propeller to cause the blade angle to load the engine and hold it at the rpm selected. This maximum propeller speed is generally about 2200 rpm. As the power lever is moved farther forward, the engine speed attempts to increase, but this increase is absorbed by the propeller seeking a higher blade angle in order to maintain a constant propeller rpm. As the propeller absorbs the additional engine power, the torque produced increases until takeoff torque limits are reached.

ALLISON 250-B17 REVERSING TURBOPROPELLER SYSTEM

Description of Propeller Operation

The Hartzell propeller shown in Fig. 21-38 is a hydraulically actuated constant-speed type utilizing oil pressure from an engine-mounted governor to supply oil under pressure through the engine shaft. Oil is used to decrease pitch and to reverse pitch. Oil pressure is opposed by the blade counterweights plus the heavy spring mounted in the propeller hub. When the oil supply is cut off and the drain opens, the action of the counterweights combined with the spring force pushes oil out of the piston chamber so that the blade pitch moves into a higher positive blade angle, either from the reverse-pitch position or from a positive-pitch position.

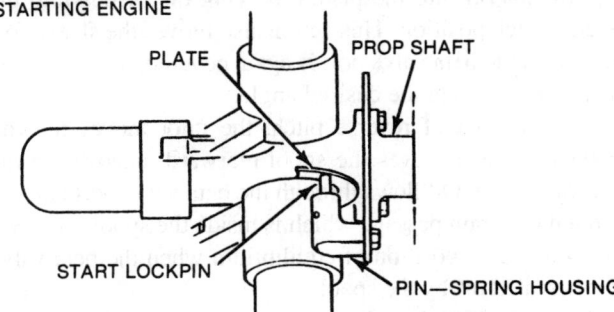

PROPELLER MUST BE AT MINIMUM BLADE ANGLE PRIOR TO STARTING ENGINE

PLATE   PROP SHAFT

START LOCKPIN   PIN—SPRING HOUSING

ROTATE PROP TO REVERSE TO DROP IN FRONT OF THE PLATE

FIGURE 21-37 Start lock system. (*Honeywell Inc.*)

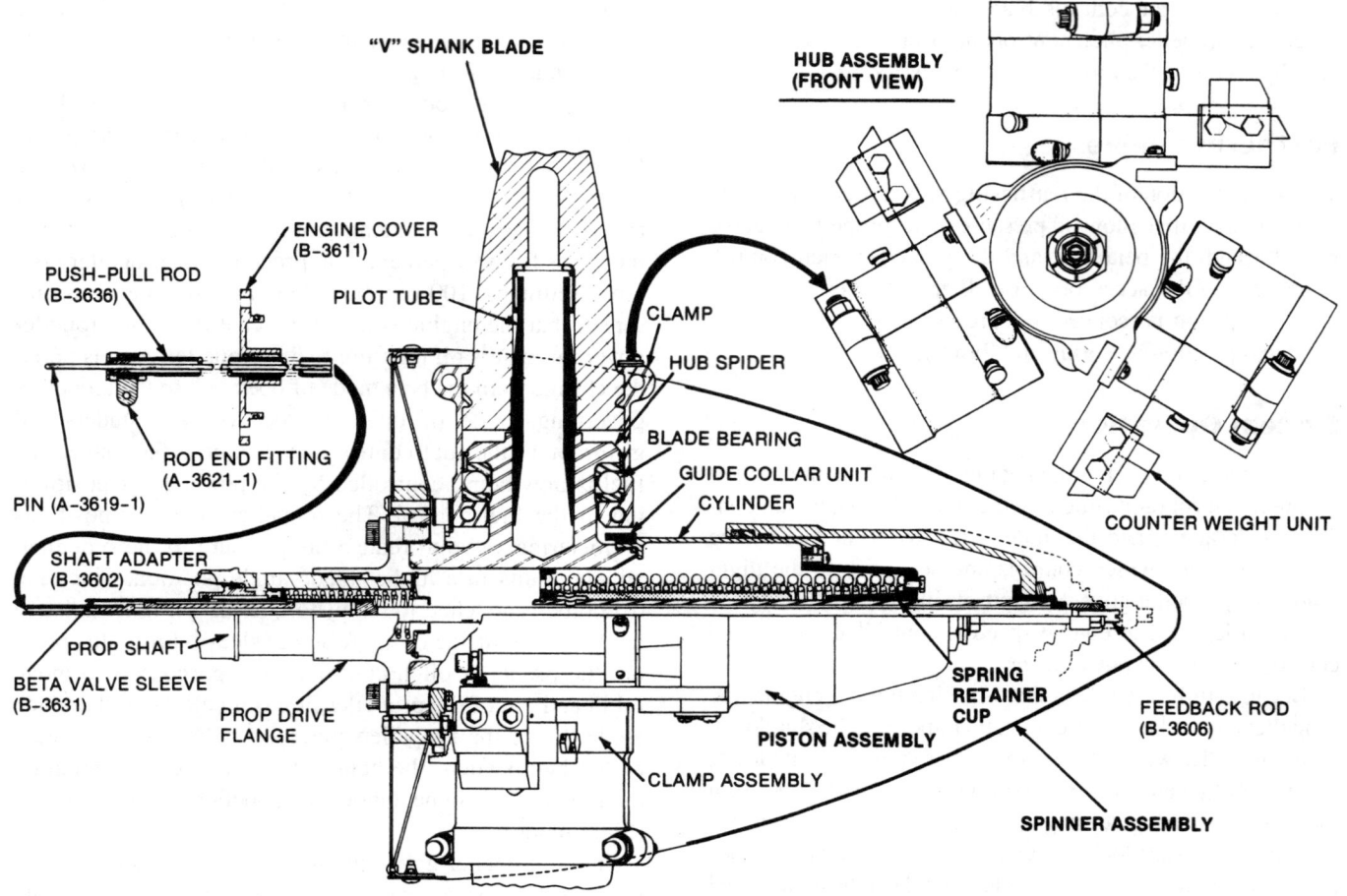

FIGURE 21-38 Hartzell HC-A3Vf-7 propeller hub assembly. (*Allison Rolls-Royce.*)

Beta Valve Operation

The beta valve is installed inside the engine drive shaft, as illustrated in Fig. 21-39. The beta valve feedback rod extends forward through the center of the propeller shaft and on out beyond the front of the piston. The beta control rod extends out through the rear of the engine gearbox, where it is connected to the engine-propeller control system.

There are three basic elements which comprise the beta valve:

1. The shaft adapter (B-3602), which is fixed inside of the engine drive shaft, rotating therein.

2. The sleeve (B-3631), which slides fore and aft with respect to the shaft adapter and also rotates with it. The sleeve is normally held in its most rearward position by the action of the springs. However, when the governor calls for reduced propeller pitch, the piston moves forward until gap X reaches zero, after which time the sleeve moves forward along with the piston.

3. The spool, which does not rotate but slides fore and aft, within the sleeve, in response to the control setting of the engine-propeller system. There is a relative rotational motion between the sleeve and the spool.

The **beta valve** is considered to be the hydraulic low-pitch stop. It cuts off the flow of oil from the governor to the piston, which is tending to reduce pitch, when the piston

reaches a predetermined low-pitch position. The path of the oil from the governor enters the shaft adapter (B-3602) and goes into the sleeve. The oil flows through the beta valve port into the engine shaft and into the propeller shaft, acting to push the piston forward (or toward low pitch or reverse pitch). When the propeller piston moves to the point where it closes gap X and picks up the rod (B-3606), it then pulls the sleeve forward until the beta valve port is closed. This shuts off the flow of oil to the piston, and the piston comes to a stop at the low-pitch position.

In order to reverse the propeller pitch, the pilot repositions the pushrod in the forward direction, which moves the spool forward, uncovering the beta valve port. The governor then pumps oil into the piston, moving the blades into the reverse-pitch position. This action also moves the sleeve forward until the beta valve port is again covered, at which time the pitch is fixed at the desired angle.

To come out of reverse pitch, the pilot moves the rod rearward, which moves the spool rearward, uncovering the beta valve port. Oil flows through the beta valve port and out through the drain passage which is inside the spool. The system comes to a position of equilibrium when the beta valve port is sealed off by the spool.

A pin (A-3619-1) is fastened to the inner end of the rod (B-3606) for the purpose of indicating the position of the blade pitch when in reverse. An electric switch is used to

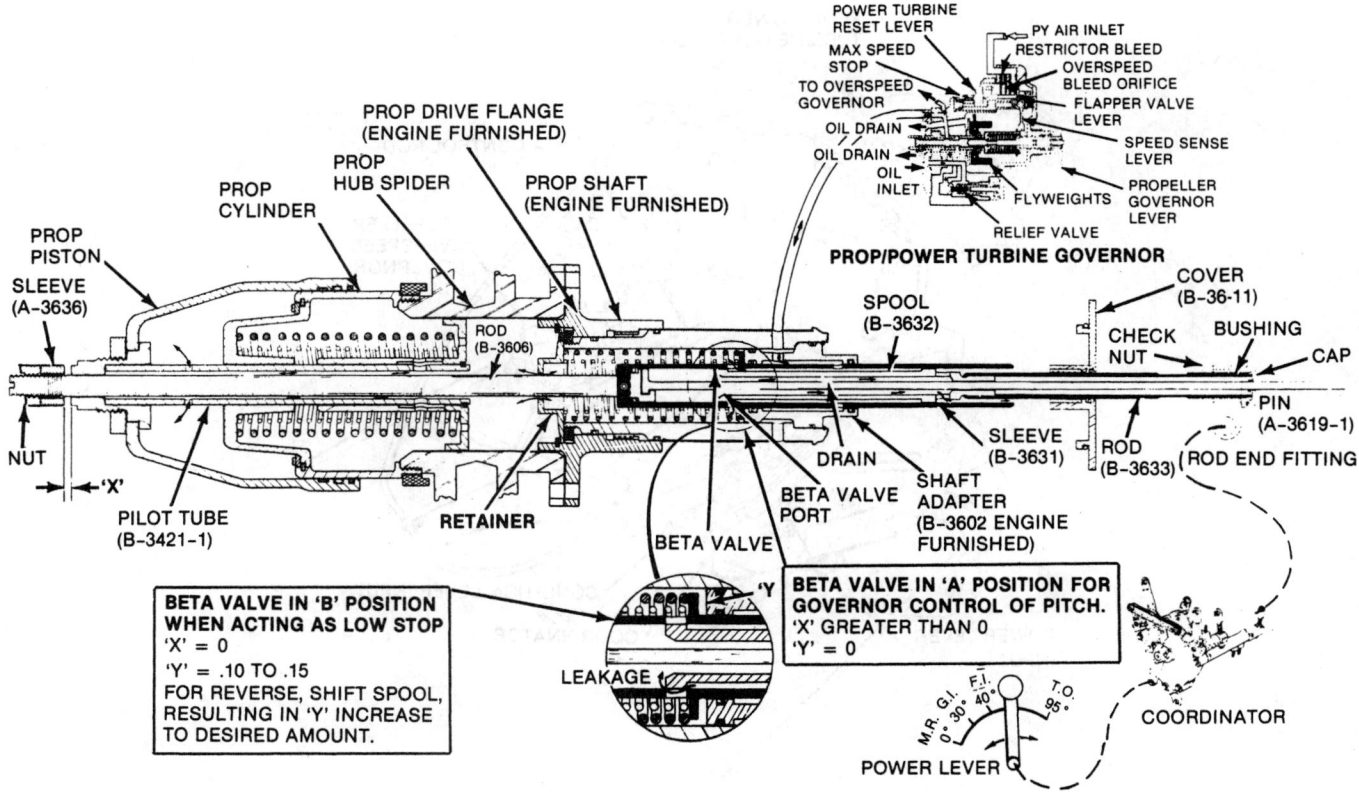

FIGURE 21-39 Beta valve schematic. (*Allison Rolls-Royce.*)

engage the end of the pin in order to illuminate a light in the cockpit when the pitch is in the reverse region. A rotating joint at the juncture of the pin and the end of the rod (B-3606) provides for zero rotation of the pin.

Overspeed Protection

The propeller power-turbine governor and the propeller overspeed governor function in the Allison 250-B17 in much the same manner as they do in the PT6A system previously described. The overspeed control system provides speed governing for the power-turbine rotor and for the gas-producer-turbine rotors. The system consists of the pneumatic N_f governor in the propeller governor and the hydraulic-mechanical overspeed governor (see Fig. 21-40).

Cockpit Controls

The **power lever**, located in the aircraft cockpit, allows engine thrust modulation from takeoff to maximum reverse. This lever has the following specific positions: MAXIMUM REVERSE, GROUND IDLE START, FLIGHT IDLE, and MAXIMUM. The power lever is connected through aircraft linkage to the input lever on the coordinator.

The **condition lever**, also located in the aircraft cockpit, allows engine start-up, engine shutdown, and propeller feathering and has the capability of varying the propeller governor setting between 82 and 104 percent of propeller speed. Fuel shutoff and propeller feathering are affected simultaneously through positioning of the condition lever. This lever has the following positions: FUEL SHUTOFF AND PROPELLER FEATHERING, MINIMUM PROPELLER SPEED, and MAXIMUM (104 percent) PROPELLER SPEED. The condition lever is connected through aircraft linkage to a lever mounted on the top right side of the coordinator.

The coordinator, illustrated in Fig. 21-40, is located on the engine just aft of the fuel control, mounted on brackets attached to the accessory gearbox. The function of the **coordinator** is to provide automatic sequencing of the multiple powerplant controls in response to input from the pilot-operated power and condition levers. The coordinator assembly includes a base, a cover, and a cam disc shaft that has three cam follower slots. The cam disc shaft is rotated by the power input lever, which is splined onto the shaft where it extends through the cover. The power input lever is connected to and rotated by movement of the aircraft power lever. The assembly thus has one input, from the power lever, and three outputs, connected to the gas producer fuel control lever, the power-turbine-governor reset lever, and the propeller beta valve control lever. Any movement by the power lever will be translated into the required coordinated movement of the three output shafts to afford the necessary engine control.

Engine and Propeller Operation

Starting

The pilot must monitor the N_1 compressor speed during engine start-up, and, upon reaching the prescribed speed for lightoff, advance the condition lever to the MAXIMUM

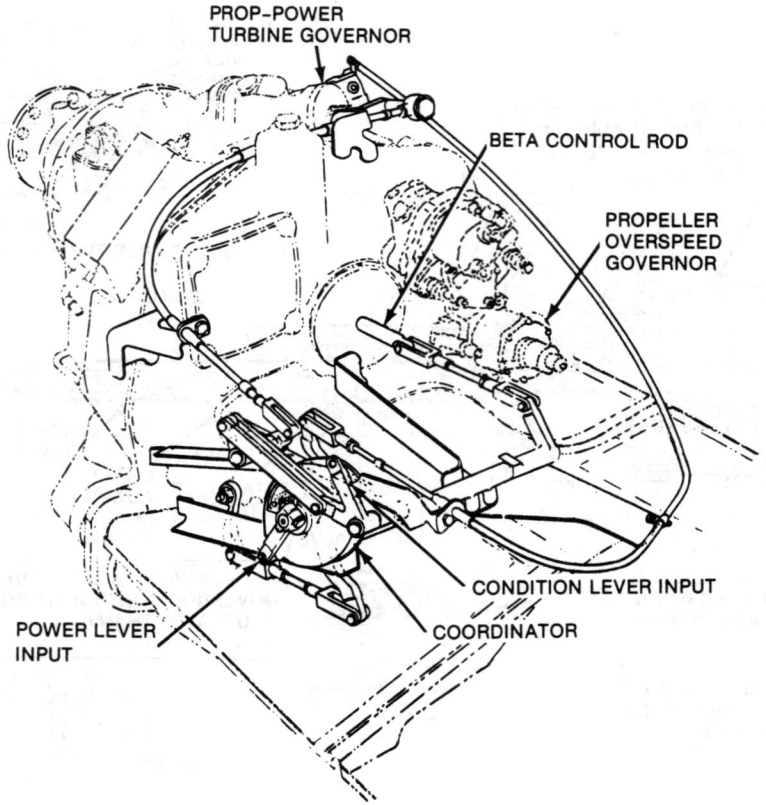

FIGURE 21-40 Allison 250 propeller control system components. (*Allison Rolls-Royce.*)

PROPELLER SPEED position to initiate fuel flow. The fuel control will automatically regulate the fuel flow during the acceleration to idle. Propeller unfeathering will automatically occur, with the propeller beta valve regulating the blade angle upon completion of the start. Both the fuel control cutoff valve and the propeller governor feathering valve are positioned by the coordinator condition lever. A ground start is accomplished with the power lever placed in the FLIGHT IDLE position.

Taxi Operation

During taxi operation, the gas producer fuel control regulates the fuel flow and the propeller beta control valve regulates the blade angle. Both the fuel control speed setting and the beta control setting are established by the position of the coordinator power input lever. The propeller governor will automatically assume control of the blade angle when high power levels are selected. During taxi operation, the condition lever should be in the MAXIMUM PROPELLER SPEED position. The power lever can be moved freely to obtain the desired thrust for taxiing.

Takeoff

During takeoff operation, the gas producer fuel control regulates the fuel flow and the propeller governor regulates the blade angle. The speed settings of the fuel control and the propeller governor are established by the position of the

power and condition levers. The condition lever should be in the position for 100 percent propeller speed, allowing the propeller governor to maintain the N_2 compressor speed control. The power setting of the engine is controlled by the power lever. Power lever movement must be controlled so as not to exceed the turbine outlet temperature (TOT) and torque limits.

Flight Operation

During flight operation, the gas producer fuel control regulates the fuel flow and the propeller governor regulates the blade angle. The speed settings of the fuel control and the propeller governor are established by the positions of the power lever and the condition lever. The propeller speed setting may be varied between 82 and 104 percent by movement of the condition lever. The power lever must not be moved below the FLIGHT IDLE position during flight. Lower lever positions may result in a total loss of engine power and in propeller feathering.

Landing

For landing, the condition lever should be positioned at 100 percent propeller speed. Landings can be made with the condition lever set for a reduced reverse-thrust potential (lower propeller speed) and possibly lower takeoff thrust in a go-around. For landing approach, the power lever may be moved to FLIGHT IDLE. When the airspeed becomes less than 100 mph, the propeller blade angle may reach the flight low-pitch stop, and the propeller speed will drop below 100 percent. Upon touchdown, the power lever may be moved below FLIGHT IDLE. Reverse-thrust selection is permitted if the aircraft airspeed is less than 90 mph. During reverse-thrust operation, the power-turbine governor regulates the fuel flow and the beta control valve regulates the propeller blade angle. The beta control valve setting is established by the position of the power lever.

Stopping

Engine stopping is effected by shutting off the fuel supply by means of the fuel control cutoff valve. Simultaneously, the propeller governor ports the oil, causing feathering of the propeller. Both the fuel control cutoff valve and the propeller governor feathering valve are positioned by the condition lever. When the engine is to be shut down, the power lever is first moved to the GROUND IDLE position, where the TOT is allowed to stabilize for a minimum of 2 min. The condition lever is then moved to FUEL SHUTOFF AND PROPELLER FEATHERING. In-flight stopping procedure is the same as ground stopping except that the power lever is moved to FLIGHT IDLE.

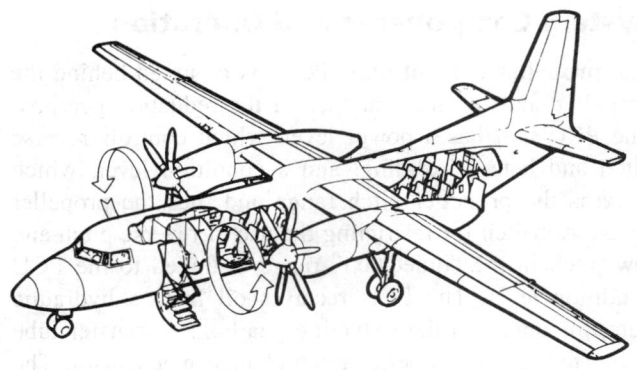

FIGURE 21-41 Fokker 50 aircraft. (*Dowty.*)

PW124 AND R352 TURBOPROPELLER ENGINE INTERFACE

The R352 propeller unit has been designed specifically for the Fokker 50 aircraft, shown in Fig. 21-41, utilizing a Pratt & Whitney 124 series engine. The R352 turbopropeller and the location of its associated equipment on the PW124 engine are illustrated in Fig. 21-42. The turbopropeller is flange-mounted on the engine gearbox drive shaft.

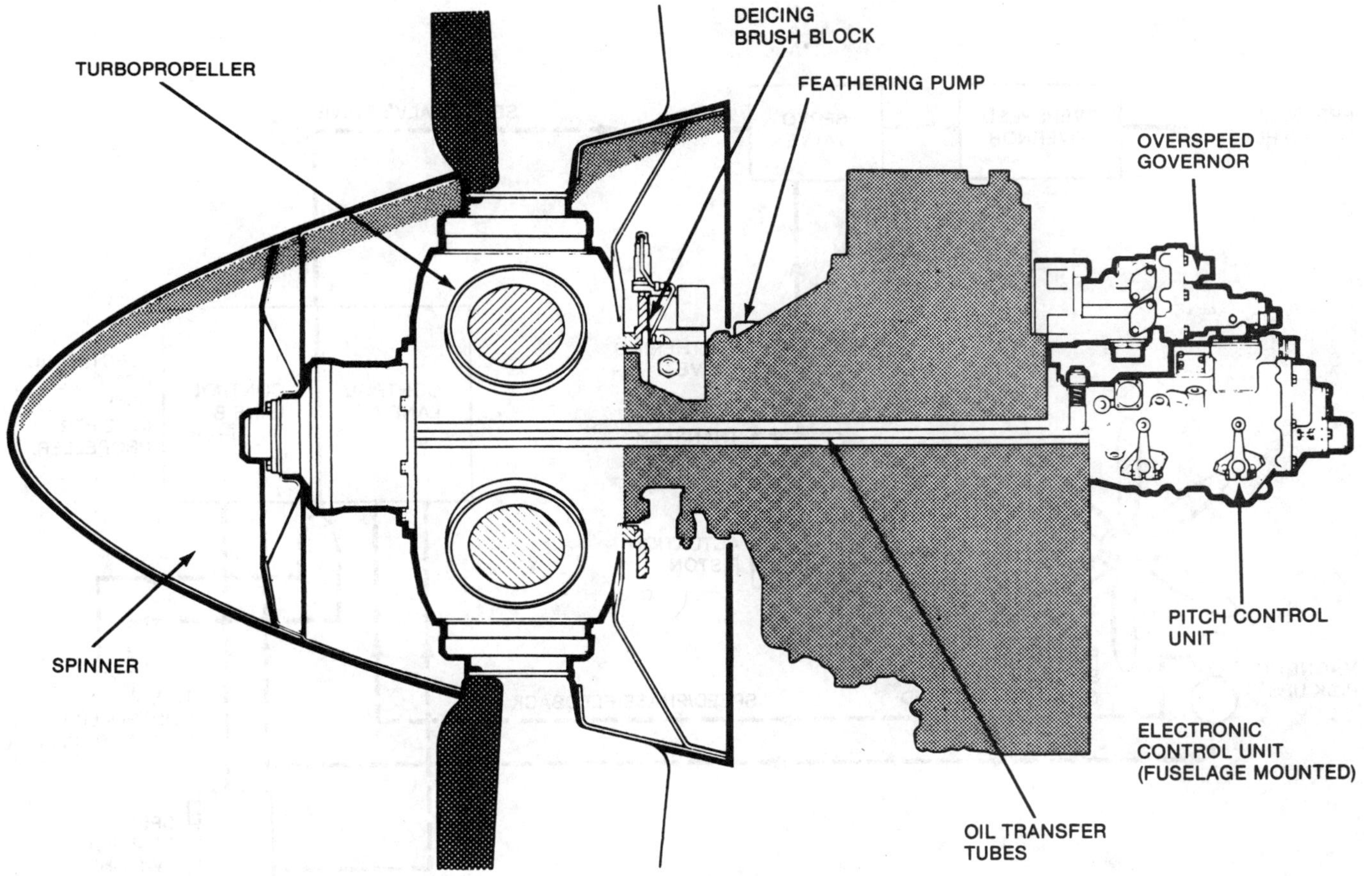

TURBOPROPELLER

DEICING BRUSH BLOCK

FEATHERING PUMP

OVERSPEED GOVERNOR

PITCH CONTROL UNIT

ELECTRONIC CONTROL UNIT (FUSELAGE MOUNTED)

OIL TRANSFER TUBES

SPINNER

FIGURE 21-42 Dowty R352 turbopropeller. (*Dowty.*)

System Components and Operation

The **propeller control unit** (PCU) is mounted behind the propeller shaft on the rear face of the reduction gearbox. The PCU unit has a power lever, which controls reverse pitch and beta scheduling, and a condition lever, which governs the propeller pitch range and thus the propeller speed. A switch for restricting the use of reverse pitch and low pitch in the "quiet taxi" range is linked to the PCU condition lever. The PCU receives oil from a hydraulic pump mounted on the reduction gearbox. A transfer tube is located in the propeller pitch-change mechanism. The PCU hydraulic pump is located on the right mounting pad of the reduction gearbox accessory drive cover under the overspeed governor. The pump boosts the pressure of oil drawn from the auxiliary tank located in the gearbox rear housing. The pump contains a pressure relief valve to regulate output and a check valve to prevent reverse flow. Rigging of the turbopropeller and control system is achieved by screwing the oil transfer tubes to a predetermined stop in the pitch control unit. The pitch control unit is signaled both mechanically and electrically.

The system, illustrated in Fig. 21-43, features a digital electronically signaled pitch control unit. The traditional hydro-mechanical flyweight-type propeller governor which is used with many other propellers has been replaced in the R352 by an extremely accurate two-stage servo valve signaled by the electronic control unit. This unit receives discrete speed demands from the pilot's engine rating select panel, and the feedback loop is provided by a magnetic turbopropeller speed and phase pickup on the gearbox.

The system achieves selected speeds to within ±2 rpm and turbopropeller phase control to within ±2°. During normal flight, blade pitch is controlled through metering of low-pitch oil by the servo valve, which acts against the centrifugal counterweights. During landing approach and ground operation, the servo valve is bypassed and the pilot's power lever directly controls blade pitch to a predetermined schedule using low- and high-pitch oil. The electronic control unit features built-in testing and fault monitoring for ease of maintenance.

Overspeed Governor

The **overspeed governor**, illustrated in Fig. 21-44, is a hydro-mechanical unit which provides overspeed protection during flight. In normal operation, it routes pressure oil to a selector valve in the PCU. The overspeed governor monitors propeller speed (N_p) and, in the event of an overspeed, bleeds pressure oil, via the PCU selector valve, from the metered side of the propeller servo-piston. This action causes the blade angle and the load on the power turbine to increase, reducing propeller speed. The overspeed governor is set to limit steady propeller rpm to an overspeed value of 103 percent. If this speed is exceeded, the force of the governor flyweight (1 in Fig. 21-44) overcomes the pressure of the spring (2). The flyweights move out, lifting the selector valve (3) and allowing oil to drain, thereby increasing propeller blade angle.

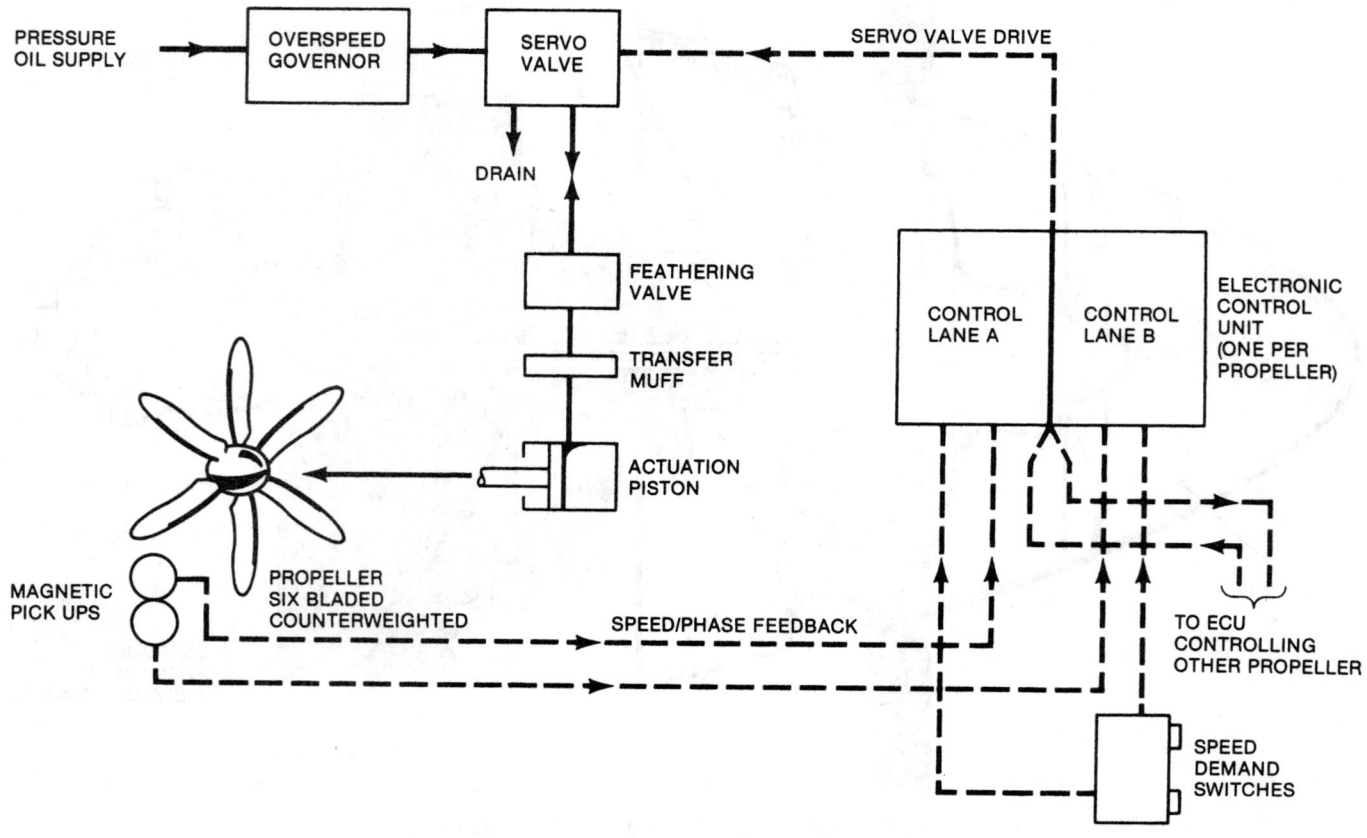

FIGURE 21-43 Propeller electronic control system. (*Dowty*.)

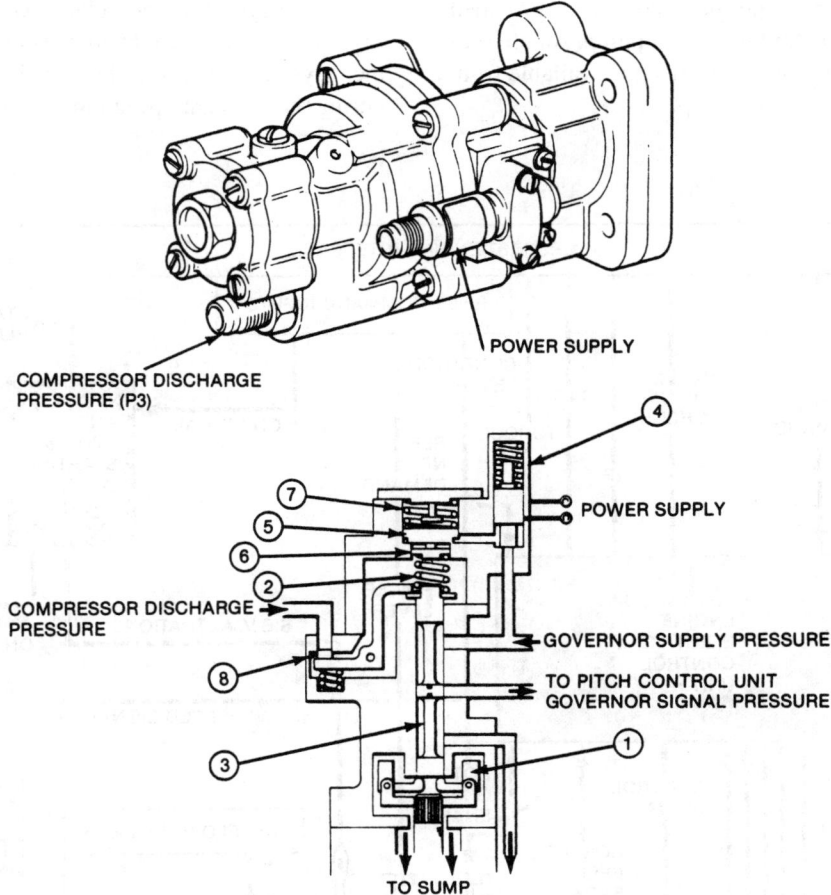

FIGURE 21-44 Propeller overspeed governor. (*Pratt & Whitney Canada.*)

When speed falls to a point at which spring pressure exceeds flyweight force, the valve moves down, restoring the flow of pressure oil to the selector valve and the PCU to normal functioning. The governor also incorporates an air-bleed orifice for additional protection against propeller overspeed (8). The orifice is opened by governor flyweight force at approximately 109 percent propeller speed and bleeds compressor discharge pressure air supplied to the hydromechanical fuel control, thereby reducing fuel flow and propeller speed. The governor is equipped with a solenoid valve (4) that enables its operation to be tested. When energized, the valve opens, allowing pressure oil to alter the position of the speed reset adjuster (5). Speeder-spring compression is reduced by the upward movement of the adjuster and integral spring seat (6), which acts against the speed reset spring (7). Reduced speeder-spring force allows the flyweights to move out at a lower speed, simulating an overspeed condition. When the solenoid is deenergized, the oil supply to the adjuster bore is cut off, allowing the governor to function normally. The overspeed governor acts independently of the electronic control unit and can override it.

Cockpit Controls, Feathering Pump, and Blade Deicing

The power lever controls the propeller in the beta mode and the engine fuel system, while the fuel lever controls the propeller feathering system and the fuel on/off valves. The feathering pump unit is signaled electrically. It provides an emergency oil supply to ensure full feathering of the blades if required.

Electric blade-deicing equipment for the turbopropeller blades and spinner uses a three-phase electric current fed by a brush gear to three slip rings on the spinner backplate. Current is then transferred through the slip rings to deicer boots on the blades, and also to elements within the spinner shell. The **deicer boot** is built into the leading edge and normally provides sufficient ice protection to the inboard leading edge of the composite blade. The remainder of the blade is protected by centrifugal action only.

The turbopropeller deicing equipment is normally switched on, together with other powerplant deicing systems, when indicated temperature is below 41°F [5°C] and visible moisture is noticed.

GENERAL ELECTRIC CT7 PROPELLER CONTROL SYSTEM

The propeller control system for the CT7 engine operates in two modes: the **constant-speed mode**, for normal flight conditions; and the **beta control mode**, for taxiing and reverse thrust.

High-pressure oil for propeller pitch changes is provided by a second stage in the main lubrication pump. An alternative supply of oil for emergency feathering is available from an electrical auxiliary feathering pump.

The design of the propeller control system, illustrated in Fig. 21-45, is integrated with the engine's two-lever management system. The power lever controls engine power output during the constant-speed mode and controls propeller blade

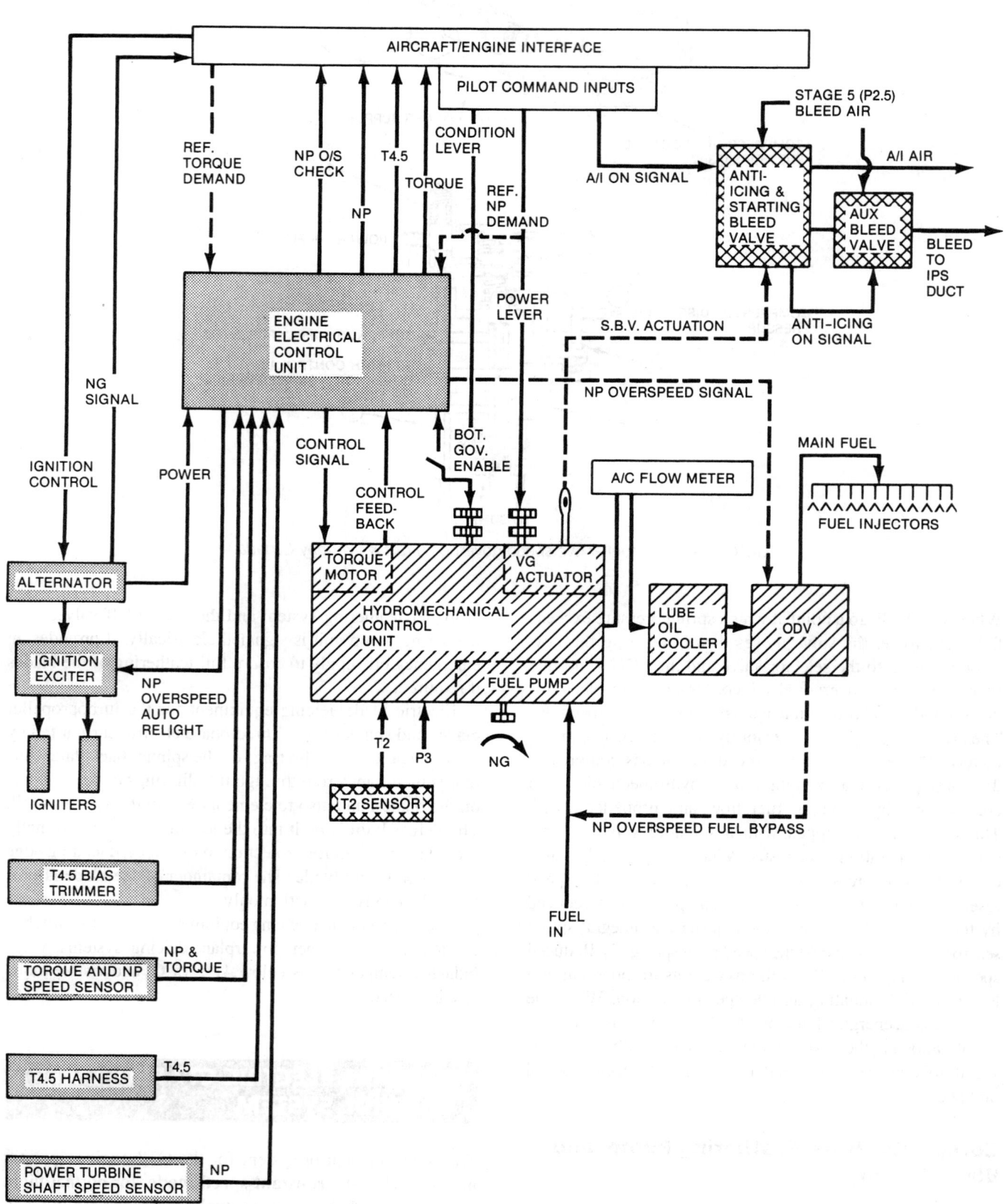

FIGURE 21-45 CT7 turbopropeller control system.

pitch in the beta control mode. Propeller rpm is selected by means of the condition lever during the constant-speed mode, and the feathering valve is controlled by the condition lever during engine shutdown. A further description of the CT7 engine control system can be found in Chap. 17.

Propeller Control

Starting

To start the engine, the power lever is set at the GROUND IDLE position and the condition lever is set at the START position. At these positions, the condition lever opens the shutoff valve in the fuel control, which allows metered fuel to flow to the engine, and the power level sets the fuel flow schedule inside the fuel control for start fuel and acceleration to ground idle. The aircraft starting system energizes the ignition circuit, and the engine automatically accelerates to ground idle. Once ground idle speed has been reached and has stabilized, the condition lever can be advanced to any position up to MAXIMUM PROP SPEED. The power lever remains in the GROUND IDLE position. Advancing the condition lever beyond unfeather actuates the feathering valve in the PCU and closes the bottoming governor enable switch located in the condition lever quadrant. The propeller will unfeather and run at a zero-thrust condition.

Once the propeller has unfeathered and reached minimum lockout speed, the bottoming governor circuit in the electronic engine control begins comparing the speed signal from the engine to a reference speed signal in the bottoming governor circuit. With the condition lever between MINIMUM PROP SPEED and MAXIMUM PROP SPEED and the power lever at GROUND IDLE, the engine control system is set for the beta control mode. Beta control mode is used to taxi the aircraft and for reverse thrust during landing, if necessary.

Taxi Operation

To taxi the aircraft, the pilot modulates the power lever between GROUND IDLE and FLIGHT IDLE. This increases propeller pitch mechanically through the PCU. As propeller pitch increases, propeller speed starts to decrease as a result of the torque load being applied to the propeller. The bottoming governor senses the drop in propeller speed and generates an error signal through the comparison of output and reference speed signals. This error signal is passed to the torque motor in the fuel control, which trims up fuel flow so as to maintain the reference propeller speed. When pitch is increased and speed is maintained, the propeller produces thrust. As the power lever is retarded back to GROUND IDLE, the pitch on the propeller is reduced and propeller speed starts to pick up. This increase in speed is sensed by the bottoming governor. The trim added by the bottoming governor will bleed off to maintain reference propeller speed.

Takeoff and Flight

After taxiing the aircraft to the runway, the pilot sets the condition lever and power lever for the constant-speed mode.

Movement of the condition lever beyond the MINIMUM PROP SPEED position will set a speed reference within the constant-speed governor from 75 percent (1038 rpm) minimum to 100 percent (1384 rpm) maximum. With the condition lever at its 100 percent position, the power lever is advanced beyond FLIGHT IDLE up to MAXIMUM POWER, as required. The power lever, as it is moved to MAXIMUM POWER, increases propeller speed equal to or beyond the condition lever constant-speed governor reference setting in the PCU. As propeller speed increases beyond the reference setting, even slightly, the PCU senses this overspeed and increases propeller pitch to maintain constant propeller speed. This results in an increase in the thrust output from the propeller. Likewise, if the power lever is moved away from maximum power, resulting in a reduction of the engine output power, propeller speed decreases. The PCU senses even slight underspeeds and schedules low pitch to maintain the constant speed set by the condition lever.

Landing and Shutdown

After landing, the condition lever is retarded to MINIMUM PROP SPEED (75 percent propeller speed). This leaves the bottoming governor enabled. The power lever is retarded to GROUND IDLE, approximately 72 percent of gas-turbine generator speed (N_g). If reverse thrust is required during landing rollout, it is achieved by retarding the power lever, which, through its connection to the PCU, creates a negative pitch change in the blades to assist in braking the aircraft. The electronic control unit, through the fuel control unit torque motor, will schedule appropriate fuel flow to maintain reference propeller speed. To taxi the aircraft, the pilot moves the power lever between GROUND IDLE and FLIGHT IDLE, which manually schedules positive pitch through the PCU.

The pilot shuts down the engine by shutting off the fuel using the condition lever, which is moved to the FUEL OFF position. However, before moving the condition lever to FUEL OFF, the pilot positions the power lever at GROUND IDLE, which sets propeller pitch for zero thrust, and positions the condition lever at FEATHER. This movement of the condition lever causes the feathering valve in the PCU to feather the propeller and opens the quadrant switch to disable the bottoming governor circuit in the electronic control unit. The engine will run at approximately 67 percent N_g with the propeller feathered. The engine must be operated at ground idle for a minimum of 2 min to stabilize engine temperatures. After the engine temperatures have stabilized, the condition lever can be moved to the FUEL OFF position and the engine will shut down.

THE ALLISON TURBOPROPELLER

The Allison (formerly Aeroproducts) A6441FN-606 and -606A turbopropeller shown in Fig. 21-46 is a hydraulically controlled four-blade unit incorporating an integral hydraulic system. The hydraulic governing system operates independently of any of the other systems and maintains precise control during all operating conditions. Electrical power is

FIGURE 21-46 The Allison A6441FN-606 turbopropeller. (*Allison Rolls-Royce.*)

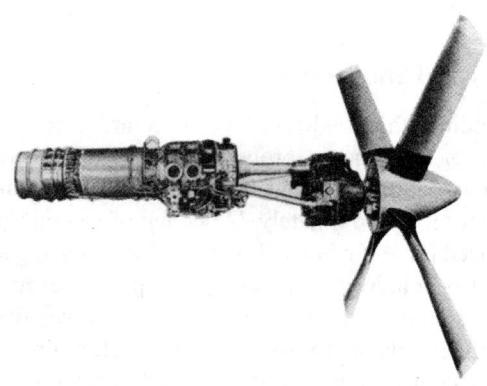

FIGURE 21-47 Turbopropeller mounted on an Allison 501-D13 engine. (*Allison Rolls-Royce.*)

supplied to the propeller for synchronization and ice control. The propeller is designed for use with the Allison Model 501-D13 turboprop (propjet) engine, as shown in Fig. 21-47. Some of the important design features of the turbopropeller are as follows:

1. Four hydraulic pumps, driven by propeller rotation, provide the required flow and pressure to maintain propeller control.

2. An electrically driven feather pump provides hydraulic pressure for static operation as well as for feathering or unfeathering.

3. Autofeathering, manual feathering, and emergency feathering systems are provided.

4. Safety devices are incorporated for protection against excessive drag in the event of engine or propeller malfunctions.

5. The propeller provides uniform variation of thrust throughout the **beta range** with power lever movement.

6. The propeller is guaranteed up to at least 4500 shp [3355.65 kW] at takeoff and 2500 shp [1864.25 kW] at maximum reverse.

7. The propeller system is designed for primary reliability and fail-safe operation.

8. The propeller can be feathered at any rotational speed up to the propeller design limitation of 142 percent of rated rpm.

9. There is no limit to the number of times a propeller can be feathered or unfeathered in flight.

10. The propeller is designed to have an increase blade-angle change rate of 15°/s and a decrease blade-angle change rate of 10°/s at 1020 propeller rpm.

11. The feather pump will provide a static blade-angle change of 3°/s.

12. When feathering is signaled during operation, the initial blade-angle change rate is well in excess of 15°/s but decreases as the rpm decreases.

Leading Particulars

Number of blades	4
Propeller diameter	13.5 ft [4.11 m]
Blade chord	18.5 in [46.99 cm]
Governing speed	1020 prop rpm
Total weight installed, including hydraulic fluid and grease	1038 lb [470.84 kg]

Aerodynamic Blade Angles at 42-in R Station

Maximum reverse	−4°
Ground idle	1.5°
Start	7°
Flight idle	20°
Feather	94.7° + 0.2, − 0.1
Beta follow-up at takeoff	31.5° + 0.5, − 1.0
Mechanical low-pitch stop	18.25 to 18.5°
Hydraulic low-pitch stop	20°

Major Assemblies and Components

The propeller consists of five major assemblies mounted on the engine propeller shaft plus the control components and assemblies mounted in the aircraft. These major assemblies are (1) hub, (2) blade and retention, (3) regulator, (4) feather reservoir, and (5) spinner. The engine-driven propeller alternator, electronic controls for synchronization and phase synchronization, ice-control accessories, and necessary switches, relays, etc., complete the installation. The major assemblies are illustrated in Fig. 21-48.

Hub Assembly

The hub assembly is the principal structural member of the propeller and consists of the **hub, torque units,** and **master gear assembly.** A cross-sectional drawing of the hub assembly is shown in Fig. 21-49. Note that each blade has a separate pitch-changing device called a **torque unit** and that the four blades are all meshed with the **master gear.** The master gear therefore ensures that all blades are at the same angle.

The hub, machined from a steel forging, provides mounting for the pitch-control mechanisms and the sockets for the retention of the blades. Splines and cone seats are machined on the inner diameter of the hub. The splines and cone seats provide the means of properly positioning the hub on the propeller shaft. The hub contains hydraulic fluid passages which deliver oil, under pressure, from the regulator to the torque units, master gear assembly, and feather reservoir. The hub is shot-peened to increase its fatigue life, and it is cadmium-plated to prevent surface corrosion.

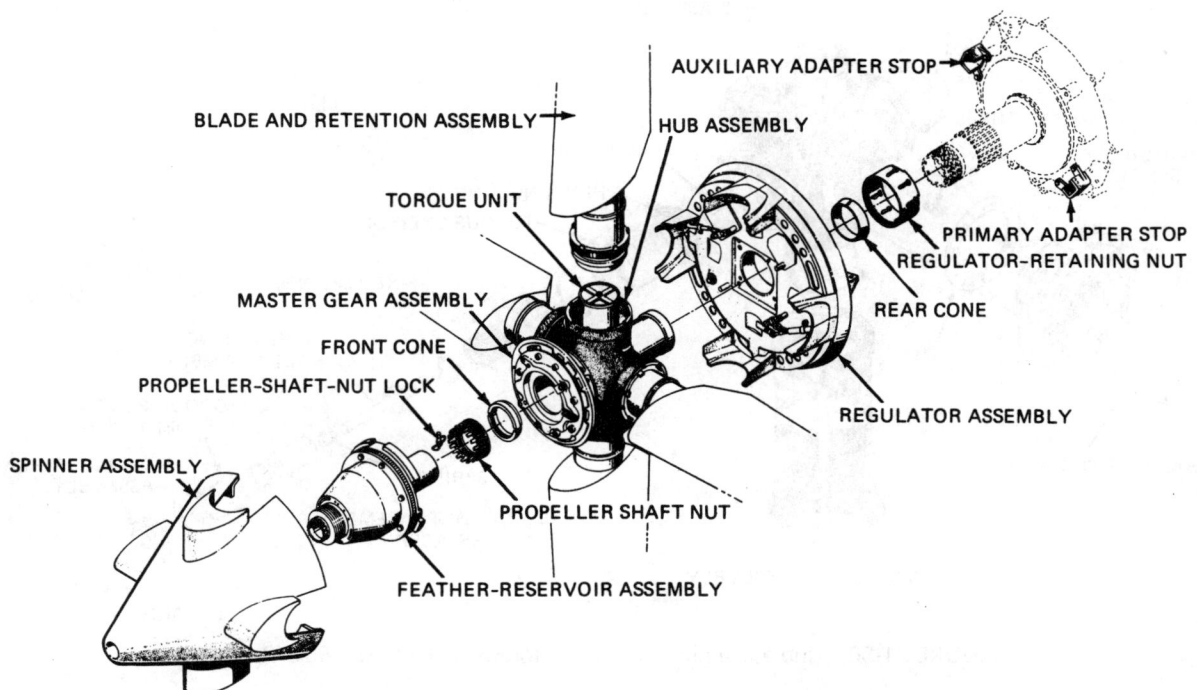

FIGURE 21-48 Major assemblies of the Allison turbopropeller. (*Allison Rolls-Royce.*)

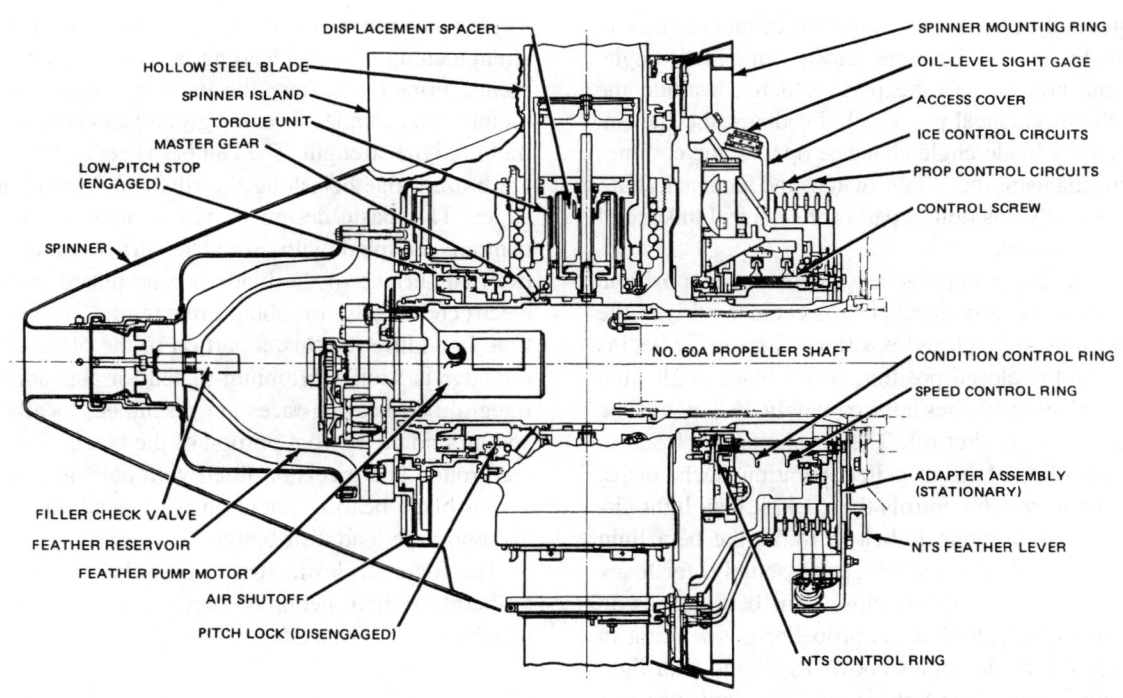

FIGURE 21-49 Cross section of the hub assembly of the Allison A6441FN-606 turbopropeller. (*Allison Rolls-Royce.*)

Each hub socket contains a **torque unit** which consists principally of a fixed spline, a piston, and a cylinder. A retaining bolt, which secures the fixed spline to the hub, incorporates a tube to provide a passage for fluid to the outboard or **decrease-pitch** side of the piston. A port in the base of the fixed spline provides an oil passage to the inboard or **increase-pitch** side of the piston. Helical splines, machined on the components of the torque unit, convert linear motion of the piston to rotation of the cylinder. Cylinder rotation is transmitted to the blade through an indexing ring and matching splines on the cylinder and in the blade root. An illustration of the hub assembly, showing the torque unit, is presented in Fig. 21-50.

The **master gear assembly** consists of a housing, master gear, mechanical low-pitch stop (MLPS), mechanical pitch lock, air shutoff, and feedback mechanism parts. The housing is splined and bolted to the front face of the hub and provides the mounting surface for the feather reservoir. The master

The Allison Turbopropeller   **645**

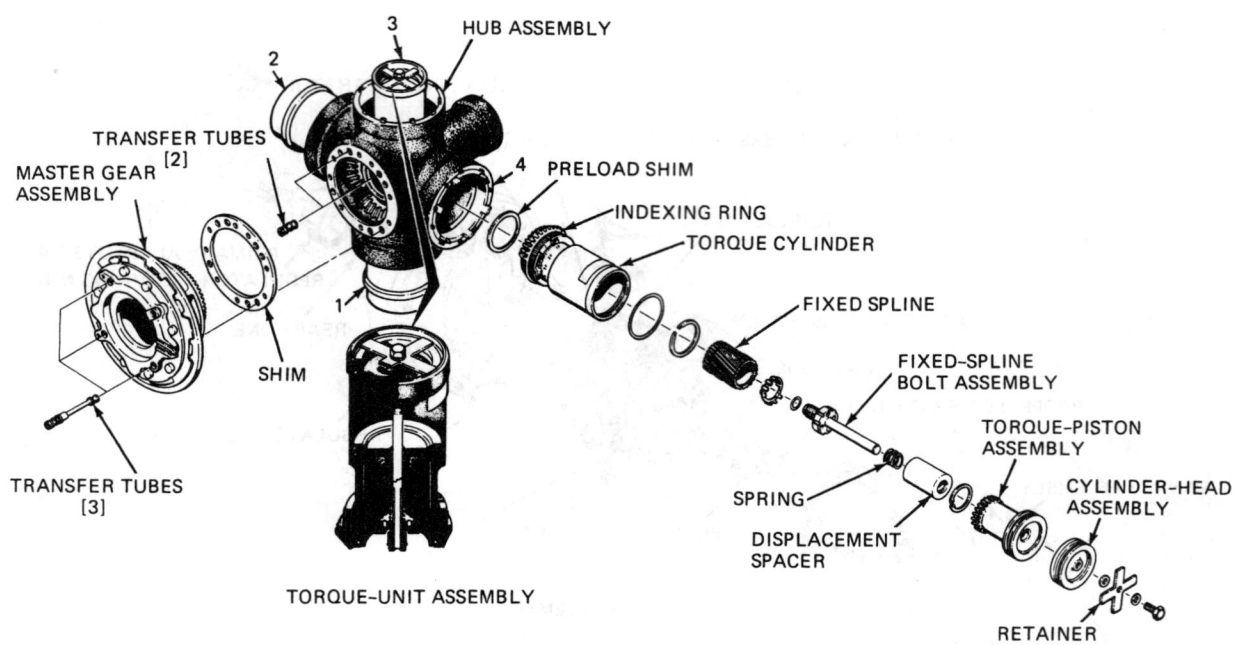

FIGURE 21-50 Hub assembly showing the torque unit. (*Allison Rolls-Royce.*)

gear coordinates the movements of the torque units so that all blades are maintained at the same aerodynamic blade angle. Splined to the housing are the parts which constitute the MLPS and the mechanical pitch lock. Feedback mechanism parts, actuated by blade-angle changes, operate the cooling-air shutoff mechanisms, beta light switch, and feedback shaft, which mechanically positions a part of the control linkage of the hydraulic governor.

As the blade angle approaches the feathered position, the master-gear-assembly feedback mechanism drives the air shutoff shutter to its closed position. The shutter begins to move toward the closed position at 60° blade angle, and when the blade angle reaches approximately 75°, all airflow through the spinner is shut off. The air shutter is closed to prevent rapid cooling of the propeller operating mechanisms.

The beta light switch controls the cockpit beta light circuit. When the blade angle is below 18.5°, the beta light switch is closed by the master-gear-assembly feedback mechanism. When the switch is closed, the beta light is on in the cockpit to indicate that the propeller is operating in the beta range. At blade angles above 18.5°, the beta light switch is open and the beta light is out. This indicates that the propeller is not operating in the beta range—that is, the governor is in control of rpm.

Blade and Retention Assemblies

Each blade and retention assembly consists of a blade cuff, deicing element, cuff deicing slip ring, and integral ball-bearing set, and is retained in its hub socket by a blade-retaining nut.

The hollow steel blade consists of a **camber sheet** and **thrust member** which are brazed together. The camber sheet, formed from sheet steel, completes the camber surface of the blade. The thrust member, machined from a steel forging, constitutes the blade shank thrust face, longitudinal strengthening ribs, and leading- and trailing-edge reinforcements. Prior to brazing, the interior surfaces of the thrust member and camber sheet are ground and polished for maximum fatigue strength. The camber sheet and thrust member are brazed together along the ribs and leading and trailing edges. This basic design provides maximum aerodynamic contours coupled with excellent structural qualities. The external surface of the blade is zinc-plated and passivated electrochemically to obtain maximum corrosion protection. In addition to this, a portion of the blade tip and leading edge is "dull" chromium-plated for abrasion resistance. Integral ball-bearing races are machined, locally hardened, and ground on the root portion of the blade. The outer races are ground and precision-fitted with balls and separators to obtain blade-bearing retention with maximum service life and optimum load distribution.

The foamed plastic structure of the cuff is covered by a fiberglass shell, neoprene fabric, and ice-control element which is flush-mounted.

Regulator Assembly

The regulator assembly, which is mounted on the rear of the hub, consists of the cover and housing assembly, pumps, valves, and hydraulic fluid required to provide controlled flow to the hub pitch-change and -control mechanisms. These parts, except for portions of the adapter assembly, are contained in the regulator cover and housing. The regulatory assembly, except for the adapter assembly, rotates with the propeller. A drawing of the regulator assembly is shown in Fig. 21-51. Observe in particular the **shoes** (speed shoe and condition shoes) which straddle the control rings mounted on the front of the engine. As the rings are moved

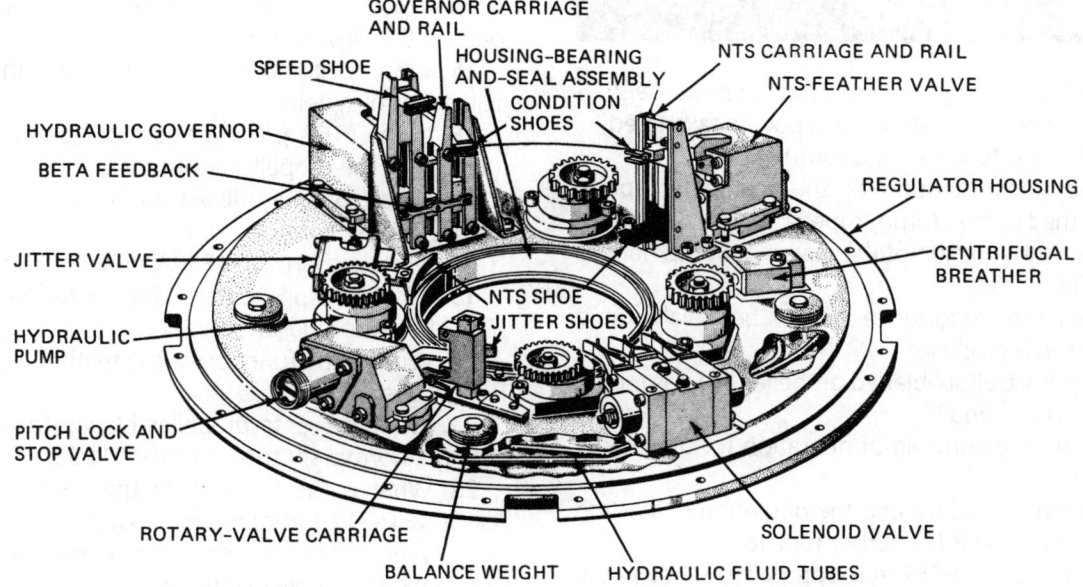

FIGURE 21-51 Propeller regulator assembly. (*Allison Rolls-Royce.*)

forward and backward in response to control from the airplane through the control screws and gears, the shoes are also moved backward and forward as they rotate around the edges of the rings. The movement of the shoes transmits signals to the components of the regulator.

When the propeller assembly is placed on the propeller shaft, two engine-mounted adapter stops engage with tangs on the adapter assembly. This prevents rotation of the adapter assembly when the propeller rotates. The adapter assembly is supported by two ball-bearing assemblies in the regulator. Seals, located in the regulator cover and housing, contact the adapter assembly to retain the hydraulic fluid.

The **adapter assembly** consists primarily of the mechanical control components which are coupled to the valves and pump power gear in the regulator and the accessory plate outside the regulator. Mechanical control movements from the engine or cockpit are transmitted to the regulator to control the hydraulic operation of the propeller. This mechanical control is accomplished by movement of the propeller condition lever or negative torque signal (NTS) feather lever.

The accessory plate, which is attached to the adapter assembly, supports the brush block assemblies, the rotary actuator, a solenoid stop, and adapter stop tangs. Two brush block assemblies are provided, one for ice control and the other for control of the solenoid valve, beta light, and feather pump. Electric power is supplied to and conducted through the brush block assemblies to the slip rings mounted on the regulator cover. Electric power is also supplied to the rotary actuator.

Feather Reservoir Assembly

The feather reservoir assembly, installed on the forward face of the hub, consists of a feather motor, pump, pressure control valve, check valve, and filler check valve mounted in a housing and cover assembly. A sufficient quantity of hydraulic fluid

is available in the feather reservoir to accomplish feathering in the event that the regulator hydraulic supply is depleted. This assembly also supplies hydraulic flow for stationary pitch-change operation and for completion of the feathering cycle. Air is admitted through the spinner nose and directed over the cover for cooling of the hydraulic fluid.

Spinner Assembly

The **spinner** is a one-piece aluminum-alloy assembly. It provides for the streamlined flow of air past the blade cuffs and over the feather reservoir, hub, and regulator. The spinner is positioned and supported by dowels on the regulator spinner mounting ring and attached to the feather reservoir by a nut in the nose of the spinner. Spray-mat-type ice-control circuits are on the external surface of the spinner. A schematic diagram of the anti-icing and deicing circuits for the spinner and cuffs is shown in Fig. 21-52.

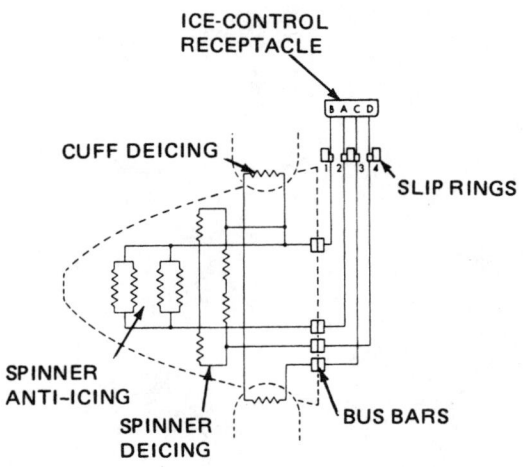

FIGURE 21-52 Schematic diagram of anti-icing and deicing circuits.

1. What is meant by the term *shaft horsepower*?
2. How is equivalent shaft horsepower calculated?
3. What is the function of a torquemeter?
4. Explain the operation of the feathering spring assembly in the Hartzell turbopropeller.
5. Discuss the function of the governor in the Hartzell turbopropeller.
6. Explain the importance of the beta valve in a Hartzell reversible propeller.
7. On the Hartzell six-bladed propeller, what is meant by the term *dual-acting*?
8. Describe the principle of operation for the Dowty R321 turbopropeller.
9. Explain the need for and the operation of starting latches on the Dowty R321 turbopropeller.
10. On the Dowty R352 turbopropeller, what is the purpose of the blade counterweights?
11. What is the function of the PCU installed on the Hamilton Model 14RF propeller?
12. What are the advantages of composite propeller blades?
13. What materials are commonly used in the construction of composite propeller blades?
14. What carries the main centrifugal and bending loads in the Dowty composite blades?
15. What cockpit controls are utilized in the PT6A propeller control systems?
16. In the PT6A propeller control system, what is the function of the propeller governor?
17. What force is utilized for blade reversing in the PT6A propeller control systems?
18. What prevents the PT6A from overspeeding?
19. What cockpit controls are utilized in the Garrett TPE331 turbopropeller operation?
20. What is the function of the feathering valve in the TPE331 system?
21. What prevents propeller blades from feathering during a normal shutdown of the TPE331?
22. What is the function of the coordinator on the Allison 250-B17 turbopropeller system?
23. What is the function of the air-bleed orifice on the Dowty R352 propeller system?
24. In what positions are the cockpit controls placed in starting up the CT7 propeller system?
25. Describe the normal engine shutdown procedures for the CT7 propeller system.

Propeller Installation, Inspection, and Maintenance 22

INTRODUCTION

Propellers are essential aircraft parts, providing the thrust necessary to move the aircraft through the air. Propellers require the highest degree of care and attention to detail in their installation, inspection, and maintenance.

The propeller installation procedures and maintenance requirements discussed in this chapter are representative of those currently in widespread use. No attempt has been made to include detailed maintenance procedures for any one particular propeller; all pressures, figures, and specifications are presented solely for the purpose of illustration and do not have specific application. *For maintenance information on a specific propeller, always refer to applicable manufacturer's instructions.*

PROPELLER INSTALLATION AND REMOVAL

Types of Hubs

The propeller is mounted on its shaft by means of several attaching parts. The types of hubs generally used to mount propellers on engine crankshafts are (1) a forged steel hub fitting a splined crankshaft, (2) a tapered forged steel hub connected to a tapered crankshaft, and (3) a hub bolted to a steel flange forged on the crankshaft.

Hubs Fitting Tapered Shafts

For some models on which the hub fits a tapered shaft, the hub is held in place by a retaining nut that screws onto the end of the shaft. A locknut safeties the retaining nut, and a puller is required for removing the propeller from the shaft. The locknut screws into the hub and bears against the retaining nut. The locknut and the retaining nut are then safetied together with either a cotter pin or a lockwire.

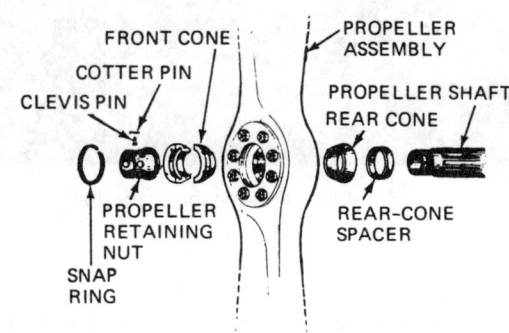

FIGURE 22-1 Installation parts for a splined-shaft propeller.

A newer design employs a snap ring instead of a locknut. When the propeller is to be removed, the retaining nut is backed off and bears against the snap ring, and the propeller is thus started from the shaft. Holes in the retaining nut and the shaft are provided for safetying.

Hubs Fitting Splined Shafts

On splined-shaft propellers, a retaining nut that screws onto the end of the shaft is used to hold a hub fitting the splined shaft, as shown in Fig. 22-1. Front and rear cones are provided to seat the propeller properly on the shaft. The **rear cone** is made of bronze and is of one-piece construction. It seats in the **rear-cone seat** of the hub. The **front cone** is a two-piece split-type steel cone. A groove around its inner surface makes it possible to fit the cone over a flange of the propeller retaining nut.

The front cone seats in the **front-cone seat** of the hub when the retaining nut is threaded into place. A snap ring is fitted into a groove in the hub forward of the front cone so that the front cone will act against the snap ring and pull the propeller from the shaft when the retaining nut is unscrewed from the propeller shaft. This snap ring must not be removed when the splined-shaft propeller is removed from its shaft, because the snap ring provides a puller for the propeller.

When a hub with a bronze bushing instead of a front cone is used, a puller may be required to start the propeller from the shaft.

FIGURE 22-2   Integral-hub flange-type crankshaft.

A **rear-cone spacer** is provided in some designs to prevent the front cone from bottoming on the forward ends of the splines. If the rear cone is too far back, the front cone will come in contact with the splines before the propeller is secure.

The principal purpose of a retaining nut is to hold the propeller firmly on its shaft. A secondary purpose, in some designs, is to function as a puller with the snap ring to aid in removing the propeller.

Integral-Hub Flange-Type Crankshaft

The integral-hub flange-type crankshaft is manufactured with the propeller mounting hub forged on the front end of the crankshaft, as shown in Fig. 22-2. The flange includes integral bushings that fit into counterbored recesses in the rear face of the propeller hub. The recesses are concentric with the bolt holes. A stubshaft on the front end of the crankshaft forward of the flange fits the propeller bore and ensures that the propeller is correctly centered.

Propeller Installation Preparation

Before installing any propeller, inspect the shaft and hub for corrosion, nicks, and other surface defects. Wipe the shaft and the inside of the hub with a clean, dry rag until they are free of dirt, grease, and other foreign substances. Small burrs or rough spots which might prevent the hub from sliding onto the shaft may be removed with a fine file or fine sandpaper. Inspect bolt holes for cleanness and thread condition, and inspect the attaching bolts for cracks and elongation.

A thin coat of light engine oil or antiseize compound is normally applied to the shaft prior to installation of the propeller.

Specific Propeller Installations, Removals, and Related Concerns

Installation of a Propeller on a Flange-Type Shaft

Flange-type shafts (see Fig. 22-2) are currently used on most opposed-type reciprocating engines. If the flanged propeller shaft has dowel pins, the propeller can be installed in only one position. In the absence of dowel pins, consult the manufacturer's maintenance manual for the proper position for installation. The propeller installation position may affect such factors as vibration, engine life, and positioning for hand propping. In the absence of dowel pins, position the propeller so that the blades are at the two o'clock and eight o'clock positions when the engine is stopped. Place the propeller on the flanged shaft with enough force to mate the recesses in the back of the propeller with the bushings on the flanged shaft. Install the propeller bolts into the holes and turn the bolts until they are finger-tight. Use a torque wrench for final tightening, and tighten in an alternating sequence so that all the bolts are pulled down evenly. Torque to the specified value, and safety the bolts.

A propeller on a flanged shaft is removed simply by unscrewing the retaining bolts and lifting the propeller from the shaft.

Installation of a Propeller on a Tapered Shaft

Before a propeller is installed on a tapered shaft, the fit of the propeller to the shaft should be checked. This may be done with Prussian blue. First, both the hub and the shaft are cleaned and all roughness is removed. Then a thin coating of Prussian blue is applied to the shaft. The propeller hub is installed, and the retaining nut is tightened to the proper torque for normal installation. The hub is then removed, and the degree of surface contact inside the hub is shown by the transfer of Prussian blue. If the contact area is 70 percent or more, the fit is satisfactory. Some authorities recommend a fit of 85 or 90 percent. If the area of contact is less than 70 percent, the fit may be improved by lapping with a fine lapping compound. When the correct fit is attained, the lapping compound must be completely removed from both the hub and the shaft, and then both surfaces should be coated with light engine oil.

The procedure used to install a propeller on a tapered shaft depends on the type of hub. If a locknut is used, lift the propeller into position. Be sure that the key on the shaft lines up with the keyway on the hub. Slide the propeller well back on the shaft. Unless there is something wrong, the hub will not bind as it slides on the tapered shaft. Screw the retaining nut onto the end of the shaft. Note that a shoulder on the retaining nut bears against a shoulder in the hub and forces the hub onto the shaft. Use the wrenches designated by the manufacturer for the final tightening, and do not apply any extra leverage.

Next, screw the locknut into the hub. Be careful in starting the nut to ensure that there is no "cross threading," because the thread on this nut is comparatively fine. Pull the locknut tight, but do not tighten it as much as the retaining nut. One of the lockwire holes must be in line with a hole in the retaining nut. Finally, use either a lockwire or a cotter pin to secure the retaining nut and the locknut.

If the propeller is designed with a snap-ring puller, simply place the propeller in the proper position on the shaft, install and tighten the retaining nut, install the snap ring, and install the safety clevis pin or bolt.

Removal of a Propeller from a Tapered Shaft

The most prevalent problem in removing a tapered-shaft propeller is that the hub often sticks to the shaft, perhaps because of overtorquing during installation or because the shaft has not been properly lubricated. If, after the retaining nut has been removed, the propeller cannot be removed from the shaft with reasonable force, it will be necessary to employ a propeller puller. Propellers that are designed with a snap ring and puller nut are easier to remove because the nut applies pulling force against the snap ring as the nut is backed off.

Removal of a Propeller from a Splined Shaft

In order to remove a propeller from a splined shaft, remove the cotter pins and clevis pin that secure the propeller retaining nut and then unscrew the propeller retaining nut. The front cone over the flange of the retaining nut presses against the snap ring in the hub and pulls the propeller away from the shaft for a short distance. When the propeller is loose, it is usually slipped off easily by hand; but if this is not possible with a reasonable amount of force, remove the snap ring, nut, and front cone. Then clean the threaded portion of the shaft and nut; lubricate the cone, nut, and shaft with clean engine oil; reassemble; and finally apply force to unscrew the nut. The rear cone and spacer are left with the engine if a new propeller is to be installed. A propeller puller is used to start the propeller from the shaft if there is a bronze bushing instead of a front cone.

Shaft and Hub Splines

The splines on the propeller shaft and inside the propeller hub should be carefully inspected for damage and wear. Wear of the splines should be checked with a single-key no-go gauge made to +0.002 in [0.05 mm] of the base drawing dimensions for spline land width. If the gauge enters more than 20 percent of the spline area, the part should be rejected.

Installation of a Propeller on a Splined Shaft

To install a propeller on a splined shaft, first install the rear-cone spacer if there is one on the assembly (see Fig. 22-1). Install the rear cone on the propeller shaft. Match the wide spline on the shaft with the wide groove in the hub, and slide the propeller well back against the rear cone. Next, assemble the front cone and the retaining nut, and screw the nut onto the propeller shaft. Before the propeller is tightened down and safetied, the front and rear centering cones must be inspected for bottoming.

Front Cones

Since hub front-cone halves are machined in pairs, the original mated halves are always used together in the same installation. If one half becomes unserviceable, both halves are rejected. Before installation and use, the two halves of a front cone are held together by a thin section of metal left over from the manufacturing process. This metal must be sawed through with a hacksaw and the two separated halves gone over carefully with a handstone to remove all rough and fine edges and to round off the sharp edges where the cones have been cut apart. After this process is completed, the two halves are always taped together when not installed.

Front-Cone Bottoming

A front cone sometimes **bottoms** against the outer ends of the propeller-shaft splines; that is, the apex of the front cone hits the ends of the splines before the cone properly seats in its cone seat in the hub. The hub is loose because it is not seated properly and held tight by the cones, even though the retaining nut may be tight.

Whenever a splined hub is found to be loose, even though the retaining nut is tight, an inspection is made for front-cone bottoming unless there is a more probable cause of the trouble. Also, this condition may be manifested by excessive propeller vibration during preflight operations.

Inspection for Front-Cone Bottoming

To check for front-cone bottoming, first apply a thin coating of Prussian blue to the apex of the front cone. Then install the propeller on the shaft and tighten the propeller retaining nut. Next, remove the retaining nut and front cone. See if the Prussian blue has been transferred to the ends of the splines of the propeller shaft. If it has not been transferred, the front cone is not bottoming and the Prussian blue can be cleaned off.

If the Prussian blue has been transferred to the ends of the shaft splines, install a steel spacer behind the rear cone to correct the condition of front-cone bottoming (see Fig. 22-3). Spacers for this purpose are generally made in any shop adjacent to the place where the work is being performed and are $\frac{1}{8}$ in [3.18 mm] thick.

The presence of the spacer moves the entire propeller assembly forward, causing the front cone to seat in the hub before its apex hits the end of the shaft spline. After the installation of the spacer, the Prussian-blue test should be made again. If bottoming is still indicated, inspect the hub-shaft end and all attaching parts for excessive wear or any other condition that might cause improper fit. Worn or defective parts should be replaced.

Rear-Cone Bottoming

Occasionally a situation will exist where the front edge of a rear cone bottoms against the ends of the splines in the propeller hub, as shown in Fig. 22-3. This condition is caused by wear of both the cone and the cone seat in the hub due to prolonged service and will prevent the cone and

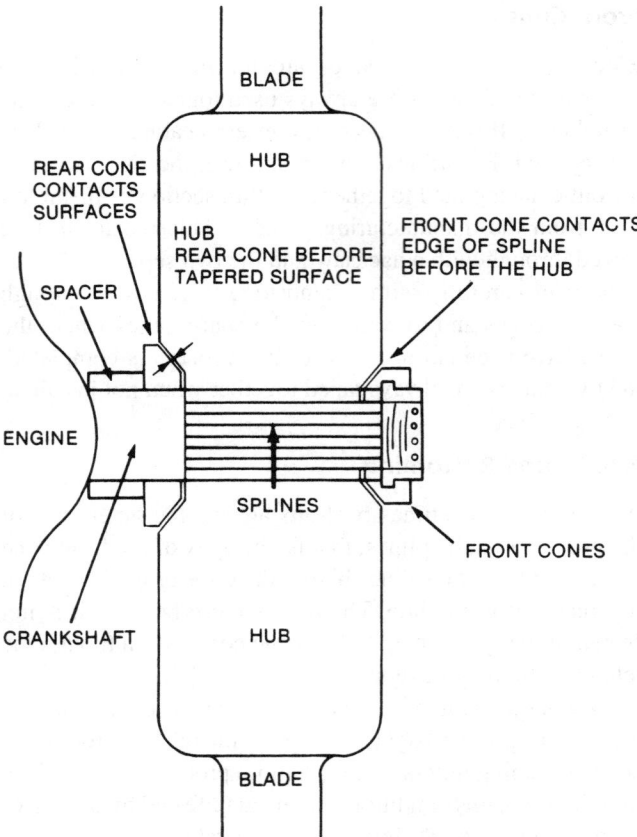

FIGURE 22-3   Cone bottoming on splined-shaft propeller.

cone seat from being firmly engaged. If inspection shows that the front of the rear cone is touching the splines, the condition can be corrected by carefully removing not more than $\frac{1}{16}$ in [1.59 mm] of material from the front edge (apex) of the cone or replacing the cone with a new one. After proper fit of the cones has been established, the propeller can be installed.

A bar about 3 ft [0.9 m] long is placed through the holes in the nut for the final tightening, as specified in the maintenance manual. It is not necessary to pound the tightening bar. The snap ring is then installed in its groove in the hub. The retaining nut is safetied with a clevis pin and a cotter pin. If the propeller retaining nuts have elongated locking holes, a washer is placed under the cotter pin. The clevis-pin head should be to the inside, and the washer and cotter pin should be to the outside.

Installation and Adjustment of Ground-Adjustable Propellers

The installation of ground-adjustable propellers follows the practices previously described for fixed-pitch propellers. The following steps may be considered typical for such an installation:

1. Make sure that the propeller being installed has been approved for the engine and aircraft on which it is being installed.

2. See that the propeller has been inspected for proper blade angle and airworthiness.

3. See that the propeller shaft and the inside of the propeller are clean and covered with a light coat of engine oil.

4. Install the rear-cone spacer (if used) and the rear cone.

5. Lift the propeller into place carefully, and slide it onto the shaft, making sure that the wide splines are aligned and that the splines are not damaged by rough handling of the propeller.

6. See that the split front cone and the retaining nut are coated with engine oil, assemble them, and install them as a unit. (This step in the procedure will vary according to the design of the retaining devices. With some propellers the front-cone halves are installed, then the retainer nut, and finally a snap ring.)

7. Tighten the retaining nut to the proper torque, as specified by the manufacturer or according to other pertinent directions. Usually a 3-ft [0.9-m] bar will enable the technician to apply adequate torque for small propeller installations.

8. Install the safety pin or other safetying device.

Adjustment of the blade angle for a ground-adjustable propeller may be done on a propeller surface table, as shown in Fig. 22-4. The propeller is mounted on a mandrel of the correct size, and the blade angle is checked with a large propeller protractor, as shown in the illustration. The blade clamps or retaining nuts are loosened so that the blades can be turned; after the correct angle is established, the blades are secured in the hub by the clamps or blade nuts. The blade angle must be checked at a specified blade station as given in the pertinent instructions.

The method for checking the blade angle when the propeller is installed on the engine is the same as that used for other propellers and is described later in this chapter.

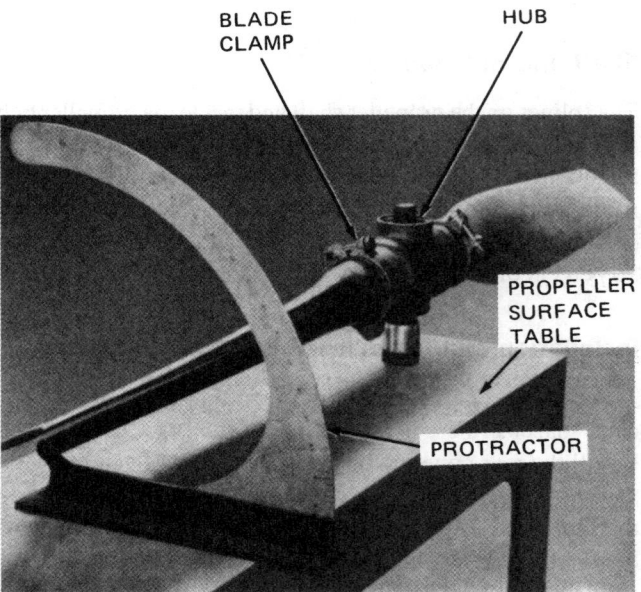

FIGURE 22-4   Adjustment of blade angle on a ground-adjustable propeller.

Removal of a Hartzell Compact Flanged Propeller

To remove a compact flanged propeller, it is usually necessary to remove the engine cowling. When the propeller is removed for overhaul, the blades should be feathered and the spinner removed.

A typical compact propeller installation is shown in Fig. 22-5. Remove the spinner nose cap by removing the attaching screws. Remove the spinner by removing the safety wire and check nut from the propeller at the forward end of the forward spinner bulkhead and also the screws that secure the spinner to the aft bulkhead. Place a drip pan under the propeller to catch oil spillage. Cut the safety wire around the propeller mounting studs, and remove the studs from the engine crankshaft flange. The nuts are frozen and pinned to the studs, so the studs should turn with the nuts. Pull the propeller from the engine shaft.

Installation of a Hartzell Compact Flanged Propeller

Clean the propeller and the engine crankshaft flange. Lubricate and install the O-ring on the engine crankshaft. Install the sleeve, spring, and thimble in the engine crankshaft, as shown in Fig. 22-5. Mount the propeller on the engine crankshaft. Screw each bolt into its mating engine flange bushing a few threads at a time until all are tight. Torque the studs to the proper specifications. Safety the bolts by inserting safety wire through the roll pins. Install the spinner, and torque the spinner screws and check nut to the proper specifications. Safety the check nut with safety wire. Charge the cylinder through the air valve with dry air or nitrogen gas to the prescribed pressure. Refer to the placard in the spinner cap or to the appropriate maintenance manual for an exact pressure value for the ambient temperature. It is very important to maintain an accurate air charge. *Note:* Do not check the pressure with the propeller in the feathered position.

Always use the amount of air pressure required for the ambient temperature, as shown by the placard or manual. If excessive pressure is used in the propeller, there is a possibility of feathering taking place at idle speed when the engine is warm and the oil is very thin. An accurate air pressure gauge is an important tool. A typical pressure check kit is shown in Fig. 22-6. **Dry air** or **nitrogen gas** should be used to recharge the propeller. It is important not to allow moisture to enter the air chamber, because this could cause the piston to freeze during cold-weather operation. A test for gas leakage may be performed by applying a soap solution or equivalent around the valve and stop adjustment nut. If the propeller is in feather on the ground, it is possible that starting the engine will cause the blades to come out of feather position. This is often an undesirable practice because of engine roughness, which will occur during unfeathering. The blades may be removed from feather without engine operation by removing the air charge, turning the blades by hand, and replacing the air charge. Install the engine cowling and spinner cap.

Installation of a McCauley Flanged Propeller

If the spinner bulkhead has been removed, position the bulkhead so that the propeller blades will emerge from the spinner with ample clearance. Install the spinner bulkhead on the

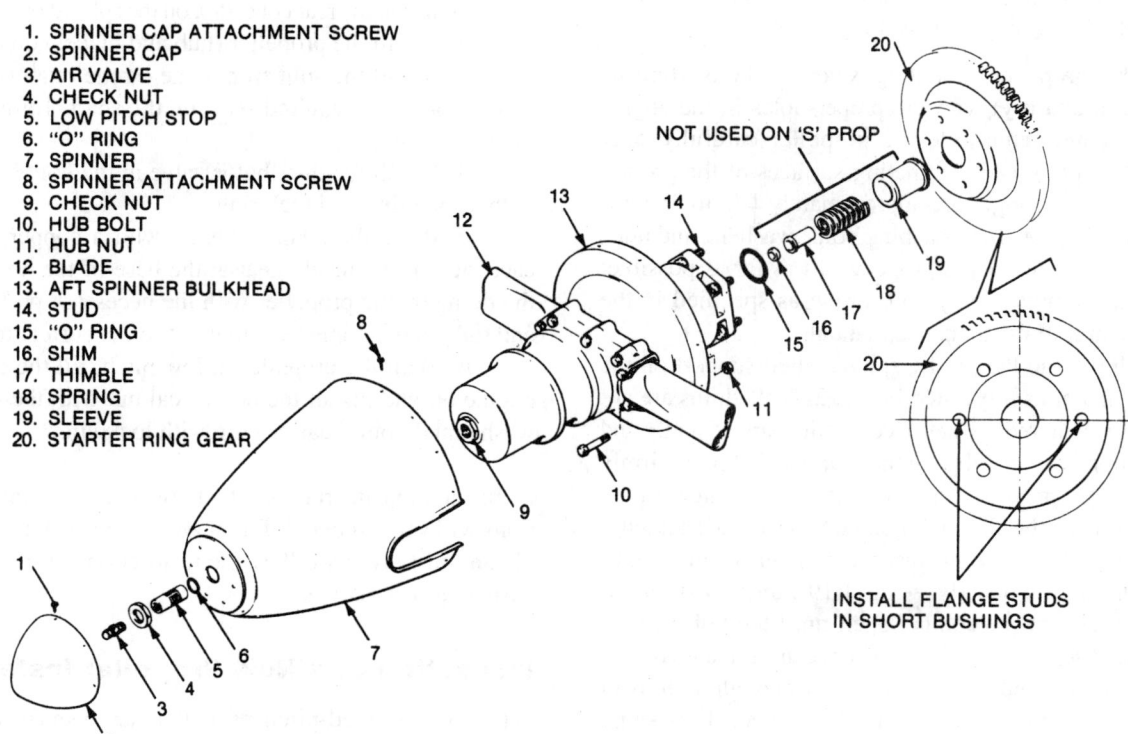

1. SPINNER CAP ATTACHMENT SCREW
2. SPINNER CAP
3. AIR VALVE
4. CHECK NUT
5. LOW PITCH STOP
6. "O" RING
7. SPINNER
8. SPINNER ATTACHMENT SCREW
9. CHECK NUT
10. HUB BOLT
11. HUB NUT
12. BLADE
13. AFT SPINNER BULKHEAD
14. STUD
15. "O" RING
16. SHIM
17. THIMBLE
18. SPRING
19. SLEEVE
20. STARTER RING GEAR

NOT USED ON 'S' PROP

INSTALL FLANGE STUDS IN SHORT BUSHINGS

FIGURE 22-5 Compact propeller installation. (*Hartzell Propeller.*)

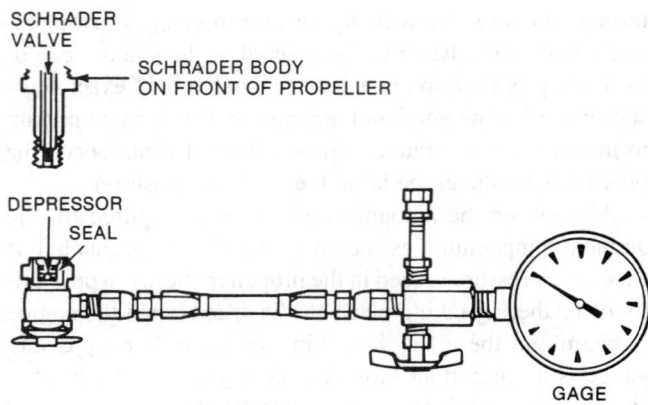

FIGURE 22-6   Air charge pressure check kit. (*Hartzell Propeller.*)

propeller or engine shaft using attaching parts (lugs, screws, etc.) as applicable. Clean the front face of the crankshaft flange, including the mounting-bolt holes, using a clean, lint-free cloth dampened in solvent. Tightly adhering dirt may be removed with fine steel wool, but be certain to wipe off any metal particles left by the steel wool. Remove any dirt and particles from the crankshaft bore. Inspect the mating flange surface and bore of the hub, and, if necessary, clean with a solvent-dampened cloth, particularly the counterbore or groove into which the O-ring packing is to be installed. Lightly lubricate a new O-ring and the crankshaft pilot with clean engine oil and install the O-ring in the propeller hub (all other surfaces should be clean and dry).

CAUTION: *The propeller must be seated against the crankshaft flange with a straight push. Rotation, cocking, or wiggling of the propeller to seat it is likely to damage the O-ring groove, and oil leakage may result.*

Align the propeller mounting studs or bolts (threads must be clean and dry) with the proper holes in the engine crankshaft flange, and slide the propeller carefully over the crankshaft pilot until the mating surfaces of the propeller and crankshaft flanges are approximately $\frac{1}{4}$ in [6.35 mm] apart. Install the propeller attaching bolts, washers, and nuts, as applicable, and work the propeller aft as far as possible; then tighten the nuts evenly and torque as specified in the appropriate aircraft maintenance manual.

Install shims and the plastic spinner shell support on the propeller cylinder as illustrated in Fig. 22-7. If shims are not mechanically centered (piloted), center the parts visually and hold them in place until the spinner support is forced firmly in place. Lightly press the shell to hold it snugly against the support and check the alignment of the holes in the shell with the holes in the bulkhead. Adjust the number of shims until the holes are approximately $\frac{3}{64}$ in [1.19 mm] out of alignment. Push hard on the spinner shell until the holes are in alignment and screws and fiber washers can be installed. The number of shims used should allow just enough alignment for you to install the screws while simultaneously pushing hard against the shell. Maintain force against the shell to hold the holes in alignment and install four screws and washers

(approximately equally spaced). Relax force and install the remaining screws and washers.

Installation of a Turbopropeller PT6A Engine (Reversing)

The turbopropeller is installed on the PT6A engine according to the instructions that follow. Place a new O-ring seal over the engine shaft. Pull the low-pitch stop collar fully forward with the puller, as shown in Fig. 22-8.

CAUTION: *To avoid damaging the propeller, make sure that the tool is not cocked. Be careful to avoid bending or otherwise damaging the spring-loaded rods and the low-pitch **stop collar** (brass ring).*

Install the propeller on the engine by inserting the two mounting studs on the propeller into the mounting holes in the drive shaft of the engine. Install the propeller mounting bolts and washers and torque to the specifications illustrated in Fig. 22-9. Secure the carbon block and arm assembly to the mounting bracket, as shown in Fig. 22-10. *Note:* To protect the low-pitch stop collar against scoring, there should be a clearance between the bottom of the collar and the head of the carbon block retaining pin, as shown in Fig. 22-10. Remove the puller and connect the propeller reversing lever to the propeller control linkage.

Installation of a Hamilton Standard Hydromatic Propeller

The Hamilton Standard hydromatic propeller is installed as follows:

1. Install the rear cone, dry, on the splined propeller shaft.
2. Install the propeller (hub and blades) on the shaft.
3. Install the split front cone on the retaining nut and install both with required seals on the shaft. Torque according to instructions.
4. Install the distributor valve in the dome and secure with a snap ring and lock ring.
5. Place the adapter flange with a copper gasket on each side of the inside gear at the base of the dome. Install the dome on the propeller with the necessary preload shims and tighten the dome retaining nut to the correct torque.
6. With the propeller in low pitch, fill the dome with engine oil and install the dome-seal nut with the dome-seal washer and dome seal. Secure with lockwire.

Installation instructions and some maintenance instructions will vary among different types of propellers. The technician must always follow the instructions provided for the particular installation.

Inspection of a New Propeller Installation

When a new fixed-pitch propeller has been installed and operated, the hub bolts should always be inspected for tightness after the first flight and after 25 h of flying.

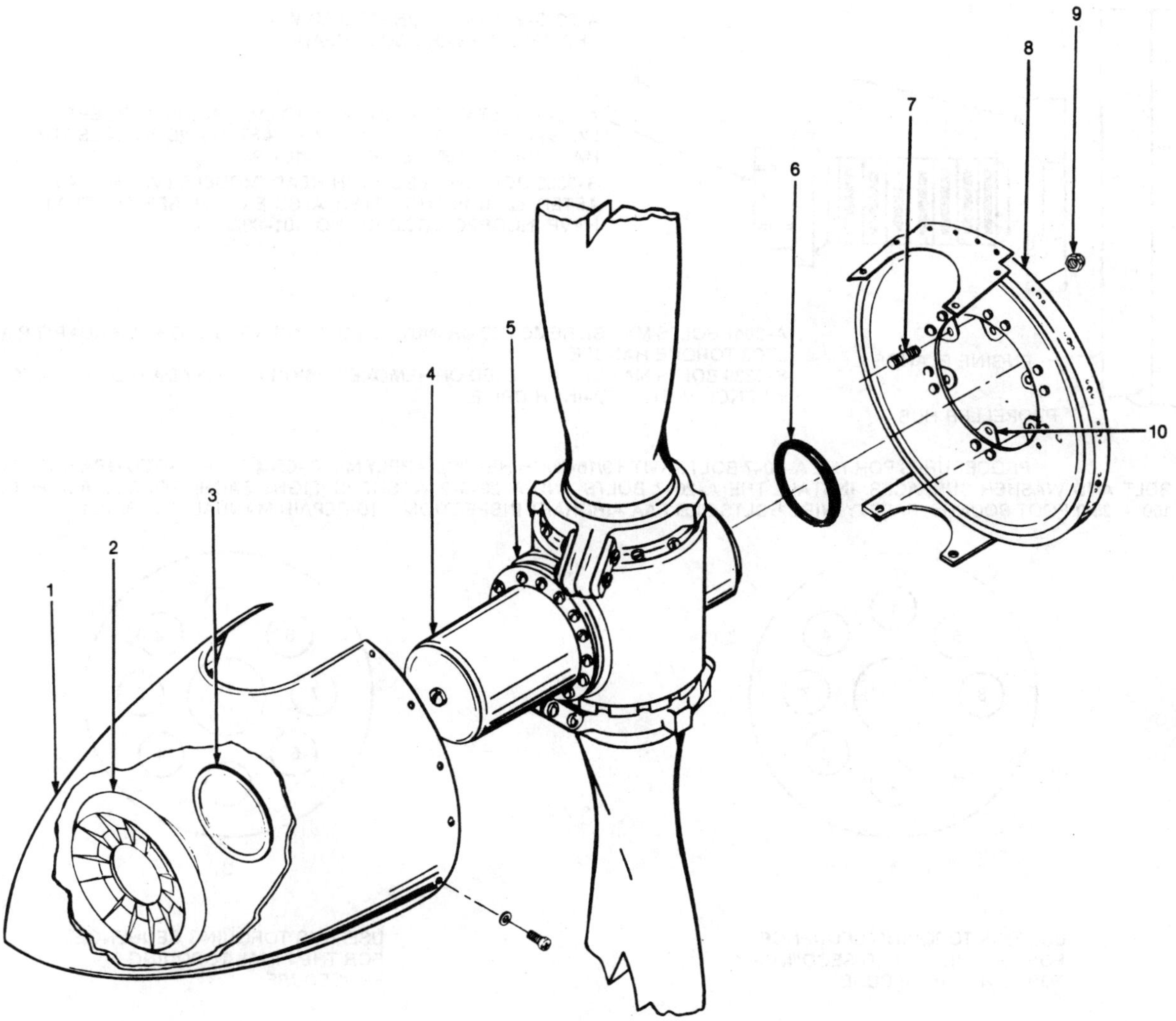

FIGURE 22-7 showing the McCauley flanged propeller installation, with the following labeled parts:

1. SPINNER
2. SUPPORT
3. SPACER
4. PROPELLER CYLINDER
5. PROPELLER HUB
6. O-RING SEAL
7. STUD
8. SPINNER BULKHEAD
9. NUT
10. LUG

FIGURE 22-7 McCauley flanged propeller installation.

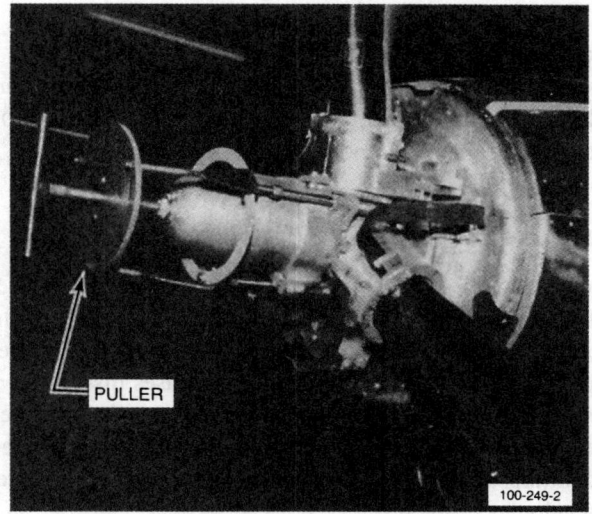

PULLER

100-249-2

FIGURE 22-8 Puller installed on PT6A propeller. (*Beech Aircraft Corp.*)

Thereafter, the bolts should be inspected and checked for tightness at least every 50 h of operation unless otherwise specified in the appropriate maintenance manual.

No definite time interval between inspections can be specified for wood propellers, since bolt tightness is affected by changes in the wood caused by the moisture content in the air where the airplane is flown and stored. During wet weather, some moisture is apt to enter the propeller wood through the drilled holes in the hub. The wood swells, but since expansion is limited by the bolts extending between the two flanges, some of the wood fibers are crushed. Later, when the propeller dries out during dry weather, a certain amount of propeller hub shrinkage takes place and the wood no longer completely fills the space between the two hub flanges. Accordingly, the shrinkage of the wood also results in loose hub bolts.

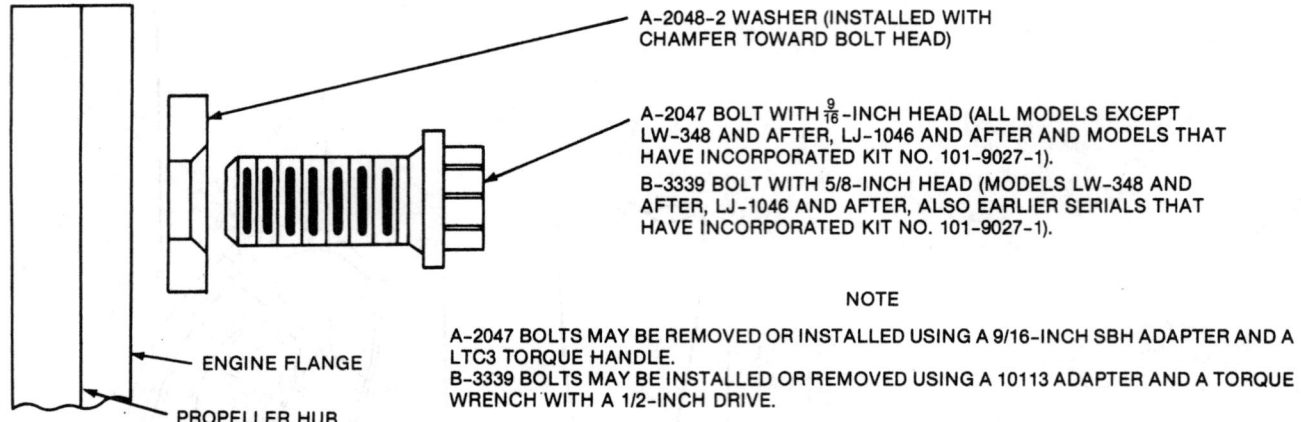

A-2048-2 WASHER (INSTALLED WITH CHAMFER TOWARD BOLT HEAD)

A-2047 BOLT WITH $\frac{9}{16}$-INCH HEAD (ALL MODELS EXCEPT LW-348 AND AFTER, LJ-1046 AND AFTER AND MODELS THAT HAVE INCORPORATED KIT NO. 101-9027-1).

B-3339 BOLT WITH 5/8-INCH HEAD (MODELS LW-348 AND AFTER, LJ-1046 AND AFTER, ALSO EARLIER SERIALS THAT HAVE INCORPORATED KIT NO. 101-9027-1).

NOTE

A-2047 BOLTS MAY BE REMOVED OR INSTALLED USING A 9/16-INCH SBH ADAPTER AND A LTC3 TORQUE HANDLE.
B-3339 BOLTS MAY BE INSTALLED OR REMOVED USING A 10113 ADAPTER AND A TORQUE WRENCH WITH A 1/2-INCH DRIVE.

ENGINE FLANGE

PROPELLER HUB

PROCEDURES FOR THE A-2047 BOLTS (WITH 9/16-INCH HEADS): APPLY MIL-T-5544 PETROLATED GRAPHITE TO THE BOLT AND WASHER SURFACES. INSTALL THE A-2047 BOLTS AND A-2048-2 WASHERS (EIGHT EACH). TORQUE ALL BOLTS TO 100 + 25 0 FOOT POUNDS. SAFETY WIRE BOLTS PER FAA AIRCRAFT INSPECTION AND REPAIR MANUAL (AC 43.13-1).

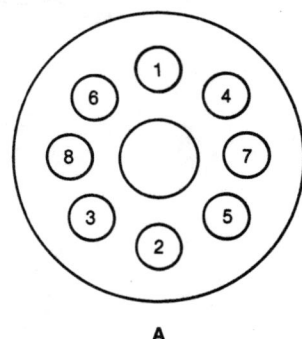

A

USE THIS TORQUING SEQUENCE FOR THE INITIAL AND SECONDARY TORQUING PROCEDURE

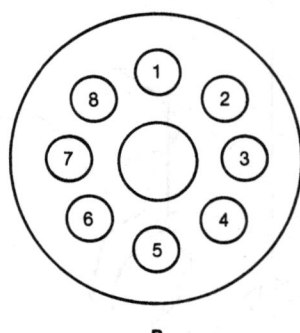

B

USE THIS TORQUING SEQUENCE FOR THE FINAL TORQUING PROCEDURE

PROCEDURES FOR THE B-3339 BOLTS (WITH 5/8-INCH HEADS): APPLY MIL-T-5544 PETROLATED GRAPHITE TO THE BOLT AND WASHER SURFACES. INSTALL THE B-3339 BOLTS AND A-2048-2 WASHERS (EIGHT EACH). TORQUE ALL BOLTS TO 40 FOOT-POUNDS, THEN TO 80 FOOT-POUNDS. USE TORQUE SEQUENCE "A" FOR THE INITIAL AND SECONDARY TORQUING PROCEDURE. FINAL TORQUE ALL BOLTS TO 100 + 5 0 FOOT-POUNDS. USE TORQUE SEQUENCE "B" FOR THE FINAL TORQUE. SAFETY WIRE ALL BOLTS PER FAA AIRCRAFT INSPECTION AND REPAIR MANUAL (AC 43.13-1).

FIGURE 22-9   Torque procedure for PT6A propeller installation (for training purposes only). (*Beech Aircraft Corp.*)

AIRCRAFT VIBRATIONS

Most aircraft vibrations are caused by rotating elements that are out of balance, and by aerodynamic forces in propellers. These vibrations may be classed as "**correctable**" and "**uncorrectable.**"

Correctable vibrations relate primarily to propeller track and balance and to the balance of shafting, accessories, etc. Track and balance disturbances are generally at the one-per-rev (once-per-revolution) rate of the propeller.

Uncorrectable vibrations are vibrations that are inherent in the aircraft, such as "*n*-per-rev" vibrations (where n = number of blades) and vibrations that are harmonically related to propeller rates. These are generally aerodynamically induced force inputs at rates that excite natural resonances in blades, airframes, mounts, etc. They are truly characteristic of the aircraft components and are changeable only by changes in design parameters by the factory. Certain worn or loose parts will aggravate these vibrations, and correction of such problems may help, but will not cure, inherent vibrations.

Purpose of Checking Track and Balance

The purpose of propeller balancing is to reduce the one-per-rev vibration induced by the out-of-balance propeller. This vibration is transmitted to the airframe to a varying degree, depending on the engine mounts, and causes discomfort to crew and passengers. It also causes shortened component life and other problems, such as premature failures in lines and fittings, cracks in structure, bearing and seal failures, and avionics problems.

The purpose of checking propeller **track** is to ensure that all blades of a propeller are flying in the same plane of rotation.

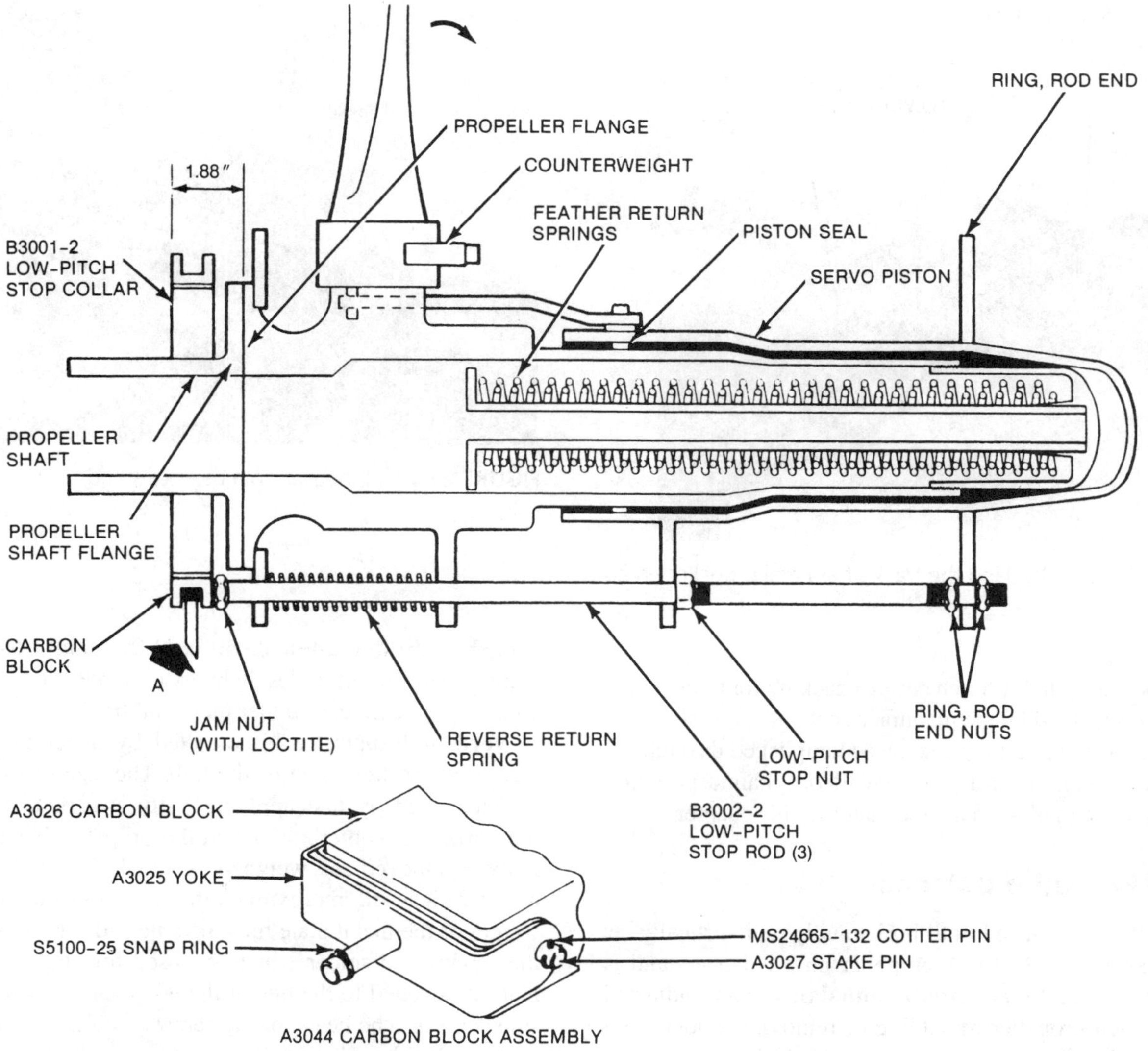

RING, ROD END

PROPELLER FLANGE

COUNTERWEIGHT

FEATHER RETURN SPRINGS

PISTON SEAL

SERVO PISTON

1.88″

B3001-2 LOW-PITCH STOP COLLAR

PROPELLER SHAFT

PROPELLER SHAFT FLANGE

CARBON BLOCK

A

JAM NUT (WITH LOCTITE)

REVERSE RETURN SPRING

LOW-PITCH STOP NUT

B3002-2 LOW-PITCH STOP ROD (3)

RING, ROD END NUTS

A3026 CARBON BLOCK

A3025 YOKE

S5100-25 SNAP RING

MS24665-132 COTTER PIN

A3027 STAKE PIN

A3044 CARBON BLOCK ASSEMBLY

VIEW A

FIGURE 22-10 PT6A propeller installation. (*Beech Aircraft Corp.*)

Track may change with power if pitch angle, blade contour, and stiffness are not exactly matched. Changing track may make it appear that balance is changing, but it is really caused by the "lift" of one blade that is different from the others. Vibration from the out-of-track condition, which changes with power and blade-angle settings, interacts with vibration from the out-of-balance condition, making it extremely difficult or impossible to achieve acceptable balance under all flight regimes.

Propeller Track

For the **track** of a propeller to be correct, corresponding points on the two blades must lie in the same plane, perpendicular to the axis of rotation. Checking the track of either a wood propeller or a metal propeller can be done on a propeller surface table, as shown in Fig. 22-11. Each corresponding point on each blade should have the same height from the table surface as indicated by the height gauge.

FIGURE 22-11 Checking propeller track on a surface table.

When a propeller is installed on an airplane engine on the aircraft, the track can be checked by rotating the tip of the propeller past a fixed reference point attached to the aircraft. This method is shown for a metal propeller in Fig. 22-12. The track of one blade should normally be within $\frac{1}{16}$ in [1.59 mm] of the other blade. Constant-speed and controllable propellers

CHECK DISTANCE HERE
FOR BOTH BLADES

REFERENCE BAR ATTACHED
TO WING STRUTS

FIGURE 22-12 Checking the track of a metal propeller on an airplane.

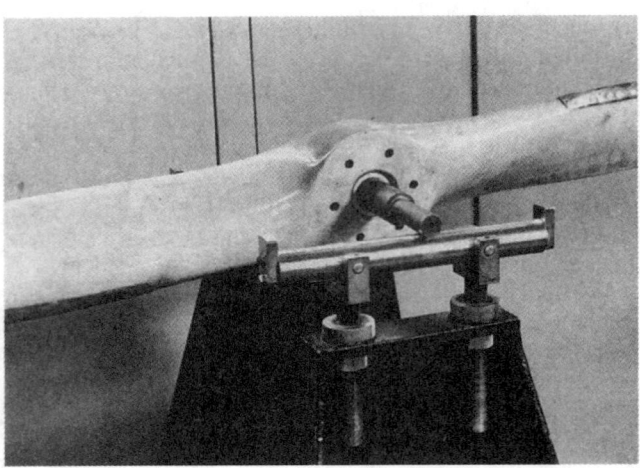

FIGURE 22-14 Propeller on a balance stand.

should be placed in low pitch before checking for track unless otherwise specified by the manufacturer.

Propeller tracking may also be accomplished through the use of a strobelight in a process called dynamic propeller tracking, which will be discussed later in this chapter.

Static Propeller Balance

A propeller is balanced both horizontally and vertically, as illustrated in Fig. 22-13. A propeller on a balance stand is shown in Fig. 22-14. **Horizontal imbalance** can be adjusted on a wooden propeller by adding or removing solder from the blade tips. When balance is achieved, the solder must blend in with the contour of the tip. The metal at the tip is

vented by drilling a few 0.040-in [1.016-mm] holes at the extreme tip. These holes help to eliminate any moisture which might condense under the metal tipping.

Vertical imbalance is corrected by attaching a metal weight to the light side of the hub. The size of the weight is determined by first applying putty at a point 90° from the horizontal centerline. When the propeller balances, the putty is removed and weighed. A metal plate is then cut to a size which will approximate the weight of the putty. The weight of the metal plate must be adjusted for the weight of the screws and solder which are used for attachment. The plate is attached to the hub at the 90° location with counter sunk screws. The heads of the screws are soldered and then smoothed with a file. Finally, varnish is applied to match the finish of the rest of the propeller.

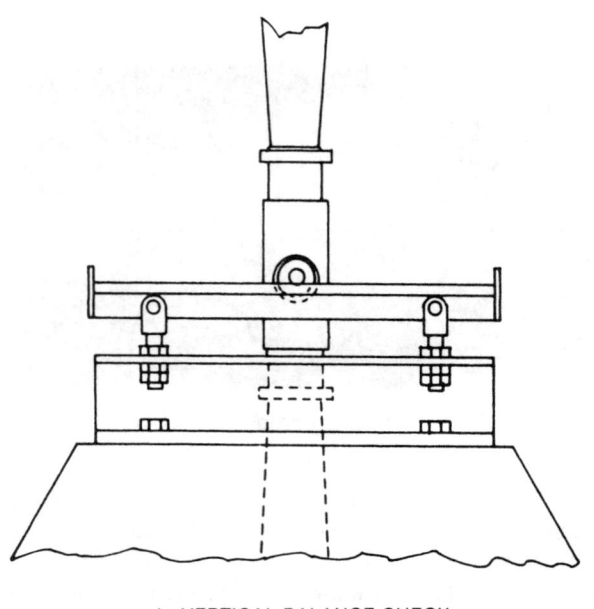

A. VERTICAL BALANCE CHECK

B. HORIZONTAL BALANCE CHECK

FIGURE 22-13 Positions of two-bladed propeller during balance check.

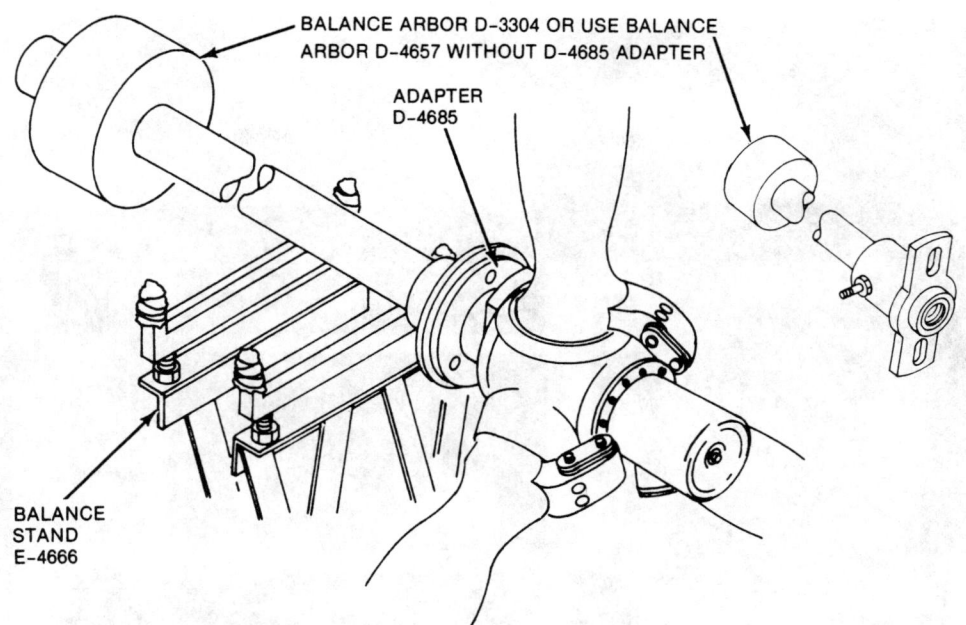

FIGURE 22-15   Balancing a three-bladed propeller.

Balancing of Controllable Propellers

Upon completion of repairs, the horizontal and vertical balance of a propeller must be checked. If any unbalanced condition is found, correction must be made according to the manufacturer's instructions. Balancing methods include the installation of weights in the shanks of the blades, packing of lead wool into holes drilled in the ends of the blades, packing of lead into hollow bolts, and others. In any event, the manufacturer's recommendations must be followed for any specific type of propeller. For some propellers, only the manufacturer is permitted to perform the balancing operations.

To balance a three-bladed constant-speed propeller, attach the propeller to a balance arbor and place the propeller and arbor on the balance stand, as shown in Fig. 22-15. Place each of the three blades alternately in the horizontal position. Locate the blade that has the greatest tendency to move up from horizontal; this is the lightest blade.

The propeller is balanced when all three blades stay in the horizontal position without any tendency to move up or down.

Dynamic Propeller Balance

If a rotating disk is perfectly balanced, no vibration will be passed on to the supporting structure. If a weight is added to the edge of the disk, the support will be forced up and down once per revolution as the disk rotates, generating a one-per-rev vibration (see Fig. 22-16). Even when the propeller is properly statically balanced and the engine is in perfect condition, tolerances allow the propeller/engine combination to produce a one-per-rev vibration. This residual imbalance can produce significant vibration energy that can stress engine mounts, seals, accessory brackets, etc. Electronic vibration-measuring

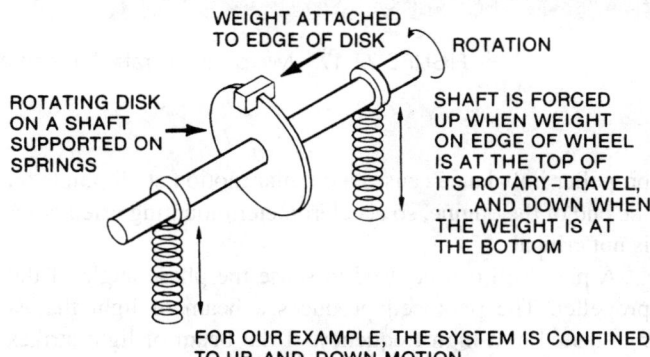

FIGURE 22-16   Up-and-down motion caused by an out-of-balance condition. (*Chadwick-Helmuth Co.*)

equipment can be used to sense this out-of-balance condition with the engine running and provide information on where and in what amount to add trim weights to eliminate the vibration.

A vibration-measuring system consists of a vibration transducer for converting the vibration into a measurable electrical signal; a photocell, magnetic pickup, or strobelight for sensing the angular position of the propeller; and an electronic instrument for filtering, measurement, and readout of the amplitude and phase of the vibration signal. The Micro-Vib aircraft vibration analyzer/balancer is shown in Fig. 22-17.

The vibration transducer most commonly used is a piezoelectric accelerometer. This type of sensor uses a crystal material that produces an electrical charge in proportion to the force applied to it. This type of sensor is sensitive to vibration in one axis only. For propeller balance, the sensor is mounted on the engine or gearbox as close as possible to the propeller, as shown in Fig. 22-18. The vibration produced by

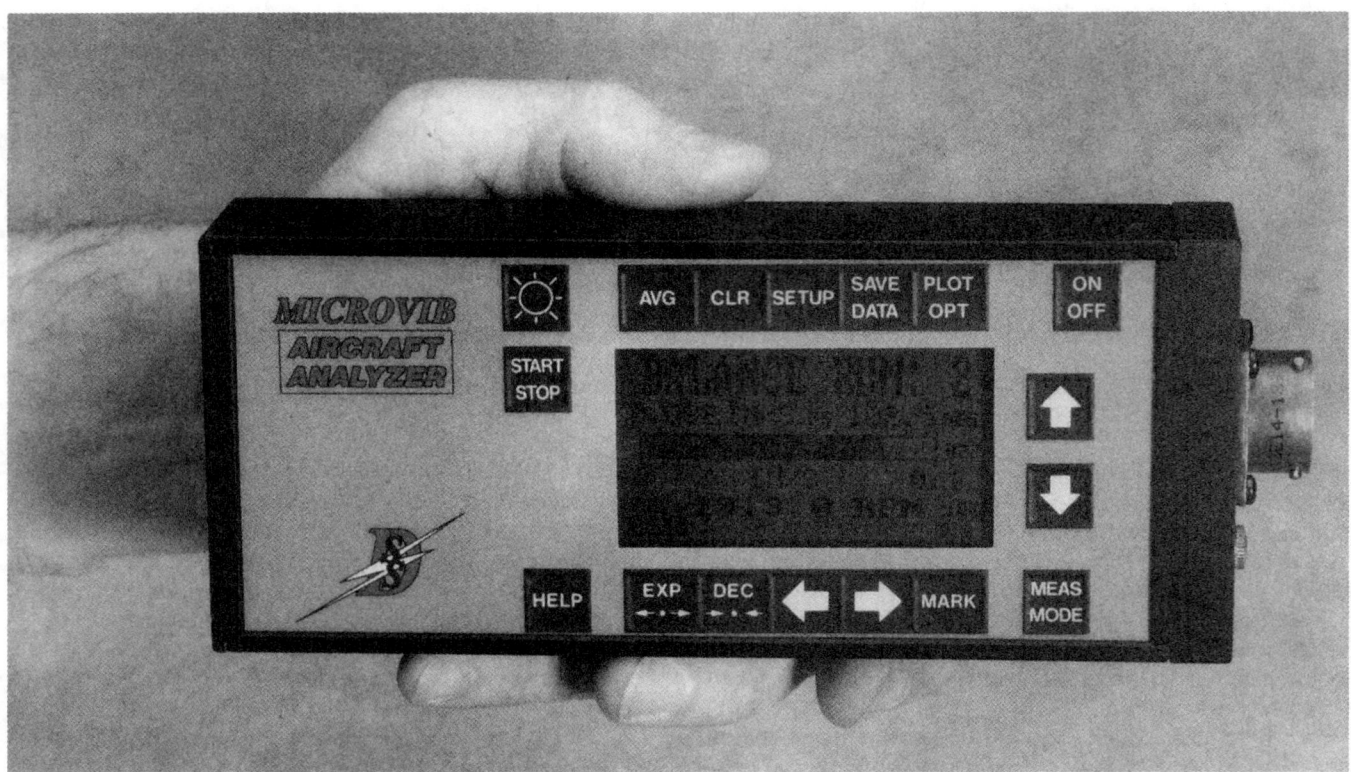

FIGURE 22-17 Micro-Vib aircraft vibration analyzer/balancer. (*Dynamic Solutions Systems, Inc.*)

propeller imbalance causes a circular motion at all points on the end of the engine, so accelerometer mounting orientation is not critical.

A photocell can be used to sense the phase angle of the propeller. The photocell produces a beam of light that is modulated at a high frequency. If the beam of light strikes a retroreflector, the return beam is sensed by the photocell and a signal pulse is generated. By attaching the photocell (photo-tach) to the engine or engine cowling (see Fig. 22-19), and placing a patch of retroreflective tape on a propeller blade, a simple phase detector is installed. The high-frequency modulation of the light beam allows the photocell to ignore other sources of ambient light. Common photocells can sense a reflector at distances from 2 to 40 in [5.08 to 101.6 cm], and universal mounting brackets provide easy alignment, making photocells easy to use.

A magnetic pickup can also be used to sense the phase angle of the propeller. A magnetic pickup consists of a bar magnet wrapped with a coil of wire. The housing of the pickup nearly completes the magnetic circuit, leaving a gap at the end of the pickup. Whenever a ferromagnetic object is introduced into this gap, the magnetic flux increases dramatically. This change in magnetic flux induces a voltage in the coil that is proportional to the rate of change of the magnetic flux. Magnetic pickups are inexpensive and very rugged, but they require a small, accurately controlled gap between the end of the pickup and the "interrupter" (blade of steel) that passes by the pickup. This controlled gap requires that special brackets and interrupters be produced for each type of aircraft.

A strobelight can also be used to measure the vibration phase angle. After the vibration signal is filtered, a zero-crossing detector circuit is used to create a pulse and trigger the strobelight. A piece of retroreflective tape is placed on the root of one of the propeller blades. Directing the strobe at the propeller will make the retroreflective tape appear to have stopped. The phase angle can be estimated visually by the user. This method has several disadvantages. It usually requires two people, one to operate the aircraft and one to stand in front of the prop and operate the strobe. Also, there is no way to average the phase angle data except by human skill level. In addition, a strobe requires a lot of power, which is usually drawn from the aircraft battery. Attachment of wires to aircraft power systems takes time and can be dangerous.

The electronic instrument that completes the dynamic balancing system must include at minimum a tuneable filter circuit for removing all but the one-per-rev signal produced by the propeller, an accurate amplitude-measuring circuit and display (similar to a voltmeter), and a phase-measuring circuit for measurement and display of the phase angle between the photocell or mag pickup signal and the accelerometer signal. The balancing process is accomplished during a ground run-up with the cables and balancer positioned as shown in Figs. 22-20 and 22-21.

Data Averaging

With older balancers, the raw amplitude and phase data were presented to the user. The user was left to manually tune the filter and watch the needle on a meter waver back and forth.

FIGURE 22-18 Accelerometer installation for balancing. (*Dynamic Solutions Systems, Inc.*)

Digital balancers can automatically tune the filter and mathematically compute the average of several readings. This method can provide repeatable, accurate measurements of data that were difficult to interpret on the older equipment. This allows dynamic balancing to much lower levels to be done with ease.

Computing a Balance Solution

Determining how much weight to add and where to add it based on the amplitude and phase angle data was commonly done with older balancers using nomographs often called "charts." The amplitude and phase readings were manually plotted on a graph of concentric circles and radial lines. An overlay of lines could be followed out to the recommended solution weight and location. More modern balancing systems compute the solution using digital computations. These balancers eliminate the need to manually plot the data on, and to read, these "charts." These "computer balancers" display the amount of weight to add and the location in degrees.

PHOTO-TACH

FIGURE 22-19 Photo-tach installation for balancing. (*Dynamic Solutions Systems, Inc.*)

Attaching Dynamic Balance Weights

Some aircraft have locations designed to accept dynamic balance weights. Many aircraft have no built-in provision for attaching these weights. Several propeller and airframe manufacturers have authorized drilling of the spinner backing plate for attachment of standard aircraft screws, nuts, and washers. The hole should be drilled slightly undersize and reamed to final size. The hole should be deburred with emery cloth to reduce the possibility of a stress concentration. The maximum amount of weight to be added at one location is strictly limited by manufacturer standards. When a stack of washers is used, at least one should be placed on either side of the bulkhead to reduce stress.

Balancers That Learn

The latest technology in computer balancers is a balancer that continuously learns during the balancing process. Nonlinear response to weight addition is commonly found in many types of aircraft. Only by learning from each step in the balancing process can the balancer reduce vibrations to a minimum in the fewest possible moves. Earlier versions of "computer balancers" learn the response of the aircraft on the first trial run only, and assume thereafter that the response is linear. This can

FIGURE 22-20   Cable and transducer installation for balancing. (*Dynamic Solutions Systems, Inc.*)

lead to a phenomenon referred to as the "Black Hole," wherein vibration apparently cannot be reduced below a moderate level.

Dynamic Propeller Tracking

A strobelight can be used to observe the visual track of the blade tips on a propeller. Small strips of retroreflective tape must be placed securely and accurately at the tip of each blade. The shape, color, or orientation of each target should be varied for identification. If the strobe has an internal oscillator, it can be set to the blade rate while the engine is running. If a mag pickup or photocell signal is available, this can be used as a synchronizing pulse if the strobe has a "sync oscillator" (or "locking oscillator").

A propeller that is out of track will usually induce a one-per-rev vibration similar to the vibration from an out-of-balance prop. This vibration will increase with thrust power (pitch), whereas an out-of-balance vibration will remain constant with increasing thrust. An out-of-track prop can also induce an axial vibration that can be sensed by reorienting the vibration sensor axis parallel with the prop shaft.

Engine Vibration Analysis

Modern computer balancers usually can provide an accurate "spectrum analysis" of the vibration of the aircraft engine. This can be done during the balancing process and requires only a few extra seconds. The engine vibration is a good indicator of the "health" of the engine. By comparing the vibrations of several engines of the same type, it is easy to see when an engine has a problem. One computer balancer maker (Dynamic Solutions Systems, Inc.) is building a large database of common engine types. These data will be shared with the aircraft industry to help improve engine reliability and flight safety.

FIGURE 22-21 Collection of balancing data during ground run-up. (*Dynamic Solutions Systems, Inc.*)

MAINTENANCE AND REPAIR OF PROPELLERS

General Nature of Propeller Repairs

When objects, such as stones, dirt, birds, etc., strike against the propeller blades and hub during flight or during takeoff and landing, they may cause bends, cuts, scars, nicks, scratches, or other defects in the blades or hub. If a defect is not repaired, local stresses are established which may cause a crack to develop, resulting eventually in the failure of the propeller or hub.

For this reason, propellers are carefully examined at frequent intervals, and any defects that are discovered are repaired immediately according to methods and procedures that will not further damage the propeller.

The terminology of propeller inspection, maintenance, and repair is very precise. Repairs and alterations are rigidly classified and assigned to certain types of repair agencies. After the work is assigned to the correct individuals or organizations, the propeller must be carefully cleaned before work is performed on it. Then the necessary inspections, repairs, alterations, and maintenance procedures may be carried out.

Authorized Repairs and Alterations

The technician contemplating repair, overhaul, or alteration should be thoroughly familiar with the approved practices and regulations governing the operation which he or she expects to perform. All repairs and alterations of propellers must be performed in accordance with the regulations set forth in Federal Aviation Regulations (FAR) and the pertinent manufacturers' manuals.

Repairs and alterations on propellers are divided into four main categories: (1) major alterations, (2) minor alterations, (3) major repairs, and (4) minor repairs.

A **major alteration** is an alteration which may cause an appreciable change in weight, balance, strength, performance, or other qualities affecting the airworthiness of a propeller. Any alteration which is not made in accordance with accepted practices or cannot be performed by means of elementary operations is also a major alteration.

A **minor alteration** is any alteration not classified as a major alteration.

A **major repair** is any repair which may adversely affect any of the qualities noted in the definition of a major alteration.

A **minor repair** is any repair other than a major repair.

Classification of Repairs and Alterations

Changes such as those in the following list are classified as major alterations unless they have been authorized in the propeller specifications issued by the Federal Aviation Administration (FAA).

1. Changes in blade design
2. Changes in hub design
3. Changes in governor or control design
4. Installation of a governor or feathering system
5. Installation of a propeller deicing system
6. Installation of parts not approved for the propeller
7. Any change in the design of a propeller or its controls

Changes classified as minor alterations are those similar to the types listed below.

1. Initial installation of a propeller spinner
2. Relocation of changes in the basic design of brackets or braces of the propeller controls
3. Changes in the basic design of propeller control rods or cables

Repairs of the types listed below are classified as propeller major repairs, since they may adversely affect the airworthiness of the propeller if they are neglected or improperly performed.

1. Any repairing or straightening of steel blades
2. Repairing or machining of steel hubs
3. Shortening of blades
4. Retipping of wood propellers
5. Replacement of outer laminations on fixed-pitch wood propellers

6. Inlay work on wood propellers
7. All repairs of composition blades
8. Replacement of tip fabric
9. Repair of elongated bolt holes in the hubs of fixed-pitch wood propellers
10. Replacement of plastic covering
11. Repair of propeller governors
12. Repair of balance propellers of rotorcraft
13. Overhaul of controllable-pitch propellers
14. Repairs involving deep dents, cuts, scars, nicks, etc., and straightening of aluminum blades
15. Repair or replacement of internal elements of blades

Propeller repairs such as those listed below are classified as propeller minor repairs.

1. Repairs of dents, cuts, scars, scratches, nicks, and leading-edge pitting of aluminum blades if the repair does not materially affect the strength, weight, balance, or performance of the propeller
2. Repairs of dents, cuts, scratches, nicks, and small cracks parallel to the grain of wood blades
3. Removal and installation of propellers
4. The assembly and disassembly of propellers to the extent necessary to permit (a) assembly of propellers partially disassembled for shipment and not requiring the use of balancing equipment, (b) routine servicing and inspection, and (c) replacement of parts other than those which normally require the use of skilled techniques, special tools, and test equipment
5. Balancing of fixed-pitch and ground-adjustable propellers
6. Refinishing of wood propellers

Persons and Organizations Authorized to Perform Repairs and Alterations on Propellers

The regulations governing the persons and organizations authorized to perform propeller repairs and alterations are subject to change, but in general, maintenance, minor repairs, or minor alterations must be done by a certificated repair station holding the appropriate ratings, an airframe and powerplant technician (A&P) or a person working under the direct supervision of such a technician, or an appropriately certificated air carrier. Major repairs or alterations on propellers may be performed only by an appropriately rated repair station, manufacturer, or air carrier in accordance with the regulations governing their respective operations.

Requirements governing persons or organizations authorized to perform maintenance and repairs on propellers are set forth in FAR Part 65.

Remember that minor repairs and alterations are those which are not likely to change the operating characteristics of the propeller or affect the airworthiness of the propeller. All other repairs and alterations are major in nature and must be performed by properly authorized agencies.

General Repair Requirements

Propellers should be repaired in accordance with the best accepted practices and the latest techniques. Manufacturers' recommendations should always be followed if such recommendations are available. It is recognized that the manuals may not be available for some of the older propellers; in such cases, the propellers should be repaired in accordance with standard practices and FAA regulations.

When a propeller is repaired or overhauled by a certificated agency, the repair station number or the name of the agency should be marked indelibly on the repaired propeller. It is recommended that a decal giving both the repair agency's name and repair station number be used for this purpose. If the original identification marks on a propeller are removed during overhaul or repair, it is necessary that they be replaced. These marks include the name of the manufacturer and the model designation.

General Inspection and Repair of Wood Propellers

Wood propellers are inspected for such defects as cracks, bruises, scars, warping, oversize holes in the hub, evidence of glue failure, evidences of separated laminations, sections broken off, and defects in the finish. The tipping should be inspected for such defects as looseness or slippage, separation of soldered joints, loose screws, loose rivets, breaks, cracks, eroded sections, and corrosion. Frequently, cracks do appear across the leading edge of the metal tipping between the front and rear slits where metal has been removed to permit easier forming of the tip curvature. These cracks are considered normal and are not cause for rejection.

The steel hub of a wood or composite propeller should be inspected for cracks and wear. When the hub is removed from the propeller, it should be magnetically inspected. Any crack in the hub is cause for rejection. The hub should also be inspected for wear of the bolt holes.

All propellers should undergo regular and careful inspection for any possible defect. Any doubtful condition such as looseness of parts, nicks, cracks, scratches, bruises, or loss of finish should be carefully investigated and the condition checked against repair and maintenance specifications for that particular type of propeller.

Causes for Rejection

Propellers worn or damaged to such an extent that it is either impossible or uneconomical to repair them and make them airworthy should be rejected and scrapped. The following conditions are deemed to render a wood propeller unairworthy and are therefore cause for rejection.

1. Cracks or deep cuts across the grain of the wood
2. Split blades
3. Separated laminations, except the outside laminations of fixed-pitch propellers
4. More screw or rivet holes, including holes filled with dowels, than are used to attach the metal leading-edge strip and tip

5. Appreciable warping
6. An appreciable portion of wood missing
7. Cracks, cuts, or other damage to the metal shanks or sleeves of blades
8. Broken lag screws which attach the metal sleeve to the blade
9. Oversize shaft holes in fixed-pitch propellers
10. Cracks between the shaft hole and the bolt holes
11. Cracked internal laminations
12. Excessively elongated bolt holes

Repair of Minor Damage

Small cracks parallel to the grain in a wood propeller should be filled with an approved glue thoroughly worked into all portions of the cracks, dried, and then sanded smooth and flush with the surface of the propeller. This treatment is also used with small cuts. Dents or scars which have rough surfaces or shapes that will hold a filler and will not induce failure may be filled with a mixture of approved glue and clean, fine sawdust, thoroughly worked and packed into the defect, dried, and then sanded smooth and flush with the surface of the propeller. It is very important that all loose or foreign matter be removed from the place to be filled so that a good bond of the glue to the wood is obtained.

Repair of Major Damage

As explained previously, the aviation technician rarely is involved with the major repair of a wood propeller. For this reason, such repairs are not described in this section. If an A&P technician should be confronted with the need to have a wood propeller reconditioned or repaired, a properly certificated propeller repair station would be able to perform the necessary work. Help and advice in such a matter can be obtained from the FAA General Aviation District Office.

General Inspection and Repair of Metal Propellers

Hollow and Solid Steel Propellers

Major damage on steel propeller blades should not be repaired except by the manufacturer. Welding or straightening is not permissible on such blades, even for very minor repairs, except by the manufacturer, because of the special process employed and the heat treatment required. A blade developing a crack of any nature in service should be returned to the manufacturer for inspection.

Inspection of Steel Blades

The inspection of steel blades may be either visual or magnetic. Visual inspection is easier to perform if the steel blades are covered with engine oil or rust-preventive compound. The full length of the leading edge, especially near the tip; the full length of the trailing edge; the grooves and shoulders on the shank; and all dents and scars should be examined with a magnifying glass to determine whether defects are scratches or cracks.

In the magnetic inspection of steel blades and propeller parts, the blade or part to be inspected is mounted in a machine, and then the blade is magnetized by passing a current through the blade or part. Either a black or a red mixture of an iron-base powder and kerosene is poured over the blade or part at the time that it is magnetized. North and south magnetic poles are established on either side of any crack in the metal. The iron filings arrange themselves in lines within the magnetic field thus created. A black or a red line, depending on the color of the mixture, will appear wherever a crack exists in the blade or part.

Repair of Minor Damage in Steel Blades

Minor injuries to the leading and trailing edges only of steel blades may be smoothed by handstoning, provided that the injury is not deep.

Aluminum-Alloy Propellers

A seriously damaged aluminum-alloy propeller blade should be repaired only by the manufacturer or by repair agencies certificated for this type of work. Such repair agencies should follow manufacturers' instructions.

Definition of Damaged Propellers

A damaged metal propeller is one that has been bent, cracked, or seriously dented. Minor surface dents, scars, nicks, etc., which are removable by field maintenance technicians are not considered sufficient to qualify the propeller as damaged.

If the model number of a damaged blade appears on the manufacturer's list of blades which cannot be repaired, the blade should be rejected.

Blades bent in face alignment. The extent of a bend in the face alignment of blades should be carefully checked by means of a protractor similar to the one illustrated in Fig. 22-22.

Manufacturers often specify the maximum bends which can be repaired by cold-straightening on specific models of propellers. Figure 22-23 is a chart which shows the maximum allowable bend for cold repair of McCauley 1A90, 1B90, and 1C90 fixed-pitch metal propellers. From the chart, for example, it can be determined that if the propeller is bent at the 16-in [40.64-cm] radius, the maximum degree of bend which can be straightened cold is 9°. At the 32-in [81.28-cm] radius, the blade can be repaired by cold-straightening if the bend is as great as 18.5°. After straightening, the affected portion of the blade must be etched and thoroughly inspected for cracks and other flaws. Blades with bends in excess of this amount require heat treatment and must be returned to the manufacturer or an authorized agent for repair.

Blades bent in edge alignment. Blades which are bent in edge alignment should be repaired by the manufacturer or a certificated repair station holding the appropriate rating.

Aluminum Blade Fatigue Failure

An investigation of a representative number of propeller blades disclosed that failures usually occurred because of **fatigue cracks** which started at mechanically formed dents, cuts, scars, scratches, nicks, or leading-edge pits. Blade material samples in most cases did not reveal evidence of failure caused by material defects or surface discontinuities existing before the blades were placed in service. Often fatigue failure occurs at a place where previous damage has been repaired, which may be a result of the failure actually having started prior to the repair or of the repair having been performed improperly.

The stresses that normally occur in a propeller blade may be envisioned as being produced by lines of force that run within the blade approximately parallel to the surface, as shown in Fig. 22-24. When a defect occurs, it tends to squeeze together the lines of force in the defect area, thereby increasing the stress. This increase in stress may be sufficient to cause a crack to start. Even a small defect such as a nick or dent may develop into a crack. The crack, in turn, results in an even greater stress concentration in the area. The resulting growth of the crack will almost inevitably result in blade failure. This condition is so common,

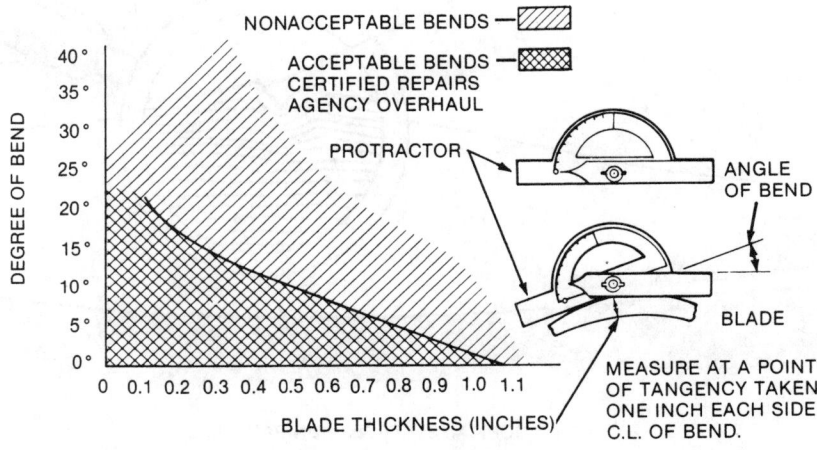

FIGURE 22-22 Blade straightening limits. (*Hartzell Propeller.*)

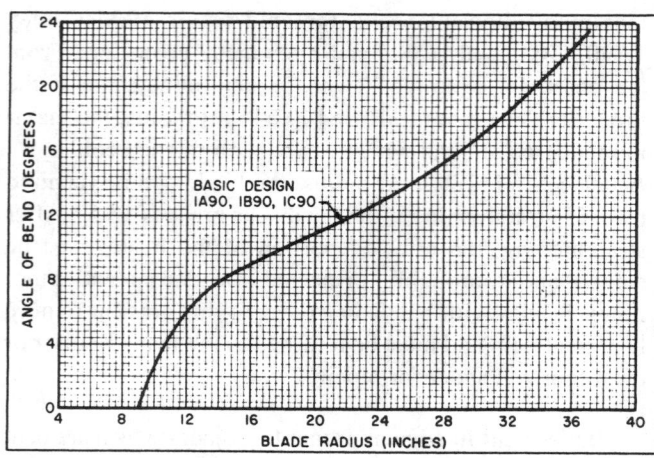

FIGURE 22-23 Chart showing the maximum allowable bend for a cold repair.

and the results are so serious, that great emphasis should be placed on the daily and preflight inspections of propeller blades for defects.

Experience indicates that fatigue failures normally occur within a few inches of the blade tip; however, failures also can occur in other portions of the blade when dents, cuts, scratches, or nicks are ignored. Failures have also been reported in blades near the shank and at the propeller hub, well out of the critical areas; therefore, no damage should be overlooked or allowed to go without correction.

When performing an inspection on the propeller, especially during the preflight inspection, inspect each complete blade—not just the leading edge—for erosion, scratches, nicks, and cracks. Regardless of how small a surface irregularity may be, consider it as a stress intensifier that makes the area subject to fatigue failure.

Propeller manufacturers' manuals, service letters, and bulletins specify methods and limits for blade maintenance,

inspection service, and repair. The proper service information should always be consulted.

Prevention and Treatment of Minor Surface Defects

To prevent propeller surface defects, avoid operating the aircraft in areas with loose stone or gravel that could be pulled into the blades and cause damage to the blade face or leading edge. When takeoff from a nonhard surface runway is initiated, blade damage can be minimized by allowing the aircraft to move prior to fully opening the throttle. Keep the blade clean of stains and foreign matter, and *do not move aircraft by pulling on propeller blades.*

Propeller blades with nicks, gouges, scratches, and leading-edge pitting can be repaired most often by a qualified technician in the field. Normally there is sufficient material available to allow a number of minor repairs to be made prior to replacement. Blades with larger nicks, gouges, etc., that may affect the structure, balance, or operation of the propeller should be referred to a qualified propeller repair station for repair or replacement.

Repair may be made using files or small air- or electric-powered equipment with suitable grinding and polishing attachments. All repairs must be made parallel to the blade axis. The manufacturer's manual should be consulted to ensure that correct information is used in the repair of specific models of propellers.

For damaged areas in the leading or trailing edge, begin with a round file and remove damaged material down to the bottom of the damaged area. Remove material from this point out on both sides, providing a smooth, faired depression, and maintaining the original airfoil concept, as shown in Fig. 22-25. The area should be smoothly faired using **emery cloth**, to remove all traces of initial filing and rework. **Crocus cloth** is used to polish the area. When all rework has been completed, inspect the reworked area with a 10× magnifying

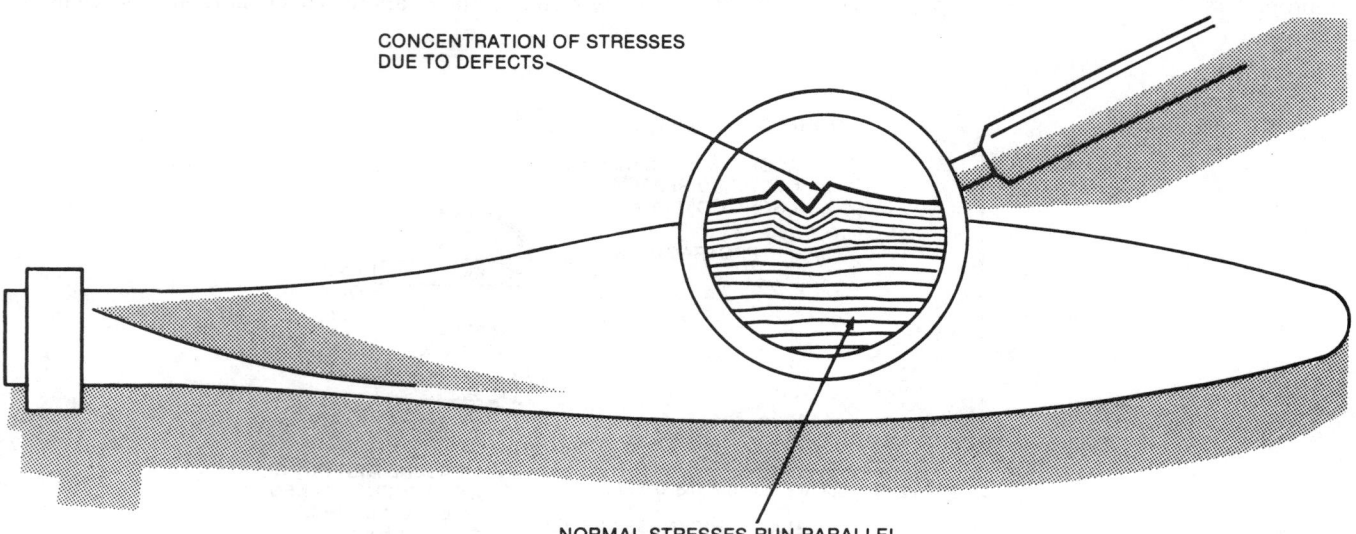

FIGURE 22-24 Fatigue stresses in a propeller.

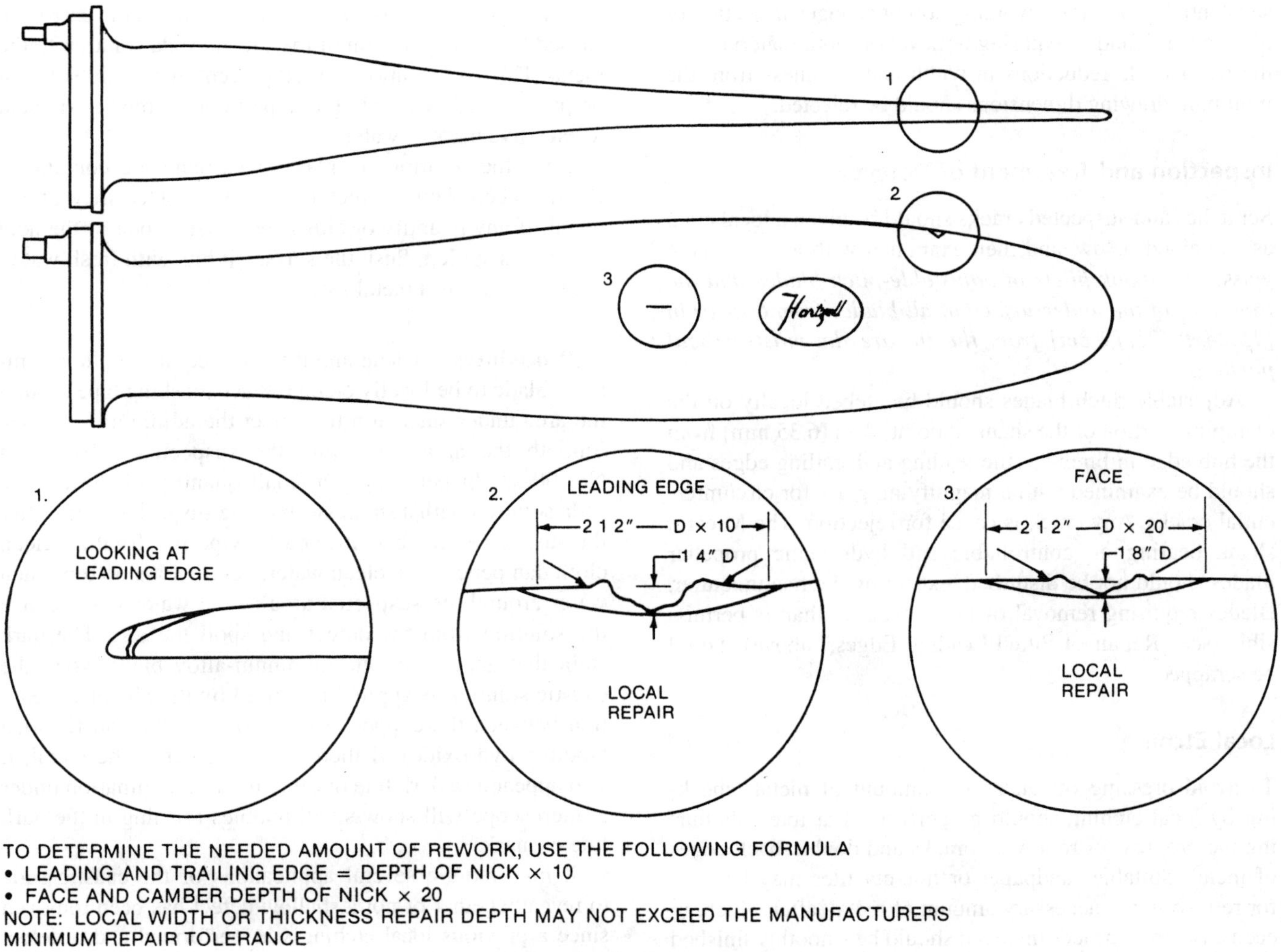

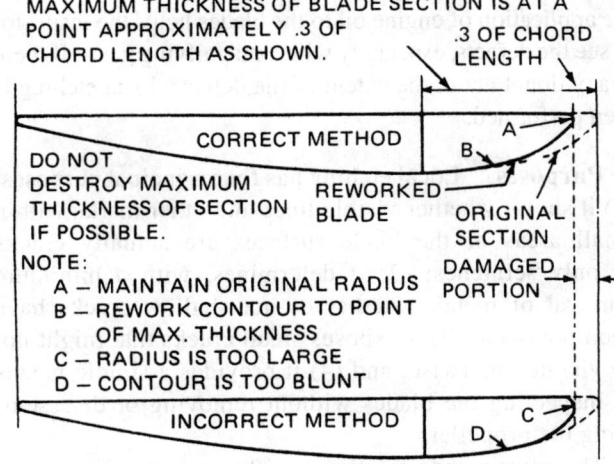

1
2
3
Hartzell
1.
LOOKING AT
LEADING EDGE
2.
LEADING EDGE
2 1 2″ — D × 10
1 4″ D
LOCAL
REPAIR
3.
FACE
2 1 2″ — D × 20
1 8″ D
LOCAL
REPAIR
TO DETERMINE THE NEEDED AMOUNT OF REWORK, USE THE FOLLOWING FORMULA
• LEADING AND TRAILING EDGE = DEPTH OF NICK × 10
• FACE AND CAMBER DEPTH OF NICK × 20
NOTE: LOCAL WIDTH OR THICKNESS REPAIR DEPTH MAY NOT EXCEED THE MANUFACTURERS
MINIMUM REPAIR TOLERANCE

FIGURE 22-25 Blade repair. (*Hartzell Propeller.*)

glass and dye penetrant to make sure that no indications of the damage or cracks remain.

Damaged areas on the face or camber sections of the blade should be reworked employing the same methods used for the leading edge (see Fig. 22-25).

All repaired areas should be chemically treated to prevent corrosion. Alodine or an approved paint should be properly applied to the repaired area prior to return of the blade to service.

Number of Defects Allowable in Blades

More than one defect falling within the above limitations is not sufficient cause alone for the rejection of a blade. A reasonable number of such defects per blade is not necessarily dangerous, if within the limits specified, unless the locations of the defects with respect to each other are such as to form a continuous line of defects that would materially weaken the blade.

Repair of Pitted Leading Edges

Blades whose leading edges are pitted from normal wear in service may be reworked by removing sufficient material

MAXIMUM THICKNESS OF BLADE SECTION IS AT A
POINT APPROXIMATELY .3 OF
CHORD LENGTH AS SHOWN.
.3 OF CHORD
LENGTH
CORRECT METHOD
A
B
DO NOT
DESTROY MAXIMUM
THICKNESS OF SECTION
IF POSSIBLE.
REWORKED
BLADE
ORIGINAL
SECTION
NOTE:
A – MAINTAIN ORIGINAL RADIUS
B – REWORK CONTOUR TO POINT
OF MAX. THICKNESS
C – RADIUS IS TOO LARGE
D – CONTOUR IS TOO BLUNT
DAMAGED
PORTION
INCORRECT METHOD
C
D

FIGURE 22-26 Rework of a propeller's leading edge.

to eliminate the defects. In this case, the metal should be removed by starting at approximately the thickest section, as shown in Fig. 22-26, and working well forward over the nose camber so that the contour of the reworked portion will remain substantially the same, avoiding abrupt changes in section or blunt edges. Blades requiring removal of more material than the permissible reductions in width and thickness from the minimum drawing dimensions should be rejected.

Inspection and Treatment of Defects

Scratches and suspected cracks should be given a local etch, as explained below, and then examined with a magnifying glass. *The shank fillets of adjustable-pitch blades and the front half of the undersurface of all blades from 6 to 10 in [15.24 to 25.40 cm] from the tip are the most critical portions.*

Adjustable-pitch blades should be etched locally on the clamping portion of the shank at points $\frac{1}{4}$ in [6.35 mm] from the hub edge in line with the leading and trailing edges and should be examined with a magnifying glass for circumferential cracks. Any crack is cause for rejection. The Micarta shank bearing on controllable and hydromatic propeller blades should not be disturbed, except by the manufacturer. Blades requiring removal of more material than is permissible (see "Repair of Pitted Leading Edges," above) should be scrapped.

Local Etching

To avoid dressing off an excess amount of metal, checking by local etching should be performed at intervals during the progress of removing cracks and double-back edges of metal. Suitable sandpaper or fine-cut files may be used for removing the necessary amount of metal, after which, in each case, the surfaces involved should be smoothly finished with No. 00 sandpaper. Each blade from which any appreciable amount of metal has been removed should be properly balanced before being used.

When aluminum-alloy blades are inspected for cracks or other failures and for bends, nicks, scratches, and corrosion, the application of engine oil to the blades helps the inspector to see the defects, especially with a magnifying glass. If there is any doubt about the extent of the defects, local etching is then performed.

Purposes. Local etching has four principal purposes: (1) it shows whether visible lines and other marks within small areas of the blade surfaces are actually cracks or only scratches; (2) it determines, with a minimum removal of metal, whether or not shallow cracks have been removed; (3) it exposes small cracks that might not be visible otherwise; and (4) it provides a simple means of inspecting the blades without removing or disassembling the propeller.

The caustic soda solution is a 20 percent solution prepared locally by adding to the required amount of water as much commercial caustic soda as the water will dissolve and then adding some soda pellets after the water has ceased to dissolve the caustic to be sure that the solution is saturated. The quantity required depends on the amount of etching to be done. This caustic soda solution should reveal the presence of any cracks.

An acid solution is used to remove the dark corrosion caused by the application of the caustic soda solution to the metal. The acid solution is a 20 percent nitric acid solution prepared locally by adding one part commercial nitric acid to each five parts of water.

Keep the solutions in glass or earthenware containers. Do not keep them in metal containers, since they attack metal. If any quantity of either the caustic soda or the acid solution is spilled, flush the surface it hits with fresh water, especially if it is a metal surface.

Procedures. Clean and dry the area of the aluminum-alloy blade to be locally etched. Place masking tape around the area under suspicion to protect the adjoining surfaces. Smooth the area containing the suspected defect with No. 00 sandpaper. Apply a small quantity of the caustic soda solution with a small swab to the suspected area. After the suspected area becomes dark, wipe it off with a clean cloth dampened with clean water, but do not slop too much water around the suspected area or the water will remove the solution from the defect and spoil the test. The dark stain that appears on an aluminum-alloy blade when the caustic solution is applied is caused by the chemical reaction between the copper in the alloy and the caustic soda (sodium hydroxide). If there is any defect in the metal, it will appear as a dark line or other mark. Examination under a microscope will show small bubbles forming in the dark line or mark.

It may require several applications of the caustic soda to reveal whether or not a shallow defect has been removed since a previous local etching was performed and a defect discovered. Immediately after the completion of the final test, all traces of caustic soda must be removed with the nitric acid solution. The blade is rinsed thoroughly with clean water, and then it is dried and coated with clean engine oil.

The inspection of aluminum-alloy propeller blades for cracks and flaws may be accomplished by means of a chromic acid anodizing process. This is superior to the caustic etching process and should therefore be used if facilities are available.

The blades should be immersed in the anodizing bath as far as possible, but all parts not made of aluminum alloy must either be kept out of the chromic acid bath or be separated from the blade by nonconductive wedges or hooks. The anodizing treatment should be followed by a rinse in clear, cold, running water for 3 to 5 min, and the blades should be dried as soon as possible after the rinse, preferably with an air blast. After the blades are dried, they should stand for at least 15 min before examination. Flaws, such as cold shuts and inclusions, will appear as fine black lines. Cracks will appear as brown stains caused by chromic acid bleeding out onto the surface.

The blades may be sealed for improved corrosion resistance by immersing them in hot water (180 to 212°F [82 to 100°C]) for $\frac{1}{2}$ h. In no case should the blades be treated with hot water before the examination for cracks, since heating expands any cracks and allows the chromic acid to be washed away.

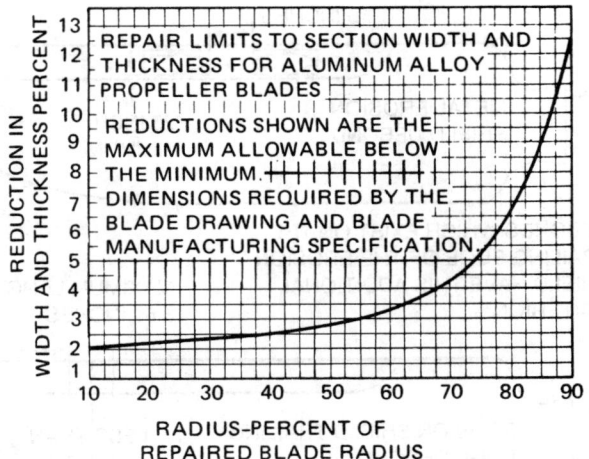

FIGURE 22-27 Propeller blade repair limits.

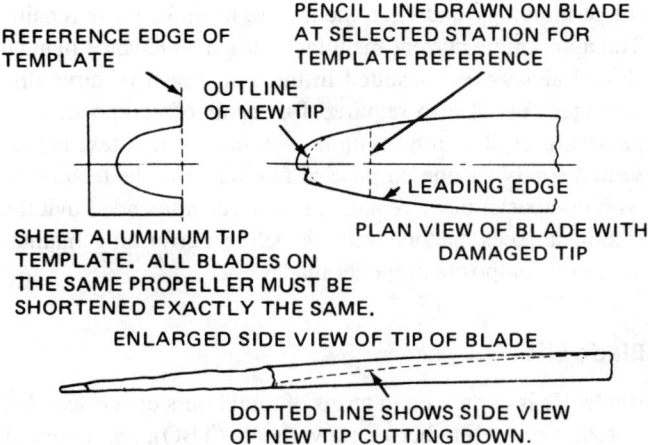

FIGURE 22-28 Repair of a damaged propeller tip.

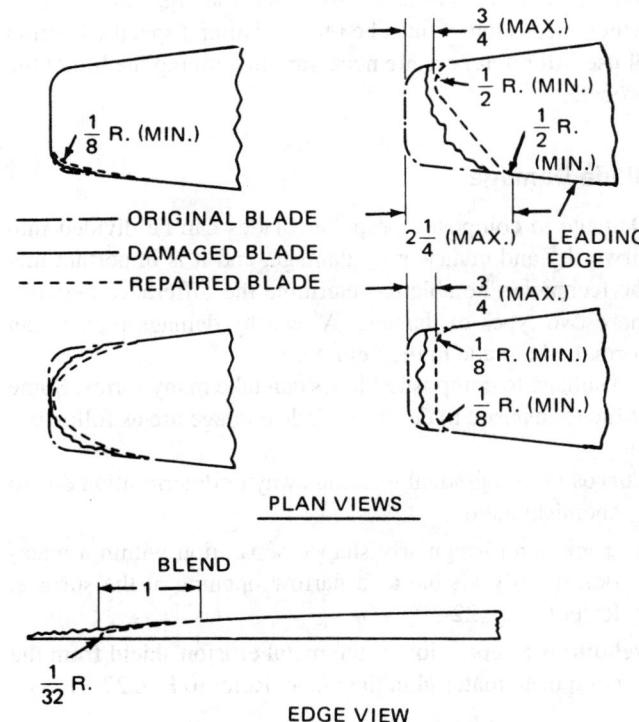

NOTE: BLADE RADII AND TIP SHAPE SHOULD BE THE SAME FOR BOTH BLADES

FIGURE 22-29 Repair of a damaged square tip.

Inspection of aluminum-alloy propeller blades for cracks and other defects may also be accomplished by means of the fluorescent penetrant process or the dye penetrant process. These methods for the inspection of nonferrous metals are explained in Chap. 10.

Tolerances Listed in Blade Manufacturing Specifications

Tolerances listed in the blade manufacturing specifications govern the width and thickness of new blades. These tolerances are to be used with the pertinent blade drawing to determine the minimum original blade dimensions to which the reductions shown in Fig. 22-27 may be applied.

For repairing blades, the permissible reductions in width and thickness from the minimum original dimensions allowed by the blade drawing and blade manufacturing specifications are shown in Fig. 22-27 for locations on the blade from the shank to 90 percent of the blade radius. In this instance, the outer 10 percent of blade length may be modified as required.

Shortening of Blades to Remove Defects

When the removal or treatment of defects on the tip necessitates shortening of a blade, each blade used with it must likewise be shortened. Such sets of blades should be kept together. Figures 22-28 and 22-29 illustrate acceptable methods of blade shortening.

With some propeller blades, the length may be reduced substantially and the propeller can then be given a new model number in accordance with the manufacturer's specifications. The reduction in length may require an increase in the blade angle, and the length must agree with the specification for the new model number.

Causes for Rejection

Unless otherwise specified in this text, a blade having any of the following defects must be rendered unserviceable:

(1) irreparable defects, such as longitudinal cracks, cuts, scratches, scars, etc., that cannot be dressed off or rounded out without materially weakening or unbalancing the blade or materially impairing its performance; (2) general unserviceability due to removal of too much stock by etching, dressing off defects, etc.; (3) slag inclusions in an excessive number or cold shuts in an excessive number, or both; and (4) transverse cracks of any size.

Composite Propeller Blade Damage Limits

To determine the damage limits and make repairs on composite propeller blades, the technician must be familiar with

Maintenance and Repair of Propellers **671**

the terminology and equipment used to make these repairs. The appropriate current manufacturer's maintenance manual should always be consulted in the assessment of airworthy damage. This is also required for repair of composite propeller blades. The information contained in this text is presented solely for the purpose of familiarizing the technician with composite blade repair. It is also recommended that the technician seek factory training before performing maintenance on composite propeller blades.

Blade Life

Blade life is expressed in terms of total hours of service (TT, or total time), time between overhauls (TBO), and hours of service since overhaul (TSO, or time since overhaul). Overhaul returns the blade assembly to zero hours TSO, but not to zero hours TT. Occasionally, a part may be "life limited," which means that it must be replaced after a specified period of use. All references are necessary in defining the life of the propeller.

Blade Damage

Damage to composite propeller blades can be divided into airworthy and unairworthy damage, and it is important that the technician be able to determine the difference between these two types of damage. Airworthy damage repairs can normally be made in the field.

Damage to composite blades can take many forms. Some of the terms used to describe blade damage are as follows:

Corrosion is a gradual wearing away or deterioration due to chemical action.

A **crack** is an irregularly shaped separation within a material, usually visible as a narrow opening at the surface. Refer to Fig. 22-30.

Debond is a separation of the metal erosion shield from the composite material in the blade. Refer to Fig. 22-31.

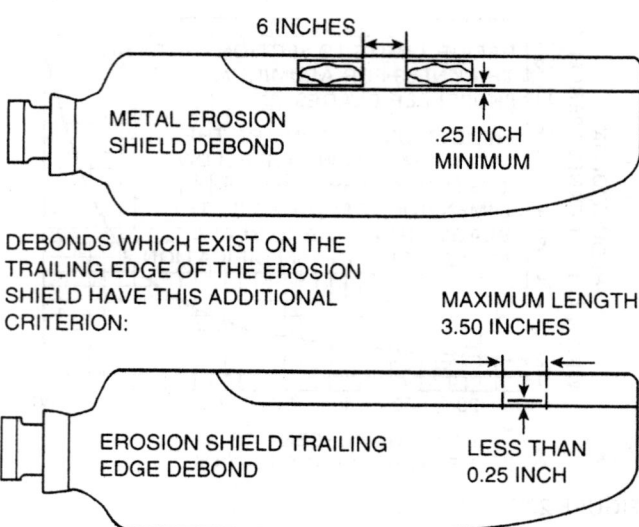

FIGURE 22-31   Limits of airworthy damage in metal erosion shield debond. (*Hartzell Propeller.*)

Delamination is an internal separation of the layers of composite material.

A **depression** is a surface area where the material has been compressed, but not removed, by contact with a sharp object.

Distortion is an alteration of the original shape or size of a component.

Erosion is a gradual wearing away or deterioration due to action of the elements.

Exposure is the condition in which material is left open to the action of the elements.

A **gouge** is a small surface area from which material has been removed by contact with a sharp object.

Impact damage occurs when the propeller blade or hub assembly strikes or is struck by an object, either in flight or on the ground.

Overspeed damage occurs when the propeller hub assembly rotates at a speed more than 10 percent in excess of the maximum for which it is designed. Overspeed damage may not produce visible indications.

A **scratch/nick** is removal of paint and possibly a small amount of the composite material not exceeding one layer (approximately 0.010 in [0.254 mm]).

A **split** is a delamination of the blade extending to the blade surface, normally found near the trailing edge or tip.

Normal airworthy damage does not affect flight safety characteristics of the blades, although areas of airworthy damage should be repaired to maintain aerodynamic efficiency. Airworthy damage should be monitored until repaired, with the repair being accomplished as soon as possible. To determine if the damage is airworthy or unairworthy, the technician should refer to the information contained in the current blade-repair manual. The following is a list of airworthy damage limits for Hartzell composite propeller blades.

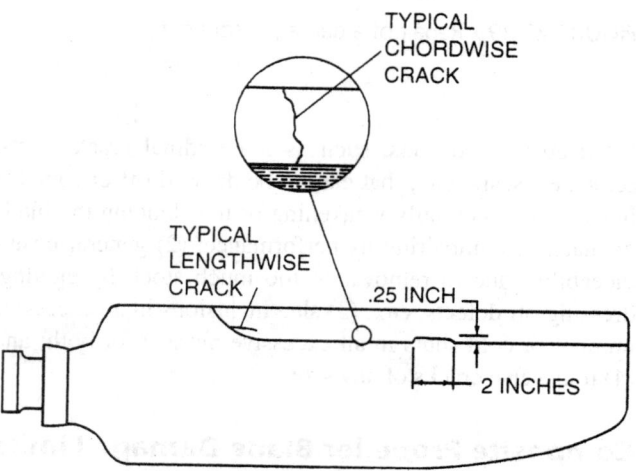

FIGURE 22-30   Missing portions of nickel erosion shield (trailing edge side) and typical cracks. (*Hartzell Propeller.*)

Erosion Shield Airworthy Damage

The following types of damage cannot be resolved without replacement of the erosion shield, but, within these limits, do not render the blade unairworthy:

Any gouge that does not penetrate through to the surface of composite material.

Any full-width chordwise crack as long as the erosion shield is not debonded within 3.5 in [8.89 cm] of the crack (Fig. 22-30).

No two full-width chordwise cracks may occur within 6 in [15.24 cm] of each other.

Chordwise cracks less than 0.5 in [1.27 cm] in length that are not debonded within 1 in [2.54 cm].

Portions of the trail side of the erosion shield may be missing as a result of erosion or removal by sanding (see Fig. 22-30 for limits).

Lengthwise cracks less than 2 in [5.08 cm] long that are not debonded within 3.5 in [8.89 cm] of the crack (see Fig. 22-30 for limits).

For blades with attached counterweight clamps, cracks within 1 in [2.54 cm] of a counterweight clamp that are not debonded.

Minor deformations due to impact damage that does not greatly affect the airfoil shape.

The following types of damage do not render the blade unairworthy but should be repaired as soon as practical to prevent degradation of the condition:

Debonds located along the trailing side of the erosion shield that together total less than 10.5 in [26.67 cm] in length. No individual debond may exceed 3.5 in [8.89 cm] in length and 0.25 in [0.64 cm] in width (Fig. 22-31).

Debond which is located at least 0.25 in [0.64 cm] from the erosion shield trail side and has a total area less than 2.5 in^2 [16.13 cm^2], and is separated by at least 6 in [15.24 cm] from any other debond area on the same blade surface. The total debonded area of all debonds may not exceed 10 in^2 [64.52 cm^2].

Blade Cuff Airworthy Damage

The types of blade cuff damage that are considered to be airworthy damage are as follows:

Nicks, scratches.

Depressions less than 1 in^2 [6.45 cm2] in area and less than 0.25 in [6.35 mm] deep.

Delaminations less than 2 in^2 [12.90 cm^2] in area.

Cracks at the root end are airworthy, but should be sealed to protect the foam from contamination (Fig. 22-32) until the next overhaul, during which these cracks can be permanently repaired.

Cracks located in the area where the cuff and blade meet must be within the limits, as shown in Fig. 22-33.

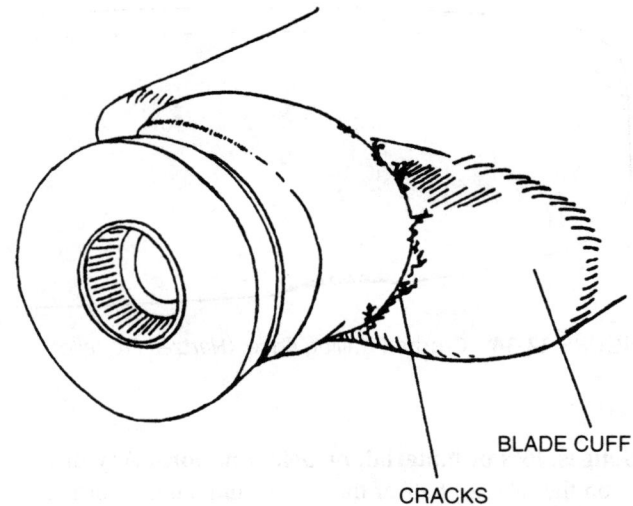

FIGURE 22-32 Blade cuff damage. (*Hartzell Propeller.*)

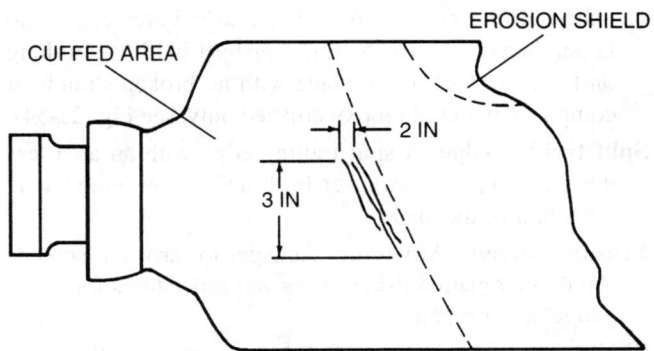

FIGURE 22-33 Cracks in the area where cuff meets blade. (*Hartzell Propeller.*)

No more than two other cracks may be located elsewhere on the cuff. These cracks must be less than 3 in [7.62 cm] in length.

No more than two damaged areas per side are permitted within 6 (linear) in [15.24 cm] of each other. Root end cracks and cracks where the blade and cuff meet are not included in this requirement.

Cuffs with no boot or erosion shield covering the leading edge may have no cracks within 2 in [5.08 cm] of leading edge.

Blade Airworthy Damage

Types of blade damage that are considered airworthy are:

Gouges or loss of material. Gouges less than 0.500 in [12.7 mm] in diameter or of equivalent area and no more than 2.5 in [6.35 cm] long and less than 0.020 in [0.508 mm] deep anywhere on the outboard half of the blade.

Delamination. Delamination on the outboard half of the blade totaling less than 2 in^2 [12.90 cm^2] in area and with no dark brown or black stain which would indicate the presence of grease.

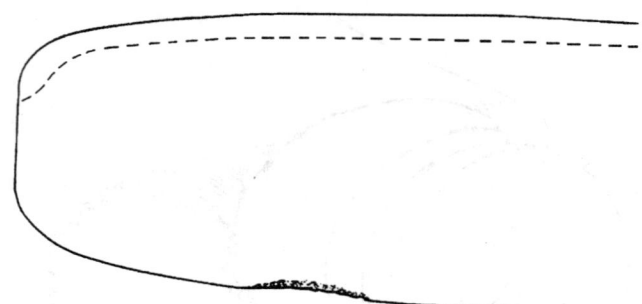

FIGURE 22-34 Crushed trailing edge. (*Hartzell Propeller.*)

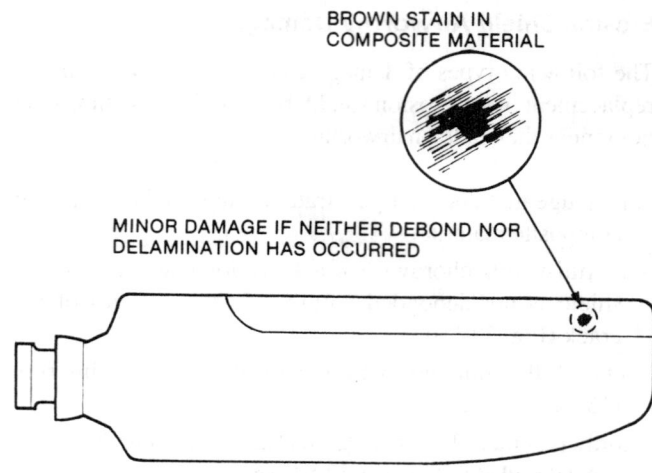

BROWN STAIN IN
COMPOSITE MATERIAL

MINOR DAMAGE IF NEITHER DEBOND NOR
DELAMINATION HAS OCCURRED

FIGURE 22-35 Evidence of lightning strike damage in a composite blade. (*Hartzell Propeller.*)

Gouges, loss of material, or delaminations. Any of these on the inboard half of the blade can be unairworthy, and the factory should be consulted.

Paint erosion. The exposure of less than 5 in² [32.26 cm²] of the composite material and/or the primer filler.

Crushed blade trailing edge. The crushed area can be no larger than 0.25 in [6.35 mm] deep by 1 in [2.54 cm] long on the outer half of the blade with no broken strands of composite material (epoxy crushed only, see Fig. 22-34).

Split trailing edge. A split trailing edge with an area less than 0.25 in [6.35 mm] deep by 1 in [2.54 cm] long on the outer half of the blade.

Erosion screen. Airworthy damage to erosion screens should be repaired using limits and procedures for blade gouge minor repair.

Blade retention windings. Cracks appearing in the paint over the blade retention windings are airworthy. These cracks should be repaired as soon as practical.

Repair of Composite Blades

Repair procedures can be categorized as either minor repair or major repair. Minor repair is the correction of damage that may be safely performed in the field by a certified aircraft technician. Major repair is the correction of damage that cannot be performed by elementary operations. Major repairs must be performed by a propeller shop that has been approved by Hartzell for the specific type of major repair.

Repair of Unairworthy Damage

Unairworthy damage is defined as any damage of the composite blade which exceeds the limits of the airworthy damage as previously described. Unairworthy damage to a composite blade must be repaired before the blade can be used on another flight. This requires returning the blade to a factory-designated facility for evaluation and repair with factory consultation.

Composite propeller blades are not subject to fatigue cracks, as are aluminum propellers, because of their composite construction. The factory or factory-approved repair station can repair many types of unairworthy damage of composite blades.

Lightning strikes usually enter a composite blade through the metal erosion shield; however, some strikes can hit the hub directly. A direct strike on the hub of a composite blade propeller results in unairworthy damage. The blades must then be overhauled according to prescribed procedures before the propeller can be used for further service. A lightning strike on the metal erosion shield leaves a darkened area and sometimes pitting near the tip of the composite blade, as shown in Fig. 22-35. If evidence of a lightning strike is found, it will require a careful debond/delamination inspection to determine the extent of the damage and whether it is airworthy or unairworthy. To determine this, perform a "coin-tap" test immediately to test for debond and/or delamination. The **"coin-tap" test** is shown in Fig. 22-36. Use of an **impactoscope flaw detector** is an approved optional method, in conjunction with a coin-tap test, for detecting delaminated areas on the blade. If only a darkened area is present on the erosion shield from the lightning strike, and all blade damage is within limits specified, the damage is considered airworthy. If the damage is determined to be unairworthy damage, return the blade to the factory.

Repair of Airworthy Damage

There are many different types of airworthy damage, and therefore there are several methods used to repair this type of damage. Repair of a composite propeller blade is usually performed by cleaning the damaged area, removing the paint, and sanding. The damaged area is then filled with laminated fiberglass cloth and epoxy. It is normally necessary to use C-clamps to apply pressure as the epoxy sets up, as shown in Fig. 22-37. After the epoxy has hardened, the area is sanded for conformance with the contour of the blade. Examples of repaired erosion shield areas are shown in Fig. 22-38. The repaired blade receives a final finish of approved primer and polyurethane paint.

Records of Repair

In the performance of repairs of composite blades, record keeping is very important. Records should indicate whether a particular instance of damage was unairworthy or airworthy,

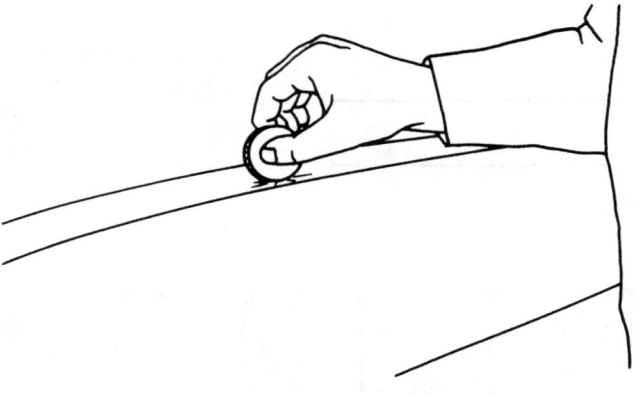

"COIN-TAP" TEST ALONG ENTIRE SURFACE
OF EROSION SHIELD CHECKS FOR DEBOND

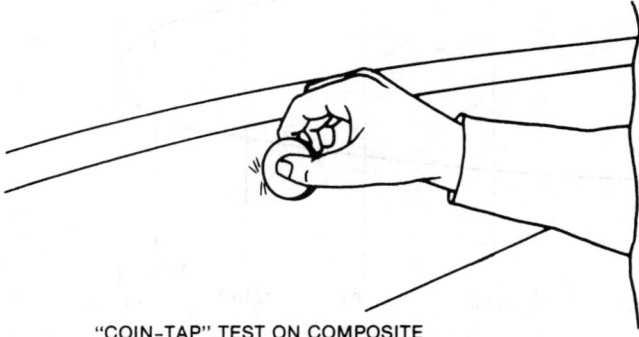

"COIN-TAP" TEST ON COMPOSITE
BLADE SURFACE CHECKS FOR DELAMINATION

NOTE: THE "COIN" USED FOR THESE TESTS
SHOULD HAVE RADIUSED EDGES AND
WEIGH AT LEAST THREE OUNCES (85 g).

FIGURE 22-36   Use of coin-tap test to check for debond and delamination. (*Hartzell Propeller.*)

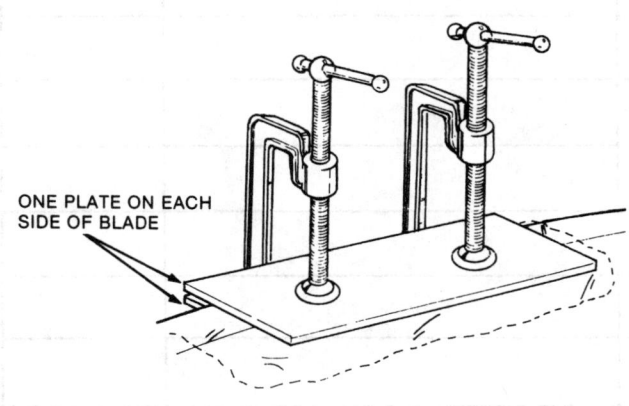

ONE PLATE ON EACH
SIDE OF BLADE

THIN PLASTIC SHEET
UNDER TOP PLATE TO
PREVENT IT FROM
BONDING TO BLADE

WOODEN OR PLASTIC PLATE (TWO REQUIRED)
.25-INCH (6 MM) THICK
THREE (3) INCHES (76 MM) WIDE
SIX (6) INCHES (150 MM) LONG

FIGURE 22–37   Use of C-clamps to apply pressure to erosion shield debond repair. (*Hartzell Propeller.*)

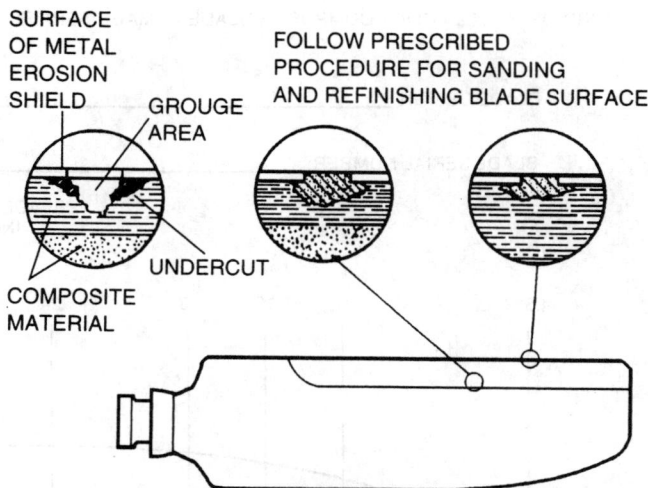

SURFACE
OF METAL
EROSION
SHIELD
GROUGE
AREA
FOLLOW PRESCRIBED
PROCEDURE FOR SANDING
AND REFINISHING BLADE SURFACE
UNDERCUT
COMPOSITE
MATERIAL

FIGURE 22-38   Field repair of airworthy (minor) damage in erosion shield. (*Hartzell Propeller.*)

and a description of the resulting repair should be recorded on the proper form. Figure 22-39 shows a typical form for recording composite blade repairs. The date, flight hours, degree of damage, description of repair, and person performing the work are recorded on the form. It is also important to show the location of the repair, because repairs may be made close to other repairs or on top of previous repairs.

CHECKING BLADE ANGLES

The blade angles of a propeller may be checked by using any precision protractor which is adjustable and is equipped with a spirit level. Such a protractor is often called a **bubble protractor**.

Universal Propeller Protractor

The blade angles of a propeller may be accurately checked by the use of a **universal propeller protractor**, which is the same instrument used to measure the throw of control surfaces. An accurate check of blade angles cannot be made by referring to the graduations on the ends of the hub barrels or on the shanks of the blades of propellers; such references are suitable only for rough routine field inspections and emergency blade settings.

A **protractor** is merely a device for measuring angles. The propeller protractor consists of an aluminum frame in which a steel ring and a disk are mounted, as shown in Fig. 22-40. The principal, or "whole-degree," scale is on the disk. The vernier, or "fractional-degree," scale is on the ring. The zeros on these two scales provide reference marks which can be set at the two sides of an angle, thereby enabling the operator to read from zero to zero to obtain the number of degrees in the angle.

Two adjusting knobs provide for the adjustment of the ring and disk. The ring adjuster is in the upper right-hand

BLADE DESIGN _____

BLADE SERIAL NUMBER _____

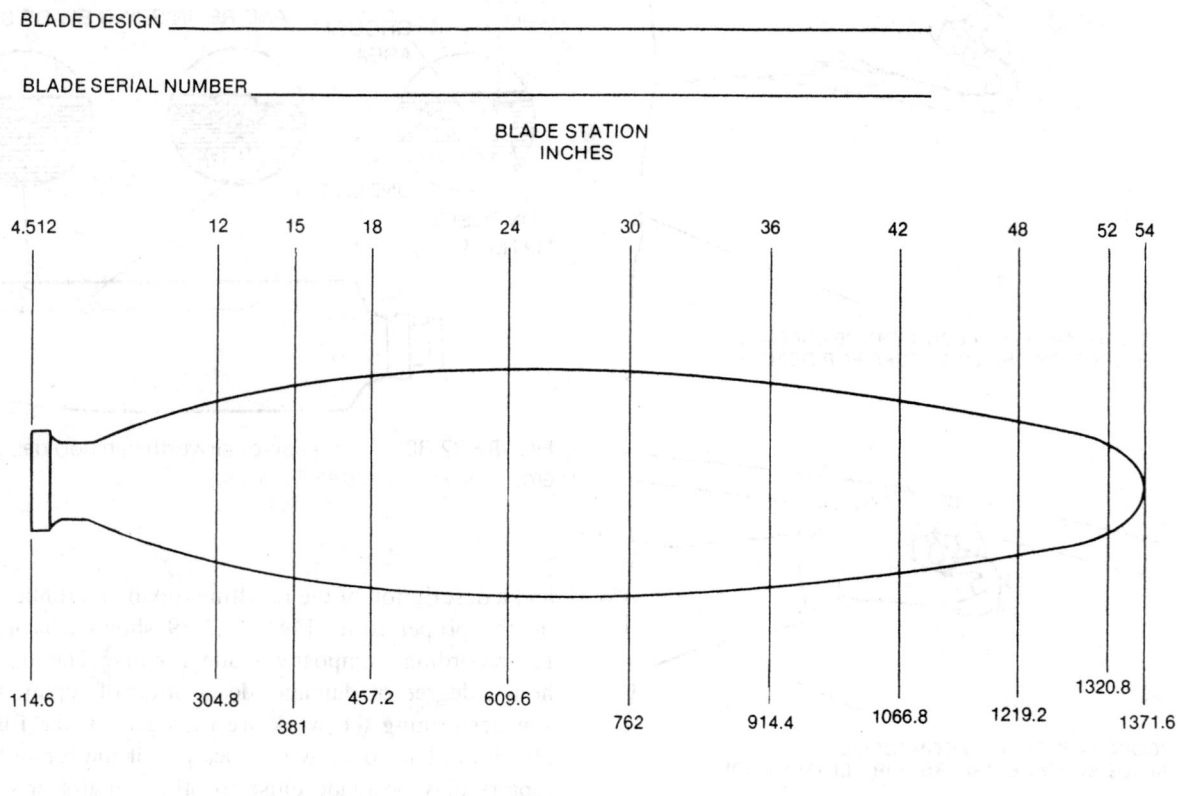

BLADE STATION
INCHES

4.512   12   15   18   24   30   36   42   48   52 54

114.6   304.8   457.2   609.6   1320.8
                 381   762   914.4   1066.8   1219.2   1371.6

BLADE STATION
MM

DATE OF ENTRY	FLIGHT HOURS	DEGREE OF DAMAGE (MAJOR OR MINOR) DESCRIPTION OF DAMAGE	DESCRIPTION OF REPAIR	REPAIRED BY

FIGURE 22-39 Composite blade damage and repair record form. (*Hartzell Propeller.*)

corner of the frame; when it is turned, the ring rotates. The disk adjuster is on the ring; when this knob is turned, the disk rotates.

There are two locks on the protractor. One is the disk-to-ring lock, located on the ring. It is a pin that is held by a spring when engaged, but it engages only when the pin is pulled out and placed in the deep slot and when the zeros on the two scales are aligned. Under these conditions, the ring and disk rotate together when the ring adjuster is turned and when the ring-to-frame lock is disengaged. To hold the spring-loaded pin of the disk-to-ring lock in the released position, it is first pulled outward and then turned 90°.

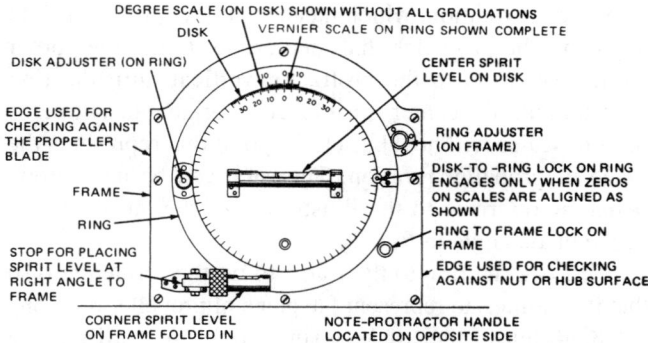

FIGURE 22-40   Propeller protractor.

The other lock, the ring-to-frame lock, is on the frame. It is a right-hand screw with a thumb nut. The disk can be turned independently of the ring by means of the disk adjuster when the ring is locked to the frame and the disk-to-ring lock is released.

There are two spirit levels on the protractor. One is the center, or disk, level. It is at right angles to the zero graduation mark on the whole-degree scale of the disk; therefore, the zero graduation mark will lie in a vertical plane through the center of the disk whenever the disk is "leveled off" in a horizontal position by means of the disk level.

The other level is the corner spirit level, located at the lower left-hand corner of the frame and mounted on a hinge. This level is swung out at right angles to the frame whenever the protractor is to be used. It is used to keep the protractor in a vertical position for the accurate checking of the blade angle.

As described earlier, the degree scale for the protractor (Fig. 22-40) is on the center disk and the vernier scale is on the ring just outside the disk. The vernier graduations have a ratio of 10 to 9 with the degree graduations; that is, ten graduations on the vernier scale will match with nine graduations on the degree scale. As shown in Fig. 22-41, the reading between the 0° point on the degree scale and the 0° point on the vernier scale is somewhat more than 15°. To find the amount in tenths of a degree, the vernier scale is read in the same direction from the 0° point on the vernier scale to the point where a vernier-scale graduation coincides approximately with a degree-scale graduation. In Fig. 22-41, this is a point eight graduations to the left of the 0° point of the vernier

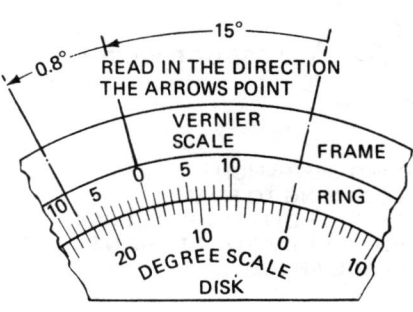

FIGURE 22-41   Reading the protractor scale.

scale and is read 0.8°. The total angle shown by the protractor is therefore 15.8°.

As pointed out above, the number of tenths of a degree in the blade angle is found by observing the number of vernier-scale spaces between the zero of the vernier scale and the vernier-scale graduation that comes closest to being in perfect alignment with a degree-scale graduation line. Always read tenths of degrees on the vernier scale in the same direction as the degrees are read on the degree scale.

How to Measure the Propeller Blade Angle

To measure the propeller blade angle, determine how much the flat side of the blade slants from the plane of rotation. If a propeller shaft is in the horizontal position when the airplane rests on the ground, the plane of propeller rotation, which is perpendicular to the axis of rotation or the propeller shaft, is vertical. Under these conditions, the blade angle is simply the number of degrees that the flat side of the blade slants from the vertical, as illustrated in Fig. 22-42. However, an airplane may rest on the ground with its propeller shaft at an angle to the horizontal. The plane of propeller rotation, being perpendicular to the propeller axis of rotation, is then at the same angle to the vertical as the propeller shaft is to the horizontal, as represented by angle A in Fig. 22-43.

Under these conditions, the number of degrees that the flat side of the propeller blade slants from the vertical is the blade angle minus the ground angle of the airplane, or angle B in

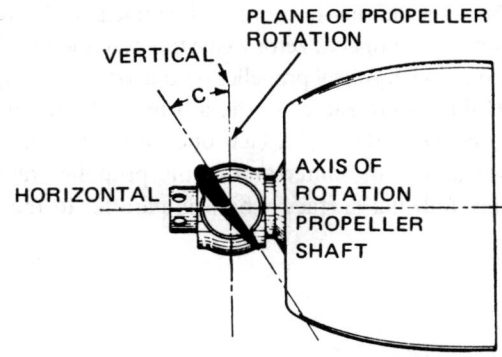

FIGURE 22-42   Measuring a propeller's blade angle.

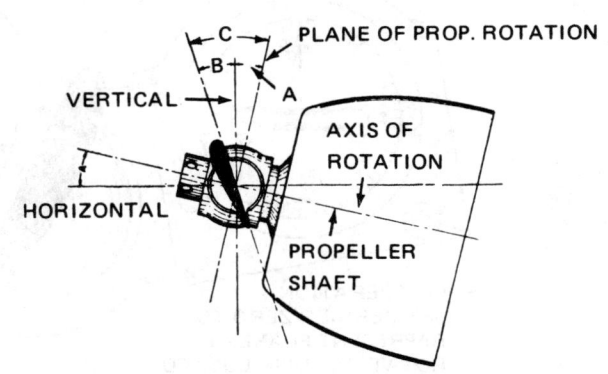

FIGURE 22-43   Effect of ground angle on the measurement of blade angle.

the same illustration. To obtain the actual blade angle, the ground angle of the airplane (which is also the angle at which the plane of rotation slants from the vertical) must be added to the angle at which the flat side of the blade slants from the vertical in the opposite direction, as represented by angle C in the same illustration.

Angles A and B are measured with a universal protractor and added in two related operations. It is then possible to read the total angle, or blade angle C, from the degree and vernier scales of the protractor.

Checking and Setting Blade Angles When the Propeller Is on the Shaft

The following steps are recommended when a universal propeller protractor is available for checking propeller blade angles while the propeller is installed on the shaft of the engine. If it is necessary to use another type of protractor, the procedure to be followed will be modified.

1. Mark the face of each blade with a lead pencil at the blade station prescribed for that particular blade.

2. Turn the propeller until the first blade to be examined is in a horizontal position with its leading edge up.

3. Using a universal propeller protractor, swing the corner spirit level out as far as it will go from the face of the protractor.

4. Turning the disk adjuster, align the zeros of both scales and lock the disk to the ring by placing the spring-loaded pin of the disk-to-ring lock in the deep slot.

5. See that the ring-to-frame lock is released. By turning the ring adjuster, turn both zeros to the top. Refer to Fig. 22-40, which shows the universal propeller protractor.

6. Hold the protractor in the left hand, by the handle, with the curved edge up. Place one vertical edge of the protractor across the outer end of the propeller retaining nut or any hub flat surface which is parallel to the plane of propeller rotation, which means that it is placed at right angles to the propeller-shaft centerline. Using the corner spirit level to keep the protractor vertical, turn the ring adjuster until the center spirit level is horizontal. The zeros of both scales will now be set at a point that represents the plane of propeller rotation. This step can be understood better by referring to the illustration of the measurement of the blade angle in Fig. 22-44.

7. Lock the ring to the frame to fix the vernier zero so that it continues to represent the plane of propeller rotation.

8. Release the disk-to-ring lock by pulling the spring-loaded pin outward and turning it to 90°. This completes what is often called the **first operation**, shown at the left in Fig. 22-44.

9. Change the protractor to the right hand, holding it in the same manner as before, and place the other vertical edge of the protractor, which is the edge opposite the first edge used, against the blade at the mark which was made with a pencil on the face of the blade. This is the beginning of the **second operation**, which is shown at the right in Fig. 22-44. Keep the protractor vertical by means of the corner spirit level. Turn the disk adjuster until the center spirit level is horizontal, as shown in the illustration. In this manner, the angle at which the flat side of the blade slants from the vertical is added to the angle at which the plane of rotation slants from the vertical in the opposite direction.

10. Read the number of whole degrees on the degree scale between the zero of the degree scale and the zero of the vernier scale. Read the tenths of a degree on the vernier scale from the vernier zero to the vernier-scale graduation that comes the closest to lining up with a degree-scale graduation. In this manner the blade angle is determined.

11. Obtain the required blade angle by making any necessary adjustments of the blade or the propeller pitch-changing mechanism.

12. Repeat this procedure for each of the remaining blades to be checked.

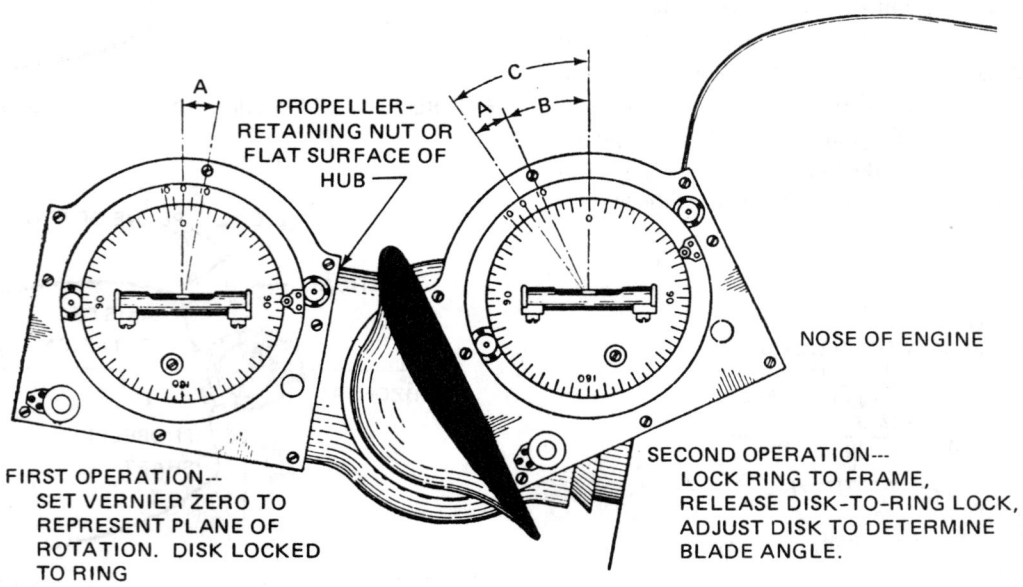

FIGURE 22-44 Measurement of the blade angle in two operations.

INSPECTIONS AND ADJUSTMENTS OF PROPELLERS

There are many types of propeller inspections, including daily inspections, 100-h inspections, and inspections performed during overhaul. Although inspection procedures will vary among the different types of propellers, some examples of inspection procedures on a compact propeller 100-h inspection are as follows:

1. Check for oil and grease leaks.

2. Clean the spinner, propeller hub interior and exterior, and blades with a noncorrosive solvent.

3. Inspect the hub parts for cracks.

4. Steel hub parts should not be permitted to rust. Use aluminum paint to touch up, if necessary, or replate during overhaul.

5. Check all visible parts for wear and safety.

6. Check blades to determine whether they turn freely on the hub pivot tube. This can be done by rocking the blades back and forth through the slight freedom allowed by the pitch-change mechanism.

7. Inspect blades for damage or cracks. Nicks in the leading edges of blades should be filed out and all edges rounded. Use fine emery cloth for finishing.

8. Check condition of propeller mounting nuts and studs.

9. For severe damage, internal repairs, or replacement of parts, transport the propeller to a certified propeller repair station.

10. Sand each blade face lightly and paint with a flat black paint to retard glare, when necessary. A light application of oil or wax may be applied to the surfaces to prevent corrosion.

11. Grease the blade hub through the zerk fittings, as shown in Fig. 22-45. Remove one of the two fittings for each propeller blade. Apply grease through the zerk fitting until fresh grease appears at the fitting hole of the removed fitting. Care should be taken to avoid blowing out hub gaskets.

12. Check for air leaks by applying soap solution around the air valve and stop adjustment nut. Internal leakage will show up as airflow through the piston rod.

13. Record the appropriate information in the propeller records.

Propeller Static RPM (Fixed-Pitch Propeller)

Normally, a fixed-pitch propeller will not develop its maximum rpm while the aircraft is sitting static, or still, because of the lack of air passing through the propeller blades. As discussed in Chap. 20, as the airspeed of the aircraft increases, the angle of attack of the propeller blades decreases, thus allowing the propeller to turn faster.

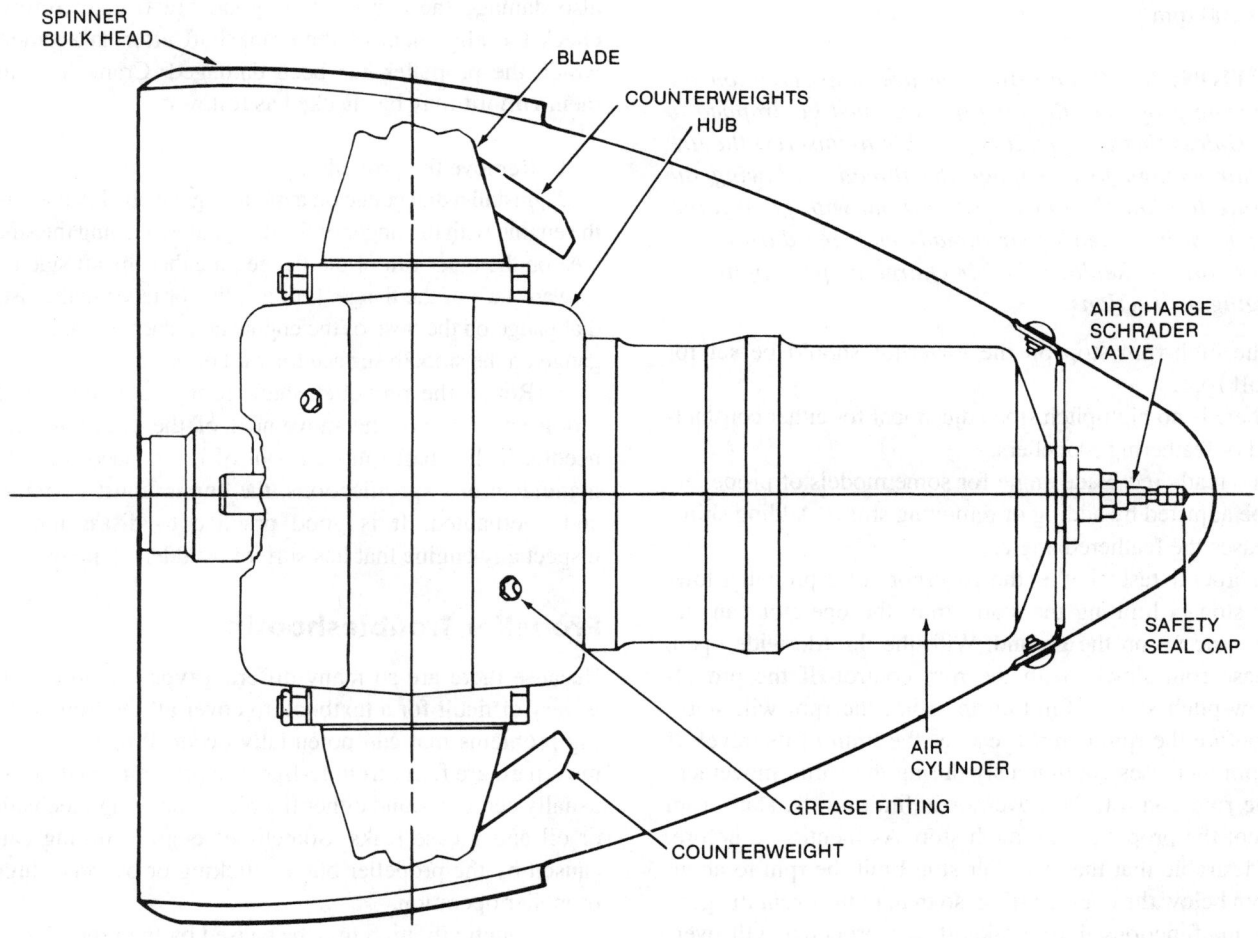

FIGURE 22-45 Grease and air charge fittings. (*Hartzell Propeller.*)

As the aircraft approaches flying speed, with throttles full open, the propeller should approach the red-line rpm as noted on the tachometer.

The **static rpm** is checked during a power run-up to determine whether the engine is developing its rated power. The static rpm is generally found in the aircraft's Type Certificate Data Sheet. For example, for an aircraft that is red-lined at 2750 rpm, static rpm may range from 2100 to 2200 rpm. If an aircraft does not produce its static rpm, this could be a signal that the engine or propeller is in need of maintenance. On a fixed-pitch propeller, there is no means of adjusting the static rpm.

Constant-Speed Propeller Adjustments

Some propeller models have items which may have to be adjusted after installation, such as low-pitch stop for static rpm, high-rpm stop, high-pitch stop, and feathering pitch stop. These adjustments are described in the proper manufacturer's manuals. The **low-pitch stop** on the Hartzell Compact propeller should be set to obtain takeoff rpm, or about 50 rpm below takeoff rpm, during engine run-up on the ground. This stop is normally set for each specific engine application at the factory. In the event that an adjustment is required, it can be made by adjusting the screw in the nose of the cylinder. Backing the screw out one-half turn will increase the static rpm by about 100 rpm; conversely, turning the screw in one-half turn will decrease the static rpm by about 100 rpm.

CAUTION: *Before adjusting the low stop screw on the feathering propeller, the air pressure must be dropped to zero. Unless this is done, it is possible to unscrew the low stop far enough to disengage the threads, allowing the pressure to blow the low stop screw out with great force. There must be at least four threads engaged during normal operation. Replace the air charge as per applicable charging instructions.*

The high-rpm stop on the governor should be set for takeoff rpm.

There is no high-pitch stop adjustment for either constant-speed or feathering propellers.

The feathered blade angle for some models of propellers can be adjusted by adding or removing shims. Adding shims increases the feathered angle.

In order to test whether the governor or the propeller low-pitch stop is limiting the static rpm, the operator can run the engine up on the ground. With the throttle wide open, increase rpm slowly with the rpm control. If the propeller low-pitch stop is limiting the rpm, the rpm will stabilize before the rpm control reaches the limit of its travel. If the rpm increases continuously during the entire movement of the rpm control, the governor is limiting the static rpm and not the propeller low-pitch stop. As mentioned before, it is desirable that the propeller stop limit the rpm to about 50 rpm below the engine rating, so that, in the event the governor malfunctions during takeoff, the propeller will overspeed a minimum amount.

Special Inspections after Accidents

A propeller strike is defined as follows:

A. Any incident, whether or not the engine is operating, that requires repair to the propeller other than minor dressing of the blades.

B. Any incident during engine operation in which the propeller impacts a solid object, which causes a drop in rpm and also requires structural repair of the propeller (incidents requiring only paint touch up are not included). This is not restricted to propeller strikes against the ground, and although the propeller may continue to rotate, damage to the engine may result, possibly progressing to engine failure.

C. A sudden rpm drop while impacting water, tall grass, or similar nonsolid medium, where propeller structural damage is not normally incurred.

If the propeller strikes or is struck by any object (sudden stoppage), examine it for damage. Disassemble any propeller that has been involved in an accident, and carefully inspect the parts for damage and misalignment before using the propeller again. Examine all steel parts and otherwise serviceable steel propeller blades for airworthy damage by means of a magnetic inspection supervised by trained personnel. Have aluminum-alloy blades which are otherwise serviceable given a general etching.

Any accident which severely damages the propeller may also damage the engine. It is good practice, therefore, to check the alignment of the crankshaft after an accident in which the propeller has been damaged. Crankshaft alignment **runout** may be checked as follows:

1. Remove the propeller.
2. Install a dial gauge on a mounting attached to the nose of the engine with the finger of the dial gauge touching the smooth area on the outer rim of the flange on either the aft side or the forward face of the flange. (For a spline or taper shaft, install a dial gauge on the nose of the engine and place the finger of the gauge on the smooth surface forward of the spline or taper.)
3. Rotate the propeller shaft through a complete revolution and observe the movement of the gauge indicating needle. If the shaft runout is out of limits according to the manufacturer's specifications, the engine must be removed and overhauled. It is good practice to disassemble and inspect any engine that has suffered sudden stoppage.

Propeller Troubleshooting

Because there are so many different types of propellers, it is very difficult for a textbook to cover all the troubleshooting problems that can potentially occur. Propellers, for the most part, are fairly trouble-free; the problems that do occur usually center around either the pitch-changing mechanisms or oil and grease leaks. Sometimes engine surging can be caused by the propeller blades sticking or by very sluggish propeller operation.

Although vibration may be caused by the propeller, there are numerous other possible sources of vibration which can

TROUBLE	PROBABLE CAUSE	CORRECTION
Propeller does not respond to movement of propeller pitch lever.	Governor speeder spring broken	Overhaul or replace governor
	Screen in governor mounting gasket clogged	Remove governor & replace gasket
	Governor drive shaft sheared	Overhaul or replace governor
Engine speed will not not stabilize.	Governor relief valve sticking	Overhaul or replace governor
	Excessive clearance in pilot valve	Overhaul or replace governor
	Excessive governor oil pump clearance	Overhaul or replace governor
Failure of propeller to go full low pitch (high rpm).	Governor arm reaches stop before maximum rpm is obtained	Adjust governor
	Defective governor	Overhaul or replace governor
Sluggish propeller movement to either high or low pitch	Excessive propeller blade friction	Grease blade bearing
Failure of propeller to feather.	Attempting to feather from too low an engine rpm	Increase rpm and attempt to feather again
	Automatic high pitch stop pin in the engaged position	Disengage stop pin. Check for freedom of operation & correct cause of pin sticking
	Feathering spring weak or broken	Overhaul propeller
Oil leaking around propeller mounting flange.	Damaged hub O-ring seal	Remove propeller and replace O-ring seal

FIGURE 22-46   Example of a troubleshooting chart for a constant-speed propeller.

make troubleshooting difficult. The dynamic track and balance procedures previously outlined should be followed in troubleshooting of propeller vibrations.

The propeller spinner can contribute to an out-of-balance condition. An indication of this would be a noticeable spinner "wobble" while the engine is running. This condition is normally caused by inadequate shimming of the spinner front support or a cracked or deformed spinner.

A typical troubleshooting chart for propeller problems is shown in Fig. 22-46.

REVIEW QUESTIONS

1. What types of hubs are generally used to mount propellers on engine crankshafts?

2. What is the purpose of the cones used in the installation of a propeller on a splined shaft?

3. How does the retaining nut for a propeller serve as a puller when the propeller is removed?

4. Why is a rear-cone spacer used with some installations?

5. What items should be checked prior to the installation of a propeller on a crankshaft?

6. What sequence should be followed during tightening of propeller attachment bolts?

7. Explain how the fit of a tapered nub and shaft is checked.

8. What special tool is needed for installation of a PT6A turbopropeller?

9. What is meant by the term *correctable vibration*?

10. What is the purpose of checking propeller track?

11. What components comprise a dynamic vibration measuring system?

12. What are three methods of measuring the phase angle of the propeller during dynamic balancing?

13. Define the term *major alteration* as it applies to propellers.

14. Who is authorized to perform major repairs and alterations on propellers?

15. Describe the repair of small cracks parallel to the grain in a wood propeller.

16. Who is authorized to repair damaged steel propeller blades?

17. How may minor damage of the leading and trailing edges of a steel blade be repaired?

18. Define a *damaged metal propeller blade*.

19. What should be done to aluminum-alloy blades that have pitted leading edges as a result of normal wear?

20. What is the purpose of local etching?

21. Define the terms *debond* and *delamination* as they apply to composite propellers.

22. How may a composite propeller blade be tested for debond and delamination?

23. What device is used for checking propeller blade angle?

24. Can static rpm be adjusted on a fixed-pitch propeller?

25. After a propeller is damaged in an accident, what inspections should be made?

Engine Indicating, Warning, and Control Systems 23

ENGINE INSTRUMENTS

The safe and efficient operation of aircraft powerplants of all types requires installation of indicating instruments for measurement of various parameters associated with the performance of the engine. Early aircraft engines usually were provided with nothing more than an oil pressure gauge, an oil temperature gauge, and a tachometer. With these basic instruments, the pilot was able to determine with reasonable accuracy whether or not the engine was performing as it should.

As engines became more complex with improved design and performance, it became necessary to measure additional factors governing performance. Accordingly, manifold pressure gauges, carburetor air temperature gauges, cylinder head temperature gauges, fuel pressure gauges, fuel flowmeters, exhaust gas temperature gauges, and a number of indicating instruments were developed and employed. Gas-turbine engines required even more gauges, including engine pressure ratio indicators, exhaust-gas temperature gauges, torquemeters, percent rpm gauges, and vibration indicators.

We will examine briefly the operating principles of engine gauges, explain how troubles are indicated by the gauges, and describe methods of maintenance and repair of indicating systems. We shall not attempt to describe the overhaul of instruments, because this type of work must be done by qualified instrument technicians in approved instrument overhaul shops. Additional information regarding the theory, construction, and maintenance of instruments and instrument systems is provided in other texts.

Instruments for Reciprocating Engines

Pressure Instruments

Pressure instruments are usually of the bourdon-tube type, diaphragm type, or bellows type. The particular type of mechanism employed to provide an indication of pressure depends on the requirements of the system and the level of

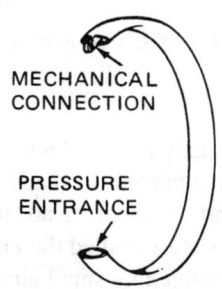

FIGURE 23-1   Bourdon tube.

pressure which must be measured. Comparatively high pressures usually require the use of bourdon-tube instruments.

The design of a typical **bourdon tube** is illustrated in Fig. 23-1. This tube is formed in a circular shape with a flattened cross section. The material of the tube is thin metal sheet, such as brass. When air or liquid pressure is applied to the open end of the bourdon tube, the tube tends to straighten out. This reaction of the tube is utilized to move a sector gear which, in turn, rotates the spur gear attached to the shaft of an indicating needle. Thus, when the bourdon tube changes its circular form because of changes in the pressure applied to it, the sector gear is moved and the indicating needle is rotated. The arrangement of a bourdon-tube indicating mechanism is shown in Fig. 23-2.

Oil pressure gauges and some oil temperature gauges are typical of bourdon-tube instruments. The face of an oil pressure gauge for a light aircraft is shown in Fig. 23-3.

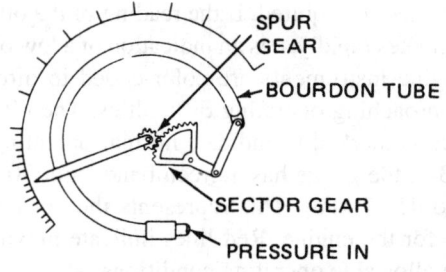

FIGURE 23-2   Bourdon-tube indicating mechanism.

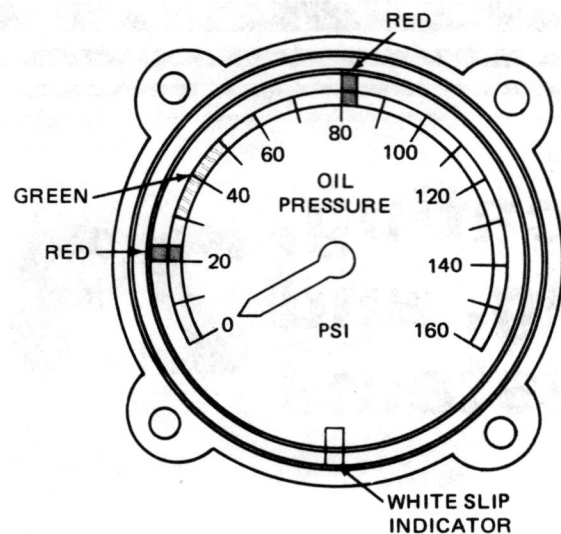

FIGURE 23-3   Oil pressure gauge with range markings.

This instrument is usually mounted with other engine instruments in a common subpanel. In many cases, oil pressure, oil temperature, and cylinder head temperature (CHT) instruments are in one case, called the **engine gauge unit**.

The oil pressure gauge in a small aircraft is directly connected by means of a tube to the engine at a point immediately downstream from the engine oil pump. The gauge therefore indicates the pressure being delivered to the engine. This instrument includes a small restriction in the inlet to prevent pressure surges from damaging the instrument. When the aircraft is operated in cold-weather conditions, the tube from the instrument to the engine is filled with light oil. The light oil allows the instrument to react to pressure changes in a normal manner, whereas engine oil would be partially congealed, thus preventing proper operation.

The oil pressure gauge is of primary importance when an engine is first started. If there is no indication of oil pressure within the first 30 s, the engine should be stopped and an investigation made to determine the cause. When an engine has been idle for an extended period of time, oil may have drained from the oil pump and it may be necessary to prime the pump by injecting oil into the inlet side. Other causes for a lack of oil pressure may be a low oil supply, a clogged oil inlet line, or a broken line.

During operation, the oil pressure gauge can provide an indication of engine condition. Below-normal oil pressure is often a sign that the engine has become worn to the extent that an overhaul is required. If the reading of the oil pressure gauge fluctuates rapidly, it is an indication of a low oil supply.

All engine instruments are color-coded to direct attention to approaching operating difficulties. The oil pressure gauge face is marked to indicate normal operating ranges. In Fig. 23-3, the gauge has a green band from 30 to 60 psi [206.85 to 413.7 kPa]. This represents the normal operating range for the engine. Red lines indicate maximum and minimum allowable operating conditions.

When the limit and range markings are placed on the glass face of an instrument, a white line is placed on the glass

and extended to the case of the instrument. This line, referred to as a **slippage mark**, will be broken and offset if the cover glass should move. When this condition is noted, the cover glass should be rotated until the white line is again in alignment. The glass should then be secured to prevent further rotation. If the cover glass rotates, the limit and range markings will not give true indications of engine operation.

Inspections of the oil gauge system include the following steps:

1. Check for oil leaks at the gauge connection and the engine fitting.
2. Check the gauge for security of mounting and make sure that there is no looseness of mounting screws.
3. Examine the routing of the oil line from the engine to the gauge. See that the line is securely attached with suitable padded clamps and that the line does not rub against any part or structure which could cause wear as a result of vibration.
4. Examine the limit and range markings to ensure that they are adequate and that the face glass has not rotated.

In some large aircraft, the oil pressure indicating system utilizes a **pressure transmitter** mounted near the engine. The transmitter consists of two chambers separated by a flexible diaphragm. One chamber is connected to the engine oil pressure outlet and the other is connected to the oil pressure gauge. The chamber on the gauge side of the transmitter is filled with a light oil, which also fills the line leading to the gauge. The advantage of this type of system is that the light oil cannot be contaminated with engine oil and will therefore continue to be effective during extreme cold-weather operation. The transmitter operates by applying engine oil pressure to the diaphragm, which, in turn, pressurizes the light oil in the instrument side of the chamber. The pressure is then applied to the oil pressure gauge through the pressure gauge tube.

Diaphragm-type gauges. Diaphragm-type gauges are generally used to measure comparatively low pressures. The construction of an instrument diaphragm is shown in Fig. 23-4. Note that the diaphragm consists of two thin metal disks concentrically corrugated and sealed together at the edges to form a cavity. One side is provided with an opening at the center

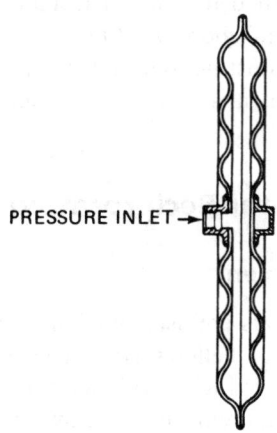

FIGURE 23-4   The construction of an instrument diaphragm.

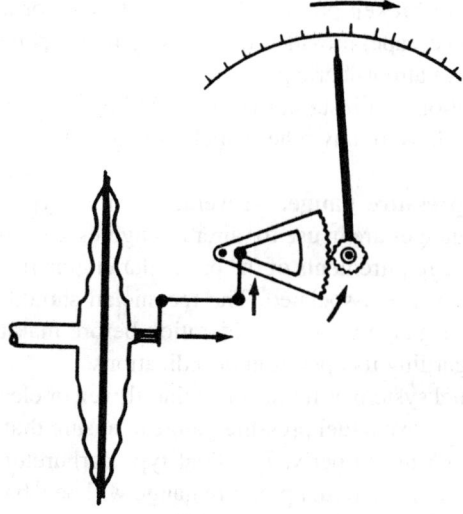

FIGURE 23-5 How a diaphragm may be linked to the indicating needle.

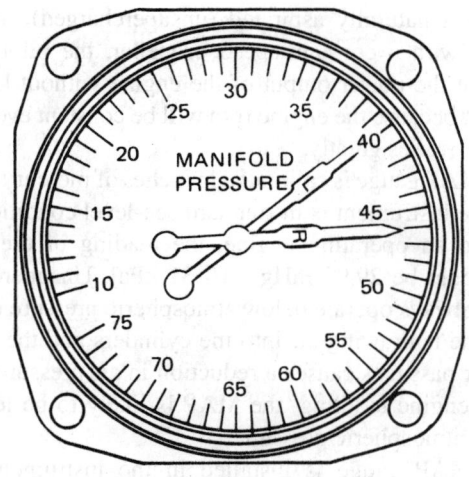

FIGURE 23-6 The face of a MAP gauge.

through which air or gas pressure can be applied to the inside of the diaphragm. As pressure is applied, the two sides spread apart, thus producing a motion which can be used to actuate an indicating needle. How a diaphragm may be linked to the indicating needle is illustrated in Fig. 23-5.

An engine instrument which utilizes a diaphragm to sense pressure is the **manifold pressure (MAP) gauge**. This gauge was mentioned in the discussions on engine performance and induction systems. The acronym MAP is used to indicate manifold pressure because manifold pressure is absolute pressure; thus, MAP stands for *manifold absolute pressure*. The manifold pressure gauge (MAP gauge) is, in effect, a barometer, because it measures atmospheric pressure when it is not connected to a running engine. When the engine is stopped, the MAP gauge should indicate the local barometric pressure. To accomplish this, the MAP gauge is provided with two diaphragms. One diaphragm is a sealed *aneroid* cell which responds to ambient atmospheric pressure. The other diaphragm is connected to the intake manifold of the engine and responds to the MAP. The effect is that the gauge provides an indication of the *absolute* pressure (pressure above zero pressure) existing in the intake manifold of the engine.

MAP gauges may be designed so that the MAP of the engine is applied to the inside of a diaphragm or to the exterior of the diaphragm in a sealed instrument case. In all cases the instrument must be designed so that the chemical constituents in the intake manifold will not get into the operating mechanisms of the instrument. Furthermore, the pressure line from the manifold to the diaphragm or instrument case must be provided with a restriction, such as a coiled capillary tube, to prevent excessive pressures from reaching the instrument, as in the case of a backfire.

The face of a typical MAP gauge is shown in Fig. 23-6, and the interior mechanism is shown in Fig. 23-7. The cover glass on the face or the dial of the instrument has limit and range markings to aid the pilot in operating the engine safely and efficiently. A range marked in blue shows the MAP values which may be used while operating in

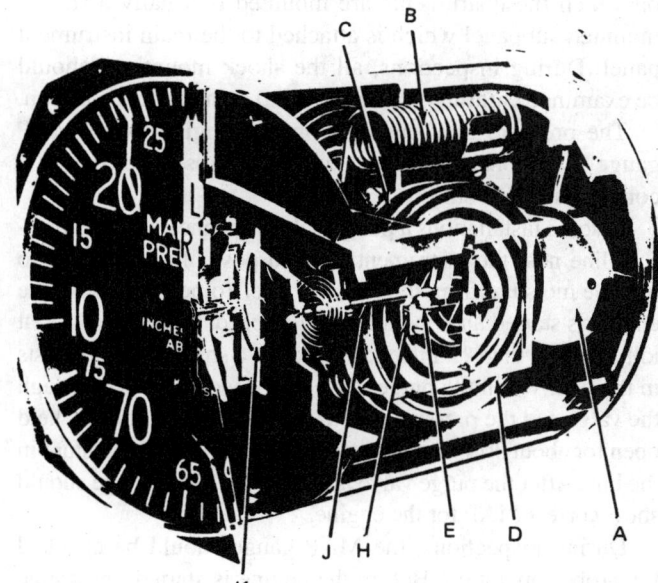

A. Pressure connection
B. Capillary coil
C. Pressure diaphragm
D. Aneroid diaphragm
E. Bimetallic temperature compensator
H. Actuating arm
J. Rocking shaft
K. Sector gear

FIGURE 23-7 Interior mechanism of a MAP gauge. (*Kollsman.*)

AUTO LEAN, and a green band shows the range of pressures which are satisfactory for AUTO RICH. A red line indicates maximum allowable MAP for takeoff power, and a white line shows slippage of the cover glass.

The MAP gauge is particularly important for use with supercharged engines and those with which constant-speed propellers are employed, because the MAP has a direct effect on mean effective pressure (mep). Excessive mep can cause severe engine damage. When an engine is being operated under supercharged MAPs, the pilot must know that safe operating pressures are not being exceeded. As explained previously, excessive MAPs cause high cylinder pressures, which result in detonation, overheating, and preignition.

When a naturally aspirated (unsupercharged) engine is operated with a constant-speed propeller, the pilot cannot determine the power output of the engine without knowing the MAP, because the engine rpm will be constant even when the power varies greatly.

The MAP gauge is calibrated in inches of mercury (inHg). When the instrument is in standard sea-level conditions and is not on an operating engine, the reading of the instrument should be 29.92 inHg [101.31 kPa]. Unsupercharged engines always operate below atmospheric pressure because the engine is drawing air into the cylinders and the friction of the air passages causes a reduction in air pressure. When such an engine is idling, the MAP is likely to be less than one-half atmospheric pressure.

The MAP gauge is installed in the instrument panel adjacent to other engine instruments. Ideally, it should be next to the tachometer so that the pilot can quickly read rpm and MAP to determine engine power output. The panel on which the instruments are mounted is usually a shock-mounted subpanel which is attached to the main instrument panel. During inspections, all the shock mountings should be examined carefully for damage and signs of deterioration.

The pressure line leading from the engine to the MAP gauge may consist of metal tubing, a pressure hose, or a combination of both.

In some installations a **purge valve** is connected to the pressure line near the instrument. The purpose of this valve is to remove moisture from the pressure line. To purge the line, the engine is started and operated at idle speed. Since the MAP at idling is less than 15 inHg [50.79 kPa], a strong suction exists in the line. When the purge valve is opened, air flows through the valve and the pressure line to the engine. The valve is held open for about 30 s, and this effectively removes all moisture in the line. After the purge valve is closed, the MAP gauge should show correct MAP for the engine.

During inspections, the MAP gauge should be checked for proper operation. Before the engine is started, the gauge should show local barometric pressure. After the engine has been started and is idling, the MAP gauge reading should drop sharply. When the engine rpm is increased, the gauge should be watched to see that it increases evenly and in proportion to power output. In the event that the reading lags or fails to register, the cause can usually be traced to one of the following discrepancies: (1) the restriction in the instrument case fitting is too small, (2) the tube leading from the engine to the gauge is clogged or leaks, or (3) the diameter of the tubing is too small for its length. If the reading is jumpy and erratic when the engine speed is increased, the restriction in the case fitting is probably too large. If there is a leak in the gauge pressure line, the MAP will read high for an unsupercharged engine. If the engine is supercharged and is operating at a MAP above atmospheric pressure, a leaking pressure line will cause the instrument to give a low reading. If the pressure line is broken or disconnected, the MAP gauge will read local barometric pressure. A leaking or broken pressure line on an unsupercharged engine will permit outside air to leak into the manifold and lean the mixture to some extent, depending on the size of the tubing and fitting. On a supercharged engine, a leaking or broken pressure line will allow air or a fuel-air mixture to escape, provided that the engine is operating at a MAP above atmospheric pressure.

Inspection and installation of the MAP gauge should be similar to those of any other panel-mounted instrument.

Fuel pressure gauge. Several different types of fuel pressure gauges are in use for aircraft engines, each designed to meet the requirements of the particular engine fuel system with which it is associated. The technician should identify the type of gauge under consideration before making judgments regarding its operation or indications.

Any fuel system utilizing an engine-driven or electric fuel pump must have a fuel pressure gauge to ensure that the system is working properly. If a float-type carburetor is used with the engine, the fuel pressure gauge will be a basic type, probably with a green range marking from 3 to 6 psi [20.69 to 41.37 kPa]. If the engine is equipped with a pressure discharge carburetor, the range marking on the fuel pressure gauge will be placed in keeping with the fuel pressure specified for the carburetor. This will probably be between 15 and 20 psi [103.43 and 137.9 kPa]. A red limit line will be placed at each end of the pressure-range band to indicate that the engine must not be operated outside the specified range.

If an engine is equipped with either a direct fuel injection system or a continuous-flow fuel injection system, the fuel pressure is a direct indication of power output. Since engine power is proportional to fuel consumption and since fuel flow through the nozzles is directly proportional to pressure, fuel pressure can be translated into engine power or fuel flow rate, or both. The fuel pressure gauge, therefore, can be calibrated in terms of percentage of power. A gauge of this type is shown in Fig. 23-8. Note that the instrument indicates a wide range of pressures at which the engine can operate. It is calibrated in pounds per square inch and also indicates percentage-of-power and altitude limitations. The face of the instrument is color-coded blue to show the normal cruise range during which the engine can be operated in AUTO LEAN and green to show the range during which the AUTO RICH mixture setting should be used.

FIGURE 23-8   A fuel pressure gauge.

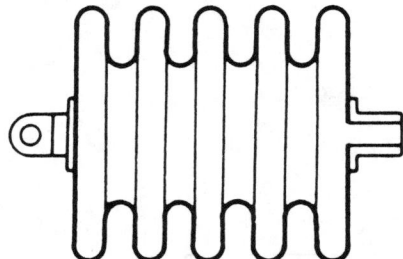

FIGURE 23-9   A bellows capsule.

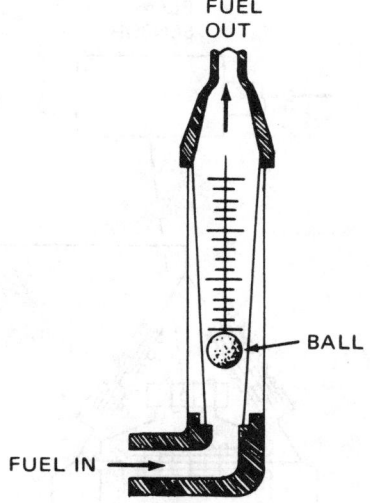

FIGURE 23-10   Tubular fuel flowmeter.

Fuel pressure gauges are similar in construction to other pressure gauges used for relatively low pressures. The actuating mechanism is either a diaphragm or a pair of **bellows**. The advantage of the bellows is that it provides a greater range of movement than does a diaphragm. In a typical fuel gauge, the mechanism includes two bellows capsules joined end to end. A bellows capsule is shown in Fig. 23-9. One capsule is connected to the fuel pressure line and the other is vented to ambient pressure in the airplane. The fuel pressure causes the fuel bellows to expand and move toward the air capsule. This movement is transmitted to the indicating needle by means of conventional linkage.

Fuel Flowmeters

Fuel flowmeters are employed on all large aircraft to provide the flight engineer or pilot with important information regarding the efficient operation of the engines. The flowmeter is the most accurate way to determine fuel consumption for all types of engines. The quantity of fuel being burned per hour, when integrated with other factors, makes it possible to adjust power settings for maximum range, maximum speed, or maximum economy of operation. If the fuel consumption of an engine is abnormal for any particular power setting, the operator is warned by the fuel flowmeter that there is something wrong (a fault) in the engine or its associated systems.

Fuel flowmeters may be scaled for pounds per hour, gallons per hour, or kilograms per hour. The most accurate units of measurement for fuel are the pound and the kilogram, because these are measures of weight. Since the weight of a gallon of fuel varies with temperature, fuel temperature would have to be considered in making a computation based on gallons per hour of fuel consumption.

In engine test cells, a tubular flowmeter is often used. This is simply a glass or plastic tube with a tapered inside diameter. The tube is mounted vertically with an indicating ball inside. Fuel enters the bottom of the tube and flows out the top and in so doing causes the indicating ball to rise a distance proportional to the rate of fuel flow. Figure 23-10 shows a tubular fuel flowmeter.

Light aircraft with continuous-flow fuel injection engines usually employ a fuel pressure gauge which also serves as a flowmeter. This is possible because the fuel flow in a direct fuel injection system is proportional to fuel pressure. The instrument can therefore be calibrated either in pounds per square inch or gallons per hour, or both. A gauge of this

FIGURE 23-11   Fuel pressure gauge also used as a flowmeter.

type is shown in Fig. 23-11. Owing to its design characteristics, this system is not as accurate as the above-mentioned fuel flowmeters. If a fuel injection nozzle were to become clogged, the gauge would indicate high fuel flow rather than the lower indication normally expected.

Fuel flow indicating systems for large aircraft are of two general types. The first, used in many older aircraft, employs a **synchro system** to transmit a measurement of fuel flow from the sensor and transmitter to the indicator. The principle of the synchro system is illustrated in Fig. 23-12. Fuel flow through the flow sensor causes rotation of the transmitter rotor to a position proportional to the fuel flow rate. The rotor, being energized electrically by 400 Hz ac, generates a three-phase current in the stator which flows through the stator of the indicator. The magnetic field thus produced reacts with the field of the rotor in the indicator and causes it to assume a position corresponding to the position of the rotor in the transmitter. The indicator needle is mounted on the shaft with the rotor; therefore, it moves to a position established

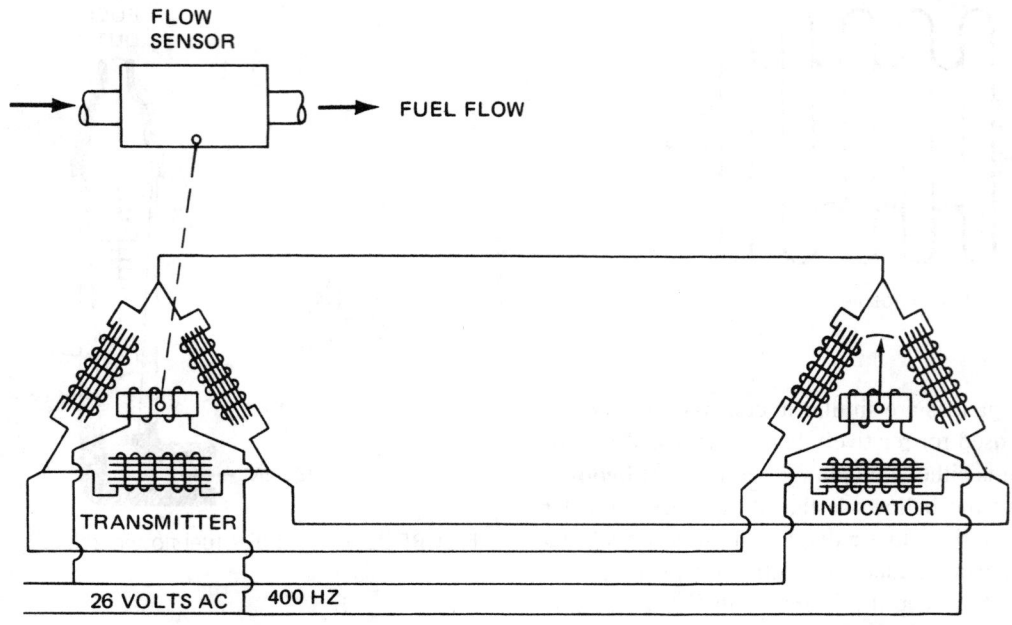

FIGURE 23-12 Flowmeter utilizing a synchro transmitting system.

by the rotor. The indicator needle thereby shows the fuel flow rate on the instrument scale.

Fuel flow indicating systems for modern jet aircraft are complex electrical and electronic circuits and devices. The transmitter is located in the fuel line between the engine-driven fuel pump and the carburetor, or fuel control unit. The sensing device may be a gate-type unit, or it may be a turbine which generates an electrical signal. In any case, the signal sent from the transmitter contains the required information regarding fuel flow rate. The electrical signal is delivered to the "receiver" of the indicator, where it is converted to a form which provides movement of the indicator needle.

Fuel Pressure Warning Systems

For aircraft equipped with multiple-tank fuel systems, it is desirable to have a means of warning the pilot or flight engineer that the fuel in an operating tank is exhausted and that the fuel selector valve must be set to draw fuel from another tank. The fuel pressure system consists of a diaphragm-operated pressure switch with one side of the diaphragm connected to the main fuel line downstream from the main fuel pressure pump, electric wiring to a warning light in the cockpit, and possibly a buzzer or other aural alarm. Fuel pressure holds the switch open at normal pressures, but the switch closes when the pressure falls. This turns on the warning light in the cockpit.

Temperature Gauges

The safe and efficient operation of an aircraft engine requires that temperatures in the engine be monitored and controlled within carefully delineated limits. The most critical temperature information needed for the operation of a reciprocating engine relates to oil temperature and cylinder head tempera-

ture (CHT). As long as these temperatures are within established limits, it is not likely that any damage to the engine will occur because of heat.

In addition to oil temperature and CHT gauges, carburetor air temperature (CAT) gauges and exhaust-gas temperature (EGT) gauges are installed in many aircraft. For reciprocating engines, the EGT gauge is the indicator for the **economy mixture indicating system**. Since the temperature of the exhaust gas increases as combustion improves, it reveals to the pilot when the engine is providing the most complete combustion, thus getting the most power for the fuel consumed. EGT gauges are also employed as exhaust-gas analyzers and may be installed with sensors in all the exhaust stacks of an engine. The purpose of the exhaust-gas analyzer is to indicate the fuel-air ratio of the mixture being burned in the cylinders. The CAT gauge is important as a means of detecting icing conditions and regulating engine performance, particularly when a float-type carburetor system is turbocharged.

Oil temperature gauge. The most common type of oil temperature gauge is operated electrically and may utilize either a Wheatstone bridge circuit or a ratiometer circuit. Since the ratiometer circuit provides a more dependable temperature indication under varying input voltage conditions, this type of instrument has become standard for most installations.

The circuit for a **Wheatstone bridge instrument** is shown in Fig. 23-13. The bridge consists of three fixed resistors of 100 Ω each and one variable resistor whose resistance is 100 Ω at a fixed point such as 0°C, or 32°F. With the bridge circuit connected to a battery power source as shown, if all four resistances are equal, current flow through each side of the bridge will be equal and there will be no differential in voltage between points A and B. When the variable

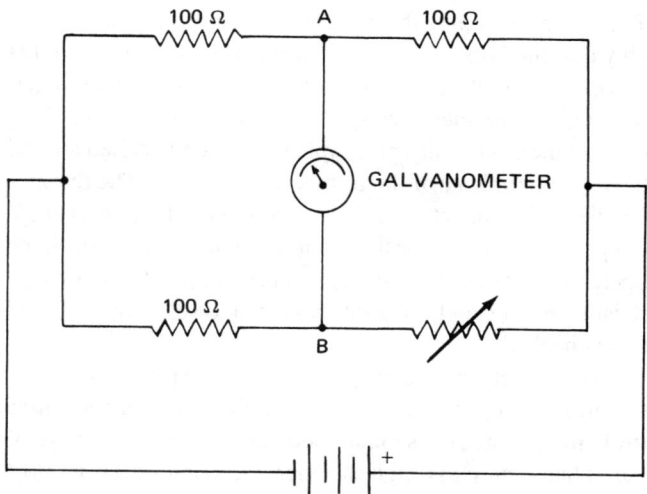

FIGURE 23-13   Circuit for a Wheatstone bridge.

resistor in the temperature-sensing bulb is exposed to heat, the resistance increases. When this happens, it can be seen that more current will flow through the top portion of the bridge than will flow through the bottom portion. This will cause a voltage differential between points A and B, and current will then flow through the galvanometer indicator. The instrument is calibrated to give a correct reading of the temperature sensed by the variable resistor in the sensing bulb. If any one segment of the bridge circuit should be broken, the indicator needle would move to one end of the scale, depending on which circuit was broken.

The circuit for a **ratiometer instrument** is shown in Fig. 23-14. This circuit has two sections, each supplied with current from the same source. As shown in the drawing, a circular iron core is located between the poles of a magnet in such a manner that the gap between the poles and the core varies in width. The magnetic flux density in the narrower parts of the gap is much greater than the flux density in the wider portions of the gap. Two coils are mounted opposite each other on the iron core, and both coils are fixed to a common shaft on which the indicating needle is mounted. The coils are wound to produce magnetic forces which oppose each other. When equal currents are flowing through the two

coils, their opposing forces will balance each other and the coils will be in the center position, as shown. If the current flowing in coil A is greater than that flowing in coil B, the force produced by coil A will be greater than the force of coil B, and coil A will move toward the wider portion of the gap, where a lower flux density exists. This will move coil B a sufficient distance into an area of higher flux density to create a force which will balance the force of coil A. At this point, the position of the indicating needle gives the appropriate temperature reading.

Like the sensing bulb for the Wheatstone bridge instrument, the bulb for the ratiometer instrument contains a coil of fine resistance wire which increases in resistance as the temperature rises. The resistance wire is sealed in a metal tube and is connected to pin connectors for the electrical contacts.

If it is found during inspection that the ratiometer instrument is giving an off-scale high reading, then either the sensing circuit is broken or there is an open circuit in the balancing coil. The circuits may be checked with an ohmmeter to locate the break. If the instrument is found to be defective, it should be removed and sent to an approved instrument repair facility.

The temperature-sensing bulb is usually located in the engine oil system at a point immediately after the oil has passed through the oil cooler. Some oil coolers have provisions for the temperature bulb to be installed near the outlet of the cooler. In any event, the temperature gauge measures the temperature of the oil entering the engine. The face of the oil temperature gauge is marked with a green band showing the range of normal operating temperatures. Red lines are used for both the minimum allowable operating temperature and the maximum safe temperature. The manufacturer usually specifies the temperature limits for engine operation. As a general rule, 40°C [104°F] is considered the minimum safe temperature for operation, 60 to 70°C [140 to 158°F] is the normal operating range, and 100°C [212°F] is the maximum allowable temperature for a reciprocating engine.

In the past, many oil temperature gauges were of the vapor-pressure type, and many of these instruments are still in service. A **vapor-pressure temperature gauge** is a simple bourdon-tube instrument connected to a liquid-filled bulb by means of a capillary tube. The bulb is filled with a highly volatile liquid which vaporizes and develops pressure inside the bulb; the pressure is transmitted to the bourdon tube through the capillary tube. The pressure of the vapor is in proportion to the temperature; therefore, the indicating needle shows the temperature of the medium in which the bulb is placed. Vapor-pressure instruments were also used for CAT indicators and coolant-temperature indicators.

In the installation and removal of a vapor-pressure instrument, the technician must take care to ensure that the capillary tube connecting the temperature bulb to the instrument is not kinked or otherwise damaged. For protection against wear and abrasion, the capillary tube is usually encased in a sheath made of fine woven wire. Since the instrument and temperature bulb for a vapor-pressure assembly form a sealed unit, the capillary tube cannot be cut to fit any particular routing. Instead, the tubing is usually longer than required and the

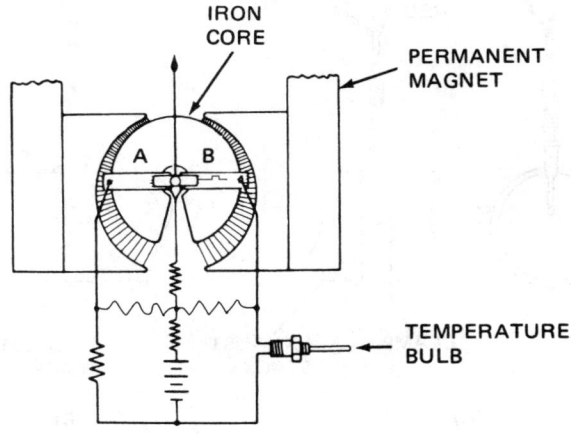

FIGURE 23-14   Circuit for a ratiometer-type instrument.

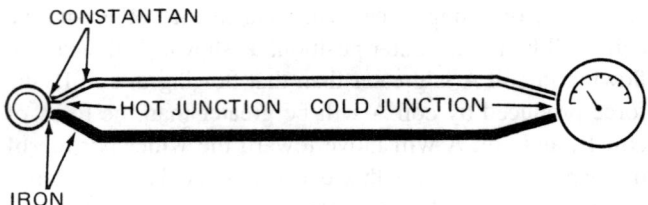

FIGURE 23-15 Thermocouple circuit.

excess length is coiled and secured with padded clamps to the aircraft structure. The entire tubing length must be adequately supported at short intervals so that it cannot be worn by rubbing against any part of the structure or other installations.

Thermocouple instruments.   The measurement of very high temperatures such as CHT and EGT requires the use of thermocouples. A **thermocouple** is the junction of two dissimilar metals which generates a small electric current that varies according to the temperature of the junction. For this reason, it does not require an external power source. To be operational, the thermocouple must be connected in a circuit, as shown in Fig. 23-15. The dissimilar metals can be **constantan and iron**, **Alumel and Chromel**, or some other combination of metals or alloys which will produce the required results. Generally, reciprocating engine CHT systems use iron and constantan, and turbojet EGT systems use Alumel and Chromel. The complete thermocouple circuit consists of the "cold" junction, the "hot" junction, electric leads made from the same material as the thermocouple, and a galvanometer-type indicating instrument.

In the thermocouple illustrated in Fig. 23-15, the dissimilar metals are constantan and iron. Constantan is chosen as one of the metals because its resistance is affected very little by changes in temperature. A constantan wire is connected from the hot junction to the cold junction at the instrument and an iron wire is connected at the other side of the circuit. When the hot junction is at a higher or lower temperature than the cold junction, a current will flow through the circuit and instrument. The value of the current will depend on the difference in temperature between the two junctions.

The indicating instrument used with a thermocouple system includes a bimetallic thermometer mechanism which provides for inclusion of the ambient temperature at the instrument. If the instrument is not connected to the thermocouple leads or if it is connected and the temperatures of the two junctions are equal, the indicator will show ambient temperature.

The hot junction of the thermocouple system may be constructed in a number of different forms. For reading CHT on reciprocating engines, the thermocouple is metallically bonded to a copper spark plug washer or is designed to be secured by a screw in a special well in the head of one of the cylinders. The cylinder to which the thermocouple is attached is determined by the manufacturer. It is the usual practice to find which cylinder operates at the highest temperature and to attach the thermocouple to this cylinder.

Thermocouples for reading EGT for either reciprocating or gas-turbine engines are constructed in the form of probes.

The probe containing the thermocouple is inserted into a suitably designed opening in the exhaust pipe or tailpipe so that it is exposed to the stream of exhaust gases. For a reciprocating engine, one thermocouple is usually adequate; but for a gas-turbine engine, thermocouple probes are installed around the entire periphery of the turbine exhaust case. The thermocouples are connected in series, thus providing an average temperature reading for the points around the exhaust. Some operators install a thermocouple on each cylinder of a reciprocating engine and use a rotary switch to select the cylinder to be checked.

The EGT gauge on a reciprocating engine is used primarily to furnish temperature readings for the **economy mixture indicating system**. As mentioned before, the temperature of the exhaust gases is highest when the combustion is most complete. The temperature therefore informs the pilot when the engine is operating at best economy.

The installation of the thermocouple probe in an exhaust stack is shown in Fig. 23-16A. The probe is mounted in a clamp and the clamp is placed around the exhaust stack. This installation does not require a special fitting to be welded in the exhaust stack. An installation with a welded fitting in the exhaust stack is shown in Fig. 23-16B.

Typical instructions for making a calibration check of the economy mixture indicating system are as follows:

1. Fly the airplane at an altitude of 7500 ft [2286 m] with an average cruising speed condition at 65 percent power.
2. Lean the mixture slowly and observe the EGT gauge. Continue leaning until the EGT reaches a peak and starts to decrease. Enrich the mixture slowly to regain the peak temperature reading.
3. Record the peak reading after the system has stabilized.
4. Lean the mixture to a setting which provides an EGT reading of not less than 25°F [13.9°C] below peak EGT.
5. Use the adjusting screw on the face of the instrument and position the pointer at $\frac{4}{5}$ full scale.

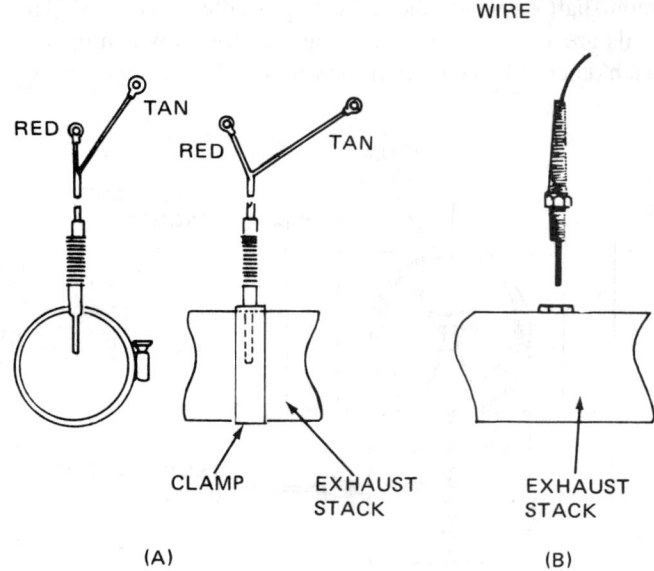

FIGURE 23-16 Thermocouple installation in an exhaust stack.

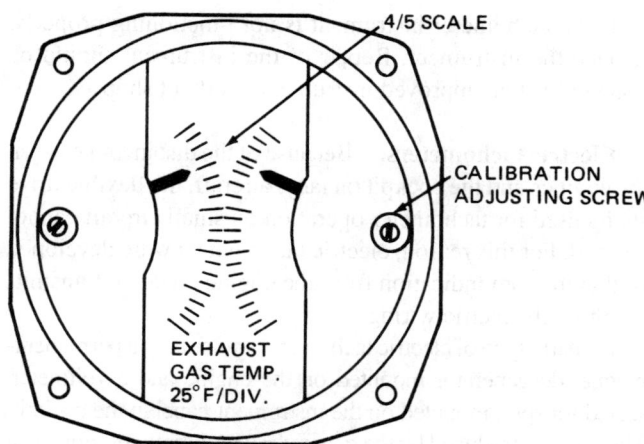

FIGURE 23-17 EGT gauge.

6. If the adjustment requires more than ±75°F [±41.7°C], adjust the calibration screw on the back of the instrument one turn clockwise for an increase in indicator reading of 25°F [13.9°C] on the indicator scale or one turn counterclockwise for a decrease of 25°F [13.9°C].

The face of a typical EGT gauge for a reciprocating engine is shown in Fig. 23-17. This is a double indicator for a two-engine airplane. Note that there are two calibrating screws on the face of the instrument. On the bottom of the back of the instrument are two additional calibrating screws for additional adjustment as required.

Installation, maintenance, and troubleshooting of thermocouple instrument systems are not difficult; however, they require careful attention to certain requirements of such systems. Among these are the following:

1. When installing a thermocouple system, make sure that the specified types of thermocouples, leads, and instruments are used.
2. Check the leads for length, resistance value, and condition. Resistance and continuity can be checked with an ohmmeter or multimeter. Set the meter for the correct range, and then connect the meter leads, one to each end of the thermocouple lead. The meter will show the resistance of the thermocouple lead. Thermocouple leads are matched to specific installations and must not be altered.
3. Check the indicating instrument to see that it indicates the ambient temperature when it is not connected to the thermocouple leads. Usually the instrument will include adjusting screws, sometimes called zero adjusters. These are used to adjust for ambient temperature.
4. Install the instrument in the instrument panel according to the manufacturer's instructions. Handle the instrument carefully to prevent damage due to shock. See that the panel is properly shock-mounted.
5. Install and route the thermocouple leads as specified by the aircraft manufacturer. Secure the leads with clamps at specified intervals and see that the leads do not rub on any structure which will cause wear or other damage. At points where there are junctions in the leads, be sure to correctly connect the terminals. *If the leads are reversed, the indicator will give a reverse reading.* See that all connections are tight and secure.
6. Install the thermocouple on the specified cylinder. See that the attachment is tightened and secured as required.

The following faults in the operation of a thermocouple system are the results of the discrepancies noted:

1. When the instrument gives a reading of ambient temperature only, there is a break in the thermocouple circuit.
2. When the reading is erratic, there is a loose connection in the circuit or in the instrument.
3. When the reading is reversed (decreases with an increase in temperature), the thermocouple leads are reversed at one of the junctions or at the instrument.

Tachometers

The tachometer is a primary engine instrument designed to provide an accurate indication of engine rpm. For reciprocating engines the reading is in rpm, but for gas-turbine engines the reading is given in percentage of maximum allowable rpm.

Mechanical tachometers. Early tachometers were of the mechanical type, employing flyweights on a rotating shaft to produce movement of the indicating needle. Figure 23-18 illustrates the use of flyweights in an instrument mechanism. Flyweight tachometers were effective and operated well; however, these instruments were subject to more wear than later types.

Magnetic tachometers. A very common and widely used tachometer mechanism is the magnetic type. This mechanism utilizes a cylindrical magnet rotating in an aluminum **drag cup** to produce movement of an indicating needle. The flexible drive shaft from the engine is connected to the cylindrical magnet. As the magnet rotates, it generates eddy currents in the aluminum drag cup, and these currents produce electromagnetic forces which cause the drag cup to rotate in the same direction as the rotating magnet. The drag cup is restrained by a coiled spring so that it can rotate only through the distance allowed by the spring. The amount of drag-cup rotation is

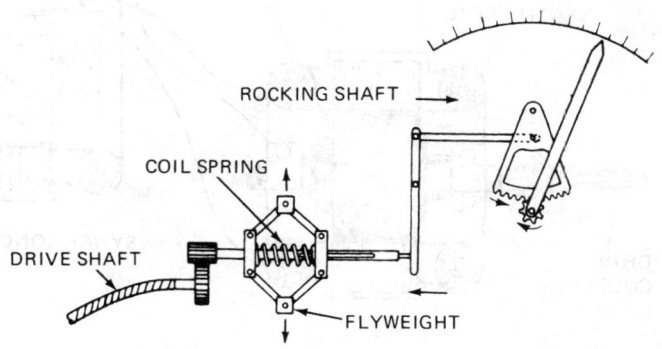

FIGURE 23-18 Use of flyweights in a tachometer.

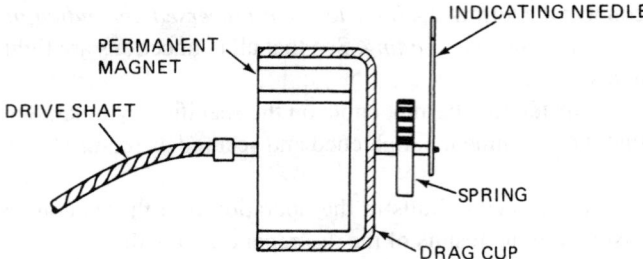

FIGURE 23-19 Illustration of the magnetic tachometer principle.

proportional to the speed of the rotating magnet; therefore, the drag-cup rotation provides a measure of engine rpm. The drag cup is mounted on the same shaft as the indicating needle of the instrument which produces the reading on the dial of the tachometer. The operation of a magnetic tachometer is illustrated in Fig. 23-19.

The flexible shaft of a mechanical drive tachometer is encased in a flexible metal housing. To operate properly, the shaft and housing must be free of kinks, dents, and sharp bends. There should be no bend with a radius of less than 6 in [15.24 cm] and no bend within 3 in [7.62 cm] of either end terminal.

If the tachometer is noisy or the pointer oscillates, check the housing and shaft for defects or improper installation. Disconnect the cable shaft from the tachometer and pull it out of the housing slowly and carefully. Check the shaft for worn spots, kinks, breaks, dirt, and lack of lubrication.

If the condition of the cable is satisfactory, clean it thoroughly and then apply Lubriplate 110, or an equivalent grease, to the lower two-thirds of the cable to provide a light coat of lubricant. Insert the shaft as far as possible into the housing, making sure the proper end is inserted. Rotate the shaft until the end terminal seats in the engine drive fitting. Insert the upper end of the shaft into the tachometer fitting and see that it seats properly. Screw the housing terminal nut onto the tachometer fitting and torque as specified. A typical instrument usually requires approximately 50 in•lb [5.65 N•m]. Safety the nut as required.

If the tachometer instrument is not functioning properly, replace the instrument. Repair of the instrument should be done only at an approved instrument overhaul shop.

Electric tachometers. Because of the distances between the engines and the cockpit on large aircraft, the flexible drive shafts used for tachometer operation on small aircraft cannot be used. For this reason, electric tachometers were developed so that the rpm indication from the engine could be transmitted through electric wiring.

An early type of electric tachometer was simply a permanent-magnet dc generator mounted on the engine and a voltmeter, scaled for rpm, mounted on the instrument panel in the cockpit. The voltage produced by the generator was directly proportional to engine rpm; therefore, the reading of the voltmeter in the cockpit was an indication of engine rpm.

A commonly employed electric tachometer system utilizes a three-phase generator mounted on the engine to develop three-phase alternating current whose frequency is determined by engine rpm. This current is transmitted to a three-phase synchronous motor in the tachometer instrument. The motor speed is determined by the frequency of the current, which, in turn, is established by the engine rpm. Thus, instead of using a flexible drive shaft to transmit engine rpm to the instrument in the cockpit, a three-phase electric current performs the same function.

Figure 23-20 illustrates the ac electric tachometer system. The alternator, mounted on the engine, produces a three-phase current which is delivered to the synchronous motor. The three-phase current creates a rotating field in the stator of the synchronous motor. The rotor of the synchronous motor, being a permanent magnet, aligns itself with the rotating field and turns at the same speed as the alternator. The synchronous motor drives a magnet in the drag cup of the instrument, thus causing the drag cup to turn and move the indicating needle. As explained previously, the drag cup is restrained by a coiled balance spring and will turn to a point where the spring force and the rotating force are balanced. The indicating needle will therefore move a distance proportional to engine speed and provide a reading of rpm.

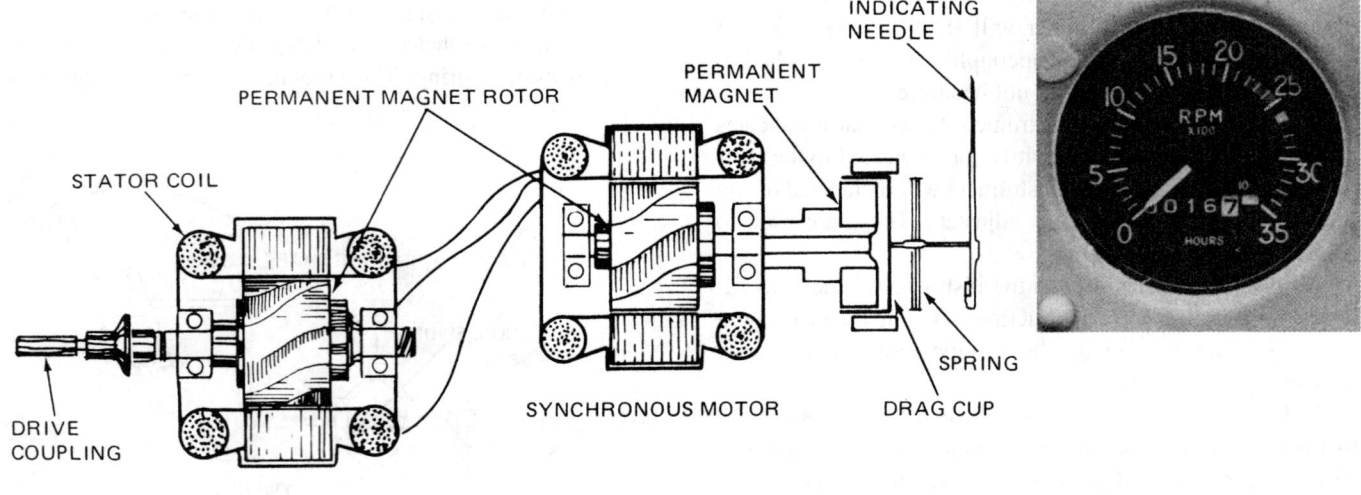

FIGURE 23-20 Alternating-current electric tachometer system.

Electronic tachometers. An electronic tachometer involves a system that utilizes electric impulses from the magneto ignition system to produce the engine rpm indication. The tachometer includes a frequency-to-voltage converting circuit that feeds into a standard meter movement. The electric impulses are obtained from a special pair of breaker points in one of the engine magnetos. The impulses, having a frequency proportional to engine rpm, produce a voltage by means of the converting circuit, which is also proportional to engine rpm. This voltage, when fed to the meter movement, produces a needle deflection that indicates engine rpm.

The simplest type of electronic tachometer has two connections. One receives the signal pulses from the magneto, and the other is connected to ground. In this case the impulses from the magneto are sufficiently strong to operate the instrument with no outside power required.

A more complex electronic tachometer operates on the principle described above; however, it includes an amplifier circuit with one or two transistors to strengthen signal impulses that otherwise might not be sufficient to operate the instrument. This type of tachometer system requires three electrical connections, one being for an input of 12 V from the aircraft electrical system.

Maintenance and inspections. Many of the older light aircraft utilize mechanically driven tachometers. The principal maintenance required for these instruments is service and lubrication of the drive shaft.

The inspection and maintenance of electrical tachometer systems is the same as for other electrical systems. If either the tachometer generator (alternator) or the instrument is defective, it should be replaced. If the instrument vibrates or shakes, it should be checked for security of mounting and the panel should be checked for integrity of the shock mounts.

During routine inspections, both the tachometer generator at the engine and the instrument in the cockpit should be checked for security of mounting. Examine the electrical plugs and wiring for condition. The plugs should be tight in the receptacles. Wiring should be properly laced or clamped to prevent vibration or fluttering which could cause wear. Usually the electrical wiring from a number of instruments will be secured together in a harness, and the harness will be supported by suitable clamps. Routing of the electrical wiring or harness should be such that the wires are not exposed to excessive heat or liquids.

Hour Meters

Many tachometers, particularly those for light aircraft, include an **hour meter** by which the number of total flight hours of the aircraft is recorded. Such an hour meter is shown in Fig. 23-20.

Hour meters are manufactured in two basic designs. For one type, the meter mechanism is geared directly to the rotating element within the tachometer. This type of mechanism records engine revolutions and provides an accurate record of time only when the engine is operating at the designed cruising rpm. For an engine designed to cruise at 2100 rpm, the hour meter will record 1 h of flight time as the engine completes 126 000 revolutions.

The Hobbs hour meter is operated by an electric clock mechanism. This system receives power from the aircraft battery or electrical system and may be connected so as to operate when the master switch is turned on, when the engine is operating, or when the aircraft is in flight. This system provides the only true indication of actual flight time; the actuating switch is operated by the landing gear. At takeoff, when the weight of the aircraft is taken off the landing gear, the switch closes and turns on the hour meter.

Another system for turning on the Hobbs meter utilizes a pressure-operated switch connected to the oil system. This switch is adjusted so that it will close when the oil pressure reaches a predetermined level.

Instruments for Gas-Turbine Engines

Gas-turbine engines often utilize some of the same engine instruments that were discussed earlier in this chapter, such as the oil pressure gauge, oil temperature gauge, and fuel flow gauge. The oil pressure and oil temperature gauges operate in the same manner as was previously described for reciprocating engines, but fuel flow gauges in gas-turbine engines are calibrated in pounds per hour instead of the gallons per hour calibration used in reciprocating engines. A typical fuel flow system consists of a fuel flow transmitter, which is fitted into the low-pressure fuel system, and an indicator that displays the rate of fuel flow. Examples of various turbofan gas-turbine engine instruments are shown in Fig. 23-21.

Gas-turbine engine instruments that are commonly used in turbine engines are percent rpm gauges and temperature gauges. A separate percent rpm gauge is used for each compressor section or fan in the engine. For example, if an engine has low- and high-pressure compressors, one of the percent rpm gauges is designated as the N_1 (low-pressure) speed indicator and the other is designated as the N_2 (high-pressure) speed indicator. This system is used on dual-spool turbofan-type engines. On turbopropeller engines, the speed of the engine (gas generator, free-turbine design) is referred to as the N_g speed, which is indicated on the percent rpm gauge. The propeller rpm is read out on a tachometer as N_p.

Turbofan engines also utilize an **engine pressure ratio (EPR) gauge** as an indication of the thrust being developed by the engine, whereas turbopropeller engines use torque gauges to measure the torque or power being developed by the propeller shaft.

EPR Indicator

The **engine pressure ratio (EPR)** measures the thrust developed by the engine. EPR is the ratio of turbine discharge pressure to engine inlet pressure. The reading is displayed on the cockpit EPR gauge, as shown in Fig. 23-22. EPR can be indicated by either electromechanical or electronic transmitters. The electronic EPR system utilizes two vibrating cylinder pressure transducers which sense the engine air pressures and vibrate at frequencies relative to these pressures.

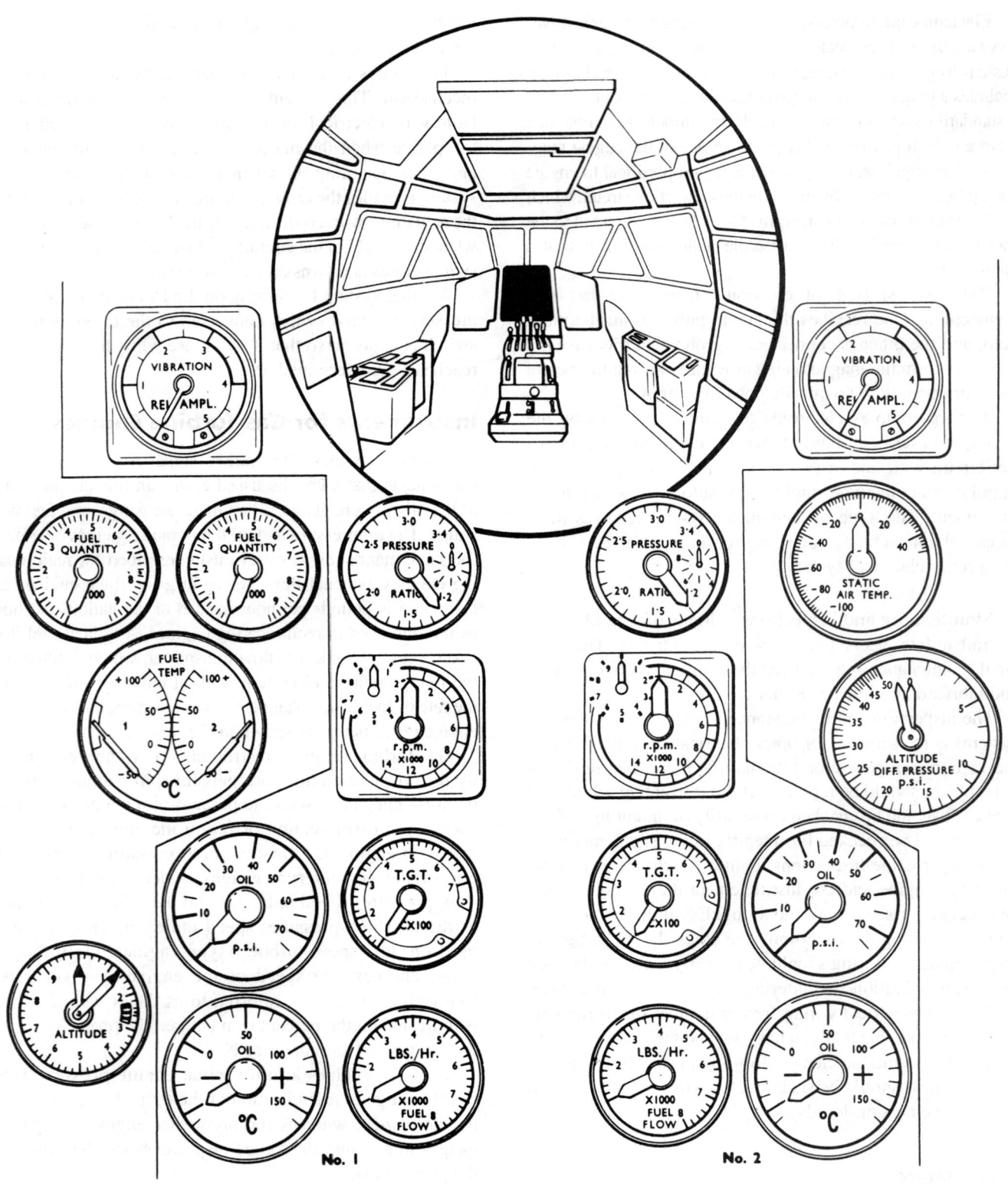

FIGURE 23-21 Typical turbine-engine instrument panel. (*Rolls-Royce.*)

From these vibration frequencies, electric signals of EPR are computed and are supplied to the EPR gauge and electronic engine control system.

EPR settings are dependent on pressure altitude and outside air temperature. The EPR setting is used for setting the correct amounts of thrust for takeoff, climb, and cruise. The amount of bleed air being used for air conditioning or anti-icing also affects the correct EPR setting. The EPR gauge is very important in the operation of a transport-category aircraft, and its use is increasing in the business-size aircraft.

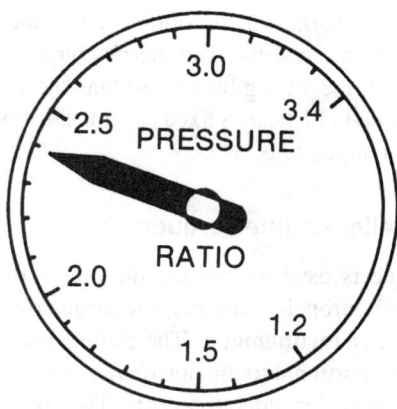

FIGURE 23-22   EPR indicator.

FIGURE 23-23   Turbine-engine tachometer.

Turbine Speed Indicators

Gas-turbine engine speed is measured in revolutions per minute (rpm). Gas-turbine engine tachometers, as shown in Fig. 23-23, are usually calibrated in **percent rpm**. For axial-compressor engines, the principal purpose of the percent rpm tachometer is to monitor rpm, especially during engine starting and to indicate an overspeed condition, if one occurs.

Engine speed indication is electrically transmitted from a small generator, driven by the engine, to an indicator that shows the actual revolutions per minute or a percentage of the maximum engine speed, as shown in Fig. 23-24. The engine speed is sometimes used to indicate engine thrust, especially if the EPR gauge is inoperative, but it does not give an absolute indication of the thrust being produced because inlet temperature and pressure conditions affect the thrust at a given engine speed.

EGT Indicator

The temperature of the exhaust section of any turbine engine must be monitored to prevent overheating of the turbine blades and other exhaust components. There are many terms used to describe the temperature of the exhaust section. Most often

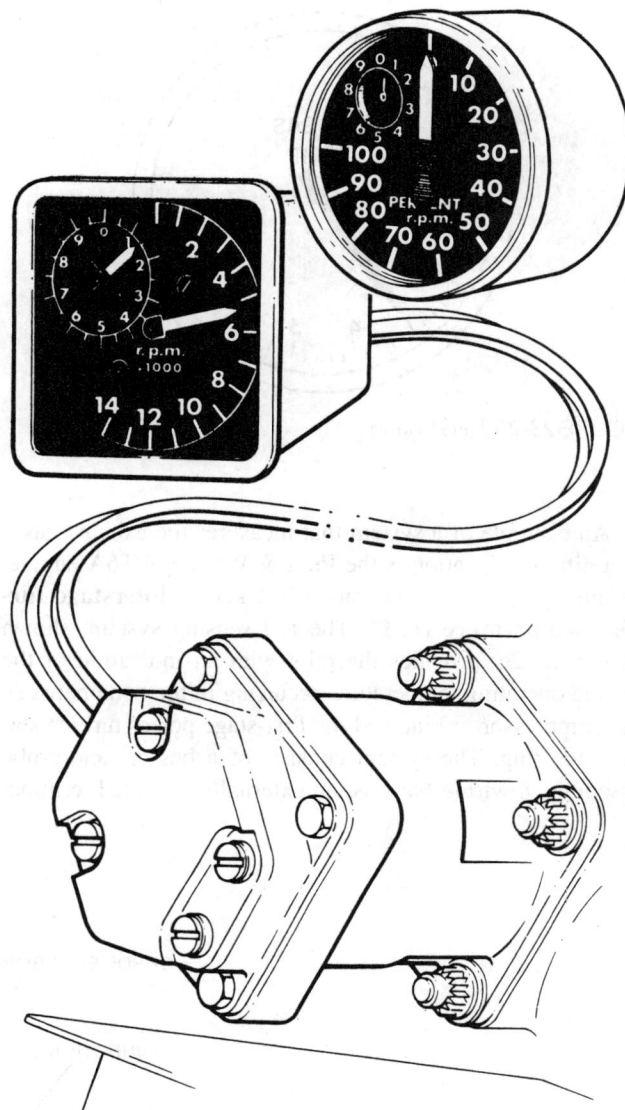

FIGURE 23-24   Engine speed indicators. (*Rolls-Royce.*)

the terms used will be determined by the location of the temperature probes in the exhaust stream. Some of the terms used are turbine-gas temperature (TGT), exhaust-gas temperature (EGT), interstage turbine temperature (ITT), turbine outlet temperature (TOT), and turbine inlet temperature (TIT). The exhaust is measured in degrees centigrade and will generally range from 350 to 500°C [662 to 932°F].

The **turbine inlet temperature** (TIT) is the most critical of all the engine variables. However, it is impractical to measure TIT in some engines, especially large models. In these engines, thermocouples are inserted at the point of turbine discharge (exhaust gas). This EGT temperature reading provides a relative indication of the temperature at the turbine inlet. In other engines, the temperature-sensing probes (thermocouples) are placed at various positions in the turbine section, as previously described. Several thermocouples are usually used, spaced at intervals around the perimeter of the engine exhaust duct near the turbine exit. The EGT indicator, shown in Fig. 23-25, displays the average temperature measured by the individual thermocouples.

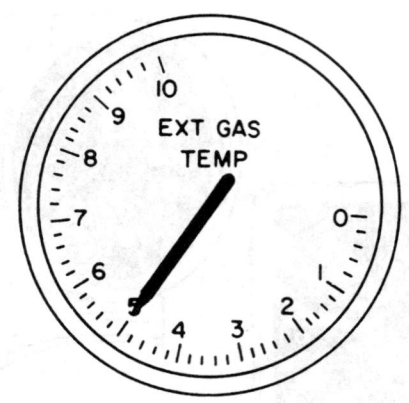

FIGURE 23-25 EGT gauge.

An example of a system that measures the exhaust gases at a different location is the Pratt & Whitney PT6A engine. In this system, the temperature is taken as **interstage turbine temperature (ITT)**. The ITT-sensing system, shown in Fig. 23-26, provides the pilot with an indication of the engine operating temperature occurring in the zone between the compressor turbine and the first-stage power-turbine stator vane ring. The system consists of a bus-bar-and-probe assembly, a wiring harness, an externally mounted terminal block, and a trim harness incorporating a thermocouple probe and preset variable resistor. The thermocouple is connected in parallel with the wiring harness to bias the ITT signal so that the indicated ITT bears a fixed relationship with the compressor inlet temperature.

Turbopropeller Engine Torquemeters

Engine torque is used to indicate the power that is developed by a turbopropeller engine. The torque display indicator is known as a **torquemeter**. The engine torque or turning moment is proportional to the horsepower and is transmitted through the propeller reduction gears. The engine operating condition, during which the torque indicating system is the most important, is called a "positive torque condition." **Positive torque** occurs when the engine is producing the power that drives the propeller. The torque indicating system is designed to measure the torque that is produced by the engine at a point between the engine and the propeller. If the propeller was driving the engine, such as during an engine out condition, then **negative torque** would be produced.

The actual horsepower at a given torque indication must be related to the effect of rpm. The formula illustrated in Fig. 23-27 indicates that horsepower is a function of a mathematical constant (K) multiplied by the rpm and then by

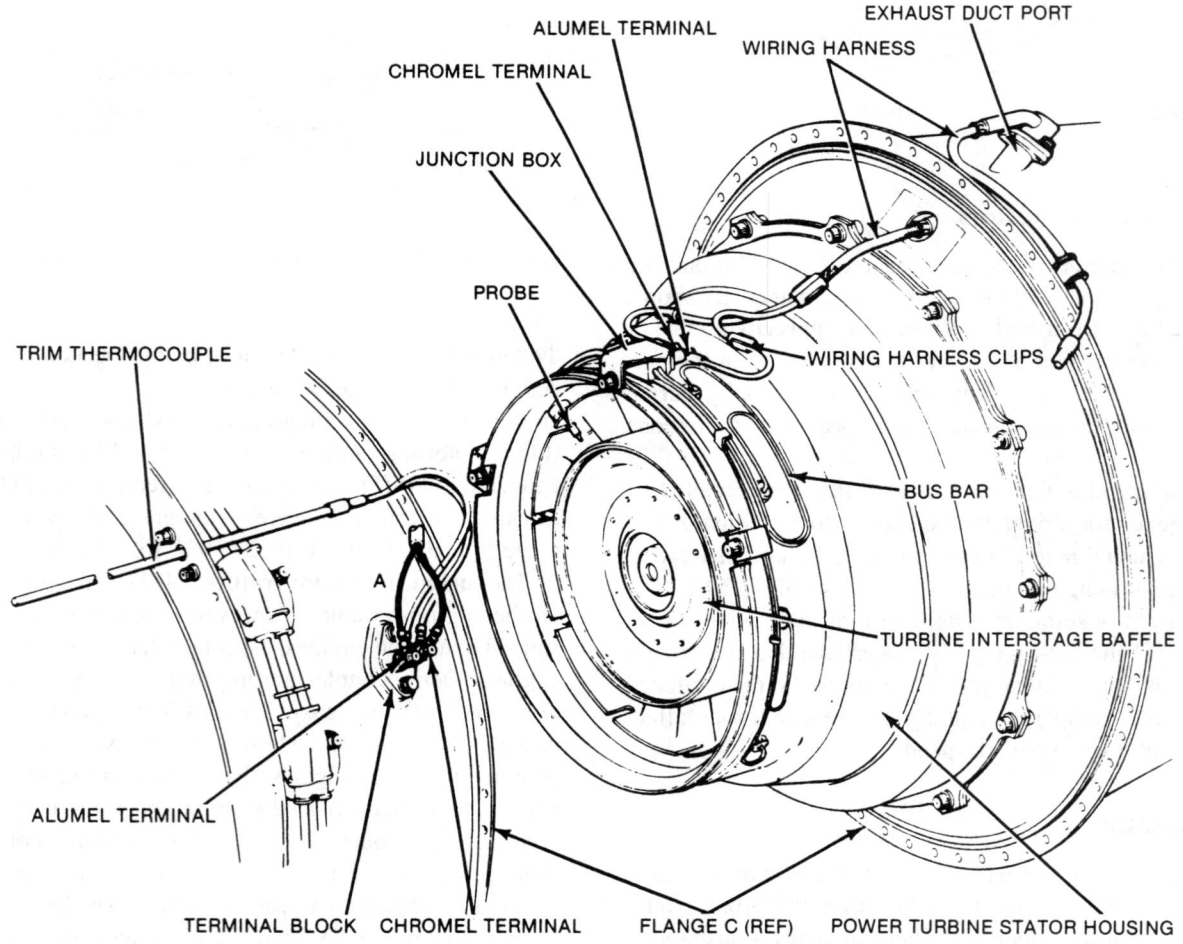

FIGURE 23-26 ITT-sensing system. (*Pratt & Whitney Canada.*)

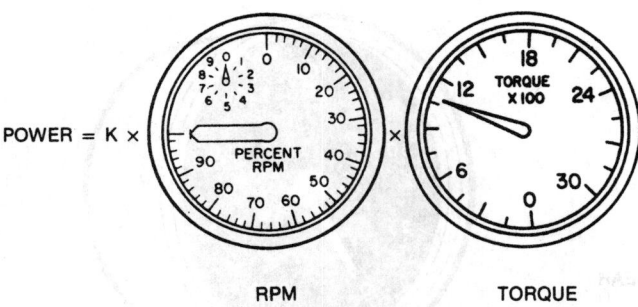

FIGURE 23-27 K × rpm × torque = horsepower. (*Honeywell Inc.*)

POWER=	K (.0001904)	×	RPM	×	TORQUE(FT LBS)
900 HP =	K	×	1591(100%) ×		2971
900 HP =	K	×	1527(96%) ×		3095

FIGURE 23-28 RPM-torque relationship. (*Honeywell Inc.*)

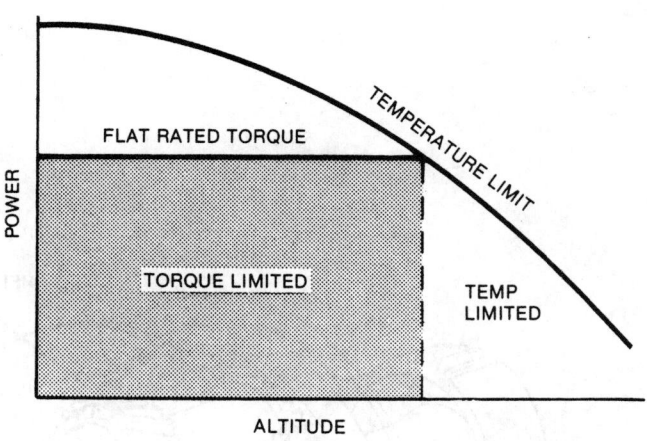

FIGURE 23-29 Torque limit operation. (*Honeywell Inc.*)

torque in foot-pounds. The K factor is a constant factor that does not change ($K = 2\pi/33\,000$). If the formula is used to determine horsepower, the effect of rpm on the torque being produced at a given horsepower can be seen in Fig. 23-28. This example assumes that 900 horsepower is desired as a limit at both 100 percent and 96 percent engine speeds. The value of K (0.0001904) times 1591 propeller rpm (100 percent) times 2971 ft•lb (torque) equals 900 horsepower. Under this condition of 900 horsepower at 100 percent rpm, the torque limit red line would be 2971 ft•lb [4029 N•m]. For 900 horsepower at a reduced rpm—representing minimum cruise of 96 percent—the propeller rpm would be 1527, and so the formula would yield a torque value of 3095 ft•lb. If the airframe manufacturer were to allow the pilot to maintain the same 900 horsepower at 96 percent rpm, there might be a second red line at the 3095 ft•lb [4197 N•m] torque limit. Thus, with decreased rpm, the torque must increase to maintain the same horsepower; this example illustrates that torque is inversely proportional to rpm.

Exceeding the engine's temperature or torque limitations may damage the aircraft or the engine. The curve in Fig. 23-29 indicates the torque/temperature relationship in the operation of an engine in a typical installation. It can be seen that at a certain point the temperature limit will take precedence over the torque limit as the aircraft reaches less dense air. That point in pressure altitude or outside air temperature is a matter of the individual torque limit ratings applied by the airframe manufacturer. Consequently, the pilot must monitor the applicable engine instruments and observe the flight manual limits.

A torquemeter system is shown in Fig. 23-30. In this hydromechanical system, the axial thrust produced moves the helical gear, which meters oil pressure acting on a piston. The position of the metering valve is controlled by the torquemeter piston, which reacts in direct proportion to engine torque. This pressure is transmitted to the transducer, which converts oil pressure into an electrical signal. The signal is passed on to the indicator in the cockpit, which is generally calibrated in foot-pounds of torque.

Another torque-indicating system which is basically an electronic system is used on the Garrett TPE331 engine. The major components of this system, illustrated in Fig. 23-31, include a torque ring containing strain gauges (sometimes referred to as the torque transducer) and an externally mounted signal conditioner, which sends a corrected signal to the cockpit torquemeter. Wiring from the electrical connector mounted on the diaphragm carries an electric signal externally from the gearbox to the signal conditioner. The signal conditioner is usually mounted somewhere in the nacelle adjacent to the engine. Electric signals from the conditioner are sent to the torquemeter in the cockpit.

In addition to providing an indication of engine power, the torquemeter system may also be used to operate a torque limiter system, if one is installed, and to signal the automatic propeller feathering system if the torquemeter oil pressure falls as a result of an engine power failure.

Engine Vibration Monitoring

A gas-turbine engine has an extremely low vibration level, and a change in vibration level, in response to an impending or partial engine failure, may pass without being noticed. Many engines are therefore fitted with vibration indicators that continually monitor the vibration level of the engine.

Airborne vibration monitoring has been used for many years on transport-category aircraft engines, although early systems were somewhat unreliable. The new generation of systems being employed on many new aircraft is made up of systems that provide improved reliability and utility.

A modern vibration monitoring system consists of three elements: vibration pickup(s), signal conditioner (monitor unit), and cockpit display or gauge. A typical airborne vibration monitoring system can be seen in Fig. 23-32. Vibration pickup locations on the engine are selected with respect to vibration sensitivity, environmental considerations, and pickup sensitivity. The most sensitive engine location may not be conducive to good reliability and pickup maintenance access. Consequently, a less sensitive location may be used in

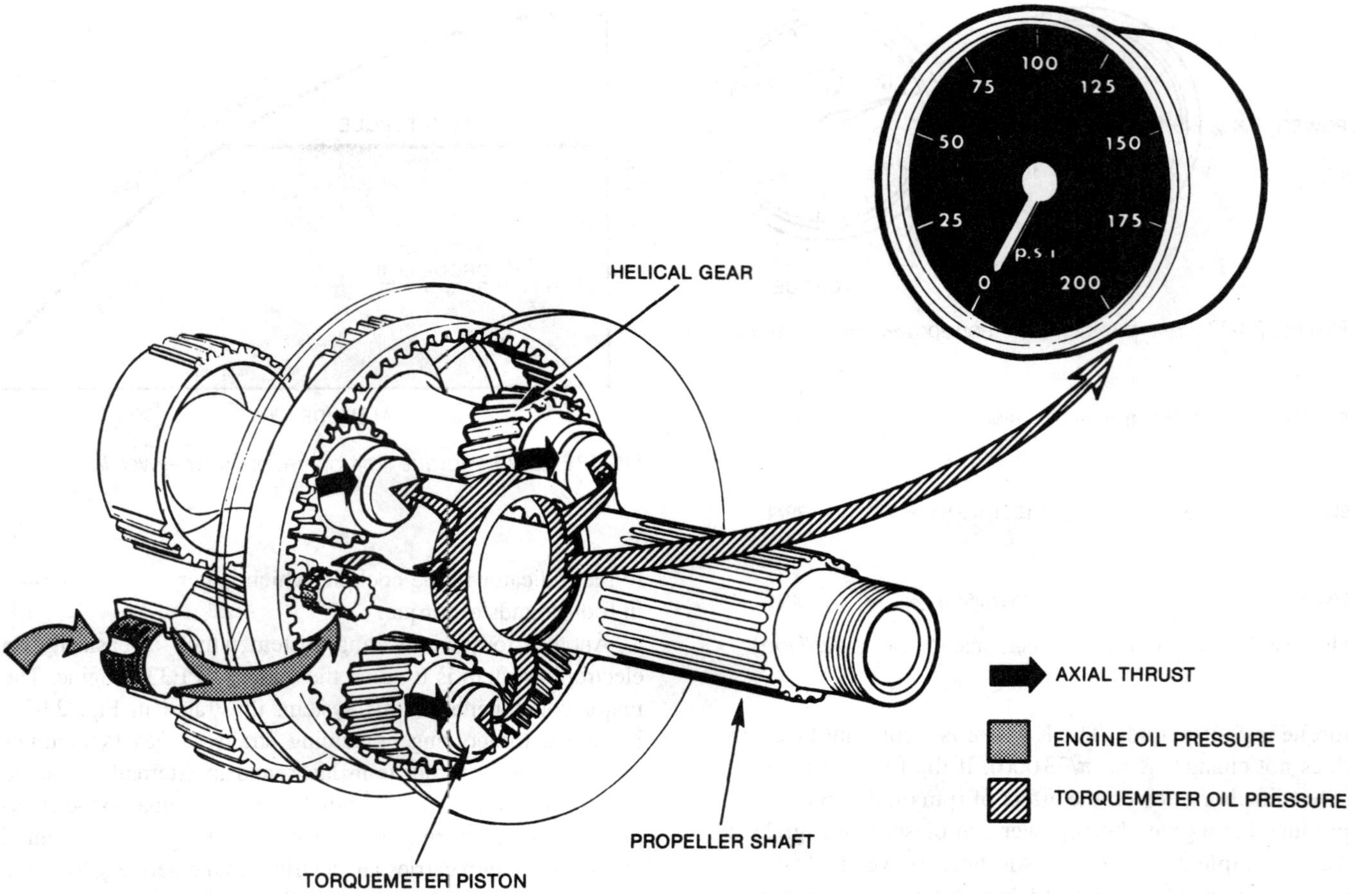

HELICAL GEAR

100
75   125
50   150
25   175
P.S.I
0   200

AXIAL THRUST

ENGINE OIL PRESSURE

TORQUEMETER OIL PRESSURE

PROPELLER SHAFT

TORQUEMETER PISTON

FIGURE 23-30 A simple torquemeter system. (*Rolls-Royce.*)

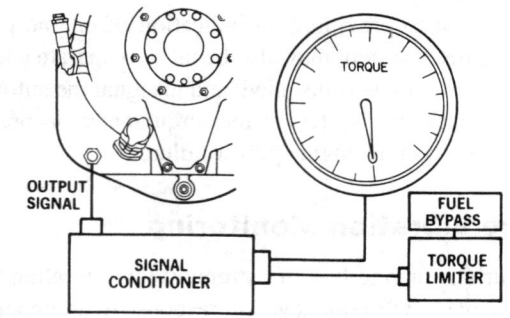

TORQUE

OUTPUT
SIGNAL

FUEL
BYPASS

SIGNAL
CONDITIONER

TORQUE
LIMITER

FIGURE 23-31 Torque indication system. (*Honeywell Inc.*)

conjunction with a more sensitive pickup. Also, the selected location should be responsive to vibration levels over the entire range of engine running speeds.

Two types of vibration pickups have been used: the **electromechanical velocity pickup** and the **piezoelectric accelerometer**. An electromechanical velocity pickup consists of a permanent magnet suspended on springs inside a coil, as shown in Fig. 23-33A. The coil is attached to the pickup case, which is attached to the engine. As the engine vibrates, the pickup case and coil move with the engine, but the magnet, being suspended, tends to remain at rest. The relative motion between the magnet and coil generates an electric signal proportional to radial velocity (measured in inches per second, or ips).

Later-generation systems employ a piezoelectric accelerometer in place of a velocity pickup. The accelerometer, shown in Fig. 23-33B, consists of a piezoelectric crystal clamped between a reference surface (accelerometer case) and a mass. Engine vibration causes the reference surface to move. Since the mass opposite the reference surface tends to remain at rest, the crystal is alternately squeezed and relaxed. This generates a charge in the crystal that is proportional to engine vibration acceleration. The accelerometer contains no moving parts, and therefore it is significantly more reliable than the velocity pickup.

Signals from velocity pickups are converted into vibration displacement, measured in thousandths of an inch, peak to peak. These units represent the physical displacement of the vibration sensor. Signals from accelerometers are converted into velocity, measured in inches per second. These units represent the peak amplitude of the velocity of the vibration sensor.

The signal conditioner converts the vibration signal from the pickup into a form suitable for cockpit display. The vibration data are then transmitted to the cockpit for display and, often, to aircraft data recording systems.

Fault Indicating and Isolating Systems

From the simple indicators used on early aircraft, engine indicating systems have developed to include (in addition to

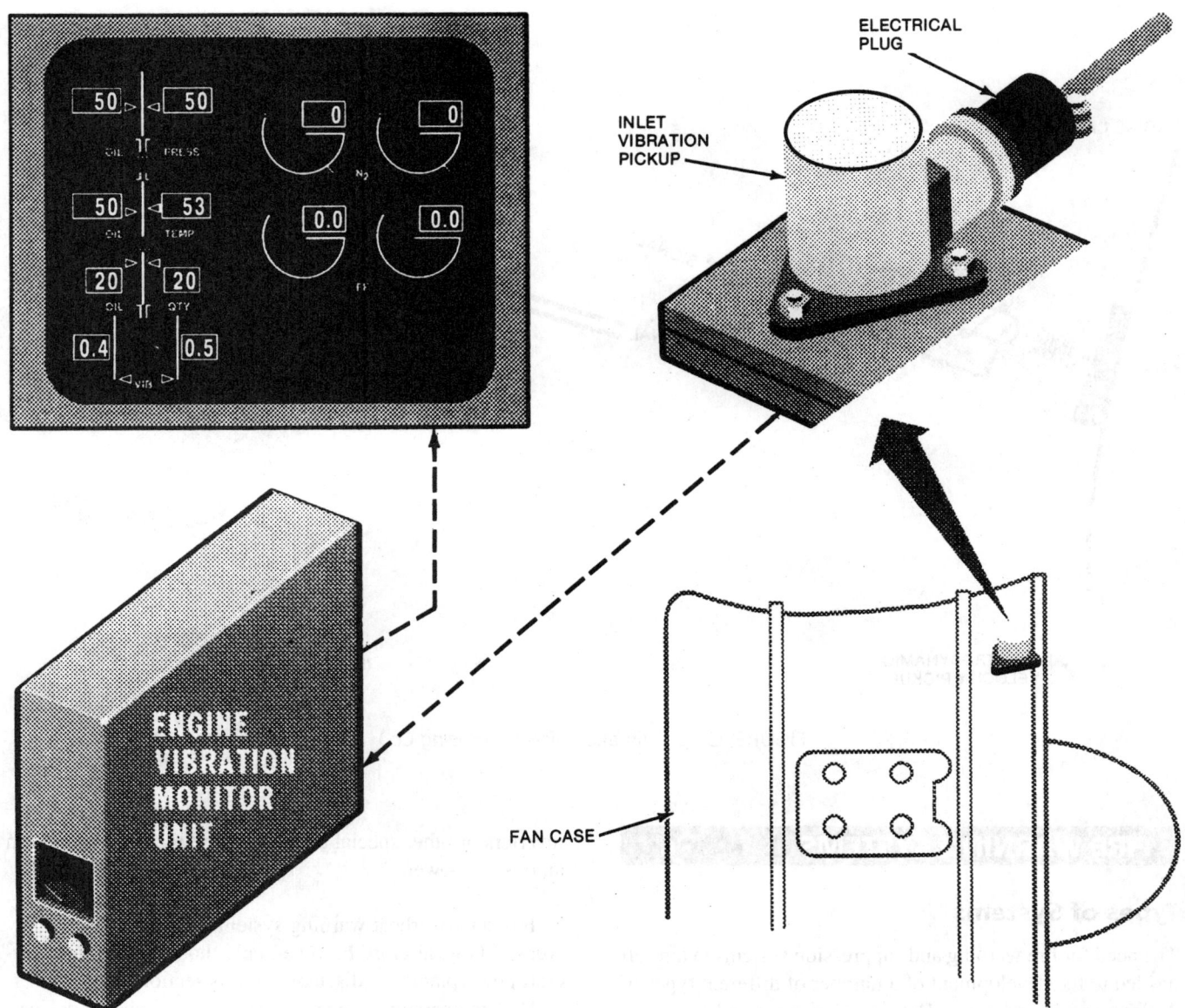

FIGURE 23-32 Airborne vibration monitoring system. (*Boeing Co.*)

basic measurements of temperatures, pressures, and speeds) a multitude of indicators that provide the flight crew of an aircraft with information about the operation of engines and powerplant systems.

An example of a fault isolating system is the FEFI/ TAFI system developed by the McDonnell Douglas Corporation for use with the DC-10 aircraft. This system incorporates two distinct programs, each set forth in a comprehensive manual. The first program deals with the collection of operational information in flight. This is called the flight engineer's fault isolation (FEFI) program. The second part of the system is called the turn around fault isolation (TAFI) program, wherein technicians employ information developed during operation to isolate faults through the use of the TAFI manual. When a fault has been isolated and identified, the corrective action is immediately indicated.

The **central maintenance computer system** (CMCS), such as that used on the Boeing 747-400 aircraft, provides a centralized location for access to maintenance data from all major electrical and electromechanical systems, including many monitored parameters of the engines. The **integrated display system** (IDS), which consists of six monitor screens (cathode-ray tubes), is used to display engine information using data transmitted from the engine and its systems. The same data are also transmitted to the CMCS. The engine information is displayed on the two center monitors. An example of this display can be seen in Fig. 23-32. System failures, engine faults, exceeding of parameters, and **engine indicating and crew alerting system** (EICAS) messages appear on the integrated display system. Additional information concerning EICAS can be found in the text *Aircraft Electricity and Electronics.*

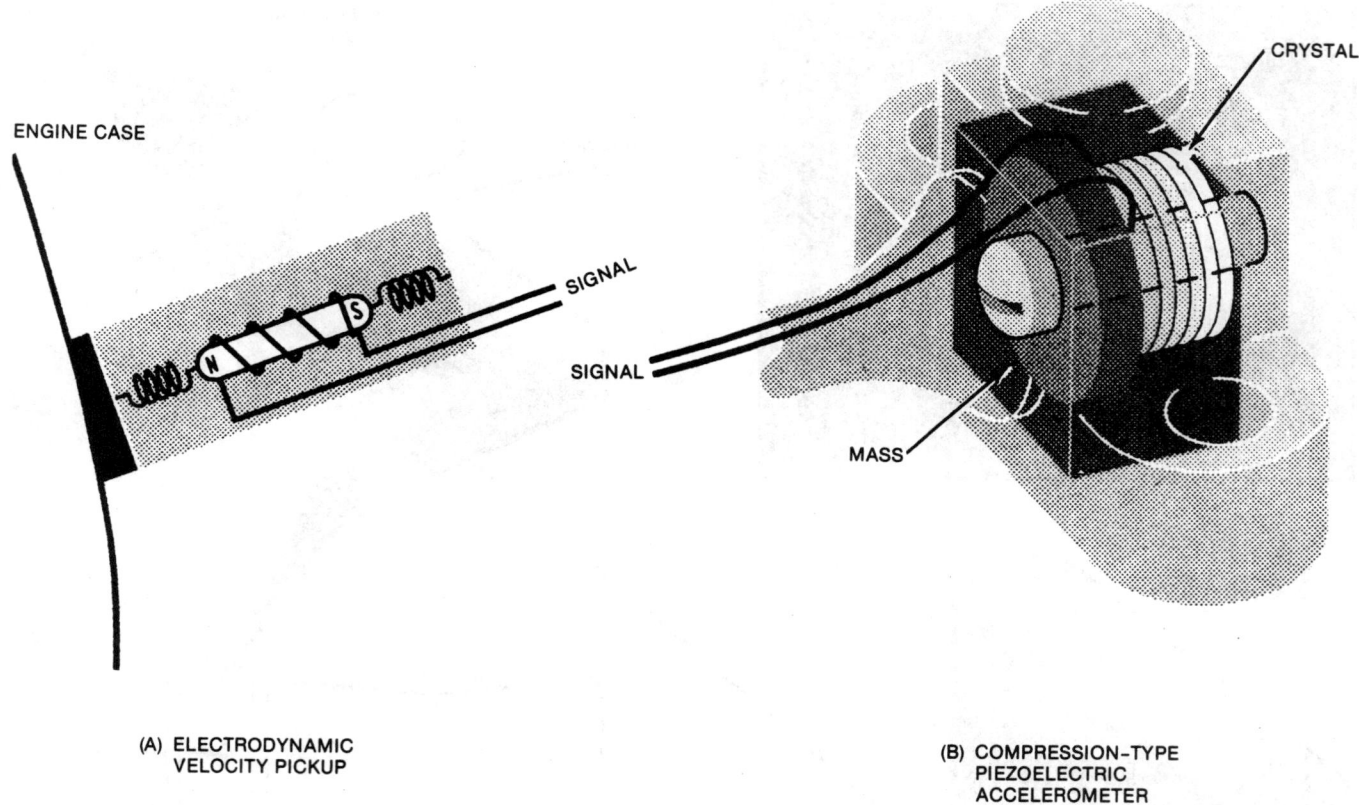

ENGINE CASE

SIGNAL

SIGNAL

(A) ELECTRODYNAMIC
VELOCITY PICKUP

CRYSTAL

MASS

(B) COMPRESSION-TYPE
PIEZOELECTRIC
ACCELEROMETER

FIGURE 23-33 Vibration pickups. (*Boeing Co.*)

FIRE WARNING SYSTEMS

Types of Systems

The need for fire warning and suppression systems in aircraft has led to the development of a number of different types of designs and installations. Extensive research and engineering studies have been done to determine which systems and devices are most effective in providing immediate warning of fire or overheat conditions so that appropriate action can be taken by flight crews. Problems encountered include the location and routing of sensing devices, false warnings due to a variety of causes, difficulties in providing means for testing of systems, and the establishment of effective maintenance and testing procedures.

The types of systems which have been developed generally meet the following requirements:

1. Provide an immediate warning of fire or overheat by means of a red light and an audible signal in the cockpit.
2. Provide an accurate indication that a fire has been extinguished, as well as an indication that a fire has reignited.
3. Incorporate durability and resistance to damage from all the environmental factors which may exist in the location where the system is installed.
4. Incorporate an accurate and effective testing system to ensure the integrity of the system.
5. Provide for easy installation and removal.
6. Operate from the aircraft electrical system without

inverters or other special equipment and require a minimum amount of power.

Fire and overheat warning systems are installed in many areas of large aircraft; however, only those applicable to aircraft powerplants are discussed in this section.

High temperatures may be detected by a number of devices. Among these are thermocouples, thermistors (resistors or materials which change resistance substantially with changes in temperature), gases in sealed elements, and bimetallic thermal switches. Fire and overheat detectors are constructed as small units a few inches in length and also as tubular units from 18 in [45.7 cm] to more than 15 ft [4.6 m] long for continuous-loop systems. Diameters of the continuous-loop sensing elements range from less than 0.060 in [1.5 mm] to more than 0.090 in [2.3 mm].

A comparatively simple fire detection system, such as that installed on light aircraft, consists of one or more thermal switches connected in parallel through an alarm circuit which includes a red fire warning light and an alarm bell. A schematic diagram of a circuit for this type of system is shown in Fig. 23-34A. The thermal switches in this system, sometimes referred to as spot detectors, are located at the points in the engine nacelle where temperatures are likely to be the highest. They are designed to close when the temperature reaches a level substantially above the normal operating temperature.

The **thermocouple** fire warning system shown in the schematic diagram in Fig. 23-34B utilizes a reference

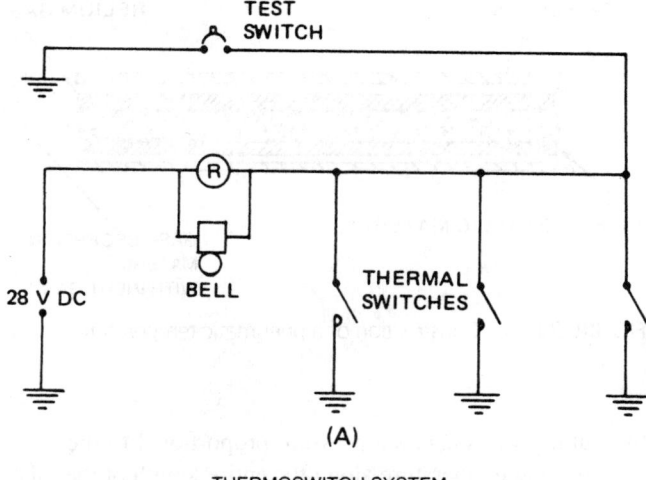

(A)

THERMOSWITCH SYSTEM

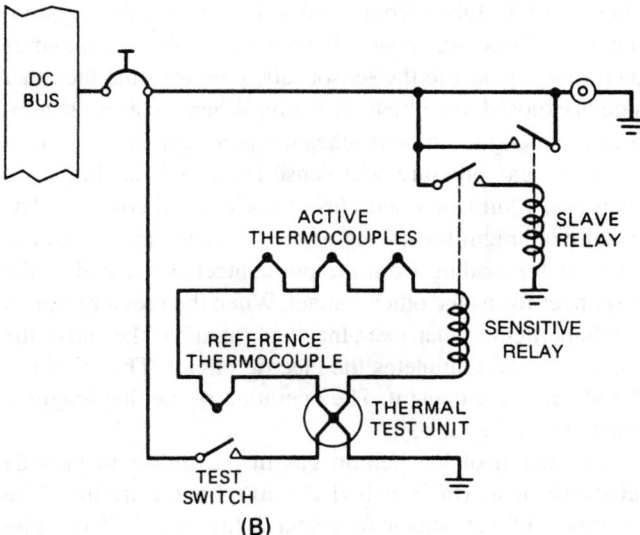

(B)

FIGURE 23-34 Fire warning system for small aircraft.

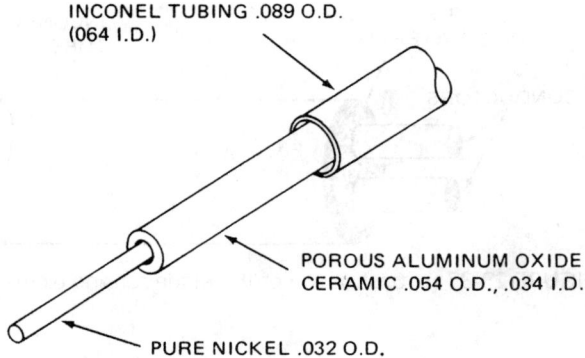

VOIDS BETWEEN TUBING, CERAMIC & WIRE AND POROSITY OF CERAMIC ARE SATURATED WITH A EUTECTIC SALT MIXTURE.

FIGURE 23-35 Section of the Fenwal sensing element. (*Fenwal.*)

consists of a small (0.089 in [2.26 mm] OD), lightweight, flexible Inconel tube with a pure nickel wire center conductor. The voids and clearances between the tubing and the conductor and the porous ceramic are saturated with a eutectic (low-melting point) salt mixture. A section of the Fenwal sensing element is illustrated in Fig. 23-35.

The tube is hermetically sealed at both ends. The nickel conductor terminates in a welded joint at the tip of a hermetically sealed terminal, thus providing an electric contact at each end of the element. The end terminals of the element are shown in Fig. 23-36.

The sensing element responds to a specified degree of elevated temperature at any point along its entire length. At the alarm temperature of the element, the resistance of the eutectic salt drops rapidly and permits increased current to flow between the inner and outer conductors. The increased current flow provides the signal which is sensed in the electronic **control unit**. The control unit produces an output signal that actuates the alarm circuit. The Fenwal system incorporates a few spot detectors wired in parallel between two separate circuits. This prevents a short circuit in either one from causing a false fire warning signal.

After corrective action has been taken and an overheat condition no longer exists, the resistance of the eutectic salt in the sensor increases to its normal level and the system returns to a standby condition. Should the fire reignite, the system would again produce an alarm.

Kidde System

The continuous fire warning system designed by Walter Kidde and Company, Inc., utilizes a **sensing element assembly** comprised of a pair of sensing elements mounted on a preshaped, rigid support tube assembly. The sensing elements

thermocouple that provides a comparison of the temperatures in the reference zone and the active zone. The reference thermocouple is isolated from the active thermocouples so that the reference signal will not be affected by engine heat. When a fire occurs, the temperature in the area of the active thermocouples will exceed the temperature of the reference thermocouple, causing a current to flow through the sensitive relay. This relay closes and activates the slave relay, which closes and directs current to the warning light in the cockpit. The test circuit of the thermocouple system includes a heating element that causes the test thermocouple in the thermal test unit to produce a current that activates the system for testing.

Fenwal System

For both reciprocating engines and turbine engines on large aircraft, continuous-loop fire warning systems are employed. The Fenwal continuous fire-detection and overheat system utilizes a **sensing element**, which may vary in length from 1.5 ft [0.46 m] to 15 ft [4.57 m], the length being selected to fit the particular engine system design. The sensing element

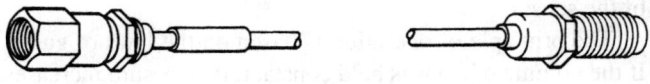

FIGURE 23-36 End terminals of the Fenwal sensing element.

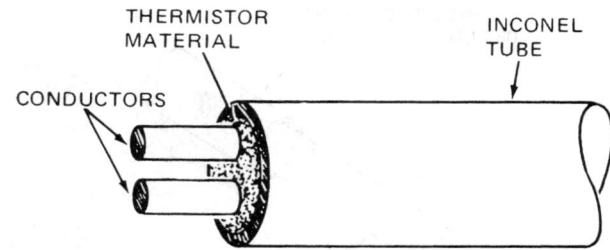

FIGURE 23-37 Construction of the Kidde sensing element.

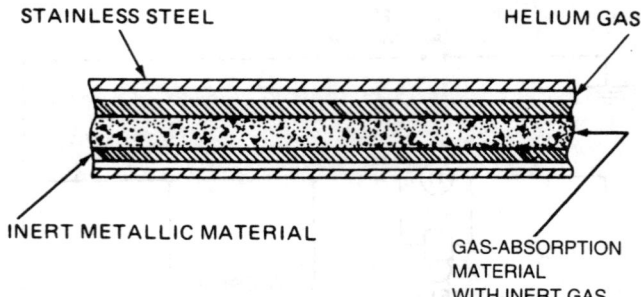

FIGURE 23-38 Construction of a pneumatic temperature sensor.

are held in place by dual clamps riveted to the tube assembly approximately every 6 in [15 cm]. Fireproof grommets are provided to cushion the sensing elements in the clamps. Four nuts per assembly are used to secure the sensing element end connectors to the bracket assemblies at either end of the support tube assembly. The nuts are secured with safety wire.

The sensing element of the Kidde system consists of an Inconel tube containing a **thermistor** (thermal resistor) material in which are embedded two electric conductors. The sensing elements terminate at both ends in electric connectors, with one of the conductors grounded to the shell of the connector.

The design of the Kidde sensing element is shown in Fig. 23-37. The resistance of the thermistor material decreases rapidly as the element is heated. When the resistance decreases to a predetermined point as the result of a fire, the control circuit monitoring the resistance of the element produces an alarm signal in the cockpit.

In the Kidde system described, each sensing loop is connected to its own electronic control circuit mounted on a separate circuit card. This arrangement provides for complete redundancy, thus ensuring operation of the system even if one side fails. Either the Fenwal system or the Kidde system will detect a fire when one sensing element is inoperative, even though the press-to-test circuit does not function.

Pneumatic System

A pneumatic fire and overheat detection system, manufactured by the Systron-Donner Corporation, utilizes a gas-filled, sealed tube to provide a signal that actuates an alarm. The tube is connected to a **responder**, which is essentially an assembly containing one or two diaphragm-type switches.

The sensor consists of two tubes, one inside the other (coaxial), containing two separate gases plus a gas-absorption material inside the inner tube. The outer tube is constructed of stainless steel. Between the inner and outer tubes, an inert gas (helium) is contained under pressure.

The inner tube is made of an inert metallic material and encloses the gas-absorption material. At temperatures below the fire alarm level, the material retains the gas. The cross-sectional drawing in Fig. 23-38 illustrates the construction of the sensor.

The principle of operation is based on the laws of gases. If the volume of a gas is held constant, its pressure increases as temperature increases; thus, the helium gas between the

two tubing walls exerts a pressure proportional to the average absolute temperature along the entire length of the tube. One end of the tube is connected to a small chamber containing a metal diaphragm switch. One side of the diaphragm is therefore exposed to the sensor tube pressure, and the other side is exposed to ambient pressure. When the average temperature along the tube reaches the predetermined overheat level, the gas pressure will cause the metal diaphragm to snap over ("oil can") and close the electrical contacts. The metal diaphragm forms one of the contacts, and the end of a conductor leading from the pin contact in the end of the responder forms the other contact. When the pressure causes the diaphragm to snap over, the diaphragm hits the end of the conductor and completes the electric circuit. This activates the alarm in the cockpit. The operation of the diaphragm is illustrated in Fig. 23-39.

The action of the helium gas in the sensor to provide an overheat alarm is called the **overheat function**. The operation of the sensor to detect a fire is called the **discrete function** and this is performed by the inert gas contained in the gas-absorption material in the inside tube of the sensor. When any portion of the sensor tube is exposed to the fire warning temperature, the gas-absorption material immediately releases a large quantity of gas, which quickly increases the pressure in the tube to a level necessary to close the diaphragm switch. This produces the fire alarm signal. When the temperature is reduced to normal levels after action has been taken to suppress the fire, the gas-absorption material reabsorbs the gas and the pressure drops, thus opening the switch and cutting off the alarm. At this time the system is ready again to produce overheat or fire warnings.

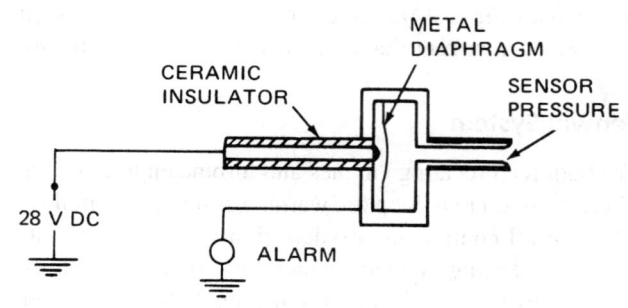

FIGURE 23-39 Diaphragm pressure switch.

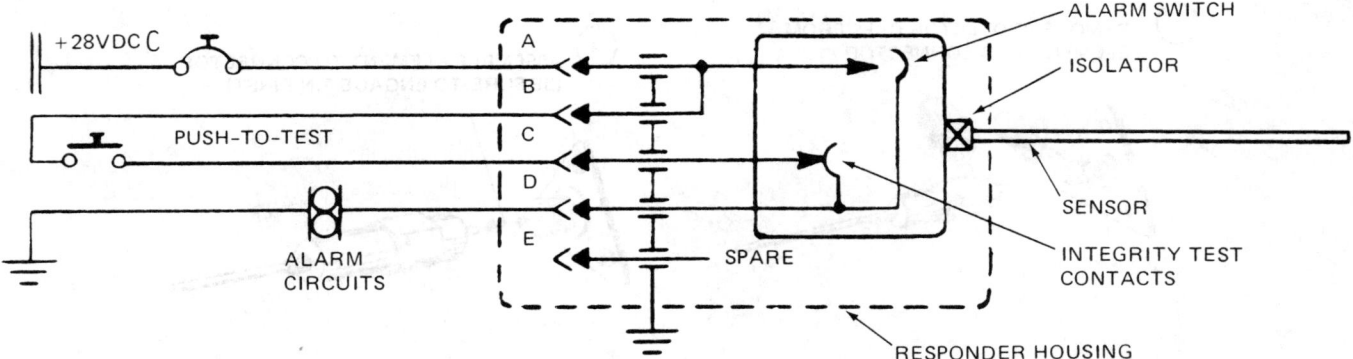

FIGURE 23-40 Schematic diagram of alarm and integrity circuits. (*Systron-Donner Corp.*)

To provide integrity monitoring for the system, an additional diaphragm switch is incorporated in the responder housing. This switch is normally held closed by the gas pressure in the sensor tube. The test circuit will function as long as the integrity switch remains closed. Figure 23-40 is a schematic diagram illustrating the operation of the **integrity monitoring system**. Note that the push-to-test switch and the integrity switch are connected in series. When both switches are closed, current can flow from the 28 V dc source through the circuit to provide an alarm signal. If there is no response when the test switch is closed, the flight engineer or pilot will know that the gas pressure in the sensor has been released as a result of some type of damage to the sensor tube.

Routing of Fire Warning Sensors

The routing of overheat and fire warning sensors has been determined by the manufacturer through extensive engineering tests. Areas in the powerplant nacelle are often divided into zones, and each zone is carefully checked to determine normal maximum operating temperatures and the areas of highest temperatures. Sensors are routed through locations where fires are most likely to occur.

Overheat conditions can occur where no fire exists; however, since overheat is an indication of a dangerous condition, the flight crew must be warned when such conditions are indicated. Overheat conditions may be caused in nacelles of large reciprocating engines by cracks or burn-through in the exhaust manifold, turbosupercharger ducts or housings, or tailpipes. Failure of engine baffling can also produce dangerous hot spots. In a turbine-engine nacelle, overheat conditions can develop as a result of a cracked compressor housing, a leaking bleed-air duct, burn-through in a combustor, and burn-through or other leaks in the exhaust system.

Remember that the routing of overheat and fire warning sensors is designed carefully to provide the most effective system possible. It is therefore essential that the replacement of sensing elements be done strictly in accordance with the original routing.

Installation of Sensors

Various manufacturers provide different types of clamps, clips, grommets, brackets, and other supporting devices for sensing elements. Typical items for installation and attachment of a Fenwal system are shown in Fig. 23-41.

The **bulkhead connector** is used at bulkheads and firewalls. It is a feed-through electrical connector which electrically joins two sections of a sensing element where it is necessary for the element to pass through the firewall.

The **wire-end fitting assembly** is an electrical connector designed according to customer specifications for connection to the electrical circuit of the aircraft.

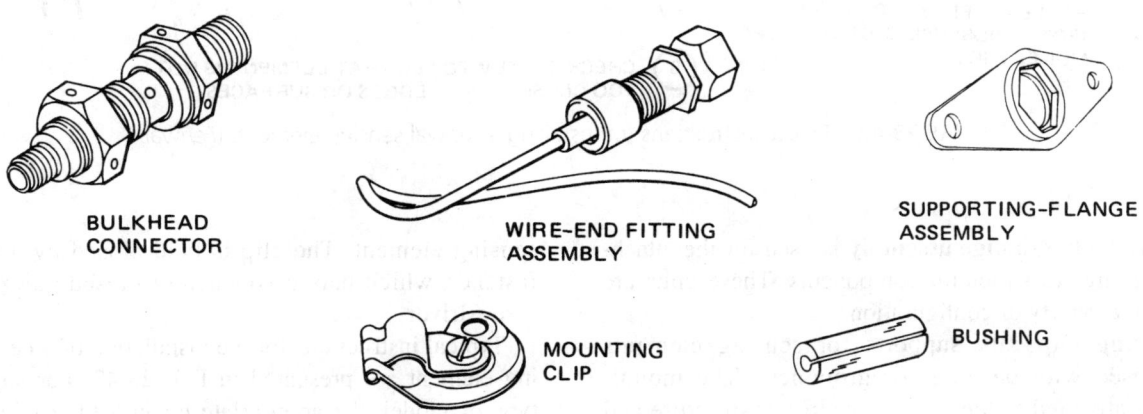

FIGURE 23-41 Installation fittings and mounting devices for attachment of a Fenwal sensor system. (*Fenwal.*)

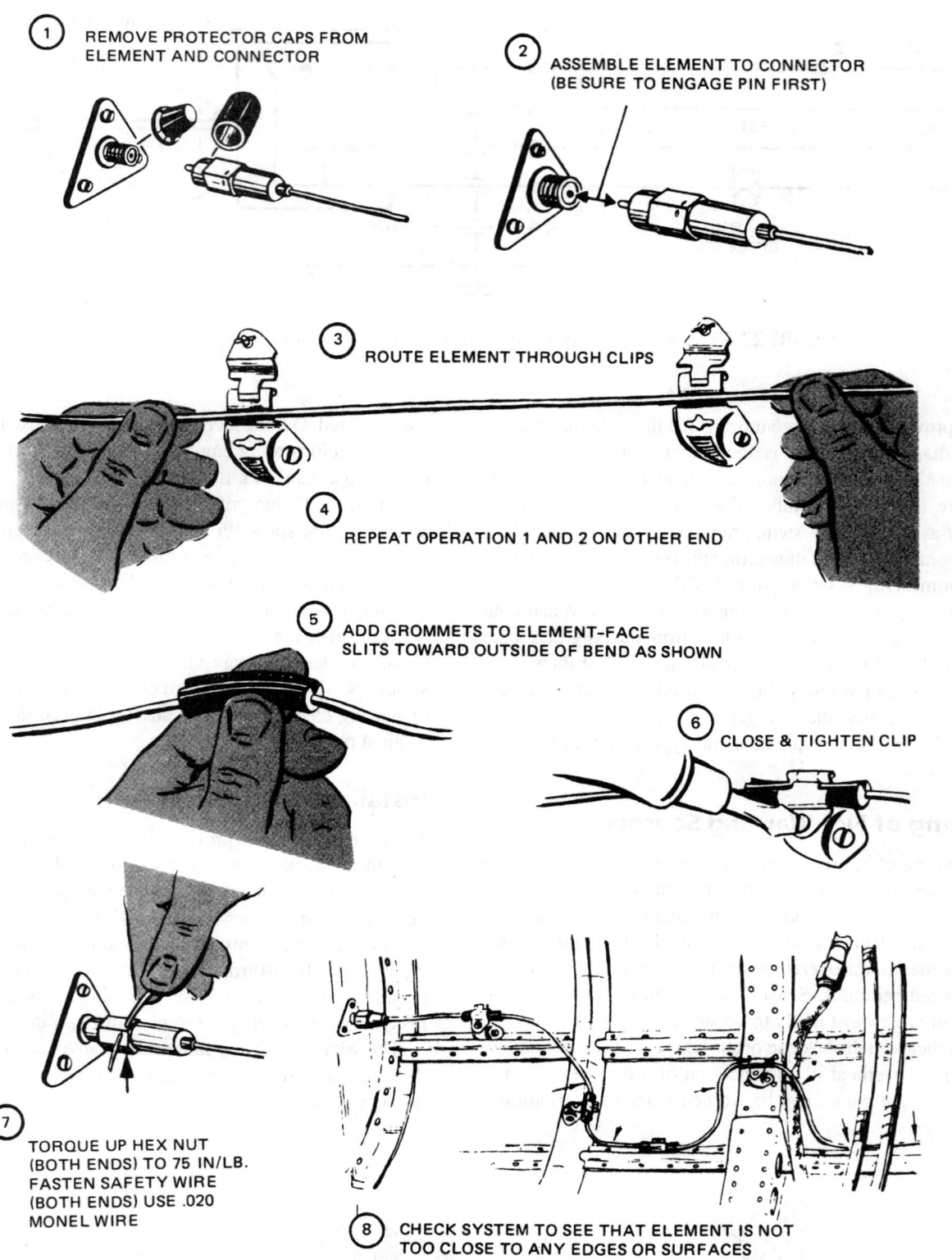

FIGURE 23-42 Typical instructions for installing a Fenwal sensing element. (*Fenwal.*)

The **supporting-flange assembly** is used for the attachment of fittings to structural components. These units are supplied in a variety of configurations.

Mounting clips are supports for sensing elements and are used with bushings or grommets. The mounting clip is attached to the engine or aircraft structure and provides a quick means of installation or removal of a sensing element. The clip is held closed by a Cam Loc fastener, which may be opened or closed quickly with a screwdriver.

Typical instructions for the installation of a Fenwal sensing element are presented in Fig. 23-42. For any specific type or model, the appropriate manufacturer's instructions should be consulted.

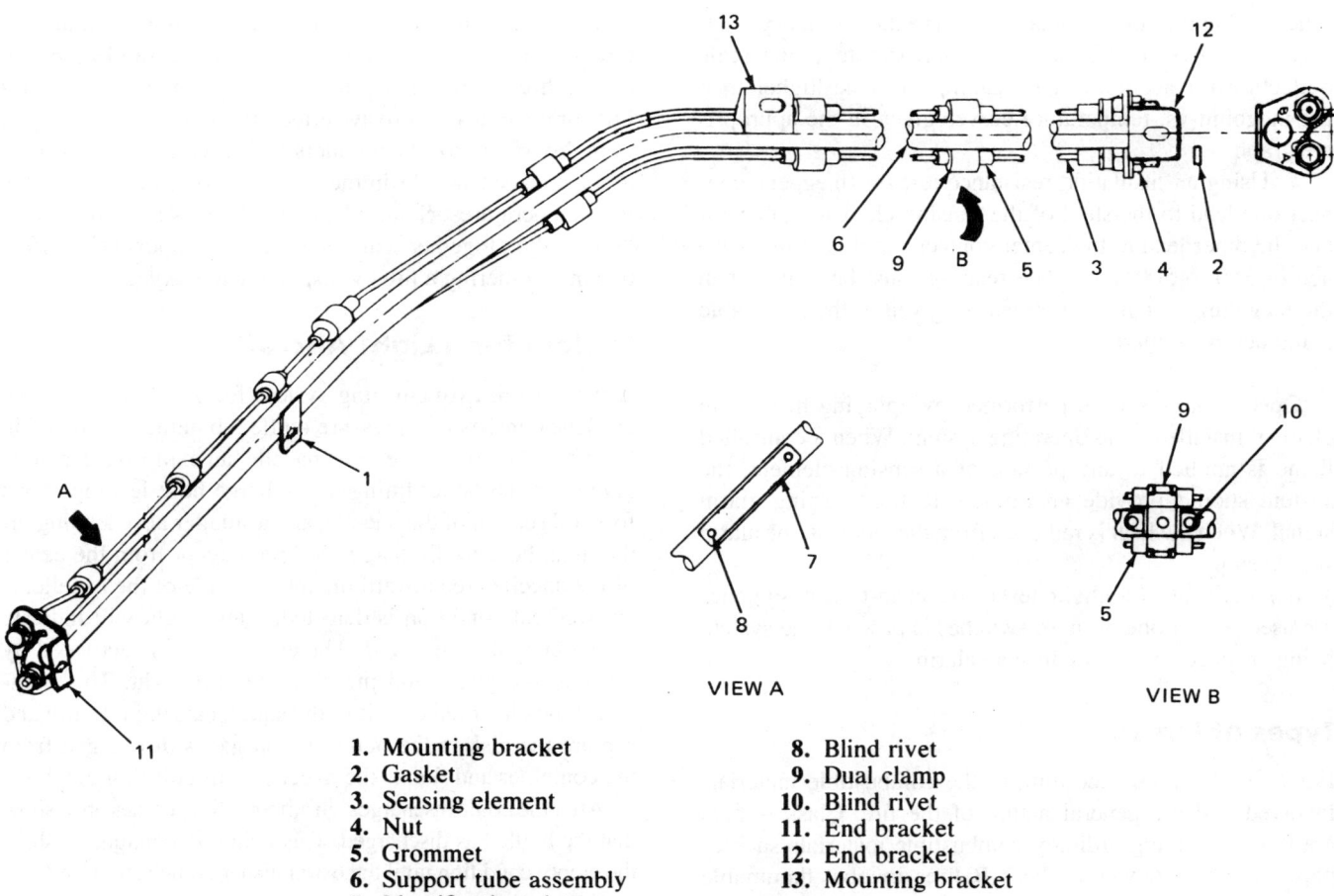

1. Mounting bracket
2. Gasket
3. Sensing element
4. Nut
5. Grommet
6. Support tube assembly
7. Identification plate
8. Blind rivet
9. Dual clamp
10. Blind rivet
11. End bracket
12. End bracket
13. Mounting bracket

FIGURE 23-43 Details of Kidde sensing-element assembly. (*Walter Kidde and Co.*)

Attachment details for a Kidde sensing-element assembly are shown in Fig. 23-43. As explained previously, this assembly consists of a preformed support tube on which two sensing elements are mounted by means of clamps and grommets. Brackets attached to the support tube provide for attachment to the nacelle or aircraft structure. At the ends of the sensing elements are pin-and-socket receptacles for electrical connections.

Inspection and Testing

Instructions for the inspection and testing of fire and over-heat warning systems are provided by the manufacturers. Because of the different types and models of systems, the technician must be sure to consult the appropriate manuals and use the correct specifications in making tests.

Typical inspection instructions are as follows:

1. Remove or open engine cowling as required to gain access to sensing elements.
2. Check end fittings and sensor connectors for tightness and proper safetying.
3. Check bulkhead fittings and supports for security.
4. Examine all clamps and grommets (bushings) and check to see that the sensing element is held securely in place.
5. Examine the entire length of each sensor for wear, sharp bends, kinks, dents, crushed sections, cuts, cracks, proximity

to structure, and security of mounting. Discard elements which do not meet the manufacturer's specifications.
6. Check the routing of sensors against the manufacturer's instructions.
7. Check the security of electronic control circuit units.

The testing of sensing elements will vary according to the design and principle of operation. Elements which produce warning signals through resistance changes (Fenwal and Kidde) are tested by means of a megger and ohmmeter as specified by the manufacturer. The sensing element is disconnected at the ends and the resistance of the conductors is checked. Insulation resistance between the conductors and outer tube is checked with a megger.

Typical testing instructions for a Kidde sensing element are as follows:

1. Allow the sensing element to come to temperature equilibrium with its surroundings in the immediate vicinity of the mercury thermometer.
2. Check the resistance of each element from the center contact at one end to the center contact at the other end. The maximum allowable resistance for each element is specified by the manufacturer. If an element exceeds the specified value, replace it.
3. Using a high-range ohmmeter, connect one lead of the meter to the shell of the sensing element connector and the

other lead to the center contact. Observe the resistance reading, using a scale that provides approximate center-scale deflection for accuracy. The reading must be higher than the megohm-vs.-temperature curves given in the appropriate graph.

4. Using an insulation resistance test set (megger), connect one lead to the shell of the sensing element connector and the other lead to the center contact. Apply the test voltage of 350 to 500 V dc. The reading must be higher than the megohm-vs.-temperature curves given in the applicable manufacturer's chart.

Operational tests are performed by applying heat to an element installed in an operating system. When a controlled flame is applied to any portion of a sensing element, the system should provide an immediate fire warning alarm signal. When the heat is removed from the element, the alarm should stop.

For daily and preflight tests, the push-to-test switches are used. When one of these switches is pressed, the system being checked should produce an alarm.

Types of Fires

Fires are classified according to the combustible materials involved and the general nature of the fire. Class A fires are those involving ordinary combustible materials such as paper, cloth, and wood. Class B fires involve flammable liquids such as gasoline, oils, kerosene, and jet fuel. Class C fires are those involving electrical wiring and equipment. Methods for extinguishing fires are determined by the nature and class of the fire. For example, water and water-based extinguishing agents must not be used for class B or class C fires. Additional information on aircraft fires and extinguishing agents will be found in the text *Aircraft Basic Science*.

FIRE SUPPRESSION SYSTEMS

A fire suppression or extinguishing system consists of one or more pressure tanks (containers) containing an effective fire extinguishing agent, an instantaneous release valve, deployment lines to carry the agent to the engine compartment (nacelle) involved with the fire, and associated controls and indicators. Some systems are designed to provide only one release of extinguishing agent per engine and others provide for two releases of agent. The two-shot feature may be attained by utilizing two separate agent containers for each area or by incorporating a cross-feed system so that the extinguishing agent can be drawn from either of two different containers for each engine.

Extinguishing agents used in engine fire extinguishing systems vary in type. Some commonly used agents are carbon dioxide (CO_2), Freon, bromotrifluoromethane (CF_3Br), and chlorobromomethane (CH_2ClBr), also known as CB. Some factors in the selection of an agent are cost, ease of handling, weight, toxicity, corrosive effect on metals, and effectiveness in extinguishing fires.

CO_2 is widely used as a fire extinguishing agent around aircraft because it can be used safely for both class B and class C fires. Furthermore, from a safety standpoint, CO_2 has little or no toxic or corrosive effect. It is contained as a liquid in highly pressurized containers and when released through a cone-shaped nozzle immediately becomes a gas at very low temperatures. Some of the gas freezes and produces a "snow" that, together with the gaseous component, is effective in smothering a fire by displacing the oxygen in the air.

System for a Light Aircraft

A typical fire extinguishing system for a light twin-engine airplane consists of a pressure bottle (container) mounted in the aft section of each engine nacelle, an explosive cartridge (squib) in the outlet fitting, a discharge hose leading to the forward section of the nacelle, and a smaller hose leading off the main hose to discharge the Freon agent from the center of the nacelle area toward the inboard side of the nacelle. If an overheat condition is detected, a FIRE light will indicate which engine is involved. The extinguisher is activated by opening the guard and pressing the FIRE light. This completes the electrical circuit to the squib, causing it to fire and rupture the disk in the outlet. Freon gas is discharged from the container and floods the nacelle, extinguishing the fire.

After the bottle discharges, an amber E light comes on to show that the bottle has discharged. This light will continue to show the empty condition until the extinguisher bottle is replaced.

Systems for a Transport Aircraft

The fire extinguishing equipment for an airliner such as the Douglas DC-10 incorporates three systems with six pressurized agent containers, two for each engine. The containers for the wing-mounted engines are located in pairs inside the leading edges of the wings between the nacelle and the fuselage on each side of the aircraft. The mounting of the containers is shown in Fig. 23-44. Note that the two containers

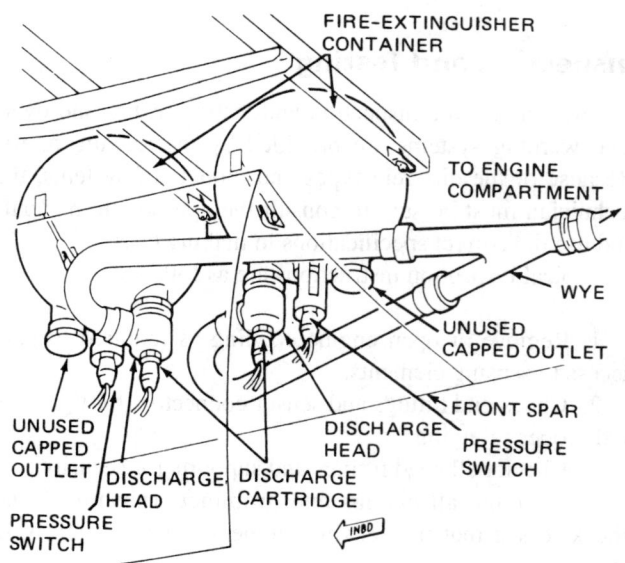

FIGURE 23-44 Mounting of extinguishing agent containers. (*McDonnell Douglas Corp.*)

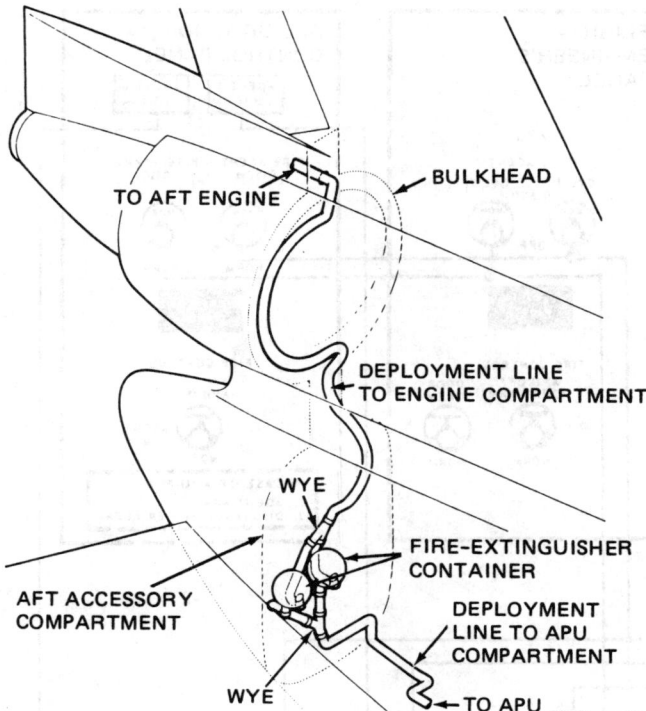

FIGURE 23-45 Deployment system for aft engine and APU compartment. (*McDonnell Douglas Corp.*)

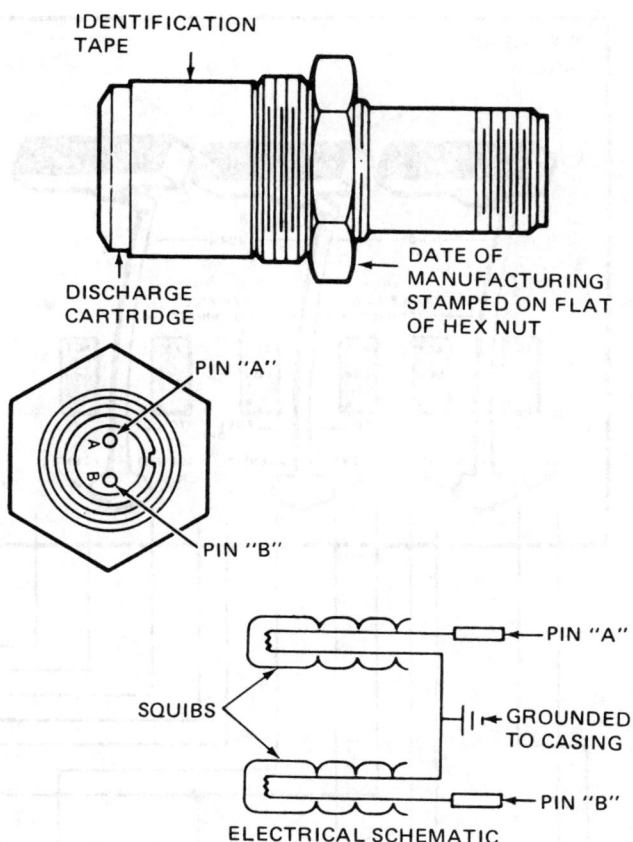

FIGURE 23-46 Discharge cartridge.

feed through a Y fitting to the common deployment line. The arrangement of the system for the aft engine and auxiliary power unit (APU) compartment is shown in Fig. 23-45.

The six agent containers are constructed of stainless steel and are hermetically sealed. They may vary in capacity in accordance with customer requirements. The containers are filled with bromotrifluoromethane (CF_3Br) and pressurized with nitrogen. Each container is provided with two outlet ports with breakable metal disks welded to each outlet neck. The disks serve as pressure relief valves that will rupture if an overpressure condition develops because of excessive heat. Depending on which disk ruptures on the wing-mounted containers, the discharge of CF_3Br will flow to the engine nacelle or out through two orifices in the cap of the unused outlet neck. For the aft-engine containers, the relief discharge will flow either to the engine nacelle or to the APU compartment.

Each container incorporates a pressure switch that remains open when the container is adequately pressurized. If the pressure drops below the specified level, the switch will close and illuminate the AGT LOW light, which is amber in color. This light remains on until the pressure is restored to normal.

The discharge cartridge is an electrically fired explosive unit that provides force to rupture the disk seal in the outlet port of the container and release the extinguishing agent to the deployment lines. One discharge cartridge is installed on each wing-mounted container and two are installed on the aft-engine container.

The discharge cartridge, shown in Fig. 23-46, contains two squibs (explosive charges) that are fired by an electrical circuit that heats the powder to the ignition point. When the fire extinguisher control handle is pulled down, the electrical circuit is closed and the squibs are fired. The force of the explosion ruptures the disk seal, thus releasing the extinguishing agent. A screen in the discharge head prevents particles of the disk from entering the deployment line.

The deployment lines and extinguisher outlets for wing-mounted engines are shown in Fig. 23-47. Note that the deployment lines terminate at five discharge nozzles in the

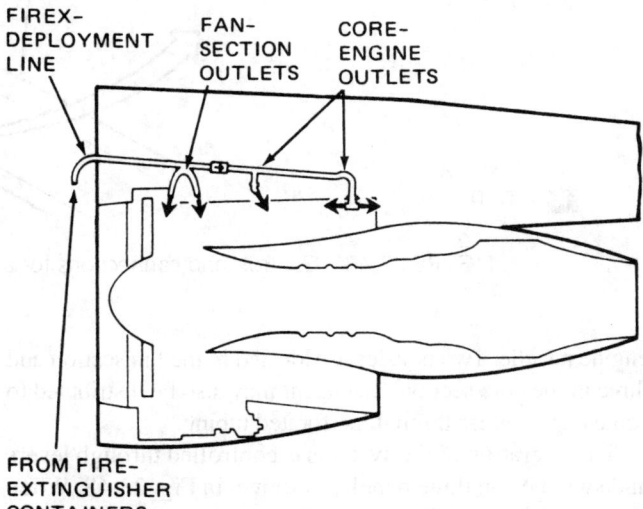

FIGURE 23-47 Deployment lines and extinguisher outlets for wing-mounted engines. (*McDonnell Douglas Corp.*)

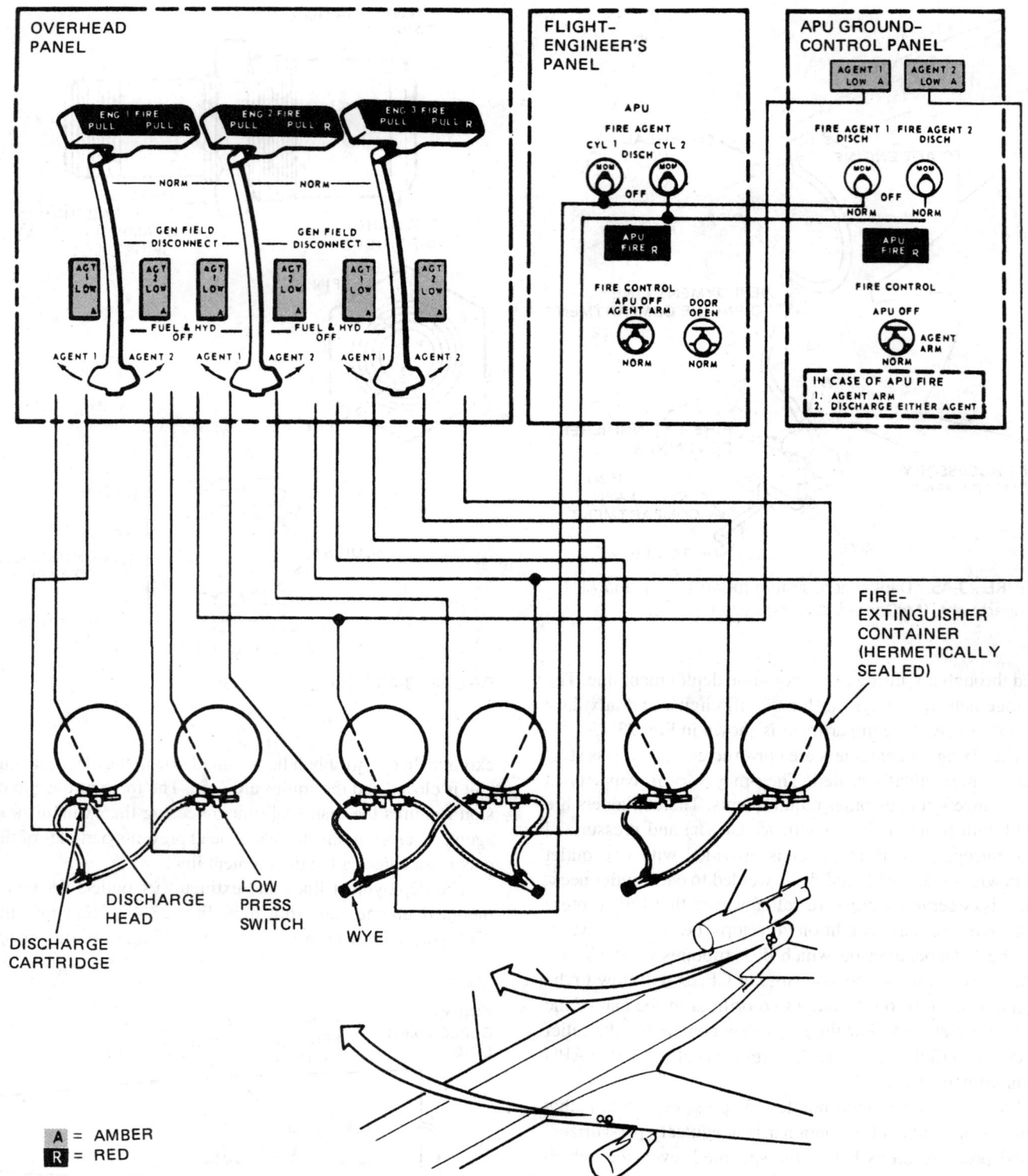

FIGURE 23-48 Controls and connections for a fire extinguishing system. (*McDonnell Douglas Corp.*)

engine nacelle. Two nozzles are located in the fan section and three in the core section. The agent may also be distributed to some engine areas through perforated tubing.

The operation of the systems is controlled through levers and switches on three panels, as shown in Fig. 23-48. When an overheat condition or fire occurs, a red light appears on the control lever handle for the engine involved. At the same time, a loud alarm sounds. The pilot pulls the illuminated handle down all the way and turns it clockwise to discharge agent 1 or counterclockwise to discharge agent 2.

When the handle is pulled down, the aural alarm is shut off, the generator field for the affected engine is disconnected, and the fuel and hydraulics for the engine are shut off. At the same time, a switch controlling the circuit to the discharge cartridge is closed, thus firing the cartridge and discharging the selected container through the deployment lines to the engine nacelle.

The two containers in the rear of the aircraft are interconnected so that the system may be used for the rear engine or the APU. If an APU fire breaks out, the red APU FIRE indicator is illuminated and an aural alarm sounds. The flight engineer arms the system with a switch on his or her panel and then closes the switch for the agent container that he or she wishes to use. This fires the discharge cartridge in the chosen container and releases the agent to the APU compartment. The APU system can also be activated from the APU ground control panel in the bottom of the fuselage.

Fire Extinguishing System Indicators

All fire extinguishing systems for aircraft are required to incorporate indicators by which the system condition is shown. The systems described above utilize amber lights to indicate when a container has been discharged or when the pressure is low. Each container includes a pressure switch that is connected to the indicator light. When pressure is low in a container, the switch closes and turns on the amber light. Indicator lights are located on the fire-control panel in the cockpit so that the pilot is warned immediately when any container is low in pressure or is discharged.

For many aircraft fire extinguishing systems, breakable colored disks are employed as condition indicators. These are mounted in the side of the fuselage, where they are readily visible to a crew member making a preflight inspection. A *yellow* disk is used to indicate when an extinguishing agent has been discharged by use of the cockpit controls, and a *red* disk is used to show when the discharge has occurred as a result of overpressure and rupture of the safety disk in the outlet of the agent container. A hose or tubing is connected from the main discharge line to the fitting behind the yellow or amber disk. Discharge of the agent by the cockpit control directs pressure to the disk and causes it to rupture.

The safety disk fitting in the agent container is connected by a hose or tubing to the red disk fitting. When the safety disk is ruptured because of overpressure, agent pressure ruptures the red disk, thus providing an indication that the container has been discharged.

Maintenance of Fire Extinguishing Systems

The maintenance of fire extinguishing systems is described in manufacturers' manuals for specific systems. There are, however, typical practices which apply to many systems. Periodic checks and tests are generally specified to ensure that a system and components are in operable condition.

The condition of lines, hoses, fittings, and components is checked visually and by feel to determine the condition and security of mounting.

Agent containers having pressure gauges are checked by observing the pressure gauges to see that the pressure is within the prescribed range for pressure vs. temperature. Containers without pressure gauges are checked by weighing them to determine whether the quantity of agent in the

container is adequate. Hydrostatic test dates must be stamped on the containers so that the technician will know when to remove them for weight checks.

Discharge cartridges are examined to ensure that the service life of the unit has not been exceeded. Service life is specified in the number of hours that a cartridge will be effective as long as it has not been exposed to temperatures above a specified limit.

When a discharge cartridge has been removed for any reason, it must be reinstalled in the same container from which it was removed. During removal and installation, the terminals of the discharge cartridge should be grounded or shorted to prevent accidental firing. Before connecting the cartridge terminals to the electrical system, the system should be checked with a voltmeter to make sure that no voltage exists at the terminals.

Because of the differences in fire extinguishing systems, instructions for inspection and maintenance will vary. For this reason it is essential that the technician follow the procedures set forth in the manufacturers' manuals.

ENGINE CONTROL SYSTEMS

Need for Engine Controls

All aircraft powerplants must be controlled precisely in order to provide the performance required for the many different modes of aircraft operation. The controls provide a means by which the pilot, copilot, or flight engineer can manipulate the engine and engine accessories. Modern aircraft engines, whether they are gas-turbine or reciprocating types, have many control systems and mechanisms, some mechanical, some electronic or electrical, some hydraulic, and others that may have a combination of actuating forces and devices. This section focuses on the controls that are mechanical in nature.

Desirable Characteristics

Each part or unit of the engine which is subject to control, and each engine accessory, must be controlled not only as an individual unit but also in consideration of its effect on all the other parts, units, and accessories. Therefore, engine control systems must be accurate in their manipulation, and positive, reliable, and effective, regardless of the distance between the controlled unit and the control in the cockpit.

Principal Types of Engine Control Systems

The three general types of mechanical engine control systems are (1) **push-pull rods** with bellcranks and levers, (2) **cable and pulley systems**, and **(3) flexible push-pull wires encased in coiled-wire sheathing**. Frequently, aircraft engine control systems embody combinations of the above systems. These are basic control systems—they may or may not be included in the hydraulic, electrical, or electronic control systems installed in modern airplanes.

It is also possible to classify engine controls as (1) manually operated, (2) semiautomatic, and (3) automatic. The rapid advancement in engine design, the use of higher power output in many powerplants, the installation of superchargers and turbosuperchargers, the introduction of highly controllable propellers, and the need for accurate fuel schedules in the operation of gas-turbine engines are some of the factors calling for the adoption of automatic or semiautomatic controls, although a method of manual control is almost always retained to provide for the possible failure of automatic or semiautomatic control mechanisms. Automatic and semiautomatic control may be accomplished by hydraulic, electrical, or electronic mechanisms.

Push-Pull Tube Characteristics

The rods or tubing used to transmit control lever movements are generally called **push-pull** rods. The amount of force to be exerted on a push-pull rod determines its diameter and its wall thickness if it is a hollow tube, since it must withstand both compression and tension. Compression tends to increase a bend, and tension tends to decrease it; therefore, bends are avoided. If bends were present, they would tend to change the length and therefore the adjustment of the mechanism under control.

Guides for Rods and Tubing

Guides are often provided for the tubing or rods to prevent flexing and to give them mechanical support where needed. The rod or tube must slide through the guides smoothly and without friction, especially where the rod or tube is long. Guides may be made of fiber, plastic, hard rubber, a composition such as Micarta, or other material of similar characteristics. The plastic called Teflon makes an excellent guide because of its almost complete freedom from friction.

Types of Attachment for Control Rods

There are four types of attachment for the ends of control rods: (1) clevis and pin, (2) ball bearing, (3) ball joint, and (4) threaded. If they are properly inspected and adjusted when necessary, these various rod ends will not be responsible for the so-called give and slack which cause much of the improper operation of control-system linkage. Clevis-rod ends may be screwed on the tube or flexible joint and locked with a jam nut. A clevis pin, washer, and cotter pin may be used to fasten the rod end to the control arm or bellcrank. When the control tube movement is not in line with the control arm movement, a ball-joint end is used on the control tube to allow an angle (usually up to 15°) between the two parts. Screw ends allow for adjustment of the length of the tube. When a tube is screwed in, it is shorter, and less thread is exposed.

Figure 23-49 shows a clevis-rod end, locknuts, a push-pull control rod cut apart in the drawing to reduce the length, the threaded end of a rod, an inspection hole, and a straight beam which is sometimes called a "walking beam."

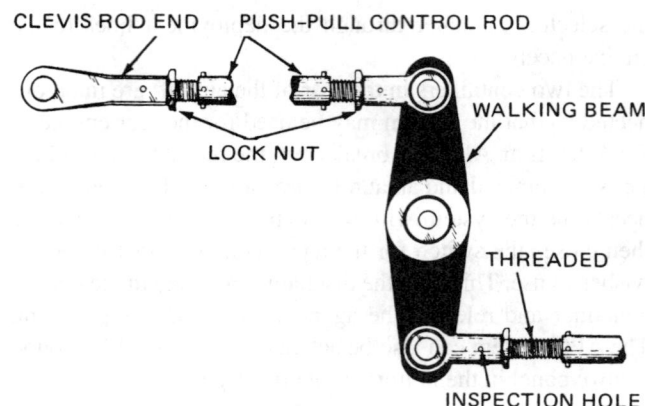

FIGURE 23-49   Beam and clevis-rod end.

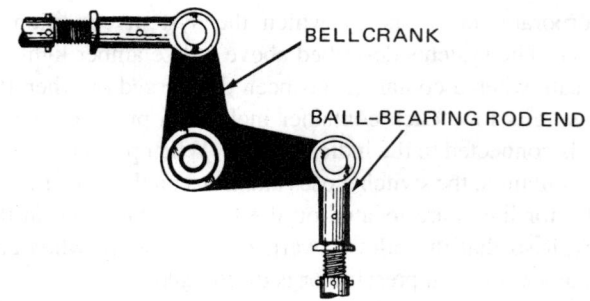

FIGURE 23-50   Bellcrank and ball-bearing rod end.

Bellcranks

A bellcrank is a double lever or crank arm in which there are two cranks approximately at right angles to each other (see Fig. 23-50). The purpose of a bellcrank is to provide a means of changing the direction of motion—that is, to transmit the motion around some obstacle. For example, it may be necessary to change a forward movement to a rearward movement or to change a horizontal movement to a vertical movement. In this manner, it is possible to obtain the desired relative movement between the engine control lever in the cockpit and the engine unit which it controls.

If the arms of a bellcrank are of equal length, the change of movement is accomplished without any gain or loss in the movement of the linkage; but if the arms are not of equal length, a gain or a loss in movement is obtained. For example, it is possible to produce a relatively great movement of an engine-unit adjustment by means of a comparatively small movement of the cockpit control. The reverse is also true. A comparatively great movement of the cockpit control may produce a relatively small movement of the engine-unit adjustment. The results obtained depend on the original design of the bellcrank, its installation, and how it is rigged to the engine units. Regardless of the transmission of movement, it must be supported by bearings that will reduce the friction to a minimum. Ball bearings are generally used for this purpose.

Cable-and-Pulley Control Systems

A **pulley** is essentially a wheel, usually grooved, mounted in a frame or block so that it can readily turn upon a fixed axis.

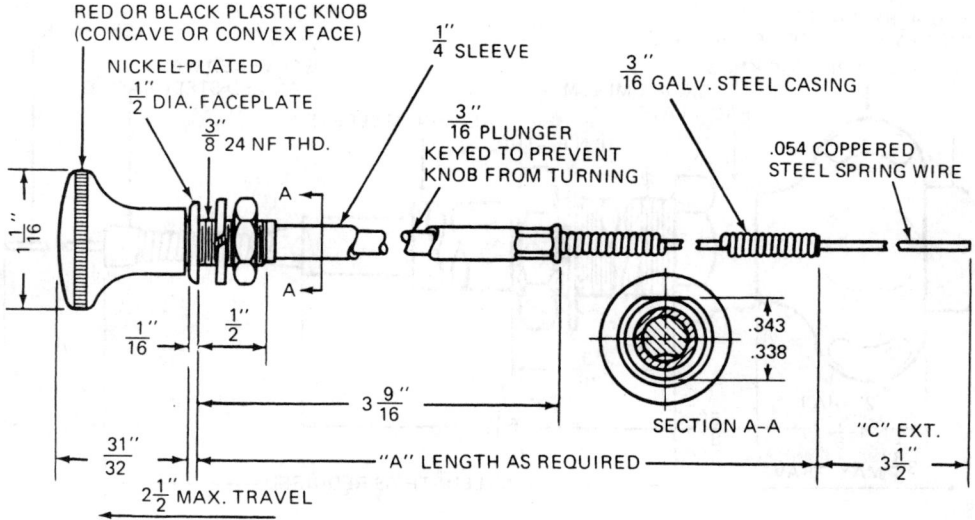

FIGURE 23-51   A push-pull control. (*Arens Controls.*)

Two or more wheels may be mounted in the same frame, either on the same axis or on different axes. When two or more wheels are mounted in the same frame, the pulley is described as having two or more **sheaves**.

The installation of cable-and-pulley systems in a multi-engine airplane is often a complicated procedure. In a single-engine airplane, where the distance between the control levers and the engine units is short, a push-pull control rod may be used; but the great distances and the need for changes of direction in a multiengine airplane make it necessary to use cables and pulleys.

Pulleys for aircraft control systems are usually made of a phenolic or Micarta composition, plastic, or aluminum alloy. They are often supported on antifriction bearings.

Engine control cables are 7 by 7, $\frac{3}{32}$-in [2.38-mm] flexible steel with standard swaged attachment fittings. The cables are usually guided and supported by **fairleads** (phenolic or Micarta blocks with holes in them to admit the cables).

In some powerplant installations, the cables extend from the control levers to the engine units; but in other installations the cables extend only from the control levers to the firewall of each engine nacelle and are then connected to push-pull control rods that transmit the movement to the engine units.

Various devices are used for adjusting the tension and the length of the cables, such as turnbuckles, adjustable pulley clusters, and adjusting links. The cables can be adjusted so that the control handles in the cockpit reach their fore-and-aft stops just after the stops in the engine sections are reached. Thus, "**springback**" is provided to ensure the complete range of control operation.

Flexible Push-Pull Controls

Flexible push-pull controls are used for a variety of remote-control situations in aircraft. This is particularly true for light aircraft, where the distance through which the control must operate is not so great that friction becomes a problem.

The construction of a simple push-pull control is shown in Fig. 23-51. This control is mounted on the instrument panel and may be used for carburetor air heat, a cabin heater, or some other similar function. The knob is attached to a **plunger** into which a steel spring wire is secured. The plunger is inserted into the **sleeve**, which serves as a guide. At the forward end of the sleeve is a threaded section and a **faceplate** to provide for mounting on the panel. The assembly is inserted through a hole in the panel and is mounted firmly by means of a locknut and washer on the rear of the panel.

The steel wire of the control is enclosed in a $\frac{3}{16}$-in [4.76-mm] galvanized-wire casing which is in the form of a coil or spiral. The casing guides the control wire to the operating mechanism.

The method by which the control should be mounted in the aircraft is shown in Fig. 23-52. Note that each end of the flexible casing is securely mounted with terminal mounting clamps so that there will be no slack, or play, in the operation. Intermediate clamps must be used on both sides of bends to prevent flexing and a resultant loss of movement. The control

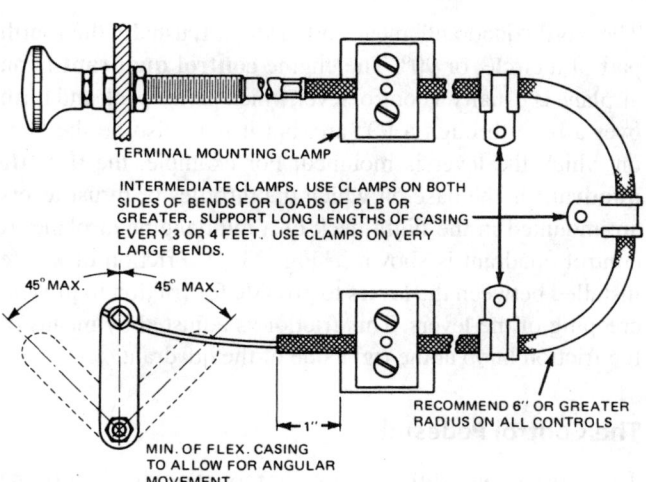

FIGURE 23-52   Mounting of a push-pull control. (*Arens Controls.*)

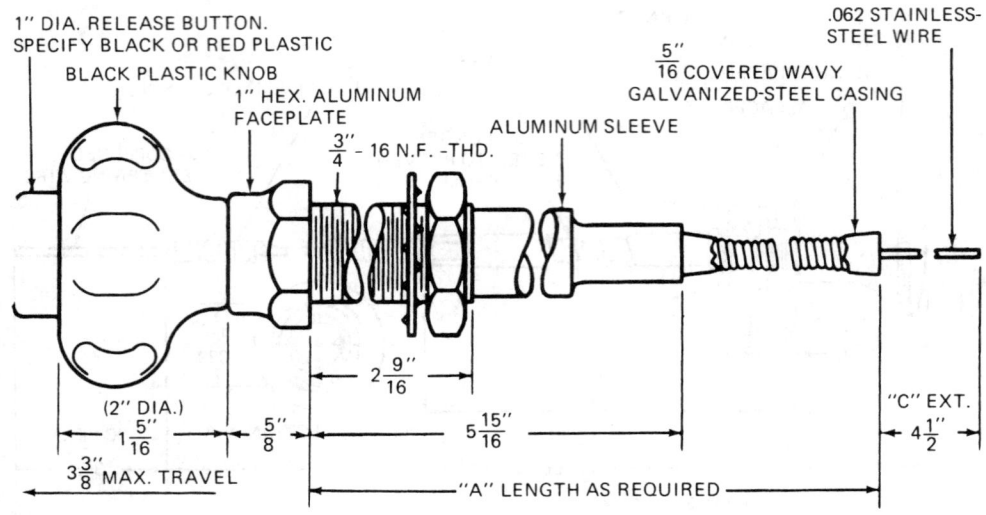

FIGURE 23-53 A vernier push-pull control. (*Arens Controls.*)

should be adjusted so that there is a small amount of spring-back at the full "in" position of the knob to ensure that the operated lever has moved through its full range.

Vernier Push-Pull Control

A vernier push-pull control is constructed to provide for a coarse adjustment and a fine (**vernier**) adjustment, especially for the throttle and propeller controls. A unit of this type is shown in Fig. 23-53. When a fine adjustment is desired, the plastic knob is rotated to the right for increase and to the left for decrease. When a coarse adjustment is desired, the lock button is depressed and the knob is pushed in or pulled out. The fine adjustment is used when precise power settings are made. The approximate engine rpm is set by moving the propeller control in or out as required. Then the final adjustment is made by rotating the knob. After the rpm is set, the manifold pressure is adjusted with the throttle control, first with the coarse adjustment and then with the vernier.

The Control Quadrant

The word "quadrant" means a fourth part, usually the fourth part of a circle, or 90°. The engine **control quadrant** in an airplane is usually a control lever which pivots back and forth over a base through a 90° arc, but it may also be the base on which the lever is mounted. For example, the **throttle quadrant** is the base on which the throttles or thrust levers are mounted in the flight deck of a multiengine airplane. A control quadrant is shown in Fig. 23-54. Friction disks are installed between the levers to provide for friction to prevent creeping of the levers. This friction is adjusted by means of the friction knob at the right side of the quadrant.

The Control Pedestal

The **control pedestal** is a frame or mount on the floor of the cockpit or flight deck in a multiengine airplane where some of the engine controls are connected. The control pedestal

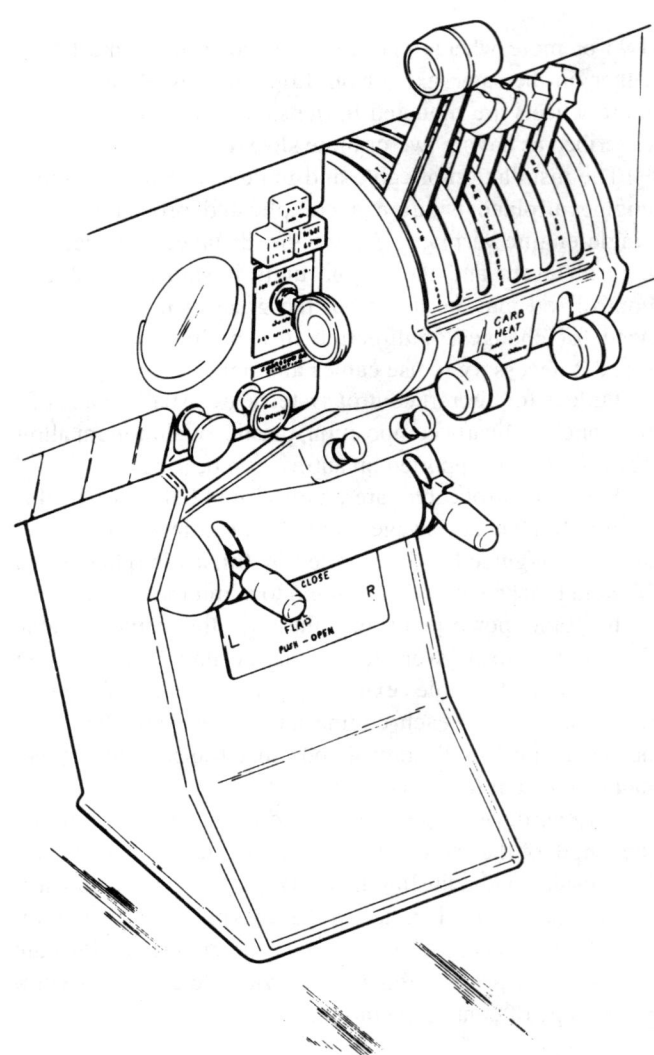

FIGURE 23-54 A control quadrant. (*Piper Aircraft Co.*)

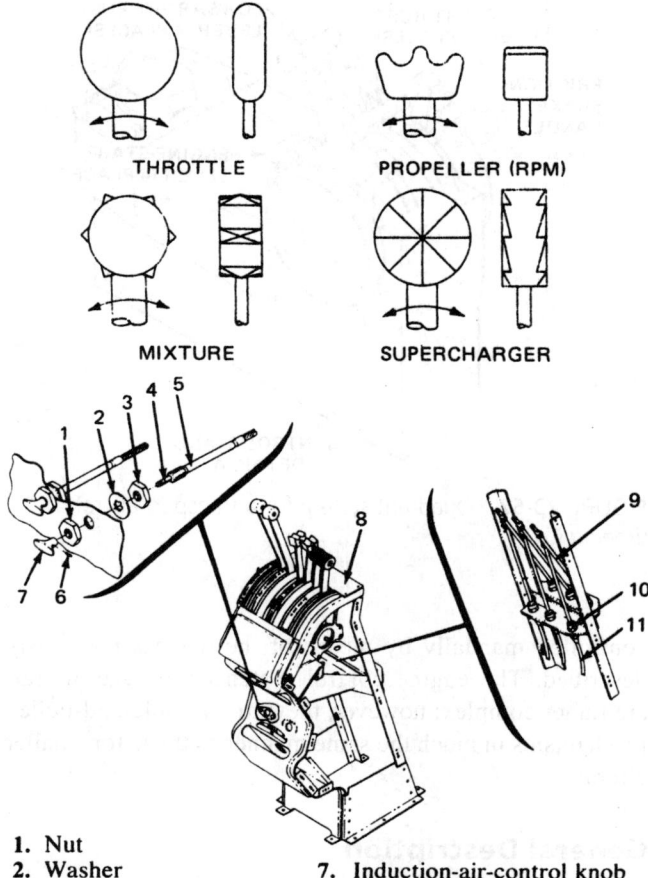

THROTTLE   PROPELLER (RPM)

MIXTURE   SUPERCHARGER

1. Nut
2. Washer
3. Nut
4. Control wire
5. Control-wire housing
6. Roll pin
7. Induction-air-control knob
8. Control pedestal
9. Teleflex push-pull unit
10. Swivel
11. Teleflex control conduit

FIGURE 23-55   Control pedestal with approved shapes for control knobs. (*Cessna Aircraft Co.*)

supports one or more control quadrants. Figure 23-55 illustrates a control pedestal for a twin-engine airplane. The control quadrant on this pedestal has controls for the throttle, engine rpm (propeller), and carburetor mixture.

Note that the quadrant levers are attached to Teleflex push-pull cable units which are mounted so that they can swivel with the movement of the lever. Note also that the operating knobs on the control levers are required to have shapes easily recognized by feel. The shapes of the knobs are illustrated in Fig. 23-55.

Controls for Use with Quadrant

Push-pull wire-type controls are often used with control quadrants on the pedestal. A simplified diagram that shows how this works is presented in Fig. 23-56. The end of the control unit is connected to the arm of a quadrant lever instead of a knob. The sleeve is secured to a bracket with a swivel fitting to allow a few degrees of side movement as the quadrant lever position is changed. The engine end of the control assembly is connected to the controlled unit with a clevis or other suitable device.

Control Assemblies

The control assemblies in the pilot's compartment or cockpit of a single-engine airplane are much simpler than those

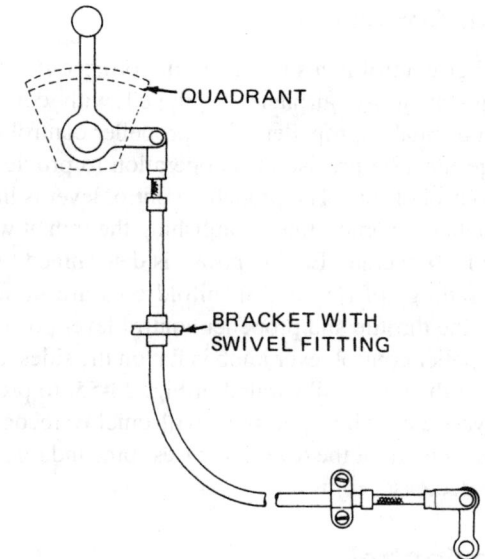

FIGURE 23-56   Lever-operated push-pull control.

installed in a multiengine airplane. A single-engine airplane may have only one control quadrant, which is used to control the throttle, the fuel-air mixture, the propeller, and possibly a supercharger. Individual control levers may be provided for such powerplant accessories as the carburetor air heater or the cowl flaps.

In a multiengine airplane, the complexity of the controls installed in the pilot's compartment may make it necessary to have two or more control quadrants mounted on a control pedestal where most of the engine controls are located to afford a central point of operation for the convenience of the operator.

A problem encountered in the operation of a control quadrant is the tendency of the controls to creep out of the positions in which they are placed. This problem is partly overcome by providing a reasonable amount of friction in the quadrant assemblies, but it is also solved by the use of mechanical locking devices, which temporarily hold the controls in the positions where they are placed by the operator until it is time to move them.

MECHANICAL ENGINE CONTROL FUNCTIONS FOR SMALL AIRCRAFT

Throttle

The throttle-control knob or lever is probably the most conspicuous of all controls, and by some persons it is considered to be the most important because it controls engine power. The throttle lever is marked with the letter T or the word "throttle," and its direction of operation is indicated by the words "open" and "closed." The throttle knob (see Fig. 23-55) is a thick disk that is flat on each side, to comply with the standard requirements of the Federal Aviation Administration (FAA). Throttle movement is always such that forward movement opens the throttle valve and rearward movement closes the valve.

Propeller Control

The propeller control is also of great importance in engine power adjustment for airplanes equipped with constant-speed or controllable propellers. The propeller control lever and linkage must be precise in its operation to provide for accurate control of rpm. The propeller control lever is linked to the propeller governor, thus establishing the rpm at which the engine is to operate. Engine power is determined by the combined settings of rpm and manifold pressure in accordance with the throttle and propeller control lever positions.

The propeller control lever knob is flat on the sides and is scalloped on the top, as illustrated in Fig. 23-55, to provide a distinctive shape which the pilot will quickly recognize. Forward movement of the lever increases rpm, and rearward movement decreases rpm.

Mixture Control

The mixture control lever on the quadrant is used to adjust the fuel-air mixture through the mixture control lever on the carburetor or fuel control unit. The lever positions are usually marked FULL RICH, LEAN, and IDLE CUTOFF. Some airplanes have markings for AUTO RICH and FULL RICH to provide two rich settings. The FULL RICH setting provides a maximum rich condition which is not affected by the automatic mixture control.

The mixture control lever is arranged so that a forward movement will provide a richer adjustment. The knob on the quadrant lever is also distinctively shaped for easy recognition. It is a thick disk like the throttle knob; however, it has a raised diamond pattern on the periphery, as shown in Fig. 23-55.

Carburetor Air Heat

The control lever or knob for the carburetor air heat may or may not be mounted on the pedestal. It is usually not a lever of the type used for the previously described controls but may be a push-pull knob on the front of the pedestal or on the instrument panel. The function of this control is to operate a gate valve in the air induction system to provide either cold air or heated air for the carburetor. Heated air is required for the engine when there is danger of carburetor icing.

Miscellaneous Engine Controls

In addition to the controls described in the foregoing paragraphs, engines may require several other controls, depending on the design of the engine and aircraft. Among the other controls an engine may require are controls for the cowl flaps, oil coolers, superchargers, and intercoolers. All such controls must be marked to show their functions and must be easily accessible to the pilot.

MECHANICAL ENGINE CONTROL SYSTEMS FOR LARGE AIRCRAFT

A variety of engine mechanisms require control from the cockpit of the airplane, and many of these can be

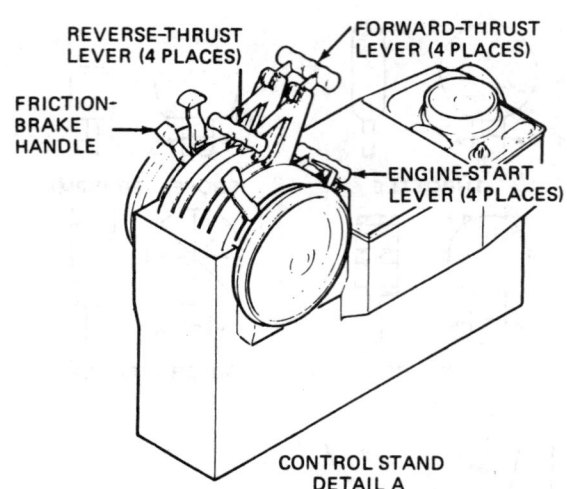

FIGURE 23-57 Quadrant system for a transport aircraft. (*Boeing Co.*)

controlled manually by means of the devices previously described. The engine control systems for large aircraft are rather complex; however, they utilize cable-and-pulley mechanisms in much the same manner as those for smaller aircraft.

General Description

A typical transport aircraft is equipped with four separate engine throttle systems (see Fig. 23-57) to provide for individual control of each engine. The starting of each engine is accomplished by the use of a single lever to energize the ignition system and to initiate the flow of fuel to the engine. Another lever assembly controls both forward and reverse thrust by regulating fuel flow and actuating the thrust reverser. An interlocking mechanism prevents simultaneous initiation of forward and reverse thrust for each engine.

The throttle system consists of an engine start lever and a thrust lever assembly for each engine, connected by a series of throttle control cables and mechanical linkages to the fuel control units on the engine. A thrust lever friction brake applies a braking force to all thrust lever assemblies during forward thrust operation.

The engine start lever is connected by cables to an engine control drum-and-shaft assembly in a nacelle strut. The arrangement of this assembly and the associated linkages is shown in Fig. 23-58. The drum-and-shaft assembly is connected by a rod-and-bellcrank arrangement on the right side of the engine to the fuel control unit. Advancing the engine start lever actuates an ignition switch that energizes the ignition system. Further movement of the start lever opens a shutoff valve in the fuel control unit.

Thrust Lever Assembly

Thrust lever assemblies on the control-stand (pedestal) quadrant control the forward and reverse thrust of the engines. Each thrust lever assembly consists of a forward-thrust lever,

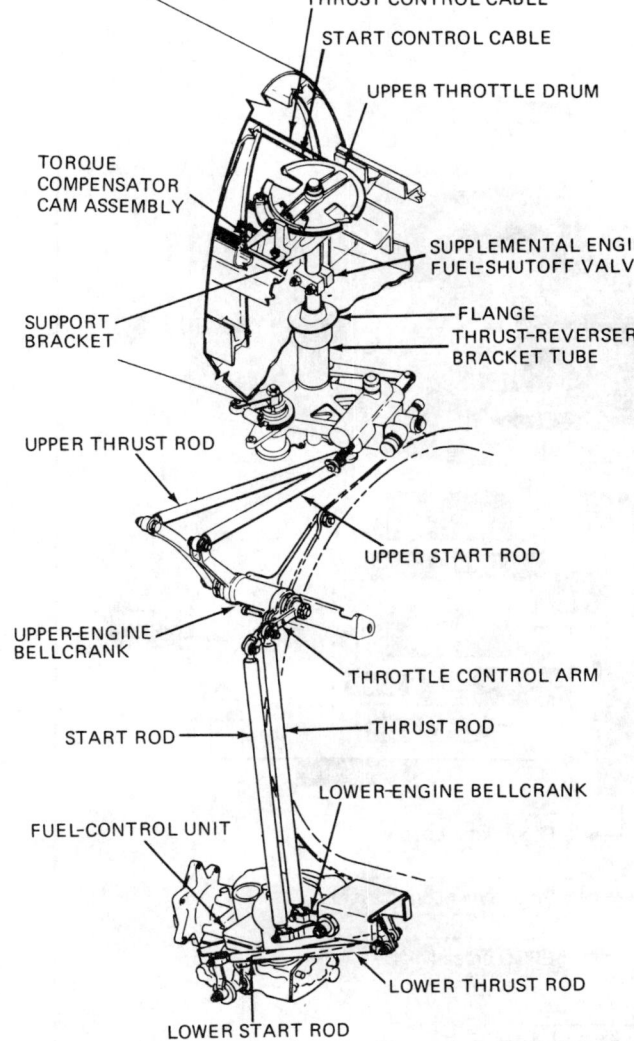

FIGURE 23-58 Drum-and-shaft assembly with linkages. (*Boeing Co.*)

a reverse-thrust lever, a reverse-thrust control link, a pawl, a brake drum, and a thrust control drum, as illustrated in Fig. 23-59. The forward-thrust lever, with the reverse-thrust lever attached to it, is mounted on the brake drum. One end of the control link is riveted to the reverse-thrust lever, and the opposite end is attached to the thrust control drum. Various positions of the assembly are shown in Fig. 23-59.

As either thrust lever is advanced from the idle position, the control link rotates the thrust control drum to actuate the fuel control unit to increase thrust. The forward-thrust idle position is against an idle stop on the quadrant, and full forward thrust is obtained before the lever is all the way forward. The reverse-thrust lever, when in the idle position, is against an idle stop on the forward-thrust lever.

An interlock mechanism prevents simultaneous actuation of the forward- and reverse-thrust levers to ensure positive forward-to-reverse-thrust control. The ability of each lever to move depends on the position of the other lever. If the forward-thrust lever is more than 2° from the idle position, the reverse-thrust lever cannot be moved more than 12° from idle. However, if the reverse-thrust lever is advanced more than 12° from idle, the forward-thrust lever cannot be moved. The interlock between the levers is a pawl riveted to the forward-thrust lever with the pawl between the thrust lever and the control link. When the forward-thrust lever is 2° or less from the idle position, the pawl is aligned with a lockout hole in the web of the thrust lever cover. As the reverse-thrust lever is returned to the idle position, the control link pushes the pawl from the hole to unlock the forward-thrust lever. When the forward-thrust lever is more than 2° from the idle position, the pawl is not aligned with the lockout hole. The web then opposes the force of the control link on the pawl so that the reverse-thrust lever cannot be moved more than 12° from idle.

Engine Start Levers

The **engine start levers** on the control-stand quadrant are used to start the engines. Each lever controls ignition and fuel, as explained previously. The start lever is provided with a spring-loaded detent catch which may be released by lifting the knob. The detent secures the lever in the CUTOFF and IDLE positions. An additional detent is provided between these two positions. This catch is provided to ensure that a throttle left insecurely in the IDLE position will not creep to the CUTOFF position and cause an unintentional engine shutdown. A stop gate and detent are provided at the START position.

Thrust Lever Friction Brake

A thrust lever friction brake on the control-stand quadrant applies a variable braking force to all thrust levers during forward-thrust operation. The friction brake is used to select manually the proper amount of friction force to prevent throttle creep during flight.

The full authority digital engine control (FADEC) permits operation of the engine control system and autothrottles for the engines to be fully integrated into the flight control and management system. A fly-by-wire (FBW) system replaces manual cable control of an aircraft engine's throttle with an electronic interface. FADEC allows maximum performance to be extracted from the aircraft without engine operational problems or high pilot workloads. The movements of engine controls are converted to electronic signals (through resolvers) transmitted by wires to the electronic fuel control unit (EFCU) control computers. From pilot or flight management inputs the computers determine how much to move the actuators at each engine's fuel actuating unit to provide the expected response from the engine. Many inputs are sent to the EFCU computers in order for the computer to determine the correct commands that are sent to the fuel actuating unit which determines the amount of fuel flowing to the engine, see Fig. 23-59. Feedback from the fuel actuating unit is constantly read by the EFCU computers.

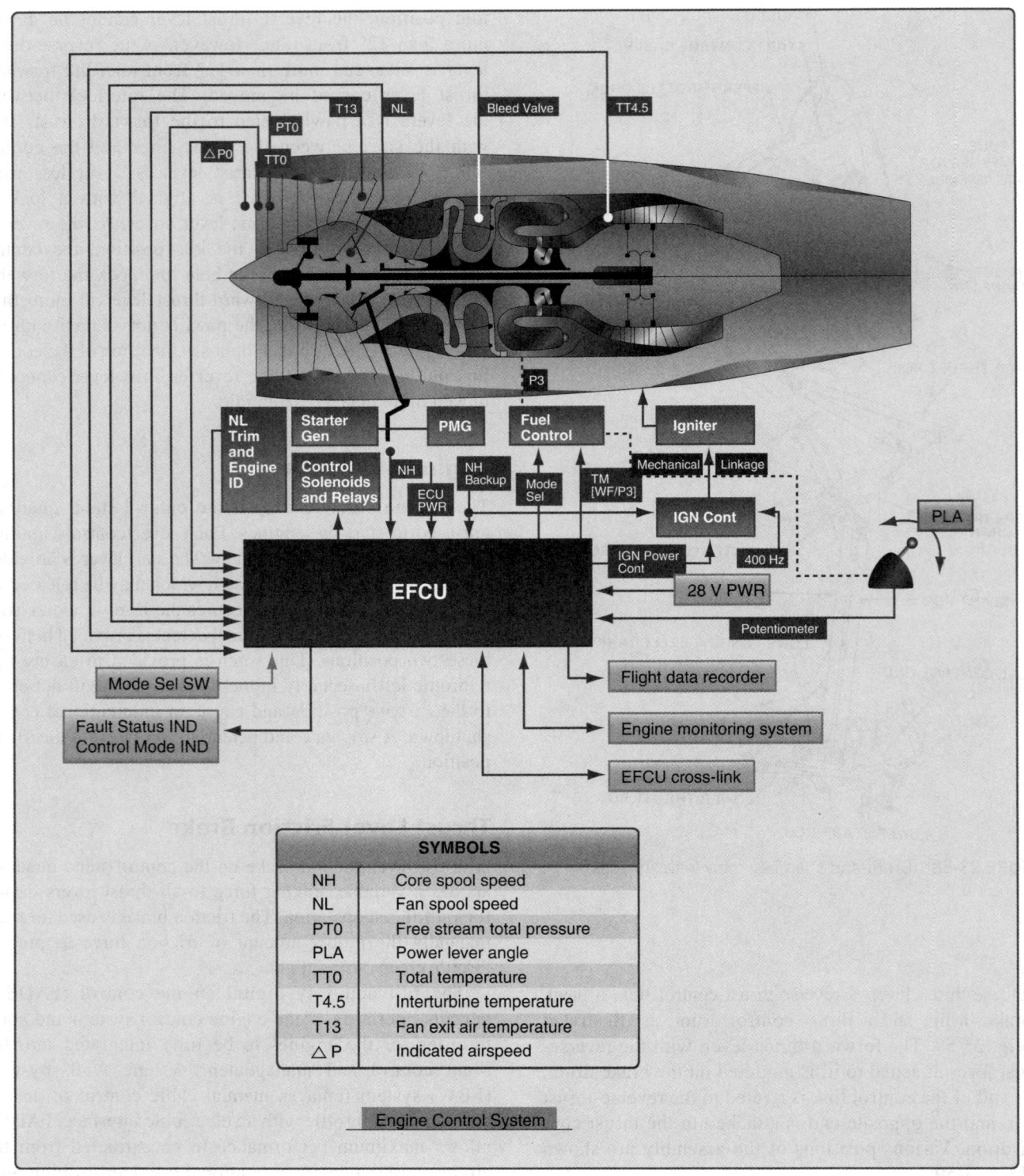

FIGURE 23-59 Engine control system.

INSPECTION AND MAINTENANCE OF CONTROL SYSTEMS

Rods and Tubing

1. Operate the system slowly, and watch for evidence of any strain on the rods or the tubing that will cause bending or twisting.

2. Examine each rod end that is threaded, and observe whether or not the rod is screwed into the socket body far enough to be seen through the inspection hole.

3. Eliminate any play by making certain that all connections are tight.

4. Examine the guides to see if the rods bind too much on the guides, but do not mistake any binding for spring-back. Replace any guides that cause binding.

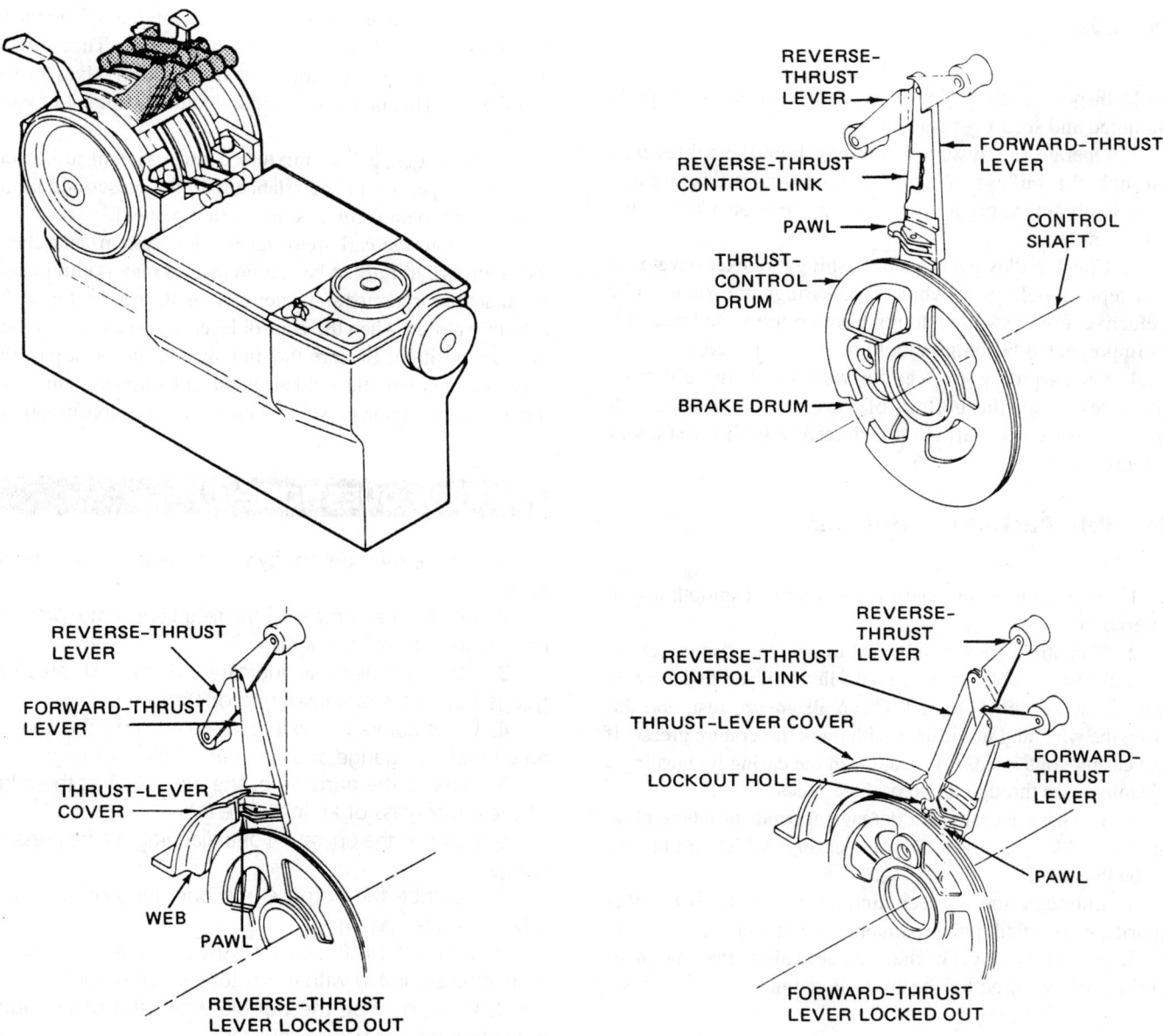

FIGURE 23-60 Thrust lever assembly. (*Boeing Co.*)

5. Adjust the lengths of screw-end rods by screwing them into the control end or backing them out. Retighten the locknuts.

6. If any rod is removed, mark it to show its location for reassembly.

7. Replace any ball-bearing rod ends that cause lost motion.

Bellcranks

1. Examine for security of mounting, for wear, and for proper lubrication.

2. Replace any bellcrank bearings that are causing lost motion.

3. Mark each part that is removed, and reassemble in the correct order.

Cables

1. Inspect for frayed or broken cables.

2. Clean the cables where they pass through fairleads or over pulleys, and then cover them with the prescribed rust preventive.

3. Maintain the correct tension constantly, not merely at the prescribed inspection intervals. Use a **tensiometer** to obtain an accurate reading of the cable tension, and check it against the approved tension given in the manufacturer's instructions.

4. Inspect all turnbuckles, adjusting links, and other devices used to maintain tension in the cables. Safety such devices after each adjustment.

5. If any cables are removed, mark them to show their locations in the system. When new cables are installed, apply any required code markings in the same manner as they were applied on the original cables.

Pulleys

1. Inspect each pulley to ensure that it is properly mounted and securely fastened.

2. Operate the system to observe how the cables pass through the pulleys. Where necessary, adjust or replace parts to obtain the proper clearance and tension. Correct any misalignment.

3. Check pulley bearings or bushings for excessive play, and replace pulleys in which the bearings are found to be defective. Pulleys which do not have prelubricated bearings or impregnated bushings should be oil-lubricated.

4. Check pulley grooves for wear occurring either on the sides or on the bottom of the groove. Replace pulleys when wear is sufficient to interfere with satisfactory operation.

Flexible Push-Pull Controls

1. Inspect push-pull controls for wear and smoothness of operation.

2. If controls stick or are hard to operate, disconnect the control wire or cable from the controlled unit and pull the wire out of the casing or conduit. Clean all grease, rust, and dirt from the wire, and lubricate it with low-temperature grease. If the casing contains dirt or rust, clean the casing by pushing a cleaning wire through it two or three times.

3. Inspect the casing for damage at bends or where abrasion may have occurred. Replace casings which are broken or badly worn.

4. Lubricate the plungers and sleeves with oil or other lubricant specified in the maintenance manual.

5. Adjust the travel of the control to allow the amount of springback required for the controlled unit.

Miscellaneous Inspections

1. Examine the control unit in the cockpit or pilot's compartment or at the flight engineer's station for free operation, for the correct travel of the control levers through their extreme ranges of movement, for security of mounting, and for lost motion.

2. Throughout all control systems, look for proper safetying, broken or misaligned pulleys, missing or loose nuts or bolts, dirty connections, and lack of lubrication. Either remedy each defective condition as it is found, or record it on a check sheet and then remedy the unsatisfactory conditions in a systematic manner; for example, perform all the lubrication jobs in one session.

3. Examine the entire engine control system for any unnecessary play. Be sure that all mounting bolts, shaft bolts, rods, check nuts, etc., are tight. Having accomplished an inspection with the engines stopped, operate the engines and check all the controls for each engine, one engine at a time. When this has been done, check the operation with all engines operating.

4. Pay particular attention to the adjustment of the throttle levers, especially in a multiengine airplane. They should be together at equal positions of the throttle valves in the carburetors. This may require individual adjustment of each throttle valve.

5. Check the fuel-air mixture control system to be certain that it operates in accordance with the prescribed limits given in the manufacturer's instruction manual.

6. Operate the carburetor air heating system, and check the controls carefully to be certain that all play is eliminated. See that the carburetor air heater valve in the carburetor intake is tightly closed when the control lever is in the OFF, CLOSED, or COLD position. Be sure that hot air does not leak past the valve and mix with the cold air in the induction system, leading to loss of engine power and increased fuel consumption.

REVIEW QUESTIONS

1. List the mechanisms by which pressure instruments are operated.

2. What is the purpose of the restriction in the pressure port of an oil pressure gauge?

3. What problem is indicated if the oil pressure gauge fluctuates rapidly during operation?

4. What colored markings are placed on the face of an oil pressure gauge and what are their meanings?

5. What is the purpose of the white line at the edge of the cover glass of an instrument?

6. Describe the operation of a diaphragm-type pressure gauge.

7. Describe the purpose and operating principles of a MAP gauge system.

8. Why is a MAP gauge essential in the operation of an engine equipped with a constant-speed propeller?

9. What may be the trouble when the MAP gauge indication lags or fails to register?

10. What is the effect of a broken or leaking MAP gauge pressure line on an unsupercharged engine? A supercharged engine?

11. Why is it possible to employ a fuel pressure gauge as a fuel flow indicator for engines with fuel injection?

12. At what point in an oil system is the temperature-sensing bulb located?

13. Explain the operating principles of thermocouple-type temperature-indicating systems.

14. How can you measure the resistance of thermocouple leads?

15. Describe a magnetic-type tachometer.

16. Describe an electric tachometer system.

17. What does an EPR gauge measure?

18. How are turbine-engine tachometers usually calibrated?

19. List the terms commonly used in measuring turbine-gas temperatures.

20. What is a torquemeter used to measure?

21. What are the basic requirements of a fire or overheat warning system?

22. Describe the different types of heat detectors.
23. List three types of fire-detection systems.
24. Describe a basic fire extinguishing system.
25. Describe a discharge cartridge.
26. What methods are used to show that an adequate extinguishing agent and pressure are available?
27. Name three basic types of engine control systems.

28. Name the types of attachments for push-pull rod systems.
29. Why are controls rigged to allow springback in the control lever?
30. What is meant by *vernier* as applied to a push-pull control?

Appendix

COMMON ABBREVIATIONS AND SYMBOLS

A	area, ft^2
ABC	after bottom center (piston)
ABDC	after bottom dead center
ac	alternating current
AGB	accessory gearbox
AIAA	American Institute of Aeronautics and Astronautics
AN	Air Force-Navy
A&P	airframe and powerplant
APU	auxiliary power unit
ASTM	American Society for Testing Materials
ATC	after top center
ATDC	after top dead center
BBC	before bottom center
BBDC	before bottom dead center
BC	bottom center
BDC	bottom dead center
bhp	brake horsepower
BITE	built-in test equipment
bmep	brake mean effective pressure
bsfc	brake specific fuel consumption
BTC	before top center
BTU	British thermal unit
CAT	carburetor air temperature
CCW	counterclockwise
CDP	compressor discharge pressure
CDT	compressor discharge temperature
CHT	cylinder head temperature
CIP	compressor inlet pressure
CIT	compressor inlet temperature
CL	condition lever
CPR	compressor discharge (pressure) ratio
CRS	certified repair station
CW	clockwise
dc	direct current
ECU	electric control unit
EGT	exhaust gas temperature
EPR	engine pressure ratio
E/S	engine speed
eshp	equivalent shaft horsepower
EVC	engine vane control
F_g	gross thrust, lb

F_n	net jet thrust, lb
F_p	ram drag of engine airflow (momentum of entering air)
FAA	Federal Aviation Administration
FAR	Federal Aviation Regulation
FSDO	Flight Standards District Office (FAA)
Fwd	forward
g	acceleration due to gravity (32.17 ft/s^2)
GADO	General Aviation District Office (FAA)
HMU	hydromechanical unit
HPC	high-pressure compressor
hp	horsepower
IGV	inlet guide vanes
ITT	interturbine temperature
Kn	knot
LPT	low-pressure turbine
M	mach number
MAP	manifold pressure (absolute)
mep	mean effective pressure
METO	maximum except takeoff (power)
MIL	military
MS	material standard
N	Newton
N_1	low-pressure compressor rotational speed, rpm
N_2	high-pressure compressor rotational speed, rpm
N_g	gas generator speed (high-speed rotor)
N_P	power turbine speed (low-speed rotor)
NASA	National Aeronautics and Space Administration
NTC	negative torque control
NTS	negative torque signal
OAT	outside air temperature
OGV	outlet guide vanes
P	pressure
P_{am}	ambient absolute pressure
P_o	standard sea-level absolute pressure
PCU	propeller control unit
PGC	propeller gear case
PL	power lever
psi	pounds per square inch
PSIA	pounds per square inch absolute
PSID	differential pressure
P/T	power turbine
PTO	power takeoff assembly
Q	torque
Q	dynamic pressure
rpm	revolutions per minute

SAE	Society of Automotive Engineers
sfc	specific fuel consumption
shp	shaft horsepower
SPEC	specification
STD	standard
Σ	the sum of
T	temperature
t_{am}	ambient temperature, °F
t_o	standard sea-level temperature, 59°F
TC	top center, Type Certificate
TCDS	Type Certificate Data Sheet
TDC	top dead center
TET	turbine exhaust temperature
T/F	turbofan

TIP	turbine inlet pressure
TIT	turbine inlet temperature
T/J	turbojet
TOP	turbine outlet pressure
TOT	turbine outlet temperature
T/S	turboshaft
tsfc	thrust specific fuel consumption
V	velocity
VG	variable geometry
VSV	variable stator vanes
W	weight
w_a	engine airflow, lb/s
w_{bl}	compressor bleed airflow, lb/s
w_f	fuel flow, lb/h

Conversion Factors

Multiply	By	To Obtain
Atmospheres (atm)	76.0	cmHg at 0°C
	29.92	inHg at 0°C
	33.90	ftH$_2$O at 4°C
	1.033	kg/cm^2
	14.696	psi
	101.33	kPa
	2116.0	psf
	1.0133	bar, hectopieze
Bars (bar)	75.01	cmHg at 0°C
	29.53	inHg at 0°C
	14.50	psi
	100.0	kPa
Barns (b)	10^{-24}	cm^2 (nuclear cross section)
British thermal units (Btu)	778.26	ft•lb
	3.930×10^{-4}	hp•h
	2.931×10^{-1}	kW•hr
	2.520×10^{-1}	kg•cal
	1.076×10^2	kg•m
	1055	J
Btu/s	1055	W
Centimeters (cm)	0.3937	in
	3.281×10^{-2}	ft
cmHg	5.354	inH$_2$O at 4°C
	4.460×10^{-1}	ftH$_2$O at 4°C
	1.934×10^{-1}	psi
	27.85	psf
	135.95	kg/m^2
	1.333	kPa
cm/s	3.2810×10^{-2}	fts
	2.237×10^{-2}	mph
	3.60×10^{-2}	km/h
cm^2	1.550×10^{-1}	in^2
	1.076×10^{-3}	ft^2
	10^4	m
cm^3	10^{-3}	L
	6.102×10^{-2}	in^3
	2.642×10^{-4}	U.S. gal
centipoise (cP)	6.72×10^{-4}	lb/s•ft
	3.60	kg/h•m
circular mils (cmil)	7.854×10^{-7}	in^2
	5.067×10^{-4}	mm^2
	7.854×10^{-1}	mil^2

Conversion Factors (*continued*)

Multiply	By	To Obtain
curies (Ci)	3.7×10^{10}	disintegration/s
degrees (arc) (°)	1.745×10^{-2}	rad
dynes (dyn)	1.020×10^{-3}	g
	2.248×10^{-6}	lb
	10^5	N
	7.233×10^{-5}	poundals
electron volts (eV)	1.602×10^{-12}	ergs
ergs (*or* dyn•cm)	9.478×10^{-11}	Btu
	6.2×10^{11}	eV
	7.376×10^{-8}	ft•lb
	1.020×10^{-3}	g•cm
	10^{-7}	Joule
	2.388×10^{-11}	kg•cal
feet (ft)	3.048×10^{-1}	m
	3.333×10^{-1}	yd
	1.894×10^{-4}	mi
	1.646×10^{-4}	nm
ft^2	929.0	cm^2
	144.0	in^2
	9.294×10^{-2}	m^2
	1.111×10^{-1}	yd^2
	2.296×10^{-5}	acres
ft^3	2.832×10^4	cm^3
	1728	in^3
	3.704×10^{-2}	yd^3
	7.481	U.S. gal
	28.32	L
	2.83×10^{-2}	m^3
ft^3/min	4.719×10^{-1}	L/s
	2.832×10^{-2}	m^3/min
ft^{3}H$_2$O	62.428	lb
ftH$_2$O at 4°C	2.950×10^{-2}	atm
	4.335×10^{-1}	psi
	62.43	psf
	3.048×10^2	kg/m^2
	2.999	kPa
	8.826×10^{-1}	inHg at 0°C
	2.240	cmHg at 0°C
ft/min	1.136×10^{-2}	mph
	1.829×10^{-2}	km/h
	5.080×10^{-1}	cm/s
ft/s	6.818×10^{-1}	mph
	1.097	km/h
	30.48	cm/s
	5.925×10^{-1}	kn

Conversion Factors (*continued*)

Multiply	By	To Obtain
ft•lb	1.383×10^{-1}	kg•m
	1.356	N•m or J.
	1.285×10^{-3}	Btu
	3.776×10^{-7}	kW•h
ft•lb/min	3.030×10^{-5}	hp
	4.06×10^{-5}	kW
ft•lb/s	1.818×10^{-3}	hp
	1.356×10^{-3}	kW
fluid ounce (fluid oz)	8	drams
	29.6	cm^3
gallon (gal), Imperial	277.4	in^3
	1.201	U.S. gal
	4.546	L
gal, U.S., dry	268.8	in^3
	1.556×10^{-1}	ft^3
	1.164	U.S. gal, liquid
		L
gal, U.S., liquid	231.0	in^3
	1.337×10^{-1}	ft^3
	3.785	L
	8.327×10^{-1}	imperial gal
	128.0	fluid oz
gigagram (Gg)	10^6	kg
grains	6.480×10^{-2}	g
grams (g)	15.43	grains
	3.527×10^{-2}	oz avdp
	2.205×10^{-3}	lb avdp
	1000	mg
	10^{-3}	kg
	980.67	dyn
g-cal	3.969×10^{-3}	Btu
g of U^{235} fissioned	23000	kW•h heat generated
g/cm	0.1	kg/m
	6.721×10^{-2}	lb/ft
	5.601×10^{-3}	lb/in
g/cm^3	1000	kg/m^3
	62.43	lb/ft^3
hectare	10^4	m^2
	2.471	acre
hectopieze	29.53	inHg
	75.01	cmHg
horsepower (hp)	33000	ft•lb/min
	550	ft•lb/s
	76.04	m•kg/s
	1.014	metric hp
	7.457×10^{-1}	kW
	745.7	W
	7.068×10^{-1}	Btu•s
hp, metric	75.0	m•kg/s
	9.863×10^{-1}	hp
	7.355×10^{-1}	kW
	6.971×10^{-1}	Btu•s
hp•h	2.545×10^3	Btu
	1.98×10^6	ft•lb
	2.737×10^5	m•kg
inch (in)	2.54	cm
	83.33×10^{-3}	ft

Conversion Factors (*continued*)

Multiply	By	To Obtain
inHg at 0°C	40.66	$inAcBr_4$
	3.342×10^{-2}	atm
	13.60	inH_2O at 4°C
	1.133	ftH_2O
	4.912×10^{-1}	psi
	70.73	psf
	3.386	kPa
	3.453×10^2	kg/m^2
inH_2O at 4°C	2.99	$AcBr_4$
	7.355×10^{-2}	inHg at 0°C
	1.868×10^{-1}	cmHg at 0°C
	3.613×10^{-2}	psi
	2.49×10^{-1}	kPa
	25.40	kg/m^2
in^2	6.452	cm^2
	6.94×10^{-3}	ft^2
	6.452×10^{-4}	m^2
in^3	16.39	cm^3
	1.639×10^{-2}	L
	4.329×10^{-3}	U.S. gal
	1.639×10^{-5}	m^3
	1.732×10^{-2}	qt
joules (J)	9.480×10^{-4}	Btu
	7.375×10^{-1}	ft•lb
	2.389×10^{-4}	kg•cal
	1.020×10^{-1}	kg•m
	2.778×10^{-4}	Watt•h
	3.725×10^{-7}	hp•h
	10^7	ergs
kilograms (kg)	2.205	lb
	9.808	N
	35.27	oz
	10^3	g
kg•cal	3.9685	Btu
	3087	ft•lb
	4.269×10^2	kg•m
	4.1859×10^3	J
kg•m	7.233	ft•lb
	9.809	J
kg/m^2	62.43×10^{-3}	lb/ft^3
	10^{-3}	g/cm^3
kg/cm^2	14.22	psi
	2.048×10^{-3}	psf
	28.96	inHg at 0°C
	32.8	ftH_2O at 4°C
	98.077	kPa
kilometers (km)	3.281×10^3	ft
	6.214×10^{-1}	mi
	5.400×10^{-1}	nmi
	10^5	cm

Multiply	By	To Obtain
km/h	9.113×10^{-1}	ft/s
	5.396×10^{-1}	kn
	6.214×10^{-1}	mph
	2.778×10^{-1}	m/s
kPa	1000	N/m²
	14.503×10^{-2}	psi
kilowatts (kW)	9.480×10^{-1}	Btu/s
	7.376×10^{2}	ft•lb/s
	1.341	hp
	2.389×10^{-1}	kg•cal/s
kW•h heat generated	4.35×10^{-5}	g U²³⁵ fissioned
knots (kn)	1.0	nmi/h
	1.688	ft/s
	1.151	mph
	1.852	km/h
	5.148×10^{-1}	m/s
liters (L)	10^{3}	cm³
	61.03	in³
	3.532×10^{-2}	ft³
	2.642×10^{-1}	U.S. gal
	2.200×10^{-1}	imperial gal
	1.057	qt
megagrams (Mg)	10^{3}	kg
megapascals (MPa)	10^{3}	kPa
meters (m)	39.37	in
	3.281	ft
	1.094	yd
	6.214×10^{-4}	mi
m/s	3.281	ft/s
	2.237	mph
	3.600	km/h
m²	10.76	ft²
	1.196	yd²
m³	35.31	ft³
	264.17	gal (U.S.)
	61023	in³
	1.308	yd³
microamperes (μA)	6.24×10^{12}	unit charges/s
microns (μm)	3.937×10^{5}	in
	10^{-6}	m
miles (mi)	5280	ft
	1.609	km
	8.690×10^{-1}	nmi
mph	1.467	ft/s
	4.470×10^{-1}	m/s
	1.609	km/h
	8.690×10^{-1}	kn

Multiply	By	To Obtain
millibars (mbar)	2.953×10^{-2}	inHg at 0°C
	100.0×10^{-3}	kPa
nautical miles (nmi)	6076.1	ft
	1.852	km
	1.151	mi
	1852	m
newtons (N)	10^{5}	dyn
	2.248×10^{-1}	lb
ounces (oz) avdp	6.250×10^{-2}	lb avdp
	28.35	g
	4.375×10^{2}	grains
oz, fluid	29.57	cm³
	1.805	in³
pascals (Pa)	1.0	N/m²
	14.503×10^{-5}	psi
	29.5247×10^{-5}	in Hg at 0°C
	10^{-1}	bar
	100	mbar
pounds (lb)	453.6	g
	7000	grains
	16.0	oz avdp
	4.448	N
	3.108×10^{-2}	slugs
lb/ft³	16.02	kg/m³
lb/in³	1728	lb/ft³
	27.68	gm/cm³
psi	2.036	inHg at 0°C
	2.307	ftH₂O at 4°C
	6.805×10^{-2}	atm
	7.031×10^{2}	kg/ m²
	6.895	kPa
radians (rad)	57.30	°(arc)
rad/s	57.30	o/s
	15.92×10^{-2}	rev/s
	9.549	rpm
revolutions (rev)	6.283	rad
rpm	1.047×10^{-1}	rad/s
slugs	14.59	kg
	32.18	lb
stokes	10^{-4}	m²/s
stones	14	lb
	6.35	kg
tons	2×10^{3}	lb
	907.2	kg
tons, metric	10^{3}	kg
	2.205×10^{3}	lb
unit charges/s	1.6×10^{-13}	μA
watts (W)	9.481×10^{-4}	Btu/s
	1.340×10^{-3}	hp
yards (yd)	3.0	ft
	36.0	in
	9.144×10^{-1}	m
yd²	9.0	ft²
	1.296×10^{3}	in²
	8.361×10^{-1}	m²
yd³	27.0	ft³
	2.022×10^{2}	gal (U.S.)
	7.646×10^{-1}	m³

	Standard Atmosphere						
	Temperature t		Pressure P			Densiity	
Altitude, ft	°F	°C	inHg	cmHg	kPa	ρ	ρ/ρ_0
0	59.0	15.0	29.92	76.00	101.33	0.002378	1.0000
1000	55.4	13.0	28.86	73.30	97.74	0.002309	0.9710
2000	51.9	11.0	27.82	70.66	94.22	0.002242	0.9428
3000	48.3	9.1	26.81	68.10	90.80	0.002176	0.9151
4000	44.7	7.1	25.84	65.63	87.51	0.002112	0.8881
5000	41.2	5.1	24.89	63.22	84.29	0.002049	0.8616
6000	37.6	3.1	23.98	60.91	81.21	0.001988	0.8358
7000	34.0	1.1	23.09	58.65	78.20	0.001928	0.8106
8000	30.5	−0.8	22.22	56.44	75.25	0.001869	0.7859
9000	26.9	−2.8	21.38	54.31	72.41	0.001812	0.7619
10000	23.3	−4.8	20.58	52.27	69.70	0.001756	0.7384
11000	19.8	−6.8	19.79	50.27	67.02	0.001702	0.7154
12000	16.2	−8.8	19.03	48.34	64.45	0.001648	0.6931
13000	12.6	−10.8	18.29	46.46	61.94	0.001596	0.6712
14000	9.1	−12.7	17.57	44.63	59.50	0.001545	0.6499
15000	5.5	−14.7	16.88	42.88	57.17	0.001496	0.6291
16000	1.9	−16.7	16.21	41.47	54.90	0.001448	0.6088
17000	−1.6	−18.7	15.56	39.52	52.70	0.001401	0.5891
18000	−5.2	−20.7	14.94	37.95	50.60	0.001355	0.5698
19000	−8.8	−22.6	14.33	36.40	48.87	0.001311	0.5509
20000	−12.3	−24.6	13.75	34.93	46.57	0.001267	0.5327
21000	−15.9	−26.6	13.18	33.48	44.64	0.001225	0.5148
22000	−19.5	−28.6	12.63	32.08	42.77	0.001183	0.4974
23000	−23.0	−30.6	12.10	30.73	40.98	0.001143	0.4805
24000	−26.6	−32.5	11.59	29.44	39.25	0.001103	0.4640
25000	−30.2	−34.5	11.10	28.19	37.59	0.001065	0.4480
26000	−33.7	−36.5	10.62	26.97	35.97	0.001028	0.4323
27000	−37.3	−38.5	10.16	25.81	34.41	0.000992	0.4171
28000	−40.9	−40.5	9.720	24.69	32.92	0.000957	0.4023
29000	−44.4	−42.5	9.293	23.60	31.47	0.000922	0.3879
30000	−48.0	−44.4	8.880	22.56	30.07	0.000889	0.3740
31000	−51.6	−46.4	8.483	21.55	28.73	0.000857	0.3603
32000	−55.1	−48.4	8.101	20.58	27.44	0.000826	0.3472
33000	−58.7	−50.4	7.732	19.64	26.19	0.000795	0.3343
34000	−62.2	−52.4	7.377	18.74	24.98	0.000765	0.3218
35000	−65.8	−54.3	7.036	17.87	23.83	0.000736	0.3098
36000	−69.4	−56.3	6.708	17.04	22.72	0.000704	0.2962
37000	−69.7	−56.5	6.395	16.24	21.62	0.000671	0.2824
38000	−69.7	−56.5	6.096	15.48	20.65	0.000640	0.2692
39000	−69.7	−56.5	5.812	14.76	19.68	0.000610	0.2566
40000	−69.7	−56.5	5.541	14.07	18.77	0.000582	0.2447

	Temperature *t*		Pressure *P*			Density	
Altitude, ft	°F	°C	inHg	cmHg	kPa	ρ	ρ/ρ_0
41 000	−69.7	−56.5	5.283	13.42	17.89	0.000554	0.2332
42 000	−69.7	−56.5	5.036	12.79	17.06	0.000529	0.2224
43 000	−69.7	−56.5	4.802	12.20	16.26	0.000504	0.2120
44 000	−69.7	−56.5	4.578	11.63	15.50	0.000481	0.2021
45 000	−69.7	−56.5	4.364	11.08	14.78	0.000459	0.1926
46 000	−69.7	−56.5	4.160	10.57	14.09	0.000437	0.1837
47 000	−69.7	−56.5	3.966	10.07	13.43	0.000417	0.1751
48 000	−69.7	−56.5	3.781	9.60	12.81	0.000397	0.1669
49 000	−69.7	−56.5	3.604	9.15	12.21	0.000379	0.1591
50 000	−69.7	−56.5	3.436	8.73	11.64	0.000361	0.1517

Standard Atmosphere (*continued*)

Glossary

Absolute. The magnitude of a pressure or temperature above a perfect vacuum or absolute zero, respectively. For temperature, absolute zero is theoretically equal to −273.18°C or −459.72°F.

Accelerating pump. A small pump in a carburetor that is used to supply momentarily rich fuel-air mixture to the engine when the throttle is suddenly opened.

Acceleration. A change in velocity per unit of time.

Accessory section. The portion of an aircraft engine crankcase on which such accessories as magnetos, carburetors, generators, fuel pumps, and hydraulic pumps are mounted.

Accumulator. As it pertains to propellers, a device used to aid in unfeathering a propeller by storing hydraulic pressure.

Additive. A material added to an oil or a fuel to change its characteristics or quality.

Aerodynamic twisting moment. An operational force on a propeller which tends to increase the propeller blade angle (center of lift).

Air (standard). Sea-level atmospheric conditions of temperature at 15°C [59.0°F] and air pressure at 14.7 psi [29.92 inHg].

Air-cooled engine. A reciprocating engine from which excess heat is removed by transferring it directly into the air that flows over the engine.

Air start. The process of starting an aircraft engine in flight.

Airworthiness. The state or quality of an aircraft or an aircraft component which will enable safe performance according to specifications.

Airworthiness Directive (AD). A directive issued by the FAA requiring that certain inspections and/or repairs be performed on specific models of aircraft, engines, propellers, rotors, or appliances, and setting forth time limits for such operations.

Alloy. A solid solution of two or more metallic constituents. One metal is usually predominant, and to it are added smaller amounts of other metals to improve strength and heat resistance.

Alpha mode of operation. The mode of operation of a propeller on a turboprop engine that includes all of the flight operations, from takeoff to landing.

Alternator. A mechanical device which produces ac current through induction.

Angle of attack (AOA). The angle between the chord line of a propeller blade section and the relative wind.

Annular combustor. A ring-shaped combustor.

Atmosphere (standard). *See* **air (standard)**.

Atomize. To break a liquid up into minute particles.

Atomizer. A device through which fuel is forced so that it enters the combustor as a fine spray.

Augmentor tube. A long, specially shaped, stainless-steel tube that is mounted around the exhaust tailpipe of an aircraft reciprocating engine.

Automatic propeller. A propeller which changes blade angles in response to operational forces.

Autorotation. A rotorcraft flight condition in which the lifting rotor is driven entirely by action of the air as the rotorcraft is in motion.

Axial-flow compressor. A type of compressor that is used in a gas-turbine engine in which the air passes through the compressor in an essentially straight flow.

Axial-flow turbine. A turbine in which the energy of flowing air is converted to shaft power while the air follows a path parallel to the turbine's axis of rotation.

Backfire. Ignition of the fuel mixture in the intake manifold caused by the flame from the cylinder combustion chamber. It can be caused by incorrect timing or too lean a mixture.

Bearing (mechanical). Part of a machine that supports a journal, pivot, or pin that rotates, oscillates, or slides.

Beta. The engine operational mode in which propeller blade pitch is hydromechanically controlled by a cockpit power lever. Used for ground operations.

Blade angle. The angle between the blade section chord line and the plane of rotation of the propeller or crankshaft.

Blade chord. A straight line through the center of a propeller blade, perpendicular to its span, between its leading edge and its trailing edge.

Blade face. The flat portion of a propeller blade.

Blade paddle. A tool used to move the blades in the propeller hub against spring force.

Blade shank. The rounded portion of the propeller blade near the butt of the blade.

Blade station. A distance, measured in inches, from the center of the propeller hub.

Blade twist. The decrease in the pitch angle of a propeller blade as the distance from the hub increases.

Blowby. Leakage or loss of pressure or compression past the piston rings.

Bore. The diameter of the inside of the cylinder of a reciprocating engine.

Borescope. An instrument that is used to examine the inside of a structure through a very small hole.

Brake horsepower (bhp). The power produced by an engine and available for work through the propeller shaft. It is usually measured as a force on a brake drum or equivalent device.

Brake mean effective pressure (bmep). The average pressure inside the cylinder of a reciprocating engine during the power stroke.

Brake specific fuel consumption (bsfc). A measure of the amount of fuel that is used in a heat engine to develop a given amount of power.

Breaker points. Two contact surfaces that are mechanically opened and closed to control current flow.

Breather. A vent line in an aircraft engine that allows the air pressure inside the engine crankcase to be the same as the pressure of the surrounding air.

British thermal unit (Btu). The quantity of heat required to raise the temperature of 1 lb of water from 62 to 63°F.

Burner, can, combustor, flame tube, liner. The sheet-metal assembly which contains the flame of a turbine engine.

Camshaft. The shaft on which there are cams, or lobes, that operate the engine valves.

Can-annular combustor. The burner section of a gas-turbine engine.

Capacitor. The component in a magneto ignition system that prevents the breaker points from burning and also aids in the rapid collapse of the current flow in the primary circuit. Also called *condenser.*

Carbon. Residues formed in the combustion chamber of an engine during the burning of fuels, which are largely composed of hydrocarbons.

Carburetor. A device for automatically metering fuel in the proper proportion with air to produce a combustible mixture.

Carburetor air temperature (CAT). The temperature of the air as it enters the carburetor.

Carburetor ice. Ice that forms inside the throat of a carburetor on a reciprocating engine.

Centering cones. Devices in a splined-shaft installation that center the propeller on the crankshaft.

Centrifugal-flow compressor. A compressor in a gas-turbine engine that is made in the form of a disk-type impeller.

Centrifugal force. The outward force an object exerts on a restraining agent when the motion of the object is circular.

Centrifugal twisting moment (CTM). The force on a propeller which tends to decrease the propeller blade angle.

Certified propeller repair station. A facility that can perform major repairs of and alterations to propellers.

Choked nozzle. A nozzle whose flow rate has reached the speed of sound.

Chord line. The imaginary line which extends from the leading edge to the trailing edge of an airfoil section.

Circuit. The completed path of electric current.

Coil. A device that is made of several turns of wire wound around some type of core.

Combustion. The process of burning the air-fuel mixture in the combustion chamber.

Combustion chamber. The section of the engine into which fuel is injected and burned and which contains the flame tube or combustor.

Compression ratio. The ratio of the pressure (or volume) of air discharged from a compressor to the pressure (or volume) of air entering it. *See also* **pressure ratio**.

Compression rings. Piston rings that normally confine the combustion gases to the combustion chamber.

Compressor. The section of the engine which acts like an air pump to increase the energy of the air received from the entrance duct and discharged into the turbine section.

Compressor stage. A compressor stage in a gas-turbine engine is made up of one set of rotor blades and the following set of stator blades.

Compressor surge. An operating region of violent pulsating airflow usually outside the operating limits of the engine.

Condenser. *See* **capacitor**.

Condition lever. An engine control used with a turboprop engine that controls the pitch of the propeller for the flight operating range (the alpha range of operation).

Connecting rod. The engine part that connects the piston to the crankshaft.

Constant-speed propeller system. A propeller system in which a governor controls the propeller blade angle to maintain the selected rpm.

Controllable-pitch propeller. A propeller whose pitch can be changed in flight.

Convergent duct. An air passage or channel of decreasing cross-sectional area. In flowing through such a duct, a gas increases in velocity and decreases in pressure.

Cooling fins. Thin ribs extending out from a surface that carry heat from the surface to the air that flows through the fins.

Crankcase. The housing to which the cylinder block is connected and within which the crankshaft and many other parts of the engine operate.

Crankshaft. The main shaft of the engine, which, in conjunction with the connecting rods, changes the linear reciprocating motion of the piston into rotary motion.

Critical rpm range. The rpm range at which destructive harmonic vibrations exist in a propeller and engine combination.

Cycle. A series of occurrences for which conditions at the end are the same as they were at the beginning.

Cylinder barrel. A steel cylinder within which a piston moves up and down. It has machined cooling fins, and its inside surface provides the seal between the piston rings and the cylinder.

Cylinder head temperature (CHT). The temperature of the cylinder head of an air-cooled reciprocating engine.

Cylinder skirt. The portion of the cylinder of an air-cooled aircraft engine that sticks into the crankcase of the engine.

Deicing system. A system on the leading edge of a wing on which ice forms and is then broken loose in cycles. This also can refer to a propeller leading edge which is deiced or anti-iced using electric boots or liquid alcohol.

Delta *P* (Δ*P*). The difference in pressure between two points.

Density altitude. Pressure altitude corrected for nonstandard temperature.

Detonation. After normal ignition, the explosion of the remaining air-fuel mixture due to above-normal combustion chamber pressure or temperature.

Diffuser. A duct of increasing cross-sectional area which is designed to convert high-speed gas flow into low-speed flow at an increased pressure.

Dispersant. An additive to oils which keeps particles of dirt and other foreign materials in suspension instead of settling out as sludge.

Divergent duct. An air passage or channel of increasing cross-sectional area. In flowing through such a duct, a gas decreases in velocity and increases in pressure.

Droop. A decrease in speed, voltage, air pressure, or other parameter which results when load is applied.

Duct. A passage or tube used for directing gases.

Dwell. The length of time that breaker points are closed.

Dynamic balance. The condition in which a mass remains free of vibration while in motion: all forces exerted on various parts of the mass are balanced by equal and opposite forces.

Dynamic damper. A heavy weight in the shape of a spool that is mounted on the crankshaft of a reciprocating engine to absorb torsional vibrations.

Effective pitch. The distance forward that a propeller actually moves in one revolution.

Efficiency. The ratio of power output to power input. (Power input equals power output plus power wasted.)

E-gap angle. The number of degrees that the magnet in a magneto rotates beyond its neutral position before the breaker points open.

Electronic fuel control unit (EFCU). A fuel control that uses electronic inputs from engine parameters feed into a computer that determines the amount of fuel to flow to the engine.

Emulsifier. A material which, when applied over a film of penetrant on the surface of a part, mixes with the penetrant and thereby enables the penetrant to be washed off the surface with water.

End play. The movement along the axis of a mounted shaft.

Energy. The capacity to do work or overcome resistance. Potential, or latent, energy is "stored work" waiting to be used; kinetic energy is associated with motion. The units of energy are the foot-pound and the joule.

Engine mount. The part of an aircraft structure that is designed for attachment of the engine.

Engine pressure ratio (EPR). The pressure measurement that is used as an indication of the amount of thrust that is being developed by an axial-flow gas-turbine engine.

Exhaust-gas temperature (EGT). The temperature of the exhaust gas at the discharge side of the turbines.

Exhaust valve. A valve which permits exhaust gas to exit the combustion chamber.

False start. An unsuccessful or aborted engine start.

Turbofan engine. A gas-turbine engine that employs a fan to accelerate a large volume of air through a bypass duct to increase thrust and engine efficiency.

Feathering propeller. A propeller that can have its blades turned so that they are facing directly into the airstream as the aircraft is moving through the air.

Fixed-pitch propeller. A propeller whose blade angles cannot be changed except by bending the blades to a new pitch.

Flameout. An unintended extinction of flame.

Flanged crankshaft. A crankshaft whose propeller mounting surface is perpendicular to the shaft centerline and to which the propeller bolts directly.

Float. A device which is ordinarily used to automatically operate a needle valve and seat for controlling the entrance of fuel into a float carburetor.

Float level. The predetermined height of the fuel in the carburetor bowl, usually regulated by means of a float.

Fluid. Any substance having elementary particles that move easily with respect to each other—i.e., liquids (incompressible fluids) and gases (compressible fluids).

Fluorescent penetrant. A penetrant incorporating a fluorescent dye to improve the visibility of indications at the flaw.

Flutter. A vibration or oscillation of definite period set up in an aileron, wing, or other surface by aerodynamic forces and maintained by a combination of those forces and the elastic inertial forces of the object itself.

Flyweights. Objects which are activated by centrifugal force.

Force. Any action which tends to produce, retard, or modify motion.

Fouling. The addition or formation of foreign material, such as carbon, to or on the electrodes of a spark plug which prevents the spark from jumping the gap.

Four-stroke-cycle engine. An engine that has one power stroke for two revolutions of the crankshaft.

Free turbine. A turbine which operates independent shafts for high- and low-pressure rotors.

Fuel-air ratio. The ratio by weight of fuel to air.

Fuel control unit (FCU). A device used to regulate the flow of fuel to the combustion chambers. It may respond to one or more of the following factors: power control lever setting, inlet air temperature and pressure, compressor rpm, and combustion chamber pressure.

Fuel nozzle. The nozzle in a gas-turbine engine combustor through which the fuel is discharged. The spray pattern from the fuel nozzle is such that the flame is always centered in the burner so that it will not overheat it.

Gas turbine. An engine, consisting of a compressor, burner, or heat exchanger, and a turbine, that uses a gaseous fluid as the working medium and produces shaft horsepower, jet thrust, or both.

Gasket. A substance placed between two metal surfaces to act as a seal.

Gear ratio. A gear relationship, usually expressed numerically, used to compare input to output speed.

Geometric pitch. The theoretical distance that a propeller will move forward in one revolution.

Governor. The speed-sensing propeller control device that adjusts and maintains system rpm by adjusting oil flow to and from certain types of constant-speed propellers.

Ground. The contact point for the completion of an electrical circuit.

Ground-adjustable propeller. A propeller which is adjusted only on the ground to change the blade angles.

Ground idle. The engine speed that is normally used for operating a gas-turbine engine on the ground so that it will produce the minimum amount of thrust.

Guide vanes. Stationary airfoil sections which direct the flow of air or gases from one major part of the engine to another.

Helicoil. The registered trade name of a special helical insert that is used to restore threads that have been stripped from a hole.

High tension. High voltage produced in the magneto secondary and then transmitted via high-tension leads to the spark plug.

Horsepower. The amount of force needed to move 33 000 lb through a distance of 1 ft in 1 min (or 550 ft•lb/s). hp = $K \times$ torque (ft•lb) $\times$ rpm; hp = $0.000\,190\,4 \times$ ft•lb $\times$ rpm.

Hot start. An engine start, or attempted start, which results in the turbine temperature exceeding the specified limits. It is caused by an excessive fuel-to air ratio.

Hub. The central portion of a propeller which is attached to the engine crankshaft.

Idle. The lowest recommended operating speed of an engine.

Idle cutoff. The position of the mixture control in a fuel system of a reciprocating engine that shuts off all of the fuel to the cylinders.

Igniter. A device, such as a spark plug, used to start the burning of the air-fuel mixture in a combustion chamber.

Ignition event. The act of igniting a combustible mixture, by means of an electric spark, on the compression stroke.

Ignition system. The means of igniting the air-fuel mixture in the cylinders; it includes spark plugs, high-tension leads, ignition switches, and magnetos.

Impeller. The main rotor of a radial compressor which increases the velocity of the air being pumped.

Impulse coupling. A spring-loaded coupling between the magneto and an aircraft reciprocating engine that causes the magneto to produce a hot and late spark for starting the engine.

Indicated horsepower. The theoretical horsepower that a reciprocating engine is developing.

Indicated mean effective pressure (IMEP). The average pressure that exists inside the cylinder of a reciprocating engine during its power stroke.

Induction system. The system, consisting of inlet ducts and an inlet plenum, that admits air to the engine.

Inertia. The opposition of a body to a change in its state of rest or motion.

Intake manifold. The tubes or housings used to conduct the air-fuel mixture to the cylinders.

Intake valve. A valve which permits the air-fuel mixture to enter the cylinder combustion chamber.

Interstage turbine temperature (ITT). Gas temperature measured at the inlet of the second-stage turbine stator assembly.

Jet. A small, tubelike device through which a fluid or gas flows.

Jet engine. Any engine that ejects a jet or stream of gas or fluid and obtains all or most of its thrust by reaction to the ejection.

Joule. A unit of work or energy equal to approximately 0.7375 ft•lb.

Journal. A shaft machined to fit a bearing.

Kinetic energy. Energy associated with motion.

Knot. A rate of speed equivalent to 1 nmi/h (6076 ft/h) [1852 m/h].

Labyrinth seal. A high-speed seal which provides interlocking passages to discourage the flow of air, oil, etc., from one area to another.

Lap. An abrasive operation undertaken to match the contours of surfaces for fit.

Lightoff. Ignition or the air-fuel mixture in the combustion chamber.

Low tension (electrical). Low voltage.

Lubrication. The process of reducing friction.

Mach number. The ratio of the velocity of a mass to the speed of sound under the same atmospheric conditions. A speed of Mach 1.0 means the speed of sound, regardless of temperature; a speed of Mach 0.7 means that the speed is seven-tenths the speed of sound.

Magneto. The source of high-voltage electrical energy that is used to furnish the spark for igniting the fuel-air mixture inside the cylinders of a reciprocating engine.

Magneto coil. Essentially a transformer which, through the action of induction, converts low voltage to high voltage.

Mass flow. Airflow measured in slugs per second.

Metering jet. A precision orifice in a fluid system that is used to restrict the flow of fluid.

Micron. A unit of length equal to 1/1000 (0.001) mm or 1/25 000 (0.000 039 37) in.

MIL spec. A specification for a material or part that ensures compliance with quality and performance standards. (Such specifications were originally for military purposes; therefore, the abbreviation MIL.)

Mixture control. A control that is installed in a reciprocating-engine-powered aircraft that allows the pilot to vary the fuel-air mixture while the engine is running.

Momentum. The tendency of a body to continue in motion after being placed in motion.

Multiviscosity index. Two SAE viscosities (e.g., 10-30) the lower of which an oil will exhibit at low temperatures and the higher of which it will exhibit at high temperatures.

Negative-torque-sensing system. A system wherein propeller torque drives the engine, which the NTS detects and automatically drives the propeller to high pitch to reduce drag on the aircraft.

Nozzle, fuel. A spray device which directs atomized fuel into a combustion chamber.

Nozzle, turbine. A convergent duct through which hot gases are directed to the turbine blades.

Octane rating. A standardized comparison of the antiknock qualities of a given fuel with those of a test fuel. The higher the octane rating, the higher the fuel's resistance to combustion due to heat and pressure.

Oil control rings. Piston rings that control the amount of oil on the cylinder walls.

Oil cooler. An air-oil heat exchanger that is used to remove excess heat from the lubricating oil that is used in a reciprocating engine.

Oil dilution. A method of preparing the oil in a reciprocating engine in order to make it possible for the engine to start when the temperature is very low.

Oil separator. A device that is used to remove oil from the air that is pumped by an oil-lubricated air pump.

Orifice. A small opening in a passage, jet, or nozzle.

Overspeed. Engine speed which exceeds the selected rpm by a set percentage.

Overtemperature. Any exhaust temperature that exceeds the minimum allowable temperature for a given operating condition.

Penetrability. The property that causes a penetrant to find its way into very fine openings, such as cracks.

Penetrant. A fluid—usually a liquid but possibly a gas—that is used to enter a discontinuity and thereby indicate the flaw.

Piston. A cylindrical part closed at one end which is connected to the crankshaft by the connecting rod and which works up and down in the cylinder barrel.

Piston displacement. The volume, in cubic inches or liters, displaced by a piston in moving from one end of its stroke to the other.

Piston head. The part of the piston above the rings that receives the thrust of combustion; the top of the piston.

Piston pin. The journal and bearings that connect an engine's connecting rod and piston together. Also known as *wrist pin*.

Piston ring. A springy split ring, usually made of gray iron, that is placed in a piston-ring groove to provide a seal between the cylinder barrel and the piston.

Piston-ring gap (end gap). The clearance between the ends of a piston ring when the piston is placed squarely in the cylinder.

Piston-ring groove. One of the grooves in a piston near its head in which a piston ring is placed.

Piston-ring lands. The ridges of metal between the piston ring grooves.

Pitch. The distance, in inches, over which a propeller theoretically moves forward in one revolution.

Pitch distribution. The twist in a propeller blade along its length.

Plane of rotation. The plane in which a propeller blade rotates.

Plenum. A duct, housing, or enclosure used to contain air under pressure.

Port. A hole through which gases may enter or exit.

Power. A measure of the rate at which work is performed—i.e., the amount of work accomplished per unit of time.

Power lever. The cockpit lever used to change propeller pitch during beta operation and also to select engine fuel flow during prop governing.

Preignition. Ignition of the air-fuel mixture before the normal firing time.

Pressure ratio. In a gas-turbine engine, the ratio of compressor discharge pressure to compressor inlet pressure.

Primary air. The portion of the compressor output air that is used for the actual combustion of fuel, usually 20 to 25 percent.

Primer. A small hand-operated pump that is used to spray gasoline into the induction system at the cylinder head.

Probe. A sensing element that is extended into the airstream or gas stream so as to measure pressure, velocity, or temperature.

Propeller. A device used for converting brake horsepower, or torque, into thrust to propel an aircraft forward.

Propeller feathering. Rotation of the propeller blades to eliminate the drag of a windmilling propeller on a multiengine aircraft in the event of engine failure.

Propeller governing. A mode of engine operation wherein the propeller governor selects the blade pitch to control engine rpm and the fuel flow is established manually.

Propeller reversing. Causing the blades of a propeller used on a turboprop engine to rotate at a negative angle to produce a reversing thrust that brakes the aircraft.

Propeller track. The plane of rotation of a propeller blade.

Psia. Pounds per square inch absolute, a measure of pressure. "Absolute" is the zero pressure in a perfect vacuum.

Psig. Pounds per square inch gauge, a measure of the pressure inside a tube, plenum, or duct compared with the pressure outside it.

Pushrod. A stiff rod or a hollow tube that is used to open the intake and exhaust valves of a reciprocating engine.

Radial engine. A form of reciprocating engine that was at one time very popular for use on aircraft. The cylinders are arranged radially around a central crankcase.

Rated horsepower. The horsepower that the manufacturer of an aircraft engine guarantees that the engine will produce under a given set of conditions.

Retard breaker points. An extra set of breaker points installed in an aircraft magneto that is equipped with the Bendix Shower of Sparks starting system.

SAE number. Any of a series of numbers established as standards by the Society of Automotive Engineers for grading materials, components, and other products.

Scavenging. Removal of hot combustion products from the cylinder through the exhaust port or valve.

Secondary air. The portion of compressor output air that is used for cooling combustion gases and engine parts.

Shaft horsepower (shp). The horsepower that is actually available at a rotating shaft.

Skirt. The lower, hollow part of a piston or cylinder.

Slip. The difference between geometric pitch and effective pitch.

Slug. The mass to which a 1-lb force can impart an acceleration of 1 ft/s/s. It is frequently used in aeronautical computations.

Spark. Voltage sufficiently high to jump through the air from one electrode to another.

Spark advance. The interval between the spark-plug-firing and top-dead-center positions of a piston.

Spark gap. The space between the electrodes of a spark plug through which the spark jumps.

Spark plug. A device inserted into the combustion chamber of an engine to deliver the spark needed for combustion.

Specific gravity. The ratio of the density of a substance to the density of water at a standard temperature. (Density is mass per unit volume.)

Specific heat. The ratio of the thermal capacity of a substance to the thermal capacity of water.

Specific weight. The ratio of the weight of a homogeneous fluid to its volume at a given temperature and pressure. Also called *weight density*.

Spectrometric oil analysis. An analysis of the contaminants in a sample of engine oil.

Splined shaft. A cylindrical crankshaft which has splines on its outer surface to prevent propeller rotation on the shaft.

Stage (turbine). A row of turbine nozzle guide vanes the following row of turbine blades, which together extract power from hot gases to drive the compressors and accessories.

Static rpm. The maximum engine rpm that can be obtained under conditions of propeller load at full throttle on the ground in a no-wind condition.

Stator. A row of stationary guide vanes.

Stroke. The distance traveled by a piston from the top to the bottom of its stroke.

Subsonic speed. Speed lower than the speed of sound.

Supersonic speed. Speed higher than the speed of sound.

Synchronization system. A system which is designed to keep all engines at the same rpm.

Synchrophasing system. A synchronization system which allows the pilot to adjust the blade phase angle (relative position) as the propellers rotate to eliminate noise or vibration.

Tapered shaft. An older crankshaft in which the propeller-mounting surface tapers to a smaller diameter for acceptance of the propeller mounting taper.

Thermocouple. A pair of jointed wires of two dissimilar metals. A dc voltage is produced at one joint when the other joint is at a higher temperature.

Throttle plate. The movable plate in the carburetor that regulates the air-fuel mixture entering the intake manifold and combustion chamber.

Throw. The distance from the center of the crankshaft main bearing to the center of the connecting-rod journal.

Thrust. A pushing force exerted by one mass against another, which tends to produce motion. In jet propulsion, the force in the direction of motion caused by the pressure of reactive forces on the inner surfaces of the engine.

Thrust bearing. A form of bearing used in a mechanism to support rotating loads that are parallel to the axis of the shaft on which the bearing is mounted.

Thrust bending force. An operational force which tends to bend the propeller blades forward.

Thrust reverser. A device used to partially reverse the flow of an engine's nozzle discharge gases and thus create a thrust force in the rearward direction.

Top dead center (TDC). The uppermost of the two positions of the piston when the crank and rod are in the same straight line.

Torque. The force which is produced by a turning effort; it is measured in foot-pounds.

Torque bending force. An operational force which tends to bend the propeller blades in the direction opposite to that of rotation.

Torquemeter. A meter for measuring torque, such as the torque of an aircraft engine shaft.

Torque nose. A mechanism or apparatus at the nose section of an engine that senses the engine torque and activates a torquemeter.

Turbine. A rotating device turned by either direct or reactive forces (or a combination of the two) and used to transform some of the kinetic energy of the exhaust gases into shaft horsepower to drive the compressor (s) and accessories.

Turbine blade. A fin mounted on the turbine disk and so shaped and positioned as to extract energy from the exhaust gases to rotate the disk.

Turbine exhaust cone.   A fixed or adjustable bullet-shaped structure over which the exhaust gases pass before converging in the exhaust section.

Turbine inlet temperature (TIT).   Temperature of hot gases as they enter the engine turbine.

Turbojet.   A gas-turbine engine whose entire propulsive output is delivered by the jet of gases through the turbine nozzle.

Turboprop.   A type of gas-turbine engine that converts heat energy into propeller shaft work and some jet thrust.

Turbulence.   An agitation of, or disturbance in, the normal flow pattern.

Two-position propeller.   A propeller which can be changed from low- to high-pitch blade angles (and vice versa) in flight.

Valve.   A device for opening and closing a passage that either admits the air-fuel mixture to, or exhausts gases from, the cylinder.

Valve clearance.   The gap allowed between the end of the valve stem and the rocker arm to compensate for expansion due to heat.

Valve overlap.   The time in the operating cycle during which both valves are off their seats or open. It increases volumetric efficiency.

Valve stem.   The part of the valve that rides in the valve guide.

Valve stem guide.   A guide that is shrunk into the cylinder head to keep the valve square with the valve seat.

Vaporize.   To change a liquid to a gaseous form.

Vector.   A line which, by scaled length, indicates the magnitude of a force and whose arrowhead shows the direction of action of the force.

Velocity.   The rate of change of distance with respect to time. The average velocity is equal to total distance divided by total time.

Viscosity.   A fluid's resistance to flow under an applied force.

Work.   A force acting through a distance.

Index

Note: Page numbers followed by *f* denote figures; page numbers followed by *t* denote tables.

Absolute pressure, 73, 137
Absolute-pressure controller, 118
AC. *See* Alternating current
ACC. *See* Active clearance control
Accelerating system, 148, 148*f*, 149*f*
Accelerating well, 148
Acceleration, 314, 543*f*
Accelerometer, 661*f*, 698
Accessory case, 55–56, 55*f*
Accessory drives and gears, 50–51, 342,
    343*f*, 414, 429, 501, 509–510, 509*f*,
    512*f*, 529
Accessory input-gear shaft, 504
Accessory section, 31, 31*f*, 32, 55–56,
    56*f*, 429
Active clearance control (ACC), 430, 434
AD. *See* Ashless dispersant
Additives, to aviation oil, 86
ADI. *See* Antidetonant injection
Adiabatic, 315
Adjustable propeller, 581
Adjusting worm, 598
Aeolipile, 307
Aerobatic (A) engine classification, 7
Aerodynamic twisting force, 587
Aeromax Aviation, 21, 23*f*, 24*f*
Aeroshell oil, 87
Aerospatiale Super Caravelle, 413
Afterburners, 339
Afterburning, 321, 341*f*
Aftercoolers, 109
Afterfiring, 142, 265
Air blast fuel nozzle, 499*f*
Air bleed, 143, 144*f*, 171, 173, 185,
    331–333, 360, 510–511, 511*f*
Air casings, 482
Air cooling, 8, 122–123, 122*f*, 123*f*,
    335*f*, 337*f*
Air density, 106, 106*f*, 142
Air-diffuser housing, 528
Air filters, 101, 254
Air impingement starting, 407, 407*f*
Air inlet icing, 326
Air inlet section, 414
Air metering force, 163
Air-port mixture control system, 150,
    151, 151*f*

Air pressure-oil propellers, 603
Air scoop, 101, 113*f*
Air seals, 431
Air spray nozzle, 359, 360*f*
Air throttle assembly, 170
Air-turbine starters, 402–406, 404*f*, 405*f*
Aircraft vibrations, 656–663
AiResearch, 405*f*
    ATS 100-350 air-turbine, 409
    Garrett-AiResearch TFE731 turbofan
        engine, 463–466
    Garrett-AiResearch TPE331 turboshaft
        engine, 616
Airflow, 315–317, 315*f*, 331, 420*f*, 472*f*
    through Allison 250-C20B engine,
        535*f*
    in combustion liner, 333*f*
    in Garrett ATF3-6 engine, 469*f*
    for Lycoming T53 engine, 526–527
Airworthiness Directives, 270
Airworthy damage, 672*f*, 673–674, 675*f*
Alcohol aviation fuels, 131–132
Allied Signal, GTCP85 Series APU,
    523, 525*f*
Allison engines, 518*f*, 643–647, 644*f*, 645*f*
    250-B17 reversing turbopropeller
        system, 635–639
    250-C20 engine, 534*f*, 536*f*, 537*f*
    250-C20B engine, 535*f*
    250 engine, 370, 515–518, 517*f*, 519*f*,
        520*f*, 532–536, 533*f*, 638, 638*f*
    501-D13, 644*f*
    A6441FN-606 turbopropeller, 644*f*, 645*f*
    GMA 2100, 499, 501*f*
    GMA 3007 engine, 464*f*
    T56/501D series III, 497, 498*f*
    T56/501D series IV, 498–499, 498*f*
    T406-AD-400, 499
Alnico, 193
Alternate air inlet heating system, 158
Alternating current (AC), 14, 20
Alternating-current electric tachometer
    system, 692*f*
Alternative air valve, 101–102, 102*f*
Altitude, 77, 77*f*, 106, 106*f*
    effects, 318, 319*f*
    turbochargers and, 113–114

Altitude compensating valve, 173
Altitude-control valve, 150, 151*f*
Aluminum-alloy propellers, 667
Aluminum blade fatigue failure, 667–668
AMC. *See* Automatic mixture control
    systems
American Petroleum Institute (API), 80,
    80*f*, 521
Angle of attack, 584–585, 585*f*
Animal lubricants, 79
Annual inspection checklist, 249*f*
Annular combustor, 334*f*, 440, 440*f*
Annulus gear, 481
Anti-corrosion agents, 83
Anti-icing, 136, 247, 425*f*, 528
    deicing circuits and, 647*f*
    of propellers, 607–609, 608*f*
Antidetonant injection (ADI), 164–167, 320
Antifriction, aviation oils and, 85
Antiknock value, 130
Antioxidants, 83
Antiwear, aviation oils and, 85
API. *See* American Petroleum Institute
APUs. *See* Auxiliary power units
Aromatic aviation fuels, 131–132
Articulated rods, 40
Ash content, 83
Ashless dispersant (AD), 86
Asymmetrical loading (P factor),
    589, 589*f*
Atmospheric pressure, 137
Atomic absorption, 392
Atomization, 358
Autofeathering systems, 612, 612*f*
Autoignition system, 397
Automatic fuel dump valve, 513
Automatic mixture control systems (AMC),
    152, 152*f*, 164, 179–183, 182*f*
Automatic pitch propeller, 591
Autorotation, 536, 539
Auxiliary ignition units, 205–207
Auxiliary power units (APUs), 26, 378–379,
    379*f*, 407, 408*f*, 521–523, 523*f*, 545*f*
Aviation gas (AVGAS), 15, 25, 131,
    132–133
Axial-flow compressor, 327–329, 327*f*,
    328*f*, 329*f*

Back-suction mixture control system, 150, 151f
Backfiring, 63, 141, 265
Baffles, 122, 122f, 123f
Balance checking, 656–657, 658f
Balance pipe, 104, 104f
Balancing, 570, 663f, 664f
  accelerometer installation for, 661f
  photo-tach installation for, 662f
  of propellers, 659, 659f
Ball-bearing rod end, 710f
Ball bearings, 33, 33f
Ball gauges, 282
Barber, John, 307
Barnett, William, 1
Base-mounted magnetos, 192, 193f
Battery ignition system, 191, 191f
Baumé scale, 80
BDC. See Bottom dead center
Bearing carrier, 405
Bearing compartment seals, 510–511, 511f
Bearing sump, 390f
Bearings, 32–33, 390, 455, 460f, 489f, 507. See also specific types
Beech Bonanza, 8
Beechcraft King Air, 616f
Bell:
  204B helicopter, 536–537, 538f
  XP-59-A airplane by, 26
Bellcranks, 710, 710f, 717
Bellmouth air inlet, 570, 571f
Bellows capsule, 687, 687f
Bels (B), 344
Bendix, 164, 165f, 235–236, 240, 370, 371f, 372f, 536f
  RSA fuel injection systems, 175–189, 176f, 177f, 179f, 182f, 183f, 184f, 185f, 186f, 187f, 188–199
  S-600 low-tension ignition system, 219f
Bendix-Stromberg, 153–154, 154f, 155f
Bentley rotary engine, 2
Benz, Karl, 1
Benzol, 132
Bernoulli's principle, 138, 336
Best economy mixture, 78, 139, 140, 140f
Best power mixture, 78, 139
Beta control mode, 497, 624, 641
Beta range, 644
Beta sleeve, 620
Beta tube, 620, 634f
Beta valve, 617, 620, 636–637, 637f
Bevel-planetary-gear, 57, 57f
bhp. See Brake horsepower
Blade angle, 583–585, 584f, 585f, 652f, 677–678, 677f, 678f
Blade butt, 582
Blade counterweights, 617, 621
Blade cuffs, 582, 582f, 673
Blade element, 582, 582f
Blade performance, 620f
Blade repair limits, 671f
Blade section, angle of attack of, 584

Blades, 560f, 568, 581f, 582, 582f, 583, 627f, 646, 667f, 669, 669f, 673–674, 673f. See also Turbine blades; specific types
  deicing of, 609f, 641
  of Dowty R321 propeller, 620
  life of, 672
  for low-pressure turbine, 445f
  of metal propellers, 593f
  relative wind and, 585f
  retainers, 437, 443
  retention windings, 674
  rod, 38
  shortening, 671
Blast tubes, 203
Bleed-off, 351
Bleeder resistor, 399
Bloom, 82
Blowby, 43
bmep. See Brake mean effective pressure
Boeing:
  727 aircraft, 387, 413
  737 aircraft, 387, 413
  747 aircraft, 313
  757 aircraft, 522f, 524f
  767 aircraft, 373
  787 aircraft, 410–411, 460–463, 461f
Boost pressure, 19, 19f
Boost pump, 133, 184
Booster coil, 206, 206f
Borescope, 548–549, 549f
Bottom dead center (BDC), 60–62, 61f
Bottoming governor system, 496
Bottoming tap, 294
Bourbon tube, 683, 683f
Boyle's law, 59–60, 105, 316
Brake horsepower (bhp), 69, 589, 615
Brake mean effective pressure (bmep), 70
Brake specific fuel consumption (bsfc), 76, 140
Brake thermal efficiency (bte), 72
Branca, Giovanni, 307
Brayton, George B., 1, 60
Brayton cycle, 60, 316, 317f
Breaker assembly, 197
Breaker points, 197
Breather system, 389
Bresson, 307
Brinelling, 271
British thermal unit (Btu), 311
Broadband noise, 345
bsfc. See Brake specific fuel consumption
bte. See Brake thermal efficiency
Bubble protractor, 675
Bulkhead connector, 703
Bullet-type AMC, 181–183
Bumbell, John, 307
Burners, 430
Burning, 271, 272f
Butterfly valves, 144, 144f
Bypass fan assembly module, 465
Bypass filter system, 91, 91f
Bypass ratio, 313, 330

Cable-and-pulley control systems, 709–711
Cables, 253, 717
  for balancing, 663f
  high-tension, 203, 213f
  for ignition, 253f, 399
Calories, 156
Cam, 49
Cam follower, 51, 197
Cam-gear backlash, 299
Cam lobes, 52
Cam plate, 50, 52, 52f, 53f
Cam rollers, 52
Cam surfaces, 273
Cam track, 52
Camber sheet, 646
Camshaft, 50–51, 50f
Can-annular combustion chamber, 333, 334f
Capacitance, 218
Capacity tester, 223
Carbon dioxide, 239
Carbon-residue test, 83
Carburetor air heat, 714
Carburetor air intake heaters, 158
Carburetor air temperature gauge (CAT), 78, 104, 157, 158
Carburetor heat valve, 102
Carburetors, 23, 136–164, 144f. See also specific types
  with ADI, 164–167
  downdraft, 153, 153f
  inspection of, 254–255
  pressure injection, 161–164
  turbochargers and, 111–113, 113f
  venturi tube and, 138, 138f, 142f
Carnot, Nicolas-Leonard-Sadi, 60
Carnot cycle, 60
Carrier ring (or cage), 57
Cartridge-type starter, 407
CAT. See Carburetor air temperature gauge
Catalytically active metals, 83
Catenary thermal shield, 443
Cecil, Reverend W., 1
CECO. See Chandler Evans Company
Central conductor, 276, 277f
Central maintenance computer system (CMCS), 699
Centrifugal breather, 385f
Centrifugal-flow compressor, 326–327
Centrifugal force, 326, 587
Centrifugal impeller, 502
Centrifugal turboprop engine, 327f
Centrifugal twisting force, 587
Centrifugal twisting moment, 607
Certification:
  of overhaul station, 269
  type certification, 270
Cessna, 8
CFM56-3, 450f, 452f
CFM56 engine, 413, 449–451, 451f

Chandler Evans Company (CECO), 536
Charles, Jacques A. C., 106
Charles' law, 59–60, 106, 316
Check valve, 166
Chokebored cylinder barrel, 45
Circlets, 44
Circular magnetization, 276, 276f
Clamping rings, 582
Cleaning:
    of engine mounts, 257
    of float-type carburetors, 158
    of reciprocating engine, 273–275
    of spark plugs, 229, 253
Clearance:
    ACC, 430, 434
    cold, 54
    cylinder-wall, 42
    for dimensional inspection, 282
    propellers, 590
    valve, 54–55
Clearance-adjusting screw, 52
Cleveland open-cup tester, 81, 81f
Cloud point, 82–83
CMCS. See Central maintenance computer
    system
Cockpit controls, 627, 637, 641
Cockpit indication signals, 496
Cockpit instruments, 590f
Coefficient of absolute viscosity, 381
Coil assembly, 196
Coin-tap test, 674, 675f
Cold clearance, 54
Cold-cylinder check, 262
Cold repair, 668f
Cold-soaked engine, 546f
Cold weather engine operation, 82, 546f
Collective-pitch increase, 538–539
Color, of lubricants, 82
Colorimeter, 82
Combination pressure-splash lubrication
    systems, 88
Combination starter-ignition switch, 209,
    209f
Combustion chamber, 333–334, 333f, 334f,
    358f, 418, 418f, 419f, 426f, 427f,
    455f, 456, 461f, 480, 482f, 501, 505,
    505f, 553, 557f
Combustion efficiency, 322
Combustion section, 414, 418–419, 420f
    for Allison 250 gas-turbine engine,
        532, 534
    for Pratt & Whitney JT9D turbofan
        engine, 426
    for Rolls-Royce Dart engine, 482
    for Rolls-Royce RB 211 engine,
        456–458
Combustion starters, 406–407
Combustor, 440–441, 441f, 528
Coming-in speed, 203
Compact propellers, 603–604, 617, 653f
Composite propeller blades, 624–626,
    671–672, 674–675, 674f, 676f
Compression event, 65, 65f

Compression pressure, 107–108, 108f
Compression ratio, 61, 67, 75–76,
    76f, 131
Compression rings, 43, 43f
Compression stroke, 61, 62, 138
Compression testing, 250–251, 250f
Compressor airflow, 331
Compressor blades, 328f, 565, 566f
Compressor bleed valve, 510, 510f
Compressor-intermediate case, 416,
    417, 417f
Compressor pressure ratio, 326
Compressor section, 414–417, 456,
    532, 534
Compressor stall, 326, 330–331
Compressor turbines, 558f, 559f, 569f
    disk and blade assembly, 553–554
    vanes of, 505
Compressors, 439f, 440f, 446, 446f,
    501–504, 504f, 528, 567f, 579f.
    See also specific types
    front-stator assembly, 436
    for gas-turbine engine, 326, 327f
    for General Electric CF6 turbofan
        engine, 436–438
    for Lycoming T53 engine, 527
    for Rolls-Royce Dart engine, 482
    in supercharger, 104, 105f
Computing system, 364–366
Concentric oil tubes, 622f
Condition lever, 489, 633, 637
Condition monitoring, 577–578
Cone bottoming, 652f
Connected-rod bearing journal, 33
Connecting rods, 37–40, 38f, 60, 60f, 283f,
    284f, 293, 293f, 297–298
Connector-well flashover, 232f
Conradson test, 83
Constant-effort spring, 179, 181f
Constant head idle spring, 179, 180f
Constant-pressure cycle, 316
Constant-speed counterweight propeller,
    604
Constant-speed mode, 497, 641
Constant speed propeller, 587, 591
Constant-speed propeller, 242, 594–598,
    596f, 680, 681f
Constant torque, 496
Contact breaker, 197
Continental, 8, 8f, 11f, 36–37, 169, 169f,
    172–175, 172f, 173f, 176f, 207–215,
    208f, 240–241. See also Teledyne
    Continental
    D-3000 magneto ignition, 215f
    IO-470-D, 96–99, 97f
    S-200 magneto, 209f, 212–213, 214
    TSIO-360, 121
    turbochargers for, 118–121, 120f
Continuous-flow injection systems,
    169–175
Continuous ignition, 397, 400
Control pedestal, 712–713, 713f
Control quadrant, 712, 712f

Control systems, 115–117, 116f, 615–647,
    716–718. See also specific types
    for Dowty R321 four-blade turbopropeller,
        620–621
    for Garrett ATF3-6 engine, 470f
    for General Electric CT7 engine, 495
    for Lycoming T53 engine fuel, 531f
    for Pratt & Whitney PW100 series
        turboprop engine, 489
    for RB 211 engine, 461f
Control units, 369, 701
Controllable-pitch propellers, 591,
    594, 659
Convergent-divergent nozzle, 315,
    315f, 339f
Cooling:
    with air, 8, 122–123, 122f, 123f
    aviation oils and, 85
    by evaporation, 156
    of fuel, 179
    with liquid, 8, 123–124, 123f
    for Lycoming T53 engine, 527
    nozzle-vane construction and, 442f
    for Pratt & Whitney JT9D turbofan
        engine, 337f
    for Rolls-Royce RB 211 turbofan
        engine, 337f
    transpiration, 335
    turbine, 430, 510–511, 511f
    of turbine nozzles vanes and blades,
        428f
Cooling systems, 8, 15, 15f, 16–17, 17f,
    21, 109, 121–123
Corrosion, 84, 85, 131, 272, 672
Corrugated-perimeter type, 345
Counterweights, 7, 33, 33f, 34, 34f, 293,
    604, 617, 621
Cowl assembly, 440
Cowl flaps, 122, 122f
Crank cheek, 34
Crankcases, 7, 29–32, 30f, 31f, 273,
    293, 298
Crankpin, 33–34, 33f, 282f
Crankpin end. See Large end
Crankshafts, 5–8, 7f, 33–37, 33f, 35f, 38,
    50–51, 60, 60f, 63–64, 63f, 285f,
    292–293
Crete function, 702
Critical altitude, 7, 111
Crocus cloth, 668
Cross-flow system, 133
Crosshead, 619
Cruise control, 243
Crushed blade trailing edge, 674
Cumene, 132
Cuno oil filter, 92
Current, 277f
Current induced in a coil, 194
Curtis, Charles G., 308
Curtiss engines:
    D-12, 3
    Jennie (JN-4), 3
    OX-5, 3, 4f

Curvic coupling, 456
Cycle:
  Brayton, 60, 316, 317*f*
  Carnot, 60
  Diesel, 60
  efficiency, 322
  engine, 60, 138–139
  flight, 546
  four-stroke-cycle engines, 61–62,
    61*f*, 191
  pressure, 71, 71*f*
  rotary-cycle engines, 66, 67*f*
  starting, 395
  two-stroke engines, 15–16, 15*f*, 17*f*
  Wankel, 66
Cycle pressure ratio, 322
Cylinder barrels, 45–46, 290–292, 291*f*
Cylinder bore gauge, 282, 283*f*
Cylinder glazing hone, 291*f*
Cylinder heads, 46–47, 141, 141*f*, 286
Cylinder honing machine, 291, 291*f*
Cylinder pressure, 70, 70*f*
Cylinder-wall clearance, 42
Cylinders, 45–47, 46*f*, 298–299, 616
  classification by, 5–8, 7*f*
  compression testing of, 250
  crankshaft and, 5–8, 7*f*
  detonation within, 75*f*
  displacement and, 7–8
  exterior finish for, 47
  fins of, 292
  flange of, 292
  in hydraulic unit assembly, 51
  in internal-combustion engine, 60, 60*f*
  of reciprocating engines, 286
  skirt of, 292

Da Vinci, Leonardo, 307
Daimler, Gottlieb, 1
Dampers, 34–35, 248
Dassault Mercure, 413
Data averaging, 660–661
DC trembler unit, 396–397, 396*f*
DCDI. *See* Dual capacitor discharge
  ignition
Deaerator tray, 484
Decarbonizing, 274
Deck pressure, 114, 115, 117
Dee Howard Co. business aircraft, 342*f*
Degreasing, 229, 273–274
Dehydrator bag, 303
Deicer boot, 641
Deicing, 607–609, 608*f*, 609*f*, 610*f*,
  641, 647*f*
Delamination, 672, 673
Delay bleed, 166
Demeshing spring, 234
Density altitude, 77
Density controller, 114–117, 114*f*
Depth gauge, 285, 285*f*
Detonation, 74–75, 74*f*, 75*f*, 109,
  140, 157
Dew point, 156

Diaphragm pressure switch, 702*f*
Diaphragm-type gauges, 684–686, 685*f*
Dielectric, 222
Diesel, Rudolf, 60
Diesel cycle, 60
Diesel engine, 66–68
Differential bellows, 513
Differential-pressure:
  compression tester, 251–252, 251*f*
  controller, 114–117, 115*f*, 177
Diffuser, 333
  air-diffuser housing, 528
  case, 417, 418*f*, 425, 427*f*
  section, 326, 417–418
  vanes of, 111, 482
Dimensional inspection, 282
Diminishing returns, theory of, 8
Direct-compression check, 251
Direct-cranking starter, 233*f*, 235*f*
Directionally solidified casting, 348
Discharge nozzle, 142, 163, 171, 183
Displacement, 7–8
Disposable oil filter cartridges, 92, 93*f*, 387
Dissimilar metals, 274
Distortion, 672
Distributor, 199–201, 210*f*, 606
Dome, 441, 606
Double-entry centrifugal-flow compressor,
  327*f*
Double-row radial engine, 3–4, 4*f*, 7*f*
Double-throw crankshaft, 36
Doubler circuit, 399
Douglas:
  DC-9, 387
  DC-10, 313, 407*f*, 409
Downdraft carburetors, 153, 153*f*
Dowty, 619*f*, 621*f*, 626, 628*f*
  R321 four-blade turbopropeller,
    619–621
  R352 engine, 621, 622*f*, 623*f*, 639–641,
    639*f*
Draining sumps, 247
Drum-and-shaft assembly, 715*f*
Dry motoring check, 572
Dry paper filters, 254
Dry-sump system, 98–99, 98*f*
Dryewiecki theory, 582
Dual-acting hydraulic pitch-control system,
  617
Dual capacitor discharge ignition (DCDI),
  14–16
Dual-element deicing system, 609, 609*f*
Ducted fan, 313
Dump valve, 489–490
Duplex nozzle, 358, 359*f*
Dyna-focal engine mount, 257*f*
Dynamic balancing, 7, 34, 35*f*, 570,
  659, 662
Dynamic propeller tracking, 663
Dynes, 381

E-gap angle, 198
Easyout, 294*f*

Economizer, 141, 148–150, 149*f*
  MAP and, 150, 150*f*
Economy mixture indicating system,
  688, 690
ECS, for propeller, 640*f*
ECU. *See* Electronic control unit
Eddy currents, 201, 281
EEC. *See* Electronic engine control
Effective pitch, 585–586, 586*f*
Efficiency:
  bte, 72
  propeller, 589
  propulsive, 321–322, 321*f*
  thermal, 72*f*, 322–323, 322*f*
  volumetric, 61, 73
EGME. *See* Ethylene glycol monomethyl
  ether
EGT. *See* Exhaust gas temperature
EICAS. *See* Engine indicating and crew
  alerting system
Ejectors, 127
Electric-discharge machining, 346
Electric starters, 401–402, 402*f*
Electrode:
  construction of, 227*f*
  erosion of, 401*f*
Electronic control system, 489
Electronic control unit (ECU), 220, 495
Electronic engine control (EEC), 373–379,
  374*f*, 376*f*, 431, 490
Electronic imaging, 548–549
Electronic tachometers, 692, 693
Electrostatic buildup, 351
Electrostatic charging, in fuel, 15
Electrostream drilling, 346
Elements, 582
Elongation, for reciprocating engine,
  272
Emulsifier, 279, 280*f*
Engagement mechanism, 405, 405*f*
Engine control systems, 240*f*, 248, 255,
  683–718, 716*f*
  for Pratt & Whitney 2037 turbofan
    engine, 434
  for Pratt & Whitney PW100 series
    engine, 490–492
  for Rolls-Royce RB 211 engine, 458
Engine cycle, 60, 138–139
Engine-driven fuel pump, 134
Engine indicating and crew alerting system
  (EICAS), 699
Engine mounts, 248, 257
Engine pressure ratio (EPR), 377, 573,
  693–694
Engine section, 247
Engine systems:
  for Garrett TPE331 engine, 514–515
  for General Electric CF6 turbofan
    engine, 447–449
  for Pratt & Whitney JT8D engine,
    422
Engine vane control (EVC), 364, 366–369,
  368*f*, 369*f*, 574

Engines. *See also specific types, components, and relevant topics*
acceleration, 543f
adjustment, 573–574
failure warning system, 539
fire, 239
gauge unit, 684
idle, 545f
indicating systems, 683–718
inspection for, 550f, 551f
instruments, 683–699
noise, 344–345
overhaul for, 267
parallel harness, 575
preheating, 247
preservation, 302–304
records, 270
speed, 319, 695f
start levers, 547f, 715
start switch, 408, 547f
storage, 302–304, 573–574
testing, 300–302, 570–573
thermocouples, 575–576
transportation, 573–574
trimming, 573–574
troubleshooting for, 260f–261f
underspeeding, 597f
vibration, 663, 697–698
warning systems, 683–718
Enrichment valve, 141
Entrained water, 136
EP. *See* Extreme-pressure lubricants
EPR. *See* Engine pressure ratio
Equivalent shp, 479
Erosion, 401f, 672, 673, 674, 675f
Ethylene glycol monomethyl ether (EGME), 136
EVC. *See* Engine vane control
Exciter units, 396, 397, 532
Exhaust:
augmentors, 78, 127, 127f
back pressure, 78
bypass-valve assembly, 114–115, 115f
diffuser, 528
duct, 509, 509f
nozzles, 339
port, 47
stacks, 248, 690f
stroke, 62, 67, 138
valves, 48
Exhaust event, 65, 65f
Exhaust-gas flow, 338
Exhaust gas temperature (EGT), 140, 246f, 574–575, 574f, 695–696
circuit error, 576
gauge, 691f, 696f
troubleshooting, 575–577
Exhaust systems, 21, 255–257, 256f, 338, 338f
for light-aircraft engines, 125, 125f
for opposed-type engines, 124–125, 125f
for radial engines, 125–127, 126f
for reciprocating engines, 124–126, 124f

Expansion chambers, 482
Expansion stroke, 62
Experimental engines, 16–26
Experimental mean pitch, 586
External-type supercharger, 109–110, 109f
Extreme-pressure lubricants (EP), 84

F/A. *See* Fuel-air ratio
FADEC. *See* Full-authority digital electronic control system
Fairing assembly, 421, 421f
False start, 544
Fan blades, 453f, 550, 551f, 564f
of gas-turbine engines, 564–565
Pratt & Whitney aircraft, 432f
Fan bypass ratio, 330
Fan-discharge front-compressor inner duct, 417
Fan-discharge rear-compressor inner duct, 417
Fan discharge section, 414
Fan drive shaft, 444
Fan frame, 445
Fan sections:
for General Electric CF6 engine, 436, 446f
for Pratt & Whitney JT8D gas-turbine engine, 414–417
for Pratt & Whitney JT9D-7R4 engine, 429
for Rolls-Royce RB 211 engine, 456
Fatigue cracks, 667
Fatigue stresses, in propeller, 668f
Fault indicators, 577, 698–699
FCU. *See* Fuel control unit
Feather edge, 289, 289f
Feather reservoir assembly, 647
Feather valve, 621
Feathered blade, 592
Feathering, 592, 599, 601–604, 617, 631
autofeathering systems, 612, 612f
full-feathering, 601–603
nonfeathering, 598–601, 603, 603f
pump, 641
valve, 635
Feathering-spring assembly, 616
Fenwal sensing element, 703f
Fenwal sensing system, 701, 701f, 703f, 704f
Fernihough, W. F., 308
fhp. *See* Friction horsepower
Fiberscope, 548–549, 549f, 550f
Film cooling, 335
Fine-wire spark plugs, 226f, 230f
Fins:
of cylinders, 292
of reciprocating engines, 286
Fir-tree method, 337f
Fire, types, 706
Fire extinguishers, 706–709
controls and connections for, 708f
mounting, 706f
for wing-mounted engines, 707f

Fire point, 81
Fire sleeve, 89, 89f
Fire wall seals, 258
Fire warning systems, 700–706, 701f
First-stage low-pressure turbine nozzle, 444, 444f
First-stage stator, 425f
First-stage turbine nozzle assembly, 426
First-stage turbine rotor, 444f
Fixed forces, 594
Fixed oils, 82
Fixed pitch propellers, 581f, 591, 592–593
Fixed shaft engine, 479
Fixed-shaft-type power conversion, 479f
Flame propagation, 141, 141f
Flame tubes, 482
Flange-mounted magnetos, 192
Flange-type propeller shafts, 37, 37f, 650, 653–654, 655f
Flanges, 292, 416f, 424f, 451f, 489f
Flash point, 80–81
Flexible push-pull controls, 709, 711–712, 718
Flexible vibration dampers, 248
Flight cycle, 546
Float level, 159–161, 160f, 161f
Float mechanism, 144–145, 145f
Float-type carburetors, 144–161, 146f, 147f, 148f, 240
AMC for, 152, 152f
by Bendix-Stromberg, 153–154, 154f, 155f
economizer for, 148–150, 149f
by Marvel-Schebler, 154, 154f, 155f
mixture control system for, 150–151, 151f
troubleshooting for, 161, 162t
Floating points, 262
Flow dividers, 183–184, 183f, 184f, 489–490
Flowmeter, 687f, 688f
Fluid pressure, 137–138
Flux leakage, 276
Flux linkage, 194
Flyweights, 205, 595, 596f, 597f, 598, 691f
Foam-type air filters, 254
FOD. *See* Foreign-object damage
Fokker 50 aircraft, 639f
Foot-pounds, 296
Foreign-object damage (FOD), 541, 549–550
Fork-and-blade connecting-rod assembly, 38–39, 38f, 39f
Forked rod, 38
Forward shaft, 436, 443
Forward turbofan engine, 313f
Four-blade propeller, 583f
Four cylinder engines, 9, 12f, 88f
Four-pole rotating magnet, 195f
Four-stroke-cycle engines, 61–62, 61f, 191
Four-throw crankshaft, 36, 36f
Fourth-stage turbine outer seal ring, 419
Fractional-distillation test, 129, 129f
Free turbine, 479, 479f, 499, 524

Freewheeling clutch, 536
Friction horsepower (fhp), 69
Front accessory drive, 414, 422
Front-compressor, 414, 421, 421*f*, 482
Front-cone bottoming, 651
Front-cone seat, 649
Front cones, 649, 651
Front section. *See* Nose section
Fuel, 129–133. *See also specific types*
　additives, 351
　cooling, 179
　electrostatic charging in, 15
　for gas-turbine engine, 349–379
　gravities and weights of, 350*f*
　microorganisms in, 351
　performance number for, 130, 130*f*
　for Pratt & Whitney PW100 series
　　turboprop engine, 489
　testing, 129–130, 129*f*
　vaporization of, 143–144
　water in, 350
Fuel-air control unit, 170–171, 170*f*
Fuel-air ratio (F/A), 78, 130, 131, 139*f*,
　140, 141, 141*f*
　ADI and, 165
　for Bendix RSA fuel injection system,
　　175
　carburetors and, 139–142
　for pressure injection carburetors, 162
Fuel boost pump, 134
Fuel control systems, 170, 359–361, 366,
　370–373
Fuel control unit (FCU), 163, 164, 360*f*,
　361–364, 369*f*, 485*f*, 513
Fuel-cooled oil cooler, 354*f*
Fuel discharge nozzle. *See* Discharge
　nozzle
Fuel drain valve, 32
Fuel filters, 134, 352*f*
Fuel flowmeters, 687–688
Fuel heater, 512
Fuel injected (I) engine classification, 7
Fuel injection systems, 7, 169–189,
　170*f*, 264*f*
　by Bendix RSA, 175–189, 176*f*
　by Continental, 169, 169*f*, 172–175,
　　172*f*, 173*f*
　inspection of, 255
　maintenance of, 255
Fuel manifold, 171, 171*f*, 513, 553, 554*f*
Fuel metering, 136, 163
Fuel metering valve, 171, 171*f*, 373*f*
Fuel nozzles, 358*f*, 418, 536, 559–564.
　See also specific types
Fuel pressure gauge, 184, 184*f*, 686–687,
　686*f*, 687*f*
Fuel pressure warning systems, 688
Fuel pump, 134, 352*f*, 353*f*, 489, 491*f*, 536
Fuel pump pressure, 174
Fuel regulator system, 529–531
Fuel spray nozzles, 357–359
Fuel strainer, 145, 146*f*
Fuel sump strainers, 134

Fuel systems, 129–136, 134*f*, 135*f*, 254–255,
　262–263, 349–379, 357*f*
　for Allison 250 gas-turbine engine, 536
　for Bendix RSA fuel injection system,
　　175
　icing of, 136
　for PT6A engine, 513
　for Rolls-Royce Dart engine, 484–485
　for Rotax opposed-type engines,
　　17–18, 17*f*
　for Rotax two-cycle engines, 15, 17*f*
　troubleshooting chart for, 263*f*
Full-authority digital electronic control
　　system (FADEC), 219–221,
　　373–379, 435
Full-authority electronic engine control, 434
Full-authority electronic fuel control, 468
Full-feathering, 601–603
Full-floating knuckle pins, 40, 40*f*
Full-floating piston pins, 44
Full-flow system, 91–92, 91*f*, 386
Full rich mixture, 152

Galvanometer, 194
Gap-setting tool, 230*f*
Garrett Turbine Engine Company,
　377*f*, 463
　ATF3-6 engine, 345, 468–469, 468*f*,
　　469*f*, 470*f*
　TFE731-5 engine, 377, 466–468, 467*f*
　TFE731 engine, 463–466, 464*f*, 465*f*,
　　466*f*, 467*f*
　TPE331-14 engine, 514*f*
　TPE331 engine, 514–515, 515*f*, 516*f*,
　　616, 631–635, 634*f*
Gas:
　supercharging and, 105–106, 105*f*, 106*f*
　temperature and, 106–107, 106*f*, 107*f*
Gas flow, 420*f*, 526–527, 535*f*
　through convergent-divergent nozzle,
　　339*f*
　in Garrett ATF3-6 engine, 469*f*
Gas generator, 479, 504–505, 572*f*
Gas-path health, 577–578
Gas producer, 479, 534, 536
Gas-producer fuel control, 370–373
Gas turbines, 26–27, 59, 307–323, 311*f*,
　312*f*, 325–348, 330*f*, 401–407,
　541–578
　advanced manufacturing processes of,
　　345–348
　air-bleed of, 331–333
　combustion chambers of, 333–334
　component maintenance of, 564–568
　compressors of, 326, 327*f*
　electronic engine controls for, 373–379
　engine bearings and seals, 342–344
　engine noise for, 344–345
　exhaust nozzles of, 339
　exhaust systems of, 338
　fan blades of, 564–565
　fuel filter of, 352*f*
　fuel requirements of, 349

Gas turbines (*Cont.*):
　fuel system components of, 352–353
　fuel systems, 354–355
　fuels and fuel systems of, 349–379
　ignition systems of, 395–411
　inlet of, 325–326
　inspection of, 546–557
　instruments for, 693–697
　internal air supply systems of, 331–333
　internal airflow pattern of, 332*f*
　jet fuel properties of, 349–359
　limits of stable airflow in, 331*f*
　lubrication for, 381–382, 390–391
　overhaul of, 568–575
　performance of, 317–321
　spur-gear fuel pump of, 352*f*
　starters for, 406, 406*f*
　starting of, 541–544
　starting systems of, 395–411
　TBO for, 5
　troubleshooting of, 577–578
　turbine nozzle diaphragm of, 334–335
　variable stator vanes of, 332*f*
Gasoline:
　AVGAS *vs.*, 132–133
　characteristics of, 129–133
Gear pump, 20–21, 94, 95*f*
Geared (G) engine classification, 7
General Electric:
　CF6, 313, 407, 413, 435–449, 438*f*,
　　439*f*, 440*f*, 441*f*, 442*f*, 446*f*, 448*f*
　CF6-6, 436
　CF6-50, 436, 437*f*, 438*f*
　CF6-80, 449
　CT7, 314*f*, 492–497, 493*f*, 494*f*, 495*f*,
　　496*f*, 497*f*, 641–643, 642*f*
　GE I-A, 26
　GE90, 27, 27*f*, 413
Geometric pitch, 586, 586*f*
Glow plugs, 397, 400–401, 400*f*,
　402*f*, 512
Gnome-Monosoupape rotary engine, 2, 2*f*
Go gauges, 282
Gouges, 272, 672, 673
Governor flyweights, 596*f*, 597*f*
Governor mode, 624
Governors:
　bottoming governor system, 496
　overspeed, 490, 631, 633*f*, 640–641,
　　641*f*
　power turbines, 373, 513, 529, 536
　propellers, 596–598, 621, 627, 630*f*,
　　632*f*
Gravesande, Willem Jako, 307
Gravity-feed fuel systems, 134, 134*f*
Grease and air charge fittings, 679*f*
Great Plains Aircraft, 23–24, 24*f*
Grit-blasting machine, 275*f*
Groove, 48, 272
Groove land, 44, 44*f*
Ground adjustable propeller, 652*f*
Ground-adjustable propellers, 581, 581*f*,
　591, 593–594, 652

Ground angle effects, 677*f*
Ground clearances, 590
Ground run-up, 664*f*

Hamilton Standard, 375*f*, 604, 606–607, 621–624, 654
    24PF propeller, 623*f*, 624*f*
    EEC system with, 376*f*
    JFC68, 374
    JFC68 FCU, 364–368, 365*f*, 367*f*
    Model 14RF propeller, 623–624
    Model 24PF propeller, 623*f*
    PCU, 624*f*
Hand cranking, 241
Harness, 215
Harness testing and inspection, 253–254
Hartzell, 599–604, 601*f*, 603*f*, 615–619, 626*f*, 627*f*, 653
    HC-82XF-2 feathering propeller hub assembly, 603*f*
    HC-A3Vf-7 propeller hub assembly, 636*f*
    HC-A6A-3 six-blade propeller, 619*f*
    HC-B3TN-2 propeller, 615, 616*f*, 617*f*
    HC-B3TN-3 propeller, 617*f*
    HV-D4N-5 four-blade propeller, 618*f*
    lightweight turbine propeller, 617
Head:
    of cylinder, 46–47, 141, 141*f*, 286
    of piston, 40–41, 41*f*
Heat engines, 59, 311
Heat rejection, 54
Heat shields, 420*f*
Heater muff, 102
Heli-coils, 46, 294, 295*f*
Helicopter:
    Allison 250 engine in, 537–539
    power trains, 536–537
Hero, 307, 307*f*
High-bypass engine, 313, 340*f*
High bypass turbofan engine, 27
High-compressor drive turbine, 427
High-energy capacitor discharge system, 395
High-output radial engine, 110
High pitch, 586*f*, 624
High-pressure air-turbine starters, 406
High-pressure compressors, 423, 426*f*, 427, 427*f*, 429, 456
High-pressure fuel system, 356*f*
High-pressure oil, 607
High-pressure turbine (HPT), 441–443, 443*f*
High-speed gear train, 481
High-tension cable, 203, 213*f*
High-tension harness, 213–214
High-tension ignition system, 209*f*
High-tension magneto, 192, 202*f*, 213*f*, 214*f*, 218
High-viscosity-index (high-VI), 86
High-voltage corona, 218
HIP. *See* Hot isostatic pressing
Hispano-Suiza engine, 2, 2*f*

HMU. *See* Hydromechanical metering unit
Honeywell, 377, 475, 477*f*
Hopper, 99, 99*f*
Horizontal engine assembly, 571*f*
Horizontal (H) engine classification, 7
Horsepower, 68, 77*f*, 108, 108*f*, 318, 615. *See also specific types*
Hot isostatic pressing (HIP), 347
Hot section inspection, 552, 553–555, 580*f*
Hot spot, 102
Hot start condition, 543, 544
Hour meters, 693
Houston Chemical Co., 351
Howmet Turbine Components Corporations, 348
HPT. *See* High-pressure turbine
Hubs, 649–650
    assemblies, 581, 599*f*, 606, 644, 645*f*, 646*f*
    splines, 651
Humidity, relative, 156
Humidity effect, 319–320, 320*f*
Hung start, 544
Hydraulic lock, 45
Hydraulic unit assembly, 51, 51*f*
Hydrobore, 293, 293*f*
Hydromatic pitch-changing mechanism, 607*f*
Hydromatic propeller assembly, 606*f*
Hydromechanical metering unit (HMU), 489
Hydromechanical torquemeter, 532*f*
Hypoid lubricants. *See* Extreme-pressure lubricants

IAE 2500 engine, 413
IAE V2500 engine, 458–460, 463*f*
Icing, 104, 136, 154–158
Idle:
    adjustments, 161, 173–174, 174*f*, 177–179, 179*f*, 188
    check, 573
    cutoff, 146, 151–152, 188
    engine, 545*f*
Idling system, 146–148, 147*f*, 153*f*
Idling valves, 163
IDS. *See* Integrated display system
iE_2. *See* Integrated Electronic Engine
IFB. *See* Integral front bearing
Igniters, for turbine engines, 399–401, 399*f*, 400*f*
Ignition, 61, 191–237
    boosters, 205–207
    cable assembly for, 399, 399*f*
    cable tester for, 253*f*
    event, 65, 65*f*, 191
    exciters, 396, 397, 532
    magnetos, 191
    malfunctions, 262
    shielding, 203–204
    switch, 203–205, 204*f*
    TDC and, 70
Ignition-current regulator, 513, 513*f*

Ignition-starter switches, 204*f*
Ignition systems, 236–237, 396*f*, 512–513
    with combination starter-ignition switch, 209*f*
    for gas-turbine engines, 395–411
    for General Electric CT7 engine, 495
    with glow plugs, 397
    in Jabiru engine, 20–21
    for JT8D, 398–399
    for Pratt & Whitney JT9D turbofan engine, 398*f*
    for Pratt & Whitney PW100 series engine, 492
    for Rolls-Royce Dart engine, 486
Ignition unit, 398*f*
ihp. *See* Indicated horsepower
imep. *See* Indicated mean effective pressure
Impact damage, 672
Impact ice, 157
Impactoscope flaw detector, 674
Impeller, 111, 502, 527
Impulse coupling, 205–206, 205*f*, 206*f*
Impulse turbines, 335, 336*f*
In-line engines, 2–3, 7*f*, 31–32, 64
Inch-pounds, 296
Indicated horsepower (ihp), 68–69, 71
Indicated mean effective pressure (imep), 69, 70
Induction systems, 101–104, 101*f*, 102*f*, 103*f*, 157
    ducting of, 254
    of reciprocating engines, 264–265
Induction vibrator, 206–207, 206*f*
Inductor-rotor magnetos, 192
Inertia bonding, 347
Injecting water, 320
Inlet:
    for gas-turbine engine, 325–326
    guide vanes, 325
    housing, 528
Inspection:
    of Bendix RSA fuel injection systems, 189
    of Continental fuel injection systems, 174–175
    of control systems, 716–718
    dimensional, 281–286
    for engine, 550*f*, 551*f*
    for engine controls, 255
    for engine fuel systems and carburetors, 254–255
    for engine mounts, 257
    of exhaust systems, 255–257, 256*f*
    of fire warning systems, 705–706
    of float-type carburetors, 158–161
    for front-cone bottoming, 651
    of fuel injection system, 255
    of gas turbines, 546–557
    of glow plugs, 400–401
    hot section, 552, 553–555, 580*f*
    of igniters, 400–401
    of induction system air filters and ducting, 254

Inspection (*Cont.*):
  of magnetos, 221–224, 252, 252*f*
  of metal propellers, 666–667
  of metering jet, 159, 159*f*
  of new propeller installation, 654–655
  of oil filters, 92–93
  of oil system lines, 250
  for overlimit operation, 551–552
  for overspeed, 552
  of propellers, 649–681
  of PT6A hot section inspection, 555
  of reciprocating engines, 247–258
  of spark plugs, 231, 231*f*, 253
  structural, 275–281
  of superchargers, 258
  of tachometers, 693
  for turbine-engines, 400, 411
  of turbochargers, 258
  of valves, 273*f*
  of wood propellers, 666
Installation, 303–304
  of Fenwal sensing element,
    703*f*, 704*f*
  of ground-adjustable propellers, 652
  of Hamilton Standard hydromatic
    propeller, 654
  of Hartzell compact flanged propeller,
    653
  of McCauley flanged propeller,
    653–654
  of Photo-tach, 662*f*
  of propellers, 649–681
  for reciprocating engine, 299–300
  of sensors, 703–705
  for Slick Series 4300, 215–217
  for Slick Series 6300, 215–217
  of spark plug, 231–232
  for splined-shaft propeller, 649*f*
  in test stand, 299–300
  of thermocouples, 690*f*
  for turbine-engine igniters, 401
Insulating blanket, 340*f*
Intake event, 65, 65*f*
Intake manifolds, 102
Intake pipes, 104, 104*f*
Intake port, 47
Intake stroke, 61, 62, 138
Intake valves, 48
Integral front bearing (IFB), 21, 23*f*, 24*f*
Integral-hub flange-type crankshaft,
  650, 650*f*
Integrated display system (IDS), 699
Integrated Electronic Engine (iE$_2$),
  10–14, 13*f*
Integrity monitoring system, 703
Intercoolers, 109, 158
Interference fit, 288, 288*f*
Intermediate casing, 416, 417, 417*f*, 424,
  455, 482
Intermediate-pressure bearing, 455
Intermediate-pressure compressor, 456
Internal air supply systems, 331–333,
  332*f*

Internal-combustion engines, 1, 59–78,
    65*f*, 67*f*. *See also specific types*
  efficiency of, 72–73
  firing order for, 64–65, 64*f*
  operating fundamentals for, 60–62
  power in, 68–72
  valve timing for, 62–64, 63*f*
Internal-type superchargers, 109–110, 109*f*
Interstage baffle, 506
Interstage bleed system, 528–529, 528*f*
Interstage seal, 442, 505
Interstage turbine temperature (ITT), 544*f*,
  696, 696*f*
Inverted in-line engine, 2, 3*f*, 96
Inverted-V-type engine, 3
Investment casting, 347, 347*f*
Isolating systems, 698–699
ITT. *See* Interstage turbine temperature

Jabiru, 20–21, 20*f*, 22*f*, 23*f*
Jet engine symbols, 317
Jet fuels, 349–359
Jet propulsion, 26, 307, 309, 310
Joule, 398

Kerosene-type jet fuels, 350
Kevlar shell, 626
Kickback, 142
Kidde sensing element, 701–702,
  702*f*, 705*f*
Kinematic viscosity, 381
Knuckle pins, 40, 40*f*

Labyrinth, 510
Labyrinth seals, 390
Land. *See* Groove land
Langen, Eugen, 1
Large end, of connecting-rod
  assembly, 38
Laser-beam welding, 346–347
Laser drilling, 346, 346*f*
Latent heat of vaporization, 129, 156
Laws of evaporation, 156
Laws of thermodynamics, 316
Lead, in aviation fuels, 130–132
Leading edges, 581, 669–670, 669*f*
Leakage test, 560
Lean best-power mixture, 139, 153
Lean die-out, 360
Leaning, 245, 246*f*
Left-hand rotation (L) engine
  classification, 7
Left-hand rule, 195
Lenoir, Joseph Étienne, 1
Lenz' Law, 195
LeRhone rotary engine, 2, 2*f*
Liberty engine, 3, 3*f*
Light aircraft engines, 111–114, 112*f*
  exhaust system for, 125, 125*f*
  fire suppression system for, 706
  low-tension ignition system for, 219
Light-sport engines, 14–26, 20*f*
Lightning, 558, 626, 674*f*

Limiting compressor discharge pressure,
  364
Link arms, 616
Link rods. *See* Articulated rods
Liquid cooling, 8, 123–124, 123*f*
Liquid-fuel rocket engine, 310*f*
Liquid lock. *See* Hydraulic lock
Liquid penetrant inspection, 279–280, 280*f*
Lock-plate assembly, 40, 40*f*
Lockheed L-1011, 313
Lockwiring, 296*f*
Longitudinal magnetization, 276
Low-bypass air mixer unit, 339*f*
Low-compressor drive turbine, 428, 428*f*
Low-intensity ignition, 397
Low pitch, 586*f*, 624
Low pitch stop, 606, 634*f*, 680
Low-pressure air-turbine starter, 402–406
Low-pressure compressors, 425*f*, 428
  for General Electric CF6 turbofan
    engine, 436
  for Pratt & Whitney 2037 turbofan
    engine, 432*f*
  for Pratt & Whitney JT9D-7R4 engine,
    429
  for Pratt & Whitney JT9D turbofan
    engine, 423–424
Low-pressure fuel-cooled oil cooler, 385*f*
Low-pressure thrust bearing, 455
Low-pressure turbines, 444–445, 444*f*,
  445*f*, 456*f*
Low-speed gear train, 481
Low-tension ignition system, 218–219,
  218*f*, 219*f*
Low-tension ignition systems, 218–219
Low-tension magneto, 192
Low-tension wiring, 203
Low-voltage electric starting system, 403*f*
Low-voltage harness, 220
Lubbock nozzle, 358, 359*f*
Lubricants, 79–87. *See also specific types*
  for turbine engines, 381–393
Lubricating oils, 381
Lubrication systems, 89, 96–99,
  382–385, 387*f*
  for Allison 250-C20 engine, 537*f*
  for Allison 250 gas-turbine engine, 536
  for Continental IO-470-D, 96–98, 97*f*
  engine design features and, 95–96
  for four cylinder engines, 88*f*
  for gas-turbine engine, 390–391
  for gas turbines, 381–382
  in inverted in-line engine, 96
  for Lycoming T53 engine, 529, 530*f*
  for opposed-type engine, 96–98, 97*f*
  for Pratt & Whitney turbofan engine,
    388*f*
  for PT6A engine, 511–512, 512*f*
  for radial engines, 96, 96*f*
  for Rotax opposed-type engines, 18, 18*f*
  for Rotax two-cycle engines, 15
  for turbine engines, 381–393
  of turbine engines, 385–391

Lubrication systems (*Cont.*):
for turbochargers, 113
for turbofan engine, 386–390
Lycoming, 9–14, 12*f*, 56
IO-390 series engines by, 10
IO-540 series engines by, 9–10, 11*f*, 12*f*
IO-580 series engines by, 10, 13*f*
IO-720 engines by, 10
O-233 engine by, 25–26, 25*f*
O-320-B, 160-hp, 244*f*
T53 engine, 524–532, 526*f*, 530*f*, 531*f*
T53-L-13B engine, 525*f*

Magnetic chip detector, 384, 384*f*
Magnetic particle testing, 276–279, 276*f*, 278*f*
Magnetic tachometers, 691, 692*f*
Magnetizing coil, 276*f*
Magnetos, 191, 192*f*, 193–203, 197*f*, 199*f*, 200*f*, 207, 214–215, 222*f*
continental ignition D-3000, 215*f*
distributor, 201*f*
flange-mounted, 192
inspection of, 223–224, 252, 252*f*
maintenance and inspection of, 221–223
outlets, 214*f*
overhaul of, 223–225
rotating magnet, 192, 196*f*
safety gap of, 203
single-type, 192
Slick Series 4300, 215–218, 216*f*, 217*f*
Slick Series 6300, 215–217
sparking order, 202–203
speed, 201–202
test stand, 224, 224*f*
timing for, 211
timing marks, 212*f*
Maintenance, 247–258, 559, 693
of Continental ignition S-200 magneto, 214
of control systems, 716–718
of ignition and starting system, 236–237
of induction system air filters and ducting, 254
of magnetos, 221–223
of propellers, 649–681
release, 302
for Slick 4300 series, 217–218
for turbine-engine starters, 411
of turbochargers, 258
Major overhaul, 268
Manifold absolute pressure (MAP), 71, 73–74, 117, 590, 685
aviation fuels and, 131
for Bendix RSA fuel injection systems, 188–199
critical altitude and, 111
economizer and, 150, 150*f*
F/A and, 141
fuel injection systems and, 184
gauge, 685*f*
horsepower and, 108, 108*f*
icing and, 156–157

Manifold absolute pressure (*Cont.*):
normalizers and, 104
relief valve, 121, 121*f*
superchargers and, 102, 104–105, 107
waste gate and, 113
Manifold adapter, 562*f*, 563*f*
MAP. *See* Manifold absolute pressure
Marvel-Schebler, 154, 154*f*, 155*f*
Mass, 105, 315
Master and articulated connecting-rod assembly, 39–40, 39*f*, 40*f*
Master gear, 644, 645
Master rod, 39, 39*f*
Master strainers, 134
Maximum allowable bend, 668*f*
Maximum compression ratio, 76
Maximum except takeoff (METO), 71
Maximum increase in rpm, 245
Maximum power, 72
McCauley, 598–599, 600*f*, 601*f*, 624, 625*f*, 653–654, 655*f*
2A36C18 propeller, 598*f*, 599*f*
Met-L-Prop propeller, 593
McDonnell Douglas, 409*f*, 410*f*, 699
DC-9, 413
DC-10, 342*f*, 407
MD-90, 458
Mean effective pressure (mep), 70–71
Mechanical control system, 489
Mechanical efficiency, 72
Mechanical energy, 59
Mechanical engine control systems:
for large aircraft, 714–715
for small aircraft, 713–714
Mechanical tachometers, 691
Megohmmeter, 253
Menasco Pirate, 2, 3*f*
mep. *See* Mean effective pressure
Mesh screen filter assembly, 353*f*
Meshing spring, 234
Metal particle concentration, 392*f*
Metal propellers, 593, 593*f*, 658*f*, 666–667
Metered fuel, 361
Metered fuel pressure, 163, 183
Metering jet, 143, 146, 146*f*, 159, 159*f*, 177
Metering pressure control valve, 166
Metering system, 145–146, 146*f*, 364
METO. *See* Maximum except takeoff
Micrometer, 134, 282*f*
MIL-L-7808C lubricant, 411
MIL-L-23699 lubricant, 409, 411
Mineral lubricants, 79, 80
Mixer unit, 338
Mixture control systems, 150–151, 151*f*, 162, 714
AMC, 152, 152*f*, 164, 179–183, 182*f*
Mixture leaning, 246*f*
Mixture thermometer. *See* Carburetor air temperature gauge
Modular construction, 458, 469*f*
Modular design, 449–451, 461*f*, 466*f*
Modules, 493, 568

Moment-weighed, 338
Moss, Sanford A., 308
Muffler system, 21
Multigrade oils, 86, 86*f*
Multiple-compressor axial-flow engines, 329–330
Multispeed supercharger, 110
Multiviscosity oils, 86–87

N_2 rpm indicator speed switch, 408
Naturally aspirated engine, 73, 74, 111
Needle-type mixture control, 150–151, 151*f*
Needle valve economizer, 149, 149*f*
Negative angle of attack, 585*f*
Negative torque, 696
Net thrust, 615
Neutralization number, 84
Newton-meter, 296
Newton's Laws of Motion, 309*f*, 314–315
Newton's steam carriage, 307, 307*f*
N_f governor, 629
90 degree adapter drive, 234–235, 235*f*
Nitriding, 45, 291
Nitrogen gas, 653
No-go gauges, 282
No-lightoff condition, 543
Noise, 344, 345, 345*f*, 346*f*
Nonfeathering, 598–601, 603, 603*f*
Nonferrous-metal plugs, 45
Normalizers, 104, 111–115, 112*f*
Nose section, 7, 29, 31*f*
Nozzle box, 482
Nozzles. *See also specific types*
for Bendix RSA fuel injection systems, 184–185, 185*f*, 186*f*
vanes of, 335*f*, 442*f*, 482

Octane, 130
Offset adapter, torque wrench with, 297*f*
Oil:
analysis, 390, 392–393, 392*f*
functions of, 85–86
grading of, 81, 81*f*
Oil breather system, 384
Oil consumption run, 301
Oil coolers, 89, 89*f*, 384
Oil damped bearing, 388
Oil dilution, 82, 95, 95*f*
Oil filters, 91–93
accessory gearbox showing, 512*f*
cartridges for, 92, 93*f*, 387
inspection of, 92–93
for turbine-engine, 383
Oil-in temperature, 82
Oil indicating and warning systems, 384–385
Oil inlet tube, 51
Oil pressure, 116, 242
gauge, 94, 684*f*
pumps, 94–95, 94*f*, 511
relief valve for, 90–91, 511, 512*f*
Oil rings, 43–44, 44*f*
Oil sample analysis report, 393, 394*f*

Oil scraper rings. *See* Oil wiper rings
Oil seal, 390, 403, 405
Oil separator, 94
Oil system, 250, 263–264
    for Pratt & Whitney PW100 series
        turboprop engine, 489
    for Rolls-Royce Dart engine, 483–484,
        483*f*
Oil tank, 99, 99*f*, 382, 382*f*
Oil temperature gauges, 94, 688–690
Oil temperature regulator unit, 89, 90*f*
Oil viscosity valve, 89, 90*f*
Oil wiper rings, 43–44
On condition, 568
On-speed condition, 597*f*, 629
Opposed cylinders (O) engine
    classification, 7
Opposed-type engines, 7*f*, 55–56, 55*f*
    camshaft for, 50*f*
    crankcase for, 29, 30*f*
    exhaust systems for, 124–125, 125*f*
    for experimental, 16–26
    firing order for, 64, 64*f*
    induction system of, 102–103, 103*f*
    for light-sport, 16–26
    lubrication system for, 96–98, 97*f*
    in post-World War I, 5, 6*f*
    by Rotax, 16–19, 17*f*, 18*f*, 19*f*
    supercharger for, 110
Optical emission spectrometry, 392
Otto, August, 1
Otto cycle. *See* Engine cycle
Out-of-balance condition, 659*f*
Overboost, 117
Overhaul, 267–305
    of gas turbines, 568–575
    of magnetos, 223–225
Overhaul shop, 268–270
Overhaul station, 269
Overheat function, 702
Overlimit operation, 551–552
Overrunning clutch, 233–234, 234*f*, 537
Overshoot, 117
Overspeed, 71, 364, 495–496, 552, 596*f*,
    637, 672
    governor, 490, 631, 633*f*, 640–641, 641*f*
Oxidation, 83, 85, 272

Packard V-12 engine, 3
Paint erosion, 674
Parsons, Sir Charles, 308
Partial-flow lubrication system, 91
Passage leading to idling system, 146
PCU. *See* Propeller control unit
Peak EGT, 245
Percent rpm, 695
Performance:
    for Allison 250 gas-turbine engine, 534
    of aviation fuels, 131
    of internal-combustion engines, 73–78
    for Pratt & Whitney JT9D turbofan
        engine, 422–429
    for Rolls-Royce RB 211 engine, 454

Performance number, for aircraft fuel,
    130, 130*f*
Permanent magnet, 193*f*, 194*f*
PGC. *See* Propeller gear case
Phase angle relationships, 610*f*
Photo-tach installation, 662*f*
Pickup, 698
Piezoelectric accelerometer, 698
Pilot valve, 598
Pin-joint attachment, 417*f*
Piston-pin end. *See* Small end
Piston-pin plugs, 45
Piston-pin retainers, 44–45, 45*f*
Piston pins, 44, 44*f*
Piston rings, 42–44, 43*f*, 44*f*, 284*f*
Piston skirt, 44, 272*f*
Piston-type economizer, 149–150, 149*f*
Pistons, 40–45, 41*f*, 43*f*, 44*f*, 292,
    298–299, 616
    crankshaft and, 63–64, 63*f*
    cylinder-wall clearance and, 42
    displacement, 68
    in internal-combustion engine, 60, 60*f*
Pitch, 585–587
    change, 484, 586–587, 620*f*
    control, 620, 633–635, 634*f*
    distribution, 583, 584*f*
    ratio, 586
Pivot-type breaker points, 197
Pivot-valve plunger, 595
Pivotless breaker assembly, 197*f*
Plain bearings, 32, 32*f*
Plain connecting rod, 38, 38*f*
Planetary gears, 57, 57*f*
Plug gauges, 282
Plunger, 51, 711
Plunger-type fuel pump, 353*f*
Pneumatic control, for fuel metering
    valve, 373*f*
Pneumatic start system components, 546*f*
Pneumatic starter, 409*f*
Pneumatic system, 702–703
Pneumatic temperature sensor, 702*f*
Point contact, 84
Poise, 381
Poppet-type valves, 47–48, 47*f*
Porsche, 26, 26*f*
Position indicating switch, 409
Positive torque, 696
Post-World War I, 3–5, 4*f*
Pounds per square inch absolute
    (psia), 137
Pounds per square inch gauge (psig),
    73, 137
Pour point, 82, 83
Powder metallurgy, 347
Power. *See also specific types and relevant*
    topics
    assurance check, 573
    control system, 518
    efficiency and, 72
    event, 65, 65*f*
    F/A and, 139*f*, 141*f*

Power (*Cont.*):
    formula for, 69
    icing and, 156–157
    in internal-combustion engines, 68–72
    lever, 489, 633, 634*f*, 637
    output, 320*f*
Power enrichment system. *See* Economizer
Power section, 31, 31*f*, 553, 556*f*
Power settings, 242–245
    for Lycoming Model O-320-B, 160-hp,
        244*f*
Power stroke, 61–62, 67, 138
Power train, 536
Power turbines, 479
    governor, 373, 513, 529, 536
    guide vanes, 506
    overspeed protection system, 495–496
    rotor, 534
    shaft and bearings, 507*f*
    shaft housing, 507
    speed, 496
    support, 509, 534
PowerLink ignition system, 220–221
Pratt & Whitney, 348
    2037 engine, 375, 431–434, 432*f*, 433*f*
    fan blades comparison, 432*f*
    Hornet engine, 3
    ignition system components, 396*f*
    JT8D-7R4 engine, 432
    JT8D engine, 386, 387, 391, 391*f*,
        413–422, 413*f*, 414*f*, 415*f*,
        416*f*, 422*f*
    JT9D-3A engine, 422
    JT9D-7R4 engine, 429–431, 430*f*,
        431*f*, 436
    JT9D-7RN engine, 373, 374
    JT9D-59A engine, 429
    JT9D engine, 313, 333*f*, 336, 337*f*, 345,
        364, 366, 398*f*, 413, 422–429, 423*f*,
        424*f*, 425*f*, 427*f*, 436, 542*f*
    JT15D engine, 470–475, 471*f*, 472*f*,
        473*f*, 474*f*
    lubrication system for, 388*f*
    PT-6 engine, 362*f*, 363*f*, 618*f*
    PT-6A-60/65 engine, 361
    PT6A-27 engine, 342
    PT6A engine, 499–513, 501*f*, 502*f*,
        503*f*, 504*f*, 511*f*, 512*f*, 553–555,
        555*f*, 559–564, 626–631, 629*f*, 654,
        655*f*, 656*f*, 657*f*, 696
    PW100 engine, 486–492, 486*f*, 487*f*
    PW120 engine, 487*f*, 488*f*, 489*f*, 490*f*,
        491*f*, 492*f*
    PW124 engine, 639–641
    PW4000 engine, 434–435, 435*f*
    R-985 engine, 56, 56*f*, 110, 111*f*
    R-2800 engine, 111
    R-4360 engine, 4–5, 6*f*
    series 2000, 373, 413
    series 4000, 373
    Wasp engine, 3
Precipitation number, 83
Precoiling, 300–301

Preheating, 247
Preignition, 74–75, 76f, 109, 157
Prelubrication, 297
Preservation run-in, 303
Pressure. *See also specific types and relevant topics*
  defined, 137
  U-shaped tubes and, 142, 142f
Pressure altitude, 77
Pressure baffles, 122, 122f
Pressure balance holes, 483
Pressure fuel systems, 134, 135f
Pressure injection carburetors, 161–164, 163f, 164f, 165f
Pressure instruments, 683–687
Pressure jet atomization, 358
Pressure lubrication systems, 87
Pressure oil filter, 383f
Pressure oil pump, 382, 383f
Pressure oil system, 474
Pressure-ratio controller, 118
Pressure ratio sensor, 369f
Pressure regulator unit, 164
Pressure relief valve system, 385–386
Pressure-testing, 231f
Pressure transmitter, 684
Pressure tube, 443
Pressurization, 527
Pressurized magnetos, 218
Pressurizing valve, 359f
Primary capacitor, 197–198, 197f
Primary circuits, 193, 196–201, 203–205
Primary winding, 196, 198
Prist, 351
Projected core nose, 226
Prony brake, 69, 69f
Propeller control systems, 589–590, 590f, 595f, 714
  of General Electric CT7 propeller control system, 643
  for PT-6 turboprop engine, 618f
Propeller control unit (PCU), 490, 640
Propeller-flange timing marks, 211f
Propeller gear case (PGC), 494
Propeller static rpm (fixed-pitch propeller), 679–680
Propeller torque reaction, 589
Propeller track, 657–658, 657f
Propeller weight-diameter comparisons, 625f
Propellers, 587f. *See also specific components and types*
  adjustments of, 679–681
  of Allison 250-B17 reversing turbopropeller system, 637–638
  alterations of, 665
  anti-icing for, 607–609, 608f
  authorized repairs of, 665
  on balance stand, 658f
  with blade cuffs, 582f
  blade element of, 582f
  checking blade angles of, 675–678
  classification of, 590–592

Propellers (*Cont.*):
  clearances, 590
  defects in, 670
  deicing systems of, 607–609
  ECS for, 640f
  efficiency, 589
  fatigue stresses in, 668f
  forces on, 587
  for General Electric CT7 turboprop engine, 496–497
  governor, 596–598, 621, 627, 630f, 632f
  hubs, 649
  inspection of, 649–681
  installation of, 649–681
  instruments, 589–590
  leading edge of, 581, 669–670, 669f
  load, 588–589, 588f
  maintenance of, 649–681
  overspeed governor of, 490, 631, 633f, 640–641, 641f
  path through air, 583f
  protractor, 677f
  reduction gear for, 56–57, 212f
  regulator assembly, 647f
  removal, 649–655
  repair of, 664–675
  shafts, 37, 37f, 481
  speed, 496
  surface defect prevention and treatment of, 668–671
  synchrophaser system, 609–612, 611f
  theory, 581–612
  thrust horsepower, 322
  troubleshooting of, 680–681
Propulsive efficiency, 321–322, 321f
Protected air, 102
Protractor scale, 675, 677f
psia. *See* Pounds per square inch absolute
psig. *See* Pounds per square inch gauge
Pulleys, 709–711, 718
Pulse jet engine, 310
Push-pull control, 709, 711–712, 711f, 712f, 718
Push-wire, 226
Pusher propellers, 591
Pushrods, 49, 51, 54

Quadrant system, 713, 714f

Radial (R) engine classification, 7
Radial engines, 7
  accessory section for, 56, 56f
  cam plate for, 52, 52f
  crankcase for, 29–31, 31f
  cylinder heads for, 47
  exhaust system for, 125–127, 126f
  firing order for, 64
  high-output, 110
  lubrication system for, 96, 96f
  in post-World War I, 3–4, 4f
  supercharger for, 110–111, 110f
  valve operating mechanism for, 51–54, 53f
Radial gradient vanes, 434

Radio shielding, 203
Radiographic inspection, 281
Ram effect, 319, 319f, 322f
Ram jet engines, 310
Ramsbottom test, 83
Range and airspeed, 243, 244f
Range markings, 684f
Rate-of-change controller, 118
Rated power, 71
Ratio lever, 364
Ratio of velocity, 588
Ratiometer instrument, 689, 689f
Reaction principles, 314–315
Reaction turbines, 335, 336f
Rear-compressor drive shaft, 416
Rear-compressor drive-turbine rotor assembly, 421f
Rear cone, 649, 650
  bottoming, 651–652
Rebarreling, 292
Rebuilt engine, 267
Reciprocating engines, 8–14, 59, 239–266. *See also specific engines*
  abrasion for, 271
  accessory section of, 55–56
  afterfiring of, 265
  air cooling for, 122–123, 122f, 123f
  backfiring of, 265
  bearings in, 32–33
  brinelling for, 271
  burning for, 271
  burr for, 271
  cam surfaces for, 273
  chafing for, 271
  chipping for, 272
  cleaning for, 273–275
  connecting-rod assemblies in, 37–40, 38f
  construction and nomenclature for, 29–57
  by Continental, 8, 8f, 11f
  cooling systems for, 121–123
  corrosion for, 272
  counterweights of, 293
  crack for, 272
  crankcase for, 29–32, 273
  crankshafts of, 33–37, 33f, 292–293
  cruise control of, 243
  cylinders of, 45–47, 46f, 286
  degreasing for, 273–274
  engine operating conditions for, 245–246
  exhaust systems for, 124–126, 124f
  fins of, 286
  fuel system troubles of, 262–263
  gears for, 273
  ignition for, 191–237
  ignition malfunctions for, 262
  ignition systems for, 236–237
  induction system problems of, 264–265
  inspection of, 247–258
  installation for, 299–300
  instruments for, 683–687

Reciprocating engines (*Cont.*):
  liquid cooling for, 123–124, 123*f*
  by Lycoming, 9–14, 12*f*, 13*f*
  maintenance of, 247–258
  oil pressure of, 242
  oil system problems of, 263–264
  opening and cleaning of, 248
  operating requirements of, 241–242
  operation for, 239–241
  overhaul of, 267–305
  oxidation for, 272
  pistons in, 40–45, 41*f*, 44*f*
  power settings and adjustments of,
    242–243
  power settings of, 243–245
  preheating for, 247
  pressure injection carburetors in, 163
  propeller reduction gears for, 56–57
  range and speed charts of, 243
  reassembly for, 295–299
  receiving inspection for, 270
  repair and replacement for, 286–295
  servicing oil screens and filters of,
    248–250
  sludge chambers of, 293
  spark plugs of, 262
  starters for, 232–236
  starting of, 239–240, 301–302
  starting systems for, 191–237
  stopping procedure for, 245
  studs for, 273, 286
  symptom recognition of, 258
  tappets for, 273
  temperature check of, 242
  troubleshooting of, 258–265
  valve guides of, 286–287
  valve seats for, 287
  valves for, 47–55, 273
  winterization procedures for, 246–247
Reduction gear, 56–57, 212*f*, 342–344,
  405, 480, 481, 481*f*
Reduction gearbox, 486, 507, 508*f*
Regulator assembly, 646–647
Regulator servo unit, 177
Reid vapor pressure bomb, 130, 130*f*
Relative humidity, 156
Relative pressure, 137
Relative wind, 585, 585*f*
Relief valve, 90–91, 121, 121*f*, 134, 511
Repair:
  of airworthy damage, 674
  for blades, 568
  composite blade damage and, 676*f*
  of composite blades, 674–675
  for compressor blades, 566*f*
  for fan blades, 564*f*
  of float-type carburetors, 159
  of leading edges, 669–670
  of propellers, 664–675, 671*f*
  for reciprocating engine, 286–295
  of superchargers, 258
  for turbine nozzle, 568
  of turbochargers, 258

Repair (*Cont.*):
  of unairworthy damage, 674
  of vanes, 568
  of wood propellers, 666
Resistor-type spark plugs, 225
Retention assemblies, 646
Reverse pitch propeller, 592
Reversible steel hub propellers, 603
Reversing operation, 631
Rich best-power mixture, 139, 152
Rich blowout, 360
Rocker arms, 49–50, 51, 51*f*, 290
Rockets, 310
Rods, 709, 710. *See also* Connecting rods
  control system, 716–717
Roller bearings, 32–33, 32*f*
Rolling friction, 84
Rolls-Royce, 343*f*
  Dart engine, 480–486, 480*f*, 481*f*, 482*f*,
    483*f*, 485*f*
  RB 211-524G/H engine, 459*f*
  RB 211 engine, 313, 336, 337*f*, 345,
    413, 452–458, 458*f*, 460*f*, 461*f*
  Tay engine, 458, 462*f*
  Trent 1000 engine, 461, 464*f*
Rotating magnet, 195–196, 195*f*, 196*f*
Rotating-magnet magnetos, 192, 193–194
Rotax:
  in-line engines by, 14*f*
  opposed-type engines by, 16–19, 17*f*,
    18*f*, 19*f*
  two-stroke engines by, 14–16, 14*f*, 15*f*,
    16*f*, 17*f*
Round wire gauges, 230
Routine operational inspections, 547–548
Rpm lever, 633
RPM-torque relationship, 697*f*
Run-in, 300–302, 301*f*
Running engagements, 409
Runout, 272

SACS1. *See* Stator vane actuator
Safety circlets, 48
Safety gap, 203
Safety wire, 296
Saturated space, 156
Saturation level, 350
Saybolt Universal viscosimeter, 81, 81*f*,
  381
Scale of sound values, 344*f*
Scavenge oil system, 382–383, 383*f*, 474
Scavenge pump, 95
Scavenging stroke, 62
SCDI. *See* Single capacitor discharge
  ignition
Scheduled maintenance, 557–558
Scoring, 272, 272*f*
Screw extractor, 294, 294*f*
Sea-level boosted engine, 114
Secondary winding, 196
Self-inductance, 198
Self-locking nuts, 296
Semifloating piston pins, 44

Semisolid lubricants, 79
Sensing element, 701
Service bulletins, 270
Servo pressure regulator, 175, 178*f*
Setting gap, 230*f*
sfc. *See* Specific fuel consumption
Shaft horsepower (shp), 479, 521, 615
Shaft splines, 37, 649–650, 649*f*, 651, 652*f*
Shaft turbine, 521
Sheaves, 711
Shell threads, 227
Shielded terminal thread designs, 227*f*
Shielding:
  ignition, 203
  for spark plugs, 225*f*
shp. *See* Shaft horsepower
Shrink-fit method, for joining cylinder
  barrel and head, 46
Shroud segment grinding, 555, 560*f*, 561*f*
Shrouded fuel nozzle, 173, 173*f*
Shutoff valve, 631
Silica abrasive, 230
Simplex spray nozzle, 358
Single can-type combustor, 333*f*
Single capacitor discharge ignition (SCDI),
  14–16
Single-crystal castings, 348
Single-crystal turbine blades, 434, 434*f*
Single-element deicing system, 609, 610*f*
Single-grade oils, 86, 86*f*
Single-plane balancing, 570
Single pressure relief valve, 90–91, 91*f*
Single-row radial engine, 3, 7*f*, 64
Single-speed supercharger, 110–111, 110*f*
Single-throw crankshaft, 35–36, 35*f*, 36*f*
Single-type magnetos, 192
Six-throw crankshaft, 36–37, 36*f*
Skirt:
  for combustor, 441
  of cylinders, 45, 292
  of piston, 44, 272*f*
Sleeve, 711
Slick Series 4300 magnetos, 215–218,
  216*f*, 217*f*
Slick Series 6300 magnetos, 215–218
Sliding friction, 84
Sliding (rotation) idle valve, 179
Slinger ring, 607
Slip, 586
Slippage mark, 684
Sludge chambers, 34, 95, 293
Small end, of connecting-rod assembly, 38
Small-hole gauge, 282, 283*f*
Solenoid, 95, 95*f*, 277*f*
Solid lubricants, 79
Solvent degreasing, 274*f*
Spalling, 272, 272*f*
Spar/shell, 625, 625*f*
Spark ignition system, 25, 395, 555*f*
Spark plugs, 214*f*, 225–232, 225*f*, 228*f*,
  230*f*, 245–246, 253, 262
  gap gauge, 230*f*
  heat range of, 227–228

Spark plugs (*Cont.*):
　inspection of, 229, 231, 253
　installation of, 231–232
　pressure-testing of, 231*f*
　reach, 227, 227*f*
　regapping of, 230
　removal of, 229
　rotation of, 231, 232*f*
　temperature ranges of, 228*f*
Specific fuel consumption (sfc), 140
Specific gravity, 80
Speed-adjusting lever, 598
Speed control, 362
Speeder spring, 595, 598
Spill-type nozzle, 359
Spill valve, 484
Spin-on oil filter, 92, 93*f*
Spinner, 647
Splash lubrication, 88
Splines, 37, 649–650, 649*f*,
　651, 652*f*
Split, 672
Split-clamp crankshaft, 35–36
Split trailing edge, 674
Sport aviation oils, 87
Sprag clutch, 536
Spring rings, 44, 48
Spring-type clutch, 235*f*
Spur-gear, 56*f*, 57, 352*f*, 405
Square engine, 60, 68
Squealer tips, 443, 505
Squeeze film bearing, 388, 389*f*
Stall control, 331
Standard engine rating, 71
Standard pressure, 137
Start lock system, 635*f*
Starter generators, 401
Starter motor, 21, 233, 236–237
Starters, 232–236, 234*f*, 236*f*
　for gas turbines, 406, 406*f*
　shutoff valve, 408, 409*f*
　turbine, 407
　for turbine engines, 411
Starting, 240–241
　cycle, 395
　of gas-turbine engines, 541–544
　ignition system and, for Rolls-Royce
　　Dart engine, 486
　latches, 620
　for reciprocating engine, 301–302
　temperature-limit chart for, 551*f*
　for turbojet engine, 395*f*
Starting systems:
　for Douglas DC-10, 407*f*
　for gas turbines, 395–411
　for reciprocating engines, 191–237
　for turbofan engine, 407–411
Starting vibrator, 215
Static balancing, 570
Static-flux curve, 196, 196*f*
Static propeller balance, 658–660
Static rpm, 680
Stationary piston pins, 44

Stator, 421, 425*f*, 436
　of axial-flow compressor, 328*f*
　housing, 506
　for PT6A turboprop engine, 502–504
Stator vane actuator (SACS1), 369
Steel hub propellers, 599–603, 604*f*
Stoichiometric mixture, 140
Stop-withdrawal, 484
Stopping procedure, 245
Straight-fluted screw extractor, 294*f*
Straight mineral oil, 86, 86*f*
Strainer-type oil filter, 92, 92*f*
Stroke, 60–61, 61*f*
Structural clearances, 590
Structural inspection, 275
Stud-and-nut joint method, 46
Studs, 247, 273, 286, 293–294
Submersible fuel boost pump, 352*f*
Sun gear, 57
Superchargers, 7, 104–111, 110*f*
　compression pressure and, 107–108,
　　108*f*
　compressor in, 104, 105*f*
　crankcase and, 31, 31*f*
　critical altitude and, 7
　gases and, 105–106, 105*f*, 106*f*
　maintenance of, 258
　MAP and, 102, 104–105, 107
　for opposed-type engine, 110
　repair of, 258
Supersonic airflow, 315*f*
Supervisory EEC, 373, 374, 374*f*
Supervisory engine control, 431
Supporting-flange assembly, 704
Swing check valve, 111
Swirl vanes, 333
Synchro system, 687, 688*f*
Synchrophaser system, 609–611, 611*f*
Synthetic lubricants, 80, 381–382

Table of limits, 270, 271*f*
Tabular fuel flowmeter, 687*f*
Tachometers, 529, 575, 577, 590, 691–693,
　691*f*, 692*f*, 695*f*
Takeoff power, 71, 165
Takeoff system, 496
Tapered propeller shafts, 37
Tapered shaft propellers, 649, 650–651
Tappets, 49, 52, 273
Taylor, Charles, 1
TBO. *See* Time between overhauls
TCP. *See* Tricresyl phosphate
TCU. *See* Turbo control unit
TDC. *See* Top dead center
TEL. *See* Tetraethyl-lead
Teledyne Continental, 6*f*, 7, 56
　O-200 engine by, 24–25, 25*f*
Temperature. *See also specific related
　topics*
　aircraft fuel and, 129, 129*f*
　Bendix RSA fuel injection systems
　　and, 188
　cockpit indication signals of, 496

Temperature (*Cont.*):
　compensator, 513
　density controller and, 117
　effect, 320
　gas and, 106–107, 106*f*, 107*f*
　gauges, 688–691
　Pratt & Whitney JT9D engine and, 542*f*
　of reciprocating engines, 242
　of spark plugs, 228*f*
　thrust and, 320*f*
　viscosity and, 85
Temperature-accelerating well, 99
Temperature-limit chart, 551*f*
Temperature-time limits, 551*f*
Test stand, 224, 224*f*, 299–300
Tetraethyl-lead (TEL), 130
Thermal efficiency, 72, 72*f*, 322–323, 322*f*
Thermocouples, 575–576, 700
　installation of, 690*f*
　instruments, 690–691
Thermostatic control valve, 89, 92
Thickness gauge, 284
Threaded-joint method, for joining
　cylinder barrel and head, 46
Three-blade constant-speed feathering
　aluminum hub propeller, 605*f*
Three-blade constant-speed feathering
　steel hub propeller, 602*f*
Three-bladed propeller, 659*f*
Throttle, 713
　body, 154
　ice, 157
　quadrant, 712
　sensitivity, 117
　unit, 163, 164
　valve, 144, 144*f*
Throw. *See* Crankpin
Thrust, 317
　afterburning and, 321
　augmentation, 320
　bending force, 587
　horsepower, 589
　lever assembly, 714–715, 717*f*
　lever friction brake, 715
　low-pressure thrust bearing, 455
　member, 646
　net, 615
　pressure thrust, 338
　propeller thrust horsepower, 322
　reversers, 341–342, 342*f*
　temperature and, 320*f*
　vectoring, 340
　zero-thrust pitch, 586
Thrust specific fuel consumption, 321
Time between overhauls (TBO), 5, 568
Timing:
　of distributor, 210*f*
　kit, 211*f*
　for magnetos, 211
　for Slick Series 4300, 215–217
　of valves, 62–64, 63*f*
Timing marks, 211–212, 211*f*, 212*f*
Timing pin, 216, 217*f*

Tip speed, 587–588
TIT. *See* Turbine inlet temperature
Toluene, 132
Top dead center (TDC), 60–62, 61*f*, 70
Torque, 69, 589, 696
    bending force, 587
    hold, 496
    indication system, 71, 698*f*
    limit operation, 697*f*
    for PT6A propeller installation, 656*f*
    sensor, 621
    units, 644, 645, 646*f*
    valves, 296
    wrench, 297*f*
Torque signal condition (TSC), 490
Torquemeters, 481, 508, 508*f*, 531–532,
    696–697, 698*f*
Torsion, 587
Total-loss oil system, 386
Tower shaft, 424
TPE331 anti-icing system diagram, 516*f*
Track checking, 656–657
Tractor propellers, 590–591
Tractor-type airplanes, 31
Trailing edge, 581
Transformer coil, 219, 219*f*
Transistor ignition exciter unit, 397
Transistorized ignition unit, 397*f*
Transpiration cooling, 335
Trend analysis, 576*f*, 577–578, 578*f*,
    579*f*, 580*f*
Tricresyl phosphate (TCP), 131
Trim curve, 574
Troubleshooting, 259*f*
    of aircraft tachometer system, 577
    for Bendix RSA fuel injection systems,
      186, 187*f*
    for constant-speed propeller, 681*f*
    for Continental fuel injection systems,
      175, 176*f*
    EGT system, 575–577
    for engines, 260*f*–261*f*
    for float-type carburetors, 161, 162*t*
    of fuel injection, 264*f*
    for fuel systems, 263*f*
    for gas-turbine engine, 577–578
    of propellers, 680–681
    for reciprocating engines, 258–265
      ignition and starting system, 236–237
    for turbines, 576*f*
True-pitch propeller, 583
TSC. *See* Torque signal condition
Tubing, 710, 716–717
Turbine blades, 567–568
    air cooling and, 337*f*
    nozzle vanes and, 335*f*
    for Pratt & Whitney JT9D-7R4 engine,
      430
Turbine engines, 322–323.
    See also specific engines
    calibration, 571–572
    igniters for, 399–401, 399*f*, 400*f*
    instrument panel, 694*f*

Turbine engines (*Cont.*):
    lubricants for, 381–393
    lubricating systems of, 385–391
    oil analysis for, 391–393
    oil bearings and seals for, 390
    oil breather system of, 384
    oil coolers of, 384
    oil filter for, 383
    oil indicating and warning systems of,
      384–385
    starters for, 411
    tachometer for, 695*f*
Turbine inlet temperature (TIT), 695
Turbine nozzles, 420, 565–567, 568
    diaphragm, 334–335, 335*f*
    inner case and seal assembly, 419–420
    vanes and blades, 428*f*
Turbine section, 420*f*, 466, 505
    for Allison 250 gas-turbine engine, 532,
      534–535
    for Pratt & Whitney JT9D turbofan
      engine, 427–429
    for Rolls-Royce Dart engine, 482–483
    for Rolls-Royce RB 211 engine, 458
Turbine-shaft heat shields, 418
Turbines, 336*f*. *See also specific types*
    cooling of, 430, 510–511, 511*f*
    exhaust case, 419, 421, 421*f*, 428, 428*f*
    fir-tree method for, 337*f*
    front case, 419
    of gas-turbine engines, 335–338
    horsepower calculations, 318
    ignition systems, 395–396
    midframe, 447, 447*f*
    rear case, 419, 428
    rear frame, 447, 447*f*
    rotor inner first-stage air seal, 420
    rotor section, 506*f*
    rotor spacer, 443
    shrouds, 443
    speed indicators, 695
    starters, 407
    troubleshooting for, 576*f*
    wash, 565
    wheels, 483
Turbo control unit (TCU), 19
Turbocharged (T) engine classification, 7
Turbochargers, 7, 104–123, 111*f*, 112*f*,
    113*f*, 116*f*
    for Continental, 118–121, 120*f*
    differential-pressure controller in,
      114–117, 115*f*
    fuel injection systems and, 173
    inspection of, 258
    lubrication system for, 113
    normalizers and, 111–115
    in Rotax opposed-type engines,
      18–19, 19*f*
    variable-pressure controllers in,
      117, 119*f*
Turbofan engines, 26, 313, 413–475.
    See also specific engines
    on Boeing 787, 460–463

Turbofan engines (*Cont.*):
    filter cartridge for, 387
    fuel systems for, 357*f*
    lubrication system for, 386–390
    noise-absorbent linings in, 346*f*
Turbojet engines, 26, 59, 311, 395*f*
Turbomachinery, 486–489
Turbonormalize, 109
Turboprop engines, 26, 313, 314*f*,
    361–364, 479–518, 615–647.
    See also specific engines
    on Allison 501-D13 engine, 644*f*
    horsepower of, 615
    torquemeters on, 696–697
Turboshaft engines, 26, 313, 521–539.
    See also specific engines
    APU location in, 521–523
    auxiliary power unit of, 521
    Bendix fuel control system for,
      371*f*, 372*f*
    fuel control system for, 370–373
Twin-engine deicing system, 608*f*
Twin-spool axial-flow gas-turbine engine,
    414*f*
Twin-spool turbofan engine, 316*f*
Twisting forces, 587
Two-bladed propeller positions, 658*f*
Two-plane balancing, 570
Two-position propellers, 586, 591, 594,
    594*f*, 604, 606*f*
Two-speed supercharger, 109, 111
Two-stroke engines, 14–16, 14*f*, 15*f*, 16*f*,
    17*f*, 65–66, 65*f*
Type certification, 270

U-shaped tubes, 142, 142*f*
Ultrasonic inspection, 280–281
Unairworthy damage, 674
Unbalanced forces, 309*f*
Uncontrolled oscillations, 203
Uncorrectable vibrations, 656
Underspeed, 364, 597*f*, 630, 633
Universal propeller protractor, 675–677
Unshielded spark plug, 226*f*
Updraft carburetor, 147
Upright-V-type engine, 3

V-type engines, 3, 7, 7*f*, 31–32,
    36–37, 64
Valve-adjusting chart, 55*f*
Valve clearance, 54–55
Valve face conditions, 290*f*
Valve guides, 48–49, 48*f*, 286–287
Valve lag, 63
Valve lap. *See* Valve overlap
Valve lead, 63
Valve lifters, 49
Valve mechanism, 299
Valve operating mechanism, 49–54, 50*f*,
    53*f*
Valve overlap, 62, 63
Valve seat grinder, 287*f*
Valve seats, 49, 49*f*, 287, 288*f*

Valve springs, 49, 285*f*
Valve stem, 48
Valve tappets, 49, 52, 273
Valve tip, 48
Valves, 47–55. *See also specific types*
  with feather edge, 289*f*
  inspection of, 273*f*
  interference fit of, 288*f*
  for reciprocating engine, 273, 288–290
  refacing of, 288*f*
  timing of, 62–64, 63*f*
Vane type oil pressure pump, 94
Vanes, 565–567
  air cooling of, 335*f*
  of compressor turbine, 505
  of diffuser, 111, 482
  of nozzles, 335*f*, 442*f*, 482
  repairs for, 568
Vapor blasting, 275
Vapor degreasing, 274
Vapor ejectors, 170
Vapor lock, 134–136
Vapor-pressure temperature gauge, 689
Vapor separators, 136, 170
Vaporization, 129, 156
  of fuel, 143–144
Variable-absolute-pressure controller, 118, 120*f*, 121
Variable-area exhaust nozzle, 339–340
Variable force, 594
Variable frequency starter-generator (VFSG), 410, 411*f*
Variable inlet guide vanes, 331
Variable-inlet guide-vanes, 529
Variable-port, 358

Variable-port fuel spray nozzle, 359*f*
Variable-pressure controllers, 117, 119*f*
Variable pump, 485*f*
Variable-stator-vane actuators, 436
Variable stator vanes, 331, 332*f*
Variable vanes, 424, 426*f*
Vegetable lubricants, 79
Ventilated-type oil control rings, 44
Venturi tube, 138, 138*f*, 142*f*, 163, 175, 177*f*
Vernier push-pull control, 712, 712*f*
Vertical engine assembly, 570*f*
Vertical (V) engine classification, 7
Vertical imbalance, 658
VFSG. *See* Variable frequency starter-generator
VI. *See* Viscosity index
Vibration, 7, 656–663, 700*f*
Vibrator, 207
Videoscope, 551*f*
Viscosity, 81–82, 85, 381
  multiviscosity oils, 86–87
Viscosity index (VI), 82, 86
Volatility, 129, 143–144, 349
Volkswagen, 21–23, 24*f*
Volumetric efficiency, 61, 73

W/A. *See* Water-alcohol
Walk-around, 547
Wankel, Felix, 66
Wankel cycle, 66
Washers, 296
Waste gate, 110, 111, 113, 121*f*
Water-alcohol (W/A), 164, 166, 166*f*
Water clearances, 590

Water injection. *See* Antidetonant injection
Water-methanol system, 485
Water pressure indicator, 166
Water-vapor pressure, 78
Watt, 68
Wear metals, 391
Wedge-shaped piston rings, 44
Weight-power ratio, 76–77
Wet motoring check, 572–573
Wet-sump system, 96–99, 97*f*
Wetted oil air filter, 254
Wheatstone bridge, 688, 689*f*
Whittle, Frank, 26, 308
Whittle W1 engine, 308*f*
Wide-cut gasolines, 350
Williams/Rolls FJ44 turbofan engine, 475, 477*f*
Wing-mounted engines, 707*f*
Winterization procedures, 246–247
Wiping friction, 84–85
Wire-end fitting assembly, 703
Wire mesh wetted oil-type air filter, 254
Wood propellers, 592, 593*f*, 666
Woodward propellers, 595*f*, 598*f*
Work, 69, 71
World War I, 1–3
World War II, 26
Wright brothers, 1, 2*f*
Wright Hisso engine, 3
Wright Whirlwind engine, 3

X-ray inspection, 281, 281*f*
Xylene, 132

Zero-thrust pitch, 586